ADX500 CAS:107-13-1 *HR: 3*

ACRYLONITRILE

DOT: UN 1093

mf: C_3H_3N mw: 53.07

PROP: Colorless, mobile liquid; mild odor. Sol in water. Mp: $-82°$, bp: $77.3°$, fp: $-83°$, flash p: $30°F$ (TCC), lel: 3.1%, uel: 17%, d: 0.806 @ $20°/4°$, autoign temp: $898°F$, vap press: 100 mm @ $22.8°$, vap d: 1.83, flash p: (of 5% aq sol): $<50°F$.

SYNS: ACRYLNITRIL (GERMAN, DUTCH) ◇ ACRYLONITRILE MONOMER ◇ AKRYLONITRYL (POLISH) ◇ CARBACRYL ◇ CIANURO di VINILE (ITALIAN) ◇ CYANOETHYLENE ◇ CYANURE de VINYLE (FRENCH) ◇ ENT 54 ◇ FUMIGRAIN ◇ NITRILE ACRILICO (ITALIAN) ◇ NITRILE ACRYLIQUE (FRENCH) ◇ PROPENENITRILE ◇ 2-PROPENENITRILE ◇ RCRA WASTE NUMBER U009 ◇ TL 314 ◇ VENTOX ◇ VINYL CYANIDE

TOXICITY DATA with REFERENCE

bfa-rat/sat 30 mg/kg TXCYAC 16,67,80

dns-rat:lvr 1 mmol/L PMRSDJ 5,371,85

slt-dmg-orl 1520 μmol/L PMRSDJ 5,325,85

skn-hmn 500 mg nse INMEAF 17,199,48

skn-rbt 10 mg/24H open JIHTAB 30,63,48

skn-rbt 500 mg MLD SCCUR* -,1,61

eye-rbt 20 mg SEV JIHTAB 30,63,48

ipr-ham TDLo:641 mg/kg (female 8D post):TER TJADAB 23,325,81

orl-rat TDLo:650 mg/kg (female 6-15D post):REP DOWCC* 03NOV76

orl-rat TDLo:18200 mg/kg/52W C:CAR FCTOD7 24,129,86

ihl-rat TCLo:5 ppm/52W-I:ETA MELAAD 68,401,77

orl-rat LD:3640 mg/kg/52W-C:NEO DOWCC* MAR77

ihl-hmn TCLo:16 ppm/20M:EYE,PUL INMEAF 17,199,48

ihl-man LCLo:1 g/m^3/1H:CNS,GIT ZAARAM 16,1,66

skn-chd LDLo:2015 mg/kg:CNS,RSP,GIT DMWOAX 75,1087,50

orl-rat LD50:78 mg/kg JOHYAY 3,106,59

ihl-rat LCLo:500 ppm/4H JIDHAN 31,343,49

skn-rat LD50:148 mg/kg GISAAA 41(10)103,76

ihl-mus LCLo:315 ppm/4H NTIS** PB280-478

ipr-mus LD50:46 mg/kg TXAPA9 59,589,81

orl-mus LD50:27 mg/kg JHEMA2 3,106,59

scu-mus LD50:35 mg/kg JHEMA2 3,106,59

ihl-dog LCLo:110 ppm/4H JIHTAB 24,27,42

CONSENSUS REPORTS: NTP Fifth Annual Report on Carcinogens. IARC Cancer Review: Group 2A IMEMDT 7,79,87; Human Limited Evidence IMEMDT 19,73,79; Animal Limited Evidence IMEMDT 19,73,79. Community Right-To-Know List. EPA Extremely Hazardous Substances List. Reported in EPA TSCA Inventory.

OSHA PEL: TWA 2 ppm; CL 10 ppm/15M; Cancer Hazard.
ACGIH TLV: Suspected Human Carcinogen, TWA 2 ppm (skin).
DFG TRK: 3 ppm (7 mg/m^3), Animal Carcinogen, Suspected Human Carcinogen.
NIOSH REL: TWA 1 ppm; CL 10 ppm/15M
DOT Classification: Flammable Liquid and Poison.

HR: Hazard Rating indicating relative hazard for toxicity, fire, and reactivity with 3 denoting the worst hazard level.
See Introduction: paragraph 3, p. xiii.

CAS: – The American Chemical Society's Chemical Abstracts Service number. A complete CAS number cross-index is located in Section 1.
See Introduction: paragraph 4, p. xiii.

Cited Reference: – A code which represents the cited reference for the toxicity data. Complete bibliographic citations in order are listed in Section 3.
See Introduction: paragraph 16, p. xxiv.

Safety Profiles: – These are text summaries of the toxicity, flammability, reactivity, incompatibility and other dangerous properties of the entry.
See Introduction: paragraph 19, p. xxvii.

SAFETY PROFILE: Confirmed human carcinogen with experimental carcinogenic, neoplastigenic, and tumorigenic data. Poison by inhalation, ingestion, skin contact, and other routes. Human systemic effects by inhalation and skin contact: conjunctival irritation, somnolence, general anesthesia, cyanosis and diarrhea. An experimental teratogen. Other experimental reproductive effects. Human mutation data reported. Dangerous fire hazard when exposed to heat, flame, or oxidizers. Moderate explosion hazard when exposed to flame. Can react vigorously with oxidizing materials (see also CYANIDES).

Sax's Dangerous Properties of Industrial Materials

Sax's
Dangerous
Properties of
Industrial Materials

Eighth Edition

Volume 1

RICHARD J. LEWIS, SR.

 VAN NOSTRAND REINHOLD
New York

DISCLAIMER

Extreme care has been taken in preparation of this work.
However, neither the publisher nor the authors shall be
held responsible or liable for any damages resulting in
connection with or arising from the use of any of the
information in this book.

Copyright © 1992 by Van Nostrand Reinhold

Library of Congress Catalog Card Number 92–3896
ISBN VOLUME I   0-442-01276-4
ISBN VOLUME II   0-442-01277-2
ISBN VOLUME III   0-442-01278-0
ISBN SET   0-442-01132-6

Manufactured in the United States of America

Published by Van Nostrand Reinhold
115 Fifth Avenue
New York, NY 10003

Chapman and Hall
2–6 Boundary Row
London, SE 1 8HN

Thomas Nelson Australia
102 Dodds Street
South Melbourne 3205
Victoria, Australia

Nelson Canada
1120 Birchmont Road
Scarborough, Ontario M1K 5G4, Canada

16 15 14 13 12 11 10 9 8 7 6 5 4 3 2 1

Library of Congress Cataloging-in-Publication Data
Lewis, Richard J., Sr.
    Sax's dangerous properties of industrial materials / Richard J.
Lewis, Sr.—8th ed.
        p.   cm.
    Rev. ed. of: Dangerous properties of industrial materials /N.
Irving Sax and Richard J. Lewis, Sr. 7th ed. ©1989.
    Includes bibliographical references and index.
    ISBN 0-442-01132-6
    1. Hazardous substance—Handbooks, manuals, etc.   I. Sax, N.
Irving (Newton Irving).   Dangerous properties of industrial
materials.   II. Title.   III. Title: Dangerous properties of
industrial materials.
T55.3.H3S3 1992
    604.7—dc20   92-3896
                                              CIP

Dedicated to Grace Ross Lewis
for her help and encouragement
which made this revision possible.

Acknowledgments

I extend thanks to Judith Joseph and Mark Licker for their encouragement. My thanks and best wishes to Alberta Gordon for her expert professional advice and assistance in converting the manuscript to this volume.

Contents

Preface

This eighth edition of *Dangerous Properties of Industrial Materials*, a three volume set, represents a major revision and updating of the seventh edition. The objective of the book, however, remains the same: to promote safety by providing the most up-to-date hazard information available. The growth in the availability of toxicological and hazard control reports continues unabated. This book cannot contain all the published data and continue to provide the accessibility for which it is known. To continue to provide complete hazard assessments for the maximum number of entries, data for each entry has been selectively reduced. In particular, carcinogenic and reproductive data lines above those required to establish the hazard of the entry have been excluded. Complete data for these entries are available in the books *Carcinogenically Active Chemicals* and *Reproductively Active Chemicals*, both available from the publisher.

Over 14,000 entries have been revised for this edition, and 1,500 new entries have been added. Some less useful entries have been eliminated or combined with related entries. Emphasis has been placed on including new carcinogenic and reproductive entries. A special effort has resulted in a significant increase in entries containing skin and eye irritation data.

All carcinogenic entries were reviewed and have been categorized as either confirmed, suspected, or questionable. This assessment was based on a detailed classification scheme discussed under Safety Profiles in the Introduction.

All reproductive entries have been updated and new data added.

Numerous synonyms have been added to assist in locating the many materials which are known under a variety of systematic and common names. The synonym cross-index contains the entry name as well as each synonym. This index should be consulted first to locate a material by name. Synonyms are given in English, as well as other major languages such as French, German, Dutch, Polish, Japanese, and Italian.

Additional physical and chemical properties have been added. Whenever available, physical descriptions, formulas, molecular weights, melting points, boiling points, explosion limits, flash points, densities, autoignition temperatures, and the like, have been supplied.

The following classes of data have been updated for all entries for which they apply as follows:

1. IARC Group 1-4 classes and recent assessments.
2. OSHA revised standards which were published in January 1989 and take effect on December 31, 1992.
3. ACGIH TLVs and BEIs reflect the latest recommendations and now include "intended Changes."
4. NTP Fifth Annual Report on Carcinogens entries are identified.
5. DOT classifications now have the corresponding guide number for entries having multiple guide numbers.
6. CAS numbers are provided for additional entries.

Each entry concludes with a safety profile, a textual summary of the hazards presented by the entry. The discussion of human exposures includes target organs and specific effects reported. Carcinogenic and reproductive assessments have been completely revised for this edition.

Fire and explosion hazards are briefly summarized in terms of conditions of flammable or reactive hazard. Where feasible, fire-fighting materials and methods are discussed. Materials which are known to be incompatible with an entry are listed here.

Also included in the safety profile are comments on disaster hazards which serve to alert users of materials to the dangers that may be encountered on entering storage premises during a fire or other emergency. Although the presence of water, steam, acid fumes, or powerful vibrations can cause the decomposition of many materials into dangerous compounds, of particular concerned are high temperatures (such as those resulting from a fire) since these can cause many otherwise mild chemicals to emit highly toxic gases or vapors such as NO_x, SO_x, acids, and so forth, or evolve vapors of antimony, arsenic, mercury, and the like.

The book, which consists of three volumes, is divided as follows:

The first volume contains a CAS number cross-index, a synonym cross-index, and the complete citations for bibliographic references given in the data section.

Section 1 contains the CAS Number cross-index for CAS numbers for the listed materials.

Section 2 contains the prime name and synonym cross-index for the listed materials

Section 3 contains the complete bibliographic references.

The main section of the book is contained in the Volumes II and III. It lists and describes approximately 20,000 materials in alphabetical order by entry name.

Please refer to the Introduction in Volume I for an explanation of the sources of data and codes used.

Every effort was made to include the most current and complete information. The author welcomes comments or corrections to the data presented.

Richard J. Lewis, Sr.

Key to Abbreviations

abs – absolute
ACGIH – American Conference of Governmental Industrial Hygienists
alc – alcohol
alk – alkaline
amorph – amorphous
anhyd – anhydrous
approx – approximately
aq – aqueous
atm – atmosphere
autoign – autoignition
aw – atomic weight
af – atomic formula
BEI – ACGIH Biological Exposure Indexes
bp – boiling point
b range – boiling range
CAS – Chemical Abstracts Service
cc – cubic centimeter
CC – closed cup
CL – ceiling concentration
COC – Cleveland open cup
conc – concentration, concentrated
compd(s) – compound(s)
contg – containing
cryst, crys – crystal(s), crystalline
d – density
D – day(s)
decomp, dec – decomposition
deliq – deliquescent
dil – dilute
DOT – U.S. Department of Transportation
EPA – U.S. Environmental Protection Agency
eth – ether
(F) – Fahrenheit
FCC – Food Chemical Codex
FDA – U.S. Food and Drug Administration
flam – flammable
flash p – flash point
fp – freezing point
g, gm – gram
glac – glacial
gran – granular, granules
hygr – hygroscopic
H, hr – hour(s)
HR: – hazard rating
htd – heated

htg – heating
IARC – International Agency for Research on Cancer
immisc – immiscible
incomp – incompatible
insol – insoluble
IU – International Unit
kg – kilogram (one thousand grams)
L,l – liter
lel – lower explosive limit
liq – liquid
M – minute(s)
m^3 – cubic meter
mf – molecular formula
mg – milligram
misc – miscible
μ, u – micron
mL, ml – milliliter
mm – millimeter
mod – moderately
mp – melting point
mppcf – million particles per cubic foot
mw – molecular weight
ng – nanogram
NIOSH – National Institute for Occupational Safety and Health
nonflam – nonflammable
NTP – National Toxicology Program
OBS – obsolete
OC – open cup
org – organic
OSHA – Occupational Safety and Health Administration
Pa – Pascals
PEL – permissible exposure level
petr – petroleum
pg – picogram (one trillionth of a gram)
Pk – peak concentration
pmole – picomole
powd – powder
ppb – parts per billion (v/v)
pph – parts per hundred (v/v)(percent)
ppm – parts per million (v/v)
ppt – parts per trillion (v/v)
prep – preparation
PROP – properties
refr – refractive
rhomb – rhombic

S,sec – second(s)
sl, slt, sltly – slightly
sol – soluble
soln – solution
solv(s) – solvent(s)
spont – spontaneous(ly)
STEL – short term exposure limit
subl – sublimes
TCC – Tag closed cup
tech – technical
temp – temperature
TLV – Threshold Limit Value
TOC – Tag open cup
TWA – time weighted average
U, unk – unknown, unreported
μ, u – micron
uel – upper explosive limit
μg, ug – microgram

ULC, ulc – Underwriters Laboratory Classification
USDA – U.S. Department of Agriculture
vac – vacuum
vap – vapor
vap d – vapor density
vap press – vapor pressure
vol – volume
visc – viscosity
vsol – very soluble
W – week(s)
Y – year(s)
% – percent(age)
> – greater than
< – less than
< = – equal to or less than
= > – equal to or greater than
° – degrees of temperature in Celsius (Centigrade)
(F),°F – temperature in Fahrenheit

Introduction

The list of potentially hazardous materials includes drugs, food additives, preservatives, ores, pesticides, dyes, detergents, lubricants, soaps, plastics, extracts from plant and animal sources, plants and animals which are toxic by contact or consumption, and industrial intermediates and waste products from production processes. Some of the information refers to materials of undefined composition. The chemicals included are assumed to exhibit the reported toxic effect in their pure state unless otherwise noted. However, even in the case of a supposedly "pure" chemical, there is usually some degree of uncertainty as to its exact composition and the impurities which may be present. This possibility must be considered in attempting to interpret the data presented since the toxic effects observed could in some cases be caused by a contaminant. Some radioactive materials are included but the effect reported is the chemically produced effect rather than the radiation effect.

For each entry the following data are provided when available: the DPIM number, hazard rating, entry name; CAS number; DOT number; molecular formula; molecular weight; line structural formula; a description of the material and physical properties; and synonyms. Following this are listed toxicity data with references for reports of primary skin and eye irritation, mutation, reproductive, carcinogenic, and acute toxic dose data. The Consensus Reports section contains, where available, NTP Fifth Annual Report on Carcinogens, IARC reviews, NTP Carcinogenesis Testing Program results, EPA Extremely Hazardous Substances List, the EPA Genetic Toxicology Program, and the Community Right to Know List. We also indicate the presence of the material on the update of the EPA TSCA inventory of chemicals in use in the United States. The next grouping consists of the U.S. Occupational Safety and Health Administration's (OSHA) permissible exposure levels, the American Conference of Governmental Industrial Hygienists' (ACGIH) Threshold Limit Values (TLVs), German Research Society's (MAK) values, National Institute for Occupational Safety and Health (NIOSH) recommended exposure levels, and U.S. Department of Transportation (DOT) classifications. Each entry concludes with a safety profile which discusses the toxic and other hazards of the entry.

1. *DPIM Entry Number* identifies each entry by a unique number consisting of three letters and three numbers, for example, AAA123. The first letter of the entry number indicates the alphabetical position of the entry. Numbers beginning with "A" are assigned to entries indexed with the A's. Each listing in the cross-indexes is referenced to its appropriate entry by the DPIM entry number.

2. *Entry Name* The name of each material is selected, where possible, to be a commonly used designation.

3. *Hazard Rating* (*HR:*) is assigned to each material in the form of a number 1, 2, or 3 that briefly identifies the level of the toxicity or hazard. The letter "D" is used where the data available are insufficient to indicate a relative rating. In most cases a "D" rating is assigned when only in vitro mutagenic or experimental reproductive data are available. Ratings are assigned on the basis of low (1), medium (2), or high (3) toxic, fire, explosive, or reactivity hazard.

The number "3" indicates an LD50 below 400 mg/kg or an LC50 below 100 ppm; or that the material is explosive, highly flammable, or highly reactive.

The number "2" indicates an LD50 of 400-4,000 mg/kg or an LC50 of 100-500 ppm; or that the material is flammable or reactive.

The number "1" indicates an LD50 of 4,000-40,000 mg/kg or an LC50 of 500-5000 ppm; or that the material is combustible or has some reactivity hazard.

4. *Chemical Abstracts Service Registry Number* (*CAS*) is a numeric designation assigned by the American Chemical Society's Chemical Abstracts Service and uniquely identifies a specific chemical compound. This entry allows one to conclusively identify a material regardless of the name or naming system used.

5. *DOT:* indicates a four digit hazard code assigned by the U.S. Department of Transportation. This code is recognized internationally and is in agreement with the

United Nations coding system. The code is used on transport documents, labels, and placards. It is also used to determine the regulations for shipping the material.

6. *Molecular Formula* (*mf:*) or atomic formula (*af:*) designates the elemental composition of the material and is structured according to the Hill System (see *Journal of the American Chemical Society*, 22(8): 478-494, 1900) in which carbon and hydrogen (if present) are listed first, followed by the other elemental symbols in alphabetical order. The formula for compounds that do not contain carbon is ordered strictly alphabetically by element symbol. Compounds such as salts or those containing waters of hydration have molecular formulas incorporating the CAS dot-disconnect convention. In this convention, the components are listed individually and separated by a period. The individual components of the formula are given in order of decreasing carbon atom count, and the component ratios given. A lower case "x" indicates that the ratio is unknown. A lower case "n" indicates a repeating, polymer like structure. The formula is obtained from one of the cited references or a chemical reference text, or derived from the name of the material.

7. *Molecular Weight* (*mw:*) or atomic weight (*aw:*) is calculated from the molecular formula using standard elemental molecular weights (carbon = 12.01).

8. *Structural formula* is a line formula indicating the the structure of a given material.

9. *Properties* (*PROP:*) are selected to be useful in evaluating the hazard of a material and designing its proper storage and use procedures. A definition of the material is included where necessary. The physical description of the material may include the form, color, and odor to aid in positive identification. When available, the boiling point, melting point, density, vapor pressure, vapor density, and refractive index are given. The flash point, autoignition temperature, and lower and upper explosive limits are included to aid in fire protection and control. An indication is given of the solubility or miscibility of the material in water and common solvents. Unless otherwise indicated temperature is given in Celsius, pressure is in millimeters of mercury.

10. *Synonyms* for the entry name are listed alphabetically. Synonyms include other chemical names, common or generic names, foreign names (with the language in parentheses), or codes. Some synonyms consist in whole or in part of registered trademarks. These trademarks are not identified as such. The reader is cautioned that some synonyms, particularly common names, may be ambiguous and refer to more than one material.

11. *Skin and Eye Irritation Data* lines include, in sequence, the tissue tested (skin or eye); the species of animal tested; the total dose and, where applicable, the du-

ration of exposure; for skin tests only, whether open or occlusive; an interpretation of the irritation response severity when noted by the author; and the reference from which the information was extracted. Only positive irritation test results are included.

Materials that are applied topically to the skin or to the mucous membranes can elicit either (a) systemic effects of an acute or chronic nature or (b) local effects, more properly termed "primary irritation." A primary irritant is a material that, if present in sufficient quantity for a sufficient period of time, will produce a nonallergic, inflammatory reaction of the skin or of the mucous membrane at the site of contact. Primary irritants are further limited to those materials that are not corrosive. Hence, concentrated sulfuric acid is not classified as a primary irritant.

a. Primary Skin Irritation. In experimental animals, a primary skin irritant is defined as a chemical that produces an irritant response on first exposure in a majority of the test subjects. However, in some instances compounds act more subtly and require either repeated contact or special environmental conditions (humidity, temperature, occlusion, etc.) to produce a response.

The most standard animal irritation test is the Draize procedure (*Journal of Pharmacology and Experimental Therapeutics*, 82: 377-419, 1944). This procedure has been modified and adopted as a regulatory test by the Consumer Product Safety Commission (CPSC) in 16 CFR 1500.41 (formerly 21 CFR 191.11). In this test a known amount (0.5 mL of a liquid or 0.5 grams of a solid or semisolid) of the test material is introduced under a one square inch gauze patch. The patch is applied to the skin (clipped free of hair) of twelve albino rabbits. Six rabbits are tested with intact skin and six with abraded skin. The abrasions are minor incisions made through the stratum corneum, but are not sufficiently deep to disturb the dermis or to produce bleeding. The patch is secured in place with adhesive tape, and the entire trunk of the animal is wrapped with an impervious material, such as rubberized cloth, for a 24-hour period. The animal is immobilized during exposure. After 24 hours the patches are removed and the resulting reaction evaluated for erythema, eschar, and edema formation. The reaction is again scored at the end of 72 hours (48 hours after the initial reading), and the two readings are averaged. A material producing any degree of positive reaction is cited as an irritant.

As the modified Draize procedure described above has become the standard test specified by the U.S. government, nearly all of the primary skin irritation data either strictly adheres to the test protocol or involves only simple modifications to it. When test procedures other than those described above are reported in the literature, appropriate codes are included in the data line to indicate those deviations.

The most common modification is the lack of occlusion of the test patch, so that the treated area is left open to the atmosphere. In such cases the notation "open" appears in the irritation data line. Another frequent modification involves immersion of the whole arm or whole body in the test material or, more commonly, in a dilute aqueous solution of the test material. This type of test is often conducted on soap and detergent solutions. Immersion data are identified by the abbreviation "imm" in the data line.

The dose reported is based first on the lowest dose producing an irritant effect and second on the latest study published. The dose is expressed as follows:

(1) Single application by the modified Draize procedure is indicated by only a dose amount. If no exposure time is given, then the data are for the standard 72-hour test. For test times other than 72 hours, the dose data are given in mg (or an appropriate unit)/duration of exposure, e.g., 10 mg/24H.

(2) Multiple applications involve administration of the dose in divided portions applied periodically. The total dose of test material is expressed in mg (or an appropriate unit)/duration of exposure, with the symbol "I" indicating intermittent exposure, e.g., 5 mg/6D-I.

The method of testing materials for primary skin irritation given in the Code of Federal Regulations does not include an interpretation of the response. However, some authors do include a subjective rating of the irritation observed. If such a severity rating is given, it is included in the data line as mild ("MLD"), moderate ("MOD"), or severe ("SEV"). The Draize procedure employs a rating scheme which is included here for informational purposes only, since other researchers may not categorize irritation response in this manner.

Category	Code	Skin Reaction (Draize)
Slight (Mild)	MLD	Well defined erythema and slight edema (edges of area well defined by definite raising)
Moderate	MOD	Moderate to severe erythema and moderate edema (area raised approximately 1 mm)
Severe	SEV	Severe erythema (beet redness) to slight eschar formation (injuries in depth) and severe edema (raised more than 1 mm and extending beyond area of exposure)

b. Primary Eye Irritation. In experimental animals, a primary eye irritant is defined as a chemical that produces an irritant response in the test subject on first exposure. Eye irritation study procedures developed by Draize have been modified and adopted as a regulatory test by CPSC in 16 CFR 1500.42. In this procedure, a known amount of the test material (0.1 mL of a liquid or 100 mg of a solid or paste) is placed in one eye of each of six albino rabbits; the other eye remains untreated, serving as a control. The eyes are not washed after instillation and are examined at 24, 48, and 72 hours for ocular reaction. After the recording of ocular reaction at 24 hours, the eyes may be further examined following the application of fluorescein. The eyes may also be washed with a sodium chloride solution (U.S.P. or equivalent) after the 24-hour reaction has been recorded.

A test is scored positive if any of the following effects are observed: (1) ulceration (besides fine stippling); (2) opacity of the cornea (other than slight dulling of normal luster); (3) inflammation of the iris (other than a slight deepening of the rugae or circumcorneal injection of the blood vessel); (4) swelling of the conjunctiva (excluding the cornea and iris) with eversion of the eyelid; or (5) a diffuse crimson-red color with individual vessels not clearly identifiable. A material is an eye irritant if four of six rabbits score positive. It is considered a nonirritant if none or only one of six animals exhibits irritation. If intermediate results are obtained, the test is performed again. Materials producing any degree of irritation in the eye are identified as irritants. When an author has designated a substance as either a mild, moderate, or severe eye irritant, this designation is also reported.

The dose reported is based first on the lowest dose producing an irritant effect and second on the latest study published. Single and multiple applications are indicated as described above under "Primary Skin Irritation." Test times other than 72 hours are noted in the dose. All eye irritant test exposures are assumed to be continuous, unless the reference states that the eyes were washed after instillation. In this case, the notation "rns" (rinsed) is included in the data line.

Species Exposed. Since Draize procedures for determining both skin and eye irritation specify rabbits as the test species, most of the animal irritation data are for rabbits, although any of the species listed in Table 2 may be used. We have endeavored to include as much human data as possible, since this information is directly applicable to occupational exposure, much of which comes from studies conducted on volunteers (for example for cosmetic or soap ingredients) or from persons accidentally exposed. When accidental exposure, such as a spill, is cited, the line includes the abbreviation "nse" (nonstandard exposure). In these cases it is often very difficult to determine the precise amount of the material to which the individual was exposed. Therefore, for accidental exposures an estimate of the concentration or strength of the material, rather than a total dose amount, is generally provided.

12. *Mutation Data* lines include, in sequence, the mutation test system utilized, the species of the tested organ-

ism (and where applicable, the route of administration or cell type), the exposure concentration or dose, and the reference from which the information was extracted.

A mutation is defined as any heritable change in genetic material. Unlike irritation, reproductive, tumorigenic, and toxic dose data which report the results of whole animal studies, mutation data also include studies on lower organisms such as bacteria, molds, yeasts, and insects, as well as in vitro mammalian cell cultures. Studies of plant mutagenesis are not included. No attempt is made to evaluate the significance of the data or to rate the relative potency of the compound as a mutagenic risk to man.

Each element of the mutation line is discussed below.

a. *Mutation Test System.* A number of test systems are used to detect genetic alterations caused by chemicals. Additional ones may be added as they are reported in the literature. Each test system is identified by the 3-letter code shown in parentheses. For additional information about mutation tests, the reader may wish to consult the *Handbook of Mutagenicity Test Procedures,* edited by B. J. Kilbey, M. Legator, W. Nichols, and C. Ramel (Amsterdam: Elsevier Scientific Publishing Company/North-Holland Biomedical Press, 1977).

(1) The Mutation in Microorganisms (mmo) System utilizes the detection of heritable genetic alterations in microorganisms which have been exposed directly to the chemical.

(2) The Microsomal Mutagenicity Assay (mma) System utilizes an in vitro technique which allows enzymatic activation of promutagens in the presence of an indicator organism in which induced mutation frequencies are determined.

(3) The Micronucleus Test (mnt) System utilizes the fact that chromosomes or chromosome fragments may not be incorporated into one or the other of the daughter nuclei during cell division.

(4) The Specific Locus Test (slt) System utilizes a method for detecting and measuring rates of mutation at any or all of several recessive loci.

(5) The DNA Damage (dnd) System detects the damage to DNA strands including strand breaks, crosslinks, and other abnormalities.

(6) The DNA Repair (dnr) System utilizes methods of monitoring DNA repair as a function of induced genetic damage.

(7) The Unscheduled DNA Synthesis (dns) System detects the synthesis of DNA during usually nonsynthetic phases.

(8) The DNA Inhibition (dni) System detects damage that inhibits the synthesis of DNA.

(9) The Gene Conversion and Mitotic Recombination (mrc) System utilizes unequal recovery of genetic markers in the region of the exchange during genetic recombination.

(10) The Cytogenetic Analysis (cyt) System utilizes cultured cells or cell lines to assay for chromosomal aberrations following the administration of the chemical.

(11) The Sister Chromatid Exchange (sce) System detects the interchange of DNA in cytological preparations of metaphase chromosomes between replication products at apparently homologous loci.

(12) The Sex Chromosome Loss and Nondisjunction (sln) System measures the nonseparation of homologous chromosomes at meiosis and mitosis.

(13) The Dominant Lethal Test (dlt). A dominant lethal is a genetic change in a gamete that kills the zygote produced by that gamete. In mammals, the dominant lethal test measures the reduction of litter size by examining the uterus and noting the number of surviving and dead implants.

(14) The Mutation in Mammalian Somatic Cells (msc) System utilizes the induction and isolation of mutants in cultured mammalian cells by identification of the gene change.

(15) The Host-Mediated Assay (hma) System uses two separate species, generally mammalian and bacterial, to detect heritable genetic alteration caused by metabolic conversion of chemical substances administered to host mammalian species in the bacterial indicator species.

(16) The Sperm Morphology (spm) System measures the departure from normal in the appearance of sperm.

(17) The Heritable Translocation Test (trn) measures the transmissibility of induced translocations to subsequent generations. In mammals, the test uses sterility and reduced fertility in the progeny of the treated parent. In addition, cytological analysis of the F1 progeny or subsequent progeny of the treated parent is carried out to prove the existence of the induced translocation. In Drosophila, heritable translocations are detected genetically using easily distinguishable phenotypic markers, and these translocations can be verified with cytogenetic techniques.

(18) The Oncogenic Transformation (otr) System utilizes morphological criteria to detect cytological differences between normal and transformed tumorigenic cells.

(19) The Phage Inhibition Capacity (pic) System utilizes a lysogenic virus to detect a change in the genetic characteristics by the transformation of the virus from noninfectious to infectious.

(20) The Body Fluid Assay (bfa) System uses two separate species, usually mammalian and bacterial. The test substance is first administered to the host, from whom body fluid (e.g., urine, blood) is subsequently taken. This body fluid is then tested in vitro, and mutations are measured in the bacterial species.

b. *Species.* Those test species that are peculiar to mutation data are designated by the 3-letter codes shown below.

	Code	Species
Bacteria	bcs	Bacillus subtilis
	esc	Escherichia coli
	hmi	Haemophilus influenzae
	klp	Klebsiella pneumoniae
	sat	Salmonella typhimurium
	srm	Serratia marcescens
Molds	asn	Aspergillus nidulans
	nsc	Neurospora crassa
Yeasts	smc	Saccharomyces cerevisiae
	ssp	Schizosaccharomyces pombe
Protozoa	clr	Chlamydomonas reinhardi
	eug	Euglena gracilis
	omi	other microorganisms
Insects	dmg	Drosophila melanogaster
	dpo	Drosophila pseudo-obscura
	grh	grasshopper
	slw	silkworm
	oin	other insects
Fish	sal	salmon
	ofs	other fish

If the test organism is a cell type from a mammalian species, the parent mammalian species is reported, followed by a colon and the cell type designation. For example, human leukocytes are coded "hmn:leu." The various cell types currently cited in this edition are as follows:

Designation	Cell Type
ast	Ascites tumor
bmr	bone marrow
emb	embryo
fbr	fibroblast
hla	HeLa cell
kdy	kidney
leu	leukocyte
lng	lung
lvr	liver
lym	lymphocyte
mmr	mammary gland
ovr	ovary
spr	sperm
tes	testis
oth	other cell types not listed above

In the case of host-mediated and body fluid assays, both the host organism and the indicator organism are given as follows: host organism/indicator organism, e.g., "ham/sat" for a test in which hamsters were exposed to the test chemical and *S. typhimurium* was used as the indicator organism.

For in vivo mutagenic studies, the route of administration is specified following the species designation, e.g., "mus-orl" for oral administration to mice. See Table 1 for a complete list of routes cited. The route of administration is not specified for in vitro data.

c. Units of Exposure. The lowest dose producing a positive effect is cited. The author's calculations are used to determine the lowest dose at which a positive effect was observed. If the author fails to state the lowest effective dose, two times the control dose will be used. Ideally, the dose should be reported in universally accepted toxicological units such as milligrams of test chemical per kilogram of test animal body weight. While this is possible in cases where the actual intake of the chemical by an organism of known weight is reported, it is not possible in many systems using insect and bacterial species. In cases where a dose is reported or where the amount can be converted to a dose unit, it is normally listed as milligrams per kilogram (mg/kg). However, micrograms (μg), nanograms (ng), or picograms (pg) per kilogram may also be used for convenience of presentation. Concentrations of gaseous materials in air are listed as parts per hundred (pph), per million (ppm), per billion (ppb), or per trillion (ppt).

Test systems using microbial organisms traditionally report exposure data as an amount of chemical per liter (L) or amount per plate, well, disc, or tube. The amount may be on a weight (g, mg, μg, ng, or pg) or molar (millimole (mmol), micromole (μmol), nanomole (nmol), or picomole (pmol)) basis. These units describe the exposure concentration rather than the dose actually taken up by the test species. Insufficient data currently exist to permit the development of dose amounts from this information. In such cases, therefore, the material concentration units used by the author are reported.

Since the exposure values reported in host-mediated and body fluid assays are doses delivered to the host organism, no attempt is made to estimate the exposure concentration to the indicator organism. The exposure values cited for host-mediated assay data are in units of milligrams (or other appropriate unit of weight) of material administered per kilogram of host body weight or in parts of vapor or gas per million (ppm) parts of air (or other appropriate concentrations) by volume.

13. *Toxicity Dose Data* lines include, in sequence, the route of exposure; the species of animal studied; the toxicity measure; the amount of material per body weight or concentration per unit of air volume and, where applicable, the duration of exposure; a descriptive notation of the type of effect reported; and the reference from which the information was extracted. Only positive toxicity test results are cited in this section.

All toxic dose data appearing in the book are derived from reports of the toxic effects produced by individual materials. For human data, a toxic effect is defined as any reversible or irreversible noxious effect on the body, any benign or malignant tumor, any teratogenic effect, or any death that has been reported to have resulted from exposure to a material via any route. For humans, a toxic effect is any effect that was reported in the source reference. There is no qualifying limitation on the duration of exposure or for the quantity or concentration of the material, nor is there a qualifying limitation on the circumstances that resulted from the exposure. Regardless of the absurdity of the circumstances that were involved in a toxic exposure, it is assumed that the same circumstances could recur. For animal data, toxic effects are limited to the production of tumors, benign (neoplastigenesis) or malignant (carcinogenesis); the production of changes in the offspring resulting from action on the fetus directly (teratogenesis); and death. There is no limitation on either the duration of exposure or on the quantity or concentration of the dose of the material reported to have caused these effects.

The report of the lowest total dose administered over the shortest time to produce the toxic effect was given preference, although some editorial liberty was taken so that additional references might be cited. No restrictions were placed on the amount of a material producing death in an experimental animal nor on the time period over which the dose was given.

Each element of the toxic dose line is discussed below:

a. Route of Exposure or Administration. Although many exposures to materials in the industrial community occur via the respiratory tract or skin, most studies in the published literature report exposures of experimental animals in which the test materials were introduced primarily through the mouth by pills, in food, in drinking water, or by intubation directly into the stomach. The abbreviations and definitions of the various routes of exposure reported are given in Table 1.

TABLE 1.   Routes of Administration to, or Exposure of, Animal Species to Toxic Substances

Abbreviation	Route	Definition
Eyes	eye	Administration directly onto the surface of the eye. Used exclusively for primary irritation data. See Ocular
Intraaural	ial	Administration into the ear
Intraarterial	iat	Administration into the artery
Intracerebral	ice	Administration into the cerebrum
Intracervical	icv	Administration into the cervix
Intradermal	idr	Administration within the dermis by hypodermic needle
Intraduodenal	idu	Administration into the duodenum
Inhalation	ihl	Inhalation in chamber, by cannulation, or through mask
Implant	imp	Placed surgically within the body location described in reference
Intramuscular	ims	Administration into the muscle by hypodermic needle
Intraplacental	ipc	Administration into the placenta
Intrapleural	ipl	Administration into the pleural cavity by hypodermic needle
Intraperitoneal	ipr	Administration into the peritoneal cavity
Intrarenal	irn	Administration into the kidney
Intraspinal	isp	Administration into the spinal canal
Intratracheal	itr	Administration into the trachea
Intratesticular	itt	Administration into the testes
Intrauterine	iut	Administration into the uterus
Intravaginal	ivg	Administration into the vagina
Intravenous	ivn	Administration directly into the vein by hypodermic needle
Multiple	mul	Administration into a single animal by more than one route
Ocular	ocu	Administration directly onto the surface of the eye or into the conjunctival sac. Used exclusively for systemic toxicity data
Oral	orl	Per os, intragastric, feeding, or introduction with drinking water
Parenteral	par	Administration into the body through the skin. Reference cited is not specific concerning the route used. Could be ipr, scu, ivn, ipl, ims, irn, or ice
Rectal	rec	Administration into the rectum or colon in the form of enema or suppository
Subcutaneous	scu	Administration under the skin
Skin	skn	Application directly onto the skin, either intact or abraded. Used for both systemic toxicity and primary irritant effects
Unreported	unr	Dose, but not route, is specified in the reference

b. Species Exposed. Since the effects of exposure of humans are of primary concern, we have indicated, when available, whether the results were observed in man, woman, child, or infant. If no such distinction was made in the reference, the abbreviation "hmn" (human) is used. However, the results of studies on rats or mice are the most frequently reported and hence provide the most useful data for comparative purposes. The species and abbreviations used in reporting toxic dose data are listed alphabetically in Table 2.

c. Description of Exposure. In order to better describe the administered dose reported in the literature, six abbreviations are used. These terms indicate whether the dose caused death (LD) or other toxic effects (TD) and whether it was administered as a lethal concentration (LC) or toxic concentration (TC) in the inhaled air. In general, the term "Lo" is used where the number of subjects studied was not a significant number from the population or the calculated percentage of subjects showing an effect was listed as 100. The definition of terms is as follows:

TDLo - Toxic Dose Low - the lowest dose of a material introduced by any route, other than inhalation, over any

TABLE 2. Species
(With Assumptions For Toxic Dose Calculation From Non-Specific Data*)

Species	Abbrev.	Approximate Age	Approximate Weight	Consumption (Approx.) Food gm/day	Consumption (Approx.) Water ml/day	1 ppm in Food Equals, in mg/kg/day	Gestation Period days
Bird—type not specified	brd		1 kg				
Bird—wild bird species	bwd		40 gm				
Cat, adult	cat		2 kg	100	100	0.05	64 (59-68)
Child	chd	1-13 Y	20 kg				
Chicken, adult	ckn	8 W	800 gm	140	200	0.175	
Cattle	ctl		500 kg	10,000		0.02	284 (279-290)
Duck, adult (domestic)	dck	8 W	2.5 kg	250	500	0.1	
Dog, adult	dog	52 W	10 kg	250	500	0.025	62 (56-68)
Domestic animals (Goat, Sheep)	dom		60 kg	2,400		0.04	G: 152 (148-156) S: 146 (144-147)
Frog, adult	frg		33 gm				
Guinea Pig, adult	gpg		500 gm	30	85	0.06	68
Gerbil	grb		100 gm	5	5	0.05	25 (24-26)
Hamster	ham	14 W	125 gm	15	10	0.12	16 (16-17)
Human	hmn	Adult	70 kg				
Horse, Donkey	hor		500 kg	10,000		0.02	H: 339 (333-345) D: 365
Infant	inf	0-1 Y	5 kg				
Mammal (species un-specified in reference)	mam		200 gm				
Man	man	Adult	70 kg				
Monkey	mky	2.5 Y	5 kg	250	500	0.05	165
Mouse	mus	8 W	25 gm	3	5	0.12	21
Non-mammalian species	nml						
Pigeon	pgn	8 W	500 gm				
Pig	pig		60 kg	2,400		0.041	114 (112-115)
Quail (laboratory)	qal		100 gm				
Rat, adult female	rat	14 W	200 gm	10	20	0.05	22
Rat, adult male	rat	14 W	250 gm	15	25	0.06	
Rat, adult	rat	14 W	200 gm	15	25		
Rat, weanling	rat	3 W	50 gm	15	25	0.3	
Rabbit, adult	rbt	12 W	2 kg	60	330	0.03	31
Squirrel	sql		500 gm		44		
Toad	tod		100 gm				
Turkey	trk	18 W	5 kg				
Woman	wmn	Adult	50 kg		270		

*Values given above in Table 2 are within reasonable limits usually found in the published literature and are selected to facilitate calculations for data from publications in which toxic dose information has not been presented for an individual animal of the study. See, for example, Association of Food and Drug Officials, Quarterly Bulletin, volume 18, page 66, 1954; Guyton, American Journal of Physiology, volume 150, page 75, 1947; The Merck Veterinary Manual, 5th Edition, Merck & Co., Inc., Rahway, N.J., 1979; and The UFAW Handbook on the Care and Management of Laboratory Animals, 4th Edition, Churchill Livingston, London, 1972. Data for lifetime exposure are calculated from the assumptions for adult animals for the entire period of exposure. For definitive dose data, the reader must review the referenced publication.

given period of time and reported to produce any toxic effect in humans or to produce carcinogenic, neoplastigenic, or teratogenic effects in animals or humans.

TCLo - Toxic Concentration Low - the lowest concentration of a material in air to which humans or animals have been exposed for any given period of time that has produced any toxic effect in humans or produced a carcinogenic, neoplastigenic, or teratogenic effect in animals or humans.

LDLo - Lethal Dose Low - the lowest dose (other than LD50) of a material introduced by any route, other than inhalation, over any given period of time in one or more divided portions and reported to have caused death in humans or animals.

LD50 - Lethal Dose Fifty - a calculated dose of a material which is expected to cause the death of 50% of an entire defined experimental animal population. It is determined from the exposure to the material by any route other than inhalation of a significant number from that population. Other lethal dose percentages, such as LD1, LD10, LD30, and LD99, may be published in the scientific literature for the specific purposes of the author. Such data would be published if these figures, in the absence of a calculated lethal dose (LD50), were the lowest found in the literature.

LCLo - Lethal Concentration Low - the lowest concentration of a material in air, other than LC50, which has been reported to have caused death in humans or animals. The reported concentrations may be entered for periods of exposure which are less than 24 hours (acute) or greater than 24 hours (subacute and chronic).

LC50 - Lethal Concentration Fifty - a calculated concentration of a material in air, exposure to which for a specified length of time is expected to cause the death of 50% of an entire defined experimental animal population. It is determined from the exposure to the material of a significant number from that population.

The following table summarizes the above information:

Category	Exposure Time	Route of Exposure	TOXIC EFFECTS Human	TOXIC EFFECTS Animal
TDLo	Acute or Chronic	All except Inhalation	Any Non-Lethal	CAR, NEO, NEO, ETA, TER, REP
TCLo	Acute or Chronic	Inhalation	Any Non-Lethal	CAR, NEO, ETA, ETA, TER, REP
LDLo	Acute or Chronic	All except Inhalation	Death	Death
LD50	Acute	All except Inhalation	Not Applicable	Death (Statistically Determined)
LCLo	Acute or Chronic	Inhalation	Death	Death
LC50	Acute	Inhalation	Not Applicable	Death (Statistically Determined)

d. Units of Dose Measurement. As in almost all experimental toxicology, the doses given are expressed in terms of the quantity administered per unit body weight, or quantity per skin surface area, or quantity per unit volume of the respired air. In addition, the duration of time over which the dose was administered is also listed, as needed. Dose amounts are generally expressed as milligrams (one thousandth of a gram) per kilogram (mg/kg). In some cases, because of dose size and its practical presentation in the file, grams per kilogram (g/kg), micrograms (one millionth of a gram) per kilogram (μg/kg), or nanograms (one billionth of a gram) per kilogram (ng/kg) are used. Volume measurements of dose were converted to weight units by appropriate calculations. Densities were obtained from standard reference texts. Where densities were not readily available, all liquids were assumed to have a density of one gram per milliliter. Twenty drops of liquid are assumed to be equal in volume to one milliliter.

All body weights have been converted to kilograms (kg) for uniformity. For those references in which the dose was reported to have been administered to an animal of unspecified weight or a given number of animals in a group (e.g., feeding studies) without weight data, the weights of the respective animal species were assumed to be those listed in Table 2 and the dose is listed on a per kilogram body weight basis. Assumptions for daily food and water intake are found in Table 2 to allow approximating doses for humans and species of experimental animals where the dose was originally reported as a concentration in food or water. The values presented are selections which are reasonable for the species and convenient for dose calculations.

Concentrations of a gaseous material in air are generally listed as parts of vapor or gas per million parts of air by volume (ppm). However, parts per hundred (pph or per cent), parts per billion (ppb), or parts per trillion (ppt) may be used for convenience of presentation. If the material is a solid or a liquid, the concentrations are listed preferably as milligrams per cubic meter (mg/m^3) but may, as applicable, be listed as micrograms per cubic meter (μg/m^3), nanograms per cubic meter (ng/m^3), or picograms (one trillionth of a gram) per cubic meter (pg/m^3) of air. For those cases in which other measurements of contaminants are used, such as the number of fibers or particles, the measurement is spelled out.

Where the duration of exposure is available, time is presented as minutes (M), hours (H), days (D), weeks (W), or years (Y). Additionally, continuous (C) indicates that the exposure was continuous over the time administered, such as ad libitum feeding studies or 24-hour, 7-day per week inhalation exposures. Intermittent (I) indicates that the dose was administered during discrete periods, such as daily, twice weekly, etc. In all cases, the total duration of exposure appears first after the kilogram body weight and slash, followed by descriptive data; e.g., 10 mg/kg/3W-I indicates ten milligrams per kilogram body weight administered over a period of three weeks, intermittently in a number of separate, discrete doses. This description is intended to provide the reader with enough information for an approximation of the experimental conditions, which can be further clarified by studying the reference cited.

e. Frequency of Exposure. Frequency of exposure to the test material varies depending on the nature of the experiment. Frequency of exposure is given in the case of an inhalation experiment, for human exposures (where applicable), or where CAR, NEO, ETA, REP, or TER is specified as the toxic effect.

f. Duration of Exposure. For assessment of tumorigenic effect, the testing period should be the life span of the animal, or until statistically valid calculations can be obtained regarding tumor incidence. In the toxic dose line, the total dose causing the tumorigenic effect is given. The duration of exposure is included to give an indication of the testing period during which the animal was exposed to this total dose. For multigenerational studies, the time during gestation when the material was administered to the mother is also provided.

g. Notations Descriptive of the Toxicology. The toxic dose line thus far has indicated the route of entry, the species involved, the description of the dose, and the amount of the dose. The next entry found on this line when a toxic exposure (TD or TC) has been listed is the toxic effect. Following a colon will be one of the notations found in Table 3. These notations indicate the

TABLE 3. Notations Descriptive of the Toxicology

Abbreviation	Definition (not limited to effects listed)
ALR	Allergic systemic reaction such as might be experienced by individuals sensitized to penicillin.
BCM	Blood clotting mechanism effects—any effect which increases or decreases clotting time.
BLD	Blood effects—effect on all blood elements, electrolytes, pH, protein, oxygen carrying or releasing capacity.
BPR	Blood pressure effects—any effect which increases or decreases any aspect of blood pressure.
CAR	Carcinogenic effects—see paragraph 9g in text.
CNS	Central nervous system effects—includes effects such as headaches, tremor, drowsiness, convulsions, hypnosis, anesthesia.
COR	Corrosive effects—burns, desquamation.
CUM	Cumulative effects—where material is retained by the body in greater quantities than is excreted, or the effect is increased in severity by repeated body insult.
CVS	Cardiovascular effects—such as an increase or decrease in the heart activity through effect on ventricle or auricle; fibrillation; constriction or dilation of the arterial or venous system.
DDP	Drug dependence effects—any indication of addiction or dependence.
ETA	Equivocal tumorigenic agent—see text.
EYE	Eye effects—irritation, diploplia, cataracts, eye ground, blindness by affecting the eye or the optic nerve.
GIT	Gastrointestinal tract effects—diarrhea, constipation, ulceration.
GLN	Glandular effects—any effect on the endocrine glandular system.
IRR	Irritant effects—any irritant effect on the skin, eye, or mucous membrane.
MLD	Mild irritation effects—used exclusively for primary irritation data.
MMI	Mucous membrane effects—irritation, hyperplasia, changes in ciliary activity.
MOD	Moderate irritation effects—used exclusively for primary irritation data.
MSK	Musculo-skeletal effects—such as osteoporosis, muscular degeneration.
NEO	Neoplastic effects—see text.
PNS	Peripheral nervous system effects.
PSY	Psychotropic effects—exerting an effect upon the mind.
PUL	Pulmonary system effects—effects on respiration and respiratory pathology.
RBC	Red blood cell effects—includes the several anemias.
REP	Reproductive effects—see text.
SEV	Severe irritation effects—used exclusively for primary irritation data.
SKN	Skin effects—such as erythema, rash, sensitization of skin, petechial hemorrhage.
SYS	Systemic effects—effects on the metabolic and excretory function of the liver or kidneys.
TER	Teratogenic effects—nontransmissible changes produced in the offspring.
UNS	Unspecified effects—the toxic effects were unspecific in the reference.
WBC	White blood cell effects—effects on any of the cellular units other than erythrocytes, including any change in number or form.

organ system affected or special effects that the material produced, e.g., TER = teratogenic effect. No attempt was made to be definitive in reporting these effects because such definition requires much detailed qualification and is beyond the scope of this book. The selection of the dose was based first on the lowest dose producing an effect and second on the latest study published.

14. *Reproductive Effects Data* lines include, in sequence, the reproductive effect reported, the route of exposure, the species of animal tested, the type of dose, the total dose amount administered, the time and duration of administration, and the reference from which the information was extracted. Only positive reproductive effects data for mammalian species are cited. Because of differences in the reproductive systems among species and the systems' varying responses to chemical exposures, no attempt is made to extrapolate animal data or to evaluate the significance of a substance as a reproductive risk to humans.

Each element of the reproductive effects data line is discussed below.

a. Reproductive Effect. For human exposure, the effects are included in the safety profile. The effects include those reported to effect the male or female reproductive systems, mating and conception success, fetal effects including abortion, transplacental carcinogenesis, and post birth effects on parents and offspring.

b. Route of Exposure or Administration. See Table 1 for a complete list of abbreviations and definitions of the various routes of exposure reported. For reproductive effects data, the specific route is listed either when the substance was administered to only one of the parents or when the substance was administered to both parents by the same route. However, if the substance was administered to each parent by a different route, the route is indicated as "mul" (multiple).

c. Species Exposed. Reproductive effects data are cited for mammalian species only. Species abbreviations are shown in Table 2. Also shown in Table 2 are approximate gestation periods.

d. Type of Exposure. Only two types of exposure, TDLo and TCLo, are used to describe the dose amounts reported for reproductive effects data.

e. Dose Amounts and Units. The total dose amount that was administered to the exposed parent is given. If the substance was administered to both parents, the individual amounts to each parent have been added together and the total amount shown. Where necessary, appropriate conversion of dose units has been made. The dose amounts listed are those for which the reported effects are statistically significant. However, human case reports are cited even when no statistical tests can be performed. The statistical test is that used by the author. If no statistic is reported, a Fisher's Exact Test is applied with significance at the 0.05 level, unless the author makes a strong case for significance at some other level.

Dose units are usually given as an amount administered per unit body weight or as parts of vapor or gas per million parts of air by volume. There is no limitation on either the quantity or concentration of the dose or the duration of exposure reported to have caused the reproductive effect(s).

f. Time and Duration of Treatment. The time when a substance is administered to either or both parents may significantly affect the results of a reproductive study, because there are differing critical periods during the reproductive cycles of each species. Therefore, to provide some indication of when the substance was administered, which should facilitate selection of specific data for analysis by the user, a series of up to four terms follows the dose amount. These terms indicate to which parent(s) and at what time the substance was administered. The terms take the general form:

(uD male/vD pre/w-xD preg/yD post)

where u = total number of days of administration to male prior to mating

v = total number of days of administration to female prior to mating

w = first day of administration to pregnant female during gestation

x = last day of administration to pregnant female during gestation

y = total number of days of administration to lactating mother after birth of offspring

If administration is to the male only, then only the first of the above four terms is shown following the total dose to the male, e.g., 10 mg/kg (5D male). If administration is to the female only, then only the second, third, or fourth term, or any combination thereof, is shown following the total dose to the female, e.g.,

10 mg/kg (3D pre)

10 mg/kg (3D pre/4-7D preg)

10 mg/kg (3D pre/4-7D preg/5D post)

10 mg/kg (3D pre/5D post)

10 mg/kg (4-7D preg)

10 mg/kg (4-7D preg/5D post)

10 mg/kg (5D post) (NOTE: This example indicates administration to the lactating mother only after birth of the offspring.)

If administration is to both parents, then the first term and any combination of the last three terms are listed, e.g., 10 mg/kg (5D male/3D pre/4-7D preg). If administration is continuous through two or more of the above periods, the above format is abbreviated by replacing the slash (/) with a dash (-). For example, 10 mg/kg (3D pre-5D post) indicates a total of 10 mg/kg administered to the female for three days prior to mating, on each day during gestation, and for five days following birth. Approximate gestation periods for various species are shown in Table 2.

g. Multigeneration Studies. Some reproductive studies entail administration of a substance to several consecutive generations, with the reproductive effects measured in the final generation. The protocols for such studies vary widely. Therefore, because of the inherent complexity and variability of these studies, they are cited in a simplified format as follows. The specific route of administration is reported if it was the same for all parents of all generations; otherwise the abbreviation "mul" is used. The total dose amount shown is that administered to the F0 generation only; doses to the Fn (where n = 1, 2, 3, etc.) generations are not reported. The time and duration of treatment for multigeneration studies are not included in the data line. Instead, the dose amount is followed by the abbreviation "(MGN)", e.g., 10 mg/kg (MGN). This code indicates a multigeneration study, and the reader must consult the cited reference for complete details of the study protocol.

15. Carcinogenic Study Result. Tumorigenic citations are classified according to the reported results of the study to aid the reader in selecting appropriate references for in-depth review and evaluation. The classification, ETA (equivocal tumorigenic agent), denotes those studies reporting uncertain, but seemingly positive, results. The criteria for the three classifications are listed below. These criteria are used to abstract the data in individual reports on a consistent basis and do not represent a comprehensive evaluation of a material's tumorigenic potential to humans.

The following nine technical criteria are used to abstract the toxicological literature and classify studies that report positive tumorigenic responses. No attempts are made either to evaluate the various test procedures or to correlate results from different experiments.

(1) A citation is coded "CAR" (carcinogenic) when review of an article reveals that all the following criteria are satisfied:

(a) A statistically significant increase in the incidence of tumors in the test animals. The statistical test is that used by the author. If no statistic is reported, a Fisher's Exact Test is applied with significance at the 0.05 level, unless the author makes a strong case for significance at some other level.

(b) A control group of animals is used and the treated and control animals are maintained under identical conditions.

(c) The sole experimental variable between the groups is the administration or nonadministration of the test material (see (9) below).

(d) The tumors consist of autonomous populations of cells of abnormal cytology capable of invading and destroying normal tissues, or the tumors metastasize as confirmed by histopathology.

(2) A citation is coded "NEO" (neoplastic) when review of an article reveals that all the following criteria are satisfied:

(a) A statistically significant increase in the incidence of tumors in the test animals. The statistical test is that used by the author. If no statistic is reported, a Fisher's Exact Test is applied with significance at the 0.05 level, unless the author makes a strong case for significance at some other level.

(b) A control group of animals is used, and the treated and control animals are maintained under identical conditions.

(c) The sole experimental variable between the groups is the administration or nonadministration of the test material (see (9) below).

(d) The tumors consist of cells that closely resemble the tissue of origin, that are not grossly abnormal cytologically, that may compress surrounding tissues, but that neither invade tissues nor metastasize; or

(e) The tumors produced cannot be classified as either benign or malignant.

(3) A citation is coded "ETA" (equivocal tumorigenic agent) when some evidence of tumorigenic activity is presented, but one or more of the criteria listed in (1) or (2) above is lacking. Thus, a report with positive pathological findings, but with no mention of control animals, is coded "ETA."

(4) Since an author may make statements or draw conclusions based on a larger context than that of the particular data reported, papers in which the author's conclusions differ substantially from the evidence presented in the paper are subject to review.

(5) All doses except those for transplacental carcinogenesis are reported in one of the following formats.

(a) For all routes of administration other than inhalation: cumulative dose in mg (or appropriate unit)/kg/duration of administration.

Whenever the dose reported in the reference is not in the units discussed herein, conversion to this format is made. The total cumulative dose is derived from the lowest dose level that produces tumors in the test group.

(b) For inhalation experiments: concentration in ppm (or mg/m^3)/total duration of exposure.

The concentration refers to the lowest concentration that produces tumors.

(6) Transplacental carcinogenic doses are reported in one of the following formats.

(a) For all routes of administration other than inhalation: cumulative dose in mg/kg/(time of administration during pregnancy).

The cumulative dose is derived from the lowest single dose that produces tumors in the offspring. The test chemical is administered to the mother.

(b) For inhalation experiments: concentration in ppm (or mg/m^3)/(time of exposure during pregnancy).

The concentration refers to the lowest concentration that produces tumors in the offspring. The mother is exposed to the test chemical.

(7) For the purposes of this listing, all test chemicals

are reported as pure, unless otherwise stated by the author. This does not rule out the possibility that unknown impurities may have been present.

(8) A mixture of compounds whose test results satisfy the criteria in (1), (2), or (3) above is included if the composition of the mixture can be clearly defined.

(9) For tests involving promoters or initiators, a study is included if the following conditions are satisfied (in addition to the criteria in (1), (2), or (3) above):

(a) The test chemical is applied first, followed by an application of a standard promoter. A positive control group in which the test animals are subjected to the same standard promoter under identical conditions is maintained throughout the duration of the experiment. The data are only used if positive and negative control groups are mentioned in the reference.

(b) A known carcinogen is first applied as an initiator, followed by application of the text chemical as a promoter. A positive control group in which the test animals are subjected to the same initiator under identical conditions is maintained throughout the duration of the experiment. The data are only used if positive and negative control groups are mentioned in the reference.

16. *Cited Reference* is the final entry of the irritation, mutation, reproductive, tumorigenic, and toxic dose data lines. This is the source from which the information was extracted. All references cited are publicly available. No governmental classified documents have been used for source information. All references have been given a unique six-letter CODEN character code (derived from the American Society for Testing and Materials *CODEN for Periodical Titles* and the CAS *Source Index*), which identifies periodicals, serial publications, and individual published works. For those references for which no CODEN was found, the corresponding six-letter code includes asterisks (*) in the last one or two positions following the first four or five letters of an acronym for the publication title. Following the CODEN designation (for most entries) is the number of the volume, followed by a comma; the page number of the first page of the article, followed by a comma; and a two-digit number, indicating the year of publication in this century. When the cited reference is a report, the report number is listed. Where contributors have provided information on their unpublished studies, the CODEN consists of the first three letters of the last name, the initials of the first and middle names, and a number sign (#). The date of the letter supplying the information is listed. All CODEN acronyms are listed in alphabetical order and defined in the CODEN Section.

17. *Consensus Reports* lines supply additional information to enable the reader to make knowledgeable evaluations of potential chemical hazards. Two types of reviews are listed: (a) International Agency for Research on Cancer (IARC) monograph reviews, which are published by the United Nations World Health Organization (WHO), and the National Toxicology Program (NTP).

a. Cancer Reviews. In the U.N. International Agency for Research on Cancer (IARC) monographs, information on suspected environmental carcinogens is examined, and summaries of available data with appropriate references are presented. Included in these reviews are synonyms, physical and chemical properties, uses and occurrence, and biological data relevant to the evaluation of carcinogenic risk to humans. The over 43 monographs in the series contain an evaluation of approximately 900 materials. Single copies of the individual monographs (specify volume number) can be ordered from WHO Publications Centre USA, 49 Sheridan Avenue, Albany, New York 12210, telephone (518) 436-9686.

The format of the IARC data line is as follows. The entry "IARC Cancer Review:" indicates that the carcinogenicity data pertaining to a compound has been reviewed by the IARC committee. The committee's conclusion are summarized in three words. The first word indicates whether the data pertains to humans or to animals. The next two words indicate the degree of carcinogenic risk as defined by IARC.

For experimental animals the evidence of carcinogenicity is assessed by IARC and judged to fall into one of four groups defined as follows:

(1) Sufficient Evidence of carcinogenicity is provided when there is an increased incidence of malignant tumors: (a) in multiple species or strains; or (b) in multiple experiments (preferably with different routes of administration or using different dose levels); or (c) to an unusual degree with regard to the incidence, site, or type of tumor, or age at onset. Additional evidence may be provided by data on dose-response effects.

(2) Limited Evidence of carcinogenicity is available when the data suggest a carcinogenic effect but are limited because: (a) the studies involve a single species, strain, or experiment; (b) the experiments are restricted by inadequate dosage levels, inadequate duration of exposure to the agent, inadequate period of follow-up, poor survival, too few animals, or inadequate reporting; or (c) the neoplasms produced often occur spontaneously and, in the past, have been difficult to classify as malignant by histological criteria alone (e.g., lung adenomas and adenocarcinomas, and liver tumors in certain strains of mice).

(3) Inadequate Evidence is available when, because of major qualitative or quantitative limitations, the studies cannot be interpreted as showing either the presence or absence of a carcinogenic effect.

(4) No Evidence applies when several adequate studies are available which show that within the limitations of the tests used, the chemical is not carcinogenic.

It should be noted that the categories Sufficient Evidence and Limited Evidence refer only to the strength of the experimental evidence that these chemicals are carcinogenic and not to the extent of their carcinogenic activity nor to the mechanism involved. The classification of any chemical may change as new information becomes available.

The evidence for carcinogenicity from studies in humans is assessed by the IARC committees and judged to fall into one of four groups defined as follows:

(1) Sufficient Evidence of carcinogenicity indicates that there is a causal relationship between the exposure and human cancer.

(2) Limited Evidence of carcinogenicity indicates that a causal relationship is credible, but that alternative explanations, such as chance, bias, or confounding, could not adequately be excluded.

(3) Inadequate Evidence, which applies to both positive and negative evidence, indicates that one of two conditions prevailed: (a) there are few pertinent data; or (b) the available studies, while showing evidence of association, do not exclude chance, bias, or confounding.

(4) No Evidence applies when several adequate studies are available which do not show evidence of carcinogenicity.

This cancer review reflects only the conclusion of the IARC committee based on the data available for the committee's evaluation. Hence, for some substances there may be a disagreement between the IARC determination and the information on the tumorigenic data lines (see paragraph 10). Also, some substances previously reviewed by IARC may be reexamined as additional data become available. These substances will contain multiple IARC review lines, each of which is referenced to the applicable IARC monograph volume.

An IARC entry indicates that some carcinogenicity data pertaining to a compound has been reviewed by the IARC committee. It indicates whether the data pertain to humans or to animals and whether the results of the determination are positive, suspected, indefinite, negative, or no data.

This cancer review reflects only the conclusion of the IARC Committee based on the data available at the time of the Committee's evaluation. Hence, for some materials there may be disagreement between the IARC determination and the tumorigenicity information in the toxicity data lines.

b. NTP Status. The notation "NTP Fifth Annual Report on Carcinogens" indicated that the entry is listed on the fifth report made to the U. S. Congress by the National Toxicology Program (NTP) as required by law. This listing implies that the entry is assumed to be a human carcinogen.

Another NTP notation indicates that the material has been tested by the NTP under its Carcinogenesis Testing Program. These entries are also identified as National Cancer Institute (NCI), which reported the studies before the NCI Carcinogenesis Testing Program was absorbed by NTP. To obtain additional information about NTP, the Carcinogenesis Testing Program, or the status of a particular material under test, contact the Toxicology Information and Scientific Evaluation Group, NTP/TRTP/NIEHS, Mail Drop 18-01, P.O. Box 12233, Research Triangle Park, NC 27709.

c. EPA Extremely Hazardous Substances List. This list was developed by the U.S. Environmental Protection Agency (EPA) as required by the Superfund Amendments and Reauthorization Act of 1986 (SARA). Title III Section 304 requires notification by facilities of a release of certain extremely hazardous substances. These 402 substances were listed by EPA in the *Federal Register* November 17, 1986.

d. Community Right to Know List. This list was developed by the EPA as required by the Superfund Amendments and Reauthorization Act of 1986 (SARA). Title III, Sections 311-312 require manufacturing facilities to prepare Material Safety Data Sheets and notify local authorities of the presence of listed chemicals. Both specific chemicals and classes of chemicals are covered by these sections.

e. EPA Genetic Toxicology Program (GENE-TOX). This status line indicates that the material has had genetic effects reported in the literature during the period 1969-1979. The test protocol in the literature is evaluated by an EPA expert panel on mutations and the positive or negative genetic effect of the substance is reported. To obtain additional information about this program, contact GENE-TOX program, USEPA, 401 M Street, SW, TS796, Washington, DC 20460, Telephone (202) 382-3513.

f. EPA TSCA Status Line. This line indicates that the material appears on the chemical inventory prepared by the Environmental Protection Agency in accordance with provisions of the Toxic Substances Control Act (TSCA). Materials reported in the inventory include those that are produced commercially in or imported into this country. The reader should note, however, that materials already regulated by EPA under FIFRA and by the Food and Drug Administration under the Food, Drug, and Cosmetic Act, as amended, are not included in the TSCA inventory. Similarly, alcohol, tobacco, and explosive materials are not regulated under TSCA. TSCA regulations should be consulted for an exact definition of reporting requirements. For additional information about TSCA, contact EPA, Office of Toxic Substances, Washington, D.C. 20402. Specific questions about the inventory can be directed to the EPA Office of Industry Assistance, telephone (800) 424-9065.

18. *Standards and Recommendations.* This section contains regulations by agencies of the United States Government or recommendations by expert groups.

"OSHA" refers to standards promulgated under Section 6 of the Occupational Safety and Health Act of 1970. "DOT" refers to materials regulated for shipment by the Department of Transportation. Because of frequent changes to and litigation of Federal regulations, it is recommended that the reader contact the applicable agency for information about the current standards for a particular material. Omission of a material or regulatory notation from this edition does not imply any relief from regulatory responsibility.

a. OSHA Air Contaminant Standards. The values given are for the revised standards which were published in January 13, 1989 and take effect on September 1, 1989 through December 31, 1992. These are noted with the entry "OSHA PEL:" followed by "TWA" or "CL" meaning either time-weighted average or ceiling value, respectively, to which workers can be exposed for a normal 8-hour day, 40-hour work week without ill effects. For some materials, TWA, CL, and Pk (peak) values are given in the standard. In those cases, all three are listed. Finally, some entries may be followed by the designation "(skin)." This designation indicates that the compound may be absorbed by the skin and, even though the air concentration may be below the standard, significant additional exposure through the skin may be possible.

b. ACGIH Threshold Limit Values. The American Conference of Governmental Industrial Hygienists (ACGIH) Threshold Limit Values are noted with the entry "ACGIH TLV:" followed by "TWA" or "CL" meaning either time-weighted average or ceiling value, respectively, to which workers can be exposed for a normal 8-hour day, 40-hour work week without ill effects. The notation "CL" indicates a ceiling limit which must not be exceeded. The notation "skin" indicates that the material penetrates intact skin, and skin contact should be avoided even though the TLV concentration is not exceeded. STEL indicates a short-term exposure limit, usually a 15-minute time-weighted average, which should not be exceeded. Biological Exposure Indices (*BEI:*) are, according to the ACGIH, set to provide a warning level ". . .of biological response to the chemical, or warning levels of that chemical or its metabolic product(s) in tissues, fluids, or exhaled air of exposed workers. . ."

The latest annual TLV list is contained in the publication *Threshold Limit Values and Biological Exposure Indices.* This publication should be consulted for future trends in recommendations. The ACGIH TLVs are adopted in whole or in part by many countries and local administrative agencies throughout the world. As a result, these recommendations have a major effect on the control of workplace contaminant concentrations. The ACGIH may be contacted for additional information at 6500 Glenway Ave., Cincinnati, Ohio 45211, USA.

c. DFG MAK. These lines contain the German Research Society's Maximum Allowable Concentration values. Those materials which are classified as to workplace hazard potential by the German Research Society are noted on this line. The MAK values are also revised annually and discussions of materials under consideration for MAK assignment are included in the annual publication together with the current values. *BAT:* indicates Biological Tolerance Value for a Working Material which is defined as, ". . .the maximum permissible quantity of a chemical compound, its metabolites, or any deviation from the norm of biological parameters induced by these substances in exposed humans." *TRK:* values are Technical Guiding Concentrations for workplace control of carcinogens. For additional information, write to Deutsche Forschungsgemeinschaft (German Research Society), Kennedyallee 40, D-5300 Bonn 2, Federal Republic of Germany. The publication *Maximum Concentrations at the Workplace and Biological Tolerance Values for Working Materials* can be obtained from Verlag Chemie GmbH, Buchauslieferung, P.O. Box 1260/1280, D-6940 Weinheim, Federal Republic of Germany, or Verlag Chemie, Deerfield Beach, Florida.

d. NIOSH REL. This line indicates that a NIOSH criteria document recommending a certain occupational exposure has been published for this compound or for a class of compounds to which this material belongs. These documents contain extensive data, analysis, and references. The more recent publications can be obtained from the National Institute for Occupational Safety and Health, U.S. Department of Health and Human Services, 4676 Columbia Pkwy., Cincinnati, Ohio 45226.

e. DOT Classification. This is the hazard classification according to the U.S. Department of Transportation (DOT) or the International Maritime Organization (IMO.) This classification gives an indication of the hazards expected in transportation, and serves as a guide to the development of proper labels, placards, and shipping instructions. The basic hazard classes include compressed gases, flammables, oxidizers, corrosives, explosives, radioactive materials, and poisons. Although a material may be designated by only one hazard class, additional hazards may be indicated by adding labels or by using other means as directed by DOT. Many materials are regulated under general headings such as "pesticides" or "combustible liquids" as defined in the regulations. These are not noted here as their specific concentration or properties must be known for proper classification. Special regulations may govern shipment by air. This information should serve *only as a guide*, since the regulation of transported materials is carefully controlled in most countries by federal and local agencies. Because of frequent changes to regulations, it is recommended that the reader contact the applicable agency for information about the current standards for a particular material. United States transportation regulations are found in 40 CFR, Parts 100 to 189. Contact the U.S. Department of Transportation, Materials Transportation Bureau, Washington, D.C. 20590.

19. *Safety Profiles* are text summaries of the reported hazards of the entry. The word "experimental" indicates that the reported effects resulted from a controlled exposure of laboratory animals to the substance. Toxic effects reported include carcinogenic, reproductive, acute lethal, human nonlethal effects, skin and eye irritation, and positive mutation study results.

Human effects are identified either by human or more specifically by man, woman, child, or infant. Specific symptoms or organ systems effects are reported when available.

Carcinogenicity potential is denoted by the words "confirmed," "suspected," or "questionable." The substance entries are grouped into three classes based on experimental evidence and the opinion of expert review groups. The OSHA, IARC, ACGIH, and DFG MAK decision schedules are not related or synchronized. Thus an entry may have had a recent review by only one group. The most stringent classification of any regulation or expert group is taken as governing.

Class I - Confirmed Carcinogens

These substances are capable of causing cancer in exposed humans. An entry was assigned to this class if it had one or more of the following data items present.

a. An OSHA regulated carcinogen.

b. AN ACGIH assignment as a human or animal carcinogen.

c. A DFG MAK assignment as a confirmed human or animal carcinogen.

d. An IARC assignment of human or animal sufficient evidence of carcinogenicity or higher.

e. NTP Fifth Annual Report on Carcinogens.

Class II - Suspected Carcinogens

These substances may be capable of causing cancer in exposed humans. The evidence is suggestive, but not sufficient to convince expert review committees. Some entries have not yet had expert review, but contain experimental reports of carcinogenic activity. In particular, an entry is included if it has positive reports of carcinogenic endpoint in two species. As more studies are published, many Class II carcinogens will have their carcinogenicity confirmed. On the other hand, some will be judged noncarcinogenic in the future. An entry was assigned to this class if it had one or more of the following data items present.

a. An ACGIH assignment of suspected carcinogen.

b. An DFG MAK assignment of suspected carcinogen.

c. An IARC assignment of human or animal limited evidence.

d. Two animal studies reported positive carcinogenic endpoint in different species.

Class III - Questionable Carcinogens

These entries have minimal published evidence of possible carcinogenic activity. The reported endpoint is often neoplastic growth with no spread or invasion characteristic of carcinogenic pathology. An even weaker endpoint is that of equivocal tumorigenic agent (ETA). Reports are assigned this designation when the study was defective. The report may have lacked control animals, may have used very small sample size, often lack complete pathology reporting, or suffer many other study design defects. Many of these were designed for other than carcinogenic evaluation, and the reported carcinogenic effect is a byproduct of the study, not the goal. The data are presented because some of these substances may be carcinogens. There is insufficient data to affirm or deny the possibility. An entry was assigned to this class if it had one or more of the following data items present.

a. An IARC assignment of inadequate or no evidence.

b. A single human report of carcinogenicity.

c. A single experimental carcinogenic report, or duplicate reports in the same species.

d. One or more experimental neoplastic or equivocal tumorigenic agent report.

Fire and explosion hazards are briefly summarized in terms of conditions of flammable or reactive hazard. Materials which are incompatible with the entry are listed here. Fire and explosion hazards are briefly summarized in terms of conditions of flammable or reactive hazard. Where feasible, fire-fighting materials and methods are discussed. A material with a flash point of 100°F or less is considered dangerous; if the flash point is from 100 to 200°F, the flammability is considered moderate; if it is above 200°F, the flammability is considered low (combustible).

Also included in the safety profile are disaster hazards comments which serve to alert users of materials, safety professionals, researchers, supervisors, and fire fighters to the dangers that may be encountered on entering storage premises during a fire or other emergency. Although the presence of water, steam, acid fumes, or powerful vibrations can cause the decomposition of many materials into dangerous compounds, we are particularly concerned with high temperatures (such as those resulting from a fire) since these can cause many otherwise mild chemicals to emit highly toxic gases or vapors such as NO_x, SO_x, acids, and so forth, or evolve vapors of antimony, arsenic, mercury, and the like.

Sax's Dangerous Properties of Industrial Materials

CAS Number Cross-Index

50-00-0 see FMV000
50-01-1 see GKY000
50-02-2 see SOW000
50-03-3 see HHQ800
50-04-4 see CNS825
50-06-6 see EOK000
50-07-7 see AHK500
50-09-9 see ERE000
50-10-2 see ORQ000
50-11-3 see DJO800
50-12-4 see MKB250
50-13-5 see DAM700
50-14-6 see VSZ100
50-18-0 see CQC650
50-19-1 see HNJ000
50-21-5 see LAG000
50-22-6 see CNS625
50-23-7 see CNS750
50-24-8 see PMA000
50-27-1 see EDU500
50-28-2 see EDO000
50-29-3 see DAD200
50-31-7 see TIK500
50-32-8 see BCS750
50-33-9 see BRF500
50-34-0 see HKR500
50-35-1 see TEH500
50-36-2 see CNE750
50-37-3 see DJO000
50-39-5 see HNY500
50-41-9 see CMX700
50-44-2 see POK000
50-47-5 see DSI709
50-48-6 see EAH500
50-49-7 see DLH600
50-50-0 see EDP000
50-52-2 see MOO250
50-53-3 see CKP250
50-54-4 see QHA000
50-55-5 see RDK000
50-56-6 see ORU500
50-59-9 see TEY000
50-60-2 see PDW400
50-62-4 see HFF500
50-63-5 see CLD250
50-65-7 see DFV400
50-67-9 see AJX500
50-70-4 see SKV200
50-71-5 see AFT750
50-76-0 see AEB000
50-78-2 see ADA725
50-79-3 see DER400
50-81-7 see ARN000
50-89-5 see TFX790
50-90-8 see CFJ750
50-91-9 see DAR400
50-98-6 see EAW500
50-98-6 see EAY000
50-99-7 see GFG000
51-02-5 see INT000
51-03-6 see PIX250
51-05-8 see AIT250
51-06-9 see AJN500
51-12-7 see BET000
51-15-0 see FNZ000
51-17-2 see BCB750
51-18-3 see TND500
51-20-7 see BOL000
51-21-8 see FMM000

51-28-5 see DUZ000
51-30-9 see IMR000
51-34-3 see SBG000
51-40-1 see NNO699
51-41-2 see NNO500
51-42-3 see AES000
51-44-5 see DER600
51-45-6 see HGD000
51-46-7 see TBJ000
51-48-9 see TFZ275
51-50-3 see DCR000
51-52-5 see PNX000
51-55-8 see ARR000
51-57-0 see MDT600
51-58-1 see NIJ400
51-60-5 see DQY909
51-61-6 see DYC400
51-62-7 see BBK750
51-63-8 see BBK500
51-64-9 see AOA500
51-66-1 see AAR250
51-67-2 see TOG250
51-68-3 see DPE000
51-71-8 see PFC500
51-73-0 see DWX600
51-74-1 see HGE000
51-75-2 see BIE250
51-78-5 see ALU500
51-79-6 see UVA000
51-80-9 see TDR750
51-83-2 see CBH250
51-84-3 see CMF250
51-85-4 see MCN500
51-98-9 see ABU000
52-01-7 see AFJ500
52-21-1 see SOV100
52-24-4 see TFQ750
52-26-6 see MRO750
52-28-8 see CNG500
52-31-3 see TDA500
52-39-1 see AFJ875
52-43-7 see AFS500
52-46-0 see AQO000
52-49-3 see BBV000
52-51-7 see BNT250
52-52-8 see AJK250
52-53-9 see ILC000
52-60-8 see DJR800
52-62-0 see PBT000
52-66-4 see PAP500
52-67-5 see MCR750
52-68-6 see TIQ250
52-76-6 see NNV000
52-78-8 see ENX600
52-85-7 see FAB600
52-86-8 see CLY500
52-88-0 see MGR500
52-89-1 see CQK250
52-90-4 see CQK000
53-03-2 see PLZ000
53-06-5 see CNS800
53-16-7 see EDV000
53-18-9 see DIG800
53-19-0 see CDN000
53-21-4 see CNF000
53-36-1 see DAZ117
53-38-3 see MNC125
53-39-4 see AOO125
53-41-8 see HJB050

53-43-0 see AOO450
53-44-1 see MBW775
53-46-3 see DJM800
53-46-3 see XCJ000
53-60-1 see PMI500
53-64-5 see BKP300
53-69-0 see DQI600
53-70-3 see DCT400
53-79-2 see AEI000
53-85-0 see DGK100
53-86-1 see IDA000
53-89-4 see BCP650
53-94-1 see HIU500
53-95-2 see MDT600
53-96-3 see FDR000
54-04-6 see MDI500
54-05-7 see CLD000
54-06-8 see AES639
54-11-5 see NDN000
54-12-6 see TNW500
54-16-0 see HLJ000
54-21-7 see SJO000
54-25-1 see RJA000
54-30-8 see NOC000
54-31-9 see CHJ750
54-36-4 see MCJ370
54-42-2 see DAS000
54-47-7 see PII100
54-49-9 see HNB875
54-62-6 see AMG750
54-64-8 see MDI000
54-71-7 see PIF250
54-77-3 see DTO000
54-80-8 see INS000
54-84-2 see CMP900
54-85-3 see ILD000
54-86-4 see NDW500
54-87-5 see SEE250
54-88-6 see TLE750
54-91-1 see BHJ250
54-92-2 see ILE000
54-95-5 see PBI500
54-96-6 see DCD000
55-03-8 see LEQ300
55-18-5 see NJW500
55-21-0 see BBB000
55-22-1 see ILC000
55-27-6 see NNP050
55-31-2 see AES500
55-37-8 see DST200
55-38-9 see FAQ900
55-43-6 see DCR200
55-48-1 see ARR500
55-51-6 see BIA750
55-52-7 see PDN000
55-55-0 see MGJ750
55-56-1 see BIM250
55-57-2 see PEO000
55-63-0 see NGY000
55-63-0 see SLD000
55-65-2 see GKQ000
55-68-5 see MCU750
55-80-1 see DUH600
55-81-2 see MFC500
55-86-7 see BIE500
55-91-4 see IRF000
55-93-6 see DSU000
55-97-0 see HEA000
55-98-1 see BOT250

56-04-2 see MPW500
56-10-0 see AJY250
56-12-2 see PIM500
56-17-7 see CQJ750
56-18-8 see AIX250
56-23-5 see CBY000
56-24-6 see TMI250
56-25-7 see CBE750
56-29-1 see ERD500
56-33-7 see DWN150
56-34-8 see TCC250
56-35-9 see BLL750
56-36-0 see TIC000
56-38-2 see PAK000
56-38-2 see PAK250
56-38-2 see PAK260
56-40-6 see GHA000
56-47-3 see DAQ800
56-49-5 see MIJ750
56-53-1 see DKA600
56-54-2 see QFS000
56-55-3 see BBC250
56-57-5 see NJF000
56-65-5 see ARQ500
56-69-9 see HOO100
56-72-4 see CNU750
56-75-7 see CDP250
56-81-5 see GGA000
56-85-9 see GFO050
56-86-0 see GFO000
56-87-1 see LJM700
56-89-3 see CQK325
56-92-8 see HGD500
56-93-9 see BFM250
56-95-1 see CDT125
56-97-3 see TLQ500
56-99-5 see CCK650
57-06-7 see AGJ250
57-09-0 see HCQ500
57-10-3 see PAE250
57-11-4 see SLK000
57-12-5 see COI500
57-13-6 see USS000
57-14-7 see DSF400
57-15-8 see ABD000
57-22-7 see LEY000
57-24-9 see SMN500
57-27-2 see MRO500
57-29-4 see NAG500
57-30-7 see SID000
57-33-0 see NBU000
57-37-4 see BCA000
57-39-6 see TNK250
57-41-0 see DKQ000
57-42-1 see DAM600
57-43-2 see AMX750
57-44-3 see BAG000
57-47-6 see PIA500
57-50-1 see SNH000
57-52-3 see BLN500
57-53-4 see MQU750
57-55-6 see PML000
57-56-7 see HGU000
57-57-8 see PMT100
57-62-5 see CMA750
57-63-6 see EEH500
57-64-7 see PIA750
57-66-9 see DWW000
57-67-0 see AHO250

57-68-1 see SNJ000
57-74-9 see CDR750
57-83-0 see PMH500
57-85-2 see TBG000
57-88-5 see CMD750
57-91-0 see EDO500
57-92-1 see SLW500
57-94-3 see TOA000
57-95-4 see TNY750
57-96-5 see DWM000
57-97-6 see DQJ200
58-00-4 see AQP250
58-08-2 see CAK500
58-13-9 see PDC875
58-14-0 see TGD000
58-15-1 see DOT000
58-18-4 see MPN500
58-20-8 see TBF600
58-22-0 see TBF500
58-25-3 see LFK000
58-27-5 see MMD500
58-28-6 see DLS600
58-32-2 see PCP250
58-33-3 see PMI750
58-34-4 see MRW000
58-36-6 see OMY850
58-38-8 see PMF500
58-39-9 see CJM250
58-40-2 see DQA600
58-46-8 see TBJ275
58-54-8 see DFP600
58-55-9 see TEP000
58-56-0 see PPK500
58-58-2 see POK300
58-60-6 see ALQ625
58-61-7 see AEH750
58-63-9 see IDE000
58-71-9 see SFQ500
58-72-0 see TMS250
58-73-1 see BBV500
58-74-2 see PAH000
58-82-2 see BML500
58-85-5 see VSU100
58-89-9 see BBQ500
58-90-2 see TBT000
58-93-5 see CFY000
58-94-6 see CLH750
58-96-8 see UVJ000
59-00-7 see DNC200
59-01-8 see KAL000
59-02-9 see VSZ450
59-05-2 see MDV500
59-14-3 see BNC750
59-23-4 see GAV000
59-26-7 see DJS200
59-30-3 see FMT000
59-31-4 see CCC250
59-32-5 see CKV625
59-33-6 see DBM800
59-35-8 see DSV400
59-40-5 see QTS000
59-42-7 see NCL500
59-43-8 see TES750
59-46-1 see AIL750
59-47-2 see GGS000
59-49-4 see BDJ000
59-50-7 see CFE250
59-51-8 see MDT740
59-52-9 see BAD750

77-07-6 see LFG000	78-85-3 see MGA250	80-43-3 see DGR600	83-88-5 see RIK000	87-10-5 see THW750
77-09-8 see PDO750	78-86-4 see CEU250	80-46-6 see AON000	83-89-6 see ARQ250	87-11-6 see ABI250
77-10-1 see PDC890	78-87-5 see PNJ400	80-47-7 see MCE000	83-93-2 see TEF775	87-13-8 see EEV200
77-20-3 see NEA500	78-88-6 see DGH400	80-48-8 see MLL250	83-98-7 see MJH900	87-17-2 see SAH500
77-21-4 see DYC800	78-89-7 see CKR500	80-49-9 see MDL000	84-01-5 see CLY750	87-19-4 see IJN000
77-26-9 see AGI750	78-90-0 see PMK250	80-50-2 see LJS000	84-02-6 see PMF250	87-20-7 see IME000
77-27-0 see AGL375	78-92-2 see BPW750	80-51-3 see OPE000	84-04-8 see CJL500	87-25-2 see EGM000
77-28-1 see BPF500	78-93-3 see MKA400	80-52-4 see MCD750	84-08-2 see PAJ750	87-29-6 see API750
77-36-1 see CLY600	78-94-4 see BOY500	80-54-6 see LFT000	84-11-7 see PCX250	87-31-0 see DUR800
77-37-2 see CPQ250	78-95-5 see CDN200	80-56-8 see PIH250	84-12-8 see PCY300	87-33-2 see CCK125
77-41-8 see MLP800	78-96-6 see AMA500	80-58-0 see BMY250	84-15-1 see TBC640	87-39-8 see AFU000
77-46-3 see SNY500	78-97-7 see LAQ000	80-59-1 see TGA700	84-16-2 see DLB400	87-42-3 see CKV500
77-47-4 see HCE500	78-98-8 see PQC000	80-62-6 see MLH750	84-17-3 see DAL600	87-44-5 see CCN000
77-49-6 see NHO500	78-99-9 see DGF400	80-63-7 see MIF800	84-19-5 see DHB500	87-47-8 see PPQ625
77-50-9 see PNA250	79-00-5 see TIN000	80-71-7 see HMB500	84-36-6 see RCA200	87-48-9 see BNL750
77-51-0 see DIK000	79-01-6 see TIO750	80-77-3 see CKF500	84-51-5 see EGL500	87-51-4 see ICN000
77-54-3 see CCR250	79-02-7 see DEM200	80-81-9 see TCR250	84-57-1 see DFQ200	87-52-5 see DYC000
77-58-7 see DDV600	79-03-8 see PMW500	80-99-9 see CMD000	84-58-2 see DEX400	87-56-9 see MRU900
77-65-8 see BNK000	79-04-9 see CEC250	81-04-9 see AQY400	84-60-6 see DMH600	87-59-2 see XMJ000
77-66-7 see ACE000	79-05-0 see PMU250	81-07-2 see BCE500	84-65-1 see APK250	87-60-5 see CLK200
77-67-8 see ENG500	79-06-1 see ADS250	81-13-0 see PAG200	84-66-2 see DJX000	87-62-7 see XNJ000
77-71-4 see DSF300	79-07-2 see CDY850	81-15-2 see TML750	84-68-4 see DEQ400	87-65-0 see DFY000
77-73-6 see DGW000	79-08-3 see BMR750	81-16-3 see ALH750	84-69-5 see DNJ400	87-66-1 see PPQ500
77-74-7 see MNJ100	79-09-4 see PMU750	81-20-9 see NMS500	84-74-2 see DEH200	87-68-3 see HCD250
77-75-8 see EQL000	79-09-4 see PMV000	81-21-0 see BGA250	84-75-3 see DKP600	87-69-4 see TAF750
77-76-9 see DOM400	79-10-7 see ADS750	81-23-2 see DAL000	84-76-4 see DVJ000	87-76-3 see TLN150
77-77-0 see DXR200	79-11-8 see CEA000	81-24-3 see TAH250	84-77-5 see DGX600	87-85-4 see HEC000
77-78-1 see DUD100	79-14-1 see GHO000	81-25-4 see CME750	84-79-7 see HLY500	87-86-5 see PAX250
77-79-2 see DMF000	79-16-3 see MFT750	81-30-1 see NAP300	84-80-4 see VTA000	87-87-6 see TBQ500
77-80-5 see BLT250	79-17-4 see TFQ000	81-45-8 see CEL750	84-86-6 see ALI000	87-90-1 see TIQ750
77-81-6 see EIF000	79-19-6 see TFQ000	81-46-9 see AIB250	84-96-8 see AFL500	87-99-0 see XPJ000
77-83-8 see ENC000	79-20-9 see MFW100	81-48-1 see HOK000	84-97-9 see PCK500	88-04-0 see CLW000
77-86-1 see TEM500	79-21-0 see PCL500	81-49-2 see AJL500	85-00-7 see DWX800	88-05-1 see TLG500
77-89-4 see ADD750	79-22-1 see MIG000	81-54-9 see TKN750	85-01-8 see PCW250	88-06-2 see TIW000
77-92-9 see CMS750	79-24-3 see NFY500	81-55-0 see DMN400	85-02-9 see BDB750	88-09-5 see DHI400
77-93-0 see TJP750	79-27-6 see ACK250	81-60-7 see TDD250	85-31-4 see TFJ500	88-10-8 see DIW400
77-98-5 see TCC500	79-29-8 see DQT400	81-61-8 see TDD000	85-34-7 see TIY500	88-12-0 see EEG000
78-00-2 see TCF000	79-31-2 see IJU000	81-64-1 see DMH000	85-36-9 see AAM875	88-14-2 see FQF000
78-06-8 see DEJ200	79-34-5 see TBQ100	81-81-2 see WAT200	85-41-6 see PHX000	88-15-3 see ABI500
78-07-9 see EQA000	79-35-6 see DFA300	81-84-5 see NAQ000	85-43-8 see TDB000	88-19-7 see TGN250
78-08-0 see TJN250	79-36-7 see DEN400	81-88-9 see FAG070	85-44-9 see PHW750	88-24-4 see MJN250
78-10-4 see EPF550	79-37-8 see OLI000	81-96-9 see BMU000	85-60-9 see BRP750	88-26-6 see IFX200
78-11-5 see PBC250	79-38-9 see CLQ750	81-98-1 see DDK000	85-68-7 see BEC500	88-29-9 see ACL750
78-11-5 see PBC500	79-39-0 see MDN500	82-02-0 see AHK750	85-70-1 see BQP750	88-32-4 see BQI010
78-12-6 see NCH500	79-40-3 see DXO200	82-05-3 see BBI250	85-71-2 see MOD000	88-35-7 see AOS500
78-13-7 see TCD250	79-41-4 see MDN250	82-08-6 see KAJ500	85-73-4 see PHY750	88-44-8 see AKQ000
78-19-3 see DXR000	79-42-5 see TFK250	82-21-3 see DVW100	85-79-0 see DDT200	88-51-7 see AJJ250
78-20-6 see DEF200	79-43-6 see DEL000	82-22-4 see IBI000	85-82-5 see FAG080	88-61-9 see XJJ000
78-27-3 see EQL500	79-44-7 see DQY950	82-24-6 see AJS000	85-83-6 see SBC500	88-62-0 see AMT000
78-28-4 see ENF000	79-46-9 see NIY000	82-28-0 see AKP750	85-84-7 see FAG130	88-63-1 see PFA250
78-30-8 see TMO600	79-55-0 see PBJ000	82-33-7 see DBY700	85-86-9 see OHA000	88-67-5 see IEE000
78-34-2 see DVQ709	79-57-2 see HOH500	82-34-8 see NET000	85-91-6 see MGQ250	88-72-2 see NMO525
78-35-3 see LGB000	79-58-3 see SKR000	82-35-9 see DUQ000	85-98-3 see DJC400	88-73-3 see CJB750
78-37-5 see LGA000	79-61-4 see DRT100	82-43-9 see DEO750	86-00-0 see NFP500	88-74-4 see NEO000
78-40-0 see TJT750	79-69-6 see IGW500	82-44-0 see CEI000	86-13-5 see BDI000	88-75-5 see NIE500
78-41-1 see TMP500	79-74-3 see DCH400	82-45-1 see AIA750	86-21-5 see TMJ750	88-82-4 see TKQ250
78-42-2 see TNI250	79-78-7 see AGI500	82-46-2 see DEO700	86-26-0 see PEG000	88-84-6 see GKO000
78-43-3 see TNG750	79-81-2 see VSP000	82-48-4 see APK635	86-29-3 see DVX200	88-85-7 see BRE500
78-44-4 see IPU000	79-90-3 see TJE880	82-50-8 see DVS200	86-30-6 see DWI000	88-89-1 see PID000
78-46-6 see DDV800	79-92-5 see CBA500	82-54-2 see CNT625	86-34-0 see MNZ000	88-89-1 see PID250
78-48-8 see BSH250	79-93-6 see CKE750	82-66-6 see DVV600	86-35-1 see EOL100	88-89-1 see PID750
78-51-3 see BPK250	79-95-8 see TBO750	82-68-8 see PAX000	86-40-8 see XAK000	88-99-3 see PHW250
78-52-4 see DJN600	79-96-9 see IPJ000	82-71-3 see SMP500	86-50-0 see ASH500	89-02-1 see DUR400
78-53-5 see DJA400	79-97-0 see IPK000	82-92-8 see EAN600	86-52-2 see CIP750	89-19-0 see BQX250
78-57-9 see ASD000	80-00-2 see CKI625	82-93-9 see CFF500	86-54-4 see HGP495	89-25-8 see NNT000
78-59-1 see IMF400	80-05-7 see BLD500	83-05-6 see BIL500	86-56-6 see DSU400	89-55-4 see BOE500
78-62-6 see DHG000	80-06-8 see BIN000	83-12-5 see PFJ750	86-57-7 see NHQ000	89-57-6 see AMM500
78-63-7 see DRJ800	80-07-9 see BIN750	83-26-1 see PIH175	86-60-2 see ALI300	89-61-2 see DFT400
78-67-1 see ASL750	80-08-0 see SOA500	83-30-7 see TKO000	86-65-7 see NBE850	89-69-0 see TIT750
78-70-6 see LFX000	80-09-1 see SOB000	83-34-1 see MKV750	86-72-6 see CBN100	89-72-5 see BSE000
78-71-7 see BIK325	80-10-4 see DFF000	83-40-9 see CNX625	86-74-8 see CBN000	89-73-6 see SAL500
78-75-1 see DDR400	80-11-5 see THE500	83-43-2 see MOR500	86-85-1 see MLH000	89-78-1 see MCF750
78-75-1 see DDR600	80-12-6 see TDX500	83-44-3 see DAQ400	86-86-2 see NAK000	89-80-5 see MCG275
78-76-2 see BMX750	80-13-7 see HAF000	83-59-0 see PNP250	86-87-3 see NAK500	89-81-6 see MCF250
78-77-3 see BNR750	80-15-9 see IOB000	83-63-6 see AKR300	86-88-4 see AQN635	89-83-8 see TFX810
78-78-4 see EIK000	80-25-1 see DME400	83-64-7 see AKH750	86-93-1 see PGJ750	89-84-9 see DMG400
78-79-5 see IMS000	80-26-2 see TBE250	83-66-9 see BRU500	86-95-3 see QNA000	89-86-1 see HOE600
78-80-8 see MHU250	80-33-1 see CJT750	83-67-0 see TEO500	86-96-4 see QEJ800	89-94-1 see MPO000
78-81-9 see IIM000	80-35-3 see AKO500	83-70-5 see AKX500	87-01-4 see DPJ800	89-98-5 see CEI500
78-82-0 see IJX000	80-38-6 see CJR500	83-73-8 see DNF600	87-02-5 see AKI000	90-00-6 see PGR250
78-83-1 see IIL000	80-40-0 see EPW500	83-79-4 see RNZ000	87-08-1 see PDT500	90-01-7 see HMK100
78-84-2 see IJS000	80-41-1 see CHI125	83-86-3 see PIB250	87-09-2 see AGK250	90-02-8 see SAG000

90-03-9 see CHW675	92-53-5 see PFS750	94-86-0 see IRY000	96-45-7 see IAQ000	98-53-3 see BQW250
90-04-0 see AOV900	92-54-6 see PFX000	94-87-1 see CHO125	96-47-9 see MPO500	98-54-4 see BSE500
90-05-1 see GKI000	92-62-6 see DBN600	94-91-7 see DWS400	96-48-0 see BOV000	98-55-5 see TBD750
90-12-0 see MMB750	92-64-8 see CPJ500	94-93-9 see DWY200	96-49-1 see GHM000	98-56-6 see CEM825
90-13-1 see CIZ000	92-67-1 see AJS100	94-96-2 see EKV000	96-50-4 see AMS250	98-57-7 see CKG750
90-15-3 see NAW500	92-69-3 see BGJ500	95-04-5 see EHP000	96-53-7 see TFS250	98-59-9 see TGO250
90-16-4 see BDH000	92-71-7 see DWI200	95-06-7 see CDO250	96-54-8 see MPB000	98-60-2 see CEK375
90-17-5 see TIT000	92-82-0 see PDB500	95-08-9 see TJQ250	96-64-0 see SKS500	98-64-6 see CEK000
90-20-0 see AKH000	92-83-1 see XAT000	95-13-6 see IBX000	96-66-2 see TFD000	98-71-5 see PFI500
90-27-7 see PEP250	92-84-2 see PDP250	95-14-7 see BDH250	96-67-3 see HMY500	98-72-6 see NIJ500
90-30-2 see PFT250	92-87-5 see BBX000	95-16-9 see BDE500	96-69-5 see TFC600	98-73-7 see BQK500
90-33-5 see MKP500	92-88-6 see BGG500	95-19-2 see HAT500	96-73-1 see CJB825	98-77-1 see PIY500
90-34-6 see PMC300	92-93-3 see NFQ000	95-21-6 see MHK500	96-75-3 see NEP500	98-80-6 see BBM000
90-39-1 see SKX500	92-94-4 see TBC750	95-24-9 see AJE750	96-76-4 see DEG000	98-82-8 see COE750
90-41-5 see BGE250	93-01-6 see HMU500	95-25-0 see CDQ750	96-80-0 see DNP000	98-83-9 see MPK250
90-43-7 see BGJ250	93-05-0 see DJV200	95-26-1 see DQO800	96-83-3 see IFY100	98-84-0 see ALW250
90-45-9 see AHS500	93-08-3 see ABC500	95-30-7 see BDF250	96-84-4 see IFZ800	98-85-1 see PDE000
90-46-0 see XBJ000	93-09-4 see NAV500	95-31-8 see BQK750	96-88-8 see SBB000	98-86-2 see ABH000
90-47-1 see XBS000	93-10-7 see QEA000	95-32-9 see BDF750	96-91-3 see DUP400	98-87-3 see BAY300
90-49-3 see PFB350	93-14-1 see RLU000	95-33-0 see CPI250	96-93-5 see AKI250	98-88-4 see BDM500
90-64-2 see MAP000	93-15-2 see AGE250	95-35-2 see BHA500	96-96-8 see MFB000	98-92-0 see NCR000
90-65-3 see PAP750	93-16-3 see IKR000	95-38-5 see AHP500	96-99-1 see CJC500	98-94-2 see DRF709
90-69-7 see LHY000	93-18-5 see EEY500	95-39-6 see BFY250	97-00-7 see CGM000	98-95-3 see NEX000
90-72-2 see TNH000	93-19-6 see IJM000	95-41-0 see HFO700	97-02-9 see DUP600	98-96-4 see POL500
90-82-4 see POH000	93-23-2 see LBW000	95-45-4 see DBH000	97-05-2 see SOC500	98-98-6 see PIB930
90-84-6 see DIP400	93-26-5 see AAR000	95-47-6 see XHJ000	97-06-3 see MMH400	99-03-6 see AHR500
90-87-9 see HII600	93-28-7 see EQS000	95-48-7 see CNX000	97-11-0 see POO000	99-04-7 see TGP750
90-89-1 see DIW000	93-29-8 see AAX750	95-49-8 see CLK100	97-16-5 see DFY400	99-05-8 see AIH500
90-94-8 see MQS500	93-45-8 see NBF500	95-50-1 see DEP600	97-17-6 see DFK600	99-06-9 see HJI100
90-98-2 see DES000	93-46-9 see NBL000	95-51-2 see CEH670	97-18-7 see TFD250	99-08-1 see NMO500
91-08-7 see TGM800	93-51-6 see MEK250	95-53-4 see TGQ750	97-23-4 see MJM500	99-09-2 see NEN000
91-10-1 see DOJ200	93-53-8 see COF000	95-54-5 see PEY250	97-39-2 see DXP200	99-11-6 see DMV400
91-15-6 see PHY000	93-54-9 see EGQ000	95-55-6 see ALT000	97-42-7 see CCM750	99-24-1 see MKI100
91-16-7 see DOA200	93-55-0 see EOL500	95-56-7 see BNV000	97-44-9 see ABX500	99-30-9 see RDP300
91-17-8 see DAE800	93-58-3 see MHA750	95-57-8 see CJK250	97-45-0 see MCD000	99-35-4 see TMK500
91-19-0 see QRJ000	93-59-4 see PCM000	95-63-6 see TLL750	97-52-9 see NEQ000	99-45-6 see MGC350
91-20-3 see NAJ500	93-60-7 see NDV000	95-64-7 see XNS000	97-53-0 see EQR500	99-48-9 see MKY250
91-21-4 see TCU500	93-62-9 see HKM500	95-65-8 see XLJ000	97-54-1 see IKQ000	99-49-0 see MCD250
91-22-5 see QMJ000	93-65-2 see CIR500	95-68-1 see XMS000	97-56-3 see AIC250	99-54-7 see DFT600
91-23-6 see NER000	93-68-5 see ABA000	95-69-2 see CLK220	97-61-0 see MQJ750	99-55-8 see NMP500
91-33-8 see BDE250	93-71-0 see CFK000	95-70-5 see TGM000	97-62-1 see ELS000	99-56-9 see ALL500
91-38-3 see CIB700	93-72-1 see TIX500	95-71-6 see MKO250	97-63-2 see EMF000	99-57-0 see NEM500
91-40-7 see PEG500	93-76-5 see TAA100	95-74-9 see CLK215	97-64-3 see LAJ000	99-59-2 see NEQ500
91-44-1 see DIL400	93-78-7 see TGF210	95-76-1 see DEO300	97-72-3 see IJW000	99-60-5 see CJC250
91-49-6 see BPU500	93-79-8 see BSQ750	95-78-3 see XNA000	97-74-5 see BJL600	99-61-6 see NEV000
91-51-0 see LFT100	93-80-1 see TIW750	95-79-4 see CLK225	97-77-8 see DXH250	99-62-7 see DNN829
91-53-2 see SAV000	93-89-0 see EGR000	95-80-7 see TGL750	97-81-4 see ENZ000	99-63-8 see IMO000
91-56-5 see ICR000	93-90-3 see MKQ250	95-82-9 see DEO400	97-84-7 see TDN000	99-65-0 see DUQ200
91-57-6 see MMC000	93-96-9 see MHN500	95-83-0 see CFK125	97-85-8 see IIW000	99-66-1 see PNR750
91-58-7 see CJA000	93-99-2 see PEL500	95-85-2 see CEH250	97-86-9 see IIY000	99-71-8 see BSE250
91-59-8 see NBE500	94-04-2 see VOU000	95-86-3 see DCA200	97-88-1 see MHU750	99-72-9 see THD750
91-60-1 see NAP500	94-07-5 see HLV500	95-87-4 see XKS000	97-90-5 see BKM250	99-73-0 see DDJ600
91-62-3 see MPF800	94-09-7 see EFX000	95-88-5 see CLD750	97-93-8 see TJN750	99-75-2 see MPX850
91-63-4 see QEJ000	94-11-1 see IOY000	95-92-1 see DJT200	97-94-9 see TJP250	99-76-3 see HJL500
91-64-5 see CNV000	94-13-3 see HNU500	95-93-2 see TDM750	97-95-0 see EGW000	99-77-4 see ENO000
91-66-7 see DIS700	94-14-4 see MOT750	95-94-3 see TBN750	97-96-1 see DHI000	99-79-6 see ELQ500
91-71-4 see TFD750	94-15-5 see DNY000	95-95-4 see TIV750	97-99-4 see TCT000	99-80-9 see MJG750
91-75-8 see PDC000	94-17-7 see BHM750	95-99-8 see BHM250	98-00-0 see FPU000	99-83-2 see MCC000
91-79-2 see DPJ200	94-20-2 see CKK000	96-05-9 see AGK500	98-01-1 see FPQ875	99-85-4 see MCB750
91-80-5 see TEO250	94-24-6 see BQA010	96-08-2 see LFV000	98-02-2 see FPM000	99-86-5 see MLA250
91-81-6 see TMP750	94-25-7 see BPZ000	96-09-3 see EBR000	98-03-3 see TFM500	99-87-6 see CQI000
91-84-9 see WAK000	94-26-8 see BSC000	96-10-6 see DHI885	98-04-4 see TMB750	99-89-8 see IQZ000
91-85-0 see NCD500	94-28-0 see FCD560	96-11-7 see GGG000	98-05-5 see BBL750	99-91-2 see CEB250
91-93-0 see DCJ400	94-30-4 see AOV000	96-12-8 see DDL800	98-06-6 see BQJ250	99-92-3 see AHR750
91-94-1 see DEQ600	94-35-9 see PFJ000	96-13-9 see DDS000	98-07-7 see BFL250	99-93-4 see HIO000
91-95-2 see BGK500	94-36-0 see BDS000	96-14-0 see MNI500	98-08-8 see BDH500	99-94-5 see TGQ250
91-97-4 see DQS000	94-41-7 see CDH000	96-17-3 see MJX500	98-09-9 see BBS750	99-96-7 see SAI500
91-99-6 see DHF400	94-46-2 see IHP100	96-18-4 see TJB600	98-10-2 see BBR500	99-98-9 see DTL600
92-04-6 see CHN500	94-47-3 see PFB750	96-19-5 see TJC000	98-11-3 see BBS250	99-99-0 see NMO550
92-06-8 see TBC620	94-51-9 see DWS800	96-20-8 see AJA250	98-12-4 see CPR250	100-00-5 see NFS525
92-09-1 see AJY750	94-52-0 see NFD500	96-21-9 see DDR800	98-13-5 see TJA750	100-01-6 see NEO500
92-13-7 see PIF000	94-58-6 see DMD600	96-22-0 see DJN750	98-14-6 see PDO250	100-02-7 see NIF000
92-15-9 see ABA500	94-59-7 see SAD000	96-23-1 see DGG400	98-16-8 see AID500	100-06-1 see MDW750
92-16-0 see BDJ750	94-62-2 see PIV600	96-24-2 see CDT750	98-17-9 see TKE750	100-07-2 see AOY250
92-23-9 see LET000	94-63-9 see POS750	96-27-5 see MRM750	98-19-1 see BSK000	100-10-7 see DOT400
92-24-0 see NAI000	94-67-7 see SAG500	96-29-7 see EMU500	98-29-3 see BSK000	100-14-1 see NFN400
92-26-2 see DBT200	94-74-6 see CIR250	96-31-1 see DUM200	98-44-2 see AIE000	100-15-2 see MMF800
92-31-9 see AJP250	94-75-7 see DAA800	96-32-2 see MHR250	98-46-4 see NFJ500	100-16-3 see NIR000
92-32-0 see PPQ750	94-78-0 see PEK250	96-33-3 see MGA500	98-47-5 see NEB500	100-17-4 see NER500
92-43-3 see PDM500	94-80-4 see BQZ000	96-34-4 see MIF775	98-50-0 see ARA250	100-20-9 see TAV250
92-48-8 see MIP750	94-81-5 see CLN750	96-37-7 see MIU500	98-51-1 see BSP500	100-21-0 see TAN750
92-52-4 see BGE000	94-82-6 see DGA000	96-40-2 see CFI625	98-52-2 see BQW000	100-22-1 see BJF500

100-25-4 see DUQ600
100-29-8 see NID000
100-32-3 see BLA500
100-33-4 see DBM000
100-34-5 see BBN500
100-35-6 see CGV500
100-36-7 see DJI400
100-37-8 see DHO500
100-38-9 see DIY600
100-39-0 see BEC000
100-40-3 see CPD750
100-41-4 see EGP500
100-42-5 see SMQ000
100-44-7 see BEE375
100-45-8 see CPC625
100-47-0 see BCQ250
100-50-5 see FNK025
100-51-6 see BDX500
100-52-7 see BAY500
100-53-8 see TGO750
100-56-1 see PFM500
100-57-2 see PFN100
100-58-3 see PFL600
100-60-7 see MIT000
100-61-8 see MGN750
100-63-0 see PFI000
100-64-1 see HLI500
100-65-2 see PFJ250
100-66-3 see AOX750
100-68-5 see TFC750
100-72-1 see MDS500
100-73-2 see ADR500
100-74-3 see ENL000
100-75-4 see NLJ500
100-79-8 see DVR600
100-85-6 see BFM500
100-86-7 see DQQ200
100-88-9 see CPQ625
100-89-0 see TKM250
100-97-0 see HEI500
100-99-2 see TKR500
101-00-8 see TKT200
101-01-9 see TMS500
101-02-0 see TMU250
101-05-3 see DEV800
101-07-5 see DJK200
101-08-6 see DVO819
101-14-4 see MJM200
101-20-2 see TIL500
101-21-3 see CKC000
101-25-7 see DVF400
101-26-8 see MDL600
101-27-9 see CEW500
101-31-5 see HOU000
101-33-7 see TJM000
101-37-1 see THN500
101-38-2 see CHR000
101-39-3 see MIO000
101-40-6 see PNN400
101-41-7 see MHA500
101-42-8 see DTP400
101-48-4 see PDX000
101-49-5 see BEN250
101-50-8 see AJS500
101-54-2 see PFU500
101-59-7 see AOS750
101-61-1 see MJN000
101-67-7 see DVK400
101-68-8 see MJP400
101-72-4 see PFL000
101-73-5 see HKF000
101-77-9 see MJQ000
101-79-1 see CEH125
101-80-4 see OPM000
101-83-7 see DGT600
101-84-8 see PFA850
101-85-9 see AOH000
101-86-0 see HFO500
101-87-1 see PET000
101-90-2 see REF000
101-92-8 see AAY250
101-93-9 see BJO500

101-96-2 see DEG200
101-97-3 see EOH000
101-99-5 see CBL750
102-01-2 see AAY000
102-06-7 see DWC600
102-07-8 see CBM250
102-08-9 see DWN800
102-09-0 see DVZ000
102-17-2 see APE000
102-19-2 see IHV000
102-20-5 see PDI000
102-28-3 see AHQ000
102-29-4 see RDZ900
102-36-3 see IKH099
102-50-1 see MGO500
102-51-2 see MFG000
102-54-5 see FBC000
102-60-3 see QAT000
102-62-5 see DBF600
102-67-0 see TMY100
102-69-2 see TMY250
102-70-5 see THN000
102-71-6 see TKP500
102-76-1 see THM500
102-77-2 see BDG000
102-79-4 see BQM000
102-81-8 see DDU600
102-82-9 see THX250
102-83-0 see DDV200
102-85-2 see TIA750
102-98-7 see PFP250
103-03-7 see CBL000
103-05-9 see BEC250
103-07-1 see MNT000
103-08-2 see ENW500
103-09-3 see OEE000
103-11-7 see ADU250
103-16-2 see AEY000
103-17-3 see CEP000
103-18-4 see AKE500
103-23-1 see AEO000
103-24-2 see BJQ500
103-26-4 see MIO500
103-27-5 see PFO000
103-28-6 see IJV000
103-29-7 see BFX500
103-33-3 see ASL250
103-34-4 see BKU500
103-36-6 see EHN000
103-37-7 see BED000
103-38-8 see ISW000
103-41-3 see BEG750
103-44-6 see ELB500
103-45-7 see PFB250
103-48-0 see PDF750
103-50-4 see BEO250
103-52-6 see PFB800
103-53-7 see BEE250
103-54-8 see CMQ730
103-56-0 see CMR850
103-58-2 see HHQ500
103-59-3 see CMR750
103-61-7 see CMQ800
103-64-0 see BOF000
103-65-1 see IKG000
103-69-5 see EGK000
103-70-8 see FNJ000
103-71-9 see PFK250
103-72-0 see ISQ000
103-73-1 see PDM000
103-75-3 see EER500
103-76-4 see HKY500
103-81-1 see PDX750
103-82-2 see PDY850
103-83-3 see DQP800
103-84-4 see AAQ500
103-85-5 see PGN250
103-89-9 see ABJ250
103-90-2 see HIM000
103-93-5 see THA250
103-95-7 see COU500
104-01-8 see MFE250

104-04-1 see NEK000
104-06-3 see FNF000
104-12-1 see CKB000
104-14-3 see AKT250
104-15-4 see TGO000
104-19-8 see DPH400
104-45-0 see PNE250
104-46-1 see PMQ750
104-47-2 see MFF000
104-50-7 see OCE000
104-51-8 see BQI750
104-53-0 see HHP000
104-54-1 see CNX500
104-55-2 see CMP969
104-57-4 see BEP250
104-61-0 see CNF250
104-62-1 see PFC250
104-64-3 see HHQ000
104-65-4 see CMR500
104-68-7 see PEQ750
104-75-6 see EKS500
104-76-7 see EKQ000
104-78-9 see DIY800
104-82-5 see MHN300
104-85-8 see TGT750
104-89-2 see END000
104-90-5 see EOS000
104-91-6 see NLF200
104-93-8 see MGP000
104-94-9 see AOW000
105-01-1 see IIR100
105-07-7 see COK250
105-08-8 see BKH325
105-10-2 see DTL800
105-11-3 see DVR200
105-13-5 see MED500
105-21-5 see HBA550
105-28-2 see GIO000
105-29-3 see MNL775
105-30-6 see AOK750
105-31-7 see HFZ000
105-34-0 see MIQ000
105-36-2 see EGV000
105-37-3 see EPB500
105-38-4 see VQK000
105-39-5 see EHG500
105-40-8 see EMQ500
105-44-2 see MKW750
105-45-3 see MFX250
105-46-4 see BPV000
105-52-2 see DKP400
105-53-3 see EMA500
105-54-4 see EHE000
105-55-5 see DKC400
105-56-6 see EHP500
105-57-7 see AAG000
105-58-8 see DIX200
105-59-9 see MKU250
105-60-2 see CBF700
105-64-6 see DNR400
105-67-9 see XKJ500
105-68-0 see ILW000
105-74-8 see LBR000
105-75-9 see DEC600
105-76-0 see DED600
105-79-3 see IIT000
105-82-8 see AAG850
105-83-9 see BGU750
105-85-1 see CMT750
105-86-2 see GCY000
105-87-3 see DTD800
105-91-9 see NCP000
105-95-3 see EJQ500
105-99-7 see AEO750
106-11-6 see HKJ000
106-14-9 see HOG000
106-19-4 see DWQ875
106-20-7 see DJA800
106-21-8 see DTE600
106-22-9 see CMT250
106-23-0 see CMS845
106-24-1 see DTD000

106-25-2 see DTD200
106-27-4 see IHP400
106-30-9 see EKN050
106-32-1 see ENY000
106-33-2 see ELY700
106-34-3 see QFJ000
106-35-4 see EHA600
106-36-5 see PNU000
106-40-1 see BMT325
106-41-2 see BNU750
106-42-3 see XHS000
106-43-4 see TGY075
106-44-5 see CNX250
106-45-6 see TGP250
106-46-7 see DEP800
106-47-8 see CEH680
106-48-9 see CJK750
106-49-0 see TGR000
106-50-3 see PEY500
106-51-4 see QQS200
106-54-7 see CEK425
106-55-8 see DTR400
106-61-6 see GGO000
106-63-8 see IIK000
106-67-2 see ELT500
106-67-2 see EMZ000
106-68-3 see ODI000
106-69-4 see HES500
106-71-8 see ADT111
106-72-9 see DSD775
106-74-1 see ADT500
106-75-2 see OPO000
106-83-2 see BRH250
106-86-5 see VNZ000
106-87-6 see VOA000
106-88-7 see BOX750
106-89-8 see EAZ500
106-90-1 see ECH500
106-91-2 see ECI000
106-92-3 see AGH150
106-93-4 see EIY500
106-94-5 see BNX750
106-95-6 see AFY000
106-96-7 see PMN500
106-97-8 see BOR000
106-99-0 see BOP500
107-00-0 see EFS500
107-02-8 see ADR000
107-03-9 see PML500
107-04-0 see CES500
107-05-1 see AGB250
107-06-2 see EIY600
107-07-3 see EIU800
107-08-4 see PNO750
107-10-8 see PND250
107-11-9 see AFW000
107-12-0 see PMV750
107-13-1 see ADX500
107-14-2 see CDN500
107-15-3 see EEA500
107-16-4 see HIM500
107-18-6 see AFV500
107-19-7 see PMN450
107-20-0 see CDY500
107-21-1 see EJC500
107-22-2 see GIK000
107-25-3 see MQL750
107-25-5 see MQM000
107-27-7 see CHC500
107-29-9 see AAH250
107-30-2 see CIO250
107-31-3 see MKG750
107-32-4 see PCM500
107-35-7 see TAG750
107-36-8 see HKI500
107-37-9 see AGU250
107-41-5 see HFP875
107-44-8 see IPX000
107-45-9 see TDN250
107-46-0 see HEE000
107-49-3 see TCF250
107-49-3 see TCF270

107-49-3 see TCF280
107-61-9 see DVR000
107-66-4 see DEG700
107-69-7 see TIO500
107-70-0 see MEX250
107-71-1 see BSC250
107-72-2 see PBY750
107-74-4 see DTE400
107-75-5 see CMS850
107-81-3 see BNU500
107-82-4 see BNP250
107-83-5 see IKS600
107-84-6 see CII000
107-87-9 see PBN250
107-88-0 see BOS500
107-89-1 see AAH750
107-91-5 see COJ250
107-92-6 see BSW000
107-94-8 see CKS500
107-96-0 see MCQ000
107-98-2 see PNL250
107-98-2 see UBA000
107-99-3 see CGW000
108-00-9 see DPC000
108-01-0 see DOY800
108-03-2 see NIX500
108-05-4 see VLU250
108-09-8 see DQU600
108-10-1 see HFG500
108-11-2 see MKW600
108-13-4 see MAO000
108-16-7 see DPT800
108-18-9 see DNM200
108-20-3 see IOZ750
108-21-4 see INE100
108-22-5 see MQK750
108-23-6 see IOL000
108-24-7 see AAX500
108-30-5 see SNC000
108-31-6 see MAM000
108-32-7 see CBW500
108-34-9 see MOX250
108-38-3 see XHA000
108-39-4 see CNW750
108-42-9 see CEH675
108-43-0 see CJK500
108-44-1 see TGQ500
108-45-2 see PEY000
108-46-3 see REA000
108-50-9 see DTU800
108-55-4 see GFU000
108-57-6 see DXQ745
108-58-7 see REA100
108-59-8 see DSM200
108-60-1 see BII250
108-62-3 see TDW500
108-64-5 see ISY000
108-67-8 see TLM050
108-68-9 see XLS000
108-69-0 see XOA000
108-73-6 see PGR000
108-77-0 see TJD750
108-78-1 see MCB000
108-80-5 see THS000
108-82-7 see DNH800
108-83-8 see DNI800
108-84-9 see HFJ000
108-86-1 see PEO500
108-87-2 see MIQ740
108-88-3 see TGK750
108-89-4 see MOY250
108-90-7 see CEJ125
108-91-8 see CPF500
108-93-0 see CPB750
108-94-1 see CPC000
108-95-2 see PDN750
108-95-2 see PDO000
108-98-5 see PFL850
108-99-6 see PIB920
109-01-3 see MOD250
109-02-4 see MMA250
109-05-7 see MOG750

109-06-8 see MOY000	110-65-6 see BST500	112-12-9 see UKS000	115-20-8 see TIN500	118-42-3 see PJB750
109-08-0 see MOW750	110-66-7 see PBM000	112-14-1 see OEG000	115-21-9 see EPY500	118-46-7 see ALJ750
109-09-1 see CKW000	110-67-8 see MFL750	112-15-2 see CBQ750	115-24-2 see ABD500	118-48-9 see IHN200
109-19-3 see ISX000	110-68-9 see MHV000	112-18-5 see DRR800	115-25-3 see CPS000	118-52-5 see DFE200
109-20-6 see GDK000	110-69-0 see BSU500	112-19-6 see UMS000	115-26-4 see BJE750	118-55-8 see PGG750
109-21-7 see BQM500	110-71-4 see DOE600	112-23-2 see HBO500	115-28-6 see CDS000	118-58-1 see BFJ750
109-27-3 see TEF500	110-73-6 see EGA500	112-24-3 see TJR000	115-29-7 see EAQ750	118-60-5 see ELB000
109-29-5 see OKU000	110-74-7 see PNM500	112-25-4 see HFT500	115-31-1 see IHZ000	118-61-6 see SAL000
109-31-9 see ASC000	110-75-8 see CHI250	112-26-5 see TKL500	115-32-2 see BIO750	118-68-3 see AJB250
109-42-2 see BSS100	110-77-0 see EPP500	112-27-6 see TJQ000	115-37-7 see TEN000	118-71-8 see MAO350
109-43-3 see DEH600	110-78-1 see PNP000	112-30-1 see DAI600	115-38-8 see ENB500	118-74-1 see HCC500
109-44-4 see BJO225	110-80-5 see EES350	112-31-2 see DAG000	115-43-5 see AGQ875	118-75-2 see TBO500
109-46-6 see DEI000	110-81-6 see DJC600	112-31-2 see DAG200	115-44-6 see AFY500	118-79-6 see THV750
109-53-5 see IJQ000	110-82-7 see CPB000	112-32-3 see OEY100	115-58-2 see PBS250	118-90-1 see TGQ000
109-55-7 see AJQ100	110-83-8 see CPC579	112-33-4 see AMC250	115-63-9 see HFG400	118-91-2 see CEL250
109-56-8 see INN400	110-85-0 see PIJ000	112-34-5 see DJF200	115-67-3 see PAH500	118-92-3 see API500
109-57-9 see AGT500	110-86-1 see POP250	112-35-6 see TJQ750	115-69-5 see ALB000	118-93-4 see HIN500
109-59-1 see INA500	110-87-2 see DMC200	112-36-7 see DIW800	115-71-9 see OHG000	118-96-7 see TMN490
109-60-4 see PNC250	110-88-3 see TMP000	112-37-8 see UKA000	115-76-4 see PMB250	118-96-7 see TMN500
109-61-5 see PNH000	110-89-4 see PIL500	112-38-9 see ULS000	115-77-5 see PBB750	119-04-0 see NCF000
109-62-6 see EMD000	110-91-8 see MRP750	112-40-3 see DXT200	115-78-6 see THY500	119-06-2 see DXQ200
109-63-7 see BMH250	110-94-1 see GFS000	112-42-5 see UNA000	115-79-7 see MSC100	119-07-3 see OEU000
109-64-8 see TLR000	110-95-2 see TDU500	112-43-6 see UMA000	115-82-2 see EKQ500	119-12-0 see POP000
109-65-9 see BMX500	110-96-3 see DNH400	112-44-7 see UJJ000	115-84-4 see BKH625	119-26-6 see DVC400
109-66-0 see PBK250	110-97-4 see DNL600	112-45-8 see ULJ000	115-86-6 see TMT750	119-27-7 see DUP800
109-69-3 see BQQ750	110-98-5 see OQM000	112-48-1 see DDW400	115-90-2 see FAQ800	119-32-4 see NMP000
109-70-6 see BNA825	111-10-4 see MEP750	112-49-2 see TKL875	115-91-3 see EAV000	119-33-5 see NFU500
109-71-7 see CHR400	111-12-6 see MND275	112-50-5 see EFL000	115-93-5 see CQL250	119-34-6 see NEM480
109-72-8 see BRR739	111-13-7 see ODG000	112-53-8 see DXV600	115-95-7 see LFY100	119-36-8 see MPI000
109-73-9 see BPX750	111-14-8 see HBE000	112-54-9 see DXT000	115-96-8 see CGO500	119-38-0 see DSK200
109-74-0 see BSX250	111-15-9 see EES400	112-55-0 see LBX000	115-98-0 see VQF000	119-39-1 see PHV750
109-75-1 see BOX500	111-16-0 see PIG000	112-56-1 see BPL250	116-01-8 see DNX600	119-41-5 see ELH600
109-76-2 see PMK500	111-17-1 see HFM500	112-57-2 see TCE500	116-02-9 see TLO500	119-47-1 see MJO500
109-77-3 see MAO250	111-20-6 see SBJ500	112-58-3 see DKO800	116-06-3 see CBM500	119-48-2 see DUO400
109-78-4 see HGP000	111-21-7 see EJB500	112-59-4 see HFN000	116-09-6 see ABC000	119-51-7 see ILI000
109-79-5 see BRR900	111-22-8 see TJQ500	112-60-7 see TCE250	116-14-3 see TCH500	119-53-9 see BCP250
109-82-0 see HHW000	111-24-0 see DDR000	112-61-8 see MJW000	116-14-3 see TCH750	119-58-4 see TDO750
109-83-1 see MGG000	111-25-1 see HFM500	112-62-9 see OHW000	116-15-4 see HDF000	119-60-8 see DGV600
109-84-2 see HHC000	111-26-2 see HFK000	112-66-3 see DXV400	116-16-5 see HCL500	119-61-9 see BCS250
109-85-3 see MEM500	111-27-3 see HFJ500	112-70-9 see TJI750	116-17-6 see TKT500	119-64-2 see TCX500
109-86-4 see EJH500	111-28-4 see HCS500	112-72-1 see TBY250	116-29-0 see CKM000	119-65-3 see IRX000
109-87-5 see MGA850	111-29-5 see PBK750	112-73-2 see DDW200	116-38-1 see EAE600	119-68-6 see MGQ000
109-89-7 see DHJ200	111-30-8 see GFQ000	112-79-8 see EAF500	116-52-9 see DGQ200	119-72-2 see ALP750
109-90-0 see ELS500	111-31-9 see HES000	112-80-1 see OHU000	116-54-1 see DEM800	119-79-9 see ALI250
109-92-2 see EQF500	111-34-2 see VMZ000	112-90-3 see OHM700	116-71-2 see DCU800	119-84-6 see HHR500
109-92-2 see EQG000	111-35-3 see EFG000	112-92-5 see OAX000	116-76-7 see APL750	119-90-4 see DCJ200
109-93-3 see VOP000	111-36-4 see BRQ500	112-96-9 see OBG000	116-80-3 see CIA000	119-93-7 see TGJ750
109-94-4 see EKL000	111-40-0 see DJG600	112-98-1 see TCE350	116-82-5 see AIY500	120-02-5 see CBI250
109-95-5 see ENN000	111-41-1 see AJW000	113-00-8 see GKW000	116-84-7 see AJH250	120-07-0 see BKD500
109-97-7 see PPS250	111-42-2 see DHF000	113-15-5 see EDC000	116-85-8 see AKE250	120-08-1 see DRS800
109-99-9 see TCR750	111-43-3 see PNM000	113-18-8 see CHG000	116-90-5 see DCV000	120-11-6 see BES750
110-00-9 see FPK000	111-44-4 see DFJ050	113-38-2 see EDR000	117-03-3 see APM000	120-12-7 see APG500
110-01-0 see TDC730	111-45-5 see AGO000	113-42-8 see PAM000	117-05-5 see BDK750	120-15-0 see TCU700
110-02-1 see TFM250	111-46-6 see DJD600	113-45-1 see MNQ000	117-10-2 see DMH400	120-18-3 see NAP000
110-05-4 see BSC750	111-48-8 see TFI500	113-48-4 see OES000	117-11-3 see AJE325	120-20-7 see DOE200
110-12-3 see MKW450	111-49-9 see HDG000	113-52-0 see DLH630	117-12-4 see DMH200	120-32-1 see CJU250
110-13-4 see HEQ500	111-55-7 see EJD759	113-53-1 see DYC875	117-14-6 see APK625	120-36-5 see DGB000
110-15-6 see SMY000	111-60-4 see EJM500	113-59-7 see TAF675	117-18-0 see TBR500	120-40-1 see BKE500
110-16-7 see MAK900	111-65-9 see OCU000	113-69-9 see BDW000	117-27-1 see BIN500	120-45-6 see PFR000
110-17-8 see FOU000	111-69-3 see AER250	113-80-4 see AQW125	117-34-0 see DVW800	120-47-8 see HJL000
110-18-9 see TDQ750	111-70-6 see HBL500	113-92-8 see TAI500	117-39-5 see QCA000	120-51-4 see BCM000
110-19-0 see IIJ000	111-71-7 see HBB500	113-98-4 see BFD000	117-51-1 see HGK500	120-53-6 see EEU500
110-20-3 see ABE250	111-75-1 see BQC000	114-03-4 see HOA575	117-52-2 see ABF500	120-54-7 see TEF750
110-22-5 see ACV500	111-76-2 see BPJ850	114-03-4 see HON800	117-55-5 see AMM125	120-55-8 see DJE000
110-22-5 see ACV750	111-77-3 see DJG000	114-07-8 see EDH500	117-61-3 see BBX500	120-57-0 see PIW250
110-26-9 see MJL500	111-79-5 see MNC000	114-26-1 see PMY300	117-62-4 see AIH000	120-58-1 see IRZ000
110-27-0 see IQN000	111-80-8 see MNC000	114-42-1 see FBY000	117-79-3 see AIB000	120-61-6 see DUE000
110-32-7 see AEQ000	111-83-1 see BNU000	114-45-4 see IRU000	117-80-6 see DFT000	120-62-7 see ISA000
110-38-3 see EHE500	111-84-2 see NMX000	114-49-8 see HOT500	117-81-7 see DVL700	120-66-1 see ABJ000
110-40-7 see DJY600	111-86-4 see OEK000	114-63-6 see HJM000	117-82-8 see DOF400	120-67-2 see DGP800
110-43-0 see MGN500	111-87-5 see OEI000	114-80-7 see POD000	117-83-9 see BHK000	120-71-8 see MGO750
110-44-1 see SKU000	111-90-0 see CBR000	114-83-0 see ACX750	117-84-0 see DVL600	120-72-9 see ICM000
110-45-2 see IHS000	111-91-1 see BID750	114-85-2 see BFW250	117-89-5 see TKE500	120-73-0 see POJ250
110-46-3 see IMB800	111-92-2 see DDT800	114-86-3 see PDF000	117-96-4 see DCK000	120-75-2 see MHI750
110-49-6 see EJJ500	111-94-4 see BIQ500	114-90-9 see BGS250	117-97-5 see BLC500	120-78-5 see BDE750
110-51-0 see POQ250	111-96-6 see BKN750	115-02-6 see ASA500	117-98-6 see AAW750	120-80-9 see CCP850
110-52-1 see DDL000	111-97-7 see DGS600	115-07-1 see PMO500	118-00-3 see GLS000	120-82-1 see TIK250
110-54-3 see HEN000	112-03-8 see TLW500	115-09-3 see MDD750	118-03-6 see ALI750	120-83-2 see DFX800
110-57-6 see BRG000	112-04-9 see OBI000	115-10-6 see MJW500	118-10-5 see CMP925	120-92-3 see CPW500
110-60-1 see BOS000	112-05-0 see NMY000	115-11-7 see IIC000	118-28-5 see AJU500	120-93-4 see IAS000
110-61-2 see SNE000	112-06-1 see HBL000	115-18-4 see MHU100	118-29-6 see HMP100	120-94-5 see MPB250
110-63-4 see BOS750	112-07-2 see BPM000	115-19-5 see MHX250	118-33-2 see ALH500	120-97-8 see DEQ200

121-00-6 see BRN000	123-11-5 see AOT530	124-97-0 see NCI600	127-95-7 see OLE000	131-91-9 see NLB000
121-02-8 see MMH500	123-12-6 see TJV500	124-98-1 see EBL000	128-03-0 see PLD500	131-99-7 see IDE200
121-03-9 see MMH250	123-17-1 see TLV000	124-99-2 see GFC000	128-04-1 see SGM500	132-17-2 see TNU000
121-14-2 see DVH000	123-19-3 see DWT600	125-04-2 see HHR000	128-08-5 see BOF500	132-18-3 see DVW700
121-17-5 see NFS700	123-20-6 see VNF000	125-10-0 see PLZ100	128-09-6 see SND500	132-19-4 see DJP500
121-19-7 see HMY000	123-23-9 see SNC500	125-28-0 see DKW800	128-13-2 see DMJ200	132-20-7 see TMK000
121-21-1 see POO050	123-25-1 see SNB000	125-29-1 see OOI000	128-37-0 see BFW750	132-27-4 see BGJ750
121-29-9 see POO100	123-28-4 see TFD500	125-30-4 see DVO700	128-42-7 see DVF600	132-32-1 see AJV000
121-32-4 see EQF000	123-29-5 see ENW000	125-33-7 see DBB200	128-44-9 see SJN700	132-35-4 see POF500
121-39-1 see EOK600	123-30-8 see ALT250	125-40-6 see BPF000	128-46-1 see DME000	132-45-6 see PGP500
121-43-7 see TLN000	123-31-9 see HIH000	125-42-8 see EMO500	128-50-7 see DTB800	132-53-6 see NLB500
121-44-8 see TJO000	123-32-0 see DTU600	125-46-2 see UWJ000	128-51-8 see DTC000	132-60-5 see PGG000
121-45-9 see TMD500	123-33-1 see DMC600	125-51-9 see PJA000	128-53-0 see MAL250	132-67-2 see SIO500
121-46-0 see NNG000	123-34-2 see AGP500	125-56-6 see MDP750	128-56-3 see DLJ800	132-69-4 see BBW500
121-47-1 see SNO000	123-35-3 see MRZ150	125-58-6 see MDO775	128-58-5 see JAT000	132-75-2 see NBD500
121-54-0 see BEN000	123-36-4 see ECO500	125-60-0 see RDA375	128-59-6 see DMJ800	132-86-5 see NAN000
121-58-4 see DUD000	123-38-6 see PMT750	125-64-4 see DNW400	128-62-1 see NOA000	132-89-8 see CDS250
121-59-5 see CBJ000	123-39-7 see MKG500	125-68-8 see LFD200	128-66-5 see DCZ000	132-93-4 see PDD350
121-66-4 see ALQ000	123-42-2 see DBF750	125-69-9 see DBE200	128-76-7 see LBO200	132-98-9 see PDT750
121-69-7 see DQF800	123-43-3 see SNU000	125-71-3 see DBE150	128-79-0 see IBJ000	133-06-2 see CBG000
121-73-3 see CJB250	123-51-3 see IHP000	125-72-4 see DYF000	128-80-3 see BLK000	133-07-3 see TIT250
121-75-5 see MAK700	123-54-6 see ABX750	125-72-4 see MLZ000	128-87-0 see IBD000	133-10-8 see SEP000
121-79-9 see PNM750	123-56-8 see SND000	125-73-5 see DBE800	128-89-2 see IBE000	133-14-2 see BIX750
121-82-4 see CPR800	123-61-5 see BBP000	125-84-8 see AKC600	128-93-8 see BNN550	133-16-4 see DHT300
121-86-8 see CJG800	123-62-6 see PMV500	125-85-9 see PET250	128-95-0 see DBP000	133-17-5 see HGB200
121-87-9 see CJA175	123-63-7 see PAI250	125-86-0 see CBG250	129-00-0 see PON250	133-18-6 see APJ500
121-88-0 see ALO000	123-66-0 see EHF000	125-88-2 see BOQ750	129-03-3 see PCI500	133-32-4 see ICP000
121-89-1 see NEL500	123-68-2 see AGA500	126-02-3 see CQH500	129-06-6 see WAT220	133-49-3 see PAY500
121-90-4 see NFK500	123-72-8 see BSU250	126-06-7 see BNA325	129-15-7 see MMG000	133-55-1 see DRO400
121-91-5 see IMJ000	123-73-9 see COB260	126-07-8 see GKE000	129-16-8 see MCV000	133-58-4 see NHK900
121-92-6 see NFG000	123-75-1 see PPS500	126-11-4 see HMJ500	129-17-9 see ADE500	133-67-5 see HII500
121-98-2 see AOV750	123-76-2 see LFH000	126-15-8 see BHJ500	129-18-0 see BOV750	133-78-8 see AMT250
122-00-9 see MFW250	123-77-3 see ASM270	126-17-0 see POJ000	129-20-4 see HNI500	133-90-4 see AJM000
122-03-2 see COE500	123-77-3 see ASM300	126-22-7 see BPG000	129-40-8 see CJA250	133-91-5 see DNH000
122-04-3 see NFK100	123-81-9 see MCN000	126-27-2 see DTL200	129-42-0 see DBP400	134-03-2 see ARN125
122-06-5 see SLS000	123-82-0 see TNX750	126-30-7 see DTG400	129-43-1 see HJE000	134-20-3 see APJ250
122-09-8 see DTJ400	123-86-4 see BPU750	126-31-8 see SHX000	129-44-2 see DBP200	134-29-2 see AOX250
122-10-1 see SOY000	123-88-6 see MEP250	126-33-0 see SNW500	129-49-9 see MQP500	134-31-6 see QPS000
122-11-2 see SNN300	123-91-1 see DVQ000	126-39-6 see EIO500	129-51-1 see EDB500	134-32-7 see NBE000
122-14-5 see DSQ000	123-92-2 see IHO850	126-52-3 see EEH000	129-56-6 see APK000	134-37-2 see ALY250
122-15-6 see DRL200	123-93-3 see MCM750	126-64-7 see LFZ000	129-63-5 see AAN000	134-49-6 see PMA750
122-19-0 see DTC600	123-99-9 see ASB750	126-68-1 see TJU000	129-66-6 see TML000	134-50-9 see AHS750
122-20-3 see NEI500	124-02-7 see DBI600	126-72-7 see TNC500	129-66-8 see TML250	134-53-2 see AOD250
122-31-6 see MAN750	124-03-8 see EKN500	126-73-8 see TIA250	129-67-9 see DXD000	134-58-7 see AJO500
122-34-9 see BJP000	124-04-9 see AEN250	126-75-0 see DAP200	129-71-5 see CDR000	134-62-3 see DKC800
122-37-2 see AOT000	124-05-0 see EIQ000	126-84-1 see ABD250	129-74-8 see BOM250	134-63-4 see LHZ000
122-39-4 see DVX800	124-06-1 see ENL850	126-85-2 see CFA500	129-77-1 see EOY000	134-71-4 see EAX500
122-40-7 see AOG500	124-07-2 see OCY000	126-91-0 see LFY000	129-79-3 see TMM250	134-72-5 see EAY500
122-42-9 see CBM000	124-07-2 see ODE200	126-92-1 see TAV750	129-99-7 see SIG000	134-80-5 see DIP600
122-43-0 see BQJ350	124-09-4 see HEO000	126-93-2 see OLU000	130-01-8 see ARS500	134-81-6 see BBY750
122-51-0 see ENY500	124-09-4 see HEO500	126-98-7 see MGA750	130-15-4 see NBA500	134-84-9 see MHF750
122-52-1 see TJT800	124-13-0 see OCO000	126-99-8 see NCI500	130-17-6 see ALX000	134-90-7 see CDP325
122-56-5 see THX500	124-16-3 see BPL500	127-00-4 see CKR750	130-26-7 see CHR500	134-96-3 see DOF600
122-57-6 see SMS500	124-17-4 see BQP500	127-06-0 see ABF000	130-61-0 see MOO500	135-02-4 see AOT525
122-59-8 see PDR100	124-18-5 see DAG400	127-07-1 see HOO500	130-80-3 see DKB000	135-07-9 see MIV500
122-60-1 see PFH000	124-19-6 see NMW500	127-08-2 see PKT750	130-83-6 see AGD500	135-12-6 see CJD600
122-62-3 see BJS250	124-20-9 see SLA000	127-09-3 see SEG500	130-85-8 see PAF100	135-14-8 see PJA120
122-66-7 see HHG000	124-22-1 see DXW000	127-17-3 see PQC100	130-86-9 see FOW000	135-16-0 see TCR400
122-67-8 see IIQ000	124-25-4 see TBX500	127-18-4 see PCF275	130-89-2 see QJS000	135-19-3 see NAX000
122-70-3 see PDK000	124-26-5 see OAR000	127-19-5 see DOO800	130-90-5 see QIS300	135-20-6 see ANO500
122-72-5 see HHP500	124-28-7 see DTC400	127-20-8 see DGI600	130-95-0 see QHJ000	135-23-9 see DPJ400
122-73-6 see BES500	124-30-1 see OBC000	127-21-9 see DGL400	131-01-1 see RDF000	135-44-4 see LEU000
122-74-7 see HHQ550	124-38-9 see CBU250	127-25-3 see MFT500	131-04-4 see PAE100	135-48-8 see PAV000
122-78-1 see BBL500	124-38-9 see CBU500	127-31-1 see FHH100	131-07-7 see SCA475	135-49-9 see DBT400
122-79-2 see PDY750	124-38-9 see CBU750	127-33-3 see MIJ500	131-08-8 see SER000	135-51-3 see ROF300
122-80-5 see AHQ250	124-40-3 see DOQ800	127-41-3 see IFW000	131-11-3 see DTR200	135-87-5 see BCI500
122-82-7 see AAZ000	124-40-3 see DOR000	127-47-9 see VSK900	131-14-6 see APK850	135-88-6 see PFT500
122-84-9 see AOV875	124-41-4 see SIK450	127-48-0 see TLP750	131-15-7 see BLB750	135-98-8 see BQJ000
122-88-3 see CJN000	124-41-4 see SIK500	127-52-6 see SFV275	131-16-8 see DWV500	136-16-3 see ORS200
122-93-0 see ADU750	124-43-6 see HIB500	127-56-0 see SNQ000	131-17-9 see DBL200	136-23-2 see BIX000
122-96-3 see PIJ750	124-47-0 see UTJ000	127-57-1 see PPO250	131-18-0 see AON300	136-25-4 see PBK000
122-97-4 see HHP050	124-48-1 see CFK500	127-58-2 see SJW475	131-27-1 see ALH250	136-30-1 see SGF500
122-98-5 see AOR750	124-58-3 see MGQ530	127-63-9 see PGI750	131-49-7 see AOO875	136-32-3 see SKK500
122-99-6 see PER000	124-64-1 see TDH750	127-65-1 see CDP000	131-52-2 see SJA000	136-35-6 see DWO800
123-00-2 see AMF250	124-65-2 see HKC500	127-65-1 see SFV550	131-55-5 see BCS325	136-36-7 see HNH500
123-01-3 see PEW500	124-68-5 see IIA000	127-68-4 see NFC500	131-56-6 see DMI600	136-40-3 see PDC250
123-03-5 see CCX000	124-70-9 see DFS800	127-69-5 see SNN500	131-57-7 see MES000	136-44-7 see GGQ000
123-04-6 see EKU000	124-76-5 see IHY000	127-71-9 see SNH800	131-70-4 see MRF525	136-45-8 see EAU500
123-05-7 see BRI000	124-85-6 see PMA450	127-79-5 see ALF250	131-73-7 see HET500	136-47-0 see TBN000
123-06-8 see EEW000	124-87-8 see PIE500	127-85-5 see ARA500	131-74-8 see ANS500	136-60-7 see BQK250
123-08-0 see FOF000	124-90-3 see DLX400	127-90-2 see OAL000	131-79-3 see FAG135	136-77-6 see HFV500
123-09-1 see CKG500	124-94-7 see AQX250	127-91-3 see POH750	131-89-5 see CPK500	136-78-7 see CNW000

136-80-1 see TGT000	140-89-6 see PLF000	143-98-6 see DBE825	150-78-7 see DOA400	192-70-1 see BCP750
136-85-6 see MHK250	140-90-9 see SHE500	143-98-6 see MMA000	150-84-5 see AAU000	192-97-2 see BCT000
136-90-3 see MFQ500	140-93-2 see SIA000	144-00-3 see EKT500	150-86-7 see PIB600	193-39-5 see IBZ000
136-92-5 see DJD400	140-95-4 see DTG700	144-02-5 see BAG250	151-00-8 see IJT000	193-40-8 see DCX400
136-95-8 see AIS500	141-00-4 see CAI750	144-11-6 see PAL500	151-05-3 see BEL750	194-59-2 see DCY000
137-03-1 see HBN500	141-01-5 see FBJ100	144-12-7 see HNM000	151-06-4 see ARW750	194-60-5 see BDA500
137-05-3 see MIQ075	141-02-6 see DVK600	144-14-9 see ALW750	151-18-8 see AMB500	194-62-7 see BDA250
137-06-4 see TGP000	141-03-7 see SNA500	144-19-4 see TLY750	151-21-3 see SIB600	194-69-4 see BCG750
137-07-5 see AIF500	141-04-8 see DNH125	144-21-8 see DXE600	151-38-2 see MEO750	195-19-7 see BCR750
137-08-6 see CAU750	141-05-9 see DJO200	144-29-6 see PIJ500	151-41-7 see MRH250	195-84-6 see TJH250
137-09-7 see AHP000	141-11-7 see RHA000	144-33-2 see DXC400	151-50-8 see PLC500	196-78-1 see BCH000
137-17-7 see TLG250	141-16-2 see DTF800	144-34-3 see SBQ000	151-50-8 see PLC750	196-79-2 see BCQ000
137-18-8 see XQJ000	141-18-4 see BHJ750	144-41-2 see MRU250	151-56-4 see EJM900	198-46-9 see BCI000
137-20-2 see SIY000	141-22-0 see RJP000	144-48-9 see IDW000	151-63-3 see AHQ750	198-55-0 see PCQ250
137-26-8 see TFS350	141-23-1 see HOG500	144-49-0 see FIC000	151-67-7 see HAG500	201-42-3 see DCR600
137-29-1 see CNL500	141-25-3 see DTF400	144-55-8 see SFC500	151-73-5 see BFV770	201-65-0 see DCX000
137-30-4 see BJK500	141-28-6 see AEP750	144-62-7 see OLA000	152-02-3 see AGI000	202-98-2 see CPU000
137-32-6 see MHS750	141-31-1 see HIT500	144-74-1 see TEX500	152-16-9 see OCM000	203-20-3 see DCR400
137-40-6 see SJL500	141-32-2 see BPW100	144-80-9 see SNP500	152-18-1 see TMH525	203-64-5 see CPX250
137-53-1 see SKJ300	141-37-7 see ECB000	144-82-1 see MPQ750	152-20-5 see DUG500	205-82-3 see BCJ500
137-58-6 see DHK400	141-38-8 see EKV500	144-83-2 see PPO000	152-43-2 see QFA250	205-99-2 see BAW250
137-86-0 see GEK200	141-43-5 see EEC600	145-41-5 see SGD500	152-47-6 see MFN500	206-00-8 see FDP000
137-89-3 see BJQ750	141-52-6 see SGR800	145-42-6 see SKB500	152-53-4 see COX325	206-44-0 see FDF000
137-97-3 see DXP600	141-53-7 see SHJ000	145-49-3 see DBP909	152-62-5 see DYF759	207-08-9 see BCJ750
138-14-7 see DAK300	141-57-1 see PNX250	145-63-1 see BAT000	152-72-7 see ABF750	207-83-0 see DDB000
138-15-8 see GFO025	141-65-1 see TBA750	145-73-3 see EAR000	152-97-6 see FDA925	207-84-1 see DCX600
138-22-7 see BRR600	141-66-2 see DGQ875	145-94-8 see CDV700	153-18-4 see RSU000	207-85-2 see BDB250
138-24-9 see TMB250	141-76-4 see IEY000	145-95-9 see MCQ250	153-32-2 see HDK000	207-88-5 see BDB000
138-31-8 see SLP500	141-78-6 see EFR000	146-22-5 see DLY000	153-34-4 see DKY000	207-89-6 see BDA000
138-59-0 see SCE000	141-79-7 see MDJ750	146-28-1 see TFP250	153-39-9 see TCN750	208-07-1 see BCE000
138-61-4 see AMB000	141-82-2 see CCC750	146-36-1 see ARZ000	153-61-7 see CCX250	208-96-8 see AAF500
138-65-8 see ARL750	141-84-4 see RGZ550	146-37-2 see AJS750	153-78-6 see FDI000	211-91-6 see BAW125
138-85-2 see SHU000	141-86-6 see DCC800	146-48-5 see YBJ000	153-87-7 see ECW600	213-46-7 see PIB750
138-86-3 see MCC250	141-90-2 see TFR250	146-54-3 see TKL000	153-94-6 see TNW300	214-17-5 see BCG500
138-89-6 see DSY600	141-91-3 see DST600	146-56-5 see FMP000	154-06-3 see MQI250	215-58-7 see BDH750
139-02-6 see SJF000	141-93-5 see DIU200	146-59-8 see ADJ875	154-17-6 see DAR600	215-64-5 see DDC600
139-05-9 see SGC000	141-95-7 see SIE500	146-77-0 see CEF100	154-21-2 see LGD000	217-59-4 see TMS000
139-06-0 see CAR000	141-97-9 see EFS000	147-20-9 see LJR000	154-23-4 see CCP875	218-01-9 see CML810
139-07-1 see BEM000	142-04-1 see BBL000	147-24-0 see BAU750	154-41-6 see PMJ500	220-42-8 see EBA500
139-08-2 see TCA500	142-16-5 see BJR000	147-47-7 see TLP500	154-42-7 see AMH250	224-41-9 see DCT600
139-10-6 see AOB500	142-18-7 see HMH300	147-61-5 see HMH300	154-69-8 see POO750	224-42-0 see DCS600
139-13-9 see AMT500	142-19-8 see AGH250	147-82-0 see THU750	154-87-0 see TET750	224-53-3 see DCS800
139-25-3 see MJN750	142-22-3 see AGD250	147-84-2 see DJC800	154-93-8 see BIF750	224-98-6 see NAZ000
139-26-4 see FMJ000	142-25-6 see DPK400	147-90-0 see SAI100	154-97-2 see PLX250	225-51-4 see BAW750
139-33-3 see EIX500	142-26-7 see HKM000	147-93-3 see MCK750	154-99-4 see DTM600	226-36-8 see DCS400
139-40-2 see PMN850	142-28-9 see DGF800	147-94-4 see AQQ750	155-00-0 see CJZ000	226-47-1 see DDC800
139-45-7 see TMY000	142-29-0 see CPX750	148-01-6 see DUP300	155-04-4 see BHA750	230-27-3 see BDC000
139-59-3 see PDR500	142-46-1 see BLJ250	148-18-5 see SGJ000	155-41-9 see SBH500	239-01-0 see BCG250
139-60-6 see BJT500	142-47-2 see MRL500	148-24-3 see QPA000	155-91-9 see SNI500	239-60-1 see DDB200
139-65-1 see TFI000	142-59-6 see DXD200	148-51-6 see DAY825	155-97-5 see PPI800	239-64-5 see DCX800
139-66-2 see PGI500	142-62-1 see HEU000	148-56-1 see TKG750	156-10-5 see NKB500	239-67-8 see BDA750
139-70-8 see CMU050	142-63-2 see PIK500	148-65-2 see CHY250	156-25-2 see POP750	240-39-1 see PPI815
139-86-6 see TEY250	142-64-3 see PIK000	148-72-1 see PIF500	156-28-5 see PDE500	240-44-8 see BCI250
139-87-7 see ELP000	142-71-2 see CNI250	148-78-7 see CCE000	156-31-0 see BBK250	244-63-3 see NNR300
139-88-8 see EMT500	142-72-3 see MAD000	148-79-8 see TEX000	156-39-8 see HNL500	257-07-8 see DDE200
139-91-3 see FPI000	142-73-4 see IBH000	148-82-3 see PED750	156-43-4 see PDD500	258-76-4 see DUP000
139-93-5 see SAP500	142-82-5 see HBC500	148-87-8 see EHQ500	156-48-9 see MNU150	260-94-6 see ADJ500
139-94-6 see ENV500	142-83-6 see SKT500	149-15-5 see BOP000	156-51-4 see PFC750	262-12-4 see DDA800
139-96-8 see SON000	142-84-7 see DWR000	149-16-6 see BOO750	156-54-7 see SFN600	262-20-4 see PDQ750
140-03-4 see MFX750	142-87-0 see SOK000	149-29-1 see CMV000	156-56-9 see MJP500	262-38-4 see TBI775
140-05-6 see MIF500	142-88-1 see HEP000	149-30-4 see BDF000	156-57-0 see MCN750	271-89-6 see BCK250
140-08-9 see PHO000	142-90-5 see DXJ000	149-44-0 see MHW000	156-59-2 see DFI200	273-53-0 see BDI500
140-11-4 see BDX000	142-91-6 see IQW000	149-45-1 see DXH300	156-60-5 see ACK000	274-09-9 see MJR000
140-18-1 see BEE500	142-92-7 see HFI500	149-57-5 see BRI250	156-62-7 see CAQ250	280-57-9 see DCK400
140-20-5 see MFC000	142-96-1 see BRH750	149-64-4 see SBG500	156-72-9 see MJQ500	283-56-7 see TJK750
140-25-0 see BEU750	143-07-7 see LBL000	149-73-5 see TLX600	156-87-6 see PMM250	283-60-3 see TMO750
140-26-1 see PDF775	143-08-8 see NNB500	149-74-6 see DFQ800	157-03-9 see DCQ400	283-66-9 see DCK700
140-29-4 see PEA750	143-13-5 see NNB400	149-91-7 see GBE000	157-22-2 see DCM800	286-20-4 see CPD000
140-31-8 see AKB000	143-16-8 see DKO600	149-95-1 see ARL500	188-96-5 see DDC400	287-23-0 see COW000
140-39-6 see MNR250	143-18-0 see OHY000	150-05-0 see AES250	189-55-9 see BCQ500	287-92-3 see CPV750
140-40-9 see ABY900	143-19-1 see OIA000	150-13-0 see AIH600	189-58-2 see APF750	288-13-1 see POM500
140-41-0 see CJY000	143-22-6 see TKL750	150-19-6 see REF050	189-64-0 see DCY200	288-32-4 see IAL000
140-49-8 see CEC000	143-24-8 see PBO500	150-38-9 see TNL250	189-92-4 see PCV500	288-94-8 see TEF650
140-56-7 see DOU600	143-27-1 see HCO500	150-39-0 see HKS000	190-03-4 see BCR500	289-14-5 see EJO500
140-57-8 see SOP500	143-28-2 see OBA000	150-46-9 see TJP500	190-07-8 see BCR250	289-95-2 see PPO750
140-64-7 see DBL800	143-29-3 see BHK750	150-49-2 see BKX750	191-07-1 see CNS250	290-87-9 see THR525
140-66-9 see TDN500	143-33-9 see SGA500	150-50-5 see TIG250	191-24-2 see BCR000	290-96-0 see TEF600
140-67-0 see AFW750	143-33-9 see SGB000	150-60-7 see DXH200	191-26-4 see APE750	291-21-4 see TLS500
140-76-1 see MQM500	143-50-0 see KEA000	150-68-5 see CJX750	191-27-5 see ADK250	291-64-5 see COX500
140-79-4 see DVF200	143-57-7 see POF000	150-69-6 see EFE000	191-30-0 see DCY400	294-93-9 see COD475
140-87-4 see COH250	143-62-4 see DMJ000	150-74-3 see DSA000	192-47-2 see DDC200	297-76-7 see EQJ500
140-88-5 see EFT000	143-81-7 see BPF250	150-76-5 see MFC700	192-65-4 see NAT500	297-78-9 see OAN000

297-88-1 see MDO760	304-28-9 see BGP250	328-74-5 see BLO250	356-27-4 see HAY000	391-70-8 see TNV625
297-90-5 see MKR250	304-43-8 see BGK250	329-01-1 see TKJ250	356-69-4 see PCH275	392-56-3 see HDB000
297-97-2 see EPC500	304-55-2 see DNV800	329-21-5 see BSG000	356-69-4 see PCH350	392-83-6 see BOJ750
297-99-4 see PGX300	304-81-4 see PDB750	329-56-6 see NNP000	357-07-3 see ORG100	393-52-2 see FGP000
298-00-0 see MNH000	304-84-7 see DKE200	329-63-5 see AES625	357-08-4 see NAH000	395-47-1 see TKF525
298-00-0 see MNH010	305-03-3 see CDO500	329-65-7 see EBB500	357-09-5 see FJF100	396-01-0 see UVJ450
298-00-0 see MNH020	305-33-9 see IGG700	329-71-5 see DVA000	357-56-2 see AFJ400	398-32-3 see FKZ000
298-02-2 see PGS000	305-53-3 see SIN000	329-89-5 see ALL250	357-57-3 see BOL750	400-44-2 see CHK750
298-03-3 see DAO500	305-80-6 see BBO250	329-99-7 see MIT600	358-21-4 see PCG760	401-78-5 see BOJ500
298-04-4 see DXH325	305-84-5 see CCK665	330-55-2 see DGD600	358-52-1 see POA250	402-26-6 see TKF530
298-06-6 see PHG500	305-85-1 see DNG000	330-64-3 see DNS200	358-74-7 see DJJ400	402-31-3 see BLO270
298-07-7 see BJR750	305-97-5 see LGU000	330-68-7 see FFH000	359-06-8 see FFR000	402-51-7 see TKF535
298-12-4 see GIQ000	306-07-0 see BEX500	331-39-5 see CAK375	359-40-0 see OLK000	402-71-1 see THH450
298-14-6 see PKX100	306-12-7 see OOK100	331-87-3 see FFI000	359-46-6 see TJX750	404-42-2 see DKF170
298-18-0 see DHB600	306-23-0 see HNG700	331-91-9 see FHQ010	359-48-8 see PCO250	404-72-8 see ISL000
298-39-5 see SLZ000	306-37-6 see DSF800	332-14-9 see PDI500	359-83-1 see DOQ400	404-82-0 see PDM250
298-45-3 see HLT000	306-40-1 see CMG250	332-54-7 see FHQ100	360-53-2 see HAY500	404-86-4 see CBF750
298-46-4 see DCV200	306-53-6 see MKU750	332-69-4 see BMN350	360-54-3 see MKK750	405-22-1 see FPE100
298-51-1 see THK600	306-67-2 see GEK000	332-97-8 see EKK500	360-68-9 see DKW000	405-86-7 see FHV000
298-57-7 see CMR100	306-83-2 see TJY500	333-18-6 see EIW000	360-70-3 see NNE550	406-20-2 see MKE000
298-59-9 see RLK000	308-48-5 see PCG755	333-20-0 see PLV750	360-89-4 see OBO000	406-23-5 see AGG750
298-81-7 see XDJ000	309-00-2 see AFK250	333-25-5 see DEW000	360-97-4 see AKK250	406-90-6 see TKB250
298-93-1 see DUG400	309-29-5 see ENL100	333-29-9 see DXN600	361-09-1 see SFW000	407-25-0 see TJX000
298-96-4 see TMV500	309-36-4 see MDU500	333-36-8 see HDC000	361-37-5 see MLD250	407-83-0 see FHF000
299-11-6 see MNO500	309-43-3 see SBN000	333-40-4 see DTV400	362-74-3 see COV625	407-98-7 see FFW000
299-26-3 see AME500	311-28-4 see TBL000	333-41-5 see DCM750	363-03-1 see PEL750	409-02-9 see MKK000
299-27-4 see PLG800	311-45-5 see NIM500	333-93-7 see POK325	363-13-3 see BEH000	409-21-2 see SCQ000
299-28-5 see CAS750	311-47-7 see CLV375	334-22-5 see BHN750	363-17-7 see FER000	420-04-2 see COH500
299-29-6 see FBK000	311-89-7 see HAS000	334-44-1 see FIY000	363-20-2 see TIH800	420-12-2 see EJP500
299-39-8 see SKX750	312-45-8 see HAQ000	334-48-5 see DAH400	363-24-6 see DVJ200	420-23-5 see DRY289
299-42-3 see EAW000	312-93-6 see BFW325	334-56-5 see FHL000	363-42-8 see TLQ000	420-52-0 see TKF775
299-45-6 see PKT000	313-06-4 see DAZ115	334-62-3 see FHN000	363-49-5 see HIK500	421-17-0 see TKB300
299-75-2 see TFU500	313-67-7 see AQY250	334-64-5 see FHM000	364-62-5 see AJH000	421-20-5 see MKG250
299-84-3 see RMA500	313-74-6 see DQJ400	334-71-4 see FIA000	364-71-6 see FFT000	421-53-4 see TJZ000
299-85-4 see DGD800	313-93-9 see EPZ000	334-88-3 see DCP800	364-98-7 see DCQ700	422-05-9 see PBE750
299-86-5 see COD850	313-94-0 see BSR500	335-57-9 see PCH000	365-26-4 see HKH500	422-61-7 see PBF300
300-08-3 see AQU000	313-95-1 see FMF000	335-76-2 see PCG725	366-18-7 see BGO500	422-63-9 see PBE500
300-37-8 see DNG400	313-96-2 see MFQ750	337-28-0 see EFA000	366-70-1 see PME500	422-64-0 see PBF000
300-42-5 see DBA800	314-03-4 see MOO750	337-47-3 see SOX500	366-71-2 see MKN750	424-40-8 see DJK100
300-54-9 see MRW250	314-04-5 see DKJ200	338-66-9 see DKH825	366-93-8 see BHN000	425-51-4 see MCB375
300-57-2 see AFX000	314-13-6 see BGT250	339-43-5 see BSM000	367-51-1 see SKH500	425-87-6 see CLR000
300-62-9 see BBK000	314-19-2 see AQP500	340-56-7 see MDT250	368-43-4 see BBT250	427-00-9 see DKX600
300-76-5 see NAG400	314-35-2 see CNR125	341-69-5 see OJW000	368-47-8 see PGO500	427-01-0 see GCG300
300-88-9 see CEI250	314-40-9 see BMM650	341-70-8 see DHF600	368-68-3 see DKG100	427-45-2 see TNG050
301-04-2 see LCV000	314-42-1 see BNM000	342-69-8 see MPU000	368-97-8 see DKI400	427-51-0 see CQJ500
301-11-1 see LBO000	315-18-4 see DOS000	342-95-0 see BHT250	369-57-3 see BBO325	429-30-1 see TJX250
301-12-2 see DAP000	315-22-0 see BHT250	343-75-9 see DUK200	370-14-9 see FMS875	431-03-8 see BOT500
302-01-2 see HGS000	315-30-0 see ZVJ000	343-89-5 see FFG000	370-81-0 see COF675	431-46-9 see MFR500
302-15-8 see MKN500	315-37-7 see TBF750	343-94-2 see AJX250	371-28-8 see CGZ000	431-97-0 see BLO300
302-17-0 see CDO000	315-72-0 see DCV800	345-78-8 see POH250	371-29-9 see FIN000	432-04-2 see TNN775
302-22-7 see CBF250	315-80-0 see DCW800	346-18-9 see PKL250	371-40-4 see FHY000	432-60-0 see AGF750
302-23-8 see PMG600	316-14-3 see MGW000	348-67-4 see MDT730	371-41-5 see FKV000	433-27-2 see EFK500
302-27-2 see ADH750	316-41-6 see BFN625	349-37-1 see DKH000	371-47-1 see SIE000	434-03-7 see GEK500
302-33-0 see DIG400	316-42-7 see EAN000	350-03-8 see ABI000	371-62-0 see FIE000	434-05-9 see PMC700
302-40-9 see DHU900	316-46-1 see FMN000	350-46-9 see FKL000	371-67-5 see TJY275	434-07-1 see PAN100
302-41-0 see PJA140	316-49-4 see MGV500	350-87-8 see DKH100	371-69-7 see EKK550	434-13-9 see LHW000
302-48-7 see BGX775	316-81-4 see TFM100	351-63-3 see DKG980	371-78-8 see BLQ325	434-16-2 see DAK600
302-49-8 see EHV500	317-34-0 see TEP500	351-65-5 see DRL000	371-86-8 see PHF750	434-22-0 see NNX400
302-66-9 see MNM500	317-52-2 see HEG000	352-21-6 see AKF375	372-09-8 see COJ500	434-64-0 see PCH500
302-70-5 see CFA750	317-64-6 see DQK000	352-32-9 see FMC000	372-18-9 see DKF800	436-30-6 see MKD750
302-79-4 see VSK950	318-03-6 see BDB500	352-93-2 see EPH000	372-48-5 see FLT100	436-40-8 see DCN000
302-83-0 see TAL490	318-22-9 see TKH000	353-03-7 see EKI000	372-64-5 see BLO325	437-38-7 see PDW500
302-95-4 see SGE000	318-98-9 see ICC000	353-13-9 see FMO000	372-91-8 see FHC000	437-74-1 see XCS000
302-96-5 see AOO400	319-84-6 see BBQ000	353-17-3 see FHD000	373-02-4 see NCX000	438-41-5 see MDQ250
303-04-8 see DFM000	319-85-7 see BBR000	353-18-4 see CON500	373-14-8 see FJA000	438-60-8 see DDA600
303-21-9 see DVG200	319-86-8 see BFW500	353-21-9 see FIZ000	373-88-6 see TKA750	438-67-5 see EDV600
303-25-3 see MAX275	320-67-2 see ARY000	353-36-6 see FIB000	373-91-1 see TKD375	439-14-5 see DCK759
303-26-4 see NNK500	320-72-9 see DGK200	353-42-4 see BMH000	375-22-4 see HAX500	439-25-8 see MKD500
303-33-3 see HAL500	321-25-5 see FHQ000	353-50-4 see CCA500	376-18-1 see HCO000	440-17-5 see TKK250
303-34-4 see LBG000	321-38-0 see FKK000	353-59-3 see BNA250	376-53-4 see PCG600	440-58-4 see AAI750
303-45-7 see GJM000	321-54-0 see CIK750	354-06-3 see TJY000	376-89-6 see HDC300	441-38-3 see BCP500
303-47-9 see CHP250	321-55-1 see DFH600	354-13-2 see TIJ175	377-38-8 see TCJ000	442-51-3 see HAI500
303-49-1 see CDU750	321-64-2 see TCJ075	354-21-2 see TIM000	378-44-9 see BFV750	443-30-1 see DOT600
303-53-7 see PMH600	324-93-6 see AKC500	354-23-4 see TJY750	379-79-3 see EDC500	443-48-1 see MMN250
303-54-8 see DPW600	325-23-5 see BCI750	354-32-5 see TJX500	381-73-7 see DKF200	444-27-9 see TEV000
303-69-5 see DYB600	325-69-9 see FLE000	354-93-8 see PCH300	382-21-8 see OBM000	445-29-4 see FGH000
303-70-8 see DPX400	326-43-2 see PGG355	355-02-2 see PCH290	382-67-2 see DBA875	446-72-0 see GCM350
303-81-1 see SMB000	326-61-4 see PIX000	355-43-1 see PCH100	383-73-3 see BLO000	446-86-6 see ASB250
303-98-0 see UAH000	327-97-9 see CHK175	355-66-8 see OBK100	388-72-7 see FFZ000	447-05-2 see PPJ900
304-06-3 see PGH000	327-98-0 see EPY000	355-80-6 see OBU000	389-08-2 see EID000	447-25-6 see DAR150
304-17-6 see IRG000	328-04-1 see CJD650	356-12-7 see FDD150	390-64-7 see PEV750	451-40-1 see PEB000
304-20-1 see HGP500	328-38-1 see LER000		391-57-1 see FEE000	452-06-2 see AMH000

452-35-7 see EEN500	475-08-1 see CBK125	494-47-3 see FPS000	506-30-9 see EAF000	518-75-2 see CMS775
452-86-8 see DNE200	475-26-3 see FHJ000	494-52-0 see AON875	506-32-1 see AQS750	518-82-1 see MQF250
453-13-4 see DKI800	475-81-0 see TDI475	494-97-3 see NNR500	506-59-2 see DOR600	519-23-3 see EAI850
453-18-9 see MKD000	475-83-2 see NOE500	494-98-4 see PQB500	506-61-6 see PLS250	519-37-9 see HLC000
454-41-1 see ALF600	476-32-4 see CDL000	495-18-1 see BCL500	506-63-8 see DQR200	519-65-3 see DVU100
455-14-1 see TKB750	476-70-0 see DNZ100	495-48-7 see ASO750	506-64-9 see SDP000	519-87-9 see PDX500
455-16-3 see TGO500	477-27-0 see ADE000	495-54-5 see DBP999	506-65-0 see GIW189	519-88-0 see DDT300
455-80-1 see HHM500	477-29-2 see DAN375	495-73-8 see BDD000	506-68-3 see COO500	520-07-0 see AQN250
456-59-7 see DNU100	477-30-5 see MIW500	496-06-0 see IGF300	506-77-4 see COO750	520-09-2 see DYC700
456-88-2 see FLD000	478-15-9 see TDI750	496-11-7 see IBR000	506-78-5 see COP000	520-18-3 see ICE000
457-60-3 see NCJ500	478-84-2 see BNM250	496-67-3 see BNP750	506-82-1 see DQW800	520-26-3 see HBU000
457-87-4 see EGI500	478-99-9 see LJI000	496-72-0 see TGM250	506-85-4 see FOS050	520-36-5 see CDH250
458-24-2 see ENJ000	479-13-0 see COF350	496-74-2 see TGN000	506-87-6 see ANE000	520-45-6 see MFW500
458-88-8 see PNT000	479-18-5 see DNC000	497-18-7 see CBS500	506-93-4 see GLA000	520-52-5 see PHU500
459-02-9 see FKG000	479-23-2 see CMC000	497-19-8 see SFO000	506-96-7 see ACD750	520-53-6 see HKE000
459-22-3 see FLC000	479-45-8 see TEG250	497-25-6 see OMM000	507-02-8 see ACO500	520-68-3 see EAC500
459-72-3 see EKG500	479-50-5 see DHU000	497-26-7 see MJH775	507-09-5 see TFA500	520-85-4 see MBZ150
459-80-3 see GCW000	479-92-5 see INY000	497-38-1 see NNK000	507-16-4 see SNT200	521-10-8 see AOO475
459-99-4 see FIM000	480-16-0 see MRN500	497-39-2 see DDX000	507-19-7 see BQM250	521-11-9 see MJE760
460-07-1 see ACB250	480-18-2 see DMD000	497-56-3 see DUT000	507-20-0 see BQR000	521-18-6 see DME500
460-12-8 see BOQ625	480-22-8 see APH250	497-76-7 see HIH100	507-25-5 see CBY500	521-24-4 see DLK000
460-19-5 see COO000	480-30-8 see SKS700	498-02-2 see HLQ500	507-28-8 see TEA300	521-35-7 see CBD625
461-56-3 see FLR000	480-54-6 see RFP000	498-23-7 see CMS320	507-40-4 see BRP500	521-74-4 see DDS600
461-58-5 see COP125	480-79-5 see IDG000	498-24-8 see MDI250	507-42-6 see THU500	521-78-8 see SOX550
461-72-3 see HGO600	480-81-9 see SBX500	498-66-8 see NNH500	507-60-8 see SBF500	522-00-9 see DIR000
461-78-9 see CLY250	481-06-1 see SAU500	499-04-7 see AQT625	507-70-0 see BMD000	522-12-3 see QCJ000
461-89-2 see THR750	481-39-0 see WAT000	499-12-7 see ADH000	508-53-2 see HJX500	522-16-7 see ARS250
462-06-6 see FGA000	481-42-5 see PJH610	499-44-5 see IRR000	508-59-8 see PAM175	522-23-6 see MDU750
462-08-8 see AMI250	481-49-2 see CCX550	499-75-2 see CCM000	508-65-6 see EBL500	522-25-8 see POL000
462-72-6 see FHA000	481-72-1 see DMU600	500-00-5 see MCE500	508-75-8 see CNH780	522-40-7 see DKA200
462-73-7 see FHB000	481-85-6 see MMC250	500-28-7 see MIJ250	508-77-0 see CQH750	522-48-5 see VRZ000
462-94-2 see PBK500	482-41-7 see MKD250	500-34-5 see EQP500	509-09-1 see PBF250	522-60-1 see HHR700
462-95-3 see EFT500	482-44-0 see IHR300	500-38-9 see NBR000	509-14-8 see TDY250	522-70-3 see BLX750
463-04-7 see AOL500	482-49-5 see DYB000	500-55-0 see AQO250	509-15-9 see GCK000	523-44-4 see FAG010
463-40-1 see OAX100	483-04-5 see AFG750	500-64-1 see GJI250	509-20-6 see ADG500	523-50-2 see FQC000
463-51-4 see KEU000	483-18-1 see EAL500	500-72-1 see OLW000	509-67-1 see TCY750	523-80-8 see AGE500
463-58-1 see CCC000	483-55-6 see HMI000	500-92-5 see CKB250	509-86-4 see COY500	523-86-4 see SMC500
463-71-8 see TFN500	483-57-8 see FBP300	501-30-4 see HLH500	510-13-4 see MAK500	523-87-5 see DYE600
463-82-1 see NCH000	483-63-6 see BEF500	501-53-1 see BEF500	511-09-1 see EDB100	524-42-5 see NBA000
463-88-7 see VQR300	484-20-8 see MFN275	501-68-8 see BEG000	511-12-6 see DLK800	524-83-4 see DWE800
464-10-8 see NMQ000	484-23-1 see OJD300	502-26-1 see SLL400	511-13-7 see CMW700	525-02-0 see BEM750
464-45-9 see NCQ820	484-47-9 see TMS750	502-37-4 see HOW100	511-46-6 see CKE000	525-05-3 see HNB000
464-48-2 see CBB000	484-78-6 see HJD000	502-39-6 see MLF250	511-55-7 see PEM750	525-26-8 see CMX840
464-49-3 see CBB250	485-19-8 see SLX500	502-42-1 see SMV000	512-13-0 see TLW000	525-64-4 see FDM000
465-16-7 see OHQ000	485-31-4 see BGB500	502-44-3 see LAP000	512-16-3 see COW700	525-66-6 see ICB000
465-19-0 see BOM655	485-35-8 see CQL500	502-47-6 see CMT125	512-24-3 see CCZ000	526-08-9 see AIF000
465-39-4 see BOM650	485-47-2 see DMV200	502-49-8 see CPS250	512-48-1 see DJU200	526-18-1 see DYE700
465-65-6 see NAG550	485-89-2 see OPK300	502-55-6 see BJU000	512-56-1 see TMD250	526-26-1 see SML500
465-69-0 see FKF100	486-17-9 see BRS000	502-56-7 see NMZ000	512-64-1 see EAD500	526-55-6 see ICS000
465-73-6 see IKO000	486-25-9 see FDO000	502-72-7 see CPU250	512-85-6 see ARM500	526-62-5 see BGW750
466-06-8 see POB500	486-84-0 see MPA050	502-85-2 see HJS500	513-10-0 see TLF500	526-73-8 see TLL500
466-11-5 see DBC550	487-10-5 see ASN500	503-01-5 see ILK000	513-12-2 see EPI000	526-75-0 see XKJ000
466-24-0 see PIC250	487-19-4 see NDX300	503-09-3 see EBU000	513-31-5 see DDS200	526-84-1 see DMW200
466-40-0 see IKZ000	487-53-6 see DHO600	503-17-3 see COC500	513-35-9 see AOI750	526-99-8 see GAR000
466-81-9 see EDG500	487-54-7 see SAN200	503-20-8 see FFJ000	513-36-0 see CIU500	527-07-1 see SHK800
466-99-9 see DLW600	487-93-4 see DPG109	503-28-6 see ASN400	513-37-1 see IKE000	527-53-7 see TDM500
467-22-1 see CBQ625	488-17-5 see DNE000	503-30-0 see OMW000	513-38-2 see IIV509	527-73-1 see NHG000
467-36-7 see TES500	488-23-3 see TDM250	503-41-3 see DXI500	513-42-8 see IMW000	528-21-2 see TKN250
467-60-7 see DWK400	488-41-5 see DDP600	503-49-1 see HMC000	513-44-0 see IIX000	528-29-0 see DUQ400
467-63-0 see TJK000	488-81-3 see RIF000	503-74-2 see ISU000	513-48-4 see IEH000	528-48-3 see FBW000
468-28-0 see LIU000	489-84-9 see DSJ800	503-80-0 see BJJ250	513-77-9 see BAJ250	528-53-0 see DAM400
468-61-1 see DHQ200	490-02-8 see ARO000	503-87-7 see TFJ750	513-81-5 see DQT150	528-58-5 see COI750
469-21-6 see DYE500	490-31-3 see RLP000	504-15-4 see MPH500	513-85-9 see BOT000	528-74-5 see DFO000
469-59-0 see JCS000	490-78-8 see DMG600	504-17-6 see MCK500	513-86-0 see ABB500	528-76-7 see DUR200
469-61-4 see CCR500	490-79-9 see GCU000	504-20-1 see PGW250	513-92-8 see TDE250	528-92-7 see IQX000
469-62-5 see DAB879	490-91-5 see IQF000	504-24-5 see AMI500	514-10-3 see AAC500	528-97-2 see BQH250
469-65-8 see HJO500	491-07-6 see IKY000	504-29-0 see AMI000	514-61-4 see MDM350	529-05-5 see DRV000
469-81-8 see MRN675	491-35-0 see LEL000	504-60-9 see PBA250	514-65-8 see BGD500	529-19-1 see TGT500
470-82-6 see CAL000	491-36-1 see QFA000	504-63-2 see PML250	514-73-8 see DJT800	529-33-9 see TDI350
470-90-6 see CDS750	491-59-8 see CML750	504-75-6 see IAT000	514-85-2 see VSK975	529-34-0 see DLX200
471-03-4 see BIH500	491-92-9 see RHZ000	504-88-1 see NIY500	515-64-0 see SNJ350	529-65-7 see EFQ500
471-25-0 see PMT275	492-08-0 see PAB250	504-90-5 see TFS500	515-83-3 see TIO000	530-31-4 see ANM250
471-29-4 see MKI750	492-17-1 see BGF109	505-14-6 see TFH600	516-21-2 see COX400	530-35-8 see EJR500
471-35-2 see TDP250	492-18-2 see SIH500	505-22-6 see DVP600	516-95-0 see EBA100	530-43-8 see CDP700
471-46-5 see OLO000	492-41-1 see NNM000	505-44-2 see MPN100	517-09-9 see ECV000	530-78-9 see TKH750
471-53-4 see GIE000	492-80-8 see IBB000	505-57-7 see HFA500	517-16-8 see EME500	530-91-6 see TCX750
471-77-2 see NBU800	492-94-4 see FPZ000	505-60-2 see BIH250	517-25-9 see TMM500	531-18-0 see HDY000
471-95-4 see BON000	493-52-7 see CCE500	505-66-8 see HGI900	517-28-2 see HAP500	531-72-6 see TFA350
472-54-8 see NNT500	493-53-8 see ADA750	505-71-5 see DUU800	517-85-1 see DCR000	531-76-0 see BHT750
473-41-6 see BSQ250	494-03-1 see BIF250	505-71-5 see DUV800	518-34-3 see TDX830	531-82-8 see AAL750
474-25-9 see CDL325	494-19-9 see DKY800	505-75-9 see CMN000	518-47-8 see FEW000	531-85-1 see BBX750
474-86-2 see ECW000	494-38-2 see BJF000	506-12-7 see HAS500		531-86-2 see BBY000

532-03-6 see GKK000	540-42-1 see PMV250	547-63-7 see MKX000	556-89-8 see NMQ500	578-54-1 see EGK500
532-11-6 see A00490	540-47-6 see CQE750	547-91-1 see IEP200	556-90-1 see IAP000	578-66-5 see AML250
532-27-4 see CEA750	540-51-2 see BNI500	547-95-5 see MIG850	556-97-8 see CLV500	578-94-9 see PDB000
532-28-5 see MAP250	540-54-5 see CKP750	548-00-5 see BKA000	557-05-1 see ZMS000	579-07-7 see PGA500
532-32-1 see SFB000	540-59-0 see DFI100	548-26-5 see BNK700	557-07-3 see ZJS000	579-10-2 see MFW000
532-33-4 see BRT000	540-61-4 see GHI000	548-42-5 see AEY375	557-09-5 see ZEJ000	580-48-3 see CDQ325
532-43-4 see TET500	540-63-6 see EEB000	548-43-6 see EAJ000	557-11-9 see AGV000	580-74-5 see DVU000
532-49-0 see DDW000	540-67-0 see EMT000	548-57-2 see SBE500	557-18-6 see DJO100	581-28-2 see AHS000
532-54-7 see ILH000	540-69-2 see ANH500	548-61-8 see THP000	557-19-7 see NDB500	581-29-3 see ADJ375
532-62-7 see AIT750	540-72-7 see SIA500	548-62-9 see AOR500	557-20-0 see DKE600	581-64-6 see AKK750
532-76-3 see COU250	540-73-8 see DSF600	548-68-5 see DHY400	557-21-1 see ZGA000	581-88-4 see IKB000
532-82-1 see PEK000	540-80-7 see BRV760	548-73-2 see DYF200	557-30-2 see EEA000	581-89-5 see NHQ500
532-94-5 see SHN500	540-84-1 see TLY500	548-83-4 see GAZ000	557-34-6 see ZBS000	582-08-1 see ASN750
533-06-2 see CBK500	540-88-5 see BPV100	548-93-6 see AKE750	557-40-4 see DBK000	582-17-2 see NAO500
533-23-3 see EHY600	541-09-3 see UPS000	549-18-8 see EAI000	557-48-2 see NMV760	582-25-2 see PKW760
533-28-8 see IJZ000	541-19-5 see BJI000	549-49-5 see QJJ100	557-66-4 see EFW000	582-61-6 see BDL750
533-31-3 see MJU000	541-22-0 see DAF600	550-24-3 see EAJ600	557-91-5 see DDN800	583-03-9 see BQJ500
533-45-9 see CHD750	541-25-3 see CLV000	550-28-7 see AKL625	557-93-7 see BOA250	583-15-3 see MCX500
533-51-7 see SDU000	541-41-3 see EHK500	550-33-4 see RJF000	557-98-2 see CKS000	583-39-1 see BCC500
533-58-4 see IEV000	541-42-4 see IQQ000	550-34-5 see NAH800	557-99-3 see ACM000	583-57-3 see DRF800
533-73-3 see BBU250	541-47-9 see MHT500	550-70-9 see TMX775	558-13-4 see CBX750	583-58-4 see LJB000
533-74-4 see DSB200	541-53-7 see DXL800	550-74-3 see PIE000	558-17-8 see TLU000	583-59-5 see MIR000
533-75-5 see TNV550	541-58-2 see DUG200	550-82-3 see HNG500	558-25-8 see MDR750	583-60-8 see MIR500
533-87-9 see TKP000	541-59-3 see MAM750	550-90-3 see LIQ800	560-53-2 see SMN002	583-63-1 see BDC250
533-96-0 see SJT750	541-64-0 see FPY000	550-99-2 see NCW000	561-27-3 see HBT500	583-75-5 see MHR500
534-07-6 see BIK250	541-66-6 see FMX000	551-06-4 see ISN000	561-43-3 see IBP200	583-80-2 see MQI000
534-13-4 see DSK900	541-69-5 see PEY750	551-08-6 see BRQ100	561-78-4 see NEB000	584-02-1 see IHP010
534-15-6 see DOO600	541-73-1 see DEP699	551-11-1 see POC500	562-09-4 see CIS000	584-03-2 see BOS250
534-17-8 see CDC750	541-85-5 see EGI750	551-16-6 see PCU500	562-10-7 see PGE775	584-08-7 see PLA000
534-22-5 see MKH000	541-91-3 see MIT625	551-36-0 see AMM750	562-74-3 see TBD825	584-26-9 see ADC750
534-33-8 see ACN250	541-95-7 see MOU500	551-58-6 see SOW500	562-95-8 see TJS500	584-79-2 see AFR250
534-52-1 see DUS700	542-46-1 see CMU850	551-74-6 see MAW750	563-12-2 see EEH600	584-84-9 see TGM750
534-76-9 see EQS500	542-54-1 see MNJ000	551-92-8 see DSV800	563-25-7 see DDY800	584-93-0 see BOL250
535-55-7 see MCU500	542-55-2 see IIR000	552-16-9 see NFG500	563-41-7 see SBW500	584-94-1 see DSE509
535-65-9 see GEW750	542-56-3 see IJD000	552-30-7 see TKV000	563-45-1 see MHT250	585-08-0 see PCX000
535-89-7 see CCP500	542-59-6 see EJI000	552-41-0 see PAC250	563-46-2 see MHT000	585-54-6 see AQZ900
536-17-4 see DOT800	542-62-1 see BAK750	552-46-5 see NBF000	563-47-3 see CIU750	586-06-1 see DMV800
536-21-0 see AKT000	542-63-2 see DIV000	552-80-7 see DUC300	563-52-0 see CEV250	586-11-8 see DVA600
536-25-4 see AKF000	542-69-8 see BRQ250	552-86-3 see FQI000	563-54-2 see DGG800	586-38-9 see AOU500
536-29-8 see DFX400	542-75-6 see DGG950	552-89-6 see NEU500	563-68-8 see TEI250	586-62-9 see TBE000
536-33-4 see EPQ000	542-76-7 see CKT250	552-94-3 see SAN000	563-71-3 see FBH100	586-77-6 see BNF250
536-43-6 see BPR500	542-78-9 see PMK000	553-24-2 see AJQ250	563-80-4 see MLA750	586-92-5 see POQ500
536-46-9 see DTM000	542-88-1 see BIK000	553-27-5 see AOQ875	564-00-1 see DHB800	586-98-1 see POR800
536-50-7 see TGZ000	542-90-5 see EPP000	553-30-0 see DBN400	564-25-0 see DYE425	587-15-5 see DGV200
536-59-4 see PCI550	542-92-7 see CPU500	553-53-7 see NDU500	564-36-3 see EDA600	587-63-3 see DLR000
536-60-7 see CQI250	543-15-7 see HBB000	553-54-8 see LGW000	565-33-3 see AKQ250	587-84-8 see DVY000
536-69-6 see BSI000	543-20-4 see SNG000	553-68-4 see IAC000	565-59-3 see DTI200	587-85-9 see DWD800
536-74-3 see PEB750	543-21-5 see ACJ250	553-69-5 see PGG350	565-80-0 see DTI600	587-98-4 see MDM775
536-80-1 see IFC000	543-38-4 see AKD500	553-84-4 see PCI750	566-09-6 see SKR500	588-16-9 see AAQ750
537-00-8 see CCY500	543-39-5 see MLO250	553-97-9 see MHI250	566-28-9 see ONO000	588-22-7 see DFY500
537-05-3 see PDN500	543-49-7 see HBE500	554-12-1 see MOT000	566-48-3 see HJB200	588-42-1 see TJL250
537-12-2 see DVV500	543-53-3 see PPB550	554-13-2 see LGZ000	567-47-5 see HMX000	588-59-0 see SLR000
537-45-1 see CHQ750	543-59-9 see PBW500	554-14-3 see MPV000	568-69-4 see TNJ750	589-16-2 see EGL000
537-92-8 see ABI750	543-63-5 see BRS750	554-18-7 see A00800	568-70-7 see HMF500	589-18-4 see MHB250
537-98-4 see FBP200	543-67-9 see PNQ750	554-35-8 see GFC100	568-75-2 see HMF000	589-38-8 see HEV500
538-02-3 see CPV609	543-80-6 see BAH500	554-70-1 see TJT775	568-81-0 see DQK200	589-41-3 see HKQ025
538-03-4 see ARL000	543-81-7 see BFP000	554-76-7 see SLP600	569-57-3 see CLO750	589-43-5 see DSE600
538-04-5 see CHO750	543-82-8 see ILM000	554-84-7 see NIE000	569-58-4 see AGW750	589-59-3 see ITA000
538-07-8 see BID250	543-86-2 see IHQ000	554-92-7 see TKW750	569-59-5 see PDD000	589-79-7 see EKJ500
538-09-9 see DBJ200	543-90-6 see CAD250	554-99-4 see MJV000	569-61-9 see RMK020	589-82-2 see HBF000
538-23-8 see TMO000	543-94-2 see SME000	555-06-6 see SEO500	569-64-2 see AFG500	589-90-2 see DRG200
538-28-3 see BEU500	544-13-8 see TLR500	555-15-7 see NGC000	569-65-3 see HGC500	589-92-4 see MIR625
538-32-9 see BFN125	544-16-1 see BRV500	555-16-8 see NEV500	569-77-7 see TDD500	589-93-5 see LJA000
538-42-1 see SFX000	544-17-2 see CAS250	555-21-5 see NIJ000	570-22-9 see IAM000	590-00-1 see PLS750
538-65-8 see BQV500	544-25-2 see COY000	555-30-6 see DNA800	571-22-2 see HJB100	590-01-2 see BSJ500
538-71-6 see DXX000	544-40-1 see BSM125	555-37-3 see BRA250	571-60-8 see NAN500	590-14-7 see BOA000
538-79-4 see MDM750	544-63-8 see MSA250	555-57-7 see MOS250	572-48-5 see DXO000	590-19-2 see BOP250
538-93-2 see IIN000	544-92-3 see CNL000	555-65-7 see BMM600	573-56-8 see DVA200	590-21-6 see PMR750
539-03-7 see CDZ100	544-97-8 see DUO200	555-77-1 see TNF250	573-58-0 see SGQ500	590-28-3 see PLC250
539-17-3 see DPO200	545-06-2 see TII750	555-84-0 see NDY000	573-83-1 see PLQ775	590-29-4 see PLG750
539-21-9 see AHI875	545-55-1 see TND250	555-89-5 see NCM700	574-25-4 see TFJ825	590-63-6 see HOA500
539-35-5 see CMP885	545-91-5 see DWH600	555-96-4 see BEQ000	576-08-9 see LAR100	590-86-3 see MHX500
539-48-0 see PEX250	545-93-7 see QCS000	556-12-7 see FPF000	576-26-1 see XLA000	590-88-5 see BOR750
539-71-9 see ELO000	546-46-3 see ZFJ250	556-22-9 see GII000	576-55-6 see TBJ500	590-92-1 see BOB250
539-88-8 see EFS600	546-48-5 see PAP110	556-24-1 see ITC000	576-68-1 see MAW500	590-96-5 see HMG000
539-90-2 see BSW500	546-71-4 see ENQ000	556-52-5 see GGW500	577-11-7 see DJL000	591-08-2 see ADD250
540-09-0 see TJF250	546-80-5 see TFW000	556-61-6 see ISE000	577-55-9 see DNN800	591-09-3 see ACS750
540-18-1 see AOG000	546-88-3 see ABB250	556-64-9 see MPT000	577-59-3 see NEL450	591-10-6 see BJJ500
540-23-8 see TGS750	546-93-0 see MAC650	556-65-0 see TFF500	577-66-2 see EEO000	591-11-7 see MKH500
540-36-3 see DKG000	547-32-0 see MRM250	556-72-9 see PBJ750	577-71-9 see DVA400	591-12-8 see A00750
540-37-4 see IEC000	547-44-4 see SNQ550	556-82-1 see MHU110	577-91-3 see PDM750	591-12-8 see MKH250
540-38-5 see IEW000	547-58-0 see MND600	556-88-7 see NHA500	578-32-5 see DRK800	591-21-9 see DRG000

```
7681-53-0 see SHV000      7772-99-8 see TGC000      7783-92-8 see SDN399      7790-03-6 see COM125      8002-65-1 see NBS300
7681-55-2 see SHV500      7773-01-5 see MAR000      7783-93-9 see SDV000      7790-07-0 see AEP250      8002-66-2 see CDH500
7681-57-4 see SII000      7773-06-0 see ANU650      7783-95-1 see SDQ500      7790-12-7 see AKD625      8002-74-2 see PAH750
7681-76-7 see MMN750      7773-34-4 see DJB200      7784-08-9 see SDM100      7790-21-8 see PLO750      8002-75-3 see PAE500
7681-82-5 see SHW000      7774-29-0 see MDD000      7784-13-6 see AGZ000      7790-28-5 see SJB500      8002-89-9 see TEQ250
7681-93-8 see PIF750      7774-29-0 see MDD250      7784-18-1 see AHB000      7790-29-6 see RQA000      8003-03-0 see ARP250
7682-90-8 see BAS000      7774-41-6 see ARC500      7784-19-2 see THQ500      7790-30-9 see TEK500      8003-05-2 see MDH500
7683-59-2 see DMV600      7775-09-9 see SFS000      7784-21-6 see AHB500      7790-47-8 see TGD750      8003-08-5 see POF275
7689-03-4 see CBB870      7775-11-3 see DXC200      7784-24-9 see AHF200      7790-59-2 see PLR750      8003-19-8 see DGG000
7693-26-7 see PLJ250      7775-14-6 see SHR500      7784-27-2 see AHD900      7790-76-3 see CAW450      8003-34-7 see POO250
7697-37-2 see NED500      7775-27-1 see SJE000      7784-30-7 see PHB500      7790-78-5 see CAE425      8004-13-5 see PFA860
7697-37-2 see NEE500      7775-41-9 see SDQ000      7784-33-0 see ARF250      7790-79-6 see CAG250      8004-87-3 see MQN025
7698-91-1 see MGL600      7776-33-2 see SBU710      7784-34-1 see ARF500      7790-84-3 see CAJ250      8006-25-5 see EDC565
7698-97-7 see FAO200      7778-18-9 see CAX500      7784-35-2 see ARI250      7790-86-5 see CCY750      8006-28-8 see SEE000
7699-31-2 see DJL600      7778-39-4 see ARB250      7784-37-4 see MDF350      7790-91-2 see CDX750      8006-38-0 see ADH500
7699-41-4 see SCL000      7778-43-0 see ARC000      7784-40-9 see LCK000      7790-92-3 see HOV000      8006-39-1 see TBD500
7699-43-6 see ZSJ000      7778-44-1 see ARB750      7784-41-0 see ARD250      7790-93-4 see CDU000      8006-61-9 see GBY000
7700-17-6 see COD000      7778-50-9 see PKX250      7784-42-1 see AHD500      7790-94-5 see CLG500      8006-64-2 see TOD750
7704-34-9 see SOD500      7778-54-3 see HOV500      7784-44-3 see DCG800      7790-98-9 see PCD500      8006-75-5 see DNU400
7704-71-4 see MAF750      7778-66-7 see PLK000      7784-45-4 see ARG750      7790-99-0 see IDS000      8006-78-8 see LBK000
7704-99-6 see ZRA000      7778-74-7 see PLO500      7784-46-5 see SEY500      7791-10-8 see SMF500      8006-80-2 see OHI000
7705-07-9 see TGG250      7778-80-5 see PLT000      7784-46-5 see SEZ000      7791-10-8 see SMG000      8006-81-3 see YAT000
7705-08-1 see FAU000      7778-83-8 see PNH250      7785-33-3 see GDO000      7791-11-9 see RPF000      8006-82-4 see BLW250
7705-08-0 see FAV000      7779-27-3 see HDW000      7785-84-4 see SKM500      7791-12-0 see TEJ250      8006-83-5 see HOX000
7705-12-6 see IHB675      7779-41-1 see AFJ700      7785-87-7 see MAU250      7791-13-1 see CNB800      8006-84-6 see FAP000
7718-54-9 see NDH000      7779-80-8 see IIS000      7786-17-6 see MJO750      7791-18-6 see MAE500      8006-90-4 see PCB250
7719-09-7 see TFL000      7779-88-6 see ZJJ000      7786-29-0 see MNC175      7791-20-0 see NDA000      8006-99-3 see CDL500
7719-12-2 see PHT275      7780-06-5 see IOO000      7786-30-3 see MAE250      7791-21-1 see DEL600      8007-00-9 see PCP750
7720-78-7 see FBN100      7782-39-0 see DBB800      7786-34-7 see MQR750      7791-23-3 see SBT500      8007-00-9 see PCQ000
7721-01-9 see TAF000      7782-41-4 see FEZ000      7786-67-6 see MCE750      7791-25-5 see SOT000      8007-01-0 see RNA000
7722-06-7 see AMN300      7782-44-7 see OQW000      7786-81-4 see NDK500      7791-26-6 see URA000      8007-01-0 see RNF000
7722-64-7 see PLP000      7782-49-2 see SBO500      7787-32-8 see BAM000      7791-27-7 see PPR500      8007-02-1 see LEH000
7722-73-8 see HGN000      7782-49-2 see SBP000      7787-36-2 see PCK000      7795-91-7 see EJJ000      8007-08-7 see GEQ000
7722-84-1 see HIB000      7782-50-5 see CDV750      7787-47-5 see BFQ000      7796-16-9 see TIV275      8007-11-2 see OJO000
7722-84-1 see HIB010      7782-61-8 see IHC000      7787-49-7 see BFR500      7803-49-8 see HLM500      8007-12-3 see OGQ100
7722-86-3 see PCN750      7782-63-0 see FBO000      7787-52-2 see BFR250      7803-51-2 see PGY000      8007-20-3 see CCQ500
7722-88-5 see TEE500      7782-64-1 see MAS750      7787-56-6 see BFU500      7803-52-3 see SLQ000      8007-45-2 see CMY800
7723-14-0 see PHO500      7782-65-2 see GEI100      7787-62-4 see BKW750      7803-55-6 see ANY250      8007-45-2 see CMY805
7723-14-0 see PHO740      7782-68-5 see IDK000      7787-69-1 see CDC500      7803-57-8 see HGU500      8007-46-3 see TFX500
7723-14-0 see PHP000      7782-75-4 see MAH775      7787-71-5 see BMQ325      7803-62-5 see SDH575      8007-46-3 see TFX750
7726-95-6 see BMP000      7782-77-6 see NMR000      7788-99-0 see CMG850      7803-68-1 see TAI750      8007-56-5 see HHM000
7727-15-3 see AGX750      7782-78-7 see NMJ000      7789-00-6 see PLB250      8000-03-1 see EDQ000      8007-70-3 see AOU250
7727-21-1 see DWQ000      7782-79-8 see HHG500      7789-04-0 see CMK300      8000-25-7 see RMU000      8007-75-8 see BFO000
7727-37-9 see NGP500      7782-87-8 see PLQ750      7789-06-2 see SMH000      8000-26-8 see PIH400      8007-80-5 see CCO750
7727-37-9 see NGP510      7782-89-0 see LGT000      7789-09-5 see ANB500      8000-26-8 see PIH500      8008-20-6 see KEK000
7727-43-7 see BAP000      7782-92-5 see SEN000      7789-12-0 see SGI500      8000-27-9 see CCR000      8008-26-2 see OGO000
7727-54-0 see ANR000      7782-94-7 see NEG000      7789-17-5 see CDE000      8000-28-0 see LCC000      8008-31-9 see TAD500
7732-18-5 see WAT259      7782-99-2 see SOO500      7789-18-6 see CDE250      8000-28-0 see LCD000      8008-45-5 see NOG500
7733-02-0 see ZNA000      7783-00-8 see SBO000      7789-20-0 see HAK000      8000-29-1 see CMT000      8008-46-6 see OGU000
7738-94-5 see CMH250      7783-06-4 see HIC500      7789-21-1 see FLZ000      8000-34-8 see CMY075      8008-51-3 see CBB500
7739-33-5 see TKU000      7783-07-5 see HIC000      7789-23-3 see PLF500      8000-42-8 see CBG500      8008-52-4 see CNR735
7745-89-3 see CLK230      7783-08-6 see SBN500      7789-24-4 see LHF000      8000-46-2 see GDA000      8008-56-8 see LEI000
7751-31-7 see BHQ750      7783-20-2 see ANU750      7789-25-5 see NMH500      8000-48-4 see EQQ000      8008-56-8 see LEJ000
7757-79-1 see PLL500      7783-28-0 see ANR500      7789-26-6 see NMU000      8000-66-6 see CCJ625      8008-57-9 see OGY000
7757-81-5 see SJV000      7783-30-4 see MDC750      7789-27-7 see TEK000      8000-68-8 see PAL750      8008-74-0 see SCB000
7757-82-6 see SJY000      7783-33-7 see NCP500      7789-29-9 see PKU250      8000-95-1 see CAK800      8008-79-5 see SKY000
7757-83-7 see SJZ000      7783-35-9 see MDG500      7789-29-9 see PKU500      8001-15-8 see EAM500      8008-94-4 see LFN300
7757-93-9 see CAW100      7783-36-0 see MDG250      7789-30-2 see BMQ000      8001-23-8 see SAC000      8011-47-0 see CBB375
7758-01-2 see PKY300      7783-40-6 see MAF500      7789-33-5 see IDN200      8001-25-0 see OIQ000      8011-76-5 see SOV500
7758-02-3 see PKY500      7783-41-7 see ORA000      7789-38-0 see SFG000      8001-26-1 see LGK000      8012-34-8 see MCV750
7758-05-6 see PLK250      7783-42-8 see TFL250      7789-40-4 see TEI750      8001-28-3 see COC250      8012-54-2 see ARI500
7758-09-0 see PLM500      7783-46-2 see LDF000      7789-41-5 see CAR375      8001-29-4 see CNU000      8012-74-6 see LIC000
7758-16-9 see DXF800      7783-47-3 see TGD100      7789-42-6 see CAD600      8001-30-7 see CNS000      8012-75-7 see CAP750
7758-19-2 see SFT500      7783-48-4 see SMI500      7789-43-7 see CNB250      8001-31-8 see CNO000      8012-89-3 see BAU000
7758-19-2 see SFU000      7783-49-5 see ZHS000      7789-46-0 see IGP000      8001-35-2 see CDV100      8012-91-7 see JEA000
7758-23-8 see CAW110      7783-50-8 see FAX000      7789-47-1 see MCY000      8001-50-1 see TBC500      8012-95-1 see MQV750
7758-88-5 see CDA750      7783-53-1 see MAW000      7789-51-7 see SBN550      8001-54-5 see AFP250      8012-96-2 see IGF000
7758-89-6 see CNK250      7783-54-2 see NGW000      7789-57-3 see THX000      8001-58-9 see CMY825      8013-10-3 see JEJ000
7758-94-3 see FBI000      7783-55-3 see PHQ500      7789-59-5 see PHU000      8001-61-4 see CNH792      8013-75-0 see FQT000
7758-95-4 see LCQ000      7783-56-4 see AQE000      7789-60-8 see PHT250      8001-64-7 see CNH775      8013-76-1 see BLV500
7758-97-6 see LCR000      7783-60-0 see SOR000      7789-61-9 see AQK000      8001-74-9 see CDK750      8013-77-2 see SAX000
7758-98-7 see CNP250      7783-61-1 see SDF650      7789-67-5 see TGB750      8001-79-4 see CCP250      8013-86-3 see CQJ000
7758-99-8 see CNP500      7783-64-4 see ZQS000      7789-69-7 see PHR250      8001-85-2 see BMA750      8013-90-9 see IFV000
7759-35-5 see HMB650      7783-66-6 see IDT000      7789-75-5 see CAS000      8001-88-5 see BGO750      8014-13-9 see COF325
7761-45-7 see MQR100      7783-70-2 see AQF250      7789-78-8 see CAT200      8001-95-4 see AGW300      8014-19-5 see PAE000
7761-88-8 see SDS000      7783-71-3 see TAF250      7789-79-9 see CAT250      8002-01-5 see AER750      8014-29-7 see OHE000
7763-77-1 see CGX000      7783-79-1 see SBS000      7789-80-2 see CAT500      8002-03-7 see PAO000      8014-95-7 see SOI520
7764-50-3 see DKV175      7783-80-4 see TAK250      7789-82-4 see CAT750      8002-05-9 see PCR250      8015-01-8 see MBU500
7765-88-0 see MGD210      7783-81-5 see UOJ000      7789-89-1 see TJB000      8002-05-9 see PCS250      8015-12-1 see EEH520
7770-47-0 see MKW000      7783-81-5 see UOS000      7789-90-4 see TJC750      8002-09-3 see PIH750      8015-14-3 see LJE000
7772-76-1 see ANR750      7783-82-6 see TOC550      7789-92-6 see TJM250      8002-26-4 see TAC000      8015-18-7 see AMK250
7772-98-7 see SKI000      7783-91-7 see SDN500      7790-01-4 see MFX000      8002-33-3 see TOD500      8015-19-8 see DNX500
```

8015-29-0 see MDL750	8054-43-1 see CDB760	9003-00-3 see ADY250	9006-42-2 see MQQ250	10022-60-3 see AEP500
8015-30-3 see EAP000	8056-51-7 see NNL500	9003-01-4 see ADW200	9007-12-9 see TFZ000	10022-68-1 see CAH250
8015-35-8 see DAB000	8056-92-6 see EQK100	9003-04-7 see SJK000	9007-13-0 see CAW500	10023-25-3 see DKS800
8015-43-8 see DVJ500	8057-51-0 see EDK600	9003-07-0 see PKI250	9007-16-3 see ADW000	10024-58-5 see DAH450
8015-54-1 see DXP800	8057-62-3 see PAM782	9003-07-0 see PMP500	9007-40-3 see COC875	10024-70-1 see MEF500
8015-55-2 see AFM375	8059-82-3 see RCA275	9003-11-6 see PJH630	9007-43-6 see CQM325	10024-74-5 see BKQ500
8015-64-3 see AOO760	8059-83-4 see BOO650	9003-13-8 see BRP250	9007-49-2 see DAZ000	10024-78-9 see EMY000
8015-73-4 see BAR250	8062-14-4 see CMP250	9003-17-2 see PJL350	9007-73-2 see FBB000	10024-89-2 see MRQ600
8015-79-0 see OGK000	8063-06-7 see COF750	9003-20-7 see AAX250	9007-81-2 see FOO600	10024-93-8 see NBY000
8015-88-1 see CCL750	8063-14-7 see CBD750	9003-22-9 see AAX175	9007-92-5 see GEW875	10024-97-2 see NGU000
8015-92-7 see CDH750	8063-18-1 see SLV500	9003-34-3 see PJR500	9008-54-2 see CMS225	10025-65-7 see PJE000
8015-97-2 see CMY100	8063-77-2 see CBV250	9003-39-8 see PKQ250	9008-57-5 see GHU000	10025-67-9 see SON510
8016-03-3 see DAC400	8064-08-2 see CCS675	9003-39-8 see PKQ500	9008-97-3 see PIB575	10025-68-0 see SBS500
8016-20-4 see GJU000	8064-18-4 see TNX375	9003-39-8 see PKQ750	9009-54-5 see PKL500	10025-69-1 see TGC275
8016-23-7 see GKM000	8064-35-5 see MLX850	9003-39-8 see PKR000	9009-65-8 see POD750	10025-70-4 see SMH525
8016-26-0 see LAC000	8064-38-8 see TGJ350	9003-39-8 see PKR250	9009-86-3 see RJK000	10025-73-7 see CMJ250
8016-31-7 see LII000	8064-66-2 see MCA500	9003-39-8 see PKR500	9010-06-4 see SEH450	10025-74-8 see DYG600
8016-38-4 see NCO000	8064-76-4 see EEH575	9003-39-8 see PKR750	9010-53-1 see UVJ475	10025-76-0 see ERA500
8016-45-3 see PIG730	8064-77-5 see BAV350	9003-39-8 see PKS000	9011-04-5 see HCV500	10025-77-1 see FAW000
8016-68-0 see SBA000	8064-79-7 see IGI000	9003-53-6 see SMQ500	9011-13-6 see SEA500	10025-78-2 see TJD500
8016-69-1 see SED500	8064-90-2 see TKX000	9003-54-7 see ADY500	9011-14-7 see PKB500	10025-82-8 see ICK000
8016-87-3 see TAE500	8065-36-9 see BTA250	9003-55-8 see SMR000	9011-18-1 see DBD750	10025-85-1 see NGQ500
8016-88-4 see TAF700	8065-48-3 see DAO600	9004-06-2 see EAG875	9011-93-2 see LJQ000	10025-87-3 see PHQ800
8017-59-2 see SLY200	8065-83-6 see CAY950	9004-07-3 see CML850	9012-59-3 see CBA000	10025-91-9 see AQC500
8018-01-7 see DXI400	8065-91-6 see CNV750	9004-10-8 see IDF300	9014-01-1 see BAC000	10025-97-5 see IGJ499
8018-15-3 see MCS000	8066-01-1 see MK000	9004-17-5 see IDF325	9014-02-2 see NBV500	10025-98-6 see PLN750
8021-27-0 see AAC250	8066-21-5 see ZJS300	9004-32-4 see SFO500	9014-92-0 see TAX500	10025-99-7 see PJD250
8021-29-2 see FBV000	8066-27-1 see FAQ930	9004-34-6 see CCU150	9015-68-3 see ARN800	10026-03-6 see SBU000
8021-39-4 see BAT850	8067-24-1 see DLL400	9004-38-0 see CCU050	9015-73-0 see DHW600	10026-04-7 see SCQ500
8022-00-2 see MIW100	8067-82-1 see AFK500	9004-51-7 see IGU000	9016-00-6 see PJR000	10026-07-0 see TAJ250
8022-15-9 see LCA000	8067-85-4 see AOC875	9004-53-9 see DBD800	9016-01-7 see OJM400	10026-08-1 see TFT000
8022-37-5 see ARL250	8068-28-8 see SFY500	9004-54-0 see DBC800	9016-45-9 see NND500	10026-10-5 see UQJ000
8022-56-8 see SAE500	8069-64-5 see TEQ000	9004-54-0 see DBD000	9016-45-9 see PKF000	10026-11-6 see ZPA000
8022-56-8 see SAE550	8069-76-9 see DVI800	9004-54-0 see DBD200	9016-45-9 see TAW250	10026-12-7 see NEA000
8022-81-9 see BMA600	8070-50-6 see EJK000	9004-54-0 see DBD400	9016-45-9 see TAW500	10026-13-8 see PHR500
8022-91-1 see HGF000	8072-20-6 see CKL500	9004-54-0 see DBD600	9016-45-9 see TAX000	10026-17-2 see CNC100
8023-53-8 see AFP750	8075-78-3 see DAP810	9004-54-0 see DBD700	9016-45-9 see TAX250	10026-18-3 see CNE250
8023-75-4 see NBP000	9000-01-5 see AQQ500	9004-64-2 see HNV000	9031-11-2 see GAV100	10028-15-6 see ORW000
8023-80-1 see GCM000	9000-02-6 see AHJ000	9004-65-3 see HNX000	9034-34-8 see HHK000	10028-18-9 see NDC500
8023-85-6 see CCQ750	9000-07-1 see CCL250	9004-66-4 see IGS000	9034-40-6 see LIU360	10028-22-5 see FBA000
8023-89-0 see EAI500	9000-21-9 see FPQ000	9004-70-0 see CCU250	9036-06-0 see PMJ100	10029-04-6 see ELI500
8023-91-4 see GBC000	9000-28-6 see GLY000	9004-70-0 see CNH000	9036-19-5 see GHS000	10031-13-7 see LCL000
8023-94-7 see HGN500	9000-29-7 see GLY100	9004-74-4 see MFJ750	9036-19-5 see IAH000	10031-18-2 see MCX750
8023-95-8 see HAK500	9000-30-0 see GLU000	9004-74-4 see MFK000	9036-19-5 see OFM000	10031-26-2 see IGQ000
8024-14-4 see DAJ800	9000-36-6 see KBK000	9004-74-4 see MFK250	9036-19-5 see OFO000	10031-27-3 see TAK600
8024-37-1 see COG000	9000-36-6 see SLO500	9004-81-3 see PJY000	9036-19-5 see OFQ000	10031-37-5 see CMK425
8025-81-8 see SLC000	9000-40-2 see LIA000	9004-82-4 see SIB500	9036-19-5 see OFS000	10031-43-3 see CNN000
8028-73-7 see ARE500	9000-55-9 see PJJ000	9004-86-8 see PKE750	9036-19-5 see OFU000	10031-53-5 see ERC000
8028-73-7 see ARE750	9000-64-0 see BAF000	9004-96-0 see PJY100	9038-95-3 see GHY000	10031-58-0 see TDS500
8028-75-9 see ARE250	9000-65-1 see THJ250	9004-96-0 see PJY250	9038-95-3 see MNE250	10031-59-1 see TEL750
8028-89-5 see CBG125	9000-70-8 see PCU360	9004-98-2 see OIG000	9038-95-3 see UBS000	10031-82-0 see EEL500
8029-29-6 see BAF250	9001-00-7 see BMO000	9004-98-2 see OIG040	9038-95-3 see UCA000	10031-96-6 see EQS100
8029-68-3 see IAD000	9001-13-2 see CMY325	9004-98-2 see OIK000	9038-95-3 see UCJ000	10032-00-5 see AAY500
8030-30-6 see NAI500	9001-33-6 see FBS000	9004-98-2 see PJW500	9038-95-3 see UDA000	10032-02-7 see GDG000
8030-55-5 see GME000	9001-37-0 see GFG100	9004-98-2 see PKE500	9038-95-3 see UDJ000	10032-15-2 see HFR200
8031-00-3 see FMS000	9001-62-1 see GGA800	9004-99-3 see PJV250	9038-95-3 see UDS000	10034-81-8 see PCE000
8031-03-6 see MQV000	9001-73-4 see PAG500	9004-99-3 see PJV500	9038-95-3 see UEA000	10034-85-2 see HHI500
8031-42-3 see DKL200	9002-01-1 see SLW450	9004-99-3 see PJV750	9038-95-3 see UEJ000	10034-93-2 see HGW500
8034-17-1 see EAL000	9002-07-7 see TNW000	9004-99-3 see PJW000	9038-95-3 see UFA000	10035-10-6 see HHJ000
8036-63-3 see FCN050	9002-18-0 see AEX250	9004-99-3 see PJW250	9039-53-6 see TCI000	10036-47-2 see TCI000
8037-19-2 see TGH750	9002-60-2 see AES650	9004-99-3 see PJW750	9039-61-6 see RDA350	10038-98-9 see GDY000
8042-47-5 see MQV875	9002-61-3 see CMG675	9005-00-9 see SLM500	9041-08-1 see HAQ550	10039-33-5 see DVM600
8044-51-7 see DAP900	9002-62-4 see PMH625	9005-00-9 see SLN000	9041-93-4 see BLY780	10039-54-0 see OLS000
8047-15-2 see SAV500	9002-67-9 see LIU300	9005-08-7 see PJU500	9045-81-2 see PKQ100	10040-45-6 see SJJ175
8047-28-7 see EDC575	9002-68-0 see FMT100	9005-08-7 see PJU750	9047-13-6 see AOM150	10042-59-8 see PNN250
8047-67-4 see IHG000	9002-70-4 see SCA750	9005-25-8 see SLJ500	9049-05-2 see CAO250	10042-76-9 see SMK000
8048-31-5 see TEO750	9002-72-6 see PJA250	9005-27-0 see HLB400	9049-76-7 see HNY000	10042-84-9 see SEP500
8048-52-0 see DBX400	9002-83-9 see KDK000	9005-32-7 see AFL000	9056-38-6 see NMB000	10042-88-3 see TAM000
8049-17-0 see FBG000	9002-84-0 see TAI250	9005-37-2 see PNJ750	9060-10-0 see BLY760	10043-01-3 see AHG750
8049-19-2 see FBF000	9002-86-2 see PKQ059	9005-38-3 see SEH000	9061-82-9 see SFP000	10043-09-1 see BJN750
8049-47-6 see PAF600	9002-88-4 see PJS750	9005-46-3 see SFQ000	9067-32-7 see HGN600	10043-18-2 see BPL750
8049-62-5 see LEK000	9002-89-5 see PKP750	9005-49-6 see HAQ500	9074-07-1 see SBI860	10043-35-3 see BMC000
8050-07-5 see OIM000	9002-92-0 see DXY000	9005-64-5 see PKG000	9076-25-9 see PCC000	10043-52-4 see CAO750
8050-07-5 see OIM025	9002-92-0 see EEU000	9005-64-5 see PKL000	9082-07-9 see SFW300	10043-67-1 see AHF100
8050-88-2 see CCU000	9002-92-0 see LBS000	9005-65-6 see PKL100	9084-06-4 see BLX000	10045-34-8 see AGM060
8051-03-4 see ZKJ000	9002-92-0 see LBT000	9005-66-7 see PKG500	9086-60-6 see CCH000	10045-94-0 see MDF000
8052-16-2 see AEA750	9002-92-0 see LBU000	9005-67-8 see PKL030	9087-70-1 see PAF550	10045-95-1 see NCB000
8052-41-3 see SLU500	9002-93-1 see PKF500	9005-70-3 see TOE250	10004-44-1 see HLM000	10048-13-2 see SLP000
8052-42-4 see ARO500	9002-98-6 see PJX800	9005-71-4 see SKV195	10008-90-9 see CLU500	10048-32-5 see PAJ500
8052-42-4 see ARO750	9002-98-6 see PJX825	9005-81-6 see CCT250	10010-36-3 see FLY000	10048-95-0 see ARC250
8052-42-4 see PCR500	9002-98-6 see PJX835	9006-00-2 see PJH500	10022-31-8 see BAN250	10049-03-3 see FFD000
8053-39-2 see TNO275	9002-98-6 see PJX845	9006-04-6 see ROH900	10022-50-1 see NMT500	10049-04-4 see CDW450

10049-06-6 see TGG750	10138-21-3 see DFK400	10294-40-3 see BAK250	10457-59-7 see IOR000	11041-12-6 see CME400
10049-07-7 see RHK000	10138-34-8 see EBK000	10294-41-4 see CDB250	10457-90-6 see BNU725	11043-98-4 see MQX000
10049-08-8 see RRZ000	10138-39-3 see AGF500	10294-46-9 see CNO350	10465-10-8 see DJI350	11043-99-5 see MQX250
10051-06-6 see PBD500	10138-41-7 see ECX500	10294-48-1 see PCF775	10465-27-7 see SAR000	11048-13-8 see NBR500
10058-07-8 see PPN000	10138-47-3 see EGJ000	10294-54-9 see CDE500	10466-65-6 see PLQ000	11048-92-3 see GME300
10058-20-5 see SMZ000	10138-52-0 see GAH000	10294-66-3 see PLW000	10467-10-4 see EMA000	11050-62-7 see IKV000
10060-12-5 see CMK450	10138-60-0 see HFF000	10294-70-9 see TGD500	10473-64-0 see DWL500	11051-88-0 see CNF159
10061-01-5 see DGH200	10138-62-2 see HGG000	10308-82-4 see AKD375	10473-70-8 see CKI500	11052-01-0 see EDC600
10061-02-6 see DGH000	10138-63-3 see PFC100	10308-83-5 see DBW100	10473-98-0 see DWL525	11052-70-3 see MRA275
10070-95-8 see BHR500	10138-79-1 see TNF000	10308-84-6 see GLC000	10476-81-0 see SMF000	11054-63-0 see TNX650
10072-24-9 see QCS875	10138-87-1 see EGJ500	10308-90-4 see DUX600	10476-85-4 see SMG500	11054-70-9 see LBF500
10072-25-0 see DFH000	10138-89-3 see TLA000	10309-37-2 see BAD625	10476-86-5 see SMJ500	11055-06-4 see FPD000
10072-50-1 see PKD050	10139-98-7 see DAZ140	10309-79-2 see MHN750	10476-95-6 see AAW250	11056-06-7 see BLY000
10075-36-2 see IHO200	10140-08-6 see DJA600	10310-38-0 see EIU000	10477-72-2 see PCU425	11056-12-5 see CMS232
10083-53-1 see MGL500	10140-75-7 see DBF875	10311-84-9 see DBI099	10482-16-3 see MKR000	11069-34-4 see MGS500
10085-81-1 see BCH750	10140-84-8 see DGC000	10312-83-1 see MDW250	10484-36-3 see AOK000	11071-15-1 see AQH000
10086-50-7 see IJA000	10140-87-1 see DFG159	10318-23-7 see DPP400	10486-00-7 see SJB350	11071-47-9 see IKF000
10087-89-5 see DWL400	10140-89-3 see DBH800	10318-26-0 see DDJ000	10488-36-5 see TGJ250	11072-93-8 see EDL500
10094-34-5 see BEL850	10140-91-7 see CGH675	10319-70-7 see AHN000	10497-05-9 see PHE750	11077-03-5 see PAF000
10097-26-4 see BSA750	10140-94-0 see CHH125	10322-73-3 see EDU600	10504-99-1 see DWM600	11078-23-2 see CNH800
10097-28-6 see SDH000	10140-97-3 see CJG750	10325-39-0 see DEU300	10510-77-7 see EOE000	11081-39-3 see PKB775
10099-57-7 see IRC000	10141-00-1 see PLB500	10325-94-7 see CAH000	10519-11-6 see DAF100	11082-38-5 see TNM850
10099-58-8 see LAX000	10141-05-6 see CNC500	10326-21-3 see MAE000	10519-12-7 see DAF150	11085-39-5 see FPA000
10099-59-9 see LBA000	10141-07-8 see COD100	10326-24-6 see ZDS000	10534-86-8 see HBX000	11094-61-4 see ECX000
10099-60-2 see LBB000	10141-15-8 see COM250	10328-51-5 see CPL100	10535-87-2 see PEF500	11096-42-7 see NND000
10099-66-8 see LIW000	10141-19-2 see VQU000	10329-95-0 see MRU080	10537-47-0 see DED000	11096-82-5 see PJN250
10099-67-9 see LIY000	10141-22-7 see DFR400	10331-57-4 see DFD000	10540-29-1 see NOA600	11097-67-9 see AEC175
10099-70-4 see BOX250	10143-20-1 see DSJ200	10335-79-2 see NAV000	10543-95-0 see HDA000	11097-69-1 see PJN000
10099-71-5 see MAL000	10143-22-3 see DTG200	10339-31-8 see NGY700	10544-63-5 see EHO200	11097-82-8 see GCO200
10099-72-6 see MAL500	10143-23-4 see DTI400	10339-55-6 see ELZ000	10544-72-6 see NGU500	11100-14-4 see PJN750
10099-73-7 see MAL750	10143-38-1 see DYH100	10347-38-3 see BLK750	10544-72-6 see NGV000	11103-86-9 see PLW500
10099-74-8 see LDO000	10143-53-0 see DJF000	10347-81-6 see MAW850	10545-99-0 see SOG500	11104-28-2 see PJM000
10099-76-0 see LDW000	10143-54-1 see DJF600	10350-81-9 see PMY000	10546-24-4 see MME500	11105-11-6 see TOD000
10101-41-4 see CAX750	10143-56-3 see DJG200	10356-76-0 see FHO000	10552-94-0 see NLQ000	11107-01-0 see TOB750
10101-50-5 see SJC000	10143-60-9 see DJK600	10356-92-0 see DAS400	10557-85-4 see IEI600	11111-23-2 see LHX350
10101-68-5 see MAU750	10143-66-5 see DOB200	10361-03-2 see SII500	10563-70-9 see TDL000	11111-49-2 see HDB500
10101-83-4 see SKC000	10143-67-6 see AAG750	10361-37-2 see BAK000	10578-16-2 see DKG600	11112-10-0 see AQB250
10101-88-9 see TNM750	10158-43-7 see DTM800	10361-44-1 see BKW250	10580-77-5 see BJN000	11113-75-0 see NDL100
10101-97-0 see NDL000	10161-84-9 see SBU900	10361-76-9 see PLP750	10584-98-2 see DDY600	11114-18-4 see LFN000
10102-06-4 see URA200	10161-85-0 see DJA325	10361-79-2 see PLX750	10588-01-9 see SGI000	11114-20-8 see CCL350
10102-17-7 see SKI500	10163-15-2 see DXD600	10361-80-5 see PLY250	10589-74-9 see PBX500	11114-46-8 see FBD000
10102-18-8 see SJT500	10165-33-0 see AKM250	10361-82-7 see SAR000	10592-13-9 see HGP550	11114-92-4 see CNA750
10102-20-2 see SKC500	10168-80-6 see ECY500	10361-83-8 see SAT200	10595-95-6 see MKB000	11115-82-5 see EAT800
10102-23-5 see SBN505	10168-81-7 see GAL000	10361-84-9 see SBC000	10598-82-0 see NFO700	11116-31-7 see BLY250
10102-40-6 see DXE875	10168-82-8 see HGH100	10361-91-8 see YDJ000	10599-90-3 see CDO750	11116-32-8 see BLY500
10102-43-9 see NEG100	10169-00-3 see ANL500	10361-92-9 see YES000	10605-21-7 see MHC750	11118-72-2 see AQM000
10102-44-0 see NGR500	10171-76-3 see BJM750	10361-95-2 see ZES000	11001-74-4 see QQS075	11119-62-3 see ADG125
10102-44-0 see NGS000	10171-78-5 see HNX800	10369-17-2 see HEJ375	11002-18-9 see CBF500	11120-29-9 see PJO250
10102-45-1 see TEK750	10176-39-3 see FND100	10371-86-5 see MEL775	11002-21-4 see CCO000	11121-08-7 see KDA025
10102-49-5 see IGN000	10182-82-8 see HOC500	10377-48-7 see LHR000	11002-90-7 see ART500	11121-57-6 see PII150
10102-50-8 see IGM000	10187-79-8 see MMK000	10377-60-3 see MAH000	11003-24-0 see ARY750	11133-78-1 see SCR000
10102-53-1 see ARB000	10187-86-7 see AKY000	10377-98-7 see LAO000	11004-30-1 see SKQ500	11133-98-5 see CNI600
10103-50-1 see ARD000	10192-29-7 see ANE250	10379-14-3 see CFG750	11005-02-0 see ORI300	11135-81-2 see PLS500
10103-60-3 see ARD600	10193-95-0 see BKM000	10380-28-6 see BLC250	11005-63-3 see SMN000	11138-49-1 see AHG000
10108-56-2 see BQW750	10196-18-6 see NEF500	10380-77-5 see COE125	11005-70-2 see CCX625	11138-87-7 see LAZ000
10108-64-2 see CAE250	10199-89-0 see NFS550	10387-13-0 see BIJ750	11005-92-8 see CCX175	11141-16-5 see PJM250
10108-73-3 see CDB000	10210-36-3 see GJG000	10389-72-7 see DRC600	11005-94-0 see CNW100	12001-26-2 see MQS250
10117-38-1 see PLT500	10210-68-1 see CNB500	10390-80-4 see EBD000	11006-22-7 see FCA000	12001-28-4 see ARM275
10118-76-0 see CAV250	10212-25-6 see COW900	10397-75-8 see TDQ230	11006-31-8 see TAB300	12001-29-5 see ARM268
10118-90-8 see MQW250	10213-09-9 see DFW800	10402-33-2 see AGL000	11006-33-0 see PGQ500	12001-47-7 see BKV250
10119-31-0 see ZPJ000	10213-15-7 see MAH250	10402-52-5 see PGB750	11006-33-0 see PGQ750	12001-65-9 see HEI650
10121-94-5 see MNS000	10213-74-8 see EGY000	10402-53-6 see ECU550	11006-64-7 see ITG000	12001-79-5 see VSZ500
10124-36-4 see CAJ000	10213-75-9 see EKZ000	10402-90-1 see PFB000	11006-70-5 see OIS000	12001-85-3 see NAT000
10124-37-5 see CAU000	10213-77-1 see TNA000	10405-02-4 see KEA300	11006-76-1 see VRF000	12001-89-7 see DGR200
10124-43-3 see CNE125	10215-25-5 see TEU000	10405-27-3 see DKF400	11006-90-9 see EAE400	12002-03-8 see COF500
10124-48-8 see MCW500	10215-33-5 see BPS500	10415-75-5 see MDE750	11006-92-1 see CKN250	12002-19-6 see MCV250
10124-50-2 see PKV500	10218-83-4 see DDQ100	10415-87-9 see PFR200	11011-73-7 see BML750	12002-43-6 see GEO000
10124-53-5 see MAK000	10222-01-2 see DDM000	10416-59-8 see TMF000	11013-97-1 see MKK600	12003-96-2 see AHH125
10124-56-8 see SHM500	10232-90-3 see DFJ500	10419-79-1 see CHG375	11014-59-8 see LAU400	12005-86-6 see SHM000
10124-65-9 see PLK750	10232-91-4 see EGY500	10420-90-3 see HCS100	11015-37-5 see MRA250	12007-25-9 see MAD250
10125-76-5 see NJM400	10232-92-5 see DMM600	10427-00-6 see FPV000	11016-29-8 see PJJ350	12007-33-9 see BMH659
10125-85-6 see DTO100	10232-93-6 see DMD000	10428-19-0 see TBL500	11016-71-0 see RQF350	12007-46-4 see SIX550
10137-69-6 see CPE500	10233-88-2 see GJE000	10429-82-0 see EAK000	11016-72-1 see RQK000	12007-56-6 see CAN250
10137-73-2 see CQB275	10235-09-3 see PHN500	10431-86-4 see BSY000	11018-93-2 see TER250	12007-97-5 see MRC650
10137-74-3 see CAO500	10238-21-8 see CEH700	10433-59-7 see EAK500	11024-24-1 see DKL400	12008-41-2 see DXF200
10137-80-1 see EKT000	10241-05-1 see MRD500	10436-39-2 see TBT500	11028-39-0 see MCE275	12008-61-6 see ANG125
10137-87-8 see EMO000	10256-92-5 see TCJ025	10443-70-6 see EHL500	11028-71-0 see CNH625	12009-21-1 see BAP750
10137-90-3 see ENX000	10262-69-8 see LIN800	10447-38-8 see DWM400	11029-61-1 see GJO025	12010-12-7 see BFQ750
10137-96-9 see EJK000	10265-92-6 see DTQ400	10448-09-6 see HBA600	11030-13-0 see ITH000	12010-53-6 see PJH775
10137-98-1 see EJL000	10290-12-7 see CNN500	10453-86-8 see BEP500	11031-48-4 see SAX500	12010-67-2 see PJK500
10138-01-9 see ERC550	10294-33-4 see BMG400	10457-58-6 see BQY000	11032-05-6 see TAK750	12011-67-5 see IGQ750
10138-17-7 see DRP000	10294-34-5 see BMG500		11032-12-5 see TOE750	12011-76-6 see DAC450

13520-96-2 see POS500	13762-26-0 see ZQJ000	13963-57-0 see TNN000	14259-66-6 see MRD000	14674-72-7 see CAP000
13523-86-9 see VSA000	13762-51-1 see PKY250	13965-73-6 see DDI800	14259-67-7 see CML000	14674-74-9 see BAK125
13529-51-6 see HHH000	13762-80-6 see HHA000	13967-90-3 see BAI750	14264-16-5 see BLS250	14679-73-3 see CBS000
13529-75-4 see LHE525	13764-35-7 see PDP500	13968-14-4 see SIX000	14264-31-4 see SFZ100	14684-25-4 see PFW500
13530-65-9 see ZFJ100	13764-49-3 see EQO500	13973-88-1 see CDW000	14268-23-6 see PLW300	14689-97-5 see DFG600
13536-84-0 see UQA000	13765-19-0 see CAP500	13977-28-1 see BMN250	14271-02-4 see EDJ000	14696-82-3 see IDN000
13536-85-1 see FFB000	13766-26-2 see PJL325	13980-04-6 see HDV500	14285-43-9 see MIB000	14698-07-8 see ARP875
13537-18-3 see TFW500	13767-90-3 see PLD250	13983-17-0 see WCJ000	14286-84-1 see BAV250	14698-29-4 see OOG000
13537-32-1 see PHJ250	13768-38-2 see OKG000	13984-07-1 see MLL000	14293-44-8 see CLF325	14708-14-6 see NDC000
13537-45-6 see HGU100	13768-67-7 see YDS800	13988-24-4 see DBJ100	14293-70-0 see PLM750	14709-62-7 see ZJJ400
13539-59-8 see AQN750	13769-43-2 see PLK900	13988-26-6 see DJD700	14301-11-2 see DUC200	14722-22-6 see IEO000
13548-38-4 see CMJ600	13770-16-6 see MAT899	13991-87-2 see CNP000	14314-27-3 see PLA525	14733-73-4 see BMU750
13551-87-6 see NHH500	13770-61-1 see ICI000	13993-65-2 see MNQ500	14323-41-2 see TBW250	14737-08-7 see CFG500
13551-92-3 see NHI000	13770-89-3 see NDK000	14003-66-8 see MMM750	14324-55-1 see BJC000	14737-91-8 see MEJ775
13552-44-8 see MJQ100	13770-96-2 see SEM500	14007-07-9 see CDT500	14332-59-3 see ZLS200	14742-53-1 see BII000
13553-79-2 see RKU000	13776-58-4 see XFS000	14007-45-5 see PKV600	14333-26-7 see DBC000	14746-03-3 see BPA250
13556-50-8 see DCL100	13779-41-4 see PHF250	14008-44-7 see MQR000	14334-41-9 see SDZ000	14751-74-7 see HIR000
13560-89-9 see DAI460	13779-41-4 see PHF500	14008-53-8 see BOG000	14354-56-4 see PFN250	14751-76-9 see HIR500
13561-08-5 see BJO000	13780-03-5 see CAN000	14008-79-8 see CMB125	14355-21-6 see BME250	14751-87-2 see FDU875
13569-63-6 see RGP000	13780-57-9 see SOF000	14018-82-7 see ZIJ000	14362-31-3 see CDR250	14751-89-4 see TND750
13569-65-8 see RHP000	13782-01-9 see PLI750	14023-01-9 see HEZ800	14362-68-6 see TNG775	14751-90-7 see HIP500
13573-18-7 see SKN000	13782-22-4 see SKL500	14023-90-6 see TMX600	14362-70-0 see PCG775	14753-13-0 see MMR250
13589-15-6 see MIH250	13783-04-5 see TGG625	14024-00-1 see HCC475	14368-49-1 see NFA625	14753-14-1 see CJF250
13590-71-1 see PGZ950	13812-39-0 see HGY000	14024-18-1 see IGL000	14370-50-4 see AKY750	14760-99-7 see DRN300
13590-82-4 see CDB400	13814-96-5 see LDE000	14024-58-9 see MAQ500	14402-89-2 see SIU500	14763-77-0 see CNL250
13590-97-1 see DXX800	13814-98-7 see SMI000	14024-63-6 see ZCJ000	14405-45-9 see ICG000	14774-78-8 see CKC325
13593-03-8 see DJY200	13815-28-6 see ANI000	14024-64-7 see BGQ750	14410-98-1 see DAQ125	14779-78-3 see AOI250
13597-95-0 see BFU000	13815-31-1 see ANI500	14024-75-0 see BKU250	14426-20-1 see DHO700	14781-32-9 see BIU125
13597-99-4 see BFT000	13816-33-6 see IOD050	14026-03-0 see NLI500	14433-76-2 see DQX000	14788-78-4 see JDA000
13598-15-7 see BFS000	13820-40-1 see ANE750	14028-44-5 see AOA095	14435-92-8 see CCA000	14798-26-6 see PID500
13598-36-2 see PGZ899	13820-41-2 see ANV800	14038-43-8 see IGY000	14436-50-1 see PCU400	14807-75-1 see DXN400
13598-47-5 see SHO000	13820-81-0 see PAR750	14040-11-0 see HCC000	14437-17-3 see CFC750	14807-96-6 see TAB750
13598-56-6 see UPA000	13820-83-2 see HBX500	14044-65-6 see BMB300	14437-41-3 see CGB250	14807-96-6 see TAB775
13600-88-9 see THQ250	13820-91-2 see PJD750	14046-99-2 see PNC925	14448-38-5 see HOW500	14808-60-7 see SCJ500
13600-98-1 see SFX750	13822-05-4 see CKA500	14047-09-7 see TBN500	14452-57-4 see MAH750	14816-18-3 see BAT750
13601-02-0 see DCJ600	13823-29-5 see TFT500	14051-55-9 see CGG600	14456-34-9 see HAD000	14816-67-2 see SKW500
13601-08-6 see TBI000	13823-36-4 see LAY499	14060-38-9 see ARJ500	14458-95-8 see HCI475	14817-09-5 see DAJ400
13603-07-1 see MMY750	13825-74-6 see TGH250	14073-00-8 see MMQ500	14459-29-1 see PHV275	14838-15-4 see NNM500
13607-48-2 see BGL000	13826-66-9 see BLA000	14075-02-6 see MRP000	14459-95-1 see PLH100	14838-45-0 see HNT075
13609-67-1 see HHQ825	13826-83-0 see ANH000	14075-53-7 see PKY000	14461-87-1 see HLE450	14842-81-0 see MAI600
13614-98-7 see MQW100	13826-86-3 see NHS500	14076-05-2 see CHM750	14464-46-1 see SCJ000	14848-01-2 see CEO000
13629-82-8 see DOV200	13826-94-3 see PLU250	14076-19-8 see CJE750	14465-96-4 see DNU850	14860-49-2 see CMW459
13637-63-3 see CDX250	13835-15-9 see CEP675	14088-71-2 see PMF550	14477-33-9 see TME250	14860-53-8 see TEC250
13637-65-5 see CHK250	13838-16-9 see EAT900	14090-22-3 see BQY500	14481-29-9 see ANH875	14861-06-4 see VNU000
13637-76-8 see LDS499	13839-85-5 see YEA000	14094-43-0 see MMQ750	14484-64-1 see FAS000	14873-10-0 see BJY000
13637-84-8 see SOT500	13840-33-0 see LHJ000	14094-45-2 see MMR500	14486-19-2 see CAG000	14881-92-6 see ZHJ000
13642-52-9 see SKW000	13845-23-3 see HID000	14094-48-5 see MMQ750	14487-05-9 see RKP400	14884-42-5 see CMK275
13647-35-3 see EBY600	13847-65-9 see NGS500	14096-51-6 see DFJ000	14488-49-4 see DQF650	14885-29-1 see IGH000
13654-91-6 see CHL250	13847-66-0 see TEI500	14097-03-1 see BAQ750	14504-15-5 see BED750	14899-36-6 see DBC525
13655-52-2 see AGW250	13851-11-1 see FAO000	14100-52-8 see CJE500	14519-10-9 see BMT250	14901-07-6 see IFX000
13655-95-3 see MJW875	13860-69-0 see MMV000	14104-20-2 see SDN000	14520-53-7 see AGM750	14901-08-7 see COU000
13669-70-0 see FAL000	13862-16-3 see TNJ250	14128-54-2 see LHR700	14521-96-1 see EQO450	14913-29-2 see PJI500
13674-84-5 see TNG000	13862-62-9 see BAL750	14144-91-3 see DQE800	14536-17-5 see FBH050	14913-33-8 see DEX000
13674-87-8 see FQU875	13863-31-5 see TGE100	14148-99-3 see DWA600	14543-76-1 see DTK200	14917-59-0 see BMV250
13682-73-0 see PLC175	13863-59-7 see BMP750	14149-43-0 see TLG000	14545-08-5 see IDZ400	14918-35-5 see DBB400
13684-56-5 see EE0500	13863-88-2 see SDM500	14152-28-4 see POC250	14545-72-3 see CDX000	14919-77-8 see SCA400
13684-63-4 see MEG250	13864-01-2 see LBC000	14167-18-1 see BLH250	14549-32-7 see DNN000	14925-39-4 see BMT000
13693-09-9 see XEJ300	13865-57-1 see DRQ000	14168-01-5 see DLO880	14550-84-6 see TEI625	14929-11-4 see SDY500
13701-67-2 see DDI600	13877-99-1 see DPA800	14168-44-6 see MJO775	14551-09-8 see MKB500	14930-96-2 see CQM125
13706-09-7 see TFO500	13889-92-4 see PNH150	14173-58-1 see BNT500	14557-50-7 see DIF600	14932-06-0 see BIT750
13706-10-0 see TFO250	13897-55-7 see DBY000	14187-32-7 see COD575	14567-61-4 see GKE900	14938-35-3 see AOM250
13706-86-0 see MKL300	13908-93-5 see CPL750	14207-78-4 see DRF000	14579-91-0 see BIL250	14938-42-2 see DLJ600
13707-88-5 see AGW000	13908-93-5 see FIJ000	14208-08-3 see TDQ325	14589-65-2 see HBW500	14940-68-2 see ZSS000
13709-32-5 see PCM250	13909-02-9 see CGW250	14211-01-9 see IOV000	14597-45-6 see TGM580	14949-00-9 see AMR750
13709-61-0 see XFA000	13909-09-6 see CHD250	14215-29-3 see CAD350	14600-07-8 see DAD000	14976-57-9 see FOS100
13710-19-5 see CLK325	13909-11-0 see CFH500	14215-30-6 see CNJ500	14604-29-6 see TGM575	14977-61-8 see CML125
13713-13-8 see DQB309	13909-12-1 see CFH750	14215-33-9 see MCX250	14611-52-0 see DAZ125	14986-60-8 see FFA000
13717-04-9 see PMX250	13909-13-2 see CHE500	14215-58-8 see CFW800	14612-92-1 see HGA500	14986-84-6 see HEY500
13718-26-8 see SKP000	13909-14-3 see CGV000	14220-17-8 see NDI000	14617-95-9 see MED750	15005-90-0 see CMG800
13721-39-6 see SIY250	13911-65-4 see HMK200	14221-47-7 see ANG925	14621-84-2 see DHA450	15016-15-6 see SPB800
13743-07-2 see HKW500	13912-77-1 see MLO750	14222-46-9 see PPQ000	14628-06-9 see MPG250	15017-02-4 see DXP400
13746-66-2 see PLF250	13927-77-0 see BIW750	14222-60-7 see PNW750	14631-45-9 see EFH500	15020-57-2 see DUC600
13746-89-9 see ZSA000	13929-01-6 see FEE100	14226-68-7 see BFC250	14634-93-6 see ZHA000	15029-32-0 see MRR750
13749-37-6 see BOI000	13929-35-6 see RKK000	14233-86-4 see FKM000	14635-33-7 see PCY400	15056-34-5 see THQ750
13754-23-9 see BLF250	13933-32-9 see TBI100	14235-86-0 see PFN000	14641-96-4 see LFC000	15061-57-1 see CNO325
13754-56-8 see DVT400	13935-78-9 see AKL100	14239-51-1 see BJB750	14642-66-1 see DIQ100	15083-53-1 see DPU800
13755-29-8 see SKE000	13939-06-5 see HCB500	14239-68-0 see BJB500	14644-61-2 see ZTJ000	15086-94-9 see BMO250
13755-38-9 see SIW500	13940-16-4 see TGC280	14244-61-2 see PLU590	14650-81-8 see BGV500	15090-10-5 see BEW750
13758-99-1 see PPT325	13943-58-3 see TEC500	14248-66-9 see DSY200	14663-23-1 see DAB840	15090-12-7 see BEW500
13759-11-6 see HAD500	13952-84-6 see BPY000	14252-80-3 see BOO000	14666-77-4 see DEB000	15090-13-8 see BEL500
13759-83-6 see SAT000	13955-12-9 see PHN000	14255-87-9 see BQK000	14666-78-5 see DJU600	15090-16-1 see BEN750
13762-14-6 see CNC250	13956-29-1 see CBD599	14255-88-0 see DGA200	14667-55-1 see TME270	15091-30-2 see DDT000

15096-52-3 see SHF000	15572-56-2 see INL000	16110-13-7 see BBH250	16587-71-6 see AOH750	16930-96-4 see HFX000
15104-03-7 see MMZ000	15589-00-1 see DOG600	16111-27-6 see NNL400	16590-41-3 see CQF099	16940-66-2 see SFF500
15105-92-7 see SJN650	15589-31-8 see TAL575	16111-62-9 see DJK800	16595-80-5 see LFA020	16940-81-1 see HDE000
15114-92-8 see DNU350	15597-43-0 see ISF000	16118-49-3 see CBL500	16607-77-5 see ODQ300	16941-12-1 see CKO750
15120-17-9 see ARD500	15598-34-2 see PPC100	16120-70-0 see BRK100	16607-80-0 see CPR500	16941-12-1 see CKP000
15121-11-6 see CKQ750	15605-28-4 see NMI000	16130-58-8 see COJ625	16634-82-5 see CEB500	16941-92-7 see HIA500
15130-85-5 see IDD000	15611-84-4 see PLN050	16136-32-6 see BBW250	16646-44-9 see TDG250	16949-15-8 see LHT000
15131-55-2 see BIS200	15622-65-8 see MRB250	16142-27-1 see MRU075	16648-69-4 see OQU000	16960-39-7 see MPS250
15131-84-7 see CML815	15625-89-5 see TLX175	16143-89-8 see MFF750	16650-10-5 see TBQ275	16961-83-4 see SCO500
15148-80-8 see BQB250	15640-93-4 see TCG500	16176-02-6 see COT000	16672-39-2 see DJF400	16962-07-5 see AHG875
15151-00-5 see AGR750	15652-38-7 see DAE600	16179-44-5 see HKQ500	16672-87-0 see CDS125	16962-40-6 see ANI250
15154-36-6 see AIN000	15660-29-4 see ANX800	16187-03-4 see PBQ300	16675-05-1 see PGU250	16967-79-6 see ECT600
15154-37-7 see AIM250	15662-33-6 see RSZ000	16187-15-8 see BOX825	16680-47-0 see ECW520	16984-48-8 see FEX875
15158-64-2 see ERB500	15663-27-1 see PJD000	16203-97-7 see APH500	16690-44-1 see MER250	16987-02-3 see CKT100
15158-67-5 see CDA500	15663-42-0 see PAS859	16219-75-3 see ELO500	16694-30-7 see KGK150	16993-94-5 see BBX250
15172-86-8 see BGP750	15669-07-5 see FBG200	16219-98-0 see NKQ000	16699-07-3 see MMR750	17003-75-7 see FHW000
15180-03-7 see DBK400	15676-16-1 see EPD500	16224-33-2 see BQT500	16699-10-8 see MMS750	17003-82-6 see FHZ000
15181-46-1 see HIC600	15686-71-2 see ALV000	16227-10-4 see BPU000	16714-23-1 see MGR800	17010-21-8 see CAG500
15191-85-2 see SCN500	15687-18-0 see CKG000	16230-71-0 see DWI400	16714-68-4 see PAY000	17010-59-2 see CIL710
15213-49-7 see PLD600	15687-27-1 see IIU000	16238-56-5 see BNR000	16719-32-7 see BOQ500	17010-61-6 see DFD400
15216-10-1 see NJL000	15699-18-0 see NCY050	16239-84-2 see ALN500	16721-80-5 see SHR000	17010-62-7 see EPU000
15219-97-3 see OLK200	15702-65-5 see BKJ275	16260-59-6 see DFL800	16731-55-8 see PLR250	17010-63-8 see EPT500
15227-42-6 see DFF500	15708-41-5 see EJA379	16268-87-4 see AJS250	16747-33-4 see EIK500	17010-64-9 see DJB400
15230-48-5 see DFL600	15709-62-3 see BLS500	16282-67-0 see DKH830	16752-77-5 see MDU600	17010-65-0 see EOI000
15233-65-5 see DON200	15712-13-7 see TBN150	16291-96-6 see CDI250	16755-07-0 see RIU000	17012-89-4 see MIK250
15237-44-2 see AEM250	15718-71-5 see EGT500	16291-96-6 see CDI500	16757-80-5 see MHH750	17012-91-8 see BBG750
15246-55-6 see TNG100	15721-02-5 see TBO000	16291-96-6 see CDJ500	16757-81-6 see MHM250	17013-01-3 see DXD800
15263-52-2 see BHL750	15721-33-2 see DNO200	16291-96-6 see CDK000	16757-82-7 see MHG500	17013-07-9 see AHB750
15267-04-6 see AMN500	15727-43-2 see DCW000	16301-26-1 see ASP000	16757-83-8 see MHH500	17013-35-3 see EMO875
15267-95-5 see CIY500	15736-98-8 see SHJ500	16310-68-2 see TCO000	16757-84-9 see DQO200	17013-37-5 see EJY000
15271-41-7 see CFF250	15742-33-3 see HBW000	16320-04-0 see ENX575	16757-85-0 see DQN000	17013-41-1 see DDV400
15284-15-8 see MDP500	15769-72-9 see EES000	16322-14-8 see CBO750	16757-86-1 see DQN200	17014-71-0 see PLP250
15284-39-6 see MQM775	15773-47-4 see DJN800	16329-92-3 see BJV635	16757-87-2 see DQN800	17018-24-5 see MPY250
15285-42-4 see BFA250	15790-54-2 see PNV800	16329-93-4 see BJV630	16757-88-3 see DQN400	17024-18-9 see NIC000
15301-48-1 see BCE825	15804-19-0 see QRS000	16338-97-9 see NJV500	16757-89-4 see DQO400	17025-30-8 see DTY200
15307-79-6 see DEO600	15805-73-9 see VNK000	16338-99-1 see HBP000	16757-90-7 see DQN600	17026-81-2 see AJT750
15308-34-6 see NNT100	15825-70-4 see MAW250	16339-01-8 see MMY250	16757-91-8 see DQO000	17029-22-0 see PLH500
15311-77-0 see CMY135	15826-16-1 see TLX175	16339-04-1 see ELX500	16757-92-9 see TLM500	17031-32-2 see CNO500
15318-45-3 see MPN000	15826-37-6 see CNX825	16339-05-2 see BRY250	16758-28-4 see EKD500	17040-19-6 see DAP600
15336-58-0 see SDJ025	15829-53-5 see MDF750	16339-07-4 see NKW500	16760-11-5 see POU250	17043-56-0 see MNM750
15336-81-9 see DKM130	15842-89-4 see CJT125	16339-12-1 see DSZ000	16760-12-6 see PPB000	17057-98-6 see DRY100
15339-36-3 see MAR750	15860-21-6 see CPH500	16339-14-3 see NLY000	16760-13-7 see AMP750	17066-89-6 see DHS400
15347-57-6 see LCG000	15861-05-9 see FFU000	16339-16-5 see CIQ500	16760-14-8 see AMQ000	17068-78-9 see ARM266
15351-05-0 see BTA325	15875-13-5 see TNH250	16339-18-7 see NKL500	16760-18-2 see AMP500	17070-44-9 see MEJ250
15356-70-4 see MCG000	15876-67-2 see DXG800	16339-21-2 see MMX750	16760-22-8 see PPA750	17074-42-9 see NFW100
15364-94-0 see MAU000	15879-93-3 see GFA000	16354-47-5 see MEU500	16760-23-9 see PPB250	17074-44-1 see NFW200
15375-94-7 see HAP000	15886-84-7 see TLR250	16354-48-6 see MJX000	16773-42-5 see OJS000	17083-63-5 see PLI500
15383-68-3 see SDN100	15904-73-1 see PDP600	16354-50-0 see EMM500	16781-80-9 see CIN000	17083-85-1 see TCB750
15387-10-7 see AQN500	15914-23-5 see DRE200	16354-52-2 see DIT800	16800-47-8 see THM250	17088-21-0 see EEF000
15388-46-2 see TNE000	15922-78-8 see HOC000	16354-53-3 see DOA000	16802-49-6 see AMP750	17088-72-1 see PAR800
15411-45-7 see DTJ159	15923-42-9 see HDQ500	16354-54-4 see MOU250	16802-50-9 see POU000	17088-73-2 see DMN200
15421-84-8 see DIO200	15930-94-6 see CMK500	16354-55-5 see EMN000	16808-85-8 see FDV000	17090-79-8 see MRE225
15438-31-0 see FBN000	15932-80-6 see MCF500	16357-59-8 see EFJ500	16813-36-8 see NJY000	17090-93-6 see SEZ355
15442-77-0 see BIX500	15932-89-5 see HME000	16361-01-6 see DKY200	16816-67-4 see PAG150	17095-24-8 see RCU000
15446-08-9 see CKK500	15942-48-0 see MIA000	16368-97-1 see BJR625	16824-81-0 see LDA500	17097-76-6 see CAT125
15457-05-3 see NIX000	15954-91-3 see CAF750	16378-21-5 see PJA190	16825-72-2 see IFS350	17099-81-9 see HIA000
15457-77-9 see PJI250	15954-94-6 see LDD000	16378-22-6 see TMK150	16825-74-4 see IFQ775	17109-49-8 see EIM000
15457-87-1 see TFT250	15954-98-0 see ZGS100	16395-80-5 see MNA250	16830-14-1 see NKO425	17124-74-2 see DAZ135
15460-48-7 see DGB800	15964-31-5 see IPR000	16409-43-1 see RNU000	16830-15-2 see ARN500	17125-80-3 see BAO750
15467-20-6 see DXF000	15972-60-8 see CFX000	16409-45-3 see MCG500	16842-03-8 see CNC230	17140-68-0 see DIH600
15468-32-3 see SCK000	15973-99-6 see DVF000	16413-88-0 see TCG250	16845-29-7 see RSA000	17140-78-2 see DYB400
15475-56-6 see MDV750	15978-93-5 see DNF000	16417-75-7 see DIM400	16846-24-5 see JDS200	17156-85-3 see TBJ475
15481-70-6 see DCE400	15980-15-1 see OLY000	16423-68-0 see FAG040	16853-85-3 see LHS000	17156-88-6 see DRJ400
15489-16-4 see AQH500	16006-09-0 see IIE000	16432-36-3 see TNN500	16870-90-9 see HOP259	17157-48-1 see BMR000
15493-35-3 see SLY000	16017-38-2 see CML325	16448-54-7 see IHC100	16871-71-9 see ZIA000	17160-71-3 see FDA880
15500-66-0 see PAF625	16032-41-0 see DPT200	16449-54-0 see AHA875	16871-90-2 see PLH750	17167-73-6 see AAH100
15500-66-0 see PAF630	16033-21-9 see MOA725	16454-60-7 see NCB500	16879-01-9 see DLS275	17168-82-0 see AQQ100
15503-86-3 see RFU000	16033-21-9 see PFS500	16456-56-7 see MRF500	16883-45-7 see TDJ500	17168-83-1 see DBY300
15521-65-0 see BJK250	16037-91-5 see AQH800	16468-98-7 see ENE000	16887-79-9 see ILF000	17168-85-3 see THP250
15529-90-5 see CLQ500	16037-91-5 see AQI250	16484-77-8 see CIR325	16893-05-3 see TBO776	17169-60-7 see FBD500
15535-69-0 see DED800	16039-55-7 see CAG750	16485-39-5 see OMS400	16893-06-4 see TBO777	17176-77-1 see DDH400
15535-79-2 see DVL200	16051-77-7 see ISC500	16488-48-5 see BQA000	16893-85-9 see DXE000	17185-68-1 see AQQ125
15538-67-7 see THU275	16053-71-7 see DLF000	16501-01-2 see MEQ000	16893-93-9 see PLI250	17210-48-9 see DNY400
15545-48-9 see CIS250	16054-41-4 see BIX125	16509-79-8 see ZEA000	16919-27-0 see PLI000	17224-08-7 see DKF600
15546-11-9 see BKO250	16056-11-4 see TMB000	16524-23-5 see TGV000	16919-58-7 see ANF250	17224-09-8 see TCH250
15546-12-0 see BJR250	16064-14-5 see CLC750	16532-79-9 see BNV750	16919-73-6 see PLA750	17230-87-4 see CLS250
15546-16-4 see BHK250	16066-38-9 see DWV400	16543-55-8 see NLD500	16921-30-5 see PLR000	17230-88-5 see DAB830
15547-17-8 see EPK000	16069-36-6 see DGV100	16550-39-3 see MBV775	16921-96-3 see IDQ000	17236-22-5 see IFG000
15565-25-0 see BEK500	16071-86-6 see CMO750	16561-29-8 see PGV000	16923-95-8 see PLG500	17242-52-3 see PKV000
15567-46-1 see MMY000	16088-56-5 see DVQ600	16566-62-4 see NAU500	16924-00-8 see PLH000	17243-64-0 see EOA500
15571-58-1 see DVM800	16091-18-2 see DVK200	16566-64-6 see NAU000	16924-32-6 see DRJ200	17243-64-0 see EOB000
15571-91-2 see PLU000	16102-24-2 see DGW875	16568-02-8 see AAH000	16925-39-6 see CAX250	17247-77-7 see DLX800

17264-01-6 see MIS750
17268-47-2 see DOX400
17278-93-2 see MJA000
17284-75-2 see BQH500
17300-62-8 see MAJ775
17309-87-4 see DSI800
17311-31-8 see DVQ800
17321-77-6 see CDV000
17333-74-3 see TGM550
17333-83-4 see CEJ500
17333-84-5 see CEJ250
17333-86-7 see BBN250
17369-59-4 see PNO500
17372-87-1 see BNH500
17381-88-3 see DBJ400
17397-89-6 see ECE500
17400-65-6 see DPQ400
17400-68-9 see DQE400
17400-69-0 see MJF500
17400-70-3 see MJF750
17401-48-8 see DQH800
17406-45-0 see THG250
17416-17-0 see DPQ800
17416-18-1 see DPQ600
17416-20-5 see MJG000
17416-21-6 see DQE600
17418-58-5 see AKI750
17427-00-8 see ABV250
17433-31-7 see ADA000
17433-39-5 see ADD000
17449-96-6 see CMW700
17455-13-9 see COD500
17455-23-1 see DGT300
17463-44-4 see AKP500
17466-45-4 see PCU350
17471-59-9 see TDU750
17471-82-8 see GKG300
17476-04-9 see LGS000
17496-59-2 see DEL800
17498-10-1 see DJN875
17501-44-9 see PBL750
17508-17-7 see DVC700
17513-40-5 see MGY500
17518-47-7 see NBN000
17523-77-2 see PLN500
17524-18-4 see THP750
17526-17-9 see DMT200
17526-24-8 see BBH500
17526-74-8 see GGY000
17548-36-6 see HCY500
17549-30-3 see DJD200
17557-23-2 see NCI300
17560-51-9 see ZAK300
17563-48-3 see BRA550
17573-21-6 see BCY250
17573-23-8 see DKU000
17575-20-1 see LAT000
17575-21-2 see LAT500
17575-22-3 see LAU000
17576-63-5 see FKO000
17576-88-4 see BNE600
17596-45-1 see BMB150
17597-95-4 see HFG700
17598-65-1 see DBH200
17599-02-9 see DPC200
17599-08-5 see AIM000
17599-09-6 see AIU000
17601-12-6 see BLE250
17605-71-9 see MJU750
17607-20-4 see BGW710
17608-59-2 see NKC000
17615-73-5 see AOT125
17617-13-9 see DID800
17617-23-1 see FMQ000
17617-45-7 see PIE510
17622-94-5 see EJM000
17639-93-9 see CKT000
17650-86-1 see AHL000
17654-88-5 see MCP500
17661-50-6 see TCB100
17667-23-1 see DVO809
17672-21-8 see HJC500

17673-25-5 see PGS250
17676-08-3 see LJM600
17689-16-6 see TEY750
17692-39-6 see PDU250
17697-55-1 see ASP500
17702-41-9 see DSI800
17702-57-7 see FNE500
17710-62-2 see CJU125
17719-22-1 see MPT300
17721-94-7 see BRZ200
17721-95-8 see DTA400
17730-82-4 see BTA000
17737-65-4 see CMX770
17750-93-5 see TCJ775
17751-20-1 see DEX600
17780-75-5 see CMY000
17784-47-3 see ADJ750
17794-13-7 see TAO750
17804-35-2 see BAV575
17804-49-8 see PMF540
17814-73-2 see MOR250
17822-71-8 see DHP550
17822-72-9 see DHP450
17822-73-0 see DHP500
17822-74-1 see DHQ800
17831-71-9 see ADT050
17861-62-0 see NDG550
17869-27-1 see AMG500
17874-34-9 see DQV000
17902-23-7 see FMB000
17918-11-5 see PEH500
17924-92-4 see ZAT000
17927-57-0 see CFK325
17959-11-4 see MDX250
17959-12-5 see MLX000
17969-20-9 see CKK250
17977-68-3 see FOE000
17982-67-1 see EML500
18046-21-4 see CKI750
18051-18-8 see MIQ400
18067-13-5 see DAB750
18109-81-4 see BOR350
18139-02-1 see FHX000
18139-03-2 see BJU500
18156-74-6 see TMF250
18172-33-3 see CQD000
18179-67-4 see MFO000
18181-70-9 see IEN000
18181-80-1 see IOS000
18186-71-5 see DXU200
18197-22-3 see FNL000
18204-79-0 see TMF625
18207-29-9 see NLD800
18230-75-6 see TMF125
18237-15-5 see AME750
18237-16-6 see AJB500
18244-91-2 see TJU500
18252-65-8 see DEU115
18264-75-0 see ALN750
18264-88-5 see DMA400
18268-70-7 see FCD512
18268-70-7 see PJC250
18273-30-8 see BJE250
18278-44-9 see MOB500
18279-20-4 see DVU600
18279-21-5 see DKP200
18283-93-7 see TBJ300
18323-44-9 see CMV675
18355-50-5 see BGM000
18355-54-9 see EKD000
18356-02-0 see MPJ000
18361-48-3 see LEV025
18365-12-3 see PNH500
18378-89-7 see MQW750
18380-68-2 see DQG400
18413-14-4 see ELC000
18417-89-5 see SAU000
18429-70-4 see DQJ600
18429-71-5 see TLJ250
18431-36-2 see HFR100
18433-84-6 see PAT799
18444-66-1 see COE250

18454-12-1 see LCS000
18461-55-7 see AAU250
18463-85-9 see DPO400
18463-86-0 see DQM200
18466-11-0 see MOB250
18472-51-0 see CDT250
18472-87-2 see ADG250
18479-58-8 see DLX000
18497-13-7 see DLO400
18501-44-5 see HDX500
18520-54-2 see TAR000
18520-57-5 see TAR250
18523-48-3 see ASD375
18523-69-8 see NGN000
18530-56-8 see HDP500
18538-45-9 see BCC100
18539-34-9 see DRR500
18559-59-6 see IDD100
18559-60-9 see TMJ800
18559-63-2 see TKX125
18559-92-7 see DPO600
18559-94-9 see BQF500
18559-95-0 see SMT000
18588-49-3 see DCA500
18588-50-6 see DBY100
18591-81-6 see ALB625
18598-63-5 see MBX800
18648-22-1 see DQP000
18649-64-4 see EQN000
18656-21-8 see IDJ600
18656-25-2 see DWE200
18662-53-8 see NEI000
18683-91-5 see AHJ250
18694-40-1 see MCH550
18705-22-1 see PNV750
18714-34-6 see NJB500
18719-22-7 see NNL000
18727-07-6 see UTU400
18760-80-0 see MNN350
18774-85-1 see HFS500
18787-40-1 see PIY000
18791-02-1 see DDS100
18791-21-4 see PPI775
18810-58-7 see BAI000
18810-58-7 see BAI250
18815-73-1 see MQP250
18820-29-6 see MAV000
18854-01-8 see DJV600
18857-59-5 see HMI500
18868-43-4 see MRD250
18868-66-1 see EGP000
18869-73-3 see ACD500
18883-66-4 see SMD000
18886-42-5 see ENC600
18897-36-4 see CAI350
18902-42-6 see NMI500
18905-29-8 see GLK100
18912-80-6 see DJF800
18917-82-3 see LDL000
18917-91-4 see AHC750
18921-70-5 see DBY500
18924-91-9 see TNI000
18928-76-2 see TDH100
18936-66-8 see BND750
18936-75-9 see AIC750
18936-78-2 see FNS000
18943-30-1 see SMJ000
18968-99-5 see MAG250
18972-56-0 see MAG250
18976-74-4 see RCZ000
18979-94-7 see CHL500
18984-80-0 see EQQ500
18987-38-7 see TME600
18996-35-5 see MRL000
18997-62-1 see DUN800
18999-28-5 see HBK500
19009-56-4 see MIW000
19010-66-3 see LCW000
19010-79-8 see CAI400
19025-95-7 see BKY500
19039-44-2 see GLO000
19042-19-4 see LAN000

19049-40-2 see BFT500
19056-01-0 see MLW250
19056-03-2 see PNV250
19056-06-5 see MLW500
19060-37-8 see PPA000
19060-39-0 see PPA250
19060-40-3 see PPD250
19060-43-6 see POV250
19060-44-7 see PPD000
19060-45-8 see POT750
19060-74-3 see POW250
19060-75-4 see POW000
19060-76-5 see POW500
19072-57-2 see DRK600
19083-81-9 see PPF250
19083-82-0 see PPG250
19089-24-8 see PQB750
19089-92-0 see HFM600
19105-63-6 see EMR500
19109-66-1 see TMA500
19115-30-1 see HKA700
19137-90-7 see EPC999
19139-31-2 see DKP000
19142-68-8 see CJL409
19142-70-2 see POT000
19142-71-3 see PPE500
19143-00-1 see CPD500
19143-28-3 see MOC500
19146-99-7 see PPH000
19155-52-3 see MFQ250
19168-23-1 see ANF000
19202-92-7 see TAL550
19210-06-1 see ZGS000
19216-56-9 see AJP000
19218-16-7 see BJP899
19237-84-4 see FPP100
19245-07-9 see FAC155
19246-24-3 see TAK800
19247-68-8 see DKK200
19267-68-6 see BIU260
19287-45-7 see DDI450
19293-56-2 see AAK250
19311-91-2 see DJY400
19315-64-1 see HIN000
19356-22-0 see PCN000
19361-41-2 see AAK400
19367-79-4 see MEP000
19381-50-1 see NAX500
19383-97-2 see PEK750
19387-91-8 see TGD250
19395-58-5 see MRN250
19395-78-9 see PCJ350
19396-06-6 see PKE100
19408-46-9 see KCK000
19416-93-4 see PLK825
19423-89-3 see LDI000
19441-09-9 see ANT300
19456-73-6 see DQF200
19456-74-7 see DQE000
19456-75-8 see DQF400
19456-77-0 see DPP709
19464-55-2 see TKT750
19471-27-3 see DPP600
19471-28-4 see DQE200
19473-49-5 see MRK500
19477-24-8 see CCJ500
19481-39-1 see DJQ800
19481-40-4 see EIL500
19482-31-6 see PAS829
19485-03-1 see BRG500
19493-75-5 see TJU750
19521-84-7 see BBO400
19525-20-3 see TER500
19526-81-9 see EAJ500
19546-20-4 see BLB500
19561-70-7 see HKW450
19562-30-2 see PAF250
19590-85-3 see DAE695
19595-66-5 see DQF600
19597-69-4 see LGV000
19598-90-4 see GAN000
19622-19-6 see EIH500

19624-22-7 see PAT750
19689-86-2 see NFN000
19704-60-0 see BJX750
19706-58-2 see DEI600
19708-47-5 see THR100
19716-21-3 see DQC600
19721-74-5 see BIY250
19750-95-9 see CJJ500
19763-77-0 see DGY000
19767-45-4 see MDK875
19780-25-7 see EHO000
19780-35-9 see EJS000
19789-69-6 see FKP000
19792-18-8 see TBO765
19794-93-5 see THK875
19816-89-8 see BMT500
19834-02-7 see CPF750
19836-78-3 see MND750
19841-73-7 see DWF865
19855-39-1 see DAD650
19864-71-2 see IBP000
19875-60-6 see LJE500
19879-06-2 see GJS000
19882-03-2 see HGI700
19895-66-0 see DCI800
19899-80-0 see HJP575
19910-65-7 see BSD000
19926-22-8 see BBD000
19932-64-0 see DCO509
19932-85-5 see BMV000
19935-86-5 see PNR500
19937-59-8 see MQR225
19952-47-7 see AJE500
19974-69-7 see BBW000
19982-87-7 see FMH000
19986-35-7 see CGF000
19992-69-9 see PPL500
19996-03-3 see CFB500
20004-62-0 see HAL000
20056-92-2 see DXU800
20057-09-4 see PCW500
20064-00-0 see NFW470
20064-38-4 see CAB125
20064-40-8 see CKJ825
20064-41-9 see MQB100
20123-68-6 see ALV750
20123-80-2 see DMI300
20153-98-4 see CNR750
20168-99-4 see CMP950
20170-20-1 see PAM500
20182-56-3 see BKD250
20187-55-7 see BAV325
20198-77-0 see DFJ400
20228-27-7 see RSU450
20228-97-1 see CJI125
20231-45-2 see VSF400
20240-62-4 see MHW000
20240-98-6 see MNT500
20241-03-6 see DSR200
20248-45-7 see TNH500
20265-97-8 see AOX500
20268-52-4 see CEJ000
20275-19-8 see BGH000
20281-00-9 see CDE325
20296-29-1 see OCY100
20300-26-9 see GJM025
20302-25-4 see POV500
20311-78-8 see EAO500
20325-40-0 see DOA800
20333-40-8 see BRF550
20344-15-4 see MCH535
20354-26-1 see BGD250
20369-63-5 see TID150
20373-56-2 see DEW400
20389-01-9 see DNG800
20398-06-5 see EEE000
20404-94-8 see DDO400
20405-19-0 see MIH750
20408-97-3 see GFK000
20427-56-9 see RSK000
20427-59-2 see CNM500
20436-27-5 see OGI100

20519-92-0 see TLT100	21083-47-6 see DWN600	21667-01-6 see BHW250	22349-59-3 see DTJ200	23129-50-2 see HEY000
20537-88-6 see AMD000	21085-56-3 see MNO775	21679-31-2 see TNN250	22373-78-0 see MRE230	23135-22-0 see DSP600
20539-85-9 see BJ0125	21087-64-9 see MQR275	21704-44-9 see ELY575	22388-72-3 see HBY000	23139-00-6 see MNA650
20548-54-3 see CAY000	21107-27-7 see TLP275	21704-46-1 see CBI675	22392-07-0 see DKB175	23139-02-8 see BNX125
20562-02-1 see SKS000	21109-95-5 see BAP250	21708-94-1 see EAF100	22432-68-4 see TBQ255	23155-02-4 see PHA550
20562-03-2 see CDG500	21124-09-4 see MRI250	21711-65-9 see NLA000	22445-73-4 see TDX835	23180-57-6 see PAC000
20570-96-1 see BEQ250	21124-13-0 see MQB250	21715-46-8 see CGQ500	22451-06-5 see TAI800	23182-46-9 see PIX800
20611-21-6 see BBT000	21140-85-2 see MFG600	21722-83-8 see EHS000	22471-42-7 see HFA000	23184-66-9 see CFW750
20624-25-3 see SGJ500	21149-87-1 see EMP500	21725-46-2 see BLW750	22480-64-4 see BNX000	23191-75-5 see FOO875
20627-28-5 see DQJ800	21172-28-1 see ALG500	21736-83-4 see SLI325	22494-42-4 see DKI600	23209-59-8 see CAX260
20627-31-0 see DQK800	21187-98-4 see DBL700	21738-42-1 see OLT000	22506-53-2 see DUW120	23210-56-2 see IAG660
20627-32-1 see TLK000	21208-99-1 see CQC250	21739-91-3 see CQK600	22514-23-4 see FMU225	23210-58-4 see IAG625
20627-33-2 see TLJ750	21209-02-9 see MIT250	21794-01-4 see RQP000	22545-60-4 see AEG250	23214-92-8 see AES750
20627-34-3 see TLK500	21224-57-7 see EOK500	21813-99-0 see FQR100	22571-95-5 see SPB500	23217-81-4 see AAP750
20645-04-9 see ZTS000	21224-77-1 see EFC000	21816-42-2 see DSY000	22575-95-7 see DOF200	23217-86-9 see AMV750
20652-39-5 see DNP600	21224-81-7 see THB250	21820-82-6 see DMB000	22578-17-2 see SJA500	23217-87-0 see AAQ000
20661-60-3 see NJH500	21230-20-6 see BBR380	21829-25-4 see AEC750	22583-29-5 see DSC800	23233-88-7 see BOL325
20667-12-3 see SDU500	21232-47-3 see TBN550	21842-58-0 see DLQ800	22585-64-4 see BOL310	23239-41-0 see SGB500
20675-51-8 see PBW400	21239-57-6 see TBN200	21848-62-4 see NNR200	22604-10-0 see TAI050	23239-51-2 see RLK700
20680-07-3 see MHS375	21243-26-5 see DLO200	21884-44-6 see LIV000	22608-53-3 see TMD625	23246-96-0 see RJZ000
20684-29-1 see FAC130	21247-98-3 see BCV250	21888-96-0 see DBE000	22609-73-0 see NDY600	23255-69-8 see FQR000
20685-78-3 see SPE000	21248-00-0 see BMU500	21892-31-9 see PBD300	22653-19-6 see TDP500	23255-93-8 see HGO500
20689-96-7 see ENS500	21248-01-1 see CEK500	21905-27-1 see AGQ775	22670-79-7 see PEJ250	23256-30-6 see NGG000
20691-83-2 see DSN200	21254-73-9 see CPU750	21905-32-8 see HNX600	22691-91-4 see DSV289	23256-50-0 see GKO750
20691-84-3 see CCE750	21255-83-4 see BMP500	21905-40-8 see CKA575	22692-30-4 see FHZ200	23257-56-9 see LFO000
20697-04-5 see CLF000	21256-18-8 see OLW600	21908-53-2 see MCT500	22713-35-5 see DQP125	23257-58-1 see LFG100
20706-25-6 see PNA225	21259-20-1 see FQS000	21916-66-5 see POB250	22722-03-8 see ELY550	23261-20-3 see DCI600
20719-22-6 see POX750	21259-76-7 see TFK270	21917-91-9 see MNO250	22722-98-1 see SGK800	23273-02-1 see DOX200
20719-23-7 see EKB000	21260-46-8 see BKW000	21923-23-9 see CLJ875	22750-53-4 see CAD325	23279-53-0 see AAP500
20719-34-0 see TAP000	21267-72-1 see CDS275	21928-82-5 see MMT250	22750-56-7 see CDC125	23279-54-1 see EFP000
20719-36-2 see TAP250	21282-96-2 see AAY750	21947-75-1 see CKS099	22750-65-8 see BGR250	23282-20-4 see NMV000
20731-44-6 see NIA700	21284-11-7 see COB000	21947-76-2 see CKS100	22750-69-2 see DCM600	23291-96-5 see DAL350
20737-02-4 see SDS500	21299-86-5 see INU200	21961-08-0 see HFY000	22750-85-2 see DPR200	23292-52-6 see TJX375
20738-78-7 see DWX200	21308-79-2 see MNF250	21962-24-3 see PNJ500	22750-86-3 see DPR400	23292-85-5 see TMW250
20740-05-0 see HFN500	21327-74-2 see BGQ325	21970-53-6 see DQD200	22750-93-2 see EOD000	23315-05-1 see EAG000
20743-57-1 see BGF000	21340-68-1 see MIO975	21987-62-2 see DAZ110	22751-18-4 see NFX500	23319-66-6 see TNI500
20762-60-1 see PKW000	21351-79-1 see CDD750	21992-92-7 see MPI225	22751-23-1 see NIJ200	23324-72-3 see MKR500
20762-98-5 see FQJ025	21361-93-3 see NDP500	22047-25-2 see ADA350	22751-24-2 see NFA500	23327-57-3 see NBS500
20770-41-6 see PLQ500	21362-69-6 see MCH600	22059-60-5 see RSZ600	22754-97-8 see RPB100	23339-04-0 see DRX600
20777-39-3 see LCA100	21368-68-3 see CBA800	22071-15-4 see BDU500	22755-01-7 see OOS000	23344-17-4 see SMB850
20777-49-5 see DKV160	21372-60-1 see TKH025	22086-53-9 see CGN325	22755-07-3 see DCW200	23362-09-6 see MPJ100
20780-49-8 see DTE800	21380-82-5 see ACI375	22089-22-1 see TNT500	22755-25-5 see SIW625	23395-20-2 see HMS500
20794-96-1 see CHT500	21393-59-9 see IGX000	22098-38-0 see NIG500	22755-34-6 see TNE275	23399-90-8 see TIY250
20816-12-0 see OKK000	21413-28-5 see DWJ300	22103-31-7 see TAG875	22755-36-8 see TLE500	23412-26-2 see THD300
20819-54-9 see TMD400	21416-67-1 see RCA375	22113-51-5 see SIM100	22760-18-5 see POB300	23414-72-4 see ZLA000
20820-80-8 see EIF500	21416-87-5 see PIK250	22120-39-4 see DST400	22771-17-1 see DGV900	23422-53-9 see DSO200
20829-66-7 see EIV700	21436-96-4 see XOJ000	22131-79-9 see AGN000	22771-18-2 see MRH212	23435-31-6 see DON400
20830-75-5 see DKN400	21436-97-5 see TLG750	22137-01-5 see JAZ000	22781-23-3 see DQM600	23436-19-3 see IIG000
20830-81-3 see DAC000	21450-81-7 see AAS500	22144-77-0 see ZUS000	22788-18-7 see CCE250	23452-05-3 see AGW500
20839-15-0 see ACZ000	21452-14-2 see AKR500	22151-75-3 see HOO875	22797-20-2 see DLG000	23456-94-2 see DSA600
20839-16-1 see PGS500	21457-22-7 see DME700	22188-15-4 see NMG000	22826-61-5 see TBJ250	23471-13-8 see EED600
20854-03-9 see TGY250	21466-07-9 see BNV500	22189-32-8 see SKY500	22830-45-1 see FBL000	23471-23-0 see MHS250
20856-57-9 see CDP750	21466-08-0 see NNU000	22199-08-2 see SNI425	22832-87-7 see MQS560	23476-83-7 see POC000
20859-73-8 see AHE750	21476-57-3 see BJI125	22199-30-0 see CMX820	22839-47-0 see ARN825	23483-74-1 see BNL275
20866-13-1 see HAA360	21482-59-7 see TDP275	22204-24-6 see POK575	22862-76-6 see AOY000	23488-38-2 see TBK250
20917-34-4 see DSL400	21498-08-8 see LIA400	22204-53-1 see MFA500	22868-13-9 see SGR500	23489-00-1 see EKN100
20917-49-1 see OBY000	21535-47-7 see BMA625	22205-30-7 see DVM400	22885-98-9 see MPQ500	23489-01-2 see ENY100
20917-50-4 see OCA000	21548-32-3 see DHH200	22205-45-4 see CNP750	22904-40-1 see LDB000	23489-02-3 see ENX100
20919-99-7 see HBD650	21554-20-1 see AJQ500	22205-57-8 see CDC250	22916-47-8 see MQS550	23489-03-4 see DAJ300
20921-41-9 see MNX850	21561-99-9 see MMY500	22223-55-8 see MQH100	22936-86-3 see CQI750	23491-45-4 see MOD500
20921-50-0 see PGQ000	21564-17-0 see BOO635	22224-92-6 see FAK000	22952-87-0 see ASG675	23495-12-7 see EJK500
20929-99-1 see BJW250	21572-61-2 see TBO778	22225-32-7 see FDU000	22953-41-9 see BHG000	23505-41-1 see DIN600
20930-00-1 see BJW500	21586-21-0 see MNI525	22232-71-9 see MBV250	22953-53-3 see BHV750	23509-16-2 see BAR750
20930-10-3 see DVY900	21590-92-1 see AHP125	22235-85-4 see AGM250	22953-54-4 see BHW000	23510-39-6 see CHG400
20941-65-5 see EPJ000	21593-23-7 see CCX500	22236-53-9 see CJC750	22958-08-3 see ASE875	23518-25-4 see POC525
20977-05-3 see EDM000	21595-62-0 see PPC250	22238-17-1 see THY000	22960-71-0 see EEX500	23521-13-3 see DUA200
20977-50-8 see FGV000	21600-42-0 see DTV200	22248-79-9 see RAF100	22966-79-6 see EDR500	23521-14-4 see DUA400
20982-36-9 see BHQ000	21600-43-1 see DJY000	22251-01-0 see FDT000	22966-83-2 see MPC250	23526-02-5 see BAC125
20982-74-5 see NMV740	21600-45-3 see HKV000	22254-24-6 see IGG000	22967-92-6 see MLF550	23535-89-9 see DEH800
20986-33-8 see IMM000	21600-51-1 see CBP250	22260-42-0 see COF825	23031-25-6 see TAN100	23537-16-8 see RRA000
20991-79-1 see TEK300	21609-90-5 see LEN000	22260-51-1 see BNB325	23031-32-5 see TAN250	23541-50-6 see DAC200
21000-42-0 see DWX000	21615-29-2 see BKB000	22262-18-6 see MPI205	23031-38-1 see POD875	23560-59-0 see HBK700
21001-46-7 see TCM400	21626-24-4 see IGX875	22262-19-7 see MPN275	23067-13-2 see EDI500	23564-05-8 see PEX500
21019-39-6 see HKB600	21638-36-8 see MMJ000	22295-11-0 see TDU250	23093-74-5 see BON400	23564-06-9 see DJV000
21035-25-6 see BCD325	21642-82-0 see PBS750	22295-99-4 see HCA000	23103-98-2 see DOX600	23581-62-6 see EAV100
21059-46-1 see CAM675	21642-83-1 see PBS500	22298-04-0 see DLO000	23107-11-1 see DAL100	23593-75-1 see MRX500
21062-28-2 see WCJ750	21644-95-1 see AKK000	22304-30-9 see ASA000	23107-12-2 see DAL400	23595-00-8 see TKF699
21064-50-6 see MHD000	21645-51-2 see AHC000	22316-47-8 see CIR750	23107-96-2 see BDS500	23605-05-2 see APN000
21070-32-6 see BRW750	21649-57-0 see CBO000	22323-45-1 see ZJA000	23109-05-9 see AHI625	23606-32-8 see SDL500
21070-33-7 see BRM750	21650-02-2 see ARX150	22326-31-4 see SNU500	23110-15-8 see FOZ000	23657-27-4 see TNO300
21075-41-2 see MNS250	21658-26-4 see NIQ500	22345-47-7 see GJS200	23111-70-8 see BEB500	23668-11-3 see PAB500
21082-50-8 see EME000	21662-09-9 see DAI360	22346-43-6 see HLX000	23115-33-5 see BLR500	23668-76-0 see DNR200

23674-86-4 see DKJ300
23679-20-1 see DBL000
23705-25-1 see PLN100
23712-05-2 see DGA800
23734-06-7 see CCL109
23734-88-5 see AKC625
23745-86-0 see PLG000
23746-34-1 see PKX500
23753-67-5 see BNU125
23757-42-8 see AMQ500
23777-55-1 see MJB300
23777-63-1 see DCF750
23777-80-2 see HCA275
23779-99-9 see TKG000
23784-96-5 see CFB825
23795-03-1 see DWW200
23828-92-4 see AHJ500
23834-96-0 see BND325
23840-95-1 see ZTK300
23844-24-8 see BCL250
23903-11-9 see DWK900
23930-19-0 see AFK875
23947-60-6 see BRI750
23950-58-5 see DTT600
23978-09-8 see LFQ000
24017-47-8 see THT750
24026-35-5 see BDT500
24027-84-7 see BCX500
24049-18-1 see AGN250
24050-16-6 see MPR000
24050-58-6 see DWK700
24089-00-7 see MND100
24094-93-7 see CMJ850
24095-80-5 see BLP325
24096-53-5 see DGF000
24124-25-2 see LGJ000
24143-08-6 see AAL500
24143-17-7 see OMK300
24166-13-0 see CMY125
24167-76-8 see SGM000
24168-96-5 see IKN200
24219-97-4 see MQS220
24220-18-6 see DPS200
24221-86-1 see EAX000
24234-06-8 see ELA600
24264-08-2 see DJN700
24268-87-9 see NFW430
24268-89-1 see NFW460
24279-91-2 see BGX750
24280-93-1 see MRX000
24283-57-6 see CFC500
24292-52-2 see HBU400
24301-86-8 see PEF750
24301-89-1 see DWB300
24305-27-9 see TNX400
24342-55-0 see HGI200
24342-56-1 see MIQ725
24345-16-2 see AQN650
24346-78-9 see HBP250
24356-60-3 see HMK000
24356-66-9 see AEH100
24356-94-3 see DAM300
24358-29-0 see CGJ250
24365-47-7 see LEX400
24365-61-5 see PGF125
24378-32-3 see PAY600
24380-21-0 see EDK000
24382-04-5 see MAN700
24397-89-5 see AEA109
24407-55-4 see BKP200
24423-68-5 see DNL200
24425-13-6 see BQJ750
24426-36-6 see IPG000
24458-48-8 see NHK800
24477-37-0 see DBE885
24519-85-5 see DLK700
24526-64-5 see NMV700
24536-75-2 see TLA525
24543-59-7 see BMA650
24549-06-2 see MJY000
24554-26-5 see NGM500
24570-10-3 see LHX495

24570-11-4 see LHX600
24570-12-5 see LHX510
24579-91-7 see CHN750
24589-78-4 see MQG750
24596-38-1 see IOT875
24596-39-2 see BRB450
24596-41-6 see BRB460
24600-36-0 see FMU000
24602-86-6 see DUJ400
24613-89-6 see CMI250
24621-17-8 see ZQB100
24623-77-6 see AHC250
24632-47-1 see NDY350
24632-48-2 see NDY370
24656-22-2 see PIC100
24684-41-1 see MJD750
24684-42-2 see DLU200
24684-49-9 see MEV250
24684-56-8 see DLP600
24684-58-0 see ABN250
24690-46-8 see DTT400
24699-40-9 see CNE375
24704-64-1 see AHE875
24729-96-2 see CMV690
24735-35-1 see ZTS100
24748-25-2 see TDH775
24771-52-6 see PBH150
24781-13-3 see SAK500
24800-44-0 see TMZ000
24815-24-5 see TLN500
24818-79-9 see AHA150
24829-11-6 see DHD400
24848-81-5 see TMO500
24853-80-3 see ASC250
24854-67-9 see DHC000
24869-88-3 see DOS300
24891-41-6 see TLK600
24909-09-9 see DLC000
24928-15-2 see PGS750
24928-17-4 see PGT250
24934-91-6 see CDY299
24937-79-9 see DKH600
24938-91-8 see TJJ250
24939-03-5 see PJY750
24951-05-1 see PIX750
24961-39-5 see BNO750
24961-49-7 see DMK400
24973-25-9 see DOJ800
24991-55-7 see DOM100
25013-15-4 see MPK500
25013-15-4 see VQK600
25013-16-5 see BQI000
25014-41-9 see PJK250
25036-33-3 see PKP250
25038-54-4 see PJK750
25038-54-4 see PJY500
25038-59-9 see PKF750
25046-79-1 see PMG000
25047-48-7 see CMV375
25056-70-6 see TIX250
25057-89-0 see MJY500
25061-59-0 see MBV710
25068-38-6 see EBF500
25068-38-6 see EBG000
25068-38-6 see EBG500
25068-38-6 see ECL000
25068-38-6 see IPM000
25068-38-6 see IPN000
25068-38-6 see IPO000
25068-38-6 see IPP000
25081-01-0 see CLF500
25090-71-5 see DAT000
25090-72-6 see DAT600
25090-73-7 see DAT200
25103-09-7 see ILR000
25103-12-2 see TKT000
25103-58-6 see DXT800
25122-46-7 see CMW300
25122-57-0 see CMW400
25134-21-8 see NAC000
25136-55-4 see DRO800
25136-85-0 see PJL600

25147-05-1 see CBT125
25148-26-9 see CIN500
25151-00-2 see BSO750
25152-84-5 see DAE450
25152-85-6 see HFE500
25154-52-3 see NNC500
25154-54-5 see DUQ180
25155-18-4 see BAC750
25155-25-3 see BHL100
25155-30-0 see DXW200
25167-31-1 see CEI325
25167-67-3 see BOW250
25167-70-8 see TMA250
25167-83-3 see TBS250
25167-93-5 see CJA950
25168-04-1 see NMS000
25168-15-4 see TGF200
25168-24-5 see BKK250
25168-26-7 see ILO000
25201-35-8 see PAV250
25230-72-2 see TNV575
25231-46-3 see TGO100
25231-47-4 see DCI000
25234-79-1 see DSF100
25238-02-2 see CLS125
25240-93-1 see DUV000
25251-03-0 see PCC750
25265-19-4 see PAN250
25265-19-4 see PAN500
25265-77-4 see TEG500
25267-15-6 see PJQ250
25267-54-6 see CNQ375
25284-83-7 see MDR775
25287-52-9 see PFF750
25295-51-6 see DXB400
25301-02-4 see TDN750
25310-48-9 see MLX300
25311-71-1 see IMF300
25316-40-9 see HKA300
25321-09-9 see DNN709
25321-14-6 see DVG600
25321-41-9 see XJA000
25322-20-7 see TBP750
25322-68-3 see PJT000
25322-68-3 see PJT200
25322-68-3 see PJT225
25322-68-3 see PJT230
25322-68-3 see PJT240
25322-68-3 see PJT250
25322-68-3 see PJT500
25322-68-3 see PJT750
25322-68-3 see PJU000
25322-68-3 see PJV000
25322-69-4 see PKI500
25322-69-4 see PKI550
25322-69-4 see PKI750
25322-69-4 see PKJ250
25322-69-4 see PKJ500
25322-69-4 see PKK000
25323-30-2 see DFH800
25324-56-5 see TGE500
25331-92-4 see AFH550
25332-09-6 see PGF000
25332-39-2 see CKJ000
25332-39-2 see THK880
25333-83-9 see PMX500
25339-17-7 see IKK000
25339-56-4 see HBK350
25340-17-4 see DIU000
25340-18-5 see TJO750
25351-18-2 see TLY250
25354-97-6 see HFP500
25355-59-3 see MFG400
25355-61-7 see MMW775
25376-45-8 see TGL500
25377-73-5 see DXV000
25382-52-9 see MJU350
25384-17-2 see AGV890
25387-67-1 see CBB875
25389-94-0 see KAM000
25389-94-0 see HLT300
25395-41-9 see PKL750

25402-50-0 see IDP000
25405-85-0 see PGT000
25410-64-4 see LHX515
25410-69-9 see LHX498
25413-64-3 see PNF750
25417-20-3 see NBS700
25425-12-1 see CMS500
25429-29-2 see PAV600
25430-97-1 see DJL200
25439-20-7 see PEB775
25450-02-6 see LHX500
25455-73-6 see SDV700
25474-92-4 see DXH400
25480-76-6 see CIQ625
25481-21-4 see EJA500
25483-10-7 see TCG450
25486-91-3 see DUF000
25486-92-4 see DLT400
25487-36-9 see HAA370
25498-49-1 see MEV500
25523-79-9 see XES000
25526-93-6 see DAR200
25535-16-4 see PMT000
25546-65-0 see XQJ650
25550-58-7 see DUY600
25550-58-7 see DUY810
25551-13-7 see TLL250
25566-92-1 see DGR400
25567-67-3 see CGL750
25596-24-1 see TMG775
25601-84-7 see DOL800
25604-70-0 see BMY825
25604-71-1 see CHR325
25606-41-1 see PNI250
25614-03-3 see BNB250
25614-78-2 see ACL250
25627-41-2 see POD800
25639-42-3 see MIQ745
25639-45-6 see DKK400
25640-78-2 see IOF200
25655-41-8 see PKE250
25677-40-1 see PBX350
25679-28-1 see PMR000
25682-07-9 see AJU875
25704-18-1 see SJK375
25704-81-8 see APK500
25710-89-8 see XEJ000
25711-26-6 see BLL500
25717-80-0 see MRN275
25723-52-8 see EIW500
25724-33-8 see DXR400
25724-50-9 see DUJ000
25724-58-7 see PHW500
25735-67-5 see AOM750
25736-79-2 see UIA000
25736-79-2 see UIJ000
25736-79-2 see UIS000
25748-74-7 see PKO750
25764-08-3 see CDB325
25791-96-2 see NCT000
25796-01-4 see NBO525
25805-16-7 see PKP000
25808-74-6 see LDG000
25812-30-0 see GCK300
25843-45-2 see ASP250
25843-64-5 see IAK100
25852-70-4 see BSS000
25854-41-5 see SGP600
25855-92-9 see BNP850
25868-47-7 see DWA400
25869-93-6 see HAE000
25870-02-4 see SDV500
25875-51-8 see RLK890
25876-47-5 see FOV000
25889-63-8 see DRY000
25898-71-9 see HGA000
25910-37-6 see NFA600
25913-34-2 see TGE165
25916-47-6 see ADW250
25928-94-3 see ECK500
25930-79-4 see TME500
25931-01-5 see PKL750

25951-54-6 see PKQ000
25952-35-6 see TCQ275
25953-06-4 see DIO300
25954-13-6 see ANG750
25956-17-6 see FAG100
25961-87-9 see HBP275
25967-29-7 see FMR100
25987-94-4 see SDX200
25991-93-9 see TIS500
26006-71-3 see SKC100
26011-83-6 see DOK400
26016-98-8 see CAW376
26016-99-9 see DXF600
26027-38-3 see NNB300
26037-72-9 see IFL000
26043-11-8 see NDD000
26045-95-4 see PAY610
26049-68-3 see HHD500
26049-69-4 see DSG400
26049-70-7 see HHE000
26049-71-8 see HHB000
26072-78-6 see DBK120
26076-87-9 see DCC125
26087-47-8 see BKS750
26095-59-0 see OKO400
26096-99-1 see CGI125
26097-80-3 see CBA100
26129-32-8 see CJP750
26134-62-3 see LHM000
26140-60-3 see TBD000
26148-68-5 see AJD750
26156-56-9 see HBV000
26157-96-0 see DCL300
26159-34-2 see NBO550
26162-66-3 see MRN600
26163-27-9 see SDP550
26166-37-0 see DAP875
26171-23-3 see TGJ850
26219-22-7 see TLA500
26225-59-2 see MBX000
26241-10-1 see PLD710
26249-01-4 see SFB100
26259-45-0 see BQC250
26270-58-6 see DEW200
26270-59-7 see DOD600
26270-60-0 see DOC800
26270-61-1 see DON000
26270-62-2 see DHP600
26283-13-6 see CDD250
26305-03-3 see PCB750
26308-28-1 see POL475
26309-95-5 see AOD000
26311-45-5 see PBW000
26328-00-7 see IRQ000
26328-53-0 see AOA050
26351-01-9 see AHB625
26354-18-7 see OIY000
26375-23-5 see PKM250
26377-04-8 see TIF600
26377-29-7 see PHI250
26388-78-3 see BIT500
26388-85-2 see PCE250
26399-36-0 see CQG250
26400-24-8 see DAL060
26401-20-7 see HFJ600
26401-97-8 see BKK750
26419-73-8 see DRR000
26427-28-1 see TCW500
26444-49-5 see TGY750
26445-05-6 see POP100
26445-82-9 see TIL300
26447-14-3 see TGZ100
26447-28-9 see FQE000
26464-99-3 see DUL550
26471-62-5 see TGM740
26488-34-6 see ABL250
26493-63-0 see BGT125
26506-47-8 see CNJ900
26509-45-5 see MKC000
26523-63-7 see HCD500
26530-20-1 see OFE000
26538-44-3 see RBF100

31751-59-4 see DWN100	33060-69-4 see CBM875	34031-32-8 see ARS150	35038-45-0 see SFA100	35743-94-3 see AQG000
31785-60-1 see DLX300	33069-62-4 see TAH775	34037-79-1 see DBK600	35038-46-1 see ASH000	35763-44-1 see BPF750
31793-07-4 see PJA220	33073-60-8 see CHF000	34044-10-5 see TAH675	35050-55-6 see THG500	35764-54-6 see MFT250
31820-22-1 see NKU570	33082-92-7 see MHI600	34099-73-5 see BMC250	35065-27-1 see HCD000	35764-59-1 see RDZ875
31828-50-9 see SBN450	33089-61-1 see MJL250	34114-98-2 see DAD600	35077-51-1 see DQH200	35764-73-9 see DMF800
31842-01-0 see IDA400	33094-66-5 see NFD700	34140-59-5 see TKU675	35080-11-6 see PNC875	35788-21-7 see AEI250
31868-18-5 see MQR760	33094-74-5 see NHN550	34148-01-1 see CMV500	35108-88-4 see AER500	35788-27-3 see AEL750
31874-15-4 see PFF000	33098-26-9 see BRB750	34149-92-3 see PKP500	35133-55-2 see BEU800	35788-28-4 see AEL000
31876-38-7 see MRE250	33098-27-0 see BRC000	34177-12-3 see DW0875	35133-58-5 see RCA435	35788-29-5 see AEL250
31895-22-4 see TFH750	33100-27-5 see PB0000	34183-22-7 see PMJ525	35133-59-6 see RCA450	35788-31-9 see AEJ750
31897-92-4 see BNO000	33125-97-2 see HOU100	34195-34-1 see DKX050	35142-05-3 see AQX825	35834-26-5 see RMF000
31897-98-0 see AMC500	33132-61-5 see BRE250	34200-96-9 see OKS150	35142-06-4 see MGQ525	35844-94-1 see ARN700
31898-11-0 see BNO250	33132-71-7 see BQS250	34202-69-2 see HDA500	35154-45-1 see ISZ000	35846-53-8 see MBU820
31927-64-7 see MGP250	33132-75-1 see AJB000	34211-26-2 see PPC000	35158-25-9 see IKM100	35849-41-3 see BHX250
31932-35-1 see MOY750	33132-85-3 see BQS000	34214-51-2 see CHJ000	35170-28-6 see AEM500	35856-82-7 see TDP325
31959-87-2 see MKI000	33132-87-5 see AGF000	34255-03-3 see CMX920	35175-75-8 see MRR100	35865-33-9 see DCI400
32006-07-8 see TGG600	33175-34-7 see DGD075	34257-95-9 see DLO875	35187-24-7 see TLJ500	35869-74-0 see HKD550
32017-76-8 see BOF250	33196-65-5 see TDO500	34289-01-5 see MFF575	35187-27-0 see TLL000	35884-45-8 see NLP480
32059-15-7 see GKO800	33204-76-1 see CMS241	34291-02-6 see BSX325	35187-28-1 see DQK400	35891-69-1 see AOP750
32061-49-7 see CNL625	33212-68-9 see DSM289	34320-82-6 see TJA500	35231-36-8 see DAD850	35906-51-5 see TAH650
32078-95-8 see PKK775	33229-34-4 see HKN875	34333-07-8 see TNG250	35249-69-5 see SIZ000	35920-39-9 see AEK250
32106-51-7 see SKE500	33232-39-2 see DPE100	34346-90-2 see TLA250	35250-53-4 see POM000	35920-40-2 see AEI750
32144-31-3 see DKM100	33237-74-0 see FBP850	34346-96-8 see BNQ750	35257-18-2 see PGN840	35941-65-2 see BPT500
32179-45-6 see MIH300	33244-00-7 see PEK675	34346-97-9 see BNQ250	35258-12-9 see BKO840	35941-71-0 see CJH750
32210-23-4 see BQW500	33245-39-5 see FDA900	34346-98-0 see BMX250	35271-57-9 see MGF250	35943-35-2 see TJE870
32222-06-3 see DMJ400	33265-79-1 see PIN275	34346-99-1 see BNQ000	35278-53-6 see HAA300	35944-73-1 see CPW250
32226-65-6 see CEL000	33286-22-5 see DNU600	34375-78-5 see DLJ000	35281-27-7 see HDR500	35944-74-2 see MQQ750
32226-69-0 see QUJ300	33330-91-5 see CCC325	34381-68-5 see AAE125	35281-29-9 see DLH800	35944-82-2 see DYD600
32228-97-0 see PFE900	33364-51-1 see BJE000	34388-29-9 see MNG525	35281-34-6 see HDF500	35944-83-3 see DYD200
32238-28-1 see PKM500	33372-39-3 see BKC250	34419-05-1 see MJJ250	35282-68-9 see HMB595	35944-84-4 see DYD800
32248-37-6 see PIV750	33372-40-6 see DNB600	34423-54-6 see NKU590	35282-69-0 see HMB600	35967-49-8 see DXG500
32266-10-7 see HFG650	33384-03-1 see PDI250	34433-31-3 see HAA345	35285-69-9 see PNO250	35991-93-6 see NAC500
32266-82-3 see CAR875	33389-33-2 see DLY200	34435-70-6 see IGF325	35287-69-5 see EDA500	36011-19-5 see CQM250
32283-21-9 see BCJ125	33389-36-5 see HKO000	34438-27-2 see EAT810	35306-34-4 see DAP825	36031-66-0 see EOQ500
32362-68-8 see DQP400	33390-21-5 see LJB700	34444-01-4 see CCX300	35317-79-4 see CMB675	36039-40-4 see DCF700
32373-17-4 see BBE000	33396-37-1 see MCI750	34461-00-2 see SGK600	35321-46-1 see HLI325	36069-45-1 see MRV750
32385-11-8 see SDY750	33400-47-4 see DAB200	34461-49-9 see FEI000	35322-07-7 see FOL100	36069-46-2 see DAD800
32388-21-9 see OMK000	33401-94-4 see TKV750	34461-68-2 see CNO750	35335-07-0 see DRI800	36104-80-0 see CFY250
32446-40-5 see AON500	33402-03-8 see HNC000	34465-46-8 see HAJ500	35335-60-5 see EON500	36105-20-1 see FMQ100
32449-92-6 see GFM000	33406-36-9 see BKO825	34481-84-0 see MJQ250	35363-12-3 see ENR000	36115-09-0 see DDB400
32458-57-4 see EQN700	33419-42-0 see EAV500	34491-04-8 see DTP800	35367-38-5 see CJV250	36133-88-7 see NGK500
32524-44-0 see NBO500	33419-68-0 see SAC875	34491-12-8 see BJD000	35375-29-2 see AOC750	36148-80-8 see MHQ775
32598-13-3 see TBO700	33421-40-8 see ALY000	34493-98-6 see DCQ800	35398-20-0 see PJQ275	36167-63-2 see HAF500
32607-15-1 see TEE750	33445-03-3 see CCO675	34494-09-2 see DCK200	35400-43-2 see SOU625	36167-69-8 see CGS500
32607-31-1 see PLU600	33447-90-4 see DSU100	34501-24-1 see BBE750	35412-68-1 see ICY000	36169-16-1 see BOU550
32617-22-4 see PLD575	33453-19-9 see CQF125	34513-77-4 see MRW275	35440-49-4 see MKR100	36170-25-9 see IOW500
32665-36-4 see EQY600	33467-74-2 see HFE650	34521-09-0 see AQ1750	35449-36-6 see TDO260	36211-73-1 see BBN650
32707-10-1 see BKW850	33499-84-2 see DUN300	34521-14-7 see TAI100	35457-80-8 see MBY150	36226-64-9 see BEA500
32740-01-5 see DQI400	33531-34-9 see CIK250	34521-15-8 see MRU775	35483-50-2 see VRP200	36236-67-6 see MBX500
32750-98-4 see TKD350	33531-59-8 see MIE250	34521-16-9 see EDW300	35506-85-5 see BFK000	36236-73-4 see CIO375
32752-13-9 see PMF535	33543-31-6 see MKC500	34522-40-2 see DUL400	35507-35-8 see DWZ000	36294-69-6 see TBG625
32755-26-3 see CJI809	33550-22-0 see TID000	34529-29-8 see ELK500	35507-78-9 see AEF250	36304-84-4 see MLP250
32764-43-5 see AJL125	33564-30-6 see CCS510	34548-72-6 see BRC250	35523-89-8 see SBA600	36323-28-1 see TBK000
32766-75-9 see HMU000	33564-31-7 see DKF125	34552-83-5 see LIH000	35543-24-9 see BOM600	36330-85-5 see BGL250
32767-68-3 see III000	33568-99-9 see BKL000	34562-99-7 see AJC500	35554-08-6 see SBA500	36333-41-2 see DDK875
32774-16-6 see HCD100	33605-67-3 see CCK550	34580-14-8 see KGK200	35554-44-0 see FPB875	36355-01-8 see HCA500
32780-64-6 see HMM500	33634-75-2 see SIZ100	34590-94-8 see DWT200	35556-06-0 see DDE000	36368-30-6 see PEI750
32784-82-0 see IBQ100	33671-46-4 see CKA000	34604-38-1 see PFO550	35572-78-2 see AJR750	36375-30-1 see MFW750
32787-44-3 see MLF520	33691-06-4 see IAG300	34624-48-1 see AFR500	35607-13-7 see IAG700	36393-56-3 see NNV500
32795-47-4 see NMV725	33693-04-8 see BQC500	34627-78-6 see ACV000	35607-66-0 see CCS500	36417-16-0 see DFN500
32795-84-9 see BMT750	33697-73-3 see TGJ625	34643-46-4 see DGC800	35619-65-9 see TNP275	36457-20-2 see HJI500
32808-51-8 see CPJ000	33755-46-3 see DBC575	34661-75-1 see USJ000	35627-29-3 see NLU000	36478-76-9 see URA100
32808-53-0 see CPJ250	33765-68-3 see CDF375	34675-84-8 see CDF375	35629-37-9 see AAJ750	36499-65-7 see DGQ400
32809-16-8 see PMF750	33765-80-9 see ELF110	34681-10-2 see MPU250	35629-38-0 see AAM500	36504-65-1 see BCV750
32838-28-1 see BPF825	33770-60-4 see DFC800	34681-23-7 see SOB500	35629-39-1 see AAK000	36504-66-2 see BCW000
32852-21-4 see FNB000	33804-48-7 see DPQ200	34691-31-1 see FAQ000	35629-40-4 see ACI629	36504-67-3 see ECQ050
32854-75-4 see LBD000	33813-20-6 see EJQ000	34692-97-2 see DCF710	35629-44-8 see EOA000	36504-68-4 see ECQ000
32865-01-3 see DNW759	33854-16-9 see DVJ100	34701-14-9 see NHJ000	35629-70-0 see AKY875	36505-84-7 see PPP250
32871-90-2 see BEK250	33855-47-9 see BMW000	34758-84-4 see MFG250	35631-27-7 see NKU875	36508-71-1 see ROZ000
32887-03-9 see MCB550	33857-26-0 see DAC800	34807-41-5 see MDJ250	35653-70-4 see DRL400	36519-25-2 see NCK000
32889-48-8 see PMF600	33868-17-6 see MMX000	34816-55-2 see MRU600	35658-65-2 see CAE500	36551-21-0 see DXM000
32891-29-5 see KBB000	33878-50-1 see EGS000	34828-67-6 see PJR750	35691-65-7 see DDM500	36556-75-9 see BOM510
32891-29-5 see MBV100	33904-55-1 see NJS500	34839-12-8 see AJI600	35695-70-6 see AMU500	36568-91-9 see CCG500
32949-37-4 see HMN000	33942-87-9 see MJA500	34839-13-9 see AJI550	35695-72-8 see AMU000	36573-63-4 see ACF000
32954-58-8 see FQL100	33942-88-0 see MHE250	34866-46-1 see BQD500	35697-34-8 see BJA500	36576-14-4 see ICA000
32976-87-7 see DQL400	33944-90-0 see SBU500	34879-34-0 see AKK625	35700-21-1 see MLO300	36576-23-5 see DTL000
32976-88-8 see ENT000	33953-73-0 see BCX750	34920-64-4 see HNL100	35700-23-3 see CCC100	36614-38-7 see DSK800
32986-56-4 see TGI250	33965-80-9 see CHU000	34929-08-3 see TKF250	35700-27-7 see MOV800	36653-82-4 see HCP000
32988-50-4 see VQZ000	33972-75-7 see IHB600	34961-28-9 see CPH250	35711-34-3 see SKJ340	36702-44-0 see NLJ000
33005-95-7 see SOX400	33979-15-6 see CMV950	34963-49-4 see DIE200	35725-30-5 see DYH000	36711-31-6 see MAD100
33017-08-2 see TKE550	34014-18-1 see BSN000	34983-45-4 see DRK500	35725-31-6 see HGH000	36734-19-7 see GIA000
33020-34-7 see IPW000	34018-28-5 see LCO000	35035-05-3 see TGB160	35725-33-8 see TFX250	36774-74-0 see PEM000

36788-39-3 see OQO000	37795-69-0 see DLQ400	38940-50-0 see BDS750	40343-32-6 see HKW345	41735-30-2 see ENN500
36791-04-5 see RJA500	37795-71-4 see MJE250	38940-51-1 see BDU250	40358-04-1 see NMO400	41753-43-9 see PAF450
36798-79-5 see DQU400	37882-31-8 see POB000	38957-41-4 see EAN700	40378-58-3 see MON500	41772-23-0 see ALK250
36885-49-1 see VQA400	37924-13-3 see TKF750	38965-69-4 see XCS680	40428-75-9 see SNF000	41814-78-2 see MQC000
36911-94-1 see DIT400	37971-99-6 see XBA000	38965-70-7 see XCS700	40487-42-1 see DRN200	41825-28-9 see LDA000
36911-95-2 see DIT600	37994-82-4 see BCY000	38998-91-3 see BJM700	40502-72-5 see BEH250	41826-92-0 see CNG835
36950-85-3 see EPC200	38001-34-2 see DXS000	39013-93-9 see CGY000	40507-23-1 see THH350	41892-01-7 see DRS400
36981-93-8 see PEV600	38005-31-1 see IDU000	39047-21-7 see BJG000	40507-78-6 see IBR200	41948-17-8 see AQS875
36993-63-2 see SHY500	38029-10-6 see POM800	39070-08-1 see MNB250	40527-42-2 see PIW500	41956-77-8 see CCI500
37032-15-8 see FNE000	38035-28-8 see DRL600	39071-30-2 see CQD250	40548-68-3 see NLT500	41956-90-5 see HFW500
37033-23-1 see AEF000	38048-87-2 see DRX800	39079-58-8 see DHC400	40560-76-7 see TGN100	41992-22-7 see SLD800
37045-40-2 see MGD000	38078-09-0 see DIR875	39108-14-0 see AGX300	40561-27-1 see COR750	42011-48-3 see NGN500
37065-29-5 see MQU525	38082-89-2 see PQC525	39133-31-8 see TKU650	40568-90-9 see MHG250	42013-48-9 see DTN775
37106-97-1 see CML835	38092-76-1 see HKM175	39156-41-7 see DBO400	40580-75-4 see DFE229	42013-69-4 see AJW750
37106-99-3 see PMI000	38094-02-9 see DRS600	39196-18-4 see DAB400	40580-89-0 see NJK500	42017-07-2 see LHF625
37132-72-2 see FOM000	38105-25-8 see ABS250	39197-62-1 see ENP000	40596-69-8 see KAJ000	42028-27-3 see DLM000
37150-27-9 see BDE000	38105-27-0 see ABO250	39202-39-6 see GLQ000	40661-97-0 see MLG500	42028-33-1 see HJE575
37169-10-1 see DFU800	38115-19-4 see PHP500	39236-46-9 see GDO800	40665-92-7 see CMX880	42135-22-8 see NGB500
37203-43-3 see OAD090	38139-15-0 see PAS879	39254-48-3 see PGH500	40666-04-4 see IAD100	42145-91-5 see EGC500
37205-85-9 see MND000	38194-50-2 see SOU550	39274-39-0 see XEJ100	40709-23-7 see AHP375	42149-31-5 see DRK400
37209-31-7 see DBB500	38198-35-5 see PPW000	39277-41-3 see VRP000	40711-41-9 see BRL750	42200-33-9 see CNR675
37221-23-1 see PJJ500	38222-35-4 see DXS200	39277-47-9 see AEX750	40713-31-3 see CGQ250	42238-29-9 see NFA000
37224-57-0 see CMK000	38232-63-2 see MCX000	39293-24-8 see ACP000	40762-15-0 see DKV400	42242-58-0 see BCL100
37226-23-6 see COE100	38234-12-7 see HIJ000	39300-45-3 see AQT500	40778-40-3 see PMC600	42242-72-8 see CBK250
37227-61-5 see NCY000	38237-74-0 see AMW500	39300-88-4 see GMA000	40816-40-8 see SKS100	42279-29-8 see PAW250
37229-14-4 see LHX360	38237-76-2 see AJD250	39301-00-3 see AFI750	40828-46-4 see TEN750	42281-59-4 see CQF079
37230-84-5 see HAC000	38241-20-2 see NLF000	39306-82-6 see PJC500	40847-64-1 see DKR400	42310-84-9 see BLD000
37231-28-0 see MCB525	38241-21-3 see NLT000	39323-48-3 see BFW000	40853-53-0 see IQH000	42397-64-8 see DVD600
37235-82-8 see DDT250	38246-95-6 see HNG100	39340-46-0 see ERC800	40853-56-3 see AAV250	42397-65-9 see DVD800
37239-28-4 see ADZ125	38252-74-3 see BQQ250	39367-89-0 see DIA800	40911-07-7 see BRN500	42436-07-7 see HFE625
37244-85-2 see TAI600	38252-75-4 see BRO500	39387-42-3 see BBA625	40942-73-2 see OOO100	42461-89-2 see BOA750
37247-90-8 see BGB250	38285-49-3 see MHX000	39389-47-4 see DXG400	40951-13-1 see DLT800	42471-28-3 see NDY800
37268-68-1 see PJL100	38293-27-5 see ONU100	39391-39-4 see APT750	40953-35-3 see DCP880	42472-93-5 see MOC750
37270-71-6 see EQM500	38304-91-5 see DCB000	39404-28-9 see SAU350	40959-16-8 see TCT250	42472-96-8 see OOM300
37280-35-6 see CBF680	38357-93-6 see ITD100	39413-47-3 see BFV250	41024-90-2 see CDM625	42489-15-6 see PLV000
37280-56-1 see LEX000	38363-32-5 see PAP230	39425-26-8 see MQN250	41029-45-2 see PMS775	42489-15-6 see SKF500
37286-64-9 see MKS250	38363-40-5 see PAP225	39426-77-2 see ACG125	41083-11-8 see THT500	42498-58-8 see EBW000
37297-87-3 see MCW350	38402-95-8 see BLT775	39456-76-3 see LIJ000	41084-90-6 see DGQ650	42509-80-8 see PHK000
37299-86-8 see RGZ100	38411-22-2 see HCD075	39456-78-5 see SAU475	41093-93-0 see DAB925	42520-97-8 see DEV300
37300-23-5 see ZFJ120	38434-77-4 see ENT500	39457-24-4 see BAS500	41096-46-2 see EQD000	42540-40-9 see FOD000
37312-62-2 see SCA625	38460-95-6 see UMJ000	39472-31-6 see KBU000	41109-80-2 see PIK075	42542-07-4 see MJR750
37317-41-2 see PJP000	38462-04-3 see ARM750	39482-21-8 see TMF600	41136-03-2 see ALY500	42579-28-2 see NKJ000
37324-23-5 see PJN500	38514-71-5 see AEK000	39491-47-9 see AEK000	41136-22-5 see CCS560	42583-55-1 see CCK575
37324-24-6 see PJO000	38524-82-2 see EPC175	39515-41-8 see DAB825	41155-82-2 see HKW380	42609-52-9 see DQQ700
37329-49-0 see TOC250	38539-23-0 see ABN500	39543-79-8 see BAU255	41191-04-2 see THR500	42615-29-2 see AFO500
37333-40-7 see BIT250	38562-01-5 see POC750	39562-70-4 see EMR600	41198-08-7 see BNA750	42713-66-6 see NJT000
37338-91-3 see TBF350	38571-73-2 see GGI000	39597-90-5 see EBQ550	41208-07-5 see MQY760	42794-63-8 see MQY125
37339-90-5 see LEK100	38589-14-9 see MOJ500	39603-53-7 see NLM500	41217-05-4 see DQG000	42794-87-6 see TCG750
37346-96-6 see PLV250	38668-83-6 see TNK400	39603-54-8 see ORS000	41226-18-0 see DHQ000	42824-34-0 see HAA310
37350-58-6 see MQR144	38726-90-8 see DCO200	39637-16-6 see DGB400	41226-20-4 see DOC000	42839-36-1 see CAY875
37353-63-2 see PJO500	38726-91-9 see DCC600	39660-55-4 see OBS800	41240-93-1 see DIP800	42840-17-5 see MFO500
37388-50-4 see PHP250	38748-32-2 see TNC175	39687-95-1 see MKX575	41262-21-9 see TLH250	42884-33-3 see ALJ250
37394-32-4 see PGW000	38753-50-3 see NIR550	39729-81-2 see TBH250	41287-56-3 see BOD500	42920-39-8 see BKO835
37394-33-5 see IDB000	38777-13-8 see PMY310	39735-49-4 see NEW550	41287-72-3 see BOD000	42924-53-8 see MFA300
37415-55-7 see PGV750	38780-35-7 see DGU709	39753-42-9 see MFR775	41288-00-0 see MPA250	42959-18-2 see SDH670
37415-56-8 see BSX750	38780-36-8 see BIS250	39754-64-8 see PDU750	41296-95-1 see BPJ000	42971-09-5 see EGM100
37415-62-6 see SIN850	38780-37-9 see DGT200	39801-14-4 see MRI750	41363-50-2 see PFR350	42978-42-7 see BDP500
37425-13-1 see AGU750	38780-39-1 see DGC600	39831-55-5 see APT000	41365-24-6 see DJC200	42978-43-8 see ACU500
37517-28-5 see APS750	38780-42-6 see DEU200	39834-38-3 see DQL600	41372-08-1 see MJE780	43033-72-3 see ACQ690
37517-30-9 see AAE100	38787-96-1 see PMP250	39884-52-1 see NLE000	41394-05-2 see ALA500	43047-59-2 see CFP750
37519-65-6 see SBA875	38802-82-3 see DNT000	39885-14-8 see MEW250	41409-50-1 see DAE625	43085-16-1 see PEE600
37557-89-4 see BEM250	38819-28-2 see AFI500	39900-38-4 see CCR524	41422-43-9 see TDQ000	43087-91-8 see ALV500
37558-17-1 see PGT750	38821-53-3 see SBN440	39920-56-4 see AAU500	41427-34-3 see COS750	43121-43-3 see CJO250
37571-88-3 see DLC600	38838-26-5 see GGE600	40036-79-1 see TCA750	41448-29-7 see EHN500	43143-11-9 see OIU850
37574-47-3 see BCV500	38860-48-9 see MED000	40054-69-1 see EQN600	41451-75-6 see BOL500	43200-80-2 see ZUA450
37574-48-4 see HJN500	38870-89-2 see MDW780	40068-20-0 see PCV775	41481-90-7 see PAR799	43210-67-9 see FAL100
37636-51-4 see DAS600	38892-09-0 see SGQ100	40112-23-0 see GJM030	41483-43-6 see BRJ000	43222-48-6 see ARW000
37640-57-6 see THO750	38911-59-0 see NNX650	40180-04-9 see TGA600	41510-23-0 see CCW500	43229-80-7 see FNE100
37661-08-8 see BAB250	38914-96-4 see CGR000	40193-47-3 see DKA000	41519-23-7 see HFE520	45173-31-7 see TKT100
37686-84-3 see DLR100	38914-97-5 see CGX750	40202-39-9 see CQB250	41542-50-1 see DHZ000	45776-10-1 see HEW000
37686-85-4 see DLR150	38915-00-3 see BHZ000	40225-02-3 see SAB800	41572-59-2 see DEX800	46061-25-0 see NJO300
37699-43-7 see DSX800	38915-14-9 see EHJ500	40237-34-1 see DUG889	41575-87-5 see DCF800	46231-41-8 see PDS500
37717-82-1 see BFV900	38915-22-9 see CGX500	40267-72-9 see GDG100	41575-94-4 see CCC075	46941-74-6 see THJ755
37723-78-7 see IGA000	38915-28-5 see IAE000	40283-68-9 see BMD250	41593-31-1 see DKW200	47897-65-4 see AEW750
37724-43-9 see MOF750	38915-40-1 see BPI500	40283-91-8 see BFC000	41598-07-6 see POC275	48145-04-6 see PER250
37724-45-1 see MLR500	38915-49-0 see CGP500	40283-92-9 see BFC500	41621-49-2 see BAR800	48163-10-6 see LBQ000
37724-47-3 see MOG000	38915-50-3 see CGR250	40284-08-0 see AEG000	41663-73-4 see AJI650	49538-98-9 see DNT200
37750-86-0 see DQM800	38915-59-2 see CHH000	40284-10-4 see AEG500	41699-09-6 see MHE500	49558-46-5 see NIW300
37753-10-9 see SOD000	38915-61-6 see CGR500	40321-76-4 see PAW000	41708-72-9 see TGI699	49561-54-8 see WBA600
37764-25-3 see DBJ600	38915-61-6 see CGR750	40334-69-8 see BIQ250	41708-76-3 see ICD100	49575-13-5 see MPL600
37764-28-6 see EBY500	38925-90-5 see CGS750	40343-30-4 see HKW350	41735-28-8 see MMU000	49642-61-7 see EFO000
37793-22-9 see MNB000	38940-46-4 see PFY750		41735-29-9 see ENS000	49720-72-1 see DBA175

49720-83-4 see PDK900
49773-64-0 see BJE600
49780-10-1 see NMV750
49830-98-0 see PIN100
49842-07-1 see TGI500
49850-29-5 see AMS625
49852-84-8 see CIH000
49852-85-9 see BNP000
50264-69-2 see DEL200
50264-86-3 see CEQ625
50264-96-5 see GGR100
50274-95-8 see CJC610
50277-65-1 see PMT300
50285-70-6 see NLX500
50285-71-7 see DRV600
50285-72-8 see DJP600
50308-82-2 see CIP000
50308-83-3 see CIO750
50308-84-4 see CIP250
50308-86-6 see DSR600
50308-87-7 see DSR800
50308-88-8 see MEW750
50308-89-9 see MEW450
50308-94-6 see QOJ250
50309-02-9 see DBW000
50309-03-0 see DCC200
50309-11-0 see MLT250
50309-16-5 see AJW500
50309-17-6 see MLT750
50309-20-1 see AAL000
50311-48-3 see TKE525
50335-03-0 see CDG750
50335-09-6 see EOP500
50355-74-3 see TJA000
50355-75-4 see DUI800
50370-12-2 see DYF700
50416-18-7 see AJE250
50424-93-6 see MMI250
50425-34-8 see DSS000
50425-35-9 see MLU750
50454-68-7 see TGJ875
50465-39-9 see TGJ000
50471-44-8 see RMA000
50475-76-8 see LGQ000
50485-03-5 see ACP750
50510-11-7 see AJI500
50510-12-8 see AJM750
50528-97-7 see XGA725
50570-59-7 see BHN250
50577-64-5 see SDK000
50588-13-1 see BGR325
50602-44-3 see THG600
50645-52-8 see IHJ000
50650-74-3 see AQQ000
50657-29-9 see OHQ300
50662-24-3 see LHM850
50679-08-8 see TAI450
50707-40-9 see EPW000
50722-38-8 see ACH075
50764-78-8 see EPM600
50765-87-2 see PMB500
50782-69-9 see EIG000
50802-21-6 see TGD125
50809-32-0 see TFZ100
50814-62-5 see EAG500
50816-18-7 see DAI450
50831-29-3 see TDI300
50832-74-1 see ALN250
50838-36-3 see MQA000
50846-45-2 see BAC325
50847-92-2 see DUG600
50864-67-0 see BAO300
50865-01-5 see DXF700
50880-57-4 see HLX600
50892-23-4 see CLW250
50892-99-4 see FPK100
50901-84-3 see ALA000
50901-87-6 see ALA250
50906-29-1 see PHY275
50908-62-8 see AEJ000
50924-49-7 see BMM000
50933-33-0 see GJK000

50935-04-1 see CCK625
50976-02-8 see DGW400
51003-81-7 see DPY600
51003-83-9 see BRC500
51012-33-0 see TGA375
51022-69-6 see COW825
51022-70-9 see SAF400
51022-74-3 see IGD100
51023-56-4 see CCW750
51025-94-6 see CQK100
51029-30-2 see FEH000
51034-39-0 see CKR000
51077-50-0 see DMD100
51104-87-1 see DEU509
51109-89-8 see NOA700
51110-01-1 see SKS600
51131-85-2 see HKH000
51138-16-0 see NIL625
51155-15-8 see TET780
51165-36-7 see SBL500
51200-87-4 see DTG750
51207-31-9 see TBO780
51218-45-2 see MQA060
51218-49-6 see PMB850
51234-28-7 see OJI750
51235-04-2 see HFA300
51249-05-9 see ALZ000
51257-84-2 see LEJ700
51264-14-3 see ADL750
51286-83-0 see PLV275
51312-42-6 see SJJ000
51317-24-9 see LDP000
51321-79-0 see PGY750
51325-35-0 see DBF400
51333-22-3 see BOM520
51340-26-2 see DAQ002
51371-34-7 see IHO700
51410-44-7 see HKI000
51460-26-5 see AER666
51473-23-5 see LEP000
51481-10-8 see VTF500
51481-61-9 see TAB250
51481-65-3 see MQS200
51503-61-8 see ANS250
51542-33-7 see NKR000
51566-62-2 see CMU000
51579-82-9 see AIU500
51622-02-7 see DKO000
51627-14-6 see APT250
51630-58-1 see FAR100
51635-81-5 see EDX000
51707-55-2 see TEX600
51753-57-2 see PDB350
51762-05-1 see CCS530
51764-33-1 see IFP800
51775-17-8 see CGX625
51775-36-1 see THH575
51781-06-7 see CCL800
51781-21-6 see MQT550
51786-53-9 see XOS000
51787-42-9 see TLI500
51787-43-0 see TLI750
51787-44-1 see TDL500
51799-29-2 see NDY360
51821-32-0 see HKT200
51877-12-4 see DDN100
51898-39-6 see AET250
51909-61-6 see GAA100
51922-16-8 see GAA120
51938-12-6 see NLN000
51938-13-7 see BRX750
51938-14-8 see BRO000
51938-15-9 see BRY000
51938-16-0 see MKP000
51938-42-2 see SKR875
51940-44-4 see PIZ000
52049-26-0 see CHA500
52061-60-6 see IQE000
52093-21-7 see MQS579
52096-16-9 see NEN300
52096-22-7 see BCD125
52098-56-3 see EPM000

52112-09-1 see BLN750
52112-66-0 see AIX750
52112-67-1 see AJL750
52112-68-2 see AJF750
52125-53-8 see EJV000
52129-71-2 see DUV400
52137-03-8 see PJU250
52152-93-9 see CCS550
52157-57-0 see CIO275
52162-18-2 see AAM000
52171-37-6 see DOD400
52171-41-2 see DOD200
52171-42-3 see DOF800
52171-92-3 see DRE400
52171-93-4 see DLI000
52171-94-5 see DUF200
52175-10-7 see POJ100
52195-07-0 see PIJ600
52205-73-9 see EDT100
52207-83-7 see AGH500
52207-87-1 see TGH690
52212-02-9 see PII250
52214-84-3 see CMS210
52217-47-7 see ACL500
52236-34-7 see DWF300
52237-03-3 see BLG100
52279-59-1 see MRU750
52292-20-3 see PKN000
52315-07-8 see RLF350
52329-60-9 see AIH000
52338-90-6 see TJK500
52340-46-2 see CHL875
52351-96-9 see MEC500
52400-55-2 see EGC000
52400-58-5 see DJU800
52400-60-9 see DJI200
52400-61-0 see DPC400
52400-65-4 see MLL500
52400-66-5 see MHD750
52400-76-7 see BDY500
52400-77-8 see CFP250
52400-80-3 see DDM800
52400-82-5 see DHT800
52400-99-4 see DHT600
52401-02-2 see DPI400
52401-04-4 see DHT400
52444-01-6 see BIL000
52463-83-9 see PIH100
52470-25-4 see GLB300
52479-14-8 see DIC600
52479-15-9 see DHU600
52479-18-2 see AKA500
52485-79-7 see TAL325
52486-78-9 see MCV500
52547-00-9 see MLB600
52549-17-4 see PLX400
52551-67-4 see BKC000
52557-97-8 see MHY600
52578-56-0 see BJE500
52581-71-2 see UHA000
52583-02-5 see BNF000
52583-06-9 see CIF000
52622-27-2 see CBF710
52623-84-4 see SFS500
52623-88-8 see MAP300
52645-53-1 see AHJ750
52663-81-7 see DXS375
52663-84-0 see FEO000
52670-52-7 see BSL450
52670-78-7 see MNW100
52673-65-1 see BJG750
52673-66-2 see BJI750
52684-23-8 see OIW000
52694-54-9 see CPK625
52712-76-2 see AJA375
52716-12-8 see CEX255
52731-39-2 see BRX000
52740-16-6 see ARK780
52740-56-4 see MEA500
52775-76-5 see MJU500
52777-39-6 see MGK750
52783-44-5 see HAT000

52794-97-5 see KCA000
52831-39-7 see FGJ000
52831-41-1 see DLN000
52831-45-5 see FGB000
52831-55-7 see DLN200
52831-56-8 see FGD000
52831-58-0 see FGO000
52831-60-4 see FJT000
52831-62-6 see FGL000
52831-65-9 see FGG000
52831-67-1 see DLN400
52831-68-2 see FGF000
52833-75-7 see TKC000
52851-26-0 see THQ600
52913-14-1 see TGO575
52918-63-5 see DAF300
52934-83-5 see RMK200
52936-25-1 see EJC000
52955-41-6 see MRY100
52968-02-2 see MOA000
52977-61-4 see NKP500
53004-03-8 see DOU000
53011-73-7 see IGF200
53025-21-1 see PAG225
53043-14-4 see AOH250
53043-29-1 see LEF800
53051-16-4 see NLS500
53067-74-6 see DWC050
53112-33-7 see MPO750
53123-88-9 see RBK000
53125-86-3 see SKO000
53127-17-6 see EEI060
53130-67-9 see HIX200
53142-01-1 see AHN625
53152-21-9 see BOO630
53153-66-5 see MNB500
53164-05-9 see AAE625
53179-09-2 see APY500
53184-19-3 see SBU950
53198-41-7 see ABT750
53198-46-2 see NKP000
53198-87-1 see EQZ000
53213-78-8 see BJP300
53214-97-4 see DNE875
53216-90-3 see GKC000
53221-79-7 see MGA000
53221-83-3 see ADO750
53221-85-5 see ADO250
53221-86-6 see ADO500
53221-88-8 see ADQ000
53222-10-9 see MFY750
53222-14-3 see AAJ000
53222-15-4 see MPM000
53222-25-6 see AHT850
53222-52-9 see ALC750
53230-00-5 see DWU800
53231-79-1 see PJF500
53296-30-3 see ERD000
53305-31-0 see PJI600
53306-53-9 see DSF200
53317-25-2 see CQK500
53365-77-8 see MPB175
53370-90-4 see HFS759
53378-72-6 see PBJ600
53384-39-7 see MJK750
53404-82-3 see TIE600
53421-36-6 see CNN750
53421-38-8 see DWD200
53459-38-4 see MAG050
53460-80-3 see CIL850
53460-81-4 see CIQ400
53469-21-9 see PJM500
53477-43-3 see MFG510
53478-39-0 see MFZ000
53499-68-6 see MLK750
53516-81-7 see AMU125
53532-37-9 see BJY125
53534-20-6 see DRI700
53555-01-4 see TIL750
53555-67-2 see ABL750
53558-25-1 see PPP750
53569-62-3 see CBV000

53569-64-5 see NGY500
53578-07-7 see IPI350
53581-53-6 see BNE250
53583-79-2 see EPD100
53597-25-4 see SAO200
53597-27-6 see FAO100
53597-29-8 see MQF750
53607-04-8 see SNL830
53609-64-6 see DNB200
53639-82-0 see PMC275
53663-14-2 see NHY500
53681-76-8 see MRA300
53697-17-9 see POC360
53716-43-1 see BAV275
53716-49-7 see CCK800
53716-50-0 see OMY500
53734-79-5 see IBQ300
53746-45-5 see FAP100
53757-28-1 see NGM400
53757-29-2 see MMJ975
53757-30-5 see FNG000
53757-31-6 see NGL000
53760-19-3 see PAM775
53762-93-9 see VRA000
53778-51-1 see DXF400
53780-34-0 see DUK000
53797-35-6 see RIP000
53847-48-6 see TEL150
53858-86-9 see FBP050
53866-33-4 see DHA425
53885-35-1 see TGA525
53902-12-8 see RLK800
53908-27-3 see DYE400
53910-25-1 see PBT100
53912-89-3 see PIT650
53935-32-3 see POE050
53940-49-1 see EBT500
53955-81-0 see MLL600
53956-04-0 see GIE100
53962-20-2 see PAM789
53988-42-4 see AFK900
53994-73-3 see PAG075
54024-22-5 see DBA750
54025-36-4 see NNX600
54063-28-4 see CBA375
54063-38-6 see FLK000
54083-22-6 see ROU800
54086-41-8 see DKN250
54096-45-6 see EBD600
54099-11-5 see AED750
54099-12-6 see AEE000
54099-13-7 see DPB400
54099-14-8 see DHT200
54099-23-9 see DHP000
54156-67-1 see IKW000
54240-36-7 see KGK300
54262-83-8 see AMH800
54301-15-4 see ADL500
54301-19-8 see HJE600
54323-85-2 see POE100
54328-07-3 see PLR500
54340-62-4 see BQD000
54350-48-0 see EMJ500
54363-49-4 see MNH500
54405-61-7 see MGT000
54448-39-4 see PLL000
54453-03-1 see CNL750
54460-46-7 see HCP500
54472-62-7 see INN500
54481-45-7 see ABQ250
54484-91-2 see CFQ000
54504-70-0 see TEQ500
54514-12-4 see TGM425
54524-31-1 see MCK300
54527-84-3 see PCG550
54531-52-1 see BCJ150
54567-24-7 see ASE500
54573-23-8 see CPW325
54579-17-8 see EDI000
54593-27-0 see BJG500
54597-56-7 see MRJ600
54605-45-7 see IDJ500

54622-43-4 see DYE550
54634-49-0 see NLP600
54643-52-6 see HOB000
54708-51-9 see CJW500
54708-68-8 see CKD250
54746-50-8 see BPD000
54749-90-5 see CLX000
54767-75-8 see SOU600
54779-53-2 see AJS875
54818-88-1 see DDD400
54824-17-8 see MQX775
54824-20-3 see MAB055
54827-17-7 see TDM800
54856-23-4 see BFV350
54889-82-6 see PNE750
54897-62-0 see ELE500
54897-63-1 see NKE000
54925-45-0 see AEL500
54927-63-8 see SCA000
54965-21-8 see VAD000
54965-24-1 see TAD175
54976-93-1 see SFV300
55011-46-6 see NMK000
55028-70-1 see AQT575
55028-71-2 see ECW550
55042-15-4 see UOA000
55044-04-7 see ELN600
55049-48-4 see SFX725
55077-30-0 see AAC875
55079-83-9 see REP400
55080-20-1 see ABO500
55090-44-3 see NKU000
55102-43-7 see ROF200
55102-44-8 see RFU600
55112-89-5 see BIB250
55118-19-9 see DPB200
55123-66-5 see AAL300
55124-14-6 see BJE325
55158-44-6 see CNK700
55165-33-8 see BET750
55216-04-1 see BIO500
55217-61-3 see AKL750
55256-53-6 see ARU250
55257-88-0 see DXY750
55268-74-1 see BGB400
55268-75-2 see CCS600
55283-68-6 see ENE500
55294-15-0 see EAE675
55297-95-5 see TET800
55297-96-6 see DAR000
55299-24-6 see CGB000
55308-37-7 see MFF650
55308-57-1 see PEU650
55308-64-0 see EFC259
55309-14-3 see MFJ110
55335-06-3 see TJE890
55365-87-2 see EPC150
55380-34-2 see DRO000
55398-24-8 see BDO199
55398-25-9 see BDQ000
55398-26-0 see BDO500
55398-27-1 see EOJ000
55398-86-2 see MRG000
55448-20-9 see MAS250
55467-31-7 see CCX725
55477-20-8 see CEV750
55477-27-5 see CHO250
55489-49-1 see CFL750
55489-90-4 see CFR000
55509-78-9 see SBN510
55514-14-2 see MLX250
55520-67-7 see VOA550
55541-30-5 see DBC500
55556-85-9 see NLK000
55556-86-0 see DTA800
55556-88-2 see DRN800
55556-91-7 see NLL000
55556-92-8 see NLU500
55556-93-9 see NLK500
55556-94-0 see MJH000
55557-00-1 see DVE600
55557-02-3 see NKH500

55566-30-8 see TDI000
55600-34-5 see CMX845
55620-97-8 see EAB100
55621-29-9 see BRN250
55636-92-5 see HKS600
55644-07-0 see DGS700
55644-07-0 see DGS800
55651-31-5 see DLP400
55651-36-0 see DLP200
55661-38-6 see ALF500
55673-54-6 see AJD375
55688-38-5 see HLK500
55689-65-1 see OMU000
55719-85-2 see PFD250
55726-47-1 see EAU075
55738-54-0 see DPL000
55764-18-6 see OOK000
55769-64-7 see CNW125
55779-06-1 see FOK000
55792-21-7 see PIO750
55798-64-6 see BCA250
55818-96-7 see TDH500
55837-18-8 see IJH000
55837-27-9 see PDW250
55837-29-1 see TGF175
55843-86-2 see IAY000
55864-39-6 see ALG375
55870-36-5 see BIT350
55870-64-9 see EBC000
55898-33-4 see LJP000
55902-04-0 see AGL750
55921-66-9 see ALA750
55936-75-9 see MEG500
55936-76-0 see MEG000
55936-77-1 see ABQ500
55936-78-2 see MEH250
55939-60-1 see SHT500
55941-39-4 see AAJ500
55981-23-2 see CEK875
55984-51-5 see NKV000
56001-43-5 see NCN800
56011-02-0 see IHV050
56047-14-4 see RHK250
56073-10-0 see TAC800
56090-02-9 see HEK550
56092-91-2 see COQ375
56124-62-0 see TJX350
56139-33-4 see BQU000
56172-46-4 see DTE000
56173-18-3 see HHE500
56179-80-7 see ECR259
56179-83-0 see DLZ000
56183-20-1 see MEB000
56187-09-8 see HCN100
56217-89-1 see NDY380
56217-90-4 see NDY390
56222-04-9 see FAJ200
56222-35-6 see NLP700
56235-95-1 see NKC500
56238-63-2 see SFQ300
56239-24-8 see CHB000
56267-87-9 see TJT500
56275-41-3 see DFB400
56280-76-3 see PHT000
56283-74-0 see DAK000
56287-19-5 see TLM250
56287-41-3 see CFO750
56287-74-2 see AEW625
56299-00-4 see POF800
56302-13-7 see SAY950
56305-04-5 see TNR625
56316-37-1 see PNG500
56370-81-1 see TBG600
56375-33-8 see NJO000
56386-98-2 see PKN750
56390-16-0 see FGQ000
56391-56-1 see SBD000
56391-57-2 see NCP550
56393-22-7 see HHD000
56395-66-5 see HLF500
56400-60-3 see DSG200
56411-66-6 see DDN700

56411-67-7 see DHC309
56420-45-2 see EBB100
56433-01-3 see CJC600
56455-90-4 see DEK000
56470-64-5 see AOO150
56480-06-9 see HCA600
56488-59-6 see BSE750
56501-30-5 see DXX100
56501-31-6 see EPJ600
56501-32-7 see EOB200
56501-33-8 see EKN600
56501-34-9 see ENX875
56501-35-0 see DXY700
56501-36-1 see DXX875
56516-72-4 see NKG500
56530-47-3 see DAU600
56530-48-4 see DXU600
56530-49-5 see DAS800
56533-30-3 see DKF620
56538-00-2 see CGQ280
56538-01-3 see CGQ400
56538-02-4 see CGX325
56573-85-4 see TIC500
56583-56-3 see KGK400
56602-09-6 see DAT400
56605-16-4 see SLD900
56606-38-3 see PPP000
56611-65-5 see PHV725
56614-97-2 see BBG200
56622-38-9 see BKO800
56631-46-0 see PKO000
56641-03-3 see CDV625
56643-49-3 see AMD250
56654-52-5 see DEF000
56654-53-6 see BRE000
56668-59-8 see NDD500
56713-63-4 see MNT750
56717-11-4 see AGE750
56726-04-6 see DAU400
56741-95-8 see AIY850
56743-33-0 see CCD625
56749-17-8 see WCB100
56764-40-0 see LCZ000
56767-15-8 see DWA700
56775-88-3 see ZBA500
56776-01-3 see BQE250
56776-25-1 see DFO600
56779-19-2 see PKN500
56795-65-4 see BRL500
56795-66-5 see PNO000
56796-20-4 see CCS350
56833-74-0 see SMC375
56856-83-8 see AAW000
56863-02-6 see BKF500
56892-30-9 see BCX000
56892-31-0 see BCY500
56892-32-1 see BCY750
56892-33-2 see BCZ000
56894-91-8 see BIJ500
56927-39-0 see DSH200
56929-36-3 see DXI200
56937-68-9 see PGV500
56943-26-1 see TJE050
56961-60-5 see BBC000
56961-62-7 see EGO000
56961-65-0 see MJY250
56970-24-2 see MEA000
56973-16-1 see TCI500
56974-46-0 see PEU000
56974-61-9 see GAD400
56986-35-7 see BPV325
56986-36-8 see BRX500
57021-61-1 see XLS300
57074-51-8 see HOT000
57109-90-7 see CDQ250
57117-24-5 see NKX000
57117-31-4 see PAW100
57164-87-1 see ADN000
57164-89-3 see ADN250
57165-71-6 see TKD000
57170-08-8 see MFJ105

57197-43-0 see SNL850
57229-41-1 see ABY150
57267-78-4 see ANL100
57281-35-3 see CPG500
57282-49-2 see LJM800
57285-10-6 see IFT100
57294-74-3 see AJK625
57296-63-6 see IBQ400
57314-55-3 see MRU757
57377-32-9 see HOT200
57383-74-1 see MCJ300
57404-88-3 see DLC800
57414-02-5 see MMD750
57420-66-3 see EQD600
57421-56-4 see SDR350
57432-60-7 see EDA875
57432-61-8 see MJV750
57449-30-6 see COC825
57455-37-5 see UJA200
57472-68-1 see DWS650
57495-14-4 see KGK100
57512-42-2 see DCL800
57524-15-9 see TGJ885
57530-25-3 see AMZ125
57541-72-7 see DFW200
57541-73-8 see DDQ800
57554-34-4 see CBG375
57558-46-0 see SBN300
57576-44-0 see APU500
57583-34-3 see MQH500
57590-20-2 see PBJ875
57598-00-2 see HJG000
57608-59-0 see FCC000
57619-29-1 see PMJ550
57629-90-0 see BPM500
57644-85-6 see MFB250
57645-49-5 see DBA250
57645-91-7 see CMV325
57647-13-9 see BJA809
57647-35-5 see BGC500
57647-53-7 see BJH000
57648-21-2 see TGB175
57651-82-8 see HJV000
57653-85-7 see HCF000
57665-49-3 see BJA750
57670-85-6 see SOI200
57707-64-9 see ASE000
57716-89-9 see MPN600
57726-65-5 see DWF700
57775-22-1 see EQO000
57775-27-6 see PIK625
57781-14-3 see HAG325
57801-81-7 see LEJ600
57808-65-8 see CFC100
57808-66-9 see DYB875
57816-08-7 see DKY400
57835-92-4 see NJA100
57846-03-4 see BNR325
57872-78-3 see AEK750
57872-80-7 see AEM000
57891-85-7 see PLJ780
57897-99-1 see SBU600
57943-81-4 see CMV250
57948-13-7 see DFQ400
57962-60-4 see MNY750
57982-77-1 see LIU420
57982-78-2 see BOM530
57998-68-2 see ASK875
58001-89-1 see MCA775
58011-68-0 see POM275
58030-91-4 see DLC400
58048-24-1 see AMH500
58048-25-2 see AEJ500
58048-26-3 see AMH750
58050-46-7 see BEB000
58050-49-0 see ENL500
58086-32-1 see ADB250
58098-08-1 see SNI000
58100-26-8 see CQK125
58128-20-4 see CMS230
58131-55-8 see NFF000
58138-08-2 see TJK100

58139-32-5 see MMW250
58139-33-6 see DTN400
58139-34-7 see MMW750
58139-35-8 see MNV500
58152-03-7 see AKJ000
58164-88-8 see AQE250
58169-97-4 see MMV250
58194-38-0 see ART000
58209-98-6 see AEU500
58229-88-2 see DVN600
58240-55-4 see THQ000
58243-85-9 see FNK200
58270-08-9 see DFE469
58302-42-4 see SDP600
58306-30-2 see BKN250
58338-59-3 see DBQ125
58344-42-6 see DSH600
58429-99-5 see DQL200
58430-00-5 see DQK600
58430-01-6 see TLI250
58430-94-7 see TLT500
58431-24-6 see ABT500
58451-82-4 see BJM500
58451-85-7 see BJM250
58451-87-9 see BIV000
58484-07-4 see CHA000
58494-43-2 see CHA750
58513-59-0 see TFQ600
58546-54-6 see SBE450
58551-69-2 see CCC110
58580-55-5 see PAG050
58581-89-8 see ASC125
58658-27-8 see AHT250
58682-45-4 see BMS750
58683-84-4 see MFB250
58695-41-3 see CPG000
58696-86-9 see ANM625
58718-68-6 see PAE875
58763-27-2 see CIU325
58763-31-8 see BPI625
58776-08-2 see PFJ780
58785-63-0 see LIF000
58786-99-5 see BPG325
58812-37-6 see CML825
58814-86-1 see AEC625
58817-05-3 see AOI500
58845-80-0 see LIG000
58886-98-9 see DLC200
58911-02-7 see TCY500
58911-30-1 see CLR825
58917-67-2 see BCU250
58917-67-2 see DMP900
58917-67-2 see DMQ600
58917-91-2 see DMR000
58933-55-4 see HAN000
58941-14-3 see DFL709
58957-92-9 see DAN000
58958-60-4 see ISC550
58970-76-6 see BFV300
58989-02-9 see HKY000
59017-64-0 see IGD200
59034-32-1 see TKB325
59034-34-3 see CFX625
59086-92-9 see PGU750
59122-46-2 see MJE775
59128-97-1 see HAG800
59130-69-7 see HCP550
59160-29-1 see LFO300
59163-97-2 see MID750
59177-62-7 see CID000
59177-64-9 see BNN250
59177-70-7 see MOW000
59177-76-3 see MOW250
59177-78-5 see BFG500
59177-85-4 see BNN500
59178-29-9 see TGC285
59182-63-7 see CPP000
59183-17-4 see DFE300
59183-18-5 see ACH125
59198-70-8 see DKF130

59209-40-4 see POF550	60479-97-2 see BTA125	61462-73-5 see CED500	62421-98-1 see MNB300	63019-82-9 see POQ000
59230-81-8 see BNR250	60494-19-1 see BKO500	61471-62-3 see DUJ800	62422-00-8 see DAS500	63019-93-2 see DOV400
59261-17-5 see DVQ759	60504-57-6 see DAY835	61477-94-9 see PJA170	62450-06-0 see TNX275	63019-97-6 see MGF500
59277-89-3 see AEC700	60504-95-2 see DAM000	61481-19-4 see BQK850	62450-07-1 see ALD500	63019-98-7 see MGE000
59297-18-6 see SLQ625	60510-57-8 see PGN400	61490-68-4 see TDD750	62536-49-6 see SLO000	63020-03-1 see MFA250
59327-98-9 see ASG625	60525-15-7 see ZBA525	61499-28-3 see HNX500	62571-86-2 see MCO750	63020-21-3 see DDD800
59333-90-3 see ERE100	60539-20-0 see DXR800	61503-59-1 see CNR250	62573-57-3 see PBX750	63020-25-7 see MIQ250
59348-49-1 see AJI530	60548-62-1 see BOU100	61514-68-9 see SDW100	62573-59-5 see NJP000	63020-27-9 see EEX000
59348-62-8 see DCP700	60550-91-6 see COW500	61556-82-9 see TDR250	62593-23-1 see NJQ500	63020-32-6 see PNF000
59355-75-8 see MFX600	60553-18-6 see FIK875	61583-30-0 see CJY250	62637-93-8 see TLE250	63020-33-7 see PNI500
59405-47-9 see POS250	60560-33-0 see COS500	61614-71-9 see TDF250	62641-66-1 see PNL500	63020-37-1 see MPO250
59413-14-8 see HNZ000	60561-17-3 see SNH100	61691-82-5 see ABN725	62641-67-2 see NJX500	63020-39-3 see TDL750
59419-71-5 see BGT150	60566-40-7 see AHT000	61695-69-0 see ONE000	62641-68-3 see ELH000	63020-45-1 see BBG500
59467-70-8 see MQT525	60573-88-8 see AKG250	61695-70-3 see VMF000	62681-13-4 see EFR500	63020-47-3 see INZ000
59483-61-3 see CJA150	60580-30-5 see DWU400	61695-72-5 see ONC000	62765-90-6 see IHK100	63020-48-4 see IOA000
59512-21-9 see BDT750	60597-20-8 see TLT750	61695-74-7 see ONG000	62765-93-9 see NCS000	63020-53-1 see IOF000
59536-65-1 see FBU000	60599-38-4 see NJN000	61702-43-0 see ALO500	62778-13-6 see BJJ000	63020-60-0 see MEN750
59544-89-7 see TKH250	60605-72-3 see BJK000	61706-44-3 see ALX500	62783-48-6 see NKR500	63020-61-1 see MEU250
59547-52-3 see LGK100	60607-34-3 see OMG000	61711-25-9 see AFJ375	62783-49-7 see NKS000	63020-69-9 see DRH800
59557-05-0 see MCG850	60616-74-2 see MAG500	61734-86-9 see NLE500	62783-50-0 see NKS500	63020-76-8 see MIV250
59558-23-5 see THB000	60633-76-3 see COK659	61734-88-1 see ELM000	62796-23-0 see EDD500	63020-91-7 see CGK500
59643-84-4 see PMY750	60634-59-5 see SNB500	61734-89-2 see BSB500	62851-48-3 see AFM000	63021-00-1 see DRU600
59652-20-9 see MIY250	60662-79-5 see DWC100	61734-90-5 see BSB750	62851-59-6 see ERC600	63021-11-4 see OIC000
59652-21-0 see MIY000	60672-60-8 see PBH500	61735-77-1 see FGI000	62861-56-7 see CKS325	63021-32-9 see BBM500
59653-73-5 see TBC450	60676-83-7 see AJK500	61735-78-2 see DKG400	62861-57-8 see MEH775	63021-33-0 see EHW000
59665-11-1 see MFX500	60676-86-0 see SCK600	61738-03-2 see MEO500	62865-26-3 see CMS205	63021-35-2 see EHW500
59680-34-1 see AEC250	60676-90-6 see ZRS000	61738-04-3 see MEV750	62893-20-3 see CCS369	63021-43-2 see MSB000
59690-88-9 see ALY750	60706-43-6 see MOE250	61738-05-4 see MEP500	62907-78-2 see SNJ400	63021-45-4 see NBC000
59703-84-3 see SJJ200	60706-49-2 see MOP500	61785-70-4 see NIV000	62912-45-2 see PBH500	63021-46-5 see NMF000
59748-51-5 see NEM000	60706-52-7 see MOQ000	61785-72-6 see MFH750	62912-47-4 see PBI000	63021-47-6 see NMF500
59748-95-7 see MDY000	60719-83-7 see CJX000	61785-73-7 see THD275	62912-51-0 see OPQ000	63021-48-7 see NMD500
59766-02-8 see DDF400	60719-84-8 see AOD375	61788-32-7 see HHW800	62928-11-4 see IGG775	63021-49-8 see NME000
59804-37-4 see TAL485	60723-51-5 see BOM000	61788-33-8 see PJP250	62929-91-3 see PME600	63021-50-1 see NME500
59863-59-1 see DFW600	60748-45-0 see ARS000	61788-72-5 see FAB920	62967-27-5 see IKI000	63021-51-2 see NNA000
59865-13-3 see CQH100	60762-50-7 see GFO200	61788-90-7 see CNF325	62973-76-6 see TJF000	63021-62-5 see SMU000
59901-90-5 see ABW250	60763-49-7 see CMS125	61789-01-3 see FAB900	62987-05-7 see BLH325	63021-67-0 see OBW000
59901-91-6 see HOE500	60764-83-2 see MCP000	61789-30-8 see CNF175	63007-70-5 see CFJ000	63035-21-2 see NCG700
59921-81-2 see FGW000	60784-40-9 see CHE325	61789-51-3 see NAR500	63018-40-6 see BBC500	63039-89-4 see EMM000
59928-80-2 see MEI500	60784-41-0 see EIP500	61790-14-5 see NAS500	63018-49-5 see BBI750	63039-90-7 see NOE550
59935-47-6 see FEP000	60784-42-1 see PNJ000	61790-53-2 see DCJ800	63018-50-8 see BBJ000	63039-95-2 see HBF500
59937-28-9 see MAO275	60784-43-2 see TDQ250	61791-14-8 see EEJ500	63018-56-4 see BBJ250	63040-01-7 see TLH500
59943-31-6 see CIT625	60784-44-3 see PBG750	61791-24-0 see EEJ500	63018-57-5 see BBH750	63040-02-8 see TLH750
59960-30-4 see NJR500	60784-46-5 see CHB750	61791-24-0 see EEK000	63018-59-7 see BBH000	63040-05-1 see TLH350
59960-90-6 see DHO000	60784-47-6 see CHC250	61791-42-2 see SFY000	63018-62-2 see BBF750	63040-09-5 see MIM000
59965-27-4 see EGS500	60784-48-7 see CHA250	61792-05-0 see HLO400	63018-63-3 see BNF315	63040-20-0 see QQA000
59978-65-3 see NDE500	60789-89-1 see MNP300	61827-74-5 see AJO000	63018-64-4 see BQI500	63040-21-1 see TBI750
59985-27-2 see AMJ750	60828-78-6 see TBB775	61840-09-3 see HDF100	63018-67-7 see CIG000	63040-24-4 see HLR500
59988-01-1 see ADM500	60842-44-6 see DJY100	61840-39-9 see OKO200	63018-68-8 see COM000	63040-25-5 see AKN250
60012-89-7 see BLG325	60864-95-1 see DML200	61848-66-6 see DAD075	63018-69-9 see BBB500	63040-27-7 see CGK750
60047-17-8 see LFY500	60883-74-1 see BMX000	61848-70-2 see DAD040	63018-70-2 see MGY000	63040-30-2 see PEB250
60050-37-5 see BNX250	60967-88-6 see BBD250	61866-12-4 see CHF250	63018-94-0 see TLH050	63040-32-4 see TMG000
60062-60-4 see PNX500	60967-89-7 see BBD500	61907-23-1 see APL250	63018-98-4 see ACC750	63040-43-7 see PGU500
60075-74-3 see EEQ100	60967-90-0 see BBF000	61912-76-3 see LNE000	63018-99-5 see AOE750	63040-44-8 see PGU000
60084-10-8 see RJF500	60968-01-6 see DKS600	61925-70-0 see BQX000	63019-08-9 see DKA800	63040-49-3 see DOB600
60086-22-8 see CMX860	60968-08-3 see DKS400	61941-56-8 see AHK625	63019-09-0 see TMF750	63040-53-9 see BCQ750
60102-37-6 see PCQ750	60996-85-2 see CAL075	61947-30-6 see DNJ000	63019-12-5 see EHD000	63040-54-0 see DCY600
60145-64-4 see AKD875	61001-31-8 see PEU600	62018-89-7 see HKS500	63019-14-7 see DSV000	63040-55-1 see FNK000
60153-49-3 see MMS200	61001-36-3 see PGE350	62018-90-0 see HJS000	63019-23-8 see DRE600	63040-56-2 see FNU000
60160-75-0 see DTM200	61001-40-9 see MFH800	62018-91-1 see BRB000	63019-25-0 see DQL800	63040-57-3 see FNV000
60166-93-0 see IFY000	61001-42-1 see PGE300	62018-92-2 see BRX250	63019-29-4 see EEV000	63040-58-4 see FNQ000
60168-88-9 see FAK100	61005-12-7 see BSF250	62037-49-4 see MIG500	63019-32-9 see HBN000	63040-63-1 see DPO800
60172-01-2 see PIG500	61034-40-0 see NJL850	62046-37-1 see DET600	63019-34-1 see HFL000	63040-64-2 see DPP000
60172-03-4 see PIG250	61036-64-4 see TAI400	62046-63-3 see PAI750	63019-42-1 see SNX000	63040-65-3 see DPP200
60172-05-6 see PPD500	61136-62-7 see CDC375	62064-66-8 see BEY800	63019-46-5 see CME000	63040-98-2 see DKM800
60172-05-6 see SMW000	61137-63-1 see DML000	62116-25-0 see BMS000	63019-50-1 see BAY250	63041-00-9 see BOI250
60172-15-8 see POT500	61177-45-5 see PLB775	62126-20-9 see CNQ250	63019-51-2 see CGW750	63041-01-0 see DKM400
60223-52-1 see TDJ000	61227-05-2 see TET250	62133-36-2 see TJP780	63019-52-3 see CGH250	63041-05-4 see MJD000
60254-95-7 see HJN875	61242-71-5 see NNG500	62169-70-4 see TAL350	63019-53-4 see EHL000	63041-07-6 see MRI500
60268-85-1 see BCU000	61318-91-0 see SNH480	62178-60-3 see NLY500	63019-57-8 see DHI800	63041-14-5 see MQD000
60268-85-1 see DMQ800	61336-70-7 see AOA100	62207-76-5 see EIS000	63019-59-0 see DQC200	63041-15-6 see MQD500
60318-52-7 see TJF350	61413-38-5 see FKE000	62209-22-7 see COG750	63019-60-3 see DQC000	63041-19-0 see ILG000
60325-46-4 see SOU650	61413-39-6 see MEJ500	62229-50-9 see EBA275	63019-65-8 see DBF000	63041-23-6 see TLP250
60345-95-1 see MDF100	61413-61-4 see HMG500	62229-70-3 see PAC500	63019-67-0 see AKC000	63041-25-8 see TIJ000
60364-26-3 see PMS825	61417-04-7 see MDX500	62258-26-8 see MIG750	63019-68-1 see FEI500	63041-30-5 see AJL250
60391-92-6 see NKI500	61417-05-8 see MDY250	62303-19-9 see BDN125	63019-69-2 see MEB250	63041-43-0 see SGF000
60397-73-1 see BKN500	61417-08-1 see CED750	62314-67-4 see DML400	63019-70-5 see TDM000	63041-44-1 see DCU600
60398-22-3 see MID250	61417-10-5 see ADL000	62316-46-5 see NMG500	63019-72-7 see MEK750	63041-48-5 see DDX200
60414-81-5 see NJM500	61422-45-5 see CCK630	62355-03-7 see COG500	63019-73-8 see SMS000	63041-49-6 see DKV800
60444-92-0 see BPG750	61424-17-7 see HJS400	62362-59-8 see SBZ000	63019-76-1 see DUM600	63041-50-9 see DLT200
60448-19-3 see BCW250	61443-57-0 see DML000	62374-53-2 see BKJ250	63019-77-2 see NIK500	63041-56-5 see DLK600
60452-14-4 see BTA500	61445-55-4 see MIF250	62375-91-1 see MMH000	63019-78-3 see MOY500	63041-61-2 see DRE000
60462-51-3 see HJA000	61447-07-2 see MMI640	62399-48-8 see APK750	63019-81-8 see AKI500	63041-62-3 see DRD800

63041-70-3 see ION000	63716-96-1 see QVJ000	63905-38-4 see AFX750	63957-36-8 see PDL000	64037-07-6 see AIX500
63041-72-5 see MEL000	63717-25-9 see AKJ500	63905-44-2 see BIE750	63957-37-9 see ALT750	64037-08-7 see AIY250
63041-77-0 see MHH000	63717-27-1 see AMC750	63905-54-4 see EEQ000	63957-38-0 see ALU000	64037-09-8 see AIY000
63041-78-1 see MIK000	63731-92-0 see AGO500	63905-60-2 see FQM000	63957-39-1 see ALU250	64037-10-1 see AJF250
63041-80-5 see MIL750	63731-93-1 see AJQ000	63905-64-6 see PAG750	63957-48-2 see DDF600	64037-11-2 see AJG750
63041-83-8 see MIY500	63732-07-0 see BLX250	63905-89-5 see MDB500	63957-59-5 see HNK550	64037-12-3 see AJM500
63041-84-9 see MIY750	63732-23-0 see MKF000	63905-98-6 see AOF250	63968-64-9 see ARL375	64037-13-4 see AKL500
63041-85-0 see MIZ000	63732-31-0 see TMR750	63906-14-9 see BGR750	63976-07-8 see ALL000	64037-14-5 see AKN000
63041-88-3 see MLN500	63732-42-3 see ALS500	63906-56-9 see TEM250	63977-49-1 see DQS600	64037-15-6 see AKQ750
63041-90-7 see NFM500	63732-43-4 see ALS750	63906-57-0 see BSM825	63978-55-2 see BHP500	64037-50-9 see EIJ000
63041-92-9 see HDH000	63732-56-9 see DVC800	63906-63-8 see PNW250	63978-73-4 see TJI500	64037-51-0 see MJJ500
63041-95-2 see MJB250	63732-62-7 see AFT250	63906-64-9 see BFA000	63978-93-8 see NBG000	64037-53-2 see CES250
63042-08-0 see CBS750	63732-63-8 see CNG000	63906-75-2 see AAJ250	63979-26-0 see ALW000	64037-56-5 see BQL500
63042-11-5 see OLG000	63732-98-9 see BKX250	63906-88-7 see DKK800	63979-37-3 see DMF400	64037-57-6 see BQL750
63042-13-7 see SNF500	63748-54-9 see PJG000	63907-04-0 see TDS250	63979-62-4 see DNL400	64037-65-6 see AGJ750
63042-19-3 see EDP500	63765-69-5 see AEE500	63907-07-3 see EGX500	63979-65-7 see BSL750	64037-73-6 see ASM000
63042-22-8 see EDQ500	63765-78-6 see FIO000	63907-29-9 see ARY500	63979-84-0 see SNG500	64038-09-1 see ALW500
63042-50-2 see DRJ000	63765-80-0 see AKS750	63907-33-5 see CCF500	63979-86-2 see SJW300	64038-10-4 see SHN275
63042-68-2 see DTY400	63765-88-8 see DWH800	63907-41-5 see NGT500	63979-95-3 see BHL000	64038-38-6 see DRB800
63059-68-7 see MED250	63766-15-4 see MCT250	63915-52-6 see DGU400	63980-13-2 see TKH500	64038-39-7 see FHR000
63074-03-3 see CHY750	63770-20-7 see NCQ000	63915-54-8 see MHQ500	63980-18-7 see TGW750	64038-40-0 see TLI000
63074-08-8 see TEF700	63815-37-2 see BHY750	63915-76-4 see SFB500	63980-19-8 see TGW500	64038-41-1 see CCG250
63077-00-9 see DLA000	63815-42-9 see EFE500	63915-77-5 see SFC000	63980-20-1 see DKD200	64038-48-8 see HOQ000
63077-09-8 see BGU000	63833-90-9 see EHF500	63915-89-9 see AGK000	63980-27-8 see TGX000	64038-49-9 see HOQ500
63089-76-9 see EBF000	63833-98-7 see COS899	63916-54-1 see PIQ750	63980-44-9 see BIG750	64038-55-7 see DCP600
63104-32-5 see MHH500	63834-20-8 see DEP400	63916-83-6 see HFH500	63980-59-6 see DPA000	64038-56-8 see DUI400
63124-33-4 see NKA500	63834-30-0 see HKS300	63916-90-5 see IJR000	63980-61-0 see PHO250	64038-57-9 see MLF000
63139-69-5 see HKB200	63834-87-7 see MKJ250	63916-96-1 see LDT000	63980-62-1 see IJB000	64039-27-6 see TFJ250
63141-79-7 see DUN400	63839-60-1 see ANX875	63916-97-2 see LDV000	63980-89-2 see PHM750	64043-53-4 see CMR250
63148-62-9 see DUB600	63843-89-0 see PMK800	63916-98-3 see TJT000	63981-09-9 see HHF500	64043-55-6 see TDE000
63148-62-9 see SCR400	63867-09-4 see EKO000	63917-01-1 see LEQ000	63981-20-4 see BGP500	64046-00-0 see ANT250
63148-65-2 see PKI000	63867-52-7 see BKB250	63917-04-4 see LIX000	63981-49-7 see MCY500	64046-01-1 see SJR500
63160-33-8 see MLI350	63867-64-1 see BNU250	63917-06-6 see DFJ200	63981-53-3 see HJZ000	64046-07-7 see TGH680
63224-44-2 see ACS000	63868-62-2 see DKE800	63917-71-5 see MGF000	63981-92-0 see HJY500	64046-11-3 see TGH670
63224-45-3 see FEM000	63868-75-7 see DKQ200	63917-76-0 see AIL500	63982-03-6 see HAU500	64046-12-4 see TGH665
63224-46-4 see MSB250	63868-82-6 see PIC750	63918-29-6 see PIT250	63982-15-0 see FIS000	64046-20-4 see TGH675
63270-67-7 see BJX800	63868-93-9 see AAR750	63918-36-5 see BIC500	63982-25-2 see PEQ000	64046-38-4 see TGH660
63307-29-9 see ALS250	63868-94-0 see AAS000	63918-38-7 see ELG000	63982-32-1 see MIC250	64046-47-5 see CJD625
63323-29-5 see DMP600	63868-95-1 see NIS000	63918-39-8 see MKQ500	63982-40-1 see MID000	64046-56-6 see DWY600
63323-30-8 see DVO175	63868-96-2 see HLP000	63918-49-0 see TDG750	63982-47-8 see DPL900	64046-59-9 see TGV500
63323-31-9 see BMK620	63868-98-4 see HLP500	63918-50-3 see FAM000	63982-49-0 see MIA775	64046-61-3 see AAT500
63339-68-4 see CML820	63869-00-1 see MDH750	63918-55-8 see EHK000	63982-52-5 see BKR000	64046-62-4 see ALQ750
63357-09-5 see DMP800	63869-01-2 see CFK750	63918-56-9 see AGE000	63989-69-5 see IGO000	64046-79-3 see QDJ000
63382-64-9 see BJJ200	63869-02-3 see CGB750	63918-66-1 see AGD750	63989-75-3 see AJV850	64046-93-1 see AQC250
63394-00-3 see WBS675	63869-04-5 see HLO500	63918-74-1 see DLP000	63989-79-7 see AJJ750	64046-96-4 see CAL500
63394-02-5 see PJR250	63869-05-6 see BLK250	63918-82-1 see DOV000	63989-82-2 see DUT200	64046-97-5 see SJG000
63394-02-5 see SPB000	63869-07-8 see MLB000	63918-83-2 see DCT000	63989-84-4 see DUU200	64046-99-7 see CBK750
63394-05-8 see PJB500	63869-08-9 see PFN750	63918-85-4 see PFI750	63989-85-5 see DUU400	64047-26-3 see MCS500
63412-06-6 see MMU500	63869-14-7 see AEE250	63918-89-8 see DFK200	63990-56-7 see DWY400	64047-30-9 see TLY000
63428-83-1 see NOH000	63869-15-8 see DNW000	63918-97-8 see LEE000	63990-88-5 see BFT750	64047-82-1 see NIH000
63438-26-6 see DMO800	63869-17-0 see DHC200	63918-98-9 see EFG500	63990-96-5 see AOF500	64047-83-2 see NIH500
63441-20-3 see CPI350	63869-87-4 see TMI500	63919-00-6 see DNI200	63991-01-0 see MIX000	64048-05-1 see SFJ500
63449-39-8 see PAH800	63869-91-0 see TFX000	63919-01-7 see FID000	63991-23-1 see AKA250	64048-06-2 see PLE750
63449-39-8 see PAH810	63884-28-6 see DAF450	63919-14-2 see ZRJ000	63991-26-4 see MGH250	64048-07-3 see SIM000
63449-87-6 see EPX000	63884-40-2 see HGW000	63919-18-6 see SJQ500	63991-31-1 see MKS000	64048-08-4 see HLQ000
63468-05-3 see MNL500	63884-67-3 see DQY000	63919-21-1 see CND000	63991-43-5 see MQC250	64048-13-1 see DJB600
63469-15-8 see TEA500	63884-71-9 see DQX800	63919-22-2 see SIW000	63991-48-0 see AHT900	64048-70-0 see BCT500
63496-48-0 see PDP100	63884-80-0 see AGB000	63934-40-7 see HJR000	63991-49-1 see AHT950	64048-90-4 see DGO600
63504-15-4 see PFV000	63884-81-1 see BED250	63934-41-8 see HJR500	63991-57-1 see ECG500	64048-94-8 see AKF750
63516-07-4 see FMR300	63884-92-4 see FIH000	63937-14-4 see MDC500	63991-70-8 see MIX000	64048-98-2 see TNO000
63521-15-3 see RCK000	63885-01-8 see ZDA000	63937-26-8 see SFR000	63992-02-9 see SJR000	64048-99-3 see DJN400
63528-82-5 see DKA400	63885-02-9 see BLN000	63937-27-9 see AGL500	63992-41-6 see SFV500	64049-00-9 see MID900
63551-77-9 see SCC550	63885-07-4 see HMC500	63937-32-6 see BQM750	64005-62-5 see AOL750	64049-02-1 see IET000
63638-90-4 see BMM625	63885-67-6 see NNC100	63937-47-3 see ODO000	64011-26-3 see TNC250	64049-03-2 see TBG710
63642-17-1 see MQY325	63885-68-7 see AOJ750	63938-10-3 see CLH000	64011-32-1 see TGR250	64049-07-6 see PER600
63643-78-7 see HJX625	63886-45-3 see DRM800	63938-16-9 see CMQ000	64011-35-4 see TGR500	64049-11-2 see CLV250
63659-19-8 see KEA350	63886-56-6 see HKR550	63938-20-5 see YFA000	64011-36-5 see TGR750	64049-21-4 see HBN600
63665-41-8 see DWP500	63886-75-9 see BKF750	63938-21-6 see DSM600	64011-37-6 see MCY250	64049-22-5 see HFP000
63673-37-0 see CFV250	63886-77-1 see TCI250	63938-24-9 see ACP500	64011-39-8 see BLQ750	64049-23-6 see IHQ100
63680-76-2 see DQY400	63886-82-8 see CGM375	63938-26-1 see LJK000	64011-44-5 see MEA250	64049-28-1 see MCS250
63680-78-4 see MID500	63887-17-2 see AGO250	63938-27-2 see EDE000	64011-46-7 see ECE550	64049-29-2 see MJM250
63681-01-6 see DME600	63887-34-3 see AJN750	63938-28-3 see EDE500	64011-62-7 see CPG625	64049-85-0 see TGH650
63681-05-0 see NBV000	63887-51-4 see AGN500	63938-82-9 see PBH075	64019-93-8 see DWP559	64050-01-7 see TGH655
63698-37-3 see BJW000	63887-52-5 see AGN750	63938-93-2 see CCH250	64024-07-3 see BIS750	64050-03-9 see HOL000
63698-38-4 see BJV750	63904-81-4 see HIQ000	63941-74-2 see IFS400	64024-08-4 see UTA000	64050-15-3 see EFI000
63710-09-8 see APV000	63904-82-5 see ZTA000	63944-10-5 see NFW400	64025-05-4 see SJQ000	64050-20-0 see DSO400
63710-10-1 see MAX000	63904-83-6 see ZCS000	63950-89-0 see BHA000	64025-06-5 see PBL250	64050-23-3 see CGH500
63710-43-0 see INW000	63904-87-0 see IOE000	63951-03-1 see ARJ900	64036-46-0 see DJV800	64050-44-8 see CCJ000
63716-10-9 see EET000	63904-99-4 see MKE750	63951-08-6 see BJN850	64036-72-2 see MGD750	64050-46-0 see BIJ000
63716-17-6 see DWT000	63905-05-5 see MIH000	63951-09-7 see BKH000	64036-79-9 see BHD000	64050-54-0 see DOL400
63716-40-5 see BPS750	63905-13-5 see IOW000	63951-11-1 see CIL700	64036-86-8 see LIP000	64050-77-7 see HNR000
63716-63-2 see QPJ000	63905-29-3 see AEO250	63951-48-4 see DTH600	64036-91-5 see BKD750	64050-79-9 see HNR500

64050-81-3 see HNS000	64502-82-5 see RHA150	65036-47-7 see BNG125	66232-30-2 see BHS750	67055-59-8 see CCX600
64050-83-5 see HNS500	64506-49-6 see IMS300	65039-20-5 see CIS325	66267-18-3 see DMK200	67057-34-5 see DEB600
64050-95-9 see TGH685	64508-90-3 see BHY625	65041-92-1 see ACS500	66267-19-4 see DLD400	67110-84-3 see CNH300
64051-06-5 see HNN000	64521-13-7 see DLE600	65043-22-3 see IBW500	66267-67-2 see HJG100	67114-26-5 see SHY000
64051-08-7 see HNO000	64521-14-8 see MBV500	65057-90-1 see TAC500	66276-87-7 see BHS500	67176-33-4 see FEG000
64051-12-3 see ABV500	64521-15-9 see DLF200	65057-91-2 see TAC750	66289-74-5 see DLJ500	67195-50-0 see BQU750
64051-16-7 see BED500	64521-16-0 see ECP500	65089-17-0 see CLW500	66309-69-1 see CCS375	67195-51-1 see ECA500
64051-18-9 see HNP500	64544-07-6 see CCS625	65094-73-7 see NFW435	66332-77-2 see IJJ000	67196-02-5 see PIU800
64051-20-3 see HNQ500	64550-80-7 see BAD000	65098-93-3 see TE0800	66357-35-5 see RBF400	67210-66-6 see DIL600
64057-51-8 see CHE250	64551-89-9 see DMO500	65141-46-0 see NDL800	66408-78-4 see CMS324	67227-20-7 see AOS000
64057-52-9 see DDD600	64552-25-6 see MLC500	65146-47-6 see AIW500	66409-97-0 see BPW000	67227-30-9 see FPX000
64057-57-4 see SJB000	64598-80-7 see DLE000	65184-10-3 see TAL560	66409-98-1 see FNP000	67230-61-9 see DET125
64057-58-5 see TIR750	64598-81-8 see DLE200	65210-28-8 see CPG250	66427-01-8 see HKF600	67230-67-5 see PQB275
64057-70-1 see MJR500	64598-82-9 see DMP200	65210-29-9 see MNP250	66471-17-8 see ALW900	67238-91-9 see MRP500
64057-75-6 see MNQ250	64598-83-0 see DMS260	65210-30-2 see DPH200	66472-85-3 see SBU100	67242-54-0 see TCP600
64057-79-0 see CBV750	64604-09-7 see TCA250	65210-31-3 see DIC800	66486-68-8 see HBI725	67255-31-6 see HHH100
64058-13-5 see AFX500	64622-45-3 see IAB000	65210-32-4 see MLP750	66499-61-4 see MCS600	67262-60-6 see MEM750
64058-14-6 see AGA000	64624-44-8 see GFU200	65210-33-5 see MOI500	66535-86-2 see CKL325	67262-61-7 see MEN000
64058-26-0 see BKE750	64653-03-8 see BIW250	65210-37-9 see BKQ250	66547-10-2 see APE529	67262-62-8 see EES500
64058-30-6 see COX250	64686-82-4 see AJD500	65216-94-6 see COM500	66552-77-0 see TDK885	67262-64-0 see EMI000
64058-54-4 see OOO000	64693-33-0 see BJP325	65229-18-7 see HMQ500	66634-53-5 see DSH700	67262-69-5 see EMG500
64058-57-7 see AIE125	64709-48-4 see DKS909	65232-69-1 see TJL775	66686-30-4 see FDN000	67262-71-9 see EMH500
64058-65-7 see AKF500	64709-49-5 see DKT000	65235-63-4 see BND250	66731-42-8 see DMB200	67262-72-0 see EMH000
64058-72-6 see ABK000	64709-50-8 see DKT200	65235-79-2 see SDJ500	66733-21-9 see EDC650	67262-74-2 see IMX000
64058-74-8 see BGQ250	64719-39-7 see QBS000	65268-91-9 see DHU400	66734-13-2 see AFI980	67262-75-3 see IPZ000
64058-92-0 see TBD250	64741-44-2 see GBW000	65271-80-9 see MQY090	66788-01-0 see DKU400	67262-78-6 see DRT400
64059-02-6 see DXJ400	64741-49-7 see MQV755	65272-47-1 see MKJ000	66788-03-2 see DNC400	67262-79-7 see DSN800
64059-26-3 see LAH000	64741-50-0 see MQV815	65277-42-1 see KFK100	66788-06-5 see DMK600	67262-80-0 see MEN250
64059-29-6 see BEA250	64741-51-1 see MQV785	65296-81-3 see CCJ350	66788-11-2 see ECQ150	67262-82-2 see MEN500
64059-42-3 see CIQ000	64741-52-2 see MQV810	65313-33-9 see CIH900	66788-41-8 see DXI800	67292-61-9 see HIF000
64059-53-6 see PPN250	64741-53-3 see MQV780	65313-34-0 see DIV400	66789-14-8 see ADG425	67292-62-0 see HKA200
64070-10-6 see AQH250	64741-58-8 see GBW025	65313-35-1 see TNR500	66793-67-7 see PCG650	67292-63-1 see HIE550
64070-11-7 see AQG500	64741-79-3 see PCS000	65313-36-2 see EIY550	66795-86-6 see EDL000	67292-68-6 see MLI750
64070-12-8 see AQG750	64741-88-4 see MQV850	65313-37-3 see TDP300	66813-55-6 see TAI725	67292-88-0 see CIQ250
64070-13-9 see DHA400	64741-89-5 see MQV855	65400-79-5 see MPR250	66826-72-0 see DAC975	67293-64-5 see CIO500
64070-14-0 see AOW750	64741-96-4 see MQV845	65400-81-9 see AAL250	66826-73-1 see DGH500	67293-75-8 see SHQ000
64070-15-1 see AOX000	64741-97-5 see MQV852	65405-73-4 see GDM100	66827-45-0 see ABN700	67293-86-1 see MEA750
64070-83-3 see ARE000	64742-03-6 see MQV860	65405-77-8 see SAJ000	66827-50-7 see AGC750	67293-88-3 see DBB600
64082-34-4 see SKQ000	64742-04-7 see MQV859	65445-59-2 see DTA600	66827-74-5 see NEP000	67298-49-1 see NCM275
64082-35-5 see LHK000	64742-05-8 see MQV862	65445-60-5 see CJG375	66839-97-2 see PPT400	67312-43-0 see SIW600
64082-43-5 see MGX750	64742-10-5 see MQV863	65445-61-6 see CJG500	66839-98-3 see MEF400	67330-25-0 see BRJ325
64083-05-2 see LCE000	64742-11-6 see MQV857	65454-27-5 see MCB100	66843-04-7 see MLP500	67335-42-6 see BBG000
64083-08-5 see MAB050	64742-17-2 see MQV872	65496-97-1 see SDX630	66859-63-0 see EPC135	67335-43-7 see BBD750
64090-82-0 see AOW500	64742-18-3 see MQV760	65520-53-8 see HOH000	66862-11-1 see HEJ350	67360-94-5 see ARK750
64091-90-3 see MMS250	64742-19-4 see MQV770	65521-60-0 see PLW275	66877-41-6 see DJI250	67360-95-5 see BCB000
64091-91-4 see MMS500	64742-20-7 see MQV765	65546-74-9 see ROA400	66902-62-3 see BHQ250	67360-95-6 see BCB250
64092-23-5 see BHB000	64742-21-8 see MQV775	65561-73-1 see DNA300	66903-23-9 see BKR250	67371-65-7 see BDP899
64092-48-4 see ZUA300	64742-44-5 see PCS260	65567-32-0 see EOL050	66922-62-6 see EOD500	67410-20-2 see TMW600
64093-79-4 see NBW000	64742-45-6 see PCS270	65573-02-6 see IBQ075	66922-79-0 see AQI500	67411-81-8 see DLF400
64140-51-8 see DWR200	64742-52-5 see MQV790	65597-24-2 see CDX800	66941-08-0 see DSP200	67418-30-8 see TGA275
64187-24-2 see CBR200	64742-53-6 see MQV800	65597-25-3 see TJX650	66941-43-3 see DSI200	67445-50-5 see TMY850
64187-25-3 see CBR175	64742-54-7 see MQV795	65654-08-2 see MEA800	66941-48-8 see AFV700	67465-04-7 see BEA000
64187-27-5 see CBR215	64742-55-8 see MQV805	65654-13-9 see TEN725	66941-49-9 see AFY300	67465-26-3 see PNV760
64187-42-4 see CBR235	64742-56-9 see MQV840	65664-23-5 see PLU575	66941-53-5 see AGA250	67465-39-8 see CHW000
64187-43-5 see CBR225	64742-63-8 see MQV820	65666-07-1 see SDX625	66941-60-4 see AGD000	67465-41-2 see BII500
64241-34-5 see CAK275	64742-64-9 see MQV835	65700-59-6 see DAU200	66941-77-3 see AGL250	67465-42-3 see EET500
64245-83-6 see SPA500	64742-65-0 see MQV825	65700-60-9 see DAU000	66941-81-9 see AGM000	67465-43-4 see INE000
64245-99-4 see MJS250	64742-68-3 see MQV865	65734-38-5 see ACN500	66955-43-9 see PCL775	67465-44-5 see MES250
64246-03-3 see BKR500	64742-69-4 see MQV867	65763-32-8 see DLD800	66955-44-0 see PCO175	67466-28-8 see EJT000
64246-07-7 see DOV800	64742-70-7 see MQV868	65792-56-5 see NLN500	66964-37-2 see MHA000	67466-58-4 see HLO300
64246-13-5 see BKR750	64742-71-8 see MQV870	65793-50-2 see AIS250	66967-60-0 see MFR250	67479-03-2 see MCR250
64246-17-9 see MJS500	64742-86-5 see GBW010	65860-38-0 see TCU000	66967-65-5 see DRW000	67523-22-2 see DLF600
64253-71-0 see BAR830	64743-05-1 see PCR750	65899-73-2 see TGF050	66967-84-8 see HNG000	67526-05-0 see ENA000
64253-73-2 see RHA125	64755-14-2 see BMK325	65928-58-7 see SMP400	66968-12-5 see PHY250	67527-71-3 see MQU000
64267-45-4 see DXA900	64771-59-1 see MCX600	65979-81-9 see ZKS000	66968-89-6 see EMW000	67536-44-1 see BKH125
64267-46-5 see NMV600	64781-77-7 see DCQ500	65986-79-0 see ABR500	66969-02-6 see BNE500	67557-56-6 see BSX500
64285-06-9 see AOO120	64808-48-6 see LHZ600	65986-80-3 see ENR500	66988-15-6 see DAI800	67557-57-7 see MMS000
64296-43-1 see HBA259	64817-78-3 see MSA700	65996-93-2 see CMZ100	66997-69-1 see ECR500	67590-46-9 see BRO750
64309-76-8 see AKL250	64838-75-1 see DMP000	65997-15-1 see PKS750	67011-39-6 see BCP000	67590-56-1 see EQD100
64314-28-9 see BAR825	64840-90-0 see EAV700	66007-89-4 see SFP500	67031-48-5 see AIU250	67590-57-2 see IRQ100
64314-52-9 see DAM800	64854-98-4 see CBD500	66009-08-3 see BIW500	67032-45-5 see BPH750	67639-45-6 see FHH025
64318-79-2 see CDB775	64854-99-5 see CBD250	66017-91-2 see PNR250	67037-37-0 see DKH875	67658-42-8 see EJS100
64365-11-3 see CDI000	64910-63-0 see BSM250	66064-11-7 see AIZ000	67049-51-8 see HFV000	67658-46-2 see PNL800
64387-78-6 see MLM600	64920-31-6 see DLD200	66085-59-4 see NDY700	67049-95-0 see OMY700	67664-94-2 see TJB750
64431-68-1 see MAB250	64925-80-0 see PGR775	66104-23-2 see MPU500	67050-00-4 see BPA750	67694-88-6 see ECQ200
64440-87-5 see CMN125	64953-12-4 see LBH200	66104-24-3 see BFP500	67050-04-8 see BPB500	67722-96-7 see ECG100
64441-42-5 see BRK250	64977-44-2 see FJY000	66147-68-0 see ADP750	67050-11-7 see BPC500	67730-10-3 see DWW700
64461-82-1 see TGH600	64977-46-4 see FKA000	66147-69-1 see ADP000	67050-26-4 see BRJ250	67730-11-4 see AKS250
64475-85-0 see PCT250	64977-47-5 see FKB000	66217-76-3 see MHB000	67050-64-0 see SFJ875	67749-11-5 see HCP600
64485-93-4 see CCR950	64977-48-6 see FKC000	66231-56-9 see CNH650	67050-97-9 see EPC000	67762-92-9 see SDF000
64490-92-2 see SKJ350	64977-49-7 see FKD000	66232-25-5 see BHS250	67051-25-6 see ILT000	67774-31-6 see NKA850
64491-74-3 see PEW750	65002-17-7 see MCO775	66232-28-8 see BHT000	67051-27-8 see ILU000	67774-32-7 see FBU509

73622-67-0 see DBH800	73927-94-3 see IEF000	74940-26-4 see HIE700	77430-23-0 see MPS300	77966-82-6 see DHK800
73637-11-3 see CIF750	73927-95-4 see TIE500	74955-23-0 see HIE570	77469-44-4 see BIF625	77966-83-7 see DIK800
73637-16-8 see ACA750	73927-96-5 see QTJ000	75016-34-1 see NKT100	77491-30-6 see SAD100	77966-84-8 see DNM600
73639-62-0 see CIL775	73927-97-6 see TDO000	75016-36-3 see NKT105	77500-04-0 see AJQ675	77966-85-9 see DTR800
73651-49-7 see DKX800	73927-98-7 see TIG000	75034-93-4 see DKQ650	77503-17-4 see FQQ400	77966-90-6 see PIN000
73665-15-3 see MRW500	73927-99-8 see TNC000	75038-71-0 see PLU500	77536-66-4 see ARM260	77966-93-9 see CGI750
73666-84-9 see TEF725	73928-00-4 see TKT850	75084-25-2 see DEN300	77536-67-5 see ARM264	77967-05-6 see BGH500
73671-86-0 see DJP700	73928-01-5 see CLE500	75198-31-1 see NGI800	77536-68-6 see ARM280	77967-24-9 see DHL200
73680-58-7 see AHB250	73928-02-6 see MFP500	75219-46-4 see BFV325	77650-28-3 see DHE800	77967-25-0 see DHL400
73684-69-2 see MRW750	73928-03-7 see MPJ250	75236-19-0 see EPC950	77650-95-4 see DJX300	77984-94-2 see CFR500
73688-63-8 see APM250	73928-04-8 see MPJ500	75318-62-6 see BGO325	77680-87-6 see SED700	77985-16-1 see CFN500
73688-85-4 see DPO275	73928-11-7 see DCJ000	75318-76-2 see CLN325	77698-19-2 see NKL300	77985-17-2 see CKU000
73693-97-7 see PIM750	73928-18-4 see TJS750	75321-19-6 see TMN000	77698-20-5 see NLE400	77985-21-8 see DHM400
73696-62-5 see ALG250	73928-21-9 see TJT250	75321-20-9 see DVD400	77791-20-9 see DHN800	77985-23-0 see DHV200
73696-64-7 see BKU000	73940-79-1 see SJP500	75348-40-2 see DHQ600	77791-27-6 see DIN400	77985-24-1 see DHV400
73696-65-8 see ENH000	73940-85-9 see DKC600	75348-49-1 see PPV750	77791-37-8 see END500	77985-25-2 see DHV600
73698-75-6 see AIK500	73940-86-0 see TMI100	75410-87-6 see ICZ100	77791-38-9 see EOU500	77985-27-4 see PIO000
73698-76-7 see AIK750	73940-87-1 see BLS900	75410-89-8 see BMK634	77791-40-3 see MDK250	77985-28-5 see PIO250
73698-77-8 see AIL000	73940-88-2 see TIE250	75411-83-5 see NKU500	77791-41-4 see MRT250	77985-29-6 see PIO500
73698-78-9 see AIL250	73940-89-3 see TIF750	75432-59-6 see IDH000	77791-42-5 see MLS250	77985-30-9 see PPW750
73713-75-4 see EAN500	73940-90-6 see CHU750	75464-10-7 see PMW760	77791-43-6 see MLS750	77985-31-0 see PPX000
73728-78-6 see TDJ250	73941-35-2 see THV500	75524-40-2 see CHJ625	77791-53-8 see BPR000	77985-32-1 see PPX250
73728-79-7 see TLF000	73953-53-4 see HHA100	75530-68-6 see NDY650	77791-55-0 see BQA500	78003-71-1 see SKS800
73728-82-2 see AKF250	73954-17-3 see BCA375	75662-22-5 see DGW450	77791-57-2 see CFR250	78109-79-2 see BKC500
73747-22-5 see BEL550	73963-72-1 see CMP825	75738-58-8 see CCS300	77791-58-3 see CFL500	78109-80-5 see BPJ500
73747-29-2 see AJO750	73973-02-1 see PDB300	75775-83-6 see ADP500	77791-63-0 see DEY000	78109-81-6 see BPJ750
73758-18-6 see TJJ750	73986-95-5 see DGA400	75841-84-8 see MBV735	77791-64-1 see DFF200	78109-87-2 see DII400
73758-56-2 see DEK400	73987-00-5 see ACH375	75881-16-2 see NKU500	77791-66-3 see DIS000	78109-88-3 see EEY000
73771-13-8 see SJJ190	73987-16-3 see TLO600	75881-17-3 see NJI850	77791-67-4 see DIC000	78109-90-7 see PGE500
73771-52-5 see BJA000	73987-51-6 see TGK500	75881-19-5 see NKU580	77791-69-6 see ONQ000	78110-10-8 see CLB250
73771-72-9 see FJU000	73987-52-7 see EEP000	75881-20-8 see NKW800	77824-42-1 see AQE300	78110-23-3 see MDK000
73771-73-0 see FJV000	73990-29-1 see AHI500	75881-22-0 see MMX200	77824-43-2 see AQE320	78110-37-9 see CEY250
73771-74-1 see FJW000	74007-80-0 see OPE100	75884-37-6 see FQQ100	77824-44-3 see AQE305	78110-38-0 see ARX875
73771-79-6 see DNC600	74011-58-8 see EAU100	75888-03-8 see FPO100	77846-96-9 see CKU250	78128-69-5 see BKD000
73771-81-0 see QIS000	74037-18-6 see DEU125	75889-62-2 see DIU500	77855-81-3 see MQT600	78128-80-0 see BRO250
73785-34-9 see DBG200	74037-31-3 see ASA250	75965-74-1 see MFB400	77879-90-4 see GEO200	78128-81-1 see CPM250
73785-40-7 see NJS300	74037-60-8 see BHB500	75965-75-2 see MFB410	77893-24-4 see MRJ700	78128-83-3 see MJF250
73790-27-9 see MOK000	74038-45-2 see DHH600	75993-65-6 see MPI800	77944-89-9 see CIK500	78128-84-4 see HND000
73791-29-4 see AIY750	74039-01-3 see AIW750	76050-49-2 see FMR700	77945-03-0 see DIA000	78128-85-5 see PGC750
73791-32-9 see TMA750	74039-02-4 see AMO750	76059-11-5 see DVS100	77945-09-6 see BPJ250	78173-90-7 see SON520
73791-39-6 see ALV100	74039-78-4 see BLT500	76059-13-7 see DVS300	77966-20-2 see BPJ250	78173-91-8 see ARS135
73791-40-9 see BRQ800	74039-79-5 see BLT750	76059-14-8 see DVS400	77966-25-7 see BPY500	78173-92-9 see ARS130
73791-41-0 see DGA425	74039-80-8 see BLU250	76069-32-4 see TAB785	77966-26-8 see DHJ800	78186-37-5 see MQO250
73791-42-1 see CKD800	74039-81-9 see BLU500	76095-16-4 see EAO100	77966-27-9 see DHK000	78186-61-5 see MLJ750
73791-43-2 see IPS100	74050-97-8 see HAG300	76145-76-1 see THG700	77966-28-0 see PPU750	78193-30-3 see SON525
73791-44-3 see MNU050	74093-43-9 see SDM550	76175-45-6 see TEY600	77966-30-4 see AFW250	78218-16-3 see CGY500
73791-45-4 see MOT800	74193-14-9 see DHA200	76180-96-6 see AKT600	77966-31-5 see BDY250	78218-37-8 see CGF500
73803-48-2 see CHK825	74195-73-6 see RGP450	76206-36-5 see NBI000	77966-32-6 see BEE750	78218-38-9 see CIE500
73806-23-2 see OMA000	74203-42-2 see POV750	76206-37-6 see NBM500	77966-34-8 see BEJ250	78218-40-3 see CKO000
73806-49-2 see THJ750	74203-45-5 see NDF000	76206-38-7 see BRY750	77966-38-2 see CEX500	78218-42-5 see CIU250
73815-11-9 see MEW800	74203-58-0 see PGA250	76263-73-5 see POM700	77966-40-6 see CFH000	78218-43-6 see BQN500
73816-43-0 see BKJ500	74203-59-1 see MOD750	76298-68-5 see ACY700	77966-41-7 see CFI000	78218-49-2 see TMU750
73816-74-7 see CHG250	74203-61-5 see SJK475	76319-15-8 see PAM800	77966-42-8 see CFL000	78219-33-7 see EOY100
73816-75-8 see DGO000	74220-04-5 see SBV500	76379-66-3 see DXU250	77966-43-9 see CFR750	78219-35-9 see MOI250
73816-77-0 see IBF000	74222-97-2 see SNW550	76379-67-4 see OAX050	77966-44-0 see CFS000	78219-38-2 see PFD500
73825-79-9 see DAQ000	74273-75-9 see DCM700	76429-97-5 see DWP950	77966-45-1 see CFS250	78219-45-1 see PFD750
73825-85-1 see SNS100	74278-22-1 see KHU000	76429-98-6 see DCM875	77966-46-2 see CFM250	78219-52-0 see PNT500
73826-58-1 see DGN600	74339-98-3 see DKX875	76487-32-6 see ALU875	77966-47-3 see CFM500	78219-53-1 see PNT750
73840-42-3 see MOA250	74340-04-8 see DKX900	76487-65-5 see DVY875	77966-48-4 see CFM750	78219-57-5 see EOV000
73870-33-4 see NLV500	74356-00-6 see CCS371	76541-72-5 see CMU875	77966-49-5 see CFN000	78219-58-6 see EOW000
73873-83-3 see OPY000	74444-58-9 see DND000	76556-13-3 see ANG625	77966-51-9 see CFU500	78219-61-1 see MON000
73926-79-1 see TGE750	74444-59-0 see ECT000	76631-42-0 see NMV480	77966-52-0 see CFW250	78219-62-2 see PDJ000
73926-80-4 see DDF000	74465-36-4 see DMR400	76648-01-6 see TAA400	77966-53-1 see CFW000	78219-63-3 see PDJ250
73926-81-5 see DDG600	74465-38-6 see DMT000	76674-14-1 see BJW600	77966-54-2 see CGC000	78246-54-5 see HLX925
73926-83-7 see THU250	74465-39-7 see DMS800	76714-88-0 see DGC100	77966-55-3 see CGD000	78265-89-1 see CAB500
73926-85-9 see BIV900	74469-00-4 see ARS125	76738-28-8 see DFQ100	77966-56-4 see CGE750	78265-90-4 see CAB750
73926-87-1 see CET000	74512-62-2 see PNR800	76749-37-6 see ACS250	77966-58-6 see CGN250	78265-91-5 see CAB250
73926-88-2 see CHL000	74578-38-4 see UNJ810	76790-18-6 see NBC500	77966-59-7 see CGP380	78265-95-9 see HLR000
73926-89-3 see CIC000	74749-73-8 see DUO800	76790-19-7 see BPQ250	77966-61-1 see CHS750	78265-97-1 see PFF500
73926-90-6 see TJU850	74758-13-7 see DPI600	76822-96-3 see VRU000	77966-62-2 see CID250	78280-29-2 see DII600
73926-91-7 see DGK400	74758-19-3 see BJH500	76824-35-6 see FAB500	77966-63-3 see CIT750	78280-31-6 see AMN000
73926-92-8 see DGK600	74764-93-5 see MNU750	76828-34-7 see ECJ100	77966-67-7 see CKN750	78281-06-8 see HLE650
73926-94-0 see CLG000	74790-08-2 see SLE875	77094-11-2 see AJQ600	77966-68-8 see CKT750	78308-37-9 see BQH750
73927-60-3 see EHX000	74816-28-7 see CFM000	77162-70-0 see NFW350	77966-70-2 see DDM600	78308-53-9 see MKS750
73927-86-3 see DEI400	74816-32-3 see CFU250	77174-66-4 see CMW550	77966-71-3 see DHS800	78313-59-4 see BKS800
73927-87-4 see DWW400	74847-35-1 see PPQ650	77234-90-3 see DAP880	77966-72-4 see DIC200	78329-75-6 see SCA550
73927-88-5 see BSP000	74886-24-1 see CNH525	77251-47-9 see CPD625	77966-73-5 see DIK600	78329-87-0 see BAN500
73927-89-6 see TMV825	74920-78-8 see EKL250	77287-90-2 see COW675	77966-75-7 see BEJ500	78329-88-1 see BSI750
73927-90-9 see QSJ800	74926-97-9 see BRR500	77327-05-0 see DHA300	77966-77-9 see BRI500	78329-97-2 see EIA000
73927-91-0 see TID750	74926-98-0 see IQD000	77337-54-3 see PNM650	77966-79-1 see DDU200	78330-02-6 see ELK000
73927-92-1 see TNB500	74927-02-9 see BKQ750	77372-67-9 see ACE250	77966-80-4 see DHT000	78338-31-5 see DTA690
73927-93-2 see TIE000	74940-23-1 see HIE600	77372-68-0 see BPQ000	77966-81-5 see DIC400	78338-32-6 see DTA700

78343-32-5 see FJI500	82423-05-0 see COL125	89947-76-2 see HLE750	100836-65-5 see MQO500	102206-99-5 see EGH500
78350-94-4 see LHM750	82464-70-8 see INY100	89957-52-8 see BFW010	100836-66-6 see MQO750	102207-73-8 see AJL875
78354-52-6 see MBU775	82508-31-4 see POH600	89958-12-3 see SAF300	100836-67-7 see MQP000	102207-75-0 see AJN375
78371-75-2 see XWS000	82508-32-5 see POH550	89997-47-7 see SAF000	100836-70-2 see PEQ250	102207-76-1 see AJO625
78371-84-3 see BIP500	82547-81-7 see TAA420	90028-00-5 see DKL300	100836-84-8 see DFV800	102207-77-2 see AJP125
78371-85-4 see BIP750	82636-28-0 see CCW375	90043-86-0 see AKD775	101052-67-9 see DAB820	102207-84-1 see DHO800
78371-90-1 see CLK500	82643-25-2 see MQS600	90466-79-8 see BJA200	101418-02-4 see NJP500	102207-85-2 see DHS600
78371-91-2 see CLK750	83053-57-0 see EER400	90566-09-9 see BGS825	101491-82-1 see EOF500	102207-86-3 see DRT600
78371-92-3 see CLL250	83053-59-2 see INA400	90584-32-0 see DFJ100	101491-83-2 see EOG000	102259-67-6 see DIL000
78371-93-4 see CLL000	83053-62-7 see DLU700	90729-15-0 see DWQ850	101491-84-3 see EPV000	102262-34-0 see SDM000
78371-94-5 see CLL750	83053-63-8 see DLO950	90742-91-9 see DFM875	101491-85-4 see EQI000	102280-81-9 see DWH500
78371-95-6 see CLL500	83195-98-6 see TDI100	91216-69-2 see BHJ625	101491-86-5 see EQI000	102280-93-3 see MLH100
78371-96-7 see CLM000	83335-32-4 see NJN300	91297-11-9 see DBL300	101516-88-5 see CNW105	102338-56-7 see PJC000
78371-97-8 see CLM500	83435-67-0 see DAM315	91315-15-0 see AFJ850	101563-89-7 see DBE600	102338-57-8 see PJC750
78371-98-9 see CLM250	83463-62-1 see BMY800	91336-54-8 see MKK500	101563-93-3 see GKA000	102338-88-5 see DMD200
78371-99-0 see CLM750	83665-55-8 see IEL700	91503-79-6 see FJT100	101564-02-7 see FCN000	102366-69-8 see PAN800
78372-00-6 see CLN000	83768-87-0 see VLU200	91682-96-1 see SLQ650	101564-54-9 see BKH800	102366-79-0 see DKV125
78372-01-7 see CLN500	83876-50-0 see DLE500	91724-16-2 see SME500	101564-67-4 see DWF875	102367-57-7 see DNH500
78372-02-8 see CLO500	83876-56-6 see MET875	91845-41-9 see PCR000	101564-69-6 see DWK500	102395-05-1 see NCQ500
78372-03-9 see DIB000	83876-62-4 see ABQ600	92145-26-1 see MGU550	101565-05-3 see HNE400	102395-10-8 see NCV500
78372-04-0 see DIJ200	83968-18-7 see LGF875	92202-07-8 see PJH615	101607-48-1 see MPO400	102395-72-2 see PAW600
78372-05-1 see DPE800	83997-16-4 see PEE100	92456-72-9 see CCO700	101607-49-2 see MPO390	102395-95-9 see TGV100
78372-06-2 see DPG400	84082-68-8 see NOG000	92760-57-1 see TEW000	101651-36-9 see MDK750	102418-04-2 see MCB250
78393-38-1 see CEP250	84371-65-3 see HNB750	93023-34-8 see EIF450	101651-44-9 see PPI750	102418-06-4 see NCK500
78393-39-2 see CLN250	84504-69-8 see MQY300	93164-88-6 see AAD250	101651-55-2 see BEI250	102418-13-3 see TAH500
78431-47-7 see POM710	84604-12-6 see RMP000	93405-68-6 see DEY400	101651-60-9 see CFQ250	102418-16-6 see TEG750
78499-27-1 see DLI650	84604-12-6 see RNK000	93407-11-5 see DPO100	101651-61-0 see CFT500	102418-17-7 see TEH000
78600-25-6 see ABX800	84775-95-1 see RGA000	93431-23-3 see BLD325	101651-62-1 see CGD750	102419-73-8 see CML822
78776-28-0 see DEC775	84777-85-5 see VRP775	93763-70-3 see PCJ400	101651-63-2 see CHG750	102420-56-4 see DLE400
78822-08-9 see KDA000	84837-04-7 see SLB500	93793-83-0 see HNT100	101651-64-3 see CIL750	102433-74-9 see NKY500
78831-88-6 see DWP900	84928-98-3 see BQI125	94110-08-4 see AJE350	101651-65-4 see CIU000	102433-83-0 see BNN125
78907-15-0 see BQG850	84928-99-4 see TNC800	94362-44-4 see DCQ650	101651-66-5 see CIU000	102488-99-3 see AEB750
78907-16-1 see BPY625	84929-34-0 see WBA000	95004-22-1 see UAG025	101651-69-8 see CNW500	102489-36-1 see THH560
78919-11-6 see DMS500	85086-83-5 see AEA625	95282-98-7 see ABL625	101651-73-4 see EPL000	102489-45-2 see BEI750
78937-12-9 see CDE400	85287-61-2 see CCS525	95524-59-7 see EMY100	101651-76-7 see AAN500	102489-46-3 see BEI500
78937-14-1 see PLO100	85303-87-3 see EOM600	95619-40-2 see DTS625	101651-77-8 see AAO500	102489-47-4 see BRA500
79127-36-9 see HIS300	85303-88-4 see BSH070	95770-03-9 see CGE250	101651-78-9 see AAO750	102489-48-5 see CFO000
79127-47-2 see MSB100	85303-89-5 see BSG100	96081-07-1 see DYB300	101651-79-0 see EOG500	102489-49-6 see CFP000
79201-85-7 see PIB700	85303-91-9 see MFJ115	96231-64-0 see OJC000	101651-86-9 see KCU000	102489-50-9 see CFQ500
79307-93-0 see ASC130	85303-98-6 see EFC600	96806-34-7 see NKJ050	101651-88-1 see BKC750	102489-51-0 see CFQ750
79448-03-6 see MEG750	85559-57-5 see DAQ100	96806-35-8 see NKJ100	101651-90-5 see CFU000	102489-52-1 see CFS500
79458-80-3 see EAJ100	85616-56-4 see DLI300	96811-96-0 see DPI750	101651-94-9 see CGE000	102489-53-2 see CFS750
79543-29-6 see MIY200	85622-95-3 see MQY110	96860-89-8 see DJX350	101651-96-1 see CGF250	102489-54-3 see CFT000
79796-14-8 see PLT750	85625-90-7 see MAQ600	97194-20-2 see DBV200	101651-97-2 see CGG750	102489-55-4 see CFT250
79796-40-0 see TFR000	85681-49-8 see BLK500	97196-24-2 see XVJ000	101651-98-3 see CHB250	102489-56-5 see CFT750
79818-59-0 see EAI800	85720-47-4 see DJI375	97702-94-8 see DIB200	101652-00-0 see CHQ000	102489-58-7 see CHK125
79867-78-0 see MRR850	85720-57-6 see DJI300	97702-95-9 see DIB800	101652-01-1 see DEY600	102489-59-8 see CHR750
80266-48-4 see DUI200	85721-24-0 see CAP250	97702-97-1 see DIE400	101652-02-2 see DFD600	102489-61-2 see DII000
80277-11-8 see AOD425	85723-21-3 see AAI125	97702-98-2 see DII800	101652-05-5 see DFM600	102489-70-3 see TCI100
80382-23-6 see LII300	85886-25-5 see SAO475	97702-99-3 see DIJ000	101652-07-7 see DFM800	102492-24-0 see SDL000
80387-97-9 see BJK600	85923-37-1 see DRD850	97703-02-1 see DIK200	101652-10-2 see DUA600	102504-44-9 see EMG000
80449-58-7 see ACF250	86045-52-5 see BLQ900	97703-04-3 see DIK400	101652-11-3 see DPG200	102504-45-0 see EMI500
80471-63-2 see EBH400	86166-58-7 see BLK700	97703-08-7 see DIM800	101670-43-3 see HGM500	102504-51-8 see AAO000
80539-34-0 see VQU500	86255-25-6 see SDT300	97703-11-2 see DIN200	101670-51-3 see DHD800	102504-52-9 see AAO250
80611-44-5 see PGQ275	86341-95-9 see PLJ775	97864-38-5 see DFE000	101692-44-8 see DHR800	102504-64-3 see CGJ000
80660-68-0 see BLN100	86425-12-9 see DGQ625	97919-22-7 see CMA600	101809-53-4 see MRE500	102504-65-4 see CIE250
80702-47-2 see RJF400	86451-37-8 see NMV450	98459-16-6 see CCW800	101809-55-6 see ITF000	102505-08-8 see EDM500
80734-02-7 see LEJ500	86539-71-1 see MEW775	99071-30-4 see HNK575	101809-59-0 see MBW250	102516-61-0 see AKI900
80748-58-9 see IKA000	86641-76-1 see SLD600	99520-58-8 see IBZ100	101809-59-0 see MBW500	102516-65-4 see THH550
80790-68-7 see MRT100	86674-51-3 see NLP375	99814-12-7 see SDY675	101831-88-3 see DPG000	102517-11-3 see PIE750
80830-42-8 see FAQ950	86886-16-0 see DNQ700	99999-42-5 see CIC500	101833-83-4 see DDF200	102534-95-2 see EDW000
81098-60-4 see CMS237	87050-74-8 see TJX600	100215-34-7 see NCT500	101834-51-9 see BER300	102571-36-8 see DPU600
81265-54-5 see IBW100	87050-95-1 see HAY600	100447-46-9 see HDO500	101913-67-1 see TKH300	102577-46-8 see TGF025
81295-38-7 see SAF500	87209-80-1 see MQY350	100466-04-4 see DKT400	101913-96-6 see HNA500	102583-64-2 see TDB750
81296-95-9 see NCG715	87425-02-3 see MGD100	100482-23-3 see CCK790	101952-86-7 see SHN150	102583-65-3 see SLN509
81412-43-3 see TJJ500	87625-62-5 see POI100	100482-34-6 see PFI600	101952-95-8 see SAY000	102583-71-1 see ZVS000
81424-67-1 see CBG075	88026-65-7 see SPD100	100620-36-8 see CGE500	101975-70-6 see NNE500	102584-01-0 see EJZ000
81486-22-8 see NEA100	88208-15-5 see NLY750	100700-12-8 see HNA000	101976-64-1 see NEC500	102584-11-2 see POX250
81781-28-4 see TMO775	88208-16-6 see NJY500	100700-20-7 see NJU000	101997-27-7 see PGQ350	102584-14-5 see POY750
81789-85-7 see IBY600	88254-07-3 see COP765	100700-23-0 see AAN250	101997-51-7 see BEA275	102584-42-9 see CKE250
81840-15-5 see DLF700	88321-09-9 see OMY925	100700-24-1 see AAN750	102070-98-4 see PAN750	102584-86-1 see XSJ000
81852-50-8 see EGN100	88338-63-0 see CDM575	100700-25-2 see AAP250	102071-30-7 see CKI175	102584-88-3 see NJU500
81866-63-9 see THC500	88746-71-8 see ARP625	100700-28-5 see FJG000	102071-76-1 see BQF825	102584-89-4 see AAP000
81910-05-6 see MHR100	88845-25-4 see AQZ100	100700-29-6 see MEI000	102071-88-5 see MRR125	102584-97-4 see CGT000
81994-68-5 see BQB825	88969-41-9 see DLX100	100700-34-3 see MFU250	102107-61-9 see NHE600	102585-37-5 see BQG500
82018-90-4 see NLX700	89022-11-7 see CBT175	100733-36-6 see MPZ000	102128-78-9 see MHD300	102585-38-6 see BQG750
82038-92-4 see DVE300	89022-12-8 see MDY300	100786-01-4 see CCP800	102128-91-6 see DIE600	102585-42-2 see CJQ500
82177-75-1 see PIE550	89194-77-4 see YGA700	100836-53-1 see MDK500	102128-92-7 see DIN000	102585-43-3 see CJQ750
82177-80-8 see SDM525	89367-92-0 see IGE100	100836-60-0 see HMZ100	102129-02-2 see CIL000	102585-52-4 see EFN500
82198-80-9 see PMW770	89591-51-5 see DUO300	100836-61-1 see MMV500	102129-03-3 see CIX000	102585-53-5 see EFM500
82219-81-6 see CCS635	89911-79-5 see NOC400	100836-62-2 see MMV750	102129-21-5 see DIA400	102585-59-1 see MFU500
82419-36-1 see OGI300	89930-55-2 see OAD100	100836-63-3 see BRV000	102129-33-9 see MKH750	102585-60-4 see MFU000
			102206-93-9 see DIR800	

102585–62–6 see SLQ500	102648–39–5 see LEW000	104931–87–5 see HAA320	109509–25–3 see CHR850
102612–93–1 see WBS860	102648–46–4 see POG000	105735–71–5 see DUW100	109651–74–3 see CHR700
102612–94–2 see WBS855	102916–22–3 see ZNS000	106440–54–4 see CDB770	110335–28–9 see CGP375
102629–86–7 see MAC600	103331–86–8 see TBA800	106440–55–5 see CDB772	112945–52–5 see SCH000
102646–51–5 see UBJ000	103416–59–7 see AAI100	107572–59–8 see BJE700	730771–71–0 see HKI075
102647–16–5 see OIU499	104639–49–8 see SIS000	108278–70–2 see NKO900	

11Asee NHG000
A-19 see BGY140
A 21 see DMV600
A-36 see DAE600
A 42 see TED500
A65 see PJA140
A 66 see PMA750
A 71 see CHG000
9AA see AHS500
A-101 see CGA500
A-139 see BDC750
A-20D see GLU000
A-310 see ALF250
A 348 see PAH500
A 350 see AJH129
A 350 see DKR200
A 361 see ARQ725
A 363 see DOR400
A 3-80 see SMQ500
A 435 see SNQ550
A 446 see LFW300
A 468 see DLS800
A-502 see SNJ000
A 585 see PIK625
593-A see PIK075
A 688 see PDT250
688A see DDG800
A 884 see DBA800
A-980 see CEW500
A 1141 see GGS000
A-1348 see PEC250
A 1390 see MNW150
A 1803 see BHJ250
A-1981 see DTO200
A 2275 see DBJ100
A 2297 see TLP750
A-2371 see MQW750
A 3322 see TEO250
A 4077 see FMP000
A-4700 see EDK875
A-4760 see EDM000
A-4828 see TNT500
A 4942 see IMH000
A 6413 see AAC000
A-8103 see BHJ250
A-8506 see TFQ275
A 10846 see DUD800
A-11025 see RDF000
A 11032 see UVA000
A-12253 see TAL350
A 19120 see BEX500
A-27053 see CBR500
A-32686 see POB500
A 3823A see MRE225
A 41304 see DBA875
A-45975 see TEF700
A-48257 see BOM600
A-91033 see DQA400
734571A see DGB500
A7301153 see HLF500
A 1 (sorbent) see AHE250
A 15 (polymer) see AAX175
A 100 (pharmaceutical) see IGS000
AA-9 see DXW200
AA-497 see TCU000
A 3733A see HAL000
AAB see PEI000
AACAPTAN see CBG000
AACIFEMINE see EDU500

A. AESTIVALIS see PCU375
AAF see FDR000
2-AAF see FDR000
AAFERTIS see FAS000
AA223 LEDERLE see AFI625
AALINDAN see BBQ500
AAMANGAN see MAS500
A. AMOREUXI VENOM see AOO250
A. AMURENSIS see PCU375
AAMX see ABA250
A. ANNUA see PCU375
9AAP see AHT000
AAPROTECT see BJK500
AARANE see CNX825
AARARRE see CNX825
AAT see AIC250, PAK000
o-AAT see AIC250
AATACK see TFS350
AATP see PAK000
AATREX see ARQ725
AATREX 4L see ARQ725
AATREX 80W see ARQ725
AATREX NINE-O see ARQ725
A. AUSTRALIS HECTOR VENOM see AOO265
AAVOLEX see BJK500
AAZIRA see BJK500
2-AB see BPY000
AB-15 see ALV750
AB-42 see COH250
AB 109 see DVU300
AB 206 see MQU525
AB 35616 see CDQ250
ABACIN see TKX000
ABACTRIM see TKX000
ABADOL see AMS250
ABADOLE see AMS250
ABAR see LEN000
ABASIN see ACE000
ABATE see TAL250
ABATHION see TAL250
ABAVIT see PFP500
ABBOCILLIN see BDY669
ABBOCORT see HHQ800
ABBOLEXIN see PGG350
ABBOMEEN E-2 see DMT400
ABBOTT-22370 see TKX250
ABBOTT-28440 see DEW400
ABBOTT-30360 see FIW000
ABBOTT 30400 see PAP000
ABBOTT-35616 see CDQ250
ABBOTT 36581 see BOR350
ABBOTT 38414 see VGU700
ABBOTT 40566 see TCM250
ABBOTT-43326 see MQT550
ABBOTT 44090 see PNR750
ABBOTT-44747 see FOL000
ABBOTT-45975 see FPP100
ABBOTT ANTIBIOTIC M259 see AAC000
ABBOTT'S A.P. 43 see DIF200
ABC 8/3 see OLM300
ABC 12/3 see TEQ175
ABCID see SNN300
ABC LANATOSIDE COMPLEX see LAU400
ABD see AID650
A. BELLADONNA see AHI635
ABELMOSCHUS MANIHOT (Linn.) Medik., extract see HGA550
ABENSANIL see HIM000
ABEREL see VSK950
ABESON NAM see DXW200

AB 50912 HEMIHYDROCHLORIDE see CCS300
ABIES ALBA OIL see AAC250
ABIETIC ACID see AAC500
ABIETIC ACID, METHYL ESTER see MFT500
ABILIT see EPD500
ABIOL see HJL500
ABIROL see DAL300
"A" BLASTING POWDER see ERF500
ABMINTHIC see DJT800
ABN-Δ⁸-THC see HNE400
ABOL see DOX600
ABOVIS see AAC875
ABRACOL S.L.G see OAV000
ABRAMYCIN see TBX000
ABRAREX see AHE250
ABRICYCLINE see TBX000
ABRIN see AAD000
ABRINS see AAD000
ABRODEN see SHX000
ABRODIL see SHX000
ABROMA AUGUSTA Linn., root extract see UIS300
ABROMEEN E-25 see CPG125
ABROVAL see BNP750
ABRUS PRECATORIUS L., seed kernel extract see AAD100
ABRUS PRECATORIUS L. (SEED) see RMK250
ABRUS PRECATORIUS OIL see AAD125
ABS see AFO500
ABS (pyrolysis products) see ADX750
ABSENTOL see TLP750
ABSETIL see TLP750
ABSIN see ACE000
ABSINTHIUM see ARL250
ABSOLUTE ETHANOL see EFU000
ABSOLUTE FRENCH ROSE see RMP000
ABSOLUTE MIMOSA see MQV000
ABSORBABLE GELATIN SPONGE see PCU360
ABSTENSIL see DXH250
ABSTINYL see DXH250
ABTS see ASH375
ABURAMYCIN see OIS000
ABURAMYCIN B see CMK650
5-AC see ARY000
AC-R-11 see BHJ500
AC 1198 see PMO250
AC 1370 see CCS525
AC 2770 see FND100
AC 3092 see BFA000
AC 3422 see EEH600
AC 5223 see DXX400
AC 5230 see ADA725
AC-12682 see DSP400
AC 18133 see EPC500
AC 18682 see IOT000
AC-18,737 see EAS000
AC 24055 see DUI000
AC-43064 see DXN600
AC 47031 see PGW750
AC 47470 see DHH400
AC 52160 see TAL250
AC 64475 see DHH200
AC 84777 see ARW000
AC 85258 see DFE469
AC 92553 see DRN200
AC 1692-40 see DKK100
AC 921000 see BSO000
ACABEL see CDG250
ACACIA see AQQ500
ACACIA DEALBATA GUM see AQQ500
ACACIA (EXTRACT) see AAD250
ACACIA FARNESIANA (Linn.) Willd., extract excluding roots see AAD500
ACACIA GUM see AQQ500
ACACIA MOLLISSIMA TANNIN see MQV250
ACACIA PALIDA (PUERTO RICO) see LED500
ACACIA SENEGAL see AQQ500
ACACIA SYRUP see AQQ500
ACACIA VILLOSA see AAD750
ACADYL see BJZ000
ACAJOU (HAITI) see MAK300
A. CALIFORNICA see HGL575
ACALMID see CKE750
ACALO see CKE750
ACAMOL see HIM000
ACAMYLOPHENINE see NOC000
ACAMYLOPHENINE DIHYDROCHLORIDE see AAD875

A. CANADENSIS see PAM780
ACANTHOPHIA ANTARCTICUS VENOM see ARU875
ACAPRIN see PJA120
ACAR see DER000
ACARABEN 4E see DER000
ACARACIDE see SOP500
ACARALATE see PNH750
ACARICYDOL E 20 see CJT750
ACARIN see BIO750
ACARITHION see TNP250
ACAROL see IOS000
ACARON see CJJ250
ACAVYL see BJZ000
ACCELERATE see DXD000
ACCELERATOR EFK see ZHA000
ACCELERATOR L see BJK500
ACCELERATOR THIURAM see TFS350
ACCELERINE see DSY600
ACCEL R see BKU500
ACCENT see MRL500
ACCO FAST RED KB BASE see CLK225
ACCONEM see DHH200
ACCOSPERSE TOLUIDINE RED XL see MMP100
ACCOTHION see DSQ000
ACCUCOL see PPN750
ACCUZOLE see SNN500
ACD 7029 see MIB000
AC-DI-SOL NF see SFO500
ACDRILE see MBX800
8:9-ACE-1:2-BENZANTHRACENE see BAW000
4,10-ACE-1,2-BENZANTHRACENE see AAE000
ACEBUTOLOL see AAE100
(±)-ACEBUTOLOL see AAE100
dl-ACEBUTOLOL see AAE100
ACEBUTOLOL HYDROCHLORIDE see AAE125
ACECARBROMAL see ACE000
ACECLIDINE see AAE250
ACECLIDIN-HCL see QUS000
ACECOLINE see ABO000, CMF250
ACEDAPSONE see SNY500
ACEDE CRESYLIQUE (FRENCH) see CNW500
ACEDIST see BNV500
ACEDOXIN see DKL800
ACEDRON see BBK500
ACEF see CCS250
ACEFEN see AAE500
ACEGLUTAMIDE ALUMINUM see AGX125
ACEITE CHINO (CUBA) see TOA275
ACELAN, combustion products see ADX750
ACEMETACIN see AAE625
ACENAPHTHALENE see AAE750
ACENAPHTHANTHRACENE see AAF000
5-ACENAPHTHENAMINE see AAF250
13H-ACENAPHTHO(1,8-ab)PHENANTHRENE see DCR600
ACENAPHTHYLENE see AAF500
ACENOCOUMARIN see ABF750
ACENOCOUMAROL see ABF750
ACENTERINE see ADA725
ACEOTHION see DSQ000
ACEPHAT (GERMAN) see DOP600
ACEPHATE see DOP600
ACEPHATE-MET see DTQ400
ACEPHENE see DPE000
ACEPRAMINE see AJD000
ACEPREVAL see AAF625
ACEPROMAZINA see ABH500
ACEPROMAZINE see ABH500
ACEPROMAZINE MALEATE see AAF750
ACEPROMIZINA see ABH500
ACEPROSOL see AOO800
ACEPYRENE see CPX500
ACEPYRYLENE see CPX500
ACESAL see ADA725
ACETAAL (DUTCH) see AAG000
ACETACID RED B see HJF500
ACETACID RED J see FMU070
ACETACID RED 2BR see FAG020
3-ACETAET 4ᴮ-PROPANOATE LEUCOMYCIN V see LEV025
ACETAGESIC see HIM000
ACETAL see AAG000, ADA725
ACETALDEHYD (GERMAN) see AAG250
ACETALDEHYDE see AAG250
ACETALDEHYDE, AMINE SALT see AAG500
ACETALDEHYDE AMMONIA see AAG500

ACETALDEHYDE BIS(2-METHOXYETHYL)ACETAL see AAG750
ACETALDEHYDE DIBUTYL ACETAL see DDT400
ACETALDEHYDE DIMETHYL ACETAL see DOO600
ACETALDEHYDE, ((3,7-DIMETHYL-2,6-OCTADIENYL)OXY)-, (E)- see GDM100
ACETALDEHYDE, DIPROPYL ACETAL see AAG850
ACETALDEHYDE-DI-n-PROPYL ACETAL see AAG850
ACETALDEHYDE-N-FORMYL-N-METHYLHYDRAZONE see AAH000
ACETALDEHYDE, (HEXYLOXY)-, DIMETHYL ACETAL see HFG700
ACETALDEHYDE-N-METHYL-N-FORMYLHYDRAZONE see AAH000
ACETALDEHYDE METHYLHYDRAZONE see AAH100
ACETALDEHYDE, N-METHYLHYDRAZONE see AAH100
ACETALDEHYDE OXIME see AAH250
ACETALDEHYDE SODIUM BISULFITE see AAH500
ACETALDEHYDE SODIUM SULFITE see AAH500
ACETALDEHYDE, TETRAMER see TDW500
ACETALDEHYDE, TRIMER see PAI250
ACETAL DIETHYLIQUE (FRENCH) see AAG000
ACETALDOL see AAH750
ACETALDOXIME see AAH250
ACETALE (ITALIAN) see AAG000
ACETALGIN see HIM000
ACETAMIDE see AAI000
ACETAMIDE, 2-AMINO-N-(2-(2,5-DIMETHOXYPHENYL)-2-HYDROXYETHYL)-,
    MONOHYDROCHLORIDE, (±)- (9CI) see MQT530
ACETAMIDE, 2-AMINO-N-(β-HYDROXY-2,5-DIMETHOXYPHENETHYL)-,
    MONOHYDROCHLORIDE, (±)- see MQT530
ACETAMIDE, 2-CHLORO-N-(2,6-DIETHYLPHENYL)-N-(2-PROPOXYETHYL)-
    see PMB850
ACETAMIDE, N,N'-(1,4-CYCLOHEXYLENEDIMETHYLENE)BIS(2-(1-
    AZIRIDINYL)- see CPL100
ACETAMIDE, 2-(DIETHYLAMINO)-N-(1,3-DIMETHYL-4-(o-FLUOROBENZOYL)-
    5-PYRAZOLYL)-, MONOHYDROCHLORIDE see AAI100
ACETAMIDE, N,N-DIETHYL-N'-(1,2,3,4-TETRAHYDRO-1-NAPHTHYL)-
    see TGI725
ACETAMIDE, N-(4-(2-FLUOROBENZOYL)-1,3-DIMETHYL-1H-PYRAZOL-5-YL)-
    2-((3-(2-METHYL-1-PIPERIDINYL)PROPYL)AMINO)-, (Z)-2-BUTENEDIO-
    ATE (1:2) see AAI125
2-ACETAMIDE-4-MERCAPTOBUTYRIC ACID γ-THIOLACTONE see TCZ000
5-ACETAMIDE-1,3,4-THIADIAZOLE-2-SULFONAMIDE see AAI250
ACETAMIDINE HYDROCHLORIDE see AAI500
3-ACETAMIDO-5-(ACETAMIDOMETHYL)-2,4,6-TRIIODOBENZOIC ACID
    see AAI750
N-(p-(9-(3-ACETAMIDOACRIDINYL)AMINO)PHENYL)METHANESULFONAMIDE
    see AAJ000
3-ACETAMIDO-5-AMINO-2,4,6-TRIIODOBENZOIC ACID see AAJ125
4-ACETAMIDOANILINE see AHQ250
p-ACETAMIDOANILINE see AHQ250
p-ACETAMIDOAZOBENZENE see PEH750
p-ACETAMIDOBENZALDEHYDE THIOSEMICARBAZONE see FNF000
ACETAMIDOBENZENE see AAQ500
p-ACETAMIDOBENZENEARSONIC ACID, SODIUM SALT, TETRAHYDRATE
    see ARA000
p-ACETAMIDOBENZENESTIBONIC ACID SODIUM SALT see SLP500
4'-ACETAMIDOBENZIDINE see ACC000
((p-ACETAMIDOBENZOYL)OXY)TRIBUTYLSTANNANE see TIB750
ACETAMIDOBIPHENYL see PDY000
4-ACETAMIDOBIPHENYL see PDY500
N-(4'-ACETAMIDOBIPHENYLYL)ACETOHYDROXAMIC ACID see HKB000
2-ACETAMIDO-4,5-BIS(ACETOXYMERCURI)THIAZOLE see AAJ250
4-ACETAMIDOBUTYRIC ACID see AAJ350
4-ACETAMIDO-4-CARBOXAMIDO-n-(N-NITROSO)BUTYLCYANAMIDE see
    AAJ500
3-ACETAMIDODIBENZFURANE see DDC000
3-ACETAMIDODIBENZOFURAN see DDC000
3-ACETAMIDODIBENZTHIOPHENE see DDD600
3-ACETAMIDODIBENZTHIOPHENE OXIDE see DDD800
7-ACETAMIDO-6,7-DIHYDRO-10-HYDROXY-1,2,3-TRIMETHOXY-
    BENZO(a)HEPTALEN-9(5H)-ONE see ADE000
7-ACETAMIDO-6,7-DIHYDRO-1,2,3,10-TETRAMETHOXY-BENZO(a)HEPTALEN-
    9(5H)-ONE see CNG830
2-ACETAMIDO-4,5-DIMETHYLOXAZOLE see AAJ750
2-ACETAMIDO-4,5-DIPHENYLOXAZOLE see AAK000
ACETAMIDOETHANE see EFM000
2-ACETAMIDOETHANOL see HKM000
1-ACETAMIDO-4-ETHOXYBENZENE see ABG750
2,2'-((5-ACETAMIDO-2-ETHOXYPHENYL)IMINO)DIETHANOL see BKB000
4-(2-ACETAMIDOETHYLDITHIO)BUTANESULFINATE SODIUM see AAK250
3-ACETAMIDOFLUORANTHENE see AAK400
2-ACETAMIDOFLUORENE see FDR000
N-(2-ACETAMIDOFLUOREN-1-YL)-N-FLUOREN-2-YL ACETAMIDE see AAK500
1-(N-ACETAMIDOFLUOROMETHYL)-NAPHTHALENE see MME809
3-ACETAMIDO-4-HYDROXY-PHENYLARSONIC ACID see ABX500
7-ACETAMIDO-10-HYDROXY-1,2,3-TRIMETHOXY-6,7-
    DIHYDROBENZO(a)HEPTALEN-9(5H)-ONE see ADE000

ACETAMIDOMALONIC ACID DIETHYL ESTER see AAK750
2-ACETAMIDO-4-MERCAPTOBUTYRIC ACID THIOLACTONE see TCZ000
l-α-ACETAMIDO-β-MERCAPTOPROPIONIC ACID see ACH000
2-(3-ACETAMIDO-5-N-METHYL-ACETAMIDO-2,4,6-TRIIODOBENZAMIDO)2-
    DEOXY-d-GLUCOSE see MQR300
6-ACETAMIDO-4-METHYL-1,2-DITHIOLO(4,3-B)PYRROL-5(4H)-ONE see ABI250
7-ACETAMIDO-1-METHYL-4-(p-(p-((1-METHYLPYRIDINIUM-4-
    YL)AMINO)BENZAMIDO)ANILINO)QUINOLINIUM DI-p-
    TOLUENESULFONATE see AAL000
5-ACETAMIDO-1-METHYL-3-(5-NITRO-2-FURYL)-s-TRIAZOLE see MMK000
3-ACETAMIDO-5-METHYLPYRROLIN-4-ONE(4,3-D)-1,2-DITHIOLE see ABI250
2-ACETAMIDO-N-(3-METHYL-2-THIAZOLIDINYLIDENE)ACETAMIDE see
    AAL250
2-(2-ACETAMIDO-4-METHYLVALERAMIDO)-N-(1-FORMYL-4-
    GUANIDINOBUTYL)-4-METHYLVALERAMIDE see LEX400
(S)-2-(2-ACETAMIDO-4-METHYLVALERAMIDO)-N-(1-FORMYL-4-
    GUANIDINOBUTYL)-4-METHYL-VALERAMIDE see AAL300
p-ACETAMIDONITROBENZENE see NEK000
5-ACETAMIDO-3-(5-NITRO-2-FURYL)-6H-1,2,4-OXADIAZINE see AAL500
2-ACETAMIDO-4-(5-NITRO-2-FURYL)THIAZOLE see AAL750
2-ACETAMIDO-5-(NITROSOCYANAMIDO)VALERAMIDE see AAM000
ACETAMIDO-5-NITROTHIAZOLE see ABY900
4-ACETAMIDO-4-PENTEN-3-ONE-1-CARBOXAMIDE see PMC750
p-ACETAMIDOPHENACYL CHLORIDE see CEC000
3-ACETAMIDOPHENANTHRENE see PCY500
9-ACETAMIDOPHENANTHRENE see PCY750
2-ACETAMIDOPHENATHRENE see AAM250
2-ACETAMIDOPHENOL see HIL000
3-ACETAMIDOPHENOL see HIL500
4-ACETAMIDOPHENOL see HIM000
m-ACETAMIDOPHENOL see HIL500
o-ACETAMIDOPHENOL see HIL000
p-ACETAMIDOPHENOL see HIM000
1-(4-ACETAMIDOPHENOXY)-3-ISOPROPYLAMINO-2-PROPANOL see ECX100
4'-(ACETAMIDO)PHENYL-2-HYDROXYETHYLSULFONE see SAN600
p-ACETAMIDOPHENYL ACETYLSALICYLATE see SAN600
1-(p-ACETAMIDOPHENYL)-3,3-DIMETHYLTRIAZENE see DUI000
3-ACETAMIDOPHENYL ISOTHIOCYANATE see ISG000
2-ACETAMIDO-4-PHENYLOXAZOLE see AAM500
4-ACETAMIDOPYRIDINE see AAM750
1-ACETAMIDO-2-PYRROLIDINONE see NNE400
4-ACETAMIDOSTILBENE see SMR500
trans-4-ACETAMIDOSTILBENE see SMR500
2-ACETAMIDO-5-SULFONAMIDO-1,3,4-THIADIAZOLE see AAI250
ACETAMIDOTHIADIAZOLESULFONAMIDE see AAI250
α-ACETAMIDO-γ-THIOBUTYROLACTONE see TCZ000
3-ACETAMIDOTOLUENE see ABI250
p-ACETAMIDOTOLUENE see ABJ250
3-ACETAMIDO-2,4,6-TRIIODOBENZOIC ACID see AAM875
3-ACETAMIDO-2,4,6-TRIIODOBENZOIC ACID SODIUM SALT see AAN000
2-(3-ACETAMIDO-2,4,6-TRIIODO-5-(N-METHYLACETAMIDO)BENAZMIDO)-2-
    DEOXY-d-GLUCOPYRANOSE see MQR300
2-(3-ACETAMIDO-2,4,6-TRIIODO-5-(N-METHYLACETAMIDO)BENZAMIDO)-2-
    DEOXY-d-GLUCOSE see MQR300
2-((4-(3-ACETAMIDO-2,4,6-TRIIODOPHENOXY)BUTOXY)METHYL)BUTYRIC
    ACID SODIUM SALT see AAN250
2-(2-(3-ACETAMIDO-2,4,6-TRIIODOPHENOXY)ETHOXY)ACETIC ACID SO-
    DIUM SALT see AAN500
2-(2-(3-ACETAMIDO-2,4,6-TRIIODOPHENOXY)ETHOXY)BUTYRIC ACID SO-
    DIUM SALT see AAN750
2-((2-(3-ACETAMIDO-2,4,6-TRIIODOPHENOXY)ETHOXY)METHYL)
    PROPIONIC ACID SODIUM SALT see AAO000
2-(2-(3-ACETAMIDO-2,4,6-TRIIODOPHENOXY)ETHOXY)PROPIONIC ACID SO-
    DIUM SALT see AAO250
2-(2-(3-ACETAMIDO-2,4,6-TRIIODOPHENOXY)ETHOXY)-2-(o-TOLYL)ACETIC
    ACID SODIUM SALT see AAO500
2-(2-(3-ACETAMIDO-2,4,6-TRIIODOPHENOXY)ETHOXY)-2-(p-TOLYL)ACETIC
    ACID SODIUM SALT see AAO750
2-(2-(3-ACETAMIDO-2,4,6-TRIIODOPHENOXY)ETHOXY)VALERIC ACID SO-
    DIUM SALT see AAP000
2-((3-(3-ACETAMIDO-2,4,6-TRIIODOPHENOXY)PROPOXY)METHYL)BUTYRIC
    ACID SODIUM SALT see AAP250
2-(3-ACETAMIDO-2,4,6-TRIIODOPHENYL)BUTYRIC ACID see AAP500
2-(3-ACETAMIDO-2,4,6-TRIIODOPHENYL)PROPIONIC ACID see AAP750
2-(3-ACETAMIDO-2,4,6-TRIIODOPHENYL)VALERIC ACID see AAQ000
ACETAMIN see SNY500
ACETAMINE DIAZO BLACK RD see DCJ200
ACETAMINE YELLOW CG see AAQ250
m-ACETAMINOANILINE see AHQ000
p-ACETAMINOBENZYLIDENETHIOSEMICARBAZONE see FNF000
3-ACETAMINODIBENZOTHIOPHENE see DDD600
ACET-o-AMINOFENOL (CZECH) see HIL000
p-ACETAMINOFENYL-2-HYDROXYETHYLSULFON see HLB500
p-ACETAMINOFENYL-β-HYDROXYETHYLSULFON see HLB500

2-ACETAMINOFLUORENE see FDR000
2-ACETAMINO-4-(5-NITRO-2-FURYL)THIAZOLE see AAL750
4-ACETAMINO-2-NITROPHENETOLE see NEL000
ACETAMINOPHEN see HIM000
2-ACETAMINOPHENANTHRENE see AAM250
3-ACETAMINOPHENANTHRENE see PCY500
9-ACETAMINOPHENANTHRENE see PCY750
2-ACETAMINOPHENOL see HIL000
p-ACETAMINOPHENOL see HIM000
1-(4-ACETAMINOPHENYL)-3,3-DIMETHYLTRIAZENE see DUI000
trans-4-ACETAMINOSTILBENE see SMR500
ACETAMOX see AAI250
ACETANIL see AAQ500
ACETANILIDE see AAQ500
ACETANILIDE, 4'-(2-HYDROXYETHYLSULFONYL)- see HLB500
m-ACETANISIDIDE see AAQ750
o-ACETANISIDIDE see AAR000
p-ACETANISIDIDE see AAR250
ACETANISOLE (FCC) see MDW750
ACETARSIN see ACN250
ACETARSOL see ABX500
ACETARSONE see ABX500
ACETARSONE DIETHYLAMINE SALT see ACN250
ACETATE d'AMYLE (FRENCH) see AOD725
ACETATE-AS see HHQ800
ACETATE BLUE G see TBG700
ACETATE BRILLIANT BLUE 4B see MGG250
ACETATE de BUTYLE (FRENCH) see BPU750
ACETATE de BUTYLE SECONDAIRE (FRENCH) see BPV000
ACETATE C-7 see HBL000
ACETATE C-8 see OEG000
ACETATE C-11 see UMS000
ACETATE C-12 see DXV400
ACETATE de CELLOSOLVE (FRENCH) see EES400
21-ACETATECORTICOSTERONE see CNS650
ACETATE CORTISONE see CNS825
ACETATE de CUIVRE (FRENCH) see CNI250
16-ACETATE DIGITOXIN see ACH375
ACETATE de l'ETHER MONOETHYLIQUE DE L'ETHYLENE-GLYCOL
  (FRENCH) see EES400
ACETATE d'ETHYLGLYCOL (FRENCH) see EES400
ACETATE FAST ORANGE R see AKP750
ACETATE of 4-(HYDROXYPHENYL)-2-BUTANONE see AAR500
ACETATE d'ISOBUTYLE (FRENCH) see IIJ000
ACETATE d'ISOPROPYLE (FRENCH) see INE100
ACETATE de L'ETHER MONOMETHYLIQUE de L'ETHYLENE-GLYCOL
  (FRENCH) see EJJ500
ACETATE of LIME see CAL750
ACETATE de METHYLE (FRENCH) see MFW100
ACETATE de METHYLE GLYCOL (FRENCH) see EJJ500
ACETATE P.A. see AGQ750
ACETATE PHENYLMERCURIQUE (FRENCH) see ABU500
ACETATE de PLOMB (FRENCH) see LCV000
1-ACETATE-1,2,3-PROPANETRIOL see GGO000
ACETATE de PROPYLE NORMAL (FRENCH) see PNC250
ACETATE RED VIOLET R see DBP000
ACETATE-REPLACING FACTOR see DXN800
ACETATE de TRIMETHYLPLOMB (FRENCH) see ABX125
ACETATE de TRIPHENYL-ETAIN (FRENCH) see ABX250
ACETATE de TRIPROPYLPLOMB (FRENCH) see ABX325
ACETATE YELLOW 6G see MEB750
ACETATO(2-AMINO-5-NITROPHENYL)MERCURY see ABQ250
(ACETATO)(p-AMINOPHENYL)MERCURY see ABQ000
(ACETATO)BIS(HEPTYLOXY)PHOSPHINYLMERCURY see AAR750
(ACETATO)BIS(HEXYLOXY)PHOSPHINYLMERCURY see AAS000
ACETATO di CELLOSOLVE (ITALIAN) see EES400
(ACETATO)(DIETHOXYPHOSPHINYL)MERCURY see AAS250
(ACETATO)ETHYLMERCURY see EMD000
(ACETATO-o)-ETHYLMERCURY see EMD000
ACETATO((5-HYDROXYMERCURI)-2-THIENYL)MERCURY see HLQ000
(ACETATO-o)HYDROXY-mu-2,5-THIOPHENEDIYLDIMERCURY see HLQ000
ACETATO(2-METHOXYETHYL)MERCURY see MEO750
ACETATO di METIL CELLOSOLVE (ITALIAN) see EJJ500
ACETATOPHENYLMERCURATE(1-) AMMONIUM SALT see PFO550
(ACETATO)PHENYLMERCURY see ABU500
(ACETATO)PHENYL-MERCURY mixed with CHLOROETHYL-MERCURY
  (19:1) see PFO500
ACETATO di STAGNO TRIFENILE (ITALIAN) see ABX250
(ACETATO)(2,3,5,6-TETRAMETHYLPHENYL)MERCURY see AAS500
(ACETATO)(8-THEOPHYLLINYL)MERCURY see TEP750
(ACETATO)(TRIMETAARSENITO)DICOPPER see COF500
ACETATOTRIPHENYLSTANNANE see ABX250
ACETAZINE see ABH500
ACETAZOLAMID see AAI250

ACETAZOLAMIDE see AAI250
ACETAZOLAMIDE SODIUM see AAS750
ACETAZOLAMIDE SODIUM SALT see AAS750
ACETAZOLEAMIDE see AAI250
ACETCARBROMAL see ACE000
ACETDIMETHYLAMIDE see DOO800
ACETDRON see AOB250
ACETEIN see ACH000
ACETENE see EIO000
ACETETHYLANILIDE see EFQ500
ACETEUGENOL see EQS000
ACETHION see DIX000
ACETHION AMIDE see AAT000
ACETHIONE see DIX000
ACETHROPAN see AES650
ACETHYDRAZIDE see ACM750
ACETHYDROXAMSAURE (GERMAN) see ABB250
ACETHYLPROMAZIN see ABH500
ACETIC ACID see AAT250
ACETIC ACID (aqueous solution) (DOT) see AAT250
ACETIC ACID-4-((ACETOXYMETHYL)NITROSAMINO)BUTYL ESTER
  see ABM500
ACETIC ACID-2-((ACETOXYMETHYL)NITROSAMINO)ETHYL ESTER
  see ABN750
ACETIC ACID-4-((ACETOXYMETHYL)NITROSAMINO)PROPYL ESTER
  see ABV750
ACETIC ACID (N-ACETYL-N-(4-BIPHENYL)AMINO) ESTER see ABJ750
ACETIC ACID (N-ACETYL-N-(2-FLUORENYL)AMINO) ESTER see ABL000
ACETIC ACID(N-ACETYL-N-(4-FLUORENYL)AMINO)ESTER see ABO500
ACETIC ACID (N-ACETYL-N-(2-PHENANTHRYL)AMINO)ESTER see ABK250
ACETIC ACID-(N-ACETYL-N-(p-STYRYLPHENYL)AMINO) ESTER see ABW500
ACETIC ACID ALLYL ESTER see AFU750
ACETIC ACID-3-ALLYLOXYALLYL ESTER see AAT500
ACETIC ACID, 4-ALLYLOXY-3-CHLOROPHENYL-, compounded with 2-
  AMINOETHANOL see MDI225
ACETIC ACID AMIDE see AAI000
ACETIC ACID, 2,2'-(((4-(AMINOCARBONYL)PHENYL)ARSINIDENE)
  BIS(THIO))BIS-(9CI) see TFA350
ACETIC ACID, AMMONIUM SALT see ANA000
ACETIC ACID, AMYL ESTER see AOD725, AOD750
ACETIC ACID, ANHYDRIDE see AAX250
ACETIC ACID, ANHYDRIDE with NITRIC ACID (1:1) see ACS750
ACETIC ACID ANILIDE see AAQ500
ACETIC ACID, BARIUM SALT see BAH500
ACETIC ACID, BENZ(a)ANTHRACENE-7,12-DIMETHANOL DIESTER
  see BBF750
ACETIC ACID, BENZ(a)ANTHRACENE-7-METHANOL ESTER see BBH500
ACETIC ACID BENZYL ESTER see BDX000
ACETIC ACID, (((3,5-BIS(1,1-DIMETHYLETHYL)-4-
  HYDROXYPHENYL)METHYL)THIO)-, 2-ETHYLHEXYL ESTER see BJK600
ACETIC ACID, 2,2'-((BIS(PHENYLMETHYL)STANNYLENE)BIS(THIO))BIS-,
  DIISOOCTYL ESTER (9CI) see EDC560
ACETIC ACID, compounded with BORON FLUORIDE (BF3) (8CI) see BMG750
ACETIC ACID-1,3-BUTADIENYL ESTER see ABM250
ACETIC ACID-2-BUTOXY ESTER see BPV000
ACETIC ACID-2-(tert-BUTYL)-4,6-DINITRO-m-TOLYL ESTER see BRU750
ACETIC ACID n-BUTYL ESTER see BPU750
ACETIC ACID-tert-BUTYL ESTER see BPV100
ACETIC ACID, 3-BUTYL-5-METHYL-TETRAHYDRO-2H-PYRAN-4-YL ESTER
  see MHX000
ACETIC ACID (sec-BUTYLNITROSAMINOMETHYL) ESTER see BPV500
ACETIC ACID-1-(BUTYLNITROSOAMINO)BUTYL ESTER see BPV325
ACETIC ACID, CADMIUM SALT see CAD250
ACETIC ACID, CADMIUM SALT, DIHYDRATE see CAD275
ACETIC ACID, CEDROL ESTER see CCR250
ACETIC ACID CHLORIDE see ACF750
ACETIC ACID, 3-CHLOROPROPYLENE ESTER see CIL900
ACETIC ACID, CHROMIUM (2+) SALT (8CI, 9CI) see CMJ000
ACETIC ACID, CINNAMYL ESTER see CMQ730
ACETIC ACID, CITRONELLYL ESTER see AAU000
ACETIC ACID, COBALT(2+) SALT see CNC000
ACETIC ACID, COBALT(2+) SALT, TETRAHYDRATE see CNA500
ACETIC ACID, CUPRIC SALT see CNI250
ACETIC ACID, CYANO-, TRIPHENYLSTANNYL ESTER see TMV825
ACETIC ACID, CYCLOHEXYLETHYL ESTER see EHS000
ACETIC ACID, 9-DECENYL ESTER see DAI450
ACETIC ACID, ((DIBENZYLSTANNYLENE)DITHIO)DI-, DIISOOCTYL ESTER
  see EDC560
ACETIC ACID, 2,2'-((DIBUTYLSTANNYLENE)BIS(THIO))BIS-, DINONYL
  ESTER see DEH650
ACETIC ACID, ((DIBUTYLSTANNYLENE)DITHIO)DI-, DINONYL ESTER
  (8CI) see DEH650
ACETIC ACID, (3,4-DICHLOROPHENOXY)- see DFY500
ACETIC ACID, (2,4-DICHLOROPHENOXY)-, ETHYL ESTER see EHY600

ACETIC ACID, ((2,3-DIHYDRO-6,7-DICHLORO-2-METHYL-1-OXO-2-PHENYL-1H-INDEN-5-YL)OXY)-, (±)- see IBQ400
ACETIC ACID DIHYDROTERPINYL ESTER see DME400
ACETIC ACID DIMETHYLAMIDE see DOO800
ACETIC ACID-α-(DIMETHYLAMINOMETHYL)BENZYL ESTER see ABN700
ACETIC ACID-1,3-DIMETHYLBUTYL ESTER see HFJ000
ACETIC ACID-2,6-DIMETHYL-m-DIOXAN-4-YL ESTER see ABC250
ACETIC ACID-1,1-DIMETHYLETHYL ESTER see BPV100
ACETIC ACID-3,7-DIMETHYL-6-OCTEN-1-YL ESTER see AAU000
ACETIC ACID-(2,4-DINITRO-6-sec-BUTYLPHENYL) ESTER see ACE500
ACETIC ACID-(4,6-DINITRO-2-sec-BUTYLPHENYL) ESTER see ACE500
ACETIC ACID-4,6-DINITRO-o-CRESYL ESTER see AAU250
ACETIC ACID, DIPHENYL-, 2-(BIS(2-HYDROXYETHYL)AMINO)ETHYL ESTER, HYDROCHLORIDE see DVW900
ACETIC ACID, DODECYL ESTER see DXV400
ACETIC ACID ESTER with N-4-BIPHENYLYLACETOHYDROXAMIC ACID see ABJ750
ACETIC ACID ESTER with N-(FLUOREN-3-YL)ACETOHYDROXAMIC ACID see ABO250
ACETIC ACID, ESTER with N-(FLUOREN-4-YL)ACETOXYHYDROXAMIC ACID see ABO500
ACETIC ACID, ESTER with N-NITROSO-2,2'-IMINODIETHANOL see NKM500
ACETIC ACID ESTER with N-(2-PHENANTHRYL)ACETOHYDROXAMIC ACID see ABK250
ACETIC ACID, ESTER with N-(p-STYRYLPHENYL)ACETOHYDROXAMIC ACID see SMT000
ACETIC ACID-ESTER with N-(p-STYRYLPHENYL)ACETOHYDROXAMIC ACID see ABW500
ACETIC ACID ETHENYL ESTER see VLU250
ACETIC ACID ETHENYL ESTER HOMOPOLYMER see AAX250
ACETIC ACID ETHENYL ESTER POLYMER with CHLORETHENE (9CI) see AAX175
ACETIC ACID, ETHOXY- see EEK500
ACETIC ACID-2-ETHOXYETHYL ESTER see EES400
ACETIC ACID, (ETHYLENEDINITRILO)TETRA-, CALCIUM (II) COMPLEX see CAR800
ACETIC ACID, (ETHYLENEDINITRILO)TETRA-, YTTRIUM complex see YFA100
ACETIC ACID, (ETHYLENEDINITRILO)TETRA-, ZINC(II) COMPLEX see ZGS100
ACETIC ACID α-ETHYLEXYL ESTER see OEE000
ACETIC ACID-1-(ETHYLNITROSAMINO)ETHYL ESTER see ABT500
ACETIC ACID GERANIOL ESTER see DTD800
ACETIC ACID, GLACIAL (DOT) see AAT250
ACETIC ACID-3-HEPTANOL ESTER see AAU500
ACETIC ACID-2,4-HEXADIEN-1-OL ESTER see AAU750
ACETIC ACID, HEXYLENE GLYCOL see HFQ000
ACETIC ACID HEXYL ESTER see HFI500
ACETIC ACID-3-HYDROXYHEPTYL ESTER see AAU500
ACETIC ACID, (o-((2-HYDROXY-3-HYDROXYMERCURI)PROPYL)CARBAM-OYL)PHENOXY- see NCM800
ACETIC ACID, IRON(2+) SALT see FBH000
ACETIC ACID, ISOBUTYL ESTER see IIJ000
ACETIC ACID, ISOPENTYL ESTER see IHO850
ACETIC ACID ISOPROPYL ESTER see INE100
ACETIC ACID-2-ISOPROPYL-5-METHYL-2-HEXEN-1-YL ESTER see AAV250
ACETIC ACID, LEAD SALT see LCG000
ACETIC ACID LEAD (2+) SALT see LCV000
ACETIC ACID, LEAD(+2) SALT TRIHYDRATE see LCJ000
ACETIC ACID, LINALOOL ESTER see LFY100
ACETIC ACID, MAGNESIUM SALT see MAD000
ACETIC ACID MANGANESE(II) SALT (2:1) see MAQ000
ACETIC ACID-p-MENTHAN-8-OL ESTER see DME400
ACETIC ACID, MERCURY(2+) SALT see MCS750
ACETIC ACID-3-METHOXYBUTYL ESTER see MHV750
ACETIC ACID, 2-(2-(2-METHOXYETHOXY)ETHOXY)ETHYL ESTER see AAV500
ACETIC ACID METHYL ESTER see MFW100
ACETIC ACID-1-METHYLETHYL ESTER (9CI) see INE100
ACETIC ACID-2-METHYL-6-METHYLENE-7-OCTEN-2-YL ESTER see AAW500
ACETIC ACID METHYLNITROSAMINOMETHYL ESTER see AAW000
ACETIC ACID, 2-METHYL-2,4-PENTANEDIOL DIESTER see HFQ000
ACETIC ACID, α-METHYL-PHENETHYL ESTER see ABU800
ACETIC ACID-4-METHYLPHENYL ESTER see MNR250
ACETIC ACID-2-METHYL-2-PROPENE-1,1-DIOL DIESTER see AAW250
ACETIC ACID-2-METHYLPROPYL ESTER see IIJ000
ACETIC ACID-1-METHYLPROPYL ESTER (9CI) see BPV000
ACETIC ACID, MONOGLYCERIDE see GGO000
ACETIC ACID MYRCENYL ESTER see AAW500
ACETIC ACID, NICKEL(2+) SALT see NCX000
ACETIC ACID, NITRILOTRI-, IRON(III) chelate see IHC100
ACETIC ACID, OCTYL ESTER see OEG000
ACETIC ACID, OXIME see ABB250
ACETIC ACID, OXO-, METHYL ESTER see MKI550

ACETIC ACID, (4-OXO-2-THIAZOLIDINYLIDENE)-, BUTYL ESTER (9CI) see BPI300
ACETIC ACID, PHENYL-, BUTYL ESTER see BQJ350
ACETIC ACID, PHENYL-, 3,7-DIMETHYL-6-OCTENYL ESTER see CMU050
ACETIC ACID-2-PHENYLETHYL ESTER see PFB250
ACETIC ACID PHENYLHYDRAZONE see ACX750
ACETIC ACID, PHENYLMERCURY DERIVATIVE see ABU500
ACETIC ACID PHENYLMETHYL ESTER see BDX000
ACETIC ACID-2-PROPENYL ESTER see AFU750
ACETIC ACID, n-PROPYL ESTER see PNC250
ACETIC ACID-1-(PROPYLNITROSAMINO)PROPYL ESTER see ABT750
ACETIC ACID, SAMARIUM SALT see SAR000
ACETIC ACID, SODIUM SALT see SEG500
ACETIC ACID, TRIANHYDRIDE with ANTIMONIC ACID see AQJ750
ACETIC ACID, TRIETHYLENE GLYCOL DIESTER see EJB500
ACETIC ACID (m-TRIMETHYLAMMONIO)PHENYL ESTER METHYLSULF-ATE see ABV500
ACETIC ACID-VETIVEROL ESTER see AAW750
ACETIC ACID VINYL ESTER see VLU250
ACETIC ACID, VINYL ESTER, CHLOROETHYLENE COPOLYMER see PKP500
ACETIC ACID, VINYL ESTER, POLYMER with CHLOROETHYLENE see AAX175
ACETIC ACID VINYL ESTER POLYMERS see AAX250
ACETIC ACID, ZINC SALT see ZBS000
ACETIC ACID, ZINC SALT, DIHYDRATE see ZCA000
ACETIC ALDEHYDE see AAG250
ACETIC ANHYDRIDE see AAX500
ACETIC CHLORIDE see ACF750
ACETIC-4-CHLOROANILIDE see CDZ100
ACETIC ETHER see EFR000
ACETIC OXIDE see AAX500
ACETICYL see ADA725
ACETIDIN see EFR000
ACETILARSANO see ACN250
3-(3-ACETIL-4-(3-tert-BUTILAMINO)-2-HIDROXIPROPOXI)FENIL)-1,1-DIETILUREA HCl (SPANISH) see SBN475
ACETILE DIAZO BLACK N see DPO200
ACETILSALICILICO see ADA725
ACETILUM ACIDULATUM see ADA725
ACETIMIDIC ACID see AAI000
ACETIN see GGO000
ACETIODONE see AAN000
ACETIROMATE see ADD875
ACETISAL see ADA725
ACETISOEUGENOL see AAX750
ACETKARBROMAL see ACE000
ACETMIN see AOB500
ACETOACETAMIDOBENZENE see AAY000
ACETOACETANILIDE see AAY000
o-ACETOACETANISIDE see ABA500
ACETOACET-o-ANISIDIN (CZECH) see ABA500
ACETOACET-p-CHLORANILIDE see AAY250
ACETOACETIC ACID ANILIDE see AAY000
ACETOACETIC ACID-o-ANISIDIDE see ABA500
ACETOACETIC ACID BUTYL ESTER see BPV250
ACETOACETIC ACID-3,7-DIMETHYL-2,6-OCTADIENYL ESTER see AAY500
ACETOACETIC ACID, ETHYL ESTER see EFS000
ACETOACETIC ACID-2-HYDROXYETHYL ESTER ACRYLATE see AAY750
ACETOACETIC ANILIDE see AAY000
ACETOACETIC ESTER see EFS000
ACETOACETIC METHYL ESTER see MFX250
ACETOACETONE see ABX750
p-ACETOACETOPHENETIDIDE see AAZ000
2-ACETOACETOXYETHYL ACRYLATE see AAY750
ACETOACET-p-PHENETIDIDE see AAZ000
ACETOACET-o-TOLUIDIDE see ABA000
ACETOACET-m-XYLIDIDE see ABA250
2-ACETOACETYLAMINOANISOLE see ABA500
((ACETOACETYL)AMINO)BENZENE see AAY000
2-ACETOACETYLAMINOTOLUENE see ABA000
ACETOACETYLANILINE see AAY000
ACETOACETYL-o-ANISIDE see ABA500
ACETOACETYL-o-ANISIDINE see ABA500
ACETOACETYL-o-ANISINE see ABA500
ACETOACETYL-4-CHLOROANILINE see AAY250
ACETOACETYL-2-METHYLANILIDE see ABA000
p-ACETOAMINOANILINE see AHQ250
ACETOAMINOFLUORENE see FDR000
ACETO-m-AMINOTOLUENE see ABI750
ACETOANILIDE see AAQ500
ACETO-m-ANISIDIDE see AAQ750
ACETO-p-ANISIDIDE see AAR250
ACETOARSENITE de CUIVRE (FRENCH) see COF500

ACETO AZIB see ASL750
ACETOBUTOLOL HYDROCHLORIDE see AAE125
ACETO-CAUSTIN see TII250
ACETOCID see SNP500
ACETO-CORT see HHQ800
ACETO DIPP see NBL000
ACETO DNPT 40 see DVF400
ACETO DNPT 80 see DVF400
ACETO DNPT 100 see DVF400
ACETOFERROCENE see ABA750
ACETOGUAIACONE see HLQ500
ACETOHEXAMIDE see ABB000
ACETO HMT see HEI500
ACETOHYDRAZIDE see ACM750
ACETOHYDROXAMIC ACID see ABB250
ACETOHYDROXAMIC ACID, FLUOREN-2-YL-o-GLUCURONIDE see HIQ500
ACETOHYDROXAMIC ACID, N-(7-IODOFLUOREN-2-YL)-O-MYRISTOYL- see MSB100
ACETOHYDROXIMIC ACID see ABB250
ACETOIN see ABB500
ACETOL see ADA725
ACETOL (1) see ABC000
ACETOMESIDIDE see TLD250
ACETOMETHOXAN see ABC250
ACETOMETHOXANE see ABC250
ACETOMETHYLANILIDE see MFW000
ACETOMORFINE see HBT500
ACETOMORPHINE see HBT500
ACETON (GERMAN, DUTCH, POLISH) see ABC750
1-ACETONAPHTHALENE see ABC475
β-ACETONAPHTHALENE see ABC500
ACETONAPHTHONE see ABC500
1-ACETONAPHTHONE see ABC475
2-ACETONAPHTHONE see ABC500
α-ACETONAPHTHONE see ABC475
β-ACETONAPHTHONE see ABC500
1'-ACETONAPHTHONE see ABC475
2'-ACETONAPHTHONE see ABC500
ACETONCIANHIDRINEI (ROUMANIAN) see MLC750
ACETONCIANIDRINA (ITALIAN) see MLC750
ACETONCYAANHYDRINE (DUTCH) see MLC750
ACETONCYANHYDRIN (GERMAN) see MLC750
ACETONE see ABC750
ACETONE ANIL see TLP500
ACETONE BIS(ETHYL SULFONE) see ABD500
ACETONE-o-CARBANILOYLOXIME see PEQ500
ACETONE CHLOROFORM see ABD000
ACETONECYANHYDRINE (FRENCH) see MLC750
ACETONE CYANOHYDRIN (DOT) see MLC750
ACETONE-1,3-DICHLORO-1,1,3,3-TETRAFLUOROACETONE see DGL400
ACETONE DIETHYL KETAL see ABD250
ACETONE DIETHYLSULFONE see ABD500
ACETONE-OXIME-N-PHENYLCARBAMATE see PEQ500
ACETONE OXIME PHENYLURETHANE see PEQ500
ACETONE PEROXIDE see ABE000
ACETONE SEMICARBAZONE see ABE250
ACETONIC ACID see LAG000
ACETONITRIL (GERMAN, DUTCH) see ABE500
ACETONITRILE see ABE500
ACETONITRILE, AMINO-, MONOHYDROCHLORIDE (9CI) see GHI100
ACETONITRILE IMIDAZOLE-5,7,7,12,14,14-HEXAMETHYL-1,4,8,11-TETRAAZA-4,11-CYCLOTETRADECA DIENE IRON (11) PERCHLORATE see ABE750
ACETONITRILETHIOL see MCK300
ACETONKYANHYDRIN (CZECH) see MLC750
ACETONOXIME see ABF000
ACETONYL see ADA725
ACETONYL ACETONE see HEQ500
p-ACETONYLANISOLE see AOV875
3-(α-ACETONYLBENZYL)-4-HYDROXYCOUMARIN see WAT200
3-(α-ACETONYLBENZYL)-4-HYDROXY-COUMARIN SODIUM SALT see WAT220
ACETONYL BROMIDE see BNZ000
ACETONYL CHLORIDE see CDN200
3-(α-ACETONYLFURFURYL)-4-HYDROXYCOUMARIN see ABF500
3-(α-ACETONYL-p-NITROBENZYL)-4-HYDROXY-COUMARIN see ABF750
p-ACETONYLOXYBENZENEARSONIC ACID SODIUM SALT see SJK475
2-ACETONYLPIPERIDINE see PAO500
ACETO PAN see PFT250
ACETO PBN see PFT500
ACETOPHEN see ADA725
ACETO-p-PHENALIDE see ABG750
ACETOPHENAZINE see ABG000
ACETOPHENAZINE MALEATE see ABG000
p-ACETOPHENETIDE see ABG750

m-ACETOPHENETIDIDE see ABG250
p-ACETOPHENETIDIDE see ABG750
ACETO-p-PHENETIDIDE see ABG750
ACETOPHENETIDIN see ABG750
ACETOPHENETIDINE see ABG750
ACETO-4-PHENETIDINE see ABG750
ACETOPHENETIN see ABG750
ACETOPHENONE see ABH000
ACETOPHENONE THIOSEMICARBAZONE see ABH250
ACETOPHOS see DIW600
ACETOPROMAZINE see ABH500
ACETOPROPIONIC ACID see LFH000
ACETOPROPYL ALCOHOL see ABH750
3-ACETOPYRIDINE see ABI000
ACETOPYRROTHINE see ABI250
ACETOQUAT CPC see CCX000
ACETOQUAT CTAB see HCQ500
ACETOQUINONE BLUE L see TBG700
ACETOQUINONE BLUE R see TBG700
ACETOQUINONE LIGHT HELIOTROPE NL see DBP000
ACETOQUINONE LIGHT ORANGE JL see AKP750
ACETOQUINONE LIGHT PURE BLUE R see MGG250
ACETOSAL see ADA725
ACETOSALIC ACID see ADA725
ACETOSALIN see ADA725
ACETO SDD 40 see SGM500
ACETOSPAN see AQX500
ACETOSULFAMIN see SNP500
ACETO TETD see TFS350
2-ACETOTHIENONE see ABI500
ACETOTHIOAMIDE see TFA000
2-ACETOTHIOPHENE see ABI500
ACETO TMTM see BJL600
ACETOTOLUIDE see ABI750
4-ACETOTOLUIDE see ABJ250
o-ACETOTOLUIDE see ABJ000
p-ACETOTOLUIDE see ABJ250
m-ACETOTOLUIDIDE see ABI750
o-ACETOTOLUIDIDE see ABJ000
p-ACETOTOLUIDIDE see ABJ250
ACETOVANILLONE see HLQ500
ACETOVANILONE see HLQ500
ACETOVANYLLON see HLQ500
ACETOXIME see ABF000
ACETOXON see DIW600
N-ACETOXY-4-ACETAMIDOBIPHENYL see ABJ750
N-ACETOXY-2-ACETAMIDOFLUORENE see ABL000
ACETOXY(2-ACETAMIDO-5-NITROPHENYL)MERCURY see ABK000
N-ACETOXY-2-ACETAMIDOPHENANTHRENE see ABK250
N-ACETOXY-4-ACETAMIDOSTILBENE see ABW500, SMT000
2-ACETOXY-4'-ACETAMINO)PHENYLBENZOATE see SAN600
N-ACETOXY-2-ACETYLAMINOFLUORENE see ABL000
N-ACETOXY-N-ACETYL-2-AMINOFLUORENE see ABL000
N-ACETOXY-2-ACETYLAMINOPHENANTHRENE see ABK250
trans-N-ACETOXY-4-ACETYL-AMINOSTILBENE see ABL250
2-ACETOXYACRYLONITRILE see ABL500
α-ACETOXYACRYLONITRILE see ABL500
19-ACETOXY-Δ$^{(1,4)}$-ANDROSTADIENE-3,17-DIONE see ABL625
2-ACETOXYBENZOIC ACID see ADA725
o-ACETOXYBENZOIC ACID see ADA725
6-ACETOXY-BENZO(a)PYRENE see ABL750
N-α-ACETOXYBENZYL-N-BENZYLNITROSAMINE see ABL875
N-ACETOXY-4-BIPHENYLACETAMIDE see ABJ750
3-β-ACETOXY-BIS NOR-Δ^5-CHOLENIC ACID see ABM000
1-ACETOXY-1,3-BUTADIENE see ABM250
N-(4-ACETOXYBUTYL)-N-(ACETOXYMETHYL)NITROSAMINE see ABM500
N-(α-ACETOXY)BUTYL-N-BUTYLNITROSAMINE see ABT250, BPV325
17-ACETOXY-6-CHLORO-6-DEHYDROPROGESTERONE see CBF250
17-α-ACETOXY-6-CHLORO-6-DEHYDROPROGESTERONE see CBF250
17-α-ACETOXY-6-CHLORO-6,7-DEHYDROPROGESTERONE see CBF250
ACETOXY-4'-CHLORO-3,5-DIIODOBENZANILIDE see CGB250
2-ACETOXY-2'-CHLORO-N-METHYL-DIETHYLAMINE see MFW750
17-α-ACETOXY-6-CHLORO-1-α,2-α-METHYLENEPREGNA-4,6-DIENE-3,20-DIONE see CQJ500
17-α-ACETOXY-6-CHLORO-4,6-PREGNADIENE-3,20-DIONE see CBF250
17-α-ACETOXY-6-CHLOROPREGNA-4,6-DIENE-3,20-DIONE see CBF250
ACETOXYCYCLOHEXIMIDE see ABN000
1'-ACETOXY-2',3'-DEHYDROESTRAGOLE see EQN225
17-α-ACETOXY-6-DEHYDRO-6-METHYLPROGESTERONE see VTF000
2-ACETOXY-3-DIETHYLCARBAMYL-9,10-DIMETHOXY-1,2,3,4,6,7-HEXAHYDRO-11B-BENZO(a)QUINOLIZINE see BCL250
ACETOXYDIETHYLPHENYLSTANNANE see DJV800
11-ACETOXY-15-DIHYDROCYCLOPENTA(a)PHENANTHRACEN-17-ONE see ABN250

1-ACETOXY-1,4-DIHYDRO-4-(HYDROXYAMINO)QUINOLINE ACETATE (ESTER) see ABN500
21-ACETOXY-7,18-DIHYDROXY-16,18-DIMETHYL-10-PHENYL-(11)CYTOCHALASA-6(12),13,19-TRIENE-1-ONE see PAM775
21-ACETOXY-11-β,17-α-DIHYDROXYPREGN-4-ENE-3,20-DIONE see HHQ800
(11-β)-21-(ACETOXY)-11,17-DIHYDROXY-PREGN-4-ENE-3,20-DIONE see HHQ800
21-ACETOXY-3,20-DIKETOPREGN-4-ENE see DAQ800
1-ACETOXYDIMERCURIO-1-PERCHLORATODIMERCURIOPROPEN-2-ONE see ABN625
6-ACETOXY-2,4-DIMETHYL-m-DIOXANE see ABC250
α-ACETOXY DIMETHYLNITROSAMINE see AAW000
3-(2-(5-ACETOXY-3,5-DIMETHYL-2-OXOCYCLOHEXYL)-2-HYDROXYETHYL)GLUTARIMIDE see ABN000
β-ACETOXY-N,N-DIMETHYLPHENETHYLAMINE see ABN700
1′-ACETOXYELEMICIN see TLC850
1′-ACETOXYESTRAGOLE see ABN725
ACETOXYETHANE see EFR000
7-α-(ACETOXYETHANE)OXYFLAVONE see ELH600
N-(2-ACETOXYETHYL)-N-(ACETOXYMETHYL)NITROSAMINE see ABN750
N-ACETOXYETHYL-N-CHLOROETHYLMETHYLAMINE see MFW750
1-(2-ACETOXYETHYL)-4-(3-(2-CHLORO-10-PHENOTHIAZINYL)PROPYL)PIPERAZIME DIHYDROCHLORIDE see TFP250
1-ACETOXYETHYLENE see VLU250
17-β-ACETOXY-16-β-ETHYLESTR-4-EN-3-ONE see ELF110
N-(α-ACETOXY)ETHYL-N-ETHYLNITROSAMINE see ABT500
1-ACETOXYETHYL 2-(2-FLUORO-4-BIPHENYLYL)PROPIONATE see FJT100
N-(α-ACETOXY)ETHYL-N-METHYLNITROSAMINE see ABR500
N-(2-ACETOXYETHYL)-N-NITROSOCARBAMIC ACID ETHYL ESTER see EFR500
2-ACETOXYETHYLTRIMETHYLAMMONIUM CHLORIDE see ABO000
N-ACETOXYFLUORENYLACETAMIDE see ABO250
N-ACETOXY-2-FLUORENYLACETAMIDE see ABL000
N-ACETOXY-3-FLUORENYLACETAMIDE see ABO250
N-ACETOXY-4-FLUORENYLACETAMIDE see ABO500
N-ACETOXY-2-FLUORENYLBENZAMIDE see ABO750
21-ACETOXY-17,α-HYDROXYPREGN-4-ENE-3,11,20-TRIONE see CNS825
21-ACETOXY-17,α-HYDROXY-3,11,20-TRIKETOPREGNENE-4 see CNS825
(2-(4-ACETOXY-2-ISOPROPYL-5-METHYLPHENOXY)ETHYL)DIMETHYLAMINE HYDROCHLORIDE see TFY000
2-ACETOXYISOSUCCINODINITRILE see MFX000
ACETOXYL see BDS000
2′,4′-ACETOXYLIDIDE see ABP750
2′,6′-ACETOXYLIDIDE see ABP250
3′,4′-ACETOXYLIDIDE see ABP500
p-(ACETOXYMERCURI)ANILINE see ABQ000
(ACETOXYMERCURI)BENZENE see ABU500
2-(ACETOXYMERCURI)-5-(HYDROXYMERCURI)THIOPHENE see HLQ000
2′-(ACETOXYMERCURI)-4′-NITROACETANILIDE see ABK000
2-(ACETOXYMERCURI)-4-NITROANILINE see ABQ250
6-ACETOXYMERCURI-5-NITROGUAIACOL and 4,6-DIACETOXYMERCURI-5-NITROGUAIACOL see GKO500
1-ACETOXYMERCURIO-1-PERCHLORATOMERCURIOPROPEN-2-ONE see ABQ375
8-(ACETOXYMERCURIO)THEOPHYLLINE see TEP750
1-ACETOXY-2-METHOXY-4-ALLYLBENZENE see EQS000
4-ACETOXY-3-METHOXY-1-PROPENYLBENZENE see AAX750
N-ACETOXY-N-METHYL-4-AMINOAZOBENZENE see ABQ500
p-((p-(ACETOXYMETHYLAMINO)PHENYL)AZO)-BENZOIC ACID METHYL ESTER see MEG000
4-ACETOXY-7-METHYLBENZ(c)ACRIDINE see ABQ600
10-ACETOXYMETHYL-1,2-BENZANTHRACENE see BBH500
6-ACETOXY METHYL BENZO(a)PYRENE see ACU500
ACETOXYMETHYLBUTYLNITROSAMINE see BRX500
N-(ACETOXY)METHYL-N,N-BUTYLNITROSAMINE see BRX500
ACETOXYMETHYLETHYLNITROSAMINE see ENR500
N-(ACETOXY)METHYL-N-ETHYLNITROSAMINE see ENR500
N-(ACETOXYMETHYL)-N-ISOBUTYLNITROSAMINE see ABR125
7-ACETOXYMETHYL-12-METHYLBENZ(a)ANTHRACENE see ABR250
ACETOXYMETHYL-METHYL-NITROSAMIN (GERMAN) see AAW000
ACETOXYMETHYL METHYLNITROSAMINE see AAW000
N-α-ACETOXYMETHYL-N-METHYLNITROSAMINE see AAW000
N-ACETOXYMETHYL-N-NITROSOETHYLAMINE see ENR500
1-ACETOXY-N-METHYL-N-NITROSOETHYLAMINE see ABR500
N-(1-ACETOXYMETHYL)-N-NITROSOETHYL AMINE see ENR500
ACETOXYMETHYLPHENYLNITROSAMINE see ABR625
17-ACETOXY-6-METHYLPREGNA-4,6-DIENE-3,20-DIONE see VTF000
17-α-ACETOXY-6-METHYLPREGNA-4,6-DIENE-3,20-DIONE see VTF000
17-α-ACETOXY-6-METHYL-4,6-PREGNADIENE-3,20-DIONE see VTF000
17-α-ACETOXY-6-α-METHYLPREGN-4-ENE-3,20-DIONE see MCA000
17-ACETOXY-6-α-METHYLPROGESTERONE see MCA000
ACETOXYMETHYLPROPYLNITROSAMINE see PNR250
N-(ACETOXY)METHYL-N-n-PROPYLNITROSAMINE see PNR250
ACETOXYMETHYLTRIDEUTEROMETHYLNITROSAMINE see MMS000

N-ACETOXY-2-MYRISTOYL-AMINOFLUORENE see FEM000
N-ACETOXY-N-(1-NAPTHYL)-ACETAMIDE see ABS250
p-ACETOXYNITROBENZENE see ABS750
α-ACETOXY-N-NITROSODIBENZYLAMINE see ABL875
1-ACETOXY-N-NITROSODIBUTYLAMINE see ABT250, BPV325
1-ACETOXY-N-NITROSODIETHYLAMINE see ABT500
1-ACETOXY-N-NITROSODIMETHYLAMINE see AAW000
1-ACETOXY-N-NITROSODIPROPYLAMINE see ABT750
1-ACETOXY-N-NITROSO-N-TRIDEUTEROMETHYLMETHYLAMINE see MMS000
17-ACETOXY-19-NOR-17-α-PREGN-4-EN-20-YN-3-ONE see ABU000
17-β-ACETOXY-19-NOR-17-α-PREGN-4-EN-20-YN-3-ONE see ABU000
2-ACETOXYPENTANE see AOD735
N-ACETOXY-4-PHENANTHRYLACETAMIDE see ABK250
3-ACETOXYPHENOL see RDZ900
4-(p-ACETOXYPHENYL)-2-BUTANONE see AAR500
ACETOXYPHENYLMERCURY see ABU500
2-ACETOXY-1-PHENYLPROPANE see ABU800
3-ACETOXYPHENYLTRIMETHYLAMMONIUM IODIDE see ABV250
3-ACETOXYPHENYL TRIMETHYLAMMONIUM METHYLSULFATE see ABV500
2-ACETOXY-N-(3-(m-(1-PIPERIDINYLMETHYL)PHENOXY)PROPYL)ACETAMIDE HYDROCHLORIDE see HNT100
3-β-ACETOXYPREGN-6-ENE-20-CARBOXYLIC ACID see ABM000
17-α-ACETOXYPREGN-4-ENE-3-β-OL-20-ONE and MESTRANOL (20:1) see ABV600
17-β-ACETOXY-5-α-17-α-PREGN-2-ENE-20-YNE see PMA450
ACETOXYPROGESTERONE see PMG600
17-ACETOXYPROGESTERONE see PMG600
17-α-ACETOXYPROGESTERONE see PMG600
1-ACETOXYPROPANE see PNC250
2-ACETOXYPROPANE see INE100
3-ACETOXYPROPENE see AFU750
N-(3-ACETOXYPROPYL)-N-(ACETOXYMETHYL)NITROSAMINE see ABV750
N-(α-ACETOXY)PROPYL-N-n-PROPYLNITROSAMINE see ABT750
3-ACETOXYQUINUCLIDINE GLAUCOSTAT see AAE250
1′-ACETOXYSAFROLE see ACV000
1′-ACETOXYSAFROLE-2′,3′-OXIDE see ABW250
N-ACETOXY-N-(4-STILBENYL) ACETAMIDE see ABW500
4-ACETOXYTOLUENE see MNR250
p-ACETOXYTOLUENE see MNR250
α-ACETOXYTOLUENE see BDX000
ACETOXYTRIBUTYLPLUMBANE see THY850
ACETOXYTRIBUTYLSTANNANE see TIC000
ACETOXYTRICYCLOHEXYLSTANNANE see ABW600
ACETOXYTRIETHYLPLUMBANE see TJU150
ACETOXYTRIETHYLSTANNANE see ABW750
ACETOXYTRIETHYLTIN see ABW750
ACETOXYTRIHEXYLSTANNANE see ABX000
ACETOXYTRIHEXYLTIN see ABX000
ACETOXYTRIISOPROPYLSTANNANE see TKT750
ACETOXYTRIMETHYLPLUMBANE see ABX125
ACETOXYTRIMETHYLSTANNANE see TMI000
ACETOXYTRIOCTYLSTANNANE see ABX150
3-ACETOXY-1,2,4-TRIOXOLANE see VLU310
ACETOXYTRIPENTYLSTANNANE see ABX175
ACETOXYTRIPHENYLLEAD see TMT000
ACETOXY-TRIPHENYL-STANNAN (GERMAN) see ABX250
ACETOXYTRIPHENYLSTANNANE see ABX250
ACETOXY-TRIPHENYLSTANNANE see ABX250
ACETOXYTRIPHENYLTIN see ABX250
ACETOXYTRIPROPYLPLUMBANE see ABX325
ACETOXYTRIPROPYLSTANNANE see TNB000
ACETOZALAMIDE see AAI250
ACETO ZDBD see BIX000
ACETO ZDED see BJK500
ACETO ZDMD see BJK500
ACET-p-PHENALIDE see ABG750
ACETPHENARSINE see ABX500
ACETPHENETIDIN see ABG750
ACET-p-PHENETIDIN see ABG750
p-ACETPHENETIDIN see ABG750
ACETRIZOATE SODIUM see AAN000
ACETRIZOIC ACID see AAM875
ACETRIZOIC ACID SODIUM SALT see AAN000
ACET-THEOCIN see TEP000
ACETYLACETANILIDE see AAY000
α-ACETYLACETANILIDE see AAY000
ACETYLACETONATE-1,5-CYCLOOCTADIENE RHODIUM see CPR840
ACETYL ACETONE see ABX750
N-(ACETYLACETYL)ANILINE see AAY000
3-ACETYLACONITINE HYDROBROMIDE see ABX800
ACETYL ADALIN see ACE000
ACETYLADRIAMYCIN see DAC000

3-(ACETYLAMINO)-5-((ACETYLAMINO)METHYL)-2,4,6-TRIIODOBENZOIC ACID see AAI750
3-(ACETYLAMINO)-5-(ACETYLMETHYLAMINO)-2,4,6-TRIIODO-BENZOIC ACID see MQR350
2-((3-(ACETYLAMINO)-5-(ACETYLMETHYLAMINO)-2,4,6-TRIIODOBENZOYL)AMINO)-2-DEOXY-d-GLUCOSE see MQR300
3-ACETYLAMINOANILINE see AHQ000
4-(ACETYLAMINO)ANILINE see AHQ250
m-(ACETYLAMINO)ANILINE see AHQ000
p-(ACETYLAMINO)ANILINE see AHQ250
4-ACETYLAMINOAZOBENZENE see PEH750
p-ACETYLAMINOBENZALDEHYDE THIOSEMICARBAZONE see FNF000
ACETYLAMINOBENZENE see AAQ500
N-ACETYL-p-AMINOBENZENEARSONIC ACID, SODIUM SALT, TETRAHYDRATE see ARA000
N-ACETYL-4-AMINOBENZENESULFONAMIDE see SNP500
4-(ACETYLAMINO)BENZOIC ACID with 2-(DIMETHYLAMINO)ETHANOL (1:1) see DOZ000
2-ACETYLAMINOBIPHENYL see PDY000
3-ACETYLAMINOBIPHENYL see PDY250
4-ACETYLAMINOBIPHENYL see PDY500
γ-ACETYLAMINOBUTYRIC ACID see AAJ350
N-((ACETYLAMINO)CARBONYL)-2-BROMO-2-ETHYLBUTANAMIDE see ACE000
N-((ACETYLAMINO)CARBONYL)-α-ETHYLBENZENEACETAMIDE see ACX500
3-ACETYLAMINODIBENZOFURAN see DDC000
2-ACETYLAMINODIBENZOTHIOPHENE see DDD400
3-ACETYLAMINODIBENZOTHIOPHENE see DDD600
3-ACETYLAMINODIBENZOTHIOPHENE-5-OXIDE see DDD800
2-ACETYLAMINO-9,10-DIHYDROPHENANTHRENE see DMA400
4-ACETYLAMINO-p-DIPHENYLBENZENE see TBD250
2-ACETYLAMINOETHANOL see HKM000
S-(2-(ACETYLAMINO)ETHYL)-O,O-DIMETHYL PHOSPHORODITHIOATE see DOP200
3-ACETYLAMINO-FLUORANTHEN see AAK400
3-ACETYLAMINOFLUORANTHENE see AAK400
2-ACETYLAMINO-FLUOREN (GERMAN) see FDR000
4-ACETYLAMINOFLUOREN (GERMAN) see ABY000
4-ACETYLAMINOFLUORENE see ABY000
N-ACETYL-2-AMINOFLUORENE see FDR000
2-ACETYLAMINOFLUORENE (OSHA) see FDR000
2-ACETYLAMINO-9-FLUORENOL see ABY150
2-ACETYLAMINOFLUORENONE see ABY250
2-ACETYLAMINO-9-FLUORENONE see ABY250
4-ACETYLAMINO-2-HYDROXYMERCURIBENZOIC ACID, SODIUM SALT see SJP500
3-ACETYLAMINO-4-HYDROXYPHENYLARSONIC ACID see ABX500
(3-(ACETYLAMINO)-4-HYDROXYPHENYL)ARSONINE (9CI) see ABX500
3-ACETYLAMINO-5-N-METHYL-ACETYLAMINO-2,4,6-TRIIODOBENZOYL GLUCOSAMINE see MQR300
6-(ACETYLAMINO)-4-METHYL-1,2-DITHIOLO(4,3-B)PYRROL-5(4H)-ONE see ABI250
5-ACETYLAMINO-N-METHYL-2,4,6-TRIIODOISOPHTHALAMIC ACID see IGD000
2-ACETYLAMINO-4-(5-NITRO-2- FURYL)THIAZOLE see AAL750
2-ACETYLAMINO-5-NITROTHIAZOLE see ABY900
5-(ACETYLAMINO)-4-OXO-5-HEXENAMIDE see PMC750
p-(ACETYLAMINO)PHENACYL CHLORIDE see CEC000
2-ACETYLAMINOPHENANTHRENE see AAM250
3-ACETYLAMINOPHENANTHRENE see PCY500
9-ACETYLAMINOPHENANTHRENE see PCY750
N-(o-(ACETYLAMINO)PHENETHYL)-1-HYDROXY-2-NAPHTHALENECARBOXAMIDE see HIJ500
2-(ACETYLAMINO)PHENOL see HIL000
3-(ACETYLAMINO)PHENOL see HIL500
m-(ACETYLAMINO)PHENOL see HIL500
o-(ACETYLAMINO)PHENOL see HIL000
p-ACETYLAMINOPHENOL see HIM000
N-ACETYL-2-AMINOPHENOL see HIL000
N-ACETYL-p-AMINOPHENOL see HIM000
p-N-ACETYLAMINOPHENYLACETYLSALICYLATE see SAN600
p-ACETYLAMINOPHENYL DERIVATIVE of NITROGEN MUSTARD see BHO500
ACETYL-p-AMINOPHENYLSTIBINSAURES NATRIUM (GERMAN) see SLP500
N(1)-ACETYL-4-AMINOPHENYLSULFONAMIDE see SNP500
2-ACETYLAMINOPYRENE see PON500
4-ACETYLAMINOPYRIDINE see AAM750
4-ACETYLAMINOSTILBENE see SMR500
trans-4-ACETYLAMINOSTILBENE see SMR500
4-ACETYLAMINO-p-TERPHENYL see TBD250
2-ACETYLAMINO-1,3,4-THIADIAZOLE-5-SULFONAMIDE see AAI250
4-(ACETYLAMINO)TOLUENE see ABJ250
3-(ACETYLAMINO)-2,4,6-TRIIODOBENZOIC ACID see AAM875

5-ACETYLAMINO-2,4,6-TRIIODO ISOPHTHALIC ACID DI-(N-METHYL-2,3-DIHYDROXYPROPYLAMIDE) see ACA125
3-(2-(3-ACETYLAMINO-2,4,6-TRIIODOPHENOXY)ETHOXY)-2-ETHYLPROPIONIC ACID see IGA000
2-((2-(3-(ACETYLAMINO)-2,4,6-TRIIODOPHENOXY)ETHOXY)METHYL)BUTANOIC ACID see IGA000
ACETYL ANHYDRIDE see AAX500
ACETYLANILINE see AAQ500
3-ACETYLANILINE see AHR500
4-ACETYLANILINE see AHR750
m-ACETYLANILINE see AHR500
N-ACETYLANILINE see AAQ500
ACETYL-p-ANISIDINE see AAR250
4-ACETYLANISOLE see MDW750
p-ACETYLANISOLE see MDW750
9-ACETYL-1,7,8-ANTHRACENETRIOL see ACA750
10-ACETYL-1,8,9-ANTHRACENETRIOL see ACA750
10-ACETYLANTHRALIN see ACA750
ACETYL-l-ARGININE, NITROSATED see AAM000
ACETYLARSAN see ACN250
N-ACETYLARSANILIC ACID, SODIUM SALT, TETRAHYDRATE see ARA000
ACETYL AZIDE see ACB000
1-ACETYLAZIRIDINE see ACB250
4-(p-(p-ACETYLBENZAMIDO)ANILINO)-6-AMINO-1-METHYLQUINOLINIUM)-p-AMIDINOHYDRAZONE-p-TOLUENESULFONATE-MONO-p-TOLUENSULFONATE see ACB750
ACETYLBENZENE see ABH000
1-(p-ACETYLBENZENESULFONYL)-3-CYCLOHEXYLUREA see ABB000
N-ACETYLBENZIDINE see ACC000
5-ACETYL BENZO(C)PHENANTHRENE see ACC750
ACETYL BENZOYL ACONINE see ADH750
ACETYL BENZOYL PEROXIDE (solid) see ACC250
ACETYL BENZOYL PEROXIDE (solution) see ACC500
2-ACETYL-3584-BENZPHENANTHRENE see ACC750
N-ACETYL-4-BIPHENYLHYDROXYLAMINE see ACD000
1-ACETYL-3,3-BIS(4-(ACETYLOXY)PHENYL)-1,3-DIHYDRO-2H-INDOL-2-ONE see ACD500
N-(N-ACETYL-3-(p-BIS(2-CHLOROETHYL)AMINO)PHENYL)ALANYL-3-PHENYLALANINE ETHYL ESTER see ACD250
1-ACETYL-3,3-BIS(p-HYDROXYPHENYL)OXINDOLE DIACETATE see ACD500
ACETYL BROMIDE see ACD750
ACETYLBROMODIETHYLACETYLCARBAMIDE see ACE000
N-ACETYL-N-BROMODIETHYLACETYLCARBAMIDE see ACE000
N-ACETYL-N-BROMODIETHYLACETYLUREA see ACE000
N-ACETYL-N'-α-BROMO-α-ETHYLBUTYRYLCARBAMIDE see ACE000
1-ACETYL-3-(2-BROMO-2-ETHYLBUTYRYL)UREA see ACE000
1-ACETYL-3-(α-BROMO-α-ETHYLBUTYRYL)UREA see ACE000
o-ACETYL-N-(p-BUTOXYPHENYLACETYL)HYDROXYLAMINE see ACE250
3-(3-ACETYL-4-(3-tert-BUTYLAMINO-2-HYDROXYPROPOXY)PHENYL)-1,1-DIETHYLHARNSTOFF HCl (GERMAN) see SBN475
3-(3-ACETYL-4-(3-tert-BUTYLAMINO-2-HYDROXYPROPOXY)PHENYL)-1,1-DIETHYLUREA HYDROCHLORIDE see SBN475
o-ACETYL-2-sec-BUTYL-4,6-DINITROPHENOL see ACE500
3-ACETYL-5-sec-BUTYL-4-HYDROXY-3-PYROLIN-2-ONE,MONOSODIUM SALT see ADC250
3-ACETYL-5-sec-BUTYL-4-HYDROXY-3-PYRROLIN-2-ONE see VTA750
3-ACETYL-5-sec-BUTYL-4-HYDROXY-3-PYRROLIN-2-ONE SODIUM SALT see ADC250
1-(2-ACETYL-4-n-BUTYRAMIDOPHENOXY)-2-HYDROXY-3-ISOPROPYLAMINOPROPANE see AAE100
dl-1-(2-ACETYL-4-BUTYRAMIDOPHENOXY)-2-HYDROXY-3-ISOPROPYLAMINOPROPANE HYDROCHLORIDE see AAE125
ACETYLBUTYRYL see HEQ200
3'-o-ACETYLCALOTROPIN see ACF000
o-ACETYL-N-CARBOXY-N-PHENYL-HYDROXYLAMINE ISOPROPYL ESTER see ING400
ACETYLCARBROMAL see ACE000
ACETYL CEDRENE see ACF250
ACETYL CHLORIDE see ACF750
N-ACETYL-N-(2-CHLORO-6-METHYLPHENYL)-1-PYRROLIDINEACETAMIDE MONOHYDROCHLORIDE see CLB250
4-ACETYL-4-(3-CHLOROPHENYL)-1-(3-(4-METHYLPIPERAZINO)-PROPYL) PI- PERIDINE TRIHYDROCHLORIDE see ACG125
1-(3-(3-ACETYL-4-(p-(CHLOROPHENYL)PIPERIDINO)PROPYL)-4-METHYLPIPERAZINE TRIHYDROCHLORIDE see ACG125
ACETYLCHOLINE see CMF250
ACETYLCHOLINE CHLORIDE see ABO000
ACETYLCHOLINE HYDROCHLORIDE see ABO000
ACETYL CHOLINE ION see CMF250
ACETYLCHOLINIUM CHLORIDE see ABO000
ACETYLCOLAMINE see HKM000
N-ACETYL COLCHINOL see ACG250
4-ACETYL-N-((CYCLOHEXYLAMINO)CARBONYL)-BENZENESULFONAMIDE see ABB000

ACETYLCYSTEINE see ACH000
N-ACETYLCYSTEINE see ACH000
N-ACETYL-l-CYSTEINE see ACH000
N-ACETYL-N-CYSTEINE see ACH000
N-ACETYL-l-CYSTEINE (9CI) see ACH000
3-ACETYLDEOXYNIVALENOL see ACH075
ACETYL-1,1-DICHLOROETHYL PEROXIDE see ACH125
1-ACETYL-3-(2,2-DICHLOROETHYL)UREA see ACH250
cis-1-ACETYL-4-(4-((2-(2,4-DICHLOROPHENYL)-2-(1H-IMIDAZOL-1-
   YLMETHYL)-1,3-DIOXOLAN-4-YL)METHOXY)PHENYL)-PIPERAZINE
   see KFK100
1-ACETYL-9,10-DIDEHYDRO-N,N-DIETHYL-6-METHYLERGOLINE-8β-CAR-
   BOXAMIDE BITARTRATE see ACP750
1-ACETYL-9,10-DIDEHYDRO-N-ETHYL-6-METHYLERGOLINE-8β-CAR-
   BOXAMIDE see ACP750
3B-o-(4-o-ACETYL-2,6-DIDEOXY-3-C-METHYL-α-l-
   ARABINOHEXOPYRANOSYL)-7-METHYL-OLIVOMYCIN D see CMK650
16-ACETYLDIGITALINUM VERUM see ACH375
ACETYLDIGITOXIN-α see ACH500
α-ACETYLDIGITOXIN see ACH500
ACETYLDIGITOXIN-β see ACH750
ACETYLDIGOXIN-α see ACI000
α-ACETYLDIGOXIN see ACI000
β-ACETYLDIGOXIN see ACH750, ACI250
ACETYLDIGOXIN-β see ACI250
ε-ACETYLDIGOXIN (GERMAN) see DKN600
3-ACETYL-1,5-DIHYDRO-4-HYDROXY-5-(1-METHYLPROPYL)-2H-PYRROL-2-
   ONE see VTA750
3-ACETYL-1,5-DIHYDRO-4-HYDROXY-5-(1-METHYLPROPYL)-2H-PYRROL-2-
   ONE SODIUM SALT see ADC750
3-ACETYL-10-(3-DIMETHYLAMINOPROPYL)PHENOTHIAZINE see ABH500
2-ACETYL-10-(3-(DIMETHYLAMINO)PROPYL)PHENOTHIAZINE, MALEATE
   see AAF750
ACETYLDIMETHYLARSINE see ACI375
3-ACETYL-2,4-DIMETHYLPYRROLE see ACI500
N-ACETYLDIPHENYLAMINE see PDX500
N-ACETYL-N-(4,5-DIPHENYL-2-OXAZOLYL)ACETAMIDE see ACI629
ACETYLEN see ACI750
ACETYLENE see ACI750
ACETYLENE, dissolved (DOT) see ACI750
ACETYLENE BLACK see CBT750
(ACETYLENECARBONYLOXY)TRIPHENYLTIN see TMW600
ACETYLENECARBOXYLIC ACID METHYL ESTER see MOS875
ACETYLENE CHLORIDE see ACJ000
ACETYLENE COMPOUNDS see ACJ125
ACETYLENEDICARBOXAMIDE see ACJ250
ACETYLENEDICARBOXYLIC ACID DIAMIDE see ACJ250
ACETYLENEDICARBOXYLIC ACID DIMETHYL ESTER see DOP400
ACETYLENEDICARBOXYLIC ACID MONOPOTASSIUM SALT see ACJ500
ACETYLENE DICHLORIDE see DFI100
trans-ACETYLENE DICHLORIDE see ACK000
ACETYLENE TETRABROMIDE see ACK250
ACETYLENE TETRACHLORIDE see TBQ100
ACETYLENE TRICHLORIDE see TIO750
N-ACETYL ETHANOLAMINE see HKM000
ACETYL ETHER see AAX500
N-ACETYL ETHYL CARBAMATE see ACL000
N-ACETYLETHYL-2-cis-CROTONYLCARBAMIDE see ACL250
ACETYL ETHYLENE see BOY500
ACETYLETHYLENEIMINE see ACB250
N'-ACETYL ETHYLNITROSOUREA see ACL500
ACETYL ETHYL TETRAMETHYL TETRALIN see ACL750
ACETYLETHYL TETRAMETHYLTETRALIN see ACL750
ACETYLEUGENOL see EQS000
N-ACETYL-p-FENYLENDIAMIN (CZECH) see AHQ250
N-ACETYL-m-FENYLENEDIAMIN (CZECH) see AHQ000
ACETYLFERROCENE see ABA750
N-ACETYL-N-9H-FLUOREN-2-YL-ACETAMIDE see DBF200
ACETYL FLUORIDE see ACM000
21-ACETYL-6-α-FLUORO-16-α-METHYLPREDNISOLONE see PAL600
ACETYLFORMALDEHYDE see PQC000
ACETYLFORMYL see PQC000
ACETYLGITOXIN-α see ACM500
16-ACETYLGITOXIN see ACM250
N-ACETYL-l-GLUTAMINE ALUMINUM SALT see AGX125
N-ACETYLHOMOCYSTEINE THIOLACTONE see TCZ000
N-ACETYLHOMOCYSTEINTHIOLAKTON (GERMAN) see TCZ000
ACETYL HYDRAZIDE see ACM750
N-ACETYLHYDRAZINE see ACM750
ACETYL HYDROPEROXIDE see PCL500
ACETYLHYDROQUINONE see DMG600
2-ACETYLHYDROQUINONE see DMG600
ACETYLHYDROXAMIC ACID see ABB250
N-ACETYL-4-HYDROXY-m-ARSANILIC ACID see ABX500

N-ACETYL-4-HYDROXY-m-ARSANILIC ACID, CALCIUM SALT see CAL500
N-ACETYL-4-HYDROXYARSANILIC ACID compounded with DIETHYLAMINE
   (1581) see ACN250
N-ACETYL-4-HYDROXY-m-ARSANILIC ACID DIETHYLAMINE SALT
   see ACN250
N-ACETYL-4-HYDROXY-m-ARSANILIC ACID SODIUM SALT see SEG000
17-β-ACETYL-17-HYDROXYESTR-4-ENE-3-ONE HEXANOATE see GEK510
2-ACETYL-10-(3-(4-(β-HYDROXYETHYL)PIPERAZINYL)PROPYL)PHENOTHI-
   AZINE see ABG000
2-ACETYL-10-(3-(4-(β-HYDROXYETHYL)PIPERIDINO)PROPYL)PHENOTHI-
   AZINE see PII500
2-ACETYL-7-((2-HYDROXY-3-ISOPROPYLAMINO)PROPOXY)BENZOFURAN
   HYDROCHLORIDE see BAU255
3'-ACETYL-4'-(2-HYDROXY-3-(ISOPROPYLAMINO)PROPOXY)BUTYRANILIDE
   see AAE100
3'-ACETYL-4'-(2-HYDROXY-3-(ISOPROPYLAMINO)PROPOXY)BUTYRANILIDE
   HYDROCHLORIDE see AAE125
(±)-N-(3-ACETYL-4-(2-HYDROXY-3-((1-METHYLETHYL)AMINO)PROPOXY)
   PHENYL)BUTANAMIDE see AAE100
N-ACETYL-N'-(p-HYDROXYMETHYL)PHENYLHYDRAZINE see ACN500
2-ACETYL-5-HYDROXY-3-OXO-4-HEXENOIC ACID Δ-LACTONE see MFW500
ACETYL HYPOBROMITE see ACN875
ACETYLIDES see ACO000
1-ACETYLIMIDAZOLE see ACO250
N-ACETYLIMIDAZOLE see ACO250
ACETYLIN see ADA725
ACETYL IODIDE see ACO500
ACETYLISOEUGENOL see AAX750
ACETYL ISONIAZID see ACO750
N-ACETYLISONIAZID see ACO750
1-ACETYL-2-ISONICOTINOYLHYDRAZINE see ACO750
N-ACETYLISONICOTINYLHYDRAZIDE see ACO750
N-ACETYL-N'-ISONICOTINYL HYDRAZIDE see ADA000
ACETYLISOPENTANOYL see MKL300
ACETYL ISOVALERYL see MKL300
3-o-ACETYLJERVINE see JDA000
ACETYLKIDAMYCIN see ACP000
ACETYL KITASAMYCIN see MRE750
(2-ACETYLLACTOYLOXYETHYL)TRIMETHYLAMMONIUM HEMI-1,5-NAPH-
   THALENEDISULFONATE see AAC875
(2-ACETYLLACTOYLOXYETHYL)TRIMETHYLAMMONIUM 1,5-NAPH-
   THALENEDISULFONATE see AAC875
ACETYL-LANATOSID A (GERMAN) see ACP250
ACETYLLANATOSIDE A see ACP250
ACETYLLANDROMEDOL see AOO375
N-ACETYL-l-LEUCYL-N-(4-((AMINOIMINOMETHYL)AMINO)-1-
   FORMYLBUTYL)-l-LEUCINAMIDE (9CI) see LEX400
N-ACETYL-l-LEUCYL-l-LEUCYL-l-ARGINAL see LEX400
1-ACETYLLYSERGIC ACID DIETHYLAMIDE BITARTRATE see ACP500
1-ACETYLLYSERGIC ACID ETHYLAMIDE see ACP750
d-1-ACETYL LYSERGIC ACID MONOETHYLAMIDE see ACP750
ACETYLMANDELIC ACID-(2-(DIMETHYLAMINO)-1-PHENYL)ETHYL ESTER
   see ACQ000
β-ACETYLMANDELOYLOXY-β-PHENYLETHYL DIMETHYLAMINE see ACQ000
ACETYL MERCAPTAN see TFA500
ACETYLMERCAPTOACETIC ACID see ACQ250
N-ACETYL-3-MERCAPTOALANINE see ACH000
α-1-ACETYLMETHADOL see ACQ666
l-α-ACETYLMETHADOL HYDROCHLORIDE see ACQ690
N-ACETYL-5-METHOXYTRYPTAMINE see MCB350
4-(N-ACETYL-N-METHYL)AMINO-4'-(N',N'-DIMETHYLAMINO)AZOBENZENE
   see DPQ200
N'-ACETYL-N'-METHYL-4'-AMINO-N,N-DIMETHYL-4-AMINOAZOBENZENE
   see DPQ200
4-(N-ACETYL-N-METHYL)AMINO-4'-N-METHYLAMINOAZOBENZENE
   see MLK750
N-ACETYL-METHYLANILINE see MFW000
ACETYL METHYL BROMIDE see BNZ000
N-ACETYL-N-(METHYLCARBAMOYLOXY)-N-METHYLUREA see CBG075
ACETYL METHYL CARBINOL see ABB500
o-ACETYL-β-METHYLCHOLINE CHLORIDE see ACR000
α-ACETYL-6-METHYLERGOLINE-8-β-PROPIONAMIDE see ACR100
N-ACETYL-N-(2-METHYL-4-((2-METHYLPHENYL)AZO)PHENYL)ACETAMIDE
   see ACR300
ACETYL-METHYL-NITROSO-HARNSTOFF (GERMAN) see ACR400
ACETYLMETHYLNITROSOUREA see ACR400
N'-ACETYL-METHYLNITROSOUREA see ACR400
3-ACETYL-6-METHYL-2,4-PYRANDIONE see MFW500
3-ACETYL-6-METHYLPYRANDIONE-2,4 see MFW500
3-ACETYL-6-METHYL-2H-PYRAN-2,4(3H)-DIONE see MFW500
N'-ACETYL-N'-MONOMETHYL-4'-AMINO-N-ACETYL-N-MONOMETHYL-4-
   AMINOAZOBENZENE see DQH200
N'-ACETYL-N'-MONOMETHYL-4'-AMINO-N-MONOMETHYL-4-
   AMINOAZOBENZENE see MLK750

4-ACETYLMORPHOLINE see ACR750
N-ACETYLMORPHOLINE see ACR750
N-ACETYL-N-MYRISTOYLOXY-2-AMINOFLUORENE see ACS000
o-ACETYL-N-(2-NAPHPTHOYL)HYDROXYLAMINE see ACS250
1-ACETYLNAPHTHALENE see ABC475
2-ACETYLNAPHTHALENE see ABC500
β-ACETYLNAPHTHALENE see ABC500
N-ACETYL-1-NAPHTHYLAMINE see NAK000
N-ACETYL-2-NAPHTHYLHYDROXYLAMINE see NBD000
N-ACETYLNEOMYCIN see ACS375
8-ACETYL NEOSOLANIOL see ACS500
ACETYL NITRATE see ACS750
ACETYL NITRITE see ACT000
2-ACETYL-5-NITROFURAN see ACT250
O-ACETYL NITROHYDROXYAMINE see NEC500
O-ACETYL-N-NITROHYDROXYLAMINE see NEC500
ACETYLNITROPEROXIDE see PCL750
1-ACETYL-4-NITROSO-2,6-DIMETHYLPIPERAZINE see NJI850
ACETYLON FAST BLUE G see TBG700
ACETYLON FAST RED VIOLET R see DBP000
ACETYL OXIDE see AAX500
7-ACETYL-5-OXO-5H-(1)BENZOPYRANO(2,3-b)PYRIDINE see ACU125
2-(ACETYLOXY)BENZOIC ACID see ADA725
2-(ACETYLOXY)BENZOIC ACID 4-(ACETYLAMINO)PHENYL ESTER
  see SAN600
2-(ACETYLOXY)BENZOIC ACID, mixed with 3,7-DIHYDRO-1,3,7-TRIMETHYL-
  1H-PURINE-2,6-DIONE and N-(4-ETHOXYPHENYL)ACETAMIDE
  see ARP250
2-(ACETYLOXY)-3-BROMO-N-(4-BROMOPHENYL)-5-CHLORO-
  BENZENECARBOTHIOAMIDE see BOL325
2-(ACETYLOXY)-N-(4-CHLOROPHENYL)-3,5-DIIODOBENZAMIDE see CGB250
17-(ACETYLOXY)-6-CHLOROPREGNA-4,6-DIENE-3,20-DIONE see CBF250
ACETYLOXYCYCLOHEXIMIDE see ABN000
(11-β,16-α)-21-(ACETYLOXY)-16,17-(CYCLOPENTYLIDENEBIS(OXY))-9-
  FLUORO-11-HYDROXYPREGNA-1,4-DIENE-3,20-DIONE see COW825
(11-β)-21-(ACETYLOXY)-11,17-DIHYDROXY-PREGN-4-ENE-3,20-DIONE (9CI)
  see HHQ800
4-ACETYLOXY-12,13-EPOXY-3,7,15-TRIHYDROXY-(3α,4-β,7-β)-TRICHOTHEC-
  9-EN-8-ONE see FQR000
(16-β,17-β)-17-(ACETYLOXY)-16-ETHYL-ESTR-4-EN-3-ONE (9CI) see ELF110
(2-(ACETYLOXY)ETHYL)NITROSOCARBAMIC ACID ETHYL ESTER
  see EFR500
N-(ACETYLOXY)-N-9H-FLUOREN-2-YL-TETRADECANAMIDE see FEM000
6-β-(ACETYLOXY)-3-β-(β-d-GLUCOPYRANOSYLOXY)-8,14-DIHYDROXYBUFA-
  4,20,22-TRIENOLIDE see SBF500
3-β,6-β-6-ACETYLOXY-3-(β-d-GLUCOPYRANOSYLOXY)-8,14-
  DIHYDROXYBUFA-4,20,22-TRIENOLIDE see SBF500
21-(ACETYLOXY)-11-β-HYDROXY-17-((1-OXOPENTYL)OXY)PREGNA-1,4-
  DIENE-3,20-DIONE see AAF625
(11-β)-21-(ACETYLOXY)-11-HYDROXY-PREGN-4-ENE-3,20-DIONE (9CI)
  see CNS650
21-(ACETYLOXY)-17-HYDROXY-PREGN-4-ENE-3,11,20-TRIONE (9CI)
  see CNS825
4-(4-(ACETYLOXY)-3-IODOPHENOXY)-3,5-DIIODO-BENZOIC ACID see
  ADD875
4-((4-((ACETYLOXY)METHYLAMINO)PHENYL)AZO)BENZOIC ACID
  METHYL ESTER (9CI) see MEG000
(6R-(6-α,7-β(Z)))-3-((ACETYLOXY)METHYL)-7-(((2-AMINO-4-
  THIAZOLYL)(METHOXYIMIO)ACETYL)AMINO)-8-OXO-5-THIA-1-
  AZABICYCLO(4,2,0)OCT-2-ENE-2-CARBOXYLIC ACID, SODIUM SALT
  see CCR950
6-ACETYLOXYMETHYLBENZO(a)PYRENE see ACU500
17-(ACETYLOXY)-6-METHYL-16-METHYLENEPREGNA-4,6-DIENE-3,20-DIONE
  (9CI) see MCB380
N-(ACETYLOXY)-N-METHYL-4-(PHENYLAZO)BENZENAMINE (9CI)
  see ABQ500
(6-α)-17-(ACETYLOXY)-6-METHYLPREG-4-ENE-3,20-DIONE see MCA000
17-(ACETYLOXY)-6-METHYLPREGN-4-ENE-3,20-DIONE, (6-α)- mixed with (17-
  α)-19-NORPREGNA-1,3,5(10)-TRIEN-20-YNE-3,17-DIOL see POF275
(17-α)-17-(ACETYLOXY)-19-NORPREGN-4-EN-20-YN-3-ONE see ABU000
17-(ACETYLOXY(17-α)-19-NORPREGN-4-ESTREN-17-β-OL-ACETATE-3-ONE
  see ABU000
2-(2-(ACETYLOXY)-1-OXOPROPOXY)-N,N,N-TRIMETHYLETHANAMINIUM
  1,5-NAPHTHALENEDISULFONATE (2:1) see AAC875
(ACETYLOXY)PHENYL-CARBAMIC ACID see ING400
4-((4-((ACETYLOXY)PHENYL)CYCLOHEXYLIDENEMETHYL)PHENOL ACE-
  TATE see FBP100
β-ACETYLOXY-β-PHENYLETHYL DIMETHYLAMINE see ABN700
4-((4-(ACETYLOXY)PHENYL)(2-METHYLCYCLOHEXYLIDENE)METHYL)
  PHENOL ACETATE see BGQ325
5-(1-ACETYLOXY-2-PROPENYL)-1,3-BENZODIOXOLE see ACV000
2-(ACETYLOXY)-N,N,N-TRIMETHYLETHANAMINIUM see CMF250
2-(ACETYLOXY)-N,N,N-TRIMETHYLETHANAMINIUM CHLORIDE
  see ABO000

(ACETYLOXY)TRIOCTYLSTANNANE see ABX150
(ACETYLOXY)TRIPHENYLPLUMBANE see TMT000
(ACETYLOXY)TRIPHENYL-STANNANE (9CI) see ABX250
ACETYL PEROXIDE see ACV500
ACETYL PEROXIDE (solution) see ACV750
ACETYL PEROXIDE SOLUTION, not over 25% peroxide (DOT) see ACV750
ACETYLPHENETIDIN see ABG750
N-ACETYL-p-PHENETIDINE see ABG750
ACETYLPHENETURIDE see ACX500
ACETYL PHENOL see PDY750
2-ACETYLPHENOL see HIN500
4-ACETYLPHENOL see HIO000
o-ACETYLPHENOL see HIN500
p-ACETYLPHENOL see HIO000
4-(N-ACETYL)PHENYLAMINO-1-ETHYL-2,2,6,6-TETRAMETHYLPIPERIDINE see
  EPL000
ACETYL-p-PHENYLENEDIAMINE see AHQ250
N-ACETYL-m-PHENYLENEDIAMINE see AHQ000
1-ACETYL-3-PHENYLETHYLACETYLUREA see ACX500
ACETYLPHENYLHYDRAZINE see ACX750
β-ACETYLPHENYLHYDRAZINE see ACX750
1-ACETYL-2-PHENYLHYDRAZINE see ACX750
1-(p-ACETYLPHENYL)-3-METHYL-3-NITROSOUREA see MFX725
cis-2-ACETYL-3-PHENYL-5-TOSYL-3a,4,5-TETRAHYDROPYRAZOLO(4,3-c)
  QUINOLINE see ACY700
1-ACETYL-2-PHENYL-1,5,5-TRIMETHYL-SEMIOXAMAZIDE see DVU100
12-O-ACETYL-PHORBOL-13-DECA-(Δ-2)-ENOATE see ACY750
12-O-ACETYL-PHORBOL-13-DECANOATE see ACZ000
ACETYLPHOSPHORAMIDOTHIOIC ACID-O,S-DIMETHYL ESTER see DOP600
ACETYL PHTHALYL CELLULOSE see CCU050
1-ACETYL-2-PICOLINOLHYDRAZINE see ADA000
1-ACETYL-2-PICOLINOYLHYDRAZINE see ADA000
N-ACETYLPIPERIDIN (GERMAN) see ADA250
1-ACETYLPIPERIDINE see ADA250
ACETYLPROMAZINE see ABH500
ACETYLPROMAZINE MALEATE (1:1) see AAF750
β-ACETYLPROPIONIC ACID see LFH000
ACETYLPROPIONYL see PBL350
2-ACETYL PYRAZINE see ADA350
3-ACETYLPYRIDINE see ABI000
β-ACETYLPYRIDINE see ABI000
ACETYLRESORCINOL see RDZ900
4-ACETYLRESORCINOL see DMG400
ACETYLSAL see ADA725
ACETYLSALICYLIC ACID see ADA725
ACETYLSALICYLIC ACID SODIUM SALT see ADA750
o-ACETYLSALICYLIC ACID, SODIUM SALT see ADA750
ACETYLSALICYLSAEURE NATRIUMSALZ (GERMAN) see ADA750
ACETYLSALICYLSAURE (GERMAN) see ADA725
N-ACETYL-SARCOLYSIL VALINE ETHYL ETHER see ARM000
2-(ACETYLSELENO)BENZOIC ACID see SBU100
ACETYLSELENO-2 BENZOIC ACID see SBU100
5-ACETYLSPIRO(BENZOFURAN-2(3H),1'-CYCLOPROPAN)-3-ONE see SLE880
o-ACETYLSTERIGMATOCYSTIN see ADB250
N-ACETYLSULFANILAMIDE see SNP500
N'-ACETYLSULFANILAMIDE see SNP500
N¹-ACETYLSULFANILAMIDE see SNP500
N(¹)-ACETYLSULFANILAMIDE SODIUM SALT see SNQ000
N-ACETYLSULFANILAMINE see SNP500
N-ACETYL-N-TETRADECANOYLOXY-2-AMINOFLUORENE see ACS000
6-ACETYL-1,1,4,4-TETRAMETHYL-7-ETHYL-1,2,3,4,-TETRALIN see ACL750
7-ACETYL-1,1,4,4-TETRAMETHYL-1,2,3,4-TETRAHYDRONAPHTHALENE
  see ACL750
3-ACETYLTETRAMIC ACID SODIUM SALT see ADC250
S-ACETYLTHIOGLYCOLIC ACID see ACQ250
1-ACETYL-2-THIOHYDANTOIN see ADC750
5-(ACETYLTHIOMETHYL)-4-AMINOMETHYL-2-METHYL-3-PYRIDINOL
  HYDROBROMIDE see ADD000
7-α-ACETYLTHIO-3-OXO-17-α-PREGN-4-ENE-21,17-β-CARBOLACTONE
  see AFJ500
7-α-ACETYLTHIO-3-OXO-17-β-PREGN-4-ENE-21,17-β-CARBOLACTONE
  see AFJ500
2-ACETYLTHIOPHENE see ABI500
ACETYL THIOUREA see ADD250
1-ACETYL-2-THIOUREA see ADD250
p-ACETYLTOLUENE see MFW250
N-ACETYL-p-TOLUIDIDE see ABJ250
ACETYL-o-TOLUIDINE see ABJ000
ACETYL-p-TOLUIDINE see ABJ250
N-ACETYL-m-TOLUIDINE see ABI750
ACETYL TRIETHYL CITRATE see ADD750
ACETYLTRIIODOTHYRONINE FORMIC ACID see ADD875
N-ACETYL TRIMETHYLCOLCHICINIC ACID see ADE000
N-ACETYL TRIMETHYLCOLCHICINIC ACID METHYLETHER see CNG830

ACETYL-l-TRP see ADE075
ACETYLTRYPTOPHAN see ADE075
ACETYL-l-TRYPTOPHAN see ADE075
N-ACETYLTRYPTOPHAN see ADE075
N-ACETYL-l-TRYPTOPHAN see ADE075
(S)-N-ACETYLTRYPTOPHAN see ADE075
ACETYLURETHANE see ACL000
ACETYLUREUM see PEC250
ACETYLZIRCONIUM, ACETONATE see PBL750
ACH see CMF250
ACHANIA, flower extract see ADE125
ACH CHLORIDE see ABO000
ACHILLEIC ACID see ADH000
ACHIOTE see APE100
ACHLESS see TKH750
ACHLETIN see HII500
ACHROCIDIN see ABG750
ACHROMYCIN see TBX000, TBX250
ACHROMYCIN HYDROCHLORIDE see TBX250
ACHROMYCIN (PURINE DERIVATIVE) see AEI000
ACHTL see TCZ000
AC 2197 HYDROCHLORIDE see DBA475
ACIBILIN see TAB250
ACICLOVIR see AEC700
ACID see DJO000
ACIDAL BRIGHT PONCEAU 3R see FMU080
ACIDAL FAST ORANGE see HGC000
ACIDAL GREEN G see FAE950
ACID ALIZARINE VIOLET N see HLI000
ACIDAL LIGHT GREEN SF see FAF000
ACIDAL PONCEAU G see FMU070
ACIDAL WOOL GREEN BS see ADF000
ACID AMARANTH see FAG020
ACID AMIDE see NCR000
ACID AMMONIUM CARBONATE see ANB250
ACID BLUE 1 see ADE500
ACID BLUE 7 see ADE675
ACID BLUE 9 see FMU059
ACID BLUE 92 see ADE750
ACID BLUE A see ADE750
ACID BLUE W see FAE100
ACID BRILLIANT GREEN BS see ADF000
ACID BRILLIANT GREEN SF see FAF000
ACID BRILLIANT PINK B see FAG070
ACID BRILLIANT RUBINE 2G see HJF500
ACID BRILLIANT SCARLET 3R see FMU080
ACID BUTYL PHOSPHATE see ADF250
ACID CALCIUM PHOSPHATE see CAW110
ACID CARBOYS, EMPTY see ADF500
ACID CHROME BLUE BA see HJF500
ACID COPPER ARSENITE see CNN500
ACIDE ACETIQUE (FRENCH) see AAT250
ACIDE ACETYLSALICYLIQUE (FRENCH) see ADA725
ACIDE ACETYL SELENO-2 BENZOIQUE see SBU100
ACIDE ANISIQUE (FRENCH) see MPI000
ACIDE ARSENIEUX (FRENCH) see ARI750
ACIDE ARSENIQUE LIQUIDE (FRENCH) see ARB250
ACIDE BENZOIQUE (FRENCH) see BCL750
ACIDE BENZOYL-2-PHENYLACETIQUE (FRENCH) see BDS500
ACIDE BROMACETIQUE (FRENCH) see BMR750
ACIDE BROMHYDRIQUE (FRENCH) see HHJ000
l'ACIDE BUCLOXIQUE (FRENCH) see CPJ000
ACIDE BUCLOXIQUE CALCIUM (FRENCH) see CPJ250
ACIDE CACODYLIQUE (FRENCH) see HKC000
ACIDE CARBOLIQUE (FRENCH) see PDN750
ACIDE CHLORACETIQUE (FRENCH) see CEA000
ACIDE CHLORHYDRIQUE (FRENCH) see HHL000
ACIDE 2-(4-CHLORO-2-METHYL-PHENOXY)PROPIONIQUE (FRENCH)
  see CIR500
ACIDE p-CHLOROPHENYL-2-THIAZOLE-ACETIQUE-4 (FRENCH) see CKK250
ACIDE CHROMIQUE (FRENCH) see CMH250
ACIDE CYANACETIQUE (FRENCH) see COJ500
ACIDE CYANHYDRIQUE (FRENCH) see HHS000
l'ACIDE (CYCLOHEXYL-4, CHLORO-3, PHENYL)-4,OXO-4, BUTYRIQUE CAL-
  CIUM (FRENCH) see CPJ250
ACIDE DEHYDROCHOLIQUE (FRENCH) see DAL000
ACIDE-2,4-DICHLORO PHENOXY-ACETIQUE (FRENCH) see DAA800
ACIDE-2-(2,4-DICHLORO-PHENOXY) PROPIONIQUE (FRENCH) see DGB000
ACIDE DIMETHYLARSINIQUE (FRENCH) see HKC000
ACIDE DIMETHYL-ETHYL-ALLENOLIQUE ETHER METHYLIQUE
  (FRENCH) see DRU600
ACIDE DIPHENYLETHOXYACETIQUE (FRENCH) see EES300
ACIDE DISELINO SALICYLIQUE see SBU150
ACIDE ETHYLENEDIAMINETETRACETIQUE (FRENCH) see EIX000

ACIDE 1-ETIL-7-METIL-1,8-NAFTIRIDIN-4-ONE-3-CARBOSSILICO (ITALIAN)
  see EID000
ACIDE FLUORHYDRIQUE (FRENCH) see HHU500
ACIDE FLUOROSILICIQUE (FRENCH) see SCO500
ACIDE FLUOSILICIQUE (FRENCH) see SCO500
ACIDE FORMIQUE (FRENCH) see FNA000
ACIDE (ISOBUTYL-4 PHENYL)-2 PROPIONIQUE (FRENCH) see IIU000
ACIDE ISO-NICOTINIQUE (FRENCH) see ILC000
ACIDE ISOPHTALIQUE (FRENCH) see IMJ000
ACIDE ISOVANILLIQUE (FRENCH) see HJC000
ACIDE o-METHOXYCINNAMIQUE (FRENCH) see MEJ775
ACIDE β-(1-METHOXY-4-NAPHTHOYL)PROPIONIQUE (FRENCH) see MEZ300
ACIDE METHYL-o-BENZOIQUE (FRENCH) see MPI000
ACIDE x-METHYLFOLIQUE (FRENCH) see MKG275
ACIDE METHYL SELENO-2-BENZOIQUE see MPI200
ACIDE METIAZINIQUE (FRENCH) see MNQ500
ACIDE MONOCHLORACETIQUE (FRENCH) see CEA000
ACIDE-MONOFLUORACETIQUE (FRENCH) see FIC000
ACIDE NALIDIXICO (ITALIAN) see EID000
ACIDE NALIDIXIQUE (FRENCH) see EID000
ACIDE NICOTINIQUE (FRENCH) see NCQ900
ACIDE NIFLUMIQUE (FRENCH) see NDX500
ACIDE NITRIQUE (FRENCH) see NED500
l'ACIDE OLEIQUE (FRENCH) see OHU000
ACIDE ORTHOVANILLIQUE see HJB500
ACIDE OXALIQUE (FRENCH) see OLA000
ACIDE PERACETIQUE (FRENCH) see PCL500
ACIDE PHENYLBORIQUE (FRENCH) see BBM000
ACIDE-α-PHENYL-o-METHOXYCINNAMIQUE (FRENCH) see MFG600
ACIDE PHOSPHORIQUE (FRENCH) see PHB250
ACIDE PHTALIQUE (FRENCH) see PHW250
ACIDE PICOLIQUE (FRENCH) see PIB930
ACIDE PICRAMIQUE (FRENCH) see DUP400
ACIDE PICRIQUE (FRENCH) see PID000
ACIDE PROPIONIQUE (FRENCH) see PMU750
l'ACIDE RICINOLEIQUE (FRENCH) see RJP000
ACIDE SULFHYDRIQUE (FRENCH) see HIC500
ACIDE SULFURIQUE (FRENCH) see SOI500
d'ACIDE TANNIQUE (FRENCH) see TAD750
ACIDE TEREPHTALIQUE (FRENCH) see TAN750
ACIDE THIOGLYCOLIQUE (FRENCH) see TFJ100
ACIDE TRICHLORACETIQUE (FRENCH) see TII250
ACIDE 2,4,5-TRICHLORO PHENOXYACETIQUE (FRENCH) see TAA100
ACIDE 2-(2,4,5-TRICHLORO-PHENOXY) PROPIONIQUE (FRENCH) see TIX500
ACIDE VANILLIQUE see VFF000
ACID FAST ORANGE EGG see HGC000
ACID FAST RED FB see HJF500
ACID FAST YELLOW AG see SGP500
ACID GREEN see FAE950
ACID GREEN 3 see FAE950
ACID GREEN 40 see CMM400
ACID GREEN A see FAF000
ACIDIC METANIL YELLOW see MDM775
ACID IV see ALH250
ACID LEAD ARSENATE see LCK000
ACID LEAD ORTHOARSENATE see LCK000
ACID LEATHER BLUE IC see FAE100
ACID LEATHER BLUE R see ADE750
ACID LEATHER BROWN 2G see XMA000
ACID LEATHER GREEN S see ADF000
ACID LEATHER LIGHT BROWN G see SGP500
ACID LEATHER ORANGE BZR see ADG000
ACID LEATHER ORANGE I see FAG010
ACID LEATHER ORANGE PGW see HGC000
ACID LEATHER RED KPR see FMU070
ACID LEATHER YELLOW PRW see MDM775
ACID LEATHER YELLOW R see MDM775
ACID LEATHER YELLOW T see FAG140
ACID LIGHT ORANGE G see HGC000
ACID METANIL YELLOW see MDM775
ACIDO ACETICO (ITALIAN) see AAT250
ACIDO 3-ACETILAMINO-2,4,6-TRIIODOBENZOICO (ITALIAN) see AAM875
ACIDO o-ACETIL-BENZOICO (ITALIAN) see ADA725
ACIDO ACETILSALICILICO (ITALIAN) see ADA725
ACIDO-3-AMINO-2,4,6-TRIIODOBENZOICO (ITALIAN) see AMU625
ACIDO BROMIDRICO (ITALIAN) see HHJ000
ACIDO CIANIDRICO (ITALIAN) see HHS000
ACIDO CLORIDRICO (ITALIAN) see HHL000
ACIDO 1-(p-CLOROBENZOIL)-5-METOSSI-2-METIL-3-IN-
  DOLILACETOIDROSSAMICO (ITALIAN) see OLM300
ACIDO 2-(4-CLORO-2-METIL-FENOSSI)-PROPIONICO (ITALIAN) see CIR500
ACIDO (2,4-DICLORO-FENOSSI)-ACETICO (ITALIAN) see DAA800
ACIDO-2-(2,4-DICLORO-FENOSSI)-PROPIONICO (ITALIAN) see DGB000
ACIDO (3,6-DICLORO-2-METOSSI)-BENZOICO (ITALIAN) see MEL500

ACIDO DOISYNOLICO (SPANISH) see DYB000
ACIDO FENCLOZICO (ITALIAN) see CKK250
ACIDO-5-FENIL-5-ALLILBARBITURICO (ITALIAN) see AGQ875
ACIDO-5-FENIL-5-ETILBARBITURICO (ITALIAN) see EOK000
ACIDO FENOFIBRICO see CJP750
ACIDO FLUFENAMICO (ITALIAN) see TKH750
ACIDO FLUORIDRICO (ITALIAN) see HHU500
ACIDO FLUOSILICICO (ITALIAN) see SCO500
ACIDO FORMICO (ITALIAN) see FNA000
ACIDO FOSFORICO (ITALIAN) see PHB250
ACIDO-m-IDROSSIBENZOICO (ITALIAN) see HJI100
ACIDO INDOXAMICO (ITALIAN) see OLM300
ACIDO-4-(ISONICOTINIL-IDRAZONE)PIMELICO (ITALIAN) see ILG000
ACIDO-1-METIL-5-(p-TOLNIL)-PIRROL-2-ACETICO (SPANISH) see TGJ850
ACIDOMONOCLOROACETICO (ITALIAN) see CEA000
ACIDO MONOFLUOROACETIO (ITALIAN) see FIC000
ACIDOMYCIN see CCI500, CMP885
ACIDO NIFLUMICO (ITALIAN) see NDX500
ACIDO NITRICO (ITALIAN) see NED500
ACIDO ORTOCRESOTINICO (ITALIAN) see CNX625
ACIDO OSSALICO (ITALIAN) see OLA000
ACIDO 3-OSSI-5-METIL-BENZOICO (ITALIAN) see CNX625
ACIDO PICRICO (ITALIAN) see PID000
ACID ORANGE 10 see HGC000
ACID ORANGE 11 see DDO200
ACID ORANGE 24 see XMA000
ACID ORANGE No. 3 see SGP500
ACIDO SALICILICO (ITALIAN) see SAI000
ACIDO SOLFORICO (ITALIAN) see SOI500
ACIDO TRICLOROACETICO (ITALIAN) see TII250
ACIDO (2,4,5-TRICLORO-FENOSSI)-ACETICO (ITALIAN) see TAA100
ACIDO 2-(2,4,5-TRICLORO-FENOSSI)-PROPIONICO (ITALIAN) see TIX500
ACIDO 1-(m-TRIFLUOROMETILFENIL)-N-NITROSO ANTRANILICO (ITAL-
IAN) see TKF699
ACID OXALATE see AMX825
ACID PHOSPHINE CL see FAG010
ACID PONCEAU 4R see FMU080
ACID PONCEAU R see FMU070
ACID POTASSIUM SULFATE see PKX750
ACID QUININE HYDROCHLORIDE see QIJ000
ACID RED see ADG125
ACID RED 18 see FMU080
ACID RED 26 see FMU070
ACID RED 92 see ADG250
ACID RED 388 see RGZ100
ACID RUBINE see HJF500
ACID SCARLET see FMU070
ACID SCARLET 3R see FMU080
ACID SKY BLUE A see FAE000
ACID, SLUDGE (DOT) see SEA000
ACID-SPAR see CAS000
ACID-TREATED HEAVY NAPHTHENIC DISTILLATE see MQV760
ACID-TREATED HEAVY PARAFFINIC DISTILLATE see MQV765
ACID-TREATED LIGHT NAPHTHENIC DISTILLATE see MQV770
ACID-TREATED LIGHT PARAFFINIC DISTILLATE see MQV775
ACID-TREATED RESIDUAL OIL see MQV872
ACIDUM ACETYLSALICYLICUM see ADA725
ACIDUM FENCLOZICUM see CKK250
ACIDUM NICOTINICUM see NCQ900
ACID VIOLET see FAG120
ACID WOOL BLUE RL see ADE750
ACID YELLOW 36 see MDM775
ACID YELLOW E see SGP500
ACID YELLOW TRA see FAG150
ACIFLOCTIN see AEN250
ACIGENA see HCL000
ACILAN FAST NAVY BLUE R see ADE750
ACILAN GREEN SFG see FAF000
ACILAN ORANGE GX see HGC000
ACILAN PONCEAU RRL see FMU070
ACILAN RED SE see FAG020
ACILAN SCARLET V3R see FMU080
ACILAN TURQUOISE BLUE AE see FMU059
ACILAN YELLOW GG see FAG140
ACILETTEN see CMS750
ACILLIN see AIV500
ACIMETION see MDT740
ACIMETTEN see ADA725
ACINETTEN see AEN250
ACINITRAZOLE see ABY900
ACINTENE A see PIH250
ACINTENE DP see MCC250
ACINTENE DP DIPENTENE see MCC250
ACINTENE O see PMQ750

ACIPEN V see PDT500
ACIPHENOCHINOLINE see PGG000
ACIPHENOCHINOLINIUM see PGG000
ACISAL see ADA725
ACKEE see ADG400
ACKET see SAH000
ACL 60 see SGG500
ACL 70 see DGN200
ACL 85 see TIQ750
ACLACINOMYCIN A see APU500
ACLACINOMYCIN Y see ADG425
ACLACINOMYCIN Y1 see ADG425
ACLATONIUM NAPADISILATE see AAC875
ACM see AAE625
ACNBC see AAJ500
ACNEGEL see BDS000
ACNESTROL see DKA600
ACNU see ALF500
ACOCANTHERIN see OKS000
ACODEEN see BOR350
ACOKANTHERA (VARIOUS SPECIES) see BOO700
ACOLEN see DAL000
A. COLUMBIANUM see MRE275
ACON see VSK600
ACONCEN see CNV750
ACONINE see ADG500
ACONITANE see ADH750
ACONITE see MRE275
ACONITIC ACID see ADH000
ACONITIN CRISTALLISAT (GERMAN) see ADH750
ACONITINE, AMORPHOUS see ADH500
ACONITINE (crystalline) see ADH750
ACONITINE HYDROCHLORIDE see ADH875
ACONITUM CARMICHAELI see ADI250
ACONITUM JAPONICUM see ADI500
ACONITUM (Various Species) see MRE275
A. CONTORTRIX CONTORTRIX VENOM see SKW775
A. CONTORTRIX MOKASEN VENOM see NNX700
A. CORDATA see TOA275
ACORN TANNIN see ADI625
ACORTAN see AES650
ACORTO see AES650
ACORUS CALAMUS Linn., oil extract see OGL020
ACORUS CALAMUS OIL see OGL020
ACP-M-728 see AJM000
AC-PC see AJJ875
ACPC see AJK250
ACP GRASS KILLER see TII500
ACQUINITE see ADR000, CKN500
ACRALDEHYDE see ADR000
ACRAMINE RED see DBN000
ACRAMINE YELLOW see AHS750
ACRANIL see ADI750
ACRANIL DIHYDROCHLORIDE see ADI750
ACRANIL HYDROCHLORIDE see ADI750
ACREX see CBW000
ACRIBEL, combustion products see ADX750
ACRICHINE see ARQ250, CFU750
ACRICID see BGB500
2-ACRIDINAMINE see AHS000
9-ACRIDINAMINE see AHS500
3-ACRIDINAMINE (9CI) see ADJ375
9-ACRIDINAMINE MONOHYDROCHLORIDE see AHS750
ACRIDINE see ADJ500
ACRIDINE, 9-(2-((2-CHLOROETHYL)AMINO)ETHYLAMINO)-6-CHLORO-2-ME-
THOXY-, DIHYDROCHLORIDE, HYDRATE see QCS875
2,6-ACRIDINEDIAMINE see DBN000
3,6-ACRIDINEDIAMINE see DBN600
3,9-ACRIDINEDIAMINE (9CI) see ADJ625
3,6-ACRIDINEDIAMINE, MONOHYDROCHLORIDE (9CI) see PMH250
3,6-ACRIDINEDIAMINE SULFATE (2:1) see DBN400
3,6-ACRIDINEDIAMINE SULPHATE see DBN400
ACRIDINE HYDROCHLORIDE see ADJ750
ACRIDINE MONOHYDROCHLORIDE see ADJ750
ACRIDINE MUSTARD see ADJ875
ACRIDINE ORANGE see BAQ250, BJF000
ACRIDINE ORANGE FREE BASE see BJF000
ACRIDINE ORANGE NO see BAQ250
ACRIDINE ORANGE R see BAQ250
ACRIDINE RED see ADK000
ACRIDINE RED 3B see ADK000
ACRIDINE RED, HYDROCHLORIDE see ADK000
ACRIDINE YELLOW see DBT400

ACRIDINE YELLOW BASE see DBT200
ACRIDINE YELLOW G see DBT400
ACRIDINIUM CHLORIDE see ADJ750
ACRIDINO(2,1,9,8-klmna)ACRIDINE see ADK250
ACRIDINO(2,1,9,8-klmna)ACRIDINE SULFATE see DCK200
4'-(9-ACRIDINYLAMINO)-2'-AMINOMETHANESULFONANILIDE see ADK750, ADL000
4'-(9-ACRIDINYLAMINO)HEXANESULFONANILIDE see ADL250
4'-(9-ACRIDINYLAMINO)METHANESULFON-m-ANISIDE MONOHYDROCHLORIDE see ADL500
4'-(9-ACRIDINYLAMINO)METHANESULPHON-m-ANISIDIDE see ADL750
4'-(9-ACRIDINYLAMINO)-2'-METHOXYMETHANESULFONANILIDE see ADM000
4'-(9-ACRIDINYLAMINO)-3'-METHOXYMETHANESULFONANILIDE see ADL750, ADM250
N-(4-(ACRIDINYL-9-AMINO)-3-METHOXYPHENYL)ETHANESULFONAMIDE METHANESULFONATE see ADM000
N-(4-(9-ACRIDINYLAMINO)-3-METHOXYPHENYL)METHANESULFONAMIDE see ADM000
N-(4-(9-ACRIDINYLAMINO)-3-METHOXYPHENYL)METHANESULFONAMIDE compounded with LACTIC ACID see AOD425
4'-(9-ACRIDINYLAMINO)-2-METHYLMETHANESULFONANILIDE see ADN000
4'-(9 ACRIDINYLAMINO)-3'-METHYLMETHANESULFONANILIDE see ADN250
4'-(9-ACRIDINYLAMINO)METHYLSULFONYL-m-ANISIDINE see ADL750
4'-(9-ACRIDINYLAMINO)-2'-NITROMETHANESULFONANILIDE see ADN500
N-(p-(ACRIDIN-9-YLAMINO)PHENYL)BUTANESULFONAMIDE, HYDROCHLORIDE see ADO250
N-(p-(9-ACRIDINYLAMINO)PHENYL)-1-ETHANESULFONAMIDE see ADO500
N-(p-(ACRIDIN-9-YLAMINO)PHENYL)-ETHANESULFONAMIDE, HYDROCHLORIDE-3 see ADO750
N-(p-(ACRIDIN-9-YLAMINO)PHENYLHEXANESULFONAMIDE HYDROCHLORIDE see ADP000
N-(p-(ACRIDIN-9-YLAMINO)PHENYL)METHANESULFONAMIDE HYDROCHLORIDE see ADP500
N-(p-(ACRIDIN-9-YLAMINO)PHENYL)PENTANESULFONAMIDE HYDROCHLORIDE see ADP750
N-(p-(9-ACRIDINYLAMINO)PHENYL)-1-PROPANESULFONAMIDE see ADQ000
N-(p-(9-ACRIDINYLAMINO)PHENYL)PROPANESULFONAMIDE HYDROCHLORIDE see ADQ250
N'-9-ACRIDINYL-N-(2-CHLOROETHYL)-N-ETHYL-1,3-PROPANEDIAMINE DIHYDROCHLORIDE see CGX750
N-(9-ACRIDINYL)-N'-(2-CHLOROETHYL)-1,3-PROPANEDIAMINE see ADQ500
N'-9-ACRIDINYL-N,N-DIMETHYL-1,4-BENZENEDIAMINE see DOS800
ACRIFLAVIN see DBX400
ACRIFLAVINE see XAK000
ACRIFLAVINE NEUTRAL see XAK000
ACRIFLAVINE mixture with PROFLAVINE see DBX400
ACRIFLAVINIUM CHLORIDE see DBX400
ACRIFLAVINIUM CHLORIDUM see DBX400
ACRIFLAVON see DBX400, XAK000
ACRILAFIL see ADY500
ACRINAMINE see ARQ250
ACRINOL see EDW500
ACRIQUINE see ARQ250
ACROFOL see SFV250
ACROLACTINE see EDW500
ACROLEIC ACID see ADS750
ACROLEIN see ADR000
ACROLEINA (ITALIAN) see ADR000
ACROLEIN CYANOHYDRIN see HJQ000
ACROLEIN DIACETATE see ADR250
ACROLEIN DIMER see ADR500
ACROLEINE (DUTCH, FRENCH) see ADR000
ACROMONA see MMN250
ACROMYCINE see ADR750
ACRONINE see ADR750
ACRONIZE see CMA750
ACRONYCINE see ADR750
ACROPOR see ADY250
ACROSTICHUM AUREUM Linn., extract see ADS150
ACRYL, combustion products see ADX750
ACRYLALDEHYD (GERMAN) see ADR000
ACRYLALDEHYDE see ADR000
ACRYLAMIDE see ADS250
ACRYLAMIDE, N-(METHOXYMETHYL)- see MEX300
ACRYLAMIDE, N-METHYL- see MGA300
ACRYLATE d'ETHYLE (FRENCH) see EFT000
ACRYLATE de METHYLE (FRENCH) see MGA500
ACRYL BRILLIANT GREEN B see AFG500
ACRYLIC ACID see ADS750
ACRYLIC ACID (ACGIH,DOT,OSHA) see ADS750
ACRYLIC ACID, inhibited (DOT) see ADS750
ACRYLIC ACID, 3-p-ANISOYL-3-BROMO-, SODIUM SALT, (E)- see CQK600
ACRYLIC ACID BUTYL ESTER see BPW100

ACRYLIC ACID n-BUTYL ESTER (MAK) see BPW100
ACRYLIC ACID CHLORIDE see ADZ000
ACRYLIC ACID-β-CHLOROETHYL ESTER see ADT000
ACRYLIC ACID-2-CYANOETHYL ESTER see ADT111
ACRYLIC ACID, DIESTER with TETRAETHYLENE GLYCOL see ADT050
ACRYLIC ACID, DIESTER with TRIETHYLENE GLYCOL see TJQ100
ACRYLIC ACID-N,N-DIETHYLAMINOETHYL ESTER see DHT125
ACRYLIC ACID ESTER with HYDRACRYLONITRILE see ADT111
ACRYLIC ACID-2-ETHOXYETHANOL DIESTER see ADT250
ACRYLIC ACID-2-ETHOXYETHANOL ESTER see ADT500
ACRYLIC ACID-2-ETHOXYETHYL ESTER see ADT500
ACRYLIC ACID, ETHYLENE ESTER see EIP000
ACRYLIC ACID, ETHYLENE GLYCOL DIESTER see EIP000
ACRYLIC ACID ETHYL ESTER see EFT000
ACRYLIC ACID-2-ETHYLHEXYL ESTER see ADU250
ACRYLIC ACID ETHYLHEXYL ESTER mixed with HYDROXYETHYL ESTER (50:50) see ADU500
ACRYLIC ACID-2-(5'-ETHYL-2-PYRIDYL)ETHYL ESTER see ADU750
ACRYLIC ACID, GLACIAL see ADS750
ACRYLIC ACID HEXYL ESTER see ADV000
ACRYLIC ACID HOMOPOLYMER see ADW200
ACRYLIC ACID-2-HYDROXYETHYL ESTER see ADV250
ACRYLIC ACID-2-HYDROXYPROPYL ESTER see HNT600
ACRYLIC ACID ISOBUTYL ESTER see IIK000
ACRYLIC ACID, ISODECYL ESTER see IKL000
ACRYLIC ACID 2-METHOXYETHOXY ESTER see MEM250
ACRYLIC ACID-2-METHOXYETHYL ESTER see MIF750
ACRYLIC ACID METHYLCARBAMYLETHYL ESTER see MID750
ACRYLIC ACID METHYL ESTER (MAK) see MGA500
ACRYLIC ACID METHYLTHIOETHYL ESTER see MPT250
ACRYLIC ACID-2-(METHYLTHIO)ETHYL ESTER see MPT250
ACRYLIC ACID-1-METHYLTRIMETHYLENE ESTER see BRG500
ACRYLIC ACID-5-NORBORNEN-2-METHYL ESTER see BFY250
ACRYLIC ACID-5-NORBORNEN-2-YLMETHYL ESTER see BFY250
ACRYLIC ACID, OXYBIS(ETHYLENEOXYETHYLENE) ESTER see ADT050
ACRYLIC ACID OXYDIETHYLENE ESTER see ADT250
ACRYLIC ACID, PENTAERITHRITOL TRIESTER see PBC750
ACRYLIC ACID, POLYMERS see ADW200
ACRYLIC ACID, POLYMER with SUCROSEPOLYALLYL ETHER see ADW000
ACRYLIC ACID POLYMER, ZINC SALT see ADW250
ACRYLIC ACID RESIN see ADW200
ACRYLIC ACID, TELOMER with TRICHLOROACETIC ACID see ADW750
ACRYLIC ACID ((3a,4,7,7a-TETRAHYDRO)-4,7-METHANOINDENYL) ESTER see DGW400
ACRYLIC ACID TRIDECYL ESTER see ADX000
ACRYLIC ALDEHYDE see ADR000
ACRYLIC AMIDE see ADS250
ACRYLIC POLYMER see ADW200
ACRYLIC RESIN see ADW200
ACRYLITE see PKB500
ACRYLNITRIL (GERMAN, DUTCH) see ADX500
ACRYLOAMIDE see ADS250
ACRYLONITRILE see ADX500
ACRYLONITRILE-BUTADIENE-STYRENE (pyrolysis products) see ADX750
ACRYLONITRILE MONOMER see ADX500
ACRYLONITRILE POLYMER with 1,3-BUTADIENE, and STYRENE, COMBUSTION PRODUCTS see ADX750
ACRYLONITRILE POLYMER with CHLOROETHYLENE see ADY250
ACRYLONITRILE POLYMER with STYRENE see ADY500
ACRYLONITRILE-STYRENE COPOLYMER see ADY500
ACRYLONITRILE-STYRENE POLYMER see ADY500
ACRYLONITRILE-STYRENE RESIN see ADY500
ACRYLOPHENONE see PMQ250
2-ACRYLOXYETHYLDIMETHYLSULFONIUM METHYL SULFATE see ADY750
ACRYLOYL CHLORIDE see ADZ000
2-(ACRYLOYLOXY)ETHANOL see ADV250
ACRYLSAEUREAETHYLESTER (GERMAN) see EFT000
ACRYLSAEUREMETHYLESTER (GERMAN) see MGA500
ACRYLYL CHLORIDE see ADZ000
ACRYPET see PKB500
ACRYSOL A 3 see ADW200
ACS see ADY500, AJD000
AC 1370 SODIUM see CCS525
ACT see AEB000
ACTAEA (VARIOUS SPECIES) see BAF325
ACTAMER see TFD250
ACTASAL see CMG000
ACTEDRON see BBK000
ACTELIC see DIN800
ACTELLIC see DIN800
ACTELLIFOG see DIN800
ACTEMIN see AOB500
ACTEROL see NHH000
ACTH see AES650

ACTHAR see AES650
ACTI-AID see CPE750
ACTICEL see SCH000
ACTI-CHLORE see CDP000
ACTIDILAT see TMX775
ACTIDIONE see CPE750
ACTIDIONE TGF see CPE750
ACTIDOL see TMX775
ACTIDONE see CPE750
ACTIHAEMYL see ADZ125
ACTILIN see NCF000
ACTINE see EHP000
ACTINIC RADIATION see AEA000
ACTINOBOLIN see AEA109
ACTINOCHRYSIN see AEA750
ACTINOGAN see AEA250
ACTINOLITE ASBESTOS see ARM260
ACTINOMYCIN see AEA500
ACTINOMYCIN 23-21 see AEA625
ACTINOMYCIN 1048A see AEC000
ACTINOMYCIN 2104L see AEB750
ACTINOMYCIN BV see AEC200
ACTINOMYCIN C see AEA750
ACTINOMYCIN D see AEB000
ACTINOMYCINDIOIC D ACID, DILACTONE see AEB000
ACTINOMYCIN DV see AEC200
ACTINOMYCIN I see AEB000
ACTINOMYCIN J1 see AEC200
ACTINOMYCIN K see AEB500
ACTINOMYCIN L see AEB750
ACTINOMYCIN S see AEC000
ACTINOMYCIN S3 see AEC175, AEC200
ACTINOMYCIN-V see AEC200
ACTINOMYCIN X2 see AEC200
ACTINOSPECTACIN DIHYDROCHLORIDE PENTAHYDRATE see SKY500
ACTINOXANTHIN see AEC250
ACTINOXANTHINE see AEC250
ACTIOL see MBX800
ACTIOQUINONE LIGHT YELLOW see AAQ250
ACTISPRAY see CPE750
ACTITHIAZIC ACID see CCI500, CMP885
ACTIVATED ALUMINUM OXIDE see AHE250
ACTIVATED CARBON see CDI000
ACTIVE ACETYL ACETATE see EFS000
ACTIVE DICUMYL PEROXIDE see DGR600
ACTIVIN see DYF450
ACTIVOL see ALT250
ACTOL see NDX500
ACTON see AES650
ACTONAR see AES650
ACTOR Q see DVR200
ACTOZINE see BCA000, PGA750
ACTRAPID see IDF300
AC-17 TRIHYDRATE see AER666
ACTRIL see HKB500
AC-TRY see ADE075
ACTYBARYTE see BAP000
ACTYLOL see LAJ000
ACULEACIN A see AEC625
ACUPAN see NBS500
ACYCLOGUANOSINE see AEC700
ACYCLOGUANOSINE SODIUM (OBS.) see AEC725
ACYCLOVIR see AEC700
ACYCLOVIR SODIUM SALT see AEC725
ACYLANID see ACH500
ACYLATE see ING400
ACYLATE-1 see ING400
ACYLPYRIN see ADA725
ACYTOL see LAJ000
5-ACZ see ARY000
AD see AEB000
AD1M see AGX000
AD 32 see TJX350
AD 122 see OEM000
AD-205 see DSH000
AD-810 see BCE750
AD-1590 see DLI650
ADAB see DPO200
ADAKANE 12 see DXT200
ADALAT see AEC750
ADALIN see BNK000
ADAM AND EVE see ITD050
1-ADAMANTAMINE see TJG250
1-ADAMANTANAMINE see TJG250

ADAMANTANAMINE HYDROCHLORIDE see AED250
1-ADAMANTANAMINE HYDROCHLORIDE see AED250
1-ADAMANTANEACETIC ACID-2-(DIETHYLAMINO)ETHYL ESTER, ETHYL
    IODIDE-11-5 see AED750
1-ADAMANTANEACETIC ACID-3-(DIMETHYLAMINO)PROPYL ESTER,
    ETHYL IODIDE-12-6 see AEE000
ADAMANTINE HYDROCHLORIDE see AED250
S-((N-1-ADAMANTYLAMIDINO)METHYL)HYDROGEN THIOSULFATE, HY-
    DRATE (4581)-14-7 see AEE250
ADAMANTYLAMINE HYDROCHLORIDE see AED250
1-ADAMANTYLAMINE HYDROCHLORIDE see AED250
1-(1-ADAMANTYLAMINO)-2,2,2-TRIFLUORO-1-(TRIFLUOROMETHYL)
    ETHANOLSESQUIHYDRATE see AEE500
5-(1-ADAMANTYL)-2,4-DIAMINO-6-ETHYLPYRIMIDINE ETHYLSULFONATE
    see AEF000
5-(1-ADAMANTYL)-2,4-DIAMINO-6-METHYLPYRIMIDINE ETHYLSULFON-
    ATE see AEF250
N-1-ADAMANTYL-N-(2-(DIMETHYLAMINO)ETHOXY)ACETAMIDE HYDRO-
    CHLORIDE-2 see AEF500
2,2'-(1,3-ADAMANTYLENE)N,N,N',N'-TETRAMETHYL-ETHYLAMINE DIHY-
    DROCHLORIDE see BJG750
N-(2-ADAMANTYL)-2-MERCAPTOACETAMIDINE HYDROCHLORIDE
    see AEG000
S-(N-(1-ADAMANTYLMETHYLAMIDINO)METHYL)PHOSPHOROTHIOATE
    MONOSODIUM SALT see AEG129
N-(1-ADAMANTYLMETHYL)-2-MERCAPTOACETAMIDINE HYDROCHLO-
    RIDE see AEG250
(2-(2-ADAMANTYLOXY)ETHYL)TRIETHYL-AMMONIUMIODIDE see DHP000
(2-(1-ADAMANTYLOXY)PROPYL)DIMETHYLETHYLAMMONIUM IODIDE
    see DPU600
3-(1-ADAMANTYL)PROPYLAMINE HYDROCHLORIDE see AMC500
N-(3-(1-ADAMANTYL)PROPYL)-2-MERCAPTOACETAMIDINE HYDROCHLO-
    RIDE HYDRATE (10:10:3) see AEG500
ADAMSITE see PDB000
ADANON see MDO750
ADANON HYDROCHLORIDE see MDP000, MDP750
ADANTON HYDROCHLORIDE see AEG625
ADAPIN see AEG750
ADAPTOL see AEG875
ADC AURAMINE O see IBA000
ADC BRILLIANT GREEN CRYSTALS see BAY750
ADC MALACHITE GREEN CRYSTALS see AFG500
ADC PERMANENT RED TONER R see CJD500
ADC RHODAMINE B see FAG070
ADC TOLUIDINE RED B see MMP100
ADDEX-THAM see TEM500
ADDISOMNOL see BNK000
ADDITIN 30 see PFT250
ADDUKT HEXACHLORCYKLOPENTADIENU S CYKLOPENTADIENEM
    (CZECH) see HCN000
A1-DEHYDROMETHYLTESTERONE see DAL300
ADELFA (PUERTO RICO) see OHM875
ADEMINE see UVJ450
ADEMOL see TKG750
ADENARON see AER666
ADENINE see AEH000
ADENINE ARABINOSIDE see AEH100, AQQ900
ADENINE ARABINOSIDE MONOPHOSPHATE see AQQ905
ADENINE ARABINOSIDE 5'-MONOPHOSPHATE see AQQ905
ADENINE, 9-BENZYL- see BDX100
ADENINE DEOXYRIBONUCLEOSIDE see DAQ200
ADENINE DEOXYRIBOSE see DAQ200
ADENINE-1-N-OXIDE see AEH250
ADENINE RIBOSIDE see AEH750
ADENINIMINE see AEH000
ADENIUM (VARIOUS SPECIES) see DBA450
ADENOCK see ZVJ000
ADENOHYPOPHYSEAL GROWTH HORMONE see PJA250
ADENOHYPOPHYSEAL LUTEOTROPIN see PMH625
ADENOSIN (GERMAN) see AEH750
ADENOSINE see AEH750
β-ADENOSINE see AEH750
β-d-ADENOSINE see AEH750
ADENOSINE-3'-(α-AMINO-p-METHOXYHYDROCINNAMAMIDO)-3'-DEOXY-
    N,N-DIMETHYL-79-2 see AEI000
ADENOSINE-5'-CARBOXAMIDE see AEI250
ADENOSINE, 2-CHLORO- CEF100
ADENOSINE CYCLIC MONOPHOSPHATE see AOA130
ADENOSINE-3',5'-CYCLIC MONOPHOSPHATE see AOA130
ADENOSINE CYCLIC-3',5'-PHOSPHATE see AOA130
ADENOSINE-5'-(N-CYCLOBUTYL)CARBOXAMIDE see AEI500
ADENOSINE-5'-(N-CYCLOPENTYL)CARBOXAMIDE see AEI750
ADENOSINE-3',5'-CYCLOPHOSPHATE see AOA130
ADENOSINE-5'-(N-CYCLOPROPYL)CARBOXAMIDE see AEJ000

ADENOSINE-5'-(N-CYCLOPROPYL)CARBOXAMIDE-N-OXIDE see AEJ250
ADENOSINE-5'-(N-CYCLOPROPYLMETHYL)CARBOXAMIDE see AEJ500
ADENOSINE-5'-(N-(2-(DIMETHYLAMINO)ETHYL))CARBOXAMIDE see AEJ750
ADENOSINE-5'-(N,N-DIMETHYL)CARBOXAMIDE HYDRATE see AEK000
ADENOSINE-5'-(N-ETHYL)CARBOXAMIDE HEMIHYDRATE see AEK250
ADENOSINE-5'-(N-ETHYL)CARBOXAMIDE-N-OXIDE see AEK500
ADENOSINE-5'-(N-HEXYL)CARBOXAMIDE HEMIHYDRATE see AEK750
ADENOSINE-5'-(N-(2-HYDROXYETHYL)CARBOXAMIDE see AEL000
ADENOSINE-5'-(N-ISOPROPYL)CARBOXAMIDE see AEL250
ADENOSINE-5'-(N-METHOXY)CARBOXAMIDE HYDRATE see AEL500
ADENOSINE-5'-(N-METHYL)CARBOXAMIDE HEMIHYDRATE see AEL750
ADENOSINE, N-METHYL-, mixed with SODIUM NITRITE (1:4) see SIS650
ADENOSINE-5'-MONOPHOSPHATE see AOA125
ADENOSINE-3',5'-MONOPHOSPHATE see AOA130
ADENOSINE-5'-MONOPHOSPHATE POTASSIUM SALT see AEM500
ADENOSINE 5'-MONOPHOSPHATE SODIUM SALT see AEM750
ADENOSINE-5-MONOPHOSPHORIC ACID see AOA125
ADENOSINE-5'-MONOPHOSPHORIC ACID see AOA125
ADENOSINE-5'-MONOPHOSPHORIC ACID POTASSIUM SALT see AEM500
ADENOSINE PHOSPHATE see AOA125
ADENOSINE-5'-PHOSPHATE see AOA125
ADENOSINE-3',5'-PHOSPHATE see AOA130
ADENOSINE-5'-PHOSPHATE POTASSIUM SALT see AEM500
ADENOSINE-5'-PHOSPHORIC ACID see AOA125
ADENOSINE-5'-PHOSPHORIC ACID POTASSIUM SALT see AEM500
ADENOSINE-5'-(N-PROPYL)CARBOXAMIDE see AEM000
ADENOSINE-5'-(TETRAHYDROGENTRIPHOSPHATE) SODIUM SALT
    see AEM250
ADENOSINE TRIPHOSPHATE see ARQ500
ADENOSINE-5'-TRIPHOSPHATE see ARQ500
ADENOSINE-5'-TRIPHOSPHORIC ACID see ARQ500
ADENOVITE see AOA125
ADENYL see AOA125
ADENYLDEOXYRIBOSIDE see DAQ200
ADENYLIC ACID see AOA125
tert-ADENYLIC ACID see AOA125
5'-ADENYLIC ACID POTASSIUM SALT see AEM500
5'-ADENYLIC ACID SODIUM SALT see AEM750
ADENYLPYROPHOSPHORIC ACID see ARQ500
ADEPHOS see ARQ500
ADEPSINE OIL see MQV750
ADERMINE see PPK250
ADERMINE HYDROCHLORIDE see PPK500
ADETOL see ARQ500
ADHERE see MIQ075
ADIABEN see CKK000
ADIAZINE see PPP500
ADILACTETTEN see AEN250
ADINOL T see SIY000
ADIPAMIDE see AEN000
ADIPAN see AOB250, BBK000
ADIPARTHROL see AOB250
ADIPEX see MDQ500, MDT600
ADIPHENIN see DHX800, THK000
ADIPHENINE see DHX800
ADIPIC ACID see AEN250
ADIPIC ACID BIS(3,4-EPOXY-6-METHYLCYCLOHEXYLMETHYL) ESTER
    see AEN750
ADIPIC ACID, BIS(2-ETHOXYETHYL) ESTER see BJO225
ADIPIC ACID BIS(2-ETHYLHEXYL) ESTER see AEO000
ADIPIC ACID-3-CYCLOHEXENYLMETHANOL DIESTER see AEO250
ADIPIC ACID DIALLYL ESTER see AEO500
ADIPIC ACID DIAMIDE see AEN000
ADIPIC ACID, DIBUTOXYETHYL ESTER see BHJ750
ADIPIC ACID DIBUTYL ESTER see AEO750
ADIPIC ACID DI-(3-CARBOXY-2,4,6-TRIIODOANILIDE) DISODIUM
    see BGB325
ADIPIC ACID DIDECYL ESTER (mixed isomers) see AEP000
ADIPIC ACID-DI-2-(2-ETHYLBUTOXY)ETHYL) ESTER see AEP250
ADIPIC ACID DI(2-ETHYLBUTYL) ESTER see AEP500
ADIPIC ACID DIETHYL ESTER see AEP750
ADIPIC ACID DI(2-HEXYLOXYETHYL) ESTER see AEQ000
ADIPIC ACID DIHYDRAZIDE see AEQ250
ADIPIC ACID DIISOPENTYL ESTER see AEQ500
ADIPIC ACID DIISOPROPYL ESTER see DNL800
ADIPIC ACID DINITRILE see AER250
ADIPIC ACID DI-2-PROPYNYL ESTER see AEQ750
ADIPIC ACID, METHYL VINYL ESTER see MQL000
ADIPIC ACID NITRILE see AER250
ADIPIC ACID, POLYMER with 1,4-BUTANEDIOL and METHYLENEDI-p-PHE-
    NYLENE ISOCYANATE see PKM250
ADIPIC ACID, POLYMER with 1,4-BUTANEDIOL, METHYLENEDI-p-PHE-
    NYLENE ISOCYANATE and 2,2'-(p-PHENYLENEDIOXY)DIETHANOL
    see PKM500

ADIPIC ACID, POLYMER with ETHYLENE GLYCOL and METHYLENEDI-p-
    PHENYLENE ISOCYANATE see PKL750
ADIPIC ACID, UREA mixed with CARBOXYMETHYLCELLULOSE ACIDS
    see AER000
ADIPIC DIAMIDE see AEN000
ADIPIC DIHYDRAZIDE see AEQ250
ADIPIC KETONE see CPW500
ADIPINIC ACID see AEN250
ADIPINSAEURE-DI-(3-CARBOXY-2,4,6-TRIJOD-ANILID) DINATRIUM
    (GERMAN) see BGB325
ADIPIODONE MEGLUMINE see BGB315
ADIPLON see PPO000
ADIPODINITRILE see AER250
ADIPOL 2EH see AEO000
5,5'-(ADIPOLYDIMINO)BIS(2,4,6-TRIIODO-N-METHYLISOPHTHALAMIC
    ACID) see IDJ500
ADIPONITRILE see AER250
ADIPOSETTIN see NNW500
ADIPRAZINE see HEP000
ADITYL see ACE000
ADJUDETS see BBK500
ADK (CZECH) see AGL500
ADM see AES750
ADMA 2 see DRR800
ADMER PB 02 see PMP500
ADMEX 741 see FAB920
ADMEX 746 see FAB920
ADM HYDROCHLORIDE see HKA300
ADMUL see OAV000
ADNEPHRINE see VGP000
2-ADO see DDB600
ADOBACILLIN see AIV500
ADOBIOL see AER500
ADOL see HCP000, OAX000, OBA000
ADOL 34 see OBA000
ADOL 68 see OAX000
ADOL 80 see OBA000
ADOL 85 see OBA000
ADOL 90 see OBA000
ADOL 320 see OBA000
ADOL 330 see OBA000
ADOL 340 see OBA000
ADONA see AER666
ADONAL see EOK000
ADONA TRIHYDRATE see AER666
ADONIDIN see AER750
ADONIS (VARIOUS SPECIES) see PCU375
ADONITOL see RIF000
ADOPON see NOC000
ADORM see TDA500
ADPHEN see DKE800
ADR see HKA300
A. DRACONTIUM see JAJ000
ADRAN see IIU000
ADRAXONE see AES639
ADRECHROS see AER666
ADRENAL see VGP000
ADRENAL CORTEX HORMONE see AES650
ADRENALEX see CNS800
1-ADRENALIN see VGP000
ADRENALIN BITARTRATE see AES000
ADRENALIN CHLORIDE see AES500
d-ADRENALINE see AES250
l-(+)-ADRENALINE see AES250
ADRENALINE ACID TARTRATE see AES000
(−)-ADRENALINE ACID TARTRATE see AES000
ADRENALINE BITARTRATE see AES000
(−)-ADRENALINE BITARTRATE see AES000
l-ADRENALINE BITARTRATE see AES000
1-ADRENALINE-d-BITARTRATE see AES000
1-ADRENALINE CHLORIDE see AES500
(−)-ADRENALINE HYDROCHLORIDE see AES500
1-ADRENALINE HYDROCHLORIDE see AES500
(±)-ADRENALINE HYDROCHLORIDE see AES625
dl-ADRENALINE HYDROCHLORIDE see AES625
ADRENALINE HYDROGEN TARTRATE see AES000
(−)-ADRENALINE HYDROGEN TARTRATE see AES000
1-ADRENALINE HYDROGEN TARTRATE see AES000
ADRENALINE TARTRATE see AES000
(−)-ADRENALINE TARTRATE see AES000
l-ADRENALINE TARTRATE see AES000
ADRENALIN HYDROCHLORIDE see AES500
ADRENALIN-MEDIHALER see VGP000
ADRENALONE see MGC350

ADRENAMINE see VGP000
ADRENAN see VGP000
ADRENAPAX see VGP000
ADRENASOL see VGP000
ADRENATRATE see VGP000
ADRENOCHROME see AES639
ADRENOCHROME SULFONATE AC 17 TRIHYDRATE see AER666
ADRENOCORTICOTROPHIC HORMONE see AES650
ADRENOCORTICOTROPHIN see AES650
ADRENOCORTICOTROPIC HORMONE see AES650
ADRENOCORTICOTROPIN see AES650
ADRENODIS see VGP000
ADRENOHORMA see VGP000
ADRENOMONE see AES650
ADRENON see MGC350
ADRENONE see MGC350
ADRENOR see NNO500
ADRENOTROPHIN see AES650
ADRENUTOL see VGP000
ADRESON see CNS800, CNS825
ADREVIL see PEU000
ADRIACIN see HKA300
ADRIAMICINA see MDO250
ADRIAMYCIN see AES750, HKA300
ADRIAMYCIN-HCl see AES750
ADRIAMYCIN, HYDROCHLORIDE see HKA300
ADRIAMYCIN-14-OCTANOATEHYDROCHLORIDE see AET250
ADRIAMYCIN SEMIQUINONE see AES750
ADRIANOL see SPC500
ADRIBLASTIN see HKA300
ADRIBLASTINA see AES750
ADRIBLASTINE see HKA300
ADRINE see VGP000
ADRIXINE see BBK500
ADROIDIN see PAN100
ADRONAL see CPB750
ADROYD see PAN100
ADRUCIL see FMM000
ADUMBRAN see CFZ000
ADVASTAB 401 see BFW750
ADVASTAB 405 see MJO500
ADVASTAB 17 MO see BKK750
ADVAWAX 140 see OAV000
ADVENTAN see FQJ100
ADYNOL see ARQ500
AEAMN see ABN750
AED see CQJ750
AENH (GERMAN) see ENV000
A 38414 (ENZYME) see VGU700
AEORLIN see BQF500
AERBRON see POF500
AERO see MCB000
AERO-CYANAMID see CAQ250
AERO CYANAMID GRANULAR see CAQ250
AERO CYANAMID SPECIAL GRADE see CAQ250
AERO CYANATE see PLC250
AEROL 1 (pesticide) see TIQ250
AERO liquid HCN see HHS000
AEROSEB-DEX see SOW000
AEROSEB-HC see CNS750
AEROSIL see SCH000
AEROSOL GPG see DJL000
AEROSOL of THERMOVACUUM CADMIUM see CAK000
AEROSPORIN see PKC500
AEROTEX GLYOXAL 40 see GIK000
AEROTHENE MM see MJP450
AEROTHENE TT see MIH275
AESCIN (GERMAN) see EDK875
α-AESCIN see EDL000
β-AESCIN see EDL500
AESCIN SODIUM SALT see EDM000
AESCIN TRIETHANOLAMINE SALT see EDM500
AESCULETIN DIMETHYL ETHER see DRS800
AESCULUS (VARIOUS SPECIES) see HGL575
AESCUSAN see EDK875
α-AESCUSAN see EDL000
β-AESCUSAN see EDL500
AESCUSAN SODIUM SALT see EDM000
AES-2Mg see MAI500
AES-MG see MAI500
AESTOCIN see DPE200
AET see AJY250
AET-2HBR see AJY250
AET BROMIDE see AJY250

AET DICHLORIDE see AJY500
AET DIHYDROBROMIDE see AJY250
AETHALDIAMIN (GERMAN) see EEA500
AETHANETHIOL (GERMAN) see EMB100
AETHANOL (GERMAN) see EFU000
AETHANOLAMIN (GERMAN) see EEC600
AETHER see EJU000
2-AETHINYLBUTANOL see EQL000
AETHIONIN see EEI000
AETHISTERON see GEK500
AETHOBROMID DES α,α-DICYCLOPENTYLESSIGSAEURE-β'-DIAETHYLAMINO AETHYLESTER (GERMAN) see DGW600
AETHON see ENY500
AETHOPHYLLINUM see HLC000
AETHOPROPROPAZIN see DIR000
AETHOSUXIMIDE (GERMAN) see ENG500
AETHOXEN see VMA000
2-AETHOXY-AETHYLACETAT (GERMAN) see EES400
1-AETHOXY-4-(1-CAETO-2-HYDROXYAETHYL)-NAPHTALAENE SUCCINATE see SNB500
3-(AETHOXYCARBONYLAMINOPHENYL)-N-PHENYL-CARBAMAT (GERMAN) see EEO500
S-AETHOXY-CARBONYLTHIAMIN HYDROCHLORID (GERMAN) see EEQ500
2-AETHOXY-6,9-DIAMINOACRIDINLACTAT (GERMAN) see EDW500
1-p-AETHOXYPHENYL-3-DIAETHYLAMINO-INDAN CITRAT (GERMAN) see DHR800
p-AETHOXYPHYLHARNSTOFF (GERMAN) see EFE000
5-AETHOXY-3-TRICHLORMETHYL-1,2,4-THIADIAZOL (GERMAN) see EFK000
AETHUSA CYNAPIUM see FMU200
AETHYLACETAT (GERMAN) see EFR000
AETHYL ACETOXYMETHYLNITROSAMIN (GERMAN) see ENR500
AETHYLACRYLAT (GERMAN) see EFT000
AETHYL-AETHANOL-NiTROSOAMIN (GERMAN) see ELG500
AETHYLALKOHOL (GERMAN) see EFU000
AETHYLAMINE (GERMAN) see EFU400
4-AETHYLAMINO-2-tert-BUTYLAMINO-6-METHYLTHIO-s-TRIAZIN (GERMAN) see BQC750
2-AETHYLAMINO-5-BUTYL-4-YL-DIMETHYLSULFAMAT (GERMAN) see BRJ000
2-AETHYLAMINO-4-CHLOR-6-ISOPROPYLAMINO-1,3,5-TRIAZIN (GERMAN) see ARQ725
2-AETHYLAMINO-6-CHLOR-4-METHYL-4-PHENYL-4H-3,1-BENZOXAZIN (GERMAN) see CGQ500
2-AETHYLAMINO-4-ISOPROPYLAMINO-6-CHLOR-1,3,5-TRIAZIN (GERMAN) see ARQ725
2-AETHYLAMINO-3-PHENYL-NOR-CAMPHAN (GERMAN) see EOM000
AETHYLANILIN (GERMAN) see EGK000
AETHYLBENZOL (GERMAN) see EGP500
AETHYLBUTYLKETON (GERMAN) see EHA600
AETHYL-N-BUTYL-NITROSOAMIN (GERMAN) see EHC000
AETHYL-tert-BUTYL-NITROSOAMIN (GERMAN) see NKD500
AETHYLCARBAMAT (GERMAN) see UVA000
AETHYLCHLORID (GERMAN) see EHH000
AETHYL-CHLORVYNOL (GERMAN) see CHG000
1-AETHYL-CYCLOHEXANOL-(1) (GERMAN) see EHR500
O-AETHYL-S-(2-DIMETHYLAMINOAETHYL)-METHYLPHOSPHONOTHIOATE (GERMAN) see EIF500
O-AETHYL-S,S-DIPHENYL-DITHIOPHOSPHAT (GERMAN) see EIM000
S-AETHYL-N,N-DIPROPYLTHIOLCARBAMAT (GERMAN) see EIN500
AETHYLEN-BIS-THIURAMMONOSULFID (GERMAN) see ISK000
AETHYLENBROMID (GERMAN) see EIY500
AETHYLENCHLORID (GERMAN) see EIY600
AETHYLENECHLORHYDRIN (GERMAN) see EIU800
AETHYLENEDIAMIN (GERMAN) see EEA500
AETHYLENGLYKOLAETHERACETAT (GERMAN) see EES400
AETHYLENGLYKOLMETHYLAETHERACETAT (GERMAN) see EJJ500
AETHYLENGLYKOL-MONOMETHYLAETHER (GERMAN) see EJH500
AETHYLENIMIN (GERMAN) see EJM900
AETHYLENIMINO-2-OXYBUTEN (GERMAN) see VMA000
AETHYLENOXID (GERMAN) see EJN500
AETHYLENSULFID (GERMAN) see EJP500
AETHYLFORMIAT (GERMAN) see EKL000
AETHYLHARNSTOFF und NATRIUMNITRIT (GERMAN) see EQE000
AETHYLHARNSTOFF und NITRIT (GERMAN) see EQE000
S-AETHYL-N-HEXAHYDRO-1H-AZEPINTHIOLCARBAMAT (GERMAN) see EKO500
1-AETHYLHEXANOL (GERMAN) see EKQ000
AETHYLIDENCHLORID (GERMAN) see DFF809
AETHYLIS see EHH000
AETHYLIS CHLORIDUM see EHH000
2-(O-AETHYL-N-ISOPROPYLAMINDOTHIOPHOSPHORYLOXY)-BENZOSAEURE-ISOPROPYLESTER (GERMAN) see IMF300
AETHYL-ISOPROPYL-NITROSOAMIN (GERMAN) see ELX500
AETHYLMERCAPTAN (GERMAN) see EMB100

AETHYLMETHYLKETON (GERMAN) see MKA400
2-AETHYL-6-METHYL-N-(1-METHYL-2-METHOXYAETHYL)-
CHLORACETANILID (GERMAN) see MQQ450
O-AETHYL-O-(3-METHYL-4-METHYLTHIOPHENYL)-ISOPROPYLAMIDO-
PHOSPHORSAEURE ESTER (GERMAN) see FAK000
3-β-AETHYL-1-METHYL-4-PHENYL-4α-PIPERIDYLPROPIONAT
HYDROCHLORID (GERMAN) see NOE550
3-β-AETHYL-1-METHYL-4-PHENYL-4α-PROPIONYLOXYPIPERIDIN
HYDROCHLORID (GERMAN) see NOE550
N-AETHYL-N'-NITRO-N-NITROSOGUANIDIN (GERMAN) see ENU000
O-AETHYL-O-n(4-NITROPHENYL)-PHENYL-MONOTHIOPHOSPHONAT (GER-
MAN) see EBD700
AETHYLNITROSO-HARNSTOFF (GERMAN) see ENV000
AETHYLNITROSOURETHAN (GERMAN) see NKE500
5-AETHYL-5-PENTYL-(2')-BARBITURSAEURE (GERMAN) see PBS250
O-AETHYL-S-PHENYL-AETHYL-DITHIOPHOSPHONAT (GERMAN) see
FMU045
S-AETHYL-N-PHENYL-DITHIOCARBAMAT see EOK550
5-AETHYL-5-PHENYL-HEXAHYDROPYRIMIDIN-4,6-DION (GERMAN) see
DBB200
4-AETHYL-1-PHOSPHA-2,6,7-TRIOXABICYCLO(2.2.2)OCTAN (GERMAN) see
TNI750
4-AETHYL-1-PHOSPHA-2,6,7-TRIOXABICYCLO(2.2.2)OCTAN-1-OXID (GER-
MAN) see ELJ500
AETHYL-4-PICOLYLNITROSAMIN (GERMAN) see NLH000
N-AETHYLPIPERIDIN (GERMAN) see EOS500
N-(1-AETHYLPROPYL)-3,4-DIMETHYL-2,6-DINITROANILIN (GERMAN) see
DRN200
N-(1-AETHYLPROPYL)-2,6-DINITRO-3,4-XYLIDIN (GERMAN) see DRN200
AETHYLPROPYLKETON (GERMAN) see HEV500
AETHYLRHODANID (GERMAN) see EPP000
AETHYLSENFOEL (GERMAN) see ISK000
1-N-AETHYLSISOMICIN see SBD000
S-2-AETHYLSULFINYL-1-METHYL AETHYL-O,O DIMETHYL-
MONOTHIOPHOSPHAT see DSK600
O-AETHYL-O-(2,4,5-TRICHLORPHENYL)-AETHYLTHIONOPHOSPHONAT
(GERMAN) see EPY000
AETHYLTRICHLORPHON (GERMAN) see EPY600
N-AETHYL-N-(2,4,6-TRIJOD-3-AMINOPHENYL)-SUCCINAMIDSAEURE (GER-
MAN) see AMV375
AETHYLURETHAN (GERMAN) see UVA000
AETHYL-VINYL-NITROSOAMIN (GERMAN) see NKF000
AETHYLZINNTRICHLORID (GERMAN) see EPS000
AETINA see EPQ000
AETIVA see EPQ000
AETM (GERMAN) see ISK000
AETT see ACL750
AF see XAK000
AF2 see FQO000
AF-45 see EQN259
AF 101 see DXQ500
AF 260 see AHC000
AF 425 see CKI000
AF 594 see CPN750
AF 864 see BBW500
AF 983 see BAV325
AF 1161 see CKJ000, THK880
AF 1890 see DEL200
AF 2071 see BAV275
AF 2259 see IJJ000
AF-2 (preservative) see FQN000
AFASTOGEN BLUE 5040 see DNE400
AFATIN see BBK500
AFAXIN see VSK600
AFBI see AEU250
A.F. BLUE No. 1 see FMU059
A.F. BLUE No. 2 see FAE100
AFCOLAC B 101 see SMR000
AFCOLENE see SMQ500
2AAF DIMER see AAK500
AFESIN see CKD500
A.F. GREEN No. 1 see FAE950
A.F. GREEN No. 2 see FAF000
AF 438 HYDROCHLORIDE see OOE100
AFIBRIN see AJD000
AFICIDA see DOX600
AFICIDE see BBQ500
A-FIL CREAM see TGG760
AFILINE see MPU250
AFI-PHYLLIN see DNC000
AFI-TIAZIN see PDP250
AFKO-HIST see WAK000
AFKO-SAL see SAH000
AFL 1081 see FFF000

AFL 1082 see FFH000
AFLATOXICOL see AEW500
AFLATOXIN see AET750
AFLATOXIN B see AEU250
AFLATOXIN B1 see AEU250
AFLATOXIN B2 see AEU750
AFLATOXIN B1-2,3-DICHLORIDE see AEU500
AFLATOXIN G1 see AEV000
AFLATOXIN G2 see AEV500
AFLATOXIN G1 mixed with AFLATOXIN B1 see AEV250
AFLATOXIN M1 see AEW000
AFLATOXIN Ro see AEW500
A. FLAVA see HGL575
AFLIX see DRR200
AFLON see TAI250
AFLOQUALONE see AEW625
A. FLOS-AQUAE TOXIN see AON825
AFLOXAN see POF550
AFLUON see MAF500
AFNOR see CJJ000
A.F. ORANGE No. 1 see FAG010
A.F.ORANGE No. 2 see TGW000
A. FORDII see TOA275
AFOS see DJI000
AFRAZINE see AEX000
A.F. RED No. 1 see FAG018
A.F. RED No. 5 see XRA000
AFRICAN COFFEE TREE see CCP000
AFRICAN LILAC TREE see CDM325
AFRIDOL BLUE see AEW750
AFRIN see AEX000
AFRIN HYDROCHLORIDE see AEX000
AFTATE see TGB475
AF 1312/TS see CEQ625
A.F. VIOLET No 1 see FAG120
A.F YELLOW No. 2 see FAG130
A.F. YELLOW No. 3 see FAG135
AG 3 see CBR500
8 AG see AJO500
AG-629 see SLE880
AG. 5895 see VGA100
AG 58107 see TAI600
AGALITE see TAB750
AGALLOL see MEP250
AGALLOLAT see MEP250
AGAR see AEX250
AGAR-AGAR see AEX250
AGAR AGAR FLAKE see AEX250
AGAR-AGAR GUM see AEX250
AGARIN see AKT750
AGASTEN see FOS100
AGATE see SCI500, SCJ500
AGC see GFA000
AGE see AGH150
AGEDAL see DPH600
AGEFLEX BGE see BRK750
AGEFLEX CGE see GGS000
AGEFLEX EGDM see BKM250
AGEFLEX FA-2 see DHT125
AGEFLEX FA-10 see IKL000
AGEFLEX FM-1 see DPG600
AGEFLEX FM-4 see BQD250
AGEFLEX FM 246 see DXY200
AGEFLEX n-HA see ADV000
AGELFLEX FM-10 see IKM000
AGENAP see NAR000
AGENT 504 see DAI600
AGENT AT 717 see PKQ250
AGENT BLUE see HKC000
AGENT ORANGE see AEX750
AGERATOCHROMENE see AEX850
AGERITE see AEY000, BLE500
AGERITE 150 see HKF000
AGERITE ALBA see AEY000
AGERITEDPPD see BLE500
AGERITE ISO see HKF000
AGERITE POWDER see PFT500
AGERITE WHITE see NBL000
AGEROPLAS see BKB250
AGGLUTININ see AAD000
AGIDOL see BFW750
AGILENE see DRK600, PJS750
AGIOLAN see VSK600
AGI TALC, BC 1615 see TAB750

AGKISTRODON ACUTUS VENOM see HGM600
AGKISTRODON CONTORTRIX CONTROTRIX VENOM see SKW775
AGKISTRODON CONTORTRIX MOKASEN VENOM see NNX700
AGKISTRODON CONTORTRIX VENOM see AEY125
AGKISTRODON PISCIVORUS PISCIVORUS VENOM see EAB200
AGKISTRODON PISCIVORUS VENOM see AEY130
AGKISTRODON RHODOSTOMA VENOM see AEY135
A. GLABRA see HGL575
AGLICID see BSQ000
AGLUMIN see DIS600
AGOFOLLIN see EDR000
AGOSTILBEN see DKA600
AGOTAN see PGG000
AGOVIRIN see TBG000
AGP see AKC625
AGRAZINE see PDP250
AGREFLAN see DUV600
AGRIA 1050 see DSQ000
AGRIBON see SNN300
AGRICIDE MAGGOT KILLER (F) see CDV100
AGRICULTURAL LIMESTONE see CAO000
AGRIDIP see CNU750
AGRIFLAN 24 see DUV600
AGRI-MYCIN see SLY500
AGRIMYCIN 17 see SLW500
AGRION see SGH500
AGRISIL see EPY000
AGRISOL G-20 see BBQ500
AGRISTREP see SLY500
AGRITAN see DAD200
AGRITOX see CIR250
AGRITOX see EPY000
AGRIYA 1050 see DSQ000
AGRIZAN see CNK559
A-GRO see MNH000
AGROCERES see HAR000
AGROCIDE see BBQ500
AGROCLAVINE see AEY375
AGROFOROTOX see TIQ250
AGROMICINA see TBX000
AGRONAA see NAK500
AGRONEXIT see BBQ500
AGROSAN see ABU500
AGROSOL see MLF250
AGROSOL S see CBG000
AGROTECT see DAA800
AGROTHION see DSQ000
AGROXONE see CIR250
AGROXONE 3 see SIL500
AGROX 2-WAY and 3-WAY see CBG000
AGRYPNAL see EOK000
AGSTONE see CAO000
AGUATHOL see DXD000
AH see DBM800
AH-42 see TEO250
AH 289 see CDR000
AH 1932 see PFY105
AH 3232 see CDQ250
AH 3365 see BQF500
AH 19065 see RBF400
AHA see ABB250
AH 5158A see HMM500
AHCOCID FAST SCARLET R see FMU070
AHCO DIRECT BLACK GX see AQP000
AHCOQUINONE BLUE IR BASE see HOK000
AHCOVAT PRINTING GOLDEN YELLOW see DCZ000
AHCTL see TCZ000
AHE POI (HAWAII) see EAI600
AH-289 HYDROCHLORIDE see CDR250
A. HIPPOCASTANUM see HGL575
AHOUAI des ANTILLES see YAK350
AHR 233 see MFD500
AHR 376 see AEY400
AHR-619 see ENL100, SLU000
AHR-1680 see DMB000
AHR-1767 see ENC600
AHR-3053 see CCH125
AHR 3219 see EQN600
AHR 5850D MONOHYDRATE see AHK625
AHYDOL (RUSSIAN) see TMJ000
A 66 HYDROCHLORIDE see MNV750
A 446 HYDROCHLORIDE see LFW300
A-HYDROCORT see HHR000
AHYGROSCOPIN-B see VGZ000

AHYPNON see MKA250
AI 318284 see MRQ750
AI3-18285 see CBF725
AI 3-22542 see DKC800
AI 3-29024 see AFR750
AI3-29158 see AHJ750
AI3-35966 see ISZ000
AI3-36175-Ga see EMY100
AI3-36329-A see MIS500
AI3-36401 see HJG100
AI3-36420 see ODO000
AI3-36537 see CPI350
AI3-36543 see MLM500
AI3-36558 see MLM600
AI3-36561 see DSP650
AI3-36563 see TCV400
AI3-36564 see MLL650
AI3-36565 see MLL655
AI3-36566 see MLL660
AI3-36570 see TCV375
AI3-37220 see CPD625
AI 3-51254 see MGE100
AI3-70087 see HKQ500
AIA see PNR800
AIB see MGB000
AIBN see ASL750
TRANS-AID see ANW750
AIGLONYL see EPD500
A-250-II see CBF680
A 11725 II see MRW750
AIL du CANADA (CANADA) see WBS850
AIMAX see MLJ500
AIMSAN see DRR400
AIP see AHE750
AIPTASIA PALLIDA VENOM see SBI800
AIPYSURUS LAEVIS VENOM see SBI880
AIPYSURUS LAEVIS VENOM (AUSTRALIA) see AFG000
AIR, compressed see AFG250
AIRBRON see ACH000
AIRDALE BLUE IN see FAE100
AIREDALE BLACK ED see AQP000
AIREDALE BLUE D see CMO500
AIREDALE BLUE 2BD see CMO000
AIREDALE BLUE FFD see CMN750
AIREDALE BLUE RL see ADE750
AIREDALE CARMOISINE see HJF500
AIREDALE VIOLET ND see CMP000
AIREDALE YELLOW E see SGP500
AIREDALE YELLOW T see FAG140
AIR-FLO GREEN see CNN500
AIRONE see ZMA000
AISELAZINE see HGP500
AISEMIDE see CHJ750
A. ITALICUM see ITD050
AITC see AGJ250
AIZEN AMARANTH see FAG020
AIZEN AURAMINE see IBA000
AIZEN BRILLIANT BLUE FCF see FMU059
AIZEN BRILLIANT SCARLET 3RH see FMU080
AIZEN CATHILON RED GTLH see BAQ750
AIZEN CHROME VIOLET BH see HL1000
AIZEN CRYSTAL VIOLET EXTRA PURE see AOR500
AIZEN DIAMOND GREEN GH see BAY750
AIZEN DIRECT BLUE 2BH see CMO000
AIZEN DIRECT DEEP BLACK GH see AQP000
AIZEN DIRECT SKY BLUE 5BH see CMO500
AIZEN EOSINE GH see BNH500, BNK700
AIZEN ERYTHROSINE see FAG040
AIZEN FOOD BLUE No. 2 see FAE000
AIZEN FOOD GREEN No. 3 see FAE950
AIZEN FOOD ORANGE No. 1 see FAG010
AIZEN FOOD ORANGE No. 2 see TGW000
AIZEN FOOD RED No. 5 see XRA000
AIZEN FOOD VIOLET No 1 see FAG120
AIZEN FOOD YELLOW No. 5 see FAG150
AIZEN MALACHITE GREEN see AFG500
AIZEN METANIL YELLOW see MDM775
AIZEN METHYLENE BLUE BH see BJI250
AIZEN NAPHTHOL ORANGE I see FAG010
AIZEN ORANGE I see FAG010
AIZEN PONCEAU RH see FMU070
AIZEN PRIMULA BROWN BRLH see CMO750
AIZEN RHODAMINE BH see FAG070
AIZEN TARTRAZINE see FAG140

AIZEN URANINE see FEW000
AIZEN VICTORIA BLUE BOH see VKA600
AJAN see NBS500
AJAX, LEMON (scouring powder) see AFG625
AJI CABALLERO (PUERTO RICO) see PCB275
AJI de GALLINA (PUERTO RICO) see PCB275
AJI GUAGUAO (CUBA) see PCB275
AJINOMOTO see MRL500
AJI PICANTE (PUERTO RICO) see PCB275
AJMALICINE see AFG750
AJMALICINE HYDROCHLORIDE see AFH000
AJMALICINE MONOHYDROCHLORIDE see AFH000
AJMALINE see AFH250
AJMALINE BIS(CHLOROACETATE) (ester) HYDROCHLORIDE see AFH275
AJMALINE HYDROCHLORIDE see AFH280
AJO see WBS850
AK-33X see MAV750
AK, flower extract see CAZ075
'AKA'AKAI (HAWAII) see WBS850
'AKA'AKAI-PILAU (HAWAII) see WBS850
AKAR see DER000
AKARITHION see TNP250
AKARITOX see CKM000
AKEE see ADG400
AKETDRIN see AOB250
AKI see ADG400
AKINETON see BGD500
AKINETON HYDROCHLORIDE see BGD750
AKINOPHYL see BGD500, BGD750
AKIRIKU RHODAMINE B see FAG070
AKLAVIN see DAY835
AKLOMIX-3 see HMY000
AKLONIN (GERMAN) see PCV750
AKLONINE see PCV750
AKOIN HYDROCHLORID (GERMAN) see PDN500
AKOTIN see NCQ900
AK PS see AFH500
AKRA, flower extract see CAZ075
AKRICHIN see ARQ250
AKROFOL see SFV250
AKROLEIN (CZECH) see ADR000
AKROLEINA (POLISH) see ADR000
AKRO-ZINC BAR 85 see ZKA000
AKRYLAMID (CZECH) see ADS250
AKRYLONITRYL (POLISH) see ADX500
AKSA, combustion products see ADX750
AKTAMIN see NNO500
AKTAMIN HYDROCHLORIDE see NNP000
AKTEDRIN see AOB250
AKTEDRON see AOB500
AKTIKON see ARQ725
AKTIKON PK see ARQ725
AKTINIT A see ARQ725
AKTINIT PK see ARQ725
AKTINIT S see BJP000
AKTIVAN see FNF000
AKTIVEX see CCX000
AKTIVIN see CDP000
AKTON see DIX600
AKULON see PJY500
AKYL DIMETHYL BENZYL AMMONIUM SACCHARINATE see BBA625
AKYPOROX O 50 see PJY100
AKYPOSAL TLS see SON000
AKZO CHEMIE MANEB see MAS500
AL-50 see RDP300
AL-100 see ZVJ000
AL-1021 see FGV000
AL 1076 see PFJ000
AL-1612 see AFH550
'ALA-AUMOE (HAWAII) see DAC500
ALABASTER see CAX750
ALACHLOR (USDA) see CFX000
ALACIL see AFW500
ALACINE see PFC750
AL-ALCHILI (ITALIAN) see DNI600
ALAMANDA MORADA FALSA (PUERTO RICO) see ROU450
ALAMINE 6 see HCO500
ALAMINE 11 see OHM700
ALAMINE 308 see DVL000
ALAMINE 336 see DVL000
ALANE see AHB500
ALANEX see CFX000
ALANINE, 3-(((3-AMINO-4-HYDROXYPHENYL)PHENYLARSINO)THIO)-
see AKI900

ALANINE, N-BENZYLOXYCARBONYL-3-PHENYL-, VINYL ESTER, l-
see CBR235
ALANINE, N-CARBOXY-3-PHENYL-, N-BENZYL l-VINYL ESTER, l-
see CBR235
ALANINE, 3-(p-CHLOROPHENYL)-, dl- see FAM100
ALANINE, 3,3'-DISELENOBIS-(9CI) see SBP600
ALANINE, 3,3'-DISELENODI-, dl- see SBU200
ALANINE, N-l-γ-GLUTAMYL-3-(METHYLENECYCLOPROPYL)- see HOW100
ALANINE, 3-(4-(4-HYDROXY-3-IODOPHENOXY)-3,5-DIIODOPHENYL)-, l-
see LGK050
ALANINE MUSTARD see BHN500
ALANINE NITROGEN MUSTARD see BHV250, PED750
β-ALANINOL see PMM250
ALANOSINE see AFH750
l-ALANOSINE see AFH750
4-N-d-ALANYL-2,4-DIAMINO-2,4-DIDEOXY-l-ARABINOSE see AFI500
β-ALANYL-l-HISTIDINE see CCK665
N-l-ALANYL-3-(5-OXO-7-OXABICYCLO(4.1.0)HEPT-2-YL)-l-ALANINE
see BAC175
ALANYL-2,6-XYLIDIDE see TGI699
ALA de PICO (MEXICO) see RBZ400
ALAR see DQD400
ALAR-85 see DQD400
ALATHON see PKF750
ALAUN (GERMAN) see AGX000
ALAZIN see PFC750
ALAZINE see PFC750
ALAZOPEPTIN see AFI625
ALBA-DOME see AEY000
ALBAGEL PREMIUM USP 4444 see BAV750
ALBALON LIQUIFILM see NCW000
ALBAMINE see SNP500
ALBAMIX see SMB000
ALBAMYCIN see NOB000, SMB000
ALBAMYCIN SODIUM see NOB000
ALBARICOQUE (SPANISH) see AQP890
ALBEGO see CFY250
ALBEMAP see BBK500
ALBENDAZOLE (USDA) see VAD000
ALBEXAN see SNM500
ALBIGEN A see PKQ250
ALBIOTIC see LGD000
ALBITOCIN see AFI750
ALBOLINE see MQV750
ALBOMYCIN A1 see GJU800
ALBON see SNN300
ALBONE see HIB000
ALBONE 35 see HIB010
ALBONE 50 see HIB010
ALBONE 70 see HIB010
ALBONE 35CG see HIB010
ALBONE 50CG see HIB010
ALBONE 70CG see HIB010
ALBORAL see DCK759
ALBOSAL see SNM500
AL BROMOHYDRATE see AGY000
ALBSAPOGENIN see GMG000
ALBUCID see SNP500
ALBUMIN see AFI850
ALBUMIN MACRO AGGREGATES see AFI850
ALBUTEROL see BQF500
ALCALASE see BAC000
ALCANFOR see CBB250
ALCHLOQUIN see CHR500
ALCIAN BLUE see AFI900
ALCIDE see CDW450
ALCLOFENAC EPOXIDE see AFI950
ALCLOFENAC SODIUM SALT see AGN250
ALCLOMETASONE DIPROPIONATE see AFI980
ALCLOPHENAC see AGN000
ALCOA 331 see AHC000
ALCOA F 1 see AHE250
ALCOA SODIUM FLUORIDE see SHF500
ALCOBAM NM see SGM500
ALCOBAM ZM see BJK500
ALCOBON see FHI000
ALCOGUM see ADW200
ALCOHOL see EFU000
ALCOHOL, anhydrous see EFU000
ALCOHOL, dehydrated see EFU000
ALCOHOL C-8 see OEI000
ALCOHOL C-9 see NNB500
ALCOHOL C-10 see DAI600
ALCOHOL C-11 see UMA000, UNA000

ALCOHOL C-12 see DXV600
ALCOHOL C-16 see HCP000
ALCOHOL, DENATURED see AFJ000
ALCOHOLS, C7-9 see LGF875
ALCOHOLS, C12-14-SECONDARY, ETHOXYLATED PROPOXYLATED
  see TBA800
ALCOHOLS, N.O.S. see AFJ250
ALCOHOL SULFATE see AFJ375
ALCOHOL SULPHATE see AFJ375
ALCOID see AFJ400
ALCOOL ALLILCO (ITALIAN) see AFV500
ALCOOL ALLYLIQUE (FRENCH) see AFV500
ALCOOL AMILICO (ITALIAN) see IHP000
ALCOOL AMYLIQUE (FRENCH) see AOE000
ALCOOL BUTYLIQUE (FRENCH) see BPW500
ALCOOL BUTYLIQUE SECONDAIRE (FRENCH) see BPW750
ALCOOL BUTYLIQUE TERTIAIRE (FRENCH) see BPX000
ALCOOL ETHYLIQUE (FRENCH) see EFU000
ALCOOL ETILICO (ITALIAN) see EFU000
l'ALCOOL n-HEPTYLIQUE PRIMAIRE (FRENCH) see HBL500
ALCOOL ISOAMYLIQUE (FRENCH) see IHP000
ALCOOL ISOBUTYLIQUE (FRENCH) see IIL000
ALCOOL ISOPROPILICO (ITALIAN) see INJ000
ALCOOL ISOPROPYLIQUE (FRENCH) see INJ000
ALCOOL METHYL AMYLIQUE (FRENCH) see MKW600
ALCOOL METHYLIQUE (FRENCH) see MGB150
ALCOOL METILICO (ITALIAN) see MGB150
ALCOOL PROPILICO (ITALIAN) see PND000
ALCOOL PROPYLIQUE (FRENCH) see PND000
ALCOPAN-250 see PAG200
ALCOPHOBIN see DXH250
ALCOPOL O see DJL000
ALCUONIUM DICHLORIDE see DBK400
ALCURONIUM CHLORIDE see DBK400
ALDABAN see TAI450
ALDACOL Q see PKQ250
ALDACTAZIDE see AFJ500
ALDACTIDE see AFJ500
ALDACTONE see AFJ500
ALDACTONE A see AFJ500
ALDANIL see FMW000
ALDECARB see CBM500
ALDECIN see AFJ625
ALDEHYDE-14 see UJJ000
ALDEHYDE ACETIQUE (FRENCH) see AAG250
ALDEHYDE ACRYLIQUE (FRENCH) see ADR000
ALDEHYDE AMMONIA see AAG500
ALDEHYDE B see COU500
ALDEHYDE BUTYRIQUE (FRENCH) see BSU250
ALDEHYDE C-6 see HEM000
ALDEHYDE C-8 see OCO000
ALDEHYDE C-9 see NMW500
ALDEHYDE C10 see DAG000
ALDEHYDE C-18 see CNF250
ALDEHYDE C-10 DIMETHYLACETAL see AFJ700
ALDEHYDECOLLIDINE see EOS000
ALDEHYDE C-14 PURE see UJA800
ALDEHYDE CROTONIQUE (FRENCH) see COB260
ALDEHYDE C-11, UNDECYLENIC see ULJ000
ALDEHYDE-2-ETHYLBUTYRIQUE (FRENCH) see DHI000
ALDEHYDE FORMIQUE (FRENCH) see FMV000
ALDEHYDE PROPIONIQUE (FRENCH) see PMT750
ALDEHYDES see AFJ800
ALDEHYDINE see EOS000
ALDEHYDODICHLOROMALEIC ACID see MRU900
ALDEIDE ACETICA (ITALIAN) see AAG250
ALDEIDE ACRILICA (ITALIAN) see ADR000
ALDEIDE BUTIRRICA (ITALIAN) see BSU250
ALDEIDE FORMICA (ITALIAN) see FMV000
ALDER BUCKTHORN see MBU825
ALDERLIN see INS000
ALDERLIN HYDROCHLORIDE see INT000
ALDICARB (USDA) see CBM500
ALDICARBE (FRENCH) see CBM500
ALDIFEN see DUZ000
AL-DIISOBUTYL see DNI600
ALDIMORPH see AFJ850
ALDINAMID see POL500
ALDO see CLY500
ALDO-28 see OAV000
ALDO-72 see OAV000
ALDOCORTEN see AFJ875
ALDOCORTENE see AFJ875
ALDOCORTIN see AFJ875

ALDO HMS see OAV000
ALDOL see AAH750
ALDOMET see DNA800, MJE780
ALDOMETIL see DNA800, MJE780
ALDOMIN see DNA800, MJE780
ALDO MS see OAV000
ALDO MSA see OAV000
ALDO MSLG see OAV000
ALDOMYCIN see NGE500
ALDOSPERSE L 9 see DXY000
(+)-ALDOSTERONE see AFJ875
d-ALDOSTERONE see AFJ875
ALDOSTERONE (8CI) see AFJ875
ALDOXIME see AAH250
2-ALDOXIME PYRIDINIUM-N-METHYL METHANESULPHONATE see PLX250
ALDOXYCARB see AFK000
ALDREX see AFK250
ALDREX 30 see AFK250
ALDRICH see DOJ200
ALDRIN see AFK250
ALDRIN, cast solid (DOT) see AFK250
ALDRINE (FRENCH) see AFK250
ALDRITE see AFK250
ALDROSOL see AFK250
ALECOR see ECU600
ALELAILA (PUERTO RICO) see CDM325
ALENTIN see BSM000
ALENTOL see AOB250
ALEPSIN see DNU000
ALERMINE see CLD250
ALERYL see BBV500
ALESTEN see SNP500
ALETAMINE HYDROCHLORIDE see AGQ250
ALEUDRIN see DMV600
ALEURITES (VARIOUS SPECIES) see TOA275
ALEURITIC ACID, tech see TKP000
ALEVAIRE see TDN750
ALEVIATIN see DKQ000
ALEXAN see COW900
ALEXANDRIAN LAUREL see MBU780
ALF see AKR500
ALFACALCIDOL see HJV000
ALFACILLIN see PDD350
ALFACRON see IEN000
ALFADIONE see AFK500
ALFALFA MEAL see AFK750
ALFAMAT see GFA000
ALFANAFTILAMINA (ITALIAN) see NBE000
ALFA-NAFTYLOAMINA (POLISH) see NBE000
ALFATESINE (FRENCH) see AFK500
ALFATIL see CCR850
ALFA-TOX see DCM750, MEI500
ALFATROFIN see AES650
ALFAXALONE see AFK875
ALFENAMIN see AHA875
ALFENOL 3 see PKF500
ALFENOL 9 see PKF500
ALFEPROL (RUSSIAN) see AGW250
ALFIBRATE see AHA150
ALFICETYN see CDP250
ALFIDE see SOX875
ALFIMID see DYC800
ALFLORONE see FHH100
ALFOCILLIN see PDD350
ALFOL 8 see OEI000
ALFOL-10 see ODE000
ALFOL 12 see DXV600
ALFONAL K see AFK900
ALFUCIN see NGE500
ALGAFAN see PNA500
ALGAMON see SAH000
ALGAROBA see LIA000
ALGERIAN IVY see AFK950
ALGERIL see PMX250
ALGESTONE ACETOPHENIDE see DAM300
ALGIAMIDA see SAH000
ALGIDON see MDP750
ALGIL see DAM700
ALGIMYCIN see ABU500
ALGIN see SEH000
ALGIN (polysaccharide) see SEH000
ALGINATE KMF see SEH000
ALGIN GUM see CAO250
ALGINIC ACID see AFL000

ALGINIC ACID HYDROGEN SULFATE SODIUM SALT (9CI) see SEH450
ALGIPON L-1168 see SEH000
ALGISTAT see DFT000
ALGLOFLON see TAI250
ALGODON de SEDA (CUBA, PUERTO RICO) see COD675
ALGOFLON SV see TAI250
ALGOFRENE TYPE 1 see TIP500
ALGOFRENE TYPE 2 see DFA600
ALGOFRENE TYPE 5 see DFL000
ALGOFRENE TYPE 6 see CFX500
ALGOFRENE TYPE 67 see ELN500
ALGOLYSIN see MDP750
ALGOSEDIV see TEH500
ALGOTROPYL see HIM000
ALGRAIN see EFU000
ALGUSI, extract see AHI630
ALGYLEN see TIO750
ALHELI EXTRANJERO (PUERTO RICO) see OHM875
ALICYANATE see PLC250
ALIDOCHLOR see CFK000
ALIMEMAZINE see AFL500
ALIMEMAZINE-S,S-DIOXIDE see AFL750
ALIMEZINE see AFL500
ALINAMIN see DXO300
ALINAMIN F see FQJ100
ALINDOR see BRF500
ALIOMYCIN see AFM000
ALIPHATIC and AROMATIC EPOXIDES see AFM250
ALIPHATIC CHLORINATED HYDROCARBONS see CDV250
ALIPORINA see TEY000
ALIPUR see AFM375
ALIPUR-O see CPT000
ALIQUAT 203 see DGX200
ALIQUAT 206 see DRK200
ALIQUAT 336 see MQH000
ALISOBUMAL see AGI750
ALITON see DXO300
ALIVAL see NMV725
ALIVAL (ANTIDEPRESSANT) see NMV725
ALIZARIN see DMG800
ALIZARINA see DMG800
ALIZARINE see DMG800
ALIZARINE CHROME ORANGE G see NEY000
ALIZARINE CYANINE GREEN BASE see BLK000
ALIZARINE LAKE RED 2P see DMG800
ALIZARINE L PASTE see DMG800
ALIZARINE NAC see DMG800
ALIZARINE ORANGE see NEY000
ALIZARINE ORANGE 2GN see NEY000
ALIZARINE ORANGE R see NEY000
ALIZARINE ORANGE RD see NEY000
ALIZARINE R-CF see NEY000
ALIZARINE RED see DMG800
ALIZARINE VIOLET 3B BASE see HOK000
ALIZARINE YELLOW P see NEY000
ALIZARINE YELLOW R see NEY000
ALIZARINE YELLOW RW see NEY000
ALIZAROL ORANGE R see NEY000
ALKALIES see AFM500
ALKALI GRASS see DAE100
ALKALI RESISTANT RED DARK see NAY000
ALKALOID C see AFG750
ALKALOID H 3, from COLCHICUM ANTUMNALE see MIW500
ALKALOID II see AFG750
ALKALOIDS see AFM750
ALKALOID SALTS see AFM750
ALKALOIDS, VERATRUM see VIZ000
ALKALOID V see CPT750
ALKAMID see PJY500
ALKANES see AFN250
ALKAPOL PEG-200 see PJT000
ALKAPOL PEG-400 see MFD500
ALKAPOL PPG-1200 see PKI500
ALKARSODYL see HKC500
ALKATHENE see PJS750
ALKATHENE RXDG33 see TAI250
ALK-AUBS see DXH250
ALKAVERVIR see VIZ000
ALKAZENE 42 see EHY000
α-ALKENESULFONIC ACID see AFN500
ALKENYL DIMETHYLETHYL AMMONIUM BROMIDE see AFN750
ALK-ENZYME see BAC000
ALKERAN see PED750
ALKERAN (RUSSIAN) see BHV000

ALKIRON see MPW500
ALKOHOL (GERMAN) see EFU000
ALKOHOLU ETYLOWEGO (POLISH) see EFU000
ALKOVERT see PIB250
3-(ALKYLAMINO)PROPIONITRILE see AFO200
ALKYL ARYL POLYETHER ALCOHOLS see ARL875
ALKYLARYLPOLYGLYKOLAETHER (GERMAN) see ARL875
ALKYL ARYL SULFONATE see AFO250
ALKYLBENZENESULFONATE see AFO500
p-n-ALKYLBENZENESULFONIC ACID DERIVATIVE, SODIUM SALT
    see AFO750
ALKYLBENZENESULFONIC ACID SODIUM SALT see AFP000
p-N-ALKYLBENZENSULFONAN SODNY (CZECH) see AFO750
ALKYL(C$_8$H$_{17}$ to C$_{18}$H$_{37}$) DIMETHYL-3,4-DICHLOROBENZYL AMMONIUM
    CHLORIDE see AFP750
ALKYL(C-13) POLYETHOXYLATES(ETHOXY-6) see TJJ250
ALKYL(C-14) POLYETHOXYLATES(ETHOXY-7) see TBY750
ALKYL(C9-15)TOLYL METHYLTRIMETHYL AMMONIUM CHLORIDE
    see DYA600
ALKYL DIMETHYL BENZALKONIUM CHLORIDE see AFP075
ALKYL DIMETHYL BENZALKONIUM SACCHARINATE see BBA625
ALKYL DIMETHYLBENZYL AMMONIUM CHLORIDE see AFP250
ALKYL(C$_8$H$_{17}$ to C$_{18}$H$_{37}$) DIMETHYL-3,4-DICHLOROBENZYLAMMONIUM
    CHLORIDE see AFP750
ALKYLDIMETHYLETHYLBENZYL AMMONIUM CHLORIDE see BBA500
ALKYLDIMETHYL(PHENYLMETHYL)QUATERNARY AMMONIUM CHLO-
    RIDES see AFP250
ALKYL((ETHYLPHENYL)METHYL)DIMETHYL QUATERNARY AMMONIUM
    CHLORIDES see BBA500
ALKYLNITRILE see AFQ000
ALKYL PHENOL POLYGLYCOL ETHERS see ARL875
ALKYL PHENOXY POLYETHOXY ETHANOLS see ARL875
ALKYL PHENYL POLYETHYLENE GLYCOL ETHER see AFQ250
ALKYL PYRIDINES R see AFQ500
ALKYROM see AFQ575
ALLACYL see AFW500
ALLAMANDA see AFQ625
ALLAMANDA CATHARTICA see AFQ625
ALLANTOXANIC ACID, POTASSIUM SALT see AFQ750
ALLBRI NATURAL COPPER see CNI000
ALLEDRYL see BBV500
(+)-ALLELRETHONYL (+)-cis,trans-CHRYSANTHEMATE see AFR250
ALLENE see AFR000
ALLEOSIDE A DIHYDRATE see HAO000
ALLERCLOR see TAI500
ALLERCUR see CMV400
ALLERCURE HYDROCHLORIDE see CMV400
ALLERGAN see CKV625
ALLERGAN 211 see DAS000
ALLERGAN B see BBV500
ALLERGEFON MALEATE see CGD500
ALLERGEN see LJR000
ALLERGEVAL see BBV500
ALLERGICAL see BBV500
ALLERGIN see BBV500, TAI500
ALLERGINA see BBV500
ALLERGISAN see TAI500
ALLERGIVAL see BBV500
ALLERON see PAK000
ALLETHRIN see AFR250
d-ALLETHRIN see AFR250
(+)-cis-ALLETHRIN see AFR500
d-trans ALLETHRIN see BGC750
trans-(+)-ALLETHRIN see AFR750
ALLETHRIN I see AFR250
ALLETHRIN RACEMIC MIXTURE see AFS000
d-ALLETHROLONE CHRYSANTHEMUMATE see AFR750
(+)-ALLETHRONYL (+)-trans-CHRYSANTHEMUMATE see AFR750
ALLICIN see AFS250
ALLIDOCHLOR see CFK000
ALLIGATOR LILY see BAR325
ALLILE (CLORURO DI) (ITALIAN) see AGB250
ALLIL-GLICIDIL-ETERE (ITALIAN) see AGH150
1-ALLILOSSI-2,3 EPOSSIPROPANO (ITALIAN) see AGH150
ALLILOWY ALKOHOL (POLISH) see AFV500
ALLIONAL see AFT000
ALLISAN see RDP300
ALLIUM (Various Species) see WBS850
ALLIUM SATIVUM Linn., powder see GBU850
ALLOBARBITAL see AFS500
ALLOBARBITONE see AFS500
ALLOCAINE see AIL750, AIT250
ALLOCLAMIDE see AFS625
ALLOCLAMIDE HYDROCHLORIDE see AFS640

ALLODAN see AFS750
ALLODENE see BBK000
ALLOFENYL see AGQ875
ALLOFERIN see DBK400
ALLOMALEIC ACID see FOU000
ALLOMALEIC ACID DIMETHYL ESTER see DSB600
ALLOMYCIN see AHL000
ALLONAL see AFT000
ALLO-OCIMENOL see LFX000
1-(p-ALLOPHANOYLBENZYL)-2-METHYLHYDRAZINE HYDROBROMIDE
　　see MKN750
ALLOPHENYLUM see AGQ875
ALLOPREGNAN-3-β-OL-20-ISONICOTINYLHYDRAZONE see AFT125
ALLOPSEUDOCODEINE HYDROCHLORIDE see AFT250
ALLOPURINOL see ZVJ000
ALLOPYDIN see AGN000
ALLORPHINE see AFT500
ALLOTROPAL see EQL000
ALLOXAN see AFT750
ALLOXAN MONOHYDRATE see MDL500
ALLOXAN-5-OXIME see AFU000
ALLOXANTIN see AFU250
2-ALLOXYETHANOL (CZECH) see AGO000
ALLOZYM see ZVJ000
ALLURAL see ZVJ000
ALLPHASEM see AGQ875
ALLSPICE see PIG740
ALLTEX see CDV100
ALLTOX see CDV100
ALLUMINIO(CLORURO DI) (ITALIAN) see AGY750
ALLUMINIO DIISOBUTIL-MONOCLORURO (ITALIAN) see CGB500
ALLURAL see ZVJ000
ALLURA RED AC see FAG100
ALLUVAL see BNP750
ALLYL ACETATE see AFU750
ALLYLACETIC ACID see PBQ750
ALLYL ADIPATE see AEO500
ALLYL AL see AFV500
ALLYL ALCOHOL see AFV500
ALLYL ALDEHYDE see ADR000
ALLYLALKOHOL (GERMAN) see AFV500
5-ALLYL-5-(1-(ALLYLTHIO)ETHYL)BARBITURIC ACID SODIUM SALT
　　see AFV750
ALLYLAMINE see AFW000
2-(ALLYLAMINO)-6'-CHLORO-o-ACETOTOLUIDIDE HYDROCHLORIDE
　　see AFW250
2-(ALLYLAMINO)-2'-CHLORO-6'-METHYLACETANILIDE HYDROCHLORIDE
　　see AFW250
1-ALLYL-6-AMINO-3-ETHYL-2,4(1H,3H)-PYRIMIDINEDIONE see AFW500
1-ALLYL-6-AMINO-3-ETHYLURACIL see AFW500
p-ALLYLANISOLE see AFW750
ALLYLBARBITAL see AGI750
ALLYLBARBITONE see AGI750
ALLYLBARBITURAL see AFS500
ALLYLBARBITURIC ACID see AGI750
ALLYLBENZENE see AFX000
ALLYL BENZENE SULFONATE see AFX250
5-ALLYL-1,3-BENZODIOXOLE see SAD000
5-ALLYL-5-BENZYL-2-THIOBARBITURIC ACID SODIUM SALT see AFX500
ALLYL-BIS(β-CHLOROETHYL)AMINE HYDROCHLORIDE see AFX750
ALLYL BROMIDE see AFY000
ALLYL BUTANOATE see AFY250
5-ALLYL-5-(2-BUTENYL)-2-THIOBARBITURIC ACID SODIUM SALT
　　see AFY300
5-ALLYL-5-sec-BUTYLBARBITURIC ACID see AFY500
ALLYL-sec-BUTYL THIOBARBITURIC ACID see AFY750
5-ALLYL-5-sec-BUTYL-2-THIOBARBITURIC ACID see AFY750
5-ALLYL-5-sec-BUTYL-2-THIOBARBITURIC ACID SODIUM SALT see AGA000
5-ALLYL-5-(1-BUTYLTHIO)ETHYL)BARBITURIC ACID SODIUM SALT
　　see AGA250
ALLYL BUTYRATE see AFY250
ALLYL CAPROATE see AGA500
ALLYL CAPRYLATE see AGM500
ALLYL CARBAMATE see AGA750
N-ALLYL CARBAMIC ACID-3-DIMETHYLAMINOPHENYL ESTER HYDRO-
　　CHLORIDE see AGB000
ALLYLCARBAMIC ESTER of m-OXYPHENYLDIMETHYLAMINE HYDRO-
　　CHLORIDE see AGB000
ALLYLCARBAMIDE see AGV000
ALLYLCATECHOL METHYLENE ETHER see SAD000
ALLYLCHLORID (GERMAN) see AGB250
ALLYL CHLORIDE see AGB250
ALLYL CHLOROCARBONATE see AGB500
ALLYL CHLOROFORMATE (DOT) see AGB500
ALLYLCHLOROHYDRIN ETHER see AGB750

ALLYL (3-CHLORO-2-HYDROXYPROPYL) ETHER see AGB750
ALLYL CINERIN see AFR250
ALLYL CINNAMATE see AGC000
ALLYL COMPOUNDS see AGC125
ALLYL CYANIDE see BOX500
6-ALLYL-α-CYANOERGOLINE-8-PROPIONAMIDE see AGC200
ALLYL CYCLOHEXANEACETATE see AGC250
ALLYL CYCLOHEXANEPROPIONATE see AGC500
5-ALLYL-5-(2-CYCLOHEXEN-1-YL)-2-THIOBARBITURIC ACID see TES500
ALLYL CYCLOHEXYLACETATE see AGC250
2-ALLYL-2-CYCLOHEXYLACETIC ACID-3-(DIETHYLAMINO)-2,2-
　　DIMETHYLPROPYLESTER HYDROCHLORIDE see AGC750
3-ALLYLCYCLOHEXYL PROPIONATE see AGC500
5-ALLYL-5-(2-CYCLOPENTENYL)-2-THIOBARBITURIC ACID see AGD000
N-ALLYL-7,8-DEHYDRO-4,5-EPOXY-3,6-DIHYDROXYMORPHINAN see AFT500
17-α-ALLYL-3-DEOXY-19-NORTESTOSTERONE see AGF750
N-ALLYL-N-DESMETHYLMORPHINE see AFT500
l-ALLYL-(6-DIAZO-5-OXO)-l-NORLEUCYL-(6-DIAZO-5-OXO)-l-NORLEUCINE
　　see AFI625
ALLYL DIGLYCOL CARBONATE see AGD250
6-ALLYL-6,7-DIHYDRO-5H-DIBENZ(c,e)AZEPINE see ARZ000
6-ALLYL-6,7-DIHYDRO-5H-DIBENZ(c,e)AZEPINE PHOSPHATE see AGD500
6-ALLYL-6,7-DIHYDRO-3,9-DICHLORO-5H-DIBENZ(c,e)AZEPINE see AGD750
l-N-ALLYL-7,8-DIHYDRO-14-HYDROXYNORMORPHINONE see NAG550
N-ALLYL-7,8-DIHYDRO-14-HYDROXYNORMORPHINONE, HYDROCHLO-
　　RIDE see NAH000
6-ALLYL-6,7-DIHYDRO-6-METHYL-5H-DIBENZ(c,e)AZEPINIUM IODIDE
　　see AGE000
N-ALLYL-2,12-DIHYDROXY-1,11-EPOXYMORPHINENE-13 HYDROCHLORIDE
　　see NAG500
1-ALLYL-3,4-DIMETHOXYBENZENE see AGE250
4-ALLYL-1,2-DIMETHOXYBENZENE see AGE250
1-ALLYL-2,5-DIMETHOXY-3,4-METHYLENEDIOXYBENZENE see AGE500
ALLYLDIMETHYLARSINE see AGE625
1-ALLYL-1-(3,7-DIMETHYLOCTYL)PIPERIDINIUM BROMIDE see AGE750
1-ALLYL-1-(3,7-DIMETHYLOCTYL)-PIPERIDIUMBROMID (GERMAN)
　　see AGE750
β-ALLYL-N,N-DIMETHYLPHENETHYLAMINE see AGF000
ALLYLDIOXYBENZENE METHYLENE ETHER see SAD000
5-ALLYL-1,3-DIPHENYLBARBITURIC ACID see AGF250
5-ALLYL-1,3-DIPHENYL-2,4,6(1H,3H,5H)-PYRIMIDINETERIONE see AGF250
ALLYLE (CHLORURE D') (FRENCH) see AGB250
ALLYL ENANTHATE see AGH250
17-ALLYL-4,5-α-EPOXY-3,14-DIHYDROXYMORPHINAN-6-ONE see NAG550
17-ALLYL-4,5-α-EPOXY-3,14-DIHYDROXYMORPHINAN-6-ONE HYDROCHLO-
　　RIDE see NAH000
ALLYL-3,4-EPOXY-6-METHYLCYCLOHEXANECARBOXYLATE see AGF500
ALLYL-2,3-EPOXYPROPYL ETHER see AGH150
ALLYL-9,10-EPOXYSTEARATE see ECO500
17-α-ALLYL-ESTRATRIENE-4,9,11,17-β-OL-3-ONE see AGW675
ALLYLESTRENOL see AGF750
17-α-ALLYL-4-ESTREN-17-β-OL see AGF750
17-α-ALLYLESTR-4-EN-17-β-OL see AGF750
ALLYLETHER see DBK000
ALLYL ETHER of PROPYLENE GLYCOL see PNK000
1-ALLYL-3-ETHYL-6-AMINOTETRAHYDROPYRIMIDINEDIONE see AFW500
ALLYL ETHYL ETHER see AGG000
2-ALLYL-5-ETHYL-2'-HYDROXY-9-METHYL-6,7-BENZOMORPHAN
　　see AGG250
ALLYL FLUORIDE see AGG500
ALLYL FLUOROACETATE see AGG750
ALLYL FORMATE see AGH000
ALLYL GLUCOSINOLATE see AGH125
α-ALLYL GLYCEROL ETHER see AGP500
ALLYLGLYCIDAETHER (GERMAN) see AGH150
ALLYL GLYCIDYL ETHER see AGH150
4-ALLYLGUAIACOL see EQR500
ALLYL HEPTANOATE see AGH250
ALLYL HEPTOATE see AGH250
ALLYL HEPTYLATE see AGH250
ALLYL HEXAHYDROPHENYLPROPIONATE see AGC500
ALLYL HEXANOATE (FCC) see AGA500
ALLYL HOMOLOG of CINERIN I see AFR250, BGC750
8-ALLYL-(±)-1-α-H,5-α-H-NORTHROPAN-3-α-OL see AGM250
ALLYLHYDRAZINE HYDROCHLORIDE see AGH500
ALLYL HYDROPEROXIDE see AGH750
17-α-ALLYL-17-β-HYDROXY-Δ⁴-ESTREN see AGF750
17-α-ALLYL-17-β-HYDROXY-4-ESTRENE see AGF750
17-α-ALLYL-17-β-HYDROXYESTR-4-EN-3-ONE see AGM200
4-ALLYL-1-HYDROXY-2-METHOXYBENZENE see EQR500
2-ALLYL-4-HYDROXY-3-METHYL-2-CYCLOPENTEN-1-ONE see AFR750
d,l-2-ALLYL-4-HYDROXY-3-METHYL-2-CYCLOPENTEN-1-ONE-d,l-
　　CHRYSANTHEMUMMONOCARBOXYLATE see AFR250
N-ALLYL-3-HYDROXYMORPHINAN see AGI000

17-α-ALLYLHYDROXY-19-NOR-4-ANTROSTENE see AGF750
1-N-ALLYL-14-HYDROXYNORDIHYDROMORPHINONE see NAG550
1-N-ALLYL-14-HYDROXYNORDIHYDROMORPHINONE HYDROCHLORIDE
  see NAH000
ALLYLHYDROXYPHENYLARSINE OXIDE see AGQ775
ALLYLIC ALCOHOL see AFV500
ALLYLIDENE DIACETATE see ADR250
ALLYL IODIDE see AGI250
ALLYL-α-IONONE see AGI500
ALLYLISOBUTYLBARBITAL see AGI750
ALLYLISOBUTYLBARBITURATE see AGI750
5-ALLYL-5-ISOBUTYLBARBITURIC ACID see AGI750
ALLYL ISOCYANATE see AGJ000
ALLYLISOPROPYLACETYLCARBAMIDE see IQX000
ALLYLISOPROPYLACETYLUREA see IQX000
5-ALLYL-5-ISOPROPYLBARBITURATE see AFT000
ALLYLISOPROPYLBARBITURIC ACID see AFT000
5-ALLYL-5-ISOPROPYLBARBITURIC ACID see AFT000
ALLYLISOPROPYLMALONYLUREA see AFT000
ALLYL ISORHODANIDE see AGJ250
ALLYL ISOSULFOCYANATE see AGJ250
ALLYL ISOTHIOCYANATE see AGJ250
ALLYL ISOTHIOCYANATE, stabilized (DOT) see AGJ250
ALLYL ISOVALERATE see ISV000
ALLYL ISOVALERIANATE see ISV000
3-ALLYL-4-KETO-2-METHYLCYCLOPENTENYL
  CHRYSANTHEMUMMONOCARBOXYLATE see AFR250
ALLYLLITHIUM see AGJ375
ALLYL MERCAPTAN see AGJ500
α-ALLYLMERCAPTO-α, α-DIETHYLACETAMIDE see AGJ750
2-ALLYLMERCAPTO-2-ETHYLBUTYRAMIDE see AGJ750
2-ALLYLMERCAPTOISOBUTYRAMIDE see AGK000
α-ALLYLMERCAPTOISOBUTYRAMIDE see AGK000
ALLYLMERCAPTOMETHYLPENICILLIN see AGK250
ALLYLMERCAPTOMETHYLPENICILLINIC ACID see AGK250
ALLYL MESYLATE see AGK750
ALLYL METHACRYLATE see AGK500
ALLYL METHANESULFONATE see AGK750
4-ALLYL-1-METHOXYBENZENE see AFW750
5-ALLYL-1-METHOXY-2,3-(METHYLENEDIOXY)BENZENE see MSA500
4-ALLYL-2-METHOXYPHENOL see EQR500
4-ALLYL-2-METHOXYPHENOL ACETATE see EQS000
4-ALLYL-2-METHOXYPHENOL FORMATE see EQS100
4-ALLYL-2-METHOXYPHENYLPHENYLACETATE see AGL000
5-ALLYL-5-(1-METHYLALLYL)-2-THIOBARBITURIC ACID SODIUM SALT
  see AGL250
5-ALLYL-5-(1-METHYLBUTYL)BARBITURIC ACID see SBM500
5-ALLYL-5-(1-METHYLBUTYL)BARBITURIC ACID SODIUM DERIVATIVE
  see SBN000
5-ALLYL-5-(1-METHYLBUTYL)BARBITURIC ACID SODIUM SALT see SBN000
5-ALLYL-5-(1-METHYLBUTYL)MALONYLUREA see SBM500
5-ALLYL-5-(1-METHYLBUTYL)MALONYLUREA SODIUM SALT see SBN000
5-ALLYL-5-(1-METHYLBUTYL)-2-THIOBARBITURATE SODIUM see SOX500
5-ALLYL-5-(1-METHYLBUTYL)-2-THIOBARBITURIC ACID see AGL375
5-ALLYL-5-(1-METHYLBUTYL)-2-THIO-BARBITURIC ACID SODIUM SALT
  see SOX500
ALLYL 3-METHYLBUTYRATE see ISV000
2-ALLYL-3-METHYL-2-CYCLOPENTEN-1-ON-4-YL-N,N-DIMETHYL-
  KARBAMAT (CZECH) see AGL500
1-ALLYL-3,4-METHYLENEDIOXYBENZENE see SAD000
4-ALLYL-1,2-METHYLENEDIOXYBENZENE see SAD000
2-ALLYL-3-METHYL-4-HYDROXY-2-CYCLOPENTEN-1-ONE
  DIMETHYLCARBAMATE-9 see AGL500
5-ALLYL-1-METHYL-5-(1-METHYL-2-PENTYNYL)BARBITURIC ACID SO-
  DIUM SALT see MDU500
N-ALLYL-3-METHYL-N-α-METHYLPHENETHYL-6-OXO-1(6H)-PYRIDAZINE
  ACETAMIDE see AGL750
3-ALLYL-2-METHYL-4-OXO-2-CYCLOPENTEN-1-YL CHRYSANTHEMATE
  see AFR250
dl-3-ALLYL-2-METHYL-4-OXOCYCLOPENT-2-ENYL-dl-cis trans CHRYSANTHE-
  MATE see AFR250
(±)-3-ALLYL-2-METHYL-4-OXO-2-CYCLOPENTENYL 2,2,3,3-
  TETRAMETHYLCYCLOPROPANE CARBOXYLATE see TAL575
(±)-5-ALLYL-5-(1-METHYL-2-PENTYNYL)-2-THIOBARBITURIC ACID
  see AGL875
dl-1-ALLYL-1-METHYL-4-PHENYL-4-PIPERIDINOL PROPIONATE HYDRO-
  CHLORIDE see AGV890
5-ALLYL-5-(1-METHYLPROPENYL)BARBITURIC ACID see AGM000
5-ALLYL-5-(1-METHYLPROPYL) BARBITURIC ACID see AFY500
5-ALLYL-5-(2'-METHYL-N-PROPYL) BARBITURIC ACID see AGI750
ALLYL MONOSULFIDE see AGS250
ALLYLMORPHINE HYDROCHLORIDE see NAG500
ALLYL MUSTARD OIL see AGJ250
ALLYLNITRILE see BOX500

1-ALLYL-2-NITROIMIDAZOLE see AGM060
4-(ALLYLNITROSAMINO)-1-BUTANOL see HJS400
3-(ALLYLNITROSAMINO)-1,2-PROPANEDIOL see NJY500
1-(ALLYLNITROSAMINO)-2-PROPANONE see AGM125
4-(ALLYLNITROSOAMINO)BUTRIC ACID see CCJ375
N-ALLYL-N-NITROSO-3-BUTENYLAMINE see BPD000
1-ALLYL-1-NITROSOUREA see NJK000
N-ALLYLNORMORPHINE see AFT500
N-ALLYLNORMORPHINE HYDROCHLORIDE see NAG500
17-ALLYL-19-NORTESTOSTERONE see AGM200
17-α-ALLYL-19-NORTESTOSTERONE see AGM200
ALLYL OCTANOATE see AGM500
17-α-ALLYL-4-OESTRENE-17-β-OL see AGF750
ALLYLOESTRENOL see AGF750
2-ALLYLOXYBENZAMIDE see AGM750
o-(ALLYLOXY)BENZAMIDE see AGM750
ALLYLOXYCARB see DBI800
2-(ALLYLOXY)-4-CHLORO-N-(2-DIETHYLAMINO)ETHYL)BENZAMIDE
  see AFS625
2-ALLYLOXY-4-CHLORO-N-(2-(DIETHYLAMINO)ETHYL)BENZAMIDE HY-
  DROCHLORIDE see AFS640
(4-ALLYLOXY-3-CHLOROPHENYL)ACETIC ACID see AGN000
(4-(ALLYLOXY)-3-CHLOROPHENYL)ACETIC ACID SODIUM SALT see
  AGN250
1-ALLYLOXY-3-CHLORO-2-PROPANOL see AGB750
(±)-1-(β-(ALLYLOXY)-2,4-DICHLOROPHENETHYL)IMIDAZOLE see FPB875
2-(ALLYLOXY)-N-(2-(DIETHYLAMINO)ETHYL)α,α,α-TRIFLUORO-p-TOLU-
  AMIDE see FDA875
o-ALLYLOXY-N,N-DIETHYLBENZAMIDE see AGN500
o-ALLYLOXY-N,N-DIMETHYLBENZAMIDE see AGN750
1-ALLYLOXY-2,3-EPOXY-PROPAAN (DUTCH) see AGH150
1-ALLYLOXY-2,3-EPOXYPROPAN (GERMAN) see AGH150
1-(ALLYLOXY)-2,3-EPOXYPROPANE see AGH150
2-ALLYLOXYETHANOL see AGO000
o-ALLYLOXY-N-(β-HYDROXYETHYL)BENZAMIDE see AGO250
6-ALLYLOXY-2-METHYLAMINO-4-(N-METHYLPIPERAZINO)-5-
  METHYLTHIOPYRIMIDINE see AGO500
2-(o-ALLYLOXY)-2-HYDROXY-N-ISOPROPYL-1-PROPYLAMINE
  HYDROCHLORIDE see THK750
1-(o-ALLYLOXY)PHENOXY)-3-(ISOPROPYLAMINO)-2-PROPANOL see CNR500
1-(o-ALLYLOXYPHENOXY)-3-ISOPROPYLAMINOPROPAN-2-OL HYDRO-
  CHLORIDE see THK750
(−)-1-(o-ALLYLOXYPHENOXY)-3-ISOPROPYLAMINO-2-
  PROPANOLHYDROCHLORIDE-73-9 see AGP000
(+)-1-(o-ALLYLOXYPHENOXY)-3-ISOPROPYLAMINO-2-
  PROPANOLHYDROCHLORIDE-23-5 see AGO750
3-(ALLYLOXYPHENOXY)-1,2-PROPANEDIOL see AGP250
5-(o-(ALLYLOXY)PHENYL)-3-(o-TOLYL)-s-TRIAZOLE see AGP400
3-ALLYLOXY-1,2-PROPANEDIOL see AGP500
3-ALLYLOXYPROPIONITRILE see AGP750
β-ALLYLOXY-PROPIONITRILE see AGP750
ALLYL PALLADIUM CHLORIDE DIMER see AGQ000
α-ALLYL PHENETHYLAMINEHYDROCHLORIDE see AGQ250
2-ALLYL PHENOL see AGQ500
o-ALLYL PHENOL see AGQ500
ALLYL PHENOXYACETATE see AGQ750
1-(o-ALLYLPHENOXY)-3-(ISOPROPYLAMINO)-2-PROPANOL see AGW250
1-(o-ALLYLPHENOXY)-3-(ISOPROPYLAMINO)-2-PROPANOL HYDROCHLO-
  RIDE see AGW000
ALLYL PHENYLACETATE see PMS500
ALLYL-3-PHENYLACRYLATE see AGC000
ALLYL PHENYL ARSINIC ACID see AGQ775
5-ALLYL-5-PHENYLBARBITURIC ACID see AGQ875
ALLYL PHENYL ETHER see AGR000
1-ALLYL-2-PHENYL-ETHYLAMINE HYDROCHLORIDE see AGQ250
2-ALLYL-2-PHENYL-4-PENTENOIC ACID 2-(DIETHYLAMINO)ETHYL ESTER
  HYDROCHLORIDE see AGR125
ALLYL PHENYL THIOUREA see AGR250
1-ALLYL-3-PHENYL-2-THIOUREA see AGR250
ALLYL PHOSPHATE see THN750
ALLYLPRODINE HYDROCHLORIDE see AGV890
ALLYL PROPYL DISULFIDE see AGR500
ALLYLPROPYMAL see AFT000
m-ALLYLPYROCATECHIN METHYLENE ETHER see SAD000
4-ALLYLPYROCATECHOL FORMALDEHYDE ACETAL see SAD000
ALLYLPYROCATECHOL METHYLENE ETHER see SAD000
1-ALLYLQUINALDINUM BROMIDE see AGR750
ALLYLRETHRONYL dl-cis-trans-CHRYSANTHEMATE see AFR250
ALLYLRHODANID (GERMAN) see AGS750
ALLYLSENFOEL (GERMAN) see AGJ250
ALLYL SEVENOLUM see AGJ250
ALLYLSUCCINIC ANHYDRIDE see AGS000
ALLYL SULFIDE see AGS250
ALLYL SULFOCYANIDE see AGS750

12-ALLYL-7,7a,8,9-TETRAHYDRO-3,7a-DIHYDROXY-4aH-8,9c-IMINO-
ETHANOPHENANTHRO(4,5-bcd)FURANONE see NAG550
ALLYL-TETRA-HYDROGERANYL-PIPERIDINIUMBROMID (GERMAN)
see AGE750
3-(ALLYL-(TETRAHYDRONAPHTHYL)AMINO)-N,N-DIETHYL-PROPIONAM-
IDE see AGS375
3-(ALLYL-(TETRAHYDRONAPHTHYL)AMINO)-N,N-DIETHYLPROPIONAM-
IDE HYDROCHLORIDE see AGS375
ALLYLTHEOBROMINE see AGS500
1-ALLYLTHEOBROMINE see AGS500
ALLYLTHIOCARBAMIDE see AGT500
ALLYL THIOCARBONIMIDE see AGJ250
ALLYL THIOCYANATE see AGS750
2-ALLYLTHIO-2-ETHYLBUTYRAMIDE see AGJ750
ALLYLTHIOMETHYLPENICILLIN see AGK250
4-ALLYLTHIOSEMICARBAZIDE see AGT000
2-(ALLYLTHIO)-2-THIAZOLINE see AGT250
1-ALLYLTHIOUREA see AGT500
N-ALLYLTHIOUREA see AGT500
1-ALLYL-2-THIOUREA see AGT500
ALLYL TRENBOLONE see AGW675
ALLYL TRI-N-BUTYLPHOSPHONIUMCHLORIDE see AGT750
ALLYL TRICHLORIDE see TJB600
ALLYL TRICHLOROSILANE see AGU250
ALLYL TRIMETHYLHEXANOATE see AGU400
ALLYL 3,5,5-TRIMETHYLHEXANOATE see AGU400
ALLYLTRIPHENYL STANNANE see AGU500
ALLYLTRIPHENYLTIN see AGU500
9-ALLYL-2-(4-(2-TRITYLETHYL)-1-PIPERAZINYL)-9H-
PURINEDIMETHANESULFONATE see AGU750
ALLYLUREA see AGV000
4-ALLYLVERATROLE see AGE250
ALLYL VINYL ETHER see AGV250
N-ALLYNORATROPINE see AGM250
ALLYXYCARB see DBI800
(ALL-Z)-5,8,11,14-EICOSATETRAENOIC ACID see AQS750
ALMAZINE see CFC250
ALMECILLIN see AGK250
ALMEDERM see HCL000
ALMEFRIN see SPC500
ALMEFROL see VSZ450
(±)-ALMINOPROFEN see MGC200
ALMITE see AHE250
ALMITRINA (SPANISH) see BGS500
ALMOCARPINE see PIF000, PIF250
ALMOND ARTIFICIAL ESSENTIAL OIL see BAY500
ALMOND OIL BITTER, FFPA (FCC) see BLV500
ALMORA see MAG000
ALNOVIN see FAO100
ALNOX see AFS500
ALNOXIN see PQC500
ALOBARBITAL see AFS500
ALOCASIA (VARIOUS SPECIES) see EAI600
ALOCHLOR see CFX000
ALODAN (GEROT) see DAM700
ALOE see AGV875
ALOE-EMODIN see DMU600
ALOGINAN see FOS100
ALON see AHE250
A. LONGIFLORA see BOO700
ALOPERIDIN see CLY500
ALOPERIDOLO see CLY500
ALOSITOL see ZVJ000
ALOTEC see MDM800
ALPEN see AIV500
ALPEN-N see SEQ000
ALPERIDINE HYDROCHLORIDE see AGV890
ALPEROX C see LBR000
ALPHA CHYMAR see CML850
ALPHA-CHYMAR OPHTH see CML850
ALPHACILINA see AOD000
ALPHACILLIN see AOD000
ALPHACROIC ORANGE R see NEY000
ALPHADIONE see AFK500
ALPHALIN see VSK600
ALPHA MEDOPA see DNA800
ALPHAMIN see FOS100
ALPHANAPHTHYL THIOUREA see AQN635
ALPHANAPHTYL THIOUREE (FRENCH) see AQN635
ALPHAPRODINE see NEA500
ALPHAPRODINE HYDROCHLORIDE see NEB000
ALPHASOL OT see DJL000
ALPHASONE ACETOPHENIDE see DAM300
ALPHASPRA see NAK500

ALPHASTEROL see VSK600
ALPHAXALONE see AFK875
ALPHAZURINE see FMU059
ALPHEBA see AGQ875
ALPHENAL see AGQ875
ALPHENATE see AGQ875
AL-PHOS see AHE750
ALPINE TALC USP, BC 127 see TAB750
ALPINE TALC USP, BC 141 see TAB750
ALPINE TALC USP, BC 662 see TAB750
ALPINYL see HIM000
ALPRAZOLAM see XAJ000
ALPRENOL HYDROCHLORIDE see AGW000
ALPRENOLOL see AGW250
ALPROSTADIL see POC350
ALPROSTADIL-α-CYCLODEXTRIN CLATHRATE see AGW275
ALQOVERIN see BRF500
ALQUEQUENJE (PUERTO RICO) see JBS100
ALRATO see AQN635
ALRHEUMAT see BDU500
ALRHEUMUM see BDU500
ALSEROXYLON see AGW300
ALSTONINE HYDROCHLORIDE see AGW375
ALTABACTINA see FPI000
ALTADIOL see EDS100
ALTAFUR see FPI000
ALTAN see DMH400
ALTAX see BDE750
ALTERNARIOL see AGW476
ALTERNARIOL and ALTERNARIOL MONOMETHYL ETHER (1:1)
see AGW550
ALTERNARIOL-9-METHYL ETHER see AGW500
ALTERNARIOL MONOMETHYL ETHER see AGW500
ALTERNARIOL MONOMETHYL ETHER and ALTERNARIOL (1:1)
see AGW550
ALTERTON see CKE750
ALTERUNGSSCHUTZMITTEL ZMB see ZIS000
ALTEZOL see AKO500
ALTHESIN see AFK500
ALTHOSE HYDROCHLORIDE see MDP000
ALTICINA see PDD350
ALTOCYL BRILLIANT BLUE B see MGG250
ALTODEL see POB000
ALTODOR see DIS600
ALTOSID see KAJ000
ALTOSID (GERMAN) see AGW625
ALTOSIDE see AGW625
ALTOSID IGR see KAJ000
ALTOSID SR 10 see KAJ000
ALTOTAL see PQC500
ALTOWHITES see KBB600
ALTOX see AFK250
ALTOZAR see EQD000
ALTRAD see EDO000
ALTRENOGEST see AGW675
ALTRETAMINE see HEJ500
ALUDRINE see DMV600
ALUFENAMINE see AHA875
ALUFIBRATE see AHA150
ALULINE see ZVJ000
ALUM see AHF200, AHG750
ALUMIGEL see AHC000
ALUMINA see AHE250, AHE250
β-ALUMINA see AHG000
γ-ALUMINA see AHE250
β″-ALUMINA see AHG000
α-ALUMINA (OSHA) see AHE250
ALUMINA FIBRE see AGX000
ALUMINA HYDRATE see AHC000
ALUMINA HYDRATED see AHC000
ALUMINA TRIHYDRATE see AHC000
α-ALUMINA TRIHYDRATE see AHC000
ALUMINIC ACID see AHC000
ALUMINON see AGW750
ALUMINOPHOSPHORIC ACID see PHB500
ALUMINUM see AGX000
ALUMINUM ACEGLUTAMIDE see AGX125
ALUMINUM ACETYLACETONATE see TNN000
ALUMINUM(III) ACETYLACETONATE see TNN000
ALUMINUM ACID PHOSPHATE see PHB500
ALUMINUM ALLOY, Al,Be see BFP250
ALUMINUM AMMONIUM SULFATE see AGX250
ALUMINUM AZIDE see AGX300
ALUMINUM BERYLLIUM ALLOY see BFP250

ALUMINUM BOROHYDRIDE see AGX500
ALUMINUM BROMHYDROXIDE see AGY000
ALUMINUM BROMIDE see AGX750
ALUMINUM BROMIDE, anhydrous see AGX750
ALUMINUM BROMIDE, solution (DOT) see AGX750
ALUMINUM BROMIDE HYDROXIDE see AGY000
ALUMINUM BROMOHYDROL see AGY000
ALUMINUM CALCIUM SILICATE see AGY100
ALUMINUM CARBIDE see AGY250
ALUMINUM CHLORATE see AGY500
ALUMINUM CHLORHYDRATE see AHA000
ALUMINUM CHLORHYDROL see AHA000
ALUMINUM CHLORHYDROXIDE see AHA000
ALUMINUMCHLORID (GERMAN) see AGY750
ALUMINUM CHLORIDE see AGY750
ALUMINUM CHLORIDE (1:3) see AGY750
ALUMINUM CHLORIDE, solution (DOT) see AGY750
ALUMINUM CHLORIDE, anhydrous (DOT) see AGY750
ALUMINUM CHLORIDE HEXAHYDRATE see AGZ000
ALUMINUM(III) CHLORIDE, HEXAHYDRATE see AGZ000
ALUMINUM CHLORIDE HYDROXIDE see AHA000
ALUMINUM CHLORIDE NITROMETHANE see AHA125
ALUMINUM CHLOROHYDROXIDE see AHA000
ALUMINUM CLOFIBRATE see AHA150
ALUMINUM COMPOUNDS see AHA175
ALUMINUM DEHYDRATED see AGX000
ALUMINUM DEXTRAN see AHA250
ALUMINUM ETHYLATE see AHA750
ALUMINUM FLAKE see AGX000
ALUMINUM FLUFENAMATE see AHA875
ALUMINUM FLUORIDE see AHB000
ALUMINUM FLUOROSULFATE, HYDRATE see AHB250
ALUMINUM FLUORURE (FRENCH) see AHB000
ALUMINUM FORMATE see AHB375
ALUMINUM FOSFIDE (DUTCH) see AHE750
ALUMINUM HEXAFLUOROSILICATE see THH000
ALUMINUM HYDRATE see AHC000
ALUMINUM HYDRIDE see AHB500
ALUMINUM HYDRIDE-DIETHYL ETHER see AHB625
ALUMINUM HYDRIDE-TRIMETHYL AMINE see AHB750
ALUMINUM HYDROXIDE see AHC000
ALUMINUM(III) HYDROXIDE see AHC000
ALUMINUM HYDROXIDE CHLORIDE see AHA000
ALUMINUM HYDROXIDE GEL see AHC000
ALUMINUM HYDROXIDE OXIDE see AHC250
ALUMINUM HYDROXYBROMIDE see AGY000
ALUMINUM HYDROXYCHLORIDE see AHA000
ALUMINUM IODIDE see AHC500
ALUMINUM LACTATE see AHC750
ALUMINUM LITHIUM HYDRIDE see LHS000
ALUMINUM MAGNESIUM PHOSPHIDE see AHD250
ALUMINUM METAHYDROXIDE see AHC250
ALUMINUM, METALLIC, POWDER (DOT) see AGX000
ALUMINUM METHYL see AHD500
ALUMINUM MONOPHOSPHATE see PHB500
ALUMINUM MONOPHOSPHIDE see AHE750
ALUMINUM MONOSTEARATE see AHA250
ALUMINUMNATRIUMLACTAT (GERMAN) see AHF750
ALUMINUM NITRATE (DOT) see AHD750
ALUMINUM(III) NITRATE (1:3) see AHD750
ALUMINUM NITRATE NONAHYDRATE see AHD900
ALUMINUM(III) NITRATE, NONAHYDRATE (1:3:9) see AHD900
ALUMINUM NITRIDE see AHE000
ALUMINUM OXIDE see AHE250, EAL100
α-ALUMINUM OXIDE see AHE250
β-ALUMINUM OXIDE see AHE250
γ-ALUMINUM OXIDE see AHE250
ALUMINUM OXIDE (2:3) see AHE250
ALUMINUM OXIDE HYDRATE see AHC000
ALUMINUM OXIDE SILICATE see AHF500
ALUMINUM OXIDE TRIHYDRATE see AHC000
ALUMINUM PHOSPHATE see PHB500
ALUMINUM PHOSPHATE (1:1) see PHB500
ALUMINUM PHOSPHATE, solution (DOT) see PHB500
ALUMINUM PHOSPHIDE see AHE750
ALUMINUM PHOSPHINATE see AHE875
ALUMINUM PICRATE see AHF000
ALUMINUM POTASSIUM SULFATE see AHF100
ALUMINUM POTASSIUM SULFATE, DODECAHYDRATE see AHF200
ALUMINUM POWDER see AGX000
ALUMINUM POWDER, UNCOATED, NON-PYROPHORIC (DOT) see AGX000
ALUMINUM SESQUIOXIDE see AHE250
ALUMINUM(III) SILICATE (2:1) see AHF500
ALUMINUM SODIUM FLUORIDE see SHF000

ALUMINUM SODIUM HYDRIDE see SEM500
ALUMINUM SODIUM LACTATE see AHF750
ALUMINUM SODIUM OXIDE see AHG000
ALUMINUM SODIUM SULFATE see AHG500
ALUMINUM STEARATE (ACGIH) see AHA250
ALUMINUM SULFATE (2:3) see AHG750
ALUMINUM TETRAHYDROBORATE see AGX500, AHG875
ALUMINUM THALLIUM SULFATE see AHH000
ALUMINUM-TITANIUM ALLOY (1:1) see AHH125
ALUMINUM TRIACETYLACETONATE see TNN000
ALUMINUM TRIBROMIDE see AGX750
ALUMINUM, TRIBROMOTRIMETHYLDI- see MGC225
ALUMINUM TRICHLORIDE see AGY750
ALUMINUM TRICHLORIDEHEXAHYDRATE see AGZ000
ALUMINUM, TRICHLOROTRIMETHYLDI- see MGC230
ALUMINUM TRIFLUORIDE see AHB000
ALUMINUM TRIHYDRAT see AHC000
ALUMINUM TRIHYDRIDE see AHB500
α-ALUMINUM TRIHYDRIDE see AHB500
ALUMINUM TRIHYDROXIDE see AHC000
ALUMINUM TRINITRATE see AHD750
ALUMINUM TRINITRATE NONAHYDRATE see AHD900
ALUMINUM, TRIPROPYL- see TMY100
ALUMINUM TRIPROPYL see AHH750
ALUMINUM TRIS(ACETYLACETONATE) see TNN000
ALUMINUM, TRIS(2-HYDROXYPROPANOATO-O^1),O^2)- (9CI) see AHC750
ALUMINUM, TRIS(LACTATO)- see AHC750
ALUMINUM TRIS(2-((3-TRIFLUOROMETHYL)PHENYL)AMINO)BENZOATO-N,o)-o-PYRIN see AHA875
ALUMINUM TRISULFATE see AHG750
ALUMITE see AHE250
ALUNDUM see AHE250
ALUNEX see TAI500
ALUPENT see MDM800
ALUPHOS see PHB500
ALURAL see BNP750
ALURATE see AFT000
ALURENE see CLH750
ALUSAL see AHC000
ALUTOR M 70 see PKB500
ALUTYL see PGG000
ALUZINE see CHJ750
ALVEDON see HIM000
ALVINOL see CHG000
ALVIT see DHB400
ALVO see OLW600
ALYPIN see AHI250
ALYPINE see AHI250
ALYSINE see SJO000
ALZODEF see CAQ250
AM-715 see BAB625
AM-2604 A see VRP775
1-A-2-MA (RUSSIAN) see AKM250
AMABEVAN see CBJ000
AMACEL BLUE BNN see MGG250
AMACEL BLUE GG see TBG700
AMACEL BRILLIANT BLUE B see MGG250
AMACEL DEVELOPED NAVY SD see DCJ200
AMACEL HELIOTROPE R see DBP000
AMACEL PURE BLUE B see TBG700
AMACEL YELLOW G see AAQ250
AMACID AMARANTH see FAG020
AMACID BLUE FG CONC see FMU059
AMACID BRILLIANT BLUE see FAE100
AMACID CHROME BLUE R see HJF500
AMACID FAST BLUE R see ADE750
AMACID GREEN G see FAF000
AMACID LAKE SCARLET 2R see FMU070
AMACID YELLOW M see MDM775
A. MACULATUM see ITD050
AMADIL see HIM000
AMAIZO W 13 see SLJ500
AMALOX see ZKA000
AMANIL BLACK GL see AQP000
AMANIL BLUE 2BX see CMO000
AMANIL FAST VIOLET N see CMP000
AMANIL SKY BLUE see CMO250
AMANIL SKY BLUE 6B see CMN750
AMANIL SUPRA BROWN LBL see CMO750
AMANITA RUBESCENS TOXIN see AHI500
α-AMANITIN (8CI, 9CI) see AHI625
α-AMANITINE see AHI625
AMANO N-AP see GGA800
AMANTADINE see TJG250

AMANTADINE HYDROCHLORIDE see AED250
AMANTHRENE GOLDEN YELLOW see DCZ000
AMAPALO AMRILLO (PUERTO RICO) see PLW800
AMAPLAST GREEN OZ see BLK000
AMAPLAST RED VIOLET P 2R see DBP000
AMARANT see FAG020
AMARANTH see FAG020
AMARANTHE USP (biological stain) see FAG020
AMARBEL EXTRACT see AHI630
AMAREX see MQQ250
AMARSAN see ABX500
AMARTHOL FAST RED TR BASE see CLK220, CLK235
AMARTHOL FAST RED TR SALT see CLK235
AMARTHOL FAST SCARLETT GG BASE see DEO400
AMARYLLIS see AHI635
AMATIN see HCC500
AMATOL see AHI750
AMAX see BDG000
AMAZE see IMF300
AMAZOLON see AED250
AMAZON YELLOW X2485 see DEU000
AMB see AOC500
AMBACAMP see BAB250
AMBAM see ANZ000
AMBATHIZON see FNF000
AMBAZON see AHI875
AMBAZONE see AHI875
AMBENONIUM CHLORIDE see MSC100
AMBENONIUM DICHLORIDE see MSC100
AMBENYL see BAU750
AMBER see AHJ000
AMBER ACID see SMY000
AMBERGRIS TINCTURE see AHJ000
AMBEROL ST 140F see AHC000
AMBESIDE see SNM500
AMBESTIGMIN CHLORIDE see MSC100
AMBIBEN see AJM000
AMBILHAR see NML000
AMBINON see CMG675
AMBISTRIN see SLY200
AMBLOSIN see AIV500
AMBOCHLORIN see CDO500
AMBOCLORIN see CDO500
AMBOFEN see CDP250
AMBOMYCIN see AFI625
AMBOX see BGB500
AMBRA see AHJ000
AMBRACYN see TBX250
AMBRAMICINA see TBX000
AMBRAMYCIN see TBX000
AMBROXOL see AHJ250
AMBROXOL HYDROCHLORIDE see AHJ500
AMBRUNATE see MNQ500
AMBUCETAMID see DDT300
AMBUCETAMIDE see DDT300
AMBUSH see AHJ750, CBM500
AMBUTYROSIN A see BSX325
AMBUYROSIN A see BSX325
6-AMC see CML800
AMCAP see AOD125
AMCHA see AJV500
trans-AMCHA see AJV500
AMCHEM 68-250 see CDS125
AMCHEM GRASS KILLER see TII250
AMCHEM R 14 see PKL750
AMCHEM 2,4,5-TP see TIX500
AMCIDE see ANU650
AMCILL see AIV500, AOD125
AMCILL-S see SEQ000
AMCINONIDE see COW825
AMCO see PMP500
AMD see DNA800, MJE780
AMDEX see BBK500
AMDON GRAZON see PIB900
AMDRAM see DBA800
AME see AGW500
AME and AOH (1:1) see AGW550
AMEBACILIN see FOZ000
AMEBAN see CBJ000
AMEBARSONE see CBJ000
AMEBICIDE see EAN000
AMEBIL see CHR500
AMECHOL see ACR000
AMEDEL see BHJ250

AMEDRINE see DBA800
AMEISENATOD see BBQ500
AMEISENMITTEL MERCK see BBQ500
AMEISENSAEURE (GERMAN) see FNA000
AMEPROMAT see MQU750
AMERCIAN CYANAMID 18133 see EPC500
AMERCIDE see CBG000
AMERFIL see PMP500
AMERICAINE see EFX000
AMERICAN ALLSPICE see CCK675
AMERICAN BITTERSWEET see AHJ875
AMERICAN CYANAMID 4,049 see MAK700
AMERICAN CYANAMID 5223 see DXX400
AMERICAN CYANAMID 12,008 see DJN600
AMERICAN CYANAMID 12,503 see POP000
AMERICAN CYANAMID 12880 see DSP400
AMERICAN CYANAMID 18682 see IOT000
AMERICAN CYANAMID 18706 see DNX600
AMERICAN CYANAMID-38023 see FAB600
AMERICAN CYANAMID-43073 see MJG500
AMERICAN CYANAMID 47031 see PGW750
AMERICAN CYANAMID AC 43,064 see DXN600
AMERICAN CYANAMID AC 43,913 see PHN250
AMERICAN CYANAMID AC 52,160 see TAL250
AMERICAN CYANAMID CL-24055 see DUI000
AMERICAN CYANAMID CL-38,023 see FAB600
AMERICAN CYANAMID CL-43913 see PHN250
AMERICAN CYANAMID CL-47,300 see DSQ000
AMERICAN CYANAMID CL-47470 see DHH400
AMERICAN CYANAMID E.I. 43,913 see PHN250
AMERICAN CYANIMID 24,055 see DUI000
AMERICAN ELDER see EAI100
AMERICAN HELLEBORE see VIZ000
AMERICAN HOLLY see HGF100
AMERICAN LAUREL see MRU359
AMERICAN MANDRAKE see MBU800
AMERICAN MEZEREON see LEF100
AMERICAN NIGHTSHADE see PJJ315
AMERICAN PENICILLIN see BFD250
AMERICAN PENNYROYAL OIL see PAR500
AMERICAN VERATRUM see VIZ000
AMERICAN VERMILION see CJD500
AMERICAN WHITE HELLEBORE see FAB100
AMERICIUM see AHK000
AMERICIUM TRICHLORIDE see AHK250
AMERIZOL see TOA000
AMERLATE P see IPS500
AMERLATE W see IPS500
AMEROL see AMY050
AMEROX OE-20 see PJW500
AMET (GERMAN) see AAI750
AMETANTRONE ACETATE see BKB300
AMETHOCAINE see BQA010
AMETHOCAINE HYDROCHLORIDE see TBN000
AMETHONE HYDROCHLORIDE see DIF200
AMETHOPTERIN see MDV500
AMETHOPTERIN SODIUM see MDV600
AMETHYST see SCI500, SCJ500
AMETOX see SKI500
AMETRIODINIC ACID see AAI750
AMETYCIN see AHK500
AMEZINIUMMETILSULFAT (GERMAN) see MQS100
AMEZINIUM METILSULFATE see MQS100
AMFENACO (SPANISH) see AIU500
AMFENAC SODIUM MONOHYDRATE see AHK625
AMFETAMINA see AOB250
AMFETAMINE see AOB250
d-AMFETASUL see BBK500
AMFETYLINE HYDROCHLORIDE see CBF825
AMFH see AAH100
AMFIPEN see AIV500
AMFOMYCIN see AOB875
AMGABA see AJC375
AMIANTHUS see ARM250
AMIBIARSON see CBJ000
AMICAR see AJD000
AMICARDINE see AHK750
AMICETIN see AHL000
AMICETIN CITRATE see AHL250
AMICHLOPHENE see CJN750
AMICIDE see ANU650
AMIDANTEL see DPF200
AMIDATE see HOU100
AMIDAZIN see EPQ000

AMIDAZOPHEN see DOT000
AMIDEFRINE MESYLATE see AHL500
AMIDE of GABA see AJC375
AMIDEPHRINE MESYLATE see AHL500
AMIDEPHRINE MONOMETHANESULFONATE see AHL500
AMIDE PP see NCR000
AMIDES see AHL750
AMIDINE BLUE 4B see CMO250
p-AMIDINOBENZOIC ACID BUTYL ESTER see AHL875
p-AMIDINOBENZOIC ACID HEXYL ESTER see AHL880
p-AMIDINOBENZOIC ACID PENTYL ESTER see AHL885
p-AMIDINOBENZOIC ACID PROPYL ESTER see AHL890
N-AMIDINO-2-(2,6-DICHLOROPHENYL)ACETAMIDE HYDROCHLORIDE
    see GKU300
N-(2-AMIDINOETHYL)-3-AMINOCYCLOPENTANOCARBOXAMIDE
    see AHN625
N''-(2-AMIDINOETHYL)-4-FORMAMIDO-1,1',1''-TRIMETHYL-(N,4':N',4''-TER-
    PYRROLE)-2-CARBOXAMIDE see SLI300
N''-(2-AMIDINOETHYL)-4-FORMAMIDO-1,1',1''-TRIMETHYL-(N,4':N',4''-TER-
    PYRROLE)-2-CARBOXAMIDE HYDROCHLORIDE see DXG600
1-AMIDINOHYDRAZONO-4-THIOSEMICARBAZONO-2,5-CYCLOHEXADIENE
    see AHI875
S-(AMIDINOMETHYL) HYDROGEN THIOSULFATE see AHN000
AMIDINOMYCIN see AHN625
3-AMIDINO-1-p-NITROPHENYLUREA HYDROCHLORIDE see NIJ400
4-AMIDINO-1-(NITROSAMINOAMIDINO)-1-TETRAZENE see TEF500
N¹-AMIDINOSULFANILAMIDE see AHO250
AMIDITHION see AHO750
AMID KYSELINY OCTOVE see AAI000
AMID KYSELINY STAVELOVE (CZECH) see OLO000
o-AMIDOAZOTOLUOL (GERMAN) see AIC250
o-AMIDOBENZOIC ACID see API500
AMIDOCYANOGEN see COH500
AMIDOFEBRIN see DOT000
AMIDOFOS see COD850
AMIDO-G-ACID see NBE850
AMIDOL see AHP000
AMIDOLINE see AHP125
AMIDON see ILD000
AMIDONE see MDO750
AMIDONE HYDROCHLORIDE see MDP000
AMIDON HYDROCHLORIDE see MDP750
AMIDOPHEN see DOT000
AMIDOPHENAZONE see DOT000
AMIDOPHOS see COD850
AMIDOPYRAZOLINE see DOT000
AMIDOPYRIN see DOT000
AMIDOSAL see SAH000
AMIDOSULFONIC ACID see SNK500
AMIDOSULFURIC ACID see SNK500
AMIDO SULFURYL AZIDE see AHP250
AMIDOTRIZOATE MEGLUMINE see AOO875
AMIDOTRIZOIC ACID see DCK000
AMIDOUREA HYDROCHLORIDE see SBW500
AMIDOX see DAA800
AMIDOXAL see SNN500
AMIDO YELLOW EA-CF see SGP500
AMIDRINE see ILM000
AMIDRYL see BBV500
AMID-SAL see SAH000
AMID-THIN see NAK000
AMIFENATSOL HYDROCHLORIDE see DCA600
AMIFUR see NGE500
AMIKACIN see APS750
AMIKACIN SULFATE see APT000
AMIKAPRON see AJV500
AMIKHELLIN HYDROCHLORIDE see AHP375
AMIKIN see APT000
AMIKLIN see APT000
AMILAN see NOH000
AMILAN CM 1001 see PJY500
AMILAR see PKF750
AMILPHENOL see AON000
AMINACRINE see AHS500
AMINACRINE HYDROCHLORIDE see AHS750
AMINARSON see CBJ000
AMINARSONE see CBJ000
AMINASINE see CKP250
AMINATE BASE see AKC750
AMINAZIN see CKP250
AMINAZINE see CKP250
AMINAZIN MONOHYDROCHLORIDE see CKP500
AMINE 220 see AHP500
AMINES see AHP750

AMINES, FATTY see AHP760
AMINE 2,4,5-T FOR RICE see TAA100
AMINIC ACID see FNA000
AMINICOTIN see NCR000
AMINITROZOLE see ABY900
5-AMINOACENAPHTHENE see AAF250
5-((2-AMINOACETAMIDO)METHYL)-1-(4-CHLORO-2-(o-
    CHLOROBENZOYL)PHENYL)-N,N-DIMETHYL-1H-s-TRIAZOLE-3-CAR-
    BOXAMIDE, HYDROCHLORIDE, DIHYDRATE see AHP875
2-AMINO-4-ACETAMINIFENETOL (CZECH) see AJT750
3-AMINOACETANILID (CZECH) see AHQ000
4'-AMINOACETANILID (CZECH) see AHQ250
4-AMINOACETANILIDE see AHQ250
m-AMINOACETANILIDE see AHQ000
p-AMINOACETANILIDE see AHQ250
3'-AMINOACETANILIDE see AHQ000
4'-AMINOACETANILIDE see AHQ250
AMINOACETIC ACID (GERMAN) see GHA000
AMINOACETONITRILE see GHI000
AMINOACETONITRILE BISULFATE see AHQ750
AMINOACETONITRILE HYDROCHLORIDE see GHI100
AMINOACETONITRILE HYDROGEN SULFATE see AHQ750
AMINOACETONITRILE HYDROSULFATE see AHQ750
AMINOACETONITRILE SULFATE see AHR000
2-AMINOACETOPHENONE see AHR250
m-AMINOACETOPHENONE see AHR500
p-AMINO ACETOPHENONE see AHR750
β-AMINOACETOPHENONE see AHR500
3'-AMINOACETOPHENONE see AHR500
4'-AMINOACETOPHENONE see AHR750
ω-AMINOACETOPHENONE see AHR250
m-AMINOACETYLBENZENE see AHR500
p-AMINOACETYLBENZENE see AHR750
2-AMINOACRIDINE see AHS000
3-AMINOACRIDINE see ADJ375
5-AMINOACRIDINE see AHS500
9-AMINOACRIDINE see AHS500
2-AMINOACRIDINE (EUROPEAN) see ADJ375
3-AMINOACRIDINE (EUROPEAN) see AHS000
AMINOACRIDINE HYDROCHLORIDE see AHS750
5-AMINOACRIDINE HYDROCHLORIDE see AHS750
9-AMINOACRIDINE MONOHYDROCHLORIDE see AHS750
9-AMINOACRIDINE PENICILLIN see AHT000
4'-(2-AMINO-9-ACRIDINYLAMINO)METHANESULFONANILIDE see AHT825
4'-((3-AMINO-9-ACRIDINYL)AMINO)METHANESULFONANILIDE see AHT250
1-AMINOADAMANTANE see TJG250
AMINOADAMANTANE HYDROCHLORIDE see AED250
1-AMINOADAMANTENE HYDROCHLORIDE see AED250
1-AMINOADAMATANE see TJG250
2-AMINOADENINE see POJ500
2-AMINOETHANOL (GERMAN) see EEC600
2-AMINOAETHYLISOSELENOURONIUMBROMID-HYDROBROMID(GER-
    MAN) see AJY000
β-AMINOAETHYLISOTHIURONIUM-CHLORID-HYDROCHLORID (GERMAN)
    see AJY500
β-AMINOAETHYL-ISOTHIURONIUM DIHYDROBROMID(GERMAN)
    see AJY250
β-AMINOAETHYL-MORPHOLIN (GERMAN) see AKA750
4-AMINO-6-(2-AMINO-1,6-DIMETHYLPYRIMIDINIUM-4-YL)AMINO)-1-
    METHYL-QUINALDINIUM BIS(METHYL SULFATE) see AQN625
6-AMINO-2-(3'-AMINOFENYL)BENZIMIDAZOL HYDROCHLORID (CZECH)
    see AHT900
6-AMINO-2-(4'-AMINOFENYL)BENZIMIDAZOL HYDROCHLORID (CZECH)
    see AHT950
3-AMINO-N-(3-AMINO-3-IMINOPROPYL)CYCLOPENTANECARBOXAMIDE
    see AHN625
6-AMINO-4-((3-AMINO-4-(((4-((1-METHYLPYRIDINIUM-4-YL)AMINO)PHENYL)
    AMINO)CARBONYL)PHENYL)AMINO)-1-METHYLQUINOLINIUM),DIIO-
    DIDE see AHT850
4-AMINO-N-(2'-AMINOPHENYL)BENZAMIDE see DBQ125
6-AMINO-2-(3'-AMINOPHENYL)BENZIMIDAZOLE, DIHYDROCHLORIDE
    see AHT900
6-AMINO-2-(4'-AMINOPHENYL)BENZIMIDAZOLE DIHYDROCHLORIDE
    see AHT950
2-AMINOANILINE see PEY250
3-AMINOANILINE see PEY000
4-AMINOANILINE see PEY500
m-AMINOANILINE see PEY000
p-AMINOANILINE see PEY500
3-AMINOANILINE DIHYDROCHLORIDE see PEY750
4-AMINOANILINE DIHYDROCHLORIDE see PEY650
m-AMINOANILINE DIHYDROCHLORIDE see PEY750
p-AMINOANILINE DIHYDROCHLORIDE see PEY650
3-AMINO-p-ANISIC ACID see AIA250

2-AMINOANISOLE see AOV900
4-AMINOANISOLE see AOW000
o-AMINOANISOLE see AOV900
p-AMINOANISOLE see AOW000
4-AMINOANISOLE-3-SULFONIC ACID see AIA500
1-AMINOANTHRACENE see APG050
2-AMINOANTHRACENE see APG100
α-AMINOANTHRACENE see APG050
β-AMINOANTHRACENE see APG100
1-AMINO-9,10-ANTHRACENEDIONE see AIA750
2-AMINO-9,10-ANTHRACENEDIONE see AIB000
1-AMINOANTHRACHINON (CZECH) see AIA750
1-AMINOANTHRAQUINONE see AIA750
2-AMINOANTHRAQUINONE see AIB000
α-AMINOANTHRAQUINONE see AIA750
β-AMINOANTHRAQUINONE see AIB000
1-AMINO-9,10-ANTHRAQUINONE see AIA750
N-(4-AMINOANTHRAQUINONYL)BENZAMIDE see AIB250
2-AMINO-9,10-ANTRAQUINONE see AIB000
4-AMINO-1-ARABINOFURANOSYL-2-OXO-1,2-DIHYDROPYRIMIDINE
    see AQQ750
4-AMINO-1-β-d-ARABINOFURANOSYL-2(1H)-PYRIMIDINONE (9CI)
    see AQQ750
2-AMINO-4-ARSENOSOPHENOL see OOK100
2-AMINO-4-ARSENOSOPHENOL HYDROCHLORIDE see ARL000
AMINOARSON see CBJ000
AMINOAZOBENZENE see PEI000
4-AMINOAZOBENZENE see PEI000
p-AMINOAZOBENZENE see PEI000
4-AMINO-1,1'-AZOBENZENE see PEI000
4-AMINOAZOBENZENE-3,4'-DISULFONIC ACID see AJS500
4-AMINOAZOBENZENE HYDROCHLORIDE see PEI250
p-AMINOAZOBENZENE HYDROCHLORIDE see PEI250
4-AMINOAZOBENZOL see PEI000
p-AMINOAZOBENZOL see PEI000
6-AMINO-3,4'-AZODI-BENZENESULFONIC ACID see AJS500
o-AMINOAZOTOLUENE (MAK) see AIC250
2-AMINO-5-AZOTOLUENE see AIC250
p-AMINO-2:3-AZOTOLUENE see AIC000
2'-AMINO-2:5'-AZOTOLUENE see TGW500
4'-AMINO-2,5'-AZOTOLUENE see AIC250
4'-AMINO-2:3'-AZOTOLUENE see AIC250
4'-AMINO-3,2'-AZOTOLUENE see AIC000
4'-AMINO-4,2'-AZOTOLUENE see AIC500
4'-AMINO-4-3'-AZOTOLUENE see TGW750
AMINOAZOTOLUENE (indicator) see AIC250
o-AMINOAZOTOLUENO (SPANISH) see AIC250
o-AMINOAZOTOLUOL see AIC250
AMINOBENZ see AKF000
10-AMINOBENZ(a)ACRIDINE see AIC750
4-AMINOBENZALDEHYDE THIOSEMICARBAZONE see AID250
p-AMINOBENZALDEHYDETHIOSEMICARBAZONE see AID250
m-AMINOBENZAL FLUORIDE see AID500
3-AMINOBENZAMIDE see AID625
m-AMINOBENZAMIDE see AID625
3-AMINO-BENZAMIDE (9CI) see AID625
4-AMINOBENZAMIDINE see AID650
p-AMINOBENZAMIDINE see AID650
1-AMINO-4-BENZAMIDOANTHRAQUINONE see AIB250
5-AMINO-1:2-BENZANTHRACENE see BBC000
10-AMINO-1,2-BENZANTHRACENE see BBB750
AMINOBENZENE see AOQ000
(±)-α-AMINO-BENZENEACETIC ACID, DECYL ESTER, HYDROCHLORIDE
    see PFF500
(±)-α-AMINO-BENZENEACETIC ACID HEPTYL ESTER HYDROCHLORIDE
    see PFF750
(±)-α-AMINOBENZENEACETIC ACID,3-METHYLBUTYL ESTER HYDRO-
    CHLORIDE (9CI) see PCV750
(±)-α-AMINO-BENZENEACETIC ACID NONYL ESTER HYDROCHLORIDE
    see PFG250
(±)-α-AMINO-BENZENEACETIC ACID OCTYL ESTER HYDROCHLORIDE
    see PFG500
(±)-α-AMINO-BENZENEACETIC ACID PENTYL ESTER HYDROCHLORIDE
    see PFG750
4-AMINOBENZENEARSONIC ACID see ARA250
p-AMINOBENZENEARSONIC ACID see ARA250
p-AMINOBENZENEAZODIMETHYLANILINE see DPO200
p-AMINO BENZENE DIAZONIUMPERCHLORATE see AID750
2-AMINO-p-BENZENEDISULFONIC ACID see AIE000
1-AMINO-2,5-BENZENEDISULFONIC ACID see AIE000
2-AMINO-1,4-BENZENEDISULFONIC ACID see AIE000
2-AMINO-BENZENE-1,4-DISULFONIC ACID see AIE000
(S)-α-AMINOBENZENEPROPANOIC ACID see PEC750
4-AMINOBENZENESTIBONIC ACID see SLP600

p-AMINOBENZENESTIBONIC ACID see SLP600
p-AMINOBENZENESTIBONIC ACID COMPOUND with UREA (3:1) see AIE125
p-AMINOBENZENESULFAMIDE see SNM500
3-(p-AMINOBENZENESULFAMIDO)-6-METHOXYPYRIDAZINE see AKO500
2-(p-AMINOBENZENESULFANAMIDE)-3-METHOXYPYRAZINE see MFN500
p-AMINOBENZENESULFONACETAMIDE see SNP500
4-AMINOBENZENESULFONAMIDE see SNM500
p-AMINOBENZENESULFONAMIDE see SNM500
6-(4-AMINOBENZENESULFONAMIDO)-4,5-DIMETHOXYPYRIMIDINE
    see AIE500
5-(p-AMINOBENZENESULFONAMIDO)-3,4-DIMETHYLISOOXALE see SNN500
5-(p-AMINOBENZENESULFONAMIDO)-3,4-DIMETHYLISOXAZOLE see SNN500
2-(p-AMINOBENZENESULFONAMIDO)-4,5-DIMETHYLOXAZOLE see AIE750
2-(p-AMINOBENZENESULFONAMIDO)-4,6-DIMETHYLPYRIMIDINE see SNJ000
6-(4-AMINOBENZENESULFONAMIDO)-2,4-DIMETHYLPYRIMIDINE see SNJ350
6-(p-AMINOBENZENESULFONAMIDO)-2,4-DIMETHYLPYRIMIDINE see SNJ350
2-(p-AMINOBENZENESULFONAMIDO)-5-METHOXYPYRIMIDINE SODIUM
    SALT see MFO000
2-(p-AMINOBENZENESULFONAMIDO)-5-METHYLTHIADIAZOLE see MPQ750
3-(p-AMINOBENZENESULFONAMIDO)-2-PHENYLPYRAZOLE see AIF000
2-p-AMINOBENZENESULFONAMIDOQUINOXALINE see QTS000
2-(p-AMINOBENZENESULFONAMIDO)THIAZOLE see TEX250
3-AMINO-BENZENESULFONIC ACID see SNO000
m-AMINOBENZENESULFONIC ACID see SNO000
1-AMINOBENZENE-3-SULFONIC ACID see SNO000
m-AMINOBENZENESULFONIC ACID SODIUM SALT see AIF250
p-AMINOBENZENESULFONYL-2-AMINO-4,5-DIMETHYLOXAZOLE see AIE750
6-(p-AMINOBENZENESULFONYL)AMINO-2,4-DIMETHYLPYRIMIDINE
    see SNJ350
N-(4-AMINOBENZENESULFONYL)-N-BUTYLUREA see BSM000
p-AMINOBENZENESULFONYLGUANIDINE see AHO250
p-AMINOBENZENESULFONYLUREA see SNQ550
1-(4-AMINOBENZENESULFONYL)UREA see SNQ550
5-(p-AMINOBENZENESULPHONAMIDE)-3,4-DIMETHYLISOXAZOLE
    see SNN500
2-p-AMINOBENZENESULPHONAMIDO-4,6-DIMETHOXYPYRIMIDINE
    see SNI500
5-(p-AMINOBENZENESULPHONAMIDO)-3,4-DIMETHYLISOXAZOLE
    see SNN500
3-p-AMINOBENZENESULPHONAMIDO-7-METHOXYPYRIDAZINE see AKO500
2-p-AMINOBENZENESULPHONAMIDOQUINOXALINE see QTS000
2-(p-AMINOBENZENESULPHONAMIDO)THIAZOLE see TEX250
N-p-AMINOBENZENESULPHONYLGUANIDINEMONOHYDRATE see AHO250
2-AMINOBENZENETHIOL see AIF500
4-AMINOBENZENETHIOL see AIF750
p-AMINOBENZENETHIOL see AIF750
2-AMINOBENZIMIDAZOLE see AIG000
2-AMINO-6-BENZIMIDAZOLYL PHENYLKETONE see AIH000
2-AMINOBENZOIC ACID see API500
4-AMINOBENZOIC ACID see AIH600
m-AMINOBENZOIC ACID see AIH500
o-AMINOBENZOIC ACID see API500
p-AMINOBENZOIC ACID see AIH600
γ-AMINOBENZOIC ACID see AIH600
p-AMINOBENZOIC ACID-2-N-AMYLAMINOETHYL ESTER see PBV750
p-AMINOBENZOIC ACID BUTYL ESTER see BPZ000
p-AMINOBENZOIC ACID-3-(DIBUTYLAMINO)PROPYL ESTER, HYDROCHLO-
    RIDE see AIT000
p-AMINOBENZOIC ACID-2-(2-(2-(2-
    (DIETHYLAMINO)ETHOXY)ETHOXY)ETHOXY)ETHYL ESTER,
    HYDROCHLORIDE see AIK500
p-AMINOBENZOIC ACID-2-(2-(2-
    (DIETHYLAMINO)ETHOXY)ETHOXY)ETHOXY)ETHYL ESTER, HYDRO-
    CHLORIDE see AIK750
p-AMINOBENZOIC ACID-2-(2-(2-(DIETHYLAMINO)ETHOXY)ETHOXY)ETHYL
    ESTER, HYDROCHLORIDE see AIL000
p-AMINOBENZOIC ACID-2-(2-(DIETHYLAMINO)ETHOXY)ETHYL ESTER, HY-
    DROCHLORIDE see AIL250
p-AMINOBENZOIC ACID-3-(β-DIETHYLAMINO)ETHOXY)PROPYL ESTER
    see AIL500
4-AMINOBENZOIC ACID DIETHYLAMINOETHYL ESTER see AIL750
p-AMINOBENZOIC ACID-2-DIETHYLAMINOETHYL ESTER see AIL750
4-AMINOBENZOIC ACID 2-(DIETHYLAMINO)ETHYL ESTER, HYDROCHLO-
    RIDE see AIT250
p-AMINOBENZOIC ACID-2-DIETHYLAMINOETHYL ESTER, HYDROCHLO-
    RIDE see AIT250
p-AMINOBENZOIC ACID-N-1-DIETHYLAMINO-1-ISOBUTYLETHANOL
    METHANESULFONATE see LEU000
p-AMINOBENZOIC ACID-β-DIETHYLAMINOISOHEXYL ESTER
    METHANESULFONATE see LEU000
p-AMINOBENZOIC ACID-3-(DIETHYLAMINO)PROPYL ESTER HYDROCHLO-
    RIDE see AIM000
p-AMINOBENZOIC ACID-N,N-DIETHYLLEUCINOL ESTER
    METHANESULFONATE see LEU000

p-AMINOBENZOIC ACID-2-(DIISOPROPYLAMINO)ETHYL ESTER, HYDRO-
CHLORIDE see AIT500
p-AMINOBENZOIC ACID-3-(DIISOPROPYLAMINO)PROPYL ESTER HYDRO-
CHLORIDE-37-7 see AIM250
p-AMINOBENZOIC ACID 3-(DIMETHYLAMINO)-1,2-DIMETHYLPROPYL
ESTER, HYDROCHLORIDE see AIT750
p-AMINOBENZOIC ACID-(2-(DIMETHYLAMINO)-1-PHENYL)ETHYL ESTER
see AIU250
p-AMINOBENZOIC ACID-2-(DIPROPYLAMINO)ETHYL ESTER HYDROCHLO-
RIDE see AIN000
p-AMINOBENZOIC ACID 3-(DIPROPYLAMINO)PROPYL ESTER, HYDRO-
CHLORIDE see AIU000
4-AMINOBENZOIC ACID ETHYL ESTER see EFX000
o-AMINOBENZOIC ACID, ETHYL ESTER see EGM000
p-AMINOBENZOIC ACID ETHYL ESTER see EFX000
3-AMINOBENZOIC ACID ETHYL ESTER METHANESULFONATE see EFX500
p-AMINO-BENZOIC ACID 1-ETHYL-4-PIPERIDYL ESTER, HYDROCHLO-
RIDE see EOV000
p-AMINOBENZOIC ACID ISOBUTYL ESTER see MOT750
2-AMINOBENZOIC ACID METHYL ESTER see APJ250
o-AMINOBENZOIC ACID METHYL ESTER see APJ250
p-AMINOBENZOIC ACID-(2-METHYL-2-(1-METHYL-HEPTYLAMINO)PROPYL
ESTER HYDROCHLORIDE see MLO750
p-AMINOBENZOIC ACID-3-(3-METHYLPIPERIDINO) PROPYL ESTER HY-
DROCHLORIDE see MOK500
p-AMINO-BENZOIC ACID-1-METHYL-4-PIPERIDYL ESTER, HYDROCHLO-
RIDE see MON000
4-AMINOBENZOIC ACID, MODOSODIUM SALT see SEO500
p-AMINOBENZOIC ACID MONOGLYCERYL ESTER see GGQ000
p-AMINO-BENZOIC ACID-1-PHENETHYL-4-PIPERIDYL ESTER HYDROCHLO-
RIDE see PDJ000
2-AMINOBENZOIC ACID-3-PHENYL-2-PROPENYL ESTER see API750
p-AMINOBENZOIC ACID PHOSPHATE see AIQ875
p-AMINOBENZOIC ACID-3-PIPERIDINOPROPYL ESTER HYDROCHLORIDE
see PIS000
m-AMINOBENZOIC ACID-2-(2-PIPERIDYL)ETHYL ESTER HYDROCHLORIDE
see AIQ880
o-AMINOBENZOIC ACID 2-(2-PIPERIDYL)ETHYL ESTER HYDROCHLORIDE
see AIQ885
p-AMINOBENZOIC ACID-2-(2-PIPERIDYL)ETHYL ESTER HYDROCHLORIDE
see AIQ890
p-AMINOBENZOIC ACID SODIUM SALT see SEO500
p-AMINOBENZOIC DIETHYLAMINOETHYLAMIDE see AJN500
2-(p-AMINOBENZOLSULFONAMIDO)-4,5-DIMETHYLOXAZOL (GERMAN)
see AIE750
6-(4'-AMINOBENZOL-SULFONAMIDO)-2,4-DIMETHYLPYRIMIDIN (GERMAN)
see SNJ000
(p-AMINOBENZOLSULFONYL)-2-AMINO-4,6-DIMETHYLPYRIMIDIN (GER-
MAN) see SNJ000
(p-AMINOBENZOLSULFONYL)-4-AMINO-2,6-DIMETHYLPYRIMIDIN (GER-
MAN) see SNJ350
(p-AMINOBENZOLSULFONYL)-4-AMINO-2-METHYL-6-METHOXY-PYRIMIDIN
(GERMAN) see MDU300
(p-AMINOBENZOLSULFONYL)-2-AMINO-4-METHYLPYRIMIDIN (GERMAN)
see ALF250
4-AMINO-BENZOLSULFONYL-METHYLCARBAMAT (GERMAN) see SNQ500
2-AMINOBENZONITRILE see APJ750
3-AMINOBENZONITRILE see AIR125
m-AMINOBENZONITRILE see AIR125
o-AMINOBENZONITRILE see APJ750
4-AMINOBENZONITRILE (9CI) see COK125
p-AMINOBENZONITRILE (8CI) see COK125
p-AMINOBENZOPHENONE see AIR250
3-AMINOBENZO-6,7-QUINAZOLINE-4-ONE see AIS250
3-AMINOBENZO(g)QUINAZOLIN-4(3H)-ONE see AIS250
2-AMINOBENZOTHIAZOLE see AIS500
6-AMINO-2-BENZOTHIAZOLETHIOL see AIS550
3-AMINOBENZOTRIFLUORIDE see AID500
m-AMINOBENZOTRIFLUORIDE see AID500
p-AMINOBENZOTRIFLUORIDE see TKB750
2-AMINOBENZOXAZOLE see AIS600
3-(p-AMINOBENZOXY)-1-DI-n-BUTYLAMINOPROPANE see BOO750
3-(p-AMINOBENZOXY)-1-DI-n-BUTYLAMINOPROPANE SULFATE see BOP000
1-AMINO-4-BENZOYLAMINOANTHRACHINON (CZECH) see AIB250
1-AMINO-4-(BENZOYLAMINO)ANTHRAQUINONE see AIB250
4-AMINO-1-BENZOYLAMINOANTHRAQUINONE see AIB250
p-AMINOBENZOYLAMINOMETHYLHYDROCOTARNINE see AIS625
2-AMINO-3-BENZOYL BENZENEACETIC ACID see AIU500
2-AMINO-3-BENZOYLBENZENEACETIC ACID SODIUM SALT HYDRATE
see AHK625
2-AMINO-5-BENZOYLBENZIMIDAZOLE see AIH000
p-AMINOBENZOYLDIBUTYLAMINOPROPANOL see BOO750
AMINOBENZOYLDIBUTYLAMINOPROPANOL HYDROCHLORIDE see AIT000
p-AMINOBENZOYLDIBUTYLAMINOPROPANOL SULFATE see BOP000

p-AMINOBENZOYLDIETHYLAMINOETHANOL see AIL750
p-AMINOBENZOYLDIETHYLAMINOETHANOL HYDROCHLORIDE see AIT250
p-AMINO BENZOYL DIETHYL AMINO PROPANOL HYDROCHLORIDE
see AIM000
o-AMINOBENZOYL DI(ISOPROPYLAMINO)ETHANOL HYDROCHLORIDE
see IJZ000
p-AMINO BENZOYL DI-ISO-PROPYL AMINO ETHANOL HYDROCHLORIDE
see AIT500
p-AMINO BENZOYL DIISOPROPYL AMINO PROPANOL HYDROCHLORIDE
see AIM250
p-AMINOBENZOYLDIMETHYLAMINO-1,2-DIMETHYLPROPANOL HYDRO-
CHLORIDE see AIT750
p-AMINO BENZOYL DI-N-PROPYL AMINO ETHANOL HYDROCHLORIDE
see AIN000
p-AMINO BENZOYL DI-N-PROPYL AMINOPROPANOL HYDROCHLORIDE
see AIU000
o-AMINOBENZOYLFORMIC ANHYDRIDE see ICR000
β-4-AMINOBENZOYLOXY-β-PHENYLETHYL DIMETHYLAMINE see AIU250
2-AMINO-3-BENZOYLPHENYLACETIC ACID see AIU500
2-AMINOBENZTHIAZOLE see AIS500
4-(4-AMINOBENZYL)ANILINE see MJQ000
α-(α-AMINOBENZYL)BENZYL ALCOHOL HYDROCHLORIDE see DWB000
2-AMINO-6-BENZYLMERCAPTOPURINE see BFL125
2-AMINO-6-BENZYL-MP see BFL125
AMINOBENZYLPENICILLIN see AIV500
d-(−)-α-AMINOBENZYLPENICILLIN see AIV500
d-(−)-α-AMINOBENZYLPENICILLIN SODIUM SALT see SEQ000
AMINOBENZYLPENICILLIN TRIHYDRATE see AOD125
α-AMINOBENZYLPENICILLIN TRIHYDRATE see AOD125
4-AMINO-N-(1-BENZYL-4-PIPERIDYL)-5-CHLORO-o-ANISAMIDE
HYDROXYSUCCINATE see AIV625
2-AMINO-6-(BENZYLTHIO)PURINE see BFL125
2-AMINOBIPHENYL see BGE250
4-AMINOBIPHENYL see AJS100
o-AMINOBIPHENYL see BGE250
p-AMINOBIPHENYL see AJS100
4-AMINOBIPHENYL DIHYDROCHLORIDE see BGE300
4-AMINO-3-BIPHENYLOL see AIV750
4'-AMINO-4-BIPHENYLOL see AIW000
3-AMINO-4-BIPHENYLOL HYDROCHLORIDE see AIW250
4-AMINO-3-BIPHENYLOL HYDROGEN SULFATE see AKF250
N-(4'-AMINO(1,1'-BIPHENYL)-4-YL)-ACETAMIDE see ACC000
2-AMINO-5-BIPHENYLYLIMIDAZOLEHYDROCHLORIDE see AIW500
2-(4'-AMINO-1,1'-BIPHENYL-4-YL)-2H-NAPHTHO(1,2-d)TRIAZOLE-6,8-DIS
ULFONIC ACID, DIPOTASSIUM SALT see AIW750
5-AMINO(3,4'-BIPYRIDIN)-6-(1H)-ONE see AOD375
2-AMINO-4-(p-(BIS(2-CHLOROETHYL)AMINO)PHENYL)BUTYRIC ACID
see AJE000
3-((AMINO(BIS(2-CHLOROETHYL)AMINO)PHOSPHINYL)OXY)PROPANOIC
ACID see CCE250
5-AMINO-1-BIS(DIMETHYLAMIDE)PHOSPHORYL-3-PHENYL-1,2,4-TRIAZOLE
see AIX000
5-AMINO-1-BIS(DIMETHYLAMIDO)PHOSPHORYL-3-PHENYL-1,2,4-TRI-
AZOLE see AIX000
5-AMINO-1-(BIS(DIMETHYLAMINO)PHOSPHINYL)-3-PHENYL-1,2,4-TRI-
AZOLE see AIX000
AMINOBIS(PROPYLAMINE) see AIX250
1-AMINO-4-BROMANTHRACHINON-2-SULFONAN SODNY (CZECH)
see DKR000
1-AMINO-2-BROM-4-HYDROXYANTHRACHINON (CZECH) see AIY500
2-AMINO-5-BROMOBENZOXAZOLE see AIX500
2-AMINO-6-BROMOBENZOXAZOLE see AIX750
2-AMINO-5-BROMO-6-CHLOROBENZOXAZOLE see AIY000
2-AMINO-6-BROMO-5-CHLOROBENZOXAZOLE see AIY250
4-AMINO-5-BROMO-N-(2-(DIETHYLAMINO)ETHYL)-o-ANISAMIDE see VCK100
1-AMINO-2-BROMO-4-HYDROXYANTHRAQUINONE see AIY500
1-AMINO-2-BROMO-4-(2-(2-HYDROXYETHYL)SULFONYL-4-
METHYLPHENYLAMINO)ANTHRAQUINONE see AIY750
2-AMINO-5-BROMO-6-PHENYL-4(1H)-PYRIMIDINONE see AIY850
N-AMINO-2-(m-BROMOPHENYL)SUCCINIMIDE see AIZ000
1-AMINO-3-BROMOPROPANE HYDROBROMIDE see AJA000
1-AMINO-BUTAAN (DUTCH) see BPX750
1-AMINOBUTAN (GERMAN) see BPX750
1-AMINOBUTANE see BPX750
2-AMINOBUTANE see BPY000
4-AMINOBUTANOIC ACID see PIM500
2-AMINOBUTAN-1-OL see AJA250
2-AMINO-1-BUTANOL see AJA250
4-AMINO-2-(4-BUTANOYLHEXAHYDRO-1H-1,4-DIAZEPIN-1-YL)-6,7-
DIMETHOXYQUINAZOLINE HYDROCHLORIDE see AJA375
4-AMINO-2-BUTOXY-BENZOIC ACID 2-(DIETHYLAMINO)ETHYL ESTER HY-
DROCHLORIDE see SPB800
3-AMINO-2-BUTOXYBENZOIC ACID-2-DIETHYLAMINOETHYL ESTER HY-
DROCHLORIDE-87-1 see AJA500

4-AMINO-2-BUTOXY-BENZOIC ACID 2-(DIETHYLAMINO)ETHYL ESTER, MONOHYDROCHLORIDE see SPB800

2-AMINO-n-BUTYL ALCOHOL see AJA250

4-AMINO-N-((BUTYLAMINO)CARBONYL)BENZENESULFONAMIDE see BSM000

dl-4-AMINO-α-(tert-BUTYLAMINO)-3-CHLORO-5-(TRIFLUOROMETHYL)PHENETHYL ALCOHOL HCl see MAB300

4-AMINO-α-((tert-BUTYLAMINO)METHYL)-3-CHLORO-5-(TRIFLUOROMETHYL)BENZYL ALCOHOL HCl see KGK300

4-AMINO-α-((tert-BUTYLAMINO)METHYL)-3,5-DICHLOROBENZYL ALCOHOL HYDROCHLORIDE see VHA350

4-AMINO-α-((tert-BUTYLAMINO)METHYL)-3,5-DICHLOROBENZYLALKOHOL-HYDROCHLORID (GERMAN) see VHA350

N¹-3-((4-AMINOBUTYL)AMINO)PROPYL)BLEOMYCINAMIDE see BLY500

2-AMINO-5-BUTYLBENZIMIDAZOLE see AJA650

(4-AMINOBUTYL)DIETHOXYMETHYLSILANE see AJA750

p-AMINO-β-sec-BUTYL-N,N-DIMETHYLPHENETHYLAMINE see AJB000

3-(2-AMINOBUTYL)INDOLE ACETATE see AJB250

3-(4-AMINOBUTYL)INDOLE HYDROCHLORIDE see AJB500

Δ-AMINOBUTYLMETHYLDIETHOXYSILANE see AJA750

4-AMINO-6-tert-BUTYL-3-METHYLTHIO-as-TRIAZIN-5-ONE see MQR275

4-AMINO-6-tert-BUTYL-3-(METHYLTHIO)-1,2,4-TRIAZIN-5-ONE see MQR275

5-AMINO-N-BUTYL-2-PROPARGYLOXYBENZAMIDE see AJC000

5-AMINO-N-BUTYL-2-(2-PROPYNYLOXY)BENZAMIDE see AJC000

(4-AMINOBUTYL)TRIETHOXYSILANE see AJC250

4-AMINOBUTYRAMIDE see AJC375

4-AMINOBUTYRIC ACID see PIM500

γ-AMINO-N-BUTYRIC ACID see PIM500

γ-AMINOBUTYRIC ACID CETYL ESTER see AJC500

4-AMINOBUTYRIC ACID METHYL ESTER see AJC625

4-AMINO-2-(4-BUTYRYLHEXAHYDRO-1H-1,4-DIAZEPIN-1-YL)-6,7-DIMETHOXYQUINAZOLINE HYDROCHLORIDE see BON350

AMINOCAINE see AIT250

AMINOCAPROIC ACID see AJD000

6-AMINOCAPROIC ACID see AJD000

omega-AMINOCAPROIC ACID see AJD000

epsilon-AMINOCAPROIC ACID see AJD000

AMINOCAPROIC LACTAM see CBF700

p-AMINO CAPROPHENONE see AJD250

AMINOCARB see DOR400

4-AMINO-4'-(2-CARBAMOYLETHYL)-1,1-DIMETHYL-N,4'-BI(PYRROLE-2-CARBOXAMIDE) see AJD375

4-AMINO-5-CARBAMYL-3-BENZYLTHIAZOLE-2(3H)-THIONE see AJD500

AMINOCARBE (FRENCH) see DOR400

3-AMINO-9-p-CARBETHOXYAMINOPHENYL-10-METHYLPHENANTHRIDINIUM ETHANESULPHONATE see CBQ125

AMINO-α-CARBOLINE see AJD750

2-AMINO-α-CARBOLINE see AJD750

(3-((AMINOCARBONYL)AMINO)-2-METHOXYPROPYL)CHLOROMERCURY see CHX250

2,2'-((4-((AMINOCARBONYL)AMINO)PHENYL)ARSINIDENE)BIS(THIO)BISACETIC ACID see CBI250

2,2'-((4-((AMINOCARBONYL)AMINO)PHENYL)ARSINIDENEBIS(THIO))BIS)BENZOIC ACID see TFD750

(4-((AMINOCARBONYL)AMINO)PHENYL)ARSONIC ACID see CBJ000

N-(AMINOCARBONYL)BENZENEACETAMIDE see PEC250

N-(AMINOCARBONYL)-2-BROMO-2-ETHYLBUTANAMIDE see BNK000

N-(AMINOCARBONYL)-2-BROMO-3-METHYLBUTANAMIDE see BNP750

(6R-(6-α,7-β(R*)))-4-(AMINOCARBONYL)-1-((2-CARBOXY-8-OXO-7-((PHENYLSULFOACETYL)AMINO)-5-THIO-1-AZABICYCLO(4.2.0)OCT-2-EN-3-YL)METHYL)-PYRIDINIUM HYDROXIDE, inner salt, MONOSODIUM SALT see CCS550

(Z)-N-(AMINOCARBONYL)-2-ETHYL-2-BUTENAMIDE see EHP000

N-(AMINOCARBONYL)HYDROXYLAMINE see HOO500

N-(AMINOCARBONYL)-2-(1-METHYLETHYL)-4-PENTENAMIDE see IQX000

N-AMINOCARBONYL-4-((2-METHYLHYDRAZINO)METHYL)BENZAMIDE MONOHYDROBROMIDE see MKN750

N-(AMINOCARBONYL)-N-NITROGLYCINE see NKI500

l-N⁵-(AMINOCARBONYL)-N⁵-NITROSOORNITHINE see NJT000

dl-N⁵-(AMINOCARBONYL)-N⁵-NITROSOORNITHINE see NJS500

N-(AMINOCARBONYL)OXY)ETHANIMIDOTHIOIC ACID, METHYL ESTER see MPS250

(6R-(6-α,7-β(Z)))-3-(((AMINOCARBONYL)OXY)METHYL)-7-((2-FURANYL(METHYOXYIMINO)ACETYL)AMINO)-8-OXO-5-THIA-1-AZABICYCLO(4.2.0)OCT-2-ENE-2-CARBOXYLIC ACID MONOSODIUM SALT see SFQ300

(6R-(6-α,7-β(Z)))-3-(((AMINOCARBONYL)OXY)METHYL)-7-((2-FURANYL(METHYOXYIMINO)ACETYL)AMINO)-8-OXO-5-THIA-1-AZABICYCLO(4.2.0)OCT-2-ENE-2-CARBOXYLIC ACID see CCS600

2-(((AMINOCARBONYL)OXY)METHYL)-2-METHYLPENTYL ESTER BUTYL CARBAMIC ACID see MOV500

(6R-cis)-3-(((AMINOCARBONYL)OXY)METHYL)-7-METHOXY-8-OXO-7-((2-THIENYLACETYL)AMINO)-5-THIA-1-AZABICYCLO(4.2.0)OCT-2-ENE-2-CARBOXYLIC ACID MONOSODIUM SALT see CCS510

2-((AMINOCARBONYL)OXY)-N,N,N-TRIMETHYLETHANAMINIUM CHLORIDE see CBH250

2-((AMINOCARBONYL)OXY)-N,N,N-TRIMETHYL-1-PROPANAMINIUM CHLORIDE see HOA500

(((4-(AMINOCARBONYL)PHENYL)ARSINIDINE)BIS(THIO))BIACETIC ACID see TFA350

1-((4-(AMINOCARBONYL)PYRIDINIO)METHOXY)METHYL)-2-((HYDROXYIMINO)METHYL) PYRIDINIUM 2Cl see HAA345

(((4-(AMINOCARBONYL)PYRIDINO)METHOXY)-2-((HYDROXYIMINO)METHYL)PYRIDINIUM DICHLORIDE see HAA345

2-AMINO-6-(p-CARBOXYAMINOPHENYL)-5-METHYLPHENANTHRIDINIUM ETHANESULFONATE ETHYL ESTER see CBQ125

1-AMINO-2-CARBOXYBENZENE see API500

1-AMINO-4-CARBOXYBENZENE see AIH600

(6R-(6-α,7-α))-7-((((2-AMINO-2-CARBOXYETHYL)THIO)ACETYL)AMINO)-7-METHOXY-3-(((1-METHYL-1H-TETRAZOL-5-YL)THIO)METHYL)-8-OXO-5-THIA-1-AZABICYCLO(4.2.0)OCT-2-ENE-2-CARBOXYLIC ACID MONOSODIUM SALT see CCS365

1-AMINO-2-CARBOXYLATE-4-NITRO-ANTHRAQUINONE see AJS000

(6R-cis)-7-(((4-(2-AMINO-1-CARBOXY-2-OXOETHYL)-1,3-DITHIETAN-2-YL)CARBONYL)AMINO)-7-METHOXY-3-(((1-METHYL-1H-TETRAZOL-5-YL)THIO)METHYL)-8-OXO-5-THIA-1-AZABICYCLO(4.2.0)OCT-2-ENE-2-CARBOXYLIC ACID MONOSODIUM SALT see CCS371

N²-(((+)-5-AMINO-5-CARBOXYPENTYLAMINO)METHYL)TETRACYCLINE see MRV250

3-AMINO-N-(α-CARBOXYPHENETHYL)SUCCINAMIC ACID N-METHYL ESTER, stereoisomer see ARN825

AMINOCARDOL see TEP500

AMINOCHLORAMBUCIL see AJE000

2-AMINO-6'-CHLORO-o-ACETOTOLUIDIDE, HYDROCHLORIDE see AJE250

(s)-N-(5-AMINO-1-(CHLOROACETYL)PENTYL)-4-METHYLBENZENESULFONAMIDE see THH500

4-AMINO-6-p-CHLOROANILINO-1,2-DWUHYDRO-2,2-DWUMETHYLO-1,3,5-TROJAZYNA see COX400

1-AMINO-5-CHLOROANTHRAQUINONE see AJE325

m-AMINOCHLOROBENZENE see CEH675

1-AMINO-2-CHLOROBENZENE see CEH670

1-AMINO-3-CHLOROBENZENE see CEH675

1-AMINO-4-CHLOROBENZENE see CEH680

4-AMINO-2-CHLOROBENZOIC ACID see CEG750

3-AMINO-4-CHLOROBENZOIC ACID 2-((DIMETHYLAMINO)ETHYL) ESTER HYDROCHLORIDE see AJE350

2-AMINO-4-CHLOROBENZOTHIAZOLE see AJE500

2-AMINO-6-CHLOROBENZOTHIAZOLE see AJE750

2-AMINO-4-CHLOROBENZOXAZOLE see AJF250

2-AMINO-5-CHLOROBENZOXAZOLE see AJF500

2-AMINO-6-CHLOROBENZOXAZOLE see AJF750

2-AMINO-7-CHLOROBENZOXAZOLE see AJG750

4-AMINO-5-CHLORO-N-(2-(DIETHYLAMINO)ETHYL)-N-ANISAMIDE see AJH000

4-AMINO-5-CHLORO-N-(2-(DIETHYLAMINO)ETHYL)-o-ANISAMIDE DIHYDROCHLORIDE MONOHYDRATE see MQQ300

4-AMINO-5-CHLORO-N-(2-(DIETHYLAMINO)ETHYL)-2-METHOXYBENZAMIDE see AJH000

5-AMINO-4-CHLORO-2,3-DIHYDRO-3-OXO-2-PHENYLPYRIDAZINE see PEE750

4-AMINO-5-CHLORO-N-(2-(ETHYLAMINOETHYL)-o-ANISAMIDE see AJH125

4-AMINO-5-CHLORO-N-(2-ETHYLAMINOETHYL)-2-METHOXYBENZAMIDE see AJH125

6-AMINO-2-(2-CHLOROETHYL)-2,3-DIHYDRO-4H-1,3-BENZOXAZIN-4-ONE see DKR200

6-AMINO-2-(2-CHLOROETHYL)-2,3-DIHYDRO-4H-1,3-BENZOXAZIN-4-ONEHYDROCHLORIDE see AJH129

4-AMINO-5-CHLORO-N-(2-ETILAMINOETIL)-2-METOSSIBENZAMIDE (ITALIAN) see AJH125

1-AMINO-4-CHLORO-4-HYDROXYANTHRAQUINONE see AJH250

3-AMINO-5-CHLORO-4-HYDROXYBENZENESULFONIC ACID see AJH500

2-AMINO-5-CHLORO-6-HYDROXYBENZOXAZOLE see AJH750

2-AMINO-5-CHLORO-6-METHOXYBENZOXAZOLE see AJI000

4-AMINO-5-CHLORO-2-METHOXY-N-(1-BENZYL-4-PIPERIDYL)BENZAMIDE MALATE see CMV325

4-AMINO-5-CHLORO-2-METHOXY-N-(4-PIPERIDYL)BENZAMIDE see DBA250

2-AMINO-2'-CHLORO-6'-METHYLACETANILIDE, HYDROCHLORIDE see AJE250

1-AMINO-2-CHLORO-6-METHYLBENZENE see CLK200

1-AMINO-3-CHLORO-2-METHYLBENZENE see CLK200

1-AMINO-3-CHLORO-4-METHYLBENZENE see CLK215

1-AMINO-3-CHLORO-6-METHYLBENZENE see CLK225

2-AMINO-3-CHLORO-1,4-NAPHTHOQUINONE see AJI250

1-AMINO-2-CHLORO-4-NITROBENZENE see CJA175

2-AMINO-4-CHLOROPHENOL (DOT) see CEH250

γ-AMINO-β-(p-CHLOROPHENYL)BUTYRIC ACID see BAC275

5-AMINO-4-CHLORO-2-PHENYL-3(2H)-PYRIDAZINONE see PEE750

2-AMINO-5-((p-CHLOROPHENYL)THIOMETHYL)-2-OXAZOLINE see AJI500

(-)-1-AMINO-3-CHLORO-2-PROPANOL see AJI520

l-1-AMINO-3-CHLORO-2-PROPANOL see AJI520
(±)-1-AMINO-3-CHLORO-2-PROPANOL see AJI530
dl-1-AMINO-3-CHLORO-2-PROPANOL see AJI530
(−)-1-AMINO-3-CHLORO-2-PROPANOL HYDROCHLORIDE see AJI550
l-AMINO-3-CHLORO-2-PROPANOL HYDROCHLORIDE see AJI550
l-1-AMINO-3-CHLORO-2-PROPANOL HYDROCHLORIDE see AJI550
(±)-1-AMINO-3-CHLORO-2-PROPANOL HYDROCHLORIDE see AJI600
dl-1-AMINO-3-CHLORO-2-PROPANOL HYDROCHLORIDE see AJI600
4-AMINO-N-(5-CHLORO-2-QUINOXALINYL)BENZENESULFONAMIDE
    see CMA600
2-AMINO-5-CHLOROTHIAZOLE see AJI650
2-AMINO-4-CHLOROTOLUENE see CLK225
2-AMINO-5-CHLOROTOLUENE see CLK220
2-AMINO-6-CHLOROTOLUENE see CLK200
4-AMINO-2-CHLOROTOLUENE see CLK215
2-AMINO-5-CHLOROTOLUENE HYDROCHLORIDE see CLK235
6-AMINO-4-CHLORO-m-TOLUENESULFONIC ACID see AJJ250
4-AMINO-3-CHLOROTRIFLUOROMETHYL-α-((tert-
    BUTYLAMINO)METHYL)BENZYLALCOHOL HYDROCHLORIDE
    see KGK300
dl-1-(4-AMINO-3-CHLORO-5-TRIFLUOROMETHYLPHENYL)-2-tert-
    BUTYLAMINOETHANOL HYDROCHLORIDE see MAB300
AMINOCHLORTHENOXAZINE see DKR200
AMINOCHLORTHENOXAZIN HYDROCHLORIDE see AJH129
6-AMINOCHRYSENE see CML800
3-AMINO-4-CLORO-BENZOATO di DIMETILAMINOETILE CLORIDRATO
    (ITALIAN) see AJE350
6-AMINOCOUMARIN COUMARIN-3-CARBOXYLIC ACID SALT see AJJ500
6-AMINOCOUMARIN HYDROCHLORIDE see AJJ750
5-AMINO-o-CRESOL see AKJ750
3-AMINO-p-CRESOL METHYL ESTER see MGO750
m-AMINO-p-CRESOL, METHYL ESTER see MGO750
4-AMINO-5-CYANO-7-(d-RIBOFURANOSYL)-7H-PYRROLO(2,3-d)PYRIMIDINE
    see VGZ000
2-AMINO-2,4,6-CYCLOHEPTATRIEN-1-ONE see AJJ800
(6R-(6-α,7-β(R*)))-7-((AMINO-1,4-CYCLOHEXADIEN-1-YLACETAL)AMINO)-3-
    METHYL-8-OXO-5-THIA-1-AZABICYCLO(4.2.0)OCT-2-ENE-2-CARBOXYLIC
    ACID see CCS530
(6R-(6-α,7-β(R*)))-7-((AMINO-1,4-CYCLOHEXADIEN-1-YLACETAL)AMINO)-3-
    METHYL-8-OXO-5-THIA-1-AZABICYCLO(4.2.0)OCT-2-ENE-2-CARBOXYLIC
    ACID DIHYDRATE see CCS535
7-(D-2-AMINO-2-(1,4-CYCLOHEXADIENYL)ACETAMIDE)-3-METHOXY-3-
    CEPHEM-4-CARBOXYLIC ACID see CCS530
d-7-(2-AMINO-2-(1,4-CYCLOHEXADIEN-1-YL)ACETAMIDO)-3-METHYL-8-OXO-
    5-THIA-1-AZABICYCLO(4.2.0)OCT-2-ENE-2-CARBOXYLIC ACID HY-
    DRATE see SBN450
AMINOCYCLOHEXANE see CPF500
6-(1-AMINOCYCLOHEXANECARBOXAMIDO)PENICILLANIC ACID see AJJ875
AMINOCYCLOHEXANE HYDROCHLORIDE see CPA775
AMINOCYCLOHEXYLPENICILLIN see AJJ875
(1-AMINOCYCLOHEXYL)PENICILLIN see AJJ875
1-AMINO-2-(o-CYCLOHEXYLPHENOXY)PROPIONALDOXIME see AJK000
1-AMINOCYCLOPENTANE-1-CARBOXYLIC ACID see AJK250
1-AMINO-1-CYCLOPENTANECARBOXYLIC ACID see AJK250
4-AMINO-N-CYCLOPROPYL-3,5-DICHLOROBENZAMIDE see AJK500
N⁴-AMINOCYTIDINE see AJK625
4-AMINO-DAB see DPO200
AMINODARONE see AJK750
23-AMINO-O⁴-DEACETYL-23-DEMETHOXYVINCALEUKOBLASTINE SUL-
    FATE see VGU750
1-AMINODECANE see DAG600
2-AMINO-2-DEOXY-l-ASCORBIC ACID see AJL125
2-AMINO-2-DEOXY-d-GALACTOSE HYDROCHLORIDE see GAT000
o-3-AMINO-3-DEOXY-α-d-GLUCOPYRANOSYL-(1-4)-o-(2,6-DIAMINO-2,6-DIDE-
    OXY-α-d-GLUCOPYRANOSYL-(1-6)-2-DEOXY-d-STREPTAMINE see BAU270
o-3-AMINO-3-DEOXY-α-d-GLUCOPYRANOSYL-(1-6)-o-(2,6-DIAMINO-2,6-
    DIDEOXY-α-d-GLUCOPYRANOSYL-(1-4))-2-DEOXY-d-STREPTAMINE SUL-
    FATE (1:1) see KBA100
o-3-AMINO-3-DEOXY-α-d-GLUCOPYRANOSYL-(1-6)-o-(2,6-DIAMINO-2,3,4,6-
    TETRADEOXY-α-d-erythro-HEXOPYRANOSYL-(1-4))-2-DEOXY-d-STREPT-
    AMINE SULFATE (SALT) see PAG050
AMINODEOXYKANAMYCIN see BAU270
2′-AMINO-2′-DEOXYKANAMYCIN see BAU270
AMINODEOXYKANAMYCIN SULFATE see KBA100
4-AMINO-4-DEOXY-N¹⁰-METHYLPTEROYLGLUTAMATE see MDV500
4-AMINO-4-DEOXY-N¹⁰-METHYLPTEROYLGLUTAMIC ACID see MDV500
4-AMINO-1-(2-DEOXY-β-d-erythro-PENTOFURANOSYL)-s-TRIAZIN-2(1H)-ONE
    see ARY125
4-AMINO-4-DEOXYPTEROYLGLUTAMATE see AMG750
2-AMINO-9-(2-DEOXY-β-d-RIBOFURANOSYL)-9H-PURINE-6-THIOL HYDRATE
    see TFJ250
β-d-5-(2-AMINO-2-DEOXY-l-XYLONAMIDO)-1,5-DIDEOXY-1-(3,4-DIHYDRO-5-
    HYDROXYMETHYL)-2,4-DIOXO-1(2H)-PYRIMIDINYL ALLOFURANURO-
    NIC ACID, MONOCARBAMATE (ester) see PKE100

7-AMINODIBENZ(a,h)ANTHRACENE see AJL250
9-AMINO-1,2,5,6-DIBENZANTHRACENE see AJL250
3-AMINODIBENZOFURAN see DDB600
1-AMINO-2,4-DIBROMANTHRACHINON (CZECH) see AJL500
1-AMINO-2,4-DIBROMOANTHRAQUINONE see AJL500
trans-4-((2-AMINO-3,5-DIBROMOBENCIL)AMINO CICLOHEXANOL (SPANISH)
    see AHJ250
2-AMINO-5,7-DIBROMOBENZOXAZOLE see AJL750
trans-4-((2-AMINO-3,5-DIBROMOBENZYL)AMINO)CYCLOHEXANOL HYDRO-
    CHLORIDE see AHJ500
N-(2-AMINO-3,4-DIBROMOCICLOHEXIL)-trans-4-AMINOCICLOHEXANOL
    (SPANISH) see AHJ250
N-(2-AMINO-3,4-DIBROMOCYCLOHEXYL)-trans-4-AMINOCYCLOHEXANOL
    see AHJ250
2-AMINO-3,5-DIBROMO-N-CYCLOHEXYL-N-METHYL-BENZENEMETHANAM
    INE MONOHYDROCHLORIDE (9CI) see BMO325
2-AMINO-4-DIBUTYLAMINOETHOXYPYRIMIDINE see AJL875
2-AMINO-4-(2-DIBUTYLAMINOETHOXY)PYRIMIDINE see AJL875
2-AMINO-4-DICHLOROARSINOPHENOL HYDROCHLORIDE see DFX400
1-AMINO-3,4-DICHLOROBENZENE see DEO300
3-AMINO-2,5-DICHLOROBENZOIC ACID see AJM000
2-AMINO-5,6-DICHLOROBENZOXAZOLE see AJM500
α-AMINO-γ-(p-DICHLOROETHYLAMINO)-PHENYLBUTYRIC ACID see AJE000
3-AMINO-1-(3,4-DICHLORO-α-METHYLBENZYL)-2-PYRAZOLIN-5-ONE
    see EAE675
5-AMINO-2-(1-(3,4-DICHLOROPHENYL)-ETHYL)-2,4-DIHYDRO-3H-PYRAZOL-3-
    ONE see EAE675
2-AMINO-5-((3,4-DICHLOROPHENYL)THIOMETHYL)-2-OXAZOLINE
    see AJM750
5-AMINO-9-(DIETHYLAMINO)BENZO(a)PHENOXAZIN-7-IUM SULFATE (2:1)
    see AJN250
2-AMINO-4-DIETHYLAMINOETHOXYPYRIMIDINE see AJN375
2-AMINO-4-(2-DIETHYLAMINOETHOXY)PYRIMIDINE see AJN375
p-AMINO-N-(2-DIETHYLAMINOETHYL)BENZAMIDE see AJN500
4-AMINO-N-(2-(DIETHYLAMINO)ETHYL)-BENZAMIDE (9CI) see AJN500
p-AMINO-N-(2-(DIETHYLAMINO)ETHYL))BENZAMIDE HYDROCHLORIDE
    see PME000
p-AMINO-N-(2-DIETHYLAMINOETHYL)BENZAMIDE SULFATE see AJN750
N-(2-AMINO-5-DIETHYLAMINOPHENETHYL)METHANE SUL-
    FONAMIDEHYDROCHLORIDE-74-5 see AJO000
1-AMINO-3-(DIETHYLAMINO)PROPANE see DIY800
2-AMINO-4-γ-DIETHYLAMINOPROPYLAMINO-5,6-DIMETHYLPYRIMIDINE
    see AJQ000
p-AMINO DIETHYLANILINE HYDROCHLORIDE see AJO250
4-AMINODIFENIL (SPANISH) see AJS100
p-AMINODIFENYLAMIN (CZECH) see PFU500
N-(4-AMINO-9,10-DIHYDRO-9,10-DIOXO-1-ANTHRACENTY)-BENZAMIDE
    see AIB250
2-AMINO-1,9-DIHYDRO-9-((2-HYDROXYETHOXY)METHYL)-6H-PURIN-6-ONE
    see AEC700
2-AMINO-1,9-DIHYDRO-9-((2-HYDROXYETHOXY)METHYL)-6H-PURIN-6-ONE
    MONOSODIUM SALT see AEC725
6-AMINO-1,2-DIHYDRO-1-HYDROXY-2-IMINO-4-PIPERIDINOPYRIMIDINE
    see DCB000
(s)-7-AMINO-6,7-DIHYDRO-10-HYDROXY-1,2,3-
    TRIMETHOXYBENZO(a)HEPTALEN-9(5H)-ONE see TLN750
4-o-(3-AMINO-3,4-DIHYDRO-6-((METHYLAMINO)METHYL)-2H-PYRAN-2-YL)-2-
    DEOXY-6-o-(3-DEOXY-4-C-METHYL-3-(METHYLAMINO)β-l-
    ARABINOPYRANOSYL)-d-STREPTAMINE (2S-cis)-, SULFATE (salt)
    see GAA120
3-AMINO-1,5-DIHYDRO-5-METHYL-1-β-d-RIBOFURANOSYL-1,4,5,6,8-PEN-
    TAAZAACENAPHTHYLENE see TJE870
α-AMINO-2,3-DIHYDRO-3-OXO-5-ISOXAZOLEACETIC ACID see AKG250
(s)-7-AMINO-6,7-DIHYDRO-1,2,3,10-TETRAMETHOXYBENZO(a)HEPTALEN-
    9(5H)-ONE see TLO000
5-AMINO-1,6-DIHYDRO-7H-v-TRIAZOLO(4,5-d)PYRIMIDIN-7-ONE see AJO500
5-AMINO-1,4-DIHYDRO-7H-1,2,3-TRIAZOLO(4,5-d)PYRIMIDIN-7-ONE (9CI)
    see AJO500
l-2-AMINO-1-(3,4-DIHYDROXYPHENYL)ETHANOL see NNO500
2-AMINO-3-(3,4-DIHYDROXYPHENYL)PROPANOIC ACID see DNA200
2-AMINO-4-DI-ISOBUTYLAMINOETHOXYPYRIMIDINE see AJO625
2-AMINO-4-(2-DIISOBUTYLAMINOETHOXY)PYRIMIDINE see AJO625
3-AMINO-2-(2-(DIISOPROPYLAMINO)ETHOXY)BUTYROPHENONE-DIHY-
    DROCHLORIDE-29-2 see AJO750
2-AMINO-1-(2,5-DIMETHOXYPHENYL)-1-PROPANOL HYDROCHLORIDE
    see MDW000
4-AMINO-N-(2,6-DIMETHOXY-4-PYRIMIDINYL)BENZENESULFONAMIDE
    see SNN300
4-AMINO-N-(4,6-DIMETHOXY-2-PYRIMIDINYL)BENZENESUL-FONAMIDE
    see SNI500
4-AMINO-N-(5,6-DIMETHOXY-4-PYRIMIDINYL)BENZENESULFONAMIDE
    see AIE500
1-(4-AMINO-6,7-DIMETHOXY-2-QUINAZOLINYL-4-(2-FURANYLCARBONYL)
    PIPERAZINE see AJP000

1-(4-AMINO-6,7-DIMETHOXY-2-QUINAZOLINYL)-4-(2-FURANYLCARBONYL)
   PIPERAZINE HYDROCHLORIDE see FPP100
1-(4-AMINO-6,7-DIMETHOXY-2-QUINAZOLINYL)-4-(2-FUROYL)PIPERAZINE
   MONOHYDROCHLORIDE see FPP100
1-(4-AMINO-6,7-DIMETHOXY-2-QUINAZOLINYL)-4-((TETRAHYDRO-2-
   FURANYL)CARBON YL)PIPERAZINE HCl 2H$_2$O see TEF700
4-AMINO-4'-DIMETHYLAMINOAZOBENZENE see DPO200
4'-AMINO-N,N-DIMETHYL-4-AMINOAZOBENZENE see DPO200
2-AMINO-4-DIMETHYLAMINOETHOXYPYRIMIDINE see AJP125
2-AMINO-4-(2-DIMETHYLAMINOETHOXY)PYRIMIDINE see AJP125
3-AMINO-7-DIMETHYLAMINO-2-METHYLPHENAZATHIONIUM CHLORIDE
   see AJP250
3-AMINO-7-DIMETHYLAMINO-2-METHYLPHENAZINE HYDROCHLORIDE
   see AJQ250
3-AMINO-7-(DIMETHYLAMINO)-2-METHYL-PHENAZINE
   MONOHYDROCHLORIDE see AJQ250
3-AMINO-7-(DIMETHYLAMINO)-2-METHYL-PHENOSELENAZIN-5-IUM CHLO-
   RIDE see SBU950
1-AMINO-3-DIMETHYLAMINOPROPANE see AJQ100
AMINODIMETHYLAMINOTOLUAMINOZINE HYDROCHLORIDE see AJQ250
p-AMINODIMETHYLANILINE see DTL800
4-AMINO-2',3-DIMETHYLAZOBENZENE see AIC250
4'-AMINO-2,3'-DIMETHYLAZOBENZENE see AIC250
AMINODIMETHYLBENZENE see XMA000
1-AMINO-2,4-DIMETHYLBENZENE see XMS000
1-AMINO-2,5-DIMETHYLBENZENE see XNA000
3-AMINO-1,4-DIMETHYLBENZENE see XNA000
4-AMINO-1,3-DIMETHYLBENZENE see XMS000
1-AMINO-2,4-DIMETHYLBENZENE HYDROCHLORIDE see XOJ000
1-AMINO-2,5-DIMETHYLBENZENE HYDROCHLORIDE see XOS000
3-AMINO-1,4-DIMETHYLBENZENE HYDROCHLORIDE see XOS000
4-AMINO-1,3-DIMETHYLBENZENE HYDROCHLORIDE see XOJ000
5-AMINO-1,4-DIMETHYLBENZENE HYDROCHLORIDE see XOS000
3-AMINO-1,4-DIMETHYL-γ-CARBOLINE see TNX275
2-AMINODIMETHYLETHANOL see IIA000
4-AMINO-6-(1,1-DIMETHYLETHYL)-3-(METHYLTHIO)-1,2,4-TRIAZIN-5(4H)-
   ONE see MQR275
4-AMINO-3',5'-DIMETHYL-4'-HYDROXYAZOBENZENE see AJQ500
2-AMINO-3,4-DIMETHYLIMIDAZO(4,5-f)QUINOLINE see AJQ600
2-AMINO-3,8-DIMETHYLIMIDAZO(4,5-f)QUINOXALINE see AJQ675
2-AMINO-3,8-DIMETHYL-3H-IMIDAZO(4,5-f)QUINOXALINE see AJQ675
4-AMINO-N-(3,4-DIMETHYL-5-ISOXAZOLYL)BENZENESULPHONAMIDE
   see SNN500
2-AMINO-6-DIMETHYL-4-(p-(p-((p-((1-METHYLPYRIDINIUM-3-YL)CARBAM-
   OYL)PHENYL)CARBABENZAMIDO)ANILINO)PYRIMIDIMIUM,DIIODIDE
   see AJQ750
4-AMINO-N-(4,5-DIMETHYL-2-OXAZOLYL)BENZENESULFONAMIDE
   see AIE750
p-AMINO-N,α-DIMETHYLPHENETHYLAMINE see AJR000
2-AMINO-4,5-DIMETHYLPHENOL see AMW750
5-AMINO-1,3-DIMETHYL-4-PYRAZOLYL o-FLUOROPHENYL KETONE
   see AJR400
(5-AMINO-1,3-DIMETHYL-1H-PYRAZOL-4-YL)(2-
   FLUOROPHENYL)METHANONE see AJR400
3-AMINO-1,4-DIMETHYL-5H-PYRIDO(4,3-b)INDOLE see TNX275
3-AMINO-1,4-DIMETHYL-5H-PYRIDO(4,3-b)INDOLE ACETATE see AJR500
4-AMINO-N-(2,6-DIMETHYL-4-PYRIMIDINYL)BENZENESULFONAMIDE (9CI)
   see SNJ350
4-AMINO-N-(4,6-DIMETHYL-2-PYRIMIDINYL)BENZENESULFONAMIDE,
   MONOSODIUM SALT see SJW500
2-AMINO-4,6-DINITROPHENOL see DUP400
2-AMINO-4,6-DINITROTOLUENE see AJR750
4-AMINO-3,5-DINITROTOLUENE see DVI100
1-AMINO-9,10-DIOXO-9,10-DIHYDRO-2-ANTHRACENECARBOXYLIC ACID
   see AJS000
2-AMINODIPHENYL see BGE250
4-AMINODIPHENYL see AJS100
o-AMINODIPHENYL see BGE250
p-AMINODIPHENYL see AJS100
p-AMINODIPHENYLAMINE see PFU500
2-AMINODIPHENYLENE OXIDE see DDB600
2-AMINODIPHENYLENOXYD (GERMAN) see DDB800
2-AMINO-1,2-DIPHENYLETHANOL HYDROCHLORIDE see DWB000
4-AMINODIPHENYL ETHER see PDR500
2-AMINO-6-((1,2-DIPHENYLETHYL)AMINO)-3-PYRIDINECARBAMIC ACID
   ETHYL ESTER, MONOHYDROCHLORIDE see DAB200
p-AMINODIPHENYLIMIDE see PEI000
2-AMINODIPYRIDO(1,2-a:3',2'-d)-IMIDAZOLE see DWW700
2-AMINODIPYRIDO(1,2-a:3',2'-d)IMIDAZOLE HYDROCHLORIDE see AJS225
2-AMINO-4,6-DIPYRROLIDINOTRIAZINE see AJS250
4-AMINO-3,4'-DISULFOAZOBENZENE see AJS500
2-AMINO-1,4-DISULFOBENZENE see AIE000
1-AMINO-3,6-DISULFO-8-NAFTYLESTER KYSELINA p-TOLUENSULFONOVE
   (CZECH) see AKH500

5-AMINO-1,2,4-DITHIAZOLE-3-THIONE see IBL000
4-AMINO-1-DODECYLQUINALDINIUM ACETATE see AJS750
AMINODUR see TEP500
9-AMINOELLIPTICINE see AJS875
2-AMINOETANOLO (ITALIAN) see EEC600
AMINOETHANE see EFU400
1-AMINOETHANE see EFU400
2-AMINO-ETHANESELENOL HYDROCHLORIDE see AJS900
2-AMINOETHANESELENOSULFURIC ACID see AJS950
2-AMINOETHANESULFONIC ACID see TAG750
2-AMINOETHANESULFONO-p-PHENETIDIDE HYDROCHLORIDE see TAG875
2-AMINOETHANETHIOL see AJT250
2-AMINO-ETHANETHIOL DIHYDROGEN PHOSPHATE(ester), MONOSODIUM
   SALT see AKB500
2-AMINOETHANETHIOSULFURIC ACID see AJT500
1-AMINOETHANOL see AAG500
2-AMINOETHANOL (MAK) see EEC600
2-AMINOETHANOL componded with 6-CYCLOHEXYL-1-HYDROXY-4-METHYL-
   2(1H)-PYRIDINONE (1:1) see BAR800
3-AMINO-4-ETHOXYACETANILIDE see AJT750
(L)-3-(2-AMINOETHOXY)ALANINE see OLK200
2-AMINO-3-ETHOXYCARBONYL-5-BENZYL-4,5,6,7-TETRAHYDROTHIENO
   (2,3-c)PYRIDINE HYDROCHLORIDE see AJU000
2-AMINO-3-ETHOXYCARBONYL-6-BENZYL-4,5,6,7-TETRAHYDROTHIENO(2,3-
   c)PYRIDINE HYDROCHLORIDE see TGE165
2-AMINOETHOXYETHANOL see AJU250
2-(2-AMINOETHOXY)ETHANOL see AJU250
5-AMINO-6-ETHOXY-2-NAPHTHALENESULFONIC ACID see AJU500
1-(1-AMINOETHYL)ADAMANTANE HYDROCHLORIDE see AJU625
α-AMINOETHYL ALCOHOL see AAG500
β-AMINOETHYL ALCOHOL see EEC600
(3-(2-AMINOETHYL)AMINOPROPYL)TRIMETHOXYSILANE see TLC500
2-AMINOETHYLAMMONIUM PERCHLORATE see AJU875
o-AMINOETHYLBENZENE see EGK500
β-AMINOETHYLBENZENE see PDE250
1-AMINO-4-ETHYLBENZENE see EGL000
4-(2-AMINOETHYL)-1,2-BENZENEDIOL HYDROCHLORIDE see DYC600
α-(1-AMINOETHYL)BENZENEMETHANOL HYDROCHLORIDE see PMJ500
4-(2-AMINOETHYL)-1,2,3-BENZENETRIOL HYDROCHLORIDE see HKG000
α-(1-AMINOETHYL)-BENZYL ALCOHOL see NNM000
dl-α-(1-AMINOETHYL)BENZYL ALCOHOL see NNM500
α-(1-AMINOETHYL)BENZYL ALCOHOL HYDROCHLORIDE see NNN000,
   PMJ500
3-(2-AMINOETHYL)-1-BENZYL-5-METHOXY-2-METHYLINDOLE HYDRO-
   CHLORIDE see BEM750
3-AMINO-9-ETHYLCARBAZOLE see AJV000
3-AMINO-N-ETHYLCARBAZOLE see AJV000
3-AMINO-9-ETHYLCARBAZOLEHYDROCHLORIDE see AJV250
trans-4-AMINOETHYLCYCLOHEXANE-1-CARBOXYLIC ACID see AJV500
4-(2-AMINOETHYL)-6-DIAZO-2,4-CYCLOHEXADIENONE HYDROCHLORIDE
   see DCQ575
α-(1-AMINOETHYL)-2,4-DIMETHOXYBENZYL ALCOHOL HYDROCHLORIDE
   see AJV850
α-(1-AMINOETHYL)-2,5-DIMETHOXYBENZYL ALCOHOL HYDROCHLORIDE
   see MDW000
2-AMINOETHYL DISULFIDE DIHYDROCHLORIDE see CQJ750
AMINOETHYLENE see EJM900, VLU400
2-AMINOETHYL ESTER CARBAMIMIDOTHIOIC ACID DIHYDROBROMIDE
   see AJY250
AMINOETHYLETHANDIAMINE see DJG600
AMINOETHYL ETHANOLAMINE see AJW000
N-AMINOETHYLETHANOLAMINE see AJW000
N-(2-AMINOETHYL)ETHYLENEDIAMINE see DJG600
6-AMINO-1-ETHYL-4-p-((p-((1-ETHYLPYRIDINIUM-4-YL)AMINO)2-
   AMINOPHENYL)CARBAMOYL)ANILINO)QUINOLINIUM DIIODIDE
   see AJW250
6-AMINO-1-ETHYL-4-(p-(p-((1-ETHYLPYRIDINIUM-4-YL)AMINO)BENZAMIDO)
   ANILINO)QUINOLINIUM DIIODIDE see AJW500
6-AMINO-1-ETHYL-4-((p-((p-(1-ETHYLPYRIDINIUM-4-YL)AMINO)PHENYL)
   CARBAMOYL)ANILINOQUINOLINIUM DIBROMIDE see AJW750
β-AMINOETHYLGLYOXALINE see HGD000
1-AMINO-2-ETHYLHEXAN (CZECH) see EKS500
1-α-(1-AMINOETHYL)-m-HYDROXYBENZYL ALCOHOL see HNB875
(−)-α-(1-AMINOETHYL)-m-HYDROXYBENZYL ALCOHOL BITARTRATE
   see HNC000
1-α-(1-AMINOETHYL)-m-HYDROXYBENZYL ALCOHOL BITARTRATE
   see HNC000
1-α-(1-AMINOETHYL)-m-HYDROXYBENZYL ALCOHOL HYDROGEN-d-TAR-
   TRATE see HNC000
3-(β-AMINOETHYL)-5-HYDROXYINDOLE see AJX500
α-(1-AMINOETHYL)-HYDROXY-4-METHYLBENZYL ALCOHOL HYDRO-
   CHLORIDE see MKS000
β-AMINOETHYLIMIDAZOLE see HGD000
4-(2-AMINOETHYL)IMIDAZOLE see HGD000

4-(2-AMINOETHYL)IMIDAZOLE BIS(DIHYDROGEN PHOSPHATE) see HGE000
4-(2-AMINOETHYL)IMIDAZOLE DI-ACID PHOSPHATE see HGE000
4-AMINOETHYLIMIDAZOLE HYDROCHLORIDE see HGE500
3-(2-AMINOETHYL)INDOLE see AJX000
(AMINO-2 ETHYL)-3-INDOLE (FRENCH) see AJX000
3-(1-AMINOETHYL)INDOLE HYDROCHLORIDE see LIU100
3-(2-AMINOETHYL)INDOLE HYDROCHLORIDE see AJX250
3-(2-AMINOETHYL)INDOL-5-OL see AJX500
3-(2-AMINOETHYL)INDOL-5-OL CREATININE SULFATE see AJX750
AMINOETHYLISOSELENOURONIUM BROMIDE HYDROCHLORIDE
    see AJY000
2-β-AMINOETHYLISOTHIOUREA see AJY250
S-β-AMINOETHYLISOTHIOURONIC DIHYDROCHLORIDE see AJY500
2-(β-AMINOETHYL)ISOTHIOURONIUM BROMIDE HYDROBROMIDE
    see AJY250
2-AMINOETHYLISOTHIOURONIUM DIBROMIDE see AJY250
2-AMINOETHYLISOTHIOURONIUMDICHLORIDE see AJY500
2-AMINOETHYLISOTHIURONIUM BROMIDE HYDROBROMIDE see AJY250
β-AMINOETHYLISOTHIURONIUM BROMIDE HYDROBROMIDE see AJY250
S-(2-AMINOETHYL)ISOTHIURONIUM BROMIDE HYDROBROMIDE
    see AJY250
S-(β-AMINOETHYL)ISOTHIURONIUM BROMIDE HYDROBROMIDE
    see AJY250
2-AMINOETHYLISOTHIURONIUM DIHYDROBROMIDE see AJY250
2-AMINOETHYL MERCAPTAN see AJT250
4-AMINO-N-ETHYL-m-(β-METHANESULFONAMIDOETHYL)-m-TOLUIDINE
    see AJY750
3-(2-AMINOETHYL)-5-METHOXYBENZOFURAN HYDROCHLORIDE
    see AJZ000
6-(2-AMINOETHYL)-5-METHOXYBENZOFURAN HYDROCHLORIDE
    see AKA000
6-(β-AMINOETHYL)-5-METHOXYBENZOFURANHYDROCHLORIDE
    see AKA000
α-(1-AMINOETHYL)-4-METHOXYBENZYLALCOHOL HYDROCHLORIDE
    see AKA250
α-(1-AMINOETHYL)-4-METHOXYBENZYL ALCOHOL HYDROCHLORIDE
    see AKA250
3-(2-AMINOETHYL)-5-METHOXYINDOLE see MFS400
3-(2-AMINOETHYL)-5-METHOXYINDOLE HYDROCHLORIDE see MFT000
3-(2-AMINOETHYL)-6-METHOXYINDOLE HYDROCHLORIDE see MFS500
3-(2-AMINOETHYL)-5-METHOXY-2-METHYLBENZO-FURAN, HYDROCHLO-
    RIDE see MGH750
2-(AMINOETHYL)-2-METHYL-1,3-BENZODIOXOLE HYDROCHLORIDE
    see AKA500
2-AMINOETHYL-2-METHYL-1,3-BENZODIOXOLE HYDROCHLORIDE
    see AKA500
N-AMINOETHYLMORPHOLINE see AKA750
p-(2-AMINOETHYL)PHENOL see TOG250
p-β-AMINOETHYLPHENOL see TOG250
4-(2-AMINOETHYL)PHENOL HYDROCHLORIDE see TOF750
p-(2-AMINOETHYL)PHENOL MONOCHLORIDE see TOF750
AMINOETHYLPIPERAZINE see AKB000
N-AMINOETHYLPIPERAZINE see AKB000
1-(2-AMINOETHYL)PIPERAZINE see AKB000
N-(2-AMINOETHYL)PIPERAZINE see AKB000
N-(β-AMINOETHYL)PIPERAZINE see AKB000
6-AMINO-3-ETHYL-1-(2-PROPENYL)-2,4(1H,3H-)-PYRIMIDINEDIONE
    see AFW500
4-(2-AMINOETHYL)PYROCATECHOL see DYC400
4-(2-AMINOETHYL)PYROCATECHOL HYDROCHLORIDE see DYC600
5-AMINO-2-ETHYLTETRAZOL see AKB125
2-AMINO-4-(ETHYLTHIO)BUTYRIC ACID see AKB250, EEI000
l-2-AMINO-4-(ETHYLTHIO)BUTYRIC ACID see AKB250
dl-2-AMINO-4-(ETHYLTHIO)BUTYRIC ACID see EEI000
S-(2-AMINOETHYL)THIOPHOSPHATEMONOSODIUM SALT see AKB500
2-AMINOETHYL-2-THIOPSEUDOUREA DICHLORIDE see AJY500
2-(2-AMINOETHYL)-2-THIOPSEUDOUREA DIHYDROCHLORIDE see AJY500
2-(2-AMINOETHYL)-2-THIOPSEUDOUREA HYDROBROMIDE see AJY250
3-AMINO-α-ETHYL-2,4,6-TRIIODOHYDROCINNAMIC ACID see IFY100
3′-AMINO-N-ETHYL-2′,4′,6′-TRIIODOSUCCINANILIC ACID see AMV375
AMINOFENAZONE (ITALIAN) see DOT000
5-AMINO-3-FENIL-1-BIS(-DIMETILAMINO)-FOSFORIL-1,2,4-TRIAZOLO(ITAL-
    IAN) see AIX000
m-AMINOFENOL (CZECH) see ALT500
p-AMINOFENOL (CZECH) see ALT250
5-AMINO-3-FENYL-1-BIS(DIMETHYL-AMINO)-FOSFORYL-1,2,4-TRIAZOOL
    (DUTCH) see AIX000
m-AMINOFENYLMOCOVINA HYDROCHLORID (CZECH) see ALY750
AMINOFILINA (SPANISH) see TEP500
AMINOFLUOREN (GERMAN) see FDI000
2-AMINOFLUORENE see FDI000
2-AMINOFLUORENONE see AKB875
4-AMINOFLUORENONE see AKB900
2-AMINO-9-FLUORENONE see AKB875

4-AMINO-9-FLUORENONE see AKB900
2-AMINO-N-FLUOREN-2-YLACETAMIDE see AKC000
2-AMINO-5-FLUOROBENZOXAZOLE see AKC250
4-AMINO-4′-FLUORODIPHENYL see AKC500
6-AMINO-2-(FLUOROMETHYL)-3-(2-METHYLPHENYL)-4(3H)-
    QUINAZOLINONE (9CI) see AEW625
6-AMINO-2-FLUOROMETHYL-3-(o-TOLYL)-4(3H)-QUINAZOLINONE
    see AEW625
2-AMINO-1-(3-FLUOROPHENYL)ETHANOL HYDROBROMIDE see AKS500
(2-AMINO-6-(((4-FLUOROPHENYL)METHYL)AMINO)-3-PYRIDINYL)CAR-
    BAMIC ACID ETHYL ESTER MALEATE see FMP100
3-AMINO-2-FLUORO-PROPANOIC ACID HYDROCHLORIDE see FFT100
4-AMINO-5-FLUORO-2(1H)-PYRIMIDINONE see FHI000
AMINOFORM see HEI500
AMINOFORMAMIDINE see GKW000
AMINOFOSTINE see AMD000
2-AMINOGLUTARAMIC ACID see GFO050
l-2-AMINOGLUTARAMIDIC ACID see GFO050
α-AMINOGLUTARIC ACID see GFO000
l-2-AMINOGLUTARIC ACID see GFO000
α-AMINOGLUTARIC ACID HYDROCHLORIDE see GFO025
l-2-AMINOGLUTARIC ACID HYDROCHLORIDE see GFO025
AMINOGLUTETHIMIDE see AKC600
p-AMINOGLUTETHIMIDE see AKC600
AMINOGLUTETHIMIDE PHOSPHATE see AKC625
1-AMINOGLYCEROL see AMA250
AMINOGLYCOL see ALB000
AMINOGUANIDINE see AKC750
AMINOGUANIDINE SULFATE see AKD250
AMINOGUANIDINE SULPHATE see AKD250
AMINO GUANIDINIUM NITRATE see AKD375
l,2-AMINO-4-(GUANIDINOOXY)BUTYRIC ACID see AKD500
2-AMINO-4-(GUANIDINOOXY)-l-BUTYRIC ACID see AKD500
dl-2-AMINOHEPTANE see TNX750
2-AMINOHEPTANE SULFATE see AKD600
7-AMINOHEPTANOIC ACID, ISOPROPYL ESTER see AKD625
3-(7-AMINOHEPTYL)INDOLE ADIPATE see AKD750
AMINOHEXAHYDROBENZENE see CPF500
AMINOHEXAHYDROBENZENE HYDROCHLORIDE see CPA775
9-AMINO-2,3,5,6,7,8-HEXAHYDRO-1H-CYCLOPENTA(b)QUINOLINE HYDRO-
    CHLORIDE HYDRATE see AKD775
1-AMINOHEXANE see HFK000
omega-AMINOHEXANOIC ACID see AJD000
6-AMINOHEXANOIC ACID CYCLIC LACTAM see CBF700
6-AMINOHEXANOIC ACID HOMOPOLYMER see PJY500
3-AMINO-4-HOMOISOTWISTANE see AKD875
α-AMINOHYDROCINNAMIC ACID see PEC750
2-AMINO-3-HYDROXYACETOPHENONE see AKE000
1-AMINO-4-HYDROXYANTHRAQUINONE see AKE250
4-AMINO-4′-HYDROXYAZOBENZENE see AKE500
2-AMINO-1-HYDROXYBENZENE see ALT000
3-AMINO-1-HYDROXYBENZENE see ALT500
4-AMINO-1-HYDROXYBENZENE see ALT250
3-AMINO-4-HYDROXYBENZENEARSONIC ACID see HJE500
4-AMINO-2-HYDROXYBENZENEARSONIC ACID see HJE400
2-AMINO-3-HYDROXYBENZOIC ACID see AKE750
4-AMINO-2-HYDROXYBENZOIC ACID see AMM250
5-AMINO-2-HYDROXYBENZOIC ACID see AMM500
4-AMINO-2-HYDROXYBENZOIC ACID, 2-(DIETHYLAMINO)ETHYL ESTER,
    HYDROCHLORIDE (9CI) see AMM750
4-AMINO-2-HYDROXYBENZOIC ACID, 2-(DIMETHYLAMINO)ETHYL ESTER,
    HYDROCHLORIDE see AMN000
2-AMINO-3-HYDROXYBENZOIC ACID, METHYL ESTER see HJC500
3-AMINO-2-HYDROXYBENZOIC ACID METHYL ESTER see AKF000
α-AMINO-p-HYDROXYBENZYLPENICILLIN TRIHYDRATE see AOA100
4-AMINO-4′-HYDROXYBIPHENOL see AIW000
4-AMINO-3-HYDROXYBIPHENYL see AIV750
4-AMINO-3-HYDROXYBIPHENYL SULFATE see AKF250
4-AMINO-3-HYDROXYBUTYRIC ACID see AKF375
γ-AMINO-β-HYDROXYBUTYRIC ACID see AKF375
l-2-AMINO-3-HYDROXYBUTYRIC ACID see TFU750
1-AMINO-4-HYDROXY-5-CHLORANTHRACHINON (CZECH) see AJH250
5-AMINO-2-HYDROXY-2,4,6-CYCLOHEPTATRIEN-1-ONE see AMV800
4-AMINO-2-HYDROXY-2,4,6-CYCLOHEPTATRIEN-1-ONE see AMV790
2-AMINO-3-HYDROXY-1,7-DIHYDRO-7-METHYL-6H-PURIN-6-ONE
    see HMD000
2-AMINO-3-HYDROXY-1,7-DIHYDRO-8-METHYL-6H-PURIN-6-ONE
    see HMD500
3-AMINO-4-HYDROXYDIPHENYL HYDROCHLORIDE see AIW250
3-AMINO-4-(2-HYDROXY)ETHOXYBENZENARSONIC ACID see AKF500
(3-AMINO-4-(2-HYDROXYETHOXY)PHENYL)ARSINE OXIDE see AKF750
(R)-4-(2-AMINO-1-HYDROXYETHYL)-1,2-BENZENEDIOL see NNO500
4-AMINO-N-(2-HYDROXYETHYL)-o-TOLUENESULFONAMIDE see AKG000
2-AMINO-1-HYDROXYHYDROINDENE HYDROCHLORIDE see AKL100

α-AMINO-3-HYDROXY-5-ISOXAZOLEACETIC ACID HYDRATE see AKG250
α-AMINO-3-HYDROXY-5-ISOXAZOLESSIGSAURE HYDRAT (GERMAN) see AKG250
AMINO-(3-HYDROXY-5-ISOXAZOLYL)ACETIC ACID see AKG250
2-AMINO-5-HYDROXYLEVULINIC ACID see AKG500
4-AMINO-5-HYDROXYMETHYL-2-METHOXYPYRIMIDINE see AKO750
2-AMINO-2-(HYDROXYMETHYL)PROPANE-1,3-DIOL see TEM500
2-AMINO-2-(HYDROXYMETHYL)-1,3-PROPANEDIOL see TEM500
4-AMINO-5-HYDROXY-2,7-NAPHTHALENEDISULFONIC ACID see AKH000
8-AMINO-7-HYDROXY-3,6-NAPHTHALENEDISULFONIC ACID, SODIUM SALT see AKH250
4-AMINO-5-HYDROXY-2,7-NAPHTHALENEDISULFONIC ACID-p-TOLUENESULFONATE (ESTER) see AKH500
4-AMINO-5-HYDROXY-1-NAPHTHALENESULFONIC ACID see AKH750
7-AMINO-4-HYDROXY-2-NAPHTHALENESULFONIC ACID see AKI000
3-AMINO-4-HYDROXYNITROBENZENE see NEM500
3-AMINO-2-HYDROXY-5-NITRO-BENZENESULFONIC ACID see HMY500
3-AMINO-4-HYDROXY-5-NITROBENZENESULFONIC ACID see AKI250
4-AMINO-3-HYDROXY-4'-NITRODIPHENYLHYDROCHLORIDE see AKI500
l-2-AMINO-4-(HYDROXYNITROSAMINO)PROPIONIC ACID see AFH750
2-AMINO-5-HYDROXY-4-OXOPENTANOIC ACID see AKG500
(S-(4*,s*))-N-(3-AMINO-2-HYDROXY-1-OXO-4-PHENYLBUTYL)-l-LEUCINE see BFV300
3-AMINO-4-HYDROXY-PHENARSINE HYDROCHLORIDE see ARL000
1-AMINO-4-HYDROXY-2-PHENOXYANTHRAQUINONE see AKI750
(2S-(2-α,5-α,6-β(S*))-6-((AMINO(4-HYDROXYPHENYL)ACETYL)AMINO)-3,3-DIMETHYL-7-OXO-4-THIA-1-AZABICYCLO(3.2.0)HEPTANE-2-CARBOXYLIC ACID TRIHYDRATE see AOA100
((5-(3-AMINO-4-HYDROXYPHENYL)ARSENO)-2-HYDROXYANILINO)METHANOL SULFOXYLATE SODIUM see NCJ500
(3-AMINO-4-HYDROXYPHENYL)ARSENOUS ACID see OOK100
3-AMINO-4-HYDROXYPHENYLARSINE OXIDE HYDROCHLORIDE see ARL000
3-AMINO-4-HYDROXYPHENYL ARSINOXIDE HYDROCHLORIDE see ARL000
(3-AMINO-4-HYDROXYPHENYL)ARSONOUS DICHLORIDE MONOHYDROCHLORIDE see DFX400
3-(R)-AMINO-2-(s)-HYDROXY-4-PHENYLBUTANOYL-(2)-LEUCINE see BFV300
3-AMINO-4-HYDROXYPHENYL DICHLORARSINE HYDROCHLORIDE see DFX400
(3-AMINO-4-HYDROXYPHENYL)DICHLOROARSINE HYDROCHLORIDE see DFX400
2-AMINO-3-HYDROXYPHENYL METHYL KETONE see AKE000
3-(((3-AMINO-4-HYDROXYPHENYL)PHENYLARSINO)THIO)ALANINE see AKI900
dl-2-AMINO-1-HYDROXY-1-PHENYLPROPANE see NNM500
1-N-(S-AMINO-2-HYDROXYPROPIONYL)BETAMYCIN see AKJ000
1-N-(S-3-AMINO-2-HYDROXYPROPIONYL)-GENTAMICIN B SULFATE see SBE800
1-N-(S-3-AMINO-2-HYDROXYPROPIONYL) GENTAMYCIN B see AKJ000
3-AMINO-4(1-(2-HYDROXY)PROPOXY)BENZENEARSONIC ACID see AKJ250
S-3-AMINO-2-HYDROXYPROPYL SODIUMHYDROGEN PHOS-PHOROTHIOATETETRAHYDRATE see AKJ500
l-N-(p-(((-2-AMINO-4-HYDROXY-6-PTERIDINYL)METHYL)AMINO)BENZOYL) GLUTAMIC ACID see FMT000
4-AMINO-2-HYDROXYTOLUENE see AKJ750
5-AMINO-7-HYDROXY-1H-v-TRIAZOLO(d)PYRIMIDINE see AJO500
4-AMINO-4'-METHYLAZO-2,3',5'-TRIMETHYLAZOBENZENE see AKK000
2-AMINOHYPOXANTHINE see GLI000
5-AMINOIMIDAZOLE-4-CARBOXAMIDE see AKK250
5-AMINO-4-IMIDAZOLECARBOXAMIDE UREIDOSUCCINATE see AKK625
5-AMINO-1H-IMIDAZOLE HYDROCHLORIDE see AKL250
α-AMINOIMIDAZOLE-4-PROPIONIC ACID, COBALT(2+) SALT see BJY000
α-AMINO-o-IMINOETHANE HYDROCHLORIDE see AAI500
9-((AMINOIMINOMETHYL)AMINO)-N-(10-((AMINOIMINOMETHYL)AMINO)-1-(3-AMINOPROPYL)-20-HYDROXYDECYL)NONANAMIDE see EQS500
4-((AMINOIMINOMETHYL)AMINO)BENZOIC ACID 2-(METHOXYCARBONYL)PHENYL ESTER see CBT175
4-((AMINOIMINOMETHYL)AMINO)BENZOIC ACID 2-METHOXY-4-(2-PROPE-NYL)PHENYL ESTER see MDY300
N-(4-((AMINOIMINOMETHYL)AMINO)-1-FORMYLBUTYL)-3-PHENYLALANYL-l-PROLINAMIDE SULFATE, (S)- see PEE100
o-((AMINOIMINOMETHYL)AMINO)-l-HOMOSERINE see AKD500
3-(((2-((AMINOIMINOMETHYL)AMINO)-4-THIAZOLYL)METHYL)THIO)-N-(AMINOSULFONYL)PROPANIMIDAMIDE see FAB500
N-(AMINOIMINOMETHYL)-2,6-DICHLOROBENZENEACETAMIDE HYDRO-CHLORIDE see GKU300
N-(AMINOIMINOMETHYL)-N'-(4-NITROPHENYL)UREA MONOHYDROCHLORIDE see NIJ400
(2-((AMINOIMINOMETHYL)THIO)ETHYL)TRIETHYLAMMONIUM BROMIDE HYDROBROMIDE see TAI100
2-((AMINOIMINOMETHYL)THIO)-N,N,N-TRIETHYL-ETHANAMINIUM BRO-MIDE, MONOHYDROBROMIDE (9CI) see TAI100
2-AMINO-4-IMINO-1(4H)-NAPHTHALENONE HYDROCHLORIDE see ALK625
7-AMINO-3-IMINO-3H-PHENOTHIAZINEMONOHYDROCHLORIDE see AKK750

2-AMINOINDANE HYDROCHLORIDE see AKL000
2-AMINOINDAN HYDROCHLORIDE see AKL000
AMINOINDANOL HYDROCHLORIDE see AKL100
5-AMINOINDAZOLE HYDROCHLORIDE see AKL250
α-AMINO-INDOLE-3-PROPRIONIC ACID see TNX000
l-α-AMINO-3-INDOLEPROPRIONIC ACID see TNX000
α'-AMINO-3-INDOLEPROPRIONIC ACID see TNX000
2-AMINO-3-INDOL-3-YL-PROPRIONIC ACID see TNX000
2-AMINOINOSINE see GLS000
m-AMINOIODOBENZENE see IEB000
2-AMINO-5-IODOBENZOXAZOLE see AKL500
2-AMINOISOBUTANE see BPY250
α-AMINOISOBUTANOIC ACID see MGB000
β-AMINOISOBUTANOL see IIA000
2-AMINOISOBUTYRIC ACID see MGB000
α-AMINOISOBUTYRIC ACID see MGB000
α-AMINOISOCAPROIC ACID see LES000
AMINOISOMETRADIN see AKL625
AMINOISOMETRADINE see AKL625
3-AMINOISONAPHTHOIC ACID see ALJ000
α-AMINOISOPROPYL ALCOHOL see AMA500
2-AMINOISOPROPYLBENZENE see INX000
o-AMINOISOPROPYLBENZENE see INX000
d-R-AMINO-3-ISOSSAZOLIDONE (ITALIAN) see CQH000
l-(+)-α-AMINOISOVALERIC ACID see VBP000
d-4-AMINO-3-ISOXAZOLIDINONE see CQH000
d-4-AMINO-3-ISOXAZOLIDONE see CQH000
AMINOKAPRON see AJD000
trans-1-AMINO-2-MERCAPTOMETHYLCYCLOBUTANEHYDROCHLORIDE see AKL750
3-AMINO-4-MERCAPTO-6-METHYLPYRIDAZIN (GERMAN) see ALB625
3-AMINO-4-MERCAPTO-6-METHYLPYRIDAZINE see ALB625
2-AMINO-6-MERCAPTOPURINE see AMH250
2-AMINO-6-MERCAPTOPURINE RIBONUCLEOSIDE see TFJ500
2-AMINO-6-MERCAPTOPURINE RIBOSIDE see TFJ500
2-AMINO-6-MERCAPTO-9(β-d-RIBOFURANOSYL)PURINE see TFJ500
2-AMINO-5-MERCAPTO-1,3,4-THIADIAZOLE see AKM000
5-AMINO-2-MERCAPTO-1,3,4-THIADIAZOLE see AKM000
AMINOMERCURIC CHLORIDE see MCW500
AMINOMESITYLENE see TLG500
2-AMINOMESITYLENE see TLG500
AMINOMESITYLENE HYDROCHLORIDE see TLH000
2-AMINOMESITYLENE HYDROCHLORIDE see TLH000
6-AMINO-1-METALLYL-3-METHYLPYRIMIDINE-2,4-DIONE see AKL625
AMINOMETHANAMIDINE see GKW000
AMINOMETHANE see MGC250
6-AMINOMETHAQUALONE see AKM125
1-AMINO-2-METHOXYANTHRAQUINONE see AKM250
4-AMINO-3-METHOXYAZOBENZENE see MFB000, MFF500
1-AMINO-2-METHOXYBENZENE see AOV900
1-AMINO-4-METHOXYBENZENE see AOW000
2-AMINO-5-METHOXY BENZENESULFONIC ACID see AIA500
3-AMINO-4-METHOXYBENZOIC ACID see AIA250
2-AMINO-4-METHOXYBENZOTHIAZOLE see AKM750
2-AMINO-5-METHOXYBENZOXAZOLE see AKN000
4-AMINO-4'-METHOXY-3-BIPHENYLOL see HLR500
4-AMINO-4'-METHOXY-3-BIPHENYLOLHYDROCHLORIDE see AKN250
2-AMINO-3-METHOXYDIPHENYLENE OXIDE see AKN500
2-AMINO-3-METHOXYDIPHENYLENOXYD (GERMAN) see MDZ000
4-AMINO-N-(2-METHOXYETHYL)-7-((2-METHOXYETHYL)AMINO-2-PHENYL-6-PTERIDINECARBOXAMIDE see AKN750
3'-(α-AMINO-p-METHOXYHYDROCINNAMAMIDO)-3'-DEOXY-N,N-DIMETHYLADENOSINE MONOHYDROCHLORIDE see POK250
3'-(α-AMINO-p-METHOXYHYDROCINNAMAMIDO)-3'-DEOXY-N,N-DIMETHYLADENOSINE DIHYDROCHLORIDE see POK300
3'-(l-α-AMINO-p-METHOXYHYDROCINNAMAMIDO)-3'-DEOXY-N,N-DIMETHYLADENOSINE see AEI000
1-AMINO-2-METHOXY-5-METHYLBENZENE see MGO750
6-AMINO-8-METHOXY-1-METHYL-4-(p-(p-((1-METHYLPYRIDINIUM-4-YL)AMINO)BENZAMIDO)ANILINOQUINOLINIUMDI-p-TOLUENESULFON-ATE see AKO000
7-AMINO-4-(2-METHOXY-p-(p-((1-METHYLPYRIDINIUM-4-YL)AMINO)BENZAMIDO)ANILINO)-1-METHYLQUINOLINIUM DIBROMIDE see AKO250
4-AMINO-N-(6-METHOXY-2-METHYL-4-PYRIMIDINYL)BENZENESULFONAM-IDE see MDU300
7-AMINO-9-α-METHOXYMITOSANE see AHK500
2-AMINO-1-METHOXYNAPHTHALENE see MFA000
3-AMINO-4-METHOXYNITROBENZENE see NEQ500
2-AMINO-1-METHOXY-4-NITROBENZENE see NEQ500
(S)-3'-((2-AMINO-3-(4-METHOXYPHENYL)-1-OXOPROPYL)AMINO)-3'-DEOXY-N,N-DIMETHYLADENOSINE see AEI000
4-AMINO-6-METHOXY-1-PHENYLPYRIDAZINIUM-METHYLSULFAT (GER-MAN) see MQS100

4-AMINO-6-METHOXY-1-PHENYLPYRIDAZINIUM METHYL SULFATE see MQS100
4-AMINO-N-(6-METHOXY-3-PYRIDAZINYL)-BENZENESULFONAMIDE see AKO500
4-AMINO-2-METHOXY-5-PYRIMIDINEMETHANOL see AKO750
4-AMINO-N-(6-METHOXY-4-PYRIMIDINYL)-BENZENESULFONAMIDE (9CI) see SNL800
4-AMINO-N-(5-METHOXY-2-PYRIMIDINYL)BENZENESULFONAMIDE SODIUM SALT see MFO000
3-AMINO-4-METHOXYTOLUEN (CZECH) see MFQ500
3-AMINO-4-METHOXYTOLUENE see MGO750
4-AMINO-1-METHYLAMINOANTHRAQUINONE see AKP250
dl-α-AMINO-β-METHYLAMINOPROPIONIC ACID see AKP500
2-AMINO-N-METHYLANILINE see MNU000
4-AMINO-2-METHYLANILINE see TGM000
2-AMINO-4-METHYLANISOLE see MGO750
1-AMINO-2-METHYL-9,10-ANTHRACENEDIONE see AKP750
1-AMINO-2-METHYLANTHRAQUINONE see AKP750
1-AMINO-2-METHYLBENZENE see TGQ750
2-AMINO-1-METHYLBENZENE see TGQ750
3-AMINO-1-METHYLBENZENE see TGQ500
4-AMINO-1-METHYLBENZENE see TGR000
1-AMINO-2-METHYLBENZENE HYDROCHLORIDE see TGS500
2-AMINO-1-METHYLBENZENE HYDROCHLORIDE see TGS500
β-(AMINOMETHYL)BENZENEPROPANOIC ACID see PEE500
β-(AMINOMETHYL)-BENZENEPROPANOIC ACID HYDROCHLORIDE see GAD000
2-AMINO-5-METHYLBENZENESULFONIC ACID see AKQ000
3-AMINO-4-METHYLBENZENESULFONYLCYCLOHEXYLUREA see AKQ250
2-AMINO-4-METHYLBENZOTHIAZOLE see AKQ500
2-AMINO-6-METHYLBENZOTHIAZOLE see MGD500
2-AMINO-4-METHYL-BENZOTHIAZOLE, HYDROCHLORIDE see MGD750
2-AMINO-5-METHYLBENZOXAZOLE see AKQ750
3-AMINO-α-METHYLBENZYL ALCOHOL see AKR000
m-AMINO-α-METHYLBENZYL ALCOHOL see AKR000
8-(4-AMINO-1-METHYLBUTYLAMINO)-6-METHOXYQUINOLINE see PMC300
8-((4-AMINO-1-METHYLBUTYL)AMINO)-6-METHOXYQUINOLINE DIPHOSPHATE see AKR250
2-AMINO-3-METHYL-α-CARBOLINE see ALD750
3-AMINO-1-METHYL-γ-CARBOLINE see ALD500
2-AMINO-4-METHYL-5-CARBOXANILIDOTHIAZOLE see AKR500
β-(AMINOMETHYL)-4-CHLOROBENZENEPROPANOIC ACID see BAC275
β-(AMINOMETHYL)-p-CHLOROHYDROCINNAMIC ACID see BAC275
trans-p-(AMINOMETHYL)CYCLOHEXANECARBOXYLIC ACID see AJV500
trans-1-AMINOMETHYLCYCLOHEXANE-4-CARBOXYLIC ACID see AJV500
trans-4-AMINOMETHYL-1-CYCLOHEXANECARBOXYLIC ACID see AJV500
trans-4-(((4-(AMINOMETHYL)CYCLOHEXYL)CARBONYL)OXY)BENZENEPROPANOIC ACID see CDF375
trans-4-(((4-(AMINOMETHYL)CYCLOHEXYL)CARBONYL)OXY)-BENZENEPROPANOIC ACID HYDROCHLORIDE see CDF380
4-AMINO-3-METHYL-N,N-DIETHYLANILINEHYDROCHLORIDE see AKR750
2-AMINOMETHYL-2,3-DIHYDRO-4H-PYRAN see AKS000
2-AMINOMETHYL-3,4-DIHYDRO-2H-PYRAN see AKS000
α-(AMINOMETHYL)-3,4-DIHYDROXYBENZYL ALCOHOL see ARL500, ARL750
l-α-(AMINOMETHYL)-3,4-DIHYDROXYBENZYL ALCOHOL see NNO500
2-AMINO-6-METHYLDIPYRIDO(1,2-a:3′,2′-d)IMIDAZOLE see AKS250
2-AMINO-6-METHYLDIPYRIDO(1,2-a:3′,2′-d)IMIDAZOLE HYDROCHLORIDE see AKS275
N-AMINOMETHYLDOPA see CBQ500
α-AMINO-2-METHYLENE-CYCLOPROPANEPROPANOIC ACID (9CI) see MJP500
α-AMINOMETHYLENECYCLOPROPANEPROPIONIC ACID see MJP500
l-α-AMINO-β-METHYLENECYCLOPROPANEPROPIONIC ACID see MJP500
α-AMINO-β-(2-METHYLENECYCLOPROPYL)PROPIONIC ACID see MJP500
2-AMINO-4,5-METHYLENEHEX-5-ENOIC ACID see MJP500
α-AMINOMETHYL-3-FLUOROBENZENYLALCOHOL HYDROBROMIDE see AKS500
4-AMINO-10-METHYLFOLIC ACID see MDV500
2-AMINO-6-METHYLHEPTANE see ILM000
6-AMINO-2-METHYLHEPTANE see ILM000
6-AMINO-2-METHYL-2-HEPTANOL HYDROCHLORIDE see HBB000
4-AMINO-2-METHYL-3-HEXANOL see AKS750
β-(AMINOMETHYL)HYDROCINNAMIC ACID see PEE500
β-(AMINOMETHYL)-HYDROCINNAMIC ACID HYDROCHLORIDE see GAD000
α-(AMINOMETHYL)-m-HYDROXYBENZYL ALCOHOL see AKT000
α-(AMINOMETHYL)-p-HYDROXYBENZYL ALCOHOL see AKT250
α-AMINOMETHYL-3-HYDROXYBENZYLALCOHOL HYDROCHLORIDE see AKT500
(±)-α-(AMINOMETHYL)-m-HYDROXYBENZYL ALCOHOL HYDROCHLORIDE see NNT100
5-AMINOMETHYL-3-HYDROXYISOXAZOLE see AKT750
2-AMINO-3-METHYLIMIDAZO(4,5-f)QUINOLINE see AKT600
2-AMINO-3-METHYLIMIDAZO(4,5-f)QUINOLINE DIHYDROCHLORIDE see AKT620

5-(AMINOMETHYL)-3-ISOXAZOLOL see AKT750
5-(AMINOMETHYL)-3(2H)-ISOXAZOLONE see AKT750
4-AMINO-N-(5-METHYL-3-ISOXAZOLYL)BENZENESULFONAMIDE see SNK000
5-AMINOMETHYL-3-ISOXYZOLE see AKT750
l-α-AMINO-γ-METHYLMERCAPTOBUTYRIC ACID see MDT750
4-AMINOMETHYL-5-MERCAPTOMETHYL-2-METHYL-3-PYRIDINOL THIO ACETATE HYDROBROMIDE see ADD000
6-AMINO-3-METHYL-1-(2-METHYLALLYL)-2,4(1H,3H)-PYRIMIDINEDIONE see AKL625
6-AMINO-3-METHYL-1-(2-METHYLALLYL)URACIL see AKL625
6-AMINO-3-METHYL-1-(2-METHYL-2-PROPENYL)-2,4(1H,3H)-PYRIMIDINEDIONE see AKL625
4-AMINO-2-METHYL-1-NAPHTHALENOL see AKX500
4-AMINO-2-METHYL-1-NAPHTHOL see AKX500
3-AMINO-4-METHYL-5-(5-NITRO-2-FURYL)-s-TRIAZOLE see AKY000
2-AMINO-4-((E)-2-(1-METHYL-5-NITRO-1H-IMIDAZOL-2-YL)ETHENYL)PYRIMIDINE see TJF000
2-AMINO-6-(1-METHYL-4-NITRO-5-IMIDAZOLYL)MERCAPTOPURINE see AKY250
2-AMINO-6-(1′-METHYL-4′-NITRO-5′-IMIDAZOLYL)MERCAPTOPURINE see AKY250
(E)-2-AMINO-4-(2-(1-METHYL-5-NITROIMIDAZOL-2-YL)VINYL)PYRIMIDINE see TJF000
2-(AMINOMETHYL)NORBORNANE see AKY750
2-AMINO-2-METHYLOL-1,3-PROPANEDIOL see TEM500
2-AMINO-4-METHYLOXAZOLE see AKX875
2-AMINO-3-METHYLPENTANOIC ACID see IKX000
2-AMINO-4-METHYLPENTANOIC ACID see LES000
4-AMINO-3-METHYLPHENOL see AKZ000
5-AMINO-2-METHYLPHENOL see AKJ750
cis-2-AMINO-5-METHYL-4-PHENYL-1-PYRROLINE see ALA000
trans-2-AMINO-5-METHYL-4-PHENYL-1-PYRROLINE see ALA250
8-AMINO-2-METHYL-4-PHENYL-1,2,3,4-TETRAHYDROISOQUINOLINE see NMV700
8-AMINO-2-METHYL-4-PHENYL-1,2,3,4-TETRAHYDROISOQUINOLINE MALEATE see NMV725
4-AMINO-3-METHYL-6-PHENYL-1,2,4-TRIAZIN-5(4H)-ONE see ALA500
2-AMINO-4-(N-METHYLPIPERAZINO)-5-METHYLTHIO-6-CHLORO-PYRIMIDINE see ALA750
1-AMINO-2-METHYLPROPANE see IIM000
2-AMINO-2-METHYLPROPANE see BPY250
2-AMINO-2-METHYL-1,3-PROPANEDIOL see ALB000
2-AMINO-2-METHYLPROPANOIC ACID see MGB000
2-AMINO-2-METHYLPROPANOL see IIA000
1-AMINO-2-METHYL-2-PROPANOL see ALB250
2-AMINO-2-METHYLPROPAN-1-OL see IIA000
2-AMINO-2-METHYL-1-PROPANOL see IIA000
S-2-AMINO-2-METHYLPROPYL DIHYDROGEN PHOSPHOROTHIOATE see ALB500
S-(2-AMINO-2-METHYLPROPYL)PHOSPHOROTHIOATE see ALB500
(−)-α-(AMINOMETHYL)PROTOCATECHUYL ALCOHOL see NNO500
4-AMINO-N^{10}-METHYLPTEROYLGLUTAMIC ACID see MDV500
4-AMINO-N^{10}-METHYLPTEROYLGLUTAMIC ACID DISODIUM SALT see MDV600
3-AMINO-6-METHYL-4-PYRIDAZINETHIOL see ALB625
2-AMINOMETHYLPYRIDINE see ALB750
2-AMINO-3-METHYLPYRIDINE see ALC000
2-AMINO-4-METHYLPYRIDINE see ALC250
2-AMINO-6-METHYLPYRIDINE see ALC500
4-((3-AMINO-4-((4-((1-METHYLPYRIDINIUM-4-YL)AMINO)BENZOYL)AMINO)PHENYL)AMINO)-1-METHYLQUINOLINIUM)DIBROMIDE-52-9 see ALC750
2-AMINO-3-METHYL-9H-PYRIDO(2,3-b)INDOLE see ALD750
3-AMINO-1-METHYL-5H-PYRIDO(4,3-b)INDOLE see ALD500
3-AMINO-1-METHYL-5H-PYRIDO(4,3-b)INDOLE ACETATE see ALE750
4-AMINO-N-(4-METHYL-2-PYRIMIDINYL)-BENZENESULFONAMIDE see ALF250
4-AMINO-N-(4-METHYL-2-PYRIMIDINYL)-BENZENESULFONAMIDE MONOSODIUM SALT see SJW475
1-(4-AMINO-2-METHYLPYRIMIDIN-5-YL)METHYL-3-(2-CHLOROETHYL)-3-NITROSOUREA see ALF500
3-(4-AMINO-2-METHYL-5-PYRIMIDINYL)METHYL-1-(2-CHLOROETHYL)-1-NITROSOUREA see NDY800
3-((4-AMINO-2-METHYL-5-PYRIMIDINYL)METHYL)-1-(2-CHLOROETHYL)-1-NITROSOUREA HYDROCHLORIDE see ALF500
N′-((4-AMINO-2-METHYL-5-PYRIMIDINYL)METHYL)-N-(2-CHLOROETHYL)-N-NITROSOUREA HCl see ALF500
3-((4-AMINO-2-METHYL-5-PYRIMIDINYL)METHYL)-5-(2-HYDROXYETHYL)-4-METHYLTHIAZOLIUM CHLORIDE see TES750
3-(4-AMINO-2-METHYLPYRIMIDYL-5-METHYL)-4-METHYL-5β-HYDROXYETHYLTHIAZOLIUM NITRATE see TET500
4-AMINO-N-METHYL-1,2,5-SELENADIAZOLE-3-CARBOXAMIDE see MGL600
2-AMINO-4-(METHYLSELENYL)BUTYRIC ACID see SBU725
2-AMINO-4-(METHYLSULFINYL)BUTYRIC ACID see ALF600

2-AMINO-4-(S-METHYLSULFONIMIDOYL)-BUTANOIC ACID (9CI)
  see MDU100
2-AMINOMETHYLTETRAHYDROPYRAN see ALF750
2-AMINO-N-(3-METHYL-2-THIAZOLIDINYLIDENE)ACETAMIDE see ALG250
2-AMINO-4-(METHYLTHIO)BUTYRIC ACID see MDT750
l(−)-AMINO-γ-METHYLTHIOBUTYRIC ACID see MDT750
5-AMINO-3-METHYLTHIO-1,2,4-OXADIAZOLE see ALG375
4-AMINO-3-METHYLTOLUENE see XMS000
2-AMINO-4-METHYLTOLUENE HYDROCHLORIDE see XOS000
4-AMINO-3-METHYLTOLUENE HYDROCHLORIDE see XOJ000
6-AMINO-2-METHYL-3-(o-TOLYL)-1(3H)-QUINAZOLINONE see AKM125
α-AMINOMETHYL-m-TRIFLUOROMETHYLBENZYL ALCOHOL see ALG500
2-AMINO-4-METHYLVALERIC ACID see LES000
α-AMINO-β-METHYLVALERIC ACID see IKX000
α-AMINO-γ-METHYLVALERIC ACID see LES000
l-2-AMINO-4-METHYLVALERIC ACID see LES000
dl-2-AMINO-4-METHYLVALERIC ACID see LER000
AMINOMETRADINE see AFW500
AMINOMETRAMIDE see AFW500
2-AMINO-6-MP see AMH250
1-AMINONAFTALEN (CZECH) see NBE000
2-AMINONAFTALEN (CZECH) see NBE500
1-AMINONAPHTHALENE see NBE000
2-AMINONAPHTHALENE see NBE500
3-AMINO-2-NAPHTHALENECARBOXYLIC ACID see ALJ000
2-AMINO-1,5-NAPHTHALENEDISULFONIC ACID see ALH000
2-AMINO-4,8-NAPHTHALENEDISULFONIC ACID see ALH250
3-AMINO-1,5-NAPHTHALENEDISULFONIC ACID see ALH250
6-AMINO-NAPHTHALENE-1,3-DISULFONIC ACID see ALH500
7-AMINO-1,5-NAPHTHALENEDISULFONIC ACID see NBE850
7-AMINO-1,5-NAPHTHALENEDISULFONIC ACID see ALH250
1-AMINONAPHTHALENE-4-SULFONIC ACID see ALI000
1-AMINO-6-NAPHTHALENESULFONIC ACID see ALI250
2-AMINO-1-NAPHTHALENESULFONIC ACID see ALH750
4-AMINO-1-NAPHTHALENESULFONIC ACID see ALI000
5-AMINO-2-NAPHTHALENESULFONIC ACID see ALI250
7-AMINO-1-NAPHTHALENESULFONIC ACID see ALI300
5-AMINO-2-NAPHTHALENESULFONIC ACID SODIUM SALT see ALI500
7-AMINO-1,3,6-NAPHTHALENETRISULFONIC ACID see ALI750
AMINONAPHTHALENOL see ALJ250
3-AMINO-2-NAPHTHOIC ACID see ALJ000
4-AMINO-1-NAPHTHOIC ACID 2-(DIETHYLAMINO)ETHYL ESTER HYDRO-
  CHLORIDE see NAH800
2-AMINO-1-NAPHTHOL see ALJ250
8-AMINO-2-NAPHTHOL see ALJ750
1-AMINO-2-NAPHTHOL-3,6-DISULPHONIC ACID SODIUM SALT
  see AKH250
1-AMINO-2-NAPHTHOL HYDROCHLORIDE see ALK000
1-AMINO-4-NAPHTHOL HYDROCHLORIDE see ALK500
2-AMINO-1-NAPHTHOL HYDROCHLORIDE see ALK250
4-AMINO-1-NAPHTHOL HYDROCHLORIDE see ALK500
2-AMINO-1-NAPHTHOL PHOSPHATE (ESTER) SODIUM SALT see BGU000
AMINONAPHTHOL SULFONIC ACID J see AKI000
AMINONAPHTHOL SULFONIC ACID S see AKH750
2-AMINO-1,4-NAPHTHOQUINONE IMINE HYDROCHLORIDE see ALK625
2-AMINO-1-NAPHTHYL ESTER SULFURIC ACID see ALK750
2-AMINO-1-NAPHTHYLGLUCOSIDURONIC ACID see ALL000
2-AMINO-1-NAPHTHYL HYDROGEN SULFATE see ALK750
2-AMINO-1-NAPHTHYL HYDROGEN SULPHATE see ALK750
AMINONICOTINAMIDE see ALL250
6-AMINONICOTINAMIDE see ALL250
6-AMINONICOTINIC ACID AMIDE see ALL250
6-AMINO-NICOTINSAEUREAMID (GERMAN) see ALL250
6-AMINONIKOTINSAEUREAMID (GERMAN) see ALL250
2-AMINO-4-NITROANILINE see ALL500
4-AMINO-2-NITROANILINE see ALL750
2-AMINO-5-NITROANISOL (CZECH) see NEQ000
2-AMINO-4-NITROANISOLE see NEQ500
3-AMINONITROBENZENE see IEB000
m-AMINONITROBENZENE see NEN500
p-AMINONITROBENZENE see NEO500
1-AMINO-2-NITROBENZENE see NEO000
1-AMINO-3-NITROBENZENE see NEN500
1-AMINO-4-NITROBENZENE see NEO500
2-AMINO-5-NITRO BENZENESULFONIC ACID see NEP500
2-AMINO-4-NITRO-BENZOIC ACID see NES500
4-AMINO-4′-NITROBIPHENYL see ALM000
4-AMINO-4′-NITRO-3-BIPHENYLOL HYDROCHLORIDE see AKI500
4-AMINO-4′-NITRODIPHENYL SULFIDE see AOS750
2-AMINO-5-(5-NITRO-2-FURYL)-1,3,4-OXADIAZOLE see ALM250
2-AMINO-5-(5-NITRO-2-FURYL)-1,3,4-THIADIAZOLE see NGI500
5-AMINO-2-(5-NITRO-2-FURYL)-1,3,4-THIADIAZOLE see NGI500
2-AMINO-4-(5-NITRO-2-FURYL)THIAZOLE see ALM500
5-AMINO-3-(5-NITRO-2-FURYL)-s-TRIAZOLE see ALM750

3-AMINO-6-(2-(5-NITRO-2-FURYL)VINYL)PYRIDAZINE HYDROCHLORIDE
  see ALN250
2-AMINO-4-(2-(5-NITRO-2-FURYL)VINYL)THIAZOLE see ALN500
3-AMINO-6-(2-(5-NITRO-2-FURYL)VINYL)-as-TRIAZINE see FPF000
3-AMINO-6-(2-(5-NITRO-2-FURYL)VINYL)-1,2,4-TRIAZINE see FPF000
1-AMINO-3-NITRO GUANIDINE see ALN750
2-AMINO-4-NITROPHENOL see NEM500
2-AMINO-5-NITROPHENOL see ALO000
4-AMINO-2-NITROPHENOL see NEM480
2-AMINO-4-NITROPHENOL SODIUM SALT see ALO500
2-((4-AMINO-2-NITROPHENYL)AMINO)ETHANOL see ALO750
2-AMINO-4-(p-NITROPHENYL)THIAZOLE see ALP000
l-2-AMINO-3-((N-NITROSO)HYDROXYLAMINO)PROPIONIC ACID see AFH750
4-AMINO-4′-NITRO-2,2′-STILBENEDISULFONIC ACID see ALP750
AMINONITROTHIAZOLE see ALQ000
2-AMINO-5-NITROTHIAZOLE see ALQ000
AMINONITROTHIAZOLUM see ALQ000
2-AMINO-4-NITROTOLUENE see NMP500
4-AMINO-2-NITROTOLUENE see NMP000
AMINONUCLEOSIDE see ALQ625
AMINONUCLEOSIDE PUROMYCIN see ALQ625
l-α-AMINO-4(OR 5)-IMIDAZOLEPROPIONIC ACID see HGE700
1-(4-AMINO-4-OXO-3,3-DIPHENYLBUTYL)-1-METHYLPIPERIDINIUM BRO-
  MIDE (9CI) see RDA375
(4-((2-AMINO-2-OXOETHYL)AMINO)PHENYL)ARSONIC ACID see CBJ750
N-(2-AMINO-2-OXOETHYL)-2-DIAZOACETAMIDE see CBK000
α-((2-AMINO-1-OXOPROPYL)AMINO)-5-OXO-7-OXABICYCLO(4.1.0)HEPTANE-
  2-PROPANOIC ACID see BAC175
N-(p-(((2-AMINO-4-OXO-6-PTERIDINYL)METHYL)-N-NITROSOAMINO)
  BENZOYL)-l-GLUTAMIC ACID see NLP000
AMINOOXYACETIC ACID see ALQ650
2-AMINOOXYACETIC ACID BUTYL ESTER, HYDROCHLORIDE see ALQ750
1-AMINO-4-OXYANTHRAQUINONE (RUSSIAN) see AKE250
α-(AMINOOXY)-6-BROMO-m-CRESOL see BMM600
AMINOOXYTRIPHENE HYDROCHLORIDE see TNJ750
AMINOPAR see AMM250
6-AMINOPENICILLANIC ACID see PCU500
d-(−)-α-AMINOPENICILLIN see AIV500
AMINOPENTAFLUOROBENZENE see PBD250
AMINOPENTAMIDE see DOY400
dl-AMINOPENTAMIDE HYDROCHLORIDE see ALR250
2-AMINOPENTANE see DHJ200
2-AMINOPENTANEDIOIC ACID see GFO000
2-AMINOPENTANEDIOIC ACID HYDROCHLORIDE see GFO025
3-(5-AMINOPENTYL)INDOLE ADIPATE see ALR500
AMINOPERIMIDINE see ALR750
4-AMINO-PGA see AMG750
AMINOPHEN see AOQ000
1-AMINOPHENANTHRENE see ALS000
2-AMINOPHENANTHRENE see PDA250
3-AMINOPHENANTHRENE see PDA500
9-AMINOPHENANTHRENE see PDA750
AMINOPHENAZONE see DOT000
AMINOPHENAZONE mixed with SODIUM NITRITE (1:1) see DOT200
17-(p-AMINOPHENETHYL)-MORPHINAN-3-OL (−)- see ALS250
(−)-17-(m-AMINOPHENETHYL)-MORPHINAN-3-OL, HYDROCHLORIDE
  see ALS500
(±)-17-(p-AMINOPHENETHYL)MORPHINAN-3-OL, HYDROCHLORIDE
  see ALS750
1-(p-AMINOPHENETHYL)-4-PHENYLISONIPECOTIC ACID, ETHYL ESTER
  see ALW750
1-(p-AMINOPHENETHYL)-4-PHENYLPIPERIDINE-4-CARBOXYLIC ACID
  ETHYL ESTER see ALW750
4-AMINOPHENETOLE see PDD500
p-AMINOPHENETOLE see PDD500
2-AMINOPHENOL see ALT000
3-AMINOPHENOL see ALT500
4-AMINOPHENOL see ALT250
m-AMINOPHENOL see ALT500
o-AMINOPHENOL see ALT000
p-AMINOPHENOL see ALT250
m-AMINOPHENOL (DOT) see ALT500
p-AMINOPHENOL (DOT) see ALT250
m-AMINOPHENOL, chlorinated see ALT550
m-AMINOPHENOL ANTIMONYL TARTRATE see ALT750
o-AMINOPHENOL ANTIMONYL TARTRATE see ALU000
p-AMINOPHENOL ANTIMONYL TARTRATE see ALU250
2-AMINOPHENOL-4-ARSONIC ACID DIETHYLAMINE SALT see ACN250
4-AMINOPHENOL HYDROCHLORIDE. see ALU500
p-AMINOPHENOL HYDROCHLORIDE see ALU500
o-AMINOPHENOL-OXO(TARTRATO)ANTIMONATE(1-)- see ALU000
p-AMINOPHENOL-OXO(TARTRATO)ANTIMONATE(1-)- see ALU250
4-AMINOPHENOL PHOSPHATE (3:1) (ester) see THO550
p-AMINOPHENOL PHOSPHATE (3:1) (ester) see THO550

4-AMINOPROPYLMORPHOLINE see AMF250
N-AMINOPROPYLMORPHOLINE (DOT) see AMF250
AMINOPROPYLON see AMF375
AMINOPROPYLONE see AMF375
m-(2-AMINOPROPYL)PHENOL see AMF500
p-AMINO-1-PROPYL-4-PIPERIDYL ESTER HYDROCHLORIDE BENZOIC
   ACID see PNT500
6-AMINO-1-PROPYL-4-(p-((p-((1-PROPYLPYRIDINIUM-4-YL)AMINO)-2-
   AMINOPHENYL)CARBAMOYL)ANILINO)QUINOLINIUM) DIIODIDE
   see AMF750
6-AMINO-1-PROPYL-4-(p-((p-((1-PROPYLPYRIDINIUM-4-YL)AMINO)PHENYL)
   CARBAMOYL)ANILINO)QUINOLINIUM)DIBROMIDE see AMG000
α-(1-AMINOPROPYL)PROTOCATECHUYL ALCOHOL HYDROCHLORIDE
   see ENX500
3-AMINOPROPYLSILATRAN (CZECH) see AMG500
(3-AMINOPROPYL)TRIETHOXYSILANE see TJN000
(γ-AMINOPROPYL)TRIETHOXYSILANE see TJN000
1-(3-AMINOPROPYL)-2,8,9-TRIOXA-5-AZA-1-SILABICYCLO(3.3.3) UNDECANE-
   27-1 see AMG500
AMINOPTERIDINE see AMG750
AMINOPTERIN see AMG750
4-AMINOPTEROYLGLUTAMIC ACID see AMG750
2-AMINOPURINE see AMH000
6-AMINOPURINE see AEH000
6-AMINO-1H-PURINE see AEH000
6-AMINO-3H-PURINE see AEH000
6-AMINO-9H-PURINE see AEH000
2-AMINOPURINE-6-THIOL see AMH250
2-AMINO-6-PURINETHIOL see AMH250
2-AMINOPURINE-6(1H)-THIONE see AMH250
1-(6-AMINO-9H-PURIN-9-YL)-N-CYCLOBUTYL-1-DEOXYRIBOFURAN-
   URONAMIDE see AEI500
1-(6-AMINO-9H-PURIN-9-YL)-N-CYCLOPROPYL-1-DEOXY-2,3-
   DIHYDROXYRIBOFURANURONAMIDE DIACETATE see AMH500
1-(6-AMINO-9H-PURIN-9-YL)-N-CYCLOPROPYL-1-DEOXYRIBOFURAN-
   URONAMIDE see AEJ000
1-(6-AMINO-9H-PURIN-9-YL)-N-CYCLOPROPYL-1-DEOXYRIBOFURAN-
   URONAMIDE-N-OXIDE see AEJ250
1-(6-AMINO-9H-PURIN-9-YL)-N-CYCLOPROPYLMETHYL-1-
   DEOXYRIBOFURANURONAMIDE see AEJ500
1-(6-AMINO-9H-PURIN-9-YL)-1-DEOXY-2,3-DIHYDROXY-N-
   ETHYLRIBOFURANURONAMIDE DIACETATE see AMH750
1-(6-AMINO-9H-PURIN-9-YL)-1-DEOXY-N,N-DIMETHYLRIBOFURANURONAM-
   IDE HYDRATE see AEK000
1-(6-AMINO-9H-PURIN-9-YL)-1-DEOXY-N-ETHYLRIBOFURANURONAMIDE
   HEMIHYDRATE see AEK250
1-(6-AMINO-9H-PURIN-9-YL)-1-DEOXY-N-ETHYLRIBOFURANURONAMIDE-N-
   OXIDE see AEK500
1-(6-AMINO-9H-PURIN-9-YL)-1-DEOXY-N-HEXYLRIBOFURANURONAMIDE
   HEMIHYDRATE see AEK750
1-(6-AMINO-9H-PURIN-9-YL)-1-DEOXY-N-(2-
   HYDROXYETHYL)RIBOFURANURONAMIDE see AEL000
1-(6-AMINO-9H-PURIN-9-YL)-1-DEOXY-N-ISOPROPYLRIBOFURANURONAM-
   IDE see AEL250
1-(6-AMINO-9H-PURIN-9-YL)-1-DEOXY-N-METHOXYRIBOFURANURONAM-
   IDE HYDRATE see AEL500
1-(6-AMINO-9H-PURIN-9-YL)-1-DEOXY-N-METHYLRIBOFURANURONAMIDE
   HEMIHYDRATE see AEL750
1-(6-AMINO-9H-PURIN-9-YL)-1-DEOXY-N-PROPYLRIBOFURANURONAMIDE
   see AEM000
β-d-1-(6-AMINO-9H-PURIN-9-YL)-1-DEOXYRIBOFURANURONAMIDE
   see AEI250
1-(6-AMINO-9H-PURIN-9-YL)-N-(2-(DIMETHYLAMINO)-ETHYL-1-
   DEOXYRIBOFURANURONAMIDE see AEJ750
(S)-3-(6-AMINO-9H-PURIN-9-YL)-1,2-PROPANEDIOL see AMH800
4-AMINOPYRAZOLOPYRIMIDINE see POM600
4-AMINOPYRAZOLO(3,4-d)PYRIMIDINE see POM600
1-AMINOPYRENE see PON000
3-AMINOPYRENE see PON000
AMINOPYRIDINE see POP100
2-AMINOPYRIDINE see AMI000
AMINO-2-PYRIDINE see AMI000
3-AMINOPYRIDINE see AMI250
AMINO-3-PYRIDINE see AMI250
4-AMINOPYRIDINE see AMI500
AMINO-4-PYRIDINE see AMI500
o-AMINOPYRIDINE see AMI000
p-AMINOPYRIDINE see AMI500
α-AMINOPYRIDINE see AMI000
γ-AMINOPYRIDINE see AMI500
m-AMINOPYRIDINE (DOT) see AMI250
3-AMINOPYRIDINE HYDROCHLORIDE see AMI750
4-AMINOPYRIDINE HYDROCHLORIDE see AMJ000
4-AMINO-PYRIDINEN-OXIDE see AMJ250

4-AMINOPYRIDINE-1-OXIDE see AMJ250
4-AMINO-N-(2-(4-(2-PYRIDINYL)-1-PIPERAZINYL)ETHYL)BENZAMIDE
   see AMJ500
5-AMINO-5-(4-PYRIDINYL)-2(1H)-PYRIDINONE see AOD375
2-AMINO-9H-PYRIDO(2,3-B)INDOLE see AJD750
3-AMINO-5H-PYRIDO(4,3-b)INDOLE see AMJ600
2-AMINO-5-(4-PYRIDYL)-1,3,4-THIADIAZOLEHYDROCHLORIDE see AMJ625
4-AMINO-N-2-PYRIMIDINYLBENZENESULFONAMIDE see PPP500
4-AMINO-N-2-PYRIMIDINYL-BENZENESULFONAMIDE MONOSILVER(1+)
   SALT see SNI425
4-AMINO-N-(2-PYRIMIDINYL)BENZENESULFONAMIDE SILVER SALT
   see SNI425
2-(2-AMINO-4-PYRIMIDINYLVINYL)QUINOXALINE-N,N-DIOXIDE see AMJ750
AMINOPYRINE see DOT000, POP100
AMINOPYRINE-BARBITAL see AMK250
AMINOPYRINE mixed with SODIUM NITRITE (1:1) see DOT200
AMINOPYRINE SODIUM SULFONATE see AMK500
AMINOQUIN see RHZ000
8-AMINOQUINOLINE see AML250
4-AMINOQUINOLINE-1-OXIDE see AML500
2-AMINO-4-((2-QUINOXALINYL-N,N-DIOXIDE)VINYL)PYRIMIDINES
   see AMJ750
AMINO REDUCTONE see AJL125
AMINOREXFUMARATE see ALX250
6-AMINO-9-β-d-RIBOFURANOSYL-9H-PURINE see AEH750
2-AMINO-9-(β-d-RIBOFURANOSYL)PURINE-6-THIOL see TFJ500
2-AMINO-9-β-d-RIBOFURANOSYL-9H-PURINE-6-THIOL see TFJ500
4-AMINO-1-β-d-RIBOFURANOSYL-2(1H)-PYRIMIDINONE see CQM500
4-AMINO-7-(β-d-RIBOFURANOSYL)-PYRROLO(2,3-D)PYRIMIDINE see TNY500
4-AMINO-7-β-d-RIBOFURANOSYL-7H-PYRROLO(2,3-D)PYRIMIDINE
   see TNY500
4-AMINO-7-β-d-RIBOFURANOSYL-7H-PYRROLO(2,3-d)PYRIMIDINE-5-CARBO-
   NITRILE see VGZ000
4-AMINO-7-β-d-RIBOFURANOSYL-7H-PYRROLO(2,3-d)PYRIMIDINE-5-CARBO-
   XAMIDE see SAU000
2-AMINO-9-β-d-RIBOFURANOSYL-6-SELENO-OH-PURIN-6(1H)-ONE see SBU700
4-AMINO-1-β-d-RIBOFURANOSYL-d-TRIAZIN-2(1H)-ONE see ARY000
5-AMINO-2-β-d-RIBOFURANOSYL-as-TRIAZIN-3(2H)-ONE see AMM000
4-AMINO-1-β-d-RIBOFURANOSYL-1,3,5-TRIAZIN-2(1H)-ONE see ARY000
AMINO-S ACID see AMM125
p-AMINOSALICYLATE SODIUM see SEP000
AMINOSALICYLIC ACID see AMM250
4-AMINOSALICYLIC ACID see AMM250
5-AMINOSALICYLIC ACID see AMM500
m-AMINOSALICYLIC ACID see AMM500
p-AMINOSALICYLIC ACID see AMM250
4-AMINOSALICYLIC ACID-2-(DIETHYLAMINO)ETHYL ESTER HYDROCHLO-
   RIDE see AMM750
p-AMINOSALICYLIC ACID, 2-(DIETHYLAMINO)ETHYL ESTER, HYDRO-
   CHLORIDE see AMM750
p-AMINOSALICYLIC ACID, 2-(DIMETHYLAMINO)ETHYL ESTER HYDRO-
   CHLORIDE-31-6 see AMN000
p-AMINOSALICYLIC ACID HYDRAZIDE see AMN250
p-AMINOSALICYLIC ACID SODIUM SALT see SEP000
p-AMINOSALICYLSAEURE (GERMAN) see AMM500
p-AMINOSALICYLSAEUREDIAETHYLAMINOAETHYLESTER-
   CHLORHYDRAT (GERMAN) see AMM750
p-AMINOSALICYLSAURES SALZ (GERMAN) see TMJ750
4-AMINO-1,2,5-SELENADIAZOLE-3-CARBOXAMIDE see AMN300
2-AMINOSELENOAZOLIN (GERMAN) see AMN500
2-AMINOSELENOAZOLINE see AMN500
4-AMINOSEMICARBAZIDE see CBS500
AMINOSIDIN see NCF500
AMINOSIDINE SULFATE see APP500
AMINOSIDINE SULPHATE see APP500
AMINOSIDIN SULFATE see APP500
AMINOSIN see AGT500
4-AMINOSTILBENE see SLQ900
p-AMINOSTILBENE see SLQ900
trans-4-AMINOSTILBENE see AMO000
4-AMINO-2-STILBENECARBONITRILE see COS750
2-(p-AMINOSTYRYL)-6-(p-ACETYLAMINOBENZOYLAMINO)QUINOLINE
   METHOACETATE see AMO250
o-AMINOSULFANILIC ACID see PFA250
2,(4'-AMINO-3'-SULFO-1,1'-BIPHENYL-4-YL)-2H-NAPHTHO(1,2-4)TRIAZOLE-
   6,8-DISULFONIC ACID, TRIPOTASSIUM SALT see AMO750
6-AMINO-5-SULFOMETHYL-2-NAPHTHALENESULFONIC ACID see AMP000
1-AMINO-4-SULFONAPHTHALENE see ALI000
1-AMINO-6-SULFONAPHTHALENE see ALI250
AMINOSULFONIC ACID see SNK500
3-(AMINOSULFONYL)-5-(BUTYLAMINO)-4-PHENOXY-3-(AMINOSULFONYL)-5-
   (BUTYLAMINO)-4-PHENOXYBENZOIC ACID see BON325
3-(AMINOSULFONYL)-4-CHLORO-N-(2,3-DIHYDRO-2-METHYL-1H-INDOL-1-
   YL)-BENZAMIDE (9CI) see IBV100

5-(AMINOSULFONYL)-4-CHLORO-N-(2,6-DIMETHYLPHENYL)-2-HYDROXY
  BENZAMIDE (9CI) see CLF325
5-(AMINOSULFONYL)-4-CHLORO-2-((2-FURNAYLMETHYL)AMINO)BENZOIC
  ACID see CHJ750
5-(AMINOSULFONYL)-N-((1-ETHYL-2-PYRROLIDINYL)METHYL)-2-
  METHOXYBENZAMIDE see EPD500
3-(AMINOSULFONYL)-4-PHENOXY-5-(1-PYRROLIDINYL)-BENZOIC ACID
  see PDW250
N-(5-(AMINOSULFONYL)-1,3,4-THIADIAZOL-2-YL)ACETAMIDE see AAI250
2-AMINO-5-((4-SULFOPHENYL)AZO)-BENZENESULFONIC ACID see AJS500
4-(4-AMINO-3-SULFOPHENYLAZO)BENZENESULFONIC ACID see AJS500
3,3′-(2-AMINOTEREPHTHALOYLBIS(IMINO-p-PHENYLENECARBONYLIMINO))
  BIS(1-ETHYLPYRIDINIUM, DI-p-TOLUENESULFONATE see AMQ000
3,3′-(2-AMINOTEREPHTHALOYLBIS(IMINO(3-AMINO-p-PHENYLENE)
  CARBONYLIMINO))BIS(1-ETHYLPYRIDINIUM, DI-p-TOLUENESULFON-
  ATE see AMP500
3,3′-(2-AMINOTEREPHTHALOYLBIS(IMINO(3-AMINO-p-PHENYLENE)
  CARBONYLIMINO))BIS(1-PROPYLPYRIDINIUM, DI-p-TOLUENESULFON-
  ATE see AMP750
3,3′-(2-AMINOTEREPHTHALOYLBIS(IMINO-p-
  PHENYLENECARBONYLIMINO))BIS(1-METHYLPYRIDINIUM, DI-p-
  TOLUENESULFONATE see AMQ250
o-2-AMINO-2,3,4,6-TETRADEOXY-6-(METHYLAMINO)α-d-glycero-HEX-4-EN-
  OPYRANOSYL-(1-4)-o-(3-DEOXY-4-C-METHYL-3-(METHYLAMINO)β-l-
  ARABINOPYRANOSYL-(1-6))-2-DEOXY-d-STREPTAMINE see GAA100
d-o-2-AMINO-2,3,4,6-TETRADEOXY-6-(METHYLAMINO)α-d-erythro-
  HEXOPYRANOSYL-(1-4)-o-(3-DEOXY-4-C-METHYL-3-(METHYLAMINO)β-l-
  ARABINOPYRANOSYL-(1-6))-2-DEOXY-STREPTAMINE SULFATE
  see MQS600
9-AMINO-1,2,3,4-TETRAHYDROACRIDINE see TCJ075
5-AMINO-6,7,8,9-TETRAHYDROACRIDINE (EUROPEAN) see TCJ075
8-AMINO-1,2,3,4-TETRAHYDRO-2-METHYL-4-PHENYLISOQUINOLINE MALE-
  ATE see NMV725
2-AMINO-1,2,3,4-TETRAHYDRONAFTALEN (CZECH) see TCY250
5-AMINO-2,2,4,4-TETRAKIS(TRIFLUOROMETHYL)IMIDAZOLIDINE
  see AMQ500
4-AMINO-2,2,5,5-TETRAKIS(TRIFLUOROMETHYL)-3-IMIDAZOLINE
  see AMQ500
2-AMINOTETRALIN see TCY250
AMINOTETRALIN (CZECH) see TCY250
1-AMINO-2,2,6,6-TETRAMETHYLPIPERIDINE see AMQ750
AMINOTETRAZOLE see AMR000
5-AMINOTETRAZOLE see AMR000
5-AMINO-1H-TETRAZOLE see AMR000
AMINOTHIADIAZOLE see AMR250
2-AMINO-1,3,4-THIADIAZOLE see AMR250
2-AMINO-1,3,4-THIADIAZOLEHYDROCHLORIDE see AMR500
2-AMINO-1,3,4-THIADIAZOLE, MONOHYDROCHLORIDE see AMR500
2-AMINO-1,3,4-THIADIAZOLE-5-SULFONAMIDE SODIUM SALT see AMR750
2-AMINO-5-1,3,4-THIADIAZOLE-5-THIOL see AKM000
5-AMINO-1,3,4-THIADIAZOLE-2-THIOL see AKM000
5-AMINO-1,3,4-THIADIAZOLINE-2-THIONE see AKM000
2-AMINO-Δ(2)-1,3,4-THIADIAZOLINE-5-THIONE see AKM000
5-AMINO-1,2,3,4-THIATRIAZOLE see AMS000
AMINOTHIAZOLE see AMS250
2-AMINOTHIAZOLE see AMS250
4-AMINO-N-2-THIAZOLYLBENZENESULFONAMIDE see TEX250
1-AMINO-2-(4-THIAZOLYL)-5-BENZIMIDAZOLECARBAMIC ACID ISOPRO-
  PYL ESTER see AMS625
(6R-(6-α,7-β(Z)))-1-((7-(((2-AMINO-4-THIAZOLYL)((1-CARBOXY-1-
  METHYLETHOXY)IMINO)ACETYL)AMINO)-2-CARBOXY-8-OXO-5-THIA-1-
  AZABICYCLO(4.2.0)OCT-2-EN-3-YL)METHYL)-PYRIDINIUM HYDROXIDE,
  inner salt, PENTAHYDRATE see CCQ200
2-(((1-(2-AMINO-4-THIAZOLYL)-2-((2-METHYL-4-OXO-1-SULFO-3-
  AZETIDINYL)AMINO)-2-OXOETHYLIDENE)AMINO)OXY)-2-
  METHYLPROPANOIC ACID, (2S-(2-α,3β(Z)))- see ARX875
1-(2-AMINOTHIAZOLYL)-2-(5-NITRO-2-FURYL)ETHYLENE see ALN500
o-AMINOTHIOFENOLAT ZINECNATY (CZECH) see BGV500
α-AMINO-2-THIOPHENEPROPANOIC ACID see TEY250
2-AMINOTHIOPHENOL see AIF500
4-AMINOTHIOPHENOL see AIF750
o-AMINOTHIOPHENOL see AIF500
p-AMINOTHIOPHENOL see AIF750
N-AMINOTHIOUREA see TFQ000
N-(4-(((AMINOTHIOXOMETHYL)HYDRAZONO)METHYL)PHENYL)
  ACETAMIDE see FNF000
3-AMINOTOLUEN (CZECH) see TGQ500
4-AMINOTOLUEN (CZECH) see TGR000
2-AMINOTOLUENE see TGQ750
3-AMINOTOLUENE see TGQ500
4-AMINOTOLUENE see TGR000
m-AMINOTOLUENE see TGQ500
o-AMINOTOLUENE see TGQ750
p-AMINOTOLUENE see TGR000

2-AMINOTOLUENE HYDROCHLORIDE see TGS500
4-AMINOTOLUENE HYDROCHLORIDE see TGS750
o-AMINOTOLUENE HYDROCHLORIDE see TGS500
α-AMINO-p-TOLUENESULFONAMIDE, MONOACETATE see MAC000
2-AMINO-p-TOLUENESULFONIC ACID see AMT000
4-AMINOTOLUENE-3-SULFONIC ACID see AKQ000
4-AMINO-o-TOLUENESULFONIC ACID see AMT250
6-AMINO-m-TOLUENESULFONIC ACID see AKQ000
3-AMINO-p-TOLUIDINE see TGL500
5-AMINO-o-TOLUIDINE see TGL750
p-((4-AMINO-m-TOLYL)AZO)BENZENESULFONAMIDE see SNX000
1-(3-AMINO-p-TOLYLSULFONYL)-3-CYCLOHEXYLUREA see AKQ250
AMINOTRATE PHOSPHATE see TJL250
AMINOTRIACETIC ACID see AMT500
2-AMINOTRIAZOLE see AMY050
3-AMINOTRIAZOLE see AMY050
3-AMINO-s-TRIAZOLE see AMY050
2-AMINO-1,3,4-TRIAZOLE see AMY050
3-AMINO-1,2,4-TRIAZOLE see AMY050
3-AMINO-1H-1,2,4-TRIAZOLE see AMY050
AMINOTRIAZOLE (plant regulator) see AMY050
AMINO TRIAZOLE WEEDKILLER 90 see AMY050
5-AMINO-1H-v-TRIAZOLO(d)PYRIMIDIN-7-OL see AJO500
5-AMINO-v-TRIAZOLO(4,5-d)PYRIMIDIN-7-OL see AJO500
AMINOTRIAZOL-SPRITZPULVER see AMY050
3-AMINO-1-TRICHLORO-2-PENTANOL see AMU000
((3-AMINO-2,4,6-TRICHLOROPHENYL)METHYLENE) HYDRAZIDE
  BENZENESULFONIC ACID see AMU125
4-AMINO-3,5,6-TRICHLOROPICOLINIC ACID see PIB900
4-AMINO-3,5,6-TRICHLORO-2-PICOLINIC ACID see PIB900
3-AMINO-1-TRICHLORO-2-PROPANOL see AMU500
4-AMINO-3,5,6-TRICHLORPICOLINSAEURE (GERMAN) see PIB900
1-AMINOTRICYCLO(3.3.1.1^{3,7})DECANE see TJG250
30-AMINO-3,14,25-TRIHYDROXY-3,9,14,20,25-PENTAAZATRIACONTANE-
  2,10,13,21,24-PENTAONE see DAK200
3-AMINO-2,4,6-TRIIODO-BENZOIC ACID see AMU625
N-(3-AMINO-2,4,6-TRIIODOBENZOYL)-N-(2-CARBOXYETHYL)ANILINE
  see AMU750
3-((3-AMINO-2,4,6-TRIIODOBENZOYL)PHENYLAMINO)PROPIONIC ACID
  see AMU750
2-(3-AMINO-2,4,6-TRIIODOBENZYL)BUTYRIC ACID see IFY100
4-((3-AMINO-2,4,6-TRIIODOPHENYL)ETHYLAMINO)-4-OXO-BUTANOIC ACID
  see AMV375
3-(3-AMINO-2,4,6-TRIIODOPHENYL)-2-ETHYLPROPANOIC ACID see IFY100
β-(3-AMINO-2,4,6-TRIIODOPHENYL)-α-ETHYLPROPIONIC ACID see IFY100
2-(3-AMINO-2,4,6-TRIIODOPHENYL)VALERIC ACID see AMV750
N-(3-AMINO-2,4,6-TRIJODBENZOYL)-N-PHENYL-β-AMINOPROPIONSAE-
  URE(GERMAN) see AMU750
1-AMINO-2,4,6-TRIMETHYLBENZEN (CZECH) see TLG500
1-AMINO-2,4,5-TRIMETHYLBENZENE see TLG250
2-AMINO-1,3,5-TRIMETHYLBENZENE see TLG500
1-AMINO-2,4,5-TRIMETHYLBENZENE HYDROCHLORIDE see TLG750
2-AMINO-1,3,5-TRIMETHYLBENZENE HYDROCHLORIDE see TLH000
4-AMINO-a,a,4-TRIMETHYLCYCLOHEXANEMETHAMINE see MCD750
AMINOTRIMETHYLOMETHANE see TEM500
AMINOTRIS(HYDROXYMETHYL)METHANE see TEM500
4-AMINOTROPOLONE see AMV790
5-AMINOTROPOLONE see AMV800
2-AMINOTROPONE see AJJ800
AMINO-TS-ACID see AMV875
AMINOUNDECANOIC ACID see AMW000
11-AMINOUNDECANOIC ACID see AMW000
11-AMINOUNDECYLIC ACID see AMW000
6-AMINOURACIL see AMW250
AMINOURACIL MUSTARD see BIA250
AMINOUREA see HGU000
AMINOUREA HYDROCHLORIDE see SBW500
p-AMINO VALEROPHENONE see AMW500
AMINOX see AMM250
2-(4′-AMINOXENYL)NAFTO-α,β-TRIAZOL-6,8-DISULFONAN DRASELNY
  (CZECH) see AIW750
2-(4′-AMINOXENYL)NAFTO-α,β-TRIAZOL-6,8,3′TRISULFONAN DRASELNY
  (CZECH) see AMO750
2-AMINO-1,4-XYLENE see XNA000
4-AMINO-1,3-XYLENE see XMS000
2-AMINO-1,4-XYLENE HYDROCHLORIDE see XOS000
4-AMINO-1,3-XYLENE HYDROCHLORIDE see XOJ000
2-AMINO-4,5-XYLENOL see AMW750
3-AMINO-4-(2-(2,6-XYLYLOXY)ETHYL)-4H-1,2,4-TRIAZOLE see AMX000
AMINOZIDE see DQD400
1,2,-AMINOZOPHENYLENE see BDH250
AMINUTRIN see LJM700
AMINZOL SOLUBLE see ALQ000
AMIODARONE see AJK750

AMIODOXYL BENZOATE see AMX250
AMIOYL see BCA000
AMIP 15m see TAI250
AMIPAN T see DUO400
AMIPAQUE see MQR300
AMIPENIX S see AIV500
AMIPHENAZOLE HYDROCHLORIDE see DCA600
AMIPHOS see DOP200
AMI-PILO see PIF250
AMIPOLNE see AAE500
AMIPROL see DCK759
AMIPTAN see AHK750
AMIPURIMYCIN HYDRATE see AMX500
AMIPYLO see AMF375
AMIRAL see CJO250
AMISEPAN see SBG500
AMISOMETRADIN see AKL625
AMISOMETRADINE see AKL625
AMISTURA P see PIF250
AMISYL see BCA000
AMITAKON see BCA000
AMITAL see AMX750
AMITHIOZONE see FNF000
AMITID see EAI000
AMITIOZON see FNF000
AMITOL see AMY050
AMITON see DJA400
AMITON OXALATE see AMX825
AMITRAZ see MJL250
AMITRAZ ESTRELLA see MJL250
AMITRENE see BBK250, BBK500
AMITRIL see AMY050, EAI000
AMITRIL T.L. see AMY050
AMITRIPTILINE see EAH500
AMITRIPTYLIN (GERMAN) see EAH500
AMITRIPTYLINE see EAH500
AMITRIPTYLINE CHLORIDE see EAI000
AMITRIPTYLINE-N-OXIDE see AMY000
AMITRIPTYLINOXIDE see AMY000
AMITROL see AMY050
AMITROL 90 see AMY050
AMITROLE see AMY050
AMITROL-T see AMY050
AMITRYPTYLINE HYDROCHLORIDE see EAI000
AMIXICOTYN see NCR000
AMIZIL HYDROCHLORIDE see BCA000
AMIZOL see AMY050
AMMAT see ANU650
AMMATE see ANU650
AMMICARDINE see AHK750
AMMIDIN see IHR300
AMMI-KHELLIN see AHK750
AMMINE PENTAHYDROXO PLATINUM see AMY250
AMMINETRICHLORO-PLATINATE(1⁻) POTASSIUM (SP-4-2) see PJD750
AMMIPURAN see AHK750
AMMISPASMIN see AHK750
AMMIVIN see AHK750
AMMIVISNAGEN see AHK750
AMMN see AAW000
AMMO see RLF350
AMMOFORM see HEI500
AMMOIDIN see XDJ000
AMMONIA see AMY500
AMMONIA, solution (DOT) see ANK250
AMMONIAC (FRENCH) see AMY500
AMMONIACA (ITALIAN) see AMY500
AMMONIA GAS see AMY500
AMMONIAK (GERMAN) see AMY500
AMMONIATED GLYCYRRHIZIN see GIE100
AMMONIATED MERCURY see MCW500
AMMONIO (DICROMATO DI) (ITALIAN) see ANB500
AMMONIOFORMALDEHYDE see HEI500
2-AMMONIOTHIAZOLE NITRATE see AMZ125
AMMONIUM ACETATE see ANA000
AMMONIUM ACID ARSENATE see DCG800
AMMONIUM-AETHYL-CARBAMOYL-PHOSPHONAT (GERMAN) see ANG750
AMMONIUM ALUMINUM FLUORIDE see THQ500
AMMONIUM AMIDOSULFONATE see ANU650
AMMONIUM AMIDOSULPHATE see ANU650
AMMONIUM AMINOFORMATE see AND750
AMMONIUM (AMINYLENIUM BIS [TRIHYDROBORATE]) see ANA500
AMMONIUM ARSENATE, solid (DOT) see DCG800
AMMONIUM AURINTRICARBOXYLATE see AGW750
AMMONIUM AZIDE see ANA750

AMMONIUM BENZAMIDOOXYACETATE see ANB000
AMMONIUM-2-(BENZAMIDOOXY)ACETATE see ANB000
AMMONIUM (3-N-BENZYLCARBAMOYLOXY)PHENYL)TRIMETHYL METHYLSULFATE see BED500
AMMONIUM BICARBONATE (1581) see ANB250
AMMONIUMBICHROMAAT (DUTCH) see ANB500
AMMONIUM BICHROMATE see ANB500
AMMONIUM BIFLUORIDE see ANJ000
AMMONIUM BISULFIDE see ANJ750
AMMONIUM BITHIOLICUM see IAD000
AMMONIUM BOROFLUORIDE see ANH000
AMMONIUM BROMATE see ANC000
AMMONIUM BROMIDE see ANC250
AMMONIUM BROMO SELENATE see ANC750
AMMONIUM CADMIUM CHLORIDE see AND250
AMMONIUM CALCIUM ARSENATE see AND500
AMMONIUM CARBAMATE see AND750
AMMONIUMCARBONAT (GERMAN) see ANE000
AMMONIUM CARBONATE see ANB250, ANE000
AMMONIUM, (3-CARBOXY-2-HYDROXYPROPYL)TRIMETHYL-, CHLORIDE, (-)- see CCK660
AMMONIUM CARBOXYMETHYL CELLULOSE see CCH000
AMMONIUM CHLORATE see ANE250
AMMONIUMCHLORID (GERMAN) see ANE500
AMMONIUM CHLORIDE see ANE500
AMMONIUM CHLOROHEPTENE ARSONATE see CEI250
AMMONIUM (2-(p-((2-CHLORO-4-NITROPHENYL)AZO)PHENETHYLAMINO)ETHYL)TRIMETHYL see BAQ750
AMMONIUM CHLOROPALLADATE(II) see ANE750
AMMONIUM CHLOROPALLADATE(IV) see ANF000
AMMONIUM CHLOROPLATINATE see ANF250
AMMONIUM CHROMATE see ANF500
AMMONIUM CHROME ALUMS see ANF625
AMMONIUM CHROMIC SULFATE see ANF750
AMMONIUM CITRATE see ANF800
AMMONIUM CITRATE, DIBASIC (DOT) see ANF800
AMMONIUM CRYOLITE see THQ500
AMMONIUM CYANIDE see ANG000
AMMONIUM DECAHYDRODECABORATE (2−) see ANG125
AMMONIUM, (4-((4,6-DIAMINO-m-TOLYL)IMINO)-2,5-CYCLOHEXADIEN-1-YLIDENE)DIMETHYL-, CHLORIDE, MONOHYDRATE see TGU500
AMMONIUMDICHROMAAT (DUTCH) see ANB500
AMMONIUMDICHROMAT (GERMAN) see ANB500
AMMONIUM DICHROMATE see ANB500
AMMONIUM DICHROMATE(VI) see ANB500
AMMONIUM DIFLUORIDE mixed with HYDROCHLORIC ACID see ANG250
AMMONIUM DIHEXADECYLDIMETHYL-, CHLORIDE see DRK200
AMMONIUM DIMETHYL DITHIOCARBAMATE see ANG500
AMMONIUM-3,5-DINITRO-1,2,4-TRIAZOLIDE see ANG625
AMMONIUM DISULFATONICKELATE(II) see NCY050
AMMONIUM DNOC see DUT800
AMMONIUM DODECYL SULFATE see SOM500
AMMONIUM-N-DODECYL SULFATE see SOM500
AMMONIUM ETHYLCARBAMOYLPHOSPHONATE see ANG750
AMMONIUM FERRIC OXALATE see ANG925
AMMONIUM FERRIOXALATE see ANG925
AMMONIUM FLUOALUMINATE see THQ500
AMMONIUM FLUOBORATE see ANH000
AMMONIUM FLUORIDE see ANH250
AMMONIUM FLUOROBORATE see ANH000
AMMONIUM FLUORURE (FRENCH) see ANH250
AMMONIUM FLUOSILICATE see COE000
AMMONIUM FORMATE see ANH500
AMMONIUMGLUTAMINAT (GERMAN) see MRF000
AMMONIUM GLYCYRRHIZINATE see GIE100
AMMONIUM HEXACHLOROPALLADATE see ANF000
AMMONIUM HEXACHLOROPLATINATE(IV) see ANF250
AMMONIUM HEXACYANOFERRATE(II) see ANH875
AMMONIUM HEXAFLUOROALUMINATE see THQ500
AMMONIUM HEXAFLUOROFERRATE see ANI000
AMMONIUM HEXAFLUOROSILICATE see COE000
AMMONIUM HEXAFLUOROTITANATE see ANI250
AMMONIUM HEXAFLUOROVANADATE see ANI500
AMMONIUM, HEXAMETHYLENEBIS((CARBOXYMETHYL)DIMETHYL-, DI-CHLORIDE, DIDODECYL ESTER see HEF200
AMMONIUM, HEXAMETHYLENEBIS(TRIMETHYL-, DIBENZENESULFON-ATE see HEG100
AMMONIUM HEXANITRO COBALTATE see ANI750
AMMONIUM HYDROGEN CARBONATE see ANB250
AMMONIUM HYDROGEN FLUORIDE see ANJ000
AMMONIUM HYDROGEN FLUORIDE, solid see ANJ000
AMMONIUM HYDROGEN FLUORIDE (solution) see ANJ250
AMMONIUM HYDROGEN FLUORIDE, solution (DOT) see ANJ250

AMMONIUM HYDROGEN SULFATE see ANJ500
AMMONIUM HYDROGEN SULFIDE see ANJ750
AMMONIUM HYDROSULFIDE see ANJ750
AMMONIUM HYDROSULFIDE, solution (DOT) see ANJ750
AMMONIUM HYDROXIDE see ANK250
AMMONIUM HYPOPHOSPHITE see ANK500
AMMONIUM ICHTHOSULFONATE see IAD000
AMMONIUM IODATE see ANK750
AMMONIUM IODIDE see ANL000
AMMONIUM ISETHIONATE see ANL100
AMMONIUM LANTHANUM NITRATE see ANL500
AMMONIUM LAURYL SULFATE see SOM500
AMMONIUM MAGNESIUM ARSENATE see ANL750
AMMONIUM MAGNESIUM CHROMATE see ANM000
AMMONIUM MANDELATE see ANM250
AMMONIUM MERCAPTAN see ANJ750
AMMONIUM MERCAPTOACETATE see ANM500
AMMONIUM METAVANADATE (DOT) see ANY250
AMMONIUM-3-METHYL-2,4,6-TRINITROPHENOXIDE see ANM625
AMMONIUM MOLYBDATE see ANM750
AMMONIUM MURIATE see ANE500
AMMONIUM NICKEL SULFATE see NCY050
AMMONIUM NITRATE see ANN000
AMMONIUM NITRATE (DOT) see ANN000
AMMONIUM(I) NITRATE(1581) see ANN000
AMMONIUM NITRITE see ANO250
AMMONIUM aci-NITROMETHANE see ANO400
AMMONIUM-N-NITROSOPHENYLHYDROXYLAMINE see ANO500
AMMONIUM ORTHOPHOSPHITE see ANS250
AMMONIUM OXALATE see ANO750
AMMONIUM OXOFLUOROMOLYBDATE see ANO875
AMMONIUM PARAMOLYBDATE see ANM750
AMMONIUM PARATUNGSTATE HEXAHYDRATE see ANO900
AMMONIUM PENTADECAFLUOROOCTANATE see ANP625
AMMONIUM PENTA PEROXODICHROMATE see ANP000
AMMONIUM PERCHLORATE see ANP250
AMMONIUM PERCHLORATE (DOT) see PCD500
AMMONIUM PERCHLORYL AMIDE see ANP500
AMMONIUM PERFLUOROCAPRILATE see ANP625
AMMONIUM PERFLUOROCAPRYLATE see ANP625
AMMONIUM PERFLUOROOCTANOATE see ANP625
AMMONIUM-m-PERIODATE see ANP750
AMMONIUM PERMANGANATE see PCJ750
AMMONIUM PEROXO BORATE see ANQ250
AMMONIUM PEROXO DISULFATE see ANQ500
AMMONIUM PEROXY CHROMATE see ANQ750
AMMONIUM PEROXYDISULFATE see ANR000
AMMONIUM PERSULFATE see ANR000
AMMONIUM PERSULFATE (DOT) see ANR000
AMMONIUM PHENYLDITHIOCARBAMATE see ANR250
AMMONIUM PHOSPHATE see ANR500
AMMONIUM PHOSPHATE DIBASIC see ANR500
AMMONIUM PHOSPHATE, MONOBASIC see ANR750
AMMONIUM PHOSPHIDE see ANS000
AMMONIUM PHOSPHITE see ANS250
AMMONIUM PICRATE see ANS500
AMMONIUM PICRATE, wet with 10% or more water (DOT) see ANS500
AMMONIUM PICRATE, wet with 10% or more water, over 16 oz in one outside packaging (DOT) see ANS500
AMMONIUM PICRONITRATE see ANS500
AMMONIUM PLATINIC CHLORIDE see ANF250
AMMONIUM POLYSULFIDE (solution) see ANT000
AMMONIUM POTASSIUM SELENIDE mixed with AMMONIUM POTASSIUM SULFIDE-00-0 see ANT250
AMMONIUM POTASSIUM SULFIDE mixed with AMMONIUM POTASSIUM SELENIDE see ANT250
AMMONIUM REINECKATE HYDRATE see ANT300
AMMONIUM RHODANATE see ANW750
AMMONIUM RHODANIDE see ANW750
AMMONIUM SACCHARIN see ANT500
AMMONIUM SALTS of PHOSPHATIDIC ACIDS see ANU000
AMMONIUMSALZ der AMIDOSULFONSAURE (GERMAN) see ANU650
AMMONIUMSALZ des α-METHYL-β-HYDROXY-Δ^α,β-BUTYLCOLACTAM (GERMAN) see MKS750
AMMONIUM SILICOFLUORIDE (DOT) see COE000
AMMONIUM SULFAMATE see ANU650
AMMONIUM SULFATE (2:1) see ANU750
AMMONIUM SULFATE, and CHROMIC SULFATE, TETRACOSAHYDRATE see ANF625
AMMONIUM SULFHYDRATE see ANJ750
AMMONIUM SULFIDE, solution, red see ANT000
AMMONIUM SULFOCYANATE see ANW750
AMMONIUM SULFOCYANIDE see ANW750
AMMONIUM SULFOICHTHYOLATE see IAD000

AMMONIUM SULPHAMATE see ANU650
AMMONIUM SULPHATE see ANU750
AMMONIUM-d-TARTRATE see DCH000
AMMONIUM TARTRATE (DOT) see DCH000
AMMONIUM TELLURATE see ANV750
AMMONIUM TETRACHLOROPALLADATE see ANE750
AMMONIUM TETRACHLOROPLATINATE see ANV800
AMMONIUM TETRAFLUOROBORATE see ANH000
AMMONIUM TETRAFLUOROBORATE(1-) see ANH000
AMMONIUM TETRANITROPLATINATE(II) see ANW250
AMMONIUM TETRAPEROXO CHROMATE see ANW500
AMMONIUM THIOCYANATE see ANW750
AMMONIUM THIOGLYCOLATE see ANM500
AMMONIUM THIOGLYCOLLATE see ANM500
AMMONIUM TRICHLOROACETATE see ANX750
AMMONIUM TRIFLUOROSTANNITE see ANX800
AMMONIUM-2,4,5-TRINITROIMIDAZOLIDE see ANX875
AMMONIUM TRIOXALATOFERRATE(III) see ANG925
AMMONIUM VANADATE see ANY250
AMMONIUM VANADI-ARSENATE see ANY500
AMMONIUM VANADO-ARSENATE see ANY750
AMMONIUMYL, DIBUTYL-, HEXACHLOROSTANNATE(2-) (2:1) see BIV900
AMMONYX see AFP250
AMMONYX 4 see DTC600
AMMONYX CA SPECIAL see DTC600
AMMONYX CPC see CCX000
AMMONYX DME see EKN500
AMMONYX LO see DRS200
AMMOPHYLLIN see TEP500
AMN see AOL000
AMNESTROGEN see ECU750
AMNICOTIN see NCR000
AMNOSED see TDA500
AMNUCOL see SEH000
AMOBAM see ANZ000
AMOBARBITAL see AMX750
AMOBEN see AJM000
AMOCO 1010 see PMP500
AMOEBAL see ABX500
AMOENOL see CHR500
AMOGLANDIN see POC500
AMOIL see AON300
AMOKIN see CLD000
AMOLANONE HYDROCHLORIDE see DIF200
AMONIAK (POLISH) see AMY500
A. MONTANA see TOA275
AMONYX AO see DRS200
AMOPYROQUIN DIHYDROCHLORIDE see PMY000
A1-MORIN see MRN500
AMORPHAN see NNW500
AMORPHOUS AESCIN see EDK875
AMORPHOUS CROCIDOLITE ASBESTOS see ARM275
AMORPHOUS FUSED SILICA see SCK600
AMORPHOUS SILICA DUST see SCH000
AMOSCANATE see AOA050
AMOSENE see MQU750
AMOSITE ASBESTOS see ARM262
AMOSITE (OBS.) see ARM250
AMOSULALOL HYDROCHLORIDE see AOA075
AMOSYT see DYE600
AMOTRIL see ARQ750
AMOTRIPHENE HYDROCHLORIDE see TNJ750
AMOX see NGS500
AMOXAPINE see AOA095
AMOXEPINE see AOA095
AMOXICILLIN mixed with POTASSIUM CLAVULANATE (2:1) see ARS125
AMOXICILLIN TRIHYDDRATE see AOA100
AMOXONE see DAA800
AMP see AOA125
5-AMP see AOA125
A5MP see AOA125
5'-AMP see AOA125
3',5'-AMP see AOA130
AMP (nucleotide) see AOA125
AMPAZINE see DQA600
AMPD see ALB000
AMPERIL see AIV500, AOD125
AMPHAETEX see BBK500
AMPHATE see AOB500
AMPHEDRINE see BBK500
AMPHEDROXY see DBA800
AMPHEDROXYN see DBA800
AMPHENAZOLE HYDROCHLORIDE see DCA600
AMPHENICOL see CDP250

AMPHENIDONE see ALY250
AMPHEPRAMONUM HYDROCHLORIDE see DIP600
AMPHEREX see BBK500
AMPHETAMINE see AOA250
(+)-AMPHETAMINE see AOA500
d-AMPHETAMINE see AOA500
dl-AMPHETAMINE see BBK000
AMPHETAMINE HYDROCHLORIDE see AOA750
AMPHETAMINE PHOSPHATE see AOB500
dl-AMPHETAMINE PHOSPHATE see AOB500
dl-AMPHETAMINE SALT with FINE RESIN see AOB000
(−)-AMPHETAMINE SULFATE see BBK750
(+)-AMPHETAMINE SULFATE see BBK500
d-AMPHETAMINE SULFATE see BBK500
l-AMPHETAMINE SULFATE see BBK750
(±)-AMPHETAMINE SULFATE see AOB250
dl-AMPHETAMINE SULFATE see AOB250
AMPHETANE PHOSPHATE see AOB500
AMPHIBOLE see ARM250
AMPHICOL see CDP250
AMPHISOL HYDROCHLORIDE see DCA600
AMPHOIDS S see BBK250
AMPHOJEL see AHC000
AMPHOMORONAL see AOC500
AMPHOMYCIN see AOB875
AMPHORDS S see BBK250
AMPHOS see AOB500
AMPHOTERGE K-2 see AOC250
AMPHOTERIC-2 see AOC250
AMPHOTERIC-17 see AOC275
AMPHOTERICIN beta see AOC500
AMPHOTERICIN B see AOC500
AMPHOTERICIN B, METHYL ESTER HYDROCHLORIDE see AOC750
AMPHOTERICINE B see AOC500
AMPHOZONE see AOC500
AMPI-BOL see AIV500
AMPICHEL see AOD125
d-AMPICILLIN see AIV500
d-(−)-AMPICILLIN see AIV500
AMPICILLIN (USDA) see AIV500
AMPICILLIN A see AIV500
AMPICILLIN ACID see AIV500
AMPICILLIN ANHYDRATE see AIV500
AMPICILLIN-OXACILLIN MIXTURE see AOC875
AMPICILLIN PIVALOYLOXYMETHYL ESTER HYDROCHLORIDE
    see AOD000
AMPICILLIN SODIUM see SEQ000
AMPICILLIN SODIUM SALT see SEQ000
AMPICILLIN TRIHYDRATE see AOD125
AMPICIN see AIV500
AMPIKEL see AIV500, AOD125
AMPIMED see AIV500
AMPINOVA see AOD125
AMPIPENIN see AIV500
AMPLIACTIL see CKP250
AMPLIACTIL MONOHYDROCHLORIDE see CKP500
AMPLICITIL see CKP250
AMPLIGRAM see TEY000
AMPLIN see AOD125
AMPLISOM see AIV500
AMPLITAL see AIV500
5′-AMP POTASSIUM SALT see AEM500
AMPROLENE see EJN500
AMPROTROPINE PHOSPHATE see AOD250
AMPY-PENYL see AIV500
AMPYROX see SBH500
AMRINONE see AOD375
AMRITAMYCIN see MQX250
AMROOD, extract see POH800
AMS see ANU650
AMSA see ADL750
m-AMSA see ADL750, ADM000
AMSACRINE see ADL750
AMSACRINE LACTATE see AOD425
m-AMSA HYDROCHLORIDE see ADL500
m-AMSA LACTATE see AOD425
m-AMSA METHANESULFONATE see ADL750
AMSCO TETRAMER see PMP750
AMSECLOR see CDP250
AMSIDINE see ADL750
AMSINCKIA INTERMEDIA see TAG250
AMSINE see ADL750
AMSTAT see AJV500
AMSUBIT see IAD000

AMSUSTAIN see AOA500, BBK500
AMTHIO see ANW750
AMUDANE see GKE000
AMUNO see IDA000
AMYDRICAINE see AHI250
AMYGDALIC ACID see MAP000
AMYGDALIN see AOD500
d,l-AMYGDALIN see IHO700
AMYGDALINIC ACID see MAP000
AMYGDALONITRILE see MAP250
n-AMYL ACETATE see AOD725
AMYL ACETATE (DOT) see AOD725
sec-AMYL ACETATE see AOD735
AMYL ACETATE (mixed isomers) see AOD750
AMYL ACETIC ESTER see AOD725
AMYL ACETIC ETHER see AOD725
AMYL ACID PHOSPHATE (DOT) see PBW750
AMYL ALCOHOL see AOE000
N-AMYL ALCOHOL see AOE000
sec-AMYL ALCOHOL (DOT) see PBM750
tert-AMYL ALCOHOL (DOT) see PBV000
AMYL ALCOHOL, NORMAL see AOE000
AMYL ALDEHYDE see VAG000
N-AMYLALKOHOL (CZECH) see AOE000
AMYLAMINE (mixed isomers) (DOT) see PBV500
2-N-AMYLAMINOETHYL-p-AMINOBENZOATE see PBV750
AMYLAZETAT (GERMAN) see AOD725
AMYL AZIDE see AOE500
AMYLBARBITONE see AMX750
5-n-AMYL-1:2-BENZANTHRACENE see AOE750
tert-AMYLBENZENE see AOF000
AMYL BENZOATE see IHP100
4-AMYL-N-BENZOHYDRYLPYRIDINIUM BROMIDE see AOF250
p-AMYLBENZOIC ACID see PBW000
AMYL BIPHENYL see AOF500
d-AMYL BROMIDE see AOF750
AMYL BUTYRATE see IHP400
n-AMYL BUTYRATE see AOG000
γ-N-AMYLBUTYROLACTONE see CNF250
AMYLCAINE see PBV750
AMYLCARBINOL see HFJ500
n-AMYL CHLORIDE see PBW500
AMYL CHLORIDE (DOT) see PBW500
α-AMYL CINNAMALDEHYDE see AOG500
AMYL CINNAMATE see AOG600
AMYL CINNAMIC ACETATE see AOG750
α-AMYLCINNAMIC ALCOHOL see AOH000
α-AMYL CINNAMIC ALDEHYDE see AOG500
α-AMYLCINNAMYL ALCOHOL see AOH000
AMYL CINNAMYLIDENE METHYL ANTHRANILATE see AOH100
6-n-AMYL-m-CRESOL see AOH250
4-tert-AMYLCYCLOHEXANONE see AOH750
AMYLCYCLOHEXYL ACETATE (mixed isomers) see AOI000
AMYLDICHLORARSINE see AOI200
N-AMYLDICHLORARSINE see AOI200
AMYL-p-DIMETHYLAMINOBENZOATE see AOI250
AMYLDIMETHYL-p-AMINO BENZOIC ACID see AOI500
AMYL DIMETHYL PABA see AOI250
AMYLEINE see AOM000
α,η-AMYLENE see AOI750
2,4-AMYLENEGLYCOL see PBL000
AMYLENE HYDRATE see PBV000
AMYLENES, MIXED see AOJ000
AMYL ETHER see PBX000
N-AMYL ETHER see PBX000
AMYLETHYLCARBINOL see OCY100
AMYL ETHYL KETONE see EGI750, ODI000
AMYLETHYLMETHYLCARBINOL see MND050
n-AMYL FORMATE see AOJ500
AMYL FORMATE (DOT) see AOJ500
AMYL HARMOL HYDROCHLORIDE see AOJ750
o-n-AMYL HARMOL HYDROCHLORIDE see AOJ750
AMYL HEXANOATE see IHU100
n-AMYLHYDRAZINE HYDROCHLORIDE see PBX250
AMYL HYDRIDE (DOT) see PBK250
tert-AMYL HYDROPEROXIDE see PBX325
AMYL HYDROSULFIDE see PBM000
3-AMYL-1-HYDROXY-6,6,9-TRIMETHYL-6H-DIBENZO(b,d)PYRAN see CBD625
AMYLISOEUGENOL see AOK000
AMYL LACTATE see AOK250
AMYL LAURATE see AOK500
n-AMYL MERCAPTAN see PBM000
AMYL MERCAPTAN (DOT) see PBM000
AMYL METHYL ALCOHOL see AOK750

AMYL METHYL CARBINOL see HBE500
AMYL-METHYL-CETONE (FRENCH) see MGN500
n-AMYL METHYL KETONE see MGN500
AMYL METHYL KETONE (DOT) see MGN500
n-AMYL-N-METHYLNITROSAMINE see AOL000
AMYL NITRATE see AOL250
n-AMYL NITRITE see AOL500
AMYL NITRITE (DOT) see AOL500
n-AMYLNITROSOUREA see PBX500
1-AMYL-1-NITROSOUREA see PBX500
n-AMYL-N-NITROSOURETHANE see AOL750
AMYLOBARBITAL see AMX750
AMYLOBARBITONE see AMX750
AMYLOCAINE see AOM000
AMYLOFENE see EOK000
AMYLOMAIZE VII see SLJ500
AMYLOPECTINE SULPHATE see AOM150
AMYLOPECTIN, HYDROGEN SULFATE see AOM150
AMYLOPECTIN SULFATE see AOM150
AMYLOPECTIN SULFATE (SN-263) see AOM150
AMYLOWY ALKOHOL (POLISH) see IHP000
AMYLOXYISOEUGENOL see AOK000
4-n-AMYLPHENOL see AOM250
AMYL PHENOL 4T see AON000
2-sec-AMYLPHENOL see AOM500
4-sec-AMYLPHENOL see AOM750
4-tert-AMYLPHENOL see AON000
p-tert-AMYLPHENOL see AON000
α-AMYL-β-PHENYLACROLEIN see AOG500
α-N-AMYL-β-PHENYLACRYL ACETATE see AOG750
3-sec-AMYLPHENYL-N-METHYLCARBAMATE see AON250
2-AMYL-3-PHENYL-2-PROPEN-1-OL see AOH000
AMYL PHTHALATE see AON300
AMYL PROPIONATE see AON350
AMYL SALICYLATE see SAK000
AMYLSINE see PBV750
AMYL SULFHYDRATE see PBM000
AMYL THIOALCOHOL see PBM000
n-AMYL THIOCYANATE see AON500
AMYL TRICHLOROSILANE see PBY750
AMYLTRIETHOXYSILANE see PBZ000
AMYLTRIMETHYLAMMONIUM IODIDE see TMA500
AMYLUM see SLJ500
AMYL-Δ-VALEROLACTONE see DAF200
AMYLVINYLCARBINOL see ODW000
AMYL ZIMATE see BJK500
AMYTAL see AMX750
AMYTAL SODIUM see AON750
AN see AAQ500
6-AN see ALL250
AN 23 see DJO800
AN-148 see MDP750
AN 1041 see LJR000
AN 1087 see EOY000
AN 1324 see GFM200
ANA see NAK500
6-ANA see ALL250
ANABACTYL see CBO250
ANABAENA FLOS-AQUAE TOXIN see AON825
ANABASIDE HYDROCHLORIDE see PIT650
ANABASIN see AON875
(−)-ANABASIN see AON875
ANABASIN CHLORIDE see PIT650
ANABASINE see AON875
ANABASINE MONOHYDROCHLORIDE see PIT650
ANABASIN HYDROCHLORIDE see PIT650
ANABAZIN see AON875
ANABET see CNR675
ANABOLEEN see DME500
ANABOLEX see DME500
ANABOLIN see DAL300
ANAC 110 see CNI000
ANACARDONE see DJS200
ANACEL see TBN000
ANACETIN see CDP250
ANACOBIN see VSZ000
ANACORDONE see DJS200
ANADOLOR see AIT250
ANADOMIS GREEN see CMJ900
ANADREX see DBB000
ANADROL see PAN100
ANADROYD see PAN100
ANAESTHETIC ETHER see EJU000
ANAFEBRINA see DOT000

ANAFLON see HIM000
ANAFRANIL see CDU750, CDV000
ANAGESTONE ACETATE mixed with MESTRANOL (10:1)
    see AOO000
ANAGIARDIL see MMN250
ANAHIST see RDU000
ANALEPTIN see HLV500
ANALEXIN see PGG350, PGG355
ANALGIZER see DFA400
O-ANALOG of DIMETHOATE see DNX800
ANALUD see PEW000
ANALUX see DPE000
ANAMENTH see TIO750
ANAMID see SAH000
ANANASE see BMO000
ANANSIOL see MNM500
ANAPAC see ABG750
A. NAPELLUS see MRE275
ANAPHRANIL see CDV000
ANAPOLON see PAN100
ANAPRAL see TLN500
ANAPREL see TLN500
ANAPRILIN see ICC000
ANAPRIME see FDD075
ANAPROTIN see DME500
ANARCON see AFT500
ANASCLEROL see VLF000
ANASTERON see PAN100
ANASTERONAL see PAN100
ANASTERONE see PAN100
ANASTRESS see MQU750
ANATENSIN see GGS000
ANATENSOL see TJW500
ANATHYLMON see MQU750
ANATOLA see VSK600
ANATOXIN-a see AOO120
ANATOXIN I see AOO120
ANATRAN see ABH500, HII500
ANATROPIN mixed with MESTRANOL (10:1) see AOO000
ANAUTINE see DYE600
ANAVAR see AOO125
ANAYODIN see IEP200, SHW000
ANAZOLENE, SODIUM see ADE750
ANC 113 see OLW400
ANCEF see CCS250
ANCHOIC ACID see ASB750
ANCHRED STANDARD see IHD000
ANCILLIN see AOD125
ANCISTRODON PISCIVORUS VENOM see AOO135
ANCITABINE see COW875
ANCITABINE HYDROCHLORIDE see COW900
ANCOBON see FHI000
ANCOLAN see HGC500
ANCOLAN DIHYDROCHLORIDE see MBX250
ANCOR EN 80@1 = 50 see IGK800
ANCORTONE see PLZ000
ANCROD see VGU700
ANCYLOL see DNG000
ANCYTABINE see COW875
ANDAKSIN see MQU750
ANDANTOL see AEG625
ANDAXIN see MQU750
ANDERE see BQL000
ANDHIST see RDU000
ANDIAMINE see HFF500
ANDORDRIN DIPROPIONATE see AOO150
ANDRACTIM see DME500
ANDRAMINE see DYE600
ANDRANE see CCR510
ANDREZ see SMR000
ANDROCTONUS AMOREUXI VENOM see AOO250
ANDROCTONUS AUSTRALIS HECTOR VENOM see AOO265
ANDRODIOL see AOO475
ANDROFLUORENE see AOO275
ANDROFLUORONE see AOO275
ANDROFURAZANOL see AOO300
ANDROGEN see TBG000
ANDROLIN see TBF500
ANDROLONE see DME500
ANDROMEDOTOXIN see AOO375
ANDROMETH see MPN500
ANDRONAQ see TBF500
ANDROSAN see TBG000, MPN500
ANDROSAN (tablets) see MPN500

ANDROSTA-2,5-DIENO(2,3-d)ISOXAZOL-17-OL, 4,4,17-TRIMETHYL-, (17-β)-
(9CI) see HOM259
ANDROSTALONE see MJE760
ANDROSTANAZOL see AOO400
ANDROSTANAZOLE see AOO400
5-α-ANDROSTAN-3α,17β-DIOL, 2β,16β-DIPIPECOLINIO-, DIBROMIDE,
DIACETATE see PAF625
5-α-ANDROSTANE-2-α-CARBONITRILE, 4-α-5-EPOXY-17-β-HYDROXY-3-OXO-
see EBY600
ANDROSTANE-2-CARBONITRILE, 4,5-EPOXY-17-HYDROXY-3-OXO-, (2-α-4-α-
5-α-17-β)- see EBY600
5-α-ANDROSTANE-17-α-METHYL-17-β-OL-3-ONE see MJE760
ANDROSTANOLONE see DME500
5-α-ANDROSTAN-17-β-OL-3-ONE see DME500
5-α-ANDROSTAN-3-α-OL-17-ONE see HJB050
ANDROSTANOLONE PROPIONATE see DME525
5-α-ANDROSTAN-17-ONE, 3-α-HYDROXY- see HJB050
ANDROSTAN-17-ONE, 3-HYDROXY-, (3-α-5-α)- see HJB050
ANDROSTAN-3-ONE, 17-HYDROXY-17-METHYL-, (5-α-17-β)-(9CI) see MJE760
ANDROSTANON-3-α-OL-17-ONE see HJB050
ANDROSTEN see MPN500
Δ-(⁴)-ANDROSTEN-3,17-DIONE see AOO425
ANDROST-2-ENE-2-CARBONITRILE, 3,17-DIHYDROXY-4,17-DIMETHYL-4,5-
EPOXY-, (4-α-5-α- 17-β)- see EBH400
ANDROST-5-ENE-3,17-DIOL, DIPROPANOATE, (3-β,17-β)- (9CI) see AOO410
ANDROSTENEDIOL DIPROPIONATE see AOO410
ANDROST-5-ENE-3-β,17-β-DIOL, DIPROPIONATE see AOO410
ANDROSTENEDIONE see AOO425
4-ANDROSTENE-3,17-DIONE see AOO425
Δ-4-ANDROSTENEDIONE see AOO425
Δ(⁴)-ANDROSTENE-3,17-DIONE see AOO425
4-ANDROSTENE-17-α-METHYL-17-β-OL-3-ONE see MPN500
Δ⁴-ANDROSTENE-17-β-PROPIONATE-3-ONE see TBG000
ANDROSTENOLONE see AOO450
ANDROST-4-EN-17β-OL-3-ONE see TBF500
Δ⁴-ANDROSTEN-17(β)-OL-3-ONE see TBF500
ANDROST-4-EN-3-ONE, 17-(3-CYCLOPENTYL-1-OXOPROPOXY)-, (17-β)- (9CI)
see TBF600
ANDROST-4-EN-3-ONE, 17-β-HYDROXY-1-α-7-α-DIMERCAPTO-17-METHYL-,
1,7-DIACETATE see TFK300
ANDROSTEROLO see AOO275
ANDROSTERONE see HJB050
cis-ANDROSTERONE see HJB050
ANDROSTESTONE-M see AOO475
ANDROSTESTON-M see AOO475
ANDROTARDYL see TBF750
ANDROTESTON see TBG000
ANDROTEST P see TBG000
ANDROTEX see AOO425
ANDRUSOL see TBF500
ANDRUSOL-P see TBG000
ANECOTAN see DFA400
ANECTINE see CMG250, HLC500
ANECTINE CHLORIDE see HLC500
ANELIX see HIM000
ANELMID see DJT800
ANEMONE see PAM780
ANEMONE (VARIOUS SPECIES) see PAM780
ANERGAN see ABH500
ANERTAN see MPN500, TBG000
ANERTAN (tablets) see MPN500
ANERVAL see BRF500
ANESTACON see DHK400
ANESTACON HYDROCHLORIDE see DHK600
ANESTHENYL see MGA850
ANESTHESIA ETHER see EJU000
ANESTHESIN see EFX000
ANESTHESOL see AIT250
ANESTHETIC COMPOUND No. 347 see EAT900
ANESTHETIC ETHER see EJU000
ANESTHONE see EFX000
ANESTIL see AIT250
ANETAIN see BQA010
ANETHAINE see TBN000
trans-ANETHOL see PMR250
cis-ANETHOLE see PMR250
ANETHOLE (FCC) see PMQ750
trans-ANETHOLE see PMR250
ANETHOLE TRITHIONE see AOO490
ANETHOLTRITHION see AOO490
ANEURAL see MQU750
ANEURIMEC see DXO300
ANEURINE see TES750
ANEUXRAL see MQU750

ANEXOL see CDP000
ANFETAMINA see AOB250
ANFLAGEN see IIU000
ANFRAM 3PB see TNC500
ANFT see ALM500
ANG 66 see AOP250
ANGECON see FQC000
ANGELICA LACTONE see MKH250
α-ANGELICA LACTONE see AOO750
β-ANGELICA LACTONE see MKH500
β,γ-ANGELICA LACTONE see MKH250
Δ¹-ANGELICA LACTONE see MKH500
Δ²-ANGELICA LACTONE see MKH250
ANGELICA OIL, root see AOO760
ANGELICA ROOT OIL see AOO760
ANGELICIN (coumarin derivative) see FQC000
ANGELIKA OEL see AOO760
ANGELI'S SULFONE see AOO800
ANGELI SULFONE see AOO800
ANGEL'S TRUMPET see AOO825
ANGEL TULIP see SLV500
ANGEL WINGS see CAL125
ANGICAP see PBC250
ANGIFLAN see DBX400
ANGIGRAFIN see AOO875
ANGINAL see PCP250
ANGININ see PPH050
ANGININE see NGY000, PPH050
ANGINON see AHI875
ANGINYL see DNU600
ANGIOCICLAN see POD750
ANGIOGRAFIN see AOO875
ANGIOKAPSUL see ARQ750
ANGIOMIN see XCS000
ANGIOPAC see VLF000
ANGIOTENSIN see AOO900
ANGIOTONIN see AOO900
ANGIOXINE see PPH050
ANGITET see PBC250
ANGITRIT see TJL250
ANGLISLITE see LDY000
ANGOLAMYCIN see AOO925
ANGORLISIN see ELH600
ANG.-STERANTHREN (GERMAN) see DCR800
ANG-STERANTHRENE see DCR800
ANGUIDIN see AOP250
ANGUIDINE see AOP250
ANGUIFUGAN see DJT800
ANHIBA see HIM000
ANHISTABS see WAK000
ANHISTAN see FOS100
ANHISTOL see WAK000
ANHYDRIDE ACETIQUE (FRENCH) see AAX500
ANHYDRIDE ARSENIEUX (FRENCH) see ARI750
ANHYDRIDE ARSENIQUE (FRENCH) see ARH500
ANHYDRIDE CARBONIQUE (FRENCH) see CBU250
ANHYDRIDE CARBONIQUE et OXYDE d'ETHYLENE MELANGES (FRENCH)
see EJO000
ANHYDRIDE CHROMIQUE (FRENCH) see CMK000
ANHYDRIDE KYSELINY 4-CHLOR-1,2,3,6-TETRAHYDROFTA-LOVE (CZECH)
see CFG500
ANHYDRIDE PHTALIQUE (FRENCH) see PHW750
ANHYDRIDES see AOP500
ANHYDRIDE VANADIQUE (FRENCH) see VDU000
ANHYDRID KYSELINY 3,6-ENDOMETHYLEN-Δ(SUP 4)-
TETRAHYDROFTALOVE (CZECH) see BFY000
ANHYDRID KYSELINY KROTONOVE see COB900
ANHYDRID KYSELINY TETRAHYDROFTALOVE (CZECH) see TDB000
ANHYDRID KYSELINY TRIFLUOROCTOVE (CZECH) see TJX000
2,2'-ANHYDRO-1-β-d-ARABINOFURANOSYLCYTOSINE HYDROCHLORIDE
see COW900
2,2'-ANHYDROARABINOSYLCYTOSINE see COW875
2,2'-ANHYDROARABINOSYLCYTOSINE HYDROCHLORIDE see COW900
ANHYDROARA C see COW875
ANHYDRO-4,4'-BIS(DIETHYLAMINO)TRIPHENYLMETHANOL-2',4''-DIS-
ULPHONIC ACID, MONOSODIUM SALT see ADE500
ANHYDRO-4-CARBAMOYL-5-HYDROXY-1-β-d-RIBOFURANOSYL-IM-
IDAZOLIUMHYDROXIDE see BMM000
2,2'-ANHYDROCYTARABINE HYDROCHLORIDE see COW900
ANHYDROCYTIDINE see COW875
2,2'-ANHYDROCYTIDINE see COW875
2,2'-ANHYDROCYTIDINE HYDROCHLORIDE see COW900
3,6-ANHYDRO-d-GALACTAN see CCL250

ANHYDROGITALIN see GEU000
ANHYDRO-d-GLUCITOL MONOOCTADECANOATE see SKV150
ANHYDROGLUCOCHLORAL see GFA000
10-(1',5'-ANHYDROGLUCOSYL)ALOE-EMODIN-9-ANTHRONE see BAF825
ANHYDROHYDROXYPROGESTERONE see GEK500
ANHYDROL see EFU000
ANHYDROMYRIOCIN see AOP750
ANHYDRONE see PCE000
ANHYDROSORBITOL STEARATE see SKV150
ANHYDRO-o-SULFAMINEBENZOIC ACID see BCE500
ANHYDROTRIMELLIC ACID see TKV000
ANHYDROUS AMMONIA see AMY500
ANHYDROUS CHLOROBUTANOL see ABD000
ANHYDROUS IRON OXIDE see IHD000
ANHYDROUS OXIDE of IRON see IHD000
ANHYDROXYPROGESTERONE see GEK500
ANI see ISN000
ANICON KOMBI see CIR250
ANICON M see CIR250
ANIDRIDE ACETICA (ITALIAN) see AAX500
ANIDRIDE CROMICA (ITALIAN) see CMK000
ANIDRIDE CROMIQUE (FRENCH) see CMJ900
ANIDRIDE FTALICA (ITALIAN) see PHW750
ANILANA, combustion products see ADX750
ANILAZIN see DEV800
ANILAZINE see DEV800
l'ANILIDE de l'ACIDE (PYRROLIDINO-N)-3-N-BUTYRIQUE (FRENCH)
     see MPB500
ANILIDE of (PYRROLIDINO-N)-3-N-BUTYRIC ACID see MPB500
ANILIN (CZECH) see AOQ000
ANILINA (ITALIAN, POLISH) see AOQ000
ANILINE see AOQ000
ANILINE ANTIMONYL TARTRATE see AOQ250
ANILINE, p-ARSENOSO- see ARJ755
ANILINE, p-ARSENOSO-N,N-BIS(2-CHLOROETHYL)- see ARJ760
ANILINE, p-ARSENOSO-N,N-BIS(2-HYDROXYETHYL)- see ARJ770
ANILINE, p-ARSENOSO-N,N-DIETHYL- see ARJ800
ANILINE, p-ARSENOSO-, DIHYDRATE see ALV100
p-ANILINEARSONIC ACID see ARA250
ANILINE, p-(7-BENZOFURYLAZO)-N,N-DIMETHYL- see BCL100
ANILINE, N,N-BIS(2-CHLOROETHYL)-2,3-DIMETHOXY- see BIC600
ANILINE, N,N-BIS(2-(2,3-EPOXYPROPOXY)ETHOXY)- see BJN850
ANILINE, N,N-BIS(2-(2,3-EPOXYPROPOXY)ETHYL)- see BJN875
ANILINE, p-((3-BROMO-4-ETHYLPHENYL)AZO)-N,N-DIMETHYL- see BNK275
ANILINE, p-((4-BROMO-3-ETHYLPHENYL)AZO)-N,N-DIMETHYL- see BNK100
ANILINE, p-(m-BROMOPHENYLAZO)-N,N-DIMETHYL- see BNE600
ANILINE, p-(3-BROMO-p-TOLYL)AZO)-N,N-DIMETHYL- see BNQ100
ANILINE, p-((4-BROMO-m-TOLYL)AZO)-N,N-DIMETHYL- see BNQ110
ANILINE, p-((p-BUTYLPHENYL)AZO)-N,N-DIMETHYL- see BRB450
ANILINE, p-((p-tert-BUTYL)PHENYL)AZO)-N,N-DIMETHYL- see BRB460
ANILINE CARMINE POWDER see FAE100
ANILINE CHLORIDE see BBL000
ANILINE, 4-CHLORO-3-NITRO- see CJA185
ANILINE, p-DICHLOROARSINO-, HYDROCHLORIDE see AOR640
ANILINE, N,N-DIMETHYL-p-(4'-CHLORO-3'-METHYLPHENYLAZO)-
     see CIL710
ANILINE, N,N-DIMETHYL-p-(3,5-DIFLUOROPHENYLAZO)- see DKH100
ANILINE-2,5-DISULFONIC ACID see AIE000
ANILINE DYES see AOQ500
ANILINE, p-((o-ETHYLPHENYL)AZO)-N,N-DIMETHYL- see EIF450
ANILINE GREEN see AFG500, BAY750
ANILINE HYDROCHLORIDE (DOT) see BBL000
ANILINE MUSTARD see AOQ875
ANILINE OIL see AOQ000
ANILINE OIL DRUMS, EMPTY see AOR000
"ANILINE SALT" see BBL000
p-ANILINESULFONAMIDE see SNM500
m-ANILINESULFONIC ACID see SNO000
ANILINE-p-SULFONIC AMIDE see SNM500
ANILINE, 2,3,4-TRIFLUORO- see TJX900
ANILINE VANADATE, DIHYDRATE see AOR250
ANILINE VIOLET see AOR500
ANILINE YELLOW see PEI000
ANILINIUM CHLORIDE see BBL000
ANILINIUM NITRATE see AOR625
ANILINIUM PERCHLORATE see AOR630
ANILINOBENZENE see DVX800
2-ANILINOBENZOIC ACID see PEG500
o-ANILINOBENZOIC ACID see PEG500
4-ANILINODICHLOROARSINE, HYDROCHLORIDE see AOR640
1-ANILINO-2,5-DISULFONIC ACID see AIE000
ANILINOETHANE see EGK000
2-ANILINOETHANOL see AOR750
(2-ANILINOETHYL)HYDRAZONE DIHYDROCHLORIDE see AOS000

ANILINOMETHANE see MGN750
ANILINONAPHTHALENE see PFT500
1-ANILINONAPHTHALENE see PFT250
2-ANILINONAPHTHALENE see PFT500
2-ANILINO-5-NITROBENZENESULFONIC ACID see AOS500
ANILINO (p-NITROPHENYL) SULFIDE see AOS750
4-ANILINOPHENOL see AOT000
p-ANILINOPHENOL see AOT000
4-((4-ANILINO-5-SULFO-1-NAPHTHYL)AZO)-5-HYDROXY-2,7-NAPHTHALENE-
     DIFULFONIC ACID TRISODIUM see ADE750
6-(p-ANILINOSULFONYL)METANILAMIDE see AOT125
ANILITE see AOT250
ANILIX see CKL500
ANIMAL CONIINE see PBK500
ANIMAL GALACTOSE FACTOR see OJV500
ANIMAL OIL see BMA750
ANIMERT see CKL750
ANIMERT V-10 see CKL750
ANIMERT V-101 see CKL750
ANIMERT V-10K see CKL750
ANIPRIME see FDD075
2-ANISALDEHYDE see AOT525
o-ANISALDEHYDE see AOT525
p-ANISALDEHYDE see AOT530
o-ANISAMIDE see AOT750
o-ANISAMIDE, N-((1-ETHYL-2-PYRROLIDINYL)METHYL)-5-
     (ETHYLSULFONYL)- see EPD100
ANISE ALCOHOL see MED500
ANISE CAMPHOR see PMQ750
ANISEED OIL see AOU250
ANISENE see CLO750
ANISE OIL see AOU250
m-ANISIC ACID see AOU500
o-ANISIC ACID see MPI000
p-ANISIC ACID, ETHYL ESTER see AOV000
ANISIC ACID HYDRAZIDE see AOV500
o-ANISIC ACID, HYDRAZIDE see AOV250
p-ANISIC ACID, HYDRAZIDE see AOV500
p-ANISIC ACID, METHYL ESTER see AOV750
ANISIC ALCOHOL see MED500
ANISIC ALDEHYDE see AOT530
ANISIC HYDRAZIDE see AOV500
ANISIC KETONE see AOV875
2-ANISIDINE see AOV900
4-ANISIDINE see AOW000
o-ANISIDINE see AOV900
p-ANISIDINE see AOW000
m-ANISIDINE ANTIMONYL TARTRATE see AOW500
o-ANISIDINE ANTIMONYL TARTRATE see AOW750
p-ANISIDINE ANTIMONYL TARTRATE see AOX000
o-ANISIDINE HYDROCHLORIDE see AOX250
p-ANISIDINE HYDROCHLORIDE see AOX500
2-ANISIDINE NITRATE see MEA600
o-ANISIDINE NITRATE see NEQ500
ANISKETONE see AOV875
ANIS OEL (GERMAN) see AOU250
p-ANISOL ALCOHOL see MED500
ANISOLE see AOX750
ANISOLE, p-(3-BROMOPROPENYL)-, (E)- see BMT300
ANISOMYCIN see AOY000
ANISOPIROL see HAH000
ANISOPYRADAMINE see DBM800
ANISOTROPINE METHOBROMIDE see LJS000
3-p-ANISOYL-3-BROMOACRYLIC ACID, SODIUM SALT see SIK000
(E)-3-p-ANISOYL-3-BROMOACRYLIC ACID SODIUM SALT see CQK600
ANISOYL CHLORIDE see AOY250
ANISOYLHYDRAZINE see AOV500
p-ANISOYLHYDRAZINE see AOV500
ANISTADIN see HII500
ANISYL ACETATE see AOY400
2-(p-ANISYL)ACETIC ACID see MFE250
ANISYLACETONE see MFF580
ANISYLACETONITRILE see MFF000
ANISYL ALCOHOL (FCC) see MED500
o-ANISYLAMINE see AOV900
p-ANISYLAMINE see AOW000
N-(o-ANISYL)-2-(p-BUTOXYPHENOXY)-N-(2(DIETHYLAMINO)ETHYL)ACET-
     AMIDE HYDROCHLORIDE see APA000
ANISYL-N-BUTYRATE see MED750
ANISYL FORMATE see MFE250
ANISYL METHYL KETONE see AOV875
ANISYL PHENYLACETATE see APE000
p-ANISYL CHLORIDE see AOY250
ANIT see ISN000

ANITSOTROPINE METHYLBROMIDE see LJS000
ANKERBIN see SLJ050
ANKILOSTIN see PCF275
ANN (GERMAN) see AAW000
ANNALINE see CAX750
ANNATTO EXTRACT see APE100
ANNONA MURICATA see SKV500
ANNUAL POINSETTIA see EQX000
(6)ANNULENE see BBL250
ANODYNON see EHH000
ANOFEX see DAD200
ANOL see CPB750
ANOPROLIN see ZVJ000
A-NORANDROSTANE-2-α-17-α-DIETHYNYL-2-β,17-β-DIOL see EQN259
ANORDRIN see AOO150
ANOREXIDE see BBK000
ANOVIGAM see EEM000
ANOVLAR 21 see EEH520
ANOZOL see DJX000
ANP 235 see DPE000
ANP 246 see CJN750
ANP 3548 see BPM750
ANP 3624 see TGA600
ANP 4364 see DFO600
ANPARTON see ARQ750
235 ANP HYDROCHLORIDE see AAE500
ANPROLENE see EJN500
ANPROLENE see EJN500
ANQI see ALK625
ANQUIL see FLK100
ANSADOL see SAH500
ANSAL see AQN250
ANSAMITOCIN P-4 see APE529
ANSAR see HKC000
ANSAR 160 see HKC500
ANSAR 170 see MRL750
ANSAR 184 see DXE600
ANSAR DSMA LIQUID see DXE600
ANSATIN see TKH750
ANSEPRON see PGA750
ANSIACAL see MDQ250
ANSIATAN see MQU750
ANSIBASE RED KB see CLK225
ANSIL see MQU750
ANSILAN see CGA000
ANSIOLISINA see CFZ000, DCK759
ANSIOWAS see MQU750
ANSIOXACEPAM see CFZ000
ANSOLYSEN see PBT000
ANSOLYSEN BITARTRATE see PBT000
ANSOLYSEN TARTRATE see PBT000
ANSUL ETHER 181AT see PBO500
ANT-1 see TKH750
ANTABUS see DXH250
ANTABUSE see DXH250
ANTADIX see DXH250
ANTADOL see BRF500
ANTAENYL see DXH250
ANTAETHAN see DXH250
ANTAETHYL see DXH250
ANTAETIL see DXH250
ANTAGE W 400 see MJO500
ANTAGONATE see TAI500
ANTAGOSAN see PAF550
ANTAGOTHYROID see TFR250
ANTAGOTHYROIL see TFR250
ANTAK see DAI600, ODE000
ANTALCOL see DXH250
ANTALERGAN see WAK000
ANTALVIC see PNA500
ANTAMINE see WAK000
ANTAN see NAH500
ANTAROX A-200 see PKF500
ANTARSIN see DNV600
ANTASTEN see PDC000
ANTAZOLINE see PDC000
ANTEMOQUA see AJX500
ANTEMOVIS see AJX500
ANTENE see BJK500
ANTEPAR see PIJ500
ANTERGAN see BEM500
ANTERGAN HYDROCHLORIDE see PEN000
ANTERGYL see SEO500
ANTERIOR PITUITARY GROWTH HORMONE see PJA250

ANTERIOR PITUITARY LUTEOTROPIN see PMH625
ANTERON see SCA750
ANTETAN see DXH250
ANTETHYL see DXH250
ANTETIL see DXH250
ANTEX-490 see SCA750
ANTEYL see DXH250
ANTHALLAN HYDROCHLORIDE see APE625
ANTHANTHREN (GERMAN) see APE750
ANTHANTHRENE see APE750
ANTHECOLE see PIJ500
ANTHELMYCIN see APF000
ANTHELONE U see UVJ475
ANTHELVET see TDX750
ANTHER see IHV050
ANTHGLUTIN see GFO100
ANTHIO see DRR200
ANTHIOLIMINE see LGU000
ANTHIOMALINE see LGU000
ANTHIOMALINE NONAHYDRATE see AQE500
ANTHION see DWQ000
ANTHISAN see WAK000
ANTHISAN MALEATE see DBM800
ANTHIUM DIOXCIDE see CDW450
ANTHON see TIQ250
ANTHOPHYLITE see ARM264
ANTHRA(9,1,2-cde)BENZO(h)CINNOLINE see APF750
ANTHRACEN (GERMAN) see APG500
1-ANTHRACENAMINE see APG050
2-ANTHRACENAMINE see APG100
ANTHRACENE see APG500
ANTHRACENE BROWN see TKN500
9,10-ANTHRACENEDIONE see APK250
9,10-ANTHRACENEDIONE, 1-BROMO-4-(METHYLAMINO)- see BNN550
9,10-ANTHRACENEDIONE, 2,6-DIAMINO- see APK850
9,10-ANTHRACENEDIONE, 1,4-DIAMINO-5-NITRO-(9CI) see DBY700
9,10-ANTHRACENEDIONE, 1,5-DICHLORO- see DEO700
9,10-ANTHRACENEDIONE, 1,8-DICHLORO- see DEO750
9,10-ANTHRACENEDIONE, 1,5-DIPHENOXY- see DVW100
1,8,9-ANTHRACENETRIOL see APH250
1,8,9-ANTHRACENETRIOL TRIACETATE see APH500
9(10H)-ANTHRACENONE, 1,8-DIHYDROXY-10-(1-OXOPROPYL)- see PMW760
ANTHRACHINON-1,8-DISULFONAN DRASELNY (CZECH) see DLJ600
ANTHRACHINON-1,5-DISULFONAN SODNY (CZECH) see DLJ700
ANTHRACHINON-1-SULFONAN SODNY (CZECH) see SER000
ANTHRACHINON-1-SULFONAN SODNY (CZECH) see DLJ800
ANTHRACIN see APG500
ANTHRACITE PARTICLES see CMY635
1-ANTHRACYLAMINE see APG050
2-ANTHRACYLAMINE see APG100
ANTHRADIONE see APK250
ANTHRAFLAVIC ACID see DMH600
ANTHRAFLAVIN see DMH600
ANTHRAGALLIC ACID see TKN500
ANTHRAGALLOL see TKN500
ANTHRALAN YELLOW RRT see SGP500
ANTHRALIN see APH250
1-ANTHRAMINE see APG050
2-ANTHRAMINE see APG100
ANTHRAMYCIN see API000
ANTHRAMYCIN METHYL ETHER see API125
ANTHRAMYCIN-11-METHYL ETHER see API125
ANTHRANILIC ACID see API500
ANTHRANILIC ACID, N-(2-BENZYLIDENEHEPTYLIDENE)-, METHYL ESTER see AOH100
ANTHRANILIC ACID, N-(3-(p-tert-BUTYLPHENYL)-2-METHYLPROPYLIDENE)-, METHYL ESTER see LFT100
ANTHRANILIC ACID, CINNAMYL ESTER see API750
ANTHRANILIC ACID, LINALYL ESTER see APJ000
ANTHRANILIC ACID, METHYL ESTER see APJ250
ANTHRANILIC ACID, PHENETHYL ESTER see APJ500
ANTHRANILONITRILE see APJ750
m-ANTHRANILONITRILE see AIR125
ANTHRANOL CHROME YELLOW R see NEY000
ANTHRANTHRENE see APE750
ANTHRAPOLE AZ see BQK250
ANTHRA(1,9-cd)PYRAZOL-6(2H)-ONE see APK000
β-ANTHRAQUINOLINE see NAZ000
ANTHRAQUINONE see APK250
9,10-ANTHRAQUINONE see APK250
ANTHRAQUINONE BRILLIANT GREEN CONCENTRATE ZH see APK500
ANTHRAQUINONE, 2-BROMO-1,5-DIAMINO-4,8-DIHYDROXY- see BNC800
ANTHRAQUINONE, 1-BROMO-4-(METHYLAMINO)- see BNN550
ANTHRAQUINONE, 2,6-DIAMINO- see APK850

1,2-ANTHRAQUINONEDIOL see DMG800
1,8-ANTHRAQUINONEDISULFINIC ACID see APK635
1,5-ANTHRAQUINONEDISULFONIC ACID see APK625
ANTHRAQUINONEDISULFONIC ACID, DIPOTASSIUM SALT see DLJ600
9,10-ANTHRAQUINONE-2-SODIUM SULFONATE see SER000
2-ANTHRAQUINONESULFONATE SODIUM see SER000
ANTHRAQUINONE-2-SULFONATE SODIUM SALT see SER000
2-ANTHRAQUINONESULFONIC ACID SODIUM SALT see SER000
α-ANTHRAQUINONYLAMINE see AIA750
β-ANTHRAQUINONYLAMINE see AIB000
ANTHRAQUINONYLAMINOANTHRAQUINONE see IBI000
((N-ANTHRAQUINON-2-YL)AMINOMETHYLENE)DIMETHYLAMMONIUM
    CHLORIDE see APK750
1,4-ANTHRAQUINONYLDIAMINE see DBP000
1,5-ANTHRAQUINONYLDIAMINE see DBP200, DBP400
1,8-ANTHRAQUINONYLDIAMINE see DBP400
2,6-ANTHRAQUINONYLDIAMINE see APK850
N,N'''-(2,6-ANTHRAQUINONYLENE)BIS(N,N-DIETHYLACETAMIDE)
    see APL250
2,2'-(1,4-ANTHRAQUINONYLENEDIIMINO)BIS(5-METHYLBENZENESULFO
    NIC ACID) DISODIUM SALT see APL500
1,1'-(ANTHRAQUINON-1,4-YLENEDIIMINO)DIANTHRAQUINONE see APL750
1,1'-(ANTHRAQUINON-1,5-YLENEDIIMINO)DIANTHRAQUINONE see APM000
4,4'-(1,4-ANTHRAQUINONYLENEDIIMINODIPHENYL-1,4-ENEDIOXO)
    BENZENESULFONIC ACID see APM250
ANTHRARUFIN see DMH200
1,8,9-ANTHRATRIOL see APH250
ANTHRAVAT GOLDEN YELLOW see DCZ000
ANTHRIMIDE see IBI000
ANTHROGON see FMT100
9-ANTHRONOL see APM750
ANTHROPODEOXYCHOLIC ACID see CDL325
ANTHROPODESOXYCHOLIC ACID see CDL325
ANTHROPODODESOXYCHOLIC ACID see CDL325
2-ANTHRYLAMINE see APG100
ANTHURIUM see APM875
ANTIAETHAN see DXH250
ANTIANGOR see CBR500
α-ANTIARBIN see APN000
ANTIB see FNF000
ANTIBASON see MPW500
ANTIBIOCIN see PDT750
ANTIBIOTIC 60-6 see CCX725
ANTIBIOTIC 899 see VRA700
ANTIBIOTIC 1037 see VGZ000
ANTIBIOTIC 1600 see APP500
ANTIBIOTIC 1719 see ASO501
ANTIBIOTIC 66-40 see SDY750
ANTIBIOTIC 833A see PHA550
ANTIBIOTIC 205T3 see NMV500
ANTIBIOTIC 29275 see CBF680
ANTIBIOTIC 33876 see APF000
ANTIBIOTIC 67-694 see RMF000
ANTIBIOTIC 281471 see MRW800
ANTIBIOTIC 6761-31 see TFQ275
ANTIBIOTIC A see API125
ANTIBIOTIC A-246 see FPC000
ANTIBIOTIC A-649 see OIU499
ANTIBIOTIC A 130A see LEJ700
ANTIBIOTIC A-5283 see PIF000
ANTIBIOTIC A 8506 see TFQ275
ANTIBIOTIC A-64922 see OIU499
ANTIBIOTIC A 3733A see HAL000
ANTIBIOTIC A 28695 A see SCA000
ANTIBIOTIC AB 206 see MQU525
ANTIBIOTIC AD 32 see TJX350
ANTIBIOTIC A-250-II see CBF680
ANTIBIOTIC AK PS see AFH500
ANTIBIOTIC AM-2604 A see VRP775
ANTIBIOTIC APM see XFS600
ANTIBIOTIC A-399-Y4 see VGZ000
ANTIBIOTIC AY 22989 see RBK000
ANTIBIOTIC B 41D see MQT600
ANTIBIOTIC B 599 see CMK650
ANTIBIOTIC B-14437 see SAU000
ANTIBIOTIC B-98891 see MQU000
ANTIBIOTIC BAY-f 1353 see MQS200
ANTIBIOTIC BB-K 8 see APS750
ANTIBIOTIC BB-K8 SULFATE see APT000
ANTIBIOTIC BL-640 see APT250
ANTIBIOTIC BL-S 640 see APT250
ANTIBIOTIC BU 2231A see TAC500
ANTIBIOTIC BU 2231B see TAC750
ANTIBIOTIC CC 1065 see APT375

ANTIBIOTIC CGP 9000 see CCS530
ANTIBIOTIC D-45 see SLW475
ANTIBIOTIC DC 11 see TEF725
ANTIBIOTIC DE 3936 see LIF000
ANTIBIOTIC E212 see VGZ000
ANTIBIOTIC 1163 F.I. see LIN000
ANTIBIOTIC FN 1636 see PEC750
ANTIBIOTIC FR 1923 see APT750
ANTIBIOTIC G-52 see GAA100
ANTIBIOTIC G-52 SULFATE see APU000
ANTIBIOTIC G-52 SULFATE see GAA120
ANTIBIOTIC HA-9 see TFC500
ANTIBIOTIC KA 66061 see SLF500
ANTIBIOTIC KM 208 see BAC175
ANTIBIOTIC KW 1062 see MQS579
ANTIBIOTIC KW-1070 see FOK000
ANTIBIOTIC LA 7017 see MQW750
ANTIBIOTIC M 4365A2 see RMF000
ANTIBIOTIC MA 144A see APU500
ANTIBIOTIC MA 144A1 see APU500
ANTIBIOTIC MA 144A2 see TAH675
ANTIBIOTIC MA 144B2 see TAH650
ANTIBIOTIC MA 144M1 see MAB250
ANTIBIOTIC MA 144S2 see APV500
ANTIBIOTIC MA 144T1 see DAY835
ANTIBIOTIC MM 14151 see CMV250
ANTIBIOTIC N-329 B see VBZ000
ANTIBIOTIC No. 899 see VRF000
ANTIBIOTIC NSC 70845 see NMV500
ANTIBIOTIC NSC-71936 see CMQ725
ANTIBIOTIC OS 3966A see RMK200
ANTIBIOTIC 20-798RP see DAC300
ANTIBIOTIC PA-93 see SMB000
ANTIBIOTIC PA-106 see AOY000
ANTIBIOTIC PA147 see APV750
ANTIBIOTIC PA 11481 see VRA700
ANTIBIOTIC PA 114 B1 see VRA700
ANTIBIOTIC Ro 21 6150 see LEJ700
ANTIBIOTIC S 15-1A see RAG300
ANTIBIOTIC 6059-S see LBH200
ANTIBIOTIC S 7481F1 see CQH100
ANTIBIOTIC SF 733 see XQJ650
ANTIBIOTIC SF 837 see MBY150
ANTIBIOTIC SF 767B see NCF500
ANTIBIOTIC SF 837 A1 see MBY150
ANTIBIOTIC SF 837 A₁ see MBY150
ANTIBIOTIC 1719 SODIUM SALT see ASO510
ANTIBIOTIC 66-40 SULFATE see APY500
ANTIBIOTIC T see COB000
ANTIBIOTIC TM 481 see LIF000
ANTIBIOTIC U 18496 see ARY000
ANTIBIOTIC U 48160 see QQS075
ANTIBIOTICUM PA147 (GERMAN) see APV750
ANTIBIOTIC 44 VI see MRW800
ANTIBIOTIC W-847-A see MCA250
ANTIBIOTIC WR 141 see GJS000
ANTIBIOTIC X 146 see TFQ275
ANTIBIOTIC X 537 see LBF500
ANTIBIOTIC X-465A see CDK250
ANTIBIOTIC XK 41C see MCA250
ANTIBIOTIC XK 62-2 see MQS579
ANTIBIOTIC XS-89 see SMM500
ANTIBIOTIC X465A SODIUM SALT see CDK500
ANTIBIOTIC YL-704 A3 see JDS200
ANTIBIOTIC YL 704 B₁ see MBY150
ANTIBIOTIC YL 704 B1 see MBY150
ANTIBIOTIC YL-704 B3 see LEV025
ANTIBIOTIQUE see NCF000
ANTIBULIT see SHF500
ANTICARIE see HCC500
ANTICHLOR see SKI500
ANTI-CHROMOTRICHIA FACTOR see AIH600
ANTIDEPRIN see DLH600
ANTIDEPRIN HYDROCHLORIDE see DLH630
ANTIDUROL see DAM700
ANTIEGENE MB see BCC500
ANTIETANOL see DXH250
ANTI-ETHYL see DXH250
ANTIETIL see DXH250
ANTIFEBRIN see AAQ500
ANTIFEEDANT 24005 see DUI000
ANTIFEEDING COMPOUND 24,055 see DUI000
ANTIFOAM FD 62 see SCR400
ANTIFOLAN see MDV500

ANTIFORMIN see SHU500
ANTI-GERM 77 see BEN000
ANTIGESTIL see DKA600
ANTIHELMYCIN see AQB000
ANTIHEMORRHAGIC VITAMIN see VTA000
ANTIHIST see DBM800
ANTIHISTAL see PDC000
ANTI-INFECTIVE VITAMIN see VSK600
ANTI-INFLAMMATORY HORMONE see CNS750
ANTIKNOCK-33 see MAV750
ANTIKOL see DXH250
ANTIKREIN see PAF550
ANTILEPSIN see DNU000
ANTILIPID see ARQ750
ANTILYSIN see PAF550
ANTILYSINE see PAF550
ANTIMALARINA see ARQ250
ANTIMICINA see ILD000
ANTIMIGRANT C 45 see SEH000
ANTIMILACE see TDW500
ANTIMIT see BIE500
ANTIMOINE FLUORURE (FRENCH) see AQE000
ANTIMOINE (TRICHLORURE d') see AQC500
ANTIMOL see SFB000
ANTIMONIAL SAFFRON see AQF500
ANTIMONIC "ACID" see AQF750
ANTIMONIC ACID, SODIUM SALT see AQB250
ANTIMONIC CHLORIDE see AQD000, AQD250
ANTIMONIC OXIDE see AQF750
ANTIMONIC SULFIDE see AQF500
ANTIMONIO (PENTACLORURO DI) (ITALIAN) see AQD000
ANTIMONIO (TRICLORURO di) see AQC500
ANTIMONIOUS OXIDE see AQF000
ANTIMONOUS CHLORIDE see AQC500
ANTIMONOUS CHLORIDE (DOT) see AQC500
ANTIMONOUS FLUORIDE see AQE000
ANTIMONOUS SULFATE see AQJ250
ANTIMONOUS SULFIDE see AQL500
ANTIMONPENTACHLORID (GERMAN) see AQD000
ANTIMONTRICHLORID see AQC500
ANTIMONWASSERSTOFFES (GERMAN) see SLQ000
ANTIMONY see AQB750
ANTIMONY(III) ACETATE see AQJ750
ANTIMONY AMMONIA TRIACETIC ACID see AQC000
ANTIMONY, BIS(TRICHLORO) compounded with 1 mole of OCTAMETHYL
    PYROPHOSPHORAMIDE see AQC250
ANTIMONY BLACK see AQB750
ANTIMONY BROMIDE see AQK000
ANTIMONY BUTTER see AQC500
ANTIMONY CHLORIDE see AQC500
ANTIMONY(III) CHLORIDE see AQC500
ANTIMONY(V) CHLORIDE see AQD000
ANTIMONY CHLORIDE (DOT) see AQC500
ANTIMONY(V) CHLORIDE (solution) see AQD250
ANTIMONY COMPOUNDS see AQD500
ANTIMONY DIMERCAPTOSUCCINATE see AQD750
ANTIMONY DIMERCAPTOSUCCINATE(IV) see AQD750
ANTIMONY EMETINE IODIDE see EAM000
ANTIMONY FLUORIDE see AQF250
ANTIMONY(III) FLUORIDE (1583) see AQE000
ANTIMONY(V) FLUORIDE see AQF250
ANTIMONY GLANCE see AQL500
ANTIMONY HYDRIDE see SLQ000
ANTIMONY LACTATE see AQE250
ANTIMONY LACTATE, solid (DOT) see AQE250
ANTIMONYL ANILINE TARTRATE see AOQ250
ANTIMONYLBRENZEATECHINDISULFOSAURES NATRIUM (GERMAN)
    see AQH500
ANTIMONYL-2,4-DIHYDROXY-5-HYDROXYMETHYL PYRIMIDINE
    see AQE300
ANTIMONYL-2,4-DIHYDROXY PYRIMIDINE see AQE305
ANTIMONYL-7-FORMYL-8-HYDROXYQUINOLINE-5-SULPHONATE
    see AQE320
ANTIMONY LITHIUM THIOMALATENONAHYDRATE see AQE500
ANTIMONYL POTASSIUM TARTRATE see AQG250
ANTIMONY, compounded with NICKEL (1:1) see NCY100
ANTIMONY NITRIDE see AQE750
ANTIMONY ORANGE see AQL500
ANTIMONY OXIDE see AQF000
ANTIMONY PENTACHLORIDE see AQD000
ANTIMONY PENTACHLORIDE (DOT) see AQD000
ANTIMONY PENTACHLORIDE, solution (DOT) see AQD250
ANTIMONY(V) PENTAFLUORIDE see AQF250
ANTIMONY PENTAOXIDE see AQF750

ANTIMONY PENTASULFIDE see AQF500
ANTIMONY PENTOXIDE see AQF750
ANTIMONY PERCHLORIDE see AQD000, AQD250
ANTIMONY PEROXIDE see AQF000
ANTIMONY POTASSIUM DIMETHYLCYSTEINOTARTRATE see AQG000
ANTIMONY POTASSIUM TARTRATE see AQG250
d-ANTIMONY POTASSIUM TARTRATE see AQG500
l-ANTIMONY POTASSIUM TARTRATE see AQH000
dl-ANTIMONY POTASSIUM TARTRATE see AQG750
meso-ANTIMONY POTASSIUM TARTRATE see AQH250
ANTIMONY PYROCATECHOL SODIUM DISULFONATE see AQH500
ANTIMONY RED see AQF500
ANTIMONY REGULUS see AQB750
ANTIMONY SESQUIOXIDE see AQF000
ANTIMONY SODIUM DIMETHYL CYSTEINO TARTRATE see AQH750
ANTIMONY SODIUM GLUCONATE see AQH800
ANTIMONY(III) SODIUM GLUCONATE see AQI000
ANTIMONY(V) SODIUM GLUCONATE see AQI250
ANTIMONY SODIUM OXIDE-l-(+)-TARTRATE see AQI750
ANTIMONY SODIUM PROPYLENE DIAMINE TETRAACETIC ACID DIHY-
    DRATE see AQI500
ANTIMONY SODIUM TARTRATE see AQI750
ANTIMONY(III) SULFATE (2:3) see AQJ250
ANTIMONY SULFIDE see AQF500, AQL500
ANTIMONY TARTRATE see AQJ500
ANTIMONY TELLURIDE see AQL750
ANTIMONY TRIACETATE see AQJ750
ANTIMONY TRIBROMIDE see AQK000
ANTIMONY TRICHLORIDE see AQC500
ANTIMONY TRICHLORIDE, solid (DOT) see AQC500
ANTIMONY TRICHLORIDE, liquid (DOT) see AQC500
ANTIMONY TRICHLORIDE, solution (DOT) see AQC500
ANTIMONY TRICHLORIDE OXIDE see AQK250
ANTIMONY TRIETHYL see AQK500
ANTIMONY TRIFLUORIDE see AQE000
ANTIMONY TRIHYDRIDE see SLQ000
ANTIMONY TRIIODIDE see AQK750
ANTIMONY TRIMETHYL see AQL000
ANTIMONY TRIOXIDE (MAK) see AQF000
ANTIMONY TRIPHENYL see AQL250
ANTIMONY TRIPHENYLDICHLORIDE see DGO800
ANTIMONY TRISULFATE see AQJ250
ANTIMONY TRISULFIDE see AQL500
ANTIMONY TRITELLURIDE see AQL750
ANTIMONY WHITE see AQF000
ANTIMOONPENTACHLORIDE (DUTCH) see AQD000
ANTIMOONTRICHLRIDE see AQC500
ANTIMOSAN see AQH500
ANTIMUCIN WDR see ABU500
ANTIMYCIN see AQM000, CMS775
ANTIMYCIN A see AQM250
ANTIMYCIN A1 see DUO350
ANTIMYCIN A3 see BLX750
ANTIMYCIN A4 see AQM260
ANTIMYCOIN see AQM500
ANTINONIN see DUS700
ANTINOSIN see TDE750
ANTIO see DRR200
ANTI OX see MJO500
ANTIOXIDANT 1 see MJO500
ANTIOXIDANT 29 see BFW750
ANTIOXIDANT 116 see PFT500
ANTIOXIDANT 330 see TMJ000
ANTIOXIDANT 425 see MJN250
ANTIOXIDANT 736 see TFD000
ANTIOXIDANT 754 see IFX200
ANTIOXIDANT D see HLI500
ANTIOXIDANT DBPC see BFW750
ANTIOXIDANT MB (CZECH) see BCC500
ANTIOXIDANT No. 33 see DEG000
ANTIOXIDANT PBN see PFT500
ANTIOXIDANT TOD (CZECH) see BLB500
ANTIOXIDANT ZMB see ZIS500
ANTIPAR see DHF600, DII200
ANTI-PELLAGRA VITAMIN see NCQ900
ANTIPERZ see TII500
ANTIPHEN see MJM500
ANTI-PICA see FDA880, HAF400
ANTIPIRICULLIN see AQM250
ANTIPRESSINE DIHYDROCHLORIDE see DQB800
ANTIPREX 461 see ADW200
ANTIPYONIN see SFF000
ANTIPYRINE see AQN000
ANTIPYRINE SALICYLATE see AQN250

N-ANTIPYRINYL-2-(DIMETHYLAMINO)PROPIONAMIDE see AMF375
N-((ANTIPYRINYLISOPROPYLAMINO)METHYL)NICOTINAMIDE see AQN500
(ANTIPYRINYLMETHYLAMINO)METHANESULFONIC ACID SODIUM SALT
   see AMK500
ANTIRAD see AJY250
ANTIRADON see AJY250
ANTIREN see PIJ000
ANTIREX see EAE600
ANTI-RUST see SIQ500
ANTISACER see DKQ000, DNU000
ANTISAL 1a see TGK750
ANTISEPTOL see BEN000
ANTISERUM against the isozyme of LACTATE DEHYDROGENASE see LAO300
ANTISERUM to LUTEINIZING HORMONE see LIU350
ANTISERUM to SPERM SPECIFIC LACTATE DEHYDROGENASE see LAO300
ANTISOL 1 see PCF275
ANTISTERILITY VITAMIN see VSZ450
ANTISTINE see PDC000
ANTISTOMINUM see BBV500
ANTISTREPT see SNM500
ANTI-STRESS see EQL000
ANTITANIL see DME300
ANTI-TETANY SUBSTANCE 10 see DME300
ANTITROMBOSIN see BJZ000
ANTITUBERKULOSUM see ILD000
ANTIULCERA MASTER see CDQ500
ANTIVERM see PDP250
ANTIVITIUM see DXH250
ANTIXEROPHTHALMIC VITAMIN see VSK600
ANTLERMICIN A see TEF725
ANTOBAN see PIJ500
ANTOFIN see AFT500
ANTOMIN see BBV500
ANTORA see PBC250
ANTORPHINE see AFT500
ANTOSTAB see SCA750
ANTOXYLIC ACID see ARA250
ANTRACOL see ZMA000
ANTRAMYCIN see API000
ANTRANCINE 12 see BQI000
ANTRAPUROL see DMH400
ANTRENIL see ORQ000
ANTRENYL see ORQ000
ANTRENYL BROMIDE see ORQ000
ANTROMBIN K see WAT209
ANTRYCIDE see AQN625
ANTRYCIDE METHYL SULFATE see AQN625
ANTRYPOL see BAT000
ANTU see AQN635
ANTUITRIN S see CMG675
ANTURAT see AQN635
ANTURIO see APM875
ANTUSSAN see DBE200
ANTX-a see AOO120
ANTYMON (POLISH) see AQB750
ANTYMONOWODOR (POLISH) see SLQ000
ANTYWYLEGACZ see CMF400
ANU see PBX500
ANURAL see MQU750
ANUSPIRAMIN see BRF500
A. NUTTALLIANA see PAM780
ANVITOFF see AJV500
ANXIETIL see MQU750
ANXINE see GGS000
ANXIOLIT see CFZ000
ANZIEF see ZVJ000
AO 29 see BFW750
AO-40 see TMJ000
AO 4K see BFW750
AO 754 see IFX200
AOAA see ALQ650
A. OBLONGIFOLIA see BOO700
AOH see AGW476
AOH and AME (1:1) see AGW550
AOM see ASP250
AOMB see BCC500
A OO see AGX000
A. OPPOSITIFOLIA see BOO700
AORAL see VSK600
AOS see AFN500
AP see PEK250
4-AP see AMI500
AP-14 see PAM500
17-AP see PMG600

AP 43 see DIF200
AP-237 see BTA000
AP 407 see AOD250
A1-0109 P see AHE250
APACHLOR see CDS750
A. PACHYPODA see BAF325
APACIL see AMM250
APADODINE see DXX400
APADON see HIM000
APADRIN see MRH209
APAETP see AMD000
A. PALAESTINUM see ITD050
APAMIDE see HIM000
APAMIDON see FAB400
APAMIN see AQN650
APAMINE see AQN650, DBA800
APAMN see ABV750
APAP see HIM000
APARKAN see BBV000
APARKAZIN see DHF600
APARSIN see BBQ500
APAS see AMM250
APASCIL see MQU750
APATATE DRAPE see TES750
A. PATENS see PAM780
APAURIN see DCK759
APAVAP see DGP000
APAVINPHOS see MQR750
APAZONE see AQN750
APAZONE DIHYDRATE see ASA000
APC see ABG750, DBI800
APC (pharmaceutical) see ARP250
APCO 2330 see PEY000
APD see DBB500
'APE (HAWAII) see EAI600
APELAGRIN see NCQ900
APESAN see IPU000
APETAIN see BBK500
APEX 462-5 see TNC500
APEXOL see VSK600
APFO see ANP625
APGA see AMG750
APH see ACX750
APHAMITE see PAK000
APHENYLBARBIT see EOK000
APHOLATE see AQO000
APHOSAL see GFA000
APHOX see DOX600
APHOXIDE see TND250
APHRODINE see YBJ000
APHRODINE HYDROCHLORIDE see YBS000
APHROSOL see YBJ000
APHTIRIA see BBQ500
APIGENIN see CDH250
APIGENINE see CDH250
APIGENOL see CDH250
APIOL see AGE500
APIRACHOL see TGJ150
APIRACOHL see TGJ150
APIROLIO 1476 C see PAV600
APL see CMG675
APL (hormone) see CMG675
APLAKIL see CFZ000
APLIDAL see BBQ500
APL-LUSTER see TEX000
APM see XFS600
APN see AQO000
β-APN see AMB750
APO see TND250
APOATROPIN see AQO250
APOATROPINE see AQO250
APOCHOLIC ACID see AQO500
APOCID ORANGE 2G see HGC000
APOCODEINE see AQO750
APOCYNAMARIN see SMM500
APOCYNINE see HLQ500
APOIDINA see CMG675
APOLAN see ARQ750
APOLON B₆ see PII100
APOMINE BLACK GX see AQP000
APOMORFIN see AQP250
APOMORPHINE see AQP250
APONORIN see HII500
APOPEN see PDT500

APORMORPHINE see AQP250
APORMORPHINE CHLORIDE see AQP500
6A-β-APORMPHINE-10,11-DIOL HYDROCHLORIDE see AQP500
6A-β-APORMPHINE-10,11-DIOL see AQP250
APOSULFATRIM see TKX000
APOTHESINE see AQP750
3-α,16-α-APOVINCAMINIC ACID ETHYL ESTER see EGM100
A-POXIDE see MDQ250
APOZEPAM see DCK759
4-APP see POM600
APPA see PHX250
APPALACHIAN TEA see HGF100
APPEX see RAF100
APPLE SEEDS see AQP875
APPLE of SODOM (extract) see AQP800
APPL-SET see NAK500
APPRESINUM see HGP500
APPRESSIN see HGP495
APRELAZINE see HGP500
APRESAZIDE see HGP500
APRESINE see HGP500
APRESOLIN see HGP495, HGP500
APRESOLINE-ESIDRIX see HGP500
APRESOLINE HYDROCHLORIDE see HGP500
APREZOLIN see HGP495, HGP500
APRICOT PITS see AQP890
APRIDOL see EQL000
APRIL FOOLS see PAM780
APRINDINE HYDROCHLORIDE see FBP850
APRINOX see BEQ625
APROBARBITAL see AFT000
APROBARBITAL SODIUM see BOQ750
APROBARBITONE see AFT000
APROBARBITONE SODIUM see BOQ750
APROBIT see DQA400
APROCARB see PMY300
APROL 160 see MND100
APROL 161 see MND050
APRONAL see IQX000
APRONALIDE see IQX000
APROTININ see PAF550
APROZAL see AFT000
APSIN VK see PDT750
APTAL see CFE250
APTIN see AGW000
APTROL SULFATE see AQQ000
A. PULSATILLA see PAM780
APURIN see ZVJ000
APURINA see DWW000
APUROL see ZVJ000
APV see CBR000
APYONINE AURAMINE BASE see IBB000
A-200 PYRINATE see AAB250
AQ 110 see TKX125
AQD see TMJ800
AQUA AMMONIA see ANK250
AQUACAT see CNA250
AQUA CERA see HKJ000
AQUACHLORAL see CDO000
AQUACIDE see DWX800
AQUACRINE see EDV000
AQUA-1,2-DIAMINOETHANE DIPEROXO CHROMIUM(IV) see AQQ100
AQUA-1,2-DIAMINOPROPANEDIPEROXOCHROMIUM(IV) DIHYDRATE
    see AQQ125
AQUAFIL see SCH000
AQUA FORTIS see NED500
AQUAKAY see MMD500
AQUA-KLEEN see DAA800
AQUALINE see ADR000
AQUA MEPHYTON see VTA000
AQUAMOLLIN see EIV000
AQUAMYCETIN see CDP250
AQUAMYCIN see ACJ250
AQUAPEL (POLYSACCHARIDE) see SLJ500
AQUAPHOR see CLF325
AQUAPLAST see SFO500
AQUA REGIA see HHM000
AQUAREX METHYL see SIB600
AQUARILLS see CFY000
AQUARIUS see CFY000
AQUASOL see VSP000
AQUASYNTH see VSK600
AQUATAG see BDE250
AQUATENSEN see MIV500

AQUATHOL see EAR000
AQUATIN see CLU000
AQUA-VEX see TIX500
AQUAVIRON see TBG000
AQUAZINE see BJP000
AQUEOUS AMMONIA see ANK250
AQUINONE see MMD500
AR-11 see HNP500
AR-12 see MID000
AR-13 see HNR000
AR-16 see TGH665
AR-17 see HNP000
AR-19 see AGB000
AR-21 see HNQ500
AR-22 see BED250
AR-23 see BED500
AR-25 see HNS500
AR-32 see DQY909
AR-33 see HNQ000
AR-35 see HNS000
AR-41 see DQX800
AR-45 see AQQ250
AR 12008 see DIO200
ARA-A see AEH100
ARA-ATP see ARQ500
ARABIC GUM see AQQ500
ARABINOCYTIDINE see AQQ750
9-β-d-ARABINO FURANOSYL ADENINE see AQQ900
9-(β-d-ARABINOFURANOSYL)ADENINE-5-(DIHYDROGEN PHOSPHATE)
    see AQQ905
9-β-d-ARABINOFURANOSYLADENINEMONOHYDRATE see AEH100
9-β-d-ARABINOFURANOSYLADENINE 5'-TRIPHOSPHATE see ARQ500
1-β-d-ARABINOFURANOSYL-4-AMINO-2(1H)PYRIMIDINONE see AQQ750
1-β-d-ARABINOFURANOSYL-2,2'-ANHYDRO-CYTOSINE HYDROCHLORIDE
    see COW900
1-ARABINOFURANOSYLCYTOSINE see AQQ750
1-β-ARABINOFURANOSYLCYTOSINE see AQQ750
1-(β-d-ARABINOFURANOSYL)CYTOSINE see AQQ750
1-β-d-ARABINOFURANOSYLCYTOSINE HYDROCHLORIDE see AQR000
1-β-d-ARABINOFURANOSYLCYTOSINE-5'-PALMITATE see AQS875
1-β-d-ARABINOFURANOSYLCYTOSINE-5'-PALMITOYL ESTER see AQS875
N-(1-β-d-ARABINOFURANOSYL-1,2-DIHYDRO-2-OXO-4-
    PYRIMIDINYL)DOCOSANAMIDE see EAU075
1-β-d-ARABINOFURANOSYL-5-FLUOROCYTOSINE see AQR250
9-β-d-ARABINOFURANOSYL-9H-PURINE-6-AMINE MONOHYDRATE
    see AEH100
1-β-d-ARABINOFURANOSYL-2',3',5'-TRIACETATE see AQR500
ARABINOSYLADENINE see AQQ900
9-ARABINOSYLADENINE see AQQ900
β-d-ARABINOSYLADENINE see AQQ900
ARABINOSYLADENINE MONOPHOSPHATE see AQQ905
5'-ARABINOSYLADENINE MONOPHOSPHATE see AQQ905
β-d-ARABINOSYLCYTOSINE see AQQ750
ARABINOSYLCYTOSINE HYDROCHLORIDE see AQR000
ARABINOSYLCYTOSINE PALMITATE see AQS875
ARAB RAT DETH see WAT200
ARACHIC ACID see EAF000
ARACHIDIC ACID see EAF000
ARACHIDONIC ACID see AQS750
ARACHIS OIL see PAO000
ARACID see CJR500
ARACIDE see SOP500
ARA-CP see AQS875
ARA-C PALMITATE see AQS875
ARACYTIDINE-5'-PALMITATE see AQS875
ARAGONITE see CAO000
ARALDITE ACCELERATOR 062 see DQP800
ARALDITE ERE 1359 see REF000
ARALDITE 6010 mixed with ERR 4205 (1:1) see OPI200
ARALDITE HARDENER HY 951 see TJR000
ARALDITE HY 951 see TJR000
ARALEN see CLD000
ARALEN DIPHOSPHATE see CLD250
ARALEN PHOSPHATE see CLD250
ARALO see PAK000
ARAMINE see HNB875, HNC000
ARAMITE see SOP500
ARAMITEARARAMITE-15W see SOP500
ARANCIO CROMO (ITALIAN) see LCS000
ARASAN see TFS350
ARATAN see MEP250
ARATEN PHOSPHATE see AQT250
ARATHANE see AQT500
ARATRON see SOP500

ARBAPROSTIL see AQT575
ARBITEX see BBQ500
ARBOGAL see DSQ000
ARBOL DEL QUITASOL (CUBA) see CDM325
ARBOL de PERU (MEXICO) see PCB300
ARBORICID see BSQ750
ARBOROL see DUS700
ARBOTECT see TEX000
ARBRE FRICASSE (HAITI) see ADG400
ARBUTIN see HIH100
ARBUZ see PAG500
ARCACIL see PDT750
ARCADINE see BCA000
ARCASIN see PDT750
ARCHIDONATE see AQS750
ARCHIDYN see RKP000
ARCOBAN see MQU750
ARCOMONOL TABLETS see MQY400
ARCOSOLV see DWT200
ARCOTRATE see PBC250
ARCTON see CBY750
ARCTON 3 see CLR250
ARCTON 4 see CFX500
ARCTON 6 see DFA600
ARCTON 7 see DFL000
ARCTON 9 see TIP500
ARCTON 33 see FOO509
ARCTON 63 see FOO000
ARCTON 114 see FOO509
ARCTON O see CBY250
ARCTUVIN see HIH000
ARDALL see SJO000
ARDAP see RLF350
ARDEX see BBK500
ARDUAN see PII250
ARECA CATECHU see BFW000
ARECA CATECHU Linn., fruit extract see BFW000
ARECA CATECHU Linn., nut extract see BFW000
ARECAIDINE see AQT625
ARECAIDINE METHYL ESTER see AQT750
ARECAINE see AQT625
ARECA NUT see AQT650
ARECHIN see CLD250
ARECOLINE see AQT750
ARECOLINE BASE see AQT750
ARECOLINE BROMIDE see AQU000
ARECOLINE HYDROBROMIDE see AQU000
ARECOLINE HYDROCHLORIDE see AQU250
AREDION see CKM000
AREGINAL see EKL000
ARELIZ see PDW250
ARESIN see CKD500
ARESKAP 100 see AQU500
ARESKET 300 see AQU750
ARESKLENE 400 see AQV000
ARESOL see RLU000
ARETAN-NIEUW see BKS810
ARETIT see ACE500, BRE500
AREZIN see CKD500
AREZINE see CKD500
ARFICIN see RKP000
ARFONAD see TKW500
ARFONAD CAMPHORSULFONATE see TKW500
ARFONAD ROCHE see TKW500
ARGAMINE see AQW000
ARGEMONE OIL mixed with MUSTARD OIL see OGS000
ARGENT FLUORURE (FRENCH) see SDQ500
ARGENTIC FLUORIDE see SDQ500
ARGENTIUM CREDE see SDI750
ARGENTOUS OXIDE see SDU500
ARGENTUM see SDI500
ARGEZIN see ARQ725
l-ARGININE, nitrosated see NJU500
ARGININE HYDROCHLORIDE see AQW000
l-ARGININE HYDROCHLORIDE see AQW000
ARGININE MONOHYDROCHLORIDE see AQW000
l-ARGININE MONOHYDROCHLORIDE see AQW000
8-l-ARGININEOXYTOCIN see AQW125
ARGININE-VASOTOCIN see AQW125
8-ARGININE VASOTOCIN see AQW125
ARGIPRESTOCIN see AQW125
ARGIVENE see AQW000
ARGO BRAND CORN STARCH see SLJ500
ARGON see AQW250

ARGONAL see PFM250
(ARG^8)OXYTOCIN see AQW125
ARGUN see AGN000
ARG-VASOTOCIN see AQW125
8-ARG-VASOTOCIN see AQW125
ARIBINE see MPA050
ARICHIN see CFU750
ARILATE see BAV575
ARINAMINE see DXO300
ARIOTOX see TDW500
ARISAEMA (VARIOUS SPECIES) see JAJ000
ARISAN see CKF750
ARISTAMID see SNJ350
ARISTAMIDE see SNJ350
ARISTOCORT see AQX250
ARISTOCORT ACETONIDE see AQX500
ARISTOCORT DIACETATE see AQX750
ARISTOCORT FORTE PARENTERAL see AQX750
ARISTOCORT SYRUP see AQX750
ARISTODERM see AQX500
ARISTOGEL see AQX500
ARISTOGYN see SNJ350
ARISTOLIC ACID see AQX825
ARISTOLIC ACID METHYL ESTER see MGQ525
ARISTOLICHIA INDICA L., ALCOHOLIC EXTRACT see AQY000
ARISTOLOCHIC ACID see AQY250
ARISTOLOCHIC ACID SODIUM SALT see AQY125
ARISTOLOCHINE see AQY250
ARISTOPHYLLIN see DNC000
ARISTOSPAN see AQY375
ARIZOLE see PMQ750
ARKITROPIN see MDL000
ARKLONE P see FOO000
ARKOFIX NG see DTG000
ARKOPAL N-090 see PKF000
ARKOTINE see DAD200
ARKOZAL see BSQ000
ARLACEL 60 see SKV150
ARLACEL 80 see SKV100
ARLACEL 161 see OAV000
ARLACEL 169 see OAV000
ARLANTHRENE GOLDEN YELLOW see DCZ000
ARLEF see TKH750
ARLIDIN HYDROCHLORIDE see DNU200
ARLOSOL GREEN B see BLK000
ARMAZAL see COH250
ARMCO IRON see IGK800
ARMEEN L-7 see TNX750
ARMEEN 16D see HCO500
ARMEEN DM-12D see DRR800
ARMEEN O see OHM700
ARMENIAN BOLE see IHD000
ARMINE see ENQ000
ARMODOUR see PKQ059
ARMOFOS see SKN000
ARMOSTAT 801 see OAV000
ARMOTAN MO see SKV100
ARMOTAN MS see SKV150
ARMOTAN PMO-20 see PKL100
ARMSTRONG'S ACID see AQY400
ARMSTRONG'S S ACID see AQY400
ARMYL see MRV250
ARNAUDON'S GREEN see CMK300
ARNAUDON'S GREEN (HEMIHEPTAHYDRATE) see CMK300
ARNICA see AQY500
ARNITE A see PKF750
ARNOSULFAN see SNN300
AROALL see SJO000
AROCHLOR 1221 see PJM000
AROCHLOR 1242 see PJM500
AROCHLOR 1254 see PJN000
AROCHLOR 1260 see PJN250
AROCLOR see PJL750
AROCLOR 54 see CLD250
AROCLOR 1016 see PJL800
AROCLOR 1232 see PJM250
AROCLOR 1242 see PJM500
AROCLOR 1248 see PJM750
AROCLOR 1254 see PJN000
AROCLOR 1260 see PJN250
AROCLOR 1262 see PJN500
AROCLOR 1268 see PJN750
AROCLOR 2565 see PJO000
AROCLOR 4465 see PJO250

AROCLOR 5442 see PJP750
AROFT see AEW625
AROFUTO see AEW625
AROMA BLANCA (CUBA, HAWAII) see LED500
AROMATIC AMINES see AQY750
AROMATIC CASTOR OIL see CCP250
AROMATIC SOLVENT see NAI500
AROMATIC SPIRITS of AMMONIA see AQZ000
AROMATOL see AQZ100
AROMOX DMMC-W see DRS200
ARON COMPOUND HW see PKQ059
AROSOL see PER000
AROTINOIC ACID see AQZ200
AROTINOIC METHANOL see AQZ300
AROTINOID ETHYL ESTER see AQZ400
AROVIT see VSP000
ARPEZINE see PIJ500
ARPHONAD see TKW500
ARQUAD DM14B-90 see TCA500
ARQUAD DM18B-90 see DTC600
ARQUAD DMMCB-75 see AFP250
ARQUEL see DGM875
ARRESIN see CKD500
ARRESTEN see MQQ050
ARRET see CMG000
ARRHENAL see DXE600
ARROWROOT STARCH see SLJ500
ARROW WOOD see MBU825
ARSACETIN see AQZ900
ARSACETIN SODIUM SALT see AQZ900
ARSACETIN SODIUM SALT, TETRAHYDRATE see ARA000
ARSACETIN TETRAHYDRATE see ARA000
ARSAMBIDE see CBJ000
ARSAMINOL see SAP500
ARSAN see HKC000
ARSANILIC ACID see ARA250
4-ARSANILIC ACID see ARA250
p-ARSANILIC ACID see ARA250
ARSANILIC ACID, N-ACETYL-, SODIUM SALT see AQZ900
p-ARSANILIC ACID, N,N-BIS(2-CHLOROETHYL)- see BIA300
p-ARSANILIC ACID, N,N-BIS(2-HYDROXYETHYL)- see BKD600
p-ARSANILIC ACID, BISMUTH, SODIUM SALT see BKX500
p-ARSANILIC ACID, N,N-DIETHYL- see DIS775
ARSANILIC ACID, MONOSODIUM SALT see ARA500
ARSANILIC ACID SODIUM SALT see ARA500
ARSAPHENAN see ACN250
ARSECLOR see DFX400
ARSECODILE see HKC500
ARSENAMIDE see TFA350
ARSENATE see ARB250
ARSENATE of IRON, FERRIC see IGN000
ARSENATE of IRON, FERROUS see IGM000
ARSENATE of LEAD see LCK000
ARSENENOUS ACID, POTASSIUM SALT see PKV500
ARSENENOUS ACID, SODIUM SALT (9CI) see SEY500
ARSENIATE de CALCIUM (FRENCH) see ARB750
ARSENIATE de MAGNESIUM (FRENCH) see ARD000
ARSENIATE de PLOMB (FRENCH) see ARC750
ARSENIC see ARA750
ARSENIC-75 see ARA750
ARSENIC, metallic (DOT) see ARA750
ARSENIC, white, solid (DOT) see ARI750
ARSENIC ACID see ARH500
m-ARSENIC ACID see ARB000
o-ARSENIC ACID see ARB250
ARSENIC ACID, solid (DOT) see ARB250, ARC500
ARSENIC ACID, liquid (DOT) see ARB250
ARSENIC ACID ANHYDRIDE see ARH500
ARSENIC ACID, CALCIUM SALT (2:3) see ARB750
ARSENIC ACID, DISODIUM SALT see ARC000
ARSENIC ACID, DISODIUM SALT, HEPTAHYDRATE see ARC250
ARSENIC ACID, HEMIHYDRATE see ARC500
ARSENIC ACID, LEAD SALT see ARC750
ARSENIC ACID, MAGNESIUM SALT see ARD000
ARSENIC ACID, METHYLPHENYL-(9CI) see HMK200
ARSENIC ACID, MONOPOTASSIUM SALT see ARD250
ARSENIC ACID, MONOSODIUM SALT see ARD500, ARD600
ARSENIC ACID, SODIUM SALT see ARD750
ARSENIC ACID, SODIUM SALT (9CI) see ARD500
ARSENIC ACID, TRICESIUM SALT see CDC375
ARSENIC(V) ACID, TRISODIUM SALT, HEPTAHYDRATE (1:3:7) see ARE000
ARSENIC ACID, ZINC SALT see ZDJ000
ARSENICAL solution see FOM050

ARSENICAL DIP see ARE250
ARSENICAL DIP, liquid (DOT) see ARE250
ARSENICAL DUST see ARE500
ARSENICAL FLUE DUST see ARE500
ARSENICAL FLUE DUST (DOT) see ARE750
ARSENICALS see ARA750, ARF750
ARSENIC ANHYDRIDE see ARH500
ARSENIC BISULFIDE see ARF000
ARSENIC BLACK see ARA750
ARSENIC BLANC (FRENCH) see ARI750
ARSENIC(III) BROMIDE see ARF250
ARSENIC BUTTER see ARF500
ARSENIC CHLORIDE see ARF500
ARSENIC(III) CHLORIDE see ARF500
ARSENIC COMPOUNDS see ARF750
ARSENIC DICHLOROETHANE see DFH200
ARSENIC DIETHYL see ARG000
ARSENIC DIMETHYL see ARG250
ARSENIC FLUORIDE see ARI250
ARSENIC HEMISELENIDE see ARG500
ARSENIC HYDRIDE see ARK250
ARSENIC IODIDE see ARG750
ARSENIC OXIDE see ARH500, ARI750
ARSENIC(III) OXIDE see ARI750
ARSENIC(V) OXIDE see ARH500
ARSENIC PENTASULFIDE see ARH250
ARSENIC PENTOXIDE see ARH500
ARSENIC PHOSPHIDE see ARH750
ARSENIC SESQUIOXIDE see ARI750
ARSENIC SESQUISULFIDE see ARI000
ARSENIC SULFIDE see ARI000
ARSENIC SULFIDE YELLOW see ARI000
ARSENIC SULPHIDE see ARI000
ARSENIC TRIBROMIDE see ARF250
ARSENIC TRIFLUORIDE see ARI250
ARSENIC TRIHYDRIDE see ARK250
ARSENIC TRIIODIDE see ARG750
ARSENIC TRIIODIDE mixed with MERCURIC IODIDE see ARI500
ARSENIC TRIOXIDE see ARI750
ARSENIC TRIOXIDE mixed with SELENIUM DIOXIDE (1:1) see ARJ000
ARSENIC TRISULFIDE see ARI000
ARSENIC YELLOW see ARI000
ARSENIDES see ARJ250
ARSENIGEN SAURE (GERMAN) see ARI750
ARSENIOUS ACID (MAK) see ARI750
ARSENIOUS ACID, CALCIUM SALT see CAM500
ARSENIOUS ACID, SODIUM SALT see ARJ500, SEY500
ARSENIOUS ACID, SODIUM SALT POLYMERS see ARJ500
ARSENIOUS ACID, STRONTIUM SALT see SME500
ARSENIOUS ACID, TRISILVER(1+) SALT see SDM100
ARSENIOUS ACID, ZINC SALT see ZDS000
ARSENIOUS CHLORIDE see ARF500
ARSENIOUS and MERCURIC IODIDE, solution (DOT) see ARI500
ARSENIOUS OXIDE see ARI750
ARSENIOUS SULPHIDE see ARI000
ARSENIOUS TRIOXIDE see ARI750
ARSENITE de POTASSIUM (FRENCH) see PKV500
ARSENITE de SODIUM (FRENCH) see SEY500
ARSENIURETTED HYDROGEN see ARK250
ARSENO 39 see ARL000
ARSENO-BISMULAK see BKX500
ARSENOMARCASITE see ARJ750
1-(p-ARSENOPHENYL)UREA see CBI500
ARSENOPYRITE see ARJ750
ARSENOSAN see ARL000
p-ARSENOSOANILINE see ARJ755
4-ARSENOSOANILINE, DIHYDRATE see ALV100
ARSENOSOBENZENE see PEG750
p-ARSENOSO-N,N-BIS(2-CHLOROETHYL)ANILINE see ARJ760
p-ARSENOSO-N,N-BIS(2-HYDROXYETHYL)ANILINE see ARJ770
p-ARSENOSO-N,N-DIETHYLANILINE see ARJ800
4-ARSENOSO-2-NITROPHENOL see NHE600
p-ARSENOSOPHENOL see HNG800
p-ARSENOSOTOLUENE see TGV100
ARSENOUS ACID see ARI750
ARSENOUS ACID ANHYDRIDE see ARI750
ARSENOUS ACID, TRISILVER(1+) SALT (9CI) see SDM100
ARSENOUS ANHYDRIDE see ARI750
ARSENOUS BROMIDE see ARF250
ARSENOUS CHLORIDE see ARF500
ARSENOUS FLUORIDE see ARI250
ARSENOUS HYDRIDE see ARK250
ARSENOUS IODIDE see ARG750
ARSENOUS OXIDE see ARI750

ARSENOUS OXIDE ANHYDRIDE see ARI750
ARSENOUS SULFIDE see ARI000
ARSENOUS TRIBROMIDE see ARF250
ARSENOUS TRICHLORIDE (9CI) see ARF500
ARSENOUS TRIIODIDE (9CI) see ARG750
ARSENOWODOR (POLISH) see ARK250
ARSENOXIDE see ARL000
ARSENOXIDE SODIUM see ARJ900
ARSENPHENOLAMINE HYDROCHLORIDE see SAP500
ARSEN (GERMAN, POLISH) see ARA750
ARSENWASSERSTOFF (GERMAN) see ARK250
ARSEVAN see NCJ500
ARSINE see ARK250
ARSINE, (p-AMINOPHENYL)DICHLORO-, HYDROCHLORIDE see AOR640
ARSINE, (p-AMINOPHENYL)OXO-, DIHYDRATE see ALV100
ARSINE, AMYLDICHLORO- see AOI200
ARSINE BORON TRIBROMIDE see ARK500
ARSINE, sec-BUTYLDICHLORO- see BQY300
ARSINE, CHLORO(2-CHLOROVINYL)PHENYL- see PER600
ARSINE, (2-CHLOROETHYL)DICHLORO- see CGV275
ARSINE, DICHLOROHEPTYL- see HBN600
ARSINE, DICHLOROHEXYL- see HFP600
ARSINE, DICHLOROISOPENTYL- see IHQ100
ARSINE, DICHLOROPENTYL- see AOI200
ARSINE, DICHLOROPROPYL- see PNH650
ARSINE, DIIODOMETHYL- see MGQ775
ARSINE, DIPHENYLHYDROXY- see DVY100
ARSINE, ETHYLENEBIS(DIPHENYL- see EIQ200
ARSINE, HYDROXYDIPHENYL- see DVY100
ARSINE OXIDE, ALLYLHYDROXYPHENYL- see AGQ775
ARSINE OXIDE, BUTYLHYDROXYISOPROPYL- see BRQ800
ARSINE OXIDE, (o-CHLOROPHENYL)(3-(2,4-DICHLOROPHENOXY)-2-
   HYDROXYPROPYL)HYDROXY- see DGA425
ARSINE OXIDE, (m-CHLOROPHENYL)HYDROXY(β-HYDROXYPHENETHYL)-
   see CKA575
ARSINE OXIDE, (p-CHLOROPHENYL)HYDROXYMETHYL- see CKD800
ARSINE OXIDE, DIBUTYLHYDROXY- see DDV250
ARSINE OXIDE, DIETHYLHYDROXY- see DIS850
ARSINE OXIDE, HYDROXYDIMETHYL-, SODIUM SALT, TRIHYDRATE
   see HKC550
ARSINE OXIDE, HYDROXY(2-HYDROXYPROPYL)PHENYL- see HNX600
ARSINE OXIDE, HYDROXYISOBUTYLISOPROPYL- see IPS100
ARSINE OXIDE, HYDROXYMETHYLPHENETHYL- see MNU050
ARSINE OXIDE, HYDROXYMETHYLPHENYL- see HMK200
ARSINE OXIDE, HYDROXYMETHYLPROPYL- see MOT800
ARSINE, OXO(4-CARBOXY)PHENYL- see CCI550
ARSINE SELENIDE, TRIMETHYL- see TLH250
ARSINE SULFIDE, DIMETHYLDI- see DQG700
ARSINE-TRI-1-PIPERIDINIUM CHLORIDE see ARK750
ARSINETTE see LCK000
ARSINIC ACID, DIBUTYL-(9CI) see DDV250
ARSINOTRIS PIPERIDINIUM TRICHLORIDE see ARK750
ARSINYL see DXE600
ARSION see CEI250
ARSONATE liquid see MRL750
ARSONIC ACID see ABX500
ARSONIC ACID, (4-(ACETYLAMINO)PHENYL)-, MONOSODIUM SALT (9CI)
   see AQZ900
ARSONIC ACID, CALCIUM SALT (1:1) see ARK780
ARSONIC ACID, COPPER(2+) SALT (1:1) (9CI) see CNN500
ARSONIC ACID, (4-HYDROXYPHENYL)-, polymer with FORMALDEHYDE
   see BCJ150
ARSONIC ACID, METHYL-, IRON SALT (9CI) see IHB680
ARSONIC ACID, POTASSIUM SALT see PKV500
ARSONIC ACID, SODIUM SALT (9CI) see ARJ500
ARSONIUM, (3-HYDROXYPHENYL)DIETHYLMETHYL-, IODIDE,
   METHYLCARBAMATE see MID900
ARSONIUM, TETRAPHENYL-, CHLORIDE see TEA300
((p-ARSONOPHENYL)CARBAMOYL)DITHIOCARBAMIC ACID see DXM100
4-ARSONOPHENYLGLYCINAMIDE see CBJ750
p-ARSONOPHENYLUREA see CBJ000
(3-(p-ARSONOPHENYL)UREIDO)DITHIOBENZOIC ACID see ARK800
ARSONOUS DICHLORIDE, (1-METHYLPROPYL)-(9CI) see BQY300
ARSONOUS DIIODIDE, METHYL-(9CI) see MGQ775
N-((p-ARSONPHENYL)CARBAMOYL)DITHIOGLYCINE see DXM100
ARSPHEN see ABX500
ARSPHENAMINE see SAP500
ARSPHENAMINE METHYLENESULFOXYLIC ACID SODIUM SALT
   see NCJ500
ARSPHENOXIDE see ARL000
ARSYCODILE see HKC500
ARSYNAL see DXE600
ARTAM see PGG000
ARTANE see BBV000

ARTANE HYDROCHLORIDE see BBV000
ARTANE TRIHEXYPHENIDYL see BBV000
ARTEANNUIN see ARL375
ARTEMETHER see ARL425
ARTEMISIA OIL see ARL250
ARTEMISIA OIL (WORMWOOD) see ARL250
ARTEMISINE see ARL375
ARTEMISININ see ARL375
ARTEMISININELACTOL METHYL ETHER see ARL425
ARTERENOL see NNO500
d-ARTERENOL see ARL500
l-ARTERENOL see NNO500
dl-ARTERENOL see ARL750
l-ARTERENOL BITARTRATE see NNO699
dl-ARTERENOL HYDROCHLORIDE see NNP050
ARTERIOFLEXIN see ARQ750
ARTERIOVINCA see VLF000
ARTEROCOLINE see ABO000, CMF250
ARTEROCYN see PLV750
ARTERODY see BBJ750
ARTEROSOL see ARQ750
ARTES see ARQ750
ARTEVIL see ARQ750
d-ARTHIN see VSZ100
ARTHODIBROM see NAG400
ARTHO LM see MLH000
ARTHROBID see SOU550
ARTHROCHIN see CLD000
ARTHROCINE see SOU550
ARTHROPAN see CMG000
ARTHRYTIN OXOATE see AMX250
ARTIC see MIF765
ARTIFICIAL ALMOND OIL see BAY500
ARTIFICIAL ANT OIL see FPQ875
ARTIFICIAL BARITE see BAP000
ARTIFICIAL CINNAMON OIL see CCO750
ARTIFICIAL GUM see DBD800
ARTIFICIAL HEAVY SPAR see BAP000
ARTIFICIAL MUSTARD OIL see AGJ250
ARTIFICIAL SWEETENING SUBSTANZ GENDORF 450 see SJN700
ARTISIL BLUE BSG see MGG250
ARTISIL BLUE SAP see TBG700
ARTISIL ORANGE 3RP see AKP750
ARTISIL VIOLET 2RP see DBP000
ARTIZIN see BRF500
ARTO-ESPASMOL see DNU100
ARTOLON see MQU750
ARTOMEY see VCK100
ARTOMYCIN see TBX250
ARTONIL see BBW750
ARTOSIN see BSQ000
ARTOZIN see BSQ000
ARTRACIN see IDA000
ARTRIBID see SOU550
ARTRIL 300 see IIU000
ARTRINOVO see IDA000
ARTRIONA see CNS825
ARTRIVIA see IDA000
ARTRIZONE see BRF500
ARTROBIONE see CMG000
ARTROFLOG see HNI500
ARTROPAN see BRF500
A. RUBRA see BAF325
ARUM (Various Species) see ITD050
ARUMEL see FMM000
ARUSAL see IPU000
ARVIN see VGU700
ARVYNOL see CHG000
ARWIN see VGU700
ARWOOD COPPER see CNI000
ARYL ALKYL POLYETHER ALCOHOL see ARL875
ARZENE see PEG750
AS see AFJ375
AS-15 see SLY500
AS-17665 see NDY500
AS (surfactant) see AFJ375
ASA see ADA725, ADA725
ASA 158-5 see PJA130
ASABAINE see XCJ000
ASABAINE see DJM800
ASA COMPOUND see ABG750
A.S.A. EMPIRIN see ADA725
ASAGRAEA OFFICINALIS see VHZ000
ASAGRAN see ADA725

ASAHISOL 1527 see AAX250
ASALIN see ARM000
ASALINE see ARM000
ASAMEDOL see DHS200
ASAMID see ENG500
ASARON see IHX400
ASARONE see IHX400
α-ASARONE see IHX400
ASARONE, trans- see IHX400
trans-ASARONE see IHX400
ASARUM CAMPHOR see IHX400
ASATARD see ADA725
ASAZOL see MRL750
ASB 516 see AAX250
ASBEST (GERMAN) see ARM250
ASBESTINE see TAB750
ASBESTOS see ARM250
7-45 ASBESTOS see ARM268
ASBESTOS (ACGIH) see ARM260, ARM262, ARM264, ARM268,
    ARM275, ARM280
ASBESTOS, ACTINOLITE see ARM260
ASBESTOS, AMOSITE see ARM262
ASBESTOS, ANTHOPHYLITE see ARM264
ASBESTOS, ANTHOPHYLLITE see ARM266
ASBESTOS, CHRYSOTILE see ARM268
ASBESTOS, CROCIDOLITE see ARM275
ASBESTOS FIBER see ARM250
ASBESTOS, TREMOLITE see ARM280
ASBESTOS, WHITE (DOT) see ARM268
ASCABIN see BCM000
ASCABIOL see BCM000
ASCAREX SYRUP see PIJ500
ASCARIDOL see ARM500
ASCARIDOLE see ARM500
ASCARISIN see ARM500
ASCARYL see HFV500
ASCENSIL see ALC250
ASCEPTICHROME see MCV000
AS 61CL see ADY500
ASCLEPIN see ACF000
ASCOFURANONE see ARM750
ASCOPHEN see ARP250
ASCORBIC ACID see ARN000
l-ASCORBIC ACID see ARN000
l(+)-ASCORBIC ACID see ARN000
ASCORBIC ACID SODIUM SALT see ARN125
l-ASCORBIC ACID SODIUM SALT see ARN125
ASCORBICIN see ARN125
ASCORBIN see ARN125
ASCORBUTINA see ARN000
ASCORPHYLLINE see HLC000
ASCOSIN see ARN250
ASCUMAR see ABF750
ASCURON see BJI000
2-ASe see AMN500
ASEBOTOXIN see AOO375
ASECRYL see GIC000
ASENDIN see AOA095
ASEPTA HERBAN see HDP500
ASEPTICHROME see MCV000
ASEPTOFORM see HJL500
ASEPTOFORM E see HJL000
ASEPTOFORM P see HNU500
ASEX see SFS000
ASHES (residues) see CMY650
ASHLENE see NOH000
ASIATICOSIDE see ARN500
ASIDON 3 see TEH500
ASILAN see SNQ500
ASIPRENOL see DMV600
ASL-603 see BMV750
AS-LH see LIU350
ASMADION see TEH500
ASMALAR see DMV600
ASMATANE MIST see AES000, VGP000
ASMAVAL see TEH500
ASM MB see BCC500
ASOZIN see MGQ750
ASP 51 see TED500
ASPALON see ADA725
ASPAMINOL HYDROCHLORIDE see ARN700
ASPARA see MAI600
ASPARAGINASE see ARN800
l-ASPARAGINASE see ARN800

l-ASPARAGINASE X see ARN800
l-ASPARAGINASI (ITALIAN) see ARN800
ASPARAGINATE CALCIUM see CAM675
l-ASPARAGINE AMIDOHYDROLASE see ARN800
ASPARA K see PKV600
ASPARTAME see ARN825
ASPARTAT see MAI750
ASPARTIC ACID DISODIUM SALT see SEZ350
l-ASPARTIC ACID, DISODIUM SALT (9CI) see SEZ350
l-ASPARTIC ACID, POTASSIUM SALT (9CI) see PKV600
l-ASPARTIC ACID, SODIUM SALT (9CI) see SEZ355
ASPARTYLPHENYLALANINE METHYL ESTER see ARN825
N-l-α-ASPARTYL-l-PHENYLALANINE 1-METHYL ESTER (9CI) see ARN825
ASPERASE see ARN875
ASPERFILLUS ALKALINE PROTEINASE see SBI860
ASPERGILLIC ACID see ARO000
ASPERGILLIN see ARO250
ASPERGILLOPEPTIDASE B see SBI860
ASPERGUM see ADA725
ASPHALT see ARO500
ASPHALT (CUT BACK) see ARO750
ASPHALT, PETROLEUM see PCR500
ASPHALTUM see ARO500
A. SPICATA see BAF325
ASPICULAMYCIN see ARP000
ASPIRDROPS see ADA725
ASPIRIN see ADA725
ASPIRIN ACETAMINOPHEN ESTER see SAN600
ASPIRINE see ADA725
ASPIRIN-ISOPROPYLANTIPYRINE see PNR800
ASPIRIN-dl-LYSINE see ARP125
ASPIRIN-NATRIUM (GERMAN) see ADA750
ASPIRIN, PHENACETIN and CAFFEINE see ARP250
ASPON see TED500
ASPON-CHLORDANE see CDR750
ASPOR see EIR000
ASPORUM see EIR000
ASPRO see ADA725
ASPRON see HBT500
ASSAM TEA see ARP500
ASSIFLAVINE see DBX400
ASSIMIL see MJE760
ASSIPRENOL see DMV600
ASSIUM PENTACARBONYL VANADATE(3⁻) see PLO100
ASSUGRIN see SGC000
ASSUGRIN FEINUSS see SGC000
ASSUGRIN VOLLSUSS see SGC000
ASTA see CQC650
ASTA 5122 see SOD000
ASTA B518 see CQC650
ASTA C see CNR750
ASTA CD 072 see CNR825
ASTARIL see SLP600
ASTA Z 4828 see TNT500
ASTA Z 4942 see IMH000
ASTA Z 7557 see ARP625
ASTEMIZOL (GERMAN) see ARP675
ASTEMIZOLE see ARP675
ASTERIC see ADA725
ASTEROMYCIN see ARP000
ASTHENTHILO see DKL800
ASTHMA METER MIST see VGP000
ASTHMA WEED see CCJ825
ASTMAHALIN see VGP000
ASTMAMASIT see DNC000
A-STOFF see CDN200
ASTOMIN see MLP250
ASTRA CHRYSOIDINE R see PEK000
ASTRACILLIN see PDD350
ASTRA DIAMOND GREEN GX see BAY750
ASTRAFER see IGU000
ASTRALON see PKQ059
ASTRANTIAGENIN D see GMG000
ASTRATONE see EJQ500
ASTRAZOLO see SNN500
ASTRESS see CFZ000
ASTRIDINE see CCK125
ASTRINGEN see AHA000
ASTROBAIN see OKS000
ASTROBOT see DGP900
ASTROCAR see DJS200
ASTROPHYLLIN see DNC000
ASTROTIA STOKESII VENOM see SBI890
ASTYN see EHP000

ASUCROL see CKK000
ASUGRYN see SGC000
ASULAM see SNQ500
ASULFIDINE see PPN750
ASULFOX F see SNQ500
ASULOX see SNQ500
ASULOX 40 see SNQ500
ASUNTHOL see CNU750
A. SUPERBA (AUSTRALIA) VENOM see ARU750
A. SUPERBA VENOM see ARV625
ASURO see DBD750
ASVERIN-C see ARP875
ASVERIN CITRATE see ARP875
ASVERINE CITRATE see ARP875
AS XVII see KEA300
ASYMMETRICAL TRIMETHYL BENZENE see TLL750
ASYMMETRIN see FNO000
AT see AGK250, AGW675, AMY050
AT 7 see HCL000
o-AT see AIC250
A.T. 10 see DME300
AT 101 see HID350
AT-290 see PED750
AT 327 see BLV000
AT-581 see ARQ000
AT 717 see PKQ250
AT-2266 see EAU100
ATA see AMY050
ATABRINE see ARQ250
ATABRINE DIHYDROCHLORIDE see CFU750
ATABRINE HYDROCHLORIDE see CFU750
ATACTIC POLY(ACRYLIC ACID) see ADW200
ATACTIC POLYPROPYLENE see PMP500
ATACTIC POLYSTYRENE see SMQ500
ATACTIC POLY(VINYL CHLORIDE) see PKQ059
ATADIOL see CKE750
ATALCO C see HCP000
ATALCO O see OBA000
ATALCO S see OAX000
ATAMASCO LILY see RBA500
ATARA see CJR909
ATARAX see CJR909
ATARAX HYDROCHLORIDE see VSF000
ATARAXOID see CJR909
ATARAZOID see CJR909
ATA-Sb see AQC000
ATAZINA see CJR909
ATAZINAX see ARQ725
ATC see TEV000
AT 327 CITRATE see ARP875
ATCOTIBINE see ILD000
ATCP see PIB900
ATDA see AMR250
ATDA HYDROCHLORIDE see AMR500
ATECULON see ARQ750
ATEM see IGG000
ATEMPOL see EQL000
ATENOLOL see TAL475
ATENSIN see GGS000
ATENSINE see DCK759
ATERAX see CJR909
ATERIAN see AHO250
ATERIOSAN see ARQ750
ATEROCYN see PLV750
ATEROID see SEH450
ATEROSAN see PPH050
ATGARD see DGP900
ATHAPROPAZINE see DIR000
ATHEBRATE see ARQ750
ATHERILINE see ARQ325
ATHEROLINE see ARQ325
ATHEROLIP see AHA150
ATHEROMIDE see ARQ750
ATHEROPRONT see ARQ750
ATHOPROPAZIN see DIR000
ATHRANID-WIRKSTOFF see ARQ750
ATHROMBIN see WAT220
ATHROMBINE-K see WAT200
ATHROMBON see PFJ750
ATHYLEN (GERMAN) see EIO000
ATHYLENGLYKOL (GERMAN) see EJC500
ATHYLENGLYKOL-MONOATHYLATHER (GERMAN) see EES350
ATHYL-GUSATHION see EKN000
ATHYMIL see BMA625

ATILEN see DCK759
ATILON see SNY500
ATIPI see ARQ500
ATIRAN see MEP250
ATIRIN see CCS250
ATIVAN see CFC250
ATLACIDE see SFS000
ATLANTIC BLACK BD see AQP000
ATLANTIC BLUE 2B see CMO000
ATLANTIC CONGO RED see SGQ500
ATLANTIC RESIN FAST BLUE see CMN750
ATLANTIC RESIN FAST BROWN BRL see CMO750
ATLANTIC VIOLET N see CMP000
ATLAS "A" see SEY500
ATLAS G 924 see SLL000
ATLAS G-2133 see DXY000
ATLAS G-2142 see PJY100
ATLAS G-2144 see PJY100
ATLAS G 2146 see HKJ000
ATLAS WHITE TITANIUM DIOXIDE see TGG760
ATLATEST see TBF750
AT LIQUID see AMY050
ATLOX 1087 see PKL100
ATMOS 150 see OAV000
ATMUL 67 see OAV000
ATMUL 84 see OAV000
ATMUL 124 see OAV000
ATOCIN see PGG000
A″-TOMATIDINE see THG250
ATOMIT see CAO000
ATONIN O see ORU500
ATOPHAN see PGG000
ATOPHAN-NATRIUM (GERMAN) see SJH000
ATOPHAN SODIUM see SJH000
ATOREL see IDE000
ATOSIL see DQA400
ATOVER see PPH050
ATOXICOCAINE see AIT250
ATOXYL see ARA500
ATOXYLIC ACID see ARA250
ATP see ARQ500
5′-ATP see ARQ500
ATP (nucleotide) see ARQ500
AT-17 PHOSPHATE see MLP250
ATP Na SALT see AEM250
ATRAL see PJA120
ATRANEX see ARQ725
ATRASINE see ARQ725
ATRATOL see SFS000
ATRATOL A see ARQ725
ATRATOL 80W see ARQ700
ATRATON see EGD000
ATRATONE see EGD000
ATRAVET see AAF750, ABH500
ATRAXINE see MQU750
ATRAZIN see ARQ725
ATRAZINE see ARQ725
ATRED see ARQ725
ATREX see ARQ725
ATRIPHOS see ARQ500
A. TRIPHYLLUM see JAJ000
A. TRISPERMA see TOA275
ATRIVYL see MMN250
ATROCHIN see SBG000
ATROL see DPA000
ATROLEN see ARQ750
ATROMID see ARQ750
ATROMIDIN see ARQ750
ATROMID S see ARQ750
ATROPA BELLADONNA see DAD880
ATROPAMIN see AQO250
ATROPAMINE see AQO250
ATROPIN (GERMAN) see ARR000
ATROPINE see ARR000
(−)-ATROPINE see HOU000
ATROPINE METHOBROMIDE see MGR250
ATROPINE METHONITRATE see MGR500
ATROPINE METHYLBROMIDE see MGR250
ATROPINE METHYL NITRATE see MGR500
ATROPINE OCTABROMIDE see OEM000
ATROPINE-N-OCTYL BROMIDE see OEM000
ATROPINE SULFATE (1:1) see ARR250
ATROPINE SULFATE (2:1) see ARR500
ATROPIN-N-OCTYLBROMID (GERMAN) see OEM000

ATROPIN SIRAN (CZECH) see ARR500
ATROPINSULFAT (GERMAN) see ARR500
ATROPYLTROPEINE see AQO250
ATROQUIN see SBG000
ATROTON see EGD000
ATROVENT see IGG000
ATROVIS see ARQ750
AT SEFEN see DPE000
ATSETOZIN see ABH500
ATTAC 6 see CDV100
ATTAC 6-3 see CDV100
ATTAPULGITE see PAE750
ATTAR ROSE TURKISH see OHC000
ATUL ACID CRYSTAL ORANGE G see HGC000
ATUL ACID SCARLET 3R see FMU080
ATUL CONGO RED see SGQ500
ATUL CRYSTAL RED F see HJF500
ATUL DIRECT BLACK E see AQP000
ATUL DIRECT BLUE 2B see CMO000
ATUL DIRECT VIOLET N see CMP000
ATUL FAST YELLOW R see DOT300
ATUL INDIGO CARMINE see FAE100
ATUL OIL ORANGE T see TGW000
ATUL ORANGE R see PEJ500
ATUL TARTRAZINE see FAG140
ATUMIN see ARR750
ATURBANE HYDROCHLORIDE see ARR875
ATX II see ARS000
ATYSMAL see ENG500
AU-95722 see AMD000
l'AUBIER de TILIA SYLVESTRIS (FRENCH) see SAW300
l'AUBIER de TILLEUL (FRENCH) see SAW300
AUBYGEL GS see CCL250
AUBYGUM DM see CCL250
AUCUBA JAPONICA see JBS050
AUGMENTIN see ARS125
AUGMENTIN (antibiotic) see ARS125
'AUKO'I (HAWAII) see CNG825
AULES see TFS350
AULIGEN see BJU000
AURAMINE (MAK) see IBA000, IBB000
AURAMINE BASE see IBB000
AURAMINE HYDROCHLORIDE see IBA000
AURAMINE O (BIOLOGICAL STAIN) see IBA000
AURAMINE YELLOW see IBA000
AURAMYCIN A see ARS130
AURAMYCIN B see ARS135
AURANETIN see ARS250
AURANILE see DKQ000, DNU000
AURANOFIN see ARS150
AURANTHINE see ARS250
AURANTICA see MRN500
AURANTIN see AEA500
AURANTINE see ARS250
AURANTIN (FLAVONE) see ARS250
AUREINE see ARS500
AUREMETINE see ARS750
AUREOCICLINA see CMB000
AUREOCINA see CMA750
AUREOCYCLINE see CMB000
AUREOFUSCIN see ART000
AUREOLIC ACID see MQW750
AUREOMYCIN see CMA750
AUREOMYCIN A-377 see CMA750
AUREOMYCIN HYDROCHLORIDE see CMB000
AUREOMYKOIN see CMA750
AUREOTAN see ART250
AUREOTHIN see DTU200
AURIC CHLORIDE see GIW176
AURICIDINE see GJG000
AURINE-TRICARBOXYLATE d'AMMONIUM (French) see AGW750
AURINTRICARBOXYLIC ACID AMMONIUM SALT see AGW750
AURLELIC ACID see MQW750
AUROCIDIN see GJG000
AUROLIN see GJG000
AUROMOMYCIN see ART125
AUROMYOSE see ART250
AUROPAN see NBU000
AUROPEX see GJG000
AUROPIN see GJG000
AURORA YELLOW see CAJ750
AUROSAN see GJG000
AUROTAN see ART250
1-AUROTHIO-d-GLUCOPYRANOSE see ART250

AUROTHIOGLUCOSE see ART250
AUROTHION see GJG000
AUROVERTIN see ART500
AURUMINE see ART250
AUSOVIT see DXO300
AUSTIOX see TGG760
AUSTOCYSTIN D see ARU250
AUSTRACIL see CDP250
AUSTRACOL see CDP250
AUSTRALIAN BROWN SNAKE VENOM see ARU500, TEG650
AUSTRALIAN COPPERHEAD SNAKE VENOM see ARU750
AUSTRALIAN DEATH ADDER SNAKE VENOM see ARU875
AUSTRALIAN GUM see AQQ500
AUSTRALIAN KING BROWN SNAKE VENOM see ARV000
AUSTRALIAN KING COBRA SNAKE VENOM see ARV125
AUSTRALIAN RED-BELLIED BLACK SNAKE VENOM see ARV250
AUSTRALIAN ROUGH SCALED SNAKE VENOM see ARV375
AUSTRALIAN TAIPAN SNAKE VENOM see ARV500
AUSTRALIAN TIGER SNAKE VENOM see ARV550
AUSTRALIS see ARV000
AUSTRALOL see IQZ000
AUSTRAPEN see AIV500
AUSTRASTAPH see SLJ050
AUSTRELAPS SUPERBA (AUSTRALIA) VENOM see ARU750
AUSTRELAPS SUPERBA VENOM see ARV625
AUSTRELAPS SUPERBUS VENOM see ARV625
AUSTRIAN CINNABAR see LCS000
AUSTROMINAL see EOK000
AUSTROVIT PP see NCR000
AUTAN see DKC800
AUTHRON see ART250
AuTM see GJC000
AUTOMIN see BBV500
AUTOMOBILE EXHAUST CONDENSATE see GCE000
AUTUMN CROCUS see CNX800
AUXEOMYCIN see CMB000
AUXILSON see DBC510
AUXINUTRIL see PBB750
AUXOBIL see DYE700
AVACAN see AAD875
AVACAN HYDROCHLORIDE see ARV750
AVADEX see DBI200
AVADEX BW see DNS600
AVADYL see NOC000
AVAGAL see DJM800, XCJ000
AVAPENA see CKV625
AVAZYME see CML850
AVELLANO (DOMINICAN REPUBLIC) see TOA275
AVENGE see ARW000
A. VERNALIS see PCU375
AVERSAN see DXH250
AVERTIN see ARW250, THV000
AVERZAN see DXH250
AVESYL see GGS000
A/VI see EBY500
AVIBEST C see ARM268
AVIBON see VSK600
AVICADE see RLF350
AVICOL see ARW750, PAX000
AVIL-RETARD see TMK000
AVIOMARIN see DYE600
AVIROL 118 CONC see SIB600
AVISUN see PMP500
AVITA see VSK600
AVITOL see VSK600
AVITROL see AMI500
AVLOCLOR see CLD000, CLD250
AVLON see DBX400, XAK000
AVLOSULPHONE see SOA500
AVLOTANE see HC1000
AVOCAN see NOC000
AVOGODRITE see PKY000
AVOLIN see DTR200
AVOMINE see DQA400
AVON GREEN A-4379 see BAY750
AVOXYL see GGS000
AVT see AQW125
AW-14-2446 see CMW600
AW-14'2333 see HOU059
AWD 19-166 see BMB125
AWELYSIN see SLW450
AWPA #1 see CMY825
AXIOM see DIX600
AXM see ABN000

AY-5406 see BCA000
AY-6108 see AIV500
AY 8682 see CQH625
AY 9944 see BHN000
AY-11483 see EDU600
AY 21011 see ECX100
AY-22,352 see CIO375
AY 22989 see RBK000
AY 24034 see LIU360
AY-25,329 see NMV750
AY-25,674 see ONQ000
AY-57,062 see ABH500
AY-61122 see MLJ500
AY 61123 see ARQ750
AY-62014 see BPT750
AY 62021 see CLX250
AY-62022 see MBZ100
AY 64043 see ICB000, ICC000
AY 136155 see MBZ100
AY 11483-16 see EDU600
AYAA see AAX250
AYAF see AAX250
AYERMATE see MQU750
AYERST 62013 see TJQ333
AYFIVIN see BAC250
AYUSH-47 see ARX125
9-AZAANTHRACENE see ADJ500
10-AZAANTHRACENE see ADJ500
5-AZA-10-ARSENAANTHRACENE CHLORIDE see PDB000
12-AZABENZ(a)ANTHRACENE see BAW750
AZABENZENE see POP250
AZABICYCLANE CITRATE see ARX150
1-AZABICYCLO(2.2.2)OCTAN-3-OL BENZILATE (9CI) see QVA000
1-AZABICYCLO(2.2.2)OCTAN-3-OL, BENZYLATE (ESTER), HYDROCHLO-
   RIDE see QWJ000
1-AZABICYCLO(3.2.1)OCTAN-6-OL DIPHENYLACETATE HYDROCHLORIDE
   see ARX500
AZABICYCLO(2.2.1)OCTAN-3-OL, DIPHENYLACETATE (ester), HYDROGEN
   SULFATE (2:1) DIHYDRATE see QVJ000
1-AZABICYCLO(3.2.1)OCTAN-6-OL-9-FLUORENECARBOXYLATE HYDRO-
   CHLORIDE-48-9 see ARX750
AZABICYCLOOCTANOL METHYL BROMIDE DIPHENYLACETATE
   see ARX770
1-(3-AZABICYCLO(3.3.0)OCT-3-YL)-3-(p-TOLYLSULFONYL)UREA see DBL700
AZACITIDINE see ARY000
AZACOSTEROL DIHYDROCHLORIDE see ARX800
AZACOSTEROL HYDROCHLORIDE see ARX800
AZACTAM see ARX875
AZACYCLOHEPTANE see HDG000
1-AZACYCLOHEPTANE see HDG000
2-AZACYCLOHEPTANONE see CBF700
AZACYCLOHEXANE see PIL500
AZACYCLONOL HYDROCHLORIDE see PIY750
β-1-AZACYCLOOCTYLETHYLGUANIDINE SULFATE see GKS000
1-AZACYCLOOCT-2-YL METHYL GUANIDINE see GKO800
1-AZA-2,4-CYCLOPENTADIENE see PPS250
AZACYCLOPENTANE see PPS500
AZACYCLOPROPANE see EJM900
AZACYTIDINE see ARY000
5-AZACYTIDINE see ARY000
6-AZACYTIDINE see AMM000
5'-AZACYTIDINE see ARY000
5-AZADEOXYCYTIDINE see ARY125
5-AZA-2'-DEOXYCYTIDINE see ARY125
7-AZADIBENZ(a,h)ANTHRACENE see DCS400
7-AZADIBENZ(a,j)ANTHRACENE see DCS600
14-AZADIBENZ(a,j)ANTHRACENE see DCS800
7-AZA-7H-DIBENZO(c,g)FLUORENE see DCY000
AZADIENO see MJL500
9-AZAFLUORENE see CBN000
AZAGUANINE see AJO500
AZAGUANINE-8 see AJO500
8-AZAGUANINE see AJO500
2-AZAHYPOXANTHINE see ARY500
1-AZAINDENE see ICM000
3-AZAINDOLE see BCB750
AZALEA see RHU500
AZALEUCINE see ARY625
4-AZALEUCINE see ARY625
AZALINE see ARM000
AZALOMYCIN F see ARY750
AZALONE see PDC000
AZAMETHONE see MKU750
AZAMETHONIUM BROMIDE see MKU750

AZAMETON see MKU750
AZAN see AJO500
1-AZANAPHTHALENE see QMJ000
2-AZANAPHTHALENE see IRX000
AZANIDAZOLE see TJF000
AZANIL RED SALT TRD see CLK235
AZANIN see ASB250
3-AZAPENTANE-1,5-DIAMINE see DJG600
AZAPERONE (USDA) see FLU000
AZAPETINE see ARZ000
AZAPETINE PHOSPHATE see AGD500
1-AZAPHENANTHRENE see BDB750
4-AZAPHENANTHRENE see BDC000
3-AZAPHTHALIMID (GERMAN) see POQ750
4-AZAPHTHALIMID (GERMAN) see POR000
AZAPICYL see ADA000
AZAPLANT see AMY050
AZAPROPAZON (GERMAN) see AQN750
AZAPROPAZON DIHYDRAT (GERMAN) see ASA000
AZAPROPAZONE see ASA000
AZAPROPAZONE (anhydrous) see AQN750
AZAPROPAZONE SODIUM see ASA250
AZAPROPAZONE NATRIUMSALZ (GERMAN) see ASA250
AZARIBINE see THM750
AZASERIN see ASA500
AZASERINE see ASA500
l-AZASERINE see ASA500
6-AZASPIRO(3,4)OCTANE-5,7-DIONE see ASA750
AZA-6-SPIRO(3,4)OCTANE-DIONE-5,7 (FRENCH) see ASA750
4-(3-AZASPIRO(5.5)UNDEC-3-YL)-4'-FLUORO-BUTYROPHENONE HYDRO-
   CHLORIDE-8 see ASA875
AZASTENE see HOM259
AZASTEROL see ARX800
AZATADINE DIMALEATE see DLV800
AZATADINE MELEATE see DLV800
AZATHIOPRINE see ASB250
8-AZATHIOXANTHINE see MLY000
AZATIOPRIN see ASB250
N-(4-AZA-endo-TRICYCLO(5.2.1.$5^{2,6}$)-DECAN-4-YL)-4-CHLORO-3-SUL-
   FAMOYLBENZAMIDE see CHK825
6-AZAURACIL see THR750
4(6)-AZAURACIL see THR750
6-AZAURACILRIBOSIDE see RJA000
6-AZAURACIL-β-d-RIBOSIDE see RJA000
AZAURIDINE see RJA000
6-AZAURIDINE see RJA000
8-AZAXANTHINE see THT350
AZBLLEN ASBESTOS see ARM266
AZBOLEN ASBESTOS see ARM264
AZDEL see PMP500
AZDID see BRF500
AZELAIC ACID see ASB750
AZELAIC ACID DI(2-ETHYLHEXYL)ESTER see BJQ500
AZELAIC ACID DIHEXYL ESTER see ASC000
AZELASTIN see ASC130
AZELASTINE see ASC125
AZELASTINE HYDROCHLORIDE see ASC130
AZEPERONE see FLU000
AZEPHEN see ASC250
AZEPINE PHOSPHATE see AGD500
AZEPROMAZINE see ABH500
AZETALDEHYDSCHWEFLIGSAUREN NATRIUMS (GERMAN) see AAH500
l-2-AZETIDINECARBOXYLIC ACID see ASC500
AZETYLAMINOFLUOREN (GERMAN) see FDR000
N-AZETYL-COLCHINOL (GERMAN) see ACG250
AZG see AJO500
AZIDE see SFA000
AZIDES see ASC750
AZIDINE BLUE 3B see CMO250
AZIDITHION see ASD000
AZIDOACETIC ACID see ASD375
AZIDOACETONE see ASD500
AZIDOACETO NITRILE see ASE000
AZIDOBENZENE see PEH000
N-AZIDO CARBONYL AZEPINE see ASE250
AZIDOCARBONYL GUANIDINE see ASE500
AZIDOCODEINE see ASE875
AZIDODIMETHYL BORANE see ASF500
2-AZIDO-3,5-DINITROFURAN see ASF625
AZIDODITHIOFORMIC ACID see ASF750
(5-α,6-β)-6-AZIDO-4,5-EPOXY-17-METHYL-MORPHINAN-3,14-DIOL see HJE600
AZIDO FLUORINE see FFA000
2-AZIDO-4-ISOPROPYLAMINO-6-METHYLTHIO-s-TRIAZINE see ASG250
2-AZIDO-4-ISOPROPYLAMINO-6-METHYLTHIO-1,3,5-TRIAZINE see ASG250

N-AZIDO METHYL AMINE see ASG500
2-AZIDOMETHYLBENZENEDIAZONIUM TETRAFLUOROBORATE see ASG625
4-AZIDO-N-(1-METHYLETHYL)-6-(METHYLTHIO)-1,3,5-TRIAZIN-2-AMINE
    see ASG250
4-AZIDO-N-(1-METHYLETHYL)-6-(METHYLTHIO)-1,3,5-TRIAZIN-2-AMINI
    see ASG250
AZIDOMETHYLPHOSPHONIC ACID, ETHYL METHYL ESTER see EML500
AZIDOMORPHINE see ASG675
5-AZIDOTETRAZOLE see ASH000
3-AZIDO-1,2,4-TRIAZOLE see ASH250
AZIDOTRIMETHYLSILANE see TMF100
AZIJNZUUR (DUTCH) see AAT250
AZIJNZUURANHYDRIDE (DUTCH) see AAX500
AZIMETHYLENE see DCP800
AZIMIDOBENZENE see BDH250
AZIMINOBENZENE see BDH250
AZINDOLE see BCB750
AZINE see POP250
AZINE DEEP BLACK EW see AQP000
AZINFOS-ETHYL (DUTCH) see EKN000
AZINFOS-METHYL (DUTCH) see ASH500
2,2'-AZINOBIS(3-ETHYL-7-BENZOTHIAZOLINESULFONIC ACID, DIAMMON-
    IUM SALT see ASH375
2,2'-AZINO-DI(3-ETHYL-BENZTHIAZOLINE SULPHONIC ACID (6), AMMO-
    NIUM SALT see ASH375
AZINOS see EKN000
AZINOTHRICIN see ASH425
AZINPHOS-AETHYL (GERMAN) see EKN000
AZINPHOS ETHYL see EKN000
AZINPHOS-ETILE (ITALIAN) see EKN000
AZINPHOS METHYL see ASH500
AZINPHOS METHYL, liquid (DOT) see ASH500
AZINPHOS-METILE (ITALIAN) see ASH500
AZIONYL see ARQ750
AZIPROTRYN see ASG250
AZIPROTRYNE see ASG250
AZIRANE see EJM900
AZIRIDIN (GERMAN) see EJM900
AZIRIDINE see EJM900
1-AZIRIDINEACETAMIDE, N,N'-(1,4-CYCLOHEXYLENEDIMETHYLENE)BIS-
    see CPL100
1-AZIRIDINECARBOXAMIDE, N,N-DIMETHYL- see EJA100
1-AZIRIDINECARBOXAMIDE, N-METHYL- see EJN400
1-AZIRIDINECARBOXANILIDE see PEH250
AZIRIDINE CARBOXYLIC ACID ETHYL ESTER see ASH750
1-AZIRIDINE ETHANOL see ASI000
1-AZIRIDINEPROPANENITRILE see ASI250
1-AZIRIDINE PROPIONITRILE see ASI250
AZIRIDINE-1,3,5,2,4,6-TRIAZATRIPHOSPHORINE DERIVATIVE see AQO000
1-AZIRIDINYL-BIS(DIMETHYLAMINO)PHOSPHINE OXIDE see ASK500
1-(1-AZIRIDINYL)-N-(m-CHLOROPHENYL)FORMAMIDE see CJR250
6-(1-AZIRIDINYL)-4-CHLORO-2-PHENYLPYRIMIDINE see CGW750
2-(1-AZIRIDINYL)ETHANOL see ASI000
4-AZIRIDINYL-3-HYDROXY-1,2-BUTENE see VMA000
1-(1-AZIRIDINYL)-N-(p-METHOXYPHENYL)FORMAMIDE see MFF250
α-(1-AZIRIDINYLMETHYL)BENZYL ALCOHOL see PEH500
2-(1-AZIRIDINYL)-1-PHENYLETHANOL see PEH500
1-AZIRIDINYL PHOSPHINE OXIDE (TRIS) (DOT) see TND250
1-AZIRIDINYLPHOSPHONITRILE TRIMER see AQO000
AZIRIDINYLQUINONE see ASK875
p-1-AZIRIDINYL-N,N,N',N'-TETRAMETHYLAMINO-PHOSPHINE OXIDE
    see ASK500
p-1-AZIRIDINYL-N,N,N',N'-TETRAMETHYLPHOSPHINIC DIAMIDE
    see ASK500
2-(1-AZIRIDINYL)-1-VINYLETHANOL see VMA000
AZIRIDYL BENZOQUINONE see BDC750
AZIRPOTRYNE see ASG250
AZIUM see SFA000, SOW000
AZO-33 see ZKA000
AZOAETHAN (GERMAN) see ASN250
AZOAMINE SCARLET see NEQ500
AZOBASE DCA see DEO400
AZOBASE MNA see NEN500
AZOBENZEEN (DUTCH) see ASL250
AZOBENZEN (CZECH) see COX250
AZOBENZENE see ASL250
AZOBENZENE OXIDE see ASO750
AZOBENZIDE see ASL250
AZOBENZOL see ASL250
AZOBISBENZENE see ASL250
1,1'-AZOBISCARBAMIDE see ASM270
AZOBISCARBONAMIDE see ASM270
AZOBISCARBOXAMIDE see ASM270
1,1'-AZOBIS(FORMAMIDE) see ASM270

AZOBISISOBUTYLONITRILE see ASL750
α,α'-AZOBISISOBUTYLONITRILE see ASL750
AZOBIS ISOBUTYRAMIDE HYDROCHLORIDE see ASM000
AZOBISISOBUTYRONITRILE see ASL750
2,2'-AZOBIS(ISOBUTYRONITRILE) see ASL750
2,2'-AZOBIS(2-METHYLPROPIONITRILE) see ASL750
AZO BLUE see ASM100
AZOCARBONITRILE see DGS300
AZOCARD BLACK EW see AQP000
AZOCARD BLUE 2B see CMO000
AZOCARD RED CONGO see SGQ500
AZOCARD VIOLET N see CMP000
AZOCHLORAMIDE see ASM250
AZOCHROMAL ORANGE R see NEY000
AZOCYCLOTIN see THT500
AZODIBENZENE see ASL250
AZODIBENZENEAZOFUME see ASL250
AZODICARBAMIDE see ASM270
AZODICARBOAMIDE see ASM270
AZODICARBONAMIDE see ASM270, ASM300
AZODICARBOXAMIDE see ASM270
AZODICARBOXYLIC ACID DIAMIDE see ASM270
AZODIISOBUTYRONITRILE see ASL750
AZODIISOBUTYRONITRILE (DOT) see ASL750
2,2'-AZODIISOBUTYRONITRILE see ASL750
α,α'-AZODIISOBUTYRONITRILE see ASL750
AZODINE see PDC250
N,N'-(AZODI-4,1-PHENYLENE)BIS(N-METHYLACETAMIDE) see DQH200
AZODIUM see PDC250
AZODOX-55 see ZKA000
AZODRIN see MRH209
"AZODRIN" see ASN000
AZO DYE No. 6945 see MIH500
AZODYNE see PDC250
AZOENE FAST BLUE BASE see DCJ200
AZOENE FAST ORANGE GR SALT see NEO000
AZOENE FAST RED KB BASE see CLK225
AZOENE FAST RED TR BASE see CLK220
AZOENE FAST RED TR SALT see CLK235
AZOEN FAST SCARLET 2G BASE see DEO400
AZO ETHANE see ASN250
AZOFENE see BDJ250
AZOFIX BLUE B SALT see DCJ200
AZOFIX RED GG SALT see NEO500
AZOFIX SCARLET G SALT see NMP500
AZOFIX SCARLET GG SALT see DEO400
AZOFORMALDOXIME see ASN375
AZOFOS see MNH000
AZO-GANTANOL see SNK000
AZO GANTRISIN see PDC250, SNN500
AZO GASTANOL see PDC250
AZOGEN DEVELOPER A see NAX000
AZOGEN DEVELOPER H see TGL750
AZOGENE ECARLATE R see NEQ500
AZOGENE FAST RED TR see CLK220, CLK235
AZOGENE FAST SCARLET G see NMP500
AZOGNE FAST BLUE B see DCJ200
AZOIC DIAZO COMPONENT 32 see CLK225
AZOIC DIAZO COMPONENT 37 see NEO500
AZOIC DIAZO COMPONENT 46 see CLK200
AZOIC DIAZO COMPONENT 11 BASE see CLK220, CLK235
AZOIC DIAZO COMPONENT 13 BASE see NEQ500
AZOIC RED 36 see MGO750
AZOIMIDE see HHG500
AZOLAN see AMY050
AZOLASTONE see MQY110
AZOLE see AMY050, PPS250
AZOLID see BRF500
AZOLMETAZIN see SNJ000
AZO-MANDELAMINE see PDC250
AZOMETHANE see ASN400
AZOMINE see PDC250
AZOMINE BLACK EWO see AQP000
AZOMINE BLUE 2B see CMO000
AZOMYCIN see NHG000
1,1'-AZONAPHTHALENE see ASN500
2,2'-AZONAPHTHALENE see ASN750
8-AZONIABICYCLO(3.2.1)OCTANE, 8-(2-FLUOROETHYL)-3-
    ((HYDROXYDIPHENYLACETYL)OXY)-8-METHYL-, BROMIDE, (endo,syn)-,
    MONOHYDRATE see FMR300
AZONIASPIRO(3-α-BENZILOYLOXY-NORTROPAN-8,1'-PYRROLIDINE)-CHLO-
    RIDE see KEA300
AZONIASPIRO COMPOUND XVII see KEA300
AZOPHENYLENE see PDB500

AZOPHOS see MNH000
AZOPYRIN see PPN750
AZO RED R see FAG020
AZORESORCIN see HNG500
AZORUBIN see HJF500
AZOSEMIDE see ASO375
AZOSEPTALE see TEX250
AZOSSIBENZENE (ITALIAN) see ASO750
AZO-STANDARD see PDC250
AZO-STAT see PDC250
AZOSULFIZIN see SNN500
AZOTA CABALLO see GIW200
AZOTE (FRENCH) see NGR500
AZOTHIOPRINE see ASB250
AZOTIC ACID see NED500
AZOTO (ITALIAN) see NGR500
2:3'-AZOTOLUENE see DQH000
AZOTOMYCIN see ASO501, ASO510
AZOTOMYCIN SODIUM see ASO510
AZOTOMYCIN SODIUM SALT see ASO510
AZOTOWY KWAS (POLISH) see NED500
AZOTOX see DAD200
AZOTOYPERITE see BIE500
AZOTREX see PDC250
AZOTURE de SODIUM (FRENCH) see SFA000
AZOTU TLENKI (POLISH) see NGT500
AZOXYAETHAN (GERMAN) see ASP000
AZOXYBENZEEN (DUTCH) see ASO750
AZOXYBENZENE see ASO750
AZOXYBENZIDE see ASO750
AZOXYBENZOL (GERMAN) see ASO750
AZOXYDIBENZENE see ASO750
AZOXYETHANE see ASP000
AZOXYISOPROPANE see ASP510
AZOXYMETHANE see ASP250
1-AZOXYPROPANE see ASP500
2-AZOXYPROPANE see ASP510
1,1'-AZOXYPROPANE see ASP500
AZQ see ASK875
AZS see ASA500
AZTEC BPO see BDS000
AZTHREONAM see ARX875
AZTREONAM see ARX875
AZUCENA de MEJICO (MEXICO) see AHI635
AZULENE, 1,2,3,4,5,6,7,8-OCTAHYDRO-1,4-DIMETHYL-7-(1-
    METHYLETHYLIDENE)-, MONOEPOXIDE see EBU100
AZULENO(5,6,7-cd)PHENALENE see ASP750
AZULFIDINE see PPN750
AZULON see DSJ800
AZUR see RJA000
6-AZUR see RJA000
AZUREN see ILD000
AZURENE see BNU725
6-AZURIDINE see RJA000
AZUROL ALIZARINOVY SW (CZECH) see DKX800
AZURRO DIRETTO 3B see CMO250

B10 see SFO500
B-12 see VSZ000
23B see DHY400
B-28 see AJO500
B32 see HCL000
B-45 see AOF250
B-71 see PKC000
B/77 see DNX600
B-314 see BEI000
B-322 see BEW750
B-324 see BEW500
B-325 see BEX250
B-329 see BEL500
B-331 see BEN750
B-343 see IHQ100
B 377 see DVC200
B 41D see MQT600
B-436 see PEV750
B 518 see CQC650
B 577 see HKK000
B586 see CKE750
B-622 see DEV800
693B see SLP600
B 995 see DQD400
B 1500 see MCA100
B-1,776 see BSH250
B-1843 see BLG500

B2734 see NIV000
B2740 see THD275
B2772 see MFH750
B2837 see NFF000
B-3015 see SAZ000
B-4130 see AAI750
B 4992 see BKS810
B 5333 see CFY250
B-5477 see EAT810
B-9002 see HMV000
B-10094 see CON300
B 11163 see DTQ800
B-11420 see IGA000
B-14437 see SAU000
B-15000 see IFY000
B 17476 see RQF350
B-66256 see USJ000
B 66347 see CLW625
B 67347 see CLW625
B 77488 see BAT750
B 90912 see SAU000
B-98891 see MQU000
B 10 (polysaccharide) see SFO500
BA see BBC250
BA 168 see LIA400
276-Ba see LIN800
BA 1355 see PDC875
BA 2650 see SNY500
BA 2666 see DBE150
Ba 2724 see DGQ400
BA 2726 see DBF800
BA 2784 see DRV000
Ba 2797 see CAY500
BA5968 see HGP495
BA 5968 see HGP500
Ba-16038 see AKC600
BA 21381 see IAN100
BA 30,803 see BCH750
BA 32644 see NML000
BA 34276 see MAW850
Ba 34647 see BAC275
Ba 36278 see SGB500
Ba-39089 see THK750
Ba 49249 see EEH575
BA 51-090278 see VQA100
BA 51-090462 see DCY400
BAAM see MJL250
BABEIRO AMARILLO (PUERTO RICO) see YAK300
BABIDIUM CHLORIDE see HGI000
BABN see BPV325, BRX000
B. ABORTUS Bang. LIPOPOLYSACCHARIDE see LGK350
BA 253 BR see ONI000
Ba-253-BR-L see ONI000
BABROCID see NGE500
Ba 598 BROMIDE HYDRATE see FMR300
BABS see BLX500
BABUL STEM BARK EXTRACT see AAD250
BABURAN see PJA120
BABY PEPPER see ROA300
B(c)AC see BAW750
BACACIL see BAB250
BACAMPICILLIN HYDROCHLORIDE see BAB250
BACARATE see DKE800
BACCIDAL see BAB625
BA 32644 CIBA see NML000
B''-ACID see LIU000
BACIFURANE see DGQ500
BACIGUENT see BAC250
BACI-JEL see BAC250
BACILIQUIN see BAC250
BACILLIN see BAC175, ILD000
BACILLOL see CNW500
BACILLOMYCIN (8CI, 9CI) see BAB750
BACILLOMYCIN R see BAB750
BACILLOPEPTIDASE A see BAC000
BACILLOPEPTIDASE B see BAC000
BACILLUS CEREUS exo-ENTEROTOXIN see BAB650
BACILLUS SPHAERICUS ALKALINE PROTEINASE see SCC550
BACILLUS SUBTILIS BPN see BAB750
BACILLUS SUBTILIS CARLSBERG see BAC000
BACILLUS THURINGIENSIS EXOTOXIN see BAC125
BACILYSIN see BAC175
BACIMETHRIN see AKO750
BACITEK OINTMENT see BAC250

BACITRACIN see BAC250
BACLOFEN see BAC275
BACLON see BAC275
BACMECILLINAM see BAC325
BACO AF 260 see AHC000
BACTERAMID see SNM500
BACTERIAL VITAMIN H1 see AIH600
BACTESULF see SNN500
BACTINE see BAC750, MHB500
B 2992 ACTIVE PRINCIPLE see MQX250
BACTOCILL see MNV250
BACTOL see CHR500
BACTOLATEX see SMQ500
BACTOPEN see SLJ000
BACTRAMIN see TKX000
BACTRAMYL see PAF250
BACTRATYCIN see TOG500
BACTRIM see SNK000, TKX000
BACTROL see BGJ750
BACTROMIN see TKX000
BACTROVET see SNN300
BACTYLAN see SEP000
BACULO (PUERTO RICO) see SBC550
BADEN ACID see ALI300
BA-1,2-DIHYDRODIOL see BBD250
BA-3,4-DIHYDRODIOL see BBD500
BA-5,6-DIHYDRODIOL see BBE250
BA-8,9-DIHYDRODIOL see BBE750
BA-10,11-DIHYDRODIOL see BBF000
BA-5,6-cis-DIHYDRODIOL see BBE000
BA-5,6-trans-DIHYDRODIOL see BBE250
BA-1,2-DIOL-3,4-EPOXIDE-1 see DMP000
BA-3,4-DIOL-1,2-EPOXIDE-1 see DLE000, DLE200, DLE000
BA-10,11-DIOL-8,9-EPOXIDE-1 see BAD000
BA-8,9-DIOL-10,11-EPOXIDE-1 see DMO600
BADISCHE ACID see ALI300
BAF see THR500
BAFHAMERITIN-M see XQS000
BAGAODRYL see BBV500
BAGASSE DUST see BAD250
BAJATEN see IBV100
BAJKAIN see BMA125
BAKELITE AYAA see AAX250
BAKELITE DYNH see PJS750
BAKELITE LP 70 see AAX175
BAKELITE LP 90 see AAX250
BAKELITE RMD 4511 see ADY500
BAKELITE SMD 3500 see SMQ500
BAKELITE VLFV see AAX175
BAKELITE VMCC see AAX175
BAKELITE VYNS see AAX175
BAKER'S ANTIFOL see DKC800
BAKER'S ANTIFOLANTE see THR500
BAKER'S ANTIFOL SOLUBLE see THR500
BAKER'S P AND S LIQUID and OINTMENT see PDN750
BAKING SODA see SFC500
BAKONTAL see BAP000
BAKTAR see TKX000
BAKTOL see CFE250
BAKTOLAN see CFE250
BAKUCHIOL see BAD625
BAL see BAD750
BALATA see BAE000
BALFON 7000 see TAI250
BALMADREN see VGP000
BALSAM APPLE see BAE325, FPD100
BALSAM CAPTIVI see CNH792
BALSAM GURJUN see GME000
BALSAM PEAR see FPD100
BALSAM of PERU see BAE750
BALSAM PERU OIL (FCC) see BAE750
BALSAMS, COPAIBA see CNH792
BALSAMS, TOLU see BAF000
BALSAM TOLU see BAF000
BAMBERMYCIN see MRA250
BAMBICORT see HHQ800
BAMD 400 see MQU750
BAMETAN SULFATE see BOV825
BAMETHANE see BQF250
BAMETHAN SULFATE see BOV825
BAMIFYLLINE HYDROCHLORIDE see BEO750, THL750
B-AMIN see TES750
BAMIPHYLLINE HYDROCHLORIDE see BEO750, THL750
BAMN see BRX500

BANABIN see CKF500
BANABIN-SINTYAL see CKF500
BANANA OIL see IHO850
BANANOTE see MDW750
BANDANE see BAF250
BANDOL see CBQ625
BANEBERRY see BAF325
BANEX see MEL500
BANGTON see CBG000
BAN-HOE see CBM000
BANISTERINE see HAI500
BANLEN see MEL500
BANMINTH see TCW750
BANOCIDE see DIW200
BANOL see CGI500
BANOL TUCO SOK see CGI500
BANOL TURF FUNGICIDE see PNI250
BANOMITE see BAF500
BANROT see PEW750
BANTHIN see DJM800, XCJ000
BANTHINE see XCJ000
BANTHINE BROMIDE see DJM800, XCJ000
BANTHIONINE see MDT740
BANTROL see HKB500
BANUCALAD (HAWAII) see TOA275
BANVEL see MEL500
BANVEL HERBICIDE see MEL500
BANVEL T see TIK000
BAP see NJM500
9-BAP see BDX100
BAPC see BAB250
2-BAP HYDROCHLORIDE see BEA000
BAPN see AMB500
BAPN FUMARATE see AMB750
BAPTITOXIN see CQL500
BAPTITOXINE see CQL500
BARACOUMIN see BJZ000
BARAMINE see BBV500
BARATOL see TLG000
BARAZAE see SNN500
BARAZAN see BAB625
BARBADOS GOOSEBERRY see JBS100
BARBADOS LILY see AHI635
BARBADOS NUT see CNR135
BARBADOS PRIDE see CAK325
BARBALLYL see AFS500
BARBALOIN see BAF825
BARBAMATE see CEW500
BARBAN see CEW500
BARBANE see CEW500
BARBAPIL see EOK000
BARBASCO see RNZ000
BARBENYL see EOK000
BARBEXACLONE see CPO500
BARBEXACLONUM see CPO500
BARBIDAL see AFS500
BARBIDORM see ERD500
BARBILEHAE (BARBILETTAE) see EOK000
BARBIMON see AMK250
BARBIPHENYL see EOK000
BARBITA see EOK000
BARBITAL see BAG000, BAG500
BARBITAL Na see BAG250
BARBITAL SODIUM see BAG250, BAG500
BARBITAL SOLUBLE see BAG250
BARBITONE see BAG000, BAG500
BARBITONE SODIUM see BAG250
BARBITURATES see BAG500
BARBITURIC ACID see BAG750
BARBITURIC ACID, 1-(1-(1-CYCLOHEXYL-N-METHYL-2-PROPANAMINE)-5-ETHYL-5-PHENYL see CPO500
BARBONAL see EOK000
BARBOPHEN see EOK000
BARBOSEC see SBM500, SBN000
BARDAC 22 see DGX200
BAR-DEX see AOB500
BARDIOL see EDO000
BARHIST see DPJ400
BARIDIUM see PDC250
BARIDOL see BAP000
B. ARIETANS VENOM see BLV075
BARIO (PEROSSIDO di) (ITALIAN) see BAO250
BARITE see BAP000
BARITOP see BAP000

BARITRATE see PBC250
BARIUM see BAH250
BARIUM ACETATE see BAH500
BARIUM ACETYLIDE see BAH750
BARIUM AZIDE see BAI000
BARIUM AZIDE (wet) see BAI250
BARIUM AZIDE, dry or containing less than 50% water (DOT) see BAI000
BARIUM BENZOATE see BAI500
BARIUM BICHROMATE see BAL500
BARIUM BINOXIDE see BAO250
BARIUM BIS(TETRAFLUOROBORATE) see BAL750
BARIUM BROMATE see BAI750
BARIUM CADMIUM STEARATE see BAI800
BARIUM CAPRYLATE see BAI825
BARIUM CARBIDE see BAJ000
BARIUM CARBONATE see BAJ250
BARIUM CARBONATE (1:1) see BAJ250
BARIUM CHLORATE see BAJ500
BARIUM CHLORATE (wet) see BAJ750
BARIUM CHLORIDE see BAK000
BARIUM CHLORITE see BAK125
BARIUM CHROMATE (1:1) see BAK250
BARIUM CHROMATE(VI) see BAK250
BARIUM CHROMATE OXIDE see BAK250
BARIUM COMPOUNDS (soluble) see BAK500
BARIUM CYANIDE see BAK750
BARIUM CYANIDE, solid (DOT) see BAK750
BARIUM CYANOPLATINITE see BAL000
BARIUM DIACETATE see BAH500
BARIUM DIAZIDE see BAL250
BARIUM DIBENZYLPHOSPHATE see BAL275
BARIUM DICHLORIDE see BAK000
BARIUM DICHROMATE see BAL500
BARIUM DICYANIDE see BAK750
BARIUM DINITRATE see BAN250
BARIUM DIOXIDE see BAO250
BARIUM DISTEARATE see BAO825
BARIUM FERRATE see BAL625
BARIUM FERRITE see BAL625
BARIUM FLUOBORATE see BAL750
BARIUM FLUORIDE see BAM000
BARIUM FLUOROSILICATE see BAO750
BARIUM FLUOSILICATE see BAO750
BARIUM HEXAFERRITE see BAL625
BARIUM HEXAFLUOROSILICATE see BAO750
BARIUM HEXAFLUOROSILICATE(2-) see BAO750
BARIUM HYDRIDE see BAM250
BARIUM HYDROXIDE see BAM500
BARIUM HYPOPHOSPHITE see BAM750
BARIUM IODATE see BAN000
BARIUM MONOXIDE see BAO000
BARIUM NITRATE (DOT) see BAN250
BARIUM(II) NITRATE (1:82) see BAN250
BARIUM NITRIDE see BAN500
BARIUM NITROSYLPENTACYANOFERRATE see PAY600
BARIUM OCTANOATE see BAI825
BARIUM OCTOATE see BAI825
BARIUM OXIDE see BAO000
BARIUM PERCHLORATE (DOT) see PCD750
BARIUM PERMANGANATE see PCK000
BARIUMPEROXID (GERMAN) see BAO250
BARIUM PEROXIDE see BAO250
BARIUMPEROXYDE (DUTCH) see BAO250
BARIUMPOLYSULFID see BAO300
BARIUM POLYSULFIDE see BAO300
BARIUM PROTOXIDE see BAO000
BARIUM RHODANIDE see BAO500
BARIUM SILICOFLUORIDE see BAO750
BARIUM SILICON FLUORIDE see BAO750
BARIUM STEARATE see BAO825
BARIUM SULFATE see BAP000
BARIUM SULFIDE see BAO300, BAP250
BARIUM SUPEROXIDE see BAO250
BARIUM TETRAFLUOROBORATE see BAL750
BARIUM THIOCYANATE see BAP500
BARIUM ZIRCONATE see BAP750
BARIUM ZIRCONIUM OXIDE see BAP750
BARIUM ZIRCONIUM(IV) OXIDE see BAP750
BARIUM ZIRCONIUM TRIOXIDE see BAP750
BARLENE 125 see DRR800
BARNETIL see EPD100
BAROS CAMPHOR see BMD000
BAROSPERSE see BAP000
BAROTRAST see BAP000

BARPENTAL see NBU000
BARQUAT MB-50 see AFP250
BARQUAT SB-25 see DTC600
BARQUINOL see CHR500
BARRICADE see RLF350
BARSEB HC see CNS750
BAR-TIME see BBK250
BARTOL see EOK000
BARYTA see BAO000
BARYTA WHITE see BAP000
BARYTA YELLOW see BAK250
BARYTES see BAP000
BARYUM FLUORURE (FRENCH) see BAM000
BAS see BEM750
BAS-3050 see MNS500
BAS 3220 see DJV000
BAS-3460 see MHC750
BAS 4239 see CEQ500
BAS 2900H see CDS275
BAS 2903H see CDS275
BAS 32500F see PEX500
BAS 36801F see MAP300
BASAGRAN see MJY500
BASALIN see FDA900
BASAMAIZE see CDS275
BASAMID see DSB200
BASAMID-FLUID see VFU000
BASAMID G see DSB200
BASAMID-GRANULAR see DSB200
BASAMID P see DSB200
BASAMID-PUDER see DSB200
BASANITE see BRE500
BASCOREZ see AAX250
BASCURAT see BOV825
BASE 661 see SMR000
BASECIL see MPW500
BASEDOL see AMS250
BASE LP 12 see DXY000
BASE OIL see PCR250
BASERGIN see LJL000
BASETHYRIN see MPW500
BAS 352 F see RMA000
BAS 2203F see TJJ500
BAS 2205-F see DUJ400
BASFAPON see DGI400
BASFAPON B see DGI400, DGI600
BASFAPON/BASFAPON N see DGI400
BASF-GRUNKUPFER see CNK559
BASF III see SMQ500
BASF-MANEB SPRITZPULVER see MAS500
BASFUNGIN see MLX850
BASFUNGINE see MLX850
BASF URSOL 3GA see ALT000
BASF URSOL D see PEY500
BASF URSOL EG see ALT500
BASF URSOL ERN see NAW500
BASF URSOL P BASE see ALT250
BASF URSOL SLA see DBO400
BAS-290-H see CDS275
BAS 351-H see MJY500
BAS 392-H see FDA900
BASIC ALUMINUM CHLORATE see AHA000
BASIC BLUE 9 see BJI250
BASIC BLUEK see VKA600
BASIC BRIGHT GREEN see BAY750
BASIC CHROMIC SULFATE see NBW000
BASIC CHROMIC SULPHATE see NBW000
BASIC CHROMIUM CARBONATE see CMJ100
BASIC CHROMIUM SULFATE see NBW000
BASIC CHROMIUM SULPHATE see NBW000
BASIC COPPER CARBONATE see CNJ750
BASIC COPPER CHLORIDE see CNK559
BASIC CUPRIC CARBONATE see CNJ750
BASIC GREEN 4 see AFG500
BASIC LEAD ACETATE see LCH000
BASIC LEAD CHROMATE see LCS000
BASIC MERCURIC SULFATE see MDG000
BASIC NICKEL CARBONATE see NCY500
BASIC ORANGE 3RN see BAQ250, BJF000
BASIC PANCREATIC TRYPSIN INHIBITOR see PAF550
BASIC PARAFUCHSINE see RMK020
BASIC RED 18 see BAQ750
BASIC VIOLET 10 see FAG070
BASIC VIOLET K see MQN025

BASIC YELLOW K see DBT400
BASIC ZINC CHROMATE see ZFJ100
BASIC ZIRCONIUM CHLORIDE see ZSJ000
BASIL OIL see BAR250
BASIL OIL, EUROPEAN TYPE (FCC) see BAR250
BASINEX see DGI400
BASKET FLOWER see BAR325
BASLE GREEN see COF500
BASOFORTINA see MJV750, PAM000
BASOLAN see MCO500
BASON see BNK350
BASORA CORRA see BAR500
B. ASPER VENOM see BMI000
BASSA see MOV000
BASSINET (CANADA) see FBS100
BASTARD ACACIA see GJU475
BASUDIN see DCM750
BASUDIN 10 G see DCM750
BATASAN see ABX250
BATAZINA see BJP000
BATEL see BFW250
BATRACHOTOXIN see BAR750
BATRAFEN see BAR800
BATRILEX see PAX000
BATROXOBIN see RDA350
B. ATROX VENOM see BMI125
BATTRE AUTOUR (HAITI) see JBS100
BAUMYCIN A1 see BAR825
BAUMYCIN A2 see BAR830
BAUXITE RESIDUE see IHD000
BAVISTIN see MHC750
BAX see BAU750
BAX 1400Z see PMO250
BAX 2793Z see BEO750
1BA-4-XA (RUSSIAN) see CEL750
BAXACOR see DHS200
BAXARYTMON see PMJ525
BAX 2793Z see THL750
BAY 1040 see AEC750
BAY 1470 see DMW000
BAY 1521 see DPH600
BAY 2353 see DFV400
BAY 3517 see AJV500
BAY 4059 see BOL325
BAY 4503 see PMX250
BAY 4934 see MGQ750
BAY 5097 see MRX500
BAY 5621 see BAT750
BAY 5821 see DJY200
BAY 9010 see PMY300
BAY 9015 see DFD000
BAY 9017 see DJR800
BAY 9026 see DST000
BAY 9027 see ASH500
BAY 10756 see DAO600
BAY 11405 see MNH000
BAY 15203 see DAO800, MIW100
BAY 15922 see TIQ250
BAY 16225 see EKN000
BAY 18436 see DAP400
BAY 19149 see DGP900
BAY 21097 see DAP000
BAY 23129 see PHI500
BAY 23323 see OQS000
BAY 23655 see DSK600
BAY 25141 see FAQ800
BAY 25634 see EAT600
BAY 29492 see LIN400
BAY 29493 see FAQ900
BAY 30130 see DGI000
BAY 32394 see TNH750
BAY 32651 see MIB250
BAY 33051 see DRR400
BAY 33819 see BIM000
BAY 34042 see ENI500
BAY 34727 see COQ399
BAY 37341 see DJR800
BAY 37342 see DST200
BAY 38156 see PHM750
BAY 41831 see DSQ000
BAY 42696 see DQE800
BAY 42903 see EOO000
BAY 44646 see DOR400
BAY 45432 see DNX800

BAY 46131 see ZMA000
BAY 47531 see DFL200
BAY 48130 see DLS800
BAY 50282 see DBI800
BAY 50519 see EOP500
BAY 61597 see MQR275
BAY 62863 see DLS800
BAY 68138 see FAK000
BAY 70143 see CBS275
BAY 70533 see CFC750
BAY 71628 see DTQ400
BAY 75546 see BAS000
BAY 77049 see DJY200
BAY 77488 see BAT750
BAY 79770 see CDP750
BAY-92114 see IMF300
BAY 105807 see MIA250
BAY A 1040 see AEC750
BAY-a 7168 see NDY600
BAY BUE 1452 see THT500
BAY BUSH (BAHAMAS) see CNH789
BAYCAIN see BMA125
BAYCAINE see BMA125
BAYCALNE see BMA125
BAYCARD see MOV000
BAYCARON see MCA100
BAYCID see FAQ900
BAYCLEAN see AFP250
BAY COE 3664 see BAS500
BAYCOVIN see DIZ100
BAY d8815 see DPF200
BAY DIC 1468 see MQR275
BAY-DRW 1139 see ALA500
BAY E-601 see MNH000
BAY E-605 see PAK000
BAY e 5009 see EMR600
BAY-E 9736 see NDY700
BAY ENE 11183 B see EAT600
BAYER 73 see DFV400, DFV600
BAYER-186 see CMW700
BAYER 205 see BAT000
BAYER 693 see SLP600
BAYER 1219 see AFL500
BAYER 1355 see PDC850
BAYER 1362 see BSZ000
BAYER 1420 see PMM000
BAYER 2353 see DFV400
BAYER 2502 see NGG000
BAYER 3231 see TND000
BAYER 4245 see DET000
BAYER 4964 see ORU000
BAYER 5072 see DOU600
BAYER 5080 see DOR400
BAYER 5081 see EPY000
BAYER 5312 see EPQ000
BAYER 5360 see MMN250
BAYER 8169 see DAO500, DAO600
BAYER 9007 see FAQ900
BAYER 9013 see DST200
BAYER 9015 see DFD000
BAYER 9051 see TDX750
BAYER 15080 see BDD000
BAYER 15922 see TIQ250
BAYER 16259 see EKN000
BAYER 17147 see ASH500
BAYER 18510 see DRR400
BAYER 19639 see DXH325
BAYER 20315 see DAP600
BAYER 23655 see DSK600
BAYER 25142 see PHM000
BAYER 25198 see EAV000
BAYER 25 634 see EAT600
BAYER 25648 see DFV600
BAYER 25820 see HMV000
BAYER 29492 see LIN400
BAYER 29952 see MOB599
BAYER 32394 see TNH750
BAYER 33172 see FQK000
BAYER 34727 see COQ399
BAYER 36205 see ORU000
BAYER 37289 see EPY000
BAYER 37341 see DJR800
BAYER 37342 see DST200
BAYER 37344 see DST000

BAYER 38819 see BIM000
BAYER 38920 see HCK000
BAYER 39007 see PMY300
BAYER 41831 see DSQ000
BAYER 42903 see EOO000
BAYER 44646 see DOR400
BAYER 45,432 see DNX800
BAYER 46131 see ZMA000
BAYER 47531 see DFL200
BAYER 52553 see PHN000
BAYER 6159H see MQR275
BAYER 62863 see DLS800
BAYER 6443H see MQR275
BAYER 70533 see CFC750
BAYER 71628 see DTQ400
BAYER 78418 see EIM000
BAYER 94337 see MQR275
BAYER 21/116 see MIW100
BAYER 21/199 see CNU750
BAYER 25/154 see DAP400
BAYER A-128 see PAF550
BAYER A 139 see BDC750
BAYER B-186 see CMW700
BAYER-E 393 see SOD100
BAYER E-605 see PAK000
BAYER G4073 see BGW750
BAYERITIAN see TGG760
BAYER 1440 L see DNA800, MJE780
BAYER L 13/59 see TIQ250
BAYER R39 SOLUBLE see BDC750
BAYER S767 see FAQ800
BAYER S 4400 see EPY000
BAYER S 5660 see DSQ000
BAYERTITAN see TGG760
BAY 6681 F see CJO250
BAY G 2821 see EAE675
BAYGON see PMY300
BAYGON, NITROSO derivative see PMY310
BAY H 4502 see BGA825
BAY HOL 0574 see EON500
BAY-HOX-1901 see EPR000
BAY LEAF OIL see BAT500, LBK000
BAYLETON see CJO250
BAYLUSCID see DFV400, DFV600
BAYLUSCIDE see DFV600
BAY-MEB-6447 see CJO250
BAYMIX 50 see CNU750
BAY NTN 8629 see DGC800
BAY-NTN-9306 see SOU625
BAY OIL see BAT500
BAYOL F see MQV750
BAYPRESOL see DNA800, MJE780
BAYRE 77488 see BAT750
BAYRITES see BAP000
BAYROGEL see HKK000
BAYRUSIL see DJY200
BAY S 2758 see MIB250
BAY-SRA-12869 see IMF300
BAYTAN see MEP250
BAYTEX see FAQ900
BAYTHION see BAT750
BAYTITAN see TGG760
BAY VA 1470 see DMW000
BAY-VA 4059 see BOL325
BAZUDEN see DCM750
BB-8 see OAH000
BBAL see NJO200
BBC see BAV575, BMW250
BBC 12 see DDL800
BBCE see BIQ500
BBH see BBQ500
BB-K8 see APT000
"B" BLASTING POWDER see ERF500
BBN see BMW250, HJQ350
BBNOH see HJQ350
BBP see BEC500
BCF-BUSHKILLER see TAA100
B-CHLORO-N,N-DIMETHYLAMINODIBORANE see CGD399
BCM see MHC750
BCM (NH) see BCD325
BCME see BIK000
BCNU see BIF750
BCP see BQW825
BCPE see BIN000

BCPE mixed with SPAS see CKL500
BCPN see BQQ250
BCS COPPER FUNGICIDE see CNP250
BCTB see DIG400
BD 40A see FNE100
BDCM see BND500
(BDH) see EQL000
BDH 312 see GGS000
BDH 1298 see VTF000
BDH 2700 see BAT795
BDH 6140 see BAT800
BDH 6146 see TKH050
BDH 29-801 see TAI250
B.D.H. 200 HYDROCHLORIDE see EEQ000
B-DIETHYLAMINOETHYLAMINOPHENYLACETIC ACID ISOAMYL ESTER
    see NOC000
BDMA see DQP800
BDM-CHLORIDE (RUSSIAN) see AFP075
BD(a,h)P see DCY200
BDU see BNC750
5-BDU see BNC750
BE see BNI500
BE 1293 see CLF325
Be 724-A see BEQ625
BEACH APPLE see MAO875
BEACILLIN see BFC750
BEAD TREE see CDM325
BEAM see MQC000
BEAMETTE see PMP500
BEAN SEED PROTECTANT see CBG000
BEAN TREE see GIW195
BEAUTYLEAF see MBU780
BEAVER POISON see WAT325
BEBUXINE see CQH325
BEC 001 see AQP800
BECAMPICILLIN see BAB250
BECANTAL see DEE600
BECANTEX see DEE600
BECANTYL see DEE600
BECAPTAN see AJT250
BECAPTAN DISULFURE (FRENCH) see MCN500
BECILAN see PPK500
BECLACIN see AFJ625
BECLAMID see BEG000
BECLAMIDE see BEG000
BECLOFORTE see AFJ625
BECLOMETASONE DIPROPIONATE see AFJ625
BECLOMETASONE-17,21-DIPROPIONATE see AFJ625
BECLOMETHASONE DIPROPIONATE see AFJ625
BECLOVAL see AFJ625
BECLOVENT see AFJ625
BECOREL see PAN100
BECOTIDE see AFJ625
BEECHWOOD CRESOATE see BAT850
BEESIX see PPK250
BEESWAX see BAU000
BEESWAX, WHITE see BAU000
BEESWAX, YELLOW see BAU000
BEET-KLEEN see CBM000, CKC000, DTP400
"BEETLE" see BAU250
BEET SUGAR see SNH000
BEFLAVINE see RIK000
BEFUNOLOL HYDROCHLORIDE see BAU255
BEHA see AEO000
BEHEN see MBU800
N⁴-BEHENOYL-1-β-d-ARABINOFURANOSYLCYTOSINE see EAU075
BEHENOYLCYTOSINE ARABINOSIDE see EAU075
N⁴-BEHENOYLCYTOSINE ARABINOSIDE see EAU075
BEHP see DVL700
BEI-1293 see CLF325
BEIENO (MEXICO) see HAQ100
BEIVON see TES750
BEJUCO AHOJA VACA (DOMINICAN REPUBLIC) see YAK300
BEJUCO DO PEO (PUERTO RICO) see CDH125
BEJUCO de LOMBRIZ (CUBA) see PGQ285
BEK see BJU000
BEKADID see OOI000
BEKANAMYCIN see BAU270
BEKANAMYCIN SULFATE see KBA100
BEKLAMID see BEG000
BELAMINE BLACK GX see AQP000
BELAMINE BLUE 2B see CMO000
BELAMINE SKY BLUE A see CMO500
BELAMINE SKY BLUE FF see CMN750

BELDAVRIN see HOT500
BELFENE see LJR000
BELGANYL see BAT000
BELGENINE see BAU325
BELLADONA (HAWAII) see AOO825
BELLADONNA see BAU500, DAD880
BELLADONNA LILY see AHI635
BELLASTHMAN see DMV600
BELL MINE see CAT225
BELL MINE PULVERIZED LIMESTONE see CAO000
BELLUAINE (CANADA) see SED550
BELLYACHE BUSH see CNR135
BELMARK see FAR100
BELOC see SBV500
BELOSIN see NOC000
BELSEREN see CDQ250
BELT see CDR750
BELUSTINE see CGV250
BEMACO see CLD000
BEMAPHATE see CLD000, CLD250
BEMASULPH see CLD000
BEMEGRIDE see MKA250
BEN-30 see BAV000
BENA see BAU750, BBV500
BENACHLOR see BBV500
BENACTIZINA (ITALIAN) see DHU900
BENACTIZINE HYDROCHLORIDE see BCA000
BENACTYZIN see DHU900
BENACTYZIN (CZECH) see BCA000
BENACTYZINE see DHU900
BENACTYZINE CHLORIDE see BCA000
BENACTYZINE HYDROCHLORIDE see BCA000
BENADON see BBV500, PPK500
BENADRIN see BBV500
BENADRYL see BAU750, BBV500
BENADRYL HYDROCHLORIDE see BAU750
BEN-A-HIST see PDC000
BENAKTIN see BCA000
BENALGIN see BBW500
BEN-ALLERGIN see BBV500
BENANSERIN HYDROCHLORIDE see BEM750
BENAPON see BBV500
BENASPIR see ADA725
BENAZALOX see BAV000
BENAZIDE see IKC000
BENAZOLIN see BAV000
BENAZOL P see HML500
BENCARBATE see DQM600
BENCHINOX see BDD000
BENCICLANE see BAV250
BENCIDAL BLACK E see AQP000
BENCIDAL BLUE 2B see CMO000
BENCIDAL BLUE 3B see CMO250
BENCIDAL FAST VIOLET N see CMP000
BENCONASE see AFJ625
BEN-CORNOX see BAV000
BENCYCLANE see BAV250
BENCYCLANE FUMARATE see BAV250
BENDACORT see BAV275
BENDAZAC see BAV325
BENDAZOL see BEA825
BENDAZOLE see BEA825
BENDAZOLIC ACID see BAV325
BENDECTIN see BAV350
BENDEX see BLU000
BENDIOCARB see DQM600, MHZ000
BENDIOXIDE see MJY500
BENDOPA see DNA200
BENDRALAN see PDD350
BENDROFLUAZIDE see BEQ625
BENDROFLUMETHIAZIDE see BEQ625
BENDYLATE see BAU750
BENECARDIN see AHK750
BENECID see DWW000
BENEMID see DWW000
BENEPEN see PAQ100
BENESAL see SAH000
BENETACIL see PAQ100
BENETHAMINE PENICILLIN see PAQ100
BENETHAMINE PENICILLIN G see PAQ100
BENETOLIN see PAQ100
BENFOS see DGP900
BENGAL GELATIN see AEX250
BENGAL ISINGLASS see AEX250

BENGUINOX see BDD000
BEN-HEX see BBQ500
BENICOT see NCR000
BENIROL see BBA500
BENKFURAN see NGE000
BENLATE 50 see BAV575
BENLATE and SODIUM NITRITE see BAV500
BENMOXINE see NCQ100
BENNIE see AOB250
BENOCTEN see BAU750
BENODAINE HYDROCHLORIDE see BCI500
BENODIN see BBV500
BENODINE see BBV500
BENOMYL see BAV575
BENOMYL 50W see BAV575
BENOPAN see BAV000
BENOQUIN see AEY000
BENORAL see SAN600
BENORILATE see SAN600
BENORTAN see SAN600
BENORTERONE see MNC150
BENORYLATE see SAN600
BENOTERONE see MNC150
BENOVOCYLIN see EDP000
BENOXAPROFEN see OJI750
BENOXIL see OPI300
BENOXINATE HYDROCHLORIDE see OPI300
BENOXYL see BDS000
BENOZIL see DAB800
BEN-P see BFC750
BENPERIDOL see FGU000, FLK100
BENPROPERINE PHOSPHATE see PJA130
BENQUINOX see BDD000
BENSECAL see BAV000
BENSERAZIDE HYDROCHLORIDE see SCA400
BENSULFOID see SOD500
BENSULIDE see DNO800
BENSYLYT see DDG800
BENSYLYTE see PDT250
BENT see MDQ250
BENTANEX see EEO500
BENTANIDOL see BFW250
BENTAZEPAM see BAV625
BENTAZON see MJY500
BENTHIOCARB see SAZ000
BENTHIOZONE see FNF000
BENTIROMIDE see CML835
BENTONE see KBB600
BENTONITE see BAV750
BENTONITE 2073 see BAV750
BENTONITE MAGMA see BAV750
BENTONYL see TJL250
BENTOX see BAU255
BENTOX 10 see BBQ500
BENTRIDE see BEQ625
BENTROL see DNF400, HKB500
BENURIDE see PFB350
BENURON see BEQ625
BEN-U-RON see HIM000
BENURYL see DWW000
BENVIL see MOV500
(5R,6R)-BENXYLPENICILLIN see BDY669
BENYLAN see BBV500
BENYLATE see BCM000
BENZABAR see TIK500
BENZAC see BDS000, PJQ000, TIK500
BENZAC 1281 see DOR800
BENZ(1)ACEANTHRENE see BAW000
BENZ(1)ACEANTHRYLENE see BAW125
BENZ(j)ACEANTHRYLENE see CMC000
BENZ(j)ACEANTHRYLENE, 1,2-DIHYDRO-3,6-DIMETHYL-(9CI) see DRD850
1,2-BENZACENAPHTHENE see FDF000
BENZ(k)ACEPHENANTHRENE see AAF000
BENZ(e)ACEPHENANTHRYLENE see BAW250
3,4-BENZ(e)ACEPHENANTHRYLENE see BAW250
BENZACILLIN see BFC750
BENZACIN see BAW500
BENZACINE see BAW500
BENZACINE HYDROCHLORIDE see BAW500
BENZACIN HYDROCHLORIDE see BAW500
BENZACONINE see PIC250
BENZ(a)ACRIDIN-10-AMINE see AIC750
BENZ(c)ACRIDINE see BAW750
3,4-BENZACRIDINE see BAW750

7,8-BENZACRIDINE (FRENCH) see BAW750
3,4-BENZACRIDINE-9-ALDEHYDE see BAX250
BENZ(c)ACRIDINE-7-CARBONITRILE see BAX000
BENZ(c)ACRIDINE-7-CARBOXALDEHYDE see BAX250
BENZ(a)ACRIDINE-5,6-DIOL, 5,6-DIHYDRO-12-METHYL-, (Z)- see DLE500
BENZ(c)ACRIDINE 3,4-DIOL-1,2-EPOXIDE-1 see EBH875
BENZ(c)ACRIDINE 3,4-DIOL-1,2-EPOXIDE-2 see EBH850
BENZ(c)ACRIDINE, 3-METHOXY-7-METHYL- see MET875
BENZ(c)ACRIDIN-4-OL, 7-METHYL-, ACETATE (ESTER) see ABQ600
N'-BENZ(c)ACRIDIN-7-YL-N-(2-CHLOROETHYL)-N-ETHYL-1,2-ETHANEDIAM
     INE DIHYDROCHLORIDE see EHI500
α-(BENZ(c)ACRIDIN-7-YL)-N-(p-(DIMETHYLAMINO)PHENYL)NITRONE
     see BAY250
α-(9-(3,4-BENZACRIDYL)-N-(p-DIMETHYLAMINO-PHENYL)-NITRONE
     see BAY250
BENZADONE GOLDEN YELLOW see DCZ000
BENZADONE GREY M see CMU475
BENZADOX see ANB000
BENZAHEX see BBQ750
BENZAIDIN see BFW250
BENZAKNEW see BDS000
BENZALACETON (GERMAN) see SMS500
BENZALACETONE see SMS500
2-BENZALACETOPHENONE see CDH000
BENZAL ALCOHOL see BDX500
BENZAL-(BENZYL-CYANID) (GERMAN) see DVX600
BENZAL CHLORIDE see BAY300
BENZALDEHYDE see BAY500
BENZALDEHYDE CYANOHYDRIN see MAP250
BENZALDEHYDE, DIMETHYL ACETAL see DOG700
BENZALDEHYDE FFC see BBM500
BENZALDEHYDE GLYCERYL ACETAL (FCC) see BBA000
BENZALDEHYDE GREEN see AFG500, BAY750
BENZALDEHYDE, 2-METHOXY-(9CI) see AOT525
BENZALDEHYDE THIOSEMICARBAZONE see BAZ000
BENZALDEHYDKYANHYDRIN (CZECH) see MAP250
BENZAL GLYCERYL ACETAL see BBA000
BENZALIN see DLY000
BENZALKONIUM CHLORIDE see AFP250, BBA500
BENZALKONIUM SACCHARINATE see BBA625
BENZALMALONONITRILE see BBA750
BENZAMIDE see BBB000
BENZAMIDE, 4-AMINO-5-BROMO-N-(2-(DIETHYLAMINO)ETHYL)-2-
     METHOXY-(9CI) see VCK100
BENZAMIDE, 4-AMINO-5-CHLORO-N-(1-(3-(4-FLUOROPHENOXY)PROPYL)-3-
     METHOXY-4-PIPERIDINYL)-2-METHOXY-, MONOHYDRATE, cis-
     see CMS237
BENZAMIDE, N-(5-CHLORO-4-((4-CHLOROPHENYL)CYANOMETHYL)-2-
     METHYLPHENYL)-2-HYDROXY-3,5- DIIODO- see CFC100
BENZAMIDE, N-((1-ETHYL-2-PYRROLIDINYL)METHYL)-5-
     (ETHYLSULFONYL)-2-METHOXY- see EPD100
BENZAMIDEPHENYLHYDRAZONE HYDROCHLORIDE see PEK675
(2-BENZAMIDO)ACETOHYDROXAMIC ACID see BBB250
N-(4-BENZAMIDO-4-CARBAMOYLBUTYL)-N-NITROSOCYANAMIDE
     see CBK250
1-BENZAMIDO-5-CHLORO-ANTHRAQUINONE see BDK750
dl-α-BENZAMIDO-p-(2-(DIETHYLAMINO)ETHOXY)-N,N-
     DIPROPYLHYDROCINNAMAMIDE see TGF175
dl-4-BENZAMIDO-N,N-DIPROPYLGLUTARAMIC ACID see BGC625
dl-4-BENZAMIDO-N,N-DIPROPYLGLUTARAMIC ACID SODIUM SALT
     see PMH575
(S)-p-(α-BENZAMIDO-p-HYDROXYHYDROCINNAMAMIDO)BENZOIC ACID
     see CML835
BENZAMIDOOXY ACETIC ACID, AMMONIUM SALT see ANB000
1-BENZAMIDO-1-PHENYL-3-PIPERIDINOPROPANE HYDROCHLORIDE
     see DKK800
N-(3-BENZAMIDO-3-PHENYL)PROPYL PIPERIDINE HYDROCHLORIDE
     see DKK800
BENZAMIL BLACK E see AQP000
BENZAMIL SUPRA BROWN BRLL see CMO750
BENZAMINE BLUE see CMO250
BENZAMPHETAMINE see AOB250
BENZANIDINE see BFW250
BENZANIL BLUE 2B see CMO000
BENZANILIDE, 2'-CHLORO-2-(2-(DIETHYLAMINO)EHOXY)- see DHP450
BENZANILIDE, 4'-CHLORO-2-(2-(DIETHYLAMINO)ETHOXY)- see DHP550
BENZANIL VIOLET N see CMP000
BENZ(a)ANTHRACEN-7-ACETIC ACID, METHYL ESTER see BBC500
BENZ(a)ANTHRACEN-7-ACETONITRILE see BBB500
BENZ(a)ANTHRACEN-7-AMINE see BBB750
BENZ(a)ANTHRACEN-8-AMINE see BBC000
BENZANTHRACENE see BBC250
BENZ(a)ANTHRACENE see BBC250
BENZ(b)ANTHRACENE see NAI000

1,2-BENZANTHRACENE see BBC250
2,3-BENZANTHRACENE see NAI000
1,2-BENZ(a)ANTHRACENE see BBC250
1,2:5,6-BENZANTHRACENE see DCT400
1,2-BENZANTHRACENE-10-ACETIC ACID, METHYL ESTER see BBC500
1,2-BENZANTHRACENE-10-ALDEHYDE see BBC750
BENZ(a)ANTHRACENE, 8-BROMO-7,12-DIMETHYL- see BNF315
BENZ(a)ANTHRACENE-7-CARBOXALDEHYDE see BBC750
BENZ(a)ANTHRACENE-7,12-DICARBOXALDEHYDE see BBD000
BENZ(a)ANTHRACENE-1,2-DIHYDRODIOL see BBD250
BENZ(a)ANTHRACENE-3,4-DIHYDRODIOL see BBD500
BENZ(a)ANTHRACENE-5,6-DIHYDRODIOL see BBE250
BENZ(a)ANTHRACENE-10,11-DIHYDRODIOL see BBF000
BENZ(a)ANTHRACENE-5,6-cis-DIHYDRODIOL see BBE000
BENZ(a)ANTHRACENE-5,6-trans-DIHYDRODIOL see BBE250
trans-BENZ(a)ANTHRACENE-8,9-DIHYDRODIOL see BBE750
(+)-(3S,4S)trans-BENZ(a)ANTHRACENE-3,4-DIHYDRODIOL see BBD750
BENZ(a)ANTHRACENE, 3,4-DIHYDROXY-1,2-EPOXY-1,2,3,4-TETRAHYDRO-,
     (Z), (+)- see DMO500
BENZ(a)ANTHRACENE-7,12-DIMETHANOL see BBF500
BENZ(a)ANTHRACENE-7,12-DIMETHANOLDIACETATE see BBF750
BENZ(a)ANTHRACENE-3,9-DIOL see BBG200
(−)(3R,4R)-trans-BENZ(a)ANTHRACENE-3,4-DIOL see BBG000
BENZ(a)ANTHRACENE 3,4-DIOL-1,2-EPOXIDE-2 see DLE000
BENZ(a)ANTHRACENE-5,6-EPOXIDE see BDJ500
BENZ(a)ANTHRACENE-7-ETHANOL see BBG500
BENZ(a)ANTHRACENE-7-METHANEDIOLDIACETATE (ester) see BBG750
BENZ(a)ANTHRACENE-7-METHANETHIOL see BBH000
BENZ(a)ANTHRACENE-7-METHANOL see BBH250
BENZ(a)ANTHRACENE-7-METHANOL ACETATE see BBH500
7H-BENZ(de)ANTHRACENE-7-ONE see BBI250
BENZ(a)ANTHRACENE-5,6-OXIDE see BDJ500, EBP000
BENZ(a)ANTHRACENE, 8-PHENYL- see PEK750
BENZ(a)ANTHRACENE-7-THIOL see BBH750
BENZ(a)ANTHRACEN-5-OL see BBI000
7H-BENZ(de)ANTHRACEN-7-ONE see BBI250
N-(BENZ(a)ANTHRACEN-5-YLCARBAMOYL)GLYCINE see BBI750
N-(BENZ(a)ANTHRACEN-7-YLCARBAMOYL)GLYCINE see BBJ000
BENZ(a)ANTHRACEN-7-YL-OXIRANE see ONC000
1-(BENZ(a)ANTHRACEN-7-YL)-2,2,2-TRICHLOROETHANONE see TIJ000
BENZ(a)ANTHRACEN-7-YL TRICHLOROMETHYL KETONE see TIJ000
BENZ(a)ANTHRA-5,6-OXIDE see EBP000
BENZ(3,4)ANTHRA(1,2-6)OXIRENE see BDJ500
1,2-BENZANTHRAZEN (GERMAN) see BBC250
BENZANTHRENE see BBC250
1,2-BENZANTHRENE see BBC250
2,3-BENZANTHRENE see NAI000
BENZANTHRENONE see BBI250
BENZANTHRONE see BBI250
1,2-BENZANTHRYL-3-CARBAMIDOACETIC ACID see BBI750
1,2-BENZANTHRYL-10-CARBAMIDOACETIC ACID see BBJ000
1,2-BENZANTHRYL-10-ISOCYANATE see BBJ250
1,2-BENZANTHRYL-10-MERCAPTAN see BBH750
1,2-BENZANTHRYL-10-METHYLMERCAPTAN see BBH000
7-BENZANTHRYLOXIRANE see ONC000
BENZANTINE see BBV500
BENZAR see BAV000
BENZARONE see BBJ500
BENZATHINE BENZYLPENICILLIN see BFC750
BENZATHINE PENICILLIN see BFC750
BENZATHINE PENICILLIN G see BFC750
BENZATROPINE METHANESULFONATE see TNU000
BENZAZIDE, BENZOIC ACID AZIDE see BDL750
BENZAZIMIDE see BDH000
BENZAZIMIDONE see BDH000
1-BENZAZINE see QMJ000
2-BENZAZINE see IRX000
1-BENZAZOLE see ICM000
BENZAZOLINE see BBW750
BENZAZOLINE HYDROCHLORIDE see BBJ750
BENZAZON VII see NGC400
BENZBROMARON see DDP200
BENZBROMARONE see DDP200
1,2-BENZCARBAZOLE see BCG250
BENZ-o-CHLOR see DER000
BENZCHLOROPROPAMIDE see BEG000
BENZCHLORPROPAMID see BEG000
3,4-BENZCHRYSENE see PIB750
BENZCURINE IODIDE see PDD300
15,16-BENZDEHYDROCHOLANTHRENE see DCR400
BENZEDREX (SKF) see PNN300
BENZEDRINE see BBK000
(±)-BENZEDRINE see BBK000
dl-BENZEDRINE see BBK000

BENZEDRINE SULFATE see BBK250
d-BENZEDRINE SULFATE see BBK500
l-BENZEDRINE SULFATE see BBK750
BENZEDRYNA see AOB250
BENZEEN (DUTCH) see BBL250
BENZEHIST see BAU750
BENZEN (POLISH) see BBL250
BENZENACETIC ACID see PDY850
BENZENAMINE see AOQ000
BENZENAMINE, 4-((3-BROMOPHENYL)AZO)-N,N-DIMETHYL-(9CI)
  see BNE600
BENZENAMINE, 2,6-DICHLORO-N-2-IMIDAZOLIDINYLIDENE- (9CI)
  see DGB500
BENZENAMINE, 4-((3,4-DICHLOROPHENYL)AZO)-N,N-DIMETHYL-(9CI)
  see DFD400
BENZENAMINE, 4-((3,4-DIETHYLPHENYL)AZO)-N,N-DIMETHYL-(9CI)
  see DJB400
BENZENAMINE, N,N-DIMETHYL-3'-BROMO-4'-ETHYL-4-(PHENYLAZO)-
  see BNK275
BENZENAMINE, N,N-DIMETHYL-4'-BROMO-3'-ETHYL-4-(PHENYLAZO)-
  see BNK100
BENZENAMINE, N,N-DIMETHYL-3'-BROMO-4'-METHYL-4-(PHENYLAZO)-
  see BNQ100
BENZENAMINE, N,N-DIMETHYL-4'-BROMO-3'-METHYL-4-(PHENYLAZO)-
  see BNQ110
BENZENAMINE, N,N-DIMETHYL-3'-CHLORO-4'-ETHYL-4-(PHENYLAZO)-
  see CGW100
BENZENAMINE, N,N-DIMETHYL-4'-CHLORO-3'-ETHYL-4-(PHENYLAZO)-
  see CGW105
BENZENAMINE, N,N-DIMETHYL-2'-ETHYL-4-(PHENYLAZO)- see EIF450
BENZENAMINE, 4-((2-ETHYLPHENYL)AZO)-N,N-DIMETHYL- see EIF450
BENZENAMINE, 4-FLUORO-(9CI) see FFY000
BENZENAMINE HYDROCHLORIDE see BBL000
BENZENAMINE, N-(2-METHYL-2-NITROPROPYL)-p-NITROSO-(9CI)
  see NHK800
BENZENAMINE, 4-STIBONO-(9CI) see SLP600
BENZENAMINE, 2,3,4-TRIFLUORO- see TJX900
BENZENAMINIUM,3-(((DIMETHYLAMINO)CARBONYL)OXY)-N,N,N-
  TRIMETHYL-, BROMIDE (9CI) see POD000
BENZENE see BBL250
BENZENEACETALDEHYDE see BBL500
BENZENEACETAMIDE (9CI) see PDX750
BENZENEACETIC ACID see PDY850
BENZENEACETIC ACID, α-AMINO-3-HYDROXY-4-(HYDROXYMETHYL)-, (S)-
  see FOJ100
BENZENEACETIC ACID, BUTYL ESTER (9CI) see BQJ350
BENZENEACETIC ACID, 2,4-DICHLORO-, ETHYL ESTER (9CI) see EHY600
BENZENEACETIC ACID, 3,7-DIMETHYL-6-OCTENYL ESTER (9CI)
  see CMU050
BENZENEACETIC ACID, ETHYL ESTER (9CI) see EOH000
BENZENEACETIC ACID, 3-HEXENYL ESTER, (Z)- see HFE625
BENZENEACETIC ACID, α-HYDROXY-, 2-(2-ETHOXYETHOXY)ETHYL
  ESTER see HJG100
BENZENEACETIC ACID-2-METHOXY-4-(2-PROPENYL)PHENYL ESTER
  see AGL000
BENZENEACETIC ACID, 3-METHYLBUTYL ESTER (9CI) see IHV000
BENZENEACETIC ACID, METHYL ESTER see MHA500
BENZENEACETIC ACID, α-METHYL-4-((2-METHYL-2-PROPENYL)AMINO)-,
  (±)- see MGC200
BENZENEACETIC ACID, α-METHYL-4-(2-METHYLPROPYL)-, SODIUM SALT
  (9CI) see IAB100
BENZENEACETIC ACID, α-PHENYL-, 2-(DIETHYLAMINO)ETHYL ESTER,
  (9CI) see DHX800
BENZENEACETIC ACID, 2-PHENYLETHYL ESTER see PDI000
BENZENEACETIC ACID, 2-PROPENYL ESTER see PMS500
BENZENEACETONITRILE see PEA750
BENZENEARSONIC ACID see BBL750
BENZENEARSONIC ACID, p-ACETONYLOXY-, SODIUM SALT see SJK475
BENZENEARSONIC ACID, 4-AMINO-2-HYDROXY- see HJE400
BENZENEARSONIC ACID, 3,4-DIFLUORO- see DKG100
BENZENEARSONIC ACID, 4-(p-DIMETHYLAMINOPHENYLAZO)-, HYDRO-
  CHLORIDE see DPO275
BENZENEARSONIC ACID, p-FLUORO- see FGA100
BENZENE AZIMIDE see BDH250
4-BENZENEAZOANILINE see PEI000
BENZENEAZO-2-ANTHROL see PEI750
BENZENEAZOBENZENE see ASL250
BENZENEAZOBENZENEAZO-β-NAPHTHOL see OHA000
BENZENEAZODIMETHYLANILINE see DOT300
BENZENEAZO-β-NAPHTHOL see PEJ500
BENZENE-1-AZO-2-NAPHTHOL see PEJ500
l-BENZENEAZO-2-NAPHTHYLAMINE see FAG130
l-BENZENE-AZO-β-NAPHTHYLAMINE see FAG130
p-BENZENEAZOPHENOL see HJF000

BENZENE-1,3-BIS(SULFONYL AZIDE) see BBL825
BENZENEBORONIC ACID see BBM000
BENZENEBUTANOIC ACID, 3-CHLORO-4-CYCLOHEXYL-α-OXO- see CPJ000
BENZENECARBALDEHYDE see BAY500
BENZENECARBINOL see BDX500
BENZENECARBONAL see BAY500
BENZENECARBONYL CHLORIDE see BDM500
BENZENECARBOPEROXOIC ACID (9CI) see PCM000
BENZENECARBOTHIOAMIDE see BBM250
BENZENECARBOXALDEHYDE see BBM500
BENZENECARBOXIMIDAMIDE, 4-AMINO- (9CI) see AID650
BENZENECARBOXYLIC ACID see BCL750
BENZENE CHLORIDE see CEJ125
BENZENE, 4-CHLORO-1-(4-CHLOROPHENOXY)-2-NITRO- see CJD600
n-BENZENE-n-CYCLOPENTADIENYL IRON(II)PERCHLORATE see BBN000
m-BENZENEDIAMINE see PEY000
o-BENZENEDIAMINE see PEY250
p-BENZENEDIAMINE see PEY500
1,2-BENZENEDIAMINE see PEY250
1,3-BENZENEDIAMINE see PEY000
1,4-BENZENEDIAMINE see PEY500
m-BENZENEDIAMINE DIHYDROCHLORIDE see PEY750
p-BENZENEDIAMINE DIHYDROCHLORIDE see PEY650
1,4-BENZENEDIAMINE DIHYDROCHLORIDE see PEY650
1,3-BENZENEDIAMINE HYDROCHLORIDE see PEY750
BENZENE, 1,4-DIAZIDO- see DCL125
BENZENE DIAZONIUM-2-CARBOXYLATE see BBN250
BENZENE DIAZONIUM CHLORIDE see BBN500
BENZENEDIAZONIUM FLUOBORATE see BBO325
BENZENEDIAZONIUM FLUOROBORATE see BBO325
BENZENEDIAZONIUM HYDROGEN SULFATE see BBN650
BENZENEDIAZONIUM, 4-(HYDROXYMETHYL)-, SULFATE (2:1) see HLX900
BENZENEDIAZONIUM, 4-(HYDROXYMETHYL)-, TETRAFLUOROBORATE(1-)
  see HLX925
BENZENE DIAZONIUM NITRATE see BBN750
BENZENEDIAZONIUM-4-OXIDE see BBN850
BENZENE DIAZONIUM SALTS see BBO000
BENZENEDIAZONIUM-2-SULFONATE see BBO125
BENZENE DIAZONIUM-4-SULFONATE see BBO250
BENZENEDIAZONIUM TETRACHLOROZINCATE see DCW000
BENZENEDIAZONIUM TETRAFLUOROBORATE see BBO325
BENZENEDIAZONIUM TRIBROMIDE see BBO400
1,3-BENZENEDICARBONITRILE see PHX550
1,3-BENZENEDICARBONYL CHLORIDE see IMO000
1,4-BENZENEDICARBONYL CHLORIDE see TAV250
1,4-BENZENEDICARBONYL DICHLORIDE see TAV250
p-BENZENEDICARBOXALDEHYDE see TAN500
1,4-BENZENEDICARBOXALDEHYDE (9CI) see TAN500
m-BENZENEDICARBOXYLIC ACID see IMJ000
o-BENZENEDICARBOXYLIC ACID see PHW250
p-BENZENEDICARBOXYLIC ACID see TAN750
BENZENE-1,2-DICARBOXYLIC ACID see PHW250
1,2-BENZENEDICARBOXYLIC ACID see PHW250
1,4-BENZENEDICARBOXYLIC ACID see TAN750
1,2-BENZENEDICARBOXYLIC ACID ANHYDRIDE see PHW750
1,2-BENZENEDICARBOXYLIC ACID BI(2-METHOXYETHYL)ESTER (9CI)
  see DOF400
1,2-BENZENEDICARBOXYLIC ACID, BUTYL PHENYLMETHYL ESTER
  see BEC500
1,2-BENZENEDICARBOXYLIC ACID, DECYL OCTYL ESTER see OEU000
o-BENZENEDICARBOXYLIC ACID, DIBUTYL ESTER see DEH200
BENZENE-o-DICARBOXYLIC ACID DI-n-BUTYL ESTER see DEH200
1,2-BENZENEDICARBOXYLIC ACID, DIETHYL ESTER see DJX000
1,2-BENZENEDICARBOXYLIC ACID, DIHEPTYL ESTER (9CI) see HBP400
1,2-BENZENEDICARBOXYLIC ACID DIHEXYL ESTER see DKP600
1,2-BENZENEDICARBOXYLIC ACID, DIISOOCTYL ESTER see ILR100
1,2-BENZENEDICARBOXYLIC ACID DIMETHYL ESTER see DTR200
1,3-BENZENEDICARBOXYLIC ACID, DIMETHYL ESTER see IML000
1,4-BENZENE DICARBOXYLIC ACID DIMETHYL ESTER (9CI) see DUE000
o-BENZENEDICARBOXYLIC ACID DIOCTYL ESTER see DVL600
1,2-BENZENEDICARBOXYLIC ACID DIOCTYL ESTER see DVL600
1,2-BENZENEDICARBOXYLIC ACID, DIPENTYL ESTER see AON300
1,2-BENZENEDICARBOXYLIC ACID, DIPROPYL ESTER see DWV500
1,2-BENZENEDICARBOXYLIC ACID, DITRIDECYL ESTER see DXQ200
1,2-BENZENEDICARBOXYLIC ACID, MONO(2-METHYLPROPYL) ESTER
  (9CI) see MRI775
BENZENE, o-DIETHOXY- see CCP900
BENZENE, 1,2-DIETHOXY-(9CI) see CCP900
BENZENE-1,3-DIISOCYANATE see BBP000
BENZENE-, 1,3-DIISOCYANATOMETHYL- see TGM740
p-BENZENEDINITRILE see BBP250
m-BENZENEDIOL see REA000
o-BENZENEDIOL see CCP850
p-BENZENEDIOL see HIH000

1,2-BENZENEDIOL see CCP850
1,3-BENZENEDIOL see REA000
1,4-BENZENEDIOL see HIH000
(1,2-BENZENEDIOLATO-O)PHENYLMERCURY see PFO750
1,3-BENZENEDIOL, DIACETATE see REA100
1,3-BENZENEDIOL, 5-(2-((1,1-DIMETHYLETHYL)AMINO)-1-HYDROXYETHYL)-
    (9CI) see TAN100
1,2-BENZENEDIOL, 4-(1-HYDROXY-2-(METHYLAMINO)ETHYL)-, HYDRO-
    CHLORIDE, (R)- (9CI) see AES500
1,3-BENZENEDIOL, MONOACETATE see RDZ900
1,4-BENZENEDIOL, 2,3,5-TRIMETHYL- (9CI) see POG300
BENZENEFORMIC ACID see BCL750
BENZENE HEXACHLORIDE see BBP750
BENZENEHEXACHLORIDE (mixed isomers) see BBQ750
α-BENZENEHEXACHLORIDE see BBQ000
BENZENE HEXACHLORIDE-α-isomer see BBQ000
β-BENZENEHEXACHLORIDE see BBR000
Γ-BENZENE HEXACHLORIDE see BBQ500
BENZENE HEXACHLORIDE-γ-isomer see BBQ500
trans-α-BENZENEHEXACHLORIDE see BBR000
Δ-BENZENEHEXACHLORIDE see BFW500
BENZENE ISOPROPYL see COE750
BENZENE-1-ISOTHIOCYANATE see ISQ000
BENZENEMETHANAMINE-N-(2-CHLOROETHYL)-2-ETHOXY-5-NITRO
    see CGX325
BENZENEMETHANAMINE, 3,4-DIMETHOXY- see VIK050
BENZENEMETHANAMINE-N-NITROSO-N-PHENYL see BFE750
BENZENEMETHANOIC ACID see BCL750
BENZENEMETHANOL see BDX500
BENZENEMETHANOL, α-(1-AMINOETHYL)-3-HYDROXY-, (R-(R*,S*))-(9CI)
    see HNB875
BENZENEMETHANOL, α-(((2-(3,4-DIMETHOXYPHENYL)ETHYL)AMINO)
    METHYL)-4-HYDROXY-, (R)- see DAP850
BENZENEMETHANOL, 2-((4-(DIMETHYLAMINO)PHENYL)AZO)-(9CI)
    see HMB595
BENZENEMETHANOL, α-ETHYNYL-4-METHOXY- see HKA700
BENZENEMETHANOL, 2-HYDROXY- (9CI) see HMK100
BENZENEMETHANOL, α-METHYL- see PDE000
BENZENEMETHANOL, 4-(2-(5,6,7,8-TETRAHYDRO-5,5,8,8-TETRAMETHYL-2-
    NAPHTHALENYL)-1-PROPENYL)-, (E)- see AQZ300
BENZENE, METHYL- see TGK750
BENZENE, (2-(3-METHYLBUTOXY)ETHYL)- see IHV050
BENZENENITRILE see BCQ250
BENZENE, 1,1'-OXYBIS-, HEXACHLORO derivatives (9CI) see CDV175
BENZENEPHOSPHONIC ACID see PFV500
BENZENEPHOSPHONIC ACID, DIOCTYL ESTER see PFV750
BENZENE PHOSPHORUS DICHLORIDE (DOT) see DGE400
BENZENE PHOSPHORUS THIODICHLORIDE see PFW500
BENZENEPROPANAL see HHP000
BENZENEPROPANAMINE, N-(1,1-DIMETHYLETHYL)-α-METHYL-Γ-PHENYL-,
    HYDROCHLORIDE (9CI) see TBC200
BENZENEPROPANENITRILE (9CI) see HHP100
3-BENZENEPROPANOL see HHP050
BENZENEPROPANOL CARBAMATE see PGA750
BENZENEPROPANOL, PROPANOATE (9CI) see HHQ550
BENZENEPROPIONITRILE see HHP100
BENZENESELENIC ACID see BBR325
BENZENESELENONIC ACID see PGH500
BENZENESULFANILIDE see BBR750
BENZENESULFINYL AZIDE see BBR380
BENZENE SULFINYL CHLORIDE see BBR390
BENZENESULFONAMIDE see BBR500
BENZENESULFONAMIDE, 4-AMINO-N-(AMINOCARBONYL)- (9CI) see SNQ550
BENZENESULFONAMIDE, 4-AMINO-N-(5-CHLORO-2-QUINOXALINYL)-
    see CMA600
BENZENESULFONAMIDE, 4-AMINO-N-(5-(1,1-DIMETHYLETHYL)-1,3,4-
    THIADIAZOL-2-YL)- (9CI) see GEW750
BENZENESULFONAMIDE, 4-AMINO-N-(5-ETHYL-1,3,4-THIADIAZOL-2-YL)-,
    MONOSODIUM SALT (9CI) see SJW300
BENZENESULFONAMIDE, 4-CHLORO-N-((CYCLOHEXYLAMINO)CAR-
    BONYL)- see CDR550
BENZENESULFONAMIDE, 2-CHLORO-5-(2,3-DIHYDRO-1-HYDROXY-3-OXO-
    1H-ISOINDOL-1-YL)- (9CI) see CLY600
BENZENESULFONAMIDE, 2-CHLORO-5-(1-HYDROXY-3-OXO-1-
    ISOINDOLINYL)- see CLY600
BENZENESULFONAMIDE, 4-CHLORO-N-(((1-METHYLETHYL)AMINO)CAR-
    BONYL)- (9CI) see CDY100
BENZENESULFONAMIDE, N-CHLORO-4-METHYL-, SODIUM SALT (9CI)
    see CDP000
BENZENESULFONAMIDE, N-(3-CHLORO-2-OXO-1-(PHENYLMETHYL)PRO-
    PYL)-4-METHYL-,(S)- see THH450
BENZENESULFONAMIDE, p-((p-(DIMETHYLAMINO)PHENYL)AZO)-
    see SNW800
2-BENZENESULFONAMIDO-5-tert-BUTYL-1,3,4-THIADIAZOLE   see GFM200

2-BENZENESULFONAMIDO-5-(β-METHOXYETHOXY)PYRIMIDINE SODIUM
    SALT see GHK200
2-BENZENESULFONAMIDO-5-TERTIOBUTYL-1-THIA-3,4-DIAZOLE
    see GFM200
BENZENESULFONANILIDE see BBR750
BENZENESULFONATE de 4-CHLOROPHENYLE (FRENCH) see CJR500
BENZENE SULFONCHLORIDE see BBS750
BENZENESULFONIC ACID see BBS250
BENZENESULFONIC ACID, ALKYL DERIVATIVES see AFO500
BENZENESULFONIC ACID, (ACID) CHLORIDE see BBS750
BENZENESULFONIC ACID, 4-CHLOROPHENYL ESTER see CJR500
BENZENESULFONIC ACID, DODECYL ESTER see DXW400
BENZENE SULFONYL AZIDE see BBS500
BENZENESULFONYL CHLORIDE see BBS750
2-(BENZENESULFONYL)ETHANOL see BBT000
BENZENESULPHONAMIDE see BBR500
BENZENE SULPHONYL CHLORIDE (DOT) see BBS750
BENZENESULPHONYL FLUORIDE see BBT250
BENZENE, 1,2,3,5-TETRACHLORO- see TBO700
BENZENETETRAHYDRIDE see CPC579
BENZENETHIOL (DOT) see PFL850
1,2,4-BENZENETRICARBOXYLIC ACID ANHYDRIDE see TKV000
1,2,4-BENZENETRICARBOXYLIC ACID, CYCLIC 1,2-ANHYDRIDE see TKV000
1,2,4-BENZENETRICARBOXYLIC ANHYDRIDE see TKV000
BENZENETRIFUROXAN see BBU125
cis-BENZENE TRIIMINE see THQ600
BENZENE, 1,2,4-TRIMETHOXY-5-PROPENYL-, (E)- see IHX400
BENZENE, 1,2,4-TRIMETHOXY-5-PROPENYL-, trans- see IHX400
BENZENE-s-TRIOL see PGR000
1,2,3-BENZENETRIOL see PPQ500
1,2,4-BENZENETRIOL see BBU250
BENZENE-1,3,5-TRIOL see PGR000
1,3,5-BENZENETRIOL see PGR000
BENZENE TRIOZONIDE see BBU500
BENZENOL see PDN750
BENZENOSULFOCHLOREK (POLISH) see BBS750
BENZENOSULPHOCHLORIDE see BBS750
BENZENYL CHLORIDE see BFL250
BENZENYL FLUORIDE see BDH500
BENZENYL TRICHLORIDE see BFL250
BENZETAMOPHYLLINE HYDROCHLORIDE see BEO750, THL750
BENZETHACIL see BFC750
BENZETHIDIN see BBU625
BENZETHIDINE see BBU625
BENZETHONIUM CHLORIDE see BEN000
BENZETHONIUM CHLORIDE MONOHYDRATE see BBU750
BENZETIMIDE see BBU800
BENZETONIUM CHLORIDE see BEN000
BENZEX see BBQ750
2,3-BENZFLUORANTHENE see BAW250
3,4-BENZFLUORANTHENE see BAW250
10,11-BENZFLUORANTHENE see BCJ500
trans-BENZ(a,e)FLUORANTHENE-12,13-DIHYDRODIOL see BBU825
trans-BENZ(a,e)FLUOROANTHENE-3,4-DIHYDRODIOL see BBU810
BENZ(j)FLUOROANTHRENE see BCJ500
BENZHEXOL see PAL500
BENZHEXOL CHLORIDE see BBV000
BENZHEXOL HYDROCHLORIDE see BBV000
BENZHORMOVARINE see EDP000
BENZHYDRAMINE see BBV500
BENZHYDRAMINE HYDROCHLORIDE see BAU750
BENZHYDRAMINUM see BBV500
BENZHYDRAZIDE see BBV250
BENZHYDRIL see BBV500
BENZHYDRYL see BBV500
o-BENZHYDRYLDIMETHYLAMINOETHANOL see BBV500
o-BENZHYDRYLDIMETHYLAMINOETHANOL-8-CHLOROTHEOPHYLLINATE
    see DYE600
1-BENZHYDRYL-4-(2-(2-HYDROXYETHOXY)ETHYL)PIPERAZINE see BBV750
2-BENZHYDRYL-3-HYDROXY-N-METHYLPIPERIDINE HYDROCHLORIDE
    see BBW000
N-BENZHYDRYL-N-METHYL PIPERAZINE see EAN600
N-BENZHYDRYL-N'-METHYLPIPERAZINE HYDROCHLORIDE see MAX275
N-BENZHYDRYL-N'-METHYLPIPERAZINE MONOHYDROCHLORIDE
    see MAX275
(N-BENZHYDRYL)(N'-METHYL)DIETHYLENEDIAMINE see EAN600
2-(BENZHYDRYLOXY)-N,N-DIMETHYLETHYLAMINE see BBV500
2-(BENZHYDRYLOXY)-N,N-DIMETHYLETHYLAMINE with 8-
    CHLOROTHEOPHYLLINE see DYE600
2-(BENZHYDRYLOXY)-N,N-DIMETHYLETHYLAMINEHYDROCHLORIDE
    see BAU750
2-(BENZHYDRYLOXYETHYL)GUANIDINE see BBW250
4-(BENZHYDRYLOXY)-1-METHYLPIPERIDINE see LJR000
BENZIDAMINE HYDROCHLORIDE see BBW500

BENZIDAZOL see BBW750
BENZIDIN (CZECH) see BBX000
BENZIDINA (ITALIAN) see BBX000
BENZIDINE see BBX000
3,3'-BENZIDINEDICARBOXYLIC ACID see BFX250
3,3'-BENZIDINE DICARBOXYLIC ACID, DISODIUM SALT see BBX250
3,3'-BENZIDINE-Γ,Γ''-DIOXYDIBUTYRIC ACID see DEK400
Γ,Γ''-,3,3'-BENZIDINE DIOXYDIBUTYRIC ACID see DEK400
2,2'-BENZIDINEDISULFONIC ACID see BBX500
BENZIDINE HYDROCHLORIDE see BBX750
BENZIDINE LACQUER YELLOW G see DEU000
BENZIDINE SULFATE see BBY000
BENZIDINE-3-SULFURIC ACID see BBY250
BENZIDINE SULPHATE and HYDRAZINE-BENZENE see BBY300
BENZIDINE-3-SULPHURIC ACID see BBY250
BENZIDINE YELLOW see DEU000
BENZIDINE YELLOW TONER YT-378 see DEU000
BENZIDIN-3-YL ESTER SULFURIC ACID see BBY500
BENZIDIN-3-YL HYDROGEN SULFATE see BBY500
BENZIES see AOB250
BENZIL see BBY750
BENZILAN see DER000
BENZILATE DU DIETHYLAMINO-ETHANOL CHLORHYDRATE (FRENCH)
   see BCA000
1-BENZILBIGUANIDE CLORIDRATO (ITALIAN) see BEA850
BENZILE (CLORURO di) (ITALIAN) see BEE375
N-β-(BENZILFENILAMINO)ETILPIPERIDINA CLORIDRATO (ITALIAN)
   see BEA275
1-(2-(2-BENZILFENOSSI)-1-METILETIL)-PIPERIDINA see MOA600
1-(2-(2-BENZILFENOSSI)-1-METILETIL)-PIPERIDINA FOSFATO (ITALIAN)
   see PJA130
BENZILIC ACID-β-DIETHYLAMINOETHYL ESTER see DHU900
BENZILIC ACID-β-DIETHYLAMINOETHYL ESTER HYDROCHLORIDE
   see BCA000
BENZILIC ACID,-3-(2,5-DIMETHYL-1-PYRROLIDINYL)PROPYL ESTER, HY-
   DROCHLORIDE see BCA250
BENZILIC ACID ESTER with 1-ETHYL-3-HYDROXY-1-METHYLPIPERIDIN-
   IUM BROMIDE see PJA000
BENZILIC ACID, ester with ETHYL (2-HYDROXYETHYL)DIMETHYLAMMON-
   IUM CHLORIDE see ELF500
BENZILIC ACID ester with 3-HYDROXY-1,1-DIMETHYLPIPERIDINIUM BRO-
   MIDE see CBF000
BENZILIC ACID ester with 2-(HYDROXYMETHYL)-1,1-DIMETHYLPIPERIDIN-
   IUM METHYL SULFATE see CDG250
BENZILIC ACID-1-METHYL-3-PIPERIDYL ESTER see MON250
BENZILIC ACID, 3-QUINUCLIDINYL ESTER, HYDROCHLORIDE see QWJ000
(2-BENZILOXYETHYL)DIMETHYLOCTYLAMMONIUM BROMIDE
   see DSH000
8-α-BENZILOYLOXY-6,10-ETHANO-5-AZONIASPIRO(4.5)DECANE CHLORIDE
   see KEA300
8-BENZILOYLOXY-6,10-ETHANO-5-AZONIASPIRO(4.5)DECDANE CHLORIDE
   see BCA375
4-BENZILOYLOXY-1,1,2,2,6-PENTAMETHYLPIPERIDINIUMCHLORIDE
   (β FORM) see BCB000
4-BENZILOYLOXY-1,1,2,6,6-PENTAMETHYLPIPERIDINIUMCHLORIDE
   (α FORM) see BCB250
BENZILSAEURE-(N,N-DIMETHYL-2-HYDROXYMETHYL-PIPERIDINIUM)-
   ESTER-METHYLSULFAT (GERMAN) see CDG250
BENZILSAEURE-DIMETHYL-OCTYL-AMMONIUM-AETHYLESTER BROMIDE
   (GERMAN) see DSH000
BENZILSAEURE-DIMETHYL-PENTYL-AMMONIUM-AETHYLESTER BRO-
   MIDE (GERMAN) see DSH200
BENZILOXYETHYLDIMETHYLETHYLAMMONIUM CHLORIDE see ELF500
4-BENZILYLOXY-1,2,2,6-TETRAMETHYLPIPERIDINE METHOCHLORIDE
   (α FORM) see BCB250
4-BENZILYLOXY-1,2,2,6-TETRAMETHYLPIPERIDINE METHOCHLORIDE
   (β FORM) see BCB000
BENZIMIDAZOLE see BCB750
o-BENZIMIDAZOLE see BCB750
1H-BENZIMIDAZOLE (9CI) see BCB750
2-BENZIMIDAZOLEACETONITRILE see BCC000
BENZIMIDAZOLE CARBAMATE see BCC100
2-BENZIMIDAZOLECARBAMIC ACID see BCC100
2-BENZIMIDAZOLECARBAMIC ACID, 5-(p-FLUOROBENZOYL)-, METHYL
   ESTER see FDA887
BENZIMIDAZOLE-2-CARBAMIC ACID, METHYL ESTER see MHC750
1H-BENZIMIDAZOLE, 5-CHLORO-6-(2,3-DICHLOROPHENOXY)-2-
   (METHYLTHIO)- see CFL200
BENZIMIDAZOLE, 1-(2-DIETHYLAMINOETHYL)-2-(p-ETHOXYBENZYL)-5-
   NITRO-, HYDROCHLORIDE see EQN750
BENZIMIDAZOLE METHYLENE MUSTARD see BCC250
BENZIMIDAZOLE MUSTARD see BCC250
2-BENZIMIDAZOLETHIOL see BCC500

1H-BENZIMIDAZOLIUM HEXAKIS-1-DODECYL-3-METHYL-2-PHENYL-
   (CYANO-C)FERRATE(1−) see TNH750
BENZIMIDAZOLIUM-1-NITROIMIDATE see BCD125
2-BENZIMIDAZOLYLACETONITRILE see BCC000
2-BENZIMIDAZOLYLCARBAMIC ACID see BCC100
1H-BENZIMIDAZOL-2-YLCARBAMIC ACID METHYL ESTER see MHC750
1-(2-BENZIMIDAZOLYL)-3-METHYLUREA see BCD325
4-(2-BENZIMIDAZOLYL)THIAZOLE see TEX000
BENZIMINAZOLE see BCB750
BENZIN see NAI500
BENZIN (OBS.) see BBL250
BENZINDAMINE see BCD750
BENZINDAMINE HYDROCHLORIDE see BBW500
1H-BENZ(6,7)INDAZOLO(2,3,4-fgh)NAPHTH(2'',3'':6',7')INDOLO
   (3',2':5,6)ANTHR A(2,1,9-mna) ACRIDINE-5,8,13,25-TETRAONE see CMU475
BENZ(e)INDENO(1,2-b)INDOLE see BCE000
BENZINDOPYRINE HYDROCHLORIDE see BCE250
BENZINE see PCT250
1-BENZINE see QMJ000
BENZINE (OBS.) see BBL250
BENZINOFORM see CBY000
BENZINOL see TIO750
3-BENZISOTHIAZOLINONE-1,1-DIOXIDE see BCE500
1,2-BENZISOTHIAZOLIN-3-ONE 1,1-DIOXIDE AMMONIUM SALT see ANT500
1,2-BENZISOTHIAZOL-3(2H)-ONE-1,1-DIOXIDE see BCE500
1,2-BENZISOTHIAZOL-3(2H)-ONE-1,1-DIOXIDE, CALCIUM SALT see CAM750
BENZISOTRIAZOLE see BDH250
1,2-BENZISOXAZOLE-3-METHANESULFONAMIDE see BCE750
BENZITRAMIDE see BCE825
3,4-BENZOACRIDINE see BAW750
BENZOANTHRACENE see BBC250
BENZO(a)ANTHRACENE see BBC250
1,2-BENZOANTHRACENE see BBC250
BENZO(a)ANTHRACENE-5,6-OXIDE see EBP000
7H-BENZO(de)ANTHRACEN-7-ONE see BBI250
BENZOANTHRONE see BBI250
BENZOATE see BCL750
17-BENZOATE-3-n-BUTYRATE d'OESTRADIOL (FRENCH) see EDP500
BENZOATE d'OESTRADIOL (FRENCH) see EDP000
BENZOATE d'OESTRONE (FRENCH) see EDV500
BENZOATE of SODA see SFB000
BENZOATE SODIUM see SFB000
1-BENZOAZO-2-NAPHTHOL see PEJ500
BENZOBARBITAL see BDS300
BENZO(f)(1)BENZOTHIENO(3,2-b)QUINOLINE see BCF500
BENZO(h)(1)BENZOTHIENO(3,2-b)QUINOLINE see BCF750
BENZO(e)(1)BENZOTHIOPYRANO(4,3-b)INDOLE see BCG000
BENZO BLUE see CMO250
BENZO BLUE GS see CMO000
BENZOCAINE see EFX000
11H-BENZO(a)CARBAZOLE see BCG250
8,9-BENZO-Γ-CARBOLINE see BDB500
BENZOCHINAMIDE see BCL250
BENZO-CHINON (GERMAN) see QQS200
BENZOCHLORPROPAMID see BEG000
BENZO(a)CHRYSENE see PIB750
BENZO(b)CHRYSENE see BCG500
BENZO(c)CHRYSENE see BCG750
BENZO(g)CHRYSENE see BCH000
2,3-BENZOCHRYSENE see BCG500
BENZO(d,e,f)CHRYSENE see BCS750
BENZO(10,11)CHRYSENO(1,2-b)OXIRENE-6β,7-α-DIHYDRO see BCV750
BENZO CONGO RED see SGQ500
N-6-(3,4-BENZOCOUMARINYL)ACETAMIDE see BCH250
BENZOCTAMINE HYDROCHLORIDE see BCH750
BENZO(de)CYCLOPENT(a)ANTHRACENE see BCI000
1H-BENZO(a)CYCLOPENT(b)ANTHRACENE see BCI250
BENZO DEEP BLACK E see AQP000
BENZODIAPIN see MDQ250
1,4-BENZODIAZINE see QRJ000
1,3-BENZODIAZOLE see BCB750
1,2-BENZODIHYDROPYRONE (FCC) see HHR500
BENZODIOXANE HYDROCHLORIDE see BCI500
2-((1,4-BENZODIOXAN-2-YLMETHYL)AMINO)ETHANOL HYDROCHLORIDE
   see HKN500
3-(((1,4-BENZODIOXAN-2-YL)METHYL)AMINO)-N-METHYLPROPIONAMIDE
   see MHD300
3-(((1,4-BENZODIOXAN-2-YL)METHYL)AMINO)-1-MORPHOLINO-1-
   PROPANONE see MRR125
1-(1,4-BENZODIOXAN-2-YLMETHYL-1-BENZYL)-HYDRAZINE TARTRATE
   see BCI750
1-(1,4-BENZODIOXAN-2-YLMETHYL)PIPERIDINEHYDROCHLORIDE
   see BCI500
2,3-BENZODIOXIN-1,4-DIONE see PHY500

3,4-BENZODIOXOLE-5-CARBOXALDEHYDE see PIW250
1,3-BENZODIOXOLE-5-METHANOL, α-(OXIRANYL)-, ACETATE (ester)
  see ABW250
1,3-BENZODIOXOLE-5-(2-PROPEN-1-OL) see BCJ000
2-(4-(1,3-BENZODIOXOL-5-YLMETHYL)-1-PIPERAZINYL)PYRIMIDINE
  see TNR485
1-(5-(1,3-BENZODIOXOL-5-YL)-1-OXO-2,4-PENTADIENYL)PIPERIDINE(E,E)-
  (9CI) see PIV600
1,3-BENZODIOXOL-5-YL-OXO-2,4-PENTADIENYL-PIPERINE see PIV600
1,3-BENZODITHIOLIUM PERCHLORATE see BCJ125
BENZODOL see BCJ150
BENZOE-DIAETHYL (GERMAN) see BQA000
BENZOEPIN see EAQ750
BENZOESAEURE (GERMAN) see BCL750
BENZOESAEURE (NA-SALZ) (GERMAN) see SFB000
BENZOESTROFOL see EDP000
BENZOFLEX 9-88 see DWS800
BENZOFLEX 9-98 see DWS800
BENZOFLEX P-600 see PKE750
BENZOFLEX S-404 see GGU000
BENZOFLEX S-552 see PBC000
BENZOFLEX P 200 see PKE750
BENZOFLEX 9-88 SG see DWS800
BENZO(l)FLUORANTHENE see BCJ500
BENZO(b)FLUORANTHENE see BAW250
BENZO(e)FLUORANTHENE see BAW250
BENZO(j)FLUORANTHENE see BCJ500
BENZO(k)FLUORANTHENE see BCJ750
2,3-BENZOFLUORANTHENE see BAW250
3,4-BENZOFLUORANTHENE see BAW250
7,8-BENZOFLUORANTHENE see BCJ500
8,9-BENZOFLUORANTHENE see BCJ750
11,12-BENZOFLUORANTHENE see BCJ750
11,12-BENZO(k)FLUORANTHENE see BCJ750
2,3-BENZOFLUORANTHRENE see BAW250
BENZO(jk)FLUORENE see FDF000
BENZOFOLINE see EDP000
BENZOFORM BLACK BCN-CF see AQP000
BENZOFURAN see BCK250
BENZO(b)FURAN see BCK250
2,3-BENZOFURAN see BCK250
2-BENZO-FURANCETONITRILE see BCK750
BENZOFUR D see PEY500
BENZOFURFURAN see BCK250
BENZOFUR GG see ALT000
BENZOFUR MT see TGL750
6H-BENZOFURO(3,2-c)(1)BENZOPYRAN-6-ONE, 3,9-DIHYDROXY- see COF350
22H-BENZOFURO(3A,3-H)(1,5,10)TRIAZACYCLOEICOSINE-3,14,22-
  TRIONE,4,5,6,7,8,9,10,11,12,13,20A,21,23,24-TETRADECAHYDRO-17,19-
  ETHENO-, HYDROCHLORIDE see LIP000
BENZOFUROLINE see BEP500
BENZOFUR P see ALT250
p-(7-BENZOFURYLAZO)-N,N-DIMETHYLANILINE see BCL100
BENZOGUANAMINE see BCL250
BENZO-GYNOESTRYL see EDP000
BENZOHEXONIUM see HEG100
BENZOHYDRAZIDE see BBV250
BENZOHYDRAZINE see BBV250
BENZOHYDROQUINONE see HIH000
BENZOHYDROXAMATE see BCL500
BENZOHYDROXAMIC ACID see BCL500
2-(BENZOHYDRYLOXY)-N,N-DIMETHYLETHYLAMINE see BBV500
BENZOIC ACID see BCL750
BENZOIC ACID (DOT) see BCL750
BENZOIC ACID, 2-(ACETYLSELENO)- see SBU100
BENZOIC ACID AMIDE see BBB000
BENZOIC ACID, 4-((AMINOIMINOMETHYL)AMINO)-, 2-
  (METHOXYCARBONYL)PHENYL ESTER see CBT175
BENZOIC ACID, 4-((AMINOIMINOMETHYL)AMINO)-, 2-METHOXY-4-(2-PRO-
  PENYL)PHENYL ESTER see MDY300
BENZOIC ACID, 5-(AMINOSULFONYL)-2,4-DICHLORO- see DGK900
BENZOIC ACID, 4-ARSENOSO- see CCI550
BENZOIC ACID, BENZYL ESTER see BCM000
BENZOIC ACID-n-BUTYL ESTER see BQK250
BENZOIC ACID, 2-((2-CARBOXYPHENYL)AMINO)-4-CHLORO-, DISODIUM
  SALT see LHZ600
BENZOIC ACID, CHLORIDE see BDM500
BENZOIC ACID, 3-CHLORO-2-HYDROXYPROPYL ESTER see CKQ500
BENZOIC ACID, 2-CHLORO-5-NITRO-, METHYL ESTER see CJC515
BENZOIC ACID, CINNAMYL ESTER see CMQ750
BENZOIC ACID, 3,5-DIACETAMIDO-2,4,6-TRIIODO-, compd. with 1-DEOXY-1-
  (METHYLAMINO)-d-GLUCITOL see AOO875
BENZOIC ACID, 2,4-DICHLORO-5-SULFAMOYL- see DGK900
BENZOIC ACID, DIESTER with DIETHYLENE GLYCOL see DJE000

BENZOIC ACID DIESTER with DIPROPYLENE GLYCOL see DWS800
BENZOIC ACID DIESTER with POLYETHYLENE GLYCOL 600 see PKE750
BENZOIC ACID DIETHYLAMIDE see BCM250
BENZOIC ACID-N,N-DIETHYLAMIDE see BCM250
BENZOIC ACID, 2,4-DIHYDROXY- (9CI) see HOE600
BENZOIC ACID,2-((3-(4-(1,1-DIMETHYLETHYL)PHENYL)-2-
  METHYLPROPYLIDENE)AMINO)-, METHYLESTER see LFT100
BENZOIC ACID, 2,6-DIMETHYL-4-PROPOXY-, 2-METHYL-2-(1-
  PYRROLIDINYL)PROPYL ESTER, HYDROCHLORIDE see DTS625
BENZOIC ACID, o-((3-(4,6-DIMETHYL-2-PYRIMIDINYL)UREIDO)SULFONYL)-,
  METHYL ESTER see SNW550
BENZOIC ACID-n-DIPROPYLENE GLYCOL DIESTER see DWS800
BENZOIC ACID, 2,2'-DISELENOBIS- see SBU150
BENZOIC ACID ESTRADIOL see EDP000
BENZOIC ACID-1-ETHYL-4-PIPERIDYL ESTER, HYDROCHLORIDE
  see EOW000
BENZOIC ACID-2-(1-ETHYL-2-PIPERIDYL)ETHYL ESTER HYDROCHLORIDE
  see EOY100
BENZOIC ACID, p-GUANIDINO-, 4-ALLYL-2-METHOXYPHENYL ESTER
  see MDY300
BENZOIC ACID, p-GUANIDINO-, 4-METHYL-2-OXO-2H-1-BENZOPYRAN-7-
  YL ESTER see GLC100
BENZOIC ACID, HEXYL ESTER see HFL500
BENZOIC ACID HYDRAZIDE, 3-HYDRAZONE with DAUNORUBICIN,
  MONOHYDROCHLORIDE see ROZ000
BENZOIC ACID(4-(HYDROXYIMINO)-2,5-CYCLOHEXADIEN-1-YLIDENE)
  HYDRAZIDE see BDD000
BENZOIC ACID, 2-HYDROXY-, 3-METHYL-2-BUTENYL ESTER see PMB600
BENZOIC ACID, 2-HYDROXY-, 4-METHYLPHENYL ESTER (9CI) see THD850
BENZOIC ACID, 2-HYDROXY-, compounded with MORPHOLINE (1:1)
  see SAI100
BENZOIC ACID-m-HYDROXYPHENYL ESTER see HNH500
BENZOIC ACID, ISOPROPYL ESTER see IOD000
BENZOIC ACID, LITHIUM SALT see LGW000
BENZOIC ACID, 3-METHYL-2-BUTENYL ESTER see MHU150
BENZOIC ACID, 1-(3-METHYL)BUTYL ESTER see IHP100
BENZOIC ACID, 4-METHYLPHENYL ESTER see TGX100
BENZOIC ACID-2-(4-METHYLPIPERIDINO)ETHYL ESTER HYDROCHLO-
  RIDE see MOI250
BENZOIC ACID-1-METHYL-4-PIPERIDYL ESTER HYDROCHLORIDE
  see MON500
BENZOIC ACID, 2-(METHYLSELENO)- see MPI200
BENZOIC ACID, o-(METHYLSELENO)- see MPI200
BENZOIC ACID, o-(METHYLSELENO)-, SODIUM SALT see MPI205
BENZOIC ACID, o-(METHYLTELLURO)-, SODIUM SALT see MPN275
BENZOIC ACID NITRILE see BCQ250
BENZOIC ACID, PEROXIDE see BDS000
BENZOIC ACID-1-PHENETHYL-4-PIPERIDYL ESTER HYDROCHLORIDE
  see PDJ250
BENZOIC ACID, 2-(2-PHENYLETHYLPIPERIDINO) ETHYL ESTER, HYDRO-
  CHLORIDE see PFD500
BENZOIC ACID-3-(2-PHENYLETHYLPIPERIDINO) PROPYL ESTER, HYDRO-
  CHLORIDE see PFD750
BENZOIC ACID, o-(PHENYLHYDROXYARSINO)- see PFI600
BENZOIC ACID, PHENYLMETHYL ESTER see BCM000
BENZOIC ACID, 2-(PHOSPHONOOXY)- (9CI) see PHA575
BENZOIC ACID, 1-PROPYL-1-PIPERIDYL ESTER HYDROCHLORIDE
  see PNT750
BENZOIC ACID, SODIUM SALT see SFB000
BENZOIC ACID, TETRAESTER with PENTAERYTHRITOL see PBC000
BENZOIC ACID, p-((E)-2-(5,6,7,8-TETRAHYDRO-5,5,8,8-TETRAMETHYL-2-
  NAPHTHYL)-1-PROPENYL)-, ETHYL ESTER see AQZ400
BENZOIC ACID, p-TOLYL ESTER see TGX100
BENZOIC ACID TRIESTER with GLYCERIN see GGU000
BENZOIC ACID, VINYL ESTER see VMK000
BENZOIC ALDEHYDE see BAY500
BENZOIC-3-CHLORO-N-ETHOXY-2,6-DIMETHOXYBENZIMIDIC ANHYDRIDE
  see BCP000
BENZOIC ETHER see EGR000
BENZOIC HYDRAZIDE see BBV250
o-BENZOIC SULPHIMIDE see BCE500
BENZOIC TRICHLORIDE see BFL250
BENZOIMIDAZOLE see BCB750
BENZOIN see BCP250
4,5-BENZO-2,3-1',2'-INDENOINDOLE (FRENCH) see BCE000
BENZOIN METHYL ETHER see MHE000
BENZOINOXIM (CZECH) see BCP500
α-BENZOIN OXIME see BCP500
BENZOKETCTRIAZINE see BDH000
BENZOL (DOT) see BBL250
BENZOLAMIDE see PGJ000
BENZOLE see BBL250
BENZOLENE see BBL250
BENZOLIN see NOF500

BENZOLINE see PCT250
BENZOLIN HYDROCHLORIDE see NOF500
BENZOLO (ITALIAN) see BBL250
BENZOMATE see BCP000
BENZOMETAN see BCP650
BENZOMETHAMINE BROMIDE see BCP685
BENZONAL see BDS300
BENZO(B)NAPHTHACENE see PAV000
BENZO(a)NAPHTHO(8,1,2-cde)NAPHTHACENE see BCP750
BENZO(h)NAPHTHO(1,2-f,s-3)QUINOLINE see BCQ000
BENZONE see BRF500
BENZONITRILE see BCQ250
BENZONITRILE (DOT) see BCQ250
BENZONITRILE, p-ISOPROPYL- see IOD050
BENZONITRILE, 4-(1-METHYLETHYL)- see IOD050
BENZOPARADIAZINE see QRJ000
BENZOPENICILLIN see BDY669
BENZO(rst)PENTAPHENE see BCQ500
BENZO(rst)PENTAPHENE-5-CARBOXALDEHYDE see BCQ750
BENZOPERIDOL see FGU000, FLK100
BENZOPEROXIDE see BDS000
1,12-BENZOPERYLENE see BCR000
BENZO(ghi)PERYLENE see BCR000
BENZO(a)PHENALENO(1,9-hi)ACRIDINE see BCR250
BENZO(a)PHENALENO(1,9-i,j)ACRIDINE see BCR500
BENZO(c)PHENALENO(1,9-l,j)ACRIDINE see BCR250
BENZO(h)PHENALENO(1,9-bc)ACRIDINE see BCR500
BENZO(1)PHENANTHRENE see TMS000
BENZO(a)PHENANTHRENE see BBC250, CML810
BENZO(b)PHENANTHRENE see BBC250
BENZO(c)PHENANTHRENE see BCR750
1,2-BENZOPHENANTHRENE see CML810
2,3-BENZOPHENANTHRENE see BBC250
3,4-BENZOPHENANTHRENE see BCR750
9,10-BENZOPHENANTHRENE see TMS000
BENZO(def)PHENANTHRENE see PON250
(±)-BENZO(c)PHENANTHRENE-3,4-DIHYDRODIOL see BCS100
(−)-BENZO(c)PHENANTHRENE-3,4-DIOL-1,2-EPOXIDE-2 see BCS110
(+)-BENZO(c)PHENANTHRENE-3,4-DIOL-1,2-EPOXIDE-1 see BCS103
(+)-BENZO(c)PHENANTHRENE-3,4-DIOL-1,2-EPOXIDE-2 see BCS105
(±)-BENZO(c)PHENANTHRENE-3,4-DIOL-1,2-EPOXIDE-1 see BMK634
BENZO(c)PHENANTHRENE-3-α-4-β-DIOL, 1,2,3,4-TETRAHYDRO-1-β,2-β-
  EPOXY-, (±)- see BMK634
BENZO(c)PHENANTHRENE-3,4-DIOL, 1,2,3,4-TETRAHYDRO-1,2-EPOXY-1,
  (E)-(+)-(1S,2R,3R,4S)- see BCS105
BENZO(c)PHENANTHRENE-3,4-DIOL, 1,2,3,4-TETRAHYDRO-1,2-EPOXY-,
  (E)-(−)-(1R,2S,3S,4R)- see BCS110
BENZO(c)PHENANTHRENE-3,4-DIOL, 1,2,3,4-TETRAHYDRO-1,2-EPOXY-,
  (Z)-(+)-(1R,2S,3R,4S)- see BCS103
BENZO(c)PHENANTHRENE-8-CARBOXALDEHYDE see BCS000
BENZOPHENONE see BCS250
BENZOPHENONE-2 see BCS325
BENZOPHENONE-3 see MES000
BENZOPHOSPHATE see BDJ250
BENZOPIPERILONE (ITALIAN) see BCP650
3,4-BENZOPIRENE (ITALIAN) see BCS750
BENZOPROPYL see AHI250
3H-2-BENZOPYRAN-7-CARBOXYLIC ACID, 4,6-DIHYDRO-8-HYDROXY-3,4,5-
  TRIMETHYL-6-OXO-, (3R-trans)- see CMS775
2H-1-BENZOPYRAN, 6,7-DIMETHOXY-2,2-DIMETHYL- see AEX850
2H-1-BENZOPYRAN-5-OL, 2-METHYL-2-(4-METHYL-3-PENTENYL)-7-PENTYL-
  see PBW400
2H-1-BENZOPYRAN-2-ONE see CNV000
4H-1-BENZOPYRAN-4-ONE, 7-((6-O-(6-DEOXY-α-l-MANNOPYRANOSYL)-β-D-
  GLUCOPYRANOSYL)OXY)-2,3-DIHYDRO-5-HYDROXY-2-(3-HYDROXY-4-
  METHOXYPHENYL)-, MONOMETHYL ETHER see MKK600
2-(5H-(1)BENZOPYRANO(2,3-b)PYRIDIN-7-YL)PROPIONIC ACID see PLX400
1-(2H-1-BENZOPYRAN-8-YLOXY)-3-ISOPROPYLAMINO-2-PROPANOL
  see HLK800
BENZO(A)PYRAZINE see QRJ000
BENZO(a)PYRENE see BCS750
BENZO(e)PYRENE see BCT000
1,2-BENZOPYRENE see BCT000
3,4-BENZOPYRENE see BCS750
4,5-BENZOPYRENE see BCT000
6,7-BENZOPYRENE see BCS750
BENZO(a)PYRENE-6-CARBOXALDEHYDE THIOSEMICARBAZONE see BCT500
BENZO(a)PYRENE-6-CARBOXYALDEHYDE see BCT250
BENZO(a)PYRENE-4,5-DIHDYRODIOL see DMK400
BENZO(a)PYRENE-7,8-DIHYDRODIOL see BCT750
(E)-BENZO(a)PYRENE-4,5-DIHYDRODIOL see DLC600
(E)-BENZO(a)PYRENE-7,8-DIHYDRODIOL see DLC800
BENZO(a)PYRENE-7,8-DIHYDRODIOL-9,10-EPOXIDE (anti) see BCU000
anti-BENZO(a)PYRENE-7,8-DIHYDRODIOL-9,10-OXIDE see BCU000

BENZO(a)PYRENE-7,8-DIHYDRO-7,8-EPOXY see BCV750
BENZO(a)PYRENE, 4,5-DIHYDROXY-4,5-DIHYDRO- see DLB800
anti-BENZO(a)PYRENE-DIOLEPOXIDE see DMQ000
anti(±)BENZO(a)PYRENE-DIOL-EPOXIDE see BCU250
BENZO(e)PYRENE, 9,10-DIOL-11,12-EPOXIDE 1 (cis) see DMR150
BENZO(a)PYRENE DIOL EPOXIDE ANTI see BCU250
1,6-BENZO(a)PYRENEDIONE see BCU500
3,6-BENZO(a)PYRENEDIONE see BCU750
BENZO(a)PYRENE-1,6-DIONE see BCU500
BENZO(a)PYRENE-3,6-DIONE see BCU750
6,12-BENZO(a)PYRENEDIONE see BCV000
BENZO(a)PYRENE-6,12-DIONE see BCV000
BENZO(a)PYRENE-4,5-EPOXIDE see BCV500
BENZO(a)PYRENE-7,8-EPOXIDE see BCV750
BENZO(a)PYRENE-4,5-IMINE see BCV125
BENZO(a)PYRENE-6-METHANOL see BCV250
BENZO(a)PYRENE MONOPICRATE see BDV750
BENZO(a)PYRENE-4,5-OXIDE see BCV500
BENZO(a)PYRENE-7,8-OXIDE see BCV750
BENZO(a)PYRENE-9,10-OXIDE see BCW000
BENZO(a)PYRENE-11,12-OXIDE see BCW250
6,12-BENZOPYRENE QUINONE see BCV000
BENZO(a)PYRENE-1,6-QUINONE see BCU500
BENZO(a)PYRENE-3,6-QUINONE see BCU750
BENZO(a)PYRENE-6,12-QUINONE see BCV000
BENZO(a)PYRENE, 7,8,9,10-TETRAHYDRO-7-β,8-α-9-α-10-α-TETRAHYDROXY-
  see TDD750
BENZO(a)PYRENE-7-β,8-α-9-α-10-α-TETRAOL see TDD750
BENZO(a)PYRENE-6-YL ACETATE see ABL750
BENZO(a)PYREN-1-OL see BCW750
BENZO(a)PYREN-2-OL see BCX000
BENZO(a)PYREN-3-OL see BCX250
BENZO(a)PYREN-4-OL see HJN500
BENZO(a)PYREN-5-OL see BCX500
BENZO(a)PYREN-6-OL see BCX750
BENZO(a)PYREN-7-OL see BCY000
BENZO(a)PYREN-9-OL see BCY250
BENZO(a)PYREN-10-OL see BCY500
BENZO(a)PYREN-11-OL see BCY750
BENZO(a)PYREN-12-OL see BCZ000
BENZO(1,2)PYRENO(4,5-b)OXIRENE-3b,4b-DIHYDRO see BCV500
BENZO(b)PYRIDINE see QMJ000
BENZO(c)PYRIDINE see IRX000
7H-BENZO(a)PYRIDO(3,2-g)CARBAZOLE see BDA000
7H-BENZO(c)PYRIDO(2,3-g)CARBAZOLE see BDA250
7H-BENZO(c)PYRIDO(3,2-g)CARBAZOLE see BDA500
13H-BENZO(g)PYRIDO(3,2-i)CARBAZOLE see BDA750
13H-BENZO(g)PYRIDO(2,3-a)CARBAZOLE see BDB000
13H-BENZO(g)PYRIDO(3,2-a)CARBAZOLE see BDB250
1,2-BENZOPYRIDO(3',2':5,6)CARBAZOLE see BDA000
3,4-BENZOPYRIDO(3',2':5,6)CARBAZOLE see BDA500
5,6-BENZOPYRIDO(2',3':1,2)CARBAZOLE see BDB000
5,6-BENZOPYRIDO(3',2':1,2)CARBAZOLE see BDB250
5,6-BENZOPYRIDO(3',2':3,4)CARBAZOLE see BDA250
7,8-BENZOPYRIDO(2',3':1,2)CARBAZOLE see BDA750
11H-BENZO(g)PYRIDO(4,3-b)INDOLE see BDB500
1,2-BENZOPYRONE see CNV000
BENZOPYRROLE see ICM000
2,3-BENZOPYRROLE see ICM000
BENZOQUIN see AEY000
BENZOQUINAMIDE see BCL250
1,4-BENZOQUINE see QQS200
BENZOQUINOL see HIH000
BENZO(b)QUINOLINE see ADJ500
BENZO(f)QUINOLINE see BDB750
BENZO(h)QUINOLINE see BDC000
α-BENZOQUINOLINE see BDC000
2,3-BENZOQUINOLINE see ADJ500
5,6-BENZOQUINOLINE see BDB750
7,8-BENZOQUINOLINE see BDC000
2H-BENZO(a)QUINOLIZIN-2-ONE, 1,3,4,6,7,11b-HEXAHYDRO-3-ISOBUTYL-
  9,10-DIMETHOXY- see TBJ275
o-BENZOQUINONE see BDC250
p-BENZOQUINONE see QQS200
1,2-BENZOQUINONE see BDC250
1,4-BENZOQUINONE see QQS200
BENZOQUINONE (DOT) see BDC250, QQS200
p-BENZOQUINONE AMIDINOHYDRAZONE THIOSEMICARBAZONE
  see AHI875
BENZOQUINONE AZIRIDINE see BDC750
1,4-BENZOQUINONE-N-BENZOYLHYDRAZONE OXIME see BDD000
BENZOQUINONE-1,4-BIS(CHLOROIMINE)(1,4-BIS(CHLORIMIDO)-2,5-
  CYCLOHEXADIENE) see BDD125
p-BENZOQUINONE DIIMINE see BDD250

1,4-BENZOQUINONE DIIMINE see BDD200
1,4-BENZOQUINONE DIOXINE see DVR200
BENZOQUINONE GUANYLHYDRAZONE THIOSEMICARBAZONE see AHI875
p-BENZOQUINONE, compounded with HYDROQUINONE see QFJ000
p-BENZOQUINONE IMINE see BDD500
p-BENZOQUINONE MONOIMINE see BDD500
1,2-BENZOQUINONE MONOXIME see NLF300
p-BENZOQUINONE OXIME BENZOYLHYDRAZONE see BDD000
p-BENZOQUINONINIME see BDD500
2,1,3-BENZOSELENADIAZOLE, 5,6-DIMETHYL- see DQO650
2,1,3-BENZOSELENADIAZOLE, 5-METHYL- see MHI300
BENZOSELENAZOLIUM, 3-ETHYL-2-(3-(3-ETHYL-2-BENZOSELENAZO-
    LINYLIDENE)-2-METHYLPROPENYL)-, IODIDE see DJQ300
BENZO SKY BLUE A-CF see CMO500
o-BENZOSULFIMIDE see BCE500
BENZOSULFONAMIDE see BBR500
BENZOSULFONAZOLE see BDE500
BENZOSULPHIMIDE see BCE500
BENZO-2-SULPHIMIDE see BCE500
3,4-BENZOTETRACENE see BCG500
BENZO(c)TETRAPHENE see BCG500
3,4-BENZOTETRAPHENE see BCG500
1H-2,1,4-BENZOTHIADIAZIN-3-YL-CARBAMIC ACID METHYL ESTER (9CI)
    see MHI600
BENZO-1,2,3-THIADIAZOLE-1,1-DIOXIDE see BDE000
BENZOTHIAMIDE see BBM250
BENZOTHIAZIDE see BDE250
BENZOTHIAZOLE see BDE500
BENZOTHIAZOLE DISULFIDE see BDE750
7-BENZOTHIAZOLESULFONIC ACID see ALX000
2-BENZOTHIAZOLETHIOL see BDF000
2-BENZOTHIAZOLETHIOL, ZINC SALT (2:1) see BHA750
BENZOTHIAZOLIUM, 3-METHYL-2-METHYLTHIO-, p-TOLUENESULFON-
    ATE see MLX250
(p-(2-BENZOTHIAZOLYL)BENZYL)PHOSPHONIC ACID DIETHYL ESTER
    see DIU500
2-BENZOTHIAZOLYL-N,N-DIETHYLTHIOCARBAMYL SULFIDE see BDF250
BENZOTHIAZOLYL DISULFIDE see BDE750
2-BENZOTHIAZOLYL DISULFIDE see BDE750
N-(2-BENZOTHIAZOLYL)-N-METHYL-N'NITROSOUREA see NKR000
N-(2-BENZOTHIAZOLYL)-N-METHYLUREA see MHM500
2-BENZOTHIAZOLYL MORPHOLINODISULFIDE see BDF750
2-BENZOTHIAZOLYL-N-MORPHOLINOSULFIDE see BDG000
2-BENZOTHIAZOLYLSULFENYL MORPHOLINE see BDG000
S-2-BENZOTHIAZOLYLTHIOGLYCOLIC ACID see CCH750
4-(2-BENZOTHIAZOLYLTHIO)MORPHOLINE see BDG000
4-BENZOTHIENYL METHYLCARBAMATE see BDG250
BENZO(b)THIEN-4-YL METHYLCARBAMATE see BDG250
BENZOTHIOAMIDE see BBM250
BENZO(b)THIOPHENE-4-OL METHYLCARBAMATE see BDG250
6-BENZOTHIOPURINE see BDG325
BENZOTHIOZANE see FNF000
BENZOTHIOZON see FNF000
BENZOTRIAZINEDITHIOPHOSPHORIC ACID DIMETHOXY ESTER
    see ASH500
BENZOTRIAZINE derivative of an ETHYL DITHIOPHOSPHATE see EKN000
BENZOTRIAZINE derivative of a METHYL DITHIOPHOSPHATE see ASH500
1,2,3-BENZOTRIAZIN-4(1H)-ONE see BDH000
3H-1,2,3-BENZOTRIAZIN-4-ONE see BDH000
1H-BENZOTRIAZOLE see BDH250
1,2,3-BENZOTRIAZOLE see BDH250
2-(2H-BENZOTRIAZOL-2-YL)-4-METHYLPHENOL see HML500
BENZOTRICHLORIDE (DOT, MAK) see BFL250
BENZOTRIFLUORIDE see BDH500
BENZOTRIFUROXAN see BBU125
2,5,8-BENZOTRIOXACYCLOUNDECIN-1,9-DIONE, 3,4,6,7-TETRAHYDRO-(9CI)
    see DJD700
BENZO(b)TRIPHENYLENE see BDH750
BENZOTRIS(c)FURAZAN-2-OXIDE see BBU125
BENZOTROPINE see BDI000
BENZOTROPINE MESYLATE see TNU000
BENZOTROPINE METHANESULFONATE see TNU000
BENZOURACIL see QEJ800
BENZO VIOLET N see CMP000
BENZOXALE see TMP750
BENZOXAMATE see BCP000
2-BENZOXAXOLOL see BDJ000
2H-3,1-BENZOXAZINE-2,4(1H)-DIONE see IHN200
BENZOXAZOLE see BDI500
2-BENZOXAZOLINONE see BDJ000
2-BENZOXAZOLINONE see BDJ000
S-((3-BENZOXAZOLINYL-6-CHLORO-2-OXO)METHYL) O,O-
    DIETHYLPHOSPHORODITHIOATE see BDJ250
BENZOXAZOLONE see BDJ000

2(3H)-BENZOXAZOLONE see BDJ000
BENZOXINE see BFW250
BENZ(a)OXIRENO(c)ANTHRACENE see BDJ500
3-BENZOXY-1-(2-METHYLPIPERIDINO)PROPANE see PIV750
3-BENZOXY-1-(2-METHYLPIPERIDINO)PROPANE HYDROCHLORIDE
    see IJZ000
dl-3-BENZOXY-1-(2-METHYLPIPERIDINO)PROPANE HYDROCHLORIDE
    see IJZ000
BENZOYL see BDS000
BENZOYLACET-o-ANISIDIDE see BDJ750
BENZOYLACONINE see PIC250
BENZOYL ALCOHOL see BDX500
BENZOYLAMIDE see BBB000
4-BENZOYLAMIDO-4-CARBOXAMIDO-n(N-NITROSO)-BUTYLCYANAMIDE
    see CBK250
2-BENZOYLAMIDOFLUORENE-2-CARBOXYLATE see FEN000
BENZOYLAMINOACETOHYDROXAMIC ACID see BBB250
1-BENZOYLAMINO-2-CHLOROANTHRAQUINONE see CEL750
5-BENZOYLAMINO-1-CHLOROANTHRAQUINONE see BDK750
(±)-α-(BENZOYLAMINO)-4-(2-(DIETHYLAMINO)ETHOXY)-N,N-
    DIPROPYLBENZENEPROPANAMIDE see TGF175
(±)-α-(BENZOYLAMINO)-5-(DIPROPYLAMINO)-5-OXO-PENTANOIC ACID
    see BGC625
2-BENZOYLAMINOFLUORENE see FDX000
2-BENZOYLAMINOFLUORENE-2-CARBOXYLATE see FEN000
4-((2-(BENZOYLAMINO)-3-(4-HYDROXYPHENYL)-1-OXOPROPYL)AMINO)
    BENZOIC ACID see CML835
(S)-4-((2-BENZOYLAMINO)-3-(4-HYDROXYPHENYL)-1-OXOPROPYL)
    AMINO)BENZOIC ACID see CML835
1-BENZOYLAMINO-4-METHOXY-5-CHLORANTHRACHINON (CZECH)
    see CIA000
BENZOYL-l-ARGININEAMIDE, NITROSATED see CBK250
BENZOYL AZIDE see BDL750
BENZOYLBENZENE see BCS500
N-2 (5-BENZOYL-BENZIMIDAZOLE) CARBAMATE de METHYLE (FRENCH)
    see MHL000
5-BENZOYL-2-BENZIMIDAZOLECARBAMIC ACID METHYL ESTER
    see MHL000
N-(BENZOYL-5-BENZIMIDAZOLYL)-2, CARBAMATE de METHYLE
    (FRENCH) see MHL000
α-BENZOYL-omega-(BENZOYLOXY)POLY(OXY-1,2-ETHANEDIYL) see PKE750
BENZOYLCARBINOL TRIMETHYLACETATE see PCV350
BENZOYLCARBINYL TRIMETHYLACETATE see PCV350
BENZOYL CHLORIDE see BDM500
BENZOYL CHLORIDE (DOT) see BDM500
BENZOYL CYANIDE-o-(DIETHOXYPHOSPHINOTHIOYL)OXIME see BAT750
N-BENZOYL-N-DEACETYL COLCHICINE see BDV250
N-BENZOYL-N-(3,4-DICHLOROPHENYL)-l-ALANINE ETHYL ESTER
    see EGS000
BENZOYLDIETHYLAMINE see BCM250
4-BENZOYL-3,5-DIMETHYL-N-NITROSOPIPERAZINE see NJL850
BENZOYLENEUREA see QEJ800
1-BENZOYL-5-ETHYL-5-PHENYLBARBITURIC ACID see BDS300
1-(2-BENZOYLETHYL)-4-PHENYLISONIPECOTIC ACID ETHYL ESTER HY-
    DROCHLORIDE see CBP325
1-BENZOYL-5-ETHYL-5-PHENYL-2,4,6-TRIOXOHEXAHYDROPYRIMIDINE
    see BDS300
N-(β-BENZOYLETHYL)PIPERIDINE HYDROCHLORIDE see PIV500
N-BENZOYLFERRIOXAMINE B see DAK200
3'-BENZOYL-2-FORMYL-1,1'-(OXYDIMETHYLENE)DIPYRIDINIUM DIIODIDE
    DIHYDRATE see HAA340
N-BENZOYLGLYCINE, MONOSODIUM SALT see SHN500
3-BENZOYLHYDRATROPIC ACID see BDU500
m-BENZOYLHYDRATROPIC ACID see BDU500
BENZOYL HYDRAZIDE see BBV250
BENZOYLHYDRAZONE DAUNORUBICIN see ROU800
2-BENZOYLHYDRAZONO-1,3-DITHIOLANE see BDN125
BENZOYLHYDROGEN PEROXIDE see PCM000
BENZOYL HYDROPEROXIDE see PCM000
BENZOYLHYDROXAMIC ACID see BCL500
BENZOYLLUMINAL see BDS300
BENZOYL METHIDE see ABH000
3-BENZOYL-α-METHYL-BENZENEACETIC ACID SODIUM SALT see KGK100
1-(BENZOYL)-2-(α-METHYLBENZYL)HYDRAZINE see NCQ100
BENZOYLMETHYLECGONINE see CNE750
o-BENZOYL-N-METHYL-N-(p-(PHENYLAZO)PHENYL)HYDROXYLAMINE
    see BDP000
BENZOYL-Γ-(2-METHYLPIPERIDINE)PROPANOL HYDROCHLORIDE
    see IJZ000
BENZOYL-Γ-(2-METHYLPIPERIDINO)PROPANOL see PIV750
5-BENZOYL-α-METHYL-2-THIOPHENEACETIC ACID see SOX400
BENZOYL NITRATE see BDN500
N-BENZOYLOXY-ACETYLAMINOFLUORENE see FDZ000
1-BENZOYLOXY-3-CHLOROPROPAN-2-OL see CKQ500

4,6-o-BENZYLIDENE-β-d-GLUCOPYRANOSIDE PODOPHYLLOTOXIN
  see BER500
BENZYLIDENE GLYCEROL see BBA000
2-BENZYLIDENE-1-HEPTANOL see AOH000
BENZYLIDENEMETHYLPHOSPHORODITHIOATE see BES250
BENZYLIDENEPHENYLACETONITRILE see DVX600
BENZYLIDYNE CHLORIDE see BFL250
BENZYLIDYNE FLUORIDE see BDH500
2-BENZYL-2-IMIDAZOLINE see BBW750
2-BENZYL-4,5-IMIDAZOLINE see BBW750
BENZYLIMIDAZOLINE HYDROCHLORIDE see BBJ750
2-BENZYL-4,5-IMIDAZOLINE HYDROCHLORIDE see BBW750
2-BENZYL-2-IMIDAZOLINE MONOHYDROCHLORIDE see BBJ750
((1-BENZYL-1H-INDAZOL-3-YL)OXY)ACETIC ACID see BAV325
4-(1-BENZYL-3-INDOLETHYL)PYRIDINE HYDROCHLORIDE see BCE250
BENZYL ISOAMYL ETHER see BES500
BENZYL ISOBUTYRATE (FCC) see IJV000
BENZYL ISOEUGENOL see BES750
BENZYL ISOEUGENOL ETHER see BES750
N-BENZYL-β-(ISONICOTINOYLHYDRAZINE)PROPIONAMIDE see BET000
N-BENZYL-β-(ISONICOTINYLHYDRAZINO)PROPIONAMIDE see BET000
BENZYL ISOPENTYL ETHER see BES500
4-(3-BENZYLISOPROPYLAMINO-2-HYDROXYPROPOXY)-9-METHOXY-7-
  METHYL-FURO(3,2-g)CHROMONE, HYDROCHLORIDE see BET750
BENZYLISOPROPYL PROPIONATE see DQQ400
BENZYL-ISOTHIOCYANATE see BEU250
BENZYLISOTHIOUREA HYDROCHLORIDE see BEU500
BENZYLISOTHIOURONIUM CHLORIDE see BEU500
2-BENZYLISOTHIOURONIUM CHLORIDE see BEU500
BENZYL ISOVALERATE (FCC) see ISW000
BENZYL LAURATE see BEU750
BENZYL MERCAPTAN see TGO750
6-BENZYLMERCAPTOPURINE see BDG325
BENZYL METHANOATE see BEP250
2-BENZYL-4-METHANOL-1,3-DIOXANE see PDX250
4-BENZYL-α-(4-METHOXYPHENYL)-β-METHYL-1-PIPERIDINEETHANOL
  see BEU800
BENZYL-2-METHOXY-4-PROPENYLPHENYL ETHER see BES750
1-BENZYL-2-METHYL-3-(2-AMINOETHYL)-5-METHOXYINDOLE HYDRO-
  CHLORIDE see BEM750
BENZYL-3-METHYLBUTANOATE see ISW000
BENZYL-3-METHYL BUTYRATE see ISW000
BENZYLMETHYLCARBINYL ACETATE see ABU800
1-BENZYL-2-METHYLHYDRAZINE see MHN750
1-BENZYL-1-(5-METHYL-3-ISOXAZOIYLCARBONYL)HYDRAZINE see IKC000
1-BENZYL-2-(5-METHYL-3-ISOXAZOIYL-CARBONYL)HYDRAZINE see IKC000
N'-BENZYL N-METHYL-5-ISOXAZOLECARBOXYLHYDRAZIDE-3 see IKC000
1-BENZYL-2-(3-METHYLISOXAZOL-5-YL)CARBONYL HYDRAZINE
  see BEW000
1-BENZYL-2-METHYL-5-METHOXYTRYPTAMINE HYDROCHLORIDE
  see BEM750
1-BENZYL-3-METHYL-5-(2-(4-METHYL-1-PIPERAZINYL)ETHOXY)PYRAZOLE
  see BEW500
1-BENZYL-3-METHYL-5-(2-(2-METHYLPIPERIDINO)ETHOXY)PYRAZOLE
  see BEW750
4-BENZYL-1-(1-METHYL-4-PIPERIDYL)-3-PHENYL-3-PYRAZOLIN-5-ONE
  see BCP650
BENZYL-2-METHYL PROPIONATE see IJV000
N-BENZYL-N-METHYL-2-PROPYNYLAMINE see MOS250
BENZYLMETHYLPROPYNYLAMINE HYDROCHLORIDE see BEX500
N-BENZYL-N-METHYL-2-PROPYNYLAMINE HYDROCHLORIDE
  see BEX500
(1-BENZYL-3-METHYL-5-PYRAZOLYLOXYETHYL)TRIMETHYLAMMONIUM
  IODIDE see BEX750
N-BENZYL-α-METHYL-3-TRIFLUOROMETHYLPHENETHYLAMINE
  see BEY800
N-BENZYL-α-METHYL-m-TRIFLUOROMETHYLPHENETHYLAMINE
  see BEY800
BENZYL MONOCHLORACETATE see BEE500
8-BENZYL-7-(1'-MORPHOLINO-2'-AMINO)ETHYLTHEOPHYLLINE HYDRO-
  CHLORIDE see BFA000
6-BENZYL-MP see BDG325
BENZYL MUSTARD OIL see BEU250, BFL000
BENZYL NITRATE see BFA250
BENZYL NITRILE see PEA750
α-(BENZYLNITROSAMINO)BENZYL ALCOHOL ACETATE (ester) see ABL875
1-BENZYL-1-NITROSOUREA see NJM000
BENZYL NORMECHLORETHAMINE see BIA750
BENZYL OXIDE (CZECH) see BEO250
BENZYLOXY ACETYLENE see BFA899
BENZYLOXYCARBONYL CHLORIDE see BEF500
BENZYLOXYCARBONYLGLYCINE see CBR125
N-BENZYLOXYCARBONYLGLYCINE see CBR125
(BENZYLOXYCARBONYL)PHENYLALANINE see CBR220

l-N-BENZYLOXYCARBONYL-3-PHENYLALANINE-1,2-DIBROMOETHYL
  ESTER see CBR225
N-BENZYLOXYCARBONYL-l-PHENYLALANINE VINYL ESTER see CBR235
S-((N-(2-BENZYLOXYETHYL)AMIDINO)METHYL) HYDROGEN THIOSUL-
  FATE see BFC000
5-BENZYLOXY-3-(1-METHYL-2-PYRROLIDINYL)INDOLE see BFC250
p-BENZYLOXYPHENOL see AEY000
1-(p-(BENZYLOXY)PHENYL)-2-(o-FLUOROPHENYL)-1-PHENYLETHYLENE
  see BFC400
2-(m-(BENZYLOXY)PHENYL)PYRAZOLO(1,5-a)QUINOLINE see BFC450
S-((N-(3-BENZYLOXYPROPYL)AMIDINO)METHYL) HYDROGEN THIOSUL-
  FATE see BFC500
BENZYLPENCILLINDIBENZYLETHYLENEDIAMINE SALT see BFC750
BENZYLPENICILLIN see BDY669
BENZYLPENICILLIN BENZATHINE see BFC750
BENZYLPENICILLIN G see BDY669
BENZYLPENICILLINIC ACID see BDY669
BENZYLPENICILLINIC ACID POTASSIUM SALT see BFD000
BENZYL PENICILLINIC ACID SODIUM SALT see BFD250
BENZYLPENICILLIN POTASSIUM see BFD000
BENZYLPENICILLIN POTASSIUM SALT see BFD000
BENZYLPENICILLIN SODIUM see BFD250
N-BENZYL-N-PHENOXYISOPROPYL-β-CHLORETHYLAMINE HYDROCHLO-
  RIDE see DDG800
1-(2-BENZYLPHENOXY)-2-PIPERIDINOPROPANE PHOSPHATE see PJA130
BENZYL PHENYLACETATE see BFD400
N-BENZYL-N'-PHENYLACETYLHYDRAZIDE see PDY870
BENZYL γ-PHENYLACRYLATE see BEG750
N-β-(BENZYL-PHENYLAMINO)ETHYLPIPERIDINE HYDROCHLORIDE
  see BEA275
BENZYL PHENYLFORMATE see BCM000
BENZYL PHENYL KETONE see PEB000
BENZYLPHENYL NITROSAMINE see BFE750
2-(4-BENZYL-PIPERIDINO)-1-(4-HYDROXYPHENYL)-1-PROPANOL TAR-
  TRATE (2:1) see IAG625
(+)-2-(1-BENZYL-4-PIPERIDYL)-2-PHENYLGLUTARIMIDE HYDROCHLO-
  RIDE see DBE000
(+)-3-(1-BENZYL-4-PIPERIDYL)-3-PHENYLPIPERIDINE-2,6-DIONE HYDRO-
  CHLORIDE see DBE000
N-BENZYL-4-PROTOADAMANTANEMETHANAMINE MALEATE see BFG500
4-BENZYLPYRIDINE see BFG750
1-BENZYLPYRIDINIUM CHLORIDE see BFH000
BENZYL-(α-PYRIDYL)-DIMETHYLAETHYLENDIAMIN (GERMAN) see TMP750
N-BENZYL-N-α-PYRIDYL-N',N'-DIMETHYL-AETHYLENDIAMIN-
  HYDROCHLORID (GERMAN) see POO750
1-BENZYL-3-(2-(4-PYRIDYL)ETHYL)INDOLE HYDROCHLORIDE see BCE250
2-(BENZYL(2-(PYRROLIDINYL)ETHYL)AMINO)-2-CHLOROACETANILIDE DI-
  HYDROCHLORIDE see BFI250
BENZYLRODIURAN see BEQ625
BENZYL SALICYLATE see BFJ750
BENZYLSENFOEL (GERMAN) see BEU250
BENZYL SILANE see BFJ825
BENZYL SODIUM see BFJ850
BENZYLSTEARYLDIMETHYLAMMONIUM CHLORIDE see DTC600
BENZYL SULFITE see BFK000
BENZYL SULFOXIDE see DDH800
3-BENZYLSYDNONE-4-ACETAMIDE see BED750
BENZYLT see PDT250
1-BENZYL-2-(3-(4,5,6,7-TETRAHYDROBENZISOXAZOYLYL)CARBONYL)
  HYDRAZINE HYDROCHLORIDE see BFK325
2-(4-BENZYL-1,2,3,6-TETRAHYDROPYRIDINO)-1-(4-METHOXYPHENYL)-1-
  PROPANOL see RCA450
D-BENZYL TG see EDC560
S-BENZYL THIOBENZOATE see BFK750
BENZYL THIOCYANATE see BFL000
BENZYLTHIOGUANINE see BFL125
6-BENZYLTHIOGUANINE see BFL125
BENZYLTHIOL see TGO750
3-((BENZYLTHIO)METHYL)-6-CHLORO-1,2,4-BENZOTHIADIAZINE-7-
  SULFONAMIDE-1,1-DIOXIDE see BDE250
3-BENZYLTHIOMETHYL-6-CHLORO-2H-1,2,4-BENZOTHIADIAZINE-7-
  SULFONAMIDE-1,1-DIOXIDE see BDE250
3-BENZYLTHIOMETHYL-6-CHLORO-7-SULFAMOYL-1,2,4-BENZOTHIADIAZ-
  INE-1,1-DIOXIDE see BDE250
3-BENZYLTHIOMETHYL-6-CHLORO-7-SULFAMYL-1,2,4-BENZOTHIADIAZINE-
  1,1-DIOXIDE see BDE250
3-BENZYLTHIOMETHYL-6-CHLORO-7-SULFAMYL-2H-1,2,4-BENZOTHIADIAZ-
  INE-1,1-DIOXIDE see BDE250
BENZYL THIOPSEUDOUREA HYDROCHLORIDE see BEU500
2-BENZYL-2-THIO-PSEUDOUREA HYDROCHLORIDE see BEU500
6-(BENZYLTHIO)PURINE see BDG325
BENZYLTHIURONIUM CHLORIDE see BEU500
S-BENZYLTHIURONIUM CHLORIDE see BEU500
BENZYL TRICHLORIDE see BFL250

3-BENZYL-6-TRIFLUOROMETHYL-7-SULFAMOYL-3,4-DIHYDRO-1,2,4-
   BENZOTHIADIAZINE-1,1-DIOXIDE see BEQ625
1-BENZYL-2-TRIMETHYLACETYLHYDRAZINE HYDROCHLORIDE
   see BFM000
BENZYLTRIMETHYLAMMONIUM CHLORIDE see BFM250
BENZYLTRIMETHYLAMMONIUM HYDROXIDE see BFM500
BENZYL TRIMETHYL AMMONIUM IODIDE see BFM750
BENZYLUREA see BFN125
1-BENZYLUREA see BFN125
N-BENZYLUREA see BFN125
BENZYL VIOLET see FAG120
BENZYL VIOLET 3B see FAG120
BENZYLYT see DDG800
3,4-BENZYPYRENE see BCS750
BENZYRIN see BBW500
BENZYTOL see CLW000
BEOSIT see EAQ750
BEP see BJP899, BKH625
BEPANTHEN see PAG200
BEPANTHENE see PAG200
BEPANTOL see PAG200
BEPARON see TCC000
BEPERIDEN see BGD500
BEPROCHINE see RHZ000
BERBERIN see BFN500
BERBERINE see BFN500
BERBERINE CHLORIDE DIHYDRATE see BFN550
BERBERINE HYDROCHLORIDE BIHYDRATE see BFN550
BERBERINE SULFATE see BFN625
BERBERINE SULFATE (2:1) see BFN625
BERBERINE SULFATE TRIHYDRATE see BFN750
BERBERIN SULFATE see BFN625
BERCEMA see EIR000
BERCEMA FERTAM 50 see FAS000
BERCULON A see FNF000
BERELEX see GEM000
BERGAMIOL see LFY100
BERGAMOT OIL rectified see BFO000
BERGAMOTTE OEL (GERMAN) see BFO000
BERGAPTEN see MFN275
BERGENIN HYERATE see BFO100
BERGIUS COAL HYDROGENATION PRODUCTS FRACTION 1 see HHW509
BERGIUS COAL HYDROGENATION PRODUCTS FRACTION 3 see HHW519
BERGIUS COAL HYDROGENATION PRODUCTS FRACTION 4 see HHW529,
   HHW549
BERGIUS COAL HYDROGENATION PRODUCTS FRACTION 7 see HHW539
BERKAZON see FNF000
BERKENDYL see CDP000
BERKFLAM B 10E see PAU500
BERKFURIN see NGE000
BERKOLOL see ICC000
BERKOMINE see DLH600, DLH630
BERKOZIDE see BEQ625
BERLISON F see HHQ800
BERMAT see CJJ250
BERNARENIN see VGP000
BERNICE see CNE750
BERNIES see CNE750
BERNOCAINE see AIT250
BERNSTEINSAEURE-2,2-DIMETHYLHYDRAZID (GERMAN) see DQD400
BERNSTEINSAURE (GERMAN) see SMY000
BERNSTEINSAURE-ANHYDRID (GERMAN) see SNC000
BEROCILLIN see AOD000
BEROL 478 see DJL000
BEROMYCIN see PDT500, PDT750
BEROMYCIN 400 see PDT750
BEROMYCIN (penicillin) see PDT750
BERONALD see CHJ750
BEROTEC see FAQ100
BEROTEC HYDROBROMIDE see FAQ100
BERRY ALDER see MBU825
BERSAMA ABYSSINICA Fres. ssp. ABYSSINICA, leaf extract see BFO125
BERSEN see DDT300
BERTHOLITE see CDV750
BERTHOLLET SALT see PLA250
BERTRANDITE see BFO250
BERUBIGEN see VSZ000
BERYL see BFO500
BERYLLIA see BFT250
BERYLLIUM see BFO750
BERYLLIUM-9 see BFO750
BERYLLIUM, metal powder (DOT) see BFO750
BERYLLIUM ACETATE see BFP000
BERYLLIUM ACETATE, BASIC see BFT500

BERYLLIUM ACETATE, NORMAL see BFP000
BERYLLIUM ALUMINOSILICATE see BFO500
BERYLLIUM ALUMINUM ALLOY see BFP250
BERYLLIUM ALUMINUM SILICATE see BFO500
BERYLLIUM CARBONATE see BFP500
BERYLLIUM CARBONATE (1:1) see BFP750
BERYLLIUM CARBONATE, BASIC see BFP500
BERYLLIUM CHLORIDE see BFQ000
BERYLLIUM COMPOUND with NIOBIUM (12:1) see BFQ750
BERYLLIUM COMPOUNDS see BFQ500
BERYLLIUM COMPOUND with TITANIUM (12:1) see BFR000
BERYLLIUM COMPOUND with VANADIUM (12:1) see BFR250
BERYLLIUM-COPPER-COBALT ALLOY see CNK700
BERYLLIUM DICHLORIDE see BFQ000
BERYLLIUM DIFLUORIDE see BFR500
BERYLLIUM DIHYDROXIDE see BFS250
BERYLLIUM DINITRATE see BFT000
BERYLLIUM FLUORIDE see BFR500
BERYLLIUM HYDRATE see BFS250
BERYLLIUM HYDRIDE see BFR750
BERYLLIUM HYDROGEN PHOSPHATE (1:1) see BFS000
BERYLLIUM HYDROXIDE see BFS250
BERYLLIUM LACTATE see LAH000
BERYLLIUM MANGANESE ZINC SILICATE see BFS750
BERYLLIUM MONOXIDE see BFT250
BERYLLIUM-NICKEL ALLOY see NCY000
BERYLLIUM NITRATE see BFT000
BERYLLIUM ORTHOSILICATE see SCN500
BERYLLIUM OXIDE see BFT250
BERYLLIUM OXIDE ACETATE see BFT500
BERYLLIUMOXIDE CARBONATE see BFP500
BERYLLIUM OXYACETATE see BFT500
BERYLLIUM OXYFLUORIDE see BFT750
BERYLLIUM PERCHLORATE see BFU000
BERYLLIUM PHOSPHATE see BFS000
BERYLLIUM SILICATE see SCN500
BERYLLIUM SILICATE HYDRATE see BFO250
BERYLLIUM SILICIC ACID see SCN500
BERYLLIUM SULFATE (1:1) see BFU250
BERYLLIUM SULFATE TETRAHYDRATE (1:1:4) see BFU500
BERYLLIUM SULPHATE TETRAHYDRATE see BFU500
BERYLLIUM TETRAHYDROBORATE see BFU750
BERYLLIUM TETRAHYDROBORATETRIMETHYLAMINE see BFV000
BERYLLIUM ZINC SILICATE see BFV250
BERYL ORE see BFO500
BESAN see POH000
BESTATIN see BFV300
BE-STILL TREE see YAK350
BESTRABUCIL see BFV325
BETABION see TES750
BETACIDE P see HNU500
BETADID see HJS850
BETADINE see PKE250
BETADRENOL see BQB250
BETADRENOL HYDROCHLORIDE see BQB250
BETAFEDRINA see BBK500
BETAFEDRINE see BBK500
BETAFEN see AOB250
BETAHISTINE MESILATE see BFV350
BETAHISTINE MESYLATE see BFV350
BETAINE CEPHALORIDINE see TEY000
BETAISODONA see PKE250
BETALGIL see DTL200
BETALIN 12 CRYSTALLINE see VSZ000
BETALING see BFW250
BETALIN S see TES750
BETALOC see SBV500
BETAMEC see DNO800
BETAMETHASONE see BFV750
BETAMETHASONE ACETATE and BETAMETHASONE PHOSPHATE
   see BFV755
BETAMETHASONE ACETATE mixed with BETAMETHASONE SODIUM PHOS-
   PHATE see CCS675
BETAMETHASONE BENZOATE see BFV760
BETAMETHASONE 17-BENZOATE see BFV760
BETAMETHASONE DIPROPIONATE see BFV765
BETAMETHASONE 17,21-DIPROPIONATE see BFV765
BETAMETHASONE DISODIUM PHOSPHATE see BFV770
BETAMETHASONE-21-DISODIUM PHOSPHATE see BFV770
BETAMETHASONE PHOSPHATE and BETAMETHASONE ACETATE
   see BFV755
BETAMETHASONE SODIUM PHOSPHATE see BFV770
BETAMETHASONE SODIUM PHOSPHATE mixed with BETAMETHASONE AC-
   ETATE see CCS675

BETAMETHASONE VALERATE see VCA000
BETAMETHASONE 17-VALERATE see VCA000
BETA METHYL DIGOXIN see MJD300
BETA-NAFTYLOAMINA (POLISH) see NBE500
BETANAL see MEG250
BETANAL AM see EEO500
BETA-NEG see ICC000
BETANEX see EEO500
BETANIDINE SULFATE see BFV900
BETANIDIN SULFATE see BFV900
BETANIDOLE see BFW250
BETAPEN see PAQ100
BETAPEN-VK see PDT750
d-BETAPHEDRINE see BBK500
BETAPRONE see PMT100
BETAPTIN see AGW000
BETAPYRIMIDUM see DJS200
BETAQUIL see CBI000
BETASAN see DNO800
BETATRON see DXO300
BETAXIN see TES750
BETAXINA see EID000
BETAXOLOL HYDROCHLORIDE see KEA350
BETAZED see BRF500
BETEL LEAVES see BFV975
BETEL NUT see AQT650, BFW000
BETEL NUT, polyphenol fraction see BFW010
BETEL NUT TANNIN see BFW050
BETEL QUID see BFW120
BETEL QUID EXTRACT see BFW125
BETEL TOBACCO EXTRACT see BFW135
BETHABARRA WOOD see HLY500
BETHAINE CHOLINE CHLORIDE see HOA500
BETHAMETHASONE 17-BENZOATE see BFV760
BETHANECHOL CHLORIDE see HOA500
BETHANID see BFW250
BETHANIDINE, HEMISULFATE see BFW250
BETHANIDINE SULFATE see BFW250
BETHIAMIN see TES750
BETNELAN see BFV750
BETNELAN PHOSPHATE see BFW325
BETNESOL see BFV770
BETNOVATE see VCA000
BETNOVATEAT see VCA000
BETRAMIN see BBV500
BETRILOL see BON400
BETSOLAN see BFV750
BETULA OIL see MPI000
BEVATINE-12 see VSZ000
BEVIDOX see VSZ000
BEVONIUM METHYL SULFATE see CDG250
BEVONIUM METILSULFATE see CDG250
BEWON see TES750
BEXANE see CLN750
BEXIDE see BJU000
BEXOL see BBQ500
BEXON see MMN250
BEXONE see CLN750
BEXT see BJU000, DKE400
BEXTON see CHS500
BEXTRENE XL 750 see SMQ500
BEZITRAMIDE see BCE825
B(b)F see BAW250
B(j)F see BCJ500
BF 121 see FOJ100
BF 5930 see OJW000
BFE 60 see BAU255
B 169-FERRICYANIDE see TNH750
BFH see BRK100
BFP see BJE750
BFPO see BJE750
BFV see FMV000
BG 5930 see OJW000
B. GABONICA VENOM see BLV080
BGE see BRK750
BH 6 see BGS250
BHA (FCC) see BQI000
3-tert-BHA see BQI010
BHANG see CBD750
BHB see BPI125
BHBN see HJQ350
BHC see BBQ500, BJZ000
α-BHC see BBQ000
β-BHC see BBR000

Γ-BHC see BBQ500
BHC (USDA) see BBP750
Δ-BHC see BFW500
BHD see BDN125
BH 2,4-D see DAA800
BH DALAPON see DGI400
BHEN see BRO000
B-HERBATOX see SFS000
BHFT see BEQ625
BHIMSAIM CAMPHOR see BMD000
BH MCPA see CIR250
BH MECOPROP see CIR500
BHP see DNB200
BHPBN see HIE600
BHT (food grade) see BFW750
BI-58 see DSP400
Bi 3411 see CDO000
4',4'''-BIACETANILIDE see BFX000
BIACETYL see BOT500
BIALAMICOL HYDROCHLORIDE see BFX125
BIALFLAVINA see DBX400
BIALLYL see HCR500
BIALLYLAMICOL DIHYDROCHLORIDE see BFX125
BIALLYLAMICOL HYDROCHLORIDE see BFX125
BIALMINAL see EOK000
BIALPIRINIA see ADA725
BIALZEPAM see DCK759
p,p-BIANILINE see BBX000
4,4'-BIANILINE see BBX000
N,N'-BIANILINE see HHG000
o,p'-BIANILINE see BGF109
BIANISIDINE see TGJ750
5,5'-BIANTHRANILIC ACID see BFX250
BIARISON see POB300
BIARSAN see POB300
1,1'-BIAZIRIDINYL see BFX325
BIAZOLINA see CCS250
BIBENZENE see BGE000
BIBENZYL see BFX500
BIBESOL see DGP900
BIC see BRQ500, IAN000
BICAM ULV see DQM600
BICA-PENICILLIN see BFC750
BICARBONATE of SODA see SFC500
BICARBURET of HYDROGEN see BBL250
BICARBURRETTED HYDROGEN see EIO000
BICEP see MQQ450
BICHE PRIETO (MEXICO) see CNG825
BICHLORACETIC ACID see DEL000
BICHLORENDO see MQW500
BICHLORIDE of MERCURY see MCY475
BICHLORURE d'ETHYLENE (FRENCH) see EIY600
BICHLORURE de MERCURE (FRENCH) see MCY475
BICHLORURE de PROPYLENE (FRENCH) see PNJ400
BICHOL see DYE700
BICHROMATE d'AMMONIUM (FRENCH) see ANB500
BICHROMATE OF POTASH see PKX250
BICHROMATE of SODA see SGI000
BICHROMATE de SODIUM (FRENCH) see SGI000
BICILLIN see BFC750
BICIRON see RGP450
BICKIE-MOL see HIM000
BiCNU see BIF750
BICOLASTIC A 75 see SMQ500
BICOLENE P see PMP500
BICORTONE see PLZ000
BICP see CEX250
BI-CURVAL see ZJS300
BICYCLO(4.4.0)DECANE see DAE800
BICYCLO(2,2,2)-1,4-DIAZAOCTANE see DCK400
BICYCLO(2.2.1)HEPTADIENE see NNG000
BICYCLO(2.2.1)HEPTANE, 2,2,5,6-TETRACHLORO-1,7,7-
    TRIS(CHLOROMETHYL)-, (5-endo,6-exo)- see THH575
(endo,endo)-BICYCLO(2.2.1)HEPT-5-ENE-2,3-DICARBOXYLIC ACID
    DIMETHYL ESTER see DRB400
cis-BICYCLO(2.2.1)HEPT-5-ENE-2,3-DICARBOXYLIC ACID, DIMETHYL
    ESTER see DRB400
BICYCLO(2.2.1)HEPTENE-2-DICARBOXYLIC ACID, 2-ETHYLHEXYLIMIDE
    see OES000
cis-BICYCLO(2,2,1-HEPTENE-2,3-DICARBOXYLIC ACID) METHYL ESTER
    see DRB400
BICYCLO(2.2.1)-HEPT-5-ENE-2,3-DICARBOXYLIC ANHYDRIDE see BFY000
BICYCLO(2.2.1)HEPT-5-ENE-2-METHYLOL ACRYLATE see BFY250
trans-BICYCLO(2.2.1)HEPT-5-ENYL-2,3-DICARBONYLCHLORIDE see NNI000

α-BICYCLO(2.2.1)HEPT-5-EN-1-YL-α-PHENYL-PIPERIDINEPROPANOL HYDROCHLORIDE see BGD750
α-(BICYCLO(2.2.1)HEPT-5-EN-2-YL)-α-PHENYL-1-PIPERIDINEPROPANOL HYDROCHLORIDE see BGD750
1-BICYCLOHEPTENYL-1-PHENYL-3-PIPERIDINO-PROPANOL-1 see BGD500
α-(BICYCLO(2.2.1)HEPT-5-EN-2-YL)-α-PHENYL-1-PIPERIDINO PROPANOL see BGD500
1-BICYCLOHEPTENYL-1-PHENYL-3-PIPERIDINOPROPANOL-1 HYDROCHLORIDE see BGD750
N'-BICYCLO(2.2.1)HEPT-2-YL-N-(2-CHLOROETHYL)-N-NITROSOUREA see CHE500
BICYCLO(4,3,0)NONA-3,7-DIENE see TCU250
BICYCLONONADIENE DIEPOXIDE see BFY750
BICYCLONONALACTONE see OBV200
3,3'-(BICYCLO(2.2.2)OCTANE-1,4-DIYLBIS(CARBONYLIMINO-4,1-PHENYLENECARBONYL IMINO))BIS(1-ETHYLPYRIDINIUM, SALT with 4-METHYLBENZENESULFONIC ACID (1582) see BFZ000
BICYCLOPENTADIENE see DGW000
BICYCLOPENTADIENE DIOXIDE see BGA250
BICYCLOPENTADIENYLBIS(TRICARBONYLIRON) see BGA500
BICYCLO(2.1.0)PENT-2-ENE see BGA650
BIDERON see DGC800
BIDIPHEN see TFD250
BIDIRL see DGQ875
BIDISIN see CFC750
BIDOCEF see DYF700
BIDRIN see DGQ875
BIETASERPINE see DIG800
BIETASERPINE BITARTRATE see DIH000
BIETHYLENE see BOP500
1,1'-BI(ETHYLENE OXIDE) see BGA750
BIETHYLXANTHOGENTRISULFIDE see BJU000
BIFEX see PMY300
BIFLORINE see FOW000
BIFLUORIDEN (DUTCH) see FEZ000
BIFLUORURE de POTASSIUM (FRENCH) see PKU250
BIFONAZOL see BGA825
BIFONAZOLE see BGA825
BIFORMAL see GIK000
BIFORMYL see GIK000
BIFORON see BQL000
BIFURON see NGG500
BIG DIPPER see DVX800
BIGITALIN see GEU000
BIG LEAF IVY see MRU359
BIGUANIDINE, CHROMATE see BJW825
BIGUMAL see CKB250
BIGUNAL see BQL000
BIHEXYL see DXT200
BIHOROMYCIN (crystalline) see BGB250
10204-BII see FBP300
BIKAVERIN see LJB700
BI-KELLINA see AHK750
BIKLIN see APT000
BILAGEN see TGZ000
BILARCIL see TIQ250
BILCOLIC see MKP500
BILENE see DYE700
BILETAN see DXN800
BILEVON see HCL000
BILEVON M see DFD000
BILHARCID see PIJ600
BILICANTE see MKP500
BILIDREN see DAL000
BILIGRAFIN FORTE see BGB315
BILIGRAFIN NATRIUM (GERMAN) see BGB325
BILIGRAFIN SODIUM see BGB325
BILIMIN see SKM000
BILIMIRO see IGA000
BILIMIRON see IGA000
BILINE-8,12-DIPROPIONIC ACID, 1,10,19,22,23,24-HEXAHYDRO-2,7,13,17-TETRAMETHYL-1,19-DIOXO- 3,18-DIVINYL- see HAO900
BILINEURINE see CMF000
BILIOGNOST see PDM750
BILI-ORAL see EQC000
BILIRON see SGD500
BILIRUBIN see HAO900
BILIRUBIN IX-α see HAO900
BILISCOPIN see IGD100
BILISELECTAN see PDM750
BILITRAST see TDE750
BILIVISTAN NATRIUM (GERMAN) see BGB350
BILIVISTAN SODIUM see BGB350
BILOBRAN see MRH209

BILOCOL see DYE700
BI-LOFT, combustion products see ADX750
BILOPAC see SKO500
BILOPAQUE see SKO500
BILOPTIN see SKM000
BILOPTINON see SKM000
BILOSTAT see DAL000
BILTRICIDE see BGB400
BIM see MQC000
6,6'-BIMETANILIC ACID see BBX500
BIMETHYL see EDZ000
BINAPACRYL see BGB500
(1,1'-BINAPHTHALENE)-2,2'-DIAMINE see BGB750
(1,2'-BINAPHTHALENE)-1,2-DIAMINE see BGC000
(2,2-BINAPHTHALENE)-8,8'-DICARBOXALDEHYDE, 1,1',6,6',7,7'-HEXAHYDROXY-3,3'-DIMETHYL-5,5'-BIS (1-METHYLETHYL)-, (±)- see GJM030
(±)-1,1-BI-2-NAPHTHOL see CDM625
(8,8'-BI-1H-NAPHTHO(2,3-c)PYRAN)-3,3'-DIACETIC ACID, 3,3',4,4'-TETRAHYDRO-9,9',10,10'-TETRAHYDRO-7,7'-DIMETHOXY-1,1'-DIOXO-, DIMETHYL ESTER see VRP200
2,3,1',8'-BINAPHTHYLENE see BCJ750
β,β-BINAPHTHYLENEETHENE see PIB750
BINAZIN see TGJ150
BINAZINE see TGJ150
BINDAN see PFJ750
BINDAZAC see BAV325
BINDON see ONY000
BINDON ATHYLATHER see BGC250
BINDON ETHYL ETHER see BGC250
BINITROBENZENE see DUQ200
BINODALINE HYDROCHLORIDE see BGC500
BINODALIN HYDROCHLORID (GERMAN) see BGC500
BINOSIDE see BGC625
BINOTAL see AIV500
BINOTAL SODIUM see SEQ000
BINOVA see DXO300
BIO 1,137 see BIT250
BIO 5,462 see EAQ750
BIO (JAPANESE) see SDY600
BIOACRIDIN see DBX400
BIOALETRINA (PORTUGUESE) see BGC750
BIOALLETHRIN see AFR250, BGC750
S-BIOALLETHRIN see AFR750
S-trans-BIOALLETHRIN see AFR750
BIOBAMAT see MQU750
BIOCAMYCIN see HGP550
BIOCETIN see CDP250
BIOCIDE see ADR000
N-1386 BIOCIDE see BLM500
BIO-CLAVE see CJU250
BIOCOLINA see CMF750
BIOCORT ACETATE see CNS825
BIOCORTAR see HHQ800
BIO-DAC 50-22 see DGX200
BIO-DES see DKA600
BIODIASTASE 1000 see BGC825
BIODOPA see DNA200
BIOEPIDERM see VSU100
BIOFANAL see NOH500
BIOFERMIN see DWX600
BIOFLAVONOID see RSU000
BIOFUREA see NGE500
BIOGASTRONE see BGD000, CBO500
BIOGRISIN-FP see GKE000
BIOMET TBTO see BLL750
BIOMIORAN see CDQ750
BIOMITSIN see CMA750
BIOMYCIN see CMA750
BIOMYDRIN see SPC500
BIONIC see NCQ900
BIOPAL CVL-10 see NND000
BIOPAL NR-20 see NND000
BIOPAL VRO 10 see NND000
BIOPAL VRO 20 see NND000
BIOPHEDRIN see EAW000
BIOPHENICOL see CDP250
BIOPHYLL see CKN000
BIO-PHYLLINE see HLC000
BIOPRASE see BAC000
BIO-QUAT 50-24 see AFP250
BIOQUIN see BLC250, QPA000
BIOQUIN 1 see BLC250
BIORAL see BGD000, CBO500

BIRCH TAR OIL see BGO750
BIRCH TAR OIL, RECTIFIED (FCC) see BGO750
BIRD of PARADISE see CAK325
BIRD PEPPER see PCB275
BIRLANE see CDS750
BIRMI, LEAF EXTRACT see KCA100
BIRNENOEL see AOD725
BIRTHWORT see AQY250
BIRUTAN see RSU000
2,7-BIS(ACETAMIDO)FLUORENE see BGP250
BIS(4-ACETAMIDOPHENYL)SULFONE see SNY500
BIS(p-ACETAMIDOPHENYL) SULFONE see SNY500
BIS-4-ACETAMINO PHENYL SELENIUMDIHYDROXIDE see BGP500
BIS(ACETATO-O)(ω-(2-(ACETYLAMINO)-5-NITRO-1,3-PHENYLENE)DI-
     MERCURY see BGQ250
BIS(ACETATO-O,O')OXOZIRCONIUM see ZTS000
BIS(ACETATO)TETRAHYDROXYTRILEAD see LCH000
BIS(ACETATO)TRIHYDROXYTRILEAD see LCJ000
BIS(ACETO)DIHYDROXYTRILEAD see LCH000
BIS(ACETO)DIOXOURANIUM DIHYDRATE see UQT700
BIS(ACETO-O)DIOXOURANIUM DIHYDRATE see UQT700
4,4'-BISACETOPHENONE-α,α'-DI(3-METHYLPYRIDINIUM) DIBROMIDE
     see BGP750
BIS-(ACETOXYAETHYL)NITROSAMIN (GERMAN) see NKM500
BIS(ACETOXY)CADMIUM see CAD250
BIS(ACETOXYDIBUTYLSTANNANE) OXIDE see BGQ000
2,6-BIS(ACETOXYMERCURI)-4-NITROACETANILIDE see BGQ250
9,10-BISACETOXYMETHYL-1,2-BENZANTHRACENE see BBF750
BIS-(p-ACETOXYPHENYL)-CYCLOHEXYLIDENEMETHANE see FBP100
BIS(p-ACETOXYPHENYL)-2-METHYLCYCLOHEXYLIDENEMETHANE
     see BGQ325
BIS(ACETYLACETONATO) TITANIUM OXIDE see BGQ750
BIS(ACETYL ACETONE)COPPER see BGR000
2,5-BIS(ACETYLAMINO)FLUORENE see BGR250
3,5-BIS(ACETYLAMINO)-2,4,6-TRIIODOBENZOIC ACID see DCK000
4,4'-BIS(N-ACETYL-N-METHYLAMINO)AZOBENZENE see DQH200
1,1'-(3,17-BIS(ACETYLOXY)ANDROSTANE-2,16-DIYL)BIS(1-
     METHYLPIPERIDINIUM) DIBROMIDE see PAF625
1,1'-((2-β,3-α,5-α,16-β,17-β)-3,17-BIS(ACETYLOXY)ANDROSTANE-2,16-
     DIYL)BIS(1-METHYLPIPERIDINIUM, DIBROMIDE see BGR325
17,21-BIS(ACETYLOXY)-2-BROMO-6-β,9-DIFLUORO-11-β-HYROXYPREGNA-1,4-
     DIEN-3,20-DIONE see HAG325
BIS(ACETYLOXY)DIBUTYLSTANNANE see DBF800
17,21-BIS(ACETYLOXY)-6,9-DIFLUORO-11-HYDROXY-16-METHYLPREGNA-1,4-
     DIENE-3,20-DIONE (6-α,11-β,16-β)- see DKF125
BIS(ACETYLOXY)MERCURY see MCS750
BIS(ACETYL-N-PHENYLCARBAMYLMETHYL)-4,4'-DISAZO-3,3'-
     DICHLOROBIPHENYL see DEU000
BIS(ACRYLONITRILE) NICKEL (O) see BGR500
S-(1,2-BIS(AETHOXY-CARBONYL)-AETHYL)-O,O-DIMETHYL-
     DITHIOPHASPHAT (GERMAN) see MAK700
2,4-BIS(AETHYLAMINO)-6-CHLOR-1,3,5-TRIAZIN (GERMAN) see BJP000
BISAKLOFEN BP see MJO500
1,4-BIS(4-ALDOXIMINOPYRIDINIUM)BUTANEDIOL-2,3-BIBROMIDE
     see BGR750
1,3-BIS(4-ALDOXIMINOPYRIDINIUM) DIMETHYL ETHER BICHLORIDE
     see BGS250
BIS-3-ALLOXY-2-HYDROXYPROPYL-1-ESTER KYSELINY FUMAROVE
     (CZECH) see BGS750
2,4-BIS(ALLYLAMINO)-6-(4-(BIS-(p-FLUOROPHENYL)METHYL)-1-
     PIPERAZINYL)-s-TRIAZINE see BGS500
1-(4,6-BISALLYLAMINO-s-TRIAZINYL)-4-(p,p'-DIFLUOROBENZHYDRYL)-
     PIPERAZINE see BGS500
BIS(ALLYL CARBONATE)DIETHYLENE GLYCOL see AGD250
BIS(ALLYLDITHIOCARBAMATO)ZINC see ZCS000
BIS(3-ALLYLOXY-2-HYDROXYPROPYL) FUMARATE see BGS750
4,5-BIS(ALLYLOXY)-2-IMIDAZOLINDINONE see BGS825
BISAMIDINE see APL250
BIS AMINE see MJM200
BIS(4-AMINO-1-ANTHRAQUINONYL)AMINE see IBD000
BIS(4-AMINO-3-CHLOROPHENYL) ETHER see BGT000
BIS(2-AMINOETHYL)AMINE see DJG600
BIS(β-AMINOETHYL)AMINE see DJG600
BIS(2-AMINOETHYL)AMINE COBALT(III) AZIDE see BGT125
BIS(2-AMINOETHYL)AMINEDIPEROXOCHROMIUM(IV) see BGT150
N,N'-BIS(2-AMINOETHYL)-1,2-DIAMINOETHANE see TJR000
BIS(β-AMINOETHYL)DISULFIDE see MCN500
N,N'-BIS(2-AMINOETHYL)ETHYLENEDIAMINE see TJR000
N,N'-BIS(2-AMINOETHYL)-1,2-ETHYLENEDIAMINE see TJR000
BIS-p-AMINOFENYLMETHAN (CZECH) see MJQ000
4,4'-BIS(1-AMINO-8-HYDROXY-2,4-DISULFO-7-NAPHTHYLAZO)-3,3'-
     BITOLYL, TETRASODIUM SALT see BGT250
4,4'-BIS(7-(1-AMINO-8-HYDROXY-2,4-DISULFO)NAPHTHYLAZO)-3,3'-
     BITOLYL, TETRASODIUM SALT see BGT250

4,4'-BIS(1-AMINO-8-HYDROXY-2,4-DISULPHO-7-NAPHTHYLAZO)-3,3'-
     BITOLYL, TETRASODIUM SALT see BGT250
1,3-BIS-AMINOMETHYLBENZEN (CZECH) see XHS800
1,4-BIS-AMINOMETHYLBENZEN (CZECH) see PEX250
1,3-BIS(AMINOMETHYL)CYCLOHEXANE see BGT500
1,4-BIS(AMINOMETHYL)CYCLOHEXANE see BGT750
BIS-4-AMINO-3-METHYLFENYLMETHAN (CZECH) see MJO250
BIS(2-AMINO-1-NAPHTHYL)SODIUM PHOSPHATE see BGU000
BIS(4-AMINOPHENOXYPHENYL)SULFONE see SNZ000
BIS(2-AMINOPHENYL)DISULFIDE see DXJ800
BIS(o-AMINOPHENYL)DISULFIDE see DXJ800
1,1'-BIS(2-AMINOPHENYL)DISULFIDE see DXJ800
BIS(4-AMINOPHENYL)ETHER see OPM000
BIS(p-AMINOPHENYL)ETHER see OPM000
BIS(4-AMINOPHENYL)METHANE see MJQ000
BIS(p-AMINOPHENYL)METHANE see MJQ000
2',4-BIS(AMINOPHENYL)METHANE see MJP750
BIS(4-AMINOPHENYL) SULFIDE see TFI000
BIS(p-AMINOPHENYL)SULFIDE see TFI000
BIS(4-AMINOPHENYL) SULFONE see SOA500
BIS(m-AMINOPHENYL) SULFONE see SOA000
BIS(p-AMINOPHENYL) SULFONE see SOA500
BIS(4-AMINOPHENYL) SULPHIDE see TFI000
BIS(p-AMINOPHENYL)SULPHIDE see TFI000
BIS(4-AMINOPHENYL)SULPHONE see SOA500
BIS(p-AMINOPHENYL)SULPHONE see SOA500
BIS(2-AMINOPHENYLTHIO)ZINC see BGV500
2,2-BIS(p-AMINOPHENYL)-1,1,1-TRICHLOROETHANE see BGU500
BIS-(3-AMINOPROPYL)AMINE see AIX250
1,4-BIS(AMINOPROPYL) BUTANEDIAMINE see DCC400
N,N'-BIS(3-AMINOPROPYL)-1,4-BUTANEDIAMINE see DCC400
1,4-BIS(AMINOPROPYL)BUTANEDIAMINE TETRAHYDROCHLORIDE
     see GEK000
N,N'-BIS(3-AMINOPROPYL)-1,4-DIAMINOBUTANE see DCC400
BIS(3-AMINOPROPYL)METHYLAMINE see BGU750
BIS(Γ-AMINOPROPYL)METHYLAMINE see BGU750
N,N-BIS(3-AMINOPROPYL)METHYLAMINE see BGU750
N,N-BIS(Γ-AMINOPROPYL)METHYLAMINE see BGU750
BIS(ω-AMINOPROPYL)METHYLAMINE see BGU750
1,4-BIS(AMINOPROPYL)PIPERAZINE see BGV000
BIS(AMINOPROPYL)PIPERAZINE (DOT) see BGV000
BIS(2-AMINOTHIOPHENOL), ZINC SALT see BGV500
1,3-BIS(5-AMINO-1,3,4-TRIAZOL-2-YL)TRIAZENE see BGV750
2,2,-BIS(p-ANILINE)-1,1,1-TRICHLOROETHANE see BGU500
4,4'-BIS((4-ANILINO-6-BIS(2-HYDROXYETHYL)AMINO-w-TRIAZIN-2-
     YL)AMINO)-2,2'-STILBENEDISULFONIC ACID DISODIUM SALT
     see BGW000
4,4'-BIS((4-ANILINO-6-METHOXY-s-TRIAZIN-2-YL)AMINO)-2,2'-
     STILBENEDISULFONIC ACID, DISODIUM SALT see BGW100
2,2-BIS(p-ANISYL)-1,1,1-TRICHLOROETHANE see MEI450
1,4-BIS-1'-ANTHRACHINONYLAMINO-ANTHRACHINON (CZECH) see APL750
1,5-BIS-1'-ANTHRACHINONYLAMINO-ANTHRACHINON (CZECH)
     see APM000
BISANTRENE HYDROCHLORIDE see BGW325
BIS(AQUO)-N,N'-(2-IMINO-5-PHENYL-4-OXAZOLIDINONE)MAGNESIUM(II)
     see PAP000
BIS-A-TDA see MJL750
2,5-BIS-ATHYLENIMINOBENZOCHINON-1,4 (GERMAN) see BGW750
BIS-o-AZIDO BENZOYL PEROXIDE see BGW500
1,2-BIS(AZIDOCARBONYL)CYCLOPROPANE see BGW650
BIS(2-AZIDOETHOXYMETHYL)NITRAMINE see BGW700
1,5-BIS-(4-AZIDOFENYL)-1,4-PENTADIEN-3-ON see BGW720
N,N-BIS(AZIDOMETHYL)NITRIC AMIDE see DCM499
3,3-BIS(AZIDOMETHYL)OXETANE see BGW710
1,5-BIS(p-AZIDOPHENYL)-1,4-PENTADIEN-3-ONE see BGW720
BIS(AZIDOTHIOCARBONYL)DISULFIDE see BGW739
N,N'-BIS(AZIRIDINEACETYL)-1,8-OCTAMETHYLENEDIAMINE see BGX500
2,5-BIS(AZIRIDINO)BENZOQUINONE see BGW750
N,N'-BIS-AZIRIDINYLACETYL-1,4-CYCLOHEXYLDIMETHYLENEDIAMINE
     see CPL100
N,N'-BIS(AZIRIDINYLACETYL)-1,8-OCTAMETHYLENE DIAMINE see BGX500
P,P-BIS(1-AZIRIDINYL)-p-(1-ADAMANTYL)-PHOSPHINE OXIDE see BJP325
BIS(1-AZIRIDINYL)AMINOPHOSPINE SULFIDE see DNU850
2,5-BIS(1-AZIRIDINYL)-3,6-BIS(2-METHOXYETHOXY)-p-BENZOQUINONE
     see BDC750
2,5-BIS(1-AZIRIDINYL)-3,6-BIS(2-METHOXYETHOXY)-2,5-CYCLOHEXADIENE-
     1,4-DIONE see BDC750
2,5-BIS(1-AZIRIDINYL)-3-(2-CARBAMOYLOXY-1-METHOXYETHYL)-6-
     METHYL-1,4-BENZOQUINONE see BGX750
2,4-BIS(1-AZIRIDINYL)-6-CHLOROPYRIMIDINE see EQI600
2,6-BIS(1-AZIRIDINYL)-4-CHLOROPYRIMIDINE see EQI600
p,p-BIS(1-AZIRIDINYL)-2,5-DIIODOBENZOYLPHOSPINIC AMIDE see DNE800
p,p-BIS(1-AZIRIDINYL)-N,N-DIMETHYLAMINOPHOSPHINE OXIDE
     see DOV600

p-(BIS-(β-CHLOROETHYL)AMINO)BENZYLIDENE MALONONITRILE
  see BHP250
1,6-BIS(CHLOROETHYLAMINO)-1,6-BIS-DEOXY-d-MANNITOL see MAW500
2-(BIS(2-CHLOROETHYL)AMINO)-3-(2-CHLOROETHYL)TETRAHYDRO-2H-
  1,3,2-OXAPHOSPHORINE-2-OXIDE see TNT500
1,6-BIS-(CHLOROETHYLAMINO)-1,6-DESOXY-d-MANNITOLDIHYDROCHLOR-
  IDE see MAW750
1,6-BIS-(CHLOROETHYLAMINO)-1,6-DIDEOXY-d-MANNITE see MAW500
1,6-BIS(CHLOROETHYLAMINO)-1,6-DIDEOXY-d-MANNITEDIHYDROCHLOR-
  IDE see MAW750
1,6-BIS(2-CHLOROETHYL)AMINO)-1,6-DIDEOXY-d-MANNITOL see MAW500
1,6-BIS((β-CHLOROETHYL)AMINO)-1,6-DIDEOXY-d-MANNITOL see MAW500
2-(p-BIS(2-CHLOROETHYL)AMINO)-N-ETHYLACETAMIDE see PCV775
9-(2-(BIS(2-CHLOROETHYL)AMINO)ETHYLAMINO)-6-CHLORO-2-
  METHOXYACRIDINE DIHYDROCHLORIDE see DFH000
4-(BIS(2-CHLOROETHYL)AMINO)-N-ETHYL-BENZENEACETAMIDE
  see PCV775
3-(2-(BIS(2-(CHLOROETHYL)AMINO)ETHYL)-1,3-DIAZASPIRO(4.5)DECANE-
  2,4-DIONE see SLD900
3-(2-(BIS(2-(CHLOROETHYL)AMINO)ETHYL)-5,5-PENTAMETHYLENE-
  HYDANTOIN see SLD900
2-(BIS(2-CHLOROETHYL)AMINO)ETHYL VINYL SULFONE see BHP500
4'-(BIS(2-CHLOROETHYL)AMINO)-2-FLUORO ACETANILIDE see BHP750
2-(BIS(2-CHLOROETHYL)AMINO)HEXAHYDRO-1,3,2-DIAZAPHOSPHORINE-
  2-OXIDE see BHQ000
2-(BIS(2-CHLOROETHYL)AMINO)-4-HYDROPEROXYTETRAHYDRO-2H-1,3,2-
  OXAZAPHOSPHORINE see HIF000
2-(BIS(2-CHLOROETHYL)AMINO)-4-HYDROXYTETRAHYDRO-2H-1,3,2-
  OXAZAPHOSPHORINE see HKA200
p-BIS(2-CHLOROETHYL)AMINO-o-METHOXYPHENYLALANINE see BHQ250
3-(4-(BIS(2-CHLOROETHYL)AMINO)-3-METHOXYPHENYL)ALANINE
  see BHQ250
3-BIS(2-CHLOROETHYL)AMINO-4-METHYLBENZOIC ACID see AFQ575
3-(BIS(2-CHLOROETHYL)AMINOMETHYL)-2-BENZOXAZOLINONE
  see BHQ750
9-(4-(BIS(2-CHLOROETHYL)AMINO)-1-METHYLBUTYLAMINO)-6-CHLORO-2-
  METHOXYACRIDINE DIHYDROCHLORIDE see QDS000
9-(4-(BIS-β-CHLOROETHYLAMINO)-1-METHYLBUTYLAMINO)-6-CHLORO-2-
  METHOXYACRIDINE see QDJ000
4-((4-(BIS(2-CHLOROETHYL)AMINO)-1-METHYLBUTYL)AMINO-7-
  CHLOROQUINOLINE, DIHYDROCHLORIDE see CLD500
2-(BIS(2-CHLOROETHYL)AMINOMETHYL)-5,5-DIMETHYLBENZIMIDAZOLE
  HYDROCHLORIDE see BCC250
4-(BIS(2-CHLOROETHYL)AMINOMETHYL)-2,3-DIMETHYL-1-PHENYL-3-
  PYRAZOLIN-5-ONE HYDROCHLORIDE see BHR500
3-(o-((BIS(2-CHLOROETHYL)AMINO)METHYL)PHENYL)ALANINE DIHYDRO-
  CHLORIDE see ARQ000
2-((BIS(2-CHLOROETHYL)AMINO)METHYL)-PHENYLALANINE DIHYDRO-
  CHLORIDE (9CI) see ARQ000
o-BIS(2-CHLOROETHYL)AMINOMETHYLPHENYLALANINE HYDROCHLO-
  RIDE see ARQ000
5-(BIS(2-CHLOROETHYL)AMINO)-6-METHYLURACIL see DYC700
2-BIS(2-CHLOROETHYL)AMINONAPHTHALENE see BIF250
2-(BIS(2-CHLOROETHYL)AMINO)-1-OXA-3-AZA-2-PHOSPHOCYCLOHEXANE
  2-OXIDE MONOHYDRATE see CQC500
2-(BIS(2-CHLOROETHYL)AMINO)-2H-1,3,2-OXAAZAPHOSPHORINE 2-OXIDE
  see CQC650
1-BIS(2-CHLOROETHYL)AMINO-1-OXO-2-AZA-5-OXAPHOSPHORIDINE MON-
  OHYDRATE see CQC500
4-(BIS(2-CHLOROETHYL)AMINO)PHENOL see BHR750
p-(BIS(2-CHLOROETHYL)AMINO)PHENOL BENZOATE see BHV500
p-(BIS(2-CHLOROETHYL)AMINO)PHENOL-p-BROMOBENZOATE see BHV750
p-(BIS(2-CHLOROETHYL)AMINO)PHENOL-m-CHLOROBENZOATE
  see BHW000
p-(BIS(2-CHLOROETHYL)AMINO)PHENOL-2,6-DIMETHYLBENZOATE
  see BHW250
(p-BIS(2-CHLOROETHYL)AMINO)PHENYL)ACETATE CHOLESTEROL
  see CME250
(p-(BIS(2-CHLOROETHYL)AMINO)PHENYL)ACETIC ACID see PCU425
2-(p-(BIS(2-CHLOROETHYL)AMINO)PHENYL)ACETIC ACID see PCU425
2-(N,N-BIS(2-CHLOROETHYL)AMINOPHENYL) ACETIC ACID BUTYL ESTER
  see BHS250
(p-(BIS(2-CHLOROETHYL)AMINO)PHENYL)ACETIC ACID CHOLESTEROL
  ESTER see CME250
(4-(BIS(2-CHLOROETHYL)AMINO)PHENYL)ACETIC ACID CHOLESTERYL
  ESTER see CME250
2-(N,N-BIS(2-CHLOROETHYL)AMINOPHENYL)ACETIC ACID DECYL ESTER
  see BHS500
2-(N,N-BIS(2-CHLOROETHYL)AMINOPHENYL)ACETIC ACID OCTADECYL
  ESTER see BHS750
2-(N,N-BIS(2-CHLOROETHYL)AMINOPHENYL)ACETIC ACID TETRADECYL
  ESTER see BHT000
3-(p-(BIS(2-CHLOROETHYL)AMINO)PHENYL)ALANINE see BHT750
4-(BIS(2-CHLOROETHYL)AMINO)-d-PHENYLALANINE see SAX200

4-(BIS(2-CHLOROETHYL)AMINO)-l-PHENYLALANINE see PED750
(+)-3-(p-(BIS(2-CHLOROETHYL)AMINO)PHENYL)ALANINE see SAX200
4-(BIS(2-CHLOROETHYL)AMINO)-dl-PHENYLALANINE see BHT750
d-3-(p-(BIS(2-CHLOROETHYL)AMINO)PHENYL)ALANINE see SAX200
l-3-(m-(BIS(2-CHLOROETHYL)AMINO)PHENYL)ALANINE see SAX210
l-3-(p-(BIS(2-CHLOROETHYL)AMINO)PHENYL)ALANINE see PED750
p-N-BIS(2-CHLOROETHYL)AMINO-l-PHENYLALANINE see PED750
3-(o-(BIS(β-CHLOROETHYL)AMINO)PHENYL)-dl-ALANINE see BHT250
3-(p-(p-(BIS(2-CHLOROETHYL)AMINO)PHENYL)-l-ALANINE see PED750
dl-3-(p-(BIS(2-CHLOROETHYL)AMINO)PHENYL)ALANINE see BHT750
3-(p-(BIS(β-CHLOROETHYL)AMINO)PHENYL)-d-ALANINE HYDROCHLO-
  RIDE see BHU750
3-(m-(BIS(β-CHLOROETHYL)AMINO)PHENYL)-dl-ALANINE HYDROCHLO-
  RIDE see BHU500
3-(p-(BIS(β-CHLOROETHYL)AMINO)PHENYL)-dl-ALANINE HYDROCHLO-
  RIDE see BHV000
4-(BIS(2-CHLOROETHYL)AMINO)-d-PHENYLALANINE MONOHYDROCHLOR-
  IDE see BHU750
4-(BIS(2-CHLOROETHYL)AMINO)-dl-PHENYLALANINE
  MONOHYDROCHLORIDE see BHV000
l-3-(p-(BIS(2-CHLOROETHYL)AMINO)PHENYL)ALANINE
  MONOHYDROCHLORIDE see BHV250
p-(BIS(2-CHLOROETHYL)AMINO)PHENYL BENZOATE see BHV500
p-(BIS(2-CHLOROETHYL)AMINO)PHENYL-p-BROMOBENZOATE see BHV750
4-(p-(BIS(2-CHLOROETHYL)AMINO)PHENYL)BUTYRIC ACID see CDO500
4-(p-(BIS(β-CHLOROETHYL)AMINOPHENYL)BUTYRIC ACID see CDO500
Γ-(p-(BIS(2-CHLOROETHYL)AMINOPHENYL)BUTYRIC ACID see CDO500
p-(BIS(2-CHLOROETHYL)AMINO)PHENYL-m-CHLOROBENZOATE
  see BHW000
p-(BIS(2-CHLOROETHYL)AMINO)PHENYL-2,6-DIMETHYLBENZOATE
  see BHW250
2-(p-BIS(2-CHLOROETHYL)AMINOPHENYL)-1,3,2-DITHIARSENOLANE
  see BHW300
4-(p-BIS(β-CHLOROETHYLAMINO)PHENYLETHYLAMINO)-7-
  CHLOROQUINOLINE MONOHYDROCHLORIDE see BHW500
l-3-(p-(BIS(2-CHLOROETHYL)AMINO)PHENYL)-N-FORMYLALANINE
  see BHX250
2-(((4-(BIS(2-CHLOROETHYL)AMINO)PHENYL)IMINO)METHYL)-1-METHYL-
  QUINOLINIUM CHLORIDE (9CI) see IAK100
5-(4-BIS(2-CHLOROETHYL)AMINOPHENYL)PENTANOIC ACID see BHY625
o-(4-(BIS(2-CHLOROETHYL)AMINO)PHENYL-dl-TYROSINE see BHY500
5-(p-(BIS(2-CHLOROETHYL)AMINO)PHENYL)VALERIC ACID see BHY625
β-(BIS(2-CHLOROETHYLAMINO))PROPIONITRILE see BHY750
9-((3-(BIS(2-CHLOROETHYL)AMINO)PROPYL)AMINO)ACRIDINE DIHYDRO-
  CHLORIDE see BHZ000
N,N-BIS(β-CHLOROETHYL)-AMINO-N'-O-PROPYLENE-PHOSPHORIC ACID
  ESTER DIAMIDE see IMH000
N,N-BIS(2-CHLOROETHYL)-o-(3-AMINOPROPYL)PHOSPHORAMIDATE,
  ZWITTERION see CQN125
5-(BIS(2-CHLOROETHYL)AMINO)-2,4(1H,3H)PYRIMIDINEDIONE see BIA250
2-(BIS(2-CHLOROETHYL)AMINO)TETRAHYDROOXAZAPHOSPHORINE
  CYCLOHEXYLAMINE SALT see CQN000
(BIS(CHLORO-2-ETHYL)AMINO)-2-TETRAHYDRO-3,4,5,6-OXAZAPHOSPHOR-
  INE-1,3,2-OXIDE-2-MONOHYDRATE see CQC500
3-BIS(2-CHLOROETHYL)AMINO-p-TOLUIC ACID see AFQ575
o-(4-BIS(β-CHLOROETHYL)AMINO-o-TOLYLAZO)BENZOIC ACID see BIA000
2-(BIS(2-CHLOROETHYL)AMINO)-6-TRIFLUOROMETHYLTETRAHYDRO-2H-
  1,3,2-OXAZAPHOSPHORINE 2-OXIDE see TKD000
5-(BIS(2-CHLOROETHYL)AMINO)URACIL see BIA250
5-N,N-BIS(2-CHLOROETHYL)AMINOURACIL see BIA250
BIS(2-CHLOROETHYL)AMMONIUM CHLORIDE see BHO250
N,N-BIS(2-CHLOROETHYL)ANILINE see AOQ875
N,N-BIS(2-CHLOROETHYL)-p-ARSANILIC ACID see BIA300
N,N-BIS(2-CHLOROETHYL)BENZENAMINE see AOQ875
N,N-BIS(2-CHLOROETHYL)BENZENEMETHANAMINE see BIA750
N,N-BIS(2-CHLOROETHYL)BENZENEMETHANAMINE HYDROCHLORIDE
  see EAK000
BIS(2-CHLOROETHYL)BENZYLAMINE see BIA750
N,N-BIS(2-CHLOROETHYL)BENZYLAMINE see BIA750
N,N-BIS(2-CHLOROETHYL)BENZYLAMINE HYDROCHLORIDE see EAK000
N-N-BIS(2-CHLOROETHYL)BUTYLAMINE see DEV300
N,N-BIS(2-CHLOROETHYL)BUTYLAMINE HYDROCHLORIDE see BIB250
N,N-BIS(2-CHLOROETHYL)-p-CHLOROBENZYLAMINE HYDROCHLORIDE
  see BIB750
O,O-BIS(2-CHLOROETHYL)-O-(3-CHLORO-4-METHYL-7-COUMARINYL) PHOS-
  PHATE see DFH600
BIS(β-CHLOROETHYL)-β-CHLOROPROPYLAMINE HYDROCHLORIDE
  see NOB700
N,N-BIS(2-CHLOROETHYL)-2-CHLOROPROPYLAMINE HYDROCHLORIDE
  see NOB700
2,3-(N,N(1)-BIS(2-CHLOROETHYL)DIAMIDO-1,3,2-OX-
  AZAPHOSPHORIDINOXY see IMH000
1,4-BIS(2-CHLOROETHYL)-1,4-DIAZONIABICYCLO(2.2.1)HEPTANE (Z)-2-
  BUTENEDIOATE (1:2) see BIC325

cis-BIS(CYCLOBUTYLAMMINE)DICHLOROPLATINUM(II) see DGT200
BISCYCLOHEXANONE OXALDIHYDRAZONE see COF675
BIS(CYCLOHEXYL)CARBOXYLIC ACID DIETHYLAMINOETHYL ESTER HY-
    DROCHLORIDE see ARR750
BIS(n-CYCLOOCTATETRANENE)URANIUM(O) see BIR000
BISCYCLOPENTADIENE see DGW000
BIS(CYCLOPENTADIENYL)BIS(PENTAFLUOROPHENYL)ZIRCONIUM see BIR250
BIS(CYCLOPENTADIENYLCHROMIUM TRICARBONYL)MERCURY see BIR500
BIS(CYCLOPENTADIENYL)COBALT see BIR529
BISCYCLOPENTADIENYLIRON see FBC000
BIS(n-CYCLOPENTADIENYL)MAGNESIUM see BIR750
BIS(2-CYCLOPENTENYL)ETHER see BIS200
cis-BIS(CYCLOPENTYLAMMINE)PLATINUM(II) see BIS250
BIS-DEAE-FLUORENONE HYDROCHLORIDE see TGB000
BIS(DECANOYLOXY)DI-n-BUTYLSTANNANE see BIS500
BIS(DECANOYLOXY)DI-N-BUTYLTIN see BIS500
BISDEHYDRODOISYNOLIC ACID 7-METHYL ETHER see BIT030
dl-cis-BISDEHYDRODOISYNOLIC ACID METHYL ETHER see BIS750
BISDEHYDROISYNOLIC ACID METHYL ESTER see BIT000
BIS(DIALKYLPHOSPHINOTHIOYL)DISULFIDE see BIT250
BIS(1,2-DIAMINOETHANE)DIAQUACOBALT(III) PERCHLORATE see BIT350
BIS-1,2-DIAMINO ETHANE DICHLORO COBALT(III) CHLORATE see BIT500
BIS-1,2-DIAMINO ETHANE DICHLORO COBALT(III) PERCHLORATE
    see BIT750
cis-BIS-1,2-DIAMINO ETHANE DINITRO COBALT(III) IODATE BIU000
BIS(1,2-DIAMINOETHANE)DINITROCOBALT(III) PERCHLORATE see BIU125
BIS(1,2-DIAMINO ETHANE)HYDROXOOXO RHENIUM(V) DIPERCHLORATE
    see BIU250
BIS(1,2-DIAMINOETHANE)HYDROXOOXORHENIUM(V) PERCHLORATE
    see BIU260
BIS-1,2-DIAMINO PROPANE-cis-DICHLORO CHROMIUM(III) PERCHLORATE
    see BIU500
BIS(DIAZOACETYL)BUTANE see BIU750
1,4-BIS(DIAZOACETYL)BUTANE see BIU750
1,8-BIS(DIAZO)-2,7-OCTANEDIONE see BIU750
BIS(DI-n-BENZENE CHROMIUM(IV)DICHROMATE: see BIU900
2,6-BIS-(DIBENZYLHYDROXYMETHYL)PIPERIDINE see BIV000
2,2-BIS-(3',5'-DIBROM-4'-GLYCIDOXYFENYL)PROPAN (CZECH) see BIV500
1,4-BIS(DIBROMMETHYL)BENZEN (CZECH) see TBK250
2,2-BIS(3,5-DIBROMO-4-(2,3-EPOXYPROPOXY)PHENYL)PROPANE see BIV500
1,3-BIS(DIBROMOMETHYL)BENZENE see TBK000
BIS(2,3-DIBROMOPROPYL)PHOSPHATE see BIV825
BIS(2,3-DIBROMOPROPYL)PHOSPHATE, MAGNESIUM SALT see MAD100
BIS(DIBUTYLACETOXYTIN)OXIDE see BGQ000
BIS(DIBUTYLAMMONIUM)HEXACHLOROSTANNATE see BIV900
BIS(DIBUTYLBORINO)ACETYLENE see BIW000
BIS(DIBUTYLDITHIOCARBAMATO)DIBENZYLSTANNANE see BIW250
BIS(DIBUTYLDITHIOCARBAMATO)DIMETHYLSTANNANE see BIW500
BIS(DIBUTYLDITHIOCARBAMATO)NICKEL see BIW750
BIS(DIBUTYLDITHIOCARBAMATO)ZINC see BIX000
BIS((DIBUTYLDITHIOCARBAMOYL)OXY)DIBENZYLSTANNANE see BIW250
BIS((DIBUTYLDITHIOCARBAMOYL)OXY)DIMETHYLSTANNANE see BIW500
2,2-BIS-DI-tert-BUTYLPEROXYBUTANE see DVT500
BIS(O,O-DIBUTYLPHOSPHORODITHIOATO-S)MERCURY see MDA750
BIS(DIBUTYLTHIOCARBAMOYL) DISULFIDE see TBM750
2,2-BIS(3',5'-DICHLOR-4'-GLYCIDOXYFENYL)PROPAN (CZECH) see BIY000
BIS(DICHLOROACETYL)DIAMINE see BIX125
N,N'-BIS(DICHLOROACETYL)-1,8-DIAMINOOCTANE see BIX250
N,N'-BIS(DICHLOROACETYL)-N,N-DIETHYL-1,6-HEXANEDIAMINE
    see HEF300
N,N'-BIS(DICHLOROACETYL)-N,N-DIETHYL-1,4-XYLYLENEDIAMINE
    see PFA600
1,2-BIS(DICHLOROACETYL)HYDRAZINE see BIX125
N,N'-BIS(DICHLOROACETYL)-1,8-OCTAMETHYLENEDIAMINE see BIX250
BIS(3,4-DICHLOROBENZOATO)NICKEL see BIX500
BIS(2,4-DICHLORO BENZOYL)PEROXIDE see BIX750
BIS(2,4-DICHLOROBENZOYL)PEROXIDE see BIX750
trans-1,4-BIS(2-DICHLOROBENZYLAMINOETHYL)CYCLOHEXANE
    DICHLORHYDRATE (FRENCH) see BHN000
2,2-BIS(3,5-DICHLORO-4-(2,3-EPOXYPROPOXY)PHENYL)PROPANE see BIY000
BIS(1,2-DICHLOROETHYL)SULFONE see BIY250
BIS(3,4-DICHLORO-2(5)-FURANONYL) ETHER see MRV000
BIS(3,4-DICHLOROFURANON-5-YL-2) ETHER see MRV000
2,2-BIS(3,5-DICHLORO-4-HYDROXYPHENYL)PROPANE see TBO750
BIS(3,4-DICHLOROPHENYL)-DIAZENE (9CI) see TBN500
BIS(3,4-DICHLOROPHENYL)-DIAZENE 1-OXIDE (9CI) see TBN550
1,3-DICHLORO-1,1,3,3-TETRAETHYLDISTANNOXANE) see BIY500
1,3-BIS(DI-n-CYCLOPENTADIENYL IRON)-2-PROPEN-1-ONE see BIZ000
1,5-BIS(4-(2,3-DIDEHYDROTRIAZIRIDINYL)PHENYL)-1,4-PENTADIEN-3-ONE
    see BJA000
2,6-BIS(DIETHANOLAMINO)-4,8-DIPIPERIDINOPYRIMIDO(5,4-d)PYRIMIDINE
    see PCP250
BIS(2,2-DIETHOXYETHYL)DISELENIDE see BJA200
BIS(S-(DIETHOXYPHOSPHINOTHIOYL)MERCAPTO)METHANE see EEH600

α,α'-BIS(DIETHYLAMINO)-5,5'-DIALLYL-m,m'-BITOLYL-4,4'-DIOL DIHYDRO-
    CHLORIDE see BFX125
2,6-BIS(2-(DIETHYLAMINO)ETHOXY)-9,10-ANTHRACENEDIONE DIHYDRO-
    CHLORIDE see BJA500
BIS(1-(2-DIETHYLAMINOETHOXYCARBONYL)-1-PHENYLCYCLOPENT-
    ANE)ETHANE DISULFONATE see CBG250
2,7-BIS(2-(DIETHYLAMINO)ETHOXY)-FLUGREN-9-ONE DIHYDROCHLORIDE
    see TGB000
2,7-BIS(2-(DIETHYLAMINO)ETHOXY)-9H-FLUOREN-9-ONE DIHYDROCHLO-
    RIDE see TGB000
1-(BIS(2-(DIETHYLAMINO)ETHYL)AMINO)-5-CHLORO-3-(p-
    CHLOROPHENYL)INDOLE DIHYDROCHLORIDE HEMIHYDRATE
    see BJA750
1-(BIS(2-(DIETHYLAMINO)ETHYL)AMINO)-3-PHENYLINDOLE DIHYDRO-
    CHLORIDE see BJA809
N,N'-BIS(2-DIETHYLAMINOETHYL)OXAMIDE BIS(2-CHLOROBENZYL CHLO-
    RIDE) see MSC100
1,3-BIS(DIETHYLAMINO)-2-(α-PHENYL-α-CYCLOHEXYLMETHYL)PROPANE
    HYDROCHLORIDE see LFK200
3,6-BIS(3-DIETHYLAMINOPROPOXY)PYRIDAZINE BISMETHIODIDE
    see BJA825
BIS(DIETHYLAMINO)THIOXOMETHYL)DISULPHIDE see DXH250
BIS(DIETHYLDITHIOCARBAMATO)CADMIUM see BJB500
BIS(DIETHYLDITHIOCARBAMATO)MERCURY see BJB750
BIS(DIETHYLDITHIOCARBAMATO)ZINC see BJC000
1,4-BIS(N,N'-DIETHYLENE PHOSPHAMIDE)PIPERAZINE see BJC250
BISDIETHYLENE TRIAMINE COBALT(III) PERCHLORATE see BJC500
BIS-O,O-DIETHYLPHOSPHORIC ANHYDRIDE see TCF250
BIS-O,O-DIETHYLPHOSPHOROTHIONIC ANHYDRIDE see SOD100
BIS(DIETHYLTHIOCARBAMOYL) DISULFIDE see DXH250
BIS(N,N-DIETHYLTHIOCARBAMOYL) DISULFIDE see DXH250
BIS(DIETHYLTHIOCARBAMOYL)DISULFIDE mixed with SODIUM NITRITE
    see SIS200
BIS(N,N-DIETHYLTHIOCARBAMOYL)DISULPHIDE see DXH250
BIS(DIETHYLTHIO)CHLORO METHYL PHOSPHONATE see BJD000
BIS(2,2-DIFLUOROAMINO)-1,3-BIS(DINITRO-FLUOROETHOXY)PROPANE
    see SPA500
BIS(DIFLUOROAMINO)DIFLUOROMETHANE see BJD250
1,1-BIS(DIFLUOROAMINO)-2,2-DIFLUORO-2-NITROETHYL METHYL ETHER
    see BJD375
1,2-BIS(DIFLUOROAMINO)ETHANOL see BJD500
1,2-BIS(DIFLUOROAMINO)ETHYL VINYL ETHER see BJD750
4,4-BIS(DIFLUOROAMINO)-3-FLUOROIMINO-1-PENTENE see BJE000
1,2-BIS(DIFLUOROAMINO)-N-NITROETHYLAMINE see BJE250
BIS(DIFLUOROBORYL)METHANE see BJE325
(T-4-(R)(R)-BIS(N-(2-DIHYDROXY-3,3-DIMETHYL-1-OXOBUTYL β-ALANIN-
    ATO)ZINC see ZKS000
2,2-BISDIHYDROXYMETHYL-1,3-PROPANEDIOL TETRANITRATE
    see PBC250
1,4-BIS(3,4-DIHYDROXYPHENYL)-2,3-DIMETHYLBUTANE see NBR000
N,N'-BIS(2-(3',4'-DIHYDROXYPHENYL)-2-
    HYDROXYETHYL)HEXAMETHYLENEDIAMINE DIHYDROCHLORIDE
    see HFG600
N,N'-BIS(2-(3',4'-DIHYDROXYPHENYL)-2-HYDROXYETHYL)
    HEXAMETHYLENEDIAMINE SULFATE see HFG650
BIS(DIHYDROXYPHENYL)SULFIDE see BJE500
2,5-BIS(3,4-DIMETHOXYPHENYL)-1,3,4-THIADIAZOLE see BJE600
3,5-BIS(3,4-DIMETHOXYPHENYL)-1H-1,2,4-TRIAZOLE see BJE700
BIS(DIMETHYLAMIDO)FLUORO PHOSPHATE see BJE750
BIS(DIMETHYLAMIDO)PHOSPHORYL FLUORIDE see BJE750
2,8-BISDIMETHYLAMINOACRIDINE see BJF000
3,6-BIS(DIMETHYLAMINO)ACRIDINE see BJF000
BIS(DIMETHYLAMINO)-3-AMINO-5-PHENYLTRIAZOLYL PHOSPHINE
    OXIDE see AIX000
p-BIS(DIMETHYLAMINO)BENZENE see BJF500
1,4-BIS(DIMETHYLAMINO)BENZENE see BJF500
4,4'-BIS(DIMETHYLAMINO)BENZHYDRYLIDENIMINE HYDROCHLORIDE
    see IBA000
4,4'-BIS(DIMETHYLAMINO)BENZOHYDROL see TDO750
4,4'-BIS(DIMETHYLAMINO)BENZOPHENONE see MQS500
p,p'-BIS(N,N-DIMETHYLAMINO)BENZOPHENONE see MQS500
4,4'-BIS(DIMETHYLAMINO)BENZOPHENONE-IMINE HYDROCHLORIDE
    see IBA000
BIS(DIMETHYLAMINOBORANE)ALUMINUM TETRAHYDROBORATE
    see BJG000
BIS(DIMETHYLAMINO)CARBONOTHIOYL) DISULPHIDE see TFS350
BIS(DIMETHYLAMINO)DIMETHYLSTANNANE see BJG125
4,4'-BIS(DIMETHYLAMINO)DIPHENYLMETHANE see MJN000
p,p'-BIS(DIMETHYLAMINO)DIPHENYLMETHANE see MJN000
3,5-BIS-DIMETHYLAMINO-1,2,4-DITHIAZOLIUM CHLORIDE see BJG150
1,2-BIS-(DIMETHYLAMINO-ETHANE (DOT) see TDQ750
BIS(2-DIMETHYLAMINOETHOXY)ETHANE see BJG250
3,6-BIS(2-(DIMETHYLAMINO)ETHOXY)-9H-XANTHEN-9-ONE DIHYDRO-
    CHLORIDE see BJG500

BIS(2-ETHYLHEXYL)-1,2-BENZENEDICARBOXYLATE see DVL700
BIS(2-ETHYLHEXYL) ESTER, PEROXYDICARBONIC ACID see DJK800
BIS(2-ETHYLHEXYL) ESTER PHOSPHORUS ACID CADMIUM SALT
    see CAD500
BIS(ETHYLHEXYL) ESTER of SODIUM SULFOSUCCINIC ACID see DJL000
BIS(2-ETHYLHEXYL)ETHER see DJK600
BIS(2-ETHYLHEXYL) FUMARATE see DVK600
BIS(2-ETHYLHEXYL)HYDROGEN PHOSPHATE see BJR750
BIS(2-ETHYLHEXYL) HYDROGEN PHOSPHITE see BJQ709
BIS(2-ETHYLHEXYL) ISOPHTHALATE see BJQ750
BIS(2-ETHYLHEXYL)MALEATE see BJR000
BIS(2-ETHYLHEXYL)ORTHOPHOSPHORIC ACID see BJR750
BIS(2-ETHYLHEXYLOXYCARBONYLMETHYLTHIO)DIBUTYLSTANNANE
    see BKK250, DDY600
BIS(2-ETHYLHEXYLOXYCARBONYLMETHYLTHIO)DIMETHYLSTANNANE
    see BKK500
BIS((2-(ETHYL)HEXYLOXY)MALEOYLOXY)DI(n-BUTYL)STANNANE
    see BJR250
BIS(2-ETHYLHEXYL)PHENYL PHOSPHATE see BJR625
BIS(2-ETHYLHEXYL)PHOSPHATE see BJR750
BIS(2-ETHYLHEXYL)PHOSPHORIC ACID see BJR750
BIS(2-ETHYLHEXYL)PHTHALATE see DVL700
BIS(2-ETHYLHEXYL)SEBACATE see BJS250
BIS(2-ETHYLHEXYL)SODIUM SULFOSUCCINATE see DJL000
BIS(2-ETHYLHEXYL)-S-SODIUM SULFOSUCCINATE see DJL000
1,4-BIS(2-ETHYLHEXYL) SODIUM SULFOSUCCINATE see DJL000
1,4-BIS(2-ETHYLHEXYL)SULFOBUTANEDIOIC ACID ESTER, SODIUM SALT
    see DJL000
BIS(2-ETHYLHEXYLTHIOGLYCOLATE)DIBUTYLTIN see DDY600
BIS(2-ETHYLHEXYLTHIOGLYCOLATE)DIOCTYLTIN see DVM800
BIS(ETHYLMERCURI)PHOSPHATE see BJT250
N,N'-BIS(1-ETHYL-3-METHYLPENTYL)-p-PHENYLENEDIAMINE see BJT500
BIS(ETHYLPHENYLCARBODITHIOATO-S,S)-(T-4)-ZINC see ZHA000
1,1-BIS(p-ETHYLPHENYL)-2,2-DICHLOROETHANE see DJC000
2,2-BIS(p-ETHYLPHENYL)-1,1-DICHLOROETHANE see DJC000
BIS(N-ETHYL-N-PHENYL)UREA see DJC400
2,2-BIS(ETHYLSULFONYL)BUTANE see BJT750
2,2-BIS(ETHYLSULFONYL)PROPANE see ABD500
BIS(ETHYLXANTHIC)DISULFIDE see BJU000
BISETHYL XANTHOGEN DISULFIDE see BJU000
BIS(ETHYLXANTHOGEN) TETRASULFIDE see BJU250
BIS(ETHYLXANTHOGEN) TRISULFIDE see DKE400
S-(1,2-BIS(ETOSSI-CARBONIL)-ETIL)-O,O-DIMETIL-DITIOFOSFATO(ITAL-
    IAN) see MAK700
BISEXOVIS see AOO410
BISEXOVISTER see AOO410
BISFEROL A (GERMAN) see BLD500
1,6-BIS(9 FLUORENYLDIMETHYL-AMMONIUM)HEXANE BROMIDE
    see HEG000
BIS(2-FLUORO-2,2-DINITROETHOXY)DIMETHYLSILANE see BJU350
BIS(2-FLUORO-2,2-DINITROETHYL)AMINE see BJU500
1,1-BIS(FLUOROOXY)HEXAFLUOROPROPANE see BJV625
2,2-BIS(FLUOROOXY)HEXAFLUOROPROPANE see BJV630
1,1-BIS(FLUOROOXY)TETRAFLUOROETHANE see BJV635
1-(4,4-BIS(p-FLUOROPHENYL)BUTYL)-4-(2-OXO-1-BENZIMIDAZOLINYL)
    PIPERIDINE see PIH000
trans-4-(4,4-BIS(p-FLUOROPHENYL)BUTYL)-1-(2-(4-
    PHENYLCYCLOHEXYLAMINO)ETHYL)PIPERAZINE TRIHYDROCHLOR-
    IDE see BJV750
8-(4,4-BIS(p-FLUOROPHENYL)BUTYL)-1-PHENYL-1,3,8-
    TRIAZASPIRO(4,5)DECAN-4-ONE see PFU250
trans-2-(4-(4,4-BIS(p-FLUOROPHENYL)BUTYL)PIPERAZINYL)-N-(4-
    PHENYLCYCLOHEXYL)ACETAMIDE DIHYDROCHLORIDE see BJW000
1-(1-(4,4-BIS(p-FLUOROPHENYL)BUTYL)-4-PIPERIDYL)-2-
    BENZIMIDAZOLINONE see PIH000
(3)-1-(BIS(p-FLUOROPHENYL)METHYL)-4-CINNAMYLPIPERAZINE DIHYDRO-
    CHLORIDE see FDD080
(E)-1-(BIS(4-FLUOROPHENYL)METHYL)-4-(3-PHENYL-2-PROPENYL)PIPERA-
    ZINE DIHYDROCHLORIDE see FDD080
1,1-BIS(4-FLUOROPHENYL)-2-PROPYNYL-N-CYCLOHEPTYLCARBAMATE
    see BJW250
1,1-BIS(4-FLUOROPHENYL)-2-PROPYNYL-N-CYCLOOCTYL CARBAMATE
    see BJW500
α-α-BIS(4-FLUOROPHENYL)-1H-1,2,4-TRIAZOLE-1-ETHANOL see BJW600
1,1-BIS(p-FLUOROPHENYL)-2,2,2-TRICHLOROETHANE see FHJ000
2,2-BIS(p-FLUOROPHENYL)-1,1,1-TRICHLOROETHANE see FHJ000
BIS(3-FLUOROSALICYLALDEHYDE)-ETHYLENEDIIMINE-COBALT see EIS000
BIS(N-FORMYL-p-AMINOPHENYL)SULFONE see BJW750
BIS(FORMYLMETHYL) MERCURY see BJW800
1,3-BIS(4-FORMYLPYRIDINIUM)-PROPANE BISOXIDE DIBROMIDE
    see TLQ500
1,3-BIS(4-FORMYLPYRIDINIUM)-PROPANE BISOXIME DICHLORIDE
    see TLQ750
2-(BIS(FURFURYLIDENAMINO))METHYLFURAN see FPS000

3,5-BIS(2-FURYL)-1H-1,2,4-TRIAZOLE see AMX000
3,5-BIS-d-GLUCONAMIDO-2,4,6-TRIIODO-N-METHYLBENZAMIDE see IFS400
m-BIS(GLYCIDYLOXY)BENZENE see REF000
2,3-BIS(GLYCIDYLOXY)-1,4-DIOXANE see BJN750
BIS(4-GLYCIDYLOXYPHENYL)DIMETHYAMETHANE see BLD750
2,2-BIS(p-GLYCIDYLOXYPHENYL)PROPANE see BLD750
BIS(GUANIDINIUM) CHROMATE see BJW825
(BIS-(HEPTYLOXY)PHOSPHINYL)MERCURY ACETATE see AAR750
BIS(HEXANOYLOXY)DI-n-BUTYLSTANNANE see BJX750
BIS(HEXANOYLOXY)DI-n-BUTYL-TIN see BJX750
BIS(2-HEXYLOXY)ETHYL)ADIPATE see AEQ000
(BIS(HEXYLOXY)PHOSPHINYL)MERCURY ACETATE see AAS000
BIS(l-HISTIDINATO)COBALT see BJY000
BIS(l-HISTIDINATO)MANGANESE TETRAHYDRATE see BJX800
BIS(l-HISTIDINE)COBALT see BJY000
BIS-HM-A-TDA see BJY125
BISHYDRAZINE NICKEL(II)PERCHLORATE see BJY250
BISHYDRAZINE TIN(II)CHLORIDE see BJY500
BIS(HYDROGEN MALEATO)DIBUTYL-TIN BIS(2-ETHYLHEXYL) ESTER
    see BJR250
BIS(HYDROGEN MALEATO)DIOCTYLTIN BIS(2-ETHYLHEXYL) ESTER
    see DVM600
BIS(1-HYDROPEROXY CYCLOHEXYL)PEROXIDE see BJY750
2,2-BIS(HYDROPEROXY)PROPANE see BJY825
1,2-BIS(HYDROXOMERCURIO)-1,1,2,2-BIS(OXYDIMERCURIO)ETHANE
    see BKH125
BIS(HYDROXYAETHYL)-AETHER-DINITRAT (GERMAN) see DJE400
BIS(β-HYDROXYAETHYL)NITROSAMIN (GERMAN) see NKM000
BIS(2-HYDROXYBENZOATO-O^1,O^2-, (T-4)-CADMIUM (9CI) see CAI400
BIS(2-HYDROXY-3-tert-BUTYL-5-ETHYLPHENYL) METHANE see MJN250
BIS(4-HYDROXY-5-tert-BUTYL-2-METHYLPHENYL) SULFIDE see TFC600
BIS-2-HYDROXY-5-CHLORFENYLMETHAN (CZECH) see MJM500
BIS(2-HYDROXY-5-CHLOROPHENYL)METHANE see MJM500
BISHYDROXYCOUMARIN see BJZ000
BIS(4-HYDROXY-3-COUMARIN) ACETIC ACID ETHYL ESTER see BKA000
BIS-3,3'-(4-HYDROXYCOUMARINYL)ACETIC ACID ETHYL ESTER
    see BKA000
BIS-(4-HYDROXY-3-COUMARINYL)ETHYL ACETATE see BKA000
BIS(4-HYDROXYCOUMARIN-3-YL)METHANE see BJZ000
BIS(1-HYDROXYCYCLOHEXYL)PEROXIDE see BKA250
2,2-BIS(4-HYDROXY-3,5-DICHLOROPHENYL)PROPANE see TBO750
BIS(2-HYDROXY-3,5-DICHLOROPHENYL) SULFIDE see TFD250
BIS(2-HYDROXY ETHYL)AMINE see DHF000
2-BIS-HYDROXYETHYLAMINO-4-ACETAMINOFENETOL (CZECH) see BKB000
3'-BIS(2-HYDROXYETHYL)AMINO)-p-ACETOPHENETIDIDE see BKB000
2-(BIS(β-HYDROXYETHYL)AMINO)-4,5-DIPHENYLOXAZOLE MONOHY-
    DRATE see BKB250
1,4-BIS((2-((2-HYDROXYETHYL)AMINO)ETHYLAMINO)-9,10-
    ANTHRACENEDIONE see BKB325
1,4-BIS((2-((2-HYDROXYETHYL)AMINO)ETHYL)AMINO)-9,10-
    ANTHRACENEDIONE DIACETATE see BKB300
1,4-BIS((2-((HYDROXYETHYL)AMINO)ETHYL)AMINO)-9,10-
    ANTHRACENEDIONE DIACETATE (SALT) (9CI) see BKB300
1,4-BIS((2-((2-HYDROXYETHYL)AMINO)ETHYL)AMINO)ANTHRAQUINONE
    see BKB325
5,8-BIS((2-((2-HYDROXYETHYL)AMINO)ETHYL)AMINO)-1,4-DIHYDROXY-9,10-
    ANTHRACENEDIONE DIACETATE see DBE875
5,8-BIS((2-((HYDROXYETHYL)AMINO)ETHYL)AMINO)-1,4-
    DIHYDROXYANTHRAQUINONE see MQY090
5,8-BIS((2-((2-HYDROXYETHYL)AMINO)ETHYL)AMINO)-1,4-
    DIHYDROXYANTHRAQUINONE 1,4-DIACETATE see DBE875
3-(BIS(2-HYDROXYETHYL)AMINO)-6-
    HYDRAZINOPYRIDAZINEDIHYDROCHLORIDE see BKB500
4-(N,N-BIS(2-HYDROXYETHYL)AMINO)-7-NITROBENZOFURAZAN-1-OXIDE
    see NFF000
4-BIS(2-HYDROXYETHYL)AMINO-2-(5-NITRO-2-FURYL)QUINAZOLINE
    see BKB750
2-(BIS(2-HYDROXYETHYL)AMINO)-5-NITROPHENOL see BKC000
4-BIS(2-HYDROXYETHYL)AMINO-2-(5-NITRO-2-THIENYL)QUINAZOLINE
    see BKC250
N-(3-(BIS(2-HYDROXYETHYL)AMINO)PROPYL)BENZAMIDE HYDROCHLO-
    RIDE see BKC500
10-(3-(BIS(2-HYDROXYETHYL)AMINO)PROPYL)-7-CHLOROISOALLOXAZINE
    SULFATE see BKC750
N-(3-(BIS(2-HYDROXYETHYL)AMINO)PROPYL)-o-PROPOXYBENZAMIDE HY-
    DROCHLORIDE see BKD000
4,4'-BIS((4-(2-HYDROXYETHYL)AMINO-6-(p-SULFOANILINO)-s-TRIAZIN-2-
    YL)AMINO)-2,2'-STILBENEDISULFONIC ACID TETRASODIUM SALT
    see BKD250
N,N-BIS(2-HYDROXYETHYL)ANILINE see BKD500
N,N-BIS(2-HYDROXYETHYL)-p-ARSANILIC ACID see BKD600
1,1-BIS(2-HYDROXYETHYL)AZIRIDINIUM CHLORIDE see BKE750
BIS(2-HYDROXYETHYL)CARBAMODITHIOIC ACID, MONOPOTASSIUM
    SALT see PKX500

BIS(2-HYDROXYETHYL)-2-(2-CHLORO ETHYL THIO)ETHYL SULFONIUM)
CHLORIDE see BKD750
BIS(2-HYDROXYETHYL)DITHIOCARBAMIC ACID, MONOPOTASSIUM SALT
see PKX500
BIS(2-HYDROXYETHYL)DITHOCARBAMIC ACID, POTASSIUM SALT
see PKX500
N,N-BIS(2-HYDROXYETHYL)DODECAN AMIDE see BKE500
BIS(2-HYDROXYETHYL) ETHER see DJD600
1,1-BIS(β-HYDROXYETHYL)ETHYLENINONIUM CHLORIDE see BKE750
N,N-BIS-2-HYDROXYETHYL-p-FENYLENDIAMIN STRAN (CZECH) see BKF750
1,1-BIS(2-HYDROXYETHYL)HYDRAZINE see HHH000
BIS(2-HYDROXYETHYL)LAURAMIDE see BKE500
N,N-BIS(HYDROXYETHYL)LAURAMIDE see BKE500
N,N-BIS(2-HYDROXYETHYL)LAURAMIDE see BKE500
N,N-BIS(β-HYDROXYETHYL)LAURAMIDE see BKE500
N,N-BIS(2-HYDROXYETHYL)-3-METHYLANILINE see DHF400
N,N-BIS(β-HYDROXYETHYL)-3-METHYLANILINE see DHF400
N',N'-BIS(2-HYDROXYETHYL)-N-METHYL-2-NITRO-p-PHENYLENEDIAMINE
see BKF250
BIS(β-HYDROXYETHYL)NITROSAMINE see NKM000
N,N-BIS(2-HYDROXYETHYL)-9,12-OCTADECADIENAMIDE see BKF500
N,N-BIS(2-HYDROXYETHYL)-p-PHENYLENEDIAMINE SULFATE (1:1) see
BKF750
1,4-BIS(2-HYDROXYETHYL)PIPERAZINE see PIJ750
N,N'-BIS(β-HYDROXYETHYL)PIPERAZINE see PIJ750
BIS(2-HYDROXYETHYL)SULFIDE see TFI500
BIS(β-HYDROXYETHYL)SULFIDE see TFI500
N,N-BIS(2-HYDROXYETHYL)-m-TOLUIDINE see DHF400
2,2-BIS-4'-HYDROXYFENYLPROPAN (CZECH) see BLD500
l-(+)-N,N'BIS(2-HYDROXY)-1-HYDROXYMETHYLETHYL)-2,4,6-TRIIODO-5-
LACTAMIDE ISOPHTHALAMIDE see IFY000
1,3-BIS(4-HYDROXYIMINOMETHYL-1-PYRIDINIO)-2-OXAPROPANE DICHLO-
RIDE see BGS250
1,3-BIS(4-HYDROXYIMINOMETHYL-1-PYRIDINIO)PROPANE DICHLORIDE
see TLQ750
1,4-BIS(4-HYDROXYIMINOMETHYL-PYRIDINIUM-(1))-BUTANEDIOL-2,3
DIBROMID (GERMAN) see BGR750
1,4-BIS(4-HYDROXYIMINOMETHYL-PYRIDINIUM-(1))BUTANEDIOL(2,3)-
DIPERCHLORAT (GERMAN) see DND400
BIS((HYDROXYIMINO-METHYL)-PYRIDINIUM-(1)-METHYL-AETHER-
DICHLORID (GERMAN) see PPC000
BIS(2-HYDROXYIMINOMETHYLPYRIDINIUM-1-METHYL)ETHER DICHLO-
RIDE see PPC000
BIS(4-HYDROXYIMINOMETHYLPYRIDINIUM-1-METHYL)ETHER DICHLO-
RIDE see BGS250
BIS(HYDROXYLAMINE) SULFATE see OLS000
BISHYDROXYL AMINE ZINC(II)CHLORIDE see BKG250
3,5-BIS(3-HYDROXYMERCURI-2-METHOXYPROPYL)BARBITURIC ACID SO-
DIUM SALT see BKG500
5,5-BIS(3-HYDROXYMERCURI-2-METHOXYPROPYL)BARBITURIC ACIDSOD-
IUM SALT see BKG750
2,6-BIS(HYDROXYMERCURI)-4-NITROANILINE see BKH000
3-BIS(HYDROXYMETHYL)AMINO-6-(5-NITRO-2-FURYLETHENYL)-1,2,4-
TRIAZINE see BKH500
2-N,N-BIS(HYDROXYMETHYL)AMINO-1,3,4-THIADIAZOLE see BJY125
9:10-BISHYDROXYMETHYL-1:2-BENZANTHRACENE see BBF500
1,4-BIS(HYDROXYMETHYL)CYCLOHEXANE see BKH325
BIS(HYDROXYMETHYL)FURATRIZINE see BKH500
3,3-BIS(HYDROXYMETHYL)HEPTANE see BKH625
1,2-BISHYDROXYMETHYL-1-METHYLPYRROLE see MPB175
2,3-BISHYDROXYMETHYL-1-METHYLPYRROLE see MPB175
3,3-BIS(HYDROXYMETHYL)PENTANE see PMB250
BIS HYDROXYMETHYL PEROXIDE see BKH750
BIS(HYDROXYMETHYL)PEROXIDE see DMN200
BIS-(1-HYDROXYMETHYL)PEROXIDE see DMN200
meso-3,4-BIS(4-HYDROXY-3-METHYLPHENYL)HEXANE see DJI350
2,2-BIS(HYDROXYMETHYL)-1,3-PROPANEDIOL see PBB750
2,2-BIS(HYDROXYMETHYL)-1,3-PROPANEDIOL TETRANITRATE see PBC250
d,N,N'-BIS(1-HYDROXYMETHYLPROPYL)ETHYLENEDIAMINE see TGA500
2,3-BIS(HYDROXYMETHYL)QUINOXALINE DI-N-OXIDE see DVQ800
1,3-BIS(HYDROXYMETHYL)UREA see DTG700
N,N'-BIS(HYDROXYMETHYL)UREA see DTG700
BIS(4-HYDROXY-3-NITROPHENYL)MERCURY see MCS600
BIS(4-HYDROXY-2-OXO-2H-1-BENZOPYRAN-3-YL)ACETIC ACID ETHYL
ESTER see BKA000
3,3-BIS(p-HYDROXYPHENOL)-PHTHALIDE see PDO750
N,N-BIS(2-HYDROXY-3-PHENOXYPROPYL)ETHYLENEDIAMINE DIHYDRO-
CHLORIDE see IAG700
BIS-(p-HYDROXYPHENYL)-CYCLOHEXYLIDENEMETHANE DIACETATE
see FBP100
BIS(4-HYDROXYPHENYL) DIMETHYLMETHANE see BLD500
BIS(4-HYDROXYPHENYL)DIMETHYLMETHANE DIGLYCIDYL ETHER
see BLD750
3,4-BIS(4-HYDROXYPHENYL)-2,4-HEXADIENE see DAL600

3,4-BIS(p-HYDROXYPHENYL)-2,4-HEXADIENE see DAL600
3,4-BIS(p-HYDROXYPHENYL)HEXANE see DLB400
meso-3,4-BIS(p-HYDROXYPHENYL)-n-HEXANE see DLB400
3,4-BIS(p-HYDROXYPHENYL)-2-HEXANONE see BKH800
3,4-BIS(p-HYDROXYPHENYL)-3-HEXENE see DKA600
BIS(4-HYDROXYPHENYL)METHANE see BKI250
BIS(p-HYDROXYPHENYL)METHANE see BKI250
p,p'-BIS(HYDROXYPHENYL)METHANE see BKI250
BIS(4-HYDROXYPHENYL)PROPANE see BLD500
2,2-BIS(4-HYDROXYPHENYL)PROPANE see BLD500
2,2-BIS(p-HYDROXYPHENYL)PROPANE see BLD500
2,2-BIS(4-HYDROXYPHENYL)PROPANE, DIGLYCIDYL ETHER see BLD750
2,2-BIS(p-HYDROXYPHENYL)PROPANE, DIGLYCIDYL ETHER see BLD750
BIS(4-HYDROXYPHENYL) SULFIDE see TFJ000
2,2-BIS(4-HYDROXYPHENYL)-1,1,1-TRICHLOROETHANE see BKI500
2,3-BIS(p-HYDROXYPHENYL)VALERONITRILE see BKI750
BIS(2-HYDROXYPROPYL)AMINE see DNL600
N,N'-(BIS-ω-HYDROXYPROPYL)HOMOPIPERAZINE 3,4,5-
TRIMETHOXYBENZOATE DIHYDROCHLORIDE see CNR750
N-BIS(2-HYDROXYPROPYL)NITROSAMINE see DNB200
2,2'-BISHYDROXYPROPYLNITROSAMINE see DNB200
BIS(3-HYDROXY-1-PROPYNYL)MERCURY see BKJ250
BIS(1-HYDROXY-2(1H)-PYRIDINETHIONATO)ZINC see ZMJ000
BIS(8-HYDROXYQUINOLINE-5-SULFONIC ACID) MANGANESE(II) see BKJ275
BIS-(5-HYDROXY-7-SULFO-2NAFTYL)AMIN (CZECH) see IBF000
1,3-BIS(1-HYDROXY-2,2,2-TRICHLOROETHYL)UREA see DGQ200
BIS(2-HYDROXY-3,5,6-TRICHLOROPHENYL)METHANE see HCL000
BISIBUTIAMINE see BKJ325
BISINA see CKF500
BIS(3-INDOLEMETHYLENEMORPHOLINIUM)HEXACHLOROSTANNATE
see BKJ500
BIS(ISOBUTYL)ALUMINUM CHLORIDE see CGB500
BIS(ISOBUTYL)HYDROALUMINUM see DNI600
BIS(4-ISOCYANATOCYCLOHEXYL)METHANE see MJM600
BIS(4-ISOCYANATOPHENYL)METHANE see MJP400
BIS(p-ISOCYANATOPHENYL)METHANE see MJP400
BIS(1,4-ISOCYANATOPHENYL)METHANE see MJP400
BIS(ISONICOTINALDOXIME 1-METHYL) ETHER DICHLORIDE see BGS250
BIS(ISOOCTYLOXYCARBONYLMETHYLTHIO)DIBUTYL STANNANE
see BKK250
BIS(ISOOCTYLOXYCARBONYLMETHYLTHIO)DIMETHYLSTANNANE
see BKK500
BIS(ISOOCTYLOXYCARBONYLMETHYLTHIO)DIOCTYL STANNANE
see BKK750
BIS(ISOOCTYLOXYMALEOYLOXY)DIOCTYLSTANNANE see BKL000
BIS(ISOPROPYLAMIDO) FLUOROPHOSPHATE see PHF750
2,4-BIS(ISOPROPYLAMINO)-6-CHLORO-s-TRIAZINE see PMN850
2,4-BIS(ISOPROPYLAMINO)-6-ETHYLTHIO-s-TRIAZINE see EPN500
2,4-BIS(ISOPROPYLAMINO)-6-METHOXY-s-TRIAZINE see MFL250
2,4-BIS(ISOPROPYLAMINO)-6-METHYLMERCAPTO-s-TRIAZINE see BKL250
4,6-BIS(ISOPROPYLAMINO)-2-METHYLMERCAPTO-s-TRIAZINE see BKL250
2,4-BIS(ISOPROPYLAMINO)-6-METHYLTHIO-s-TRIAZINE see BKL250
2,4-BIS(ISOPROPYLAMINO)-6-METHYLTHIO-1,3,5-TRIAZINE see BKL250
2,4-BIS(ISOPROPYLAMINO)-6-(METHYLTHIO)-s-TRIAZINE mixed with
METHANEARSONIC ACID MONOSODIUM SALT (1:4) see BKL500
BIS(ISOPROPYLBENZENE)CHROMIUM see DGR200
BIS(LAUROYLOXYCARBONYLMETHYLTHIO)DIOCTYLSTANNANE
see DVN000
BIS(LAUROYLOXY)DIBUTYLSTANNANE see DDV600
BIS(LAUROYLOXY)DI(n-BUTYL)STANNANE see DDV600
BIS(LAUROYLOXY)DIOCTYLSTANNANE see DVJ800
1,3-BISMALEIMIDO BENZENE see BKL750
BISMARSEN see BKV250
BISMATE see BKW000
BIS(MERCAPTOACETATE)-1,4-BUTANEDIOL see BKM000
BIS(MERCAPTOACETATE)DIOCTYLTIN BIS(BUTYL) ESTER see DVM200
BIS(MERCAPTOACETATE)DIOCTYLTIN BIS(2-ETHYLHEXYL) ESTER
see DVM800
BIS(MERCAPTOACETATE)DIOCTYL-TIN BIS(ISOOCTYL) ESTER see BKK750
BIS(MERCAPTOBENZIMIDAZOLATO)ZINC see ZIS000
BIS(MERCAPTOBENZOTHIAZOLATO)ZINC see BHA750
BIS(MERCAPTO)DIOCTYLTIN BIS(DODECYL) ESTER see DVM400
BIS(MESCALINIUM)TETRACHLOROMANGANATE(II) see BKM100
1,2-BIS(MESYLOXY)ETHANE see BKM125
1,4-BIS(2'-MESYLOXYETHYLAMINO)-1,4-DIDEOXYMESOERYTHRITOL
see LJD500
1,2-BIS(METHACRYLOYLOXY)ETHANE see BKM250
1,4-BIS(METHANESULFONOXY)BUTANE see BOT250
BIS(METHANE SULFONYL)-d-MANNITOL see BKM500
(1,4-BIS(METHANESULFONYLOXY)BUTANE) see BOT250
BISMETHIN see PEV750
3,4-BIS(METHOXY)BENZYL CHLORIDE see BKM750
3,5-BIS(METHOXYCARBONYL)BENZENESULFONIC ACID, SODIUM SALT
see BKN000

N-(2-(2,3-BIS-(METHOXYCARBONYL)-GUANIDINO)-5-(PHENYLTHIO)-PHENYL)-2-METHOXYACETAMIDE see BKN250
1,2-BIS(METHOXYCARBONYLTHIOUREIDO)BENZENE see PEX500
o-BIS(3-METHOXYCARBONYL-2-THIOUREIDO)BENZENE see PEX500
1,2-BIS(3-(METHOXYCARBONYL)-2-THIOUREIDO)BENZENE see PEX500
2,5-BISMETHOXYETHOXY-3,6-BISETHYLENEIMINO-1,4-BENZOQUINONE see BDC750
3,6-BIS(β-METHOXYETHOXY)-2,5-BIS(ETHYLENEIMINO)-p-BENZOQUINONE see BDC750
3,6-BIS(β-METHOXYETHOXY)-2,5-BIS(ETHYLENIMINO)-p-BENZOQUINONE see BDC750
BIS(2-(2-METHOXYETHOXY)ETHYL) ETHER see PBO500
4,4'-BIS((4-(2-METHOXYETHOXY)-6-(N-METHYL-N-2-SULFOETHYL)AMINO-s-TRIAZIN-2-YL)AMINO)-2,2'-STILBENEDISULFONIC ACID see BKN500
BIS(2-METHOXYETHYL)ESTER MALEIC ACID see DOF000
BIS(2-METHOXY ETHYL)ETHER see BKN750
BIS(2-METHOXYETHYL)ETHER see PBO500
BIS(2-METHOXYETHYL)NITROSOAMINE see BKO000
BIS(2-METHOXYETHYL) PHTHALATE see DOF400
BIS(2-METHOXYETHYL) PHTHALATE see DOF400
BIS(METHOXYMALEOYLOXY)DIBUTYLSTANNANE see BKO250
BIS(METHOXYMALEOYLOXY)DIOCTYLSTANNANE see BKO500
2,6-BIS(p-METHOXYPHENETHYL)-1-METHYLPIPERIDINE ETHANESULFONATE see MJE800
4,4'-BIS(4-METHOXY-6-PHENYLAMINO-2-s-TRIAZINYLAMINO)-2,2'-STILBENEDISULFONIC ACID see BKO750
3,4-BIS(p-METHOXYPHENYL)-3-BUTEN-2-ONE see BKO800
2-(4-(1,2-BIS(4-METHOXYPHENYL)-1-BUTENYL)PHENOXY)-N,N-DIETHYL-ETHANAMINE (9CI) see BKO825
trans-1-(2-(p-(1,2-BIS(p-METHOXYPHENYL)-1-BUTENYL)PHENOXY)ETHYL)PYRROLIDINE HYDROCHLORIDE see HAA300
2-(p-(1,2-BIS(p-METHOXYPHENYL)-1-BUTENYL)PHENOXY)TRIETHYLAMINE see BKO825
2-(p-(1,2-BIS(p-METHOXYPHENYL)-1-BUTENYL)PHENOXY)TRIETHYLAMINE CITRATE see BKO835
trans-2-(p-(1,2-BIS(p-METHOXYPHENYL)-1-BUTENYL)PHENOXY)TRIETHYLMA- INE HYDROCHLORIDE see BKO840
1,5-BIS(o-METHOXYPHENYL)-3,7-DIAZAADMANTAN-9-ONE see BKP000
trans-1,2-BIS(p-METHOXYPHENYL)-1-(p-(2-(N,N-DIETHYLAMINO)ETHOXY)PHENYL)BUT-1-ENE HCl see BKO840
3,4-BIS(p-METHOXYPHENYL)-3-HEXENE see DJB200
N,N,N'-BIS(4-METHOXYPHENYL)-N'-(4-ETHOXYPHENYL)GUANIDINE HYDRO-CHLORIDE see PDN500
2,3-BIS(4-METHOXYPHENYL)PENT-2-ENENITRILE see BKP300
2,3-BIS(p-METHOXYPHENYL)-2-PENTENONITRILE see BKP300
2-(p-(1,2-BIS(p-METHOXYPHENYL)PROPENYL)PHENOXY)TRIETHYLAMINE, CITRATE, MONOHYDRATE see BKP325
2-(p-(1,2-BIS(p-METHOXYPHENYL)PROPENYL)PHENOXY)TRIETHYLAMINE CITRATE HYDRATE see BKP325
trans-1,2-BIS(p-METHOXYPHENYL)-1-(p-(2-N-PYRROLIDINOETHOXY)PHENYL)BUT-1-ENE HYDROCHLORIDE see HAA300
1,1-BIS(p-METHOXYPHENYL)-2,2,2-TRICHLOROETHANE see MEI450
2,2-BIS(p-METHOXYPHENYL)-1,1,1-TRICHLOROETHANE see MEI450
1,10-BIS(N-METHYL-N-(1'-ADAMANTYL)AMINO)DECANE DIIODOMETHYL-ATE see DAE500
BIS-(2-METHYL-4-AMINO-6-QUINOLYLOXY)ETHANE DIHYDROCHLORIDE see EJB100
BIS(N-METHYLANILINE)METHANE see MJO000
BIS(N-METHYLANILINO)METHAN (GERMAN) see MJO000
N,N'-(BIS(2-(2-METHYL-1,3-BENZODIOXOL-2-YL)ETHYL))ETHYLENEDIAM-INE DIHYDROCHLORIDE see BKQ250
BIS(α-METHYLBENZYL)AMINE see BKQ500
N,N'-BIS(α-METHYLBENZYL)ETHYLENEDIAMINE see DQQ600
3,6-BIS(5-(3-METHYL-2-BUTENYL)INDOL-3-YL)-2,5-DIHYDROXY-p-BENZOQUINONE see CNF159
BIS(3-METHYLBUTYL) ADIPATE see AEQ500
2,6-BIS(1-METHYLBUTYL)PHENOL see BKQ750
1,4-BIS(METHYLCARBAMYLOXY)-2-ISOPROPYL-5-METHYLBENZENE see BKR000
BIS(3-METHYLCYCLOHEXYL PEROXIDE) see BKR250
1,1-BIS((3,4-METHYLENEDIOXYPHENOXY)METHYL)-N,N-DIMETHYL-1-BUTANOL CITRATE see BKR500
α,α-BIS(3,4-(METHYLENEDIOXY)PHENOXY)METHYL)-1-PIPERIDINEBUTANOLACETATE CITRATE see BKR750
1-(1,3-BIS(3,4-(METHYLENEDIOXY)PHENOXY)-2-PROPYL)PYRROLIDINE CI-TRATE see BKS500
α-(2-BIS(1-METHYLETHYL)AMINO)ETHYL)-α-PHENYL-2-PYRIDINEACETAM-IDE PHOSPHATE (9CI) see RSZ600
BIS(1-METHYLETHYL)DIAZENE 1-OXIDE see ASP510
N,N'-BIS(1-METHYLETHYL)-6-METHYL-THIO-1,3,5-TRIAZINE-2,4-DIAMINE see BKL250
3,5-BIS(1-METHYLETHYL)PHENOL METHYLCARBAMATE see DNS200

3,5-BIS(1-METHYLETHYL)PHENYL ESTER METHYL CARBAMIC ACID see DNS200
3,5-BIS(1-METHYLETHYL)PHENYL METHYLCARBAMATE see DNS200
O,O-BIS(1-METHYLETHYL)-S-(PHENYLMETHYL)PHOSPHOROTHIOATE see BKS750
O,O-BIS(1-METHYLETHYL)-S-(2-((PHENYLSULFONYL)AMINO)ETHYL)PHEOSPHORODITHIOATE see DNO800
BIS(2-METHYLGLYCIDYL) ETHER see BJN500
BIS(6-METHYLHEPTYL)ESTER of PHTHALIC ACID see ILR100
N,N-BIS(5-METHYL-3-HEPTYL)-p-PHENYLENEDIAMINE see BJT500
BIS(3-METHYLHEXYL)PHTHALIC ACID ESTER see DSF200
1,1-BIS(2-METHYL-4-HYDROXY-5-tert-BUTYLPHENYL)BUTANE see BRP750
BIS(2-METHYL-3-HYDROXY-4-METHOXYMETHYL-5-METHYLPYRIDYL) DISULFIDE DIHYDROCHLORIDE see BKS800
3,5-BIS-METHYLKARBOXY-BENZENSULFONAN SODNY (CZECH) see BKN000
BIS(METHYLMERCURIC)SULFATE see BKS810
BIS-(METHYLMERCURY)-SULFATE see BKS810
BIS-(METHYLMERKURI)SULFAT see BKS810
N,N-BIS(N-METHYL-N-PHENYL-tert-BUTYLACETAMIDO)β-HYDROXYETHYLAMINE see DTL200
N,N'-BIS((1-METHYL-4-PHENYL-4-PIPERIDINYL)METHYL)-DECANEDIAM-IDE (9CI) see BKS825
N,N'-BIS(1-METHYL-4-PHENYL-4-PIPERIDYLMETHYL)SEBACAMIDE see BKS825
N,N'-BIS(2-METHYLPHENYLTHIOUREA see DXP600
N,N'-BIS(2-METHYLPROPYL)CARBAMOTHIOIC ACID-S-ETHYL ESTER see EID500
N-BISMETHYLPTEROYLGLUTAMIC ACID see MDV500
BIS(2-METHYL PYRIDINE)SODIUM see BKT250
N,N-BIS(METHYLQUECKSILBER)-p-TOLUOL-SULFAMID see MLH100
N,N-BIS(METHYLSULFONEPROPOXY)AMINE HYDROCHLORIDE see YCJ000
1,6-BIS-o-METHYLSULFONYL-d-MANNITOL see BKM500
BIS(METHYLSULFONYL)OXY)DIBUTYLSTANNANE see DEI400
BIS((METHYLSULFONYL)OXY)DIPROPYLSTANNANE see DWW400
BIS(3-METHYLSULFONYLOXYPROPYL)AMINE p-TOLUENESULFONATE see IBQ100
N,N'-BIS(3-METHYL-2-THIAZOLIDINYLIDENE)UREA see BKU000
BIS(METHYLXANTHOGEN) DISULFIDE see DUN600
1,3-BIS(3,4-METILENDIOSSIFENOSSI)-2-AMINOPROPANO (ITALIAN) see MJS250
1,3-BIS-(3,4-METILENDIOSSIFENOSSI)-2-(3-DIMETILAMINOPROPIL)PROPAN-2-OLO CITRATO (ITALIAN) see BKR500
1,3-BIS-(3,4-METILENDIOSSIFENOSSI)-2-PIRROLIDINOPROPANO CITRATO (ITALIAN) see BKS500
BIS(MONOISOPROPYLAMINO)FLUOROPHOSPHATE see PHF750
BIS(MONOISOPROPYLAMINO)FLUOROPHOSPHINE OXIDE see PHF750
BIS(4-MORPHOLINECARBODITHIOATO)MERCURY see BKU250
N,N'-BISMORPHOLINE DISULFIDE see BKU500
BISMORPHOLINO DISULFIDE see BKU500
BIS(MORPHOLINO-)METHAN (GERMAN) see MJQ750
BISMORPHOLINO METHANE see MJQ750
5,7-BIS(MORPHOLINOMETHYL)-2-HYDROXY-3-ISOPROPYL-2,4,6-CYCLOHEPTATRIEN-1-ONE DIHYDROCHLORIDE see IOW500
BISMUTH see BKU750
BISMUTH-209 see BKU750
BISMUTH AMIDE OXIDE see BKV000
BISMUTH ARSPHENAMINE SULFONATE see BKV250
BISMUTH CHROMATE see DDT250
BISMUTH COMPOUNDS see BKV750
BISMUTH DIMETHYL DITHIOCARBAMATE see BKW000
BISMUTH EMETINE IODIDE see EAM500
BISMUTH NITRATE see BKW250
BISMUTH NITRIDE see BKW500
BISMUTH PENTAFLUORIDE see BKW750
BISMUTH PERCHLORATE see BKW850
BISMUTH PLUTONIDE see BKX000
BISMUTH POTASSIUM SODIUM TARTRATE(SOLUBLE) see BKX250
BISMUTH SESQUITELLURIDE see BKY000
BISMUTH SODIUM-p-AMINOPHENYLARSONATE see BKX500
BISMUTH SODIUM THIOGLYCOLLATE see BKX750
BISMUTH STANNATE PENTAHYDRATE see BKY250
BISMUTH TELLURIDE see BKY000
BISMUTH TELLURIDE, UNDOPED see BKY000
BISMUTH TIN OXIDE see BKY250
BISMUTH TRISODIUM THIOGLYCOLLATE see BKY500
BIS(NITRATO-O,O')DIOXO URANIUM (solid) see URA200
BIS(NITRATO)DIOXOURANIUM HEXAHYDRATE see URS000
2,5-BIS-(NITRATOMERCURIMETHYL)-1,4-DIOXANE see BKZ000
BIS(NITRATO-O)OXOZIRCONIUM see BLA000
BIS-p-NITRO BENZENE DIAZO SULFIDE see BLA250
1,5-BIS(5-NITRO-2-FURANYL)-1,4-PENTADIEN-3-ONE, (AMINOIMINOMETHYL)HYDRAZONE see PAF500
BIS(5-NITROFURFURYLIDENE)ACETONE GUANYLHYDRAZONE see PAF500

sym-BIS(5-NITRO-2-FURFURYLIDENE) ACETONE GUANYLHYDRAZONE
see PAF500
1,5-BIS(5-NITRO-2-FURYL)-1,4-PENTADIENE-3-AMINOHYDRAZONE HYDRO-
CHLORIDE see DKK100
1,5-BIS(5-NITRO-2-FURYL)-3-PENTADIENONE AMIDINONHYDRAZONE
see PAF500
1,5-BIS(5-NITRO-2-FURYL)-3-PENTADIENONE GUANYLHYDRAZONE
see PAF500
BIS(4-NITROPHENYL)DISULFIDE see BLA500
BIS(p-NITROPHENYL)DISULFIDE see BLA500
BIS(p-NITROPHENYL)SULFIDE see BLA750
BIS(NONYLOXYMALEOYLOXY)DIOCTYLSTANNANE see DEI800
BIS(OCTANOYLOXY)DI-n-BUTYL STANNANE see BLB250
BIS(OCTANOYLOXY)DI-n-BUTYLTIN see BLB250
2,2-BIS(3'-tert-OCTYL-4'-HYDROXYPHENYLPROPANE see BLB500
BIS(2-OCTYL)PHTHALATE see BLB750
BISODIUM TARTRATE see BLC000
BISOFLEX 81 see DVL700
BISOFLEX 91 see DVJ000
BISOFLEX DOA see AEO000
BISOFLEX DOP see DVL700
BISOFLEX DOS see BJS250
BIS(OLEOYLOXY)DIBUTYLSTANNANE see DEJ000
BISOLVOMYCIN see HOI000
BISOLVON see BMO325
BISOLVON HYDROCHLORIDE see BMO325
BISOMER 2HEA see ADV250
BIS(OXIRANYLMETHYL)BENZENAMINE (9CI) see DKM100
BIS(2-OXO-9-BORNANESULFONIC ACID) DIETHYLSTANNYL ESTER
see DKC600
BIS(1-OXODODECYL)PEROXIDE see LBR000
BIS-(2-OXOPROPYL)-N-NITROSAMINE see NJN000
BIS(1-OXOPROPYL)PEROXIDE see DWQ800
BIS(8-OXYQUINOLINE)COPPER see BLC250
BIS(PANTOTHENAMIDOETHYL) DISULFIDE see PAG150
BIS(PENTACHLOR-2,4-CYCLOPENTADIEN-1-YL) see DAE425
BIS(PENTACHLOROCYCLOPENTADIENYL) see DAE425
BIS(PENTACHLORO-2,4-CYCLOPENTADIEN-1-YL) see DAE425
BIS(PENTACHLOROPHENOL), ZINC SALT see BLC500
BIS(PENTAFLUOROETHYL)ETHER see PCG760
BIS(PENTA FLUORO PHENYL)ALUMINUM BROMIDE see BLC750
BISPENTAFLUOROSULFUR OXIDE see BLD000
BIS(PENTAMETHYLENETHIURAM)-TETRASULFIDE see TEF750
BIS(2,4-PENTANEDIONATO)CHROMIUM see BLD250
BIS(2,4-PENTANEDIONATO)COPPER see BGR000
BIS(2,4-PENTANEDIONATO)TITANIUM OXIDE see BGQ750
BIS(2,4-PENTANEDIONATO-O,O)ZINC see ZCJ000
4,5-BIS(4-PENTENYLOXY)-2-IMIDAZOLIDINONE see BLD325
BISPHENOL A see BLD500
BISPHENOL A DIGLYCIDYL ETHER see BLD750
BISPHENOL DIGLYCIDYL ETHER, MODIFIED see BLE000
BIS(PHENOXARSIN-10-YL) ETHER see OMY850
BIS(10-PHENOXARSYL) OXIDE see OMY850
BIS(10-PHENOXYARSINYL) OXIDE see OMY850
10,10'-BIS(PHENOXYARSINYL) OXIDE see OMY850
BIS(p-PHENOXYPHENYL)DIPHENYLSTANNANE see BLE250
BIS(p-PHENOXYPHENYL)DIPHENYLTIN see BLE250
1,4-BIS(PHENYL AMINO)BENZENE see BLE500
BISPHENYL-(2-CHLORPHENYL)-1-IMIDAZOLYL-METHAN (GERMAN)
see MRX500
N,N'-BIS(PHENYLISOPROPYL)PIPERAZINE DIHYDROCHLORIDE see BLF250
BIS(PHENYLMERCURI)METHYLENEDINAPHTHALENESULFONATE
see PFN000
BIS(PHENYLMERCURYLAURYL)SULFIDE see PFN750
BIS-N,N'-(3-PHENYLPROPYL-2)-PIPERAZINE DIHYDROCHLORIDE
see BLF250
BIS(PHENYLSELENIDE) see BLF500
BIS(PHENYLTHIO)DIMETHYLTIN see BLF750
1,3-BIS(PHENYL)TRIAZENO)BENZENE see BLG000
4,4'-BIS(4-PHENYL-2H-1,2,3-TRIAZOL-2-YL)-2,2'-STILBENEDISULFONIC ACID
DIPOTASSIUM SALT see BLG100
N,N-BIS(PHOSPHONOMETHYL)GLYCINE see BLG250
2,6-BIS(PICRYLAMINO)-3,5-DINITROPYRIDINE see PQC525
BIS(PIPERIDINOTHIOCARBONYL) TETRASULFIDE see TEF750
3,8-BIS(1-PIPERIDINYLMETHYL)-2,7-DIOXASPIRO(4.4)NONANE-1,6-DIONE
see BLG325
2,6-BIS(1-PIPERIDYLMETHYL)-4-(α,α-DIMETHYLBENZYL)PHENOL
DIHYDROBROMIDE see BHR750
BIS(2-PROPANOL)AMINE see DNL600
BIS(2-PROPOXYETHYL)-1,4-DIHYDRO-2,6-DIMETHYL-4-(3-NITROPHENYL)-
3,5-PYRIDINEDICARBOXYLATE see NDY600
2,4-BIS(PROPYLAMINO)-6-CHLOR-1,3,5-TRIAZIN (GERMAN) see PMN850
BIS(2-PROPYLOXY)DIAZENE see DNQ700
trans-1,2-BIS(n-PROPYLSULFONYL)ETHYLENE see BLG500

2,2-BIS(((3-PYRIDINYLCARBONYL)OXY)METHYL)-1,3-PROPANEDIYL ESTER
of 3-PYRIDINECARBOXYLIC ACID see NCW300
BIS(2-PYRIDYLTHIO)ZINC, 1,1'-DIOXIDE see ZMJ000
BIS(8-QUINOLINATO)COPPER see BLC250
BIS(8-QUINOLINOLATO)COPPER see BLC250
BIS(8-QUINOLINOLATO-N^1),O(^{8}) see BLC250
BIS(SALICYLALDEHYDE)ETHYLENEDIIMINE COBALT(II) see BLH250
BIS(SUCCINYLDICHLOROCHOLINE) see HLC500
BIS(5-SULFO-8-QUINOLINOLATO-N^1,O^8) MANGANESE(II) see BKJ275
2,2'-BIS-6-TERC.BUTYL-p-KRESYLMETHAN (CZECH) see MJO500
2,2-BIS-3'-TERC. OKTYL-4'-HYDROXYFENYLPROPAN (CZECH) see BLB500
BISTERIL see PDC250
BIS(2,3,3,3-TETRACHLOROPROPYL) ETHER see OAL000
BIS(TETRADECANOYLOXY)DIBUTYLSTANNANE see BLH309
1,3-BIS(TETRAHYDRO-2-FURANYL)-5-FLUORO-2,4-PYRIMIDINEDIONE
see BLH325
1,3-BIS(TETRAHYDRO-2-FURYL)-5-FLUOROURACIL see BLH325
BIS-N,N,N',N'-TETRAMETHYLPHOSPHORODIAMIDIC ANHYDRIDE
see OCM000
1,6-BIS(5-TETRAZOLYL)HEXAAZ-1,5-DIENE see BLI000
3,4-BIS(1,2,3,4-THIATRIAZOL-5-YL THIO) MALEIMIDE see BLI250
BIS(1,2,3,4-THIATRIAZOL-5-YL THIO)METHANE see BLI500
1-3,3-BIS(3'-THIENYL)-2-PROPENYL-(3-HYDROXY-3-PHENYLPROPYL-2)
AMINE see DXI800
BIS(THIOCARBAMOYL)DISULFIDE see TFS500
BISTHIOCARBAMYL HYDRAZINE see BLJ250
BIS(THIOCYANATO)-MERCURY see MCU250
p,p'-BIS(α-THIOL CARBAMYLACETAMIDO)BIPHENYL see MCL500
BIS(THIOUREA) see BLJ250
BISTOLUENE DIAZO OXIDE see BLJ500
1,4-BIS(p-TOLYLAMINO)ANTHRAQUINONE see BLK000
BIS-1,4-p-TOLYLAMINOANTHRCHINON (CZECH) see BLK000
N-BIS(p-TOLYLSULFONYL)AMIDOMETHYL MERCURY see BLK250
1,3-BIS(o-TOLYL)-2-THIOUREA see DXP600
3,5-BIS(o-TOLYL)-s-TRIAZOLE see BLK500
BISTON see DCV200
BIS(TRIBENZYLSTANNYL)SULFIDE see BLK750
BIS(TRIBENZYLTIN) SULFIDE see BLK750
BIS-(TRI-N-BUTYLCIN)OXID (CZECH) see BLL750
BIS(TRIBUTYLOXIDE) of TIN see BLL750
BIS(TRI-N-BUTYLPHOSPHINE)DICHLORONICKEL see BLS250
BIS(TRIBUTYL(SEBACOYLDIOXY))TIN see BLL000
BIS((TRI-n-BUTYL)CYCLOPENTADIENYL)IRON see BLL250
1,1'-BIS(TRIBUTYLSTANNYL)FERROCENE see BLL250
BIS(TRIBUTYLSTANNYL)OXIDE see BLL750
BIS(TRIBUTYLTIN) ITACONATE see BLL500
BIS(TRIBUTYL TIN)OXIDE see BLL750
BIS(TRIBUTYLTIN)SULFIDE see HCA700
BIS(TRI-N-BUTYLZINN)-OXYD (GERMAN) see BLL750
BIS-2,3,5-TRICHLOR-6-HYDROXYFENYLMETHAN (CZECH) see HCL000
m-BIS(TRICHLORMETHYL)BENZENE see BLL825
BIS(TRICHLOROACETYL)PEROXIDE see BLM000
BIS-2,4,5-TRICHLORO BENZENE DIAZO OXIDE see BLM250
1,3-BIS(2,2,2-TRICHLORO-1-HYDROXYETHYL)UREA see DGQ200
BIS(3,5,6-TRICHLORO-2-HYDROXYPHENYL)METHANE see HCL000
m-BIS(TRICHLOROMETHYL)BENZENE see BLL825
1,3-BIS(TRICHLOROMETHYL)BENZENE see BLL825
1:4-BIS-TRICHLOROMETHYL BENZENE see HCM500
BIS(TRICHLOROMETHYL)SULFONE see BLM500
BISTRICHLOROMETHYLTRISULFID (CZECH) see BLM750
BIS(TRICHLORO METHYL)TRISULFIDE see BLM750
1,3-BIS(2,4,5-TRICHLOROPHENOXY)-1,1,3,3-TETRABUTYLDISTANNOXANE
see OPE100
BIS(2,3,5-TRICHLOROPHENYLTHIO)ZINC see BLN000
BIS(TRIETHYLENETETRAMINE)TUNGSTATONICKEL see BLN100
BIS(TRIETHYL TIN)ACETYLENE see BLN250
BIS(TRIETHYLTIN) SULFATE see BLN500
BIS(TRIFLUOROACETIC) ANHYDRIDE see TJX000
BIS(TRIFLUOROACETOXY)DIBUTYLTIN see BLN750
BIS(TRIFLUOROACETYL)PEROXIDE see BLO000
BIS(TRIFLUOROETHYL)ETHER see HDC000
BIS(2,2,2-TRIFLUOROETHYL)ETHER see HDC000
3,5-BIS(TRIFLUOROMETHYL)ANILINE see BLO250
1,3-BIS(TRIFLUOROMETHYL)BENZENE see BLO270
BIS(TRIFLUOROMETHYL)CHLOROPHOSPHINE see BLO280
BIS(TRIFLUOROMETHYL)CYANOPHOSPHINE see BLO300
BIS(TRIFLUOROMETHYL)DISULFIDE see BLO325
2,2-BIS(TRIFLUOROMETHYL)-4-METHYL-5-PHENYLOXAZOLIDINE
HYDRATE see BLP250
BIS(TRIFLUOROMETHYL)NITROXIDE see BLP300
2-(3,5-BIS(TRIFLUOROMETHYL)PHENYL)-N-METHYL-
HYDRAZINECARBOTHIOAMIDE (9CI) see BLP325
1-((3,5-BIS-TRIFLUOROMETHYL)PHENYL)-4-METHYL-THIOSEMICARBAZIDE
see BLP325

BIS(TRIFLUOROMETHYL)PHOSPHORUS(III) AZIDE see BLP500
α,α-BIS(TRIFLUOROMETHYL)-1-PIPERIDINEMETHANOL HYDRATE
   see BLQ250
BIS(TRIFLUOROMETHYL)SULFIDE see BLQ325
2,2-BIS(TRIFLUOROMETHYL)THIAZOLIDINE HYDRATE see BLQ500
2,2'-BIS(1,6,7-TRIHYDROXY-3-METHYL-5-ISOPROPYL-8-AL-
   DEHYDONAPHTHALENE see GJM000
BIS(TRIISOBUTYLSTANNANE) see HDY100
BISTRIMATE see BKX750
N,N'-BIS(3-(3,4,5-TRIMETHOXYBENZOYLOXY)PROPYL)HOMOPIPERAZINE
   DIHYDROCHLORIDE see CNR750
1,4-BIS(3-(3,4,5-TRIMETHOXYBENZOYLOXY)-PROPYL)PERHYDRO-1,4-
   DIAZEPINE DIHYDROCHLORIDE see CNR750
BIS-3,4,5-TRIMETHOXY-β-PHENETHYLAMMONIUM
   TETRACHLOROMANGANATE(II) see BKM100
2,3-BISTRIMETHYLACETOXYMETHYL-1-METHYLPYRROLE see BLQ600
α,omega-BIS(TRIMETHYL AMMONIUM)HEXANE DIBROMIDE see HEA000
α,omega-BIS(TRIMETHYLAMMONIUM)HEXANE DICHLORIDE see HEA500
BIS(TRIMETHYLHEXYL)TIN DICHLORIDE see BLQ750
BIS(TRIMETHYLSILYL)ACETAMIDE see TMF000
N,o-BIS(TRIMETHYLSILYL)ACETAMIDE see TMF000
BIS(TRIMETHYLSILYL)AMINE see HED500
N,N'-BIS(TRIMETHYLSILYL)AMINOBORANE see BLQ850
cis-BIS(TRIMETHYLSILYLAMINO)TELLURIUM TETRAFLUORIDE see BLQ900
BIS(TRIMETHYLSILYL)CHROMATE see BLR000
1,2-BIS(TRIMETHYLSILYL)HYDRAZINE see BLR125
BIS(TRIMETHYLSILYL)MERCURY see BLR140
BISTRIMETHYL SILYL OXIDE see BLR250
BIS(TRIMETHYLSILYL)PEROXOMONOSULFATE see BLR500
BIS(1,3,7-TRIMETHYL-8-XANTHINYL)MERCURY see MCT000
N,N'-BIS(2,2,2-TRINITROETHYL)UREA see BLR625
BIS(2,4,6-TRINITRO-PHENYL)-AMIN (GERMAN) see HET500
BIS(TRINITROPHENYL)SULFIDE see BLR750
BISTRIPERCHLORATO SILICON OXIDE see BLS000
BIS(TRIPHENYLPHOSPHINE)DICHLORONICKEL see BLS250
BIS(TRIPHENYL PHOSPHINE)NICKEL DITHIOCYANATE see BLS500
BIS(TRIPHENYL SILYL)CHROMATE see BLS750
BIS(TRIPHENYLTIN)ACETYLENEDICARBOXYLATE see BLS900
BIS(TRIPHENYLTIN)SULFATE see BLT000
BIS(TRIPHENYLTIN)SULFIDE see BLT250
BIS(TRIPROPYLTIN)OXIDE see BLT300
BIS(TRIS(p-CHLOROPHENYL)PHOSPHINE)MERCURIC CHLORIDE COM-
   PLEX see BLT500
BIS(TRIS(p-DIMETHYLAMINOPHENYL)PHOSPHINE)MERCURIC CHLORIDE
   COMPLEX see BLT750
BIS(TRIS(p-DIMETHYLAMINOPHENYL)PHOSPHINE OXIDE)STANNIC CHLO-
   RIDE COMPLEX see BLT775
BIS(TRIS(β,β-DIMETHYLPHENETHYL)TIN)OXIDE see BLU000
BIS(TRIS(p-METHOXYPHENYL)PHOSPHINE)MERCURIC CHLORIDE COM-
   PLEX see BLU250
BIS(TRIS(2-METHYL-2-PHENYLPROPYL)TIN)OXIDE see BLU000
BIS(TRIS(p-METHYLTHIOPHENYL)PHOSPHINE)MERCURIC CHLORIDE
   COMPLEX see BLU500
BISTRIUM CHLORIDE see HEA500
BISULFAN see BOT250
BISULFITE see HIC600, SOH500
BISULFITE de SODIUM (FRENCH) see SFE000
BISULPHANE see BOT250
BISULPHITE see HIC600
BIS(2-VINYLOXYETHYL)ETHER see DJE600
BITEMOL see BJP000
BITEMOL S 50 see BJP000
BITHIODINE see ARP875
BITHION see TAL250
BITHIONOL see TFD250
BITHIONOL SULFIDE see TFD250
BITIN see TFD250
BITIODIN see BLV000
BITIRAZINE see DIW000
BITIS ARIETANS VENOM see BLV075
BITIS GABONICA VENOM see BLV080
BITOLTEROL MESILATE see BLV125
BITOLTEROL MESYLATE see BLV125
4,4'-BI-o-TOLUIDINE see TGJ750
(m,o'-BITOLYL)-4-AMINE see BLV250
BITOSCANATE see PFA500
β-BITTER ACID see LIU000
BITTER ALMOND OIL see BLV500
BITTER ALMOND OIL CAMPHOR see BCP250
BITTER CUCUMBER see FPD250
BITTER FENNEL OIL see FAP000
BITTER GOURD see FPD100
BITTER ORANGE OIL see BLV750
BITTER SALTS see MAJ500

BITTERSWEET see AHJ875
BITUMEN (MAK) see ARO500
Δ(1,1')-BIUREA see ASM270
BIVERM see PDP250
BIVINYL see BOP500
BIXA ORELLANA see APE100
BIZMUTHIOL II (CZECH) see MCP500
BK see BML500
BK8 see TOC000
BK 15 see TOB750
BKF see MJO500
B-K LIQUID see SHU500
B-K POWDER see HOV500
BL 9 see DXY000
Γ-BL see BOV000
BL 139 see DOY400
BL 191 see PBU100
BLA see LCH000
BLACAR 1716 see PKQ059
1743 BLACK see BMA000
BLACK ACACIA see GJU475
BLACK AND WHITE BLEACHING CREAM see HIH000
BLACK BIRCH OIL see SOY100
BLACK BLASTING POWDER see ERF500
BLACK CALLA see ITD050
BLACK DOGWOOD see MBU825
BLACK 2EMBL see AQP000
BLACK EYED SUSAN see RMK250
BLACK LEAF see NDN000
BLACK LOCUST see GJU475
BLACK MANGANESE OXIDE see MAS000
BLACK NIGHTSHADE see DAD880
BLACK OXIDE of IRON see IHD000
BLACK PEARLS see CBT500
BLACK PEPPER OIL see BLW250
BLACK PN see BMA000
BLACK POWDER (DOT) see PLL750
BLACK POWDER, compressed (DOT) see PLL750
BLACK POWDER, granular or as meal (DOT) see PLL750
BLACK WIDOW SPIDER VENOM see BLW500
BLACOSOLV see TIO750
BLADAFUME see SOD100
BLADAFUN see SOD100
BLADAN see EEH600, HCY000, PAK000, TCF250
BLADAN BASE see HCY000
BLADAN-M see MNH000
BLADDERON see FCB100
BLADDERPOD LOBELIA see CCJ825
BLADEX see BLW750
BLADEX G see DFY800
BLADEX H see TAH900
BLADEX 80WP see BLW750
BLANC FIXE see BAP000
BLANCOL see BLX000
BLANCOL DISPERSANT see BLX000
BLANDLUBE see MQV750
BLANOSE BWM see SFO500
BLA-S see BLX500
BLASCORID see MOA600, PJA130
BLASTICIDEN-S-LAURYLSULFONATE see BLX250
BLASTICIDIN see BLX500
BLASTICIDIN S see BLX500
BLASTING GELATIN (DOT) see NGY000
BLASTING OIL see NGY000
BLASTING POWDER (DOT) see PLL750
BLASTMYCIN see BLX750
BLASTOESTIMULINA see ARN500
BLASTOMYCIN see BLX750
BLATTANEX see PMY300
BLATTERALKOHOL see HFE000
L-BLAU 2 (GERMAN) see FAE100
BLAUSAEURE (GERMAN) see HHS000
BLAUWZUUR (DUTCH) see HHS000
BLAZING RED see CJD500
BLEACHING POWDER see HOV500
BLEACHING POWDER, containing 39% or less chlorine (DOT) see HOV500
BLEDO CARBONERO (CUBA) see PJJ315
BLEIACETAT (GERMAN) see LCV000
BLEIAZETAT (GERMAN) see LCJ000
BLEIPHOSPHAT (GERMAN) see LDU000
BLEISTEARAT (GERMAN) see LDX000
BLEISTIFTBAUMS (GERMAN) see EQY000
BLEISULFAT (GERMAN) see LDY000
BLEKIT EVANSA (POLISH) see BGT250

BLEMINOL see ZVJ000
BLENDED RED OXIDES of IRON see IHD000
BLENOXANE see BLY000, BLY780
BLEO see BLY000
BLEOCIN see BLY000
BLEOMYCETIN see BLY500
BLEOMYCIN see BLY000
BLEOMYCIN A2 see BLY250
BLEOMYCIN A5 see BLY500
BLEOMYCIN A COMPLEX see BLY750
BLEOMYCIN B2 see BLY760
BLEOMYCIN PEP see BLY770
BLEOMYCIN SULFATE see BLY780
BLEOMYCIN, SULFATE (salt) (9CI) see BLY780
BLEPH-10 see SNP500
BLEU BRILLIANT FCF see FMU059
BLEU DIAMINE see CMO250
BLEX see DIN800
BLEXANE see BLY780
BL H368 see BEQ625
BLIGHIA SAPIDA see ADG400
BLIGHTOX see EIR000
BLISTER FLOWER see FBS100
BLISTERING BEETLES see CBE250
BLISTERING FLIES see CBE250
BLISTER WORT see FBS100
BLITEX see EIR000
BLITOX see CNK559
BLITOX 50 see CNK559
BLIZENE see EIR000
BLM see BLY000
BLM-PEP see BLY770
BLO see BOV000
BLOC see FAK100
BLOCADREN see DDG800
BLOCAN see SBH500
BLON see BOV000
BLOODBERRY see ROA300
BLOOD STONE see HAO875
BLOTIC see MKA000
BLOXANTH see ZVJ000
BL P 152 see PDD350
BLP-1011 see DGE200
BLP 1322 see HMK000
BL-S 578 see DYF700
BLS 640 see APT250
BLUE 2B see CMO000
BLUE 1084 see ADE500
1206 BLUE see FAE000
1311 BLUE see FAE100
11388 BLUE see FMU059
12070 BLUE see FAE100
BLUE 'APE (HAWAII) see XCS800
BLUE ASBESTOS (DOT) see ARM275
BLUEBELL see CMV390
BLUEBERRY ROOT see BMA150
BLUE BLACK BN see BMA000
BLUE BN BALSE see DCJ200
BLUECAIN see BMA125
BLUE CARDINAL FLOWER see CCJ825
BLUE COHOSH see BMA150
BLUE COPPER see CNK559, CNP250
BLUE COPPER-50 see CNK559
BLUE COPPERAS see CNP500
BLUE CROSS see CGN000
BLUE DEVIL WEED see VQZ675
BLUE EMB see CMO250
BLUE GINSENG see BMA150
BLUE JESSAMINE see CMV390
BLUE OIL see AOQ000, COD750
BLUE-OX see ZLS000
BLUE POWDER see ZBJ000
BLUE STONE see CNP250
BLUESTONE see CNP500
BLUE TARO (HAWAII) see XCS800
BLUE VITRIOL see CNP250, CNP500
BLUTENE see AJP250
BLUTENE CHLORIDE see AJP250
BLUTON see IIU000
BM 1 see HNI500
BM 3055 see IBP200
7-BMBA see BNO750
BMC see BRS750, MHC750
BMIH see IKC000

BMOO see BRT000
BN see BFW000
B-NINE see DQD400
BNM see BAV575
B. N. MEXICANUS VENOM see BMJ500
BNP see NIY500
BNP 30 see BRE500
BNU see BSA250
BO 714 see TCZ000
BO-ANA see FAB600
BOB see NKL300
BOEA see BKA000
B.O.E.A. see BKA000
BOG MANGANESE see MAS000
BOG ONION see JAJ000
BOH see HHC000
BOILER COMPOUND, (liquid) see BMA500
BOIS D'ARC (FRENCH) see MRN500
BOIS GENTIL (CANADA) see LAR500
BOIS d'INDE see BAT500
BOIS JAMBETTE (HAITI) see GIW200
BOIS JOLI (CANADA) see LAR500
BOIS de PLOMB (CANADA) see LEF100
BOL see BNM250
BOL-148 see BNM250
BOLATRON see PKQ059
BOLDIN see DNZ100
BOLDINE see DNZ100
(+)-BOLDINE see DNZ100
(S)-BOLDINE see DNZ100
(+)-(S)-BOLDINE see DNZ100
BOLDINE DIMETHYL ETHER see TDI475
BOLDO LEAF OIL see BMA600
BOLERO see SAZ000
BOLETIC ACID see FOU000
BOLETIC ACID DIMETHYL ESTER see DSB600
BOLINAN see PKQ250
BOLLS-EYE see HKC000, HKC500
BOLSTAR see SOU625
BOLVIDON see BMA625
BOMBITA see DBA800
BOMT see BMA650
BOMYL see SOY000
BONABOL see XQS000
BONADETTES see HGC500
BONADOXIN see HGC500
BONAMID see PJY500
BONAMINE see HGC500
BONAPAR see PGG350
BONAPHTHON see BNS750
BONAPICILLIN see AIV500
BONARE see CFZ000
BONAZEN see ZNA000
BONBONNIER (HAITI) see LAU600
BONBRAIN see TEH500
BOND CH 18 see AAX250
BONE OIL see BMA750
BONGAY see HGL575
BONIBAL see DXH250
BONICOR see HNY500
BONIDE BLUE DEATH RAT KILLER see PHO740
BONIDE KRAB CRABGRASS KILLER see PLC250
BONIDE RYATOX see RSZ000
BONIDE TOPZOL RAT BAITS and KILLING SYRUP see RCF000
BONIFEN see BMB000
BONINE see MBX500
BONITON see AEH750
BONLAM see CBA100
BONLOID see PKQ059
BONNECOR see BMB125
BONOFORM see TBQ100
BONOMOLD OE see HJL000
BONOMOLD OP see HNU500
BOOKSAVER see AAX250
BOOTS BTS 27419 see MJL250
BOP see NJN000
BORACIC ACID see BMC000
BORACSU see SFF000
BORANE-AMMONIA see BMB150
BORANE, COMPOUND with N,N-DIMETHYLMETHANAMINE (1:1) see BMB250
BORANE, COMPOUND with TRIMETHYLAMINE (1:1) see BMB250
BORANE with DIMETHYLAMINE (1:1) see DOR200
BORANE, compound with DIMETHYLSULFIDE see MPL250

BORANE-HYDRAZINE see BMB260
BORANE, compounded with MORPHOLINE see MRQ250
BORANE-PHOSPHORUS TRIFLUORIDE see BMB270
BORANE-PYRIDINE see POQ250
BORANES see BMB280
BORANE-TETRAHYDROFURAN see BMB300
BORASSUS FLABELLIFER Linn., extract see BMB325
BORATES, TETRA, SODIUM SALT, anhydrous (OSHA, ACGIH) see SFE500, SFF000
BORAX (8CI) see SFF000
BORAX DECAHYDRATE see SFF000
BORAZINE see BMB500
BORAZOLE see BMB500
BORDEAU ARSENITE, liquid or solid (DOT) see BMB750
BORDEAUX see FAG020
BORDEAUX ARSENITE see BMB750
BORDEN 2123 see AAX250
BORDERMASTER see CIR250
BOREA see BMM650
BORER SOL see EIY600
BORESTER 2 see THX750
BORESTER O see TLN000
BORIC ACID see BMC000
BORIC ACID, ETHYL ESTER see BMC250
BORIC ACID, PHENYLMERCURY SILVER derivative see PFP250
BORIC ACID, SODIUM SALT see SJD000
BORIC ACID, TRI-n-AMYL ESTER see TMQ000
BORIC ACID, TRIBUTYL ESTER see THX500
BORIC ACID, TRI-sec-BUTYL ESTER see THX750, THY000
BORIC ACID, TRI-o-CHLOROPHENYL ESTER see TIY750
BORIC ACID, TRI-o-CRESYL ESTER see TJF500
BORIC ACID, TRIETHYL ESTER see TJK250, TJP500
BORIC ACID, TRIHEXYL ESTER see TKM000
BORIC ACID, TRIISOBUTYL ESTER see TKR750
BORIC ACID, TRIOCTADECYL ESTER see TMN750
BORIC ACID, TRI-n-OCTYL ESTER see TMO250
BORIC ACID, TRIOLEYL ESTER see BMC500
BORIC ACID, TRI-n-PENTYL ESTER see TMQ000
BORIC ACID, TRIS(2-AMINOETHYL) ESTER see TJK750
BORIC ACID, TRIS(1-AMINO-2-PROPYL) ESTER see TKT200
BORIC ACID, TRIS(2-ETHYLHEXYL) ESTER see TJR500
BORIC ACID, TRIS(1-METHYLHEPTYL) ESTER see TMO500
BORIC ACID, TRIS(4-METHYL-2-PENTYL) ESTER see BMC750
BORIC ACID, TRIS(PHENYLCYCLOHEXYL) ESTER see TMR750
BORIC ACID, TRISTEARYL ESTER see TMN750
BORIC ANHYDRIDE see BMG000
BORICIN see SFF000
BOR-IND see IDA400
BORNANE, 2,2,5-endo,6-exo,8,9,10-HEPTACHLORO- see THH575
1-2-BORNANOL see NCQ820
2-BORNANONE see CBA750
(+)-2-BORNANONE see CBB250
d-2-BORNANONE see CBB250
BORNATE see IHZ000
BORNEO CAMPHOR see BMD000
BORNEOL see BMD000
(−)-BORNEOL see NCQ820
BORNEOL (DOT) see BMD000
trans-BORNEOL see BMD000
(1S,2R,4S)-(−)-1-BORNEOL see NCQ820
BORNYL ACETATE see BMD100
l-BORNYL ACETATE see BMD100
BORNYL ALCOHOL see BMD000
1-BORNYL ALCOHOL see NCQ820
S-((N-BORNYLAMIDIN)METHYL) HYDROGEN THIOSULFATE see BMD250
BOROETHANE see DDI450
BOROFAX see BMC000
BOROHYDRURE de POTASSIUM (FRENCH) see PKY250
BOROHYDRURE de SODIUM (FRENCH) see SFF500
BOROLIN see PIB900
BORON see BMD500
BORON AZIDE DICHLORIDE see BMD750
BORON AZIDE DIIODIDE see BMD825
BORON BROMIDE see BMG400
BORON BROMIDE DIIODIDE see BME250
BORON CHLORIDE see BMG500
BORON COMPOUNDS see BME500
BORON DIBROMIDE IODIDE see BME750
BORON FLUORIDE see BMG700
BORON HYDRIDE see DDI450
BORON OXIDE see BMG000
BORON PHOSPHIDE see BMG250
BORON SESQUIOXIDE see BMG000
BORON TRIAZIDE see BMG325

BORON TRIBROMIDE see BMG400
BORON TRICHLORIDE see BMG500
BORON TRIFLUORIDE see BMG700
BORON TRIFLUORIDE-ACETIC ACID COMPLEX see BMG750
BORON TRIFLUORIDE-ACETIC ACID COMPLEX (DOT) see BMG750
BORON TRIFLUORIDE DIETHYL ETHERATE see BMH250
BORON TRIFLUORIDE-DIMETHYL ETHER see BMH000
BORON TRIFLUORIDE DIMETHYL ETHERATE (DOT) see BMH000
BORON TRIFLUORIDE ETHERATE see BMH250
BORONTRIFLUORIDE MONOETHYLAMINE see EFU500
BORON TRIIODIDE see BMH500
BORON TRIOXIDE see BMG000
BORON TRISULFIDE see BMH659
BOROPHENYLIC ACID see BBM000
BORRELIDIN see BMH750
BORSAURE (GERMAN) see BMC000
BORTRAN see RDP300
BORTRYSAN see DEV800
BOSAN SUPRA see DAD200
BOSMIN see VGP000
BOTHROPS ASPER VENOM see BMI000
BOTHROPS ATROX VENOM see BMI125
BOTHROPS COLOMIBIENSIS VENOM see BMI250
BOTHROPS GODMANI VENOM see BMI500
BOTHROPS LATERALIS VENOM see BMI750
BOTHROPS NASUTUS VENOM see BMJ000
BOTHROPS NIGROVIRIDIS NEGROVIRIDIS VENOM see BMJ250
BOTHROPS NUMMIFER MEXICANUS VENOM see BMJ500
BOTHROPS OPHYOMEGA VENOM see BMJ750
BOTHROPS PICADOI VENOM see BMK000
BOTHROPS SCHLEGLII VENOM see BMK250
BOTHROPS VENOM PROTEINASE see RDA350
BOTRAN see RDP300
BOTROPASE see RDA350
BOTRYODIPLODIN see BMK290
(−)-BOTRYODIPLODIN see BMK290
BOURBONAL see EQF000
BOURREAU DES ARBRES (CANADA) see AHJ875
BOUTON d'OR (CANADA) see FBS100
BOUVARDIN see BMK325
BOV see SOI500
BOVIDAM see CBA100
BOVIDERMOL see DAD200
BOVINE LACTOGENIC HORMONE see PMH625
BOVINE PINEAL GLAND EXTRACT see BMK400
BOVINE PROLACTIN see PMH625
BOVINOCIDIN see NIY500
BOVINOX see TIQ250
BOVITROL see DAM300
BOVIZOLE see TEX000
BOVOFLAVIN see DBX400
BOVOLIDE see BMK500
BOY see HBT500
BOYGON see PMY300
BP2 see AFJ625
B(a)P see BCS750
B(e)P see BCT000
BP 400 see MOO750, MOP000
BPDE see BCU250
BPDE-syn see DMR000
anti-BPDE see BCU250, DMQ600
BP-4,5-DIHYDRODIOL see DLB800
BP-7,8-DIHYDRODIOL see BCT750, DML000, DML200
BP-9,10-DIHYDRODIOL see DLC400
B(e)P-4,5-DIHYDRODIOL see DMK400
B(E)P 9,10-DIHYDRODIOL see DMK600
trans-BP-7,8-DIHYDRODIOL DIACETATE see DBG200
BP-7,8-DIHYDRODIOL-9,10-EPOXIDE (anti) see BCU000
anti-BP-7,8-DIHYDRODIOL-9,10-OXIDE see BCU000
B(e)P DIOL EPOXIDE-1 see DMR150, DMR400
B(E)P DIOL EPOXIDE-2 see DMR200
anti-BP-DIOLEPOXIDE see DMQ000
BP 7,8-DIOL-9,10-EPOXIDE 2 see DMP900
B(e)P 9,10-DIOL-11,12-EPOXIDE-1 see DMR150
(−)-BP 7-α,8-β-DIOL-9-β,10-β-EPOXIDE 2 see DVO175
(−)-BP-7,β,8,α-DIOL-9,β,10,β-EPOXIDE 1 see DMP800
(+)-BP-7,α,8-β-DIOL-9,α,10,α-EPOXIDE 1 see DMP600
(+)-BP-7-β,8-α-DIOL-9-α,10-α-EPOXIDE 2 see BMK620
BP DIOL EPOXIDE ANTI see BCU250
BP-4,5-EPOXIDE see BCV500
BP 7,8-EPOXIDE see BCV750
B(a)P EPOXIDE I see DMR000
B(a)P EPOXIDE II see BMK630
BPG 400 see PKK500

BPG 800 see PKK750
B(E)P H4-9,10-DIOL see DNC400
B(c)PH DIOL EPOXIDE-1 see BMK634
B(c)PH DIOL EPOXIDE-2 see BMK635
B(e)P H4-9,10-EPOXIDE see ECQ150
BP-3-HYDROXY see BCX250
BPL see PMT100
BPMC see MOV000
BP 4,5-OXIDE see BCV500
BP 7,8-OXIDE see BCV750
BP-9,10-OXIDE see BCW000
BP-11,12-OXIDE see BCW250
BPPS see SOP000
BP-3,6-QUINONE see BCU750
BP-6,12-QUINONE see BCV000
BPZ see FLL000
BR 700 see CKI750
BR 750 see GKO750
BR-931 see CLW500
BRACKEN FERN, CHLOROFORM FRACTION see BMK750
BRACKEN FERN, DRIED see BML000
BRACKEN FERN TANNIN see BML250
BRACKEN FERN, TANNIN-FREE see TAE250
BRACKEN FERN TOXIC COMPONENT see SCE000
BRADILAN see TDX860
BRADYKININ see BML500
BRADYKININ (synthetic) see BML500
BRADYL see NAC500
BRAMYCIN see BML750
BRAN ABSOLUTE see WBJ700
BRASILAMINA BLACK GN see AQP000
BRASILAMINA BLUE 2B see CMO000
BRASILAMINA BLUE 3B see CMO250
BRASILAMINA CONGO 4B see SGQ500
BRASILAMINA VIOLET 3R see CMP000
BRASILAN AZO RUBINE 2NS see HJF500
BRASILAN METANIL YELLOW see MDM775
BRASILAN ORANGE 2G see HGC000
BRASILAZET BLUE GR see TBG700
BRASILAZINA OIL RED B see SBC500
BRASILAZINA OIL SCARLET 6G see XRA000
BRASILAZINA OIL YELLOW G see PEI000
BRASILAZINA OIL YELLOW R see AIC250
BRASILAZINA ORANGE Y see PEK000
BRASIL (CUBA) see CAK325
BRASILETTO (BAHAMAS) see CAK325
BRASILIAN CHROME ORANGE R see NEY000
BRASORAN see ASG250
BRASSICOL see PAX000
BRAUNOSAN H see PKE250
BRAUNSTEIN (GERMAN) see MAS000
BRAVO see TBQ750
BRAVO 6F see TBQ750
BRAVO-W-75 see TBQ750
BRAXORONE see BML825
BRAZILIAN PEPPER TREE see PCB300
BREADFRUIT VINE see SLE890
BRECHWEINSTEIN see AQJ500
BRECOLANE NDG see DJD600
BREDININ see BMM000
BREDININE see BMM000
BREK see LIH000
BRELLIN see GEM000
BREMFOL see BMM125
BREMIL see CFY000
BRENAL see AAE500
BRENDIL see VGK000
BRENOL see ART250
BRENTAMINE FAST BLUE B BASE see DCJ200
BRENTAMINE FAST ORANGE GR BASE see NEO000
BRENTAMINE FAST RED TR BASE see CLK220
BRENTAMINE FAST RED TR SALT see CLK235
BREON see PKQ059
BREON 351 see AAX175
BRESIT see ARQ750
BRESTAN see ABX250
BRESTANOL see CLU000
BRETHINE see TAN250
BRETOL see EKN500
BRETYLAN see BMV750
BRETYLATE see BMV750
BRETYLIUM-p-TOLUENESULFONATE see BMV750
BRETYLIUM TOSYLATE see BMV750
BRETYLOL see BMV750

BREVIMYTAL see MDU500
BREVINYL see DGP900
BREVIRENIN see VGP000
BREVITAL SODIUM see MDU500
BRIANIL see IPU000
BRICAN see TAN100
BRICANYL see TAN100, TAN250
BRICAR see TAN100
BRICARIL see TAN100
BRICK OIL see CMY825
BRICYN see TAN100
BRIDAL see BEM500
BRIER (BAHAMAS) see CAK325
BRIETAL SODIUM see MDU500
BRIGHT RED see CHP500
BRIJ 30 see DXY000
BRIJ 92 see OIG000
BRIJ 98 see PJW500
BRIJ 92((2)-OLEYL) see OIG000
BRIJ 96((10) OLEYL) see OIG040
BRILLIANT ACID BLACK BNA EXPORT see BMA000
BRILLIANT ACID BLACK BN EXTRA PURE A see BMA000
BRILLIANT ACRIDINE ORANGE E see BJF000
BRILLIANT BLACK see BMA000
BRILLIANT BLACK A see BMA000
BRILLIANT BLACK BN see BMA000
BRILLIANT BLACK NAF see BMA000
BRILLIANT BLACK N.FQ see BMA000
BRILLIANT BLUE see FMU059
BRILLIANT BLUE FCD No. 1 see FAE000
BRILLIANT BLUE FCF see FAE000
BRILLIANT BLUE R see BMM500
BRILLIANT CRIMSON RED see HJF500
BRILLIANT FAST YELLOW see DOT300
BRILLIANT GREEN 3EMBL see FAE950
BRILLIANT GREEN SULFATE see BAY750
BRILLIANT OIL ORANGE R see PEJ500
BRILLIANT OIL ORANGE Y BASE see PEK000
BRILLIANT OIL SCARLET B see XRA000
BRILLIANT OIL YELLOW see IBB000
BRILLIANT PINK B see FAG070
BRILLIANT PONCEAU 3R see FMU080
BRILLIANT PONCEAU G see FMU070
BRILLIANT RED see CHP500
BRILLIANT RED 5SKH see PMF540
BRILLIANTSAEURE GRUEN BS (GERMAN) see ADF000
BRILLIANT SCARLET see CHP500, FMU080
BRILLIANTSCHWARZ BN (GERMAN) see BMA000
BRILLIANT SULFAFLAVINE see CMM750
BRILLIANT TANGERINE 13030 see DVB800
BRILLIANT TONER Z see CHP500
BRILLIANT TONING RED AMINE see AJJ250
BRILLIANT YELLOW SLURRY see DEU000
BRIMSTONE see SOD500
BRIPADON see FLG000
BRISPEN see DGE200
BRISTACICLIN α see TBX000
BRISTACIN see PPY250
BRISTACYCLINE see TBX000, TBX250
BRISTAMIN HYDROCHLORIDE see DTO800
BRISTAMYCIN see EDJ500
BRISTOL A-649 see OIU499
BRISTOL LABORATORIES BC 2605 see CQF079
BRISTOPHEN see MNV250
BRISTURIC see BEQ625
BRISTURON see BEQ625
BRITACIL see AIV500
BRITAI see CFH825, CMV500
BRITISH ALUMINUM AF 260 see AHC000
BRITISH ANTILEWISITE see BAD750
BRITISH EAST INDIAN LEMONGRASS OIL see LEG000
BRITON see TIQ250
BRITTEN see TIQ250
BRITTOX see DDP000
BRL see AIV500
BRL 152 see PDD350
BRL 556 see IBP200
BRL 1341 see AIV500
BRL 1383 see SGS500
BRL 1400 see DSQ800, MNV250
BRL-1621 see SLJ000
BRL-1702 see DGE200
BRL-2064 see CBO250
BRL 3475 see CBO000

BRL 25000 see ARS125
BRL 147777 see MFA300
BRL 14151K see PLB775
BRL-1621 SODIUM SALT see SLJ050
BRL 2333 TRIHYDRATE see AOA100
BR 55N see PAU500
BROBAMATE see MQU750
BROCADISIPAL see MJH900, OJW000
BROCADOPA see DNA200
BROCASIPAL see MJH900, OJW000
BROCIDE see EIY600
BROCKMANN, ALUMINUM OXIDE see AHE250
BROCRESIN see BMM600
BROCRESINE see BMM600
BROCSIL see PDD350
BRODAN see CMA100
BRODIAR see DDS600
BRODIFACOUM see TAC800
BROFAREMINE HYDROCHLORIDE see BMM625
BROFENE see BNL250
BROGDEX 555 see SGM500
BROM (GERMAN) see BMP000
BROMACETOCARBAMIDE see BNK000
BROMACETYLENE see BMS500
BROMACIL see BMM650
BROMADAL see BNK000
BROMADEL see BNK000
BROMADIALONE see BMN000
BROMADIOLONE see BMN000
BROMADRYL see BMN250
BROMAL HYDRATE see THU500
BROMALLYLENE see AFY000
5-(2'-BROMALLYL)-5-ISOPROPYLBARBITURIC ACID see QCS000
BROMAMID see BMN350
BROMAMIDE see BMN350, BMT250
BROMAMIDE (pharmaceutical) see BMN350
4-BROMANILINU (CZECH) see BMT325
BROMANMINAN SODNY (CZECH) see DKR000
BROMANYLPROMIDE see BMN350
BROMARAL see BNP750
BROMAT see HCQ500
BROMATES see BMN500
BROMATE de SODIUM (FRENCH) see SFG000
BROMAZEPAM see BMN750
BROMAZIL see BMM650
10-BROM-1,2-BENZANTHRACEN (GERMAN) see BMT750
3-BROMBENZANTHRONE see BMU000
2-BROMBENZOTRIFLUORID (CZECH) see BOJ750
3-BROMBENZOTRIFLUORID (CZECH) see BOJ500
BROMBENZYL CYANIDE see BMW250
N-p-BROMBENZYL-N-α-PYRIDYL-N',N'-DIMETHYL-AETHYLENDIAMIN-
    HYDROCHLORID (GERMAN) see HGA500
N-p-BROMBENZYL-N-α-PYRIDYL-N'-METHYL-N'-AETHYL-
    AETHYLENDIAMIN-MALEINAT (GERMAN) see BMW000
BROMCARBAMIDE see BNP750
BROMCHLOPHOS see NAG400
BROMCHLORENONE see BMZ000
BROMCHOLITIN see TDI475
BROMDEFENURON see MHS375
O-(4-BROM-2,5-DICHLOR-PHENYL)-O,O-DIMETHYL-MONOTHIOPHOSPHAT
    (GERMAN) see BNL250
d-2-BROM-DIETHYLAMIDE of LYSERGIC ACID see BNM250
BROME (FRENCH) see BMP000
BROMEK DWUMETYLOLAURYLOBENZYLOAMONIOWY (POLISH) see
    BEO000
BROMELAIN see BMO000
BROMELAINS see BMO000
BROMELIA see EEY500
BROMELIN see BMO000
BROMEOSIN see BMO250
BROMETHOL see ARW250, THV000
BROMEX see CES750, DFK600, NAG400
BROMFENOFOS see BNV500
BROMFENPHOS see BNV500
1-p-BROMFENYL-3,3-DIMETHYLTRIAZEN (CZECH) see BNW250
2-BROMFLUORBENZEN (CZECH) see FGX000
3-BROMFLUORBENZEN (CZECH) see FGY000
BROMHEXINE CHLORIDE see BMO325
BROMHEXINE HYDROCHLORIDE see BMO325
BROMIC ACID, POTASSIUM SALT see PKY300
BROMIC ACID, SODIUM SALT see SFG000
BROMIDES see BMO750
BROMIDE SALT OF POTASSIUM see PKY500
BROMIDE SALT of SODIUM see SFG500

BROMINAL see DDP000
BROMINAL M & PLUS see CIR250
BROMINATED VEGETABLE (SOYBEAN) OIL see BMO825
BROMINE see BMP000
BROMINE, solution (DOT) see BMP000
BROMINE AZIDE see BMP250
BROMINE CYANIDE see COO500
BROMINE DIOXIDE see BMP500
BROMINE FLUORIDE see BMP750
BROMINE PENTAFLUORIDE see BMQ000
BROMINE PERCHLORATE see BMQ250
BROMINE TRIFLUORIDE see BMQ325
BROMINE(1) TRIFLUOROMETHANESULFONATE see BMQ500
BROMINE TRIOXIDE see BMQ750
BROMINEX see DDP000
BROMINIL see DDP000
5-BROMISATIN (CZECH) see BNL750
5-BROM-3-ISOPROPYL-6-METHYL-URACIL (GERMAN) see BNM000
BROMISOVAL see BNP750
BROMISOVALERYLUREA see BNP750
α-BROMISOVALERYLUREA see BNP750
BROMISOVALUM see BNP750
BROMIZOVAL see BNP750
BROMKAL 80 see OAH000
BROMKAL 80-9D see NMV735
BROMKAL 83-10DE see PAU500
BROMKAL 82-ODE see PAU500
BROMKAL P 67-6HP see TNC500
BROM LSD see BNM250
BROMLYSERGAMIDE see BNM250
2-BROM-d-LYSERGIC ACID DIETHYLAMINE see BNM250
BROM-METHAN (GERMAN) see MHR200
BROMNATRIUM (GERMAN) see SFG500
5-BROM-5-NITRO-1,3-DIOXAN (GERMAN) see BNT000
BROMO (ITALIAN) see BMP000
BROMOACETALDEHYDE see BMR000
2-BROMOACETALDEHYDE see BMR000
α-BROMOACETALDEHYDE see BMR000
α-BROMOACETIC ACID see BMR750
BROMOACETIC ACID, solid (DOT) see BMR750
BROMOACETIC ACID, solution (DOT) see BMR750
BROMOACETIC ACID ETHYLENE ESTER see BHD250
BROMOACETIC ACID, ETHYL ESTER see EGV000
BROMOACETIC ACID METHYL ESTER see MHR250
BROMOACETONE see BNZ000
BROMOACETONE (DOT) see BNZ000
BROMOACETONE, liquid (DOT) see BNZ000
BROMOACETONE OXIME see BMS000
1-BROMOACETOXY-2-PROPANOL see BMS250
BROMOACETYLENE see BMS500
BROMOACETYLENYLETHYLMETHYLCARBINOL see BNK350
BROMO ACID see BNH500, BNK700
4'-(3-BROMO-9-ACRIDINYLAMINO)METHANESULFONANILIDE see BMS750
2-BROMOACROLEIN see BMT000
1-(2-BROMO-1-ADAMANTYL)-N-METHYL-2-PROPYLAMINE HYDROCHLO-
    RIDE see BNO000
1-(3-BROMO-1-ADAMANTYL)-N-METHYL-2-PROPYLAMINE HYDROCHLO-
    RIDE see BNO250
5-(2-BROMOALLYL)-5-sec-BUTYLBARBITURIC ACID see BOR000
1'-BROMOALLYLENE see PMN500
5-(2'-BROMOALLYL)-5-(1'-METHYL-N-PROPYL)BARBITURIC ACID
    see BOR000
BROMOAMINE see BMT250
3'-BROMO-trans-ANETHOLE see BMT300
4-BROMOANILINE see BMT325
p-BROMOANILINE see BMT325
BROMOAPROBARBITAL see QCS000
BROMOAZIDE see BMP250
1-BROMOAZIRIDINE see BMT500
BROMO B see BNK700
10-BROMO-1,2-BENZANTHRACENE see BMT750
3-BROMOBENZ(d,e)ANTHRONE see BMU000
3-BROMO-7H-BENZ(DE)ANTHRACEN-7-ONE see BMU000
4-BROMO-BENZENAMINE (9CI) see BMT325
BROMOBENZENE (DOT) see PEO500
4-BROMOBENZENEACETONITRILE see BNV750
β-(p-BROMOBENZHYDRYLOXY)ETHYLDIMETHYLAMINE HYDROCHLO-
    RIDE see BNW500
2-(4-BROMOBENZOHYDRYLOXY)ETHYLDIMETHYLAMINE HYDROCHLO-
    RIDE see BNW500
6-BROMOBENZO(a)PYRENE see BMU500
2-BROMOBENZOTRIFLUORIDE see BOJ750
3-BROMOBENZOTRIFLUORIDE see BOJ500
m-BROMOBENZOTRIFLUORIDE see BOJ500

o-BROMOBENZOTRIFLUORIDE see BOJ750
5-BROMO-2-BENZOXAZOLINONE see BMU750
6-BROMO-2-BENZOXAZOLINONE see BMV000
p-BROMOBENZOYL AZIDE see BMV250
p-BROMOBENZOYLTHIOHYDROXIMIC ACID-5-DIETHYLAMINOETHYL
    ESTER HYDROCHLORIDE see DHZ000
4-BROMOBENZYLCYANIDE see BNV750
p-BROMOBENZYL CYANIDE see BNV750
α-BROMOBENZYL CYANIDE see BMW250
2-((p-BROMOBENZYL)(2-(DIMETHYLAMINO)ETHYL)AMINO)PYRIDINE
    HYDROCHLORIDE see HGA500
N-p-BROMOBENZYL-N',N'-DIMETHYL-N-2-PYRIDYLETHYLENE-DIAMINE
    HYDROCHLORIDE see HGA500
(o-BROMOBENZYL)ETHYLDIMETHYLAMMONIUM-p-TOLUENESULFONATE
    see BMV750
N-p-BROMOBENZYL-N'-ETHYL-N'-METHYL-N-2-PYRIDYLETHYLENEDIAM-
    INE MALEATE see BMW000
BROMOBENZYLNITRILE see BMW250
α-BROMOBENZYLNITRILE see BMW250
3-BROMOBENZYLTRIFLUORIDE see BOJ500
o-BROMOBENZYLTRIFLUORIDE see BOJ750
3-(3-(4'-BROMO(1,1'-BIPHENYL)-4-YL)3-HYDROXY-1-PHENYLPROPYL)-4-
    HYDROXY-2H-1-BENZOPYRAN-2-ONE see BMN000
3-(3-(4'-BROMOBIPHENYL-4-YL)-1,2,3,4-TETRAHYDRONAPHTH-1-YL)-4-
    HYDROXYCOUMARIN see TAC800
3-(3-(4'-BROMO-1,1'-BIPHENYL-4-YL)-1,2,3,4-TETRAHYDRO-1-NAPHTHYL)-4-
    HYDROXYCOUMARIN see TAC800
α-BROMO-β,β-BIS(p-ETHOXYPHENYL)STYRENE see BMX000
4-BROMO-7-BROMOMETHYLBENZ(a)ANTHRACENE see BMX250
2-BROMO-2-(BROMOMETHYL)GLUTARONITRILE see DDM500
4-BROMO-α-(4-BROMOPHENYL)-α-HYDROXYBENZENEACETIC ACID-1-
    METHYLETHYL ESTER see IOS000
2-BROMO-6-(N-(p-BROMOPHENYL)THIOCARBAMOYL)-4-CHLORO-BENZOIC
    ACID see BOL325
1-BROMOBUTANE see BMX500
2-BROMOBUTANE see BMX750
4-BROMO-1-BUTENE see BMX825
5-BROMO-3-sec-BUTYL-6-METHYLURACIL see BMM650
2-BROMOBUTYRIC ACID see BMY250
α-BROMOBUTYRIC ACID see BMY250
4-BROMOBUTYRONITRILE see BMY500
BROMOCARBAMIDE see BNP750
BROMOCHLOROACETONITRILE see BMY800
BROMOCHLOROACETYLENE see BMY825
6-BROMO-5-CHLORO-2-BENZOXAZOLINONE see BMZ000
6-BROMOCHLOROBENZOXAZOLONE see BMZ000
1-BROMO-1-CHLORO-2,2-DIFLUOROETHENE see BNA000
1-BROMO-1-CHLORO-2,2-DIFLUOROETHYLENE see BNA000
2-BROMO-2-CHLORO-1,1-DIFLUOROETHYLENE see BNA000
BROMOCHLORODIFLUOROMETHANE see BNA250
3-BROMO-1-CHLORO-5,5-DIMETHYLHYDANTOIN see BNA325
3-BROMO-1-CHLORO-5,5-DIMETHYL-2,4-IMIDAZOLIDINEDIONE see BNA325
3-BROMO-N-(2-CHLOROMERCURICYCLOHEXYL)PROPIONAMIDE see CET000
BROMOCHLOROMETHANE see CES650
BROMOCHLOROMETHYL CYANIDE see BMY800
1-BROMO-1-(p-CHLOROPHENYL)-2,2-DIPHENYLETHYLENE see BNA500
O-(4-BROMO-2-CHLOROPHENYL)-O-ETHYL-S-PROPYL PHOSPHOROTHIO-
    ATE see BNA750
3-(4-BROMO-3-CHLOROPHENYL)-1-METHOXY-1-METHYLUREA see CES750
N'-(4-BROMO-3-CHLOROPHENYL)-N-METHOXY-N-METHYLUREA see CES750
N-(4-BROMO-3-CHLOROPHENYL)-N'-METHOXY-N'-METHYLUREA see CES750
2-BROMO-4-(2-CHLOROPHENYL)-9-METHYL-6H-THIENO(3,2-f)(1,2,4)
    TRIAZOLO(4,3-a)(1,4)DIAZEPINE see LEJ600
1-BROMO-3-CHLOROPROPANE see BNA825
BROMOCHLOROTRIFLUOROETHANE see HAG500
2-BROMO-2-CHLORO-1,1,1-TRIFLUOROETHANE see HAG500
BROMOCRIPTIN see BNB250
BROMOCRIPTINE see BNB250
BROMOCRIPTINE MESILATE see BNB325
BROMOCYAN see COO500
BROMOCYANOGEN see COO500
1-BROMO-12-CYCLOTRIDECADIEN-4,8,10-TRIYNE see BNB750
BROMODEOXYGLYCEROL see MRF275
BROMODEOXYURIDINE see BNC750
5-BROMODEOXYURIDINE see BNC750
5-BROMO-2-DEOXYURIDINE see BNC750
5-BROMO-2'-DEOXY URIDINE see BNC750
5-BROMODESOXYURIDINE see BNC750
2-BROMO-1,5-DIAMINO-4,8-DIHYDROXYANTHRAQUINONE see BNC800
2-BROMO-1,8-DIAMINO-4,5-DIHYDROXYANTHRAQUINONE see BND250
BROMODIBORANE see BND325
BROMODICHLOROMETHANE see BND500
4-BROMO-2,5-DICHLOROPHENOL-o-ESTER with O,O-DIETHYL PHOS-
    PHOROTHIOATE see EGV500

O-(4-BROMO-2,5-DICHLOROPHENYL)-O,O-DIETHYL PHOSPHOROTHIOATE
    see EGV500
O-(4-BROMO-2,5 DICHLOROPHENYL)-O,O-DIETHYLPHOSPHOROTHIONATE
    see EGV500
4-BROMO-2,5-DICHLOROPHENYL DIMETHYL PHOSPHOROTHIONATE
    see BNL250
o-(4-BROMO-2,5-DICHLOROPHENYL)-o-ETHYL PHENYLPHOSPHONOTHIO-
    ATE see BND750
O-(4-BROMO-2,5-DICHLOROPHENYL)-O-METHYL
    PHENYLPHOSPHONOTHIOATE see LEN000
O-(4-BROMO-2,5-DICLORO-FENIL)-O,O-DIMETIL-MONOTIOFOSFATO (ITAL-
    IAN) see BNL250
2-BROMO-9,10-DIDEHYDRO-N,N-DIETHYL-6-METHYLERGOLINE-8β-CAR-
    BOXAMIDE see BNM250
BROMODIETHYLACETYLCARBAMIDE see BNK000
BROMODIETHYLACETYLUREA see BNK000
5-BROMO-2-(2-(DIETHYLAMINO)ETHOXY)BENZANILIDE see DHP400
7-BROMO-1,3-DIHYDRO-5-(2-PYRIDYL)-2H-1,4-BENZDIAZEPIN-2-ONE
    see BMN750
dl-4-BROMO-2,5-DIMETHOXYAMPHETAMINE HYDROBROMIDE see BNE250
2-BROMO-3,5-DIMETHOXYANILINE see BNE325
dl-4-BROMO-2,5-DIMETHOXY-α-METHYLPHENETHYLAMINE HYDROBROM-
    IDE see BNE250
21-BROMO-3,17-DIMETHOXY-19-NOR-17-α-PREGNA-1,3,5(10)-TRIEN-20-YNE
    see BAT800
2-BROMO-N,N-DIMETHYL-1-ADAMANATANEMETHANAMINE
    HYDROCHLORIDEHEMIHYDRATE see BNE500
2-BROMO-N,N-DIMETHYL-1-ADAMANTANEPROPANAMINE HYDROCHLO-
    RIDE see BNF000
3'-BROMO-4-DIMETHYLAMINOAZOBENZENE see BNE600
2-(p-BROMO-α-(2-DIMETHYLAMINO)ETHYL)BENZYL)PYRIDINE BIMALE-
    ATE see BNE750
(+)-2-(p-BROMO-α-(2-DIMETHYLAMINO)ETHYL)BENZYL)PYRIDINE MALE-
    ATE see DXG100
(±)-2-(p-BROMO-α-(2-DIMETHYLAMINO)ETHYL)BENZYL)PYRIDINE MALE-
    ATE see DNW759
2-(p-BROMO-α-(2-(DIMETHYLAMINO)ETHYL)BENZYL)PYRIDINE MALEATE
    (1581) see BNE750
2-BROMO-1-(N,N-DIMETHYLAMINOMETHYL)ADAMANTANE
    HYDROCHLORIDEHEMIHYDRATE see BNE500
2-BROMO-1-(3-DIMETHYLAMINOPROPYL)ADAMANTANE HYDROCHLO-
    RIDE see BNF000
4-BROMODIMETHYLANILINE see BNF250
p-BROMO-N,N-DIMETHYL ANILINE see BNF250
3-BROMO-7,12-DIMETHYLBENZ(a)ANTHRACENE see BNF300
4-BROMO-7,12-DIMETHYLBENZ(a)ANTHRACENE see BNF310
5-BROMO-9,10-DIMETHYL-1,2-BENZANTHRACENE see BNF315
p-BROMO-α,α-DIMETHYLPHENETHYLAMINE HYDROCHLORIDE see BNF750
α-BROMO-β-DIMETHYLPROPANOYLUREA see BNP750
3-BROMO-5,7-DIMETHYL PYRAZOLYL-2-PYRIMIDINEPHOSPHOROTHIOIC
    ACID-O,O-DIETHYL ESTER. see BAS000
6-BROMO-2,4-DINITROBENZENEDIAZONIUM HYDROGEN SULFATE
    see BNG125
3-BROMO-2,7-DINITRO-5-BENZO(b)-THIOPHENEDIAZONIUM-4-OLATE
    see BNG250
2-(1-(4-BROMODIPHENYL)ETHOXY)-N,N-DIMETHYLETHYLAMINE
    HYDROCHLORIDE see BMN250
BROMODIPHENYLMETHANE see BNG750
BROMODIPHENYLMETHANE (solution) see BNH000
3-BROMO-DMBA see BNF300
4-BROMO-DMBA see BNF310
BROMOEOSIN see BMO250
BROMOEOSINE see BNH500, BNK700
1-BROMO-4-(EPOXYETHYL)BENZENE see BOF250
3-BROMO-1,2-EPOXYPROPANE see BNI000
α-BROMOERGOCRIPTINE see BNB250
BROMOERGOCRYPTINE see BNB250
2-BROMOERGOCRYPTINE see BNB250
2-BROMO-α-ERGOCRYPTINE METHANESULFONATE see BNB325
2-BROMO-α-ERGOKRYPTIN see BNB250
2-BROMO-α-ERGOKRYPTINE-MESILATE (GERMAN) see BNB325
BROMOETHANE see EGV400
2-BROMO-ETHANEPHOSPHORIC ACID BIS(2-BROMOETHYL) ESTER
    see TNE600
BROMOETHANIOC ACID see BMR750
α-BROMOETHANIOC ACID see BMR750
BROMOETHANOL see BNI500
2-BROMO ETHANOL see BNI500
BROMOETHENE see VMP500
2-BROMO-2-ETHYLBUTYRLUREA see BNK000
(α-BROMO-α-ETHYLBUTYRYL)CARBAMIDE see BNK000
1-BROMO-ETHYL-BUTYRYL-UREA see BNK000
2-BROMO-2-ETHYLBUTYRYLUREA see BNK000
(α-BROMO-α-ETHYLBUTYRYL)UREA see BNK000

3'-BROMO-4'-ETHYL-4-DIMETHYLAMINOAZOBENZENE see BNK275
4'-BROMO-3'-ETHYL-4-DIMETHYLAMINOAZOBENZENE see BNK100
2-BROMO-N-ETHYL-N,N-DIMETHYLBENZENEMETHANAMINIUM 4-
   METHYLBENZENESULFONATE see BMV750
BROMOETHYLENE see VMP000
BROMOETHYLENE POLYMER see PKQ000
2-BROMO ETHYL ETHYL ETHER see BNK250
2-BROMOETHYL-3-NITROANISOLE see NFN000
1-(2-BROMOETHYL)-1-NITROSOUREA see NJN500
p-((3-BROMO-4-ETHYLPHENYL)AZO)-N,N-DIMETHYLANILINE see BNK275
p-((4-BROMO-3-ETHYLPHENYL)AZO)-N,N-DIMETHYLANILINE see BNK100
(2-BROMOETHYL)TRIMETHYLAMMONIUM BROMIDE see BNK325
BROMOETHYNE see BMS500
2-BROMOETHYNYL-2-BUTANOL see BNK350
BROMOETHYNYLETHYLMETHYLCARBINOL see BNK350
4-BROMO-2-FENILINDAN-1,3-DIONE (ITALIAN) see BNW750
BROMOFENOFOS see BNV500
BROMOFLOR see CDS125
9-BROMOFLUORENE see BNK500
BROMOFLUORESCEIC ACID see BMO250, BNH500, BNK700
BROMO FLUORESCEIN see BNH500, BNK700
BROMOFLUOROFORM see TJY100
1-BROMO-8-FLUOROOCTANE see FKS000
10-BROMO-11b-(2-FLUOROPHENYL)2,3,7,11b-TETRAHYDROOXAZOLO
   (3,2-d)(1,4)BENZODIAZEPIN-6(5H)-ONE see HAG800
BROMOFORM see BNL000
BROMOFORME (FRENCH) see BNL000
BROMOFORMIO (ITALIAN) see BNL000
BROMOFOS see BNL250
BROMOFOS-ETHYL see EGV500
BROMOFOSMETHYL see BNL250
BROMOFUME see EIY500
BROMO-O-GAS see MHR200
2-BROMO-2,3,3,4,4,4-HEXAFLUOROBUTYRIC ACID METHYL ESTER
   see EKO000
BROMOHEXANE see HFM500
BROMOHEXYLMERCURY see HFR100
α-BROMOHYDRIN see MRF275
BROMO(2-HYDROXYETHYL)MERCURY see EED600
BROMO(2-HYDROXYETHYL)MERCURY AMMONIA SALT see BNL275
2-BROMO-12'-HYDROXY-2'-(1-METHYLETHYL)-5'-α-(2-
   METHYLPROPYL)ERGOTAMIN-3',6',18-TRIONE see BNB250
6-α-BROMO-17-β-HYDROXY-17-α-METHYL-4-OXA-5-α-ANDROSTAN-3-ONE
   see BMA650
5-BROMOINDOLE-2,3-DIONE see BNL750
2-BROMOISOBUTANE see BQM250
5-BROMO-3-ISOPROPYL-6-METHYL, 2,4-PYRIMIDINEDIONE (FRENCH)
   see BNM000
5-BROMO-3-ISOPROPYL-6-METHYLURACIL see BNM000
5-BROMO-3-ISOPROPYL-6-METIL-URACIL (ITALIAN) see BNM000
3'-BROMOISOSAFROLE see BOA750
α-BROMOISOVALERIC ACID UREIDE see BNP750
α-BROMOISOVALEROYLUREA see BNP750
(α-BROMOISOVALERYL)UREA see BNP750
9-α-BROMO-11-KETOPROGESTERONE see BML825
BROMOL see THV750
2-BROMO-d-LYSERGIC ACID DIETHYLAMIDE see BNM250
BROMOLYSERGIDE see BNM250
2-(BROMOMERCURI)ETHANOL see EED600
2-(BROMOMERCURI) ETHANOL-AMMONIA (1:0.8 moles) compound
   see BNL275
7-BROMOMESOBENZANTHRONE see BMU000
BROMOMETANO (ITALIAN) see MHR200
BROMO METHANE see MHR200
BROMOMETHANE mixed with DIBROMOETHANE see BNM750
4-(7-BROMO-5-METHOXY-2-BENZOFURANYL)PIPERIDINE HYDROCHLO-
   RIDE see BMM625
3-BROMO-3-(4-METHOXYBENZOYL)ACRYLIC ACID SODIUM SALT
   see SIK000
2-(5-BROMO-2-METHOXYBENZYLOXY)TRIETHYLAMINE see BNN125
BROMO(METHOXYCARBONYL) MERCURY see MHS250
(Z)-3-BROMO-4-(4-METHOXYPHENYL)-4-OXO-2-BUTENOIC ACID, SODIUM
   SALT see SIK000
2-BROMO-N-METHYL-1-ADAMANTANEETHYLAMINE MALEATE see BNN250
2-BROMO-N-METHYL-1-ADAMANTANEMETHANAMINE HYDROCHLORIDE
   see BNN500
1-BROMO-4-(METHYLAMINO)ANTHRAQUINONE see BNN550
2-BROMO-1-(2-METHYLAMINOPROPYL)ADAMANTANE HYDROCHLORIDE
   see BNO000
3-BROMO-1-(2-METHYLAMINOPROPYL)ADAMANTANE HYDROCHLORIDE
   see BNO250
9-BROMOMETHYLANTHRACENE see BNO500
7-BROMO METHYL BENZ(a)ANTHRACENE see BNO750
(BROMOMETHYL)BENZENE see BEC000

p-BROMO-α-METHYLBENZHYDRYL-2-DIMETHYLAMINOETHYL ETHER HY-
   DROCHLORIDE see BMN250
6-BROMOMETHYLBENZO(a)PYRENE see BNP000
1-BROMO-3-METHYL BUTANE see BNP250
2-BROMO-3-METHYLBUTYRYLUREA see BNP750
10-BROMOMETHYL-9-CHLOROANTHRACENE see BNP850
9-(BROMOMETHYL)-10-CHLOROANTHRACENE see BNP850
7-BROMOMETHYL-4-CHLOROBENZ(a)ANTHRACENE see BNQ000
3'-BROMO-4'-METHYL-4-DIMETHYLAMINOAZOBENZENE see BNQ100
4'-BROMO-3'-METHYL-4-DIMETHYLAMINOAZOBENZENE see BNQ110
7-BROMOMETHYL-6-FLUOROBENZ(a)ANTHRACENE see BNQ250
2-BROMO METHYL FURAN see BNQ500
7-BROMO METHYL-1-METHYLBENZ(a)ANTHRACENE see BNQ750
12-BROMOMETHYL-7-METHYLBENZ(a)ANTHRACENE see BNR250
7-BROMO METHYL-12-METHYLBENZ(a)ANTHRACENE see BNR000
2-BROMOMETHYL-5-METHYLFURAN see BNR325
BROMOMETHYL METHYL KETONE see BNZ000
5-BROMO-6-METHYL-3-(1-METHYLPROPYL)-2,4(1H,3H)-PYRIMIDINEDIONE
   see BMM650
5-BROMO-6-METHYL-3-(1-METHYLPROPYL)URACIL see BMM650
p-(BROMOMETHYL)NITROBENZENE see BEC000
1-(BROMOMETHYL)-4-NITROBENZENE see NFN000
6-α-BROMO-17-β-METHYL-4-OXA-5-α-ANDROSTAN-3-ONE see BMA650
1-BROMO-3-METHYLPENTIN-3-OL see BNK350
1-BROMO-3-METHYL-1-PENTYN-3-OL see BNK350
2-(p-BROMO-α-METHYL-α-PHENYLBENZYL)OXY)-N,N-DIMETHYLETHYLAM-
   INE HYDROCHLORIDE see BMN250
1-BROMO-2-METHYL PROPANE see BNR750
2-BROMO-2-METHYL PROPANE see BQM250
2-BROMO-2-METHYLPROPANE (DOT) see BQM250
6-BROMO-1,2-NAPHTHOQUINONE see BNS750
BROMONE see BMN000
8-β-((5-BROMONICOTINOYLOXY)METHYL)-1,6-DIMETHYL-10-
   α-METHOXYERGOLINE see NDM000
5-BROMO-5-NITRO-m-DIOXANE see BNT000
5-BROMO-5-NITRO-1,3-DIOXANE see BNT000
2-BROMO-2-NITROPANE-1,3-DIOL see BNT250
2-BROMO-2-NITROPROPAN-1,3-DIOL see BNT250
2-BROMO-2-NITRO-1,3-PROPANEDIOL see BNT250
3-BROMO-4-NITROQUINOLINE-1-OXIDE see BNT500
α-BROMO-p-NITROTOLUENE see NFN000
β-BROMO-β-NITROTRIMETHYLENEGLYCOL see BNT250
1-BROMOOCTANE see BNU000
1-BROMO-2-OXIMINOPROPANE see BMS000
α-BROMOPARANITROTOLUENE see NFN000
4-BROMO-PDMT see BNW250
1-BROMOPENTABORANE (9) see BNU125
4-BROMO-1,2,2,6,6-PENTAMETHYLPIPERIDINE see BNU250
2-BROMOPENTANE see BNU500
4-(2-(5-BROMO-2-PENTYLOXYBENZYLOXY)ETHYL)MORPHOLINE see BNU660
2-(5-BROMO-2-PENTYLOXYBENZYLOXY)TRIETHYLAMINE see BNU700
BROMOPERIDOL see BNU725
4-BROMOPHENACYL BROMIDE see DDJ600
p-BROMOPHENACYL BROMIDE see DDJ600
BROMOPHENIRAMINE MALEATE see BNE750
dl-BROMOPHENIRAMINE MALEATE see DNW759
4-BROMOPHENOL see BNU750
o-BROMOPHENOL see BNV000
p-BROMOPHENOL see BNU750
BROMO PHENOLS see BNV250
BROMOPHENPHOS see BNV500
4-BROMOPHENYLACETONITRILE see BNV750
p-BROMOPHENYLACETONITRILE see BNV750
α-BROMOPHENYLACETONITRILE see BMW250
2-(4-BROMOPHENYL)ACETONITRILE see BNV750
p-BROMOPHENYLAMINE see BMT325
3-((BROMOPHENYL)AMINO)-N,N-DIMETHYL-PROPANAMIDE (9CI)
   see BMN350
p-(m-BROMOPHENYLAZO)-N,N-DIMETHYLANILINE see BNE600
4-BROMOPHENYL CHLOROMETHYL SULFONE see BNV850
(S)-Γ-(4-BROMOPHENYL)-N,N-DIMETHYL-2-PYRIDINEPROPANAMINE (Z)-2-
   BUTENEDIOATE (1:1) see DXG100
(±)-(Z)-Γ-(4-BROMOPHENYL)-N,N-DIMETHYL-2-PYRIDINEPROPANAMINE 2-
   BUTENEDIOATE (1:1) see DNW759
3-(4-BROMOPHENYL)-N,N-DIMETHYL-3-(3-PYRIDINYL)-2-PROPEN-1-AMINE
   see ZBA500
(Z)-3-(4-BROMOPHENYL)-N,N-DIMETHYL-3-(3-PYRIDINYL)-2-PROPEN-1-
   AMINE DIHYDROCHLORIDE see ZBA525
3-(p-BROMOPHENYL)-N,N-DIMETHYL-3-(3-PYRIDYL)ALLYLAMINE
   see ZBA500
1-(4-BROMOPHENYL)-3,3-DIMETHYLTRIAZENE see BNW250
p-BROMOPHENYL ESTER ISOTHIOCYANIC ACID see BNW825
α-BROMO-β-PHENYLETHYLENE see BOF000
BROMO PHENYL HYDRAMINE HYDROCHLORIDE see BNW500

3-(α-(p-(p-BROMOPHENYL)-β-HYDROXYPHENETHYL)BENZYL)-4-HYDROXY-COUMARIN see BMN000
4-(4-(4-BROMOPHENYL)-4-HYDROXYPIPERIDINO)-4'-FLUORO-BUTYROPHENONE see BNU725
4-(4-(4-(p-BROMOPHENYL)-4-HYDROXYPIPERIDINO)-4'-FLUORO-BUTYROPHENONE see BNU725
4-(4-(p-BROMOPHENYL)-4-HYDROXYPIPERIDINOL)-4'-FLUORO-BUTYROPHENONE see BNU725
4-(4-(4-BROMOPHENYL)-4-HYDROXY-1-PIPERIDINYL)-1-(4-FLUOROPHENYL)-1-BUTANONE see BNU725
2-(p-BROMOPHENYL)IMIDAZO(2,1-a)ISOQUINOLINE see BNW625
4-BROMO-2-PHENYL-1,3-INDANDIONE see BNW750
5-BROMO-2-PHENYLINDAN-1,3-DIONE see UVJ400
5-BROMO-2-PHENYL-1,3-INDANDIONE see UVJ400
5-BROMO-2-PHENYL-1H-INDENE-1,3(2H)-DIONE see UVJ400
p-BROMOPHENYL ISOTHIOCYANATE see BNW825
p-BROMO PHENYL LITHIUM see BNX000
BROMOPHENYLMETHANE see BEC000
3-(p-BROMOPHENYL)-1-METHOXY-1-METHYLUREA see PAM785
N'-(4-BROMOPHENYL)-N-METHOXY-N-METHYLUREA see PAM785
3-(p-BROMOPHENYL)-1-METHYL-1-METHOXYUREA see PAM785
3-(p-BROMOPHENYL)-1-METHYL-1-NITROSOUREA see BNX125
1-(p-BROMOPHENYL)-3-METHYLUREA see MHS375
1-(p-BROMOPHENYL)-3-METHYLUREA mixed with SODIUM NITRITE see SIQ675
2-(m-BROMOPHENYL)-N-(4-MORPHOLINOMETHYL)SUCCINIMIDE see BNX250
(p-BROMOPHENYL)OXIRANE see BOF250
(4-BROMOPHENYLOXIRANE) (9CI) see BOF250
1-(p-BROMOPHENYL)-1-PHENYL-1-(2-DIMETHYLAMINOETHOXY)ETHANE HYDROCHLORIDE see BMN250
2-(1-(4-BROMOPHENYL)-1-PHENYLETHOXY)-N,N-DIMETHYLETHANAMINE HYDROCHLORIDE see BMN250
(2-(1-p-BROMOPHENYL-1-PHENYLETHOXY)ETHYL)DIMETHYLETHYLAM-INE HYDROCHLORIDE see BMN250
(Z)-3-(4'-BROMOPHENYL)-3-(3''-PYRIDYL)DIMETHYLALLYLAMINE see ZBA500
BROMOPHOS see BNL250
BROMOPHOSETHYL see EGV500
BROMOPICRIN see NMQ000
9-BROMOPREGN-4-ENE-3,11,20-TRIONE see BML825
9-α-BROMOPREGN-4-ENE-3,11,20-TRIONE see BML825
BROMOPRIDA see VCK100
BROMOPRIDE see VCK100
3-BROMO-1-PROPANAMINE HYDROBROMIDE see AJA000
1-BROMOPROPANE see BNX750
2-BROMOPROPANE see BNY000
1-BROMOPROPANE (DOT) see BNX750
3-BROMO-1,2-PROPANEDIOL see MRF275
3-BROMOPROPANOL see BNY750
3-BROMO-1-PROPANOL see BNY750
BROMO-2-PROPANONE see BNZ000
1-BROMO-2-PROPANONE see BNZ000
2-BROMOPROPENALDEHYDE see BMT000
1-BROMOPROPENE see BOA000
2-BROMOPROPENE see BOA250
3-BROMOPROPENE see AFY000
1-BROMO-1-PROPENE see BOA000
(E)-p-(3-BROMOPROPENYL)ANISOLE see BMT300
5-(3-BROMO-1-PROPENYL)-1,3-BENZODIOXOLE see BOA750
5-(2-BROMO-2-PROPENYL)-5-(1-METHYLETHYL)-2,4,6(1H,3H,5H)-PYRIMIDINETRIONE see QCS000
3-BROMOPROPIONIC ACID see BOB250
α-BROMOPROPIONIC ACID see BOB000
β-BROMOPROPIONIC ACID see BOB250
3-BROMO PROPIONITRILE see BOB510
3-BROMOPROPYLAMINE HYDROBROMIDE see AJA000
BROMOPROPYLATE see IOS000
3-BROMOPROPYL CHLORIDE see BNA825
2-BROMOPROPYLENE see BOA250
3-BROMOPROPYLENE see AFY000
3-BROMO-1-PROPYNE see PMN500
3-BROMOPROPYNE (DOT) see PMN500
3-BROMOPYRIDINE see BOC510
7-BROMO-5-(2-PYRIDYL)-3H-1,4-BENZODIAZEPIN-2(1H)-ONE see BMN750
3-(2-(5-BROMO-2-PYRIDYLOXY)ETHYL)THIAZOLIDINE HYDROCHLORIDE see BOD000
2-(6-(5-BROMO-2-PYRIDYL OXY)HEXYL)AMINOETHANE THIOL HYDRO-CHLORIDE see BOD500
1-BROMO-2,5-PYRROLIDINEDIONE see BOF500
UINE see QJJ100
5-BROMOSALICYLIC ACID see BOE500
BROMO SELTZER see ABG750
BROMOSILANE see BOE750

β-BROMOSTYRENE see BOF000
ω-BROMOSTYRENE see BOF000
4-BROMOSTYRENE OXIDE see BOF250
p-BROMOSTYRENE OXIDE see BOF250
4'-BROMOSTYRENE OXIDE see BOF250
p-BROMOSTYRENE-7,8-OXIDE see BOF250
BROMOSTYROL see BOF000
BROMOSTYROLENE see BOF000
N-BROMOSUCCIMIDE see BOF500
N-BROMO SUCCINIMIDE see BOF500
3-BROMO-1,1,2,2-TETRAFLUOROPROPANE see BOF750
3-BROMOTETRAHYDROTHIOPHENE-1,1-DIOXIDE see BOG000
N-BROMO TETRAMETHYL GUANIDINE see BOG250
5-BROMO-2-(2-(3-THIAZOLIDINYL)ETHOXY)PYRIDINE HYDROCHLORIDE see BOD000
p-BROMOTHIOBENZOHYDROXIMIC ACID-S-DIETHYLAMINOETHYL ESTER HYDROCHLORIDE see DHZ000
α-BROMOTOLUENE (DOT) see BEC000
ω-BROMOTOLUENE see BEC000
α-BROMO-α-TOLUNITRILE see BMW250
p-((3-BROMO-p-TOLYL)AZO)-N,N-DIMETHYLANILINE see BNQ100
p-((4-BROMO-m-TOLYL)AZO)-N,N-DIMETHYLANILINE see BNQ110
BROMOTRIBUTYLSTANNANE see TIC250
BROMOTRICHLOROMETHANE see BOH750
3-BROMO-1,1,1-TRICHLORO PROPANE see BOI000
3-BROMOTRICYCLOQUINAZOLINE see BOI250
BROMOTRIETHYLSTANNANE see BOI750
BROMOTRIETHYLSTANNANE compounded with 2-PIPECOLINE (1:1) see TJU850
BROMOTRIFLUOROETHENE see BOJ000
BROMO TRIFLUOROETHYLENE see BOJ000
BROMOTRIFLUOROMETHANE see TJY100
3-BROMOTRIFLUOROMETHYLBENZENE see BOJ500
m-BROMO(TRIFLUOROMETHYL)BENZENE see BOJ500
m-BROMO-α,α,α-TRIFLUOROTOLUENE see BOJ500
o-BROMO-α,α,α-TRIFLUOROTOLUENE see BOJ750
BROMOTRIPENTYLSTANNANE see BOK250
BROMOTRIPHENYLETHYLENE see BOK500
BROMOTRIPROPYLSTANNANE see BOK750
5-BROMOURACIL see BOL000
BROMOURACIL DEOXYRIBOSIDE see BNC750
5-BROMOURACIL DEOXYRIBOSIDE see BNC750
5-BROMOURACIL-2-DEOXYRIBOSIDE see BNC750
BROMOVAL see BNP750
α-BROMOVALERIC ACID see BOL250
BROMOVALEROCARBAMIDE see BNP750
BROMOVALERYLUREA see BNP750
(E)-5-(2-BROMOVINYL)-2'-DEOXYURIDINE see BOL300
trans-5-(2-BROMOVINYL)-2'-DEOXYURIDINE see BOL300
BROMOWODOR (POLISH) see HHJ000
BROMOXIL see BNP750
BROMOXYNIL see DDP000
BROMOXYNIL OCTANOATE see DDM200
BROMPERIDOL see BNU725
p-BROMPHENACYL-8 see DDJ600
d-BROMPHENIRAMINE MALEATE see DXG100
(±)-BROMPHENIRAMINE MALEATE see DNW759
dl-BROMPHENIRAMINE MALEATE see DNW759
BROMPHENPHOS see BNV500
3-(4-BROMPHENYL)-1-METHOXYHARNSTOFF (GERMAN) see PAM785
β-BROMSTYROL see BOF000
BROMURAL see BNP750
BROMURE de CYANOGEN (FRENCH) see COO500
BROMURE d'ETHYLE see EGV400
BROMURE de METHYLE (FRENCH) see MHR200
BROMURE de VINYLE (FRENCH) see VMP000
BROMURE de XYLYLE (FRENCH) see XRS000
BROMURO di ETILE (ITALIAN) see EIY500
BROMURO di METILE (ITALIAN) see MHR200
BROMURO de OXITROPIO(SPANISH) see ONI000
BROMUVAN see BNP750
BROMVALERYLUREA see BNP750
BROMVALETONE see BNP750
BROMVALETONUM see BNP750
BROMVALUREA see BNP750
BROMWAUREA see BNP750
BROMWASSERSTOFF (GERMAN) see HHJ000
BROMYL see BNP750
BROMYL FLUORIDE see BOL310
BRONCHIOCAIN see BQH250
BRONCHOCAIN see BQH250
BRONCHOCAINE see BQH250
BRONCHODIL see DNA600
BRONCHOLYSIN see ACH000
BRONCHOSELECTAN see AAN000

BRONCHOSPASMIN see DNA600
BRONCOVALEAS see BQF500
BRONKAID MIST see VGP000
BRONKEPHRINE see DMV600
BRONKEPHRINE HYDROCHLORIDE see ENX500
BRONOCOT see BNT250
BRONOPOL see BNT250
BRONOSOL see BNT250
BRONTYL see HOA000
BRONZE BROMO see BNH500, BNK700
BRONZE GREEN TONER A-8002 see AFG500
BRONZE POWDER see CNI000
BRONZE RED RO see CHP500
BRONZE SCARLET see CHP500
BROOM (DUTCH) see BMP000
BROOM ABSOLUTE see GCM000
O-(4-BROOM-2,5-DICHLOOR-FENYL)-O,O-DIMETHYL-MONOTHIOFOSFAAT
    (DUTCH) see BNL250
5-BROOM-3-ISOPROPYL-6-METHYL-URACIL DUTCH) see BNM000
BROOMMETHAAN (DUTCH) see MHR200
BROOMWATERSTOF (DUTCH) see HHJ000
BROPIRAMINE see AIY850
BROPIRIMINE see AIY850
BROTIANIDE see BOL325
BROTIZOLAM see LEJ600
BROTOPON see CLY500
BROVALIN see BNP750
BROVALUREA see BNP750
BROVARIN see BNP750
BROVEL see ECU600
1545 BROWN see CMP250
11460 BROWN see XMA000
11660 BROWN see CMB750
BROWN ACETATE see CAL750
BROWN COPPER OXIDE see CNO000
BROWN DRAGON see JAJ000
BROWN FK see CMP250
BROWN HEMATITE see LFW000
BROWN IRON ORE see LFW000
BROWN IRONSTONE CLAY see LFW000
BROWN SALT NV see MIH500
BROXIL see PDD350
BROXURIDINE see BNC750
BROXYKINOLIN see DDS600
BROXYNIL see DDP000
BROXYQUINOLINE see DDS600
BRS 640 see BML500
BRUCEANTIN see BOL500
BRUCIN (GERMAN) see BOL750
BRUCINA (ITALIAN) see BOL750
BRUCINE see BOL750
(−)-BRUCINE see BOL750
BRUCINE, solid (DOT) see BOL750
BRUCINE ALKALOID see BOL750
BRUCINE IODOMETHYLATE see BOM000
BRUCINE IODOMETHYLE (FRENCH) see BOM000
BRUCINE METHIODIDE see BOM000
BRUCITE see NBT000
BRUDR see BNC750
BRUFANEUXOL see DOT000
BRUFANIC see IIU000
BRUFEN see IIU000
BRUGMANSIA ARBOREA see AOO825
BRUGMANSIA SANGUINEA see AOO825
BRUGMANSIA SUAVEOLENS see AOO825
BRUGMANSIA X CANDIDA see AOO825
BRUINSTEEN (DUTCH) see MAS000
BRULAN see BSN000
BRUMIN see WAT200
BRUNEOMYCIN see SMA000
BRUOMOPHOS (RUSSIAN) see BNL250
BRUSH BUSTER see MEL500
BRUSH-OFF 445 MLD VOLATILE BRUSH KILLER see TAA100
BRUSH RHAP see TAA100
BRUSHTOX see TAA100
BRYAMYCIN see TFQ275
BS 572 see DNU100
BS 4231 see PMG000
BS 5930 see OJW000
BS 5933 see DRR500
BS 6825 see TGX500
BS 7029 see CQH625
BS 7051 see HBT000
BS 7331 see TGJ250

BS 7020a see DAZ140
BS100-141 see GKU300
BSA see BBR500, TMF000
BSC-REFINE D see BBS750
B-SELEKTONON M see CIR250
BT see BPG325, BPU000
BT 621 see CBS000
BTC see AFP250
BTC 471 see BBA500
BTC 1010 see DGX200
BTF see BBU125
621-BT HYDROCHLORIDE HYDRATE see TGJ150
BTKH see BEU500
BTO see BLL750
BTPABA see CML835
B-1,3,5-TRICHLOROBORAZINE see TIL300
B-TRIMETHYLBORAZINE see TLN100
BTS 18322 see FLG100
BTT see BSO750
BTZ see DKU875
BU-6 see BGS250
BU 533 see PIN100
BU 2231A see TAC500
BUBAN 37 see DUV400
BUBBIE BLOSSOMS see CCK675
BUBBY BUSH see CCK675
BUBURONE see IIU000
BUCACID AZURE BLUE see FMU059
BUCACID BRILLIANT SCARLET 3R see FMU080
BUCACID FAST ORANGE G see HGC000
BUCACID FAST WOOL BLUE R see ADE750
BUCACID GUINEA GREEN BA see FAE950
BUCACID INDIGOTINE B see FAE100
BUCACID METANIL YELLOW see MDM775
BUCACID ORANGE R see ADG000
BUCACID TARTRAZINE see FAG140
BUCARBAN see BSM000
BUCB see DJF200
BUCCALSONE see HHR000
BUCETIN see HJS850
BUCK BRUSH see SED550
BUCKEYE see HGL575
BUCKTHORN see BOM125, MBU825
BUCLADESINE see DEJ300
BUCLIZINE DIHYDROCHLORIDE see BOM250
BUCLODIN see BOM250
BUCLOSINSAEURE (GERMAN) see CPJ000
BUCLOXIC ACID see CPJ000
BUCLOXIC ACID CALCIUM see CPJ250
BUCLOXIC ACID CALCIUM SALT see CPJ250
BUCLOXINSAEURE KALZIUM (FERMAN) see CPJ250
BUCLOXONIC ACID see CPJ000
BUCLOXONIC ACID CALCIUM SALT see CPJ250
BUCOLOM see BQW825
BUCOLOME see BQW825
BUCROL see BSM000
BUCS see BPJ850
BUCTRIL see DDP000
BUCTRIL INDUSTRIAL see DDP000
BUCUMOLOL HYDROCHLORIDE see BOM510
dl-BUCUMOLOL HYDROCHLORIDE see BOM510
BUDESONIDE see BOM520
BUDIPIN (GERMAN) see BOM530
BUDIPINE see BOM530
BUD-NIP see CKC000
BUDOFORM see CHR500
BUDORM see BPF500
BUDR see BNC750
5-BUDR see BNC750
BUDRALAZINE see DQU400
BU2AE see DDU600
BUENO see MRL750
BUFEDIL see BOM600
BUFEMID see BGL250
BUFEN see ABU500
BUFENCARB see BTA250
BUFETOLOL HYDROCHLORIDE see AER500
BUFEXAMIC ACID see BPP750
BUFF-A-COMP see ABG750
BUFLOMEDIL see BOM600
BUFLOMEDIL HYDROCHLORIDE see BOM600
BUFOGENIN see BOM650
BUFOGENIN B see BOM655
BUFON see DKA600

BUFONAMIN see BOM750, BQL000
BUFOPTO ZINC SULFATE see ZNA000
BUFORMIN see BRA625
BUFORMINE see BRA625
BUFORMIN HYDROCHLORIDE see BOM750, BQL000
BUFOTALIN see BON000
BUFOTALINE see BON000
BUFOTENIN see DPG109
BUFURALOL see BQD000
BUHACH see POO250
BUIS de SAPIA (CANADA) see YAK500
BUKARBAN see BSM000
BUKS see BFW750
BUKSAMIN see AKF375
BULANA, combustion products see ADX750
BULAN and PROLAN MIXTURE (2581) see BON250
BULBOCAPNINE see HLT000
d-BULBOCAPNINE see HLT000
BULBONIN see BQL000
BULGARIAN see RNF000
BULKOSOL see BON300
BULL FLOWER see MBU550
BULPUR see PLC250
BUMADIZON CALCIUM SALT HEMIHYDRATE see CCD750
BUMETANIDE see BON325
BUMEX see BON325
BUNAIOD see EQC000
BUNAMIODYL see EQC000
BUNAZOCINE HYDROCHLORIDE see BON350
BUNAZOSIN HYDROCHLORIDE see BON350
BUNGARUS CAERULEUS VENOM see BON365
BUNGARUS FASCIATUS VENOM see BON367
BUNGARUS MULTICINCTUS VENOM see BON370
BUNIODYL see EQC000
BUNITROLOL HYDROCHLORIDE see BON400
BUNK see PJJ300
BUNSENITE see NDF500
BUNT-CURE see HCC500
BUNT-NO-MORE see HCC500
BUPATOL see BOV825
BUPHENINE HYDROCHLORIDE see DNU200
BUPICAINE HYDROCHLORIDE (+) see BON750
BUPICAINE HYDROCHLORIDE (±) see BOO000
BUPIVACAINE see BSI250
l(−)-BUPIVACAINE see BOO500
d(+)-BUPIVACAINE see BOO250
dl-BUPIVACAINE see BSI250
BUPIVACAINE HYDROCHLORIDE see BOO000
BUPLEURUM FALCATUM L see SAF300
BUPLEURUM MARGINATUM WALL. EX. DC., EXTRACT see BOO625
BUPRANOLOL HYDROCHLORIDE see BQB250
BUPRENORPHINE see TAL325
BUPRENORPHINE HYDROCHLORIDE see BOO630
BUPROPION HYDROCHLORIDE see WBJ500
BUR see EHA100
BURCOL see BSM000
BURESE see CNE750
BUREX (CZECH) see PEE750
BURFOR see NCW300
BURGODIN see BCE825
BURINE see BON325
BURINEX see BON325
BURITAL SODIUM see SOX500
BURLEY TOBACCO see TGH725, TGH750
BURMA GREEN B see AFG500
BURN BEAN see NBR800
BURNISH GOLD see GIS000
BURNOL see DBX400, XAK000
BURNTISLAND RED see IHD000
BURNT LIME see CAU500
BURNT SIENNA see IHD000
BURNT UMBER see IHD000
BUROFLAVIN see DBX400
BURONIL see FKI000
BURPLEURUM FALCATUM LINN. VAR. MARGINATUM (WALL. EX. DC.)
   CL., EXTRACT see BOO625
BURTONITE 44 see FPQ000
BURTONITE V-7-E see GLU000
BURTONITE-V-40-E see CCL250
BUSAN 72A see BOO635
BUSCAPINA see SBG500
BUSCAPINE see SBG500
BUSCOL see SBG500
BUSCOLAMIN see SBG500

BUSCOLYSINE see SBG500
BUSCOPAN see SBG500
BUSCOPAN COMPOSITUM see BOO650
BUSERELIN see LIU420
BUSHMAN'S POISON see BOO700
BUSONE see BRF500
BUSPIRONE see PPP250
BUSTREN see SMQ500
BUTABARB see BPF000
BUTABARBITAL see BPF000
BUTABARBITAL SODIUM see BPF250
BUTABARBITONE see BPF000
BUTABITAL see AFY500
BUTACAINE see BOO750
BUTACAINE SULFATE see BOP000
BUTACARB see DEG400
BUTACARBE (FRENCH) see DEG400
BUTACHLOR see CFW750
BUTACIDE see PIX250
BUTACOMPREN see BRF500
BUTACOTE see BRF500
BUTADIEEN (DUTCH) see BOP500
BUTA-1,3-DIEEN (DUTCH) see BOP500
BUTADIEN (POLISH) see BOP500
BUTA-1,3-DIEN (GERMAN) see BOP500
BUTADIENDIOXYD (GERMAN) see BGA750
1,2-BUTADIENE see BOP250
1,3-BUTADIENE see BOP500
BUTA-1,3-DIENE see BOP500
α-Γ-BUTADIENE see BOP500
BUTADIENE DIEPOXIDE see BGA750
l-BUTADIENE DIEPOXIDE see BOP750
1,3-BUTADIENE DIEPOXIDE see BGA750
BUTADIENE DIMER see CPD750
BUTADIENE DIOXIDE see BGA750
dl-BUTADIENE DIOXIDE see DHB600
BUTADIENE MONOXIDE see EBJ500
BUTADIENE PEROXIDE see BOQ250
1,3-BUTADIENE-STYRENE COPOLYMER see SMR000
BUTADIENE-STYRENE POLYMER see SMR000
1,3-BUTADIENE-STYRENE POLYMER see SMR000
BUTADIENE-STYRENE RESIN see SMR000
BUTADIENE-STYRENE RUBBER (FCC) see SMR000
BUTADIENE SULFONE see DMF000
BUTADIEN-FURFURAL COPOLYMER see BHJ500
N-2,3-BUTADIENYL-N-METHYLBENZYLAMINE HYDROCHLORIDE
   see BOQ500
1,3-BUTADIYNE see BOQ625
BUTAFLOGIN see HNI500
BUTAFUME see BPY000
BUTAKON 85-71 see SMR000
BUTAL see BSU250
BUTALAMINE HYDROCHLORIDE see PEU000
BUTALAN see BRF500
BUTALBARBITAL see AGI750
BUTALBITAL see AGI750
BUTALBITAL SODIUM see BOQ750
BUTALDEHYDE see BSU250
BUTALGIN see MDP750
BUTALGINA see BRF500
BUTALIDON see BRF500
BUTALLYLONAL see BOR000
BUTALLYLONAL SODIUM see BOR250
BUTALYDE see BSU250
BUTAMBEN see BPZ000
BUTAMID see BSQ000
BUTAMIN see AIT750
BUTAMIRATE CITRATE see BOR350
BUTAMYRATE CITRATE see BOR350
BUTANAL see BSU250
n-BUTANAL (CZECH) see BSU250
BUTANAL OXIME see BSU500
BUTANAMIDE, N-(4-ETHOXYPHENYL)-3-HYDROXY-
   see HJS850
1-BUTANAMINE see BPX750
2-BUTANAMINE see BPY000
1,3-BUTANDIOL (GERMAN) see BOS500
BUTANE see BOR500
n-BUTANE (DOT) see BOR500
BUTANECARBOXYLIC ACID see VAQ000
1-BUTANECARBOXYLIC ACID see VAQ000
1,3-BUTANEDIAMINE see BOR750
1,4-BUTANEDIAMINE see BOS000
1,4-BUTANEDIAMINE, 2-METHYL-, POLYMER with α-HYDRO-omega-

HYDROXYPOLY(OXY-1,4-BUTANEDIYL) and 1,1'-METHYLENEBIS(4-ISOCYANATOCYCLOHEXANE) see PKN500
1,4-BUTANEDICARBOXAMIDE see AEN000
1,4-BUTANEDICARBOXYLIC ACID see AEN250
BUTANE DIEPOXIDE see BGA750
1,4-BUTANEDINITRILE see SNE000
BUTANEDIOIC ACID see SMY000
BUTANEDIOIC ACID, 2,3-BIS(BENZOYLOXY)-, (R-(R*,R*))- see DDE300
BUTANEDIOIC ACID, DIETHYL ESTER see SNB000
BUTANEDIOIC ACID DIPROPYL ESTER see DWV800
BUTANEDIOIC ACID MONO(7-CHLORO-2,3-DIHYDRO-2-OXO-5-PHENYL-1H-1,4-BENZODIAZEPIN-3-YL) ESTER see CFY500
BUTANEDIOIC ACID MONO(2,2-DIMETHYLHYDRAZIDE) see DQD400
BUTANEDIOIC ACID MONO(3-((2-ETHYLHEXYL)AMINO)-1-METHYL-3-OX-OPROPYL) ESTER (9CI) see BPF825
BUTANEDIOIC ANHYDRIDE see SNC000
1,2-BUTANEDIOL see BOS250
1,3-BUTANEDIOL see BOS500
BUTANE-1,3-DIOL see BOS500
1,4-BUTANEDIOL see BOS750
BUTANE-1,4-DIOL see BOS750
2,3-BUTANEDIOL see BOT000
1,4-BUTANEDIOL, BISSULFAMATE (ester) see BOU100
1,2-BUTANEDIOL, CYCLIC CARBONATE see BOT200
1,3-BUTANEDIOL, CYCLIC SULFITE see MQG500
1,3-BUTANEDIOL DIACRYLATE see BRG500
1,4-BUTANEDIOL DIMETHANESULPHONATE see BOT250
1,4-BUTANEDIOL DIMETHYL SULFONATE see BOT250
1,4-BUTANEDIOL, POLYMER with 1,6-DIISOCYANATOHEXANE see PKO750
1,4-BUTANEDIOL POLYMER with 1,1'-METHYLENEBIS(4-ISOCYANATOBENZENE) see PKP000
2,3-BUTANEDIONE see BOT500
N,N'-(1,4-BUTANEDIYL)BIS-1-AZIRIDIENCARBOXAMIDE see TDQ225
1,4-BUTANEDIYL SULFAMATE see BOU100
BUTANEFRINE HYDROCHLORIDE see ENX500
BUTANEN (DUTCH) see BOR500
BUTANENITRILE see BSX250
n-BUTANENITRILE see BSX250
BUTANESULFONE see BOU250
BUTANE SULTONE see BOU250
1,4-BUTANESULTONE (MAK) see BOU250
Δ-BUTANE SULTONE see BOU250
BUTANETETRACARBOXYLIC ACID see BOU500
1,2,3,4-BUTANETETRACARBOXYLIC ACID see BOU500
5H,6H-6,5A,13A,14-(1,2,3,4)BUTANETETRAYCYCLOOCTA(1,2-B:5,6-B')DINAPHTHALENE see LIV000
BUTANETHIOL see BRR900
n-BUTANETHIOL see BRR900
tert-BUTANETHIOL see MOS000
1-BUTANETHIOL, TIN(2+) SALT see BOU550
BUTANEX see CFW750
BUTANI (ITALIAN) see BOR500
BUTANILICAINE HYDROCHLORIDE see BQA750
BUTANIMIDE see SND000
BUTANOIC ACID see BSW000
BUTANOIC ACID, 2-AMINO-4-(METHYLSULFINYL)-(9CI) see ALF600
BUTANOIC ACID-2-BUTOXY-1-METHYL-2-OXOETHYL ESTER (9CI) see BQP000
BUTANOIC ACID, 4,4'-DISELENOBIS(2-AMINO-(9CI) see SBU710
BUTANOIC ACID ETHYL ESTER see EHE000
BUTANOIC ACID, 3-OXO-, 5-METHYL-2-(1-METHYLETHYL)CYCLOHEXYL ESTER, (1R-(1-α-2-β, 5α-)- see MCG850
BUTANOIC ACID, 1,2,3-PROPANETRIYL ESTER see TIG750
BUTANOIC ACID 2,2,2-TRICHLORO-1-(DIMETHOXYPHOSPHINYL)ETHYL ESTER see BPG000
1-BUTANOL see BPW500
BUTAN-1-OL see BPW500
BUTAN-2-OL see BPW750
2-BUTANOL see BPW750
n-BUTANOL see BPW500
BUTANOL (DOT) see BPW500
tert-BUTANOL see BPX000
BUTANOL (FRENCH) see BPW500
sec-BUTANOL (DOT) see BPW750
2-BUTANOL ACETATE see BPV000
3-BUTANOL see AAH750
BUTANOL-2-AMINE see AJA250
BUTANOL (4)-BUTYL-NITROSAMINE see HJQ350
BUTANOLEN (DUTCH) see BPW500
4-BUTANOLIDE see BOV000
1-BUTANOL, 4-(NITROSO-2-PROPENYLAMINO)- see HJS400
BUTANOLO (ITALIAN) see BPW500
2-BUTANOL-3-ONE see ABB500
BUTANOL SECONDAIRE (FRENCH) see BPW750

BUTANOL TERTIAIRE (FRENCH) see BPX000
2-BUTANONE (OSHA) see MKA400
BUTANONE 2 (FRENCH) see MKA400
2-BUTANONE, 4-(4-HYDROXY-3-METHOXYPHENYL)- see VFP100
2-BUTANONE, 4-(6-METHOXY-2-NAPHTHALENYL)- see MFA300
2-BUTANONE OXIME see EMU000
2-BUTANONE OXIME HYDROCHLORIDE see BOV625
2-BUTANONE, SEMICARBAZONE see MKA750
BUTANOVA see HNI500
BUTAPERAZINE DIMALEATE see BSZ000
BUTAPHENE see BRE500
BUTAPIRAZOL see BRF500
BUTAPIRONE see HNI500
BUTAPYRAZOLE see BRF500
BUTARECBON see BRF500
n-BUTARSAMIDE see CBK750
BUTARTRINA see BRF500
BUTATAB see BPF000
BUTATAL see BPF000
BUTATENSIN see MBW750
BUTAZATE see BIX000
BUTAZATE 50-D see BIX000
BUTAZINA see BRF500
BUTAZOLIDINE SODIUM see BOV750
BUTAZONA see BRF500
BUTAZONE see BRF500
BUTEA FRONDOSA, seed extract see BOV800
BUTEDRIN see BOV825
BUTEDRINE see BQF250
BUTELLINE see BOP000
2-BUTENAL see COB250
(E)-2-BUTENAL see COB260
trans-2-BUTENAL see COB260
1-BUTENE see BOW250
cis-2-BUTENE see BOW500
trans-2-BUTENE see BOW750
1-BUTENE, 3,4-DICHLORO- see DEV100
(±)-(Z)-2-BUTENEDIOATE-1H-INDEN-5-OL, 2,3-DIHYDRO-6-((2-METHYL-1-PIPERIDINYL)METHYL)- see PJI600
(Z)-2-BUTENEDIOATE (1:1) 1,2,3,4-TETRAHYDRO-2-METHYL-4-PHENYL-8-ISOQUINOLINAMINE see NMV725
(E)-BUTENEDIOIC ACID see FOU000
(Z)-2-BUTENEDIOIC ACID see MAK900
cis-BUTENEDIOIC ACID see MAK900
trans-BUTENEDIOIC ACID see FOU000
2-BUTENEDIOIC ACID BIS(1,3-DIMETHYLBUTYL) ESTER see DKP400
2-BUTENEDIOIC ACID BIS(2-ETHYLHEXYL) ESTER see DVK600
2-BUTENEDIOIC ACID BIS(1-METHYLETHYL) ESTER see BOX250
2-BUTENEDIOIC ACID, DIBUTYL ESTER see DED600
(Z)-2-BUTENEDIOIC ACID DIETHYL ESTER see DJO200
trans-2-BUTENEDIOIC ACID DIMETHYL ESTER see DSB600
2-BUTENEDIOIC ACID, MONO(2-(2-HYDROXYETHOXY)ETHYL) ESTER see MAL500
2-BUTENEDIOIC ACID, MONO(2-HYDROXYPROPYL) ESTER see MAL750
cis-BUTENEDIOIC ANHYDRIDE see MAM000
2-BUTENE-1,4-DIOL, DIMETHANESULFONATE, (E)- see DNX000
2-BUTENENITRILE see COQ750
3-BUTENE NITRILE see BOX500
1-BUTENE-4-NITRILE see BOX500
3-BUTENE-2-ONE see BOY500
1-BUTENE OXIDE see BOX750
1,2-BUTENE OXIDE see BOX750
trans-2-BUTENE OZONIDE see BOX825
1-BUTENE, 2,3,4-TRICHLORO- see TIL360
2-BUTENOIC ACID see COB500
α-BUTENOIC ACID see COB500
trans-2-BUTENOIC ACID see ERE000
2-BUTENOIC ACID, ANHYDRIDE (9CI) see COB900
2-BUTENOIC ACID, 3-BROMO-4-(4-METHOXYPHENYL)-4-OXO-, SODIUM SALT, (E)- (9CI) see CQK600
2-BUTENOIC ACID, 2-BUTENYLIDENE ESTER, (E,E,E)-(8CI,9CI) see COD100
2-BUTENOIC ACID, 2,3-DICHLOR-4-OXO-, (Z)-(9CI) see MRU900
2-BUTENOIC ACID-3,7-DIMETHYL-6-OCTENYL ESTER see CMT500
2-BUTENOIC ACID, ETHENYL ESTER see VNU000
2-BUTENOIC ACID, 3-(((ETHYLAMINO)METHOXYPHOSPHINOTHIOYL)OXY)-, 1-METHYLETHYL, (E)-, mixt. with 2,2-DICHLOROETHENYL DIMETHYL PHOSPHATE see SAD100
2-BUTENOIC ACID, ETHYL ESTER see EHO200
2-BUTENOIC ACID, HEXYL ESTER see HFM600
2-BUTENOIC ACID, 2-METHYL-, (E)-(9CI) see TGA700
trans-2-BUTENOIC ACID METHYL ESTER see COB825
2-BUTENOIC ACID, 2-METHYL, 1-ISOPROPYL ESTER (E)- see IRN100
2-BUTENOL see BOY000
2-BUTEN-1-OL see BOY000

BUTOXYETHYL 2,4,5-T see TAH900
2-BUTOXYETHYLVINYL ETHER see VMU000
(3-n-BUTOXY-2-HYDROXYPROPYL)PHENYL ETHER see BPP250
BUTOXYL see MHV750
N-BUTOXYMETHYL-2-CHLORO-2',6'-DIETHYLACETANILIDE see CFW750
N-(BUTOXYMETHYL)-2-CHLORO-N-(2,6-DIETHYLPHENYL)ACETAMIDE
  see CFW750
(BUTOXYMETHYL)NITROSOMETHYLAMINE see BPM500
2-(p-BUTOXYPHENOXY)-N-(2-(DIETHYLAMINO)ETHYL)-2,5-DIETHOXYAC-
  ETANILIDE MONOHYDROCHLORIDE see BPM750
2-(p-BUTOXYPHENOXY)-N-(2-(DIETHYLAMINO)ETHYL)-N-(2,4-
  DIMETHOXYPHENYL)ACETAMIDE HYDROCHLORIDE see BPN000
2-(p-BUTOXYPHENOXY)-N-(2-(DIETHYLAMINO)ETHYL)-N-(2,5-
  DIMETHOXYPHENYL)ACETAMIDE HYDROCHLORIDE see BPN250
2-(p-BUTOXYPHENOXY)-N-(2(DIMETHYLAMINO)ETHYL)-N-(2,6-
  DIMETHYLPHENYL)ACETAMIDE HYDROCHLORIDE see BPO250
1-BUTOXY-3-PHENOXY-2-PROPANOL see BPP250
3-n-BUTOXY-1-PHENOXY-2-PROPANOL see BPP250
BUTOXYPHENYL see BSF750
4-BUTOXYPHENYLACETOHYDROXAMIC ACID see BPP750
p-BUTOXY PHENYL ACETOHYDROXAMIC ACID see BPP750
4-N-BUTOXYPHENYLACETOHYDROXAMIC ACID-o-ACETATE ESTER
  see ACE250
4-N-BUTOXYPHENYLACETOHYDROXAMIC ACID-o-FORMATE ESTER
  see BPQ250
4-N-BUTOXYPHENYLACETOHYDROXAMIC ACID-o-PROPIONATE ESTER
  see BPQ000
N-(p-BUTOXYPHENYL ACETYL)-o-FORMYLHYDROXYLAMINE see BPQ250
4'-BUTOXY-2-PIPERIDINOACETANILIDE HYDROCHLORIDE see BPR000
4-BUTOXY-3-(PIPERIDINO)PROPIOPHENONE HYDROCHLORIDE see BPR250
4'-BUTOXY-3-PIPERIDINO PROPIOPHENONE HYDROCHLORIDE see BPR500
4-n-BUTOXY-β-(1-PIPERIDYL)PROPIOPHENONE HYDROCHLORIDE
  see BPR500
BUTOXYPOLYPROPYLENE GLYCOL see BRP250
BUTOXYPROPANEDIOL POLYMER see BRP250
3-BUTOXYPROPANENITRILE see BPT000
3-BUTOXY PROPANOIC ACID see BPS000
BUTOXYPROPANOL (mixed isomers) see BPS750
n-BUTOXYPROPANOL (mixed isomers) see BPS750
1-BUTOXY-2-PROPANOL see BPS250
3-BUTOXY-1-PROPANOL see BPS500
3-BUTOXYPROPIONIC ACID see BPS000
3-BUTOXYPROPIONITRILE see BPT000
4'-BUTOXY-2-PYRROLIDINYLACETANILIDE HYDROCHLORIDE see BPT250
2-BUTOXYQUINOLINE-4-CARBOXYLIC ACID DIETHYLAMINOETHYLAM-
  IDE see DDT200
BUTOXYRHODANODIETHYL ETHER see BPL250
1-BUTOXY-2-(2-THIOCYANATOETHXY)ENTHANE see BPL250
2-BUTOXY-2'-THIOCYANODIETHYL ETHER see BPL250
β-BUTOXY-β'-THIOCYANODIETHYL ETHER see BPL250
1-BUTOXY-2-(2-THIOCYANOETHOXY)ETHANE see BPL250
BUTOXYTRIETHYLENE GLYCOL see TKL750
BUTOXYTRIGLYCOL see TKL750
1-BUTOXY-2-(VINYLOXY)ETHANE see VMU000
BUTOZ see BRF500
BUTRATE see BPF000
BUTRIPTYLINE see BPT500
BUTRIPTYLINE HYDROCHLORIDE see BPT750
BUTRIZOL see BPU000
BUTROPIPAZON see FLL000
BUTROPIPAZONE see FLL000
BUTROPIUM BROMIDE see BPI125
BUTTER of ANTIMONY see AQC500, AQD000
BUTTER CRESS see FBS100
BUTTERCUP see FBS100
BUTTERCUP YELLOW see CMK500, PLW500, ZFJ100
BUTTER DAISY see FBS100
BUTTERSAEURE (GERMAN) see BSW000
BUTTER YELLOW see AIC250, DOT300
BUTTER of ZINC see ZFA000
BUTURON see CKF750
N-BUTYLACETANILIDE see BPU500
BUTYLACETAT (GERMAN) see BPU750
BUTYL ACETATE see BPU750
1-BUTYL ACETATE see BPU750
2-BUTYL ACETATE see BPV000
n-BUTYL ACETATE see BPU750
sec-BUTYL ACETATE see BPV000
tert-BUTYL ACETATE see BPV100
BUTYLACETATEN (DUTCH) see BPU750
BUTYLACETIC ACID see HEU000
BUTYL ACETOACETATE see BPV250
N-BUTYL-N-(1-ACETOXYBUTYL)NITROSAMINE see BPV325, BRX000
BUTYL ACETOXYMETHYLNITROSAMINE see BRX500

N-BUTYL-N-(ACETOXYMETHYL)NITROSAMINE see BRX500
sec-BUTYL ACETOXYMETHYL NITROSAMINE see BPV500
N-sec-BUTYL-N-(ACETOXYMETHYL)NITROSOAMINE see BPV500
n-BUTYL-3,o-ACETYL-12-β-13-α-DIHYDROJERVINE see BPW000
n-BUTYL ACID PHOSPHATE see ADF250
BUTYL ACRYLATE see BPW100
n-BUTYL ACRYLATE see BPW100
BUTYLACRYLATE, INHIBITED (DOT) see BPW100
tert-BUTYL-1-ADAMANTANE PEROXYCARBOXYLATE see BPW250
BUTYL ADIPATE see AEO750
BUTYLAETHYLMALONSAEURE-AETHYL-DIAETHYLAMINOAETHYL-DI-
  ESTER (GERMAN) see BRJ125
2-BUTYL ALCOHOL see BPW750
n-BUTYL ALCOHOL see BPW500
BUTYL ALCOHOL (DOT) see BPW500
sec-BUTYL ALCOHOL see BPW750
tert-BUTYL ALCOHOL see BPX000
sec-BUTYL ALCOHOL ACETATE see BPV000
BUTYL ALCOHOL HYDROGEN PHOSPHITE see DEG800
n-BUTYL ALDEHYDE see BSU250
sec-BUTYL ALLYL BARBITURIC ACID see AFY500
BUTYLALYLONAL see BOR000
n-BUTYL AMIDO SULFURYL AZIDE see BPX500
n-BUTYLAMIN (GERMAN) see BPX750
n-BUTYLAMINE see BPX750
sec-BUTYLAMINE see BPY000
tert-BUTYLAMINE see BPY250
BUTYLAMINE, tertiary see BPY250
2-(BUTYLAMINO)-p-ACETOPHENETIDIDE HYDROCHLORIDE
  see BPY500
3-(tert-BUTYLAMINO)ACETYLINDOLE HYDROCHLORIDE HYDRATE
  see BPY625
3-((tert-BUTYLAMINO)ACETYL)INDOLE HYDROCHLORIDE HYDRATE
  see BPY625
2-tert-BUTYLAMINO-4-AETHYLAMINO-6-CHLOR-1,3,5-TRIAZIN (GERMAN)
  see BQB000
BUTYL-p-AMINOBENZOATE see BPZ000
p-BUTYLAMINOBENZOIC ACID-2-(DIETHYLAMINO)ETHYL ESTER
  MONOHYDROCHLORIDE see BQA000
p-(BUTYLAMINO)BENZOIC ACID-2-(DIMETHYLAMINO)ETHYL ESTER
  see BQA010
p-(BUTYLAMINO)BENZOIC ACID, 2-(DIMETHYLAMINO)ETHYL ESTER, HY-
  DROCHLORIDE see TBN000
p-BUTYLAMINOBENZOYL-2-DIMETHYLAMINOETHANOL see BQA010
p-BUTYLAMINOBENZOYL-2-DIMETHYLAMINOETHANOL HYDROCHLO-
  RIDE see TBN000
N-((BUTYLAMINO)CARBONYL)-4-METHYLBENZENESULFONAMIDE
  see BSQ000
2-(BUTYLAMINO)-2'-CHLOROACETANILIDE HYDROCHLORIDE see BQA500
2-(BUTYLAMINO)-6'-CHLORO-o-ACETOTOLUIDIDE MONOHYDROCHLOR-
  IDE see BQA750
2-tert-BUTYLAMINO-4-CHLORO-6-ETHYLAMINO-s-TRIAZINE see BQB000
2-(BUTYLAMINO)-6'-CHLORO-o-HEXANOTOLUIDIDE HYDROCHLORIDE
  see CAB500
1-(tert-BUTYLAMINO)-3-(2-CHLORO-5-METHYLPHENOXY)-2-PROPANOL HY-
  DROCHLORIDE see BQB500
2-(BUTYLAMINO)-N-(2-CHLORO-6-METHYLPHENYL)ACETAMIDE HYDRO-
  CHLORIDE see BQA750
(±)-α-tert-BUTYLAMINO-3-CHLOROPROPIOPHENONE HYDROCHLORIDE
  see WBJ500
1-(BUTYLAMINO)CYCLOHEXYLPHOSPHONIC ACID DIBUTYL ESTER
  see ALZ000
(−)-1-(tert-BUTYLAMINO)-3-(o-CYCLOPENTYLPHENOXY)-2-PROPANOL
  SULFATE see PAP230
5-BUTYLAMINO-2-(2-DIETHYLAMINOETHYL)-1H-ISOINDOLE-1,3(2H)-DIONE
  HYDROCHLORIDE see BQB825
4-BUTYLAMINO-N-(2-(DIETHYLAMINO)ETHYL)PHTHALIMIDE HYDRO-
  CHLORIDE see BQB825
2-BUTYLAMINOETHANOL see BQC000
2-sec-BUTYLAMINO-4-ETHYLAMINO-6-METHOXY-s-TRIAZINE see BQC250
2-tert-BUTYLAMINO-4-ETHYLAMINO-6-METHOXY-s-TRIAZINE
  see BQC500
2-sec-BUTYLAMINO-4-ETHYLAMINO-6-METHOXY-1,3,5-TRIAZINE see BQC250
2-tert-BUTYLAMINO-4-ETHYLAMINO-6-METHOXY-1,3,5-TRIAZINE
  see BQC500
2-tert-BUTYLAMINO-4-ETHYLAMINO-6-METHYLMERCAPTO-s-TRIAZINE
  see BQC750
2-tert-BUTYLAMINO-4-ETHYLAMINO-6-METHYLTHIO-s-TRIAZINE see BQC750
2-tert-BUTYLAMINO-1-(7-ETHYL-2-BENZOFURANYL)ETHANOL HYDRO-
  CHLORIDE see BQD000
3-(2-(tert-BUTYLAMINO)ETHYL)-6-HYDROXYBENZYL ALCOHOL SULFATE
  (2581) see BQD125
tert-BUTYL AMINO ETHYL METHACRYLATE see BQD250
2-(tert-BUTYLAMINO)ETHYL METHACRYLATE see BQD250

(5-(2-(tert-BUTYLAMINO)-1-HYDROXYETHYL)-2-HYDROXYPHENYL)UREA HYDROCHLORIDE see BQD500

6-(2-(tert-BUTYLAMINO)-1-HYDROXYETHYL)-3-HYDROXY-2-PYRIDINEMETHANOL DIHYDROCHLORIDE see POM800

4-(2-(tert-BUTYLAMINO)-1-HYDROXYETHYL)-o-PHENYLENE DI-p-TOLUATE MESILATE see BLV125

2-(tert-BUTYLAMINO)-1-(4-HYDROXY-3-HYDROXYMETHYLPHENYL) ETHANOL see BQF500

2-BUTYLAMINO-1-p-HYDROXYPHENYLETHANOL see BQF250

o-(3-tert-BUTYLAMINO-2-HYDROXYPROPOXY)BENZONITRILE HYDROCHLORIDE see BON400

5-(3-tert-BUTYLAMINO-2-HYDROXY)PROPOXY-3,4-DIHYDROCARBOSTYRIL HYDROCHLORIDE see MQT550

5-(3-tert-BUTYLAMINO-2-HYDROXY-PROPOXY)-3,4-DIHYDRO-(2(1H)-CHINOLINON-HYDROCHLORID (GERMAN) see MQT550

8-(3-tert-BUTYLAMINO-2-HYDROXY)PROPOXY-5-METHYLCOUMARIN HYDROCHLORIDE see BOM510

(±)-2-(3'-tert-BUTYLAMINO-2'-HYDROXYPROPYLTHIO)-4-(5'-CARBAMOYL-2'-THIENYL)THIAZOLE HYDROCHLORIDE see BQE000

2-(tert-BUTYLAMINO)-1-(3-INDOLYL)-1-PROPANONE HYDROCHLORIDE HYDRATE see BQG850

2-(tert-BUTYLAMINO)-1-(3-INDOLYL)-1-PROPANONE MONOHYDROCHLORIDE, MONOHYDRATE see BQG850

α-(tert-BUTYLAMINO)METHYL-2-CHLOROBENZYL ALCOHOL HYDROCHLORIDE see BQE250

α-((tert-BUTYLAMINO)METHYL)-o-CHLOROBENZYL ALCOHOL HYDROCHLORIDE see BRA250

α-(BUTYLAMINO)METHYL)-3,5-DIHYDROXYBENZYL ALCOHOL, SULFATE (2:1) see TAN250

α-((BUTYLAMINO)METHYL)-4-HYDROXYBENZENEMETHANOL see BQF250

α-((BUTYLAMINO)METHYL)-p-HYDROXYBENZYL ALCOHOL see BQF250

α-((BUTYLAMINO)METHYL)-p-HYDROXYBENZYL ALCOHOL SULFATE see BOV825

α-1-(tert-BUTYLAMINO)METHYL)-4-HYDROXY-m-XYLENE-α,α-DIOL see BQF500

α'-(tert-BUTYL AMINO)METHYL)-4-HYDROXY-m-XYLENE-α,α'-DIOL see BQF500

1-(tert-BUTYLAMINO)3-(3-METHYL-2-NITROPHENOXY)-2-PROPANOL see BQF750

1-(BUTYLAMINO)-3-((4-METHYLPHENYL)AMINO)-2-PROPANOL (9CI) see BQH800

2-(BUTYLAMINO)-2-METHYL-1-PROPANOL BENZOATE HYDROCHLORIDE see BQF825

2-(BUTYLAMINO)-2-METHYL-1-PROPANOL BENZOATE (ester) HYDROCHLORIDE see BQF825

2-(BUTYLAMINO)-N-METHYL-N-(1-(2,6-XYLYLOXY)-2-PROPYL) ACETAMIDE HYDROCHLORIDE see BQG250

2-(sec-BUTYLAMINO)-N-METHYL-N-(1-(2,4-XYLYLOXY)-2-PROPYL) ACETAMIDE HYDROCHLORIDE see BQG500

(−)-1-(tert-BUTYLAMINO)-3-((4-MORPHOLINO-1,2,5-THIADIAZOL-3-YL)OXY)-2-PROPANOL MALEATE see TGB185

2-(BUTYLAMINO)-N-(1-PHENOXY-2-PROPYL)ACETAMIDE HYDROCHLORIDE see BQG750

3-(BUTYLAMINO)-4-PHENOXY-5-SULFAMOYLBENZOIC ACID see BON325

3-(tert-BUTYLAMINO)PROPIONYLINDOLE HYDROCHLORIDE HYDRATE see BQG850

4-(BUTYLAMINO)SALICYLIC ACID 2-(DIETHYLAMINO)ETHYL ESTER HYDROCHLORIDE see BQH250

p-BUTYLAMINO SALICYLIC ACID-2-(DIETHYLAMINO)ETHYL ESTER HYDROCHLORIDE see BQH250

4-(BUTYLAMINO)-SALICYLIC ACID 2-(DIETHYLAMINO)ETHYL ESTER MONOHYDROCHLORIDE see BQH250

p-BUTYLAMINOSALICYLIC ACID-2-(DIMETHYLAMINO)ETHYL ESTER HYDROCHLORIDE see BQH500

p-BUTYLAMINOSALICYLIC ACID-1-ETHYL-4-PIPERIDYL ESTER HYDROCHLORIDE see BQH750

1-(tert-BUTYLAMINO)-3-((5,6,7,8-TETRAHYDRO-cis-6,7-DIHYDROXY-1-NAPHTHYL)OXY)-2-PROPANOL see CNR675

1-(tert-BUTYLAMINO)-3-(o-((TETRAHYDROFURFURYL)OXY)PHENOXY)-2-PROPANOL HYDROCHLORIDE see AER500

1-(BUTYLAMINO)-3-p-TOLUIDINO-2-PROPANOL see BQH800

BUTYLAMYLNITROSAMIN (GERMAN) see BRY250

N-BUTYLANILINE see BQH850

N-(n-BUTYL)ANILINE see BQH850

N-n-BUTYLANILINE (DOT) see BQH850

BUTYLATE see EID500

BUTYLATED HYDROXYANISOLE see BQI000

3-tert-BUTYLATED HYDROXYANISOLE see BQI010

BUTYLATED HYDROXYTOLUENE see BFW750

BUTYLATE-2,4,5-T see BSQ750

N-BUTYL-N-2-AZIDOETHYLNITRAMINE see BQI125

tert-BUTYL AZIDO FORMATE see BQI250

8-BUTYLBENZ(a)ANTHRACENE see BQI500

5-n-BUTYL-1,2-BENZANTHRACENE see BQI500

N-BUTYLBENZENAMINE (9CI) see BQH850

n-BUTYLBENZENE see BQI750

sec-BUTYLBENZENE see BQJ000

tert-BUTYLBENZENE see BQJ250

BUTYLBENZENEACETATE see BQJ350

α-BUTYLBENZENEMETHANOL see BQJ500

2-tert-BUTYLBENZIMIDAZOLE see BQJ750

5-BUTYL-2-BENZIMIDAZOLECARBAMIC ACID METHYL ESTER see BQK000

N-(BUTYL-5-BENZIMIDAZOLYL)-2-CARBAMATE de METHYLE (FRENCH) see BQK000

(4-BUTYL-1H-BENZIMIDAZOL-2-YL)-CARBAMIC ACID METHYL ESTER see BQK000

BUTYL BENZOATE see BQK250

n-BUTYL BENZOATE see BQK250

2-BUTYL-3-BENZOFURANYL p-((2-DIETHYLAMINO)ETHOXY)-m,m-DIIODOPHENYL KETONE see AJK750

p-tert-BUTYL BENZOIC ACID see BQK500

N-tert-BUTYL-2-BENZOTHIAZOLESULFENAMIDE see BQK750

α-BUTYLBENZYL ALCOHOL see BQJ500

1-(p-tert-BUTYLBENZYL)-4-(p-CHLORODIPHENYLMETHYL)PIPERAZINE DIHYDROCHLORIDE see BOM250

1-(p-tert-BUTYLBENZYL-4-p-CHLORO-α-PHENYLBENZYL)PIPERAZINE DIHYDROCHLORIDE see BOM250

BUTYL BENZYL PHTHALATE see BEC500

n-BUTYL BENZYL PHTHALATE see BEC500

tert-BUTYL-BICYCLOPHOSPHATE see BQK850

BUTYLBIGUANIDE see BRA625

1-BUTYLBIGUANIDE HYDROCHLORIDE see BQL000

N-BUTYLBIGUANIDE HYDROCHLORIDE see BQL000

BUTYLBIS(β-CHLOROETHYL)AMINE HYDROCHLORIDE see BIB250

N-BUTYL-BIS(2-CHLOROETHYLAMINE) HYDROCHLORIDE see BIB250

sec-BUTYLBIS(2-CHLOROETHYL)AMINE HYDROCHLORIDE see BQL500

sec-BUTYL-BIS(β-CHLOROETHYL)AMINE HYDROCHLORIDE see BQL500

tert-BUTYLBIS(2-CHLOROETHYL)AMINE HYDROCHLORIDE see BQL750

tert-BUTYLBIS(β-CHLOROETHYL)AMINE HYDROCHLORIDE see BQL750

N-BUTYL-N,N-BIS(HYDROXY ETHYL)AMINE see BQM000

BUTYL BORATE see THX750

n-BUTYL BORATE see THX750

sec-BUTYL-BROM-ALLYL BARBITURIC ACID SODIUM SALT see BOR250

1-BUTYL BROMIDE see BNR750

i-BUTYL BROMIDE see BNR750

N-BUTYL BROMIDE see BMX500

BUTYL BROMIDE (DOT) see BMX500

sec-BUTYL BROMIDE see BMX750

tert-BUTYL BROMIDE see BQM250

BUTYL BROMIDE, NORMAL (DOT) see BMX500

3-sek.BUTYL-5-BROM-6-METHYLURACIL (GERMAN) see BMM650

5-sec-BUTYL-5-(β-BROMOALLYL)BARBITURIC ACID see BOR000

N-BUTYL-1-BUTANAMINE see DDT800

N-tert-BUTYL-1,4-BUTANEDIAMINE DIHYDROCHLORIDE see BQM309

BUTYL BUTANEDIOATE see SNA500

n-BUTYL BUTANOATE see BQM500

BUTYL-BUTANOL(4)-NITROSAMIN see HJQ350

BUTYL-BUTANOL-NITROSAMINE see HJQ350

BUTYL-2-BUTOXYCYCLOPROPANE-1-CARBOXYLATE see BQM750

p-(N-BUTYL-2-(BUTYLAMINO)ACETAMIDO)BENZOIC ACID BUTYL ESTER HYDROCHLORIDE see BQN250

N-BUTYL-2-(BUTYLAMINO)-2',6'-PROPIONOXYLIDIDE HYDROCHLORIDE see BQN500

BUTYL BUTYRATE (FCC) see BQM500

n-BUTYL BUTYRATE see BQM500

n-BUTYL n-BUTYRATE see BQM500

Γ-n-BUTYL-Γ-BUTYROLACTONE see OCE000

N-n-BUTYL-N-4-(1,4-BUTYROLACTONE)NITROSAMINE see NJO200

BUTYL BUTYROLLACTATE see BQP000

BUTYL BUTYRYL LACTATE see BQP000

BUTYL CAPROATE see BRK900

BUTYLCAPTAX see BSN325

BUTYL CARBAMATE see BQP250

tert-BUTYLCARBAMIC ACID ESTER with 3-(m-HYDROXYPHENYL-1,1-DIMETHYLUREA see DUM800

1-(BUTYLCARBAMOYL)-2-BENZIMIDAZOLECARBAMIC ACID, METHYL ESTER see BAV575

1-(BUTYLCARBAMOYL)-2-BENZIMIDAZOLECARBAMIC ACID METHYL ESTER and SODIUM NITRITE (1:6) see BAV500

1-(BUTYLCARBAMOYL)-2-BENZIMIDAZOL-METHYLCARBAMAT (GERMAN) see BAV575

1-(N-BUTYLCARBAMOYL)-2-(METHOXY-CARBOXAMIDO)-BENZIMIDAZOL (GERMAN) see BAV575

N'-(BUTYLCARBAMOYL)SULFANILAMIDE see BSM000

N¹-(BUTYLCARBAMOYL)SULFANILAMIDE see BSM000

N-BUTYLCARBINOL see AOE000

dl-sec-BUTYLCARBINOL see MHS750

BUTYL CARBITOL see DJF200

BUTYL CARBITOL ACETATE see BQP500
BUTYLCARBITOL FORMAL see BHK750
BUTYL CARBITOL 6-PROPYLPIPERONYL ETHER see PIX250
BUTYL CARBITOL RHODANATE see BPL250
BUTYL CARBITOL THIOCYANATE see BPL250
BUTYL CARBITYL (6-PROPYLPIPERONYL) ETHER see PIX250
BUTYL CARBOBUTOXYMETHYL PHTHALATE see BQP750
5-BUTYL-2-(CARBOMETHOXYAMINO)BENZIMIDAZOLE see BQK000
1-((tert-BUTYLCARBONYL-4-CHLOROPHENOXY)METHYL)-1H-1,2,4-
   TRIAZOLE see CJO250
N-BUTYL-N-(2-CARBOXYETHYL)NITROSAMINE see BRX250
4-BUTYL-4-(β-CARBOXYPROPIONYL-OXYMETHYL)-1,2-DIPHENYL-3,5-
   PYRAZOLIDINEDIONE see SOX875
N-BUTYL-(3-CARBOXY PROPYL)NITROSAMINE see BQQ250
4-tert-BUTYLCATECHOL see BSK000
BUTYL CELLOSOLVE see BPJ850
BUTYL CELLOSOLVE ACETATE see BPM000
BUTYL CELLOSOLVE ACRYLATE see BPK500
BUTYL "CELLOSOLVE" ADIPATE see BHJ750
BUTYL "CELLOSOLVE" PHTHALATE see BHK000
BUTYL CHEMOSEPT see BSC000
o-(4-tert BUTYL-2-CHLOOR-FENYL)-o-METHYL-FOSFORZUUR-N-METHYL-
   AMIDE (DUTCH) see COD850
BUTYL-CHLORHYDRINETHER (CZECH) see BQT500
n-BUTYL CHLORIDE see BQQ750
BUTYL CHLORIDE (DOT) see BQQ750
sec-BUTYL CHLORIDE see CEU250
tert-BUTYL CHLORIDE see BQR000
3-tert-BUTYL-5-CHLOR-6-METHYLURACIL (GERMAN) see BQT750
α-BUTYL-p-CHLOROBENZYL ESTER of SUCCINIC ACID see CKI000
b-BUTYL-3-CHLORO-N,N-DIMETHYL-4-ETHOXYPHENETHYLAMINE
   see BQR250
β-sec-BUTYL-3-CHLORO-N,N-DIMETHYL-4-METHOXYPHENETHYLAMINE
   see BQR750
β-sec-BUTYL-5-CHLORO-N,N-DIMETHYL-2-METHOXYPHENETHYLAMINE
   see BQS000
β-sec-BUTYL-p-CHLORO-N,N-DIMETHYLPHENETHYLAMINE see BQS250
β-sec-BUTYL-5-CHLORO-2-ETHOXY-N,N-DIISOPROPYLPHENETHYLAMINE
   see BQT000
1-(β-sec-BUTYL-5-CHLORO-2-ETHOXYPHENETHYL)PIPERIDINE
   see BQT250
BUTYL (3-CHLORO-2-HYDROXYPROPYL) ETHER see BQT500
3-tert-BUTYL-5-CHLORO-6-METHYLURACIL see BQT750
tert-BUTYL CHLOROPEROXYFORMATE see BQU000
4-tert-BUTYL-2-CHLORO PHENYL METHYL METHYL PHOSPHORAMIDATE
   see COD850
4-tert.-BUTYL 2-CHLOROPHENYL METHYLPHOSPHORAMIDATE de
   METHYLE (FRENCH) see COD850
o-(4-tert-BUTYL-2-CHLOROPHENYL)-o-METHYL PHOSPHORAMIDOTHION-
   ATE see BQU500
o-(4-tert-BUTYL-2-CHLOR-PHENYL)-o-METHYL-PHOSPHORSAEURE-N-
   METHYL AMID (GERMAN) see COD850
3-tert-BUTYLCHOLANTHRENE see BQU750
tert-20-BUTYLCHOLANTHRENE see BQU750
tert-BUTYL CHROMATE see BQV000
α-BUTYLCINNAMALDEHYDE see BQV250
n-BUTYL CINNAMATE see BQV500
BUTYL CINNAMIC ALDEHYDE see BQV250
α-BUTYLCINNAMIC ALDEHYDE see BQV250
2-tert-BUTYL-p-CRESOL see BQV750
4-tert-BUTYLCYCLOHEXANOL see BQW000
p-tert-BUTYLCYCLOHEXANONE see BQW250
4-tert-BUTYLCYCLOHEXYL ACETATE see BQW500
p-tert-BUTYLCYCLOHEXYL ACETATE see BQW500
N-BUTYL CYCLOHEXYL AMINE see BQW750
5-BUTYL-1-CYCLOHEXYLBARBITURIC ACID see BQW825
N-(4-tert-BUTYL CYCLOHEXYL)-3,3-DIPHENYL PROPYLAMINE HYDRO-
   CHLORIDE see BQX000
5-BUTYL-1-CYCLOHEXYL-2,4,6(1H,3H,5H)-PYRIMIDINETRIONE
   see BQW825
5-n-BUTYL-1-CYCLOHEXYL-2,4,6-TRIOXOPERHYDROPYRIMIDINE
   see BQW825
BUTYL 2,4-D see BQZ000
BUTYL DECYL PHTHALATE see BQX250
N-tert-BUTYL-1,4-DIAMINOBUTANE DIHYDROCHLORIDE see BQM309
tert-BUTYL DIAZOACETATE see BQX750
14-n-BUTYL DIBENZ(a,h)ACRIDINE see BQY000
10-n-BUTYL-1,2,5,6-DIBENZACRIDINE (FRENCH) see BQY000
n-BUTYL-2-DIBUTYLTHIOUREA see BQY250
sec-BUTYLDICHLORARSINE see BQY300
sec-BUTYLDICHLOROARSINE see BQY300
BUTYLDICHLOROBORANE see BQY500
N-BUTYL-2,2'-DICHLORODIETHYLAMINE see DEV300
N-sec-BUTYL-2,2'-DICHLORODIETHYLAMINE, HYDROCHLORIDE see BQL500

N-tert-BUTYL-2,2'-DICHLORO-DIETHYLAMINE HYDROCHLORIDE
   see BQL750
BUTYL DICHLOROPHENOXYACETATE see BQZ000
BUTYL (2,4-DICHLOROPHENOXY)ACETATE see BQZ000
1-BUTYL-3-(3,4-DICHLOROPHENYL)-1-METHYLUREA see BRA250
N-BUTYL-N'-(3,4-DICHLOROPHENYL)-N-METHYLUREA see BRA250
N-BUTYLDIETHANOLAMINE see BQM500
2-(BUTYL(2-(DIETHYLAMINO)ETHYL)AMINO)-6-CHLORO-o-ACETOTOLUID-
   IDE HYDROCHLORIDE see BRA500
o-BUTYL DIETHYLENE GLYCOL see DJF200
n-BUTYLDIETHYLTIN IODIDE see BRA550
tert-BUTYLDIFLUOROPHOSPHINE see BRA600
BUTYL DIGLYME see DDW200
BUTYLDIGUANIDE see BRA625
1-BUTYLDIGUANIDE see BRA625
1-BUTYLDIGUANIDE HYDROCHLORIDE see BQL000
BUTYL-3,4-DIHYDRO-2,2-DIMETHYL-4-OXO-2H-PYRAN-6-CARBOXYLATE
   see BRT000
n-BUTYL-12-β-13-α-DIHYDROJERVINE-3-ACETATE see BPW000
9-BUTYL-1,9-DIHYDRO-6H-PURINE-6-THIONE see BRS500
N-BUTYL-N-(2,4-DIHYDROXYBUTYL)NITROSAMINE see BRB000
2-BUTYL-3-(3,5-DIIODO-4-(2-DIETHYLAMINOETHOXY)BENZOYL)
   BENZOFURAN see AJK750
2-N-BUTYL-3',5'-DIIODO-4'-N-DIETHYLAMINOETHOXY-3-
   BENZOYLBENZOFURAN see AJK750
4'-n-BUTYL-4-DIMETHYLAMINOAZOBENZENE see BRB450
4'-tert-BUTYL-4-DIMETHYLAMINOAZOBENZENE see BRB460
3-BUTYL-1-(2-(DIMETHYLAMINO)ETHOXY)ISOQUINOLINE HYDROCHLO-
   RIDE see DNX400
4-(1-sec-BUTYL-2-(DIMETHYLAMINO)ETHYL)PHENOL see BRB500
2-(1-sec-BUTYL-2-(DIMETHYLAMINO)ETHYL)QUINOLINE see BRB750
2-(1-sec-BUTYL-2-(DIMETHYLAMINO)ETHYL)QUINOXALINE see BRC000
2-(1-sec-BUTYL-2-(DIMETHYLAMINO)ETHYL)THIOPHENE see BRC250
5-n-BUTYL-2-DIMETHYLAMINO-4-HYDROXY-6-METHYLPYRIMIDINE
   see BRD000
6-BUTYL-5-DIMETHYLAMINO-5H-INDENO(5,6-d)-1,3-DIOXOLE HYDROCHLO-
   RIDE see BRC500
2-n-BUTYL-3-DIMETHYLAMINO-5,6-METHYLENEDIOXYINDENE HYDRO-
   CHLORIDE see BRC500
BUTYL-3-((DIMETHYLAMINO)METHYL)-4-HYDROXYBENZOATE see BRC750
5-BUTYL-2-(DIMETHYLAMINO)-6-METHYL-4-PYRIMIDINOL see BRD000
5-BUTYL-2-(DIMETHYLAMINO)-6-METHYL-4(1H)-PYRIMIDINONE see BRD000
2-(4-tert-BUTYL-2,6-DIMETHYLBENZYL)-2-IMIDAZOLINE HYDROCHLORIDE
   see OKO500
2-(4-tert-BUTYL-2,6-DIMETHYLBENZYL)-2-IMIDAZOLINE
   MONOHYDROCHLORIDE see OKO500
8-tert-BUTYL-4,6-DIMETHYLCOUMARIN see DQV000
N-(7-BUTYL-4,9-DIMETHYL-2,6-DIOXO-8-HYDROXY-1,5-DIOXONAN-3-YL)-3-
   FORMAMIDOSALICYLAMIDE see DAM000
β-sec-BUTYL-N,N-DIMETHYL-2-ETHOXY-5-FLUOROPHENETHYLAMINE
   see BRD500
β-sec-BUTYL-N,N-DIMETHYL-5-FLUORO-2-METHOXYPHENETHYLAMINE
   see BRD750
2-(4-tert-BUTYL-2,6-DIMETHYL-3-HYDROXYBENZYL)-2-IMIDAZOLINIUM
   CHLORIDE see AEX000
1-BUTYL-3,3-DIMETHYL-1-NITROSOUREA see BRE000
β-sec-BUTYL-N,N-DIMETHYLPHENETHYLAMINE see BRE250
β-sec-BUTYL-N,N-DIMETHYLPHENETHYLAMINE HYDROCHLORIDE
   see BRE255
6-tert-BUTYL-2,4-DIMETHYLPHENOL see BST000
1-tert-BUTYL-3,5-DIMETHYL-2,4,6-TRINITROBENZENE see TML750
2-sec-BUTYL-4,6-DINITROPHENOL see BRE500
o-tert-BUTYL-4,6-DINITROPHENOL see DRV200
2-sec-BUTYL-4,6-DINITROPHENOL AMMONIUM SALT see BPG250
2-sec-BUTYL-4,5-DINITROPHENOL ISOPROPYL CARBONATE see CBW000
2-sec-BUTYL-4,6-DINITROPHENOL- 2,2',2''-NITRILOTRIETHANOL SALT
   see BRE750
o-sec-BUTYL-4,6-DINITROPHENOLTRIETHANOLAMINE SALT see BRE750
2-sec-BUTYL-4,6-DINITROPHENYLACETATE see ACE500
6-sec-BUTYL-2,4-DINITROPHENYLACETATE see ACE500
2-tert-BUTYL-4,6-DINITROPHENYL ACETATE see DVJ400
2-sec-BUTYL-4,6-DINITROPHENYL-3,3-DIMETHYLACRYLATE see BGB500
2-sec-BUTYL-4,6-DINITROPHENYL ISOPROPYL CARBONATE see CBW000
2-sec-BUTYL-4,6-DINITROPHENYL-3-METHYL-2-BUTENOATE see BGB500
2-sec-BUTYL-4,6-DINITROPHENYL-3-METHYLCROTONATE see BGB500
2-sec-BUTYL-4,6-DINITROPHENYL SENECIOATE see BGB500
BUTYL DIOXITOL see DJF200
4-BUTYL-1,2-DIPHENYL-3,5-DIOXO PYRAZOLIDINE see BRF500
4-BUTYL-1,2-DIPHENYLPYRAZOLIDINE-3,5-DIONE see BRF500
4-BUTYL-1-1,2-DIPHENYL-3,5-PYRAZOLIDINEDIONE with 4-
   (DIMETHYLAMINO)-1,2-DIHYDRO-1,5-DIMETHYL-2-PHENYL-3H-
   PYRAZOL-3-ONE see IGI000
4-BUTYL-1,2-DIPHENYL-3,5-PYRAZOLIDINEDIONE SODIUM SALT
   see BOV750

BUTYL DISELENIDE see BRF550
6-BUTYLDODECAHYDRO-7,14-METHANO-2H,6H-DIPYRIDO:(1,2-α1',2'-e)(1,5)DIAZOCINE see BSL450
BUTYLE (ACETATE de) (FRENCH) see BPU750
BUTYLENE see BOW250
α-BUTYLENE see BOW250
Γ-BUTYLENE see IIC000
1,2-BUTYLENE CARBONATE see BOT200
1,3-BUTYLENE DIACRYLATE see BRG500
BUTYLENEDIAMINE see BOS000
1,4-BUTYLENEDIAMINE see BOS000
2-BUTYLENE DICHLORIDE see BRG000
β-BUTYLENE GLYCOL see BOS500
1,2-BUTYLENE GLYCOL see BOS250
1,3-BUTYLENE GLYCOL (FCC) see BOS500
1,4-BUTYLENE GLYCOL see BOS750
2,3-BUTYLENE GLYCOL see BOT000
BUTYLENE GLYCOL BIS(MERCAPTOACETATE) see BKM000
1,3-BUTYLENE GLYCOL DIACRYLATE see BRG500
BUTYLENE HYDRATE see BPW750
BUTYLENE OXIDE see TCR750
1,2-BUTYLENE OXIDE see BOX750
1,4-BUTYLENE SULFONE see BOU250
1,2-BUTYLENIMINE see EGN500
BUTYLENIN see IIU000
BUTYL-9,10-EPOXYSTEARATE see BRH250
2,4,5-T-N-BUTYL ESTER see BSQ750
2,4-d,n-BUTYL ESTER mixed with 2,4,5-T,n-BUTYL ESTER (1:1) see AEX750
2,4,5-T,n-BUTYL ESTER mixed with 2,4-d,n-BUTYL ESTER see AEX750
BUTYL ESTER 2,4-D see BQZ000
n-BUTYL ESTER of 3,4-DIHYDRO-2,2-DIMETHYL-4-OXO-2H-PYRAN-6-CARBOXYLIC ACID see BRT000
N-BUTYLESTER KYSELINI-2,4,5-TRICHLORFENOXYOCTOVE (CZECH) see BSQ750
BUTYLESTER KYSELINY MRAVENCI see BRK000
BUTYL ETHANOATE see BPU750
n-BUTYL ETHER see BRH750
BUTYL ETHER (DOT) see BRH750
BUTYL ETHYL ACETALDEHYDE see BRI000
BUTYL ETHYL ACETIC ACID see BRI250
n-BUTYL-2-(ETHYLAMINO)-2',6'-ACETOXYLIDIDE HYDROCHLORIDE see BRI500
5-n-BUTYL-2-ETHYLAMINO-4-HYDROXY-6-METHYL-PYRIMIDINE see BRI750
5-BUTYL-2-(ETHYLAMINO)-6-METHYL-4(1H)-PYRIMIDINONE see BRI750
5-BUTYL-2-ETHYLAMINO-6-METHYLPYRIMIDIN-4-YL DIMETHYLSULPHAMATE see BRJ000
5-BUTYL-5-ETHYLBARBITURIC ACID see BPF500
5-sec-BUTYL-5-ETHYLBARBITURIC ACID see BPF000
5-sec-BUTYL-5-ETHYLBARBITURIC ACID SODIUM SALT see BPF250
BUTYL ETHYLENE see HFB000
o-BUTYL ETHYLENE GLYCOL see BPJ850
n-BUTYL ETHYL KETONE see EHA600
BUTYLETHYLMALONIC ACID-2-(DIETHYLAMINO)ETHYL ETHYL ESTER see BRJ125
5-sec-BUTYL-5-ETHYLMALONYL UREA see BPF000
5-sec-BUTYL-5-ETHYL-1-METHYLBARBITURIC ACID see BRJ250
BUTYLETHYL-PROPANEDIOIC ACID-2-(DIETHYLAMINO)ETHYL ETHYL ESTER (9CI) see BRJ125
2-BUTYL-2-ETHYL-1,3-PROPANEDIOL see BKH625
5-BUTYL-5-ETHYL-2,4,6(1H,3H,5H)-PYRIMIDINETRIONE (9CI) see BPF500
5-BUTYL-5-ETHYL-2,4,6(1H,3H,5H)-PYRIMIDINETRIONE MONOSODIUM SALT (9CI) see BPF750
BUTYLETHYLTHIOCARBAMIC ACID S-PROPYL ESTER see PNF500
2-sec.-BUTYLFENOL (CZECH) see BSE000
p-tert-BUTYLFENOL (CZECH) see BSE500
BUTYL FLUFENAMATE see BRJ325
BUTYL FORMAL see VAG000
tert-BUTYL FORMAMIDE see BRJ750
n-BUTYL FORMATE see BRK000
BUTYL FORMATE (DOT) see BRK000
N-n-BUTYL-N-FORMYLHYDRAZINE see BRK100
n-BUTYL-N'-(2-FUROYL) see BRK250
1-BUTYL-3-(2-FUROYL)UREA see BRK250
n-BUTYL GLYCIDYL ETHER see BRK750
BUTYL GLYCOL see BPJ850
BUTYLGLYCOL (FRENCH, GERMAN) see BPJ850
BUTYL GLYCOL PHTHALATE see BHK000
4-tert-BUTYLHEXAHYDROPHENYL ACETATE see BQW500
BUTYL HEXANOATE see BRK900
n-BUTYL HEXANOATE see BRK900
BUTYLHYDRAZINE ETHANEDIOATE see BRL750
n-BUTYLHYDRAZINE HYDROCHLORIDE see BRL500
BUTYLHYDRAZINE OXALATE see BRL750
O,O-tert-BUTYL HYDROGEN MONOPEROXY MALEATE see BRM000

O-O-tert-BUTYLHYDROGEN MONOPEROXY MALEATE see PCN250
BUTYL HYDROGEN PHTHALATE see MRF525
terc. BUTYLHYDROPEROXID (CZECH) see BRM250
tert-BUTYLHYDROPEROXIDE see BRM250
N-BUTYL-N-(1-HYDROPEROXYBUTYL)NITROSAMINE see HIE600
tert-BUTYLHYDROQUINONE see BRM500
BUTYL HYDROXIDE see BPW500
tert-BUTYL HYDROXIDE see BPX000
6-BUTYL-4-HYDROXYAMINOQUINOLINE-1-OXIDE see BRM750
BUTYLHYDROXYANISOLE see BQI000
tert-BUTYLHYDROXYANISOLE see BQI000
tert-BUTYL-4-HYDROXYANISOLE see BQI000
3-tert-BUTYL-4-HYDROXYANISOLE see BRN000
2(3)-tert-BUTYL-4-HYDROXYANISOLE see BQI000
BUTYL-o-HYDROXYBENZOATE see BSL250
BUTYL-p-HYDROXYBENZOATE see BSC000
BUTYL p-HYDROXYBENZOATE see DTC800
n-BUTYL-o-HYDROXYBENZOATE see BSL250
α-n-BUTYL-β-HYDROXY-Δ^α,β-BUTENOLID (GERMAN) see BRO250
n-BUTYL-(4-HYDROXYBUTYL)NITROSAMINE see HJQ350
n-BUTYL-N-(2-HYDROXYBUTYL)NITROSAMINE see BRN250
n-BUTYL-N-(3-HYDROXYBUTYL)NITROSAMINE see BRN500
N-BUTYL-N-(4-HYDROXYBUTYL)NITROSAMINE see HJQ350
N-(7-BUTYL-8-HYDROXY-4,9-DIMETHYL-2,6-DIOXO-1,5-DIOXONAN-3-YL)-3-FORMAMIDOSALICYLAMIDE see DAM000
BUTYL(2-HYDROXYETHYL)NITROSOAMINE see BRO000
3-BUTYL-4-HYDROXY-2(5H)FURANONE see BRO250
BUTYLHYDROXYISOPROPYLARSINE OXIDE see BQ800
n-BUTYL-N-(2-HYDROXYL-3-CARBOXYPROPYL)NITROSAMINE see BRO500
4-BUTYL-4-HYDROXYMETHYL-1,2-DIPHENYL-3,5-PYRAZOLIDINEDIONE HYDROGEN SUCCINATE see SOX875
2-(tert-BUTYL)-2-(HYDROXYMETHYL)-1,3-PROPANEDIOL, CYCLIC PHOSPHITE (1581) see BRO750
BUTYLHYDROXYOXOSTANNANE see BSL500
4-BUTYL-2-(4-HYDROXYPHENYL)-1-PHENYL-3,5-DIOXOPYRAZOLIDINE see HNI500
4-BUTYL-1-(4-HYDROXYPHENYL)-2-PHENYL-3,5-PYRAZOLIDINEDIONE see HNI500
4-BUTYL-1-(p-HYDROXYPHENYL)-2-PHENYL-3,5-PYRAZOLIDINEDIONE see HNI500
4-BUTYL-2-(p-HYDROXYPHENYL)-1-PHENYL-3,5-PYRAZOLIDINEDIONE see HNI500
α-BUTYL-omega-HYDROXYPOLY(OXY(METHYL-1,2-ETHANEDIYL)) see BRP250
BUTYL α-HYDROXYPROPIONATE see BRR600
BUTYL-(3-HYDROXYPROPYL)NITROSAMINE see BRX750
BUTYLHYDROXYTOLUENE see BFW750
2-(tert-BUTYL)-2-(HYEROXYMETHYL)-1,3-PROPANEDIOL, CYCLIC PHOSPHATE (1:1) see BQK850
BUTYLHYOSCINE see SBG500
N-BUTYLHYOSCINE BROMIDE see SBG500
N-BUTYLHYOSCINIUM BROMIDE see SBG500
tert-BUTYL HYPOCHLORITE see BRP500
4,4'-BUTYLIDENEBIS(6-tert-BUTYL-m-CRESOL) see BRP750
4,4'-BUTYLIDENEBIS(6-tert-BUTYL-3-METHYLPHENYL) see BRP750
4,4'-BUTYLIDENEBIS(3-METHYL-6-tert-BUTYLPHENOL) see BRP750
(11-β,16-α)-16,17-(BUTYLIDENEBIS(OXY))-11,21-DIHYDROXYPREGNA-1,4-DIENE-3,20-DIONE see BOM520
6,6'-BUTYLIDENEBIS(2,4-XYLENOL) see BRQ000
16-α,17-α-BUTYLIDENEDIOXY-11-β,21-DIHYDROXY-1,4-PREGNADIENE-3,20-DIONE see BOM520
BUTYLIDENE PHTHALIDE see BRQ100
3-BUTYLIDENE PHTHALIDE see BRQ100
n-BUTYLIDENE PHTHALIDE see BRQ100
6-t-BUTYL-3-(2-IMIDAZOLIN-2-YLMETHYL)-2,4-DIMETHYLPHENOL see ORA100
6-tert-BUTYL-3-(2-IMIDAZOLIN-2-YLMETHYL)-2,4-DIMETHYLPHENOL HYDROCHLORIDE see AEX000
N-BUTYLIMIDODICARBONIMIDIC DIAMIDE MONOHYDROCHLORIDE (9CI) see BQL000
N-BUTYL-2,2'-IMINODIETHANOL see BQM000
n-BUTYL IODIDE see BRQ250
sec-BUTYL IODIDE see IEH000
tert-BUTYL IODIDE see TLU000
BUTYL ISOBUTYRATE see BRQ350
n-BUTYL ISOCYANATE see BRQ500
tert-BUTYL ISOCYANIDE see BRQ750
tert-BUTYLISONITRILE see BRQ750
n-BUTYL ISOPENTANOATE see ISX000
BUTYL(ISOPROPYL)ARSINIC ACID see BRQ800
tert-BUTYL ISOPROPYL BENZENE HYDROPEROXIDE see BRR250
tert-BUTYL ISOPROPYL BENZENE HYDROPEROXIDE (DOT) see BRR250
2-sec-BUTYL-6-ISOPROPYLPHENOL see BRR500

2-((3-BUTYL-1-ISOQUINOLINYL)OXY)-N,N-DIMETHYLETHANAMINE
   MONOHYDROCHLORIDE see DNX400
1-BUTYL ISOVALERATE see ISX000
n-BUTYL ISOVALERATE see ISX000
BUTYL ISOVALERIANATE see ISX000
BUTYL KETONE see NMZ000
2-tert-BUTYL-p-KRESOL (CZECH) see BQV750
BUTYL LACTATE see BRR600
n-BUTYL LACTATE see BRR600
BUTYL LITHIUM see BRR739
tert-BUTYL LITHIUM see BRR750
BUTYLMALONIC ACID MONO(1,2-DIPHENYLHYDRAZIDE) CALCIUM SALT
   HEMIHYDRATE see CCD750
BUTYL-MALONSAEURE-MONO-(1,2-DIPHENYL-HYDRAZID)-CALCIUM-
   SEMIHYDRAT (German) see CCD750
BUTYL MERCAPTAN see BRR900
n-BUTYL MERCAPTAN see BRR900
BUTYL MERCAPTAN (DOT) see BRR900
tert-BUTYL MERCAPTAN see MOS000
p-BUTYLMERCAPTOBENZHYDRYL-β-DIMETHYLAMINOETHYLSULPHIDE
   see BRS000
2-(BUTYLMERCAPTO)ETHYL VINYL ETHER see VNA000
BUTYLMERCAPTOMETHYLPENICILLIN see BRS250
9-BUTYL-6-MERCAPTOPURINE see BRS500
n-BUTYLMERCURIC CHLORIDE see BRS750
S-(BUTYLMERCURIC)-THIOGLYCOLIC ACID, SODIUM SALT see SFJ500
n-BUTYL MESITYL OXIDE OXALATE see BRT000
n-BUTYLMESITYLOXID OXALATE see BRT000
BUTYL MESYLATE see BRT250
BUTYLMETHACRYLAAT (DUTCH) see MHU750
BUTYL-2-METHACRYLATE see MHU750
N-BUTYL METHACRYLATE see MHU750
BUTYL METHANESULFONATE see BRT250
n-BUTYL METHANESULFONATE see BRT250
BUTYL METHOXYMETHYLNITROSAMINE see BRT750
sec-BUTYL METHOXYMETHYLNITROSAMINE see BRU000
2-tert-BUTYL-4-METHOXYPHENOL see BRN000
3-tert-BUTYL-4-METHOXYPHENOL see BQI010
BUTYL-p-METHYLBENZENESULFONATE see BSP750
n-BUTYL-α-METHYLBENZYLAMINE see BRU250
BUTYL 3-METHYLBUTYRATE see ISX000
6-tert-BUTYL-3-METHYL-2,4-DINITRO ANISOLE see BRU500
2-tert-BUTYL-5-METHYL-4,6-DINITROPHENYL ACETATE see BRU750
1-BUTYL-2-METHYL-HYDRAZINE DIHYDROCHLORIDE see MHW250
p-tert-BUTYL-α-METHYLHYDROCINNAMALDEHYDE see LFT000
p-tert-BUTYL-α-METHYLHYDROCINNAMIC ALDEHYDE see LFT000
BUTYL METHYL KETONE see HEV000
n-BUTYL METHYL KETONE see HEV000
tert-BUTYL METHYL KETONE see DQU000
2-tert-BUTYL-4-METHYLPHENOL see BQV750
1-BUTYL-3-(p-METHYLPHENYLSULFONYL)UREA see BSQ000
2-sec-BUTYL-2-METHYL-1,3-PROPANEDIOL DICARBAMATE see MBW750
BUTYL-2-METHYL-2-PROPENOATE see MHU750
N-BUTYL-2-METHYL-2-PROPYL-1,3-PROPANEDIOL DICARBAMATE
   see MOV500
N-N-BUTYL-2-METHYL-2-PROPYL-1,3-PROPANEDIOL DICARBAMATE
   see MOV500
3-BUTYL-5-METHYL-TETRAHYDRO-2H-PYRAN-4-YL ACETATE see MHX000
tert-BUTYL-N-(3-METHYL-2-THIAZOLIDINYLIDENE)CARBAMATE see BRV000
2-sec-BUTYL-2-METHYLTRIMETHYLENE DICARBAMATE see MBW750
BUTYLMIN see SBG500
BUTYL MONOSULFIDE see BSM125
4-BUTYLMORPHOLINE see BRV100
N-BUTYLMORPHOLINE see BRV100
N-(n-BUTYL)MORPHOLINE see BRV100
9-BUTYL-6-MP see BRS500
BUTYL NAMATE see SGF500
BUTYL NITRATE see BRV325
n-BUTYL NITRITE see BRV500
BUTYL NITRITE (DOT) see BRV500
sec-BUTYL NITRITE see BRV750
tert-BUTYL NITRITE see BRV760
tert-BUTYLNITROACETATE see DSV289
tert-BUTYL NITROACETYLENE see BRW000
N-BUTYL-N'-NITRO-N-NITROSOGUANIDINE see NLC000
tert-BUTYL-p-NITRO PEROXY BENZOATE see BRW250
BUTYL-p-NITROPHENYL ESTER of ETHYLPHOSPHONIC ACID see BRW500
6-BUTYL-4-NITROQUINOLINE-1-OXIDE see BRW750
BUTYLNITROSAMINE see NJO000
4-(BUTYLNITROSAMINO)-1-BUTANOL see HJQ350
4-(n-BUTYLNITROSAMINO)-1-BUTANOL see HJQ350
1-(BUTYLNITROSAMINO)BUTYL ACETATE see ABT250
4-(BUTYLNITROSAMINO)BUTYL ACETATE see BRX000
2-(BUTYLNITROSAMINO)ETHANOL see BRO000

4-(N-BUTYLNITROSAMINO)-4-HYDROXYBUTYRIC ACID LACTONE
   see NJO200
N-BUTYL-N-NITROSO-β-ALANINE see BRX250
4-(BUTYLNITROSOAMINO)-1,3-BUTANEDIOL see BRB000
4-(BUTYLNITROSOAMINO)BUTANOIC ACID see BQQ250
1-(BUTYLNITROSOAMINO)-2-BUTANOL see BRN250
4-(BUTYLNITROSOAMINO)-2-BUTANOL see BRN500
1-(BUTYLNITROSOAMINO)BUTYL ACETATE see BPV325
4-(BUTYLNITROSOAMINO)-3-HYDROXYBUTYRIC ACID see BRO500
BUTYLNITROSOAMINOMETHYL ACETATE see BRX500
3-(BUTYLNITROSOAMINO)-1-PROPANOL see BRX750
1-(BUTYLNITROSOAMINO)-2-PROPANONE see BRY000
N-BUTYL-N-NITROSO AMYL AMINE see BRY250
n-BUTYL-N-NITROSO-1-BUTAMINE see BRY500
N-BUTYL-N-NITROSOBUTYRAMIDE see NJO150
N-BUTYL-N-NITROSOCARBAMIC ACID-1-NAPHTHYL ESTER see BRY750
N-BUTYL-N-NITROSO ETHYL CARBAMATE see BRZ000
BUTYLNITROSOHARNSTOFF (GERMAN) see BSA250
N-BUTYL-N-NITROSOPENTYLAMINE see BRY250
4-tert-BUTYL-1-NITROSOPIPERIDINE see BRZ200, NJO300
N-BUTYL-N-NITROSOSUCCINAMIC ACID ETHYL ESTER see EHC800
2-BUTYL-3-NITROSOTHIAZOLIDINE see BSA000
n-BUTYLNITROSOUREA see BSA250
1-BUTYL-1-NITROSOUREA see BSA250
N-n-BUTYL-N-NITROSOUREA see BSA250
1-sec-BUTYL-1-NITROSOUREA see NJO500
1-BUTYL-1-NITROSOURETHAN see BRZ000
N-BUTYL-N-NITROSOURETHAN see BRZ000
n-BUTYLNORSYMPATHOL see BQF250
BUTYLNORSYMPATOL see BOV825
BUTYL-NOR-SYMPATOL see BQF250
n-BUTYLNORSYNEPHRINE see BQF250
BUTYLOCAINE see TBN000
2-BUTYL-1-OCTANOL see BSA500
2-BUTYLOCTYL ALCOHOL see BSA500
2-BUTYLOCTYL ESTER METHACRYLIC ACID see BSA750
BUTYLOHYDROKSYANIZOL (POLISH) see BQI000
BUTYL OLEATE see BSB000
BUTYLONE see NBU000
BUTYLOWY ALKOHOL (POLISH) see BPW500
BUTYL OXITOL see BPJ850
N-BUTYL-N-(1-OXOBUTYL)NITROSAMINE see NJO150
N-BUTYL-N-(2-OXOBUTYL)NITROSAMINE see BSB500
N-BUTYL-N-(3-OXOBUTYL)NITROSAMINE see BSB750
4-tert-BUTYL-1-OXO-1-PHOSPHA-2,6,7-TRIOXABICYCLO(2.2.2)OCTANE
   see BQK850
BUTYL(2-OXOPROPYL)NITROSOAMINE see BRY000
tert-BUTYLOXYCARBONYL AZIDE see BQI250
2-(n-BUTYLOXYCARBONYLMETHYLENE)THIAZOLID-4-ONE see BPI300
α-BUTYLOXYCINCHONINIC ACID DIETHYLETHYLENEDIAMIDE
   see DDT200
N-(BUTYLOXY)METHYL-N-METHYLNITROSAMINE see BPM500
BUTYL PARABEN see BSC000
n-BUTYL PARAHYDROXYBENZOATE see BSC000
BUTYL PARASEPT see BSC000
N-BUTYL-N-PENTYLNITROSAMINE see BRY250
tert-BUTYL PERACETATE see BSC250
tert-BUTYLPERBENZOAN (CZECH) see BSC500
tert-BUTYL PERBENZOATE see BSC500
tert-BUTYL PEROXIDE see BSC750
tert-BUTYL PEROXYACETATE see BSC250
tert-BUTYL PEROXYACETATE, more than 76% in solution (DOT) see BSC250
tert-BUTYL PEROXY BENZOATE see BSC500
tert-BUTYL PEROXYBENZOATE, technical pure or in concentration of more
   than 75% (DOT) see BSC500
sec-BUTYL PEROXYDICARBONATE see BSD000
tert-BUTYL PEROXYPIVALATE see BSD250
BUTYLPHEN see BSE500
2-n-BUTYLPHENOL see BSD500
4-n-BUTYLPHENOL see BSD750
4-sec BUTYL PHENOL see BSE250
o-sec-BUTYLPHENOL see BSE000
p-sec-BUTYL PHENOL see BSE250
4-tert-BUTYLPHENOL see BSE500
p-tert-BUTYLPHENOL (MAK) see BSE500
o-BUTYLPHENOL, solid (DOT) see BSD500
o-BUTYLPHENOL, liquid (DOT) see BSD500
2-(p-tert-BUTYLPHENOXY)CYCLOHEXYL PROPARGYL SULFITE see SOP000
2-(p-tert-BUTYLPHENOXY)CYCLOHEXYL 2-PROPYNYL SULFITE see SOP000
4'-(3-(4'-tert-BUTYLPHENOXY)-2-HYDROXYPROPOXY)BENZOIC ACID
   see BSE750
BUTYLPHENOXYISOPROPYL CHLOROETHYL SULFITE see SOP500
2-(p-BUTYLPHENOXY)ISOPROPYL 2-CHLOROETHYL SULFITE see SOP500
2-(4-tert-BUTYLPHENOXY)ISOPROPYL-2-CHLOROETHYL SULFITE see SOP500

2-(p-tert-BUTYLPHENOXY)ISOPROPYL 2'-CHLOROETHYL SULPHITE see SOP500
2-(p-tert-BUTYLPHENOXY)-1-METHYLETHYL 2-CHLOROETHYL ESTER of SULPHUROUS ACID see SOP500
2-(p-BUTYLPHENOXY)-1-METHYLETHYL 2-CHLOROETHYL SULFITE see SOP500
2-(p-tert-BUTYLPHENOXY)-1-METHYLETHYL-2-CHLOROETHYL SULFITE ESTER see SOP500
2-(p-tert-BUTYLPHENOXY)-1-METHYLETHYL 2'-CHLOROETHYL SULPHITE see SOP500
2-(p-tert-BUTYLPHENOXY)-1-METHYLETHYL SULPHITE of 2-CHLORO- ETHANOL see SOP500
1-(o-SEC-BUTYL PHENOXY)-2-PROPANOL see PNK500
1-(p-tert-BUTYLPHENOXY)-2-PROPANOL-2-CHLOROETHYL SULFITE see SOP500
BUTYL PHENYL ACETATE see BBA000, BQJ350
n-BUTYL PHENYLACETATE see BQJ350
α-n-BUTYL-β-PHENYLACROLEIN see BQV250
n-BUTYL PHENYLACRYLATE see BQV500
p-((p-BUTYLPHENYL)AZO)-N,N-DIMETHYLANILINE see BRB450
p-((p-tert-BUTYLPHENYL)AZO)-N,N-DIMETHYLANILINE see BRB460
o-sec-BUTYLPHENYL CARBAMATE see BSF250
4-BUTYL-1-PHENYL-3,5-DIOXOPYRAZOLIDINE see MQY400
BUTYL PHENYL ETHER see BSF750
S-p-tert-BUTYLPHENYL-o-ETHYL ETHYLPHOSPHONODITHIOATE see BSG000
3-(o-BUTYLPHENYL)-5-(m-METHOXYPHENYL)-s-TRIAZOLE see BSG100
o-sec-BUTYLPHENYL METHYLCARBAMATE see MOV000
2-sec-BUTYLPHENYL-N-METHYLCARBAMATE see MOV000
3-sec-BUTYLPHENYL-N-METHYLCARBAMATE see BSG250
m-sec-BUTYLPHENYL-N-METHYLCARBAMATE see BSG250
β-(4-tert-BUTYLPHENYL)-α-METHYLPROPIONALDEHYDE see LFT000
2-tert-BUTYL-3-PHENYL OXAZIRANE see BSH000
3-(o-BUTYLPHENYL)-5-PHENYL-s-TRIAZOLE see BSH075
α-sec-BUTYL-α-PHENYL-1-PIPERIDINEBUTYRONITRILE HYDROCHLORIDE see EQZ000
4-BUTYL-1-PHENYL-3,5-PYRAZOLIDINEDIONE see MQY400
BUTYL PHOSPHITE see MRF500
BUTYL PHOSPHORIC ACID see ADF250
BUTYL PHOSPHOROTRITHIOATE see BSH250
n-BUTYL PHTHALATE (DOT) see DEH200
BUTYL PHTHALATE BUTYL GLYCOLATE see BQP750
BUTYLPHTHALIDE see BSH500
3-BUTYLPHTHALIDE see BSH500
3-n-BUTYLPHTHALIDE see BSH500
BUTYL PHTHALYL BUTYL GLYCOLATE see BQP750
5-BUTYL PICOLINIC ACID see BSI000
5-BUTYLPICOLINIC ACID CALCIUM SALT HYDRATE see FQR100
1-BUTYL-2',6'-PIPECOLOXYLIDIDE see BSI250
1-BUTYL-2',6'-PIPECOLOXYLIDIDE (±) see BOO000
l-(-)-1-BUTYL-2',6'-PIPECOLOXYLIDIDE see BOO500
d-(+)-1-BUTYL-2',6'-PIPECOLOXYLIDIDE see BOO250
1-BUTYL-2',6'-PIPECOLOXYLIDIDE HYDROCHLORIDE (+) see BON750
(±)-1-BUTYL-2',6'-PIPECOLOXYLIDIDE MONOHYDROCHLORIDE, MONOHYDRATE see BOO000
p-(N-BUTYL-2-(PIPERIDINO)ACETAMIDO)BENZOIC ACID BUTYL ESTER HYDROCHLORIDE see BSI750
BUTYL POTASSIUM XANTHATE see PKY850
BUTYLPROPANEDIOIC ACID MONO(1,2-DIPHENYLHYDRAZIDE) CALCIUM SALT HEMIHYDRATE see CCD750
BUTYL PROPANOATE see BSJ500
BUTYL-2-PROPENOATE see BPW100
BUTYL PROPIONATE see BSJ500
n-BUTYL PROPIONATE see BSJ500
9-BUTYL-9H-PURINE-6-THIOL see BRS500
5-BUTYL-2-PYRIDINECARBOXYLIC ACID see BSI000
5-BUTYL-2-PYRIDINECARBOXYLIC ACID CALCIUM SALT HYDRATE see FQR100
BUTYLPYRIN see BRF500
4-tert-BUTYLPYROCATECHOL see BSK000
p-BUTYLPYROCATECHOL see BSK000
4-tert-BUTYLPYROKATECHIN (CZECH) see BSK000
n-BUTYLPYRROLIDINE see BSK250
N-BUTYL-α-PYRROLIDINE-CARBOXY-MESIDIDE HYDROCHLORIDE see PQB750
n-BUTYL RHODANATE see BSN500
BUTYL SALICYLATE see BSL250
n-BUTYL SALICYLATE see BSL250
BUTYLSCOPOLAMINE BROMIDE see SBG500
N-BUTYLSCOPOLAMINE BROMIDE see SBG500
n-BUTYLSCOPOLAMINE TANNATE see BSL325
N-BUTYLSCOPOLAMINIUM BROMIDE see SBG500
BUTYLSCOPOLAMMONIUM BROMIDE see SBG500
N-BUTYLSCOPOLAMMONIUM BROMIDE see SBG500
N-BUTYLSCOPOLAMMONIUM BROMIDE combined with SODIUM SULPYRINE (1:25) see BOO650

17-BUTYLSPARTEIN see BSL450
BUTYL STANNOIC ACID see BSL500
n-BUTYL-k-STROPHANTHIDIN see BSL750
N-BUTYLSULFANILYLUREA see BSM000
1-BUTYL-3-SULFANILYL UREA see BSM000
BUTYL SULFIDE see BSM125
n-BUTYL-SULFIDE see BSM125
1-BUTYLSULFONIMIDOCYCLOHEXAMETHYLENE see BSM250
BUTYLSYMPATHOL see BQF250
BUTYL-2,4,5-T see BSQ750
BUTYL TEGOSEPT see BSC000
1-BUTYL THEOBROMINE see BSM825
N-(5-tert-BUTYL-1,3,4-THIADIAZOL-2-YL)BENZENESULFONAMIDE see GFM200
1-(5-tert-BUTYL-1,3,4-THIADIAZOL-2-YL)-3-DIMETHYLHARNSTOFF (GERMAN) see BSN000
1-(5-(tert-BUTYL)-1,3,4-THIADIAZOL-2-YL)-1,3-DIMETHYLUREA see BSN000
2-BUTYLTHIOBENZOTHIAZOLE see BSN325
BUTYLTHIOBUTANE see BSM125
n-BUTYL THIOCYANATE see BSN500
5-(1-(BUTYLTHIO)ETHYL)-5-ETHYLBARBITURIC ACID SODIUM SALT see SFJ875
2-(BUTYLTHIO)ETHYL VINYL ETHER see VNA000
S-((tert-BUTYLTHIO)METHYL)-O,O-DIETHYLPHOSPHORODITHIOATE see BSO000
n-BUTYLTHIOMETHYLPENICILLIN see BRS250
2-((p-(BUTYLTHIO)-α-PHENYLBENZYL)THIO)-N,N-DIMETHYLETHYLAMINE see BRS000
(BUTYLTHIO)TRIOCTYLSTANNANE see BSO200
(BUTYLTHIO)TRIPROPYLSTANNANE see TMY850
n-BUTYL THIOUREA see BSO500
n-BUTYLTIN TRICHLORIDE see BSO750
BUTYLTIN TRI(DODECANOATE) see BSO750
BUTYLTIN TRILAURATE see BSO750
n-BUTYLTIN TRIS(DIBUTYLDITHIOCARBAMATE) see BSP000
BUTYL TITANATE see BSP250
p-tert-BUTYLTOLUENE see BSP500
BUTYL-p-TOLUENESULFONATE see BSP750
n-BUTYL-p-TOLUENESULFONATE see BSP750
n-BUTYL-N'-p-TOLUENESULFONYLUREA see BSQ000
1-BUTYL-3-(p-TOLYL SULFONYL)UREA see BSQ000
1-BUTYL-3-(p-TOLYLSULFONYL)UREA, SODIUM SALT see BSQ250
BUTYL TOSYLATE see BSP750
1-BUTYL-3-TOSYLUREA see BSQ000
N-n-BUTYL-N'-TOSYLUREA see BSQ000
4-BUTYL-s-TRIAZOLE see BPU000
4-N-BUTYL-4H-1,2,4-TRIAZOLE see BPU000
BUTYLTRICHLOROGERMANE see BSQ500
BUTYLTRICHLOROGERMANE see BSQ500
BUTYL-2,4,5-TRICHLOROPHENOXYACETATE see BSQ750
N-BUTYL (2,4,5-TRICHLOROPHENOXY)ACETATE see BSQ750
BUTYL TRICHLORO SILANE see BSR000
BUTYL TRICHLORO SILANE see BSR250
3-tert-BUTYLTRICYCLOQUINAZOLINE see BSR500
BUTYL-2-((3-(TRIFLUOROMETHYL)PHENYL)AMINO)BENZOATE see BRJ325
BUTYL-o-((m-(TRIFLUOROMETHYL)PHENYL)AMINO)BENZOATE see BRJ325
BUTYLTRI(LAUROYLOXY)STANNANE see BSO750
1-BUTYL-N-(2,4,6-TRIMETHYLPHENYL)-2-PYRROLIDINECARBOXAMIDE MONOHYDROCHLORIDE see PQB750
N-tert-BUTYL-N-TRIMETHYLSILYLAMINOBORANE see BSR825
5-tert-BUTYL-2,4,6-TRINITRO-m-XYLENE see TML750
4-(tert-BUTYL)-2,6,7-TRIOXA-1-PHOSPHABICYCLO(2.2.2)OCTANE see BRO750
4-(tert-BUTYL)-2,6,7-TRIOXA-1-PHOSPHABICYCLO(2.2.2)OCTAN-1-ONE see BQK850
BUTYLTRIS(DIBUTYLDITHIOCARBAMATO)STANNANE see BSP000
BUTYLTRIS(2-ETHYLHEXYLOXYCARBONYLMETHYLTHIO)STANNANE see BSS000
BUTYLTRIS(ISOOCTYLOXYCARBONYLMETHYLTHIO)STANNANE see BSS000
BUTYL 10-UNDECENOATE see BSS100
BUTYL UNDECYLENATE see BSS100
N-BUTYLUREA see BSS250
1-BUTYLUREA and SODIUM NITRITE (2581) see BSS500
1-BUTYLURETHAN see EHA100
BUTYLURETHANE see EHA100
1-BUTYLURETHANE see EHA100
N-BUTYLURETHANE see EHA100
BUTYL VINYL ETHER see VMZ000
BUTYL VINYL ETHER (inhibited) see VMZ000
BUTYL-XANTHIC ACID POTASSIUM SALT see PKY850
sec-BUTYLXANTHIC ACID SODIUM SALT see DXM000
5-tert-BUTYL-m-XYLENE see DQU800
6-tert-BUTYL-2,4-XYLENOL see BST000
BUTYL ZIMATE see BIX000
BUTYL ZIRAM see BIX000
BUTYN see BOO750

1-BUTYNE see EFS500
2-BUTYNE see COC500
2-BUTYNEDIAMIDE see ACJ250
2-BUTYNEDINITRILE see DGS000
2-BUTYNE-1,4-DIOL see BST500
1,4-BUTYNEDIOL (DOT) see BST500
2-BUTYNE-1-THIOL see BST750
BUTYNOIC ACID, 3-PHENYL-2-PROPENYL ESTER see CMQ800
BUTYN-1-OL-3-ESTER of m-CHLOROPHENYLCARBAMIC ACID see CEX250
BUTYNORATE see DDV600
BUTYN SULFATE see BOP000
3-BUTYNYL-m-CHLOROCARBANILATE see CEX250
2-BUTYNYL-4-CHLORO-m-CHLOROCARBANILATE see CEW500
1-BUTYN-3-YL-m-CHLOROPHENYLCARBAMATE see CEX250
BUTYNYL-3N-3-CHLOROPHENYLCARBAMATE mixed with 3-CYCLOOCTYL-
   1,1-DIMETHYL UREA see AFM375
1,1'-(2-BUTYNYLENEDIOXY)BIS(3-CHLORO-2-PROPANOL) see BST900
1,1'-(2-BUTYNYLENE)DIPYRROLIDINE see DWX600
17-α-(1-BUTYNYL)-17-β-HYDROXYESTR-4-EN-3-ONE see EKF550
3-BUTYN-1-YL-p-TOLUENE SULFONATE see BSU000
2-BUTYOXY-N-(2-(DIETHYLAMINO)ETHYL)-4-QUINOLINECARBOXAMIDE
   see DDT200
BUTYRAC see DGA000, EAK500
BUTYRAC ESTER see DGA000
BUTYRAL see BSU250
BUTYRALDEHYD (GERMAN) see BSU250
n-BUTYRALDEHYDE see BSU250
BUTYRALDEHYDE (CZECH) see BSU250
m-BUTYRALDEHYDE OXIME see BSU500
N-BUTYRALDOXIME see BSU500
BUTYRALDOXIME (DOT) see BSU500
3-BUTYRAMIDO-α-ETHYL-2,4,6-TRIIODOCINNAMIC ACID SODIUM SALT
   see EQC000
3-BUTYRAMIDO-α-ETHYL-2,4,6-TRIIODOHYDROCINNAMIC ACID SODIUM
   SALT see SKO500
5'-BUTYRAMIDO-2'-(2-HYDROXY-3-
   ISOPROPYLAMINOPROPOXY)ACETOPHENONE see AAE100
2-(3-BUTYRAMIDO-2,4,6-TRIIODOPHENYLMETHYLENE)BUTYRIC ACID SO-
   DIUM SALT see EQC000
2-(3-BUTYRAMIDO-2,4,6-TRIIODOPHENYL)PROPIONIC ACID see BSV250
n-BUTYRANILIDE see BSV500
BUTYRANILIDE, 4'-ETHOXY-3-HYDROXY- see HJS850
BUTYRATE SODIUM see SFN600
(BUTYRATO)PHENYLMERCURY see BSV750
BUTYRHODANID (GERMAN) see BSN500
n-BUTYRIC ACID see BSW000
BUTYRIC ACID, 4-ACETAMIDO- see AAJ350
BUTYRIC ACID, 2-AMINO-4-(METHYLSULFINYL)- see ALF600
BUTYRIC ACID, CINNAMYL ESTER see CMQ800
BUTYRIC ACID-3,7-DIMETHYL-6-OCTENYL ESTER see DTF800
BUTYRIC ACID, α-α-DIMETHYLPHENETHYL ESTER see BEL850
BUTYRIC ACID, 4,4'-DISELENOBIS(2-AMINO- see SBU710
BUTYRIC ACID ESTER with BUTYL LACTATE see BQP000
BUTYRIC ACID-γ-FLUORO-β-HDYROXY-THIOL-METHYL ESTER see FJG000
BUTYRIC ACID, HEXYL ESTER see HFM700
BUTYRIC ACID ISOBUTYL ESTER see BSW500
BUTYRIC ACID LACTONE see BOV000
BUTYRIC ACID NITRILE see BSX250
BUTYRIC ACID TRIESTER with GLYCERIN see TIG750
BUTYRIC ACID, VINYL ESTER see VNF000
BUTYRIC ALDEHYDE see BSU250
BUTYRIC ETHER see EHE000
BUTYRIC or NORMAL PRIMARY BUTYL ALCOHOL see BPW500
BUTYRINASE see GGA800
17-BUTYRLOXY-11-β-HYDROXY-21-PROPIONYLOXY-4-PREGNENE-3,20-
   DIONE see HHQ850
α-BUTYROLACTONE see BOV000
β-BUTYROLACTONE see BSX000
γ-BUTYROLACTONE (FCC) see BOV000
BUTYRON see CKF750
BUTYRONE (DOT) see DWT600
BUTYRONITRILE see BSX250
BUTYRONITRILE (DOT) see BSX250
BUTYRONITRILE, 4-(DIETHOXYMETHYLSILYL)- see COR500
BUTYRONITRILE, 4-(TRIETHOXYSILYL)- see COS800
BUTYROPHENONE, 3-(p-CHLOROPHENYL)-2-PHENYL-4'-(2-(1-
   PYRROLIDINYL)ETHOXY)-, erythro- see CKI600
BUTYROSIN A see BSX325
N-(1-BUTYROXYMETHYL)METHYLNITROSAMINE see BSX500
N-(1-BUTYROXYMETHYL)-N-NITROSOMETHYLAMINE see BSX500
12-o-BUTYROYL-PHORBOLDODECANOATE see BSX750
3-(3-BUTYRYLAMINO-2,4,6-TRIIODOPHENYL)-2-ETHYLACRYLIC ACID SO-
   DIUM SALT see EQC000
1-BUTYRYLAZIRIDINE see BSY000

1-n-BUTYRYLAZIRIDINE see BSY000
BUTYRYL CHLORIDE see BSY250
1-BUTYRYL-4-CINNAMYLPIPERAZINE HYDROCHLORIDE see BTA000
1-N-BUTYRYL-4-CINNAMYL PIPERAZINE HYDROCHLORIDE see BTA000
2-BUTYRYL-β-(N,N-DIISOPROPYL)PHENOXYETHYLAMINE HYDROCHLO-
   RIDE see DNN000
2-BUTYRYL-10-(3-DIMETHYLAMINOPROPYL)PHENOTHIAZINE MALEATE
   see BTA125
BUTYRYLETHYLENEIMINE see BSY000
BUTYRYLETHYLENIMINE see BSY000
BUTYRYL LACTONE see BOV000
BUTYRYL NITRATE see BSY750
BUTYRYLPERAZINE DIMALEATE see BSZ000
1-BUTYRYL-4-(PHENYLALLYL)PIPERAZINE HYDROCHLORIDE see BTA000
BUTYRYLPROMAZINE MALEATE see BTA125
BUTYRYL TRIGLYCERIDE see TIG750
BUVETZONE see BRF500
BUX see BTA250
BUX-TEN see BTA250
2-n-BUYTLAMINOETHANOL see BQC000
BUZEPIDE METHIODIDE see BTA325
BUZON see BRF500
BUZULFAN see BOT250
BVU see BNP750
B-W see SJU000
BW-197U see MQR100
BW 47-83 see EAN600
BW 5071 see AMH250
BW 56-72 see TKZ000
BW 33-T-57 see MKW250
BW 50-197 see MQR100
BW 56-158 see ZVJ000
B.W. 57-233 see SBE500
BW 57-322 see ASB250
BW 57-323 see AKY250
BW 58-271 see RLZ000
BW 58-283 see DBX875
BW 57-323H see AKY250
BW 58-283b see DBX875
B.W. 356-C-61 see KFA100
BW 467-C-60 see BFW250
BW 248U see AEC700
BW 283U see DBX875
BW-21-Z see AHJ750
B-2847-Y see THD300
BYKOMYCIN see NCD550
BYLADOCE see VSZ000
2,2'-BYPYRIDIN see BGO500
BZ see BBU800
BZ 55 see BSQ000
BZCF see BEF500
6-Bz-1-DIBROMBENZANTHRON (CZECH) see DDK000
BZF-60 see BDS000
B-3-Zh see CMS205
BZI see BCB750
BZL see BTA500
BZQ see BCL250
BZT see BEN000, CBF825

C 6 see HEA000
C-10 see BPH750
C 45 see CGG500
C-56 see HCE500
C 78 see BQE250
C 172 see PAQ060
C-272 see BLG500
C 283 see LEF300
C 283 see NFW500
C-299 see OHI875
C-410 see NFW200
4-C-32 see TGA525
C-492 see NFW460
C-516 see NFW100
C-541 see NFW430
C 570 see FAB400
C 609 see NFW450
C 661 see MNX260
C-666 see CAB125
C-684 see NFW470
C-702 see NFW400, NFW425
C 709 see DGQ875
C-776 see DTP600
C-829 see NFW435
C-835 see NFW350

C-847 see CEW500
C-854 see CJT750
C 1,006 see CJT750
C 1120 see DLS800
C-1228 see OIU499
C 1686 see MDK000
C 1739 see DOK600
C 1863 see DIN000
C 1983 see CJQ000
C 2039 see DII800
C 2046 see DIB200
C 2047 see DIK200
C 2048 see MDK750
C 2052 see DIE600
C 2053 see DIE400
C 2054 see EOG500
C 2057 see DIJ000
C 2059 see DUK800
C 2060 see DIM000
C 2061 see MDK250
C 2085 see MRT250
C 2094 see MNX250
C 2095 see CJQ750
C 2096 see CJQ500
C 2097 see DIK400
C 2098 see DIB800
C 2102 see EOU500
C 2103 see END500
C 2126 see EOG000
C 2127 see EQI500
C 2136 see MDK500
C 2137 see EOF500
C 2138 see EPV000
C 2140 see MQO500
C 2141 see MQO750
C 2142 see EQI000
C 2242 see CIS250
C 2446 see AHO750
C 3037 see CFM250
C 3039 see DDM600
C 3049 see CFU250
C 3053 see DEX600
C 3054 see DHO800
C 3057 see DFF200
C 3058 see CKT750
C 3059 see CHS750
C 3061 see CFM000
C 3062 see PIN000
C 3063 see CGP380
C 3065 see DHK800
C 3067 see CIK500
C 3068 see CFO000
C 3069 see CIJ250
C 3070 see CFL500
C 3071 see CGN250
C 3072 see CFL000
C 3073 see CLC125
C 3074 see CFW250
C 3078 see CFW000, CLC100
C 3080 see DHK200
C 3085 see PIM750
C 3087 see PPU750
C 3089 see DHL400
C 3094 see DHK000
C 3095 see DHJ800
C 3101 see CFM500
C 3102 see DHL200
C 3103 see DDU200
C 3104 see AJE250
C 3115 see CFH000
C 3117 see BDY250
C 3120 see CFI000
C 3121 see BPJ000
C 3124 see AFW250
C 3125 see BPR000
C-3126 see PAM785
C 3127 see DIP800
C 3130 see BPT250
C 3133 see CGE750
C 3134 see MQO250
C 3135 see DHN800
C 3136 see BEE750
C 3137 see DHT000
C 3138 see CKO000

C 3139 see CIU250
C 3140 see CGJ000
C 3141 see DHX600
C 3144 see DII600
C 3145 see DIJ200
C 3150 see DHS600
C 3152 see CFM750
C 3155 see DWC000
C 3156 see CHR750
C 3158 see CIE250
C 3160 see BQN500
C 3162 see CIE500
C 3164 see BRI500
C 3167 see CID250
C 3172 see TNK250
C 3173 see CFP000
C 3181 see BSI750
C 3182 see CLL250
C 3183 see BIP500
C 3184 see CLL500
C 3186 see CLL750
C 3187 see BPI750
C 3189 see CKU000
C 3191 see CFL750
C 3192 see BQN250
C 3193 see CLO500
C 3199 see CLB250
C 3201 see CFN000
C 3205 see CGT000
C 3206 see CAB250
C 3207 see CAB500
C 3208 see CAB750
C 3209 see DHO000
C 3211 see CHK125
C 3213 see CLM500
C 3214 see CLM000
C 3215 see DPG400
C 3218 see BIP750
C 3219 see XWS000
C 3221 see DHV600
C 3222 see DIA000
C 3223 see DHS400
C 3229 see CLM750
C 3230 see DIA800
C 3234 see DIB000
C 3235 see DII400
C 3246 see CLM250
C 3247 see CLL000
C 3249 see CFS500
C 3253 see CFQ500
C 4200 see AMN000
C 4201 see AMM750
C 4207 see BQH500
C 4208 see BQH250
C 4211 see BQH750
C 4910 see DEY200
C 4920 see CFT500
C 4924 see DII000
C 4926 see CFT750
C 4928 see DIH800
C-5068 see HGP495
C 5123 see DHS800
C 5124 see EPC875
C 5125 see DHM400
C 5126 see CFN500
C 5290 see CFT000
C 5296 see BEI750
C 5307 see DHV200
C 5308 see DHV400
C 5309 see PIO000
C 5310 see PIO500
53-11 C see EQQ100
C 5311 see PIO250
C 5312 see PPW750
C 5318 see PPX250
C 5319 see PPX000
C 5320 see CLN000
C 5323 see CEP250
C 5324 see CLN250
C 5326 see CLN500
53-32C see TGA525
C 5334 see DIN400
C 5342 see DIC400
C 5343 see CEX750

C 5346 see DIC200
C 5347 see CFS750
C 5348 see BEI250
C 5351 see BEI500
C 5352. see CER250
C 5353 see BEJ250
C 5354 see BEJ500
C 5364 see CEX500
C 5365 see DEY000
C 5366 see CFR000
C 5384 see CFQ750
C 5385 see CFT250
C 5388 see BRA500
C 5397 see CFR500
C 5398 see CFS250
C 5399 see BFI250
C 5400 see CFR250
C 5401 see CFR750
C 5402 see CFS000
C 5405 see CIL000
C 5406 see CIU000
C 5407 see CIX000
C 5410 see CHG750
C 5412 see CFQ250
C 5413 see BQA500
C 5414 see BPY500
C 5415 see CIL750
C 5416 see DIC000
C 5417 see CGD750
C 5420 see CIW250
C 5422 see BPR250
C 5458 see CGE500
C 5501 see CGE250
C 5968 see HGP495
C 6005 see MQP000
C 6257 see BQG750
C 6259 see BQG250
C 6260 see BQG500
C 6304 see PEQ250
C-6313 see CES750
C 6379 see DBA800
C 6575 see MLS250
C 6583 see MLS750
C 6606 see DXY400
C 6608 see DIM800
C 6610 see DIN200
C 6866 see BIE500
C-6989 see NIX000
C 7019 see ASG250
C 7239 see CFN750
C-7441 see DKQ200
C 7441 see OJD300
74C48 see CBQ125
75-20 C see NCQ100
C 8514 see CJJ250
C 9295 see MKU750
C-9491 see IEN000
C-10015 see CDS750
C-10015 see DRP600
11925 C see PCY300
C 20684 see EQN750
295 C 51 see TMX775
356C61 see KFA100
467-C-60 see BFW250
C 45 (pharmaceutical) see CGG500
CA see BMW250
CAA see COJ500
CABADON M see VSZ000
CABELLOS de ANGEL (CUBA, PUERTO RICO) see CMV390
CABEZA de BURRO see EAI600
CABEZA de VIEJO (MEXICO) see CMV390
CAB-O-GRIP see AHE250
CAB-O-GRIP II see SCH000
CAB-O-LITE 100 see WCJ000
CAB-O-LITE 130 see WCJ000
CAB-O-LITE 160 see WCJ000
CAB-O-LITE F 1 see WCJ000
CAB-O-LITE P 4 see WCJ000
CABLONGA see YAK350
CAB-O-SIL see SCH000
CAB-O-SPERSE see SCH000
CABRAL see PGG350
CABRONAL see EOK000
CACHALOT O-1 see OBA000

CACHALOT O-3 see OBA000
CACHALOT O-8 see OBA000
CACHALOT O-15 see OBA000
CACHALOT C-50 see HCP000
CACHALOT L-50 see DXV600
C ACID see ALH250
C-8 ACID see OCY000
CACODYLATE de SODIUM (FRENCH) see HKC500
CACODYL HYDRIDE see DQG600
CACODYLIC ACID (DOT) see HKC000
CACODYLIC ACID SODIUM SALT see HKC500
CACODYL NEW see DXE600
CACODYL SULFIDE see CAC250
CACP see PJD000
CACTINOMYCIN see AEA750
C. ADAMANTEUS VENOM see EAB225
CADAVERIN see PBK500
CADAVERINE see PBK500
CADCO 0115 see SMQ500
CADDY see CAE250
CADE OIL RECTIFIED see JEJ000
CADET see BDS000
CADIA DEL PERRO see CAC500
CADIZEM see DNU600
CADMINATE see CAI750
CADMIUM see CAD000
CADMIUM(II) ACETATE see CAD250
CADMIUM ACETATE (DOT) see CAD250
CADMIUM ACETATE DIHYDRATE see CAD275
CADMIUM AMIDE see CAD325
CADMIUM AZIDE see CAD350
CADMIUM BARIUM STEARATE see BAI800
CADMIUM BIS(2-ETHYLHEXYL) PHOSPHITE see CAD500
CADMIUM, BIS(1-HYDROXY-2(1H)-PYRIDINETHIONATO)- see CAI350
CADMIUM, BIS(SALICYLATO)- see CAI400
CADMIUM BROMIDE see CAD600
CADMIUM CAPRYLATE see CAD750
CADMIUM CHLORATE see CAE000
CADMIUM CHLORIDE see CAE250
CADMIUM CHLORIDE, DIHYDRATE see CAE375
CADMIUM CHLORIDE, HYDRATE (2:5) see CAE425
CADMIUM CHLORIDE, MONOHYDRATE see CAE500
CADMIUM COMPOUNDS see CAE750
CADMIUM DIACETATE see CAD250
CADMIUM DIACETATE DIHYDRATE see CAD275
CADMIUM DIAMIDE see CAD325
CADMIUM DIAZIDE see CAD350
CADMIUM DIBROMIDE see CAD600
CADMIUM DICHLORIDE see CAE250
CADMIUM DICYANIDE see CAF500
CADMIUM DIETHYL DITHIOCARBAMATE see BJB500
CADMIUM DILAURATE see CAG775
CADMIUM DINITRATE see CAH000
CADMIUM DODECANOATE see CAG775
CADMIUM(II) EDTA COMPLEX see CAF750
CADMIUM FLUOBORATE see CAG000
CADMIUM FLUORIDE see CAG250
CADMIUM FLUOROBORATE see CAG000
CADMIUM FLUORURE (FRENCH) see CAG250
CADMIUM FLUOSILICATE see CAG500
CADMIUM FUME see CAH750
CADMIUM GOLDEN see CMS205
CADMIUM GOLDEN 366 see CAJ750
CADMIUM LACTATE see CAG750
CADMIUM LAURATE see CAG775
CADMIUM LEMON see CMS205
CADMIUM LEMON YELLOW 527 see CAJ750
CADMIUM NITRATE see CAH000
CADMIUM(II) NITRATE TETRAHYDRATE (1:2:4) see CAH250
CADMIUM NITRIDE see TIH000
CADMIUM ORANGE see CAJ750
CADMIUM OXIDE see CAH500
CADMIUM OXIDE FUME see CAH750
CADMIUM PHOSPHATE see CAI000
CADMIUM PHOSPHIDE see CAI125
CADMIUM PRIMROSE see CMS205
CADMIUM PRIMROSE 819 see CAJ750
CADMIUM PROPIONATE see CAI250
CADMIUM PT see CAI350
CADMIUM 2-PYRIDINETHIONE see CAI350
CADMIUM SALICYLATE see CAI400
CADMIUM SELENIDE see CAI500
CADMIUM STEARATE see OAT000
CADMIUM SUCCINATE see CAI750

CADMIUM SULFATE see CAJ000
CADMIUM SULFATE (1:1) see CAJ000
CADMIUM SULFATE (1:1) HYDRATE (3:8) see CAJ250
CADMIUM SULFATE OCTAHYDRATE see CAJ250
CADMIUM SULFATE TETRAHYDRATE see CAJ500
CADMIUM SULFIDE see CAJ750
CADMIUM SULFIDE mixed with ZINC SULFIDE (1:1) see CMS205
CADMIUM SULPHATE see CAJ000
CADMIUM SULPHIDE see CAJ750
CADMIUM THERMOVACUUM AEROSOL see CAK000
CADMIUM-THIONEIN see CAK250
CADMIUM salt of 2,4-5-TRIBROMOIMIDAZOLE see THV500
CADMIUM YELLOW see CAJ750
CADMOPUR YELLOW see CAJ750
CADOX see BDS000, BSC750
CADOX TBH see BRM250
CADOX TS see BIX750
CADOX TS 40,50 see BIX750
CADPX PS see BHM750
CADRALAZINE see CAK275
CADUCID see FLG000
CAESALPINIA (various species) see CAK325
CAF see CDP250, CEA750
CAFFEIC ACID see CAK375
CAFFEIN see CAK500
CAFFEINE see CAK500
CAFFEINE BROMIDE see CAK750
CAFFEINE HYDROBROMIDE see CAK750
CAFFEINE and SODIUM BENZOATE see CAK800
3-CAFFEOYLQUINIC ACID see CHK175
3-o-CAFFEOYLQUINIC ACID see CHK175
CAFRON see BCA000
CAID see CJJ000
CAIMONICILLO (DOMINICAN REPUBLIC) see ROA300
CAIN'S QUINOLINIUM see QOJ250
CAIROX see PLP000
CAJEPUTENE see MCC250
CAJEPUTOL see CAL000
CAKE ALUM see AHG750
CALACIDOL see CAL075
CALADIO (PUERTO RICO) see CAL125
CALADIUM see CAL125
CALAMINE (spray) see ZKA000
CALAMUS OIL see OGK000
CALAN see VHA450
CALAR see CAM000
CALCAMINE see DME300
CAL CHEM 5655 see MOU750
CALCIA see CAU500
CALCIATE(2-), ((ETHYLENEDINITRILO)TETRAACETATO)-, DIHYDROGEN (8CI) see CAR800
CALCICAT see CAL250
CALCIC LIVER of SULFUR see CAY000
CALCIFEROL see VSZ100
CALCIFERON 2 see VSZ100
CALCINED BARYTA see BAO000
CALCINED BRUCITE see MAH500
CALCINED DIATOMITE see SCJ000
CALCINED MAGNESIA see MAH500
CALCINED MAGNESITE see MAH500
CALCINED SODA see SIN500
CALCIOPEN K see PDT750
CALCIRETARD see CAM675
CALCITE see CAO000
CALCITRIOL see DMJ400
CALCIUM see CAL250
CALCIUM, pyrophoric (DOT) see CAL250
CALCIUM, non-pyrophoric (DOT) see CAL250
CALCIUM ACETARSONE see CAL500
CALCIUM ACETATE see CAL750
CALCIUM ACETYLIDE see CAN750
CALCIUM ACID METHANEARSONATE see CAM000
CALCIUM ACID METHYL ARSONATE see CAM000
CALCIUM ALUMINUM SILICATE see AGY100
CALCIUMARSENAT see ARB750
CALCIUM ARSENATE (MAK) see ARB750
CALCIUM ARSENITE see CAM500
CALCIUM ARSENITE, solid (DOT) see CAM500
CALCIUM ASPARTATE see CAM675
CALCIUM-l-ASPARTATE see CAM675
CALCIUM BENZOATE see CAM680
CALCIUM-o-BENZOSULFIMIDE see CAM750
CALCIUM-2-BENZOSULPHIMIDE see CAM750
CALCIUM-o-BENZOSULPHIMIDE see CAM750

CALCIUM BIPHOSPHATE see CAW110
CALCIUM BISULFITE (solution) see CAN000
CALCIUM BISULFITE, solution (DOT) see CAN000
CALCIUM BORATE see CAN250
CALCIUM BROMATE see CAN400
CALCIUM BROMIDE see CAR375
CALCIUM BUCLOXATE see CPJ250
CALCIUM 5-BUTYLPICOLINATE HYDRATE see FQR100
CALCIUM CARAGEENIN see CAO250
CALCIUM CARBIDE see CAN750
CALCIUM CARBIMIDE see CAQ250
CALCIUM CARBONATE see CAO000
CALCIUM CARRAGEENAN see CAO250
CALCIUM CARRAGHEENATE see CAO250
CALCIUM CHEL-330 see CAY500
CALCIUM CHLORATE see CAO500
CALCIUM CHLORATE, aqueous solution (DOT) see CAO500
CALCIUM CHLORIDE see CAO750
CALCIUM CHLORIDE, anhydrous see CAO750
CALCIUM CHLORIDE with STREPTOMYCIN (1:1) see SLY000
CALCIUM CHLORITE see CAP000
CALCIUM CHLOROHYDROCHLORITE see HOV500
CALCIUM-4-(p-CHLOROPHENYL)-2-PHENYL-5-THIAZOLEACETATE see CAP250
CALCIUM CHROMATE see CAP500
CALCIUM CHROMATE (VI) see CAP500
CALCIUM CHROMATE(VI) DIHYDRATE see CAP750
CALCIUM CHROME YELLOW see CAP500, CAP750
CALCIUM CHROMIUM OXIDE (CaCrO$_4$) see CAP500
CALCIUM COMPOUNDS see CAQ000
CALCIUM CYANAMID see CAQ250
CALCIUM CYANAMIDE see CAQ250
CALCIUM CYANIDE see CAQ500
CALCIUM CYANIDE (mixture) see CAQ750
CALCIUM CYANIDE, solid (DOT) see CAQ500
CALCIUM CYANIDE MIXTURE, solid (DOT) see CAQ750
CALCIUM CYCLAMATE see CAR000
CALCIUM CYCLOHEXANESULFAMATE see CAR000
CALCIUM CYCLOHEXANE SULPHAMATE see CAR000
CALCIUM CYCLOHEXYLSULFAMATE see CAR000
CALCIUM CYCLOHEXYLSULPHAMATE see CAR000
CALCIUM DIACETATE see CAL750
CALCIUM DIBROMIDE see CAR375
CALCIUM DIFLUORIDE see CAS000
CALCIUM DIHYDRIDE see CAT200
CALCIUM-d-(+)-4-(2,4-DIHYDROXY-3,3-DIMETHYLBUTYRAMIDE)BUTY-RATE HEMIHDYRATE see CAT175
CALCIUM d(+)-N-(α,Γ-DIHYDROXY-β,β-DIMETHYLBUTYRYL)-β-ALANIN-ATE see CAU750
CALCIUM DIOXIDE see CAV500
CALCIUM DISILICIDE see CAR750
CALCIUM DOBESILATE see DMI300
CALCIUM-DTPA see CAY500
CALCIUM EDTA COMPLEX see CAR800
CALCIUM (−)-(1R,2S)-(1,2-EPOXYPROPYL)PHOSPHONATE HYDRATE see CAW376
CALCIUM ESFAR see CPJ250
CALCIUM-N-2-ETHYLHEXYL-β-OXYBUTYRAMIDE SEMISUCCINATE see CAR875
CALCIUM FLUORIDE see CAS000
CALCIUM FLUOSILICATE see CAX250
CALCIUM FORMATE see CAS250
CALCIUM FOSFOMYCIN HYDRATE see CAW376
CALCIUM FUSARATE see FQR100
CALCIUM GLUCONATE see CAS750
CALCIUM HEXAFLUOROSILICATE see CAX250
CALCIUM HEXAMETAPHOSPHATE see CAS825
CALCIUM HOMOPANTOTHENATE see CAT125
CALCIUM-d-HOMOPANTOTHENATE see CAT125
CALCIUM HOPANTENATE see CAT125
CALCIUM HOPANTENATE HEMIHYDRATE see CAT175
CALCIUM HYDRATE see CAT225
CALCIUM HYDRIDE see CAT200
CALCIUM HYDROGEN METHANEARSONATE see CAM000
CALCIUM HYDROGEN SULFITE, solution (DOT) see CAN000
CALCIUM HYDROXIDE see CAT225
CALCIUM HYPOCHLORIDE see HOV500
CALCIUM HYPOCHLORITE see HOV500
CALCIUM HYPOCHLORITE MIXTURE, dry (DOT) see HOW000
CALCIUM HYPOPHOSPHITE see CAT250
CALCIUM IODATE see CAT500
CALCIUM, METAL (DOT) see CAL250
CALCIUM, METAL, CRYSTALLINE (DOT) see CAL250
CALCIUM METHANEARSONATE see CAM000
CALCIUM METHIONATE see CAT700

CALCIUM MOLYBDATE see CAT750
CALCIUM MONOCHROMATE see CAP500
CALCIUM NEMBUTAL see CAV000
CALCIUM NITRATE (DOT) see CAU000
CALCIUM(II) NITRATE (1:2) see CAU000
CALCIUM(II) NITRATE TETRAHYDRATE (1:2:4) see CAU250
CALCIUM NITRIDE see TIH250
CALCIUM ORTHOARSENATE see ARB750
CALCIUM OXIDE see CAU500
CALCIUM OXYCHLORIDE see HOV500
CALCIUM PANTHOTHENATE (FCC) see CAU750
CALCIUM PANTOTHENATE see CAU750
CALCIUM-d-PANTOTHENATE see CAU750
d-CALCIUM PANTOTHENATE see CAU750
CALCIUM PANTOTHENATE, CALCIUM CHLORIDE DOUBLE SALT
   see CAU780
CALCIUM PENTOBARBITAL see CAV000
CALCIUM PERMANGANATE see CAV250
CALCIUM PEROXIDE see CAV500
CALCIUM PEROXOCHROMATE see CAV750
CALCIUM PEROXODISULPHATE see CAW000
CALCIUM-2-(m-PHENOXYPHENYL)PROPIONATE DIHYDRATE see FAP100
CALCIUM-2-PHENYL-4-(p-CHLOROPHENYL)-5-THIAZOLEACETATE see CAP250
CALCIUM PHOSPHATE, DIBASIC see CAW100
CALCIUM PHOSPHATE, MONOBASIC see CAW110
CALCIUM PHOSPHATE, TRIBASIC see CAW120
CALCIUM PHOSPHIDE see CAW250
CALCIUM PHOSPHINATE see CAT250
CALCIUM PHOSPHONOMYCIN HYDRATE see CAW376
CALCIUM PYROPHOSPHATE see CAW450
CALCIUM RESINATE see CAW500
CALCIUM RESINATE, fused (DOT) see CAW500
CALCIUM RESINATE, technically pure (DOT) see CAW500
CALCIUM RHODANID (GERMAN) see CAY250
CALCIUM SACCHARIN see CAM750
CALCIUM SACCHARINA see CAM750
CALCIUM SACCHARINATE see CAM750
CALCIUM SILICATE see CAW850
CALCIUM SILICIDE see CAX000
CALCIUM SILICOFLUORIDE see CAX250
CALCIUM SODIUM METAPHOSPHATE see CAX260
CALCIUM SULFATE see CAX500
CALCIUM(II) SULFATE DIHYDRATE (1:1:2) see CAX750
CALCIUM SULFIDE see CAY000
CALCIUM SUPEROXIDE see CAV500
CALCIUM THIOCYANATE see CAY250
CALCIUM TRISODIUM CHEL 330 see CAY500
CALCIUM TRISODIUM DIETHYLENE TRIAMINE PENTAACETATE
   see CAY500
CALCIUM TRISODIUM DTPA see CAY500
CALCIUM TRISODIUM PENTETATE see CAY500
CALCIUM TRISODIUM SALT of DIETHYLENETRIAMINEPENTAACETIC
   ACID see CAY500
CALCIUM VALPROATE see CAY675
CALCO 2246 see MJO500
CALCOCHROME ORANGE R see NEY000
CALCOCID AMARANTH see FAG020
CALCOCID BLUE EG see FMU059
CALCOCID BRILLIANT SCARLET 3RN see FMU080
CALCOCID ERYTHROSINE N see FAG040
CALCOCID FAST BLUE SR see ADE750
CALCOCID FAST LIGHT ORANGE 2G see HGC000
CALCOCID 2RIL see FMU070
CALCOCID URANINE B4315 see FEW000
CALCOCID VIOLET 4BNS see FAG120
CALCOCID YELLOW MXXX see MDM775
CALCOCID YELLOW XX see FAG140
CALCODUR BROWN BRL see CMO750
CALCODUR RESIN FAST BLUE see CMN750
CALCOGAS ORANGE NC see PEJ500
C 10 ALCOHOL see DAI600
CALCOLAKE SCARLET 2R see FMU070
CALCOLOID GOLDEN YELLOW see DCZ000
CALCOLOID NAVY BLUE 2GC see CMU500
CALCOMINE BLACK see AQP000
CALCOMINE BLUE 2B see CMO000
CALCOMINE VIOLET N see CMP000
CALCO OIL ORANGE 7078 see PEJ500
CALCO OIL RED D see SBC500
CALCO OIL SCARLET BL see XRA000
CALCOSYN SAPPHIRE BLUE R see MGG250
CALCOSYN YELLOW GC see AAQ250
CALCOTONE ORANGE 2R see DVB800
CALCOTONE RED see IHD000

CALCOTONE RED 3B see NAY000
CALCOTONE TOLUIDINE RED YP see MMP100
CALCOTONE WHITE T see TGG760
CALCOZINE BLUE ZF see BJI250
CALCOZINE BRILLIANT GREEN G see BAY750
CALCOZINE CHRYSOIDINE Y see PEK000
CALCOZINE MAGENTA N see RMK020
CALCOZINE ORANGE YS see PEK000
CALCOZINE RED BX see FAG070
CALCOZINE RHODAMINE BX see FAG070
CALCOZINE YELLOW OX see IBA000
CALCYANIDE see CAQ500
CALDAN see BHL750
C-8 ALDEHYDE see OCO000
C-9 ALDEHYDE see NMW500
C-10 ALDEHYDE see DAG000
C-16 ALDEHYDE see ENC000
C-12 ALDEHYDE, LAURIC see DXT000
CALDON see BRE500
CALEDON DARK BLUE G see CMU500
CALEDON GOLDEN YELLOW see DCZ000
CALEDON GREY M see CMU475
CALEDON JADE GREEN 2G see APK500
CALEDON PRINTING NAVY G see CMU500
CALEDON PRINTING YELLOW see DCZ000
CALF KILL see MRU359
CALF THYMUS DNA see DAZ000
CALGON see SHM500
CALIBENE see SOX875
CALICO BUSH see MRU359
CALICO YELLOW see MRN500
CALIDRIA RG 100 see ARM268
CALIDRIA RG 144 see ARM268
CALIDRIA RG 600 see ARM268
CALIFORNIA CHEMICAL COMPANY RE5305 see BSG250
CALIFORNIA FERN see PJJ300
CALIFORNIA PEPPER TREE see PCB300
CALIGRAN M see MAP300
CALIXIN see TJJ500
CALLA see CAY800
CALLA LILY see CAY800
CALLA PALUSTRIS see WAT300
CALMADIN see MQU750
CALMATHION see MAK700
CALMAX see MQU750
CALMDAY see CGA500
CALMINAL see EOK000
CALMINOL see MNM500
CALMIREN see MQU750
CALMIXENE see MOO750
CALMOCITENE see DCK759
CALMODEN see MDQ250
CALMONAL see HGC500
CALMORE see TEH500
CALMOREX see TEH500
CALMOSINE see EJA379
CALMOTIN see BNP750
CALNEGYT see CAY875
CALOCAIN see MNQ000
CALOCHLOR see MCY475
CALO-CLOR see CAY950
CALOGREEN see MCW000
CALOMEL see MCW000
CALOMELANO (ITALIAN) see MCW000
CALOMEL and MAGNESIUM SULFATE (5:8) see CAZ000
CALOPHYLLUM INOPHYLLUM see MBU780
CALOSAN see MCW000
CALOTROPIS PROCERA (Ait.) R.Br., flower extract see CAZ075
CALOTROPIS (VARIOUS SPECIES) see COD675
CALPANATE see CAU750
CALPLUS see CAO750
CALPOL see HIM000
CALPURNINE see CAZ125
CALSMIN see DLY000
CALSOFT F-90 see DXW200
CALSOL see EIV000
CALTAC see CAO750
CALTHA (VARIOUS SPECIES) see MBU550
CALTHOR see AJJ875
CALVACIN see CBA000
CALVISKEN see VSA000
CALX see CAU500
CALYCANTH see CCK675
CALYCANTHINE, HYDROCHLORIDE see CBA075

CALYCANTHUS (VARIOUS SPECIES) see CCK675
CALYSTIGINE see PAE100
CAM see CDP250
CAMA see CAM000
CAMATROPINE see MDL000
CAMAZEPAM see CFY250
CAMBAXIN see BAB250
CAMBENDAZOLE see CBA100
CAMBENZOLE see CBA100
CAMBET see CBA100
CAMBOGIC ACID see CBA125
CAMCOLIT see LGZ000
CAMELLIA SINENSIS see ARP500
CAMFOSULFONATO del d-3-4-(1'DIBENZIL-2-CHETO-IMIDAZOLIDO)-1,2-
	TRIMETILTHIOPHANIUM (ITALIAN) see TKW500
CAMILAN see SNY500
CAMILICHIGUI (MEXICO) see DHB309
CAMITE see BMW250
CAMIVERINE see CBA375
CAMOFORM HYDROCHLORIDE see BFX125
CAMOMILE OIL, ENGLISH TYPE (FCC) see CDH750
CAMOMILE OIL GERMAN see CDH500
cAMP see AOA130
CAMPANA (CUBA, PUERTO RICO) see AOO825
CAMPAPRIM A 1544 see AMY050
CAMPBELLINE OIL ORANGE see PEJ500
CAMPECHE (PUERTO RICO) see LED500
2-CAMPHANOL see BMD000
1-2-CAMPHANOL see NCQ820
2-CAMPHANONE see CBA750
d-2-CAMPHANONE see CBB250
CAMPHECHLOR see CDV100
CAMPHENE see CBA500
CAMPHIDONIUM see TLG000
CAMPHOCHLOR see CDV100
CAMPHOCLOR see CDV100
CAMPHOFENE HUILEUX see CDV100
CAMPHOGEN see CQI000
CAMPHOPHYLINE see CNR125
CAMPHOR see CBA750
(+)-CAMPHOR see CBB250
d-CAMPHOR see CBB250
l-(−)-CAMPHOR see CBB000
l-CAMPHOR see CBB000
(±)-CAMPHOR see CBA800
d-(+)-CAMPHOR see CBB250
dl-CAMPHOR see CBA800
(1R,4R)-(+)-CAMPHOR see CBB250
CAMPHOR-natural see CBA750
CAMPHOR, synthetic (ACGIH, DOT) see CBA750
CAMPHORATED OIL see CBB375
CAMPHOR LINIMENT see CBB375
CAMPHOR OIL see CBB500
CAMPHOR OIL, RECTIFIED see CBB500
CAMPHOR OIL WHITE see CBB500
CAMPHOR OIL YELLOW see CBB500
CAMPHOR TAR see NAJ500
CAMPHOR USP see CBB250
CAMPHOZONE see DJS200
CAMPILIT see COO500
CAMPOSAN see CDS125
CAMPOVITON 6 see PPK500
CAMPTOTHECIN see CBB870
CAMPTOTHECINE see CBB870
20(S)-CAMPTOTHECINE see CBB870
CAMPTOTHECIN, SODIUM SALT see CBB875
CAMUZULENE see DRV000
CAMYLOFINE see NOC000
CAMYLOFINE DIHYDROCHLORIDE see AAD875
CAMYLOFINE HYDROCHLORIDE see AAD875
CAMYLOFIN HYDROCHLORIDE see AAD875
CAMYLOPIN see NOC000
CANACERT AMARANTH see FAG020
CANACERT BRILLIANT BLUE FCF see FAE000
CANACERT ERYTHROSINE BS see FAG040
CANACERT INDIGO CARMINE see FAE100
CANACERT SUNSET YELLOW FCF see FAG150
CANACERT TARTRAZINE see FAG140
CANADA MOONSEED see MRN100
CANADIEN 2000 see BMN000
CANADINE see TCJ800
(−)-CANADINE see TCJ800
α-CANADINE see TCJ800
CANADOL see PCT250

CANAFISTOLA (CUBA) see GIW300
CANANGA see CAL125
CANARIO (PUERTO RICO) see AFQ625
CANARY CHROME YELLOW 40-2250 see LCR000
CANARY IVY see AFK950
CANAVANIN see AKD500
l-CANAVANINE see AKD500
CANCER JALAP see PJJ315
CANDAMIDE see LGZ000
CANDASETPIC see CFE250
CANDELILLA (MEXICO) see SDZ475
CANDEPTIN see LFF000
CANDEREL see ARN825
CANDEX see ARQ725, NOH500
CANDIDA ALBICANS GLYCOPROTEINS see CBC375
CANDIDIN see CBC500
CANDIDIN B see CBC750
CANDIDINE see CBC500
CANDIMON see LFF000
CANDIO-HERMAL see NOH500
CANDLEBERRY see TOA275
CANDLENUT see TOA275
CANDLETOXIN A see CBD250
CANDLETOXIN B see CBD500
CANESCINE see RDF000
CANESCINE 10-METHOXYDERIVATIVE see MEK700
CANESTEN see MRX500
CANE SUGAR see SNH000
CANNABICHROME see PBW400
CANNABICHROMENE see PBW400
CANNABIDIOL see CBD599
(−)-CANNABIDIOL see CBD599
(−)-trans-CANNABIDIOL see CBD599
CANNABINOL see CBD625
CANNABIS see CBD750
CANNABIS RESIN see CBD750
CANNABIS SMOKE RESIDUE see CBD760
CANNANBICHROMENE see PBW400
CANNE-A-GRATTER (HAITI) see DHB309
CANNE-MADERE (HAITI) see DHB309
CANNOGENIN-α-l-THEVETOSIDE see EAQ050
C. ANNUUM see PCB275
CANOCENTA see CKL325
CANOGARD see DGP900
CANQUIL-400 see MQU750
CANTABILINE see MKP500
CANTABILINE SODIUM see HMB000
CANTHARIDES see CBE250
CANTHARIDES CAMPHOR see CBE750
CANTHARIDIN see CBE750
CANTHARIDINE see CBE750
CANTHARONE see CBE750
CANTIL see CBF000
CANTREX see KAL000, KAV000
CANTRIL see CBF000
CAN-TROL see CLN750
CANTROL see CLO000
CAO 1 see BFW750
CAO 3 see BFW750
CAOBA (CUBA, DOMINICAN REPUBLIC, PUERTO RICO) see MAK300
CAOUTCHOUC see ROH900
CAOUTCHOUC (HAITI) see ROU450
CAP see CBF250, CDP250, CEA750
CAPACIDIN see CBF500
CAPARISIDE see TFA350
CAPAROL see BKL250
CAPARSOLATE see TFA350
CAPATHYN see DPJ400
CAPAURIDINE see TDI750
CAPAURINE, (±)- see TDI750
CAPAURINE, dl- see TDI750
CAPE BELLADONNA see AHI635
CAPE GOOSEBERRY see JBS100
CAPEN see MCI375
CAPER SPURGE see EQW000
CAPILAN see DNU100
CAPISTEN see BDU500
CAPITUS see ENG500
CAPLENAL see ZVJ000
CAPMUL see PKL030
CAPMUL POE-O see PKL100
CAPOBENATE see CBF625
CAPOBENATE SODIUM see CBF625
CAPORIT see HOV500

CAPOTEN see MCO750
CAPOTILLO (MEXICO) see CAL125
CAP-P see CDP700
CAP-PALMITATE see CDP700
CAPRALDEHYDE see DAG000
CAPRALENSE see AJD000
CAPRAMOL see AJD000
CAPRAN 80 see PJY500
CAPREOMYCIN DISULFATE see CBF675
CAPREOMYCIN 1A see CBF680
CAPRIC ACID see DAH400
n-CAPRIC ACID see DAH400
CAPRIC ACID ETHYL ESTER see EHE500
CAPRIC ALCOHOL see DAI600
CAPRIN see ADA725
CAPRINIC ACID see DAH400
CAPRINIC ALCOHOL see DAI600
CAPROALDEHYDE see HEM000
CAPROAMIDE see HEM500
CAPROAMIDE POLYMER see PJY500
CAPROCID see AJD000
CAPROCIN see CBF675
CAPRODAT see IPU000
CAPROIC ACID see HEU000
n-CAPROIC ACID see HEU000
CAPROIC ALDEHYDE see HEM000
CAPROKOL see HFV500
CAPROLACTAM see CBF700
6-CAPROLACTAM see CBF700
omega-CAPROLACTAM (MAK) see CBF700
CAPROLACTAM OLIGOMER see PJY500
epsilon-CAPROLACTAM POLYMERE (GERMAN) see PJY500
CAPROLACTONE see LAP000
epsilon-CAPROLACTONE see LAP000
CAPROLATTAME (FRENCH) see CBF700
CAPROLISIN see AJD000
CAPROLON see NOH000
CAPROMYCIN see CBF680
CAPRON see HNT500
CAPRON see PJY500
CAPRONALDEHYDE see HEM000
CAPRONAMIDE see HEM500
CAPRONIC ACID see HEU000
CAPRONITRILE see HER500
CAPROYL ALCOHOL see HFJ500
n-CAPROYLALDEHYDE see HEM000
1-CAPROYLAZIRIDINE see HEW000
CAPROYLETHYLENEIMINE see HEW000
CAPRYL ALCOHOL see OEI000
CAPRYLAMINE see OEK000
CAPRYLATE d'OESTRADIOL (FRENCH) see EDQ500
CAPRYLDINITROPHENYL CROTONATE see AQT500
2-CAPRYL-4,6-DINITROPHENYL CROTONATE see AQT500
CAPRYLIC ACID see OCY000
n-CAPRYLIC ACID see OCY000
CAPRYLIC ACID TRIGLYCERIDE see TMO000
CAPRYLIC ALCOHOL see OEI000
CAPRYLIC/CAPRIC TRIGLYCERIDE see CBF710
CAPRYLIC ETHER see OEY000
4-CAPRYLMORPHOLINE see CBF725
CAPRYL-o-PHTHALATE see BLB750
CAPRYLYL ACETATE see OEG000
CAPRYLYLAMINE see OEK000
CAPRYLYL PEROXIDE, solution (DOT) see OFI000
CAPRYNIC ACID see DAH400
CAPSAICIN see CBF750
CAPSAICINE see CBF750
CAPSEBON see CAJ750
CAPSICUM (VARIOUS SPECIES) see PCB275
CAPSINE see DUS700
CAPSTAT see CBF680
CAPTAF see CBG000
CAPTAFOL see CBF800
CAPTAGON HYDROCHLORIDE see CBF825
CAPTAMINE HYDROCHLORIDE see DOY600
CAPTAN see CBG000
CAPTANCAPTENEET 26,538 see CBG000
CAPTANE see CBG000
CAPTAN-STREPTOMYCIN 7.5-0.1 POTATO SEED PIECE PROTECTANT
    see CBG000
CAPTAX see BDF000
CAPTEX see CBG000
CAPTEX 300 see CBF710
CAPTODIAME see BRS000

CAPTODIAMIN see BRS000
CAPTODIAMINE see BRS000
CAPTOFOL see CBF800
CAP-O-TRAN see MQU750
CAPUT MORTUUM see IHD000
CAPVAL see NOA000
CAPVAL HYDROCHLORIDE see NOA500
CAP-WAKO see CCU050
CARACEMIDE see CBG075
CARACHOL see SGD500
CARADATE 30 see MJP400
CARADRIN see POB500
CARAGARD see BQC500
CARAIBE (HAITI) see EAI600, XCS800
CARAMEL see CBG125
CARAMEL COLOR see CBG125
CARAMIFENE (ITALIAN) see CBG375
CARAMIPHEN see PET250
CARAMIPHENE HYDROCHLORIDE see PET250
CARAMIPHEN ETHANE DISULFONATE see CBG250
CARAMIPHEN HYDROCHLORIDE see CBG375
CARASTAY see CCL250
CARASTAY G see CCL250
CARAWAY OIL see CBG500
CARBACHOL see CBH250
CARBACHOL CHLORIDE see CBH250
CARBACHOLIN see CBH250
CARBACHOLINE CHLORIDE see CBH250
CARBACOLINA see CBH250
CARBACRYL see ADX500
CARBADINE see EIR000
CARBADIPIMIDINE see CBH500
CARBADIPIMIDINE MALEATE see CCK790
CARBADOX (USDA) see FOI000
CARBAETHOXYDIGOXIN (GERMAN) see EEP000
CARBAICA see AKK625
CARBAM see VFU000
CARBAMALDEHYDE see FMY000
CARBAMAMIDINE see GKW000
CARBAMATE see FAS000
CARBAMATE de l'ETHINYLCYCLOHEXANOL (FRENCH) see EEH000
CARBAMATE (ISOAMYL) see IHQ000
CARBAMATE de METHYLPENTINOL (FRENCH) see MNM500
CARBAMATE du PROPINYLCYCLOHEXANOL (FRENCH) see POA250
CARBAMATES see CBH750
CARBAMAZEPEN see DCV200
CARBAMAZEPINE see DCV200
CARBAMAZINE see DIW000
CARBAMEZEPINE see DCV200
CARBAMIC ACID, ALLYL ESTER see AGA750
CARBAMIC ACID, 1H-BENZIMIDAZOL-2-YL- see BCC100
CARBAMIC ACID, BUTYL ESTER see BQP250
CARBAMIC ACID-2-sec-BUTYL-2-METHYLTRIMETHYLENE ESTER see MBW750
CARBAMIC ACID, 1-(4-CHLOROPHENYL)-1-PHENYL-2-PROPYNYL ESTER
    see CKI500
CARBAMIC ACID-3-DIMETHYLAMINOPHENYL ESTER, METHOSULFATE
    see HNP500
CARBAMIC ACID, DIMETHYLDITHIO-, ANHYDROSULFIDE see BJL600
CARBAMIC ACID, DIMETHYLDITHIO-, ZINC SALT (2:) see BJK500
CARBAMIC ACID, DIMETHYL-, ester with (m-
    HYDROXYPHENYL)TRIMETHYLAMMONIUM BROMIDE see POD000
CARBAMIC ACID, N,N-DIMETHYL-, m-ISOPROPYLPHENYL ESTER
    see DQX300
CARBAMIC ACID, ESTER with CHOLINE CHLORIDE see CBH250
CARBAMIC ACID, ESTER with 2-(HDYROXYMETHYL)-1-
    METHYLPENTYLISOPROPYLCARBAMATE see IPU000
CARBAMIC ACID, ESTER with 2-(HYDROXYMETHYL)-2-METHYLPENTYL
    BUTYLCARBAMATE see MOV500
CARBAMIC ACID, ESTER with 2-METHYL-2-PROPYL-1,3-PROPANEDIOL
    BUTYLCARBAMATE see MOV500
CARBAMIC ACID, ESTER with 2-METHYL-2-PROPYL-1,3-PROPANEDIOL
    ISOPROPYLCARBAMATE see IPU000
CARBAMIC ACID, ETHYL ESTER see UVA000
CARBAMIC ACID-1-ETHYL-1-METHYL-2-PROPYNYL ESTER see MNM500
CARBAMIC ACID-2-ETHYNYL-2-BUTYL ESTER see MNM500
CARBAMIC ACID, (5-(4-FLUOROBENZOYL)-1H-BENZIMIDAZOL-2-YL)-,
    METHYL ESTER (9CI) see FDA887
CARBAMIC ACID HYDRAZIDE see HGU000
CARBAMIC ACID-2-HYDROXYETHYL ESTER see HKQ000
CARBAMIC ACID-β-HYDROXYPHENETHYL ESTER see PFJ000
CARBAMIC ACID (2-ISOBUTOXYETHYL) ESTER see IIE000
CARBAMIC ACID, ISOPROPYL ESTER see IOJ000
CARBAMIC ACID, N-METHYL-, 3-DIETHYLARSINOPHENYL ESTER,
    METHIODIDE see MID900

CARBAMIC ACID-N-METHYL-3-DIMETHYLAMINOPHENYL ESTER METHIODIDE see HNO500
CARBAMIC ACID-1-METHYLETHYL ESTER see IOJ000
CARBAMIC ACID, METHYL-, NAPHTHALENYL ESTER see MME800
CARBAMIC ACID, METHYLNITROSO-, o-ISOPROPOXYPHENYL ESTER see PMY310
CARBAMIC ACID, METHYLNITROSO-, 2-(1-METHYLETHOXY)PHENYL ESTER (9CI) see PMY310
CARBAMIC ACID-α-METHYLPHENETHYL ESTER see CBI000
CARBAMIC ACID, (3-METHYLPHENYL)-3-((METHOXYCARBONYL)AMINO)PHENYL ESTER (9CI) see MEG250
CARBAMIC ACID-3-PHENYLPROPYL ESTER see PGA750
CARBAMIC ACID, PROPYL ESTER see PNG250
CARBAMIC ACID, THIO, S-ESTER with 2-MERCAPTOACETANILIDE see MCL500
CARBAMIC ACID, 2,2,2-TRICHLOROETHYL ESTER see TIO500
CARBAMIC ACID, (m-TRIMETHYLAMMONIO)PHENYL ESTER, METHYLSULFATE see HNP500
CARBAMIC ACID, VINYL ESTER see VNK000
CARBAMIC ESTER of 3-OXYPHENYLTRIMETHYLAMMONIUM METHYLSULFATE see HNP500
CARBAMIDAL see DJS200
CARBAMIDE see USS000
CARBAMIDE PEROXIDE see HIB500
CARBAMIDE PHENYLACETATE see PEC250
CARBAMIDE RESIN see USS000
CARBAMIDINE see GKW000
p-CARBAMIDOBENZENEARSONIC ACID see CBJ000
β-CARBAMIDOCARBOMETHYL-O,O-DIETHYLDITHIOPHOSPHATE see AAT000
p-CARBAMIDOPHENYL ARSENOUS ACID see CBI500
p-CARBAMIDOPHENYL ARSENOUS OXIDE see CBI500
4-CARBAMIDOPHENYL BIS(CARBOXYMETHYLTHIO)ARSENITE see CBI250
p-CARBAMIDOPHENYL-BIS(2-CARBOXYPHENYLMERCAPTO)ARSINE see TFD750
4-CARBAMIDOPHENYL BIS(o-CARBOXYPHENYLTHIO)ARSENITE see TFD750
p-CARBAMIDOPHENYL-DI(1'-CARBOXYPHENYL-2') THIOARSENITE see TFD750
4-CARBAMIDOPHENYLOXOARSINE see CBI500
CARBAMIDSAEURE-AETHYLESTER (GERMAN) see UVA000
CARBAMIMIDIC ACID see USS000
CARBAMIMIDOTHIOIC ACID-2-(DIMETHYLAMINO)ETHYL ESTER DIHYDROCHLORIDE see NNL400
CARBAMIMIDOTHIOIC ACID, ETHYL ESTER with METAPHOSPHORIC ACID (1:1) see ELY575
CARBAMIMIDOTHIOIC ACID, ETHYL ESTER, MONO(DIETHYL PHOSPHATE) see CBI675
CARBAMINOCHOLINE CHLORIDE see CBH250
p-CARBAMINO PHENYL ARSONIC ACID see CBJ000
CARBAMINOPHENYL-p-ARSONIC ACID see CBJ000
CARBAMINOTHIOGLYCOLIC ACID ANILIDE see MCL500
CARBAMINOYLCHOLINE CHLORIDE see CBH250
CARBAMIOTIN see CBH250
CARBAMODITHIOIC ACID, PHENYL-, ETHYL ESTER see EOK550
CARBAMOHYDROXAMIC ACID see HOO500
CARBAMOHYDROXIMIC ACID see HOO500
CARBAMOHYDROXYAMIC ACID see HOO500
CARBAMONITRILE see COH500
CARBAMOTHIOIC ACID-S,S'-(2-(DIMETHYLAMINO)-1,3-PROPANEDIYL) ESTER, MONOHYDROCHLORIDE (9CI) see BHL750
(p-CARBAMOYLAMINO)PHENYLARSINOBIS(2-THIO-ACETIC ACID) see CBI250
N-CARBAMOYLARSANILIC ACID see CBJ000
8-CARBAMOYL-3-(2-CHLOROETHYL)IMIDAZO(5,1-d)-1,2,3,5-TETRAZIN-4(3H)-ONE see MQY110
CARBAMOYLCHOLINE CHLORIDE see CBH250
γ-CARBAMOYL CHOLINE CHLORIDE see CBH250
5-CARBAMOYL-5H-DIBENZ(b,f)AZEPINE see DCV200
5-CARBAMOYLDIBENZO(b,f)AZEPINE see DCV200
5-CARBAMOYL-5H-DIBENZO(b,f)AZEPINE see DCV200
N-CARBAMOYL-2-(2,6-DICHLOROPHENYL)ACETAMIDINE HYDROCHLORIDE see PFV000
N-(2-CARBAMOYLETHYL)ARSANILIC ACID SODIUM SALT see SJG000
1-CARBAMOYLFORMIMIDIC ACID see OLO000
4'-CARBAMOYL-2-FORMYL-1,1'-(OXYDIMETHYLENE)DI-PYRIDINIUM-DICHLORIDE-2-OXIME see HAA345
CARBAMOYLHYDRAZINE see HGU000
N-CARBAMOYLHYDROXYLAMINE see HOO500
p-((CARBAMOYLMETHYL)AMINO)-BENXENEARSONIC ACID see CBJ750
N-(CARBAMOYLMETHYL)ARSANILIC ACID see CBJ750
N-(CARBAMOYLMETHYL)-2-DIAZOACETAMIDE see CBK000
1-p-CARBAMOYLMETHYLPHENOXY-3-ISOPROPYLAMINO-2-PROPANOL see TAL475
2-CARBAMOYL-2-NITROACETONITRILE see CBK125
N-(1-CARBAMOYL-4-(NITROSOCYANAMIDO)BUTYL)BENZAMIDE see CBK250

N-CARBAMOYL-N-NITROSOGLYCINE see NKI500
l-N^5-CARBAMOYL-N^5-NITROSOORNITHINE see NJT000
dl-N^5-CARBAMOYL-N^5-NITROSOORNITHINE see NJS500
CARBAMOYL OXIME see HOO500
1-CARBAMOYLOXY-2-HYDROXY-3-(o-METHYLPHENOXY)PROPANE see CBK500
3-CARBAMOYLOXY-3-METHYL-4-PENTYNE see MNM500
(3-(N-CARBAMOYLOXY)PHENYL)DIETHYLMETHYL-AMMONIUM METHOSULFATE see HNK550
1-CARBAMOYLOXY-3-PHENYLPROPANE see PGA750
((m-CARBAMOYLOXY)PHENYL)TRIMETHYLAMMONIUM METHYLSULFATE see HNP500
2-CARBAMOYLOXYPROPYLTRIMETHYLAMMONIUM CHLORIDE see HOA500
1-CARBAMOYLOXY-1-(2-PROPYNYL)CYCLOHEXANE see POA250
4-CARBAMOYL-4-PHTHALYLBUTYRIC ACID see PIA250
10-(3-(4-CARBAMOYLPIPERIDINE)PROPYL)-2-(METHANESULFONYL)PHENO-THIAZINE see MQR000
N-(1-CARBAMOYLPROPYL)ARSANILIC ACID see CBK750
4-CARBAMOYL-1-β-d-RIBOFURANOSYL-IMIDAZOLIUM-5-OLATE see BMM000
CARBAMULT see CQI500
4-CARBAMYLAMINOPHENYLARSONIC ACID see CBJ000
N-CARBAMYL ARSANILIC ACID see CBJ000
CARBAMYLCHOLINE CHLORIDE see CBH250
5-CARBAMYLDIBENZO(b,f)AZEPINE see DCV200
5-CARBAMYL-5H-DIBENZO(b,f)AZEPINE see DCV200
CARBAMYLHYDRAZINE see HGU000
CARBAMYLHYDRAZINE HYDROCHLORIDE see SBW500
CARBAMYL HYDROXAMATE see HOO500
CARBAMYLMETHYLCHOLINE CHLORIDE see HOA500
4-CARBAMYLPHENYL BIS(CARBOXYMETHYLTHIO)ARSENITE see TFA350
1-CARBAMYL-2-PHENYLHYDRAZINE see CBL000
2-CARBAMYL PYRAZINE see POL500
CARBANIL see PFK250
CARBANILALDEHYDE see FNJ000
CARBANILIC ACID, DITHIO-, ETHYL ESTER see EOK550
d-(−)-CARBANILIC ACID (1-ETHYLCARBAMOYL)ETHYL ESTER see CBL500
CARBANILIC ACID ETHYL ESTER see CBL750
CARBANILIC ACID ISOPROPYL ESTER see CBM000
CARBANILIDE see CBM250
m-CARBANILOYLOXYCARBANILIC ACID ETHYL ESTER see EEO500
CARBANOCHLORIDIC ACID, NAPHTHYL ESTER see NBH200
CARBANOLATE see CBM500, CGI500
CARBANTHRENE GOLDEN YELLOW see DCZ000
CARBANTHRENE NAVY BLUE G see CMU500
CARBANTHRENE RED BROWN 5R see CMU800
CARBARSONE (USDA) see CBJ000
CARBARSONE OXIDE see CBI500
CARBARYL see CBM750
CARBASED see ACE000
CARBASONE see CBJ000
CARBATENE see MQQ250
CARBATHIONE see VFU000
CARBATOX-60 see CBM750
CARBAVINE see CBM875
CARBAX see BIO750
CARBAZALDEHYDE see FNN000
CARBAZAMIDE see HGU000
CARBAZEPINE see DCV200
CARBAZIC ACID, ETHYL ESTER see EHG000
CARBAZIC ACID HYDRAZIDE see CBS500
CARBAZIDE see CBS500
CARBAZIDE (DOT) see CBS500
CARBAZILQUINONE see BGX750
CARBAZINC see BJK500
CARBAZOCHROME SODIUM SULFONATE see AER666
CARBAZOCHROME SODIUM SULFONATE TRIHYDRATE see AER666
CARBAZOLE see CBN000
9H-CARBAZOLE see CBN000
CARBAZOLE, 3-(p-HYDROXYANILINO)- see CBN100
4-(3-CARBAZOLYLAMINO)PHENOL see CBN100
CARBAZON see AER666
CARBAZOTIC ACID see PID000
CARBECIN see CBO250
CARBENDAZIM see MHC750
CARBENDAZIME and SODIUM NITRITE (1:1) see SIQ700
CARBENDAZIM and SODIUM NITRITE (5:1) see CBN375
CARBENDAZOLE see MHC750
CARBENDAZYM see MHC750
CARBENICILLIN DISODIUM SALT see CBO250
CARBENICILLIN PHENYL see CBN750
CARBENICILLIN PHENYL ESTER see CBN750
CARBENICILLIN PHENYL SODIUM see CBO000
CARBENICILLIN SODIUM see CBO250
CARBENOXALONE, DISODIUM SALT see CBO500

CARBENOXOLONE see BGD000
CARBENOXOLONE, DISODIUM SALT see CBO500
CARBENOXOLONE SODIUM see CBO500
CARBESTROL see CBO625
CARBETAMEX see CBL500
CARBETAMID (GERMAN) see CBL500
CARBETAMIDE see CBL500
CARBETHOXYACETIC ESTER see EMA500
3-CARBETHOXYAMINO-5-DIMETHYLAMINOACETYL-10,11-
 DIHYDRODIBENZ(b,f)AZEPINE HYDROCHLORIDE see BMB125
α-CARBETHOXY-β,β-BISCYCLOPROPYL ACRYLONITRILE see DGX000
1-CARBETHOXY-1,2-DIHYDROQUINOLINE see CBO750
N-CARBETHOXYETHYLENIMINE see ASH750
CARBETHOXYHYDRAZINE see EHG000
1-CARBETHOXY HYDRAZINE see EHG000
N-(CARBETHOXY)HYDRAZINE see EHG000
CARBETHOXY MALAOXON see OPK250
CARBETHOXY MALATHION see MAK700
2-CARBETHOXYMETHYLENE-3-METHYL-5-PIPERIDINO-4-THIAZOLIDONE
 see MNG000
p-CARBETHOXYPHENOL see HJL000
1(4-CARBETHOXYPHENYL)-3,3-DIMETHYLTRIAZENE see CBP250
2-(N-(4-CARBETHOXY-4-PHENYL)PIPERIDINO)PROPIOPHENONE HYDRO-
 CHLORIDE see CBP325
CARBETHOXYSYRINGOYL METHYLRESERPATE see RCA200
S-CARBETHOXYTHIAMINE HYDROCHLORIDE see EEQ500
CARBETOVUR see MAK700
CARBETOX see MAK700
CARBICRON see DGQ875
CARBIDE 6-12 see EKV000
CARBIDE BLACK E see AQP000
CARBIDIUM ETHANESULFONATE see CBQ125
CARBIDIUM ETHANESULPHONATE see CBQ125
CARBIDOPA see CBQ500
CARBIDOPA MONOHYDRATE see CBQ529
CARBILAZINE see DIW000
CARBIMIDE see COH500
CARBIN see CEW500
CARBINAMINE see MGC250
CARBINOL see MGB150
CARBINOLBASE DES KRISTALLVIOLETT (GERMAN) see TJK000
CARBINOLBASE des MALACHITGRUEN (GERMAN) see MAK500
CARBINOLBASE des METHYLVIOLETT (GERMAN) see MQN250
CARBINOXAMINE MALEATE see TAI500
CARBINOXAMINE DIPHENYLDISULFONATE see CBQ575
CARBINOXAMINE MALEATE see CGD500
p-CARBINOXAMINE MALEATE see CGD500
CARBIPHENE HYDROCHLORIDE see CBQ625
CARBITOL see CBR000, DJD600
CARBITOL ACETATE see CBQ750
CARBITOL CELLOSOLVE see CBR000
CARBITOL SOLVENT see CBR000
CARBOBENZOXY CHLORIDE see BEF500
N-CARBOBENZOXYGLYCINE-1,2-DIBROMOETHYL ESTER see CBR175
N-CARBOBENZOXYGLYCINE VINYL ESTER see CBR200
N-CARBOBENZOXY-l-LEUCINE-1,2-DIBROMOETHYL ESTER see CBR210
N-CARBOBENZOXY-l-LEUCINE VINYL ESTER see CBR215
CARBOBENZOXYLGLYCINE see CBR125
CARBOBENZOXYPHENYLALANINE see CBR220
CARBOBENZOXY-l-PHENYLALANINE see CBR220
N-CARBOBENZOXY-l-PHENYLALANINE see CBR220
N-CARBOBENZOXY-l-PHENYLALANINE-1,2-DIBROMOETHYL ESTER see
 CBR225
N-CARBOBENZOXY-l-PHENYLALANINE VINYL ESTER see CBR235
N-CARBOBENZOXY-l-PROLINE-1,2-DIBROMOETHYL ESTER see CBR245
N-CARBOBENZOXY-l-PROLINE VINYL ESTER see CBR247
CARBOBENZOYL GLYCINE see CBR125
N-CARBOBENZOYLGLYCINE see CBR125
CARBOBENZYLOXY CHLORIDE see BEF500
CARBOBENZYLOXYGLYCINE see CBR125
N-CARBOBENZYLOXYGLYCINE see CBR125
2-CARBO-n-BUTOXY-6,6-DIMETHYL-5,6-DIHYDRO-1,4-PYRONE see BRT000
CARBOCAINE see SBB000
CARBOCAINE HYDROCHLORIDE see CBR250
CARBOCHOL see CBH250
CARBOCHOLIN see CBH250
CARBOCHROMENE HYDROCHLORIDE see CBR500
CARBOCISTEINE see CBR675
CARBOCIT see CBR675
CARBO-CORT see CMY800
CARBOCROMENE see CBR500
CARBOCYSTEINE see CBR675
CARBODIHYDRAZIDE see CBS500
CARBOETHOXYHYDRAZINE see EHG000

2-CARBOETHOXY-1-METHYLVINYL-DIETHYLPHOSPHATE see CBR750
N¹-CARBOETHOXY-N²-PHTHALAZINO HYDRAZINE see CBS000
CARBOETHOXYPHTHALAZINO HYDRAZINE see CBS000
CARBOFENOTHION (DUTCH) see TNP250
CARBOFLUORENE AMINO ESTER see CBS250
CARBOFOS see MAK700
CARBOFURAN see CBS275
CARBOHYDRAZIDE see CBS500
N-CARBOISOPROPOXY-o-ACETYL-N-PHENYL CARBAMATE see ING400
CARBOLIC ACID see PDN750
CARBOLIC ACID, LIQUID (DOT) see PDO000
CARBOLITH see LGZ000
CARBOLON see SCQ000
CARBOLSAURE (GERMAN) see PDN750
CARBOMAL see BNK000
CARBOMER 934 see ADW000
CARBOMETHENE see KEU000
2-CARBOMETHOXYANILINE see APJ250
o-CARBOMETHOXYANILINE see APJ250
2-β-CARBOMETHOXY-3-β-BENZOXYTROPANE see CNE750
4'-CARBOMETHOXY-2,3'-DIMETHYLAZOBENZENE see CBS750
4'-CARBOMETHOXY-2,3'-DIMETHYLAZOBENZOL see CBS750
N-CARBOMETHOXYMETHYLIMINOPHOSPHORYL CHLORIDE see CBT125
α-2-CARBOMETHOXY-1-METHYLVINYL DIMETHYL PHOSPHATE
 see MQR750
2'-CARBOMETHOXYPHENYL 4-GUANIDINOBENZOATE see CBT175
2-CARBOMETHOXY-1-PROPEN-2-YL DIMETHYL PHOSPHATE see MQR750
N-(3-CARBOMETHOXYPROPYL)-N-(1-ACETOXYBUTYL)NITROSAMINE
 see MEI000
p-CARBOMETHOXYTOLUENE see MPX850
CARBOMYCIN see CBT250
CARBOMYCIN A see CBT250
CARBON see CBT500
CARBONA see CBY000
CARBON, ACTIVATED see CDI000
CARBONATE MAGNESIUM see MAC650
(CARBONATO)DIHYDROXYDICOPPER see CNJ750
CARBONAZIDIC ACID, 1,1-DIMETHYLETHYL ESTER see BQI250
CARBON BICHLORIDE see PCF275
CARBON BISULFIDE (DOT) see CBV500
CARBON BISULPHIDE see CBV500
CARBON BLACK see CBT750
CARBON BROMIDE see CBX750
CARBON CHLORIDE see CBY000
CARBON CHLOROSULFIDE see TFN500
CARBON D see DXD200
CARBON DICHLORIDE see PCF275
CARBON DIFLUORIDE OXIDE see CCA500
CARBON DIOXIDE see CBU250
CARBON DIOXIDE (solid) see CBU750
CARBON DIOXIDE, solid (DOT) see CBU750
CARBON DIOXIDE (liquefied) see CBU500
CARBON DIOXIDE, liquefied (DOT) see CBU500
CARBON DIOXIDE, refrigerated liquid (DOT) see CBU500
CARBON DIOXIDE and ETHYLENE OXIDE MIXTURES, with more than 6%
 ETHYLENE OXIDE (DOT) see EJO000
CARBON DIOXIDE, mixture with NITROGEN OXIDE (N₂O) see CBV000
CARBON DIOXIDE mixed with NITROUS OXIDE see CBV000
CARBON DIOXIDE-NITROUS OXIDE mixture (DOT) see CBV000
CARBON DIOXIDE mixed with OXYGEN see CBV250
CARBON DIOXIDE-OXYGEN mixture (DOT) see CBV250
CARBON DISULFIDE see CBV500
CARBON DISULPHIDE see CBV500
CARBONE (ITALIAN) see CBT500
CARBONE (OXYCHLORURE de) (FRENCH) see PGX000
CARBONE (OXYDE de) (FRENCH) see CBW750
CARBONE (SUFURE de) (FRENCH) see CBV500
CARBON FERROCHROMIUM see FBD000
CARBON FLUORIDE see CBY250
CARBON FLUORIDE OXIDE see CCA500
CARBON HEXACHLORIDE see HCI000
CARBONIC ACID, ALLYL ESTER, DIESTER with DIETHYLENE GLYCOL
 see AGD250
CARBONIC ACID, AMMONIUM SALT see ANE000
CARBONIC ACID, BARIUM SALT (1:1) see BAJ250
CARBONIC ACID BERYLLIUM SALT (1:1) see BFP750
CARBONIC ACID BIS(2-METHYLALLYL) ESTER see CBV750
CARBONIC ACID-2-sec-BUTYL-4,6-DINITROPHENYL-2,4-DINITROPHENYL
 ESTER (8CI) see DVC200
CARBONIC ACID-2-sec-BUTYL-4,6-DINITROPHENYL ISOPROPYL ESTER
 see CBW750
CARBONIC ACID, CALCIUM SALT (1:1) see CAO000
CARBONIC ACID, CHROMIUM SALT see CMJ100
CARBONIC ACID, CYCLIC 3-CHLOROPROPYLENE ESTER see CBW400

CARBONIC ACID, CYCLIC ETHYLENE ESTER see GHM000
CARBONIC ACID, CYCLIC ETHYLETHYLENE ESTER see BOT200
CARBONIC ACID CYCLIC PROPYLENE ESTER see CBW500
CARBONIC ACID, DIAMMONIUM SALT see ANE000
CARBONIC ACID, DICESIUM SALT see CDC750
CARBONIC ACID, DIESTER with 1-(2,3-DIMETHYLPHENYL)-3-(2-HYDROXYETHYL)UREA see PLK580
CARBONIC ACID DIHYDRAZIDE see CBS500
CARBONIC ACID, DILITHIUM SALT see LGZ000
CARBONIC ACID, DIPHENYL ESTER see DVZ000
CARBONIC ACID, DIPOTASSIUM SALT see PLA000
CARBONIC ACID, DISODIUM SALT see SFO000
CARBONIC ACID, DITHALLIUM(1+) SALT see TEJ000
CARBONIC ACID GAS see CBU250
CARBONIC ACID, LEAD(2+) SALT (1:1) see LCP000
CARBONIC ACID LITHIUM SALT see LGZ000
CARBONIC ACID, MAGNESIUM SALT see MAC650
CARBONIC ACID METHYL-4-(o-TOLYLAZO)-o-TOLYL ESTER see CBS750
CARBONIC ACID, MONOAMMONIUM SALT see ANB250
CARBONIC ACID MONOSODIUM SALT see SFC500
CARBONIC ACID, NICKEL SALT (1:1) see NCY500
CARBONIC ACID, cyclic VINYLENE ESTER see VOK000
CARBONIC ACID, ZINC SALT (1:1) see ZEJ050
CARBONIC ANHYDRASE INHIBITOR NO. 6063 see AAI250
CARBONIC ANHYDRIDE see CBU250
CARBONIC DIAZIDE see CCA000
CARBONIC DIFLUORIDE see CCA500
CARBONIC DIHYDRAZIDE see CBS500
CARBON ICE (DOT) see CBU750
CARBONIC OXIDE see CBW750
CARBONIMIDIC DIHYDRAZIDE, BIS((4-CHLOROPHENYL)METHYLENE)-
  see RLK890
4,4'-CARBONIMIDOYLBIS(N,N-DIMETHYLBENZENAMINE) see IBB000
4,4'-CARBONIMIDOYLBIS(N,N-DIMETHYLBENZENAM-
  INE)MONOHYDROCHLORIDE see IBA000
CARBON IODIDE see CBY500
CARBONIO (OSSICLORURO di) (ITALIAN) see PGX000
CARBONIO (OSSIDO di) (ITALIAN) see CBW750
CARBONIO (SOLFURO di) (ITALIAN) see CBV500
CARBON MONOXIDE see CBW750
CARBON MONOXIDE, CRYOGENIC liquid (DOT) see CBW750
CARBON NITRIDE see COO000
CARBON NITRIDE ION (CN1-) see COI500
CARBONOCHLORIDE ACID-1-METHYL ESTER see IOL000
CARBONOCHLORIDIC ACID-2-CHLOROETHYL ESTER see CGU199
CARBONOCHLORIDIC ACID, 1-NAPHTHALENYL ESTER see NBH200
CARBONOCHLORIDIC ACID PHENYL ESTER see CBX109
CARBONOCHLORIDIC ACID, PROPYL ESTER see PNH000
CARBONODITHIOIC ACID, O-ETHYL ESTER, POTASSIUM SALT see PLF000
CARBONOHYDRAZIDE see CBS500
CARBONOHYDRAZONIC DIHYDRAZIDE, MONONITRATE (9CI) see THN800
CARBON OIL see BBL250
CARBONOTHIOIC DICHLORIDE see TFN500
CARBONOTHIOIC DIHYDRAZIDE see TFE250
CARBONOTRITHIOIC ACID, METHYL TRICHLOROMETHYL ESTER see TIS500
CARBONOTRITHIOIC ACID-2-PROPENYLTRICHLOROMETHYL ESTER
  see TIR750
CARBON OXIDE SULFIDE see CCC000
CARBON OXYCHLORIDE see PGX000
CARBON OXYFLUORIDE see CCA500
CARBON OXYSULFIDE see CCC000
CARBON REMOVER (liquid) see CBX250
CARBON S see SGM500
CARBON SILICIDE see SCQ000
CARBON SULFIDE see CBV500
CARBON SULPHIDE (DOT) see CBV500
CARBON TET see CBY000
CARBON TETRABROMIDE see CBX750
CARBON TETRACHLORIDE see CBY000
CARBON TETRAFLUORIDE see CBY250
CARBON TETRAIODIDE see CBY500
CARBON TRIFLUORIDE see CBY750
CARBONYL AZIDE see CCA000
CARBONYLCHLORID (GERMAN) see PGX000
CARBONYL CHLORIDE see PGX000
CARBONYL DIAMIDE see USS000
CARBONYLDIAMINE see USS000
CARBONYL DIAZIDE see CCA000
CARBONYL DICYANIDE see OOM400
CARBONYL DIFLUORIDE see CCA500
CARBONYLDIHYDRAZINE see CBS500
CARBONYL DIISOTHIOCYANATE see CCA125
CARBONYL FLUORIDE see CCA500
CARBONYL IRON see IGK800

CARBONYL LITHIUM see CCB250
CARBONYL POTASSIUM see CCB500
CARBONYLS see CCB609
CARBONYL SODIUM see CCB750
CARBONYL SULFIDE see CCC000
CARBONYL SULFIDE-^{32}S see CCC000
CARBOPHOS see MAK700
CARBOPLATIN see CCC075
CARBOPOL 934 see ADW000, ADW200
CARBOPROST see CCC100
CARBOPROST TROMETHAMINE see CCC110
CARBOQUONE see BGX750
CARBORAFFIN see CDI000
CARBORAFINE see CDI000
CARBORANYLMETHYLETHYL SULFIDE see DEJ600
CARBORANYLMETHYLPROPYL SULFIDE see DEJ800
CARBORUNDEUM see SCQ000
CARBORUNDUM see SCQ000
CARBOSPOL see AGJ250
CARBOSTESIN see BOO000
CARBOSTYRIL see CCC250
2-CARBOSYPYRIDINE see PIB930
CARBOTHIALDIN see DSB200
CARBOTHIALDINE see DSB200
CARBOWAX see PJT000, PJT200
CARBOWAX 1000 see PJT250
CARBOWAX 1500 see PJT500
CARBOWAX 4000 see PJT750
CARBOWAX 6000 see PJU000
CARBOWAX 1000 DISTEARATE see PJU750
CARBOWAX 1000 MONOSTEARATE see PJW000
16-β-CARBOXAMIDE-3-β-ACETOXY-Δ^5-(17-α)-ISOPREGNENE-20-ONE
  see HND800
CARBOXAMIDOACETAMIDE see MAO000
1-p-CARBOXAMIDOPHENYL)-3,3-DIMETHYLTRIAZINE see CCC325
5-CARBOXANILIDO-2,3-DIHYDRO-6-METHYL-1,4-OXATHIIN see CCC500
CARBOXIN (USDA) see CCC500
CARBOXINE see CCC500
CARBOXYACETIC ACID see CCC750
1-(p-CARBOXYAETHYLPHENYL)-3,3-DIMETHYLTRIAZEN (GERMAN)
  see CBP250
CARBOXYANILINE see API500
2-CARBOXYANILINE see API500
4-CARBOXYANILINE see AIH600
o-CARBOXYANILINE see API500
(3-(α-CARBOXY-o-ANISAMIDO)-2-(2-HYDROXYETHOXY)PROPYL)HYDROXY-
  MERCURY MONOSODIUM SALT see HLO300
(3-(α-CARBOXY-p-ANISAMIDO)-2-HYDROXYPROPYL)HYDROXYMERCURY
  see HLP500
3-(α-CARBOXY-o-ANISAMIDO)-2-METHOXYPROPYL HYDROXYMERCURY,
  MONOSODIUM SALT see SIH500
CARBOXYBENZENE see BCL750
p-CARBOXYBENZENESULFONDICHLOROAMIDE see HAF000
CARBOXYBENZENESULFONYL AZIDE see CCD625
(o-CARBOXYBENZOYL)-p-AMINOPHENYLSULFONAMIDOTHIAZOLE
  see PHY750
N⁴-(o-CARBOXYBENZOYL)-N'-2-THIAZOLYL-SULFANILAMIDE
  SULFAPHIHALAZOLE see PHY750
CARBOXYBENZYLPENICILLIN PHENYL ESTER SODIUM SALT see CBO000
CARBOXYBENZYLPENICILLIN SODIUM see CBO250
α-CARBOXYCAPROYL-N,N'-DIPHENYLHYDRAZINE CALCIUM SALT HEMI-
  HYDRATE see CCD750
N-(2-CARBOXYCAPROYL)HYDRAZOBENZENE CALCIUM SALT HEMIHY-
  DRATE see CCD750
p-CARBOXYCARBANILIC ACID-4-BIS(2-CHLOROETHYLAMINO)PHENYL
  ESTER see CCE000
(6R-(6-α,7-β(R*)))-1-((2-CARBOXY-7-(((((5-CARBOXY-1H-IMIDAZOL-4-YL)
  CARBONYL)AMINO)PHENYLACETYLE)AMINO)-8-OXO-5-THIA-1-
  AZABICYCLO(4.2.0)OCT-2-EN-3-YL)METHYL)-4-(2-SULFOETHYL)-PYRIDIN-
  IUM HYDROXODIE, inner salt, MONOSODIUM SALT see CCS525
CARBOXYCYCLOPHOSPHAMIDE see CCE250
N-(1-((1-CARBOXY-5-DIAZO-4-OXOPENTYL)CARBAMOYL)-5-DIAZO-4-
  OXOPENTYL)-GLUTAMINE SODIUM SALT see ASO510
2-CARBOXY-1-((2,6-DICARBOXY-2,3-DIHYDRO-4(1H)-
  PYRIDINYLIDENE)ETHYLIDENE)5,6-DIHYDROXY-1H-INDOLIUM, HY-
  DROXIDE, INNER SALT, SULFATE (SALT) see BFV900
2-CARBOXY-4'-(DIMETHYLAMINO)AZOBENZENE see CCE500
3'-CARBOXY-4-DIMETHYLAMINOAZOBENZENE see CCE750
N-(2-CARBOXY-3,3-DIMETHYL-7-OXO-4-THIA-1-AZABICYCLO(3.2.0)HEPT-6-
  YL)-2-PHENYL-MALONAMIC ACID DISODIUM SALT see CBO250
2-CARBOXYDIPHENYLAMINE see PEG500
2-CARBOXYDIPHENYLARSINOUS ACID see PFI600
CARBOXYETHANE see PMU750
S-2-CARBOXYETHYL-l-CYSTEINE see CCF250

CARBOXYETHYLGERMANIUM SESQUIOXIDE see CCF125
2-CARBOXYETHYLGERMASESQUIOXANE see CCF125
3-CARBOXY-1-ETHYL-7-METHYL-1,8-NAPHTHIDIN-4-ONE see EID000
4'-(2-CARBOXYETHYL)PHENYL-trans-4-AMINOMETHYLCYCLOHEXANE
    CARBOXYLATE HYDROCHLORIDE see CDF380
2-(4-(1-CARBOXYETHYL)PHENYL)-1-ISOINDOLINONE see IDA400
3-((2-CARBOXYETHYL)THIO)ALANINE see CCF250
12-CARBOXYEUDESMA-3,11(13)-DIENE see EQR000
2-CARBOXYFURAN see FQF000
3-CARBOXY-4-HYDROXYBENZENESULFONIC ACID see SOC500
2'-CARBOXY-2-HYDROXY-4-METHOXYBENZOPHENONE(o-(2-HYDROXY-p-
    ANISOYL)BENZOIC ACID) see HLS500
3-(3-CARBOXY-4-HYDROXYPHENYL)-2-PHENYL-4,5-DIHYDRO-3H-BENZ(e)
    INDOLE see FAO100
(3-CARBOXY-2-HYDROXYPROPYL)TRIMETHYLAMMONIUM CHLORIDE
    see CCK650
(-)-(3-CARBOXY-2-HYDROXYPROPYL)TRIMETHYLAMMONIUM CHLORIDE
    see CCK660
l-(3-CARBOXY-2-HYDROXYPROPYL)TRIMETHYLAMMONIUM CHLORIDE
    see CCK660
3-CARBOXY-5-HYDROXY-1-p-SULFOPHENYL-4-o-SUL-
    FOPHENYLAZOPYRAZOLE TRISODIUM SALT see FAG140
(R)-3-CARBOXY-2-HYDROXY-N,N,N-TRIMETHYL-1-PROPANAMINIUM CHLO-
    RIDE see CCK660
3-CARBOXY-2-HYDROXY-N,N,N-TRIMETHYL-1-PROPANAMINIUM CHLO-
    RIDE (9CI) see CCK650
1-(4'-CARBOXYLAMIDOPHENYL)-3,3-DIMETHYLTRIAZINE see CCC325
(3-(4-CARBOXYLATOMETHOXY)PHENYL)-2-HYDROXYPROPYL)HYDROXY-
    MERCURATE(1-), SODIUM see CCF500
l-N-CARBOXYLEUCINE-N-BENZYL-1-(1,2-DIBROMOETHYL) ESTER
    see CBR210
l-N-CARBOXYLEUCINE N-BENZYL 1-VINYL ESTER see CBR215
6-CARBOXYL-4-HYDROXYLAMINOQUINOLINE-1-OXIDE see CCF750
6-CARBOXYL-4-NITROQUINOLINE-1-OXIDE see CCG000
(CARBOXYMETHOXY)AMINE see ALQ650
2-(CARBOXY-METHOXY)BENZALDEHYDE SODIUM SALT see CCG250
(3-(o-(CARBOXYMETHOXY)BENZAMIDO)-2-METHOXYPROPYL)HYDROXY
    MERCURY, MONOSODIUM SALT compounded with THEOPHYLLINE
    see SAQ000
2'-CARBOXYMETHOXY-4,4'-BIS(3-METHYL-2-BUTENYLOXY)CHALCONE
    see IMS300
(4-(CARBOXY METHOXY)-3-CHLOROPHENYL)(5,5-DIETHYL-2,4,6(1H,3H,5H)-
    PYRIMIDINETRIONATO-O²-MERCURY, MONOSODIUM SALT see CCG500
4-CARBOXYMETHYLBIPHENYL see BGE125
CARBOXYMETHYL CARBAMIMONIOTHIOATE CHLORIDE see CCH199
1-CARBOXYMETHYL-1-CARBOXYETHOXYETHYL-2-COCO-IMIDAZOLINIUM
    BETAINE see AOC250
CARBOXYMETHYL CELLULOSE see SFO500
CARBOXYMETHYL CELLULOSE, AMMONIUM SALT see CCH000
CARBOXYMETHYLCELLULOSE NORDIC see CCH000
CARBOXYMETHYL CELLULOSE, SODIUM see SFO500
CARBOXYMETHYL CELLULOSE, SODIUM SALT see SFO500
l-CARBOXYMETHYLCYSTEINE see CBR675
S-(CARBOXYMETHYL)CYSTEINE see CBR675
S-CARBOXYMETHYLCYSTEINE see CCH125
4-o-(CARBOXYMETHYL)-1-DEOXY-1,4-DIHYDRO-4-HYDROXY-1-OXO-
    RIFAMYCIN γ-LACTONE see RKP400
N-(CARBOXYMETHYL)-N-(2-(2,6-DIMETHYLPHENYL)AMINO)-2-OXOETHYL)-
    GLYCINE (9CI) see LFO300
3,3'-(CARBOXYMETHYLENE)BIS(4-HYDROXYCOUMARIN) ETHYL ESTER
    see BKA000
N-(CARBOXYMETHYL)GLYCINE see IBH000
N-(CARBOXYMETHYL)-N'-(2-HYDROXYETHYL)-N,N'-ETHYLENEDIGLYCINE
    see HKS000
(o-CARBOXYMETHYL)HYDROXYLAMINE see ALQ650
((CARBOXYMETHYLIMINO)BIS(ETHYLENENITRILO))TETRAACETIC ACID
    see DJG800
2-CARBOXYMETHYLISOTHIOURONIUM CHLORIDE see CCH199
N-(γ-CARBOXYMETHYLMERCAPTOMERCURI-β-METHOXY)
    PROPYLCAMPHORAMIC ACID DISODIUM SALT see TFK270
2-(CARBOXYMETHYLMERCAPTO)PHENYLSTIBONIC ACID see CCH250
2-(CARBOXYMETHYLMERCAPTO)PHENYL-STIBONSAEURE (GERMAN)
    see CCH250
5-CARBOXYMETHYL-3-METHYL-2H-1,3,5-THIADIAZINE-2-THIONE
    see MPP750
CARBOXYMETHYLNITROSOUREA see CCH500, NKI500
1-(CARBOXYMETHYL)-1-NITROSOUREA see NKI500
6-(5-CARBOXY-3-METHYL-2-PENTENYL)-7-HYDROXY-5-METHOXY-4-
    METHYLPHTHALIDE see MRX000
4-o-(CARBOXYMETHYL)RIFAMYCIN see RKK000
(CARBOXYMETHYLTHIO)ACETIC ACID see MCM750
3-(CARBOXYMETHYLTHIO)ALANINE see CBR675
3-((CARBOXYMETHYL)THIO)ALANINE see CCH125
l-3-((CARBOXYMETHYL)THIO)ALANINE see CBR675

2-CARBOXYMETHYLTHIOBENZOTHIAZOLE see CCH750
5-CARBOXYMETHYL-3-p-TOLYL-THIAZOLIDINE-2,4-DIONE-2-
    ACETOPHENONE HYDRAZONE see CCH800
(CARBOXYMETHYL)TRIMETHYLAMMONIUM IODIDE-4-
    (DIMETHYLAMINO)-3-ISOPROPYLPHENYL ESTER see IOU200
(CARBOXYMETHYL)TRIMETHYLAMMONIUM IODIDE-5-
    (DIMETHYLAMINO)-4-ISOPROPYL-o-TOLYL ESTER see DOW875
(CARBOXYMETHYL)TRIMETHYLAMMONIUM IODIDE-6-
    (DIMETHYLAMINO)-4-ISOPROPYL-m-TOLYL ESTER see DQD500
N-CARBOXY-3-MORPHOLINOSYDNONIMINE ETHYL ESTER see MRN275
1-CARBOXY-4-NITROBENZENE see CCI250
6-CARBOXY-4-NITROQUINOLINE-1-OXIDE see CCG000
3-(3-CARBOXY-1-OXOPROPOXY)-11-OXOOLEAN-12-EN-29-OIC ACID, DISOD-
    IUM SALT (3-β,20-β) see CBO500
21-(3-CARBOXY-1-OXOPROPOXY)-5-β-PREGNANE-3,20-DIONE SODIUM
    SALT see VJZ000
3-CARBOXY-2,4-PENTADIENALLACTOL see APV750
2-(5-CARBOXYPENTYL)-4-THIAZOLIDONE see CCI500
3-CARBOXYPHENOL see HJI100
4-CARBOXYPHENOL see SAI500
o-CARBOXYPHENYL ACETATE see ADA725
l-N-CARBOXY-3-PHENYLALANINE-N-BENZYL ESTER see CBR220
l-N-CARBOXY-3-PHENYLALANINE N-BENZYL 1-VINYL ESTER see CBR235
p-CARBOXYPHENYLAMINE see AIH600
2-((2-CARBOXYPHENYL)AMINO)-4-CHLOROBENZOIC ACID DISODIUM
    SALT see LHZ600
p-CARBOXY PHENYLARSENOXIDE see CCI550
(p-CARBOXYPHENYL)CHLOROMERCURY see CHU500
9-o-CARBOXYPHENYL-6-DIETHYLAMINO-3-ETHYLIMINO-3-ISOXANTHENE,
    3-ETHOCHLORIDE see FAG070
(9-(o-CARBOXYPHENYL)-6-(DIETHYLAMINO)-3H-XANTHEN-3-YLIDENE)
    DIETHYLAMMONIUM CHLORIDE see FAG070
9-(o-CARBOXYPHENYL)-6-HYDROXY-3-ISOXANTHENONE see FEV000
9-o-CARBOXYPHENYL-6-HYDROXY-3-ISOXANTHONE, DISODIUM SALT
    see FEW000
(o-CARBOXYPHENYL)HYDROXY-MERCURY SODIUM SALT see SHT500
9-(o-CARBOXYPHENYL)-6-HYDROXY-2,4,5,7-TETRAIODO-3-ISOXANTHONE
    see FAG040
9-(o-CARBOXYPHENYL)-6-HYDROXY-3H-XANTHEN-3-ONE see FEV000
4'-CARBOXYPHENYLMETHANESULFONANILIDE, SODIUM SALT see CCJ000
o-CARBOXYPHENYL PHOSPHATE see PHA575
((o-CARBOXYPHENYL)THIO)ETHYLMERCURY SODIUM SALT see MDI000
CARBOXYPHOSPHAMIDE see CCE250
4-CARBOXYPHTHALATO(1,2-DIAMINOCYCLOHEXANE)PLATINUM(II)
    see CCJ350
4-CARBOXYPHTHALIC ANHYDRIDE see TKV000
3-o-(β-CARBOXYPROPIONYL)-11-OXO-18-β-OLEAN-12-EN-30-OIC ACID, DIS-
    ODIUM SALT see CBO500
3-β-(3-CARBOXYPROPIONYLOXY)-11-OXO-OLEAN-12-EN-30-OIC ACID
    see BGD000
3-CARBOXYPROPYL(2-PROPENYL)NITROSAMINE see CCJ375
3-CARBOXYPYRIDINE see NCQ900
4-CARBOXYPYRIDINE see ILC000
4-CARBOXYRESORCINOL see HOE600
6-CARBOXYURACIL see OJV500
4-CARBOXYTHIAZOLIDINE see TEV000
CARBOXY VINYL POLYMER see CCJ400
CARBRITAL see NBU000
CARBUTAMID see BSM000
CARBUTAMIDE see BSM000
CARBUTEN see MBW750
CARBUTEROL HYDROCHLORIDE see BQD500
CARBYL see CBH250
CARBYL SULFATE see DXI500
CARBYNE see CEW500
CARCHOLIN see CBH250
CARCINOCIDIN see CCN250
CARCINOLIPIN see CCJ500
CARDAMINE see DJS200
CARDAMON see CCJ625
CARDAMON OIL see CCJ625
CARDELMYCIN see SMB000
CARDENAL de MACETA (MEXICO) see CCJ825
CARDIAGEN see DJS200
CARDIAMID see DJS200
CARDIAMINA see DJS200
CARDIAMINE see DJS200
CARDIDIGIN see DKL800
CARDIEM see DNU600
CARDIGIN see DKL800
CARDILATE see VSA000
CARDIMON see DJS200
CARDINAL FLOWER see CCJ825
CARDINOPHILLIN see CCN500

CARDINOPHYLLIN see CCN500
CARDIO see CCK125
CARDIOFILINA see TEP500
CARDIOGRAFIN see AOO875
CARDIO-GREEN see CCK000
CARDIO-KHELLIN see AHK750
CARDIOLANATA see LAU400
CARDIOLIPOL see NCW300
CARDIOMIN see TEP500
CARDIOMONE see AOA125
CARDION see POB500
CARDIOQUIN see GAX000
CARDIORYTHMINE see AFH250
CARDIOTRAST see DNG400
CARDIOVITE see POB500
CARDIS see CCK125
CARDITIN see PEV750
CARDITOXIN see DKL800
CARDOGENEN-(20:22)-DIOL-(3-β,14) (GERMAN) see DMJ000
CARDOGENEN-(20:22)-TRIOL-(3-β,12,14) (GERMAN) see DKN300
β-CARDONE see CCK250
CARDOPHYLIN see TEP500
CARDOPHYLLIN see TEP500
CARDOVERINA see PAH250
CARDOXIN see PCP250
CARENA see TEP500
3-CARENE see CCK500
S-3-CARENE see CCK500
Δ3-CARENE see CCK500
CARFECILLIN see CBN750
CARFECILLIN SODIUM see CBO000
CARFENE see ASH500
CARFIMAT see PGE000
CARFLOC D 1000 see CNH125
CARGUTOCIN see CCK550
CARIAQUILLO (PUERTO RICO) see LAU600
CARICIDE see DIW000, DIW200
CARIDOROL see ICC000
CARINA see PKQ059
CARINEX GP see SMQ500
CARIOMIN see TEP500
CARIOTA (CUBA) see FBW100
CARISOL see IPU000
CARISOMA see IPU000
CARISOPRODATE see IPU000
CARISOPRODATUM see IPU000
CARISOPRODOL see IPU000
CARITROL see DIW200
CARLONA P see PMP500
CARLSODAL see IPU000
CARLSOMA see IPU000
CARLYTENE see TFY000
CARMAZINE see DXI400
CARMAZON see POB500
CARMETHOSE see SFO500
CARMETIZIDE see CCK575
CARMINAPH see PEJ500
CARMINE BLUE (BIOLOGICAL STAIN) see FAE100
CARMINOMICIN I see CCK625
CARMINOMYCIN see KBU000
CARMINOMYCIN HYDROCHLORIDE see KCA000
CARMINOMYCIN I see CCK625
CARMIN (PUERTO RICO) see ROA300
CARMOFUR see CCK630
CARMOISIN (GERMAN) see HJF500
CARMOISINE ALUMINUM LAKE see HJF500
CARMOISINE SUPRA see HJF500
CARMOL HC see HHQ800
CARMUBRIS see BIF750
CARMUSTIN see BIF750
CARMUSTINE see BIF750
CARNACID-COR see GEW000
CARNATION RED TONER B see NAY000
CARNELIO HELIO RED see MMP100
CARNELIO RED 2G see DVB800
CARNELIO RED R see CJD500
CARNELIO YELLOW GX see DEU000
CARNITINE CHLORIDE see CCK650
l-CARNITINE CHLORIDE see CCK660
l-CARNITINE HYDROCHLORIDE see CCK660
(R)-CARNITINE HYDROCHLORIDE see CCK660
CARNOSINE see CCK665
l-CARNOSINE see CCK665
CAROB BEAN GUM see LIA000

CAROB FLOUR see LIA000
CAROID see PAG500
CAROLINA ALLSPICE see CCK675
CAROLINA JASMINE see YAK100
CAROLINA PINK see PIH800
CAROLINA TEA see HGF100
CAROLINA WILD WOODBINE see YAK100
CAROLINA YELLOW JASMINE see YAK100
CAROLYSINE see BIE500
CARPENE see DXX400
CARPIPRAMINE see CBH500
CARPIPRAMINE DIHYDROCHLORIDE see CCK775
CARPIPRAMINE DIHYDROCHLORIDE MONOHYDRATE see CCK780
CARPIPRAMINE HYDROCHLORIDE see CCK775
CARPIPRAMINE MALEATE see CCK790
CARPROFEN see CCK800
CARQUEJOL see CCL109
CARRAGEEN see CCL250
kappa-CARRAGEEN see CCL350
CARRAGEENAN (FCC) see CCL250
kappa-CARRAGEENAN see CCL350
CARRAGEENAN, CALCIUM(II) SALT see CAO250
CARRAGEENAN, DEGRADED see CCL500
CARRAGEENAN GUM see CCL250
CARRAGEENAN, SODIUM SALT see SFP000
kappa-CARRAGEENIN see CCL350
CARRAGHEANIN see CCL250
CARRAGHEEN see CCL250
CARRAGHEENAN see CCL250
CARREL-DAKIN SOLUTION see SHU500
CARRIOMYCIN, SODIUM SALT see SFP500
CARROT, SEED EXTRACT see GAP500
CARROT SEED OIL see CCL750
CARRTIME see BBK500
CARSIL see SDX625
CARSODOL see IPU000
CARSONOL SLS see SIB600
CARSONON N-9 see PKF000
CARSONON PEG-4000 see PJT750
CARSOQUAT SDQ-25 see DTC600
CARSORON see DER800
CARTAGYL see ARQ750
CARTAP HYDROCHLORIDE see BHL750
CARTEOLOL see CCL800
CARTEOLOL HYDROCHLORIDE see MQT550
CARTOSE see GFG000
CARTWHEELS see AOB250
CARUBICIN see CCK625
CARVACROL see CCM000
CARVACRON see HII500
CARVANIL see CCK125
CARVASEPT see CEX275
CARVASIN see CCK125
l-CARVEOL see MKY250
CARVIL see MOV000
4-CARVOMENTHENOL see TBD825
3-CARVOMENTHENONE see MCF250
(−)-CARVONE see CCM120
CARVONE see MCD250
(+)-CARVONE see CCM100
l-CARVONE see CCM120
d-CARVONE see CCM100
l(−)-CARVONE see CCM120
(R)-CARVONE see CCM120
(S)-CARVONE see CCM100
d(+)-CARVONE see CCM100
(S)-(+)-CARVONE see CCM100
l-CARVYL ACETATE see CCM750
l-CARVYL PROPIONATE see MCD000
CARYNE see CEW500
CARYOLYSIN see BIE250
CARYOLYSINE see BIE500
CARYOLYSINE HYDROCHLORIDE see BIE500
CARYOPHYLENE OXIDE see CCN100
CARYOPHYLLENE see CCN000
β-CARYOPHYLLENE (FCC) see CCN000
CARYOPHYLLENE EPOXIDE see CCN100
β-CARYOPHYLLENE EPOXIDE see CCN100
CARYOPHYLLENE OXIDE see CCN100
(-)-CARYOPHYLLENE OXIDE see CCN100
β-CARYOPHYLLENE OXIDE see CCN100
CARYOPHYLLIC ACID see EQR500
CARYOTA (VARIOUS SPECIES) see FBW100
CARZAZO (DOMINICAN REPUBLIC) see CAK325

CARZINOCIDIN see CCN250
CARZINOPHILIN see CCN500
CARZINOPHILIN A see CCN750
CARZINOSTATININ see CCO000
CARZOL see CJJ250
CARZOL SP see DSO200
CARZONAL see FMB000, FMM000
CASALIS GREEN see CMJ900
CASANTIN see DHF600, DII200
CASCABELILLO (PUERTO RICO) see RBZ400
CASCARA see MBU825
CASCARITA see CMV390
CASEIN and CASEINATE SALTS (FCC) see SFQ000
CASEIN-SODIUM see SFQ000
CASEIN, SODIUM COMPLEX see SFQ000
CASEINS, SODIUM COMPLEXES see SFQ000
CASHES see PJJ300
CASHMILON, combustion products see ADX750
CASIFLUX VP 413-004 see WCJ000
CASIMAN (MEXICO, PUERTO RICO) see SLE890
CASING HEAD GASOLINE (DOT) see GBY000
CASORON 133 see DER800
CASPAN see MDD750
CASSAINE HYDROCHLORIDE see CCO675
CASSAPPRET SR see PKF750
CASSAVA see CCO680, CCO700
CASSAVA, MANIHOT UTILISSIMA see CCO700
CASSAVA MEAL see CCO700
CASSAVA POWDER see CCO700
CASSE (HAITI) see GIW300
CASSEL BROWN see MAT500
CASSEL GREEN see MAT250
CASSELLA 532 see OJD300
CASSELLA 4489 see CBR500
CASSENA see HGF100
CASSIA ALDEHYDE see CMP969
CASSIA FISTULA see GIW300
CASSIA OCCIDENTALIS see CNG825
CASSIA OIL see CCO750
CASSIAR AK see ARM268
CASSIA TORA Linn., leaf extract see CCO800
CASSURIT LR see DTG000
CASTANEA SATIVA MILL TANNIN see CDM250
CASTOR BEAN see CCP000
CASTOR BEANS (DOT) see CCP000
CASTOR FLAKE (DOT) see CCP000
CASTOR MEAL (DOT) see CCP000
CASTOR OIL see CCP250
CASTOR OIL AROMATIC see CCP250
CASTOR OIL PLANT see CCP000
CASTOR POMACE (DOT) see CCP000
CASTRIX see CCP500
CASTRON see PDN000
CAT (herbicide) see BJP000
CATACIDE see DIW000
CATALIN CAO-3 see BFW750
CATALOID see SCH000
CATALYTIC-DEWAXED HEAVY NAPHTHENIC DISTILLATE see MQV865
CATALYTIC-DEWAXED HEAVY PARAFFINIC DISTILLATE see MQV868
CATALYTIC-DEWAXED LIGHT NAPHTHENIC DISTILLATE see MQV867
CATALYTIC-DEWAXED LIGHT PARAFFINIC DISTILLATE see MQV870
CATAMINE AB see AFP250
CATANAC SP see CCP675
CATANAC SP ANTISTATIC AGENT see CCP675
CATANIL see CKK000
CATAPRES see CMX760
CATAPRESAN see CMX760
CATAPYRIN see AFW500
CATATOXIC STEROID No. 1 see CCP750
CATECHIN see CCP800, CCP850, CCP875
(+)-CATECHIN see CCP875
d-CATECHIN see CCP875
d-(+)-CATECHIN see CCP875
CATECHIN (FLAVAN) see CCP875
CATECHIN HYDRATE see DMD000
CATECHINIC ACID see CCP875
CATECHOL see CCP850, CCP875
(+)-CATECHOL see CCP875
d-CATECHOL see CCP875
CATECHOL DIETHYL ETHER see CCP900
CATECHOL (FLAVAN) see CCP875
CATECHU see CCP800
CATECHUIC ACID see CCP875
CATENULIN see NCF500

CATERGEN see CCP875
CATESBY'S VINE (BAHAMAS) see YAK300
CATEUDYL see QAK000
CATHA EDULIS Forsk, leaf extract see KGK350
CATHOCIN see SMB000
CATHOMYCIN see SMB000
CATHOMYCIN SODIUM see NOB000
CATHOMYCIN SODIUM LYOVAC see NOB000
CATILAN see CDP250
CATIONIC SP see CCP675
CATOLIN 14 see MJO500
CATOVITAN see PNS000
CATRAL see PDN000
CATRAN see PDN000
CATRON HYDROCHLORIDE see PDN250
CATRONIACID see PDN250
CATRONIAZIDE see PDN000
C. ATROX VENOM see WBJ600
CAT'S BLOOD see ROA300
CATS EYES see PAM780
CAULOPHYLLUM THALICTROIDES see BMA150
CAULOPHYLLUM THALICTROIDES, glycoside extract see CCQ125
CAURITE see DTG700
CAUSOIN see DKQ000
CAUSTIC BARLEY see VHZ000
CAUSTIC POTASH see PLJ500
CAUSTIC POTASH, dry, solid, flake, bead, or granular (DOT) see PLJ500
CAUSTIC POTASH, liquid or solution (DOT) see PLJ500
CAUSTIC SODA see SHS000
CAUSTIC SODA, bead (DOT) see SHS000
CAUSTIC SODA, dry (DOT) see SHS000
CAUSTIC SODA, flake (DOT) see SHS000
CAUSTIC SODA, granular (DOT) see SHS000
CAUSTIC SODA, liquid (DOT) see SHS000
CAUSTIC SODA, solid (DOT) see SHS000
CAUSTIC SODA, solution see SHS500
CAUSTIC SODA, solution (DOT) see SHS000
CAUTIVA (PUERTO RICO) see AFQ625
CAVALITE BRILLIANT BLUE R see BMM500
CAV-ECOL see WBJ700
CAVINTON see EGM100
CAVI-TROL see SHF500
CAVODIL see PDN000
CAVONLY see TDA500
CAVUMBREN see BGB315
CAYENNE PEPPER see PCB275
CAZ PENTAHYDRATE see CCQ200
4-CB see TBO700
CB-154 see BNB250, BNB325
CB 304 see THM750
CB-337 see HMC000
804 CB see CPJ000
CB 1331 see PCU425
CB 1348 see CDO500
CB 1356 see BHY625
CB-1385 see AJE000
CB 1506 see CHC750
1522 CB see ABH500
1613-CB see BTA125
CB 1639 see AJK250
1678 CB see IDA500
1678 CB see PMX500
CB 1729 see BHT250
C.B. 2041 see BOT250
CB2058 see DNX200
CB 2095 see DNX000
CB 2511 see BKM500
CB 2562 see TFU500
CB 3008 see BHV000
CB 3025 see BHV250, PED750
3026 C.B. see SAX200
CB-3026 see SAX200
CB-3307 see BHT750
CB 4261 see CFG750
CB 4306 see CDQ250
4311 CB see DKV700
4361 CB see CFG750
CB-4564 see CQC500, CQC650
CB-4835 see BIA250
CB 8000 see DEQ200
CB 8019 see IPU000
8022 C.B. see ENX600
8065 C.B. see MPN000
8089 CB see FGU000

8089 C.B. see FLK100
CB 10286 see CCC325
2-CBA see CEL250
C 34647Ba see BAC275
C-39089-Ba see THK750
C 49249Ba see EEH575
CBBP see THY500
CBC 806495 see TFQ750
CBC 900139 see SBE500
CBC 906288 see TND250
CB 804 CALCIUM see CPJ250
CBD see CBD599
CBD 90 see TIQ750
CBDCA see CCC075
CBDZ see CBA100
8102 CB HYDROCHLORIDE see BEO750
C. BICOLOR see CAL125
CBN see CBD625, CEW500
C. BONDUC see CAK325
CBPC see CBO250
CBS-1114 see PEK675
CB 1348 SODIUM SALT see CDO625
(CBZ)GLY see CBR125
CC 914 see CBI250
CC-1065 see APT375
CC 11511 see DYC800
CCA see LHZ600
"C" CARRIE see CNE750
CCC see CAQ250
CCC PLANT GROWTH REGULATOR see CMF400
C. CERASTES VENOM see CCX620
CCH see HOV500
CCHO see CPD000
C 7337 CIBA see PDW400
CCK 179 see DLL400
CCL see PAG075
CCN52 see RLF350
C.C. No. 914 see CBI250
CCNU see CGV250
CCRG 81010 see MQY110
CCS see CJT750
CCS 203 see BPW500
CCS 301 see BPW750
CCUCOL see ASB250
CD see TGD000
CD 2 see LFK000
CD 68 see CDR750
CD-3400 see MKR100
CDA 101 see CNI000
CDA 102 see CNI000
CDA 110 see CNI000
CDA 122 see CNI000
CDAA see CFK000
CDAAT see CFK000
CDA: CETYLCIDE see EKN500
CDB 63 see SGG500
CDBM see CFK500
CDC see CDL325
CDCA see CDL325
CDDP see PJD000
CDEC see CDO250
2-CDF see PBT100
6CDG see CFJ375
CDHA see DGT600
C. DIURNUM see DAC500
CDM see CJJ250
CDNA see RDP300
CDP see LFK000
CDP-CHOLIN see CMF350
CDP-CHOLINE see CMF350
CDP-COLINA see CMF350
CdPT see CAI350
C. DRUMMONDII see CAK325
CDT see BJP000
CDT see COW935
114 C.E. see HNM000
746 CE see ECU550
305CE see FDA875
CE 3624 see TGA600
C-14919 E-2 see MAB400
CEBESINE see OPI300
CEBETOX see MIW250
CEBITATE see ARN125
CEBOLLA see WBS850

CEBOLLEJA (MEXICO) see GJU460
CEBROGEN see GFO050
CEBRUM see MDQ250
CECALGINE TBV see SEH000
CE CE CE see CMF400
CECENU see CGV250
CECIL see CNE750
CECLOR see CCR850
CECOLENE see TIO750
CEDAD see BCA000
CEDAR LEAF OIL see CCQ500
CEDARWOOD OIL ATLAS see CCQ750
CEDARWOOD OIL MOROCCAN see CCQ750
CEDARWOOD OIL (VIRGINIA) see CCR000
CEDILANID see LAU000
CEDOCARD see CCK125
CEDRAMBER see CCR525
CEDRANE, 8,9-EPOXIDE see CCR510
8-β-H-CEDRAN-8-OL ACETATE see CCR250
CEDRANYL ACETATE see CCR250
CEDR-8-ENE see CCR500
α-CEDRENE see CCR500
CEDR-8-ENE EPOXIDE see CCR510
CEDROL FORMATE see CCR524
CEDROL METHYL ETHER see CCR525
CEDRO OIL see LEI000
CEDRUS ATLANTICA OIL see CCQ750
CEDRYL ACETATE see CCR250
CEDRYL FORMATE see CCR524
CEE see PMB000
CEE DEE see HCQ500
CEENU see CGV250
CEEPRYN see CCX000, CDF750
CEEPRYN CHLORIDE see CCX000
CEFACETRILE SODIUM see SGB500
CEFACIDAL see CCS250
CEFACLOR see CCR850, PAG075
CEFACLOR HYDRATE see CCR850
CEFADOL see CCR875
CEFADOLE see CCX300
CEFADYL see HMK000
CEFA-ISKIA see ALV000
CEFALOGLYCIN see CCR890
CEFALOJECT see HMK000
CEFALORIDIN see TEY000
CEFALORIDINE see TEY000
CEFALORIZIN see TEY000
CEFALOTHINE SODIUM see SFQ500
CEFALOTIN see CCX250
CEFALOTINA SODICA (SPANISH) see SFQ500
CEFALOTO see ALV000
CEFAMANDOL see CCX300
CEFAMANDOLE see CCX300
l-CEFAMANDOLE see CCX300
CEFAMANDOLE NAFATE see FOD000
CEFAMANDOLE SODIUM see CCR925
CEFAMEDIN see CCS250
CEFAMEZIN see CCS250
CEFAPIRIN (GERMAN) see CCX500
CEFAPIRIN SODIUM see HMK000
CEFAPRIN SODIUM see HMK000
CEFATIN see OAV000
CEFATOXIME SODIUM see CCR950
CEFATREXYL see HMK000
CEFATRIZINE see APT250
CEFAZEDONE SODIUM SALT see RCK000
CEFAZIL see CCS250
CEFAZINA see CCS250
CEFAZOLIN see CCS250
CEFAZOLINE SODIUM see CCS250
CEFAZOLIN SODIUM SALT see CCS250
CEFEDRIN see CCX600
CEFLORIN see TEY000
CEFMENOXIME HEMIHYDROCHLORIDE see CCS300
CEFMETAZOLE see CCS350
CEFMETAZOLE SODIUM see CCS360
CEFMINOX see CCS365
CEFOPERAZONE SODIUM see CCS369
CEFOTAN see CCS371
CEFOTAXIME SODIUM see CCR950
CEFOTETAN see CCS373
CEFOTETAN DISODIUM SALT see CCS371
CEFOTIAM DIHYDROCHLORIDE see CCS375
CEFOTIAM HYDROCHLORIDE see CCS375

CEFOXITIN see CCS500
CEFOXITIN SODIUM SALT see CCS510
CEFOXOTIN SODIUM see CCS510
CEFPIMIZOLE SODIUM see CCS525
CEFRACYCLINE SUSPENSION see TBX000
CEFRACYCLINE TABLETS see TBX250
CEFRADINE see SBN440
CEFROXADIN see CCS530
CEFROXADIN DIHYDRATE see CCS535
CEFROXADINE see CCS530
CEFSULODIN SODIUM see CCS550
CEFTAZIDIME see CCQ200
CEFTAZIDIME PENTAHYDRATE see CCQ200
CEFTEZOLE SODIUM see CCS560
CEFTIZOXIME SODIUM see EBE100
CEFTIZOXIM-NATRIUM (GERMAN) see EBE100
CEFUROXIM see CCS600
CEFUROXIME see CCS600
CEFUROXIME AXETIL see CCS625
CEFUROXIME SODIUM see SFQ300
CEFUROXIME SODIUM SALT see SFQ300
CEFZONAME SODIUM see CCS635
CEGLUTION see LGZ000
CEKIURON see DXQ500
CEKUDIFOL see BIO750
CEKUFON see TIQ250
CEKUGIB see GEM000
CEKUMETA see TDW500
CEKUMETHION see MNH000
CEKUQUAT see PAJ000
CEKUSAN see BJP000
CEKUSAN see DGP900
CEKUSIL see ABU500
CEKUSIL UNIVERSAL A see MEO750
CEKUSIL UNIVERSAL C see MEP250
CEKUTHOATE see DSP400
CEKUTROTHION see DSQ000
CEKUZINA-S see BJP000
CEKUZINA-T see ARQ725
CELA S-2225 see EGV500
CELA S-2957 see CLJ875
CELA A-36 see DAE600
CELAMERCK S-2957 see CLJ875
CELANAR see PKF750
CELANDINE see CCS650
CELANEX see BBQ500
CELANOL DOS 75 see DJL000
CELANTHRENE BRILLIANT BLUE see MGG250
CELANTHRENE PURE BLUE BRS see TBG700
CELANTHRENE RED VIOLET R see DBP000
CELA S 1942 see BNL250
CELASTROL-METHYLETHER see PMD525
CELASTRUS SCANDENS see AHJ875
CELATHION see CLJ875
CELATOX see TNX375
CELESTAN-DEPOT see CCS675
CELESTODERM see VCA000
CELESTONE see BFV750
CELESTONE CHRONODOSE see CCS675
CELESTONE SOLOSPAN see CCS675
CELESTONE SOLUSPAN see BFV755
CELESTONE SOLUSPAN see CCS675
CELGARD 2500 see PMP500
CELINHOL -A see OAV000
CELIOMYCIN see VQZ000
CELIPROLOL HYDROCHLORID (GERMAN) see SBN475
CELIPROLOL HYDROCHLORIDE see SBN475
CELLACETATE see CCU050
CELLATIVE see AAE500
CELLIDRIN see ZVJ000
CELLITAZOL B see DCJ200
CELLITON BLUE FFR see MGG250
CELLITON BLUE G see TBG700
CELLITON BRILLIANT YELLOW 8G see MEB750
CELLITON FAST RED VIOLET R see DBP000
CELLITON FAST YELLOW G see AAQ250
CELLITON ORANGE R see AKP750
CELLMIC S see OPE000
CELLOCIDIN see ACJ250
CELLOFAS see SFO500
CELLOFOR (CZECH) see DWO800
CELLOGEL C see SFO500
CELLOIDIN see CCU250
CELLON see TBQ100

CELLOPHANE see CCT250
CELLOSOLVE (DOT) see EES350
CELLOSOLVE ACETATE (DOT) see EES400
CELLOSOLVE ACRYLATE see ADT500
CELLOSOLVE SOLVENT see EES350
CELLPRO see SFO500
CELLRYL see CCT825
CELLUFIX FF 100 see SFO500
CELLUFLEX see CGO500
CELLUFLEX 179C see TNP500
CELLUFLEX DOP see DVL600
CELLUFLEX DPB see DEH200
CELLUFLEX FR-2 see TNG750
CELLUFLEX TPP see TMT750
CELLUGEL see SFO500
CELLULASE AP3 see CCT900
"CELLULOID" see CCU000
CELLULOID, in blocks, rods, rolls, sheets, tubes (DOT) see CCU000
CELLULOID SCRAP (DOT) see CCU000
CELLULOSE, ACETATE HYDROGEN 1,2-BENZENEDICARBOXYLATE (9CI)
    see CCU050
CELLULOSE ACETATE MONOPHTHALATE see CCU050
CELLULOSE, ACETATE PHTHALATE see CCU050
CELLULOSE ACETOPHTHALATE see CCU050
CELLULOSE ACETYLPHTHALATE see CCU050
CELLULOSE GEL see CCU100
CELLULOSE GLYCOLIC ACID, SODIUM SALT see SFO500
CELLULOSE GUM see SFO500
CELLULOSE, MICROCRYSTALLINE see CCU100
CELLULOSE NITRATE see CCU250
CELLULOSE, POWDERED see CCU150
CELLULOSE SODIUM GLYCOLATE see SFO500
CELLULOSE TETRANITRATE see CCU250
CELLUPHOS 4 see TIA250
CELLU-QUIN see BLC250
CELLUTATE RED VIOLET RH see DBP000
CELMER see ABU500, MEP250
CELMIDE see EIY500
CELMIDOL see CCR875
CELMONE see NAK500
CELOCURINE see BJI000
CELOGEN OT see OPE000
CELON A see EIX000
CELON ATH see EIX000
CELON E see EIV000
CELON H see EIV000
CELON IS see EIV000
CELONTIN see MLP800
CELOSEN AZ see ASM270
CELOSPOR see SGB500
CELPHIDE see AHE750
CELPHOS see AHE750, PGY000
CELTHIGN see MAK700
CELUTATE BLUE BLT see MGG250
CEMENT, adhesive (DOT) see CCV250
CEMENT, leather see CCV000
CEMENT (liquid) see CCV250
CEMENT (pyroxylin) see CCV750
CEMENT (roofing liquid) see CCW000
CEMENT (rubber) see CCW250
CEMENT BLACK see MAS000
CEMENT, PORTLAND see PKS750
CEMENT, PYROXYLIN (DOT) see CCV750
CEMENT, ROOFING, liquid (DOT) see CCW000
CEMENT, RUBBER (DOT) see CCW250
CEMIDON see ILD000
CEMULSOL D-8 see PJY100
CEMULSOL 1050 see PJY100
CEMULSOL A see PJY100
CEMULSOL C 105 see PJY100
CENESTIL see PMS825
CENITRON OB see OPE000
CENOLATE see ARN125
CENOL GARDEN DUST see RNZ000
CENOMYCIN see CCS510
CENSTIM see DLH630
CENSTIN see DLH600, DLH630
CENTBUCRIDINE HYDROCHLORIDE see CCW375
CENTBUTINDOLE see CCW500
CENTCHROMAN see CCW725
CENTCHROMAN HYDROCHLORIDE see CCW750
CENTEDEIN see MNQ000
CENTEDRIN see RLK000
CENTELASE see ARN500

CENTIMIDE see HCQ500
CENTPHENAQUIN see CCW800
CENTRALGIN see DAM700
CENTRALINE BLUE 3B see CMO250
CENTRAX see DAP700
CENTREDIN see MNQ000
CENTRINE see DOY400
CENTROFENOXINA see DPE000
CENTROPHENOXINE see AAE500
CENTRUROIDES SUFFUSUS SUFFUSUS VENOM see CCW925
CENTURINA see AOD000
CENTURY 1240 see SLK000
CENTURY CD FATTY ACID see OHU000
CENTYL see BEQ625
CEP see CDS125
2-CEPA see CDS125
CEPA CABALLERO (CUBA) see MQW525
CEPACILINA see BFC750
CEPACILLINA see BFC750
CEPACOL see CDF750
CEPACOL CHLORIDE see CCX000
CEPALORIDIN see TEY000
CEPALORIN see TEY000
CEPH 87/4 see TEY000
CEPHA see CDS125
CEPHACETRILE SODIUM see SGB500
CEPHADOLE see CCX300
(−)-CEPHAELINE DIHYDROCHLORIDE see CCX125
CEPHAELINE HYDROCHLORIDE see CCX125
CEPHAELINE METHYL ETHER see EAL500
CEPHALEXIN see ALV000
CEPHALOGLYCIN see CCR890
CEPHALOGLYCINE see CCR890
d-CEPHALOGLYCINE see CCR890
CEPHALOMYCIN see CCX175
CEPHALORIDIN see TEY000
CEPHALORIDINE see TEY000
(3(R))-CEPHALOTAXINE-4-METHYL-2-HYDROXY-2-(4-HYDROXY-4-
  METHYLPENTYL)BUTANEDIOATE (ESTER) see HGI575
CEPHALOTHIN see CCX250
CEPHALOTHIN SODIUM see HMK000
CEPHALOTHIN SODIUM see SFQ500
CEPHALOTIN see CCX250
CEPHAMANDOLE see CCX300
CEPHAMYCIN see CCS365
CEPHAOGLYCIN ACID see CCR890
CEPHAPIRIN see CCX500
CEPHARANTHIN see CCX550
CEPHARANTHINE see CCX550
CEPHA 10LS see CDS125
CEPHATREXYL see HMK000
CEPHEDRINE see CCX600
CEPHOXITIN see CCS500
CEPHRADIN see SBN440
CEPHRADINE see SBN440
CEPHROL see CMT250
CEPHUROXIME see CCS600
CEPORAN see TEY000
CEPOREX see ALV000
CEPOREXIN see ALV000
CEPOREXINE see ALV000
CEPORINE see TEY000
CEPOVENIN see SFQ500
CEPRIM see CCX000
CEQUARTYL see BBA500
CERAMIC FIBRE see AHF500
CERAPHYL 230 see DNL800
CERAPHYL 368 see OFE100
CERAPHYL 375 see ISC550
CERASINE YELLOW GG see DOT300
CERASINROT see OHA000
CERASTES CERASTES VENOM see CCX620
CERASYNT see HKJ000
CERASYNT 1000-D see OAV000
CERASYNT PA see SLL000
CERASYNT PN see SLL000
CERASYNT S see OAV000
CERASYNT SD see OAV000
CERASYNT SE see OAV000
CERASYNT WM see OAV000
CERAZOL (suspension) see TEX250
CERBERIGENIN see DMJ000
CERBEROSID (GERMAN) see CCX625
CERBEROSIDE see CCX625

CERBROSIDE see CCX625
CER-o-CILLIN see AGK250
CERCINE see DCK759
CERCOBIN see DJV000
CERCOBIN METHYL see PEX500
CEREB see CMF350
CEREBON see DPE000
CEREBROFORTE see NNE400
CEREDON see BDD000
CERELINE see BDD000
CERELOSE see GFG000
CERENOX see BDD000
CEREPAP see DNA200
CEREPAX see CFY750
CERESAN see ABU500
CERESAN see CHC500
CERESAN M see EME500
CERESAN UNIVERSAL-FEUCHTBEIZE see BKS810
CERESAN UNIVERSAL NAZBEIZE see MEP250
CERES ORANGE R see PEJ500
CERES ORANGES RR see XRA000
CERESPAN see PAH250
CERES RED 7B see EOJ500
CERES YELLOW GGN see OHI875
CERES YELLOW R see PEI000
CEREWET see BKS810
CEREXIN A see CCX725
CEREZA (SPANISH) see AQP890
CERFA 114 see HNM000
CERIC DISULFATE see CDB400
CERIC OXIDE see CCY000
CERIC SULFATE see CDB400
CERIC SULPHATE see CDB400
CERIMAN see SLE890
CERIMAN de MEJICO (CUBA) see SLE890
CERISE TONER X1127 see FAG070
CERISOL SCARLET G see XRA000
CERISOL YELLOW AB see FAG130
CERISOL YELLOW TB see FAG135
CERIUM see CCY250
CERIUM ACETATE see CCY500
CERIUM AZIDE see CCY699
CERIUM CHLORIDE see CCY750
CERIUM(III) CHLORIDE see CCY750
CERIUM CITRATE see CCZ000
CERIUM(III) CITRATE see CCZ000
CERIUM COMPOUNDS see CDA250
CERIUM DIOXIDE see CCY000
CERIUM DISULFATE see CDB400
CERIUM EDETATE see CDA500
CERIUM FLUORIDE see CDA750
CERIUM FLUORURE (FRENCH) see CDA750
CERIUM NITRATE see CDB000
CERIUM(3+) NITRATE see CDB000
CERIUM(III) NITRATE see CDB000
CERIUM NITRATE, HEXAHYDRATE see CDB250
CERIUM(III) NITRATE, HEXAHYDRATE (1583586) see CDB250
CERIUM NITRIDE see CDB325
CERIUM SULFATE see CDB400
CERIUM(4+) SULFATE see CDB400
CERIUM(IV) SULFATE see CDB400
CERIUM(III) TETRAHYDROALUMINATE see CDB500
CERIUM TRIACETATE see CCY500
CERIUM TRICHLORIDE see CCY750
CERIUM TRIFLUORIDE see CDA750
CERIUM TRIHYDRIDE see CDB750
CERIUM TRINITRATE see CDB000
CERIUM TRINITRATE HEXAHYDRATE see CDB250
CERIUM TRISULFIDE see DEK600
CERM-1766 see IRP000
CERM-1841 see TKJ500
3024 CERM see MFG250
CERM 10,137 see TGJ885
CERNILTON see CDB760
CERNITIN GBX see CDB770
CERNITIN T-60 see CDB772
CERN KYPOVA 8 see CMU475
CERN OSTAZINOVA H-N (CZECH) see DGN600
CERN PRIMA 38 see AQP000
CERN REAKTIVNI 8 see CMS227
CEROTINORANGE G see PEJ500
CEROTINSCHARLACH G see XRA000
CEROUS ACETATE see CCY500
CEROUS CHLORIDE see CCY750

CEROUS CITRATE see CCZ000
CEROUS FLUORIDE see CDA750
CEROUS NITRATE see CDB000
CEROUS NITRATE HEXAHYDRATE see CDB250
CEROXONE see BJK750
CERTICOL BLACK PNW see BMA000
CERTICOL CARMOISINE S see HJF500
CERTICOL ORANGE GS see HGC000
CERTICOL PONCEAU MXS see FMU070
CERTICOL PONCEAU 4RS see FMU080
CERTICOL PONCEAU SXS see FAG050
CERTINAL see ALT250
CERTIQUAL ALIZARINE see DMG800
CERTIQUAL EOSINE see BNH500, BNK700
CERTIQUAL FLUORESCEINE see FEW000
CERTIQUAL ORANGE 1 see FAG010
CERTIQUAL RHODAMIEN see FAG070
CERTOL see DNF400
CERTOMYCIN see NCP550
CERTOX see SMN500
CERTROL see HKB500
CERUBIDIN see DAC000
CERUBIDINE see DAC200
CERULENIN see ECE500
CERULIGNOL see MFM750
CERUSSETE see LCP000
CERUTIL see AAE500
CERVAGEM see CDB775
CERVEN BRILANTNI OSTACETOVA F-LB (CZECH) see AKI750
CERVEN BRILANTNI OSTAZINOVA H-3B (CZECH) see CHG250
CERVEN BRILANTNI OSTAZINOVA S-5B (CZECH) see DGN800
CERVEN KOSENILOVA A see FMU080
CERVEN KUMIDINOVA see FAG018
CERVEN KYSELA 26 see FMU070
CERVEN KYSELA 27 see FAG020
CERVEN POTRAVINARSKA 1 see FAG050
CERVEN POTRAVINARSKA 9 see FAG020
CERVICUNDIN see EQJ500
CERVOLIDE see OKW110
CERVOXAN see DOZ000
CES see ECU750
CES see SOP500
CESALIN see CDB800
CESIUM see CDC000
CESIUM-133 see CDC000
CESIUM ACETYLIDE see CDC125
CESIUM AMIDE see CDC250
CESIUM ARSENATE see CDC375
CESIUM BROMIDE see CDC500
CESIUM BROMOXENATE see CDC699
CESIUM CARBONATE see CDC750
CESIUM CHLORIDE see CDD000
CESIUM CHLOROXENATE see CDD250
CESIUM CYANOTRIDECAHYDRODECABORATE (2-) see CDD325
CESIUM FLUORIDE see CDD500
CESIUM GRAPHITE see CDD625
CESIUM HYDRATE see CDD750
CESIUM HYDROXIDE see CDD750
CESIUM HYDROXIDE, solid (DOT) see CDD750
CESIUM HYDROXIDE, solution (DOT) see CDD750
CESIUM HYDROXIDE DIMER see CDD750
CESIUM IODIDE see CDE000
CESIUM LITHIUM TRIDECAHYDRONONABORATE see CDE125
CESIUM METAL (DOT) see CDC000
CESIUM MONOCHLORIDE see CDD000
CESIUM MONOFLUORIDE see CDD500
CESIUM NITRATE (DOT) see CDE250
CESIUM(I) NITRATE (1581) see CDE250
CESIUM NITRIDE see TIH750
CESIUM OXIDE see CDE325
CESIUM OZONIDE see CDF000
CESIUM PENTACARBONYLVANADATE (3-) see CDE400
CESIUM, POWDERED (DOT) see CDC000
CESIUM SELENIDE see DEJ400
CESIUM SULFATE see CDE500
CESIUM TRIOXIDE ("OZONATE") see CDF000
CESOL see BGB400
CESTRUM (VARIOUS SPECIES) see DAC500
CET see BJP000
CET see CCX250
CETAB see HCQ500
CETACORT see CNS750
CETADOL see HIM000
CETAFFINE see HCP000

CETAIN see AIT250
CETAL see HCP000
CETALOL CA see HCP000
CETAMIUM see CCX000
CETARIN see MKR250
CETAROL see HCQ500
CETAVLON see HCQ500
CETIL LIGHT ORANGE GG see HGC000
CETOCYLINE see CDF250
CETONE V see AGI500
CETRAMIN see CNH125
CETRAXATE see CDF375
CETRAXATE HYDROCHLORIDE see CDF380
CETRIMIDE see HCQ500
CETRIMONIUM BROMIDE see HCQ500
CETYL ALCOHOL see HCP000
CETYLAMIN (GERMAN) see HCO500
CETYLAMINE see HCO500, HCQ500
CETYLAMINE-HF see CDF400
CETYLAMINE HYDROFLUORIDE see CDF400
CETYLAMINHYDROFLUORID (GERMAN) see CDF400
CETYL-γ-AMINOBUTYRATE see AJC500
CETYLDIETHYLETHYLAMMONIUM BROMIDE see CDF500
CETYL DIMETHYL ETHYL AMMONIUM BROMIDE see EKN500
CETYL ETHYL DIMETHYLAMMONIUM BROMIDE see EKN500
CETYL 2-ETHYLHEXANOATE see HCP550
CETYL GABA see AJC500
CETYLIC ACID see PAE250
CETYLIC ALCOHOL see HCP000
CETYLOL see HCP000
CETYLPYRIDINIUM CHLORIDE see CCX000
1-CETYLPYRIDINIUM CHLORIDE see CCX000
N-CETYLPYRIDINIUM CHLORIDE see CCX000
CETYLPYRIDINIUM CHLORIDE MONOHYDRATE see CDF750
CETYLTRIETHYLAMMONIUM BROMIDE see CDF500
CETYLTRIMETHYLAMMONIUM BROMIDE see HCQ500
N-CETYLTRIMETHYLAMMONIUM BROMIDE see HCQ500
CETYLUREUM see PEC250
CEVADENE see CDG000
CEVADIC ACID see TGA700
CEVADILLA see VHZ000
CEVADIN see CDG000
CEVADINE see CDG000
CEVADINE see VHZ000
CEVANE-3-β,4-β,7-α,14,15-α,16-β,20-HEPTOL,4,9-EPOXY-, 15-((+)-2-HYDROXY-
  2-METHYLBUTYRATE) 3-((-)-2-METHYLBUTYRATE) see VHF000
CEVANOL see BCA000
CEVIAN A 678 see AAX250
CEVIAN HL see ADY500
CEVIN see EBL000
CEVINE see EBL000
CEVITAMIC ACID see ARN000
CEVITAMIN see ARN000
CEX see ALV000
CEYLON ISINGLASS see AEX250
CEZ SODIUM see CCS250
CF 125 see CDT000
CFC see PGE000
CFC 31 see CHI900
CFC 133a see TJY175
C. FERTILIS see CCK675
C. FLORIDUS see CCK675
CFNU see CPL750, FIJ000
C. FRUTESCENS see PCB275
C.F.S. see TEM000
CFT 1201 see AGR125
CFV see CDS750
CFX see CCS500
CG 113 see PMB850
CG-120 see SFX725
CG 201 see CDG250
CG 315 see THJ500
CG 601 see MRU080
CG-1283 see MQW500
CGA 10832 see CQG250
CGA-12223 see PHK000
CGA 15324 see BNA750
CGA-18762 see PMF600
CGA-24705 see MQQ450
CGA 26351 see CDS750
CGA 26423 see PMB850
CGA 89317 see CFL200
CG 10213 GO see SAY950
C. GIGANTEA see COD675

C. GILLIESII see CAK325
C 9333 GO see AOA050
CGP 2175 see MQR144
CGP 4540 see AOA050
CGP 9000 see CCS530
CGP 71743 see CCS550
CGP-11305A see BMM625
CGP-9000 DIHYDRATE see CCS535
C-GREEN 10 see BLK000
CGS 10787B see CDG300
CGT see COX400
CH see SBW500
CH 800 see CKI750
CH 3565 see TIQ000
CHA see CPF500
α-CHACONINE see CDG500
CHAETOGLOBOSIN A see CDG750
CHALCEDONY see SCI500, SCJ500
CHALCONE see CDH000
CHALCONE, 2',3,4'-TRIHYDROXY-4,6'-DIMETHOXY-, 4'-(6-O-(6-DEOXY-α-l-MANNOPYRANOSYL)-β-d-GLUCOPYRANOSIDE) see HBU400
CHALICE VINE see CDH125
CHALK see CAO000
CHALOTHANE see HAG500
CHAMAZULEN see DRV000
CHAMAZULENE see DRV000
CHAMBER CRYSTALS see NMJ000
CHAMELEON MINERAL see PLP000
CHAMICO BEJUCO (CUBA) see CDH125
CHAMOMILE see CDH250
CHAMOMILE-GERMAN OIL see CDH500
CHAMOMILE OIL see CDH500
CHAMOMILE OIL (ROMAN) see CDH750
CHANNEL BLACK see CBT750
CHANNING'S SOLUTION see NCP500
CHARAS see CBD750
CHARCOAL see CDI250
CHARCOAL (wood, lump) see CDJ750
CHARCOAL (wood, ground, crushed, granulated or pulverized) see CDJ500
CHARCOAL, ACTIVATED (DOT) see CDI000
CHARCOAL BLACK see CBT500
CHARCOAL (BRIQUETTES) see CDI250
CHARCOAL SCREENINGS, MADE from "PINON" WOOD (DOT) see CDI500
CHARCOAL (SHELL) see CDJ000
CHARCOAL, SHELL (DOT) see CDJ000
CHARCOAL WOOD SCREENINGS, OTHER THAN ""PINON" WOOD SCREENINGS (DOT) see CDK000
CHARGER E see GHS000
CHARTREUSIN see CDK250
CHARTREUSIN, SODIUM SALT see CDK500
CHA-SULFATE see CPF750
CHAULMOOGRA OIL see CDK750
CHAULMOOGRIC ACID, SODIUM SALT see SFR000
CHAVICOL METHYL ETHER see AFW750
1,3-CHBP see BNA825
CHEBUTAN see KGK000
CHEELOX BF see EIV000
CHEELOX BF ACID see EIX000
CHEELOX BR-33 see EIV000
CHEL 330 see DJG800
CHEL 330 ACID see DJG800
CHELADRATE see EIX500
CHELAFER see FBC100
CHELAFRIN see VGP000
CHELAPLEX III see EIX500
CHELATON III see EIX500
CHEL DTPA see DJG800
CHELEN see EHH000
CHELIDONINE see CDL000
CHELIDONIUM MAJUS L. see CCS650
CHEL-IRON see FBC100
CHELLIN see AHK750
CHELLINA (ITALIAN) see AHK750
β-CHELOCARDIN see CDF250
CHELON 100 see EIV000
CHEMAGRO 1,776 see BSH250
CHEMAGRO 2353 see DFV400
CHEMAGRO 5461 see BJD000
CHEMAGRO 25141 see FAQ800
CHEMAGRO 37289 see EPY000
CHEMAGRO B-1776 see BSH250
CHEMAGRO B-1776 see TIG250
CHEMAGRO B-1843 see BLG500
CHEMAGRO B-9002 see HMV000

CHEMAGRO D-113 see DFS200
CHEMAGRO R-5461 see BJD000
CHEMAID see HKC500
CHEMANOX 11 see BFW750
CHEMANOX 21 see MJO500
CHEMATHION see MAK700
CHEMAX NP SERIES see NND500
CHEM BAM see DXD200
CHEMCOCCIDE see RLK890
CHEMCOLOX 200 see EIV000
CHEMCOLOX 340 see EIX000
CHEM DM ACID see HKS000
CHEMESTER 300-OC see PJY100
CHEM FISH see RNZ000
CHEMFORM see DKC800, MEI450
CHEMFORM see SLW500
CHEM-FROST see SFS500
CHEM-HOE see CBM000
CHEMIAZID see ILD000
CHEMICAL 109 see AQN635
CHEMICAL MACE see CEA750
CHEMICETIN see CDP250
CHEMICETINA see CDP250
CHEMI-CHARL see SHM500
CHEMICTIVE BRILLIANT RED 5B see PMF540
CHEMIFLUOR see SHF500
CHEMIOCHIN see CFU750
CHEMIOFURAN see NGE000
CHEMIPEN see PDD350
CHEMIPEN-C see PDD350
CHEMITRIM see TKX000
CHEMLON see PJY500
CHEM-MITE see RNZ000
CHEM NEB see MAS500
CHEMOCCIDE see RLK890
CHEMOCHIN see CLD000
CHEMOCIN see CNK559
CHEMOFURAN see NGE500
CHEMOSAN see MGC350
CHEMOSEPT see TEX250
CHEMOTHERAPY CENTER No. 606 see CBI500
CHEMOUAG see SNN500
CHEMOX GENERAL see BRE500
CHEMOX P.E. see BRE500
CHEMOX PE see DUZ000
CHEMOX SELECTIVE see BPG250
CHEMPAR see CNK559
CHEM PELS C see SEY500
CHEM-PHENE see CDV100
CHEMRAT see PIH175
CHEM RICE see DGI000
CHEMSECT DNOC see DUS700
CHEM-SEN 56 see SEY500
CHEM-TOL see PAX250
CHEMYSONE see HHQ800
CHEM ZINEB see EIR000
CHENDAL see CDL325
CHENDOL see CDL325
CHENIC ACID see CDL325
CHENIX see CDL325
CHENOCEDON see CDL325
CHENODEOXYCHOLIC ACID see CDL325
CHENODEOXYCHOLIC ACID SODIUM SALT see CDL375
CHENODESOXYCHOLIC ACID see CDL325
CHENODESOXYCHOLIC ACID SODIUM SALT see CDL375
CHENODESOXYCHOLSAEURE (GERMAN) see CDL325
CHENODEX see CDL325
CHENODIOL see CDL325
CHENOFALK see CDL325
CHENOPODIUM AMBROSIOIDES see SAF000
CHENOPODIUM OIL see CDL500
CHENOSAURE see CDL325
CHENOSSIL see CDL325
CHEPIROL see KGK000
CHEQUE see MQS225
CHERRY see AQP890
CHERRY BARK OAK see CDL750
CHERRY LAUREL OIL see CDM000
CHERRY PEPPER see PCB275
CHERTS see SCI500, SCJ500
CHESTNUT COMPOUND see CNJ750
CHESTNUT TANNIN see CDM250
CHETAZOLIDIN see KGK000
CHETIL see KGK000

17-CHETOVIS see AOO450
CHEVREFEUILLE (CANADA) see HGK700
CHEVRON 9006 see DTQ400
CHEVRON ORTHO 9006 see DTQ400
CHEVRON RE5305 see BSG250
CHEVRON RE 5655 see MOU750
CHEVRON RE 12,420 see DOP600
CHEWING TOBACCO see SED400
CHEXMATE see HKC000
CHFB see CHK750
CHICAGO ACID S see AKH750
CHICAGO BLUE 6B see CMN750
CHICLIDA see HGC500
1,4-CHIDM see BKH325
CHILDREN'S BANE see WAT325
CHILE PEPPER see PCB275
CHILE SALTPETER see SIO900
CHIMCOCCIDE see RLK890
CHIMIPAL AE 3 see DXY000
CHIMOREPTIN see DLH630
CHINABERRY see CDM325
CHINACRIN HYDROCHLORIDE see CFU750
CHINA GREEN (BIOLOGICAL STAIN) see AFG500
CHINALDINE see QEJ000
CHINALPHOS see DJY200
CHINA TREE see CDM325
CHINAWOOD OIL see TOA510
CHINAWOOD OIL TREE see TOA275
CHINE APE see EAI600
CHINESE INKBERRY see DAC500
CHINESE ISINGLASS see AEX250
CHINESE LANTERN PLANT see JBS100
CHINESE RED see LCS000
CHINESE SEASONING see MRL500
CHINESE WHITE see ZKA000
CHINGAMIN see CLD000, CLD250
CHINIDIN (GERMAN) see QFS000
CHININ (GERMAN) see QHJ000
CHININDIHYDROCHLORID (GERMAN) see QIJ000
CHININ HYDROBROMID (GERMAN) see QJJ100
CHINIOFON see IEP200
CHINOFER see IGS000
CHINOFORM see CHR500
CHINOFUNGIN see TGB475
CHINOIN see EID000
CHINOIN 103 see CPG500
CHINOIN-127 see CDM500
CHINOIN-170 see CDM575
CHINOLEINE see QMJ000
CHINOLIN (CZECH) see QMJ000
CHINOLINE see QMJ000
CHINOMETHIONATE see ORU000
p-CHINON (GERMAN) see QQS200
CHINON (DUTCH, GERMAN) see QQS200
CHINONE see QQS200
CHINON I (GERMAN) see BGW750
CHINONOXIM-BENZOYLHYDRAZON (GERMAN) see BDD000
CHINONOXIME-BENZOYLHYDRAZONE see BDD000
CHINORTA see NIM500
CHINOTILIN see KGK400
CHINOXONE see OPK300
3-CHINUCLIDYLBENZILATE see QVA000
CHIP see IGG775
CHIPCO 26019 see GIA000
CHIPCO BUCTRIL see DDP000
CHIPCO CRAB-KLEEN see DDP000
CHIPCO CRAB KLEEN see DXE600
CHIPCO THIRAM 75 see TFS350
CHIPCO TURF HERBICIDE "D" see DAA800
CHIPCO TURF HERBICIDE MCPP see CIR500
CHIPMAN 3,142 see TBR750
CHIPMAN 6199 see AMX825
CHIPMAN 6200 see DJA400
CHIPMAN 11974 see BDJ250
CHIPMAN R-6, 199 see AMX825
CHIPTOX see CIR250
CHIRAL BINAPHTHOL see CDM625
CHIRONEX FLECKERI TOXIN see CDM700
CHISSO 507B see PMP500
CHISSONOX 201 see ECB000
CHISSONOX 206 see VOA000
CHITA ROOT EXTRACT see PJH615
CHITIN see CDM750
CHITINA (ITALIAN) see CDM750

CHITRAKA ROOT EXTRACT see PJH615
CHKHZ 18 see DVF400
ChKhZ 21 see ASM270
ChKhZ 21R see ASM270
CHLODITAN see CDN000
CHLODITHANE see CDN000
CHLOFENVINPHOS see CDS750
CHLOFEXAMIDE see CJN750
CHLOMAPHENE see CMX500
CHLOMIN see CDP250
CHLOMYCOL see CDP250
CHLONIXIN see CMX770
CHLOOR (DUTCH) see CDV750
3-CHLOORANILINEN (DUTCH) see CEH675
2-CHLOORBENZALDEHYDE (DUTCH) see CEI500
o-CHLOORBENZALDEHYDE (DUTCH) see CEI500
CHLOORBENZEEN (DUTCH) see CEJ125
CHLOORBENZIDE (DUTCH) see CEP000
(4-CHLOOR-BENZYL)-(4-CHLOOR-FENYL)-SULFIDE (DUTCH) see CEP000
2-CHLOOR-1,3-BUTADIEEN (DUTCH) see NCI500
(4-CHLOOR-BUT-2-YN-YL)-N-(3-CHLOOR-FENYL)-CARBAMAAT (DUTCH) see CEW500
CHLOORDAAN (DUTCH) see CDR750
O-2-CHLOOR-1-(2,4-DICHLOOR-FENYL)-VINYL-O,O-DIETHYLFOSFAAT (DUTCH) see CDS750
(2-CHLOOR-3-DIETHYLAMINO-1-METHYL-3-OXO-PROP-1-EN-YL)-DIMETHYL-FOSFAAT see FAB400
2-CHLOOR-4-DIMETHYLAMINO-6-METHYL-PYRIMIDINE (DUTCH) see CCP500
1-CHLOOR-2,4-DINITROBENZEEN (DUTCH) see CGM000
1-CHLOOR-2,3-EPOXY-PROPAAN (DUTCH) see EAZ500
CHLOORETHAAN (DUTCH) see EHH000
2-CHLOORETHANOL (DUTCH) see EIU800
CHLOORFACINON (DUTCH) see CJJ000
3-(4-(4-CHLOOR-FENOXY)-FENOXY)-FENYL)-1,1-DIMETHYLUREUM (DUTCH) see CJQ000
CHLOORFENSON (DUTCH) see CJT750
(4-CHLOOR-FENYL)-BENZEEN-SULFONAAT (DUTCH) see CJR500
(4-CHLOOR-FENYL)-4-CHLOOR-BENZEEN-SULFONAAT (DUTCH) see CJT750
3-(4-CHLOOR-FENYL)-1,1-DIMETHYLUREUM (DUTCH) see CJX750
2(2-(4-CHLOOR-FENYL-2-FENYL)-ACETYL)-INDAAN-1,3-DION (DUTCH) see CJJ000
N-(3-CHLOOR-FENYL)-ISOPROPYL CARBAMAAT (DUTCH) see CKC000
CHLOOR-HEXAVIET see HEA500
CHLOOR-METHAAN (DUTCH) see MIF765
2-(4-CHLOOR-2-METHYL-FENOXY)-PROPIONZUUR (DUTCH) see CIR500
1-CHLOOR-4-NITROBENZEEN (DUTCH) see NFS525
O-(3-CHLOOR-4-NITRO-FENYL)-O,O-DIMETHYL-MONOTHIOFOSFAAT (DUTCH) see MIJ250
O-(4-CHLOOR-3-NITRO-FENYL)-O,O-DIMETHYLMONOTHIOFOSFAAT (DUTCH) see NFT000
CHLOORPIKRINE (DUTCH) see CKN500
CHLOORTHION (DUTCH) see MIJ250
CHLOORWATERSTOF (DUTCH) see HHL000
CHLOPHEDIANOL HYDROCHLORIDE see CMW700
CHLOPHEN see PJL750
CHLOR (GERMAN) see CDV750
CHLORACETAMID (GERMAN) see CDY850
CHLORACETIC ACID see CEA000
CHLORACETONE see CDN200, CDN200
CHLORACETONITRILE see CDN500
CHLORACETYL CHLORIDE see CEC250
CHLORACON see BEG000
CHLORACTIL see CKP500
2-CHLORAETHANOL (GERMAN) see EIU800
N-(2-CHLORAETHYL)-N'-(2 CHLOROETHYL)-N'-o-PROPYLEN-PHOSPHORSAUREESTER-DIAMID (GERMAN) see IMH000
α-CHLOR-6'-AETHYL-n-(2-METHOXY-1-METHYLAETHYL)-ACET-o-TOLUIDIN (GERMAN) see MQQ450
2-CHLORAETHYL-PHOSPHONSAURE (GERMAN) see CDS125
2-CHLORAETHYL-TRIMETHYLAMMONIUMCHLORID see CMF400
CHLORAK see TIQ250
CHLORAKON see BEG000
CHLORAL ALCOHOLATE see TIO000
CHLORALDEHYDE see DEM200
CHLORALDURAT see CDO000
CHLORAL ETHYLALCOHOLATE see TIO000
CHLORAL, ETHYL HEMIACETAL see TIO000
CHLORAL HYDRATE see CDO000
CHLORALLYL DIETHYLDITHIOCARBAMATE see CDO250
2-CHLORALLYL DIETHYLDITHIOCARBAMATE see CDO250
CHLORALLYLENE see AGB250
CHLORALONE see CDP000
CHLORALOSANE see GFA000
α-CHLORALOSE see GFA000

CHLORAMBEN see AJM000
CHLORAMBUCIL see CDO500
CHLORAMBUCIL SODIUM SALT see CDO625
CHLORAMEISENSAEURE METHYLESTER (GERMAN) see MIG000
CHLORAMEX see CDP250
CHLORAMFICIN see CDP250
CHLORAMFILIN see CDP250
CHLORAMIDE see CDO750
CHLORAMIFENE see CMX500
CHLORAMIN see BIE500
CHLORAMINE see BIE500
CHLORAMINE see CDO750
CHLORAMINE (inorganic compound) see CDO750
CHLORAMINE B see SFV275
CHLORAMINE BLACK C see AQP000
CHLORAMINE BLUE see CMO250
CHLORAMINE BLUE 2B see CMO000
CHLORAMINE FAST BROWN BRL see CMO750
CHLORAMINE SKY BLUE 4B see CMO500
CHLORAMINE T see CDP000
CHLORAMINE-T see SFV550
CHLORAMIN HYDROCHLORIDE see BIE500
1-CHLOR-5-AMINOANTHRACHINON (CZECH) see AJE325
CHLORAMINOPHEN see CDO500
CHLORAMINOPHENE see CDO500
CHLORAMIPHENE see CMX500, CMX700
CHLORAMIPHENE CITRATE see CMX700
CHLORAMP (RUSSIAN) see PIB900
CHLORAMPHENICOL see CDP250
d-CHLORAMPHENICOL see CDP250
d-threo-CHLORAMPHENICOL see CDP250
l(+)-threo-CHLORAMPHENICOL see CDP325
CHLORAMPHENICOL ACID SUCCINATE see CDP725
CHLORAMPHENICOL HEMISUCCINATE see CDP725
CHLORAMPHENICOL HYDROGEN SUCCINATE see CDP725
CHLORAMPHENICOL MONOPALMITATE see CDP700
CHLORAMPHENICOL MONOSUCCINATE see CDP725
CHLORAMPHENICOL MONOSUCCINATE SODIUM SALT see CDP500
CHLORAMPHENICOL PALMITATE see CDP700
CHLORAMPHENICOL SODIUM MONOSUCCINATE see CDP500
CHLORAMPHENICOL SODIUM SUCCINATE see CDP500
CHLORAMPHENICOL SUCCINATE see CDP725
CHLORAMPHENICOL SUCCINATE SODIUM see CDP500
CHLORAMPHENICOL-SUKZINAT-NATRIUM (GERMAN) see CDP500
CHLORAMSAAR see CDP250
CHLORANAUTINE see DYE600
CHLORANIFORMETHAN see CDP750
CHLORANIFORMETHANE see CDP750
CHLORANIL see TBO500
4-CHLORANILIN (CZECH) see CEH680
2-(2-CHLORANILIN)-4,6-DICHLOR-1,3,5-TRIAZIN (GERMAN) see DEV800
m-CHLORANILINE see CEH675
o-CHLORANILINE see CEH670
p-CHLORANILINE see CEH680
CHLORANOCRYL see DFO800
1-CHLORANTHRACHINON (CZECH) see CEI000
CHLORAQUINE see CLD000
CHLORARSENOL see CEI250
CHLORARSOL see DFX400
CHLORASAN see CDP000
CHLORASEN see DFX400
CHLORASEPTINE see CDP000
CHLORASOL see CDP250
CHLORA-TABS see CDP250
CHLORATE de CALCIUM (FRENCH) see CAO500
CHLORATE of POTASH (DOT) see PLA250
CHLORATE de POTASSIUM (FRENCH) see PLA250
CHLORATES see CDQ000
CHLORATE SALT of MAGNESIUM see MAE000
CHLORATE SALT of SODIUM see SFS000
CHLORATE of SODA (DOT) see SFS000
2-(3-(2-CHLORATHYL)-3-NITROSOUREIDO)ATHYLMETHANSULFONAT (GERMAN) see CHF250
CHLORAX see SFS000
CHLORAZAN see CDP000
CHLORAZENE see CDP000
CHLORAZEPAM see CDQ250
CHLORAZEPATE DIPOTASSIUM see CDQ250
CHLORAZIN see CKP500
CHLORAZINE see CDQ325
CHLORAZOL BLACK E (biological stain) see AQP000
CHLORAZOL BLACK EA see AQP000
CHLORAZOL BLACK EN see AQP000
CHLORAZOL BLUE 3B see CMO250

CHLORAZOL BLUE B see CMO000
CHLORAZOL SKY BLUE FF see BGT250
CHLORAZOL SKY BLUE FF see CMN750
CHLORAZOL VIOLET N see CMP000
CHLORAZONE see CDP000
CHLORBENSID (GERMAN) see CEP000
CHLORBENSIDE see CEP000
CHLORBENXIDE see CEP000
2-CHLORBENZALDEHYD (GERMAN) see CEI500
CHLORBENZENE see CEJ125
p-CHLORBENZENESULFOCHLORID (CZECH) see CEK375
p-CHLORBENZENSULFONAN SODNY (CZECH) see CEK250
1-p-CHLORBENZHYDRYL-m-METHYLBENZYLPIPERAZINE DIHYDROCHLORIDE see MBX250
CHLORBENZIDE see CEP000
CHLORBENZILATE see DER000
CHLORBENZOL see CEJ125
o-CHLORBENZONITRIL (CZECH) see CEM000
CHLORBENZOSAMINE DIHYDROCHLORIDE see CDQ500
CHLORBENZOXAMINE DIHYDROCHLORIDE see CDQ500
5-CHLORBENZOXAZOLIN-2-ON see CDQ750
CHLORBENZOXYETHAMINE DIHYDROCHLORIDE see CDQ500
1-CHLOR-5-BENZOYLAMINOANTHRACHINON (CZECH) see BDK750
N-(p-CHLORBENZOYL)-γ-(2,6-DIMETHYLANILINO)-BUTTERSAEURE (GERMAN) see CLW625
1-(p-CHLORBENZOYL)-5-METHOXY-2-METHYLINDOL-3-ACETOXY) ESSIGSAEURE (GERMAN) see AAE625
N-p-CHLORBENZOYL-5-METHOXY-2-METHYLINDOLE-3-ACETIC ACID see IDA000
5-CHLORBENZOZAZOLIN-2-ON see CDQ750
(4-CHLOR-BENZYL)-(4-CHLOR-PHENYL)-SULFID (GERMAN) see CEP000
1-p-CHLORBENZYL-2-METHYL-BENZIMIDAZOL (GERMAN) see CDY325
p-CHLORBENZYL-α-PYRIDYL-DIMETHYL-AETHYLENDIAMIN (GERMAN) see CKV625
CHLORBICYCLENE (FRENCH) see DAM700
CHLORBISAN see DMN000
CHLORBROMURON see CES750
CHLORBUFAM see CEX250
CHLORBUFAN mixed with CYCEURON see AFM375
CHLORBUPHAM see CEX250
2-CHLOR-1,3-BUTADIEN (GERMAN) see NCI500
CHLORBUTANOL see ABD000
4-CHLORBUTAN-1-OL (GERMAN) see CEU500
(4-CHLOR-BUT-2-IN-YL)-N-(3-CHLOR-PHENYL)-CARBAMAT (GERMAN) see CEW500
CHLORBUTOL see ABD000
CHLORCARVACROL see CEX275
CHLORCHOLINCHLORID see CMF400
CHLORCHOLINE CHLORIDE see CMF400
p-CHLOR-m-CRESOL see CFE250
CHLORCYAN see COO750
CHLORCYCLINE see CFF500
CHLORCYCLIZINE see CFF500
CHLORCYCLIZINE DIHYDROCHLORIDE see CDR000
CHLORCYCLIZINE HYDROCHLORIDE see CDR250
CHLORCYCLIZINE HYDROCHLORIDE A see CDR500
CHLORCYCLIZINIUM CHLORIDE see CDR250
CHLORCYCLOHEXAMIDE see CDR550
4-(3-CHLOR-4-CYCLOHEXYL-PHENYL)-4-OXO-BUTTERSAEURE KALZIUM (GERMAN) see CPJ250
6-CHLOR-N-CYCLOPROPYL-N'-(1-METHYLETHYL)-1,3,5-TRIZAINE-2,4-DIAMINE see CQI750
CHLORDAN see CDR750, CDR675
γ-CHLORDAN see CDR575, CDR750
cis-CHLORDAN see CDR675
trans-CHLORDAN see CDR575
CHLORDANE see CDR750
α-CHLORDANE see CDR675
cis-CHLORDANE see CDR675
α(cis)-CHLORDANE see CDR675
γ(trans)-CHLORDANE see CDR575
CHLORDANE, liquid (DOT) see CDR750
CHLORDECONE see KEA000
CHLORDENE see HCN000
7-CHLOR-4-(4-(DIAETHYLAMINO)-1-METHYLBUTYLAMINO)-CHINOLINDIPHOSPHAT (GERMAN) see CLD250
(2-CHLOR-3-DIAETHYLAMINO-1-METHYL-3-OXO-PROP-1-EN-YL)-DIMETHYLPHOSPHAT see FAB400
CHLORDIAZACHEL see MDQ250
CHLORDIAZEPOXIDE see LFK000
CHLORDIAZEPOXIDE HYDROCHLORIDE see MDQ250
CHLORDIAZEPOXIDE MONOHYDROCHLORIDE see MDQ250
CHLORDIAZEPOXIDE, NITROSATED see SIS000
CHLORDIAZEPOXIDE mixed with SODIUM NITRITE (1:1) see SIS000

O-2-CHLOR-1-(2,4-DICHLOR-PHENYL)-VINYL-O,O-DIAETHYLPHOSPHAT (GERMAN) see CDS750
7-CHLOR-2,3-DIHYDRO-1-METHYL-5-PHENYL-1H-1,4-BENZODIAZEPIN HYDROCHLORID (GERMAN) see MBY000
CHLORDIMEFORM see CJJ250
CHLORDIMEFORM HYDROCHLORIDE see CJJ500
2-CHLOR-11-(2-DIMETHYAMINOAETHOXY)-DIBENZO(b,f)-THIEPIN (GERMAN) see ZUJ000
2-CHLOR-4-DIMETHYLAMINO-6-METHYLPYRIMIDIN (GERMAN) see CCP500
CHLORDIMETHYLETHER (CZECH) see CIO250
1-CHLOR-2,4-DINITROBENZENE see CGM000
CHLORE (FRENCH) see CDV750
CHLOREFENIZON (FRENCH) see CJT750
CHLORENDIC ACID see CDS000
CHLORENDIC IMIDE see CDS100
CHLOREPIN see CIR750
1-CHLOR-2,3-EPOXY-PROPAN (GERMAN) see EAZ500
CHLORESENE see BBQ500
CHLORESSIGSAEURE-N-ISOBUTINYLANILID (GERMAN) see CDS275
CHLORESSIGSAEURE-N-ISOPROPYLANILID (GERMAN) see CHS500
CHLORESSIGSAEURE-N-(METHOXYMETHYL)-2,6-DIAETHYLANILID (GERMAN) see CFX000
CHLORESTROLO see CLO750
CHLORETHAMINACIL see BIA250
CHLORETHAMINE see BIE500
CHLOR-ETHAMINE see EIW000
2-CHLORETHANOL (GERMAN) see EIU800
CHLORETHAZINE see BIE500
CHLORETHENE see VNP000
CHLORETHEPHON see CDS125
CHLORETHIAZOL see CHD750
2-(2-CHLOROETHOXY)ETHYL 2'-CHLOROETHYL ETHER see TKL500
CHLORETHYL see EHH000
CHLORETHYLBENZMETHOXAZONE see CDS250
CHLORETHYLENE see VNP000
2-CHLOROETHYLPHOSPHONIC ACID see CDS125
2-CHLOROETHYL VINYL ETHER see CHI250
CHLORETIN see CDS275
CHLORETONE see ABD000
CHLOREX see DFJ050
CHLOREXTOL see PJL750
CHLORFACINON (GERMAN) see CJJ000
CHLORFENAC see TIY500
CHLORFENAMIDINE see CJJ250
CHLORFENETHOL see BIN000
CHLORFENIDIM see CJX750
p-CHLORFENOL (CZECH) see CJK750
2-(4'-CHLORFENOXY)ETHANOL (CZECH) see CJO500
CHLORFENPROP-METHYL see CFC750
CHLORFENSON see CJT750
CHLORFENSONE see CJT750
CHLORFENSULFID (GERMAN) see CDS500
CHLORFENSULFIDE see CDS500
CHLORFENVINFOS see CDS750, CDS750
CHLORFENVINPHOS see CDS750
1-p-CHLORFENYL-3,3-DIMETHYLTRIAZEN (CZECH) see CJI100
3-CHLOR-p-FENYLENDIAMIN (CZECH) see CEG600
p-CHLORFENYLISOKYANAT (CZECH) see CKB000
p-CHLORFENYLMERKAPTOMETHYLCHLORID (CZECH) see CFB750
p-CHLORFENYLMONOGLYKOLETHER (CZECH) see CJO500
p-CHLORFENYLSILATRAN (CZECH) see CKM750
CHLORFLURAZOLE see DGO400
CHLORFLURECOL see CDT000
CHLORFLURECOL-METHYL see CDT000
CHLORFLURECOL-METHYL ESTER see CDT000
CHLORFLURENOL see CDT000
CHLORFLURENOL METHYL ESTER see CDT000
CHLORFONIUM see THY500
CHLORFOS see TIQ250
CHLOR-N-(2-FURYLMETHYL)-5-SULFAMYLANTHRANILSAEURE (GERMAN) see CHJ750
CHLORGUANIDE see CKB250
CHLORGUANIDE HYDROCHLORIDE see CKB500
CHLORGUANIDE TRIAZINE see COX400
CHLORHEXAMIDE see CDR550
CHLORHEXIDIN (CZECH) see BIM250
CHLORHEXIDINE see BIM250
CHLORHEXIDINE ACETATE see CDT125
CHLORHEXIDINE DIACETATE see CDT125
CHLORHEXIDINE DIGLUCONATE see CDT250
CHLORHEXIDINE GLUCONATE see CDT500
CHLORHEXIDIN GLUKONATU (CZECH) see CDT500
CHLORHYDRATE de ACETOXY-THYMOXY-ETHYL-DIMETHYLAMINE (FRENCH) see TFY000

CHLORHYDRATE d'AMIKHELLINE (FRENCH) see AHP375
CHLORHYDRATE d'ANILINE (FRENCH) see BBL000
CHLORHYDRATE de 4-CHLOROORTHOTOLUIDINE (FRENCH) see CLK235
CHLORHYDRATE de CONESSINE (FRENCH) see CNH660
CHLORHYDRATE de N-(DIETHOXY-2,5-PHENYL)-N-DIETHYLAMINO-2-ETHYL BUTOXY-4-PHENOXYACETAMIDE see BPM750
CHLORHYDRATE de DIETHYLAMINOETHYLTHEOPHYLLINE (FRENCH) see DIH600
CHLORHYDRATE de (N-ETHYL,N,β-CHLORETHYL)AMINO-METHYLBENZODIOXANE (FRENCH) see CGX625
CHLORHYDRATE d'HISTAMINE (FRENCH) see HGE500
CHLORHYDRATE de (NAPHTHYLOXY-1)-4 HYDROXY-3 BUTYRAMIDOXIME (FRENCH) see NAC500
CHLORHYDRATE de NICOTINE (FRENCH) see NDP400
CHLORHYDRATE de PAPAVERINE (FRENCH) see PAH250
CHLORHYDRATE de PHENETHYL-8-OXA-1-DIAZA-3,8-SPIRO(4,5)DECANONE-2 (FRENCH) see DAI200
CHLORHYDRATE de α-PHENYL-α-(β'-DIETHYLAMINOETHYL) GLUTARIMIDE (FRENCH) see ARR875
CHLORHYDRATE de PIPERIDINOMETHYLCYCLOHEXANE (FRENCH) see PIR000
CHLORHYDRATE de (PIPERONYL-4-PIPERAZINO)-1)-(PHENYL-1-PYRROLIDONE-2-CARBOXAMIDE-4) (FRENCH) see PGA250
CHLORHYDRATE de RAUGALLINE (FRENCH) see AFH280
CHLORHYDRATE de TETRACYCLINE (FRENCH) see TBX250
CHLORHYDRATE de (TRIMETHOXY-2-4-6) PHENYL-(PYRROLIDINE-3) PROPYLACETONE (FRENCH) see BOM600
CHLORHYDRIN see CDT750
α-CHLORHYDRIN see CDT750
CHLORHYDROL see AHA000
3-CHLOR-4-HYDROXYBIFENYL (CZECH) see CHN500
2-CHLOR-9-HYDROXYFLUOREN-CARBONSAEURE-(9)-METHYLESTER (GERMAN) see CDT000
CHLORIAZID see CLH750
CHLORIC ACID see CDU000
CHLORIC ACID, solution, containing not more than 10% acid (DOT) see CDU000
CHLORIC ACID, BARIUM SALT see BAJ500
CHLORIC ACID, BARIUM SALT (wet) see BAJ750
CHLORIC ACID, COPPER SALT see CNJ900
CHLORIC ACID, STRONTIUM SALT see SMF500
CHLORIC ACID, THALLIUM(1+) SALT see TEJ100
CHLORICOL see CDP250
CHLORID AMONNY (CZECH) see ANE500
CHLORID ANILINU (CZECH) see BBL000
CHLORID ANTIMONITY see AQC500
CHLORIDAZON see PEE750
CHLORID-N-BUTYLCINICITY (CZECH) see BSR250
CHLORID CHROMITY HEXAHYDRAT see CMK450
CHLORID DI-n-BUTYLCINICITY (CZECH) see DDY200
CHLORID DRASELNY (CZECH) see PLA500
CHLORIDEAZEPOXIDE HYDROCHLORIDE see MDQ250
CHLORIDE de CHOLINE (FRENCH) see CMF750
CHLORIDE of DIAMINOMETHYLPHENYLDIMETHYL-p-BENZOQUINONE-DIIMINE see DCE800
CHLORIDE of LIME (DOT) see HOV500
CHLORIDE of PHOSPHORUS see PHT275
CHLORIDES see CDU250
CHLORIDE of SULFUR (DOT) see SOG500, SON510
CHLORID FENYLRTUTNATY (CZECH) see PFM500
CHLORIDIAZEPIDE see LFK000
CHLORIDIAZEPOXIDE see LFK000
CHLORIDIN see TGD000
CHLORIDINE see TGD000
CHLORID KREMICITY (CZECH) see SCQ500
CHLORID KYSELINY-p-CHLORBENSULFONOVE (CZECH) see CEK375
CHLORID KYSELINY CHLORMETHANSULFONOVE (CZECH) see CHY000
CHLORID MEDNY (CZECH) see CNK250
CHLORID RTUTNATY (CZECH) see MCY475
CHLORID TRIBENZYLCINICITY (CZECH) see CLP000
CHLORID TRI-n-BUTYLCINICITY (CZECH) see CLP500
CHLORIDUM see EHH000
CHLORIERTE BIPHENYLE, CHLORGEHALT 42% (GERMAN) see PJM500
CHLORIERTE BIPHENYLE, CHLORGEHALT 54% (GERMAN) see PJN000
CHLORIERTES CAMPHEN see CDU325
CHLOR-IFC see CKC000
CHLORIMIPRAMINE see CDU750
CHLORIMIPRAMINE HYDROCHLORIDE see CDV000
CHLORINAT see CEW500
CHLORINATED BIPHENYL see PJL750
CHLORINATED CAMPHENE see CDV100
CHLORINATED DIBENZO DIOXINS see CDV125
CHLORINATED DIPHENYL see PJL750
CHLORINATED DIPHENYLENE see PJL750
CHLORINATED DIPHENYL OXIDE see CDV175

CHLORINATED HC, ALIPHATIC see CDV250
CHLORINATED HC AROMATIC see CDV500
CHLORINATED HYDROCARBONS, ALIPHATIC see CDV250
CHLORINATED HYDROCARBONS, AROMATIC see CDV500
CHLORINATED HYDROCHLORIC ETHER see DFF809
CHLORINATED LIME (DOT) see HOV500
CHLORINATED NAPHTHALENES see CDV575
CHLORINATED PARAFFINS (C12, 60% CHLORINE) see PAH800
CHLORINATED PARAFFINS (C23, 43% CHLORINE) see PAH810
CHLORINATED POLYETHER POLYURETHAN see CDV625
CHLORINDAN see CDR750
CHLORINDANOL see CDV700
CHLORINE see CDV750
CHLORINE AZIDE see CDW000
CHLORINE CYANIDE see COO750
CHLORINE DIOXIDE see CDW450
CHLORINE DIOXIDE, not hydrated (DOT) see CDW450
CHLORINE DIOXYGEN TRIFLUORIDE see CDW500
CHLORINE FLUORIDE see CDX750
CHLORINE FLUORIDE (ClF$_5$) see CDX250
CHLORINE FLUORIDE OXIDE see PCF750
CHLORINE MOL. see CDV750
CHLORINE NITRATE see CDX000
CHLORINE NITRIDE (NITROGEN) TRICHLORIDE see NGQ500
CHLORINE OXIDE see CDW450
CHLORINE(IV) OXIDE see CDW450
CHLORINE OXYFLUORIDE see PCF750
CHLORINE PENTAFLUORIDE see CDX250
CHLORINE PENTAFLUORIDE (DOT) see CDX250
CHLORINE PERCHLORATE see CDX500
CHLORINE PEROXIDE see CDW450
CHLORINE SULFIDE see SOG500
CHLORINE TETROXYFLUORIDE see FFD000
CHLORINE TRIFLUORIDE see CDX750
CHLORINE(1)TRIFLUOROMETHANESULFONATE see CDX800
CHLOR-IPC see CKC000
CHLORISONDAMINE see CDY000
CHLORISONDAMINE CHLORIDE see CDY000
CHLORISONDAMINE DIMETHOCHLORIDE see CDY000
CHLORISOPROPAMIDE see CDY100
CHLORITES see CDY250
5-CHLOR-7-JOD-8-8HYDROXY-CHINOLIN (GERMAN) see CHR500
CHLOR KIL see CDR750
CHLORKU LITU (POLISH) see LHB000
CHLORMADINON see CDY275
CHLORMADINON ACETATE see CBF250
CHLORMADINONE see CDY275
CHLORMADINONE ACETATE see CBF250
CHLORMADINONE ACETATE mixed with MESTRANOL see CNV750
CHLORMADINONU (POLISH) see CBF250
CHLORMENE see TAI500
CHLORMEPHOS see CDY299
CHLORMEQUAT see CMF400
CHLORMEQUAT CHLORIDE see CMF400
CHLORMEROPRIN see CHX250
CHLOR-METHAN (GERMAN) see MIF765
CHLORMETHANSULFOCHLORID (CZECH) see CHY000
CHLORMETHAZANONE see CKF500
CHLORMETHAZONE see CKF500
CHLORMETHIAZOLE see CHD750
CHLORMETHINE see BIE250
CHLORMETHINE HYDROCHLORIDE see BIE500
CHLORMETHINE-N-OXIDE HYDROCHLORIDE see CFA750
CHLORMETHINUM see BIE500
3-(3-CHLOR-4-METHOXYPHENYL)-1,1-DIMETHYLHARNSTOFF (GERMAN) see MQR225
N'-(3-CHLOR-4-METHOXY-PHENYL)-N,N-DIMETHYLHARNSTOFF (GERMAN) see MQR225
CHLORMETHYL-METHYL-DIETHOXYSILAN (CZECH) see CIO000
α-(CHLORMETHYL)-2-METHYL-5-NITRO-IMIDAZOL-1-AETHANOL (GERMAN) see OJS000
4-(4-CHLOR-2-METHYLPHENOXY)-BUETTERSAEURE (GERMAN) see CLN750
4-(4-CHLOR-2-METHYLPHENOXY)-BUTTERSAEURE (GERMAN) see CLN750
4-(4-CHLOR-2-METHYL-PHENOXY)-BUTTERSAEURE NATRIUMSALZ (GERMAN) see CLO000
2-(4-CHLOR-2-METHYL-PHENOXY)-PROPIONSAEURE (GERMAN) see CIR500
3-(3-CHLOR-4-METHYLPHENYL)-1,1-DIMETHYLHARNSTOFF (GERMAN) see CIS250
N-(3-CHLOR-METHYLPHENYL)-2-METHYLPENTANAMID (GERMAN) see SKQ400
3-CHLOR-2-METHYL-PROP-I-EN (GERMAN) see CIU750
CHLORMETHYL-TRIETHOXYSILAN (CZECH) see CIY500
CHLORMEZANONE see CKF500

CHLORMIDAZOLE see CDY325
CHLORMITE see PNH750
CHLORNAFTINA see BIF250
CHLORNAPHAZIN see BIF250
α-CHLORNAPHTHALENE see CIZ000
CHLORNAPHTHIN see BIF250
1-CHLOR-5-NITROANTHRACHINON (CZECH) see CJA250
1-CHLOR-4-NITROBENZOL (GERMAN) see NFS525
CHLORNITROFEN see NIW500
CHLORNITROMYCIN see CDP250
O-(3-CHLOR-4-NITRO-PHENYL)-O,O-DIMETHYL-MONOTHIOPHOSPHAT (GERMAN) see MIJ250
O-(4-CHLOR-3-NITRO-PHENYL)-O,O-DIMETHYL-MONOTHIOPHOSPHAT (GERMAN) see NFT000
CHLOROACETALDEHYDE see CDY500
2-CHLOROACETALDEHYDE see CDY500
CHLOROACETALDEHYDE MONOMER see CDY500
CHLOROACETAMIDE see CDY850
2-CHLORO ACETAMIDE see CDY850
N-CHLOROACETAMIDE see CDY825
α-CHLOROACETAMIDE see CDY850
CHLOROACETAMIDE OXIME see CDZ000
4'-CHLOROACETANILIDE see CDZ100
CHLOROACETIC ACID see CEA000
α-CHLOROACETIC ACID see CEA000
CHLOROACETIC ACID, solid (DOT) see CEA000
CHLOROACETIC ACID, liquid (DOT) see CEA000
CHLOROACETIC ACID BENZYL ESTER see BEE500
CHLOROACETIC ACID CHLORIDE see CEC250
CHLOROACETIC ACID, ETHYL ESTER see EHG500
CHLORO-ACETIC ACID, PHENETHYL ESTER see PDF500
CHLOROACETIC ACID SODIUM SALT see SFU500
CHLOROACETIC CHLORIDE see CEC250
p-CHLOROACETO ACETANILIDE see AAY250
4'-CHLOROACETO ACETANILIDE see AAY250
CHLOROACETONE see CDN200
CHLOROACETONE, stabilized (DOT) see CDN200
2-CHLOROACETONITRILE see CDN500
α-CHLOROACETONITRILE see CDN500
CHLOROACETONITRILE (DOT) see CDN500
1-CHLOROACETOPHENONE see CEA750
4-CHLOROACETOPHENONE see CEB250
p-CHLOROACETOPHENONE see CEB250
α-CHLOROACETOPHENONE see CEA750
2'-CHLOROACETOPHENONE see CEB000
4'-CHLOROACETOPHENONE see CEB250
omega-CHLOROACETOPHENONE see CEA750
CHLOROACETOPHENONE, gas, liquid or solid (DOT) see CEA750
2-CHLORO-4-ACETOTOLUIDIDE see CEB500
2-CHLOROACETO-p-TOLUIDIDE see CEB500
3'-CHLORO-p-ACETOTOLUIDIDE see CEB750
2-CHLORO-10-(3-(4-(2-ACETOXYETHYL)PIPERAZINYL)PROPYL) PHENOTHIAZINE see TFP250
6-CHLORO-17-α-ACETOXY-4,6-PREGNADIENE-3,20-DIONE see CBF250
6-α-CHLORO-17-α-ACETOXYPROGESTERONE see CEB875
6-CHLORO-Δ^6-17-ACETOXYPROGESTERONE see CBF250
Δ^6-6-CHLORO-17-α-ACETOXYPROGESTERONE see CBF250
6-CHLORO-Δ^6-(17-α)ACETOXYPROGESTERONE see CBF250
(CHLOROACETOXY)TRIBUTYLSTANNANE see TIC750
4'-CHLOROACETYL ACETANILIDE see CEC000
4'-(CHLOROACETYL)ACETANILIDE see CEC000
CHLOROACETYL CHLORIDE see CEC250
N-CHLOROACETYLDIETHYLAMINE see DIX400
CHLOROACETYLENE see CEC500
I-N-(α-(CHLOROACETYL)PHENETHYL)-p-TOLUENESULFONAMIDE see THH450
N-(CHLOROACETYL)-3-PHENYL-N-(p-TOLYLSULFONYL)ALANINE see THH550
N-(CHLOROACETYL)-p-TOLUIDINE see CEB500
4'-(2-CHLORO-9-ACRIDINYLAMINO)METHANESULFONANILIDE see CED500
4'-(3-CHLORO-9-ACRIDINYLAMINO)METHANESULFONANILIDE see CED750
N-(4-((2-CHLORO-9-ACRIDINYL)AMINO)PHENYL)METHANESULFONAMIDE see CED500
CHLOROACRYLIC ACID see CEE500
2-CHLOROACRYLIC ACID see CEE500
α-CHLOROACRYLIC ACID see CEE500
2-CHLOROACRYLIC ACID, METHYL ESTER see MIF800
cis-β-CHLOROACRYLIC ACID SODIUM SALT see SFV250
CHLOROACRYLONITRILE see CEE750
2-CHLOROACRYLONITRILE see CEE750
α-CHLOROACRYLONITRILE see CEE750
2-CHLOROADENOSINE see CEF100
2-CHLOROADENOSINE-5'-SULFAMATE see CEF125
CHLOROAETHAN (GERMAN) see EHH000

β-CHLORO ALLYL ALCOHOL see CEF250
pi-CHLORO ALLYL ALCOHOL see CEF500
α-CHLOROALLYL CHLORIDE see DGG950
γ-CHLOROALLYL CHLORIDE see DGG950
2-CHLOROALLYL DIETHYLDITHIOCARBAMATE see CDO250
2-CHLOROALLYL-N,N-DIETHYLDITHIOCARBAMATE see CDO250
CHLOROALLYLENE see AGB250
1-(3-CHLOROALLYL)-3,5,7-TRIAZA-1-AZONIAADAMANTANE CHLORIDE
    see CEG550
CHLOROALONIL see TBQ750
CHLOROALOSANE see GFA000
CHLOROAMBUCIL see CDO500
CHLOROAMINE see CDO750
3-CHLORO-4-AMINOANILINE see CEG600
3-CHLORO-4-AMINOANILINE SULFATE see CEG625
5-CHLORO-1-AMINOANTHRAQUINONE see AJE325
2-CHLORO-4-AMINOBENZOIC ACID see CEG750
4'-CHLORO-4-AMINOBIPHENYL ETHER see CEH125
3-CHLORO-4-AMINODIPHENYL see CEH000
4-CHLORO-4'-AMINODIPHENYL ETHER see CEH125
2-CHLORO-3-AMINO-1,4-NAPHTHOQUINONE see AJI250
p-CHLORO-o-AMINOPHENOL see CEH250
3-CHLORO-4-AMINOSTILBENE see CLE500
2-CHLORO-4-AMINOTOLUENE see CLK215
4-CHLORO-2-AMINOTOLUENE see CLK225
5-CHLORO-2-AMINOTOLUENE see CLK220
5-CHLORO-2-AMINOTOLUENE HYDROCHLORIDE see CLK235
CHLOROAMITRIPTYLINE HYDROCHLORIDE see CEH500
2-CHLOROANILINE see CEH670
3-CHLOROANILINE see CEH675
4-CHLOROANILINE see CEH680
m-CHLOROANILINE see CEH675
o-CHLOROANILINE see CEH670
p-CHLOROANILINE see CEH680
3-CHLOROANILINE (ITALIAN) see CEH675
m-CHLOROANILINE, solid (DOT) see CEH675
o-CHLOROANILINE, solid (DOT) see CEH670
p-CHLOROANILINE, solid (DOT) see CEH680
m-CHLOROANILINE, liquid (DOT) see CEH675
o-CHLOROANILINE, liquid (DOT) see CEH670
p-CHLOROANILINE, liquid (DOT) see CEH680
(o-CHLOROANILINO)DICHLOROTRIAZINE see DEV800
1-((p-(2-(CHLORO-o-ANISAMIDO)ETHYL)PHENYL)SULFONYL)-3-
    CYCLOHEXYL UREA see CEH700
3-CHLOROANISIDINE see CEH750
1-CHLORO-9,10-ANTHRACENEDIONE see CEI000
1-CHLOROANTHRAQUINONE see CEI000
α-CHLOROANTHRAQUINONE see CEI000
1-CHLORO-9,10-ANTHRAQUINONE see CEI000
CHLOROARSENOL see CEI250
CHLOR(O)AZIDE see CDW000
1-CHLOROAZIRIDINE see CEI325
CHLOROBEN see DEP600
2-CHLOROBENZALDEHYDE see CEI500
o-CHLOROBENZALDEHYDE see CEI500
α-CHLOROBENZALDEHYDE see BDM500
2-CHLOROBENZAL MALONONITRILE see CEQ600
o-CHLOROBENZAL MALONONITRILE see CEQ600
1-CHLORO-5-BENZAMIDO-ANTHRAQUINONE see BDK750
7-CHLOROBENZ(a)ANTHRACENE see CEJ000
10-CHLORO-1,2-BENZANTHRACENE see CEJ000
CHLOROBENZEN (POLISH) see CEJ125
3-CHLOROBENZENAMINE see CEH675
4-CHLOROBENZENAMINE see CEH680
2-CHLORO-BENZENAMINE (9CI) see CEH670
CHLOROBENZENE see CEJ125
4-CHLORO BENZENAMINE see CEH680
3-CHLORO-BENZENECARBOPEROXOIC ACID (9CI) see CJI750
o-CHLOROBENZENECARBOXALDEHYDE see CEI500
2-CHLORO-1,4-BENZENEDIAMINE see CEG600
4-CHLORO-1,3-BENZENEDIAMINE see CJY120
2-CHLORO-1,4-BENZENEDIAMINE SULFATE see CEG625
2-CHLOROBENZENEDIAZONIUM SALTS see CEJ500
m-CHLOROBENZENEDIAZONIUM SALTS see CEJ250
o-CHLOROBENZENEDIAZONIUM SALTS see CEJ500
p-CHLOROBENZENESULFONAMIDE see CEK000
4-CHLOROBENZENESULFONATE de 4-CHLOROPHENYLE (FRENCH) see
    CJT750
p-CHLOROBENZENESULFONIC ACID-p-CHLOROPHENYL ESTER see CJT750
p-CHLOROBENZENESULFONIC ACID, SODIUM SALT see CEK250
p-CHLOROBENZENESULFONYL CHLORIDE see CEK375
1-(p-CHLOROBENZENESULFONYL)-3-PROPYLUREA see CKK000
N-(p-CHLOROBENZENESULFONYL)-N-PROPYLUREA see CKK000
4-CHLOROBENZENETHIOL see CEK425

6-CHLOROBENZENO(a)PYRENE see CEK500
1-(p-CHLOROBENZHYDRYL)-4-(p-tert-BUTYLBENZYL)DIETHYLENEDIAMINE
    DIHYDROCHLORIDE see BOM250
1-p-CHLOROBENZHYDRYL-4-p-(tert)-BUTYLBENZYLPIPERAZINE DIHYDRO-
    CHLORIDE see BOM250
1-(p-CHLOROBENZHYDRYL)-4-(2-(2-
    HYDROXYETHOXY)ETHYL)DIETHYLENEDIAMINE see CJR909
1-(p-CHLOROBENZHYDRYL)-4-(2-(2-
    HYDROXYETHOXY)ETHYL)DIETHYLENEDIAMINE HYDROCHLORIDE
    see VSF000
N-(4-CHLOROBENZHYDRYL)-N'-(HYDROXYETHOXYETHYL)PIPERAZINE
    see CJR909
1-(p-CHLOROBENZHYDRYL)-4-(2-(2-HYDROXYETHOXY)ETHYL)PIPERAZINE
    see CJR909
1-(p-CHLOROBENZHYDRYL)-4-(m-METHYLBENZYL)DIETHYLENEDIAMINE
    see HGC500
1-p-CHLOROBENZHYDRYL-4-m-METHYLBENZYLPIPERAZINE
    see HGC500
1-(4-CHLOROBENZHYDRYL)-4-METHYLPIPERAZINE see CFF500
1-(4-CHLOROBENZHYDRYL)-4-METHYLPIPERAZINE DIHYDROCHLORIDE
    see CDR000
1-(p-CHLOROBENZHYDRYL)-4-METHYLPIPERAZINE HYDROCHLORIDE
    see CDR250
N¹-(4'-CHLOROBENZHYDRYL)-N⁴-SPIROMORPHOLINO-PIPERAZINIUM
    CHLORIDE HYDROCHLORIDE see CEK875
1-(4-CHLOROBENZHYDRYL)PIPERAZINE see NNK500
4-(4-CHLOROBENZHYDRYL)PIPERAZINE see NNK500
N-(p-CHLOROBENZHYDRYL)PIPERAZINE see NNK500
1-(CHLOROBENZIL)-2-PIRROLIDIL-METIL-BENZIMIDAZOLOCLORIDATO
    (ITALIAN) see CMV400
2-CHLOROBENZO(e)(1)BENZOTHIOPYRANO(4,3-b)INDOLE see CEL000
2-CHLOROBENZOIC ACID see CEL250
o-CHLOROBENZOIC ACID see CEL250
4-CHLOROBENZOIC ACID-3-ETHYL-7-METHYL-3,7-
    DIAZABICYCLO(3.3.1)NON-9-YL ESTER HYDROCHLORIDE
    see YGA700
o-CHLOROBENZOIC ACID NICKEL(II) SALT see CEL500
CHLOROBENZOL (DOT) see CEJ125
CHLOROBENZONE see CEL750
o-CHLOROBENZONITRILE see CEM000
p-CHLOROBENZONITRILE see CEM250
6-CHLORO-2H-1,2,4-BENZOTHIADIAZINE-7-SULFONAMIDE-1,1-DIOXIDE
    see CLH750
2-CHLOROBENZOTHIAZOLE see CEM500
1-CHLOROBENZOTRIAZOL see CEM625
p-CHLOROBENZOTRIFLUORIDE see CEM825
5-CHLORO-2-BENZOXAZOLAMINE see AJF500
5-CHLOROBENZOXAZOLIDONE see CDQ750
5-CHLORO-2-BENZOXAZOLINONE see CDQ750
6-CHLORO-2-BENZOXAZOLINONE see CDQ750
5-CHLOROBENZOXAZOL-2-ONE see CDQ750
5-CHLORO-3(H)-2-BENZOXAZOLONE see CDQ750
p-CHLOROBENZOYL AZIDE see CEO000
5-(4-CHLOROBENZOYL)-1,4-DIMETHYL-1H-PYRROLE-2-ACETIC ACID SO-
    DIUM SALT DIHYDRATE see ZUA300
m-CHLOROBENZOYL HYDROPEROXIDE see CJI750
1-(4-CHLOROBENZOYL)-N-HYDROXY-5-METHOXY-2-METHYL-1H-INDOLE-3-
    ACETAMIDE see OLM300
1-(p-CHLOROBENZOYL)-5-METHOXY-2-METHYLINDOLE-3-ACETIC ACID
    see IDA000
1-(4-CHLOROBENZOYL)-5-METHOXY-2-METHYL-1H-INDOLE-3-ACETIC
    ACID CARBOXYMETHYL ESTER see AAE625
1-(p-CHLOROBENZOYL)-5-METHOXY-2-METHYLINDOLE-3-
    ACETOHYDROXAMIC ACID see OLM300
((1-(4-CHLOROBENZOYL)-5-METHOXY-2-METHYLINDOLE-3-
    YL)ACETOXY)ACETIC ACID see AAE625
1-(p-CHLOROBENZOYL)-5-METHOXY-2-METHYL-3-IN-
    DOLYLACETOHYDROXAMIC ACID see OLM300
1-(p-CHLOROBENZOYL)-2-METHYL-5-METHOXYINDOLE-3-ACETIC ACID
    see IDA000
1-(p-CHLOROBENZOYL)-2-METHYL-5-METHOXY-3-INDOLE-ACETIC ACID
    see IDA000
α-(1-(p-CHLOROBENZOYL)-2-METHYL-5-METHOXY-3-INDOLYL)ACETIC
    ACID see IDA000
(3-((p-CHLOROBENZOYLMETHYL)-N-METHYLAMINO)PROPYL)-5H-
    DIBENZ(b,f)AZEPINE see IFZ900
1-(2-(2-CHLOROBENZOYL)-4-NITROPHENYL)-2-(DIETHYLAMINOMETHYL)
    IMIDAZOLE FUMARATE see NMV400
p-CHLOROBENZOYL PEROXIDE see BHM750
p-CHLOROBENZOYL PEROXIDE (DOT) see BHM750
CHLOROBENZYLATE see DER000
p-CHLOROBENZYL-p-CHLOROPHENYL SULFIDE see CEP000
4-CHLOROBENZYL-4-CHLOROPHENYL SULPHIDE see CEP000
p-CHLOROBENZYL-p-CHLOROPHENYL SULPHIDE see CEP000

1-(4-CHLOROBENZYL)-3-(6-CHLORO-o-TOLYL)-1-(2-PYRROLIDINYLETHYL) UREA HYDROCHLORIDE see CEP250

α-(p-CHLOROBENZYL)-4-DIETHYLAMINOETHOXY-4-METHYLBENZHYDROL see TMP500

S-(4-CHLOROBENZYL)-N,N-DIETHYLTHIOCARBAMATE see SAZ000

2-((p-CHLOROBENZYL)(2-(DIMETHYLAMINO)ETHYL)AMINO)PYRIDINE see CKV625

4-(p-CHLOROBENZYL)-2-(2-DIMETHYLAMINO)ETHYL)-1(2H)-PHTHALAZINONE HYDROCHLORIDE see CEP675

2-(p-CHLOROBENZYL)-3-DIMETHYLAMINOMETHYL-2-BUTANOL HYDRO-CHLORIDE see CMW500

N-(p-CHLOROBENZYL)-N',N'-DIMETHYL-N-(2-PYRIDYL)ETHYLENEDIAMINE see CKV625

N-3'-CHLOROBENZYL-N'-ETHYLUREA see LII400

4-(p-CHLOROBENZYL)-2-(HEXAHYDRO-1-METHYL-1H-AZEPIN-4-YL)-1-(2H)-PHTHALAZINONE see ASC125

4-(p-CHLOROBENZYL)-2-(HEXAHYDRO-1-METHYL-1H-AZEPIN-4-YL)-1-(2H)-PHTHALAZINONE HCl see ASC130

p-CHLOROBENZYL-3-HYDROXYCROTONATE DIMETHYL PHOSPHATE see CEQ500

o-CHLOROBENZYLIDENE MALONITRILE see CEQ600

2-CHLOROBENZYLIDENE MALONONITRILE see CEQ600

o-CHLOROBENZYLIDENE MALONONITRILE see CEQ600

1-(4-CHLOROBENZYL)-1H-INDAZOLE-3-CARBOXYLIC ACID see CEQ625

1-p-CHLOROBENZYL-1H-INDAZOLE-3-CARBOXYLIC ACID see CEQ625

1-(p-CHLOROBENZYL)-1H-INDAZOLE-3-CARBOXYLIC ACID 1,3-DIHY-DROXY-2-PROPYL ESTER see GGR100

4-CHLOROBENZYL ISOTHIOCYANATE see CEQ750

3-(p-CHLOROBENZYL)OCTAHYDRO-QUINOLIZINE TARTRATE (1:1) see CMX920

1-(p-CHLOROBENZYLOXYCARBONYL)-1-PROPEN-2-YL-DIMETHYLPHOSPHATE see CEQ500

1-(1-(2-((3-CHLOROBENZYL)OXY)PHENYL)VINYL)-1H-IMIDAZOLE HYDRO-CHLORIDE see CMW550

2-(p-CHLOROBENZYL(2-(PYRROLIDINYL)ETHYL)AMINO)-o-ACETOTOLUID-IDE DIHYDROCHLORIDE see CER250

1-(p-CHLOROBENZYL)-2-(1-PYRROLIDINYLMETHYL)BENZIMIDAZOLE HY-DROCHLORIDE see CMV400

1-p-CHLOROBENZYL-PYRROLIDYL-METHYLENE-BENZIMIDAZOLE HYDRO-CHLORIDE see CMV400

3-(p-CHLOROBENZYL)QUINOLIZIDINE TARTRATE see CMX920

5-(o-CHLOROBENZYL)-4,5,6,7-TETRAHYDROTHIENO(3,2-c)PYRIDINE HY-DROCHLORIDE see TGA525

7-CHLOROBICYCLO(3.2.0)HEPTA-2,6-DIEN-6-YL DIMETHYL PHOSPHATE see HBK700

CHLORO BIPHENYL see PJL750

CHLORO-1,1-BIPHENYL see PJL750

2-CHLORO-1,1'-BIPHENYL see CGM750

3-CHLOROBIPHENYLAMINE see CEH000

2-CHLORO-N,N-BIS(2-CHLOROETHYL)-1-PROPYLAMINE HYDROCHLORIDE see NOB700

2-CHLORO-4,6-BIS(DIETHYLAMINO)-s-TRIAZINE see CDQ325

2-CHLORO-4,6-BIS(ETHYLAMINO)-s-TRIAZINE see BJP000

1-CHLORO-3,5-BISETHYLAMINO-2,4,6-TRIAZINE see BJP000

2-CHLORO-4,6-BIS(ETHYLAMINO)-1,3,5-TRIAZINE see BJP000

4-CHLORO-2,6-BIS-ETHYLENEIMINOPYRIMIDINE see EQI600

2-CHLORO-1,1-BIS(FLUOROOXY)TRIFLUOROETHANE see CER825

CHLOROBIS(2-METHYLPROPYL)ALUMINUM see CGB500

CHLOROBLE M see MAS500

2-CHLOROBMN see CEQ600

1-CHLORO-3-BROMO-BUTENE-1 see CES250

1-CHLORO-2-BROMOETHANE see CES500

sym-CHLOROBROMOETHANE see CES500

CHLOROBROMOMETHANE see CES650

4-CHLORO-7-BROMOMETHYLBENZ(a)ANTHRACENE see BNQ000

1-(1-CHLORO-4-BROMOPHENYL)-3-METHYL-3-METHOXYUREA see CES750

1-CHLORO-3-BROMOPROPANE (DOT) see BNA825

omega-CHLOROBROMOPROPANE see BNA825

trans-CHLORO(2-(3-BROMOPROPIONAMIDO)CYCLOHEXYL)MERCURY see CET000

CHLOROBRUMURON see CES750

CHLOROBUFAM see CEX250

CHLOROBUTADIENE see NCI500

1-CHLOROBUTADIENE see CET250

1-CHLORO-1,3-BUTADIENE see CET250

2-CHLOROBUTA-1,3-DIENE see NCI500

2-CHLORO-1,3-BUTADIENE see NCI500

1-CHLOROBUTANE see CEU000

2-CHLOROBUTANE see CEU250

1-CHLOROBUTANE (DOT) see BQQ750

4-CHLORO-1-BUTANE-OL see CEU500

3-CHLOROBUTANOIC ACID see CEW000

CHLOROBUTANOL see ABD000

4-CHLOROBUTANOL see CEU500

4-CHLORO-1-BUTANOL see CEU500

1-CHLORO-2-BUTANONE see CEU750

1-CHLORO-2-BUTENE see CEU825

2-CHLORO-2-BUTENE see CEV000

3-CHLORO-1-BUTENE see CEV250

1-CHLORO-1-BUTEN-3-ONE see CEV500

CHLOROBUTIN see CDO500

CHLOROBUTINE see CDO500

o-CHLORO-α-((tert-BUTYLAMINO)METHYL)BENZYLALCOHOL HYDROCHLO-RIDE see BQE250

4-CHLORO-2-(tert-BUTYLAMINO)-6-(4-METHYLPIPERAZINO)-5-METHYLTHIOPYRIMIDINE see CEV750

5-CHLORO-3-tert-BUTYL-6-METHYLURACIL see BQT750

4-CHLORO-2-BUTYNOL see CEV800

CHLORO-2-BUTYNYL-m-CHLOROCARBAMATE see CEW500

4-CHLOROBUT-2-YNYL-m-CHLOROCARBANILATE see CEW500

4-CHLORO-2-BUTYNYL-m-CHLOROCARBANILATE see CEW500

4-CHLOROBUT-2-YNYL-3-CHLOROPHENYLCARBAMATE see CEW500

4-CHLORO-2-BUTYNYL-N-(3-CHLOROPHENYL)CARBAMATE see CEW500

3-CHLOROBUTYRIC ACID see CEW000

β-CHLOROBUTYRIC ACID see CEW000

4-CHLOROBUTYRIC ACID TRIBUTYLSTANNYL ESTER see TID000

CHLOROCAIN see CBR250

CHLOROCAINE see AIT250

CHLOROCAMPHENE see CDV100

CHLOROCAPS see CDP250

m-CHLORO CARBANILIC ACID-4-CHLORO-2-BUTYNYL ESTER see CEW500

3-CHLOROCARBANILIC ACID, ISOPROPYL ESTER see CKC000

m-CHLOROCARBANILIC ACID, ISOPROPYL ESTER see CKC000

m-CHLOROCARBANILIC ACID-1-METHYL-2-PROPYNYL ESTER see CEX250

CHLOROCARBONATE D'ETHYLE (FRENCH) see EHK500

CHLOROCARBONATE de METHYLE (FRENCH) see MIG000

CHLOROCARBONIC ACID METHYL ESTER see MIG000

N-(CHLOROCARBONYLOXY)TRIMETHYLUREA see CEX255

3-CHLOROCARPIPRAMINE DIHYDROCHLORIDE see DKV309

5-CHLOROCARVACROL see CEX275

CHLOROCHIN see CLD000

3-CHLOROCHLORDENE see HAR000

6'-CHLORO-2-(p-CHLOROBENZYL(2-(DIETHYLAMINO)ETHYL)AMINO)-o-ACETOLUIDIDE DIHYDROCHLORIDE see CEX500

6'-CHLORO-2-(p-CHLOROBENZYL(2-(PYRROLIDINYL)ETHYL)AMINO)-o-ACETOTOLUIDIDE DIHYDROCHLORIDE see CEX500

1-CHLORO-2-(β-CHLOROETHOXY)ETHANE see DFJ050

6-CHLORO-9-((2-((2-CHLOROETHYL)AMINO)ETHYL)AMINO)-2-METHOXYACRIDINE 2-HYDROCHLORIDE SESQUIHYDRATE see CEY250

2-CHLORO-N-(2-CHLOROETHYL)-N,N-DIMETHYLETHANAMINIUM CHLO-RIDE see DQS600

13-CHLORO-N-(2-CHLOROETHYL)-N,11-DINITROSO-10-OXO-5,6-DITHIA-2,9,11-TRIAZATRIDECANAMIDE see BIF625

2-CHLORO-N-(2-CHLOROETHYL)ETHANAMINE HYDROCHLORIDE see BHO250

7-CHLORO-10-(3-(N-(2-CHLOROETHYL)-N-ETHYL)AMINOPROPYLAMINO)-2-METHOXY-BENZO(B)(1,5)NAPHTHYRIDINE DIHYDROCHLORIDE see IAE000

6-CHLORO-9-(3-(2-CHLOROETHYL)MERCAPTOPROPYLAMINO)-2-METHOXYACRIDINE HYDROCHLORIDE see CFA250

2-CHLORO-N-(2-CHLOROETHYL)-N-METHYLETHANAMINE HYDROCHLO-RIDE see BIE500

2-CHLORO-N-(2-CHLOROETHYL)-N-METHYL ETHANAMINE-N-OXIDE see CFA500

2-CHLORO-N-(2-CHLOROETHYL)-N-METHYLETHANAMINE-N-OXIDE HY-DROCHLORIDE see CFA750

1-CHLORO-2-(β-CHLOROETHYLTHIO)ETHANE see BIH250

CHLORO(CHLOROMETHOXY)METHANE see BIK000

9-CHLORO-10-CHLOROMETHYL ANTHRACENE see CFB500

cis-2-CHLORO-3-(CHLOROMETHYL)OXIRANE see DAC975

trans-2-CHLORO-3-(CHLOROMETHYL)OXIRANE see DGH500

1-CHLORO-4-(CHLOROMETHYLTHIO)BENZENE see CFB750

2-CHLORO-5-CHLOROMETHYLTHIOPHENE see CFB825

5-CHLORO-N-(2-CHLORO-4-NITROPHENYL)-2-HYDROXYBENZAMIDE see DFV400

5-CHLORO-N-(2-CHLORO-4-NITROPHENYL)-2-HYDROXYBENZAMIDE with 2-AMINOETHANOL (1:1) see DFV600

3-CHLORO-4-(3-CHLORO-2-NITROPHENYL)PYRROLE see CFC000

5-CHLORO-2'-CHLORO-4'-NITROSALICYLANILIDE see DFV400

N-(5-CHLORO-4-((4-CHLOROPHENYL)CYANOMETHYL)-2-METHYLPHENYL)-2-HYDROXY-3, 5-DIIODOBENZAMIDE see CFC100

4-CHLORO-α-(4-CHLOROPHENYL)-α-CYCLOPROPYLBENZENEMETHANOL (9CI) see PMF550

7-CHLORO-5-(2-CHLOROPHENYL)-1,3-DIHYDRO-3-HYDROXY-2H-1,4-BENZODIAZEPIN-2-ONE see CFC250

7-CHLORO-5-(o-CHLOROPHENYL)-1,3-DIHYDRO-3-HYDROXY-2H-1,4-BENZODIAZEPIN-2-ONE see CFC250

7-CHLORO-5-(2-CHLOROPHENYL)-1,3-DIHYDRO-3-HYDROXY-1-METHYL-2H-
    1,4-BENZODIAZEPIN-2-ONE see MLD100
5-CHLORO-3-(4-CHLOROPHENYL)-4'-FLUORO-2'-METHYLSALICYLANILIDE
    see CFC500
7-CHLORO-5-(2-CHLOROPHENYL)-3-HYDROXY-1H-1,4-BENZODIAZEPIN-
    2(3H)-ONE see CFC250
7-CHLORO-5-(2-CHLOROPHENYL)-3-HYDROXY-1-METHYL-2,3-DIHYDRO-1H-
    1,4-BENZODIAZEPIN-2-ONE see MLD100
2-CHLORO-3-(4-CHLOROPHENYL)METHYLPROPIONATE see CFC750
1-CHLORO-4-(((4-CHLOROPHENYL)METHYL)THIO)BENZENE see CEP000
8-CHLORO-6-(o-CHLOROPHENYL)-1-METHYL-4H-s-TRIAZOLO(4,3-
    a)(1,4)BENZODIAZEPINE see THS800
8-CHLORO-6-(2-CHLOROPHENYL)-1-METHYL-4H-(1,2,4)TRIAZOLO(4,3-
    a)(1,4)BENZODIAZEPINE see THS800
2-CHLORO-3-(4-CHLOROPHENYL)PROPIONIC ACID METHYL ESTER
    see CFC750
4-CHLORO-α-(4-CHLOROPHENYL)-α-
    (TRICHLOROMETHYL)BENZENEMETHANOL see BIO750
CHLORO(2-CHLOROVINYL)MERCURY see CFD250
CHLOROCHOLINE CHLORIDE see CMF400
CHLOROCID see CDP250
CHLOROCIDE see CEP000
CHLOROCIDIN C TETRAN see CDP250
CHLOROCOL see CDP250
CHLOROCRESOL see CFE250
p-CHLOROCRESOL see CFE250
4-CHLORO-m-CRESOL see CFE250
4-CHLORO-o-CRESOL see CFE000
6-CHLORO-m-CRESOL see CFE250
p-CHLORO-m-CRESOL see CFE250
4-CHLORO-o-CRESOXYACETIC ACID see CIR250
CHLOROCTAN SODNY (CZECH) see SFU500
CHLOROCYAN see COO750
CHLOROCYANIDE see COO750
CHLOROCYANOACETYLENE see CFE750
2-CHLORO-1-CYANOETHANOL see CHU000
CHLOROCYANOGEN see COO750
2-CHLORO-α-CYANO-6-METHYLERGOLINE-8-PROPIONAMIDE see CFF100
2-CHLORO-4-(1-CYANO-1-METHYLETHYLAMINO)-6-ETHYLAMINO-1,3,5-
    TRIAZINE see BLW750
3-CHLORO-6-CYANO-2-NORBORNANONE-o-(METHYLCARBAMOYL)OXIME
    see CFF250
endo-3-CHLORO-exo-6-CYANO-2-NORBORNANONE-o-
    (METHYLCARBAMOYL)OXIME see CFF250
2-exo-CHLORO-6-endo-CYANO-2-NORBORNANONE-o-
    (METHYLCARBAMOYL)OXIME2-CARBONITRILE see CFF250
3-CHLORO-6-CYANONORBORNANONE-2-OXIME-o,N-METHYLCARBAMATE
    see CFF250
CHLOROCYCLAMIDE-R see CDR550
CHLOROCYCLINE see CFF500
CHLOROCYCLIZINE see CFF500
CHLOROCYCLIZINE HYDROCHLORIDE see CDR500
2-CHLOROCYCLOHEXANONE see CFG250
α-CHLOROCYCLOHEXANONE see CFG250
4-CHLORO-4-CYCLOHEXENE-1,2-DICARBOXYLIC ANHYDRIDE see CFG500
7-CHLORO-5-(CYCLOHEXEN-1-YL)-1,3-DIHYDRO-1-METHYL-2H-1,4-
    BENZODIAZEPIN-2-ONE see CFG750
7-CHLORO-5-(1-CYCLOHEXENYL)-1-METHYL-2-OXO-2,3-DIHYDRO-1H-(1,4)-
    BENZO(f)DIAZEPINE see CFG750
6'-CHLORO-2-(CYCLOHEXYLAMINO)-o-ACETOTOLUIDIDE HYDROCHLO-
    RIDE see CFH000
5-CHLORO-N-(2-(4-((((CYCLOHEXYLAMINO) CARBONYL)AMINO)SULFO-
    NYL)PHENYL) ETHYL)-2-METHOXYBENZAMIDE see CEH700
3-(3-CHLORO-4-CYCLOHEXYLBENZOYL)PROPIONIC ACID see CPJ000
3-(3-CHLORO-4-CYCLOHEXYLBENZOYL)PROPIONIC ACID CALCIUM SALT
    see CPJ250
(Z)-3-(2-CHLOROCYCLOHEXYL)-1-(2-CHLOROETHYL)-1-NITROSOUREA
    see CFH500
cis-3-(2-CHLOROCYCLOHEXYL)-1-(2-CHLOROETHYL)-1-NITROSOUREA
    see CFH500
cis-N'-(2-CHLOROCYCLOHEXYL)-N-(2-CHLOROETHYL)-N-NITROSOUREA
    see CFH500
trans-3-(2-CHLOROCYCLOHEXYL)-1-(2-CHLOROETHYL)-1-NITROSOUREA
    see CFH750
trans-N'-(2-CHLOROCYCLOHEXYL)-N-(2-CHLOROETHYL)-N-NITROSOUREA
    see CFH750
6-CHLORO-5-CYCLOHEXYL-2,3-DIHYDRO-1H-INDENE-1-CARBOXYLIC ACID
    (9CI) see CMV500
(±)-6-CHLORO-5-CYCLOHEXYL-2,3-DIHYDRO-1H-INDENE-1-CARBOXYLIC
    ACID (9CI) see CFH825
6-CHLORO-5-CYCLOHEXYL-1-INDANCARBOXYLIC ACID see CFH825,
    CMV500
(±)-6-CHLORO-5-CYCLOHEXYLINDAN-1-CARBOXYLIC ACID see CFH825
(±)-6-CHLORO-5-CYCLOHEXYL-1-INDANCARBOXYLIC ACID see CFH825

6'-CHLORO-2-(N-CYCLOHEXYL-N-METHYLAMINO)-o-ACETOTOLUIDIDE HY-
    DROCHLORIDE see CFI000
3-CHLORO-4-CYCLOHEXYL-α-OXOBENZENEBUTANOIC ACID see CPJ000
3-CHLORO-4-CYCLOHEXYL-α-OXO-BENZENEBUTANOIC ACID see CPJ250
3-CHLORO-4-CYCLOHEXYL-α-OXOBENZENEBUTANOIC ACID CALCIUM
    SALT see CPJ250
4-(3-CHLORO-4-CYCLOHEXYLPHENYL)-4-OXO-BUTYRIC ACID see CPJ000
4-(3-CHLORO-4-CYCLOHEXYLPHENYL)-4-OXOBUTYRIC ACID CALCIUM
    SALT see CPJ250
CHLOROCYCLOPENTANE see CFI250
2-CHLOROCYCLOPENTANONE see CFI500
α-CHLOROCYCLOPENTANONE see CFI500
3-CHLOROCYCLOPENTENE see CFI625
4-CHLORO-2-CYCLOPENTYL PHENOL see CFI750
2-CHLORO-4-CYCLOPROPYLAMINO-6-ISOPROPYLAMINO-1,3,5-TRIAZINE
    see CQI750
2-CHLORO-4-CYCLOPROPYLAMINO-6-ISOPROPYLAMINO-sec-TRIAZINE
    see CQI750
2-((4-CHLORO-6-(CYCLOPROPYLAMINO)-1,3,5-TRIAZIN-2-YL)AMINO)-2-
    METHYLPROPANENITRILE see PMF600
2-(4-CHLORO-6-(CYCLOPROPYLAMINO)-s-TRIAZIN-2-YL)AMINO-2-
    METHYLPROPIONITRILE see PMF600
3-CHLORO-4-CYCLO-PROPYLMETHOXYPHENYLACETIC ACID LYSINE
    SALT (d,l) see CFJ000
2-(3-CHLORO-4-CYCLOPROPYLMETHOXYPHENYL)ACETIC ACID LYSINE
    SALT (d,l) see CFJ000
7-CHLORO-1-CYCLOPROPYLMETHYL-1,3-DIHYDRO-5-(2-FLUOROPHENYL)-
    2H-1,4-BENZODIAZEPIN-2-ONE see FMR100
7-CHLORO-1-(CYCLOPROPYLMETHYL)-1,3-DIHYDRO-5-PHENYL-2H-1,4-
    BENZODIAZEPIN-2-ONE see DAP700
7-CHLORO-1-CYCLOPROPYLMETHYL-5-PHENYL-1H-1,4-BENZODIAZEPIN-
    2(3H)-ONE see DAP700
6-CHLORO-1-(CYCLOPROPYLMETHYL)-4-PHENYL-2(1H)-QUINAZOLINONE
    see CQF125
CHLORODANE see CDR750
CHLORODECANE see DAJ200
6-CHLORO-6-DEHYDRO-17-α-ACETOXYPROGESTERONE see CBF250
6-CHLORO-Δ⁶-DEHYDRO-17-ACETOXYPROGESTERONE see CBF250
6-CHLORO-6-DEHYDRO-17-α-ACETOXYPROGESTERONE mixed with
    MESTRENOL see CNV750
6-CHLORO-6-DEHYDRO-17-α-HYDROXYPROGESTERONE ACETATE
    see CBF250
7-CHLORO-6-DEMETHYLTETRACYCLINE DEMETHYLCHLOROTETRACY-
    CLINE see MIJ500
7-CHLORO-6-DEMETHYLTETRACYCLINE HYDROCHLORIDE see DAI485
CHLORODEN see DEP600
6-CHLORO-6-DEOXYGLUCOSE see CFJ375
6-CHLORO-6-DEOXY-d-GLUCOSE see CFJ375
CHLORODEOXYGLYCEROL see CDT750
CHLORODEOXYGLYCEROL DIACETATE see CIL900
7(S)-CHLORO-7-DEOXYLINCOMYCIN see CMV675
7(S)-CHLORO-7-DEOXYLINCOMYCIN-2-PHOSPHATE see CMV690
5-CHLORODEOXYURIDINE see CFJ750
5-CHLORO-2'-DEOXYURIDINE see CFJ750
CHLORODIABINA see CKK000
2-CHLORO-N,N-DIALLYLACETAMIDE see CFK000
α-CHLORO-N,N-DIALLYLACETAMIDE see CFK000
1-CHLORO-2,4-DIAMINOBENZENE see CJY120
4-CHLORO-1,2-DIAMINOBENZENE see CFK125
α-(2-CHLORO-4-(4,6-DIAMINO-2,2-DIMETHYL-S-TRIAZINE-1(2H)-
    YL)PHENOXY)-N,N-DIMETHYL-m-TOLUAMIDE ETHANESULFONATE
    see THR500
CHLORODIAZEPOXIDE see LFK000
1-CHLORODIBENZO-p-DIOXIN see CDV125
2-((8-CHLORODIBENZO(b,f)THIEPIN-10-YL)OXY-N,N-DIMETHYLETHANAM-
    INE see ZUJ000
CHLORODIBORANE see CFK325
CHLORODIBROMOMETHANE see CFK500
1-CHLORO-2,3-DIBROMOPROPANE see DDL800
3-CHLORO-1,2-DIBROMOPROPANE see DDL800
CHLORO(DIBUTOXYPHOSPHINYL)MERCURY see CFK750
6'-CHLORO-2-(DIBUTYLAMINO)-o-ACETOTOLUIDIDE, HYDROCHLORIDE
    see CFL000
1-CHLORO-2-(2,2-DICHLORO-1-(4-CHLOROPHENYL)ETHYL)BENZENE
    see CDN000
p-CHLORO-DI-(2-CHLOROETHYL)BENZYLAMINE HYDROCHLORIDE
    see BIB750
1-CHLORO-2,2-DICHLOROETHYLENE see TIO750
6-CHLORO-3-(DICHLOROMETHYL)-3,4-DIHYDRO-2H-1,2,4-BENZOTHIADIAZ-
    INE-7-SULFONAMIDE-1,1-DIOXIDE see HII500
6-CHLORO-3-(DICHLOROMETHYL)3,4-DIHYDRO-7-SULFAMYL-1,2,4-
    BENZOTHIADIAZINE-1,1-DIOXIDE see HII500
6-CHLORO-5-(2,3-DICHLOROPHENOXY)-2-METHYLTHIO-BENZIMIDAZOLE
    see CFL200

7-CHLORO-1,3-DIHYDRO-3-HYDROXY-5-PHENYL-2H-1,4-BENZODIAZEPINE-2-ONE see CFZ000

7-CHLORO-2,3-DIHYDRO-1-METHYL-5-PHENYL-1H-1,4-BENZODIAZEPINE see CGA000

7-CHLORO-2,3-DIHYDRO-1-METHYL-5-PHENYL-1H-1,4-BENZODIAZEPINE, HYDROCHLORIDE see MBY000

7-CHLORO-1,3-DIHYDRO-1-METHYL-5-PHENYL-2H-1,4-BENZODIAZEPIN-2-ONE see DCK759

6-CHLORO-3,4-DIHYDRO-2-METHYL-3-(((2,2,2-TRIFLUOROETHYL)THIO)METHYL)2H-1,2,4-BENZOTHIADIAZINE-7-SULFONAMIDE, 1,1-DIOXIDE see PKL250

10-CHLORO-5,10-DIHYDROPHENARSAZINE see PDB000

7-CHLORO-1,3-DIHYDRO-5-PHENYL-2H-1,4-BENZODIAZEPIN-2-ONE see CGA500

7-CHLORO-1,3-DIHYDRO-5-PHENYL-1-(2-PROPYNYL)-2H-1,4-BENZODIAZEPIN-2-ONE see PIH100

7-CHLORO-1,3-DIHYDRO-5-PHENYL-1-TRIMETHYLSILYL-2H-1,4-BENZODIAZEPIN-2-ONE see CGB000

6-CHLORO-3,4-DIHYDRO-7-SULFAMOYL-2H-1,2,4-BENZOTHIADIAZINE-1,1-DIOXIDE see CFY000

(S-(e,e))-3-CHLORO-4,6-DIHYDROXY-2-METHYL-5-(3-METHYL-7-(TETRAHYDRO-5,5-DIMETHYL-4-OXO-2-FURANYL)-2,6-OCTADIENYL)-BENZALDEHYDE see ARM750

1-CHLORO-2,3-DIHYDROXYPROPANE see CDT750

3-CHLORO-1,2-DIHYDROXYPROPANE see CDT750

4′-CHLORO-3,5-DIIODOSALICYLANILIDE ACETATE see CGB250

CHLORO DIISOBUTYL ALUMINUM see CGB500

CHLORO(DIISOPROPOXYPHOSPHINYL)MERCURY see CGB750

6′-CHLORO-2-(DIISOPROPYLAMINO)-o-ACETOTOLUIDIDE HYDROCHLORIDE see CGC000

21-CHLORO-3,17-DIMETHOXY-19-NOR-17α-PREGN-1,3,5(10)-TRIEN-20-YNE see BAT795

N-CHLORODIMETHYLAMINE see CGC200

6′-CHLORO-2-(DIMETHYLAMINO)-o-ACETOTOLUIDIDE HYDROCHLORIDE see CGD000

p-CHLORO DIMETHYLAMINOAZOBENZENE see CGD250

2′-CHLORO-4-DIMETHYLAMINOAZOBENZENE see DRD000

3′-CHLORO-4-DIMETHYLAMINOAZOBENZENE see DRC800

4′-CHLORO-4-DIMETHYLAMINOAZOBENZENE see CGD250

β-CHLORODIMETHYLAMINO DIBORANE see CGD399

CHLORO(DIMETHYLAMINO)ETHANE see CGW000

2-(p-CHLORO-α-(2-(DIMETHYLAMINO)ETHOXY)BENZYL)PYRIDINE BIMALE-ATE see CGD500

2-(p-CHLORO-α-(2-(DIMETHYLAMINO)ETHOXY)BENZYL)-PYRIDINE DIPHENYLDISULFONATE see CBQ575

2-(p-CHLORO-α-(2-(DIMETHYLAMINO)ETHOXY)BENZYL)PYRIDINE MALE-ATE see CGD500

2-CHLORO-11-(2-(DIMETHYLAMINO)ETHOXY)DIBENZO(b,f)THIEPIN see ZUJ000

2-CHLORO-11-(2-DIMETHYLAMINOETHOXY)DIBENZO(b,f)THIEPINE see ZUJ000

2-CHLORO-α-(2-(DIMETHYLAMINO)-ETHYL)BENZHYDROL HYDROCHLORIDE see CMW700

dl-2-(p-CHLORO-α-2-(DIMETHYLAMINO)ETHYLBENZYL)PYRIDINE BIMALE-ATE see TAI500

(+)-2-(p-CHLORO-α-(2-(DIMETHYLAMINO)ETHYL)BENZYL)PYRIDINE MALE-ATE see PJJ325

2-(p-CHLORO-α-(2-(DIMETHYLAMINO)ETHYL)BENZYL)PYRIDINE MALEATE (1:1) see CLD250

2′-CHLORO-2-((2-(DIMETHYLAMINO)ETHYL)ETHYLAMINO)ACETANILIDE DIHYDROCHLORIDE see CGD750

7-CHLORO-10-(2-(DIMETHYLAMINO)ETHYL)ISOALLOXAZINE SULFATE see CGE000

2-CHLORO-α-(2-(DIMETHYLAMINO)ETHYL)-α-PHENYL-BENZENEMETHANOL HYDROCHLORIDE (9CI) see CMW700

2′-CHLORO-2-(2-(DIMETHYLAMINO)ETHYLTHIO)ACETANILIDE HYDRO-CHLORIDE see CGE250

6′-CHLORO-2-(2-(DIMETHYLAMINO)ETHYLTHIO)-o-ACETOTOLUIDIDE see CGE500

6′-CHLORO-2-(DIMETHYLAMINO)-N-METHYL-o-ACETOTOLUIDIDE HYDRO-CHLORIDE see CGE750

5-CHLORO-3-(DIMETHYLAMINOMETHYL)-2-BENZOXAZOLINONE see CGF000

7-CHLORO-10-(4-(DIMETHYLAMINO)-1-METHYLBUTYL)ISOALLOXAZINE SULFATE see CGF250

4-CHLORO-α-(2-(DIMETHYLAMINO)-1-METHYLETHYL)-α-METHYLBENZENEETHANOL HYDROCHLORIDE see CMW500

p-CHLORO-α-(2-(DIMETHYLAMINO)-1-METHYLETHYL)-α-METHYL-PHENETHYL ALCOHOL see CMW459

p-CHLORO-α-(2-(DIMETHYLAMINO)-1-METHYLETHYL)-α-METHYLPHENETHYL ALCOHOL HYDROCHLORIDE see CMW500

2-CHLORO-4-DIMETHYLAMINO-6-METHYL-PYRIMIDINE see CCP500

7-CHLORO-4-(DIMETHYLAMINO)-1,4,4a,5,5a,6,11,12a-OCTAHYDRO-2-NAPHTHACENECARBOXAMIDE see CMA750

p-CHLORO-α,α-DIMETHYL-2-AMINOPROPIONATE-PHENETHYL ALCOHOL HYDROCHLORIDE see CJX000

6′-CHLORO-3-(DIMETHYLAMINO)-o-PROPIONOTOLUIDIDE HYDROCHLO-RIDE see CGF500

(α-2-CHLORO-9-omega-DIMETHYLAMINO-PROPYLAMINE)THIOXANTHENE see TAF675

7-CHLORO-10-(3-DIMETHYLAMINOPROPYL)-BENZO-(b)(1,8)-5(10H)-NAPHTHAPYRIDONE HYDROCHLORIDE see CGG500

3-CHLORO-5-(3-(DIMETHYLAMINO)PROPYL)-10,11-DIHYDRO-5H-DIBENZ(b,f)AZEPINE see CDU750

3-CHLORO-5-(3-(DIMETHYLAMINO)PROPYL)-10,11-DIHYDRO-5H-DIBENZ(b,f)AZEPINE HYDROCHLORIDE see CGG600

3-CHLORO-5-(3-(DIMETHYLAMINO)PROPYL)-10,11-DIHYDRO-5H-DIBENZ(b,f)AZEPINE MONOHYDROCHLORIDE see CDV000

2-CHLORO-9-(3-(DIMETHYLAMINO)PROPYLIDENE)-THIOXANTHENE see TAF675

2-CHLORO-9-(omega-DI-METHYLAMINOPROPYLIDENE)THIOXANTHENE see TAF675

7-CHLORO-10-(3-(DIMETHYLAMINO)PROPYL)ISOALLOXAZINE HYDRO-CHLORIDE see CGG750

2-CHLORO-10-(3-(DIMETHYLAMINO)PROPYL)PHENOTHIAZINE see CKP250

CHLORO-3-(DIMETHYLAMINO-3-PROPYL)-10 PHENOTHIAZINE (FRENCH) see CKP250

2-CHLORO-10-(3-DIMETHYLAMINOPROPYL) PHENOTHIAZINE MONOHYDROCHLORIDE see CKP500

3′-CHLORO-N,N-DIMETHYLAMINOSTIBEN (GERMAN) see CGK750

4′-CHLORO-N,N-DIMETHYLAMINOSTIBEN (GERMAN) see CGL000

2′-CHLORO-4-DIMETHYLAMINOSTILBENE see CGK500

3′-CHLORO-4-DIMETHYLAMINOSTILBENE see CGK750

4′-CHLORO-4-DIMETHYLAMINOSTILBENE see CGL000

9-CHLORO-8,12-DIMETHYLBENZ(a)ACRIDINE see CGH250

2-CHLORO-1,10-DIMETHYL-5,6-BENZACRIDINE (FRENCH) see CGH250

2-CHLORO-1,10-DIMETHYL-7,8-BENZACRIDINE (FRENCH) see DRB800

2-CHLORO-α,α-DIMETHYLBENZENEETHANIAMINE HYDROCHLORIDE see DRC600

10-CHLORO-6,9-DIMETHYL-5,10-DIHYDRO-3,4-BENZOPHENARSAZINE see CGH500

p-CHLORO-5,10-DIMETHYL-2,4-DIOXA-p-THIONO-3-PHOSPHABICYCLO(4.4.0)DECANE see CGH675

CHLORO((3-(5,5-DIMETHYL-2,4-DIOXO-3-IMIDAZOLIDINYL)-2-METHOXY)PROPYL)MERCURY see CHV250

2-CHLORO-N,N-DIMETHYLETHYLAMINE HYDROCHLORIDE see DRC000

5-CHLORO-3-(1,1-DIMETHYLETHYL)-6-METHYL-2,4(1H,3H)-PYRIMIDINEDIONE see BQT750

1-(2-CHLORO-5,7-DIMETHYL-3-METHYL-1-ADAMANTYL)-N-METHYL-2-PROPYL AMINE HYDROCHLORIDE see CIF000

p-CHLORO-N-α-DIMETHYLPHENETHYLAMINE see CIF250

p-CHLORO-α,α-DIMETHYLPHENETHYLAMINE see CLY250

4-CHLORO-α,α-DIMETHYLPHENETHYLAMINE HYDROCHLORIDE see ARW750

O-CHLORO-α,α-DIMETHYLPHENETHYLAMINE HYDROCHLORIDE see DRC600

p-CHLORO-α,α-DIMETHYLPHENETHYLAMINE HYDROCHLORIDE see ARW750

N-(p-CHLORO-α,α-DIMETHYLPHENETHYL)-2-(DIETHLAMINO)PROPIONAM-IDE HYDROCHLORIDE see CGI125

4-CHLORO-3,5-DIMETHYLPHENOL see CLW000

2-CHLORO-4,5-DIMETHYLPHENOL, METHYL CARBAMATE see CGI500

((4-CHLORO-6-((2,3-DIMETHYLPHENYL)AMINO)-2-PYRIMIDINYL)THIO)ACETIC ACID see CLW250

(2-CHLORO-4,5-DIMETHYL)PHENYL ESTER, CARBAMIC ACID see CGI500

2-CHLORO-4,5-DIMETHYLPHENYL METHYLCARBAMATE see CGI500

CHLORODIMETHYLPHOSPHINE see CGI625

7-CHLORO-1-((DIMETHYLPHOSPHINYL)METHYL)-1,3-DIHYDRO-5-PHENYL-2H-1,4-BENZODIAZEPINE-2-ONE see FOL100

6′-CHLORO-2-(2,6-DIMETHYLPIPERIDINO)-o-ACETOTOLUIDIDE HYDRO-CHLORIDE see CGI750

6′-CHLORO-3-(2,6-DIMETHYLPIPERIDINO)-o-PROPIONOTOLUIDIDE HYDRO-CHLORIDE see CGJ000

2-CHLORO-5-(3,5-DIMETHYLPIPERIDINO SULPHONYL)BENZOIC ACID see CGJ250

2-CHLORO-4-(1,1-DIMETHYLPROPYL)PHENYL METHYL METHYLPHOSPHORAMIDATE see DYE200

2′-CHLORO-N,N-DIMETHYL-4-STILBENAMINE see CGK500

3′-CHLORO-N,N-DIMETHYL-4-STILBENAMINE see CGK750

4′-CHLORO-N,N-DIMETHYL-4-STILBENAMINE see CGL000

2-CHLORO-N,N-DIMETHYLTHIOXANTHENE-Δ⁹-γ-PROPYLAMINE see TAF675

N-CHLORO-4,5-DIMETHYLTRIAZOLE see CGL125

4′-CHLORO-2,2-DIMETHYLVALERANILIDE see CGL250

2-CHLORO-4,6-DINITROANILINE see CGL325

4-CHLORO-2,6-DINITROANILINE see CGL500

CHLORODINITROBENZENE see CGL750

1-CHLORO-2,4-DINITROBENZENE see CGM000

4-CHLORO-1,3-DINITROBENZENE see CGM000

6-CHLORO-1,3-DINITROBENZENE see CGM000
CHLORODINITROBENZENE (DOT) see CGL750
CHLORODINITRO BENZENE (mixed isomers) see CGL750
4-CHLORO-2,5-DINITROBENZENE DIAZONIUM-6-OXIDE see CGM199
1-CHLORO-2,4-DINITROBENZOL (GERMAN) see CGM000
1-CHLORO-2,4-DINITRONAPHTHALENE see DUS600
p-CHLORO-2,4-DIOXA-5-ETHYL-p-THIONO-3-PHOSPHABICYCLO(4.4.0)
　　DECANE see CGM375
p-CHLORO-2,4-DIOXA-5-METHYL-p-THIONO-3-PHOSPHABICYCLO(4.4.0)
　　DECANE see CGM400
CHLORO((3-(2,4-DIOXO-3-IMIDAZOLIDINYL)-2-METHOXY)PROPYL)
　　MERCURY see CHV750
CHLORO((3-(2,4-DIOXO-5-IMIDAZOLIDINYL)-2-METHOXY)PROPYL)
　　MERCURY see CGM450
6-CHLORO-1,3-DIOXO-5-ISOINDOLINESULFONAMIDE see CGM500,
　　CLG825
CHLORO(3-(2,4-DIOXO-1-METHYL-3-IMIDAZOLIDINYL)-2-METHOXY)
　　PROPYL)MERCURY see CHW250
CHLORO((3-(2,4-DIOXO-3-METHYL-1-IMIDAZOLIDINYL)-2-METHOXY)
　　PROPYL)MERCURY see CHW000
CHLORO((3-(2,4-DIOXO-3-METHYL-5-IMIDAZOLIDINYL)-2-METHOXY)
　　PROPYL)MERCURY see CHW500
2-CHLORODIPHENYL see CGM750
o-CHLORODIPHENYL see CGM750
2-CHLORO-2,2-DIPHENYLACETIC ACID-2-(DIETHYLAMINO)ETHYL ESTER
　　HYDROCHLORIDE see DHW200
CHLORODIPHENYLARSINE see CGN000
1-(o-CHLORO-α,α-DIPHENYLBENZYL)IMIDAZOLE see MRX500
CHLORODIPHENYL (21% Cl) see PJM000
CHLORODIPHENYL (32% Cl) see PJM250
CHLORODIPHENYL (41% Cl) see PJL800
CHLORODIPHENYL (48% Cl) see PJM750
CHLORODIPHENYL (60% Cl) see PJN250
CHLORODIPHENYL (62% Cl) see PJN500
CHLORODIPHENYL (68% Cl) see PJN750
CHLORODIPHENYL (42% Cl) (OSHA) see PJM500
CHLORODIPHENYL (54% Cl) (OSHA) see PJN000
2-(4-(2-CHLORO-1,2-DIPHENYLETHENYL)PHENOXY)-N,N-
　　DIETHYLETHANAMINE see CMX500
1-(p-CHLORODIPHENYLMETHYL)-4-(2-(2-HYDROXYETHOXY)ETHYL)
　　PIPERAZINE see CJR909
4-CHLORODIPHENYL SULFONE see CKI625
(E)-2-(p-(2-CHLORO-1,2-DIPHENYLVINYL)PHENOXY)TRIETHYLAMINE, CI-
　　TRATE see CMX750
2-(p-(2-CHLORO-1,2-DIPHENYL VINYL)PHENOXY)TRIETHYLAMINE CI-
　　TRATE (1:1) see CMX700
2-CHLORO-N,N-DI-2-PROPENYLACETAMIDE see CFK000
6'-CHLORO-2-(DIPROPYLAMINO)-o-ACETOTOLUIDIDE HYDROCHLORIDE
　　see CGN250
CHLORODIPROPYLBORANE see CGN325
CHLOROEPOXYETHANE see CGX000
1-CHLORO-2,3-EPOXYPROPANE see EAZ500
3-CHLORO-1,2-EPOXYPROPANE see EAZ500
cis-1-CHLORO-1,2-EPOXYPROPANE see CKS099
trans-1-CHLORO-1,2-EPOXYPROPANE see CKS100
CHLOROETENE see MIH275
2-CHLORO-1-ETHANAL see CDY500
2-CHLOROETHANAMIDE see CDY850
CHLOROETHANE see EHH000
2-CHLOROETHANEPHOSPHONIC ACID see CDS125
2-CHLOROETHANE SULFOCHLORIDE see CGO125
2-CHLOROETHANESULFONYL CHLORIDE see CGO125
β-CHLOROETHANESULFONYL CHLORIDE see CGO125
CHLOROETHANOIC ACID see CEA000
2-CHLOROETHANOL (MAK) see EIU800
Δ-CHLOROETHANOL see EIU800
2-CHLOROETHANOL ACRYLATE see ADT000
2-CHLOROETHANOL-2-(p-tert-BUTYLPHENOXY)-1-METHYLETHYL SULFITE
　　see SOP500
2-CHLOROETHANOL ESTER with 2-(p-tert-BUTYLPHENOXY)-1-
　　METHYLETHYL SULFITE see SOP500
2-CHLOROETHANOL HYDROGEN PHOSPHATE ESTER with 3-CHLORO-7-
　　HYDROXY-4-METHYLCOUMARIN see DFH600
2-CHLORO-ETHANOL-4-METHYLBENZENESULFONATE (9CI) see CHI125
2-CHLOROETHANOL PHOSPHATE see CGO125
2-CHLOROETHANOL PHOSPHATE DIESTER ESTER with 3-CHLORO-7-
　　HYDROXY-4-METHYLCOUMARIN see DFH600
2-CHLOROETHANOL PHOSPHITE (3:1) see PHO000
2-CHLORO-ETHANOL, PHOSPHOROTHIOATE (3:1) see PHN500
2-CHLORO-ETHANOL-p-TOLUENESULFONATE (8CI) see CHI125
CHLOROETHENE see MIH275
CHLOROETHENE see VNP000
CHLOROETHENE HOMOPOLYMER see PKQ059
(2-CHLOROETHENYL) ARSONOUS DICHLORIDE see CLV000

1,1',1''-(1-CHLORO-1-ETHENYL-2-YLIDENE)-TRIS(4-METHOXYBENZENE)
　　see CLO750
17-α-CHLOROETHINYL-17-β-HYDROXYESTRA-4,9-DIEN-3-ONE see CHP750
(2-CHLOROETHOXY)CARBONYL CHLORIDE see CGU199
2-(2-CHLOROETHOXY)ETHANOL see DKN000
(2-CHLOROETHOXY)ETHENE see CHI250
2-(2-CHLOROETHOXY)ETHYL 2'-CHLOROETHYL ETHER see TKL500
CHLOROETHYL ACRYLATE see ADT000
2-CHLOROETHYL ACRYLATE see ADT000
β-CHLOROETHYL ACRYLATE see ADT000
2-CHLOROETHYL ALCOHOL see EIU800
β-CHLOROETHYL ALCOHOL see EIU800
2-CHLOROETHYLAMINE see CGP125
2-CHLOROETHYLAMINE HYDROCHLORIDE see CGP250
2-CHLORO-4-ETHYLAMINEISOPROPYLAMINE-s-TRIAZINE see ARQ725
3'-CHLORO-2-ETHYLAMINO-o-ACETOTOLUIDIDE HYDROCHLORIDE
　　see CGP375
6'-CHLORO-2-(ETHYLAMINO)-o-ACETOTOLUIDIDE HYDROCHLORIDE
　　see CGP380
2-CHLORO-4-ETHYLAMINO-6-(1-CYANO-1-METHYL)ETHYLAMINO-s-
　　TRIAZINE see BLW750
7-((2-((2-CHLOROETHYL)AMINO)ETHYL)AMINO)BENZ(c)ACRIDINE DIHY-
　　DROCHLORIDE HYDRATE see CGP500
9-(2-((2-CHLOROETHYL)AMINO)ETHYLAMINO)-6-CHLORO-2-
　　METHOXYACRIDINE, DIHYDROCHLORIDE see QCS875
2-CHLOROETHYLAMINOETHYL DEHYDROABIETATE HYDROCHLORIDE
　　see CGQ250
N-(2-CHLOROETHYL)AMINOETHYL-4-ETHOXYNITROBENZENE see CGX325
6'-CHLORO-2-(ETHYLAMINO)-o-HEXANOTOLUIDIDE HYDROCHLORIDE
　　see CAB750
1-CHLORO-3-ETHYLAMINO-5-ISOPROPYLAMINO-s-TRIAZINE see ARQ725
2-CHLORO-4-ETHYLAMINO-6-ISOPROPYLAMINO-s-TRIAZINE see ARQ725
1-CHLORO-3-ETHYLAMINO-5-ISOPROPYLAMINO-2,4,6-TRIAZINE
　　see ARQ725
2-CHLORO-4-ETHYLAMINO-6-ISOPROPYLAMINO-1,3,5-TRIAZINE see ARQ725
2'-CHLORO-2-(ETHYLAMINO)-6'-METHYLACETANILIDE, HYDROCHLORIDE
　　see CGP380
N-(2-CHLOROETHYL)AMINOMETHYL-4-HYDROXYNITROBENZENE
　　see CGQ280
N-(2-CHLOROETHYL)AMINOMETHYL-4-METHOXYNITROBENZENE
　　see CGQ400
2-(((2-CHLOROETHYL)AMINO)METHYL)-4-NITROPHENOL see CGQ280
2-CHLORO-3-(ETHYLAMINO)-1-METHYL-3-OXO-1-PROPENYL DIMETHYL
　　ESTER PHOSPHORIC ACID see DTP600
2-CHLORO-3-(ETHYLAMINO)-1-METHYL-3-OXO-1-PROPENYL DIMETHYL
　　PHOSPHATE see DTP600
6-CHLORO-2-ETHYLAMINO-4-METHYL-4-PHENYL-4H-3,1-BENZOXAZINE
　　see CGQ500
9-((3-((2-CHLOROETHYL)AMINO)PROPYL)AMINO)ACRIDINE DIHYDRO-
　　CHLORIDE HYDRATE see CGR000
7-((3-((2-CHLOROETHYL)AMINO)PROPYL)AMINO)BENZ(c)ACRIDINE DIHY-
　　DROCHLORIDE SESQUIHYDRATE see CGR250
7-((3-((2-CHLOROETHYL)AMINO)PRO-
　　PYL)AMINO)BENZO(b)(1,8)PHENANTHROLINE DIHYDROCHLORIDE HY-
　　DRATE see CGR750
7-((3-((2-CHLOROETHYL)AMINO)PRO-
　　PYL)AMINO)BENZO(b)(1,10)PHENANTHROLINE DIHYDROCHLORIDE
　　see CGR500
10-((2-CHLOROETHYLAMINO)PROPYLAMINO)-2-METHOXY-7-
　　CHLOROBENZO(b)-(1,5)-NAPHTHYRIDINE see CGS500
4-((3-((2-CHLOROETHYL)AMINO)PROPYL)AMINO)-6-METHOXYQUINOLINE
　　HYDROCHLORIDE see CGS750
2-(4-CHLORO-6-ETHYLAMINO-s-TRIAZINE-2-YLAMINO)-2-METHYL-
　　PROPIONITRILE see BLW750
2-(4-CHLORO-6-ETHYLAMINO-1,3,5-TRIAZINE-2-YLAMINO)-2-
　　METHYLPROPIONITRILE see BLW750
2-((4-CHLORO-6-(ETHYLAMINO)-1,3,5-TRIAZIN-2-YL)AMINO)-2-METHYL-
　　PROPANENITRILE see BLW750
2-((4-CHLORO-6-(ETHYLAMINO)-s-TRIAZIN-2-YL)AMINO)-2-
　　METHYLPROPIONITRILE see BLW750
CHLOROETHYLAMINOURACIL see DYC700
6'-CHLORO-2-(ETHYLAMINO)-o-VALEROTOLUIDIDE HYDROCHLORIDE
　　see CGT000
2-CHLOROETHYLAMMONIUM CHLORIDE see CGP250
2-(2-CHLOROETHYL)-3-AZA-4-CHROMANONE see CDS250
CHLOROETHYLBENZENE see EHH500
3-(2-CHLOROETHYL)-2-(BIS(2-CHLOROETHYL)AMINO)PERHYDRO-2H-1,3,2-
　　OXAZAPHOSPHORINE-2-OXIDE see TNT500
β-CHLOROETHYL-β-(BIS(β-HYDROXYETHYL)SULFONIUM)ETHYL SULFIDE
　　CHLORIDE see BKD750
β-CHLOROETHYL-β'-(p-tert-BUTYLPHENOXY)-α'- METHYLETHYL SULFITE
　　see SOP500
β-CHLOROETHYL-β-(p-tert-BUTYLPHENOXY)-α-METHYLETHYL SULPHITE
　　see SOP500

6-CHLORO-N-ETHYL-N'-(1-METHYLETHYL)-1,3,5-TRIAZINE-2,4-DIAMINE (9CI) see ARQ725

2-(2-CHLOROETHYL)-2-METHYLHYDRAZONE BENZALDEHYDE see MIG500

α-CHLORO-2'-ETHYL-6'-METHYL-N-(1-METHYL-2-METHOXYETHYL)-ACETANILIDE see MQQ450

5-(2-CHLOROETHYL)-4-METHYLOXAZOLE-1,2-ETHANEDISULFONATE (2:1) see CHD675

N-(2-CHLOROETHYL)-N-(1-METHYL-2-PHENOXYETHYL)BENZENEMETHANAMINE see PDT250

N-(2-CHLOROETHYL)-N-(1-METHYL-2-PHENOXYETHYL)BENZENEMETHANAMINE HYDROCHLORIDE see DDG800

N-(2-CHLOROETHYL)-N-(1-METHYL-2-PHENOXYETHYL)BENZYLAMINE see PDT250

N-(2-CHLOROETHYL)-N-(1-METHYL-2-PHENOXYETHYL)BENZYLAMINE HYDROCHLORIDE see DDG800

6-CHLORO-N-ETHYL-4-METHYL-4-PHENYL-4H-3,1-BENZOXAZIN-2-AMINE see CGQ500

2-CHLORO-N-(2-ETHYL-6-METHYLPHENYL)-N-(2-METHOXY-1-METHYLETHYL) ACETAMIDE see MQQ450

3-(2-CHLOROETHYL)-2-METHYLPYRIDINE HYDROCHLORIDE see CHD700

5-(2-CHLOROETHYL)-4-METHYLTHIAZOLE see CHD750

5-(2-CHLOROETHYL)-4-METHYLTHIAZOLE ETHANE DISULFONATE see CHD800

4-(2-CHLOROETHYL)MORPHOLINE see CHD875

4-(2-CHLOROETHYL)MORPHOLINE HYDROCHLORIDE see CHE000

1-(2-CHLOROETHYL)-3-MORPHOLINO-1-NITROSOUREA see MRR775

N-(2-CHLOROETHYL)-N-NITROSOACETAMIDE see CHE250

N-(β-CHLOROETHYL)-N-NITROSOACETAMIDE see CHE250

4-((((2-CHLOROETHYL)NITROSOAMINO)CARBONYL)AMINO)CYCLOHEXANE CARBOXYLIC ACID, ETHYL ESTER see CHF000

2-((((2-CHLOROETHYL)NITROSOAMINO)CARBONYL)AMINO)-2-DEOXY-d-GLUCOPYRANOSE see CLX000

2-((((2-CHLOROETHYL)NITROSOAMINO)CARBONYL)AMINO)-2-DEOXY-d-GLUCOSE see CLX000

(17-β)-17-((((2-CHLOROETHYL)NITROSOAMINO)CARBONYL)OXY)ESTR-4-EN-3-ONE see NNX600

(2-CHLOROETHYL)NITROSOCARBAMIC ACID-2-FLUOROETHYL ESTER see FIH000

N-2-CHLOROETHYL-N-NITROSO-CARBAMIC ACID METHYLESTER see MIH250

N-(2-CHLOROETHYL)-N-NITROSOCARBOMOYL AZIDE see CHE325

N-(2-CHLOROETHYL)-N-NITROSOETHYLCARBAMATE see CHF500

1-(2-CHLOROETHYL)-1-NITROSO-3-(2-NORBORNYL)UREA see CHE500

1-(2-CHLOROETHYL)-1-NITROSO-3-RIBOPYRANOSYLUREA-2,3',4'-TRIACETATE see ROF200

1-(2-CHLOROETHYL)-1-NITROSOUREA see CHE750

N-(2-CHLOROETHYL)-N-NITROSOUREA see CHE750

trans-4-(3-(2-CHLOROETHYL)-3-NITROSOUREIDOCYCLOHEXANE CARBOXYLIC ACID ETHYL ESTER see CHF000

2-(3-(2-CHLOROETHYL)-3-NITROSOUREIDO)-2-DEOXY-d-GLUCOSOPYRANOSE see CLX000

2-(3-(2-CHLOROETHYL)3-NITROSOUREIDO)ETHYL METHANE SULFONATE see CHF250

2-(3-(2-CHLOROETHYL)-3-NITROSOUREIDO)-d-GLUCO-PYRANOSE see CLX000

N-(β-CHLOROETHYL)-N-NITROSOURETHAN see CHF500

2-CHLOROETHYL-N-NITROSOURETHANE see CHF500

1-(2-CHLOROETHYL)-3-(2-NORBORNYL)-1-NITROSOUREA see CHE500

CHLOROETHYLOWY ALKOHOL (POLISH) see EIU800

1-CHLORO-3-ETHYL-1-PENTEN-4-YN-3-OL see CHG000

CHLOROETHYLPHENAMIDE see BEG000

4-(4-CHLORO-6-ETHYLPHENYLAMINO-2-s-TRIAZINYLAMINO-5-HYDROXY-6-(4-METHYL-2-SULFOPHENYLAZO)-2,7-NAPHTHALENE DISULFONIC ACID TRISODIUM SALT see CHG250

p-((3-CHLORO-4-ETHYLPHENYL)AZO)-N,N-DIMETHYLANILINE see CGW100

p-((4-CHLORO-3-ETHYLPHENYL)AZO)-N,N-DIMETHYLANILINE see CGW105

(2-CHLOROETHYL)PHOSPHONIC ACID DIETHYL ESTER see CHG375

(2-CHLOROETHYL)PHOSPHONIC ACID MONOETHYL ESTER see CHG400

3-(2-CHLOROETHYL)-2-PICOLINE HYDROCHLORIDE see CHD700

β-CHLOROETHYLPIPERIDINE HYDROCHLORIDE see CHG500

1-(2-CHLOROETHYL)PIPERIDINE HYDROCHLORIDE see CHG500

2'-CHLORO-2-(ETHYL(2-PIPERIDINOETHYL)AMINO)ACETANILIDE DIHYDROCHLORIDE see CHG750

7-(3-(2-CHLOROETHYL-n-PROPYLAMINO)PROPYLAMINO)BENZO(b)(1,10)-PHENATHROLINE HYDROCHLORIDE see CHH000

2-CHLORO-5-ETHYL-4-PROPYL-2-THIONO-1,3,2-DIOXAPHOSPHORINANE see CHH125

(CHLORO-2-ETHYL)-1-(RIBOFURANOSYLISOPROPYLIDENE-2'-3'-PARANITROBENZOATE-5')-3-NITROSOUREA see RFU600

(CHLORO-2-ETHYL)-1-(RIBOPYRANOSYLTRIACETATE-2',3',4')-3-NITROSOUREA see ROF200

2-CHLOROETHYLSULFONYL CHLORIDE see CGO125

2-CHLOROETHYL SULFUROUS ACID-2-(4-(1,1-DIMETHYLETHYL)PHENOXY)-1-METHYLETHYL ESTER see SOP500

2-CHLOROETHYL SULPHITE of 1-(p-tert-BUTYLPHENOXY)-2-PROPANOL see SOP500

2-((3-(2-CHLOROETHYL)TETRAHYDRO-2H-1,3,2-OXAZAPHOSPHORIN-2-YL)AMINO)-ETHANOL, METHANESULFONATE (ester), p-OXIDE see SOD000

1-CHLORO-2-(ETHYLTHIO)ETHANE see CGY750

2-((2-CHLOROETHYL)THIO)ETHANOL see CHC000

2-(2-CHLOROETHYL)THIOETHYLBIS(2-HYDROXYETHYL)-CHLORIDE see BKD750

2-CHLORO-N-(6-ETHYL-o-TOLYL)-N-(2-METHOXY-1-METHYLETHYL)-ACETAMIDE see MQQ450

2-CHLOROETHYL TOSYLATE see CHI125

N-2-CHLOROETHYL)-α,α,α-TRIFLUORO-2,6-DINITRO-N-PROPYL-p-TOLUIDINE see FDA900

(2-CHLOROETHYL)TRIMETHYLAMMONIUM CHLORIDE see CMF400

(β-CHLOROETHYL)TRIMETHYLAMMONIUM CHLORIDE see CMF400

2-CHLOROETHYL VINYL ETHER see CHI250

CHLOROETHYNE see ACJ000

17-α-CHLOROETHYNLY-19-NOR-4,9-ANDROSTADIEN-17β-OL-3-ONE see CHP750

17-α-CHLOROETHYNYL-3,17-β-DIMETHOXY-OESTRA-1,3,5(10)-TRIENE see BAT795

17-α-CHLOROETHYNYL-17-β-HYDROXY-19-NOR-4,9-ANDROSTADIEN-3-ONE see CHP750

CHLOROETHYNYL NORGESTREL mixed with MESTRANOL (20:1) see CHI750

4-CHLORO-α-ETHYNYL-α-PHENYLBENZENEMETHANOL CARBAMATE see CKI250, CKI500

(CHLORO-2-ETIL)-1-(RIBOFURANOSILISOPROPILIDENE-2,3'-PARANITROBENZOATO)-3-NITROSOUREA see RFU600

CHLOROFENIZON see CJT750

3-(4-(4-CHLORO-FENOSSIL)-1,1-DIMETIL-UREA (ITALIAN) see CJQ000

CHLOROFENSULPHIDE see CKL500

CHLOROFENVINPHOS see CDS750

p-CHLOROFENYLESTER KYSELINY BENZENSULFONOVE (CZECH) see CJR500

N-2-(7-CHLORO)FLUORENYLACETAMIDE see CHI825

N-(7-CHLORO-2-FLUORENYL)ACETAMIDE see CHI825

2-CHLORO-N-(9-FLUORENYL)DIETHYLAMINE HYDROCHLORIDE see FEE100

1-CHLORO-10-FLUORO-DECANE see FHN000

21-CHLORO-9-FLUORO-11-β,17-DIHYDROXY-16-β-METHYLPREGNA-1,4-DIENE-3,20-DIONE-17-PROPIONATE see CMW300

21-CHLORO-9-FLUORO-17-HYDROXY-16-β-METHYLPREGNA-1,4-DIENE-3,11,20-TRIONE BUTYRATE see CMW400

CHLOROFLUOROMETHANE see CHI900

1-CHLORO-8-FLUOROOCTANE see FKT000

1-CHLORO-5-FLUOROPENTANE see FFW000

7-CHLORO-5-(o-FLUOROPHENYL)-1,3-DIHYDRO-1-METHYL-2H-1,4-BENZODIAZEPIN-2-ONE see FDB100

7-CHLORO-5-(2-FLUOROPHENYL)-1-METHYL-1H-1,4-BENZODIAZEPIN-2(3H)-ONE see FDB100

8-CHLORO-6-(2-FLUOROPHENYL)-1-METHYL-4H-IMIDAZO(1,5-a)(1,4)BENZODIAZEPINE see MQT525

6-(3-(2-CHLORO-6-FLUOROPHENYL)-5-METHYL-4-ISOXAZOLECARBOXAMIDO)PENICILLANIC ACID SODIUM SALT see CHJ000

3-(2-CHLORO-6-FLUOROPHENYL)-5-METHYL-4-ISOXAZOLYLPENICILLIN SODIUM MONOHYDRATE see CHJ000

3-CHLORO-2-FLUOROPROPENE see CHJ250

3-CHLORO-2-FLUORO-1-PROPENE see CHJ250

CHLORO FLUORO SULFONE see SOT500

21-CHLORO-9-FLUORO-11-β,16-α-17-TRIHYDROXYPREGN-4-ENE-3,20-DIONE cyclic 16,17-ACETAL with ACETONE see HAF300

CHLOROFLURAZOLE see DGO400

CHLOROFLURENOL-METHYL ESTER see CDT000

CHLOROFORM see CHJ500

CHLOROFORMAMIDINIUM CHLORIDE see CHJ599

CHLOROFORMAMIDINIUM NITRATE see CHJ625

CHLOROFORME (FRENCH) see CHJ500

CHLOROFORMIC ACID BENZYL ESTER see BEF500

CHLOROFORMIC ACID-2-CHLOROETHYL ESTER see CGU199

CHLOROFORMIC ACID DIMETHYLAMIDE see DQY950

CHLOROFORMIC ACID, ESTER with 1-NAPHTHOL see NBH200

CHLOROFORMIC ACID ETHYL ESTER see EHK500

CHLOROFORMIC ACID ISOPROPYL ESTER see IOL000

CHLOROFORMIC ACID 1-MAPHTHYL ESTER see NBH200

CHLOROFORMIC ACID METHYL ESTER see MIG000

CHLOROFORMIC ACID PHENYL ESTER see CBX109

CHLOROFORMIC ACID PROPYL ESTER see PNH000

CHLOROFORMIC DIGITALIN see DKN400

CHLOROFORMYL CHLORIDE see PGX000

CHLOROFOS see TIQ250

CHLOROFTALM see TIQ250
CHLORO-2-FURANYL MERCURY see CHK000
4-CHLORO-N-FURFURYL-5-SULFAMOYLANTHRANILIC ACID see CHJ750
CHLORO(2-FURYL)MERCURY see CHK000
6'-CHLORO-2-(2-FURYLMETHYL)AMINO-o-ACETOTOLUIDIDE HYDROCHLO-
    RIDE see CHK125
4-CHLORO-N-(2-FURYLMETHYL)-5-SULFAMOYLANTHRANILIC ACID
    see CHJ750
CHLOROGENIC ACID see CHK175
CHLOROGERMANE see CHK250
CHLOROGUANIDE see CKB250
CHLOROGUANIDE HYDROCHLORIDE see CKB500
CHLOROGUANIDINE HYDROCHLORIDE see CKB500
(2-CHLORO-1-HEPTENYL)ARSONIC ACID MONOAMMONIUM SALT
    see CEI250
2-CHLORO-1,1,1,4,4,4-HEXAFLUOROBUTENE-2 see CHK750
endo-4-CHLORO-N-(HEXAHYDRO-4,7-METHANOISOINDOL-2-YL)-3-
    SULFAMOYLBENZAMIDE see CHK825
2-CHLORO-10-(3-(HEXAHYDROPYRROLO(1,2-a)PYRAZIN-2(1H)-YL)-
    PROPIONYL)PHENOTHIAZINE 2HCl see NMV750
CHLORO(2-HEXANAMIDOCYCLOHEXYL)MERCURY, (E)- see CHL000
trans-CHLORO(2-HEXANAMIDOCYCLOHEXYL)MERCURY see CHL000
CHLOROHEXYL ISOCYANATE see CHL250
4-CHLORO-2-HEXYLPHENOL see CHL500
CHLOROHYDRIC ACID see HHL000
α-CHLOROHYDRIN see CDT750
dl-α-CHLOROHYDRIN see CHL875
epi-CHLOROHYDRIN see EAZ500
α-CHLOROHYDRIN DIACETATE see CIL900
CHLOROHYDROQUINONE see CHM000
5-CHLORO-4-(HYDROXYAMINO)QUINOLINE-1-OXIDE see CHM500
6-CHLORO-4-(HYDROXYAMINO)QUINOLINE-1-OXIDE see CHM750
7-CHLORO-4-(HYDROXYAMINO)QUINOLINE-1-OXIDE see CHN000
3-CHLORO-4-HYDROXYBIPHENYL see CHN500
CHLORO(2-HYDROXY-3,5-DINITROPHENYL)MERCURY see CHN750
3-CHLORO-4-HYDROXYDIPHENYL see CHN500
5-CHLORO-2-HYDROXYDIPHENYLMETHANE see CJU250
2-CHLORO-N-(2-HYDROXYETHYL)ANILINE see CHO125
4-CHLORO-6-(2-HYDROXYETHYLPIPERAZINO-2-METHYLAMINO-5-
    METHYLTHIOPYRIMIDINE see CHO250
5-CHLORO-3-(4-(2-HYDROXYETHYL)-1-PIPERAZINYL)CARBONYLMETHYL-2-
    BENZOTHIAZOLINONE see CJH750
2-CHLORO-10-3-(1-(2-HYDROXYETHYL)-4-PIPERAZINYL)PROPYL PHENO-
    THIAZINE see CJM250
5-CHLORO-8-HYDROXY-7-IODOQUINOLINE see CHR500
2-CHLORO-4-(HYDROXY MERCURI)PHENOL see CHO750
5-CHLORO-4-HYDROXYMETANILIC ACID see AJH500
2-CHLORO-9-HYDROXY-9-METHYLCARBOXYLATEFLUORENE see CDT000
3-CHLORO-7-HYDROXY-4-METHYLCOUMARIN BIS(2-CHLOROETHYL)PHOS-
    PHATE see DFH600
3-CHLORO-7-HYDROXY-4-METHYL-COUMARIN-O,O-DIETHYL PHOS-
    PHOROTHIOATE see CNU750
3-CHLORO-7-HYDROXY-4-METHYL-COUMARIN-O-ESTER with O,O-DI-
    ETHYL PHOSPHOROTHIOATE see CNU750
(−)-N-((5-CHLORO-8-HYDROXY-3-METHYL-1-OXO-7-ISOCHROMANYL)
    CARBONYL)-3-PHENYLALANINE see CHP250
6-CHLORO-17-α-HYDROXY-16-α-METHYLPREGNA-4,6-DIENE-3,20-DIONE
    see CHP375
5-CHLORO-2-((2-HYDROXY-1-NAPHTHALENYL)AZO)-4-METHYLBENZENE
    SULFONIC ACID, BARIUM SALT (2:1) see CHP500
5-CHLORO-2-((2-HYDROXY-1-NAPHTHALENYL)AZO)-4-METHYLBENZENE
    SULPHONIC ACID, BARIUM SALT see CHP500
5-CHLORO-2-((2-HYDROXY-1-NAPHTHYL)AZO)-p-TOLUENE SULFONIC
    ACID, BARIUM SALT see CHP500
21-CHLORO-17-HYDROXY-19-NOR-17-α-PREGNA-4,9-DIEN-20-YN-3-ONE
    see CHP750
7-CHLORO-3-HYDROXY-5-PHENYL-1,3-DIHYDRO-2H-1,4-BENZODIAZEPIN-2-
    ONE see CFZ000
(3-CHLORO-4-HYDROXYPHENYL)HYDROXYMERCURY see CHO750
CHLORO(o-HYDROXYPHENYL)MERCURY see CHW675
CHLORO(p-HYDROXYPHENYL)MERCURY see CHW750
7-CHLORO-10-(2-HYDROXY-3-PIPERIDINOPROPYL)ISOALLOXAZINE SUL-
    FATE see CHQ000
6-CHLORO-17-HYDROXYPREGNA-4,6-DIENE-3,20-DIONE see CDY275
6-CHLORO-17-α-HYDROXYPREGNA-4,6-DIENE-3,20-DIONE ACETATE see
    CBF250
6-α-CHLORO-17-α-HYDROXYPREGN-4-ENE-3,20-DIONE ACETATE see CEB875
6-CHLORO-17-α-HYDROXY-Δ⁶-PROGESTERONE ACETATE see CBF250
1-(3-CHLORO-2-HYDROXYPROPYL)-5-IODO-2-METHYL-4-NITROIMIDAZOLE
    see CIN000
1-(3-CHLORO-2-HYDROXYPROPYL)-2-METHYL-5-NITROIMIDAZOLE
    see OJS000
1-(3-CHLORO-2-HYDROXYPROPYL)-2-NITROIMIDAZOLE see CIQ250
3-CHLORO-2-HYDROXYPROPYL PERCHLORATE see CHQ250

2-CHLORO-HYDROXYTOLUENE see CFE250
6-CHLORO-3-HYDROXYTOLUENE see CFE250
5-CHLORO-4-(2-IMIDAZOLIN-2-YLAMINO)-2,1,3-BENZOTHIADIAZOLE
    HYDROCHLORIDE see TGH600
4-CHLOROIMINO-2,5-CYCLOHEXADIENE-1-ONE see CHQ500
4-CHLOROIMINO-2,6-DIBROMO-2,5-CYCLOHEXADIENE-1-ONE see CHQ750
4-CHLOROIMINO-2,6-DICHLORO-2,5-CYCLOHEXADIENE-1-ONE see CHR000
3-CHLOROIMIPRAMINE see CDU750
3-CHLOROIMIPRAMINE HYDROCHLORIDE see CDV000
CHLOROIMIPRAMINE MONOHYDROCHLORIDE see CDV000
CHLOROIN see CLD250
7-CHLORO-4-INDANOL see CDV700
CHLOROIODOACETYLENE see CHR325
CHLOROIODOETHYNE see CHR325
5-CHLORO-7-IODO-8-HYDROXYQUINOLINE see CHR500
3-CHLORO-1-IODOPROPYNE see CHR400
CHLOROIODOQUINE see CHR500
5-CHLORO-7-IODO-8-QUINOLINOL see CHR500
2-CHLOROISOBUTANE see BQR000
3'-CHLORO-2-ISOBUTYLAMINO-p-ACETOTOLUIDIDE HYDROCHLORIDE
    see CHR700
6'-CHLORO-2-(ISOBUTYLAMINO)-o-ACETOTOLUIDIDE HYDROCHLORIDE
    see CHR700
3'-CHLORO-3-ISOBUTYLAMINO-o-PROPIONOTOLUIDIDE HYDROCHLORIDE
    see CHR850
α-CHLOROISOBUTYLENE see IKE000
γ-CHLOROISOBUTYLENE see CIU750
1-CHLORO-2-ISOPROPOXY-2-PROPANOL see CHS250
2-CHLORO-N-ISOPROPYLACETANILIDE see CHS500
α-CHLORO-N-ISOPROPYLACETANILIDE see CHS500
2-(α-CHLORO-β-ISOPROPYLAMINE)ETHYLNAPHTHALENE HYDROCHLO-
    RIDE see CHT500
6'-CHLORO-2-(ISOPROPYLAMINO)-o-ACETOTOLUIDIDE HYDROCHLORIDE
    see CHS750
2'-CHLORO-2-(ISOPROPYLAMINO)-6'-METHYLACETANILIDE HYDROCHLO-
    RIDE see CHS750
β-CHLORO-N-ISOPROPYL-2-NAPHTHALENEETHYLAMINE HYDROCHLO-
    RIDE see CHT500
2-CHLORO-N-ISOPROPYL-N-PHENYLACETAMIDE see CHS500
CHLOROJECT L see CDP250
3-CHLORO-LACTONITRILE see CHU000
CHLOROMADINONE ACETATE see CBF250
CHLOROMAX see CDP250
S-(6-CHLORO-3-(MERCAPTOMETHYL)-2-BENZOXAZOLINONE)-O,O-DIETHYL
    PHOSPHORODITHIOATE see BDJ250
(CHLOROMERCURI)BENZENE see PFM500
p-(CHLOROMERCURI)BENZOIC ACID see CHU500
p-CHLOROMERCURIC BENZOIC ACID see CHU500
N-(2-CHLOROMERCURICYCLOHEXYL) HEXANAMIDE, (E)- see CHL000
N-(2-CHLOROMERCURICYCLOHEXYL)PROPIONAMIDE see CIC000
2-(CHLOROMERCURI)-4,6-DINITROPHENOL see CHN750
N-(CHLOROMERCURI)FORMANILIDE see CHU750
2-CHLOROMERCURIFURAN see CHK000
3-(3-CHLOROMERCURI-2-METHOXY-1-PROPYL)-5,5-DIMETHYLHYDANTOIN
    see CHV250
1-(3-CHLOROMERCURI-2-METHOXY)PROPYLHYDANTOIN see CHV500
1-(3-CHLOROMERCURI-2-METHOXY-1-PROPYL)-HYDANTOIN
    see CHV500
3-(3-CHLOROMERCURI-2-METHOXY-1-PROPYL)HYDANTOIN see CHV750
3-(3-(CHLOROMERCURI)-2-METHOXYPROPYL)-1-METHYLHYDANTOIN
    see CHW250
1-(3-CHLOROMERCURI-2-METHOXY-1-PROPYL)-3-METHYLHYDANTOIN
    see CHW000
3-(3-CHLOROMERCURI-2-METHOXY-1-PROPYL)-1-METHYLHYDANTOIN
    see CHW250
5-(3-CHLOROMERCURI-2-METHOXY-1-PROPYL)-3-METHYLHYDANTOIN
    see CHW500
(3-(CHLOROMERCURI)-2-METHOXYPROPYL)UREA see CHX250
1-(3-(CHLOROMERCURI)-2-METHOXYPROPYL)UREA see CHX250
o-CHLOROMERCURIPHENOL see CHW675
p-CHLOROMERCURIPHENOL see CHW750
p-(CHLOROMERCURI)PHENOL see CHW750
3-(CHLOROMERCURI)PYRIDINE see CKW500
1-(3-(CHLOROMERCURY)-2-METHOXYPROPYL)HYDANTOIN
    see CHV500
CHLOROMERIDIN see CHX250
CHLOROMERODRIN see CHX250, CHX250
CHLOROMETHANE see MIF765
CHLOROMETHANE mixed with DICHLOROMETHANE see CHX750
CHLOROMETHANE SULFONATE d'ETHYLE (FRENCH) see CHC750
CHLOROMETHANE SULFONYL CHLORIDE see CHY000
CHLOROMETHAPYRILENE see CHY250
2-((3-((6-CHLORO-2-METHOXY-9-ACRIDINYL)AMINO))PROPYL)
    ETHYLAMINOETHANOL DIHYDROCHLORIDE see CHY750

1-((6-CHLORO-2-METHOXY-9-ACRIDYL)-AMINO)-3-(DIETHYLAMINO)-2-
PROPANOL DIHYDROCHLORIDE see ADI750
N-(5-CHLORO-4-METHOXYANTHRAQUINONYL)BENZAMIDE see CIA000
N-(4-(2-(5-CHLORO-2-METHOXYBENZAMIDO)ETHYL)PHENYLSULFONYL)-N-
CYCLOHEXYLUREA see CEH700
3-CHLORO-4-METHOXY-BENZENAMINE (9CI) see CEH750
CHLORO(trans-2-METHOXYCYCLOOCTYL)MERCURY see CIB500
3-CHLORO-3-METHOXYDIAZIRINE see CIB625
5-CHLORO-4-METHOXYDIPHENYLAMINE-2-CARBOXYLIC ACID see CIB700
CHLOROMETHOXY ETHANE see CIM000
CHLORO(2-METHOXYETHYL)MERCURY see MEP250
3-CHLORO-7-METHOXY-9-(1-METHYL-4-DIETHYLAMINOBUTYLAMINO)
ACRIDINE see ARQ250
3-CHLORO-7-METHOXY-9-(1-METHYL-4-DIETHYLAMINOBUTYLAMINO)
ACRIDINE DIHYDROCHLORIDE see CFU750
2-CHLORO-1-METHOXY-4-NITROBENZENE (9CI) see CJA200
4-CHLORO-N-(p-METHOXYPHENYL)ANTHRANILIC ACID see CIB700
CHLORO-(4-METHOXYPHENYL)DIAZIRINE see CIB725
3-(3-CHLORO-4-METHOXYPHENYL)-1,1-DIMETHYL-3-NITROSOUREA
see NKY500
3-(3-CHLORO-4-METHOXYPHENYL)-1,1-DIMETHYL-UREA see MQR225
N-(3-CHLORO-4-METHOXYPHENYL)-N',N'-DIMETHYLUREA see MQR225
5-CHLORO-2-METHOXYPROCAINAMIDE see AJH000
CHLORO(2-(3-METHOXYPROPIONAMIDO)CYCLOHEXYL)MERCURY
see CIC000
7-CHLORO-8-METHOXY-10-(2-PYRROLIDINYLETHYL)ISOALLOXAZINE ACE-
TATE see CIC500
2-CHLORO-4'-METHYLACETANILIDE see CEB500
2-CHLORO-N-METHYL-1-ADAMANTANE METHANAMINE HYDROCHLO-
RIDE see CID000
6'-CHLORO-2-(METHYLAMINO)-o-ACETOTOLUIDIDE HYDROCHLORIDE
see CID250
5-CHLORO-6-(((((METHYLAMINO)CAR-
BONYL)OXY)IMINO)BICYCLO(2.2.1)HEPTANE see CFF250
4'-CHLORO-2-((METHYLAMINO)METHYL)BENZHYDROL HYDROCHLORIDE
see CID825
7-CHLORO-2-METHYLAMINO-5-PHENYL-3H-1,4-BENZODIAZEPINE 4-OXIDE
see LFK000
7-CHLORO-2-METHYLAMINO-5-PHENYL-3H-1,4-BENZODIAZEPINE-4-OXIDE,
mixed with SODIUM NITRITE (1:1) see SIS000
7-CHLORO-2-METHYLAMINO-5-PHENYL-3H-1,4-BENZODIAZEPIN 4-OXIDE
see LFK000
7-CHLORO-2-METHYLAMINO-5-PHENYL-3H-1,4-BENZODIAZEPIN, 4-OXIDE,
HYDROCHLORIDE see MDQ250
6'-CHLORO-2-(METHYLAMINO)-o-PROPIONOTOLUIDIDE HYDROCHLORIDE
see CIE250
6'-CHLORO-3-(METHYLAMINO)-o-PROPIONOTOLUIDIDE HYDROCHLORIDE
see CIE500
2-CHLORO-1-(2-METHYLAMINOPROPYL)-3,5,7-TRIMETHYLADAMANTANE
HYDROCHLORIDE see CIF000
p-CHLORO-N-METHYLAMPHETAMINE see CIF250
d-1-p-CHLORO-METHYLAMPHETAMINE (FRENCH) see CIF250
3-CHLORO-2-METHYLANILINE see CLK200
3-CHLORO-4-METHYLANILINE see CLK215
3-CHLORO-6-METHYLANILINE see CLK225
4-CHLORO-2-METHYLANILINE see CLK220
4-CHLORO-6-METHYLANILINE see CLK220
5-CHLORO-2-METHYLANILINE see CLK225
4-CHLORO-2-METHYLANILINE HYDROCHLORIDE see CLK235
4-CHLORO-6-METHYLANILINE HYDROCHLORIDE see CLK235
o-CHLOROMETHYLANISALDEHYDE see CIF750
2-CHLOROMETHYL-p-ANISALDEHYDE see CIF750
7-CHLOROMETHYL BENZ(a)ANTHRACENE see CIG250
8-CHLORO-7-METHYLBENZ(a)ANTHRACENE see CIG000
10-CHLORO-7-METHYLBENZ(a)ANTHRACENE see CIG500
5-CHLORO-10-METHYL-1,2-BENZANTHRACENE see CIG000
7-CHLORO-10-METHYL-1,2-BENZANTHRACENE see CIG500
CHLOROMETHYLBENZENE see BEE375
4-CHLORO-1-METHYLBENZENE see TGY075
2-CHLORO-1-METHYLBENZENE (9CI) see CLK100
4-CHLORO-2-METHYLBENZENEAMINE see CLK220
4-CHLORO-2-METHYLBENZENEAMINE HYDROCHLORIDE see CLK235
4-CHLORO-2-METHYLBENZENEDIAZONIUM SALTS see CIG750
7-CHLORO-1-METHYL-5-3H-1,4-BENZODIAZEPIN-2(1H)-ONE see DCK759
2-(2-(5-CHLORO-2-METHYL-1,3-BENZODIOXOL-2-YL)ETHOXY)-N,N-
DIETHYLETHANAMINE see CFO750
6-CHLOROMETHYL BENZO(a)PYRENE see CIH000
7-CHLORO-3-METHYL-2H-1,2,4-BENZOTHIADIAZINE-1,1-DIOXIDE
see DCQ700
1-(4-CHLORO-2-METHYLBENZYL)-1H-INDAZOLE-3-CARBOXYLIC ACID
see TGJ875
4-CHLOROMETHYLBIPHENYL see CIH825
CHLOROMETHYL BISMUTHINE see CIH900
1-CHLORO-3-METHYLBUTANE see CII000

2-CHLORO-2-METHYLBUTANE see CII250
2-CHLORO-6-METHYLCARBANILIC ACID-2-(DIETHYLAMINO)ETHYL
ESTER, HYDROCHLORIDE see CIJ250
2-CHLORO-6-METHYLCARBANILIC ACID-N-METHYL-4-PIPERIDINYL
ESTER see CIK250
2-CHLORO-6-METHYLCARBANILIC ACID-2-(PYRROLIDINYL)ETHYL ESTER
HYDROCHLORIDE see CIK500
dl-6-CHLORO-α-METHYLCARBAZOLE-2-ACETIC ACID see CCK800
10-CHLOROMETHYL-9-CHLOROANTHRACENE see CFB500
1-(CHLOROMETHYL)-N-((CHLOROMETHYL)DIMETHYLSILYL)-1,1-
DIMETHYL-SILANAMINE see BIL250
3-CHLORO-4-METHYL-7-COUMARINYL DIETHYLPHOSPHATE see CIK750
3-CHLORO-4-METHYL-7-COUMARINYL DIETHYL PHOSPHOROTHIOATE
see CNU750
O-3-CHLORO-4-METHYL-7-COUMARINYL-O,O-DIETHYL PHOSPHOROTHIO-
ATE see CNU750
CHLOROMETHYL CYANIDE see CDN500
6-CHLORO-16-α-METHYL-Δ(6)-DEHYDRO-17-α-ACETOXYPROGESTERONE
see CHP375
3-CHLORO-3-METHYLDIAZIRINE see CIK825
o-CHLORO-2-(METHYL(2-(DIETHYLAMINO)ETHYL)AMINO)PROPIONANIL-
IDE DIHYDROCHLORIDE see CIL000
S-(CHLOROMETHYL)-O,O-DIETHYL PHOSPHORODITHIOATE see CDY299
S-CHLOROMETHYL-O,O-DIETHYL PHOSPHOROTHIOLOTHIOATE
see CDY299
S-CHLOROMETHYL-O,O-DIETHYL PHOSPHOROTHIOLOTHIONATE
see CDY299
7-CHLORO-2-METHYL-3,3a-DIHYDRO-2H,9H-ISOXAZOLO(3,2-
b)(1,3)BENZOXAZIN-9-ONE see CIL500
12-(CHLOROMETHYL)-2-β,3-β-DIHYDROXY-27-NORO-13-ENE-23,28-DIOIC
ACID see SBY000
3'-CHLORO-4'-METHYL-4-DIMETHYLAMINOAZOBENZENE see CIL700
4'-CHLORO-3'-METHYL-4-DIMETHYLAMINOAZOBENZENE see CIL710
o-CHLORO-2-(METHYL(2-(DIMETHYLAMINO)ETHYL)AMINO)ACETANILIDE
DIHYDROCHLORIDE see CIL750
2-CHLORO-4-METHYL-6-DIMETHYLAMINOPYRIMIDINE see CCP500
4-(CHLOROMETHYL)-2,2-DIMETHYL-1,3-DIOXA-2-SILACYCLOPENTANE
see CIL775
4-(CHLOROMETHYL)-2,2-DIMETHYL-1,3-DIOXOLANE see CIL800
S-(CHLOROMETHYL)-O,O-DIMETHYL PHOSPHORODITHIOIC ACID, ESTER
see CDY299
1-(4-CHLOROMETHYL-1,3-DIOXOLAN-2-YL)-2-PROPANONE see CIL850
6-CHLORO-1,2-α-METHYLENE-6-DEHYDRO-17α-HYDROXYPROGESTERONE
ACETATE see CQJ500
6-CHLORO-Δ⁶-1,2-α-METHYLENE-17-α-HYDROXYPROGESTERONE ACETATE
see CQJ500
6-CHLORO-1,2-α-METHYLENE-17-α-HYDROXY-Δ⁶-PROGESTERONE ACE-
TATE see CQJ500
d-2-CHLORO-6-METHYLERGOLINE-8-β-ACETONITRILE METHANESULFO-
NIC ACID SALT see LEP000
1-CHLOROMETHYL-1,2-ETHANEDIOL DIACETATE see CIL900
2-CHLORO-N-METHYL-ETHYLAMINE HYDROCHLORIDE see MIG250
2-CHLORO-1-METHYL-10-ETHYL-7,8-BENZACRIDINE (FRENCH) see EHL000
1-CHLOROMETHYLETHYLENE GLYCOL CYCLIC SULFITE see CKQ750
(CHLOROMETHYL)ETHYLENE OXIDE see EAZ500
CHLOROMETHYL ETHYL ETHER see CIM000
(2-CHLORO-1-METHYLETHYL) ETHER see BII250
2-CHLORO-N-(1-METHYLETHYL)-N-PHENYLACETAMIDE see CHS500
O-(5-CHLORO-1-(METHYLETHYL)-1H 1,2,4-TRIAZOL-3-YL) O,O-DIETHYL
PHOSPHOROTHIOATE see PHK500
7-(CHLOROMETHYL)-5-FLUORO-12-METHYLBENZ(a)ANTHRACENE
see FHH025
3-CHLOROMETHYLFURAN see CIM300
3-CHLORO-METHYLHEPTANE see EKU000
3-CHLORO-4-METHYL-7-HYDROXYCOUMARIN DIETHYL THIOPHOSPHO-
RIC ACID ESTER see CNU750
N-CHLORO-4-METHYL-2-IMIDAZOLINONE see CIM399
4-CHLORO-1-METHYLIMIDAZOLIUM NITRATE see CIM500
4-CHLORO-N-(2-METHYL-1-INDOLINYL)-3-SULFAMOYLBENZAMIDE
see IBV100
α-(CHLOROMETHYL)-5-IODO-2-METHYL-4-NITROIMIDAZOLE-2-ETHANOL
see CIN000
CHLOROMETHYLMERCURY see MDD750
2'-CHLORO-6'-METHYL-2-(METHYLAMINO)ACETANILIDE HYDROCHLO-
RIDE see CID250
4-CHLORO-N-METHYL-3-((METHYLAMINO)SULFONYL)BENZAMIDE
see CIP500
10-CHLOROMETHYL-9-METHYLANTHRACENE see CIN500
7-CHLOROMETHYL-12-METHYL BENZ(a)ANTHRACENE see CIN750
CHLOROMETHYLMETHYLDIETHOXY SILANE see CIO000
CHLOROMETHYL METHYL ETHER see CIO250
2-CHLOROMETHYL-5-METHYLFURAN see CIO275
α-(CHLOROMETHYL)-2-METHYL-5-NITRO-1H-IMIDAZOLE-1-ETHANOL
see OJS000

4-(CHLOROMETHYL)-2-METHYL-2-PENTYL-1,3-DIOXOLANE see CIO375

2-CHLORO-10-((2-METHYL-3-(4-METHYL-1-PIPERAZINYL)PROPYL-PHENOTHIAZINE see CIO500

7-CHLORO-1-METHYL-4-(p-((1-METHYLPYRIDINIUM-4-YL)AMINO)PHENYL)CARBAMOYL)ANILINO)QUINOLINIUM DIBROMIDE see CIO750

6-CHLORO-1-METHYL-4-(p-((p-((1-METHYLPYRIDINIUM-4-YL)AMINO)PHENYL)CARBAMOYL)ANILINO)QUINOLINIUM, DI-p-TOLUENESULFON-ATE see CIP000

8-CHLORO-1-METHYL-4-(p-((p-((1-METHYLPYRIDINIUM-4-YL)AMINO)PHENYL)CARBAMOYL)ANILINO)QUINOLINIUM) DI-p-TOLUENESULFON-ATE see CIP250

4-CHLORO-N-METHYL-3-(METHYLSULFAMOYL)BENZAMIDE see CIP500

3'-CHLORO-α-(METHYL((MORPHOLINOCARBONYL)METHYL)AMINO)-o-BENZOTOLUIDIDE HYDROCHLORIDE see FMU000

3'-CHLORO-α-(N-METHYL-N-((MORPHOLINOCARBONYL)METHYL)AMINOMETHYL)BENZANILIDE HYDROCHLORIDE see FMU000

3'-CHLORO-2'-(N-METHYL-N-((MORPHOLINOCARBONYL)METHYL)AMINOMETHYL)BENZANILIDE HY-DROCHLORIDE see FMU000

1-CHLOROMETHYL NAPHTHALENE see CIP750

α-CHLOROMETHYLNAPHTHALENE see CIP750

2-(8-CHLOROMETHYL-1-NAPHTHYLTHIO)ACETIC ACID see CIQ000

1-(CHLOROMETHYL)-2-NITROBENZENE see NFN500

5-CHLORO-3-METHYL-2-NITROBENZOFURAN see NHN550

α-(CHLOROMETHYL)-2-NITROIMIDAZOLE-2-ETHANOL see CIQ250

4-(CHLOROMETHYL)-2-(o-NITROPHENYL)-1,3-DIOXOLANE see CIQ400

2-CHLORO-2-METHYL-N-NITROSOETHANAMINE see CIQ500

2-CHLORO-N-METHYL-N-NITROSOETHYLAMINE see CIQ500

N-CHLORO-5-METHYL-2-OXAZOLIDINONE see CIQ625

CHLOROMETHYLOXIRANE see EAZ500

2-(CHLOROMETHYL)OXIRANE see EAZ500

cis-CHLORO-3-METHYLOXIRANE see CKS099

trans-2-CHLORO-3-METHYLOXIRANE see CKS100

7-CHLORO-1-METHYL-2-OXO-5-PHENYL-3H-1,4-BENZODIAZEPINE see DCK759

4-CHLORO-2-METHYLPHENOL see CIR000

4-CHLORO-3-METHYLPHENOL see CFE250

(4-CHLORO-2-METHYLPHENOXY)ACETIC ACID see CIR250

4-CHLORO-2-METHYLPHENOXYACETIC ACID, ETHYL ESTER see EMR000

4-CHLORO-2-METHYLPHENOXYACETIC ACID SODIUM SALT see SIL500

4-(4-CHLORO-2-METHYLPHENOXY)BUTANOIC ACID see CLN750

4-(4-CHLORO-2-METHYLPHENOXY)BUTANOIC ACID, SODIUM SALT see CLO000

4-(4-CHLORO-2-METHYLPHENOXY)BUTYRIC ACID see CLN750

γ-(4-CHLORO-2-METHYLPHENOXY)BUTYRIC ACID see CLN750

CHLOROMETHYLPHENOXYBUTYRIC ACID SODIUM SALT see CLO000

4-(4-CHLORO-2-METHYLPHENOXY)BUTYRIC ACID SODIUM SALT see CLO000

1-(2-CHLORO-5-METHYLPHENOXY)-3-((1,1-DIMETHYLETHYL)AMINO)-2-PROPANOL HYDROCHLORIDE see BQB250

2-(4-CHLORO-2-METHYLPHENOXY)PROPANOIC ACID (R) (9CI) see CIR325

2-(4-CHLORO-2-METHYLPHENOXY)PROPIONIC ACID see CIR500

4-CHLORO-2-METHYLPHENOXY-α-PROPIONIC ACID see CIR500

(+)-α-(4-CHLORO-2-METHYLPHENOXY) PROPIONIC ACID see CIR500

N-(3-CHLORO-2-METHYLPHENYL)ANTHRANILIC ACID see CLK325

7-CHLORO-N-METHYL-5-PHENYL-3H-1,4-BENZODIAZEPIN-2-AMINE-4-OXIDE see LFK000

7-CHLORO-1-METHYL-5-PHENYL-1H-1,5-BENZODIAZEPINE-2,4(3H,5H)-DIONE see CIR750

7-CHLORO-1-METHYL-5-PHENYL-2H-1,4-BENZODIAZEPIN-2-ONE see DCK759

2-((p-CHLORO-α-METHYL-α-PHENYLBENZYL)OXY)-N,N-DIMETHYLAMINE HYDROCHLORIDE see CIS000

(+)-2-(2-((p-CHLORO-α-METHYL-α-PHENYLBENZYL)OXY)ETHYL)-1-METHYL PYRROLIDINE FUMARATE see FOS100

7-CHLORO-1-METHYL-5-PHENYL-1,3-DIHYDRO-2H-1,4-BENZODIAZEPIN-2-ONE see DCK759

N'-(4-CHLORO-2-METHYLPHENYL)-N,N-DIMETHYLMETHANIMIDAMIDE see CJJ250

3-(3-CHLORO-4-METHYLPHENYL)-1,1-DIMETHYL-UREA see CIS250

N-(3-CHLORO-4-METHYLPHENYL)-N,N'-DIMETHYLUREA see CIS250

7-CHLORO-N-METHYL-5-PHENYL-EH-1,4-BENZODIAZEPIN-2-AMINE-4-OXIDE, MONOHYDROCHLORIDE see MDQ250

2-CHLORO-5-METHYLPHENYLHYDROXYLAMINE see CIS325

5-CHLORO-1-METHYL-3-PHENYL-1H-IMIDAZO(4,5-b)PYRIDIN-2(3H)-ONE see MNT100

CHLOROMETHYL PHENYL KETONE see CEA750

N'-(4-CHLORO-2-METHYLPHENYL)-METHANIMIDAMIDE MONOHYDROCHLORIDE see CJJ500

N-(3-CHLORO-4-METHYLPHENYL)-2-METHYLPENTANAMIDE see SKQ400

CHLOROMETHYLPHENYLSILANE see CIS625

8-CHLORO-1-METHYL-6-PHENYL-4H-s-TRIAZOLO(4,3-a)(1,4)BENZODIAZEP-INE see XAJ000

2-CHLORO-11-(4-METHYLPIPERAZINO)DIBENZO(b,f)(1,4)THIAZEPINE see CIS750

2-CHLORO-11-(4'-METHYL)PIPERAZINO-DIBENZO(b,f)(1,4)THIAZEPINE HYDROCHLORIDE see CIT000

7-CHLORO-3-(4-METHYL-1-PIPERAZINYL)-4H-1,2,4-BENZOTHIADIAZINE-1,1-DIOXIDE see CIT625

8-CHLORO-11-(4-METHYL-1-PIPERAZINYL)-5H-DIBENZO(b,e)(1,4)DIAZEPINE see CMY250

2-CHLORO-11-(4-METHYL-1-PIPERAZINYL)-DIBENZO(b,f)(1,4)OXAZEPINE see DCS200

2-CHLORO-11-(4-METHYL-1-PIPERAZINYL)-DIBENZO(b,f)(1,4)OXOAZEPINE see DCS200

2-CHLORO-11-(4-METHYL-1-PIPERAZINYL)DIBENZO(b,f)(1,4)THIAZEPINE see CIS750

2-CHLORO-11-(4-METHYL-1-PIPERAZINYL)DIBENZO(b,f)(1,4)THIAZEPINE HYDROCHLORIDE see CIT000

2-CHLORO-10-(3-(1-METHYL-4-PIPERAZINYL)-PROPYL)-PHENOTHIAZINE see PMF500

2-CHLORO-10-(3-(1-METHYL-1-PIPERAZINYL)PROPYL)PHENOTHIAZINE see PMF500

3-CHLORO-10-(3-(1-METHYL-4-PIPERAZINYL)PROPYL)PHENOTHIAZINE see PMF500

CHLORO-3 (N-METHYLPIPERAZINYL-3 PROPYL)-10 PHENOTHIAZINE (FRENCH) see PMF500

2-CHLORO-10-(3-(4-METHYL-1-PIPERAZINYL)-PROPYL-10H-PHENOTHI-AZINE-(Z)-2-BUTENEDIOATE (1:2) see PMF250

2-CHLORO-10-(3-(4-METHYL-1-PIPERAZINYL)PROPYL)PHENOTHIAZINE, DIMALEATE see PMF250

2-CHLORO-10-(3-(4-METHYL-1-PIPERAZINYL)PROPYL)PHENOTHIAZINE DIMALEATE see PMF250

2-CHLORO-10-(3-(1-METHYL-4-PIPERAZINYL)PROPYL)PHENOTHIAZINE EDISYLATE see PME700

2-CHLORO-10-(3-(4-METHYL-1-PIPERAZINYL)PROPYL)PHENOTHIAZINE 1,2-ETHANEDISULFONATE (1:1) see PME700

2-CHLORO-10-(3-(4-METHYL-1-PIPERAZINYL)PROPYL)PHENOTHIAZINE MA-LEATE see PMF250

6'-CHLORO-2-(2-METHYLPIPERIDINO)-o-ACETOTOLUIDIDE HYDROCHLO-RIDE see CIT750

o-CHLORO-2-(METHYL(2-(PIPERIDINO)ETHYL)AMINO)ACETANILIDE DIHY-DROCHLORIDE see CIU000

6'-CHLORO-3-(2-METHYLPIPERIDINO)-o-PROPIONOTOLUIDIDE HYDRO-CHLORIDE see CIU250

3'-CHLORO-5'-METHYL-3-PIPERIDINO-4'-PROPOXY-PROPIOPHENONE HY-DROCHLORIDE see CIU325

1-CHLORO-2-METHYLPROPANE see CIU500

2-CHLORO-2-METHYLPROPANE see BQR000

1-CHLORO-2-METHYLPROPENE see IKE000

3-CHLORO-2-METHYLPROPENE see CIU750

1-CHLORO-2-METHYL-1-PROPENE see IKE000

3-CHLORO-2-METHYL-1-PROPENE see CIU750

1-(3-CHLORO-5-METHYL-4-PROPOXYPHENYL)-3-(1-PIPERIDINYL)1-PROPANONE HYDROCHLORIDE (9CI) see CIU325

2'-CHLORO-6'-METHYL-2-(PROPYLAMINO)ACETANILIDE HYDROCHLO-RIDE see CKT750

2-CHLORO-5-(1-METHYLPROPYL)PHENYL METHYLCARBAMATE see MOU750

2-CHLORO-N-(1-METHYL-2-PROPYNYL)ACETANILIDE see CDS275

2-CHLORO-N-(1-METHYL-2-PROPYNYL)-ACETANILIDE (8CI) see CDS275

2-CHLORO-N-(1-METHYL-2-PROPYNYL)-N-PHENYLACETAMIDE see CDS275

2-(CHLOROMETHYL) PYRIDINE HYDROCHLORIDE see PIC000

3-(CHLOROMETHYL) PYRIDINE HYDROCHLORIDE see CIV000

5'-CHLORO-2-(METHYL(2-(PYRROLIDINYL)ETHYL)AMINO)-O-ACETOTOLUIDIDE DIHYDROCHLORIDE see CIW250

o-CHLORO-2-(METHYL(2-(PYRROLIDINYL)ETHYL)AMINO)PROPIONANILIDE DIHYDROCHLORIDE see CIX000

6-CHLORO-2-METHYL-4-QUINAZOLINONE see MII750

2-CHLOROMETHYL TETRAHYDROFURAN see CIX750

2-CHLORO-5-METHYL-6,7,9,10-TETRAHYDRO-5H-ISOQUINO(2,1-D)(1,4)BENZODIAZEPIN-6 ONE see IRW000

2-CHLOROMETHYLTHIOPHENE see CIY250

(CHLOROMETHYL)TRICHLOROSILANE see CIY325

CHLOROMETHYL(TRICHLORO)SILANE see CIY325

(CHLOROMETHYL)TRIETHOXYSILANE see CIY500

CHLOROMETHYL-TRIMETHOXYSILAN (CZECH) see CIY750

CHLOROMETHYL TRIMETHOXYSILANE see CIY750

3-CHLORO-4-METHYL-UMBELLIFERONE BIS(2-CHLOROETHYL)PHOS-PHATE see DFH600

3-CHLORO-4-METHYLUMBELLIFERONE-O-ESTER with O,O-DIETHYL PHOS-PHOROTHIOATE see CNU750

3'-CHLORO-2-METHYL-p-VALEROTOLUIDIDE see SKQ400

N-CHLORO-3-MORPHOLINONE see CIY899

CHLOROMROWCZAN 1-NAFTYLU (CZECH) see NBH200

CHLOROMYCETIN see CDP250

CHLOROMYCETIN SUCCINATE see CDP725

CHLORONAFTINA see BIF250
1-CHLORONAPHTHALENE see CIZ000
2-CHLORONAPHTHALENE see CJA000
α-CHLORONAPHTHALENE see CIZ000
β-CHLORONAPHTHALENE see CJA000
CHLORONAPHTHINE see BIF250
CHLORONASE see CKK000
CHLORONEB see CJA100
CHLORONEBE (FRENCH) see CJA100
CHLORONEOANTERGAN see CKV625
CHLORONITRIN see CDP250
2-CHLORO-4-NITROANILINE see CJA175
4-CHLORO-3-NITROANILINE see CJA185
N-CHLORO-4-NITROANILINE see CJA150
o-CHLORO-p-NITROANILINE see CJA175
CHLORONITROANISOLE see CJA200
2-CHLORO-4-NITRO-ANISOLE see CJA200
o-CHLORO-p-NITROANISOLE see CJA200
1-CHLORO-5-NITRO-9,10-ANTHRACENEDIONE see CJA250
1-CHLORO-5-NITROANTHRAQUINONE see CJA250
5-CHLORO-1-NITROANTHRAQUINONE see CJA250
CHLORONITROBENZENE see CJA950
2-CHLORONITROBENZENE see CJB750
4-CHLORONITROBENZENE see NFS525
CHLORO-m-NITROBENZENE see CJB250
m-CHLORONITROBENZENE see CJB250
CHLORO-o-NITROBENZENE see CJB750
o-CHLORONITROBENZENE see CJB750
p-CHLORONITROBENZENE see NFS525
1-CHLORO-2-NITROBENZENE see CJB750
1-CHLORO-3-NITROBENZENE see CJB250
1-CHLORO-4-NITROBENZENE see NFS525
2-CHLORO-1-NITROBENZENE see CJB750
4-CHLORO-1-NITROBENZENE see NFS525
m-CHLORONITROBENZENE (DOT) see CJB250
o-CHLORONITROBENZENE (DOT) see CJB750
2-CHLORO-4-NITROBENZENEAZO-2'-AMINO-4'-METHOXY-5'-METHYLBEN-
    ZENE see MIH500
2-CHLORO-5-NITROBENZENESULFONIC ACID see CJB825
2-CHLORO-3,5-NITROBENZENESULFONIC ACID, SODIUM SALT see CJC000
2-CHLORO-4-NITROBENZOIC ACID see CJC250
4-CHLORO-3-NITROBENZOIC ACID see CJC500
2-CHLORO-5-NITROBENZOIC ACID METHYL ESTER see CJC515
4-CHLORO-7-NITROBENZO-2-OXA-1,3-DIAZOLE see NFS550
2-CHLORO-5-NITROBENZYL ALCOHOL see CJC549
6-CHLORO-2-NITROBENZYL BROMIDE see CJC600
2-CHLORO-4-NITROBENZYL CHLORIDE see CJC610
4-CHLORO-2-NITROBENZYL CHLORIDE see CJC625
2-CHLORO-2-NITROBUTANE see CJC750
3-CHLORO-4-(2'-NITRO-3'-CHLOROPHENYL)PYRROLE see CFC000
1-CHLORO-1-NITROETHANE see CJC800
2-((o-CHLORO-α-(NITROMETHYL)BENZYL)THIO)ETHYLAMINE HYDRO-
    CHLORIDE see NEC000
2-CHLORO-4-NITROPHENOL see CJD250
2-CHLORO-4-NITROPHENYLAMIDE-6-CHLOROSALICYLIC ACID see DFV400
1-((2-CHLORO-4-NITROPHENYL)AZO)-2-NAPHTHOL see CJD500
4-CHLORO-2-NITROPHENYL p-CHLOROPHENYL ETHER see CJD600
N-(2-CHLORO-4-NITROPHENYL)-5-CHLOROSALICYLAMIDE see DFV400
O-(2-CHLORO-4-NITROPHENYL) O,O-DIMETHYL PHOSPHOROTHIOATE
    see NFT000
O-(3-CHLORO-4-NITROPHENYL) O,O-DIMETHYL PHOSPHOROTHIOATE
    see MIJ250
2-CHLORO-5-NITROPHENYL ESTER ACETIC ACID see CJD625
o-(2-CHLORO-4-NITROPHENYL)-o-ISOPROPYL ETHYLPHOSPHONOTHIO-
    ATE see CJD650
CHLORONITROPROPAN (POLISH) see CJD750
CHLORONITROPROPANE see CJD750, CJE000
1-CHLORO-1-NITROPROPANE see CJE000
1-CHLORO-2-NITROPROPANE see CJD750
2-CHLORO-2-NITROPROPANE see CJE250
3-CHLORO-4-NITROQUINOLINE-1-OXIDE see CJE500
5-CHLORO-4-NITROQUINOLINE-1-OXIDE see CJE750
6-CHLORO-4-NITROQUINOLINE-1-OXIDE see CJF000
7-CHLORO-4-NITROQUINOLINE-1-OXIDE see CJF250
3'-CHLORO-5-NITROSALICYLANILIDE see CJF500
1-CHLORO-1-NITROSOCYCLOHEXANE see CJF825
2-CHLORO-1-NITROSO-2-PHENYLPROPANE see CJG000
3-CHLORONITROSOPIPERIDINE see CJG375
4-CHLORONITROSOPIPERIDINE see CJG500
3-CHLORO-1-NITROSOPIPERIDINE see CJG375
4-CHLORO-1-NITROSOPIPERIDINE see CJG500
α-CHLORO-p-NITROSTYRENE see CJG750
2-CHLORO-4-NITROTOLUENE see CJG800
α-CHLORO-o-NITROTOLUENE see NFN500

α-CHLORO-p-NITROTOLUENE see NFN400
4-CHLORO-3-NITRO-α,α,α-TRIFLUOROTOLUENE see NFS700
CHLORONIUM PERCHLORATE see CJH250
9-CHLORONONANOIC ACID see CJH500
CHLOROOXIRANE see CGX000
(3-CHLORO-4-(OXIRANYLMETHOXY)PHENYL)ACETIC ACID
    see AFI950
5-CHLORO-1-(1-(3-(2-OXO-1-BENZIMIDAZOLINYL)PROPYL)-4-PIPERIDYL)-2-
    BENZIMIDAZOLINONE see DYB875
4-CHLORO-2-OXO-3(2H)-BENZOTHIAZOLEACETIC ACID see BAV000
4-CHLORO-2-OXOBENZOTHIAZOLIN-3-YL ACETIC ACID see BAV000
4-((5-CHLORO-2-OXO-3(2H)-BENZOTHIAZOLYL)ACETYL)-1-
    PIPERAZINEETHANOL see CJH750
3-(6-CHLORO-2-OXOBENZOXAZOLIN-3-YL)METHYL-O,O-DIETHYL PHOS-
    PHOROTHIOLOTHIONATE see BDJ250
exo-5-CHLORO-6-OXO-endo-2-NORBORNANECARBONITRILE-o-
    (METHYLCARBAMOYL)OXIME see CFF250
9-CHLORO-5-OXO-7-(1H-TETRAZOL-5-YL)-5H-1-BENZOPYRANO(2,3-b)
    PYRIDINE SODIUM SALT PENTAHYDRATE see THK850
CHLOROPARACIDE see CEP000
CHLOROPARAFFIN XP-470 see CJI000
CHLORO-PDMT see CJI100
CHLOROPENTAFLUOROACETONE HYDRATE see CJI250
CHLOROPENTAFLUOROETHANE see CJI500
CHLOROPENTAHYDROXYDIALUMINUM see AHA000
1-CHLOROPENTANE see PBW500
1-CHLORO-3-(PENTYLOXY)-2-PROPANOL see CHS250
CHLOROPEPTIDE see CJI609
CHLOROPERALGONIC ACID see CJH500
3-CHLOROPERBENZOIC ACID see CJI750
m-CHLOROPERBENZOIC ACID see CJI750
3-CHLOROPEROXYBENZOIC ACID see CJI750
m-CHLOROPEROXYBENZOIC ACID see CJI750
CHLOROPEROXYL see CDW450
CHLOROPEROXYTRIFLUOROMETHANE see CJI809
CHLOROPHACINONE see CJJ000
CHLOROPHEN see PAX250
CHLOROPHENAMADIN see CJJ250
CHLOROPHENAMIDINE see CJJ250
CHLOROPHENAMIDINE HYDROCHLORIDE see CJJ500
CHLOROPHENE see CJU250
4-CHLOROPHENE-1,3-DIAMINE see CJY120
1-(p-CHLOROPHENETHYL)-6,7-DIMETHOXY-2-METHYL-1,2,3,4-
    TETRAHYDROISOQUINOLINE see MDV000
1-(p-CHLOROPHENETHYL)HYDRAZINE HYDROGEN SULFATE see CJK000
1-(p-CHLOROPHENETHYL)-2-METHYL-6,7-DIMETHOXY-1,2,3,4-
    TETRAHYDROISOQUINOLINE see MDV000
1-(p-CHLOROPHENETHYL)-1,2,3,4-TETRAHYDRO-6,7-DIMETHOXY-2-
    METHYLISOQUINOLINE see MDV000
2-(o-CHLOROPHENETHYL)-3-THIOSEMICARBAZIDE see CJK100
2-CHLOROPHENOL see CJK250
3-CHLOROPHENOL see CJK500
4-CHLOROPHENOL see CJK750
m-CHLOROPHENOL see CJK500
o-CHLOROPHENOL see CJK250
p-CHLOROPHENOL see CJK750
m-CHLOROPHENOL, solid (DOT) see CJK500
o-CHLOROPHENOL, solid (DOT) see CJK250
p-CHLOROPHENOL, solid (DOT) see CJK750
m-CHLOROPHENOL, liquid (DOT) see CJK500
o-CHLOROPHENOL, liquid (DOT) see CJK250
p-CHLOROPHENOL, liquid (DOT) see CJK750
CHLOROPHENOLS see CJL000
CHLOROPHENOTHAN see DAD200
CHLOROPHENOTHANE see DAD200
2-(2-(4-(2-((2-CHLORO-10-PHENOTHIAZINYL)METHYL)PROPYL)-1-
    PIPERAZINYL)ETHOXY)ETHANOL see CJL409
1-(3-(2-CHLOROPHENOTHIAZIN-10-YL)PROPYL)-ISONIPECOTAMIDE
    see CJL500
4-(3-(2-CHLOROPHENOTHIAZIN-10-YL)PROPYL)-1-PIPERAZINEETHANOL
    see CJM250
4-(3-(2-CHLOROPHENOTHIAZIN-10-YL)PROPYL)-1-PIPERAZINEETHANOL DI-
    HYDROCHLORIDE see PCO500
4-(3-(2-CHLOROPHENOTHIAZIN-10-YL)PROPYL)-1-PIPERAZINEETHANOL
    MALEATE see PCO850
8-(3-(2-CHLOROPHENOTHIAZIN-10-YL)PROPYL-1-THIA-4,8-
    DIAZASPIRO(4.5)DECAN-3-ONE HYDROCHLORIDE see SLB000
CHLOROPHENOTOXUM see DAD200
10-CHLOROPHENOXARSINE see CJM750
10-CHLORO-10H-PHENOXARSINE see CJM750
(4-CHLOROPHENOXY)ACETIC ACID see CJN000
p-CHLOROPHENOXYACETIC ACID see CJN000
p-CHLOROPHENOXYACETIC ACID-β-DIMETHYLAMINOETHYL ESTER
    see DPE000

(p-CHLOROPHENOXY)ACETIC ACID 2-(DIMETHYLAMINO)ETHYL ESTER HYDROCHLORIDE see AAE500

p-CHLOROPHENOXYACETIC ACID-2-ISOPROPYLHYDRAZIDE see CJN250

1-(p-CHLOROPHENOXYACETYL)-2-ISOPROPYL HYDRAZINE see CJN250

4-(4-CHLOROPHENOXY)ANILINE see CEH125

p-(p-CHLOROPHENOXY)ANILINE see CEH125

10-CHLOROPHENOXYARSINE see CJM750

4-(4-CHLOROPHENOXY)-BENZENAMINE (9CI) see CEH125

2-(4-CHLOROPHENOXY)-N-(2-(DIETHYLAMINO)ETHYL)ACETAMIDE see CJN750

2-(p-CHLOROPHENOXY)-N-(2-(DIETHYLAMINO)ETHYL)ACETAMIDE see CJN750

2-(p-CHLOROPHENOXY)-N-(2-(DIETHYLAMINO)ETHYL) ACETAMIDE COMPOUND with 4-BUTYL-1,2-DIPHENYL-3,5-PYRAZOLIDINEDIONE (1:1) see CMW750

(p-CHLOROPHENOXY)DIMETHYL-ACETIC ACID see CMX000

1-(4-CHLOROPHENOXY)-3,3-DIMETHYL-1-(1,2,4-TRIAZOL-1-YL)-2-BUTAN-2-ONE see CJO250

1-(4-CHLOROPHENOXY)-3,3-DIMETHYL-1-(1H-1,2,4-TRIAZOL-1-YL)-2-BUTANONE see CJO250

2-(p-CHLOROPHENOXY)ETHANOL see CJO500

2-(p-CHLOROPHENOXY)ETHANOL see CJO500

(2-(p-CHLOROPHENOXY)ETHYL)HYDRAZINE HYDROCHLORIDE see CJP250

1-(2-(o-CHLOROPHENOXY)ETHYL)HYDRAZINE HYDROGEN SULFATE see CJP500

3-p-CHLOROPHENOXY)-2-HYDROXYPROPYL CARBAMATE see CJQ250

CHLORO-4 PHENOXYISOBUTYRATE D'HYDROXY-4 N-DIMETHYLBUTYRAMIDE (FRENCH) see LGK200

α-(p-CHLOROPHENOXY)ISOBUTYRIC ACID see CMX000

α-p-CHLOROPHENOXYISOBUTYRYL ETHYL ESTER see ARQ750

N-2(p-CHLOROPHENOXY)ISOBUTYRYL-N-MORPHOLINOMETHYLUREA see PJB500

2-(4-CHLOROPHENOXY)-2-METHYLPROPANOIC ACID see CMX000

2-(4-CHLOROPHENOXY)-2-METHYLPROPANOIC ACID, 3,4-DIHYDRO-2,5,7,8-TETRAMETHYL-2-(4,8,12-TRIMETHYLTRIDECYL)-2H-1-BENZOPYRAN-6-YL-ESTER, (2r(4r,8r)) see TGJ000

2-(4-CHLOROPHENOXY)-2-METHYLPROPANOIC ACID ETHYL ESTER see ARQ750

2-(4-CHLOROPHENOXY)-2-METHYLPROPANOIC ACID-1,3-PROPANEDIYL ESTER see SDY500

2-(4-CHLOROPHENOXY-2-METHYL)PROPIONIC ACID see CIR500

2-(p-CHLOROPHENOXY)-2-METHYLPROPIONIC ACID see CMX000

2-(4-CHLOROPHENOXY)-2-METHYL-PROPIONIC ACID 4-(DIMETHYLAMINO)-4-OXOBUTYL ESTER (9CI) see LGK200

2-(p-CHLOROPHENOXY)-2-METHYLPROPIONIC ACID ETHYL ESTER see ARQ750

2-(p-CHLOROPHENOXY)-2-METHYLPROPIONIC ACID TRIMETHYLENE ESTER see SDY500

N-2(p-CHLOROPHENOXY)-2-METHYLPROPIONYL-N-MORPHOLINOMETHYLUREA see PJB500

2-(4-(4-CHLOROPHENOXY)PHENOXY)PROPIONIC ACID see CJP750

3-(p-(p-CHLOROPHENOXY)PHENYL-1,1-DIMETHYLUREA see CJQ000

N'-4-(4-CHLOROPHENOXY)PHENYL-N,N-DIMETHYLUREA see CJQ000

1-(4-(4-CHLORO-PHENOXY)PHENYL)-3,3-DMETHYLUREE (FRENCH) see CJQ000

3-(4-CHLOROPHENOXY)-1,2-PROPANEDIOL-1-CARBAMATE see CJQ250

3-(p-CHLOROPHENOXY)-1,2-PROPANEDIOL-1-CARBAMATE see CJQ250

N-(1-(o-CHLOROPHENOXY)-2-PROPYL)-2-(DIETHYLAMINO)-N-ETHYLACETAMIDE HYDROCHLORIDE see CJQ500

N-(1-(o-CHLOROPHENOXY)-2-PROPYL)-2-(DIETHYLAMINO)-N-METHYLACETAMIDE HYDROCHLORIDE see CJQ750

CHLOROPHENTERMINE see CLY250

CHLOROPHENTERMINE HYDROCHLORIDE see ARW750

N-(4-CHLOROPHENYL)ACETAMIDE see CDZ100

2-(α-p-CHLOROPHENYLACETYL)INDANE-1,3-DIONE see CJJ000

4-CHLOROPHENYLALANINE see CJR125

p-CHLOROPHENYLALANINE see CJR125

3-(p-CHLOROPHENYL)ALANINE see CJR125

dl-4-CHLOROPHENYLALANINE see FAM100

dl-p-CHLOROPHENYLALANINE see FAM100

p-CHLORO-dl-PHENYLALANINE see CJR125

3-CHLOROPHENYLAMINE see CEH675

4-CHLOROPHENYLAMINE see CEH680

m-CHLOROPHENYLAMINE see CEH675

β-(p-CHLOROPHENYL)-γ-AMINOBUTYRIC ACID see BAC275

N-(((4-CHLOROPHENYL)AMINO)CARBONYL)-2,6-DIFLUOROBENZAMIDE see CJV250

N-(3-CHLOROPHENYL)-1-AZIRIDINECARBOXAMIDE see CJR250

4-CHLOROPHENYL BENZENESULFONATE see CJR500

p-CHLOROPHENYL BENZENESULFONATE see CJR500

4-CHLOROPHENYL BENZENESULPHONATE see CJR500

p-CHLOROPHENYL BENZENESULPHONATE see CJR500

1-(α-(2-CHLOROPHENYL)BENZHYDRYL)IMIDAZOLE see MRX500

1-(p-CHLORO-α-PHENYLBENZYL)HEXAHYDRO-4-METHYL-1H-1,4-DIAZEPINE DIHYDROCHLORIDE see CJR809

1-(p-CHLORO-α-PHENYLBENZYL)HEXAHYDRO-4-METHYL)-1H-1,4-DIAZEPINE HYDROCHLORIDE see HGI200

1-(p-CHLORO-α-PHENYLBENZYL)-4-(2-((2-HYDROXYETHOXY)ETHYL) PIPERAZINE see CJR909

1-(p-CHLORO-α-PHENYLBENZYL)-4-(m-METHYLBENZYL)PIPERAZINE see HGC500

1-(p-CHLORO-α-PHENYLBENZYL)-4-(m-METHYLBENZYL)PIPERAZINE HYDROCHLORIDE see MBX500

1-(p-CHLORO-α-PHENYLBENZYL)-4-METHYLPIPERAZINE see CFF500

1-(p-CHLORO-α-PHENYLBENZYL)-4-METHYL-PIPERAZINE DIHYDROCHLORIDE see CDR000

1-(p-CHLORO-α-PHENYLBENZYL)-4-METHYLPIPERAZINE HYDROCHLORIDE see CDR500

2-(α-(p-CHLOROPHENYL)BENZYLOXY)-N,N-DIMETHYLETHYLAMINE HYDROCHLORIDE see CJR959

3-α-((p-CHLORO-α-PHENYLBENZYL)OXY)-1-α-H,5-α-H-TROPANE HYDROCHLORIDE see CMB125

1-(p-CHLORO-α-PHENYLBENZYL)PIPERAZINE see NNK500

1-(p-CHLORO-α-PHENYLBENZYL)-PIPERAZINE HYDROCHLORIDE see NNL000

1-(α-(4-CHLOROPHENYL)BENZYL)-PIPERAZINE HYDROCHLORIDE see NNL000

2-(2-(4-(p-CHLORO-α-PHENYLBENZYL)-1-PIPERAZINYL)ETHOXY)ETHANOL see CJR909

2-(2-(2-(4-(p-CHLORO-α-PHENYLBENZYL)-1-PIPERAZINYL)ETHOXY)ETHOXY)ETHANOL DIMALEATE see HHK100

1-(4-CHLOROPHENYL)BIGUANIDINIUM HYDROGEN DICHROMATE see CJT125

1-(o-CHLOROPHENYL)-2-tert-BUTYLAMINO ETHANOL HYDROCHLORIDE see BQE250

N-(3-CHLORO PHENYL) CARBAMATE de 4-CHLORO 2-BUTYNYLE (FRENCH) see CEW500

N-(3-CHLORO PHENYL) CARBAMATE D'ISOPROPYLE (FRENCH) see CKC000

(3-CHLOROPHENYL)CARBAMIC ACID 4-CHLORO-2-BUTYNYL ESTER see CEW500

N-(3-CHLOROPHENYL)CARBAMIC ACID, ISOPROPYL ESTER see CKC000

(3-CHLOROPHENYL)CARBAMIC ACID, 1-METHYLETHYL ESTER see CKC000

3-CHLOROPHENYLCARBAMIC ACID-1-METHYLPROPYNYL ESTER see CEX250

3-CHLOROPHENYL-N-CARBAMOYLAZIRIDINE see CJR250

p-CHLOROPHENYL CHLORIDE see DEP800

4-CHLOROPHENYL-4-CHLOROBENZENESULFONATE see CJT750

p-CHLOROPHENYL-p-CHLOROBENZENE SULFONATE see CJT750

4-CHLOROPHENYL-4-CHLOROBENZENESULPHONATE see CJT750

4-CHLOROPHENYL-4'-CHLOROBENZYL SULFIDE see CEP000

2-(o-CHLOROPHENYL)-2-(p-CHLOROPHENYL)-1,1-DICHLOROETHANE see CDN000

(2-CHLOROPHENYL)-α-(4-CHLOROPHENYL)-5-PYRIMIDINEMETHANOL see FAK100

α-(2-CHLOROPHENYL)-α-(4-CHLOROPHENYL)-5-PYRIMIDINEMETHANOL see FAK100

p-CHLOROPHENYL-N-(4'-CHLOROPHENYL)THIOCARBAMATE see CJU125

4-CHLORO-α-PHENYLCRESOL see BEF750

4-CHLORO-α-PHENYL-o-CRESOL see CJU250

1-(p-CHLOROPHENYL)-4,6-DIAMINO-2,2-DIMETHYL-1,2-DIHYDRO-s-TRIAZINE see COX400

5-(4'-CHLOROPHENYL)-2,4-DIAMINO-6-ETHYLPYRIMIDINE see TGD000

(o-CHLOROPHENYL)(3-(2,4-DICHLOROPHENOXY)-2-HYDROXYPROPYL)HYDROXYARSINEOXID E see DGA425

3-(4-CHLOROPHENYL)-4',5-DICHLOROSALICYLANILIDE see TIP750

2-p-CHLOROPHENYL-1-(p-(2-DIETHYLAMINOETHOXY)PHENYL)-1-p-TOLYLETHANOL see TMP500

2-(p-CHLOROPHENYL)-1-(p-(β-DIETHYLAMINOETHOXY)PHENYL)-1-(p-TOLYL)ETHANOL see TMP500

1-(4-CHLOROPHENYL)-3-(2,6-DIFLUOROBENZOYL)UREA see CJV250

1-p-CHLOROPHENYL-1,2-DIHYDRO-2,2-DIMETHYL-4,6-DIAMINO-s-TRIAZINE see COX400

1-(4-CHLOROPHENYL)-1,6-DIHYDRO-6,6-DIMETHYL-1,3,5-TRIAZINE-2,4-DIAMINE MONOHYDROCHLORIDE see COX325

5-p-CHLOROPHENYL-2,3-DIHYDRO-5H-IMIDAZO(2,1-A)ISOINDOL-5-OL see MBV250

5-(o-CHLOROPHENYL)-1,3-DIHYDRO-7-NITRO-2H-1,4-BENZODIAZEPIN-2-ONE see CMW000

(4-CHLOROPHENYL)(DIMETHOXYPHOSPHINYL)METHYL PHOSPHORIC ACID DIMETHYL ESTER see CMU875

1-(3-CHLOROPHENYL)-3,N,N-DIMETHYLCARBAMOYL-5-METHOXYPYRAZOLE see CJW500

1-(m-CHLOROPHENYL)-3-N,N-DIMETHYLCARBAMOYL-5-METHOXYPYRAZOLE see CJW500

1-(4-CHLOROPHENYL)-2,3-DIMETHYL-4-DIMETHYLAMINO-2-BUTANOL see CMW459

2-CHLORO-1-PROPANOL see CKR500
1-CHLORO-2-PROPANOL with 2-CHLORO-1-PROPANOL see CKR750
CHLOROPROPANONE see CDN200
1-CHLORO-2-PROPANONE see CDN200
3-CHLOROPROPANONITRILE see CKT250
7-CHLORO-1-PROPARGYL-5-PHENYL-2H-1,4-BENZODIAZEPIN-2-ONE
　see PIH100
1-CHLOROPROPENE see PMR750
3-CHLOROPROPENE see AGB250
1-CHLORO-1-PROPENE see PMR750
1-CHLORO PROPENE-2 see AGB250
1-CHLORO-2-PROPENE see AGB250
2-CHLORO-1-PROPENE see CKS000
3-CHLORO-1-PROPENE see AGB250
2-CHLOROPROPENE (DOT) see CKS000
cis-1-CHLOROPROPENE OXIDE see CKS099
trans-1-CHLOROPROPENE OXIDE see CKS100
2-CHLORO-2-PROPENE-1-THIOL DIETHYLDITHIOCARBAMATE see CDO250
2-CHLORO-2-PROPENOIC ACID METHYL ESTER (9CI) see MIF800
2-CHLORO-2-PROPEN-1-OL see CEF250
3-CHLORO-2-PROPEN-1-OL see CEF500
2-CHLORO-2-PROPENYL DIETHYLCARBAMODITHIOATE see CDO250
2-CHLORO-4-(2-PROPENYLOXY)BENZENEACETIC ACID see AGN000
2-CHLORO-2-PROPENYL TRIFLUOROMETHANE SULFONATE see CKS325
CHLOROPROPHAM see CKC000
CHLOROPROPHENYPYRIDAMINE MALEATE see TAI500
2-CHLOROPROPIONALDEHYDE see CKP700
α-CHLOROPROPIONALDEHYDE see CKP700
2-CHLOROPROPIONATE SODIUM SALT see CKT100
3-CHLOROPROPIONIC ACID see CKS500
α-CHLOROPROPIONIC ACID see CKS750
β-CHLOROPROPIONIC ACID see CKS500
2-CHLOROPROPIONIC ACID METHYL ESTER see CKT000
2-CHLOROPROPIONIC ACID SODIUM SALT see CKT100
α-CHLOROPROPIONIC ACID SODIUM SALT see CKT100
3-CHLOROPROPIONITRILE see CKT250
β-CHLOROPROPIONITRILE see CKT250
N-(3-CHLOROPROPIONYL)BENZYLAMINE see BEG000
p-CHLOROPROPIOPHENONE see CKT500
2-CHLOROPROPYL ALCOHOL see CKR500
α-CHLOROPROPYLALDEHYDE see CKP700
6'-CHLORO-2-(PROPYLAMINO)-o-ACETOTOLUIDIDE HYDROCHLORIDE
　see CKT750
6'-CHLORO-2-(PROPYLAMINO)-o-BUTYROTOLUIDIDE HYDROCHLORIDE
　see CKU000
4-CHLORO-4-((PROPYLAMINO)CARBONYL)BENZENESULFONAMIDE
　see CKK000
9-((2-((2-CHLOROPROPYL)AMINO)ETHYL)AMINO)-2-METHOXYACRIDINE DI-
　HYDROCHLORIDE HEMIHYDRATE see CKU250
2-CHLORO-4-(2-PROPYLAMINO)-6-ETHYLAMINO-s-TRIAZINE see ARQ725
CHLOROPROPYLATE see PNH750
3-CHLOROPROPYL BROMIDE see BNA825
3-CHLOROPROPYLENE see AGB250
α-CHLOROPROPYLENE see AGB250
3-CHLORO-1-PROPYLENE see AGB250
1-CHLORO-2,3-PROPYLENE DINITRATE see CKU625
3-CHLOROPROPYLENE GYLCOL see CDT750
CHLOROPROPYLENE OXIDE see EAZ500
γ-CHLOROPROPYLENE OXIDE see EAZ500
3-CHLORO-1,2-PROPYLENE OXIDE see EAZ500
N-(3-CHLOROPROPYL)-α-METHYLPHENETHYLAMINE HYDROCHLORIDE
　see PKS500
3-CHLOROPROPYL-n-OCTYLSULFOXIDE see CKU750
3-CHLOROPROPYNE see CKV275
1-CHLORO-2-PROPYNE see CKV250
CHLOROPROPYNENITRILE see CFE750
CHLOROPROTHIXENE see TAF675
CHLOROPTIC see CDP250
6-CHLOROPURINE see CKV500
6-CHLORO-9H-PURINE see CKV500
6-CHLORO-1H-PURINE (9CI) see CKV500
CHLOROPYRAMINE see CKV625
CHLOROPYRIBENZAMINE see CKV625
2-CHLOROPYRIDINE see CKW000
3-CHLOROPYRIDINE see CKW250
m-CHLOROPYRIDINE see CKW250
o-CHLOROPYRIDINE see CKW000
α-CHLOROPYRIDINE see CKW000
6-CHLORO-2-PYRIDINECARBOXYLIC ACID see CKN375
2-CHLOROPYRIDINE-N-OXIDE see CKW325
2-(4-(3-(3-CHLORO-10H-PYRIDO(3,2-b)-1,4-BENZOTHIAZINE-10-YL)PROPYL)-1-
　PIPERAZINYLETHANOL see CMY135
2-(4-(3-(3-CHLORO-10H-PYRIDO(3,2-b)(1,4)BENZOTHIAZIN-1-OYL)PROPYL)-1-
　PIPERAZINYL) ETHANOL see CMY135

4-(3-(3-CHLORO-10H-PYRIDO(3,2-b)(1,4)-BENZOTHIAZIN-10-YL)PROPYL)-1-
　PIPERAZINE ETHANOL see CMY135
CHLORO-3-PYRIDYLMERCURY see CKW500
4-(2-((6-CHLORO-2-PYRIDYL)THIO)ETHYL)MORPHOLINE
　MONOHYDROCHLORIDE see FMU225
CHLOROPYRILENE see CHY250
1-CHLORO-2,5-PYRROLIDINEDIONE see SND500
4-CHLORO-N-((1-PYRROLIDINYLAMINO)CARBONYL)BENZENESULFONAM-
　IDE (9CI) see GHR609
6'-CHLORO-2-(PYRROLIDINYL)-o-DIACETOTOLUIDIDE HYDROCHLORIDE
　see CLB250
6'-CHLORO-2-PYRROLIDINYL-o-HEXANOTOLUIDIDE HYDROCHLORIDE
　see CAB250
4'-CHLORO-2-PYRROLIDINYL-α,α,α-TRIFLUORO-m-ACETOTOLUIDIDE,
　HYDROCHLORIDE see CLC125
6'-CHLORO-2-PYRROLIDINYL-α,α,α-TRIFLUORO-m-ACETOTOLUIDINE,
　HYDROCHLORIDE see CLC100
2'-CHLORO-2-PYRROLIDINYL-5'-TRIFLUOROMETHYLACETANILIDE
　HYDROCHLORIDE see CLC100
4'-CHLORO-2-PYRROLIDINYL-3'-TRIFLUOROMETHYLACETANILIDE
　HYDROCHLORIDE see CLC125
3-CHLORO-4-(3-PYRROLIN-1-YL)HYDRATROPIC ACID see PJA220
CHLOROQUINALDOL see CLC500
6-CHLORO-4-QUINAZOLINONE see CLC750
6-CHLORO-4(3H)-QUINAZOLINONE see CLC750
CHLOROQUINE see CLD000
CHLOROQUINE DIPHOSPHATE see CLD250
CHLOROQUINE MUSTARD see CLD500
CHLOROQUINE PHOSPHATE see AQT250, CLD250
CHLOROQUINIUM see CLD000
4-((7-CHLORO-4-QUINOLINYL)AMINO)-2-(1-PYRROLIDINYLMETHYL)
　PHENOL DIHYDROCHLORIDE see PMY000
N⁴-(7-CHLORO-4-QUINOLINYL)-N¹,N¹-DIETHYL-1,4-PENTANEDIAMINE
　see CLD000
2-((4-(7-CHLORO-4-QUINOLYL)AMINO)PENTYL)-ETHYLAMINO)ETHANOL
　see PJB750
4-((7-CHLORO-4-QUINOLYL)AMINO)-α-1-PYRROLIDINYL-o-CRESOL DIHY-
　DROCHLORIDE see PMY000
4-CHLORORESORCINOL see CLD750
7-CHLORO-3-β-d-RIBOFURANOSYL-3H-IMIDAZO(4,5-b)PYRIDINE see AEB500
CHLOROS see SHU500
CHLORO-S.C.T.Z. see CHD750
CHLOROSILANES see CLE250
3-CHLORO-4-STILBENAMINE see CLE500
2'-CHLORO-4-STILBENYL-N,N-DIMETHYLAMINE see CGK500
3'-CHLORO-4-STILBENYL-N,N-DIMETHYLAMINE see CGK750
4'-CHLORO-4-STILBENYL-N,N-DIMETHYLAMINE see CGL000
CHLOROSTOP see PKQ059
o-CHLOROSTYRENE see CLE750
3-CHLOROSTYRENE OXIDE see CLF000
m-CHLOROSTYRENE OXIDE see CLF000
CHLOROSULFACIDE see CEP000
N-(4'-CHLORO-3'-SULFAMOYLBENZENESULFONYL)-N-METHYL-2-
　AMINOMETHYL-2-METHYLTETRAHYDROFURAN see MCA100
6-CHLORO-7-SULFAMOYL-2H-1,2,4-BENZOTHIADIAZINE-1,1-DIOXIDE
　see CLH750
6-CHLORO-7-SULFAMOYL-3,4-DIHYDRO-2H-1,2,4-BENZOTHIADIAZINE-1,1-
　DIOXIDE see CFY000
4-CHLORO-5-SULFAMOYL-2',6'-SALICYLOXYLIDIDE see CLF325
N-CHLOROSULFINYLIMIDE see CLF500
4-CHLORO-4'-(6-SULFO-2H-NAPHTHO(1,2-d)TRIAZOL-2-YL)-2,2'-
　STILBENEDISULFONIC ACID TRISODIUM SALT see CLG000
CHLOROSULFONIC ACID (DOT) see CLG500
CHLOROSULFONIC ACID-SULFUR DIOXIDE MIXTURE (DOT) see CLG750
CHLOROSULFONIC ANHYDRIDE see PPR500
5-(CHLOROSULFONYL)-2,4-DICHLOROBENZOIC ACID see CLG200
1-CHLOROSULFONYL-5-DIMETHYLAMINONAPHTHALENE see DPN200
CHLOROSULFONYL FLUORIDE see SOT500
CHLOROSULFONYLISOCYANATE see CLG250
1-(4-CHLORO-o-SULFO-5-TOLYLAZO)-2-NAPHTHOL,BARIUM SALT
　see CHP500
CHLOROSULFURIC ACID see CLG500
CHLOROSULFURIC ACID, mixed with SULFUR DIOXIDE see CLG750
4-CHLORO-5-SULPHAMOYLPHTHALIMIDE see CGM500, CLG825
CHLOROSULTHIADIL see CFY000
7-CHLOROTETRACYCLINE see CMA750
CHLOROTETRACYCLINE HYDROCHLORIDE see CMB000
6-CHLORO-N,N,N',N'-TETRAETHYL-1,3,5-TRIAZINE-2,4-DIAMINE see CDQ325
CHLOROTETRAFLUOROETHANE see CLH000
7-CHLORO-1,2,3,4-TETRAHYDRO-2-METHYL-3-(2-METHYLPHENYL)-4-OXO-6-
　QUINAZOLINESULFONAMIDE see ZAK300
N-CHLOROTETRAMETHYLGUANIDINE see CLH500
9-CHLORO-7-(1H-TETRAZOL-5-YL)-5H-1-BENZOPYRANO(2,3-b)PYRIDIN-5-
　ONE SODIUM PENTAHYDRATE see THK850

2-CHLORO-5-(1H-TETRAZOL-5-YL)-N(sup 4)-2-THENYLSULFANILAMIDE
  see ASO375
CHLOROTHALIDONE see CLY600
CHLOROTHALONIL see TBQ750
CHLOROTHANE NU see MIH275
CHLOROTHEN see CHY250
CHLOROTHENE see MIH275
CHLOROTHENE (inhibited) see MIH275
CHLOROTHENE NU see MIH275
CHLOROTHENE VG see MIH275
2-((5-CHLORO-2-THENYL)(2-DIMETHYLAMINOETHYL)AMINO)PYRIDINE
  see CHY250
CHLOROTHENYLPYRAMINE see CHY250
8-CHLORO-THEOPHYLLINE compounded with 4-(DIPHENYLMETHOXY)-1-
  METHYLPIPERIDINE (1:1) see PIZ250
5-CHLORO-1,2,3-THIADIAZOLE see CLH625
CHLOROTHIAMIDE see DGM600
CHLOROTHIAZID see CLH750
CHLOROTHIAZIDE see CLH750
4-CHLOROTHIOANISOLE see CKG500
p-CHLOROTHIOANISOLE see CKG500
p-CHLOROTHIOCARBANILIC ACID-o-(p-CHLOROPHENYL) ESTER
  see CJU125
CHLOROTHIOFORMIC ACID ETHYL ESTER see CLJ750
2,3,4,5-CHLOROTHIOPHENE see TBV750
4-CHLOROTHIOPHENOL see CEK425
p-CHLOROTHIOPHENOL see CEK425
(Z)-2-(2-CHLORO-9H-THIOXANTHEN-9-YLIDENE)-N,N-DIMETHYL-1-
  PROPANAMINE see TAF675
4-(3-(2-CHLOROTHIOXANTHEN-9-YLIDENE)PROPYL)-1-
  PIPERAZINEETHANOL DIHYDROCHLORIDE see CLX250
4-CHLORO-o-TOLOXYACETIC ACID see CIR250
2-CHLOROTOLUENE see CLK100
4-CHLOROTOLUENE see TGY075
o-CHLOROTOLUENE see CLK100
α-CHLOROTOLUENE see BEE375
p-CHLOROTOLUENE (DOT) see TGY075
ω-CHLOROTOLUENE see BEE375
2-CHLORO-p-TOLUIDINE see CLK210
3-CHLORO-p-TOLUIDINE see CLK200
3-CHLORO-p-TOLUIDINE see CLK215
4-CHLORO-2-TOLUIDINE see CLK220
4-CHLORO-o-TOLUIDINE see CLK220
5-CHLORO-o-TOLUIDINE see CLK225
3-CHLORO-p-TOLUIDINE HYDROCHLORIDE see CLK230
4-CHLORO-2-TOLUIDINE HYDROCHLORIDE see CLK235
4-CHLORO-o-TOLUIDINE HYDROCHLORIDE see CLK235
4-CHLORO-o-TOLUIDINE HYDROCHLORIDE (DOT) see CLK235
2-(2-CHLORO-p-TOLUIDINO)-2-IMIDAZOLINE NITRATE see TGJ885
CHLOROTOLURON see CIS250
N-(3-CHLORO-o-TOLYL)ANTHRANILIC ACID see CLK325
p-((3-CHLORO-p-TOLYL)AZO)-N,N-DIMETHYLANILINE see CIL700
p-((4-CHLORO-m-TOLYL)AZO)-N,N-DIMETHYLANILINE see CIL710
1-(6-CHLORO-o-TOLYL)-3-CYCLOHEXYL-3-(2-
  (DIETHYLAMINO)ETHYL)UREA HYDROCHLORIDE see CLK500
1-(6-CHLORO-o-TOLYL)-3-(3-(DIBUTYLAMINO)PROPYL)UREA HYDROCHLO-
  RIDE see CLK750
1-(6-CHLORO-o-TOLYL)-3-(2-(DIETHYLAMINO)ETHYL)-3-METHYLUREA
  see CLL000
1-(6-CHLORO-o-TOLYL)-3-(2-(DIETHYLAMINO)ETHYL)UREA HYDROCHLO-
  RIDE see CLL250
1-(6-CHLORO-o-TOLYL)-1-(2-(DIETHYLAMINO)ETHYL)-3-(2,6-XYLYL)UREA
  HYDROCHLORIDE see CLL750
1-(6-CHLORO-o-TOLYL)-3-(2-(DIETHYLAMINO)ETHYL)-3-(2,6-XYLYL)UREA
  HYDROCHLORIDE see CLL500
1-(6-CHLORO-o-TOLYL)-3-(3-(DIETHYLAMINO)PROPYL)UREA see CLM000
1-(6-CHLORO-o-TOLYL)-3-(2-(DIMETHYLAMINO)ETHYL)-3-ISOPROPYLUREA
  HYDROCHLORIDE see CLM250
1-(6-CHLORO-o-TOLYL)-3-(2-(DIMETHYLAMINO)ETHYL)UREA HYDROCHLO-
  RIDE see CLM500
1-(6-CHLORO-o-TOLYL)-3-(3-(DIMETHYLAMINO)PROPYL)UREA HYDRO-
  CHLORIDE see CLM750
N'-(4-CHLORO-o-TOLYL)-N,N-DIMETHYLFORMAMIDINE see CJJ250
N'-(4-CHLORO-o-TOLYL)-N,N-DIMETHYLFORMAMIDINE HYDROCHLORIDE
  see CJJ500
1-(6-CHLORO-o-TOLYL)-3-(4-METHOXYBENZYL)-3-(2-
  PIPERIDINOETHYL)UREA see CLN000
1-(6-CHLORO-o-TOLYL)-3-(4-METHOXYBENZYL)-3-(2-
  (PYRROLIDINYL)ETHYL)UREA HYDROCHLORIDE see CLN250
3-(4-CHLORO-o-TOLYL)-5-(m-METHOXYPHENYL)-s-TRIAZOLE see CLN325
1-(4-CHLORO-o-TOLYL)-3-(3-PHENYL-PROPYL)-3-(2-
  PYRROLIDINYLETHYL)UREA HYDROCHLORIDE see CLN500
((4-CHLORO-o-TOLYL)OXY)ACETIC ACID see CIR250
((4-CHLORO-o-TOLYL)OXY)ACETIC ACID, ETHYL ESTER see EMR000

((4-CHLORO-O-TOLYL)OXY)-ACETIC ACID with 2,2'-IMINODIETHANOL (1:1)
  see MIH750
(p-CHLORO-o-TOLYLOXY)ACETIC ACID SODIUM SALT see SIL500
(4-CHLORO-o-TOLYLOXY)BUTYRIC ACID see CLN750
4-((4-CHLORO-o-TOLYL)OXY)BUTYRIC ACID see CLN750
(4-CHLORO-o-TOLYLOXY)BUTYRIC ACID SODIUM SALT see CLO000
2-(p-CHLORO-o-TOLYLOXY)PROPIONIC ACID see CIR500
2-((4-CHLORO-o-TOLYL)OXY)PROPIONIC ACID POTASSIUM SALT
  see CLO200
1-(6-CHLORO-o-TOLYL)-3-(2-PYRROLIDINYLETHYL)UREA HYDROCHLO-
  RIDE see CLO500
CHLOROTRIANISENE see CLO750
CHLOROTRIANIZEN see CLO750
CHLOROTRIAZINE see TJD750
CHLOROTRIBENZYLSTANNANE see CLP000
CHLOROTRIBUTYLGERMANIUM see CLP250
CHLOROTRIBUTYLSTANNANE see CLP500
3-CHLORO-3-TRICHLOROMETHYLDIAZIRINE see CLP625
2-CHLORO-6-(TRICHLOROMETHYL)PYRIDINE see CLP750
2-CHLORO-1-(2,4,5-TRICHLOROPHENYL)VINYL DIMETHYL PHOSPHATE
  see TBW100
(Z)-2-CHLORO-1-(2,4,5-TRICHLOROPHENYL)VINYL DIMETHYL PHOS-
  PHATE see RAF100
2-CHLORO-1-(2,4,5-TRICHLOROPHENYL(VINYL PHOSPHORIC ACID
  DIMETHYL ESTER see TBW100
2-CHLOROTRIETHYLAMINE see CGV500
β-CHLOROTRIETHYLAMINE see CGV500
2-CHLOROTRIETHYLAMINE HYDROCHLORIDE see CLQ250
CHLORO(TRIETHYLPHOSPHINE)GOLD see CLQ500
CHLOROTRIETHYLSTANNANE see TJV000
CHLOROTRIETHYLTIN see TJV000
6-CHLORO-n,n,n'-TRIETHYL-1,3,5-TRIAZINE-2,4-DIAMINE see TJL500
CHLOROTRIFLUORIDE see CDX750
1-CHLORO-2,2,2-TRIFLUOROETHANE see TJY175
2-CHLORO-1,1,1-TRIFLUOROETHANE see TJY175
2-CHLORO-1,1,2-TRIFLUOROETHYL DIFLUOROMETHYL ETHER see EAT900
CHLOROTRIFLUOROETHYLENE see CLQ750
1-CHLORO-1,2,2-TRIFLUOROETHYLENE see CLQ750
2-CHLORO-1,1,2-TRIFLUOROETHYLENE see CLQ750
2-CHLORO-1,1,2-TRIFLUOROETHYL METHYL ETHER see CLR000
CHLOROTRIFLUOROMETHANE see CLR250
4-CHLOROTRIFLUOROMETHYLBENZENE see CEM825
p-CHLOROTRIFLUOROMETHYLBENZENE see CEM825
3-CHLORO-3-TRIFLUOROMETHYLDIAZIRINE see CLR825
4-CHLORO-3-TRIFLUOROMETHYLPHENOL see CLS000
p-CHLORO-m-TRIFLUOROMETHYLPHENOL see CLS000
3-CHLORO-1,1,1-TRIFLUOROPROPANE see TJY200
2-CHLORO-N,N,N'-TRIFLUOROPROPIONAMIDE see CLS125
4-(4-CHLORO-α,α,α-TRIFLUORO-m-TOLYL)-1-(4,4-BIS(p-
  FLUOROPHENYL)BUTYL)-4-PIPERIDINOL see PAP250
4-(4-(4-CHLORO-α,α,α-TRIFLUORO-m-TOLYL)-4-
  HYDROXYPIPERIDINO)BUTYROPHENONE-4-FLUOROHYDROCHLORIDE
  see CLS250
CHLORO(TRIISOBUTYL)STANNANE see CLS500
7-CHLORO-4,6,2'-TRIMETHOXY-6'-METHYLGRIS-2'-EN-3,4'-DIONE see GKE000
1-CHLORO-4-(TRIMETHYL)-BENZENE (9CI) see CEM825
2-CHLORO-N,N,N-TRIMETHYLETHANAMINIUM CHLORIDE see CMF400
CHLOROTRIMETHYLPLUMBANE see TLU175
CHLOROTRIMETHYLSILANE see CLS750
CHLOROTRIMETHYLSILICANE see TLN250
CHLOROTRIMETHYLSTANNANE see CLT000
CHLOROTRIMETHYLTIN see CLT000
CHLOROTRINITROMETHANE see CLT250
CHLOROTRIPHENYLMETHANE see CLT500
CHLOROTRIPHENYLSTANNANE see CLU000
CHLOROTRIPHENYLTIN see CLU000
CHLOROTRIPROPYLPLUMBANE see TNA750
CHLOROTRIPROPYLSTANNANE see CLU250
CHLOROTRISIN see CLO750
CHLOROTRIS(p-METHOXYPHENYL)ETHYLENE see CLO750
(CHLOROTRITYL)IMIDAZOLE see MRX500
1-(o-CHLOROTRITYL)IMIDAZOLE see MRX500
CHLORO(TRIVINYL)STANNANE see CLU500
CHLOROUS ACID see SDN500
CHLOROVINYLARSINE DICHLORIDE see CLV000
β-CHLOROVINYLBICHLOROARSINE see CLV000
2-CHLOROVINYLDICHLOROARSINE see CLV000
(2-CHLOROVINYL)DICHLOROARSINE see CLV000
(2-CHLOROVINYL)DIETHOXYARSINE see CLV250
2-CHLOROVINYL DIETHYL PHOSPHATE see CLV375
β-CHLOROVINYL ETHYLETHYNYL CARBINOL see CHG000
(2-CHLOROVINYL)MERCURIC CHLORIDE see CFD250
3-(β-CHLOROVINYL)-1-PENTYN-3-OL see CHG000
CHLOROVULES see CDP250

CHLOROWODOR (POLISH) see HHL000
CHLOROWODORKU 10-γ-DWUMETYLOAMINOPROPYLO-7-
  CHLOROBENZO(b)-(1,8)-NAFTYRYDONU-5 (POLISH) see CGG500
CHLOROX see SHU500
4(2-CHLORO-9H-XANTHEN-9-YLIDENE)-1-METHYLPIPERIDINE
  METHANESULFONATE see CMX860
CHLOROXAZONE see CDQ750
CHLOROXIFENIDIM see CJQ000
CHLOROXONE see DAA800
CHLOROXURON see CJQ000
CHLOROXYLAM see CGI500
5-CHLORO-m-XYLENE see CLV500
1-CHLORO-3,5-XYLENE see CLV500
5-CHLORO-1,3-XYLENE see CLV500
CHLORO-XYLENOL see CLW000
p-CHLORO-m-XYLENOL see CLW000
4-CHLORO-3,5-XYLENOL see CLW000
6-CHLORO-3,4-XYLENYL N-METHYLCARBAMATE see CGI500
(4-CHLORO-6-(2,3-XYLIDINO)-2-PYRIMIDINYLTHIO)ACETIC ACID
  see CLW250
2-((4-CHLORO-6-(2,3-XYLIDINO)-2-PYRIMIDINYL)THIO)-N-(2-
  HYDROXYETHYL)ACETAMIDE see CLW500
4-(p-CHLORO-N-2,6-XYLYLBENZAMIDO)BUTYRIC ACID see CLW625
2-CHLORO-4,5-XYLYL ESTER, CARBAMIC ACID see CGI500
6-CHLORO-3,4-XYLYL N-METHYLCARBAMATE see CGI500
CHLOROXYPHOS see TIQ250
CHLOROZIRCONYL see ZSJ000
CHLOROZODIN see ASM250
CHLOROZONE see CDP000
CHLOROZOTOCIN see CLX000
CHLORPARACIDE see CEP000
CHLORPARANITRANILINE RED see CJD500
CHLORPENTHIXOL DIHYDROCHLORIDE see CLX250
CHLORPERALGONIC see CJH500
CHLORPERAZINE see PMF500
CHLORPERPHENTHIXENE DIHYDROCHLORIDE see CLX250
CHLORPHACINON (ITALIAN) see CJJ000
CHLORPHEDIANOL HYDROCHLORIDE see CMW700
CHLORPHENAMIDINE see CJJ250
CHLORPHENESIN CARBAMATE see CJQ250
CHLORPHENIRAMINE MALEATE see TAI500
(+)-CHLORPHENIRAMINE MALEATE see PJJ325
d-CHLORPHENIRAMINE MALEATE see PJJ325
S-(+)-CHLORPHENIRAMINE MALEATE see PJJ325
o-CHLORPHENOL (GERMAN) see CJK250
CHLORPHENOXAMINE HYDROCHLORIDE see CIS000
2-(4-(4'-CHLORPHENOXY)-PHENOXY)-PROPIONSAEURE (GERMAN)
  see CJP750
3-(4-(4-CHLOR-PHENOXY)-PHENYL)-1,1-DIMETHYLHARNSTOFF (GERMAN)
  see CJQ000
CHLORPHENPROP-METHYL see CFC750
CHLORPHENTERAMINE see CLY250
CHLORPHENTERMINE see CLY250
CHLORPHENTERMINE HYDROCHLORIDE see ARW750
CHLORPHENVINFOS see CDS750
CHLORPHENVINPHOS see CDS750
6-(o-CHLORPHENYL)-8-AETYL-1-METHYL-4H-sec-TRIAZOLO(3,4-
  c)THIENO(2,3-e)(1,4)DIAZEPIN (GERMAN) see EQN600
(±)-p-CHLORPHENYLALANINE see FAM100
(4-CHLOR-PHENYL)-BENZOLSULFONAT (GERMAN) see CJR500
3-CHLORPHENYL-CARBAMIDSAURE-BUTIN-(1)-YL(3)-ESTER (GERMAN)
  see CEX250
4-CHLORPHENYL-4'-CHLORBENZOLSULFONAT (GERMAN) see CJT750
(4-CHLOR-PHENYL)-4-CHLOR-BENZOL-SULFONATE (GERMAN)
  see CJT750
3-(4-CHLORPHENYL)-2-CHLORPROPIONSAEUREMETHYLESTER (GERMAN)
  see CFC750
3-(4-CHLOR-PHENYL)-1,1-DIMETHYL-HARNSTOFF (GERMAN) see CJX750
N-(4-CHLORPHENYL)-2,2-DIMETHYLPENTAMID (GERMAN) see CGL250
1-(p-CHLOR-PHENYL)-3,3-DIMETHYL-TRIAZEN (GERMAN) see CJI100
N-(4-CHLOR-PHENYL)-2,2-DIMETHYL-VALERIANSAEUREAMID (GERMAN)
  see CGL250
4-(2-CHLORPHENYL)-2-ETHYL-9-METHYL-6H-THIENO(3,2-f)(1,2,4)TRIAZOLO
  (4,3-a)(1,4)DIAZEPINE see EQN600
γ-(4-(p-CHLORPHENYL)-4-HYDROXPIPERIDINO)-p-
  FLUORBUTYROPHENONE see CLY500
N-(3-CHLORPHENYL)-ISOPROPYL-CARBAMAT (GERMAN) see CKC000
4-CHLOR-PHENYL-ISOTHIOCYANAT (GERMAN) see ISH000
3-(4-CHLORPHENYL)-1-METHOXY-1-METHYLHARNSTOFF (GERMAN)
  see CKD500
3-(4-CHLORPHENYL)-1-METHYL-1-ISOBUTINYLHARNSTOFF (GERMAN)
  see CKF750
N-(4-CHLORPHENYL)-N'-METHYL-N'-ISOBUTINYLHARNSTOFF (GERMAN)
  see CKF750

2-(p-CHLORPHENYL)-3-METHYL-1,3-PERHYDROTHIAZIN-4-ON-1,1-DIOXIDE
  see CKF500
((4-CHLORPHENYL)-1-PHENYL)-ACETYL-1,3-INDANDION (GERMAN) see CJJ000
1-(4-CHLORPHENYL)-1-PHENYL-ACETYL-INDAN-1,3-DION (GERMAN)
  see CJJ000
2(2-(4-CHLOR-PHENYL-2-PHENYL)ACETYL)INDAN-1,3-DION (GERMAN)
  see CJJ000
4-(p-CHLORPHENYLTHIO)-BUTANOL (GERMAN) see CKK500
4-CHLORPHENYL-2',4',5'-TRICHLORPHENYLAZOSULFID (GERMAN)
  see CDS500
CHLORPHONIUM CHLORIDE see THY500
CHLORPHTHALIDOLONE see CLY600
CHLORPHTHALIDONE see CLY600
CHLOR-O-PIC see CKN500
CHLORPIKRIN (GERMAN) see CKN500
CHLORPROETHAZINE see CLY750
CHLORPROETHAZINE HYDROCHLORIDE see CLZ000
CHLORPROHEPTADIEN see CEH500
CHLORPROHEPTADIENE HYDROCHLORIDE see CEH500
CHLORPROHEPTATRIEN see CMA000
CHLORPROMAZIN see CKP250
CHLORPROMAZINE see CKP250
CHLORPROPAMID see CKK000
CHLORPROPAMIDE see CKK000
3-CHLORPROPEN (GERMAN) see AGB250
CHLORPROPHAM see CKC000
CHLORPROPHAME (FRENCH) see CKC000
N-(3-CHLORPROPYL)-1-METHYL-2-PHENYL-AETHYLAMIN-
  HYDROCHLORID (GERMAN) see PKS500
CHLORPROTHIXEN see TAF675
CHLORPROTHIXENE see TAF675
α-CHLORPROTHIXENE see TAF675
cis-CHLORPROTHIXENE see TAF675
CHLORPROTIXEN see TAF675
CHLORPROTIXENE see TAF675
CHLORPROTIXINE see TAF675
CHLORPYRIFOS see CMA100
CHLORPYRIFOS-METHYL see CMA250
CHLOR-PZ see CKP250
CHLORQUINALDOL see CLC500
CHLORQUINOX see CMA500
CHLORSAL see CLH750
CHLORSAURE (GERMAN) see SFS000
CHLORSEPTOL see CDP000
2-(4''-CHLOR-4'-STILBYL)NAFTOTRIAZOL-6,2',2''-TRISULFONAN SODNY
  (CZECH) see CLG000
N-CHLORSUCCINIMIDE see SND500
CHLORSUCCINYLCHOLIN (GERMAN) see HLC500
4-CHLOR-5-SULFAMOYL-2',6'-SALICYLOXYLIDID (GERMAN) see CLF325
CHLORSULFAQUINOXALINE see CMA600
CHLORSULFONAMIDO DIHYDROBENZOTHIADIAZINE DIOXIDE see CFY000
CHLORSULPHACIDE see CEP000
CHLORTALIDONE see CLY600
CHLORTEN see MIH275
CHLORTETRACYCLINE see CMA750
CHLORTETRACYCLINE HYDROCHLORIDE see CMB000
4-CHLORTETRAHYDROFTALANHYDRID (CZECH) see CFG500
CHLORTETRIN see DAI485
CHLORTHAL-DIMETHYL see TBV250
CHLORTHALIDON see CLY600
CHLORTHALIDONE see CLY600
CHLORTHAL-METHYL see TBV250
CHLORTHALONIL (GERMAN) see TBQ750
CHLORTHIAZIDE see CLH750
CHLORTHIEPIN see EAQ750
p-CHLORTHIOFENOL (CZECH) see CEK425
CHLORTHION METHYL see MIJ250
CHLORTHIOPHOS see CLJ875
CHLORTION (CZECH) see MIJ250
2-CHLOR-4-TOLUIDIN (CZECH) see CLK210
3-CHLOR-2-TOLUIDIN (CZECH) see CLK200
α-CHLORTOLUOL (GERMAN) see BEE375
CHLORTOLURON see CIS250
N'-(4-CHLOR-o-TOLYL)-N,N-DIMETHYLFORMAMIDIN (GERMAN) see CJJ250
CHLORTOX see CDR750
CHLORTRIANISEN see CLO750
CHLORTRIFLUORAETHYLEN (GERMAN) see CLQ750
CHLOR-TRIMETON see CLD250
CHLOR-TRIMETON see TAI500
CHLOR-TRIMETON MALEATE see TAI500
CHLOR-TRIPOLON see TAI500
CHLORTROPBENZYL see CMB125
CHLORURE d'ALUMINUM (FRENCH) see AGY750
CHLORURE ANTIMONIEUX see AQC500

CHLORURE d'ARSENIC (FRENCH) see ARF500
CHLORURE ARSENIEUX (FRENCH) see ARF500
CHLORURE de BENZENYLE (FRENCH) see BFL250
CHLORURE de 1-BENZYL-3-BENZYL-CARBOXY-PYRIDINIUM (FRENCH)
  see SAA000
CHLORURE de BENZYLE (FRENCH) see BEE375
CHLORURE de BENZYLIDENE (FRENCH) see BAY300
CHLORURE de BORE (FRENCH) see BMG500
CHLORURE de BUTYLE (FRENCH) see BQQ750
CHLORURE de CHLORACETYLE (FRENCH) see CEC250
CHLORURE de CHROMYLE (FRENCH) see CML125
CHLORURE de CYANOGENE (FRENCH) see COO750
CHLORURE de DICHLORACETYLE (FRENCH) see DEN400
CHLORURE de l'ETHYLAL TRIMETHYLAMMONIUM PROPANEDIOL
  (FRENCH) see MJH800
CHLORURE d'ETHYLE (FRENCH) see EHH000
CHLORURE d'ETHYLENE (FRENCH) see EIY600
CHLORURE d'ETHYLIDENE (FRENCH) see DFF809
CHLORURE de FUMARYLE (FRENCH) see FOY000
CHLORURE de LITHIUM (FRENCH) see LHB000
CHLORURE de MAGNESIUM HYDRATE (FRENCH) see MAE500
CHLORURE MERCUREUX (FRENCH) see MCW000
CHLORURE MERCURIQUE (FRENCH) see MCY475
CHLORURE de METHALLYLE (FRENCH) see CIU750
CHLORURE de METHYLE (FRENCH) see MIF765
CHLORURE de METHYLENE (FRENCH) see MJP450
CHLORURE PERRIQUE see FAU000
CHLORURE de SUCCINILCOLINE (FRENCH) see HLC500
CHLORURE de VINYLE (FRENCH) see VNP000
CHLORURE de VINYLIDENE (FRENCH) see VPK000
CHLORURE de ZINC (FRENCH) see ZFA000
CHLORURIT see CLH750
CHLORVINPHOS see DGP900
CHLORWASSERSTOFF (GERMAN) see HHL000
CHLORXYLAM see DET600
CHLORYL see EHH000
CHLORYL ANESTHETIC see EHH000
CHLORYLEA see TIO750
CHLORYL HYPOFLUORITE see CMB250
CHLORYL PERCHLORATE see CMB500
CHLORYL RADICAL see CDW450
CHLORZIDE see CFY000
CHLORZOXAZONE see CDQ750
CHLOTAZOLE see CMB675
CHLOTHIXEN see TAF675
CHLOTRIDE see CLH750
CHLOTRIMAZOLE see MRX500
CHLZ see CLX000
CHNU-1 see NKJ050
CHOCOLA A see VSK600
CHOCOLATE BROWN FB see CMB750
CHOKE CHERRY see AQP890
CHOLAGON see DAL000
CHOLAIC ACID see TAH250
CHOLALIN see CME750
CHOLAN DH see DAL000
CHOLANORM see CDL325
CHOLANTHRENE see CMC000
CHOLAXINE see SKV200
CHOLECALCIFEROL see CMC750
CHOLEDYL see CMF500
CHOLEGYL see CMF500
CHOLEIC ACID see DAQ400
CHOLEPULVIS see TDE750
CHOLERA ENTERO-EXOTOXIN see CMC800
CHOLERA ENTEROTOXIN see CMC800
CHOLERAGEN see CMC800
CHOLEREBIC see DAQ400
CHOLESOLVIN see SDY500
5,7-CHOLESTADIEN-3-β-OL see DAK600
(3-β)CHOLESTA-5,7-DIEN-3-OL see DAK600
CHOLESTA-5,7-DIEN-3-β-OL ACETATE see DAK800
CHOLESTAN-3-α-OL see EBA100
3-β-CHOLESTANOL see DKW000
(3-β,5-β)-CHOLESTAN-3-OL see DKW000
CHOLEST-5-EN-3-β-OL see CMD750
5-CHOLESTEN-3-β-OL see CMD750
7-CHOLESTEN-3-β-OL see CMD000
5:6-CHOLESTEN-3-β-OL see CMD750
5-α-CHOLEST-7-EN-3-β-OL see CMD000
Δ⁷-CHOLESTENOL see CMD000
Δ⁵-CHOLESTEN-3-β-OL see CMD750
5-CHOLESTEN-3-β-OL 3-(p-(BIS(2-CHLOROETHYL)AMINO)PHENYL)ACE-
  TATE see CME250

CHOLEST-6-EN-3-β-OL-5-α-HYDROPEROXIDE see CMD250
Δ⁶-CHOLESTEN-3-β-OL-5-α-HYDROPEROXIDE see CMD250
Δ⁶-CHOLESTEN-3-β-OL-5-α-HYDROPEROXYD (GERMAN) see CMD250
CHOLESTENONE see CMD500
CHOLEST-5-EN-3-ONE see CMD500
5-CHOLESTEN-3-ONE see CMD500
Δ⁽⁵⁾-CHOLESTENONE see CMD500
CHOLESTERIN see CMD750
CHOLESTERIN (GERMAN) see CMD000
CHOLESTEROL see CMD750
Δ⁷-CHOLESTEROL see DAK600
Δ⁵,⁷-CHOLESTEROL see DAK600
CHOLESTEROL BASE H see CMD750
CHOLESTEROL-α-EPOXIDE see EBM000
CHOLESTEROL-5-α,6-EPOXIDE see EBM000
CHOLESTEROL-5-α-HYDROPEROXIDE see CMD250
CHOLESTEROL ISOHEPTYLATE see CME000
CHOLESTEROL-5-METHYL-1-HEXANOATE see CME000
CHOLESTEROL mixed with OROTIC ACID mixed with CHOLIC ACID (2:2:1)
  see OJV525
CHOLESTEROL OXIDE see EBM000
CHOLESTEROL-α-OXIDE see EBM000
CHOLESTERONE see CMD500
CHOLESTERYL ALCOHOL see CMD750
CHOLESTERYL-p-BIS(2-CHLOROETHYL)AMINO PHENYLACETATE
  see CME250
CHOLESTERYL-14-METHYLHEXADECANOATE see CCJ500
CHOLESTRIN see CMD750
CHOLESTROL see CMD750
CHOLESTYRAMINE see CME400
CHOLESTYRAMINE CHLORIDE see CME400
CHOLESTYRAMINE RESIN see CME400
CHOLEXAMIN see CME675
CHOLEXAMINE see CME675
CHOLIBIL see CNG835
CHOLIC ACID see CME750
CHOLIC ACID mixed with CHOLESTEROL mixed with OROTIC ACID (1:2:2)
  see OJV525
CHOLIC ACID, MONOSODIUM SALT see SFW000
CHOLIC ACID, SODIUM SALT see SFW000
CHOLIFLAVIN see DBX400
CHOLIMED see DAL000
CHOLINE see CMF000
CHOLINE ACETATE see CMF250
CHOLINE ACETATE (ESTER) see CMF250
CHOLINE CARBAMATE CHLORIDE see CBH250
CHOLINE CHLORHYDRATE see CMF750
CHOLINE CHLORIDE (FCC) see CMF750
CHOLINE CHLORIDE ACETATE see ABO000
CHOLINE, CHLORIDE CARBAMATE(ESTER) see CBH250
CHOLINE CHLORINE CARBAMATE see CBH250
CHOLINE CYTIDINE DIPHOSPHATE see CMF350
CHOLINE 5'-CYTIDINE DIPHOSPHATE see CMF350
CHOLINE DICHLORIDE see CMF400
CHOLINE HYDROCHLORIDE see CMF750
CHOLINE, HYDROXIDE, 5'-ESTER with CYTIDINE 5'-(TRIHYDROGEN PYRO-
  PHOSPHATE), inner salt see CMF350
CHOLINE, INNER SALT, METHYLPHOSPHONOFLUORIDATE see MKF250
CHOLINE, IODIDE, PROPIONATE see PMW750
CHOLINE IODIDE SUCCINATE (2:1) see BJI000
CHOLINE ION see CMF000
CHOLINE 1,5-NAPHTHALENEDISULFONATE (2:1), DILACTATE, DIACET-
  ATE see AAC875
CHOLINE SALICYLATE see CMG000
CHOLINE SALICYLATE B see CMG000
CHOLINE, SALICYLATE (SALT) see CMG000
CHOLINE SALICYLIC ACID SALT see CMG000
CHOLINE SUCCINATE (ester) see CMG250
CHOLINE SUCCINATE DICHLORIDE see HLC500
CHOLINE SUCCINATE (2581) (ESTER) see CMG250
CHOLINE THEOPHYLLINATE see CMF500
CHOLINE, with THEOPHYLLINE (1581) see CMF500
CHOLINE THEOPHYLLINE SALT see CMF500
CHOLINE-2,6-XYLYL ETHER BROMIDE see XSS900
CHOLINIUM CHLORIDE see CMF750
CHOLINOPHYLLINE see CMF500
CHOLIT-URSAN see DMJ200
CHOLLY see CNE750
CHOLOGON see DAL000
CHOLOGRAF-N-METHYLGLUCAMINE see BGB315
CHOLOLIN see DAL000
CHOLOREBIC see DAQ400
CHOLOVUE see IFP800
CHOLOXIN see SKJ300

CHOLSAEURE (GERMAN) see CME750
CHOLUMBRIN see TDE750
CHOLYLTAURINE see TAH250
CHONDROITIN POLYSULFATE SODIUM see SFW300
CHONDRON see SFW300
CHONDRUS see CCL250
CHONDRUS EXTRACT see CCL250
CHONGRASS see PJJ315
CHOPSUI POTATO see YAG000
CHORAFURONE see AHK750
CHORIGON see CMG675
CHORIGONADOTROPIN see CMG675
CHORIGONIN see CMG675
CHORIONIC GONADOTROPHIN see CMG675
CHORIONIC GONADOTROPIC HORMONE see CMG675
CHORIONIC GONADOTROPIN see CMG675
C. HORRIDUS HORRIDUS VENOM see TGB150
CHORULON see CMG675
CHORYLEN see TIO750
CHOT see PBC250
CHOU PUANT (CANADA) see SDZ450
CHP see PNN300
CHP-PHENOBARBITALAT (GERMAN) see CPO500
CHRISTMAS BERRY TREE see PCB300
CHRISTMAS CANDLE see SDZ475
CHRISTMAS FLOWER see EQX000
CHRISTMAS ROSE see CMG700
CHROMACID FAST RED 3B see CMG750
CHROMALUM HEXAHYDRATE see CMG800
2-CHROMANONE see HHR500
CHROMAR see XHS000
CHROMARGYRE see MCV000
CHROMATE(1-), DIAMMINETETRAKIS(ISOTHIOCYANATO)-, AMMONIUM,
    HYDRATE see ANT300
CHROMATE OF POTASSIUM see PLB250
CHROMATE de PLOMB (FRENCH) see LCR000
CHROMATE of SODA see DXC200
CHROMATEX ORANGE R see DVB800
CHROMATEX RED J see MMP100
CHROME see CMI750
CHROME ALUM see CMG850, PLB500
CHROME ALUM (DODECAHYDRATE) see CMG850
CHROME FAST BLUE 2R see HJF500
CHROME FAST ORANGE RW see NEY000
CHROME FERROALLOY see FBD000
CHROME GREEN see CMJ900, LCR000
CHROME LEATHER BLACK EM see AQP000
CHROME LEATHER BLUE 2B see CMO000
CHROME LEATHER BLUE 3B see CMO250
CHROME LEATHER BROWN BRLL see CMO750
CHROME LEATHER SKY BLUE see CMN750
CHROME LEMON see LCR000
CHROME OCHER see CMJ900
CHROME ORANGE see LCS000
CHROME ORANGE MR see NEY000
CHROME ORANGE R see NEY000
CHROME ORANGE RLE see NEY000
CHROME ORE see CMI500
CHROME OXIDE see CMJ900
CHROME OXIDE GREEN see CMJ900
CHROME POTASH ALUM see PLB500
CHROME (TRIOXYDE de) (FRENCH) see CMK000
CHROME VERMILION see MRC000
CHROME YELLOW see LCR000
CHROME YELLOW 3RN see NEY000
CHROMIA see CMJ900
CHROMIC ACETATE see CMH000
CHROMIC ACETATE(III) see CMH000
CHROMIC ACETYLACETONATE see TNN250
CHROMIC ACID see CMH250, CMJ900, CMK000
CHROMIC(VI) ACID see CMH250, CMK000
CHROMIC ACID (mixture) see CMH500
CHROMIC ACID, solid (DOT) see CMK000
CHROMIC ACID (solution) see CMH750
CHROMIC ACID, solution (DOT) see CMK000
CHROMIC ACID, BARIUM SALT (1:1) see BAK250
CHROMIC ACID, BIS(TRIPHENYLSILYL) ESTER see BLS750
CHROMIC ACID, CALCIUM SALT (1:1) see CAP500
CHROMIC ACID, CALCIUM SALT (1:1), DIHYDRATE see CAP750
CHROMIC ACID, CHROMIUM(3+) SALT (3:2) see CMI250
CHROMIC ACID, COPPER-ZINC-COMPLEX see CNQ750
CHROMIC ACID, DI-tert-BUTYL ESTER see BQV000
CHROMIC ACID, DILITHIUM SALT see LHD000
CHROMIC ACID, DIPOTASSIUM SALT see PKX250

CHROMIC ACID, DISODIUM SALT see SGI000
CHROMIC ACID, DISODIUM SALT, DECAHYDRATE see SFW500
CHROMIC ACID GREEN see CMJ900
CHROMIC ACID, LEAD and MOLYBDENUM SALT see LDM000
CHROMIC ACID, LEAD(2+) SALT (1:1) see LCR000
CHROMIC ACID LEAD SALT with LEAD MOLYBDATE see LDM000
CHROMIC ACID, MERCURY ZINC COMPLEX see ZJA000
CHROMIC ACID MIXTURE, DRY (DOT) see CMH500
CHROMIC ACID, POTASSIUM ZINC SALT (2:2:1) see PLW500
CHROMIC ACID, STRONTIUM SALT (1:1) see SMH000
CHROMIC ACID, ZINC SALT see ZFJ100
CHROMIC ACID, ZINC SALT (1:2) see CMK500
CHROMIC AMMONIUM SULFATE see ANF625
CHROMIC ANHYDRIDE (DOT) see CMK000
CHROMIC CHLORIDE see CMJ250
CHROMIC CHLORIDE HEXAHYDRATE see CMK450
CHROMIC CHROMATE see CMI250
CHROMIC(III) HYDROXIDE see CMH750
CHROMIC NITRATE see CMJ600
CHROMIC OXIDE see CMJ900
CHROMIC OXYCHLORIDE see CML125
CHROMIC PHOSPHATE see CMK300
CHROMIC POTASSIUM SULFATE see PLB500
CHROMIC POTASSIUM SULPHATE see PLB500
CHROMIC TRIOXIDE (DOT) see CMK000
CHROMIS ACID ($H_2Cr_2O_7$), DISODIUM SALT, DIHYDRATE (9CI) see SGI500
CHROMITE see CMI500
CHROMITE (mineral) see CMI500
CHROMITE ORE see CMI500
CHROMIUM see CMI750
CHROMIUM ACETATE see CMH000
CHROMIUM(2+) ACETATE see CMJ000
CHROMIUM(II) ACETATE see CMJ000
CHROMIUM(III) ACETATE see CMH000
CHROMIUM ACETATE HYDRATE see CMJ000
CHROMIUM ACETYLACETONATE see TNN250
CHROMIUM(3+) ACETYLACETONATE see TNN250
CHROMIUM(III) ACETYLACETONATE see TNN250
CHROMIUM ALLOY, BASE, Cr,C,Fe,N,Si (FERROCHROMIUM) see FBD000
CHROMIUM ALLOY, Cr,C,Fe,N,Si see FBD000
CHROMIUM, BIS(BENZENE)-(8CI) see BGY700
CHROMIUM, BIS(eta^6)-BENZENE)-(9CI) see BGY700
CHROMIUM(1+), BIS(BENZENE)-, IODIDE (8CI) see BGY720
CHROMIUM(1+), BIS(eta^6-BENZENE)-, IODIDE (9CI) see BGY720
CHROMIUM, BIS(BENZENE)IODO- see BGY720
CHROMIUM CARBONATE see CMJ100
CHROMIUM CARBONYL (MAK) see HCB000
CHROMIUM CARBONYL (OC-6-11) (9CI) see HCB000
CHROMIUM CHLORIDE see CMJ250
CHROMIUM(III) CHLORIDE (1:3) see CMJ250
CHROMIUM CHLORIDE, anhydrous see CMJ250
CHROMIUM(III) CHLORIDE, HEXAHYDRATE (1:3:6) see CMK450
CHROMIUM CHLORIDE, HEXAHYDRATE (8CI,9CI) see CMK450
CHROMIUM CHLORIDE, HEXAUREA see HEZ800
CHROMIUM CHLORIDE OXIDE see CML125
CHROMIUM CHROMATE (MAK) see CMI250
CHROMIUM-COBALT ALLOY see CNA750
CHROMIUM-COBALT-MOLYBDENUM ALLOY see VSK000
CHROMIUM COMPOUNDS see CMJ500
CHROMIUM DIACETATE see CMJ000
CHROMIUM DICHLORIDE DIOXIDE see CML125
CHROMIUM DIOXIDE DICHLORIDE see CML125
CHROMIUM(VI) DIOXYCHLORIDE see CML125
CHROMIUM(II), DIPHENYL- see BGY700
CHROMIUM (III), DIPHENYL-, IODIDE see BGY720
CHROMIUM DISODIUM OXIDE see DXC200
CHROMIUM HEXACARBONYL see HCB000
CHROMIUM(3+), HEXAKIS(UREA-O)-, TRICHLORIDE, (OC-6-11)-(9CI)
    see HEZ800
CHROMIUM(3+), HEXAKIS(UREA)-, TRICHLORIDE (8CI) see HEZ800
CHROMIUM(III) HEXA-UREA CHLORIDE see HEZ800
CHROMIUM HYDROXIDE SULFATE see NBW000
CHROMIUM LEAD OXIDE see LCS000
CHROMIUM LITHIUM OXIDE see LHD000
CHROMIUM MONOPHOSPHATE see CMK300
CHROMIUM NITRATE see CMJ600
CHROMIUM (3+) NITRATE see CMJ600
CHROMIUM NITRATE (DOT) see CMJ600
CHROMIUM(III) NITRATE see CMJ600
CHROMIUM NITRIDE see CMJ850
CHROMIUM ORTHOPHOSPHATE see CMK300
CHROMIUM OXIDE see CMJ900, CMK000
CHROMIUM(3+) OXIDE see CMJ900
CHROMIUM(VI) OXIDE see CMK000

CHROMIUM(III) OXIDE see CMJ900
CHROMIUM(VI) OXIDE (1583) see CMK000
CHROMIUM(III) OXIDE (2583) see CMJ900
CHROMIUM OXIDE, aerosols see CMJ910
CHROMIUM OXIDE, NICKEL OXIDE, and IRON OXIDE FUME see IHE000
CHROMIUM OXYCHLORIDE see CML125
CHROMIUM PENTAFLUORIDE see CMK275
CHROMIUM PHOSPHATE see CMK300
CHROMIUM POTASSIUM SULFATE (1:1:2) see PLB500
CHROMIUM POTASSIUM SULPHATE see PLB500
CHROMIUM POTASSIUM ZINC OXIDE see CMK400
CHROMIUM SESQUICHLORIDE see CMK450
CHROMIUM SESQUIOXIDE see CMJ900
CHROMIUM SODIUM OXIDE see DXC200, SGI000
CHROMIUM SULFATE see NBW000
CHROMIUM SULFATE, BASIC see NBW000
CHROMIUM(III) SULFATE, HEXAHYDRATE (2:3:6) see CMG800
CHROMIUM SULFATE, PENTADECAHYDRATE see CMK425
CHROMIUM SULPHATE see NBW000
CHROMIUM TRIACETATE see CMH000
CHROMIUM TRIACETYLACETONATE see TNN250
CHROMIUM TRICHLORIDE see CMJ250
CHROMIUM TRICHLORIDE HEXAHYDRATE see CMK450
CHROMIUM TRINITRATE see CMJ600
CHROMIUM TRIOXIDE see CMK000
CHROMIUM(3+) TRIOXIDE see CMJ900
CHROMIUM(6+) TRIOXIDE see CMK000
CHROMIUM TRIOXIDE, anhydrous (DOT) see CMK000
CHROMIUM TRIS(ACETYLACETONATE) see TNN250
CHROMIUM TRIS(BENZOYLACETONATE) see TNN500
CHROMIUM TRIS(2,4-PENTANEDIONATE) see TNN250
CHROMIUM YELLOW see LCR000
CHROMIUM ZINC OXIDE see ZFJ100
CHROMIUM(6+)ZINC OXIDE HYDRATE (1582586581) see CMK500
CHROMOFLAVINE see DBX400
CHROMOFLAVINE see XAK000
CHROMOL ORANGE R. EXTRA see NEY000
CHROMOMYCIN see OIS000
CHROMOMYCIN A3 see CMK650
CHROMOMYCIN SODIUM see CMK750
CHROMOMYSIN A₃ see CMK650
CHROMONAR HYDROCHLORIDE see CBR500
CHROMOSMON see BJI250
CHROMOSORB T see TAI250
CHROMOTRICHIA FACTOR see AIH600
CHROMO (TRIOSSIDO di) (ITALIAN) see CMK000
CHROMOUS ACETATE see CMJ000
CHROMOUS ACETATE MONOHYDRATE see CMJ000
CHROMOXYCHLORID (GERMAN) see CML125
CHROMSAEUREANHYDRID (GERMAN) see CMK000
CHROMTRIOXID (GERMAN) see CMK000
CHROMYL AZIDE CHLORIDE see CML000
CHROMYLCHLORID (GERMAN) see CML125
CHROMYL CHLORIDE see CML125
[(CHROMYLDIOXY)IODO]BENZENE see PFJ775
CHROMYL NITRATE see CML325
CHROMYL PERCHLORATE see CML500
CHRONICIN FOAM see CDP725
CHRONOGYN see DAB830
CHROOMOXYLCHLORIDE (DUTCH) see CML125
CHROOMTRIOXYDE (DUTCH) see CMK000
CHROOMZUURANHYDRIDE (DUTCH) see CMK000
CHRYSANTHEMUM CINERAREAEFOLIUM see POO250
CHRYSANTHEMUMDICARBOXYLIC ACID MONOMETHYL ESTER
    PYRETHROLONE ESTER see POO100
(+)-trans-CHRYSANTHEMUMIC ACID ESTER of (+−)-ALLETHROLONE
    see BGC750
CHRYSANTHEMUM MONOCARBOXYLIC ACID PYRETHROLONE ESTER
    see POO050
CHRYSAROBIN see CML750
CHRYSAZIN see DMH400
6-CHRYSENAMINE see CML800
CHRYSENE see CML810
CHRYSENE-5,6-EPOXIDE see CML815
CHRYSENE-K-REGION EPOXIDE see CML815
CHRYSENE-5,6-OXIDE see CML815
CHRYSENEX see CML800
α-CHRYSIDINE see BAW750
CHRYSOIDIN see PEK000
CHRYSOIDIN A see DBP999
CHRYSOIDINE see PEK000
CHRYSOIDINE(II) see PEK000
CHRYSOIDINE A see PEK000
CHRYSOIDINE B see PEK000

CHRYSOIDINE C CRYSTALS see PEK000
CHRYSOIDINE G see PEK000
CHRYSOIDINE GN see PEK000
CHRYSOIDINE HR see PEK000
CHRYSOIDINE J see PEK000
CHRYSOIDINE M see PEK000
CHRYSOIDINE ORANGE see PEK000
CHRYSOIDINE PRL see PEK000
CHRYSOIDINE PRR see PEK000
CHRYSOIDINE SL see PEK000
CHRYSOIDINE SPECIAL (biological stain and indicator) see PEK000
CHRYSOIDINE SS see PEK000
CHRYSOIDINE Y see PEK000
CHRYSOIDINE Y BASE NEW see PEK000
CHRYSOIDINE Y CRYSTALS see PEK000
CHRYSOIDINE Y EX see PEK000
CHRYSOIDINE YGH see PEK000
CHRYSOIDINE YL see PEK000
CHRYSOIDINE YN see PEK000
CHRYSOIDINE Y SPECIAL see PEK000
CHRYSOIDIN FB see PEK000
CHRYSOIDIN Y see PEK000
CHRYSOIDIN YN see PEK000
CHRYSOMYKINE see CMA750
CHRYSON see BEP500
CHRYSONEX see CML800
CHRYSOPHANIC ACID ANTHRANOL see CML750
CHRYSOTILE (DOT) see ARM268
CHRYSOTILE ASBESTOS see ARM268
CHRYSRON see BEP500
CHRYTEMIN see DLH630
CHRYZOIDYNA F.B. (POLISH) see PEK000
CHS see CPF750
CHUANGHSINMYCIN see CML820
CHUANGHSINMYCIN SODIUM see CML822
CHUANGXIMYCIN SODIUM see CML822
CHUANGXINMYCIN see CML820
CHUANLIANSU see CML825
CHWASTOKS see SIL500
CHWASTOX see CIR250
CHWASTOX see SIL500
CHYMAR see CML850
CHYMEX see CML835
CHYMOTEST see CML850
α-CHYMOTRYPSIN see CML850
CHYMOTRYPSIN A see CML850
CHYMOTRYPSIN B see CML850
CI-2 see MAV750
C.I. 27 see HGC000
C.I. 79 see FMU070
C.I. 184 see FAG020
C.I. 185 see FMU080
C.I. 258 see SBC500
CI-337 see ASA500
CI 366 see ENG500
CI1395 see AOO500
CI-406 see PAN100
C.I. 440 see TKH750
C.I. 456 see CIP500
CI-473 see XQS000
CI-505 see BQM309
C.I. 515 see GLS700
C.I. 556 see SNY500
CI 581 see CKD750
CI-588 see BGO500
CI 624 see MGD200
CI-628 see NHP500
C.I. 633 see CGB250
C.I. 671 see FMU059
CI-683 see POL475
CI-705 see QAK000
CI-719 see GCK300
CI 720 see TDO260
C.I. 749 see FAG070
C.I. 766 see FEW000
CI 881 see BKB300
CI-914 see CML890
C.I. 925 see AJP250
C.I. 1956 see CKN000
C.I. 7581 see FAE100
C.I. 10305 see PID000
C.I. 10355 see DVX800
C.I. 10385 see SGP500
C.I. 11000 see PEI000

C.I. 11020 see DOT300
C.I. 11021 see OHI875
C.I. 11025 see DPO200
C.I. 11050 see DHM500
C.I. 11160 see AIC250
C.I. 11270 see DBP999
C.I. 11270 see PEK000
C.I. 11285 see NBG500
C.I. 11380 see FAG130
C.I. 11390 see FAG135
C.I. 11855 see AAQ250
C.I. 11860 see OHK000
C.I. 12055 see PEJ500
C.I. 12075 see DVB800
C.I. 12085 see CJD500
C.I. 12100 see TGW000
C.I. 12120 see MMP100
C.I. 12140 see XRA000
C.I. 12150 see CMS242
C.I. 12156 see DOK200
C.I. 12355 see NAY000
C.I. 13020 see CCE500
C.I. 13025 see MND600
C.I. 13065 see MDM775
C.I. 13390 see ADE750
C.I. 14030 see NEY000
C.I. 14600 see FAG010
C.I. 14700 see FAG050
C.I. 14720 see HJF500
C.I. 15670 see HLI000
C.I. 15985 see FAG150
C.I. 16035 see FAG100
C.I. 16105 see CMG750
C.I. 16150 see FMU070
C.I. 16155 see FAG018
C.I. 16185 see FAG020
C.I. 16255 see FMU080
C.I. 19140 see FAG140
C.I. 20285 see CMP500
C.I. 21090 see DEU000
C.I. 22120 see SGQ500
C.I. 22195 see ADG000
C.I. 22570 see CMP000
C.I. 22610 see CMO000
C.I. 23060 see DEQ600
C.I. 23685 see ASM100
C.I. 23850 see CMO250
C.I. 23860 see BGT250
C.I. 24110 see DCJ200
C.I. 24400 see CMO500
C.I. 24410 see CMN750
C.I. 26050 see EOJ500
C.I. 28440 see BMA000
C.I. 30145 see CMO750
C.I. 30235 see AQP000
C.I. 35570 see AKH000
C.I. 37010 see DEO400
C.I. 37020 see DBR400
C.I. 37025 see NEO000
C.I. 37030 see NEN500
C.I. 37035 see NEO500
C.I. 37077 see TGQ750
C.I. 37085 see CLK235
C.I. 37105 see NMP500
C.I. 37107 see TGR000
C.I. 37115 see AOX250
C.I. 37125 see NEQ000
C.I. 37130 see NEQ500
C.I. 37200 see MIH500
C.I. 37210 see MPY750
C.I. 37225 see BBX000
C.I. 37230 see TGJ750
C.I. 37240 see PFU500
C.I. 37270 see NBE500
C.I. 37275 see AIA750
C.I. 37500 see NAX000
C.I. 38480 see AJU500
C.I. 40645 see TGE155
C.I. 41000 see IBA000
C.I. 42000 see AFG500
C.I. 42040 see BAY750
C.I. 42053 see FAG000
C.I. 42080 see ADE675
C.I. 42090 see FAE000, FMU059

C.I. 42095 see FAF000
C.I. 42500 see RMK020
C.I. 42535 see MQN025
C.I. 42581 see VQU500
C.I. 42640 see FAG120
C.I. 44040 see VKA600
C.I. 44050 see PFT250
C.I. 44090 see ADF000
C.I. 45005 see PPQ750
C.I. 45010 see DIS200
C.I. 45160 see RGW000
C.I. 45330 see FEV000
C.I. 45380 see BNH500, BNK700
C.I. 45405 see CMM000
C.I. 45430 see FAG040
C.I. 46000 see XAK000
C.I. 46005 see BAQ250, BJF000
C.I. 47031 see PGW750
C.I. 50040 see AJQ250
C.I. 50411 see DCE800
C.I. 50435 see DCE800
C.I. 52040 see AJP250
C.I. 56205 see CMM750
C.I. 57000 see TKN250
C.I. 58000 see DMG800
C.I. 58205 see TKN750
C.I. 58900 see MEB750
C.I. 59040 see TNM000
C.I. 59100 see DCZ000
C.I. 59815 (CZECH) see CMU750
C.I. 59820 see CMU750
C.I. 59825 see JAT000
C.I. 59830 see APK500
C.I. 60700 see AKP750
C.I. 60710 see AKE250
C.I. 61100 see DBP000
C.I. 61105 see AKP250
C.I. 61200 see BMM500
C.I. 61505 see MGG250
C.I. 61565 see BLK000
C.I. 62015 see DBX000
C.I. 63340 see TGS000
C.I. 64500 see TBG700
C.I. 65010 see IBJ000
C.I. 69020 see CMU800
C.I. 70300 see APK000
C.I. 71000 see CMU475
C.I. 71200 see CMU500
C.I. 73015 see FAE100
C.I. 75300 see COG000
C.I. 75410 see TKN750
C.I. 75440 see MQF250
C.I. 75490 see HLY500
C.I. 75500 see WAT000
C.I. 75670 see QCA000
C.I. 75720 see QCJ000
C.I. 75730 see RSU000
C.I. 76000 see AOQ000
C.I. 76005 see AHQ250
C.I. 76010 see PEY250
C.I. 76020 see ALL500
C.I. 76025 see PEY000
C.I. 76027 see CJY120
C.I. 76035 see TGL750
C.I. 76042 see TGM000
C.I. 76043 see DCE600
C.I. 76043 see TGM400
C.I. 76050 see DBO000
C.I. 76051 see DBO400
C.I. 76060 see PEY500
C.I. 76061 see PEY650
C.I. 76065 see CEG600
C.I. 76066 see CEG625
C.I. 76070 see ALL750
C.I. 76075 see DTL800
C.I. 76085 see PFU500
C.I. 76500 see CCP850
C.I. 76505 see REA000
C.I. 76515 see PPQ500
C.I. 76520 see ALT000
C.I. 76535 see ALO000
C.I. 76545 see ALT500
C.I. 76555 see NEM480
C.I. 76605 see NAW500

C.I. 76645 see NAO500
C.I. 77000 see AGX000
C.I. 77002 see AHC000
C.I. 77050 see AQB750
C.I. 77056 see AQC500
C.I. 77060 see AQL500
C.I. 77061 see AQF500
C.I. 77086 see ARI000
C.I. 77099 see BAJ250
C.I. 77103 see BAK250
C.I. 77120 see BAP000
C.I. 77180 see CAD000
C.I. 77185 see CAD250
C.I. 77199 see CAJ750
C.I. 77205 see CMS205
C.I. 77223 see CAP500, CAP750
C.I. 77231 see CAX750
C.I. 77266 see CBT500
C.I. 77288 see CMJ900
C.I. 77295 see CMJ250
C.I. 77320 see CNA250
C.I. 77322 see CND125
C.I. 77400 see CNI000
C.I. 77402 see CNO000
C.I. 77410 see COF500
C.I. 77450 see CNQ000
C.I. 77491 see IHD000
C.I. 77575 see LCF000
C.I. 77577 see LDN000
C.I. 77578 see LDS000
C.I. 77580 see LCX000
C.I. 77600 see LCR000
C.I. 77601 see LCS000
C.I. 77605 see MRC000
C.I. 77610 see LCU000
C.I. 77622 see LDU000
C.I. 77630 see LDY000
C.I. 77640 see LDZ000
C.I. 77713 see MAC650
C.I. 77718 see TAB750
C.I. 77726 see MAT250
C.I. 77727 see MAT500
C.I. 77728 see MAS000
C.I. 77755 see PLP000
C.I. 77760 see MCT500
C.I. 77764 see MCW000
C.I. 77775 see NCW500
C.I. 77777 see NDF500
C.I. 77779 see NCY500
C.I. 77795 see PJD500
C.I. 77805 see SBO500
C.I. 77820 see SDI500
C.I. 77847 see SMM000
C.I. 77864 see TGC000
C.I. 77891 see TGG760
C.I. 77901 see TOC750
C.I. 77938 see VDU000
C.I. 77940 see VEZ000
C.I. 77945 see ZBJ000
C.I. 77947 see ZKA000
C.I. 77955 see ZFJ100
C.I. 11160B see AIC250
C.I. 45370:1 see DDO200
C.I. 45380:2 see BMO250
C.I. 52 015 (CZECH) see BJI250
C.I. 61 570 (CZECH) see APL500
C.I. ACID BLUE 74 see FAE100
C.I. ACID BLUE 92 see ADE750
C.I. ACID BLUE 9, DIAMMONIUM SALT see FMU059
C.I. ACID BLUE 9, DISODIUM SALT see FAE000
C.I. ACID BLUE 1, SODIUM SALT see ADE500
C.I. ACID BLUE 92, TRISODIUM SALT see ADE750
C.I. ACID GREEN 1 see NAX500
C.I. ACID GREEN 5 see FAF000
C.I. ACID GREEN 40 see CMM400
C.I. ACID GREEN 5, DISODIUM SALT see FAF000
C.I. ACID GREEN 3, MONOSODIUM SALT see FAE950
C.I. ACID GREEN 50, MONOSODIUM SALT see ADF000
C.I. ACID ORANGE 3 see SGP500
C.I. ACID ORANGE 10 see HGC000
C.I. ACID ORANGE 20 see FAG010
C.I. ACID ORANGE 52 see MND600
C.I. ACID ORANGE 45, DISODIUM SALT see ADG000
C.I. ACID ORANGE 20, MONOSODIUM SALT see FAG010

C.I. ACID RED 2 see CCE500
C.I. ACID RED 18 see FMU080
C.I. ACID RED 26 see FMU070
C.I. ACID RED 27 see FAG020
C.I. ACID RED 51 see FAG040
C.I. ACID RED 87 see BNK700
C.I. ACID RED 92 see ADG250
C.I. ACID RED 98 see CMM000
C.I. ACID RED 14, DISODIUM SALT see HJF500
C.I. ACID RED 26, DISODIUM SALT see FMU070
C.I. ACID YELLOW 7 see CMM750
C.I. ACID YELLOW 36 see MDM775
C.I. ACID YELLOW 73 see FEW000
C.I. ACID YELLOW 36 MONOSODIUM SALT see MDM775
CIAFOS see COQ399
CIANATIL MALEATE see COS899
CIANAZIL see COH250
CIANIDANOL see CCP875
CIANURO di SODIO (ITALIAN) see SGA500
CIANURO di VINILE (ITALIAN) see ADX500
CIATYL see CLX250
C.I. AZOIC COUPLING COMPONENT 1 see NAX000
C.I. AZOIC COUPLING COMPONENT 107 see TGR000
C.I. AZOIC DIAZO COMPONENT 3 see DEO400
C.I. AZOIC DIAZO COMPONENT 6 see NEO000
C.I. AZOIC DIAZO COMPONENT 7 see NEN500
C.I. AZOIC DIAZO COMPONENT 11 see CLK235
C.I. AZOIC DIAZO COMPONENT 12 see NMP500
C.I. AZOIC DIAZO COMPONENT 13 see NEQ500
C.I. AZOIC DIAZO COMPONENT 21 see MIH500
C.I. AZOIC DIAZO COMPONENT 37 see BIN500, NEO500
C.I. AZOIC DIAZO COMPONENT 48 see DCJ200
C.I. AZOIC DIAZO COMPONENT 112 see BBX000
C.I. AZOIC DIAZO COMPONENT 113 see TGJ750
C.I. AZOIC DIAZO COMPONENT 114 see NBE000
C.I. AZOIC RED 83 see MGO750
C.I. 41000B see IBB000
C.I. 42555B see TJK000
CIBA 34 see MAW850
CIBA 570 see FAB400
CIBA 709 see DGQ875
CIBA 1983 see CJQ000
CIBA 2059 see DUK800
CIBA 2446 see AHO750
CIBA-3126 see PAM785
CIBA 5968 see HGP495, HGP500
CIBA 6313 see CES750
CIBA 8353 see DVS000
CIBA 8514 see CJJ250
CIBA 8514 see KEA000
CIBA 9295 see MKU000
CIBA 9491 see IEN000
CIBA 11925 see PCY300
CIBA 32644 see NML000
CIBA 12669A see MIW500
CIBA 17309 BA see DAL300
CIBA 32644-BA see NML000
CIBA 34276 BA see MAW850
CIBA 34,647-Ba see BAC275
CIBA 36278-BA see SGB500
CIBA 39089-Ba see THK750
CIBA 42155-BA see AGO750
CIBA 42244-BA see AGP000
CIBA C-768 see DTP800
CIBA C-776 see DTP600
CIBA C-2307 see DOL800
CIBA C 7019 see ASG250
CIBA C-7824 see MPG250
CIBA C-9491 see IEN000
CIBACET BRILLIANT BLUE BG NEW see MGG250
CIBACETE DIAZO NAVY BLUE 2B see DCJ200
CIBACET SAPPHIRE BLUE G see TBG700
CIBACET VIOLET 2R see DBP000
CIBACET YELLOW GBA see AAQ250
CIBA CO. 2825 see PIT250
CIBACRON BLACK B-D see CMS227
CIBACTHEN see AES650
CIBA-GEIGY C-9491 see IEN000
CIBA-GEIGY C-10015 see DRP600
CIBA-GEIGY GS 13005 see DSO000
CIBA-GEIGY GS 19851 see IOS000
CIBA GO.4350 see CMV375
CIBA 9333 GO see AOA050
CIBA 34276 HYDROCHLORIDE see MAW850

CIBANONE GOLDEN YELLOW see DCZ000
CIBA 2696GO see BLP325
C.I. BASIC BLUE 9 see BJI250
C.I. BASIC BLUE 17 see AJP250
C.I. BASIC GREEN 4 see AFG500
C.I. BASIC GREEN 1, SULFATE (1:1) see BAY750
C.I. BASIC ORANGE 2 see PEK000
C.I. BASIC ORANGE 3 see PEK000
C.I. BASIC ORANGE 14 see BAQ250, BJF000
C.I. BASIC ORANGE 2, MONOHYDROCHLORIDE see PEK000
C.I. BASIC RED 5 see AJQ250
C.I. BASIC RED 1, MONOHYDROCHLORIDE see RGW000
C.I. BASIC RED 5, MONOHYDROCHLORIDE see AJQ250
C.I. BASIC RED 9, MONOHYDROCHLORIDE see RMK020
C.I. BASIC VIOLET 1 see MQN025
C.I. BASIC VIOLET 10 see FAG070
C.I. BASIC YELLOW 2 see IBA000
C.I. BASIC YELLOW 2, FREE BASE see IBB000
C.I. BASIC YELLOW 2, MONOHYDROCHLORIDE see IBA000
CIBA THIOCRON see AHO750
CIC see BAR800
CICHORIUM INTYBUS, ETHANOL EXTRACT see CMM800
CI-628 CITRATE see NHP500
CICLACILLIN see AJJ875
CICLACILLUM see AJJ875
CICLIZINA see EAN600
CICLOBIOTIC see MDO250
CICLOESANO (ITALIAN) see CPB000
CICLOESANOLO (ITALIAN) see CPB750
CICLOESANONE (ITALIAN) see CPC000
6-CICLOESIL-2,4-DINITR-FENOLO (ITALIAN) see CPK500
1-CICLOESIL-3-p-TOLILSOLFONILUREA (ITALIAN) see CPR000
CICLONIUM IODIDE see OLW400
CICLOPIROX ETHANOLAMINE SALT (1:1) see BAR800
CICLOPIROXOLAMIN see BAR800
CICLOPIROXOLAMINE see BAR800
CICLORAL see BSM000
CICLOSERINA (ITALIAN) see CQH000
CICLOSOM see TIQ250
CICLOSPASMOL see DNU100
CICLOSPORIN see CQH100
CICP see CKC000
CICUTA BULBIFERA L. see WAT325
CICUTA DOUGLASII see WAT325
CICUTAIRE (CANADA) see WAT325
CICUTA MACULATA see WAT325
CICUTIN see PNT000
CICUTINE see PNT000
CICUTOXIN see CMN000
CIDAL see SAH000
CIDALON see IHZ000
CIDAMEX see AAI250
CIDANCHIN see CLD000
CIDANDOPA see DNA200
CIDEFERRON see CMN125
CIDEMUL see DRR400
C.I. DEVELOPER 4 see REA000
C.I. DEVELOPER 5 see NAX000
C.I. DEVELOPER 13 see PEY500
C.I. DEVELOPER 17 see NEO500
CIDEX see GFQ000
CIDIAL see DRR400
C.I. DIRECT BLACK 38 see AQP000
C.I. DIRECT BLUE 1 see CMN750
C.I. DIRECT BLUE 14 see CMO250
C.I. DIRECT BLUE 53 see BGT250
C.I. DIRECT BLUE 1, TETRASODIUM SALT see CMN750
C.I. DIRECT BLUE 6, TETRASODIUM SALT see CMO000
C.I. DIRECT BLUE 14, TETRASODIUM SALT see CMO250
C.I. DIRECT BLUE 15, TETRASODIUM SALT see CMO500
C.I. DIRECT BROWN see CMO750
C.I. DIRECT BROWN 78, DIAMMONIUM SALT see FMU059
C.I. DIRECT RED 28 see SGQ500
C.I. DIRECT RED 28, DISODIUM SALT see SGQ500
C.I. DIRECT VIOLET 28 see ASM100
C.I. DIRECT VIOLET 1, DISODIUM SALT see CMP000
C.I. DIRECT VIOLET 28, DISODIUM SALT see ASM100
C.I. 45350 DISODIUM SALT see FEW000
C.I. DISPERSE BLACK 3 see DPO200
C.I. DISPERSE BLACK 6 see DCJ200
C.I. DISPERSE BLACK-6-DIHYDROCHLORIDE see DOA800
C.I. DISPERSE BLUE 1 see TBG700
C.I. DISPERSE BLUE 3 see MGG250
C.I. DISPERSE ORANGE 11 see AKP750

C.I. DISPERSE RED 11 see DBX000
C.I. DISPERSE VIOLET 1 see DBP000
C.I. DISPERSE VIOLET 4 see AKP250
C.I. DISPERSE YELLOW 3 see AAQ250
C.I. DISPERSE YELLOW 13 see MEB750
CIDOCETINE see CDP250
CIDOXEPIN HYDROCHLORIDE see AEG750
CIDREX see CFY000
C.I. FLUORESCENT BRIGHTENER 46 see TGE155
C.I. FLUORESCENT BRIGHTENING AGENT 46, SODIUM SALT see TGE155
C.I. FOOD BLACK 1, TETRASODIUM SALT see BMA000
C.I. FOOD BLUE 1 see FAE100
C.I. FOOD BLUE 2 see FAE000, FMU059
C.I. FOOD BROWN 1 see CMP250
C.I. FOOD BROWN 2 see CMB750
C.I. FOOD BROWN 3, DISODIUM SALT see CMP500
C.I. FOOD GREEN 1 see FAE950
C.I. FOOD GREEN 2 see FAF000
C.I. FOOD GREEN 3 see FAG000
C.I. FOOD GREEN 4 see ADF000
C.I. FOOD ORANGE 4 see HGC000
C.I. FOOD RED 1 see FAG050
C.I. FOOD RED 3 see HJF500
C.I. FOOD RED 5 see FMU070
C.I. FOOD RED 6 see FAG018
C.I. FOOD RED 7 see FMU080
C.I. FOOD RED 9 see FAG020
C.I. FOOD RED 15 see FAG070
C.I. FOOD RED 1, DISODIUM SALT see FAG050
C.I. FOOD RED 6, DISODIUM SALT see FAG018
C.I. FOOD VIOLET 2 see FAG120
C.I. FOOD VIOLET 3 see VQU500
C.I. FOOD YELLOW 4 see FAG140
C.I. FOOD YELLOW 10 see FAG130
C.I. FOOD YELLOW 11 see FAG135
C.I. 45350 (FREE ACID) see FEV000
CIGARETTE REFINED TAR see CMP800
CIGARETTE SMOKE CONDENSATE see SEC000
CIGARETTE TAR see CMP800
CIGUE (CANADA) see PJJ300
CI 624 HYDROCHLORIDE see MGD210
CI-IPC see CKC000
CILAG 61 see HDY000
CILEFA BLACK B see BMA000
CILEFA PINK B see FAG040
CILEFA PONCEAU 4R see FMU080
CILEFA RUBINE 2B see FAG020
CILLA BLUE EXTRA see TBG700
CILLA FAST BLUE FFR see MGG250
CILLA FAST RED VIOLET RN see DBP000
CILLA ORANGE R see AKP750
CILLENTA see BFC750
CILLORAL see BDY669, BFD000
CILOPEN see BDY669
CILOSTAZOL see CMP825
CIM see CDU750
CIMAGEL see DXY000
CIMETIDINE see TAB250
CIMETIDINE mixed with SODIUM NITRITE (4581) see CMP875
CIMEXAN see MAK700
C.I. MORDANT ORANGE 1 see NEY000
C.I. MORDANT ORANGE 1, MONOSODIUM SALT see SIU000
C.I. MORDANT VIOLET 39, TRIAMMONIUM SALT (8CI) see AGW750
CI-583 NA see SIF425
CINAMINE see TLN500
CINAMONIN see CMP885
CINANSERIN HYDROCHLORIDE see CMP900
CINATABS see TLN500
C.I. NATURAL BROWN 3 see CCP800
C.I. NATURAL BROWN 7 see WAT000
C.I. NATURAL BROWN 8 see MAT500
C.I. NATURAL RED 1 see QCA000
C.I. NATURAL YELLOW 1 see CDH250
C.I. NATURAL YELLOW 8 see MRN500
C.I. NATURAL YELLOW 10 see QCA000
C.I. NATURAL YELLOW 11 see MQF250
C.I. NATURAL YELLOW 16 see HLY500
CINCAINE HYDROCHLORIDE see NOF500
CINCHOCAINE see DDT200
CINCHOCAINE HYDROCHLORIDE see NOF500
CINCHOCAINIUM CHLORIDE see NOF500
CINCHOLEPIDINE see LEL000
CINCHOMERONIC ACID IMIDE see POR000
CINCHONAN-9-OL, 6'-METHOXY-, DIHYDROCHLORIDE, (8-α-9R)-, mixt.

with 2-(ACETYLOXY) BENZOIC ACID and 2-HYDROXY-1,2,3-PRO-
PANETRICARBOXYLIC ACID, TRILITHIUM SALT see TGJ350
d-CINCHONINE see CMP925
CINCHOPHENE see PGG000
CINCHOPHENIC ACID see PGG000
CINCHOPHEN SODIUM see SJH000
CINCHOPHEN, SODIUM SALT see SJH000
CINCO NEGRITOS (MEXICO) see LAU600
CINCOPHEN see PGG000
CINDOMET see CMP950
CINEB see EIR000
CINENE see MCC250
1,8-CINEOL see CAL000
CINEOLE see CAL000
1,8-CINEOLE see CAL000
CINEPAZIDE MALEATE see VGK000
CINERIN I ALLYL HOMOLOG see AFR250
CINERIN 1 or 11 see POO250
CINERUBIN A see TAH675
CINERUBIN B see TAH650
CINERUBINE A see TAH675
CINERUBINE B see TAH650
CINMETACIN see CMP950
CINMETHACIN see CMP950
CINNAMAL see CMP969
CINNAMALDEHYDE see CMP969
CINNAMEIN see BEG750
CINNAMENE see SMQ000
CINNAMENOL see SMQ000
CINNAMIC ACID see CMP975
trans-CINNAMIC ACID BENZYL ESTER see BEG750
CINNAMIC ACID-n-BUTYL ESTER see BQV500
CINNAMIC ACID-3-(DIETHYLAMINO) PROPYL ESTER see AQP750
CINNAMIC ACID-1,5-DIMETHYL-1-VINYL-4-HEXENYL ESTER see LGA000
CINNAMIC ACID-1,5-DIMETHYL-1-VINYL-4-HEXEN-1-YL ESTER see LGA000
CINNAMIC ACID, p-(1H-IMIDAZOL-1-YLMETHYL)-, SODIUM SALT, (E)-
  see SHV100
CINNAMIC ACID, ISOBUTYL ESTER see IIQ000
CINNAMIC ACID, ISOPROPYL ESTER see IOO000
CINNAMIC ACID, LINALYL ESTER see LGA000
CINNAMIC ACID, NICKEL(II) SALT see CMQ000
CINNAMIC ACID, PROPYL ESTER see PNH250
CINNAMIC ACID, SODIUM SALT see SFX000
CINNAMIC ALCOHOL see CMQ740
CINNAMIN see AQN750
CINNAMOHYDROXAMIC ACID see CMQ475
CINNAMON BARK OIL see CCO750
CINNAMON BARK OIL, CEYLON TYPE (FCC) see CCO750
CINNAMONIN see CCI500
CINNAMONITRILE see CMQ500
CINNAMON OIL see CCO750
CINNAMOPHENONE see CDH000
CINNAMOYLHYDROXAMIC ACID see CMQ475
1-CINNAMOYL-2-METHOXY-5-METHOXY-3-INDOLYLACETIC ACID
  see CMP950
1-CINNAMOYL-5-METHOXY-2-METHYLINDOLE-3-ACETIC ACID see CMP950
14-CINNAMOYLOXYCODEINONE see CMQ625
CINNAMYCIN see CMQ725
CINNAMYL ACETATE see CMQ730
CINNAMYL ALCOHOL see CMQ740
CINNAMYL ALCOHOL ANTHRANILATE see API750
CINNAMYL ALCOHOL, BENZOATE see CMQ750
CINNAMYL ALCOHOL, FORMATE see CMR500
CINNAMYL ALCOHOL, SYNTHETIC see CMQ740
CINNAMYL ALDEHYDE see CMP969
CINNAMYL-2-AMINOBENZOATE see API750
CINNAMYL-o-AMINOBENZOATE see API750
CINNAMYL ANTHRANILATE (FCC) see API750
CINNAMYL BENZOATE see CMQ750
CINNAMYL BUTYRATE see CMQ800
1-CINNAMYL-4-(DIPHENYLMETHYL)PIPERAZINE see CMR100
trans-1-CINNAMYL-(4-DIPHENYLMETHYL)PIPERAZINE see CMR100
d-CINNAMYLEPHEDRINE HYDROCHLORIDE see CMR250
CINNAMYLEPHEDRINE HYDROCHLORIDE, DEXTRO see CMR250
CINNAMYL FORMATE see CMR500
CINNAMYL ISOBUTYRATE see CMR750
CINNAMYL ISOVALERATE see CMR800
CINNAMYL METHANOATE see CMR500
CINNAMYL NITRILE see CMQ500
CINNAMYL PROPIONATE see CMR850
CINNARIZIN see ARQ750
CINNARIZINE see CMR100
CINNARIZINE CLOFIBRATE see CMS125
CINNIMIC ALDEHYDE see CMP969

CINNOPROPAZONE see AQN750
C.I. No. 77278 see CMJ900
C.I. No. 46005:1 see BJF000
CINOBAC see CMS200
CINOPAL see BGL250
CINOPOP see BGL250
CINOXACIN see CMS200
CIN-QUIN see QFS000, QHA000
CINU see CGV250
CINX see CMS200
CIODRIN see COD000
CIODRIN VINYL PHOSPHATE see COD000
CIOVAP see COD000
C.I. OXIDATION BASE see TGL750
C.I. OXIDATION BASE 4 see TGM400
C.I. OXIDATION BASE 7 see ALT500
C.I. OXIDATION BASE 10 see PEY500
C.I. OXIDATION BASE 12 see DBO000
C.I. OXIDATION BASE 16 see PEY250
C.I. OXIDATION BASE 17 see ALT000
C.I. OXIDATION BASE 19 see AHQ250
C.I. OXIDATION BASE 21 see DUP400
C.I. OXIDATION BASE 22 see ALL750
C.I. OXIDATION BASE 26 see CCP850
C.I. OXIDATION BASE 31 see REA000
C.I. OXIDATION BASE 32 see PPQ500
C.I. OXIDATION BASE 33 see NAW500
C.I. OXIDATION BASE 6A see ALT250
C.I. OXIDATION BASE 10A see PEY650
C.I. OXIDATION BASE 12A see DBO400
C.I. OXIDATION BASE 13A see CEG625
CIPC see CKC000
C.I. PIGMENT BLACK 13 see CND125
C.I. PIGMENT BLACK 14 see MAS000
C.I. PIGMENT BLACK 16 see ZBJ000
C.I. PIGMENT BLUE 34 see CNQ000
C.I. PIGMENT BROWN 8 see MAS000
C.I. PIGMENT GREEN 17 see CMJ900
C.I. PIGMENT GREEN 21 (9CI) see COF500
C.I. PIGMENT METAL 2 see CNI000
C.I. PIGMENT METAL 4 see LCF000
C.I. PIGMENT METAL 6 see ZBJ000
C.I. PIGMENT ORANGE 5 see DVB800
C.I. PIGMENT ORANGE 20 see CAJ750
C.I. PIGMENT ORANGE 21 see LCS000
C.I. PIGMENT RED see CHP500
C.I. PIGMENT RED see LCS000
C.I. PIGMENT RED 3 see MMP100
C.I. PIGMENT RED 4 see CJD500
C.I. PIGMENT RED 23 see NAY000
C.I. PIGMENT RED 101 see IHD000
C.I. PIGMENT RED 104 see LDM000
C.I. PIGMENT RED 104 see MRC000
C.I. PIGMENT RED 105 see LDS000
C.I. PIGMENT WHITE 3 see LDY000
C.I. PIGMENT WHITE 4 see ZKA000
C.I. PIGMENT WHITE 6 see TGG760
C.I. PIGMENT WHITE 10 see BAJ250
C.I. PIGMENT WHITE 11 see AQF000
C.I. PIGMENT WHITE 21 see BAP000
C.I. PIGMENT WHITE 25 see CAX750
C.I. PIGMENT YELLOW 12 see DEU000
C.I. PIGMENT YELLOW 31 see BAK250
C.I. PIGMENT YELLOW 32 see SMH000
C.I. PIGMENT YELLOW 33 see CAP500, CAP750
C.I. PIGMENT YELLOW 34 see LCR000
C.I. PIGMENT YELLOW 35 see CMS205
C.I. PIGMENT YELLOW 36 see ZFJ100
C.I. PIGMENT YELLOW 37 see CAJ750
C.I. PIGMENT YELLOW 46 see LDN000
C.I. PIGMENT YELLOW 48 see LCU000
CIPLAMYCETIN see CDP250
CIPRIL (PUERTO RICO) see SLJ650
CIPROFIBRATE see CMS210
CIPROMID see CQJ250
CIRAM see BJK500
CIRANTIN see HBU000
CIRCAIN see DNC000
CIRCAIR see DNC000
CIRCANOL see DLL400
CIRCOSOLV see TIO750
CIRCULIN see CMS225
C.I. REACTIVE BLACK 8 see CMS227
C.I. REACTIVE BLUE 19 see BMM500

C.I. REACTIVE BLUE 19, DISODIUM SALT see BMM500
C.I. REACTIVE RED 2 see PMF540
C.I. REACTIVE YELLOW 14 see RCZ000
C.I. REACTIVE YELLOW 73 see CMS230
CIRENE BRILLIANT BLUE R see ADE750
CIROLEMYCIN see CMS232
CIRPONYL see MQU750
CIRRASOL 185A see NMY000
CIRRASOL-OD see HCQ500
CISAPRIDE see CMS237
CISCLOMIPHENE see CMX500
CISMETHRIN see RDZ875
CISOBITAN see CMS241
C.I. SOLVENT BLUE 7 see PEI000
C.I. SOLVENT BLUE 18 see TBG700
C.I. SOLVENT GREEN 3 see BLK000
C.I. SOLVENT GREEN 7 see TNM000
C.I. SOLVENT ORANGE 2 see TGW000
C.I. SOLVENT ORANGE 3 see PEK000
C.I. SOLVENT ORANGE 7 see XRA000
C.I. SOLVENT ORANGE 15 see BJF000
C.I. SOLVENT RED see CMS242
C.I. SOLVENT RED 19 see EOJ500
C.I. SOLVENT RED 23 see OHA000
C.I. SOLVENT RED 24 see SBC500
C.I. SOLVENT RED 41 see MAC500
C.I. SOLVENT RED 43 see BMO250
C.I. SOLVENT RED 72 see DDO200
C.I. SOLVENT RED 80 see DOK200
C.I. SOLVENT VIOLET 9 see TJK000
C.I. SOLVENT VIOLET 11 see DBP000
C.I. SOLVENT VIOLET 12 see AKP250
C.I. SOLVENT VIOLET 13 see HOK000
C.I. SOLVENT VIOLET 26 see DBX000
C.I. SOLVENT YELLOW 1 see PEI000
C.I. SOLVENT YELLOW 2 see DOT300
C.I. SOLVENT YELLOW 3 see AIC250
C.I. SOLVENT YELLOW 5 see FAG130
C.I. SOLVENT YELLOW 7 see HJF000
C.I. SOLVENT YELLOW 12 see OHK000
C.I. SOLVENT YELLOW 14 see PEJ500
C.I. SOLVENT YELLOW 34 see IBB000
C.I. SOLVENT YELLOW 56 see OHI875
C.I. SOLVENT YELLOW 94 see FEV000
C.I. SOLVENT YELLOW 1, MONOHYDROCHLORIDE see PEI250
C.I. SOLVENT YELLOW 3 MONOHYDROCHLORIDE see MPY750
CISPLATINO (SPANISH) see PJD000
CISPLATYL see PJD000
CISTANCHE TUBULOSA Wight (extract) see CMS248
CISTAPHOS see AKB500
CISTEAMINA (ITALIAN) see AJT250
CITANEST see CMS250
CITANEST HYDROCHLORIDE see CMS250
CITARIN see TDX750
CITARIN L see LFA020
CITEXAL see QAK000
CITGRENILE see MNB500
CITHROL PO see PJY100
CITICHOLINE see CMF350
CITICOLINE see CMF350
CITIDIN DIFOSFATO de COLINA see CMF350
CITIDOLINE see CMF350
CITIFLUS see ARQ750
CITILAT see AEC750
CITIOLASE see TCZ000
CITIOLONE see TCZ000
CITIREUMA see SOU550
CITOBARYUM see BAP000
CITOCOR see DJS200
CITODON see ERD500
CITOFUR see FMB000
CITOL see ALT250
CITOMULGAN M see OAV000
CITOPAN see ERD500
CITOSARIN see AJJ875
CITOSULFAN see BOT250
CITOX see DAD200
CITRACETAL see CMS324
CITRACONIC ACID see CMS320
CITRACONIC ACID ANHYDRIDE see CMS322
CITRACONIC ANHYDRIDE see CMS322
CITRA-FORT see ABG750
CITRAL (FCC) see DTC800
CITRAL DIMETHYL ACETAL see DOE000

CITRAL ETHYLENE GLYCOL ACETAL see CMS324
CITRAL METHYLANTHRANILATE, SCHIFF'S BASE see CMS325
CITRAM see AMX825, DJA400
CITRAMON see ARP250
CITRATE de ACETOXY-THYMOXY-ETHYL-DIMETHYLAMINE (FRENCH) see MRN600
CITRAZINIC ACID see DMV400
CITRAZON see BCP000
CITREOVIRIDIN see CMS500
CITREOVIRIDINE see CMS500
CITRETTEN see CMS750
CITRIC ACID see CMS750
CITRIC ACID, anhydrous see CMS750
CITRIC ACID, ACETYL TRIETHYL ESTER see ADD750
CITRIC ACID, AMMONIUM SALT see ANF800
CITRIC ACID, COPPER(2+) SALT (8CI) see CNK625
CITRIC ACID, DYSPROSIUM(3+) salt (1:1) see DYG800
CITRIC ACID, GALLIUM SALT (1:1) see GBO000
CITRIC ACID, MONOSODIUM SALT see MRL000
CITRIC ACID, SAMARIUM SALT see SAS000
CITRIC ACID, SODIUM SALT see MRL000
CITRIC ACID, TRILITHIUM SALT see TKU500
CITRIC ACID, TRIPOTASSIUM SALT see PLB750
CITRIC ACID, YTTRIUM SALT (3:1) see YFA000
CITRIC ACID, ZINC SALT (2:3) see ZFJ250
CITRIDIC ACID see ADH000
CITRININ see CMS775
CITRO see CMS750
CITROFLEX 2 see TJP750
CITROFLUYL see MRL000
CITRONELLAL see CMS845
CITRONELLAL HYDRATE see CMS850
CITRONELLA OIL see CMT000
CITRONELLENE see CMT050
CITRONELLIC ACID see CMT125
CITRONELLOL see CMT250
α-CITRONELLOL see DTF400
α-CITRONELLYL ACETATE see RHA000
CITRONELLYL ACETATE (FCC) see AAU000
CITRONELLYL-2-BUTENOATE see CMT500
CITRONELLYL BUTYRATE see CMT600
CITRONELLYL-α-CROTONATE see CMT500
CITRONELLYL ETHYL ETHER see EES100
CITRONELLYL FORMATE see CMT750
CITRONELLYL ISOBUTYRATE see CMT900
CITRONELLYL NITRILE see CMU000
CITRONELLYL PHENYLACETATE see CMU050
CITRONELLYL PROPIONATE see CMU100
CITRON YELLOW see PLW500
CITRON YELLOW see ZFJ100
CITROSODINE see TNL000
CITROVIOL see DTC000
CITRULLAMON see DKQ000, DNU000
l-CITRULLINE and SODIUM NITRITE (2:1) see SIS100
CITRUS AURANTIUM see LBE000
CITRUS HYSTRIX DC., fruit peel extract see CMU300
CITRUS RED No. 2 see DOK200
CITTERAL see SEQ000
C.I. VAT BLACK 8 see CMU475
C.I. VAT BLACK 28 see IBJ000
C.I. VAT BLUE 16 see CMU500
C.I. VAT BLUE 22 see CMU750
C.I. VAT BROWN 25 see CMU800
C.I. VAT GREEN 2 see APK500
C.I. VAT YELLOW see DCZ000
CIVETONE see CMU850
cis-CIVETONE see CMU850
CIZARON see TMP750
CL 337 see ASA500
CL 369 see CKD750
CL-395 see PDC890
CL-845 see PJA170
CL-911C see DVP400
CL 912C see LFG100
CL 10304 see AJD000
CL 11366 see PGJ000
CL 12503 see POP000
CL 12,625 see PIF750
CL 12880 see DSP400
CL 13494 see AKO500
CL 13,900 see AEI000
CL-14377 see MDV500
CL 16,536 see POK300
CL 18133 see EPC500

CL 24055 see DUI000
CL 27,319 see MFD500
CL 34433 see AQY375
CL-34699 see COW825
CL-38023 see FAB600
CL 40881 see EDW875
CL-43,064 see DXN600
CL 47300 see DSQ000
CL-47,470 see DHH400
CL 52160 see TAL250
CL 54998 see BMM600
CL 59806 see MQW250
CL-62362 see DCS200
CL 64475 see DHH200
CL 65336 see AJV500
CL 67772 see AOA095
CL 69049 see PMF550
CL-71563 see DCS200
CL82204 see BGL250
CL 88236 see AJI550
CL 216942 see BGW325
CL 227193 see SJJ200
CL 232315 see MQY100
CL 67310465 see PBT100
CL 1950675526 see TES000
CL 19217 4090L 7-5525 see EGI000
Cl-ADO see CEF100
CLAFEN see CQC500, CQC650
CLAIRFORMIN see CMV000
CLAIRSIT see PCF300
CLAM POISON DIHYDROCHLORIDE see SBA500
CLANDILON see DNU100
CLANICLOR see CMU875
CLAODICAL see CCK125
CLAPHENE see CQC650
CLARIPEX see ARQ750
CLARK 1 see CGN000
CLARO 5591 see SLJ500
CLAUDELITE see ARI750
CLAUDETITE see ARI750
CLAVACIN see CMV000
CLAVELLINA (PUERTO RICO) see CAK325
CLAVULANIC ACID SODIUM SALT see CMV250
3',4'-Cl2-DAB see DFD400
CLEARASIL BENZOYL PEROXIDE LOTION see BDS000
CLEARASIL BP ACNE TREATMENT see BDS000
CLEARJEL see SLJ500
CLEARJREL see SLJ500
CLEARTUF see PKF750
CLEARY 3336 see DJV000
CLEBOPRIDE HYDROGEN MALATE see CMV325
CLEBOPRIDE MALATE see CMV325
CLEISTANTHIN A see CMV375
CLELAND'S REAGENT see DXO775, DXO800
CLEMANIL see FOS100
CLEMASTINE FUMARATE see FOS100
CLEMASTINE HYDROGEN FUMARATE see FOS100
CLEMATIS see CMV390
CLEMATITE AUX GEAUX (CANADA) see CMV390
CLEMIZOLE HDYROCHLORIDE see CMV400
CLENBUTEROL HYDROCHLORIDE see VHA350
CLENIL-A see AFJ625
CLEOCIN see CMV675
CLEOCIN PHOSPHATE see CMV690
CLEP see CMV475
C. LEPTOSEPALA see MBU550
CLERA see NCW000
CLERIDIUM 150 see PCP250
CLESTOL see DJL000
CLEVE'S ACID-1,6 see ALI250
CLEVE'S BETA-ACID see ALI250
CLHORAMEISENSAEUREAETHYLESTER (GERMAN) see EHK500
CLIACIL see PDT750
CLIDANAC see CFH825, CMV500
CLIFT see MCI750
CLIMATERINE see DKA600
CLIMBING BITTERSWEET see AHJ875
CLIMBING LILY see GEW800
CLIMBING ORANGE ROOT see AHJ875
CLIMESTRONE see ECU750
CLIN see SJO000
CLINDAMYCIN see CMV675
CLINDAMYCINE (FRENCH) see CMV675
CLINDAMYCIN-2-PALMITATE MONOHYDROCHLORIDE see CMV680

CLINDAMYCIN PHOSPHATE see CMV690
CLINDAMYCIN-2-PHOSPHATE see CMV690
CLINDROL 101CG see BKE500
CLINDROL LT 15-73-1 see BKF500
CLINDROL SDG see HKJ000
CLINDROL SEG see EJM500
CLINDROL SUPERAMIDE 100L see BKE500
CLINESTROL see DKB000
CLINOFIBRATE see CMV700
CLINOPTILOLITE see CMV850
CLINORIL see SOU550
CLINOXAN see CFG750
CLIOQUINOL see CHR500
CLIOXANIDE see CGB250
CLIQUINOL see CHR500
CLIRADON HYDROCHLORIDE see KFK000
CLISTIN see CGD500
CLISTINE MALEATE see CGD500
CLISTIN MALEATE see CGD500
CLIVIA (VARIOUS SPECIES) see KAJ100
CLIVORINE see CMV950
CLIXODYNE see HIM000
CLM see SFY500
4-Cl-M-PD see CJY120
CLOAZEPAM see CMW000, CMW000
CLOBAZAM see CIR750
CLOBBER see CQJ250
CLOBENZEPAM HYDROCHLORIDE see CMW250
CLOBERAT see ARQ750
CLOBETASOL PROPIONATE see CMW300
CLOBETASOL-17-PROPIONATE see CMW300
CLOBETASONE BUTYRATE see CMW400
CLOBETASONE-17-BUTYRATE see CMW400
CLOBRAT see ARQ750
CLOBREN-SF see ARQ750
CLOBUTINOL see CMW459
CLOBUTINOL HYDROCHLORIDE see CMW500
CLOCAPRAMINE DIHYDROCHLORIDE see DKV309
CLOCARPRAMINE DIHYDROCHLORIDE see DKV309
CLOCETE see AAE500
CLOCONAZOLE HYDROCHLORIDE see CMW550
CLODAZONE see CMW600
CLOFAR see ARQ750
CLOFEDANOL HYDROCHLORIDE see CMW700
CLOFENOTANE see DAD200
CLOFENOXIN see DPE000
CLOFEXAMIDE see CJN750
CLOFEXAMIDE PHENYLBUTAZONE see CMW750
CLOFEXAMIDE-PHENYLBUTAZONE MIXTURE see CMW750
CLOFEZON see CMW750
CLOFEZONE see CMW750
CLOFIBRAM see ARQ750
CLOFIBRAT see ARQ750
CLOFIBRATO (SPANISH) see ARQ750
CLOFIBRATO de CINARIZINA (SPANISH) see CMS125
CLOFIBRIC ACID see CMX000
CLOFIBRINIC ACID see CMX000
CLOFIBRINSAEURE (GERMAN) see CMX000
CLOFINIT see ARQ750
CLOFIPRONT see ARQ750
CLOFLUPEROL HYDROCHLORIDE see CLS250
CLOMEPHENE B see CMX500
CLOMETHIAZOLE see CHD750
CLOMETHIAZOLUM see CHD750
CLOMID see CMX700
CLOMIDAZOLE see CDY325
CLOMIFEN CITRATE see CMX700
CLOMIFENE see CMX500
trans-CLOMIFENE CITRATE see CMX750
CLOMIFENO see CMX700
CLOMEPHENE see CMX500
CLOMIPHENE CITRATE see CMX700
trans-CLOMIPHENE CITRATE see CMX750
racemic-CLOMIPHENE CITRATE see CMX700
CLOMIPHENE DIHYDROGEN CITRATE see CMX700
CLOMIPHENE-R see CMX700
CLOMIPHINE see CMX700
CLOMIPRAMINE see CDU750
CLOMIPRAMINE HYDROCHLORIDE see CDV000
CLOMIVID see CMX700
CLOMPHID see CMX700
CLONAZEPAM see CMW000
CLONIDIN see DGB500
CLONIDINE see DGB500

CLONIDINE HYDROCHLORIDE see CMX760
CLONITARLID see DFV600
CLONITRALID see DFV400
CLONIXIC ACID see CMX770
CLONIXIN see CMX770
CLONIXINE see CMX770
CLONT see MMN250
CLOPANE HYDROCHLORIDE see CPV609
CLOPENTHIXOL DIHYDROCHLORIDE see CLX250
CLOPEPRAMINE HYDROCHLORIDE see IFZ900
CLOPERASTINA CLORIDRATO (ITALIAN) see CMX820
CLOPERASTINE see CMX800
CLOPERASTINE HYDROCHLORIDE see CMX820
CLOPERIDONE HYDROCHLORIDE see CMX840
CLOPHEDIANOL HYDROCHLORIDE see CMW700
CLOPHEN see PJL750
CLOPHEN A-30 see CMX845
CLOPHEN A60 see PJN250
CLOPHENOXATE see DPE000
CLOPIDOL see CMX850
CLOPIPAZAN MESYLATE see CMX860
CLOPIXOL see CLX250
CLOPOXIDE see LFK000
CLOPRADONE see EQO000
CLOPROMAZINA (ITALIAN) see CKP250
CLOPROSTENOL see CMX880
CLOPYRALID see DGJ100
CLOQUINOZINE TARTRATE see CMX920
CLORAMIDINA see CDP250
CLORAMIN see BIE250
CLORARSEN see DFX400
CLORAZEPATE DIPOTASSIUM see CDQ250
CLOR CHEM T-590 see CDV100
CLORDAN (ITALIAN) see CDR750
CLORDIAZEPOSSIDO (ITALIAN) see LFK000
CLORDION see CBF250
CLOREPIN see CIR750
CLORESTROLO see CLO750
CLOREX see DFJ050
CLORGYLINE HYDROCHLORIDE see CMY000
CLORHIDRATO de CELIPROLOL (SPANISH) see SBN475
CLORIDRATO DI-2-BENZIL-4,5-IMIDAZOLINA (ITALIAN) see BBW750
CLORILAX see CKF500
CLORINA see CDP000
CLORINDANOL see CDV700
CLORMETAZANONE see CKF500
CLORMETHAZON see CKF500
CLORNAPHAZINE see BIF250
CLORO (ITALIAN) see CDV750
CLOROAMFENICOLO (ITALIAN) see CDP250
4-CLORO-3-AMINOBENZOATO di DIMETILAMINOETILE CLORIDRATO (ITALIAN) see AJE350
CLOROBEN see DEP600
2-CLOROBENZALDEIDE (ITALIAN) see CEI500
CLOROBENZENE (ITALIAN) see CEJ125
(4-CLORO-BENZIL)-(4-CLORO-FENIL)-SOLFURO (ITALIAN) see CEP000
1-p-CLORO-BENZOIL-5-METOXI-2-METILINDOL-3-ACIDO ACETICO (SPANISH) see IDA000
2-CLORO-1,3-BUTADIENE (ITALIAN) see NCI500
(4-CLORO-BUT-2-IN-IL)-N-(3-CLORO-FENIL)-CARBAMMATO (ITALIAN) see CEW500
CLOROCHINA see CLD000
O-2-CLORO-1-(2,4-DICLORO-FENIL)-VINYL-O,O-DIETILFOSFATO (ITALIAN) see CDS750
(2-CLORO-3-DIETILAMINO-1-METIL-3-OXO-PROP-1-EN-IL)-DIMETIL-FOSFATO see FAB400
CLORODIFENILI, CLORO 42% (ITALIAN) see PJM500
CLORODIFENILI, CLORO 54% (ITALIAN) see PJN000
p-CLORO-α-(2-DIMETILAMINO)-1-METILETIL)α-METIL FENETIL ALCOOL (ITALIAN) see CMW459
p-CLORO-α-(2-DIMETILAMINO)-1-METILETIL)-α-METIL FENETIL ALCOOL CLORIDRATO (ITALIAN) see CMW500
2-CLORO-4-DIMETILAMINO-6-METIL-PIRIMIDINA (ITALIAN) see CCP500
2-CLORO-10-(3-DIMETILAMINOPROPIL)FENOTIAZINA (ITALIAN) see CKP250
1-CLORO-2,4-DINITROBENZENE (ITALIAN) see CGM000
1-CLORO-2,3-EPOSSIPROPANO (ITALIAN) see EAZ500
CLOROETANO (ITALIAN) see EHH000
2-CLOROETANOLO (ITALIAN) see EIU800
(CLORO-2-ETIL)-1-CICLOESIL-3-NITROSOUREA (ITALIAN) see CGV250
(CLORO-2-ETIL)-1-(RIBOPIRANOSILTRIACETATO-2,3',4')-3-NITROSOUREA (ITALIAN) see ROF200
1-(2-(p-CLORO-α-FENILBENZILOSSI)ETIL)PIPERIDINA CLORIDRATO (ITALIAN) see CMX820
(4-CLORO-FENIL)-BENZOL-SOLFONATO (ITALIAN) see CJR500

(4-CLORO-FENIL)-4-CLORO-VENZOL-SOLFONATO (ITALIAN) see CJT750
3-(4-CLORO-FENIL)-1,1-DIMETIL-UREA (ITALIAN) see CJX750
2(2-(4-CLORO-FENIL-2-FENIL)-ACETIL)INDAN-1,3-DIONE (ITALIAN) see CJJ000
α-(p-CLOROFENIL)-α-FENIL-2-PIPERIDILMETANOLO CLORIDRATO (ITALIAN) see CKI175
N-(3-CLORO-FENIL)-ISOPROPIL-CARBAMMATO (ITALIAN) see CKC000
CLOROFORMIO (ITALIAN) see CHJ500
CLOROFOS (RUSSIAN) see TIQ250
CLOROMETANO (ITALIAN) see MIF765
7-CLORO-2-METILAMINO-5-FENIL-3H-1,4-BENZOIDIAZEPINA 4-OSSIDO (ITALIAN) see LFK000
7-CLORO-3-METIL-2H-1,2,4-BENZOTIODIAZINE-1,1-DIOSSIDO (ITALIAN) see DCQ250
3-CLORO-2-METIL-PROP-1-ENE (ITALIAN) see CIU750
CLOROMISAN see CDP250
1-CLORO-4-NITROBENZENE (ITALIAN) see NFS525
O-(4-CLORO-3-NITRO-FENIL)-O,O-DIMETIL-MONOIIOFOSFATO (ITALIAN) see NFT000
O-(3-CLORO-4-NITRO-FENIL)-O,O-DIMETIL-MONOTIOFOSFATO (ITALIAN) see MIJ250
CLOROPHENE see CJU250
CLOROPICRINA (ITALIAN) see CKN500
CLOROPIRIL see TAI500
CLOROPRENE (ITALIAN) see NCI500
CLOROSAN see CDP000
CLOROSINTEX see CDP250
CLOROTEPINE see ODY100
CLOROTETRACICLINA CLORIDRATO (ITALIAN) see CMB000
CLOROTRISIN see CLO750
CLOROX see SHU500
CLORPROPAMIDE (ITALIAN) see CKK000
CLORTERMINE HYDROCHLORIDE see DRC600
CLORTOKEM see CIS250
CLORTRAN see ABD000
CLORURO DI ETILE (ITALIAN) see EHH000
CLORURO di ETHENE (ITALIAN) see EIY600
CLORURO di ETILIDENE (ITALIAN) see DFF809
CLORURO di MERCURIO (ITALIAN) see MCY475
CLORURO MERCUROSO (ITALIAN) see MCW000
CLORURO di METALLILE (ITALIAN) see CIU750
CLORURO di METILE (ITALIAN) see MIF765
CLORURO di SUCCINILCOLINA (ITALIAN) see HLC500
CLORURO di VINILE (ITALIAN) see VNP000
CLOSANTEL see CFC100
CLOTAM see CLK325
CLOTEPIN see ODY100
CLOTHEPIN see ODY100
CLOTIAZEPAM see CKA000
CLOTRIDE see CLH750
CLOTRIMAZOL see MRX500
CLOUT see DXE600
CLOVE BUD OIL see CMY075
CLOVE LEAF OIL see CMY100
CLOVE LEAF OIL MADAGASCAR see CMY100
CLOWN TREACLE see WBS850
CLOXACILLIN SODIUM MONOHYDRATE see SLJ000
CLOXACILLIN SODIUM SALT see SLJ050
CLOXAPEN see SLJ000, SLJ050
CLOXAZEPINE see DCS200
CLOXAZOLAM see CMY125
CLOXAZOLAZEPAM see CMY125
CLOXYPEN see SLJ000
CLOXYPENDYL see CMY135
CLOZAPIN see CMY250
CLOZAPINE see CMY250
CIP see CKV500
4-Cl-o-PD see CFK125
2-Cl-P-PD see CEG625
CL 13,850 SODIUM see DRU875
CL 5343 SODIUM SALT see AMR750
CL 251931 SODIUM SALT see CCS635
CLUDR see CFJ750
CLUSIA ROSEA see BAE325
CLY-503 see SDY500
CLYSAR see PMP500
CM 6912 see EKF600
CMA see CBF250, CIF250
4-CMB see CIH825
CMC see SFO500
S-CMC see CCH125
CMC 7H see SFO500
CM-CELLULOSE Na SALT see SFO500
CMC SODIUM SALT see SFO500

CMDP see MQR750
CME see CBD750
C-METON see TAI500
C. MEXICANA see CAK325
CMH see MAE500
C. MITIS see FBW100
CMME see CIO250
CMMP see SKQ400
CMPABN see MEI000
CMPF see MIT600
CMPP see CIR500
CM S 2957 see CLJ875
CMU see CJX750
CMW BONE CEMENT see PKB500
CMZ SODIUM see CCS360
CN see CEA750
CN 009 see CDB760
CN 447 see DEJ000
CN 3123 see SNL850
CN 8676 see EEI000
CN-15,757 see ASA500
CN-27,554 see TKH750
CN-35355 see XQS000
CN-36337 see CIP500
CN 38703 see QAK000
CN 59,567 see CGB250
CN-25,253-2 see AOO500
CN-52,372-2 see CKD750
CN-55945-27 see NHP500
CNA see RDP300
CNCC see BIF625
C. NOCTURNUM see DAC500
CNP see NIW500
CNP 1032 see NIW500
CNU see CHE750
CNUEMS see CHF250
CNU-ETHANOL see CHB750
CO 12 see DXV600
CO-1214 see DXV600
CO-1670 see HCP000
CO-1895 see OAX000
CO-1897 see OAX000
COAGULASE see CMY325
COAKUM see PJJ315
COAL ASH see CMY650
COAL CONVERSION MATERIALS, SRC-II HEAVY DISTILLATE see CMY625
COAL DUST see CMY635
COAL DUST, EXTRACT, NITROSATED see NJH750
COAL FACINGS see CMY635
COAL FLY ASH see CMY650
COAL GAS see HHJ500
COAL, GROUND BITUMINOUS (DOT) see CMY635
COAL LIQUID see PCR250
COAL-MILLED see CMY635
COAL NAPHTHA see BBL250
COAL OIL see KEK000, PCR250
COAL OIL (export shipment only) (DOT) see KEK000
COAL SLAG-MILLED see CMY635
COAL TAR see CMY800
COAL TAR, AEROSOL see CMY805
COAL TAR CREOSOTE see CMY825
COAL TAR NAPHTHA see NAI500
COAL TAR OIL see CMY825
COAL TAR OIL (DOT) see CMY825
COAL TAR PITCH VOLATILES see CMZ100
COAPT see MIQ075
COATHYLENE PF 0548 see PMP500
COBADEX see CNS750
COBADOCE FORTE see VSZ000
COBALIN see VSZ000
COBALT see CNA250
COBALT-59 see CNA250
COBALT ACETATE see CNC000
COBALT(2+) ACETATE see CNC000
COBALT(II) ACETATE see CNC000
COBALT ACETATE TETRAHYDRATE see CNA500
COBALT ALLOY, Co,Cr see CNA750
COBALT(III) AMIDE see CNB000
COBALTATE(3-), HEXAKIS(NITRITO-N)-, TRISODIUM (OC-6-11)- (9CI) see SFX750
COBALTATE(3-), HEXANITRO-, TRISODIUM see SFX750
COBALT(II) AZIDE see CNB099
COBALT, BIS(N-9H-FLUOREN-2-YL-N-HYDROXYACETAMIDATO-O,O')-(9CI) see FDU875

COBALT BLACK see CND125
COBALT(II) BROMIDE see CNB250
COBALT CARBONYL see CNB500
COBALT(II) CHLORIDE see CNB599
COBALT(III) CHLORIDE see CNB750
COBALT(2+) CHLORIDE HEXAHYDRATE see CNB800
COBALT(II) CHLORIDE HEXAHYDRATE see CNB800
COBALT CHLORIDE, HEXAHYDRATE (8CI, 9CI) see CNB800
COBALT-CHROMIUM ALLOY see CNA750
COBALT-CHROMIUM-MOLYBDENUM ALLOY see VSK000
COBALT COMPOUNDS see CNB850
COBALT DIACETATE see CNC000
COBALT DIACETATE TETRAHYDRATE see CNA500
COBALT DICHLORIDE see CNB599
COBALT DICHLORIDE HEXAHYDRATE see CNB800
COBALT DIFLUORIDE see CNC100
COBALT DINITRATE see CNC500
COBALT(2)-EDATHAMIL see DGQ400
COBALT N-FLUOREN-2-YLACETOHYDROXAMATE see FDU875
COBALT(II) FLUORIDE see CNC100
COBALT-HISTIDINE see BJY000
COBALT HYDROCARBONYL see CNC230
COBALT-METHYLCOBALAMIN see VSZ050
COBALT MOLYBDATE see CNC250
COBALT(2+) MOLYBDATE see CNC250
COBALT MOLYBDENUM OXIDE see CNC250
COBALT MONOOXIDE see CND125
COBALT MONOSULFIDE see CNE200
COBALT MONOXIDE see CND125
COBALT MURIATE see CNB599
COBALT NAPHTHENATE, POWDER (DOT) see NAR500
COBALT(II) NITRATE see CNC500
COBALT(II) NITRIDE see CNC750
COBALT NITROPRUSSIDE see CND000
COBALT NITROSOPENTACYANOFERRATE(3) see CND000
COBALT NITROSYLPENTACYANOFERRATE see PAY610
COBALTOCENE see BIR529
COBALT OCTACARBONYL see CNB500
COBALTOUS ACETATE TETRAHYDRATE see CNA500
COBALTOUS CHLORIDE see CNB599
COBALTOUS CHLORIDE, HEXAHYDRATE see CNB800
COBALTOUS DIACETATE see CNC000
COBALTOUS DICHLORIDE see CNB599
COBALTOUS FLUORIDE see CNC100
COBALTOUS MOLYBDATE see CNC250
COBALTOUS NITRATE see CNC500
COBALTOUS OXIDE see CND125
COBALTOUS SULFATE see CNE125
COBALTOUS SULFIDE see CNE200
COBALT OXIDE see CND125
COBALT(2+) OXIDE see CND125
COBALT(II) OXIDE see CND125
COBALT(III) OXIDE see CND825
COBALT RESINATE, precipitated see CNE000
COBALT SALTS see FDU875
COBALT SULFATE see CNE125
COBALT SULFATE (1:1) see CNE125
COBALT (2+) SULFATE see CNE125
COBALT(II) SULFATE (1581) see CNE125
COBALT SULFIDE see CNE200
COBALT(II) SULFIDE see CNE200
COBALT SULFIDE (amorphous) see CNE200
COBALT(II) SULPHATE see CNE125
COBALT TETRACARBONYL see CNB500
COBALT TETRACARBONYL DIMER see CNB500
COBALT TRIFLUORIDE see CNE250
COBAMIN see VSZ000
COBAN see MRE230
COBEN see CNE375
COBEN P see PIC100
COBEX see CNE500
COBEX (polymer) see PKQ059
COBEXO see CNE500
COBH see BDD000
COBINAMIDE, COBALT-METHYL derivative, HYDROXIDE, DIHYDROGEN PHOSPHATE (ester), inner salt, 3'-ESTER with 5,6-DIMETHYL-1-α-D-RIBOFURANOSYLBENZIMIDAZOLE see VSZ050
COBIONE see VSZ000
COBOX see CNK559
COBRATEC #99 see BDH250
COBRATEC TT 100 see MHK000
COCAFURIN see NGE500
COCAIN-CHLORHYDRAT (GERMAN) see CNF000
COCAINE see CNE750

(−)-COCAINE see CNE750
l-COCAINE see CNE750
β-COCAINE see CNE750
COCAINE CHLORIDE see CNF000
COCAINE HYDROCHLORIDE see CNF000
(−)-COCAINE HYDROCHLORIDE see CNF000
l-COCAINE HYDROCHLORIDE see CNF000
COCAINE MURIATE see CNF000
CO CAP IMIPRAMINE 25 see DLH630
COCARTRIT see CLD000
C. OCCIDENTALIS see CCK675
COCCIDINE A see DUP300
COCCIDIOSTAT C see CMX850
COCCIDOT see DUP300
COCCINE see FMU080
COCCOCLASE see PPO000
COCCULIN see PIE500
COCCULUS see PIE500
COCCULUS solid (DOT) see PIE500
COCHENILLEROT A see FMU080
COCHIN see LEG000
COCHINEAL RED A see FMU080
COCHLIOBOLIN see CNF109
COCHLIOBOLIN A see CNF109
COCHLIODINOL see CNF159
COCOA FATTY ACIDS, POTASSIUM SALTS see CNF175
COCO-DIAZINE see PPP500
COCO DIETHANOLAMIDE see BKE500
COCONUT ALDEHYDE see CNF250
COCONUT BUTTER see CNR000
COCONUT DIMETHYL AMINE OXIDE see CNF325
COCONUT MEAL PELLETS, containing 6-13% moisture and no more than 10% residual fat (DOT) see CNR000
COCONUT OIL (FCC) see CNR000
COCONUT OIL AMIDE of DIETHANOLAMINE see BKE500
COCONUT PALM OIL see CNR000
COCUM see PJJ315
COD see CPR825
CODAL see MQQ450
CODECARBOXYLASE see PII100
CODECHINE see BBQ500
CODE H 133 see DER800
CODEINE see CNF500
CODEINE HYDROCHLORIDE see CNF750
CODEINE METHOCHLORIDE see CNG000
CODEINE NICOTINATE (ESTER) see CNG250
CODEINE PHOSPHATE see CNG500
CODEINE PHOSPHATE SESQUIHYDRATE see CNG675
CODEINE SULFATE see CNG750
CODEINONE, DIHYDRO-, TARTRATE see DKX050
CODELCORTONE see PMA000
CODEMPIRAL see ABG750
CO-DERGOCRINE MESYLATE see DLL400
CODETHYLINE see ENK000
CODETHYLINE HYDROCHLORIDE see DVO700
CODHYDRINE see DKW800
CODIAZINE see PPP500
CODIBARBITA see EOK000
CODYLIN see TCY750
COENZYME Q₁₀ see UAH000
COENZYME R see VSU100
CO-ESTRO see ECU750
COFFEBERRY see MBU825
COFFEE see CNG775
COFFEE SENNA see CNG825
COFFEIN (GERMAN) see CAK500
COFFEINE see CAK500
CO-FRAM see SNL850
COGESIC see DTO200
COGILOR BLUE 512.12 see FAE000
COGILOR RED 321.10 see FAG070
COGOMYCIN see FPC000
COHASAL-1H see SEH000
COHOSH see BAF325
CO-HYDELTRA see PMA000
COHYDRIN see DKW800
COIR DEEP BLACK C see AQP000
COKAN see PJJ315
COKE see CNE750
COLACE see DJL000
COLACID BLUE A see ADE750
COLACID PONCEAU 4R see FMU080
COLACID PONCEAU SPECIAL see FMU070
COLALIN see CME750

COLAMINE see EEC600
COLCEMIDE see MIW500
COLCHAMINE see MIW500
COLCHICIN (GERMAN) see CNG830
COLCHICINA (ITALIAN) see CNG830
COLCHICINE see CNG830
7-α-H-COLCHICINE see CNG830
COLCHICOSIDE see DAN375
COLCHICUM AUTUMNALE see CNX800
COLCHICUM SPECIOSUM see CNX800
COLCHICUM VERNUM see CNX800
COLCHINE, N-DEACETYL-N-METHYL see MIW500
COLCHINEOS see CNG830
COLCHINIC ACID TRIMETHYL see TLO000
COLCHISOL see CNG830
COLCIN see CNG830
COLCOTHAR see IHD000
COLDAN see NCW000
COLDRIN see CMW700
COLEBENZ see BCM000
COLECALCIFEROL see CMC750
COLEMANITE see CAN250
COLEMID see MIW500
COLEP see MOB699
COLEPAX see IFY100
COLEPUR see DDS600
COLESTERINEX see PPH050
COLESTYRAMIN see CME400
COL-EVAC see SFC500
COLEYTL see CBH250
COLFARIT see ADA725
COLIBIL see CNG835
COLIMYCIN see PKD250
COLIMYCIN M see SFY500
COLIOPAN see BPI125
COLIPAR see DDS600
COLISONE see PLZ000
COLISTICINA see PKD250
COLISTIMETHATE SODIUM see SFY500
COLISTIN see PKD250
COLISTINASE see BAC000
COLISTIN SODIUM METHANESULFONATE see SFY500
COLISTIN SULFOMETHAT see SFY500
COLISTIN SULFOMETHATE SODIUM see SFY500
COLISTRIMETHATE SODIUM see SFY500
COLITE see CMF350
COLLARGOL see SDI750
COLLIDINE, ALDEHYDECOLLIDINE see EOS000
COLLIRON I.V. see IHG000
COLLODION see CNH000
COLLODION COTTON see CCU250
COLLOID 775 see CCL250
COLLOIDAL ARSENIC see ARA750
COLLOIDAL CADMIUM see CAD000
COLLOIDAL FERRIC OXIDE see IHD000
COLLOIDAL GOLD see GIS000
COLLOIDAL MANGANESE see MAP750
COLLOIDAL MERCURY see MCW250
COLLOIDAL SELENIUM see SBO500
COLLOIDAL SILICA see SCH000
COLLOIDAL SILICON DIOXIDE see SCH000
COLLOIDAL SULFUR see SOD500
COLLOIDOX see CNK559
COLLOKIT see SOD500
COLLOMIDE see SNM500
COLLOWELL see SFO500
COLLOXYLIN see CCU250
COLLUNOSOL see TIV750
COLLUNOVAR see NCJ500
COLLUNOVER see NCJ500
COLLUSUL-HC see HHQ800
COLOCASIA ESCULENTA see EAI600
COLOCASIA GIGANTEA see EAI600
COLOGNE EARTH see MAT500
COLOGNE SPIRIT see EFU000
COLOGNE SPIRITS (ALCOHOL) (DOT) see EFU000
COLOGNE UMBER see MAT500
COLOGNE YELLOW see LCR000
COLOMBIAN BLACK TOBACCO CIGARETTE REFINED TAR see CMP800
COLONATRAST see BAP000
COLONIAL SPIRIT see MGB150
COLORADO RIVER HEMP see SBC550
COLORFIX see CNH125
COLORINES (MEXICO) see NBR800

COLOR-SET see TIX500
COLPOVISTER see EDU500
COLPRO see MBZ100
COLPRONE see MBZ100
COLSALOID see CNG830
COLSUL see SOD500
COLSULANYDE see SNM500
COLTIROT see TOG500
COLTSFOOT see CNH250
COLTS FOOT see PCR000
COLUMBIA BLACK EP see AQP000
COLUMBIAN CARBON see CBT500
COLUMBIAN SPIRITS (DOT) see MGB150
COLUMBIUM PENTACHLORIDE see NEA000
COLUMBIUM POTASSIUM FLUORIDE see PLN500
COLUTOID see GEK500
COLY-MYCIN see PKD250
COLY-MYCIN INJECTABLE see SFY500
COLYMYSIN S see PKD250
COLYONAL see DBD750
COLYSTINMETHANSULFONAT (GERMAN) see SFY500
COMAC see CNM500
COMBANTRIN see POK575
COMBETIN see SMN000
COMBINAL K1 see VTA000
COMBOT EQUINE see TIQ250
COMBRETODENDRON AFRICANUM (Welw), extract see CNH275
COMBUSTION IMPROVER -2 see MAV750
COMELIAN see CNR750
COMESA see EQL000
COMESA see MNM500
COMESTROL see DKA600
COMESTROL ESTROBENE see DKA600
COMFREY, RUSSIAN see RRK000, RRP000
COMITAL see DKQ000
COMITE see SOP000
COMITIADONE see PEC250
COMMISTERONE see HKG500
COMMON GROUNDSEL see RBA400
COMMON SALT see SFT000
COMMOTIONAL see ABG750
COMPALOX see AHE250
COMPAZINE see PMF250, PMF500
COMPENDIUM see BMN750
COMPERLAN LD see BKE500
COMPITOX see CIR500
COMPLAMEX see XCS000
COMPLAMIN see XCS000
COMPLEMIX see DJL000
COMPLEXONE see EIV000
COMPLEXON II see EIX000
COMPLEXON III see EIX500
COMPOCILLIN G see BDY669
COMPOCILLIN-VK see PDT750
COMPOSE 134 P (FRENCH) see MLK800
COMPOUND 42 see WAT200
COMPOUND 118 see AFK250
COMPOUND 269 see EAT500
COMPOUND 338 see DER000
COMPOUND 347 see EAT900
COMPOUND 497 see DHB400
COMPOUND 604 see DFT000
COMPOUND-666 see BBP750
COMPODOL 711 see IKO000
COMPOUND 864 see YCJ000
COMPOUND 889 see DVL700
COMPOUND 88R see SOP500
COMPOUND 923 see DFY400
COMPOUND 1081 see FFF000
COMPOUND 1189 see KEA000
COMPOUND 1275 see PDV700
COMPOUND 1836 see CLV375
COMPOUND 2046 see MQR750
COMPOUND 3422 see PAK000
COMPOUND-3916 see DRB400
COMPOUND 3956 see CDV100
COMPOUND-4018 see CNL500
COMPOUND 4049 see MAK700
COMPOUND 4072 see CDS750
COMPOUND 47-83 see EAN600
COMPOUND-4992 see BKS810
COMPOUND 593A see PIK075
COMPOUND 6515 see DJA300
COMPOUND 6890 see TLQ000

COMPOUND 7215 see FBP300
COMPOUND 8958 see CQH500
COMPOUND 01748 see DJT800
COMPOUND 10854 see COF250
COMPOUND 17309 see DAL300
COMPOUND 20-438 see CNH300
COMPOUND 33355 see MKB750
COMPOUND 33,828 see MLJ500
COMPOUND 33T57 see MKW250
COMPOUND 38,174 see INS000
COMPOUND 42339 see ADR750
COMPOUND 48/80 see CNH375
COMPOUND 64716 see CMS200
COMPOUND 67/20 see CCW725
COMPOUND 74-637 see ALU875
COMPOUND 90459 see OJI750
COMPOUND S-6,999 see NNF000
COMPOUND 69@1 = 83 see CNH500
COMPOUND 78/702 see CNH525
COMPOUND B see CNS625
COMPOUND B DICAMBA see MEL500
COMPOUND C-9491 see IEN000
COMPOUND E see CNS800
COMPOUND E ACETATE see CNS825
COMPOUND F see CNS750
COMPOUND F-2 see ZAT000
COMPOUND-1452-F see EME500
COMPOUND F ACETATE see HHQ800
COMPOUND G-11 see HCL000
COMPOUND HP1275 see PDV700
COMPOUND 26539 HYDROCHLORIDE see EPL600
COMPOUND 6-12 INSECT REPELLENT see EKV000
COMPOUND M-81 see PHI500
COMPOUND 14045 METHIODIDE see CNH550
COMPOUND 14045 METHOCHLORIDE see EAI875
COMPOUND 14045 METHSULFATE see TJG225
COMPOUND No. 1080 see SHG500
COMPOUND R-242 see CKI625
COMPOUND R-25788 see DBJ500
COMPOUND 2339 RP see PEN000
COMPOUND SN see SAY000
COMPOUND UC-20047 A see CFF250
COMPOUNE 732 see BQT750
COMPTIE see CNH789
COMYCETIN see CDP250
d-CON see WAT200
CON A see CNH625
CONCANAVALIN A see CNH625
CONCHININ see QFS000
CONCILIUM see FGU000
CONCO AAS-35 see DXW200
CONCOGEL 2 CONCENTRATE see SIY000
CONCO NI-90 see PKF000
CONCO NIX-100 see PKF500
CONCO SULFATE WA see SIB600
CONCO XAL see DRS200
CONCTASE C see CNH650
CONDACAPS see VSZ100
CONDENSATE PL see BKE500
CONDENSATES (PETROLEUM), VACUUM TOWER (9CI) see MQV755
CONDITION see DCK759
CONDITIONER 1 see OBA000
CONDOCAPS see VSZ100
CONDOL see VSZ100
CONDYLON see CNG830
CONDY'S CRYSTALS see PLP000
CONESSINE DIHYDROBROMIDE see DOX000
CONESSINE HYDROCHLORIDE see CNH660
CONEST see ECU750
CONESTORAL see EDV600
CONESTRON see ECU750
CONFECTIONER'S SUGAR see SNH000
CONFORTID see IDA000
CONGOBLAU 3B see CMO250
CONGO BLUE see CMO250
CONGOCIDINE DIHYDROCHLORIDE see NCQ000
CONGO RED see SGQ500
CONGO RED R-138 see NAY000
Γ-CONICEIN see CNH730
Γ-CONICEINE see CNH730
d-CONICINE see PNT000
CONIGON BC see EIV000
CONIIN see PNT000
CONIINE see PNT000

(+)-CONIINE see PNT000
CONINE see PNT000
α-CONINE see PNT000
CONIUM MACULATUM see CNH750, PJJ300
CONJES see ECU750
CONJUGATED EQUINE ESTROGEN see PMB000
CONJUGATED ESTROGENS see ECU750
CONJUNCAIN see OPI300
CONJUTABS see ECU750
7-CON-o-METHYLNOGAROL see MCB600
CONOCO C-50 see DXW200
CONOCO DBCL see DXW600
CONOTRANE see PFN000
CONOVA 30 see DAP810
CONOVID see EAP000
CONOVID E see EAP000
CONQUERORS see HGL575
CONQUININE see QFS000
CONRAXIN H see IDJ600
CONRAY see IGC000
CONRAY 30 see IGC000
CONRAY 60 see IGC000
CONRAY 280 see IGC000
CONRAY MEGLUMIN see IGC000
CONRAY MEGLUMINE 282 see IGC000
CONSDRIN see SPC500
CONSDRIN HYDROCHLORIDE see SPC500
CONSTAPHYL see DGE200
CONSTONATE see DJL000
CONT see MMN250
CONTAVERM see PDP250
CONTEBEN see FNF000
CONTERGAN see TEH500
CONTIMET 30 see TGF250
CONTINAL see NBU000
CONTINENTAL see KBB600
CONTIZELL see PKQ059
CONTRAC see BMN000
CONTRA CREME see ABU500
CONTRADOL see ABG750
CONTRALGIN see BQA010
CONTRALIN see DXH250
CONTRAPAR see KFA100
CONTRAPOT see DXH250
CONTRATHION see PLX250
CONTRHEUMA RETARD see ADA725
CONTRISTAMINE HYDROCHLORIDE see CIS000
CONTRIX 28 see IGC000
CONTROL see MFD500
CONTROVLAR see EEH520
CONTUREX see SHX000
CONVALLAOTOXIN see CNH780
CONVALLARIA MAJALIS see LFT700
CONVALLARIN see CNH775
CONVALLATON see CNH780
CONVALLATOXIGENIN see SMM500
CONVALLATOXIN see CNH780
CONVALLATOXOL see CNH785
CONVALLATOXOSIDE see CNH780
CONVAL LILY see LFT700
CONVALLOTOXIN see CNH780
CONVALLOTOXOL see CNH785
CONVALOTOXOL see CNH785
CONVENIXA see TLP750
CONVUL see DKQ000
CONVULEX see PNX750
COOLSPAN see EPD500
COOMASSIE BLUE see ADE750
COOMASSIE BLUE MEDICINAL see ADE750
COOMASSIE BLUE RL see ADE750
COOMASSIE VIOLET see FAG120
COONTIE see CNH789
CO-OP HEXA see HCC500
COPAGEL PB 25 see SFO500
COPAIBA BALSAM see CNH792
COPAIBA OIL see CNH792
COPAIBA OLEORESIN see CNH792
COPAL (CUBA) see PCB300
COPAL Z see SMQ500
COPANOIC see IFY100
COPAROGIN see FMB000
COPELLIDIN see END000
COPEY see BAE325
COPHARCILIN see AIV500

COPIAMYCIN see CNH800
COPIRENE see KGK000
COPOX see CNO000
COPPER see CNI000
COPPER-8 see BLC250
COPPER ACETATE see CNI250
COPPER(2+) ACETATE see CNI250
COPPER(II) ACETATE see CNI250
COPPER(2+) ACETATE, MONOHYDRATE see CNI325
COPPER(II) ACETATE MONOHYDRATE see CNI325
COPPER ACETOARSENITE (DOT) see COF500
COPPER ACETOARSENITE, solid (DOT) see COF500
COPPER(II) ACETYLACETONATE see BGR000
COPPER(II) ACETYLIDE see CNI500
COPPER-AIRBORNE see CNI000
COPPER ALLOY, Cu, Be see CNI600
COPPER ALLOY, Cu, Be, Co see CNK700
COPPER ARSENATE (BASIC) see CNI900
COPPER ARSENATE HYDROXIDE see CNI900
COPPER ARSENITE, solid (DOT) see CNN500
COPPERAS see FBN100, FBO000
COPPER ASCORBATE see CNJ325
COPPER(II) AZIDE see CNJ500
COPPER-BERYLLIUM ALLOY see CNI600
COPPER BIS(ACETYLACETONATE) see BGR000
COPPER BIS(ACETYLACETONE) see BGR000
COPPER, BIS(ETHYLENEDIAMINE)(MERCURICTETRATHIOCYANATO)-
  see BJP425
COPPER(2+), BIS(ETHYLENEDIAMINE)-, TETRAKIS(THIOCYANATO)MER-
  CURATE(2-), POLYMERS see BJP425
COPPER BIS(2,4-PENTANEDIONATE) see BGR000
COPPER BLUE see CNQ000
COPPER BRONZE see CNI000
COPPER CARBONATE HYDROXIDE see CNJ750
COPPER(II) CARBONATE HYDROXIDE (2581582) see CNJ750
COPPER CHELATE of N-HYDROXY-2-ACETYLAMINOFLUORENE see HIP500
COPPER CHLORATE see CNJ900
COPPER CHLORATE (DOT) see CNJ900
COPPER(I) CHLORIDE see CNK250
COPPER CHLORIDE (DOT) see CNK500
COPPER(II) CHLORIDE (1582) see CNK500
COPPER CHLORIDE, BASIC see CNK559
COPPER CHLORIDE, mixed with COPPER OXIDE, HYDRATE see CNK559
COPPER CHLORIDE OXIDE see CNK559
COPPER(I) CHLOROACETYLIDE see CNK599
COPPER CHROMATE see CNK609
COPPER CITRATE see CNK625
COPPER(I) CITRATE see CNK625
COPPER-COBALT-BERYLLIUM see CNK700
COPPER COMPOUNDS see CNK750
COPPER CYANIDE see CNL000
COPPER(II) CYANIDE see CNL250
COPPER CYANIDE (DOT) see CNL250
COPPER CYNANAMIDE see CNL250
COPPER DIACETATE see CNI250
COPPER(2+) DIACETATE see CNI250
COPPER DIACETATE MONOHYDRATE see CNI325
COPPER DIACETYLACETONATE see BGR000
COPPER DIHYDROXIDE see CNM500
COPPER DIMETHYLDITHIOCARBAMATE see CNL500
COPPER DINITRATE see CNM750
COPPER DINITRATE TRIHYDRATE see CNN000
COPPER(II)-1,3-DI(5-TETRAZOLYL)TRIAZENIDE see CNL625
COPPER EDTA COMPLEX see CNL750
COPPERFINE-ZINC see CNP500
COPPER FUME see CNM000
COPPER(I) HYDRIDE see CNM250
COPPER HYDROXIDE see CNM500
COPPER(2+) HYDROXIDE see CNM500
COPPER HYDROXYQUINOLATE see BLC250
COPPER-8-HYDROXYQUINOLATE see BLC250
COPPER-8-HYDROXYQUINOLINATE see BLC250
COPPER-8-HYDROXYQUINOLINE see BLC250
COPPER LONACOL see ZJS300
COPPER-MILLED see CNI000
COPPER MONOCHLORIDE see CNK250
COPPER MONOSULFATE see CNP250
COPPER MONOSULFIDE see CNQ000
COPPER NAPHTHENATE see NAS000
COPPER(2+) NITRATE see CNM750
COPPER(II) NITRATE see CNM750
COPPER(II) NITRATE, TRIHYDRATE (1582583) see CNN000
COPPER(I) NITRIDE see CNN250
COPPER NORDOX see CNO000

COPPER OC FUNGICIDE see CNK559
COPPER 1,3,5-OCTATRIEN-7-YNIDE see CNN399
COPPER ORTHOARSENITE see CNN500
COPPER(I) OXALATE see CNN750
COPPER(I) OXIDE see CNO000
COPPER(II) OXIDE see CNO250
COPPER OXINATE see BLC250
COPPER (2+) OXINATE see BLC250
COPPER OXINE see BLC250
COPPER OXYCHLORIDE see CNK559
COPPER OXYCHLORIDE-ZINEB mixture see ZJS300
COPPER OXYQUINOLATE see BLC250
COPPER OXYQUINOLINE see BLC250
COPPER(I) PERCHLORATE see CNO325
COPPER(II) PERCHLORATE see CNO350
COPPER(II) PERCHLORATE, DIHYDRATE see CNO500
COPPER(II) PHOSPHINATE see CNO750
COPPER(I) POTASSIUM CYANIDE see PLC175
COPPER QUINOLATE see BLC250
COPPER-8-QUINOLATE see BLC250
COPPER-8-QUINOLINOL see BLC250
COPPER QUINOLINOLATE see BLC250
COPPER-8-QUINOLINOLATE see BLC250
COPPER SALT 2-HYDROXY-1,2,3-PROPANETRICARBOXYLIC ACID (1:2)
   see CNK625
COPPERSAN see CNK559
COPPER SARDEX see CNO000
COPPER SLAG-AIRBORNE see CNI000
COPPER SLAG-MILLED see CNI000
COPPER SODIUM CYANIDE see SFZ100
COPPER SORBATE see CNP000
COPPER complex with trans-N-(p-STYRYLPHENYL)ACETOHYDROXAMIC
   ACID see SMU000
COPPER SULFATE see CNP250
COPPER(II) SULFATE (1:1) see CNP250
COPPER(II) SULFATE PENTAHYDRATE (1:1:5) see CNP500
COPPER SULFIDE see CNP750, CNQ000
COPPER(I) SULFIDE see CNP750
COPPER(2+) SULFIDE see CNQ000
COPPER (II) SULFIDE see CNQ000
COPPER(I) TETRAHYDROALUMINATE see CNQ250
COPPER TRICHLOROPHENOLATE see CNQ375
COPPER-2,4,5-TRICHLOROPHENOLATE see CNQ375
COPPER UVERSOL see NAS000
COPPER-ZINC ALLOYS see CNQ500
COPPER-ZINC CHROMATE COMPLEX see CNQ750
COPRA (DOT) see CNR000
COPRAMAT see ZJS300
COPRA (OIL) see CNR000
COPRA PELLETS (DOT) see CNR000
COPREN see GFW000
COPROL see DJL000
COPROSTANOL see DKW000
COPROSTAN-3-β-OL see DKW000
COPROSTEROL see DKW000
COPSAMINE see WAK000
COPTICIDE see SNM500
COP-TOX see CNK559
CoQ₁₀ see UAH000
COQUE MOLLE (HAITI) see JBS100
COQUERET (CANADA) see JBS100
COQUES DU LEVANT (FRENCH) see PIE500
CORACON see DJS200
CORADON see WAK000
CORAETHAMIDE see DJS200
CORAETHAMIDUM see DJS200
CORAFIL see CNR125
CORAL BEAD PLANT see RMK250
CORAL BEAN see NBR800
CORAL BERRY see ROA300
CORALEPT see DJS200
CORALITOS (CUBA) see ROA300
CORAL PLANT see CNR135
CORAL SNAKE VENOM see CNR150
CORAL VEGETAL (CUBA) see CNR135
CORALYNE SULFOACETATE see CNR250
CORAMINE see DJS200
CORATOL see POB500
CORAVITA see DJS200
CORAX see MDQ250
CO-RAX see WAT200
CORAZON de CABRITO (CUBA) see CAL125
CORAZONE see DJS200
CORCAT see PJX000

CORCHORGENIN see SMM500
CORCHORIN see SMM500
CORCHOSIDE A AGLYCON see SMM500
CORDABROMIN see HNY500
CORDALEROMIN see HNY500
CORDALIN see HLC000
CORDIAMID see DJS200
CORDIAMIN see DJS200
CORDIAMINE see DJS200
CORDILAN see LAU400
CORDILOX see VHA450
CORDIPIN see AEC750
CORDITON see DJS200
CORDOVAL see GEW000
CORDULAN see CMD750
CORDYCEPIN see DAQ225
CORDYCEPINE see DAQ225
9-CORDYCEPOSIDOADENINE see DAQ225
CORDYNIL see DJS200
COREDIOL see DJS200
COREINE see CCL250
CORESPIN see DJS200
CORETAL see CNR500
CORETAL see THK750
CORETHAMIDE see DJS200
CORETONE see DJS200
CORFLEX 880 see ILR100
CORGARD see CNR675
CORGLYCON see CNH780
CORGLYCONE see CNH780
CORGLYKON see CNH780
CORIAMYRTIN see CNR725
CORIAMYRTINE see CNR725
CORIAMYRTIONE see CNR725
CORIANDER OIL see CNR735
CORIANTIN see CMG675
CORIARI MYRTIFOLIA see CNR740
CORICIDIN see ABG750
CORIFORTE see ABG750
CORIL see ELH600
CORINE see CNE750
CORINTH FLOUR see AIB250
CORISOL see VGP000
CORIZIUM see NGG500
CORLAN see HHR000
CORLIN see CNS800, ELH600
Δ-CORLIN see PLZ100
CORLUTIN see PMH500
CORLUTIN L.A. see HNT500
CORLUVITE see PMH500
CORMALONE see SOV100
CORMED see DJS200
CORMELIAN see CNR750
CORMELIAN-DIGOTAB see CNR825
CORMID see DJS200
CORMOGRIZIN see GJU800
CORMOTYL see DJS200
CORNE CABRITE (HAITI) see YAK300
CORN LILY see FAB100
CORNMINT OIL, PARTIALLY DEMENTHOLIZED see MCB625
CORNOCENTIN see EDB500, LJL000
CORN OIL see CNS000
CORNOTONE see DJS200
CORNOX CWK see BAV000
CORNOX-M see CIR250
CORNOX RD see DGB000
CORNOX RK see DGB000
CORN PRODUCTS see SLJ500
CORN SUGAR see GFG000
CORNUCOPIA see AOO825
CORODANE see CDR750
CORODANE see OPC000
CORODIL see HNY500
CORODILAN see DHS200
CORODINOC see DUU600
CORONA COROZATE see BJK500
CORONAL see DNC000
CORONARIDINE HYDROCHLORIDE see CNS200
(-)-CORONARIDINE MONOHYDROCHLORIDE see CNS200
CORONARIN see DNC000
CORONARINE see PCP250
CORONENE see CNS250
CORONIN see AHK750

COROPHYLLIN-N see HLC000
COROSANIN see ELH600
COROSORBIDE see CCK125
COROSUL D AND S see SOD500
COROTHION see PAK000
COROTONIN see DJS200
COROTRAN see CJT750
COROVIT see DJS200
COROVLISS see CCK125
COROXON see CIK750
COROZATE see BJK500
CORPAX see PEV750
CORPHOS see HHQ875
CORPHYLLIN see DNC000
CORPORIN see PMH500
CORPS PRALINE see MAO350
CORPS R. 261 see DBA200
CORPS 2339 R P (FRENCH) see PEN000
CORPUS LUTEUM HORMONE see PMH500
CORRIGAST see HKR500
CORRIGEN see OHQ000
CORRONAROBETIN see TEH500
CORROSIVE MERCURY CHLORIDE see MCY475
CORROSIVE SUBLIMATE see MCY475
CORRY'S SLUG DEATH see TDW500
CORSONE see SOW000
CORSTILINE see AES650
CORT A see CNS650
CORTACET see DAQ800
CORTACREAM see HHQ800
CORTADREN see CNS800, CNS825
CORTAID see HHQ800
CORTAN see PLZ000
CORTANCYL see PLZ000, PLZ100
CORTATE see DAQ800
CORT-DOME see CNS750
CORTEF ACETATE see HHQ800
CORTELAN see CNS825
Δ-CORTELAN see PLZ000
CORTELL see HHQ800
CORTENIL see DAQ800
CORTES see HHQ800
CORTESAN see DAQ800
CORTEX ALDEHYDE see PDR000
CORTEXILAR see FDD075
CORTEXONE see DAQ600
CORTEXONE ACETATE see DAQ800
COR-THEOPHYLLINE see DNC000
CORTHION see PAK000
CORTHIONE see PAK000
CORTICOSTERON see CNS625
CORTICOSTERONE see CNS625
CORTICOSTERONE ACETATE see CNS650
CORTICOTROPHIN see AES650
CORTICOTROPIN see AES650
CORTICOTROPIN-LIKE SUBSTANCES see AES650
CORTIDELT see PLZ000
CORTIFAR see DAQ800
CORTIFOAM see HHQ800
CORTIGEN see DAQ800
CORTILAN-NEU see CDR750
CORTINAQ see DAQ800
CORTINAZINE see ILD000
CORTINELLUS SHIITAKE EXTRACT (JAPANESE) see JDJ100
CORTIPHATE INJECTABLE see HHQ875
CORTIPHYSON see AES650
CORTIPRED see SOV100
CORTIRON see DAQ800
CORTISAL see CNS800, CNS825
CORTISATE see CNS800, CNS825
CORTISOL see CNS750
Δ¹-CORTISOL see PMA000
CORTISOL ACETATE see HHQ800
CORTISOL ALCOHOL see CNS750
CORTISOL HEMISUCCINATE SODIUM SALT see HHR000
CORTISOL PHOSPHATE see HHQ875
CORTISOL-21-PHOSPHATE see HHQ875
CORTISOL SODIUM HEMISUCCINATE see HHR000
CORTISOL SODIUM SUCCINATE see HHR000
CORTISOL-21-SODIUM SUCCINATE see HHR000
CORTISOL SUCCINATE, SODIUM SALT see HHR000
CORTISONE see CNS800
Δ-CORTISONE see PLZ000
Δ¹-CORTISONE see PLZ000

CORTISONE ACETATE see CNS825
CORTISONE-21-ACETATE see CNS825
CORTISONE MONOACETATE see CNS825
CORTISPRAY see CNS750
CORTISTAB see CNS825
CORTISTAL see CNS800
CORTISYL see CNS825
CORTIVIS see DAQ800
CORTIVITE see CNS800, CNS825
CORTIXYL see DAQ800
CORTOCIN-F see FDB000
CORTOGEN see CNS800, CNS825
CORTOGEN ACETATE see CNS825
CORTONE see CNS800, CNS825
Δ-CORTONE see PLZ000
CORTONE ACETATE see CNS825
CORTRIL ACETATE see HHQ800
CORTRIL ACETATE-AS see HHQ800
CORTROPHIN see AES650
CORTROPHYSON see AES650
CORTUSSIN see RLU000
CORUNDUM see EAL100
CORUNDUM FUME see CNT250
CORVASAL see MRN275
CORVASYMTON see SPD000
CORVATON see MRN275
CORVIC 55/9 see PKQ059
CORVIC 236581 see AAX175
CORVITAN see DJS200
CORVITIN see DBA800
CORVITOL see DJS200
CORVITONE see DJS200
CORWIN see XAH000
CORYBAN-D see ABG750
CORYDALOID see CNT325
CORYDININE see FOW000
CORYLON see HMB500
CORYLONE see HMB500
CORYLOPHYLINE see GFG100
CORYNINE see YBJ000
CORYSTIBIN see AQH500
CORYWAS see DJS200
CORYZOL see DPJ400
COSAN see SOD500
COSCOPIN see NBP275, NOA000
COSCOPIN HYDROCHLORIDE see NOA500
COSCOTABS see NOA000
COSCOTABS HYDROCHLORIDE see NOA500
COSDEN 550 see SMQ500
COSLAN see XQS000
COSMEGEN see AEB000
COSMETIC BLUE LAKE see FAE000
COSMETIC BRILLIANT PINK BLUISH D CONC see FAG070
COSMETIC CORAL RED KO BLUISH see CHP500
COSMETIC GREEN BLUE R25396 see ADE500
COSMETIC WHITE C47-5175 see TGG760
COSMETOL see CCP250
COSMIC see TJJ500
COSMOPEN see BDY669, BFD000
COTALMON see AMK250
COTARNIN see CNT625
COTARNINE see CNT625
COTEL see VSZ000
COTINAZIN see ILD000
COTNION-ETHYL see EKN000
COTNION METHYL see ASH500
COTOFILM see HCL000
COTOFOR see EPN500
COTONE see PLZ000
COTORAN see DUK800
COTORAN MULTI see MQQ450
COTORAN MULTI 50WP see DUK800
CO-TRIFAMOLE see SNL850
CO-TRIMOXAZOLE see SNK000, TKX000
COTTON DUST see CNT750
COTTONEX see DUK800
COTTON RED L see SGQ500
COTTONSEED OIL (unhydrogenated) see CNU000
COTTON VIOLET R see CMP000
COUER SAIGNANT (HAITI) see CAL125
COUMADIN see WAT200
COUMADIN SODIUM see WAT220
COUMAFENE see WAT200
COUMAFURYL see ABF500

COUMAMYCIN see CNV500
COUMAPHOS see CNU750
COUMAPHOS-O-ANALOG see CIK750
COUMAPHOS OXYGEN ANALOG (USDA) see CIK750
4-COUMARIC ACID see CNU825
p-COUMARIC ACID see CNU825
COUMARIN see CNV000
COUMARIN 1 see DIL400
COUMARIN 4 see MKP500
COUMARIN, 7-HYDROXY-3-(p-HYDROXYPHENYL)-4-PHENYL- see HNK600
cis-o-COUMARINIC ACID LACTONE see CNV000
COUMARINIC ANHYDRIDE see CNV000
COUMARIN, 4-METHYL-7-HYDROXY-, p-GUANIDINOBENZOATE see GLC100
COUMARONE see BCK250
COUMATETRALYL see EAT600
COUMERMYCIN AL see CNV500
COUMESTROL see COF350
COUNTER see BSO000
COUNTER 15G SOIL INSECTICIDE see BSO000
COUNTER 15G SOIL INSECTICIDE-NEMATICIDE see BSO000
COUNTRY WALNUT see TOA275
COURLOSE A 590 see SFO500
COVALLATOXOL see CNH785
COVATIN see BRS000
COVATIX see BRS000
CO-VIDARABINE see PBT100
COVI-OX see VSZ450
COVIT see VSZ000
COWBUSH (BAHAMAS) see LED500
COW GARLIC see WBS850
COWSLIP see MBU550
COXIGON see OJI750
COXISTAT see NGE500
COYDEN see CMX850
COYOTILLO see BOM125
COZYME see PAG200
CP see CCJ400, CQC650
4-CP see CJN000
CP 34 see TFM250
CP 105 see ECI000
CP 3438 see TFD250
CP 4572 see CDO250
CP 556S see SOU600
CP 6,343 see CFK000
CP 10,188 see FAM100
CP 14,957 see OAN000
CP 15,336 see DBI200
CP 16171 see FAJ100
CP 19699 see CON300
CP 23426 see DNS600
CP 25017 see NHK800
CP 31393 see CHS500
CP 34089 see SOU650
CP 40294 see MOB699
CP-40507 see MOB750
CP 43858 see TIP750
CP 47114 see DSQ000
CP 48985 see CFC500
CP 49674 see DOP200
CP 50144 see CFX000
CP 53619 see CFW750
CP 53926 see DRR200
CP-12,252-1 see NBP500
CP-15-639-2 see CBO250
CP-16533-1 see IRV000
CP 10423-18 see TCW750
CP-15467-61 see LGZ000
CPA see CJN000, CQC650, CQJ500
6-CPA see CKN375
C-PAL see FAM100
C. PALUSTRIS see MBU550
CPAS see CDS500
CPAS mixed with BCPE see CKL500
CPB see CJR500
CP BASIC SULFATE see CNP250
CPBS see CJR500
CPBU 7 see GHR609
CPC 3005 see SLJ500
CPC 6448 see SLJ500
CPCA see BIO750
CPCBS see CJT750
C.P. CHROME LIGHT 2010 see LCS000
C.P. CHROME ORANGE DARK 2030 see LCS000
C.P. CHROME ORANGE MEDIUM 2020 see LCS000

C.P. CHROME YELLOW LIGHT see LCR000
CPDC see PJD000
CPDD see PJD000
CPH see CBL000, CDP250
CPIB see ARQ750
CPIRON see FBJ100
CP 1044 J3 see BPP750
CPMC see CKF000
CPNU-I see NKJ100
cis-CPO see CKS099
trans-CPO see CKS100
CPP-3,4-OXIDE see CPX625
C15-PREDIOXIN see TIL750
C. PROCERA see COD675
CPS see MIG850
CPSA see CDP725
CP 45899 SODIUM SALT see PAP600
CPT see CLK215
CPT see TAF675
C.P. TITANIUM see TGF250
CP 49952 p-TOLUENESULFONATE see SOU675
C.P. TOLUIDINE TONER A-2989 see MMP100
C. PULCHERRIMA see CAK325
CPX see TAF675
CPZ see CCS369, CKP250, CKP500
C.P. ZINC YELLOW X-883 see ZFJ100
CQ see CLD250
C-QUENS see CBF250, CNV750
CR see DDE200
CR 39 see AGD250
CR 242 see BGC625
CR 409 see BJE750
CR-604 see POF550
CR-605 see TGF175
CR/662 see BLV000
CR 3029 see MAS500
CRAB-E-RAD see DXE600
CRAB'S EYES see AAD000
CRAB'S EYES see RMK250
CRADEX see BBK500
CRAG 341 see GII000
CRAG 85W see DSB200
CRAG 974 see DSB200
CRAG DCU-73w see DGQ200
CRAG EXPERIMENTAL HERBICIDE 2 see DGQ200
CRAG FLY REPELLENT see BRP250
CRAG FRUIT FUNGICIDE 34 see GIO000
CRAG FRUIT FUNGICIDE 341 see GII000
CRAG FUNGICIDE 658 see CNQ750
CRAG FUNGICIDE 974 see DSB200
CRAG HERBICIDE see CNW000
CRAG HERBICIDE 1 see CNW000
CRAG HERBICIDE 2 see DGQ200
CRAG NEMACIDE see DSB200
CRAG SESONE see CNW000
CRAG SEVIN see CBM750
CRAIN see FBS100
CRALO-E-RAD see DXE600
CRAMCILLIN-S see PDD350
CRAMPOL see ACX500
CRAMPOLE see ACX500
CRANOMYCIN see CNW100
CRANOMYCIN HYDROCHLORIDE see CNW105
CRASTIN S 330 see PKF750
CRATAEGUS, EXTRACT see EDK600
CRATECIL see EOK000
CRAVITEN see CNW125
CRAWHASPOL see TIO750
CRC 7001 see CEK875
CRD-401 see DNU600
CREAM of TARTER see PKU600
CREATININE SULFATE compounded with 3-(2-AMINOETHYL)INDOLE-5-OL
(1:1:1), MONOHYDRATE see AJX750
CREATININE SULFATE compounded with 3-(2-AMINOETHYL)INDOL-5-OL
(1:1:1) see AJX750
CREDO see SHF500
CREIN see DAL300
CREMODIAZINE see PPP500
CREMOMERAZINE see ALF250
CREMOMETHAZINE see SNJ000
p-CRESOL see MEK250
CREOSOTE see CMY825
CREOSOTE, from COAL TAR see CMY825
CREOSOTE OIL see CMY825

CREOSOTE P1 see CMY825
CREOSOTUM see CMY825
CRESAN UNIVERSAL TROCKENBEIZE see MEP000
CRESIDINE see MGO750
m-CRESIDINE see MGO500
p-CRESIDINE see MGO750
CRESOATE, WOOD see BAT850
CRESODIOL see GGS000
CRESOL see CNW500
2-CRESOL see CNX000
3-CRESOL see CNW750
4-CRESOL see CNX250
m-CRESOL see CNW750
o-CRESOL see CNX000
p-CRESOL see CNX250
p-CRESOL ACETATE see MNR250
m-CRESOL, α-(AMINOOXY)-6-BROMO- see BMM600
m-CRESOL, 4,4′-(1,2-DIETHYLETHYLENE)DI- see DJI300
o-CRESOL, 4,4′-(1,2-DIETHYLETHYLENE)DI- see DJI350
CRESOL DIPHENYL PHOSPHATE see TGY750
CRESOL FLYCIDYL ETHER see TGZ100
o-CRESOL GLYCERYL ETHER see GGS000
CRESOLI (ITALIAN) see CNW500
p-CRESOL METHYL ETHER see MGP000
CRESON see RLU000
CRESOPUR see BAV000
CRESORCINOL DIISOCYANATE see TGM750
CRESOSSIDIOLO see GGS000
CRESOSSIPROPANDIOLO see GGS000
CRESOTIC ACID see CNX625
o-CRESOTIC ACID see CNX625
2,3-CRESOTIC ACID see CNX625
CRESOTINE BLUE 2B see CMO000
CRESOTINE BLUE 3B see CMO250
CRESOTINIC ACID see CNX625
o-CRESOTINIC ACID see CNX625
β-CRESOTINIC ACID see CNX625
2,3-CRESOTINIC ACID see CNX625
CRESOTOL see DUU600
CRESOXYDIOL see GGS000
CRESOXYPROPANEDIOL see GGS000
CRESSA CRETICA Linn., extract see CNX700
CRESTABOLIC see AOO475
CRESTANIL see MQU750
CRESTOMYCIN see NCF500
CRESTOXO see CDV100
p-CRESYL ACETATE (FCC) see MNR250
p-CRESYL BENZOATE see TGX100
p-CRESYL CAPRYLATE see THB000
CRESYL DIPHENYL PHOSPHATE see TGY750
m-CRESYL ESTER of N-METHYLCARBAMIC ACID see MIB750
o-CRESYL-α-GLYCERYL ETHER see GGS000
CRESYLGLYCIDE ETHER see TGZ100
CRESYL GLYCIDYL ETHER see TGZ100
CRESYLIC ACID see CNW500
m-CRESYLIC ACID see CNW750
o-CRESYLIC ACID see CNX000
p-CRESYLIC ACID see CNX250
CRESYLIC CREOSOTE see CMY825
p-CRESYL ISOBUTYRATE see THA250
CRESYLITE see TML500
m-CRESYL METHYLCARBAMATE see MIB750
p-CRESYL METHYL ETHER see MGP000
p-CRESYL OCTANOATE see THB000
CRESYL PHOSPHATE see TNP500
o-CRESYL PHOSPHATE see TMO600
p-CRESYL SALICYLATE see THD850
CRILL 3 see SKV150
CRILL 10 see PKL100
CRILL 26 see SLL000
CRILL K 3 see SKV150
CRILLON L.D.E. see BKE500
CRIMIDIN (GERMAN) see CCP500
CRIMIDINA (ITALIAN) see CCP500
CRIMIDINE see CCP500
CRIMSON ANTIMONY see AQL500
CRIMSON EMBL see HJF500
CRIMSON SX see FMU080
CRINODORA see CFU750
CRINOTHENE see PKB500
CRINOVARYL see EDV000
CRINUM (VARIOUS SPECIES) see SLB250
CRINURYL see DFP600
CRISALBINE see GJG000

CRISALIN see DUV600
CRISAPON see DGI400
CRISATRINA see ARQ725
CRISAZINE see ARQ725
CRISEOCICLINE see TBX000
CRISEOCIL see MCH525
CRISODIN see MRH209
CRISODRIN see MRH209
CRISONAR see CFY750
CRISPATINE see FOT000
CRISPIN see THJ750
CRISQUAT see PAJ000
CRISTALLOSE see SJN700
CRISTALLOVAR see EDV000
CRISTALOMICINA see KAV000
CRISTAPEN see BFD000
CRISTAPURAT see DKL800
CRISTERONA MB see DME500
CRISTERONE T see TBF500
CRISTOBALITE see SCI500, SCJ000
CRISTOXO 90 see CDV100
CRISULFAN see EAQ750
CRISURON see DXQ500
CRITTOX see EIR000
CRNGDP 3,167,82 see NID000
CROCEOMYCIN see HAL000
CROCIDOLITE (DOT) see ARM275
CROCIDOLITE ASBESTOS see ARM275
CROCOITE see LCR000
CROCUS see CNX800
CROCUS MARTIS ADSTRINGENS see IHD000
CRODACID see MSA250
CRODACOL A.10 see OBA000
CRODACOL-CAS see HCP000
CRODACOL-O see OBA000
CRODACOL-S see OAX000
CRODAMOL IPP see IQW000
CRODET O 6 see PJY100
CROLEAN see ADR000
CROMILE, CLORURO di (ITALIAN) see CML125
CROMOCI see CQM325
CROMOGLYCATE see CNX825
CROMOGLYCATE DISODIUM see CNX825
CROMOLYN SODIUM see CNX825
CROMOLYN SODIUM SALT see CNX825
CROMO, OSSICLORURO di (ITALIAN) see CML125
CRONETAL see DXH250
CRONETON see EPR000
CRONIL see EHP000
CROP RIDER see DAA800
CROTALARIA BERTEROANA see RBZ400
CROTALARIA INCANA see RBZ400
CROTALARIA JUNCEA see RBZ400
CROTALARIA JUNCEA Linn., seed extract see CNX827
CROTALARIA RETUSA see RBZ400
CROTALARIA SPECTABILIS see RBZ400
CROTALINE see MRH000
CROTALUS ADAMANTEUS VENOM see EAB225
CROTALUS ATROX VENOM see WBJ600
CROTALUS CERASTES VENOM see CNX830
CROTALUS DURISSUS DURISSUS VENOM see CNX835
CROTALUS DURISSUS TERRIFICUS VENOM see CNY000
CROTALUS HORRIDUS HORRIDUS VENOM see TGB150
CROTALUS HORRIDUS VENOM see CNY300
CROTALUS RUBER RUBER VENOM see CNY325
CROTALUS SCUTULATUS SCUTULATUS (CALIF) VENOM
    see CNY339
CROTALUS SCUTULATUS SCUTULATUS VENOM see CNY350
CROTALUS SCUTULATUS VENOM see CNY375
CROTALUS VIRIDIS CONCOLOR VENOM see CNY750
CROTALUS VIRIDIS HELLERI VENOM see COA000
CROTALUS VIRIDIS LUTOSUS VENOM see COA250
CROTALUS VIRIDIS OREGANUS VENOM see COA500
CROTALUS VIRIDIS VENOM see COA750
CROTALUS VIRIDIS VIRIDIS TOXIN see VRP200
CROTALUS VIRIDIS VIRIDIS VENOM see PLX100
CROTALUS VIRIDUS CERERUS VENOM see CNZ000
CROTILIN see DAA800
CROTOCIN see COB000
CROTONAL see COB260
CROTONALDEHYDE see COB250, COB260
(E)-CROTONALDEHYDE see COB260
CROTONAMIDE, 2-CHLORO-N,N-DIETHYL-3-HYDROXY-, DIMETHYL PHOS-
    PHATE see FAB400

CROTONATE de 2,4-DINITRO 6-(1-METHYL-HEPTYL)-PHENYLE (FRENCH) see AQT500
CROTONATE d'ETHYLE (FRENCH) see COB750
CROTONIC ACID see COB500
α-CROTONIC ACID see COB500
CROTONIC ACID, solid see COB500
CROTONIC ACID ANHYDRIDE see COB900
CROTONIC ACID, 2-BUTENYLIDENE ESTER see COD100
CROTONIC ACID, ETHYL ESTER see EHO200
(E)-CROTONIC ACID, ETHYL ESTER see COB750
α-CROTONIC ACID ETHYL ESTER see COB750
CROTONIC ACID GERNAIOL ESTER see DTE000
CROTONIC ACID, 2-METHYL-, (E)- see TGA700
(E)-CROTONIC ACID METHYL ESTER see COB825
CROTONIC ACID, VINYL ESTER see VNU000
CROTONIC ALDEHYDE see COB250, COB260
CROTONIC ANHYDRIDE see COB900
CROTONOEL (GERMAN) see COC250
CROTON OIL see COC250
CROTON RESIN see COC250
CROTON TIGLIUM L. OIL see COC250
CROTONYL ALCOHOL see BOY000
CROTONYLENE see COC500
CROTONYL-N-ETHYL-o-TOLUIDINE see EHO500
CROTONYLOXYMETHYL-4,5,6-TRIHYDROOXYCYCLOHEX-2-ENONE see COC825
CROTOXIN see COC875
CROTOXYPHOS see COD000
CROTURAL see EHP000
CROTYL ALCOHOL see BOY000
CROTYL-2,4-DICHLOROPHENOXYACETATE see DAD000
CROTYLIDENE ACETIC ACID see SKU000
CROTYLIDENE DICROTONATE see COD100
1-CROTYL THEOBROMINE see BSM825
CROVARIL see HNI500
CROW BERRY see PJJ315
CROWFOOD see FBS100
CROWN 18 see COD575
12-CROWN-4 see COD475
15-CROWN-5 see PBO000
18-CROWN-6 see COD500
CROWN BEAUTY see BAR325
CROWN FLOWER see COD675
CROYSULFONE see SOA500
CRP-401 see DNU600
CRTRON see VSZ100
CRUDE ARSENIC see ARI750
CRUDE COAL TAR see CMY800
CRUDE ERGOT see EDB500
CRUDE OIL see PCR250
CRUDE OIL, synthetic see COD725
CRUDE SAIKOSIDE see SAF300
CRUDE SHALE OILS see COD750
CRUFOMATE see COD850
CRUFOMATE A see COD850
CRUFORMATE see COD850
CRUMERON, combustion products see ADX750
CRUSTECDYSON see HKG500
CRYOFLEX see BHK750
CRYOFLUORAN see FOO509
CRYOFLUORANE see FOO509
CRYOGENINE see CBL000
CRYOLITE see SHF000
CRYPTOCILLIN see MNV250
CRYPTOGIL OL see PAX250
CRYPTOGYL NA (ITALIAN) see PAX750
CRYPTOHALITE see COE000
CRYPTOSTEGIA GRANDIFLORA see ROU450
CRYPTOSTEGIA MADAGASCARIENSIS see ROU450
CRYSTAL CHROME ALUM see PLB500
CRYSTALETS see VSK900
CRYSTALLINA see VSZ100
CRYSTALLINE DEHYDROXY SODIUM ALUMINUM, CARBONATE see DAC450
CRYSTALLINE DIGITALIN see DKL800
CRYSTALLINE LIENOMYCIN see LFP000
CRYSTALLINIC ACID see SMB000
CRYSTALLIZED VERDIGRIS see CNI250
CRYSTALLOMYCIN see COE100
CRYSTALLOSE see SJN700
CRYSTAL O see CCP250
CRYSTAL ORANGE 2G see HGC000
CRYSTAL PROPANIL-4 see DGI000
CRYSTALS of VENUS see CNI250

CRYSTAMET see SJU000
CRYSTAMIN see VSZ000
CRYSTAPEN see BFD000, BFD250
CRYSTAR see ADA725
CRYSTEX see SOD500
CRYSTHION 2L see ASH500
CRYSTHYON see ASH500
CRYSTODIGIN see DKL800
CRYSTOGEN see EDV000
CRYSTOIDS see HFV500
CRYSTOL CARBONATE see SFO000
CRYSTOSOL see MQV750
CRYSTWEL see VSZ000
CS see CEQ600
CS-1 see CCP750
CS-61 see DON700
CS 359 see BOM510
CS 370 see CMY125
CS 386 see MQR760
CS-430 see HAG800
CS-439 see ALF500
CS-600 see LII300
CS 708 see BON250
CS-847 see CEW500
CS 1170 see CCS350
50-CS-46 see EME050
60-CS-16 see CMF400
CS 645A see BIN500
CSAC see EES400
CSC see SEC000
CSF-GIFTWEIZEN see TEM000
CSI PASTE see DTG700
C-Sn-9 see BLL750
C-3 SODIUM SALT see CBF625
CSP see CNP500
CS 1170 SODIUM see CCS360
C. SUFFUSUS SUFFUSUS VENOM see CCW925
CT see CCX250, CLH750
CT-1341 see AFK500
CT 3318 see COE125
4365 CT see MLI750
CT 4436 see MFO250
CTA see CLO750
CTAB see HCQ500
CTC see CMA750
CTCP see CNQ375
CTFE see CLQ750
CTH see CLK230
C-TOXIFERINE 1 see THI000
CTR 6669 see MHC750
CTT see CCS373
CTX see CCR950, CQC650
CU-56 see CNK559
CUBAN LILLY see SLH200
CUBE see RNZ000
CUBE EXTRACT see RNZ000
CUBE-PULVER see RNZ000
CUBE ROOT see RNZ000
CUBES see DJO000
CUBIC NITER see SIO900
CUBOR see RNZ000
CUCKOOPINT see ITD050
CUCKOO PLANT see JAJ000
CUCUMBER ALCOHOL see NMV780
CUCUMBER ALDEHYDE see NMV760
CUCURBITACIN E see COE250
CUCURBITACINE (D) see EAH100
CUCURBITACINE-E see COE250
CUEMID see CME400
CUENTAS de ORO (PUERTO RICO) see GIW200
CUIPU (MEXICO) see CNR135
CULLEN EARTH see MAT500
CULPEN see CHJ000
CUM see COE750
CUMA see BJZ000
CUMAFOS (DUTCH) see CNU750
CUMAFURYL (GERMAN) see ABF500
CUMALDEHYDE see COE500
CUMAN see BJK500
CUMAN L see BJK500
p-CUMARIC ACID see CNU825
CUMATE see CNL500
CUMATETRALYL (GERMAN, DUTCH) see EAT600
CUMEEN (DUTCH) see COE750

CUMEENHYDROPEROXYDE (DUTCH) see IOB000
CUMENE see COE750
psi-CUMENE see TLL750
CUMENE ALDEHYDE see COF000
CUMENE HYDROPEROXIDE (DOT) see IOB000
CUMENE HYDROPEROXIDE, TECHNICALLY PURE (DOT) see IOB000
CUMENE PEROXIDE see DGR600
p-CUMENOL see IQZ000
m-CUMENOL METHYLCARBAMATE see COF250
CUMENT HYDROPEROXIDE see IOB000
p-(p-CUMENYLAZO)-N,N-DIMETHYLANILINE see IOT875
CUMENYL HYDROPEROXIDE see IOB000
m-CUMENYL METHYLCARBAMATE see COF250
CUMERTILIN SODIUM see MCS000
CUMIC ALCOHOL see CQI250
p-CUMIC ALDEHYDE see COE500
CUMID see BJZ000
o-CUMIDINE see INX000
psi-CUMIDINE see TLG250
psi-CUMIDINE HYDROCHLORIDE see TLG750
CUMINALDEHYDE see COE500
CUMINIC ACETALDEHYDE see IRA000
CUMINIC ALCOHOL see CQI250
CUMINIC ALDEHYDE (FCC) see COE500
CUMIN OIL see COF325
CUMINOL see CQI250
CUMINYL ALCOHOL see CQI250
CUMINYL ALDEHYDE see COE500
CUMINYL NITRILE see IOD050
CUMMIN see COF325
CUMOESTEROL see COF350
psi-CUMOHYDROQUINONE see POG300
CUMOLHYDROPEROXID (GERMAN) see IOB000
CUMOSTROL see COF350
CUMYL ACETALDEHYDE see IRA000
CUMYL ALCOHOL see CQI250
α-CUMYL ALCOHOL see DTN100
CUMYL HYDROPEROXIDE see IOB000
α-CUMYL HYDROPEROXIDE see IOB000
CUMYL HYDROPEROXIDE, TECHNICAL PURE (DOT) see IOB000
CUMYL PEROXIDE see DGR600
CUNDEAMOR (CUBA, PUERTO RICO) see FPD100
CUNILATE see BLC250
CUNILATE 2472 see BLC250
CUPEY see BAE325
CUPFERRON see ANO500
CUP-OF-GOLD see CDH125
CUPPER OXIDE (RUSSIAN) see CNO000
CUPRAL see SGJ000
CUPRASULFIDE see CNP750
CUPRATE(1-), DICYANO-, POTASSIUM see PLC175
CUPRATE(2-), TRIS(CYANO-C)-, DISODIUM see SFZ100
CUPRAVIT BLAU see CNM500
CUPRAVIT BLUE see CNM500
CUPRENIL see MCR750
CUPRIC ACETATE see CNI250
CUPRIC ACETATE MONOHYDRATE see CNI325
CUPRIC ACETOARSENITE see COF500
CUPRIC ACETYLACETONATE see BGR000
CUPRIC ARSENITE see CNN500
CUPRIC CARBONATE see CNJ750
CUPRIC CHELATE of 2-N-HYDROXYFLUORENYL ACETAMIDE see HIP500
CUPRIC CHLORIDE see CNK500
CUPRIC CITRATE see CNK625
CUPRIC CYANIDE (DOT) see CNL250
CUPRIC DIACETATE see CNI250
CUPRIC DINITRATE see CNM750
CUPRIC DIPERCHLORATE TETRAHYDRATE see CNO500
CUPRIC GREEN see CNN500
CUPRIC HYDROXIDE see CNM500
CUPRIC-8-HYDROXYQUINOLATE see BLC250
CUPRICIN see CNL000
CUPRIC NITRATE (DOT) see CNM750
CUPRIC NITRATE TRIHYDRATE see CNN000
CUPRIC OXALATE see CNN750
CUPRIC OXIDE see CNO250
CUPRIC-8-QUINOLINOLATE see BLC250
CUPRIC SULFATE see CNP250
CUPRIC SULFATE PENTAHYDRATE see CNP500
CUPRIC SULFIDE see CNQ000
CUPRIETHYLENE DIAMINE see DBU800
CUPRIETHYLENEDIAMINE, solution (DOT) see DBU800
CUPRIMINE see MCR750
CUPRINOL see NAS000

CUPRIZANE see COF675
CUPRIZONE see COF675
CUPROCIN see ZJS300
CUPROCITROL see CNK625
CUPRON (CZECH) see BCP500
CUPRONE see BCP500
CUPROUS ARSENATE, BASIC see CNI900
CUPROUS CHLORIDE see CNK250
CUPROUS CYANIDE see CNL000
CUPROUS DICHLORIDE see CNK250
CUPROUS OXIDE see CNO000
CUPROUS POTASSIUM CYANIDE see PLC175
CUPROUS SULFIDE see CNP750
CUPROZAN see ZJS300
CURACIT see BJ1000
CURACRON see BNA750
CURALIN M see MJM200
CURAMAGUEY (CUBA) see YAK300
CURANTYL see PCP250
CURARE see COF750
CURARIL see GGS000
CURARIN see COF825
CURARINE see COF825
(+)-CURARINE see COF825
CURARIN-HAF see TOA000
CURARYTHAN see GGS000
CURATERR see CBS275
CURATIN see AEG750
CURBISET see CDT000
CURCUMA LONGA Linn., rhizome extract see COF850
CURCUMA OIL see COG000
CURCUMIN see COG000
CURCUMINE see COG000
CURENE 442 see MJM200
C. URENS see FBW100
CURETARD see COG250
CURETARD A see DWI000
CUREX FLEA DUSTER see RNZ000
CURITAN see DXX400
CURITHANE see MJQ000
CURITHANE 103 see MGA500
CURITHANE C126 see DEQ600
CURL FLOWER see CMV390
CURLY HEADS see CMV390
CUROL BRIGHT RED 4R see FMU080
CURON FAST YELLOW 5G see FAG140
CURRAL see AFS500
CURTACAIN see TBN000
CUSCOHYGRIN DIMETHIODIDE see COG750
CUSCOHYGRINE BIS(METHYL BENZENESULFONATE) see COG500
CUSCOHYGRINE DIMETHYLIODIDE see COG750
CUSCUTA REFLEXA Roxb., extract excluding roots see AHI630
CUTCH (DYE) see CCP800
CUTICURA ACNE CREAM see BDS000
CUTISTEROL see FDB000
CUT LEAF PHILODENDRON see SLE890
CUTTING OILS see COH000
CUVALIT see LJE500
CV 1006 see CDF380
CV 3317 see DAM315
C. VESICARIA see CAK325
C. VIRIDIS VIRIDIS TOXIN see VRP200
C. VIRIDIS VIRIDIS VENOM see PLX100
CVK see PDD350
CVMP see RAF100
CVP see CDS750
C-WEISS 7 (GERMAN) see TGG760
CX-59 see HKE000
CXA-DPS see CBQ575
CXD see CCS530
CXM see CCS600
CXM see SFQ300
CXM-AX see CCS625
CY see CQC650
CY-39 see PHU500
CY 116 see AJD000
CYAANWATERSTOF (DUTCH) see HHS000
CYACETACID see COH250
CYACETACIDE see COH250
CYACETAZID see COH250
CYACETAZIDE see COH250
CYALANE see DXN600
CYAMEMAZINE MALEATE see COS899
CYAMEPROMAZINE MALEATE see COS899

CYAMOPSIS GUM see GLU000
CYANACETAMIDE see COJ250
CYANACETATE ETHYLE (GERMAN) see EHP500
CYANACETHYDRAZIDE see COH250
CYANACETIC ACID HYDRAZIDE see COH250, COH250
CYANACETOHYDRAZIDE see COH250
CYANACETYLHYDRAZIDE see COH250
CYANAMID 24055 see DUI000
CYANAMIDE see CAQ250, COH500
CYANAMIDE CALCIQUE (FRENCH) see CAQ250
CYANAMIDE, CALCIUM SALT (1:1) see CAQ250
CYANAMID GRANULAR see CAQ250
CYANAMID SPECIAL GRADE see CAQ250
CYANASET see MJM200
CYANATES see COH750
CYANATOTRIBUTYLSTANNANE see COI000
CYANAZIDE see COH250
CYANAZINE see BLW750
CYANEA CAPILLATA TOXIN see COI125
CYANESSIGSAEURE (GERMAN) see COJ500
CYANHYDRINE d'ACETONE (FRENCH) see MLC750
CYANIC ACID, POTASSIUM SALT see PLC250
CYANIC ACID, SODIUM SALT see COI250
CYANIC ACID, TRIMETHYLSTANNYL ESTER see TMI100
CYANIDE see COI500
CYANIDE, solution (DOT) see HHT000
CYANIDE ANION see COI500
CYANIDE or CYANIDE mixture, dry (DOT) see HHT000
CYANIDELONON 1522 see QCA000
CYANIDE of POTASSIUM see PLC500
CYANIDES see HHT000
CYANIDE of SODIUM see SGA500
CYANIDOL see COI750
CYANINE ACID BLUE R see ADE750
CYANINE ACID BLUE R NEW see ADE750
CYANINE GREEN G BASE see BLK000
CYANITE see AHF500
CYANIZIDE see COH250
CYANOACETAMIDE see COJ250
2-CYANOACETAMIDE see COJ250
CYANOACETHYDRAZIDE see COH250
CYANOACETIC ACID see COJ500
CYANOACETIC ACID ETHYL ESTER see EHP500
CYANOACETIC ACID HYDRAZIDE see COH250
CYANOACETIC ACID METHYL ESTER see MIQ000
CYANOACETIC ESTER see EHP500
CYANOACETOHYDRAZIDE see COH250
α-CYANOACETOHYDRAZIDE see COH250
CYANOACETONITRILE see MAO250
CYANOACETYL CHLORIDE see COJ625
N-CYANOACETYL ETHYL CARBAMATE see COJ750
CYANOACETYLHYDRAZIDE see COH250
1-(CYANOACETYL)MORPHOLINE see MRR750
4-CYANOACETYLMORPHOLINE see MRR750
2-CYANOACRYLATE ACID METHYL ESTER see MIQ075
α-CYANOACRYLATE ACID METHYL ESTER see MIQ075
CYANOAMINE see COH500
2-CYANO-4-AMINOSTILBENE see COS750
2-CYANOANILINE see APJ750
3-CYANOANILINE see AIR125
4-CYANOANILINE see COK125
m-CYANOANILINE see AIR125
o-CYANOANILINE see APJ750
p-CYANOANILINE see COK125
CYANO-B12 see VSZ000
7-CYANOBENZ(c)ACRIDINE see BAX000
4-CYANOBENZALDEHYDE see COK250
p-CYANOBENZALDEHYDE see COK250
7-CYANOBENZ(a) ANTHRACENE see COK500
10-CYANO-1,2-BENZANTHRACENE see COK500
CYANOBENZENE see BCQ250
p-CYANOBENZENECARBOXALDEHYDE see COK250
7-CYANOBENZO(c)ACRIDINE see BAX000
4-CYANOBENZONITRILE see BBP250
5-o-CYANOBENZYL-4,5,6,7-TETRAHYDROTHIENO(3,2-c)PYRIDINE
   METHANESULFONATE see TDC725
CYANOBORANE OLIGOMER see COK659
CYANOBRIK see SGA500
CYANOBROMIDE see COO500
N-CYANO-2-BROMOETHYLBUTYLAMINE see COK750
N-CYANO-2-BROMOETHYLCYCLOHEXYLAMINE see COL000
CYANO-CARBAMIC ACID METHYL ESTER, DIMER see MIQ350
2-CYANOCETYLCOUMARONE see BCK750
CYANOCOBALAMIN see VSZ000

p-CYANOCUMENE see IOD050
2-CYANO-N-(2-CYANOETHYL)ETHANAMINE see BIQ500
CYANOCYCLINE A see COL125
1-(1-CYANOCYCLOHEXYL)PIPERIDINE see PIN225
N-CYANODIALLYLAMINE see DBJ200
5-CYANO-10,11-DIHYDRO-5-(3-DIMETHYLAMINOPROPYL)-5H-DIBENZO
   (a,d)CYCLOHEPTENE HYDROCHLORIDE see COL250
4-CYANO-2,6-DIIODOPHENOL see HKB500
4-CYANO-2,6-DIJODPHENOL (GERMAN) see HKB500
4-CYANO-2,6-DIJODPHENOL CAPRYSAEUREESTER (GERMAN)
   see DNG200
4-CYANO-2,6-DIJODPHENOL LITHIUMSALZ (GERMAN) see DNF400
CYANO-3-(DIMETHYLAMINO-3-METHYL-2-PROPYL)-10-PHENOTHIAZINE
   MALEATE see COS899
CYANODIMETHYLARSINE see COL750
5-CYANO-9,10-DIMETHYL-1,2-BENZANTHRACENE see COM000
α-CYANO-2,6-DIMETHYLERGOLINE-8-PROPIONAMIDE see COM075
2-CYANO-3,3-DIPHENYLACRYLIC ACID, ETHYL ESTER see DGX000
α-CYANODIPHENYLMETHANE see DVX200
2-CYANO-3,3-DIPHENYL-2-PROPENOIC ACID, ETHYL ESTER see DGX000
1'-(3-CYANO-3,3-DIPHENYLPROPYL)(1,4'-BIPIPERIDINE)-4'-CARBOXAMIDE
   see PJA140
1-(3-CYANO-3,3-DIPHENYLPROPYL)-4-(2-OXO-3-PROPIONYL-1-
   BENZIMIDAZOLINYL)PIPERIDINE see BCE825
1-(3-CYANO-3,3-DIPHENYLPROPYL)-4-PHENYLISONIPECOTIC ACID ETHYL
   ESTER HYDROCHLORIDE see LIB000
1-(1-(3-CYANO-3,3-DIPHENYLPROPYL)-4-PIPERIDYL)-3-PROPIONYL-2-
   BENZIMIDAZOLINONE see BCE825
1-CYANO-3,4-EPITHIOBUTANE see EBD600
CYANOETHANE see PMV750
2-CYANOETHANOL see HGP000
2-(2-CYANOETHOXY)ETHYL ESTER ACRYLIC ACID see COM125
4-CYANOETHOXY-2-METHYL-2-PENTANOL see COM250
CYANOETHYDRAZIDE see COH250
CYANOETHYL ACRYLATE see ADT111
2-CYANOETHYL ACRYLATE see ADT111
2-CYANOETHYL ALCOHOL see HGP000
β-CYANOETHYLAMINE see AMB500
1-(2-CYANOETHYL)AZIRIDINE see ASI250
N-(2-CYANOETHYL)AZIRIDINE see ASI250
(2-CYANOETHYL)BENZENE see HHP100
2-(N-(2-CYANOETHYL)-N-CYCLOHEXYL)AMINO-ETHANOL see CPJ500
N-(CYANOETHYL)DIETHYLENETRIAMINE see COM500
CYANOETHYLENE see ADX500
1-(2-CYANOETHYL)ETHYLIMINE see ASI250
N-(β-CYANOETHYL)ETHYLENIMINE see ASI250
2-CYANOETHYL-2'-FLUOROETHYLETHER see CON500
N-(2-CYANOETHYL)-N-(β-HYDROXYETHY)-ANILINE see CPJ500
β-CYANOETHYLMERCAPTAN see COM750
2-CYANOETHYL PROPENOATE see ADT111
2-CYANOETHYLTRICHLOROSILANE see CON000
β-CYANOETHYLTRICHLOROSILANE see CON000
(2-CYANOETHYL)TRIETHOXYSILANE see CON250
β-CYANOETHYLTRIETHOXYSILANE see CON250
CYANOFENPHOS see CON300
2-CYANO-2'-FLUORODIETHYL ETHER see CON500
CYANOFORMYL CHLORIDE see CON825
CYANOGAS see CAQ500
CYANOGEN see COO000
CYANOGENAMIDE see COH500
CYANOGEN AZIDE see COO250
CYANOGEN BROMIDE see COO500
CYANOGEN CHLORIDE see COO750
CYANOGEN CHLORIDE, inhibited (DOT) see COO750
CYANOGEN CHLORIDE, containing less than 0.9% water (DOT) see COO750
CYANOGENE (FRENCH) see COO000
CYANOGEN FLUORIDE see COO825
CYANOGEN GAS (DOT) see COO000
CYANOGEN IODIDE see COP000
CYANOGEN MONOBROMIDE see COO500
CYANOGEN NITRIDE see COH500
CYANOGRAN see SGA500
CYANOGUANIDINE see COP125
CYANOGUANIDINE METHYLMERCURY DERIV. see MLF250
S-1-CYANO-2-HYDROXY-3-BUTENE see COP400
CYANOHYDROXYMERCURY see COP500
2-CYANO-10-(3-(4-HYDROXYPIPERIDINO)PROPYL)PHENOTHIAZINE
   see PIW000
2-CYANO-10-(3-(4-HYDROXY-1-PIPERIDYL)PROPYL)PHENOTHIAZINE
   see PIW000
CYANO-3-((HYDROXY-4 PIPERIDYL-1)-3 PROPYL)-10-PHENOTHIAZINE
   (FRENCH) see PIW000
CYANOIMINOACETIC ACID see COJ250
α-CYANO-6-ISOBUTYLERGOLINE-8-PROPIONAMIDE see COP600

α-CYANOISOPROPYLAMIDE OF THE O,O-DIETHYLTHIOPHOSPHORYL
ACETIC ACID see PHK250
CYANOKETONE see COS909
CYANOLYT see MIQ075
CYANOMETHANE see ABE500
CYANOMETHANOL see HIM500
O-(4-CYANO-2-METHOXYPHENYL)-O,O-DIMETHYL PHOSPHOROTHIOATE
see DTQ800
CYANOMETHYL ACETATE see COP750
CYANOMETHYLAMINE see GHI000
10-CYANOMETHYL-1,2-BENZANTHRACENE see BBB500
7-CYANO-12-METHYL-BENZ(a)ANTHRACENE see MIQ250
5-CYANO-10-METHYL-1,2-BENZANTHRACENE see MGY000
7-CYANO-10-METHYL-1,2-BENZANTHRACENE see MGY250
8-CYANO-7-METHYLBENZ(a)ANTHRACINE see MGY000
(CYANOMETHYL)BENZENE see PEA750
N-(CYANOMETHYL)DIMETHYLAMINE see DOS200
S-N-(1-CYANO-1-METHYLETHYL)CARBAMOYLMETHYL DIETHYL PHOS-
PHOROTHIOLATE see PHK250
S-(((1-CYANO-1-METHYL-ETHYL)CARBAMOYL)METHYL) O,O-DIETHYL
PHOSPHOROTHIOATE see PHK250
17-α-CYANOMETHYL-17-β-HYDROXY-ESTRA-4,9(10)-DIEN-3-ONE see SMP400
3-(CYANOMETHYL)INDOLE see ICW000
CYANO(METHYLMERCURI)GUANIDINE see MLF250
N-CYANO-N'-METHYL-N''-(2-(((5-METHYL-1H-IMIDAZOL-4-
YL)METHYL)THIO)ETHYL)GUANIDINE see TAB250
1-CYANO-2-METHYL-3-(2-(((5-METHYL-4-IM-
IDAZOLYL)METHYL)THIO)ETHYL)GUANIDINE see TAB250
2-CYANO-1-METHYL-3-(2-(((5-METHYLIMIDAZOL-4-
YL)METHYL)THIO)ETHYL)GUANIDINE see TAB250
α-CYANO-1-METHYL-β-OXO-PYRROLE-2-PROPIONANILIDE compd. with
2,2',2''-NITRILOTRIETHANOL see CDG300
5-CYANO-5-METHYLTETRAZOLE see COP759
CYANOMORPHOLINOADRIAMYCIN see COP765
CYANONITRENE see COP775
3-CYANONITROBENZENE see NFH500
m-CYANONITROBENZENE see NFH500
2-CYANO-4-NITROBENZENEDIAZONIUM HYDROGEN SULFATE see COQ325
N-CYANO-N-NITROSOETHYLAMINE see ENT500
3-(3-CYANO-1,2,4-OXADIAZOL-5-YL)-4-CYANO FURAZAN-2-(5-) OXIDE
see COQ375
α-CYANO-3-PHENOXYBENZYL-2-(4-CHLOROPHENYL)ISOVALERATE
PYDRIN see FAR100
α-CYANO-3-PHENOXYBENZYL-2-(4-CHLOROPHENYL)-3-METHYLBUTYRATE
see FAR100
(±)-α-CYANO-3-PHENOXYBENZYL 2,2-DIMETHYL-3-(2,2-
DICHLOROVINYL)CYCLOPROPANE CARBOXYLATE see RLF350
α-CYANO-3-PHENOXYBENZYL 2,2,3,3-TETRAMETHYL-1-
CYCLOPROPANECARBOXYLATE see DAB825
CYANO(3-PHENOXYPHENYL)METHYL 4-CHLORO-α-(1-
METHYLETHYL)BENZENEACETATE see FAR100
CYANOPHENPHOS see CON300
O-(4-CYANOPHENYL) O,O-DIMETHYL PHOSPHOROTHIOATE see COQ399
O-p-CYANOPHENYL O,O-DIMETHYL PHOSPHOROTHIOATE see COQ399
o-(4-CYANOPHENYL)-o-ETHYL PHENYLPHOSPHONOTHIOATE see CON300
o-p-CYANOPHENYL-o-ETHYL PHENYLPHOSPHONOTHIOATE see CON300
4-CYANOPHENYL ISOTHIOCYANATE see ISI000
p-CYANOPHENYL ISOTHIOCYANATE see ISI000
CYANOPHENYLMETHYL-β-d-GLUCOPYRANOSIDURONIC ACID see LAS000
CYANOPHOS see COQ399
CYANOPHOS see DGP900
1-CYANOPROPANE see BSX250
1-CYANOPROPENE see COQ750
2-CYANOPROPENE-1 see MGA750
1-CYANO-2-PROPEN-1-OL see HJQ000
3-CYANOPROPYLDICHLOROMETHYLSILANE see COR325
(3-CYANOPROPYL)DIETHOXY(METHYL) SILANE see COR500
2-CYANO-2-PROPYL NITRATE see COR750
(3-CYANOPROPYL) TRIETHOXYSILANE see COS800
2-CYANO-3-(4-PYRIDYL)-1-(1,2,3,TRIMETHYLPROPYL)GUANIDINE
see COS500
2-CYANO-4-STILBENAMINE see COS750
α-CYANOSTILBENE see DVX600
2-(5-CYANOTETRAZOLE)PENTAMMINECOBALT(III) PERCHLORATE
see COS825
2-CYANOTOLUENE see TGT500
4-CYANOTOLUENE see TGT750
o-CYANOTOLUENE see TGT500
p-CYANOTOLUENE see TGT750
α-CYANOTOLUENE see PEA750
ω-CYANOTOLUENE see PEA750
CYANOTOXIN see SMM500
CYANOTRICHLOROMETHANE see TII750
CYANOTRIMEPRAZINE MALEATE see COS899

CYANOTRIMETHYLANDROSTENOLONE see COS909
2-α-CYANO-4,4,17-α-TRIMETHYLANDROST-5-EN-17-β-OL-3-ONE see COS909
2-CYANO-1,2,3-TRIS(DIFLUOROAMINO)PROPANE see COT000
CYANOTUBERICIDIN see VGZ000
α-CYANOVINYL ACETATE see ABL500
CYANOX see COQ399
CYANSAN see COI250
CYANTIN see NGE000
CYANURAMIDE see MCB000
CYANURCHLORIDE see TJD750
CYANURE (FRENCH) see COI500
CYANURE d'ARGENT (FRENCH) see SDP000
CYANURE de CALCIUM (FRENCH) see CAQ500
CYANURE de CUIVRE (FRENCH) see CNL250
CYANURE de MERCURE (FRENCH) see MDA250
CYANURE de METHYL (FRENCH) see ABE500
CYANURE de PLOMB (FRENCH) see LCU000
CYANURE de POTASSIUM (FRENCH) see PLC500
CYANURE de SODIUM (FRENCH) see SGA500
CYANURE de VINYLE (FRENCH) see ADX500
CYANURE de ZINC (FRENCH) see ZGA000
CYANURIC ACID see THS000
CYANURIC ACID CHLORIDE see TJD750
CYANURIC ACID TRIGLYCIDYL ESTER see TKL250
CYANURIC CHLORIDE (DOT) see TJD750
CYANURIC FLUORIDE see TKK000
CYANURIC TRIAZIDE see THR250
CYANURIC TRIAZIDE (DOT) see THR250
CYANURIC TRICHLORIDE (DOT) see TJD750
CYANUROTRIAMIDE see MCB000
CYANUROTRIAMINE see MCB000
CYANURYL CHLORIDE see TJD750
CYANWASSERSTOFF (GERMAN) see HHS000
CYAP see COQ399
CYASORB UV 9 see MES000
CYAZID see COH250
CYAZIDE see COH250
CYAZIN see ARQ725
CYBIS see EID000
CYCAD HUSK see COT500
CYCAD MEAL see COT750
CYCAD NUT, aqueous extract see COT750
CYCAS CIRCINALIS HUSK see COT500
CYCASIN see COU000
CYCAS REVOLUTA GLUCOSIDE see COU000
CYCEURON plus CHLORBUFAN see AFM375
CYCHLORAL see CPR000
CYCLACILLIN see AJJ875
CYCLADIENE see DAL600
CYCLAINE see COU250
CYCLAINE HYDROCHLORIDE see COU250
CYCLAL CETYL ALCOHOL see HCP000
CYCLALIA see CMS850
CYCLAMAL see COU500
CYCLAMATE see CPQ625, SGC000
CYCLAMATE CALCIUM see CAR000
CYCLAMATE, CALCIUM SALT see CAR000
CYCLAMATE SODIUM see SGC000
CYCLAMEN ALDEHYDE see COU500
CYCLAMEN ALDEHYDE DIETHYL ACETAL see COU510
CYCLAMEN ALDEHYDE DIMETHYL ACETAL see COU525
CYCLAMIC ACID see CPQ625
CYCLAMIC ACID SODIUM SALT see SGC000
CYCLAMID see CPR000
CYCLAMIDE see ABB000
CYCLAMIDE see CPR000
CYCLAMIDOMYCIN see COV125
CYCLAN see CAR000
CYCLANDELATE see DNU100
CYCLAPEN see AJJ875
CYCLATE see BOV825
CYCLAZOCINE see COV500
CYCLE see PMF600
CYCLERGINE see DNU100
CYCLIC ADENOSINE-3',5'-PHOSPHATE see AOA130
CYCLIC AMP see AOA130
CYCLIC-3',5'-AMP see AOA130
CYCLIC AMP DIBUTYRATE see COV625
3',5'-CYCLIC AMP DIBUTYRATE see COV625
CYCLIC AMP N6,2'-DIBUTYRYL cAMP see COV625
CYCLIC DIBUTYRYL AMP see COV625
CYCLIC-2,6-cis-DIPHENYLHEXAMETHYLCYCLOTETRASILOXANE
see CMS241
CYCLIC ETHYLENE ACETAL-2-BUTANONE see EIO500

CYCLIC ETHYLENE CARBONATE see GHM000
CYCLIC ETHYLENE
 (DIETHOXYPHOSPHINOTHIOYL)DITHIOIMIDOCARBONATE see DXN600
CYCLICETHYLENE(DIETHOXYPHOSPHINOTHIOYL)DITHIOIMIDOCARBON-
 ATE see PGW750
CYCLIC ETHYLENE P,P-DIETHYL PHOSPHONODITHIOIMIDOCARBONATE
 see PGW750
CYCLIC ETHYLENE ESTER of
 (DIETHOXYPHOSPHINOTHIOYL)DITHIOIMIDOCARBONIC ACID
 see DXN600
CYCLIC ETHYLENE SULFITE see COV750
CYCLIC (HYDROXYMETHYL)ETHYLENE ACETAL ACETONE see DVR600
CYCLIC METHYLETHYLENE CARBONATE see CBW500
CYCLIC-S,S-(6-METHYL-2,3-QUINOXALINEDIYL) DITHIOCARBONATE
 see ORU000
CYCLIC N',O-PROPYLENE ESTER of N,N-BIS(2-CHLOROETHYL)PHOS-
 PHORODIAMIDIC ACID MONOHYDRATE see CQC500
CYCLIC PROPYLENE CARBONATE see CBW500
CYCLIC-1,2-PROPYLENE CARBONATE see CBW500
CYCLIC PROPYLENE (DIETHOXYPHOSPHINYL)DITHIOIMIDOCARBONATE
 see DHH400
CYCLIC SODIUM TRIMETAPHOSPHATE see TKP750
CYCLIC TETRAMETHYLENE SULFONE see SNW500
CYCLIC N,N'-TRIMETHYLENE-N''-BIS(2-CHLOROETHYL)-PHOSPHORIC
 TRIAMIDE see BHQ000
CYCLISCHES TRINATRIUMMETAPHOSPHAT (GERMAN) see TKP750
CYCLIZINE see EAN600
CYCLIZINE CHLORIDE see MAX275
CYCLIZINE HYDROCHLORIDE see MAX275
α-CYCLOAMYLOSE see COW925
CYCLOATE see EHT500
CYCLOBARBITAL see TDA500
CYCLOBARBITAL-SALICYLAMIDE COMPLEX see COV825
CYCLOBARBITOL see TDA500
CYCLOBARBITONE see TDA500
CYCLOBENDAZOLE see CQE325
CYCLOBENZAPRINE see PMH600
CYCLOBENZAPRINE HYDROCHLORIDE see DPX800
CYCLOBRAL see DNU100
CYCLOBUTANE see COW000
CYCLOBUTANECARBOXAMIDE-N-(2-FLUORENYL) see COW500
cis-(1,1-CYCLOBUTANEDICARBOSYLATO)DIAMMINEPLATINUM(II)
 see CCC075
1,1-CYCLOBUTANEDICARBOXYLATE DIAMMINE PLATINUM(II) see CCC075
cis-(1,1-CYCLOBUTANEDICARBOXYLATO)DIAMMINEPLATINUM(II)
 see CCC075
CYCLOBUTENE see COW250
CYCLOBUTYLENE see COW250
CYCLOBUTYL-N-(2-FLUORENYL)FORMAMIDE see COW500
17-CYCLOBUTYLMETHYL-3-HYDROXY-6-METHYLENE-8β-
 METHYLMORPHINAN see COW675
(8-β)-17-(CYCLOBUTYLMETHYL)-6-METHYLENEMORPHINAN-3-OL
 METHANESULFONATE see COW675
CYCLOBUTYROL see COW700
CYCLOCAPRON see AJV500
CYCLOCEL see CMF400
CYCLOCHEM GMS see OAV000
CYCLOCHEM INEO see ISC550
CYCLOCHLOROTINE see COW750
α-CYCLOCITRYLIDENEACETONE see IFW000
β-CYCLOCITRYLIDENEACETONE see IFX000
α-CYCLOCITRYLIDENE-4-METHYLBUTAN-3-ONE see COW780
CYCLO-CMP HYDROCHLORIDE see COW900
CYCLOCORT see COW825
CYCLOCYTIDINE see COW875, COW900
2,2'-CYCLOCYTIDINE see COW875
2,2'-o-CYCLOCYTIDINE see COW875
o-2,2'-CYCLOCYTIDINE see COW875
CYCLOCYTIDINE HYDROCHLORIDE see COW900
2,2'-CYCLOCYTIDINE HYDROCHLORIDE see COW900
2,2'-o-CYCLOCYTIDINE HYDROCHLORIDE see COW900
o-2,2'-CYCLOCYTIDINE MONOHYDROCHLORIDE see COW900
CYCLODAN see EAQ750
β-CYCLODEXTRIN see COW925
cis,trans,trans-CYCLODODECA-1,5,9-TRIENE see COW935
1,5,9-CYCLODODECATRIENE (Z,E,E) see COW935
CYCLODODECYL-2,6-DIMETHYLMORPHOLINE ACETATE see COX000
N-CYCLODODECYL-2,6-DIMETHYLMORPHOLINIUM ACETATE see COX000
CYCLODOL see BBV000
CYCLODORM see TDA500
CYCLOESTROL see DLB400
4-N-CYCLOETHYLENEUREIDOAZOBENZENE see COX250
p-N-CYCLO-ETHYLENEUREIDOAZOBENZENE see COX250
CYCLOFENIL see FBP100

CYCLOFENYL see FBP100
CYCLOGEST see PMH500
CYCLOGUANIL see COX400
CYCLOGUANIL HYDROCHLORIDE see COX325
CYCLOGUANYL see COX400
CYCLOGYL see CPZ125
3-CYCLOHENENYL CYANIDE see CPC625
CYCLOHEPTAAMYLOSE see COW925
β-CYCLOHEPTAAMYLOSE see COW925
9-CYCLOHEPTADECEN-1-ONE see CMU850
9-CYCLOHEPTADECEN-1-ONE, (Z)-(8CI,9CI) see CMU850
CYCLOHEPTAGLUCOSAN see COW925
CYCLOHEPTANE see COX500
CYCLOHEPTANECARBAMIC ACID-1,1-BIS(p-FLUOROPHENYL)-2-PRO-
 PYNYL ESTER see BJW250
CYCLOHEPTANONE see SMV000
1,3,5-CYCLOHEPTATRIENE see COY000
CYCLOHEPTATRIENE (DOT) see COY000
CYCLOHEPTATRIENE MOLYBDENUM TRICARBONYL see COY100
CYCLOHEPTENE see COY250
CYCLOHEPTENYL ETHYLBARBITURIC ACID see COY500
5-(1-CYCLOHEPTEN-1-YL)-5-ETHYLBARBITURIC ACID see COY500
CYCLOHEPTENYLETHYLMALONYLUREA see COY500
5-(1-CYCLOHEPTEN-1-YL)-5-ETHYL-2,4,6(1H,3H,5H)-PYRIMIDINETRIONE
 (9CI) see COY500
CYCLOHEXAAN (DUTCH) see CPB000
CYCLOHEXADECANOLIDE see OKU000
CYCLOHEXADEINEDIONE see QQS200
1,3-CYCLOHEXADIENE see CPA500
1,4-CYCLOHEXADIENE see CPA750
1,4-CYCLOHEXADIENEDIONE see QQS200
2,5-CYCLOHEXADIENE-1,4-DIONE see QQS200
3,5-CYCLOHEXADIENE-1,2-DIONE see BDC250
2,5-CYCLOHEXADIENE-1,4-DIONE DIOXIME see DVR200
1,4-CYCLOHEXADIENE DIOXIDE see QQS200
CYCLOHEXADIENOL-4-ONE-1-SULFONATE de DIETHYLAMINE (FRENCH)
 see DIS600
2,5-CYCLOHEXADIEN-1-ONE, 4-IMINO- see BDD500
CYCLOHEXAMETHYLENE CARBAMIDE see CPB050
CYCLOHEXAMETHYLENE CARBAMIDE see CPB050
CYCLOHEXAMETHYLENIMINE see HDG000
CYCLOHEXAMINE SULFATE see CPF750
CYCLOHEXAN (GERMAN) see CPB000
CYCLOHEXANAMIDE see CPB050
CYCLOHEXANAMINE see CPF500
CYCLOHEXANAMINE HYDROCHLORIDE see CPA775
CYCLOHEXANE see CPB000
CYCLOHEXANEACETIC ACID, 1-(HYDROXYMETHYL)-, MONOSODIUM
 SALT (9CI) see SHL500
CYCLOHEXANECARBAMIC ACID, 1,1-DIPHENYL-2-BUTYNYL ESTER
 see DVY900
CYCLOHEXANECARBAMIC ACID, 1-METHYL-1-PHENYL-2-PROPYNYL
 ESTER see MNX850
CYCLOHEXANECARBOXAMIDE see CPB050
CYCLOHEXANECARBOXYLIC ACID-(2-HYDROXYETHYL) ESTER
 see HKQ500
CYCLOHEXANECARBOXYLIC ACID, LEAD SALT see NAS500
CYCLOHEXANECARBOXYLIC ACID, TRIBUTYLSTANNYL ESTER see TID100
1,2-CYCLOHEXANEDIAMINE see CPB100
1,3-CYCLOHEXANEDIAMINE see DBQ800
(CYCLOHEXANE-1,2-DIAMMINE)(4-CARBOXYPHTHLATO)PLATINUM(II)
 see CCJ350
(Z)-(CYCLOHEXANE-1,2-DIAMMINE)ISOCITRATOPLATINUM(II) see PGQ275
2,2'-CYCLOHEXANE-1,1-DIYLBIS(p-PHENYLENEOXY)BIS(2-METHYLBUTY-
 RIC ACID) see CMV700
CYCLOHEXANEETHANOL, ACETATE see EHS000
CYCLOHEXANE ETHYL ACETATE see EHS000
CYCLOHEXANEETHYLAMINE see CPB500
CYCLOHEXANEFORMAMIDE see CPB050
CYCLOHEXANE OXIDE see CPD000
CYCLOHEXANE, PIPERIDINOMETHYL-, CAMPHOSULFATE see PIQ750
CYCLOHEXANE, PIPERIDINOMETHYL-, HYDROCHLORIDE see PIR000
CYCLOHEXANESPIRO-5'-HYDANTOIN see DVO600
CYCLOHEXANESULFAMIC ACID, CALCIUM SALT see CAR000
CYCLOHEXANESULFAMIC ACID, MONOSODIUM SALT see SGC000
CYCLOHEXANESULPHAMIC ACID see CPQ625
CYCLOHEXANESULPHAMIC ACID, MONOSODIUM SALT see SGC000
CYCLOHEXANETHIOL see CPB625
1,2,3-CYCLOHEXANETRIONE TRIOXIME see CPB650
CYCLOHEXANOL see CPB750
CYCLOHEXANOL ACETATE see CPF000
CYCLOHEXANOLAZETAT (GERMAN) see CPF000
1-CYCLOHEXANOL-α-BUTYRIC ACID see COW700
trans-(±)-CYCLOHEXANOL-2-((DIMETHYLAMINO)METHYL)-1-(3-
 METHOXYPHENYL) see THJ755

CYCLOHEXANOL, p-ISOPROPYL- see IOO300
CYCLOHEXANOL, 2-METHYL-5-(1-METHYLETHENYL)-, ACETATE,(1-α-2-
β,5α—(9CI) see DKV160
CYCLOHEXANON (DUTCH) see CPC000
CYCLOHEXANONE see CPC000
CYCLOHEXANONE-Δ see CPC250
CYCLOHEXANONE ISO-OXIME see CBF700
CYCLOHEXANONE OXIME see HLI500
CYCLOHEXANONE PEROXIDE and BIS(1-HYDROXYCYCLOHEXYL)PEROX-
IDE MIXTURE see CPC500
CYCLOHEXANYL ACETATE see CPF000
CYCLOHEXATRIENE see BBL250
CYCLOHEXENE see CPC579
3-CYCLOHEXENE-1-CARBONITRILE see CPC625
CYCLOHEXENECARBOXALDEHYDE see TCJ100
3-CYCLOHEXENE-1-CARBOXALDEHYDE see FNK025
3-CYCLOHEXENE-1-CARBOXYLIC ACID see CPC650
1-CYCLOHEXENE-1-CARBOXYLIC ACID, 3,4,5 see BML000
4-CYCLOHEXENE-1,2-DICARBOXIMIDE, N-(2,6-DIOXO-3-PIPERIDYL)-
see TDB200
1-CYCLOHEXENE-1,2-DICARBOXYLIC ACID DIMETHYL ESTER see DUF400
CYCLOHEXENE EPOXIDE see CPD000
CYCLOHEX-1-ENE-1-METHANOL, 4-(1-METHYLETHENYL)- PCI550
CYCLOHEXENE OXIDE see CPD000
CYCLOHEXENE-1-OXIDE see CPD000
1,2-CYCLOHEXENE OXIDE see CPD000
CYCLOHEXENONE see CPD250
2-CYCLOHEXEN-1-ONE see CPD250
5-Δ2:3-CYCLOHEXENYL-5-ALLYL-2-THIOBARBITURIC ACID see TES500
2-((4-CYCLOHEXEN-3-YLBUTYL)AMINO)ETHANETHIOL HYDROGEN SUL-
FATE (ESTER) see CPD500
S-2-((4-CYCLOHEXEN-3-YLBUTYL)AMINO)ETHYL THIOSULFATE see CPD500
1-(2-CYCLOHEXEN-1-YLCARBONYL)-2-METHYLPIPERIDINE see CPD625
5-(2-CYCLOHEXEN-1-YL)DIHYDRO-5-(2-PROPENYL)-2-THIOXO-4,6(1H,5H)-
PYRIMIDINEDIONE (9CI) see TES500
5-(1-CYCLOHEXEN-1-YL)-1,5-DIMETHYLBARBITURIC ACID see ERD500
5-(1-CYCLOHEXEN-1-YL)-1,5-DIMETHYLBARBITURIC ACID SODIUM SALT
see ERE000
5-(1-CYCLOHEXEN-1-YL)-1,5-DIMETHYL-2,4,6(1H,3H,5H)-
PYRIMIDINETRIONE see ERD500
5-(1-CYCLOHEXEN-1-YL)-1,5-DIMETHYL-2,4,6(1H,3H,5H,)-
PYRIMIDINETRIONE MONOSODIUM SALT see ERE000
5-(1-CYCLOHEXEN-1-YL)-1,5-DIMETHYL-2,4,6(1H,3H,5H)-
PYRIMIDINETRIONE SODIUM SALT (9CI) see ERE000
CYCLOHEXENYL-ETHYL BARBITURIC ACID see TDA500
5-(1-CYCLOHEXENYL)-5-ETHYLBARBITURIC ACID see TDA500
5-(1-CYCLOHEXEN-1-YL)-5-ETHYLBARBITURIC ACID see TDA500
CYCLOHEXENYLETHYLENE see CPD750
5-(1-CYCLOHEXEN-1-YL)-5-ETHYL-2,4,6(1H,3H,5H)-PYRIMIDINETRIONE
see TDA500
2-CYCLOHEXENYL HYDROPEROXIDE see CPE125
5-(1-CYCLOHEXENYL-1)-1-METHYL-5-METHYLBARBITURIC ACID
see ERD500
5-(Δ-1,2-CYCLOHEXENYL)-5-METHYL-N-METHYL-BARBITURSAEURE (GER-
MAN) see ERD500
p-(2-CYCLOHEXEN-1-YLOXY)BENZOIC ACID, 3-(2-METHYL-1-
PYRROLIDINYL)PROPYL ESTER see UAG025
1-(2-(1-CYCLOHEXEN-1-YL)PHENOXY)-3-((1-METHYLETHYL)AMINO)-2-
PROPANOL HYDROCHLORIDE see ERE100
CYCLOHEXENYL TRICHLOROSILANE see CPE500
CYCLOHEXIMIDE see CPE750
CYCLOHEXYL ACETATE see CPF000
CYCLOHEXYLACETIC ACID ALLYL ESTER see AGC250
CYCLOHEXYL ALCOHOL see CPB750
CYCLOHEXYLALLYL-ESSIGSAEUREESTER DES 3-DIAETHYLAMINO-2,2-
DIMETHYL-1-PROPANOL (GERMAN) see AGC750
CYCLOHEXYLAMIDOSULPHURIC ACID see CPQ625
CYCLOHEXYLAMINE see CPF500
CYCLOHEXYLAMINE SULFATE see CPF750
CYCLOHEXYLAMINESULPHONIC ACID see CPQ625
CYCLOHEXYLAMINO ACETIC ACID see CPG000
2-(CYCLOHEXYLAMINO)ETHANOL see CPG125
2-(2-(CYCLOHEXYLAMINO)ETHYL-2-METHYL-1,3-BENZODIOXOLE HYDRO-
CHLORIDE see CPG250
3-(2-(CYCLOHEXYLAMINO)ETHYL)-2-(3,4-METHYLENEDIOXYPHENYL)-4-
THIAZOLIDINONE HYDROCHLORIDE see WBS860
N-CYCLOHEXYL-N'-(3-AMINO-4-METHYLBENZENESULFONYL)UREA
see AKQ250
4-(CYCLOHEXYLAMINO)-1-(NAPHTHALENYLOXY)-2-BUTANOL see CPG500
dl-1-CYCLOHEXYL-2-AMINOPROPANE HYDROCHLORIDE see CPG625
1-(CYCLOHEXYLAMINO)-2-PROPANOL BENZOATE (ESTER) HYDROCHLO-
RIDE see COU250
CYCLOHEXYLAMMONIUM FORMATE see CPH250
CYCLOHEXYLAMMONIUM STEARATE see CPH500

3-(5-CYCLOHEXYL-o-ANISOYL)-PROPIONIC ACID SODIUM SALT
see MEK350
N-CYCLOHEXYL-1-AZIRIDINECARBOXAMIDE see CPI000
CYCLOHEXYLBENZENE see PER750
N-CYCLOHEXYL-2-BENZOTHIAZOLESULFENAMIDE see CPI250
2-(α-CYCLOHEXYLBENZYL)-N,N,N',N'-TETRAETHYL-1,3-PROPANEDIAMINE
HYDROCHLORIDE see LFK200
N-CYCLOHEXYL-1-BUTANESULFONAMIDE see BSM250
N-CYCLOHEXYLCARBAMIC ACID 1-PHENYL-1-(3,4-XYLYL)-2-PROPYNYL
ESTER see PGQ000
CYCLOHEXYL-N-CARBAMOYLAZIRIDINE see CPI000
N-CYCLOHEXYL-N-CARBAMOYLAZIRIDINE see CPI000
2-CYCLOHEXYLCARBONYL-1,2,3,6,7,11b-HEXAHYDRO-4H-PYRAZINO(2,1-
a)ISOQUINOLIN-4-ONE see BGB400
1-(CYCLOHEXYLCARBONYL)-3-METHYLPIPERIDINE see CPI350
((CYCLOHEXYLCARBONYL)OXY)TRIBUTYLSTANNANE see TID100
CYCLOHEXYLCARBOXAMIDE see CPB050
CYCLOHEXYL CARBOXYAMIDE see CPB050
4-(4-CYCLOHEXYL-3-CHLOROPHENYL)-4-OXOBUTYRIC ACID see CPJ000
4-(4-CYCLOHEXYL-3-CHLOROPHENYL)-4-OXOBUTYRIC ACID CALCIUM
SALT see CPJ250
CYCLOHEXYLCYANOETHYLETHANOLAMINE see CPJ500
N-CYCLOHEXYLCYCLOHEXANAMINE see DGT600
N-CYCLOHEXYL-N-DIETHYLTHIOCARBONYL SULFONAMIDE see CPK000
CYCLOHEXYLDIMETHYLAMINE see DRF709
N-CYCLOHEXYLDIMETHYLAMINE see DRF709
3-CYCLOHEXYL-6-(DIMETHYLAMINO)-1-METHYL-s-TRIAZINE-2,4(1H,3H)-
DIONE see HFA300
3-CYCLOHEXYL-6-(DIMETHYLAMINO)-1-METHYL-1,3,5-TRIAZINE-2,4(1H,3H)-
DIONE see HFA300
2-CYCLOHEXYL-4,6-DINITROFENOL (DUTCH) see CPK500
2-CYCLOHEXYL-4,6-DINITROPHENOL see CPK500
6-CYCLOHEXYL-2,4-DINITROPHENOL see CPK500
(+)-1-CYCLOHEXYL-4-(1,2-DIPHENYLETHYL)PIPERAZINE DIHYDROCHLO-
RIDE see CPK625
(S)-1-CYCLOHEXYL-4-(1,2-DIPHENYLETHYL)-PIPERAZINE DIHYDROCHLO-
RIDE see CPK625
(±)-1-CYCLOHEXYL-4-(1,2-DIPHENYLETHYL)PIPERAZINE DIHYDROCHLO-
RIDE see MRU757
N,N'-(1,4-CYCLOHEXYLENEDIMETHYLENE)BIS(2-(1-AZIRIDINYL)
ACETAMIDE) see CPL100
trans-N,N'-(1,4-CYCLOHEXYLENEDIMETHYLENE)BIS(2-CHLOROBENZYLAM-
INE) DIHYDROCHLORIDE see BHN000
CYCLOHEXYLENE OXIDE see CPD000
2-CYCLOHEXYLETHANOL see CPL250
CYCLOHEXYLETHYL ACETATE see EHS000
CYCLOHEXYLETHYL ALCOHOL see CPL250
CYCLOHEXYLETHYLCARBAMOTHIOIC ACID-S-ETHYL ESTER see EHT500
1-CYCLOHEXYL-4-(ETHYL-p-METHOXY-α-METHYLPHENETHYL)AMINO)-1-
BUTANONE HYDROCHLORIDE see SBN300
CYCLOHEXYLETHYLTHIOCARBAMIC ACID-S-ETHYL ESTER see EHT500
CYCLOHEXYL FLUOROETHYL NITROSOUREA see CPL750
CYCLOHEXYL FLUOROETHYL NITROSOUREA see FIJ000
3-CYCLOHEXYL-1-(2-FLUOROETHYL)-1-NITROSOUREA see CPL750
N'-CYCLOHEXYL-N-(2-FLUOROETHYL)-N-NITROSOUREA see CPL750
(+)-α-CYCLOHEXYL-α-HYDROXY-BENZENEACETIC ACID-1-METHYL-3-
PIPERIDINYL ESTER, HCl see PMS800
α-CYCLOHEXYL-β-HYDROXY-Δα,β-BUTENOLID (GERMAN) see CPM250
3-CYCLOHEXYL-4-HYDROXY-2(5H)FURANONE see CPM250
N-CYCLOHEXYLHYDROXYLAMINE see HKA109
6-CYCLOHEXYL-1-HYDROXY-4-METHYL-2(1H)-PYRIDINONE compounded
with 2-AMINOETHANOL (1:1) see BAR800
6-CYCLOHEXYL-1-HYDROXY-4-METHYL-2(1H)-PYRIDON, 2-
AMINOETHANOL-SALZ (GERMAN) see BAR800
6-CYCLOHEXYL-1-HYDROXY-4-METHYL-2(1H)-PYRIDONE, 2-
AMINOETHANOL-SALT see BAR800
6-CYCLOHEXYL-1-HYDROXY-4-METHYL-2(1H)-PYRIDONE ETHANOLAMINE
SALT see BAR800
4-(β-CYCLOHEXYL-β-HYDROXYPHENETHYL)-1,1-DIMETHYLPIPERAZIN-
IUM METHYLSULFATE see HFG400
4-(β-CYCLOHEXYL-β-HYDROXYPHENETHYL)-1,1-DIMETHYL PIPERAZIN-
IUM SULFATE see HFG000
N-(β-CYCLOHEXYL-β-HYDROXY-β-PHENYLETHYL)-N'-METHYLPIPERAZ-
INE DIMETHYLSULFATE see HFG400
1-(3-CYCLOHEXYL-3-HYDROXY-3-PHENYLPROPYL)-1-METHYL-PIPERIDIN-
IUM IODIDE see CPM750
1-(3-CYCLOHEXYL-3-HYDROXY-3-PHENYLPROPYL)-1-METHYL-
PYRROLIDINIUM CHLORIDE see EAI875
(±)-N-((3-CYCLOHEXYL-3-HYDROXY-3-PHENYL)PROPYL)-N-
METHYLPYRROLIDINIUM CHLORIDE see EAI875
1-(3-CYCLOHEXYL-3-HYDROXY-3-PHENYLPROPYL)-1-METHYL-
PYRROLIDINIUM IODIDE see CNH550
1-(3-CYCLOHEXYL-3-HYDROXY-3-PHENYLPROPYL)-1-METHYL-
PYRROLIDINIUM METHYL SULFATE see TJG225

2,2'-(CYCLOHEXYLIDENEBIS(4,1-PHENYLENEOXY)BIS(2-
    METHYLBUTANOIC ACID) see CMV700
2,2'-(4,4'-CYCLOHEXYLIDENEDIPHENOXY)-2,2-DIMETHYLDIBUTYRIC
    ACID see CMV700
2,2'-CYCLOHEXYLIMINODIETHANOL see DMT400
CYCLOHEXYL ISOCYANATE see CPN500
2-(N-CYCLOHEXYL-N-ISOPROPYLAMINOMETHYL)-1,3,4-OXADIAZOLE
    see CPN750
CYCLOHEXYLISOPROPYLMETHYLAMINE HYDROCHLORIDE see PNN300
CYCLOHEXYL-ISOTHIOCYANAT (GERMAN) see ISJ000
CYCLOHEXYLMETHANE see MIQ740
CYCLOHEXYL METHYL AMINE see MIT000
1-CYCLOHEXYL-2-METHYLAMINOPROPAN (GERMAN) see PNN400
1-CYCLOHEXYL-2-METHYLAMINOPROPANE HYDROCHLORIDE see PNN300
1,1-CYCLOHEXYL-2-METHYLAMINOPROPANE-5,5-
    PHENYLETHYLBARBITURATE see CPO500
N-(2-CYCLOHEXYL-1-METHYLETHYL)-3,3-DIPHENYLPROPYLAMINE HY-
    DROCHLORIDE see CPP000
1-CYCLOHEXYL-3-((p-(2-(5-METHYL-3-
    ISOXAZOLECARBOXAMIDO)ETHYL)PHENYL)SULFONYL)UREA
    see DBE885
4-(CYCLOHEXYLMETHYL)-α-(4-METHOXYPHENYL)-β-METHYL-1-
    PIPERIDINEETHANOL see RCA435
CYCLOHEXYL METHYLPHOSPHONOFLUORIDATE see MIT600
1-(CYCLOHEXYLMETHYL)PIPERIDINE HYDROCHLORIDE see PIR000
2-(4-CYCLOHEXYLMETHYLPIPERIDINO)-1-(4-METHOXYPHENYL)-1-
    PROPANOL see RCA435
1-CYCLOHEXYL-N-METHYL-2-PROPANAMINE see PNN400
1-CYCLOHEXYL-1-NITROSOUREA see NJV000
1-(2-CYCLOHEXYLPHENOXY)-1-(2-IMIDAZOLINYL)ETHANE HYDROCHLO-
    RIDE see CPP750
α-CYCLOHEXYL-α-PHENYL-1-PIPERIDINEPROPANOL see PAL500
α-CYCLOHEXYL-α-PHENYL-1-PIPERIDINEPROPANOL HYDROCHLORIDE
    see BBV000
2-CYCLOHEXYL-2-PHENYL-4-PIPERIDINOMETHYL-DIOXOLANE-1,3
    METHIODIDE see OLW400
2-CYCLOHEXYL-2-PHENYL-1-PIPERIDINO-1-PROPANOL see PAL500
1-CYCLOHEXYL-1-PHENYL-3-PIPERIDINO-PROPANOL, METHYLIODIDE
    see CPM750
1-CYCLOHEXYL-1-PHENYL-3-PYRROLIDINO-1-PROPANOL see CPQ250
1-CYCLOHEXYL-1-PHENYL-3-PYRROLIDINO-1-PROPANOL METHSULFATE
    see TJG225
1-CYCLOHEXYL-1-PHENYL-3-PYRROLIDINO-1-PROPANOL METHYL CHLO-
    RIDE see EAI875
1-CYCLOHEXYL-1-PHENYL-3-(1-PYRROLIDINYL)-1-PROPANOL see CPQ250
α-CYCLOHEXYL-α-(2-(PIPERIDINO)ETHYL)-BENZYLALCOHOL METHYLIOD-
    IDE see CPM750
N-CYCLOHEXYLPYRROLIDINONE see CPQ275
1-CYCLOHEXYL-2-PYRROLIDINONE see CPQ275
N-CYCLOHEXYLPYRROLIDONE see CPQ275
CYCLOHEXYLSULFAMIC ACID (9CI) see CPQ625
CYCLOHEXYL SULPHAMATE SODIUM see SGC000
CYCLOHEXYLSULPHAMIC ACID see CPQ625
N-CYCLOHEXYLSULPHAMIC ACID see CPQ625
CYCLOHEXYLSULPHAMIC ACID, CALCIUM SALT see CAR000
3-CYCLOHEXYLSYDNONE IMINE MONOHYDROCHLORIDE see CPQ650
6-(4-(1-CYCLOHEXYL-1H-TETRAZOL-5-YL)BUTOXY)-3,4-DIHYDRO-2(1H)-
    QUINOLINONE see CMP825
1-CYCLOHEXYL-3-p-TOLUENESULFONYLUREA see CPR000
1-CYCLOHEXYL-3-p-TOLYSULFONYLUREA see CPR000
CYCLOHEXYLTRICHLOROSILANE see CPR250
1-CYCLOHEXYLTRIMETHYLAMINE see CPR500
CYCLOL ACRYLATE see BFY250
CYCLOLEUCINE see AJK250
CYCLOLYT see DNU100
CYCLOMALTOHEPTAOSE see COW925
CYCLOMANDOL see DNU100
CYCLOMEN see DAB830
CYCLOMETHIAZIDE see CPR750
CYCLOMIDE DIN 295/S see BKF500
CYCLOMORPH see COX000
CYCLOMYCIN see CQH000
CYCLOMYCIN see TBX000
CYCLON see HHS000
CYCLONAL see ERD500
CYCLONAL SODIUM see ERE000
CYCLONAMINE see DIS600
CYCLONE B see HHS000
CYCLONITE see CPR800
CYCLONIUM IODIDE see OLW400
CYCLONOL see TLO500
cis,cis-CYCLOOCTA-1,5-DIENE see CPR825
(1,5-CYCLOOCTADIENE)(2,4-PENTANEDIONATO)RHODIUM see CPR840
1,5-CYCLOOCTADIENE (Z,Z) see CPR825

CYCLOOCTAFLUOROBUTANE see CPS000
CYCLOOCTANECARBAMIC ACID-1,1-BIS(p-FLUOROPHENYL)-2-PROPYNYL
    ESTER see BJW500
CYCLOOCTANONE see CPS250
1,3,5,7-CYCLOOCTATETRAENE see CPS500
3-CYCLOOCTYL-1,1-DIMETHYLHARNSTOFF (GERMAN) see CPT000
3-CYCLOOCTYL-1,1-DIMETHYLUREA see CPT000
N-CYCLOOCTYL-N',N'-DIMETHYLUREA see CPT000
3-CYCLOOCTYL-1,1-DIMETHYL UREA mixed with BUTYNYL-3N-3-
    CHLOROPHENYLCARBAMATE see AFM375
CYCLOPAMINE see CPT750
CYCLOPAN see ERD500
CYCLOPAR see TBX250
CYCLOPENIL see FBP100
4H-CYCLOPENTA(def)CHRYSENE see CPU000
CYCLOPENTADECANONE see CPU250
CYCLOPENTADECANONE, 3-METHYL- see MIT625
CYCLOPENTADIENE see CPU500
1,3-CYCLOPENTADIENE see CPU500
1,3-CYCLOPENTADIENE, DIMER see DGW000
pi-CYCLOPENTADIENYL COMPOUND with NICKEL see NDA500
CYCLOPENTADIENYL GOLD(1) see CPU750
CYCLOPENTADIENYLMANGANESE TRICARBONYL see CPV000
CYCLOPENTADIENYL SILVER PERCHLORATE see CPV250
CYCLOPENTADIENYL SODIUM see CPV500
α,β-CYCLOPENTAMETHYLENETETRAZOLE see PBI500
CYCLOPENTAMINE HYDROCHLORIDE see CPV609
CYCLOPENTA(de)NAPHTHALENE see AAF500
CYCLOPENTANE see CPV750
1,3-CYCLOPENTANEDISULFONYL DIFLUORIDE see CPW250
4,5-CYCLOPENTANOFURAZAN-N-OXIDE see CPW325
CYCLOPENTANONE see CPW500
CYCLOPENTANONE-2-α,3-α-EPITHIO-5-α-ANDROSTAN-17-β-YL METHYL AC-
    ETAL see MCH600
CYCLOPENTANONE OXIME see CPW750
CYCLOPENTANONE, 3-(2-OXOPROPYL)-2-PENTYL- see OOO100
CYCLOPENTAPHENANTHRENE see CPX250
4H-CYCLOPENTA(def)PHENANTHRENE see CPX250
CYCLOPENTA(cd)PYRENE see CPX500
CYCLOPENTA(cd)PYRENE-3,4-OXIDE see CPX625
CYCLOPENTENE see CPX750
2-CYCLOPENTENE-1-OL see CPY000
2-CYCLOPENTENE-1-TRIDECANOIC ACID, SODIUM SALT see SFR000
1,2-CYCLOPENTENO-5,10-ACEANTHRENE see CPY500
5586-CYCLOPENTENO-1582-BENZANTHRACENE see CPY750
6,7-CYCLOPENTENO-1,2-BENZANTHRACENE see BCI250
1-CYCLOPENTEN-3-OL see CPY000
2-CYCLOPENTEN-1-ONE, 2-HEXYL- see HFO700
CYCLOPENTENO(c,d)PYRENE see CPX500
2-CYCLOPENTENYL-4-HYDROXY-3-METHYL-2-CYCLOPENTEN-1-ONE
    CHRYSANTHEMATE see POO000
3-(2-CYCLOPENTEN-1-YL)-2-METHYL-4-OXO-2-CYCLOPENTEN-1-YL
    CHRYSANTHEMUMATE see POO000
3-(2-CYCLOPENTENYL)-2-METHYL-4-OXO-2-CYCLOPENTENYL
    CHRYSANTHEMUMMONOCARBOXYLATE see POO000
CYCLOPENTENYLRETHONYL CHRYSANTHEMATE see POO000
CYCLOPENTHIAZIDE see CPR750
CYCLOPENTIMINE see PIL500
CYCLOPENTOLATE HYDROCHLORIDE see CPZ125
CYCLOPENTYLAMINE see CQA000
3-(α-CYCLOPENTYL-4,6-DIMETHOXY-m-TOLUOYL)-PROPIONIC ACID SO-
    DIUM SALT see DOB325
2-CYCLOPENTYL-4,6-DINITROPHENOL see CQB250
3-CYCLOPENTYL ENOL ETHER of NORETHINDRONE ACETATE see QFA275
CYCLOPENTYL ETHER see CQB275
α-CYCLOPENTYLMANDELIC ACID (1-ETHYL-2-PYRROLIDINYL)METHYL
    ESTER HYDROCHLORIDE see PJI575
α-CYCLOPENTYLMANDELIC ACID-1-METHYL-3-PYRROLIDINYL ESTER HY-
    DROCHLORIDE see AEY400
3-CYCLOPENTYLMETHYL HYDROCHLOROTHIAZIDE DERIV see CPR750
3-(CYCLOPENTYLOXY)-19-NOR-17-α-PREGNA-3,5-DIEN-20-YN-17-OL ACE-
    TATE (ester) see QFA275
3-(CYCLOPENTYLOXY)-19-NOR-17-α-PREGNA-1,3,5(10)-TRIEN-20-YN-17-OL
    see QFA250
S-2-((5-CYCLOPENTYLPENTYL)AMINO)ETHYL THIOSULFATE see CQC250
1-(2-CYCLOPENTYLPHENOXY)-3-((1,1-DIMETHYLETHYL)AMINO)-2-PRO
    PANOL, (S)- see PAP225
α-CYCLOPENTYL-α-PHENYL-1-PIPERIDINEPROPANOL HYDROCHLORIDE
    see CQH500
α-CYCLOPENTYL-2-THIOPHENEGLYCOLATE DIETHYL(2-
    HYDROXYETHYL)METHYLAMMONIUM BROMIDE see PBS000
α-CYCLOPENTYL-2-THIOPHENEGLYCOLIC ACID-2-
    (DIETHYLAMINO)ETHYL ESTER HYDROCHLORIDE see DHW400
CYCLOPHENYL see FBP100

CYCLOHEPTAGLUCAN see COW925
CYCLOPHOSPHAMIDE see CQC650
CYCLOPHOSPHAMIDE HYDRATE see CQC500
CYCLOPHOSPHAMIDE and MNU (1582) see CQC600
CYCLOPHOSPHAMIDE-N-MONOCHLOROETHYL derivative see TNT500
CYCLOPHOSPHAMIDE MONOHYDRATE see CQC500
CYCLOPHOSPHAMIDUM see CQC500, CQC650
CYCLOPHOSPHAN see CQC500, CQC650
CYCLOPHOSPHANE see CQC500
CYCLOPHOSPHANUM see CQC500
CYCLOPHOSPHORAMIDE see CQC650
CYCLOPIAZONIC ACID see CQD000
α-CYCLOPIAZONIC ACID see CQD000
CYCLOPRATE see HCP500
5H-CYCLOPROPA(3,4)BENZ(1,2-e)AZULEN-5-ONE, 1,1a,1b,4,4a,7a,7b,8,9,9a-
  DECAHYDRO-4a,7-β, 9,9a-TETRAHYDROXY-3-(HYDROXYMETHYL)-1,1,6,8-
  TETRAMETHYL-, 9-ACETATE 9a-LAURATE see PGS500
5H-CYCLOPROPA(3,4)BENZ(1,2-e)AZULEN-5-ONE,1,1a,1b-β,4,4a,7a-
  α,7b,8,9,9a-DECAHYDRO-4a-α,7b-α,9a-α-TRIHYDROXY-3-
  HYDROXYMETHYL-1,6,8α-TRIMETHYL-1-ACETOXYMETHYL-,9a-(2-
  METHYLBUT-2-ENOATE) see CQD250
CYCLOPROPANAMINE, 2-PHENYL-, trans-(±)-, SULFATE (2:1) see PET500
CYCLOPROPANE see CQD750
CYCLOPROPANE, liquefied (DOT) see CQD750
CYCLOPROPANECARBOXYLIC ACID, 2,2-DIMETHYL-3-(2-
  METHYLPROPENYL)-, p-(METHOXYMETHYL) BENZYL ESTER
  see MNG525
CYCLOPROPANECARBOXYLIC ACID, 2,2-DIMETHYL-3-(2-
  METHYLPROPENYL)-, (2-METHYL-5-(2-PROPYNYL)-3-FURYL)METHYL
  ESTER see PMN700
CYCLOPROPANECARBOXYLIC ACID, HEXADECYL ESTER see HCP500
CYCLOPROPYLAMINE see CQE250
N-CYCLOPROPYL-4-AMINO-3,5-DICHLOROBENZAMIDE see AJK500
5-(CYCLOPROPYLCARBONYL)-2-BENZIMIDAZOLECARBAMIC ACID
  METHYL ESTER see CQE325
N-CYCLOPROPYL-3,5-DICHLORO-4-AMINOBENZAMIDE see AJK500
1-(4-(2-(CYCLOPROPYLMETHOXY)ETHYL)PHENOXY)-3-
  ISOPROPYLAMINOPROPAN-2-OL HYDROCHLORIDE
  see KEA350
1-N-CYCLOPROPYLMETHYL-3,14-DIHYDROXYMORPHINAN see CQF079
2-CYCLOPROPYLMETHYL-5,9-DIMETHYL-2'-HYDROXY-6,7-
  BENEOMORPHAN see COV500
3-CYCLOPROPYLMETHYL-6(eq),11(ax)-DIMETHYL-2,6-METHANO-3-
  BENZAZOCIN-8-OL see COV500
(5-α)-17-(CYCLOPROPYLMETHYL-4,5-EPOXY-3,14-DIHYDROXY-MORPHINAN-
  6-ONE (9CI) see CQF099
CYCLOPROPYL METHYL ETHER see CQE750
3-(CYCLOPROPYLMETHYL)1-1,2,3,4,5,6-HEXAHYDRO-6,11-DIMETHYL-2,6-
  METHANO-3-BENZAZOCIN-8-OL see COV500
N-CYCLOPROPYLMETHYL-14-HYDROXYDIHYDROMORPHINONE
  see CQF099
2-CYCLOPROPYLMETHYL-2'-HYDROXY-5,9-DIMETHYL-6,7-
  BENZOMORPHAN see COV500
CYCLOPROPYL METHYL KETONE see CQF059
(−)-17-CYCLOPROPYLMETHYLMORPHINAN-3,4-DIOL see CQF079
17-(CYCLOPROPYLMETHYL)MORPHINAN-3-OL see CQG750
N-CYCLOPROPYLMETHYLNOROXYMORPHONE see CQF099
1-CYCLOPROPYLMETHYL-4-PHENYL-6-CHLORO-2(1H)-QUINAZOLINONE
  see CQF125
N-(CYCLOPROPYLMETHYL)-α,α,α-TRIFLUORO-2,6-DINITRO-N-PROPYL-p-
  TOLUIDINE see CQG250
1-(o-CYCLOPROPYLPHENOXY)-3-(ISOPROPYLAMINO)-2-PROPANOL HYDRO-
  CHLORIDE see PMF525
dl-1-(o-CYCLOPROPYLPHENOXY)-3-ISOPROPYLAMINO-2-PROPANOL HY-
  DROCHLORIDE see PMF535
CYCLORPHAN see CQG750
CYCLORYL 21 see SIB600
CYCLORYL TAWF see SON000
CYCLORYL WAT see SON000
CYCLOSAN see MCW000
CYCLOSERINE see CQH000
CYCLO-d-SERINE see CQH000
CYCLOSIA see CMS850
CYCLOSPASMOL see DNU100
CYCLOSPORIN see CQH100
CYCLOSPORIN A see CQH100
CYCLOSPORINE see CQH100
CYCLOSPORINE A see CQH100
CYCLOSTIN see CQC650
CYCLOTEN see HMB500
CYCLOTETRAMETHYLENE OXIDE see TCR750
CYCLOTETRAMETHYLENE SULFONE see SNW500
CYCLOTETRAMETHYLENE TETRANITRAMINE see CQH250
CYCLOTETRAMETHYLENE TETRANITRAMINE, dry (DOT) see CQH250

CYCLOTETRASILOXANE, 2,6-DIPHENYL-2,4,4,6,8,8-HEXAMETHYL-
  see DWC650
CYCLOTON V see HCQ500
CYCLOTRIMETHYLENENITRAMINE see CPR800
CYCLOTRIMETHYLENETRINITRAMINE see CPR800
CYCLOTRIMETHYLENETRINITRAMINE, containing at least 10%-25% water
  (DOT) see CPR800
CYCLOTRIMETHYLENETRINITRAMINE, desensitized (DOT) see CPR800
CYCLOTRISILOXANE, 2,4-DIPHENYL-2,4,6,6-TETRAMETHYL-, (E)-
  see DWN100
CYCLOURON see CPT000
CYCLOVIROBUXIN D see CQH325
CYCLOVIROBUXINE see CQH325
CYCLOVIROBUXINE D see CQH325
CYCLURON see CPT000
CYCOCEL see CMF400
CYCOCEL-EXTRA see CMF400
CYCOGAN see CMF400
CYCOGAN EXTRA see CMF400
CYCOLAMIN see VSZ000
CYCRIMINE HYDROCHLORIDE see CQH500
CYCTEINAMINE see AJT250
dCYD see DAQ850
CYDRIN see CQK500
CYFEN see DSQ000
CYFLEE see FAB600
CYFOS see IMH000
CYGON see DSP400
CYGON INSECTICIDE see DSP400
CYHEPTAMIDE see CQH625
CYHEPTAMINE see CQH625
CYHEXATIN see CQH650
CYJANOWODOR (POLISH) see HHS000
CYKAZINE see COU000
CYKLOHEKSAN (POLISH) see CPB000
CYKLOHEKSANOL (POLISH) see CPB750
CYKLOHEKSANON (POLISH) see CPC000
CYKLOHEKSEN (POLISH) see CPC579
CYKLOHEXANTHIOL see CPB625
CYKLOHEXYLAMINOACETAT (CZECH) see CPG000
CYKLOHEXYLESTER KYSELINY THIOKYANOOCTOVE (CZECH) see TFF000
CYKLOHEXYLMERKATPAN (CZECH) see CPB625
CYKLOHEXYLTHIOKYANOACETAT (CZECH) see TFF000
CYKOBEMINET see VSZ000
CY-L 500 see CAQ250
CYLAN see CAR000, DXN600, PGW750
CYLERT see PAP000
CYLOCIDE see AQR000
CYLPHENICOL see CDP250
CYMAG see SGA500
CYMARIGENIN see SMM500
CYMARIN see CQH750
CYMARINE see CQH750
3-β-(β-d-CYMAROSYLOXY)-5,14-DIHYDROXY-19-OXO-5-β-CARD-20(22)-ENOL-
  IDE see CQH750
CYMATE see BJK500
CYMBI see AIV500, AOD125
CYMBUSH see RLF350
CYMEL see MCB000
CYMENE see CQI000
p-CYMENE see CQI000
p-CYMENE-7-CARBOXALDEHYDE see IRA000
2-p-CYMENOL see CCM000
3-p-CYMENOL see TFX810
p-CYMEN-3-OL see TFX810
p-CYMEN-7-OL see CQI250
CYMETHION see MDT750
CYMETOX see MIW250
CYMOL see CQI000
CYMONIC ACID see FIC000
m-CYM-5-YL METHYLCARBAMATE see CQI500
CYNARON see MDT740
CYNEM see EPC500
CYNKOMIEDZIAN see ZJS300
CYNKOTOX see EIR000
CYNKU TLENEK (POLISH) see ZKA000
CYNOGAN see BMM650
CYNOTOXIN see SMM500
CYOCEL see CMF400
CYODRIN see COD000
CYOLAN see DXN600
CYOLANE see PGW750
CYOLANE INSECTICIDE see DXN600
CYOLANE INSECTICIDE see PGW750

CYOLANE INSECTICIDE see DXN600
CYOLANE INSECTICIDE see PGW750
CYP see CON300
CYPENTIL see PIL500
CYPERKILL see RLF350
CYPERMETHRIN see RLF350
CYPERUS SCARIOSUS OIL see NAE505
CYPIP see DIW000
CYPONA see DGP900
CYPRAZINE see CQI750
CYPRESS OIL see CQJ000
CYPREX see DXX400
CYPREX 65W see DXX400
CYPROHEPTADIENE HYDROCHLORIDE see PCI250
CYPROHEPTADINE see PCI500
CYPROHEPTADINE HYDROCHLORIDE see PCI250
CYPROMID see CQJ250
CYPRON see MQU750
CYPRONIC ETHER see CQE750
CYPROSTERONE ACETATE see CQJ500
CYPROTERONE ACETATE see CQJ500
CYPROTERON-R ACETATE see CQJ500
CYRAL see DBB200
CYREDIN see VSZ000
CYREN see DKA600
CYREN B see DKB000
CYRSTHION see EKN000
CYSTAMIN see HEI500
CYSTAMINE see MCN500
CYSTAMINE DIHYDROCHLORIDE see CQJ750
CYSTAMINE "MCCLUNG" see PDC250
CYSTAPHOS see AKB500
CYSTAPHOS SODIUM SALT see AKB500
CYSTEAMIDE see AJT250
CYSTEAMINE see AJT250
CYSTEAMINE HYDROCHLORIDE see MCN750
CYSTEAMINHYDROCHLORID (GERMAN) see MCN750
CYSTEIN see CQK000
CYSTEINAMINE DISULFIDE see MCN500
CYSTEINE see CQK000
l-CYSTEINE see CQK000
l-(+)-CYSTEINE see CQK000
CYSTEINE CHLORHYDRATE see CQK250
CYSTEINE DISULFIDE see CQK325
CYSTEINE ETHYL ESTER HYDROCHLORIDE see EHU600
CYSTEINE-GERMANIC ACID see CQK100
CYSTEINE HYDRAZIDE see CQK125
CYSTEINE HYDROCHLORIDE see CQK250
l-CYSTEINE HYDROCHLORIDE see CQK250
l-CYSTEINE MONOHYDROCHLORIDE (FCC) see CQK250
(l-CYSTEINE)TETRAHYDROXYGERMANIUM see CQK100
l-CYSTEIN HYDROCHLORIDE see CQK250
CYSTIN see CQK325
CYSTINAMIN (GERMAN) see MCN500
(−)-CYSTINE see CQK325
l-CYSTINE see CQK325
CYSTINE ACID see CQK325
CYSTINEAMINE see MCN500
l-CYSTINE-BIS(N,N-β-CHLOROETHYL)HYDRAZIDEHYDROBROMIDE
    see CQK500
CYSTISINE see CQL500
CYSTO-CONRAY see IGC000
CYSTOGEN see HEI500
CYSTOGRAFIN see AOO875
CYSTOIDS ANTHELMINTIC see HFV500
CYSTOKON see AAN000
CYSTOPYRIN see PDC250
CYSTORELIN see LIU360
CYSTURAL see PDC250
CYTACON see VSZ000
CYTADREN see AKC600
CYTAMEN see VSZ000
CYTARABINE HYDROCHLORIDE see AQR000
CYTEL see DSQ000
CYTEMBENA see CQK600
CYTEMBENA see SIK000
CYTEN see DSQ000
CYTHIOATE see CQL250
CYTHION see MAK700
CYTIDINDIPHOSPHOCHOLIN see CMF350
CYTIDINE see CQM500
CYTIDINE CHOLINE DIPHOSPHATE see CMF350
CYTIDINE 5'-(CHOLINE DIPHOSPHATE) see CMF350
CYTIDINE DIPHOSPHATE CHOLINE see CMF350

CYTIDINE 5'-DIPHOSPHATE CHOLINE see CMF350
CYTIDINE DIPHOSPHATE CHOLINE ESTER see CMF350
CYTIDINE DIPHOSPHATE CHOLIN ESTER see CMF350
CYTIDINE DIPHOSPHOCHOLINE see CMF350
CYTIDINE 5'-DIPHOSPHOCHOLINE see CMF350
CYTIDINE DIPHOSPHORYLCHOLINE see CMF350
CYTIDOLINE see CMF350
CYTISINE see CQL500
CYTISINE HYDROCHLORIDE see CQL750
(−)7R:9S-CYTISINE HYDROCHLORIDE see CQL750
CYTITONE see CQL500
CYTOBION see VSZ000
CYTOCHALASIN B see CQM125
CYTOCHALASIN D see ZUS000
CYTOCHALASIN E see CQM250
CYTOCHALASIN-H see PAM775
CYTOCHROME C see CQM325
CYTOPHOSPHAN see CQC500, CQC650
CYTOREST see CQM325
CYTOSAR HYDROCHLORIDE see AQR000
CYTOSINE-β-ARABINOSIDE see AQQ750
CYTOSINE β-d-ARABINOSIDE see AQQ750
CYTOSINE ARABINOSIDE HYDROCHLORIDE see AQR000
CYTOSINE ARABINOSIDE PALMITATE see AQS875
CYTOSINE DEOXYRIBOSIDE see DAQ850
CYTOSINE RIBOSIDE see CQM500
CYTOSTASAN see CQM750
CYTOVIRIN see BLX500
CYTOXAL ALCOHOL see CQN000
CYTOXAN see CQC500, CQC650
CYTOXYL ALCOHOL CYCLOHEXYLAMMONIUM SALT see CQN000
CYTOXYL AMINE see CQN125
CYTROL see AMY050
CYTROLANE see DHH400
CYURAM DS see TFS350
CYZONE see PFL000
CZON see CCS635
CZT see CLX000
CZTEROCHLOREK WEGLA (POLISH) see CBY000
2,3,7,8-CZTEROCHLORODWUBENZO-p-DWUOKSYNY (POLISH) see TAI000
1,1,2,2-CZTEROCHLOROETAN (POLISH) see TBQ100
CZTEROCHLOROETYLEN (POLISH) see PCF275
CZTEROETHLEK OLOWIU (POLISH) see TCF000

D₂ see DBB800
D-13 see AHL000
2,4-D see DAA800
3,4-D see DFY500
D-40 see AFO250
D-50 see POL500
D 50 see AAX250, DAA800
D 109 see COU250
D 206 see DYB600
D 268 see DRW000
D-365 see IRV000
D-638 see BPJ750
D-649 see BPJ500
D-695 see BKC500
D-701 see BKD000
D-703 see EEY000
D 735 see CCC500
838-D see MHQ775
D 854 see CJT750
D 860 see BSQ000
D-1126 see FMU225
D 1221 see CBS275
D 1308 see EOS100
D-1410 see DSP600
D 1593 see CIP500
D-10,242 see DAB200
D-13,312 see TAL560
DA see CGN000, DNA200
3,4-DA see DFY500
D-90-A see CGL250
DA-241 see PEO750
DA 339 see DWA500
DA-398 see MCH550
DA79P see LGF875
DA-1773 see SJJ175
DA 2370 see PEW000
2,4-DAA see DBO000
1,2-DAA (RUSSIAN) see DBO800
DAAB see DWO800
DAAE see DCN800

2,4-DAA SULFATE see DBO400
DAB see BIU750, DOT300
DABCO see DCK400
DABCO S-25 see DCK400
DABCO CRYSTAL see DCK400
DABCO EG see DCK400
DABCO R-8020 see DCK400
DABCO 33LV see DCK400
DABI see DOT600
DAB-O-LITE P 4 see WCJ000
DAB-N-OXIDE see DTK600
DABROSIN see ZVJ000
DABYLEN see BAU750, BBV500
DAC 2797 see TBQ750
DACAMINE see DAA800, TAA100
DACAMOX see DAB400
DACARBAZINE see DAB600
2,4-D ACETATE see DFY800
2,4-D ACID see DAA800
DACONATE 6 see MRL750
DACONIL see TBQ750
DACONIL 2787 FLOWABLE FUNGICIDE see TBQ750
DACORENE HYDROCHLORIDE see BGO000
DACORTIN see PLZ000
DACOSOIL see TBQ750
DACOVIN see PKQ059
DAC PRO see DGL200
DACTHAL see TBV250
DACTIL see EOY000
DACTIL HYDROCHLORIDE see EOY000
DACTIN see DFE200
DACTINOL see RNZ000
DACTINOMYCIN see AEB000
DACTINOMYCIN (10%), ACTINOMYCIN C2 (45%), and ACTINOMYCIN C3
    (45%) mixture see AEA750
DAD see DCI600
DADDS see SNY500
DADEX see BBK500
DADIBUTOL see TGA500
DADOX d-CITRAMINE see BBK500
DADPE see OPM000
DADPS see SOA500
DAEP see DOP200
DAEP-ES see AEF000
DAF 68 see DVL700
DAFEN see LJR000
DAFF see BJR625
DAFFODIL see DAB700
DAFTAZOL HYDROCHLORIDE see DCA600
DAG see DCI600
DAGADIP see TNP250
DAGC see AGD250
DAGENAN see PPO000
DAGUTAN see SJN700
1,1-DAH see DBK100
1,2-DAH HYDROCHLORIDE see DBK120
DAI CARI XBN see BQK250
DAICEL 1150 see SFO500
DAIFLON see CLQ750
DAIFLON S 3 see FOO000
DAILON see DXQ500
DAIMETON see SNL800
DAINICHI BENZIDINE YELLOW GRT see DEU000
DAINICHI CHROME ORANGE R see LCS000
DAINICHI CHROME YELLOW G see LCR000
DAINICHI FAST SCARLET G BASE see NMP500
DAINICHI LAKE RED C see CHP500
DAINICHI PERMANENT RED GG see DVB800
DAINICHI PERMANENT RED 4 R see MMP100
DAINICHI PERMANENT RED RX see CJD500
DAIPIN see DAB750
DAIRYLIDE YELLOW AAA see DEU000
DAISEN see EIR000
DAISHIKI AMARANTH see FAG020
DAISHIKI BRILLIANT SCARLET 3R see FMU080
DAITO ORANGE BASE R see NEN500
DAITO RED BASE TR see CLK220
DAITO RED SALT TR see CLK235
DAITO SCARLET BASE G see NMP500
DAIYA FOIL see PKF750
DAKINS SOLUTION see SHU500
DAKTIN see DFE200
DAKURON see SKQ400
DALACIN C see CMV675

DALAPON see DGI600
DALAPON 85 see DGI400
DALAPON (USDA) see DGI400
DALAPON SODIUM see DGI600
DALAPON SODIUM SALT see DGI600
DAL-E-RAD see MRL750
DAL-E-RAD 100 see DXE600
DALF see MNH000
DALGOL see EQL000
DALMADORM see DAB800
DALMADORM HYDROCHLORIDE see DAB800
DALMANE see DAB800
DALMATE see DAB800
DALMATIAN SAGE OIL see SAE500
DALMATION INSECT FLOWERS see POO250
DALTOGEN see TKP500
DALTOLITE FAST YELLOW GT see DEU000
DALYSEP see MFN500
DALZIC see ECX100
DAM-57 see LJH000
DAMA de DIA (PUERTO RICO) see DAC500
DAMA de NOCHE (PUERTO RICO) see DAC500
DAMILAN see EAH500
DAMILEN HYDROCHLORIDE see EAI000
2,4-D AMINE SALT see DFY800
DAMINOZIDE (USDA) see DQD400
2,4-D AMMONIUM SALT see DAB020
1,4-DA-2-MOA (RUSSIAN) see DBX000
DAMPA D see DAB820
DAMP-ES see AEF250
DAN see DSU600
DA-2-N see DSU800
DANA see NJW500
DANABOL see DAL300
DANAMID see PJY500
DANAMINE see DJS200
DANANTIZOL see MCO500
DANAZOL see DAB830
DANDELION (JAMAICA) see CNG825
DANERAL see TMK000
DANEX see TIQ250
DANFIRM see AAX250
DANIFOS see DIX800
DANILON see SOX875
DANILONE see PFJ750
DANINON see CKL500
DANITOL see DAB825
DANIZOL see MMN250
DANOCRINE see DAB830
DANOL see DAB830
DANSYL see DPN200
DANSYL CHLORIDE see DPN200
DANTAFUR see NGE000
DANTEN see DKQ000, DNU000
DANTHION see PAK000
DANTHRON see DMH400
DANTINAL see DKQ000
DANTOIN see DFE200, DNU000
DANTOINAL KLINOS see DKQ000
DANTOINE see DKQ000
DANTOROLENE SODIUM see DAB840
DANTRIUM see DAB840
DANTRIUM HEMIHEPTAHYDRATE see DAB840
DANTROLENE see DAB845
DANTROLENE SODIUM see DAB840
DANTRON see DMH400
DANU see DPN400
DANUVIL 70 see PKQ059
DAONIL see CEH700
DAP see DOT000, POJ500
DAPA see DOU600
DAPACRYL see BGB500
DAPAZ see MQU750
DAPHENE see DSP400
DAPHNE MEZEREUM see LAR500
DAPHNETOXIN see DAB850
DAPLEN AD see PMP500
DAPON 35 see DBL200
DAPON R see DBL200
DAPRISAL see ABG750
DAPSONE see SOA500
DAPTAZILE HYDROCHLORIDE see DCA600
DAPTAZOLE HYDROCHLORIDE see DCA600
DARACLOR see TGD000

DARAL see VSZ100
DARAMIN see ANT500, CAM750
DARANIDE see DEQ200
DARAPRAM see TGD000
DARAPRIME see TGD000
DARATAK see AAX250
DARBID see DAB875
DAR-CHEM 14 see SLK000
DARCIL see PDD350
DARENTHIN see BMV750
DARID QH see SEH000
DARILOID QH see SEH000
DAROLON see ACE000
DAROPERVAMIN see DBA800
DAROTOL see CKN000
DARVIC 110 see PKQ059
DARVIS CLEAR 025 see PKQ059
DARVON see DAB879
DARVON COMPOUND see ABG750
DARVON HYDROCHLORIDE see PNA500
DARVON-N see DAB880, DYB400
DAS see AOP250, DOU600
DASANIDE see DEQ200
DASANIT see FAQ800
DASEN see SCA625
DASERD see GGS000
DASEROL see GGS000
DASHEEN see EAI600
DASIKON see ABG750
DASKIL see NCQ900
DATC see DBI200
DATHROID see LEQ300
DATRIL see HIM000
DATURALACTONE see DAB925
DATURA STRAMONIUM see SLV500
DATURINE see HOU000
DAUCUS CAROTA LINN., SEED EXTRACT see GAP500
DAUNAMYCIN see DAC000
DAUNOBLASTIN see DAC200
DAUNOBLASTINA see DAC200
DAUNOMYCIN see DAC000
DAUNOMYCIN BENZOYLHYDRAZONE see ROU800
DAUNOMYCIN CHLOROHYDRATE see DAC200
DAUNOMYCIN HYDROCHLORIDE see DAC200
DAUNOMYCINOL see DAC300
DAUNORUBICIN see DAC000
DAUNORUBICIN, BENZOYLHYDRAZONE, MONOHYDROCHLORIDE
    see ROZ000
DAUNORUBICINE see DAC000
DAUNORUBICIN HYDROCHLORIDE see DAC200
DAUNORUBICINOL see DAC300
DAURAN see AFJ400
DAVA see VGU750
DAVANA OIL see DAC400
DAVISON SG-67 see SCH000
DAVITAMON D see VSZ100
DAVITAMON PP see NCQ900
DAVITIN see VSZ100
DAVOSIN see AKO500
DAWE'S DESTROL see DKA600
DAWSON 100 see MHR200
DAWSONITE see DAC450
DAY BLOOMING JESSAMINE see DAC500
DAYFEN see DOZ000, LJR000
DAZOMET see DSB200
DAZZEL see DCM750
DB see CQK100
2,4-DB see DGA000
2NDB see ALL750
4NDB see ALL500
DB-905 see TBS000
DB 2041 see IDJ500
DBA see DCT400, DQJ200
DB(a,c)A see BDH750
DB(a,h)A see DCT400
1,2,5,6-DBA see DCT400
DB(a,h)AC see DCS400
DB(a,j)AC see DCS600
DBA-1,2-DIHYDRODIOL see DMK200
trans-DBA-3,4-DIHYDRODIOL see DLD400
DB(a,e)P see NAT500
DBA-5,6-EPOXIDE see EBP500
DBB see DDL000

DBCP see DDL800
DBD see ASH500, DDJ000
DBDPO see PAU500
DBE see EIY500
DBED DIHYDROCHLORIDE see DDG400
DBED DIPENCILLIN G see BFC750
DBED PENICILLIN see BFC750
DBF see DEC400, DJY100
DBH see BBP750, BBQ500, DDO800
1,1-DBH see DEC725
DBHMD see DEC699
DBI see PDF000
DBI-TD see PDF250
DBM see DDP600, DED600
DBMP see BFW750
DBN see BRY500
2,6-DBN see DER800
DBN (the herbicide) see DER800
DBNA see BRY500
DBNPA see DDM000
DBOT see DEF400
DBP see DEH200, DES000
DB(a,i)P see BCQ500
DB(a,l)P see DCY400
DBPC (technical grade) see BFW750
2,4-DB SODIUM SALT see EAK500
D.B.T.C. see DDY200
DBTL see DDV600
2,4-D BUTOXYETHANOL ESTER see DFY709
2,4-D BUTOXYETHYL ESTER see DFY709
2,4-D 2-BUTOXYETHYL ESTER see DFY709
2,4-D BUTYL ESTER see BQZ000
2,4-D BUTYRIC see DGA000
DBV see BRA625
DC-11 see TEF725
DC 360 see SCR400
DC 0572 see AHI875
DCA see DAQ800, DEL000, DEO300, DFE200
3,4-DCA see DEO300
DC-38-A see GEO200
DCA 70 see AAX250
DCAA see AFH275
DCA-ETHER (1:9) see DEN800
DCB see COC750, DEP600, DEQ600, DER800, DES000, DEV000
1,4-DCB see DEV000
DC-45-B2 see TMO775
DCBA see BIA750
D&C BLUE No. 4 see FAE000, FMU059
D&C BLUE NUMBER 1 see BJI250
DCBN see DGM600
DCDB see DEU375
DCDD see DAC800
2,3-DCDT see DBI200
1-1-DCE see VPK000
1,2-DCE see EIY600
DCEE see DFJ050
D&C GREEN 1 see NAX500
D&C GREEN No. 4 see FAF000
D&C GREEN No. 6 see BLK000
D&C GREEN No. 8 see TNM000
DCH 21 see ERE200
DCHFB see DFM000
DCI see DFN400
DCI LIGHT MAGNESIUM CARBONATE see MAC650
DCL see DFN500
DCM see DFO000, DFO800, DGQ200, MJP450
DCMA see DFO800
DCMC see DAI000
DCMO see CCC500
DCMOD see DLV200
DCMU see DXQ500
DCNA see RDP500
DCNA (fungicide) see RDP300
DCNB see DFT600
DCNU see CLX000
D&C ORANGE No. 2 see TGW000
D&C ORANGE No. 3 see FAG010, HGC000
D&C ORANGE NO. 5 see DDO200
D&C ORANGE No. 17 see DVB800
D&C ORANGE NUMBER 15 see DMG800
DCP see DFX800
2,4-DCP see DFX800
DCPA see DGI000
DCPC see BIN000

DCPE see BIN000
DCPM see NCM700
cis-DCPO see DAC975
trans-DCPO see DGH500
D&C RED 2 see FAG020
D&C RED No. 3 see FAG040
D&C RED No. 5 see FMU070
D&C RED No. 9 see CHP500
D&C RED No. 14 see TGX000
D&C RED No. 17 see OHA000
D&C RED No. 19 see FAG070
D&C RED No. 21 see BMO250
D&C RED No. 22 see BNH500
D&C RED No. 28 see ADG250
D&C RED No. 35 see MMP100
D&C RED No. 36 see CJD500
2,4-D CROTYL ESTER see DAD000
D.C.S. see BGJ750
DCU see DGQ200
DC-38-V see GEO200
D+C VIOLET No. 2 see HOK000
D&C YELLOW No. 5 see FAG140
D&C YELLOW No. 7 see FEV000
D&C YELLOW No. 8 see FEW000
D-D see DGG000
DD 234 see DAB750
DDA see DRR800
DDC see DQY950, SGJ000
cis-DDCP see DAD040
trans(−)-DDCP see DAD075
trans(+)-DDCP see DAD050
DDD see BIM500
2,4′-DDD see CDN000
o,p′-DDD see CDN000
p,p′-DDD see BIM500
DDE see BIM750
p,p′-DDE see BIM750
DDETA see HMQ500
2,4-D DIMETHYLAMINE SALT see DFY800
DDM see DSU000, MJQ000
DD-METHYL ISOTHIOCYANATE MIXTURE see MLC000
DD MIXTURE see DGG000
DDMP see MQR100
DDNO see DRS200
DDNP see DUR800
DDOA see ABC250
DDP see PJD000
cis-DDP see PJD000
DDS see DJC875, SOA500
DD SOIL FUMIGANT see DGG000
DDT see DAD200
o,p′-DDT see BIO625
p,p′-DDT see DAD200
DDT DEHYDROCHLORIDE see BIM750
DDVF see DGP900
DDVP see DGP900
D.E. see DCJ800
DE 83R see PAU500
DEA see DHF000
DEACETYLCHOLCHICEINE see TLN750
N-DEACETYLCHOLCHICEINE see TLN750
DEACETYLCOLCHICINE see TLO000
N-DEACETYLCOLCHICINE see TLO000
DEACETYLCOLCHICINE l-TARTRATE see DBA175
N-DEACETYLCOLCHICINE l-TARTRATE(1:1), HYDRATE see DBA175
DEACETYLDEMETHYLTHYMOXAMINE see DAD500
DEACETYL-HT-2 TOXIN see DAD600
DEACETYLLANATOSIDE B see DAD650
DEACETYL-LANATOSIDE B (8CI) see DAD650
DEACETYLLANATOSIDE C see DBH200
3-DE(2-(ACETYLMETHYLAMINO)PROPIONYLOXY)-3-HYDROXYMAYTANS-
INE ISOVALERATE (ESTER) see APE529
DEACETYLMETHYLCOLCHICINE see MIW500
DEACETYL-N-METHYLCOLCHICINE see MIW500
N-DEACETYL-N-METHYLCOLCHICINE see MIW500
N-DEACETYLMETHYLTHIOCOLCHICINE see DBA200
DEACETYLMULDAMINE see DAD800
N-DEACETYLTHIOCOLCHICINE see DBA200
N-DEACETYL-10-THIOCOLCHICINE see DBA200
DEACETYLTHYMOXAMINE see DAD850
DEACTIVATOR E see DJD600
DEACTIVATOR H see DJD600
DEADLY NIGHTSHADE see BAU500, DAD880
DEAD MEN'S FINGERS see WAT315

DEADOPA see DNA200
DEAE see DHO500
DEAE-D see DHW600
DEALCA TP1 see GLU000
DEALKYLPRAZEPAM see CGA500
DEAMELIN S see GHR609
3′-DEAMINO-3′-(3-CYANO-4-MORPHOLINYL)DOXORUBICIN see COP765
DEAMINO-DICARBA-(GLY^7)-OXYTOCIN see CCK550
DEAMINOHYDROXYTUBERCIDIN see DAE200
3′-DEAMINO-3′-MORPHOLINO-ADRIAMYCIN see MRT100
3′-DEAMINO-3′-(4-MORPHOLINYL)DAUNORUBICIN see MRT100
DEANER see DOZ000
DEANOL see DOY800
DEANOL ACETAMIDOBENZOATE see DOZ000
DEANOL-p-ACETAMIDOBENZOATE see DOZ000
DEANOL-p-CHLOROPHENOXYACETATE see DPE000
DEANOLESTERE see DPE000
DEANOX see IHD000
DEA OXO-5 see DBA800
DEAPASIL see AMM250
DEASERPYL see MEK700
DEATH CAMAS see DAE100
DEATH-OF-MAN see WAT325
7-DEAZAADENOSINE see TNY500
7-DEAZAADENOSINE-7-CARBOXAMIDE see SAU000
7-DEAZAINOSINE see DAE200
DEB see BGA750, DKA600
DEBA see BAG000
DEBANTIC see RAF100
DEBECACIN see DCQ800
DEBECACIN SULFATE see PAG050
DEBECILLIN see BFC750
DEBECYLINA see BFC750
DEBENAL see PPP500
DEBENAL-M see ALF250
DEBENDOX see BAV350
DEBENDRIN see BBV500
DEBETROL see SKJ300
DEBRICIN see FBS000
DEBRIDAT see TKU675
DEBRISOQUIN HYDROBROMIDE see DAI475
DEBRISOQUIN SULFATE see IKB000
DEBROUSSAILLANT 600 see DAA800
DEBROUSSAILLANT CONCENTRE see TAA100
DEBROXIDE see BDS000
DEC see DAE800, DIX200
DECABANE see DER800
DECABORANE see DAE400
DECABORANE(14) see DAE400
DECABROMOBIPHENYL ETHER see PAU500
DECABROMOBIPHENYL OXIDE see PAU500
DECABROMODIPHENYL OXIDE see PAU500
DECABROMOPHENYL ETHER see PAU500
DECACHLOR see DAE425
DECACHLOROBI-2,4-CYCLOPENTADIEN-1-YL see DAE425
1,1′,2,2′,3,3′,4,4′,5,5′-DECACHLOROBI-2,4-CYCLOPENTADIEN-1-YL see DAE425
1,2,3,5,6,7,8,9,10,10-DECACHLORO(5.2.1.0$^{(2,6)}$.0$^{(3,9)}$.0$^{(5,8)}$)DECANO-4-ONE
  see KEA000
DECACHLOROKETONE see KEA000
DECACHLORO-1,3,4-METHENO-2H-CYCLOBUTA(cd)PENTALEN-2-ONE
  see KEA000
DECACHLOROOCTAHYDROKEPONE-2-ONE see KEA000
DECACHLOROOCTAHYDRO-1,3,4-METHENO-2H-CYCLOBUTA(cd)PENTALEN-
  2-ONE see KEA000
1,1a,3,3a,4,5,5,5a,5b,6-DECACHLOROOCTAHYDRO-1,3,4-METHENO-2H-
  CYCLOBUTA(cd)PENTALEN-2-ONE see KEA000
DECACHLOROPENTACYCLO(5.2.1.0$^{(2,6)}$.0$^{(3,9)}$.0$^{(5,8)}$)DECAN-4-ONE
  see KEA000
DECACHLOROPENTACYCLO(5.3.0.0$^{(2,6)}$.0$^{(4,10)}$.0$^{(5,9)}$)DECAN-3-ONE
  see KEA000
DECACHLOROTETRACYCLODECANONE see KEA000
DECACHLOROTETRAHYDRO-4,7-METHANOINDENEONE see KEA000
DECACIL see LFK000
DECACURAN see DAF600
DECADERM see SOW000
trans,trans-2,4-DECADIENAL see DAE450
DECADONIUM DIIODIDE see DAE500
DECADRON see SOW000
DECADRON-LA see DBC400
DECADRON PHOSPHATE see DAE525
DECA-DURABOL see NNE550
DECA-DURABOLIN see NNE550
DECAETHOXY OLEYL ETHER see OIG040
DECAFENTIN see DAE600

DECAFLUOROBUTYRAMIDINE see DAE625
cis-N-(DECAHYDRO-2-METHYL-5-ISOQUINOLYL)-3,4,5-
   TRIMETHOXYBENZAMIDE see DAE695
trans-N-(DECAHYDRO-2-METHYL-5-ISOQUINOLYL)-3,4,5-
   TRIMETHOXYBENZAMIDE see DAE700
DECAHYDRONAPHTHALENE see DAE800
DECAHYDRO-2-NAPHTHALENOL see DAF000
DECAHYDRONAPHTHALEN-2-OL see DAF000
DECAHYDRO-β-NAPHTHOL see DAF000
trans-DECAHYDRO-β-NAPHTHOL see DAF000
DECAHYDRO-β-NAPHTHYL ACETATE see DAF100
DECAHYDRO-β-NAPHTHYL FORMATE see DAF150
DECAHYDRONAPTHOL-2 see DAF000
DECAHYDRO-4a,7,9-TRIHYDROXY-2-METHYL-6,8-BIS(METHYLAMINO)-4H-
   PYRANO(2,3-b)(1,4)BENZODIOXIN-4-ONE DIHYDROCHLORIDE, (2R-(2-
   α,4a-β,5a-β,6-β,7-β,8-β,9-α,9a-α,10a-β))- see SLI325
Γ-N-DECALACTONE see HKA500
Δ-DECALACTONE see DAF200
DECALIN see DAE800
DECALIN (DOT) see DAE800
2-DECALINOL see DAF000
DECALIN SOLVENT see DAE800
DECALINYL FORMATE see DAF150
2-DECALOL see DAF000
DECAMETHONIUM IODIDE see DAF800
DECAMETHONIUM see DAF600
DECAMETHONIUM BROMIDE see DAF600
DECAMETHONIUM DIBROMIDE see DAF600
DECAMETHONIUM DIIODIDE see DAF800
DECAMETHONIUM IODIDE see DAF800
DECAMETHRIN see DAF300
DECAMETHRINE see DAF300
N,N'-DECAMETHYLENEBIS((1-ADAMANTYL)DIMETHYLAMMONIUM, DIIOD-
   IDE see DAE500
1,1'-DECAMETHYLENEBIS(1-METHYLPIPERIDINIUM IODIDE) see DAF450
DECAMETHYLENEBIS(TRIMETHYLAMMONIUM BROMIDE) see DAF600
DECAMETHYLENE-1,10-BISTRIMETHYLAMMONIUM DIBROMIDE
   see DAF600
DECAMETHYLENEBIS(TRIMETHYLAMMONIUM DIIODIDE) see DAF800
DECAMETHYLENEBIS(TRIMETHYLAMMONIUM IODIDE) see DAF800
DECAMINE see DAA800
DECAMINE 4T see TAA100
1-DECANAL see DAG000
1-DECANAL (mixed isomers) see DAG200
DECANAL, DIMETHYLACETAL see AFJ700
DECANAL DIMETHYL ACETAL see DAI600
DECANE see DAG400
n-DECANE (DOT) see DAG400
1-DECANEAMINE see DAG600
1-DECANECARBOXYLIC ACID see UKA000
DECANEDIOIC ACID see SBJ500
DECANEDIOIC ACID, BIS(2-ETHYLHEXYL) ESTER see BJS250
DECANEDIOIC ACID, DIBUTYL ESTER see DEH600
1,10-DECANEDIOL, 2,2,9,9-TETRAMETHYL- see TDO260
DECANE, 1-METHOXY- see MIW075
DECANOIC ACID see DAH400
n-DECANOIC ACID see DAH400
DECANOIC ACID-4-(4-CHLOROPHENYL)-1-(4-(4-FLUOROPHENYL)-4-OXY-
   BUTYL)-4-PIPERIDINYL ESTER see HAG300
DECANOIC ACID, DIESTER with TRIETHYLENE GLYCOL (mixed isomers)
   see DAH450
DECANOIC ACID, ETHYL ESTER see EHE500
DECANOIC ACID, 4-HYDROXY-4-METHYL-, Γ-LACTONE see MIW050
DECANOL see DAI600
n-DECANOL see DAI600
1-DECANOL (FCC) see DAI600
DECANOL (mixed isomers) see DAI800
DECANOLIDE-1,4 see HKA500
DECANOLIDE-1,5 see DAF200
4-DECANOYLMORPHOLINE see CBF725
9-DECAOCTENOIC ACID, TRIBUTYLSTANNYL ESTER see TIA000
DECAPRYN see PGE775
DECAPRYN SUCCINATE see PGE775
DECAPS see VSZ100
DECARBAMOYLMITOMYCIN C see DAI000
10-DECARBAMOYLMITOMYCIN C see DAI000
DECARBAMYLMITOMYCIN C see DAI000
DECARBOFURAN see DLS800
2-DECARBOXAMIDO-2-ACETYL-4-DESDIMETHYLAMINO-4-AMINO-9-
   METHYL-5A,6-ANHYDROTETRACYCLINE see CDF250
DECARBOXYCYSTEINE see AJT250
DECARBOXYCYSTINE see MCN500
DECARIS see LFA020
DECARPYN SUCCINATE (1:1) see PGE775

DECASERPIL see MEK700
DECASERPINE see MEK700
DECASERPYL see MEK700
DECASERPYL PLUS see MEK700
DECASONE see SOW000
DECASPIRIDE HYDROCHLORIDE see DAI200
DECASPRAY see SOW000
n-DECATYL ALCOHOL see DAI600
DECCOTANE see BPY000
DECELITH H see PKQ059
DECEMTHION P-6 see PHX250
2-DECENAL see DAI350
cis-4-DECENAL see DAI360
trans-2-DECEN-1-AL see DAI350
cis-4-DECEN-1-AL (FCC) see DAI360
DECENALDEHYDE see DAI350
9-DECEN-1-OL see DAI400
1-DECEN-10-OL see DAI400
omega-DECENOL see DAI400
9-DECEN-1-OL, ACETATE see DAI450
DECENTAN see CJM250
DECENYL ACETATE see DAI450
9-DECENYL ACETATE see DAI450
DECHAN see DGU200
DECHLORANE 605 see DAI460
DECHLORANE 4070 see MQW500
DECHLORANE-A-O see AQF000
DECHLORANE PLUS see DAI460
DECHLORANE PLUS 515 see DAI460
DECHLORANE PLUS 2520 see DAI460
2,4-DECHLOROPHENYL-p-NITROPHENYL ETHER see DFT800
DECHOLIN see DAL000
DECHOLIN SODIUM SALT see SGD500
DECICAINE see TBN000
DECIMEMIDE see DAJ400
DECINCAN see VLF000
DECIS see DAF300
DECLINAX see DAI475, IKB000
DECLOMYCIN see MIJ500
DECLOMYCIN HYDROCHLORIDE see DAI485
DECLOXIZINE see BBV750
DECOFOL see BIO750
n-DECOIC ACID see DAH400
DECONTRACTIL see GGS000
DECORPA see GLU000
DECORTANCYL see PLZ000
DECORTIN see DAQ800, PLZ000
DECORTIN H see PMA000
DECORTISYL see PLZ000
DECORTON see DAQ800
DECOSERPYL see MEK700
DECOSTERONE see DAQ800
DECOSTRATE see DAQ800
DECROTOX see COD000
DECTAN see DBC400
DECTANCYL see SOW000
DECURVON see IDF300
DECYL ACRYLATE see DAI500
n-DECYL ACRYLATE see DAI500
DECYL ALCOHOL see DAI600
n-DECYL ALCOHOL see DAI600
DECYL ALCOHOL (mixed isomers) see DAI800
1-DECYL ALDEHYDE see DAG000, UJJ000
DECYLALDEHYDE DMA see AFJ700
DECYLAMINE see DAG600
DECYL BENZENE SODIUM SULFONATE see DAJ000
DECYL BUTYL PHTHALATE see BQX250
DECYL CHLORIDE (mixed isomers) see DAJ200
N-DECYL-N,N-DIMETHYL-1-DECANAMINIUM CHLORIDE (CI) see DGX200
DECYLENIC ALCOHOL see DAI400
1-DECYL-1-ETHYLPIPERIDINIUM BROMIDE see DAJ300
DECYLIC ACID see DAH400
n-DECYLIC ACID see DAH400
DECYLIC ALCOHOL see DAI600
DECYL METHYL ETHER see MIW075
DECYL OCTYL ALCOHOL see OAX000
DECYL OCTYL PHTHALATE see OEU000
N-DECYL-N-OCTYL PHTHALATE see OEU000
4-(DECYLOXY)-3,5-DIMETHOXYBENZAMIDE see DAJ400
4-n-DECYLOXY-3,5-DIMETHOXYBENZOIC ACID AMIDE see DAJ400
DECYLTRIPHENYLPHOSPHONIUM BROMOCHLOROTRIPHENYLSTAN-
   NATE see DAE600
(DECYL-TRIPHENYL-PHOSPHONIUM)-TRIPHENYL-BROM-CHLOR-
   STANNAT (GERMAN) see DAE600

DEDC see SGJ000
DEDELO see DAD200
DEDEVAP see DGP900
DEDK see SGJ000
DEDORAN see MCI500
DED-WEED see CIR250, DAA800, DGI400, TIX500
DED-WEED BRUSH KILLER see TAA100
DED-WEED CRABGRASS KILLER see PLC250
DED-WEED LV-69 see DAA800
DED-WEED LV-6 BRUSH KIL and T-5 BRUSH KIL see TAA100
DEE-OSTEROL see VSZ100
DEEP CRIMSON MADDER 10821 see DMG800
DEEP FASTONA RED see MMP100
DEEP LEMON YELLOW see SMH000
DEER BERRY see HGF100
DEE-RON see VSZ100
DEE-RONAL see VSZ100
DEE-ROUAL see VSZ100
DEER'S TONGUE see DAJ800
DEERTONGUE INCOLORE see DAJ800
DEET see DKC800
16-DEETHYL-3-o-DEMETHYL-16-METHYL-3-o-(1-OXOPROPYL)MONENSIN see DAK000
DEETILATO METOCLOPRAMIDE (ITALIAN) see AJH125
DEETILMETOCLOPRAMIDE (ITALIAN) see AJH125
DEF see BSH250
DEF DEFOLIANT see BSH250
DEFEKTON see CCK775, CCK780
DE-FEND see DSP400
DEFEROXAMINE see DAK200
DEFEROXAMINE MESILATE see DAK300
DEFEROXAMINE MESYLATE see DAK300
DEFEROXAMINE METHANESULFONATE see DAK300
DEFEROXAMINUM see DAK200
DEFERRIOXAMINE see DAK200
DEFERRIOXAMINE B see DAK200
DEFIBRASE see RDA350
DEFIBRASE R see RDA350
DEFILIN see DJL000
DEFILTRAN see AAI250
DEFLAMENE see FDB000
DEFLAMON-WIRKSTOFF see MMN250
DEFLEXOL see AJF500
DEFLOGIN see HNI500
DEFLORIN see TEY000
DE-FOL-ATE see MAE000, SFS000
DEFOLIANT 713 see DKE400
DEFOLIANT 2929 RP see TFH500
DEFOLIT see TEX600
DEFONIN see ILD000
DEFRADIN HYDRATE see SBN450
DEFTOR see MQR225
DEFY see DFY800
DEG see DJD600
DEGALAN S 85 see PKB500
DEGALOL see DAQ400
DEGLYCOSYLATED HCG see DAK325
DEGLYCOSYLATED HUMAN CHORIONIC GONADOTROPIN see DAK325
DEGMVE (RUSSIAN) see DJG400
DEGRANOL see MAW500
DEGRASSAN see DUS700
DE-GREEN see BSH250
DEGUELIA ROOT see DBA000
DEH see HHH000
D.E.H. 20 see DJG600
DEH 24 see TJR000
D.E.H. 26 see TCE500
DEHA see AEO000, DJN000
DEHACODIN see DKW800
DEHIDROBENZPERIDOL see DYF200
DEHISTIN see TMP750
DEHISTIN HYDROCHLORIDE see POO750
DEHP see DVL700
DEHPA EXTRACTANT see BJR750
DEHYCHOL see DAL000
DEHYDOL LS 4 see DXY000
DEHYDRACETIC ACID see MFW500
DEHYDRATIN see AAI250
DEHYDRITE see PCE000
DEHYDROABIETIC ACID see DAK400
DEHYDRO-ABIETIC ACID-2-(2-(CHLOROETHYL)AMINO)ETHYL ESTER see CGQ250
DEHYDROACETIC ACID (FCC) see MFW500
DEHYDROACETIC ACID, SODIUM SALT see SGD000

Δ⁶-DEHYDRO-17-ACETOXYPROGESTERONE see MCB375
Δ⁶-DEHYDRO-17-α-ACETOXYPROGESTERONE see MCB375
trans-DEHYDROANDROSTERONE see AOO450
DEHYDROBENZPERIDOL see DYF200
6-DEHYDRO-6-CHLORO-17-α-ACETOXYPROGESTERONE see CBF250
DEHYDROCHOLATE SODIUM see SGD500
7-DEHYDROCHOLESTERIN see DAK600
DEHYDROCHOLESTERIN (GERMAN) see DAK600
DEHYDROCHOLESTEROL see DAK600
7-DEHYDROCHOLESTEROL see DAK600
7-DEHYDROCHOLESTEROL ACETATE see DAK800
7-DEHYDROCHOLESTERYL ACETATE see DAK800
7-DEHYDROCHOLESTROL, ACTIVATED see CMC750
DEHYDROCHOLIC ACID see DAL000
DEHYDROCHOLIC ACID, SODIUM SALT see SGD500
DEHYDROCHOLSAEURE (GERMAN) see DAL000
Δ¹-DEHYDROCORTISOL see PMA000
1-DEHYDROCORTISONE see PLZ000
Δ-1-DEHYDROCORTISONE see PLZ000
Δ'-DEHYDROCORTISONE ACETATE see PLZ100
DEHYDROEPIANDROSTERONE see AOO450
5-DEHYDROEPIANDROSTERONE see AOO450
DEHYDROEPIANDROSTERONE SODIUM SULFATE DIHYDRATE see DAL030
DEHYDROEPIANDROSTERONE SULFATE SODIUM see DAL040
DEHYDROERGOTAMINE see DLK800
DEHYDROFOLLICULINIC ACID see BIT000
DEHYDROHELIOTRIDINE see DAL060
DEHYDROHELIOTRINE see DAL100
1-DEHYDROHYDROCORTISONE see PMA000
Δ¹-DEHYDROHYDROCORTISONE see PMA000
11-DEHYDRO-17-HYDROXYCORTICOSTERONE see CNS800
11-DEHYDRO-17-HYDROXYCORTICOSTERONE ACETATE see CNS825
11-DEHYDRO-17-HYDROXYCORTICOSTERONE-21-ACETATE see CNS825
DEHYDROISOANDROSTERONE see AOO450
5,6-DEHYDROISOANDROSTERONE see AOO450
6-DEHYDRO-6-METHYL-17-α-ACETOXYPROGESTERONE see VTF000
DEHYDRO-3-METHYLCHOLANTHRENE see DAL200
1,2-DEHYDRO-3-METHYLCHOLANTHRENE see DAL200
6-DEHYDRO-16-METHYLENE-6-METHYL-17-ACETOXYPROGESTERONE see MCB380
1-DEHYDRO-16-α-METHYL-9-α-FLUOROHYDROCORTISONE see SOW000
DEHYDROMETHYLTESTERONE see DAL300
1-DEHYDRO-17-α-METHYLTESTOSTERONE see DAL300
DEHYDROMONOCROTALINE see DAL350
DEHYDRONIVALENOL see VTF500
DEHYDRONIVALENOL MONOACETATE see ACH075
DEHYDROPHELOMYCIN D1 see BLY760
11-DEHYDROPROSTAGLANDIN F2-α see POC275
DEHYDRORETRONECINE see DAL400
6-DEHYDRO-RETRO-PROGESTERONE see DYF759
DEHYDROSTILBESTROL see DAL600
Δ³-DEHYDRO-3,4-TRIMETHYLENE-ISOBENZANTHRENE-2 see BCI000
DEHYQUART STC-25 see DTC600
DEHYSTOLIN see DAL000
DEIDROBENZPERIDOLO see DYF200
DEIDROCOLICO VITA see DAL000
DEINAIT see CJR909
DEIQUAT see DWX800
DEISOVALERYL BLASTMYCIN see DAM000
DEJO see DJT800
DEK see DJN750
DE-KALIN see DAE800
DEKALINA (POLISH) see DAE800
DEKAMETRIN (HUNGARIAN) see DAF300
DEKORTIN see PLZ000
DEKRYSIL see DUS700
DEKSONAL see DOU600
DEL see AJH000
DELAC J see DWI000
DELACURARINE see TOA000
DELADIOL see EDS100
DELADROXONE see DAM300
DELAGIL see CLD000, CLD250
DELAHORMONE UNIMATIC see EDS100
DELALUTIN see HNT500
DELAN see DLK200
DELAN-COL see DLK200
DELAPRIL HYDROCHLORIDE see DAM315
DELATESTRYL see MPN500, TBF750
DELATESTRYL and DEPO-MEDROXYPROGESTERONE ACETATE see DAM325
DELATESTRYL and DEPO-PROVERA see DAM325
DELAVAN see MHB500
DELCORTIN see PLZ100

DELCORTOL see PMA000
DELEAF DEFOLIANT see TIG250
DELESTROGEN see EDS100
DELESTROGEN 4X see EDS100
DELGESIC see ADA725
DELICIA see AHE750, PGY000
DELIVA see ARQ750
DELLIPSOIDS see BBK500
DELMOFULVINA see GKE000
DELMONEURINA see BET000
DELNAV see DVQ709
DELONIN AMIDE see NCR000
DELOWAS S see TBD000
DELOWAX OM see TBD000
DELPET 50M see PKB500
DELPHENE see DKC800
m-DELPHENE see DKC800
DELPHINIC ACID see ISU000
DELPHINIDOL see DAM400
DELPHINIUM see LBF000
DELSTEROL see CMC750
DELTA see CJJ000
DELTA-CORTEF see PMA000
DELTACORTELAN see PLZ000
DELTACORTENOL see PMA000
DELTACORTISONE see PLZ000
DELTACORTONE see PLZ000
DELTACORTRIL see PMA000
DELTA-DOME see PLZ000
DELTA F see PMA000
DELTAFLUORENE see SOW000
DELTALIN see VSZ100
DELTALONE see PLZ100
DELTAMETHRIN see DAF300
DELTA-MVE see MKB750
DELTAMYCIN A see CBT250
DELTAN see DUD800
DELTA-STAB see PMA000
DELTATHIONE see GFW000
DELTAZINA see PPP500
DELTILEN see SOV100
DELTISONE see PLZ000
DELTOIN see ENC500
DELTOSIDE see AHK750
DELTRATE-20 see PBC250
DELTYL see IQW000
DELTYL PRIME see IQW000
DELURSAN see DMJ200
DELVEX see DJT800
DELVINAL SODIUM see VKP000
DELYSID see DJO000
DEM see AJH125
DEMA see BIE500
DEMAROL see DAM600
DEMASORB see DUD800
DEMAVET see DUD800
DEMECLOCYCLINE see MIJ500
DEMECLOCYCLINE HYDROCHLORIDE see DAI485
DEMECOLCINE see MIW500
DEMEFLINE see DNV200
DEMEPHION see MIW250
DEMEROL see DAM600, DAM700
DEMEROL HYDROCHLORIDE see DAM700
DEMESO see DUD800
DEMETHON-METHYL (MAK) see MIW100
4-DEMETHOXYADRIAMYCIN see DAM800
4-DEMETHOXYDAUNOMYCIN see DAN000
4-DEMETHOXYDAUNORUBICIN see DAN000
11-DEMETHOXYRESERPINE see RDF000
N-DEMETHYLACLACINOMYCIN A see DAN200
DEMETHYLAMITRIPTYLENE see NNY000
DEMETHYLCHLOROTETRACYCLIN see MIJ500
DEMETHYLCHLOROTETRACYCLINE see MIJ500
6-DEMETHYLCHLOROTETRACYCLINE see MIJ500
6-DEMETHYL-7-CHLOROTETRACYCLINE DEMETHYLCHLORTETRACY-
   CLINE see MIJ500
DEMETHYLCHLOROTETRACYCLINE HYDROCHLORIDE see DAI485
DEMETHYLCHLORTETRACYCLINE see MIJ500
6-DEMETHYLCHLORTETRACYCLINE see MIJ500
6-DEMETHYL-7-CHLORTETRACYCLINE see MIJ500
DEMETHYLCHLORTETRACYCLINE, BASE see MIJ500
DEMETHYLCHLORTETRACYCLINE HYDROCHLORIDE see DAI485
2-DEMETHYLCOLCHICINE see DAN300
O^2-DEMETHYLCOLCHICINE see DAN300

O^{10}-DEMETHYLCOLCHICINE see ADE000
3-DEMETHYLCOLCHICINE GLUCOSIDE see DAN375
o-DEMETHYLDAUNOMYCIN see KBU000
(6R,25)-5-o-DEMETHYL-28-DEOXY-6,28-EPOXY-25-(1-
   METHYLETHYL)MILBEMYCIN B see MQT600
(11R(2R,5S,6R),12R)-10-DEMETHYL-19-DE((TETRAHYDRO-5-METHOXY-6-
   METHYL-2H-PYRAN-2-YL)OXY)-12-METHYL-11-o-(TETRAHYDRO-5-
   METHOXY-6-METHYL-2H-PYRAN-2-YL)-DIANEMYCIN see LEJ700
DEMETHYLDIAZEPAM see CGA500
1-DEMETHYLDIAZEPAM see CGA500
N-DEMETHYLDIAZEPAM see CGA500
DEMETHYLDOPAN see BIA250
DEMETHYL-EPIODOPHYLLOTOXIN ETHYLIDENE GLUCOSIDE see EAV500
4-DEMETHYLEPIODOPHYLLOTOXIN-β,d-ETHYLIDENEGLUCOSIDE
   see EAV500
4'-DEMETHYLEPIPODOPHYLLOTOXIN-9-(4,6-O-ETHYLIDENE-β-d-
   GLUCOPYRANOSIDE see EAV500
4'-DEMETHYLEPIPODOPHYLLOTOXIN ETHYLIDENE-β,d-GLUCOSIDE
   see EAV500
4-DEMETHYL-EPIPODOPHYLLOTOXIN-β,d-ETHYLIDEN-GLUCOSIDE
   see EAV500
4'-DEMETHYLEPIPODOPHYLLOTOXIN-9-(4,6-O-2-THENYLIDENE-β-d-
   GLUCOPYRANOSIDE see EQP000
4'-DEMETHYL-EPIPODOPHYLLOTOXIN-β-d-THENYLIDENE-GLUCOSIDE
   see EQP000
4'-O-DEMETHYL-1-O-(4,6-O-ETHYLIDENE-β,d-
   GLUCOPYRANOSYL)EPIPODOPHYLLOTOXIN see EAV500
DEMETHYLIMIPRAMINE see DSI709
DEMETHYLMISONIDAZOLE see NHI000
1'-DEMETHYL-(s)-NICOTINE see NNR500
N-DEMETHYLORPHENADRINE HYDROCHLORIDE see TGJ250
4'-DEMETHYL-1-O-(4,6-O,O-(2-THENYLIDENE)-β-d-
   GLUCOPYRANOSYL)EPIPODOPHYLLOTOXIN see EQP000
DEMETON see DAO500, DAO600
DEMETON METHYL see MIW100
DEMETON-O-METHYL see DAO800
DEMETON-S-METHYL see DAP400
DEMETON-S-METHYLSULFON (GERMAN) see DAP600
DEMETON-S-METHYLSULFONE see DAP600
DEMETON-S-METHYL-SULFOXID (GERMAN) see DAP000
DEMETON-O-METHYL SULFOXIDE see DAP000
DEMETON-S-METHYL SULFOXIDE see DAP000
DEMETON-METHYL SULPHOXIDE see DAP000
DEMETON-O-METILE (ITALIAN) see DAO800
DEMETON-S-METILE (ITALIAN) see DAP400
DEMETON-O see DAO500
DEMETON-O + DEMETON-S see DAO600
DEMETON-S see DAP200
DEMETRACICLINA see DAI485
DEMETRIN see DAP700
DEMISE see DFY800
DEMOCRACIN see TBX000
DEMOLOX see AOA095
DEMORPHAN see DBE200
DEMOSAN see CJA100
DEMOS-L40 see DSP400
DEMOTIL see DAP800
DEMOX see DAO600
DEMSODROX see DUD800
DEMULEN see DAP810
DEN see NJW500
DENA see NJW500
DENAMONE see VSZ450
DENAPON, NITROSATED (JAPANESE) see NBJ500
DENATURED SPIRITS see AFJ000
DENDREPAR see PEM750
DENDRID see DAS000
DENDRITIS see SFT000
DENDROASPIS ANGUSTICEPS VENOM see DAP815
DENDROASPIS JAMESONI VENOM see DAP820
DENDROASPIS VIRIDIS VENOM see GJU500
DENDROBAN-12-ONE HYDROCHLORIDE see DAP825
DENDROBINE HYDROCHLORIDE see DAP825
DENDROCALAMUS MEMBRANACEUS Munro, extract excluding roots
   see DAP840
DENEGYT see DAJ400
DENKALAC 61 see AAX175
DENKA QP3 see SMQ500
DENKA VINYL SS 80 see PKQ059
DENOPAMINE see DAP850
DENSINFLUAT see TIO750
DENUDATINE see DAP875
DENVER RESEARCH CENTER No. DRC-4575 see AMU125

DEPAMID see PNX600
DEPAMIDE see PNX600
DEPARAL see CMC750
DEPARKIN see DHF600
DEPAS see EQN600
DEPC see DIZ100
DEPC and AMMONIA see DJY050
DEPEN see MCR750
DEPHADREN see AOA500, BBK500
DEPHENIDOL HYDROCHLORIDE see CCR875
DEPHOSPHATE BROMOFENOFOS see DAZ110
DEPHRADINE HYDRATE see SBN450
DEPIGMAN see AEY000
DEPINAR see VSZ000
DEPOCID see AIF000
DEPOESTRADIOL see DAZ115
DEPOESTRADIOL CYPIONATE see DAZ115
DEPOFEMIN see DAZ115
DEPO-HEPARIN see HAQ550
DEPO-INSULIN see IDF325
DEPO-MEDRATE see DAZ117
DEPO-MEDROL see DAZ117
DEPO-MEDRONE see DAZ117
DEPO-MEDROXYPROGESTERONE ACETATE and DELATESTRYL
    see DAM325
DEPO-MEDROXYPROGESTERONE ACETATE and TESTOSTERONE ENANTH-
    ATE see DAM325
DEPO-METHYLPREDNISOLONE see DAZ117
DEPO-METHYLPREDNISOLONE ACETATE see DAZ117
DEPOMIDE see AIE750
DEPO-PROLUTON see HNT500
DEPO-PROVERA see MCA000
DEPO-PROVERA and DELATESTRYL see DAM325
DEPO-PROVERA and TESTOSTERONE ENANTHATE see DAM325
DEPOSTAT see GEK510
DEPOSUL see SNN300
DEPOSULIN see IDF325
DEPO-TESTOSTERONE see TBF600
DEPOT-MEDROL see DAZ117
DEPOT-OESTROMENINE see DJB200
DEPOT-OESTROMON see DJB200
DEPOVERNIL see AKO500
DEPOVIRIN see TBF600
DEPOXIN see DBA800
DEPP see BJR625
DEPRANCOL see PNA500
DEPRELIN see SNE000
DEPREMOL G see DTG000
(+)-DEPRENIL HYDROCHLORIDE see DAZ120
1-DEPRENIL HYDROCHLORIDE see DAZ125
(±)-DEPRENIL HYDROCHLORIDE see DAZ118
(−)-DEPRENYL HYDROCHLORIDE see DAZ125
(+)-DEPRENYL HYDROCHLORIDE see DAZ120
DEPRESSIN see HEA500
DEPREX see EAI000
DEPRIDOL see MDP750
DEPRINOL see DLH630
DEPROMIC see PNA500
DEPT see DAZ135
DEPTHON see TIQ250
DEPTOSULFONAMIDE see AIF000
DEPTROPINE METHOBROMIDE see DAZ140
DEPUALONE see AOB500
DEQUEST 2010 see HKS780
DEQUEST 2015 see HKS780
DEQUEST Z 010 see HKS780
D.E.R. 332 see BLD750
DERACIL see TFR250
DERALIN see ICC000
DERATOL see VSZ100
DERESPERINE see RDF000
DEREUMA see DOT000
DERGRAMIN see SOW000
DERIBAN see DGP900
DERIL see RNZ000
DERIZENE see DNU000, SPC500
DERMACAINE see DDT200
DERMACORT see CNS750
DERMADEX see HCL000
DERMA FAST BROWN W-GL see CMO750
DERMAFFINE see OBA000
DERMAFOSU (POLISH) see RMA500
DERMAGAN see ACR300
DERMAGEN see ACR300

DERMAGINE see OAV000
DERMALAR see SPD500
DERMAPHOS see RMA500
DERMAPLUS see FDD150
DERMASORB see DUD800
DERMATOLOGICO see ARN500
DERMATON see CDS750
DERMA YELLOW P see SGP500
DERMISTINE see BBV500
DERMODRIN see BBV500
DERMOFURAL see NGE500
DERMOGLANCIN see MEY750
DERMOXIN see TGB475
DERONIL see SOW000
DEROSAL see MHC750
DERRIBANTE see DGP900
DERRIN see RNZ000
DERRIS see RNZ000
DERRIS ELLIPTICA, root see DBA000
DERRIS RESINS see DBA000
DERRIS ROOT see DBA000
DES (synthetic estrogen) see DKA600
DESACE see DBH200
DESACETYLBUFOTALIN see BOM655
DESACETYLCHOLCHICEINE see TLN750
DESACETYLCOLCHICINE see TLO000
N-DESACETYLCOLCHICINE see TLO000
N-DESACETYLCOLCHICINE-d-TARTRATE see DBA175
DESACETYLLANATOSIDE B see DAD650
DESACETYLLANATOSIDE C see DBH200
DESACETYL-LANTOSID B (GERMAN) see DAD650
N-DESACETYL-N-METHYLCOLCHICINE see MIW500
N-DESACETYLTHIOCOLCHICINE see DBA200
DESACETYLTHYMOXAMINE see DAD850
DESACETYLVINBLASTINE AMIDE SULFATE see VGU750
DESACI see DBH200
DESADRENE see SOW000
DESAGLYBUZOLE see GFM200
DESALKYLPRAZEPAM see CGA500
DESAMETASONE see SOW000
DESAMINE see DBA800
DESBENZYL CLEBOPRIDE see DBA250
DESCETYLDIGILANIDE C see DBH200
DESCHLOROBIOMYCIN see TBX000
DESCINOLONE ACETONIDE see BFV760
DESCORTERONE see DAQ800
DESCOTONE see DAQ800
N-DESCYCLOPROPYLMETHYLPRAZEPAM see CGA500
DESD see DKB000
DESDANINE see COV125
DESDEMIN see CHJ750
DES DISODIUM SALT see DKA400
DESdp see DKA200
DESENEX see ULS000
DESENTOL see BBV500
DESERIL see MLD250
DESERNYL see MLD250
DESERPIDINE, 10-METHOXY- see MEK700
DESERPINE see RDF000
DESERTALC 57 see TAB750
DESERTOMYCIN see DBA400
DESERT RED see CHP500
DESERT ROSE see DBA450
DESERYL see MLD250
DESETHYLAPRINDINE HYDROCHLORIDE see DBA475
DESFEDRIN see DBA800
DESFERAL see DAK200
DESFERAL METHANESULFONATE see DAK300
DESFERRAL see DAK200
DESFERRIN see DAK200
DESFERRIOXAMINE see DAK200
DESFERRIOXAMINE B see DAK200
DESFERRIOXAMINE B MESYLATE see DAK300
DESFERRIOXAMINE B METHANESULFONATE see DAK300
DESFERRIOXAMINE METHANESULFONATE see DAK300
DESGLUCO-DIGITALINUM VERUM (GERMAN) see SMN275
DESGLUCODIGITONIN see DBA500
DESGLUCO-TRANSVAALIN see POB500
DES-I-CATE see DXD000
DESICCANT L-10 see ARB250
DESIMIPRAMINE see DSI709
DESINFECT see CDP000
DESIPRAMIN see DSI709
DESIPRAMINE (D4) see DSI709

DESIPRAMINE HYDROCHLORIDE see DLS600
DESLANATOSIDE see DBH200
DESLANOSIDE see DBH200
DESMA see DKA600
DESMECOLCINE see MIW500
DESMEDIPHAM see EEO500
DESMETHOXYRESERPINE see RDF000
11-DESMETHOXYRESERPINE see RDF000
DESMETHYLAMITRIPTYLINE see NNY000
DESMETHYLDIAZEPAM see CGA500
N-DESMETHYLDIAZEPAM see CGA500
DESMETHYLDOPAN see BIA250
DESMETHYLDOXEPIN see DBA600
DESMETHYLIMIPRAMINE see DSI709
DESMETHYLIMIPRAMINE HYDROCHLORIDE see DLS600
DESMETHYLMISONIDAZOLE see DBA700, NHI000
DESMETRYN (GERMAN, DUTCH) see INR000
DESMETRYNE see INR000
DESMODUR 44 see MJP400
DESMODUR T80 see TGM750
DESMODUR T100 see TGM740
2,4-DES-Na see CNW000
2,4-DES-NATRIUM (GERMAN) see CNW000
DESO see EPI500
DESOGESTREL see DBA750
DESOLET see SFS000
DESOMORPHINE see DKX600
DESORMONE see DAA800, DGB000
14-DESOSSI-14-((2-DIETILAMINOETIL)MERCAPTO-ACETOSSI)MUTILIN
    IDROGENO FUMARATO (ITALIAN) see TET800
DESOSSIEFEDRINA see DBA800
DES-OXA-D see DBA800
DESOXEDRINE see DBA800
DES-3,4-OXIDE see DKB100
DES-α,β-OXIDE see DKB100
DESOXIMETASONE see DBA875
17-DESOXIMETHASONE see DBA875
DESOXIN see DBA800
DESOXO-5 see DBA800, MDT600
DESOXYADENOSINE see DAQ200
DESOXYBENZOIN see PEB000
DESOXYCHOLIC ACID see DAQ400
DESOXYCHOLSAEURE (GERMAN) see DAQ400
DESOXYCORTICOSTERONE see DAQ600
DESOXYCORTICOSTERONE ACETATE see DAQ800
DESOXYCORTONE see DAQ600
DESOXYCORTONE ACETATE see DAQ800
DESOXYCYTIDIN (GERMAN) see DAQ850
14-DESOXY-14-((DIETHYLAMINOETHYL)-MERCAPTO ACETOXYL)-MUTILIN
    HYDROGEN FUMARATE see TET800
DESOXYEPHEDRINE see DBB000
DESOXYEPHEDRINE HYDROCHLORIDE see DBA800
d-DESOXYEPHEDRINE HYDROCHLORIDE see MDT600
l-DESOXYEPHEDRINE HYDROCHLORIDE see MDQ500
dl-DESOXYEPHEDRINE HYDROCHLORIDE see DAR100
DESOXYFED see DBA800, MDT600
2-DESOXY-d-GLUCOSE (FRENCH) see DAR600
DESOXYMETASONE see DBA875
DESOXYMETHASONE see DBA875
DESOXYN see BBK500, DBA800, DBB000, MDT600
DESOXYNE see MDT600
DESOXYNIVALENOL see VTF500
DESOXYNOREPHEDRINE see AOA250, AOB250
(±)-DESOXYNOREPHEDRINE see BBK000
racemic-DESOXYNOR-EPHEDRINE see BBK000
3-DESOXYNORLUTIN see NNV000
DESOXYPHED see DBA800
2-DESOXYPHENOBARBITAL see DBB200
DESOXYPHENOBARBITONE see DBB200
DESOXYPYRIDOXIME HYDROCHLORIDE see DAY825
DESOXYPYRIDOXINE see DAY800
DESPHEN see CDP250
DESSIN see CBW000
DESSON see CLW000
DESTENDO see BCA000
DESTIM see DBA800, MDT600
DESTOLIT see DMJ200
DESTOMYCIN A see DBB400
DESTONATE 20 see DBB400
DESTRIOL see EDU500
DESTROL see DKA600
DESTRONE see EDV000
DESTRUXOL APPLEX see DXE000
DESTRUXOL BORER-SOL see EIY600

DESTRUXOL ORCHID SPRAY see NDN000
DESTUN see TKF750
DESURIC see DDP200
DESYPHED see MDT600
DESYREL see CKJ000, THK880, THK875
DET see DKC800
m-DET see DKC800
DETA see DJG600
m-DETA see DKC800
DETAL see DUS700
DETALUP see VSZ100
DETAMIDE see DKC800
DETARIL see DWK400
DETERGENT 66 see SIB600
DETERGENT ALKYLATE see PEW500
DETERGENT HD-90 see DXW200
DETERGENTS, LIQUID containing AES see DBB450
DETERGENTS, LIQUID containing LAS see DBB460
DETF see TIQ250
DETHMORE see WAT200
DETHYLANDIAMINE see TEO000
2,4-D ETHYL ESTER see EHY600
DETHYRONA see SKJ300
DETIA GAS EX-B see AHE750, PGY000
DETICENE see DAB600
DETIGON see CMW700
DETIGON-BAYER see CMW700
DETMOL-EXTRAKT see BBQ500
DETMOL MA see MAK700
DETMOL MA 96% see MAK700
DETOX see DAD200
DETOX 25 see BBQ500
DETOXAN see DAD200
DETOXARGIN see AQW000
DETRALFATE see DBB500
DETRAVIS see DAI485
DETREOMYCINE see CDP250
DETREOPAL see CDP700
DETREX see DBA800
DETTOL see CLW000
DETYROXIN see SKJ300
DEURSIL see DMJ200
DEUSLON-A see EDU500
DEUTERIOMORPHINE see DBB600
DEUTERIUM see DBB800
DEUTERIUM FLUORIDE see DBC000
DEUTERIUM OXIDE see HAK000
DEVAL RED K see CLK220
DEVAL RED TR see CLK220
DEVEGAN see ABX500
DEVELIN see PNA500
DEVELOPER 11 see PEY000
DEVELOPER 13 see PEY500
DEVELOPER A see NAX000
DEVELOPER AMS see NAX000
DEVELOPER BN see NAX000
DEVELOPER H see TGL750
DEVELOPER P see NEO500
DEVELOPER PF see PEY500
DEVELOPER R see REA000
DEVELOPER SODIUM see NAX000
DEVICORAN see HID350
DEVIGON see DSP400
DEVIKOL see DGP900
DEVIL'S APPLE see MBU800, SLV500
DEVIL'S BACKBONE see SDZ475
DEVIL'S CLAWS see FBS100
DEVILS HAIR see CMV390
DEVIL'S IVY see PLW800
DEVILS THREAD see CMV390
DEVINCAN see VLF000
DEVIPON see DGI400
DEVISULPHAN see EAQ750
DEVITHION see MNH000
DEVOL ORANGE B see NEO000
DEVOL ORANGE R see NEN500
DEVOL RED K see CLK235
DEVOL RED TA SALT see CLK235
DEVOL RED TR see CLK235
DEVOL SCARLET A (FREE BASE) see DEO400
DEVOL SCARLET B see NMP500
DEVONIUM see AOD000
DEVORAN see BBQ500
DEVOTON see MFW100

DEX see BJU000
DEXA see SOW000
DEXACORT see DAE525, SOW000
DEXA-CORTIDELT see SOW000
DEXA-CORTIDELT HOSTACORTIN H see PMA000
DEXADELTONE see SOW000
DEXADRESON see DAE525
DEXAGRO see DAE525
DEXAIME see BBK500
DEXALINE see BBK500
DEXALME see BBK500
DEXAMBUTOL see EDW875
DEXAMED see BBK500
DEXAMETH see SOW000
DEXAMETHASONE ACETATE see DBC400
DEXAMETHASONE ALCOHOL see SOW000
DEXAMETHASONE DIPROPIONATE see DBC500
DEXAMETHASONE 17,21-DIPROPIONATE see DBC500
DEXAMETHASONE DISODIUM PHOSPHATE see DAE525
DEXAMETHASONE ISONICOTINATE see DBC510
DEXAMETHASONE-21-ISONICOTINATE see DBC510
DEXAMETHASONE-21-ORTHOPHOSPHATE see BFW325
DEXAMETHASONE PALMITATE see DBC525
DEXAMETHASONE-21-PALMITATE see DBC525
DEXAMETHASONE PHOSPHATE see BFW325
DEXAMETHASONE-21-PHOSPHATE see BFW325
DEXAMETHASONE SODIUM HEMISULFATE see DBC550
DEXAMETHASONE SODIUM PHOSPHATE see DAE525
DEXAMETHASONE SODIUM SULFATE see DBC550
DEXAMETHASONE VALERATE see DBC575
DEXAMETHASONE-17-VALERATE see DBC575
DEXAMETHAZONE SODIUM PHOSPHATE see DAE525
DEXAMINE see BBK500
DEXAMPHAMINE see BBK500
DEXAMPHETAMINE see AOA500, BBK500
DEXAMPHETAMINE SULFATE see BBK500
DEXAMYL see BBK500
DEXA-SCHEROSON (INJECTABLE) see DBC550
DEXBENZETIMIDE HYDROCHLORIDE see DBE000
DEXBROMPHENIRAMINE MALEATE see DXG100
DEXCHLOROPHENIRAMINE MALEATE see PJJ325
DEXCHLORPHENIRAMINE MALEATE see PJJ325
DEXEDRINA see BBK500
DEXEDRINE see AOA500
DEXEDRINE SULFATE see BBK500
DEXETIMIDE HYDROCHLORIDE see DBE000
DEXIES see BBK500
DEXIUM see DMI300
DEXON see DOU600
DEXONE see SOW000
DEXON E 117 see PMP500
DEXOPHRINE see DBA800
DEXOVAL see DBA800, MDT600
DEXOXADROL HYDROCHLORIDE see DVP400
DEXPANTHENOL (FCC) see PAG200
DEXTELAN see SOW000
DEXTIM see MDT600
DEXTRAN see DBD700
DEXTRAN 1 see DBC800
DEXTRAN 2 see DBD000
DEXTRAN 5 see DBD200
DEXTRAN 10 see DBD400
DEXTRAN 11 see DBD600
DEXTRAN 70 see DBD700
DEXTRAN ION COMPLEX see IGS000
DEXTRANS see DBD800
DEXTRAN SULFATE SODIUM see DBD750
DEXTRAN SULFATE SODIUM ALUMINIUM see DBB500
DEXTRARINE see DBD750
DEXTRAVEN see DBD700
DEXTRIFERRON see IGU000
DEXTRIFERRON INJECTION see IGU000
β-DEXTRIN see COW925
DEXTRINS see DBD800
DEXTROAMPHETAMINE SULFATE see BBK500
DEXTROBENZETIMIDE HYDROCHLORIDE see DBE000
DEXTRO CALCIUM PANTOTHENATE see CAU750
DEXTROCHLORPHENIRAMINE MALEATE see PJJ325
DEXTROFER 75 see IGS000
DEXTROID see SKJ300
DEXTROMETHADONE see DBE100
DEXTROMETHORPHAN see DBE150
DEXTROMETHORPHAN BROMIDE see DBE200
DEXTROMETHORPHAN HYDROBROMIDE see DBE200

DEXTRO-α-METHYLPHENETHYLAMINE SULFATE see BBK500
DEXTROMETORPHAN HYDROBROMIDE see DBE200
DEXTROMORAMIDE see AFJ400
DEXTROMYCETIN see CDP250
DEXTROMYCIN HYDROCHLORIDE see DBE600
DEXTRONE see DWX800, PAJ000
DEXTRO-1-PHENYL-2-AMINOPROPANE SULFATE see BBK500
DEXTRO-β-PHENYLISOPROPYLAMINE SULFATE see BBK500
DEXTROPROPOXYPHENE see DAB879
DEXTROPROPOXYPHENE HYDROCHLORIDE see PNA500
DEXTROPROPOXYPHENE NAPSYLATE see DBE625
DEXTROPROXYPHEN HYDROCHLORIDE see PNA500
DEXTROPUR see GFG000
DEXTRORPHAN see DBE800
DEXTRORPHAN TARTRATE see DBE825
DEXTROSE (FCC) see GFG000
DEXTROSE, anhydrous see GFG000
DEXTROSOL see GFG000
DEXTROSULPHENIDOL see MPN000
DEXTROTHYROXINE SODIUM see SKJ300
DEXTROTUBOCURARINE CHLORIDE see TOA000
DEXTROXIN see SKJ300
DEXULATE see DBD750
DEZIBARBITUR see EOK000
DEZONE see SOW000
DF 118 see DKW800, DKX000
DF 468 see PIM500
DF 469 see AAJ350
DF-521 see RDA350
DFA see DVX800
Df B see DAK200
DFM see DHE800
α-DFMO see DBE835, DKH875
DFO see DAK200
DFOA see DAK200
DFOM see DAK200
DFP see IRF000
DFT see DWN800
5'-DFUR see DYE415
DFV see DKF130
2-DG see DAR600
DGE see DKM200
DG-HCG see DAK325
DH 245 see AOO300
DH-524 see DGA800
DHA see AOO450, DAK400, MFW500
DHAQ see MQY090
DHAQ DIACETATE see DBE875
DHA-SODIUM see SGD000
DHA-S SODIUM see DAL040
2,4-DHBA see HOE600
2,5-DHBA see GCU000
DHBP see DYF200
DHC see DAL000
β-DHC see DLO880
DHK see DLR000
DHMS see DME000
DHNT see BKH500
(S)-DHPA see AMH800
DHPN see DNB200
DHS see MFW500
DHT see DME500
DHT₂ see DME300
5-β-DHT see HJB100
DHTP see DME525
DHUTRA see SLV500
D 201 HYDROCHLORIDE see AEG625
D 58SI see DLU900
DIABARIL see CKK000
DIABASE SCARLET G see NMP500
DIABASIC MALACHITE GREEN see AFG500
DIABASIC RHODAMINE B see FAG070
DIABECHLOR see CKK000
DIABEFAGOS see DQR800
DIABEN see BSQ000
DIABENAL see CKK000
DIABENESE see CKK000
DIABENEZA see CKK000
DIABENOR see DBE885
DIABENYL see BBV500
DIABETA see CEH700
DIABETAMID see BSQ000
DIABETOL see BSQ000
DIABETORAL see CKK000

DIABET-PAGES see CKK000
DIABINESE see CKK000
DIABORAL see BSM000, CPR000
DIABRIN see BOM750, BQL000
DIABUTAL see NBU000
DIABUTON see BSQ000
DIABYLEN see BBV500
DIACARB see AAI250
DIACELLITON FAST BLUE R see TBG700
DIACELLITON FAST BRILLIANT BLUE B see MGG250
DIACELLITON FAST GREY G see DCJ200
DIACELLITON FAST VIOLET 5R see DBP000
DIACEL NAVY DC see DCJ200
DIACEPAN see DCK759
DIACEPHIN see HBT500
4,4'-DIACETAMIDODIPHENYL SULFONE see SNY500
2-DIACETAMIDOFLUORENE see DBF200
2,7-DIACETAMIDOFLUORENE see BGP250
N-1-DIACETAMIDOFLUORENE see DBF000
2,4-DIACETAMIDO-6-(5-NITRO-2-FURYL)-s-TRIAZINE see DBF400
3,5-DIACETAMIDO-2,4,6-TRIIODOBENZOIC ACID see DCK000
3,5-DIACETAMIDO-2,4,6-TRIIODOBENZOIC ACID, SODIUM SALT
    see SEN500
α-5-DIACETAMIDO-2,4,6-TRIIODO-m-TOLUIC ACID see AAI750
10,040 DIACETATE see CDT125
2',4'-DIACETATE-DIS-NOGAMYCIN see DBF500
1,3-DIACETATE GLYCEROL see DBF600
1,2-DIACETATE 1,2,3-PROPANETRIOL see DBF600
DIACETATOZIRCONIC ACID see ZTS000
DIACETAZOTOL see ACR300
DIACETIC ETHER see EFS000
DIACETIN see DBF600
1,2-DI-ACETIN see DBF600
1,3-DIACETIN see DBF600
2,3-DIACETIN see DBF600
DIACETONALCOHOL (DUTCH) see DBF750
DIACETONALCOOL (ITALIAN) see DBF750
DIACETONALKOHOL (GERMAN) see DBF750
DIACETONE see DBF750
DIACETONE ACRYLAMIDE see DTH200
DIACETONE ALCOHOL see DBF750
DIACETONE-ALCOOL (FRENCH) see DBF750
DIACETOTOLUIDE see ACR300
o-DIACETOTOLUIDIDE, 4''-(o-TOLYLAZO)-(8CI) see ACR300
1,3-DIACETOXYBENZENE see REA100
DIACETOXYBUTYLTIN see DBF500
1,2-DIACETOXY-3-CHLOROPROPANE see CIL900
DIACETOXYDIBUTYLPLUMBANE see DED400
DIACETOXYDIBUTYL STANNANE see DBF800
DIACETOXYDIBUTYLTIN see DBF800
1,1-DIACETOXY-2,3-DICHLOROPROPANE see DBF875
trans-7,8-DIACETOXY-7,8-DIHYDROBENZO(a)PYRENE see DBG200
4-β,15-DIACETOXY-3-α,8-α-DIHYDROXY-12,13-EPOXYTRICHOTHEC-9-ENE
    see NCK000
DIACETOXYDIMETHYLPLUMBANE see DSL400
3-α,17-β-DIACETOXY-2-β,16-β-DIPIPERIDINO-5-α-ANDROSTANE
    DIMETHOBROMIDE see PAF625
3-β,17-β-DIACETOXY-17-α-ETHINYL-19-NOR-Δ3,5-ANDROSTADIENE
    see DBG300
3-β,17-β-DIACETOXY-17-α-ETHYNYL-4-OESTRENE see EQJ500
4-β,15-DIACETOXY-3-α-HYDROXY-12,13-EPOXYTRICHOTHEC-9-ENE
    see AOP250
DIACETOXYMERCURIPHENOL see HNK575
DIACETOXYMERCURY see MCS750
7-DIACETOXYMETHYLBENZ(a)ANTHRACENE see BBG750
4,15-DIACETOXY-8-(3-METHYLBUTYRYLOXY)-12,13-EPOXY-Δ-9-
    TRICHOTHECEN-3-OL see FQS000
4-β,15-DIACETOXY-8-α-(3-METHYLBUTYRYLOXY)-3-α-HYDROXY-12,13-
    EPOXYTRICHOTHEC-9-ENE see FQS000
3-β,17-β-DIACETOXY-19-NOR-17-α-PREGN-4-EN-20-YNE see EQJ500
DIACETOXYPROPENE see ADR250
1,3-DIACETOXYPROPENE see PMO800
3,3-DIACETOXYPROPENE see ADR250
1,1-DIACETOXYPROPENE-2 see ADR250
DIACETOXYSCIRPENOL see AOP250
4,15-DIACETOXYSCIRPEN-3-OL see AOP250
DIACETOXYTETRABUTYLDISTANNOXANE see BGQ000
DIACETYLAMINOAZOTOLUENE see ACR300
4,4'-DIACETYLAMINOBIPHENYL see BFX000
2-DIACETYLAMINOFLUORENE see DBF200
2,7-DIACETYLAMINOFLUORENE see BGP250
N-DIACETYL-2-AMINOFLUORENE see DBF200
N,N-DIACETYL-2-AMINOFLUORENE see DBF200
DI-(p-ACETYLAMINOPHENYL)SULFONE see SNY500

3,5-DIACETYLAMINO-2,4,6-TRIJODBENZOSAEURE NATRIUM (GERMAN)
    see SEN500
4,4'-DIACETYLBENZIDINE see BFX000
N,N'-DIACETYL BENZIDINE see BFX000
2-((2,4-DIACETYL-5-BENZOFURANYL)OXY)TRIETHYLAMINE HYDROCHLO-
    RIDE see DBG499
DIACETYLCHOLINE see CMG250
DIACETYLCHOLINE CHLORIDE see HLC500
DIACETYLCHOLINE DICHLORIDE see HLC500
DIACETYLCHOLINE DIIODIDE see BJI000
DIACETYLDAPSONE see SNY500
N,N'-DIACETYLDAPSONE see SNY500
4,4'-DIACETYLDIAMINODIPHENYL SULFONE see SNY500
N,N'-DIACETYL-4,4'-DIAMINODIPHENYL SULFONE see SNY500
2,6-DIACETYL-7,9-DIHYDROXY-8,9b-DIMETHYL-1,3(2H,9bH)-
    DIBENZOFURANDIONE see UWJ000
N,N'-DIACETYL-3,3'-DIMETHYLBENZIDINE see DRI400
N,N'-DIACETYL-N,N'-DINITRO-1,2-DIAMINOETHANE see DBG900
DIACETYL DIOXIME see DBH000
1,2-DIACETYLETHANE see HEQ500
α,β-DIACETYLETHANE see HEQ500
DIACETYL (FCC) see BOT750
N,N-DIACETYL-2-FLUORENAMINE see DBF200
DIACETYL GLYCERINE see DBF600
O,O'-DIACETYL 4-HYDROXYAMINOQUINOLINE-1-OXIDE see ABN500
DIACETYLLANATOSID C (GERMAN) see DBH200
DIACETYLLANATOSIDE see DBH200
DIACETYLMANGANESE see MAQ000
DIACETYLMETHANE see ABX750
DIACETYLMORFIN see HBT500
DIACETYLMORPHINE see HBT500
DIACETYLMORPHINE HYDROCHLORIDE see DBH400
N,O-DIACETYL-N-(1-NAPHTHYL)HYDROXYLAMINE see ABS250
1,3-DIACETYLOXYPROPENE see PMO800
DIACETYL PEROXIDE (MAK) see ACV500
DIACETYL PEROXIDE (solution) see ACV750
N,O-DIACETYL-N-(p-STYRYLPHENYL)HYDROXYLAMINE see ABW500
trans-N,o-DIACETYL-N-(p-STYRYLPHENYL)HYDROXYLAMINE
    see ABL250
3,4-DI(ACETYLTHIOMETHYL)-5-HYDROXY-6-METHYLPYRIDINE
    HYDROBROMIDE see DBH800
N,N-DIACETYL-o-TOLYLAZO-o-TOLUIDINE see ACR300
DIACID see BNK000
DIACID METANIL YELLOW see MDM775
DIACOTTON BLUE BB see CMO000
DIACOTTON CONGO RED see SGQ000
DIACOTTON DEEP BLACK see AQP000
DIACOTTON SKY BLUE 5B see CMO500
DIACOTTON SKY BLUE 6B see CMN750
DIACRID see DBX400
DIACTOL see VSZ100
DIACYCINE see TBX250
DIADEM CHROME BLUE R see HJF500
DIADILAN see PFJ750
DIADOL see AFS500
DI-ADRESON F see PMA000
DI-ADRESON-F-AQUOSUM see PMA100
DIAETHANOLAMIN (GERMAN) see DHF000
DIAETHANOLAMIN-3,5-DIJODPYRIDON-(4)-ESSIGSAEURE (GERMAN)
    see DNG400
DIAETHANOLNITROSAMIN (GERMAN) see NKM000
1,1-DIAETHOXY-AETHAN (GERMAN) see AAG000
2-DIAETHOXYPHOSPHINYL-THIOAETHYL-TRIMETHYL-AMMONIUM-
    JODID (GERMAN) see TLF500
DIAETHYLACETAL (GERMAN) see AAG000
DIAETHYLAETHER (GERMAN) see EJU000
O,O-DIAETHYL-S-(2-AETHYLTHIO-AETHYL)-DITHIOPHOSPHAT (GERMAN)
    see DXH325
O,O-DIAETHYL-S-(2-AETHYLTHIO-AETHYL)-MONOTHIOPHOSPHAT (GER-
    MAN) see DAP200
O,O-DIAETHYL-S-(AETHYLTHIO-METHYL)-DITHIOPHOSPHAT (GERMAN)
    see PGS000
DIAETHYLALLYLACETAMID (GERMAN) see DJU200
DIAETHYLAMIN (GERMAN) see DHJ200
DIAETHYLAMINOAETHANOL (GERMAN) see DHO500
o-DIAETHYLAMINOAETHOXY-BENZANILID (GERMAN) see DHP200
o-DIAETHYLAMINOAETHOXY-5-CHLOR-BENZANILID (GERMAN)
    see CFO250
1-DIAETHYLAMINO-AETHYLAMINO-4-METHYL-
    THIOXANTHONHYDROCHLORID (GERMAN) see SBE500
DIAETHYLAMINOAETHYL DIPHENYLCARBAMAT HYDROCHLORID (GER-
    MAN) see DHY200
N-DIAETHYLAMINOAETHYL-RESERPIN (GERMAN) see DIG800
DIAETHYLAMINOAETHYL-THEOPHYLLIN (GERMAN) see CNR125

7-(β-DIAETHYLAMINO-AETHYL)-THEOPHYLLIN-HYDROCHLORID (GER-
MAN) see DIH600
DIAETHYLANILIN (GERMAN) see DIS700
O,O-DIAETHYL-O-(4-BROM-2,5-DICHLOR)-PHENYL-MONOTHIOPHOSPHAT
(GERMAN) see EGV500
DIAETHYLCARBONAT (GERMAN) see DIX200
O,O-DIAETHYL-O-(CHINOXALYL-(2))-MONOTHIOPHOSPHAT (GERMAN)
see DJY200
O,O-DIAETHYL-O-(3-CHLOR-4-METHYL-CUMARIN-7-YL)-
MONOTHIOPHOSPHAT (GERMAN) see CNU750
O,O-DIAETHYL-S-(6-CHLOR-2-OXO-BEN(b)-1,3-OXALIN-3-YL)-METHYL-DIT
HIOPHOSPHAT (GERMAN) see BDJ250
O,O-DIAETHYL-S-((4-CHLOR-PHENYL-THIO)-METHYL)DITHIOPHOSPHAT
(GERMAN) see TNP250
O,O-DIAETHYL-o-(α-CYANBENZYLIDEN-AMINO)-THIONPHOSPHAT (GER-
MAN) see BAT750
O,O-DIAETHYL-o-(α-CYANO-BENZYLIDENAMINO)-MONOTHIOPHOSPHAT
(GERMAN) see BAT750
N-DIAETHYL CYSTEAMIN (GERMAN) see DIY600
O,O-DIAETHYL-O-(2,5-DICHLOR-4-BROMPHENYL)-THIONOPHOSPHAT
(GERMAN) see EGV500
O,O-DIAETHYL-O-1-(4,5-DICHLORPHENYL)-2-CHLOR-VINYL-PHOSPHAT
(GERMAN) see CDS750
O,O-DIAETHYL-O-2,4-DICHLOR-PHENYL-MONOTHIOPHOSPHAT (GER-
MAN) see DFK600
O,O-DIAETHYL-S-((2,5-DICHLOR-PHENYL-THIO)-METHYL)-
DITHIOPHOSPHAT (GERMAN) see PDC750
O,O-DIAETHYL-O-2,4-DICHLORPHENYL-THIONOPHOSPHAT (GERMAN)
see DFK600
1,2-DIAETHYLHYDRAZINE (GERMAN) see DJL400
O,O-DIAETHYL-O-(2-ISOPROPYL-4-METHYL-PYRIMIDIN-6-YL)-
MONOTHIOPHOSPHAT (GERMAN) see DCM750
O,O-DIAETHYL-O-(2-ISOPROPYL-4-METHYL)-6-PYRIMIDYL-
THIONOPHOSPHAT (GERMAN) see DCM750
O,O-DIAETHYL-S-1-METHYL)AETHYL)-CARBAMOYL-METHYL-
MONOTHIOPHOSPHAT (GERMAN) see PHK250
O,O-DIAETHYL-O-(4-METHYL-COUMARIN-7-YL)-MONOTHIOPHOSPHAT
(GERMAN) see PKT000
O,O-DIAETHYL-S-(3-METHYL-2,4-DIOXO-5-OXA-3-AZA-HEPTYL)-
DITHIOPHOSPHAT (GERMAN) see DJI000
O,O-DIAETHYL-O-(3-METHYL-1H-PYRAZOL-5-YL)-PHOSPHAT (GERMAN)
see MOX250
O,O-DIAETHYL-O-4-METHYLSULFINYL-PHENYL-MONOTHIOPHOSPHAT
(GERMAN) see FAQ800
DIAETHYL-NICOTINAMID (GERMAN) see DJS200
O,O-DIAETHYL-O-(4-NITROPHENYL)-MONOTHIOPHOSPHAT (GERMAN)
see PAK000
O,O'-DIAETHYL-p-NITROPHENYLPHOSPHAT (GERMAN) see NIM500
DIAETHYL-p-NITROPHENYLPHOSPHORSAEUREESTER (GERMAN)
see NIM500
DIAETHYLNITROSAMIN (GERMAN) see NJW500
O,O-DIAETHYL-S-(4-OXOBENZOTRIAZIN-3-METHYL)-DITHIOPHOSPHAT
(GERMAN) see EKN000
O,O-DIAETHYL-S-((4-OXO-3H-1,2,3-BENZOTRIAZIN-3-YL)-METHYL)-
DITHIOPHOSPHAT (GERMAN) see EKN000
O,O-DIAETHYL-N-PHTALIMIDOTHIOPHOSPHAT (GERMAN) see DJX200
DI-AETHYL-PROPANEDIOL (GERMAN) see PMB250
O,O-DIAETHYL-O-(PYRAZIN-2YL)-MONOTHIOPHOSPHAT (GERMAN)
see EPC500
O,O-DIAETHYL-O-(2-PYRAZINYL)-THIONOPHOSPHAT (GERMAN)
see EPC500
DIAETHYLSULFAT (GERMAN) see DKB110
O,O-DIAETHYL-S-(3-THIA-PENTYL)-DITHIOPHOSPHAT (GERMAN)
see DXH325, DAO500
DIAETHYLTHIOPHOSPHORSAEUREESTER des AETHYLTHIOGLYKOL (GER-
MAN) see DAP200
O,O-DIAETHYL-O-3,5,6-TRICHLOR-2-PYRIDYLMONOTHIOPHOSPHAT (GER-
MAN) see CMA100
DIAETHYLZINNDICHLORID (GERMAN) see DEZ000
DIAFEN see DVW700, LJR000
DIAFEN (antihistamine) see DVW700
DIAFEN HYDROCHLORIDE see DVW700
DIAFURON see NGG500
DIAGINOL see AAN000
DIAGNORENOL see SHX000
DIAGRABROMYL see BNP750
DIAKARB see AAI250
DIAKARMON see SKV200
DIAKON see MLH750, PKB500
DIAL see AFS500, CFU750
DIAL-A-GESIC see HIM000
DIALFERIN see DBK400
DIALICOR see DHS200
DIALIFOR see DBI099

DIALKYL 79 PHTHALATE see LGF875
DIALKYLPHTHALATE C7-C9 see LGF875
DIALKYLZINCS see DBI159
DIALLAAT (DUTCH) see DBI200
DIALLAT (GERMAN) see DBI200
DIALLATE see DBI200
N,N-DIALLYDICHLOROACETAMIDE see DBJ600
DIALLYINITROSAMIN (GERMAN) see NJV500
DIALLYL see HCR500
DIALLYLAMINE see DBI600
4-DIALLYLAMINO-3,5-DIMETHYLPHENYL-N-METHYLCARBAMATE
see DBI800
4-(DIALLYLAMINO)-3,5-XYLENOL METHYLCARBAMATE (ester) see DBI800
4-DIALLYL-AMINO-3,5-XYLYL N-METHYLCARBAMATE see DBI800
DIALLYLBARBITAL see AFS500
DIALLYLBARBITURIC ACID see AFS500
5,5-DIALLYLBARBITURIC ACID see AFS500
1,1-DIALLYL-3-(1,4-BENZODIOXAN-2-YLMETHYL)-3-METHYLUREA
see DBJ100
2-(N,N-DIALLYLCARBAMYLMETHYL)AMINOMETHYL-1,4-BENZODIOXAN
see DBJ100
DIALLYLCHLOROACETAMIDE see CFK000
N,N-DIALLYLCHLOROACETAMIDE see CFK000
N,N-DIALLYL-2-CHLOROACETAMIDE see CFK000
N,N-DIALLYL-α-CHLOROACETAMIDE see CFK000
DIALLYLCYANAMIDE see DBJ200
DIALLYLDIBROMO STANNANE see DBJ400
N,N-DIALLYL-2,2-DICHLOROACETAMIDE see DBJ600
DIALLYL DIGLYCOL CARBONATE see AGD250
DIALLYL ETHER see DBK000
DIALLYLETHER ETHYLENGLYKOLU (CZECH) see EJE000
DIALLYLHYDRAZINE see DBK100
1,1-DIALLYLHYDRAZINE see DBK100
1,2-DIALLYLHYDRAZINE DIHYDROCHLORIDE see DBK120
DIALLYLMAL see AFS500
DIALLYL MALEATE see DBK200
DIALLYL MONOSULFIDE see AGS250
DIALLYLNITROSAMINE see NJV500
DIALLYLNORTOXIFERINE DICHLORIDE see DBK400
N,N'-DIALLYLNORTOXIFERINIUM DICHLORIDE see DBK400
DIALLYL PEROXYDICARBONATE see DBK600
2,6-DIALLYLPHENOL see DBK800
DIALLYL PHOSPHITE see DBL000
DIALLYL PHTHALATE see DBL200
DIALLYL SELENIDE see DBL300
DIALLYL SULFATE see DBL400
DIALLYL SULFIDE see AGS250
DIALLYL THIOETHER see AGS250
DIALLYL THIOUREA see DBL600
DIALLYLTIN DIBROMIDE see DBJ400
DIALUMINUM OCTAVANADIUM TRIDECASILICIDE see DBL649
DIALUMINUM SULPHATE see AHG750
DIALUMINUM TRIOXIDE see AHE250
DIALUMINUM TRISULFATE see AHG750
DIALUX see ADY500
DIAMARIN see DYE600
DIAMAZO see ACR300
DIAMET see SIL500
DIAMFEN HYDROCHLORIDE see DHY200
DIAMICRON see DBL700
DIAMIDAFOS see PEV500
DIAMIDE see DRP800, HGS000
DIAMIDFOS see PEV500
DIAMIDINE see DBL800
N,N'-DIAMIDINO-9-AZA-1,17-HEPTADECANEDIAMINE HYDROGEN SUL-
FATE see GLQ000
4,4'-DIAMIDINODIPHENOXYPENTANE see DBM000
4,4'-DIAMIDINO-α,omega-DIPHENOXYPENTANE see DBM000
4,4'-DIAMIDINODIPHENOXYPENTANE DI(β-HYDROXYETHANESULFON-
ATE see DBL800
4,4'-DIAMIDINO-α,omega-DIPHENOXYPENTANE ISETHIONATE see DBL800
4,4'-DIAMIDINO-1,3-DIPHENOXYPROPANE DIHYDROCHLORIDE
see DBM400
4,4'-DIAMIDINO-α,Γ'-DIPHENOXYPROPANE DIHYDROCHLORIDE
see DBM400
DIAMIDINO STILBENE see SLS000
4,4'-DIAMIDINOSTILBENE see SLS000
4,4'-DIAMIDINOSTILBENE DIHYDROCHLORIDE see SLS500
DIAMINE see HGS000
2,4-DIAMINEANISOLE see DBO000
DIAMINE BLUE 2B see CMO000
DIAMINE BLUE 3B see CMO250
DIAMINE DEEP BLACK EC see AQP000
cis-DIAMINEDICHLOROPLATINUM see PJD000

3,6-DIAMINO-2,7-DIMETHYLACRIDINE see DBT200
3,6-DIAMINO-2,7-DIMETHYLACRIDINE HYDROCHLORIDE see DBT400
4,4'-DIAMINO-3,3'-DIMETHYLBIPHENYL see TGJ750
4,4'-DIAMINO-3,3'-DIMETHYLDIPHENYL see TGJ750
1:2'-DIAMINO-1':2-DINAPHTHYL see BGC000
2,2'-DIAMINO-1,1'-DINAPHTHYL see BGB750
p-DIAMINODIPHENYL see BBX000
2,4'-DIAMINODIPHENYL see BGF109
4,4'-DIAMINODIPHENYL see BBX000
O,O'-DIAMINO DIPHENYL DISULFIDE see DXJ800
4,4'-DIAMINODIPHENYL-2,2'-DISULFONIC ACID see BBX500
DIAMINODIPHENYL ETHER see OPM000
4,4-DIAMINODIPHENYL ETHER see OPM000
p,p'-DIAMINODIPHENYL ETHER see OPM000
2,4'-DIAMINODIPHENYLMETHAN (GERMAN) see MJP750
4,4'-DIAMINODIPHENYLMETHAN (GERMAN) see MJQ000
DIAMINODIPHENYLMETHANE see MJQ000
2,4'-DIAMINODIPHENYLMETHANE see MJP750
4,4'-DIAMINODIPHENYLMETHANE see MJQ000
o,p'-DIAMINODIPHENYLMETHANE see MJP750
p,p'-DIAMINODIPHENYLMETHANE see MJQ000
4,4'-DIAMINODIPHENYL OXIDE see OPM000
4,4'-DIAMINODIPHENYL SULFIDE see TFI000
p,p'-DIAMINODIPHENYL SULFIDE see TFI000
3,3'-DIAMINODIPHENYL SULFONE see SOA000
DIAMINO-4,4'-DIPHENYL SULFONE see SOA500
4,4'-DIAMINODIPHENYL SULFONE see SOA500
p,p'-DIAMINODIPHENYL SULFONE see SOA500
p,p'-DIAMINODIPHENYLSULFONE-N,N-DI(DEXTROSE SODIUM SULFO-
  NATE) see AOO800
p,p'-DIAMINODIPHENYL SULPHIDE see TFI000
p,p-DIAMINODIPHENYL SULPHONE see SOA500
DIAMINO-4,4'-DIPHENYL SULPHONE see SOA500
p,p'-DIAMINODIPHENYLTRICHLOROETHANE see BGU500
4:4'-DIAMINO-3-DIPHENYLYL HYDROGEN SULFATE see BBY250
4,4'-DIAMINO-3-DIPHENYLYL HYDROGEN SULFATE see BBY500
3,3-DIAMINODIPROPYLAMINE see AIX250
3,3'-DIAMINODIPROPYLAMINE see AIX250
DIAMINODITOLYL see TGJ750
1,2-DIAMINO-ETHAAN (DUTCH) see EEA500
1,2-DIAMINOETHANE see EEA500
1,2-DIAMINOETHANEBISTRIMETHYLGOLD see DBU600
1,2-DIAMINOETHANE COPPER COMPLEX see DBU800
1,2-DIAMINO-ETHANO (ITALIAN) see EEA500
2,5-DIAMINO-7-ETHOXYACRIDINE LACTATE see EDW500
6,9-DIAMINO-2-ETHOXYACRIDINE LACTATE MONOHYDRATE see EDW500
4,6-DIAMINO-1-(2-ETHYLPHENYL)-2-METHYL-2-PROPYL-s-TRIAZINE HY-
  DROCHLORIDE see DBV200
3,8-DIAMINO-5-ETHYL-6-PHENYLPHENANTHRIDINIUM BROMIDE
  see DBV400
2,7-DIAMINO-10-ETHYL-9-PHENYLPHENANTHRIDINIUM BROMIDE
  see DBV400
3,8-DIAMINO-5-ETHYL-6-PHENYLPHENANTHRIDINIUM CHLORIDE
  see HGI000
2,4-DIAMINO-6-ETHYL-5-PHENYLPYRIMIDINE see DCA300
4-((4-(2,4-DIAMINO-1-ETHYLPYRIMIDINIUM-5-YL)PHENYL)AMINO)CAR-
  BONYL)PHENYL)AMINO)-1-ETHYLQUINOLINIUM) DIIODIDE see DBW000
2,7-DIAMINOFLUORENE see FDM000
DIAMINOGUANIDINIUM NITRATE see DBW100
1,6-DIAMINOHEXANE see HEO000
(l)-2,6-DIAMINO-1-HEXANETHIOL DIHYDROCHLORIDE see DBW200
2,6-DIAMINOHEXANOIC ACID see LJM700
2,6-DIAMINOHEXANOIC ACID HYDROCHLORIDE see LJO000
S-2,6-DIAMINOHEXYL DIHYDROGEN PHOSPHOROTHIOATE DIHYDRATE
  see DBW600
4,6-DIAMINO-2-HYDROXY-1,3-CYCLOHEXANE-3,6-DIAMINO-3,6'-
  DIDEOXYDI-α-d-GLUCOSIDE see KAL000
4,6-DIAMINO-1,3-HYDROXY-1,3-CYCLOHEXYLENE 3,6'-DIAMINO-3,6'-
  DIDEOXYDI-d-GLUCOPYRANOSIDE see KAL000
α,2-DIAMINO-3-HYDROXY-γ-OXOBENZENEBUTANOIC ACID see HJD000
4,4'-DIAMINO-IMINO-1,1'BIANTHRAQUINONE see IBD000
4,4'-DIAMINO-1,1'-IMINOBISANTHRAQUINONE see IBD000
1,8-DIAMINO-p-MENTHANE see MCD750
1,4-DIAMINO-2-METHOXYANTHRAQUINONE see DBX000
2,4-DIAMINO-1-METHOXYBENZENE see DBO000, DBO400
1,3-DIAMINO-4-METHOXYBENZENE see DBO400
2,4-DIAMINO-1-METHOXYBENZENE SULPHATE see DBO400
2,8-DIAMINO-10-METHYLACRIDINIUM CHLORIDE see XAK000
3,6-DIAMINO-10-METHYLACRIDINIUM CHLORIDE see XAK000
3,6-DIAMINO-10-METHYLACRIDINIUM CHLORIDE with 3,6-ACRIDINEDIAM-
  INE see DBX400
2,8-DIAMINO-10-METHYLACRIDINIUM CHLORIDE mixture with 2,8-
  DIAMINOACRIDINE see DBX400
1,3-DIAMINO-4-METHYLBENZENE see TGL750

2,4-DIAMINO-1-METHYLBENZENE see TGL750
2,4-DIAMINO-5-METHYL-6-(BUT-2-YL)PYRIDO(2,3-d)PYRIMIDINE see DBX875
2,4-DIAMINO-5-METHYL-6-sec-BUTYLPYRIDO(2,3-d)PYRIMIDINE see DBX875
2,4-DIAMINOMETHYLCYCLOHEXANE see DBY000
2,4-DIAMINO-1-METHYLCYCLOHEXANE see DBY000
3,7'-DIAMINO-N-METHYLDIPROPYLAMINE see BGU750
3-(((2-((DIAMINOMETHYLENE)AMINO)-4-THIAZOLYL)METHYL)THIO)-N²-
  SULFAMOYLPROPIONAMIDINE see FAB500
2,4-DIAMINO-2-METHYLPENTANE see MNI525
2,4-DIAMINO-6-METHYL-5-PHENYLPYRIMIDINE see DBY100
1,2-DIAMINO-2-METHYLPROPANE AQUADIPEROXO CHROMIUM(IV)
  see DBY300
N-(4-(((2,4-DIAMINO-5-METHYL-6-QUINAZOLINYL)METHYL)AMINO)
  BENZOYL)-l-ASPARTIC ACID see DBY500
N-(p-(((2,4-DIAMINO-5-METHYL-6-QUINAZOLINYL)METHYL)AMINO)
  BENZOYL)-l-ASPARTIC ACID see DBY500
DIAMINON see MDO750
1,5-DIAMINONAPHTHALENE see NAM000
1,7-DIAMINO-8-NAPHTHOL-3,6-DISULPHONIC ACID see DBY600
DIAMINON HYDROCHLORIDE see MDP000, MDP750
1,4-DIAMINO-5-NITRO ANTHRAQUINONE see DBY700
1,2-DIAMINO-4-NITROBENZENE see ALL500
1,4-DIAMINO-2-NITROBENZENE see ALL750
4,6-DIAMINO-2-(5-NITRO-2-FURYL)-s-TRIAZINE see DBY800
1,5-DIAMINOPENTANE see PBK500
DIAMINOPHEN see DHW200
2,4-DIAMINOPHENOL see DCA200
2,4-DIAMINOPHENOL HYDROCHLORIDE see AHP000
3,7-DIAMINOPHENOTHIAZIN-5-IUM CHLORIDE see AKK750
2,6-DIAMINO-3-PHENYLAZOPYRIDINE see PEK250
2,6-DIAMINO-3-PHENYLAZOPYRIDINE HYDROCHLORIDE see PDC250
2,6-DIAMINO-3-(PHENYLAZO)PYRIDINE MONOHYDROCHLORIDE
  see PDC250
4,4'-((4,6-DIAMINO-m-PHENYLENE)BIS(AZO))DIBENZENESULFONIC ACID,
  DISODIUM SALT mixed with p-((4,6-DIAMINO-m-TOLYL)AZO)-
  BENZENESULFONIC ACID, SODIUM SALT see CMP250
4,4'-DIAMINOPHENYL ETHER see OPM000
2,7-DIAMINO-9-PHENYL-10-ETHYLPHENANTHRIDINIUM BROMIDE
  see DBV400
2,7-DIAMINO-9-PHENYL-10-ETHYLPHENANTHRIDINIUM CHLORIDE
  see HGI000
2,4-DIAMINO-5-PHENYL-6-ETHYLPYRIMIDINE see DCA300
DI-(4-AMINOPHENYL)METHANE see MJQ000
2,4-DIAMINO-6-PHENYL-5-METHYLPYRIMIDINE see DBY100
2,7-DIAMINO-9-PHENYLPHENANTHRIDINE ETHOBROMIDE see DBV400
2,4-DIAMINO-5-PHENYL-6-PROPYLPYRIMIDINE see DCA450
2,4-DIAMINO-5-PHENYLPYRIMIDINE see DCA500
DI(p-AMINOPHENYL) SULFIDE see TFI000
DI(4-AMINOPHENYL)SULFONE see SOA500
DI(p-AMINOPHENYL) SULFONE see SOA500
3,3'-DIAMINOPHENYL SULFONE see SOA000
DI(p-AMINOPHENYL)SULPHIDE see TFI000
DI(4-AMINOPHENYL)SULPHONE see SOA500
DI(p-AMINOPHENYL)SULPHONE see SOA500
2,4-DIAMINO 5-PHENYLTHIAZOL CHLORHYDRATE (FRENCH) see DCA600
2,4-DIAMINO-5-PHENYLTHIAZOLE HYDROCHLORIDE see DCA600
2,4-DIAMINO-5-PHENYLTHIAZOLE MONOHYDROCHLORIDE see DCA600
2,4-DIAMINO-6-PIPERIDINILPIRIMIDINA-3-OSSIDO (ITALIAN) see DCB000
2,4-DIAMINO-6-PIPERIDINOPYRIMIDINE-3-OXIDE see DCB000
1,2-DIAMINOPROPANE see PMK250
1,3-DIAMINOPROPANE see PMK500
DI-β-AMINOPROPIONITRILE FUMARATE see AMB750
m-DI-(2-AMINOPROPYL)BENZENE DIHYDROCHLORIDE see DCC100
p-DI-(2-AMINOPROPYL)BENZENE DIHYDROCHLORIDE see DCC125
DI(3-AMINOPROPYL) ETHER of DIETHYLENE GLYCOL see DJD800
4-((4-(((4-(2,4-DIAMINO-1-PROPYLPYRIMIDINIUM-5-YL)PHENYL)AMINO)
  CARBONYL)PHENYL)AMINO)-1-PROPYLQUINOLINIUM) DIIODIDE
  see DCC200
DIAMINOPROPYLTETRAMETHYLENEDIAMINE see DCC400
l-(+)-N-(((2,4-DIAMINO-6-PTERIDINYL)METHYL)METHYLAMINO)
  BENZOYL)GLUTAMIC ACID see MDV500
2,6-DIAMINOPURINE see POJ500
2,6-DIAMINOPURINE SULFATE see DCC600
2,6-DIAMINOPYRIDINE see DCC800
3,4-DIAMINOPYRIDINE see DCD000
DIAMINO-3,4-PYRIDINE see DCD000
DIAMINOPYRITAMIN see TGD000
2,4-DIAMINOSOLE SULPHATE see DBO400
4:4'-DIAMINOSTILBENE see SLR500
2,4-DIAMINOTOLUEN (CZECH) see TGL750
DIAMINOTOLUENE see TGL500, TGL750
2,4-DIAMINOTOLUENE see TGL750
2,5-DIAMINOTOLUENE see TGM000
2,6-DIAMINOTOLUENE see TGM100

3,4-DIAMINOTOLUENE see TGM250
2,4-DIAMINO-1-TOLUENE see TGL750
2,4-DIAMINOTOLUENE DIHYDROCHLORIDE see DCE000
2,5-DIAMINOTOLUENE DIHYDROCHLORIDE see DCE200
2,6-DIAMINOTOLUENE DIHYDROCHLORIDE see DCE400
p-DIAMINOTOLUENE SULFATE see DCE600
2,5-DIAMINOTOLUENE SULFATE see DCE600
2,5-DIAMINOTOLUENE SULPHATE see DCE600
2,4-DIAMINOTOLUOL see TGL750
(4-((4,6-DIAMINO-m-TOLYL)IMINO)-2,5-CYCLOHEXADIEN-1-YLIDENE)
    DIMETHYLAMMONIUM CHLORIDE see DCE800
(4-((4,6-DIAMINO-m-TOLYL)IMINO)-2,5-CYCLOHEXADIEN-1-
    YLIDENE)DIMETHYLAMMONIUM CHLORIDE H$_2$O see TGU500
4,6-DIAMINO-s-TRIAZINE-2-METHANETHIOL S-ESTER with O,O-
    DIMETHYLPHOSPHORODITHIOATE see ASD000
4,6-DIAMINO-s-TRIAZINE-2-THIONE see DCF000
4,6-DIAMINO-1,3,5-TRIAZINE-2-THIONE see DCF000
4,6-DIAMINO-1,3,5-TRIAZINE-2(1H)-THIONE see DCF000
S-(4,6-DIAMINO-1,3,5-TRIAZIN-2-YL)-METHYL)-O,O-DIMETHYL-
    DITHIOFOSFAAT (DUTCH) see ASD000
S-((4,6-DIAMINO-1,3,5-TRIAZIN-2-YL)-METHYL)-O,O-DIMETHYL-
    DITHIOPHOSPHAT (GERMAN) see ASD000
S-((4,6-DIAMINO-s-TRIAZIN-2-YL)METHYL)-O,O-DIMETHYL PHOS-
    PHORODITHIOATE see ASD000
4,6-DIAMINO-1,3,5-TRIAZINE-2-YLMETHYL-O,O-DIMETHYL PHOS-
    PHORODITHIOATE see ASD000
S-(4,6-DIAMINO-1,3,5-TRIAZIN-2-YLMETHYL)-O,O-DIMETHYL PHOS-
    PHORODITHIOATE see ASD000
S-(4,6-DIAMINO-1,3,5-TRIAZIN-2-YLMETHYL) DIMETHYL PHOS-
    PHOROTHIOLOTHIONATE see ASD000
3,5-DIAMINO-s-TRIAZOLE see DCF200
3,10-DIAMINOTRICYCLO(5.2.1.0^{2,6})DECANE see DCF600
2,4-DIAMINO-5-(3,4,5-TRIMETHOXYBENZYL)PYRIMIDINE see TKZ000
2,5-DIAMINOTROPONE see DCF700
2,5-DIAMINOTROPONE HYDROCHLORIDE see DCF710
1,3-DIAMINOUREA see CBS500
DIAMMIDE SEBACICA della 4-FENIL-4-AMMINOMETIL-N-
    METILPIPERIDINA (ITALIAN) see BKS825
DIAMMINEBORONIUM HEPTAHYDROTETRABORATE see DCF725
DIAMMINEBORONIUM TETRAHYDROBORATE see DCF750
DIAMMINE(1,1-CYCLOBUTANEDICARBOXYLATO)PLATINUM (II)
    see CCC075
cis-DIAMMINE(1,1-CYCLOBUTANEDICARBOXYLATO)PLATINUM(II)
    see CCC075
trans-DIAMMINEDICHLOROPLATINUM(II) see DEX000
cis-DIAMMINEDINITRATO PLATINUM (II) see DCF800
DIAMMINEMALONATO PLATINUM (II) see DCG000
DIAMMINEPALLADIUM (II) NITRATE see DCG600
DIAMMINETETRACHLOROPLATINUM (OC-6-22) (9CI) see TBO776
cis-DIAMMINOTETRACHLOROPLATINUM see TBO776
1,2-DIAMMONIOETHANE NITRATE see EIV700
DIAMMONIUM ARSENATE see DCG800
DIAMMONIUM CARBONATE see ANE000
DIAMMONIUM CITRATE see ANF800
DIAMMONIUM HEXACHLOROPALLADATE see ANF000
DIAMMONIUM HEXACHLOROPLATINATE (2-) see ANF250
DIAMMONIUM HEXAFLUOROSILICATE see COE000
DIAMMONIUM HYDROGEN ARSENATE see DCG800
DIAMMONIUM HYDROGEN PHOSPHATE see ANR500
DIAMMONIUM MOLYBDATE see ANM750
DIAMMONIUM MONOHYDROGEN ARSENATE see DCG800
DIAMMONIUM SULFATE see ANU750
DIAMMONIUM TARTRATE see DCH000
DIAMMONIUM TETRACHLOROPALLADATE see ANE750
DIAMOND CHROME YELLOW 3R see NEY000
DIAMOND GREEN B see AFG500
DIAMOND GREEN G see BAY750
DIAMOND SHAMROCK 40 see PKQ059
DIAMOND SHAMROCK 744 see AAX175
DIAMOND SHAMROCK DS-15647 see DAB400
DIAMORFINA see HBT500
DIAMORPHINE see HBT500
DIAMORPHINE HYDROCHLORIDE see DBH400
DIAMOX see AAI250
DIAMPHETAMINE SULFATE see BBK250
DIAMYCELINE see CDY325
DIAMYL AMINE see DCH200
DI-n-AMYLAMINE (DOT) see DCH200
2,5-DI-tert-AMYLHYDROQUINONE see DCH400
DIAMYL KETONE see ULA000
DIAMYLNITROSAMIN (GERMAN) see DCH600
DIAMYLPHENOL see DCH800
DI-tert-AMYLPHENOL see DCI000
2,4-DI-tert-AMYL PHENOL see DCI000

DIAMYL PHTHALATE see AON300
DIAN see BLD500
DIANABOL see DAL300, MPN500
DIANABOLE see DAL300
DIANALINEMETHANE see MJQ000
DIANAT (RUSSIAN) see MEL500
DIANATE see MEL500
DIANCINA see AOD000
DIANDRON see AOO450
DIANDRONE see AOO450
DIANEMYCIN see DCI400
DIANHYDROCULCITOL see DCI600
1,2:5,6-DIANHYDRODULCITOL see DCI600
1,2:3,4-DIANHYDROERYTHRITOL see DHB800
DIANHYDROGALACTITOL see DCI600
1,2:5,6-DIANHYDROGALACTITOL see DCI600
d-1,4:3,6-DIANHYDROGLUCITOL see HID350
DIANHYDROMANNITOL see DCI800
1:2:5:6-DIANHYDRO-d-MANNITOL see DCI800
1,4:3,6-DIANHYDROSORBITOL see HID350
1,4:3,6-DIANHYDROSORBITOL-2,5-DINITRATE see CCK125
1,2:3,4-DIANHYDRO-dl-THREITOL see DHB600
DIANILBLAU see CMO250
DIANIL BLUE see CMO250
o,p'-DIANILINE see BGF109
p,p'-DIANILINE see BBX000
DIANILINOMERCURY see DCJ000
o-DIANISIDIN (CZECH, GERMAN) see DCJ200
o-DIANISIDINA (ITALIAN) see DCJ200
o-DIANISIDINE see DCJ200
3,3'-DIANISIDINE see DCJ200
O,O'DIANISIDINE see DCJ200
o-DIANISIDINE DIHYDROCHLORIDE see DOA800
DIANISIDINE DIISOCYANATE see DCJ400
α,Γ-DI-p-ANISYL-β-(ETHOXYPHENYL)GUANIDINE HYDROCHLORIDE
    see PDN500
3,4-DIANISYL-3-HEXENE see DJB200
DIANISYL-MONOPHENETHYLGUANIDINE HYDROCHLORIDE see PDN500
DIANISYLTRICHLORETHANE see MEI450
2,2-DI-p-ANISYL-1,1,1-TRICHLOROETHANE see MEI450
DIANON see DCM750
DI-1,1'-ANTHRACHINONYLAMIN (CZECH) see IBI000
DIANTHRAQUINONYLAMINE see IBI000
1,1'-DIANTHRAQUINONYLAMINE see IBI000
1,1-DIANTHRIMID (CZECH) see IBI000
DIANTHRIMIDE see IBI000
1,1'-DIANTHRIMIDE see IBI000
DIANTHUS SUPERBUS L., extract see DCJ450
DIANTIMONY PENTOXIDE see AQF750
DIANTIMONY TRIOXIDE see AQF000
DIANTIMONY TRISULFATE see AQJ250
DIAPADRIN see DGQ875
DIAPAM see DCK759
DIAPAMIDE see CIP500
DIAPARENE see MHB500
DIAPARENE CHLORIDE see MHB500
DIAPHEN see DVW700
DIAPHEN (NEUROPLEGIC) see DHW200
DIAPHENYLSULFONE see SOA500
DIAPHENYLSULPHON see SOA500
DIAPHENYLSULPHONE see SOA500
DIAPHORM see HBT500
DIAPHTAMINE BLACK V see AQP000
DIAPHTAMINE BLUE BB see CMO000
DIAPHTAMINE PURE BLUE see CMO500
DIAPHTAMINE VIOLET N see CMP000
DIAPHYLLINE see TEP500
DIAPP see BEN000
DIAQUODIAMMINEPLATINUM DINITRATE see DCJ600
cis-DIAQUODIAMMINEPLATINUM(II) DINITRATE see DCJ600
DIAQUONE see DMH400
DIAREX 43G see SMQ500
DIAREX 600 see SMR000
DIAREX HF 77 see SMQ000
DIARLIDAN see DGQ500
DIARSEN see DXE600
(1,2-DIARSENEDIYLBIS((6-HYDROXY-3,1-PHENYLENE)IMINO))
    BISMETHANESULFONIC ACID DISODIUM SALT see SNR000
DIARSENIC PENTOXIDE see ARH500
DIARSENIC TRIOXIDE see ARI750
DIARSENIC TRISULFIDE see ARI000
DIARYLANILIDE YELLOW see DEU000
DIASAN see FNF000
DIASETIELMORFIEN see HBT500

DIASETILMORFIN see HBT500
DIASETYLMORFIIMI see HBT500
DIASONE HYDROCHLORIDE see MDP000
DIASTATIN see NOH500
DIASTYL see DKA600
DIASULFA see SNN300
DIASULFYL see SNN300
DIAT (GERMAN) see DCK000
DIATER see DXQ500
DIATERR-FOS see DCM750
DIATHESIN see HMK100
DIATO BLUE BASE B see DCJ200
DIATOMACEOUS EARTH see DCJ800
DIATOMACEOUS SILICA see DCJ800
DIATOMITE see DCJ800
DIATRAST see DNG400
DIATRIN HYDROCHLORIDE see DCJ850
DIATRIZOATE MEGLUMINE see AOO875
DIATRIZOATE METHYLGLUCAMINE see AOO875
DIATRIZOATE SODIUM SALT see SEN500
DIATRIZOESAURE (GERMAN) see DCK000
DIATRIZOIC ACID see DCK000
DIA-TUSS see TCY750
6,12-DIAZAANTHANTHRENE see ADK250
6,12-DIAZAANTHANTHRENE SULFATE see DCK200
6,12-DIAZAANTHANTHRENE SULPHATE see DCK200
1,3-DIAZABENZENE see PPO750
10'-(β-(1,4-DIAZABICYCLO(4.3.0)NONANYL-4)-PROPIONYL)-2'-CHLORO-
    PHENTHIAZIN DICHLORIDE see NMV750
1,4-DIAZABICYCLO(2,2,2)OCTANE see DCK400
1,4-DIAZABICYCLO(2.2.2)OCTANE HYDROGEN PEROXIDATE see DCK500
DIAZACHEL (OBS.) see MDQ250
DIAZACOSTEROL HYDROCHLORIDE see ARX800
1,4-DIAZACYCLOHEPTANE see HGI900
1,3-DIAZA-2,4-CYCLOPENTADIENE see IAL000
7,14-DIAZADIBENZ(a,h)ANTHRACENE see DDC800
1,12-DIAZADIBENZO(a,i)PYRENE see NAU000
4,11-DIAZADIBENZO(a,h)PYRENE see NAU500
1,2-DIAZA-3,4:9,10-DIBENZPYRENE see APF750
1,6-DIAZA-3,4,8,9,12,13-HEXAOXABICYCLO(4.4.4)TETRADECANE see DCK700
1,3-DIAZAINDENE see BCB750
2,3-DIAZAINDOLE see BDH250
DIAZAJET see DCM750
DIAZALE see THE500
DIAZAN see FNF000
1,4-DIAZANAPHTHALENE see QRJ000
3,6-DIAZAOCTANE-1,8-DIAMINE see TJR000
4,5-DIAZAPHENANTHRENE see PCY250
3,5-DIAZAPHTHALIMID (GERMAN) see PPP000
3,6-DIAZAPHTHALIMID (GERMAN) see POL750
DIAZATOL see DCM750
DIAZENEDICARBOXAMIDE see ASM270
DIAZEPAM see DCK759
1H-1,4-DIAZEPINE, HEXAHYDRO- see HGI900
DIAZETARD see DCK759
DIAZETOXYSKIRPENOL (GERMAN) see AOP250
DIAZETYLMORPHINE see HBT500
DIAZIDE see DCM750
1,3-DIAZIDOBENZENE see DCL100
1,4-DIAZIDOBENZENE see DCL125, DCL125
p-DIAZIDOBENZENE (DOT) see DCL125
2,2-DIAZIDOBUTANE see DCL159
1,2-DIAZIDOCARBONYL HYDRAZINE see DCL200
DIAZIDODIBENZALACETONE see BGW720
DIAZIDODIBENZALACETON (CZECH) see BJA000
2,5-DIAZIDO-3,6-DICHLOROBENZOQUINONE see DCL300
DIAZIDODICYANOMETHANE see DCM000
DIAZIDODIMETHYLSILANE see DCL350
1,1-DIAZIDOETHANE see DCL400
1,2-DIAZIDOETHANE see DCL600
DIAZIDO ETHIDIUM see DCL800
3,8-DIAZIDO-5-ETHYL-6-PHENYLPHENANTHRIDINIUM BROMIDE
    see DCL800
DIAZIDOMALONONITRILE see DCM000
DIAZIDOMETHYLENEAZINE see DCM200
DIAZIDOMETHYLENECYANAMIDE see DCM400
1,3-DIAZIDO-2-NITROAZAPROPANE see DCM499
1,3-DIAZIDOPROPENE see DCM600
2,6-DIAZIDOPYRAZINE see DCM700
DIAZIL see DHU900
m-DIAZINE see PPO750
DIAZINE BLACK E see AQP000
DIAZINE BLUE 2B see CMO000
DIAZINE BLUE 3B see CMO250

DIAZINE VIOLET N see CMP000
DIAZINON see DCM750
DIAZINONE see DCM750
DIAZINON mixed with METHOXYCHLOR see MEI500
DIAZIQUONE see ASK875
3,6-DIAZIRIDINYL-2,5-BIS(CARBOETHOXYAMINO)-1,4-BENZOQUINONE
    see ASK875
DIAZIRINE see DCM800, DCP800
DIAZIRINE-3,3-DICARBOXYLIC ACID see DCM875
2,5-DIAZIRINO-3,6-DIPROPOXY-p-BENZOQUINONE see DCN000
DIAZITOL see DCM750
DI-AZO see PDC250
DIAZO see DUR800
DIAZOACETALDEHYDE see DCN200
2-(DIAZOACETAMINO)-N-ETHYLACETAMIDE see DCN600
DIAZOACETATE (ESTER)-l-SERINE see ASA500
l-DIAZOACETATE (ESTER) SERINE see ASA500
DIAZO-ACETIC ACID ESTER with SERINE see ASA500
DIAZOACETIC ACID, ETHYL ESTER see DCN800
DIAZOACETIC ESTER see DCN800
N-DIAZOACETILGLICINA-AMIDE (ITALIAN) see CBK000
N-DIAZOACETILGLICINA-IDRAZIDE (ITALIAN) see DCO800
DIAZOACETONITRILE see DCN875
2-((DIAZOACETYL)AMINO)-N-ETHYLACETAMIDE see DCN600
2-((DIAZOACETYL)AMINO)-N-METHYLACETAMIDE see DCO200
DIAZOACETYL AZIDE see DCO509
DIAZOACETYLGLYCINAMIDE see CBK000
N-(DIAZOACETYL)GLYCINAMIDE see CBK000
DIAZOACETYLGLYCINE AMIDE see CBK000
N-DIAZOACETYLGLYCINE AMIDE see CBK000
N-DIAZOACETYLGLYCINEETHYLAMIDE see DCN600
DIAZOACETYLGLYCINE ETHYL ESTER see DCO600
N-DIAZOACETYLGLYCINE ETHYL ESTER see DCO600
DIAZOACETYLGLYCINE HYDRAZIDE see DCO800
N-(DIAZOACETYL)GLYCINE HYDRAZINE see DCO800
N-DIAZOACETYLGLYCINE METHYLAMIDE see DCO200
N-DIAZOACETYL GLYCYLHYDRAZIDE see DCO800
o-DIAZOACETYL-l-SERINE see ASA500
DIAZOAMINOBENZEN (CZECH) see DWO800
DIAZOAMINOBENZENE see DWO800
p-DIAZOAMINOBENZENE see DWO800
DIAZOAMINOBENZOL (GERMAN) see DWO800
6-DIAZO-2-(2-(4-AMINO-4-CARBOXYBUTYRAMIDO)-6-DIAZO-5-
    OXOHEXANAMIDO)-HEXANOIC ACID SODIUM see ASO510
DIAZOAMINOMETHANE see DUI709
DIAZOBEN see MND600
DIAZOBENZENE see ASL250
DIAZOBLEU see BGT250
DIAZOCARD CHRYSOIDINE G see PEK000
20,25-DIAZOCHOLESTEROL DIHYDROCHLORIDE see ARX800
DIAZOCYCLOPENTADIENE see DCP200
DIAZODICYANOIMIDAZOLE see DCP880
DIAZODICYANOMETHANE see DCP775
2-DIAZO-4,6-DINITROBENZENE-1-OXIDE see DUR800
DIAZODINITROPHENOL (DOT) see DUR800
DIAZODINITROPHENOL, dry (DOT) see DUR800
DIAZODINITROPHENOL, containing, by weight, at least 40% water (DOT)
    see DUR800
DIAZOESSIGSAEURE-AETHYLESTER (GERMAN) see DCN800
DIAZO FAST ORANGE R see NEN500
DIAZO FAST RED AL see AIA750
DIAZO FAST RED GG see NEO500
DIAZO FAST RED TR see CLK235
DIAZO FAST RED TRA see CLK220, CLK235
DIAZO FAST SCARLET G see NMP500
DIAZO-ICA see DCP400
DIAZOIMIDAZOLE-4-CARBOXAMIDE see DCP400
5-DIAZOIMIDAZOLE-4-CARBOXAMIDE see DCP400
5-DIAZOIMIDAZOLE-4-CARBOXAMIDE HYDROCHLORIDE see DCP600
DIAZOIMIDE see HHG500
DIAZOL see DCM750
DIAZOL BLACK 2V see AQP000
DIAZOL BLUE 2B see CMO000
DIAZOL-C see AKK250
1,2-DIAZOLE see POM500
1,3-DIAZOLE see IAL000
DIAZOLONE see PPP500
DIAZOL PURE BLUE FF see BGT250
DIAZOL VIOLET N see CMP000
DIAZOMALONIC ACID see DCP700
DIAZOMALONONITRILE see DCP775
DIAZOMETHANE see DCP800
DIAZOMYCIN B see ASO510
DIAZO NERO MICROSETILE G see DPO200

2-DIAZONIO-4,5-DICYANOIMIDAZOLIDE see DCP880
5-DIAZONIOTETRAZOLIDE see DCQ200
DIAZO-OXO-NORLEUCINE see DCQ400
6-DIAZO-5-OXONORLEUCINE see DCQ400
6-DIAZO-5-OXO-l-NORLEUCINE see DCQ400
4-DIAZO-5-PHENYL-1,2,3-TRIAZOLE see DCQ500
3-DIAZOPROPENE see DCQ550
5-DIAZOPYRIMIDINE-2,4(3H)-DIONE see DCQ600
5-DIAZO-2,4(1H,3H)-PYRIMIDINEDIONE see DCQ600
DIAZO RED RD see DCQ560
DIAZO RESIN V see DCQ650
DIAZORESORCINOL see HNG500
DIAZOSSIDO (ITALIAN) see DCQ700
DIAZOTIZING SALTS see SIQ500
3-DIAZOTYRAMINE HYDROCHLORIDE see DCQ575
DIAZOURACIL see DCQ600
5-DIAZOURACIL see DCQ600
DIAZO V see DCQ650
DIAZOXIDE see DCQ700
DIAZYL see PPP500, SNJ000
DIBA see DNH125
DIBAM see SGM500
DI-BAPN FUMARATE see AMB750
DIBAR see DTP400
DIBASIC AMMONIUM ARSENATE see DCG800
DIBASIC AMMONIUM PHOSPHATE see ANR500
DIBASIC LEAD ACETATE see LCV000
DIBASIC LEAD ARSENATE see LCK000
DIBASIC LEAD CARBONATE see LCP000
DIBASIC SODIUM PHOSPHATE see SJH090
DIBASIC ZINC STEARATE see ZMS000
DIBASOL see BEA825
DIBAZOL see BEA825
DIBAZOLE see BEA825
DIBEKACIN see DCQ800
DIBEKACIN SULFATE see PAG050
DIBENAMINE see DCR000, DCR200
DIBENAMINE HYDROCHLORIDE see DCR200
DIBENCIL see BFC750
DIBENCILLIN see BFC750
DIBENYLIN see PDT250
DIBENYLINE see PDT250
DIBENZ(a,e)ACEANTHRYLENE see DCR300
DIBENZ(a,j)ACEANTHRYLENE see DCR400
13H-DIBENZ(bc,j)ACEANTHRYLENE see DCR600
4H-DIBENZ(f,g,j)ACEANTHRYLENE, 5,5a,6,7-TETRAHYDRO- see DCR800
DIBENZACEPIN see DCS200
DIBENZ(a,d)ACRIDINE see DCS400
DIBENZ(a,f)ACRIDINE see DCS600
DIBENZ(a,h)ACRIDINE see DCS400
DIBENZ(a,j)ACRIDINE see DCS600
DIBENZ(c,h)ACRIDINE see DCS800
1,2,5,6-DIBENZACRIDINE see DCS400
1,2,7,8-DIBENZACRIDINE see DCS600
3,4,5,6-DIBENZACRIDINE see DCS600
3,4:5,6-DIBENZACRIDINE see DCS800
1,2,7,8-DIBENZACRIDINE (FRENCH) see DCS800
DIBENZ(a,c)ACRIDINE METHOSULFATE see DCT000
3,4:5,6-DIBENZACRIDINE METHOSULFATE see DCT000
4,4′-DIBENZAMIDO-1,1′-DIANTHRIMIDE see IBJ000
1,2,5,6-DIBENZANTHRACEEN (DUTCH) see DCT400
DIBENZ(a,c)ANTHRACENE see BDH750
DIBENZ(a,h)ANTHRACENE see DCT400
DIBENZ(a,j)ANTHRACENE see DCT600
1,2:3,4-DIBENZANTHRACENE see BDH750
1,2:5,6-DIBENZANTHRACENE see DCT400
1,2:7,8-DIBENZANTHRACENE see DCT600
2,3:6,7-DIBENZANTHRACENE see PAV000
DIBENZ(de,kl)ANTHRACENE see PCQ250
1,2:5,6-DIBENZ(a)ANTHRACENE see DCT400
1:2:5,6-DIBENZANTHRACENE-9-CARBAMIDO-ACETIC ACID see DCU600
1,2,5,6-DIBENZANTHRACENECHOLEIC ACID see DCT800
DIBENZ(a,h)ANTHRACENE-5,6-OXIDE see EBP500
1,2:5,6-DIBENZANTHRACENE-9,10-endo-α,β-SUCCINIC ACID see DCU200
1,2,5,6-DIBENZANTHRACENE-9,10-endo-α,β-SUCCINIC ACID, SODIUM SALT
   see SGF000
DIBENZ(a,h)ANTHRACEN-5-OL see DCU400
4,4′-DI-7H-BENZ(de)ANTHRACEN-7-ONE see DCV000
N-(DIBENZ(a,h)ANTHRACEN-7-YLCARBAMOYL)GLYCINE see DCU600
DIBENZANTHRANYL GLYCINE COMPLEX see DCU600
DIBENZANTHRONE see DCU800
4,4′-DIBENZANTHRONIL see DCV000
4,4′-DIBENZANTHRONYL see DCV000
5H-DIBENZ(b,f)AZEPINE-5-CARBOXAMIDE see DCV200

5H-DIBENZ(b,f)AZEPINE, 3-CHLORO-5-(3-(4-CARBAMOYL-4-
   PIPERIDINOPIPERIDINO)PROPYL)-10,11-DIHYDRO-, DIHYDROCHLO-
   RIDE, MONOHYDRATE see DCV400
4-(3-(5H-DIBENZ(b,f)AZEPIN-5-YL)PROPYL)-1-PIPERAZINEETHANOL
   see DCV800
(3-(5H-DIBENZ(b,f)AZEPIN-5-YL)PROPYL)-1-PIPERAZINEETHANOL DIHY-
   DROCHLORIDE see IDF000
3,4,5,6-DIBENZCARBAZOL see DCY000
1,2,5,6-DIBENZCARBAZOLE see DCX600
1,2,7,8-DIBENZCARBAZOLE see DCX800
3,4,5,6-DIBENZCARBAZOLE see DCY000
DIBENZENECHROMIUM see BGY700
DIBENZENECHROMIUM IODIDE see BGY720
DI(BENZENEDIAZONIUM)ZINC TETRACHLORIDE see DCW000
DI(BENZENEDIAZO)SULFIDE see DCW200
DIBENZENESULFONYL PEROXIDE see DCW400
DIBENZEPIN see DCW600
DIBENZEPINE see DCW600
DIBENZEPINE HYDROCHLORIDE see DCW800
DIBENZEPIN HYDROCHLORIDE see DCW800
1,2,3,4-DIBENZFLUORENE see DCX000
1,2,7,8-DIBENZFLUORENE see DDB200
DIBENZ(c,f)INDENO(1,2,3-ij)(2,7)NAPHTHYRIDINE see DCX400
1,2,3,4-DIBENZNAPHTHALENE see TMS000
DIBENZO(a,j)ACRIDINE see DCS600
1,2,5,6-DIBENZOACRIDINE see DCS400
DIBENZO(a,c)ANTHRACENE see BDH750
DIBENZO(a,h)ANTHRACENE see DCT400
1,2:3,4-DIBENZOANTHRACENE see BDH750
1,2:5,6-DIBENZOANTHRACENE see DCT400
DIBENZOAZEPINE see DCS200
3,4,5,6-DIBENZOCARBAZOLE see DCY000
7H-DIBENZO(a,g)CARBAZOLE see DCX600
7H-DIBENZO(a,i)CARBAZOLE see DCX800
7H-DIBENZO(c,g)CARBAZOLE see DCY000
DIBENZO(b,def)CHRYSENE see DCY200
DIBENZO(def,p)CHRYSENE see DCY400
DIBENZO-(drf,mno)CHRYSENE see APE750
DIBENZO(b,def)CHRYSENE-7-CARBOXALDEHYDE see DCY600
DIBENZO(def,p)CHRYSENE-10-CARBOXALDEHYDE see DCY800
DIBENZO(def,mno)CHRYSENE-12-CARBOXALDEHYDE see FNK000
DIBENZO(b,def)CHRYSENE-7,14-DIONE see DCZ000
DIBENZO-18-CROWN-6 see COD575
DIBENZO(a,d)CYCLOHEPTADIENE-5-CARBOXAMIDE see CQH625
DIBENZO(a,d)(1,4)-CYCLOHEPTADIENE-5-CARBOXAMIDE see CQH625
4-(5-DIBENZO(a,e)CYCLOHEPTATRIENYLIDENE)PIPERIDINE HYDROCHLO-
   RIDE see PCI250
5H-DIBENZO(a,d)CYCLOHEPTENE-5-PROPANAMINE, 10-11-DIHYDRO-N,N,β-
   TRIMETHYL-, (±)- see BPT500
(2-((5H-DIBENZO(a,d)CYCLOHEPTEN-5-YLIDENEAMINO)OXY)ETHYL)
   TRIMETHYLAMMONIUM IODIDE see LHX495
(2-((5H-DIBENZO(a,d)CYCLOHEPTEN-5-YLIDENEAMINO)OXY)-1-
   METHYLETHYL)ETHYLDIMETHYL AMMONIUM see LHX600
(2-((5H-DIBENZO(a,d)CYCLOHEPTEN-5-YLIDENEAMINO)OXY)-1-
   METHYLETHYL)TRIMETHYLAMMONIUM see LHX510
3-(5H-DIBENZO(a,d)CYCLOHEPTEN-5-YLIDENE)-N,N-DIMETHYL-1-
   PROPANAMINE HYDROCHLORIDE see DPX800
4-(5H-DIBENZO(a,d)CYCLOHEPTEN-5-YLIDENE)-1-METHYLPIPERIDINE
   see PCI500
4-(5-DIBENZO(a,d)CYCLOHEPTEN-5-YLIDINE)-1-METHYLPIPERIDINE
   see PCI500
N-3-(5H-DIBENZO(a,d)CYCLOHEPTEN-5-YL)PROPYL-N-METHYLAMINE
   see DDA600
DIBENZODIOXIN see DDA800
DIBENZO-p-DIOXIN see CDV125, DDA800
DIBENZO(1,4)DIOXIN see DDA800
DIBENZO(b.e)(1,4)DIOXIN see DDA800
DIBENZO(a,e)FLUORANTHENE see DCR300
2,3,5,6-DIBENZOFLUORANTHENE see DCR300
DIBENZO(a,jk)FLUORENE see BCJ500
DIBENZO(b,jk)FLUORENE see BCJ750
1,2,5,6-DIBENZOFLUORENE see DDB000
13H-DIBENZO(a,g)FLUORENE see DDB000
13H-DIBENZO(a,i)FLUORENE see DDB200
1,1′-(2,8-DIBENZOFURADIYL)BIS(2-(DIMETHYLAMINOETHANONE) DIHY-
   DROCHLORIDE HYDRATE (2:5) see DDB400
2-DIBENZOFURANAMINE see DDB600
3-DIBENZOFURANAMINE see DDB800
1,3(2H,9bH)-DIBENZOFURANDIONE, 2,6-DIACETYL-7,9-DIHYDROXY-8,9b-
   DIMETHYL-, (9bR)- see UWJ100
DIBENZOFURAN, 1,2,3,4,7,8-HEXACHLORO- see HCH400
3(9bH)-DIBENZOFURANONE, 2,6-DIACETYL-8,9b-DIMETHYL-1,7,9-TRIHY-
   DROXY-, D- see UWJ100

3(9bH)-DIBENZOFURANONE, 2,6-DIACETYL-1,7,9-TRIHYDROXY-8,9b-
    DIMETHYL-, D- see UWJ100
DIBENZOFURAN, 2,3,4,7,8-PENTACHLORO- see PAW100
3-DIBENZOFURANYLACETAMIDE see DDC000
N-3-DIBENZOFURANYLACETAMIDE see DDC000
DIBENZOFURANYLAMINE see DDB800
1,2,5,6-DIBENZONAPHTHALENE see CML810
DIBENZOPARADIAZINE see PDB500
DIBENZOPARATHIAZINE see PDP250
DIBENZO(h,rst)PENTAPHENE see DDC200
DIBENZO(cd,lm)PERYLENE see DDC400
DIBENZO(a,i)PHENANTHRENE see PIB750
1,2,3,4-DIBENZOPHENANTHRENE see BCH000
1,2:6,7-DIBENZOPHENANTHRENE see BCG500
1,2:7,8-DIBENZOPHENANTHRENE see PIB750
2,3:7,8-DIBENZOPHENANTHRENE see BCG500
DIBENZO-2,3,7,8-PHENANTHRENE see BCG500
DIBENZO(a,c)PHENAZINE see DDC600
DIBENZO(a,h)PHENAZINE see DDC800
DIBENZO PQD see DVR200
6H-DIBENZO(b,d)PYRAN-6-ONE, 1-METHYL-3,7,9-TRIHYDROXY-and 3,9-DI-
    HYDROXY-7-METHOXY-1-METHYL-DIBENZO(b,d)PYRAN-6-ONE (1:1)
    see AGW550
DIBENZOPYRAZINE see PDB500
DIBENZO(c,f)PYRAZINO(1,2-a)AZEPINE, 1,2,3,4,10,14b-HEXAHYDRO-2-
    METHYL- see MQS220
DIBENZO(a,d)PYRENE see DCY400
DIBENZO(a,e)PYRENE see NAT500
DIBENZO(a,h)PYRENE see DCY200
DIBENZO(a,i)PYRENE see BCQ500
DIBENZO(a,l)PYRENE see DCY400
DIBENZO(b,h)PYRENE see BCQ500
1,2:3,4-DIBENZOPYRENE see DCY400
1,2:4,5-DIBENZOPYRENE see NAT500
1,2:6,7-DIBENZOPYRENE see DCY200
1,2:7,8-DIBENZOPYRENE see BCQ500
2,3:4,5-DIBENZOPYRENE see DCY400
3,4:8,9-DIBENZOPYRENE see DCY200
DIBENZO(cd,mk)PYRENE see APE750
1,2:9,10-DIBENZOPYRENE see DCY400
3,4:9,10-DIBENZOPYRENE see BCQ500
DIBENZO(a,b)PYRENE-7,14-DIONE see DCZ000
2,3:7,8-DIBENZOPYRENE-1,6-QUINONE see DCZ000
DIBENZO(b,e)PYRIDINE see ADJ500
DIBENZOPYRROLE see CBN000
DIBENZO(b,d)PYRROLE see CBN000
DIBENZO(a,g)QUINOLIZINIUM, 5,6-DIHYDRO-2,3,9,10-TETRAMETHOXY-,
    HYDROXIDE see PAE100
DIBENZOSUBERONE OXIME see DDD000
DIBENZOTHIAZEPINE see CIS750
DIBENZOTHIAZINE see PDP250
DIBENZO-1,4-THIAZINE see PDP250
DI-2-BENZOTHIAZOLYLDISULFIDE see BDE750
DIBENZOTHIAZYL DISULFIDE see BDE750
2,2'-DIBENZOTHIAZYLDISULFIDE see BDE750
N-2-DIBENZOTHIENYLACETAMIDE see DDD400
N-3-DIBENZOTHIENYLACETAMIDE see DDD600
N-3-DIBENZOTHIENYLACETAMIDE-5-OXIDE see DDD800
DIBENZO(b,e)THIEPIN₁₀MN¹¹(6H),Γ-PROPYLAMINE, N,N-DIMETHYL-, 5-
    OXIDE, MALEATE (1:1) see POD800
3-DIBENZO(b,e)THIEPIN-11(6H)-YLIDENE-N,N-DIMETHYL-1-PROPAMINE
    see DYC875
3-DIBENZO(b,e)THIEPIN-11(6H)-YLIDENE-N,N-DIMETHYL-1-PROPANAMINE,
    HYDROCHLORIDE see DPY250
1,1'-(2,8-DIBENZOTHIOPHENEDIYL)BIS(2-(DIMETHYLAMINO)ETHANONE)
    DIHYDROCHLORIDE TRIHYDRATE see DDE000
DIBENZOTHIOXIN see PDQ750
1,4-DIBENZOTHIOXINE see PDQ750
DIBENZ(b,f)(1,4)OXAZEPINE see DDE200
DIBENZ(b,f)OXEPIN-2-ACETIC ACID, 10,11-DIHYDRO-α-8-DIMETHYL-11-
    OXO- see DLI650
DIBENZ(b,e)OXEPIN-Δ¹¹(6H),Γ-PROPYLAMINE see DBA600
DIBENZOYL see BBY750
DIBENZOYLDIETHYLENEGLYCOL ESTER see DJE000
DIBENZOYL DIPROPYLENE GLYCOL ESTER see DWS800
DIBENZOYLPEROXID (GERMAN) see BDS000
DIBENZOYL PEROXIDE (MAK) see BDS000
DIBENZOYLPEROXYDE (DUTCH) see BDS000
DIBENZOYLTARTARIC ACID see DDE300
DIBENZOYLTHIAZYL DISULFIDE see BDE750
1,2,3,4-DIBENZPHENANTHRENE see BCH000
1,2,5,6-DIBENZPHENANTHRENE see BCG750
DIBENZ(a,h)PHENAZINE see DDC800
1,2,3,4-DIBENZPHENAZINE see DDC600

1,2:5,6-DIBENZPHENAZINE see DDC800
DIBENZ(a,i)PYRENE see BCQ500
1,2,3,4-DIBENZPYRENE see DCY400
1,2:7,8-DIBENZPYRENE see BCQ500
3,4,8,9-DIBENZPYRENE see DCY200
4,5,6,7-DIBENZPYRENE see DCY400
3,4:9,10-DIBENZPYRENE see BCQ500
1',2',6',7'-DIBENZPYRENE-7,14-QUINONE see DCZ000
DIBENZTHIAZYL DISULFIDE see BDE750
DIBENZYL see BFX500
N,N-DIBENZYLAMINOETHYL CHLORIDE HYDROCHLORIDE
    see DCR200
DIBENZYLBUTYLSULFONIUM IODIDE MERCURIC IODIDE see DDF000
DIBENZYLBUTYLSULFONIUM IODIDE with MERCURY IODIDE (1:1)
    see DDF000
DIBENZYLCHLORETHAMINE HYDROCHLORIDE see DCR200
DIBENZYL CHLORETHYLAMINE see DCR000
DIBENZYLCHLORETHYLAMINE HYDROCHLORIDE see DCR200
N,N-DIBENZYL-β-CHLOROETHYLAMINE see DCR000
N,N-DIBENZYL-2-CHLOROETHYLAMINE HYDROCHLORIDE see DCR200
N,N-DIBENZYL-β-CHLOROETHYLAMINE HYDROCHLORIDE see DCR200
1,3-DIBENZYLDECAHYDRO-2-OXOIMIDAZO(4,5-c)THIENO(1,2-a)THIOLIUM
    10-CAMPHORSULFONATE see TKW500
1,3-DIBENZYLDECAHYDRO-2-OXO-IMIDAZO(4,5-c)THIENO(1,2-a)THIOLIUM-
    2-OXO-10-BORANESULFONATE see TKW500
1-1,3-DIBENZYLDECAHYDRO-2-OXOIMIDAZO(4,5-c)THIENO(1,2-a)THIOLIUM-
    2-OXO-10-BORANESULFONATE see DDF200
7,14-DIBENZYLDIBENZ(a,h)ANTHRACENE see DDF400
9,10-DIBENZYL-1,2,5,6-DIBENZANTHRACENE see DDF400
DIBENZYL(5-DIBENZYLAMINO-2,4-PENTADIENYLIDENE)AMMONIUM
    CHLORIDE SESQUIHYDRATE see DDF600
DIBENZYLDISULFID (CZECH) see DXH200
DIBENZYLENE see DDG800
DIBENZYLETHER (CZECH) see BEO250
N,N'-DIBENZYLETHYLENEDIAMINE BIS(BENZYL PENICILLIN) see BFC750
N,N'-DIBENZYLETHYLENEDIAMINE DIHYDROCHLORIDE see DDG400
DIBENZYLETHYLENEDIAMINE-DI-PENICILLIN G see BFC750
N,N'-DIBENZYLETHYLENEDIAMINE, compounded with PENICILLIN G (1:2)
    see BFC750
DIBENZYLETHYLSULFONIUM IODIDE MERCURIC IODIDE see DDG600
DIBENZYLETHYLSULFONIUM IODIDE with MERCURY IODIDE (1:1)
    see DDG600
DIBENZYLIN see DDG800
DIBENZYLINE see PDT250
DIBENZYLINE HYDROCHLORIDE see DDG800
1-3,4-(1',3'-DIBENZYL-2'-KETO-IMIDAZOLIDO)-1,2-TRIMETHYLENE
    THIOPHANIUM CAMPHOR SULFONATE see DDF200
d-3,4-(1',3'-DIBENZYL-2'-KETOIMIDAZOLIDO)-1,2-
    TRIMETHYLENETHIOPHANIUM-d-CAMPHORSULFONATE see TKW500
DIBENZYLMERCURY see DDH000
DIBENZYL PHOSPHITE see DDH400
2,2-DIBENZYL-4-(2-PIPERIDYL)-1,3-DIOXOLANE HYDROCHLORIDE
    see DDH600
DIBENZYLSULFOXIDE see DDH800
DIBENZYL SULPHOXIDE see DDH800
DIBENZYLTIN BIS(DIBUTYLDITHIOCARBAMATE) see BIW250
DIBENZYLTIN S,S'-BIS(ISOOCTYLMERCAPTOACETATE) see EDC560
DIBENZYRAN see DDG800
DIBERAL see DDH900
DIBESTIL see DKB000
DIBESTROL see DKA600
DIBETOS see BQL000
DI-1,2-BIS(DIFLUOROAMINO)ETHYL ETHER see DDI000
DIBISMUTH TRISULFIDE see DDI200
DI-(BISTRIFLUOROMETHYLPHOSFIDO)MERCURY see DDI400
DIBONDRIN see BBV500
DIBORANE see DDI450
DIBORANE(6) see DDI450
DIBORON HEXAHYDRIDE see DDI450
DIBORON OXIDE see DDI500
DIBORON TETRACHLORIDE see DDI600
DIBORON TETRAFLUORIDE see DDI800
DIBOVAN see DAD200
DIBP see DNJ400
DIBROLUUR see BNP750
DIBROM see NAG400
1,2-DIBROMAETHAN (GERMAN) see EIY500
DIBROMANNIT see DDP600
DIBROMANNITOL see DDP600
d-DIBROMANNITOL see DDP600
3,9-DIBROMBENZANTHRONE see DDK000
1,4-DIBROMBUTAN (GERMAN) see DDL000
DIBROMCHLORPROPAN (GERMAN) see DDL800
1,2-DIBROM-3-CHLOR-PROPAN (GERMAN) see DDL800

O-(1,2-DIBROM-2,2-DICHLORAETHYL)-O,O-DIMETHYL-PHOSPHAT (GER-MAN) see NAG400
DIBROMDULCITOL see DDJ000
DIBROMOACETONITRILE see DDJ400
α,p-DIBROMOACETOPHENONE see DDJ600
2,4′-DIBROMOACETOPHENONE see DDJ600
DIBROMOACETYLENE see DDJ800
2,4-DIBROMO-1-ANTHRAQUINONYLAMINE see AJL500
3,9-DIBROMO-7H-BENZ(de)ANTHRACEN-7-ONE see DDK000
4,4′-DIBROMOBENZILIC ACID ISOPROPYL ESTER see IOS000
2,2′-DIBROMOBIACETYL see DDK600
α,α′-DIBROMOBIACETYL see DDK600
DIBROMOBICYCLOHEPTANE see DDK800
DIBROMOBICYCLOHEPTANE (mixed isomers) see DDK800
1,4-DIBROMO-1,3-BUTADIYNE see DDK875
1,4-DIBROMOBUTANE see DDL000
DIBROMOBUTENE see DDL600
1,4-DIBROMO-2-BUTENE see DDL400
1,4-trans-DIBROMOBUTENE-2 see DDL600
trans-1,4-DIBROMOBUT-2-ENE see DDL600
DIBROMOCHLOROMETHANE see CFK500
1,2-DIBROMO-3-CHLORO-2-METHYLPROPANE see MBV720
DIBROMOCHLOROPROPANE see DDL800
1,2-DIBROMO-3-CHLOROPROPANE see DDL800
3,4′-DIBROMO-5-CHLOROTHIOSALICYLANILIDE ACETATE (ESTER) see BOL325
1,2-DIBROMO-3-CLORO-PROPANO (ITALIAN) see DDL800
DIBROMOCYANOACETAMIDE see DDM000
α,α-DIBROMO-α-CYANOACETAMIDE see DDM000
2,6-DIBROMO-4-CYANOPHENOL see DDP000
2,6-DIBROMO-4-CYANOPHENYL OCTANOATE see DDM200
DIBROMODIBUTYLSTANNANE see DDM400
DIBROMODIBUTYLTIN see DDM400
1,2-DIBROMO-2,2-DICHLOROETHYL DIMETHYL PHOSPHATE see NAG400
O-(1,2-DIBROMO-2,2-DICLORO-ETIL)-O,O-DIMETIL-FOSTATO (ITALIAN) see NAG400
1,2-DIBROMO-2,4-DICYANOBUTANE see DDM500
1,6-DIBROMODIDEOXYDULCITOL see DDJ000
1,6-DIBROMO-1,6-DIDEOXYDULCITOL see DDJ000
1,6-DIBROMO-1,6-DIDEOXYGALACTITOL see DDJ000
1,6-DIBROMO-1,6-DIDEOXY-d-GALACTITOL see DDJ000
1,6-DIBROMO-1,6-DIDEOXY-d-MANNITOL see DDP600
1,6-DIBROMO-1,6-d-DIDESOXYMANNITOL see DDP600
2′,6′-DIBROMO-2-(DIETHYLAMINO)-p-ACETOTOLUIIDE HYDROCHLORIDE see DDM600
5,6-DIBROMO-2-(2-(2-(DIETHYLAMINO)ETHYLAMINO)ETHYL)-2-METHYL-1,3-BENZODIOXOLE DIHYDROCHLORIDE see DDM800
2′,6′-DIBROMO-2-(DIETHYLAMINO)-4′-METHYLACETANILIDE HYDROCHLO-RIDE see DDM600
DIBROMODIFLUOROMETHANE see DKG850
1,2-DIBROMO-1,2-DIISOCYANATOETHANE POLYMERS see DDN100
2,2-DIBROMO-1,3-DIMETHYLCYCLOPROPANOIC ACID see DDN150
DIBROMODIMETHYL STANNANE see DUG800
DIBROMODIPHENYLSTANNANE see DDN200
DIBROMODULCITOL see DDJ000
1,6-DIBROMODULCITOL see DDJ000
2,3-DIBROMO-5,6-EPOXY-7,8-DIOXABICYCLO(2.2.2)OCTANE see DDN700
1,2-DIBROMOETANO (ITALIAN) see EIY500
1,1-DIBROMOETHANE see DDN800
1,2-DIBROMOETHANE (MAK) see EIY500
α,β-DIBROMOETHANE see EIY500
sym-DIBROMOETHANE see EIY500
DIBROMOFLUORESCEIN see DDO200
4′,5′-DIBROMOFLUORORESCEIN see DDO200
1,2-DIBROMOHEPTAFLUOROISOBUTYL METHYL ETHER see DDO400
1,6-DIBROMOHEXAN (GERMAN) see DDO800
1,6-DIBROMOHEXANE see DDO800
3,5-DIBROMO-4-HYDROXYBENZONITRILE see DDP000
3-(3,5-DIBROMO-4-HYDROXYBENZOYL-2-ETHYLBENZOFURAN see DDP200
2,7-DIBROMO-4-HYDROXYMERCURIFLUORESCEINE DISODIUM SALT see MCV000
3,5-DIBROMO-4-HYDROXYPHENYLCYANIDE see DDP000
3,5-DIBROMO-4-HYDROXYPHENYL-2-ETHYL-3-BENZOFURANYL KETONE see DDP200
(3,5-DIBROMO-4-HYDROXYPHENYL)(2-ETHYL-3-BENZOFURANYL)METHANONE see DDP200
5,7-DIBROMO-8-HYDROXYQUINOLINE see DDS600
DIBROMOMALEINIMIDE see DDP400
1,6-DIBROMOMANNITOL see DDP600
2,7-DIBROMOMESOBENZANTHRONE see DDK000
DIBROMOMETHANE see DDP800
N,N-DIBROMOMETHYLAMINE see DDQ100
DIBROMOMETHYLBORANE see DDQ125
DIBROMONEOPENTYL GLYCOL see DDQ400, DDQ400

2,2-DIBROMO-3-NITRILOPROPIONAMIDE see DDM000
3,4-DIBROMONITROSOPIPERIDINE see DDQ800
DIBROMONORBORNANE see DDK800
3,5-DIBROMO-4-OCTANOYLOXYBENZONITRILE see DDM200
DIBROMOPENTAERYTHRITOL see DDQ400
1,5-DIBROMOPENTANE see DDR000
2-(3,5-DIBROMO-2-PENTYLOXYBENZYLOXY)TRIETHYLAMINE see DDR100
DIBROMOPHENYLARSINE see DDR200
DIBROMOPROPANAL see DDS400
1,2-DIBROMOPROPANE see DDR400
1,3-DIBROMOPROPANE see TLR000
2,3-DIBROMOPROPANE see DDR600
α,Γ-DIBROMOPROPANE see TLR000
ω,ω′-DIBROMOPROPANE see TLR000
1,3-DIBROMOPROPANE polymer with N,N,N′,N′-TETRAMETHYL-1,6-HEXANEDIAMINE see HCV500
2,3-DIBROMOPROPANOL see DDS000
1,3-DIBROMO-2-PROPANOL see DDR800
2,3-DIBROMO-1-PROPANOL see DDS000
2,3-DIBROMO-1-PROPANOL HYDROGEN PHOSPHATE see BIV825
2,3-DIBROMO-1-PROPANOL HYDROGEN PHOSPHATE, MAGNESIUM SALT see MAD100
2,3-DIBROMO-1-PROPANOL PHOSPHATE see TNC500
2,3-DIBROMO-1-PROPANOL, PHOSPHATE (3:1) see TNC500
2,3-DIBROMOPROPANOYL CHLORIDE see DDS100
2,3-DIBROMOPROPENE see DDS200
2,3-DIBROMOPROPIONALDEHYDE see DDS400
2,3-DIBROMOPROPIONYL CHLORIDE see DDS100
α,β-DIBROMOPROPIONYL CHLORIDE see DDS100
N,N³-DI(β-BROMOPROPIONYL)-N¹,N²-DISPIROTRIPIPERAZINIUM DICHLO-RIDE see SLD600
(2,3-DIBROMOPROPYL) PHOSPHATE see TNC500
5,7-DIBROMO-8-QUINOLINOL see DDS600
3,4-DIBROMOSULFOLANE see DDT000
3,4-DIBROMOTETRAHYDROTHIOPHENE-1,1-DIOXIDE see DDT000
DIBROMOXYQUINOLINE see DDS600
DIBROMURE d′ETHYLENE (FRENCH) see EIY500
1,2-DIBROOM-3-CHLOORPROPAAN (DUTCH) see DDL800
O-(1,2-DIBROOM-2,2-DICHLOOR-ETHYL)-O,O-DIMETHYL-FOSFAAT (DUTCH) see NAG400
1,2-DIBROOMETHAAN (DUTCH) see EIY500
DIBUCAIN see NOF500
DIBUCAINE see DDT200
DIBUCAINE HYDROCHLORIDE see NOF500
DIBULINESULFAT see DDW000
DIBULINE SULFATE see DDW000
DIBUSMUTH DICHROMIUM NONAOXIDE see DDT250
DIBUTADIAMIN DIHYDROCHLORIDE see BQM309
DIBUTAMID (GERMAN) see DDT300
DIBUTAMIDE see DDT300
2,3:4,5-DI(2-BUTENYL)TETRAHYDROFURFURAL see BHJ500
DIBUTIL see DIR000
DIBUTIN see ILD000
DIBUTOLINE see DDW000
DIBUTOLINE SULFATE see DDW000
1,1-DIBUTOXYETHANE see DDT400
1,2-DIBUTOXYETHANE see DDW400
DIBUTOXYETHYL ADIPATE see BHJ750
DI(2-BUTOXYETHYL) ADIPATE see BHJ750
2,2′-DIBUTOXYETHYL ETHER see DDW200
DI(BUTOXYETHYL)PHTHALATE see BHK000
(DIBUTOXYPHOSPHINYL)MERCURY CHLORIDE see CFK750
DIBUTYL ACETAL see DDT400
DIBUTYL ACID PHOSPHATE see DEG700
DIBUTYL ADIPATE see AEO750
DI-N-BUTYL ADIPATE see AEO750
DIBUTYL ADIPINATE see AEO750
n-DIBUTYLAMINE see DDT800
DI-n-BUTYLAMINE see DDT800
DI-sec-BUTYLAMINE see DDT600
DI(n-BUTYL)AMINE (DOT) see DDT800
DIBUTYLAMINE, HEXACHLOROSTANNANE (2:1) see BIV900
DIBUTYLAMINE, HEXAFLUOROARSENATE(1-) see DDV225
DIBUTYLAMINE, 4-HYDROXY-N-NITROSO- see HJQ350
DIBUTYLAMINE TETRAFLUOROBORATE see DDU000
2-(DIBUTYLAMINO)-2′,6′-ACETOXYLIDIDE HYDROCHLORIDE see DDU200
1-(((DIBUTYLAMINO)CARBONYL)OXY)-N-ETHYL-N,N-DIMETHYLETHANAMINIUM SULFATE (2:1) see DDW000
DIBUTYLAMINOETHANOL see DDU600
2-DIBUTYLAMINOETHANOL see DDU600
2-N-DIBUTYLAMINOETHANOL see DDU600
2-DI-n-BUTYLAMINOETHANOL see DDU600
N,N-DI-n-BUTYLAMINOETHANOL (DOT) see DDU600
β-N-DIBUTYLAMINOETHYL ALCOHOL see DDU600

5-((2-(DIBUTYLAMINO)-ETHYL)AMINO)-3-PHENYL-1,2,4-OXADIAZOLE HY-
DROCHLORIDE see PEU000
α-DIBUTYL-AMINO-4-METHOXYBENZENEACETAMIDE (9CI) see DDT300
α-DIBUTYL-AMINO-p-METHOXYPHENYLACETAMIDE see DDT300
2-DIBUTYLAMINO-2-(p-METHOXYPHENYL)ACETAMIDE see DDT300
α-DIBUTYLAMINO-α-(p-METHOXYPHENYL)ACETAMIDE see DDT300
3-((DIBUTYLAMINO)METHYL)-4,5,6-TRIHYDROXYPHTHALIDE HYDRO-
CHLORIDE see APE625
3-(DIBUTYLAMINO)-1-PROPANOL-p-AMINOBENZOATE see BOO750
3-(DIBUTYLAMINO)-1-PROPANOL-p-AMINOBENZOATE (ESTER) SULFATE
(2:1) see BOP000
3-DIBUTYLAMINO-1-PROPANOL-4-AMINOBENZOATE (ESTER) SULFATE
(SALT) (2:1) see BOP000
3-(DIBUTYLAMINO)PROPYLAMINE see DDV200
3-DIBUTYLAMINOPROPYL-p-AMINOBENZOATE see BOO750
DIBUTYLAMINOPROPYL-p-AMINOBENZOATE SULFATE see BOP000
3'-DIBUTYLAMINOPROPYL-4-AMINOBENZOATE SULFATE see BOP000
DI-N-BUTYLAMMONIUM HEXAFLUOROARSENATE see DDV225
DI-n-BUTYLAMMONIUM TETRAFLUOROBORATE see DDU000
DIBUTYLARSINIC ACID see DDV250
DIBUTYLATED HYDROXYTOLUENE see BFW750
DIBUTYLBARBITURIC ACID see DDV400
5,5-DIBUTYLBARBITURIC ACID see DDV400
DIBUTYL-1,2-BENZENEDICARBOXYLATE see DEH200
DIBUTYLBIS((3-CARBOXYACRYLOYL)OXY)-STANNANE DIMETHYL ESTER
(Z,Z) (8CI) see BKO250
DIBUTYLBIS((2-ETHYLHEXANOYL)OXY)-STANNANE see BJQ250
DIBUTYLBIS((2-ETHYL-1-OXOHEXYL)OXY)-STANNANE (9CI) see BJQ250
DIBUTYLBIS(LAUROYLOXY)STANNANE see DDV600
DIBUTYLBIS(LAUROYLOXY)TIN see DDV600
DIBUTYLBIS(OCTANOYLOXY)STANNANE see BLB250
DIBUTYLBIS(OLEOYLOXY)STANNANE see DEJ000
DIBUTYLBIS((1-OXO-9-OCTADECENYL)OXY)STANNANE (Z,Z) see DEJ000
DIBUTYLBIS((1-OXOOCTYL)OXY)STANNANE see BLB250
DIBUTYLBIS(TRIFLUOROACETOXY)STANNANE see BLN750
DIBUTYL BUTANEPHOSPHONATE see DDV800
O,O-DIBUTYL-1-BUTYLAMINO-CYCLOHEXYLPHOSPHONATE see ALZ000
DIBUTYL BUTYLPHOSPHONATE see DDV800
DI-n-BUTYL-CARBAMYLCHOLINE SULPHATE see DDW000
(2-DIBUTYLCARBAMYLOXYETHYL)-DIMETHYLETHYLAMMONIUM SUL-
FATE see DDW000
DIBUTYL CARBITOL see DDW200
DIBUTYLCARBITOLFORMAL see BHK750
DIBUTYL-o-(o-CARBOXYBENZOYL) GLYCOLATE see BQP750
DIBUTYL-o-CARBOXYBENZOYLOXYACETATE see BQP750
DIBUTYL CELLOSOLVE see DDW400
DIBUTYL CELLOSOLVE ADIPATE see BHJ750
DIBUTYL CELLOSOLVE PHTHALATE see BHK000
4,6-DI-tert-BUTYL-m-CRESOL see DDX000
2,6-DI-tert-BUTYL-p-CRESOL (OSHA, ACGIH) see BFW750
DI-(4-tert-BUTYL-m-CRESOL)SULFIDE see BHL000
9,10-DI-n-BUTYL-1,2,5,6-DIBENZANTHRACENE see DDX200
DI-n-BUTYL(DIBUTYRYLOXY)STANNANE see DDX600
N,N'-DIBUTYL-N,N'-DICARBOXYETHYLENE DIAMINEMORPHOLIDE
see DUO400
N,N'-DIBUTYL-N,N'-DICARBOXYMORPHOLIDE-ETHYLENEDIAMINE
see DUO400
DIBUTYLDICHLOROGERMANE see DDY000
DIBUTYLDICHLOROSTANNANE see DDY200
DIBUTYLDICHLOROTIN see DDY200
DIBUTYL (DIETHYLENE GLYCOL BISPHTHALATE) see DDY400
DIBUTYLDI(2-ETHYLHEXYLOXYCARBONYLMETHYLTHIO)STANNANE
see DDY600
DIBUTYLDIFLUOROSTANNANE see DDY800
DIBUTYL(DIFORMYLOXY)STANNANE see DDZ000
2,2-DIBUTYLDIHYDRO-6H-1,3,2-OXATHIASTANNIN-6-ONE see DEJ200
DIBUTYLDIHYDROXYSTANNANE-3,3'-THIODIPROPIONATE see DEA400
DIBUTYLDIIODOSTANNANE see DEA000
2,2-DIBUTYL-1,3-DIOXA-7,9-DITHIA-2-STANNACYCLODODECAN see DEA200
2,2-DIBUTYL-1,3-DIOXA-2-STANNA-7,9-DITHIACYCLODODECAN-4,12-DIONE
see DEA200
2,2-DIBUTYL-1,3-DIOXA-2-STANNA-7-THIACYCLODECAN-4,10-DIONE
see DEA400
2,2-DIBUTYL-1,3,7,2-DIOXATHIASTANNECANE-4,10-DIONE see DEA400
DIBUTYLDIPENTANOYLOXYSTANNANE see DEA600
DI-tert-BUTYL DIPEROXYCARBONATE see DEA800
DI-tert-BUTYL DIPEROXYOXALATE see DEB000
DI-tert-BUTYL DIPEROXYPHTHALATE see DEB200
DIBUTYLDIPROPIONYLOXYSTANNANE see DEB400
DIBUTYL DISELENIDE see BRF550
DIBUTYLDISELENIUM see BRF550
DI-n-BUTYL-DISELINIDE see BRF550
DIBUTYLDITHIOCARBAMIC ACID, NICKEL SALT see BIW750
DIBUTYLDITHIOCARBAMIC ACID SODIUM SALT see SGF500

DIBUTYLDITHIOCARBAMIC ACID-S-TRIBUTYLSTANNYL ESTER see DEB600
DIBUTYLDITHIO-CARBAMIC ACID ZINC COMPLEX see BIX000
DIBUTYLDITHIOCARBAMIC ACID ZINC SALT see BIX000
((DIBUTYLDITHIOCARBAMOYL)OXY)TRIBUTYLSTANNANE see DEB600
N,N-DI-sec-BUTYL DITHIOOXAMIDE see DEB800
DI-sec-BUTYL ESTER PHOSPHOROFLUORIDIC ACID see DEC200
DIBUTYL ESTER SULFURIC ACID see DEC000
N,N-DIBUTYLETHANOLAMINE see DDU600
DI-n-BUTYL ETHER (DOT) see BRH750
N,N'-DI-n-BUTYLETHYLENEDIAMINE-N,N-DICARBOXYBISMORPHOLIDE
see DUO400
2,6-DI-sec-BUTYLFENOL (CZECH) see DEF800
DI-sec-BUTYLFLUOROPHOSPHATE see DEC200
DI-sec-BUTYL FLUOROPHOSPHONATE see DEC200
N,N-DI-n-BUTYLFORMAMIDE see DEC400
DIBUTYL FUMARATE see DEC600
DI-n-BUTYLGERMANEDICHLORIDE see DDY000
DIBUTYLGLYCOL PHTHALATE see BHK000
DIBUTYLHEXAMETHYLENEDIAMINE see DEC699
N,N'-DIBUTYLHEXAMETHYLENEDIAMINE see DEC699
1,6-N,N'-DIBUTYLHEXANEDIAMINE see DEC699
N,N'-DIBUTYL-1,6-HEXANEDIAMINE see DEC699
DIBUTYL HEXANEDIOATE see AEO750
1,1-DIBUTYLHYDRAZINE see DEC725
N,N-DIBUTYLHYDRAZINE see DEC725
1,1-DI-n-BUTYLHYDRAZINE see DEC725
1,2-DI-n-BUTYLHYDRAZINE DIHYDROCHLORIDE see DEC775
DIBUTYL HYDROGEN PHOSPHATE see DEG700
DIBUTYL HYDROGEN PHOSPHITE see DEG800
3,5-DI-tert-BUTYL-4-HYDROXYBENZYL ALCOHOL see IFX200
(3,5-DI-tert-BUTYL-4-HYDROXYBENZYLIDENE)MALONONITRILE see DED000
N,N-DIBUTYL-N-(2-HYDROXYETHYL)AMINE see DDU600
2,6-DI-tert-BUTYL-1-HYDROXY-4-METHYLBENZENE see BFW750
2,6-DI-tert-BUTYL-4-HYDROXYMETHYLPHENOL see IFX200
N,N-DIBUTYL(2-HYDROXYPROPYL)AMINE see DED200
3,5-DI-tert-BUTYL-4-HYDROXYTOLUENE see BFW750
DIBUTYL KETONE see NMZ000
2,6-DI-terc. BUTYL-p-KRESOL (CZECH) see BFW750
DIBUTYL LEAD DIACETATE see DED400
DIBUTYL MALEATE see DED600
DIBUTYLMALOYLOXYSTANNANE see DED800
DIBUTYL(3-MERCAPTOPROPIONATO(2-))TIN see DEJ200
DIBUTYLMERCURY see DEE000
DI-sec-BUTYLMERCURY see DEE200
N,N-DIBUTYLMETHYLAMINE see DEE400
2,4-DI-tert-BUTYL-5-METHYLPHENOL see DDX000
2,6-DI-tert-BUTYL-4-METHYLPHENOL see BFW750
2,6-DI-tert-BUTYL-p-METHYLPHENOL see BFW750
2,6-DI-tert-BUTYL NAPHTALENE SULFONATE SODIQUE (FRENCH)
see DEE600
DIBUTYL-NAPHTHALENE SULFATE, SODIUM SALT see NBS700
2,6-DI-tert-BUTYLNAPHTHALENESULFONIC ACID SODIUM SALT
see DEE600
2,6-DI-tert-BUTYL-4-NITROPHENOL see DEE800
DI-n-BUTYLNITROSAMIN (GERMAN) see BRY500
DI-n-BUTYLNITROSAMINE see BRY500
N,N-DI-n-BUTYLNITROSAMINE see BRY500
DIBUTYLNITROSOAMINE see BRY500
N,N-DIBUTYLNITROSOAMINE see BRY500
1,3-DIBUTYL-3-NITROSOUREA see DEF000
N,N'-DIBUTYL-N-NITROSOUREA see DEF000
2,2-DIBUTYL-1-OXA-2-STANNA-3-THIACYCLOHEXAN-6-ONE see DEJ200
2,2-DIBUTYL-1,3,2-OXATHIASTANNOLANE see DEF150
2,2-DIBUTYL-1,3,2-OXATHIASTANNOLANE-5-OXIDE see DEF200
DIBUTYL OXIDE see BRH750
DIBUTYLOXOSTANNANE see DEF400
DIBUTYLOXIDE of TIN see DEF400
DIBUTYLOXOTIN see DEF400
DI-tert-BUTYLPEROXID (GERMAN) see BSC750
DI-tert-BUTYL PEROXIDE (MAK) see BSC750
2,2-DI(tert-BUTYLPEROXY)BUTANE see DEF600
DI-tert-BUTYL PEROXYDE (DUTCH) see BSC750
DI-sec-BUTYL PEROXYDICARBONATE see BSD000
DI-sec-BUTYL PEROXYDICARBONATE, not more than 52% in solution (DOT)
see BSD000
DI-sec-BUTYL PEROXYDICARBONATE, technically pure (DOT) see BSD000
2,6-DI-tert-BUTYLPHENOL see DEF800
2,4-DI-tert-BUTYLPHENOL see DEG000
N,N'-DI-sec-BUTYL-p-PHENYLENEDIAMINE see DEG200
3,5-DI-tert-BUTYLPHENYLMETHYLCARBAMATE see DEG400
N,N-DIBUTYL-N'-(3-PHENYL-1,2,4-OXADIAZOL-5-YL)-1,2-ETHANEDIAMINE
HYDROCHLORIDE see PEU000
DIBUTYLPHENYL-PHENOL SODIUM DISULFONATE see AQV000
DIBUTYL PHENYL PHOSPHATE see DEG600

DIBUTYL PHOSPHATE see DEG700, DEG700
DI-n-BUTYL PHOSPHATE see DEG700
DIBUTYL-PHOSPHINIC ACID, 4-NITROPHENYL ESTER see NIM000
DIBUTYL PHOSPHITE see DEG800
DIBUTYL PHTHALATE see DEH200
DI-n-BUTYL PHTHALATE see DEH200
5,5-DIBUTYL-2,4,6(1H,3H,5H)-PYRIMIDINETRIONE see DDV400
DIBUTYL SEBACATE see DEH600
DI-n-BUTYL SEBACATE see DEH600
DIBUTYLSTANNANE OXIDE see DEF400
2,2'-((DIBUTYLSTANNYLENE)BIS(THIO))BISACETIC ACID DINONYL ESTER
  see DEH650
DIBUTYL SUCCINATE see SNA500
DI-N-BUTYLSUCCINATE see SNA500
DI-n-BUTYLSULFAT (GERMAN) see DEC000
DIBUTYL SULFATE see DEC000
n-DIBUTYL SULFIDE see BSM125
DI-n-BUTYLSULFIDE see BSM125
DIBUTYL SULPHIDE see BSM125
3,3-DIBUTYL-6,7,8,9-TETRACHLORO-2,4,3-BENZODIOXASTANNEPIN-1,5-
  DIONE see DEH800
DIBUTYL(TETRACHLOROPHTHALATO)STANNANE see DEH800
DIBUTYL(THIOACETOXY)STANNANE see DEF200
DIBUTYL THIOETHER see BSM125
1,3-DIBUTYLTHIOUREA see DEI000
1,3-DIBUTYL-2-THIOUREA see DEI000
N,N'-DIBUTYLTHIOUREA see DEI000
1,3-DI-n-BUTYL-2-THIOUREA see DEI000
DIBUTYLTHIOXOSTANNANE see DEI200
DIBUTYLTIN BIS(2-ETHYLHEXANOATE) see BJQ250
DIBUTYLTIN BIS(α-ETHYLHEXANOATE) see BJQ250
DIBUTYL-TIN BIS(ISOOCTYLTHIOGLYCOLLATE) see BKK250
DI-n-BUTYLTIN BISMETHANESULFONATE see DEI400
DIBUTYLTIN BIS(METHYL MALEATE) see BKO250
DIBUTYLTIN BIS(MONOMETHYL MALEATE) see BKO250
DIBUTYLTIN BIS(TRIFLUOROACETATE) see BLN750
DIBUTYLTIN CHLORIDE see DDY200
DIBUTYL TIN DIACETATE see DBF800
DIBUTYL TIN DIBROMIDE see DDM400
DI-n-BUTYLTIN DIBUTYRATE see DDX600
DIBUTYLTIN DICAPRYLATE see BLB250
DIBUTYLTIN DICHLORIDE see DDY200
DI-n-BUTYLTIN DICHLORIDE see DDY200
DI-n-BUTYLTIN DI(DODECANOATE) see DDV600
DIBUTYLTIN DI(2-ETHYLHEXANOATE) see BJQ250
DI-n-BUTYLTIN DI-2-ETHYLHEXANOATE see BJQ250
DIBUTYLTIN DI(2-ETHYLHEXOATE) see BJQ250
DI-n-BUTYLTIN DI-2-ETHYLHEXYLTHIOGLYCOLATE see DDY600
DIBUTYLTIN DIFLUORIDE see DDY800
DI-n-BUTYLTIN DIFORMATE see DDZ000
DI-n-BUTYL TIN DI(HEXADECYLMALEATE) see DEI600
DIBUTYLTIN DIIODIDE see DEA000
DIBUTYLTIN DILAURATE (USDA) see DDV600
DI-N-BUTYLTIN DI(MONOBUTYL)MALEATE see BHK250
DI-n-BUTYLTIN DI(MONONONYL)MALEATE see DEI800
DIBUTYLTIN DIOCTANOATE see BLB250
DIBUTYLTIN DIOCTATE see BLB250
DIBUTYLTIN DIOLEATE see DEJ000
DI-n-BUTYLTIN DIPENTANOATE see DEA600
DI-n-BUTYLTIN DIPROPIONATE see DEB400
DI-n-BUTYL-TIN DI(TETRADECANOATE) see BLH309
DIBUTYLTIN LAURATE see DDV600
DIBUTYLTIN MALATE see DED800
DIBUTYLTIN MERCAPTOPROPIONATE see DEJ200
DIBUTYLTIN-O,S-MERCAPTOPROPIONATE see DEJ200
DIBUTYLTIN-S,O-3-MERCAPTOPROPIONATE see DEJ200
DIBUTYLTIN-S,O-β-MERCAPTOPROPIONATE see DEJ200
DIBUTYLTIN METHYL MALEATE see BKO250
DIBUTYLTIN OCTANOATE see BLB250
DIBUTYLTIN OXIDE see DEF400
DI-n-BUTYLTIN OXIDE see DEF400
DI-n-BUTYLTIN SULFIDE see DEI200
DIBUTYLTIN TETRACHLOROPHTHALATE see DEH800
DIBUTYLTIN 3,3'-THIODIPROPIONATE see DEA400
6,6-DIBUTYL-4,8,11-TRIOXO-5,7,12-TRIOXA-6-STANNATRIDECA-2,9-DIENOIC
  ACID METHYL ESTER see BKO250
DIBUTYLZINN-S,S'-BIS(ISOOCTYLTHIOGLYCOLAT) (GERMAN) see BKK250
DI-n-BUTYL-ZINN DI-2-AETHYLHEXYL THIOGLYKOLAT (GERMAN) see
  DDY600
DI-n-BUTYL-ZINN-DICHLORID (GERMAN) see DDY200
DIBUTYL-ZINN-DILAURAT (GERMAN) see DDV600
DI-N-BUTYL-ZINN-DI(MONOBUTYL)MALEINAT (GERMAN) see BHK250
DI-n-BUTYLZINN-DIMONOMETHYLMALEINAT (GERMAN) see BKO250
DI-n-BUTYL-ZINN-DI(MONONONYL)MALEINAT (GERMAN) see DEI800

DI-n-BUTYL-ZINN-OXYD (GERMAN) see DEF400
DI-n-BUTYLZINN THIOGLYKOLAT (GERMAN) see DEF200
N^6,2'-o-DIBUTYRYL cAMP see COV625
N^6,$O^{2'}$-DIBUTYRYL cAMP see COV625
DIBUTYRYL CYCLIC AMP see COV625, DEJ300
DIBUTYRYL-3',5'-CYCLIC AMP see COV625
DIBUTYRYL CYCLIC-3',5'-AMP see COV625
N^6,2'-o-DIBUTYRYL CYCLIC AMP see COV625
N^6,$O^{2'}$-DIBUYTYRL CYCLIC AMP see COV625
DIBUYTYRYL cAMP see COV625
DIC see DAB600
DIC 1468 see MQR275
DICA see DEL200
DICACODYL SULFIDE see CAC250
DICAESIUM SELENIDE see DEJ400
DICAIN see BQA010
DICAINE see BQA010
DICALCIUM PHOSPHATE see CAW100
DICALITE see SCH000
DICAMBA (DOT) see MEL500
DICAMOYLMETHTANE see MBW750
DICANDIOL see MQU750
DICAPRYL-1,2-BENZENEDICARBOXYLATE see BLB750
DICAPRYL PHTHALATE see BLB750
DICAPTOL see BAD750
DICARBADODECABORANYLMETHYLETHYL SULFIDE see DEJ600
DICARBADODECABORANYLMETHYLPROPYL SULFIDE see DEJ800
2,2-DI(CARBAMOYLOXYMETHYL)PENTANE see MQU750
2,2-DICARBAMYLOXYMETHYL-3-METHYLPENTANE see MBW750
DICARBAZAMIDE see IBC000
S-(1,2-DICARBETHOXYETHYL)-O,O-DIMETHYLDITHIOPHOSPHATE
  see MAK700
DICARBETHOXYMETHANE see EMA500
DICARBOETHOXYETHYL-O,O-DIMETHYL PHOSPHORODITHIOATE
  see MAK700
DICARBOMETHOXYZINC see ZBS000
DICARBONIC ACID DIETHYL ESTER see DIZ100
2,3-DICARBONITRILO-1,4-DIATHIAANTHRACHINON (GERMAN)
  see DLK200
DI-mu-CARBONYLHEXACARBONYLDICOBALT see CNB500
DICARBONYL MOLYBDENUM DIAZIDE see DEJ849
DICARBONYLPYRAZINE RHODIUM(1) PERCHLORATE see DEJ859
DICARBONYLTUNGSTEN DIAZIDE see DEJ880
DICARBOXIDINE HYDROCHLORIDE see DEK000
o-DICARBOXYBENZENE see PHW250
3,5-DICARBOXYBENZENESULFONIC ACID, SODIUM SALT see DEK200
3,3'-DICARBOXYBENZIDINE see BFX250
DICARBOXYDINE see DEK400
((1,2-DICARBOXYETHYL)THIO)GOLD DISODIUM SALT see GJC000
4,5-DICARBOXYIMIDAZOLE see IAM000
DICARBOXYMETHANE see CCC750
DICAROCIDE see DIW200
DICARZOL see DSO200
DICATRON see PDN000
DICERIUM TRISULFIDE see DEK600
DICESIUM CARBONATE see CDC750
DICESIUM DICHLORIDE see CDD000
DICESIUM DIFLUORIDE see CDD500
DICESIUM SELENIDE see DEJ400
DICESIUM SULFATE see CDE500
DICESTAL see MJM500
DICETYLDIMETHYLAMMONIUM CHLORIDE see DRK200
DICHA (CUBA) see DHB309
DICHAN (CZECH) see DGU200
DICHAPETULUM CYMOSUM (HOOK) ENGL see PLG000
DICHINALEX see CLD000
DICHLOBENIL (DOT) see DER800
DICHLOFENAMIDE see DEQ200
DICHLOFENTHION see DFK600
DICHLOFENTION see DFK600
DICHLOFLUANID see DFL200
DICHLOFLUANIDE see DFL200
DICHLONE (DOT) see DFT000
p-DICHLOORBENZEEN (DUTCH) see DEP800
1,4-DICHLOORBENZEEN (DUTCH) see DEP800
1,1-DICHLOOR-2,2-BIS(4-CHLOOR FENYL)-ETHAAN (DUTCH) see BIM500
1,1-DICHLOORETHAAN (DUTCH) see DFF809
1,2-DICHLOORETHAAN (DUTCH) see EIY600
2,2'-DICHLOORETHYLETHER (DUTCH) see DFJ050
DICHLOORFEEN (DUTCH) see MJM500
(2,4-DICHLOOR-FENOXY)-AZIJNZUUR (DUTCH) see DAA800
2-(2,4-DICHLOOR-FENOXY)-PROPIONZUUR (DUTCH) see DGB000
(3,4-DICHLOOR-FENYL-AZO)-THIOUREUM (DUTCH) see DEQ000
3-(3,4-DICHLOOR-FENYL)-1,1-DIMETHYLUREUM (DUTCH) see DXQ500

3-(3,4-DICHLOOR-FENYL)-1-METHOXY-1-METHYLUREUM (DUTCH) see DGD600
3,6-DICHLOOR-2-METHOXY-BENZOEIZUUR (DUTCH) see MEL500
1,1-DICHLOOR-1-NITROETHAAN (DUTCH) see DFU000
(2,2-DICHLOOR-VINYL)-DIMETHYL-FOSFAAT (DUTCH) see DGP900
DICHLOORVO (DUTCH) see DGP900
2,6-DICHLOQUINONE see DES400
DICHLOR see DGG800
DICHLORACETIC ACID see DEL000
DICHLORACETYL CHLORIDE see DEN400
1,1-DICHLORAETHAN (GERMAN) see DFF809
1,2-DICHLOR-AETHAN (GERMAN) see EIY600
1,2-DICHLOR-AETHEN (GERMAN) see DFI100
p-DI-(2-CHLORAETHYL)-AMINO-dl-PHENYL-ALANIN (GERMAN) see BHT750
o-(p-DI(2-CHLORAETHYL)-AMINOPHENYL)-dl-TYROSIN-DIHYDROCHLORID (GERMAN) see DFH100
DICHLORALANTIPYRIN see SKS700
DICHLORALANTIPYRINE see SKS700
S-(2,3-DICHLOR-ALLYL)-N,N-DIISOPROPYL-MONOTHIOCARBAMAAT (DUTCH) see DBI200
2,3-DICHLORALLYL-N,N-(DIISOPROPYL)-THIOCARBAMAT (GERMAN) see DBI200
DICHLORALPHENAZONE see SKS700
DICHLORAL UREA see DGQ200
DICHLOR AMINE see BIE250
DICHLORAN see RDP300
DICHLORAN (amine fungicide) see RDP300
3,4-DICHLORANILIN see DEO300
3,4-DICHLORANILINE see DEO300
1-(3,4-DICHLORANILINO)-1-FORMYLAMINO-2,2,2-TRICHLORAETHAN (GERMAN) see CDP750
1,5-DICHLORANTHRACHINON see DEO700
1,8-DICHLORANTHRACHINON see DEO750
DICHLORANTIN see DFE200
o-DICHLORBENZENE see DEP600
3,3'-DICHLORBENZIDIN (CZECH) see DEQ600
4,4'-DICHLORBENZILSAEUREAETHYLESTER (GERMAN) see DER000
o-DICHLOR BENZOL see DEP600
p-DICHLORBENZOL (GERMAN) see DEP800
1,4-DICHLOR-BENZOL (GERMAN) see DEP800
2,6-DICHLORBENZONITRIL (GERMAN) see DER800
1-(2,4-DICHLORBENZYL)INDAZOLE-3-CARBOXYLIC ACID see DEL200
2,2'-DICHLORBIPHENYL (GERMAN) see DET800
1,1-DICHLOR-2,2-BIS(4-CHLOR-PHENYL)-AETHAN (GERMAN) see BIM500
2,3-DICHLOR-1,3-BUTADIEN (CZECH) see DEU400
2,2'-DICHLOR-DIAETHYLAETHER (GERMAN) see DFJ050
3,3'-DICHLOR-4,4'-DIAMINO-DIPHENYLAETHER (GERMAN) see BGT000
3,3'-DICHLOR-4,4'-DIAMINODIPHENYLMETHAN (GERMAN) see MJM200
DICHLORDIMETHYLAETHER (GERMAN) see BIK000
DICHLOREMULSION see EIY600
DICHLOREN see BIE500
DICHLOREN (GERMAN) see BIE250
DICHLOREN HYDROCHLORIDE see BIE500
DICHLORETHANOIC ACID see DEL000
2,2'-DICHLORETHYL ETHER see DFJ050
2-(α,β-DICHLORETHYL)PYRIDINE HYDROCHLORIDE see CMV475
β,β-DICHLOR-ETHYL-SULPHIDE see BIH250
DICHLORFENIDIM see DXQ500
2,6-DICHLORFENOL (CZECH) see DFY000
N-DICHLORFLUORMETHYLTHIO-N',N'-
    DIMETHYLAMINOSULFONSAEUREANILID (GERMAN) see DFL200
N-(DICHLOR-FLUOR-METHYL-THIO)-N',N'-DIMETHYL-N-PHENYL-
    SCHWEFEL-SAEUREDIAMID (GERMAN) see DFL200
DICHLORFOS (POLISH) see DGP900
DICHLORHYDRATE de DIMETHOXY-3,4 BENZYL PIPERAZINE (FRENCH) see VIK200
DI-CHLORICIDE see DEP800
DICHLORICIDE MOTHPROOFER see TBC500
DICHLORID KYSELINY FUMAROVE (CZECH) see FOY000
DICHLORINE OXIDE see DEL600
DICHLORINE TRIOXIDE see DEL800
DICHLORISOPRENALINE (GERMAN) see DFN400
DICHLORISOPROTERENOL see DFN400
3,4-DICHLOR-ISOPROTERENOL (GERMAN) see DFN400
DICHLORMETHAZANONE see DEM000
3,6-DICHLOR-3-METHOXY-BENZOESAEURE (GERMAN) see MEL500
DICHLORMEZANONE see DEM000
DI-CHLOR-MULSION see EIY600
2,3-DICHLOR-1,4-NAPHTHOCHINON (GERMAN) see DFT000
1,1-DICHLOR-1-NITROAETHAN (GERMAN) see DFU000
2,6-DICHLOR-4-NITROANILIN (CZECH) see RDP300
2,5-DICHLORNITROBENZEN (CZECH) see DFT400
3,4-DICHLORNITROBENZEN (CZECH) see DFT600
4,4'-DICHLOR-2-NITRODIFENYLETHER see CJD600

2,4-DICHLOR-6-NITROFENOL (CZECH) see DFU600
2,4-DICHLOR-6-NITROFENYLESTER KYSELINY OCTIVE (CZECH) see DFU800
2',5-DICHLOR-4'-NITRO-SALIZYLSAEUREANILID (GERMAN) see DFV400
DICHLOROACETALDEHYDE see DEM200
2,2-DICHLOROACETALDEHYDE see DEM200
α,α-DICHLOROACETALDEHYDE see DEM200
d-(−)-threo-2-DICHLOROACETAMIDO-1-p-NITROPHENYL-1,3-PROPANEDIOL see CDP250
DICHLOROACETATE SODIUM SALT see SGG000
2,2-DICHLOROACETIC ACID see DEL000
DICHLOROACETIC ACID, DIISOPROPYLAMINE SALT see DNM400
DICHLOROACETIC ACID METHYL ESTER see DEM800
DICHLOROACETIC ACID SODIUM SALT see SGG000
DICHLOROACETIC ANHYDRIDE see DEM825
1,3-DICHLOROACETONE see BIK250
α,Γ-DICHLOROACETONE see BIK250
sym-DICHLOROACETONE see BIK250
α,α'-DICHLOROACETONE see BIK250
1,3-DICHLOROACETONE (DOT) see BIK250
DICHLOROACETONITRILE see DEN000
2,2-DICHLOROACETOPHENONE see DEN200
α,α-DICHLOROACETOPHENONE see DEN200
ω,ω-DICHLOROACETOPHENONE see DEN200
8-DICHLOROACETOXY-9-HYDROXY-8,9-DIHYDRO-AFLATOXIN B1 see DEN300
DICHLOROACETYL CHLORIDE see DEN400
2,2-DICHLOROACETYL CHLORIDE see DEN400
α,α-DICHLOROACETYL CHLORIDE see DEN400
DICHLOROACETYL CHLORIDE (DOT) see DEN400
DICHLOROACETYLENE see DEN600
DICHLOROACETYLENE mixed with ETHER (1589) see DEN800
d-threo-N-DICHLOROACETYL-1-p-NITROPHENYL-2-AMINO-1,3-PROPANEDIOL see CDP250
S-(2,3-DICHLORO-ALLIL)-N,N-DIISOPROPIL-MONOTIOCARBAMMATO(ITALIAN) see DBI200
DICHLOROALLYL DIISOPROPYLTHIOCARBAMATE see DBI200
S-2,3-DICHLOROALLYL DIISOPROPYLTHIOCARBAMATE see DBI200
2,3-DICHLOROALLYL-N,N-DIISOPROPYLTHIOLCARBAMATE see DBI200
DICHLOROALLYL LAWSONE see DFN500
DICHLOROAMETHOPTERIN see DFO000
3',5'-DICHLOROAMETHOPTERIN see DFO000
2,5-DICHLORO-3-AMINOBENZOIC ACID see AJM000
3',5'-DICHLORO-4-AMINO-4-DEOXY-N₁₀-METHYLPTEROGLUTAMIC ACID see DFO000
1-DICHLOROAMINOTETRAZOLE see DEO200
3,4-DICHLOROANILIDE-α-METHYLACRYLIC ACID see DFO800
2,5-DICHLOROANILIN (CZECH) see DEO400
2,5-DICHLOROANILINE see DEO400
3,4-DICHLOROANILINE see DEO300
4,5-DICHLOROANILINE see DEO300
N,N-DICHLOROANILINE see DEO500
2-(2,6-DICHLOROANILINO)-2-IMIDAZOLINE HYDROCHLORIDE see CMX760
(o-(2,6-DICHLOROANILINO)PHENYL)ACETIC ACID MONOSODIUM SALT see DEO600
(o-((2,6-DICHLOROANILINO)PHENYL)ACETIC ACID SODIUM SALT see DEO600
3,6-DICHLORO-o-ANISIC ACID see MEL500
1,5-DICHLOROANTHRAQUINONE see DEO700
1,8-DICHLOROANTHRAQUINONE see DEO750
1,5-DICHLORO-9,10-ANTHRAQUINONE see DEO700
1,8-DICHLORO-9,10-ANTHRAQUINONE see DEO750
DICHLOROANTHRARUFIN see DFC600
p-DICHLOROARSINOANILINE HYDROCHLORIDE see AOR640
2-DICHLOROARSINOPHENOXATHIIN see DEP400
DICHLOROBENZALKONIUM CHLORIDE see AFP750
3,4-DICHLOROBENZENAMINE (9CI) see DEO300
m-DICHLOROBENZENE see DEP699
o-DICHLOROBENZENE see DEP600
p-DICHLOROBENZENE see DEP800
1,3-DICHLOROBENZENE see DEP699
1,2-DICHLOROBENZENE (MAK) see DEP600
1,4-DICHLOROBENZENE (MAK) see DEP800
2,5-DICHLOROBENZENEAMINE see DEO400
2,6-DICHLOROBENZENECARBOTHIOAMIDE see DGM600
1-(3',4'-DICHLOROBENZENEDIAZOL)-2-THIOUREA see DEQ000
3,4-DICHLOROBENZENE DIAZOTHIOCARBAMID see DEQ000
3,4-DICHLOROBENZENE DIAZOTHIOUREA see DEQ000
4,5-DICHLORO-1,2-BENZENEDIOL see DGJ200
4,5-DICHLORO-m-BENZENEDISULFONAMIDE see DEQ200
4,5-DICHLORO-1,3-BENZENEDISULFONAMIDE see DEQ200
3,4-DICHLOROBENZENEMETHANOL METHYLCARBAMATE see DET400
2,3(or 3,4)-DICHLOROBENZENEMETHANOL METHYL CARBAMATE see DET600
DICHLOROBENZENE, ORTHO, liquid (DOT) see DEP600

DICHLOROBENZENE, PARA, solid (DOT) see DEP800
3,3'-DICHLOROBENZIDENE see DEQ600
3,3'-DICHLOROBENZIDINA (SPANISH) see DEQ600
DICHLOROBENZIDINE see DEQ600
2,2'-DICHLOROBENZIDINE see DEQ400
o,o'-DICHLOROBENZIDINE see DEQ600
3',3'-DICHLOROBENZIDINE see DEQ600
DICHLOROBENZIDINE BASE see DEQ600
3,3'-DICHLOROBENZIDINE DIHYDROCHLORIDE see DEQ800
4,4'-DICHLOROBENZILATE see DER000
4,4'-DICHLOROBENZILIC ACID ETHYL ESTER see DER000
2,5-DICHLOROBENZOIC ACID see DER400
3,4-DICHLOROBENZOIC ACID see DER600
p-DICHLOROBENZOL see DEP800
2,6-DICHLOROBENZONITRILE see DER800
4,4'-DICHLOROBENZOPHENONE see DES000
p,p'-DICHLOROBENZOPHENONE see DES000
2,6-DICHLORO-p-BENZOQUINONE see DES400
4-(2,4-DICHLOROBENZOYL)-1,3-DIMETHYL-5-PYRAZOLYL-p-TOLUENE SUL-
   FONATE see POM275
4-(2,4-DICHLOROBENZOYL)-1,3-DIMETHYL-p-TOLUENE SULFONATE-1H-
   PYRAZOL-4-OL see POM275
DI-(4-CHLOROBENZOYL) PEROXIDE see BHM750
p,p'-DICHLOROBENZOYL PEROXIDE see BHM750
2,4-DICHLOROBENZOYL PEROXIDE (DOT) see BIX750
DICHLOROBENZYL ALCOHOL see DET000
7-((3,4-DICHLOROBENZYL)AMINO)ACTINOMYCIN D see DET125
1-(2,6-DICHLOROBENZYLFORMIMIDOYL)UREA HYDROCHLORIDE
   see PFV000
((2,6-DICHLOROBENZYLIDENE)AMINO)GUANIDINE ACETATE see GKO750
1-(2,4-DICHLOROBENZYL)-1H-INDAZOLE-3-CARBOXYLIC ACID see DEL200
3,4-DICHLOROBENZYL METHYLCARBAMATE see DET400
3,4-DICHLOROBENZYL METHYLCARBAMATE with 2,3-DICHLOROBENZYL
   METHYLCARBAMATE (80:20) see DET600
1-(2-((2,6-DICHLOROBENZYL)OXY)-2-(2,4-DICHLOROPHENYL)ETHYL)
   IMIDAZOLE NITRATE see IKN200
2,4-DICHLOROBENZYLTRIBUTYLPHOSPHONIUM CHLORIDE see THY500
2,2'-DICHLOROBIPHENYL see DET800
2,2'-DICHLORO-1,1'-BIPHENYL see DET800
3,3'-DICHLORO-4,4'-BIPHENYLDIAMINE see DEQ600
3,3'-DICHLOROBIPHENYL-4,4'-DIAMINE see DEQ600
2,2'-DICHLORO-(1,1'-BIPHENYL)-4,4'-DIAMINE see DEQ400
3,3'-DICHLORO-(1,1'-BIPHENYL)-4,4'-DIAMINE DIHYDROCHLORIDE
   see DEQ800
2,2'-((3,3'-DICHLORO(1,1'-BIPHENYL)-4,4'-DIYL)-BIS(AZO))BIS(3-OXO-N-
   PHENYL)BUTANAMIDE see DEU000
DICHLOROBIS(2-CHLOROCYCLOHEXYL)SELENIUM see DEU100
1,1-DICHLORO-2,2-BIS(p-CHLOROPHENYL)ETHANE see BIM500
1,1-DICHLORO-2,2-BIS(p-CHLOROPHENYL)ETHANE (DOT) see BIM500
1,1-DICHLORO-2,2-BIS(4-CHLOROPHENYL)-ETHANE (FRENCH) see BIM500
1,1-DICHLORO-2,2-BIS(p-CHLOROPHENYL)ETHYLENE see BIM750
cis-DICHLOROBIS(CYCLOBUTYLAMMINE)PLATINUM(II) see DGT200
DICHLOROBIS(eta-CYCLOPENTADIENYL)HAFNIUM see HAE500
cis-DICHLOROBIS(CYCLOPENTYLAMMINE)PLATINUM(II) see BIS250
1,1-DICHLORO-2,2-BIS(2,4'-DICHLOROPHENYL)ETHANE see CDN000
cis-DICHLOROBIS(DIMETHYLSELENIDE)PLATINUM(II) see DEU115
DICHLOROBIS(ETA⁵-2,4-CYCLOPENTADIEN-1-YL-TITANIUM (9CI)
   see DGW200
DICHLOROBIS(2-ETHOXYCYCLOHEXYL)SELENIUM see DEU125
1,1-DICHLORO-2,2-BIS(4-ETHYLPHENYL)ETHANE see DJC000
1,1-DICHLORO-2,2-BIS(p-ETHYLPHENYL)ETHANE see DJC000
2,2-DICHLORO-1,1-BIS(p-ETHYLPHENYL)ETHANE see DJC000
α,α-DICHLORO-2,2-BIS(p-ETHYLPHENYL)ETHANE see DJC000
1,1-DICHLORO-2,2-BIS(PARACHLOROPHENYL)ETHANE (DOT) see BIM500
cis-DICHLOROBIS(PYRROLIDINE)PLATINUM(II) see DEU200
N,N'-DICHLOROBIS(2,4,6-TRICHLOROPHENYL) UREA see DEU259
DICHLOROBORANE see DEU300
DICHLOROBROMOMETHANE see BND500
O-(2,5-DICHLORO-4-BROMOPHENYL)-O-METHYL PHENYLTHIOPHOSPHON-
   ATE see LEN000
DICHLOROBUTADIENE see DEU375
DICHLORO-1,3-BUTADIENE see DEU375
2,3-DICHLORO-1,3-BUTADIENE see DEU400
1,4-DICHLORO-1,3-BUTADIYNE see DEU509
mixo-DICHLOROBUTANE see DEU600
1,4-DICHLORO-2-BUTENE see BRG000, DEV000
3,4-DICHLORO-1-BUTENE see DEV100
1,4-DICHLOROBUTENE-2 (MAK) see DEV000
1,4-DICHLOROBUTENE-2 (trans) see BRG000
2,2'-DICHLORO-N-BUTYLDIETHYLAMINE see DEV300
1,4-DICHLOROBUTYNE see DEV400
1,4-DICHLORO-2-BUTYNE see DEV400
3,4-DICHLOROCARBANILIC ACID METHYL ESTER see DEV600
4,5-DICHLOROCATECHOL see DGJ200

2,3-DICHLOROCHINOXALIN-6-KARBONYLCHLORID (CZECH) see DGK000
2,4-DICHLORO-6-o-CHLORANILINO-s-TRIAZINE see DEV800
DICHLOROCHLORDENE see CDR750
2,4-DICHLORO-6-(o-CHLOROANILINO)-s-TRIAZINE see DEV800
2,4-DICHLORO-6-(2-CHLOROANILINO)-1,3,5-TRIAZINE see DEV800
1-(2,4-DICHLORO-β-((p-CHLOROBENZYL)OXY)-PHENETHYL)IMIDAZOLE NI-
   TRATE see EAE000
(±)-1-(2,4-DICHLORO-β-((4-CHLOROBENZYL)THIO)PHENETHYL)IMIDAZOLE
   NITRATE see SNH480
(±)-1-(2,4-DICHLORO-β-((2-CHLORO-3-ETHENYL)OXY)PHENETHYL)IMIDAZ-
   OLE see TGF050
1,1-DICHLORO-2-CHLOROETHYLENE see TIO750
2,4-DICHLORO-α-(CHLOROMETHYLENE)BENZYL ALCOHOL DIETHYL
   PHOSPHATE see CDS750
1,1-DICHLORO-2-(o-CHLOROPHENYL)-2-(p-CHLOROPHENYL)ETHANE
   see CDN000
4',5-DICHLORO-N-(4-CHLOROPHENYL)-2-HYDROXY-(1,1'-BIPHENYL)-3-CAR-
   BOXAMIDE see TIP750
4,6-DICHLORO-N-(2-CHLOROPHENYL)-1,3,5-TRIAZIN-2-AMINE see DEV800
2,4-DICHLORO-5-CHLOROSULPHONYLBENZOIC ACID see CLG200
DICHLORO(2-CHLOROVINYL)ARSINE see CLV000
DICHLORO(2-CHLOROVINYL)ARSINE OXIDE see DEW000
5-(3,4-DICHLOROCINNAMOYL)-4,7-DIMETHOXY-6-(2-
   DIMETHYLAMINOETHOXY)BENZOFURAN MALEATE see DEW200
DICHLOROCTAN SODNY (CZECH) see SGG000
2,6-DICHLORO-2,5-CYCLOHEXADIENE-1,4-DIONE see DES400
(SP-4-2)-trans(+)-DICHLORO(1,2-CYCLOHEXANEDIAMINE-N,N)- (9CI)
   see DAD050
DICHLORO(1,2-CYCLOHEXANEDIAMINE)PLATINUM see DAD040
3,4'-DICHLOROCYCLOPROPANECARBOXANILIDE see CQJ250
2,6-DICHLORO-N-CYCLOPROPYL-N-ETHYL ISONICOTINAMIDE
   see DEW400
2-(p-(2,2-DICHLOROCYCLOPROPYL)PHENOXY)-2-METHYL PROPIONIC
   ACID see CMS210
3,3'-DICHLORO-4,4'-DIAMINOBIPHENYL see DEQ600
3,3'-DICHLORO-4,4'-DIAMINO(1,1-BIPHENYL) see DEQ600
DICHLORO(1,2-DIAMINOCYCLOHEXANE)PLATINUM see DAD040
DICHLORO(1,2-DIAMINOCYCLOHEXANE)PLATINUM(II) see DAD040
cis-DICHLORO-1,2-DIAMINOCYCLOHEXANE PLATINUM(II) see DAD040
trans(−)-DICHLORO-1,2-DIAMINOCYCLOHEXANEPLATINUM(II) see DAD075
3,3'-DICHLORO-4,4'-DIAMINODIPHENYL ETHER see BGT000
3,3'-DICHLORO-4,4'-DIAMINODIPHENYLMETHANE see MJM200
N-(3,5-DICHLORO-4-((2,4-DIAMINO-6-PTERIDINYL
   METHYL)METHYLAMINO)BENZOYL)GLUTAMIC ACID see DFO000
cis-DICHLORODIAMMINE PLATINUM(II) see PJD000
trans-DICHLORODIAMMINEPLATINUM(II) see DEX000
2,7-DICHLORODIBENZODIOXIN see DAC800
2,7-DICHLORODIBENZO-p-DIOXIN see DAC800
2,7-DICHLORODIBENZO(b,e)(1,4)DIOXIN see DAC800
DICHLORODIBUTYLGERMANE see DDY000
DICHLORODIBUTYLSTANNANE see DDY200
DICHLORODIBUTYLTIN see DDY200
1-(2,4-DICHLORO-β-((2,4-DICHLOROBENZYL)OXY)PHENETHYL)-IMIDAZOLE
   NITRATE see MQS560
1-(2,4-DICHLORO-β-(2,6-DICHLOROBENZYLOXY)PHENETHYL)IMIDAZOLE
   NITRATE see IKN200
1,1-DICHLORO-2,2-DICHLOROETHANE see TBQ100
1,1-DICHLORO-2,2-DI(4-CHLOROPHENYL)ETHANE see BIM500
2,3-DICHLORO-5,6-DICYANOBENZOQUINONE see DEX400
DICHLORODICYCLOPENTADIENYLHAFNIUM see HAE500
DICHLORODI-pi-CYCLOPENTADIENYLHAFNIUM see HAE500
DICHLORODICYCLOPENTADIENYLTITANIUM see DGW200
DICHLORODI-pi-CYCLOPENTADIENYLTITANIUM see DGW200
2,2'-DICHLORO DIETHYLAMINE HYDROCHLORIDE see BHO250
β,β'-DICHLORODIETHYLAMINE HYDROCHLORIDE see BHO250
2',6'-DICHLORO-2-(DIETHYLAMINO)ACETANILIDE HYDROCHLORIDE
   see DEX600
7,8-DICHLORO-10-(2-(DIETHYLAMINO)ETHYL)ISOALLOXAZINE HYDRO-
   CHLORIDE see DEX800
2',6'-DICHLORO-2-(2-(DIETHYLAMINO)ETHYL)METHYLAMINOACETANIL-
   IDE DIHYDROCHLORIDE see DEY000
2',6'-DICHLORO-2-(2-(DIETHYLAMINO)ETHYL)THIOACETANILIDE HYDRO-
   CHLORIDE see DEY200
7,8-DICHLORO-10-(3-(DIETHYLAMINO)-2-HYDROXYPROPYL)ISOALLOXA-
   INE SULFATE see DEY400
7,8-DICHLORO-10-(4-(DIETHYLAMINO)-1-METHYLBUTYL)ISOALLOXAZINE
   HYDROCHLORIDE see DEY600
β,β'-DICHLORODIETHYLANILINE see AOQ875
β,β'-DICHLORODIETHYL ETHER see DFJ050
DICHLORODI(2-ETHYLHEXYL)STANNANE see DJL200
β,β-DICHLORODIETHYL-N-METHYLAMINE see BIE250
β,β'-DICHLORODIETHYL-N-METHYLAMINE HYDROCHLORIDE see BIE500
DICHLORODIETHYLSILANE see DEY800
DICHLORODIETHYLSTANNANE see DEZ000

DI-o-(CHLORODIETHYLSTANNYLOXO)BIS(CHLORO-DIETHYLTIN) see BIY500
2,2'-DICHLORODIETHYL SULFIDE see BIH250
DICHLORODIETHYLTIN see DEZ000
1,2-DICHLORO-1,1-DIFLUOROETHANE see DFA000
DICHLORODIFLUOROETHYLENE see DFA200
1,1-DICHLORO-2,2-DIFLUOROETHYLENE see DFA300
2,2-DICHLORO-1,1-DIFLUOROETHYL METHYL ETHER see DFA400
DICHLORODIFLUOROMETHANE see DFA600
DICHLORODIFLUOROMETHANE mixed with CHLORODIFLUOROMETHANE see DFB000
DICHLORODIFLUOROMETHANE-CHLORODIFLUOROMETHANE MIXTURE (DOT) see DFB000
DICHLORODIFLUOROMETHANE with 1,1-DIFLUOROETHANE see DFB400
DICHLORODIFLUOROMETHANE and DIFLUOROETHANE mixture (constant boiling mixture) (DOT) see DFB400
DICHLORODIFLUOROMETHANE mixed with TRICHLOROFLUOROMETH-ANE (1581) see DFB800
DICHLORODIFLUOROMETHANE with TRICHLOROTRIFLUOROETHANE see DFC000
DICHLORODIFLUOROMETHANE-TRICHLOROTRIFLUOROETHANE MIX-TURE (DOT) see DFC000
2,2-DICHLORO-1,1-DIFLUORO-1-METHOXYETHANE see DFA400
DICHLORODIHEXYLSTANNANE see DFC200
2,5-DICHLORO-4-(4,5-DIHYDRO-3-METHYL-5-OXO-1H-PYRAZOL-1-YL)BENZENESULFONIC ACID see DFQ200
4,8-DICHLORO-1,5-DIHYDROXYANTHRAQUINONE see DFC600
(2,5-DICHLORO-3,6-DIHYDROXY-p-BENZOQUINOLATO)MERCURY see DFC800
2,5-DICHLORO-3,6-DIHYDROXY-p-BENZOQUINONE, MERCURY SALT see DFC800
(2,5-DICHLORO-3,6-DIHYDROXY-p-BENZOQUINONE), MERCURY SALT see DFC800
cis-DICHLORO-trans-DIHYDROXYBISISOPROPYLAMINE PLATINUM (IV) see IGG775
5,5'-DICHLORO-2,2'-DIHYDROXY-3,3'-DINITROBIPHENYL see DFD000
5,5'-DICHLORO-2,2'-DIHYDROXYDIPHENYLMETHANE see MJM500
DICHLORODIISONONYL STANNANE see BLQ750
DICHLORODIISOPROPYL ETHER see BII250
DICHLORODIISOPROPYLSTANNANE see DNT000
3,4-DICHLORO-2,5-DILITHIOTHIOPHENE see DFD200
1,4-DICHLORO-2,5-DIMETHOXYBENZENE see CJA100
3',4'-DICHLORO-4-DIMETHYLAMINOAZOBENZENE see DFD400
7,8-DICHLORO-10-(2-(DIMETHYLAMINO)ETHYL)ISOALLOXAZINE SULFATE see DFD600
7,8-DICHLORO-10-(3-(DIMETHYLAMINO)PROPYL)ISOALLOXAZINE HYDRO-CHLORIDE see DFE000
1,1-DICHLORO-N-((DIMETHYLAMINO)SULFONYL)-1-FLUORO-N-PHENYL-METHANE SULFENAMIDE see DFL200
2,2-DICHLORO-3,3-DIMETHYLBUTANE see DFE100
sym-DICHLORODIMETHYL ETHER (DOT) see BIK000
DICHLORODIMETHYLHYDANTOIN see DFE200
1,3-DICHLORO-5,5-DIMETHYL HYDANTOIN see DFE200
1,3-DICHLORO-5,5-DIMETHYL-2,4-IMIDAZOLIDINEDIONE see DFE200
DICHLORO(4,5-DIMETHYL-o-PHENYLENEDIAMINE)PLATINUM(II) see DFE229
cis-DICHLORO(4,5-DIMETHYL-O-PHENYLENEDIAMINE)PLATINUM(II) see DFE229
3,5-DICHLORO-N-(1,1-DIMETHYL-2-PROPYNYL)BENZAMIDE see DTT600
3,5-DICHLORO-2,6-DIMETHYL-4-PYRIDINOL see CMX850
DICHLORODIMETHYLSILANE see DFE259
3,6-DICHLORO-3,6-DIMETHYLTETRAOXANE see DFE300
(trans-4)-DICHLORO(4,4-DIMETHYLZINC 5((((METHYLAMINO)CAR-BONYL)OXY)IMINO)PENTANENITRILE) see DFE469
4,4'-DICHLORO-6,6'-DINITRO-O,O'-BIPHENOL see DFD000
3,3'-DICHLORO-5,5'-DINITRO-O,O'-BIPHENOL (FRENCH) see DFD000
5,5'-DICHLORO-3,3'-DINITRO(1,1'-BIPHENYL)-2,2'-DIOL see DFD000
DICHLORODINITROMETHANE see DFE550
trans-2,3-DICHLORO-p-DIOXANE see DFE600
trans-2,3-DICHLORO-1,4-DIOXANE see DFE600
DICHLORODIOXOCHROMIUM see CML125
4,5-DICHLORO-3,6-DIOXO-1,4-CYCLOHEXADIENE-1,2-DICARBONITRILE see DEX400
DICHLORODIPENTYLSTANNANE see DVV200
DI-p-CHLORODIPHENOXYMETHANE see NCM700
DICHLORODIPHENYLACETIC ACID see BIL500
p,p'-DICHLORODIPHENYLACETIC ACID see BIL500
2,3-DICHLORO-6,12-DIPHENYL-DIBENZO(b,f)(1,5)DIAZOCINE see DFE700
2,8-DICHLORO-6,12-DIPHENYL-DIBENZO(b,f)(1,5)DIAZOCINE see DFE700
DICHLORODIPHENYL DICHLOROETHANE see BIM500
o,p'-DICHLORODIPHENYLDICHLOROETHANE see CDN000
p,p'-DICHLORODIPHENYLDICHLOROETHANE see BIM500
p,p'-DICHLORODIPHENYLDICHLOROETHYLENE see BIM750
2,2'-((3,3'-DICHLORO(1,1'-DIPHENYL)-4,4'-DIYL)BIS(AZO)BIS(3-OXO-N-PHENYLBUTANAMIDE see DEU000

DICHLORODIPHENYLETHANOL see BIN000
p,p'-DICHLORODIPHENYLMETHYLCARBINOL see BIN000
DICHLORO DIPHENYL OXIDE see DFE800
DICHLORO DIPHENYLSILANE see DFF000
DICHLORODIPHENYLSTANNANE see DWO400
DICHLORODIPHENYLSTANNANE complex with PYRIDINE (1:2) see DWA400
p,p'-DICHLORODIPHENYL SULFIDE see CEP000
DICHLORODIPHENYLTRICHLOROETHANE see DAD200
4,4'-DICHLORODIPHENYLTRICHLOROETHANE see DAD200
DICHLORODIPHENYLTRICHLOROETHANE (DOT) see DAD200
p,p'-DICHLORODIPHENYLTRICHLOROETHANE see DAD200
2,2-DICHLORO-N,N-DI-2-PROPENYLACETAMIDE see DBJ600
2',6'-DICHLORO-2-(DIPROPYLAMINO)ACETANILIDE HYDROCHLORIDE see DFF200
DICHLORODIPROPYLSTANNANE see DFF400
DICHLORODIPROPYLTIN see DFF400
DICHLORODIPYRIDINEPLATINUM(II) (Z) see DFF500
cis-DICHLORO(DIPYRIDINE)PLATINUM(II) see DFF500
4,5-DICHLORO-1,3-DISULFAMOYLBENZENE see DEQ200
1,4-DICHLORO-2,3-EPOXYBUTANE see DFF600
cis-1,3-DICHLORO-1,2-EPOXYPROPANE see DAC975
trans-1,3-DICHLORO-1,2-EPOXYPROPANE see DGH500
DICHLOROETHANE see DFF800
1,1-DICHLOROETHANE see DFF809
1,2-DICHLOROETHANE see EIY600
α,β-DICHLOROETHANE see EIY600
sym-DICHLOROETHANE see EIY600
DICHLORO-1,2-ETHANE (FRENCH) see EIY600
DICHLOROETHANOIC ACID see DEL000
2,2-DICHLOROETHANOL see DFG000
1,2-DICHLOROETHANOL ACETATE see DFG159
DICHLOROETHANOYL CHLORIDE see DEN400
1,1-DICHLOROETHENE see VPK000
2,2-DICHLOROETHENOL DIMETHYL PHOSPHATE see DGP900
(E)-S-(1,2-DICHLOROETHENYL)-l-CYSTEINE (9CI) see DGP125
2,2-DICHLOROETHENYL DIETHYL PHOSPHATE see DFG200
2,2-DICHLOROETHENYL DIMETHYL PHOSPHATE see DGP900
1,1'-DICHLOROETHENYLIDENE)BIS(4-CHLOROBENZENE) see BIM750
2,2-DICHLOROETHENYL PHOSPHORIC ACID DIMETHYL ESTER see DGP900
DICHLOROETHER see DFJ050
DICHLORO(4-ETHOXY-o-PHENYLENEDIAMMINE)PLATINUM(II) see DFG400
DI-(2-CHLOROETHYL) ACETAL see DFG600
1,2-DICHLOROETHYL ACETATE see DFG159
2,2-DICHLOROETHYLAMINE see DFG700
DI-2-CHLOROETHYLAMINE HYDROCHLORIDE see BHO250
p-(DI-2-CHLOROETHYLAMINE)PHENYL BUTYRIC ACID SODIUM SALT see CDO625
9-(2-(DI(2-CHLOROETHYL)AMINO)ETHYLAMINO)-6-CHLORO-2-METHOXYACRIDINE see DFH000
2-(DI-(2-CHLOROETHYL)AMINOMETHYL-5,6-DIMETHYLBENZIMIDAZOLE see BCC250
N,N-DI(2-CHLOROETHYL)AMINO-N,O-PROPYLENE PHOSPHORIC ACID ESTER DIAMIDE MONOHYDRATE see CQC500
2-(DI(2-CHLOROETHYL)AMINO)-1-OXA-3-AZA-2-PHOSPHACYCLOHEXANE-2-OXIDE MONOHYDRATE see CQC500
p-N,N-DI-(2-CHLOROETHYL)AMINOPHENYL ACETIC ACID see PCU425
p-N-DI(CHLOROETHYL)AMINOPHENYLALANINE see PED750
p-DI-(2-CHLOROETHYL)-AMINO-d-PHENYLALANINE see SAX200
p-DI(2-CHLOROETHYL)AMINO-d-PHENYLALANINE see SAX200
p-DI(2-CHLOROETHYL)AMINO-l-PHENYLALANINE see PED750
3-p-(DI(2-CHLOROETHYL)AMINO)-PHENYL-l-ALANINE see PED750
o-DI-2-CHLOROETHYLAMINO-dl-PHENYLALANINE see BHT250
p-DI(2-CHLOROETHYL)AMINO-dl-PHENYLALANINE see BHT750
Γ-(p-DI(2-CHLOROETHYL)AMINOPHENYL)BUTYRIC ACID see CDO500
p-(N,N-DI-2-CHLOROETHYL)AMINOPHENYL BUTYRIC ACID see CDO500
p-N,N-DI-(β-CHLOROETHYL)AMINOPHENYL BUTYRIC ACID see CDO500
N,N-DI-(2-CHLOROETHYL-Γ-p-AMINOPHENYLBUTYRIC ACID see CDO500
p-(N,N-DI-2-CHLOROETHYLAMINO)PHENYL-N-(p-CARBOXYPHENYL) CARBAMATE see CCE000
o-(p-DI-(2-CHLOROETHYL)AMINOPHENYL)-dl-TYROSINE DIHYDROCHLO-RIDE see DFH100
p-N,N-DI-(2-CHLOROETHYL)AMINOPHENYLVALERIC ACID see BHY625
5-(DI-2-CHLOROETHYL)AMINOURACIL see BIA250
5-(DI-(β-CHLOROETHYL)AMINO)URACIL see BIA250
N,N-DI(2-CHLOROETHYL)ANILINE see AOQ875
DICHLOROETHYLARSINE see DFH200
DICHLOROETHYLBENZENE see EHY500
DI-(2-CHLOROETHYL)BENZYLAMINE see BIA750
DICHLOROETHYLBORANE see DFH300
DI-(2-CHLOROETHYL)-3-CHLORO-4-METHYL-7-COUMARINYL PHOSPHATE see DFH600
DI-(2-CHLOROETHYL)-3-CHLORO-4-METHYLCOUMARIN-7-YL PHOSPHATE see DFH600

O,O-DI(2-CHLOROETHYL)-7-(3-CHLORO-4-METHYLCOUMARINYL)PHOSPHATE see CIK750

O,O-DI(2-CHLOROETHYL)-O-(3-CHLORO-4-METHYLCOUMARIN-7-YL) PHOSPHATE see DFH600

DICHLOROETHYLENE see DFH800, EIY600

1,1-DICHLOROETHYLENE see VPK000

1,2-DICHLOROETHYLENE see DFI100, DFI200

cis-DICHLOROETHYLENE see DFI200

sym-DICHLOROETHYLENE see DFI100

trans-DICHLOROETHYLENE see ACK000

trans-1,2-DICHLOROETHYLENE (MAK) see ACK000

DICHLORO-1,2-ETHYLENE (FRENCH) see DFI100

1,2-DICHLOROETHYLENE CARBONATE see DFI800

DICHLORO(ETHYLENEDIAMMINE)PLATINUM(II) see DFJ000

S-(1,2-DICHLOROETHYLENEYL)-l-CYSTEINE see DGP000

DI(2-CHLOROETHYL) ESTER, MALEIC ACID see DFJ200

DICHLOROETHYL ETHER see DFJ050

DI(β-CHLOROETHYL)ETHER see DFJ050

sym-DICHLOROETHYL ETHER see DFJ050

β,β'-DICHLOROETHYL ETHER see DFJ050

2,2'-DICHLOROETHYL ETHER (MAK) see DFJ050

DICHLOROETHYL FORMAL see BID750

DI-2-CHLOROETHYL FORMAL see BID750

1,2-DICHLOROETHYL HYDROPEROXIDE see DFJ100

DI-2-CHLOROETHYL MALEATE see DFJ200

2,3-DICHLORO-N-ETHYLMALEINIMIDE see DFJ400

DI(2-CHLOROETHYL)METHYLAMINE see BIE250

DI(2-CHLOROETHYL)METHYLAMINE HYDROCHLORIDE see BIE500

2-(1,2-DICHLOROETHYL)-4-METHYL-1,3-DIOXOLANE see DFJ500

DICHLOROETHYL-β-NAPHTHYLAMINE see BIF250

DI(2-CHLOROETHYL)-β-NAPHTHYLAMINE see BIF250

2-N,N-DI(2-CHLOROETHYL)NAPHTHYLAMINE see BIF250

N,N-DI(2-CHLOROETHYL)-β-NAPHTHYLAMINE see BIF250

DI((CHLORO-2-ETHYL)-2-N-NITROSO-N-CARBAMOYL)-N,N-CYSTAMINE see BIF625

DICHLOROETHYL OXIDE see DFJ050

DICHLOROETHYLPHENYLSILANE see DFJ800

DICHLOROETHYLPHOSPHINE see EOQ000

N,N-DI(2-CHLOROETHYL)-N,o-PROPYLENE-PHOSPHORIC ACID ESTER DIAMIDE see CQC650

2-(1,2-DICHLOROETHYL)PYRIDINE HYDROCHLORIDE see CMV475

DICHLOROETHYLSILANE see DFK000

DI-2-CHLOROETHYL SULFIDE see BIH250

β,β'-DICHLOROETHYL SULFIDE see BIH250

2,2'-DICHLOROETHYL SULPHIDE (MAK) see BIH250

DI-(2-CHLOROETHYL)THENYLAMINE HYDROCHLORIDE see BII000

2-2'-DI(3-CHLOROETHYLTHIO)DIETHYL ETHER see DFK200

DICHLOROETHYLVINYLSILANE see DFK400

DICHLOROETHYNE see DEN600

3-(3,4-DICHLORO-FENIL)-1-METOSSI-1-METIL-UREA (ITALIAN) see DGD600

DICHLOROFENTHION see DFK600

DICHLOROFLUOROMETHANE see DFL000

DICHLOROFLUOROMETHANE-TRICHLOROFLUOROMETHANE (DOT) see DFB800

4',5-DICHLORO-N-(4-FLUORO-2-METHYLPHENYL)-2-HYDROXY-(1,1'-BIPHE-NYL)-3-CARBOXAMIDE see CFC500

N-((DICHLOROFLUOROMETHYL)THIO)-N-((DIMETHYLAMINO)SULFONYL) ANILINE see DFL200

N-(DICHLOROFLUOROMETHYLTHIO)-N,N'-DIMETHYL-N-PHENYLSULFAM-IDE see DFL200

N-(DICHLOROFLUOROMETHYLTHIO)-N-(DIMETHYLSULFAMOYL)ANILINE see DFL200

N'-DICHLOROFLUOROMETHYLTHIO-N,N-DIMETHYL-N-(4-TOLYL)SULFAM-IDE see DFL400

α-β-DICHLORO-β-FORMYL ACRYLIC ACID see MRU900

2,2'-DICHLORO-N-FURFURYLDIETHYLAMINE HYDROCHLORIDE see FPX000

DICHLOROGERMANE see DFL600

N,N-DICHLOROGLYCINE see DFL709

1,6-DICHLORO-2,4-HEXADIYNE see DFL800

2,3-DICHLOROHEXAFLUOROBUTENE-2 see DFM000

2,3-DICHLOROHEXAFLUORO-2-BUTENE see DFM000

2,3-DICHLORO-1,1,1,4,4,4-HEXAFLUOROBUTENE-2 see DFM000

4,5-DICHLORO-3,3,4,5,6,6-HEXAFLUORO-1,2-DIOXANE see DFM099

DICHLOROHYDRIN see DGG400

α-DICHLOROHYDRIN see DGG400

6,7-DICHLORO-4-(HYDROXYAMINO)QUINOLINE-1-OXIDE see DFM200

3,4-DICHLORO-2-HYDROXYCROTONOLACTONE see MRU900

3,4-DICHLORO-2-HYDROXYCROTONOLACTONIC ACID see MRU900

6,7-DICHLORO-10-(3-(N-(2-HYDROXYETHYL)ETHYLAMINO)ISOALLOXA-INE SULFATE see DFM600

6,7-DICHLORO-10-(3-(N-2-HYDROXYETHYL)METHYLAMINO)PROPYL) ISO-ALLOXAZINE SULFATE see DFM800

d-(−)-threo-2,2-DICHLORO-N-(β-HYDROXY-α-(HYDROXYMETHYL))-p-NITROPHENETHYLACETAMIDE see CDP250

d-(−)-2,2-DICHLORO-N-(β-HYDROXY-α-(HYDROXYMETHYL)-p-NITROPHENYLETHYL)ACETAMIDE see CDP250

DI-(5-CHLORO-2-HYDROXYPHENYL)METHANE see MJM500

5,7-DICHLORO-8-HYDROXYQUINALDINE see CLC500

2,6-DICHLORO-N-2-IMIDAZOLIDINYLIDENE-BENZENAMINE HYDROCHLO-RIDE see CMX760

1-(2,5-DICHLORO-6-(1-(1H-IMIDAZOL-1-YL)VINYL)PHENOXY)-3-(ISOPROPYLAMINO)-2-PROPANOL HYDROCHLORIDE see DFM875

O-(2,5-DICHLORO-4-IODOPHENYL) O,O-DIMETHYL PHOSPHOROTHIOATE see IEN000

2,2'-DICHLORO-N-ISOBUTYL-DIETHYLAMINE HYDROCHLORIDE see BIE750

DICHLOROISOCYANURIC ACID see DGN200

DICHLOROISOCYANURIC ACID, dry (DOT) see DGN200

DICHLOROISOCYANURIC ACID POTASSIUM SALT see PLD000

DICHLOROISOCYANURIC ACID SODIUM SALT (DOT) see SGG500

sym-DICHLOROISOPROPYL ALCOHOL see DGG400

3,4-DICHLORO-α-(ISOPROPYLAMINO)METHYL)BENZYL ALCOHOL see DFN400

2,2'-DICHLORO-N-ISOPROPYLDIETHYLAMINE HYDROCHLORIDE see IPG000

2,2'-DICHLOROISOPROPYL ETHER see BII250

DICHLOROISOPROPYL ETHER (DOT) see BII250

DICHLOROKELTHANE see BIO750

DICHLOROLAWSONE see DFN500

DICHLOROMALEALDEHYDIC ACID see MRU900

2,3-DICHLOROMALEIC ALDEHYDE ACID see MRU900

DICHLOROMALEIC ANHYDRIDE see DFN700

DICHLOROMALEIMIDE see DFN800

DICHLOROMALEINIMIDE see DFN800

DICHLOROMAPHARSEN see DFX400

3',4'-DICHLORO-2-METHACRYLANILIDE see DFO800

DICHLOROMETHANE (MAK, DOT) see MJP450

DICHLOROMETHANETHIOSULFONIC ACID-S-TRICHLOROMETHYL ESTER see DFS600

DICHLOROMETHAZANONE see CKF500

DICHLOROMETHOTREXATE see DFO000

3'5'-DICHLOROMETHOTREXATE see DFO000

2,5-DICHLORO-6-METHOXYBENZOIC ACID see MEL500

3,6-DICHLORO-2-METHOXYBENZOIC ACID see MEL500

DICHLORO(4-METHOXYCARBONYL-O-PHENYLENEDIAMMINE) PLATINUM(II) see DFO200

DICHLORO(4-METHOXY-O-PHENYLENEDIAMMINE)PLATINUM(II) see DFO400

(2,3-DICHLORO-4-METHOXYPHENYL)-2-FURANYLMETHANONE)-O-(2-(DIETHYLAMINO)ETHYL) OXIME,MONOMETHANE SULFONATE see DFO600

(DICHLORO-2,3-METHOXY-4) PHENYL FURYL-2-O-(DIETHYLAMINOETHYL)-CETONE-OXIME (FRENCH) see DFO600

3',4'-DICHLORO-2-METHYLACRYLANILIDE see DFO800

N,N-DICHLOROMETHYLAMINE see DFO900

9,10-DI(CHLOROMETHYL)ANTHRACENE see BIJ750

DICHLOROMETHYLARSINE see DFP200

1,5-DICHLORO-3-METHYL-3-AZAPENTANE HYDROCHLORIDE see BIE500

4,4'-DICHLORO(METHYL BENZHYDROL) see BIN000

4,4'-DICHLORO-α-METHYLBENZHYDROL see BIN000

4,4'-DICHLORO-α-METHYLBENZOHYDROL see BIN000

1-(2,4-DICHLORO-β-(p-METHYLBENZYLOXY)PHENETHYL)IMIDAZOLE NI-TRATE see DFP500

3-DICHLOROMETHYL-6-CHLORO-7-SULFAMOYL-3,4-DIHYDRO-1,2,4-BENZOTHIADIAZINE-1,1-DIOXIDE see HII500

3-DICHLOROMETHYL-6-CHLORO-7-SULFAMY-3,4-DIHYDRO-1,2,4-BENZOTHIADIAZINE-1,1-DIOXIDE see HII500

DICHLOROMETHYL CYANIDE see DEN000

2,2'-DICHLORO-N-METHYLDIETHYLAMINE see BIE250

2,2'-DICHLORO-N-METHYLDIETHYLAMINE HYDROCHLORIDE see BIE500

2,2'-DICHLORO-N-METHYLDIETHYLAMINE-N-OXIDE see CFA500

2,2'-DICHLORO-N-METHYLDIETHYLAMINE N-OXIDE HYDROCHLORIDE see CFA750

1-(DICHLOROMETHYLDIMETHYLSILYL)-1-HEXYN-3-OL see HGA000

2,3-DICHLORO-4-(2-METHYLENEBUTYRL)PHENOXY ACETIC ACID see DFP600

(2,3-DICHLORO-4-(2-METHYLENEBUTYRYL)PHENOXY)ACETIC ACID see DFP600

(2,3-DICHLORO-4-(2-METHYLENE-1-OXOBUTYL)PHENOXY)ACETIC ACID see DFP600

α,α-DICHLOROMETHYL ETHER see DFQ000

sym-DICHLOROMETHYL ETHER see BIK000

3,4-DICHLORO-α-((((1-METHYLETHYL)AMINO)METHYL) BENZENEMETHANOL see DFN400

1,3-DICHLORO-5,5'-METHYLHYDANTOIN see DFE200

5,7-DICHLORO-2-METHYL-8-HYDROXYQUINOLINE see CLC500

DICHLORO-N-METHYLMALEIMIDE see DFP800

2,3-DICHLORO-N-METHYLMALEIMIDE see DFP800

α,α-DICHLOROMETHYL METHYL ETHER see DFQ000

d-threo-2-(DICHLOROMETHYL)-α-(p-NITROPHENYL)-2-OXAZOLINE-4-METHA-NOL see DFQ100

6,7-DICHLORO-2-METHYL-1-OXO-2-PHENYL-5-INDANYLOXYACETIC ACID see IBQ400

2,5-DICHLORO-4-(3-METHYL-5-OXO-2-PYRAZOLIN-1-YL) BENZENESULFO-NIC ACID see DFQ200

3,3-DICHLOROMETHYLOXYCYCLOBUTANE see BIK325

2-((2,6-DICHLORO-3-METHYLPHENYL)AMINO)-BENZOIC ACID (9CI) see DGM875

2-((2,6-DICHLORO-3-METHYLPHENYL)AMINO)BENZOIC ACID ETHOXYMETHYL ESTER see DGN000

2-((2,6-DICHLORO-3-METHYLPHENYL)AMINO-BENZOIC ACID MONOSO-DIUM SALT see SIF425

DICHLORO(4-METHYL-o-PHENYLENEDIAMMINE)PLATINUM(II) see DFQ400

DICHLOROMETHYLPHENYLSILANE see DFQ800

DICHLOROMETHYL PHOSPHINE see EOQ000

1,2-DICHLORO-2-METHYLPROPENE see IIQ200

2,3-DICHLORO-2-METHYLPROPIONALDEHYDE see DFR400

DICHLORO(1-METHYLPROPYL)ARSINE see BQY300

5,7-DICHLORO-2-METHYL-8-QUINOLINOL see CLC500

DICHLOROMETHYLSILANE see DFS000

4-(DICHLOROMETHYLSILYL)BUTYRONITRILE see COR325

1,2-DICHLORO-1-(METHYLSULFONYL)ETHYLENE see DFS200

O-(DICHLORO(METHYLTHIO)PHENYL) O,O-DIETHYL PHOSPHOROTHIO-ATE (3 isomers) see CLJ875

DICHLOROMETHYL TRICHLOROMETHYLTHIOSULFONE see DFS600

2,2'-DICHLORO-1''-METHYLTRIETHYLAMINE see IOF300

2,2'-DICHLORO-1''-METHYLTRIETHYLAMINE HYDROCHLORIDE see IPG000

DICHLOROMETHYLVINYLSILANE see DFS800

DICHLOROMONOFLUOROMETHANE (OSHA, DOT) see DFL000

2,3-DICHLORO-1,4-NAPHTHALENEDIONE see DFT000

2,3-DICHLORO-1,4-NAPHTHAQUINONE see DFT000

DICHLORONAPHTHOQUINONE see DFT000

2,3-DICHLORONAPHTHOQUINONE see DFT000

2,3-DICHLORO-α-NAPHTHOQUINONE see DFT000

2,3-DICHLORO-1,4-NAPHTHOQUINONE see DFT000

2,3-DICHLORONAPHTHOQUINONE-1,4 see DFT000

2,6-DICHLORO-4-NITROANILINE see RDP300

2,6-DICHLORO-4-NITROBENZENAMINE (9CI) see RDP300

2,5-DICHLORONITROBENZENE see DFT400

3,4-DICHLORONITROBENZENE see DFT600

1,2-DICHLORO-4-NITROBENZENE see DFT600

1,4-DICHLORO-2-NITROBENZENE see DFT400

2',4'-DICHLORO-4-NITROBIPHENYL ETHER see DFT800

2,4-DICHLORO-4'-NITRODIPHENYL ETHER see DFT800

DICHLORONITROETHANE see DFU000

1,1-DICHLORO-1-NITROETHANE see DFU000

1,2-DICHLORO-3-NITRONAPHTHALENE see DFU400

2,4-DICHLORO-6-NITROPHENOL see DFU600

2,4-DICHLORO-6-NITROPHENOL ACETATE see DFU800

2,4-DICHLORO-1-(4-NITROPHENOXY)BENZENE see DFT800

DICHLORO(4-NITRO-o-PHENYLENEDIAMMINE)PLATINUM(II) see DFV000

6,7-DICHLORO-4-NITROQUINOLINE-1-OXIDE see DFV200

2',5-DICHLORO-4'-NITROSALICYLANILIDE see DFV400

2',5-DICHLORO-4'-NITROSALICYLANILIDE-2-AMINOETHANOL SALT see DFV600

5,2'-DICHLORO-4'-NITROSALICYLANILIDE ETHANOLAMINE SALT see DFV600

5,2-DICHLORO-4-NITROSALICYLIC ANILIDE-2-AMINOETHANOL SALT see DFV600

2',5-DICHLORO-4'-NITROSALICYLOYLANILIDE ETHANOLAMINE SALT see DFV600

3,4-DICHLORO-N-NITROSOCARBANILIC ACID METHYL ESTER see DFV800

2,2'-DICHLORO-N-NITROSODIPROPYLAMINE see DFW000

3,4-DICHLORONITROSOPIPERIDINE see DFW200

3,4-DICHLORO-N-NITROSOPYRROLIDINE see DFW600

2,3-DICHLORO-4-OXO-2-BUTENOIC ACID see MRU900

4,5-DICHLORO-2-OXO-1,3-DIOXOLANE see DFI800

DICHLOROOXOVANADIUM see DFW800

DICHLOROOXOZIRCONIUM see ZSJ000

DICHLOROPENTANE see DFX000

1,5-DICHLOROPENTANE see DFX200

DICHLOROPENTANES (DOT) see DFX000

DICHLOROPENTYLARSINE see AOI200

DICHLOROPHENAMIDE see DEQ200

DICHLOROPHENARSINE HYDROCHLORIDE see DFX400

2,4-DICHLOROPHENOL see DFX800

2,6-DICHLOROPHENOL see DFY000

2,4-DICHLOROPHENOL BENZENESULFONATE see DFY400

3-(3,4-DICHLOROPHENOL)-1,1-DIMETHYLUREA see DXQ500

2,4-DICHLORO-PHENOL-O-ESTER with O,O-DIETHYL PHOSPHOROTHIO-ATE see DFK600

3,4-DICHLOROPHENOL, O-ESTER with O-METHYL METHYLPHOSPHORAMIDOTHIOATE see IEN000

(2,4-DICHLOROPHENOXY)ACETATE DIMETHYLAMINE see DFY800

DICHLOROPHENOXYACETIC ACID see DAA800

3,4-DICHLOROPHENOXYACETIC ACID see DFY500

2,4-DICHLOROPHENOXYACETIC ACID (DOT) see DAA800

(2,4-DICHLOROPHENOXY)ACETIC ACID BUTOXYETHYL ESTER see DFY709

(2,4-DICHLOROPHENOXY)ACETIC ACID, BUTYL ESTER see BQZ000

2,4-DICHLOROPHENOXYACETIC ACID BUTYL ESTER and 2,4,5-TRICHLOROPHENOXYACETIC ACID (45.5%:48.2%) see AEX750

(2,4-DICHLOROPHENOXY)ACETIC ACID DIMETHYLAMINE see DFY800

(2,4-DICHLOROPHENOXY)ACETIC ACID ETHYL ESTER see EHY600

2,4-DICHLOROPHENOXYACETIC ACID ISOOCTYL ESTER see ILO000

(2,4-DICHLOROPHENOXY)ACETIC ACID, ISOPROPYL ESTER see IOY000

(2-4-DICHLOROPHENOXY)ACETIC ACID-1-METHYLETHYL ESTER (9CI) see IOY000

2,4-DICHLOROPHENOXYACETIC ACID PROPYLENE GLYCOL BUTYL ETHER ESTER see DFZ000

2,4-DICHLOROPHENOXYACETIC ACID, SODIUM SALT see SGH500

4-(2,4-DICHLOROPHENOXY)BUTYRIC ACID see DGA000

γ-(2,4-DICHLOROPHENOXY)BUTYRIC ACID see DGA000

2,4-DICHLOROPHENOXYBUTYRIC ACID, SODIUM SALT see EAK500

γ-(2,4-DICHLOROPHENOXY)BUTYRIC ACID, SODIUM SALT see EAK500

5,6-DICHLORO-1-PHENOXYCARBONYL-2-TRIFLUOROMETHYLBENZIMIDAZOLE see DGA200

2,4-DICHLOROPHENOXY ETHANEDIOL see DGA400

2,4-DICHLOROPHENOXY-1,2-ETHANEDIOL see DGA400

2-(2,4-DICHLOROPHENOXY)ETHANOL see DGP800

2-(2,4-DICHLOROPHENOXY)ETHANOL HYDROGEN SULFATE SODIUM SALT see CNW000

2-(1-(2,6-DICHLOROPHENOXY)ETHYL)-4,5-DIHYDRO-1H-IMIDAZOLE MONOHYDROCHLORIDE see LIA400

2-(1-(2,6-DICHLOROPHENOXY)ETHYL)-2-IMIDAZOLINE HYDROCHLORIDE see LIA400

2,4-DICHLOROPHENOXYETHYL SULFATE, SODIUM SALT see CNW000

3-(2,4-DICHLOROPHENOXY)-2-HYDROXYPROPYL-o-CHLOROPHENYL ARSI-NIC ACID see DGA425

DI-(4-CHLOROPHENOXY)METHANE see NCM700

DI-(p-CHLOROPHENOXY)METHANE see NCM700

2-((3,4-DICHLOROPHENOXY)METHYL)-2-IMIDAZOLINE HYDROCHLORIDE see DGA800

2-((3,4-DICHLOROPHENOXY)METHYL-2-IMIDAZOLINE MONOHYDROCHLORIDE see DGA800

4-(2,4-DICHLOROPHENOXY)NITROBENZENE see DFT800

2-(2,4-DICHLOROPHENOXY) PROPIONIC ACID see DGB000

2-(2,5-DICHLOROPHENOXY)PROPIONIC ACID see DGB200

α-(2,4-DICHLOROPHENOXY) PROPIONIC ACID see DGB000

α-(2,5-DICHLOROPHENOXY)PROPIONIC ACID see DGB200

(+)-2-(2,4-DICHLOROPHENOXY)PROPIONIC ACID see DGB100

(2,4-DICHLOROPHENOXY)TRIBUTYLSTANNANE see DGB400

2-(2,4-DICHLOROPHENOXY)-4,5,6-TRICHLOROPHENOL see TIL750

6-(2,4-DICHLOROPHENOXY)-2,3,4-TRICHLOROPHENOL see TIL750

DI(p-CHLOROPHENYL)ACETIC ACID see BIL500

2-(2,6-DICHLOROPHENYL)AMINO)BENZENEACETIC ACID MONOSODIUM SALT see DEO600

2-(2,6-DICHLOROPHENYLAMINO)-2-IMIDAZOLINE see DGB500

2-(2,6-DICHLOROPHENYLAMINO)-2-IMIDAZOLIN HYDROCHLORID (GER-MAN) see CMX760

DICHLOROPHENYLARSINE see DGB600

N-(3,4-DICHLOROPHENYL)-1-AZIRIDINECARBOXAMIDE see DGB800

p-(3,4-DICHLOROPHENYL)AZO)-N,N-DIMETHYLANILINE see DFD400

3,4-DICHLOROPHENYLAZOTHIOUREA see DEQ000

3,4-DICHLOROPHENYL-AZOTHIOUREE (FRENCH) see DEQ000

2,4-DICHLOROPHENYL BENZENESULFONATE see DFY400

2,4-DICHLOROPHENYL BENZENESULPHONATE see DFY400

DICHLOROPHENYLBORANE see DGB875

(3,4-DICHLOROPHENYL)CARBAMIC ACID METHYL ESTER see DEV600

3,4-DICHLOROPHENYL-N-CARBAMOYLAZIRIDINE see DGB800

2,4-DICHLOROPHENYL "CELLOSOLVE" see DGC000

N-(3,4-DICHLOROPHENYL)-N'-(4-CHLOROPHENYL)UREA see TIL500

N-(3,4-DICHLOROPHENYL)CYCLOPROPANECARBOXAMIDE see CQJ250

2,4'-DICHLOROPHENYLDICHLOROETHANE see CDN000

1-(2-(2,4-DICHLOROPHENYL)-2-((2,4-DICHLOROPHENYL)METHOXY)ETHYL-IMIDAZOLE (9CI) see MQS550

1-(2-(2,4-DICHLOROPHENYL)-2-(2,6-DICHLOROPHENYL)METHOXY)ETHYL)-1H-IMIDAZOLE MONONITRATE see IKN200

O-2,4-DICHLOROPHENYL-O,O-DIETHYL PHOSPHOROTHIOATE see DFK600

2,4-DICHLORO-PHENYL DIETHYL PHOSPHOROTHIONATE see DFK600

1,6-DI(4'-CHLOROPHENYLDIGUANIDINO)HEXANE DIACETATE see CDT125

1,6-DI(4'-CHLOROPHENYLDIGUANIDO)HEXANE see BIM250

3-(3,5-DICHLOROPHENYL)-1,5-DIMETHYL-3-AZABICYCLO(3.1.0)HEXANE-2,4-DIONE see PMF750

N-(3',5'-DICHLOROPHENYL)-1,2-DIMETHYLCYCLOPROPANE-1,2-DICARBOXIMIDE see PMF750

3-(3,4-DICHLOROPHENYL)-1,1-DIMETHYL-3-NITROSOUREA see NKO500

2,4-DICHLORO-5-SULPHAMOYLBENZOIC ACID see DGK900
4,6-DICHLORO-2',4',5',7'-TETRABROMOFLUORESCEIN DIPOTASSIUM SALT
　　see CMM000
DICHLOROTETRAFLUOROACETONE see DGL400
sym-DICHLOROTETRAFLUOROACETONE see DGL400
DICHLOROTETRAFLUOROETHANE see DGL600
sym-DICHLOROTETRAFLUOROETHANE see FOO509
1,2-DICHLORO-1,1,2,2-TETRAFLUOROETHANE (MAK) see FOO509
DICHLOROTETRAFLUOROETHANE (OSHA, ACGIH) see FOO509
1,3-DICHLORO-1,1,3,3-TETRAFLUORO-2-PROPANONE see DGL400
2,3-DICHLOROTETRAHYDROFURAN see DGL800
3,4-DICHLOROTETRAHYDROTHIOPHENE-1,1-DIOXIDE see DGL200
(2,3-DICHLORO-4-(2-THENOYL)PHENOXY)ACETIC ACID see TGA600
endo-2,5-DICHLORO-7-THIABICYCLO(2.2.1) HEPTANE see DGL875
(2,3-DICHLORO-4-(2-THIENYLCARBONYL)PHENOXY)ACETIC ACID
　　see TGA600
2,6-DICHLOROTHIOBENZAMIDE see DGM600
DICHLOROTHIOCARBONYL see TFN500
DICHLOROTHIOLANE DIOXIDE see DGL200
(2,3-DICHLORO-4-(2-THIOPHENECARBONYL)PHENOXY)ACETIC ACID
　　see TGA600
DICHLOROTITANOCENE see DGW200
α,α-DICHLOROTOLUENE see BAY300
N-(2,6-DICHLORO-m-TOLYL)ANTHRANILIC ACID see DGM875
N-(2,6-DICHLORO-m-TOLYL)ANTHRANILIC ACID ETHOXYMETHYL ESTER
　　see DGN000
1,3-DICHLORO-s-TRIAZINE-2,4,6(1H,3H,5H)-TRIONE see DGN200
DICHLORO-s-TRIAZINE-2,4,6(1H,3H,5H)-TRIONE POTASSIUM DERIV
　　see PLD000
2-(4,6-DICHLORO-s-TRIAZIN-2-YLAMINO)-4-(4-AMINO-3-SULFO-1-ANTHRA-
　　QUINONYLAMINO)BENZENESULFONIC ACID, DISODIUM SALT
　　see DGN400
4-(4,6-DICHLORO-s-TRIAZIN-2-YLAMINO)-5-HYDROXY-6-(2-HYDROXY-5-
　　NITROPHENYLAZO)-2,7-NAPHTHALENEDISULFONIC ACID
　　see DGN600
5-(3,5-DICHLORO-s-TRIAZINYLAMINO)-4-HYDROXY-3-PHENYLAZO-2,7-
　　NAPHTHALENEDISULFONIC ACID see DGN800
2-(6-(4,6-DICHLORO-s-TRIAZINYL)METHYLAMINO-1-HYDROXY-3-SUL-
　　FONAPHTHYLAZO)-1,5-NAPHTHALENEDISULFONIC ACID see DGO000
4,4'-DICHLORO-α-(TRICHLOROMETHYL)BENZHYDROL see BIO750
DICHLORO(4,5,6-TRICHLORO-o-PHENYLENEDIAMMINE)PLATINUM(II)
　　see DGO200
2,2'-DICHLOROTRIETHYLAMINE see BID250
4,5-DICHLORO-2-TRIFLUOROMETHYLBENZIMIDAZOLE see DGO400
5,6-DICHLORO-2-TRIFLUOROMETHYLBENZIMIDAZOLE-1-CARBOXYLATE
　　see DGA200
5,6-DICHLORO-2-(TRIFLUOROMETHYL)-1H-BENZIMIDAZOLE-1-CARBOX-
　　YLIC ACID PHENYL ESTER see DGA200
1,3-DICHLORO-6-TRIFLUOROMETHYL-9-(3-(DIBUTYLAMINO)-1-
　　HYDROXYPROPYL)PHENANTHRENE HCl see HAF500
1-(1,3-DICHLORO-6-TRIFLUOROMETHYL-9-PHENANTHRYL)-3-(DI-N-
　　BUTYLAMINO)PROPANOL HYDROCHLORIDE see HAF500
DICHLORO(m-TRIFLUOROMETHYLPHENYL)ARSINE see DGO600
DICHLOROTRIPHENYLANTIMONY see DGO800
DICHLOROTRIPHENYLSTIBINE see DGO800
DICHLOROVAS see DGP900
(2,2-DICHLORO-VINIL)DIMETILFOSFATO (ITALIAN) see DGP900
2,2-DICHLOROVINYL ALCOHOL, DIMETHYL PHOSPHATE see DGP900
S-DICHLOROVINYL-l-CYSTEINE see DGP000
S-(trans-1,2-DICHLOROVINYL)-l-CYSTEINE see DGP125
2,2-DICHLOROVINYL DIETHYL PHOSPHATE see DFG200
2,2-DICHLOROVINYL DIMETHYL PHOSPHATE see DGP900
2,2-DICHLOROVINYL DIMETHYL PHOSPHORIC ACID ESTER see DGP900
l-3-((1,2-DICHLOROVINYL)THIO)ALANINE see DGP000
DICHLOROVOS see DGP900
DICHLOROXYLENE see XSS250
α,α'-DICHLOROXYLENE see XSS250
α,α'-DICHLORO-m-XYLENE see DGP200
α,α'-DICHLORO-o-XYLENE see DGP400
α,α'-DICHLORO-p-XYLENE see DGP600
DICHLORPHENAMIDE see DEQ200
2,4-DICHLORPHENOXYACETIC ACID see DAA800
2-(1-(2,6-DICHLORPHENOXY)AETHYL)-2-IMIDAZOLIN-HYDROCHLORID (GER-
　　MAN) see LIA400
(2,4-DICHLOR-PHENOXY)-ESSIGSAEURE (GERMAN) see DAA800
2-(2,4-DICHLOR-PHENOXY)-PROPIONSAEURE (GERMAN) see DGB000
(+)-2-(2,4-DICHLORPHENOXY)PROPIONSAFEURE (GERMAN) see DGB100
(3,4-DICHLOR-PHENYL-AZO)-THIOHARNSTOFF (GERMAN) see DEQ000
3-(3,4-DICHLORPHENYL)-1-N-BUTYL-HARNSTOFF (GERMAN) see BRA250
2,4-DICHLORPHENYL "CELLOSOLVE" see DGP800
3-(3,4-DICHLORPHENYL)-1,1-DIMETHYL-HARNSTOFF (GERMAN)
　　see DXQ500
3-(3,4-DICHLOR-PHENYL)-1-METHOXY-1-METHYL-HARNSTOFF (GERMAN)
　　see DGD600

3-(4,5-DICHLORPHENYL)-1-METHOXY-1-METHYLHARNSTOFF (GERMAN)
　　see DGD600
2,4,-DICHLORPHENYL-4-NITROPHENYLAETHER (GERMAN) see DFT800
1-(2-(2,4-DICHLORPHENYL)-2-PROPENYLOXY)AETHYL)-1H-IMIDAZOLE
　　see FPB875
DICHLORPHOS see DGP900
DICHLORPROP see DGB000
DICHLORPROPAN-DICHLORPROPENGEMISCH (GERMAN) see DGG000
DICHLORPROPEN-GEMISCH (GERMAN) see DGG800
DICHLOR STAPENOR see DGE200
DICHLORSULFOFENYL-METHYLPYRAZOLON (CZECH) see DFQ200
DICHLOR-s-TRIAZIN-2,4,6(1H,3H,5H)TRIONE POTASSIUM see PLD000
DICHLORURE de TRIMETHYLAMMONIUM-1-(β-N-METHYLINDOYL-3")
　　ETHYL-4'-PYRIDINIUM)-3 PROPANE see MKW100
O-(2,2-DICHLORVINYL)-O,O-DIETHYLPHOSPHAT (GERMAN) see DFG200
(2,2-DICHLOR-VINYL)-DIMETHYL-PHOSPHAT (GERMAN) see DGP900
O-(2,2-DICHLORVINYL)-O,O-DIMETHYLPHOSPHAT (GERMAN) see DGP900
DICHLORVOS see DGP900
DICHLORVOS-ETHYL see DFG200
DICHLOSALE see DFV400
DICHLOTIAZID see CFY000
DICHLOTRIDE see CFY000
DICHOLINE SUCCINATE see CMG250
DICHROMIUM TRIOXIDE see CMJ900
DICHRONIC see DEO600
DICHYSTROLUM see DME300
DI(2-CIANOETIL)AMMINA (ITALIAN) see BIQ500
DICK (GERMAN) see DFH200
DICLOCIL see DGE200
DICLOFENAC SODIUM see DEO600
DICLONDAZOLIC ACID see DEL200
DICLONIA see BPR500
DICLOPHENAC SODIUM see DEO600
DICLORALUREA see DGQ200
DICLORAN see RDP300
p-DICLOROBENZENE (ITALIAN) see DEP800
1,4-DICLOROBENZENE (ITALIAN) see DEP800
1,1-DICLORO-2,2-BIS(4-CLORO-FENIL)-ETANO (ITALIAN) see BIM500
3,3'-DICLORO-4,4'-DIAMINODIFENILMETANO (ITALIAN) see MJM200
1,1-DICLOROETANO (ITALIAN) see DFF809
1,2-DICLOROETANO (ITALIAN) see EIY600
2,2'-DICLOROETILETERE (ITALIAN) see DFJ050
(3,4-DICLORO-FENIL-AZO)-TIOUREA (ITALIAN) see DEQ000
3-(3,4-DICLORO-FENYL)-1,1-DIMETIL-UREA (ITALIAN) see DXQ500
d-threo-2-DICLOROMETIL-4-((4'-NITROFENIL)-OSSIMETIL)-2-OSSAZOLINA
　　(ITALIAN) see DFQ100
1,1-DICLORO-1-NITROETANO (ITALIAN) see DFU000
DICLOTRIDE see CFY000
DICLOXACILLIN SODIUM MONOHYDRATE see DGE200
DICLOXACILLIN SODIUM SALT see DGE200
DICO see OOI000
DICOBALT BORIDE see DGQ300
DICOBALT CARBONYL see CNB500
DICOBALT EDETATE see DGQ400
DICOBALT EDTA see DGQ400
DICOBALT OCTACARBONYL see CNB500
DICODID see OOI000
DICOFERIN see DGQ500
DICOFOL see BIO750
DICOL see DJD600
DICONIRT D see SGH500
DICOPHANE see DAD200
DICOPPER(I) ACETYLIDE see DGQ600
DICOPPER DICHLORIDE see CNK250
DICOPPER DIHYDROXYCARBONATE see CNJ750
DICOPPER(I)-1,5-HEXADIYNIDE see DGQ625
DICOPPER(I) KETENIDE see DGQ650
DICOPPER MONOSULFIDE see CNP750
DICOPPER MONOXIDE see CNO000
DICOPPER SULFIDE see CNP750
DICOPUR see DAA800
DICOPUR-M see CIR250
DICORANTIL see DNN600
DICORTOL see PMA000
DICORVIN see DKA600
DICOTEX see CIR250
DICOTEX 80 see SIL500
DICOTOX see DAA800, EHY600
DICOUMARIN see BJZ000
DICOUMAROL see BJZ000
DICRESOL see DGQ700
DICRESYL see MIB750
DICRODEN see MQH250
DICROTALIC ACID see HMC000

DICROTOFOS (DUTCH) see DGQ875
DICROTONYL PEROXIDE see DGQ859
DICROTOPHOS see DGQ875
DICRYL see DFO800
DICTYCIDE see COH250
DICTYZIDE see COH250
DICUMACYL see BKA000
DICUMAN see BJZ000
DICUMARINE see BJZ000
DICUMENE CHROMIUM see DGR200
DICUMENYLCHROMIUM see DGR200
DICUMYLMETHANE see DGR400
DI-α-CUMYL PEROXIDE see DGR600
DICUMYL PEROXIDE (DOT) see DGR600
DI-CUP see DGR600
DI-CUP 40 KF see DGR600
DI-CUPR see DGR600
DICUPRAL see DXH250
DICURAN see CIS250
DICURONE see GFM000
DICYANDIAMIDE-FORMALDEHYDE ADDUCT see CNH125
DICYANDIAMIDE-FORMALDEHYDE POLYMER see CNH125
DICYANDIAMIDE-FORMALDEHYDE RESIN see CNH125
DICYANOACETYLENE see DGS000
2,2'-DICYANO-2,2'-AZOPROPANE see ASL750
m-DICYANOBENZENE see PHX550
o-DICYANOBENZENE see PHY000
p-DICYANOBENZENE see BBP250
1,2-DICYANOBENZENE see PHY000
1,3-DICYANOBENZENE see PHX550
1,4-DICYANOBENZENE see BBP250
1,4-DICYANOBUTANE see AER250
1,4-DICYANO-2-BUTENE see DGS200
β,β-DICYANO-o-CHLOROSTYRENE see CEQ600
DICYANODIAMIDE see COP125
DICYANODIAZENE see DGS300
2,2'-DICYANODIETHYLAMINE see BIQ500
β,β'-DICYANODIETHYL ETHER see OQQ000
β,β'-DICYANODIETHYL SULFIDE see DGS600
2,3-DICYANO-1,4-DITHIA-ANTHRAQUINONE see DLK200
s-DICYANOETHANE see SNE000
1,2-DICYANOETHANE see SNE000
DI-(2-CYANOETHYL)AMINE see BIQ500
DI(2-CYANOETHYL)SULFIDE see DGS600
DICYANOFURAZAN see DGS700
DICYANOFURAZAN-N-OXIDE see DGS800
DICYANOFUROXAN see DGS800
DICYANOGEN see COO000
DICYANOGEN-N,N-DIOXIDE see DGT000
DICYANOMETHANE see MAO250
1,5-DICYANOPENTANE see HBD000
1,3-DICYANOPROPANE see TLR500
1,3-DICYANOTETRACHLOROBENZENE see TBQ750
cis-DICYCLOBUTYLAMINEDICHLOROPLATINUM(II) see DGT200
DICYCLOHEXANO-18-CROWN-6 see DGV100
DICYCLOHEXANO-24-CROWN-8 see DGT300
3,9,DI-(3-CYCLOHEXENYL)-2,4,8,10-TETRAOXASPIRO(5,5)UNDECANE
    see DGT400
DICYCLOHEXYL ADIPATE see DGT500
N,N-DICYCLOHEXYLAMINE see DGT600
DICYCLOHEXYLAMINE (DOT) see DGT600
DICYCLOHEXYLAMINE NITRITE see DGU200
DICYCLOHEXYLAMINE PENTANOATE see DGU400
1-(2-(DICYCLOHEXYLAMINO)ETHYL)-1-METHYL-PIPERIDINIUM BROMIDE
    see MJC775
1-(2-(DICYCLOHEXYLAMINO)ETHYL)-1-METHYL-PIPERIDINIUM CHLO-
    RIDE see CAL075
DICYCLOHEXYLAMINONITRITE see DGU200
cis-DICYCLOHEXYLAMMINEDICHLOROPLATINUM(II) see DGU709
DICYCLOHEXYLAMMONIUM NITRITE see DGU200
N,N-DICYCLOHEXYL-2-BENZOTHIAZOLESULFENAMIDE see DGU800
DICYCLOHEXYLCARBONYL PEROXIDE see DGV000
DICYCLOHEXYL-18-CROWN-6 see DGV100
2-(2,2-DICYCLOHEXYLETHYL)PIPERIDINE MALEATE see PCH800
DICYCLOHEXYLFLUOROPHOSPHATE see DGV200
DICYCLOHEXYL FLUOROPHOSPHONATE see DGV200
DICYCLOHEXYL KETONE see DGV600
DICYCLOHEXYL THIOUREA see DGV800
N,N'-DICYCLOHEXYLTHIOUREA see DGV800
DICYCLOHEXYLTIN OXIDE see DGV900
DICYCLOMINE HYDROCHLORIDE see ARR750
DICYCLOPENTADIENE see DGW000
DICYCLOPENTADIENE DIEPOXIDE see BGA250
DICYCLOPENTADIENE DIOXIDE see BGA250

DICYCLOPENTADIENYLCOBALT see BIR529
DICYCLOPENTADIENYLDICHLOROTITANIUM see DGW200
DICYCLOPENTADIENYLHAFNIUM DICHLORIDE see HAE500
DI-2,4-CYCLOPENTADIEN-1-YL IRON see FBC000
DICYCLOPENTADIENYL IRON (OSHA, ACGIH) see FBC000
DI-pi-CYCLOPENTADIENYLNICKEL see NDA500
DI-pi-CYCLOPENTADIENYLTITANIUM see TGH500
DICYCLOPENTADIENYLTITANIUMDICHLORIDE see DGW200
DICYCLOPENTA(c,lmn)PHENANTHREN-1(9H)-ONE, 2,3-DIHYDRO-
    see DGW300
DICYCLOPENTENYL ACRYLATE see DGW400
DICYCLOPENTENYLOXYETHYL METHACRYLATE see DGW450
α,α-DICYCLOPENTYL-ACETIC ACID-DIETHYLAMINO-ETHYLESTER
    BROMOCTYLATE see PAR600
DICYCLOPENTYLACETIC ACID-β-DIETHYLAMINOETHYL ESTER
    ETHOBROMIDE see DGW600
(2-(DICYCLOPENTYLACETOXY)ETHYL)TRIETHYLAMMONIUM BROMIDE
    see DGW600
N-(2-((DICYCLOPENTYLACETYL)OXY)ETHYL)-N,N-DIETHYL-1-OCTANAMIN-
    IUM BROMIDE (9CI) see PAR600
2-((DICYCLOPENTYLACETYL)OXY)-N,N,N-TRIETHYL-ETHANAMINIUM BRO-
    MIDE see DGW600
cis-DICYCLOPENTYLAMMINEDICHLOROPLATINUM(II) see BIS250
α,α-DICYCLOPENTYLESSIGSAURE-DIAETHYLAMINO-AETHYLESTER-
    BROMOCTYLAT (GERMAN) see PAR600
DICYCLOPROPYLDIAZOMETHANE see DGW875
3,3-DICYCLOPROPYL-2-(ETHOXYCARBONYL)ACRYLONITRILE see DGX000
DICYKLOHEXYLAMIN (CZECH) see DGT600
DICYKLOHEXYLAMINKAPRONAT (CZECH) see DGU400
DICYKLOHEXYLAMIN NITRIT (CZECH) see DGU200
N,N-DICYKLOHEXYLBENZTHIAZOLSULFENAMID (CZECH) see DGU800
DICYKLOPENTADIEN (CZECH) see DGW000
DICYNENE see DIS600
DICYNIT (CZECH) see DGU200
DICYNONE see DIS600
DICYSTEINE see CQK325
DID 47 see OMY850
DID 95 see CJM750
DIDAKENE see PCF275
DIDANDIN see DVV600
DIDAN-TDC-250 see DKQ000
DIDECANOYLTRIETHYLENE GLYCOL ESTER (mixed isomers) see DAH450
DIDECYL DIMETHYL AMMONIUM CHLORIDE see DGX200
DIDECYL PHTHALATE see DGX600
DI-N-DECYL PHTHALATE see DGX600
7,8-DIDEHYDROCHOLESTEROL see DAK600
7,8-DIDEHYDRO-3-(2-(DIETHYLAMINO)ETHOXY)-4,5-α-EPOXY-17-
    METHYLMORPHINAN-6-α-OL see DIE300
9,10-DIDEHYDRO-N,N-DIETHYL-2-BROMO-6-METHYLERGOLINE-8β-CAR-
    BOXAMIDE see BNM250
9,10-DIDEHYDRO-N,N-DIETHYL-6-METHYL-ERGOLINE-8β-CARBOXAMIDE
    see DJO000
9,10-DIDEHYDRO-N,N-DIETHYL-6-METHYL-ERGOLINE-8β-CARBOXAMIDE-
    d- TARTRATE with METHANOL (1:2) see LJG000
12,13-DIDEHYDRO-13,14-DIHYDRO-α-ERYTHROIDINE see EDG500
8,9-DIDEHYDRO-6,8-DIMETHYLERGOLINE see AEY375
(5-α,6-α)-7,8-DIDEHYDRO-4,5-EPOXY-3-ETHOXY-17-METHYLMORPHINAN-6-
    OL see ENK000
7,8-DIDEHYDRO-4,5-α-EPOXY-3-ETHOXY-17-METHYLMORPHINAN-6-α-OL
    HYDROCHLORIDE DIHYDRATE see ENK500
7,8-DIDEHYDRO-4,5-α-EPOXY-14-HYDROXY-3-METHOXY-17-
    METHYLMORPHINAN-6-ONE see HJX500
7,8-DIDEHYDRO-4,5-α-EPOXY-14-HYDROXY-3-METHOXY-17-
    METHYLMORPHINAN-5-α-6-ONE see HJX500
7,8-DIDEHYDRO-4,5-α-EPOXY-14-HYDROXY-3-METHOXY-17-
    METHYLMORPHINAN-6-ONE-N-OXIDE see DGY000
7,8-DIDEHYDRO-4,5-α-EPOXY-3-METHOXY-17-METHYL-MORPHINAN-6-α-OL
    PHOSPHATE SESQUIHYDRATE (3:3:2) see CNG675
7,8-DIDEHYDRO-4,5-α-EPOXY-17-METHYLMORPHINAN-3,6-α-DIOL
    see MRP000
7,8-DIDEHYDRO-4,5-α-EPOXY-17-METHYLMORPHINAN-3,6-α-DIOL HYDRO-
    CHLORIDE see MRO750
7,8-DIDEHYDRO-4,5-α-EPOXY-17-METHYLMORPHINE HYDROCHLORIDE
    see MRO750
7,8-DIDEHYDRO-4,5-α-EPOXY-17-METHYL-3-(2-
    MORPHOLINOETHOXY)MORPHINAN-6-α-OL see TCY750
7,8-DIDEHYDRO-4,5-α-EPOXY-17-METHYL-3-(2-
    PIPERIDINOETHOXY)MORPHINAN-6-α-OL see PIT600
9,10-DIDEHYDRO-N-ETHYL-1,6-DIMETHYLERGOLINE-8β-CARBOXAMIDE
    see MLD500
16,17-DIDEHYDRO-21-ETHYL-4-METHYL-7,20-CYCLOATIDANE-11β,15-β-
    DIOL see DAP875
9,10-DIDEHYDRO-N-ETHYL-6-METHYLERGOLINE-8β-CARBOXAMIDE, N-
    ETHYLLYSERGAMIDE see LJI000

3,8-DIDEHYDRO-HELIOTRIDINE see DAL060
9,10-DIDEHYDRO-N-(α-(HYDROXYMETHYL)ETHYL)-6-METHYLERGOLINE-8-β-CARBOXAMIDE see LJL000
9,10-DIDEHYDRO-N-(1-HYDROXYMETHYL)PROPYL)-1,6-DIMETHYLERGOL-INE-8-β-CARBOXAMIDE see MLD250
9,10-DIDEHYDRO-N-(α-(HYDROXYMETHYL)PROPYL)-6-METHYL-ERGOLINE-8-β-CARBOXAMIDE see PAM000
13,19-DIDEHYDRO-12-HYDROXY-SENECIONAN-11,16-DIONE see SBX500
13,19-DIDEHYDRO-12-HYDROXY-SENECIONAN-11,16-DIONE HYDROCHLO-RIDE see SBX525
5,6-DIDEHYDROISOANDROSTERONE see AOO450
13,14-DIDEHYDROMATRIDIN-15-ONE HYDROBROMIDE see SKS800
13,14-DIDEHYDRO-MATRIDIN-15-ONE MONOHYDROBROMIDE see SKS800
9,10-DIDEHYDRO-6-METHYL ERGOLINE-8-β-CARBOXYLIC ACID MORPHOL-IDE, TARTARIC ACID SALT see LJK000
N'-((8-α)-9,10-DIDEHYDRO-6-METHYLERGOLIN-8-YL)-N,N-DIETHYL-UREA (Z)-2-BETENEDIOATE see LJE500
3-(9,10-DIDEHYDRO-6-METHYLERGOLIN-8-YL)-1,1-DIETHYLUREA HYDRO-GEN MALEATE see LJE500
3-(9,10-DIDEHYDRO-6-METHYLERGOLIN-8-α-YL)-1,1-DIETHYLUREA MALE-ATE (1:1) see LJE500
16,17-DIDEHYDRO-19-METHYLOXAYOHIMBAN-16-CARBOXYLIC ACID METHYL ESTER see AFG750
9,10-DIDEHYDRO-N,N,6-TRIMETHYLERGOLINE-8β-CARBOXAMIDE see LJH000
7,8-DIDEHYDRORETINOIC ACID see DHA200
3,8-DIDEHYDRORETRONECINE see DAL400
4,4'-DIDEMETHYL-4,4'-DI-2-PROPENYLTOXIFERINE I DICHLORIDE see DBK400
DIDEMNIN B see DHA300
2',3'-DIDEOXYADENOSINE see DHA325
1,4-DIDEOXY-1,4-BIS((2-HYDROXYETHYL)AMINO)ERYTHRITOL 1,4-DIMETHANESULFONATE (ESTER) see LJD500
1,6-DIDEOXY-1,6-DI-(2-CHLOROETHYLAMINO)-d-MANNITOLDIHYDROCHLORIDE see MAW750
1,4-DIDEOXY-1,4-DIHYDRO-1,4-DIOXORIFAMYCIN see RKU000
DIDEOXYKANAMYCIN B see DCQ800
3',4'-DIDEOXYKANAMYCIN B see DCQ800
3',4'-DIDEOXYKANAMYCIN B SULFATE see DHA400
7-((4,6-DIDEOXY-3-(METHYL-4-(METHYLAMINO)-β-ALTROPYRANOSYL)OXY)-10,11-DIHYDRO-9,11-DIHYDROXY-8-METHYL-2-(1-PROPENYL)-1-PYRROLO(2,1-C)(1,4)BENZODIAZEPIN-5-ONE see SCF500
2,4-DIDEUTERIOESTRADIOL see DHA425
DIDEUTERIUM OXIDE see HAK000
DIDEUTERODIAZOMETHANE see DHA450
DI-2,4-DICHLOROBENZOYL PEROXIDE (DOT) see BIX750
4,4'-DI(DIETHYLAMINO)-4',6'-DISULPHOTRIPHENYLMETHANOL ANHY-DRIDE, SODIUM SALT see ADE500
DIDIGAM see DAD200
DIDIMAC see DAD200
DI-N,N'-DIMETHYLAMIDE ETHYLENEIMIDO PHOSPHATE see ASK500
3,6-DI(DIMETHYLAMINO)ACRIDINE see BJF000
1,2-DI-(DIMETHYLAMINO)ETHANE (DOT) see TDQ750
N,N-DI(1,4-DIMETHYLPENTYL)-p-PHENYLDIAMINE see BJL000
1,5-DI(2,4-DIMETHYLPHENYL-3-METHYL-1,3,5-TRIAZAPENTA-1,4-DIENE see MJL250
DIDOC see AAI250
DIDOCOL see DAL000
DIDODECANOYLOXYDIOCTYLSTANNANE see DVJ800
DIDODECYL-3,3'-THIODIPROPIONATE see TFD500
DIDRATE see DKW800
DIDROCOLO see DAL000
DIDRONEL R see DXD400
DIEFFENBACHIA MACUALTA see DHB309
DIEFFENBACHIA SEQUINE see DHB309
DIEFFENBACHIA (VARIOUS SPECIES) see DHB309
DIELDREX see DHB400
DIELDRIN see DHB400
DIELDRINE (FRENCH) see DHB400
DIELDRITE see DHB400
DIELTAMID see DKC800
DIEMAL see BAG000
DIENESTROL see DAL600
(E,E)-DIENESTROL see DHB550
DIENESTROL DIACETATE see DHB500
α-DIENESTROLPHENOL, 4,4'-(DIETHYLIDENEETHYLENE)DI-, trans-, (E,E)- see DHB550
DIENOCHLOR see DAE425
DIENOESTROL see DAL600
β-DIENOESTROL see DAL600
DIENOESTROL DIACETATE see DHB500
DIENOL see DAL600
DIENOL S see SMR000
DIENPAX see DCK759

DIEPIN see CGA000
DIEPOXYBUTANE see BGA750
l-DIEPOXYBUTANE see BOP750
2,4-DIEPOXYBUTANE see BGA750
dl-DIEPOXYBUTANE see DHB600
1,2:3,4-DIEPOXYBUTANE see BGA750
(2S,3S)-DIEPOXYBUTANE see BOP750
meso-DIEPOXYBUTANE see DHB800
(R*,S*)-DIEPOXYBUTANE see DHB800
l-1,2:3,4-DIEPOXYBUTANE see BOP750
(±)-1,2:3,4-DIEPOXYBUTANE see DHB600
dl-1,2:3,4-DIEPOXYBUTANE see DHB600
(2S,3S)-1,2:3,4-DIEPOXYBUTANE see BOP750
meso-1,2,3,4-DIEPOXYBUTANE see DHB800
trans-1,2,3,4-DIEPOXYCYCLOHEXANE see DHB875
1,2,9,10-DIEPOXYDECANE see DHC000
DIEPOXYDIHYDRO-7-METHYL-3-METHYLENE-1,6-OCTADIENE see DHC200
DIEPOXYDIHYDROMYRCENE see DHC200
2,3585,6-DIEPOXY-7,8-DIOXABICYCLO[2.2.2]OCTANE see DHC309
1,2:5,6-DIEPOXYDULCITOL see DCI600
2,5-DI(1,2-EPOXYETHYL)TETRAHYDRO-2H-PYRAN see DHC400
1,2,6,7-DIEPOXYHEPTANE see DHC600
1,2:5,6-DIEPOXYHEXAHYDROINDAN see BFY750
1,2:5,6-DIEPOXYHEXAHYDRO-4,7-METHANOINDAN see BGA250
1,2:5,6-DIEPOXY-3a,4,5,6,7,7a-HEXAHYDRO-4,7-METHANOINDAN see BGA250
1,2585,6-DIEPOXYHEXANE see DHC800
1,2,8,9-DIEPOXYLIMONENE see LFV000
1,2:8,9-DIEPOXYMENTHANE see LFV000
1,2:8,9-DIEPOXY-p-MENTHANE see LFV000
1,2,3,4-DIEPOXY-2-METHYLBUTANE see DHD200
DI(3,4-EPOXY-6-METHYLCYCLOHEXYLMETHYL)ADIPATE see AEN750
1,2588,9-DIEPOXYNONANE see DHD400
9,105812,13-DIEPOXYOCTADECANOIC ACID see DHD600
1,2,7,8-DIEPOXYOCTANE see DHD800
1,2,4,5-DIEPOXYPENTANE see DHE000
DIEPOXYPIPERAZINE see DHE100
DI(2,3-EPOXYPROPYL) ETHER see DKM200
9,10:12,13-DIEPOXYSTEARIC ACID see DHD600
1,2,15,16-DIEPOXY-4,7,10,13-TETRAOXAHEXADECANE see TJQ333
7-β,8-β:12,13-DIEPOXY-TRICHOTHEC-9-EN-6-β-OL see COB000
(4-β(Z),7-β,8-β)-7,8:12,13-DIEPOXY-TRICHOTHEC-9-EN-4-OL 2-BUTENOATE see COB000
DIESEL EXHAUST see DHE485
DIESEL EXHAUST EXTRACT see DHE500
DIESEL EXHAUST PARTICLES see DHE700
DIESEL FUEL (DOT) see FOP000
DIESEL FUEL MARINE see DHE800
S,S-DIESTER with DITHIO-p-UREIDOBENZENEARSONOUS ACID o-MERCAPTOBENZOIC ACID see TFD750
DIESTER with o-MERCAPTOBENZOIC ACID DITHIO-p-URE-IDOBENZENEARSONOUS ACID see TFD750
DI-ESTRYL see DKA600
DIETADIONE (ITALIAN) see DJT400
DIETAMINE see AOB500
DIETELMIN see HEP000
DIETHADION see DJT400
DIETHADIONE see DJT400
DIETHAMPHENAZOL MONOHYDRATE see BKB250
DIETHANOLAMINE (CZECH) see DHF000
DIETHANOLAMINE see DHF000
DIETHANOLAMINE-3,5-DIIODO-4-PYRIDONE-N-ACETATE see DNG400
DIETHANOLAMINOBENZENE see BKD500
DIETHANOLAMMONIUM MALEIC HYDRAZIDE see DHF200
DIETHANOLANILINE see BKD500
N,N-DIETHANOLANILINE see BKD500
1,1-DIETHANOLHYDRAZINE see HHH000
DIETHANOLLAURAMIDE see BKE500
N,N-DIETHANOLLAURAMIDE see BKE500
N,N-DIETHANOLLAURIC ACID AMIDE see BKE500
DIETHANOLNITRAMINE DINITRATE see NFW000
DIETHANOL-N-NITRAMINE DINITRATE see NFW000
DIETHANOLNITROSOAMINE see NKM000
DIETHANOL-m-TOLUIDINE see DHF400
DIETHAZIN see DII200
DIETHAZINE see DII200
DIETHAZINE HYDROCHLORIDE see DHF600
DIETHIBUTIN HYDROCHLORIDE see DJP500
DIETHION see EEH600
o-DIETHOXYBENZENE see CCP900
1,2-DIETHOXYBENZENE see CCP900
1,2-DIETHOXYCARBONYLDIAZENE see DIT300
1,2-DI(ETHOXYCARBONYL)ETHYL-O,O-DIMETHYL PHOSPHORODITHIO-ATE see MAK700

S-(1,2-DIETHOXYCARBONYL)ETHYL O,O-DIMETHYL PHOSPHOROTHIO-ATE see OPK250
S-(1,2-DI(ETHOXYCARBONYL)ETHYL DIMETHYL PHOS-PHOROTHIOLOTHIONATE see MAK700
DIETHOXYCHLOROSILANE see DHF800
DIETHOXYDIMETHYLSILANE see DHG000
1,1-DIETHOXY-ETHAAN (DUTCH) see AAG000
1,1-DIETHOXYETHANE see AAG000
1,2-DIETHOXYETHANE see EJE500
DIETHOXY ETHYL ADIPATE see BJO225
DIETHOXY-3-KYANPROPYL-METHYLSILAN see COR500
DIETHOXYMETHANE (DOT) see EFT500
α,α-DI(p-ETHOXYPHENYL)-β-BROMO-β-PHENYLETHYLENE see BMX000
α-(((DIETHOXYPHOSPHINOTHIOYL)OXY)IMINO)BENZENEACETONITRILE see BAT750
((DIETHOXYPHOSPHINOTHIOYL)THIO)ACETIC ACID, ETHYL ESTER see DIX000
(2-((DIETHOXYPHOSPHINOTHIOYL)THIO)ETHYL)CARBAMIC ACID, ETHYL ESTER see EMC000
(DIETHOXYPHOSPHINYL)DITHIOIMIDOCARBONIC ACID CYCLIC ETHYL-ENE ESTER see PGW750
(DIETHOXYPHOSPHINYLIMINO)-1,3-DITHIETANE see DHH200
DIETHOXYPHOSPHINYLIMINO-2-DITHIETANNE-1,3 (FRENCH) see DHH200
2-(DIETHOXYPHOSPHINYLIMINO)-1,3-DITHIOLANE see DXN600, PGW750
2-(DIETHOXYPHOSPHINYLIMINO)-4-METHYL-1,3-DITHIOLANE see DHH400
(DIETHOXY-PHOSPHINYL)MERCURY ACETATE see AAS250
(α-(DIETHOXYPHOSPHINYL)-p-METHOXYBENZYL)BIS(2-CHLOROPROPYL)ANTIMONITE see DHH600
2-((DIETHOXYPHOSPHINYL)OXY)-1H-BENZ(de)ISOQUINOLINE-1,3(2H)-DIONE see HMV000
2-DIETHOXY-PHOSPHINYLTHIOETHYL-TRIMETHYLAMMONIUM IODIDE see TLF500
N-(2-(DIETHOXYPHOSPHINYLTHIO)ETHYL)TRIMETHYLAMMONIUM IO-DIDE see TLF500
DIETHOXYPHOSPHORUS OXYCHLORIDE see DIY000
DIETHOXYPHOSPHORYL CYANIDE see DJW800
DIETHOXYPHOSPHORYL-THIOCHOLINE IODIDE see TLF500
2,2-DIETHOXYPROPANE see ABD250
3,3-DIETHOXYPROPENE see DHH800
DIETHOXYTETRAETHYLENE GLYCOL see PBO250
DI-ETHOXYTHIOKARBONYL-TRISULFID see DKE400
DIETHOXY THIOPHOSPHORIC ACID ESTER of 2-ETHYLMERCAPTOETHANOL see DAO600
DIETHOXY THIOPHOSPHORIC ACID ESTER OF 7-HYDROXY-4-METHYL COUMARIN see PKT000
(DIETHOXY-THIOPHOSPHORYLOXYIMINO)-PHENYL ACETONITRILE see BAT750
DIETHQUINALPHION see DJY200
DIETHQUINALPHIONE see DJY200
7,12-DIETHYENYL-3,8,13,17-TETRAMETHYL-21H,23H-PORPHINE-2,18-DIPROPANOIC ACID DISODIUM SALT see DXF700
DIETHYL see BOR500
DIETHYL ACETAL see AAG000
DIETHYL ACETALDEHYDE see DHI000
N,N-DIETHYLACETAMIDE see DHI200
DIETHYLACETIC ACID see DHI400
DIETHYLACETOACETAMIDE see DHI600
N,N-DIETHYLACETOACETAMIDE see DHI600
1-DIETHYLACETYLAZIRIDINE see DHI800
DIETHYL ACETYLENE DICARBOXYLATE see DHI850
DIETHYLACETYLETHYLENEIMINE see DHI800
O,O-DIETHYL-S-N-(A-CYANOISOPROPYL)CARBOMOYLMETHYL PHOS-PHOROTHIOATE see PHK250
DIETHYL ADIPATE see AEP750
l-N,N-DIETHYLALANINE-6-CHLORO-o-TOLYL ESTER HYDROCHLORIDE see DHI875
l-N,N-DIETHYLALANINE-2,6-DIETHYLPHENYL ESTER HYDROCHLORIDE see FAC157
l-N,N-DIETHYLALANINE-2,6-DIMETHOXYPHENYL ESTER HYDROCHLO-RIDE see FAC163
l-N,N-DIETHYLALANINE MESITYL ESTER HYDROCHLORIDE see FAC179
l-N,N-DIETHYLALANINE-2,6-XYLYL ESTER HYDROCHLORIDE see FAC100
DIETHYLALUMINUM BROMIDE see DHI880
DIETHYLALUMINUM CHLORIDE see DHI885
DIETHYLAMIDE de VANILLIQUE see DKE200
DIETHYLAMINE see DHJ200
N,N-DIETHYLAMINE see DHJ200
DIETHYLAMINE ACETARSONE see ACN250
DIETHYLAMINE-3-ACETYLAMINO-4-HYDROXYPHENYLARSONATE see ACN250
DIETHYLAMINE, 2-CHLORO-N-(9-FLUORENYL)-, HYDROCHLORIDE see FEE100
DIETHYLAMINE, 2,2'-DICHLORO-N-METHYL-, OXIDE see CFA500
2-(DIETHYLAMINO)ACETANILIDE see DHJ400

(DIETHYLAMINO)ACETONITRILE see DHJ600
N,N-DIETHYLAMINOACETONITRILE see DHJ600
2-(DIETHYLAMINO)-o-ACETOPHENETIDIDE, HYDROCHLORIDE see DHJ800
2-(DIETHYLAMINO)-p-ACETOPHENETIDIDE, HYDROCHLORIDE see DHK000
2-(DIETHYLAMINO)-o-ACETOTOLUIDIDE HYDROCHLORIDE see DHK200
DIETHYLAMINOACETO-2,6-XYLIDIDE see DHK400
α-DIETHYLAMINOACETO-2,6-XYLIDIDE see DHK400
α-DIETHYLAMINO-2,6-ACETOXYLIDIDE see DHK400
2-(DIETHYLAMINO)-2',6'-ACETOXYLIDIDE see DHK400
2-DIETHYLAMINO-2',6'-ACETOXYLIDIDE HYDROCHLORIDE see DHK600
2-DIETHYLAMINO-3',5'-ACETOXYLIDIDE HYDROCHLORIDE see DHK800
2-(DIETHYLAMINO)-2',6'-ACETOXYLIDIDE MONOHYDROCHLORIDE see DHK600
α-DIETHYLAMINO-2,5-ACETOXYLIDINE HYDROCHLORIDE see DHK600
DIETHYLAMINOACET-2,6-XYLIDIDE see DHK400
2-(((DIETHYLAMINO)ACETYL)AMINO)-3-METHYL-BENZOIC ACID METHYL ESTER, MONOHYDROCHLORIDE see BMA125
N-((DIETHYLAMINO)ACETYL)ANTHRANILIC ACID, ETHYL ESTER, HY-DROCHLORIDE see DHL200
N-(2-DIETHYLAMINO)ACETYLANTHRANILIC ACID, ETHYL ESTER HYDRO-CHLORIDE see DHL200
N-((DIETHYLAMINO)ACETYL)ANTHRANILIC ACID, METHYL ESTER, HY-DROCHLORIDE see DHL400
N-(2-DIETHYLAMINO)ACETYLANTHRANILIC ACID, METHYL ESTER, HY-DROCHLORIDE see DHL400
β-4-(1-DIETHYLAMINOACETYL-2-PIPERIDYL)-2,2-DIPHENYL-1,3-DIOXOL-ANE HYDROCHLORIDE see DHL600
DIETHYLAMINOACETYL-2,4,6-TRIMETHYLANILINE HYDROCHLORIDE see DHL800
(DIETHYLAMINO)ACYLANILIDE see DHJ400
4-(DIETHYLAMINO)AZOBENZENE see OHI875
p-(DIETHYLAMINO)AZOBENZENE see OHI875
N,N-DIETHYL-4-AMINOAZOBENZENE see OHI875
N,N-DIETHYLAMINOBENZENE see DIS700
3-(DIETHYLAMINO)BENZENESULFONIC ACID, SODIUM SALT see DHM000
1-DIETHYLAMINO-1-BUTEN-3-YNE see DHM200
4-DIETHYLAMINO-2-BUTYNYLPHENYL(CYCLOHEXYL)GLYCOLATE HY-DROCHLORIDE see OPK000
2-(DIETHYLAMINO)BUTYRIC ACID-2,6-XYLYL ESTER HYDROCHLORIDE see DHM309
3-(DIETHYLAMINO)-2',6'-BUTYROXYLIDIDE HYDROCHLORIDE see DHM400
DIETHYLAMINOCARBETHOXYBICYCLOHEXYL HYDROCHLORIDE see ARR750
2-(DIETHYLAMINO)CHLOROETHANE see CGV500
α-DIETHYLAMINO-2,6-DIMETHYLACETANILIDE see DHK400
omega-DIETHYLAMINO-2,6-DIMETHYLACETANILIDE see DHK400
2-(DIETHYLAMINO)-3',5'-DIMETHYLACETANILIDE HYDROCHLORIDE see DHK800
omega-DIETHYLAMINO-2,6-DIMETHYLACETANILIDE HYDROCHLORIDE see DHK600
3-(DIETHYLAMINO)-7-((p-(DIMETHYLAMINO)PHENYL)AZO)-5-PHENYLPHENAZINIUM CHLORIDE see DHM500
2-(DIETHYLAMINO)-N-(1,3-DIMETHYL-4-(o-FLUOROBENZOYL)-5-PYRAZOLYL)ACETAMI DE HYDROCHLORIDE see AAI100
2-(DIETHYLAMINO)-N-(2,6-DIMETHYLPHENYL)ACETAMIDE MONOHYDROCHLORIDE see DHK600
3-(DIETHYLAMINO)-2,2-DIMETHYL-1-PROPANOL-p-AMINOBENZOATE see DNY000
3-(DIETHYLAMINO)-2,2-DIMETHYLPROPYL TROPATE PHOSPHATE see AOD250
2-(DIETHYLAMINO)-N,N-DIPHENYLACETAMIDE HYDROCHLORIDE see DHN800
2-(DIETHYLAMINO)-N-(DIPHENYLMETHYL)ACETAMIDE HYDROCHLO-RIDE see DHO000
3-DIETHYLAMINO-1,1-DI(2'-THIENYL)BUT-1-ENE HYDROCHLORIDE see DJP500
3-DIETHYLAMINO-1,2-EPOXYPROPANE see GGW800
(DIETHYLAMINO)ETHANE see TJO000
DIETHYLAMINOETHANETHIOL see DIY600
2-(DIETHYLAMINO)ETHANETHIOL see DIY600
DIETHYLAMINOETHANETHIOL HYDROCHLORIDE see DHO400
DIETHYLAMINOETHANOL see DHO500
2-DIETHYLAMINOETHANOL see DHO500
2-(DIETHYLAMINO)ETHANOL see DHO500
N-DIETHYLAMINOETHANOL see DHO500
β-DIETHYLAMINOETHANOL see DHO500
2-N-DIETHYLAMINOETHANOL see DHO500
DIETHYLAMINOETHANOL (DOT) see DHO500
DIETHYLAMINOETHANOL-4-AMINOBENZOATE HYDROCHLORIDE see AIT250
DIETHYLAMINOETHANOL-p-AMINOSALICYLATE see DHO600
DIETHYLAMINOETHANOL ESTER OF DIPHENYLPROPYLACETIC ACID HYDROCHLORIDE see PBM500
2-DIETHYLAMINOETHANOL HYDROCHLORIDE see DHO700

7-(2-(N,N-DIETHYLAMINO)ETHYL)THEOPHYLLINE HYDROCHLORIDE
   see DIH600
2-(2-DIETHYLAMINO)ETHYL)THIOACETANILIDE HYDROCHLORIDE
   see DIH800
2-(2-(DIETHYLAMINO)ETHYL)THIO-o-ACETOTOLUIDIDE HYDROCHLORIDE
   see DII000
N-(DIETHYLAMINOETHYL)THIODIPHENYLAMINE see DII200
N-(2-(DIETHYLAMINO)ETHYL)-2,4,6-TRIMETHYLBENZAMIDE HYDROCHLO-
   RIDE see DII400
β-DIETHYLAMINOETHYL XANTHENE-9-CARBOXYLATE METHOBROMIDE
   see XCJ000
β-DIETHYLAMINOETHYL 9-XANTHENECARBOXYLATE METHOBROMIDE
   see XCJ000
β-DIETHYLAMINOETHYL XANTHENE-9-CARBOXYLATE METHOBROMIDE
   see DJM800
β-DIETHYLAMINOETHYL-9-XANTHENECARBOXYLATE METHOBROMIDE
   see DJM800
N-(2-(DIETHYLAMINO)ETHYL)-2,6-XYLIDINE DIHYDROCHLORIDE see DII600
2-(DIETHYLAMINO)-N-ETHYL-N-(1-(2,4-XYLYLOXY)-2-PROPYL)ACETAMIDE
   HYDROCHLORIDE see DII800
2-(DIETHYLAMINO)-N-ETHYL-N-(1-(3,5-XYLYLOXY)-2-PROPYL)ACET-
   AMIDE HYDROCHLORIDE see DIJ000
1-(2-(DIETHYLAMINO)ETHYL)-3-(2,6-XYLYL)UREA HYDROCHLORIDE
   see DIJ200
2-(DIETHYLAMINO)-N-(4-(2-FLUOROBENZOYL)-1,3-DIMETHYL-1H-PYRAZOL-
   5-YL)ACE TAMIDE HYDROCHLORIDE see AAI100
5-((Γ-DIETHYLAMINO-β-HYDROXYPROPYL)AMINO)-3-METHOXY-8-
   CHLOROACRIDINE DIHYDROCHLORIDE see ADI750
4-(DIETHYLAMINO)-2-ISOPROPYL-2-PHENYLVALERONITRILE see DIK000
2-(DIETHYLAMINO)-N-(1-MESITYLOXY-2-PROPYL)-N-METHYLACETAMIDE
   HYDROCHLORIDE see DIK200
2-(DIETHYLAMINO)-N-(1-(p-METHOXYPHENOXY)-2-PROPYL)-N-
   METHYLACETAMIDE HYDROCHLORIDE see DIK400
2-(DIETHYLAMINO)-N-METHYL-o-ACETOTOLUIDIDE HYDROCHLORIDE
   see DIK600
2-(DIETHYLAMINO)-N-METHYL-2′,6′-ACETOXYLIDIDE HYDROCHLORIDE
   see DIK800
7-(DIETHYLAMINO)-4-METHYL-2H-1-BENZOPYRAN-2-ONE see DIL400
p-(2-((DIETHYLAMINO)METHYL)BUTOXY)BENZOIC ACID, p-
   METHOXYPHENYL ESTER HYDROCHLORIDE see UAG000
8-((4-(DIETHYLAMINO)-1-METHYLBUTYL)AMINO)-6-METHOXYQUINOLINE
   see RHZ000
6-((4-(DIETHYLAMINO)-1-METHYLBUTYL)AMINO)-2-METHYL-4,5,8-
   TRIMETHOXYQUINOLINE-1,5-NAPHTHALENE DISULFONATE see DIL000
6((-4-(DIETHYLAMINO)-1-METHYLBUTYL)-5,8-DIMETHOXY-2,4-
   DIMETHYLQUINOLINE-1,5-NAPHTHALENE DISULFONATE see DIL200
7-DIETHYLAMINO-4-METHYLCOUMARIN see DIL400
7-DIETHYLAMINO-4-METHYLCOUMARIN, HYDROGEN SULFATE see DIL600
10-(2-DIETHYLAMINO-2-METHYLETHYL)PHENOTHIAZINE see DIR000
3-((DIETHYLAMINO)METHYL)-4-HYDROXYBENZOIC ACID ETHYL ESTER
   see EIA000
2-(DIETHYLAMINO)-N-METHYL-N-(2-MESITYLOXYETHYL)ACETAMIDE HY-
   DROCHLORIDE see DIM000
5-(DIETHYLAMINO)METHYL-3-(1-METHYL-5-NITROIMIDAZOL-2-
   YLMETHYLENE AMINO)-2-OXAZOLIDINONE HYDROCHLORIDE
   see DIM200
2-(DIETHYLAMINO)-N-(2-METHYL-1-NAPHTHYL)ACETAMIDE HYDROCHLO-
   RIDE see DIM400
4-DIETHYLAMINOMETHYL-2-(5-NITRO-2-THIENYL)THIAZOLE HYDRO-
   CHLORIDE see DIM600
2-(DIETHYLAMINO)-4-METHYL-1-PENTANOL-p-AMINOBENZOATE (ESTER)
   see LET000
2-(DIETHYLAMINO)-4-METHYL-1-PENTANOL, p-AMINOBENZOATE ESTER,
   METHANESULFONATE see LEU000
2-(DIETHYLAMINO)-N-METHYL-N-(2-PHENETHYL)ACETAMIDE HYDRO-
   CHLORIDE see DIM800
3-(DIETHYLAMINO)-N-METHYL-N-(1-PHENOXY-2-PROPYL)PROPIONAMIDE
   HYDROCHLORIDE see DIN000
2-(DIETHYLAMINO)-N-METHYL-N-(3-PHENYLPROPYL)ACETAMIDE HYDRO-
   CHLORIDE see DIN200
2-(DIETHYLAMINO)-N-(4-METHYL-2-PYRIDYL)ACETAMIDE DIHYDROCHLO-
   RIDE see DIN400
O-(2-(DIETHYLAMINO)-6-METHYL-4-PYRIMIDINYL)-O,O-DIETHYL PHOS-
   PHOROTHIOATE see DIN600
2-DIETHYLAMINO-6-METHYLPYRIMIDIN-4-YL
   DIETHYLPHOSPHOROTHIONATE see DIN600
2-DIETHYLAMINO-6-METHYLPYRIMIDIN-4-YL DIMETHYL PHOS-
   PHOROTHIONATE see DIN800
O-(2-(DIETHYLAMINO)-6-METHYL-4-PYRIMIDINYL)-O,O-DIMETHYL PHOS-
   PHOROTHIOATE see DIN800
O-(2-DIETHYLAMINO-6-METHYLPYRIMIDIN-4-YL)-O,O-DIMETHYL PHOS-
   PHOROTHIOATE see DIN800
7-DIETHYLAMINO-5-METHYL-s-TRIAZOLO(1,5-a)PYRIMIDINE see DIO200
5-(DIETHYLAMINO)-2-NITROSOPHENOL HYDROCHLORIDE see DIO300

4-(2-(DIETHYLAMINO)-2-OXOETHOXY)-3-METHOXYBENZENEACETIC
   ACID, PROPYL ESTER see PMM000
4-o-(2-(DIETHYLAMINO)-2-OXOETHYL)RIFAMYCIN see RKA000
DIETHYL-m-AMINO-PHENOLPHTHALEIN HYDROCHLORIDE see FAG070
2-(DIETHYLAMINO)-N-PHENYLACETAMIDE see DHJ400
4-((4-(DIETHYLAMINO)PHENYL)AZO)PYRIDINE-1-OXIDE see DIP000
2-(p-(DIETHYLAMINOPHENYL)-1,3,2-DITHIARSENOLANE see DIP100
2-(DIETHYLAMINO)-1-PHENYL-1-PROPANONE HYDROCHLORIDE see DIP600
10-DIETHYLAMINOPROPIONYL-3-TRIFLUOROMETHYL PHENOTHIAZINE
   HYDROCHLORIDE see FDE000
2-(DIETHYLAMINO)PROPIOPHENONE see DIP400
α-DIETHYLAMINOPROPIOPHENONE see DIP400
2-DIETHYLAMINOPROPIOPHENONE HYDROCHLORIDE see DIP600
2-(DIETHYLAMINO)-2′-PROPOXYACETANILIDE HYDROCHLORIDE
   see DIP800
3-DIETHYLAMINOPROPYLAMINE see DIQ100
N-(3-DIETHYLAMINOPROPYL)AMINE see DIY800
N,N-DIETHYLAMINOPROPYLAMINE see DIY800
3-(DIETHYLAMINO)PROPYLAMINE (DOT) see DIY800
2-DIETHYLAMINO-1-PROPYL-N-DIBENZOPARATHIAZINE see DIR000
10-(2-DIETHYLAMINOPROPYL)PHENOTHIAZINE see DIR000
3-DIETHYLAMINO-5H-PYRIDO(4,3-b)INDOLE see DIR800
DIETHYLAMINO STILBENE see DKA000
DIETHYLAMINOSULFUR TRIFLUORIDE see DIR875
2-(DIETHYLAMINO)-2′,4′,6′-TRICHLOROACETANILIDE HYDROCHLORIDE
   see DIS000
2-DIETHYLAMINO-2′,4′,6′-TRIMETHYLACETANILIDE HYDROCHLORIDE
   see DHL800
2-(DIETHYLAMINO)-2′,4′,6′-TRIMETHYLACETANILIDE
   MONOHYDROCHLORIDE see DHL800
DIETHYLAMINOTRIMETHYLENAMINE see DIY800
2-(DIETHYLAMINO)-N-(2,4,6-TRIMETHYLPHENYL)ACETAMIDE
   MONOHYDROCHLORIDE see DHL800
(6-(DIETHYLAMINO)-3H-XANTEN-3-YLIDENE)DIETHYLAMMONIUM CHLO-
   RIDE see DIS200
N-(6-(DIETHYLAMINO)-3H-XANTHEN-3-YLIDINE)-N-ETHYLETHANAMIN-
   IUM CHLORIDE see DIS200
2-DIETHYLAMMONIOETHYL NITRATE see DIS400
DIETHYLAMMONIUM CHLORIDE see DIS500
DIETHYLAMMONIUM CYCLOHEXADIEN-4-OL-1-ONE-4-SULFONATE
   see DIS600
DIETHYLAMMONIUM-2,5-DIHYDROXYBENZENE SULFONATE see DIS600
N,N-DIETHYLANILIN (CZECH) see DIS700
DIETHYLANILINE see DIS700
N,N-DIETHYLANILINE see DIS700
2-α-17-α-DIETHYL-A-NOR-5-α-ANDROSTANE-2-β,17-β-DIOL see EQN259
N,N-DIETHYL-p-ARSANILIC ACID see DIS775
DIETHYL ARSINE see DIS800
DIETHYL ARSINIC ACID see DIS850
DIETHYL AZOFORMATE see DIT300
DIETHYL AZOMALONATE see DIT350
DIETHYLBARBITONE see BAG000
DIETHYLBARBITURATE MONOSODIUM see BAG250
DIETHYL-BARBITURIC ACID see BAG000
5,5-DIETHYLBARBITURIC ACID see BAG000
5,5-DIETHYLBARBITURIC ACID SODIUM deriv. see BAG250
N,N-DIETHYLBENZAMIDE see BCM250
6,8-DIETHYLBENZ(a)ANTHRACENE see DIT400
7,12-DIETHYLBENZ(a)ANTHRACENE see DIT800
8,12-DIETHYLBENZ(a)ANTHRACENE see DIT600
9,10-DIETHYL-1,2-BENZANTHRACENE see DIT800
DIETHYL BENZENE see DIU000
m-DIETHYLBENZENE see DIU200
α,α-DIETHYLBENZENEACETIC ACID 2-(2-(DIETHYLAMINO)ETHOXY)
   ETHYL ESTER see DHQ200
N,N-DIETHYLBENZENESULFONAMIDE see DIU400
DIETHYL-4-(BENZOTHIAZOL-2-YL)BENZYLPHOSPHONATE see DIU500
DIETHYL BENZYLPHOSPHONATE see DIU600
O,O-DIETHYL-S-BENZYL THIOPHOSPHATE see DIU800
DIETHYLBERYLLIUM see DIV000
meso-α,α′-DIETHYLBIBENZYL-4,4′-DISULFONIC ACID DIPOTASSIUM SALT
   see SPA650
sym-DIETHYL BIS(DIMETHYLAMIDO)PYROPHOSPHATE see DIV200
unsym-DIETHYL BIS(DIMETHYLAMIDO)PYROPHOSPHATE see DJA300
DIETHYL BIS-DIMETHYLPYROPHOSPHORADIAMIDE (symmetrical)
   see DIV200
DIETHYL BIS-DIMETHYL PYROPHOSPHORDIAMIDE asym see DJA300
1,2-DIETHYL-1,3-BIS-(p-METHOXYPHENYL)-1-PROPENE see CNH525
DIETHYLBISMUTH CHLORIDE see DIV400
DIETHYLBIS(OCTANOYLOXY)STANNANE see DIV600
DIETHYLBIS(1-OXOOCTYL)OXY)STANNANE see DIV600
O,O-DIETHYL-O-(4-BROOM-2,5-DICHLOOR-FENYL)-MONOTHIOFOSFAAT
   (DUTCH) see EGV500
DI-2-(2-ETHYLBUTOXY)ETHYL ADIPATE see AEP250

O,O-DIETHYL-S-(β-DIETHYLAMINO)ETHYL PHOSPHOROTHIOLATE HYDROGEN OXALATE see AMX825

O,O-DIETHYL-S-(2-DIETHYLAMINOETHYL) THIOPHOSPHATE see DJA400

O,O-DIETHYL O-(2-DIETHYLAMINO-6-METHYL-4-PYRIMIDINYL)PHOSPHOROTHIOATE see DIN600

1-(2-(2-(2,6-DIETHYL-α-(2,6-DIETHYLPHENYL)BENZYLOXY)ETHOXY)ETHYL)-4-METHYLPIPERAZINE see DJA600

2,2'-DIETHYLDIHEXYLAMINE see DJA800

O,O-DIETHYL-(1,2-DIHYDRO-1,3-DIOXO-2H-ISOINDOL-2-YL)PHOSPHONOTHIOATE see DJX000

5,5-DIETHYLDIHYDRO-2H-1,3-OXAZINE-2,4(3H)-DIONE see DJT400

O,O-DIETHYL O-(2,3-DIHYDRO-3-OXO-2-PHENYL-6-PYRIDAZINYL) PHOSPHOROTHIOATE see POP000

DIETHYLDIIODOSTANNANE see DJB000

DIETHYL (DIMETHOXYPHOSPHINOTHIOYLTHIO) BUTANEDIOATE see MAK700

DIETHYL (DIMETHOXYPHOSPHINOTHIOYLTHIO)SUCCINATE see MAK700

α,α'-DIETHYL-4,4'-DIMETHOXYSTILBENE see DJB200

trans-α,α'-DIETHYL-4,4'-DIMETHOXYSTILBENE see DJB200

3',4'-DIETHYL-4-DIMETHYLAMINOAZOBENZENE see DJB400

8,8-DIETHYL-N,N-DIMETHYL-2-AZA-8-GERMASPIRO(4.5)DECANE-2-PROPANAMINE DIHYDROCHLORIDE see SLD800

DIETHYLDIMETHYLMETHANE see DTI000

p'-DIETHYL-p-DIMETHYL THIOPYROPHOSPHATE see DJB600

p-DIETHYL-p'-DIMETHYLTHIOPYROPHOSPHATE see DJB600

N,N'-DIETHYL-N,N'-DINITROSOETHYLENEDIAMINE see DJB800

N³,N³-DIETHYL-2,4-DINITRO-6-(TRIFLUOROMETHYL)-1,3-BENZENEDIAMINE see CNE500

3,3-DIETHYL-2,4-DIOXO-5-METHYLPIPERIDINE see DNW400

DIETHYLDIPHENYL DICHLOROETHANE see DJC000

DIETHYLDIPHENYLTHIURAM DISULFIDE see DJC200

N,N-DIETHYL-N,N-DIPHENYLTHIURAMDISULFIDE see DJC200

sym-DIETHYLDIPHENYLUREA see DJC400

1,3-DIETHYL-1,3-DIPHENYLUREA see DJC400

N,N'-DIETHYL-N,N'-DIPHENYLUREA see DJC400

DIETHYLDISULFID (CZECH) see DJC600

DIETHYLDISULFIDE see DJC600

N,N-DIETHYL-4,4-DI-2-THIENYL-3-BUTEN-2-AMINE HYDROCHLORIDE see DJP500

N,N-DIETHYL-3,3-DI-2-THIENYL-1-METHYLALLYLAMINE HYDROCHLORIDE see DJP500

DIETHYLDITHIO BIS(THIONOFORMATE) see BJU000

DIETHYLDITHIOCARBAMATE SODIUM see SGJ000

DIETHYLDITHIOCARBAMIC ACID see DJC800

DIETHYLDITHIOCARBAMIC ACID ANHYDROSULFIDE with DIMETHYLTHIOCARBAMIC ACID see DJC875

DIETHYLDITHIOCARBAMIC ACID-2-CHLOROALLYL ESTER see CDO250

DIETHYLDITHIOCARBAMIC ACID DIETHYLAMINE SALT see DJD000

DIETHYLDITHIOCARBAMIC ACID LEAD(II) SALT see DJD200

DIETHYLDITHIOCARBAMIC ACID SELENIUM(II) SALT see DJD400

DIETHYLDITHIOCARBAMIC ACID SODIUM see SGJ000

DIETHYLDITHIOCARBAMIC ACID, SODIUM SALT see SGJ000

DIETHYLDITHIOCARBAMIC ACID SODIUM SALT TRIHYDRATE see SGJ500

DIETHYLDITHIO CARBAMIC ACID TELLURIUM SALT see EPJ000

DIETHYLDITHIOCARBAMIC ACID ZINC SALT see BJC000

DIETHYLDITHIOCARBAMIC ANHYDRIDE of O,O-DIISOPROPYL THIONOPHOSPHORIC ACID see DKB600

DIETHYLDITHIOCARBAMIC ANHYDROSULFIDE see DKB600

DIETHYLDITHIOCARBAMIC SODIUM TRIHYDRATE see SGJ500

DIETHYLDITHIOCARBAMINIC ACID see DJC800

2-(N,N-DIETHYLDITHIOCARBAMYL)BENZOATHIAZOLE see BDF250

O,O-DIETHYLDITHIOFOSFORECNAN SODNY (CZECH) see PHG750

O,O-DIETHYLDITHIO 1,3-DITHIOLAN-2-YLIDENEPHOSPHORAMIDOTHIOATE see DXN600

DIETHYL-N-1,3-DITHIOLANYL-2-IMINO PHOSPHATE see DXN600

DIETHYLDITHIONE see DJC800

O,O-DIETHYL-DITHIOPHOSPHORIC ACID, p-CHLOROPHENYLTHIOMETHYL ESTER see TNP250

O,O-DIETHYLDITHIOPHOSPHORYLACETIC ACID-N-MONOISOPROPYLAMIDE see IOT000

3-DIETHYLDITHIOPHOSPHORYLMETHYL-6-CHLOROBENZOXAZOLONE-2 see BDJ250

DIETHYL DIXANTHOGEN see BJU000

DIETHYL EMME see EEV200

DIETHYLENDIAMINE see DPJ200

1,4-DIETHYLENEDIAMINE see PIJ000

N,N-DIETHYLENE DIAMINE (DOT) see PIJ000

DIETHYLENE DIOXIDE see DVQ000

1,4-DIETHYLENE DIOXIDE see DVQ000

DIETHYLENE ETHER see DVQ000

DIETHYLENE GLYCOL see DJD600

DIETHYLENE GLYCOL BISPHTHALATE see DJD700

DIETHYLENE GLYCOL-n-BUTYL ETHER see DJF200

DIETHYLENE GLYCOL BUTYL ETHER ACETATE see BQP500

DIETHYLENE GLYCOL DIACRYLATE see ADT250

DIETHYLENE GLYCOL DI(3-AMINOPROPYL) ETHER see DJD800

DIETHYLENE GLYCOL DIBENZOATE see DJE000

DIETHYLENEGLYCOL DIBUTYL ETHER see DDW200

DIETHYLENEGLYCOL DI-n-BUTYL ETHER see DDW200

DIETHYLENE GLYCOL, DIESTER with BUTYLPHTHALATE see DDY400

DIETHYLENE GLYCOL DIETHYL ETHER see DIW800

DIETHYLENE GLYCOL DIGLYCIDYL ETHER see DJE200

DIETHYLENE GLYCOL DIMETHYL ETHER see BKN750

DIETHYLENE GLYCOL DINITRATE see DJE400

DIETHYLENEGLYCOL DINITRATE, containing at least 25% phlegmatizer (DOT) see DJE400

DIETHYLENE GLYCOL DIVINYL ETHER see DJE600

DIETHYLENE GLYCOL ETHYL ETHER see CBR000

DIETHYLENE GLYCOL ETHYL METHYL ETHER see DJE800

DIETHYLENE GLYCOL ETHYLVINYL ETHER see DJF000

DIETHYLENE GLYCOL-n-HEXYL ETHER see HFN000

DIETHYLENE GLYCOL METHYL ETHER see DJG000

DIETHYLENE GLYCOL MONOBUTYL ETHER see DJF200

DI(ETHYLENE GLYCOL MONOBUTYL ETHER)PHTHALATE see DJF400

DIETHYLENE GLYCOL MONO-2-CYANOETHYL ETHER see DJF600

DIETHYLENE GLYCOL, MONOESTER with STEARIC ACID see HKJ000

DIETHYLENE GLYCOL MONOETHYL ETHER see CBR000

DIETHYLENE GLYCOL MONOETHYL ETHER ACETATE see CBQ750

DIETHYLENE GLYCOL MONOHEPTYL ETHER see HBP275

DIETHYLENE GLYCOL MONOHEXYL ETHER see HFN000

DIETHYLENE GLYCOL, MONO(HYDROGEN MALEATE) see MAL500

DIETHYLENE GLYCOL MONOISOBUTYL ETHER see DJF800

DIETHYLENE GLYCOL MONOMETHYL ETHER see DJG000

DIETHYLENE GLYCOL MONOMETHYL ETHER ACETATE see MIE750

DIETHYLENE GLYCOL MONOMETHYLPENTYL ETHER see DJG200

DIETHYLENEGLYCOL-MONO-2-METHYLPENTYL ETHER see DJG200

DIETHYLENE GLYCOL MONOPHENYL ETHER see PEQ750

DIETHYLENE GLYCOL MONOSTEARATE see HKJ000

DIETHYLENE GLYCOL MONOVINYL ETHER see DJG400

DIETHYLENE GLYCOL PHENYL ETHER see PEQ750

DIETHYLENE GLYCOL STEARATE see HKJ000

DIETHYLENEGLYCOL VINYL ETHER see DJG400

DIETHYLENEIMIDE OXIDE see MRP750

DIETHYLENE IMIDOXIDE see MRP750

DIETHYLENEIMINEAMIDOTHIOPHOSPHORIC ACID see DNU850

2,6-DIETHYLENEIMINO-4-CHLOROPYRIMIDINE see EQI600

DIETHYLENE OXIDE see TCR750

DI(ETHYLENE OXIDE) see DVQ000

DIETHYLENE OXIMIDE see MRP750

1,3,-DI(ETHYLENESULPHAMOYL)PROPANE see BJP899

DIETHYLENETRIAMINE see DJG600

DIETHYLENETRIAMINEPENTAACETIC ACID see DJG800

1,1,4,7,7-DIETHYLENETRIAMINEPENTAACETIC ACID see DJG800

DIETHYLENETRIAMINE PENTAACETIC ACID, CALCIUM TRISODIUM SALT see CAY500

(DIETHYLENETRINITRILO)PENTAACETIC ACID see DJG800

DIETHYLEN-GLYCOL MONOVINYL ESTER see DJG400

DIETHYLENGLYKOLDINITRAT (CZECH) see DJE400

DIETHYLENIMIDE OXIDE see MRP750

α,α'-DIETHYL-α,α'-EPOXYBIBENZYL-4,4'-DIOL see DKB100

DIETHYL-β,Γ-EPOXYPROPYLPHOSPHONATE see DJH200

DIETHYLESTER KYSELINY ACETYLAMINOMALONOVE (CZECH) see AAK750

DIETHYL ESTER of PYROCARBONIC ACID see DIZ100

DIETHYL ESTER SULFURIC ACID see DKB110

N,N-DIETHYLETHANAMINE see TJO000

N,N-DIETHYL-1,2-ETHANEDIAMINE see DJI400

DIETHYL ETHANEDIOATE see DJT200

(R*,S*)-4,4'-(1,2-DIETHYL-1,2-ETHANEDIYL)BIS-BENZENESULFONIC ACID DIPOTASSIUM SALT see SPA650

DIETHYL ETHANE PHOSPHONITE see DJH500

DIETHYLETHANOLAMINE see DHO500

N,N-DIETHYLETHANOLAMINE see DHO500

(E)-1,1'-(1,2-DIETHYL-1,2-ETHENE-DIYL)BIS(4-METHOXYBENZENE) see DJB200

4,4'-(1,2-DIETHYL-1,2-ETHENEDIYL)BIS-PHENOL see DKA600

trans-4,4'-(1,2-DIETHYL-1,2-ETHENEDIYL)BISPHENOL see DKA600

4,4'-(1,2-DIETHYL-1,2-ETHENEDIYL)BISPHENOL-(E)-BIS(DIHYDROGEN PHOSPHATE) see DKA200

trans-4,4'-(1,2-DIETHYL-1,2-ETHENEDIYL)BISPHENOL DIPROPIONATE see DKB000

DIETHYL ETHER (DOT) see EJU000

DIETHYLETHEROXODIPEROXOCHROMIUM(VI) see DJH800

DIETHYL-S-(2-ETHIOETHYL)THIOPHOSPHATE see DAP200

DIETHYLETHOXYALUMINUM see EER000

O,O-DIETHYL-S-(N-ETHOXYCARBONYL-N-METHYLCARBAMOYLMETHYL) PHOSPHORODITHIOATE see DJI000

N,N-DIETHYLLEUCINON-p-AMINOBENZOIC ACID METHANESULFONATE see LEU000
N,N-DIETHYLLYSERGAMIDE see DJO000
DIETHYL MAGNESIUM see DJO100
DIETHYL MALEATE see DJO200
DIETHYL MALONATE (FCC) see EMA500
DIETHYLMALONYLUREA see BAG000
DIETHYLMALONYLUREA SODIUM see BAG250
DIETHYL(2-MERCAPTOETHYL)AMINE see DIY600
DIETHYL MERCAPTOSUCCINATE-O,O-DIMETHYL DITHIOPHOSPHATE, S-ESTER see MAK700
DIETHYL MERCAPTOSUCCINATE-O,O-DIMETHYL PHOSPHORODITHIOATE see MAK700
DIETHYL MERCAPTOSUCCINATE-O,O-DIMETHYL THIOPHOSPHATE see MAK700
DIETHYL MERCAPTOSUCCINATE-S-ESTER with O,O-DIMETHYLPHOSPHORODITHIOATE see MAK700
DIETHYL MERCAPTOSUCCINIC ACID O,O-DIMETHYL PHOSPHORODITHIOATE see MAK700
DIETHYL MERCURY see DJO400
N,N-DIETHYLMETANILAN SODNY (CZECH) see DHM000
N,N-DIETHYL-2-(4-(6-METHOXY-2-PHENYL-1H-INDEN-3-YL)PHENOXY)-ETHANAMINE HYDROCHLORIDE see MFG260
N,N-DIETHYL-3-(4-METHOXYPHENYL)-1,2,4-OXADIAZOLE-5-ETHANAMINE see CPN750
N,N-DIETHYL-2-(4-(2-(4-METHOXYPHENYL)-1-PHENYLETHYL)PHENOXY)-ETHANAMINE (9CI) see MFG525
(((2-(DIETHYLMETHYLAMMONIO)-1-METHYL)ETHOXY)ETHYL) TRIMETHYLAMMONIUM DIIODIDE see MQF750
5,5-DIETHYL-1-METHYLBARBITURIC ACID see DJO800
N,N-DIETHYL-3-METHYLBENZAMIDE see DKC800
1,3-DIETHYL-5-METHYLBENZENE see DJP000
DIETHYL METHYL CARBINOLURETHAN see ENF000
O,O-DIETHYL S-(N-METHYL-N-CARBOETHOXYCARBAMOYLMETHYL) DITHIOPHOSPHATE see DJI000
O,O-DIETHYL-O-(4-METHYLCOUMARIN-7-YL)-MONOTHIOFOSFAAT (DUTCH) see PKT000
O,O-DIETHYL-O-(4-METHYL-7-COUMARINYL) PHOSPHOROTHIOATE see PKT000
O,O-DIETHYL-O-(4-METHYL-7-COUMARINYL) THIONOPHOSPHATE see PKT000
O,O-DIETHYL-O-(4-METHYLCOUMARINYL-7) THIOPHOSPHATE see PKT000
O,O-DIETHYL-S-(3-METHYL-2,4-DIOXO-5-OXA-3-AZA-HEPTYL)-DITHIOFOSFAAT (DUTCH) see DJI000
1,1-DIETHYL-2-METHYL-3-DIPHENYLMETHYLENEPYRROLIDINIUM BROMIDE see PAB750
N,N-DIETHYL-1-METHYL-3,3-DI-2-THIENYLALLYLAMINE HYDROCHLORIDE see DJP500
DIETHYL (4-METHYL-1,3-DITHIOLAN-2-YLIDENE)PHOSPHOROAMIDATE see DHH400
N,N-DIETHYL-N'-((8-α)-6-METHYLERGOLIN-8-YL)UREA (Z)-2-BUTENEDIOATE see DLR150
O,O-DIETHYL-O-6-METHYL-2-ISOPROPYL-4-PYRIMIDINYL PHOSPHOROTHIOATE see DCM750
O,O-DIETHYL-O-(4-METHYL-7-KUMARINYL) ESTER KYSELINY THIOFOSFORESCNE (CZECH) see PKT000
DIETHYLMETHYL METHANE see MNI500
DIETHYLMETHYL2-(N-METHYLBENZILAMIDO)ETHYL)AMMONIUM BROMIDE see BCP685
O,O-DIETHYL-O-(3-METHYL-4-(METHYLTHIO)PHENYL)PHOSPHOROTHIOATE see LIN400
1,1-DIETHYL-3-METHYL-3-NITROSOUREA see DJP600
N,N-DIETHYL-4-METHYL-3-OXO-5-α-4-AZAANDROSTANE-17-β-CARBOXAMIDE see DJP700
O,O-DIETHYL-O-(4-METHYL-2-OXO-2H-1-PHOSPHOROTHIOIC ACID BENZOPYRAN-7-YL)ESTER (9CI) see PKT000
N,N'-DI(1-ETHYL-3-METHYLPENTYL)-p-PHENYLENEDIAMINE see BJT500
N,N-DIETHYL-α-METHYL-10H-PHENOTHIAZINE-10-ETHANAMINE see DIR000
N,N-DIETHYL-2-(2-(2-METHYL-5-PHENYL-1H-PYRROL-1-YL)PHENOXY)-ETHANAMINE see LEF400
DIETHYLMETHYLPHOSPHINE see DJQ200
O,S-DIETHYL METHYLPHOSPHONOTHIOATE see DJR700
N,N-DIETHYL-4-METHYL-1-PIPERAZINECARBOXAMIDE see DIW000
N,N-DIETHYL-4-METHYL-1-PIPERAZINE CARBOXAMIDE CITRATE see DIW200
N,N-DIETHYL-4-METHYL-1-PIPERAZINECARBOXAMIDE DIHYDROGEN CITRATE see DIW200
N,N-DIETHYL-4-METHYL-1-PIPERAZINECARBOXAMIDE-2-HYDROXY-1,2,3-PROPANETIRCARBOXYLATE see DIW200
3,3-DIETHYL-5-METHYL-2,4-PIPERIDINEDIONE see DNW400
3,3-DIETHYL-5-METHYLPIPERIDINE-2,4-DIONE see DNW400
O,O-DIETHYL-O-(3-METHYL-1H-PYRAZOL-5-YL)-FOSFAAT (DUTCH) see MOX250

DIETHYL-3-METHYL-5-PYRAZOLYL PHOSPHATE see MOX250
O,O-DIETHYL-O-(3-METHYL-5-PYRAZOLYL) PHOSPHATE see MOX250
5,5-DIETHYL-1-METHYL-2,4,6(1H,3H,5H)-PYRIMIDINETRIONE see DJO800
3,3'-DIETHYL-9-METHYLSELENOCARBOCYANINE IODIDE see DJQ300
O,O-DIETHYL-O-(p-(METHYLSULFINYL)PHENYL) PHOSPHOROTHIOATE see FAQ800
O,O-DIETHYL-O-p-(METHYLSULFINYL)PHENYL THIOPHOSPHATE see FAQ800
DIETHYLMETHYLSULFONIUM IODIDEMERCURIC IODIDE (ADDITION COMPOUND) see DJQ800
DIETHYLMETHYL SULFONIUM IODINE with MERCURY IODIDE (1:1) see DJQ800
3,3'-DIETHYL-9-METHYLTHIACARBOCYANINE IODIDE see DJR200
O,S-DIETHYL METHYLTHIOPHOSPHONATE see DJR700
O,O-DIETHYL-O-(4-(METHYLTHIO)-3,5-XYLYL)PHOSPHOROTHIOATE see DJR800
N,N-DIETHYL-5-METHYL-(1,2,4)TRIAZOLO(1,5-a)PYRIMIDINE-7-AMINE see DIO200
N,N-DIETHYL-N-METHYL-2-(2-(TRIMETHYLAMMONIO)ETHOXY)-1-PROPANAMINIUM DIIODIDE see MQF750
O,O-DIETHYL-O-(4-METHYLUMBELLIFERONE) ESTER OF THIOPHOSPHORIC ACID see PKT000
O,O-DIETHYL-O-(4-METHYLUMBELLIFERONE) PHOSPHOROTHIOATE see PKT000
DIETHYL (4-METHYLUMBELLIFERYL) THIONOPHOSPHATE see PKT000
N,N-DIETHYL-N-METHYL-2-((9H-XANTHEN-9-YLCARBONYL)OXY)ETHANAMINIUM BROMIDE see DJM800
α,α-DIETHYL-1-NAPHTHALENEACETIC ACID SODIUM SALT see DJS100
O,O-DIETHYL-o-NAPHTHALIMIDE PHOSPHOROTHIOATE see NAQ500
O,O-DIETHYL-o-NAPHTHALOXIMIDO PHOSPHOROTHIOATE see NAQ500
O,O-DIETHYL-o-NAPHTHALOXIMIDOPHOSPHOROTHIONATE see NAQ500
O,O-DIETHYL-o-NAPHTHYLAMIDOPHOSPHOROTHIOATE see NAQ500
DIETHYL-NICOTAMIDE see DJS200
N,N-DIETHYLNICOTINAMIDE see DJS200
DIETHYLNITRAMINE see DJS500
N,N-DIETHYL-N'-(1-NITRO-9-ACRIDINYL)-1,2-ETHANEDIAMINE DIHYDROCHLORIDE (9CI) see NFW100
N,N-DIETHYL-N'-(1-NITRO-9-ACRIDINYL)-1,3-PROPANEDIAMINE DIHYDROCHLORIDE (9CI) see NFW200
N,N-DIETHYL-2-(4-(2-NITRO-1,2-DIPHENYLETHENYL)PHENOXY)ETHANAMINE CITRATE see EAF100
O,O-DIETHYL-O-4-NITRO-FENIL)-MONOTHIOFOSFAAT (DUTCH) see PAK000
DIETHYL-p-NITROFENYL ESTER KYSELINY FOSFORECNE (CZECH) see NIM500
O,O-DIETHYL-O-p-NITROFENYLESTER KYSELINYTHIOFOSFORECNE (CZECH) see PAK000
O,O-DIETHYL-S-p-NITROFENYLESTER KYSELINY THIOFOSFORECNE (CZECH) see DJS800
O,O-DIETHYL-O-p-NITROFENYLTIOFOSFAT (CZECH) see PAK000
O,O-DIETHYL-O-(p-NITROPHENYL) ESTER (DRY MIXTURE) PHOSPHOROTHIOIC ACID see PAK250
DIETHYL p-NITROPHENYL PHOSPHATE see NIM500
O,O-DIETHYL O-p-NITROPHENYL PHOSPHATE see NIM500
O,O-DIETHYL-O-4-NITROPHENYLPHOSPHOROTHIOATE see PAK000
O,O-DIETHYL-O-(4-NITROPHENYL) PHOSPHOROTHIOATE see PAK000
O,O-DIETHYL-O-(p-NITROPHENYL) PHOSPHOROTHIOATE see PAK000
O,O-DIETHYL-S-(4-NITROPHENYL) PHOSPHOROTHIOATE see DJS800
O,S-DIETHYL-O-(4-NITROPHENYL)PHOSPHOROTHIOATE see DJT000
O,S-DIETHYL-O-(p-NITROPHENYL) PHOSPHOROTHIOATE see DJT000
O,O-DIETHYL-S-(4-NITROPHENYL)PHOSPHOROTHIOIC ACID ESTER see DJS800
O,S-DIETHYL-O-(4-NITROPHENYL)PHOSPHOROTHIOIC ACID ESTER see DJT000
O,S-DIETHYL-O-(p-NITROPHENYL)PHOSPHOROTHIOIC ACID ESTER see DJT000
DIETHYL-4-NITROPHENYL PHOSPHOROTHIONATE see PAK000
DIETHYL-p-NITROPHENYLTHIONOPHOSPHATE see PAK000
O,O-DIETHYL-O-(p-NITROPHENYL)THIONOPHOSPHATE see PAK000
DIETHYL-p-NITROPHENYLTHIOPHOSPHATE see PAK000
O,O-DIETHYL-O-4-NITROPHENYL THIOPHOSPHATE see PAK000
O,O-DIETHYL-O-(p-NITROPHENYL) THIOPHOSPHATE see PAK000
O,O-DIETHYL-S-(4-NITROPHENYL)THIOPHOSPHATE see DJS800
O,S-DIETHYL-O-(4-NITROPHENYL)THIOPHOSPHATE see DJT000
DIETHYLNITROSAMINE see NJW500
N,N-DIETHYLNITROSAMINE see NJW500
DIETHYLNITROSOAMINE see NJW500
O,N-DIETHYL-N-NITROSOHYDROXYLAMINE see NKC500
DIETHYLOLAMINE see DHF000
O,O-DIETHYL-O-2-PYRAZINYL PHOSPHOROTHIOATE see EPC500
DIETHYL OXALATE see DJT200
5,5-DIETHYL-1,3-OXAZIN-2,4-DIONE see DJT400
5,5-DIETHYL-1,3-OXAZINE-2,4-DIONE see DJT400
DIETHYL OXIDE see EJU000

O,O-DIETHYL-S-(4-OXO-3H-1,2,3-BENZOTRIAZINE-3-YL)-METHYL-
DITHIOPHOSPHATE see EKN000
O,O-DIETHYL-S-(4-OXOBENZOTRIAZINO-3-METHYL)PHOSPHORODITHIO-
ATE see EKN000
O,O-DIETHYL-S-((4-OXO-3H-1,2,3-BENZOTRIAZIN-3-YL)-METHYL)-DITHIO
FOSFAAT (DUTCH) see EKN000
N,N-DIETHYL-3-OXO-BUTANAMIDE (9CI) see DHI600
DIETHYL OXYDIFORMATE see DIZ100
DIETHYL PARAOXON see NIM500
DIETHYLPARATHION see PAK000
3,3'-DIETHYLPENTAMETHINETHIACYANINE IODIDE see DJT800
3,3-DIETHYLPENTANE see DJU000
2,2-DIETHYL-4-PENTENAMIDE see DJU200
DIETHYL PERFLUOROGLUTARATE see DJK100
DIETHYL PEROXIDE see DJU400
DIETHYL PEROXYDICARBONATE see DJU600
DIETHYL PEROXYDIFORMATE see DJU600
DIETHYLPHENYLAMINE see DIS700
N,N-DIETHYL-p-(PHENYLAZO)ANILINE see OHI875
N,N-DIETHYL-4-(PHENYLAZO)BENZENAMINE see OHI875
p-((3,4-DIETHYLPHENYL)AZO)-N,N-DIMETHYLANILINE see DJB400
N,N-DIETHYL-N'-(2-(2-PHENYL-1,3-BENZODIOXOL-2-
YL)ETHYL)ETHYLENEDIAMINE DIMALEATE see DJU800
DI(p-ETHYLPHENYL)DICHLOROETHANE see DJC000
DIETHYL-4,4'-o-PHENYLENEBIS(3-THIOALLOPHANATE) see DJV000
DIETHYL-p-PHENYLENEDIAMINE see DJV200
N,N-DIETHYL-p-PHENYLENEDIAMINE see DJV200
N,N-DIETHYL-p-PHENYLENEDIAMINE HYDROCHLORIDE see AJO250
N,N'-DIETHYL-p-PHENYLENEDIAMINE SULFATE see DJV250
N,N-DIETHYL-2-PHENYL-GLYCINE-2,6-XYLYL ESTER HYDROCHLORIDE
see FAC150
O,O-DIETHYL-O-(5-PHENYL-3-ISOXAZOLYL) PHOSPHOROTHIOATE
see DJV600
O,O-DIETHYL-O-(3-(5-PHENYL)-1,2-ISOXAZOLYL)PHOSPHOROTHIOATE
see DJV600
O,O-DIETHYL-O-(5-PHENYL-3-ISOXAZOLYL)PHOSPHOROTHIOIC ACID
ESTER see DJV600
N,N-DIETHYL-3-PHENYL-1,2,4-OXADIAZOLE-5-ETHANAMINE see OOC000
N,N-DIETHYL-3-(1-PHENYLPROPYL)-1,2,4-OXADIAZOLE-5-ETHANAMINE CI-
TRATE see POF500
DIETHYL PHENYLTIN ACETATE see DJV800
3,3-DIETHYL-1-PHENYLTRIAZENE see PEU500
O,O-DIETHYL O-(1-PHENYL-1H-1,2,4-TRIAZOL-3-YL)PHOSPHOROTHIOATE
see THT750
DIETHYL PHOSPHINE see DJW000
DIETHYLPHOSPHINIC ACID-p-NITROPHENYL ESTER see DJW200
DIETHYL PHOSPHITE see DJW400
O,O-DIETHYL PHOSPHORIC ACID O-p-NITROPHENYL ESTER see NIM500
O,O-DIETHYLPHOSPHOROCHLORIDOTHIOATE see DJW600
DIETHYL PHOSPHOROCYANIDATE see DJW800
O,O-DIETHYL PHOSPHORODITHIOATE AMMONIUM see DJW875
O,O-DIETHYL PHOSPHORODITHIOATE S-ester with 3-(MERCAPTOMETHYL)-
1,2,3-BENZOTRIAZIN-4(3H)-ONE see EKN000
O,O-DIETHYLPHOSPHOROTHIOATE, O-ESTER with 6-HYDROXY-2-PHENYL-
3(2H)-PYRIDAZINONE see POP000
O,O-DIETHYL PHOSPHOROTHIOATE, o-ESTER with
PHENYLGLYOXYLONITRILE OXIME see BAT750
(2-(O,O-DIETHYLPHOSPHOROTHIO)ETHYL)TRIMETHYLAMMONIUM, IO-
DIDE see TLF500
DIETHYL PHTHALATE see DJX000
DIETHYL-o-PHTHALATE see DJX000
O,O-DIETHYLPHTHALIMIDOPHOSPHONOTHIOATE see DJX200
O,O-DIETHYL PHTHALIMIDOTHIOPHOSPHATE see DJX200
DIETHYL PROPANEDIOATE see EMA500
2,2-DIETHYLPROPANEDIOL-1,3 see PMB250
2,2-DIETHYLPROPANE-1,3-DIOL see PMB250
2,2-DIETHYL-1,3-PROPANEDIOL see PMB250
DIETHYLPROPIONE HYDROCHLORIDE see DIP600
DIETHYLPROPION HYDROCHLORIDE see DIP600
N,N-DIETHYL-N'-((8-α)-6-PROPYLERGOLIN-8-YL)UREA see DJX300
N,N-DIETHYL-N'-((8-α)-6-PROPYLERGOLIN-8-YL)UREA (Z)-2-BUTENEDIO-
ATE see DJX350
O,O-DIETHYL-O-(2-PROPYL-4-METHYLPYRIMIDINYL-6)PHOSPHOROTHIO-
ATE see DJX400
O,O-DIETHYL-O-(2-N-PROPYL-4-METHYL-PYRIMIDYL-6)PHOSPHOROTHIO-
ATE see DJX400
O,O-DIETHYL-O-(2-PROPYL-4-METHYL-6-PYRIMIDYL)PHOSPHOROTHIOIC
ACID ESTER see DJX400
DIETHYL PROPYLMETHYLPYRIMIDYL THIOPHOSPHATE see DJX400
N,N-DIETHYL-1-PROPYNYLAMINE see DJX600
N,N-DIETHYL-2-PROPYNYLAMINE see DJX800
DIETHYL-O-2-PYRAZINYL PHOSPHOROTHIONATE see EPC500
O,O-DIETHYL-O-2-PYRAZINYL PHOSPHOTHIONATE see EPC500
O,O-DIETHYL-O-PYRAZINYL THIOPHOSPHATE see EPC500

N,N-DIETHYL-3-PYRIDINECARBOXAMIDE see DJS200
N,N-DIETHYL-4-(4'-(PYRIDYL-1'-OXIDE)AZO)ANILINE see DIP000
3,3-DIETHYL-1-(m-PYRIDYL)TRIAZENE see DJY000
5,5-DIETHYL-2,4,6(1H,3H,5H)-PYRIMIDINETRIONE see BAG000
DIETHYL PYROCARBONATE see DIZ100
DIETHYL PYROCARBONATE mixed with AMMONIA see DJY050
DIETHYL PYROCARBONIC ACID see DIZ100
3-(2-(DIETHYLPYRROLIDINO)ETHOXY)-6-METHOXY-2-
PHENYLBENZOFURAN HYDROCHLORIDE see DJY100
O,O-DIETHYL-O-QUINOXALIN-2-YL PHOSPHOROTHIOATE see DJY200
O,O-DIETHYL-O-(2-QUINOXALINYL) PHOSPHOROTHIOATE see DJY200
O,O-DIETHYL-O-(2-QUINOXALYL) PHOSPHOROTHIOATE see DJY200
O,O-DIETHYL-O-2-QUINOXALYLTHIOPHOSPHATE see DJY200
N,N-DIETHYLRIFAMYCIN B AMIDE see RKA000
N,N-DIETHYLSALICYLAMIDE see DJY400
DIETHYL SEBACATE see DJY600
O,O-DIETHYL Se-(2-DIETHYLAMINOETHYL)PHOSPHOROSELENOATE
see DJA325
N,N-DIETHYLSELENOUREA see DJY800
1,1-DIETHYL-2-SELENOUREA see DJY800
DIETHYL SODIUM DITHIOCARBAMATE see SGJ000
DIETHYLSTANNYL DICHLORIDE see DEZ000
N,N-DIETHYL-4-STILBENAMINE see DKA000
α,α'-DIETHYLSTILBENEDIOL see DKA600
2,2'-DIETHYL-4,4'-STILBENEDIOL see DKA600
α,α'-DIETHYL-4,4'-STILBENEDIOL see DKA600
α,α'-DIETHYL-(E)-4,4'-STILBENEDIOL see DKA600
trans-α,α'-DIETHYL-4,4'-STILBENEDIOL see DKA600
α,α'-DIETHYL-4,4'-STILBENEDIOL BIS(DIHYDROGEN PHOSPHATE)
see TEE300
α,α'-DIETHYL-(E)-4,4'-STILBENEDIOL BIS(DIHYDROGEN PHOSPHATE)
see DKA200
α,α'-DIETHYL-4,4'-STILBENEDIOL DIPALMITATE see DKA800
α,α'-DIETHYL-4,4'-STILBENEDIOL, DIPROPIONATE see DKB000
trans-α,α'-DIETHYL-4,4'-STILBENEDIOL DIPROPIONATE see DKB000
α,α'-DIETHYL-4,4'-STILBENEDIOL trans-DIPROPIONATE see DKB000
α,α'-DIETHYL-4,4'-STILBENEDIOL DIPROPIONYL ESTER see DKB000
α,α'-DIETHYL-4,4'-STILBENEDIOL DISODIUM SALT see DKA400
DIETHYLSTILBENE DIPROPIONATE see DKB000
DIETHYLSTILBESTEROL see DKA600
trans-DIETHYLSTILBESTEROL see DKA600
DIETHYLSTILBESTEROL DIPHOSPHATE see DKA200
DIETHYLSTILBESTEROL DIPROPIONATE see DKB000
DIETHYLSTILBESTROL see DKA600
trans-DIETHYLSTILBESTROL see DKA600
DIETHYLSTILBESTROL DIMETHYL ETHER see DJB200
DIETHYLSTILBESTROL DIPALMITATE see DKA800
DIETHYLSTILBESTROL DIPHOSPHATE see DKA200
DIETHYLSTILBESTROL DIPHOSPHATE TETRASODIUM see TEE300
DIETHYLSTILBESTROL DIPROPIONATE see DKB000
DIETHYLSTILBESTROL DISODIUM SALT see DKA400
DIETHYLSTILBESTROL and ETHISTERONE see EEI050
DIETHYLSTILBESTROL, IODINE DERIVATIVE see TDE000
DIETHYLSTILBESTROL PHOSPHATE see DKA200
DIETHYLSTILBESTROL PHOSPHATE TETRASODIUM see TEE300
DIETHYLSTILBESTROL and PRANONE see EEI050
DIETHYLSTILBESTROL PROPIONATE see DKB000
DIETHYLSTILBESTRYL DIPHOSPHATE see DKA200
DIETHYLSTILBOESTEROL see DKA600
trans-DIETHYLSTILBOESTEROL see DKA600
DIETHYLSTILBOESTROL-3,4-OXIDE see DKB100
DIETHYLSTILBOESTROL-α,β-OXIDE see DKB100
DIETHYL SUCCINATE (FCC) see SNB000
DIETHYL SULFATE see DKB110
DIETHYLSULFID (CZECH) see EPH000
DIETHYL SULFIDE (DOT) see EPH000
DIETHYL SULFIDE-2,2'-DICARBOXYLIC ACID see BHM000
DIETHYL SULFITE see DKB119
DIETHYLSULFONATE see DKB139
DIETHYLSULFONDIMETHYLMETHANE see ABD500
DIETHYLSULFONMETHYLETHYLMETHANE see BJT750
DIETHYL SULPHOXIDE see EPI500
DIETHYL TELLURIDE see DKB150
α,α'-DIETHYL-3,3',5,5'-TETRAFLUORO-4,4'-STILBENEDIOL (E)- see TCH325
trans-α,α'-DIETHYL-3,3',5,5'-TETRAFLUORO-4,4'-STILBENEDIOL see TCH325
N,N-DIETHYL-N'-(1,2,3,4-TETRAHYDRO-1-NAPHTHYL)ACETAMIDE
see TGI725
N,N-DIETHYL-N'-2-(TETRAHYDRO-1,2,3,4-NAPHTHYL)-GLYCINAMIDE
see TGI725
5,5-DIETHYLTETRAHYDRO-2H-1,3-OXAZINE-2,4(3H)-DIONE see DJT400
O,O-DIETHYL-O-(7,8,9,10-TETRAHYDRO-6-OXOBENZO(C)CHROMAN-3-
YL)PHOSPHOROTHIOATE see DXO000
O,O-DIETHYL-O-(7,8,9,10-TETRAHYDRO-6-OXO-6H-DIBENZO(b,d)PYRAN-3-
YL)PHOSPHOROTHIOATE see DXO000

N,N-DIETHYL-N'-2-(TETRALYL)-GLYCINAMIDE see TGI725
O,O-DIETHYL-O-(3,4-TETRAMETHYLENECOUMARINYL-7) THIOPHOS-
    PHATE see DXO000
N,N-DIETHYL-N,N',N',N'-TETRAMETHYL-N,N'-(2-METHYL-3-OX-
    APENTAMETHYLENE)BIS(AMMONIUM IODIDE) see MQF750
DIETHYL THALLIUM PERCHLORATE see DKB175
DIETHYLTHIADICARBOCYANINE IODIDE see DJT800
3,3'-DIETHYLTHIADICARBOCYANINE IODIDE see DJT800
DIETHYLTHIAMBUTENE HYDROCHLORIDE see DJP500
N,N'-DIETHYLTHIOCARBAMIDE see DKC400
N,N-DIETHYLTHIOCARBAMYL-O,O-DIISOPROPYLDITHIOPHOSPHATE
    see DKB600
DIETHYLTHIOETHER see EPH000
2,2-DIETHYL-3-THIOMORPHOLINONE see DKC200
DIETHYL THIOPHOSPHORIC ACIDESTER of 3-CHLORO-4-METHYL-7-
    HYDROXYCOUMARIN see CNU750
DIETHYLTHIOPHOSPHORYL CHLORIDE (DOT) see DJW600
1,3-DIETHYLTHIOUREA see DKC400
1,3-DIETHYL-2-THIOUREA see DKC400
N,N'-DIETHYLTHIOUREA see DKC400
DIETHYLTIN CHLORIDE see DEZ000
DIETHYLTIN DI(10-CAMPHORSULFONATE) see DKC600
DIETHYLTIN DICAPRYLATE see DIV600
DIETHYLTIN DICHLORIDE see DEZ000
DIETHYLTIN DIIODIDE see DJB000
DIETHYLTIN DIOCTANOATE see DIV600
DIETHYLTOLUAMIDE see DKC800
DIETHYL-m-TOLUAMIDE see DKC800
N,N-DIETHYL-m-TOLUAMIDE see DKC800
N,N-DIETHYL-o-TOLUAMIDE see DKD000
DIETHYL TRIAZENE see DKD200
O,O-DIETHYL-O-3,5,6-TRICHLORO-2-PYRIDYL PHOSPHOROTHIOATE
    see CMA100
3,9-DIETHYLTRIDECYL-6-SULFATE see DKD400
N^4,N^4-DIETHYL-α,α,α-TRIFLUORO-3,5-DINITRO-TOLUENE-2,4-DIAMINE
    see CNE500
N,N-DIETHYL-4-(α-(α,α,α-TRIFLUORO-o-TOLYL) BENZYLOXY)PENTYLAM-
    INE CITRATE see DKD600
O,O-DIETHYL-S-2-TRIMETHYLAMMONIUM ETHYLPHOSPHONOTHIOLATE
    IODIDE see TLF500
N,N-DIETHYLVANILLAMIDE see DKE200
DIETHYL XANTHOGENATE see BJU000
DIETHYLXANTHOGEN DISULFIDE see BJU000
DI(ETHYLXANTHOGEN)TRISULFIDE see DKE400
DIETHYL YELLOW see OHI875
DIETHYLZINC see DKE600
17-β-2-epsilon,17-α-DIETHYNYL, A-NOR-ANDROSTANE-2-epsilon,
    DIHYDROXYDIPROPINATE see AOO150
2-α,17-α-DIETHYNYL-A-NOR-5-α-ANDROSTANE-2-β,17-β-DIOL
    DIHEMISUCCINATE see SDY675
2-α-17-α-DIETHYNYL-A-NOR-5-α-ANDROSTANE-2-β,17-β-DIOL DIPROPION-
    ATE see AOO150
DIETHYXIME see DHZ000
DIETIL see CAR000
DIETILAMIDE-CARBOPIRIDINA see DJS200
DIETILAMINA (ITALIAN) see DHJ200
α-DIETILAMINO-2,6-DIMETILACETANILIDE (ITALIAN) see DHK400
5-(2-DIETILAMINOETIL)-3-FENIL-1,2,4-OXADIEZOLO CLORIDRATO (ITAL-
    IAN) see OOE100
O,O-DIETIL-O-(4-BROMO-2,5 DICLORO-FENIL)-MONOTIOFOSFATO (ITAL-
    IAN) see EGV500
O,O-DIETIL-S-((2-CIAN-2-METIL-ETIL)-CARBAMOIL)-METIL-
    MONOTIOFOSFATO (ITALIAN) see PHK250
O,O-DIETIL-S-((4-CLORO-FENIL-TIO)-METILE)-DITIOFOSFATO (ITALIAN)
    see TNP250
O,O-DIETIL-O-(3-CLORO-4-METIL-CUMARIN-7-IL-MONOTIOFOSFATO)(ITAL-
    IAN) see CNU750
O,O-DIETIL-S-((6-CLORO-2-OXO-BENZOSSAZOLIN-3-IL)-METIL)-
    DITIOFOSFATO (ITALIAN) see BDJ250
O,O-DIETIL-O-(2,4-DICLORO-FENIL)-MONOTIOFOSFATO (ITALIAN)
    see DFK600
5,5-DIETILDIIDRO-1,3-OSSAZIN-2,4-DIONE (ITALIAN) see DJT400
DIETILESTILBESTROL (SPANISH) see DKA600
O,O-DIETIL-S-(2-ETILTIO-ETIL)-DITIOFOSFATO (ITALIAN) see DXH325
O,O-DIETIL-S-(2-ETILTIO-ETIL)-MONOTIOFOSFATO (ITALIAN) see DAP200
O,O-DIETIL-S-(ETILTIO-METIL)-DITIOFOSFATO (ITALIAN) see PGS000
O,O-DIETIL-S-(N-ETOSSI-CARBONIL-N-METIL-CARBAMOIL-METIL)-
    DITIOFOSFATO (ITALIAN) see DJI000
O,O-DIETIL-O-(2-ISOPROPIL-4-METIL-PIRIMIDIN-6-IL)-MONOTIOFOSFATO
    (ITALIAN) see DCM750
O,O-DIETIL-O-(4-METILCUMARIN-7-IL)-MONOTIOFOSFATO (ITALIAN)
    see PKT000
O,O-DIETIL-O-(3-METIL-1H-PIRAZOL-5-IL)-FOSFATO (ITALIAN) see MOX250
O,O-DIETIL-O-(4-NITRO-FENIL)-MONOTIOFOSFATO (ITALIAN) see PAK000

O,O-DIETIL-S-((4-OXO-3H-1,2,3-BENZOTRIAZIN-3-IL)-METIL)-DITIOFOSFATO
    (ITALIAN) see EKN000
DIETILPROPANDIOLO see PMB250
1,1-DIETOSSIETANO (ITALIAN) see AAG000
DIETREEN see RAF100
DIETROL see DKE800
DIETROXINE see DJT400
O,O-DIETYL-S-2-ETYLMERKAPTOETYLTIOFOSFAT (CZECH) see DAP200
O,O-DIETYL-O-4-METHYLKUMARINYL(7)TIOFOSFAT (CZECH) see PKT000
O,O-DIETYL-o-p-NITROFENYLFOSFAT (CZECH) see NIM500
DIF 4 see DRP800
DIFACIL see DHX800, THK000
DIFEDRYL see BBV500
DIFENAMIZOLE see PAM500
DIFENHYDRAMIN see BBV500
DIFENHYDRAMINE HYDROCHLORIDE see BAU750
DIFENIDOL see DWK200
DIFENIDOL HYDROCHLORIDE see CCR875
DIFENIDOLIN see CCR875
DIFENIDRAMINA (ITALIAN) see BBV500
DIFENILDICHETOPIRAZOLIDINA (ITALIAN) see DWA500
DIFENILHIDANTOINA (SPANISH) see DKQ000
DIFENIL-METAN-DIISOCIANATO (ITALIAN) see MJP400
DIFENIN see DKQ000, DNU000
1,5-DIFENOXYANTHRACHINON see DVW100
DIFENSON see CJT750
DIFENTHOS see TAL250
N,N'-DIFENYL-p-FENYLENDIAMIN (CZECH) see BLE500
2-(DIFENYL-HYDROXYACETOXY)ETHYL-DIETHYLAMMONIUMCHLORID
    (CZECH) see BCA000
DIFENYLIN see BGF109
DIFENYLMETHAAN-DISSOCYANAAT (DUTCH) see MJP400
DIFENYLSULFON (CZECH) see PGI750
DIFENZOQUAT METHYL SULFATE see ARW000
DIFETOIN see DNU000
DIFEXAMIDE METHIODIDE see BTA325
DIFFLAM see BBW500
DIFFOLLISTEROL see EDP000
DIFHYDAN see DKQ000, DNU000
DIFLAVINE (ACRIDINE) see DBN000
DIFLORASONE DIACETATE see DKF125
DIFLUBENZURON see CJV250
DIFLUCORTOLONE VALERATE see DKF130
DIFLUCORTOLONE 21-VALERATE see DKF130
DIFLUCORTOLONVALERIANAT (GERMAN) see DKF130
DIFLUNISAL see DKI600
DIFLUOR see MJD275
2,4'-DIFLUOROACETANILIDE see DKF170
DIFLUOROACETIC ACID see DKF200
DIFLUOROAMINE see DKF400
3-DIFLUOROAMINO-1,2,3-TRIFLUORODIAZIRIDINE see DKF600
DIFLUOROAMMONIUM HEXAFLUOROARSENATE see DKF620
m-DIFLUOROBENZENE see DKF800
p-DIFLUOROBENZENE see DKG000
3,4-DIFLUOROBENZENEARSONIC ACID see DKG100
2,10-DIFLUOROBENZO(rst)PENTAPHENE see DKG400
1,1-DIFLUORO-1-CHLOROETHANE see CFX250
DIFLUOROCHLOROMETHANE see CFX500
DIFLUORODIAZENE see DKG600
DIFLUORODIAZIRINE see DKG700
2,10-DIFLUORODIBENZO(a,i)PYRENE see DKG400
DIFLUORODIBROMOETHANE see DKG800
1,1-DIFLUORO-1,2-DIBROMOETHANE see DKG800
DIFLUORODIBROMOMETHANE see DKG850
1,1-DIFLUORO-2,2-DICHLOROETHYLENE see DFA300
DIFLUORODICHLOROMETHANE see DFA600
2',4'-DIFLUORO-4-DIMETHYLAMINOAZOBENZENE see DKG980
2',5'-DIFLUORO-4-DIMETHYLAMINOAZOBENZENE see DKH000
3',4'-DIFLUORO-4-DIMETHYLAMINOAZOBENZENE see DRL000
3',5'-DIFLUORO-4-DIMETHYLAMINOAZOBENZENE see DKH100
DIFLUORODIMETHYLSTANNANE see DKH200
DIFLUORODIPHENYLTRICHLOROETHANE see FHJ000
DIFLUOROETHANE see ELN500
1,1-DIFLUOROETHANE (DOT) see ELN500
1,1-DIFLUOROETHYLENE (DOT, MAK) see VPP000
1,1-DIFLUOROETHYLENE POLYMERS (PYROLYSIS) see DKH600
DIFLUORO-N-FLUOROMETHANIMINE see DKH825
DIFLUOROFORMALDEHYDE see CCA500
2',4'-DIFLUORO-4-HYDROXY-3-BIPHENYLCARBOXYLIC ACID
    see DKI600
2',4'-DIFLUORO-4-HYDROXY-(1,1'-BIPHENYL)-3-CARBOXYLIC ACID
    see DKI600
2',4'-DIFLUORO-4-HYDROXY-(1',1-DIPHENYL)-3-CARBOXYLIC ACID
    see DKI600

6-α,9-DIFLUORO-11-β-HYDROXY-16-α-METHYL-21-VALERYLOXY-1,4-PREG-NADIENE-3,20-DIONE see DKF130
6-α,9-α-DIFLUORO-16-α-HYDROXYPREDNISOLONE-16,17-ACETONIDE see SPD500
6-α,2-DIFLUORO-11-β-HYDROXY-21-VALERYLOXY-16-α-METHYL-1,4-PREG-NADIENE-3,20-DIONE see DKF130
DIFLUOROMETHYLENE DIHYPOFLUORITE see DKH830
2-(DIFLUOROMETHYL)ORNITHINE see DBE835
α-DIFLUOROMETHYLORNITHINE see DBE835
dl-α-DIFLUOROMETHYLORNITHINE see DKH875
α-DIFLUOROMETHYLORNITHINE HYDROCHLORIDE see EAE775
2-(DIFLUOROMETHYL)-dl-ORNITHINE HYDROCHLORIDE see EAE775
DIFLUOROMETHYLPHOSPHINE OXIDE see MJD275
DIFLUOROMONOCHLOROETHANE (DOT) see CFX250
DIFLUOROMONOCHLOROMETHANE see CFX500
3,4-DIFLUORO-2-NITROBENZENEDIAZONIUM-6-OXIDE see DKI200
3,6-DIFLUORO-2-NITROBENZENEDIAZONIUM-4-OXIDE see DKI289
DIFLUOROPHENYLARSINE see DKI400
p-((3,5-DIFLUOROPHENYL)AZO)-N,N-DIMETHYLANILINE see DKH100
5-(2,4-DIFLUOROPHENYL)SALICYLIC ACID see DKI600
1,1-DI(4-FLUOROPHENYL)-2-(1,2,4-TRIAZOLE-1-YL)-ETHANOL see BJW600
DIFLUOROPHOSPHORIC ACID, anhydrous (DOT) see PHF500
6-α,9-α-DIFLUOROPREDNISOLONE 17-BUTYRATE 21-ACETATE see DKJ300
1,3-DIFLUORO-2-PROPANOL see DKI800
1,2-DIFLUORO-1,1,2,2-TETRACHLOROETHANE see TBP050
6-α,9-DIFLUORO-11-β,16-α,17,21-TETRAHYDROXY-PREGNA-1,4-DIENE-3,20-DIONE cyclic 17-ACETAL with ACETONE-21-ACETATE see FDD150
1,1-DIFLUORO-1,2,2-TRICHLOROETHANE see TIM000
3,8-DIFLUOROTRICYCLOQUINAZOLINE see DKJ200
3,3-DIFLUORO-2-(TRIFLUOROMETHYL)ACRYLIC ACID, METHYL ESTER see MNN000
3,3-DIFLUORO-2-(TRIFLUOROMETHYL)-2-PROPENOIC ACID, METHYL ESTER see MNN000
6-α,9-DIFLUORO-11-β,17,21-TRIHYDROXYPREGNA-1,4-DIENE-3,20-DIONE-21-ACETATE-17-BUTYRATE see DKJ300
1,1-DIFLUOROUREA see DKJ225
DIFLUPREDNATE see DKJ300
DIFLUPYL see IRF000
DIFLUREX see TGA600
DIFLURON see CJV250
DIFLUROPHATE see IRF000
DIFO see BJE750
DIFOLATAN see CBF800
DIFOLLICULINE see EDP000
DIFONATE see FMU045
DIFORENE see DOZ000
DIFORMAL see GIK000
7,12-DIFORMYLBENZ(a)ANTHRACENE see BBD000
1,4-DIFORMYLBENZENE see TAN500
2,2'-DIFORMYLBIPHENYL see DVV800
N,N'-DIFORMYL-p,p'-DIAMINODIPHENYLSULFONE see BJW750
1,2-DIFORMYLHYDRAZIN (GERMAN) see DKJ600
1,2-DIFORMYLHYDRAZINE see DKJ600
2,4'-DIFORMYL-1,1'-(OXYDIMETHYLENE) DIPYRIDINIUM DICHLORIDE, DIOXIME see HAA370
DIFOSAN see CBF800
DIFURAN see DKK100, PAF500
DI-2-FURANYLETHANEDIONE see FPZ000
DIFURAZONE see PAF500
N,N-DIFURFURAL-n-PHENYLENEDIAMINE see DKK200
DI-2-FUROYL PEROXIDE see DKK400
DI-2-FURYLGLYOXAL see FPZ000
DI-2-FURYLGLYOXAL MONOXIME see FQB000
DIGACIN see DKN400
DIGALLIUM TRISULFATE see GBS100
DIGAMMACAINE see DKK800
DIGENEA SIMPLEX MUCILAGE see AEX250
11,11-DIGERANYLOXY-1-UNDECENE see UNA100
DIGERMANE see DKL000
DIGERMIN see DUV600
DIGIBUTINA see BRF500
DIGILANID A see LAT000
DIGILANID B see LAT500
DIGILANID C see LAU000
DIGILANIDE B see LAT500
DIGILANIDES see LAU400
DIGILONG see DKL800
DIGIMED see DKL800, LAU400
DIGIMERCK see DKL800
DIGISIDIN see DKL800
DIGITALIN see DKL800
DIGITALINE (FRENCH) see DKL800
DIGITALINE CRISTALLISEE see DKL800
DIGITALINE NATIVELLE see DKL800

DIGITALINUM VERUM see DKL800
DIGITALIS see DKL200, FOM100
DIGITALIS GLYCOSIDE see DKN400
DIGITALIS LANATA STANDARD see DKL300
DIGITALIS PURPUREA see FOM100
DIGITALIS PURPUREA, LEAF see DKL200
DIGITANNOID see DKL200
DIGITIN see DKL400
DIGITONIN see DKL400
DIGITOPHYLLIN see DKL800
DIGITOXIGENIN see DMJ000
DIGITOXIGENIN + 2 DIGITOXOSE + ACETYL-DIGILANIDOBOSE (GERMAN) see LAT000
DIGITOXIGENIN + 2-DIGITOXOSE + 1-ACETYL-(4)-DIGITOSE (GERMAN) see ACH750
DIGITOXIGENIN + 2-DIGITOXOSE + ACETYL-(3)-DIGITOXOSE (GERMAN) see ACH750
DIGITOXIGENINE see DMJ000
DIGITOXIGENIN-TRIDIGITOXOSID (GERMAN) see DKL800
DIGITOXIGENIN TRIDIGITOXOSIDE see DKL800
DIGITOXIN see DKL800
DIGITOXOSIDE see DKL875
DIGLYCERIDE ACETIC ACID see DBF600
N,N-DIGLYCIDYLANILIN (CZECH) see DKM120
DIGLYCIDYLANILINE see DKM100
N-N-DIGLYCIDYLANILINE see DKM120
DIGLYCIDYL BISPHENOL A ETHER see BLD750
N,N'-DIGLYCIDYL-5,5-DIMETHYLHYDANTOIN see DKM130
DIGLYCIDYL ETHER see DKM200
DIGLYCIDYL ETHER of N,N-BIS(2-HYDROXYETHOXYETHYL)ANILINE see BJN850
DIGLYCIDYL ETHER of 2,2-BIS(4-HYDROXYPHENYL)PROPANE see BLD750
DIGLYCIDYL ETHER of 2,2-BIS(p-HYDROXYPHENYL)PROPANE see BLD750
DIGLYCIDYL ETHER of N,N-BIS(2-HYDROXYPROPYL)-tert-BUTYLAMINE see DKM400
DIGLYCIDYL ETHER of BISPHENOL A see BLD750
DIGLYCIDYL ETHER of BISPHENOL A mixed with BIS(2,3-EPOXYCYCLOPENTYL) ETHER (1:1) see OPI200
DIGLYCIDYL ETHER of 4,4'-ISOPROPYLIDENEDIPHENOL see BLD750
DIGLYCIDYL ETHER of NEOPENTYL GLYCOL see NCI300
DIGLYCIDYL ETHER of PHENYLDIETHANOLAMINE see BJN875
1,3-DIGLYCIDYLOXYBENZENE see REF000
N-N-DIGLYCIDYLPHENYLAMINE see DKM120
DIGLYCIDYL PHTHALATE see DKM600
DIGLYCIDYL PIPERAZINE see DHE100
DIGLYCIDYL RESORCINOL ETHER see REF000
N,N-DIGLYCIDYL-p-TOLUENESULFONAMIDE see DKM800
N,N-DIGLYCIDYL-p-TOLUENESULPHONAMIDE see DKM800
DIGLYCIDYLTRIETHYLENE GLYCOL see TJQ333
DIGLYCIN see IBH000
DIGLYCINE see IBH000
DIGLYCOL see DJD600
DIGLYCOLAMINE see AJU250
DIGLYCOL CHLORHYDRIN see DKN000
DIGLYCOL DIMETHACRYLATE see BKM250
DIGLYCOLDINITRAAT (DUTCH) see DJE400
DIGLYCOL (DINITRATE de) (FRENCH) see DJE400
DIGLYCOLIC ACID DI-(3-CARBOXY-2,4,6-TRIIODOANILIDE) DISODIUM see BGB350
DIGLYCOL MONOBUTYL ETHER see DJF200
DIGLYCOL MONOBUTYL ETHER ACETATE see BQP500
DIGLYCOL MONOETHYL ETHER see CBR000
DIGLYCOL MONOETHYL ETHER ACETATE see CBQ750
DIGLYCOL MONOMETHYL ETHER see DJG000
DIGLYCOL MONOSTEARATE see HKJ000
DIGLYCOLSAEURE-DI-(3-CARBOXY-2,4,6-TRIJOD-ANILID) DINATRIUM (GERMAN) see BGB350
DIGLYCOL STEARATE see HKJ000
DIGLYKOKOLL see IBH000
DIGLYKOLDINITRAT (GERMAN) see DJE400
DIGLYME see BKN750
DIGOLD(I) KETENIDE see DKN250
DIGORID A see ACI000
DIGORID B see ACI250
DIGOXIGENIN see DKN300
DIGOXIGENINE see DKN300
DIGOXIGENIN-TRIDIGITOXOSID (GERMAN) see DKN400
DIGOXIGENIN + ZUCKERKETTE WIE BEI ACETYL-DIGITOXIN-α (GERMAN) see ACI250
DIGOXIGENIN + ZUCKERKETTE WIE BIE ACETYL-DIGITOXIN A (GERMAN) see ACI000
DIGOXIN see DKN400
ε-DIGOXIN ACETATE see DKN600
DIGOXINE see DKN400

DIGOXIN PENTAFORMATE see DKN875
DI-(8-GUANIDINO-OCTYL)AMINE SULFATE see GLQ000
DIGUANYL see CKB500
DIHDYROPYRONE see BRT000
((DIHDYROXYPROPYL)THIO)METHYLMERCURY see MLG750
DIHEPTYL ETHER see HBO000
DIHEPTYLMERCURY see DKO000
DIHEPTYL PHTHALATE see HBP400
DI-n-HEPTYL PHTHALATE see HBP400
DIHEXANOYL PEROXIDE see DKO400
DIHEXYL see DXT200
DIHEXYLAMINE see DKO600
DI-N-HEXYLAMINE see DKO600
DI-N-HEXYL AZELATE see ASC000
DIHEXYL trans-BUTENEDIOATE see DKP000
DIHEXYL ETHER see DKO800
DIHEXYL FUMARATE see DKP000
DIHEXYL LEAD DIACETATE see DKP200
DIHEXYL MALEATE see DKP400
DIHEXYLOXYETHYL ADIPATE see AEQ000
DIHEXYL PHTHALATE see DKP600
DI-n-HEXYL PHTHALATE see DKP600
DIHEXYL SODIUM SULFOSUCCINATE see DKP800
DIHEXYLTIN DICHLORIDE see DFC200
DIHIDRAL see BBV500
DIHIDROBENZPERIDOL see DYF200
DIHIDROCLORURO de BENZIDINA (SPANISH) see BBX750
DIHYCON see DKQ000
DI-HYDAN see DKQ000, DNU000
DIHYDANTOIN see DKQ000, DNU000
DIHYDRALAZIN see OJD300
DIHYDRALAZINE see OJD300
DIHYDRALAZINE HYDROCHLORIDE see DKQ200
DIHYDRALAZINE SULFATE see DKQ600
DIHYDRALLAZIN see OJD300
DIHYDRAZINECOBALT(II) CHLORATE see DKQ400
1,4-DIHYDRAZINONAPHTHALAZINE see OJD300
DIHYDRAZINOPHTHALAZINE see OJD300
1,4-DIHYDRAZINOPHTHALAZINE see OJD300
1,4-DIHYDRAZINOPHTHALAZINE HYDROCHLORIDE see DKQ200
1,4-DIHYDRAZINOPHTHALAZINE SULFATE see DKQ600
DIHYDREL see DKQ650
DIHYDRIN see DKW800
1,2-DIHYDRO-5-ACENAPHTHYLENAMINE see AAF250
DIHYDROAFLATOXIN B1 see AEU750
DIHYDROAMBRETTOLIDE see OKU000
9,10-DIHYDRO-1-AMINO-4-BROMO-9,10-DIOXO-2-ANTHRACENE SULFONIC
  ACID SODIUM SALT see DKR000
2,3-DIHYDRO-6-AMINO-2-(2-CHLOROETHYL)-4H-1,3-BENZOXAZIN-4-ONE
  see DKR200
2,4-DIHYDRO-5-AMINO-2-(1-(3,4-DICHLOROPHENYL)ETHYL)-3H-PYRAZOL-3-
  ONE see EAE675
2,3-DIHYDRO-6-AMINO-1,3-DIOXO-2-(p-TOLYL)1H-BENZ(de)ISOQUINOLINE-
  5-SULFONIC ACID, MONOSODIUM SALT see CMM750
1,9-DIHYDRO-2-AMINO-9-((2-HYDROXYETHOXY)METHYL)-6H-PURIN-6-ONE
  SODIUM SALT see AEC725
9,10-DIHYDRO-1-AMINO-4-(3-(2-HYDROXYETHYL)AMINOSULFONYL-4-
  METHYLPHENYLAMINO)-9,10-DIOXO-2-ANTHRACENE SULFONIC ACID
  SODIUM SALT see DKR400
DIHYDROANETHOLE see PNE250
22,23-DIHYDROAVERMECTIN B1 see ITD875
DIHYDROAZIRENE see EJM900
DIHYDRO-1H-AZIRINE see EJM900
1,2-DIHYDROBENZ(l)ACEANTHRYLENE see BAW000
1,2-DIHYDROBENZ(e)ACEANTHRYLENE see AAE000
1,2-DIHYDRO-BENZ(j)ACEANTHRYLENE see CMC000
4,5-DIHYDROBENZ(k)ACEPHENANTHRYLENE see AAF000
5,6-DIHYDROBENZ(c)ACRIDINE-7-CARBOXYLIC ACID see TEF775
1,2-DIHYDROBENZ(a)ANTHRACENE see DKS400
3,4-DIHYDROBENZ(a)ANTHRACENE see DKS600
1a,11b-DIHYDROBENZ(3,4)ANTHRA(1,2-b)OXIRENE see EBP000
5,6-DIHYDROBENZENE(e)ACEANTHRYLENE see AAE000
1,2-DIHYDROBENZO(a)ANTHRACENE see DKS400
3,4-DIHYDROBENZO(a)ANTHRACENE see DKS600
6,13-DIHYDROBENZO(e)(1)BENZOTHIOPYRANO(4,3-b)INDOLE see DKS800
5,6-DIHYDRO-7H-BENZO(c)CARBAZOLE-8-CARBOXYLIC ACID-2-
  (DIETHYLAMINO)ETHYL ESTER HYDROCHLORIDE see DKS909
5,6-DIHYDRO-7H-BENZO(c)CARBAZOLE-9-CARBOXYLIC ACID-2-
  (DIETHYLAMINO) ETHYL ESTER HYDROCHLORIDE see DKT000
5,6-DIHYDRO-7H-BENZO(c)CARBAZOLE-10-CARBOXYLIC ACID-2-
  (DIETHYLAMINO)ETHYL ESTER HYDROCHLORIDE see DKT200
2,3-DIHYDRO-1H-BENZO(h,i)CHRYSENE see DKT400
6-β,7-α-DIHYDROBENZO(10,11)CHRYSENO(1,2-b)OXIRENE see BCV750
2,3-DIHYDRO-1H-BENZO(a)CYCLOPENT(b)ANTHRACENE see CPY750

N-((2,3-DIHYDRO-1,4-BENZODIOXIN-2-YL)METHYL)-N-METHYL-N,N'-DI-2-
  PROPENYL-UREA (9CI) see DBJ100
7,8-DIHYDROBENZO(a)PYRENE see DKU000
9,10-DIHYDROBENZO(e)PYRENE see DKU400
(E)-7,8-DIHYDROBENZO(a)PYRENE-7,8-DIOL see DLC800
9,10-DIHYDROBENZO(a)PYRENE-9,10-DIOL see DLC000
trans-4,5-DIHYDROBENZO(a)PYRENE-4,5-DIOL see DLC600
3b,4a-DIHYDRO-4H-BENZO(1,2)PYRENO(4,5-b)AZIRINE see BCV125
3,4-DIHYDRO-2H-1,4-BENZOTHIAZINE HYDROCHLORIDE see DKU875
7,8-DIHYDRO-N-BENZYLADENINE see DKV125
1,3-DIHYDRO-7-BROMO-5-(2-PYRIDYL)-2H-1,4-BENZODIAZEPIN-2-ONE
  see BMN750
DIHYDRO BUTADIENE SULFONE see SNW500
2,3-DIHYDRO-5-CARBOXANILIDO-6-METHYL-1,4-OXATHIIN see CCC500
2,3-DIHYDRO-5-CARBOXANILIDO-6-METHYL-1,4-OXATHIIN-4,4-DIOXIDE
  see DLV200
DIHYDROCARVEOL see DKV150
1,6-DIHYDROCARVEOL see DKV150
DIHYDROCARVEOL ACETATE see DKV160
DIHYDROCARVEYL ACETATE see DKV160
d-DIHYDROCARVONE see DKV175
DIHYDROCARVYL ACETATE see DKV160
5,10-DIHYDRO-7-CHLOR-10-(2-(DIMETHYLAMINO)ETHYL)-11H-
  DIBENZO(b,e)(1,4)DIAZEPIN-11-ONE HCl see CMW250
DIHYDROCHLORIDE-1-NITRO-9-((2-DIMETHYLAMINO)-1-
  METHYLETHYLAMINO)-ACRIDINE see NFW435
DIHYDROCHLORIDE SALT of DIETHYLENEDIAMINE see PIK000
2,3-DIHYDRO-6-CHLORO-2-(2-CHLOROETHYL)-4H-1,3-BENZOXAZIN-4-ONE
  see DKV200
(±)-2,3-DIHYDRO-6-CHLORO-5-CYCLOHEXYL-1H-INDENE-1-CARBOXYLIC
  ACID see CFH825
1'-(3-(10,11-DIHYDRO-3-CHLORO-5H-DIBENZ(b,f)AZEPIN-5-YL)PROPYL)-(1,4-
  BIPIPERIDINE)-4'-CARBOXAMIDE DIHYDROCHLORIDE see DKV309
1,3-DIHYDRO-7-CHLORO-5-(o-FLUOROPHENYL)-3-HYDROXY-1-(2-
  HYDROXYETHYL)-2H-1,4-BENZODIAZEPIN-2-ONE see DKV400
1,3-DIHYDRO-7-CHLORO-3-HYDROXY-1-METHYL-5-PHENYL-2H-1,4-
  BENZODIAZEPIN-2-ONE see CFY750
2,3-DIHYDRO-7-CHLORO-1H-INDEN-4-OL (9CI) see CDV700
3,3a-DIHYDRO-7-CHLORO-2-METHYL-2H,9H-ISOXAZOLO(3,2-
  b)(1,3)BENZOXAZIN-9-ONE see CIL500
2,3-DIHYDRO-7-CHLORO-1-METHYL-5-PHENYL-1H-1,4-BENZODIAZEPINE
  see CGA000
4,5-DIHYDRO-N-(2-CHLORO-4-METHYLPHENYL)-1H-IMIDAZOL-2-AMINE
  MONONITRATE see TGJ885
2,3-DIHYDRO-7-CHLORO-2-OXO-5-PHENYL-1H-1,4-BENZODIAZEPINE-3-
  CARBOXYLIC ACID MONOPOTASSIUM SALT see DKV700
4,5-DIHYDRO-2-((p-CHLOROPHENYL)THIOMETHYL)OXAZOLAMINE
  see AJI500
3,4-DIHYDRO-6-CHLORO-7-SULFAMYL-1,2,4-BENZOTHIADIAZINE-1,1-
  DIOXIDE see CFY000
DIHYDROCHLOROTHIAZID see CFY000
DIHYDROCHLOROTHIAZIDE see CFY000
3,4-DIHYDROCHLOROTHIAZIDE see CFY000
6,12,b-DIHYDROCHOLANTHRENE see DKV800
meso-DIHYDROCHOLANTHRENE see DKV800
DIHYDROCHOLESTEROL see DKW000
DIHYDROCHRYSENE see DKW200
1,2-DIHYDROCHRYSENE see DKW200
3,4-DIHYDROCHRYSENE see DKW400
trans-1,2-DIHYDROCHRYSENE-1,2-DIOL see DLD200
(E)-1,2-DIHYDRO-1,2-CHRYSENEDIOL see DLD200
DIHYDROCINNAMALDEHYDE see HHP000
DIHYDROCITRONELLOL see DTE600
DIHYDROCITRONELLYL ACETATE see DTE800
DIHYDROCODEINE see DKW800
7,8-DIHYDROCODEINE see DKW800
DIHYDROCODEINE ACID TARTRATE see DKX000
DIHYDROCODEINE BITARTRATE see DKX000
DIHYDROCODEINE TARTRATE see DKX000
DIHYDROCODEINE TARTRATE (1:1) see DKX000
DIHYDROCODEINONE see OOI000
DIHYDROCODEINONE BITARTRATE see DKX050
DIHYDROCOUMARIN see HHR500
3,4-DIHYDROCOUMARIN see HHR500
DIHYDROCUMINYL ALCOHOL see PCI550
DIHYDROCUMINYL ALDEHYDE see DKX100
10,11-DIHYDRO-5-CYANO-N,N-DIMETHYL-5H-DIBENZO(a,d)CYCLOHEPTENE-
  5-PROPYLAMINE HCl see COL250
3,4-DIHYDRO-6-(4-(1-CYCLOHEXYL-1H-TETRAZOL-5-YL)BUTOXY)-2(1H)-
  QUINOLINONE see CMP825
DIHYDRODAUNOMYCIN see DAC300
13-DIHYDRODAUNOMYCIN see DAC300
13-DIHYDRODAUNORUBICIN see DAC300
12-β,13-α-DIHYDRO-11-DEOXO-11-β-HYDROXYJERVINE see DAQ002

DIHYDRODEOXYMORPHINE see DKX600
DIHYDRODESOXYMORPHINE-D see DKX600
9,10-DIHYDRO-4,5-DIAMINO-1-HYDROXY-2,7-ANTHRACENE DISULFONIC
   ACID DISODIUM SALT see DKX800
9,10-DIHYDRO-8a,10,-DIAZONIAPHENANTHRENE DIBROMIDE see DWX800
9,10-DIHYDRO-8a,10a-DIAZONIAPHENANTHRENE(1,1'-ETHYLENE-2,2'-
   BIPYRIDYLIUM)DIBROMIDE see DWX800
trans-1,2-DIHYDRODIBENZ(a,e)ACEANTHRYLENE-1,2-DIOL see DKX875
trans-10,11-DIHYDRODIBENZ(a,e)ACEANTHRYLENE-10,11-DIOL see DKX900
5,6-DIHYDRODIBENZ(a,h)ANTHRACENE see DKY000
5,6-DIHYDRODIBENZ(a,j)ANTHRACENE see DKY200
7,14-DIHYDRODIBENZ(a,h)ANTHRACENE see DKY400
9,10-DIHYDRO-1,2,5,6-DIBENZANTHRACENE see DKY400
10,11-DIHYDRO-5-DIBENZ(b,f)AZEPINE see DKY800
3,4-DIHYDRO-1,2,5,6-DIBENZCARBAZOLE see DLA000
12,13-DIHYDRO-7H-DIBENZO(a,g)CARBAZOLE see DLA000
5,8-DIHYDRODIBENZO(a,def)CHRYSENE see DLA100
7,14-DIHYDRODIBENZO(b,def)CHRYSENE see DLA120
10,11-DIHYDRO-5H-DIBENZO(a,d)CYCLOHEPTENE-5-CARBOXAMIDE
   see CQH625
10,11-DIHYDRO-5H-DIBENZO(a,d)CYCLOHEPTEN-5-ONE OXIME see DDD000
3,10-DIHYDRO-5H-DIBENZO(a,d)CYCLOHEPTEN-5-YLIDENE-N,N-DIMETHYL-
   1-PROPANAMINE see EAH500
3-(10,11-DIHYDRO-5H-DIBENZO(a,d)CYCLOHEPTEN-5-YLIDENE)-1-ETHYL-2-
   METHYLPYRROLIDINE see PJA190
3-(10,11-DIHYDRO-5H-DIBENZO(a,d)CYCLOHEPTEN-5-YLIDENE)-1-ETHYL-2-
   METHYL-PYRROLIDINE HYDROCHLORIDE see TMK150
4-((10,11-DIHYDRO-5H-DIBENZO(a,d)CYCLOHEPTEN-5-YL)OXY)-1-
   METHYLPIPERIDINE HYDROGEN MALEATE see HBT000
4-((10,11-DIHYDRO-5H-DIBENZO(a,d)CYCLOHEPTEN-5-YL)OXY)-1-
   METHYLPIPERIDINE, MALEATE (1:1) see HBT000
3-α-((10,11-DIHYDRO-5H-DIBENZO(a,d)CYCLOHEPTEN-5-YL)OXY)-8-
   METHYLTROPANIUM BROMIDE see DAZ140
5,10-DIHYDRO-3,4:8,9-DIBENZOPYRENE see DLA120
5,8-DIHYDRO-3,4:9,10-DIBENZOPYRENE see DLA100
2,3-DIHYDRODICYCLOPENTA(c,lmn)PHENANTHREN-1(9H)-ONE see DGW300
2,3-DIHYDRO-1-DIETHYLAMINOETHYL-2-METHYL-3-PHENYL-4(3H)-
   QUINAZOLINONE OXALATE see DID800
2,3-DIHYDRO-1-DIETHYLAMINOETHYL-2-METHYL-3-(o-TOLYL)-4(3H)-
   QUINAZOLINONE OXALATE see DIE200
DIHYDRO-5,5-DIETHYL-2H-1,3-OXAZINE-2,4(3H)-DIONE see DJT400
DIHYDRODIETHYLSTILBESTROL see DLB400
1,4-DIHYDRO-4-(2-(DIFLUOROMETHOXY)PHENYL)-2,6-DIMETHYL-3,5-
   PYRIDINEDICARBOXYLIC ACID DIMETHYL ESTER see RSZ375
cis-5,6-DIHYDRO-5,6-DIHYDROXYBENZ(a)ANTHRACENE see BBE000
(−)(3R,4R)-trans-3,4-DIHYDRO-3,4-DIHYDROXYBENZ(a)ANTHRACENE
   see BBG000
(+)-(3S,4S)-trans-3,4-DIHYDRO-3,4-DIHYDROXYBENZ(a)ANTHRACENE
   see BBD750
trans-3,4-DIHYDRO-3,4-DIHYDROXYBENZO(a)ANTHRACENE see BBD500
(−)(3R,4R)trans-3,4-DIHYDRO-3,4-DIHYDROXYBENZO(a)ANTHRACENE
   see BBG000
(+)-(3S,4S)-trans-3,4-DIHYDRO-3,4-DIHYDROXYBENZO(a)ANTHRACENE
   see BBD750
4,5-DIHYDRO-4,5-DIHYDROXYBENZO(a)PYRENE see DLB800
4,5-DIHYDRO-4,5-DIHYDROXYBENZO(e)PYRENE see DMK400
9,10-DIHYDRO-9,10-DIHYDROXYBENZO(a)PYRENE see DLC000
trans-4,5-DIHYDRO-4,5-DIHYDROXYBENZO(a)PYRENE see DLC600
trans-7,8-DIHYDRO-7,8-DIHYDROXYBENZO(a)PYRENE see DLC800
trans-9,10-DIHYDRO-9,10-DIHYDROXYBENZO(a)PYRENE see DLC200
(±)-trans-9,10-DIHYDRO-9,10-DIHYDROXYBENZO(a)PYRENE see DLC400
7,8-DIHYDRO-7,8-DIHYDROXYBENZO(a)PYRENE-9,10-OXIDE see DLD000
trans-1,2-DIHYDRO-1,2-DIHYDROXYCHRYSENE see DLD200
(E)-1,2-DIHYDRO-1,2-DIHYDROXYDIBENZ(a,h)ANTHRACENE see DMK200
trans-3,4-DIHYDRO-3,4-DIHYDROXYDIBENZ(a,h)ANTHRACENE see DLD400
trans-3,4-DIHYDRO-3,4-DIHYDROXYDIBENZO(a,h)ANTHRACENE see DLD400
trans-3,4-DIHYDRO-3,4-DIHYDROXYDIBENZO(a,e)FLUORANTHENE
   see BBU810, DKX900
trans-12,13-DIHYDRO-12,13-DIHYDROXYDIBENZO(a,e)FLUORANTHENE
   see BBU825, DKX875
trans-3,4-DIHYDRO-3,4-DIHYDROXY-7,12-DIMETHYLBENZ(a)ANTHRACENE
   see DLD600
trans-8,9-DIHYDRO-8,9-DIHYDROXY-7,12-DIMETHYLBENZ(a)ANTHRACENE
   see DLD800
(E)-8,9-DIHYDRO-8,9-DIHYDROXY-7,12-DIMETHYLBENZ(a)ANTHRACENE
   see DLD800
trans-10,11-DIHYDRO-10,11-DIHYDROXY-7,12-DIMETHYLBENZ(a)ANTHRA-
   CENE see DLD875
9,10-DIHYDRO-9,10-DIHYDROXY-9,10-DI-n-PROPYL-1,25,6-
   DIBENZANTHRACENE see DLK600
trans-3,4-DIHYDRO-3,4-DIHYDROXY DMBA see DLD600
trans-8,9-DIHYDRO-8,9-DIHYDROXY DMBA see DLD800
(±)-(1R,2S,3R,4R)-3,4-DIHYDRO-3,4-DIHYDROXY-1,2-EPOXYBENZ(a)ANTHRA-
   CENE see DLE000

(±)-(1S,2R,3R,4R)-3,4-DIHYDRO-3,4-DIHYDROXY-1,2-EPOXYBENZ(a)ANTHRA-
   CENE see DLE200
trans-1,2-DIHYDRO-1,2-DIHYDROXYINDENO(1,2,3-cd)PYRENE see DLE400
cis-5,6-DIHYDRO-5,6-DIHYDROXY-12-METHYLBENZ(a)ACRIDINE see DLE500
trans-3,4-DIHYDRO-3,4-DIHYDROXY-7-METHYLBENZ(c)ACRIDINE
   see MGU550
(E)-1,2-DIHYDRO-1,2-DIHYDROXY-7-METHYLBENZ(a)ANTHRACENE
   see DLE600
(E)-5,6-DIHYDRO-5,6-DIHYDROXY-7-METHYL-BENZ(a)ANTHRACENE
   see DLF000
trans-1,2-DIHYDRO-1,2-DIHYDROXY-7-METHYLBENZ(a)ANTHRACENE
   see DLE600
trans-3,4-DIHYDRO-3,4-DIHYDROXY-7-METHYLBENZ(a)ANTHRACENE
   see MBV500
trans-5,6-DIHYDRO-5,6-DIHYDROXY-7-METHYLBENZ(a)ANTHRACENE
   see DLF000
trans-8,9-DIHYDRO-8,9-DIHYDROXY-7-METHYLBENZ(a)ANTHRACENE
   see DLF200
(+)-3,4-DIHYDRO-3,8-DIHYDROXY-3-METHYLBENZ(a)ANTHRACENE-
   1,7,12(2H)-TRIONE see TDX840
cis-2-α,3-DIHYDRODIHYDROXY-3-METHYLCHOLANTHRENE see DLF300
trans-7,8-DIHYDRO-7,8-DIHYDROXY-3-METHYLCHOLANTHRENE see MJD600
trans-9,10-DIHYDRO-9,10-DIHYDROXY-3-METHYLCHOLANTHRENE
   see MJD610
trans-11,12-DIHYDRO-11,12-DIHYDROXY-3-METHYLCHOLANTHRENE
   see DML800
1,2-DIHYDRO-1,2-DIHYDROXY-5-METHYLCHRYSENE see DLF400
7,8-DIHYDRO-7,8-DIHYDROXY-5-METHYLCHRYSENE see DLF600
14,19-DIHYDRO-12,13-DIHYDROXY(13-α,14-α)-20-NORCROTALANAN-11,15-
   DIONE see MRH000
(E)-1,2-DIHYDRO-1,2-DIHYDROXYTRIPHENYLENE see DMM200
1,4-DIHYDRO-1,4-DIKETONAPHTHALENE see NBA500
5,6-DIHYDRO-9,10-DIMETHOXYBENZO(g)-1,3-BENZODIOXOLO(5,6-
   a)QUINOLIZINIUM CHLORIDE DIHYDRATE see BFN550
5,6-DIHYDRO-9,10-DIMETHOXYBENZO(g)-1,3-BENZODIOXOLO(5,6-
   a)QUINOLIZINIUM SULFATE TRIHYDRATE see BFN750
5,6-DIHYDRO-9,10-DIMETHOXY-BENZO(g)-1,3-BENZODIOXOLO(5,6-
   a)QUINOLIZINIUM SULFATE (2:1) see BFN625
3,4-DIHYDRO-6-(4-(3,4-DIMETHOXYBENZOYL)-1-PIPERAZINYL)-2(1H)-
   QUINOLINONE see DLF700
3,4-DIHYDRO-6,7-DIMETHOXY-N-(2-(2-METHOXYPHENOXY)ETHYL)-2(1H)-
   ISOQUINOLINECARBOXIMIDAMIDE see SBA875
5,10-DIHYDRO-10-(2-(DIMETHYLAMINO)ETHYL)-8-ETHYLSULFONYL-5-
   METHYL-11H-DIBENZO(b,e) (1,4)DIAZEPIN-11-ONE see DLG000
5,10-DIHYDRO-10-(2-(DIMETHYLAMINO)ETHYL)-5-METHYL-11H-
   DIBENZO(b,e)(1,4)DIAZEPIN-11-ONE see DCW600
10,11-DIHYDRO-5-(3-DIMETHYLAMINO-2-METHYLPROPYL)-5H-DIBENZ
   (b,f)AZEPINE see DLH200
10,11-DIHYDRO-5-(3-(DIMETHYLAMINO)-2-METHYLPROPYL)-5H-
   DIBENZ(b,f)AZEPINE MALEATE (1:1) see SOX550
1,2-DIHYDRO-3-DIMETHYLAMINO-7-METHYL-1,2-(PROPYLMALONYL)-1,2,4-
   BENZOTRIAZINE see AQN750
1,2-DIHYDRO-3-DIMETHYLAMINO-7-METHYL-1,2-(PROPYLMALONYL)-1,2,4-
   BENZOTRIAZINE DIHYDRATE see ASA000
1,2-DIHYDRO-3-DIMETHYLAMINO-7-METHYL-1,2-(PROPYLMALONYL)-1,2,4-
   BENZOTRIAZINE SODIUM SALT see ASA250
5,6-DIHYDRO-N-(3-(DIMETHYLAMINO)PROPYL)-11H-DIBENZ(b,e)AZEPINE
   see DLH600
10,11-DIHYDRO-5-(3-(DIMETHYLAMINO)PROPYL)-5H-DIBENZ(b,f)AZEPINE
   see DLH600
10,11-DIHYDRO-5-(3-(DIMETHYLAMINO)PROPYL)-5H-DIBENZ(b,f)AZEPINE
   HYDROCHLORIDE see DLH630
10,11-DIHYDRO-5-(γ-DIMETHYLAMINOPROPYLIDENE)-5H-
   DIBENZO(a,d)CYCLOHEPTENE see EAH500
5,6-DIHYDRO-7,12-DIMETHYLBENZ(a)ANTHRACENE see DLH800
2,3-DIHYDRO-2,2-DIMETHYL-7-BENZOFURANYL METHYLCARBAMATE
   see CBS275
2,3-DIHYDRO-2,2-DIMETHYLBENZOFURANYL-7-N-METHYLCARBAMATE
   see CBS275
2,3-DIHYDRO-2,2-DIMETHYL-7-BENZOFURANYL NITROSO-METHYL CARBA-
   MATE see NJQ500
3,4-DIHYDRO-1,11-DIMETHYLCHRYSENE see DLI000
16,17-DIHYDRO-11,17-DIMETHYLCYCLOPENTA(a)PHENANTHRENE
   see DLI200
16,17-DIHYDRO-11,12-DIMETHYL-15H-CYCLOPENTA(a)PHENANTHRENE
   see DRH800
15,16-DIHYDRO-11,12-DIMETHYLCYCLOPENTA(a)PHENANTHREN-17-ONE
   see DLI400
15,16-DIHYDRO-7,11-DIMETHYL-17H-CYCLOPENTA(a)PHENANTHREN-17-
   ONE see DLI300
10,11-DIHYDRO-N,N-DIMETHYL-5H-DIBENZ(b,f)AZEPINE-5-PROPANAMINE
   MONOHYDROCHLORIDE see DLH630
10,11-DIHYDRO-N,N-DIMETHYL-5H-DIBENZO(a,d)-CYCLOHEPTENE-Δ$^{5,\Gamma}$-
   PROPYLAMINE HCL see EAI000

5,6-DIHYDRO-2-METHYL-1,4-OXATHIIN-3-CARBOXANILIDE-4,4-DIOXIDE
see DLV200

N-(4,5-DIHYDRO-4-METHYL-5-OXO-1,2-DITHIOLO(4,3-B)PYRROL-6-YL)
ACETAMIDE see ABI250

2-(1,2-DIHYDRO-1-METHYL-2-OXO-3H-INDOLE-3-
YLIDENE)HYDRZAINECARBOTHIOAMIDE see MKW250

3,4-DIHYDRO-2-METHYL-4-OXO-3-o-TOLYLQUINAZOLINE see QAK000

(±)-DIHYDRO-5-(1-METHYL-2-PENTYNYL)-5-(2-PROPENYL)-2-THIOXO-
4,6(1H,5H)-PYRIMIDINEDIONE see AGL875

3,6-DIHYDRO-α-((2-METHYLPHENOXY)METHYL)-1(2H)-PYRIDINEETHANOL
HYDROCHLORIDE see TGK225

2,3-DIHYDRO-2-METHYL-9-PHENYL-1H-INDENO(2,1-c)PYRIDINE
HYDROBROMIDE see NOE525

5,6-DIHYDRO-2-METHYL-N-PHENYL-1,4-OXATHIIN-3-CARBOXAMIDE
see CCC500

5,6-DIHYDRO-2-METHYL-N-PHENYL-1,4-OXATHIIN-3-CARBOXAMIDE-4,4-
DIOXIDE see DLV200

5,11-DIHYDRO-11-((4-METHYL-1-PIPERAZINYL)ACETYL)-6H-PYRIDO(2,3-
b)(1,4)BENZODIAZEPIN-6-ONE DIHYDROCHLORIDE see GCE500

10,11-DIHYDRO-2-(4-METHYL-1-PIPERAZINYL)-11-(2-ATHIAZOLYL)-
PYRIDAZINO(3,4-b)(1,4)BENZOXAZEPINE see DLV400

10,11-DIHYDRO-2-(4-METHYL-1-PIPERAZINYL)-11-(3,4-
XYLYL)PYRIDAZINO(3,4-b)(1,4)BENZOXAZEPINE MALEATE see DLV600

9,10-DIHYDRO-10-(1-METHYL-4-PIPERIDINYLIDENE)-9-ANTHRACENOL HY-
DROCHLORIDE see WAJ000

3,4-DIHYDRO-4-METHYL-2-PIPERIDINYL-1(2H)-NAPTHALENONE HYDRO-
CHLORIDE see PIR100

1,2-DIHYDRO-2-(1-METHYL-4-PIPERIDINYL)-5-PHENYL-4-(PHENYLMETHYL)-
3H-PYRAZOL-3-ONE (9CI) see BCP650

6,11-DIHYDRO-11-(1-METHYL-4-PIPERIDYLIDENE)-5H-
BENZO(5,6)CYCLOHEPTA (1,2-b) PYRIDINE DIMALEATE see DLV800

3,4-DIHYDROMETHYL-2H-PYRAN see MJE750

(S)-(+)-5,6-DIHYDRO-6-METHYL-2H-PYRAN-2-ONE see PAJ500

4,9-DIHYDRO-1-METHYL-3H-PYRIDO(3,4-b)INDOL-7-OL
MONOHYDROCHLORIDE see HAI300

1,5-DIHYDRO-5-METHYL-1-β-d-RIBOFURANOSYL-1,4,5,6,8-PEN-
TAAZAACENAPHTHYLEN-3-AMINE see TJE870

cis-(−)-3,5-DIHYDRO-3-METHYL-2H-THIOPYRANO(4,3,2-cd)INDOLE-2-
CARBOXYLIC ACID see CML820

2,3-DIHYDRO-6-METHYL-2-THIOXO-4(1H)-PYRIMIDINONE see MPW500

DIHYDROMORPHINE HYDROCHLORIDE see DNU310

DIHYDROMORPHINONE see DLW600

DIHYDROMORPHINONE HYDROCHLORIDE see DNU300

DIHYDROMORPHINON-N-OXYD-DITARTARAT (GERMAN) see EBY500

3,4-DIHYDROMORPHOL see PCW500

2,3-DIHYDRO-1-(MORPHOLINOACETYL)-3-PHENYL-4(1H)-QUINAZOLINONE
see MRN250

2,3-DIHYDRO-1-(MORPHOLINOACETYL)-3-PHENYL-4(1H)-QUINAZOLINONE
HYDROCHLORIDE see PCJ350

DIHYDROMYRCENE see CMT050

DIHYDROMYRCENOL see DLX000

DIHYDROMYRCENYL ACETATE see DLX100

4,5-DIHYDRONAPHTH(1,2-k)ACEPHENANTHRYLENE see PCW000

3,4-DIHYDRO-1(2H)-NAPHTHALENONE see DLX200

4,5-DIHYDRO-2-(1-NAPHTHALENYLMETHYL)-1H-IMIDAZOLE
MONOHYDROCHLORIDE see NCW000

4,5-DIHYDRO-2-(1-NAPHTHALENYLMETHYL)-1H-IMIDAZOLE MONONI-
TRATE (9CI) see NAH550

2,3-DIHYDRO-2-(1-NAPHTHALENYL)-4(1H)-QUINAZOLINONE see DLX300

2,3-DIHYDRO-2-(1-NAPHTHYL)-4(1H)-QUINAZOLINONE see DLX300

DIHYDRONE HYDROCHLORIDE see DLX400

DIHYDRONEOPINE see DKW800

1,2-DIHYDRO-5-NITRO-ACENAPHTHYLENE see NEJ500

4H-2,3-DIHYDRO-2-(5'-NITRO-2'-FURYL)-3-HYDROXY-1,3-BENZOXALINE-4-
ONE see HMY100

1,2-DIHYDRO-2-(5'-NITROFURYL)-4-HYDROXY-CHINAZOLIN-3-OXID (GER-
MAN) see DLX800

1,2-DIHYDRO-2-(5'-NITROFURYL)-4-HYDROXYQUINAZOLINE-3-OXIDE
see DLX800

4,5-DIHYDRO-N-NITRO-1-NITROSO-1H-IMIDAZOL-2-AMINE see NLB700

1,3-DIHYDRO-7-NITRO-5-PHENYL-2H-1,4-BENZODIAZEPIN-2-ONE see DLY000

3,6-DIHYDRO-2-NITROSO-2H-1,2-OXAZINE see NJX000

DIHYDRO-1-NITROSO-2,4(1H,3H)-PYRIMIDINEDIONE see NJY000

2,5-DIHYDRO-1-NITROSO-1H-PYRROLE see NLQ000

5,6-DIHYDRO-1-NITROSOTHYMINE see NJX500

5,6-DIHYDRO-1-NITROSOURACIL see NJY000

1,2-DIHYDRO-2-(5-NITRO-2-THIENYL)QUINAZOLIN-4(3H)-ONE see DLY200

1,2-DIHYDRO-2-(5-NITRO-2'-THIENYL)-4(3H)-QUINAZOLINONE see DLY200

DIHYDRONORDICYCLOPENTADIENYL ACETATE see DLY400

DIHYDRONORGUAIARETIC ACID see NBR000

10,11-DIHYDRO-N,N,β-TRIMETHYL-5H-DIBENZ(b,f)AZEPINE-5-PROPANAM-
INE see DLH200

(±)-10,11-DIHYDRO-N,N,β-TRIMETHYL- 5H-DIBENZO(a,d),CYCLOHEPTENE-
5-PROPANAMINE HCl see BPT750

d,l-10,11-DIHYDRO-N,N,β-TRIMETHYL-5H-DIBENZO(a,d)-CYCLOHEPTENE-5-
PROPYLAMINE see BPT500

(±)-10,11-DIHYDRO-N,N,β-TRIMETHYL-5H-DIBENZO(a,d)-CYCLOHEPTENE-5-
PROPYLAMINE HCl see BPT750

(+ −)-9,10-DIHYDRO-N,N,10-TRIMETHYL-2-(TRIFLUORMETHYL)-9-
ANTHRACENPROPANAMIN (GERMAN) see DMF800

cis-(±)-9,10-DIHYDRO-N,N,10-TRIMETHYL-2-(TRIFLUOROMETHYL)-9-
ANTHRACENE PROPANAMINE see DMF800

3,6-DIHYDRO-1,2,2H-OXAZINE see DLY700

DIHYDROOXIRENE see EJN500

2,3-DIHYDRO-3-OXOBENZISOSULFONAZOLE see BCE500

2,3-DIHYDRO-3-OXOBENZISOSULPHONAZOLE see BCE500

5,13-DIHYDRO-5-OXOBENZO(e)(2)BENZOPYRANO(4,3-b)INDOLE see DLY800

3,4-DIHYDRO-4-OXO-3-BENZOTRIAZINYLMETHYL O,O-DIETHYL PHOS-
PHORODITHIOATE see EKN000

S-(3,4-DIHYDRO-4-OXO-1,2,3-BENZOTRIAZIN-3-YLMETHYL) O,O-DIETHYL
PHOSPHORODITHIOATE EKN000

S-(3,4-DIHYDRO-4-OXO-1,2,3-BENZOTRIAZIN-3-YLMETHYL)- O,O-
DIMETHYL PHOSPHORODITHIOATE see ASH500

S-(3,4-DIHYDRO-4-OXO-BENZO(α)(1,2,3)TRIAZIN-3-YLMETHYL)-O,O-
DIMETHYL PHOSPHORODITHIOATE see ASH500

6,11-DIHYDRO-11-OXO-DIBENZ(b,e)OXEPIN-3-ACETIC ACID see OMU000

3-(1,3-DIHYDRO-1-OXO-2H-ISOINDOL-2-YL)-2,6-DIOXOPIPERIDINE
see DVS600

4-(1,3-DIHYDRO-1-OXO-2H-ISOINDOL-2-YL)-α-METHYLBENZENEACETIC
ACID see IDA400

3-(1,3-DIHYDRO-1-OXO-2H-ISOINDOL-2-YL)-2-OXOPIPERIDINE see EAJ100

O-(1,6)-DIHYDRO-6-OXO-1-PHENYLPYRIDAZIN-3-LY), O,O-DIETHYL PHOS-
PHOROTHIOATE see POP000

10-((3,6-DIHYDRO-6-OXO-2H-PYRAN-2-YL)HYDROXYMETHYL)-5,9,10,11-tert-
TRAHYDRO-4-HYDROXY-5-(1-HYDROXYHEPTYL)-1H-CYCLONONA
(1,2-C :5,6-C')DIFURAN-1,3,6,8(4H)-TETRONE see RQP000

DIHYDROOXYCODEINONE HYDROCHLORIDE see DLX400

DIHYDROOXYCODEINON-N-OXYD (GERMAN) see DGY000

2,3-DIHYDRO-3,3',4',5,7-PENTAHYDROXYFLAVONE see DMD000

N-(1,1-DIHYDROPERFLUOROOCTYL)PYRIDINIUM
TRIFLUOROMETHANESULFONATE see MBV710

1,2-DIHYDROPHENANTHRENE see DLZ000

1,2-DIHYDRO-1,2-PHENANTHRENEDIOL see DMA000

3,4-DIHYDRO-3,4-PHENANTHRENEDIOL see PCW500

1a,9b-DIHYDROPHENANTHRO(9,10-B)OXIRENE (9CI) see PCX000

N-(9,10-DIHYDRO-2-PHENANTHRYL)ACETAMIDE see DMA400

5-(4,5-DIHYDRO-2-PHENYL-3H-BENZ(e)INDOL-3-YL)-2-HYDROXYBENZOIC
ACID see FAO100

5-(4,5-DIHYDRO-2-PHENYL-3H-BENZ(e)INDOL-3-YL)SALICYLIC ACID
see FAO100

4,5-DIHYDRO-N-PHENYL-N-PHENYLMETHYL-1H-IMIDAZOLE-2-
METHANAMINE see PDC000

5,6-DIHYDRO-2-PHENYLPYRAZOLO(5,1-a)ISOQUINOLINE see PEU600

5-(2-(3,6-DIHYDRO-4-PHENYL-1(2H)-PYRIDYL)ETHYL)-3-METHYL-2-OXAZOL
IDINONE see DMB000

5,6-DIHYDRO-2-PHENYL-(1,2,4)TRIAZOLO(5,1-a)ISOQUINOLINE (9CI)
see PEU650

2,3-DIHYDROPHORBOL ACETATE MYRISTATE see DMB200

2,3-DIHYDROPHORBOL MYRISTATE ACETATE see DMB200

10,11-DIHYDRO-5-(3-(4-PIPERIDINO-4-CARBAMOYLPIPERIDINO)PROPYL-
(b,f)AZEPINE see CBH500

6,7-DIHYDRO-6-(2-PROPENYL)-5H-DIBENZ(c,e)AZEPINE PHOSPHATE
see AGD500

4,5-DIHYDRO-2-(2-PROPENYLTHIO)THIAZOLE (9CI) see AGT250

(S)-2-(1,2-DIHYDRO-5-PROPYL-2(3H)-FURYLIDENE)-1,3-CYCLOPENTANEDIONE
see OKS100

2,3-DIHYDRO-6-PROPYL-2-THIOXO-4(1H)-PYRIMIDINONE see PNX000

DIHYDROPSEUDOIONONE see GDE400

α-β-DIHYDROPSEUDOIONONE see GDE400

3,7-DIHYDRO-1H-PURINE-2,6-DIONE see XCA000

1,7-DIHYDRO-6H-PURINE-6-THIONE see POK000

1,7-DIHYDRO-6H-PURINE-6-THIONE MONOHYDRATE (9CI) see MCQ100

1,7-DIHYDRO-6H-PURIN-6-ONE see DMC000

DIHYDROPYRAN see DMC200

3,4-DIHYDROPYRAN see DMC200

2H-3,4-DIHYDROPYRAN see DMC200

Δ²-DIHYDROPYRAN see DMC200

3,4-DIHYDRO-2H-PYRAN-2-CARBOXALDEHYDE see ADR500

2,3-DIHYDRO-1H-PYRAZOLO(2,3-a)IMIDAZOLE see IAN100

1,5-DIHYDRO-4H-PYRAZOLO(3,4-d)PYRIMIDIN-4-ONE see ZVJ000

1,2-DIHYDROPYRIDAZINE-3,6-DIONE see DMC600

1,2-DIHYDRO-3,6-PYRIDAZINEDIONE see DMC600

6,7-DIHYDROPYRIDO(1,2a;2',1'-C)PYRAZINEDIUM DIBROMIDE see DWX800

1,2-DIHYDROPYRIDO(2,1,e)TETRAZOLE see DMC800

3,4-DIHYDROPYRROLIDINE see SND000

DIHYDROQUERCETIN see DMD000

(+)-DIHYDROQUERCETIN see DMD000

2,3-DIHYDROQUERCETIN see DMD000

(2R,3R)-DIHYDROQUERCETIN see DMD000
DIHYDROQUINIDINE see HIG500
10,11-DIHYDROQUINIDINE see HIG500
7,8-DIHYDRORETINOIC ACID see DMD100
4a,5-DIHYDRO-RIBOFLAVIN-5'-PHOSPHATE SODIUM SALT see DMD200
DIHYDROSAFROLE see DMD600
DIHYDROSAMIDIN see DUO350
DIHYDROSTILBESTROL see DLB400
DIHYDROSTREPTOMYCIN see DME000
DIHYDROSTREPTOMYCIN and STREPTOMYCIN see SLY200
DIHYDROSTREPTOMYCIN SULFATE see DME200
2,3-DIHYDROSUCCINIC ACID see TAF750
DIHYDROTACHYSTEROL see DME300
DIHYDROTACHY STEROL see IGF000
DIHYDROTACHYSTEROL₂ see DME300
DIHYDROTERPINYL ACETATE see DME400
DIHYDROTESTOSTERONE see DME500
4-DIHYDROTESTOSTERONE see DME500
5-α-DIHYDROTESTOSTERONE see DME500
5-β-DIHYDROTESTOSTERONE see HJB100
4,5-α-DIHYDROTESTOSTERONE see DME500
DIHYDROTESTOSTERONE PROPIONATE see DME525
5-α-DIHYDROTESTOSTERONE PROPIONATE see DME525
DIHYDROTESTOSTERONE-17-β-PROPIONATE see DME525
DIHYDRO-5-TETRADECYL-2(3H)-FURANONE see SLL400
4,5-DIHYDRO-N-(4,6,7,8-TETRAHYDRO-1-NAPHTHALENYL)-1H-IMIDAZOL-2-
  AMINE HYDROCHLORIDE HYDRATE see RGP450
4,5-DIHYDRO-N-(4,6,7,8-TETRAHYDRO-1-NAPHTHALENYL)-1H-IMIDAZOL-2-
  AMINE MONOHYDROCHLORIDE see THJ825
4,5-DIHYDRO-2-(1,2,3,4-TETRAHYDRO-1-NAPHTHALENYL)-1H-IMIDAZOLE
  MONOHYDROCHLORIDE see VRZ000
5,6-DIHYDRO-2,3,9,10-TETRAMETHOXYDIBENZO(a,g)QUINOLIZINIUM HY-
  DROXIDE see PAE100
6,7-DIHYDRO-1,2,3,10-TETRAMETHOXY-7-(METHYLAMINO)-
  BENZO(α)HEPTALEN-9(5H)-ONE see MIW500
(S)-6,7-DIHYDRO-1,2,3,10-TETRAMETHOXY-7-(METHYLAMINO)-
  BENZO(a)HEPTALEN-9(5H)-ONE see MIW500
1,2-DIHYDRO-2,2,4,6-TETRAMETHYLPYRIDINE see DME600
6,7-DIHYDRO-3,5,5,7-TETRAMETHYL-5H-THIAZOLO(3,2-a)PYRIMIDIN-7-OL
  HYDROCHLORIDE see DME700
DIHYDROTHEELIN see EDO000
N-(5,6-DIHYDRO-4H-1,3-THIAZINYL)-2,6-XYLIDINE see DMW000
2,3-DIHYDROTHIIRENE see EJP500
2,5-DIHYDROTHIOPHENE DIOXIDE see DMF000
2,5-DIHYDROTHIOPHENE-1,1-DIOXIDE see DMF000
2,5-DIHYDROTHIOPHENE SULFONE see DMF000
2,3-DIHYDRO-2-THIOXO-4(1H)-PYRIMIDINONE see TFR250
1,2-DIHYDRO-s-TRIAZINE-4,6-DIAMINO-2,2-DIMETHYL-1-PHENYL-2,4,5-
  TRICHLOROPHENOXYACETATE see DMF400
16,17-DIHYDRO-11,12,17-TRIMETHYLCYCLOPENTA(a)PHENANTHRENE
  see DMF600
1,2-DIHYDRO-2,2,4-TRIMETHYL-6-ETHOXYQUINOLINE see SAV000
3,7-DIHYDRO-1,3,7-TRIMETHYL-1H-PURINE-2,6-DIONE see CAK500
3,7-DIHYDRO-1,3,7-TRIMETHYL-1H-PURINE-2,6-DIONE MONOHYDROBRO-
  MIDE see CAK750
1,2-DIHYDRO-2,2,4-TRIMETHYLQUINOLINE see TLP500
1,2-DIHYDROTRIPHENYLENE see DMG000
2,4-DIHYDROXYACETOPHENONE see DMG400
2,5-DIHYDROXYACETOPHENONE see DMG600
2',4'-DIHYDROXYACETOPHENONE see DMG400
2',5'-DIHYDROXYACETOPHENONE see DMG600
11-β,17-α-DIHYDROXY-21-ACETOXYPREGESTERONE see HHQ800
1,8-DIHYDROXY-10-ACETYL-9-ANTHRONE see ACA750
3-α,17-β-DIHYDROXY-5-α-ANDROSTANE see DMG700
1,8-DIHYDROXY-9,10-ANTHRACENEDIONE see DMH400
1,2-DIHYDROXYANTHRACHINON (CZECH) see DMG800
1,4-DIHYDROXYANTHRACHINON (CZECH) see DMH000
1,5-DIHYDROXYANTHRACHINON (CZECH) see DMH200
1,8-DIHYDROXYANTHRACHINON (CZECH) see DMH400
DIHYDROXYANTHRANOL see APH250
1,8-DIHYDROXYANTHRANOL see APH250
1,8-DIHYDROXY-9-ANTHRANOL see APH250
DIHYDROXYANTHRAQUINONE see MQY090
1,2-DIHYDROXYANTHRAQUINONE see DMG800
1,4-DIHYDROXYANTHRAQUINONE see DMH000
1,5-DIHYDROXYANTHRAQUINONE see DMH200
1,8-DIHYDROXYANTHRAQUINONE see DMH400
2,6-DIHYDROXYANTHRAQUINONE see DMH600
1,2-DIHYDROXY-9,10-ANTHRAQUINONE see DMG800
1,4-DIHYDROXY-9,10-ANTHRAQUINONE see DMH000
1,5-DIHYDROXY-9,10-ANTHRAQUINONE see DMH200
6,6'-((4,8-DIHYDROXY-1,5-ANTHRAQUINONNONYLENE)DIIMINO) DI-m-TOLU-
  ENE SULFONIC ACID DISODIUM SALT see TGS000
1,8-DIHYDROXY-9-ANTHRONE see APH250

3,4-DIHYDROXYBENZALDEHYDE METHYLENE KETAL see PIW250
N,2-DIHYDROXYBENZAMIDE see SAL500
3,9-DIHYDROXYBENZ(a)ANTHRACENE see BBG200
1,4-DIHYDROXY-BENZEEN (DUTCH) see HIH000
1,4-DIHYDROXYBENZEN (CZECH) see HIH000
DIHYDROXYBENZENE see HIH000
m-DIHYDROXYBENZENE see REA000
o-DIHYDROXYBENZENE see CCP850
p-DIHYDROXYBENZENE see HIH000
1,2-DIHYDROXYBENZENE see CCP850
1,3-DIHYDROXYBENZENE see REA000
1,4-DIHYDROXYBENZENE see HIH000
3,4-DIHYDROXYBENZENEACRYLIC ACID see CAK375
1,3-DIHYDROXYBENZENE DIACETATE see REA100
4,5-DIHYDROXY-1,3-BENZENEDISULFONIC ACID DISODIUM SALT
  see DXH300
2,5-DIHYDROXYBENZENESULFONIC ACID CALCIUM SALT see DMI300
2,5-DIHYDROXYBENZENESULFONIC ACID with N-ETHYLETHANAMINE
  see DIS600
3,3'-DIHYDROXYBENZIDINE see DMI400
2,4-DIHYDROXYBENZOFENON (CZECH) see DMI600
2,4-DIHYDROXYBENZOIC ACID see HOE600
2,5-DIHYDROXYBENZOIC ACID see GCU000
1,4-DIHYDROXY-BENZOL (GERMAN) see HIH000
2,4-DIHYDROXYBENZOPHENONE see DMI600
2,5-DIHYDROXYBIPHENYL see BGG250
2,2'-DIHYDROXYBIPHENYL see BGG000
2,6-DIHYDROXY-5-BIS(2-CHLOROETHYL)AMINOPYRAMIDINE see BIA250
1,4-DIHYDROXY-5,8-BIS(2-((2-HYDROXYETHYL)AMINO)ETHYLAMINO)-9,10-
  ANTHRACENEDIONE DIACETATE see DBE875
1,4-DIHYDROXY-5,8-BIS((2-((HYDROXYETHYL)AMINO)ETHYL)AMINO)-9,10-
  ANTHRACENEDIONE (9CI) see MQY090
(4,5-DIHYDROXY-1,3-BIS(HYDROXYMETHYL)-2-IMIDAZOLIDINONE
  see DTG000
2,5-DIHYDROXY-3,6-BIS(5-(3-METHYL-2-BUTENYL)-1H-INDOL-3-YL)-2,5-
  CYCLOHEXADIENE-1,4-DIONE see CNF159
3-β,14-β-DIHYDROXYBUFA-4,20,22-TRIENOLIDE 3-RHAMNOSIDE see POB500
1,3-DIHYDROXYBUTANE see BOS500
1,4-DIHYDROXYBUTANE see BOS750
2,3-DIHYDROXYBUTANE see BOT000
2,3-DIHYDROXYBUTANEDIOC ACID see TAF750
2,3-DIHYDROXYBUTANEDIOIC ACID, DIAMMONIUM SALT see DCH000
2,3-DIHYDROXY-(R-(R*,R*))-BUTANEDIOIC ACID DISODIUM SALT (9CI)
  see BLC000
(R*,S*)-2,3-DIHYDROXY-BUTANEDIOIC ACID ION(2-) (9CI) see TAF775
1,4-DIHYDROXY-1,4-BUTANEDISULFONIC ACID, DISODIUM SALT
  see SMX000
DIHYDROXYBUTENEDIOIC ACID see DMW200
2,6-DIHYDROXY-4-CARBOXYPYRIDINE see DMV400
3,β,14-DIHYDROXY-5, β-CARD-20(22)ENOLIDE see DMJ000
(3-β,5-β)-3,14-DIHYDROXY-CARD-20(22)-ENOLIDE see DMJ000
3-β,14-DIHYDROXY-5-β-CARD-20(22)-ENOLIDE-3-FORMATE see FNK050
(±)-2,3-DIHYDROXYCHLOROPROPANE see CHL875
DIHYDROXYCHLOROTHIAZIDUM see CFY000
DIHYDROXY 3-12 CHOLANATE de Na (FRENCH) see SGE000
3,12-DIHYDROXYCHOLANIC ACID see DAQ400
3-α,7-α-DIHYDROXYCHOLANIC ACID see CDL325
3-α,7-β-DIHYDROXYCHOLANIC ACID see DMJ200
3-α,12-α-DIHYDROXYCHOLANIC ACID see DAQ400
3,7-DIHYDROXYCHOLAN-24-OIC ACID see DMJ200
3-α,7-α-DIHYDROXY-5-β-CHOLANOIC ACID see DMJ200
3-α,12-α-DIHYDROXY-5-β-CHOLANOIC ACID see DAQ400
3-α,7-α-DIHYDROXY-5-β-CHOLAN-24-OIC ACID see CDL325
3-α,7-β-DIHYDROXY-6-β-CHOLAN-24-OIC ACID see DMJ200
3-α,12-α-DIHYDROXY-5-β-CHOLAN-24-OIC ACID see DAQ400
(3-α,5-β,7-β)-3,7-DIHYDROXYCHOLAN-24-OIC ACID see DMJ200
3-α-12-α-DIHYDROXY-5-β-CHOLAN-24-OIC ACID with DIBENZ(a,h)ANTHRA-
  CENE see DCT800
(3-α,5-β,12-α)-3,12-DIHYDROXY-CHOLAN-24-OIC ACID MONOSODIUM SALT
  see SGE000
3-α,12-α-DIHYDROXY-5-β-CHOLAN-24-OIC ACID SODIUM SALT see SGE000
3-α,7-α-DIHYDROXYCHOLANSAEURE (GERMAN) see DMJ200
3-α,12-α-DIHYDROXYCHOLANSAEURE (GERMAN) see DAQ400
1,25-DIHYDROXYCHOLECALCIFEROL see DMJ400
1a,25-DIHYDROXYCHOLECALCIFEROL see DMJ400
1-α,25-DIHYDROXYCHOLECALCIFEROL see DMJ400
3-α,12-α-DIHYDROXY-5-β-CHOL-8(14)-EN-24-OIC ACID see AQO500
3,4-DIHYDROXYCINNAMIC ACID see CAK375
DIHYDROXYCODEINONE HYDROCHLORIDE see DLX400
DI-(4-HYDROXY-3-COUMARINYL)METHANE see BJZ000
DIHYDROXYCYCLOBUTENEDIONE see DMJ600
3,4-DIHYDROXYCYCLOBUTENE-1,2-DIONE see DMJ600
3,4-DIHYDROXY-3-CYCLOBUTENE-1,2-DIONE see DMJ600
(-)-3,14-DIHYDROXY-N-(CYCLOBUTYLMETHYL)MORPHINAN see CQF079

1,3-DIHYDROXY-2-ETHOXYMETHYLANTHRAQUINONE see DMT200
DI(2-HYDROXYETHYL)AMINE see DHF000
DIHYDROXYETHYLANILINE see BKD500
N,N-DI(2-HYDROXYETHYL)ANILINE see BKD500
N,N-DI(β-HYDROXYETHYL)ANILINE see BKD500
N,N-DI(2-HYDROXYETHYL)CYCLOHEXYLAMINE see DMT400
2,2'-DIHYDROXYETHYL ETHER see DJD600
DI(HYDROXYETHYL) ETHER DINITRATE see DJE400
1,4-DI(2-HYDROXYETHYL)PIPERAZINE see PIJ750
N,N'-DI(2-HYDROXYETHYL)PIPERAZINE see PIJ750
β,β'-DIHYDROXYETHYL SULFIDE see TFI500
N,N-DIHYDROXYETHYL-m-TOLUIDINE see DHF400
DI-(HYDROXYETHYL)-o-TOLYLAMINE see DMT800
3,17-β-DIHYDROXY-17-α-ETHYNYL-1,3,5(10)-ESTRATRIENE see EEH500
3,17-β-DIHYDROXY-17-α-ETHYNYL-1,3,5(10)-OESTRATRIENE see EEH500
3',6'-DIHYDROXYFLUORAN see FEV500
DIHYDROXYFLUORANE see FEV000
11-β,17-β-DIHYDROXY-9-α-FLUORO-17-α-METHYL-4-ANDROSTER-3-ONE
   see AOO275
2,2'-DIHYDROXY-3,3',5,5',6,6'-HEXACHLORODIPHENYLMETHANE
   see HCL000
2,2'-DIHYDROXY-3,5,6,3',5',6'-HEXACHLORODIPHENYLMETHANE
   see HCL000
3,4-DIHYDROXY-1,5-HEXADIENE see DMU000
1-6,10a-β-DIHYDROXY-1,2,3,9,10,10a-HEXAHYDRO-4H(10),4a-IMINOETHANO-
   PHENANTHRENE TARTRATE see NNJ600
1,8-DIHYDROXY-3-(HYDROXYMETHYL)-9,10-ANTHRACENEDIONE
   see DMU600
1,8-DIHYDROXY-3-HYDROXYMETHYLANTHRAQUINONE see DMU600
1,8-DIHYDROXY-3-HYDROXYMETHYL-10-(6-HYDROXYMETHYL-3,4,5-TRIHY-
   DROXY-2-PYRANYL)ANTHRONE see BAF825
(2'S,3'R,6'R)-DIHYDROXY-2'-(HYDROXYMETHYL)-2',4',6'-TRIMETHYL-
   SPIRO(CYCLOPROPANE-1,5'-(5H)INDEN)-7'(6'H)-ONE, 2',3'-DIHYDRO-3',6'-
   see LIO600
7-(3,5-DIHYDROXY-2-(3-HYDROXY-1-OCTENYL)CYCLOPENTYL)-5-
   HEPTENOIC ACID see POC500
dl-7-(3,5-DIHYDROXY-2-(3-HYDROXY-1-OCTENYL)CYCLOPENTYL)-5-
   HEPTENOIC ACID see POC525
7-(3,5-DIHYDROXY-2-(3-HYDROXY-1-OCTENYL)CYCLOPENTYL)-5-
   HEPTENOIC ACID, THAM see POC750
7-(3,5-DIHYDROXY-2-(3-HYDROXY-1-OCTENYL)CYCLOPENTYL)-5-
   HEPTENOIC ACID, TRIMETHAMINE SALT see POC750
5,7-DIHYDROXY-2-(4-HYDROXYPHENYL)-4H-1-BENZOPYRAN-4-ONE
   see CDH250
2,3-DIHYDROXY-2-((4-HYDROXYPHENYL)METHYL)BUTANEDIOIC ACID
   see HJO500
dl-3,4-DIHYDROXY-N-3-(4-HYDROXYPHENYL)-1-METHYL-n-PROPYL
   PHENETHYLAMINE HYDROCHLORIDE see DXS375
d-(+)-2,4-DIHYDROXY-N-(3-HYDROXYPROPYL)-3,3-DIMETHYLBUTYRAM-
   IDE see PAG200
2,2-DIHYDROXY-1,3-INDANDIONE see DMV200
2,2-DIHYDROXY-1H-INDENE-1,3(2H)-DIONE see DMV200
2,6-DIHYDROXYISONICOTINIC ACID see DMV400
3,4-DIHYDROXY-α-((ISOPROPYLAMINO)METHYL)BENZYL ALCOHOL
   see DMV600
3,5-DIHYDROXY-α-((ISOPROPYLAMINO)METHYL)BENZYL ALCOHOL
   see DMV800
3,4-DIHYDROXY-α-((ISOPROPYLAMINO)METHYL)BENZYL ALCOHOL
   HYDROCHLORIDE see IMR000
(±)-3,4-DIHYDROXY-α-((ISOPROPYLAMINO)METHYL)BENZYL ALCOHOL
   HYDROCHLORIDE see IQS500
3,5-DIHYDROXY-α-((ISOPROPYLAMINO)METHYL)BENZYL ALCOHOL SUL-
   FATE see MDM800
β,β'-DIHYDROXYISOPROPYL CHLORIDE see CDT750
2,2'-DIHYDROXYISOPROPYL ETHER see OQM000
3,5-DIHYDROXY-4-ISOVALERYL-2,6,6-TRIS(3-METHYL-2-BUTENYL)-2,4-
   CYCLOHEXADIEN-1-ONE see LIU000
5,6-DIHYDRO-2-(2,6-XYLIDINO)-4H-1,3-THIAZINE see DMW000
DIHYDROXYMALEIC ACID see DMW200
3,7-DIHYDROXY-9-METHOXY-1-METHYL-6H-DIBENZO(b,d)PYRAN-6-ONE
   see AGW500
1,2-DIHYDROXY-3-(2-METHOXYPHENOXY)PROPANE see RLU000
2-(3,4-DIHYDROXY-5-METHOXYPHENYL)-3,5,7-TRIHYDROXYBENZOPYRYL-
   IUM, ACID ANION see PCU000
3,4-DIHYDROXY-α-METHYLAMINOACETOPHENONE see MGC350
3,4'-DIHYDROXY-2-(METHYLAMINO)ACETOPHENONE see MGC350
3,4-DIHYDROXY-α-((METHYLAMINO)METHYL)BENZYL ALCOHOL
   see VGP000
(±)-3,4-DIHYDROXY-α-((METHYLAMINO)METHYL)BENZYL ALCOHOL
   HYDROCHLORIDE see AES625
(−)-3,4-DIHYDROXY-α-((METHYLAMINO)METHYL)BENZYL) ALCOHOL (+)-
   TARTRATE (1:1) SALT see AES000
3-DI(HYDROXYMETHYL)AMINO-6-(5-NITRO-2-FURYLETHNEYL)-1,2,4-
   TRIAZINE see BKH500

3-DI(HYDROXYMETHYL)AMINO-6-(2-(5-NITRO-2-FURYL)VINYL)-1,2,4-TRI-
   AZINE see BKH500
7:12-DIHYDROXYMETHYLBENZ(a)ANTHRACENE see BBF500
1,3-DIHYDROXY-5-METHYLBENZENE see MPH500
2,6-DIHYDROXY-4-METHYL-5-BIS(2-CHLOROETHYL)AMINOPYRIMIDINE
   see DYC700
2,2-(DIHYDROXYMETHYL)-1-BUTANOL, MONOALLYL ETHER see TLX100
cis-1,2-DIHYDROXY-3-METHYLCHOLANTHRENE see MIK750
(E)-3,4-DIHYDROXY-7-METHYL-3,4-DIHYDROBENZ(a)ANTHRACENE-12-
   METHANOL see DMX000
2,12-DIHYDROXY-4-METHYL-11,16-DIOXOSENECIONANIUM see DMX200
DI-4-HYDROXY-3,3'-METHYLENEDICOUMARIN see BJZ000
DIHYDROXYMETHYL FURATRIZINE see BKH500
11-α,17-β-DIHYDROXY-17-METHYL-3-OXOANDROSTA-1,4-DIENE-2-CAR-
   BOXALDEHYDE see FNK040
14,16-DIHYDROXY-3-METHYL-7-OXO-trans-BENZOXACYCLOTETRADEC-11-
   EN-1-ONE see ZAT000
(5Z,11-α,13E,15S,17Z)-11,15-DIHYDROXY-15-METHYL-9-OXO-PROSTA-5,13-
   DIEN-1-OIC ACID see MOV800
(11-α-13E)-(±)-11,16-DIHYDROXY-16-METHYL-9-OXOPROST-13-EN-1-OIC
   ACID METHYL ESTER see MJE775
2,4-DIHYDROXY-2-METHYLPENTANE see HFP875
DIHYDROXYMETHYL PEROXIDE see DMN200
1,2-DIHYDROXY-3-(2-METHYLPHENOXY)PROPANE see GGS000
α,β-DIHYDROXY-γ-(2-METHYLPHENOXY)PROPANE see GGS000
3,4-DIHYDROXY-3-METHYL-4-PHENYL-1-BUTYNE see DMX800
d-N,N'-DI(1-HYDROXYMETHYLPROPYL)ETHYLENEDIAMINE DIHYDRO-
   CHLORIDE mixed with SODIUM NITRITE see SIS500
N,N'-DIHYDROXYMETHYLUREA see DTG700
1,8-DIHYDROXY-10-MYRISTOYL-9-ANTHRONE see MSA750
2,7-DIHYDROXYNAPHTHALENE see NAO500
N,3-DIHYDROXY-4-(1-NAPHTHALENYLOXY)BUTANINIDAMIDE HYDRO-
   CHLORIDE see NAC500
d-threo-N-(1,1'-DIHYDROXY-1-p-
   NITROPHENYLISOPROPYL)DICHLOROACETAMIDE see CDP250
2,2'-DIHYDROXY-N-NITROSODIETHYLAMINE see NKM000
3,4-DIHYDROXYNOREPHEDRINE HYDROCHLORIDE see AMB000
3,17-β-DIHYDROXYOESTRA-1,3,5-TRIENE see EDO000
3,17-α-DIHYDROXYOESTRA-1,3,5(10)-TRIENE see EDO500
3,17-β-DIHYDROXY-1,3,5(10)-OESTRATRIENE see EDO000
DIHYDROXYOESTRIN see EDO000
(11-β)-11,21-DIHYDROXY-17-(1-OXOBUTOXY)-PREGN-4-ENE-3,20-DIONE
   see HHQ825
(5Z,11-α,13E,15S)-11,15-DIHYDROXY-9-OXOPROSTA-5,13-DIEN-1-OIC ACID
   see DVJ200
9,15-DIHYDROXY-11-OXO-PROSTA-5,13-DIEN-1-OIC ACID, (5Z,9-α,13E,15S)-
   see POC275
(5Z,11-α,13E,15S)-11,15-DIHYDROXY-9-OXOPROSTA-5,13-DIEN-1-OIC ACID
   MONOSODIUM SALT see POC360
(11-α,13E,15S)-11,15-DIHYDROXY-9-OXO-PROST-13-EN-1-OIC ACID (9CI)
   see POC350
11,15-DIHYDROXY-9-OXO-PROST-13-EN-1-OIC ACID, (11-α,13E,15S)-, and
   α-CYCLODEXTRIN see AGW275
1,8-DIHYDROXY-10-(1-OXOTETRADECYL)-9(10H)-ANTHRACENONE
   see MSA750
1,2-DIHYDROXY-2-OXO-N-(2,6-XYLYL)-3-PYRIDINECARBOXAMIDE
   see XLS300
R-7,t-8-DIHYDROXY-t-9,10-OXY-7,8,9,10-TETRAHYDROBENZO(a)PYRENE
   see DMQ800
anti-r-7,trans-8-DIHYDROXY-trans-9,10-OXY-7,8,9,10-
   TETRAHYDROBENZO(a)PYRENE see BCU250
trans-7,8-DIHYDROXY-9,10-OXY-7,8,9,10-TETRAHYDROBENZO(a)PYRENE
   see BCU250
2,2'-DIHYDROXY-3,3',5,5',6-PENTACHLOROBENZANILIDE see DMZ000
1,5-DIHYDROXYPENTANE see PBK750
3,5-DIHYDROXYPHENOL see PGR000
DIHYDROXY-l-PHENYLALANINE see DNA200
3,4-DIHYDROXYPHENYLALANINE see DNA200
(−)-3,4-DIHYDROXYPHENYLALANINE see DNA200
l-DIHYDROXYPHENYL-l-ALANINE see DNA200
l-α-DIHYDROXYPHENYLALANINE see DNA200
3,4-DIHYDROXYPHENYL-l-ALANINE see DNA200
3,4-DIHYDROXYPHENYL-l-ALANINE see DNA200
l-3,4-DIHYDROXYPHENYLALANINE see DNA200
(−)-3-(3,4-DIHYDROXYPHENYL)-l-ALANINE see DNA200
3-(3,4-DIHYDROXYPHENYL)-l-ALANINE see DNA200
l-3-(3,4-DIHYDROXYPHENYL)ALANINE see DYC200
l-3,4-DIHYDROXYPHENYL-α-ALANINE see DNA200
l-β-(3,4-DIHYDROXYPHENYL)-l-ALANINE see DNA200
β-(3,4-DIHYDROXYPHENYL)-l-ALANINE see DNA200
β-(3,4-DIHYDROXYPHENYL)-α-ALANINE see DNA200
l-3,4-DIHYDROXYPHENYLALANINE METHYL ESTER see DYC300
1-(3,4-DIHYDROXYPHENYL)-2-AMINO-1-BUTANOL HYDROCHLORIDE
   see ENX500

l-1-(3,4-DIHYDROXYPHENYL)-2-AMINOETHANOL see NNO500
3,4-DIHYDROXYPHENYL)-1-AMINO-2-ETHANOL-1-HYDROCHLORIDE see NNP050
3,4-DIHYDROXYPHENYLAMINOPROPANOL HYDROCHLORIDE see AMB000
1-(3,5-DIHYDROXYPHENYL)-2-tert-BUTYLAMINOETHANOL SULPHATE see TAN250
2-(3,4-DIHYDROXYPHENYL)-2,3-DIHYDRO-3,5,7-TRIHYDROXY-4H-1-BENZOPYRAN-4-ONE see DMD000
(2R-trans)-2-(3,4-DIHYDROXYPHENYL)-2,3-DIHYDRO-3,5,7-TRIHYDROXY-4H-1-BENZOPYRAN-4-ONE see DMD000
α-(3,4-DIHYDROXYPHENYL)-β-DIMETHYLAMINOETHANOL see MJV000
l-3,4-DIHYDROXYPHENYLETHANOLAMINE see NNO500
DIHYDROXYPHENYLETHANOLISOPROPYLAMINE see DMV600
1-(2,4-DIHYDROXYPHENYL)ETHANONE see DMG400
3,4-DIHYDROXYPHENYLETHYLMETHYLAMINE HYDROCHLORIDE see EAZ000
3,4-DIHYDROXYPHENYLGLYOXIME see DNA300
meso-3,4-DI(p-HYDROXYPHENYL)-n-HEXANE see DLB400
γ,Δ-DI(p-HYDROXYPHENYL)-HEXANE see DLB400
3,4′(4,4′-DIHYDROXYPHENYL)HEX-3-ENE see DKA600
1-(3,4-DIHYDROXYPHENYL)-1-HYDROXY-2-AMINOBUTANE HYDROCHLORIDE see ENX500
1-(3,5-DIHYDROXY-PHENYL-2-((1-(4-HYDROXYBENZYL)ETHYL)AMINO)-ETHANOL HYDROBROMIDE see FAQ100
α-(3,4-DIHYDROXYPHENYL)-α-HYDROXY-β-DIMETHYLAMINOETHANE see MJV000
7-(3-(2-(3,5-DIHYDROXYPHENYL)-2-HYDROXY-ETHYLAMINO)PROPYL) THEOPHYLLINE HYDROCHLORIDE see DNA600
1-(3,4-DIHYDROXYPHENYL)-2-ISOPROPYLAMINOETHANOL see DMV600
(±)1-(3,4-DIHYDROXYPHENYL)-2-ISOPROPYLAMINOETHANOL HYDROCHLORIDE see IQS500
1-(3,5-DIHYDROXYPHENYL)-2-(ISOPROPYLAMINO)ETHANOL SULFATE see MDM800
dl-α-3,4-DIHYDROXYPHENYL-β-ISOPROPYLAMINOETHANOL SULFATE see IRU000
l-(−)-3-(3,4-DIHYDROXYPHENYL)-2-METHYLALANINE see DNA800
l(−)-β-(3,4-DIHYDROXYPHENYL)-α-METHYLALANINE see DNA800
3,4-DIHYDROXYPHENYL-1-METHYLAMINO-2-ETHANE HYDROCHLORIDE see EAZ000
1-1-(3,4-DIHYDROXYPHENYL)-2-METHYLAMINOETHANOL see VGP000
1-1-(3,4-DIHYDROXYPHENYL)-2-METHYLAMINO-1-ETHANOL HYDROCHLORIDE see AES500
1-(3,4-DIHYDROXYPHENYL)-2-(METHYLAMINO)-ETHANONE (9CI) see MGC350
11-β,17-DIHYDROXY-21-((((1-PHENYLMETHYL)-1H-INDAZOL-3-YL)OXY)ACETYLOXY)PREGN-4-ENE-3,20-DIONE see BAV275
2-(3,4-DIHYDROXYPHENYL)-2,3,4,5,7-PENTAHYDROXY-1-BENZOPYRAN see HBA259
(DIHYDROXYPHENYL)PHENYL MERCURY see PFO250
β-DI-p-HYDROXYPHENYLPROPANE see BLD500
2,2-DI(4-HYDROXYPHENYL)PROPANE see BLD500
3,4-DIHYDROXYPHENYLPROPANOLAMINE HYDROCHLORIDE see AMB000
3-(3,4-DIHYDROXYPHENYL)-2-PROPENOIC ACID (9CI) see CAK375
2-(3,4-DIHYDROXYPHENYL)-3,5,7-TRIHYDROXY-4H-1-BENZOPYRAN-4-ONE see QCA000
2-2-(2,4-DIHYDROXYPHENYL)-3,5,7-TRIHYDROXY-4H-1-BENZOPYRAN-4-ONE see MRN500
2-(3,4-DIHYDROXYPHENYL)-3,5,7-TRIHYDROXY-BENZOPYRYLIUM ACID ANION see COI750
(11-β)-11,17-DIHYDROXY-21-(PHOSPHONOOXY)-PREGN-4-ENE-3,20-DIONE (9CI) see HHQ875
17,21-DIHYDROXYPREGNA-1,4-DIENE-3,11,20-TRIONE see PLZ000
11-β,21-DIHYDROXYPREGN-3,20-DIONE see CNS625
11-β,21-DIHYDROXY-PREGN-4-ENE-3,20-DIONE see CNS625
(11-β)-11,21-DIHYDROXY-PREGN-4-ENE-3,20-DIONE (9CI) see CNS625
11-β,21-DIHYDROXYPREGN-4-ENE-3,20-DIONE ACETATE see CNS650
11-β,21-DIHYDROXY-PREGN-4-ENE-3,20-DIONE-21-ACETATE see CNS650
16-α,17-DIHYDROXYPREGN-4-ENE-3,20-DIONE CYCLIC ACETAL with ACETOPHENONE see DAM300
17α,21-DIHYDROXY-4-PREGNENE-3,11,20-TRIONE see CNS800
17,21-DIHYDROXYPREGN-4-ENE-3,11,20-TRIONE ACETATE see CNS825
17,21-DIHYDROXY-PREGN-4-ENE-3,11,20-TRIONE 21-ACETATE see CNS825
11,12-DIHYDROXYPROGESTERONE see CNS625
11-β,21-DIHYDROXYPROGESTERONE see CNS625
DIHYDROXYPROGESTERONE ACETOPHENIDE see DAM300
1,2-DIHYDROXYPROPANE see PML000
1,3-DIHYDROXYPROPANE see PML250
2,3-DIHYDROXYPROPYL ACETATE see GGO000
(S)-9-(2,3-DIHYDROXYPROPYL)ADENINE see AMH800
(17R,21-α)-17,21-DIHYDROXY-4-PROPYLAJMALANIUM see PNC875
17R,21-α-DIHYDROXY-4-PROPYLAJMALANIUM HYDROGEN TARTRATE see DNB000
(17R,21-α)-17,21-DIHYDROXY-4-PROPYLAJMALINIUM BROMIDE see PNC925
2,3-DIHYDROXYPROPYLAMINE see AMA250

DI(2-HYDROXY-n-PROPYL)AMINE see DNB200
4-(2,3-DIHYDROXYPROPYLAMINO)-2-(5-NITRO-2-THIENYL)QUINAZOLINE see DNB600
2,3-DIHYDROXYPROPYL CHLORIDE see CDT750
7-(2,3-DIHYDROXYPROPYL)-3,7-DIHYDRO-1,3-DIMETHYL-1H-PURINE-2,5-DIONE see DNC000
N,N-DI-(2-HYDROXYPROPYL)NITROSAMINE see DNB200
1-((2,3-DIHYDROXYPROPYL)NITROSAMINO)-2-PROPANONE see NJY550
4-(o-(2′,3′-DIHYDROXYPROPYLOXYCARBONYL)PHENYL)-AMINO-8-TRIFLUOROMETHYLQUINOLINE see TKG000
DIHYDROXYPROPYL THEOPHYLLINE see DNC000
7-(2,3-DIHYDROXYPROPYL)THEOPHYLLINE see DNC000
DIHYDROXYPROPYL THEOPYLIN (GERMAN) see DNC000
(1,2-DIHYDROXY-3-PROPYL)THIOPHYLLIN see DNC000
2,3-DIHYDROXYPROPYL-N-(8-(TRIFLUOROMETHYL)-4-QUINOLYL) ANTHRANILATE see TKG000
2,4-DIHYDROXY-2H-PYRAN-Δ-3(6H),α-ACETIC ACID-3,4-LACTONE see CMV000
(2,4-DIHYDROXY-2H-PYRAN-3(6H)-YLIDENE)ACETIC ACID-3,4-LACTONE see CMV000
5,7-DIHYDROXY-PYRIDOTETRAZOLE-6-CARBONITRILE see DND900
2,4-DIHYDROXYPYRIMIDINE see UNJ800
4,8-DIHYDROXYQUINALDIC ACID see DNC200
4,8-DIHYDROXYQUINALDINIC ACID see DNC200
2,4-DIHYDROXYQUINAZOLINE see QEJ800
4,8-DIHYDROXYQUINOLINE-2-CARBOXYLIC ACID see DNC200
2,3-DIHYDROXYQUINOXALINE see QRS000
8,8′-DIHYDROXY-RUGULOSIN see LIV000
12,18-DIHYDROXY-SENECIONAN-11,16-DIONE see RFP000
3′,6′-DIHYDROXYSPIRO(ISOBENZOFURAN-1(3H),9(9H)-XANTHEN)-3-ONE see FEV000
2,2′-DIHYDROXY-3,3′,5,5′-TETRACHLORODIPHENYLSULFIDE see TFD250
trans-9,10-DIHYDROXY-9,10,11,12-TETRAHYDROBENZO(e)PYRENE see DNC400
trans-1,2-DIHYDROXY-1,2,3,4-TETRAHYDROCHRYSENE see DNC600
trans-3,4-DIHYDROXY-1,2,3,4-TETRAHYDRODIBENZ(a,h)ANTHRACENE see DNC800
trans-3,4-DIHYDROXY-1,2,3,4-TETRAHYDRODIBENZO(a,h)ANTHRACENE see DNC800
trans-1,2-DIHYDROXY-1,2,3,4-TETRAHYDROTRIPHENYLENE see DND000
1,1′-(2,3-DIHYDROXYTETRAMETHYLENE)BIS(4-FORMYLPYRIDINIUM) DIPERCHLORATE, DIOXIME see DND400
5,7-DIHYDROXYTETRAZOLO(1,5-a)PYRIDINE-6-CARBONITRILE see DND900
2,3-DIHYDROXYTOLUENE see DNE100
2,5-DIHYDROXYTOLUENE see MKO250
3,4-DIHYDROXYTOLUENE see DNE200
3,5-DIHYDROXYTOLUENE see MPH500
α-2-DIHYDROXYTOLUENE see HMK100
1,3-DIHYDROXY-2,4,6-TRINITROBENZENE see SMP500
2,4-DIHYDROXY-1,3,5-TRINITROBENZENE see SMP500
3,5-DIHYDROXY-2,6,6-TRIS(3-METHYL-2-BUTENYL)-4-(3-METHYL-1-OX-OB-UTYL)-2,4-CYCLOHEXADIEN-1-ONE see LIU000
2,5-DIHYDROXY-3-UNDECYL-1,4-BENZOQUINONE see EAJ600
2,5-DIHYDROXY-3-UNDECYL-2,5-CYCLOHEXADIENE-1,4-DIONE (9CI) see EAJ600
6-(2,10-DIHYDROXYUNDECYL)-β-RESORCYLIC ACID-mu-LACTONE see RBF100
DIHYDROXYVIOLANTHRON (CZECH) see DMJ800
16,17-DIHYDROXYVIOLANTHRONE see DMJ800
DIHYDROXYVITAMIN D3 see DMJ400
1-α-DIHYDROXYVITAMIN D3 see HJV000
1-α,25-DIHYDROXYVITAMIN D3 see DMJ400
1,4-DIIDROBENZENE (ITALIAN) see HIH000
DIIDROBENZO(1-4)TIAZINA CLORIDRATO (ITALIAN) see DKU875
DIIDRO-5,5-DIETIL-2H-1,3-OSSAZIN-2,4(3H)-DIONE (ITALIAN) see DJT400
DIIDROXI-1,4-BENZENESULFONATO-3-DI-ETILAMMONIUM (ITALIAN) see DIS600
1,4-DIIMIDO-2,5-CYCLOHEXADIENE see BDD200
1,3-DIIMINOISOINDOLIN (CZECH) see DNE400
1,3-DIIMINOISOINDOLINE see DNE400
DIIODAMINE see DNE600
DIIODBENZOTEPH see DNE800
DIIODOACETYLENE see DNE500
DIIODOAMINE see DNE600
1,2-DIIODOBENZENE see DNE700
2,6-DIIODO-1,4-BENZENEDIOL see DNG200
2,6-DIIODO-p-BENZOQUINONE see DNG800
DIIODOBENZOTEF see DNE800
N-2,5-DIIODOBENZOYL-N′,N′,N″,N″-DIETHYLENEPHOSPHORTRIAMIDE see DNE800
1,4-DIIODO-1,3-BUTADIYNE see DNE875
cis-DIIODODIAMMINEPLATIUM (II) see DNF000
DIIODOETHYNE see DNE500
2,6-DIIODOHYDROQUINONE see DNF200

3,5-DIIODO-4-HYDROXYBENZONITRILE see HKB500
3,5-DIIODO-4-HYDROXYBENZONITRILE, LITHIUM SALT see DNF400
3,5-DIIODO-4-HYDROXYBENZONITRILE OCTANOATE see DNG200
β-(3,5-DIIODO-4-HYDROXYPHENYL)-α-PHENYLPROPIONIC ACID
  see PDM750
DIIODOHYDROXYQUIN see DNF600
DIIODOHYDROXYQUINOLINE see DNF600
5,7-DIIODO-8-HYDROXYQUINOLINE see DNF600
3,5-DIIODO-4-(3′-IODO-4′-ACETOXYPHENOXY)BENZOIC ACID see TKP850
DIIODOMETHANE see DNF800
DIIODOMETHYLARSINE see MGQ775
DIIODOMETHYLATE de la BIS(PIPERIDINOMETHYL-COUMARANYL-
  5)CETONE (FRENCH) see COE125
DIIODOMETILATO del BISPIPERIDINOMETILCUMARANIL-5-CHETONE
  (ITALIAN) see COE125
2,6-DIIODO-4-NITROPHENOL see DNG000
3,5-DIIODO-4-OCTANOYLOXYBENZONITRILE see DNG200
5,7-DIIODO-OXINE see DNF600
3,5-DIIODO-4-OXO-1(4H)PYRIDINEACETIC ACID-2,2′-IMINODIETHANOL
  SALT see DNG400
3,5-DIIODO-α-PHENYLPHLORETIC ACID see PDM750
3,5-DIIODO-4-PYRIDONE-N-ACETATE BIS(HYDROXYETHYL)AMMONIUM
  see DNG400
3,5-DIIODO-4-PYRIDONE-N-ACETIC ACID, DIETHANOLAMINE SALT
  see DNG400
2,6-DIIODOQUINOL see DNF200
5,7-DIIODO-8-QUINOLINOL see DNF600
DIIODOQUINONE see DNG800
3,5-DIIODOSALICYLIC ACID see DNH000
O,O-DIIOSPROPYL DITHIOPHOSPHORIC ACID ESTER of-N,N-S-
  DIETHYLTHIOCARBAMOYL-O,O-DIISOPROPYL PHOSPHOROTHIOATE
  see DKB600
DIIRON TRISULFATE see FBA000
DIISOAMYL ADIPATE see AEQ500
DIISOAMYLMERCURY see DNL200
DIISOBUTENE see TMA250
DIISOBUTILCHETONE (ITALIAN) see DNI800
DIISOBUTYL ADIPATE see DNH125
DIISOBUTYLALUMINIUM HYDRIDE see DNI600
DIISOBUTYLALUMINUM CHLORIDE see CGB500
DIISOBUTYLALUMINUM HYDRIDE see DNI600
DIISOBUTYLALUMINUM MONOCHLORIDE see CGB500
DIISOBUTYLAMINE see DNH400
DIISOBUTYLAMINOBENZOYLOXYPROPYL THEOPHYLLINE see DNH500
α-((DIISOBUTYLAMINO)METHYL)THEOPHYLLINE-8-ETHANOL BENZOATE
  (ester) see DNH500
DIISOBUTYL CARBINOL see DNH800
DI-ISOBUTYLCETONE (FRENCH) see DNI800
DIISOBUTYLCHLOROALUMINUM see CGB500
p-DIISOBUTYLCRESOXYETHYLDIMETHYLBENZYLAMMONIUM CHLORIDE
  MONOHYDRATE see MHB500
DIISOBUTYLENE see TMA250
DIISOBUTYLENE OXIDE see DNI200
DIISOBUTYL FUMARATE see DNI400
DIISOBUTYLHYDROALUMINUM see DNI600
DIISOBUTYLKETON (DUTCH, GERMAN) see DNI800
DIISOBUTYL KETONE see DNI800
DI-ISO-BUTYLNITROSAMINE see DRQ200
DIISOBUTYLOXOSTANNANE see DNJ000
DIISOBUTYLPHENOXYETHOXYETHYLDIMETHYL BENZYL AMMONIUM
  CHLORIDE see BEN000
p-DIISOBUTYLPHENOXYETHOXYETHYLDIMETHYLBENZYLAMMONIUM
  CHLORIDE MONOHYDRATE see BBU750
DIISOBUTYL PHTHALATE see DNJ400
DIISOBUTYLSULFIDE HYDRATE see IJO000
DIISOBUTYLTHIOCARBAMIC ACID-S-ETHYL ESTER see EID500
DIISOBUTYLTIN OXIDE see DNJ000
DIISOBUTYRYL PEROXIDE see DNJ600
o,o′-DIISOBUTYRYLTHIAMINE DISULFIDE see BKJ325
DIISOCARB see EID500
4-4′-DIISOCYANATE de DIPHENYLMETHANE (FRENCH) see MJP400
DI-ISOCYANATE de TOLUYLENE see TGM750
1,3-DIISOCYANATOBENZENE see BBP000
4,4′-DIISOCYANATO-3,3′-DIMETHOXY-1,1′-BIPHENYL see DCJ400
4,4′-DIISOCYANATO-3,3′-DIMETHYL-1,1′-BIPHENYL see DQS000
4,4′-DIISOCYANATODIPHENYLMETHANE see MJP400
1,6-DIISOCYANATOHEXANE see DNJ800
DI-ISO-CYANATOTOLUENE see TGM750
DIISOCYANATOMETHANE see DNK100
DIISOCYANATOMETHYLBENZENE see DNK200, TGM740
2,6-DIISOCYANATO-1-METHYLBENZENE see TGM800
2,4-DIISOCYANATO-1-METHYLBENZENE (9CI) see TGM750
1,5-DIISOCYANATONAPHTHALENE see NAM500
DIISOCYANATOTOLUENE see TGM740

2,4-DIISOCYANATOTOLUENE see TGM750
2,6-DIISOCYANATOTOLUENE see TGM800
DIISOCYANAT-TOLUOL see TGM750
4,4′-(2,3-DIISOCYANO-1,3-BUTADIENE-1,4-DIYL)BIS-1,2-BENZENEDIOL
  see XCS700
2,3-DIISONITROSOBUTANE see DBH000
DIISONONYLTIN DICHLORIDE see BLQ750
DIISOOCTYL ACID PHOSPHATE see DNK800
DIISOOCTYL ((DIOCTYLSTANNYLENE)DITHIO)DIACETATE see BKK750
DIISOOCTYL PHOSPHATE (DOT) see DNK800
DIISOOCTYL PHTHALATE see ILR100
DIISOPENTYLMERCURY see DNL200
DIISOPENTYLOXOSTANNANE see DNL400
DIISOPENTYLTIN OXIDE see DNL400
DIISOPHENOL see DNG000
DIISOPROPANOLAMINE see DNL600
DIISOPROPANOLNITROSAMINE see DNB200
N,N′-DIISOPROPIL-FOSFORODIAMMIDO-FLUORURO (ITALIAN) see PHF750
DIISOPROPOXYPHOSPHORYL FLUORIDE see IRF000
((DIISOPROPROXYPHOSPHINOTHIOYL)THIO)TRICYCLOHEXYLSTAN-
  NANE see DNT200
s-DIISOPROPYLACETONE see DNI800
DIISOPROPYL ADIPATE see DNL800
DI(ISOPROPYLAMIDO)PHOSPHORYLFLUORIDE see PHF750
DIISOPROPYLAMINE see DNM200
DIISOPROPYLAMINE DICHLORACETATE see DNM400
DIISOPROPYLAMINE with DICHLOROACETIC ACID (1:1) see DNM400
DIISOPROPYLAMINE DICHLOROETHANOATE see DNM400
2-(DIISOPROPYLAMINO)-2′,6′-ACETOXYLIDIDE HYDROCHLORIDE
  see DNM600
2-DIISOPROPYLAMINOETHANOL see DNP000
2-(2-(DIISOPROPYLAMINO)ETHOXY)BUTYROPHENONE HYDROCHLORIDE
  see DNN000
S-(2-DIISOPROPYLAMINOETHYL)-O-ETHYL METHYL PHOSPHONOTHIOL-
  ATE see EIG000
α-(2-(DIISOPROPYLAMINO)ETHYL)-α-PHENYL-2-PYRIDINEACETAMIDE
  see DNN600
α-(2-DIISOPROPYLAMINOETHYL)-α-PHENYL-2-PYRIDINEACETAMIDE PHOS-
  PHATE see RSZ600
β-DIISOPROPYLAMINOETHYL-9-XANTHENECARBOXYLATE METHOBROM-
  IDE see HKR500
2,6-DIISOPROPYLAMINO-4-METHOXYTRIAZINE see MFL250
Γ-DIISOPROPYLAMINO-α-PHENYL-α-(2-PYRIDYL)BUTYRAMIDE see DNN600
DIISOPROPYLAMMINE-trans-DIHYDROXYMALONATOPLATINUM(IV)
  see IGG775
DIISOPROPYLAMMONIUM DICHLOROACETATE see DNM400
DIISOPROPYLAMMONIUM DICHLOROETHANOATE see DNM400
DIISOPROPYLBENZENE see DNN709
m-DIISOPROPYLBENZENE see DNN829
o-DIISOPROPYLBENZENE see DNN800
1,3-DIISOPROPYLBENZENE see DNN829
DIISOPROPYLBENZENE HYDROPEROXIDE, not more than 72% in solution
  (DOT) see DNS000
DIISOPROPYLBENZENE PEROXIDE see DGR600
1,3-DIISOPROPYLBENZENE SODIUM SALT, DIHYDROPEROXIDE
  see DNN840
1,4-DIISOPROPYLBENZENE SODIUM SALT, DIISOPEROXIDE see DNN850
O,O-DIISOPROPYL-S-BENZYL PHOSPHOROTHIOLATE see BKS750
O,O-DIISOPROPYL-S-BENZYL THIOPHOSPHATE see BKS750
DIISOPROPYLBERYLLIUM see DNO200
DIISOPROPYLCARBAMIC ACID, ETHYL ESTER see DNP600
DIISOPROPYLCARBODIIMIDE see DNO400
N,N′-DIISOPROPYL-DIAMIDO-FOSFORZUUR-FLUORIDE (DUTCH) see PHF750
N,N′-DIISOPROPYL-DIAMIDO-PHOSPHORSAEURE-FLUORID (GERMAN)
  see PHF750
N,N′-DIISOPROPYLDIAMIDOPHOSPHORYL FLUORIDE see PHF750
O,O-DIISOPROPYL-S-DIETHYLDITHIOCARBAMOYLPHOSPHORODITHIO-
  ATE see DKB600
DIISOPROPYL-1,3-DITHIOL-2-YLIDENEMALONATE see MAO275
N-(2-(O,O-DIISOPROPYLDITHIOPHOSPHORYL)ETHYL)BENZENESULFONAM-
  IDE see DNO800
N-(β-O,O-DIISOPROPYLDITHIOPHOSPHORYLETHYL)BEZENESULFONAM-
  IDE see DNO800
DIISOPROPYL ESTER of DITHIOCARBAMYL PHOSPHOROTHIOIC ACID
  see DKB600
DIISOPROPYL ESTER SULFURIC ACID see DNO900
DIISOPROPYL ETHANOLAMINE see DNP000
N,N-DIISOPROPYL ETHANOLAMINE see DNP000
DIISOPROPYL ETHER see IOZ750
DIISOPROPYL ETHYL CARBAMATE see DNP600
N,N-DIISOPROPYL ETHYL CARBAMATE see DNP600
DIISOPROPYL FLUOROPHOSPHATE see IRF000
O,O-DIISOPROPYL FLUOROPHOSPHATE see IRF000
DIISOPROPYL FLUOROPHOSPHONATE see IRF000

DIISOPROPYLFLUOROPHOSPHORIC ACID ESTER see IRF000
DIISOPROPYLFLUORPHOSPHORSAEUREESTER (GERMAN) see IRF000
DIISOPROPYL FUMARATE see DNQ200
DIISOPROPYL HYDROGEN PHOSPHITE see DNQ600
DIISOPROPYL(2-HYDROXYETHYL)METHYLAMMONIUMBROMIDE with
 XANTHENE-9-CARBOXYLATE see HKR500
DIISOPROPYL HYPONITRITE see DNQ700
DIISOPROPYLIDENE ACETONE see PGW250
sym-DIISOPROPYLIDENE ACETONE see PGW250
DIISOPROPYLMERCURY see DNQ800
DIISOPROPYL METHANEPHOSPHONATE see DNQ875
DIISOPROPYL METHYLPHOSPHONATE see DNQ875
DIISOPROPYL-p-NITROPHENYL PHOSPHATE see DNR309
O,O-DIISOPROPYL-o,p-NITROPHENYL PHOSPHATE see DNR309
DIISOPROPYLNITROSAMIN (GERMAN) see NKA000
DIISOPROPYL OXIDE see IOZ750
DIISOPROPYLOXOSTANNANE see DNR200
DIISOPROPYL PARAOXON see DNR309
DIISOPROPYL PERDICARBONATE see DNR400
DIISOPROPYL PEROXYDICARBONATE see DNR400
2,6-DIISOPROPYLPHENOL see DNR800
3,5-DIISOPROPYLPHENOL METHYLCARBAMATE see DNS200
DIISOPROPYLPHENYLHYDROPEROXIDE (solution) see DNS000
3,5-DIISOPROPYLPHENYL METHYLCARBAMATE see DNS200
3,5-DIISOPROPYLPHENYL-N-METHYLCARBAMATE see DNS200
DIISOPROPYL PHOSPHITE see DNQ600
DIISOPROPYL PHOSPHOFLUORIDATE see IRF000
DIISOPROPYLPHOSPHONATE see DNQ600
O,O-DIISOPROPYL PHOSPHONATE see DNQ600
N,N'-DIISOPROPYLPHOSPHORODIAMIDIC FLUORIDE see PHF750
S-(O,O-DIISOPROPYL PHOSPHORODITHIOATE) ESTER of N-(2-
 MERCAPTOETHYL)BENZENESULFONAMIDE see DNO800
DIISOPROPYL PHOSPHOROFLUORIDATE see IRF000
O,O'-DIISOPROPYL PHOSPHORYL FLUORIDE see IRF000
DI-ISOPROPYLSULFAT (GERMAN) see DNO900
DI-ISOPROPYLSULFATE see DNO900
N-DIISOPROPYLTHIOCARBAMIC ACID-S-2,3,3-TRICHLOROALLYL ESTER
 see DNS600
N-DIISOPROPYLTHIOCARBAMIC ACID S-2,3,3-TRICHLORO-2-PROPENYL
 ESTER see DNS600
DI-ISOPROPYLTHIOLOCARBAMATE de S-(2,3-DICHLOROALLYLE)
 (FRENCH) see DBI200
DIISOPROPYL THIOUREA see DNS800
1,3-DIISOPROPYL-2-THIOUREA see DNS800
DIISOPROPYLTIN DICHLORIDE see DNT000
DIISOPROPYLTIN OXIDE see DNR200
N,N-DIISOPROPYL-2,3,3-TRICHLORALLYL-THIOLCARBAMAT (GERMAN)
 see DNS600
DIISOPROPYLTRICHLOROALLYLTHIOCARBAMATE see DNS600
O,O-DIISOPROPYL-S-TRICYCLOHEXYLTIN PHOSPHORODITHIOATE
 see DNT200
DIISOPYRAMIDE PHOSPHATE see RSZ600
1,4-DIISOTHIOCYANATOBENZENE see PFA500
1,2-DIISOTHIOCYANATOETHANE see ISK000
3,5-DIJOD-4-HYDROXY-BENZONITRIL (GERMAN) see HKB500
3,5-DIJOD-4-HYDROXY-BENZONITRIL CAPRYSAEUREESTER (GERMAN)
 see DNG200
3,5-DIJOD-4-HYDROXY-BENZONITRILE LITHIUMSALZ (GERMAN)
 see DNF400
DIKAIN see BQA010
DIKAIN HYDROCHLORIDE see TBN000
3,5-DIKARBOXYBENZENSULFONAN SODNY (CZECH) see DEK200
DIKETENE see KFA000
DIKETENE, inhibited (DOT) see KFA000
2,3-DIKETOBUTANE see BOT500
2,5-DIKETOHEXANE see HEQ500
1,3-DIKETOHYDRINDENE see IBS000
2,3-DIKETOINDOLINE see ICR000
DIKETONE ALCOHOL see DBF750
2,5-DIKETOPYRROLIDINE see SND000
2,5-DIKETOTETRAHYDROFURAN see SNC000
DIKOL see PDM750
DIKONIT see SGG500
DIKOTEKS see SIL500
DIKOTEX 30 see SIL500
DI-KU-SHUANG see MJL750
DILABIL see DAL000
DILABIL SODIUM see SGD500
DILACORAN see IRV000
DILACTONE ACTINOMYCINDIOIC D ACID see AEB000
DILAHIL see DAL000
DILAN see BON250
DILANGIL see MAW250
DILANGIO see POD750

DILANTIN see DKQ000, DNU000
DILANTIN DB see DEP600
DILANTINE see DKQ000
DILANTIN SODIUM see DNU000
DILAPHYLLIN see HLC000
DILATAN KORE see ELH600
DILATIN see DNU100
DILATIN DB see DEP600
DILATOL HYDROCHLORIDE see DNU200
DILATYL see DNU200
DILAUDID see DNU300
DILAUDID HYDROCHLORIDE see DNU300, DNU310
DILAUROYL PEROXIDE see LBR000
DILAUROYL PEROXIDE, TECHNICAL PURE (DOT) see LBR000
DILAURYLESTER KYSELINY β',β'-THIODIPROPIONOVE (CZECH) see TFD500
DILAURYL THIODIPROPIONATE see TFD500
DILAURYL-β-THIODIPROPIONATE see TFD500
DILAURYL-3,3'-THIODIPROPIONATE see TFD500
DILAURYL-β',β'-THIODIPROPIONATE see TFD500
DILAVASE see VGA300, VGF000
DILAZEP/β-ACETYLDIGOXIN see CNR825
DILAZEP DIHYDROCHLORIDE see CNR750
DILCIT see HFG550
DILEAD(II) LEAD(IV) OXIDE see LDS000
DI-LEN see DNU000
DILENE see BIM500
DILEXPAL see HFG550
DILIC see HKC000
1,3-DILITHIOBENZENE see DNU325
DILITHIUM-1,1-BIS(TRIMETHYLSILYL)HYDRAZIDE see DNU350
DILITHIUM CARBONATE see LGZ000
DILITHIUM CHROMATE see LHD000
DILL FRUIT OIL see DNU400
DILL HERB OIL see DNU400
DILL OIL see DNU400
DILL SEED OIL see DNU400
DILL SEED OIL, EUROPEAN TYPE see DNU400
DILL WEED OIL see DNU400
DILOMBRIN see DJT800
DILOR see DLO880, DNC000
DILOSPAN S see PGR000
DILOSYN see MDT500, MPE250
DILOXOL see GGS000
DILTIAZEM HYDROCHLORIDE see DNU600
DILURAN see AAI250
DILVASENE see FMX000
DILYN see RLU000
DILZEM see DNU600
DIMAGNESIUM PHOSPHATE see MAH775
1,3-DIMALEIMIDOBENZENE see BKL750
DIMALONE see DRB400
DIMANGANESE TRIOXIDE see MAT500
DIMANIN C see SGG500
DIMAPP see DQA400
DIMAPYRIN see DOT000
DIMAS see DQD400
DIMATE 267 see DSP400
DIMATIF see DNU850
DIMAVAL see DNU860
DIMAYAL see DNU860
DIMAZ see DXH325
DIMAZINE see DSF400
DIMAZON see ACR300
DIMEBOLIN see TCQ260
DIMEBOLINE see TCQ260
DIMEBON see TCQ260
DIMEBON DIHYDROCHLORIDE see DNU875
DIMEBONE see TCQ260
DIMECRON see FAB400
DIMECRON 100 see FAB400
DIMECROTIC ACID MAGNESIUM SALT see DOK400
DIMEDROL see BBV500
DIMEDRYL see BBV500
DIMEFADANE see DRX400
DIMEFLINE see DNV000
DIMEFLINE HYDROCHLORIDE see DNV200
DIMEFOX see BJE750
DIMEGLUMINE IOCARMATE see IDJ500
DIMELIN see ABB000
DIMELONE see DRB400
DIMELOR see ABB000
DIMEMORFAN PHOSPHATE see MLP250
DIMENFORMON see EDO000
DIMENFORMON BENZOATE see EDP000

DIMENFORMON DIPROPIONATE see EDR000
DIMENFORMONE see EDP000
DIMENFORMON PROLONGATUM see EDO000
DIMENHYDRINATE see DYE600
DIMENOXADOL HYDROCHLORIDE see DPE200
DIMEPHENTHIOATE see DRR400
DIMEPHENTHOATE see DRR400
DIMERAY see IDJ500
DIMERCAPROL PROPANOL see BAD750
(R*,S*)-2,3-DIMERCAPTOBUTANEDIOIC ACID see DNV800
(R*,R*)-(±)-1,4-DIMERCAPTO-2,3-BUTANEDIOL (9CI) see DXO775
dl-threo-DIMERCAPTO-2,3-BUTANEDIOL see DXO775
d-threo-1,4-DIMERCAPTO-2,3-BUTANEDIOL see DXO800
1,2-DIMERCAPTOETHANE see EEB000
DIMERCAPTOL see BAD750
2,3-DIMERCAPTOL-1-PROPANOL see BAD750
1,2-DIMERCAPTO-4-METHYLBENZENE see TGN000
4,5-DI(MERCAPTOMETHYL)-2-METHYL-3-PYRIDINOL DITHIOACETATE
   HYDROBROMIDE see DBH800
2,3-DIMERCAPTOPROPANE SODIUM SULPHONATE see DNU860
2,3-DIMERCAPTOPROPANESULFONIC ACID SODIUM SALT see DNU860
2,3-DIMERCAPTO-1-PROPANESULFONIC ACID SODIUM SALT see DNU860
DIMERCAPTOPROPANOL see BAD750
2,3-DIMERCAPTOPROPANOL see BAD750
2,3-DIMERCAPTOPROPAN-1-OL see BAD750
2,3-DIMERCAPTOPROPYL-p-TOLYSULFIDE see DNV600
4,5-DIMERCAPTOPYRIDOXINDI-THIOACETAT HYDROBROMID (GERMAN)
   see DBH800
meso-DIMERCAPTOSUCCINIC ACID see DNV800
meso-2,3-DIMERCAPTOSUCCINIC ACID see DNV800
meso-DIMERCAPTOSUCCINIC ACID SODIUM SALT see DNU860
2,5-DIMERCAPTO-1,3,4-THIADIAZOLE see TES250
DIMERCUROUS METHANE ARSONATE see DNW000
DIMERCURY IMIDE OXIDE see DNW200
DIMER CYKLOPENTADIENU (CZECH) see DGW000
DIMERIN see DNW400
DIMER X see IDJ500
DIMESTROL see DJB200
1,6-DIMESYL-d-MANNITOL see BKM500
1,4-DIMESYLOXYBUTANE see BOT250
1,4-DI(MESYLOXYETHYLAMINO)ERYTHRITOL see LJD500
DIMET see DXE600
DIMETACRINE see DNW700
DIMETACRINE BITARTRATE see DRM000
DIMETACRIN HYDROGENTARTRATE see DRM000
DIMETAN see DRL200
dl-DIMETANE MALEATE see DNW759
DIMETATE see DSP400
DIMETAZINA see SNN300
DIMETHACHLON see DGF000
DIMETHACIN see DNW700
DIMETHACINE see DNW700
DIMETHACRINE TARTRATE see DRM000
DIMETHADIONE see PMO250
DIMETHAEN see DPA000
2,5-DIMETHANESULFOMYLOXYHEXANE see DSU000
(R*,S*)-DIMETHANESULFONATE-meso-2,5-HEXANEDIOL (9CI) see DSU100
1,6-DIMETHANESULFONATE-d-MANNITOL see BKM500
1,4-DIMETHANESULFONATE THREITOL see TFU500
(2s,3s)-1,4-DIMETHANESULFONATE TREITOL see TFU500
1,4-DIMETHANESULFONOXYBUTANE see BOT250
cis-1,4-DIMETHANE SULFONOXY-2-BUTENE see DNW800
trans-1,4-DIMETHANE SULFONOXY-2-BUTENE see DNX000
1,4-DIMETHANESULFONOXY-2-BUTYNE see DNX200
1,4-DIMETHANESULFONOXY-1,4-DIMETHYLBUTANE see DSU000
1,6-DIMETHANE-SULFONOXY-d-MANNITOL see BKM500
1:3-DIMETHANESULFONOXYPROPANE see TLR250
1,4-DI(METHANESULFONYLOXY)BUTANE see BOT250
DIMETHANESULFONYL PEROXIDE see DNX300
1,6-DIMETHANESULPHONOXY-1,6-DIDEOXY-d-MANNITOL see BKM500
1,3-DIMETHANESULPHONOXYPROPANE see TLR250
1,4-DIMETHANESULPHONYLOXYBUTANE see BOT250
DIMETHESTERONE see DRT200
DIMETHICONE 350 see PJR000
DIMETHINDENE MALEATE see FMU409
DIMETHINDEN MALEATE see FMU409
DIMETHIOTAZINE see FMU039
DIMETHIRIMOL see BRD000
DIMETHISOQUIN HYDROCHLORIDE see DNX400
DIMETHISTERON see DRT200
DIMETHISTERONE see DRT200
DIMETHISTERONE and ETHINYL ESTRADIOL see DNX500
DIMETHOAT (DUTCH) see DSP400
DIMETHOAT (GERMAN) see DSP400

DIMETHOATE (USDA) see DSP400
DIMETHOATE O-ANALOG see DNX800
DIMETHOATE-ETHYL see DNX600
DIMETHOATE OXYGEN ANALOG see DNX800
DIMETHOATE PO ISOLOGUE see DNX800
DIMETHOAT TECHNISCH 95% see DSP400
DIMETHOCAINE see DNY000
DIMETHOGEN see DSP400
DIMETHOTHIAZINE see DUC400
DIMETHOTHIAZINE MESYLATE see FMU039
DIMETHOTHIAZINE METHANESULFONATE see FMU039
DIMETHOXANE see ABC250
DIMETHOXON see DNX800
1,2-DIMETHOXY-4-ALLYLBENZENE see AGE250
3,4'-DIMETHOXY-4-AMINOAZOBENZENE see DNY400
2,6-DIMETHOXY-4-(p-AMINOBENZENESULFONAMIDO)PYRIMIDINE
   see SNN300
(trans)-2,5-DIMETHOXY-4'-AMINOSTILBENE see DON400
2,5-DIMETHOXYAMPHETAMINE HYDROCHLORIDE see DOJ800
3,4-DIMETHOXYAMPHETAMINE HYDROCHLORIDE see DOK000
2,4-DIMETHOXYANILINE see DNY500
2,3-DIMETHOXYANILINE MUSTARD see BIC600
1,5-DIMETHOXY-9,10-ANTHRACENEDIONE see DNY800
1,5-DIMETHOXYANTHRACHINON (CZECH) see DNY800
1,5-DIMETHOXYANTHRAQUINONE see DNY800
1-5,6-DIMETHOXYAPORPHINE see NOE500
(R)-1,2-DIMETHOXYAPORPHINE see NOE500
1,2-DIMETHOXY-6a-β-APORPHINE see NOE500
1,10-DIMETHOXY-6a-α-APORPHINE-2,9-DIOL see DNZ100
3,4-DIMETHOXYBENZALDEHYDE see VHK000
7,12-DIMETHOXYBENZ(a)ANTHRACENE see DOA000
o-DIMETHOXYBENZENE see DOA200
p-DIMETHOXYBENZENE see DOA400
1,2-DIMETHOXYBENZENE see DOA200
3,4-DIMETHOXY-BENZENEACETONITRILE (9CI) see VIK100
2,5-DIMETHOXYBENZENEAZO-β-NAPHTHOL see DOK200
3,4-DIMETHOXYBENZENECARBONAL see VHK000
2,6-DIMETHOXY-1,4-BENZENEDIOL see DON200
3,3'-DIMETHOXYBENZIDIN (CZECH) see DCJ200
3,3'-DIMETHOXYBENZIDINE see DCJ200
3,3'-DIMETHOXYBENZIDINE DIHYDROCHLORIDE see DOA800
3,3'-DIMETHOXYBENZIDINE-4,4'-DIISOCYANATE see DCJ400
6,7-DIMETHOXYBENZOPYRAN-2-ONE see DRS800
1-(3,4-DIMETHOXYBENZOYL)-4-(1,2,3,4-TETRAHYDRO-2-OXO-6-
   QUINOLINYL)PIPERA ZINE see DLF700
3,4-DIMETHOXYBENZYLAMINE see VIK050
β-(2,4-DIMETHOXY-5-BENZYLBENZOYL)PROPIONIC ACID SODIUM SALT
   see DOA875
3,4-DIMETHOXYBENZYL CHLORIDE see BKM750
3,4-DIMETHOXYBENZYL CYANIDE see VIK100
1-(3,4)-DIMETHOXYBENZYL-6,7-DIMETHOXYISOQUINOLINE-3-CARBOX-
   YLIC ACID, SODIUM SALT see PAG750
3,4-DIMETHOXYBENZYLHYDRAZINE see VIK150
3,3-DIMETHOXY-(1,1'-BIPHENYL)-4,4'-DIAMINE DIHYDROCHLORIDE
   see DOA800
3,3'-DIMETHOXY-4,4'-BIPHENYLENE DIISOCYANATE see DCJ400
1,3-DIMETHOXYBUTANE see DOB200
1,1-DIMETHOXY-2-BUTENE see PNJ500
N-(3,4-DIMETHOXYCINNAMOYL)ANTHRANILIC ACID see RLK800
N-(3',4'-DIMETHOXYCINNAMOYL)ANTHRANILIC ACID see RLK800
6,7-DIMETHOXYCOUMARIN see DRS800
β-(2,4-DIMETHOXY-5-CYCLOHEXYLBENZOYL)PROPIONIC ACID
   see DOB275
β-(2,4-DIMETHOXY-5-CYCLOHEXYLBENZOYL)PROPIONIC ACID SODIUM
   SALT see DOB300
β-(2,4-DIMETHOXY-5-CYCLOPENTYLMETHYLBENZOYL)PROPIONIC ACID
   SODIUM SALT see DOB325
3,4-DIMETHOXY-DBA see DOB600
DIMETHOXY-DDT see MEI450
1,1-DIMETHOXYDECANE see AFJ700
10,10-DIMETHOXYDECANE see AFJ700
DIMETHOXYDIAZENE see DSI489
5,6-DIMETHOXYDIBENZ(a,h)ANTHRACENE see DOB600
3,4-DIMETHOXY-1,2:5,6-DIBENZANTHRACENE see DOB600
4,4'-DIMETHOXY-α,β-DIETHYLSTILBENE see DJB200
DIMETHOXY-1,10 DIHYDROXY-2,9 NOR-APORPHINE (FRENCH) see LBO100
4,7-DIMETHOXY-6-(2-DIISOPROPYLAMINOETHOXY)-5-(p-
   METHOXYCINNAMOYL) BENZOFURAN OXALATE see DOC000
4,7-DIMETHOXY-6-(2-DIMETHYLAMINOETHOXY)-5-(p-
   FLUOROCINNAMOYL)BENZOFURAN MALEATE see DOC800
4-,7-DIMETHOXY-6-(2-DIMETHYLAMINOETHOXY)-5-(p-
   HYDROXYCINNAMOYL)BENZOFURAN OXALATE see DOD200
4,7-DIMETHOXY-6-(2-DIMETHYLAMINOETHOXY)-5-(p-
   ISOPROPOXYCINNAMOYL)BENZOFURAN MALEATE see DOD400

4,7-DIMETHOXY-6-(2-DIMETHYLAMINOETHOXY)-5-(p-
  METHOXYCINNAMOYL)BENZOFURAN MALEATE see DOD600
6,7-DIMETHOXY-2,2-DIMETHYL-2H-BENZO(b)PYRAN see AEX850
o,o'-DIMETHOXY-α-α'-DIMETHYL-DIPHENETHYLAMINE compounded with
  LACTICACID see BKP200
6',12'-DIMETHOXY-2,2'-DIMETHYL-6,7-(METHYLENEBIS(OXY)OXY-
  ACANTHAN see CCX550
1,1-DIMETHOXY-3,7-DIMETHYL-2,6-OCTADIENE see DOE000
1,1-DIMETHOXY-3,7-DIMETHYL-2,6-OCTADIENE (cis and trans) see DOE000
2,5-DIMETHOXY-α,4-DIMETHYLPHENETHYLAMINE HYDROCHLORIDE
  see DOG600
p,p'-DIMETHOXYDIPHENYLTRICHLOROETHANE see MEI450
3,4-DIMETHOXYDOPAMINE see DOE200
DIMETHOXY-DT see MEI450
DIMETHOXYETHANE see DOE600
1,2-DIMETHOXYETHANE see DOE600
α,β-DIMETHOXYETHANE see DOE600
1,1-DIMETHOXYETHANE (DOT) see DOO600
1,2-DIMETHOXYETHANE (DOT) see DOE600
1,1-DI-(2-METHOXYETHOXY)ETHANE see AAG750
(2,2-DIMETHOXYETHYL)-BENZENE (9CI) see PDX000
DI(2-METHOXYETHYL) MALEATE see DOF000
DI(2-METHOXYETHYL)PEROXYDICARBONATE see DOF200
DIMETHOXY ETHYL PHTHALATE see DOF400
DI(2-METHOXYETHYL)PHTHALATE see DOF400
2,6-DIMETHOXYHYDROQUINONE see DON200
3,5-DIMETHOXYHYDROQUINONE see DON200
3,5-DIMETHOXY-4-HYDROXYBENZALDEHYDE see DOF600
4,7-DIMETHOXY-5-(p-HYDROXYCINNAMOYL)-6-(2-
  PYRROLIDINYLETHOXY)BENZOFURAN MALEATE see DOF800
DIMETHOXYMETHANE see MGA850
2,5-DIMETHOXY-4-METHYLAMPHETAMINE HYDROCHLORIDE see DOG600
DIMETHOXYMETHYLBENZENE see DOG700
2,5-DIMETHOXY-α-METHYLBENZENEETHANAMINE HYDROCHLORIDE
  see DOJ800
2,4-DIMETHOXY-β-METHYLCINNAMIC ACID MAGNESIUM SALT
  see DOK400
9,10-DIMETHOXY-2,3-(METHYLENEDIOXY)BERBINE see TCJ800
9,10-DIMETHOXY-2,3-(METHYLENEDIOXY)-7,8,13,13A-TETRAHYDROBERBIN-
  IUM see BFN500
5,8-DIMETHOXY-2-METHYL-6,7-FURANOCHROMONE see AHK750
5,8-DIMETHOXY-2-METHYL-4',5'-FURANO-6,7-CHROMONE see AHK750
5,8-DIMETHOXY-2-METHYL-4',5'-FURO-6,7-CHROMONE see AHK750
4,9-DIMETHOXY-7-METHYL-5H-FURO(3,2-G)(1)BENZOPYRAN-5-ONE
  see AHK750
3,4-DIMETHOXY-17-METHYLMORPHINAN-6β,14-DIOL see ORE000
4,9-DIMETHOXY-7-METHYL-5-OXO-1,8-DIOXABENZ-(F)INDENE see AHK750
4,9-DIMETHOXY-7-METHYL-5-OXOFURO(3,2-G)(1)BENZOPYRAN see AHK750
4,9-DIMETHOXY-7-METHYL-5-OXOFURO(3,2-G)-1,2-CHROMENE see AHK750
2,5-DIMETHOXY-α-METHYLPHENETHYLAMINE HYDROCHLORIDE see
  DOJ800
1-(2,5-DIMETHOXY-4-METHYLPHENYL)-2-AMINOPROPANE see DOG600
2,5-DIMETHOXY-α-METHYL-β-PHENYLETHYLAMINE HYDROCHLORIDE
  see DOJ800
3,4-DIMETHOXY-α-METHYL-β-PHENYLETHYLAMINEHYDROCHLORIDE
  see DOK000
5,6-DIMETHOXY-2-METHYL-3-(2-(4-PHENYL-1-PIPERAZINYL)ETHYL)-1H-IN-
  DOLE see ECW600
DIMETHOXYMETHYLPHENYLSILANE see DOH400
1,10-DIMETHOXY-6a-α-NORAPORPHINE-2,9-DIOL see LBO100
3,4-DIMETHOXYPHENETHYLAMINE see DOE200
3,4-DIMETHOXY-β-PHENETHYLAMINE see DOE200
3,4-DIMETHOXYPHENETHYLAMINE HYDROCHLORIDE see DOI400
4-(2,5-DIMETHOXYPHENETHYL)ANILINE see DON400
5-((3,4-DIMETHOXYPHENETHYL)METHYLAMINO)-2-(3,4-
  DIMETHOXYPHENYL)-2-ISOPROPYLVALERONITRILE see IRV000
2,6-DIMETHOXYPHENOL see DOJ200
(2-(2,5-DIMETHOXYPHENOXY)ETHYL)HYDRAZINE HYDROCHLORIDE
  see DOJ400
(2-(3,4-DIMETHOXYPHENOXY)ETHYL)HYDRAZINE HYDROCHLORIDE
  see DOJ600
N,N'-DI(3-(p-METHOXYPHENOXY)-2-HYDROXYPROPYL)ETHYLENEDIAM-
  INE DIMETHANESULPHOANTE see MQY125
3-(3,5-DIMETHOXYPHENOXY)-1,2-PROPANEDIOL see DOJ700
3-(3',5'-DIMETHOXYPHENOXY)PROPANEDIOL-(1,2) see DOJ700
3,4-DIMETHOXYPHENYLACETONITRILE see VIK100
1-(2,5-DIMETHOXYPHENYL)-2-AMINOPROPANE see DOJ800
1-(3,4-DIMETHOXYPHENYL)-2-AMINOPROPANE see DOK000
1-((2,5-DIMETHOXYPHENYL)AZO)-2-NAPHTHALENOL see DOK200
1-((2,5-DIMETHOXYPHENYL)AZO)-2-NAPHTHOL see DOK200
2,5-DIMETHOXY-1-(PHENYLAZO)-2-NAPHTHOL see DOK200
1-(1-(2,5-DIMETHOXYPHENYL)AZO)-2-NAPHTHOL see DOK200
3-(2,4-DIMETHOXYPHENYL)CROTONIC ACID MAGNESIUM SALT
  see DOK400

1,1-DIMETHOXY-2-PHENYLETHANE see PDX000
DIMETHOXYPHENYLETHYLAMINE see DOE200
3,4-DIMETHOXYPHENYLETHYLAMINE see DOE200
2-(3,4-DIMETHOXYPHENYL)ETHYLAMINE see DOE200
3,4-DIMETHOXY-β-PHENYLETHYLAMINE see DOE200
β-(3,4-DIMETHOXYPHENYL)ETHYLAMINE see DOE200
3,4-DIMETHOXYPHENYLETHYLAMINE (base) see DOE200
3,4-DIMETHOXY-β-PHENYLETHYLAMINE HYDROCHLORIDE see DOI400
4-(2-(2,5-DIMETHOXYPHENYL)ETHYL)BENZENAMINE see DON400
1-(3,4-DIMETHOXYPHENYL)-5-ETHYL-7,8-DIMETHOXY-4-METHYL-5H-2,3-
  BENZODIAZEPINE see GJS000
2-(3,4-DIMETHOXYPHENYL)-5-ETHYLTHIAZOLIDIN-4-ONE see KGU100
1-(2',5'-DIMETHOXYPHENYL)-2-GLYCINAMIDOETHANOL HYDROCHLO-
  RIDE see MQT530
β-(2,5-DIMETHOXYPHENYL)-β-HYDROXYISOPROPYLAMINE HYDROCHLO-
  RIDE see MDW000
2-(2,5-DIMETHOXYPHENYL)ISOPROPYLAMINE see DOK600
β-(2,5-DIMETHOXYPHENYL)ISOPROPYLAMINE HYDROCHLORIDE
  see DOJ800
DIMETHOXYPHENYLMETHANE see DOG700
1-((3,4-DIMETHOXYPHENYL)METHYL)-6,7-DIMETHOXYISOQUINOLINE
  see PAH000
1-((3,4-DIMETHOXYPHENYL)METHYL)-6,7-DIMETHOXYISOQUINOLINE HY-
  DROCHLORIDE see PAH250
((3,4-DIMETHOXYPHENYL)METHYL)HYDRAZINE see VIK150
DIMETHOXYPHENYLMETHYLSILANE see DOH400
1,1-DIMETHOXY-2-PHENYLPROPANE see HI1600
1-(3,4-DIMETHOXYPHENYL)-2-PROPENE see AGE250
3-(4,6-DIMETHOXY-α-PHENYL-m-TOLUOYL)-PROPIONIC ACID SODIUM
  SALT see DOA875
2,2-DI-(p-METHOXYPHENYL)-1,1,1-TRICHLOROETHANE see MEI450
DI(p-METHOXYPHENYL)-TRICHLOROMETHYL METHANE see MEI450
(DIMETHOXYPHOSPHINOTHIOYL)THIO)BUTANEDIOIC ACID DIETHYL
  ESTER see MAK700
2-DIMETHOXYPHOSPHINOTHIOYLTHIOMETHYL-4,6-DIAMINO-s-TRIAZINE
  see ASD000
2-((DIMETHOXYPHOSPHINYL)OXY)-1H-BENZ(d,e)ISOQUINOLINE-1,3(2H)-
  DIONE see DOL400
3-((DIMETHOXYPHOSPHINYL)OXY)-2-BUTENOIC ACID METHYL ESTER
  see MQR750
(E)-3-((DIMETHOXYPHOSPHINYL)OXY)-2-BUTENOIC ACID 1-
  PHENYLETHYL ESTER (9CI) see COD000
3-(DIMETHOXYPHOSPHINYLOXY)-N,N-DIMETHYL-cis-CROTONAMIDE
  see DGQ875
3-(DIMETHOXYPHOSPHINYLOXY)-N,N-DIMETHYLISOCROTONAMIDE
  see DGQ875
3-(DIMETHOXYPHOSPHINYLOXY)-N-METHYL-N-METHOXY-cis-CROTONAM-
  IDE see DOL800
3-(DIMETHOXYPHOSPHINYLOXY)N-METHYL-cis-CROTONAMIDE
  see MRH209
((DIMETHOXYPHOSPHINYL)THIO)ACETIC ACID ETHYL ESTER
  see DRB600
((DIMETHOXYPHOSPHINYL)THIO)-BUTANEDIOIC ACID DIETHYL ESTER
  (9CI) see OPK250
DIMETHOXY POLYETHYLENE GLYCOL see DOM100
1,1-DIMETHOXYPROPANE see DOM200
2,2-DIMETHOXYPROPANE see DOM400
3,3-DIMETHOXYPROPENE see DOM600
1,2-DIMETHOXY-4-PROPENYLBENZENE see IKR000
N-(3,6-DIMETHOXY-4-PYRIDAZINYL)SULFANILAMIDE see DON700
N¹-(2,6-DIMETHOXY-4-PYRIMIDINYL)SULFANILAMIDE see SNN300
N¹-(4,6-DIMETHOXYPYRIMIDIN-2-YL)SULFANILAMIDE see SNI500
N'-(5,6-DIMETHOXY-4-PYRIMIDYL)SULFANILAMIDE see AIE500
4,7-DIMETHOXY-6-(2-PYRROLIDINYLETHOXY)-5-
  CINNAMOYLBENZOFURAN MALEATE see DON000
2,6-DIMETHOXYQUINOL see DON200
4-(2,5-DIMETHOXY)STILBENAMINE see DON400
2',5'-DIMETHOXYSTILBENAMINE see DON400
2,5-DIMETHOXY-4'-STILBENAMINE see DON400
2,3-DIMETHOXYSTRYCHNINE see BOL750
DIMETHOXY STRYCHNINE (DOT) see BOL750
DIMETHOXYSULFADIAZINE see SNN300
2,4-DIMETHOXY-6-SULFANILAMIDO-1,3-DIAZINE see SNN300
3,6-DIMETHOXY-4-SULFANILAMIDOPYRIDAZINE see DON700
2,6-DIMETHOXY-4-SULFANILAMIDOPYRIMIDINE see SNN300
DIMETHOXYTETRAETHYLENE GLYCOL see PBO500
DIMETHOXYTETRAGLYCOL see PBO500
2,5-DIMETHOXYTETRAHYDROFURAN see DON800
(s-(4*,S*))-6,7-DIMETHOXY-3-(5,6,7,8-TETRAHYDRO-4-METHOXY-6-METHYL-
  1,3-DIOXOLO(4,5-g)ISOQUINOLIN-5-YL)-1(3H)-ISOBENZOFURANONE, N-
  OXIDE, HYDROCHLORIDE see NBP300
3',5'-DIMETHOXY-3,4',5,7-TETRAHYDROXYFLAVYLIUM ACID ANION
  see MAO750
DI(METHOXYTHIOCARBONYL) DISULFIDE see DUN600

DIMETHOXY-2,2,2-TRICHLORO-1-N-BUTYRYLOXY-ETHYLPHOSPHINE OXIDE see BPG000

DIMETHOXY-2,2,2-TRICHLORO-1-HYDROXY-ETHYL-PHOSPHINE OXIDE see TIQ250

3,3'-DIMETHOXYTRIPHENYLMETHANE-4,4'-BIS(1''-AZO-2''-NAPHTHOL) see DOO400

6,7-DIMETHOXY-1-VERATRYLISOQUINOLINE see PAH000

6,7-DIMETHOXY-1-VERATRYLISOQUINOLINE-3-CARBOXYLIC ACID SODIUM SALT see PAG750

DIMETHOXYVIOLANTHRONE see JAT000

16,17-DIMETHOXYVIOLANTHRONE see JAT000

DIMETHPRAMIDE see DUO300

DIMETHPYRINDENE MALEATE see FMU409

DIMETHULENE see DRV000

DIMETHWLEN see DRV000

DIMETHYL see EDZ000, SDF000

DIMETHYLACETAL see DOO600

DIMETHYLACETAMIDE see DOO800

N,N-DIMETHYLACETAMIDE see DOO800

O,O-DIMETHYL-S-(2-ACETAMIDOETHYL) ESTER PHOSPHORODITHIOIC ACID see DOP200

1,1-DIMETHYL-3-(p-ACETAMIDOPHENYL)TRIAZENE see DUI000

2,4-DIMETHYLACETANILIDE see ABP750

2,6-DIMETHYLACETANILIDE see ABP250

3,4-DIMETHYLACETANILIDE see ABP500

2',4'-DIMETHYLACETANILIDE see ABP750

3',4'-DIMETHYLACETANILIDE see ABP500

DIMETHYLACETIC ACID see IJU000

N,N-DIMETHYLACETOACETAMIDE see DOP000

DIMETHYLACETONE see DJN750

DIMETHYLACETONE AMIDE see DOO800

N,N-DIMETHYL-β-ACETOXY β-PHENYLETHYLAMINE see ABN700

O,O-DIMETHYL-S-(2-ACETYLAMINO)ETHYL) DITHIOPHOSPHATE see DOP200

O,O-DIMETHYL-S-(2-ACETYLAMINOETHYL) PHOSPHORODITHIOATE see DOP200

DIMETHYLACETYLENE see COC500

DIMETHYLACETYLENECARBINOL see MHX250

DIMETHYL ACETYLENEDICARBOXYLIC ACID see DOP400

O,S-DIMETHYLACETYLPHOSPHOROAMIDOTHIOATE see DOP600

2,7-DIMETHYL-3,6-ACRIDINEDIAMINE MONOHYDROCHLORIDE see DBT400

N,N-DIMETHYLACRYLAMIDE see DOP800

3,3-DIMETHYL-ACRYLATE de 2,4-DINITRO-6-(1-METHYLPROPYLE)PHENYLE (FRENCH) see BGB500

3,3-DIMETHYLACRYLIC ACID see MHT500

β,β-DIMETHYLACRYLIC ACID see MHT500

(E)-2,3-DIMETHYLACRYLIC ACID see TGA700

trans-2,3-DIMETHYLACRYLIC ACID see TGA700

trans-α-β-DIMETHYLACRYLIC ACID see TGA700

3,3-DIMETHYLACRYLIC ACID 2-sec-BUTYL-4,5-DINITROPHENYL ESTER see BGB500

cis-α,β-DIMETHYL ACRYLIC ACID, GERANIOL ESTER see GDO000

DIMETHYL ADIPATE see DOQ300

DIMETHYLAETHANOLAMIN (GERMAN) see DOY800

O,O-DIMETHYL-S-(2-AETHYLSULFINYL-AETHYL)-THIOLPHOSPHAT (GERMAN) see DAP000

O,O-DIMETHYL-S-(2-AETHYLSULFONYL-AETHYL)-THIOLPHOSPHAT (GERMAN) see DAP600

N-(5-(1,1-DIMETHYLAETHYL)-1,3,4-THIADIAZOL-2-YL)-N,N-DIMETHYLHARNSTOFF (GERMAN) see BSN000

O,O-DIMETHYL-S-(2-AETHYLTHIO-AETHYL)-DITHIO PHOSPHAT (GERMAN) see PHI500

O,O-DIMETHYL-O-(2-AETHYLTHIO-AETHYL MONOTHIOPHOSPHAT (GERMAN) see DAO800

O,O-DIMETHYL-S-(2-AETHYLTHIO-AETHYL)-MONOTHIOPHOSPHAT (GERMAN) see DAP400

DIMETHYL ALDEHYDE see DOO600

DIMETHYLALLYL ACETATE see DOQ350

3,3-DIMETHYLALLYL ACETATE see DOQ350

Γ,Γ-DIMETHYLALLYL ACETATE see DOQ350

DIMETHYLALLYL ALCOHOL see MHU110

3,3-DIMETHYLALLYL ALCOHOL see MHU110

Γ,Γ-DIMETHYLALLYL ALCOHOL see MHU110

2-(3,3-DIMETHYLALLYL)CYCLAZOCINE see DOQ400

2-DIMETHYLALLYL-5,9-DIMETHYL-2'-HYDORXYBENZOMORPHAN see DOQ400

2-(3,3-DIMETHYLALLYL)-5-ETHYL-2'-HYDROXY-9-METHYL-6,7-BENZOMORPHAN see DOQ600

2-(3,3-DIMETHYLALLYL)-2',2'-HYDROXY-5,9-DIMETHYL-6,7-BENZOMORPHAN see DOQ400

DIMETHYLALUMINUM CHLORIDE see DOQ700

DIMETHYLALUMINUM HYDRIDE see DOQ750

DIMETHYLAMIDE ACETATE see DOO800

DIMETHYLAMIDE DIETHYLENEIMIDE PHOSPHORIC ACID see DOV600

DIMETHYLAMIDOETHOXYPHOSPHORYL CYANIDE see EIF000

DIMETHYLAMINE see DOQ800

DIMETHYLAMINE (anhydrous) see DOR000

DIMETHYLAMINE, solution (DOT) see DOQ800

DIMETHYLAMINE, anhydrous (DOT) see DOQ800

DIMETHYLAMINE, aqueous solution (DOT) see DOQ800

DIMETHYLAMINE BENZHYDRYL ESTER HYDROCHLORIDE see BAU750

DIMETHYLAMINE BORANE see DOR200

4-DIMETHYLAMINE m-CRESYL METHYLCARBAMATE see DOR400

DIMETHYLAMINE with DIBORANE (1:1) see DOX200

DIMETHYLAMINE HYDROCHLORIDE see DOR600

4-DIMETHYLAMINEPYRIDINE see DQB600

DIMETHYLAMINE SALT of 2,4-D see DFY800

DIMETHYLAMINE SALTS of mixed POLYCHLOROBENZOIC ACIDS see PJQ000

DIMETHYLAMINE-2,3,6-TRICHLOROBENZOATE see DOR800

4-(DIMETHYLAMINE)-3,5-XYLYL-N-METHYLCARBAMATE see DOS000

DIMETHYLAMINOACETONITRILE see DOS200

(DIMETHYLAMINO)ACETYLENE see DOS300

N',N'-DIMETHYL-4'-AMINO-N-ACETYL-N-MONOMETHYL-4-AMINOAZOBENZENE see DPQ200

DIMETHYLAMINOAETHANOL (GERMAN) see DOY800

β-DIMETHYLAMINO-AETHYL-BENZHYDRYL-AETHER (GERMAN) see BBV500

N-(2'-DIMETHYLAMINOAETHYL)-(o-BENZYLPHENOL)-AETHER HYDROCHLORID (GERMAN) see DPD400

N-(4-((1-(DIMETHYLAMINO)-AETHYLIDEN)AMINO)PHENYL)-2-METHOXYACETAMID-HYDROCHLORID (GERMAN) see DPF200

N,N-DIMETHYL-β-AMINOAETHYL-ISOTHIURONIUM DIHYDROCHLORID (GERMAN) see NNL400

N-DIMETHYLAMINO-AETHYL-N-p-METHOXY-BENZYL-α-AMINO-PYRIDIN-MALEAT (GERMAN) see WAK000

5-(DIMETHYLAMINOAETHYL-OXYIMINO)-5H-DIBENZO(a,d)CYCLOHEPTA-1,4-DIENHYDROCHLORID (GERMAN) see DPH600

3-(DIMETHYLAMINO)-ALANINE (9CI) see ARY625

DIMETHYLAMINO-ANALGESINE see DOT000

p-DIMETHYLAMINOANILINE DIHYDROCHLORIDE see DTM000

9-(p-DIMETHYLAMINOANILINO)ACRIDINE see DOS800

DIMETHYLAMINOANTIPYRINE see DOT000

4-(DIMETHYLAMINO)ANTIPYRINE see DOT000

4-(DIMETHYLAMINO)ANTIPYRINE mixed with SODIUM NITRITE (1:1) see DOT200

p-DIMETHYLAMINOAZOBENZEN (CZECH) see DOT300

DIMETHYLAMINOAZOBENZENE see DOT300

4-DIMETHYLAMINOAZOBENZENE see DOT300

p-DIMETHYLAMINOAZOBENZENE see DOT300

4-(N,N-DIMETHYLAMINO)AZOBENZENE see DOT300

N,N-DIMETHYL-4-AMINOAZOBENZENE see DOT300

N,N-DIMETHYL-p-AMINOAZOBENZENE see DOT300

2',3-DIMETHYL-4-AMINOAZOBENZENE see AIC250

4-DIMETHYLAMINOAZOBENZENE AMINE-N-OXIDE see DTK600

p-(DIMETHYLAMINO)AZOBENZENE-o-CARBOXYLIC ACID see CCE500

4'-DIMETHYLAMINOAZOBENZENE-2-CARBOXYLIC ACID see CCE500

N,N-DIMETHYLAMINOAZOBENZENE-N-OXIDE see DTK600

4-DIMETHYLAMINOAZOBENZENE-4'-SULPHONIC ACID SODIUM SALT see MND600

DIMETHYLAMINOAZOBENZOL see DOT300

4-DIMETHYLAMINOAZOBENZOL see DOT300

p-DIMETHYLAMINO-AZOBENZOL (GERMAN) see DOT300

DIMETHYLAMINOAZOPHENE see DOT000

4-(DIMETHYLAMINO) BENZALDEHYDE see DOT400

p-(DIMETHYLAMINO)BENZALDEHYDE see DOT400

p-DIMETHYLAMINOBENZALDEHYDE(2-(2-CHLOROETHYL)-2-METHYL)HYDRAZONE see MIG750

1-(4-DIMETHYLAMINOBENZAL)INDENE see DOT600

p-DIMETHYLAMINOBENZALRHODANINE see DOT800

5-(p-DIMETHYLAMINOBENZAL)RHODANINE see DOT800

p-DIMETHYLAMINOBENZAL-5-RHODANINE see DOT800

(DIMETHYLAMINO)BENZENE see DQF800

3,4-DIMETHYLAMINOBENZENE see XNS000

p-DIMETHYLAMINOBENZENEAZO-1-NAPHTHALENE see DSU600

p-DIMETHYLAMINOBENZENE-1-AZO-1-NAPHTHALENE see DSU600

p-DIMETHYLAMINOBENZENE-1-AZO-2-NAPHTHALENE see DSU800

5(4-DIMETHYLAMINOBENZENEAZO)TETRAZOLE see DOU000

4-DIMETHYLAMINOBENZENECARBONAL see DOT400

p-DIMETHYLAMINOBENZENE DIAZO SODIUM SULFONATE see DOU600

p-DIMETHYLAMINOBENZENEDIAZOSODIUM SULPHONATE see DOU600

p-(DIMETHYLAMINO)BENZENEDIAZOSULFONATE see DOU600

4-DIMETHYLAMINOBENZENEDIAZOSULFONIC ACID, SODIUM SALT see DOU600

p-DIMETHYLAMINOBENZENEDIAZOSULFONIC ACID, SODIUM SALT see DOU600

p-(DIMETHYLAMINO)BENZENEDIAZOSULPHONATE see DOU600

4-DIMETHYLAMINOBENZENEDIAZOSULPHONIC ACID, SODIUM SALT
see DOU600
p-(DIMETHYLAMINO)BENZENEDIAZOSULPHONIC ACID, SODIUM SALT
see DOU600
p-DIMETHYLAMINOBENZOIC ACID, OCTYL ESTER see AOI500
p-DIMETHYLAMINOBENZOIC ACID, PENTYL ESTER see AOI500
p-DIMETHYLAMINOBENZOLDIAZOSULFONAT (NATRIUMSALZ) (GERMAN)
see DOU600
4,4'-DIMETHYLAMINOBENZOPHENONIMIDE see IBB000
5-(p-DIMETHYLAMINOBENZOYLIDENE)RHODANINE see DOT800
p-DIMETHYLAMINOBENZYLIDEN-1,2-BENZ-9-METHYL-ACRIDINE
see DQC200
p-DIMETHYLAMINOBENZYLIDEN-3,4-BENZ-9-METHYLACRIDINE
see DQC000
p-DIMETHYLAMINOBENZYLIDENE-3,4,5,6-DIBENZ-9-METHYLACRIDINE
see DOV000
(4-DIMETHYLAMINOBENZYLIDENE)INDENE see DOT600
p-DIMETHYLAMINOBENZYLIDENE RHODAMINE see DOT800
4-DIMETHYLAMINOBIPHENYL see BGF899
2',3-DIMETHYL-4-AMINOBIPHENYL see BLV250
3,2'-DIMETHYL-4-AMINOBIPHENYL see BLV250
3,3'-DIMETHYL-4-AMINOBIPHENYL see DOV200
4-(DIMETHYLAMINO)-3-BIPHENYLOL see DOV400
DIMETHYLAMINO-BIS(1-AZIRIDINYL)PHOSPHINE OXIDE see DOV600
4-(DIMETHYLAMINO)-α,α-BIS(4-(DIMETHYLAMINO)PHENYL)-
BENZENEMETHANOL (9CI) see TJK000
4-DIMETHYLAMINO-1,1-BIS[(3,4-(METHYLENEDIOXY)PHENOXY)METHYL]-1-
BUTANOL, METHYLCARBAMATE (ester), CITRATE see DOV800
N-DIMETHYL AMINO-β-CARBAMYL PROPIONIC ACID see DQD400
3-(((DIMETHYLAMINO)CARBONYL)AMINO)PHENYL-1,1-
DIMETHYLETHYL)CARBAMATE see DUM800
(DIMETHYLAMINO)CARBONYL CHLORIDE see DQY950
3-(((DIMETHYLAMINO)CARBONYL)OXY)-1-METHYL-PYRIDINIUM (9CI)
see PPI800
3-((DIMETHYLAMINO)CARBONYL)OXY)-1-METHYL-PYRIDINIUM BROMIDE
see MDL600
p-DIMETHYLAMINO-CARVACROLDIMETHYLURETHANE METHIODIDE
see DOW875
1-(N,N-DIMETHYLAMINO)-3-(p-CHLOROPHENYL-3α-PYRIDYL)PROPANE
MALEATE see TAI500
3-β-(DIMETHYLAMINO)CON-5-ENINE-DIHYDROBROMIDE see DOX000
3-β-(DIMETHYLAMINO)CON-5-ENINE HYDROCHLORIDE see CNH660
4-DIMETHYLAMINO-3-CRESYL METHYLCARBAMATE see DOR400
1-DIMETHYLAMINO-3-CYANO-3-PHENYL-4-METHYLHEXANE HYDROCHLO-
RIDE see DOX100
DIMETHYLAMINOCYANPHOSPHORSAEUREAETHYLESTER (GERMAN)
see EIF000
(DIMETHYLAMINO)CYCLOHEXANE see DRF709
N,N-DIMETHYLAMINOCYCLOHEXANE see DRF709
7-DIMETHYLAMINO-6-DEMETHYL-6-DEOXYTETRACYCLINE see MQW250
DIMETHYLAMINODIBORANE see DOX200
4-(DIMETHYLAMINO)-1,2-DIHYDRO-1,5-DIMETHYL-2-PHENYL-3H-PYRAZOL-
3-ONE see DOT000
1-(DIMETHYLAMINO)-2-((DIMETHYLAMINO)METHYL)-2-BUTANOL BENZO-
ATE,(ESTER) see AHI250
4-(DIMETHYLAMINO)-α-(4-(DIMETHYLAMINO)PHENYL)-α-PHENYL-
BENZENEMETHANOL see MAK500
4-(DIMETHYLAMINO)-3,5-DIMETHYLPHENOL METHYLCARBAMATE
(ESTER) see DOS000
4-(DIMETHYLAMINO)-3,5-DIMETHYLPHENYL ESTER, METHYLCARBAMIC
ACID see DOS000
4-(DIMETHYLAMINO)-3,5-DIMETHYLPHENYL-N-METHYLCARBAMATE
see DOS000
4-DIMETHYLAMINO-2,3-DIMETHYL-1-PHENYL-3-PYRAZOLIN-5-ONE
see DOT000
4-DIMETHYLAMINO-2,3-DIMETHYL-1-PHENYL-5-PYRAZOLONE see DOT000
3-DIMETHYLAMINO-N,N-DIMETHYLPROPIONAMIDE see DOX400
3-DIMETHYLAMINO-1,2-DIMETHYLPROPYL p-AMINOBENZOATE HYDRO-
CHLORIDE see AIT750
2-(DIMETHYLAMINO)-5,6-DIMETHYL-4-PYRIMIDINYLDIMETHYLCARBA-
MATE see DOX600
3,2'-DIMETHYL-4-AMINODIPHENYL see BLV250
3,3'-DIMETHYL-4-AMINODIPHENYL see DOV200
(3S,6S)-(-)-6-(DIMETHYLAMINO)-4,4-DIPHENYL-3-HEPTANOL ACETATE
(ester) HYDROCHLORIDE see ACQ690
d-6-(DIMETHYLAMINO)-4,4-DIPHENYL-3-HEPTANONE see DBE100
l-6-(DIMETHYLAMINO)-4,4-DIPHENYL-3-HEPTANONE see MDO775
(s)-6-DIMETHYLAMINO-4,4-DIPHENYL-3-HEPTANONE see DBE100
6-DIMETHYLAMINO-4,4-DIPHENYL-3-HEPTANONE HYDROCHLORIDE
see MDP000
1-6-(DIMETHYLAMINO)-4,4-DIPHENYL-3-HEPTANONE HYDROCHLORIDE
see MDP250
dl-6-DIMETHYLAMINO-4,4-DIPHENYL-3-HEPTANONE HYDROCHLORIDE
see MDP750

6-(DIMETHYLAMINO)-4,4-DIPHENYL-3-HEPTANONE dl-MIXTURE
see MDO760
p,p-DIMETHYLAMINODIPHENYLMETHANE see MJN000
α-4-DIMETHYLAMINO-1,2-DIPHENYL-3-METHYL-2-BUTANOL PROPIONATE
see PNA250
α-(+)-4-DIMETHYLAMINO-1,2-DIPHENYL-3-METHYL-2-BUTANOL PROPIO-
NATE ESTER see DAB879
2-(DIMETHYLAMINO)-N-(1,3-DIPHENYL-1H-PYRAZOL-5-YL) PROPANAMIDE
see PAM500
4-(DIMETHYLAMINO)-2,2-DIPHENYLVALERAMIDE see DOY400
2-DIMETHYLAMINO ETHANETHIOL HYDROCHLORIDE see DOY600
DIMETHYLAMINOETHANOL see DOY800
2-(DIMETHYLAMINO)ETHANOL see DOY800
N-DIMETHYLAMINOETHANOL see DOY800
β-DIMETHYLAMINOETHANOL see DOY800
N,N-DIMETHYLAMINOETHANOL see DOY800
2-DIMETHYLAMINOETHANOL-p-ACETAMIDOBENZOATE see DOZ000
2-(DIMETHYLAMINO)ETHANOL BITARTRATE see DPA500
2-DIMETHYLAMINOETHANOL-4-N-BUTYLAMINOBENZOATE HYDROCHLO-
RIDE see TBN000
β-DIMETHYLAMINOETHANOL DIPHENYLMETHYL ETHER see BBV500
2-(DIMETHYLAMINO)ETHANOL METHACRYLATE see DPG600
2-(DIMETHYLAMINO)ETHANOL TARTRATE see DPA200
1-(DIMETHYLAMINOETHOXYACETAMIDO)ADAMANTANE HYDROCHLO-
RIDE see AEF500
N-(p-(2-(DIMETHYLAMINO)ETHOXY)BENZYL)-3,4,5-TRIMETHOXYBENZAM-
IDE HYDROCHLORIDE see TKW750
N-(p-(2-(DIMETHYLAMINO)ETHOXY)-BENZYL)-3,4,5-TRIMETHOXYBENZAM-
IDE MONOHYDROCHLORIDE see TKW750
1-(β-DIMETHYLAMINOETHOXY)-3-N-BUTYLISOQUINOLINE HYDROCHLO-
RIDE see DNX400
1-(β-DIMETHYLAMINOETHOXY)-3-N-BUTYLISOQUINOLINE
MONOHYDROCHLORIDE see DNX400
5-(2-(N,N-DIMETHYLAMINO)ETHOXY)CARVACROL ACETATE CITRATE
see MRN600
5-(2-(N,N-DIMETHYLAMINO)ETHOXY)CARVACROL ACETATE HYDRO-
CHLORIDE see TFY000
2-(2-DIMETHYLAMINOETHOXY)CHALCONE CITRATE see DPA500
2-(2-DIMETHYLAMINOETHOXY)-N,N-DIETHYLPROPYLAMINE DIMETHIOD-
IDE see MQF750
α-(2-DIMETHYLAMINOETHOXY)DIPHENYLMETHANE see BBV500
2-(2-DIMETHYLAMINOETHOXY)ETHANOL see DPA600
2-(2-DIMETHYLAMINOETHOXY)ETHANOL-1-PHENYL-
CYCLOPENTYLCARBOXYLATE see DPA800
2-(2-(DIMETHYLAMINO)ETHOXY)ETHYL-1-
PHENYLCYCLOPENTANECARBOXYLATE see DPA800
4-(2-(DIMETHYLAMINO)ETHOXY)-5-ISOPROPYL-2-METHYLPHENOL
see DAD850
2-(α-(2-(DIMETHYLAMINO)ETHOXY)-α-METHYLBENZYL)PYRIDINE
see DYE500
DIMETHYLAMINOETHOXY-METHYL-BENZYL-PYRIDINE SUCCINATE
see PGE775
2-(α-(2-DIMETHYLAMINOETHOXY)-α-METHYLBENZYL)PYRIDINE SUCCI-
NATE see PGE775
4-(2-(DIMETHYLAMINO)ETHOXY)-2-METHYL-5-(1-METHYLETHYL)PHENOL
see DAD850
6-(2-DIMETHYLAMINOETHOXY)-2-(5-NITRO-1-METHYL-2-IMIDAZOLYL)-
METHYLENE)-1-TETRALON SULFATE see DPB200
cis-1-(p-2-(N,N-DIMETHYLAMINO)ETHOXY)PHENYL)-1,2-DIPHENYLBUT-1-
ENE see NOA600
trans-1-(p-β-DIMETHYLAMINOETHOXYPHENYL)-1,2-DIPHENYLBUT-1-ENE
CITRATE see TAD175
2-DIMETHYLAMINOETHOXYPHENYLMETHYL-2-PICOLINE see DYE500
2-DIMETHYLAMINOETHOXYPHENYLMETHYL-2-PICOLINE SUCCINATE
see PGE775
4-(2-DIMETHYLAMINOETHOXY)-N-(3,4,5-
TRIMETHOXYBENZOYL)BENZYLAMINE HYDROCHLORIDE see TKW750
N-(DIMETHYLAMINOETHYL)-1-ADAMANTANEACETAMIDE ETHYL IO-
DIDE see DPB400
β-DIMETHYLAMINOETHYL ALCOHOL see DOY800
2-DIMETHYLAMINOETHYLAMINE see DPC000
2-(DIMETHYLAMINO)ETHYL-p-AMINOBENZOATE HYDROCHLORIDE
see DPC200
2-(2-(2-(DIMETHYLAMINO)ETHYLAMINO)ETHYL)-2-METHYL-1,3-
BENZODIOXOLEDI HYDROCHLORIDE see DPC400
2-(DIMETHYLAMINO)ETHYL-p-AMINOSALICYLATE see AMN000
β-DIMETHYLAMINOETHYLBENZHYDRYLETHER see BBV500
β-DIMETHYLAMINOETHYL BENZHYDRYL ETHER HYDROCHLORIDE
see BAU750
DIMETHYLAMINOETHYL BENZILATE, HYDROCHLORIDE see BAW500
2-(DIMETHYLAMINO)ETHYL BENZILATE HYDROCHLORIDE see BAW500
β-DIMETHYLAMINOETHYL BENZILATE HYDROCHLORIDE see BAW500
N,N-DIMETHYL-β-3-AMINOETHYLBENZOTHIOPHENE HYDROCHLORIDE
see DQP000

α-(1-(DIMETHYLAMINO)ETHYL)BENZYL ALCOHOL HYDROCHLORIDE see DPD200
N-DIMETHYLAMINOETHYLBENZYLANILINE HYDROCHLORIDE see PEN000
DIMETHYLAMINOETHYL BENZYLATE HYDROCHLORIDE see BAW500
(2-(DIMETHYLAMINO)ETHYL)(o-BENZYLPHENOXY)ETHERHYDROCHLORIDE see DPD400
2-(α-(2-DIMETHYLAMINOETHYL)BENZYL)PYRIDINE see TMJ750
2-(α-(2-DIMETHYLAMINO)ETHYL)BENZYL)PYRIDINE, BIMALEATE see TMK000
2-(α-(2-DIMETHYLAMINO)ETHYL)BENZYL)PYRIDINE, MALEATE see TMK000
β-DIMETHYLAMINOETHYL-p-BROMO-α-METHYLBENZHYDRYL ETHER HYDROCHLORIDE see BMN250
DIMETHYLAMINOETHYL-p-BUTYL-AMINOBENZOATE see BQA010
2-DIMETHYLAMINOETHYL-p-BUTYLAMINOBENZOATE see BQA010
2-(DIMETHYLAMINO)ETHYL-p-(BUTYLAMINO)BENZOATE HYDROCHLORIDE see TBN000
DIMETHYLAMINOETHYL-p-N-BUTYLAMINOBENZOATE HYDROCHLORIDE see TBN000
DIMETHYLAMINOETHYL CHLORIDE see CGW000
2-DIMETHYLAMINOETHYLCHLORIDE see CGW000
β-(DIMETHYLAMINO)ETHYL CHLORIDE see CGW000
α-(2-DIMETHYLAMINOETHYL)-o-CHLOROBENZHYDROL HYDROCHLORIDE see CMW700
β-DIMETHYLAMINOETHYL (p-CHLORO-α-METHYLBENZHYDRYL) ETHER HYDROCHLORIDE see CIS000
DIMETHYLAMINOETHYL-p-CHLOROPHENOXYACETATE see DPE000
DIMETHYLAMINOETHYL 4-CHLOROPHENOXYACETATE HYDROCHLORIDE see AAE500
DIMETHYLAMINOETHYL p-CHLOROPHENOXYACETATE HYDROCHLORIDE see AAE500
DIMETHYLAMINOETHYL-4-CHLOROPHENOXYACETIC ACID see DPE000
2-DIMETHYLAMINOETHYL 2'-DIETHYLAMINOISOPROPYL ETHER BISMETHIODIDE see MQF750
β-DIMETHYLAMINOETHYL β'-DIETHYLAMINO-α'-METHYLETHYL ETHER DIMETHIODIDE see MQF750
S-(2-DIMETHYLAMINOETHYL)-O,O-DIETHYLPHOSPHORITHIOATE METHIODIDE see TLF500
5-(2-DIMETHYLAMINO)ETHYL)-2,3-DIHYDRO-3-HYDROXY-2-(p-METHOXYPHENYL)1,5-BENZOTHIAZEPIN-4(5H)-ONE-ACETATE (ESTER) HYDROCHLORIDE see DPE100
10-(2-(DIMETHYLAMINO)ETHYL)-5,10-DIHYDRO-5-METHYL-11H-DIBENZO(B,E)(1,4)DIAZEPIN-11-ONE see DCW600
5-(2-DIMETHYLAMINO)ETHYL)-2,3-DIHYDRO-2-PHENYL-1,5-BENZOTHIAZEPIN-4-(5H)-ONE HYDROCHLORIDE see TEU250
DIMETHYLAMINOETHYLDIPHENYLETHOXY ACETATE HYDROCHLORIDE see DPE200
β'-DIMETHYLAMINOETHYL-α,α-DIPHENYL-α-ETHOXYACETATEHYDROCHLORIDE see DPE200
DIMETHYLAMINOETHYL DIPHENYLHYDROXYACETATE HYDROCHLORIDE see BAW500
2-DIMETHYLAMINOETHYL ESTER-p-AMINOBENZOIC ACID HYDROCHLORIDE see DPC200
DIMETHYLAMINOETHYL ESTER of p-CHLOROPHENOXYACETIC ACID HYDROCHLORIDE see AAE500
2-DIMETHYLAMINO)ETHYL ESTER HYDROCHLORIDE ETHOXYDIPHENYLACETIC ACID see DPE200
2-(DIMETHYLAMINO)ETHYL ESTER METHACRYLIC ACID see DPG600
1-(2-(DIMETHYLAMINO)ETHYL)-1-ETHYL-3-MESITYLUREA HYDROCHLORIDE see DPE800
2-((2-(DIMETHYLAMINO)ETHYL)FURFURYLAMINO)PYRIDINE FUMARATE see MDP800
β-DIMETHYLAMINOETHYL (1-HYDROXYCYCLOPENTYL)PHENYLACETATE HYDROCHLORIDE see CPZ125
3-(β-DIMETHYLAMINOETHYL)-5-HYDROXYINDOLE see DPG109
2-(DIMETHYLAMINO)ETHYL 1-HYDROXY-α-PHENYLCYCLOPENTANEACETATE HYDROCHLORIDE see CPZ125
N-(4-(1-(DIMETHYLAMINO)ETHYLIDENE)AMINO)PHENYL)-2-METHOXYACETAMIDE HYDROCHLORIDE see DPF200
3-(2-(DIMETHYLAMINO)ETHYL)INDOLE see DPF600
3-(2-(DIMETHYLAMINO)ETHYL)INDOLESULFOSALICYLATE see DPG000
3-(2-(DIMETHYLAMINO)ETHYL)INDOL-4-OL see HKE000
3-(2-DIMETHYLAMINOETHYL)-5-INDOLOL see DPG109
3-(2-(DIMETHYLAMINO)ETHYL)-1H-INDOL-4-OL DIHYDROGEN PHOSPHATE ESTER see PHU500
3-2'-DIMETHYLAMINOETHYLINDOL-4-PHOSPHATE see PHU500
3-(2-DIMETHYLAMINOETHYL)INDOL-4-YL DIHYDROGEN PHOSPHATE see PHU500
10-(2-(DIMETHYLAMINO)ETHYL)ISOALLOXAZINE SULFATE see DPG200
1-(2-(DIMETHYLAMINO)ETHYL)-1-ISOPROPYL-3-(2,6-XYLYL)UREA HYDROCHLORIDE see DPG400
2-DIMETHYLAMINOETHYLISOTHIURONIUM CHLORIDE HYDROCHLORIDE see NNL400
DIMETHYLAMINOETHYL METHACRYLATE see DPG600

2-(DIMETHYLAMINO)ETHYL METHACRYLATE see DPG600
β-DIMETHYLAMINOETHYL METHACRYLATE see DPG600
N,N-DIMETHYLAMINOETHYL METHACRYLATE see DPG600
N-DIMETHYLAMINOETHYL-N-p-METHOXYα-AMINOPYRIDINE MALEATE see DBM800
2-((2-DIMETHYLAMINO)ETHYL)-(p-METHOXYBENZYL)AMINO)PYRIDINE see WAK000
2-((2-DIMETHYLAMINO)ETHYL)(p-METHOXYBENZYL)AMINO)PYRIDINE BIMALEATE see DBM800
2-((2-DIMETHYLAMINOETHYL)(p-METHOXYBENZYL)AMINO)PYRIDINE HYDROCHLORIDE see MCJ250
2-((2-DIMETHYLAMINO)ETHYL)(p-METHOXYBENZYL)AMINO)PYRIDINE MALEATE see DBM800
2-((2-DIMETHYLAMINO)ETHYL)(p-METHOXYBENZYL)AMINO)PYRIMIDINE see NCD500
2-((2-DIMETHYLAMINO)ETHYL)(p-METHOXY-BENZYL)AMINO)-PYRIMIDINE HYDROCHLORIDE see RDU000
2-((2-DIMETHYLAMINO)ETHYL)(p-METHOXYBENZYL)AMINO)THIAZOLE HYDROCHLORIDE see ZUA000
3-(2-(DIMETHYLAMINO)ETHYL)-5-METHOXY-2-METHYLINDOLE see MLI750
1-(DIMETHYLAMINOETHYL-METHYL)AMINO-3-PHENYLINDOLE HYDROCHLORIDE see BGC500
1-(omega-DIMETHYLAMINOETHYLMETHYL)AMINO-3-PHENYLINDOLE HYDROCHLORIDE see BGC500
2-DIMETHYLAMINOETHYL-2-METHYL-BENZHYDRYL ETHER CITRATE see DPH000
2-DIMETHYLAMINOETHYL-2-METHYLBENZHYDRYL ETHERHYDROCHLORIDE see OJW000
2-(2-(DIMETHYLAMINO)ETHYL)-2-METHYL-1,3-BENZODIOXOLE HYDROCHLORIDE see DPH200
α-(2-DIMETHYLAMINOETHYL)-α-(3-METHYL-2-BUTENYL)-1-NAPHTHALENEACETAMIDE see PMB500
2-(2-DIMETHYLAMINO)ETHYL)-2-(3-METHYL-2-BUTENYL)-2-(1-NAPHTHYL)ACETAMIDE see PMB500
10-(2-(DIMETHYLAMINO)ETHYL)-5-METHYL-5H-DIBENZO(b,e)(1,4)DIAZEPIN-11(10H)-ONE see DCW600
3-(2-(DIMETHYLAMINO)ETHYL)-5-METHYLINDOLE see DST400
2-(2-(DIMETHYLAMINO)ETHYL)-3-METHYL-2-PHENYLVALERONITRILE HYDROCHLORIDE see DOX100
1-(2-(DIMETHYLAMINO)ETHYL)-4-METHYLPIPERAZINE see DPH400
10-(2-(DIMETHYLAMINO)ETHYL)-1-NITRO-9(10H)-ACRIDINONE MONOHYDROCHLORIDE (9CI) see NFW430
5-DIMETHYLAMINOETHYLOXYIMINO-5H-DIBENZO(a,d)CYCLOHEPTA-1,4-DIENE HYDROCHLORIDE see DPH600
N-DIMETHYLAMINOETHYLPHENOTHIAZINE HYDROCHLORIDE see DPI000
N-(β-DIMETHYLAMINOETHYL)-PHENOTHIAZINEHYDROCHLORIDE see DPI000
2-(2-(DIMETHYLAMINO)ETHYL)-2-PHENYL-1,3-BENZODIOXOLE HYDROCHLORIDE see DPI400
N-(2-(DIMETHYLAMINO)ETHYL)-N-(3-PHENYL-1-INDOLYL)ACETAMIDE HYDROCHLORIDE see DPI600
β-DIMETHYLAMINOETHYL-2-PHENYLTETRAHYDROBENZOATE HYDROCHLORIDE see DPI700
s-(2-(DIMETHYLAMINO)ETHYL)PSEUDOTHIOUREA DIHYDROCHLORIDE see NNL400
β-DIMETHYLAMINO ETHYL-2-PYRIDYLAMINOTOLUENE see TMP750
β-DIMETHYLAMINOETHYL-2-PYRIDYLBENZYLAMINE see TMP750
N-(2-(DIMETHYLAMINO)ETHYL)-N-2-PYRIDYL-3-THENYLAMINE see DPJ200
2-((2-(DIMETHYLAMINO)ETHYL)(SELENOPHENE-2-YLMETHYL)AMINO)PYRIDINE see DPI750
2-DIMETHYLAMINOETHYL SUCCINATE DIMETHOCHLORIDE see HLC500
β-DIMETHYLAMINOETHYL-2,6,2',6'-TETRAMETHYLBENZHYDRYL ETHER HYDROCHLORIDE see DRR500
2-((2-(DIMETHYLAMINO)ETHYL)-2-THENYLAMINO)PYRIDINE see TEO250
2-((2-DIMETHYLAMINO)ETHYL)-3-THENYLAMINO)PYRIDINE see DPJ200
2-((2-(DIMETHYLAMINO)ETHYL)-2-THENYL-AMINO)PYRIDINE HYDROCHLORIDE see DPJ400
2-((2-(DIMETHYLAMINO)ETHYL)-3-THENYL-AMINO)-PYRIDINE HYDROCHLORIDE see TEO000
2-(2-(DIMETHYLAMINO)ETHYL)-2-THIOPSEUDOUREA DIHYDROCHLORIDE see NNL400
N-(2-(DIMETHYLAMINO)ETHYL)-N,N',N'-TRIMETHYL-1,2-ETHANEDIAMINE, (9CI) see PBG500
2-(N,N-DIMETHYLAMINO)ETHYL VINYL ETHER see VOF000
2-DIMETHYLAMINO-FLUREN (GERMAN) see DPJ600
2-DIMETHYLAMINOFLUORENE see DPJ600
N,N-DIMETHYL-2-AMINOFLUORENE see DPJ600
4-(DIMETHYLAMINO)-4'-FLUOROAZOBENZENE see DSA000
DIMETHYLAMINO HEXOSE REDUCTIONE see DXS200
2-DIMETHYLAMINO-4-HYDROXY-5-n-BUTYL-6-METHYLPYRIMIDINE see BRD000
4-DIMETHYLAMINO-3-HYDROXYDIPHENYL see DOV400
10-(3-DIMETHYLAMINOISOPROPYL)PHENOTHIAZINE HYDROCHLORIDE see PMI750

p-((p-(DIMETHYLAMINO)PHENYL)AZO)BENZENESULFONIC ACID SODIUM SALT see MND600

4-((p-(DIMETHYLAMINO)PHENYL)AZO)BENZIMIDAZOLE see DQM200

2-((4-DIMETHYLAMINO)PHENYLAZO)BENZOIC ACID see CCE500

3-((p-(DIMETHYLAMINO)PHENYL)AZO)BENZOIC ACID see CCE750

o-((p-(DIMETHYLAMINO)PHENYL)AZO)BENZOIC ACID see CCE500

6-DIMETHYLAMINOPHENYLAZOBENZOTHIAZOLE see DPO400

6-((p-(DIMETHYLAMINO)PHENYL)AZO)BENZOTHIAZOLE see DPO400

7-((p-(DIMETHYLAMINO)PHENYL)AZO)BENZOTHIAZOLE see DPO600

6-DIMETHYLAMINOPHENYLAZOBENZTHIAZOLE see DPO400

3-((p-DIMETHYLAMINOPHENYL)AZO)BENZYL ALCOHOL see HMB600

m-((p-DIMETHYLAMINOPHENYL)AZO)BENZYL ALCOHOL see HMB600

o-((p-DIMETHYLAMINOPHENYL)AZO)BENZYL ALCOHOL see HMB595

6-((p-(DIMETHYLAMINO)PHENYL)AZO)-1H-INDAZOLE see DSI800

4-((p-DIMETHYLAMINO)PHENYL)AZO)ISOQUINOLINE see DPO800

5-((p-(DIMETHYLAMINO)PHENYL)AZO)ISOQUINOLINE see DPP000

7-(p-(DIMETHYLAMINO)PHENYL)AZO)ISOQUINOLINE see DPP200

5-((p-(DIMETHYLAMINO)PHENYL)AZO)ISOQUINOLINE-2-OXIDE see DPP400

4-((4-DIMETHYLAMINO)PHENYL)AZO)-2,6-LUTIDINE-1-OXIDE see DPP800

4-((p-DIMETHYLAMINO)PHENYL)AZO)-2,5-LUTIDINE-1-OXIDE see DPP600

4-((p-DIMETHYLAMINO)PHENYL)AZO)-3,5-LUTIDINE-1-OXIDE
see DPP709

4-((p-DIMETHYLAMINO)PHENYL)AZO)-N-METHYLACETANILIDE
see DPQ200

5-((p-(DIMETHYLAMINO)PHENYL)AZO)-3-METHYLQUINOLINE see MJF500

5-((p-(DIMETHYLAMINO)PHENYL)AZO)-6-METHYLQUINOLINE see MJF750

5-((p-(DIMETHYLAMINO)PHENYL)AZO)-7-METHYLQUINOLINE see DPQ400

5-((p-(DIMETHYLAMINO)PHENYL)AZO)-8-METHYLQUINOLINE see MJG000

2-(4-DIMETHYLAMINOPHENYLAZO)NAPHTHALENE see DSU800

N-(4-((4-(DIMETHYLAMINO)PHENYL)AZO)PHENYL)-N-METHYLACETAMIDE
see DPQ200

4-((4-(DIMETHYLAMINO)PHENYL)AZO)-2-PICOLINE-1-OXIDE see DSS200

3'-(4-DIMETHYLAMINOPHENYL)AZOPYRIDINE see POP750

5-((p-(DIMETHYLAMINO)PHENYL)AZO)QUINALDINE see DPQ600

4-((p-(DIMETHYLAMINO)PHENYL)AZO)QUINOLINE see DTY200

5-((p-DIMETHYLAMINO)PHENYL)AZO)QUINOLINE see DPQ800

6-((p-(DIMETHYLAMINO)PHENYL)AZO)QUINOLINE see DPR000

4-((p-DIMETHYLAMINO)PHENYL)AZO)QUINOLINE-1-OXIDE see DTY400

5-((p-(DIMETHYLAMINO)PHENYL)AZO)QUINOLINE-1-OXIDE see DPR200

6-((p-(DIMETHYLAMINO)PHENYL)AZO)QUINOLINE-1-OXIDE see DPR400

5-((p-(DIMETHYLAMINO)PHENYL)AZO)QUINOXALINE see DUA200

6-((p-(DIMETHYLAMINO)PHENYL)AZO)QUINOXALINE see DUA400

(±)-2-(DIMETHYLAMINO)-2-PHENYLBUTYL-3,4,5-TRIMETHOXYBENZOATE
see TKU650

dl-trans-2-DIMETHYLAMINO-1-PHENYL-CYCLOHEX-3-EN-trans-1-CAR-
BONSAEUREAETHYLESTER HCl (GERMAN) see EIH000

dl-trans-2-DIMETHYLAMINO-1-PHENYL-CYCLOHEX-3-ENE-trans-CARBONIC
ACID ETHYL ESTER HCl see EIH000

(4-(DIMETHYLAMINO)PHENYL)DIAZENESULFONIC ACID, SODIUM SALT
see DOU600

4-((DIMETHYLAMINO)PHENYL)DIAZENESULFONIC ACID, SODIUM SALT
see DOU600

p-(DIMETHYLAMINO)-PHENYLDIAZO-NATRIUMSULFONAT (GERMAN)
see DOU600

DIMETHYLAMINOPHENYLDIMETHYLPYRAZOLIN see DOT000

4-DIMETHYLAMINO-1-PHENYL-2,3-DIMETHYLPYRAZOLONE see DOT000

2-(p-DIMETHYLAMINOPHENYL)-1,6-DIMETHYLQUINOLINIUM CHLORIDE
see DPS200

4-(DIMETHYLAMINO)PHENYL ESTER THIOCYANIC ACID see TFH500

4-(p-DIMETHYLAMINOPHENYL)IMINO-2,5-CYCLOHEXADIENE-1-ONE
see DPS600

1-DIMETHYLAMINO-2-PHENYL-3-METHYLPENTANE HYDROCHLORIDE
see BRE255

1-(4'-DIMETHYLAMINOPHENYL)-2-(1'-NAPHTHYL)ETHYLENE see DSV000

2-(3-DIMETHYLAMINO-1-PHENYLPROPYL)PYRIDINE see TMJ750

1-(N,N-DIMETHYLAMINO)-3-(PHENYL-3α-PYRIDYL)PROPANE MALEATE
see TMK000

2-(4-DIMETHYLAMINOPHENYL)QUINOLINE see DPT200

2-(p-DIMETHYLAMINOPHENYL)QUINOLINE see DPT200

p-(DIMETHYLAMINO)PHENYLTHIOCYANATE see TFH500

1,1-DIMETHYLAMINOPROPANOL-2 see DPT800

1,1-DIMETHYLAMINOPROPAN-2-OL see DPT800

4-(2-(DIMETHYLAMINO)PROPIONAMIDO)ANTIPYRINE see AMF375

3-(DIMETHYLAMINO)PROPIONITRILE see DPU000

β-DIMETHYLAMINOPROPIONITRILE see DPU000

3-(DIMETHYLAMINO)PROPIOPHENONE HYDROCHLORIDE see DPU400

β-DIMETHYLAMINOPROPIOPHENONE HYDROCHLORIDE see DPU400

1-(3-(DIMETHYLAMINO)PROPOXY)ADAMANTANE ETHYL IODIDE
see DPU600

5-(3-(DIMETHYLAMINO)PROPOXY)-3-METHYL-1-PHENYLPYRAZOLE
see DPU800

3-(DIMETHYLAMINO)PROPYLAMINE see AJQ100

N,N-DIMETHYL-N-(3-AMINOPROPYL)AMINE see AJQ100

2,2'-(3-DIMETHYLAMINOPROPYLAMINO)BIBENZYL see DLH600

10-(2-DIMETHYLAMINOPROPYL)-1-AZAPHENOTHIAZINE HYDROCHLO-
RIDE see AEG625

10-(3-DIMETHYLAMINOPROPYL)-1-AZAPHENOTHIAZINE HYDROCHLO-
RIDE see DYB800

10-(2-DIMETHYLAMINOPROPYL-(1)-4-AZAPHENTHIAZIN HYDROCHLORID
(GERMAN) see AEG625

10-(3-DIMETHYLAMINOPROPYL)-2-CHLOROPHENOTHIAZINE
MONOHYDROCHLORIDE see CKP500

5-(3-(DIMETHYLAMINO)PROPYL)-5H-DIBENZ(b,f)AZEPINE see DPW600

1-(3-DIMETHYLAMINOPROPYL)-4,5-DIHYDRO-2,3,6,7-DIBENZAZEPINE
see DLH600

5-(3-(DIMETHYLAMINO)PROPYL)-10,11-DIHYDRO-5H-DIBENZ(b,f)AZEPINE
HYDROCHLORIDE see DLH630

(3-(DIMETHYLAMINO)PROPYL)-10,11-DIHYDRO-5H-DIBENZ(b,f)AZEPINE-5-
OXIDE see IBP309

(3-DIMETHYLAMINO) PROPYL-10,11-DIHYDRO 5H-DIBENZ(b,f)AZEPINE, 5-
OXIDE MONOHYDROCHLORIDE see IBP000

5-(3-DIMETHYLAMINOPROPYL)-10,11-DIHYDRO-5H-DIBENZO(b,f)AZEPINE
see DLH600

dl-11-(3-DIMETHYLAMINOPROPYL)-6,11-DIHYDRODIBENZO(b,d)THIEPIN
see HII000

10-(3-(DIMETHYLAMINO)PROPYL)-9,9-DIMETHYLACRIDAN TARTRATE
(1:1) see DRM000

10-(2-(DIMETHYLAMINO)PROPYL)-N,N-DIMETHYLPHENOTHIAZINE-2-
SULFONAMIDE see DUC400

5-(3-(DIMETHYLAMINO)PROPYL)-6,7,8,9,10,11-HEXAHYDRO-5H-
CYCLOOCT(b)INDOLE see DPX200

5-(3-(DIMETHYLAMINO)PROPYL)-2-HYDROXY-10,11-DIHYDRO-5H-
DIBENZ(b,f)AZEPINE see DPX400

5-(3'-DIMETHYLAMINOPROPYLIDENE)-DIBENZO-(a,d)(1,4)-
CYCLOHEPTADIENE see EAH500

3-(3-DIMETHYLAMINOPROPYLIDENE)-1:2-4:5-DIBENZOCYCLOHEPTA-1:4-
DIENE see EAI000

5-(3-DIMETHYLAMINOPROPYLIDENE)DIBENZO(a,d)(1,4)
CYCLOHEPTADIENE HYDROCHLORIDE see EAI000

5-(3-(DIMETHYLAMINO)PROPYL)-5H-DIBENZO-(a,d)CYCLOHEPTENE
HYDROCHLORIDE see DPX800

5-(Γ-DIMETHYLAMINOPROPYLIDENE)-5H-DIBENZO(a,d)-10,11-
DIHYDROCYCLOHEPTENE see EAH500

11-DIMETHYLAMINO PROPYLIDENE-6H-DIBENZ(b,e)OXEPIN see AEG750

5-(3-DIMETHYLAMINOPROPYLIDENE)-10,11-DIHYDRO-5H-
DIBENZO(a,d)CYCLOHEPTENE see EAH500

5-(Γ-DIMETHYLAMINOPROPYLIDENE)-10,11-DIHYDRO-5H-
DIBENZO(A,D)CYCLOHEPTENE see EAH500

11-(3-DIMETHYLAMINOPROPYLIDENE)-6,11-
DIHYDRODIBENZO(b,e)THIEPIN see DYC875

11-(3-DIMETHYLAMINOPROPYLIDENE-6,11-DIHYDRODIBENZO(b,e)THIEP-
INE HYDROCHLORIDE see DPY900

cis-11-(3-DIMETHYLAMINOPROPYLIDENE)-6,11-
DIHYDRODIBENZO(b,e)THIEPIN 5-OXIDE HYDROGEN MALEATE
see POD800

11-(3-DIMETHYLAMINOPROPYLIDENE)-6,11-DIHYDRODIBENZ(b,e)OXEPIN
HYDROCHLORIDE see AEG750

11-(3-DIMETHYLAMINOPROPYLIDENE)-6,11-DIHYDRODIBENZ(b,e)OXIPIN
see DYE409

9-(3-DIMETHYLAMINOPROPYLIDENE)-10,10-DIMETHYL-9,10-
DIHYDROANTHRACENE HYDROCHLORIDE see TDL000

N-(Γ-DIMETHYLAMINOPROPYL)IMINODIBENZYL see DLH600

2,2'-(3-DIMETHYLAMINOPROPYLIMINO)DIBENZYL see DLH600

N-(3-DIMETHYLAMINOPROPYL)IMINODIBENZYL HYDROCHLORIDE
see DLH630

5-DIMETHYLAMINO-6-PROPYL-5H-INDENO(5,6-d)-1,3-DIOXOLE HYDRO-
CHLORIDE see DPY600

10-(3-DIMETHYLAMINOPROPYL)-2-METHOXYPHENOTHIAZINE see MFK500

10-(3-(DIMETHYLAMINO)PROPYL)-2-METHOXY)PHENOTHIAZINE, MALE-
ATE see MFK750

17-β-((3-(DIMETHYLAMINO)-PROPYL)METHYLAMINO)ANDROST-5-EN-3β-
OL DIHYDROCHLORIDE see ARX800

10-(3-(DIMETHYLAMINO)PROPYL)-1-NITRO-9-ACRIDANONE HYDROCHLO-
RIDE see NFW460

10-(3-(DIMETHYLAMINO)PROPYL)-1-NITRO-9(10H)-ACRIDINONE
MONOHYDROCHLORIDE (9CI) see NFW460

o-(3-(DIMETHYLAMINO)PROPYL)OXIME-5H-DIBENZO(a,d)CYCLOHEPTEN-5-
ONE MONOHYDROCHLORIDE see LHX498

10-(2-DIMETHYLAMINOPROPYL)PHENOTHIAZINE see DQA400

10-(3-DIMETHYLAMINOPROPYL)PHENOTHIAZINE see DQA600

10-(3-DIMETHYLAMINOPROPYL)PHENOTHIAZINE-3-ETHYLONE
see ABH500

10-(2-DIMETHYLAMINOPROPYL)PHENOTHIAZINE HYDROCHLORIDE
see PMI750

10-(3-DIMETHYLAMINOPROPYL)PHENOTHIAZINE HYDROCHLORIDE
see PMI500

N-(2-DIMETHYLAMINOPROPYL-1)PHENOTHIAZINE HYDROCHLORIDE
see PMI750

10-(Γ-DIMETHYLAMINO-N-PROPYL)PHENOTHIAZINE HYDROCHLORIDE see PMI500
(DIMETHYLAMINO-2-PROPYL-10-PHENOTHIAZINE HYDROCHLORIDE (FRENCH) see DQA400
10-(2-(DIMETHYLAMINO)PROPYL)PHENOTHIAZINE MONOHYDROCHLORIDE see PMI750
1-(10-(3-(DIMETHYLAMINO)PROPYL)PHENOTHIAZIN-2-YL)-1-BUTANONE MALEATE see BTA125
1-(10-(3-(DIMETHYLAMINO)PROPYL)-10H-PHENOTHIAZIN-2-YL)ETHANONE see ABH500
10-(3-DIMETHYLAMINOPROPYL)PHENOTHIAZIN-3-YLMETHYL KETONE see ABH500
10-(3-(DIMETHYLAMINO)PROPYL)PHENOTHIAZIN-2-YL METHYL KETONE MALEATE (1:1) see AAF750
1-(10-(2-DIMETHYLAMINOPROPYL)-PHENOTHIAZIN-2-YL)-1-PROPANONE MALEATE see IDA500
1-(10-(3-DIMETHYLAMINO)PROPYL)PHENOTHIAZIN-2-YL)-1-PROPANONE MALEATE see PMX500
α-(2-(DIMETHYLAMINO)PROPYL)-α-PHENYLBENZENEACETAMIDE see DOY400
10-(2-DIMETHYLAMINOPROPYL)-2-PROPIONYLPHENOTHIAZINE MALEATE see IDA500
((4-DIMETHYLAMINOPROPYLPYRIDO (3,2b) BENZOTHIAZINE)) HYDROCHLORIDE see DYB800
2-(3-DIMETHYLAMINOPROPYL)-3a,4,7,7a-TETRAHYDRO-4,7-ETHANOISOINDOLINE DIMETHIODIDE see DQB309
10-(3-DIMETHYLAMINOPROPYL)-9-THIA-1,10-DIAZAANTHRACENE HYDROCHLORIDE see AEG625
N-(3-DIMETHYLAMINOPROPYL)THIOCARBAMINSAEURE-S-AETHYLESTER-HYDROCHLORID (GERMAN) see EIH500
2'-((3-(DIMETHYLAMINO)PROPYL)THIO)CINNAMANILIDE HYDROCHLORIDE see CMP900
n-(2-(((3-(DIMETHYLAMINO)PROPYL)THIO)PHENYL)-3-PHENYL-2-PRO-PENAMIDE MONOHYDROCHLORIDE see CMP900
DIMETHYLAMINO-N-PROPYL-THIOPHENYLPYRIDYLAMINE see DYB600
10-(3-(DIMETHYLAMINO)PROPYL-2-(TRIFLUOROMETHYL) PHENOTHIAZINE see TKL000
4-DIMETHYLAMINOPYRIDINE see DQB600
p-DIMETHYLAMINOPYRIDINE see DQB600
Γ-(DIMETHYLAMINO)PYRIDINE see DQB600
2-(DIMETHYLAMINO) RESERPILINATE see DQB800
2-(DIMETHYLAMINO) RESERPILIN-24-OIC ACID ETHYL ESTER see DQB800
DIMETHYLAMINORHODANBENZOL see TFH500
4-DIMETHYLAMINOSTILBEN (GERMAN) see DUB800
N,N-DIMETHYL-4-AMINOSTILBENE see DUB800
cis-4-DIMETHYLAMINOSTILBENE see DUC200
4-DIMETHYLAMINO-trans-STILBENE see DUC000
trans-4-DIMETHYLAMINOSTILBENE see DUC000
trans-p-(DIMETHYLAMINO)STILBENE see DUC000
DIMETHYLAMINOSTOVAINE see AHI250
7-(p-DIMETHYLAMINO)STYRYL)BENZ(c)ACRIDINE see DQC000
12-(p-DIMETHYLAMINO)STYRYLBENZ(a)ACRIDINE see DQC200
2-(4-DIMETHYLAMINOSTYRYL)BENZOTHIAZOLE see DQC400
2-(p-DIMETHYLAMINO)STYRYL)BENZOTHIAZOLE see DQC400
14-(p-DIMETHYLAMINO)STYRYL)DIBENZ(a,j)ACRIDINE see DOV000
4-(p-(DIMETHYLAMINO)STYRYL)-6,8-DIMETHYLQUINOLINE see DQC600
4-(4-DIMETHYLAMINOSTYRYL)QUINOLINE see DQD000
4-(p-(DIMETHYLAMINO)STYRYL)QUINOLINE see DQD000
2-(4-N,N-DIMETHYLAMINOSTYRYL)QUINOLINE see DQD000
4-(p-(DIMETHYLAMINO)STYRYL)QUINOLINE MONOHYDROCHLORIDE see DQD200
DIMETHYLAMINOSUCCINAMIC ACID see DQD400
N-(DIMETHYLAMINO)SUCCINAMIC ACID see DQD400
N-DIMETHYLAMINO-SUCCINAMIDSAEURE (GERMAN) see DQD400
O-(4-((DIMETHYLAMINO)SULFONYL)PHENYL) O,O-DIMETHYL PHOSPHOROTHIOATE see FAB600
(DIMETHYLAMINO)-TERMINATED see SDF000
DIMETHYLAMINO-4-THIOCYANOBENZENE see TFH500
4-DIMETHYLAMINOTHIOCYANOBENZENE see TFH500
p-(DIMETHYLAMINO)THIOCYANOBENZENE see TFH500
p-N,N-DIMETHYLAMINOTHIOCYANOBENZENE see TFH500
p-DIMETHYLAMINOTHYMOLDIMETHYLURETHANE METHIODIDE see DQD500
4-(4-(DIMETHYLAMINO)-m-TOLYL)AZO)-2-PICOLINE-1-OXIDE see DQD600
4-((4-(DIMETHYLAMINO)-m-TOLYL)AZO)-3-PICOLINE-1-OXIDE see DQE000
4-((4-(DIMETHYLAMINO)-o-TOLYL)AZO)-2-PICOLINE-1-OXIDE see DQD800
4-((4-(DIMETHYLAMINO)-o-TOLYL)AZO)-3-PICOLINE-1-OXIDE see DQE200
5-((4-(DIMETHYLAMINO)-m-TOLYL)AZO)QUINOLINE see DQE400
5-((4-(DIMETHYLAMINO)-o-TOLYL)AZO)QUINOLINE see DQE600
4-(DIMETHYLAMINO)-m-TOLYL METHYLCARBAMATE see DOR400
5-DIMETHYLAMINO-4-TOLYL METHYLCARBAMATE see DQE800
S,S'-(2-(DIMETHYLAMINO)TRIMETHYLENE)BIS(THIOCARBAMATE) HYDROCHLORIDE see BHL750
DIMETHYLAMINOTRIMETHYLSILANE see DQE900

4-DIMETHYLAMINOTRIPHENYLMETHAN (GERMAN) see DRQ000
4-DIMETHYLAMINOTRIPHENYLMETHANE see DRQ000
3-DIMETHYLAMINO-1,1,2-TRIS(4-METHOXYPHENYL)-1-PROPENE HYDROCHLORIDE see TNJ750
5-DIMETHYLAMINO-1,2,3-TRITHIANE HYDROGENOXALATE see TFH750
N-(6-(DIMETHYLAMINO-3H-XANTHEN-3-YLIDENE)-N-METHYLMETHANAMINIUM CHLORIDE see PPQ750
4-DIMETHYLAMINO-3,5-XYLENOL see DQF000
4-(DIMETHYLAMINO)-3,5-XYLENOL METHYLCARBAMATE (ESTER) see DOS000
4-((4-(DIMETHYLAMINO)-2,3-XYLYL)AZO)PYRIDINE-1-OXIDE see DQF200
4-((4-(DIMETHYLAMINO)-2,5-XYLYL)AZO)PYRIDINE-1-OXIDE see DQF400
4-((4-(DIMETHYLAMINO)-3,5-XYLYL)AZO)PYRIDINE-1-OXIDE see DQF600
4-(DIMETHYLAMINO)-3,5-XYLYL ESTER METHYLCARBAMIC ACID see DOS000
4-DIMETHYLAMINO-3,5-XYLYL METHYLCARBAMATE see DOS000
4-DIMETHYLAMINO-3,5-XYLYL-N-METHYLCARBAMATE see DOS000
4-(N,N-DIMETHYLAMINO)-3,5-XYLYL N-METHYLCARBAMATE see DOS000
DIMETHYLAMMONIUM CHLORIDE see DOR600
DIMETHYLAMMONIUM 2,4-DICHLOROPHENOXYACETATE see DFY800
N-((DIMETHYLAMMONIUM)ETHYL)-4,5,6,7-TETRACHLOROISOINDOLINIUM DIMETHOCHLORIDE see CDY000
DIMETHYLAMMONIUM PERCHLORATE see DQF650
DI(4-METHYL-2-AMYL) MALEATE see DKP400
4,5'-DIMETHYL ANGELICIN see DQF700
DIMETHYLANILINE see XMA000
2,3-DIMETHYLANILINE see XMJ000
2,4-DIMETHYLANILINE see XMS000
2,5-DIMETHYLANILINE see XNA000
2,6-DIMETHYLANILINE see XNJ000
3,4-DIMETHYLANILINE see XNS000
3,5-DIMETHYLANILINE see XOA000
N,N-DIMETHYLANILINE see DQF800
N,N-DIMETHYL-p-ANILINEDIAZOSULFONIC ACID SODIUM SALT see DOU600
2,4-DIMETHYLANILINE HYDROCHLORIDE see XOJ000
2,5-DIMETHYLANILINE HYDROCHLORIDE see XOS000
N,N-DIMETHYLANILINE METHIODIDE see TMB750
2-(2,6-DIMETHYLANILINO)-5,6-DIHYDRO-4H-1,3-THIAZINE see DMW000
6,12-DIMETHYLANTHANTHRENE see DQG000
9,10-DIMETHYLANTHRACENE see DQG200
3-(10,10-DIMETHYL(10H)-ANTHRACENYLIDENE)-N,N-DIMETHYL-1-PROPANAMINE (9CI) see AEG875
3-(10,10-DIMETHYL-9(10H)-ANTHRACENYLIDENE)-N,N-DIMETHYL-1-PROPANAMINE HYDROCHLORIDE see TDL000
N,N-DIMETHYL-N'-(1-ANTHRACHINONYL)FORMAMIDINIUMCHLORID (GERMAN) see APK750
DIMETHYL ANTHRANILATE (FCC) see MGQ250
DIMETHYLANTIMONY CHLORIDE see DQG400
DIMETHYLARSENIC ACID see HKC000
DIMETHYLARSENIC ACID SODIUM SALT TRIHYDRATE see HKC550
DIMETHYLARSINE see DQG600
DIMETHYLARSINIC ACID see HKC000
DIMETHYLARSINIC ACID SODIUM SALT TRIHYDRATE see HKC550
DIMETHYL ARSINIC SULFIDE see DQG700
((DIMETHYLARSINO)OXY)SODIUM-As-OXIDE see HKC500
N,N-DIMETHYL-1-AZIRIDINECARBOXAMIDE see EJA100
N,N-DIMETHYL-p-AZOANILINE see DOT300
2,3'-DIMETHYLAZOBENZENE see DQH000
3,6'-DIMETHYLAZOBENZENE see DQH000
2,3'-DIMETHYLAZOBENZENE-4'-METHYLCARBONATE see CBS750
N,N'-DIMETHYL-4,4'-AZODIACETANILIDE see DQH200
DIMETHYL AZODIFORMATE see DQH509
DIMETHYL AZOFORMATE see DQH509
2,2'-DIMETHYL-BEBEERINIUM (8CI) see COF825
1,3-DIMETHYLBENZ(e)ACEPHENANTHRYLENE see DQH550
7,9-DIMETHYLBENZ(c)ACRIDINE see DQI200
5,7-DIMETHYL-1,2-BENZACRIDINE see DQI600
6,9-DIMETHYL-1,2-BENZACRIDINE see DQI800
7,10-DIMETHYLBENZ(c)ACRIDINE see DQI800
7,11-DIMETHYLBENZ(c)ACRIDINE see DQI400
8,10-DIMETHYL-BENZ(a)ACRIDINE see DQI600
8,12-DIMETHYLBENZ(a)ACRIDINE see DQH600
9,12-DIMETHYLBENZ(a)ACRIDINE see DQH800
1,10-DIMETHYL-5,6-BENZACRIDINE see DQH600
2,10-DIMETHYL-5,6-BENZACRIDINE see DQH800
1,10-DIMETHYL-7,8-BENZACRIDINE (FRENCH) see DQI400
2,10-DIMETHYL-7,8-BENZACRIDINE (FRENCH) see DQI800
3,10-DIMETHYL-7,8-BENZACRIDINE (FRENCH) see DQI200
N,N-DIMETHYLBENZAMIDE see DQJ000
DIMETHYLBENZANTHRACENE see DQJ200
DIMETHYLBENZ(a)ANTHRACENE see DQJ200
4,5-DIMETHYLBENZ(a)ANTHRACENE see DQJ600
6,7-DIMETHYLBENZ(a)ANTHRACENE see DQJ800

6,8-DIMETHYLBENZ(a)ANTHRACENE see DQK000
7,12-DIMETHYLBENZANTHRACENE see DQJ200
7,8-DIMETHYLBENZ(a)ANTHRACENE see DQL000
8,9-DIMETHYLBENZ(a)ANTHRACENE see DQK600
9,10-DIMETHYL-BENZANTHRACENE see DQJ200
1,12-DIMETHYLBENZ(a)ANTHRACENE see DQJ400
4,9-DIMETHYL-1,2-BENZANTHRACENE see DQK200
5,6-DIMETHYL-1,2-BENZANTHRACENE see DQK600
5,9-DIMETHYL-1,2-BENZANTHRACENE see DQK800
5:9-DIMETHYL-1:2-BENZANTHRACENE see DQK800
6,12-DIMETHYLBENZ(a)ANTHRACENE see DQK200
6,7-DIMETHYL-1,2-BENZANTHRACENE see DQL200
6,8-DIMETHYL-1,2-BENZANTHRACENE see DQK000
7,11-DIMETHYLBENZ(a)ANTHRACENE see DQK400
7,12-DIMETHYLBENZ(a)ANTHRACENE see DQJ200
8,12-DIMETHYLBENZ(a)ANTHRACENE see DQK800
9,10-DIMETHYLBENZ(a)ANTHRACENE see DQJ200, DQL200
1',9-DIMETHYL-1,2-BENZANTHRACENE see DQJ400
3,4'-DIMETHYL-1,2-BENZANTHRACENE see DQJ600
4,10-DIMETHYL-1,2-BENZANTHRACENE see DQJ800
5,10-DIMETHYL-1,2-BENZANTHRACENE see DQL000
8,10-DIMETHYL-1,2-BENZANTHRACENE see DQK400
9,10-DIMETHYL-1,2-BENZANTHRACENE see DQJ200
7,12-DIMETHYLBENZ(a)ANTHRACENE-8-CARBONITRILE see COM000
7,12-DIMETHYLBENZ(a)ANTHRACENE-D16 see DQL400
7,12-DIMETHYLBENZ(a)ANTHRACENE, DEUTERATED see DQL400
7,12-DIMETHYLBENZ(a)ANTHRACENE-3,4-DIOL see DQK900
7,12-DIMETHYLBENZ(a)ANTHRACENE-5,6-OXIDE see DQL600
95810-DIMETHYL-1582-BENZANTHRACENE-95810-OXIDE see DQL800
7,12-DIMETHYLBENZANTHRACENE-7,12-endo-α,β-SUCCINIC ACID
    see DRT000
9,10-DIMETHYL-1,2-BENZANTHRAZEN (GERMAN) see DQJ200
DIMETHYLBENZANTHRENE see DQJ200
O,O-DIMETHYL-S-(BENZAZIMINOMETHYL) DITHIOPHOSPHATE see ASH500
2,2-DIMETHYL-1,3-BENZDIOXOL-4-YL-N-METHYLCARBAMATE see DQM600
α,α-DIMETHYLBENZEETHANAMINE see DTJ400
2,3-DIMETHYLBENZENAMINE see XMJ000
2,4-DIMETHYLBENZENAMINE see XMS000
2,5-DIMETHYLBENZENAMINE see XNA000
2,6-DIMETHYLBENZENAMINE see XNJ000
3,5-DIMETHYLBENZENAMINE see XOA000
2,4-DIMETHYLBENZENAMINE HYDROCHLORIDE see XOJ000
2,5-DIMETHYLBENZENAMINE HYDROCHLORIDE see XOS000
DIMETHYLBENZENE see XGS000
m-DIMETHYLBENZENE see XHA000
o-DIMETHYLBENZENE see XHJ000
p-DIMETHYLBENZENE see XHS000
1,2-DIMETHYLBENZENE see XHJ000
1,3-DIMETHYLBENZENE see XHA000
1,4-DIMETHYLBENZENE see XHS000
N,N-DIMETHYLBENZENEAMINE see DQF800
α,α'-DIMETHYL-m-BENZENEBIS(ETHYLAMINE) DIHYDROCHLORIDE
    see DCC100
α,α'-DIMETHYL-p-BENZENEBIS(ETHYLAMINE) DIHYDROCHLORIDE
    see DCC125
N,N-DIMETHYL-1,4-BENZENEDIAMINE see DTL800
3,5-DIMETHYLBENZENEDIAZONIUM-2-CARBOXYLATE see DQL899
4,6-DIMETHYLBENZENEDIAZONIUM-2-CARBOXYLATE see DQL959
DIMETHYL-1,2-BENZENEDICARBOXYLATE see DTR200
DIMETHYL-1,4-BENZENE DICARBOXYLATE see DUE000
2,6-DIMETHYL-1,4-BENZENEDIOL (9CI) see DSG700
N,N-DIMETHYLBENZENEMETHANAMINE see DQP800
α,4-DIMETHYLBENZENEMETHANOL see TGZ000
α,α-DIMETHYLBENZENEMETHANOL see DTN100
DIMETHYL BENZENEORTHODICARBOXYLATE see DTR200
α,α-DIMETHYLBENZENEPROPANOL ACETATE see MNT000
2,4-DIMETHYLBENZENESULFONIC ACID see XJJ000
3,3'-DIMETHYLBENZIDIN see TGJ750
3,3'-DIMETHYLBENZIDINE see TGJ750
3,3'-DIMETHYLBENZI DINEDIHYDROCHLORIDE see DQM000
N,N-DIMETHYL-p-(4-BENZIMIDAZOLYAZO)ANILINE see DQM200
DIMETHYLBENZIMIDAZOLYCOBAMIDE see VSZ000
5,6-DIMETHYLBENZIMIDAZOLYCOBAMIDE CYANIDE see VSZ000
N,N-DIMETHYL-4(4'-BENZIMIDAZOLYLAZO)ANILINE see DQM200
DIMETHYL BENZMIDE see DQJ000
7,12-DIMETHYLBENZO(a)ANTHRACENE see DQJ200
6,12-DIMETHYLBENZO(1,2-b585,4-b')BIS(1)BENZOTHIOPHENE see DQM400
2,2-DIMETHYL-1,3-BENZODIOXOL-4-OL METHYLCARBAMATE see DQM600
2,2-DIMETHYLBENZO-1,3-DIOXOL-4-YL METHYLCARBAMATE see DQM600
6,12-DIMETHYLBENZO(1,2-b584,5-b')DITHIONAPHTHENE see DQM800
1,3-DIMETHYLBENZO(b)FLUORANTHENE see DQH550
N,N-DIMETHYL-p-(7-BENZOFURYLAZO)ANILINE see BCL100
7,13-DIMETHYLBENZO(b)PHENANTHRO(3,2-d)THIOPHENE see DRJ000
AR,AR-DIMETHYLBENZOPHENONE see PGP750

1,2-DIMETHYLBENZO(a)PYRENE see DQN000
1,3-DIMETHYLBENZO(a)PYRENE see DQN200
1,4-DIMETHYLBENZO(a)PYRENE see DQN400
1,6-DIMETHYLBENZO(a)PYRENE see DQN600
2,3-DIMETHYLBENZO(a)PYRENE see DQN800
3,6-DIMETHYLBENZO(a)PYRENE see DQO000
4,5-DIMETHYLBENZO(a)PYRENE see DQO400
3,12-DIMETHYLBENZO(a)PYRENE see DQO200
2,5-DIMETHYL-p-BENZOQUINONE see XQJ000
2,5-DIMETHYLBENZOSELENAZOLE see DQO600
5,6-DIMETHYL-2,1,3-BENZOSELENODIAZOLE see DQO650
2,5-DIMETHYLBENZOTHIAZOLE see DQO800
N,N-DIMETHYLBENZO(b)THIOPHENE-3-ETHYLAMINE HYDROCHLORIDE
    see DQP000
O,O-DIMETHYL-S-(1,2,3-BENZOTRIAZINYL-4-KETO)METHYL PHOS-
    PHORODITHIOATE see ASH500
1,2-DIMETHYL-1H-BENZOTRIAZOLIUM IODIDE see DQP100
1,3-DIMETHYL-3H-BENZOTRIAZOLIUM IODIDE see DQP125
1,2-DIMETHYLBENZOTRIAZOLIUM JODID (GERMAN) see DQP100
1,3-DIMETHYLBENZOTRIAZOLIUM JODID (GERMAN) see DQP125
1,4-DIMETHYL-2,3-BENZPHENANTHRENE see DQJ200
2,5-DIMETHYLBENZSELENAZOL (CZECH) see DQO600
2,5-DIMETHYLBENZTHIAZOL (CZECH) see DQO800
N,N-DIMETHYL-p-(6-BENZTHIAZOLYLAZO)ANILINE see DPO400
N,N-DIMETHYL-p-(7-BENZTHIAZOLYLAZO)ANILINE see DPO600
N,N-DIMETHYL-4-(6'-BENZTHIAZOLYLAZO)ANILINE see DPO400
N,N-DIMETHYL-4-(7'-BENZTHIAZOLYLAZO)ANILINE see DPO600
4,9-DIMETHYL-2,3-BENZTHIOPHANTHRENE see DQP400
p,α-DIMETHYLBENZYL ALCOHOL see TGZ000
α,α-DIMETHYLBENZYL ALCOHOL see DTN100
N,N-DIMETHYLBENZYLAMINE see DQP800
N,N-DIMETHYLBENZYLAMINE HEXAFLUOROARSENATE see BEL550
DIMETHYLBENZYLAMINE HYDROCHLORIDE see DQQ000
DIMETHYLBENZYLAMMONIUM CHLORIDE see DQQ000
DIMETHYL BENZYL CARBINOL see DQQ200
DIMETHYLBENZYL CARBINOLACETATE see BEL750
DIMETHYL BENZYL CARBINYL ACETATE see DQQ375
DIMETHYLBENZYLCARBINYL BUTYRATE see BEL850
DIMETHYL BENZYL CARBINYL BUTYRATE see DQQ380
DIMETHYL BENZYL CARBINYL PROPIONATE see DQQ400
4-α,α-DIMETHYLBENZYL-α,α'-DIPIPERIDINO-2,6-XYLENOL DIHYDROBROM-
    IDE see BHR750
N,N'-DI(α-METHYLBENZYL)ETHYLENEDIAMINE see DQQ600
α,α-DIMETHYLBENZYL HYDROPEROXIDE (MAK) see IOB000
1-(α,α-DIMETHYLBENZYL)-3-METHYL-3-PHENYLUREA see DQQ700
DIMETHYLBENZYLOCTADECYLAMMONIUM CHLORIDE see DTC600
N,N-DIMETHYL-N'-BENZYL-N'-(α-PYRIDYL)ETHYLENEDIAMINE see TMP750
DIMETHYL BERYLLIUM see DQR200
DIMETHYLBERYLLIUM-1,2-DIMETHOXYETHANE see DQR289
3',3'''-DIMETHYL-4',4'''-BIACETANILIDE see DRI400
6,12-DIMETHYL-BIBENZO(def,mno)CHRYSENE see DQG000
2,2-DIMETHYLBICYCLO(2.2.1)HEPTANE-3-CARBOXYLIC ACID, METHYL
    ESTER see MHY600
DIMETHYL cis-BICYCLO(2,2,1)-5-HEPTENE-2,3-DICARBOXYLATE
    see DRB400
6,6-DIMETHYLBICYCLO-(3.1.1)-2-HEPTENE-2-ETHANOL see DTB800
6,6-DIMETHYLBICYCLO(3.1.1)-2-HEPTENE-2-ETHYL ACETATE see DTC000
1,1-DIMETHYLBIGUANIDE see DQR600
N,N-DIMETHYLBIGUANIDE see DQR600
DIMETHYLBIGUANIDE HYDROCHLORIDE see DQR800
1,1-DIMETHYLBIGUANIDE HYDROCHLORIDE see DQR800
6,6'-DIMETHYL-(2,2'-BINAPHTHALENE)-1,1',8,8'-TETROL see DVO809
3,2'-DIMETHYL-4-BIPHENYLAMINE see BLV250
3,3'-DIMETHYL-4-BIPHENYLAMINE see DOV200
3,3'-DIMETHYL-4,4'-BIPHENYLDIAMINE see TGJ750
3,3'-DIMETHYLBIPHENYL-4,4'-DIAMINE see TGJ750
3,3'-DIMETHYL-(1,1'-BIPHENYL)-4,4'-DIAMINE see TGJ750
ar,ar'-DIMETHYL-(1,1'-BIPHENYL)-ar,ar'-DIOL (9CI) see DGQ700
3,3'-DIMETHYL-4,4'-BIPHENYLENE DIISOCYANATE see DQS000
N,N'-DIMETHYL-4,4'-BIPYRIDINIUM DICHLORIDE see PAJ000
1,1'-DIMETHYL-4,4'-BIPYRIDINIUM DIHYDRATE see PAI995
1,1'-DIMETHYL-4,4'-BIPYRIDYNIUM DIMETHYLSULFATE see PAJ250
DIMETHYL 1,3-BIS(CARBOMETHOXY)-1-PROPEN-2-YL PHOSPHATE see
    SOY000
DIMETHYL-BIS(β-CHLOROETHYL)AMMONIUM CHLORIDE see DQS600
β,Γ-DIMETHYL-α,Δ-BIS(3,4-DIHYDROXYPHENYL)BUTANE see NBR000
O,O-DIMETHYL-S-(1,2-BIS(ETHOXYCARBONYL)ETHYL)DITHIOPHOSPHATE
    see MAK700
O,O-DIMETHYL-S-1,2-BIS(ETHOXYCARBONYL)ETHYL PHOSPHOROTHIO-
    ATE see OPK250
DIMETHYL BIS(p-HYDROXYPHENYL)METHANE see BLD500
DIMETHYLBISMUTH CHLORIDE see DQT000
5,5-DIMETHYL-1,3-BIS(OXIRANYLMETHYL)-2,4-IMIDAZOLIDINEDIONE
    see DKM130

5,5-DIMETHYL-1,3-BIS(OXIRANYLMETHYL)-2,4-IMIDAZOLIDINEDIONE (9CI) see DKM130

DIMETHYLBIS(PHENYLTHIO)STANNANE see BLF750

N,N'-DIMETHYL-N,N'-BIS(3-(3',4',5'-TRIMETHOXYBENZOXY)PROPYL)ETHYLENEDIAMINE DIHYDROCHLORIDE see HFF500

N,N-DIMETHYL-(2-BROMAETHYL)-HYDRAZINIUMBROMID (GERMAN) see DQT100

N,N-DIMETHYL-β-(p-BROMOANILINO)PROPIONAMIDE see BMN350

O,O-DIMETHYL-O-(4-BROMO-2,5-DICHLOROPHENYL) PHOSPHOROTHIOATE see BNL250

N,N-DIMETHYL-(2-BROMOETHYL)HYDRAZINIUM BROMIDE see DQT100

2,3-DIMETHYL-1,3-BUTADIENE see DQT150

2,2-DIMETHYLBUTANE see DQT200

2,3-DIMETHYLBUTANE see DQT400

2,3-DIMETHYL-2,3-BUTANEDIOL see TDR000

1,3-DIMETHYL BUTANOL see AOK750

3,3-DIMETHYL-2-BUTANOL METHYLPHOSPHONOFLUORIDATE see SKS500

2,2-DIMETHYL-3-BUTANONE see DQT800

3,3-DIMETHYL-2-BUTANONE see DQU000

5-(1,3-DIMETHYL-2-BUTENYL)-5-ETHYL BARBITURIC ACID see DQU200

5-(1,3-DIMETHYL-2-BUTENYL)-5-ETHYL-2,4,6(1H,3H,5H)PYRIMIDINETRIONE see DQU200

1-(2-(1,3-DIMETHYL-2-BUTENYLIDENE)HYDRAZINO)PHTHALAZINE see DQU400

1,3-DIMETHYLBUTYL ACETATE see HFJ000

DI(3-METHYLBUTYL)ADIPATE see AEQ500

1,3-DIMETHYL BUTYLAMINE see DQU600

1,3-DIMETHYL-5-tert-BUTYLBENZENE see DQU800

1,1-DIMETHYL-3-(3-N-tert-BUTYLCARBAMYLOXY)-PHENYL)UREA see DUM800

4,6-DIMETHYL-8-tert-BUTYLCOUMARIN see DQV000

5-(1,3-DIMETHYLBUTYL)-5-ETHYLBARBITURIC ACID see DDH900

5-(1,3-DIMETHYLBUTYL)-5-ETHYL BARBITURIC ACID, SODIUM SALT see DQV200

5-(1,3-DIMETHYLBUTYL)-5-ETHYL-2,4,6(1H,3H,5H)-PYRIMIDINETRIONE (9CI) see DDH900

3,3-DIMETHYL-2-BUTYL METHYLPHOSPHONOFLUORIDATE see SKS500

3,3-DIMETHYL-n-BUT-2-YL METHYLPHOSPHONOFLUORIDATE see SKS500

DIMETHYL-2-BUTYNEDIOATE see DOP400

O,O-DIMETHYL-(1-BUTYRYLOXY-2,2,2-TRICHLOROETHYL) PHOSPHONATE see BPG000

DI-2-METHYLBUTYRYL PEROXIDE see DQW600

DIMETHYLCADMIUM see DQW800

N,N-DIMETHYLCAPRAMIDE see DQX000

N,N-DIMETHYLCAPROAMIDE see DQX200

N,N-DIMETHYLCAPRYLAMIDE see DTE200

DIMETHYLCARBAMATE de 5,5-DIMETHYL DIHYDRORESORCINOL (FRENCH) see DRL200

DIMETHYLCARBAMATE-d'l-ISOPROPYL-3-METHYL-5-PYRAZOLYLE (FRENCH) see DSK200

DIMETHYLCARBAMIC ACID CHLORIDE see DQY950

N,N-DIMETHYLCARBAMIC ACID-3-DIETHYLAMINOPHENYL ESTER METHIODIDE see HNK500

N,N-DIMETHYLCARBAMIC ACID-3-DIETHYLAMINOPHENYL ESTER, METOCHLORIDE see TGH670

DIMETHYLCARBAMIC ACID-(α-(DIETHYLAMINO))-o-TOLYL ESTER, HYDROCHLORIDE see DQY000

DIMETHYLCARBAMIC ACID-m-(DIETHYLMETHYLAMINO)PHENYL ESTER IODIDE see HNK500

DIMETHYLCARBAMIC ACID-m-(DIETHYLMETHYLAMMONIO)PHENYL ESTER, CHLORIDE see TGH670

DIMETHYLCARBAMIC ACID-1-((DIMETHYLAMINO)CARBONYL)-5-METHYL-1H-PYRAZOL-3-YL ESTER see DQZ000

DIMETHYLCARBAMIC ACID 2-(DIMETHYLAMINO)-5,6-DIMETHYL-4-PYRIMIDINYL ESTER see DOX600

N,N-DIMETHYLCARBAMIC ACID-4-(β-DIMETHYLAMINOETHYL)PHENYL ESTER, HYDROCHLORIDE see DQX800

N,N-DIMETHYLCARBAMIC ACID-p-(β-DIMETHYLAMINOETHYL)PHENYL ESTER, HYDROCHLORIDE see DQX800

N,N-DIMETHYLCARBAMIC ACID-3-DIMETHYLAMINOPHENYL ESTER METHOSULFATE see DQY909

DIMETHYLCARBAMIC ACID ESTER of 3-HYDROXY-1-METHYLPYRIDINIUM BROMIDE see MDL600

DIMETHYLCARBAMIC ACID ESTER with 3-HYDROXY-N,N,5-TRIMETHYLPYRAZOLE-1-CARBOXAMIDE see DQZ000

DIMETHYLCARBAMIC ACID ESTER with (m-HYDROXYPHENYL)TRIMETHYLAMMONIUM METHYL SULFATE see DQY909

DIMETHYLCARBAMIC ACID ETHYL ESTER see EII500

DIMETHYLCARBAMIC ACID ester with 3-HYDROXY-5,5-DIMETHYL-2-CYCLOHEXEN-1-ONE see DRL200

N,N-DIMETHYLCARBAMIC ACID, m-ISOPROPYL PHENYL ESTER see DQX300

DIMETHYLCARBAMIC ACID 3-METHYL-1-(1-METHYLETHYL)-1H-PYRAZOL-5-YL ESTER see DSK200

DIMETHYL CARBAMIC ACID-3-METHYL-1-PHENYL PYRAZOL-5-YL ESTER see PPQ625

DIMETHYLCARBAMIC ACID-5-METHYL-1H-PYRAZOL-3-YL ESTER see DQZ000

N,N-DIMETHYLCARBAMIC ACID-8-QUINOLINYL ESTER METHOSULFATE see DQY400

N,N-DIMETHYLCARBAMIC ACID-3-(TRIMETHYLAMMONIO)PHENYL ESTER METHYLSULFATE see DQY909

DIMETHYLCARBAMIC CHLORIDE see DQY950

DIMETHYLCARBAMIC ESTER of HORDENINE HYDROCHLORIDE see DQX800

DIMETHYLCARBAMIC ESTER of 2-OXYBENZYLDIETHYLAMINE HYDROCHLORIDE see DQY000

DIMETHYLCARBAMIC ESTER of 8-OXYMETHYLQUINOLINIUM METHYLSULFATE see DQY400

DIMETHYLCARBAMIC ESTER of 3-OXYPHENYLTRIMETHYLAMMONIUM METHYLSULFATE see DQY909

DIMETHYLCARBAMIDOYL CHLORIDE see DQY950

DIMETHYLCARBAMODITHIOIC ACID, IRON COMPLEX see FAS000

DIMETHYLCARBAMODITHIOIC ACID, IRON(3+) SALT see FAS000

DIMETHYLCARBAMODITHIOIC ACID, ZINC COMPLEX see BJK500

DIMETHYLCARBAMODITHIOIC ACID, ZINC SALT see BJK500

3-(DIMETHYLCARBAMOXY)PHENYL TRIMETHYLAMMONIUM METHYL SULFATE see DQY909

N-(DIMETHYLCARBAMOYL)AZIRIDINE see EJA100

DIMETHYLCARBAMOYL CHLORIDE see DQY950

N,N-DIMETHYLCARBAMOYL CHLORIDE (DOT) see DQY950

2-DIMETHYLCARBAMOYL-3-METHYLPYRAZOLYL-(5)-N,N-DIMETHYLCARBAMAT see DQZ000

1-DIMETHYLCARBAMOYL-5-METHYL-3-PYRAZOLYL DIMETHYLCARBAMATE see DQZ000

2-DIMETHYLCARBAMOYL-3-METHYL-5-PYRAZOLYL DIMETHYLCARBAMATE see DQZ000

cis-2-DIMETHYLCARBAMOYL-1-METHYLVINYL DIMETHYLPHOSPHATE see DGQ875

((4-(N,N-DIMETHYLCARBAMOYLOXY)-2-ISOPROPYL)PHENYL)TRIMETHYLAMMONIUM IODIDE see HJZ000

(3-(DIMETHYLCARBAMOYLOXY)PHENYL)DIETHYLMETHYLAMMONIUM CHLORIDE see TGH670

(3-(DIMETHYLCARBAMOYLOXY)PHENYL)DIETHYLMETHYLAMMONIUM IODIDE see HNK500

(3-(DIMETHYLCARBAMOYLOXY)PHENYL)TRIMETHYLAMMONIUM METHYLSULFATE see DQY909

1-(DIMETHYLCARBAMYL)AZIRIDINE see EJA100

DIMETHYLCARBAMYL CHLORIDE see DQY950

N,N-DIMETHYLCARBAMYL CHLORIDE see DQY950

DIMETHYLCARBAMYL DIETHYLTHIOCARBAMYL SULFIDE see DJC875

2-(N,N-DIMETHYLCARBAMYL)-3-METHYLPYRAZOLYL-5 N,N-DIMETHYLCARBAMATE see DQZ000

DIMETHYL 2-CARBAMYL-3-METHYLPYRAZOLYLDIMETHYLCARBAMATE (GERMAN) see DQZ000

N,N'-DIMETHYL CARBANILIDE see DRB200

DIMETHYL CARBATE see DRB400

O,O-DIMETHYL-S-(CARBETHOXY)METHYL PHOSPHOROTHIOLATE see DRB600

DIMETHYLCARBINOL see INJ000

α,α-DIMETHYL-α'-CARBOBUTOXY-DIHYDRO-γ-PYRONE see BRT000

2,2-DIMETHYL-6-CARBOBUTOXY-2,3-DIHYDRO-4-PYRONE see BRT000

O,O-DIMETHYL-S-(1-CARBOETHOXYBENZYL) DITHIOPHOSPHATE see DRR400

O,O-DIMETHYL-S-CARBOETHOXYMETHYL THIOPHOSPHATE see DRB600

O,O-DIMETHYL-O-(2-CARBOMETHOXY-1-METHYLVINYL) PHOSPHATE see MQR750

DIMETHYL-1-CARBOMETHOXY-1-PROPEN-2-YL PHOSPHATE see MQR750

DIMETHYL CARBONATE see MIF000

O,O-DIMETHYL-S-(CARBONYLMETHYLMORPHOLINO) PHOSPHORODITHIOATE see PHI500

DIMETHYLCELLOSOLVE see DOE600

DIMETHYL-CHLORMETHYL-ETHOXYSILAN (CZECH) see DRC400

O,O-DIMETHYL-O-3-CHLOR-4-NITROFENYLTIOFOSFAT (CZECH) see MIJ250

O,O-DIMETHYL-O-(3-CHLOR-4-NITROPHENYL)-MONOTHIOPHOSPHAT (GERMAN) see MIJ250

8,12-DIMETHYL-9-CHLOROBENZ(a)ACRIDINE see CGH250

1,10-DIMETHYL-2-CHLORO-5,6-BENZACRIDINE see CGH250

7,11-DIMETHYL-10-CHLOROBENZ(c)ACRIDINE see DRB800

1,10-DIMETHYL-2-CHLORO-7,8-BENZACRIDINE (FRENCH) see DRB800

O,O-DIMETHYL-O-(6-CHLOROBICYCLO(3.2.0)HEPTADIEN-1,5-YL)PHOSPHATE see HBK700

DIMETHYL 2-CHLORO-2-DIETHYLCARBAMOYL-1-METHYLVINYL PHOSPHATE see FAB400

O,O-DIMETHYL O-(2-CHLORO-2-(N,N-DIETHYLCARBAMOYL)-1-METHYLVINYL) PHOSPHATE see FAB400

DIMETHYLCHLOROETHER see CIO250

DIMETHYL(2-CHLOROETHYL)AMINE see CGW000

O,O-DIMETHYL S-(2,5-DICHLOROPHENYLTHIO)METHYL PHOS-
PHORODITHIOATE see MNO750
1,1-DIMETHYL-3-(3,4-DICHLOROPHENYL)UREA see DXQ500
DIMETHYL DICHLOROVINYL PHOSPHATE see DGP900
DIMETHYL-2,2-DICHLOROVINYL PHOSPHATE see DGP900
O,O-DIMETHYL DICHLOROVINYL PHOSPHATE see DGP900
O,O-DIMETHYL-O-2,2-DICHLOROVINYL PHOSPHATE see DGP900
O,O-DIMETHYL-O-(2,2-DICHLOR-VINYL)-PHOSPHAT (GERMAN)
see DGP900
DIMETHYLDIDECYLAMMONIUM CHLORIDE see DGX200
2,5-DIMETHYL-1,2,5,6-DIEPOXYHEX-3-YNE see DRK400
O,O-DIMETHYL-S-1,2-DI(ETHOXYCARBAMYL)ETHYL PHOSPHORODITHIO-
ATE see MAK700
DIMETHYL-DIETHOXYSILAN (CZECH) see DHG000
DIMETHYLDIETHOXYSILANE (DOT) see DHG000
DIMETHYL DIETHYLAMIDO-1-CHLOROCROTONYL (2) PHOSPHATE
see FAB400
N,N-DIMETHYL-p-((3,4-DIETHYLPHENYL)AZO)ANILINE see DJB400
2,6-DIMETHYL-1,1-DIETHYLPIPERIDINIUM BROMIDE see DRK600
trans-4,4'-DIMETHYL-α-α'-DIETHYLSTILBENE see DRK500
N,N-DIMETHYL-2,5-DIFLUORO-p-(2,5-DIFLUOROPHENYLAZO)ANILINE
see DRK800
N,N-DIMETHYL-p-(2,5-DIFLUOROPHENYLAZO)ANILINE see DKH000
N,N-DIMETHYL-p-(3,4-DIFLUOROPHENYLAZO)ANILINE see DRL000
N,N-DIMETHYL-p-(3,5-DIFLUOROPHENYLAZO)ANILINE see DKH100
N,N-DIMETHYL-3',4'-DIFLUORO-4-(PHENYLAZO)BENZENEAMINE
see DRL000
N,N-DIMETHYLDIGUANIDE see DQR600
9:10-DIMETHYL-9-10-DIHYDRO-1,2-BENZANTHRACENE-9,10-OXIDE
see DQL800
2,2-DIMETHYL-2,3-DIHYDROBENZOFURAN-7-YL ESTER, METHYLCARBA-
MIC ACID see CBS275
2,2-DIMETHYL-2,3-DIHYDRO-7-BENZOFURANYL-N-METHYLCARBAMATE
see CBS275
11,17-DIMETHYL-16,17-DIHYDRO-15H-CYCLOPENTA(a)PHENANTHRENE
see DLI200
7,11-DIMETHYL-15,16-DIHYDROCYCLOPENTA(a)PHENANTHREN-17-ONE
see DLI300
2,5-DIMETHYL-2,5-DIHYDROFURAN-2,5-ENDO PEROXIDE see DUL589
O,O-DIMETHYL-S-(3,4-DIHYDRO-4-KETO-1,2,3-BENZOTRIAZINYL-3-
METHYL) DITHIOPHOSPHATE see ASH500
5,5-DIMETHYL-DIHYDRORESORCINOL-N,N-DIMETHYLCARBAMAT
(GERMAN) see DRL200
5,5-DIMETHYLDIHYDRORESORCINOL DIMETHYLCARBAMATE see DRL200
5,5-DIMETHYL-4,5-DIHYDRO-3-RESORCYL-DIMETHYL-CARBAMAT
(GERMAN) see DRL200
1,3-DIMETHYL-7-(2,3-DIHYDROXYPROPYL)XANTHINE see DNC000
O,O-DIMETHYL-S-1,2-DIKARBETOXYLETHYLDITIOFOSFAT (CZECH)
see MAK700
DIMETHYL DIKETONE see BOT500
2,6-DIMETHYL-3,5-DIMETHOXYCARBONYL-4-(o-DIFLUORO-
METHOXYPHENYL)-1,4-DIHYDROPYRIDINE see RSZ375
DIMETHYL-α-(DIMETHOXYPHOSPHINYL)-p-CHLOROBENZYL PHOSPHATE
see CMU875
DIMETHYL 3-(DIMETHOXYPHOSPHINYLOXY)GLUTACONATE see SOY000
2,4'-DIMETHYL-4-DIMETHYLAMINOAZOBENZENE see DRL400
2',3'-DIMETHYL-4-DIMETHYLAMINOAZOBENZENE see DUN800
N,N-DIMETHYL-4-DIMETHYLAMINO-3-ISOPROPYLPHENYL ESTER
METHIODIDE, CARBAMIC ACID see HJZ000
2,3-DIMETHYL-8-(DIMETHYLAMINOMETHYL)-7-METHOXYCHROMONE
HYDROCHLORIDE see DRL600
3-keto-1,5-DIMETHYL-4-DIMETHYLAMINO-2-PHENYL-2,3-
DIHYDROPYRAZOLE see DOT000
1,5-DIMETHYL-4-DIMETHYLAMINO-2-PHENYL-3-PYRAZOLONE see DOT000
2,3-DIMETHYL-4-DIMETHYLAMINO-1-PHENYL-5-PYRAZOLONE see DOT000
9,9-DIMETHYL-10-(3-(DIMETHYLAMINO)PROPYL)ACRIDAN see DNW700
9,9-DIMETHYL-10-DIMETHYLAMINOPROPYLACRIDAN HYDROGEN TAR-
TRATE see DRM000
9,9-DIMETHYL-10-(3-DIMETHYLAMINO)PROPYLACRIDINE TARTRATE
see DRM000
5,6-DIMETHYL-2-DIMETHYLAMINO-4-PYRIMIDINYLDIMETHYLCARBA-
MATE see DOX600
6,8-DIMETHYL-(4-p-(DIMETHYLAMINO)STYRYL)QUINOLINE see DQC600
3',4'-DIMETHYL-4-DIMETHYLAMINOZOBENZENE see DUO000
O,O-DIMETHYL-O-(2-DIMETHYL-CARBAMOYL-1-METHYL-VINYL)PHOS-
PHAT (GERMAN) see DGQ875
O,O-DIMETHYL-O-(N,N-DIMETHYLCARBAMOYL-1-METHYLVINYL) PHOS-
PHATE see DGQ875
N,N-DIMETHYL-p-(N',N'-DIMETHYLCARBAMOYLOXY)PHENETHYLAMINE,
HYDROCHLORIDE see DQX800
3,4-DIMETHYL-4-(3,4-DIMETHYL-5-ISOXAZOLYAZO)-ISOXAZOLIN-5-ONE
see DRM600
O,O-DIMETHYL-O-(3,5-DIMETHYL-4-METHYLTHIOPHENYL) PHOS-
PHOROTHIOATE see DST200

O,O-DIMETHYL-O-(1,4-DIMETHYL-3-OXO-4-AZA-PENT-1-ENYL)FOSFAAT
(DUTCH) see DGQ875
O,O-DIMETHYL-O-(1,4-DIMETHYL-3-OXO-4-AZA-PENT-1-ENYL)PHOSPHATE
see DGQ875
N,N-DIMETHYL-p-(2',3'-DIMETHYLPHENYLAZO)ANILINE see DUN800
N,N-DIMETHYL-p-(3',4'-DIMETHYLPHENYLAZO)ANILINE see DUO000
3-DIMETHYL-1,2-DIMETHYLPROPYL p-AMINOBENZOATE HYDROCHLO-
RIDE see AIT750
N,N-DIMETHYL-4-(4'-(2',5'-DIMETHYLPYRIDYL-1'-OXIDE)AZO)ANILINE
see DPP600
N,N-DIMETHYL-4-(4'-(2',6'-DIMETHYLPYRIDYL-1'-OXIDE)AZO)ANILINE
see DPP800
N,N-DIMETHYL-4-(4'-(3',5'-DIMETHYLPYRIDYL-1'-OXIDE)AZO)ANILINE
see DPP709
2,5-DIMETHYL-1-(5-(2,5-DIMETHYLPYRROLIDINO)-2,4-PENTADIENYL-
IDENE) PYRROLIDINIUM CHLORIDE SESQUIHYDRATE see DRM800
O,O-DIMETHYL-O-(p-N,N-DIMETHYLSULFAMOYL)PHENYL)PHOS-
PHOROTHIOATE see FAB600
3,4-DIMETHYL-2,6-DINITRO-N-(1-ETHYLPROPYL)ANILINE see DRN200
N,N'-DIMETHYL-N,N'-DINITROOXAMIDE see DRN300
DIMETHYL-DI-NITROSO-AETHYLENDIAMIN (GERMAN) see DVE400
N,N'-DIMETHYL-N,N'-DINITROSO-1,4-BENZENEDICARBOXAMIDE
see DRO400
1,6-DIMETHYL-1,6-DINITROSOBIUREA see DRN400
DIMETHYLDINITROSOETHYLENEDIAMINE see DVE400
N,N'-DIMETHYL-N,N'-DINITROSOETHYLENEDIAMINE see DVE400
N,N'-DIMETHYL-N,N'-DINITROSO-1,2-HYDRAZINEDICARBOXAMIDE
see DRN400
DIMETHYLDINITROSOOXAMID (GERMAN) see DRN600
N,N'-DIMETHYL-N,N'-DINITROSOOXAMIDE see DRN600
2,5-DIMETHYLDINITROSOPIPERAZINE see DRN800
2,6-DIMETHYLDINITROSOPIPERAZINE see DRO000
2,5-DIMETHYL-1,4-DINITROSOPIPERAZINE see DRN800
N,N'-DIMETHYL-N,N'-DINITROSO-1,3-PROPANEDIAMINE see DRO200
N,N'-DIMETHYL-N,N'-DINITROSOTEREPHTHALAMIDE see DRO400
2,2-DIMETHYL-1,3-DIOXA-6-AZA-2-SILACYCLOOCTANE-6-ETHANOL
see SDH670
DIMETHYL-1,1-DIOXA-2,8-HYDROXYETHYL-5 SILA-1 AZA-5 CYCLOOCT-
ANE (FRENCH) see SDH670
DIMETHYL DIOXANE see DRO800
2,6-DIMETHYL-1,4-DIOXANE see DRP000
4,4-DIMETHYLDIOXANE-1,3 see DVQ400
DIMETHYL-p-DIOXANE (DOT) see DRO800
2,2-DIMETHYL-m-DIOXANE-4,6-DIONE see DRP200
2,2-DIMETHYL-1,3-DIOXANE-4,6-DIONE see DRP200
2,6-DIMETHYL-m-DIOXAN-4-OL ACETATE see ABC250
2,6-DIMETHYL-m-DIOXAN-4-YL ACETATE see ABC250
2,2-DIMETHYL-4,6-DIOXO-m-DIOXANE see DRP200
2,2-DIMETHYL-1,3-DIOXOLAN see DRP400
2,2-DIMETHYL-1,3-DIOXOLANE-4-METHANOL see DVR600
2-(4,5-DIMETHYL-1,3-DIOXOLAN-2-YL)PHENYL-N-METHYLCARBAMATE
see DRP600
N,N-DIMETHYLDIPHENYLACETAMIDE see DRP800
N,N-DIMETHYL-2,2-DIPHENYLACETAMIDE see DRP800
N,N-DIMETHYL-α,α-DIPHENYLACETAMIDE see DRP800
2',3'-DIMETHYL-2-DIPHENYLAMINECARBOXYLIC ACID see XQS000
cis-N,N-DIMETHYL-2-(p-(1,2-DIPHENYL-1-BUTENYL)PHENOXY)ETHYLAM-
INE see NOA600
N,N-DIMETHYL-2-(p-(1,2-DIPHENYL-1-BUTENYL)PHENOXY)ETHYLAMINE
CITRATE see DRP875
3,3'-DIMETHYL-4,4'-DIPHENYLDIAMINE see TGJ750
3,3'-DIMETHYLDIPHENYL-4,4'-DIAMINE see TGJ750
(R) (−)-N,N-DIMETHYL-1,2-DIPHENYLETHYLAMINE HYDROCHLORIDE
see DWA600
3,3'-DIMETHYLDIPHENYLMETHANE-4,4'-DIISOCYANATE
see MJN750
N,N-DIMETHYL-4-(DIPHENYLMETHYL)ANILINE see DRQ000
α-N,N-DIMETHYL-3,4-DIPHENYL-2-METHYL-3-PROPIONOXY-1-BUTYLAM-
INE see PNA250
1,2-DIMETHYL-3,5-DIPHENYL-1-H-PYRAZOLIUM METHYL SULFATE
see ARW000
2,2'-DIMETHYLDIPROPYLINITROSOAMINE see DRQ200
1,1'-DIMETHYL-4,4'-DIPYRIDINIUM-DICHLORID (GERMAN)
see PAJ000
1'-DIMETHYL-4,4'-DIPYRIDINIUM DI(METHYLSULFATE) see PAJ250
1,1'-DIMETHYL-4,4'-DIPYRIDYLIUM CHLORIDE see PAJ000
DIMETHYLDISULFIDE see DRQ400
N,N-DIMETHYL-4,4-DI-2-THIENYL-3-BUTEN-2-AMINE HYDROCHLORIDE
see TLQ250
N,1-DIMETHYL-3,3-DI-2-THIENYL-N-ETHYLALLYLAMINE HYDROCHLO-
RIDE see EIJ000
o,o-DIMETHYL DITHIOBIS(THIOFORMATE) see DUN600
DIMETHYLDITHIOCARBAMATE ZINC SALT see BJK500
DIMETHYLDITHIOCARBAMIC ACID COPPER SALT see CNL500

DIMETHYLDITHIOCARBAMIC ACID with DIMETHYLAMINE (1:1)
see DRQ600
DIMETHYLDITHIOCARBAMIC ACID DIMETHYL AMINE SALT see DRQ600
DIMETHYLDITHIOCARBAMIC ACID DIMETHYLAMMONIUM SALT
see DRQ600
DIMETHYLDITHIOCARBAMIC ACID, IRON SALT see FAS000
DIMETHYLDITHIOCARBAMIC ACID, IRON(3+) SALT see FAS000
DIMETHYLDITHIOCARBAMIC ACID, LEAD SALT see LCW000
DIMETHYLDITHIOCARBAMIC ACID, SODIUM SALT see SGM500
N,N-DIMETHYLDITHIOCARBAMIC ACID S-TRIBUTYLSTANNYL ESTER
see TID150
DIMETHYLDITHIOCARBAMIC ACID, ZINC SALT see BJK500
O,O-DIMETHYLDITHIOFOSFORECNAN SODNY (CZECH) see PHI250
2,4-DIMETHYL-1,3-DITHIOLANE-2-CARBOXALDEHYDE O-
((METHYLAMINO)CARBONYL)OXIME see DRR000
2,4-DIMETHYL-1,3-DITHIOLANE-2-CARBOXALDEHYDE O-
(METHYLCARBAMOYL)OXIME see DRR000
O,O-DIMETHYLDITHIOPHOSPHATE DIETHYLMERCAPTOSUCCINATE
see MAK700
DIMETHYLDITHIOPHOSPHORIC-ACID N-METHYLBENZAZIMIDE ESTER
see ASH500
O,O-DIMETHYL DITHIOPHOSPHORYLACETIC ACID-N-METHYL-N-
FORMYLAMIDE see DRR200
O,O-DIMETHYLDITHIOPHOSPHORYLACETIC ACID-N-MONOMETHYLAM-
IDE SALT see DSP400
O,O-DIMETHYL-DITHIOPHOSPHORYLESSIGSAEURE MONOMETHYLAMID
(GERMAN) see DSP400
(O,O-DIMETHYLDITHIOPHOSPHORYLPHENYL)ACETIC ACID ETHYL
ESTER see DRR400
DIMETHYL DIXANTHOGEN see DUN600
N,N-DIMETHYL-2-(DI-2,6-XYLYLMETHOXY)ETHYLAMINE HYDROCHLO-
RIDE see DRR500
DIMETHYL DIZENEDICARBOXYLATE see DQH509
2,5-DIMETHYL-DNPZ see DRN800
2,6-DIMETHYL-DNPZ see DRO000
N,N-DIMETHYLDODECANAMIDE see DRR600
N,N-DIMETHYL-1-DODECANAMINE see DRR800
N,N-DIMETHYLDODECYLAMINE see DRR800
DIMETHYLDODECYLAMINE ACETATE see DRS000
N,N-DIMETHYLDODECYLAMINE ACETATE see DRS000
DIMETHYLDODECYLAMINE HYDROCHLORIDE mixed with SODIUM NI-
TRITE (7:8) see SIS150
DIMETHYLDODECYLAMINE-N-OXIDE see DRS200
N,N-DIMETHYLDODECYLAMINE OXIDE see DRS200
N,N-DIMETHYL-DODECYLAMINOXID (CZECH) see DRS200
N,N-DIMETHYL-n-DODECYL(2-HYDROXY-3-CHLOROPROPYL)AMMONIUM
CHLORIDE see DRS400
N,N-DIMETHYL-n-DODECYL(3-HYDROXYPROPENYL) AMMONIUM CHLO-
RIDE see DRS600
7,8-DIMETHYLENEBENZ(a)ANTHRACENE see CMC000
3:4-DIMETHYLENE-1:2-BENZANTHRACENE see AAF000
8:9-DIMETHYLENE-1:2-BENZANTHRACENE see BAW000
DIMETHYLENEDIAMINE see EEA500
DIMETHYLENE DiISOTHIOCYANATE see ISK000
DIMETHYLENE GLYCOL see BOT000
DIMETHYLENEIMINE see EJM900
DIMETHYLENE OXIDE see EJN500
N,N-DIMETHYLENEOXIDEBIS(PYRIDINIUM-4-ALDOXIME) DICHLORIDE
see BGS250
DIMETHYLENIMINE see EJM900
N,N-DIMETHYLENOXID-BIS-(PYRIDINIUM-4-ALDOXIM)-DICHLORID
(GERMAN) see BGS250
7,12-DIMETHYLENE-5,6-EPOXY-5,6-DIHYDROBENZ(a)ANTHRACENE see DQL600
exo-1,2-cis-DIMETHYL-3,6-EPOXYHEXAHYDROPHTHALIC ANHYDRIDE
see CBE750
6,7-DIMETHYLESCULETIN see DRS800
O,S-DIMETHYL ESTER AMIDE of AMIDOTHIOATE see DTQ400
O,O-DIMETHYLESTER KYSELINY CHLORTHIOFOSFORECNE (CZECH)
see DTQ600
DIMETHYLESTER KYSELINY FOSFORITE (CZECH) see DSG600
DIMETHYLESTER KYSELINY SIROVE (CZECH) see DUD100
DIMETHYLESTER KYSELINY TEREFTALOVE (CZECH) see IML000
DIMETHYL ESTER PHOSPHORIC ACID ESTER with METHYL 3-
HYDROXYCROTONATE see MQR750
O,O-DIMETHYL ESTER PHOSPHOROTHIOIC ACID-S-ESTER with 1,2-
BIS(METHOXYCARBONYL)ETHANETHIOL see OPK250
O,O-DIMETHYL ESTER PHOSPHOROTHIOIC ACID-S-ESTER with ETHYL
MERCAPTOACETATE see DRB600
7,14-DIMETHYL-7,14-ETHANODIBENZ(a,b)ANTHRACENE-15,16-DICARBOXYL-
IC ACID see DRT000
1,1-DIMETHYLETHANOL see BPX000
DIMETHYLETHANOLAMINE see DOY800
N,N-DIMETHYLETHANOLAMINE see DOY800
DIMETHYLETHANOLAMINE (DOT) see DOY800

(E)-4,4'-(1,2-DIMETHYL-1,2-ETHENEDIYL)BIS-PHENOL (9CI) see DUC300
DIMETHYL ETHER (DOT) see MJW500
DIMETHYL ETHER HYDROQUINONE see DOA400
(DIMETHYL ETHER)OXODIPEROXO CHROMIUM(VI) see DRT089
DIMETHYLETHER of d-TUBOCURARINE IODIDE see DUM000
6-α,21-DIMETHYLETHISTERONE see DRT200
O,O-DIMETHYL-S-α-ETHOXY-CARBONYLBENZYL PHOSPHORODITHIOATE
see DRR400
2',6'-DIMETHYL-2-(2-ETHOXYETHYLAMINO)ACETANILIDE see DRT400
2',6'-DIMETHYL-2-(2-ETHOXYETHYLAMINO)ACETANILIDE HYDROCHLO-
RIDE see DRT600
N,N-DIMETHYL-p-((3-ETHOXYPHENYL)AZO)ANILINE see DRU000
DIMETHYLETHOXYPHENYLSILANE see DRU200
O,O-DIMETHYL-S-(5-ETHOXY-1,3,4-THIADIAZOLINYL-3-
METHYL)DITHIOPHOSPHATE see DRU400
O,O-DIMETHYL-S-(5-ETHOXY-1,3,4-THIADIAZOL-2(3H)-ONYL-(3)-
METHYL)DITHIOPHOSPHATE see DRU400
O,O-DIMETHYL-S-(5-ETHOXY-1,3,4-THIADIAZOL-2(3H)-ONYL-(3)-
METHYL)PHOSPHORODITHIOATE see DRU400
O,O-DIMETHYL-S-(2-ETHSULFONYLETHYL)PHOSPHOROTHIOATE
see DAP600
DIMETHYL-S-(2-ETHSULFONYLETHYL)THIOPHOSPHATE see DAP600
O,O-DIMETHYL-S-(2-ETHTHIOETHYL)PHOSPHOROTHIOATE see DAP400
DIMETHYL-S-(2-ETHTHIOETHYL)THIOPHOSPHATE see DAP400
O,O-DIMETHYL-S-(2-ETHTHIONYLETHYL) PHOSPHOROTHIOATE
see DAP000
DIMETHYL-S-(2-ETHTHIONYLETHYL) THIOPHOSPHATE see DAP000
DIMETHYL ETHYL ALLENOLIC ACID METHYL ETHER see DRU600
1,1-DIMETHYLETHYLAMINE see BPY250
N,N-DIMETHYL-4'-ETHYL-4-AMINOAZOBENZENE see EOI500
(±)-1-((1,1-DIMETHYLETHYL)AMINO)-3-(2,3-DIMETHYLPHENOXY)-2-PRO-
PANOL HYDROCHLORIDE see XGA500
5-(2-((1,1-DIMETHYLETHYL)AMINO)-1-HYDROXYETHYL)-1,3-BENZENEDIOL
see TAN100
5-(2-((1,1-DIMETHYLETHYL)AMINO)-1-HYDROXYETHYL)-1,3-BENZENEDIOL,
SULFATE (2:1) (SALT) see TAN250
2-(3-((1,1-DIMETHYLETHYL)AMINO)-2-HYDROXYPROPOXY)-BENZONITRILE
HYDROCHLORIDE see BON400
5-(3-((1,1-DIMETHYLETHYL)AMINO)-2-HYDROXYPROPOXY)-3,4-DIHYDRO-
2(1H)-QUINOLINONE see CCL800
5-(3-((1,1-DIMETHYLETHYL)AMINO)-2-HYDROXYPROPOXY)-1,2,3,4-
TETRAHYDRO-2,3-NAPHTHALENEDIOL see CNR675
α-1-((1,1-DIMETHYLETHYL)AMINO)METHYL)-4-HYDROXY-1,3-
BENZENEDIMETHANOL see BQF500
α⁶-(((1,1-DIMETHYLETHYL)AMINO)METHYL)-3-HYDROXY-2,6-
PYRIDINEDIMETHANOL DIHYDROCHLORIDE see POM800
N-5-((1,1-DIMETHYLETHYL)AMINO)SULFONYL)-1,3,4-THIADIAZOL-2-
YL)ACETAMIDE MONOSODIUM SALT see DRU875
1,4-DIMETHYL-7-ETHYLAZULENE see DRV000
4-(1,1-DIMETHYLETHYL)-1,2-BENZENEDIOL see BSK000
DI-N-METHYL ETHYL CARBAMATE see EII500
O,O-DIMETHYL-S-(N-ETHYLCARBAMOYLMETHYL) DITHIOPHOSPHATE
see DNX600
O,O-DIMETHYL-S-(N-ETHYLCARBAMOYLMETHYL) PHOSPHORODITHIO-
ATE see DNX600
DIMETHYLETHYLCARBINOL see PBV000
2-(1,1-DIMETHYLETHYL)-4,6-DINITROPHENOL see DRV200
2-(1,1-DIMETHYLETHYL)-4,6-DINITROPHENOL ACETATE see DVJ400
1-(1,1-DIMETHYLETHYL)-4,4-DIPHENYLPIPERIDINE see BOM530
DIMETHYLETHYLENE see BOW500
N,N-DIMETHYLETHYLENEUREA see EJA100
DIMETHYL ETHYL HEXADECYL AMMONIUM BROMIDE see EKN500
1,1-DIMETHYLETHYL HYDROPEROXIDE see BRM250
DIMETHYL-ETHYL-β-HYDROXYETHYL-AMMONIUM-SULFATE-DI-n-
BUTYLCARBAMATE see DDW000
DIMETHYLETHYL-β-HYDROXYETHYLAMMONIUM SULFATE
DIBUTYLURETHAN see DDW000
DIMETHYLETHYL(m-HYDROXYPHENYL)AMMONIUM CHLORIDE
see EAE600
O,O-DIMETHYL-S-(2-ETHYLMERCAPTOETHYL) DITHIOPHOSPHATE
see PHI500
O,O-DIMETHYL-2-ETHYLMERCAPTOETHYL THIOPHOSPHATE see TIR250
O,O-DIMETHYL-O-ETHYLMERCAPTOETHYL THIOPHOSPHATE see DAO800
O,O-DIMETHYL-S-ETHYLMERCAPTOETHYL THIOPHOSPHATE see DAP400
O,O-DIMETHYL-S-ETHYLMERCAPTOETHYL THIOPHOSPHATE, THIOLO
ISOMER see DAP400
O,O-DIMETHYL 2-ETHYLMERCAPTOETHYL THIOPHOSPHATE, THIONO
ISOMER see DAO800
O,O-DIMETHYL-S-2-ETHYLMERKAPTOETHYLESTER KYSELINY
DITHIOFOSFORECNE (CZECH) see PHI500
4-(1,1-DIMETHYLETHYL)-α-METHYLBENZENEPROPANAL see LFT000
6-(1,1-DIMETHYLETHYL)-3-METHYL-2,4-DINITROPHENYL ACETATE
see BRU750
N,N-DIMETHYL-p-(3'-ETHYL-4'-METHYLPHENYLAZO)ANILINE see EPT500

N,N-DIMETHYL-p-(4'-ETHYL-3'-METHYLPHENYLAZO)ANILINE see EPU000
N,N-DIMETHYL-N'-ETHYL-N'-1-NAPHTHYLETHYLENEDIAMINE see DRV500
N,N-DIMETHYL-N'-ETHYL-N'-2-NAPHTHYLETHYLENEDIAMINE see DRV550
α,α-DIMETHYLETHYL NITRITE see BRV760
1,1-DIMETHYL-3-ETHYL-3-NITROSOUREA see DRV600
3,5-DIMETHYL-5-ETHYLOXAZOLIDINE-2,4-DIONE see PAH500
4-(1,1-DIMETHYLETHYL)PHENOL see BSE500
2-(4-(1,1-DIMETHYLETHYL)PHENOXY)CYCLOHEXYL 2-PROPYNYL ESTER,
   SULFUROUS ACID see SOP000
2-(4-(1,1-DIMETHYLETHYL)PHENOXY)CYCLOHEXYL 2-PROPYNYL SUL-
   FITE see SOP000
4-(3-(4-(1,1-DIMETHYLETHYL)PHENOXY)-2-HYDROXYPROPOXY)BENZOIC
   ACID see BSE750
N,N-DIMETHYL-p-((4-ETHYLPHENYL)AZO)ANILINE see EOI500
N,N-DIMETHYL-p((m-ETHYLPHENYL)AZO)ANILINE see EOI000
N,N-DIMETHYL-p-((o-ETHYLPHENYL)AZO)ANILINE see EIF450
N,N-DIMETHYL-p-(3'-ETHYLPHENYLAZO)ANILINE see EOI000
N,N-DIMETHYL-3'-ETHYL-4-(PHENYLAZO)BENZENAMINE see EOI000
N,N-DIMETHYL-N'-ETHYL-N'-PHENYLETHYLENEDIAMINE see DRV850
S-(4-(1,1-DIMETHYLETHYL)PHENYL)-o-ETHYL ETHYLPHOSPHONODITHIO-
   ATE see BSG000
α-(4-(1,1-DIMETHYLETHYL)PHENYL)-4-(HYDROXYDIPHENYLMETHYL)-1-
   PIPERIDINEBUTANOL see TAI450
DIMETHYLETHYL(3-(10H-PYRIDO(3,2-b)(1,4)BENZOTHIAZIN-10-YL)
   PROPYLAMMONIUM ETHYL SULFATE see DRW000
O,O-DIMETHYL-S-(2-ETHYLSULFINYL-ETHYL)-MONOTHIOFOSFAAT
   (DUTCH) see DAP000
O,O-DIMETHYL-S-(2-(ETHYLSULFINYL)ETHYL) PHOSPHOROTHIOATE
   see DAP000
O,O-DIMETHYL-S-(2-ETHYLSULFINYL)ETHYL THIOPHOSPHATE see DAP000
DIMETHYLETHYL SULFONIUM IODIDE with MERCURY IODIDE (1:1)
   see EIL500
O,O-DIMETHYL-S-ETHYL-2-SULFONYLETHYL PHOSPHOROTHIOLATE
   see DAP600
O,O-DIMETHYL-S-ETHYLSULPHINYLETHYL PHOSPHOROTHIOLATE
   see DAP000
O,O-DIMETHYL-S-ETHYLSULPHONYLETHYL PHOSPHOROTHIOLATE
   see DAP600
O,O-DIMETHYL-S-(2-ETHYLTHIO-ETHYL)-DITHIOFOSFAAT (DUTCH)
   see PHI500
O,O-DIMETHYL-O-(2-ETHYL-THIO-ETHYL)-MONOTHIOFOSFAAT (DUTCH)
   see DAO800
O,O-DIMETHYL-S-(2-ETHYLTHIO-ETHYL)-MONOTHIOFOSFAAT (DUTCH)
   see DAP400
O,O-DIMETHYL S-(2-(ETHYLTHIO)ETHYL) PHOSPHORODITHIOATE
   see PHI500
O,O-DIMETHYL-O-2-(ETHYLTHIO)ETHYL PHOSPHOROTHIOATE
   see DAO800
O,O-DIMETHYL-S-(2-(ETHYLTHIO)ETHYL)PHOSPHOROTHIOATE see DAP400
S-(((1,1-DIMETHYLETHYL)THIO)METHYL)-O,O-DIETHYL PHOS-
   PHORODITHIOATE see BSO000
N,N-DIMETHYL-p-((3-ETHYL-p-TOLYL)AZO)ANILINE see EPT500
N,N-DIMETHYL-p-((4-ETHYL-m-TOLYL)AZO) ANILINE see EPU000
N,N-DIMETHYL-α-ETHYNYL-α-PHENYLBENZENEACETAMIDE see DWA700
DIMETHYL FANDANE see DRX400
DIMETHYL-FENYL-ETHOXYSILAN (CZECH) see DRU200
2,3-DIMETHYLFLUORANTHENE see DRX600
7,8-DIMETHYLFLUORANTHENE see DRX800
8,9-DIMETHYLFLUORANTHENE see DRY000
1,9-DIMETHYLFLUORENE see DRY100
DIMETHYLFLUOROARSINE see DRY289
2,10-DIMETHYL-3-FLUORO-5,6-BENZACRIDINE see FHR000
9,12-DIMETHYL-10-FLUOROBENZ(a)ACRIDINE see FHR000
7,12-DIMETHYL-1-FLUOROBENZ(a)ANTHRACENE see FHS000
7,12-DIMETHYL-4-FLUOROBENZ(a)ANTHRACENE see DRY400
7,12-DIMETHYL-5-FLUOROBENZ(a)ANTHRACENE see DRY600
7,12-DIMETHYL-8-FLUOROBENZ(a)ANTHRACENE see DRY800
7,12-DIMETHYL-11-FLUOROBENZ(a)ANTHRACENE see DRZ000
N,N-DIMETHYL-2-FLUORO-4-PHENYLAZOANILINE see FHQ000
N,N-DIMETHYL-p-(2-FLUOROPHENYLAZO)ANILINE see FHQ010
N,N-DIMETHYL-p-((p-FLUOROPHENYL)AZO)ANILINE see DSA000
3,3-DIMETHYL-1-(p-FLUOROPHENYL)TRIAZENE see DSA600
DIMETHYL FLUOROPHOSPHATE see DSA800
DIMETHYL FORMAL see MGA850
DIMETHYLFORMALDEHYDE see ABC750
DIMETHYLFORMAMID (GERMAN) see DSB000
DIMETHYLFORMAMIDE see DSB000
N,N-DIMETHYL FORMAMIDE see DSB000
N,N-DIMETHYLFORMAMIDE (DOT) see DSB000
DIMETHYLFORMOCARBOTHIALDINE see DSB200
2,4-DIMETHYL-2-FORMYL-1,3-DITHIOLANE OXIME METHYLCARBAMATE
   see DRR000
O,O-DIMETHYL-S-(N-FORMYL-N-METHYLCARBAMOYLMETHYL)
   PHOSPHORODITHIOATE see DRR200

6,6-DIMETHYLFULVENE see DSB400
DIMETHYL FUMARATE see DSB600
DIMETHYL FURAN see DSB800
DIMETHYL FURANE see DSB800
2,5-DIMETHYL FURANE see DSC000
4,8-DIMETHYL-2H-FURO(2,3-h)-1-BENZOPYRAN-2-ONE see DQF700
α,α-DIMETHYLGLYCINE see MGB000
DIMETHYLGLYCINE HYDROCHLORIDE mixed with SODIUM NITRITE
   (3581) see DSC200
N,N-DIMETHYLGLYCINONITRILE see DOS200
DIMETHYLGLYOXAL see BOT500
DIMETHYLGLYOXIME see DBH000
DIMETHYLGOLD SELENOCYANATE see DSC400
as-DIMETHYLGUANIDINE HYDROCHLORIDE see DSC800
N,N'-DIMETHYLHARNSTOFF (GERMAN) see DUM200
DIMETHYLHARNSTOFF and NATRIUMNITRIT (GERMAN)
   see DUM400
m-(3,3-DIMETHYLHARNSTOFF)-PHENYL-tert-BUTYLCARBAMAT (GERMAN)
   see DUM800
2,6-DIMETHYL-2,5-HEPTADIEN-4-ONE see PGW250
2,6-DIMETHYL-2,5-HEPTADIEN-4-ONE DIOZONIDE see DSD000
2,5-DIMETHYLHEPTANE see DSD200
3,5-DIMETHYLHEPTANE see DSD400
4,4-DIMETHYLHEPTANE see DSD600
2,6-DIMETHYL-4-HEPTANOL see DNH800
2,6-DIMETHYL HEPTANOL-4 see DNH800
2,6-DIMETHYL-HEPTAN-4-ON (DUTCH, GERMAN) see DNI800
2,6-DIMETHYLHEPTAN-4-ONE see DNI800
2,6-DIMETHYL-4-HEPTANONE see DNI800
2,6-DIMETHYL-5-HEPTENAL see DSD775
2,6-DIMETHYL-3-HEPTENE see DSD800
2,6-DIMETHYL-4-HEPTYLPHENOL, (o and p) see NNC500
1,3-DIMETHYLHEXAHYDROPYRIMIDONE see DSE489
N,N-DIMETHYLHEXANAMIDE see DQX200
2,3-DIMETHYLHEXANE see DSE509
2,4-DIMETHYLHEXANE see DSE600
DIMETHYLHEXANE DIHYDROPEROXIDE (dry) see DSE800
DIMETHYLHEXANE DIHYDROPEROXIDE (with 18% or more water) (DOT)
   see DSE800
DIMETHYL HEXANEDIOATE see DOQ300
3,4-DIMETHYL-2,5-HEXANEDIONE see DSF100
1,5-DIMETHYLHEXYLAMINE see ILM000
α,ε-DIMETHYLHEXYLAMINE see ILM000
N,1-DIMETHYLHEXYLAMINE HYDROCHLORIDE see NCL300
DI(3-METHYLHEXYL)PHTHALATE see DSF200
5,5-DIMETHYLHYDANTOIN see DSF300
1,2-DIMETHYLHYDRAZIN (GERMAN) see DSF600
DIMETHYLHYDRAZINE see DSF400
1,1-DIMETHYLHYDRAZINE see DSF400
1,2-DIMETHYLHYDRAZINE see DSF600
N,N-DIMETHYLHYDRAZINE see DSF400
N,N'-DIMETHYLHYDRAZINE see DSF600
uns-DIMETHYLHYDRAZINE see DSF400
asym-DIMETHYLHYDRAZINE see DSF400
sym-DIMETHYLHYDRAZINE see DSF600
unsym-DIMETHYLHYDRAZINE see DSF400
1,1-DIMETHYLHYDRAZINE (GERMAN) see DSF400
DIMETHYLHYDRAZINE, symmetrical (DOT) see DSF600
DIMETHYLHYDRAZINE, unsymmetrical (DOT) see DSF400
1,2-DIMETHYLHYDRAZINE DIHYDROCHLORIDE see DSF800
N,N'-DIMETHYLHYDRAZINE DIHYDROCHLORIDE see DSF800
sym-DIMETHYLHYDRAZINE DIHYDROCHLORIDE see DSF800
1,1-DIMETHYLHYDRAZINE HYDROCHLORIDE see DSG000
1,2-DIMETHYLHYDRAZINE HYDROCHLORIDE see DSG200
sym-DIMETHYLHYDRAZINE HYDROCHLORIDE see DSG200
DIMETHYLHYDRAZINIUM derivative of 2-CHLOROETHYLPHOSPHONIC
   ACID see DKQ650
2-(2,2-DIMETHYLHYDRAZINO)-4-(5-NITRO-2-FURYL)THIAZOLE see DSG400
DIMETHYLHYDROGENPHOSPHITE see DSG600
DIMETHYLHYDROQUINONE see DOA400
2,6-DIMETHYLHYDROQUINONE see DSG700
DIMETHYLHYDROQUINONE ETHER see DOA400
N,N-DIMETHYL-HYDROXYBENZENESULFONAMIDE see DUC600
N,N-DIMETHYL-2-HYDROXYETHYLAMINE see DOY800
N,N-DIMETHYL-N-(2-HYDROXYETHYL)AMINE see DOY800
DIMETHYL(2-HYDROXYETHYL)OCTYLAMMONIUM BROMIDE BENZILATE
   see DSH000
DIMETHYL(2-HYDROXYETHYL)PENTYLAMMONIUM BROMIDE BENZIL-
   ATE see DSH200
1,3-DIMETHYL-7-(2-HYDROXYETHYL)XANTHINE see HLC000
DIMETHYL 3-HYDROXYGLUTACONATE DIMETHYL PHOSPHATE
   see SOY000
1-1,4-DIMETHYL-10-HYDROXY-2,3,4,5,6,7-HEXAHYDRO-1,6-METHANO-1H-4-
   BENZAZONINE HYDROBROMIDE see SAA040

DIMETHYL-S-(2-(1-METHYLCARBAMOYLETHYLTHIO ETHYL) PHOSPHOROTHIOLATE see MJG500

O,O-DIMETHYL-S-(2-(1-METHYLCARBAMOYLETHYLTHIO)ETHYL) PHOSPHOROTHIOATE see MJG500

O,O-DIMETHYL-S-(N-METHYL-CARBAMOYL)-METHYL-DITHIOFOSFAAT (DUTCH) see DSP400

(O,O-DIMETHYL-S-(N-METHYL-CARBAMOYL-METHYL)-DITHIOPHOSPHAT) (GERMAN) see DSP400

O,O-DIMETHYL-S-(N-METHYLCARBAMOYLMETHYL) DITHIOPHOSPHATE see DSP400

O,O-DIMETHYL-S-((N-METHYL-CARBAMOYL)-METHYL)MONOTHIOFOSFAAT (DUTCH) see DNX800

O,O-DIMETHYL-S-(N-METHYL-CARBAMOYL)-METHYL-MONOTHIOPHOSPHAT (GERMAN) see DNX800

O,O-DIMETHYL METHYLCARBAMOYLMETHYL PHOSPHORODITHIOATE see DSP400

O,O-DIMETHYL-S-(N-METHYLCARBAMOYLMETHYL) PHOSPHORODITHIOATE see DSP400

O,O-DIMETHYL-S-((METHYLCARBAMOYL)METHYL)PHOSPHOROTHIOATE see DNX800

O,O-DIMETHYL-S-(N-METHYLCARBAMOYLMETHYL)PHOSPHOROTHIOATE see DNX800

DIMETHYL-S-(N-METHYL-CARBAMOYL-METHYL)PHOSPHOROTHIOLATE see DNX800

O,O-DIMETHYL-S-(N-METHYLCARBAMOYLMETHYL) PHOSPHOROTHIOLATE see DNX800

O,O-DIMETHYL-S-(N-METHYLCARBAMOYLMETHYL) THIOPHOSPHATE see DNX800

O,O-DIMETHYL-O-(2-N-METHYLCARBAMOYL-1-METHYL-VINYL)-FOSFAAT (DUTCH) see MRH209

O,O-DIMETHYL-O-(2-N-METHYLCARBAMOYL-1-METHYL)-VINYL-PHOSPHAT (GERMAN) see MRH209

O,O-DIMETHYL-O-(2-N-METHYLCARBAMOYL-1-METHYL) PHOSPHATE see MRH209

N,N-DIMETHYL-α-METHYLCARBAMOYLOXYIMINO-α-(METHYLTHIO) ACETAMIDE see DSP600

N',N-DIMETHYL-N-((METHYLCARBAMOYL)OXY)-1-METHYLTHIOOXAMIMIDIC ACID see DSP600

N',N-DIMETHYL-N-((METHYLCARBAMOYL)OXY)-1-THIOOXAMIMIDIC ACID METHYL ESTER see DSP600

O,O-DIMETHYL-S-(N-METHYLCARBAMYLMETHYL) THIOTHIONOPHOSPHATE see DSP400

O,O-DIMETHYL-O-(1-METHYL-2-CARBOXY-α-PHENYLETHYL)VINYL PHOSPHATE see COD000

O,O-DIMETHYL O-(1-METHYL-2-CARBOXYVINYL) PHOSPHATE see MQR750

O,O-DIMETHYL-O-(1-METHYL-2-CHLOR-2-N,N-DIAETHYL-CARBAMOYL)-VINYL-PHOSPHAT see FAB400

(O,O-DIMETHYL-O-(1-METHYL-2-CHLORO-2-DIETHYLCARBAMOYL-VINYL) PHOSPHATE) see FAB400

N,N-DIMETHYL-N'-(2-METHYL-4-CHLOROPHENYL)-FORMAMIDINE see CJJ250

N,N-DIMETHYL-N'-(2-METHYL-4-CHLOROPHENYL)-FORMAMIDINE HYDRO-CHLORIDE see CJJ500

N,N-DIMETHYL-N'-(2-METHYL-4-CHLORPHENYL)-FORMADIN (GERMAN) see CJJ250

2,6-DIMETHYL-1-((2-METHYLCYCLOHEXYL)CARBONYL)PIPERIDINE see DSP650

O,O-DIMETHYL-S-(3-METHYL-2,4-DIOXO-3-AZA-BUTYL)-DITHIOFOSFAAT (DUTCH) see DRR200

O,O-DIMETHYL-S-(3-METHYL-2,4-DIOXO-3-AZA-BUTYL)-DITHIOPHOSPHAT (GERMAN) see DRR200

6,6-DIMETHYL-2-METHYLENEBICYCLO(3.1.1)HEPTANE see POH750

DIMETHYLMETHYLENE-p,p'-DIPHENOL see BLD500

O,O-DIMETHYL-S-(N-METHYL-N-FORMYL-CARBAMOYLMETHYL)-DITHIOPHOSPHAT see DRR200

O,O-DIMETHYL-S-(N-METHYL-N-FORMYLCARBAMOYLMETHYL) PHOSPHORODITHIOATE see DRR200

DIMETHYL METHYL MALEATE see DRF200

O,O-DIMETHYL-O-4-(METHYLMERCAPTO)-3-METHYLPHENYL PHOS-PHOROTHIOATE see FAQ900

O,O-DIMETHYL-p-4-(METHYLMERCAPTO)-3-METHYLPHENYL THIOPHOS-PHATE see FAQ900

O,O-DIMETHYL O-(4-METHYLMERCAPTOPHENYL)PHOSPHATE see PHD250

N,N-DIMETHYL-N'-(2-METHYL-4-(((METHYLAMINO)CARBONYL)OXY) PHENYL)METHANIMIDAMIDE see FNE500

(E)-DIMETHYL 1-METHYL-3-(METHYLAMINO)-3-OXO-1-PROPENYL PHOS-PHATE see MRH209

DIMETHYL-1-METHYL-2-(METHYLCARBAMOYL)VINYLPHOSPHATE, cis see MRH209

O,O-DIMETHYL-O-(3-METHYL-4-METHYLMERCAPTOPHENYL)PHOS PHOROTHIOATE see FAQ900

N,N-DIMETHYL-2-METHYL-4-(4'-(2'-METHYLPYRIDYL-1'-OXIDE)AZO) ANILINE see DQD600

O,O-DIMETHYL-O-(3-METHYL-4-METHYLTHIO-FENYL)-MONOTHIOFOSFAAT (DUTCH) see FAQ900

O,O-DIMETHYL-O-(3-METHYL-4-METHYLTHIOPHENYL)-MONOTHIOPHOSPHAT (GERMAN) see FAQ900

O,O-DIMETHYL-O-3-METHYL-4-METHYLTHIOPHENYL PHOSPHOROTHIO-ATE see FAQ900

O,O-DIMETHYL-O-(3-METHYL-4-METHYLTHIO-PHENYL)-THIONOPHOSPHAT (GERMAN) see FAQ900

N-DIMETHYL-1-METHYL-N'-(1-NITRO-9-ACRIDINYL)-1,2-ETHANEDIAMINE DIHYDROCHLORIDE see NFW435

O,O-DIMETHYL-O-(3-METHYL-4-NITROFENYL)-MONOTHIOFOSFAAT (DUTCH) see DSQ000

O,O-DIMETHYL-O-(3-METHYL-4-NITRO-PHENYL)-MONOTHIOPHOSPHAT (GERMAN) see DSQ000

O,O-DIMETHYL-O-(3-METHYL-4-NITROPHENYL)PHOSPHORATE see PHD750

O,O-DIMETHYL-O-(3-METHYL-4-NITROPHENYL) PHOSPHOROTHIOATE see DSQ000

DIMETHYL-3-METHYL-4-NITROPHENYLPHOSPHOROTHIONATE see DSQ000

O,O-DIMETHYL-O-(3-METHYL-4-NITROPHENYL) THIOPHOSPHATE see DSQ000

N,N-DIMETHYL-p-(2'-METHYLPHENYLAZO)ANILINE see DUH800

N,N-DIMETHYL-p-(3'-METHYLPHENYLAZO)ANILINE see DUH600

N,N-DIMETHYL-4-((2-METHYLPHENYL)AZO)BENZENAMINE see DUH800

N,N-DIMETHYL-4-((3-METHYLPHENYL)AZO)BENZENAMINE see DUH600

N,N-DIMETHYL-4-((4-METHYLPHENYL)AZO)BENZENAMINE see DUH400

N,N-DIMETHYL-2-(α-METHYL-α-PHENYLBENZYLOXY)ETHYLAMINE see DSQ600

N,N-DIMETHYL-2-((α-METHYL-α-PHENYLBENZYL)OXY)ETHYLAMINE see DSQ600

N,N-DIMETHYL-2-((o-METHYL-α-PHENYL-BENZYL)OXY)-ETHYLAMINE CITRATE see DPH000

N,N-DIMETHYL-2-(o-METHYL-α-PHENYLBENZYLOXY)ETHYLAMINE HYDROCHLORIDE see OJW000

DIMETHYL-cis-1-METHYL-2-(1-PHENYLETHOXYCARBONYL)VINYL PHOS-PHATE see COD000

3,3-DIMETHYL-6-(((5-METHYL-3-PHENYL-4-ISOXAZOLECARBOXAMIDE-7-OXO-4-THIA-1-AZABICYCLO(3.2.0)HEPTANE-2-CARBOXYLIC ACID see DSQ800

DIMETHYL-5-(3-METHYL-1-PHENYLPYRAZOLYL) CARBAMATE see PPQ625

3,3-DIMETHYL-1-(m-METHYLPHENYL)TRIAZENE see DSR200

3,3-DIMETHYL-1-(o-METHYLPHENYL)TRIAZENE see MNT500

DIMETHYL METHYLPHOSPHONATE see DSR400

O,O-DIMETHYL-O-(3-METHYL) PHOSPHOROTHIOATE see DSQ000

N,N-DIMETHYL-9-(3-(4-METHYL-1-PIPERANIZYL)PROPYLIDENE)-9H-THIOXANTHENE-2-SULFONAMIDE, (Z)- see NBP500

N,N-DIMETHYL-9-(3-(4-METHYL-1-PIPERAZINYL)PROPYLIDENE) THIAXANTHENE-2-SULFONAMIDE see NBP500

N,N-DIMETHYL-9-(3-(4-METHYL-1-PIPERAZINYL)PROPYLIDENE) THIOXANTHENE-2-SULFONAMIDE see NBP500

DIMETHYL-3-(2-METHYL-1-PROPENYL)CYCLOPROPANECARBOXYLATE see BEP500

(+)-2,2-DIMETHYL-3-(2-METHYLPROPENYL)-CYCLOPROPANECARBOXYLIC ACID-(E)-,ESTER with (+)- see AFR750

(+)-(Z)-2,2-DIMETHYL-3-(2-METHYLPROPENYL)-CYCLOPROPANECARBOXY-LIC ACID ESTER with 2-ALLYL-4-HYDROXY-3-METHYL-2-CYCLOPENTEN-ONE see AFR500

2,2-DIMETHYL-3-(2-METHYLPROPYL)CYCLOPROPANECARBOXYLIC ACID-p-(METHOXYMETHYL)BENZYL ESTER see MBV700, MNG525

1,3-DIMETHYL-4-(p-(p-((1-METHYLPYRIDINIUM-4-YL)AMINO)PHENYL) CARBAMOYL)ANILINOQUINOLINIUM), DIBROMIDE see DSR600

1,6-DIMETHYL-4-(p-(p-((1-METHYLPYRIDINIUM-4-YL)AMINO)PHENYL) CARBAMOYL)ANILINO)QUINOLINIUM) DI-p-TOLUENESULFONATE see DSR800

1,8-DIMETHYL-4-(p-(p-((1-METHYLPYRIDINIUM-4-YL)AMINO)PHENYL) CARBAMOYL)ANILINO)QUINOLINIUM)DI-p-TOLUENESULFONATE see DSS000

N,N-DIMETHYL-4-((2-METHYL-4-PYRIDINYL)AZO)BENZENAMINE-N-OXIDE see DSS200

N,N-DIMETHYL-4-(2-METHYL-4-PYRIDYLAZO)ANILINE-N-OXIDE see DSS200

N,N-DIMETHYL-4-(4'-(2'-METHYLPYRIDYL-1'-OXIDE)AZO)ANILINE see DSS200

N,N-DIMETHYL-4-(4'-(3'-METHYLPYRIDYL-1'-OXIDE)AZO)ANILINE see MOY750

N,N'-DIMETHYL-4-(4'-(2'-METHYLPYRIDYL-1-OXIDE)AZO)-o-TOLUIDINE see DQD800

N,N-DIMETHYL-4-(5-(3-METHYLQUINOLYL)AZO)ANILINE see MJF500

N,N-DIMETHYL-4-(5'-(6'-METHYLQUINOLYL)AZO)ANILINE see MJF750

N,N-DIMETHYL-4-(5'-(7'-METHYLQUINOLYL)AZO)ANILINE see DPQ400

N,N-DIMETHYL-4-(5'-(8'-METHYLQUINOLYL)AZO)ANILINE see MJG000

N,N-DIMETHYL-2'-METHYLSTILBENAMINE see TMF750

O,O-DIMETHYL-O-(P-METHYLSULFINYLPHENYL)PHOSPHOROTHIOATE see EAV000

O,O-DIMETHYL-O-(4-(METHYLSULFINYL)-m-TOLYL) PHOSPHOROTHIOATE see DSS400

3,7-DIMETHYL-(E)-2,6-OCTADIEN-1-OL see DTD000
3,7-DIMETHYL-(Z)-2,6-OCTADIEN-1-OL see DTD200
2-cis-3,7-DIMETHYL-2,6-OCTADIEN-1-OL see DTD200
2,6-3,7-DIMETHYL-trans-2,6-OCTADIEN-8-OL see DTD000
3,7-DIMETHYL-trans-2,6-OCTADIEN-1-OL see DTD000
3,7-DIMETHYL-1,6-OCTADIEN-3-OL ACETATE see LFY100
trans-3,7-DIMETHYL-2,6-OCTADIEN-1-OL ACETATE see DTD800
3,7-DIMETHYL-1,6-OCTADIEN-3-OL BENZOATE see LFZ000
trans-3,7-DIMETHYL-2,6-OCTADIEN-1-OL-2-BUTENOATE see DTE000
3,7-DIMETHYL-1,6-OCTADIEN-3-OL CINNAMATE see LGA000
3,7-DIMETHYL-2,6-OCTADIEN-1-OL, FORMATE (cis) see FNC000
trans-3,7-DIMETHYL-2,6-OCTADIEN-1-OL FORMATE see GCY000
3,7-DIMETHYL-1,6-OCTADIEN-3-OL ISOBUTYRATE see LGB000
trans-3,7-DIMETHYL-2,6-OCTADIEN-1-OL ISOBUTYRATE see GDI000
4,7-DIMETHYL-1,6-OCTADIEN-3-OL ISOVALERATE see LGC000
(Z)-3,7-DIMETHYL-2,6-OCTADIEN-1-OL PROPIONATE see NCP000
cis-3,7-DIMETHYL-2,6-OCTADIEN-1-OL PROPIONATE see NCP000
3,7-DIMETHYL-1,6-OCTADIEN-3-YL ACETATE see LFY100
3,7-DIMETHYL-2-trans-6-OCTADIENYL ACETATE see DTD800
trans-3,7-DIMETHYL-2,6-OCTADIEN-1-YL ACETATE see DTD800
3,7-DIMETHYL-1,6-OCTADIEN-3-YL-o-AMINOBENZOATE see APJ000
3,7-DIMETHYL-1,6-OCTADIEN-3-YL BENZOATE see LFZ000
3,7-DIMETHYL-2,6-OCTADIEN-1-YL BENZOATE see GDE800
3,7-DIMETHYL-2-trans-6-OCTADIENYL CROTONATE see DTE000
trans-3,7-DIMETHYL-2,6-OCTADIEN-1-YL cis-α,β-DIMETHYL ACRYLATE
    see GDO000
3-7-DIMETHYL-2,6-OCTADIENYL ESTER-2-BUTENOIC ACID see DTE000
3,7-DIMETHYL-2,6-OCTADIENYL ESTER FORMIC ACID (E) see GCY000
(E)-3,7-DIMETHYLOCTA-2,6-DIEN-1-YL ESTER, HEXANOIC ACID see
    GDG000
trans-3,7-DIMETHYL-2,6-OCTADIENYL ESTER ISOBUTYRIC ACID see GDI000
(Z)-3,7-DIMETHYL-2,6-OCTADIENYL ESTER ISOVALERIC ACID see NCO500
(E,E,E)-3,7-DIMETHYL-2,6-OCTADIENYL ESTER-2-METHYL-2-BUTENOIC
    ACID see GDO000
trans-2,6-DIMETHYL-2,6-OCTADIEN-8-YL ETHANOATE see DTD800
3,7-DIMETHYL-1,6-OCTADIEN-3-YL FORMATE see LGA050
trans-3,7-DIMETHYL-2,6-OCTADIEN-1-YL FORMATE see GCY000
(E)-3,7-DIMETHYLOCTA-2,6-DIEN-1-YL-n-HEXANOATE see GDG000
N-(3,7-DIMETHYL-2,6-OCTADIENYLIDENE)ANTHRANILIC ACID METHYL
    ESTER see CMS325
3,7-DIMETHYL-1,6-OCTADIEN-3-YL ISOBUTYRATE see LGB000
trans-3-7-DIMETHYL-2,6-OCTADIENYL ISOBUTYRATE see GDI000
trans-3,7-DIMETHYL-2,6-OCTADIENYL ISOPENTANOATE see GDK000
3,7-DIMETHYL-1,6-OCTADIEN-3-YL ISOVALERATE see LGC000
3,7-DIMETHYL-2-cis-6-OCTADIEN-1-YL ISOVALERATE see NCO500
3,7-DIMETHYL-2,6-OCTADIEN-1-YL PHENYLACETATE see GDM400
N,N-DIMETHYLOCTANAMIDE see DTE200
3,7-DIMETHYL-1,2-OCTANEDIOL see DTE400
4,4-DIMETHYLOCTANOIC ACID, TRIBUTYLSTANNYL ESTER see TIF250
DIMETHYLOCTANOL see DTE600
2,6-DIMETHYL-8-OCTANOL see DTE600
3,7-DIMETHYL-3-OCTANOL see TCU600
3,7-DIMETHYL-1-OCTANOL (FCC) see DTE600
(4,4-DIMETHYLOCTANOYLOXY)TRIBUTYLSTANNANE see TIF250
3,7-DIMETHYLOCTANYL ACETATE see DTE800
3,7-DIMETHYLOCTANYL BUTYRATE see DTF000
DIMETHYLOCTATRIENE see DTF200
DIMETHYLOCTATRIENE (mixed isomer) see DTF200
3,7-DIMETHYL-6-OCTENAL see CMS845
3,7-DIMETHYL-6-OCTENENITRILE see CMU000
3,7-DIMETHYL-6-OCTENOIC ACID see CMT125
2,6-DIMETHYL-1-OCTEN-8-OL see DTF400
2,6-DIMETHYL-7-OCTEN-8-OL see CMT250
2,6-DIMETHYL-7-OCTEN-2-OL see DLX000
3,7-DIMETHYL-6-OCTEN-1-OL see CMT250
3,7-DIMETHYL-7-OCTEN-1-OL see DTF400
2,6-DIMETHYL-2-OCTEN-8-OL ACETATE see AAU000
3,7-DIMETHYL-7-OCTEN-1-OL ACETATE see RHA000
2,6-DIMETHYL-2-OCTEN-8-OL-BUTYRATE see DTF800
3,7-DIMETHYL-6-OCTEN-1-OL BUTYRATE see DTF800
3,7-DIMETHYL-6-OCTEN-1-OL CROTONATE see CMT500
3,7-DIMETHYL-6-OCTEN-1-OL FORMATE see CMT750
3,7-DIMETHYL-6-OCTEN-1-YL ACETATE see AAU000
2,6-DIMETHYL-2-OCTEN-8-YL BUTYRATE see DTF800
3,7-DIMETHYL-6-OCTEN-1-YL BUTYRATE see CMT600
2,6-DIMETHYL-2-OCTEN-8-YL FORMATE see CMT750
3,7-DIMETHYL-6-OCTEN-1-YL FORMATE see CMT750
3,7-DIMETHYL-6-OCTEN-1-YL ISOBUTYRATE see CMT900
3,7-DIMETHYL-6-OCTEN-1-YL PHENYLACETATE see CMU050
3,7-DIMETHYLOCTYL ACETATE see DTE800
N,N-DIMETHYL-N-OCTYLBENZENEMETHANAMINIUM CHLORIDE
    see OEW000
3,7-DIMETHYLOCTYL ESTER BUTANOIC ACID see DTF000
N,N-DIMETHYLOKTADECYLAMIN (CZECH) see DTC400

DIMETHYLOL DIHYDROXYETHYLENE UREA see DTG000
DIMETHYLOLGLYOXALUREA see DTG000
N,N-DIMETHYLOL-2-METHOXYETHYL CARBAMATE see DTG200
1,1-DIMETHYLOL-1-NITROETHANE see NHO500
DIMETHYLOLPROPANE see DTG400
DIMETHYLOLPROPANE DIACRYLATE see DUL200
DIMETHYLOL-TETRAKIS-BUTOXYMETHYLMELAMIN (CZECH) see BHB500
DIMETHYLOL THIOUREA see DTG600
1,3-DIMETHYLOLUREA see DTG700
2,3-DIMETHYL-7-OXABICYCLO(2.2.1)HEPTANE-2,3-DICARBOXYLIC ANHY-
    DRIDE see CBE750
DIMETHYL OXAZOLIDINE see DTG750
4,4-DIMETHYLOXAZOLIDINE see DTG750
DIMETHYLOXAZOLIDINEDIONE see PMO250
5,5-DIMETHYLOXAZOLIDINE-2,4-DIONE see PMO250
5,5-DIMETHYL-2,4-OXAZOLIDINEDIONE see PMO250
N^1-(4,5-DIMETHYL-2-OXAZOLYL)-SULFANILAMIDE see AIE750
3,3-DIMETHYLOXETANE see EBQ500
3,3-DIMETHYL-2-OXETANONE see DTH000
3,3-DIMETHYL-2-OXETHANONE see DTH000
O,O-DIMETHYL-S-(2-OXO-3-AZA-BUTYL)-DITHIOPHOSPHAT (GERMAN)
    see DSP400
O,O-DIMETHYL-S-(2-OXO-3-AZABUTYL)-MONOTHIOPHOSPHATE
    see DNX800
O,O-DIMETHYL-S-(4-OXOBENZOTRIAZINO-3-METHYL)PHOSPHORODITHIO
    ATE see ASH500
O,O-DIMETHYL-S-(4-OXO-1,2,3-BENZOTRIAZINO(3)-METHYL)
    THIOTHIONOPHOSPHATE see ASH500
O,O-DIMETHYL-S-((4-OXO-3H-1,2,3-BENZOTRIAZIN-3-YL)-METHYL)-
    DITHIOFOSFAAT (DUTCH) see ASH500
O,O-DIMETHYL-S-((4-OXO-3H-1,2,3-BENZOTRIAZIN-3-YL)-METHYL)-
    DITHIOPHOSPHAT (GERMAN) see ASH500
O,O-DIMETHYL-S-4-OXO-1,2,3-BENZOTRIAZIN-3(4H)-YLMETHYL PHOS-
    PHORODITHIOATE see ASH500
O,O-DIMETHYL-S-(4-OXO-3H-1,2,3-BENZOTRIZIANE-3-METHYL)PHOS-PHO-
    RODITHIOATE see ASH500
N,N-DIMETHYL-3-OXOBUTANAMIDE see DOP000
N-(1,1-DIMETHYL-3-OXOBUTYL)ACRYLAMIDE see DTH200
N-(1,1-DIMETHYL-3-OXOBUTYL)-2-PROPENAMIDE see DTH200
(5,5-DIMETHYL-3-OXO-CYCLOHEX-1-EN-YL)-N,N-DIMETHYL-CARBAMAAT
    (DUTCH) see DRL200
(5,5-DIMETHYL-3-OXO-CYCLOHEX-1-EN-YL)-N,N-DIMETHYL-CARBAMAT
    (GERMAN) see DRL200
5,5-DIMETHYL-3-OXOCYCLOHEX-1-ENYL DIMETHYLCARBAMATE
    see DRL200
5,5-DIMETHYL-3-OXO-1-CYCLOHEXEN-1-YL DIMETHYLCARBAMATE
    see DRL200
3-(2-(3,5-DIMETHYL-2-OXOCYCLOHEXYL)-2-HYDROXYETHYL)GLUTARIM-
    IDE see CPE750
1,6-DIMETHYL-4-OXO-1,6,7,8,9,9a-HEXAHYDRO-4H-PYRIDO(1,2-a)PYRIMI-
    DINE-3-CARBOXAMIDE see CDM500
3,7-DIMETHYL-1-(5-OXOHEXYL)-1H,3H-PURIN-2,6-DIONE see PBU100
DIMETHYLOXOHEXYLXANTHINE see PBU100
3,7-DIMETHYL-1-(5-OXOHEXYL)XANTHINE see PBU100
3,3-DIMETHYL-7-OXO-6-(2-PHENYLACETAMIDO)-4-THIA-1-
    AZABICYCLO(3.2.0)HEPTANE-2-CARBOXYLIC ACID compounded with
    EPHEDRINE (1:1) see PAQ120
2-(2,2-DIMETHYL-1-OXOPROPYL)-1H-INDENE-1,3(2H)-DIONE see PIH175
endo-8,8-DIMETHYL-3-((1-OXO-2-PROPYLPENTYL)OXY)-8-
    AZONIABICYCLO(3.2.1)OCTANE BROMIDE see LJS000
DIMETHYLOXOSTANNANE see DTH400
O,O-DIMETHYL-S-(3-OXO-3-THIA-PENTYL)-MONOTHIOPHOSPHAT
    (GERMAN) see DAP000
2,2-DIMETHYL-4-OXYMETHYL-1,3-DIOXOLANE see DVR600
α,Γ-DIMETHYL-α-OXYMETHYL GLUTARALDEHYDE see DTH600
1-(2,5-DIMETHYLOXYPHENYLAZO)-2-NAPHTHOL see DOK200
DIMETHYLOXYQUINAZINE see AQN000
O,O-DIMETHYL-1-OXY-2,2,2-TRICHLOROETHYL PHOSPHONATE see TIQ250
DIMETHYL PARANITROPHENYL THIONOPHOSPHATE see DTH800
DIMETHYL PARAOXON see PHD500
DIMETHYL PARATHION see MNH000
N,N-DIMETHYLPENTANAMIDE see DUN200
2,3-DIMETHYLPENTANE see DTI000
2,4-DIMETHYLPENTANE see DTI200
2,3-DIMETHYLPENTANOL see DTI400
2,3-DIMETHYL-1-PENTANOL see DTI400
2,4-DIMETHYL-3-PENTANONE see DTI600
S,S-DIMETHYLPENTASULFUR HEXANITRIDE see DTI709
DI(4-METHYL-2-PENTYL) MALEATE see DKP400
3,5-DIMETHYLPERHYDRO-1,3,5-THIADIAZIN-2-THION (CZECH, GERMAN)
    see DSB200
DIMETHYL PEROXIDE see DTJ000
DIMETHYLPEROXYCARBONATE see DTJ159
16,16-DIMETHYL-trans-Δ^2-PGE1 METHYL ESTER see CDB775

N,N¹-DIMETHYLPHAEANTHINE DIIODIDE see TDX835
1,4-DIMETHYLPHENANTHRENE see DTJ200
α,α-DIMETHYLPHENETHYL ACETATE see BEL750, DQQ375
α,α-DIMETHYLPHENETHYL ALCOHOL see DQQ200
α,α-DIMETHYLPHENETHYL ALCOHOL ACETATE see BEL750
α,α-DIMETHYLPHENETHYL ALCOHOL PROPIONATE see DQQ400
α,α-DIMETHYLPHENETHYLAMINE see DTJ400
(−)-N-α-DIMETHYLPHENETHYLAMINE HYDROCHLORIDE see MDQ500
N,α-DIMETHYLPHENETHYLAMINE HYDROCHLORIDE see DBA800
α-α-DIMETHYLPHENETHYL BUTYRATE see BEL850
3,4-DIMETHYLPHENISOPROPYLAMINE SULFATE see DTK200
DIMETHYLPHENOL see XKA000
2,3-DIMETHYLPHENOL see XKJ000
2,4-DIMETHYLPHENOL see XKJ500
2,5-DIMETHYLPHENOL see XKS000
2,6-DIMETHYLPHENOL see XLA000
3,4-DIMETHYLPHENOL see XLJ000
3,5-DIMETHYLPHENOL see XLS000
3,6-DIMETHYLPHENOL see XKS000
4,5-DIMETHYLPHENOL see XLJ000
4,6-DIMETHYLPHENOL see XKJ500
3,4-DIMETHYLPHENOL METHYLCARBAMATE see XTJ000
N,N-DIMETHYL-10H-PHENOTHIAZINE-10-PROPANAMINE see DQA600
N,N-DIMETHYL-3-PHENOTHIAZINESULFONAMIDE see DTK300
5-(2,5-DIMETHYLPHENOXY)-2,2-DIMETHYLPENTANOIC ACID (9CI)
    see GCK300
5-((3,5-DIMETHYLPHENOXY)METHYL)-2-OXAZOLIDINONE see XVS000
(2-(2,6-DIMETHYLPHENOXY)PROPYL)TRIMETHYLAMMONIUM CHLORIDE
    MONOHYDRATE see TLQ000
2-(2,6-DIMETHYLPHENOXY)-N,N,N-TRIMETHYL-ETHANAMINIUM BRO-
    MIDE (9CI) see XSS900
2-(2,6-DIMETHYLPHENOXY)-N,N,N-TRIMETHYL-1-PROPANAMINIUM HY-
    DRATE see TLQ000
α,α-DIMETHYLPHENRTHYL BUTYRATE see DQQ380
N,N-DIMETHYL-2-(2-PHENYLACETAMIDO)ACETAMIDE see PEB775
O,O-DIMETHYL-S-(PHENYLACETIC ACID ETHYL ESTER) PHOS-
    PHORODITHIOATE see DRR400
1-(5,7-DIMETHYL-2-PHENYL-1-ADAMANTYL)-N-METHYL-2-PROPYLAMINE
    HYDROCHLORIDE see DSO400
DIMETHYLPHENYLAMINE see XMA000, DQF800
2,3-DIMETHYLPHENYLAMINE see XMJ000
2,4-DIMETHYLPHENYLAMINE see XMS000
2,5-DIMETHYLPHENYLAMINE see XNA000
3,4-DIMETHYLPHENYLAMINE see XNS000
3,5-DIMETHYLPHENYLAMINE see XOA000
N,N-DIMETHYLPHENYLAMINE see DQF800
2-((2,3-DIMETHYLPHENYL)AMINO)BENZOIC ACID see XQS000
2-(2,6-DIMETHYLPHENYLAMINO)-4H-5,6-DIHYDRO-1,3-THIAZINE
    see DMW000
N-(2,6-DIMETHYLPHENYL)ANTHRANILIC ACID see XQS000
N-(2,6-DIMETHYLPHENYL)-2-AZABICYCLO(2.2.2)OCTANE-3-CARBOXAMIDE
    MONOHYDROCHLORIDE (9CI) see EAV100
2,3-DIMETHYL-4-PHENYLAZOANILINE see DTL000
N,N-DIMETHYL-p-PHENYLAZOANILINE see DOT300
N,N-DIMETHYL-p-PHENYLAZOANILINE-N-OXIDE see DTK600
N,N-DIMETHYL-4-PHENYLAZO-o-ANISIDINE see DTK800
N,N-DIMETHYL-4-(PHENYLAZO)BENZAMINE see DOT300
2,3-DIMETHYL-4-(PHENYLAZO)BENZENAMINE see DTL000
N,N-DIMETHYL-4-(PHENYLAZO)BENZENAMINE see DOT300
4-((2,4-DIMETHYLPHENYL)AZO)-3-HYDROXY-2,7-NAPHTHALENEDISULFO-
    NIC ACID, DISODIUM SALT see FMU070
4-((2,4-DIMETHYLPHENYL)AZO)-3-HYDROXY-2,7-NAPHTHALENEDISULPHO-
    NIC ACID, DISODIUM SALT see FMU070
1-((2,4-DIMETHYLPHENYL)AZO)-2-NAPHTHALENOL see XRA000
N,N-DIMETHYL-4-(PHENYLAZO)-m-TOLUIDINE see TLE750
N,N-DIMETHYL-4-(PHENYLAZO)-o-TOLUIDINE see MJF000
N,N-DIMETHYL-α-PHENYLBENZENEACETAMIDE see DRP800
(R)-N,N-DIMETHYL-α-PHENYLBENZENEETHANAMINE, HYDROCHLORIDE
    see DWA600
2-DI(N-METHYL-N-PHENYL-tert-BUTYL-CARBAMOYLMETHYL)
    AMINOETHANOL see DTL200
N-(2,6-DIMETHYLPHENYLCARBAMOYLMETHYL)-IMINODIACETIC ACID
    see LFO300
N-(N'-(2,6-DIMETHYLPHENYL)CARBAMOYLMETHYL)IMINODIACETIC
    ACID see LFO300
DIMETHYLPHENYLCARBINOL see DTN100
O,O-DIMETHYL-S-(PHENYL)(CARBOETHOXY)METHYL PHOS-
    PHORODITHIOATE see DRR400
N-(2,6-DIMETHYLPHENYL)-5,6-DIHYDRO-4H-1,3-THIAZIN-2-AMINE
    see DMW000
N-(2,6-DIMETHYLPHENYL)-5,6-DIHYDRO-4H-1,3-THIAZINE-2-AMINE (9CI)
    see DMW000
N'-(2,6-DIMETHYLPHENYL)-N-(((2,4-DIMETHYLPHENYL)IMINO)METHYL)-N-
    METHYLMETHANIMIDAMIDE see MJL250

1,3-DIMETHYL-3-PHENYL-2,5-DIOXOPYRROLIDINE see MLP800
DIMETHYL-4,4'-o-PHENYLENE-BIS-(3-THIOALLOPHANATE) see PEX500
DIMETHYL-p-PHENYLENEDIAMINE see DTL600, DTL800
N,N-DIMETHYL-p-PHENYLENEDIAMINE see DTL600, DTL800
N,N-DIMETHYL-p-PHENYLENEDIAMINE DIHYDROCHLORIDE see DTM000
N,N-DIMETHYL-p-PHENYLENEDIAMINE HEMISULFATE see DTM200
DIMETHYL-p-PHENYLENEDIAMINE HYDROCHLORIDE see DTM000
N,N-DIMETHYL-p-PHENYLENEDIAMINE MONOHYDROCHLORIDE
    see DTM400
1,1-DIMETHYL-2-PHENYLETHANAMINE see DTJ400
1,1-DIMETHYL-2-PHENYLETHANOL see DQQ200
(DIMETHYL-S-(PHENYLETHOXYCARBONYLMETHYL)PHOS-
    PHOROTHIOLOTHIONATE) see DRR400
α,α-DIMETHYL-β-PHENYLETHYLAMINE see DTJ400
DIMETHYLPHENYLETHYL CARBINOL see BEC250
DIMETHYLPHENYLETHYLCARBINYL ACETATE see MNT000
o,p-DIMETHYL-β-PHENYLETHYLHYDRAZINE DIHYDROGEN SULFATE
    see DTM600
β-(2,4-DIMETHYLPHENYL)ETHYLHYDRAZINE DIHYDROGEN SULPHATE
    see DTM600
DIMETHYLPHENYLETHYNYLTHALLIUM see DTM800
N,N-DIMETHYL-N'-PHENYL-N'-FLUORODICHLOROMETHYLTHIOSULFAM-
    IDE see DFL200
N,N-DIMETHYL-3-PHENYL-1-INDANAMINE see DRX400
2,4-DIMETHYLPHENYLMALEIMIDE see DTN000
2,4-DIMETHYL-N-PHENYLMALEIMIDE see DTN000
DIMETHYLPHENYLMETHANOL see DTN100
3,4-DIMETHYLPHENYL-N-METHYLCARBAMATE see XTJ000
3,5-DIMETHYLPHENYL-N-METHYLCARBAMATE see DTN200
N,N-DIMETHYL''-(PHENYLMETHYL)GUANIDINE SULPHATE (2:1)
    see BFW250
N,N-DIMETHYL-3((1-PHENYLMETHYL)-1H-INDAZOL-3-YL)OXY)-1-
    PROPANAMINE HYDROCHLORIDE see BBW500
3,4-DIMETHYLPHENYL-N-METHYL-N-NITROSOCARBAMATE see DTN400
N-(2,6-DIMETHYLPHENYL)-1-METHYL-2-PIPERIDINECARBOXAMIDE-
    MONOHYDROCHLORIDE see CBR250
N-(2,6-DIMETHYLPHENYL)-N'-(1-METHYL-2-PYRROLIDINYLIDENE)UREA
    see XGA725
5,5-DIMETHYL-2-PHENYLMORPHOLINE see DTN775
3,4-DIMETHYL-2-PHENYLMORPHOLINE BITARTRATE see DKE800
3,4-DIMETHYL-2-PHENYLMORPHOLINEHYDROCHLORIDE see DTN800
1,3-DIMETHYL-3-PHENYL-1-NITROSOUREA see DTN875
(3,4-DIMETHYLPHENYL)OXIRENE see EBT000
N,N-DIMETHYL-N'-PHENYL-N'-(PHENYLMETHYL)-1,2-ETHANEDIAMINE
    (9CI) see BEM500
DIMETHYLPHENYLPHOSPHINE see DTN896
1,1-DIMETHYL-4-PHENYLPIPERAZINE IODIDE see DTO000
1,1-DIMETHYL-4-PHENYLPIPERAZINIUM IODIDE see DTO000
1,1-DIMETHYL-4-PHENYLPIPERIDINIUM IODIDE see DTO100
1,3-DIMETHYL-4-PHENYL-4-PIPERIDINOL PROPIONATE (ESTER)
    see NEA500
1,3-DIMETHYL-4-PHENYL-4-PIPERIDINOL, PROPIONATE, HYDROCHLO-
    RIDE see NEB000
dl-1,3-DIMETHYL-4-PHENYL-4-PIPERIDINOL PROPIONATE HYDROCHLO-
    RIDE see NEB000
α-1,3-DIMETHYL-4-PHENYL-4-PIPERIDINYL PROPIONATE see NEA500
(±)-1,3-DIMETHYL-4-PHENYL-4-PIPERIDYL ESTER PROPIONIC ACID HY-
    DROCHLORIDE see NEB000
1,3-DIMETHYL-4-PHENYL-4-PIPERIDYL PROPIONATE HYDROCHLORIDE
    see NEB000
1,1-DIMETHYL-3-PHENYLPROPANOL see BEC250
1,1-DIMETHYL-3-PHENYL-1-PROPANOL see BEC250
1,3-DIMETHYL-4-PHENYL-4-PROPIONOXYPIPERIDINE see NEA500
α-1,3-DIMETHYL-4-PHENYL-4-PROPIONOXYPIPERIDINE see NEA500
(±)-α-1,3-DIMETHYL-4-PHENYL-4-PROPIONOXYPIPERIDINE HYDROCHLO-
    RIDE see NEB000
α,α-DIMETHYL-Δ-PHENYLPROPYL ALCOHOL see BEC250
(1,1-DIMETHYL-3-PHENYLPROPYL)ESTER ACETIC ACID see MNT000
2,3-DIMETHYL-1-PHENYL-3-PYRAZOLIN-5-ONE see AQN000
2,3-DIMETHYL-1-PHENYL-5-PYRAZOLONE see AQN000
N,N-DIMETHYL-2-(1-PHENYL-1-(2-PYRIDINYL)ETHOXY)ETHANAMINE (9CI)
    see DYE500
Z-(±)-2,6-DIMETHYL-α-PHENYL-α-(2-PYRIDYL)-1-PIPERIDINEBUTANOL
    HYDROCHLORIDE see PJA170
(±)-cis-2,6-DIMETHYL-α-PHENYL-α-2-PYRIDYL-1-PIPERIDINEBUTANOL
    MONOHYDROCHLORIDE see PJA170
N,N-DIMETHYL-3-PHENYL-3-(2-PYRIDYL)PROPYLAMINE see TMJ750
1,3-DIMETHYL-3-PHENYL-PYRROLIDIN-2,5-DIONE see MLP800
trans-3,4-DIMETHYL-3-PHENYL-4-(p-(β-PYRROLIDINOETHOXY)PHENYL)-7-
    METHOXYCHROMAN see CCW725
3,4-trans-2,2-DIMETHYL-3-PHENYL-4-(p-(β-PYRROLIDINOETHOXY)PHENYL)-
    7-METHOXYCHROMAN see CCW725
3,4-trans-2,2-DIMETHYL-3-PHENYL-4-p-(β-PYRROLIDINOETHOXY)PHENYL-7-
    METHOXYCHROMAN HCl see CCW750

1,2-DIMETHYL-3-PHENYL-3-PYRROLIDINOL PROPIONATE (ester) see DTO200
1,2-DIMETHYL-3-PHENYL-3-PYRROLIDYL PROPIONATE see DTO200
N,2-DIMETHYL-2-PHENYLSUCCINIMIDE see MLP800
N,N-DIMETHYL-2-(α-PHENYL-o-TOLOXY)ETHYLAMINE DIHYDROGEN CITRATE see DTO600
N,N-DIMETHYL-2-(α-PHENYL-o-TOLOXY)ETHYLAMINE HYDROCHLORIDE see DTO800
3,3-DIMETHYL-1-PHENYLTRIAZENE see DTP000
3,3-DIMETHYL-1-PHENYL-1-TRIAZENE see DTP000
1,1-DIMETHYL-3-PHENYLUREA see DTP400
N,N-DIMETHYL-N'-PHENYLUREA see DTP400
DIMETHYL PHOSPHATE of 2-CHLORO-N,N-DIETHYL-3-HYDROXYCROTONAMIDE see FAB400
DIMETHYL PHOSPHATE ESTER with 2-CHLORO-N-ETHYL-3-HYDROXYCROTONAMIDE see DTP600
DIMETHYL PHOSPHATE ESTER with 2-CHLORO-N-METHYL-3-HYDROXYCROTONAMIDE see DTP800
DIMETHYLPHOSPHATE ESTER with 3-HYDROXY-N,N-DIMETHYL-cis-CROTONAMIDE see DGQ875
DIMETHYL PHOSPHATE ESTER of 3-HYDROXY-N-METHYL-cis-CROTONAMIDE see MRH209
DIMETHYL PHOSPHATE-3-HYDROXY-CROTONIC ACID, p-CHLOROBENZYL ESTER see CEQ500
DIMETHYL PHOSPHATE of 3-HYDROXY-N,N-DIMETHYL-cis-CROTONAMIDE see DGQ875
DIMETHYL PHOSPHATE of α-METHYLBENZYL-3-HYDROXY-cis-CROTONATE see COD000
DIMETHYL PHOSPHATE OF 3-HYDROXY-N-METHYL-cis-CROTONAMINE see MRH209
DIMETHYL PHOSPHINE see DTQ089
DIMETHYLPHOSPHORAMIDOCYANIDIC ACID, ETHYL ESTER see EIF000
O,S-DIMETHYL PHOSPHORAMIDOTHIOATE see DTQ400
O,O-DIMETHYLPHOSPHOROCHLORIDOTHIOATE see DTQ600
DIMETHYL PHOSPHOROCHLORIDOTHIOATE (DOT) see DTQ600
O,O-DIMETHYL PHOSPHORODITHIOATE N-FORMYL-2-MERCAPTO-N-METHYLACETAMIDE-S-ESTER see DRR200
S-(O,O-DIMETHYLPHOSPHORODITHIOATE) of N-(2-MERCAPTOETHYL)ETHYLCARBAMATE see EMC000
N-((O,O-DIMETHYLPHOSPHORODITHIOYL)ETHYL)ACETAMIDE see DOP200
O,O-DIMETHYL PHOSPHOROTHIOATE-O,O-DIESTER with 4,4'-THIODIPHENOL see TAL250
O,O-DIMETHYL PHOSPHOROTHIOATE-O-ESTER with 4-HYDROXY-m-ANISONITRILE see DTQ800
5-(O,O-DIMETHYLPHOSPHORYL)-6-CHLOROBICYCLO(3.2.0)HEPTA-1,5-DIEN see HBK700
DIMETHYL PHTHALATE see DTR200
(O,O-DIMETHYL-PHTHALIMIDIOMETHYL-DITHIOPHOSPHATE) see PHX250
2,5-DIMETHYLPIPERAZINE see DTR400
α,4-DIMETHYL-1-PIPERAZINEACETIC ACID-6-CHLORO-o-TOLYL ESTER DIHYDROCHLORIDE see FAC185
α,4-DIMETHYL-1-PIPERAZINEACETIC ACID-2,6-DIISOPROPYLPHENYL ESTER DIHYDROCHLORIDE see FAC160
2-β,16-β-(4'-DIMETHYL-1'-PIPERAZINO)-3-α,17-β-DIACETOXY-5-α-ANDROSTANE 2BR see PII250
2-(2,6-DIMETHYLPIPERIDINO)-2',6'-ACETOXYLIDIDE HYDROCHLORIDE see DTR800
2,4'-DIMETHYL-3-PIPERIDINOPROPIOPHENONE see TGK200
2,4'-DIMETHYL-3-PIPERIDINOPROPIOPHENONE HYDROCHLORIDE see MRW125
N,N-DIMETHYL-4-PIPERIDYLIDENE-1,1-DIPHENYLMETHANE METHYLSULFATE see DAP800
DIMETHYLPOLYSILOXANE see DTR850
6,17-DIMETHYLPREGNA-4,6-DIENE-3,20-DIONE see MBZ100
2,2-DIMETHYLPROPANE (DOT) see NCH000
N,N-DIMETHYL-1,3-PROPANEDIAMINE see AJQ100
DIMETHYL PROPANEDIOATE see DSM200
2,2-DIMETHYL-1,3-PROPANEDIOL see DTG400
2,2-DIMETHYL-1,3-PROPANEDIOL DIACRYLATE see DUL200
2,2'-((2,2-DIMETHYL-1,3-PROPANEDIYL)BIS(OXYMETHYLENE))BISOXIRANE see NCI300
2,2-DIMETHYLPROPANOIC ACID see PJA500
2,2-DIMETHYLPROPANOIC ACID-3-(2-(ETHYLAMINO)-1-HYDROXYETHYL)PHENYL ESTER HYDROCHLORIDE see EGC500
(±)-2,2-DIMETHYL-PROPANOIC ACID-4-(1-HYDROXY-2-(METHYLAMINO)ETHYL)-1,2-PHENYLENE ESTER, HYDROCHLORIDE see DWP559
2,2-DIMETHYLPROPANOIC ACID ISOOCTADECYL ESTER see ISC550
2,2-DIMETHYL-PROPANOIC ACID-2-OXO-2-PHENYLETHYL ESTER (9CI) see PCV350
2,2-DIMETHYLPROPANOYL CHLORIDE see DTS400
1,1-DIMETHYLPROPARGYL ALCOHOL see MHX250
α,α-DIMETHYLPROPARGYL ALCOHOL see MHX250
N,N-DIMETHYL-2-PROPENAMIDE see DOP800
DIMETHYL PROPIOLACTONE see DTH000

3,3-DIMETHYL-β-PROPIOLACTONE see DTH000
N,N-DIMETHYLPROPIONAMIDE see DTS600
2,2-DIMETHYLPROPIONIC ACID see PJA500
α,α-DIMETHYLPROPIONIC ACID see PJA500
2,2-DIMETHYLPROPIONYL CHLORIDE see DTS400
2,6-DIMETHYL-4-PROPOXY-BENZOIC ACID 2-METHYL-2-(1-PYRROLIDINYL)PROPYL ESTER see UAG050
2,6-DIMETHYL-4-PROPOXY-BENZOIC ACID 2-METHYL-2-(1-PYRROLIDINYL)PROPYLESTER see DTS625
2,6-DIMETHYL-4-PROPOXY-BENZOIC ACID 2-(1-PYRROLIDINYL)PROPYL ESTER HYDROCHLORIDE see UAG075
7,12-DIMETHYL-8-PROPYL-BENZ(a)ANTHRACENE see PNI500
4-(1,1-DIMETHYLPROPYL)CYCLOHEXANONE see AOH750
N,N-DIMETHYL-1,3-PROPYLENEDIAMINE see AJQ100
DIMETHYLPROPYLENEUREA see DSE489
1,1-DIMETHYLPROPYL HYDROPEROXIDE see PBX325
5-(3-DIMETHYLPROPYLIDENE)DIBENZO(a,d)(1,4)CYCLOHEPTADIENE see EAH500
p-(1,1-DIMETHYLPROPYL)PHENOL see AON000
p-(α,α-DIMETHYLPROPYL)PHENOL see AON000
N,N-DIMETHYL-p-((p-PROPYLPHENYL)AZO)ANILINE see DTT400
(±)-3-(1,3-DIMETHYL-4-PROPYL-4-PIPERIDINYL)PHENOL HYDROCHLORIDE see PIB700
trans-(±)-3-(1,3-α-DIMETHYL-4-α-PROPYL-4-β-PIPERIDINYL)PHENOL HYDROCHLORIDE see PIB700
m-(1,2-DIMETHYL-3-PROPYL-3-PYRROLIDINYL)PHENOL see DSI000
(−)-N,α-DIMETHYL-N-2-PROPYNYLBENZENEETHANAMINE HYDROCHLORIDE see DAZ125
N-(1,1-DIMETHYLPROPYNYL)-3,5-DICHLOROBENZAMIDE see DTT600
(+)-N,α-DIMETHYL-N-2-PROPYNYLPHENETHYLAMINE HYDROCHLORIDE see DAZ120
(±)-N,α-DIMETHYL-N-2-PROPYNYLPHENETHYLAMINE HYDROCHLORIDE see DAZ118
DIMETHYL-1-PROPYNYLTHALLIUM see DTT800
16,16-DIMETHYL-trans-Δ2-PROSTAGLANDIN E1 METHYL ESTER see CDB775
3,5-DIMETHYL-4H-PYRAN-4-ONE-2-METHOXY-6-(TETRAHYDRO-4-β-METHYL-p-NITROCINNAMYLIDENE)-2-FURYL) see DTU200
2,3-DIMETHYLPYRAZINE see DTU400
2,5-DIMETHYLPYRAZINE see DTU600
2,6-DIMETHYLPYRAZINE see DTU800
2,5-DIMETHYLPYRIDINE see LJA000
3,4-DIMETHYLPYRIDINE see LJB000
2,6-DIMETHYLPYRIDINE-N-OXIDE see DTV089
2,6-DIMETHYLPYRIDINE-1-OXIDE-4-AZO-p-DIMETHYLANILINE see DPP800
N,N-DIMETHYL-3-(1-(2-PYRIDINYL)ETHYL)-1H-INDENE-2-ETHANAMINE (Z)-2-BUTENEDIOATE (1:1) see FMU409
N1-(4,6-DIMETHYL-2-PYRIDINYL)SULFANILAMIDE, MONOSODIUM SALT see SJW500
N,N-DIMETHYL-N'-2-PYRIDINYL-N'-(2-THIENYLMETHYL)-1,2-ETHANEDIAMIDE see TEO250
N,N-DIMETHYL-N'-2-PYRIDINYL-N'-(2-THIENYLMETHYL)-1,2-ETHANEDIAMINE MONOHYDROCHLORIDE see DPJ400
5,11-DIMETHYL-6H-PYRIDO(4,3-b)CARBAZOL-9-AMINE see AJS875
5,11-DIMETHYL-6H-PYRIDO(4,3-b)CARBAZOLE see EAI850
5,11-DIMETHYL-6H-PYRIDO(4,3-b)CARBAZYL-9-OL see HKH000
1,4-DIMETHYL-5H-PYRIDO(4,3-b)INDOL-3-AMINE see TNX275
1,4-DIMETHYL-5H-PYRIDO(4,3-b)INDOL-3-AMINE ACETATE see AJR500
1,4-DIMETHYL-5H-PYRIDO(4,3-b)INDOL-3-AMINE MONOACETATE see AJR500
N,N-DIMETHYL-p-(3-PYRIDYLAZO)ANILINE see POP750
N,N-DIMETHYL-4-(3'-PYRIDYLAZO)ANILINE see POP750
N,N-DIMETHYL-N'-(2-PYRIDYL)-N'-BENZYLETHYLENEDIAMINE HYDROCHLORIDE see POO750
N,N-DIMETHYL-N'-(2-PYRIDYL)-N'-(5-CHLORO-2-THENYL)ETHYLENEDIAMINE see CHY250
(3,3-DIMETHYL-1-(m-PYRIDYL-N-OXIDE))TRIAZENE see DTV200
N,N-DIMETHYL-N'-PYRID-2-YL-N'-2-THENYLETHYLENEDIAMINE see TEO250
N,N-DIMETHYL-N'-(2-PYRIDYL)-N'-THENYLETHYLENEDIAMINE HYDROCHLORIDE see DPJ400
S-(4,6-DIMETHYL-2-PYRIMIDINYL)-O,O-DIETHYL PHOSPHORODITHIOATE see DTV400
N1-(2,6-DIMETHYL-4-PYRIMIDINYL)SULFANILAMIDE see SNJ350
N1-(4,6-DIMETHYL-2-PYRIMIDINYL)SULFANILAMIDE see SNJ000
(N1-(4,6-DIMETHYL-2-PYRIMIDINYL)SULFANILAMIDO) SODIUM see SJW500
6,8-DIMETHYLPYRIMIDO(5,4-e)-as-TRIAZINE-5,7(6H,8H)-DIONE see FBP300
6,8-DIMETHYL-PYRIMIDO(5,4-e)-1,2,4-TRIAZINE-5,7(6H,8H)-DIONE see FBP300
N-(4,6-DIMETHYL-2-PYRIMIDYL)SULFANILAMIDE see SNJ000
1,3-DIMETHYL PYROGALLATE see DOJ200
20-(2,4-DIMETHYL-1H-PYRROLE-3-CARBOXYLATE) BATRACHOTOXININ A see BAR750
20-α-(2,4-DIMETHYL-1H-PYRROLE-3-CARBOXYLATE) BETRACHOTOXININ A see BAR750
3,4-DIMETHYLPYRROLIDINE ETHANOL see HKR550
2,4-DIMETHYLPYRROL-3-YL METHYL KETONE see ACI500

N-(2,5-DIMETHYL-1H-PYRROL-1-YL)-6-(4-MORPHOLINYL)-3-PYRIDAZINAM-
INE HYDROCHLORIDE see MBV735
N,N-DIMETHYL-4-(4'-QUINOLYLAZO)ANILINE see DTY200
N,N-DIMETHYL-4-(5'-QUINOLYLAZO)ANILINE see DPQ800
N,N-DIMETHYL-4-(6'-QUINOLYLAZO)ANILINE see DPR000
N,N-DIMETHYL-p-(5'-QUINOLYLAZO)ANILINE see DPQ800
N,N-DIMETHYL-4-(5'-QUINOLYLAZO)-m-TOLUIDINE see DQE400
DIMETHYLQUINOLYL METHYLSULFATE UREA see PJA120
N,N-DIMETHYL-4-((5'-QUINOLYL-1'-OXIDE)AZO)ANILINE
see DPR200
N,N-DIMETHYL-4-((4'-QUINOLYL-1'-OXIDE)AZO)ANILINE
see DTY400
N,N'-DIMETHYL-4-((6'-QUINOLYL-1'-OXIDE)AZO)ANILINE see DPR400
3,3-DIMETHYL-1(3-QUINOLYL)TRIAZENE see DTY600
N,N-DIMETHYL-p-(6-QUINOXALINYLAZO)ANILINE see DUA400
N,N-DIMETHYL-p-(6-QUINOXALYAZO)ANILINE see DUA400
N,N-DIMETHYL-p-(5-QUINOXALYLAZO)ANILINE see DUA200
6,7-DIMETHYL-9-d-RIBITYLISOALLOXAZINE see RIK000
7,8-DIMETHYL-10-d-RIBITYLISOALLOXAZINE see RIK000
7,8-DIMETHYL-10-(d-RIBO-2,3,4,5-TETRAHYDROXYPENTYL)-4a,5-
DIHYDROISOALLOXAZINE see DUA600
7,8-DIMETHYL-10-(d-RIBO-2,3,4,5-TETRAHYDROXYPENTYL)ISOALLOXA-
INE see RIK000
N,N-DIMETHYLSALICYLAMIDE see DUA800
DIMETHYL SELENATE see DUB000
DIMETHYL SELENIDE see DUB200
DIMETHYLSELENIUM see DUB200
N,N-DIMETHYLSEROTONIN see DPG109
DIMETHYLSILBOESTROL see DUC300
DIMETHYL SILICONE see DTR850
DIMETHYL SILOXANE see DUB600
(DIMETHYL SILYLMETHYL)TRIMETHYL LEAD see DUB689
N,N-DIMETHYLSTEARAMIDE see DTC200
N,N-DIMETHYL-4-STILBENAMINE see DUB800
(E)-N,N-DIMETHYL-4-STILBENAMINE see DUC000
(Z)-N,N-DIMETHYL-4-STILBENAMINE see DUC200
cis-N,N-DIMETHYL-4-STILBENAMINE see DUC200
trans-N,N-DIMETHYL-4-STILBENAMINE see DUC000
(E)-α,α'-DIMETHYL-4,4'-STILBENEDIOL see DUC300
(E)-α,α'-DIMETHYL-4,4'-STILBENEDIOL DIACETATE (ester) see DXS300
DIMETHYLSTILBESTROL see DUC300
trans-DIMETHYLSTILBESTROL DIACETATE see DXS300
trans-DIMETHYLSTILBOESTROL DIACETATE see DXS300
3,4-DIMETHYL STYRENE OXIDE see EBT000
N,N-DIMETHYL-p-STYRYLANILINE see DUB800
DIMETHYLSULFAAT (DUTCH) see DUD100
DIMETHYLSULFAMIDO-3-(DIMETHYLAMINO-2-PROPYL)-10-PHENOTHI-
AZINE see DUC400
2-(DIMETHYLSULFAMOYL)-(9-(4-METHYL-1-PIPERAZINYL)
PROPYLIDENE)THIOXANTHENE see NBP500
p-(N,N-DIMETHYLSULFAMOYL)PHENOL see DUC600
O,O-DIMETHYL-O,p-SULFAMOYLPHENYL PHOSPHOROTHIOATE
see CQL250
3,4-DIMETHYL-5-SULFANILAMIDOISOXAZOLE see SNN500
4,5-DIMETHYL-2-SULFANILAMIDOOXAZOLE see AIE750
2,4-DIMETHYL-6-SULFANILAMIDOPYRIMIDINE see SNJ350
2,6-DIMETHYL-4-SULFANILAMIDOPYRIMIDINE see SNJ350
4,6-DIMETHYL-2-SULFANILAMIDOPYRIMIDINE see SNJ000
N,N-DIMETHYLSULFANILIC ACID see DUD000
DIMETHYLSULFAT (CZECH) see DUD100
DIMETHYL SULFATE see DUD100
DIMETHYLSULFID (CZECH) see TFP000
DIMETHYL SULFIDE (DOT) see TFP000
DI-METHYLSULFIDE BORANE see MPL250
DIMETHYLSULFIDE-α,α'-DICARBOXYLIC ACID see MCM750
2-(7-(1,1-DIMETHYL-3-(4-SULFOBUTYL)BENZ(e)INDOLIN-2-YLIDENE)-1,3,5-
HEPTATRIENYL)-1,1-DIMETHYL-3-(4-SULFOBUTYL)1H-BENZ(e)INDOL-
IUM IODIDE, INNER SALT, SODIUM SALT see ICL000
2,4-DIMETHYL SULFOLANE see DUD400
3-DIMETHYLSULFONAMIDO-10-(2-DIMETHYLAMINOPROPYL)
PHENOTHIAZINE see DUC400
1,4-DIMETHYLSULFONOXYBUTANE see BOT250
3-((2,4-DIMETHYL-5-SULFOPHENYL)AZO)-4-HYDROXY-1-NAPH-
THALENESULFONIC ACID, DISODIUM SALT see FAG050
DIMETHYL SULFOXIDE see DUD800
3-DIMETHYLSULPHAMIDOPHENOTHIAZINE see DTK300
3,4-DIMETHYL-5-SULPHANILAMIDOISOXAZOLE see SNN500
as-DIMETHYL SULPHATE see MLH500
DIMETHYL SULPHIDE see TFP000
3,4-DIMETHYL-5-SULPHONAMIDOISOXAZOLE see SNN500
3-((2,4-DIMETHYL-5-SULPHOPHENYL)AZO)-4-HYDROXY-1-NAPH-
THALENESULPHONIC ACID, DISODIUM SALT see FAG050
DIMETHYL SULPHOXIDE see DUD800
DIMETHYL TEREPHTHALATE see DUE000

2,6-DIMETHYL-4-TERTIARYBUTYL-3-HYDROXYPHENYL)
METHYLIMIDAZOLINE HYDROCHLORIDE see AEX000
DIMETHYL-1,2,2,2-TETRACHLOROETHYL PHOSPHATE see DUE600
DIMETHYL TETRACHLOROTEREPHTHALATE see TBV250
DIMETHYL 2,3,5,6-TETRACHLOROTEREPHTHALATE see TBV250
N,N-DIMETHYLTETRADECANAMIDE see DSU200
N,N-DIMETHYL-N-TETRADECYLBENZENEMETHANAMINIUM, CHLORIDE
(9CI) see TCA500
7,12-DIMETHYL-1,2,3,4-TETRAHYDROBENZ(a)ANTHRACENE see TCP600
7,12-DIMETHYL-8,9,10,11-TETRAHYDROBENZ(a)ANTHRACENE see DUF000
1,11-DIMETHYL-1,2,3,4-TETRAHYDROCHRYSENE see DUF200
2,6-DIMETHYL-2,3,5,6-TETRAHYDRO-4H-1,4-OXAZINE see DST600
DIMETHYL TETRAHYDROPHTHALATE see DUF400
DIMETHYLTETRAHYDROPYRONE see TCQ350
2,6-DIMETHYL TETRAHYDRO-1,4-PYRONE see TCQ350
3,5-DIMETHYLTETRAHYDRO-1,3,5-THIADIAZINE-2-THIONE see DSB200
3,5-DIMETHYLTETRAHYDRO-1,3,5-2H-THIADIAZINE-2-THIONE see DSB200
3,5-DIMETHYL-1,3,5-2H-TETRAHYDROTHIADIAZINE-2-THIONE see DSB200
3,5-DIMETHYLTETRAHYDRO-2H-1,3,5-THIADIAZINE-2-THIONE see DSB200
3,5-DIMETHYL-1,2,3,5-TETRAHYDRO-1,3,5-THIADIAZINETHIONE-2
see DSB200
4,4'-(2,3-DIMETHYLTETRAMETHYLENE)DIPYROCATECHOL see NBR000
2,2'-DIMETHYLTETRANDRINIUM DIIODIDE see TDX835
3,6-DIMETHYL-1,2,4,5-TETRAOXANE see DUF800
DIMETHYLTHALLIUM FULMINATE see DUG000
DIMETHYLTHALLIUM-N-METHYLACETOHYDROXAMATE see DUG089
N,N-DIMETHYL-N'-(2-THENYL)-N'-(2-PYRIDYL-ETHYLENE-DIAMINE
HYDROCHLORIDE) see DPJ400
N,N-DIMETHYL-N'-(3-THENYL)-N'-(2-PYRIDYL) ETHYLENEDIAMINE
HYDROCHLORIDE see TEO000
DIMETHYLTHIAMBUTENE HYDROCHLORIDE see TLQ250
O,O-DIMETHYL-S-(3-THIA-PENTYL)-MONOTHIOPHOSPHAT (GERMAN)
see DAP400
2,4-DIMETHYLTHIAZOLE see DUG200
3-(4,5-DIMETHYLTHIAZOLYL-2)-2,5-DIPHENYLTETRAZOLIUM BROMIDE
see DUG400
DIMETHYLTHIOCARBAMIDE see DSK900
N,N'-DIMETHYLTHIOCARBAMIDE see DSK900
m,N-DIMETHYLTHIOCARBANILIC ACID-o-2 NAPHTHYL ESTER see TGB475
2,2'-DIMETHYLTHIOCARBANILIDE see DXP600
DIMETHYLTHIOMETHYLPHOSPHATE see DUG500
2,2-DIMETHYL-3-THIOMORPHOLINONE see DUG600
2,2-DIMETHYL-3-THIOMORPHOLONE see DUG600
3,5-DIMETHYL-2-THIONOTETRAHYDRO-1,3,5-THIADIAZINE
see DSB200
O,O-DIMETHYLTHIOPHOSPHORIC ACID, p-CHLOROPHENYL ESTER
see MQH750
1,3-DIMETHYLTHIOUREA see DSK900
sym-DIMETHYLTHIOUREA see DSK900
DIMETHYLTIN BIS(DIBUTYLDITHIOCARBAMATE) see BIW500
DIMETHYL-TIN BIS(ISOOCTYLTHIOGLYCOLLATE) see BKK500
DIMETHYLTIN DIBROMIDE see DUG800
DIMETHYLTIN DIFLUORIDE see DKH200
DIMETHYLTIN DINITRATE see DUG889
DIMETHYLTIN FLUORIDE see DKH200
DIMETHYLTIN OXIDE see DTH400
N,N-DIMETHYL-p-TOLUENESULFONAMIDE see DUH000
DIMETHYL-o-TOLUIDINE see DUH200
N,N-DIMETHYL-o-TOLUIDINE see DUH200
DIMETHYLTOLUTHIONINE CHLORIDE see AJP250
N,N-DIMETHYL-4-(p-TOLYLAZO)ANILINE see DUH400
N,N-DIMETHYL-p-(m-TOLYLAZO)ANILINE see DUH600
N,N-DIMETHYL-p-((o-TOLYL)AZO)ANILINE see DUH800
N,N-DIMETHYL-2-(α-(p-TOLYL)BENZYLOXY)ETHYLAMINE HYDROCHLO-
RIDE see TGJ475
N,N-DIMETHYL-N'-(4-TOLYL)-N'-(DICHLORFLUORMETHYLTHIO)
SULFAMID (GERMAN) see DFL400
N,N-DIMETHYL-N-(4-TOLYL)-N-(DICHLOROFLUOR-METHYLTHIO)
SULFAMIDE see DFL400
3,3-DIMETHYL-1-(m-TOLYL)TRIAZENE see DSR200
3,3-DIMETHYL-1-(o-TOLYL)TRIAZENE see MNT500
4'-DIMETHYLTRIAZENOACETANILIDE see DUI000
4'-(3,3-DIMETHYL-1-TRIAZENO)ACETANILIDE see DUI000
4'-(3-(3,3-DIMETHYL-1-TRIAZENO)-9-ACRIDINYLAMINO)
METHANESULFONANILIDE see DUI200
p-(3,3-DIMETHYLTRIAZENO)BENZAMIDE see CCC325
(DIMETHYLTRIAZENO)IMIDAZOLECARBOXAMIDE see DAB600
4-(DIMETHYLTRIAZENO)IMIDAZOLE-5-CARBOXAMIDE see DAB600
5-(DIMETHYLTRIAZENO)IMIDAZOLE-4-CARBOXAMIDE see DAB600
5-(3,3-DIMETHYLTRIAZENO)IMIDAZOLE-4-CARBOXAMIDE see DAB600
4-(3,3-DIMETHYL-1-TRIAZENO)IMIDAZOLE-5-CARBOXAMIDE see DAB600
5-(3,3-DIMETHYL-1-TRIAZENO)IMIDAZOLE-4-CARBOXAMIDE see DAB600
4-(5)-(3,3-DIMETHYL-1-TRIAZENO)IMIDAZOLE-5(4)-CARBOXAMIDE
see DAB600

5-(3,3-DIMETHYL-1-TRIAZENO)IMIDAZOLE-4-CARBOXAMIDE CITRATE see DUI400
p-(3,3-DIMETHYLTRIAZENO)PHENOL see DUI600
3-(3′,3′-DIMETHYLTRIAZENO)PYRIDINE-N-OXIDE see DTV200
3-(3′,3′-DIMETHYLTRIAZENO)-PYRIDIN-N-OXID (GERMAN) see DTV200
4-(3,3-DIMETHYL-1-TRIAZENYL)BENZAMIDE see CCC325
p-(3,3-DIMETHYL-1-TRIAZENYL)BENZAMIDE see CCC325
5-(3,3-DIMETHYL-1-TRIAZENYL)-1H-IMIDAZOLE-4-CARBOXAMIDE see DAB600
4-(3,3-DIMETHYL-1-TRIAZENYL)PHENOL see DUI600
N-(4-(3,3-DIMETHYL-1-TRIAZENYL)PHENYL)ACETAMIDE see DUI000
1,3-DIMETHYLTRIAZINE see DUI709
3,3-DIMETHYL-1-(2,4,6-TRIBROMOPHENYL)TRIAZENE see DUI800
O,O-DIMETHYL-(2,2,2-TRICHLOOR-1-HYDROXY-ETHYL)-FOSFONAAT (DUTCH) see TIQ250
O,O-DIMETHYL-(2,2,2-TRICHLOR-1-HYDROXY-AETHYL)PHOSPHONAT (GERMAN) see TIQ250
O,O-DIMETHYL 2,2,2-TRICHLORO-1-(N-BUTYRYLOXY)ETHYLPHOSPHONATE see BPG000
DIMETHYLTRICHLOROHYDROXYETHYL PHOSPHONATE see TIQ250
DIMETHYL-2,2,2-TRICHLORO-1-HYDROXYETHYLPHOSPHONATE see TIQ250
O,O-DIMETHYL-2,2,2-TRICHLORO-1-HYDROXYETHYL PHOSPHONATE see TIQ250
3,5-DIMETHYL-1-(TRICHLOROMETHYLMERCAPTO)PYRAZOLE see DUJ000
O,O-DIMETHYL-O-2,4,5-TRICHLOROPHENYL PHOSPHOROTHIOATE see RMA500
DIMETHYL TRICHLOROPHENYL THIOPHOSPHATE see RMA500
O,O-DIMETHYL-O-(2,4,5-TRICHLOROPHENYL)THIOPHOSPHATE see RMA500
3,3-DIMETHYL-1-(2,4,6-TRICHLOROPHENYL)-TRIAZINE see TJA000
DIMETHYL-3,5,6-TRICHLOROPYRIDYL PHOSPHATE see PHE250
DIMETHYL-3,5,6-TRICHLORO-2-PYRIDYL PHOSPHATE see PHE250
O,O-DIMETHYL-O-(3,5,6-TRICHLORO-2-PYRIDYL)PHOSPHOROTHIOATE see CMA250
O,O-DIMETHYL-O-(2,4,5-TRICHLORPHENYL)-THIONOPHOSPHAT(GERMAN) see RMA500
5-(2,3-DIMETHYLTRICYCLO(2.2.1.0^{2,6})HEPT-3-YL)-2-METHYL-2-PENTEN-1-OL see OHG000
2,6-DIMETHYL-4-TRIDECYLMORPHOLINE see DUJ400
N,N-DIMETHYL-2-(TRIFLUOROMETHYL)-10H-PHENOTHIAZINE-10-PROPANAMINE see TKL500
1,1-DIMETHYL-3-(3-TRIFLUOROMETHYLPHENYL)UREA see DUK800
N,N-DIMETHYL-N′-(3-TRIFLUOROMETHYLPHENYL)UREA see DUK800
2′,4′-DIMETHYL-5-((TRIFLUOROMETHYL)SULFONAMIDO)ACETANILIDE see DUK000
N-(2,4-DIMETHYL-5-(((TRIFLUOROMETHYL)SULFONYL)AMINO)PHENYL)ACETAMIDE see DUK000
N,N-DIMETHYL-p-(2,4,6-TRIFLUOROPHENYLAZO)ANILINE see DUK200
1,1-DIMETHYL-3-(α,α,α-TRIFLUORO-m-TOLYL) UREA see DUK800
9-cis-3,7-DIMETHYL-9-(2,6,6-TRIMETHYL-1-CYCLOHEXEN-1-YL)-2,4,6,8-NONATETRA ENAL see VSK975
3,7-DIMETHYL-9-(2,6,6-TRIMETHYL-1-CYCLOHEXEN-1-YL-2,4,6,8-NONATETRAENOIC ACID see VSK950
3,7-DIMETHYL-9-(2,6,6-TRIMETHYL-1-CYCLOHEXEN-1-YL)-2,4,6,8-NONATETRAEN-1-OL see VSK600
trans-3,7-DIMETHYL-9-(2,6,6-TRIMETHYL-1-CYCLOHEXEN-1-YL)-2,4,6-NONATRIEN OIC ACID see DMD100
trans-3,7-DIMETHYL-9-(2,6,6-TRIMETHYL-1-CYCLOHEXEN-1-YL)-7-YNE-2,4,6-NON ATRIENOIC ACID see DHA200
2,2-DIMETHYLTRIMETHYLENE ACRYLATE see DUL200
2,2-DIMETHYLTRIMETHYLENE ESTER ACRYLIC ACID see DUL200
DIMETHYLTRIMETHYLENE GLYCOL see DTG400
3,3-DIMETHYLTRIMETHYLENE OXIDE see EBQ500
β,β-DIMETHYLTRIMETHYLENE OXIDE see EBQ500
N,N-DIMETHYL-4-(3,4,5-TRIMETHYLPHENYL)AZOANILINE see DUL400
N,N-DIMETHYL-4-((3,4,5-TRIMETHYLPHENYL)AZO)BENZENAMINE see DUL400
1,2-DIMETHYL-2-TRIMETHYLSILYLHYDRAZINE see DUL500
DIMETHYLTRIMETHYLSILYLPHOSPHINE see DUL550
1,4-DIMETHYL-2,3,7-TRIOXABICYCLO[2.2.1]HEPT-5-ENE see DUL589
(3,5-DIMETHYL-1,2,4-TRIOXOLANE) see BOX825
N,N-DIMETHYL-1,2,3-TRITHIAN-5-AMINE, ETHANEDIOATE (1:1) see TFH750
N,N-DIMETHYL-1,2,3-TRITHIAN-5-AMINE HYDROGENOXALATE see TFH750
N,N-DIMETHYL-1,2,3-TRITHIAN-5-YLAMMONIUM HYDROGEN OXALATE see TFH750
N,N-DIMETHYLTRYPTAMINE see DPF600
DIMETHYL TUBOCURARINE see DUL800
o,o-DIMETHYLTUBOCURARINE see DUL800
o,o′-DIMETHYLTUBOCURARINE see DUL800
DIMETHYL TUBOCURARINE IODIDE see DUM000
α,3-DIMETHYLTYROSINE METHYL ESTER HYDROCHLORIDE see DUM100
6,10-DIMETHYL-UNDECA-5,9-DIEN-2-ONE see GDE400
1,3-DIMETHYLUREA see DUM200
N,N′-DIMETHYLUREA see DUM200
sym-DIMETHYLUREA see DUM200

SYMMETRIC DIMETHYLUREA see DUM200
DIMETHYLUREA and SODIUM NITRITE see DUM400
p-N,N-DIMETHYLUREIDOAZOBENZENE see DUM600
m-(3,3-DIMETHYLUREIDO)PHENYL-tert-BUTYL CARBAMATE see DUM800
N,N-DIMETHYLVALERAMIDE see DUN200
6,8-o-DIMETHYLVERSICOLORIN A see DUN300
6,8-o-DIMETHYLVERSICOLORIN B see DUN310
DIMETHYL VICLOGEN CHLORIDE see PAJ000
β,β-DIMETHYLVINYL CHLORIDE see IKE000
DIMETHYLVINYLETHINYL-p-HYDROXYPHENYLMETHANE see DUN400
DIMETHYL(VINYL)ETHYNYLCARBINOL see MKM300
α-(2,2-DIMETHYLVINYL)-α-ETHYNYL-p-CRESOL see DUN400
1,5-DIMETHYL-1-VINYL-4-HEXEN-1-OL BENZOATE see LFZ000
1,5-DIMETHYL-1-VINYL-4-HEXEN-1-OL CINNAMATE see LGA000
1,5-DIMETHYL-1-VINYL-4-HEXEN-1-YL-o-AMINOBENZOATE see APJ000
1,5-DIMETHYL-1-VINYL-4-HEXEN-1-YL BENZOATE see LFZ000
1,5-DIMETHYL-1-VINYL-4-HEXEN-1-YL CINNAMATE see LGA000
1,5-DIMETHYL-1-VINYL-4-HEXENYL ESTER, ISOBUTYRIC ACID see LGB000
DIMETHYL VIOLOGEN see PAI990
DIMETHYL XANTHIC DISULFIDE see DUN600
1,3-DIMETHYLXANTHINE see TEP000
1,7-DIMETHYLXANTHINE see PAK300
3,7-DIMETHYLXANTHINE see TEO500
3-((1,3-DIMETHYLXANTHIN-7-YL)METHYL)-5-METHYL-1,2,4-OXADIAZOLE see CDM575
DIMETHYLXANTHOGEN DISULFIDE see DUN600
3,3-DIMETHYL-1-XENYL-TRIAZENE see BGL500
N,N-DIMETHYL-p-(2,3-XYLYLAZO)ANILINE see DUN800
N,N-DIMETHYL-p-(3,4-XYLYLAZO)ANILINE see DUO000
2,2-DIMETHYL-5-(2,5-XYLYLOXY)VALERIC ACID see GCK300
DIMETHYL YELLOW see DOT300
DIMETHYL YELLOW-N,N-DIMETHYLANILINE see DOT300
DIMETHYLZINC see DUO200
DIMETHYLZINN-S,S′-BIS(ISOOCTYLTHIOGLYCOLAT) (GERMAN) see BKK500
DIMETHYOXYDOPAMINE see DOE200
10,11-DIMETHYSTRYCHNINE see BOL750
5-(DIMETILAMINOETILOXIMINO-5H-DIBENZO(a,d)CICLOEPTA-1,4-DIENE) CLORIDRATO (ITALIAN) see DPH600
(4-DIMETILAMINO-3-METIL-FENIL)-N-METIL-CARBAMMATO (ITALIAN) see DOR400
5-(3-DIMETILAMINOPROPILIDEN)-5H-DIBENZO-(a,d)-CICLOPENTENE(ITALIAN) see PMH600
N-(Γ-DIMETILAMINOPROPIL)-IMINODIBENZILE CLORIDRATO (ITALIAN) see DLH630
9-(3-DIMETILAMINOPROPYLIDEN)-10,10-DIMETIL-9,10-DIIDROANTHRACENE (ITALIAN) see AEG875
DIMETILAN see DQZ000
DIMETILANE see DQZ000
2,5-DIMETILBENZOCHINONE (1:4) (ITALIAN) see XQJ000
O,O-DIMETIL-O-(1,4-DIMETIL-3-OXO-4-AZA-PENT-1-ENIL)-FOSFATO(ITALIAN) see DGQ875
2,6-DIMETIL-EPTAN-4-ONE (ITALIAN) see DNI800
O,O-DIMETIL-S-(2-ETILITIO-ETIL)-MONOTIOFOSFATO (ITALIAN) see DAP400
O,O-DIMETIL-S-(2-ETIL-SOLFINIL-ETIL)-MONOTIOFOSFATO (ITALIAN) see DAP000
O,O-DIMETIL-S-(ETILTIO-ETIL)-DITIOFOSFATO (ITALIAN) see PHI500
O,O-DIMETIL-O-(2-ETILTIO-ETIL)-MONOTIOFOSFATO (ITALIAN) see DAO800
2,6-DIMETILFENILICO DELL'ACIDO α-N-METILPIPERAZINOBUTIRRICO IDOCLORIDRAT (ITALIAN) see FAC130
DIMETILFORMAMIDE (ITALIAN) see DSB000
O,O-DIMETIL-S-(N-FORMIL-N-METIL-CARBAMOIL-METIL)-DITIOFOSFATO (ITALIAN) see DRR200
O,O-DIMETIL-S-(N-METIL-CARBAMOIL-METIL)-DITIOFOSFATO (ITALIAN) see DSP400
O,O-DIMETIL-S-(N-METIL-CARBAMOIL)-METIL-MONOTIOFOSFATO(ITALIAN) see DNX800
O,O-DIMETIL-O-(2-N-METILCARBAMOIL-1-METIL-VINIL)-FOSFATO(ITALIAN) see MRH209
O,O-DIMETIL-O-(3-METIL-4-METILTIO-FENIL)-MONOTIOFOSFATO(ITALIAN) see FAQ900
O,O-DIMETIL-O-(3-METIL-4-NITRO-FENIL)-MONOTIOFOSFATO (ITALIAN) see DSQ000
O,O-DIMETIL-S-((2-METOSSI-1,3,4-(4H)-TIADIZAOL-5-ON-4-IL)-METIL)-DITIFOSFATO (ITALIAN) see DSO000
O,O-DIMETIL-S-((MORFOLINO-CARBONIL)-METIL)-DITIOFOSFATO(ITALIAN) see MRU250
O,O-DIMETIL-O-(4-NITRO-FENIL)-MONOTIOFOSFATO (ITALIAN) see MNH000
O,O-DIMETIL-S-((4-OXO-3H-1,2,3-BENZOTRIAZIN-3-IL)-METIL)-DITIOFOSFATO (ITALIAN) see ASH500
(5,5-DIMETIL-3-OXO-CICLOES-1-EN-IL)-N,N-DIMETIL-CARBAMMATO(ITALIAN) see DRL200
3,5-DIMETIL-PERIDRO-1,3,5-TIHADIAZIN-2-TIONE (ITALIAN) see DSB200
DIMETILSOLFATO (ITALIAN) see DUD100

O,O-DIMETIL-(2,2,2-TRICLORO-1-IDROSSI-ETIL)-FOSFONATO (ITALIAN)
    see TIQ250
DIMETINA see BEM500
DIMETINDENE MALEATE see FMU409
DIMETON see DSP400
3,3'-DIMETOSSIBENZODINA (ITALIAN) see DCJ200
DIMETOX see TIQ250
DIMETPRAMIDE see DUO300
DIMETRIDAZOLE see DSV800
DIMETYLFORMAMIDU (CZECH) see DSB000
O,O-DIMETYL-O-p-NITROFENYLFOSFAT (CZECH) see PHD500
DIMEVAMIDE see DOY400
DIMEVUR see DSP400
DIMEXAN see DUN600
DIMEXANO see DUN600
DIMEXIDE see DUD800
DIMEZATHINE see SNJ000
DIMID see DRP800
DIMIDIN see DUO350
DIMILIN see CJV250
DIMIPRESSIN see DLH600, DLH630
DIMITAN see BIE500
DIMITE see BIN000
DIMITRON see CMR100
DIMITRONAL see CMR100
DIMO see DLW600
DIMONOCLOROACETILAJMALINA CLORIDRATO (ITALIAN) see AFH275
DIMORLIN see AFJ400
DIMORPHOLAMINE see DUO400
DIMORPHOLINE DISULFIDE see BKU500
DIMORPHOLINIUM HEXACHLOROSTANNATE see DUO500
DIMORPHOLINO DISULFIDE see BKU500
1,5-DIMORPHOLINO-3-(1-NAPHTHYL)-PENTANE see DUO600
DIMP see DNQ875
DIMPEA see DOE200
DIMPYLATE see DCM750
DIM-SA see DNV800
DIMYRCETOL see DUO800
DIN 2.4602 see CNA750
DIN 2.4964 see CNA750
DINA see NFW000
DINACORYL see DJS200
DINACRIN see ILD000
DINAPACRYL see BGB500
DINAPHTAZIN (GERMAN) see DUP000
3,4,5,6-DINAPHTHACARBAZOLE see DCY000
1,2,5,6-DINAPHTHACRIDINE see DCS400
3,4,6,7-DINAPHTHACRIDINE see DCS600
DINAPHTHAZINE see DUP000
16H-DINAPHTHO(2,3-a:2',3'-i)CARBAZOLE-5,10,15,17-TETRAONE, 6,9-
    DIBENZAMIDO-1-METHOXY- see CMU800
DINAPHTHO(1,2,3-cd:3',2',1'-lm)PERYLENE-5,10-DIONE see DCU800
DI-(1-NAPHTHOYL)PEROXIDE (DOT) see DUP200
DI-β-NAPHTHYLDIIMIDE see ASN750
DI-β-NAPHTHYL-p-PHENYLDIAMINE see NBL000
DI-β-NAPHTHYL-p-PHENYLENEDIAMINE see NBL000
N,N'-DI-β-NAPHTHYL-p-PHENYLENEDIAMINE see NBL000
sym-DI-β-NAPHTHYL-p-PHENYLENEDIAMINE see NBL000
N,N'-DI(α-(1-NAPHTHYL)PROPIONYLOXY-2-ETHYL)PIPERAZINE DIHYDRO-
    CHLORIDE see NAD000
DINARKON see DLX400
DINATE see DXE600
DINATRIUM-AETHYLENBISDITHIOCARBAMAT (GERMAN) see DXD200
DINATRIUM-(N,N'-AETHYLEN-BIS(DITHIOCARBAMAT)) (GERMAN)
    see DXD200
DINATRIUM-(3,6-EPOXY-CYCLOHEXAAN-1,2-DICARBOXYLAAT) (DUTCH)
    see DXD000
DINATRIUM-(3,6-EPOXY-CYCLOHEXAN-1,2-DICARBOXYLAT) (GERMAN)
    see DXD000
DINATRIUM-(N,N'-ETHYLEEN-BIS(DITHIOCARBAMAAT)) (DUTCH)
    see DXD200
DINATRIUMPYROPHOSPHAT (GERMAN) see DXF800
DINDEVAN see PFJ750
DINEVAL see PFJ750
DINEX see CPK500
DINEZIN see DII200
DINGSABLCH, LEAF EXTRACT see KCA100
DINICKEL TRIOXIDE see NDH500
DINIL see PFA860
DINILE see SNE000
DINITOLMID see DUP300
DINITOLMIDE see DUP300
DINITRAMINE see CNE500
2,4-DINITRANILINE see DUP600

DINITRANILINE ORANGE see DVB800
DINITRATE de DIETHYLENE-GLYCOL (FRENCH) see DJE400
1,3-DINITRATO-2,2-BIS(NITRATOMETHYL)PROPANE see PBC250
DINITRATODIOXOURANIUM, HEXAHYDRATE see URS000
2,2'-DINITRATO-N-NITRODI-ETHYLAMINE see NFW000
DINITRILE of ISOPHTHALIC ACID see PHX550
2,3-DINITRILO-1,4-DITHIA-ANTHRAQUINONE see DLK200
2,3-DINITRILO-1,4-DITHIOANTHRACHINON (GERMAN) see DLK200
DINITRO see BRE500
DINITRO-3 see BRE500
DINITROAMINE see CNE500
4,6-DINITRO-2-AMINOPHENOL see DUP400
2,4-DINITROANILIN (GERMAN) see DUP600
2,4-DINITROANILINA (ITALIAN) see DUP600
2,4-DINITROANILINE see DUP600
DINITROANILINE ORANGE ND-204 see DVB800
DINITROANILINE RED see DVB800
2,4-DINITROANISOL see DUP800
α-DINITROANISOLE see DUP800
2,4-DINITROANISOLE see DUP800
1,5-DINITRO-9,10-ANTHRACENEDIONE see DUQ000
1,5-DINITROANTHRACHINON (CZECH) see DUQ000
1,5-DINITROANTHRAQUINONE see DUQ000
2,4-DINITROBENZENAMIME see DUP600
DINITROBENZENE see DUQ180
m-DINITROBENZENE see DUQ200
o-DINITROBENZENE see DUQ400
p-DINITROBENZENE see DUQ600
1,2-DINITROBENZENE see DUQ400
1,3-DINITROBENZENE see DUQ200
2,4-DINITROBENZENE see DUQ200
DINITROBENZENE, solution (DOT) see DUQ180
4,6-DINITROBENZENEDIAZONIUM-2-OXIDE see DUQ800
2,4-DINITROBENZENESULFENYL CHLORIDE see DUR200
2,4-DINITROBENZENESULFONIC ACID see DUR400
4,6-DINITROBENZOFURAZAN-N-OXIDE see DUR500
1,3-DINITROBENZOL see DUQ200
DINITROBENZOL, solid (DOT) see DUQ180
5,7-DINITRO-1,2,3-BENZOXADIAZOLE see DUR800
4,4'-DINITROBIFENYL (CZECH) see DUS000
4,4'-DINITROBIPHENYL see DUS000
3,5-DINITRO-N,N'-BIS(2,4,6-TRINITROPHENYL)-2,6-PYRIDINEDIAMINE
    see PQC525
2,4-DINITRO-6-BROMANILIN (CZECH) see DUS200
2,4-DINITRO-6-BROMOANILINE see DUS200
2,3-DINITRO-2-BUTENE see DUS400
4,6-DINITRO-2-sec.BUTYLFENOL (CZECH) see BRE500
4,6-DINITRO-2-sec.BUTYLFENOLATE AMMONY (CZECH) see BPG250
2,4-DINITRO-6-sec-BUTYLFENYLESTER KYSELINY OCTOVE (CZECH)
    see ACE500
2,4-DINITRO-6-sek.BUTYL-ISOPROPYLPHENYLCARBONAT (GERMAN)
    see CBW000
DINITROBUTYLPHENOL see BRE500
2,4-DINITRO-6-sec-BUTYLPHENOL see BRE500
4,6-DINITRO-2-sec-BUTYLPHENOL see BRE500
4,6-DINITRO-o-sec-BUTYLPHENOL see BRE500
2,4-DINITRO-6-tert-BUTYLPHENOL see DRV200
4,6-DINITRO-2-sec-BUTYLPHENOL AMMONIUM SALT see BPG250
4,6-DINITRO-o-sec-BUTYLPHENOL AMMONIUM SALT see BPG250
DINITROBUTYLPHENOL-2,2',2''-NITRILOTRIETHANOL SALT see BRE750
2,4-DINITRO-6-sek.BUTYL-PHENYLACETAT (GERMAN) see ACE500
4,6-DINITRO-2-sec-BUTYLPHENYL ACETATE see ACE500
4,6-DINITRO-2-sec-BUTYLPHENYL β,β-DIMETHYLACRYLATE see BGB500
2,4-DINITRO-6-sec-BUTYLPHENYL ISOPROPYL CARBONATE see CBW000
2,4-DINITRO-6-tert-BUTYLPHENYL METHANESULFONATE see DUS500
2,4-DINITRO-6-sec-BUTYLPHENYL-2-METHYLCROTONATE see BGB500
4,6-DINITRO-2-CAPRYLPHENYL CROTONATE see AQT500
4,6-DINITRO-2-(2-CAPRYL)PHENYL CROTONATE see AQT500
2,4-DINITROCHLORBENZEN-6-SULFONAN SODNY (CZECH) see CJC000
DINITROCHLOROBENZENE see CGL750
2,4-DINITROCHLOROBENZENE see CGM000
1,3-DINITRO-4-CHLOROBENZENE see CGM000
2,4-DINITRO-1-CHLOROBENZENE see CGM000
DINITROCHLOROBENZENE (DOT) see CGL750
DINITROCHLOROBENZOL see CGM000
DINITROCHLOROBENZOL (DOT) see CGM000
2,4-DINITRO-1-CHLORO-NAPHTHALENE see DUS600
DINITROCRESOL see DUS700
DINITRO-o-CRESOL see DUS700
DINITRO-p-CRESOL see DUT600
2,4-DINITRO-o-CRESOL see DUS700
2,6-DINITRO-p-CRESOL see DUT600
3,5-DINITRO-o-CRESOL see DUT000
3,5-DINITRO-p-CRESOL see DUT200

4,6-DINITRO-o-CRESOL see DUS700
4,6-DINITRO-o-CRESOL AMMONIUM SALT see DUT800
4,6-DINITRO-o-CRESOL DIETHYLAMINE SALT see DUU000
4,6-DINITRO-o-CRESOL METHYLAMINE (1581) see DUU200
4,6-DINITRO-o-CRESOL MORPHOLINE (1581) see DUU400
4,6-DINITRO-o-CRESOLO (ITALIAN) see DUS700
DINITRO-o-CRESOL SODIUM SALT see DUU600
3,5-DINITRO-o-CRESOL SODIUM SALT see DUU600
4,6-DINITRO-o-CRESOL SODIUM SALT see DUU600
DINITROCYCLOHEXYLPHENOL see CPK500
DINITRO-o-CYCLOHEXYLPHENOL see CPK500
2,4-DINITRO-6-CYCLOHEXYLPHENOL see CPK500
4,6-DINITRO-o-CYCLOHEXYLPHENOL see CPK500
DINITROCYCLOHEXYLPHENOL (DOT) see CPK500
DINITROCYCLOPENTYLPHENOL see CQB250
DINITRODENDTROXAL see DUS700
N,N'-DINITRO-1,2-DIAMINOETHANE see DUU800
DINITRODIAZOMETHANE see DUV000
DINITRODIGLICOL (ITALIAN) see DJE400
DINITRODIGLYKOL (CZECH) see DJE400
5,6-DINITRO-2-DIMETHYLAMINOPYRIMIDINONE see DUV089
4,4'-DINITRODIPHENYL DISULFIDE see BLA500
p,p'-DINITRODIPHENYL DISULFIDE see BLA500
3',5'-DINITRO-4'-(DI-n-PROPYLAMINO)ACETOPHENONE see DUV400
2,6-DINITRO-N,N-DIPROPYL-4-(TRIFLUOROMETHYL)BENZENAMINE
   see DUV600
2,6-DINITRO-N,N-DI-N-PROPYL-α,α,α-TRIFLURO-p-TOLUIDINE see DUV600
N,N'-DINITROETHYLENEDIAMINE see DUV800
2,5-DINITRO-N-(1-ETHYLPROPYL)-3,4-XYLIDINE see DRN200
2,4-DINITROFENOL (DUTCH) see DUZ000
DINITROFENOLO (ITALIAN) see DUZ000
2,4-DINITROFENYLHYDRAZIN (CZECH) see DVC400
3,7-DINITROFLUORANTHENE see DUW100
3,9-DINITROFLUORANTHENE see DUW120
4,12-DINITROFLUORANTHENE see DUW120
2,7-DINITROFLUORENE see DUW200
2,4-DINITROFLUOROBENZENE see DUW400
2,4-DINITRO-1-FLUOROBENZENE see DUW400
DINITROGEN DIOXIDE see NGU500
DINITROGEN MONOXIDE see NGU000
DINITROGEN TETRAFLUORIDE see TCI000
DINITROGEN TETROXIDE (DOT) see NGU500
DINITROGLICOL (ITALIAN) see EJG000
DINITROGLYCOL see EJG000
3,5-DINITRO-2-HYDROXYTOLUENE see DUS700
2,4-DINITRO-6-ISOBROPYL-m-CRESOL see DVG200
4,6-DINITROKRESOL (DUTCH) see DUS700
4,6-DINITRO-o-KRESOL (CZECH) see DUS700
4,6-DINITRO-o-KRESYLESTER KYSELINY OCTOVE (CZECH) see AAU250
DINITROL see DUS700
2,6-DINITRO-3-METHOXY-4-tert-BUTYLTOLUENE see BRU500
2,6-DINITRO-4-METHYLANILINE see DVI100
3,5-DINITRO-2-METHYLBENZENEDIZAONIUM-4-OXIDE see DUX509
2,5-DINITRO-3-METHYLBENZOIC ACID see DUX560
2,4-DINITRO-3-METHYL-6-tert-BUTYLPHENYL ACETATE see BRU750
2,4-DINITRO-3-METHYL-6-tert-BUTYLPHNYLACETAT (GERMAN) see BRU750
DINITROMETHYL CYCLOHEXYLTRIENOL see DUS700
N,N'-DINITRO-N-METHYL-1,2-DIAMINOETHANE see DUX600
DINITRO(1-METHYLHEPTYL)PHENYL CROTONATE see AQT500
2,4-DINITRO-6-(1-METHYLHEPTYL) CROTONATE see AQT500
2,4-DINITRO-6-METHYLPHENOL see DUS700
2,4-DINITRO-6-METHYLPHENOL SODIUM SALT see DUU600
4,6-DINITRO-2-(1-METHYL-N-PROPYL)PHENOL see BRE500
2,4-DINITRO-6-(1-METHYL-PROPYL)PHENOL (FRENCH) see BRE500
1,5-DINITRONAPHTHALENE see DUX700
2,4-DINITRO-1-NAPHTHOL see DUX800
2-4 DINITRO-α-NAPHTOL (FRENCH) see DUX800
2,4-DINITRO-6-(2-OCTYL)PHENYL CROTONATE see AQT500
2,6-DINITRO-4-PERCHLORYLPHENOL see DUY200
2,4-DINITROPHENETOLE see DUY400
DINITROPHENOL see DUY600
α-DINITROPHENOL see DUZ000
β-DINITROPHENOL see DVA200
Γ-DINITROPHENOL see DVA000
2,3-DINITROPHENOL see DUY800
2,4-DINITROPHENOL see DUZ000
2,5-DINITROPHENOL see DVA000
2,6-DINITROPHENOL see DVA200
3,4-DINITROPHENOL see DVA400
3,5-DINITROPHENOL see DVA600
DINITROPHENOL (solution) see DUY810
DINITROPHENOL, solution (DOT) see DUY810
2,4-DINITROPHENOL SODIUM SALT see DVA800
2,4-DINITROPHENYLACETYL CHLORIDE see DVB200

1-((2,4-DINITROPHENYL)AZO)-2-NAPHTHOL see DVB800
4,6-DINITROPHENYL-2-sec-BUTYL-3-METHYL-2-BUTENONATE see BGB500
2,4-DINITROPHENYL-2,4-DINITRO-6-sec-BUTYLPHENYL CARBONATE
   see DVC200
DI-4-NITROPHENYL DISULFIDE see BLA500
2,4-DINITROPHENYL ETHER of MORPHINE see DVC800
2,4-DINITROPHENYLHYDRAZINE see DVC400
2,4-DINITROPHENYLHYDRAZINIUMPERCHLORATE see DVC600
o-(2,4-DINITROPHENYL)HYDROXYLAMINE see DVC700
DINITROPHENYLMETHANE see DVG600
2,4-DINITROPHENYLMETHYL ETHER see DUP800
2,4-DINITROPHENYLMORPHINE HYDROCHLORIDE see DVC800
2,4-DINITROPHENYL THIOCYANATE see DVF800
2,2-DINITRO-1,3-PROPANEDIOL see DVD000
2,2-DINITROPROPANOL see DVD200
2,2-DINITRO-1-PROPANOL see DVD200
DINITROPYRENE see DVD400, DVD600, DVD800
1,3-DINITROPYRENE see DVD400
1,6-DINITROPYRENE see DVD600
1,8-DINITROPYRENE see DVD800
4,6-DINITROQUINOLINE-1-OXIDE see DVE000
4,7-DINITROQUINOLINE-1-OXIDE see DVE200
2,4-DINITRO-RHODANBENZOL (GERMAN) see DVF800
1,4-DINITROSOBENZENE HOMOPOLYMER see PJR500
p-DINITROSOBENZENE POLYMERS see PJR500
DINITROSOCIMETIDINE see DVE300
N,N'-DINITROSO-N,N'-DIETHYLETHYLENEDIAMINE see DJB800
N,N'-DINITROSO-N,N'-DIMETHYLETHYLENEDIAMINE see DVE400
N,N'-DINITROSO-N,N'-DIMETHYLOXAMID (GERMAN) see DRN600
DINITROSO-2,5-DIMETHYLPIPERAZINE see DRN800
DINITROSO-2,6-DIMETHYLPIPERAZINE see DRO000
1,4-DINITROSO-2,6-DIMETHYLPIPERAZINE see DRO000
N,N'-DINITROSO-2,6-DIMETHYLPIPERAZINE see DRO000
DINITROSODIMETHYLPROPANEDIAMINE see DRO200
N,N'-DINITROSO-N,N'-DIMETHYL-1,3-PROPANEDIAMINE see DRO200
N,N'-DINITROSO-N,N'-DIMETHYLTEREPHTALSAUREAMID (GERMAN)
   see DRO400
DINITROSOHOMOPIPERAZINE see DVE600
N,4-DINITROSO-N-METHYLANILINE see MJG750
DINITROSOPENTAMETHYLENETETRAMINE see DVF400
N,N-DINITROSOPENTAMETHYLENETETRAMINE see DVF400
3,4-DI-N-NITROSOPENTAMETHYLENETETRAMINE see DVF400
3,7-DI-N-NITROSOPENTAMETHYLENETETRAMINE see DVF400
N^1,N^3-DINITROSOPENTAMETHYLENETETRAMINE see DVF400
DI(N-NITROSO)-PERHYDROPYRIMIDINE see DVF000
DINITROSOPIPERAZIN (GERMAN) see DVF200
DINITROSOPIPERAZINE see DVF200
1,4-DINITROSOPIPERAZINE see DVF200
N,N'-DINITROSOPIPERAZINE see DVF200
DINITROSOPRODECTIN see PPH100
N-DINITROSOPYRIDINOLCARBAMATE see PPH100
N,N'-DINITROSOPYRIDINOL CARBAMATE see PPH100
DINITROSORBIDE see CCK125
3,7-DINITROSO-1,3,5,7-TETRAAZABICYCLO[3.3.1]NONANE see DVF400
DINITROSTILBENEDISULFONIC ACID see DVF600
4,4'-DINITRO-2,2'-STILBENEDISULFONIC ACID see DVF600
2,4-DINITROTHIOCYANATOBENZENE see DVF800
2,4-DINITROTHIOCYANOBENZENE see DVF800
2,4-DINITRO-1-THIOCYANOBENZENE see DVF800
2,4-DINITROTHIOPHENE see DVG000
2,6-DINITROTHYMOL see DVG200
DINITROTHYMOL 1-2-4 (FRENCH) see DVG200
3,5-DINITRO-o-TOLUAMIDE see DUP300
DINITROTOLUENE see DVG600
2,3-DINITROTOLUENE see DVG800
2,4-DINITROTOLUENE see DVH000
2,5-DINITROTOLUENE see DVH200
2,6-DINITROTOLUENE see DVH400
3,4-DINITROTOLUENE see DVH600
3,5-DINITROTOLUENE see DVH800
ar,ar-DINITROTOLUENE see DVG600
DINITROTOLUENE, liquid (DOT) see DVG600
DINITROTOLUENE, molten (DOT) see DVG600
DINITROTOLUENE, solid (DOT) see DVG600
2,6-DINITRO-p-TOLUIDINE see DVI100
3,5-DINITRO-o-TOLUIDINE see AJR750
2,4-DINITROTOLUOL see DVH000
4,6-DINITRO-1,2,3-TRICHLOROBENZENE see DVI600
2,6-DINITRO-4-TRIFLUORMETHYL-N,N-DIPROPYLANILIN (GERMAN)
   see DUV600
2,4'-DINITRO-4-TRIFLUOROMETHYL-DIPHENYL ETHER see NIX000
sym-DINITROXYDIETHYLNITRAMINE see NFW000
N-DINITROZO-PIRIDINOLKARBAMAT (HUNGARIAN) see PPH100
DINKUM OIL see EQQ000

DINOBUTON see CBW000
DINOC see DUS700, DUU600
DINOCTON-6 see DVI800
DINOCTON-O see DVI800
DINOFEN see CBW000
DINONYL-1,2-BENZENEDICARBOXYLATE see DVJ000
DI-n-NONYL PHTHALATE see DVJ000
DINOPOL NOP see DVL600
DINOPROST see POC500
DINOPROST METHYL ESTER see DVJ100
DINOPROSTONE see DVJ200
DINOPROST TROMETHAMINE (USDA) see POC750
18,19-DINOR-17-α-PREGN-4-EN-3-ONE, 13-ETHYL-17-HYDROXY- see NNE600
DINOSEB see BRE500
DINOSEB-ACETATE see ACE500
DINOSEB (AMINE) see BPG250
DINOSEBE (FRENCH) see BRE500
DINOSEB METHACRYLATE see BGB500
DINOSOL see SNN300
DINOTERB see DRV200
DINOTERB ACETATE see DVJ400
DINOVEX see DAL600
DINOXOL see DAA800, TAA100
DINOZOL see DUT800
DINOZOL 50 see DUT800
DINTOIN see DKQ000
DINTOINA see DNU000
DINULCID see OLM300
DINURANIA see DUS700
DINYL see PFA860
DIOCID see DVJ500
DIOCIDE see DVJ500
DIOCTLYN see DJL000
DIOCTYL ADIPATE see AEO000
DIOCTYLAL see DJL000
DIOCTYLAMINE see DVJ600
DIOCTYL AZELATE see BJQ500
DIOCTYL-o-BENZENEDICARBOXYLATE see DVL600
DIOCTYLBIS(NONYLOXYMALEOYLOXY)STANNANE see DEI800
DIOCTYLDIDODECANOYLOXYSTANNANE see DVJ800
DIOCTYLDI(LAUROYLOXY)STANNANE see DVJ800
2,2-DIOCTYL-1,3-DIOXA-2-STANNA-7-THIADECAN-4,10-DIONE see DVN909
2,2-DIOCTYL-1,3,2-DIOXASTANNEPIN-4,7-DIONE see DVK200
4,4'-DIOCTYLDIPHENYLAMINE see DVK400
DIOCTYL ESTER of SODIUM SULFOSUCCINATE see DJL000
DIOCTYL ESTER of SODIUM SULFOSUCCINIC ACID see DJL000
DIOCTYL ETHER see OEY000
DIOCTYL(ETHYLENEDIOXYBIS(CARBONYLMETHYLTHIO))STANNANE
    see DVN400
DIOCTYL FUMARATE see DVK600
DIOCTYLISOPENTYLPHOSPHINE OXIDE see DVK709
DIOCTYL ISOPHTHALATE see BJQ750
DIOCTYL MALEATE see DVK800
DI-N-OCTYL MALEATE see DVK800
"DIOCTYL" MALEATE see BJR000
DIOCTYL-MEDO FORTE see DJL000
N,N-DIOCTYL-1-OCTANAMINE see DVL000
2,2-DIOCTYL-1,3,2-OXATHIASTANNOLANE-5-OXIDE see DVL200
DIOCTYLOXOSTANNANE see DVL400
DIOCTYL PHTHALATE see DVL600, DVL700
n-DIOCTYL PHTHALATE see DVL600
DI-sec-OCTYL PHTHALATE see DVL700
DIOCTYL(1,2-PROPYLENEDIOXYBIS(MALEOYLDIOXY))STANNANE
    see DVL800
DIOCTYL SEBACATE see BJS250
DIOCTYL SODIUM SULFOSUCCINATE (FCC) see DJL000
(Z,Z)-4,4'-(DIOCTYLSTANNYLENE)BIS((OXY))BIS(4-OXO-2-BUTANOIC ACID
    DIISOOCTYL ESTER see BKL000
DIOCTYL SULFOSUCCINATE SODIUM SALT see DJL000
DIOCTYLTHIOACETOXYSTANNANE see DVL200
DIOCTYLTHIOXOSTANNANE see DVM000
DI-N-OCTYLTIN BIS(BUTYL MALEATE) see BHK500
DI-n-OCTYLTIN BIS(BUTYL MERCAPTOACETATE) see DVM200
DI-n-OCTYLTIN BIS(DODECYL MERCAPTIDE) see DVM400
DI-n-OCTYLTIN BIS(2-ETHYLHEXYL MALEATE) see DVM600
DI-n-OCTYLTIN BIS(2-ETHYLHEXYL) MERCAPTOACETATE see DVM800
DIOCTYLTINBIS(ISOOCTYL MALEATE) see BKL000
DIOCTYLTIN BIS(ISOOCTYL MERCAPTOACETATE) see BKK750
DIOCTYLTIN-S,S'-BIS(ISOOCTYL MERCAPTOACETATE) see BKK750
DIOCTYLTIN BIS(ISOOCTYL THIOGLYCOLATE) see BKK750
DIOCTYL-TIN BIS(ISOOCTYLTHIOGLYCOLLATE) see BKK750
DI-n-OCTYLTIN BIS(LAURYLTHIOGLYCOLATE) see DVN000
DI-n-OCTYLTIN-1,4-BUTANEDIOL-BIS-MERCAPTOACETATE see DVN200
DI-n-OCTYLTIN DIISOOCTYL THIOGLYCOLATE see BKK750

DIOCTYLTIN DILAURATE see DVJ800
DI-n-OCTYLTIN DILAURATE see DVJ800
DI-N-OCTYLTIN DIMONOBUTYLMALEATE see BHK500
DI-n-OCTYLTIN DI(1,2-PROPYLENEGLYCOLMALEATE) see DVL800
DI-n-OCTYLTIN ETHYLENEGLYCOL DITHIOGLYCOLATE see DVN400
DI-n-OCTYLTIN-2-ETHYLHEXYLDIMERCAPTOETHANOATE see DVM800
DI-N-OCTYLTIN-ITHIOGLYCOLIC ACID 2-ETHYLHEXYL ESTER see DVM800
DIOCTYLTIN MERCAPTIDE see DVN600
DI-n-OCTYLTIN MERCAPTIDE see DVN600
DIOCTYLTIN-β-MERCAPTOPROPIONATE see DVN800
DI-n-OCTYLTIN β-MERCAPTOPROPIONATE see DVN800
DIOCTYLTIN OXIDE see DVL400
DI-n-OCTYLTIN OXIDE see DVL400
DI-n-OCTYLTIN SULFIDE see DVM000
DIOCTYLTIN-3,3'-THIODIPROPIONATE see DVN909
DIOCTYLTIN THIOGLYCOLATE see DVL200
DI-n-OCTYLTIN THIOGLYCOLATE see DVL200
DI-n-OCTYL-ZINN AETHYLENGLYKOL-DITHIOGLYKOLAT (GERMAN)
    see DVN400
DI-n-OCTYL-ZINN-BIS(2-AETHYLHEXYLMALEINAT) (GERMAN) see DVM600
DI-n-OCTYL-ZINN-BIS(LAURYL-THIOGLYKOLAT) (GERMAN) see DVN000
DI-n-OCTYL-ZINN-1,4-BUTANDIOL-BIS-MERCAPTOACETAT (GERMAN)
    see DVN200
DI-n-OCTYL-ZINN-DI-ISOOCTYLTHIOGLYKOLAT (GERMAN) see BKK750
DI-n-OCTYL-ZINN DILAURAT (GERMAN) see DVJ800
DI-N-OCTYLZINN-DIMONOBUTYLMALEINAT (GERMAN) see BHK500
DI-n-OCTYLZINN-DIMONOMETHYLMALEINAT (GERMAN) see BKO500
DI-n-OCTYL-ZINN-DI-(1,2-PROPYLENGLYKOLMALEINAT)(GERMAN)
    see DVL800
DI-n-OCTYL-ZINN β-MERCAPTOPROPIONAT (GERMAN) see DVN800
DI-n-OCTYL-ZINN OXYD (GERMAN) see DVL400
DI-n-OCTYL-ZINN THIOGLYKOLAT (GERMAN) see DVL200
DIOCYDE see DVO000
DIOCYTLBIS(LAUROYLOXY)STANNANE see DVJ800
DIODOHYDROXYQUIN see DNF600
DIODON see DNG400
DIODONE see DNG400
DIODRAST see DNG400
DIOFORM see DFI100
DIOGYN see EDO000
DIOGYN B see EDP000
DIOGYNETS see EDO000
DIOKAN see DVQ000
DIOKSAN (POLISH) see DVQ000
DIOKSYNY (POLISH) see TAI000
DIOLAMINE see DHF000
DIOLANDRONE see AOO475
DIOLANE see HFP875
DIOLENE see IPU000
DIOL-EPOXIDE-1 see DMO500
DIOL-EPOXIDE 2 see DMO800
anti-DIOLEPOXIDE see DVO175
DIOLICE see CNU750
DIOLOSTENE see AOO475
DIOMEDICONE see DJL000
DI-ON see DXQ500
DIONE 21-ACETATE see RKP000
3,17-DIONE-19-ACETOXY-Δ(1,3)-ANDROSTADIENE see ABL625
2-4-DIONE-1,3-DIAZASPIRO(4.5)DECANE see DVO600
DIONIN see ENK000, ENK500
DIONINE see ENK000
DIONINE HYDROCHLORIDE see DVO700
DIONIN HYDROCHLORIDE see DVO700
DIONONE see DMH400
DIONONYL PHTHALATE see DVJ000
DIOPAL see SBH500
DIOPHYLLIN see TEP500
DIORTHOTOLYLGUANIDINE see DXP200
DIOSPYROL see DVO809
DIOSPYROS VIRGINIANA see PCP500
1,4-DIOSSAN-2,3-DIYL-BIS(O,O-DIETIL-DITIOFOSFATO) (ITALIAN)
    see DVQ709
DIOSSANO-1,4 (ITALIAN) see DVQ000
1,4-DIOSSIBENZENE (ITALIAN) see QQS200
2,4-DIOSSI-5-DIAZOPIRIMIDINA (ITALIAN) see DCQ600
DIOSSIDONE see BRF500
DIOSUCCIN see DJL000
DIOTHANE see DVO819
DIOTHANE HYDROCHLORIDE see DVV500
DIOTHENE see PJS750
DIOTILAN see DJL000
DIOVAC see DJL000
DIOVOCYCLIN see EDR000
DIOVOCYLIN see EDR000

DIOXAAN-1,4 (DUTCH) see DVQ000
1,4-DIOXAAN-2,3-DIYL-BIS(O,O-DIETHYL-DITHIOFOSFAAT) (DUTCH) see DVQ709
5-(1,4-DIOXA-8-AZASPIRO(4.5)DEC-8-YLMETHYL)-3-ETHYL-6,7-DIHYDRO-2-METHYL-INDOL-4(5H)-ONE see AFH550
DIOXACARB see DVS000
1,6-DIOXACYCLOHEPTADECAN-17-ONE see OKW110
1,7-DIOXACYCLOHEPTADECAN-17-ONE see OKW100
1,4-DIOXACYCLOHEXANE see DVQ000
1,3-DIOXACYCLOPENTANE see DVR800
(R*,S*)-3,14-DIOXA-2,15-DITHIA-6,11-DIAZEHEXADECANE-8,9-DIOL, 2,2,15,15-TETRAOXIDE (9CI) see LJD900
3,6-DIOXADODECANOL-1 see HFN000
1-DIOXADROL HYDROCHLORIDE see LFG100
d-DIOXADROL HYDROCHLORIDE see DVP400
2,5-DIOXAHEXANE see DOE600
m-DIOXAN see DVP600
p-DIOXAN (CZECH) see DVQ000
DIOXAN-1,4 (GERMAN) see DVQ000
2,3-p-DIOXANDITHIOL S,S-BIS(O,O-DIETHYL PHOSPHORODITHIOATE) see DVQ709
1,4-DIOXAN-2,3-DIYL-BIS(O,O-DIAETHYL-DITHIOPHOSPHAT) (GERMAN) see DVQ709
1,4-DIOXAN-2,3-DIYL-BIS(O,O-DIETHYLPHOSPHOROTHIOLOTHIONATE) see DVQ709
1,4-DIOXAN-2,3-DIYL-O,O,O',O'-TETRAETHYL DI(PHOSPHOROMITHIOATE) see DVQ709
DIOXANE see DVQ000
p-DIOXANE see DVQ000
1,3-DIOXANE see DVP600
1,4-DIOXANE (MAK) see DVQ000
2,3-p-DIOXANE-S,S-BIS(O,O-DIETHYLPHOSPHOROITHIOATE) see DVQ709
m-DIOXANE-4,4-DIMETHYL see DVQ400
cis-2,3-p-DIOXANEDITHIOL-S,S-BIS(O,O-DIETHYLPHOSPHORODITHIOATE) see DVQ600
p-DIOXANE-2,3-DITHIOL-S,S-DIESTER with O,O-DIETHYL PHOSPHORODITHIOATE see DVQ709
p-DIOXANE-2,3-DIYL ETHYL PHOSPHORODITHIOATE see DVQ709
DIOXANNE (FRENCH) see DVQ000
3,6-DIOXAOCTANE-1,8-DIOL see TJQ000
1,3-DIOXA-2-SILACYCLOPENTANE, 4-(CHLOROMETHYL)-2,2-DIMETHYL- see CIL775
4,9-DIOXATETRACYCLO(5.4.0.0^{3,5}.0^{8,10})UNDECANE see BFY750
4,10-DIOXATETRACYCLO(5.4.0^{3,5}).0^{1,7}).0^{9,11})UNDECANE see BFY750
1,3,2-DIOXATHIANE-2,2-DIOXIDE (9CI) see TLR750
1,3,2-DIOXATHIOLANE-2-OXIDE (9CI) see COV750
DIOXATHION see DVQ709
(E)-3,8-DIOXATRICYCLO(5.1.0.0^{2,4})OCTANE see DHB875
DIOXATRINE see BBU800
cis-1,4-DIOXENEDIOXETANE see DVQ759
1,4-DI-N-OXIDE 2,3-BIS(OXYMETHYL)QUINOXLINE see DVQ800
1,4-DI-N-OXIDE of DIHYDROXYMETHYLQUINOXALINE see DVQ800
2,2-DIOXIDE-1,3,2-DIOXATHIOLANE see EJP000
4,4-DIOXIDE-1,4-OXATHIANE see DVR000
1,4-DIOXIDE-2,3-QUINOXALINEDIMETHANOL see DVQ800
1,1-DIOXIDE TETRA HYDROTHIOFURAN see SNW500
DIOXIDE of VITAVAX see DLV200
DIOXIDIN see DVQ800
DIOXIDINE see DVQ800
DIOXIME-p-BENZOQUINONE see DVR200
DIOXIME-1,4-CYCLOHEXADIENEDIONE see DVR200
DIOXIME-2,5-CYCLOHEXADIENE-1,4-DIONE see DVR200
DIOXIN (herbicide contaminant) see TAI000
DIOXINE see TAI000
DIOXIN (bactericide) (OBS.) see ABC250
DIOXITOL see CBR000
DIOXOAMINOPYRINE see DVU100
9,10-DIOXOANTHRACENE see APK250
p-DIOXOBENZENE see HIH000
2,2'-((1,4-DIOXO-1,4-BUTANEDIYL)BIS(OXY))BIS(N,N,N-TRIMETHYLETHANAMINIUM see CMG250
2,2'-((1,4-DIOXO-1,4-BUTANEDIYL)BIS(OXY)BIS(N,N,N-TRIMETHYLETHANAMINIUM DICHLORIDE see HLC500
1,1'-((1,4-DIOXO-1,4-BUTANEDIYL)BIS(OXY-2,1-ETHANEDIYL))BIS-QUINOLINIUM DIIODIDE see KGK400
1,1',1''-(3,6-DIOXO-1,4-CYCLOHEXADIENE-1,2,4-TRIYL)TRISAZIRIDINE see TND000
2,6-DIOXO-5-DIAZOPYRIMIDINE see DCQ600
DIOXODICHLOROCHROMIUM see CML125
3,3'-(DIOXODIGERMOXANYLENE) DIPROPANOIC ACID see CCF125
9,10-DIOXO-9,10-DIHYDRO-1-NITRO-6-ANTHRACENESULFONIC ACID see DVR400
DIOXO-9,9-(DIMETHYLAMINO-3-METHYL-2-PROPYL)-10-PHENOTHIAZINE (FRENCH) see AFL750

5,5-DIOXO-10-(2-(DIMETHYLAMINO)PROPYL)PHENOTHIAZINE HYDROCHLORIDE see DVT400
3,5-DIOXO-1,2-DIPHENYL-4-N-BUTYLPYRAZOLIDENE see BRF500
3,5-DIOXO-1,2-DIPHENYL-4-N-BUTYLPYRAZOLIDIN SODIUM see BOV750
DI-OXO-DI-N-PROPYLNITROSAMINE see NJN000
2,2'-DIOXO-DI-N-PROPYLNITROSAMINE see NJN000
2,4-DIOXO-5-FLUORO-N-HEXYL-3,4-DIHYDRO-1(2H)-PYRIMIDINECARBOXAMIDME see CCK630
2,4-DIOXO-5-FLUORO-N-HEXYL-1,2,3,4-TETRAHYDRO-1-PYRIMIDINECARBOXAMIDE see CCK630
2,3-DIOXOINDOLINE see ICR000
DIOXOLAN see DVR600
1,3-DIOXOLAN see DVR800
1,4-DIOXOLAN-2,5-DIYLDIMETHYLENEBIS(NITROMERCURY) see BKZ000
1,3-DIOXOLANE see DVR800
DIOXOLANE (DOT) see DVR600
1,3-DIOXOLANE, 4-(CHLOROMETHYL)-2-(o-NITROPHENYL)- see CIQ400
1,3-DIOXOLANE, 2-(1,2-DICHLOROETHYL)-4-METHYL- see DFJ500
1,3-DIOXOLANE, 2-(2,6-DIMETHYL-1,5-HEPTADIENYL)- see CMS324
1,3-DIOXOLANE-4-METHANOL see DVR909
2-(1,3-DIOXOLANE-2-YL)PHENYL N-METHYLCARBAMATE see DVS000
1,3-DIOXOLAN-2-ONE see GHM000
1,3-DIOXOLAN-2-ONE, 4-(CHLOROMETHYL)- see CBW400
1,3-DIOXOLAN-2-ONE, 4-ETHYL- see BOT200
7-(1,3-DIOXOLAN-2-YLMETHYL)THEOPHYLLINE see TEQ175
((1,3-DIOXOLAN-4-YL)METHYL)TRIMETHYLAMMONIUM IODIDE see FMX000
2-(1,3-DIOXOLAN-2-YL)PHENYL-N-METHYLCARBAMAT see DVS000
o-(1,3-DIOXOLAN-2-YL)PHENYL METHYLCARBAMATE see DVS000
1,3-DIOXOL-4-EN-2-ONE see VOK000
DIOXOLONE-2 see GHM000
1,3-DIOXOL-2-ONE see VOK000
(1,3)DIOXOLO(4,5-j)PYRROLO(3,2,1-de)PHENANTHRIDINIUM, 4,5-DIHYDRO-2-HYDROXY-(9CI) see LJB800
2,6-DIOXO-4-METHYL-4-ETHYLPIPERIDINE see MKA250
3-(2-(1,3-DIOXO-2-METHYLINDANYL))GLUTARIMIDE see DVS100
DIOXONE see DJT400
9,10-DIOXO-1-NITRO-9,10-DIHYDRO-5-ANTHRACENESULFONIC ACID see DVS200
2,2'-DIOXO-N-NITROSODIPROPYLAMINE see NJN000
3,5-DIOXO-1-PHENYL-2-(p-HYDROXYPHENYL)-4-N-BUTYLPYRAZOLIDENE see HNI500
3-(2-(1,3-DIOXO-2-PHENYLINDANYL))GLUTARIMIDE see DVS300
3-(2-(1,3-DIOXO-2-PHENYL-4,5,6,7-TETRAHYDRO-4,7-DITHIAINDANYL))GLUTARIMIDE see DVS400
1,3-DIOXOPHTHALAN see PHW750
1,3-DIOXO-5-PHTHALANCARBOXYLIC ACID see TKV000
2,6-DIOXO-3-PHTHALIMIDOPIPERIDINE see TEH500
2-(2,6-DIOXOPIPERIDEN-3-YL) PHTHALIMIDINE see DVS600
2-(2-6-DIOXO-3-PIPERIDINYL)1H-ISOINDOLE-1,3(2H)-DIONE see TEH500
(s)-2-(2,6-DIOXO-3-PIPERIDINYL)-1H-ISOINDOLE-1,3(2H)-DIONE see TEH520
N-(2,6-DIOXO-3-PIPERIDYL)PHTHALIMIDE see TEH500
(±)-N-(2,6-DIOXO-3-PIPERIDYL)PHTHALIMIDE see DVT200
DIOXOPROMETHAZINE HYDROCHLORIDE see DVT400
N,N-DI(2-OXOPROPYL)NITROSAMINE see NJN000
2,2'-DIOXOPROPYL-N-PROPYLNITROSAMINE see NJN000
2,6-DIOXOPURINE see XCA000
1,3-DIOXO-2-(3-PYRIDYLMETHYLENE)INDAN see DVT459
2,4-DIOXOPYRIMIDINE see UNJ800
2,4-DIOXOTETRAHYDROQUINAZOLINE see QEJ800
3,5-DIOXO-2,3,4,5-TETRAHYDRO-1,2,4-TRIAZINE RIBOSIDE see RJA000
2,4-DIOXOTHIAZOLIDINE see TEV500
DIOXOTHIOLAN see SNW500
DIOXYAMINOPYRINE see DVU100
DIOXYANTHRANOL see APH250
1,4-DIOXYANTHRAQUINONE (RUSSIAN) see DMH000
2,6-DIOXY-8-AZAPURINE see THT350
m-DIOXYBENZENE see REA000
o-DIOXYBENZENE see CCP850
1,4-DIOXYBENZENE see QQS200
3,3'-DIOXYBENZIDINE see DMI400
1,4-DIOXY-BENZOL (GERMAN) see QQS200
DIOXYBIS(2,2'-DI-tert-BUTYLBUTANE see DVT500
DIOXYBIS METHANOL see DMN200
DIOXYBUTADIENE see BGA750
3-β,14-DIOXY-CARDEN-(20:22)-OLID (GERMAN) see DMJ000
3-α,7-β-DIOXYCHOLANIC ACID see DMJ200
DIOXYDE de BARYUM (FRENCH) see BAO250
DIOXYDEMETON-S-METHYL see DAP600
3,3'-(DIOXYDICARBONYL)DIPROPIONIC ACID see SNC500
2,4-DIOXY-3,3-DIETHYL-5-METHYLPIPERIDINE see DNW400
3-β,14-DIOXY-DIGEN-(20:22)-OLID (GERMAN) see DMJ000
DIOXYDIMETHANOL see DMN200
DIOXYDINE see DVQ800

4,4'-DIOXYDIPHENYLSULFIDE see TFJ000
1,4-DIOXYETHYLAMINO-5,8-DIOXYANTHRAQUINONE (RUSSIAN) see DMM400
N,N-DIOXYETHYLANILINE see BKD500
DIOXYETHYLENE ETHER see DVQ000
DIOXYGENYL TETRAFLUOROBORATE see DVT800
DIOXYMETHYLENE-PROTOCATECHUIC ALDEHYDE see PIW250
1,3-DIOXY-2-NICOTINSAEURENITRIL-TETRAZOL (GERMAN) see DND900
1,4-DI-p-OXYPHENYL-2,3-DI-ISONITRILO-1,3-BUTADIENE see DVU000
DI(p-OXYPHENYL)-2,4-HEXADIENE see DAL600
2-(3,4-DIOXYPHENYL)TETRAHYDRO-1,4-OXAZIN (GERMAN) see MRU100
7-(2,3-DIOXYPROPYL)THEOPHYLLINE see DNC000
DIOXYPYRAMIDON see DVU100
DIOZOL see BRF500
DIP see DNS200
DIPA see DNL600, DNM200
DIPALLADIUM TRIOXIDE see DVU200
DIPAM see DCK759
DIPAN see DVX200
DIPANE see WAK000
DIPANOL see MCC250
DIPAR see PDF250
DIPARAANISYL-MONOPHENETHYL-GUANIDIN-HYDROCHLORID (GERMAN) see PDN500
DI-PARALEN see CFF500
DIPARALENE see CFF500
DIPARALENE HYDROCHLORIDE see CDR250
DI-PARALENE-2-HYDROCHLORIDE see CDR000
DIPARCOL see DHF600, DII200
DIPAXIN see DVV600
DIPEGYL see NCR000
DIPENICILLINA-G-ALLUMINIO-SULFAMETOSSIPIRIDAZINA (ITALIAN) see DVU300
DIPENICILLIN-G-ALUMINIUM-SULPHAMETHOXYPYRIDAZINE see DVU300
DIPENINBROMID (GERMAN) see DGW600
DIPENTAMETHYLENETHIURAM TETRASULFIDE see TEF750
DI(PENTANOYLOXY)DIBUTYLSTANNANE see DEA600
DIPENTENE see MCC250
DIPENTENE DIOXIDE see LFV000
DIPENTYLAMINE see DCH200
DIPENTYL ETHER see PBX000
2,5-DI-tert-PENTYLHYDROQUINONE see DCH400
DIPENTYL KETONE see ULA000
DIPENTYL LEAD DIACETATE see DVU600
DIPENTYLNITROSAMINE see DCH600
DI-n-PENTYLNITROSAMINE see DCH600
DIPENTYLOXOSTANNANE see DVV000
DIPENTYL PHENOL see DCH800
DI-tert-PENTYLPHENOL see DCI000
2-(2,4-DI-tert-PENTYLPHENOXY)BUTYRIC ACID see DVV109
DIPENTYL PHTHALATE see AON300
DI-n-PENTYLPHTHALATE see AON300
DIPENTYLTIN DICHLORIDE see DVV200
DIPENTYLTIN OXIDE see DVV000
DIPEPTIDE SWEETENER see ARN825
2,6-DIPERCHLORYL-4,4'-DIPHENOQUINONE see DVV400
DIPERDON HYDROCHLORIDE see DVV500
DIPERFLUOROBUTYL ETHER see PCG755
DIPERODON HYDROCHLORIDE see DVV500
DIPEROXYTEREPHTHALIC ACID see DVV550
DIPHACIL see DHX800
DIPHACIN see DVV600
DIPHACINONE see DVV600
DIPHACYL see DHX800
DIPHANTINE see BBV500
DIPHANTOIN see DKQ000
DIPHANTOINE SODIUM see DNU000
DIPHASTON see DYF759
DIPHEBUZOL see BRF500
DIPHEDAL see DKQ000
DIPHEDAN see DNU000
DIPHEMANIL see DAP800
DIPHEMANIL METHYLSULFATE see DAP800
DIPHENACIN see DVV600
DIPHENADIONE see DVV600
DIPHENALDEHYDE see DVW800
DIPHENAMID see DRP800
DIPHENAMIDE see DRP800
DIPHENAMIZOLE see PAM500
DIPHENATE see DNU000
DIPHENATIL see DAP800
DIPHENATRILE see DVX200
DIPHENAZINE DIHYDROCHLORIDE see BLF250
DIPHENCHLOXAZINE HYDROCHLORIDE see DVW000

DIPHENETHYLAMINE, o,o'-DIMETHOXY-α-α'-DIMETHYL-, compounded with LACTIC ACID see BKP200
DIPHENHYDRINATE see DYE600
DIPHENICILLIN see BGK250
DIPHENIDOL see DWK200
DIPHENIN see DNU000
DIPHENINE see DKQ000
DIPHENINE SODIUM see DNU000
DIPHENMANIL METHYLSULFATE see DAP800
DIPHENMETHANIL see DAP800
DIPHENMETHANIL METHYLSULFATE see DAP800
o-DIPHENOL see CCP850
1,5-DIPHENOXYANTHRAQUINONE see DVW100
N,N'-DI-((3-PHENOXY-2-HYDROXYPROPYL)ETHYLENEDIAMINE DIHYDRO-CHLORIDE see IAG700
DIPHENOXYLATE HYDROCHLORIDE see LIB000
1,3-DIPHENOXY-2-PROPANOL see DVW600
DIPHENPYRALINE HYDROCHLORIDE see DVW700
DIPHENTOIN see DKQ000, DNU000
DIPHENYL (OSHA) see BGE000
DIPHENYLACETAMIDE see PDX500
DIPHENYLACETIC ACID see DVW800
α,α-DIPHENYLACETIC ACID see DVW800
DIPHENYLACETIC ACID 2-(BIS(2-HYDROXYETHYL)AMINO)ETHYL ESTER HYDROCHLORIDE see DVW900
DIPHENYLACETIC ACID DIETHYLAMINOETHYL ESTER see DHX800, THK000
DIPHENYLACETIC ACID, 2-(DIETHYLAMINO)ETHYL ESTER see DHX800, THK000
DIPHENYLACETIC ACID, ESTER with 3-QUINUCLIDINOL, HYDROGEN SULFATE (2:1) DIHYDRATE see QVJ000
DIPHENYLACETIC ACID-1-ETHYL-3-PIPERIDYL ESTER HYDROCHLORIDE see EOY000
DIPHENYLACETONITRILE see DVX200
6-DIPHENYLACETOXY-1-AZABICYCLO(3.2.1)OCTANE HYDROCHLORIDE see ARX500
DIPHENYLACETYLDIETHYLAMINOETHANOL see DHX800, THK000
2-DIPHENYLACETYL-1,3-DIKETOHYDRINDENE see DVV600
2-DIPHENYLACETYL-1,3-INDANDIONE see DVV600
2-(DIPHENYLACETYL)INDAN-1,3-DIONE see DVV600
2-(DIPHENYLACETYL)-1H-INDENE-1,3(2H)-DIONE see DVV600
2,3-DIPHENYLACRYLONITRILE see DVX600
α,β-DIPHENYLACRYLONITRILE see DVX600
3-(2,2-DIPHENYLAETHYL)-5-(2-PIPERIDINOAETHYL)-1,2,4-OXADIAZOL (GERMAN) see LFJ000
DIPHENYLAMIDE see DRP800
DIPHENYLAMIN-(β-DIAETHYLAMINOAETHYL)CARBAMIDTHIOESTER (GERMAN) see PDC875
DIPHENYLAMINE see DVX800
N,N-DIPHENYLAMINE see DVX800
DIPHENYLAMINE-2-CARBOXYLIC ACID see PEG500
DIPHENYLAMINECHLORARSINE see PDB000
DIPHENYLAMINECHLOROARSINE (DOT) see PDB000
DIPHENYLAMINE HYDROGEN SULFATE see DVY000
DIPHENYLAMINE SULFATE see DVY000
DIPHENYLAN see DKQ000
N,N-DIPHENYLANILINE see TMQ500
DIPHENYLAN SODIUM see DNU000
DIPHENYLARSINOUS ACID see DVY100
DIPHENYLARSINOUS CHLORIDE see CGN000
DIPHENYLBENZENE see TBD000
m-DIPHENYLBENZENE see TBC620
p-DIPHENYLBENZENE see TBC750
1,2-DIPHENYLBENZENE see TBC640
1,4-DIPHENYLBENZENE see TBC750
DIPHENYLBIS(PHENYLTHIO)STANNANE see DVY800
DIPHENYLBIS(PHENYLTHIO)TIN see DVY800
DIPHENYL BLUE 2B see CMO000
DIPHENYL BLUE 3B see CMO250
DIPHENYL BRILLIANT BLUE FF see CMN750
DIPHENYLBUTAZONE see BRF500
1,3-DIPHENYL-2-BUTEN-1-ONE see MPL000
(Z)-2-(p-(1,2-DIPHENYL-1-BUTENYL)-PHENOXY)-N,N-DIMETHYLETHYLAM-INE see NOA600
trans-2-(p-(1,2-DIPHENYL-1-BUTENYL)PHENOXY)-N,N-DIMETHYLETHYLAM-INE CITRATE see DVY875
1,2-DIPHENYL-4-BUTYL-3,5-DIKETOPYRAZOLIDINE CALCIUM SALT see PEO750
1,2-DIPHENYL-4-BUTYL-3,5-DIOXOPYRAZOLIDINE see BRF500
1,1-DIPHENYL-2-BUTYNYL-N-CYCLOHEXYLCARBAMATE see DVY900
DIPHENYLCARBAMIC ACID-2-(DIETHYLAMINO)ETHYL ESTER HYDRO-CHLORIDE see DHY200
DIPHENYLCARBAMOTHIOIC ACID-S-(2-(DIETHYLAMINO)ETHYL) ESTER HYDROCHLORIDE see PDC875

2-(3,3-DIPHENYL-3-(5-METHYL-1,3,4-OXADIAZOL-2-YL)PROPYL)-2-
AZABICYCLO(2.2.2)OCTANE see DWF700
2-(2-((4-DIPHENYLMETHYL)-1-PIPERAZINYL)ETHOXY)ETHANOL
HYDROXYDIETHYLPHENAMINE see BBV750
1-(3-(4-(DIPHENYLMETHYL)-1-PIPERAZINYL)PROPYL)-2-
BENZIMIDAZOLINONE see OMG000
1-(3-(4-(DIPHENYLMETHYL)-1-PIPERAZINYL)PROPYL)-1,3-DIHYDRO-2H-
BENZ IMIDAZOL-2-ONE see OMG000
4-DIPHENYLMETHYLPIPERIDINE see DWF865
4,4-DIPHENYL-1-METHYLPIPERIDINE MALEATE see DWF869
5,5-DIPHENYL-3-(3-(2-METHYLPIPERIDINO)PROPYL)-2-THIOHYDANTOIN
HYDROCHLORIDE see DWF875
5,5-DIPHENYL-3-(3-(2-METHYLPIPERIDINO)PROPYL)-2-THIOHYDANTOIN
MONOHYDROCHLORIDE see DWF875
α-2,2-DIPHENYL-4-(1-METHYL-2-PIPERIDYL)-1,3-DIOXOLANE HYDROCHLO-
RIDE see DWG600
β-2,2-DIPHENYL-4-(1-METHYL-2-PIPERIDYL)-1,3-DIOXOLANE HYDROCHLO-
RIDE see DWH000
α-2,2-DIPHENYL-4-(1-METHYL-2-PIPERIDYL)-1,3-DIOXOLANE METHYLIOD-
IDE see DWH200
3,3-DIPHENYL-2-METHYL-1-PYRROLINE see DWH500
DIPHENYLMETHYLSILANOL see DWH550
4-DIPHENYLMETHYLTROPYLTROPINIUM BROMIDE see PEM750
4-DIPHENYLMETHYL-dl-TROPYLTROPINIUM BROMIDE see PEM750
4,4-DIPHENYL-6-MORPHOLINO-3-HEPTANONE HYDROCHLORIDE see
DWH600
4,4-DIPHENYL-6-MORPHOLINO-3-HEXANONE HYDROCHLORIDE see
DWH800
2,3-DIPHENYL-3H-NAPHTHO(1,2-d)TRIAZOLIUM CHLORIDE see DWH875
DIPHENYLNITROSAMIN (GERMAN) see DWI000
DIPHENYLNITROSAMINE see DWI000
N,N-DIPHENYLNITROSAMINE see DWI000
DIPHENYL N-NITROSOAMINE see DWI000
o-DIPHENYLOL see BGJ250
2,2-DI(4-PHENYLOL)PROPANE see BLD500
2,5-DIPHENYLOXAZOLE see DWI200
4,5-DIPHENYL-2-OXAZOLEPROPANOIC ACID see OLW600
4,5-DIPHENYL-2-OXAZOLEPROPIONIC ACID see OLW600
N-(4,5-DIPHENYL-2-OXAZOLYL)DIACETAMIDE see ACI629
N-(4,5-DIPHENYLOXAZOL-2-YL)DIETHANOLAMINE MONOHYDRATE
see BKB250
2,2'-((4,5-DIPHENYL-2-OXAZOLYL)IMINO)-DIETHANOLMONOHYDRATE
see BKB250
3,3-DIPHENYL-2-OXETANONE see DWI400
DIPHENYL OXIDE see PFA850
DIPHENYL-4-Γ-OXO-Γ-BUTRIC ACID see BGL250
1,2-DIPHENYL-4-(3'-OXOBUTYL)-3,5-DIOXOPYRAZOLIDINE see KGK000
DIPHENYLOXOSTANNANE, POLYMER see DWO600
DIPHENYL PENTACHLORIDE see PAV600
1,5-DIPHENYL-1,4-PENTAZDIENE see DWI800
DIPHENYL-p-PHENYLENEDIAMINE see BLE500
N,N'-DIPHENYL-p-PHENYLENEDIAMINE see BLE500
1,2-DIPHENYL-4-(2'-PHENYLSULFINETHYL)-3,5-PYRAZOLIDINEDIONE
see DWM000
5,5-DIPHENYL-1-PHENYLSULFONYLHYDANTOIN see DWJ300
5,5-DIPHENYL-1-(PHENYLSULFONYL)-2,4-IMIDAZOLIDINEDIONE (9CI)
see DWJ300
1,2-DIPHENYL-4-PHENYLTHIOETHYL-3,5-PYRAZOLIDINEDIONE
see DWJ400
α,α-DIPHENYL-1-PIPERIDINEBUTANOL see DWK200
α,α-DIPHENYL-1-PIPERIDINEBUTANOL HYDROCHLORIDE see CCR875
α,α-DIPHENYL-2-PIPERIDINEMETHANOL see DWK400
α,α-DIPHENYL-2-PIPERIDINEMETHANOL HYDROCHLORIDE see PII750
α,α-DIPHENYL-4-PIPERIDINEMETHANOL HYDROCHLORIDE see PIY750
α,α-DIPHENYL-1-PIPERIDINEPROPANOL HYDROCHLORIDE see PMC250
α,α-DIPHENYL-1-PIPERIDINEPROPANOL METHANESULFONATE (ester)
see PMC275
DIPHENYL-PIPERIDINO-AETHYL-ACETAMID-BROMMETHYLAT (GERMAN)
see RDA375
1,1-DIPHENYL-3-N-PIPERIDINOBUTANOL-1 HYDROCHLORIDE see ARN700
2,2-DIPHENYL-4-(4-PIPERIDINO-4-CARBAMOYLPIPERIDINO)BUTYRONI-
TRILE see PJA140
1,1-DIPHENYL-3-PIPERIDINO-1-PROPANOL HYDROCHLORIDE see PMC250
5,5-DIPHENYL-3-(3-PIPERIDINOPROPYL)-2-THIOHYDANTOIN HYDROCHLO-
RIDE see DWK500
l-2,2-DIPHENYL-4-(2-PIPERIDYL)-1,3-DIOXOLANE HYDROCHLORIDE
see LFG100
d-2,2-DIPHENYL-4-(2-PIPERIDYL)-1,3-DIOXOLANE HYDROCHLORIDE
see DVP400
1,1-DIPHENYL-3-(1-PIPERIDYL)-1-PROPANOL HYDROCHLORIDE see PMC250
1,3-DIPHENYL-1-PROPEN-3-ONE see CDH000
α,α-DIPHENYL-β-PROPIOLACTONE see DWI400
1,2-DIPHENYL-2-PROPIONOXY-3-METHYL-4-DIMETHYLAMINOBUTANE
see PNA250

(+)-1,2-DIPHENYL-2-PROPIONOXY-3-METHYL-4-DIMETHYLAMINOBUTANE
HYDROCHLORIDE see PNA500
3-(3,3-DIPHENYLPROPYLAMINO)PROPYL-3',4',5'-TRIMETHOXYBENZOATE
HYDROCHLORIDE see DWK700
3-(N-d,d-DIPHENYLPROPYL-N-METHYL)AMINOPROPAN-1-OL HYDROCHLO-
RIDE see DWK900
N-(3,3-DIPHENYLPROPYL)-α-METHYLPHENETHYLAMINE see PEV750
N-(3,3-DIPHENYLPROPYL)-α-METHYLPHENETHYLAMINE LACTATE see
DWL200
1,1-DIPHENYL-2-PROPYN-1-OL CYCLOHEXANECARBAMATE see DWL400
1,1-DIPHENYL-2-PROPYNYL-N-CYCLOHEXYLCARBAMATE see DWL400
1,1-DIPHENYL-2-PROPYNYL ESTER CYCLOHEXANECARBAMIC ACID see
DWL400
1,1-DIPHENYL-2-PROPYNYL-N-ETHYLCARBAMATE see DWL500
1,1-DIPHENYL-2-PROPYNYL 1-PYRROLIDINECARBOXYLATE see DWL525
DIPHENYLPYRALIN-8-CHLOR-THEOPHYLLINAT (GERMAN) see PIZ250
DIPHENYLPYRALINE see LJR000
DIPHENYLPYRALINE HYDROCHLORIDE see DVW700
DIPHENYLPYRALINE TEOCLATE see PIZ250
1,2-DIPHENYL-3,5-PYRAZOLIDINEDIONE see DWA500
1,4-DIPHENYL-3,5-PYRAZOLIDINEDIONE see DWL600
1,3-DIPHENYL-5-PYRAZOLONE see DWL800
DIPHENYLPYRAZONE see DWM000
DIPHENYLPYRILENE see LJR000
3,3-DIPHENYL-3-(PYRROLIDINE-CARBONYLOXY)-1-PROPYNE see DWL525
α,α-DIPHENYL-3-QUINUCLIDINEMETHANOL HYDROCHLORIDE see
DWM400
DIPHENYL SELENIDE see PGH250
DIPHENYLSELENONE see DWM600
DIPHENYL SKY BLUE 6B see CMO500
DIPHENYLSTANNANE see DWM800
DIPHENYL SULFIDE see PGI500
DIPHENYL SULFONE see PGI750
DIPHENYL SULFOXIDE see DWN000
1-((DIPHENYLSULFOXIMIDO)METHYL)-1-METHYLPYRROLIDINIUM BRO-
MIDE see HAA325
trans-2,4-DIPHENYL-2,4,6,6-TETRAMETHYLCYCLOTRISILOXANE see DWN100
1,3-DIPHENYL-1,1,3,3-TETRAMETHYLDISILOXANE see DWN150
3,3',4,4'-DIPHENYLTETRAMINE see BGK500
DIPHENYLTHIOCARBAMIC ACID-S-(2-(DIETHYLAMINO)ETHYL) ESTER
see PDC850
DIPHENYLTHIOCARBAMIC ACID-S-(2-(DIETHYLAMINO)ETHYL) ESTER
HYDROCHLORIDE see PDC875
N,N'-DIPHENYLTHIOCARBAMIDE see DWN800
sym-DIPHENYLTHIOCARBAMIDE see DWN800
DIPHENYL THIOCARBAZIDE see DWN400
DIPHENYLTHIOCARBAZONE see DWN200
1,5-DIPHENYL-3-THIOCARBOHYDRAZIDE see DWN400
DIPHENYL THIOETHER see PGI500
5,5-DIPHENYL-2-THIOHYDANTOIN see DWN600
DIPHENYLTHIOLACETIC ACID-2-DIETHYLAMINOETHYL ESTER HYDRO-
CHLORIDE see DHY400
DIPHENYLTHIOUREA see DWN800
1,3-DIPHENYLTHIOUREA see DWN800
1,1-DIPHENYL-2-THIOUREA see DWO000
1,3-DIPHENYL-2-THIOUREA see DWN800
N,N'-DIPHENYLTHIOUREA see DWN800
sym-DIPHENYLTHIOUREA see DWN800
DIPHENYLTIN see DWM800
DIPHENYLTIN DIBROMIDE see DDN200
DIPHENYLTIN DICHLORIDE see DWO400
DIPHENYLTIN DIHYDRIDE see DWM800
DIPHENYLTIN OXIDE POLYMER see DWO600
DIPHENYL TOLYL PHOSPHATE see TGY750
1,3-DIPHENYLTRIAZENE see DWO800
5,6-DIPHENYL-as-TRIAZIN-3-OL see DWO875
3,5-DIPHENYL-s-TRIAZOLE see DWO950
3,5-DIPHENYL-1,2,4-TRIAZOLE see DWO950
3,5-DIPHENYL-1H-1,2,4-TRIAZOLE see DWO950
DIPHENYLTRICHLOROETHANE see DAD200
2,2-DIPHENYL-1,1,1-TRICHLOROETHANE see DWP000
1,3-DIPHENYLUREA see CBM250
N,N'-DIPHENYLUREA see CBM250
sym-DIPHENYLUREA see CBM250
2,2-DIPHENYL-VALERIC ACID-2-(DIETHYLAMINO)ETHYL)ESTER see DIG400
DIPHER see EIR000
DIPHERGAN see PMI750
DI-PHETINE see DKQ000, DNU000
DIPHEXAMIDE METHIODIDE see BTA325
DIPHONE see SOA500
DIPHOSGENE see PGX000
DIPHOSPHANE see DWP229
1,2-DIPHOSPHINOETHANE see DWP250
DIPHOSPHORIC ACID, DISODIUM SALT see DXF800

DIPHOSPHORIC ACID, TETRAMETHYL ESTER see TDV000
DIPHOSPHORIC ACID THETHRAETHYL ESTER see TCF250
DIPHOSPHORUS PENTOXIDE see PHS250
DIPHOSPHORUS TRIOXIDE see PHT500
DIPHTHERIA TOXIN see DWP300
DIPHYL see PFA860
DIPHYLLIN see DNC000
DIPHYLLIN-3,4-o-DIMETHYL XYLOSIDE see CMV375
DIPIDOLOR see PJA140
DIPIGYL see NCR000
DIPIN see BJC250
DIPINE see BJC250
DIPIPERAL see FHG000
2-β,16-β-DIPIPERIDINO-5-α-ANDROSTAN-3-α,17-β-DIOL DIPIVALATE HYDRO-
   CHLORIDE see DWP500
2,2′,2″,2‴-(4,8-DIPIPERIDINOPYRIMIDO(5,4-d)PYRIMIDINE-2,6-
   DIYLDINITRILO)TETRAETHANOL see PCP250
DIPIPERON see FHG000
DIPIPERONE see FHG000
DIPIRARTRIL-TROPICO see DUD800
DIPIRIN see DOT000
DIPIRITRAMIDE see PJA140
DIPIVEFRINE HYDROCHLORIDE see DWP559
DIPIVEFRIN HYDROCHLORIDE see DWP559
DIPLIN see DMZ000
DIPN see DNB200
DIPOFENE see DCM750
DIPO-SAFT see BFC750
DIPOTASSIUM4,4′-BIS(4-PHENYL-1,2,3-TRIAZOL-2-YL)STILBENE-2,2′-DIS-
   ULFONATE see BLG100
DIPOTASSIUM 4,4′-BIS(4-PHENYL-1,2,3-TRIAZOL-2-YL)STILBENE-2,2′-SULFO-
   NATE see BLG100
DIPOTASSIUM CHLORAZEPATE see CDQ250
DIPOTASSIUM CHROMATE see PLB250
DIPOTASSIUM CLORAZEPATE see CDQ250
DIPOTASSIUM CYCLOOCTATETRAENE see DWP900
DIPOTASSIUM DIAZIRINE-3,3-DICARBOXYLATE see DWP950
DIPOTASSIUM DICHLORIDE see PLA500
DIPOTASSIUM DICHROMATE see PKX250
DIPOTASSIUM-meso-N,N-DISULFO-3,4-DIPHENYLHEXANE see SPA650
DIPOTASSIUM ETHYLENEDIAMINETETRAACETATE see EJA250
DIPOTASSIUM MONOCHROMATE see PLB250
DIPOTASSIUM MONOPHOSPHATE see PLQ400
DIPOTASSIUM NICKEL TETRACYANIDE see NDI000
DIPOTASSIUM PERSULFATE see DWQ000
DIPOTASSIUM TETRACHLOPALLADATE see PLN750
DIPOTASSIUM TETRACYANONICKELATE see NDI000
DIPOTASSIUM TRICHLORONITROPLATINATE see PLN050
DIPPEL'S OIL see BMA750
DIPPING ACID see SOI500
DIPRAM see DGI000
DIPRAMID see DAB875
DIPRAMIDE see DAB875
DIPRAZINE see DQA400
DIPROFILLIN see DNC000
DIPROFILLINE see DNC000
DIPRON see SNM500
DIPROPANEDIOL DIBENZOATE see DWS800
DIPROPANOIC ACID GERMANIUM SEQUIOXIDE see CCF125
DIPROPARGYL ETHER see POA500
DI-2-PROPENYLAMINE see DBI600
DI-2-PROPENYL ESTER, 1,2-BENZENEDICARBOXYLIC ACID see DBL200
DI-2-PROPENYL PHOSPHONITE see DBL000
N-N-DI-2-PROPENYL-2-PROPEN-1-AMINE see THN000
5,5-DI-2-PROPENYL-2,4,6(1H,3H,5H)-PYRIMIDINETRIONE (9CI) see AFS500
DIPROPETRYN see EPN500
DIPROPETRYNE see EPN500
DIPROPHYLLIN see DNC000
DIPROPHYLLINE see DNC000
DIPROPIONATE BECLOMETHASONE see AFJ625
DIPROPIONATE d'OESTRADIOL (FRENCH) see EDR000
DIPROPIONATO de ESTILBENE (SPANISH) see DKB000
p,p′-DIPROPIONOXY-trans-α,β-DIETHYLSTILBENE see DKB000
DIPROPIONYL PEROXIDE see DWQ800
1,1-DIPROPOXYETHANE see AAG850
4,5-DIPROPOXY-2-IMIDAZOLIDINONE see DWQ850
DIPROPYL ACETAL see AAG850
DIPROPYLACETAMIDE see PNX600
DIPROPYLACETATE SODIUM see PNX750
DIPROPYLACETIC ACID see PNR750
N-DIPROPYLACETIC ACID see PNR750
DIPROPYLACETIC ACID CALCIUM SALT see CAY675
DIPROPYL ADIPATE see DWQ875
DI-n-PROPYL ADIPATE see DWQ875

DIPROPYLAMINE see DWR000
DI-n-PROPYLAMINE see DWR000
n-DIPROPYLAMINE see DWR000
1-DIPROPYLAMINOACETYLINDOLINE see DWR200
4-(DI-N-PROPYLAMINO)-3,5-DINITRO-1-TRIFLUOROMETHYLBENZENE
   see DUV600
4-((DIPROPYLAMINO)SULFONYL)BENZOIC ACID see DWW000
DIPROPYLCARBAMIC ACID ETHYL ESTER see DWT400
DIPROPYLCARBAMOTHIOIC ACID-S-ETHYL ESTER see EIN500
DIPROPYLDIAZENE 1-OXIDE see ASP500
DIPROPYL-2,2′-DIHYDROXYAMINE see DNL600
9,10-DI-n-PROPYL-9-10-DIHYDROXY-9,10-DIHYDRO-1,2,5,6-
   DIBENZANTHRACENE see DLK600
N,N-DI-N-PROPYL-2,6-DINITRO-4-TRIFLUOROMETHYLANILINE see DUV600
N³,N³-DIPROPYL-2,4-DINITRO-6-TRIFLUOROMETHYL-m-PHENYLENEDIAM-
   INE see DWS200
DI-n-PROPYL DISELENIDE see PNI850
α,α′-DIPROPYLENEDINITRILODI-o-CRESOL see DWS400
DIPROPYLENE GLYCOL see OQM000
DIPROPYLENE GLYCOL BUTYL ETHER see DWS600
DIPROPYLENE GLYCOL DIACRYLATE see DWS650
DIPROPYLENE GLYCOL DIBENZOATE see DWS800
DIPROPYLENE GLYCOL DIPELARGONATE see DWT000
DIPROPYLENE GLYCOL METHYL ETHER see DWT200
DIPROPYLENE GLYCOL MONOMETHYL ETHER see DWT200
DIPROPYLENE GLYCOL, 3,3,5-TRIMETHYLCYCLOHEXYL ETHER
   see TLO600
DIPROPYLENETRIAMINE see AIX250
DI-n-PROPYLESSIGSAURE (GERMAN) see PNR750
DIPROPYL ETHER see PNM000
DI-n-PROPYLETHER see PNM000
N,N-DI-n-PROPYL ETHYL CARBAMATE see DWT400
1-(N,N-DIPROPYLGLYCYL)INDOLINE see DWR200
DIPROPYL ISOCINCHOMERONATE see EAU500
DI-N-PROPYL-ISOCINCHOMERONATE (GERMAN) see EAU500
DIPROPYL KETONE see DWT600
DI-n-PROPYL MALEATE-ISOSAFROLE CONDENSATE see PNP250
DIPROPYL MERCURY see DWU000
DIPROPYL METHANE see HBC500
DI-n-PROPYL 6,7-METHYLENEDIOXY-3-METHYL-1,2,3,4-
   TETRAHYDRONAPHTHALENE see PNP250
DI-n-PROPYL-3-METHYL-6,7-METHYLENEDIOXY-1,2,3,4-
   TETRAHYDRONAPHTHALENE-1,2-DICARBOXYLATE see PNP250
S,S-DIPROPYL METHYLPHOSPHONOTRITHIOATE see DWU200
O,O-DI-n-PROPYL-O-(4-METHYLTHIOPHENYL)PHOSPHATE see DWU400
DI-n-PROPYLNITROSAMINE see NKB700
α-DIPROPYLNITROSAMINE METHYL ETHER see DWU800
DIPROPYLNITROSOAMINE see NKB700
DIPROPYL OXIDE see PNM000
DIPROPYLOXOSTANNANE see DWV000
DIPROPYL PEROXIDE see DWV200
DI-n-PROPYL PEROXYDICARBONATE see DWV400
DIPROPYL PHTHALATE see DWV500
DI-n-PROPYL PHTHALATE see DWV500
N,N-DIPROPYL-1-PROPANAMINE see TMY250
DIPROPYL 2,5-PYRIDINEDICARBOXYLATE see EAU500
DIPROPYL SUCCINATE see DWV800
DI-N-PROPYL SUCCINATE see DWV800
4-(DIPROPYLSULFAMOYL)BENZOIC ACID see DWW000
p-(DIPROPYLSULFAMOYL)BENZOIC ACID see DWW000
p-(DIPROPYLSULFAMOYL)BENZOIC ACID SODIUM SALT see DWW200
p-(DIPROPYLSULFAMYL)BENZOIC ACID see DWW000
p-(DI-N-PROPYLSULFAMYL)BENZOIC ACID SODIUM SALT see DWW200
DIPROPYL-5,6,7,8-TETRAHYDRO-7-METHYLNAPHTHO(2,3-d)-1,3-DIOXOLE-
   5,6-DICARBOXYLATE see PNP250
N,N-DIPROPYLTHIOCARBAMIC ACID-S-ETHYL ESTER see EIN500
DIPROPYLTHIOCARBAMIC ACID-S-PROPYL ESTER see PNI750
DI-n-PROPYLTIN BISMETHANESULFONATE see DWW400
DIPROPYLTIN CHLORIDE see DFF400
DIPROPYLTIN DICHLORIDE see DFF400
DI-n-PROPYLTIN DICHLORIDE see DFF400
DIPROPYLTIN OXIDE see DWV000
N,N-DIPROPYL-4-TRIFLUOROMETHYL-2,6-DINITROANILINE see DUV600
DIPROPYL ZINC see DWW500
DI(2-PROPYNYL) ETHER see POA500
DIPROSONE see BFV765
DIPROSTRON see EDR000
DIPROZIN see DQA400
DIPTERAX see TIQ250
DIPTEREX see TIQ250
DIPTEREX 50 see TIQ250
DIPTEVUR see TIQ250
DIPTHAL see DNS600
DIPYRIDAMINE see PCP250

DIPYRIDAMOL see PCP250
DIPYRIDAMOLE see PCP250
DIPYRIDAN see PCP250
DIPYRIDINESODIUM see DWW600
DIPYRIDO(2,3-D,2,3-K)PYRENE see NAU500
DIPYRIDO(1,2-a:3',2'-d)IMIDAZOL-2-AMINE see DWW700
DIPYRIDO(1,2-a:3',2'-d)IMIDAZOLE, 2-AMINO-, HYDROCHLORIDE see AJS225
DIPYRIDO(1,2-a:3',2'-d)IMIDAZOLE, 2-AMINO-6-METHYL-, HYDROCHLO-
 RIDE see AKS275
DIPYRIDO(2,3-d,2,3-1)PYRENE see NAU000
2,2'-DIPYRIDYL see BGO500
α,α'-DIPYRIDYL see BGO500
DIPYRIDYLDIHYDRATE see PAI995
DIPYRIDYL HYDROGEN PHOSPHATE see DWX000
DI-3-PYRIDYLMERCURY see DWX200
1,2-DI-3-PYRIDYL-2-METHYL-1-PROPANONE see MCJ370
DIPYRIDYL PHOSPHATE see DWX000
DIPYRIN see DOT000
DIPYROXIME see TLQ500
cis-DIPYRROLIDINEDICHLOROPLATINUM(II) see DEU200
1,4-DIPYRROLIDINYL-2-BUTYNE see DWX600, DWX600
DIPYUDAMINE see PCP250
DIQUAT see DWX800
DIQUAT DIBROMIDE see DWX800
DIQUAT DICHLORIDE see DWY000
1,3-DIQUINOLIN-6-YLUREA BISMETHOSULFATE see PJA120
DIRALGAN see TKG000
DIRAM A see ANG500
DIRAME see PMX250
DIRAX see AQN635, IDJ500
DIRCA PALUSTRIS see LEF100
DIRECT BLACK A see AQP000
DIRECT BLACK META see AQP000
DIRECT BLUE 6 see CMO000
DIRECT BLUE 14 see CMO250
DIRECT BRILLIANT BLUE FF see CMN750
DIRECT BROWN 95 see CMO750
DIRECT BROWN BR see PEY000
DIRECT FAST VIOLET N see CMP000
DIRECT PURE BLUE see CMO500
DIRECT RED 28 see SGQ500
DIRECT VIOLET C see CMP000
DIREKTAN see NCQ900
DIREMA see CFY000
DIREN see UVJ450
DIRESORCYL SULFIDE see BJE500
DIREX 4L see DXQ500
DIREZ see DEV800
DIRIAN see BOL325
DIRIDONE see PDC250, PEK250
DIRONYL see DLR150
DIROX see HIM000
DISADINE see PKE250
N,N'-DISALICYLIDENE-1,2-PROPANEDIAMINE see DWS400
DISALICYLALPROPYLENEDIIMINE see DWS400
DISALICYLIC ACID see SAN000
N,N'-DISALICYLIDENE-1,2-DIAMINOPROPANE see DWS400
N,N'-DISALICYLIDENE ETHYLENEDIAMINE see DWY200
DISALUNIL see CFY000
DISALYL see SAN000
DISAN see DNO800
DISATABS TABS see VSK600
DISCOLITE see FMW000
DISDOLEN see PHA575
DISELENIDE, BIS(2,2-DIETHOXYETHYL)- see BJA200
DISELENIDE, DIBUTYL-(9CI) see BRF550
DISELENIDE, DIPROPYL-(9CI) see PNI850
DISELENIUM DICHLORIDE see SBS500
α,α'-DISELENOBIS-o-ACETOTOLUIDIDE see DWY400
4,4'-DISELENOBIS(2-AMINOBUTYRIC ACID) see SBU710
2,2'-DISELENOBIS(N-PHENYLACETAMIDE) see DWY600
3,3'-DISELENODIALANINE see DWY800, SBP600
p,p'-DISELENODIANILINE see DWZ000
β,β'-DISELENODIPROPIONIC ACID, SODIUM SALT see DWZ100
DISELENO SALICYLIC ACID see SBU150
DI-SEPTON see DAO500
DISETIL see DXH250
DISFLAMOLL TKP see TNP500
DISFLAMOLL TOF see TNI250
DISILANE see DXA000
DISILOXANE, 1,3-DIPHENYL-1,1,3,3-TETRAMETHYL- see DWN150
DISILVER ACETYLIDE SILVER NITRATE see SDJ025
DISILVER CYANAMIDE see DXA500
DISILVER KETENIDE see DXA600

DISILVER OXIDE see SDU500
DISILVER PENTATIN UNDECAOXIDE see DXA800
DISILYN see BEN000
DISIPAL see MJH900
DISIPAL HYDROCHLORIDE see OJW000
DI-SIPIDIN see ORU500
DISNOGALAMYCINIC ACID see DXA900
DISNOGAMYCIN see NMV600
DISODIUM ANTHRAQUINONE-1,5-DISULFONATE see DLJ700
DISODIUM ARSENATE see ARC000
DISODIUM ARSENATE, HEPTAHYDRATE see ARC250
DISODIUM ARSENIC ACID see ARC000
DISODIUM AUROTHIOMALATE see GJC000
DISODIUM-4,4'-BIS((4-AMINO-6-(2-HYDROXYETHYL)AMINO-s-TRIAZIN-2-
 YL)AMINO)-2,2'-STILBENDISULFONIC ACID see DXB400
DISODIUM-4,4'-BIS((4-ANILINO-6-METHOXY-s-TRIAZIN-2-YL)AMINO)
 STILBENE-2,2'-D ISULFONATE see BGW100
DISODIUM-4,4'-BIS(2-SULFOSTYRYL)BIPHENYL see TGE150
DISODIUM CARBONATE see SFO000
DISODIUM-N-(3-(CARBOXYMETHYLTHIOMERCURI)-2-METHOXYPROPYL)-
 α-CAMPHORAMATE see TFK270
DISODIUM CHROMATE see DXC200
DISODIUM CHROMOGLYCATE see CNX825
DISODIUM CINNAMYLIDENE BISULFITE derivative of SULFAPYRIDINE
 see DXF400
DISODIUM CITRATE see DXC400
DISODIUM CROMOGLICATE see CNX825
DISODIUM CROMOGLYCATE see CNX825
DISODIUM DEXAMETHASONE PHOSPHATE see DAE525
DISODIUM DIACID ETHYLENEDIAMINETETRAACETATE see EIX500
DISODIUM-3,3'-DIAMINO-4,4'-DIHYDROXYARSENOBENZENE-N-
 DIMETHYLENESULFONATE see SNR000
DISODIUM-3,3'-DIAMINO-4,4'-DIHYDROXYARSENOBENZENE-N,N'-
 DIMETHYLENEBISULFITE see SNR000
DISODIUM p,p'-DIAMINODIPHENYLSULFONE-N,N-DIGLUCOSE SULFO-
 NATE see AOO800
DISODIUM-2,7-DIBROM-4-HYDROXY-MERCURI-FLUORESCEIN see MCV000
DISODIUM-2',7'-DIBROMO-4'-(HYDROXYMERCURY)FLUORESCEIN
 see MCV000
DISODIUM DICHROMATE see SGI000
DISODIUM DIFLUORIDE see SHF500
DISODIUM DIHYDROGEN ETHYLENEDIAMINETETRAACETATE see EIX500
DISODIUM DIHYDROGEN(ETHYLENEDINITRILO)TETRAACETATE
 see EIX500
DISODIUM DIHYDROGEN-(1-HYDROXYETHYLIDENE)DIPHOSPHONATE
 see DXD400
DISODIUM DIHYDROGEN PYROPHOSPHATE see DXF800
DISODIUM-1,3-DIHYDROXY-1,3-BIS-(aci-NITROMETHYL)-2,2,4,4-
 TETRAMETHYLCYCLO BUTANE see DXC600
N,N'-DISODIUM N,N'-DIMETHOXYSULFONYLDIAMIDE see DXC800
DISODIUM(2,4-DIMETHYLPHENYLAZO)-2-HYDROXYNAPHTHALENE-3,6-
 DISULFONATE see FMU070
DISODIUM(2,4-DIMETHYLPHENYLAZO)-2-HYDROXYNAPHTHALENE-3,6-
 DISULPHONATE see FMU070
DISODIUM DIOXIDE see SJC500
DISODIUM DIPHOSPHATE see DXF800
DISODIUM-4,4'-DISULFOXYDIPHENYL-(2-PYRIDYL)METHANE see SJJ175
DISODIUM EDATHAMIL see EIX500
DISODIUM EDETATE see EIX500
DISODIUM EDTA (FCC) see EIX500
DISODIUM-3,6-ENDOXOHEXAHYDROPHTHALATE see DXD000
DISODIUM EOSIN see BNH500
DISODIUM-3,6-EPOXYCYCLOHEXANE-1,2-DICARBOXYLATE see DXD000
DISODIUM (−)-(1R,2S)-(1,2-EPOXYPROPYL)PHOSPHONATE HYDRATE see
 FOL200
DISODIUM ETHANOL-1,1-DIPHOSPHONATE see DXD400
DISODIUM ETHYDRONATE see DXD400
DISODIUM ETHYLENEBIS(DITHIOCARBAMATE) see DXD200
DISODIUM ETHYLENE-1,2-BISDITHIOCARBAMATE see DXD200
DISODIUM ETHYLENEDIAMINETETRAACETATE see EIX500
DISODIUM ETHYLENEDIAMINETETRAACETIC ACID see EIX500
DISODIUM (ETHYLENEDINITRILO)TETRAACETATE see EIX500
DISODIUM (ETHYLENEDINITRILO)TETRAACETIC ACID see EIX500
DISODIUM ETIDRONATE see DXD400
DISODIUM FLUOROPHOSPHATE see DXD600
DISODIUM FOSFOMYCIN see DXF600
DISODIUM FOSFOMYCIN HYDRATE see FOL200
DISODIUM FUMARATE see DXD800
DISODIUM GLYCYRRHIZIN see DXD875
DISODIUM GLYCYRRHIZINATE see DXD875
DISODIUM GMP see GLS800
DISODIUM-5'-GMP see GLS800
DISODIUM-5'-GUANYLATE see GLS800
DISODIUM GUANYLATE (FCC) see GLS800

DISODIUM-5'-GUANYLATE mixed with DISODIUM 5'-INOSINATE (1:1) see RJF400
DISODIUM HEXAFLUOROSILICATE see DXE000
(2-)-DISODIUM HEXAFLUOROSILICATE see DXE000
DISODIUM HYDROGEN ARSENATE see ARC000
DISODIUM HYDROGEN CITRATE see DXC400
DISODIUM HYDROGEN ORTHOARSENATE see ARC000
DISODIUM HYDROGEN PHOSPHATE see SJH090
DISODIUM-6-HYDROXY-3-OXO-9-XANTHENE-o-BENZOATE see FEW000
DISODIUM-5,5'-((2-HYDROXYTRIMETHYLENE)DIOXY)-BIS(4-OXO-4H-1-BENZOPYRAN-2-CARBOXYLATE) see CNX825
DISODIUM-3-HYDROXY-4-((2,4,5-TRIMETHYLPHENYL)AZO)-2,7-NAPH-THALENEDISULFONATE see FAG018
DISODIUM-3-HYDROXY-4-((2,4,5-TRIMETHYLPHENYL)AZO)-2,7-NAPH-THALENEDISULFONIC ACID see FAG018
DISODIUM-3-HYDROXY-4-((2,4,5-TRIMETHYLPHENYL)AZO)-2,7-NAPH-THALENEDISULPHONATE see FAG018
DISODIUM-3-HYDROXY-4-((2,4,5-TRIMETHYLPHENYL)AZO)-2,7-NAPH-THALENEDISULPHONIC ACID see FAG018
DISODIUM IMINODIACETATE see DXE200
DISODIUM IMP see DXE500
DISODIUM INDIGO-5,5-DISULFONATE see FAE100
DISODIUM INOSINATE see DXE500
DISODIUM-5'-INOSINATE see DXE500
DISODIUM-5'-INOSINATE mixed with DISODIUM 5'-GUANYLATE (1:1) see RJF400
DISODIUM INOSINE-5'-MONOPHOSPHATE see DXE500
DISODIUM INOSINE-5'-PHOSPHATE see DXE500
DISODIUM LATAMOXEF see LBH200
DISODIUM METASILICATE see SJU000
DISODIUM METHANEARSENATE see DXE600
DISODIUM METHANEARSONATE see DXE600
DISODIUM METHOTREXATE see MDV600
DISODIUM METHYLARSENATE see DXE600
DISODIUM METHYLARSONATE see DXE600
DISODIUM MOLYBDATE see DXE800
DISODIUM MOLYBDATE DIHYDRATE see DXE875
DISODIUM MONOFLUOROPHOSPHATE see DXD600
DISODIUM MONOHYDROGEN ARSENATE see ARC000
DISODIUM MONOHYDROGEN PHOSPHATE see SJH090
DISODIUM MONOMETHYLARSONATE see DXE600
DISODIUM MONOSILICATE see SJU000
DISODIUM MONOXIDE see SIN500
DISODIUM NITRILOTRIACETATE see DXF000
DISODIUM NITROPRUSSIDE DIHYDRATE see SIW500
DISODIUM NITROSYLPENTACYANOFERRATE see SIU500
DISODIUM OCTABORATE, TETRAHYDRATE see DXF200
DISODIUM ORTHOPHOSPHATE see SJH090
DISODIUM-7-OXABICYCLO(2.2.1)HEPTANE-2,3-DICARBOXYLATE see DXD000
DISODIUM OXIDE see SIN500
DISODIUM PEROXIDE see SJC500
DISODIUM-2-(p-(1'-PHENYLPROPYLAMINO)BENZENESULFONAMIDO)PYRIDINE see DXF400
DISODIUM PHOSPHATE see SJH090
DISODIUM PHOSPHONOMYCIN see DXF600
DISODIUM PHOSPHONOMYCIN HYDRATE see FOL200
DISODIUM PHOSPHORIC ACID see SJH090
DISODIUM PHOSPHOROFLUORIDATE see DXD600
DISODIUM PROTOPORPHYRIN see DXF700
DISODIUM PYROPHOSPHATE see DXF800
DISODIUM PYROSULFITE see SII000
DISODIUM-5'-RIBONUCLEOTIDE see RJF400
DISODIUM SALT of EDTA see EIX500
DISODIUM SALT of ENDOTHALL see DXD000
DISODIUM SALT of 1-INDIGOTIN-S,S'-DISULPHONIC ACID see FAE100
DISODIUM SALT of 7-OXABICYCLO(2.2.1)HEPTANE-2,3-DICARBOXYLIC ACID see DXD000
DISODIUM SALT of 2-(4-SULPHO-1-NAPHTHYLAZO)-1-NAPHTHOL-4-SULPHONIC ACID see HJF500
DISODIUM SALT of 1-(2,4-XYLYLAZO)-2-NAPHTHOL-3,6-DISULFONIC ACID see FMU070
DISODIUM SALT of 1-(2,4-XYLYLAZO)-2-NAPHTHOL-3,6-DISULPHONIC ACID see FMU070
DISODIUM SELENATE see DXG000
DISODIUM SELENITE see SJT500
DISODIUM SEQUESTRENE see EIX500
DISODIUM SILICOFLUORIDE see DXE000
DISODIUM 2-(4-STYRYL-3-SULFOPHENYL)-7-SULFO-2H-NAPHTHO(1,2-d)TRIAZOLE see DXG025
DISODIUM SULBENICILLIN see SNV000
DISODIUM SULFATE see SJY000
DISODIUM SULFITE see SJZ000
DISODIUM SULFOBENZYLPENICILLIN see SNV000

DISODIUM α-SULFOBENZYLPENICILLIN see SNV000
DISODIUM-2-(4-SULFO-1-NAPHTHYLAZO)-1-NAPHTHOL-4-SULFONATE see HJF500
DISODIUM-2-(4-SULPHO-1-NAPHTHYLAZO)-1-NAPHTHOL-4-SULPHONATE see HJF500
DISODIUM TARTRATE see BLC000
DISODIUM l-(+)-TARTRATE see BLC000
DISODIUM TETRACEMATE see EIX500
DISODIUM-5-TETRAZOLAZOCARBOXYLATE see DXG050
DISODIUM VERSENATE see EIX500
DISODIUM VERSENE see EIX500
DISOFEN see DNG000
DISOLFURO DI TETRAMETILTIOURAME (ITALIAN) see TFS350
DISOMAR see DXE600
DISOMER MALEATE see DXG100
2,4-D ISOOCTYL ESTER see ILO000
DISOPHENOL see DNG000
2,4-D ISOPROPYL ESTER see IOY000
DISOPYRAMIDE see DNN600
DISORLON see CCK125
DISPADOL see DAM700
DISPAL see AHE250
DISPAMIL see PAH250
DISPARICIDA see ABX500
DISPASOL M see PKB500
DISPERMINE see PIJ000
DISPERSE BLUE K see MGG250
DISPERSE BLUE NO 1 see TBG700
DISPERSED BLUE 12195 see FAE000
DISPERSED VIOLET 12197 see FAG120
DISPERSE MB-61 see TFC600
DISPERSE ORANGE see AKP750
DISPERSE VIOLET K see DBP000
DISPERSE YELLOW 6Z see MEB750
DISPERSOL YELLOW PP see PEJ500
DISPEX C40 see ADW200
DISPHEX see PKE250
DISPHOLIDUS TYPHUS VENOM see DXG150
3,6-DI(SPIROCYCLOHEXANE)TETRAOXANE see DXG200
DISPRANOL see BIK500
DISSENTEN see LIH000
DISSOLVANT APV see DJD600
DISTACLOR see CCR850
DISTAKAPS V-K see PDT750
DISTAMICINA A (ITALIAN) see SLI300
DISTAMINE see PAP550
DISTAMYCIN see DXG400
DISTAMYCIN A see SLI300
DISTAMYCIN A/5 see DXG500
DISTAMYCIN A HYDROCHLORIDE see DXG600
DISTANNANE, HEXAISOPROPYL- see HDY100
DISTANNATHIANE, HEXABUTYL-(9CI) see HCA700
DISTANNATHIANE, HEXAKIS(PHENYLMETHYL)-(9CI) see BLK750
DISTANNOXANE, BIS(1,3-DITHIOCYANATO-1,1,3,3-TETRABUTYL- see BJM700
DISTANNOXANE, 1,3-BIS(2,4,5-TRICHLOROPHENOXY)-1,1,3,3-TETRABUTYL- see OPE100
DISTANNOXANE, HEXAETHYL- see HCX050
DISTANNOXANE, HEXAOCTYL- see HEW100
DISTANNOXANE, 1,1,1,3,3,3-HEXAPROPYL- see BLT300
DISTANNTHIANE, HEXABUTYL- see HCA700
DISTANNTHIANE, HEXAETHYL- see HCX100
DISTANNTHIANE, HEXAOCTYL- see HEW150
DISTANNTHIANE, HEXAPROPYL- see HEW200
DISTAQUAINE V see PDT500
DISTAQUAINE V-K see PDT750
DISTAVAL see TEH500
DISTAXAL see TEH500
DISTEARIN see OAV000
DISTESOL see EHP000
DISTESSOL see EHP000
DISTHENE see AHF500
DISTIGMINE BROMIDE see DXG800
DISTILBENE see DKA600, DKB000
DISTILLATES (PETROLEUM), ACID-TREATED HEAVY NAPHTHENIC (9CI) see MQV760
DISTILLATES (PETROLEUM), ACID-TREATED HEAVY PARAFFINIC (9CI) see MQV765
DISTILLATES (PETROLEUM), ACID-TREATED LIGHT NAPHTHENIC (9CI) see MQV770
DISTILLATES (PETROLEUM), ACID-TREATED LIGHT PARAFFINIC (9CI) see MQV775
DISTILLATES (PETROLEUM), HEAVY NAPHTHENIC (9CI) see MQV780
DISTILLATES (PETROLEUM), HEAVY PARAFFINIC (9CI) see MQV785

DISTILLATES (PETROLEUM), HYDROTREATED HEAVY NAPHTHENIC (9CI) see MQV790

DISTILLATES (PETROLEUM), HYDROTREATED HEAVY PARAFFINIC (9CI) see MQV795

DISTILLATES (PETROLEUM), HYDROTREATED LIGHT NAPHTHENIC (9CI) see MQV800

DISTILLATES (PETROLEUM), HYDROTREATED LIGHT PARAFFINIC (9CI) see MQV805

DISTILLATES (PETROLEUM), LIGHT NAPHTHENIC (9CI) see MQV810

DISTILLATES (PETROLEUM), LIGHT PARAFFINIC (9CI) see MQV815

DISTILLATES (PETROLEUM), SOLVENT-DEWAXED HEAVY NAPHTHENIC (9CI) see MQV820

DISTILLATES (PETROLEUM), SOLVENT-DEWAXED HEAVY PARAFFINIC (9CI) see MQV825

DISTILLATES (PETROLEUM), SOLVENT-DEWAXED LIGHT NAPHTHENIC (9CI) see MQV835

DISTILLATES (PETROLEUM), SOLVENT-DEWAXED LIGHT PARAFFINIC (9CI) see MQV840

DISTILLATES (PETROLEUM), SOLVENT-REFINED HEAVY NAPHTHENIC (9CI) see MQV845

DISTILLATES (PETROLEUM), SOLVENT-REFINED HEAVY PARAFFINIC (9CI) see MQV850

DISTILLATES (PETROLEUM), SOLVENT-REFINED LIGHT NAPHTHENIC (9CI) see MQV852

DISTILLATES (PETROLEUM), SOLVENT-REFINED LIGHT PARAFFINIC (9CI) see MQV855

DISTILLED LIME OIL see OGO000

DISTILLED MUSTARD see BIH250

DISTIVIT (B12 PEPTIDE) see VSZ000

DISTOBRAM see TGI250

DISTOKAL see HCI000

DISTOL 8 see EIV000

DISTOPAN see HCI000

DISTOPIN see HCI000

DISTOVAL see TEH500

DISTRANEURIN see CHD750

DISTYLIN see DMD000

DISUL see CNW000

1,3-DI(4-SULFAMOYLPHENYL)TRIAZENE see THQ900

1,3-DISULFAMYL-4,5-DICHLOROBENZENE see DEQ200

DISULFAN see DXH250

DISULFATON see DXH325

DISULFATOZIRCONIC ACID see ZTJ000

DISULFIDE DIBENZYL see DXH200

DISULFIDE DIPHENYL see PEW250

DISULFIRAM see DXH250

DISULFIRAM mixed with SODIUM NITRITE see SIS200

2,5-DISULFO-1-AMINOBENZENE see AIE000

2,5-DISULFOANILINE see AIE000

1,5-DISULFOANTHRAQUINONE see APK625

1,8-DISULFOANTHRAQUINONE see APK635

2,2'-DISULFOBENZIDINE see BBX500

3,5-DISULFOCATECHOL DISODIUM SALT see DXH300

DISULFO-meso-4,4-DIPHENYLHEXANE DIPOTASSIUM see SPA650

4,8-DISULFO-2-NAPHTHALAMINE see ALH250

DISULFOTON see DXH325

DISULFOTON DISULIDE see OQS000

DISULFOTON SULFOXIDE see OQS000

DISULFURAM see DXH250

DISULFUR DIBROMIDE see DXH350

DISULFUR DICHLORIDE see SON510

DISULFUR DINITRIDE see DXH400

DISULFURE de TETRAMETHYLTHIOURAME (FRENCH) see TFS350

DISULFUR HEPTAOXIDE see DXH600

DISULFUR PENTOXYDICHLORIDE see PPR500

DISULFURYL CHLORIDE see PPR500

DISULFURYL DIAZIDE see DXH800

DISULFURYL DICHLORIDE see PPR500

DISUL-Na see CNW000

DISULONE see SOA500

DISULPHINE LAKE BLUE EG see FMU059

DISULPHURAM see DXH250

DISULPHURIC ACID see SOI520

DISUL-SODIUM see CNW000

DISYNCRAM see MPE250

DISYNCRAN see MDT500, MPE250

DISYNFORMON see EDV000

DI-SYSTON see DXH325

DISYSTON SULFOXIDE see OQS000

DISYSTOX see DXH325

DI-TAC see DXE600

DITAK see UVJ450

DITAVEN see DKL800

DITAZOL MONOHYDRATE see BKB250

DITEFTIN see FQJ100

1,2-DI(5-TETRAZOLYL)HYDRAZINE see DXI000

1,3-DI(5-TETRAZOYL)TRIAZENE see DXI200

DITHALLIUM CARBONATE see TEJ000

DITHALLIUM SULFATE see TEM000

DITHALLIUM(1+) SULFATE see TEM000

DITHALLIUM TRIOXIDE see TEL050

DITHANE A-4 see DUQ600

DITHANE A-40 see DXD200

DITHANE D-14 see DXD200

DITHANE M-45 see DXI400

DITHANE M 22 SPECIAL see MAS500

DITHANE R-24 see BPU000

DITHANE S60 see DXI400

DITHANE SPC see DXI400

DITHANE STAINLESS see ANZ000

DITHANE ULTRA see DXI400

DITHANE Z see EIR000

1,4-DITHIAANTHRAQUINONE-2,3-DICARBONITRILE see DLK200

1,4-DITHIAANTHRAQUINONE-2,3-DINITRILE see DLK200

2,4-DITHIA-1,3-DIOXANE-2,2,4,4-TETRAOXIDE see DXI500

DITHIANON see DLK200

DITHIANONE see DLK200

1,3,2-DITHIARSENOLANE, 2-(p-BIS(2-CHLOROETHYL)AMINOPHENYL)- see BHW300

1,3,2-DITHIARSENOLANE, 2-(p-(DIETHYLAMINO)PHENYL)- see DIP100

3,6-DITHIA-3,4,5,6-TETRAHYDROPHTHALIMIDE see DLK700

2,6-DITHIA-1,3,5,7-TETRAZATRICYCLO(3.3.1.1^{3,7})DECANE-2,2,6,6-TETROXIDE see TDX500

DITHIAZANINE see DXI600

DITHIAZANINE IODIDE see DJT800

DITHIAZANIN IODIDE see DJT800

DITHIAZININE see DJT800

(R,S)-α-(1-((3,3-DI-3-THIENYLALLYL)AMINO)ETHYL)BENZYL ALCOHOL (+)-(α)-HYDROCHLORIDE see DXI800

(+)-α-(1-(93,3-DI-3-THIENYLALLYL)AMINO)ETHYL)BENZYL ALCOHOL HYDROCHLORIDE see DXI800

1-((3,3-DI-2-THIENYL-1-METHYL)ALLYL)PYRROLIDINE HYDROCHLORIDE see DXJ100

3-(DI(2-THIENYL)METHYLENE)-5-METHYLDECAHYDROQUINOLIZANIUM BROMIDE see TGF075

3-(DI-2-THIENYLMETHYLENE)-1-METHYLPIPERIDINE see BLV000

3-(DI-2-THIENYLMETHYLENE)-1-METHYLPIPERIDINE CITRATE see ARP875

3-(DI-2-THIENYLMETHYLENE)-5-METHYL-trans-QUINOLIZIDINIUM BROMIDE see TGF075

3,3-DI-2-THIENYL-N,N,1-TRIMETHYLALLYLAMINE HYDROCHLORIDE see TLQ250

1,3-DITHIETAN-2-YLIDENE PHOSPHORAMIDIC ACID DIETHYL ESTER see DHH250

DITHIO see SOD100

2,2'-DITHIOBIS(N-(1-ADAMANTYLMETHYL)ACETAMIDINE) DIHYDROCHLORIDE HEMIHYDRATE see DXJ400

O,O-DITHIO-BIS-ANILINE see DXJ800

2,2'-DITHIOBISANILINE see DXJ800

2,2'-DITHIOBIS(BENZOTHIAZOLE) see BDE750

1,1'-DITHIOBIS(N,N-DIETHYLTHIOFORMAMIDE) see DXH250

α,α'-DITHIOBIS(DIMETHYLTHIO)FORMAMIDE see TFS350

1,1'-DITHIOBIS(N,N-DIMETHYLTHIO)FORMAMIDE see TFS350

2,2'-DITHIOBIS(ETHYLAMINE) see MCN500

2,2'-DITHIO-BIS-(ETHYLAMINE) DIHYDROCHLORIDE see CQJ750

3,3'-DITHIOBIS(METHYLENE)BIS(5-HYDROXY-6-METHYL-4-PYRIDINEMETHANOL) DIHYDROCHLORIDE see BMB000

DITHIOBISMORPHOLINE see BKU500

4,4'-DITHIOBIS(MORPHOLINE) see BKU500

2,2'-DITHIOBIS(5-NITROPYRIDINE) see DXL200

2,2'-DITHIOBIS(PYRIDINE-1-OXIDE)MAGNESIUM SULFATE TRIHYDRATE see DXL400

DITHIOBIS(THIOFORMIC ACID)-o,o-DIETHYL ESTER see BJU000

2,5-DITHIOBIUREA see BLJ250

DITHIOBIURET see DXL800

DITHIOCARB see SGJ000, SGJ500

DITHIOCARBAMATE see SGJ000

DITHIOCARBAMOYLHYDRAZINE see MLJ500

DITHIOCARBANILIC ACID ETHYL ESTER see EOK550

DITHIOCARBONIC ACID-o-BUTYL ESTER POTASSIUM SALT see PKY850

DITHIOCARBONIC ACID-o-sec-BUTYL ESTER SODIUM SALT see DXM000

DITHIOCARBONIC ANHYDRIDE see CBV500

DITHIOCARBOXYMETHYL-p-CARBAMIDOPHENYLARSENOUS OXIDE see DXM100

DITHIOCARBOXYPHENYL-p-CARBAMIDOPHENYLARSENOUS OXIDE see ARK800

DI-mu-(THIOCYANATODI-n-BUTYLSTANNYLOXO)BIS(THIOCYANATODI-n-BUTYLTIN) see BJM700

DITHIODEMETON see DXH325

β,β'-DITHIODIALANINE see CQK325
2,2'-DITHIODIANILINE see DXJ800
N,N'-(DITHIODICARBONOTHIOYL)BIS(N-METHYLMETHANAMINE)
  see TFS350
2,2-DITHIODIETHANOL see DXM600
1,1'-DITHIODIETHYLENEBIS(3-(2-(CHLOROETHYL)-3-NITROSOUREA
  see BIF625
DITHIODIGLYCOL see DXM600
3,3'-DITHIODIMETHYLENEBIS(5-HYDROXY-6-METHYL-4-
  PYRIDINEMETHANOL) DIHYDROCHLORIDE HYDRATE see BMB000
5,5'-DITHIODIMETHYLENEBIS(2-METHYL-3-HYDROXY-4-
  HYDROXYMETHYLPYRIDINE)DIHYDROCHLORIDE HYDRATE
  see BMB000
N,N-DITHIODIMORPHOLINE see BKU500
4,4'-DITHIODIMORPHOLINE see BKU500
DITHIODIPHOSPHORIC ACID, TETRAETHYL ESTER see SOD100
2,2'-DITHIODIPYRIDINE-1,1'-DIOXIDE see DXN300
DITHIOETHYLENEGLYCOL see EEB000
1,1'-DITHIOFORMAMIDINE DIHYDROCHLORIDE see DXN400
DITHIOFOS see SOD100
DITHIOGLYCEROL see BAD750
1,2-DITHIOGLYCEROL see BAD750
DITHIOGLYCOL see EEB000
DITHIOGLYCOLYL p-ARSENOBENZAMIDE see TFA350
DITHIOHYDANTOIN see IAT100
DITHIOLANE see DXN600
DITHIOLANE IMINOPHOSPHATE see DXN600
1,2-DITHIOLANE-3-PENTANAMIDE (9CI) see DXN709
1,2-DITHIOLANE-3-VALERAMIDE see DXN709
1,2-DITHIOLANE-3-VALERIC ACID see DXN800
1,3-DITHIOLAN-2-YLIDENE-PHOSPHORAMIDOTHIOIC ACID DIETHYL
  ESTER see DXN600
1,3-DITHIOLAN-2-YLIDENE-PHOSPHORAMIDOTHIOIC ACID-O,O-DIETHYL
  ESTER see DXN600
4-((5-(1,2-DITHIOLAN-3-YL)-1-OXOPENTYL)AMINO)BUTANOIC ACID
  see LGK100
5-(1,2-DITHIOLAN-3-YL)VALERIC ACID see DXN800
1,3-DITHIOLIUM PERCHLORATE see DXN850
1,3-DITHIOL-2-YLIDENE-PROPANEDIOIC ACID BIS(1-METHYLETHYL)
  ESTER see MAO275
DITHIO-METHANEARSONOUS ACID BIS(ANHYDROSULFIDE) with
  DIMETHYLDITHIOCARBAMIC ACID see USJ075
DITHIOMETON (FRENCH) see PHI500
4,4'-DITHIOMORPHOLINE see BKU500
DITHION see DXO000
DITHIONE see DXO000, SOD100
DITHIONIC ACID see SOI520
DI(THIONOCARBOMETHOXY) DISULFIDE see DUN600
6,8-DITHIOOCTANOIC ACID see DXN800
DITHIOOXALDIIMIDIC ACID see DXO200
DITHIOOXAMIDE see DXO200
DITHIOPHOSPHATE de O,O-DIETHYLE et de (4-CHLORO-PHENYL)
  THIOMETHYLE (FRENCH) see TNP250
DITHIOPHOSPHATE de O,O-DIETHYLE et de S-(2-ETHYLTHIO-ETHYLE)
  (FRENCH) see DXH325
DITHIOPHOSPHATE de O,O-DIETHYLE et d'ETHYLTHIOMETHYLE
  (FRENCH) see PGS000
DITHIOPHOSPHATE de O,O-DIETHYLE et de S-N-METHYL-N-CAR-
  BOETHOXY CARBAMOYLMETHYLE (FRENCH) see DJI000
DITHIOPHOSPHATE de-O,O-DIETHYLE et de S(2,5-DICHLOROPHENYL)
  THIOMETHYLE (FRENCH) see PDC750
DITHIOPHOSPHATE de O,O-DIMETHYLE et de S-(4,6-DIAMINO-1,3,5-TRI-
  AZINE-2-YL)-METHYLE) (FRENCH) see ASD000
DITHIOPHOSPHATE de O,O-DIMETHYLE et de S-(1,2-
  DICARBOETHOXYETHYLE) (FRENCH) see MAK700
DITHIOPHOSPHATE de O,O-DIMETHYLE et de S-(2-ETHYLTHIO-ETHYLE)
  (FRENCH) see PHI500
DITHIOPHOSPHATE de O,O-DIMETHYLE et de S-
  ((MORPHOLINOCARBONYL)-METHYLE) (FRENCH) see MRU250
DITHIOPHOSPHATE de O,O-DIMETHYLE et de S-(N-METHYLCARBAMOYL-
  METHYLE) (FRENCH) see DSP400
DI(THIOPHOSPHORIC) ACID, TETRAETHYL ESTER see SOD100
DITHIOPHOSPHORSAEURE-O-AETHYL-S,S-DIPHENYLESTER (GERMAN)
  see EIM000
2,3-DITHIOPROPANOL see BAD750
DITHIOPROPYLTHIAMINE see DXO300
DITHIOPROPYLTHIAMINE HYDROCHLORIDE see DXO400
DITHIOPYROPHOSPHATE de TETRAETHYLE (FRENCH) see SOD100
DITHIOSYSTOX see DXH325
DITHIOTEP see SOD100
DITHIOTEREPHTHALIC ACID see DXO600
DITHIOTHREITOL see DXO775
1,4-DITHIOTHREITOL see DXO800
dl-DITHIOTHREITOL see DXO775

d-1,4-DITHIOTHREITOL see DXO800
rac-DITHIOTHREITOL see DXO775
dl-1,4-DITHIOTHREITOL see DXO775
DITHIOXAMIDE see DXO200
DITHIZON see DWN200
DITHIZONE see DWN200
DITHRANOL, 10-PROPIONYL- see PMW760
DITIAMINA see EIR000
DITILIN see CMG250, HLC500
DITILINE see CMG250, HLC500
DITILIN IODIDE see BJI000
DITIOVIT see DXO300
DITOIN see DNU000
DITOINATE see DKQ000
4,4'-DI-o-TOLUIDINE see TGJ750
1,4-DI-p-TOLUIDINOANTHRAQUINONE see BLK000
DI-o-TOLUYLTHIOUREA see DXP600
DITOLYLETHANE see DXP000
DI-o-TOLYLGUANIDINE see DXP200
1,3-DI-o-TOLYLGUANIDINE see DXP200
N,N'-DI-o-TOLYL-p-PHENYLENE DIAMINE see DXP400
DI-o-TOLYLTHIOUREA see DXP600
DITRAN see DXP800
DITRANIL see RDP300
DI-TRAPEX see MLC000
DITRAZIN see DIW200
DITRAZIN CITRATE see DIW200
DITRAZINE see DIW200
DITRAZINE BASE see DIW000
DITRAZINE CITRATE see DIW200
1,3-DI(TRICHLOROMETHYL)BENZENE see BLL825
DITRIDECYLAMINE see DXQ000
DITRIDECYL PHTHALATE see DXQ200
DI-(TRI-(2,2-DIMETHYL-2-PHENYLETHYL)TIN)OXIDE see BLU000
DITRIFON see TIQ250
DITRIPENTAT see CAY500
DI[TRIS-1,2-DIAMINOETHANECHROMIUM(III)]TRIPEROXODISULFATE
  see DXQ339
DI[TRIS-1,2-DIAMINOETHANECOBALT(III)]TRIPEROXODISULFATE
  see DXQ369
DITRIZOATE METHYLGLUCAMINE see AOO875
DITROPAN see OPK000
DITROSOL see DUS700
DITUBIN see ILD000
DIUCARDYN SODIUM see TFK270
DIULO see ZAK300
DI-n-UNDECYL KETONE see TJF250
DIUNDECYL PHTHALATE see DXQ400
DIURAL see CHJ750
DIURAMID see AAI250
DIURAPID see ASO375
1,1-DIUREIDISOBUTANE see IIV000
DIUREIDOISOBUTANE see IIV000
DIURESAL see CLH750
DIURESE see HII500
DIURETIC SALT see PKT750
DIURETICUM-HOLZINGER see AAI250
DIURETIN see SJO000
DIUREX see DXQ500
DIUREXAN see CLF325
DIURIL see CLH750
DIURILIX see CLH750
DIURITE see CLH750
DIUROBROMINE see TEO500
DIUROL see AMY050, DXQ500
DIURON see DXQ500
DIURON 4L see DXQ500
DIURONE see CHX250
DIUTAZOL see AAI250
DIUTRID see CLH750
DIUXANTHINE see TEP500
DIVASCOL see BBW750
DIVERCILLIN see AIV500, AOD125
DIVERON see MQU750
DIVINYL see BOP500
DIVINYL ACETYLENE see HCU500
m-DIVINYLBENZEN see DXQ745
DIVINYLBENZENE see DXQ745
m-DIVINYLBENZENE see DXQ745
DIVINYLENE OXIDE see FPK000
DIVINYLENE SULFIDE see TFM250
DIVINYLENIMINE see PPS250
DIVINYL ETHER (DOT) see VOP000
DIVINYL ETHER, inhibited (DOT) see VOP000

DIVINYLETHYLENE see HEZ000
DIVINYL MAGNESIUM see DXQ850
3,9-DIVINYLSPIROBI(m-DIOXANE) see DXR000
DIVINYL SULFONE see DXR200
2,5-DIVINYLTETRAHYDROPYRAN see DXR400
2,5-DIVINYLTETRAHYDRO-2H-PYRAN see DXR400
3,9-DIVINYL-2,4,8,10-TETRAOXASPIRO(5.5)UNDECANE see DXR000
DIVINYL ZINC see DXR500
DIVIPAN see DGP900
DIVIT URTO see VSZ100
DIVULSAN see DNU000
DIVYNYL OXIDE see VOP000
DIXANTHOGEN see BJU000
DIXARIT see CMX760
DIXERAN see AEG875, TDL000
DIXIBEN see EID000
DIXIE see KBB600
DIXON see FAB400
DIXON 164 see TAI250
DIXYRAZINE see MKQ000
DIXYRAZINE DIHYDROCHLORIDE see DXR800
(α-DIYLENE)POLY(p-AMINOBENZALDEHYDE-N) see DXS000
DIZENE see DEP600
DIZINCBIS(DIMETHYLDITHIOCARBAMATE)ETHYLENEBIS(DITHIOCARBA-
    MATE) see BJK000
DIZINON see DCM750
DIZOXIDE see DCQ700
DJ-1461 see DQU400
DJ-1550 see SNL800
D. JAMESONI VENOM see DAP820
DKB see DCQ800
DKB SULFATE see DHA400
DKC 1347 see CLK215
DKD see DIZ100
DKhM see DGQ200
DKM (RUSSIAN) see DGR400
DL 111 see EOL600
DL 152 see DIG800
D. L. 152 see DIH000
242 DL see BGC625
DL-820 see OGI300
DL-8280 see OGI300
DL 204-IT see EFC259
DL 717-IT see CKL325
DLX-6000 see TAI250
DM see DAC000, PDB000
1 DM 10 see XSS900
DMA see DOO800, DOQ800, DXE600, DXS200
DMA-4 see DAA800
DMAA see HKC000
DMAB see DOR200, DOT300
3,2'-DMAB see BLV250
DMAC see DOO800
DMAE see DOY800
DMAE p-ACETAMIDOBENZOATE see DOZ000
DMAE TARTRATE see DPA200
D 9998 MALEATE see FMP100
DMAM see DKP400
DMAMP see DPM400
DMAP see VTF000
DMASA see DQD400
DMB see DOA400
DMBA see DQJ200
7,12-DMBA see DQJ200
DMBC see DQQ200
DMBCA see BEL750
DMBEB see DDH900
DMC see BIN000
DMCC see DQY950
DMCT see MIJ500
DMDHEU see DTG000
DMDK see SGM500
DMDPN see DRQ200
DMDT see MEI450
p,p'-DMDT see MEI450
DMDZ see CGA500
DMEP see DOF400
DMES see DRK500
DMF see BJE750, DSB000
DMFA see DSB000
DMH see DSF300, DSF400, DSF600, DSF800, DSG200
DMHQ see DSG700
DMI see DSI709
DMI 50475 see DSI709

DMI HYDROCHLORIDE see DLS600
DMM see BKM500
DMMP see DSR400
DMMPA see DST800
DMN see NKA600
DMNA see NKA600
DMNM see DSV200, DTA000
DMNO see DSV200
DMN-OAC see AAW000
DMNT see DSG400
DMO see PMO250
DMP see DTR200
2,5-DMP see XKS000
2,6-DMP see XLA000
3,4-DMP see XLJ000
3,5-DMP see XLS000
DMPA see DGD800
DMPD see DTL800
DMPE see DOE200
DMPEA see DOE200
DMPP see DTO000
DMPP IODIDE see DTO000
DMPS see DNU860
DMPT see DTP000
DMS see DNV800, DUC300, DUD100, DUD400, TFP000
DMS-70 see DUD800
DMS-90 see DUD800
DMSA see DNV800, DQD400
trans-DMS-DIACETATE see DXS300
DMS(METHYL SULFATE) see DUD100
DMSO see DUD800
DMSP see FAQ800
DMT see DPF600
DMTP see FAQ900
DMTP (JAPAN) see DSO000
DMTT see DSB200
DMU see DTG700, DXQ500
DN 289 see BRE500
DNA see DAZ000, DUP600
DNBP see BRE500
DNBP AMMONIUM SALT see BPG250
DNCB see CGM000
DNDMP see DRO000
DN DRY MIX No. 1 see CPK500
DN-DRY MIX No.2 see DUS700
DN DUST No. 12 see CPK500
2,4-DNFB see DUW400
DNOC AMMONIUM SALT see DUT800
DNOCHP see CPK500
DNOC SOLDIUM SALT see DUU600
DNOK (CZECH) see DUS700
DNOK-ACETAT (CZECH) see AAU250
DNOP see DVL600
DNOSBP see BRE500
DNP see BIN500, DNG000
2,4-DNP see DUZ000
2,5-DNP see DVA000
DNPC see DUT600
DNPD see NBL000
2,4-DNPH see DVC400
DNPMT see DVF400
DNPOH see DVD200
DNPT see DVF400
DNPZ see DVF200
DNRB see DVF800
DNSBP see BRE500
DNT see DVH000
2,3-DNT see DVG800
2,4-DNT see DVH000
2,5-DNT see DVH200
2,6-DNT see DVH400
3,4-DNT see DVH600
3,5-DNT see DVH800
DNTB see DVF800
DNTBP see DRV200
DNTP see PAK000
DO 9 see DXY000
DO 14 see SOP000
DOA see AEO000, AOO500
DOBENDAN see CCX000
DOBESILATE CALCIUM see DMI300
DOBETIN see VSZ000
DOBO see TEP500
DOBREN see EPD500

DOBUTAMINE HYDROCHLORIDE see DXS375
DOBUTREX see DXS375
DOCA see DAQ800
DOCA ACETATE see DAQ800
DOC-AC see DAQ800
DOC ACETATE see DAQ800
DOCEMINE see VSZ000
DOCIBIN see VSZ000
DOCIGRAM see VSZ000
DOCITON see ICB000, ICC000
DOCTAMICINA see CDP250
DOCUSATE SODIUM see DJL000
DODAT see DAD200
DODDLE-DO (PUERTO RICO) see CAK325
DODECABEE see VSZ000
DODECACARBONYLDIVANADIUM see DXS400
DODECACARBONYLTRIIRON see DXS600
DODECACHLOROOCTAHYDRO-1,3,4-METHENO-2H-CYCLOBUTA(c,d)PEN-
TALENE see MQW500
1,1a,2,2,3,3a,4,5,5,5a,5b,6-DODECACHLOROOCTAHYDRO-1,3,4-METHENO-1H-
CYCLOBUTA(c,d)PENTALENE see MQW500
DODECACHLOROPENTACYCLODECANE see MQW500
DODECACHLOROPENTACYCLO(3,2,2,0^{2,6},0^{3,9},0^{5,10})DECANE see MQW500
DODECAHYDRODIPHENYLAMINE see DGT600
DODECAHYDRO-7,14-METHANO-2H,6H-DIPYRIDO(1,2-a:1',2'-e)(1,5)DIAZOC-
INE see SKX500
DODECAHYDROPHENYLAMINE NITRITE see DGU200
Δ-DODECALACTONE see DXS700
Δ-DODECALACTONE see HBP450
1,12'-DODECAMETHYLENEDIAMINE see DXW800
n-DODECAN (GERMAN) see DXT200
1-DODECANAL see DXT000
DODECANE see DXT200
1,12-DODECANEDIAMINE see DXW800
1,12'-DODECANEDIAMINE see DXW800
DODECANENITRILE see DXT400
1-DODECANETHIOL see LBX000
tert-DODECANETHIOL see DXT800
DODECANOIC ACID see LBL000
DODECANOIC ACID BENZYL ESTER see BEU750
DODECANOIC ACID, CADMIUM SALT (9CI) see CAG775
DODECANOIC ACID-2,3-DIHYDROXYPROPYL ESTER see MRJ000
DODECANOIC ACID, 2-THIOCYANATOETHYL ESTER see LBO000
1-DODECANOL see DXV600
n-DODECANOL see DXV600
DODECANOL ACETATE see DXV400
1-DODECANOL ACETATE see DXV400
DODECANOL, ETHOXYLATE see DXY000
DODECANOL ETHOXYLATED see EEU000
DODECANOL-ETHYLENE OXIDE (9.5 moles) CONDENSATE see DXY000
DODECANOL, POLYETHOXYLATED see DXY000
1-((n-DODECANOYLOXY)METHYL)-3-METHYLIMIDAZOLIUM CHLORIDE
see HMG500
3-(DODECANOYLOXYMETHY)-1-METHYL-1H-IMIDAZOLIUM CHLORIDE
see MKR150
DODECANOYL PEROXIDE see LBR000
DODECAN-1-YL ACETATE see DXV400
1,6,10-DODECATRIEN-3-OL, 3,7,11-TRIMETHYL-, ACETATE, (S-(Z))-
see NCN800
DODECATRIETHYLAMMONIUM BROMIDE see DXU200
5,7,11-DODECATRIYN-1-OL see DXU250
DODECAVITE see VSZ000
2-DODECENAL see DXU280
DODECENE see PMP750
DODECENE EPOXIDE see DXU400
9a-DODECENOATE see DXU600
(Z)-7-DODECEN-1-OL see DXU800
DODECENYLSUCCINIC ANHYDRIDE see DXV000
DODECOIC ACID see LBL000
DODECONIUM see HEF200
DODECYL ACETATE see DXV400
n-DODECYL ACETATE see DXV400
DODECYL ALCOHOL see DXV600
n-DODECYL ALCOHOL see DXV600
DODECYL ALCOHOL ACETATE see DXV400
DODECYL ALCOHOL CONDENSED with 4 MOLES ETHYLENE OXIDE
see LBS000
DODECYL ALCOHOL CONDENSED with 7 MOLES ETHYLENE OXIDE
see LBT000
DODECYL ALCOHOL CONDENSED with 23 MOLES ETHYLENE OXIDE
see LBU000
DODECYL ALCOHOL, ETHOXYLATED see DXY000
DODECYL ALCOHOL, HYDROGEN SULFATE, SODIUM SALT see SIB600
1-DODECYL ALDEHYDE see DXT000

DODECYLAMINE see DXW000
1-DODECYL-4-AMINOQUINALDINIUM ACETATE see AJS750
N-DODECYL-4-AMINOQUINALDINIUM ACETATE see AJS750
DODECYL AMMONIUM SULFATE see SOM500
DODECYLBENZENE see PEW500
DODECYL BENZENE SODIUM SULFONATE see DXW200
DODECYL BENZENESULFONATE see DXW400
DODECYLBENZENESULFONIC ACID SODIUM SALT see DXW200
DODECYLBENZENESULPHONATE, SODIUM SALT see DXW200
DODECYLBENZENSULFONAN SODNY (CZECH) see DXW200
DODECYLBENZYL CHLORIDE see DXW600
DODECYLBIS(AMINOETHYL)GLYCINE HYDROCHLORIDE see DYA850
DODECYLDIAMINE see DXW800
DODECYLDIMETHYLAMINE see DRR800
N-DODECYLDIMETHYLAMINE see DRR800
DODECYLDIMETHYLAMINE OXIDE see DRS200
N-DODECYLDIMETHYLAMINE OXIDE see DRS200
DODECYL DIMETHYL BENZYLAMMONIUM CHLORIDE see BEM000
DODECYLDIMETHYL(2-PHENOXYETHYL)AMMONIUM BROMIDE
see DXX000
DODECYLENE see PMP750
1,12'-DODECYLENEDIAMINE see DXW800
1-DODECYL-1-ETHYLPIPERIDINIUM BROMIDE see DXX100
DODECYL GALLATE see DXX200
N-DODECYLGUANIDINACETAT (GERMAN) see DXX400
DODECYLGUANIDINE ACETATE see DXX400
N-DODECYLGUANIDINE ACETATE see DXX400
DODECYLGUANIDINE ACETATE with SODIUM NITRITE (3585) see DXX600
DODECYLGUANIDINE HYDROCHLORIDE see DXX800
1-DODECYLHEXAHYDRO-1H-AZEPINE-1-OXIDE see DXX875
1-DODECYLHEXAMETHYLENIMINE-N-OXIDE see DXX875
α-DODECYL-ω-HYDROXY-POLYOXYETHYLENE see DXY000
2-DODECYLISOQUINOLINIUM BROMIDE see LBW000
DODECYL MERCAPTAN see LBX000
1-DODECYL MERCAPTAN see LBX000
m-DODECYL MERCAPTAN see LBX000
tert-DODECYLMERCAPTAN see DXT800
terc.DODECYLMERKAPTAN (CZECH) see DXT800
DODECYL METHACRYLATE see DXY200
3-DODECYL-1-METHYL-2-PHENYLBENZIMIDAZOLIUM FERRICYANIDE
see TNH750
1-DODECYL-3-METHYL-2-PHENYL-1H-BENZIMIDAZOLIUM,
HEXACYANOFERRATE(III) see TNH750
DODECYL-2-METHYL-2-PROPENOATE see DXY200
4-DODECYLMORPHOLINE-4-OXIDE see DXY300
4-DODECYLMORPHOLINE-N-OXIDE see DXY300
DODECYL NITRATE see NEE000
2-(DODECYLOXY)ETHANOL HYDROGEN SULFATE SODIUM SALT
see DYA000
2-(2-(2-(DODECYLOXY)ETHOXY)ETHOXY)ETHANESULFONIC ACID, SO-
DIUM SALT see SIC000
2-(2-(2-(DODECYLOXY)ETHOXY)ETHOXY)ETHANOL HYDROGEN SULFATE
SODIUM SALT see SIC000
N-(2-DODECYLOXYETHYL)-N-METHYL-2-(PYRROLIDINYL)ACETAMIDE HY-
DROCHLORIDE see DXY400
DODECYLPHENOL see DXY600
DODECYL PHENYLMERCURI SULFIDE see DYA400
1-DODECYLPIPERIDINE-1-OXIDE see DXY700
1-DODECYLPIPERIDINE-N-OXIDE see DXY700
DODECYL-POLYAETHYLENOXYD-AETHER (GERMAN) see DXY000
DODECYL POLY(OXYETHYLENE)ETHER see DXY000
N-DODECYLPYRROLIDINONE see DXY750
1-DODECYL-2-PYRROLIDINONE see DXY750
N-DODECYLSARCOSINE SODIUM SALT see DXZ000
DODECYL SODIUM ETHOXYSULFATE see DYA000
DODECYL SODIUM SULFATE see SIB600
DODECYL SULFATE see MRH250
DODECYL SULFATE, SODIUM SALT see SIB600
DODECYLSULFURIC ACID see MRH250
n-DODECYL THIOCYANATE see DYA200
tert-DODECYLTHIOL see DXT800
(DODECYLTHIO)PHENYLMERCURY see DYA400
DODECYL-p-TOLYL TRIMETHYL AMMONIUM CHLORIDE see DYA600
DODECYLTRICHLOROSILANE see DYA800
DODEMORFE (FRANCE) see COX000
DODEX see VSZ000
DODGUADINE see DXX400
DODICIN HYDROCHLORIDE see DYA850
DODINE see DXX400
DODINE ACETATE see DXX400
DODINE, mixture with GLYODIN see DXX400
DODINE with SODIUM NITRITE (3:5) see DXX600
1,4-DOEA-5,8-DAPFA (RUSSIAN) see DMM400
DOF see DVK600

DOFSOL see VSK600
DOG HOBBLE see DYA875
DOG LAUREL see DYA875
DOGMATIL see EPD500
DOGMATYL see EPD500
DOG PARSLEY see FMU200
DOG POISON see FMU200
DOGQUADINE see DXX400
DOISYNOESTROL see BIT000
DOISYNOLIC ACID see DYB000
DOJYOPICRIN see CKN500
DOKIRIN see BLC250
DOKTACILLIN see AIV500
DOL see BBQ750
DOLADENE see DJM800, XCJ000
DOLAN, combustion products see ADX750
DOLANTAL see DAM700
DOLANTIN see DAM700
DOLANTIN HYDROCHLORIDE see DAM700
DOLANTIN-N-OXIDE HYDROCHLORIDE see DYB250
DOLANTOL see DAM700
DOLAREN see DAM700
DOLARGAN see DAM700
DOLCOL see PIZ000
DOLCO MOUSE CEREAL see SMN500
DOLCONTRAL see DAM600, DAM700
DOLCYMENE see CQI000
DOLEAN pH 8 see ADA725
DOLENAL see DAM700
DOLENE see DAB879, PNA500
DOLENOL see DAM700
DOLEN-PUR see HCD250
DOLESTAN see BAU750
DOLESTINE see DAM700
DOLGIN see IIU000
DOL GRANULE see BBQ500
DOLICUR see DUD800
DOLIGUR see DUD800
DOLIN see DAM700
DOLIPOL see BSQ000
DOLIPRANE see HIM000
DOLISINA see DII200
DOLKWAL AMARANTH see FAG020
DOLKWAL BRILLIANT BLUE see FAE000
DOLKWAL ERYTHROSINE see FAG040
DOLKWAL INDIGO CARMINE see FAE100
DOLKWAL ORANGE SS see TGW000
DOLKWAL PONCEAU 3R see FAG018
DOLKWAL TARTRAZINE see FAG140
DOLKWAL YELLOW AB see FAG130
DOLKWAL YELLOW OB see FAG135
DOLLS EYES see BAF325
DOLMIX see BBQ750
DOLOBID see DKI600
DOLOBIL see DKI600
DOLOBIS see DKI600
DOLOCAP see PNA500
DOLOCHLOR see CKN500
DOLOCONEURASE see DYB300
DOLOGAL see DAM700
DOLOMIDE see SAH000
DOLOMITE see CAO000
DOLONEURINE see DAM700
DOLONIL see PDC250
DOLOPETHIN see DAM700
DOLOPHINE see MDO750, MDP000
DOLOPHINE HYDROCHLORIDE see MDP000, MDP750
d-DOLOPHINE HYDROCHLORIDE see MDP500
DOLOPHIN HYDROCHLORIDE see MDP750
DOLOSAL see DAM600, DAM700
DOLOVIN see IDA000
DOLOXENE see DAB879, DYB400, PNA500
DOLPHINE see MDP000
DOLSIMA see DII200
DOLSIN see DAM600
DOLVANOL see DAM700
DOM see BJR000
DOMAGK'S T.B.1 CONTEBEN see FNF000
DOMAIN see SNJ350
DOMAKOL see FNF000
DOMALIUM see DCK759
DOMAR see PIH100
DOMARAX see IPU000
DOMATOL see AMY050

DOMESTROL see DKA600
DOMF see MCV000
DOMICAL see EAI000
DOMICILLIN see SEQ000
DOMINAL see DYB600
DOMINAL HYDROCHLORIDE see DYB800
DOMOSO see DUD800
DOMPERIDONE see DYB875
DOMUCOR see ELH600
DON see DCQ400
DONASEVEN see AER666
DONAXINE see DYC000
DONMOX see AAI250
DONOPON see SBG500
DONOREST see CKI750
DONOVAN'S SOLUTION see ARI500
DOOJE see HBT500
DOP see DVL700
(−)-DOPA see DNA200
l-DOPA see DNA200
DOPAFLEX see DNA200
l-DOPA HYDROCHLORIDE see DYC200
DOPAL see DNA200
DOPAMET see DNA800, MJE780
l-DOPA METHYL ESTER see DYC300
DOPAMINE see DYC400
DOPAMINE CHLORIDE see DYC600
DOPAMINE HYDROCHLORIDE see DYC600
DOPAN see DYC700
DOPANE see DYC700
DOPARKINE see DNA200
DOPASOL see DNA200
DOPATEC see MJE780
DOPEGYT see DNA800, MJE780
DOPIDRIN see DBA800
DOPN see NJN000
DOPRAM see ENL100, SLU000
DOPRIN see DNA200
DOPTAEC see DNA800
DORAL see VSZ100
DORANTAMIN see WAK000
DORAPHEN see PNA500
DORBANE see DMH400
DORBANEX see DMH400
DORCOSTRIN see DAQ800
DOREVANE see IDA500
DORICO see ERD500
DORICO SOLUBLE see ERE000
DORIDEN see DYC800
DORIDEN-SED see DYC800
DORINAMIN see BBW500
DORISUL see SNN300
DORLYL see PKQ059
DORM see AFS500
DORMABROL see MQU750
DORMAL see CDO000
DORMALLYL see AFS500
DORMATE see MBW750
DORME see PMI750
DORMETHAN see DBE200
DORMIDIN see EQL000
DORMIGENE see BNP750
DORMIGOA see QAK000
DORMIN see TEO250
DORMIPHEN see EQL000
DORMIRAL see EOK000
DORMITURIN see BNK000
DORMODOR see DAB800
DORMOGEN see QAK000
DORMONAL see BAG000
DORMONE see DAA800
DORMOSAN see EQL000
DORMUPHAR see SKS700
DORMUTIL see QAK000
DORMWELL see SKS700
DORNWAL see ALY250
DORNWALL see ALY250
DORSACAINE see OPI300
DORSACAINE HYDROCHLORIDE see OPI300
DORSEDIN see QAK000
DORSIFLEX see MFD500
DORSILON see MFD500
DORSITAL see NBT500
DORSULFAN see SNN500

DORVICIDE A see BGJ750
DORVON see SMQ500
DORYL (PHARMACEUTICAL) see CBH250
DOS see BJS250
DOSAFLO see MQR225
DOSAGRAN see MQR225
DOSANEX see MQR225
DOSANEX FL see MQR225
DOSANEX MG see MQR225
DOSEGRAN see AFL750
DOSULEPIN see DYC875
DOSULEPIN CHLORIDE see DPY200
DOSULEPIN HYDROCHLORIDE see DPY200
D.O.T. see DUP300
DOTAN see CDY299
DOTG see BKK750
DOTG ACCELERATOR see DXP200
DOTHEIPIN HYDROCHLORIDE see DPY200
DOTHIEPIN see DYC875
DOTMENT 324 see AHE250
DOTYCIN see EDH500
DOUBLE STRENGTH see TIX500
DOVENIX see HLJ500
DOVIP see FAB600
DOW 209 see SMR000
DOW 860 see SMQ500
DOW 1329 see DGD800
DOWANOL see CBR000
DOWANOL 33B see PNL250
DOWANOL-50B see DWT200
DOWANOL 62B see TNA000
DOWANOL DB see DJF200
DOWANOL DE see CBR000
DOWANOL DM see DJG000
DOWANOL DPM see DWT200
DOWANOL EB see BPJ850
DOWANOL EE see EES350
DOWANOL EIPAT see INA500
DOWANOL EM see EJH500
DOWANOL EP see PER000
DOWANOL EPH see PER000
DOWANOL TE see EFL000
DOWANOL TMAT see TJQ750
DOWANOL TPM see MEV500
DOWCC 132 see COD850
DOWCHLOR see CDR750
DOWCIDE 1 see BGJ250
DOWCIDE 7 see PAX250
DOWCIDE 1 ANTIMICROBIAL see BGJ250
DOWCO 109 see BQU500
DOWCO 118 see DGD800
DOWCO 133 see DYD200
DOWCO 139 see DOS000
DOWCO 159 see DYD400
DOWCO 160 see DYD600
DOWCO 161 see EOQ500
DOWCO-163 see CLP750
DOWCO 169 see PEV500
DOWCO 177 see DYD800
DOWCO 179 see CMA100
DOWCO-183 see DYE200
DOWCO 184 see CEG550
DOWCO 186 see HON000
DOWCO 187 see AGU500
DOWCO-213 see CQH650
DOWCO 217 see CMA250
DOWCO 233 see TJE890
DOWCO 290 see DGJ100
DOWCO 356 see TJK100
DOW CORNING 200 see HEE000
DOW CORNING 346 see PJR000
DOW-CORNING 200 FLUID-LOT No. AA-4163 see DUB600
DOW CORNING SILICONE FLUID and FLUOROHYDROCARBON see XGA000
DOW DEFOLIANT see SFU500
DOW DORMANT FUNGICIDE see SJA000
DOW ET 14 see RMA500
DOW ET 57 see RMA500
DOWFLAKE see CAO750
DOWFROST see PML000
DOWFROTH 250 see MKS250
DOWFUME see MHR200
DOWFUME 40 see EIY500
DOWFUME EB-5 see DYE400
DOWFUME EDB see EIY500

DOWFUME MC-2 SOIL FUMIGANT see MHR200
DOWFUME N see DGG000
DOWFUME W-8 see EIY500
DOW GENERAL see BRE500
DOW GENERAL WEED KILLER see BRE500
DOWICIDE see BGJ750
DOWICIDE 2 see TIV750
DOWICIDE 4 see CHN500
DOWICIDE 6 see TBT000
DOWICIDE 7 see PAX250
DOWICIDE 9 see CFI750
DOWICIDE 2S see TIW000
DOWICIDE 31 see SFV500
DOWICIDE B see SKK500, TIV750
DOWICIDE EC-7 see PAX250
DOWICIDE G see PAX250
DOWICIDE G-ST see SJA000
DOWICIDE Q see CEG550
DOWICIL 75 see CEG550
DOWICIL 100 see CEG550
DOW LATEX 612 see SMR000
DOWLEX see PKF750
DOW MCP AMINE WEED KILLER see CIR250
DOWMYCIN E see EDJ500
DOW PENTACHLOROPHENOL DP-2 ANTIMICROBIAL see PAX250
DOWPEN V-K see PDT750
DOW-PER see PCF275
DOWPON see DGI400, DGI600
DOWPON M see DGI400
DOW SEED DISINFECTANT No. 5 see TBO500
DOW SELECTIVE see BPG250
DOW SELECTIVE WEED KILLER see BRE500
DOW SODIUM TCA INHIBITED see TII250, TII500
DOWSPRAY 17 see CPK500
DOWTHERM see PFA860
DOWTHERM 209 see PNL250
DOWTHERM A see PFA860
DOWTHERM E see DEP600
DOWTHERM SR 1 see EJC500
DOW-TRI see TIO750
DOWZENE DHC see PIK000
D-OX see SHR500
DOXAPRAM see ENL100
DOXAPRAM HYDROCHLORIDE HYDRATE see SLU000
DOXCIDE 50 see CDW450
DOXEPHRIN see DBA800
DOXEPIN see DYE409
DOXEPIN HYDROCHLORIDE see AEG750
DOXERGAN see AFL750
DOX HYDROCHLORIDE see HKA300
DOXICICLINA (ITALIAN) see DYE425
DOXIFLURIDINE see DYE415
DOXIGALUMICINA see HGP550
DOXINATE see DJL000
DOXIUM see DMI300
DOXO see DAQ800
DOXOL see DJL000
DOXORUBICIN see AES750, HKA300
DOXORUBICIN HYDROCHLORIDE see HKA300
DOXYCYCLINE see DYE425
DOXYCYCLINE HYCLATE see HGP550
DOXYCYCLINE HYDROCHLORIDE see HGP550
DOXYFED see DBA800, MDT600
DOXY-II see HGP550
DOXYLAMINE see DYE500
DOXYLAMINE SUCCINATE see PGE775
DOXYLAMINE SUCCINATE (1:1) see PGE775
DOXY-TABLINEN see HGP550
DOZAR see DPJ400
2,4-DP see DGB000
2-(2,4-DP) see DGB000
DPA see DGI000, DVX800, HET500
n-DPA see PNR750
2,2-DPA see DGI600
D-P-A INJECTION see PAG200
DPA SODIUM see PNX750
DPBS see DFY400
D & P DOUBLE O CRABGRASS KILLER see PLC250
DPE-HCl see DWD200
DPF see DYE550
DPG see DNA300, DWC600
DPG ACCELERATOR see DWC600
2,4-D PGBE see DFZ000
DPH see DKQ000, DNU000

DPID see DLH600
DPMA see DMB200
DPN see NKB700
DPNA see NKB700
DPP see AON300, PAK000, PEK250
DPPD see BLE500
2,4-D PROPYLENE GLYCOL BUTYL ETHER ESTER see DFZ000
DPT see DCA600
3,5-DPT see DWO950
DP X 1410 see MME809
DPX 1410 see DSP600
DPX 3654 see PMF550
DPX 3674 see HFA300
DQUIGARD see DGP900
DQV-K see PDT750
DR-15771 see SBE500
DRABET see BSQ000
DRACYLIC ACID see BCL750
DRAGIL-P see SLL000
DRAGON ARUM see JAJ000
DRAGON ROOT see JAJ000
DRAGONS HEAD see JAJ000
DRAGON TAIL see JAJ000
DRAKEOL see MQV750
DRALZINE see HGP500
DRAMAMIN see DYE600
DRAMAMINE see DYE600
DRAMARIN see DYE600
DRAMYL see DYE600
DRAPEX 4.4 see FAB920
DRAPOLENE see AFP250
DRAPOLEX see BBA500
DRASIL 507 see CNH125
DRAT see CJJ000
DRAWIN 755 see MPU250
DRAWINOL see CBW000
DRAZA see DST000
DRAZINE see PDV700
DRAZOXOLON see MLC250
DRAZOXOLONE see MLC250
DRB see DGK100, DVF800
DRC-714 see BIM000
DRC 1339 see CLK215
DRC-1,339 see CLK230
DRC 3340 see DTN200
DRC 3341 see MIB750
DRC-4575 see AMU125
DRC 6246 see AGW675
DREFT see SIB600
DRENAMIST see VGP000
DRENE see SON000
DRENOBYL see DAL000
DRENOL see CFY000
DREWMULSE POE-SMO see PKL100
DREWMULSE TP see OAV000
DREWMULSE V see OAV000
DREWSORB 60 see SKV150
DREXEL see DXQ500
DREXEL DEFOL see SFS000
DREXEL DIURON 4L see DXQ500
DREXEL DSMA LIQUID see DXE600
DREXEL METHYL PARATHION 4E see MNH000
DREXEL PARATHION 8E see PAK000
67/20CDRI see CCW750
DRIBAZIL see DYE700
DRICOL see AHL500
DRI-DIE INSECTICIDE 67 see SCH000
DRIDOL see DYF200
DRI-KIL see RNZ000
DRILL TOX-SPEZIAL AGLUKON see BBQ500
DRINALFA see DBA800, MDT600
DRINOX see AFK250, HAR000
DRINUPAL HYDROCHLORIDE see CKB500
DRIOL see DYE700
DRIOL-LABAZ see DYE700
DRISDOL see VSZ100
DRISTAN INHALER see PNN400
DRI-TRI see SJH200
DROCODE see DKW800
DROCTIL see ERE200
DROGENIL see FMR050
DROLEPTAN see DYF200
DROMETRIZOLE see HML500
DROMILAC see DBV400

DROMISOL see DUD800
DROMORAN see MKR250
levo-DROMORAN see LFG000
racemic DROMORAN see MKR250
DROMORAN HYDROBROMIDE see MDV250
DROMORAN-HYDROBROMIDE see HMH500
l-DROMORAN TARTRATE see DYF000
DROMOSTAT see TKU675
DROMYL see DYE600
DRONACTIN see PCI500
DRONCIT see BGB400
DROPCILLIN see BDY669
DROPERIDOL see DYF200
DROP LEAF see SFS000
DROPP see TEX600
DROPRENILAMINE HYDROCHLORIDE see CPP000
DROPSPRIN see SAH000
DROTEBANOL see ORE000
DROXAROL see BPP750
DROXARYL see BPP750
DROXOL see MDQ250
DROXOLAN see DAQ400
DRP 859025 see TND500
DRUMULSE AA see OAV000
DRUPINA 90 see BJK500
DRW 1139 see ALA500
DRY AND CLEAR see BDS000
DRY ICE (DOT) see CBU750
DRYISTAN see BBV500
DRYLIN see TKX000
DRYLISTAN see BBV500
DRY MIX No. 1 see CPK500
DRYOBALANOPS CAMPHOR see BMD000
DRYPTAL see CHJ750
DS see DXN300
DS-36 see SNL800
DS-M-1 see DBD750
DS-15647 see DAB400
DS 18302 see CBW000
DSDP see DJA400
DSE see DXD200
DSMA LIQUID see DXE600
DS-Na see SFW000
2,4-D SODIUM SALT see SGH500
DSP see SJH090
DSPT see THQ900
DSS see DJL000, SOA500
DS SUBSTANDE see AJX500
DST see DME000
DST 50 see SMR000
DT see DNC000, DWP000, TFX790
D 65MT see XAJ000
2,4-D and 2,4,5-T (2581) see DAB000
DTA see DLK200
D-40TA see CKL250
DTAS see MJK750
DTB see DXL800
DTBP see BSC750
DTBT see BLN750
DTDP see DXQ200
DTHYD see TFX790
DTIC see DAB600
DTIC CITRATE see DUI400
DTIC-DOME see DAB600
DTMC see BIO750
DTPA see DJG800
DTPA CALCIUM TRISODIUM SALT see CAY500
DTP HYDROCHLORIDE see DXO400
N-D1-TRIMETHYL-3,3-DI-2-THIENYLALLYLAMINE HYDROCHLORIDE
    see TLQ250
DTS see DNV800
DTT see DXO775, DXO800
D-DTT see DXO800
DU see DCQ600
DU-717 see CIT625
DU-5747 see CCK575
DU 21220 see RLK700
DU 112307 see CJV250
DUAL see MQQ450
DUAMINE see MCH525
DUATOK see TEX250
DUAZOMYCIN see DYF400
DUAZOMYCIN B see ASO501, ASO510
DUBIMAX see NAE000

DUBORIMYCIN see DAC300
DUBOS CRUDE CRYSTALS see TOG500
DUBRONAX see SOA500
DUCKALGIN see SEH000
DUCOBEE see VSZ000
DUFALONE see BJZ000
DUFASTON see DYF759
DUGERASE see AIT250
DUGRO see MMN750
DUKERON see TIO750
DUKSEN see DCK759
DULCIDOR see GFA000
DULCINE see EFE000
DULCITOLDIEPOXIDE see DCI600
DUL-DUL (PUERTO RICO) see CAK325
DULL 704 see PJY500
DULSIVAC see DJL000
DULZOR-ETAS see SGC000
DUMASIN see CPW500
DUMBCANE see DHB309
DUMB PLANT see DHB309
DUMITONE see SOA500
DUMOCYCIN see TBX250
DUMOGRAN see MPN500
DUNCAINE see DHK400
DUNCAINE HYDROCHLORIDE see DHK600
DUNERYL see EOK000
DUNKELGELB see PEJ500
DUODECANE see DXT200
DUODECIBIN see VSZ000
DUODECYL ALCOHOL see DXV600
DUODECYLIC ACID see LBL000
DUODECYLIC ALDEHYDE see DXT000
DUOGASTRONE see CBO500
DUO-KILL see COD000, DGP900
DUOLAX see DMH400
DUOLIP see TEQ500
DUOLUTON see NNL500
DUOMYCIN see CMA750
DUOSAN see MAP300
DUOSAN (pesticide) see MAP300
DUOSOL see DJL000
DUO-STREPTOMYCIN see SLY200
DUOTRATE see PBC250
DUPHAR see CKM000
DUPHASTON see DYF759
DUPLEX PERMATON RED L 20-7022 see CJD500
DUPLEX TOLUIDINE RED L 20-3140 see MMP100
DUPONOL see SIB600
DU PONT 326 see DGD600
DU PONT 732 see BQT750
DU PONT 1991 see BAV575
DUPONT DPX 3654 see PMF550
DUPONT HERBICIDE 326 see DGD600
DU PONT HERBICIDE 732 see BQT750
DU PONT HERBICIDE 976 see BMM650
DU PONT INSECTICIDE 1179 see MDU600
DU PONT INSECTICIDE 1519 see DVS000
DUPONT INSECTICIDE 1642 see MPS250
DUPONT PC CRABGRASS KILLER see PLC250
DU PONT WK see DXY000
DURABIOTIC see BFC750
DURABOL see DYF450
DURABOLIN see DYF450
DURABOLIN-O see EJT600
DURABORAL see EJT600
DURA CLOFIBRAT see ARQ750
DURAD see TNP500
DURA-ESTRADIOL see EDS100
DURAFUR BLACK R see PEY500
DURAFUR BLACK RC see PEY650
DURAFUR BROWN see ALL750
DURAFUR BROWN 2R see ALL750
DURAFUR BROWN MN see DBO400
DURAFUR BROWN RB see ALT250
DURAFUR DEVELOPER C see CCP850
DURAFUR DEVELOPER D see NAW500
DURAFUR DEVELOPER G see REA000
DURALUTON see HNT500
DURAMAX see ADA725
DURAN see DXQ500
DURANATE EXP-D 101 see PKO750
DURANIT see SMR000
DURANITRAT see CCK125

DURANOL BRILLIANT BLUE B see MGG250
DURANOL BRILLIANT BLUE CB see TBG700
DURANOL BRILLIANT YELLOW G see MEB750
DURANOL ORANGE G see AKP750
DURANOL VIOLET WR see DBP000
DURANTA REPENS see GIW200
DURA-PENITA see BFC750
DURAPHOS see MQR750
DURAPROST see OLW600
DURASORB see DUD800
DURA-TAB S.M. AMINOPHYLLINE see TEP500
DURATION see AEX000
DURATOX see DAP400, MIW100
DURAVOS see DGP900
DURAX see CPI250
DURETHAN BK see PJY500
DURETTER see FBN100
DURFAX 80 see PKL100
DURGASOL SCARLET GG SALT see DEO400
DURICEF see DYF700
DUROCHROME YELLOW 2RN see NEY000
DUROFERON see FBN100
DUROFOL P see PKQ059
DUROID 5870 see TAI250
DUROMINE see DTJ400
DURONITRIN see TJL250
DUROPENIN see BFC750
DUROPROCIN see MDU300
DUROTOX see PAX250
DUROX see AKO500
DURSBAN see CMA100
DURSBAN F see CMA100
DURSBAN METHYL see CMA250
DURTAN 60 see SKV150
DUSICNAN BARNATY (CZECH) see BAN250
DUSICNAN CERITY (CZECH) see CDB000
DUSICNAN HORECHATY (CZECH) see MAH250
DUSICNAN KADEMNATY (CZECH) see CAH250
DUSICNAN VAPENATY (CZECH) see CAU250
DUSICNAN ZINECNATY (CZECH) see NEF500
DUSICNAN ZiRKONICITY (CZECH) see ZSA000
DUSITAN DICYKLOHEXYLAMINU (CZECH) see DGU200
DUSITAN SODNY (CZECH) see SIQ500
DUSOLINE see CMD750
DUSORAN see CMD750
DUSPAR 125B see HKS400
DU-SPREX see DER800
DUST M see RAF100
DUS-TOP see MAE250
DUTCH LIQUID see EIY600
DUTCH OIL see EIY600
DUTCH-TREAT see HKC500
DU-TER see HON000
DUTOM see SKQ400
DUVADILAN see VGA300, VGF000
DUVALINE see PPH050
DUVARON see DYF759
DUVILAX BD 20 see AAX250
DUVOID see HOA500
DV see DAL600
DV-17 see DBC575
DV-79 see AJV500
D 33LV see DCK400
DV 400 see PKB500
DV 714 see LEF400
DVEDEG (RUSSIAN) see DJE600
D3-VIGANTOL see CMC750
DW-61 see FCB100
DW 62 see DNV000, DNV200
DW3418 see BLW750
DWARF BAY see LAR500
DWARF LAUREL see MRU359
DWARF PINE NEEDLE OIL see PIH400
DWARF POINCIANA see CAK325
DWELL see EFK000
DWUBROMOETAN (POLISH) see EIY500
DWUCHLOROCZTEROFLUOROETAN (POLISH) see DGL600
DWUCHLORODWUETYLOWY ETER (POLISH) see DFJ050
DWUCHLORODWUFLUOROMETAN (POLISH) see DFA600
2,4-DWUCHLOROFENOKSYOCTOSY KWAS (POLISH) see DAA800
DWUCHLOROFLUOROMETAN (POLISH) see DFL000
DWUCHLOROPROPAN (POLISH) see PNJ400
DWUCHLOROSTYREN (POLISH) see DGK800
DWUETYLOAMINA (POLISH) see DHJ200

DWUETYLOWY ETER (POLISH) see EJU000
DWUFENYLOGUANIDYNA (POLISH) see DWC600
DWUMETHYLOFORMAMID (POLISH) see DSB000
DWUMETYLOANILINA (POLISH) see DQF800
symetryczna DWUMETYLOHYDRAZYNA (POLISH) see DSF600
DWUMETYLOSULFOTLENKU (POLISH) see MAO250
DWUMETYLOWY SIARCZAN (POLISH) see DUD100
DWU-β-NAFTYLO-p-FENYLODWUAMINA (POLISH) see NBL000
DWUNITROBENZEN (POLISH) see DUQ200
DWUNITRO-o-KREZOL (POLISH) see DUS700
3,3'-DWUOKSYBENZYDYNA (POLISH) see DMI400
DWUSIARCZEK DWUBENZOTIAZYLU (POLISH) see BDE750
DX see AES750
DXMS see SOW000
DYANACIDE see ABU500
DYAZIDE see CFY000
DYBAR see DTP400
DYCARB see DQM600
DYCHOLIUM see SGD500
DYCILL see DGE200
DYCLOCAINUM see BPR500
DYCLONE HYDROCHLORIDE see BPR500
DYCLONINE HYDROCLORIDE see BPR500
DYCLOTHANE see BPR500
DYDELTRONE see PMA000
DYDROGESTERONE see DYF759
DYDROGESTERONE and HYDROXYPROGESTERONE (9:10) see PMA250
DYE C see DYF800
DYE EVANS BLUE see BGT250
DYE FD AND C RED No. 4 see FAG050
DYE FDC RED 2 see FAG020
DYE FD&C RED No. 3 see FAG040
DYE FD & C RED No. 4 see FAG050
DYE GS see ALL750
DYE ORANGE No. 1 see FAG010
DYE RED RASPBERRY see FAG020
DYESTROL see DKA600
DYETONE see SFG000
DYFLOS see IRF000
DYFONATE see FMU045
DYGRATYL see DME300
DYKANOL see PJL750
DYKOL see DAD200
DYLAMON see BBV500
DYLENE see SMQ500
DYLEPHRIN see VGP000
DYLITE F 40 see SMQ500
DYLOX see TIQ250
DYLOX-METASYSTOX-R see TIQ250
DYMADON see HIM000
DYMELOR see ABB000
DYMEX see ABH000
DYMID see DRP800
DYNACORYL see DJS200
DYNADUR see PKQ059
DYNALIN INJECTABLE see TET800
DYNAMICARDE see DJS200
DYNAMITE see DYG000
DYNAMONE see AES650
DYNAMUTILIN see TET800
DYNAPEN see DGE200
DYNAPHENIL see AOB500
DYNAPRIN see DLH600
DYNARSAN see ABX500
DYNASTEN see PAN100
DYNA-ZINA see DLH600, DLH630
DYNAZONE see NGE500
DYNEL see ADY250
DYNERIC see CMX700
DYNEX see DXQ500
DYNONE see EIH500
DYNOSOL see DUU600
DYNOTHEL see SKJ300
DYP-97 F see LBR000
DYPHONATE see FMU045
DYPHYLLINE see DNC000
DYPNONE see MPL000
DYPRIN see MDT740
DYREN see UVJ450
DYRENE see DEV800
DYRENE 50W see DEV800
DYRENIUM see UVJ450
DYREX see TIQ250
DYSEDON see AFL750

DYSPNE-INHAL see VGP000
DYSPROSIUM see DYG400
DYSPROSIUM CHLORIDE see DYG600
DYSPROSIUM CITRATE see DYG800
DYSPROSIUM NITRATE see DYH100
DYSPROSIUM(III) NITRATE HEXAHYDRATE (1583586) see DYH000
DYSPROSIUM TRINITRATE see DYH100
DYTAC see UVJ450
DYTHOL see CMD750
DYTOL E-46 see OAX000
DYTOL F-11 see HCP000
DYTOL J-68 see DXV600
DYTOL M-83 see OEI000
DYTOL S-91 see DAI600
DYTRANSIN see IJG000
DYVON see TIQ250
DYZOL see DCM750

E^1 see EDV000
E2 see PJY100
E^2 see EDO000
E-3 see ELF500
E6 see PKO500
E-48 see DJV600
E 62 see PKQ059
E-103 see BSN000
E-111 see PNN300
E 127 see FAG040
E 132 see FAE100
E 140 see CKN000
E 141 see DIS600
E 151 see BMA000
E 158 see DAP600
E-212 see VGZ000
E-236 see DUJ400
(−)-E-250 see DAZ125
E-298 see MRU080
E393 see SOD100
583E see POC750
E 600 see NIM500
E-646 see EAV700
E 736 see TFD000
E 785 see EAB100
(+)-E-250 see DAZ120
E1001 see IPM000
E 1004 see IPP000
E-1059 see DAO500
E 1059 see DAO600
264CE see AFS625
E-2663 see CML835
E 3314 see HAR000
(±)-E-250 see DAZ118
EA 166 see GLS700
E-733-A see CQH000
EA 3547 see DDE200
EAA see EFS000
EAB see EOH500
EACA see AJD000
EACA KABI see AJD000
E-73 ACETATE see ABN000
EACS see AJD000
EA-1 HYDROCHLORIDE see NCL300
EAK see OD1000
EAMN see ENR500
EARTHCIDE see PAX000
EARTH GALL see FAB100
EARTHNUT OIL see PAO000
EASEPTOL see HJL000
EASTBOND M 5 see PMP500
EASTER FLOWER see EQX000
EASTERN COTTONMOUTH VENOM see EAB200
EASTERN DIAMOND-BACK RATTLESNAKE VENOM see EAB225
EASTERN STATES DUOCIDE see WAT200
EAST INDIAN COPAIBA see GME000
EAST INDIAN LEMONGRASS OIL see LEG000
EASTMAN 910 see MIQ075
EASTMAN 7663 see DJT800
EASTMAN BLUE BNN see MGG250
EASTMAN INHIBITOR DHPB see DMI600
EASTMAN INHIBITOR RMB see HNH500
EASTOZONE see BJL000
EASTOZONE 31 see BJT500
EASTOZONE 33 see BJL000
EASY OFF-D see TIG250

EATAN see DLY000
EAU de BROUTS ABSOLUTE see EAB500
EAZAMINE see DII200
EB see BGT250, EGP500
EB-382 see MGC200
EBI see ISK000
EBIS see ISK000
EBNA see NKD500
EBNS see EHC800
EBRANTIL see USJ000
EBS see FAG040
EBZ see EDP000
E.C. 1.1.3.4 see GFG100
E.C.3.1.1.3. see GGA800
E.C. 3.4.4.5 see CML850
E.C. 3.4.4.6 see CML850
E.C. 3.4.4.7 see EAG875
E.C. 3.2.1.23 see GAV100
E.C. 3.4.21.1 see CML850
E.C. 3.4.4.16 see BAC000
E.C. 3.4.4.24 see BMO000
E.C. 3.4.2.1.11 see EAG875
E.C. 3.4.21.14 see BAC000
E.C. 3.4.21.15 see SBI860
ECARAZINE see TGJ150
ECARAZINE HYDROCHLORIDE see TGJ150
E. CARINATUS VENOM see EAD600
ECATOX see PAK000
ECBOLINE see EDC565
ECBOLINE ETHANESULFONATE see EDC575
ECCOTHAL see TEM000
β-ECDYSONE see HKG500
ECDYSTERONE see HKG500
β-ECDYSTERONE see HKG500
ECF see EHK500
ECGONINE, METHYL ESTER, BENZOATE (ESTER) see CNE750
ECH see EAZ500
ECHIMIDINE see EAC500
ECHINOMYCIN see EAD500
ECHINOMYCIN A see EAD500
ECHIS CARINATUS VENOM see EAD600
ECHIS COLORATA VENOM see EAD650
ECHIS COLORATUS VENOM see EAD650
ECHIUM PLANTAGINEUM see VQZ675
ECHIUM VULGARE see VQZ675
ECHLOMEZOL see EFK000
ECHODIDE see TLF500
ECHOTHIOPHATE see TLF500
ECHOTHIOPHATE IODIDE see TLF500
ECHUJETIN see DMJ000
ECIPHIN see EAW000
ECLERIN see BOV825
ECLORIL see CDO500
ECLORION see HBT500
ECM see ADA725
ECODOX see HGP550
ECOLID see CDY000
ECOLID CHLORIDE see CDY000
E. COLI ENDOTOXIN see EDK700
ECONAZOLE NITRATE see EAE000
ECONOCHLOR see CDP250
ECOPRO see TAL250
ECOSTATIN see EAE000
ECOTHIOPATE IODIDE see TLF500
ECOTHIOPHATE IODIDE see TLF500
ECOTRIN see ADA725
ECP see DAZ115, DFK600
ECPN see NKE000
ECR see EDB125
ECTIBAN see AHJ750
ECTIDA see EHP000
ECTILURAN see TEH500
ECTILUREA see EHP000
ECTON see EHP000
ECTORAL see RMA500
ECTRIN see FAR100
ECTYDA see EHP000
ECTYLCARBAMIDE see EHP000
ECTYLUREA see EHO700, EHP000
ECTYN see EHP000
ECUANIL see MQU750
ECYLERT see PAP000
ECZECIDIN see CHR500
ED see DFH200, EQJ100

EDA see DCN800
EDA 200 see EIV750
EDATHAMIL see EIX000
EDATHAMIL DISODIUM see EIX500
EDATHAMIL MONOSODIUM FERRIC SALT see EJA379
EDATHANIL TETRASODIUM see EIV000
EDB see EIY500
EDB-85 see EIY500
E-D-BEE see EIY500
EDC see EIY600
EDCO see MHR200
EDDO see EAI600
EDDP see EIM000
EDECRIL see DFP600
EDECRIN see DFP600
EDECRINA see DFP600
EDEINE see EAE400
EDEMEX see BDE250
EDEMOX see AAI250
EDEN see LFK000
EDENAL see MQU750
EDETATE DISODIUM see EIX500
EDETATE SODIUM see EIV000
EDETATE TRISODIUM see TNL250
EDETIC ACID see EIX000
EDETIC ACID TETRASODIUM SALT see EIV000
EDICOL AMARANTH see FAG020
EDICOL BLUE CL 2 see FAE000
EDICOL PONCEAU RS see FMU070
EDICOL SUPRA BLACK BN see BMA000
EDICOL SUPRA BLUE E6 see FMU059
EDICOL SUPRA CARMOISINE WS see HJF500
EDICOL SUPRA ERYTHROSINE A see FAG040
EDICOL SUPRA PONCEAU 4R see FMU080
EDICOL SUPRA PONCEAU SX see FAG050
EDICOL SUPRA ROSE B see FAG070
EDICOL SUPRA TARTRAZINE N see FAG140
EDIFENPHOS see EIM000
EDION see TLP750
EDIPHENPHOS see EIM000
EDIPOSIN see VBK000
EDISTIR RB 268 see SMR000
EDIWAL see NNL500
EDMBA see DQL600
EDPA HYDROCHLORIDE see DWF200
EDROFURADENE see EAE500
EDROPHONIUM BROMIDE see TAL490
EDROPHONIUM CHLORIDE see EAE600
EDRUL see EAE675
EDTA (chelating agent) see EIX000
EDTA ACID see EIX000
d'E.D.T.A. DISODIQUE (FRENCH) see EIX500
EDTA, DISODIUM SALT see EIX500
EDTA FERRIC SODIUM SALT see IGX875
EDTA, SODIUM SALT see EIV000
EDTA TETRASODIUM SALT see EIV000
EDTA TRISODIUM SALT see TNL250
EDTA TRISODIUM SALT (TRIHYDRATE) see TNL500
E-103-E see MQQ050
EECPE see QFA250
EEC SERIAL No. 124 see CMP250
EEDDKK see EIJ500
EENA see ELG500
EENKAPTON (DUTCH) see PDC750
EEREX GRANULAR WEED KILLER see BMM650
EEREX WATER SOLUBLE CONCENTRATE WEED KILLER see BMM650
EES see DKB139
EFACIN see NCQ900
EF CORLIN see CNS750
EFED see TMU250
EFEDRIN see EAW000
EFERON see PEC250
EFEROX see LEQ300
EFFEMOLL DOA see AEO000
EFFISAX see MOV500
EFFLUDERM (free base) see FMM000
EFFORTIL see EGE500
EFFROXINE see DBA800
EFFUSAN see DUS700
EFH see EKL250
EFLORAN see MMN250
EFLORNITHINE see DKH875
EFLORNITHINE HYDROCHLORIDE see EAE775
EFLOXATE see ELH600

EFO-DINE see PKE250
EFRICEL see SPC500
EFROXINE see MDT600
EFSIOMYCIN see FMR500
EFTAPAN see ECU550
EFTOLON see AIF000
EFUDEX see FMM000
EFUDIX see FMM000
EFURANOL see DLH630
EGACID ORANGE GG see FAG010
EGDME see DOE600
EGDN see EJG000
EGF see EBA275
EGF-UROGASTRONE see UVJ475
EGGOBESIN see PNN300
EGG YELLOW A see FAG140
EGITOL see HCI000
EGLONYL see EPD500
EGM see EJH500
EGME see EJH500
EGPEA see PNA225
EGYPTIAN RATTLEPOD see SBC550
EGYT 201 see BAV250
EGYT 341 see GJS200
EGYT 739 see CAY875
EGYT-1050 see DAJ400
EH2 see DGQ200
EH 121 see TNX000
EHBN see ELE500
EHDP see HKS780
EHEN see ELG500
EHRLICH 5 see ARL000
EHRLICH 594 see ABX500
EHRLICH 606 see SAP500
EHRLICH'S REAGENT see DOT400
EI see EJM900
EI-103 see BSN000
EI-1642 see MPS250
EI-12880 see DSP400
EI-18706 see DNX600
EI 38,555 see CMF400
EI 47031 see PGW750
EI 47300 see DSQ000
EI-47470 see DHH400
EI 52160 see TAL250
EICOSAFLUOROUNDECANOIC ACID see EAE875
2,2,3,3,4,4,5,5,6,6,7,7,8,8,9,9,10,10,11,11-EICOSAFLUOROUNDECANOIC ACID
  see EAE875
11-H-EICOSAFLUORUNDEKANSAEURE (GERMAN) see EAE875
omega-H-EICOSAFLUORUNDEKANSAEURE (GERMAN) see EAE875
EICOSAHYDRO DIBENZO(b,k)(1,4,7,10,13,16)HEXAOXACYCLOOCTADECIN
  see DGV100
EICOSANOIC ACID see EAF000
EICOSANYL DIMETHYL BENZYLAMMONIUM CHLORIDE see BEM250
EINALON S see CLY500
EIPW 111 CITRATE see EAF100
EIRENAL see PGA750
EISENDEXTRAN (GERMAN) see IGS000
EISENDIMETHYLDITHIOCARBAMAT (GERMAN) see FAS000
EISEN-III-HYDROXID-POLYMALTOSE (GERMAN) see IHA000
EISENOXYD see IHD000
EISEN(III)-TRIS(N,N-DIMETHYLDITHIOCARBAMAT) (GERMAN) see FAS000
EITDRONATE DISODIUM see DXD400
EJIBIL see EGQ000
EK 54 see DUU600
EK 1700 see PEY250
EKAGOM TB see TFS350
EKAGOM TEDS see DXH250
EKAGON TE see DJC200
EKALUX see DJY200
EKATIN see PHI500
EKATIN AEROSOL see PHI500
EKATINE-25 see PHI500
EKATIN ULV see PHI500
EKATOX see PAK000
EKAVYL SD 2 see PKQ059
EKILAN see MFD500
EKKO CAPSULES see DKQ000
EKOA (HAWAII) see LED500
EKOMINE see MGR500
EKTAFOS see DGQ875
EKTASOLVE de ACETATE see CBQ750
EKTASOLVE DB see DJF200
EKTASOLVE DB ACETATE see BQP500

EKTASOLVE DIB see DJF800
EKTASOLVE EB see BPJ850
EKTASOLVE EB ACETATE see BPM000
EKTASOLVE EE see EES350
EKTASOLVE EE ACETATE SOLVENT see EES400
EKTASOLVE EIB see IIP000
EKTASOLVE EP see PNG750
EKTEBIN see PNW750
EKTYLCARBAMID see EHP000
EKVACILLIN see SLJ000, SLJ050
EK 1108GY-A see TAI250
EL-103 see BSN000
EL 222 see FAK100
EL-291 see MQC000
EL 400 see BNL250
EL-620 see PKE500
EL-719 see PKE500
EL 4049 see MAK700
ELAIDIC ACID see EAF500
ELAIOMYCIN see EAG000
ELALDEHYDE see PAI250
ELAMOL see TGJ250
ELAN see PGG350
ELANCOBAN see MRE225
ELANCOLAN see DUV600
ELANIL see EAH500
ELAOL see DEH200
ELASIOMYCIN see EAG500
ELASTASE see EAG875
ELASTONON see AOA250, BBK000
ELASTOPAR see MJG750
ELASTOPAX see MJG750
ELASTOZONE 31 see BJT500
ELASTOZONE 34 see PFL000
ELASZYM see EAG875
ELATERICIN A see EAH100
α-ELATERIN see COE250
ELAVIL see EAH500, EAI000
ELAVIL HYDROCHLORIDE see EAI000
ELAYL see EIO000
ELBANIL see CKC000
ELBRUS see CGA000
ELCIDE 75 see MDI000
ELCORIL see CDO500
EL-CORTELAN SOLUBLE see HHR000
ELCOSINE see SNJ350
ELCOZINE CHRYSOIDINE Y see PEK000
ELCOZINE RHODAMINE B see FAG070
ELDADRYL see BAU750
ELDEPRYL see DAZ125
ELDERBERRY see EAI100
ELDERFIELD PYRIMIDINE MUSTARD see DYC700
ELDESINE see VGU750
ELDEZOL see NGE500
ELDIATRIC C see CAK500
ELDODRAM see DYE600
ELDOPAL see DNA200
ELDOPAQUE see HIH000
ELDOQUIN see HIH000
ELDRIN see RSU000
ELEAGOL see SCA750
ELECOR see PEV750
ELECTRO-CF 11 see TIP500
ELECTRO-CF 12 see DFA600
ELECTRO-CF 22 see CFX500
ELECTROCORTIN see AFJ875
ELECTROCORUNDUM see EAL100
ELECTRONIC E-2 see HIC000
ELEKTROCORTIN see AFJ875
ELEMENTAL SELENIUM see SBO500
ELEMI see EAI500
ELEMI OIL see EAI500
ELEN see IBW500
ELENIUM see LFK000, MDQ250
ELEPHANT'S EAR see CAL125, EAI600
ELEPHANT TRANQUILIZER see AOO500
ELEPSINDON see DKQ000
ELESTOL see CLD000
ELEUDRON see TEX250
ELEVAN see DOZ000
ELEX 334 see EAI800
ELFAN 4240 T see SON000
ELGACID ORANGE 2G see FAG010
ELGETOL see BRE500, DUS700, DUU600

ELGETOL 318 see BRE500
ELICIDE see MDI000
ELIMIN see DLV000
ELIMOCLAVIN see EAJ000
ELIPOL see DUS700
ELIPTEN see AKC600
ELITONE see DJS200
ELIXICON see TEP000
ELIXIR of VITRIOL see SOI510
ELIXOPHYLLIN see TEP000
ELIXOPHYLLINE see TEP000
ELJON FAST SCARLET RN see MMP100
ELJON LAKE RED C see CHP500
ELJON MADDER see DMG800
ELJON YELLOW BG see DEU000
ELKAPIN see MNG000
ELKOSIL see SNJ350
ELKOSIN see SNJ350
ELKOSINE see SNJ350
ELLIPTICINE see EAI850
ELLIPTISINE see EAI850
ELLSYL see MHJ500
ELMASIL see AMY050
ELMEDAL see BRF500
ELMER'S GLUE ALL see AAX250
ELOBROMOL see DDJ000
ELOCRON see DVS000
ELON see MGJ750
ELON WORT see CCS650
ELORINE see CPQ250
ELORINE CHLORIDE see EAI875
ELORINE SULFATE see TJG225
EL PETN see PBC250
ELPI see ARQ750
ELPON see PMP500
ELRODORM see DYC800
ELSAN see DRR400
ELSYL see MHJ500
ELTRIANYL see TKX000
ELTROXIN see LEQ300
ELVACITE see PKB500
ELVANOL see PKP750
ELVARON see DFL200
ELYMOCLAVIN see EAJ000
ELYMOCLAVINE see EAJ000
ELYSION see AOO500
ELYZOL see MMN250
ELZOGRAM see CCS250
EM see EDH500
EM 12 see DVS600
EM 136 see OOM300
EM 255 see EAJ100
EM 923 see DFY400
EMAFORM see CHR500
EMAGRIN see SPC500
EMAL T see SON000
EMANAY ATOMIZED ALUMINUM POWDER see AGX000
EMANAY ZINC DUST see ZBJ000
EMANAY ZINC OXIDE see ZKA000
EMANDIONE see PFJ750
EMANIL see DAS000
EMANON 4115 see PJY100
EMAR see ZKA000
EMATHLITE see KBB600
EMAZOL RED B see EAJ500
EMB see TGA500
EMBACETIN see CDP250
EMBADOL see TFK300
EMBAFUME see MHR200
EMBANOX see BQI000
EMBARIN see ZVJ000
EMBARK see DUK000
EMBARK PLANT GROWTH REGULATOR see DUK000
EMBATHION see EEH600
EMBAY 8440 see BGB400
EMBECHINE see BIE500
EMBELIC ACID see EAJ600
EMBELIN see EAJ600
EMBERLINE see EAJ600
EMB-FATOL see EDW875
EMBICHIN see BIE250, BIE500
EMBICHIN 7 see NOB700
EMBICHIN HYDROCHLORIDE see BIE500
EMBIKHINE see BIE500

EMBINAL see BAG250
EMBIOL see VSZ000
EMBITOL see EAK000
EMBONIC ACID see PAF100
EMBRAMINE HYDROCHLORIDE see BMN250
EMBUTAL see NBU000
EMBUTOX see DGA000, EAK500
EMBUTOX KLEAN-UP see DGA000
EMC see CHC500
EMCEPAN see CIR250
EMCOL 888 see EAL000
EMCOL CA see OAV000
EMCOL DS-50 CAD see HKJ000
EMCOL H-2A see PJY100
EMCOL H 31A see PJY100
EMCOL-IP see IQW000
EMCOL MSK see OAV000
EMCOL PS-50 RHP see SLL000
EMD 9806 see SDZ000
EMDABOL see TFK300
EMDABOLIN see TFK300
EMEDAN see BSM000
EMERALD GREEN see BAY750, COF500
EMERESSENCE 1150 see EJQ500
EMERESSENCE 1160 see PER000
EMEREST 2301 see OHW000
EMEREST 2316 see IQW000
EMEREST 2350 see EJM500
EMEREST 2381 see SLL000
EMEREST 2400 see OAV000
EMEREST 2401 see OAV000
EMEREST 2646 see PJY100
EMEREST 2660 see PJY100
EMEREST 2801 see OHW000
EMERGIL see FMO129
EMERICID see LIF000
EMERSAL 6400 see SIB600
EMERSAL 6434 see SON000
EMERSAL 6465 see TAV750
EMERSOL 120 see SLK000
EMERSOL 140 see PAE250
EMERSOL 143 see PAE250
EMERSOL 210 see OHU000
EMERSOL 213 see OHU000
EMERSOL 6321 see OHU000
EMERSOL 233LL see OHU000
EMERSOL 221 LOW TITER WHITE OLEIC ACID see OHU000
EMERSOL 220 WHITE OLEIC ACID see OHU000
EMERY see EAL100
EMERY 655 see MSA250
EMERY 2218 see MJW000
EMERY 2219 see OHW000
EMERY 2310 see OHW000
EMERY 5703 see BKD500
EMERY 5709 see DHF400
EMERY 5711 see TGT000
EMERY 5712 see DMT800
EMERY 5791 see MCN250
EMERY 6705 see PER000
EMERY 6802 see PJW500
EMERY OLEIC ACID ESTER 2301 see OHW000
EMERY X-88-R see DWT000
EMESIDE see ENG500
EMETE-CON see BDW000
EMETHIBUTIN HYDROCHLORIDE see EIJ000
EMETIC HOLLY see HGF100
EMETICON see BDW000
EMETIC WEED see CCJ825
EMETINE see EAL500
(−)-EMETINE see EAL500
EMETINE ANTIMONY IODIDE see EAM000
EMETINE BISMUTH IODIDE see EAM500
EMETINE with BISMUTH(III) TRIIODIDE see EAM500
EMETINE, DIHYDROCHLORIDE see EAN000
(−)-EMETINE DIHYDROCHLORIDE see EAN000
l-EMETINE DIHYDROCHLORIDE see EAN000
EMETINE DIHYDROCHLORIDE TETRAHYDRATE see EAN500
EMETINE HYDROCHLORIDE see EAN000
EMETINE TRIIODOBISMUTH(III) see EAM500
EMETIQUE (FRENCH) see AQG250
EMETIRAL see PMF250
EMETREN see CDP250
EMEX see AER666
EMFAC 1202 see NMY000

EMI-CORLIN see HHR000
EMID 6511 see BKE500
EMID 6541 see BKE500
EMINEURINA see CHD750
EMIPHEROL see VSZ450
EMISAN 6 see MEP250
EMISOL see AMY050
EMMATOS see MAK700
EMMATOS EXTRA see MAK700
EMMI see EME050
EMOCICLINA see VSZ000
EMODIN see IIU000, MQF250
EMODOL see MQF250
EMO-NIK see NDN000
EMOQUIL see EAN600
EMOREN see DTL200
EMORFAZONE see EAN700
EMORHALT see AJV500
EMOTIVAL see CFC250
EMP see EDT100, EME100
EMPAL see CIR250
EMPECID see MRX500
EMPG see ENC000
EMPILAN 2848 see EJM500
EMPILAN BP 100 see PJY100
EMPILAN BQ 100 see PJY100
EMPIRIN see ADA725
EMPIRIN COMPOUND see ABG750, ARP250
EMPLETS POTASSIUM CHLORIDE see PLA500
EMPP see EAV700
EMQ see SAV000
EMS see EMF500
EMSORB 2500 see SKV100
EMSORB 2505 see SKV150
EMSORB 2515 see SKV000
EMSORB 6900 see PKL100
EMSORB 6907 see SKV195
EMT 25,299 see MDV500
EMTAL 596 see TAB750
EMTEXATE see MDV500
EMTRYL see DSV800
EMTRYLVET see DSV800
EMTRYMIX see DSV800
EMTS see EME500
EMULGEN 100 see DXY000
EMUL P.7 see OAV000
EMULPHOR see PKE500
EMULPHOR A see PJY100
EMULPHOR ON-870 see PJW500
EMULPHOR SURFACTANTS see PKE500
EMULPHOR UN-430 see PJY100
EMULPHOR VN 430 see PJY100
EMULSAMINE BK see DAA800
EMULSAMINE E-3 see DAA800
EMULSIFIER No. 104 see SIB600
EMULSION 212 see FQU875
EMULSIPHOS 440/660 see SJH200
EMULSOV O EXTRA P see EAN800
E-MYCIN see EDH500
EMYRENIL see OOG000
EN 237 see EAL100
EN 313 see EEI025
EN-1530 see NAH000
EN 1627 see DPE000
EN 1639 see CQF099
EN 1939 see CQF099
EN-15304 see NAH000
EN 18133 see EPC500
EN-28,450 see AGT250
ENADEL see CMY125
ENALAPRIL MALEATE see EAO100
ENALLYNYMAL SODIUM see MDU500
ENAMEL WHITE see BAP000
ENANTHAL see HBB500
ENANTHALDEHYDE see HBB500
ENANTHIC ACID see HBE000
ENANTHIC ALCOHOL see HBL500
ENANTHOLE see HBB500
ENANTHOTOXIN see EAO500
ENANTHYLIC ACID see HBE000
ENARMON see TBG000
ENAVEN see EGS000
ENAVID see EAP000
ENBU see ENT000

ENC see ENT500
ENCEPHALARTOS HILDEBRANDTII see EAQ000
ENCETROP see NNE400
ENCORDIN see EAQ050
ENCORTON see PLZ000
ENCYPRATE see EGT000
ENDAK see MQT550
ENDAK MITE see MQT550
ENDECRIL see DFP600
ENDEP see EAI000
ENDIEMALUM see DJO800
ENDISON see DCV800
ENDOBIL see IFP800
ENDOBION see NCR000
ENDOCEL see EAQ750
ENDOCID see EAS000
ENDOCIDE see EAS000
ENDOCISTOBIL see BGB315
ENDODAN see EJQ000
ENDOD, EXTRACT see EAQ100
ENDO E see VSZ450
6,14-ENDOETHENO-7-(2-HYDROXY-2-PENTYL)-TETRAHYDRO-ORIPAVINE
    HYDROCHLORIDE see EQO500
ENDOFOLLICOLINA D.P. see EDR000
ENDOFOLLICULINA see EDV000
ENDOGRAFIN see BGB315
ENDOGRAPHIN see BGB315
ENDOLAT see DAM700
2,5-ENDOMETHYLENE CYCLOHEXANECARBOXYLIC ACID, ETHYL ESTER
    (mixed formyl isomers) see NNG500
(2,5-ENDOMETHYLENECYCLOHEXYLMETHYL)AMINE see AKY750
cis-3,6-ENDOMETHYLENE-Δ⁴-TETRAHYDROPHTHALIC ACID DIMETHYL
    ESTER see DRB400
ENDOMETHYLENETETRAHYDROPHTHALIC ACID, N-2-ETHYLHEXYL
    IMIDE see OES000
ENDOMYCIN see EAQ500
3,6-ENDOOXOHEXAHYDROPHTHALIC ACID see EAR000
ENDOPANCRINE see IDF300
ENDOPITUITRINA see ORU500
ENDOSAN see BGB500
ENDOSOL see EAQ750
ENDOSULFAN see EAQ750
ENDOSULPHAN see EAQ750
ENDOTAL see DXD000
ENDOTHAL see DXD000, EAR000
ENDOTHAL COMBINED with IPC (1581) see EAR500
ENDOTHALL see EAR000
ENDOTHAL-NATRIUM (DUTCH) see DXD000
ENDOTHAL-SODIUM see DXD000
ENDOTHAL TECHNICAL see EAR000
ENDOTHAL WEED KILLER see DXD000
ENDOTHION see EAS000
ENDOTOXIN see EAS100
ENDOTOXIN, ESCHERICHIA COLI see EDK700
ENDOX see EAT600
ENDOXAN see CQC650
ENDOXANA see CQC500
ENDOXANAL see CQC650
ENDOXAN-ASTA see CQC500
ENDOXAN MONOHYDRATE see CQC500
ENDOXAN R see CQC500
3,6-ENDOXOHEXAHYDROPHTHALIC ACID see EAR000
3,6-ENDOXOHEXAHYDROPHTHALIC ACID DISODIUM SALT see DXD000
ENDRATE see EIX000
ENDRATE DISODIUM see EIX500
ENDRATE TETRASODIUM see EIV000
ENDREX see EAT500
ENDRIN see EAT500
ENDRINE (FRENCH) see EAT500
ENDROCID see EAT600
ENDROCIDE see EAT600
ENDURACIDIN see EAT800
ENDURACIDIN A see EAT810
ENDURON see MIV500
ENDURONYL see RDF000
ENDUXAN see CQC500
ENDYDOL see ADA725
ENDYL see TNP250
E.N.E. see ENX500
ENE 11183 B see EAT600
ENELFA see HIM000
ENERIL see HIM000
ENERZER see IKC000
ENFENEMAL see ENB500

ENFLURANE see EAT900
ENGLISH HOLLY see HGF100
ENGLISH IVY see AFK950
ENGLISH RED see IHD000
ENHEPTIN see ALQ000
ENHEPTIN-A see ABY900
ENHEXYMAL see ERD500, ERE000
ENHEXYMAL NFN see ERE000
ENHYDRINA SCHISTOSA VENOM see EAU000
ENIACID BRILLIANT RUBINE 3B see HJF500
ENIACID LIGHT ORANGE G see HGC000
ENIACID METANIL YELLOW GN see MDM775
ENIACID ORANGE I see FAG010
ENIACROMO ORANGE R see NEY000
ENIALIT LIGHT RED RL see MMP100
ENIAL ORANGE I see PEJ500
ENIAL YELLOW 2G see DOT300
ENIAMETHYL ORANGE see MND600
ENIANIL BLACK CN see AQP000
ENIANIL BLUE 2BN see CMO000
ENIANIL BRILLIANT BLUE FF see CMN750
ENIANIL PURE BLUE AN see CMO500
ENICOL see CDP250
ENIDE see DRP800
ENIDRAN see DYE700
ENIDREL see CFZ000, EAP000
ENILOCONAZOL (SP) see FPB875
ENJAY CD 460 see PMP500
ENKALON see NOH000
ENKEFAL see DNU000
ENKELFEL see DKQ000
ENNG see ENU000
ENOCITABINE see EAU075
ENORDEN see MQU750
ENOVID see EAP000
ENOVID-E see EAP000
ENOVIT see DJV000
ENOVIT M see PEX500
ENOXACIN see EAU100
ENOXACIN HYDRATE (2:3) see EAU150
ENPHENEMAL see ENB500
ENPROMATE see DWL400
ENRADIN see EAT800
ENRADINE see EAT810
ENRAMYCIN see EAT800, EAT810
ENRIOCHROME ORANGE AOR see NEY000
ENRUMAY see WAK000
E.N.S. see ENX500
ENS see EPI300
ENSEAL see PLA500
ENSIGN see CMF350
ENSODORM see EOK000
ENSTAMINE HYDROCHLORIDE see DCJ850
ENSTAR see POB000
ENSURE see EAQ750
ENS-ZEM WEEVIL BAIT see DXE000
ENT see ENU500
ENT 5 see LBO000
ENT 6 see BPL250
ENT 9 see BRT000
17-ENT see ABU000, ENX600
ENT 38 see PDP250
ENT 54 see ADX500
ENT 92 see IHZ000
ENT 114 see DYA200
ENT 123 see VHZ000
ENT 133 see RNZ000
ENT 154 see DUS700
ENT 157 see CPK500
ENT 262 see DTR200
ENT 375 see EKV000
ENT 884 see COF500
ENT 988 see BJK500
ENT 1,122 see BRE500
ENT 1,501 see DXE000
ENT 1,506 see DAD200
ENT 1,656 see EIY600
ENT 1,716 see MEI450
ENT 1,860 see PCF275
ENT 3,424 see NDN000
ENT 3,776 see DFT000
ENT 3,797 see TBO500
ENT 4,225 see BIM500
ENT 4,504 see DFJ050

ENT 4,585 see CJR500
ENT 4,705 see CBY000
ENT 7,543 see POO100
ENT 7,796 see BBQ500
ENT 8,184 see OES000
ENT 8286 see BRP250
ENT 8,420 see DGG000
ENT 8,538 see DAA800
ENT 8,601 see BBP750
ENT 9,232 see BBQ000
ENT 9,233 see BBR000
ENT 9,234 see BFW500
ENT 9,624 see BIN000
ENT 9,735 see CDV100
ENT 9,932 see CDR750
ENT 14,250 see PIX250
ENT 14,611 see ASL250
ENT 14,689 see FAS000
ENT 14,874 see EIR000
ENT 14,875 see MAS500
ENT 15,108 see PAK000
ENT 15,152 see HAR000
ENT 15,208 see CKE750, NCM700
ENT 15,266 see PNP250
ENT 15,349 see EIY500
ENT 15,406 see PNJ400
ENT 15,949 see AFK250
ENT 16,087 see NIM500
ENT 16,225 see DHB400
ENT 16,273 see SOD100
ENT 16275 see BGC750
ENT 16,358 see CJT750
ENT 16,391 see KEA000
ENT 16,436 see DXX400
ENT 16,519 see SOP500
ENT 16,634 see ISA000
ENT 16,894 see TED500
ENT 17,034 see MAK700
ENT 17,035 see NFT000
ENT 17,251 see EAT500
ENT 17,291 see OCM000
ENT 17,292 see MNH000
ENT 17,295 see DAO600
ENT 17,470 see DFK600
ENT 17,510 see AFR250
ENT 17,588 see PPQ625
ENT 17591 see EAU500
ENT 17,596 see BHJ500
ENT 17,798 see EBD700
ENT 17,941 see CKI625
ENT 17,956 see CNU750
ENT 18,060 see CKC000
ENT 18,066 see BON250
ENT 18,544 see PDK000
ENT 18,596 see DER000
ENT 18,771 see TCF250
ENT 18,861 see MIJ250
ENT 18,862 see DAO800, MIW100
ENT 18,870 see DMC600
ENT 19,059 see PNQ250
ENT 19,060 see DSK200
ENT 19,109 see BJE750
ENT 19,244 see IKO000
ENT 19,442 see TBC500
ENT 19,507 see DCM750
ENT 19,763 see TIQ250
ENT 20,218 see DKC800
ENT 20,696 see CEP000
ENT 20,738 see DGP900
ENT 20,852 see BPG000, DDP000
ENT 20,993 see AMX825
ENT 21,040 see AGE250
ENT 22,014 see EKN000
ENT 22,335 see TBR250
ENT 22,374 see MQR750
ENT 22,542 see DKC800
ENT 22,784 see BIN500
ENT 22,865 see DJN600
ENT 22,897 see DVQ709
ENT 22,952 see POO000
ENT 23,233 see ASH500
ENT 23,284 see RMA500
ENT 23,437 see DXH325
ENT 23,438 see DRR400

ENT 27,474 see BEP500
ENT 27,488 see BAT750
ENT 27,520 see CMA250
ENT 27,521 see PHE250
ENT 27,552 see IOS000
ENT 27,566 see DSO200
ENT 27,567 see CJJ250, CJJ500
ENT 27,572 see FAK000
ENT 27,625 see DOL800
ENT 27,635 see CLJ875
ENT 27,696 see DRR000
ENT 27,738 see BLU000
ENT 27,766 see DOX600
ENT 27,822 see DOP600
ENT 27,851 see DAB400
ENT 27,910 see DED000
ENT 27,967 see MJL250
ENT 27,989 see MKA000
ENT 28,009 see HON000
ENT 28,344 see PIZ499
ENT 29,054 see CJV250
ENT 29,118 see EPC175
ENT 31,472 see MJK500
ENT 32,833 see AAR500
ENT 33,335 see HFX000
ENT 33348 see HFE520
ENT 50,003 see TNK250
ENT 50,107 see BJC250
ENT 50,146 see RDK000
ENT 50,324 see EJM900
ENT 50,434 see AQG250
ENT 50,439 see BIA250
ENT 50,698 see DYC700
ENT 50787 see BGX775
ENT 50,825 see MLY000
ENT-50,838 see TDQ225
ENT 50,852 see HEJ500
ENT 50,882 see HEK000
ENT 50,909 see AGU500
ENT 50,990 see DOV600
ENT 50,991 see ASK500
ENT 51253 see BGY500
ENT 51254 see MGE100
ENT 51256 see BGX850
ENT 51,762 see NNF000
ENT 51,799 see MJQ500
ENT 51,904 see TLR250
ENT 61,241 see ACM750
ENT 61,969 see DNU850
ENT 70,459 see EQD000
ENT 70,460 see KAJ000
ENT 70,531 see POB000
ENT 25,832-a see CON300
ENT 27,300-A see CQI500
ENTACYL see HEP000
ENT A13-29261 see AFK000
ENT 27,386GC see DRR400
ENT 27,699GC see DIN800
ENTEPAS see AMM250
ENTERAMINE see AJX500
ENTERICIN see ADA725
ENTERO-BIO FORM see CHR500
ENTERO-EXOTOXIN, CHOLERA see CMC800
ENTEROMYCETIN see CDP250
ENTEROPHEN see ADA725
ENTEROQUINOL see CHR500
ENTEROSALICYL see SJO000
ENTEROSALIL see SJO000
ENTEROSARINE see ADA725
ENTEROSEDIV see TEH500
ENTEROSEPTOL see CHR500
ENTEROTOXIN, CHOLERA see CMC800
ENTEROTOXON see NGG500
ENTERO-VIOFORM see CHR500
ENTEROZOL see CHR500
ENTERUM LOCORTEN see CHR500
ENTEX see FAQ900
ENTHOHEX see PCY300
ENTIZOL see MMN250
ENTOBEX see PCY300
ENTOMOXAN see BBQ500
ENTPROL see QAT000
ENTRA see TMX775
ENTRAMIN see ALQ000

ENTROKIN see CHR500
ENTRONON see PCY300
ENTROPHEN see ADA725
ENTSUFON see TMN490
ENTUSIL see SNN500
ENT 24,980-X see DJA400
ENT 25,545-X see OAN000
ENT 25,552-X see CDR750
ENT 25,554-X see MNO750
ENT 25,555-X see DJA200
ENT 25595-X see DQZ000
ENT 25,602-X see COD850
ENT 25,700-X see HCK000
ENT 27,395-X see CQH650
ENTYDERMA see AFJ625
ENU see ENV000, NKE500
ENUCLEN see BBA500
ENVERT 171 see DAA800
ENVERT DT see DAA800
ENVERT-T see TAA100
ENVIOMYCIN SULFATE see TNY250, VQZ100
ENZACTIN see THM500
ENZAMIN see BBW500
ENZAPROST see POC500
ENZAPROST F see POC500
ENZENE-1,3-DICARBOXYLIC ACID see IMJ000
ENZEON see CML850
ENZOSE see DBI800
E.O. see EJN500
EO 122 see EAV100
EOCT see COD475
EOSIN see BMO250, BNK700
EOSIN BLUE see ADG250
EOSINE see BMO250, BNH500
EOSINE B see BNK700
EOSINE BLUE see ADG250
EOSINE BLUISH see ADG250
EOSINE FA see BNK700
EOSINE LAKE RED Y see BNK700
EOSINE SODIUM SALT see BNH500
EOSINE YELLOWISH see BNH500
EOSIN GELBLICH (GERMAN) see BNH500
EP 30 see PAX250
E 66P see PKQ059
EP-145 see ECO500
EP-185 see DNI200
EP 201 see ECB000
EP-205 see BJN250
EP-206 see VOA000
EP 316 see CQI500
EP-332 see DSO200
EP-333 see CJJ250, CJJ500
EP-411 see PFC750
EP-452 see MEG250
EP 453 see OMY925
EP-475 see EEO500
EP-1086 see TKS000
EP 1463 see AAX250
EP-161E see ISE000
EPAL 6 see HFJ500
EPAL 8 see OEI000
EPAL 10 see DAI600
EPAL 12 see DXV600
EPAL 16NF see HCP000
E-PAM see DCK759
EPAMIN see DKQ000, DNU000
EPANUTIN see DKQ000, DNU000
EPAREN see DFL200
EPASMIR "5" see DKQ000
EPATIOL see MCI375
EPC (the plant regulator) see CBL750
EPDANTOINE SIMPLE see DKQ000
EPE see EAV500
EPELIN see DKQ000, DNU000
EPERISONE HYDROCHLORIDE see EAV700
EPHEDRAL see EAW000
EPHEDRATE see EAW000
EPHEDREMAL see EAW000
EPHEDRIN see EAW000
EPHEDRINE see EAW000
l-EPHEDRINE see EAW000
l(−)-EPHEDRINE see EAW000
psi-EPHEDRINE see POH000
d-psi-EPHEDRINE see POH000

EPHEDRINE HYDROCHLORIDE see EAW500, EAY000
(−)-EPHEDRINE HYDROCHLORIDE see EAY000
d-EPHEDRINE HYDROCHLORIDE see EAX000
l-EPHEDRINE HYDROCHLORIDE see EAY000
dl-EPHEDRINE HYDROCHLORIDE see EAX500
EPHEDRINE PENICILLIN see PAQ120
d-EPHEDRINE PHOSPHATE (ESTER) see EAY150
l-EPHEDRINE PHOSPHATE (ESTER) see EAY075
dl-EPHEDRINE PHOSPHATE (ESTER) see EAY175
l-EPHEDRINE SULFATE see EAY500
EPHEDRINHYDROCHLORID (GERMAN) see MGH250
EPHEDRITAL see EAW000
EPHEDROL see EAW000
EPHEDROSAN see EAW000
EPHEDROTAL see EAW000
EPHEDSOL see EAW000
EPHENDRONAL see EAW000
EPHERON see PEC250
EPHETONIN see EAX500
EPHETONINE see EAX500
EPHININE HYDROCHLORIDE see EAZ000
EPHIRSULPHONATE see CJT750
EPHORRAN see DXH250
EPHOXAMIN see EAW000
EPHYNAL see VSZ450
EPIB see ARQ750
EPIBENZALIN see DLY000
EPIBLOC see CDT750
EPIBROMHYDRIN see BNI000
EPIBROMOHYDRIN (DOT) see BNI000
EPIBROMOHYDRINE see BNI000
EPICHLOORHYDRINE (DUTCH) see EAZ500
EPICHLORHYDRIN (GERMAN) see EAZ500
EPICHLORHYDRINE (FRENCH) see EAZ500
EPICHLOROHYDRIN see EAZ500
α-EPICHLOROHYDRIN see EAZ500
(dl)-α-EPICHLOROHYDRIN see EAZ500
EPICHLOROHYDRYNA (POLISH) see EAZ500
EPICHLOROPHYDRIN see EAZ500
EPICLASE see PEC250
EPI-CLEAR see BDS000
EPICLORIDRINA (ITALIAN) see EAZ500
EPICUR see MQU750
EPICURE DDM see MJQ000
EPIDEHYDROCHOLESTERIN see EBA100
EPIDERMAL GROWTH FACTOR see EBA275
EPIDERMOL see ACR300
EPIDIAN 5 see IPO000
3,17-EPIDIHYDROXYESTRATRIENE see EDO000
3,17-EPIDIHYDROXYOESTRATRIENE see EDO000
EPIDIONE see TLP750
9,10-EPIDIOXY ANTHRACENE see EBA500
1,4-EPIDIOXY-1,4-DIHYDRO-6,6-DIMETHYLFULVENE see EBA600
EPIDONE see TLP750
EPIDORM see EOK000
EPIDOSIN see VBK000
4'-EPIDOXORUBICIN see EBB100
EPIDOZIN see VBK000
EPIDROPAL see ZVJ000
EPI-DX see EBB100
EPIFEN see FMS875
EPIFENYL see DKQ000, DNU000
EPIFLUOROHYDRIN see EBU000
EPIFOAM see HHQ800
EPIFRIN see VGP000
EPIHYDAN see DKQ000, DNU000
EPIHYDRIN ALCOHOL see GGW500
EPIHYDRINALDEHYDE see GGW000
EPIHYDRINAMINE, N,N-DIETHYL- see GGW800
EPIHYDRINE ALDEHYDE see GGW000
3-EPIHYDROXYETIOALLOCHOLAN-17-ONE see HJB050
EPIKOTE 828 see IPN000
EPIKOTE 1001 see IPM000
EPIKOTE 1004 see IPP000
EPILAN see DKQ000, MKB250
EPILAN-D see DNU000
EPILANTIN see DKQ000, DNU000
EPILEO PETIT MAL see ENG500
EPILIM see PNR750
EPILIN see PNX750
EPIMID see MNZ000
EPINAL see AGN000
EPINAT see DKQ000, DNU000
EPINELBON see DLY000

EPINEPHRAN see VGP000
EPINEPHRINE see VGP000
(−)-EPINEPHRINE see VGP000
l-EPINEPHRINE see VGP000
d-EPINEPHRINE see AES250
(R)-EPINEPHRINE see VGP000
dl-EPINEPHRINE see EBB500
EPINEPHRINE racemic see EBB500
l-EPINEPHRINE (synthetic) see VGP000
EPINEPHRINE BITARTRATE see AES000
(−)-EPINEPHRINE BITARTRATE see AES000
EPINEPHRINE-d-BITARTRATE see AES000
l-EPINEPHRINE BITARTRATE see AES000
l-EPINEPHRINE-d-BITARTRATE see AES000
EPINEPHRINE CHLORIDE see AES500
l-EPINEPHRINE CHLORIDE see AES500
(−)-EPINEPHRINE HYDROCHLORIDE see AES500
l-EPINEPHRINE HYDROCHLORIDE see AES500
(±)-EPINEPHRINE HYDROCHLORIDE see AES625
dl-EPINEPHRINE HYDROCHLORIDE see AES625
EPINEPHRINE HYDROGEN TARTRATE see AES000
EPINEPHRINE ISOPROPYL HOMOLOG see DMV600
l-EPINEPHRINE TARTRATE see AES000
EPINOVAL see DJU200
EPI-PEVARYL see EAE000
EPIPODOPHYLLOTOXIN see EBB600
EPIPREMNUM AUREUM see PLW800
EPIRENAMINE see VGP000
EPIRENAN see VGP000
EPI-REZ 508 see BLD750
EPI-REZ 510 see BLD750
EPI-REZ 508 mixed with ERR 4205 (1:1) see OPI200
EPIRIZOLE see MCH550
EPIROTIN see BBW500
EPIRUBICIN see EBB100
EPISED see DKQ000
EPISEDAL see EOK000
5-EPISISOMICIN see EBC000
EPISOL TARTRATE see EBD000
EPITELIOL see VSK600
EPITHELONE see ACR300
2,3-EPITHIOANDROSTAN-17-OL see EBD500
2-α,3-α-EPITHIO-5-α-ANDROSTAN-17-β-OL see EBD500
2,2-EPITHIO-17-((1-METHOXYCYCLOPENTYL)OXY)-ANDROSTANE (2-α,3-α,5-
    α,17-β) see MCH600
4,5-EPITHIOVALERONITRILE see EBD600
2-α,3-α-EPITHIO-17-β-YL 1-METHOXYCYCLOPENTYL ETHER see MCH600
EPITIOSTANOL see EBD500
EPITOPIC see DKJ300
EPITRATE see VGP000
EPL see EDN000
EPN see EBD700
EPOBRON see IIU000
EPOCAN see BMB000
EPOCELIN see EBE100
EPODYL see TJQ333
EPOLENE M 5K see PMP500
EPON 562 see EBF000
EPON 820 see EBF500
EPON 828 see BLD750, IPO000
EPON 1001 see EBG000
EPON 1007 see EBG500
EPON 828 mixed with ERR 4205 (1:1) see OPI200
EPONTHOL see PMM000
EPORAL see SOA500
(3,6-EPOSSI-CICLOESAN-1,2-DICARBOSSILATO) DISODICO (ITALIAN)
    see DXD000
EPOSTANE see EBH400
EPOXIDE-201 see ECB000
EPOXIDE 269 see LFV000
EPOXIDE A see BLD750
EPOXIDE ERLA-0510 see EBH500
1,2-EPOXYAETHAN (GERMAN) see EJN500
(E)-1-α-2-α-EPOXYBENZ(c)ACRIDINE-3-α-4-β-DIOL see EBH850
(Z)-1-β,2-β-EPOXYBENZ(c)ACRIDINE-3-α-4-β-DIOL see EBH875
1,2-EPOXYBUTANE see BOX750
1,4-EPOXYBUTANE see TCR750
2,3-EPOXYBUTANE see EBJ100
1,2-EPOXYBUTENE-3 see EBJ500
3,4-EPOXY-1-BUTENE see EBJ500
2,3-EPOXYBUTYRIC ACID BUTYL ESTER see EBK000
2,3-EPOXYBUTYRIC ACID, ETHYL ESTER see EJS000
1,2-EPOXYBUTYRONITRILE see EBK500
3,4-EPOXYBUTYRONITRILE see EBK500

(2,3-EPOXYPROPYL)TRIMETHYLAMMONIUM CHLORIDE see GGY200
3,12-EPOXY-12H-PYRANO(4,3-j)-1,2-BENZODIOXEPIN, DECAHYDRO-10-ME-
THOXY-3,6,9-TRIMETHYL-, (3-α-5a-β,6-β,8a-β,9-α-12-β,12aR)-, (+)-
see ARL425
EPOXY RESIN (EPICHLOROHYDRIN and DIETHYLENE GLYCOL)
see ECK500
EPOXY RESIN ERL-2795 see ECL000
EPOXY RESINS, CURED see ECL500
EPOXY RESINS, UNCURED see ECM500
9,10-EPOXYSTEARIC ACID see ECD500
cis-9,10-EPOXYSTEARIC ACID see ECD500
9,10-EPOXYSTEARIC ACID ALLYL ESTER see ECO500
9,10-EPOXYSTEARIC ACID-2-ETHYLHEXYL ESTER see EKV500
EPOXYSTYRENE see EBR000
α,β-EPOXYSTYRENE see EBR000
3,4-EPOXYSULFOLANE see ECP000
1,2-EPOXY-1,2,3,4-TETRAHYDROBENZ(a)ANTHRACENE see ECP500
(±)-3-α,4-α-EPOXY-1,2,3,4-TETRAHYDROBENZO(b,def)CHRYSENE-1-β,2-α-
DIOL see DMS500
cis-1-β,2-β-EPOXY-1,2,3,4-TETRAHYDROBENZO(c)PHENANTHRENE-3-α-4-β-
DIOL see BCS103
trans-1-α-2-α-EPOXY-1,2,3,4-TETRAHYDROBENZO(c)PHENANTHRENE-3-α,4-β-
DIOL see BCS105
trans-1-β,2-β-EPOXY-1,2,3,4-TETRAHYDROBENZO(c)PHENANTHRENE-3-β, 4-α-
DIOL see BCS110
7,8-EPOXY-7,8,9,10-TETRAHYDROBENZO(a)PYRENE see ECQ050
9,10-EPOXY-7,8,9,10-TETRAHYDROBENZO(a)PYRENE see ECQ100
9,10-EPOXY-9,10,11,12-TETRAHYDROBENZO(e)PYRENE see ECQ150
3,4-EPOXY-1,2,3,4-TETRAHYDROCHRYSENE see ECQ200
(+)-(E)-3,4-EPOXY-1,2,3,4-TETRAHYDRO-CHRYSENEDIOL see DMS000
1,2-EPOXY-1,2,3,4-TETRAHYDROPHENANTHRENE see ECR259
3,4-EPOXY-1,2,3,4-TETRAHYDROPHENANTHRENE see ECR500
(±)-3-α,4-α-EPOXY-1,2,3,4-TETRAHYDRO-1-β,2-α-PHENANTHRENEDIOL
see ECS000
(±)-3-β,4-β-EPOXY-1,2,3,4-TETRAHYDRO-1-β,2-α-PHENANTHRENEDIOL
see ECS500
1,2-EPOXY-1,2,3,4-TETRAHYDROTRIPHENYLENE see ECT000
12,13-EPOXY-3-α,4-β,8-α,15-TETRAHYDROXYTRICHOTHEC-9-ENE-8-
ISOVALERATE see DAD600
12,13-EPOXY-3,4,7,15-TETRAHYDROXYTRICHOTHEC-9-EN-8-ONE
see NMV000
(3-α,4-β,7-α)-12,13-EPOXY-3,4,7,15-TETRAHYDROXYTRICHOTHEC-9-EN-8-
ONE see NMV000
2,3-EPOXYTETRAMETHYLENE SULFONE see ECP000
1,2-EPOXY-3-(TOLYLOXY)PROPANE see TGZ100
12,13-EPOXY-4-β,8-α,15-TRIACETOXY-3-α-HYDROXY TRICHOTHEC-9-ENE
see ACS500
1,2-EPOXY-4,4,4-TRICHLOROBUTANE see ECT500
EPOXY-1,1,2-TRICHLOROETHANE see ECT600
1,2-EPOXY-3,3,3-TRICHLOROPROPANE see TJC250
12,13-EPOXY-TRICHOTHEC-9-ENE-3-α,15-TETROL 15-ACETATE, 8-ISOVALER-
ATE see THI250
6-β,7-β-EPOXY-3-α-TROPANYL S-(−)-TROPATE see SBG000
EPOXYTROPINE TROPATE see SBG000
EPOXYTROPINE TROPATE METHYLBROMIDE see SBH500
EPOXYTROPINE TROPATE METHYLNITRATE see SBH000
1,2-EPOXY-4-VINYLCYCLOHEXANE see VNZ000
EPRAZIN see POL500
EPRAZINONE DIHYDROCHLORIDE see ECU550
EPRAZINONE HYDROCHLORIDE see ECU550
EPROFIL see TEX000
EPROLIN see VSZ450
EPROZINOL see EQY600
EPROZINOL DIHYDROCHLORIDE see ECU600
EPSAMON see AJD000
EPSICAPRON see AJD000
EPSILAN see VSZ450
EPSOM SALTS see MAJ250, MAJ500
EPSYLON KAPROLAKTAM (POLISH) see CBF700
EPT see EQP000
EPTAC 1 see BJK500
EPTACLORO (ITALIAN) see HAR000
1,4,5,6,7,8,8-EPTACLORO-3a,4,7,7a-TETRAIDRO-4,7-endo-METANO-INDENE
(ITALIAN) see HAR000
EPTAL see DKQ000
EPTAM see EIN500, EPC150
EPTANI (ITALIAN) see HBC500
EPTAN-3-ONE (ITALIAN) see EHA600
EPTAPUR see CKF750
EPTC see EIN500
EPTOIN see DKQ000, DNU000
E-PVC see PKQ059
EQ see SAV000
EQUAL see ARN825

EQUANIL SUSPENSION see MQU750
EQUIBEN see CBA100
EQUIBRAL see MDQ250
EQUI BUTE see BRF500
EQUIGEL see DGP900
EQUIGYNE see ECU750, ECU750
EQUILENIN see ECV000
EQUILENINA (SPANISH) see ECV000
EQUILENIN BENZOATE see ECV500
EQUILENINE see ECV000
EQUILIN see ECW000
EQUILIN BENZOATE see ECW500
EQUILIN SODIUM SULFATE see ECW520
EQUILIN, SULFATE, SODIUM SALT (6CI) see ECW520
EQUILIUM see MQU750, PGE000
EQUIMATE see ECW550
EQUINE CYONIN see SCA750
EQUINE GONADOTROPHIN see SCA750
EQUINE GONADOTROPIN see SCA750
EQUINIL see MQU750
EQUINO-ACID see TIQ250
EQUINO-AID see TIQ250
EQUIPERTINE see ECW600
EQUIPOISE see CJR909
EQUIPROXEN see MFA500
EQUIPUR see VLF000
EQUISETIC ACID see ADH000
EQUIZOLE see TEX000
ER 115 see CFY750
ER5461 see CQG250
ERABUTOXINA see ECX000
ERADE see ORU000
ERADEX see CMA100
ERALDIN see ECX100
ERALON see ILD000
ERAMIDE see CDR250
ERAMIN see HGD000
ERANTIN see PNA500
ERASE see HKC000
ERASOL see BIE500
ERASOL HYDROCHLORIDE see BIE500
ERASOL-IDO see BIE500
ERAZIDON see ORU000
ERBAPLAST see CDP250
ERBAPRELINA see TGD000
ERBITOX see DUT800
ERBIUM CHLORIDE see ECX500
ERBIUM CITRATE see ECY000
ERBIUM(III) NITRATE (1:3) see ECY500
ERBIUM(III) NITRATE, HEXAHYDRATE (1:3:6) see ECZ000
ERBIUM TRICHLORIDE see ECX500
ERBN see PBK000
ERBOCAIN see PDU250
ERBON see PBK000
ERCEFUROL see DGQ500
ERCEFURYL see DGQ500
ERCO-FER see FBJ100
ERCOFERRO see FBJ100
ERCORAX see HKR500
ERCOTINA see HKR500
ERE 1359 see REF000
EREBILE see DAL000
ERGADENYLIC ACID see AOA125
ERGAM see EDC500
ERGAMINE see HGD000
ERGATE see EDC500
ERGENYL see PNX750
ERGOATETRINE see LJL000
ERGOBASINE see LJL000
ERGOCALCIFEROL see VSZ100
ERGOCHROME AA (2,2')-5-β,6-α,10-β-5',6'-α,10'-β see EDA500
ERGOCORNIN see EDA600
ERGOCORNINE see EDA600
ERGOCORNINE HYDROGEN MALEATE see EDA875
ERGOCORNINE HYDROGEN MALEINATE see EDA875
ERGOCORNINE MALEATE see EDA875
ERGOCORNINE MESYLATE see EDB000
ERGOCORNINE METHANESULFONATE (SALT) see EDB000
ERGOCORNINE METHANESULPHONATE see EDB000
ERGOCRYPTINE see EDB100
α-ERGOCRYPTINE see EDB100
ERGOCRYPTINE MESYLATE see EDB125
ERGOCRYPTINE METHANESULFONATE see EDB125
ERGOCRYPTINE METHANESULPHONATE see EDB125

ERGOKLININE see LJL000
α-ERGOKRYPTINE see EDB100
ERGOKRYPTINE METHANESULFONATE see EDB125
ERGOLINE-8-ACETAMIDE, 6-ETHYL-, (8-β)-, (R-(R*,R*))-2,3-
    DIHYDROXYBUTANEDIOATE (1:1) see EJS100
ERGOLINE-8-ACETAMIDE, 6-METHYL-, (8-β)-, (R-(R*,R*))-2,3-
    DIHYDROXYBUTANEDIOATE (2:1) see MJV500
ERGOLINE-8-ACETAMIDE, 6-(2-PROPENYL)-, (8-β)-, (R-(R*,R*))-2,3-
    DIHYDROXYBUTANEDIOATE see PMR800
ERGOLINE-8-ACETAMIDE, 6-PROPYL-, (8-β)-, (R-(R*,R*))-2,3-
    DIHYDROXYBUTANEDIOATE (2:1) see PNL800
ERGOLINE-8-CARBAMIC ACID, 6-PROPYL-, ETHYL ESTER, (8S) see EPC115
ERGOLINE-8-β-CARBOXAMIDE, 9,10-DIDEHYDRO-N,N-DIETHYL-6-METHYL-,
    TARTRATE (2:1) see LJM600
ERGOLINE-8-β-PROPIONAMIDE, α-ACETYL-6-METHYL- see ACR100
ERGOLINE-8-PROPIONAMIDE, 6-ALLYL-α-CYANO- see AGC200
ERGOLINE-8-PROPIONAMIDE, 2-CHLORO-α-CYANO-6-METHYL- see CFF100
ERGOLINE-8-PROPIONAMIDE, α-CYANO-2,6-DIMETHYL- see COM075
ERGOLINE-8-PROPIONAMIDE, α-CYANO-6-ISOBUTYL- see COP600
ERGOLINE-8-β-PROPIONAMIDE, N-ETHYL-6-METHYL-α-
    (METHYLSULFONYL)- see EMW100
ERGOLINE-8-β-PROPIONITRILE, 6-METHYL-α-(4-METHYL-1-
    PIPERAZINYLCARBONYL)- see MLR400
ERGOLINE-8-β-PROPIONITRILE, 6-METHYL-α-(1-
    PYRROLIDINYLCARBONYL)- see MPC300
ERGOMAR see EDC500
ERGOMETRINE see LJL000
ERGOMETRINE ACID MALEATE see EDB500
ERGOMETRINE MALEATE see EDB500
ERGONOVINE see LJL000
ERGONOVINE, MALEATE (1:1) (SALT) see EDB500
ERGOPLAST ADC see DGT500
ERGOPLAST AdDO see AEO000
ERGOPLAST FDO see DVL700
ERGORONE see VSZ100
ERGOSINE METHANESULFONATE see EDB200
α-ERGOSINE METHANESULFONATE see EDB200
ERGOSINE MONOMETHANESULFONATE see EDB200
ERGOSTAT see EDC500
ERGOSTEROL, activated see VSZ100
ERGOSTEROL, irradiated see VSZ100
ERGOT see EDB500
ERGOTAMAN-3',6',18-TRIONE, 12'-HYDROXY-2',5'-BIS(1-METHYLETHYL)-,
    (5'-α)- see EDA600
ERGOTAMAN-3',6',18-TRIONE, 12'-HYDROXY-2'-METHYL-5'-(2-
    METHYLPROPYL)-, (5'-α)-, MONOMETHANESULFONATE (salt) see EDB200
ERGOTAMINE see EDC000
ERGOTAMINE BITARTRATE see EDC500
ERGOTAMINE TARTRATE see EDC500
ERGOTARTRATE see EDC500
ERGOTERM OTGO see DVN600
ERGOTERM TGO see EDC560
ERGOTIDINE see HGD000
ERGOTOCINE see LJL000
ERGOTOXIN see EDC565
ERGOTOXINE see EDC565
ERGOTOXINE ETHANESULFONATE see EDC575
ERGOTOXINE ETHANESULPHONATE see EDC575
ERGOTOXINE ETHANSULFONATE see EDC575
ERGOTRATE see EDB500, LJL000
ERGOTRATE MALEATE see EDB500
ERIAMYCIN see EDC600
ERIBUTAZONE see BRF500
ERIDAN see DCK759
ERIE BLACK B see AQP000
ERIE CONGO 4B see SGQ500
ERIE VIOLET 3R see CMP000
ERINA see MQU750
ERINIT see PBC250
ERINITRIT see SIQ500
ERIOBOTRYA JAPONICA see LIH200
ERIO CHROME RED PE see CMG750
ERIO FAST ORANGE AS see HGC000
ERIO FAST YELLOW AEN see SGP500
ERIOGLAUCINE see FMU059
ERIOGLAUCINE G see FAE000
ERION see ARQ250, CFU750
ERIONITE see EDC650
ERIOSIN RHODAMINE B see FAG070
ERIOSKY BLUE see FMU059
ERISIMIN DIHYDRATE see HAO000
ERISPAN see FDB100
ERITRONE see VSZ000
ERITROXILINA see CNE750

ERIZIMIN see HAN800
ERIZOMYCIN see PPI775
ERL-2774 see BLD750
ERL-2795 see ECL000
ERLA-2270 see VOA000
ERLA-2271 see VOA000
ERMALONE see MJE760
ERMETRINE see LJL000
EROCYANINE 540 see EDD500
EROINA see HBT500
ERR 4205 see BJN250
ERR 4205 mixed with ARALDITE 6010 (1:1) see OPI200
ERR 4205 mixed with EPI-REZ 508 (1:1) see OPI200
ERR 4205 mixed with EPON 828 (1:1) see OPI200
ERROLON see CHJ750
ERSERINE see PIA500
ERTALON 6SA see PJY500
ERTILEN see CDP250
ERTRON see VSZ100
ERTUBAN see ILD000
ERYCIN see EDH500
ERYCYTOL see VSZ000
ERYPAR see EDJ500
ERYSAN see BIF250
ERYSIMIN see HAN800
ERYSIMIN DIHYDRATE see HAO000
ERYSIMOTOXIN see HAN800
ERYSIMUPICRONE see SMM500
ERYSIMUPIKRON see SMM500
ERYSODINE HYDROCHLORIDE see EDE000
ERYSOPINE HYDROCHLORIDE see EDE500
ERYTHRALINE HYDROBROMIDE see EDF000
ERYTHRENE see BOP500
ERYTHRITOL ANHYDRIDE see BGA750, DHB800
ERYTHROCIN see EDH500
ERYTHROCIN STEARATE see EDJ500
ERYTHROGRAN see EDH500
ERYTHROGUENT see EDH500
ERYTHROHYCIN GLUCEPTATE see EDI500
β-ERYTHROIDINE see EDG500
β-ERYTHROIDINE HYDROCHLORIDE see EDH000
ERYTHROMYCIN see EDH500
ERYTHROMYCIN A see EDH500
ERYTHROMYCIN CARBONATE see EDI000
ERYTHROMYCIN GLUCOHEPTONATE (1581) see EDI500
ERYTHROMYCIN HYDROCHLORIDE see EDJ000
ERYTHROMYCIN OCTADECANOATE (salt) see EDJ500
ERYTHROMYCIN STEARATE see EDJ500
ERYTHROMYCIN STEARIC ACID SALT see EDJ500
ERYTHROMYCIN SULFAMATE see EDK000
ERYTHROMYCIN SULFAMATE (SALT) see EDK000
ERYTHROSIN see FAG040
ERYTHROSINE B-FO (BIOLOGICAL STAIN) see FAG040
ERYTHROTIN see VSZ000
ERYTROXYLIN see CNE750
ES 902 see PPN000
ESACHLOROBENZENE (ITALIAN) see HCC500
ESACICLONATO see SHL500
ESAIDRO-1,3,5-TRINITRO-1,3,5-TRIAZINA (ITALIAN) see CPR800
ESAMETILENTETRAMINA (ITALIAN) see HEI500
ESAMETINA see HEA000
ESAMETONIO IODURO (ITALIAN) see HEB000
ESANI (ITALIAN) see HEN000
ESANITRODIFENILAMINA (ITALIAN) see HET500
ESANTENE see HFG550
ESAPROPIMATO see POA250
ESBATAL see BFW250
ESBECYTHRIN see DAF300
ESBERICARD see EDK600
ESBIOL see AFR750
ESBIOL CONCENTRATE 90% see AFR750
ESBRITE see SMQ500
ESCAMBIA 2160 see PKQ059
ESCHERICHIA COLI ENDOTOXIN see EDK700
ESCIN see EDK875
α-ESCIN see EDL000
β-ESCIN see EDL500
ESCINA (ITALIAN) see EDK875
β-ESCINIC ACID see EDL000
ESCIN, SODIUM SALT see EDM000
ESCIN TRIETHANOLAMINE SALT see EDM500
ESCLAMA see NHH000
ESCOPARONE see DRS800
ESCOREZ 7404 see SMQ500

ESCORPAL see PDC850, PDC875
ESCULETIN DIMETHYL ETHER see DRS800
ESDRAGOL see AFW750
ESE see EPI000
ESELIN see DIS600
ESEN see PHW750
ESERINE see PIA500
ESERINE SALICYLATE see PIA750
ESERINE SULFATE see PIB000
ESERINE SULPHATE see PIB000
ESEROLEIN, METHYLCARBAMATE (ESTER) see PIA500
ESFAR see CPJ000
ESFAR CALCIUM see CPJ250
ESGRAM see PAJ000
ESICLENE see FNK040
ESIDREX see CFY000
ESIDRIX see CFY000
ESILGAN see CKL250
ESJAYDIOL see AOO475
ESKABARB see EOK000
ESKACILLIAN V see PDT500
ESKACILLIN see BFD000
ESKADIAZINE see PPP500
ESKALIN V see VRA700, VRF000
ESKALITH see LGZ000
ESKAZINE see TKK250
ESKAZINE DIHYDROCHLORIDE see TKK250
ESKEL see AHK750
ESKIMON 11 see TIP500
ESKIMON 12 see DFA600
ESKIMON 22 see CFX500
ESMAIL see CGA000
ESMARIN see HII500
ESOBARBITALE (ITALIAN) see ERD500
E 39 SOLUBLE see BDC750
ESOMID CHLORIDE see HEA500
ESOPHOTRAST see BAP000
ESOPIN see MDL000
ESORB see VSZ450
ESP see DSK600
ESPADOL see CLW000
ESPARIN see DQA600
ESPASMO GASIUM see DAB750
ESPECTINOMICINA DIHYDROCHLORIDE PENTAHYDRATE see SKY500
ESPENAL see DXH250
ESPERAL see DXH250
ESPERAN see OLW400
ESPEROX 10 see BSC500
ESPEROX 31M see BSD250
ESPERSON see DBA875
ESPHYGMOGENINA see VGP000
ESPIGA de AMOR (PUERTO RICO) see CAK325
ESPIGELIA (CUBA) see PIH800
ESPINOMYCIN A see MBY150
ESPIRAN see DAI200
ESPRIL see BET000
ESQUINON see BGX750
ESSENCE of MIRBANE see NEX000
ESSENCE of MYRBANE see NEX000
ESSENCE of NIOBE see EGR000
ESSENTIAL OIL of ACORUS CALAMUS Linn. see OGL020
ESSENTIAL OIL of CYMBOPOGON NARDUS see CMT000
ESSENTIAL OIL from MYRTLE see OGU000
ESSENTIAL PHOSPHOLIPIDS see EDN000
ESSEX 1360 see HLB400
ESSEX GUM 1360 see HLB400
ESSIGESTER (GERMAN) see EFR000
ESSIGSAEURE (GERMAN) see AAT250
ESSIGSAEUREANHYDRID (GERMAN) see AAX500
ESSO FUNGICIDE 406 see CBG000
ESSO HERBICIDE 10 see BQZ000
EST see OMY925
ESTABAL see BFW250
ESTABEX S see BOU550
ESTABEX U 18 see DVK200
ESTANE 5703 see UVA000
ESTAR see CMY800, PKF750
ESTASIL see MQU750
ESTAZOLAM see CKL250
ESTER 25 see NIM500
ESTER d'ACIDE BENZILIQUE et DU-1-METHYLSULFATE de 1,1-DIMETHYL-(2-HYDROXY-METHYL)PIPERIDINIUM see CDG250
ESTER del ACIDO BENCILICO del-1,1-DIMETIL-2-OXIMETIL-PIPERIDINIO-METILSULFATO (SPANISH) see CDG250

ESTERCIDE T-2 and T-245 see TAA100
S-ESTER with O,O-DIMETHYL PHOSPHOROTHIOATE see MAK700
ESTER DWETYLOAMINOETYLOWSKY KWASU DWUFENYLOOCTOWEGO (POLISH) see THK000
ESTER DWUETYLOAMINOETYLOWY KWASU DWUFENYLOOCTOWEGO see DHX800
ESTERE ETOSSIMETILICO dell' ACIDO N-(2,6-DICLORO-m-TOLIL)AN-TRANILICO (ITALIAN) see DGN000
ESTERE ISOAMILICO dell'ACIDO α-(N-(PIRROLIDINOETIL))-AMINOFENILACETICO (ITALIAN) see CBA375
S-ESTER of (2-MERCAPTOETHYL)TRIMETHYLAMMONIUM IODIDE with O,O-DIETHYL PHOSPHOROTHIOATE see TLF500
O-ESTER-p-NITROPHENOL with O-ETHYL PHENYL PHOSPHONOTHIOATE see EBD700
ESTERON see DAA800
ESTERON 44 see IOY000
ESTERON 99 see DAA800
ESTERON 76 BE see DAA800
ESTERON 245 BE see TAA100
ESTERON BRUSH KILLER see DAA800, TAA100
ESTERON 99 CONCENTRATE see DAA800
ESTERONE see EDV000
ESTERONE FOUR see DAA800
ESTERON 44 WEED KILLER see DAA800
α-ESTER PALMITIC ACID with D-threo-(−)-2,2-DICHLORO-N-(β-HYDROXY-α-(HYDROXYMETHYL)-p-NITROPHENETHYL)ACETAMIDE see CDP700
ESTERS see EDN500
ESTER SULFONATE see CJT750
ESTERTRICHLOROSTANNANE see TIM500
ESTEVE see BRF500
ESTIBOGLUCONATO SODICO see AQH800, AQI250
ESTILBEN see DKA600, DKB000
ESTILBIN see DKB000
ESTIMULEX see DBA800
ESTINERVAL see PFC750
ESTOCINE see DPE200
ESTOL 103 see IQW000
ESTOL 603 see OAV000
ESTOL 1550 see DJX000
ESTOMYCIN see NCF500
ESTON see DSK600, QFA250
ESTONATE see DAD200
ESTON-B see EDP000
ESTONMITE see CJT750
ESTONOX see CDV100
ESTOTSIN see DPE200
ESTOX see DSK600
ESTRACYT see EDT100
ESTRACYT HYDRATE see EDN600
ESTRADEP see DAZ115
ESTRA-4,9-DIEN-3-ONE, 11-(4-(DIMETHYLAMINO)PHENYL)-17-HYDROXY-17-(1-PROPYNYL)-, (11-β, 17-β)- see HKB700
ESTRADIOL see EDO000
d-ESTRADIOL see EDO000
α-ESTRADIOL see EDO000
β-ESTRADIOL see EDO000
17-α-ESTRADIOL see EDO500
ESTRADIOL-17-β see EDO000
17-β-ESTRADIOL see EDO000
cis-ESTRADIOL see EDO000
3,17-β-ESTRADIOL see EDO000
d-3,17-β-ESTRADIOL see EDO000
ESTRADIOL BENZOATE see EDP000
ESTRADIOL-3-BENZOATE see EDP000
β-ESTRADIOL BENZOATE see EDP000
β-ESTRADIOL-3-BENZOATE see EDP000
ESTRADIOL-17-β-BENZOATE see EDP000
17-β-ESTRADIOL BENZOATE see EDP000
ESTRADIOL-17-β-3-BENZOATE see EDP000
17-β-ESTRADIOL-3-BENZOATE see EDP000
ESTRADIOL-17-BENZOATE-3,n-BUTYRATE see EDP500
ESTRADIOL BENZOATE mixed with PROGESTERONE (1:14 moles) see EDQ000
ESTRADIOL-3-BENZOATE mixed with PROGESTERONE (1:14 moles) see EDQ000
ESTRADIOL-17-CAPRYLATE see EDQ500
ESTRADIOL-17-β 3-CYCLOPENTYL ETHER see QFA250
ESTRADIOL CYCLOPENTYLPROPIONATE see DAZ115
ESTRADIOL-17-CYCLOPENTYLPROPIONATE see DAZ115
ESTRADIOL-17-β-CYCLOPENTYLPROPIONATE see DAZ115
ESTRADIOL-CYPINATE see DAZ115
ESTRADIOL-17-CYPIONATE see DAZ115
ESTRADIOL-17-β-CYPIONATE see DAZ115
ESTRADIOL DIPROPIONATE see EDR000
β-ESTRADIOL DIPROPIONATE see EDR000

17-β-ESTRADIOL DIPROPIONATE see EDR000
ESTRADIOL-3,17-DIPROPIONATE see EDR000
3,17-β-ESTRADIOL DIPROPIONATE see EDR000
β-ESTRADIOL-3,17-DIPROPIONATE see EDR000
ESTRADIOL 3-METHYL ETHER see MFB775
17-β-ESTRADIOL 3-METHYL ETHER see MFB775
ESTRADIOL MONOBENZOATE see EDP000
17-β-ESTRADIOL MONOBENZOATE see EDP000
ESTRADIOL MUSTARD see EDR500
ESTRADIOL PHOSPHATE POLYMER see EDS000
ESTRADIOL POLYESTER with PHOSPHORIC ACID see EDS000
ESTRADIOL VALERATE see EDS100
ESTRADIOL-17-VALERATE see EDS100
ESTRADIOL 17-β-VALERATE see EDS100
ESTRADIOL VALERIANATE see EDS100
ESTRADURIN see EDS000, PJR750
ESTRAGARD see DAL600
ESTRAGON OIL see TAF700
ESTRALDINE see EDO000
ESTRALUTIN see HNT500
ESTRAMUSTINE PHOSPHATE DISODIUM see EDT100
ESTRAMUSTINE PHOSPHATE DISODIUM HYDRATE see EDN600
ESTRAMUSTINE PHOSPHATE SODIUM see EDT100
ESTRAMUSTINE PHOSPHATE SODIUM HYDRATE see EDN600
α-ESTRA-1,3,5,7,9-PENTANE-3,17-DIOL see EDT500
β-ESTRA-1,3,5,7,9-PENTANE-3,17-DIOL see EDU000
ESTRATAB see ECU750
1,3,5,7-ESTRATETRAEN-3-OL-17-ONE see ECW000
ESTRA-1,3,5(10),7-TETRAEN-17-ONE, 3-HYDROXY-, HYDROGEN SULFATE
    SODIUMSALT (8CI) see ECW520
ESTRA-1,3,5(10),7-TETRAEN-17-ONE, 3-(SULFOOXY)-, SODIUM SALT
    see ECW520
(17-β)-ESTRA-1,3,5(10)-TRIEN-3,17-DIOL, 3-METHOXY-17-(3,3,3-TRIFLUORO-1-
    PROPYNYL) see TKH050
ESTRA-1,3,5(10)-TRIENE-2,4-D2-3,17-DIOL, (17-β)- see DHA425
1,3,5-ESTRATRIENE-3,17-α-DIOL see EDO500
1,3,5-ESTRATRIENE-3,17-β-DIOL see EDO000
ESTRA-1,3,5(10)-TRIENE-3,17-α-DIOL see EDO500
ESTRA-1,3,5(10)-TRIENE-3,17-β-DIOL see EDO000
17-β-ESTRA-1,3,5(10)-TRIENE-3,17-DIOL see EDO000
ESTRA-1,3,5(10)-TRIENE-3,17-β-DIOL, 3-BENZOATE see EDP000
1,3,5(10)-ESTRATRIENE-3,17-β-DIOL 3-BENZOATE see EDP000
ESTRA-1,3,5(10)-TRIENE-3,17-DIOL (17-β)-3-BENZOATE see EDP000
ESTRA-1,3,5(10)-TRIENE-3,17-β-DIOL-17-BENZOATE-3-n-BUTYRATE
    see EDP500
ESTRA-1,3,5(10)-TRIENE-3,17-β-DIOL 3-(BIS(2-CHLOROETHYL)CARBA-
    MATE)17-DISODIUM PHOSPHATE see EDT100
(17-β)-ESTRA-1,3,5(10)-TRIENE-3,17-DIOL 17-CYCLOPENTANEPROPANOATE
    (9CI) see DAZ115
1,3,5(10)-ESTRATRIENE-3,17-β-DIOL DIPROPIONATE see EDR000
ESTRA-1,3,5(10)-TRIENE-3,17-DIOL (17-β)-DIPROPIONATE see EDR000
ESTRA-1,3,5(10)-TRIENE-3,17-DIOL, 11-ETHYL- see EJT575
ESTRA-1,3,5(10)-TRIENE-3,17-β-DIOL, 4-FLUORO-, (17-β)-(9CI) see FIA500
ESTRA-1,3,5(10)-TRIENE-3,17-β-DIOL-17-OCTANOATE see EDQ500
(17-β)-ESTRA-1,3,5(10)-TRIENE-3,17-DIOL-17-PENTANOATE (9CI) see EDS100
(17-β)-ESTRA-1,3,5(10)-TRIENE-3,17-DIOL POLYMER with PHOSPHORIC
    ACID see EDS000
ESTRA-1,3,5(10)-TRIENE-17-β-DIOL-17-TETRAHYDROPYRANYL ETHER
    see EDU100
1,3,5-ESTRATRIENE-3-β,16-α,17-β-TRIOL see EDU500
1,3,5(10)-TRIENE-3,16-α,17-β-TRIOL see EDU500
(16-α,17-β)-ESTRA-1,3,5(10)-TRIENE-3,16,17-TRIOL see EDU500
1,3,5-ESTRATRIEN-3-OL-17-ONE see EDV000
1,3,5(10)-ESTRATRIEN-3-OL-17-ONE see EDV000
Δ-1,3,5-ESTRATRIEN-3-β-OL-17-ONE see EDV000
ESTRA-4,9,11-TRIEN-3-ONE, 17-α-ALLYL-17-HYDROXY- see AGW675
ESTRA-4,9,11-TRIEN-3-ONE, 17-HYDROXY-17-(2-PROPENYL)-, (17-β)-(9CI)
    see AGW675
ESTRA-1,3,5(10)-TRIEN-17-ONE, 3-(SULFOOXY)-, SODIUM SALT (9CI)
    see EDV600
ESTRA-1,3,5(10)-TRIETNE-3,17-β-DIOL-3-BENZOATE mixed with PROGESTER-
    ONE (1:14 moles) see EDQ000
ESTRATRIOL see EDU500
ESTRAVEL see EDS100
ESTRELLA DEL NORTE (CUBA) see ROU450
ESTR-4-EN-3-ONE, 17-β-HYDROXY-, DECANOATE see NNE550
ESTR-4-EN-3-ONE, 17-((1-OXODECYL)OXY)-, (17-β)- (9CI) see NNE550
ESTREPTOCIDA see SNM500
ESTRIFOL see ECU750
ESTRIL see DKA600
ESTRIN see EDV000
ESTRIOL see EDU500
16-α,17-β-ESTRIOL see EDU500
3,16-α,17-β-ESTRIOL see EDU500

ESTRIOLO (ITALIAN) see EDU500
ESTROATE see ECU750
ESTROBEN see DKB000
ESTROBENE see DKA600, DKB000
ESTROCON see ECU750
ESTRODIENOL see DAL600
ESTROFOL see PKF750
ESTROFURATE see EDU600
ESTROGEN see DKA600, EEH500
ESTROGENIN see DKB000
ESTROGENS, CONJUGATES see PMB000
ESTROICI see EDR000
ESTROL see EDV000
ESTROMED see ECU750
ESTROMENIN see DKA600
ESTRON see EDV000
ESTRONA (SPANISH) see EDV000
ESTRONE see EDV000
ESTRONE-A see EDV000
ESTRONE BENZOATE see EDV500
ESTRONE, HYDROGEN SULFATE, SODIUM SALT see EDV600
ESTRONE SODIUM SULFATE see EDV600
ESTRONE SULFATE SODIUM see EDV600
ESTRONE SULFATE SODIUM SALT see EDV600
ESTRONE-3-SULFATE SODIUM SALT see EDV600
ESTRONEX see EDR000
ESTROPAN see ECU750
ESTRORAL see DAL600
ESTROSEL see DGP900
ESTROSOL see DGP900
ESTROSTILBEN see DKB000
ESTROSYN see DKA600
ESTROVIS see QFA250
ESTROVIS 4000 see QFA250
ESTROVISTER see QFA250
ESTROVITE see EDO000
ESTRUGENONE see EDV000
ESTRUMATE see CMX880
ESTRUSOL see EDV000
ESTULIC see GKU300
ESTYRENE AS see ADY500
ESTYRENE G 20 see SMQ500
ESUCOS see MKQ000
ESZ see IHL000
ET 14 see RMA500
ET 57 see RMA500
ET 495 see TNR485
ETABETACIN see EDW000
ETABUS see DXH250
ETACRINIC ACID see DFP600
ETAFENONE HYDROCHLORIDE see DHS200
ETAIN (TETRACHLORURE d') (FRENCH) see TGC250
ETAKRINIC ACID see DFP600
ETAMBRO see TCC000
ETAMBUTOL see EDW875
ETAMICAN see VSZ450
ETAMIDE see EEM000
ETAMINAL SODIUM see NBU000
ETAMINOPHYLLINE see CNR125
ETAMIPHYLLIN see CNR125
ETAMIPHYLLINE see CNR125
ETAMIPHYLLINE HYDROCHLORIDE see DIH600
ETAMIPHYLLIN HYDROCHLORIDE see DIH600
ETAMON CHLORIDE see TCC250
ETAMSYLATE see DIS600
ETANAUTINE see BBV500
ETANOLAMINA (ITALIAN) see EEC600
ETANOLO (ITALIAN) see EFU000
ETANTIOLO (ITALIAN) see EMB100
ETAPERAZINE see CJM250
ETAVIT see VSZ450
ETAZIN see BQC250
ETAZINE see BQC250
ETC see EPY600
ETCHLORVINOLO see CHG000
3′,4′-Et2-DAB see DJB400
ETEAI see EDW300
ETEM see EJQ000
ETERE ETILICO (ITALIAN) see EJU000
E TETRAETHYLPYRONIN see DIS200
ETH see EEI000, EPQ000
ETHAANTHIOL (DUTCH) see EMB100
ETHACRIDINE LACTATE see EDW500
ETHACRYNIC ACID see DFP600

ETHAL see HCP000
ETHALFLURALIN see ENE500
ETHAL LA-X see DXY000
ETHAMBUTOL see TGA500
ETHAMBUTOL DIHYDROCHLORIDE see EDW875
ETHAMBUTOL HYDROCHLORIDE see EDW875
ETHAMBUTOL mixed with SODIUM NITRITE (1:1) see SIS500
ETHAMIDE see DWW000
ETHAMINAL see NBT500
ETHAMINAL SODIUM see NBU000
ETHAMON DS see EDX000
ETHAMOXYTRIPHETOL see DHS000
ETHAMSYLATE see DIS600
ETHAN see PII500
ETHANAL see AAG250
ETHANAL OXIME see AAH250
ETHANAMIDE see AAI000
ETHANAMIDINE HYDROCHLORIDE see AAI500
ETHANAMINE see EFU400
ETHANAMINIUM, 2-CHLORO-N,N,N-TRIMETHYL-, CHLORIDE (9CI)
    see CMF400
ETHANAMINIUM, 2-((CYCLOHEXYLHYDROXYPHENYLACETYL)OXY)-N,N-
    DIETHYL-N-METHYL-, BROMIDE (9CI) see ORQ000
ETHANAMINIUM, N,N-DIETHYL-N-METHYL-2-((9H-XANTHEN-9-
    YLCARBONYL)OXY)-, BROMIDE (9CI) see XCJ000
ETHANDIAL see GIK000
ETHANDROSTATE see EDY600
ETHANE see EDZ000
ETHANE, compressed (DOT) see EDZ000
ETHANE, refrigerated liquid (DOT) see EDZ000
ETHANECARBOXYLIC ACID see PMU750
ETHANE, 2-CHLORO-2-(DIFLUOROMETHOXY)-1,1,1-TRIFLUORO- (9CI)
    see IKS400
ETHANEDIAL see GIK000
ETHANEDIAL DIOXIME see EEA000
ETHANEDIAMIDE see OLO000
1,2-ETHANEDIAMINE see EEA500
1,2-ETHANEDICARBOXYLIC ACID see SMY000
ETHANE DICHLORIDE see EIY600
ETHANEDINITRILE see COO000
ETHANEDIOIC ACID see OLA000
ETHANEDIOIC ACID BIS(CYCLOHEXYLIDENE HYDRAZIDE)
    see COF675
ETHANEDIOIC ACID, DISODIUM SALT see SIY500
ETHANEDIOIC ACID, TIN(2+) SALT (1:1) (9CI) see TGE250
1,2-ETHANEDIOL see EJC500
1,2-ETHANEDIOL DIACETATE see EJD759
ETHANEDIOL DIMETHACRYLATE see BKM250
1,2-ETHANEDIOL DIMETHACRYLATE see BKM250
1,2-ETHANEDIOL DIMETHANESULFONATE (9CI) see BKM125
ETHANEDIOL DINITRATE see EJG000
1,2-ETHANEDIOL DIPROPANOATE (9CI) see COB260
1,2-ETHANEDIOL, MONOACETATE see EJI000
1,2-ETHANEDIONE see GIK000
ETHANEDIONIC ACID see OLA000
ETHANEDITHIOAMIDE see DXO200
1,2-ETHANEDITHIOL see EEB000
1,2-ETHANEDITHIOL, CYCLIC ESTER with P,P-DIETHYL PHOS-
    PHONODITHIOIMIDOCARBONATE see PGW750
1,2-ETHANEDITHIOL, CYCLIC S,S-ESTER with PHOS-
    PHONODITHIOIMIDOCARBONIC ACID P,P-DIETHYL ESTER see PGW750
N,N'-1,2-ETHANEDIYLBIS(N-BUTYL-4-MORPHOLINECARBOXAMIDE)
    see DUO400
1,2-ETHANEDIYLBIS(CARBAMODITHIOATO)(2−)-MANGANESE see MAS500
((1,2-ETHANEDIYLBIS(CARBAMODITHIOATO))(2-)ZINC see EIR000
1,2-ETHANEDIYLBIS(CARBAMODITHIOATO) (2-)-S,S'-ZINC see EIR000
1,2-ETHANEDIYLBISCARBAMODITHIOIC ACID DISODIUM SALT
    see DXD200
1,2-ETHANEDIYLBISCARBAMODITHIOIC ACID MANGANESE COMPLEX
    see MAS500
1,2-ETHANEDIYLBISCARBAMODITHIOIC ACID, MANGANESE(2+) SALT
    (1:1) see MAS500
1,2-ETHANEDIYLBISCARBAMODITHIOIC ACID, ZINC COMPLEX see EIR000
1,2-ETHANEDIYLBISCARBAMOTHIOIC ACID, ZINC SALT see EIR000
N,N'-1,2-ETHANEDIYLBIS(N-(CARBOXYMETHYL)GLYCINE see EIX000
N,N'-1,2-ETHANEDIYLBIS(N-(CARBOXYMETHYL)GLYCINE) DISODIUM
    SALT see EIX500
N,N'-1,2-ETHANEDIYLBIS(N-(CARBOXYMETHYL)GLYCINE TETRASODIUM
    SALT see EIV000
N,N'-1,2-ETHANEDIYLBIS(N-(CARBOXYMETHYL)GLYCINE, TRISODIUM
    SALT see TNL250
1,2-ETHANEDIYLBISMANEB, MANGANESE (2+) SALT (1:1) see MAS500
1,1'-(1,2-ETHANEDIYLBIS(OXY))BIS-BUTANE see DDW400
2,2'-(1,2-ETHANEDIYLBIS(OXY))BISETHANOL see TJQ000

1,2-ETHANEDIYLBIS(TRIS(2-CYANOETHYL)PHOSPHONIUM DIBROMIDE
    see EIU000
(R)-2,2'-(1,2-ETHANEDIYLDIIMINO)BIS-1-BUTANOL see TGA500
1,1'-(1,2-ETHANEDIYLDIIMINO)BIS(3-(4-METHOXYPHENOXY)-2-PROPANOL,
    DIMETHANESULFONATE (salt) see MQY125
1,2-ETHANEDIYL DIMETHANESULFONATE see BKM125
1,2-ETHANEDIYL ESTER CARBAMIMIDOTHIOIC ACID DIHYDROBROMIDE
    see EJA000
ETHANE HEXACHLORIDE see HCI000
ETHANE HEXAMERCARBIDE see BKH125, EEB500
ETHANEHYDRAZONIC ACID see ACM750
ETHANE-1-HYDROXY-1,1-DIPHOSPHONATE see HKS780
ETHANE-1-HYDROXY-1,1-DIPHOSPHONIC ACID DISODIUM SALT
    see DXD400
ETHANE-1-HYDROXY-1,1-DIPHOSPHONIC ACID, TETRAPOTASSIUM SALT
    see TEC250
ETHANE-1-HYDROXY-1,1-DIPHOSPHONIC ACID, TRATRASODIUM SALT
    see TEE250
ETHANE-1-HYDROXY-1,1-DIPHOSPHONIC ACID, TRISODIUM SALT
    see TNL750
ETHANENITRILE see ABE500
ETHANE PENTACHLORIDE see PAW500
ETHANEPEROXOIC ACID see PCL500
ETHANEPEROXOIC ACID-1,1-DIMETHYLETHYL ESTER see BSC250
ETHANESELENOL, 2-AMINO-, HYDROCHLORIDE see AJS900
ETHANE-1,1'-SULFINYLBIS see EPI500
ETHANESULFONIC ACID, 2-HYDROXY-, AMMONIUM SALT see ANL100
ETHANESULFONYL CHLORIDE see EEC000
ETHANETHIOAMIDE see TFA000
ETHANETHIOIC ACID see TFA500
ETHANETHIOL see EMB100
ETHANETHIOLIC ACID see TFA500
ETHANE TRICHLORIDE see TIN000
1,1,1-ETHANETRIOL DIPHOSPHONATE see HKS780
ETHANIMINIUM-N-(6-DIETHYLAMINO)-3H-XANTHEN-3-YLIDENE)-N-ETHYL
    CHLORIDE see DIS200
6,10-ETHANO-5-AZONIASPIRO(4.5)DECAN-8-OL CHLORIDE BENZILATE
    see BCA375
1H-4,9a-ETHANOCYCLOHEPTA(c)PYRAN-7-CARBOXYLIC ACID, 4a-
    (ACETYLOXY)-3-(4-CARBOXY-1,3-PENTADIENYL)3,4,4a,5,6,9-
    HEXAHYDRO-3-METHYL-1-OXO-, 7-METHYL ESTER, (3-α(1E,3E),4-α-4a-α-
    9a-α)-(-)- (9CI) see POH600
ETHANOIC ACID see AAT250
ETHANOIC ANHYDRATE see AAX500
ETHANOL (MAK) see EFU000
ETHANOL, solution (DOT) see EFU000
N-ETHANOLACETAMIDE see HKM000
ETHANOLAMINE see EEC600
β-ETHANOLAMINE see EEC600
ETHANOLAMINE, solution (DOT) see EEC600
ETHANOLAMINE-N,N-DIACETIC ACID see HKM500
ETHANOLAMINE PERCHLORATE see HKM175
ETHANOLAMINE PHOSPHATE see EED000
ETHANOLAMINE SALT of 5,2'-DICHLORO-4'-NITROSALICYLCLICANILIDE
    see DFV600
ETHANOL-2-(2-BUTOXYETHOXY) THIOCYANATE see BPL250
ETHANOL, 2-(2-(2-(4-(p-CHLORO-α-PHENYLBENZYL)-1-
    PIPERAZINYL)ETHOXY)ETHOXY)-, DIMALEATE see HHK100
ETHANOLETHYLENE DIAMINE see AJW000
ETHANOL, 2-(HEPTYLOXY)- see HBN250
ETHANOL, 2-(2-(HEPTYLOXY)ETHOXY)- see HBP275
ETHANOL,2,2'-IMINODI-,3,5-DIIODO-4-OXO-1(4H)-PYRIDINEACETATE (salt)
    see DNG400
ETHANOL,2,2'-IMINODI- with 3,5-DIIODO-4-OXO-1(4H)-PYRIDINEACETIC
    ACID (1:1) see DNG400
ETHANOLISOPROPYLAMINE see INN400
ETHANOL, 2-(ISOPROPYLAMINO)- see INN400
ETHANOLMERCURY BROMIDE see EED600
ETHANOL, 2-METHOXY-, PHOSPHATE (3:1) see TLA600
ETHANOL, 2-((1-METHYLETHYL)AMINO)- (9CI) see INN400
ETHANOL, 2,2',2''-NITRILOTRIS-, TRINITRATE (ester), PHOSPHATE
    (1:2)(SALT) (9CI) see TJL250
ETHANOL, 2-(((5-NITRO-2-FURANYL)METHYLENE)AMINO)-, N-OXIDE (9CI)
    see HKW450
ETHANOL, 2-((5-NITROFURFURYLIDENE)AMINO)-, N-OXIDE see HKW450
ETHANOL, 1-PHENYL- see PDE000
ETHANOL 200 PROOF see EFU000
4-ETHANOLPYRIDINE see POR500
ETHANOL, 2,2',2'',2'''-SILANETETRAYLTETRAKIS- see TDH100
ETHANOLSULFONIC ACID see HKI500
ETHANOL THALLIUM (1+) SALT see EEE000
1-ETHANOL-2-THIOL see MCN250
ETHANOL-2-(2,4,5-TRICHLOROPHENOXY)-, 2,2-DICHLOROPROPIONATE
    see PBK000

ETHIODAN see ELQ500
ETHIOFENCARB see EPR000
ETHIOFOS see AMD000
ETHIOL see EEH600
ETHIOLACAR see MAK700
ETHION see EEH600
ETHIONIAMIDE see EPQ000
ETHIONIN see EEI000
ETHIONINE see AKB250, EEI000
l-ETHIONINE see AKB250
(±)-ETHIONINE see EEI000
dl-ETHIONINE see EEI000
ETHIOPHENCARP see EPR000
ETHIRIMOL see BRI750
ETHISTERONE see GEK500
ETHISTERONE and DIETHYLSTILBESTROL see EEI050
ETHLON see PAK000
ETHMOSINE see EEI025
ETHMOZINE see EEI025
ETHNINE see TCY750
ETHOATE METHYL see DNX600
ETHOBROM see ARW250, THV000
ETHOCAINE see AIT250
ETHOCHLORVYNOL see CHG000
ETHODAN see EEH600
ETHODIN see EDW500
ETHODRYL see DIW000
ETHODRYL CITRATE see DIW200
ETHODUOMEEN see EEI060
ETHODUOMEEN, HYDROFLUORIDE see EEI100
ETHOFAT O see PJY100
ETHOFAT O 15 see PJY100
ETHOGLUCID see TJQ333
ETHOGLUCIDE see TJQ333
ETHOHEPTAZINE CITRATE see MIE600
ETHOHEXADIOL see EKV000
ETHOL see HCP000
ETHOMEEN C/15 see EEJ000
ETHOMEEN S/12 see EEJ500
ETHOMEEN S/15 see EEK000
ETHONE see ENY500
ETHOPHYLLINE see TEP500
ETHOPIAN see EDW875
ETHOPROMAZINE see DIR000
ETHOPROP see EIN000
ETHOPROPHOS see EIN000
ETHOSALICYL see EEM000
ETHOSPERSE LA-4 see DXY000
ETHOSUCCIMIDE see ENG500
ETHOSUCCINIMIDE see ENG500
ETHOSUXIDE see ENG500
ETHOSUXIMIDE see ENG500
ETHOTOIN see EOL100
ETHOVAN see EQF000
ETHOXAZOLAMIDE see EEN500
ETHOXENE see VMA000
ETHOXOL 20 see PJW500
3-ETHOXYACETANILIDE see ABG250
4-ETHOXYACETANILIDE see ABG750
m-ETHOXYACETANILIDE see ABG250
p-ETHOXYACETANILIDE see ABG750
3'-ETHOXYACETANILIDE see ABG250
ETHOXY ACETATE see EES400
ETHOXYACETIC ACID see EEK500
2-ETHOXYACETIC ACID see EEK500
4-ETHOXYACETOACETANILIDE see AAZ000
4'-ETHOXYACETOACETANILIDE see AAZ000
ETHOXY ACETYLENE see EEL000
3-ETHOXY ACROLEIN DIETHYL ACETAL see TJM500
4-ETHOXYANILINE see PDD500
p-ETHOXYANILINE see PDD500
6-ETHOXY-m-ANOL see IRY000
ETHOXYBENZALDEHYDE see EEL500
4-ETHOXYBENZALDEHYDE see EEL500
p-ETHOXYBENZALDEHYDE see EEL500
2-ETHOXYBENZAMIDE see EEM000
o-ETHOXYBENZAMIDE see EEM000
ETHOXYBENZENE see PDM000
o-ETHOXYBENZOIC ACID (1-CARBOXYETHYLIDENE)HYDRAZIDE
    see RSU450
o-ETHOXY-BENZOIL-IDRAZONE DELL'ACIDO PIRUVICO (ITALIAN)
    see RSU450
6-ETHOXY-2-BENZOTHIAZOLESULFONAMIDE see EEN500
o-ETHOXY-BENZOYL-HYDRAZONE of PYRUVIC ACID see RSU450

8-ETHOXYCAFFEINE see EEO000
3-ETHOXYCARBONYLAMINOPHENYL-N-PHENYLCARBAMATE see EEO500
N-(ETHOXYCARBONYL)AZIRIDINE see ASH750
S-α-ETHOXYCARBONYLBENZYL-O,O-DIMETHYL PHOSPHORODITHIOATE
    see DRR400
S-α-ETHOXYCARBONYLBENZYL DIMETHYL PHOSPHOROTHIOLOTHION-
    ATE see DRR400
ETHOXYCARBONYLDIAZOMETHANE see DCN800
ETHOXY CARBONYL DIGOXIN see EEP000
ETHOXYCARBONYLETHYLENE see EFT000
N-ETHOXYCARBONYLETHYLENEIMINE see ASH750
ETHOXYCARBONYL-1-ETHYLENIMINE see ASH750
ETHOXYCARBONYL HYDRAZIDE see EHG000
(ETHOXYCARBONYL) HYDRAZINE see EHG000
1-(ETHOXYCARBONYL) HYDRAZINE see EHG000
N-(ETHOXYCARBONYL) HYDRAZINE see EHG000
7-ETHOXYCARBONYL-4-HYDROXYMETHYL-6,8-DIMETHYL-1(2H)-
    PHTHALAZINONE see PHV725
4-ETHOXYCARBONYL-1-(2-HYDROXY-3-PHENOXYPROPYL) 4-
    PHENYLPIPERIDINE HYDROCHLORIDE see EEQ000
ETHOXYCARBONYLMETHYL BROMIDE see EGV000
S-(N-ETHOXYCARBONYL-N-METHYLCARBAMOYLMETHYL)-DIETHYL
    PHOSPHORODITHIOATE see DJI000
S-((N-ETHOXYCARBONYL)METHYLCARBAMOYL)METHYL-O,O-DIETHYL
    PHOSPHORODITHIOATE see DJI000
N-ETHOXYCARBONYL-N-METHYLCARBAMOYLMETHYL-O,O-DIETHYL
    PHOSPHORODITHIOATE see DJI000
(2-(ETHOXYCARBONYL)-1-METHYL)ETHYL CARBONIC ACID-p-
    IODOBENZYL ESTER see EEQ250
2-ETHOXYCARBONYL-1-METHYLVINYL DIETHYL PHOSPHATE see CBR750
N-(ETHOXYCARBONYL)-3-(4-MORPHOLINYL)SYDNONE IMINE see MRN275
N-(N-((S)-1-ETHOXYCARBONYL-3-PHENYLPROPYL)-l-ALANYL)-N-(INDAN-2-
    YL)GLYCINE HYDROCHLORIDE see DAM315
N-((S)-1-ETHOXYCARBONYL-3-PHENYLPROPYL)-l-ALANYL-l-PROLINE MA-
    LEATE see EAO100
(S)-1-(N-(1-(ETHOXYCARBONYL)-3-PHENYLPROPYL)-l-ALANYL)-l-PROLINE
    MALEATE see EAO100
S-ETHOXYCARBONYLTHIAMINE HYDROCHLORIDE see EEQ500
ETHOXY CHLOROMETHANE see CIM000
ETHOXY CLEVE'S ACID see AJU500
2-ETHOXY-6,9-DIAMINOACRIDINE LACTATE see EDW500
2-ETHOXY-6,9-DIAMINOACRIDINE LACTATE HYDRATE see EDW500
2-ETHOXY-6,9-DIAMINOACRIDINIUM LACTATE see EDW500
ETHOXY DIETHYL ALUMINUM see EER000
ETHOXY DIGLYCOL see CBR000
11-ETHOXY-15,16-DIHYDRO-17-CYCLOPENTA(a)PHENANTHREN-17-ONE
    see EER400
2-ETHOXY DIHYDROPYRAN see EER500
2-ETHOXY-2,3-DIHYDRO-γ-PYRAN see EER500
2-ETHOXY-3,4-DIHYDRO-1,2-PYRAN see EER500
2-ETHOXY-3,4-DIHYDRO-2H-PYRAN see EER500
6-ETHOXY-1,2-DIHYDRO-2,2,4-TRIMETHYLQUINOLINE see SAV000
ETHOXYDIISOBUTYLALUMINUM see EES000
3'-ETHOXY-4-DIMETHYLAMINOAZOBENZENE see DRU000
1-ETHOXY-3,7-DIMETHYL-2,6-OCTADIENE see GDG100
8-ETHOXY-2,6-DIMETHYLOCTENE-2 see EES100
ETHOXYDIPHENYLACETIC ACID see EES300
ETHOXYETHANE see EJU000
2-ETHOXYETHANOL see EES350
2-ETHOXYETHANOL ACETATE see EES400
2-ETHOXYETHANOL, ESTER with ACETIC ACID see EES400
ETHOXY ETHENE see EQF000
1-(2-ETHOXYETHOXY)-BUTANE see BPK750
2-(2-ETHOXYETHOXY)ETHANOL see CBR000
2-(2-ETHOXYETHOXY)ETHANOL ACETATE see CBQ750
1-ETHOXY-2-(β-ETHOXYETHOXY)ETHANE see DIW800
2-(2-(2-ETHOXYETHOXY)ETHOXY)ETHANOL see EFL000
3-(2-ETHOXY)ETHOXY-2-PROPANOL see EET000
2-ETHOXY-ETHYLACETAAT (DUTCH) see EES400
ETHOXYETHYL ACETATE see EES400
2-ETHOXYETHYL ACETATE see EES400
β-ETHOXYETHYL ACETATE see EES400
ETHOXYETHYL ACRYLATE see ADT500
2-ETHOXYETHYL ACRYLATE see ADT500
2-(2-ETHOXYETHYLAMINO)-2',6'-ACETOXYLIDIDE see DRT400
2-(2-ETHOXYETHYLAMINO)-2',6'-ACETOXYLIDIDE HYDROCHLORIDE
    see DRT600
2-(2-ETHOXYETHYLAMINO)-2'-METHYL-PROPIONANILIDE see EES500
2-(2-ETHOXYETHYLAMINO)-o-PROPIONOTOLUIDIDE see EES500
2-ETHOXYETHYLE, ACETATE de (FRENCH) see EES400
1,1'-((1-ETHOXYETHYL)ETHANEDIYLIDENE)BIS(3-THIOSEMICARBAZIDE)
    see KFA100
ETHOXYETHYL ETHER of PROPYLENE GLYCOL see EET000
(1-ETHOXYETHYL)GLYOXAL BIS(THIOSEMICARBAZONE) see KFA100

2-ETHOXYETHYL-2-METHOXYETHYLETHER see DJE800
2-ETHOXYETHYL-2-PROPENOATE see ADT500
2-ETHOXYETHYL-2-(VINYLOXY)ETHYL ETHER see DJF000
ETHOXYETHYNE see EEL000
ETHOXYFORMIC ANHYDRIDE see DIX200
3-ETHOXYHEXANAL DIETHYL ACETAL see TJM000
3-ETHOXY-4-HYDROXYBENZALDEHYDE see EQF000
4-ETHOXY-N-HYDROXY-BENZENAMINE see HNG100
4′-ETHOXY-3-HYDROXYBUTYRANILIDE see HJS850
N-(2-ETHOXY-3-HYDROXYMERCURIPROPYL)BARBITAL see EET500
1-ETHOXY-2-HYDROXY-4-PROPENYLBENZENE see IRY000
2-(3-ETHOXY-1-INDANYLIDENE)-1,3-DINDANDIONE see BGC250
1-ETHOXY-3-ISOPROPOXYPROPAN-2-OL see EET600
1-ETHOXY-3-ISOPROPOXY-2-PROPANOL see EET600
1-ETHOXY-4-(1-KETO-2-HYDROXYETHYL)-NAPHTHALENE SUCCINATE
    see SNB500
ETHOXYLATED LAURYL ALCOHOL see DXY000, EEU000
ETHOXYLATED OCTYL PHENOL see GHS000
ETHOXYLATED SORBITAN MONOOLEATE see PKL100
6-ETHOXY-2-MERCAPTOBENZOTHIAZOLE see EEU500
ETHOXYMETHANE see EMT000
7-(ETHOXYMETHYL)BENZ(a)ANTHRACENE see EEV000
10-ETHOXYMETHYL-1582-BENZANTHRACENE see EEV000
7-ETHOXY-12-METHYLBENZ(a)ANTHRACENE see MJX000
4-ETHOXY-2-METHYL-3-BUTYN-2-OL see EEV100
ETHOXY METHYL CHLORIDE see CIM000
ETHOXYMETHYL-N-(2,6-DICHLORO-m-TOLYL)ANTHRANILATE see DGN000
2-(ETHOXYMETHYL)-1,3-DIHYDROXY-9,10-ANTHRACENEDIONE see DMT200
ETHOXYMETHYLENEMALONIC ACID, ETHYL ESTER see EEV200
ETHOXYMETHYLENE MALONONITRILE see EEW000
2-((ETHOXY((1-METHYLETHYL)AMINO)PHOSPHINOTHIOYL)OXY)BENZOIC
    ACID 1-METHYLETHYL ESTER see IMF300
7-ETHOXY METHYL-12-METHYL BENZ(a)ANTHRACENE see EEX000
2-ETHOXY-N-METHYL-N-(2-(METHYLPHENETHYLAMINO)ETHYL)-2,2-
    DIPHENYLACETAMIDE HYDROCHLORIDE see CBQ625
4-ETHOXY-2-METHYL-5-MORPHOLINO-3(2H)-PYRIDAZINONE see EAN700
4-ETHOXY-2-METHYL-5-(4-MORPHOLINYL)-3(2H)-PYRIDAZINONE
    see EAN700
N-ETHOXYMORPHOLINO DIAZENIUM FLUOROBORATE see EEX500
O-ETHOXY-N-(3-MORPHOLINOPROPYL)BENZAMIDE see EEY000
2-ETHOXYNAPHTHALENE see EEY500
6-(2-ETHOXY-1-NAPHTHAMIDO)PENICILLIN SODIUM see SGS500
1-ETHOXY-4-NITROBENZENE see NID000
ETHOXY-4-NITROPHENOXYPHENYLPHOSPHINE SULFIDE see EBD700
N-(4-ETHOXY-3-NITRO)PHENYLACETAMIDE see NEL000
ETHOXY-4-NITROPHENYLOXYETHYLPHOSPHINEOXIDE see ENQ000
3-ETHOXY-2-OXOBUTYRALDEHYDE BIS(THIOSEMICARBAZONE) see KFA100
1-ETHOXY-2,2,3,3,3-PENTAFLUORO-1-PROPANOL see EFA000
ETHOXYPHAS see DIX000
3-(4-(β-ETHOXYPHENETHYL)-1-PIPERAZINYL)-2-METHYL-1-PHENYL-1-PRO-
    PANONE DIHYDROCHLORIDE see ECU550
3-(4-(β-ETHOXYPHENETHYL)-1-PIPERAZINYL)-2-METHYLPROPIOPHENONE
    see PFB000
2-(4-(β-ETHOXYPHENETHYL)-1-PIPERAZINYLMETHYL)PROPIOPHENONE
    DIHYDROCHLORIDE see ECU550
4-ETHOXYPHENOL see EFA100
p-ETHOXYPHENOL see EFA100
2-((2-ETHOXYPHENOXY)METHYL)MORPHOLINE see VKA875
2-((2-ETHOXYPHENOXY)METHYL)MORPHOLINE HYDROCHLORIDE
    see VKF000
2-((o-ETHOXYPHENOXY)METHYL)MORPHOLINE HYDROCHLORIDE
    see VKF000
2-(2-ETHOXYPHENOXYMETHYL)TETRAHYDRO-1,4-OXAZINE see VKA875
2-(2-ETHOXYPHENOXYMETHYL)TETRAHYDRO-1,4-OXAZINE HYDROCHLO-
    RIDE see VKF000
N-(4-ETHOXYPHENYL)ACETAMIDE see ABG750
N-p-ETHOXYPHENYLACETAMIDE see ABG750
N-(3-ETHOXYPHENYL)ACETAMIDE (9CI) see ABG250
N-(4-ETHOXYPHENYL)ACETOHYDROXAMIC ACID see HIN000
α-ETHOXY-α-PHENYL-BENZENEACETIC ACID (9CI) see EES300
2-(4-ETHOXYPHENYL)-1,3-BIS(4-METHOXYPHENYL)GUANIDINE HYDRO-
    CHLORIDE see PDN500
S-2-((4-(p-ETHOXYPHENYL)BUTYL)AMINO)ETHYL THIOSULFATE see EFC000
1-(p-ETHOXYPHENYL)-1-DIETHYLAMINO-3-METHYL-3-PHENYLPROPANE
    HYDROCHLORIDE see PEF500
2-(3-ETHOXYPHENYL)-5,6-DIHYDRO-s-TRIAZOLO(5,1-a)ISOQUINOLINE
    see EFC259
4-ETHOXY-7-PHENYL-3,5-DIOXA-6-AZA-4-PHOSPHAOCT-6-ENE-8-NITRILE 4
    SULFIDE see BAT750
5-(m-ETHOXYPHENYL)-3-(o-ETHYLPHENYL)-s-TRIAZOLE see EFC600
3-(4-(2-ETHOXY-2-PHENYLETHYL)-1-PIPERAZINYL)-2-METHYL-1-PHENYL-1-
    PROPANONE see PFB000
3-(4-(2-ETHOXY-2-PHENYLETHYL)-1-PIPERAZINYL)-2-METHYL-1-PHENYL-1-
    PROPANONE DIHYDROCHLORIDE see ECU550

N-(4-ETHOXYPHENYL)-N-HYDROXYACETAMIDE see HIN000
(p-ETHOXYPHENYL)HYDROXYLAMINE see HNG100
N-(p-ETHOXYPHENYL)HYDROXYLAMINE see HNG100
3-(4-ETHOXYPHENYL)-2-METHYL-4(3H)-QUINAZOLINONE see LID000
N-(4-ETHOXYPHENYL)-3′-NITROACETAMIDE see NEL000
4-ETHOXYPHENYLUREA see EFE000
p-ETHOXYPHENYLUREA see EFE000
N-(4-ETHOXYPHENYL)UREA see EFE000
ETHOXYPHOS see DIX000
4-ETHOXY-β-(1-PIPERIDYL)PROPIOPHENONE HYDROCHLORIDE see EFE500
1-ETHOXYPROPANE see EPC125
3-ETHOXY-1,2-PROPANEDIOL see EFF000
1-ETHOXY-2-PROPANOL see EFF500
3-ETHOXY-1-PROPANOL see EFG000
ETHOXY PROPIONALDEHYDE see EFG500
ETHOXYPROPIONIC ACID see EFH000
ETHOXYPROPIONIC ACID, ETHYL ESTER see EJV500
3-ETHOXYPROPIONIC ACID, ETHYL ESTER see EJV500
β-ETHOXYPROPIONITRILE see EFH500
ETHOXYPROPYLACRYLATE see EFI000
ETHOXYPROPYL ESTER ACRYLIC ACID see EFI000
1-ETHOXY-2-PROPYNE see EFJ000
4′-ETHOXY-2-PYRROLIDINYLACETANILIDE HYDROCHLORIDE see PPU750
ETHOXYQUIN (FCC) see SAV000
ETHOXYQUINE see SAV000
2-ETHOXY-1(2H)-QUINOLINECARBOXYLIC ACID, ETHYL ESTER see EFJ500
5-ETHOXY-3-TRICHLOROMETHYL-1,2,4-THIADIAZOLE see EFK000
ETHOXYTRIETHYLENE GLYCOL see EFL000
1-ETHOXY-2,2,2-TRIFLUOROETHANOL see EFK500
ETHOXYTRIGLYCOL see EFL000
6-ETHOXY-2,2,4-TRIMETHYL-1,2-DIHYDROQUINOLINE see SAV000
ETHOXYTRIMETHYLSILANE see EFL500
ETHOXYZOLAMIDE see EEN500
ETHRANE see EAT900
ETHREL see CDS125
ETHRIL see EDJ500
ETHYBENZTROPINE see DWE800
ETHYL 736 see TFD000
ETHYLAC see BDF250
ETHYLACETAAT (DUTCH) see EFR000
ETHYLACETAMIDE see EFM000
N-ETHYLACETAMIDE see EFM000
d-N-ETHYLACETAMIDE CARBANILATE see CBL500
2-(2-(3-(N-ETHYLACETAMIDO)-2,4,6-TRIIODOPHENOXY)ETHOXY)ACETIC
    ACID SODIUM SALT see EFM500
2-(2-(3-(N-ETHYLACETAMIDO)-2,4,6-TRIIODOPHENOXY)ETHOXY)-2-
    PHENYL ACETIC ACID SODIUM SALT see EFN500
2-(2-(3-(N-ETHYLACETAMIDO)-2,4,6-TRIIODOPHENOXY)ETHOXY)PROPI-
    ONIC ACID SODIUM SALT see EFO000
2-(3-(N-ETHYLACETAMIDO)-2,4,6-TRIIODOPHENYL)BUTYRIC ACID see
    EFP000
ETHYLACETANILIDE see EFQ500
N-ETHYLACETANILIDE see EFQ500
ETHYL ACETATE see EFR000
ETHYLACETIC ACID see BSW000
ETHYL ACETIC ESTER see EFR000
ETHYL ACETOACETATE (FCC) see EFS000
ETHYL ACETONE see PBN250
N-ETHYL-p-ACETOPHENETIDIDE see EOD500
ETHYL-N-(2-ACETOXYETHYL)-N-NITROSOCARBAMATE
    see EFR500
ETHYL ACETOXYMETHYLNITROSAMINE see ENR500
N-ETHYL-N-(ACETOXYMETHYL)NITROSAMINE see ENR500
ETHYL ACETYL ACETATE see EFS000
ETHYL ACETYLACETONATE see EFS000
ETHYL ACETYLENE see EFS500
ETHYL ACETYLENE, INHIBITED (DOT) see EFS500
ETHYL 3-ACETYLPROPIONATE see EFS600
ETHYLACRYLAAT (DUTCH) see EFT000
ETHYL ACRYLATE see EFT000
ETHYL ADIPATE see AEP750
ETHYLADRIANOL see EGE500
ETHYLAENE GLYCOL FORMAL see DVR800
ETHYLAKRYLAT (CZECH) see EFT000
ETHYLAL see EFT500
ETHYL ALCOHOL see EFU000
ETHYLALCOHOL (DUTCH) see EFU000
ETHYL ALCOHOL, anhydrous see EFU000
ETHYL ALCOHOL THALLIUM (I) see EEE000
ETHYL ALDEHYDE see AAG250
1-ETHYL-3-ALLYL-6-AMINOURACIL see AFW500
ETHYLALUMINUM DIIODIDE see EFU100
ETHYLAMINE see EFU400
ETHYLAMINE with BORON FLUORIDE (1:1) see EFU500

ETHYLAMINE, 2-(α-(p-CHLOROPHENYL)-α-METHYLBENZYLOXY)-N,N-DIETHYL see CKE000
ETHYLAMINE-2-(DIPHENYLMETHOXY)-N,N-DIMETHYL, compound with 8-CHLOROTHEOPHYLLINE (1:1) see DYE600
ETHYLAMINE HYDROCHLORIDE see EFW000
4-(ETHYLAMINOAZOBENZENE) see EOH500
N-ETHYL-4-AMINOAZOBENZENE see EOH500
N-ETHYLAMINOBENZENE see EGK000
ETHYL AMINOBENZOATE see EFX000
ETHYL-4-AMINOBENZOATE see EFX000
ETHYL-o-AMINOBENZOATE see EGM000
ETHYL-p-AMINOBENZOATE see EFX000
ETHYL-m-AMINOBENZOATE METHANESULFONATE see EFX500
ETHYL 2-AMINO-6-BENZYL-4,5,6,7-TETRAHYDROTHIENO(2,3-c)PYRIDINE-3-CARBOXYLATE HYDROCHLORIDE see TGE165
ETHYL-2-AMINO-6-BENZYL-6-THIENO(2,3-c)PYRIDINECARBOXYLATE HYDROCHLORIDE see AJU000
2-ETHYLAMINOETHANOL see EGA500
2-(ETHYLAMINO)ETHANOL see EGA500
2-(2-(ETHYLAMINO)ETHYL)-2-METHYL-1,3-BENZODIOXOLE HYDROCHLORIDE see EGC000
ETHYL-N-(2-AMINO-6-(4-FLUOROPHENYLMETHYLAMINO)PYRIDIN-3-YL)CARBAMATE MALEATE see FMP100
ETHYL-N-(2-AMINO-6-(4-FLUOR-PHENYLMETHYLAMINO)PYRIDIN-3-YL)CARBAMAT MALEAT (GERMAN) see FMP100
dl-(±)-3-(2-ETHYLAMINO-1-HYDROXYETHYL)PHENYL PIVALATE HYDROCHLORIDE see EGC500
(2-(((ETHYLAMINO)IMINOMETHYL)THIO)ETHYL)TRIMETHYL AMMONIUM BROMIDE HYDROBROMIDE see EQN700
2-(((ETHYLAMINO)IMINOMETHYL)THIO)-N,N,N-TRIMETHYLETHANAMIN IUM BROMIDE, MONOHYDROBROMIDE see EQN700
2-ETHYLAMINO-4-ISOPROPYLAMINO-6-METHOXY-s-TRIAZINE see EGD000
4-ETHYLAMINO-6-ISOPROPYLAMINO-2-METHOXY-s-TRIAZINE see EGD000
6-ETHYLAMINO-4-ISOPROPYLAMINO-2-METHOXY-1,3,5-TRIAZINE see EGD000
2-ETHYLAMINO-4-ISOPROPYLAMINO-6-METHYLMERCARPO-s-TRIAZINE see MPT500
2-ETHYLAMINO-4-ISOPROPYLAMINO-6-METHYLTHIO-s-TRIAZINE see MPT500
2-ETHYLAMINO-4-ISOPROPYLAMINO-6-METHYLTHIO-1,3,5-TRIAZINE see MPT500
2-ETHYLAMINO-4-METHYL-5-n-BUTYL-6-HYDROXYPYRIMIDINE see BRI750
α-((ETHYLAMINO)METHYL)-m-HYDROXYBENZYL ALCOHOL see EGE500
α-((ETHYLAMINO)METHYL)-m-HYDROXYBENZYL ALCOHOL 2,2-DIMETHYLPROPIONATE HYDROCHLORIDE see EGC500
dl-(±)α-(ETHYLAMINOMETHYL)-3'-HYDROXYBENZYL ALCOHOL 3-(2,2-DIMETHYLPROPIONATE)HCl see EGC500
α-((ETHYLAMINO)METHYL)-m-HYDROXYBENZYL ALCOHOL HYDROCHLORIDE see EGF000
S-(2-(ETHYLAMINO-2-OXOETHYL)-O,O-DIMETHYL PHOSPHORODITHIO-ATE see DNX600
ETHYL-1-(p-AMINOPHENETHYL)-4-PHENYLISONIPECOTATE see ALW750
3-ETHYL-3-(p-AMINOPHENYL)-2,6-DIOXOPIPERIDINE see AKC600
ETHYL-p-AMINOPHENYL KETONE see AMC000
2-ETHYLAMINO-3-PHENYL-NORCAMPHANE HYDROCHLORIDE see EOM000
N-((3-(ETHYLAMINO)PROPOXY)METHYL)DIPHENYLAMINE see EGG000
ETHYLAMINOPROPYLDIPHENYLAMINOCARBINOL HYDROCHLORIDE see EGG000
3-ETHYLAMINO-5H-PYRIDO(4,3-b)INDOLE see EGH500
2-ETHYLAMINOTHIADIAZOLE see EGI000
2-ETHYLAMINO-1,3,4-THIADIAZOLE see EGI000
2-ETHYL-3-(3-AMINO-2,4,6-TRIIODOPHENYL)PROPIONIC ACID see IFY100
ETHYL AMMONIUM CHLORIDE see EFW000
ETHYLAMPHETAMINE see EGI500
ETHYLAMYLCARBINOL see OCY100
ETHYL-n-AMYLCARBINOL see OCY100
ETHYL AMYL KETONE see EGI750, ODI000
2-(1-ETHYLAMYLOXY)ETHANOL see EGJ000
2-(2-(1-ETHYLAMYLOXY)ETHOXY)ETHANOL see EGJ500
ETHYLAN see DJC000, IPS500
ETHYLAN A3 see PJY100
ETHYLAN A6 see PJY100
ETHYLAN CP see GHS000
ETHYLANILINE see EGK000
2-ETHYLANILINE see EGK500
4-ETHYLANILINE see EGL000
N-ETHYLANILINE see EGK000
o-ETHYLANILINE see EGK500
p-ETHYLANILINE see EGL000
ETHYL ANISATE see AOV000
ETHYL-p-ANISATE (FCC) see AOV000
ETHYLAN MLD see BKE500
2-ETHYL-9,10-ANTHRACENEDIONE see EGL500
ETHYL ANTHRANILATE see EGM000

2-ETHYLANTHRAQUINONE see EGL500
2-ETHYL-9,10-ANTHRAQUINONE see EGL500
ETHYL ANTIOXIDANT 736 see TFD000
ETHYL APOVINCAMINATE see EGM100
ETHYL APOVINCAMIN-22-OATE see EGM100
ETHYLARSONOUS DICHLORIDE see DFH200
ETHYL AZIDE see EGM500
ETHYL AZIDOFORMATE see EGN000
ETHYL-2-AZIDO-2-PROPENOATE see EGN100
2-ETHYLAZIRIDINE see EGN500
ETHYL AZIRIDINECARBOXYLATE see ASH750
ETHYL-1-AZIRIDINECARBOXYLATE see ASH750
ETHYL AZIRIDINOCARBOXYLATE see ASH750
ETHYL-1-AZIRIDINYLCARBOXYLATE see ASH750
ETHYL AZIRIDINYLFORMATE see ASH750
ETHYLBARBITAL see BAG000
ETHYLBENATROPINE see DWE800
7-ETHYLBENZ(c)ACRIDINE see BBM500
9-ETHYL-3,4-BENZACRIDINE see BBM500
7-ETHYLBENZ(a)ANTHRACENE see EGO500
8-ETHYLBENZ(a)ANTHRACENE see EGO000
12-ETHYLBENZ(a)ANTHRACENE see EGP000
5-ETHYL-1,2-BENZANTHRACENE see EGO000
10-ETHYL-1,2-BENZANTHRACENE see EGO500
ETHYLBENZEEN (DUTCH) see EGP500
2-ETHYLBENZENAMINE see EGK500
N-ETHYLBENZENAMINE see EGK000
N-ETHYLBENZENAMINO see EGK000
ETHYL BENZENE see EGP500
ETHYL BENZENEACETATE see EOH000
α-ETHYLBENZENEACETIC ACID-2-(2-DIETHYLAMINO)ETHOXY)ETHYL ESTER CITRATE see BOR350
α-ETHYLBENZENEMETHANOL see EGQ000
ETHYL BENZOATE see EGR000
1-(7-ETHYLBENZOFURAN-2-YL)-2-tert-BUTYLAMINO-1-HYDROXYETHANE HYDROCHLORIDE see BQD000
2-ETHYL-3-BENZOFURANYL-p-HYDROXYPHENYL KETONE see BBJ500
ETHYLBENZOL see EGP500
5-ETHYLBENZO(c)PHENANTHRENE see EGS500
2-ETHYLBENZOXAZOLE see EGR500
ETHYL-o-BENZOYL-3-CHLORO-2,6-DIMETHOXY-BENZOHYDROXIMATE see BCP000
ETHYL-N-BENZOYL-N-(3,4-DICHLOROPHENYL)-2-AMINOPROPIONATE see EGS000
2-ETHYL-3584-BENZPHENANTHRENE see EGS500
ETHYLBENZTROPINE see DWE800
ETHYL BENZYL ACETOACETATE see EFS000
α-ETHYLBENZYL ALCOHOL see EGQ000
ETHYLBENZYLBARBITURIC ACID see BEA500
ETHYL-N-BENZYLCYCLOPROPANECARBAMATE see EGT000
ETHYL-N-BENZYL-N-CYCLOPROPYLCARBAMATE see EGT000
ETHYL-1-(2-BENZYLOXYETHYL)-4-PHENYLPIPERIDINE-4-CARBOXYLATE see BBU625
1,2-ETHYL BIS-AMMONIUM PERCHLORATE see EGT500
ETHYL (BIS(1-AZIRIDINYL)PHOSPHINYL)CARBAMATE see EHV500
ETHYLBIS(2-CHLOROETHYL)AMINE see BID250
ETHYLBIS(β-CHLOROETHYL)AMINE see BID250
ETHYLBIS(β-CHLOROETHYL)AMINE HYDROCHLORIDE see EGU000
ETHYL BISCOUMACETATE see BKA000
2-ETHYL-1,3-BIS(DISMETHYLAMINO)-2-PROPANOL BENZOATE see AHI250
ETHYL BIS(4-HYDROXYCOUMARINYL)ACETATE see BKA000
ETHYL BIS(4-HYDROXY-3-COUMARINYL)ACETATE see BKA000
S-ETHYL BIS(2-METHYLPROPYL)CARBAMOTHIOATE see EID500
ETHYL BORATE (DOT) see BMC250
ETHYL BROMACETATE see EGV000
ETHYL BROMIDE see EGV400
ETHYL BROMOACETATE see EGV000
ETHYL-α-BROMOACETATE see EGV000
N-ETHYL-N-o-BROMOBENZYL-N,N-DIMETHYLAMMONIUM TOSYLATE see BMV750
ETHYL BROMOPHOS see EGV500
2-ETHYLBUTANAL see DHI000
2-ETHYL-1-BUTANAMINE see EHA000
ETHYL BUTANOATE see EHE000
2-ETHYL BUTANOIC ACID see DHI400
2-ETHYLBUTANOL see EGW000
2-ETHYLBUTANOL-1 see EGW000
2-ETHYL-1-BUTANOL see EGW000
2-ETHYL-1-BUTANOL, SILICATE see TCD250
2-ETHYL-2-BUTENAL see EHO000
2-ETHYL-1-BUTENE see EGW500
2-ETHYL-1-BUTENE-1-ONE see DJN700
3-ETHYLBUTINOL see EQL000
2-(2-ETHYLBUTOXY)ETHANOL see EGX000

ETHYL DIAZOACETATE see DCN800
N-ETHYLDIAZOACETYLGLYCINE AMIDE see DCN600
1-ETHYLDIBENZ(a,h)ACRIDINE see EHW000
1-ETHYL-DIBENZ(a,j)ACRIDINE see EHW500
8-ETHYL DIBENZ(a,h)ACRIDINE see EHX000
1''-ETHYLDIBENZ(a,h)ACRIDINE see EHX000
1'-ETHYL-1,2,5,6-DIBENZACRIDINE (FRENCH) see EHW000
1'-ETHYL-3,4,5,6-DIBENZACRIDINE (FRENCH) see EHW500
ETHYL DIBROMOBENZENE see EHY000
ETHYLDICHLOROBENZENE see EHY500
ETHYL-4,4'-DICHLOROBENZILATE see DER000
ETHYL-p,p'-DICHLOROBENZILATE see DER000
ETHYL-4,4'-DICHLORODIPHENYL GLYCOLLATE see DER000
N-ETHYL-DICHLOROMALEINIMIDE see DFJ400
ETHYL (2,4-DICHLOROPHENOXY)ACETATE see EHY600
ETHYL-4,4'-DICHLOROPHENYL GLYCOLLATE see DER000
O-ETHYL-O-(2,4-DICHLOROPHENYL)-S-n-PROPYL-DITHIOPHOSPHATE
   see DGC800
O-ETHYL-O-2,4-DICHLOROPHENYL THIONOBENZENEPHOSPHONATE
   see SCC000
ETHYL DICHLOROSILANE (DOT) see DFK000
ETHYLDICHLORTHIOFOSFAT (CZECH) see MRI000
ETHYLDICOUMAROL see BKA000
ETHYLDICOUMAROL ACETATE see BKA000
ETHYL ((DIETHOXYPHOSPHINOTHIOYL)THIO)ACETATE see DIX000
ETHYL (2-((DIETHOXYPHOSPHINOTHIOYL)THIO)ETHYL)CARBAMATE
   see EMC000
ETHYL (DIETHOXYPHOSPHINYL)ACETATE see EIC000
ETHYL DIETHOXYPHOSPHORYL ACETATE see EIC000
ETHYL-3-((DIETHYLAMINO)METHYL)-4-HYDROXYBENZOATE see EIA000
ETHYL-N,N-DIETHYL CARBAMATE see EIB000
ETHYL DIETHYLENE GLYCOL see CBR000
4'-ETHYL-N,N-DIETHYL-p-(PHENYLAZO)ANILINE see EIB500
ETHYL (DIETHYLPHOSPHONO)ACETATE see EIC000
ETHYL DIGLYME see DIW800
5-ETHYLDIHYDRO-5-(1-METHYLBUTYL)-2-THIOXO-4,6,(1H,5H)-
   PYRIMIDINEDIONE (9CI) see PBT250
5-ETHYLDIHYDRO-5-(1-METHYLBUTYL)-2-THIOXO-4,6(1H,5H)-
   PYRIMIDINEDIONE, MONOSODIUM SALT see PBT500
3-ETHYL-2,3-DIHYDRO-6-METHYL-1H-CYCLOPENT(a)ANTHRACENE
   see DLM600
1-ETHYL-1,4-DIHYDRO-6,7-METHYLENEDIOXY-4-OXO-3-
   QUINOLINECARBOXYLIC ACID see OOG000
(3S-cis)-3-ETHYLDIHYDRO-4-((1-METHYL-1H-IMIDAZOL-5-YL)METHYL)-2(3H)-
   FURANONE see PIF000
3-ETHYL-6,7-DIHYDRO-2-METHYL-5-MORPHOLINOMETHYLINDOLE-4(5H)-
   ONE HYDROCHLORIDE see MRB250
3-ETHYL-6,7-DIHYDRO-2-METHYL-5-MORPHOLINOMETHYLINDOL-4(5H)-
   ONE HYDROCHLORIDE see MRB250
1-ETHYL-1,4-DIHYDRO-7-METHYL-4-OXO-1,8-NAPHTHYRIDINE-3-CARBOX-
   YLIC ACID see EID000
1-ETHYL-4,6-DIHYDRO-3-METHYL-8-PHENYLPYRAZOLO(4,3-e)(1,4)DIAZEP-
   INE see POL475
1-ETHYL-1,4-DIHYDRO-4-OXO(1,3)DIOXOLO(4,5-g)CINNOLINE-3-CARBOX-
   YLIC ACID see CMS200
5-ETHYL-5,8-DIHYDRO-8-OXO-1,3-DIOXOLO(4,5-g)QUINOLINE-7-CARBOX-
   YLIC ACID see OOG000
8-ETHYL-5,8-DIHYDRO-5-OXO-2-(1-PIPERAZINYL)PYRIDO(2,3-d)PYRAMID
   INE-6-CARBOXYLIC ACID 3H₂O see PII350
5-ETHYLDIHYDRO-5-PHENYL-4,6(1H,5H)-PYRIMIDINEDIONE see DBB200
2-ETHYL-2,3-DIHYDRO-3-((4-(2-(1-PIPERIDINYL)ETHOXY)PHENYL)AMINO)-
   1H-ISOINDOL-1-ONE see AHP125
9-ETHYL-1,9-DIHYDRO-6H-PURINE-6-THIONE see EMC500
ETHYL-3,4-DIHYDROXYBENZENE SULFONATE see EID100
ETHYL-4,4'-DIHYDROXYDICOUMARINYL-3,3'-ACETATE see BKA000
S-ETHYLDIISOBUTYL THIOCARBAMATE see EID500
ETHYL-N,N-DIISOBUTYLTHIOCARBAMATE see EID500
S-ETHYL N,N-DIISOBUTYLTHIOCARBAMATE see EID500
ETHYL-N,N-DIISOBUTYL THIOLCARBAMATE see EID500
O-ETHYL-S-2-DIISOPROPYLAMINOETHYL METHYLPHOSPHONOTHIOTE
   see EIG000
ETHYL-S-DIISOPROPYLAMINOETHYL METHYLTHIOPHOSPHONATE
   see EIG000
ETHYL-α-((DIMETHOXYPHOSPHENOTHIOYL)THIO)BENZENEACETATE
   see DRR400
ETHYL DIMETHYLACRYLATE see MIP800
ETHYL 3,3-DIMETHYLACRYLATE see MIP800
ETHYL β,β-DIMETHYLACRYLATE see MIP800
ETHYL DIMETHYLAMIDOCYANOPHOSPHATE see EIF000
2'-ETHYL-4-DIMETHYLAMINOAZOBENZENE see EIF450
3'-ETHYL-4-DIMETHYLAMINOAZOBENZENE see EOI000
4'-ETHYL-4-DIMETHYLAMINOAZOBENZENE see EOI500
4'-ETHYL-N,N-DIMETHYL-4-AMINOAZOBENZENE see EIB500
ETHYL N,N-DIMETHYLAMINO CYANOPHOSPHATE see EIF000

O-ETHYL-S-(2-DIMETHYL AMINO ETHYL)-METHYLPHOSPHONOTHIOATE
   see EIF500
ETHYL-S-DIMETHYLAMINOETHYL METHYLPHOSPHONOTHIOLATE
   see EIG000
ETHYL-dl-trans-2-DIMETHYLAMINO-1-PHENYL-3-CYCLOHEXENE-1-CARBOX-
   YLATE see EIH000
(±)-ETHYL-trans-2-2(DIMETHYLAMINO)-1-PHENYL-3-CYCLOHEXENE-1-CAR-
   BOXYLATE HYDROCHLORIDE see EIH000
S-ETHYL N-(3-DIMETHYLAMINOPROPYL)THIOL CARBAMATE HYDRO-
   CHLORIDE see EIH500
7-ETHYL-1,4-DIMETHYLAZULENE see DRV000
ETHYL-N,N-DIMETHYL CARBAMATE see EII500
N-ETHYL-N-1-DIMETHYL-3,3-DI-2-THIENYLALLYLAMINE HYDROCHLO-
   RIDE see EIJ000
N-ETHYL-N-1-DIMETHYL-3,3-DI-2-THIENYL-2-PROPENAMINE HYDROCHLO-
   RIDE see EIJ000
ETHYL DIMETHYLDITHIOCARBAMATE see EIJ500
ETHYLDIMETHYLMETHANE see EIK000
5-ETHYL-3,5-DIMETHYLOXAZOLIDINE-2,4-DIONE see PAH500
3-ETHYL-2,3-DIMETHYL PENTANE see EIK500
4'-ETHYL-N,N-DIMETHYL-4-(PHENYLAZO)-m-TOLUIDINE see EMS000
ETHYLDIMETHYLPHOSPHINE see EIL000
ETHYL DIMETHYLPHOSPHORAMIDOCYANIDATE see EIF000
ETHYL-N,N-DIMETHYLPHOSPHORAMIDOCYANIDATE see EIF000
ETHYL-N-(2-(O,O-DIMETHYLPHOSPHORODITHIOYL)ETHYL)CARBAMATE
   see EMC000
ETHYL-O,O-DIMETHYL PHOSPHORODITHIOYLPHENYL ACETATE
   see DRR400
2-ETHYL-3,5(6)-DIMETHYLPYRAZINE see EIL100
ETHYLDIMETHYL SULFONIUM IODIDE MERCURIC IODIDE ADDITION
   COMPOUND see EIL500
10-ETHYL-4,4-DIOCTYL-7-OXO-8-OXA-3,5-DITHIA-4-
   STANNATETRADECANOIC ACID-2-ETHYLHEXYL ESTER see DVM800
ETHYLDIOL ACRILATE (RUSSIAN) see EIP000
ETHYLDIOL METACRYLATE see BKM250
1-ETHYL-2,5-DIOXO-4-PHENYLIMIDAZOLIDINE see EOL100
O-ETHYL-S,S-DIPHENYL DITHIOPHOSPHATE see EIM000
O-ETHYL-S,S-DIPHENYL PHOSPHORODITHIOATE see EIM000
2-ETHYL-3,3-DIPHENYL-2-PROPENYLAMINE HYDROCHLORIDE
   see DWF200
O-ETHYL-S,S-DIPROPYL ESTER, PHOSPHORODITHIOIC ACID see EIN000
O-ETHYL-S,S-DIPROPYLPHOSPHORODITHIOATE see EIN000
S-ETHYL-N,N-DIPROPYLTHIOCARBAMATE see EIN500
S-ETHYL-N,N-DI-N-PROPYLTHIOCARBAMATE see EIN500
ETHYL DI-N-PROPYLTHIOLCARBAMATE see EIN500
ETHYL-N,N-DIPROPYLTHIOLCARBAMATE see EIN500
ETHYL-N,N-DI-N-PROPYLTHIOLCARBAMATE see EIN500
(O-ETHYL DITHIOCARBONATO)POTASSIUM see PLF000
ETHYLDITHIOURAME see DXH250
ETHYLDITHIURAME see DXH250
ETHYL DODECANOATE see ELY700
ETHYLE (ACETATE d') (FRENCH) see EFR000
ETHYLE, CHLOROFORMIAT D' (FRENCH) see EHK500
ETHYLEEN-CHLOORHYDRINE (DUTCH) see EIU800
ETHYLEENDIAMINE (DUTCH) see EEA500
ETHYLEENDICHLORIDE (DUTCH) see EIY600
ETHYLEENIMINE (DUTCH) see EJM900
ETHYLEENOXIDE (DUTCH) see EJN500
ETHYLE (FORMIATE d') (FRENCH) see EKL000
ETHYLENAMINE see VLU400
ETHYLENE see EIO000
ETHYLENE, compressed (DOT) see EIO000
ETHYLENE, refrigerated liquid (DOT) see EIO000
ETHYLENE ACETATE see EJD759
ETHYLENEACETIC ACID see EIO500
ETHYLENE ACRYLATE see EIP000
ETHYLENE ALCOHOL see EJC500
ETHYLENE ALDEHYDE see ADR000
ETHYLENEAMINE see VLU400
ETHYLENE, 1-(p-(BENZYLOXY)PHENYL)-2-(o-FLUOROPHENYL)-1-PHENYL-
   see BFC400
1,1'-ETHYLENE-2,2'-BIPYRIDYLIUM DIBROMIDE see DWX800
ETHYLENE BIS(BROMOACETATE) see BHD250
1,1'-ETHYLENEBIS(5-BUTOXY-3,7-DIMETHYL-1,5-AZABOROCINE-4,6-DIONE
   see BJJ750
N,N'-ETHYLENEBIS(N-BUTYL-4-MORPHOLINECARBOXAMIDE) see DUO400
1,1'-ETHYLENEBIS(3-(2-CHLOROETHYL)-3-NITROSOUREA) see EIP500
ETHYLENE BIS(CHLOROFORMATE) see EIQ000
1,1'-ETHYLENEBIS-CNU see EIP500
ETHYLENEBIS-(DIPHENYLARSINE) see EIQ200
ETHYLENEBIS(DITHIOCARBAMATE) DISODIUM SALT see DXD200
ETHYLENEBISDITHIOCARBAMATE MANGANESE see MAS500
N,N'-ETHYLENE BIS(DITHIOCARBAMATE MANGANEUX) (FRENCH)
   see MAS500

N,N'-ETHYLENE BIS(DITHIOCARBAMATE de SODIUM) (FRENCH)
    see DXD200
ETHYLENEBIS(DITHIOCARBAMATO) MANGANESE see MAS500
ETHYLENEBIS(DITHIOCARBAMATO)MANGANESE and ZINC ACETATE
    (50581) see EIQ500
ETHYLENE BIS(DITHIOCARBAMATO)ZINC see EIR000
ETHYLENEBIS(DITHIOCARBAMIC ACID) DISODIUM SALT see DXD200
ETHYLENEBIS(DITHIOCARBAMIC ACID) MANGANESE SALT see MAS500
ETHYLENEBIS(DITHIOCARBAMIC ACID MANGANESE ZINC COMPLEX
    (8CI) see DXI400
ETHYLENEBIS(DITHIOCARBAMIC ACID) MANGANOUS SALT see MAS500
ETHYLENEBIS(DITHIOCARBAMIC ACID) NICKEL(II) SALT see EIR500
ETHYLENEBIS(DITHIOCARBAMIC ACID), ZINC SALT see EIR000
N,N'-ETHYLENE BIS(3-FLUOROSALICYLIDENEIMINATO)COBALT(II)
    see EIS000
ETHYLENEBIS(IMINODIACETIC ACID) DISODIUM SALT see EIX500
ETHYLENEBIS(IMINODIACETIC ACID) TETRASODIUM SALT see EIV000
ETHYLENEBISISOTHIOCYANATE see ISK000
ETHYLENE BIS(METHANESULFONATE) see BKM125
2,2'-(ETHYLENEBIS(NITROSOIMINO))BISBUTANOL see HMQ500
(ETHYLENEBIS(OXYETHYLENENITRILO))TETRAACETIC ACID see EIT000
N,N'-ETHYLENEBIS(SALICYLIDENEIMINATO)COBALT(II) see BLH250
2,2'-ETHYLENE-BIS-(2-THIOPSEUDOUREA), DIHYDROBROMIDE see EJA000
ETHYLENE BISTHIURAM MONOSULFIDE see EJQ000
ETHYLENE-BIS-THIURAMMONO-SULFIDE see ISK000
ETHYLENEBIS(TRIS(2-CYANOETHYL)PHOSPHONIUM BROMIDE) see EIU000
ETHYLENE BRASSYLATE see EJQ500
ETHYLENE BROMIDE see EIY500
ETHYLENE BROMOACETATE see BHD250
ETHYLENE, 1-BROMO-1-(p-CHLOROPHENYL)-2,2-DIPHENYL- see BNA500
ETHYLENEBROMOHYDRIN see BNI500
ETHYLENE, BROMOTRIPHENYL- see BOK500
ETHYLENE CARBONATE see GHM000
ETHYLENE CARBONIC ACID see GHM000
ETHYLENECARBOXAMIDE see ADS250
ETHYLENECARBOXYLIC ACID see ADS750
ETHYLENE CHLORIDE see EIY600
ETHYLENE CHLOROBROMIDE see CES500
ETHYLENE CHLOROFORMATE see EIQ000
ETHYLENE CHLOROHYDRIN see EIU800
ETHYLENE CYANIDE see SNE000
ETHYLENE CYANOHYDRIN see HGP000
ETHYLENE DIACRYLATE see EIP000
1,2-ETHYLENEDIAMINE see EEA500
ETHYLENEDIAMINE (OSHA) see EEA500
ETHYLENE-DIAMINE (FRENCH) see EEA500
ETHYLENEDIAMINEACETIC ACID TRISODIUM SALT see TNL250
ETHYLENEDIAMINE, N-(5-CHLORO-2-THENYL)-N',N'-DIMETHYL-N-2-
    PYRIDYL- see CHY250
N,N'-ETHYLENEDIAMINEDIACETIC ACID TETRASODIUM SALT see EIV000
ETHYLENEDIAMINEDICHLORIDE PLATINUM (II) see DFJ000
ETHYLENEDIAMINE DIHYDROCHLORIDE see EIW000
ETHYLENE DIAMINEDINITRATE see EIV700
ETHYLENEDIAMINE ETHOXYLATE see EIV750
ETHYLENEDIAMINE ETHYLENE OXIDE ADDUCT see EIV750
ETHYLENEDIAMINE HYDROCHLORIDE see EIW000
ETHYLENEDIAMINE SULFATE see EIW500
ETHYLENEDIAMINETETRAACETATE see EIX000
ETHYLENEDIAMINETETRAACETATE DISODIUM SALT see EIX500
ETHYLENEDIAMINETETRAACETIC ACID see EIX000
ETHYLENEDIAMINE-N,N,N',N'-TETRAACETIC ACID see EIX000
ETHYLENEDIAMINETETRAACETIC ACID, DISODIUM SALT see EIX500
ETHYLENEDIAMINE TETRAACETIC ACID, IRON(III) SALT see HIA000
ETHYLENEDIAMINETETRAACETIC ACID, TETRASODIUM SALT see EIV000
ETHYLENEDIAMINETETRAACETICACID, TRISODIUM SALT see TNL250
ETHYLENEDIAMINETETRAACETONITRILE see EJB000
ETHYLENEDIAMINE, compounded with THEOPHYLLINE (1:2) see TEP500
1,2-ETHYLENE DIBROMIDE see EIY500
(E)1,2-ETHYLENEDICARBOXYLIC ACID see FOU000
cis-1,2-ETHYLENEDICARBOXYLIC ACID see MAK900
trans-1,2-ETHYLENEDICARBOXYLIC ACID see FOU000
trans-1,2-ETHYLENEDICARBOXYLIC ACID DIMETHYL ESTER see DSB600
ETHYLENEDICESIUM see EIY550
ETHYLENE DICHLORIDE see EIY600
1,2-ETHYLENE DICHLORIDE see EIY600
ETHYLENE DICYANIDE see SNE000
ETHYLENE DIGLYCOL see DJD600
ETHYLENE DIGLYCOL MONOETHYL ETHER see CBR000
ETHYLENE DIGLYCOL MONOMETHYL ETHER see DJG000
ETHYLENE DIHYDRATE see EJC500
(+)-2,2'-(ETHYLENEDIIMINO)DI-1-BUTANOL see TGA500
N,N'-ETHYLENE DIIMINO DI(o-CRESOL) see DWY200
ETHYLENE DIISOTHIOCYANATE see ISK000
ETHYLENE DIISOTHIOUREA DIHYDROBROMIDE see EJA000

ETHYLENE DIISOTHIOURONIUM DIBROMIDE see EJA000
ETHYLENE DIMERCAPTAN see EEB000
α-ETHYLENE DIMERCAPTAN see EEB000
ETHYLENE DIMETHANESULFONATE see BKM125
ETHYLENE DIMETHANESULPHONATE see BKM125
ETHYLENE DIMETHYL ETHER see DOE600
N,N-ETHYLENE-N',N'-DIMETHYLUREA see EJA100
ETHYLENEDINITRAMINE see DUV800
ETHYLENE DINITRATE see EJG000
(ETHYLENEDINITRILO)TETRAACETATE DIPOTASSIUM SALT see EJA250
((ETHYLENEDINITRILO)TETRAACETATO(2-))-COBALTATE(2-) CO-
    BALT(2+) SALT see DGQ400
((ETHYLENEDINITRILO)TETRAACETATO)-FERATE(1-), SODIUM see EJA379
(ETHYLENEDINITRILO)-TETRAACETIC ACID see TNL500
ETHYLENEDINITRILOTETRAACETIC ACID see EIX000
(ETHYLENEDINITRILO)TETRAACETIC ACID CADMIUM(II) COMPLEX
    see CAF750
(ETHYLENEDINITRILO)TETRAACETIC ACID COPPER(II) COMPLEX
    see CNL750
(ETHYLENEDINITRILO)-TETRAACETIC ACID DISODIUM SALT see EIX500
(ETHYLENEDINITRILO)TETRA ACETIC ACID, LEAD(II) COMPLEX
    see LDD000
(ETHYLENEDINITRILO)TETRA ACETIC ACID, MERCURY(II) COMPLEX
    see MDB250
(ETHYLENEDINITRILO)TETRA ACETIC ACID NICKEL(II) COMPLEX
    see EJA500
(ETHYLENEDINITRILO)TETRAACETONITRILE see EJB000
1,1',1'',1'''-(ETHYLENEDINITRIOLO)TETRA-2-PROPANOL see QAT000
ETHYLENEDINITROAMINE see DUV800
6,6'-(ETHYLENEDIOXY)BIS(4-AMINOQUINALDINE) DIHYDROCHLORIDE
    see EJB100
ETHYLENEDIOXYBIS(ETHYLENEAMINO)TETRAACETIC ACID see EIT000
2,2'-ETHYLENEDIOXYDIETHANOL see TJQ000
2,2'-ETHYLENEDIOXYDIETHANOL DIACETATE see EJB500
2,2'-(ETHYLENEDIOXY)DI(ETHYL ACETATE) see EJB500
2,2'-(ETHYLENEDIOXY)DI(ETHYL 2-ETHYLBUTYRATE) see TJQ250
2,2'-ETHYLENEDIOXYETHANOL see TJQ000
16,17-ETHYLENEDIOXYVIOLANTHRONE see CMU500
ETHYLENE DIPERCHLORATE see EJC000
1,1'-ETHYLENE-2,2'-DIPYRIDINIUM DICHLORIDE see DWY000
ETHYLENE DIPYRIDYLIUM DIBROMIDE see DWX800
1,1-ETHYLENE 2,2-DIPYRIDYLIUM DIBROMIDE see DWX800
1,1'-ETHYLENE-2,2'-DIPYRIDYLIUM DIBROMIDE see DWX800
2,2-ETHYLENEDITHIODIPSEUDOUREA DIHYDROBROMIDE see EJA000
ETHYLENE DITHIOGLYCOL see EEB000
ETHYLENEDITHIOL see EEB000
1,2-ETHYLENEDIYLBIS(CARBAMODITHIOATO)MANGANESE see MAS500
ETHYLENE EPISULFIDE see EJP500
ETHYLENE EPISULPHIDE see EJP500
ETHYLENE FLUORIDE see ELN500
ETHYLENE FORMATE see EJF000
ETHYLENE GLYCOL see EJC500
ETHYLENE GLYCOL ACETATE see EJD759, EJI000
ETHYLENE GLYCOL ACRYLATE see ADV250
ETHYLENE GLYCOL BIS(AMINOETHYL ETHER)TETRAACETATE see EIT000
ETHYLENE GLYCOL BIS(β-AMINOETHYL ETHER)TETRAACETATE
    see EIT000
ETHYLENE GLYCOL BIS(2-AMINOETHYL ETHER)TETRAACETIC ACID
    see EIT000
ETHYLENE GLYCOL BIS(β-AMINOETHYL ETHER)-N,N'-TETRAACETIC
    ACID see EIT000
ETHYLENE GLYCOL BIS(2-AMINOETHYL ETHER)-N,N,N',N'-TETRAACETIC
    ACID see EIT000
ETHYLENE GLYCOL BIS(BROMOACETATE) see BHD250
ETHYLENE GLYCOL BIS(CHLOROMETHYL)ETHER see BIJ250
ETHYLENE GLYCOL BIS(2,3-EPOXY-2-METHYLPROPYL) ETHER see EJD000
ETHYLENE GLYCOL-BIS-(2-HYDROXYETHYL ETHER) see TJQ000
ETHYLENE GLYCOL BIS(METHACRYLATE) see BKM250
ETHYLENE GLYCOL-n-BUTYL ETHER see BPJ850
ETHYLENE GLYCOL CARBONATE see GHM000
ETHYLENE GLYCOL, CHLOROHYDRIN see EIU800
ETHYLENE GLYCOL, CYCLIC CARBONATE see GHM000
ETHYLENE GLYCOL, CYCLIC SULFATE see EJP000
ETHYLENE GLYCOL DIACETATE see EJD759
ETHYLENE GLYCOL DIACRYLATE see EIP000
ETHYLENE GLYCOL DIALLYL ETHER see EJE000
ETHYLENE GLYCOL, DIBUTOXYTETRA see TCE350
ETHYLENE GLYCOL DIBUTYL see DDW400
ETHYLENE GLYCOL DI(CHLOROFORMATE) see EIQ000
ETHYLENE GLYCOL DI(2,3-EPOXY-2-METHYLPROPYL)ETHER see EJD000
ETHYLENE GLYCOL DIETHYL ETHER see EJE500
ETHYLENE GLYCOL DIFORMATE see EJF000
ETHYLENE GLYCOL DIHYDROXYDIETHYL ETHER see TJQ000
ETHYLENE GLYCOL DIMETHACRYLATE see BKM250

ETHYLENE GLYCOL DIMETHYL ETHER see DOE600
ETHYLENE GLYCOL DINITRATE see EJG000
ETHYLENE GLYCOL DINITRATE mixed with NITROGLYCERIN (1:1)
  see NGY500
ETHYLENE GLYCOL DIPROPIONATE (8CI) see COB260
ETHYLENE GLYCOL ETHYL ETHER see EES350
ETHYLENE GLYCOL ETHYL ETHER ACETATE see EES400
ETHYLENE GLYCOL-N-HEXYL ETHER see HFT500
ETHYLENE GLYCOLIDE (2,3-EPOXY-2-METHYLPROPYL)ETHER see EJD000
ETHYLENE GLYCOL ISOPROPYL ETHER see INA500
ETHYLENE GLYCOL MALEATE see EJG500
ETHYLENE GLYCOL METHACRYLATE see EJH000
ETHYLENE GLYCOL METHYL ETHER see EJH500
ETHYLENE GLYCOL METHYL ETHER ACETATE see EJJ500
ETHYLENE GLYCOL MONOACETATE see EJI000
ETHYLENE GLYCOL MONOACRYLATE see ADV250
ETHYLENE GLYCOL MONOBENZYL ETHER see EJI500
ETHYLENE GLYCOL MONO-sec-BUTYL ETHER see EJJ000
ETHYLENE GLYCOL MONOBUTYL ETHER (MAK, DOT) see BPJ850
ETHYLENE GLYCOL MONOBUTYL ETHER ACETATE (MAK) see BPM000
ETHYLENE GLYCOL MONOETHYL ETHER see EES350
ETHYLENE GLYCOL MONOETHYL ETHER (DOT) see EES350
ETHYLENE GLYCOL MONOETHYL ETHER ACETATE (MAK, DOT)
  see EES400
ETHYLENE GLYCOL MONOETHYL ETHER ACRYLATE see ADT500
ETHYLENE GLYCOL MONOETHYL ETHER PROPENOATE see ADT500
ETHYLENE GLYCOL MONOHEPTYL ETHER see HBN250
ETHYLENE GLYCOL, MONO-2,4-HEXADIENE ETHER see HCT500
ETHYLENE GLYCOL, MONOHEXYL ETHER see HFT500
ETHYLENE GLYCOL, MONO(HYDROGEN MALEATE) see EJG500
ETHYLENE GLYCOL MONOISOBUTYL ETHER see IIP000
ETHYLENE GLYCOL, MONOISOPROPYL ETHER see INA500
ETHYLENE GLYCOL, MONOMETHACRYLATE see EJH000
ETHYLENE GLYCOL MONOMETHYL ETHER (MAK, DOT) see EJH500
ETHYLENE GLYCOL MONOMETHYL ETHER ACETATE see EJJ500
ETHYLENE GLYCOL MONOMETHYL ETHER ACETYLRICINOLEATE
  see MIF500
ETHYLENE GLYCOL MONOMETHYL ETHER ACRYLATE see MIF750
ETHYLENE GLYCOL MONOMETHYL ETHER OLEATE see MEP750
ETHYLENE GLYCOL MONOMETHYLPENTYL ETHER see EJK000
ETHYLENE GLYCOL MONO-2-METHYLPENTYL ETHER see EJK000
ETHYLENE GLYCOL MONOPHENYL ETHER see PER000
ETHYLENEGLYCOL MONOPHENYL ETHER PROPIONATE see EJK500
ETHYLENE GLYCOL-MONO-PROPYL ETHER see PNG750
ETHYLENE GLYCOL-MONO-n-PROPYL ETHER see PNG750
ETHYLENE GLYCOL MONOPROPYL ETHER ACETATE see PNA225
ETHYLENE GLYCOL, MONOSTEARATE see EJM500
ETHYLENE GLYCOL MONO-2,6,8-TRIMETHYL-4-NONYL ETHER see EJL000
ETHYLENE GLYCOL MONOVINYL ETHER see EJL500
ETHYLENE GLYCOL PHENYL ETHER see PER000
ETHYLENE GLYCOL SILICATE see EJM000
ETHYLENE GLYCOL STEARATE see EJM500
ETHYLENE GLYCOL VINYL ETHER see EJL500
ETHYLENE HEXACHLORIDE see HCI000
ETHYLENE HOMOPOLYMER see PJS750
ETHYLENEIMINE see EJM900
ETHYLENE IMINE, INHIBITED (DOT) see EJM900
1-ETHYLENEIMINO-2-HYDROXY-3-BUTENE see VMA000
ETHYLENE IODOHYDRIN see IEL000
ETHYLENE METHACRYLATE see BKM250
N,N-ETHYLENE-N'-METHYLUREA see EJN400
ETHYLENE MONOCHLORIDE see VNP000
1,8-ETHYLENE NAPHTHALENE. see AAE750
ETHYLENE NITRATE see EJG000
ETHYLENENITROSOUREA see NKL000
ETHYLENE OXIDE see EJN500
ETHYLENE OXIDE, mixed with CARBON DIOXIDE see EJO000
ETHYLENE OXIDE and CARBON DIOXIDE MIXTURES (DOT) see EJO000
ETHYLENE OXIDE CYCLIC TETRAMER see COD475
ETHYLENE (OXYDE d') (FRENCH) see EJN500
1-ETHYLENEOXY-3,4-EPOXYCYCLOHEXANE see VOA000
ETHYLENE OZONIDE see EJO500
1,4-ETHYLENEPIPERAZINE see DCK400
ETHYLENE POLYMERS see PJS750
ETHYLENE PROPIONATE see COB260
ETHYLENESUCCINIC ACID see SMY000
ETHYLENE SULFATE see EJP000
ETHYLENE SULFIDE see EJP500
ETHYLENE SULFITE see COV750
1,2-ETHYLENE SULFITE see COV750
ETHYLENE SULPHIDE see EJP500
ETHYLENE TEREPHTHALATE POLYMER see PKF750
ETHYLENE TETRACHLORIDE see PCF275
ETHYLENETHIOCARBAMYL SULFIDE see EJQ000

ETHYLENE THIOUREA see IAQ000
1,3-ETHYLENE-2-THIOUREA see IAQ000
N,N'-ETHYLENETHIOUREA see IAQ000
ETHYLENETHIOUREA mixed with SODIUM NITRITE see IAR000
l'ETHYLENE THIOUREE (FRENCH) see IAQ000
ETHYLENE THIURAM MONOSULFIDE see EJQ000
ETHYLENE THIURAM MONOSULPHIDE see EJQ000
ETHYLENE TRICHLORIDE see TIO750
ETHYLENE UNDECANE DICARBOXYLATE see EJQ500
ETHYLENE UREA see IAS000
1,3-ETHYLENE UREA see IAS000
ETHYLENGLYCOL MONOVINYL ESTER (RUSSIAN) see EJL500
ETHYLENGLYKOLDINITRAT (CZECH) see EJG000
ETHYLENIMINE see EJM900
1-ETHYLENIMINO-2-HYDROXYBUTENE see VMA000
1-N-ETHYLEPHEDRINE HYDROCHLORIDE see EJR500
ETHYL-2,3-EPOXYBUTYRATE see EJS000
ETHYL-α,β-EPOXYHYDROCINNAMATE see EOK600
ETHYL α,β-EPOXY-β-METHYLHYDROCINNAMATE see ENC000
ETHYL 2,3-EPOXY-3-METHYL-3-PHENYLPROPIONATE see ENC000
ETHYL-α,β-EPOXY-α-PHENYLPROPIONATE see EOK600
(8-β)-6-ETHYLERGOLINE-8-ACETAMIDE TARTRATE see EJS100
ETHYL ESTER of N-ACETYL-dl-SARCOLYSYL-l-PHENYLALANINE
  see ACD250
ETHYL ESTER of N-ACETYL-dl-SARCOSYLYL-dl-VALINE see ARM000
ETHYL ESTER-l-CYSTEINE HYDROCHLORIDE (9CI) see EHU600
ETHYL ESTER l-CYSTEINE HYDROCHLORIDE (9CI) see MBX800
ETHYL ESTER of 1,2,5,6-DIBENZANTHRACENE-endo-α,β-SUCCINO GLY-
  CINE see EJT000
ETHYL ESTER of 4,4'-DICHLOROBENZILIC ACID see DER000
ETHYL ESTER of DIMETHYLDITHIOCARBAMIC ACID see EIJ500
ETHYL ESTER of O,O-DIMETHYLDITHIOPHOSPHORYL α-PHENYL ACE-
  TATE ACID see DRR400
ETHYL ESTER of 2,3-EPOXY-3-PHENYLBUTANOIC ACID see ENC000
O-ETHYLESTER KYSELINY DICHLORTHIOFOSFORECNE (CZECH)
  see MRI000
ETHYLESTER KYSELINY KROTONOVE see EHO200
ETHYLESTER KYSELINY ORTHOMRAVENCI (CZECH) see ENY500
ETHYL ESTER of METHANESULFONIC ACID see EMF500
ETHYL ESTER of 3-METHYLCHOLANTHRENE-endo-α,β-SUCCINOGLYCINE
  see EJT500
ETHYL ESTER of METHYLNITROSO-CARBAMIC ACID see MMX250
ETHYL ESTER of METHYLSULFONIC ACID see EMF500
ETHYL ESTER of METHYLSULPHONIC ACID see EMF500
11-β-ETHYLESTRADIOL see EJT575
11-β-ETHYLESTRA-1,3,5(10)-TRIENE-3,17-β-DIOL see EJT575
ETHYLESTRENOL see EJT600
N-ETHYL-ETHANAMINE see DHJ200
N-ETHYL-ETHANAMINE HYDROCHLORIDE (9CI) see DIS500
ETHYL ETHANE SULFONATE see DKB139
ETHYL ETHANOATE see EFR000
ETHYL ETHER see EJU000
ETHYL ETHER of 10-(β-
  MORPHOLYLPROPIONYL)PHENTHIAZINECARBAMINO ACID HYDRO-
  CHLORIDE see EEI025
ETHYL ETHER of PROPYLENE GLYCOL see EJV000
11-β-ETHYL-17-α-ETHINYLESTRADIOL see EJV400
13-ETHYL-17-α-ETHINYL-17-HYDROXYGON-4,9,11-TRIEN-3-ONE see ENX575
ETHYL-β-ETHOXYPROPIONATE see EJV500
3-ETHYL-2-(5-(3-ETHYL-2-BENZOTHIAZOLINYLIDENE)-1,3-PEN
  TADIENYL)BENZOTHIAZOLIUM IODIDE see DJT800
ETHYL-N-ETHYL CARBAMATE see EJW500
S-ETHYL-N-ETHYL-N-CYCLOHEXYLTHIOLCARBAMATE see EHT500
3-ETHYL-5-(4,4-ETHYLENEDIOXYPIPERIDINO-1-METHYL)-6,7-DIHYDRO-2-
  METHYLINDOL-4(5H)-ONE see AFH550
ETHYL ETHYLENE OXIDE see BOX750
2-ETHYLETHYLENIMINE see EGN500
2-ETHYL-N-(2-ETHYLHEXYL)-1-HEXANAMINE see DJA800
ETHYL-6-(ETHYL(2-HYDROXYPROPYL)AMINO)-3-PYRIDAZINECARBAZATE
  see CAK275
ETHYL-2-(6(ETHYL(2-HYDROXYPROPYL)AMINO)-3-
  PYRIDAZINYL)HYDRAZINECARBOXYLATE see CAK275
ETHYL 3-ETHYL-4-OXO-5-PIPERIDINO-Δ²,α-THIAZOLIDINEACETATE
  see EOA500
ETHYL(Z)-(3-ETHYL-4-OXO-5-PIPERIDINOTHIAZOLIDIN-2-YLIDENE)ACE-
  TATE see EOA500
5-ETHYL-5-(1-ETHYLPROPYL)BARBITURIC ACID see EJY000
5-ETHYL-5-(1-ETHYLPROPYL)2,4,6(1H,3H,5H)-PYRIMIDINETRIONE
  see EJY000
1-ETHYL-4-(p-(p-((p-((1-ETHYLPYRIDINIUM-4-YL)AMINO)-2-
  AMINOPHENYL)CARBAMOYL)CINNAMAMIDO)ANILINO)PYRIDINIUM,
  DIBROMIDE see EJZ000
1-ETHYL-4-(p-(p-((1-ETHYLPYRIDINIUM-4-YL)AMINO)BENZAMIDO)
  ANILINO)QUINOLINIUM DIBROMIDE see EKA000

1-ETHYL-4-(p-((p-((1-ETHYLPYRIDINIUM-4-YL)AMINO)PHENYL)CARBAM-
OYL)ANILINO)QUINOLINIUM, DIBROMIDE see EKA500
1-ETHYL-4-(p-(p-((1-ETHYLPYRIDINIUM-4-YL)AMINO)PHENYL)CARBAM-
OYL)CINNAMAMIDO)ANILINO)PYRIDINIUM, DI-p-TOLUENE SULFO-
NATE see EKB000
1-ETHYL-4-(p-((p-((1-ETHYLPYRIDINIUM-4-YL)PHENYL)CARBAMOYL)
ANILINO)QUINOLINIUM), DI-p-TOLUENE SULFONATE see EKC500
1-ETHYL-6-(p-(p-((1-ETHYLQUINOLINIUM-6-YL)CARBAM-
OYL)BENZAMIDO)BENZAMIDO)QUINOLINIUM, DI-p-TOLUENE SULFO-
NATE see EKD500
1-ETHYL-7-(p-(p-((1-ETHYLQUINOLINIUM-7-YL)CARBAM-
OYL)BENZAMIDO)BENZAMIDO)QUINOLINIUM), DI-p-TOLUENE SULFO-
NATE see EKD000
O,O-ETHYL-S-2(ETHYLTHIO)ETHYL PHOSPHORODITHIOATE see DXH325
ETHYLETHYNE see EFS500
13-ETHYL-17-α-ETHYNYLGON-4-EN-17-β-OL-3-ONE see NNQ500, NNQ520
13-ETHYL-17-α-ETHYNYL-17-β-HYDROXY-4-GONEN-3-ONE see NNQ500,
NNQ520
(±)-13-ETHYL-17-α-ETHYNYL-17-β-HYDROXYGON-4-EN-3-ONE see NNQ500
dl-13-β-ETHYL-17-α-ETHYNYL-19-NORTESTOSTERONE see NNQ500
17-α-ETHYLETHYNYL-19-NORTESTOSTERONE see EKF550
dl-13-β-ETHYL-17-α-ETHYNYL-19-NORTESTOSTERONE see NNQ500
ETHYLEX GUM 2020 see HLB400
ETHYL FLAVONE-7-OXYACETATE see ELH600
ETHYL-7-FLAVONOXYACETATE see ELH600
ETHYL FLAVON-7-YLOXYACETATE see ELH600
ETHYL FLAVONYL-7-OXYACETATE see ELH600
ETHYL FLUCLOZEPATE see EKF600
ETHYL FLUORIDE (DOT) see FIB000
ETHYL FLUOROACETATE see EKG500
ETHYL-10-FLUORODECANOATE see EKI000
ETHYL-ω-FLUORODECANOATE see EKI000
1-ETHYL-6-FLUORO-1,4-DIHYDRO-4-OXO-7-(1-PIPERAZINYL)-1,8-NAPH-
THYRIDINE-3-CARBOXYLIC ACID see EAU100
1-ETHYL-6-FLUORO-1,4-DIHYDRO-4-OXO-7-(1-PIPERAZINYL)-3-
QUINOLINECARBOXYLIC ACID see BAB625
ETHYL-6-FLUOROHEXANOATE see EKJ500
ETHYL-ω-FLUOROHEXANOATE see EKJ500
ETHYL-9-FLUORONONANECARBOXYLATE see EKI000
ETHYL-8-FLUORO OCTANOATE see EKK500
ETHYL-ω-FLUOROOCTANOATE see EKK500
ETHYL-5-FLUOROPENTANECARBOXYLATE see EKJ500
ETHYL-p-FLUOROPHENYL SULFONE see FLG000
ETHYL FLUOROSULFATE see EKK550
ETHYL FORMATE see EKL000
ETHYLFORMIAAT (DUTCH) see EKL000
ETHYLFORMIC ACID see PMU750
ETHYL FORMIC ESTER see EKL000
1-ETHYL-1-FORMYLHYDRAZINE see EKL250
N-ETHYL-N-FORMYLHYDRAZINE see EKL250
6-ETHYL-7-FORMYL-1,1,4,4-TETRAMETHYL-1,2,3,4-
TETRAHYDRONAPHTHALENE see FNK200
ETHYL FUMARATE see DJJ800
5-ETHYL-2(5H)-FURANONE see EKL500
4-ETHYL-3-FURAZANONE see ELK500
ETHYL FUROATE see EKM000
ETHYL GERANYL ETHER see GDG100
α-ETHYL GLYCEROL ETHER see EFF000
ETHYL GLYCOLATE see EKM500
ETHYLGLYKOLACETAT (GERMAN) see EES400
ETHYL GLYME see EJE500
ETHYL GREEN see BAY750
ETHYL-p-(6-GUANIDINOHEXANOYLOXY) BENZOATE METHANESULFON-
ATE see GAD400
ETHYL GUSATHION see EKN000
ETHYL GUTHION see EKN000
ETHYL HEPTANOATE see EKN050
ETHYL HEPTOATE see EKN050
(3-ETHYL-N-HEPTYL)METHYLCARBINOL see ENW500
1-ETHYL-1-HEPTYLPIPERIDINIUM BROMIDE see EKN100
ETHYLHEXABITAL see TDA500
ETHYL HEXADECYL DIMETHYL AMMONIUM BROMIDE see EKN500
1-ETHYL-1-HEXADECYLPIPERIDINIUM BROMIDE see EKN600
ETHYL HEXAFLUORO-2-BROMOBUTYRATE see EKO000
sec-ETHYL HEXAHYDRO-1H-AZEPINE-1-CARBOTHIOATE see EKO500
5-ETHYLHEXAHYDRO-4,6-DIOXO-5-PHENYLPHRIMIDINE see DBB200
5-ETHYLHEXAHYDRO-5-PHENYLPYRIMIDINE-4,6-DIONE see DBB200
2-ETHYLHEXALDEHYDE see BRI000
ETHYLHEXALDEHYDE (DOT) see BRI000
ETHYL-1-HEXAMETHYLENEIMINECARBOTHIOLATE see EKO500
S-ETHYL-1-HEXAMETHYLENEIMINOTHIOCARBAMATE see EKO500
S-ETHYL-N-HEXAMETHYLENETHIOCARBAMATE see EKO500
2-ETHYLHEXANAL see BRI000
ETHYL HEXANEDIOL see EKV000

2-ETHYL-1,3-HEXANEDIOL see EKV000
2-ETHYLHEXANE-1,3-DIOL see EKV000
2-ETHYLHEXANEDIOL-1,3 see EKV000
ETHYL HEXANOATE (FCC) see EHF000
2-ETHYLHEXANOIC ACID see BRI250
2-ETHYLHEXANOIC ACID, VINYL ESTER see VOU000
2-ETHYLHEXANOL see EKQ000
2-ETHYL-1-HEXANOL see EKQ000
2-ETHYL-1-HEXANOL ESTER with DIPHENYL PHOSPHATE see DWB800
2-ETHYL-1-HEXANOL HYDROGEN PHOSPHATE see BJR750
2-ETHYL-1-HEXANOL HYDROGEN SULFATE, SODIUM SALT see TAV750
2-ETHYL-1-HEXANOL PHOSPHATE see TNI250
2-ETHYL-1-HEXANOL SILICATE see EKQ500
2-ETHYL-1-HEXANOL SULFATE SODIUM SALT see TAV750
((2-ETHYLHEXANOYL)OXY)TRIBUTYLSTANNANE see TID250
2-ETHYLHEXANYL ACETATE see OEE000
2-ETHYLHEXENAL see EKR000
2-ETHYL-2-HEXENAL see EKR000
2-ETHYL-1-HEXENE see EKR500
2-ETHYL HEXENE-1 see EKR500
2-ETHYL-2-HEXENOIC ACID see EKS000
2-ETHYLHEXOIC ACID see BRI250
2-ETHYLHEXOIC ACID, VINYL ESTER see VOU000
2-ETHYLHEXYL ACETATE see OEE000
β-ETHYLHEXYL ACETATE see OEE000
2-ETHYLHEXYL ACRYLATE see ADU250
ETHYLHEXYL ACRYLATE 50:50 MIXTURE see ADU500
2-ETHYLHEXYL ALCOHOL see EKQ000
2-ETHYL HEXYLAMINE see EKS500
2-ETHYLHEXYL-3-AMINOPROPYL ETHER see ELA000
N-(2-ETHYLHEXYL)ANILINE see EKT000
5-ETHYL-5-HEXYLBARBITURIC ACID SODIUM SALT see EKT500
N-(2-ETHYLHEXYL)BICYCLO-(2,2,1)-HEPT-5-ENE-2,3-DICARBOXIMIDE
see OES000
2-ETHYLHEXYL-1-CHLORIDE see EKU000
N-(2-ETHYLHEXYL)CYCLOHEXYLAMINE see EKU500
2-ETHYLHEXYL DIPHENYL ESTER PHOSPHORIC ACID see DWB800
2-ETHYLHEXYL DIPHENYLPHOSPHATE see DWB800
ETHYL HEXYLENE GLYCOL see EKV000
2-ETHYLHEXYL-9,10-EPOXYOCTADECANOATE see EKV500
2-ETHYLHEXYL EPOXYSTEARATE see EKV500
2-ETHYLHEXYL ETHANOATE see OEE000
2-ETHYLHEXYL-2-ETHYLHEXANOATE see EKW000
2-ETHYLHEXYL FUMARATE see DVK600
N-(2-ETHYLHEXYL)-3-HYDROXYBUTYRAMIDE HYDROGEN SUCCINATE
see BPF825
N-2-ETHYLHEXYLIMIDEENDOMETHYLENETETRAHYDROPHTHALIC ACID
see OES000
2-ETHYLHEXYLMALEINAN DI-N-BUTYLCINICITY (CZECH) see BJR250
2-ETHYLHEXYL MERCAPTOACETATE see EKW300
2-ETHYLHEXYL METHACRYLATE see EKW500
2-ETHYL-1-HEXYL METHACRYLATE see EKW500
N-(2-ETHYLHEXYL)-5-NORBORNENE-2,3-DICARBOXIMIDE see OES000
2-ETHYLHEXYL OCTYLPHENYLPHOSPHITE see EKX000
N-2-ETHYLHEXYL-β-OXYBUTYRAMIDE SEMISUCCINATE see BPF825
2-(2-ETHYLHEXYLOXY)ETHANOL see EKX500
2-((2-ETHYLHEXYL)OXY)ETHANOL see EKX500
4-(2-ETHYLHEXYLOXY)-2-HYDROXYBENZOPHENONE see EKY000
3-((2-ETHYLHEXYL)OXY)PROPANENITRILE see EKZ000
3-(2-ETHYLHEXYLOXY)PROPIONITRILE see EKZ000
2-ETHYLHEXYLOXYPROPYLAMINE see ELA000
3-((2-ETHYLHEXYL)OXY)PROPYLAMINE see ELA000
2-ETHYLHEXYL PALMITATE see OFE100
ETHYLHEXYL PHTHALATE see DVL700
2-ETHYLHEXYL PHTHALATE see DVL700
1-ETHYL-1-HEXYLPIPERIDINIUM BROMIDE see ELA600
2-ETHYLHEXYL-2-PROPENOATE see ADU250
5-ETHYL-5-HEXYL-2,4,6-(1H,3H,5H)-PYRIMIDINETRIONE MONOSODIUM
SALT see EKT500
2-ETHYLHEXYL SALICYLATE see ELB000
2-ETHYLHEXYL SEBACATE see BJS250
2-ETHYLHEXYL SODIUM SULFATE see TAV750
2-ETHYLHEXYL SULFATE see ELB400
2-ETHYLHEXYL SULFOSUCCINATE SODIUM see DJL000
2-(2-ETHYLHEXYL)-3a,4,7,7a-TETRAHYDRO-4,7-METHANO-1H-ISOINDOLE-
1,3(2H)-DIONE see OES000
5-ETHYL-5-HEXYL-2-THIOBARBITURIC ACID SODIUM SALT see SHN275
2-ETHYLHEXYL THIOGLYCOLATE see EKW300
2-ETHYLHEXYL VINYL ETHER see ELB500
S-ETHYL-HOMOCYSTEINE see EEI000
S-ETHYL-l-HOMOCYSTEINE see AKB250
S-ETHYL-dl-HOMOCYSTEINE see EEI000
ETHYL HYDRATE see EFU000
ETHYLHYDRAZINE HYDROCHLORIDE see ELC000

ETHYL HYDRIDE see EDZ000
ETHYLHYDROCUPREINE see HHR700
ETHYLHYDROCUPREINE HYDROCHLORIDE see ELC500
ETHYL(HYDROGEN CYSTEINATO)MERCURY see EME000
ETHYL HYDROGEN PEROXIDE see ELD000
ETHYL HYDROPEROXIDE see ELD000
ETHYL HYDROPERSULFIDE see EEB000
ETHYL HYDROSULFIDE see EMB100
ETHYL HYDROXIDE see EFU000
4'-ETHYL-4-HYDROXYAZOBENZENE see ELD500
ETHYL-o-HYDROXYBENZOATE see SAL000
ETHYL-p-HYDROXYBENZOATE see HJL000
2-ETHYL-3-(p-HYDROXYBENZOYL)BENZOFURAN see BBJ500
2-ETHYL-4'-HYDROXY-3-BENZOYLBENZOFURAN see BBJ500
ETHYL-2-HYDROXY-2,2-BIS(4-CHLOROPHENYL)ACETATE see DER000
N-ETHYL-N-(4-HYDROXYBUTYL)NITROSOAMINE see ELE500
ETHYL-N-HYDROXYCARBAMATE see HKQ025
ETHYL-m-HYDROXYCARBANILATE CARBANILATE (ESTER)
    see EEO500
α-ETHYL-1-HYDROXYCYCLOHEXANEACETIC ACID see COW700
ETHYLHYDROXYCYCLOHEXANECARBONITRILE ACETATE (ESTER)
    see EHQ000
N-ETHYL-3-HYDROXY-N,N-DIMETHYL-BENZENAMINIUM BROMIDE (9CI)
    see TAL490
N-ETHYL-3-HYDROXY-N,N-DIMETHYLBENZENAMINIUM CHLORIDE (9CI)
    see EAE600
13-ETHYL-17-HYDROXY-18,19-DINOR-17-α-PREGNA-4,9,11-TRIEN-20-YN-3-
    ONE see ENX575
(±)-13-ETHYL-17-HYDROXY-18,19-DINOR-17-α-PREGN-4-EN-20-YN-3-ONE
    see NNQ500
N-ETHYL-2-((HYDROXYDIPHENYLACETYL)OXY)-N,N-
    DIMETHYLETHANAMINIUM CHLORIDE see ELF500
16-ETHYL-17-HYDROXYESTER-4-EN-3-ONE see ELF100
16-β-ETHYL-17-β-HYDROXYESTER-4-EN-3-ONE ACETATE see ELF110
16-β-ETHYL-17-β-HYDROXYESTR-4-EN-3-ONE see ELF100
16-β-ETHYL-17-β-HYDROXY-4-ESTREN-3-ONE see ELF100
1-ETHYL-1-(2-HYDROXYETHYL)AZIRIDINIUM SALT with 2,4,6-
    TRINITROBENZENESULFONIC ACID see ELG000
1-ETHYL-1-(2-HYDROXYETHYL)AZIRIDINIUM-2,4,6-
    TRINITROBENZENESULFONATE see ELG000
ETHYL (2-HYDROXYETHYL)DIMETHYLAMMONIUM BENZILATE CHLO-
    RIDE see ELF500
ETHYL(2-HYDROXYETHYL)DIMETHYLAMMONIUM CHLORIDE BENZIL-
    ATE see ELF500
ETHYL(2-HYDROXYETHYL)DIMETHYL-AMMONIUM SULFATE (SALT),
    BIS(DIBUTYLCARBAMATE) see DDW000
ETHYL(2-HYDROXYETHYL)ETHYLENIMONIUM PICRYLSULFONATE
    see ELG000
ETHYL-β-HYDROXYETHYLETHYLENIMONIUM PICRYLSULFONATE
    see ELG000
1-ETHYL-1-(β-HYDROXYETHYL)ETHYLENIMONIUM PICRYLSULFONATE
    see ELG000
ETHYL-2-HYDROXYETHYLNITROSAMINE see ELG500
N-ETHYL-N-HYDROXYETHYLNITROSAMINE see ELG500
N-ETHYL-N-(2-HYDROXYETHYL)-N-NITROSOCARBAMATE see ELH000
ETHYL-2-HYDROXYETHYL SULFIDE see EPP500
ETHYL-2-HYDROXYETHYL THIOETHER see EPP500
ETHYL-7-HYDROXYFLAVONE see ELH600
ETHYL-4-HYDROXY-3-METHOXYBENZOATE see EQE500
(8-α,9R)-1-ETHYL-9-HYDROXY-6'-METHOXYCINCHONAN-1-IUM IODIDE
    see QIS000
ETHYL-2-(HYDROXYMETHYL)ACRYLATE see ELI500
ETHYL-α-(HYDROXYMETHYL)ACRYLATE see ELI500
α-ETHYL-β-(HYDROXYMETHYL)-1-METHYL-IMIDAZOLE-5-BUTYRIC ACID,
    Γ-LACTONE see PIF000
1-ETHYL-7-HYDROXY-2-METHYL-1,2,3,4,4a,9,10,10a-OC-
    TAHYDROPHENANTHRENE-2-CARBOXYLIC ACID see DYB000
1-ETHYL-3-HYDROXY-1-METHYL-PIPERIDINIUM BROMIDE BENZILATE
    see PJA000
2-ETHYL-2-(HYDROXYMETHYL)-1,3-PROPANEDIOL, CYCLIC PHOSPHATE
    (1:1) see ELJ500
2-ETHYL-2-(HYDROXYMETHYL)-1,3-PROPANEDIOL TRIACRYLATE
    see TLX175
2-ETHYL-2-HYDROXYMETHYL-1,3-PROPANEDIOL TRIMETHACRYLATE
    see TLX250
ETHYL-4-HYDROXY-3-MORPHOLINOMETHYLBENZOATE see ELK000
5-ETHYL-2'-HYDROXY-2(N)-(3-METHYL-2-BUTENYL)-9-METHYL-6,7-
    BENZOMORPHAN see DOQ600
17-α-ETHYL-17-HYDROXYNORANDROSTENONE see ENX600
17-α-ETHYL-17-HYDROXY-4-NORANDROSTEN-3-ONE see ENX600
17-α-ETHYL-17-HYDROXY-19-NORANDROST-4-EN-3-ONE see ENX600
3-ETHYL-4-HYDROXY-1,2,5-OXADIAZOLE see ELK500
β-ETHYL-β-HYDROXYPHENETHYL CARBAMATE see HNJ000
β-ETHYL-β-HYDROXYPHENETHYL CARBAMIC ACID ESTER see HNJ000

ETHYL(m-HYDROXYPHENYL)DIMETHYLAMMONIUM BROMIDE (8CI)
    see TAL490
ETHYL(m-HYDROXYPHENYL)DIMETHYLAMMONIUM CHLORIDE
    see EAE600
ETHYL-p-HYDROXYPHENYL KETONE see ELL500
ETHYL 2-HYDROXYPROPIONATE see LAJ000
ETHYL α-HYDROXYPROPIONATE see LAJ000
3-(6-(ETHYL-(2-HYDROXYPROPYL)AMINO)PYRIDAZIN-3-YL)CARBAZIC
    ACID ETHYL ESTER see CAK275
2-(6-ETHYL(2-HYDROXYPROPYL)AMINO)-3-PYRIDAZINYL)-
    HYDRAZINECARBOXYLIC ACID ETHYL ESTER see CAK275
N-ETHYL-N-(3-HYDROXYPROPYL)NITROSAMINE see ELM000
(S)-4-ETHYL-4-HYDROXY-1H-PYRANO(3',4':6,7)INDOLIZINO(1,2-b)QUINO-
    LINE-3,14(4H,12H)-DIONE see CBB870
2-ETHYL-3-HYDROXY-4H-PYRAN-4-ONE see EMA600
α-ETHYL-3-HYDROXY-2,4,6-TRIIODOHYDROCINNAMIC ACID see IFZ800
α-ETHYL-β-(3-HYDROXY-2,4,6-TRIIODOPHENYL)PROPIONIC ACID see IFZ800
ETHYL HYPOCHLORITE see ELM500
ETHYLIC ACID see AAT250
ETHYLIDENE ACETONE see PBR500
5-ETHYLIDENEBICYCLO(2.2.1)HEPT-2-ENE see ELO500
N,N'-ETHYLIDENE-BIS(ETHYL CARBAMATE) see ELO000
1,1'-(ETHYLIDENEBIS(OXY))BISBUTANE see DDT400
1,1'-(ETHYLIDENE)BIS(OXY)BIS(2-CHLOROETHANE) see DFG600
ETHYLIDENE BROMIDE see DDN800
ETHYLIDENE CHLORIDE see DFF809
3-ETHYLIDENE-3,4,5,6,9,11,13,14,14A,14B-DECAHYDRO-6-HYDROXY-5,6-
    DIMETHYL(1,6)DIOXACYCLODODECINO(2,3,4-GH)-PYRROLIZINE-2,7-
    DIONE see IDG000
ETHYLIDENE DIBROMIDE see DDN800
ETHYLIDENEDICARBAMIC ACID, DIETHYL ESTER see ELO000
ETHYLIDENE DICHLORIDE see DFF809
ETHYLIDENE DIETHYL ETHER see AAG000
ETHYLIDENE DIFLUORIDE see ELN500
5-ETHYLIDENEDIHYDRO-2(3H)-FURANONE see HLF000
ETHYLIDENE DIMETHYL ETHER see DOO600
ETHYLIDENE DINITRATE see ELN600
ETHYLIDENE DIURETHAN see ELO000
ETHYLIDENE FLUORIDE see ELN500
ETHYLIDENE-2(5H)-FURANONE see MOW500
ETHYLIDENE GYROMITRIN see AAH000
ETHYLIDENEHYDROXYLAMINE see AAH250
trans-15-ETHYLIDENE-12-β-HYDROXY-4,12-α,13-β-TRIMETHYL 8-OXO-4,8
    SECOSENEC-1-ENINE see DMX200
ETHYLIDENELACTIC ACID see LAG000
ETHYLIDENE NORBORNENE see ELO500
5-ETHYLIDENE-2-NORBORNENE see ELO500
ETHYLIDICHLORARSINE see DFH200
ETHYLIDICHLOROARSINE (DOT) see DFH200
ETHYLIMINE see EJM900
2,2'-(ETHYLIMINO)DIETHANOL see ELP000
N-ETHYL-2,2'-IMINODIETHANOL see ELP000
N-ETHYL-N'-2-INDANYL-N'-PHENYL-1,3-PROPANEDIAMINE HYDROCHLO-
    RIDE see DBA475
ETHYL IODIDE see ELP500
ETHYL IODOACETATE see ELQ000
ETHYLIODOMETHYLARSINE see ELQ100
ETHYL-10-(p-IODOPHENYL)HENDECANOATE see ELQ500
ETHYL-10-(p-IODOPHENYL)UNDECYLATE see ELQ500
5-ETHYL-5-ISOAMYLBARBITURIC ACID see AMX750
5-ETHYL-5-ISOAMYLMALONYL UREA see AMX750
ETHYL ISOBUTANOATE see ELS000
ETHYL ISOBUTENOATE see MIP800
ETHYLISOBUTYLMETHANE, ISOHEPTANE see MKL250
ETHYL ISOBUTYRATE see ELS000
ETHYLISOBUTYRATE (DOT) see ELS000
ETHYL ISOCYANATE see ELS500
ETHYL ISOCYANATE (DOT) see ELS500
ETHYL ISOCYANIDE see ELT000
2-ETHYLISOHEXANOL see ELT500, EMZ000
2-ETHYLISOHEXYL ALCOHOL see ELT500
ETHYL ISONICOTINATE see ELU000
2-ETHYLISONICOTINIC ACID THIOAMIDE see EPQ000
α-ETHYLISONICOTINIC ACID THIOAMIDE see EPQ000
2-ETHYLISONICOTINIC THIOAMIDE see EPQ000
α-ETHYLISONICOTINOYLTHIOAMIDE see EPQ000
ETHYL ISONITRILE see ELT000
ETHYLISOPENTYLBARBITURIC ACID see AMX750
5-ETHYL-5-ISOPENTYLBARBITURIC ACID see AMX750
5-ETHYL-5-ISOPENTYLBARBITURIC ACID SODIUM SALT see AON750
O-ETHYL-O-(2-ISOPROPOXY-CARBONYL)-PHENYL
    ISOPROPYLPHOSPHORAMIDOTHIOATE see IMF300
ETHYLISOPROPYLBARBITURIC ACID see ELX000
5-ETHYL-5-ISOPROPYLBARBITURIC ACID see ELX000

ETHYL ISOPROPYL FLUOROPHOSPHONATE see ELX100
ETHYL ISOPROPYLIDENE ACETATE see MIP800
4-ETHYL-3-ISOPROPYL-4-METHYL-1-OXACYCLOBUTAN-2-ONE see DSH400
ETHYLISOPROPYLNITROSOAMINE see ELX500
5-ETHYL-5-ISOPROPYL-2-THIOBARBITURIC ACID SODIUM SALT
   see SHY500
ETHYLISOTHIAMIDE see EPQ000
2-ETHYLISOTHIONICOTINAMIDE see EPQ000
α-ETHYLISOTHIONICOTINAMIDE see EPQ000
S-ETHYLISOTHIOURONIUM HYDROGEN SULFATE see ELY550
S-ETHYLISOTHIURONIUM DIETHYL PHOSPHATE see CBI675
S-ETHYLISOTHIURONIUM METAPHOSPHATE see ELY575
ETHYL ISOVALERATE (FCC) see ISY000
ETHYL KETOVALERATE see EFS600
ETHYL 4-KETOVALERATE see EFS600
d-N-ETHYLLACTAMIDE CARBANILATE (ESTER) see CBL500
ETHYL LACTATE (DOT,FCC) see LAJ000
ETHYL LAEVULINATE see EFS600
ETHYL LAURATE see ELY700
ETHYL LEVULATE see EFS600
ETHYL LINALOOL see ELZ000
ETHYLLITHIUM see ELZ100
ETHYL LOFLAZEPATE see EKF600
ETHYL MAGNESIUM IODIDE see EMA000
ETHYL MALEATE see DJO200
N-ETHYLMALEIMIDE see MAL250
ETHYL MALONATE see EMA500
ETHYL MALTOL see EMA600
ETHYL MANDELATE see EMB000
ETHYLMERCAPTAAN (DUTCH) see EMB100
ETHYL MERCAPTAN see EMB100
ETHYL MERCAPTOACETATE see EMB200
ETHYL-2-MERCAPTOACETATE see EMB200
ETHYL-α-MERCAPTOACETATE see EMB200
ETHYL(MERCAPTOACETATO(2−)-O,S)-MERCURATE(1−)-POTASSIUM
   see PLE750
ETHYL MERCAPTOACETIC ACID see EMB200
ETHYL (2-MERCAPTOETHYL) CARBAMATE S-ESTER with O,O-DIMETHYL
   PHOSPHORODITHIOATE see EMC000
β-ETHYLMERCAPTOETHYL DIMETHYL THIONOPHOSPHATE see DAO800
2-ETHYLMERCAPTOMETHYLPHENYL-N-METHYLCARBAMATE see EPR000
2-ETHYLMERCAPTO-10-(3-(1-METHYL-4-PIPERAZINYL)PROPYL)PHENOTHI-
   AZINE DIMALEATE see TEZ000
3-ETHYLMERCAPTO-10-(1'-METHYLPIPERAZINYL-4-PROPYL)PHENOTHI-
   AZINE DIMALEATE see TEZ000
ETHYL MERCAPTOPHENYLACETATE-O,O-DIMETHYL PHOS-
   PHOROCITHIOATE see DRR400
9-ETHYL-6-MERCAPTOPURINE see EMC500
ETHYLMERCURIC ACETATE see EMD000
ETHYLMERCURIC CHLORIDE see CHC500
ETHYLMERCURIC CYSTEINE see EME000
ETHYLMERCURICHLORENDIMIDE see EME050
ETHYLMERCURIC PHOSPHATE see BJT250, EME100
N-(ETHYLMERCURI)-1,4,5,6,7,7-HEXACHLOROBICYCLO(2.2.1)HEPT-5-ENE-
   2,3-DICARBOXIMIDE see EME050
N-ETHYLMERCURI-3,4,5,6,7,7-HEXACHLORO-3,6-ENDOMETHYLENE-1,2,3,6-
   TETRAHYDROPHTHALIMIDE see EME050
N-ETHYLMERCURI-N-PHENYL-p-TOLUENESULFONAMIDE see EME500
N-ETHYLMERCURI-1,2,3,6-TETRAHYDRO-3,6-ENDOMETHANO-3,4,5,6,7,7-
   HEXACHLOROPHTHALIMIDE see EME050
o-(ETHYLMERCURITHIO)BENZOIC ACID SODIUM SALT see MDI000
ETHYLMERCURITHIOSALICYLIC ACID SODIUM SALT see MDI000
N-(ETHYLMERCURI)-p-TOLUENESULFONANILIDE see EME500
N-(ETHYLMERCURI)-p-TOLUENESULPHONANILIDE see EME500
ETHYLMERCURY CHLORIDE see CHC500
ETHYLMERCURY PHOSPHATE see BJT250, EME100
ETHYLMERCURY p-TOLUENESULFANILIDE see EME500
ETHYLMERCURY-p-TOLUENE SULFONAMIDE see EME500
ETHYLMERCURY-p-TOLUENESULFONANILIDE see EME500
ETHYLMERKAPTAN (CZECH) see EMB100
β-ETHYLMERKAPTOETHANOL (CZECH) see EPP500
β-ETHYLMERKAPTOETHYLCHLORID (CZECH) see CGY750
ETHYL METHACRYLATE see EMF000
ETHYL METHACRYLATE, INHIBITED (DOT) see EMF000
ETHYL METHANESULFONATE see EMF500
ETHYLMETHANESULFONATO-CNU see CHF250
ETHYL METHANESULPHONATE see EMF500
ETHYL METHANOATE see EKL000
ETHYL METHANSULFONATE see EMF500
ETHYL METHANSULPHONATE see EMF500
ETHYLMETHIAMBUTENE HYDROCHLORIDE see EIJ000
7-ETHYL-5-METHOXY-BENZ(a)ANTHRACENE see MEN750
ETHYL-4-METHOXYBENZOATE see AOV000
ETHYL-p-METHOXYBENZOATE see AOV000

2'-ETHYL-2-(2-METHOXY BUTYLAMINO) PROPIONANILIDE see EMG000
ETHYL o-(o-(METHOXYCARBONYL)BENZOYL)GLYCOLATE see MOD000
ETHYL o-(METHOXYCARBONYL)BENZOYLOXYACETATE see MOD000
2'-ETHYL-3-(2-METHOXYETHYL)AMINOBUTYRANILIDE HYDROCHLORIDE
   see EMG500
2'-ETHYL-4-(2-METHOXYETHYL)AMINOBUTYRANILIDE HYDROCHLORIDE
   see EMH000
2'-ETHYL-3-(2-METHOXYETHYL)AMINO-3-METHYLBUTYRANILIDE CYCLA-
   MATE see EMH500
2'-ETHYL-3-(2-METHOXYETHYL)AMINO-3-METHYLBUTYRANILIDE CYCLO-
   HEXANE SULFAMATE see EMH500
2'-ETHYL-2-(2-METHOXYETHYLAMINO)PROPIONANILIDE see EMI000
2'-ETHYL-2-(2-METHOXYETHYLAMINO)-PROPIONANILIDE HYDROCHLO-
   RIDE see EMI500
N-ETHYL-6-METHOXY-N'-(1-METHYLETHYL)-1,3,5-TRIAZINE-2,4-DIAMINE
   see EGD000
3-ETHYL-4-(p-METHOXYPHENYL)-2-METHYL-3-CYCLOHEXENE-1-CARBOX-
   YLIC ACID see CBO625
ETHYL all-trans-9-(4-METHOXY-2,3,6-TRIMETHYLPHENYL)-3,7-DIMETHYL-
   2,4,6,8-NONATETRAENOATE see EMJ500
ETHYL-2-METHYLACRYLATE see EMF000
ETHYL-α-METHYL ACRYLATE see EMF000
4-ETHYLMETHYLAMINOAZOBENZENE see ENB000
p-ETHYLMETHYLAMINOAZOBENZENE see ENB000
N-ETHYL-N-METHYL-p-AMINOAZOBENZENE see ENB000
4'-ETHYL-N-METHYL-4-AMINOAZOBENZENE see EOJ000
3-ETHYLMETHYLAMINO-1,1-DI(2'-THIENYL)BUT-1-ENE HYDROCHLORIDE
   see EIJ000
α-(1-(ETHYLMETHYLAMINO)ETHYL)BENZYL ALCOHOL HYDROCHLORIDE
   (−) see EJR500
4-(((((4-ETHYL-4-METHYL)AMINO)PHENYL)AZO)PYRIDINE 1-OXIDE
   see MKB500
ETHYL METHYL ARSINE see EMK600
ETHYL METHYL AZIDOMETHYL PHOSPHONATE see EML500
7-ETHYL-9-METHYLBENZ(c)ACRIDINE see EMM000
12-ETHYL-7-METHYLBENZ(a)ANTHRACENE see EMN000
7-ETHYL-12-METHYLBENZ(a)ANTHRACENE see EMM500
2-ETHYL-6-METHYL-BENZENAMINE see MJY000
o-ETHYL METHYLBENZENE see EPS500
p-ETHYLMETHYLBENZENE see EPT000
1-ETHYL-2-METHYLBENZENE see EPS500
1-ETHYL-4-METHYLBENZENE see EPT000
ETHYL-p-METHYL BENZENESULFONATE see EPW500
N-ETHYL(α-METHYLBENZYL)AMINE see EMO000
N-ETHYL-α-METHYLBENZYLAMINE see EMO000
5-ETHYL-5-(1-METHYL-1-BUTENYL)BARBITURATE see EMO500
5-ETHYL-5-(1-METHYL-1-BUTENYL)BARBITURIC ACID see EMO500
5-ETHYL-5-(1-METHYL-2-BUTENYL)BARBITURIC ACID see EMO875
5-ETHYL-5-(1-METHYL-1-BUTENYL)BARBITURIC ACID SODIUM SALT
   see VKP000
5-ETHYL-5-(3-METHYL-2-BUTENYL)BARBITURIC ACID SODIUM SALT
   see EMP500
5-ETHYL-5-(1-METHYL-1-BUTENYL)-2,4,6(1H,3H,5H)-PYRIMIDINETRIONE
   see EMO500
5-ETHYL-5-(1-METHYL-1-BUTENYL)-2,4,6(1H,3H,5H)-PYRIMIDINETRIONE
   SODIUM SALT see VKP000
5-ETHYL-5-(1-METHYLBUTYL)BARBITURIC ACID see NBT500
5-ETHYL-5-(3-METHYLBUTYL)BARBITURIC ACID see AMX750
5-ETHYL-5-(3-METHYLBUTYL)BARBITURIC ACID SODIUM DERIVATIVE
   see AON750
5-ETHYL-5-(1-METHYLBUTYL)BARBITURIC ACID SODIUM SALT see NBU000
S(−)-5-ETHYL-5-(1-METHYLBUTYL)BARBITURIC ACID SODIUM SALT
   see PBS750
R(+)-5-ETHYL-5-(1-METHYLBUTYL)BARBITURIC ACID SODIUM SALT
   see PBS500
5-ETHYL-5-(1-METHYLBUTYL)MALONYLUREA see NBT500
5-ETHYL-5-(1-METHYLBUTYL)-2,4,6(1H,3H,5H)-PYRIMIDINETRIONE (9CI)
   see NBT500
5-ETHYL-5-(1-METHYLBUTYL)-2,4,6(1H,3H,5H)-PYRIMIDINETRIONE MONO-
   SODIUM SALT (9CI) see NBU000
5-ETHYL-5-(1-METHYLBUTYL)-2-THIOBARBITURIC ACID see PBT250
5-ETHYL-5-(1-METHYLBUTYL)-2-THIOBARBITURIC ACID MONOSODIUM
   see PBT500
ETHYL 2-METHYLBUTYRATE see EMP600
ETHYL-N-METHYL CARBAMATE see EMQ500
ETHYLMETHYL CARBINOL see BPW750
ETHYL METHYL CETONE (FRENCH) see MKA400
3-ETHYL-7-METHYL-9-α-(4'-CHLOROBENZOYLOXY)-3,7-
   DIAZABICYCLO(3.3.1)NONANE HYDROCHLORIDE see YGA700
ETHYL-2-METHYL-4-CHLOROPHENOXYACETATE see EMR000
ETHYL 3-METHYLCROTONATE see MIP800
ETHYL α-METHYLCROTONATE see MIP800
2-ETHYL-8-METHYL-2,8-DIAZASPIRO(4,5)DECANE-1,3-DIONE HYDROBROM-
   IDE see EMR500

ETHYL NITRITE (DOT) see ENN000
ETHYL NITRITE, solution (DOT) see ENN000
5-(N-ETHYL-N-NITRO)AMINO-3-(5-NITRO-2-FURYL)-s-TRIAZOLE see ENN500
ETHYL-p-NITROBENZOATE see ENO000
ETHYL NITROBENZOATE, PARA ESTER see ENO000
N-ETHYL-N-NITROETHANAMINE (9CI) see DJS500
O-ETHYL-O-((4-NITROFENYL)-FENYL)-MONOTHIOFOSFONAAT (DUTCH)
    see EBD700
1-ETHYL-3-NITROGUANIDINE see ENP000
N-ETHYL-N'-NITROGUANIDINE see ENP000
ETHYLNITROLIC ACID see NHY100
N-ETHYL-N'-NITRO-N-NITROSOGUANIDINE see ENU000
ETHYL-p-NITROPHENYL BENZENETHIONOPHOSPHONATE see EBD700
O-ETHYL O-(4-NITROPHENYL)BENZENETHIONOPHOSPHONATE see EBD700
ETHYL-p-NITROPHENYL BENZENETHIOPHOSPHATE see EBD700
ETHYL-p-NITROPHENYL BENZENETHIOPHOSPHONATE see EBD700
ETHYL p-NITROPHENYL ETHYLPHOSPHATE see NIM500
ETHYL-4-NITROPHENYL ETHYLPHOSPHONATE see ENQ000
ETHYL-p-NITROPHENYLPENTYLPHOSPHONATE see ENQ500
ETHYL-p-NITROPHENYL PHENYLPHOSPHONOTHIOATE see EBD700
O-ETHYL-O-(4-NITROPHENYL) PHENYLPHOSPHONOTHIOATE see EBD700
O-ETHYL-O-p-NITROPHENYL PHENYLPHOSPHONOTHIOLATE see EBD700
O-ETHYL-O-p-NITROPHENYL PHENYLPHOSPHOROTHIOATE see EBD700
ETHYL-p-NITROPHENYL THIONOBENZENEPHOSPHATE see EBD700
ETHYL-p-NITROPHENYL THIONOBENZENEPHOSPHONATE see EBD700
3-ETHYL-4-NITROPYRIDINE-1-OXIDE see ENR000
2-(ETHYLNITROSAMINO)ETHANOL see ELG500
1-(ETHYLNITROSAMINO)ETHYL ACETATE see ABT500
(ETHYLNITROSAMINO)METHYL ACETATE see ENR500
4-((ETHYLNITROSAMINO)METHYL)PYRIDINE see NLH000
4-(ETHYLNITROSOAMINO)-1-BUTANOL see ELE500
4-(ETHYLNITROSOAMINO)BUTYRIC ACID see NKE000
5-(N-ETHYL-N-NITROSO)AMINO-3-(5-NITRO-2-FURYL)-s-TRIAZOLE
    see ENS000
ETHYLNITROSOANILINE see NKD000
N-ETHYL-N-NITROSOBENZENAMINE see NKD000
N-ETHYL-N-NITROSOBENZYLAMINE see ENS500
ETHYLNITROSOBIURET see ENT000
N-ETHYL-N-NITROSOBIURET see ENT000
N-ETHYL-N-NITROSO-tert-BUTANAMINE see NKD500
N-ETHYL-N-NITROSOBUTYLAMINE see EHC000
ETHYLNITROSOCARBAMIC ACID, ETHYL ESTER see NKE500
N-ETHYL-N-NITROSOCARBAMIC ACID ETHYL ESTER see NKE500
N-ETHYL-N-NITROSOCARBAMIC ACID 1-NAPHTHYL ESTER see NBI000
N-ETHYL-N-NITROSOCARBAMIDE see ENV000
ETHYLNITROSOCYANAMIDE see ENT500
N-ETHYL-N-NITROSO-ETHANAMINE see NJW500
N-ETHYL-N-NITROSOETHENAMINE see NKF000
N-ETHYL-N-NITROSOETHENYLAMINE see NKF000
N-ETHYL-N-NITROSO-N'-NITROGUANIDINE see ENU000
ETHYL 4-NITROSO-1-PIPERAZINECARBOXYLATE see NJQ000
2-ETHYL-3-NITROSOTHIAZOLIDINE see ENU500
2-ETHYL-N-NITROSOTHIAZOLIDINE see ENU500
ETHYLNITROSOUREA see ENV000
1-ETHYL-1-NITROSOUREA see ENV000
N-ETHYL-N-NITROSO-UREA see ENV000
N-ETHYL-N-NITROSOURETHANE see NKE500
N-ETHYL-N-NITROSOVINYLAMINE see NKF000
1-ETHYL-3-(5-NITRO-2-THIAZOLYL) UREA see ENV500
N-ETHYL-N'-(5-NITRO-2-THIAZOLYL)UREA see ENV500
ETHYL NONANOATE see ENW000
5-ETHYL-2-NONANOL see ENW500
5-ETHYL-3-NONEN-2-ONE see ENX000
ETHYL NONYLATE see ENW000
1-ETHYL-1-NONYLPIPERIDINIUM BROMIDE see ENX100
ETHYLNORADRENALINE HYDROCHLORIDE see ENX500
ETHYL NORADRIANOL see EGE500
ETHYL NOREPINEPHRINE HYDROCHLORIDE see ENX500
α-ETHYLNOREPINEPHRINE HYDROCHLORIDE see ENX500
ETHYLNORGESTRIENONE see ENX575
(−)-N-ETHYL-NORHYOSCINE-METHOBROMIDE see ONI000
N-ETHYLNORPHENYLEPHRINE see EGE500
11-β-ETHYL-19-NOR-17-α-PREGNA-1,3,5(10)-TRIEN-20-YNE-3,17-DIOL
    see EJV400
N-ETHYL-NORSCOPOLAMINEMETHOBROMIDE see ONI000
ETHYLNORSUPRARENIN HYDROCHLORIDE see ENX500
17-ETHYL-19-NORTESTOSTERONE see ENX600
16-β-ETHYL-19-NORTESTOSTERONE see ELF100
17-α-ETHYL-19-NORTESTOSTERONE see ENX600
N-ETHYLNORTROPINE BENZHYDRYL ETHER see DWE800
1-ETHYL-1-OCTADECYLPIPERIDINIUM BROMIDE see ENX875
1-ETHYL-1,2,3,4,4a,9,10,10a-OCTAHYDRO-7-HYDROXY-2-METHYL-2-
    PHENANTHRENECARBOXYLIC ACID see DYB000
ETHYL OCTANOATE see ENY000

ETHYL OCTYLATE see ENY000
1-ETHYL-1-OCTYLPIPERIDINIUM BROMIDE see ENY100
ETHYLOLAMINE see EEC600
1-(β-ETHYLOL)-2-METHYL-5-NITRO-3-AZAPYRROLE see MMN250
ETHYL ORTHOFORMATE see ENY500
ETHYL ORTHOSILICATE see EPF550
ETHYL OXALATE see DJT200
ETHYL OXALATE (DOT) see DJT200
ETHYL-3-OXATRICYCLO-(3.2.1.0^{2,4})OCTANE-6-CARBOXYLATE see ENZ000
1-ETHYL-3-(2-OXAZOLYL)UREA see EOA000
ETHYLOXIRANE see BOX750
ETHYL-3-OXOBUTANOATE see EFS000
ETHYL-3-OXOBUTYRATE see EFS000
ETHYL 4-OXOPENTANOATE see EFS600
(Z)-2-(3-ETHYL-4-OXO-5-PIPERIDINO-2-THIAZOLIDINYLIDENE, ACETIC
    ACID see EOA500
cis-2-(3-ETHYL-4-OXO-5-PIPERIDINO-2-THIAZOLIDINYLIDENE)ACETIC
    ACID see EOA500
2-(3-ETHYL-4-OXO-5-PIPERIDINO-2-THIAZOLIDINYLIDENE)ACETIC ACID
    ETHYL ESTER see EOB000
(3-ETHYL-4-OXO-5-(1-PIPERIDINYL)-2-THIAZOLIDINYLIDENE)ACETIC
    ACID ETHYL ESTER see EOA500
ETHYL 4-OXOVALERATE see EFS600
4-ETHYLOXYPHENOL see EFA100
ETHYL PARABEN see HJL000
ETHYL PARAOXON see NIM500
ETHYL PARASEPT see HJL000
ETHYL PARATHION see PAK000
S-ETHYL PARATHION see DJT000
ETHYL PELARGONATE see ENW000
ETHYL PENTABORANE (9) see EOB100
1-ETHYL-1-PENTADECYLPIPERIDINIUM BROMIDE see EOB200
5-ETHYL-5-PENTYLBARBITURIC ACID see PBS250
2-((1-ETHYLPENTYL)OXY)ETHANOL see EGJ000
2-(((1-ETHYLPENTYL)OXY)ETHOXY)ETHANOL see EGJ500
ETHYL PERCHLORATE see EOD000
ETHYL PEROXYCARBONATE see DJU600
20-ETHYL-PGF2-α see EPC200
ETHYLPHENACEMIDE see PFB350
ETHYL PHENACETATE see EOH000
ETHYLPHENACETIN see EOD500
5-ETHYLPHENAZINIUM ETHYLSULFATE see EOE000
N-ETHYLPHENAZONIUM ETHOSULFATE see EOE000
2-ETHYLPHENOL see PGR250
o-ETHYLPHENOL see PGR250
N-ETHYL-N-(1-PHENOXY-2-PROPYL)CARBAMIC ACID-2-
    (DIETHYLAMINO)ETHYL ESTER HYDROCHLORIDE see EOF500
N-ETHYL-N-(1-PHENOXY-2-PROPYL)CARBAMIC ACID-2-(2-
    METHYLPIPERIDINO)ETHYL ESTER HYDROCHLORIDE see EOG000
N-ETHYL-N-(1-PHENOXY-2-PROPYL)-2-(2-METHYLPIPERIDINO) ACET-
    AMIDE HYDROCHLORIDE see EOG500
ETHYL PHENYLACETATE see EOH000
N-α-ETHYLPHENYLACETYL-N-ACETYL UREA see ACX500
1-((ETHYL)PHENYLACETYL)UREA see PFB350
ETHYL-β-PHENYLACRYLATE see EHN000
ETHYLPHENYLAMINE see EGK000
(R)-N-ETHYL-2-((PHENYLAMINO)CARBONYL)OXY)PROPANAMIDE
    see CBL500
N-ETHYL-p-(PHENYLAZO)ANILINE see EOH500
N-ETHYL-4-(PHENYLAZO)BENZENAMINE see EOH500
p-((m-ETHYLPHENYL)AZO)-N,N-DIMETHYLANILINE see EOI000
p-((p-ETHYLPHENYL)AZO)-N,N-DIMETHYLANILINE see EOI500
p-(4-ETHYLPHENYLAZO)-N-METHYLANILINE see EOJ000
N-ETHYL-1-((4-(PHENYLAZO)PHENYL)AZO)-2-NAPHTHALENAMINE
    see EOJ500
N-ETHYL-1-((p-(PHENYLAZO)PHENYL)AZO)-2-NAPHTHALENAMINE
    see EOJ500
N-ETHYL-1-((4-(PHENYLAZO)PHENYL)AZO)-2-NAPHTHYLAMINE see EOJ500
N-ETHYL-1-((p-(PHENYLAZO)PHENYL)AZO)-2-NAPHTHYLAMINE see EOJ500
N-ETHYL-N-(p-(PHENYLAZO)PHENYL)HYDROXYLAMINE see HKY000
5-ETHYL-5-PHENYLBARBITURIC ACID see EOK000
5-ETHYL-5-PHENYLBARBITURIC ACID SODIUM see SID000
5-ETHYL-5-PHENYLBARBITURIC ACID SODIUM SALT see SID000
S-2-((4-(p-ETHYLPHENYL)BUTYL)AMINO)ETHYL THIOSULFATE see EOK500
2-ETHYL-2-PHENYLBUTYRIC ACID 2-(2-DIETHYLAMINOETHOXY)ETHYL
    ESTER see DHQ200
ETHYL-N-PHENYLCARBAMATE see CBL750
ETHYL PHENYLCARBAMOYLOXYPHENYLCARBAMATE see EEO500
ETHYL PHENYL CARBINOL see EGQ000
ETHYL PHENYL DICHLOROSILANE (DOT) see DFJ800
3-ETHYL-3-PHENYL-2,6-DIKETOPIPERIDINE see DYC800
3-ETHYL-3-PHENYL-2,6-DIOXOPIPERIDINE see DYC800
ETHYL PHENYLDITHIOCARBAMATE see EOK550
ETHYL N-PHENYLDITHIOCARBAMATE see EOK550

ETHYL SULFIDE see EPH000
S-(2-(ETHYLSULFINYL)ETHYL)-O,O-DIMETHYL PHOSPHOROTHIOATE
　　see DAP000
S-2-ETHYL-SULFINYL-1-METHYL-ETHYL-O,O-DIMETHYL-
　　MONOTHIOFOSFAAT see DSK600
ETHYLSULFOCHLORIDE see EEC000
ETHYL SULFOCYANATE see EPP000
ETHYLSULFONAL see BJT750
ETHYLSULFONYLETHANOL see EPI000
2-(ETHYLSULFONYL)ETHANOL see EPI000
ETHYLSULFONYLETHYL ALCOHOL see EPI000
1-(2-(ETHYLSULFONYL)-ETHYL)-2-METHYL-5-NITROIMIDAZOLE see TGD250
1-(2-(ETHYLSULFONYL)-ETHYL)-2-METHYL-5-NITRO-1H-IMIDAZOLE
　　see TGD250
1-(ETHYLSULFONYL)-4-FLUOROBENZENE see FLG000
4-(ETHYLSULFONYL)-1-NAPHTHALENE SULFONAMIDE see EPI300
Γ-(ETHYLSULFONYL)-N,N,α-TRIMETHYL-γ-PHENYLBENZENEPROPANAM-
　　INE HYDROCHLORIDE see DWC100
ETHYL SULFOXIDE see EPI500
S-2-ETHYL-SULPHINYL-1-METHYL-ETHYL-O,O-DIMETHYL PHOS-
　　PHOROTHIOLATE see DSK600
4-ETHYLSULPHONYLNAPHTHALENE-1-SULFONAMIDE see EPI300
4-ETHYLSULPHONYLNAPHTHALENE-1-SULPHONAMIDE see EPI300
ETHYL TELLURAC see EPJ000
1-ETHYL-1-TETRADECYLPIPERIDINIUM BROMIDE see EPJ600
6-ETHYL-1,2,3,4-TETRAHYDRO-9,10-ANTHRACENEDIONE see EPK000
2-ETHYL-5,6,7,8-TETRAHYDROANTHRAQUINONE see EPK000
3'-ETHYL-5',6',7',8'-TETRAHYDRO-5',6',8',8'-TETRAMETHYL-2'-
　　ACETONAPHTHONE see ACL750
1-(3-ETHYL-5,6,7,8-TETRAHYDRO-5,5,8,8-TETRAMETHYL-2-NAPH-
　　THALENYL)-ETHANONE see ACL750
ETHYL-p-((E)-2-(5,6,7,8-TETRAHYDRO-5,5,8,8-TETRAMETHYL-2-NAPHTHYL)-
　　1-PRO PENYL)BENZOATE see AQZ400
1-ETHYL-2,2,6,6-TETRAMETHYL-4-(N-ACETYL-N-PHENYL)PIPERIDINE
　　see EPL000
1-ETHYL-2,2,6,6-TETRAMETHYLPIPERIDINE HYDROCHLORIDE see EPL600
1-ETHYL-2,2,6,6-TETRAMETHYLPIPERIDINE HYDROGEN TARTRATE
　　see EPL700
1-ETHYL-2,2,6,6-TETRAMETHYL-PIPERIDINE TARTRATE see EPL700
N-(1-ETHYL-2,2,6,6-TETRAMETHYLPIPERIDIN-4-YL)ACETANILIDE
　　see EPL000
1-ETHYL-2,2,6,6-TETRAMETHYL-4-(N-PROPIONYL-N-BENZYLAMINO)PIPERI-
　　DINE see EPM000
1-ETHYL-1,1,3,3-TETRAMETHYLTETRAZENIUM see EPM550
ETHYL TETRAPHOSPHATE see HCY000
2-ETHYLTETRAZOLE see EPM590
5-ETHYLTETRAZOLE see EPM600
ETHYL THIOALCOHOL see EMB100
2-ETHYL-4-THIOAMIDYLPYRIDINE see EPQ000
2-ETHYLTHIO-4,6-BIS(ISOPROPYLAMINO)-s-TRIAZINE see EPN500
6-(ETHYLTHIO)N,N'-BIS(1-METHYLETHYL)-1,3,5-TRIAZINE-2,4-DIAMINE
　　see EPN500
2-ETHYL-4-THIOCARBAMOYLPYRIDINE see EPQ000
2-(ETHYLTHIO)CHLOROETHANE see CGY750
ETHYL THIOCYANATE see EPP000
ETHYL THIOCYANATOACETATE see TFF100
ETHYL THIOCYANOACETATE see TFF100
ETHYLTHIOETHANE see EPH000
2-(ETHYLTHIO)-ETHANETHIOL S-ESTER with O,O-DIETHYL PHOS-
　　PHOROTHIOATE see DAP200
2-(ETHYLTHIO)ETHANOL see EPP500
ETHYL THIOETHER see EPH000
2-ETHYLTHIOETHYL CHLORIDE see CGY750
S-2-(ETHYLTHIO)ETHYL O,O-DIETHYL ESTER of PHOSPHORODITHIOIC
　　ACID see DXH325
2-ETHYLTHIOETHYL O,O-DIMETHYL PHOSPHORODITHIOATE see PHI500
S-(2-(ETHYLTHIO)ETHYL) O,O-DIMETHYLPHOSPHORODITHIONATE
　　see PHI500
O-(2-(ETHYLTHIO)ETHYL)-O,O-DIMETHYL PHOSPHOROTHIOATE
　　see DAO800
S-(2-(ETHYLTHIO)ETHYL)-O,O-DIMETHYL PHOSPHOROTHIOATE
　　see DAP400
S(and O)-2-(ETHYLTHIO)ETHYL-O,O-DIMETHYL PHOSPHOROTHIOATE
　　see MIW100
S-(2-(ETHYLTHIO)ETHYL)DIMETHYL PHOSPHOROTHIOLATE see DAP400
S-(2-(ETHYLTHIO)ETHYL)DIMETHYL PHOSPHOROTHIOLOTHIONATE
　　see PHI500
2-(ETHYLTHIO)ETHYL DIMETHYL PHOSPHOROTHIONATE see DAO800
S-(2-(ETHYLTHIO)ETHYL)-O,O-DIMETHYL THIOPHOSPHATE see DAP400
ETHYL THIOGLYCOLATE see EMB200
2-ETHYLTHIOISONICOTINAMIDE see EPQ000
α-ETHYLTHIOISONICOTINAMIDE see EPQ000
2-((ETHYLTHIO)METHYL)PHENOL METHYLCARBAMATE see EPR000
2-((ETHYLTHIO)METHYL)PHENYL METHYLCARBAMATE see EPR000

(2-ETHYLTHIOMETHYLPHENYL)-N-METHYLCARBAMATE see EPR000
2-(ETHYLTHIO)-10-(3-(4-METHYL-1-PIPERAZINYL)PROPYL)-10-PHENOTHI-
　　AZINE-(Z)-2-BUTENEDIOATE see TEZ000
2-(ETHYLTHIO)-10-(3-(4-METHYL-1-PIPERAZINYL)PROPYL)PHENOTHI-
　　AZINE DIMALEATE see TEZ000
ETHYLTHIOMETON see DAO500
ETHYLTHIOMETON SULFOXIDE see OQS000
ETHYL THIOPHANATE see DJV000
2-ETHYL-2-THIO-PSEUDOUREA with METAPHOSPHORIC ACID (1:1)
　　see ELY575
ETHYL THIOPYROPHOSPHATE see SOD100
(ETHYLTHIO)TRIOCTYLSTANNANE see EPR200
ETHYL THIOUREA see EPR600
1-ETHYLTHIOUREA see EPR600
ETHYL THIRAM see DXH250
ETHYL THIUDAD see DXH250
ETHYL THIURAD see DXH250
ETHYLTIN TRICHLORIDE see EPS000
ETHYL-α-TOLUATE see EOH000
2-ETHYLTOLUENE see EPS500
4-ETHYLTOLUENE see EPT000
o-ETHYLTOLUENE see EPS500
p-ETHYLTOLUENE see EPT000
ETHYL(p-TOLUENESULFONANILIDATO)MERCURY see EME500
ETHYL-p-TOLUENESULFONATE see EPW500
p-((ETHYL-p-TOLYL)AZO)-N,N-DIMETHYLANILINE see EPT500
p-((4-ETHYL-m-TOLYL)AZO)-N,N-DIMETHYLANILINE see EPU000
N-ETHYL-N-(1-(o-TOLYLOXY-2-PROPYL)CARBAMIC ACID-2-(2-
　　METHYLPIPERIDINO)ETHYL ESTER HYDROCHLORIDE see EPV000
1-ETHYL-3-p-TOLYLTRIAZENE see EPW000
ETHYL TOSYLATE see EPW500
ETHYL-p-TOSYLATE see EPW500
ETHYL TRI-n-BUTYLAMMONIUM HYDROXIDE see EPX000
ETHYL TRICHLOROPHENYLETHYLPHOSPHONOTHIOATE see EPY000
O-ETHYL-O-2,4,5-TRICHLOROPHENYL ETHYLPHOSPHONOTHIOATE
　　see EPY000
ETHYL TRICHLOROSILANE see EPY500
ETHYLTRICHLOROSTANNANE see EPS000
ETHYLTRICHLOROTIN see EPS000
ETHYLTRICHLORPHON see EPY600
3-ETHYLTRICYCLOQUINAZOLINE see EPZ000
ETHYLTRIETHOXYSILANE see EQA000
ETHYL-2,2,3-TRIFLUORO PROPIONATE see EQB500
α-ETHYL-β-(2,4,6-TRIIODO-3-BUTYRAMIDOPHENYL)ACRYLIC ACID
　　SODIUM SALT see EQC000
α-ETHYL-β-(2,4,6-TRIIODO-3-BUTYRAMIDOPHENYL)PROPIONIC ACID
　　SODIUM SALT see SKO500
ETHYLTRIIODOGERMANE see EQC500
ETHYL-Γ-TRIMETHYLAMMONIUM PROPANEDIOL IODIDE see MJH250
ETHYL-3,7,11-TRIMETHYLDODECA-2,4-DIENOATE see EQD000
ETHYL(2E,4E)-3,7,11-TRIMETHYL-2,4-DODECADIENOATE see EQD000
ETHYL (2E,4E)-3,7,11-TRIMETHYL-DODECA-2-4-DIENOATE see EQD000
4-ETHYL-2,6,7-TRIOXA-1-ARSABICYCLO(2.2.2)OCTANE see EQD100
4-ETHYL-2,6,7-TRIOXA-1-PHOSPHABICYCLO(2.2.2)OCTANE see TNI750
4-ETHYL-2,6,7-TRIOXA-1-PHOSPHABICYCLO(2.2.2)OCTANE-1-OXIDE
　　see ELJ500
α-ETHYLTRYPTAMINE ACETATE see AJB250
dl-α-ETHYLTRYPTAMINE ACETATE see AJB250
ETHYL TUADS see BJB500, DXH250
ETHYL TUEX see DXH250
1-ETHYL-1-UNDECYLPIPERIDINIUM BROMIDE see EQD600
ETHYLUREA see EQD875
1-ETHYLUREA see EQD875
N-ETHYLUREA see EQD875
ETHYLUREA and SODIUM NITRITE (1:1) see EQD900
ETHYLUREA and SODIUM NITRITE (2:1) see EQE000
ETHYLURETHAN see UVA000
ETHYL URETHANE see UVA000
o-ETHYLURETHANE see UVA000
ETHYL VANILLATE see EQE500
ETHYL VANILLIN see EQF000
N-ETHYL-N-VINYLACETAMIDE see EQF200
m-ETHYL VINYLBENZEN (CZECH) see EPG000
ETHYLVINYLDICHLOROSILANE see DFK400
ETHYL VINYL ETHER see EQF500
ETHYL VINYL ETHER (inhibited) see EQG000
ETHYL VINYL KETONE see PBR250
ETHYLVINYLNITROSAMINE see NKF000
5-ETHYL-2-VINYLPYRIDINE see EQG500
ETHYLXANTHIC ACID ANHYDROSULFIDE with O-ETHYLTHIOLCARBON-
　　ATE see EQH000
ETHYLXANTHIC ACID POTASSIUM SALT see PLF000
ETHYLXANTHIC ACID SODIUM SALT see SHE500
ETHYL XANTHOGEN DISULFIDE see BJU000

ETM (heterocycle) see EJQ000
ETMA see EQN700
ETMOZIN see EEI025
ETMT see EFK000
ETO see EJN500
ETOCIL see EEM000
ETOCLOFENE see DGN000
ETODROXIZINE DIMALEATE see HHK100
ETODROXYZINE DIMALEATE see HHK100
ETOFEN see DGN000
ETOFENAMATE see HKK000
ETOFYLLINCLOFIBRAT (GERMAN) see TEQ500
ETOFYLLINE see HLC000
ETOFYLLINE CLOFIBRATE see TEQ500
ETOGLUCID see TJQ333
ETOGYN see MBZ100
ETOKSYETYLOWY ALKOHOL (POLISH) see EES350
ETOMAL see ENG500
ETOMIDATE see HOU100
ETOMIDE HYDROCHLORIDE see CBQ625
ETOMIDOLINE see AHP125
ETONITAZENE HYDROCHLORIDE see EQN750
ETOPERIDONE see EQO000
ETOPOSIDE see EAV500
ETOPROPEZINA see DIR000
ETOPSIDE see EBB600
ETORPHINE see EQO450
(-)-ETORPHINE see EQO450
7-α-ETORPHINE see EQO450
ETORPHINE HYDROCHLORIDE see EQO500
ETOSALICIL see EEM000
ETOSALICYL see EEM000
2-ETOSSIETIL-ACETATO (ITALIAN) see EES400
ETOSUXIMIDA see ENG500
ETOVAL see BPF500
ETOZOLINE see MNG000
ETP see EPQ000, EQP000
ETRENOL see HGO500
ETRETIN see REP400
ETRETINATE see EMJ500
ETRIDIAZOLE see EFK000
ETROFOL see CKF000
ETROFOLAN see MIA250
ETROLENE see RMA500
ETROPRES see RCA200
ETROZOLIDINA see HNI500
ETRUSCOMICINA see LIN000
ETRUSCOMYCIN see LIN000
ETRYPTAMINE ACETATE see AJB250
ETU see IAQ000
ETYBENZATROPINE see DWE800
ETYDION see TLP750
ETYLENU TLENEK (POLISH) see EJN500
ETYLOAMINA (POLISH) see EFU400
ETYLOBENZEN (POLISH) see EGP500
ETYLON see TCC000
ETYLOWY ALKOHOL (POLISH) see EFU000
ETYLU BROMEK (POLISH) see EGV400
ETYLU CHLOREK (POLISH) see EHH000
ETYLU KRZEMIAN (POLISH) see EPF550
E. TYPHOSA LIPOPOLYSACCHARIDE see EQP100
EU-1806 see NAE100
EU 4200 see TNR485
EUBASIN see PPO000
EUBASINUM see PPO000
EUBINE see DLX400
β-EUCAINE see EQP500
EUCAINE B see EQP500
EUCALMYL see FLU000
EUCALYPTOL (FCC) see CAL000
EUCALYPTOLE see CAL000
EUCALYPTUS OIL see EQQ000
EUCALYPTUS REDUNCA TANNIN see MSC000
EUCANINE GB see TGL750
EUCAST see EQQ100
EUCHEUMA SPINOSUM GUM see CCL250
EUCHRYSINE see BJF000
EUCISTEN see EID000
EUCISTIN see PDC250
EUCLIDAN see EQQ100
EUCLORINA see CDP000
EUCODAL see DLX400
EUCORAN see DJS200
EUCTAN see TGJ885

EUCUPIN DIHYRDOCHLORIDE see EQQ500
EUDATINE see BEX500
EUDEMINE INJECTION see DCQ700
EUDESMA-3,11(13)-DIEN-12-OIC ACID see EQR000
EUFIN see DIX200
EUFLAVINE see DBX400, XAK000
EUFODRIANL see MDT600
EUGENIA JAMBOS Linn., extract excluding roots see SPF200
EUGENIC ACID see EQR500
EUGENOL see EQR500
EUGENOL ACETATE see EQS000
1,3,4-EUGENOL ACETATE see EQS000
EUGENOL FORMATE see EQS100
1,3,4-EUGENOL METHYL ETHER see AGE250
EUGENOL PHENYLACETATE see AGL000
EUGENYL ACETATE see EQS000
EUGENYL FORMATE see EQS100
EUGENYL METHYL ETHER see AGE250
EUGENYL PHENYLACETATE see AGL000
EUGLUCAN see CEH700
EUGLUCON see CEH700
EUGLUCON 5 see CEH700
EUGLYCIN see AKQ250
EUGLYKON see CEH700
EUGYON see NNL500
EUHAEMON see VSZ000
EUHYPNOS see CFY750
EUKAIN B see EQP500
EUKALYPTUS OEL (GERMAN) see EQQ000
EUKODAL see DLX400
EUKRATON see MKA250
EUKUPIN DIHYDROCHLORIDE see EQQ500
EUKYSTOL see CLY500
EULICIN see EQS500
EULISSIN A see DAF800
EULIXINE see DAF800
EUMICTON see AAI250
EUMIDRINA see MGR500
EUMIN see MMN250
EUMOTOL see CCD750
EUMOVATE see CMW400
EUMYCETIN see EQT000
EUMYDRIN see MGR500
EUNASIN see BAV000
EUNATROL see OIA000
EUNERPAN see FKI000
EUNOCTIN see DLY000
EUONYMUS EUROPAEUS see ERA309
EUPAREN see DFL200
EUPARENE see DFL200
EUPHODRIN see DBA800
EUPHORBIA ABYSSINICA LATEX see EQT500
EUPHORBIA CANARIENSIS LATEX see EQU000
EUPHORBIA CANDELABRIUM LATEX see EQU500
EUPHORBIA ESULA LATEX see EQV000
EUPHORBIA GRANDIDENS LATEX see EQV500
EUPHORBIA LATHYRIS LATEX see EQW000
EUPHORBIA OBOVALIFOLIA LATEX see EQW500
EUPHORBIA POINSETTIS BUIST see EQX000
EUPHORBIA PULCHERRIMA WILLD. see EQX000
EUPHORBIA SERRATA LATEX see EQX500
EUPHORBIA TIRUCALLI LATEX see EQY000
EUPHORIA WULFENII LATEX see EQY500
EUPHORIN see AHK750, CBL750
EUPHOZID see ILE000
EUPHYLLIN see TEP500
EUPHYLLINE see TEP500
EUPLACID see EHP000
EUPNERON see ECU600, EQY600
EUPRACTONE see PMO250
EUPRAMIN see DLH600, DLH630
EUPREX see TLG000
EURAZYL see EQZ000
EURECEPTOR see TAB250
EURECOR see CCK125
EUREKENE see PNX750
EURESOL see RDZ900
EUREX see EHT500
EURINOL see HII500
EUROCERT AMARANTH see FAG020
EUROCERT AZORUBINE see HJF500
EUROCERT COCHINEAL RED A see FMU080
EUROCERT TARTRAZINE see FAG140
EURODIN see CKL250

EURODOPA see DNA200
EUROGALE see MKP500
EUROPEAN HOLLY see HGF100
EUROPEAN MISTLETOE see ERA100
EUROPEAN SPINDLE TREE see ERA309
EUROPEN see MGR500
EUROPHAN see PKQ059
EUROPIC CHLORIDE see ERA500
EUROPIUM CHLORIDE see ERA500
EUROPIUM CITRATE see ERB000
EUROPIUM EDETATE see ERB500
EUROPIUM NITRATE see ERC550
EUROPIUM(III) NITRATE, HEXAHYDRATE (1:3:6) see ERC000
EUROPIUM(II)SULFIDE see ERC550
EUROPIUM TRINITRATE see ERC550
EUROTIN (A) see ERC600
EURPHYLLIN see TEP500
EUSAL see EEM000
EUSAPRIM see SNK000, TKX000
EUSCOPOL see HOT500
EUSMANID see BFW250
EUSPIRAN see IMR000
EUSTIDIL see DFH600
EUSTIGMIN see NCL100
EUSTIGMIN BROMIDE see POD000
EUSTIGMINE see NCL100
EUSTIGMIN METHYLSULFATE see DQY909
EUSTROPHINUM see SMN000
EUTAGEN see DLX400
EUTENSIN see CHJ750
EUTHATAL see NBU000
EUTHYROX see LEQ300
EUTIMOX see PMI250
EUTONYL see MOS250
EUUFILIN see TEP500
EUVERNIL see SNQ550
EUVESTIN see DKB000
EUVIFOR see NNE400
EVABLIN see BGT250
EVADYNE see BPT750
EVALON see ILD000
EVANOL see SJJ175
EVANS BLUE DYE see BGT250
EVASPIRINE see PGG350
EVASPRINE see PGG350
EVAU-SUPER see SFS000
EVE see EQF500
EVENING TRUMPET FLOWER see YAK100
EVENTIN see PNN300
EVENTIN HYDROCHLORIDE see PNN300
EVERGREEN CASSENA see HGF100
EVERLASTING FLOWER OIL see HAK500
EVERNINOMICIN D see ERC800
EVERNINOMYCIN-B see ERD000
EVEX see ECU750, EDV600
EVION see VSZ450
EVIPAL see ERD500
EVIPAL SODIUM see ERE000
EVIPAN see ERD500
EVIPAN SODIUM see ERE000
EVIPLAST 80 see DVL700
EVIPLAST 81 see DVL700
EVISECT see TFH750
EVISEKT see TFH750
EVITAMINUM see VSZ450
EVOLA see DEP800
EVONOGENIN see DMJ000
EVRONAL see SBM500
EVRONAL SODIUM see SBN000
EVRRONAL see SBN000
EWEISS see BAP000
EX 4355 see DLS600
EX 10-781 see MHJ500
EX10781 see BAV000
EXACTHIN see AES650
EXACT-S see TFP000
EXACYL see AJV500
EXADRIN see VGP000
EXAGAMA see BBQ500
EXAL see VLA000
EXALAMIDE see HFS759
EXAPROLOL HYDROCHLORIDE see ERE100
EXD see BJU000
EXDOL see HIM000

EXELMIN see PIJ500
EXHAUST GAS see CBW750
EXHORAN see DXH250
EXHORRAN see DXH250
EXIPROBEN SODIUM see ERE200
EXLAN, combustion products see ADX750
EXLUTEN see NNV000
EXLUTION see NNV000
EXLUTON see NNV000
EXLUTONA see NNV000
EXMIGRA see EDC500
EXMIN see AHJ750
EXNA see BDE250
EXOFENE see HCL000
EXOLAN see APH500
EXON 450 see AAX175
EXON 454 see AAX175
EXON 605 see PKQ059
EXONAL see FMB000
EXOSALT see BDE250
EXOTHERM see TBQ750
EXOTHERMIC SILICON CHROME (DOT) see SCR000
EXOTHERM TERMIL see TBQ750
EXOTHION see EAS000
EXP 126 see AJU625
EXP 338 see AMQ500
EXP 999 see MQR000
EXP-105-1 see TJG250
EXP 105-1 see AED250
EXPANDEX see DBD700
EXPANSIN see CMV000
EXPERIMENTAL CHEMOTHERAPEUTANT 1,207 see NNH000
EXPERIMENTAL FUNGICIDE 341 see GII000
EXPERIMENTAL FUNGICIDE 5223 see DXX400
EXPERIMENTAL FUNGICIDE 224 (UNION CARBIDE)
    see ZJA000
EXPERIMENTAL HERBICIDE 2 see DGQ200
EXPERIMENTAL HERBICIDE 732 see BQT750
EXPERIMENTAL INSECTICIDE 711 see IKO000
EXPERIMENTAL INSECTICIDE 4049 see MAK700
EXPERIMENTAL INSECTICIDE 4124 see NFT000
EXPERIMENTAL INSECTICIDE 7744 see CBM750
EXPERIMENTAL INSECTICIDE 12008 see DJN600
EXPERIMENTAL INSECTICIDE 12,880 see DSP400
EXPERIMENTAL INSECTICIDE 52160 see TAL250
EXPERIMENTAL INSECTICIDE S-4087 see CON300
EXPERIMENTAL NEMATOCIDE 18,133 see EPC500
EXPERIMENTAL TICK REPELLENT 3 see AEO750
EXPLOSIVE D see ANS500
EXPLOSIVES, HIGH see ERF000
EXPLOSIVES, LOW see ERF500
EXPLOSIVES, PERMITTED see ERG000
EXP. MITICIDE No. 7 see BRP250
EXPONCIT see NNW500
EXPORSAN see DNO800
EXSEL see SBR000
EXSICATED SODIUM SULFITE see SJZ000
EXSICCATED FERROUS SULFATE see FBN100
EXSICCATED FERROUS SULPHATE see FBN100
EXSICCATED SODIUM PHOSPHATE see SJH090
EXTACOL see PGA750
EXT D&C GREEN No. 1 see NAX500
EXT. D&C ORANGE No.3 see FAG010
EXT. D&C RED No. 2 see CMG750
EXT. D&C RED No. 15 see FAG018
EXT D&C YELLOW No. 1 see MDM775
EXT. D&C YELLOW No. 9 see FAG130
EXT. D&C YELLOW No. 10 see FAG135
EXTENCILLINE see BFC750
EXTENDED ZINC INSULIN SUSPENSION see LEK000
EXTENICILLINE see BFC750
EXTERMATHION see MAK700
EXTERNAL BLUE 1 see BJI250
EXTHRIN see AFR250
EXTRACT D&C ORANGE No. 4 see TGW000
EXTRACT D&C RED No. 10 see HJF500
EXTRACT D&C RED No. 14 see XRA000
EXTRACT of JAMAICA GINGER see JBA000
EXTRACTS (PETROLEUM), HEAVY NAPHTHENIC DISTILLATE SOLVENT
    (9CI) see MQV857
EXTRACTS (PETROLEUM), HEAVY PARAFFINIC DISTILLATE SOLVENT
    (9CI) see MQV859
EXTRACTS (PETROLEUM), LIGHT NAPHTHENIC DISTILLATE SOLVENT
    (9CI) see MQV860

EXTRACTS (PETROLEUM), LIGHT PARAFFINIC DISTILLATE SOLVENT (9CI) see MQV862
EXTRACTS (PETROLEUM), RESIDUAL OIL SOLVENT (9CI) see MQV863
EXTRA FINE 200 SALT see SFT000
EXTRA FINE 325 SALT see SFT000
EXTRAMYCIN see APY500
EXTRANASE see BMO000
EXTRA-PLEX see DLB400
EXTRAR see DUS700
EXTRAX see RNZ000
EXTREMA see EPF550, SCQ500
EXTREMA see GDY000
EXTREN see ADA725
EXTREX P 60 see PJY100
EXTRINSIC FACTOR see VSZ000
EXURATE see DDP200
EYE BRIGHT see CCJ825
EYE-CORT see HHQ800
EYEULES see ARR000
E-Z-OFF see MAE000
E-Z-OFF D see BSH250
E-Z-PAQUE
see BAP000

F 1 see WCJ000
F10 see FHL000
F 12 see DFA600
F 13 see CLR250
16 F see ARM266
F 22 see CFX500
F-33 see FLJ000
F-53 see AOO150
F-112 see TBP050
F 114 see FOO509
F-115 see CJI500
F-139 see BPG000
F-150 see FPI000
F 156 see HEL500
190 F see ABX500
F 190 see ABX500
F-400 see DAB840
F461 see DLV200
F 735 see CCC500
F 849 see AKR500
F 933 see BCI500
F 1162 see SNM500
1262 F see BGO000
F 1262 see BGO000
1358F see SOA500
F 1358 see SOA500
1399 F see SNY500
2249F see FMX000
F 2387 see DVX600
F 2559 see PDD300
F 2966 see DXI400
F 6066 see FBP100
F 6103 see BGQ325
F 1 (complexon) see EIX500
F 10 (pesticide) see MAS500
FA see FMV000, FPQ900
Fa 100 see EQR500
FAA see FDR000, FFF000, FIC000
2-FAA see FDR000
2,7-FAA see BGP250
FABANTOL see PMM000
FABIANOL see APT000
FABT (CZECH) see ALV500
FAC see IOT000
FAC 20 see IOT000
FAC 5273 see PIX250
FA-Ca see FQR100
FACTITIOUS AIR see NGU000
FACTOR II (VITAMIN) see VSZ000
FACTOR PP see NCR000
FACTOR S see VSU100
FACTOR S (vitamin) see VSU100
FADORMIR see QAK000
FAFT see FQQ400
FAIR 85 see DAI800, ODE000
FAIRY BELLS see FOM100
FAIRY CAP see FOM100
FAIRY GLOVE see FOM100
FAIRY LILY see RBA500
FAIRY THIMBLES see FOM100

FALAPEN see BFD000
FALICAIN see PNB250
FALICAINE HYDROCHLORIDE see PNB250
FALISAN see MEP250
FALITHION see DSQ000
FALITIRAM see TFS350
FALKITOL see HCI000
FALL see SFS000
FALL CROCUS see CNX800
FALSE ACACIA see GJU475
FALSE BITTERSWEET see AHJ875
FALSE HELLEBORE see FAB100
FALSE KOA (HAWAII) see LED500
FALSE MISTLETOE see MQW525
FALSE PARSLEY see FMU200
FALSE SYCAMORE see CDM325
FAM see MME809
FAMFOS see FAB400, FAB600
FAMID see DVS000
FAMODIL see FAB500
FA MONOMER see FPQ900
FAMOPHOS see FAB600
FAMOPHOS WARBEX see FAB600
FAMOTIDINE see FAB500
FAMPHOS see FAB600
FAMPHUR see FAB600
FANASIL see AIE500
FANFOS see FAB600
FANFT see NGM500
FANNOFORM see FMV000
FANODORMO see TDA500
FANYLINE see FFK000
FANZIL see AIE500
FAREDINA see TEY000
FARGAN see DQA400, PMI750
FAR-GO see DNS600
FARIAL see IBR200
FARINEX 100 see SLJ500
FARINGOSEPT see AHI875
FARLUTAL see MBZ150
FARLUTIN see MCA000
FARMA see BAT000
FARMA 939 see BAT000
FARMACYROL see DHB500
FARMCO see DAA800
FARMCO ATRAZINE see ARQ725
FARMCO DIURON see DXQ500
FARMCO FENCE RIDER see TAA100
FARMCO PROPANIL see DGI000
FARMICETINA see CDP250
FARMIGLUCIN see APP500
FARMINOSIDIN see APP500
FARMISERINE see CQH000
FARMITALIA 204/122 see MFN500
FARMORUBICIN see EBB100
FARMOTAL see PBT250, PBT500
FARNESOL see FAB800
FARNESYL ALCOHOL see FAB800
FAROLITO (CUBA) see JBS100
FARTOX see PAX000
FAS-CILE see MQU750
FASCIOLIN see CBY000, HCI000
FASCO-TERPENE see CDV100
FASCO WY-HOE see CKC000
FASERTON see AHE250
FASINEX see CFL200
FAST ACID BLUE RL see ADE750
FAST ACID GREEN N see FAF000
FASTBALLS see BBK500
FAST BLUE B BASE see DCJ200
FAST BROWN SALT RR see MIH500
FAST CORINTH BASE B see BBX000
FAST DARK BLUE BASE R see TGJ750
FAST GARNET GBC BASE see AIC250
FAST GREEN see AFG500
FAST GREEN FCF see FAG000
FAST GREEN JJO see BAY750
FAST LIGHT ORANGE GA see HGC000
FAST LIGHT YELLOW E see SGP500
FASTOAN RED 2G see DVB800
FASTOGEN BLUE FP-3100 see DNE400
FASTOGEN BLUE SH-100 see DNE400
FAST OIL ORANGE see PEJ500
FAST OIL ORANGE II see XRA000

FAST OIL RED B see SBC500
FAST OIL YELLOW see AIC250
FAST OIL YELLOW B see DOT300
FASTONA RED B see MMP100
FASTONA RED R see CJD500
FAST ORANGE see PEJ500
FAST ORANGE 3R see CJD500
FAST ORANGE BASE GR see NEO000
FAST ORANGE GC BASE see CEH675
FAST ORANGE R SALT see NEN500
FAST RED A see MMP100
FAST RED BASE GG see NEO500
FAST RED BASE TR see CLK220
FAST RED KB AMINE see CLK225
FAST RED KB BASE see CLK225
FAST RED KB SALT see CLK225
FAST RED KB SALT SUPRA see CLK225
FAST RED KBS SALT see CLK225
FAST RED 2G SALT see NEO500
FAST RED SALT TR see CLK235
FAST RED SALT TRA see CLK235
FAST RED SALT TRN see CLK235
FAST RED SG BASE see NMP500
FAST RED SGG BASE see DEO400
FAST RED 5CT BASE see CLK220
FAST RED TR see CLK220
FAST RED TR11 see CLK220
FAST RED TR BASE see CLK220
FAST RED TRO BASE see CLK220
FAST RED TR SALT see CLK235
FAST RED 5CT SALT see CLK235
FAST SCARLET BASE B see NBE500
FAST SCARLET G see NMP500
FAST SCARLET R see NEQ500
FAST SCARLET TR BASE see CLK200
FAST SPIRIT YELLOW AAB see PEI000
FASTUM see BDU500
FAST WHITE see LDY000
FAST WOOL BLUE R see ADE750
FAST YELLOW AT see AIC250
FAST YELLOW B see AIC250
FAST YELLOW GC BASE see CEH670
FATAL see TBV250
FAT BROWN see NBG500
FATOLIAMID see EPQ000
FAT ORANGE II see TGW000
FAT RED B see SBC500
FAT RED 7B see EOJ500
FAT RED (YELLOWISH) see XRA000
FAT SCARLET 2G see XRA000
FATSCO ANT POISON see ARD750
FAT SOLUBLE GREEN ANTHRAQUINONE see BLK000
FATTY ACID, SOAP (C12-14) see FAB880
FATTY ACID, TALL OIL, EPOXIDIZED-2-ETHYLHEXYL ESTER see FAB900
FATTY ACID, TALL OIL, EPOXIDIZED, OCTYL ESTER see FAB920
FAT YELLOW see DOT300
FAUSTAN see DCK759
FAVISTAN see MCO500
FB/2 see DWX800
F-13B1 see TJY100
FB 5097 see MRX500
FBA 1420 see PMM000
FBA 1500 see MCA100
FBA 4503 see PMX250
FBC CMPP see CIR500
FBHC see BBQ750
FC-1 see CMB125
5-FC see FHI000
FC 12 see DFA600
FC 14 see CBY250
FC 31 see CHI900
FC 114 see FOO509
FC-143 see ANP625
FC 402 see FAC050
FC 403 see XSS375
FC-410 see FAC060
FC 448 see MMA525
FC 455 see DHM309
FC 457 see FAC100
FC 480 see FAC130
FC 590 see FAC150
FC 591 see FAC155
FC 642 see FAC157
FC 646 see FAC160

FC 650 see FAC163
FC 651 see FAC165
FC 652 see FAC166
FC 657 see FAC170
FC 659 see FAC175
FC 660 see FAC179
FC 668 see FAC185
FC 676 see DHI875
FC 681 see FAC195
FC-1318 see OBO000
FC 133a see TJY175
FC142b see CFX250
FC 3001 see SOX400
FC 4/58 see CBA375
FC 4648 see PKQ059
FC 152a see ELN500
FC-C 318 see CPS000
FCDR see FHO000
FCdR see FHO000
FCNU see CPL750
F-COL see FHH100
F-CORTEF see FHH100
FD-1 see BLH325
FDA see DGB600
FDA 0101 see SHF500
FDA 0109 see BHT250
FDA 0121 see ALL250
FDA 0345 see BIF750
FDA 1446 see AFR250
FDA 1541 see EIN500
FDA 1725 see TNX400
FDA 1902 see MCA100
2 FDBP see MQY400
FD&C ACID RED 32 see FAG080
FD&C BLUE No. 1 see FAE000
FD&C BLUE No. 2 see FAE100
FD&C GREEN No. 1 see FAE950
FD&C GREEN No..2 see FAF000
FD&C GREEN No. 3 see FAG000
FD&C GREEN No. 2-ALUMINUM LAKE see FAF000
FDC ORANGE 1 see FAG010
FD&C ORANGE No. 1 see FAG010
FD&C RED No. 1 see FAG018
FD&C RED No. 2 see FAG020
FD&C RED No. 3 see FAG040
FD&C RED No. 4 see FAG050
FD&C RED No. 19 see FAG070
FD&C RED No. 32 see FAG080
FD&C RED No. 40 see FAG100
FD&C RED No. 2-ALUMINIUM LAKE see FAG020
FD & C RED No. 4-ALUMINIUM LAKE see FAG050
FD&C VIOLET No. 1 see FAG120
FD&C YELLOW No. 3 see FAG130
FD&C YELLOW No. 4 see FAG135
FD&C YELLOW No. 5 see FAG140
FD&C YELLOW No. 6 see FAG150
F-diAA see DBF200
FDN see DRP800
F3DThd see TKH325
FDUR see DAR400
FEBANTEL see BKN250
FEBRILIX see HIM000
FEBRININA see DOT000
FEBRO-GESIC see HIM000
FEBROLIN see HIM000
FEBRON see DOT000
FEBRUARY DAPHNE see LAR500
FEBUPROL see BPP250
FECAMA see DGP900
FECTRIM see SNK000, TKX000
FEDACIN see NGE500
FEDAL-UN see PCF275
Fe-DEXTRAN see IGS000
FEDRIN see EAW000
FEENO see PDP250
FEGLOX see DWX800
FEINALMIN see DLH630
FEKABIT see PLA250
FELACRINOS see DAL000
FELAN see EKO500
FELAZINE see PFC750
FELBEN see BAU750
FELDENE see FAJ100
FELICAIN (GERMAN) see PNB250

FELICUR see EGQ000
FELISON see DAB800
FELITROPE see EGQ000
FELIXYN see CKE750
α-FELLANDRENE see MCC000
FELLING ZINC OXIDE see ZKA000
FELLOZINE see PMI750
FELMANE see FMQ000
FELONWORT see CCS650
FELSULES see CDO000
FELUREA see PEC250
FELVITEN see AOO490
FEMA 3230 see POM000
FEMACOID see ECU750
FEMADOL see PNA500
FEMALE WATER DRAGON see WAT300
FEMAMIDE see FAJ150
FEMA No. 2003 see AAG250
FEMA No. 2005 see MDW750
FEMA No. 2006 see AAT250
FEMA No. 2007 see THM500
FEMA No. 2008 see ABB500
FEMA No. 2009 see ABH000
FEMA No. 2011 see AEN250
FEMA No. 2026 see AGC500
FEMA No. 2031 see AGH250
FEMA No. 2032 see AGA500
FEMA No. 2033 see AGI500
FEMA No. 2034 see AGJ250
FEMA No. 2037 see AGM500
FEMA No. 2045 see ISV000
FEMA No. 2055 see IHO850
FEMA No. 2058 see IHP100
FEMA No. 2060 see IHP400
FEMA No. 2061 see AOG500
FEMA No. 2063 see AOG600
FEMA No. 2069 see IHS000
FEMA No. 2075 see IHU100
FEMA No. 2082 see AON350
FEMA No. 2084 see IME000
FEMA No. 2085 see ITB000
FEMA No. 2086 see PMQ750
FEMA No. 2097 see AOX750
FEMA No. 2098 see AOY400
FEMA No. 2099 see MED500
FEMA No. 2109 see ARN000
FEMA No. 2127 see BAY500
FEMA No. 2134 see BCS250
FEMA No. 2135 see BDX000
FEMA No. 2137 see BDX500
FEMA No. 2138 see BCM000
FEMA No. 2140 see BED000
FEMA No. 2141 see IJV000
FEMA No. 2142 see BEG750
FEMA No. 2149 see BFD400
FEMA No. 2151 see BFJ750
FEMA No. 2152 see ISW000
FEMA No. 2159 see BMD100
FEMA No. 2160 see IHX600
FEMA No. 2170 see MKA400
FEMA No. 2174 see BPU750
FEMA No. 2175 see IIJ000
FEMA No. 2178 see BPW500
FEMA No. 2179 see IIL000
FEMA No. 2183 see BQI000
FEMA No. 2184 see BFW750
FEMA No. 2186 see BQM500
FEMA No. 2187 see BSW500
FEMA No. 2188 see BRQ350
FEMA No. 2190 see BQP000
FEMA No. 2193 see IIQ000
FEMA No. 2203 see DTC800
FEMA No. 2209 see BBA000
FEMA No. 2210 see IJF400
FEMA No. 2213 see IJN000
FEMA No. 2218 see ISX000
FEMA No. 2219 see BSU250
FEMA No. 2220 see IJS000
FEMA No. 2221 see BSW000
FEMA No. 2222 see IJU000
FEMA No. 2223 see TIG750
FEMA No. 2224 see CAK500
FEMA No. 2229 see CBA500
FEMA No. 2245 see CCM000

FEMA No. 2249 see CCM100, CCM120
FEMA No. 2252 see CCN000
FEMA No. 2286 see CMP969
FEMA No. 2288 see CMP975
FEMA No. 2293 see CMQ730
FEMA No. 2294 see CMQ740
FEMA No. 2295 see API750
FEMA No. 2299 see CMR500
FEMA No. 2301 see CMR850
FEMA No. 2302 see CMR800
FEMA No. 2306 see CMS750
FEMA No. 2307 see CMS845
FEMA No. 2309 see CMT250
FEMA No. 2311 see AAU000
FEMA No. 2312 see CMT600
FEMA No. 2313 see CMT900
FEMA No. 2314 see CMT750
FEMA No. 2316 see CMU100
FEMA No. 2341 see COE500
FEMA No. 2356 see CQI000
FEMA No. 2361 see DAF200
FEMA No. 2362 see DAG000, DAG200
FEMA No. 2365 see DAI600
FEMA No. 2366 see DAI350
FEMA No. 2370 see BOT500
FEMA No. 2371 see BEO250
FEMA No. 2375 see EMA500
FEMA No. 2376 see DJY600
FEMA No. 2377 see SNB000
FEMA No. 2379 see DKV150
FEMA No. 2381 see HHR500
FEMA No. 2391 see DTE600
FEMA No. 2392 see DQQ375
FEMA No. 2393 see DQQ200
FEMA No. 2394 see DQQ380
FEMA No. 2401 see DXS700
FEMA No. 2414 see EFR000
FEMA No. 2415 see EFS000
FEMA No. 2418 see EFT000
FEMA No. 2420 see AOV000
FEMA No. 2421 see EGM000
FEMA No. 2422 see EGR000
FEMA No. 2426 see DHI000
FEMA No. 2427 see EHE000
FEMA No. 2428 see ELS000
FEMA No. 2429 see DHI400
FEMA No. 2430 see EHN000
FEMA No. 2432 see EHE500
FEMA No. 2433 see EJN500
FEMA No. 2434 see EKL000
FEMA No. 2437 see EKN050
FEMA No. 2439 see EHF000
FEMA No. 2440 see LAJ000
FEMA No. 2441 see ELY700
FEMA No. 2443 see EMP600
FEMA No. 2444 see ENC000
FEMA No. 2445 see ENL850
FEMA No. 2447 see ENW000
FEMA No. 2449 see ENY000
FEMA No. 2452 see EOH000
FEMA No. 2454 see EOK600
FEMA No. 2456 see EPB500
FEMA No. 2458 see SAL000
FEMA No. 2463 see ISY000
FEMA No. 2464 see EQF000
FEMA No. 2465 see CAL000
FEMA No. 2467 see EQR500
FEMA No. 2468 see IKQ000
FEMA No. 2469 see EQS000
FEMA No. 2470 see AAX750
FEMA No. 2475 see AGE250
FEMA No. 2476 see IKR000
FEMA No. 2478 see FAB800
FEMA No. 2489 see FPQ875
FEMA No. 2497 see DSD775, FQT000
FEMA No. 2507 see DTD000
FEMA No. 2509 see DTD800
FEMA No. 2511 see GDE800
FEMA No. 2512 see GDE825
FEMA No. 2514 see GCY000
FEMA No. 2516 see GDM400
FEMA No. 2517 see GDM450
FEMA No. 2539 see HBA550
FEMA No. 2540 see HBB500

FEMA No. 2544 see MGN500
FEMA No. 2545 see EHA600
FEMA No. 2548 see HBL500
FEMA No. 2557 see HEM000
FEMA No. 2559 see HEU000
FEMA No. 2560 see HFA525
FEMA No. 2562 see HFD500
FEMA No. 2563 see HFE000
FEMA No. 2565 see HFI500
FEMA No. 2567 see HFJ500
FEMA No. 2569 see HFO500
FEMA No. 2583 see CMS850
FEMA No. 2585 see HJV700
FEMA No. 2588 see RBU000
FEMA No. 2593 see ICM000
FEMA No. 2594 see IFW000
FEMA No. 2595 see IFX000
FEMA No. 2615 see DXT000
FEMA No. 2617 see DXV600
FEMA No. 2633 see LFU000
FEMA No. 2635 see LFX000
FEMA No. 2636 see LFY100
FEMA No. 2638 see LFZ000
FEMA No. 2640 see LGB000
FEMA No. 2642 see LGA050
FEMA No. 2645 see LGC100
FEMA No. 2665 see MCF750, MCG000, MCG250
FEMA No. 2667 see MCE250, MCG275
FEMA No. 2668 see MCG500, MCG750
FEMA No. 2670 see AOT530
FEMA No. 2672 see MFF580
FEMA No. 2677 see MFW250
FEMA No. 2681 see MGP000
FEMA No. 2682 see APJ250
FEMA No. 2683 see MHA750
FEMA No. 2684 see MNT075
FEMA No. 2685 see PDE000
FEMA No. 2690 see MIP750
FEMA No. 2697 see MIO000
FEMA No. 2698 see MIO500
FEMA No. 2700 see HMB500
FEMA No. 2707 see MKK000
FEMA No. 2718 see MGQ250
FEMA No. 2719 see MLL600
FEMA No. 2723 see ABC500
FEMA No. 2729 see MND275
FEMA No. 2731 see HFG500
FEMA No. 2733 see MHA500
FEMA No. 2743 see COU500
FEMA No. 2745 see MPI000
FEMA No. 2762 see MRZ150
FEMA No. 2770 see DTD200
FEMA No. 2780 see NMV780
FEMA No. 2781 see CNF250
FEMA No. 2782 see NMW500
FEMA No. 2788 see NNB400
FEMA No. 2789 see NNB500
FEMA No. 2797 see OCO000
FEMA No. 2798 see OCE000
FEMA No. 2800 see OEI000
FEMA No. 2802 see ODG000
FEMA No. 2806 see OEG000
FEMA No. 2809 see OEY100
FEMA No. 2841 see ABX750
FEMA No. 2842 see PBN250
FEMA No. 2856 see MCC000
FEMA No. 2857 see PFB250
FEMA No. 2858 see PDD750
FEMA No. 2862 see PDF750
FEMA No. 2866 see PDI000
FEMA No. 2868 see PDK200
FEMA No. 2871 see PDF775
FEMA No. 2873 see PDS900
FEMA No. 2874 see BBL500
FEMA No. 2876 see PDX000
FEMA No. 2878 see PDY850
FEMA No. 2885 see HHP050
FEMA No. 2886 see COF000
FEMA No. 2887 see HHP000
FEMA No. 2890 see HHP500
FEMA No. 2902 see PIH250
FEMA No. 2903 see POH750
FEMA No. 2911 see PIW250
FEMA No. 2922 see IRY000

FEMA No. 2923 see PMT750
FEMA No. 2926 see INE100
FEMA No. 2930 see PNE250
FEMA No. 2962 see MCE750
FEMA No. 2980 see CMT250
FEMA No. 2981 see DTF400, RHA000
FEMA No. 3006 see OHG000
FEMA No. 3007 see SAU400
FEMA No. 3045 see TBD500
FEMA No. 3047 see TBE250
FEMA No. 3053 see TBE600
FEMA No. 3060 see TCU600
FEMA No. 3073 see MNR250
FEMA No. 3075 see THA250
FEMA No. 3091 see UJA800
FEMA No. 3092 see UJJ000
FEMA No. 3095 see ULJ000
FEMA No. 3097 see UNA000
FEMA No. 3101 see VAQ000
FEMA No. 3102 see ISU000
FEMA No. 3103 see VAV000
FEMA No. 3107 see VFK000
FEMA No. 3126 see ADA350
FEMA No. 3135 see DAE450
FEMA No. 3149 see EIL100
FEMA No. 3164 see HAV450
FEMA No. 3183 see MEX350
FEMA No. 3213 see NNA300
FEMA No. 3237 see TDV725
FEMA No. 3244 see TME270
FEMA No. 3264 see DAI360
FEMA No. 3271 see DTU400
FEMA No. 3272 see DTU600
FEMA No. 3273 see DTU800
FEMA No. 3289 see HBI800
FEMA No. 3291 see BOV000
FEMA No. 3302 see MFN285
FEMA No. 3309 see MOW750
FEMA No. 3317 see NMV760
FEMA No. 3326 see ABC750
FEMA No. 3354 see HFM600
FEMA No. 3386 see PPS250
FEMA No. 3497 see HFE550
FEMA No. 3498 see ISZ000
FEMA No. 3499 see HFR200
FEMA No. 3558 see MLA250
FEMA No. 3559 see MCB750
FEMA No. 3565 see DKV175
FEMA No. 3581 see OCY100
FEMA No. 3583 see OEG100
FEMA No. 3632 see PDF790
FEMANTHREN GOLDEN YELLOW see DCZ000
FEMERGIN see EDC500
FEMEST see ECU750
FEMESTRAL see EDO000
FEMESTRONE see EDP000
FEMESTRONE INJECTION see EDV000
FEM H see ECU750
FEMIDYN see EDV000
FEMMA see ABU500
FEMOGEN see ECU750, EDO000
FEMOGEX see EDS100
FEMOXETINE HYDROCHLORIDE see FAJ200
FEMPROPAZINE see DIR000
FEMULEN see EQJ500
FENAB see TIY500
FENAC see TIY500
FENACAINE see BJO500
FENACEMID see PEC250
FENACEMIDE see PEC250
FENACET AST BLUE FF see MGG250
FENACET BLUE G see TBG700
FENACETEAMIDE see PEC250
FENACET FAST VIOLET 5R see DBP000
FENACETIL-KARBAMIDE see PEC250
FENACETINA see ABG750
FENACILIN see PDT500
FENACTIL see CKP250
FENADONE see MDP750
FENAGLICODOLO see CKE750
FENAKROM ORANGE R see NEY000
FENAKTYL see CKP250
FENALAC RED FKB EXTRA see NAY000
FENALAMIDE see FAJ150

FENALAN YELLOW E see SGP500
FEN-ALL see TIK500
FENALLYMAL see AGQ875
FENAM see DRP800
FENAMIC ACID see PEG500
FENAMIDE see FAJ150
FENAMIN see ARQ725
FENAMIN BLACK E see AQP000
FENAMIN BLUE 2B see CMO000
FENAMINE see AMY050, ARQ725
FENAMINOSULF see DOU600
FENAMIN SKY BLUE see CMO500
FENAMIN SKY BLUE 3F see CMN750
FENAMIPHOS see FAK000
FENAMIZOL HYDROCHLORIDE see DCA600
FENANTHRENE RED BROWN 5R see CMU800
FENANTOIN see DKQ000, DNU000
FENAPERONA (SPANISH) see FLK000
FENAPERONE see FLK000
FENARIMOL see FAK100
FENAROL see CJR909, CKF500
FENARSONE see CBJ000
FENARTIL see BRF500
FENASAL see DFV400
FENASPARATE see SAN600
FENATE see IGS000
FENATROL see ARQ725, TIY500
FENAVAR see AMY050
FENAZAFLOR see DGA200
FENAZEPAM see PDB350
FENAZIL see DQA400, PMI750
FENAZO BLUE SR see ADE750
FENAZO BLUE XI see FAE000
FENAZO BLUE XR see FMU059
FENAZO EOSINE XG see BNH500, BNK700
FENAZO GREEN 7G see FAF000
FENAZO RED C see HJF500
FENAZO SCARLET 2R see FMU070
FENAZO SCARLET 3R see FMU080
FENAZOXINE see FAL000
FENAZOXINE HYDROCHLORIDE see NBS500
FENAZO YELLOW M see MDM775
FENAZO YELLOW T see FAG140
FENAZO YELLOW XX see CMM750
FENBENDAZOL see FAL100
FENBENDAZOLE see FAL100
FENBITAL see EOK000
FENBUFEN see BGL250
FENBUTATIN OXIDE see BLU000
FENCAMFAMINE HYDROCHLORIDE see EOM000
FENCAMINE HYDROCHLORIDE see FAM000
FENCARBAMIDE see PDC850
FENCARBAMIDE HYDROCHLORIDE see PDC875
FENCAROL see DWM400
FENCHEL OEL (GERMAN) see FAP000
FENCHLONINE see FAM100
FENCHLOORFOS (DUTCH) see RMA500
FENCHLORFOS see RMA500
FENCHLORFOSU (POLISH) see RMA500
FENCHLOROPHOS see RMA500
FENCHLORPHOS see RMA500
α-FENCHOL see TLW000
endo-FENCHOL see TLW000
FENCHON (GERMAN) see TLW250
FENCHONE see TLW250
FENCHYL ACETATE see FAO000
α-FENCHYL ALCOHOL see TLW000
FENCLONIN see FAM100
FENCLONINE see FAM100
FENCLOR see PJL750
FENCLOZIC ACID see CKK250
FENDON see HIM000
FENDOSAL see FAO100
FENDOZAL see FAO100
FENELZIN see PFC750
FENERGAN see DQA400, PMI750
FENESTERIN see CME250
FENESTREL see FAO200
FENESTRIN see CME250
FENETAZINA see DQA400
FENETHAZINE HYDROCHLORIDE see DPI000
FENETHYLLINE HYDROCHLORIDE see CBF825
FENETICILLINE see PDD350
1-FENETILBIGUANIDE CLORIDRATO (ITALIAN) see PDF250

N'-β-FENETILFORMAMIDINILIMINOUREA (ITALIAN) see PDF000
8-N-FENETIL-1-OXA-2-OXO-3,8-DIAZASPIRO-(4,5)-DECANO CLORIDRATO
    (ITALIAN) see DAI000
FENFLURAMINE see ENJ000
FENFLURAMINE HYDROCHLORIDE see PDM250
FENFORMINA see PDF000
FENHYDREN see PFJ750
FENIBUT see PEE500
FENIBUTAZONA see BRF500
FENIBUT HYDROCHLORIDE see GAD000
FENIBUTOL see BRF500
FENICOL see CDP250, EGQ000
FENIDANTOIN "S" see DKQ000
FENIDIN see DTP400
FENIDRONE see OPK300
FENIGAM HYDROCHLORIDE see GAD000
2-FENIL-2-(p-AMINOFENIL)PROPIONAMMIDE (ITALIAN) see ALX500
2-FENIL-5-BROMO-INDANDIONE (ITALIAN) see UVJ400
FENILBUTINE see BRF500
FENILDICLOROARSINA (ITALIAN) see DGB600
FENILEP see PEC250
FENILFAR see SPC500
4-FENIL-4-FORMILPIPERIDINA (ITALIAN) see PFY100
FENILIDINA see BRF500
FENILIDRAZINA (ITALIAN) see PFI000
FENILIN see PFJ750
FENILISOPROPILIDRAZINA see PDN000
4-FENIL-α-METILFENILACETATO-γ-PROPILSOLFONATO SALE SODICO
    (ITALIAN) see PFR350
FENILOR see DDS600
FENILPRENAZONE see PEW000
2-FENILPROPANO (ITALIAN) see COE750
FENILPROPANOLAMINA (ITALIAN) see NNM000
FENIPENTOL see BQJ500
FENISED see PEC250
FENISTIL see FMU409
FENISTIL-RETARD see FMU409
FENITOIN see DNU000
FENITOX see DSQ000
FENITROTHION see DSQ000
FENITROTION (HUNGARIAN) see DSQ000
FENITROXON see PHD750
FENIZON (FRENCH) see CJR500
FENKAROL see DWM400
FENNEL OIL see FAP000
FENNOSAN see QPA000
FENNOSAN B 100 see DSB200
FENOBARBITAL see EOK000
FENOBOLIN see DYF450
FENOCYCLIN see BIT000
FENOCYCLINE see BIT000
FENOFIBRIC ACID see CJP750
FENOFLURAZOLE see DGA200
FENOL (DUTCH, POLISH) see PDN750
FENOLO (ITALIAN) see PDN750
FENOLOVO see HON000
FENOLOVO ACETATE see ABX250
FENOPHOSPHON see EPY000
FENOPRAIN see PMJ525
FENOPROFEN CALCIUM DIHYDRATE see FAP100
FENOPROFEN CALCIUM SALT DIHYDRATE see FAP100
FENOPROFEN SODIUM see FAQ000
FENOPROMIN see AOA250
FENOPRON see FAP100
FENOPROP see TIX500
FENORMONE see TIX500
FENOSMOLIN see PDP100
FENOSPEN see PDT500
FENOSTENYL see PEC250
FENOSTIL see FMU409
FENOSUCCIMIDE see MNZ000
FENOTEROL BROMIDE see FAQ100
FENOTEROL HYDROBROMIDE see FAQ100
FENOTHIAZINE (DUTCH) see PDP250
FENOTIAZINA (ITALIAN) see PDP250
FENOTONE see BRF500
FENOVERM see PDP250
FENOX see SPC500
FENOXEDIL see BPM750
FENOXEDIL HYDROCHLORIDE see BPM750
FENOXYBENZAMIN see DDG800
2-FENOXYETHANOL (CZECH) see PER000
FENOXYL CARBON N see DUZ000
FENOXYPEN see PDT500, PDT750

FENOZAFLOR see DGA200
FENPENTADIOL see CKG000
FENPIVERIMIUM BROMIDE see RDA375
FENPROBAMATO see PGA750
FENPROPANAGE see DAB825
FENPROPATHRIN see DAB825
FENPROPAZINA see DIR000
FENSON see CJR500
FENSPIRIDE see DAI200
FENSPIRIDE HYDROCHLORIDE see DAI200
FENSULFOTHION see FAQ800
FENTAL see FMB000
FENTANEST see PDW500, PDW750
FENTANIL see PDW500
FENTANYL see PDW500
FENTANYL CITRATE see PDW750
FENTAZIN see CJM250
FENTHION see FAQ900
FENTHION SULFONE see DSS800
FENTHIURAM see FAQ930
FENTHOATE see DRR400
FENTIAPRIL see FAQ950
FENTIAZAC see CKI750
FENTIAZAC CALCIUM SALT see CAP250
FENTIAZIN see PDP250
FENTIN ACETAAT (DUTCH) see ABX250
FENTIN ACETAT (GERMAN) see ABX250
FENTIN ACETATE see ABX250
FENTIN CHLORIDE see CLU000
FENTINE ACETATE (FRENCH) see ABX250
FENTIN HYDROXIDE see HON000
FENTIURAM see FAQ930
FENTRINOL see AHL500
FENUGREEK ABSOLUTE see FAR000
FENULON see DTP400
FENURAL see PEC250
FENUREA see PEC250
FENURON see DTP400
FENURONE see PEC250
FENVALERATE see FAR100
2-FENYL-5-AMINOBENZTHIAZOL (CZECH) see ALV500
1-FENYL-4-AMINO-5-CHLOR-6-PYRIDAZINON (CZECH) see PEE750
1-FENYL-3-AMINOPYRAZOL (CZECH) see ALX750
FENYLBUTAZON see BRF500
FENYL-CELLOSOLVE (CZECH) see PER000
N-FENYL-N'-CYKLOHEXYL-p-FENYLENDIAMIN (CZECH) see PET000
1-FENYL-4,5-DICHLOR-6-PYRIDAZINON (CZECH) see DGE800
1-FENYL-3,3-DIETHYLTRIAZEN (CZECH) see PEU500
1-FENYL-3,3-DIMETHYLTRIAZIN see DTP000
m-FENYLENDIAMIN (CZECH) see PEY000
FENYLENODWUAMINA (POLISH) see PEY500
FENYLEPSIN see DKQ000
FENYLESTER KYSELINY CHLORMRAVENCI (CZECH) see CBX109
1-FENYLETHANOL see PDE000
FENYLETTAE see EOK000
FENYL-GLYCIDYLETHER (CZECH) see PFH000
FENYLHIST see BAU750
FENYLHYDRAZINE (DUTCH) see PFI000
FENYL-β-HYDROXYETHYLSULFON (CZECH) see BBT000
FENYLMERCURIACETAT (CZECH) see ABU500
FENYLMERCURICHLORID (CZECH) see PFM500
FENYL-METHYLKARBINOL see PDE000
2-FENYLOTIOMOCZNIK (POLISH) see DWN800
2-FENYL-PROPAAN (DUTCH) see COE750
1-FENYL-3-PYRAZOLON (CZECH) see PGE250
FENYLSILATRAN (CZECH) see PGH750
FENYL-TRIFLUORSILAN (CZECH) see PGO500
FENYPRIN see DBA800
FENYRAMIDOL see PGG350
FENYRAMIDOL HYDROCHLORIDE see PGG355
FENYRIPOL see PGG350
FENYRIPOL HYDROCHLORIDE see FAR200
FENYTAN see PEC250
FENYTOINE see DKQ000, DNU000
FENZAFLOR see DGA200
FENZEN (CZECH) see BBL250
FEOJECTIN see IHG000
FEOSOL see FBN100, FBO000
FEOSPAN see FBN100
FEOSTAT see FBJ100
FEPRAZONE see PEW000
FEPRONA see FAP100
FERBAM see FAS000
FERBAM 50 see FAS000

FERBAM, IRON SALT see FAS000
FERBECK see FAS000
FERDEX 100 see IGS000
FERGON see FBK000
FERGON PREPARATIONS see FBK000
FER-IN-SOL see FBN100, FBO000
FERISAN see EJA379
FERKETHION see DSP400
FERLUCON see FBK000
FERMATE FERBAM FUNGICIDE see FAS000
FERMENICIDE LIQUID see SOH500
FERMENICIDE POWDER see SOH500
FERMENTATION ALCOHOL see EFU000
FERMENTATION AMYL ALCOHOL see IHP000
FERMENTATION BUTYL ALCOHOL see IIL000
FERMIDE see TFS350
FERMINE see DTR200
FERMOCIDE see FAS000
FERNACOL see TFS350
FERNASAN see TFS350
FERNESTA see BQZ000, DAA800
FERNEX see DIN600
FERNIDE see TFS350
FERNIMINE see DAA800
FERNISOLONE see PMA000
FERNISONE see HHQ800
FERNOS see DOX600
FERNOXENE see SGH500
FERNOXONE see DAA800
FERO-GRADUMET see FBN100, FBO000
FEROTON see FBJ100
FER PENTACARBONYLE (FRENCH) see IHG500
FERRADOW see FAS000
FERRALYN see FBN100
FERRATE (Fe$_{12}$O$_{19}$($^{2-}$)) BARIUM (1:1) (9CI) see BAL625
FERRATE(1-), (GLYCINATO-N,O)(SULFATO(2-)-O',O')-, HYDROGEN, (T-4)-(9CI) see FBD500
FERRATE(4-), HEXACYANO-, TETRAPOTASSIUM see TEC500
FERRATE(4-), HEXAKIS(CYANO-C)-, TETRAPOTASSIUM, (OC-6-11)- see TEC500
FERRATE(2-), PENTAKIS(CYANO-C)NITROSYL-, DISODIUM, DIHYDRATE (OC-6-22)-, (9CI) see SIW500
FERRIAMICIDE see MQW500
FERRIC ACETYLACETONATE see IGL000
FERRIC AMMONIUM OXALATE see ANG925
FERRIC AMMONIUM OXALATE (DOT) see ANG925
FERRIC ARSENATE, solid (DOT) see IGN000
FERRIC ARSENITE, solid (DOT) see IGO000
FERRIC ARSENITE, BASIC see IGO000
FERRIC CHLORIDE see FAU000
FERRIC CHLORIDE, anhydrous (DOT) see FAU000
FERRIC CHLORIDE, solid, anhydrous (DOT) see FAU000
FERRIC CHLORIDE, solid (DOT) see FAU000
FERRIC CHLORIDE (solution) see FAV000
FERRIC CHLORIDE, solution (DOT) see FAU000
FERRIC CHLORIDE HEXAHYDRATE see FAW000
FERRIC CHOLINE CITRATE see FBC100
FERRIC DEXTRAN see IGS000
FERRIC DIMETHYLDITHIOCARBAMATE see FAS000
FERRIC FLUORIDE see FAX000
FERRIC HYDROXIDE NITRILOTRIPROPIONIC ACID COMPLEX see FAY000
FERRIC NITRATE, NONAHYDRATE see IHC000
FERRIC NITRILOTRIACETATE see IHC100
FERRIC NITROSODIMETHYL DITHIOCARBAMATE and TETRAMETHYL THIURAM DISULFIDE see FAZ000
FERRICON see ROF300
FERRIC OXIDE see IHD000
FERRIC OXIDE, SACCHARATED see IHG000
FERRIC SACCHARATE IRON OXIDE (MIX.) see IHG000
FERRIC SODIUM EDETATE see EJA379
FERRIC SODIUM EDTA see EJA379
FERRIC SODIUM GLUCONATE COMPLEX see IHK000
FERRIC SODIUM PYROPHOSPHATE see SHE700
FERRIC SULFATE see FBA000
FERRIC TRIACETYLACETONATE see IGL000
FERRIC TRICHLORIDE HEXAHYDRATE see FAW000
FERRICYANURE de TRI(1-DODECYL-2-PHENYL-3-METHYL-1,3-BENZIMIDAZOLIUM (FRENCH) see TNH750
FERRICYTOCHROME C see CQM325
FERRIDEXTRAN see IGS000
FERRIGEN see IGU000
FERRITIN see FBB000
FERRIVENIN see IHG000
FERRLECIT see FBB100

FERROACTINOLITE see FBG200
FERROANTHOPHYLLITE see ARM264
FERROCENE see FBC000
FERROCHOLINATE see FBC100
FERROCHROME see FBD000
FERROCHROME (exothermic) see FBD000
exothermic FERROCHROME (DOT) see FBD000
FERROCHROME, exothermic (DOT) see FBD000
FERROCHROMIUM see FBD000
FERROCYANIDES see FBD100
FERROCYTOCHROME C see CQM325
FERRODEXTRAN see IGS000
FERROFLUKIN 75 see IGS000
FERROFOS 510 see HKS780
FERROFUME see FBJ100
FERROGLUCIN see IGS000
FERROGLUKIN 75 see IGS000
FERROGLYCINE SULFATE see FBD500, FBD500
FERROGLYCINE SULFATE COMPLEX see FBD500
FERRO-GRADUMET see FBN100
FERROLIP see FBC100
FERROMANGANESE (exothermic) see FBE000
exothermic FERROMANGANESE (DOT) see FBE000
FERRON see IEP200
FERRONAT see FBJ100
FERRONE see FBJ100
FERRONICUM see FBK000
FERRONORD see FBD500
FERROPHOSPHORUS see FBF000
FERROSAN see PLZ100
FERROSANOL see FBD500
FERROSILICON see FBG000, IHJ000
FERROSILICON, containing more than 30% but less than 90% SILICON (DOT)
  see FBG000
FERROSULFAT (GERMAN) see FBN100
FERROSULFATE see FBN100
FERROTEMP see FBJ100
FERRO-THERON see FBN100
FERROTREMOLITE see FBG200
FERROUS see FBN000
FERROUS ACETATE see FBH000
FERROUS ARSENATE (DOT) see IGM000
FERROUS ARSENATE, solid (DOT) see IGM000
FERROUS ASCORBATE see FBH050
FERROUS CARBONATE see FBH100
FERROUS CHLORIDE see FBI000
FERROUS CHLORIDE TETRAHYDRATE see FBJ000
FERROUS FUMARATE see FBJ100
FERROUS GLUCONATE see FBK000
FERROUS GLUCONATE DIHYDRATE see FBL000
FERROUS GLUTAMATE see FBM000
FERROUS ION see FBN000
FERROUS LACTATE see LAL000
FERROUS METAL BORINGS, SHAVINGS TURNINGS OR CUTTINGS (DOT)
  see FBN000
FERROUS SULFATE see FBN100
FERROUS SULFATE (FCC) see FBO000
FERROUS SULFATE HEPTAHYDRATE see FBO000
FERROVANADIUM DUST see FBP000
FERROXDURE see BAL625
FERRO YELLOW see CAJ750
FERRUGO see IHD000
FERRUM see FBJ100, FBP050
FERSAMAL see FBJ100
FERSOLATE see FBN100
FERTILYSIN see BIX250
FERTINORM see FMT100
FERTIRAL see LIU360
FERTODUR see FBP100
FERULA JAESCHKEANA VATKE, EXTRACT see FBP175
FERULIC ACID see FBP200
trans-FERULIC ACID see FBP200
FERVENULIN see FBP300
FERVENULINE see FBP300
FES see ZAT000
FESOFOR see FBO000
FESOTYME see FBO000
FETID NIGHTSHADE see HAQ100
FETT see FNK200
FETTER BUSH see DYA875
FETTERBUSH see FBP520
FETTORANGE B see XRA000
FETTORANGE R see PEJ500
FETTSCHARLACH see OHA000

FEUILLES CRABE (HAITI) see SLJ650
FEVER TWIG see AHJ875
FF see FQN000, FQO000
FF 106 see IJH000
FFB 32 see PHB500
F-5-FU see FMB000
F 151 FUMARATE see MDP800
FG 4963 see FAJ200
FG 5111 see FK1000
FH 099 see DTL200
FHA see FJF100
FH 122-A see NNQ500
FHCH see BBQ750
FHD-3 see BOF750
F.I 106 see AES750
FI 106 see HKA300
FI 1163 see LIN000
F.I. 58-30 see EPQ000
Fi 5853 see APP500
FI 6120 see DLH200
F.I. 6145 see PIW000
FI 6146 see BTA325
FI6339 see DAC000
FI 6341 see FDB000
FI 6714 see NDM000
FI 6804 see HKA300
FIBERGLASS see FBQ000
FIBERS, REFRACTORY CERAMIC see RCK725
FIBER V see PKF750
FIBORAN see FBP850
FIBRALEM see ARQ750
FIBRE BLACK VF see AQP000
FIBRENE C 400 see TAB750
FIBROTAN see PFN000
FIBROUS CROCIDOLITE ASBESTOS see ARM275
FIBROUS GLASS see FBQ000
FIBROUS GLASS DUST (ACGIH) see FBQ000
FIBROUS GRUNERITE see ARM250
FIBROUS TREMOLITE see ARM280
FICAM see DQM600
FICHLOR 91 see TIQ750
FICIN see FBS000
FI CLOR 91 see TIQ750
FI CLOR 60S see SGG500
FICUSIN see FQD000
FICUS PROTEASE see FBS000
FICUS PROTEINASE see FBS000
FIDDLE FLOWER see SDZ475
FIDDLE-NECK see TAG250
FIELD GARLIC see WBS850
FIGUIER MAUDIT MARRON (HAITI) see BAE325
FIGWORT see FBS100
F III (sugar fraction) see FAB000
FILARIOL see EGV500
FILARSEN see DFX400
FILIGRANA (CUBA) see LAU600
FILMERINE see SIQ500
FILORAL see CMF500
FIMALENE see ILD000
FINA, combustion products see ADX750
FINAVEN see ARW000
FINDOLAR see GGS000
FINE GUM HES see SFO500
FINEMEAL see BAP000
FINIMAL see HIM000
FINISH EN see DTG700
FINLEPSIN see DCV200
FINTIN ACETATO (ITALIAN) see ABX250
FINTINE HYDROXYDE (FRENCH) see HON000
FINTIN HYDROXID (GERMAN) see HON000
FINTIN HYDROXYDE (DUTCH) see HON000
FINTIN IDROSSIDO (ITALIAN) see HON000
FINTROL see AQM000
FIORINAL see ABG750
FIRE DAMP see MDQ750
FIREMASTER BP-6 see FBU000
FIREMASTER FF-1 see FBU509
FIREMASTER T23P-LV see TNC500
FIRMACEF see CCS250
FIRMATEX RK see DTG000
FIRMAZOLO see AIF000
FIRMOTOX see POO250
FIR NEEDLE OIL, SIBERIAN see FBV000
FIRON see FBJ100

FISCHER'S SOLUTION see FBV100
FISETIN see FBW000
FISH BERRY see PIE500
FISH POISON see HGL575
FISHTAIL PALM see FBW100
FISH-TOX see RNZ000
FISIOQUENS see EEH575
FISONS B25 see CEW500
FISONS NC 2964 see DSO000
FISONS NC 5016 see DGA200
FISSUCAIN see BQA010
FITIOS see DNX600
FITIOS B/77 see DNX600
FITOHEMAGLUTYNINA (POLISH) see PIB575
FITROL see KBB600
FITROL DESICCITE 25 see KBB600
FIXANOL BLACK E see AQP000
FIXANOL BLUE 2B see CMO000
FIXANOL VIOLET N see CMP000
FIXAPRET CP see DTG000
FIXATIVE IS see CNH125
FIXER IS see CNH125
FIXOL see CMS850
FK 235 see NDY650
FK 749 see EBE100
FK 1160 see CJH750
F KLOT see AJP250
FL see MQR225
FL-1039 see MCB550
FLACAVON R see TNC500
FLACETHYLE see ELH600
FLAGECIDIN see AOY000
FLAGEMONA see MMN250
FLAGESOL see MMN250
FLAGIL see MMN250
FLAGYL see MMN250
FLAMARIL see HNI500
FLAMAZINE see SNI425
FLAMENCO see TGG760
FLAME TONES see CJD500
FLAMINGO FLOWER see APM875
FLAMINGO LILY see APM875
FLAMING RED see CJD500
FLAMMEX AP see TNC500
FLAMULA (CUBA) see CMV390
FLAMYCIN see CMA750
FLANARIL see HNI500
FLANOGEN ELA see CCL250
FLAROXATE HYDROCHLORIDE see FCB100
FLAVACRIDINUM HYDROCHLORICUM see DBX400
FLAVASPIDIC ACID see FBY000
FLAVASPIDSAEURE (GERMAN) see FBY000
FLAVAXIN see RIK000
FLAVAZONE see NGE500
FLAVENSOMYCIN see FBZ000
FLAVIN see XAK000
FLAVINE see DBN400, DBX400, XAK000
FLAVIN SULPHATE see DBN400
FLAVIOFORM see DBX400
FLAVIPIN see DBX400
FLAVISEPT see DBX400
FLAVISPIDIC ACID BB see FBY000
FLAVITROL see EDW500
FLAVOFUNGIN (15810) see FCA000
FLAVOMYCELIN see LIV000
FLAVOMYCIN see MRA250
7-FLAVONE ETHYL HYDROXYACETATE see ELH600
FLAVONE-7-ETHYLOXYACETATE see ELH600
FLAVONOID AGLUCONE see FCB000
7-FLAVONOXYACETIC ACID ETHYL ESTER see ELH600
FLAVOPHOSPHOLIPOL see MRA250
FLAVOSAN see XAK000
FLAVOXATE HYDROCHLORIDE see FCB100
FLAVUMYCIN B see FCC000
FLAVUROL see MCV000
FLAVYLIUM, 3,4',5,7-TETRAHYDROXY-3',5'-DIMETHOXY-, CHLORIDE
    see MAO600
FLAXEDIL see PDD300
FLAX OLIVE see LAR500
FLEBOCORTID see HHR000
FLECK-FLIP see TIO750
FLECTOL H see TLP500
FLEET-X see TLM050
FLEET-X-DV-99 see TLM000

FLEUR SUREAU (CANADA, HAITI) see EAI100
FLEXAL see IPU000
FLEXAMINE G see BLE500
FLEXARTAL see IPU000
FLEXARTEL see IPU000
FLEXAZONE see BRF500
FLEXERIL see DPX800
FLEXIBAN see DPX800
FLEXILON see AJF500
FLEXIMEL see DVL700
FLEXIN see AJF500
FLEXOL A 26 see AEO000
FLEXOL DOP see DVL700
FLEXOL EP-8 see FAB900
FLEXOL 4GO see FCD512
FLEXOL PLASTICIZER 810 see FCN050
FLEXOL PLASTICIZER CC-55 see FCN000
FLEXOL PLASTICIZER DIP see ILR100
FLEXOL PLASTICIZER DOP see DVL700
FLEXOL PLASTICIZER 3GO see FCD560
FLEXOL PLASTICIZER TCP see TNP500
FLEXOL TOF see TNI250
FLEXZONE 3C see PFL000
FLIBOL E see TIQ250
FLIEGENTELLER see TIQ250
FLINDIX see CCK125
FLINT see SCI500, SCJ500
FLIT 406 see CBG000
FLOCOOL 180 see SJC500
FLOCOR see PKQ059
FLOCTAFENINE see TKG000
FLOGAR see OLM300
FLO-GARD see SCH000
FLOGENE see CKI750
FLOGHENE see HNI500
FLOGICID see BPP750
FLOGINAX see MFA500
FLOGISTIN see HNI500
FLOGITOLO see HNI500
FLOGOCID N PLASTIGEL see BPP750
FLOGODIN see HNI500
FLOGORIL see HNI500
FLOGOS see SOX875
FLOGOSTOP see HNI500
FLO-MOR see PAI000
FLOMORE see BSQ750
FLOMOXEF see FCN100
FLOMOXEF SODIUM see FCN100
FLOPIRINA see HNI500
FLO PRO T SEED PROTECTANT see TFS350
FLO PRO V SEED PROTECTANT see CCC500
FLOR de ADONIS (CUBA) see PCU375
FLORALTONE see GEM000, TKQ250
FLOR de BARBERO (CUBA) see AFQ625
FLOR de CAMARON (MEXICO) see CAK325
FLOR de CULEBRA (PUERTO RICO) see APM875
FLORDIMEX see CDS125
FLOREL see CDS125
FLORES MARTIS see FAU000
FLORIDA ARROWROOT see CNH789
FLORIDA HOLLY see PCB300
FLORIDIN see TEY000
FLORIDINE see SHF500
FLORIMYCIN see VQZ000
FLORINEF see FHH100
FLORIPONDIO (PUERTO RICO) see AOO825
FLOROCID see SHF500
FLOROMYCIN see VQZ000
FLORONE see DKF125
FLORONE (ITALIAN) see XQJ000
FLOROPIPAMIDE see FHG000
FLOROPIPETON see FLN000
FLOROPRYL see IRF000
FLOROXENE see TKB250
FLOR del PERU see YAK350
FLOSIN see IDA400
FLOSINT see IDA400
FLOU see POF500
FLOUVE OIL see FDA000
FLOVACIL see DKI600
FLOWER FENCE see CAK325
FLOWERS of ANTIMONY see AQF000
FLOWERS of SULPHUR (DOT) see SOD500
FLOWERS of ZINC see ZKA000

FLOXACILLIN SODIUM see FDA100
FLOXACILLIN SODIUM MONOHYDRATE see CHJ000
FLOXACIN see BAB625
FLOXAPEN see CHJ000
FLOXAPEN SODIUM see FDA100
FLOXURIDIN see DAR400
FLOXURIDINE see DAR400
FLOZENGES see SHF500
FLUALAMIDE see FDA875
FLUANISON see HAF400
FLUANISONE see HAF400
FLUANISONE HYDROCHLORIDE see FDA880
FLUANXOL see FMO129
FLUATE see TIO750
FLUBENDAZOLE see FDA887
FLUCHLORALIN see FDA900
FLUCINAR see SPD500
FLUCLOXACILLIN SODIUM see FDA100
FLUCLOXACILLIN SODIUM MONOHYDRATE see CHJ000
FLUCLOXACILLIN SODIUM SALT see FDA100
FLUCORT see FDD075, SPD500
FLUCORTICIN see FDD075
FLUCORTOLONE see FDA925
FLUCYTOSINE see FHI000
FLUDERMA see FDB000
FLUDEX see IBV100
FLUDIAZEPAM see FDB100
FLUDILAT see BAV250
FLUDROCORTISONE see FHH100
FLUDROCORTONE see FHH100
FLUE DUST, ARSENIC containing see ARE500
FLUE GAS see CBW750
FLUENETIL see FDB200
FLUENYL see FDB200
FLUFENAMIC ACID see TKH750
FLUFENAMINSAURE (GERMAN) see TKH750
FLUGERIL see FMR050
FLUIBIL see CDL325
FLUID-EXTRACT of JAMAICA GINGER U.S.P. see JBA000
FLUIFORT see CBR675
FLUIMUCETIN see ACH000
FLUIMUCIL see ACH000
FLUITRAN see HII500
FLUKOIDS see CBY000
FLUMAMINE see DQR600
FLUMARK see EAU100
FLUMEN see CLH750
FLUMESIL see BEQ625
FLUMETHASONE see FDD075
FLUMETHIAZIDE see TKG750
FLUMICIL see ACH000
FLUMOPERONE HYDROCHLORIDE see TKK750
FLUNARIZINE DIHYDROCHLORIDE see FDD080
FLUNARIZINE HYDROCHLORIDE see FDD080
FLUNIGET see DKI600
FLUNISOLIDE see FDD085
FLUNITRAZEPAM see FDD100
FLUOCINOLIDE see FDD150
FLUOCINOLONE ACETONIDE see SPD500
FLUOCINOLONE 16,17-ACETONIDE see SPD500
FLUOCINOLONE ACETONIDE ACETATE see FDD150
FLUOCINOLONE ACETONIDE-21-ACETATE see FDD150
FLUOCINONIDE see FDD150
FLUOCORTOLON see FDA925
FLUOCORTOLONE see FDA925
FLUODROCORTISONE see FHH100
FLUOHYDRISONE see FHH100
FLUOHYDROCORTISONE see FHH100
FLUO-KEM see TAI250
FLUOMETURON see DUK800
FLUOMINE see EIS000
FLUOMINE DUST see EIS000
FLUON see TAI250
FLUOOXENE see TKB250
FLUOPERAZINE see TKE500, TKK250
FLUOPERIDOL see FLU000
FLUOPHOSGENE see CCA500
FLUOPHOSPHORIC ACID DI(DIMETHYLAMIDE) see BJE750
FLUOPHOSPHORIC ACID, DIETHYL ESTER see DJJ400
FLUOPHOSPHORIC ACID, DIISOPROPYL ESTER see IRF000
FLUOPHOSPHORIC ACID, DIMETHYL ESTER see DSA800
FLUOR (DUTCH, FRENCH, GERMAN, POLISH) see FEZ000
FLUORACETATO di (ITALIAN) see SHG500
5-FLUORACIL (GERMAN) see FMM000

FLUORACIZINE see FDE000
FLUORAKIL 100 see FFF000
FLUORAL see SHF500
FLUORAL HYDRATE see TJZ000
FLUORAMIDE see FFU000
3,6-FLUORANDIOL see FEV000
3′,6′-FLUORANDIOL see FEV000
FLUORANE 114 see FOO509
4-FLUORANILIN see FFY000
FLUORANTHENE see FDF000
N-FLUORANTHEN-3-YLACETAMIDE see AAK400
N-3-FLUORANTHENYLACETAMIDE see AAK400
FLUORAPATITE see FDH000
o-FLUORBENZOESAEURE (GERMAN) see FGH000
FLUORCORTOLONE see FDA925
5-FLUOR-DESOXYCYTIDIN (GERMAN) see FHO000
FLUOREN-2-AMINE see FDI000
2-FLUORENAMINE see FDI000
FLUOREN-9-AMINE, N-(2-CHLOROETHYL)-N-ETHYL-, HYDROCHLORIDE
   see FEE100
9H-FLUOREN-9-AMINE, N-(2-CHLOROETHYL)-N-ETHYL-, HYDROCHLO-
   RIDE (9CI) see FEE100
2-FLUORENEAMINE see FDI000
9-FLUORENECARBOXYLATE-3-QUINUCLIDINOL HYDROCHLORIDE
   see FDK000
FLUORENE-9-CARBOXYLIC ACID-2-(DIETHYLAMINO)ETHYL ESTER
   see CBS250
FLUORENE-9-CARBOXYLIC ACID-3-QUINUCLIDINYL ESTER see FDK000
FLUORENE-2,7-DIAMINE see FDM000
2,7-FLUORENEDIAMINE see FDM000
9H-FLUORENE, 1,9-DIMETHYL- see DRY100
1,1′-(9H-FLUORENE-2,7-DIYL)BIS(2-(DIETHYLAMINO)ETHANONE) DIHYDRO-
   CHLORIDE TRIHYDRATE see FDN000
FLUOREN-9-ONE see FDO000
9-FLUORENONE see FDO000
9H-FLUOREN-9-ONE see FDO000
FLUORENO(9,1-gh)QUINOLINE see FDP000
1-FLUORENYLACETAMIDE see FDQ000
2-FLUORENYLACETAMIDE see FDR000
3-FLUORENYL ACETAMIDE see FDS000
N-FLUOREN-1-YL ACETAMIDE see FDQ000
N-1-FLUORENYLACETAMIDE see FDQ000
N-FLUOREN-2-YL ACETAMIDE see FDR000
N-2-FLUORENYLACETAMIDE see FDR000
N-3-FLUORENYL ACETAMIDE see FDS000
N-FLUOREN-3-YL ACETAMIDE see FDS000
N-FLUOREN-4-YLACETAMIDE see ABY000
N-4-FLUORENYLACETAMIDE see ABY000
N-9H-FLUOREN-3-YL ACETAMIDE see FDS000
1-(N-2′-FLUORENYLACETAMIDO-2-ACETYLAMINO)FLUORENE see AAK500
1-FLUORENYL ACETHYDROXAMIC ACID see FDT000
FLUORENYL-2-ACETHYDROXAMIC ACID see HIP000
3-FLUORENYL ACETHYDROXAMIC ACID see FDU000
N-(FLUOREN-2-YL)ACETOHYDROXAMIC ACETAMIDE see ABL000
N-(FLUOREN-3-YL)ACETOHYDROXAMIC ACETATE see ABO250
N-(FLUOREN-4-YL)ACETOHYDROXAMIC ACETATE see ABO500
N-FLUOREN-1-YL ACETOHYDROXAMIC ACID see FDT000
N-FLUOREN-2-YL ACETOHYDROXAMIC ACID see HIP000
N-2-FLUORENYL ACETOHYDROXAMIC ACID see HIP000
N-FLUOREN-3-YL ACETOHYDROXAMIC ACID see FDU000
N-FLUOREN-2-YLACETOHYDROXAMIC ACID, COBALT(2+) COMPLEX
   see FDU875
N-FLUOREN-2-YL ACETOHYDROXAMIC ACID, COPPER(2+) COMPLEX
   see HIP500
N-FLUOREN-2-YL ACETOHYDROXAMIC ACID, IRON(3+) COMPLEX
   see HIQ000
N-FLUOREN-2-YL ACETOHYDROXAMIC ACID, MANGANESE(2+) COM-
   PLEX see HIR000
N-FLUOREN-2-YL ACETOHYDROXAMIC ACID, NICKEL(2+) COMPLEX
   see HIR500
N-FLUOREN-2-YL ACETOHYDROXAMIC ACID, POTASSIUM SALT
   see HIS000
N-FLUOREN-2-YL ACETOHYDROXAMIC ACID SULFATE see FDV000
N-FLUOREN-2-YLACETOHYDROXAMIC ACID, ZINC COMPLEX see ZHJ000
N-(2-FLUORENYL)BENZAMIDE see FDX000
N-FLUOREN-2-YL BENZAMIDE see FDX000
N-9H-FLUOREN-2-YL-BENZAMIDE (9CI) see FDX000
N-FLUOREN-1-YL BENZOHYDROXAMIC ACID see FDY000
N-FLUOREN-2-YL BENZOHYDROXAMIC ACID see FDZ000
N-(2-FLUORENYL)BENZOHYDROXAMIC ACID see FDZ000
N-FLUOREN-2-YL BENZOHYDROXAMIC ACID ACETATE see ABO750
2,7-FLUORENYLBISACETAMIDE see BGP250
N,N′-FLUOREN-2,7-YLBISACETAMIDE see BGP250
N-FLUORENYLCYCLOBUTANECARBOXAMIDE see COW500

2-FLUORENYLDIACETAMIDE see DBF200
N-FLUOREN-1-YLDIACETAMIDE see DBF000
N-1-FLUORENYLDIACETAMIDE see DBF000
N-FLUOREN-2-YLDIACETAMIDE see DBF200
N-2-FLUORENYLDIACETAMIDE see DBF200
2-FLUORENYLDIMETHYLAMINE see DPJ600
2,5-FLUORENYLENEBISACETAMIDE see BGR250
N,N'-FLUOREN-2,5-YLENEBISACETAMIDE see BGR250
N,N'-FLUOREN-2,7-YLENEBISACETAMIDE see BGP250
N,N'-2,7-FLUORENYLENEBISACETAMIDE see BGP250
N,N'-(FLUOREN-2,7-YLENE)BIS(ACETYLAMINE) see BGP250
N,N'-FLUOREN-2,7-YLENE BIS(TRIFLUOROACETAMIDE) see FEE000
N,N'-2,7-FLUORENYLENEDIACETAMIDE see BGP250
N-(9-FLUORENYL)-N-ETHYL-β-CHLOROETHYLAMINE HYDROCHLORIDE
    see FEE100
N-FLUOREN-2-YL FORMAMIDE see FEF000
N,2-FLUORENYL FORMAMIDE see FEF000
N-(2-FLUORENYL)FORMOHYDROXAMIC ACID see FEG000
N-9H-FLUOREN-2-YL-N-HYDROXYBENZAMIDE see FDZ000
2-FLUORENYL HYDROXYLAMINE see HIU500
3-FLUORENYLHYDROXYLAMINE see FEH000
N-FLUOREN-2-YLHYDROXYLAMINE see HIU500
N-FLUOREN-3-YL HYDROXYLAMINE see FEH000
N-FLUOREN-2-YLHYDROXYLAMINE-o-GLUCURONIDE see FEI000
N-2-FLUORENYLHYDROXYLAMINE-o-GLUCURONIDE see FEI000
2-FLUORENYLMONOMETHYLAMINE see FEI500
N-(2-FLUORENYL)MYRISTOHYDROXAMIC ACID ACETATE see FEM000
N-(2-FLUORENYL)PHTHALAMIC ACID see FEN000
N-FLUORENYL-2-PHTHALIMIC ACID see FEN000
N-(2-FLUORENYL)PROPIONOHYDROXAMIC ACID see FEO000
N-2-FLUORENYL SUCCINAMIC ACID see FEP000
N-(FLUOREN-2-YL)-o-TETRADECANOYLACETOHYDROXAMIC ACID
    see ACS000
N-FLUOREN-2-YL-N-TETRADECANOYLHYDROXAMIC ACID see HMU000
N-FLUOREN-2-YL-2,2,2-TRIFLUOROACETAMIDE see FER000
N-(2-FLUORENYL)-2,2,2-TRIFLUOROACETAMIDE see FER000
FLUORESCEIN see FEV000
FLUORESCEIN, soluble see FEW000
FLUORESCEIN, 4',5'-DIBROMO- see DDO200
FLUORESCEINE see FEV000
FLUORESCEIN MERCURIACETATE see FEV100
FLUORESCEIN MERCURIC ACETATE see FEV100
FLUORESCEIN MERCURY ACETATE see FEV100
FLUORESCEIN SODIUM see FEW000
FLUORESCEIN SODIUM B.P see FEW000
FLUORESCEIN, 2',4',5',7'-TETRABROMO-, DISODIUM SALT see BNK700
FLUORESCENT BRIGHTENER 46 see TGE155
FLUORESONE see FLG000
FLUORESSIGAEURE (GERMAN) see SHG500
1-p-FLUORFENYL-3,3-DIMETHYLTRIAZEN (CZECH) see DSA600
FLUORIDE see FEX875
FLUORIDE(1-) see FEX875
FLUORIDE ION see FEX875
FLUORIDE ION(1-) see FEX875
FLUORIDENT see SHF500
FLUORIDES see FEY000
FLUORID HLINITY (CZECH) see AHB000
FLUORID SODNY (CZECH) see SHF500
FLUORIGARD see SHF500
FLUORIMIDE see DKF400
FLUORINE see FEZ000
FLUORINE, compressed (DOT) see FEZ000
FLUORINE AZIDE see FFA000
FLUORINEED see SHF500
FLUORINE FLUORO SULFATE see FFB000
FLUORINE MONOXIDE see ORA000
FLUORINE NITRATE see NMU000
FLUORINE OXIDE see ORA000
FLUORINE PERCHLORATE see FFD000
FLUORINSE see SHF500
FLUORISTAN see TGD100
FLUOR-I-STRIP A.T. see FEW000
FLUORITAB see SHF500
FLUORITE see CAS000
1-FLUOR-2-JODETHAN see FIQ000
FLUOR-O-KOTE see SHF500
FLUORO (ITALIAN) see FEZ000
FLUOROACETALDEHYDE see FFE000
FLUOROACETAMIDE see FFF000
2-FLUOROACETAMIDE see FFF000
7-FLUORO-2-ACETAMIDO-FLUORENE see FFG000
FLUOROACETANILIDE see FFH000
2-FLUOROACETANILIDE see FFH000
FLUOROACETATE see FIC000

FLUOROACETIC ACID see FIC000
2-FLUOROACETIC ACID see FIC000
FLUOROACETIC ACID (DOT) see FIC000
FLUOROACETIC ACID AMIDE see FFF000
FLUOROACETIC ACID (2-ETHYLHEXYL) ESTER see FFI000
FLUOROACETIC ACID, MERCURY(II) SALT see MDB500
FLUOROACETIC ACID METHYL ESTER see MKD000
FLUOROACETIC ACID, SODIUM SALT see SHG500
FLUOROACETIC ACID, TRIETHYLLEAD SALT see TJS500
FLUOROACETONITRILE see FFJ000
FLUOROACETPHENYLHYDRAZIDE see FFK000
3-FLUORO-4-ACETYLAMINOBIPHENYL see FKX000
3'-FLUORO-4-ACETYLAMINOBIPHENYL see FKY000
4'-FLUORO-4-ACETYLAMINOBIPHENYL see FKZ000
1-FLUORO-2-ACETYLAMINOFLUORENE see FFL000
3-FLUORO-2-ACETYLAMINOFLUORENE see FFM000
4-FLUORO-2-ACETYLAMINOFLUORENE see FFN000
5-FLUORO-2-ACETYLAMINOFLUORENE see FFO000
6-FLUORO-2-ACETYLAMINOFLUORENE see FFP000
7-FLUORO-2-ACETYLAMINOFLUORENE see FFG000
8-FLUORO-2-ACETYLAMINOFLUORENE see FFQ000
p-FLUOROACETYLAMINOPHENYL DERIVATIVE of NITROGEN MUSTARD
    see BHP750
FLUOROACETYL CHLORIDE see FFR000
FLUORO ACETYLENE see FFS000
o-(FLUOROACETYL)SALICYLIC ACID see FFT000
FLUORO-β-ALANINE HYDROCHLORIDE see FFT100
α-FLUORO-β-ALANINE HYDROCHLORIDE see FFT100
FLUOROAMINE see FFU000
4'-FLUORO-4-AMINODIPHENYL see AKC500
5-FLUORO AMYLAMINE see FFV000
5-FLUOROAMYL CHLORIDE see FFW000
5-FLUOROAMYL THIOCYANATE see FFX000
4-FLUOROANILINE see FFY000
p-FLUOROANILINE see FFY000
4-FLUOROBENZANTHRACENE see FFZ000
4-FLUOROBENZ(a)ANTHRACENE see FFZ000
4'-FLUORO-1,2-BENZANTHRACENE see FFZ000
4-FLUOROBENZENAMINE see FFY000
FLUOROBENZENE see FGA000
4-FLUOROBENZENEACETONITRILE see FLC000
4-FLUOROBENZENEARSONIC ACID see FGA100
2-FLUORO-BENZO(e)(1)BENZOTHIOPYRANO(4,3-b)INDOLE see FGB000
3-FLUORO-BENZO(e)(1)BENZOTHIOPYRANO(4,3-b)INDOLE see FGD000
4-FLUORO-BENZO(e)(1)BENZOTHIOPYRANO(4,3-b)INDOLE see FGF000
4-FLUORO-BENZO(g)(1)BENZOTHIOPYRANO(4,3-b)INDOLE see FGG000
o-FLUOROBENZOIC ACID see FGH000
1-(4'-FLUOROBENZOIL)-3-PIRROLIDINOPROPANO MALEATO (ITALIAN)
    see FGW000
3-FLUOROBENZO(rst)PENTAPHENE see FGI000
6-FLUOROBENZO(a)PYRENE see FGI100
2-FLUORO-(1)BENZOTHIOPYRANO(4,3-b)INDOLE see FGJ000
4-FLUORO-(1)BENZOTHIOPYRANO(4,3-b)INDOLE see FGL000
4-FLUORO-6H-(1)BENZOTHIOPYRANO(4,3-b)QUINOLINE see FGO000
(5-(4-FLUOROBENZOYL)-1H-BENZIMIDAZOLE-2-YL)CARBAMIC ACID
    METHYL ESTER see FDA887
2-FLUOROBENZOYL CHLORIDE see FGP000
o-FLUOROBENZOYL CHLORIDE see FGP000
1'-(3-(p-FLUOROBENZOYL)PROPYL)(1,4'-BIPIPERIDINE 1-4'-CARBOXAMIDE
    see FHG000
1-(3-p-FLUOROBENZOYLPROPYL)-4-p-CHLOROPHENYL-4-
    HYDROXYPIPERIDINE see CLY500
2-(3-(p-FLUOROBENZOYL)-1-PROPYL)-5α,9-α-DIMETHYL-2'-HYDROXY-6,7-
    BENZOMORPHAN see FGQ000
3-(γ-(p-FLUOROBENZOYL)PROPYL)-2,3,4,4a,5,6-HEXAHYDRO-1(H)-
    PYRAZINO(1,2A)QUINOLINE HCl see CNH500
8-(3-p-FLUOROBENZOYL-1-PROPYL)-4-OXO-1-PHENYL-1,3,8-
    TRIAZASPIRO(4,5)DECANE see SLE500
8-(3-(p-FLUOROBENZOYL)PROPYL)-1-PHENYL-1,3,8-
    TRIAZASPIRO(4.5)DECAN-4- ONE see SLE500
4-(3-(p-FLUOROBENZOYL)PROPYL)-1-PIPERAZINOCARBOXYLIC ACID
    CYCLOHEXYLESTER see FLK000
1-(3-(p-FLUOROBENZOYL)PROPYL)-4-PIPERIDINOISONIPACOTAMIDE
    see FHG000
1-(1-(3-(p-FLUOROBENZOYL)PROPYL)-4-PIPERIDYL)-2-
    BENZIMIDAZOLINETHIONE see TGB175
1-(1-(3-(p-FLUOROBENZOYL)PROPYL)-4-PIPERIDYL)-2-
    BENZIMIDAZOLINONE see FLK100
1-(1-(3-(p-FLUOROBENZOYL)PROPYL)-4-PIPERIDYL)-2-
    BENZIMIDAZOLINONE, HYDROCHLORIDE MONOHYDRATE
    see FGU000
1-(3-(4-FLUOROBENZOYL)PROPYL)-4-PIPERIDYL-N-ISOPROPYL CARBA-
    MATE see FGV000
1-(3-(4-FLUOROBENZOYL)PROPYL)-4-(2-PYRIDYL)PIPERAZINE see FLU000

1-(1-(3-(p-FLUOROBENZOYL)PROPYL)-1,2,3,6-TETRAHYDRO-4-PYRIDYL)-2-BENZIMIDAZOLINONE see DYF200

1-(4'-FLUOROBENZOYL)-3-PYRROLIDINYLPROPANE MALEATE see FGW000

4-FLUOROBENZYLCYANIDE see FLC000

p-FLUOROBENZYL CYANIDE see FLC000

1-(p-FLUOROBENZYL)-2-((1-(2-(p-METHOXYPHENYL)ETHYL)PIPERID-4-YL)AMINO)BENZIMIDAZOLE see ARP675

N-4-(4'-FLUORO)BIPHENYLACETAMIDE see FKZ000

4'-FLUORO-4-BIPHENYLAMINE see AKC500

N-(4'-FLUORO-4-BIPHENYLYL)ACETAMIDE see FKZ000

2-(2-FLUORO-4-BIPHENYLYL)PROPIONIC ACID see FLG100

FLUOROBISISOPROPYLAMINO- PHOSPHINE OXIDE see PHF750

5-FLUORO-1,3-BIS(TETRAHYDRO-2-FURANYL)-2,4(1H,3H)-PYRIMIDINEDIONE see BLH325

FLUOROBIS(TRIFLUOROMETHYL)PHOSPHINE see FGW100

FLUOROBLASTIN see FMM000

1-FLUORO-2-BROMOBENZENE see FGX000

1-FLUORO-3-BROMOBENZENE see FGY000

6-FLUORO-7-BROMOMETHYLBENZ(a)ANTHRACENE see BNQ250

4-FLUOROBUTYL BROMIDE see FHA000

4-FLUOROBUTYL CHLORIDE see FHB000

4-FLUOROBUTYL IODIDE see FHC000

4-FLUOROBUTYL THIOCYANATE see FHD000

4-FLUORO-BUTYRIC ACID-2-CHLOROETHYL ESTER see CGZ000

4-FLUOROBUTYRIC ACID METHYL ESTER see MKE000

4-FLUOROBUTYRONITRILE see FHF000

γ-FLUOROBUTYRONITRILE see FHF000

FLUOROBUTYROPHENONE see FHG000

FLUOROCARBON-12 see DFA600

FLUOROCARBON-22 see CFX500

FLUOROCARBON 113 see FOO000

FLUOROCARBON 114 see FOO509

FLUOROCARBON-115 see CJI500

FLUOROCARBON FC142b see CFX250

FLUOROCARBON No. 11 see TIP500

FLUOROCHLOROCARBON LIQUID see FHH000

5-FLUORO-7-CHLOROMETHYL-12-METHYLBENZ(a)ANTHRACENE see FHH025

FLUOROCHROME see MCV000

FLUOROCORTISONE see FHH100

4-FLUORO-CROTONIC ACID METHYL ESTER see MKE250

2-FLUORO-2'-CYANODIETHYL ETHER see CON500

5-FLUOROCYSTOSINE see FHI000

5-FLUOROCYTOSINE see FHI000

FLUORO-DDT see FHJ000

1-FLUORODECANE see FHL000

10-FLUORODECANOL see FHM000

ω-FLUORODECANOL see FHM000

10-FLUORODECYL CHLORIDE see FHN000

5-FLUORODEOXYCYTIDINE see FHO000

5-FLUORO-2'-DEOXYCYTIDINE see FHO000

3'-FLUORO-3'-DEOXYTHYMIDINE see DAR200

FLUORODEOXYURIDINE see DAR400

5-FLUORODEOXYURIDINE see DAR400

5-FLUORO-2-DEOXYURIDINE see DAR400

5-FLUORO-2'-DEOXYURIDINE see DAR400

β-5-FLUORO-2'-DEOXYURIDINE see DAR400

6-α-FLUORODEXAMETHASONE see FDD075

6-FLUORODIBENZ(a,h)ANTHRACENE see FHP000

4-FLUORO-1,2:5,6-DIBENZANTHRACENE see FHP000

FLUORODICHLOROMETHANE see DFL000

FLUORODIFEN see NIX000

9-α-FLUORO-11-β,21-DIHYDROXY-16-α-ISOPROYLIDENEDIOXY-1,4-PREGNADIENE, 3,20-DIONE see AQX500

9-α-FLUORO-11-β,17-β-DIHYDROXY-17-α-METHYL-4-ANDROSTENE-3-ONE see AOO275

FLUORO-9-α DIHYDROXY-11-β,17-β METHYL-17-α ANDROSTENE-4 ONE-3 (FRENCH) see AOO275

9-FLUORO-11-β-,17-β-DIHYDROXY-17-METHYLANDROST-4-EN-3-ONE see AOO275

9-FLUORO-11-β,21-DIHYDROXY-16-α-METHYLPREGNA-1,4-DIENE-3,20-DIONE see DBA875

6-α-FLUORO-11-β,21-DIHYDROXY-16-α-METHYLPREGNA-1,4-DIENE-3,20-DIONE see FDA925

9-α-FLUORO-11-β,17-DIHYDROXY-3-OXO-4-ANDROSTENE-17-α-PROPIONIC ACID POTASSIUM see CCP750

FLUORODIISOPROPYL PHOSPHATE see IRF000

m-FLUORODIMETHYLAMINOAZOBENZENE see FHQ100

2-FLUORO-4-DIMETHYLAMINOAZOBENZENE see FHQ000

2'-FLUORO-4-DIMETHYLAMINOAZOBENZENE see FHQ100

3'-FLUORO-4-DIMETHYLAMINOAZOBENZENE see FHQ100

4'-FLUORO-4-DIMETHYLAMINOAZOBENZENE see DSA000

4'-FLUORO-p-DIMETHYLAMINOAZOBENZENE see DSA000

4'-FLUORO-N,N-DIMETHYL-4-AMINOAZOBENZENE see DSA000

2'-FLUORO-4-DIMETHYLAMINOSTILBENE see FHU000

4'-FLUORO-4-DIMETHYLAMINOSTILBENE see FHV000

10-FLUORO-9,12-DIMETHYLBENZ(a)ACRIDINE see FHR000

3-FLUORO-2,10-DIMETHYL-5,6-BENZACRIDINE see FHR000

1-FLUORO-7,12-DIMETHYLBENZ(a)ANTHRACENE see FHS000

4-FLUORO-7,12-DIMETHYLBENZ(a)ANTHRACENE see DRY400

5-FLUORO-7,12-DIMETHYLBENZ(a)ANTHRACENE see DRY600

8-FLUORO-7,12-DIMETHYLBENZ(a)ANTHRACENE see DRY800

11-FLUORO-7,12-DIMETHYLBENZ(a)ANTHRACENE see DRZ000

4'-FLUORO-N,N-DIMETHYL-p-PHENYLAZOANILINE see DSA000

2'-FLUORO-N,N-DIMETHYL-4-STILBENAMINE see FHU000

4'-FLUORO-N,N-DIMETHYL-4-STILBENAMINE see FHV000

1,2,4-FLUORODINITROBENZENE see DUW400

1-FLUORO-2,4-DINITROBENZENE see DUW400

1-FLUORO-1,1-DINITRO-2-BUTENE see FHV300

2-FLUORO-1,1-DINITROETHANE see FHV800

2-FLUORO-2,2-DINITROETHANOL see FHW000

2-FLUORO-2,2-DINITROETHYLAMINE see FHX000

FLUORO DINITROMETHANE see FHY000

FLUORO DINITROMETHYL AZIDE see FHZ000

1-FLUORO-1,1-DINITRO-2-PHENYLETHANE see FHZ200

12-FLUORO DODECANO NITRILE see FIA000

2,7-FLUOROENEDIAMINE see FDM000

4-FLUOROESTRADIOL see FIA500

4-FLUOROESTRA-1,3,5-(10)-TRIENE-3,17-β-DIOL see FIA500

FLUOROETHANE see FIB000

FLUOROETHANOIC ACID see FIC000

FLUOROETHANOL see FID000

2-FLUOROETHANOL see FIE000

β-FLUOROETHANOL see FIE000

2-FLUOROETHANOL, PHOSPHITE (3:1) see PHO250

FLUOROETHENE see VPA000

FLUOROETHYL see HDC000

β-FLUOROETHYL-N-(β-CHLOROETHYL)-N-NITROSOCARBAMATE see FIH000

1-FLUOROETHYL-3-CYCLOHEXYL-1-NITROSOUREA see CPL750

1-(2-FLUOROETHYL)-3-CYCLOHEXYL-1-NITROSOUREA see FIJ000

FLUOROETHYL-O,O-DIETHYLDITHIOPHOSPHORYL-1-PHENYLACETATE see FIK000

FLUOROETHYLENE see VPA000

FLUOROETHYLENE OZONIDE see FIK875

2-FLUOROETHYL ESTER DIPHENYLACETIC ACID see FIP999

2-FLUOROETHYL FLUOROACETATE see FIM000

β-FLUOROETHYL FLUOROACETATE see FIM000

2-FLUORO ETHYL-γ-FLUORO BUTYRATE see FIN000

β-FLUOROETHYL-γ-FLUOROBUTYRATE see FIN000

2'-FLUOROETHYL-6-FLUOROHEXANOATE see EKJ500

2-FLUOROETHYL-5-FLUOROHEXOATE see FIO000

(8R)-8-(2-FLUOROETHYL)-3-α-HYDROXY-1-α-H,5-α-H-TROPANIUM BROMIDE BENZILATE H2O see FMR300

β-FLUOROETHYLIC ESTER of XENYLACETIC ACID see FIP999

2-FLUOROETHYL IODIDE see FIQ000

2-FLUOROETHYL MERCAPTOPHENYLACETATE-O,O-DIETHYL PHOSPHORODITHIOATE see FIK000

2-FLUOROETHYL-N-METHYL-N-NITROSOCARBAMATE see FIS000

1-(2-FLUOROETHYL)-1-NITROSO-UREA see NKG000

1-FLUORO-2-FAA see FFL000

3-FLUORO-2-FAA see FFM000

4-FLUORO-2-FAA see FFN000

5-FLUORO-2-FAA see FFO000

6-FLUORO-2-FAA see FFP000

7-FLUORO-2-FAA see FIT200

8-FLUORO-2-FAA see FFQ000

FLUOROFLEX see TAI250

N-(7-FLUOROFLUORENE-2-YL)ACETAMIDE see FFG000

N-(1-FLUOROFLUOREN-2-YL)ACETAMIDE see FFL000

N-(3-FLUOROFLUOREN-2-YL)ACETAMIDE see FFM000

N-(4-FLUOROFLUOREN-2-YL)ACETAMIDE see FFN000

N-(5-FLUOROFLUOREN-2-YL)ACETAMIDE see FFO000

N-(6-FLUOROFLUOREN-2-YL)ACETAMIDE see FFP000

N-(8-FLUOROFLUOREN-2-YL)ACETAMIDE see FFQ000

7-FLUORO-2-N-(FLUORENYL)ACETHYDROXAMIC ACID see FIT200

7-FLUORO-N-(FLUOREN-2-YL)ACETOHYDROXAMIC ACID see FIT200

N-(7-FLUOROFLUOREN-2-YL)ACETOHYDROXAMIC ACID see FIT200

4'-FLUORO-4-(8-FLUORO-2,3,4,5-TETRAHYDRO-1H-PYRIDO(4,3-b)INDOL-2-YL)BUTYROPHENONE HYDROCHLORIDE see FIW000

FLUOROFORM see CBY750

FLUOROFORMYL FLUORIDE see CCA500

FLUOROFORMYLON see FDB000

FLUOROFUR see FMB000

FLUOROGESAROL see FHJ000

1-FLUOROHEPTANE see FIX000

7-FLUOROHEPTANONITRILE see FIY000

7-FLUOROHEPTYLAMINE see FIZ000

FLUOROHEXANE see FJA000

1-(1-(4-(p-FLUOROPHENYL-4-OXOBUTYL)-1,2,3,6-TETRAHYDRO-4-PYRIDYL)-2-BENZIMIDAZOLINONE see DYF200
1-(4-FLUOROPHENYL)-4-(4-PHENYL-1-PIPERAZINYL)-1-BUTANONE see FLL000
4'-FLUORO-4-(1-(4-PHENYL)PIPERAZINO)BUTYROPHENONE see FLL000
1-(4-FLUOROPHENYL)-4-(4-(2-PYRIDINYL)-1-PIPERAZINYL)-1-BUTANONE see FLU000
FLUOROPHOSGENE see CCA500
FLUOROPHOSPHORIC ACID, anhydrous see PHJ250
4'-FLUORO-4-N-PIPERIDINO-4-CARBAMIDOPIPERIDINO)BUTYROPHENONE see FHG000
p-FLUORO-γ-(4-PIPERIDINO-4-CARBAMOYLPIPERIDINO)BUTYROPHENONE see FHG000
4'-FLUORO-4-(4-PIPERIDINO-4-PROPIONYLPIPERIDINO)BUTYROPHENONE see FLN000
FLUOROPLAST 3 see CLQ750
FLUOROPLAST 4 see TCH500
FLUOROPLEX see FMM000
3-FLUOROPROPENE see AGG500
2-FLUORO-2-PROPEN-1-OL see FLQ000
3-FLUOROPROPIONIC ACID see FLR000
ω-FLUOROPROPIONIC ACID see FLR000
FLUOROPRYL see IRF000
2-FLUOROPYRIDINE see FLT100
4'-FLUORO-4-(4-(2-PYRIDYL)-1-PIPERAZINYL)BUTYROPHENONE see FLU000
5-FLUORO-2,4-PYRIMIDINEDIONE see FMM000
5-FLUORO-2,4(1H,3H)-PYRIMIDINEDIONE see FMM000
5-FLUORO-4(1H)-PYRIMIDINONE see FMO100
4'-FLUORO-4-(n-(4-PYRROLIDINAMIDO-4-m-TOLYPIPERIDINO)BUTYROPHENONE see FLV000
4'-FLUORO-4-(1-PYRROLIDINYL)BUTYROPHENONE MALEATE see FGW000
FLUOROSILICIC ACID see SCO500
4'-FLUORO-4-STILBENAMINE see FLY000
2-FLUORO-4-STILBENYL-N,N-DIMETHYLAMINE see FHU000
4'-FLUORO-4-STILBENYL-N,N-DIMETHYLAMINE see FHV000
FLUOROSUFONIC ACID (DOT) see FLZ000
FLUOROSULFONATES see FLY100
FLUOROSULFONYL CHLORIDE see SOT500
FLUOROSULFURIC ACID see FLZ000
FLUOROSULFURYL HYPOFLUORITE see FFB000
FLUOROTANE see HAG500
mu-FLUOROTETRAFLUORODISTANNATE(1-), SODIUM see SJA500
5-FLUORO-1,2,3,6-TETRAHYDRO-2,6-DIOXO-4-PYRIMIDINECARBOXYLIC ACID see TCQ500
5-FLUORO-1-(TETRAHYDRO-2-FURANYL)-2,4-PYRIMIDINEDIONE see FMB000
5-FLUORO-1-(TETRAHYDRO-2-FURANYL)-2,4(1H,3H)-PYRIMIDINEDIONE see FMB000
5-FLUORO-1-(TETRAHYDROFURAN-2-YL)URACIL see FMB000
5-FLUORO-1-(TETRAHYDRO-3-FURYL)URACIL see FMB000
9-α-FLUORO-11-β,16-α,17,21-TETRAHYDROXYPREGNA-1,4-DIENE-3,20-DIONE see AQX250
9-α-FLUORO-11-β,16-α,17,21-TETRAHYDROXY-1,4-PREGNADIENE-3,20-DIONE see AQX250
9-α-FLUORO-11-β,16-α,17-α,21-TETRAHYDROXYPREGNA-1,4-DIENE-3,20-DIONE see AQX250
6-α-FLUORO-11-β,16-α,17,21-TETRAHYDROXYPREGNA-1,4-DIENE,-3,20-DIONE, CYCLIC 16,17-ACETAL with ACETONE see FDD085
9-FLUORO-11-β,16-α,17,21-TETRAHYDROXYPREGNA-1,4-DIENE-3,20-DIONE, CYCLIC 16,17-ACETAL with 21-(3,3-DIMETHYLBUTYRATE)ACETONE see AQY375
9-FLUORO-11-β,16-α,17,21-TETRAHYDROXYPREGNA-1,4-DIENE-3,20-DIONE-16,21-DIACETATE see AQX750
2-FLUOROTHIOPYRANO(4,3-b)BENZ(e)INDOLE see FGJ000
4-FLUOROTHIOPYRANO(4,3-b)BENZ(e)INDOLE see FGL000
4'-FLUORO-4-(4-(2-THIOXOBENZIMIDAZOL-1-YL)PIPERIDINO)-BUTYROPHENONE see TGB175
FLUOROTHYL see HDC000
p-FLUOROTOLUENE see FMC000
FLUOROTRIBUTYLSTANNANE see FME000
FLUOROTRICHLOROMETHANE (OSHA) see TIP500
2-FLUOROTRICYCLOQUINAZOLINE see FMF000
3-FLUORO-TRICYCLOQUINAZOLINE see FMG000
4-FLUORO-4'-TRIFLUOROMETHYLBENZOPHENONE GUANYLHYDRAZONE HYDROCHLORIDE see FMH000
9-FLUORO-11-β,17,21-TRIHYDROXY-16-α-METHYLPREGNA-1,4-DIENE-3,20-DIONE see SOW000
9-FLUORO-11-β,17,21-TRIHYDROXY-16-β-METHYLPREGNA-1,4-DIENE-3,20-DIONE see BFV750
9-α-FLUORO-11-β,17,21-TRIHYDROXY-16-β-METHYLPREGNA-1,4-DIENE- 3,20-DIONE see BFV750
9-α-FLUORO-11-β,17-α,21-TRIHYDROXY-16-α-METHYLPREGNA-1,4-DIENE-3,20-DIONE see SOW000
9-FLUORO-11-β,17,21-TRIHYDROXY-16-α-METHYLPREGNA-1,4-DIENE-3,20-DIONE ACETATE see DBC400

9-FLUORO-11-β,17,21-TRIHYDROXY-16-α-METHYLPREGNA-1,4-DIENE-3,20-DIONE-21-(DIHYDROGEN PHOSPHATE) DISODIUM SALT see DAE525
9-FLUORO-11-β,17,21-TRIHYDROXY-16-β-METHYLPREGNA-1,4-DIENE-3,20-DIONE, 21-(DIHYDROGEN PHOSPHATE), DISODIUM SALT see BFV770
9-FLUORO-11-β,17,21-TRIHYDROXY-16-α-METHYLPREGNA-1,4-DIENE-3,20-DIONE-17,21-DIPROPIONATE see DBC500
9-FLUORO-11-β,17,21-TRIHYDROXY-16-β-METHYLPREGNA-1,4-DIENE-3,20-DIONE, 17,21-DIPROPIONATE see BFV765
9-FLUORO-11-β,17,21-TRIHYDROXY-16-α-METHYLPREGNA-1,4-DIENE-3,20-DIONE-21-(HYDROGEN SULFATE), MONOSODIUM SALT see DBC550
9-FLUORO-11-β,17,21-TRIHYDROXY-16-α-METHYLPREGNA-1,4-DIENE-3,20-DIONE, 21-ISONICOTINATE see DBC510
9-FLUORO-11-β,17,21-TRIHYDROXY-16-α-METHYLPREGNA-1,4-DIENE-3,20-DIONE-21-PALMITATE see DBC525
9-FLUORO-11-β,17,21-TRIHYDROXY-16-α-METHYLPREGNA-1,4-DIENE-3,20-DIONE-17-VALERATE see DBC575
9-FLUORO-11-β,17,21-TRIHYDROXY-16-β-METHYLPREGNA-1,4-DIENE-3,20,DIONE-17-VALERATE see VCA000
9-FLUORO-11-β,17,21-TRIHYDROXYPREGN-4-ENE-3,20-DIONE see FHH100
9-α-FLUORO-11-β,17-α,21-TRIHYDROXY-4-PREGNENE-3,20-DIONE see FHH100
FLUOROTRINITROMETHANE see FMI000
3-FLUORO-1,2,4-TRIOXOLANE see FIK875
FLUOROTROJCHLOROMETAN (POLISH) see TIP500
3-FLUOROTYROSIN see FMJ000
3-FLUOROTYROSINE see FMJ000
m-FLUOROTYROSINE see FMJ000
FLUOROURACIL see FMM000
5-FLUOROURACIL see FMM000
5-FLUOROURACIL DEOXYRIBOSIDE see DAR400
5-FLUOROURACIL-2'-DEOXYRIBOSIDE see DAR400
5-FLUOROURIDINE see FMN000
5-FLUOROVALERONITRILE see FMO000
FLUOROWODOR (POLISH) see HHU500
FLUOROXENE see TKB250
4-FLUORPHENYL-1-ISOPROPYL-7-METHYL-2(1H)-CHINAZOLINON (GERMAN) see THH350
FLUORPLAST 4 see TAI250
5-FLUORPROPYRIMIDINE-2,4-DIONE see FMM000
FLUORSPAR see CAS000
FLUORTHYRIN see FMJ000
3-FLUORTYROSIN (GERMAN) see FMJ000
5-FLUORURACIL (GERMAN) see FMM000
FLUORURE de BORE (FRENCH) see BMG700
FLUORURE de N,N'-DIISOPROPYLE PHOSPHORODIAMIDE (FRENCH) see PHF750
FLUORURE de POTASSIUM (FRENCH) see PLF500
FLUORURES ACIDE (FRENCH) see FEZ000
FLUORURE de SODIUM (FRENCH) see SHF500
FLUORURE de SULFURYLE (FRENCH) see SOU500
FLUORURE de N,N,N',N'-TETRAMETHYLE PHOSPHORO-DIAMIDE (FRENCH) see BJE750
FLUORURE de THIONYLE (FRENCH) see TFL250
FLUORURI ACIDI (ITALIAN) see FEZ000
FLUORURIDINE DEOXYRIBOSE see DAR400
FLUORWASSERSTOFF (GERMAN) see HHU500
FLUORWATERSTOF (DUTCH) see HHU500
FLUORXENE see TKB250
FLUORYL see CBY750
FLUOSILICATE de ALUMINUM (FRENCH) see THH000
FLUOSILICATE de AMMONIUM (FRENCH) see COE000
FLUOSILICATE de MAGNESIUM (FRENCH) see MAG250
FLUOSILICATE de SODIUM see DXE000
FLUOSILICATE de ZINC (FRENCH) see ZIA000
FLUOSILICIC ACID see SCO500
FLUOSOL-DA 20% see FMO050
FLUOSTIGMINE see IRF000
FLUOSULFONIC ACID (DOT) see FLZ000
FLUOTESTIN see AOO275
FLUOTHANE see HAG500
FLUOTITANATE de POTASSIUM (FRENCH) see PLI000
FLUOTRACEN see DMF800
FLUOVITIF see SPD500
FLUOXIDINE see FMO100
FLUOXIMESTERONE see AOO275
FLUOXYDINE see FMO100
FLUOXYMESTERONE see AOO275
FLUOXYMESTRONE see AOO275
FLUOXYPREDNISOLONE see AQX250
FLUPENTHIXOL see FMO129
(α,β)-FLUPENTHIXOL see FMO129
cis-(Z)-FLUPENTHIXOL see FMO129
FLUPENTHIXOLE see FMO129
FLUPENTIXOL see FMO129
FLUPENTIXOL DIHYDROCHLORIDE see FMO150

FLUPENTIXOL HYDROCHLORIDE see FMO150
FLUPHENAMIC ACID see TKH750
FLUPHENAZINE see TJW500
FLUPHENAZINE DIHYDROCHLORIDE see FMP000
FLUPHENAZINE ENANTHATE see PMI250
FLUPHENAZINE HYDROCHLORIDE see FMP000
FLUPIRTINE MALEATE see FMP100
FLUPIRTIN-MALEAT (GERMAN) see FMP100
FLUPROQUAZONE see THH350
FLUPROSTENOL see ECW550
FLURACIL see FMM000
FLURA-GEL see SHF500
FLURAZEPAM see FMQ000
FLURAZEPAM HYDROCHLORIDE see DAB800
FLURAZEPAM MONOHYDROCHLORIDE see FMQ100
FLURAZEPAN DIHYDROCHLORIDE see DAB800
FLURBIPROFEN see FLG100
FLURCARE see SHF500
FLURENTIXOL see FMO129
FLURI see FMM000
FLURIL see FMM000
4'-FLURO-4-(1,2,4,4a,5,6-HEXAHYDRO-3H-PYRANZINO(1,2-A)QUINOLIN-3-YL)-
 BUTYROPHENONE 2HCl see CNH500
FLUROTHYL see HDC000
FLUROXENE see TKB250
FLUSTERON see AOO275
FLUTAMIDE see FMR050
FLUTAZOLAM see FMR075
FLUTESTOS see AOO275
FLUTONE see AQX500
FLUTOPRAZEPAM see FMR100
FLUTRA see HII500
FLUTROPIUM BROMIDE HYDRATE see FMR300
FLUVET see FDD075
FLUVIN see CFY000
FLUVOMYCIN see FMR500
FLUXANXOL see FMO129
FLUXEMA see POD750
FLUX MAAG see NDN000
FLY BAIT GRITS see SOY000
FLY-DIE see DGP900
FLY FIGHTER see DGP900
FLYPEL see DKC800
FLY TATARIA see HGK700
FM 100 see LFN000
FMA see ABU500, FEV100
FMA (analytical reagent) see FEV100
FMC 249 see AFR250
FMC-1240 see EEH600
FMC 5273 see PIX250
FMC 5462 see EAQ750
FMC 5488 see CKM000
FMC 9044 see BGB500
FMC-9102 see MQQ250
FMC 10242 see CBS275
FMC 11092 see DUM800
FMC-16388 see PNV750
FMC 17370 see BEP500
FMC 30980 see RLF350
FMC 33297 see AHJ750
FMC 41655 see AHJ750
FMC 45497 see RLF350
FMC 45806 see RLF350
F-6-NDBA see NJN300
F-NORSTEARANTHRENE see TCJ500
FNT see NDY500
FO see TCQ500
FOA see TCQ500
FOBEX see BCA000
FOCUSAN see TGB475
FOGARD see FMR700
FOGARD S see FMR700
FOIN ABSOLUTE see FMS000
FOIN COUPE see FMS000
FOLACIN see FMT000
FOLATE see FMT000
FOLBEX see DER000
FOLBEX SMOKE-STRIPS see DER000
FOLCID see CBF800
FOLCODAL see CMR100
FOLCODINE see TCY750
FOLCYSTEINE see FMT000
FOLEDRIN see FMS875
FOLETHION see DSQ000

FOLEX see TIG250
FOLIANDRIN see OHQ000
FOLIC ACID see FMT000
FOLIC ACID, 4-AMINO- see AMG750
FOLIDOL see PAK000
FOLIDOL M see MNH000
FOLIGAN see ZVJ000
FOLIKRIN see EDV000
FOLIMAT see DNX800
FOLINERIN see OHQ000
FOLINEVIN see OHQ000
FOLIONE see MND275
FOLIPEX see EDV000
FOLISAN see EDV000
FOLKS GLOVE see FOM100
FOLLESTRINE see EDV000
FOLLICLE-STIMULATING HORMONE see FMT100
FOLLICORMON see EDP000
FOLLICULAR HORMONE see EDV000
FOLLICULAR HORMONE HYDRATE see EDU500
FOLLICULIN see EDV000
FOLLICULINE BENZOATE see EDV000
FOLLICUNODIS see EDV000
FOLLICYCLIN P see EDR000
FOLLIDIENE see DAL600, DKA600
FOLLIDRIN see EDP000, EDV000
FOLLINYL see NNL500
FOLLITROPIN see FMT100
FOLLORMON see DAL600
FOLLUTEIN see CMG675
FOLOSAN see PAX000, TBR750
FOLOSAN DB-905 FUMITE see TBS000
FOLPAN see TIT250
FOLPET see TIT250
FOLSAN see TBS000
FOMAC see HCL000
FOMAC 2 see PAX000
FOM-Ca HYDRATE see CAW376
FOMINOBEN HYDROCHLORIDE see FMU000
FOM-Na see DXF600
FOMOCAINE see PDU250
p-FOMOCAINE see PDU250
FOMREZ SUL-3 see DBF800
FOMREZ SUL-4 see DDV600
FONATOL see DKA600
FONAZINE MESYLATE see FMU039
FONDAREN see DRP600
FONOFOS see FMU045
FONOLINE see MQV750
FONTARSAN see ARL000
FONTARSOL see DFX400
FONTEGO see BON325
FONTILIX see MQQ050
FONTILIZ see MQQ050
FONURIT see AAI250
FONZYLANE see BOM600
FOOD BLUE 1 see FMU059
FOOD BLUE 2 see FAE000
FOOD BLUE 3 see ADE500
FOOD BLUE DYE No. 1 see FAE000
FOOD DYE RED No. 104 see ADG250, CMM000
FOOD GREEN 2 see FAF000
FOOD RED 2 see FAG020
FOOD RED 4 see FAG050
FOOD RED 5 see HJF500
FOOD RED 6 see FMU080
FOOD RED 7 see FMU080
FOOD RED 9 see FAG020
FOOD RED 14 see FAG040
FOOD RED 15 see FAG070
FOOD RED COLOR No. 105, SODIUM SALT see RMP175
FOOD RED No. 101 see FMU070
FOOD RED No. 102 see FMU080
FOOD RED No. 104 see ADG250
FOOD RED No. 105, SODIUM SALT see RMP175
FOOD YELLOW No. 4 see FAG140
FOOL'S CICELY see FMU200
FOOL'S PARSLEY see FMU200
FOPIRTOLINA (SPANISH) see FMU225
FOPIRTOLINE HYDROCHLORIDE see FMU225
FOR see TEY000
FORAAT (DUTCH) see PGS000
FORALAMIN FUMARATE see MDP800
FORALMINE FUMARATE see MDP800

FORANE see IKS400
FORAPIN see MCB525
FORDIURAN see BON325
FORE see DXI400
FOREDEX 75 see DAA800
FORENOL see NDX500
FORHISTAL MALEATE see FMU409
FORIOD see TDE750
FORIT see ECW600
FORLEX see MLC000
FORLIN see BBQ500
FORMAGENE see PAI000
FORMAL see MAK700, MGA850
FORMALDEHYD (CZECH, POLISH) see FMV000
FORMALDEHYDE see FMV000
FORMALDEHYDE, solution (DOT) see FMV000
FORMALDEHYDE BIS(β-CHLOROETHYL) ACETAL see BID750
FORMALDEHYDE CYANOHYDRIN see HIM500
FORMALDEHYDE DIMETHYLACETAL see MGA850
FORMALDEHYDE HYDROSULFITE see FMW000
FORMALDEHYDE OXIDE POLYMER see FMW300
FORMALDEHYDE SODIUM BISULFITE ADDUCT see FMW000
FORMALDEHYDE SODIUM SULFOXYLATE see FMW000
FORMALDEHYDESULFOXYLIC ACID SODIUM SALT see FMW000
FORMAL GLYCOL see DVR800
FORMAL HYDRAZINE see FNN000
FORMALIN see FMV000
FORMALIN 40 see FMV000
FORMALIN (DOT) see FMV000
FORMALINA (ITALIAN) see FMV000
FORMALINE (GERMAN) see FMV000
FORMALINE BLACK C see AQP000
FORMALIN-LOESUNGEN (GERMAN) see FMV000
FORMALITH see FMV000
FORMAL-γ-TRIMETHYLAMMONIUM PROPANEDIOL see FMX000
FORMAMIDE see FMY000
FORMAMIDE, N-(1,1'-BIPHENYL)-4-YL-N-HYDROXY- see HLE650
FORMAMIDOBENZENE see FNJ000
FORMAMINE see HEI500
FORMANILIDE see FNJ000
FORMARIN see MNM500
FORMATRIX see ECU750
FORMEBOLONE see FNK040
FORMETANATE HYDROCHLORIDE see DSO200
FORMHYDRAZID (GERMAN) see FNN000
FORMHYDRAZIDE see FNN000
FORMHYDROXAMIC ACID see FMZ000
FORMHYDROXAMSAEURE (GERMAN) see FMZ000
FORMIATE de METHYLE (FRENCH) see MKG750
FORMIATE de PROPYLE (FRENCH) see PNM500
FORMIC ACID see FNA000
FORMIC ACID, ALLYL ESTER see AGH000
FORMIC ACID AMMONIUM SALT see ANH500
FORMIC ACID, 1-BUTYLHYDRAZIDE see BRK100
FORMIC ACID, CALCIUM SALT see CAS250
FORMIC ACID, CINNAMYL ESTER see CMR500
FORMIC ACID, CITRONELLYL ESTER see CMT750
FORMIC ACID-3,7-DIMETHYL-6-OCTEN-1-YL ESTER see CMT750
FORMIC ACID, ETHYL ESTER see EKL000
FORMIC ACID, 1-ETHYLHYDRAZIDE see EKL250
FORMIC ACID, GERANIOL ESTER see GCY000
FORMIC ACID, HEPTYL ESTER see HBO500
FORMIC ACID, HYDRAZIDE see FNN000
FORMIC ACID, ISOBUTYL ESTER see IIR000
FORMIC ACID, ISOPENTYL ESTER see IHS000
FORMIC ACID, ISOPROPYL ESTER see IPC000
FORMIC ACID, METHYLHYDRAZIDE see FNW000
FORMIC ACID, METHYLPENTYLIDENEHYDRAZIDE see PBJ875
FORMIC ACID (2-(4-METHYL-2-THIAZOLYL)HYDRAZIDE see FNB000
FORMIC ACID, NERYL ESTER see FNC000
FORMIC ACID, OCTYL ESTER see OEY100
FORMIC ACID, 1-PROPYLHYDRAZIDE see PNM650
FORMIC ACID, compounded with QUININE (1:1) see QIS300
FORMIC ALDEHYDE see FMV000
FORMIC BLACK C see AQP000
FORMIC ETHER see EKL000
FORMIC HYDRAZIDE see FNN000
FORMIC 2-(4-(5-NITROFURYL)-2-THIAZOLYL)HYDRAZIDE see NDY500
FORMILOXIN see FND100
FORMILOXINE see FND100
FORMIN see HEI500
FORMOCARBAM see FNE000
FORMOCORTAL see FDB000
FORMOHYDRAZIDE see FNN000

FORMOL see FMV000
FORMOLA 40 see DAA800
FORMOMALENIC THALLIUM see TEM399
FORMOPAN see FMW000
FORMOSA CAMPHOR see CBA750
FORMOSA CAMPHOR OIL see CBB500
FORMOSE OIL of CAMPHOR see CBB500
FORMOSULFACETAMIDE see SNP500
FORMOSULFATHIAZOLE see TEX250
FORMOTEROL FUMARATE DIHYDRATE see FNE100
FORMOTHION see DRR200
FORMPARANATE see FNE500
FORMULA 40 see DFY800
FORMVAR 1285 see AAX250
4'-FORMYLACETANILIDE THIOSEMICARBAZONE see FNF000
2-FORMYLAMINOFLUORENE see FEF000
2-FORMYLAMINO-4-(5-NITRO-2-FURYL)THIAZOLE see NGM500
2-FORMYLAMINO-4-(2-5-NITRO-2-FURYL)VINYL)-1,3-THIAZOLE see FNG000
FORMYLANILINE see FNJ000
N-FORMYLANILINE see FNJ000
6-FORMYLANTHANTHRENE see FNK000
7-FORMYLBENZ(c)ACRIDINE see BAX250
4-FORMYLBENZALDEHYDE see TAN500
p-FORMYLBENZALDEHYDE see TAN500
7-FORMYLBENZO(c)ACRIDINE see BAX250
4-FORMYLBENZONITRILE see COK250
p-FORMYLBENZONITRILE see COK250
6-FORMYLBENZO(a)PYRENE see BCT250
2-FORMYL-3:4-BENZPHENANTHRENE see BCS000
o-FORMYLCEFAMANDOLE SODIUM see FOD000
4-FORMYLCYCLOHEXENE see FNK025
5-FORMYL-1,2:3,4-DIBENZOPYRENE see DCY800
5-FORMYL-3,4:8,9-DIBENZOPYRENE see DCY600
5-FORMYL-3,4:9,10-DIBENZOPYRENE see BCQ750
N-FORMYL-N'-(3',4'-DICHLORPHENYL)-2,2,2-TRICHLORACETALDEHYDAM (GERMAN) see CDP750
FORMYLDIENOLONE see FNK040
3-FORMYL-DIGITOXIGENIN see FNK050
3-12-FORMYL-DIGOXIGENIN see FNK075
2-FORMYL-3,4-DIHYDRO-2H-PYRAN see ADR500
N-FORMYLDIMETHYLAMINE see DSB000
3'-FORMYL-N,N-DIMETHYL-4-AMINOAZOBENZENE see FNK100
p-FORMYLDIMETHYLANILINE see DOT400
α-FORMYLETHYLBENZENE see COF000
FORMYLETHYLTETRAMETHYLTETRALIN see FNK200
N-FORMYL-N-2-FLUORENYLHYDROXYLAMINE see FEG000
N-FORMYLFORMAMIDE see FNL000
16-FORMYL-GITOXIN see GES100
5-FORMYLGUAIACOL see FNM000
6-FORMYLGUAIACOL see VFP000
FORMYLHYDRAZIDE see FNN000
FORMYLHYDRAZINE see FNN000
N-FORMYLHYDRAZINE see FNN000
2-(2-FORMYLHYDRAZINO)-4-(5-NITRO-2-FURYL)THIAZOLE see NDY500
N-FORMYL HYDROXYAMINOACETIC ACID see FNO000
N-FORMYL-N-HYDROXYGLYCINE see FNO000
N-FORMYLHYDROXYLAMINE see FMZ000
2-FORMYL-11-α-HYDROXY-Δ¹-METHYLTESTOSTERONE see FNK040
FORMYLIC ACID see FNA000
1-FORMYLISOQUINOLINE THIOSEMICARBAZONE see IRV300
N-FORMYLJERVINE see FNP000
S-(2-(FORMYLMETHYLAMINO)-2-OXOETHYL)-O,O-DIMETHYLPHOSPHORODITHIOATE see DRR200
2-FORMYL-17-α-METHYLANDROSTA-1,4-DIENE-11-α,17-β-DIOL-3-ONE see FNK040
6-FORMYL-12-METHYLANTHANTHRENE see FNQ000
7-FORMYL-9-METHYLBENZ(c)ACRIDINE see FNR000
7-FORMYL-11-METHYLBENZ(c)ACRIDINE see FNS000
12-FORMYL-7-METHYLBENZ(a)ANTHRACENE see MGY500
7-FORMYL-12-METHYLBENZ(a)ANTHRACENE see FNT000
N-FORMYL-N-METHYLCARBAMOYLMETHYL-O,O-DIMETHYL PHOSPHORODITHIOATE see DRR200
S-(N-FORMYL-N-METHYLCARBAMOYLMETHYL)-O,O-DIMETHYL PHOSPHORODITHIOATE see DRR200
S-(N-FORMYL-N-METHYLCARBAMOYLMETHYL) DIMETHYL PHOSPHOROTHIOLOTHIONATE see DRR200
5-FORMYL-10-METHYL-3,4,588,9-DIBENZOPYRENE see FNV000
5-FORMYL-8-METHYL-3,4589,10-DIBENZOPYRENE see FNU000
1-FORMYL-1-METHYLHYDRAZINE see FNW000
N-FORMYL-N-METHYLHYDRAZINE see FNW000
4-FORMYL-4'-METHYL-1,1'-(OXYDIMETHYLENE)DIPYRIDINIUM, DICHLORIDE OXIME see MBZ000
N-FORMYL-N-METHYL-p-(PHENYLAZO)ANILINE see FNX000
2-FORMYL-1-METHYLPYRIDINIUM CHLORIDE OXIME see FNZ000

FOURRINE EG see ALT500
FOURRINE ERN see NAW500
FOURRINE M see TGL750
FOURRINE P BASE see ALT250
FOURRINE PG see PPQ500
FOURRINE SLA see DBO400
FOURRINE SO see CEG625
FOUR THOUSAND FORTY-NINE see MAK700
FOVANE see BDE250
FOWLER'S SOLUTION see FOM050
FOXGLOVE see DKL200, FOM100
FOY see GAD400
FOZALON see BDJ250
FP 70 see FLG100
FP-83 see FJT100
FPA see FHG000
FPF 1002 see TAB250
p-FPHE see FLF000
FPL see PIV650
FPL 670 see CNX825
FR-33 see FLJ000
FR 300 see PAU500
FR 1138 see DDQ400
FR-1923 see NMV480
FR 3068 see TGA600
FR-13,479 see EBE100
FR 34235 see NDY650
FRABEL see HNI500
FRACINE see NGE500
FRACTION AB see CKP250
FRADEMICINA see LGC200
FRADIOMYCIN SULFATE see NCG000
FRAESEOL see ENC000
FRAGIVIX see BBJ500
FRAILECILLO (CUBA) see CNR135
FRAMBINONE see RBU000
FRAMED see BJP000
FRAMYCETIN see NCF000
FRAMYCETIN SULFATE see NCD550
FRAMYCIN SULFATE see NCD550
FRANCILLADE (HAITI) see CAK325
FRANGULA EMODIN see MQF250
FRANKINCENSE GUM see OIM000
FRANKINCENSE OIL see OIM025
FRANKLIN see CAO000
FRANOCIDE see DIW200
FRANOZAN see DIW200
FRANROZE see FMB000
FRAQUINOL see NCD550
FRATOL see SHG500
FRAXINELLONE see FOM200
FRAXINUS JAPONICA Blume, bark extract see FON100
FRAZALON see AOO300
FREE ACID see SLW475
FREE BENZYLPENICILLIN see BDY669
FREE COCONUT OIL see CNR000
FREE HISTAMINE see HGD000
FREEMANS WHITE LEAD see LDY000
FREEURIL see BDE250
FREKAPHYLLIN see HLC000
FREKVEN see ICC000
FRENACTIL see FGU000, FLK100
FRENACTYL see FGU000, FLK100
FRENANTOL see ELL500
FRENASMA see CNX825
FRENCH GREEN see COF500
FRENCH JASMINE see COD675
FRENCH ROSE ABSOLUTE see RMP000
FRENOGASTRICO see DJM800, XCJ000
FRENOHYPON see ELL500
FRENOLON DIFUMARATE see MDU750
FRENOLYSE see AJV500
FRENQUEL HYDROCHLORIDE see PIY750
FRENTIROX see MCO500
FREON see CFX500
FREON 11 see TIP500
FREON 13 see CLR250
FREON 14 see CBY250
FREON 21 see DFL000
FREON 22 see CFX500
FREON 23 see CBY750
FREON 30 see MJP450
FREON 31 see CHI900
FREON 41 see FJK000

FREON 112 see TBP050
FREON 113 see FOO000
FREON 114 see FOO509
FREON 115 see CJI500
FREON 142 see CFX250
FREON 152 see ELN500
FREON 253 see TJY200
FREON 500 see DFB400
FREON 133a see TJY175
FREON 13B1 see TJY100
FREON 142b see CFX250
FREON 12-B2 see DKG850
FREON C-318 see CPS000
FREON F-12 see DFA600
FREON F-23 see CBY750
FREON MF see TIP500
FREON 113TR-T see FOO000
FRESMIN see VSZ000
FREUND'S ADJUVANT see FOO600
FRIAR'S CAP see MRE275
FRIDERON see RBF100
FRIGEN see CFX500
FRIGEN 11 see TIP500
FRIGEN 12 see DFA600
FRIGEN 114 see FOO509
FRIGEN 113a see FOO000
FRIGIDERM see FOO509
FRIJOLILLO (MEXICO) see NBR800
FRISIUM see CIR750
FROBEN see FLG100
FRP 53 see PAU500
FRUCOTE see BPY000
FRUCTOFURANOSE, TETRANICOTINATE see TDX860
FRUCTOSE (FCC) see LFI000
FRUCTUS PIPERIS LONGI see PIV650
FRUITDO see BLC250
FRUITONE see NAK000, NAK500
FRUITONE A see TAA100
FRUITONE T see TIX500
FRUIT RED A EXTRA YELLOWISH GEIGY see HJF500
FRUIT RED A GEIGY see FAG020
FRUIT SALAD PLANT see SLE890
FRUIT SUGAR see LFI000
FRUMIN AL see DXH325
FRUSEMIDE see CHJ750
FRUSEMIN see CHJ750
FRUSID see CHJ750
FRUSTAN see DCK759
FRUTABS see LFI000
FSH see FMT100
FSH-P see FMT100
FT see FPY000
F3T see TKH325
FT 8 see BEQ625
FTA see FQQ100
FTAALZUURANHYDRIDE (DUTCH) see PHW750
F1-TABS see SHF500
FTAFLEX DIBA see DNH125
FTALAN see TIT250
FTALOPHOS see PHX250
FTALOWY BEZWODNIK (POLISH) see PHW750
FTBG see FMH000
F3TDR see TKH325
FTIVAZID see VEZ925
FTIVAZIDE see VEZ925
FTORAFUR see FMB000
FTORIN see FOO875
FTORIN (PHARMACEUTICAL) see FOO875
FTORLON 4 see TAI250
FTOROPLAST 4 see TAI250
FTOROTAN (RUSSIAN) see HAG500
F-2 TOXIN see ZAT000
FT mixture with URACIL (1:4) see UNJ810
5-FU see FMM000
FUBERIDATOL see FQK000
FUBERIDAZOLE see FQK000
FUBERISAZOL see FQK000
FUBRIDAZOLE see FQK000
FUCHSIN see MAC250
p-FUCHSIN see RMK020
FUCHSINE BASE see MAC500
FUCHSINE DR-001 see RMK020
FUCHSINE SPC see RMK020
FUCIDINA see SHK000

FUCIDINE see SHK000
FUCLASIN see BJK500
FUCLASIN ULTRA see BJK500
FUDR see DAR400
5-FUDR see DAR400
FUEL OIL see FOP000
FUEL OIL #2 see DHE800
FUEL OIL, pyrolyzate see FOP100
FUGACILLIN see CBO250
FUGEREL see FMR050
FUGILIN see FOZ000
FUGOA see NNW500
FUGU POISON see FOQ000
FUJITHION see FOR000
FUKI-NO-TOH (JAPANESE) see PCR000
FUKINOTOXIN see PCQ750
FUKLASIN see BJK500
FUKLASIN ULTRA see FAS000
FULAID see FMB000
FULCIN see GKE000
FULCINE see GKE000
FULDAZIN see MQU525
FULFEEL see FMB000
FUL-GLO see FEW000
FULGRAM see BAB625
FULLSAFE see TKH750
FULMINATE of MERCURY, DRY (DOT) see MDC000
FULMINATE of MERCURY, WET (DOT) see MDC250
FULMINATES see FOS000
FULMINIC ACID see FOS050
FULSIX see CHJ750
FULUMINOL see FOS100
FULUVAMIDE see CHJ750
6-FULVENOSELONE see FOS300
FULVICAN GRISACTIN see GKE000
FULVICIN see GKE000
FULVINA see GKE000
FULVINE see FOT000
FULVISTATIN see GKE000
FUMADIL B see FOZ000
FUMAFER see FBJ100
FUMAGILLIN see FOZ000
FUMAGON see DDL800
FUMAR-F see FBJ100
FUMARIC ACID see FOU000
FUMARIC ACID, DIBUTYL ESTER see DEC600
FUMARIC ACID DIHEXYL ESTER see DKP000
FUMARIC ACID, DIISOPROPYL ESTER see DNQ200
FUMARIC ACID, DIMETHYL ESTER see DSB600
FUMARIC ACID ETHYL-2,3-EPOXYPROPYL ESTER see FOV000
FUMARINE see FOW000
FUMARONITRILE see FOX000
FUMAROYL CHLORIDE see FOY000
FUMARYLCHLORID (CZECH) see FOY000
FUMARYL CHLORIDE see FOY000
FUMAZONE see DDL800
FUMED SILICA see SCH000
FUMED SILICON DIOXIDE see SCH000
FUMETOBAC see NDN000
FUMETTE see MDR750
FUMIDIL see FOZ000
FUMIGACHLORIN see FPA000
FUMIGANT-1 (OBS.) see MHR200
FUMIGRAIN see ADX500
FUMING LIQUID ARSENIC see ARF500
FUMING SULFURIC ACID (DOT) see SOI520
FUMIRON see FBJ100
FUMITOXIN see AHE750
FUMO-GAS see EIY500
FUNDAL see CJJ250
FUNDAL 500 see CJJ250
FUNDAL SP see CJJ500
FUNDASOL see BAV575
FUNDEX see CJJ250
FUNDUSCEIN see FEW000
FUNGACETIN see THM500
FUNGAFLOR see FPB875
FUNGICHROMIN see FPC000
FUNGICHROMIN, HYDRATE see FPC000
FUNGICHTHOL see IAD000
FUNGICIDE 1991 see BAV575
FUNGICLOR see PAX000
FUNGIFEN see PAX250
FUNGIFOS see MIE250

FUNGILIN see AOC500
FUNGILON see TNH750
FUNGIMAR see CNO000
FUNGINON see FPC100
FUNGISONE see AOC500
FUNGISTOP see TGB475
FUNGITOX see PEX500
FUNGITOX OR see ABU500
FUNGIVIN see GKE000
FUNGIZONE see AOC500
FUNGOCIN see BAB750
FUNGOL B see SHF500
FUNGO-POLYCID see CDY325
FUNGOSTOP see BJK500
FUNGUS BAN TYPE II see CBG000
FUNICOLOSIN see FPD000
FUQUA see FPD100
FUR see FMN000
5-FUR see FMN000
FURACILLIN see NGE500
FURACIN see SPC500
FURACINETTEN see NGE500
FURACOCCID see NGE500
FURACORT see NGE500
FURACYCLINE see NGE500
FURADAN see CBS275
FURADANTIN see NGE000
FURADONIN see NGE000
FURADROXYL see FPE100
FURAFLUOR see FMB000
FURAL see FPQ875
FURALAZIN see FPF000
2-FURALDEHYDE see FPQ875
2-FURALDEHYDE AZINE see FPH000
2-FURALDEHYDE, 2,3:4,5-BIS(2-BUTENYLENE)TETRAHYDRO- see BHJ500
FURALDON see NGE500
FURALE see FPQ875
FURALTADONE see FPI000
FURALTADONE HYDROCHLORIDE see FPI100
l-FURALTADONE HYDROCHLORIDE see FPI150
FURAMETHRIN see POD875
FURAMON see FPY000
FURAMON IODIDE see FPY000
FURAN see FPK000
FURANACE see NDY400
FURANACE-10 see NDY400
2-FURANALDEHYDE see FPQ875
FURAN-2-AMIDOXIME see FPK100
2-FURANCARBINOL see FPU000
2-FURANCARBONAL see FPQ875
2-FURANCARBOXALDEHYDE see FPQ875
α-FURANCARBOXYLIC ACID see FQF000
FURAN-α-CARBOXYLIC ACID METHYL ESTER see MKH600
2,5-FURANDIONE see MAM000
FURANIDINE see TCR750
FURANIUM see FEW000
2-FURANMETHANETHIOL see FPM000
2-FURANMETHANOL see FPU000
2-FURANMETHYLAMINE see FPW000
FURAN-OFTENO see NGE500
FURANOL see FPY000
2-FURANPROPIONIC ACID, TETRAHYDRO-α-(1-NAPHTHYLMETHYL)-, 2-(DIETHYLAMINO)ETHYL ESTER, OXALATE (1:1) see NAE100
FURANTHRIL see CHJ750
FURANTHRYL see CHJ750
FURANTOIN see NGE000
FURANTRIL see CHJ750
2-(2-FURANYL)-1H-BENZIMIDAZOLE see FQK000
5-(3-FURANYL)-5-HYDROXY-2-PENTANONE see IGF325
N-(2-FURANYLMETHYL)-N',N'-DIMETHYL-N-2-PYRIDINYL-1,2-ETHANEDIAMINE FUMARATE see MDP800
1-(3-FURANYL)-4-METHYL-1-PENTANONE see PCI750
1-(3-FURANYL)-2,4-PENTANEDIOL see IGF200
1-(3-FURANYL)-1,4-PENTANEDIONE see IGF300
FURAPLAST see NGE500
FURAPROMIDIUM see FPO000
FURAPYRIMIDONE see FPO100
FURASEPTYL see NGE500
FURATOL see SHG500
FURATONE see BKH500
FURATONE-S see BKH500
FURAXONE see NGG500
FURAZABOL see AOO300
FURAZOL see NGG500

FURAZOLIDON see NGG500
FURAZOLIDONE (USDA) see NGG500
FURAZOLIN see FPI000
FURAZOLINE see FPI000
FURAZON see NGG500
FURAZONE see NGE500
FURAZOSIN see AJP000
FURAZOSIN HYDROCHLORIDE see FPP100
FUR BLACK 41867 see PEY500
FUR BROWN 41866 see PEY500
FURCELLERAN GUM see FPQ000
FURESIS see CHJ750
FURESOL see NGE500
FURETHIDINE see FPQ100
FURFURAL see FPQ875
2-FURFURAL see FPQ875
FURFURAL-ACETONE ADDUCT see FPQ900
FURFURAL-ACETONE MONOMER see FPQ900
1:1 FURFURAL-ACETONE MONOMER see FPQ900
FURFURAL ACETONE MONOMER FA see FPQ900
FURFURAL ALCOHOL see FPU000
FURFURALDEHYDE see FPQ875
FURFURALE (ITALIAN) see FPQ875
FURFURAL OXIME see FPR000
FURFURAMIDE see FPS000
FURFURAN see FPK000
FURFURIN see NGE500
FURFUROL see FPQ875
FURFUROLATSETONOVYI MONOMER FA see FPQ900
FURFUROLE see FPQ875
FURFURYLACETONE see FPT000
FURFURYL ALCOHOL see FPU000
FURFURYL ALCOHOL PHOSPHATE (3581) see FPV000
2-FURFURYLALKOHOL (CZECH) see FPU000
FURFURYLAMINE see FPW000
FURFURYL-BIS(2-CHLOROETHYL)AMINE HYDROCHLORIDE see FPX000
FURFURYL-BIS(β-CHLOROETHYL)AMINE HYDROCHLORIDE see FPX000
FURFURYL MERCAPTAN see FPM000
N-(2-FURFURYL)-N-(2-PYRIDYL)-N',N'-DIMETHYLETHYLENEDIAMINE
    FUMARATE see MDP800
FURFURYLTRIMETHYLAMMONIUM IODIDE see FPY000
FURIDAZOL see FQK000
FURIDAZOLE see FQK000
FURIDIAZINE see NGI500
FURIDON see NGG500
FURIL see FPZ000
α-FURIL see FPZ000
2,2'-FURIL see FPZ000
2-FURIL-METANALE (ITALIAN) see FPQ875
α-FURILMONOXIME see FQB000
FURITON see NGB700
FURLOE see CKC000
FURLOE 4EC see CKC000
FURMETHANOL see FPI000
FURMETHIDE see FPY000
FURMETHONOL see FPI000, FPI150
FURMETONOL see FPI000
FURMITHIDE IODIDE see FPY000
FURNACE BLACK see CBT750
FUROBACTINA see NGE000
2H-FURO(2,3-h)(1)BENZOPYRAN-2-ONE see FQC000
7H-FURO(3,2-g)(1)BENZOPYRAN-7-ONE see FQD000
FUROCOUMARIN see FQD000
FURO(2',3',7,6)COUMARIN see FQD000
FURO(4',5',6,7)COUMARIN see FQD000
FURO(5',4',7,8)COUMARIN see FQC000
FURODAN see CBS275
FUROFUTRAN see FMB000
FUROIC ACID see FQE000
2-FUROIC ACID see FQF000
α-FUROIC ACID see FQF000
2-FUROIC ACID, METHYL ESTER see MKH600
FUROIN see FQI000
FUROLE see FPQ875
α-FUROLE see FPQ875
FURO(3',4':6,7)NAPHTHO(2,3-d)-1,3-DIOXOL-6(5aH)-ONE, 5,8,8a,9-
    TETRAHYDRO-9-HYDROXY-5-(3,4,5- TRIMETHOXYPHENYL)-, (5R-(5-α-5a-
    β,8a-α-9-β))- see EBB600
FUROSEDON see CHJ750
FUROSEMID see CHJ750
FUROSEMIDE see CHJ750
FUROSEMIDE "MITA" see CHJ750
FUROTHIAZOLE see FQJ000
FUROVAG see NGG500

FUROX see NGG500
FUROXAL see NGG500
FUROXANE see NGG500
FUROXONE SWINE MIX see NGG500
2-FUROYL AZIDE see FQJ025
N-FUROYL-N'-n-BUTYLHARNSTOFF (GERMAN) see BRK250
FUROYL CHLORIDE see FQJ050
(FUROYLOXY)TRIETHYL PLUMBANE see TJS750
2-(4-(2-FUROYL)PIPERAZIN-1-YL)-4-AMINO-6,7-DIMETHOXYQUINAZOLINE
    see AJP000
2-(4-(2-FUROYL)PIPERAZIN-1-YL)-4-AMINO-6,7-DIMETHOXYQUINAZOLINE
    HYDROCHLORIDE see FPP100
FUROZOLIDINE see NGG500
FURPIRINOL see NDY400
FURPYRINOL see NDY400
FURRO 4R see DUP400
FURRO D see PEY500
FURRO EG see ALT500
FURRO ER see NAW500
FURRO L see DBO000
FURRO P BASE see ALT250
FURRO SLA see DBO400
FURSEMID see CHJ750
FURSEMIDE see CHJ750
FURSULTIAMIN see FQJ100
FURSULTIAMINE see FQJ100
FURTHRETHONIUM IODIDE see FPY000
FURTRETHONIUM IODIDE see FPY000
FURTRIMETHONIUM IODIDE see FPY000
FUR YELLOW see PEY500
3-(α-FURYL-β-ACETYLAETHYL)-4-HYDROXYCUMARIN (GERMAN)
    see ABF500
3-(1-FURYL-3-ACETYLETHYL)-4-HYDROXYCOUMARIN see ABF500
FURYL ALCOHOL see FPU000
FURYLAMIDE see FQN000
2-(2-FURYL)BENZIMIDAZOLE see FQK000
2-(2'-FURYL)-BENZIMIDAZOLE see FQK000
4-(2-FURYL)-2-BUTANONE see FPT000
2-FURYLCARBINOL see FPU000
α-FURYLCARBINOL see FPU000
17-α-(3-FURYL)ESTRA-1,3,5(10),7-TETRAENE-3,17-DIOL-3-ACETATE
    see EDU600
FURYLFURAMIDE see FQN000, FQO000
2-FURYL α-HYDROXYFURFURYL KETONE see FQI000
1-(3-FURYL)-4-HYDROXYPENTANONE see FQL100
1-(β-FURYL)-4-HYDROXYPENTANONE see FQL100
5-(3-FURYL)-5-HYDROXY-2-PENTANONE see IGF325
β-FURYL ISOAMYL KETONE see PCI750
2-FURYLISOPROPYLAMINE SULFATE see FQM000
β-(2-FURYL)ISOPROPYLAMINE SULFATE see FQM000
2-FURYLMERCURIC CHLORIDE see CHK000
2-FURYLMERCURY CHLORIDE see CHK000
2-FURYL-METHANAL see FPQ875
(2-FURYL)METHANOL see FPU000
1-(2-FURYL)METHYLAMINE see FPW000
1-(3-FURYL)-4-METHYL-1-PENTANONE see PCI750
α-2-FURYL-5-NITRO-2-FURANACYRLAMIDE see FQN000
2-(2-FURYL)-3-(5-NITRO-2-FURYL)ACRYLAMIDE see FQN000
(E)-2-(2-FURYL)-3-(5-NITRO-2-FURYL)ACRYLAMIDE see FQO000
trans-2-(2-FURYL)-3-(5-NITRO-2-FURYL)ACRYLAMIDE see FQO000
2-(2-FURYL)-3-(5-NITRO-2-FURYL)ACRYLIC ACID AMIDE see FQN000
α-(FURYL)-β-(5-NITRO-2-FURYL)ACRYLIC AMIDE see FQN000
1-(3-FURYL)-1,4-PENTANEDIONE see IGF300
N-(4-(2-FURYL)-2-THIAZOLYL)ACETAMIDE see FQQ100
N-(4-(2-FURYL)-2-THIAZOLYL)FORMAMIDE see FQQ400
FUSARENONE X see FQR000
FUSAREX see TBR750, TBS000
FUSARIC ACID see BSI000
FUSARIC ACID-Ca see FQR100
FUSARIC ACID CALCIUM SALT see FQR100
FUSARINIC ACID see BSI000
FUSARIOTOXIN T 2 see FQS000
FUSARIUM TOXIN see ZAT000
FUSED BORIC ACID see BMG000
FUSED QUARTZ see SCK600
FUSED SILICA (ACGIH) see SCK600
FUSELOEL (GERMAN) see FQT000
FUSEL OIL see FQT000
FUSEL OIL, REFINED (FCC) see FQT000
FUSID see CHJ750
FUSIDATE SODIUM see SHK000
FUSIDIC ACID see FQU000
FUSIDIN see SHK000
FUSIDINE see FQU000

FUSIN see SHK000
FUSSOL see FFF000
FUTRAFUL see FMB000
FUTRAMINE D see PEY500
FUTRAMINE EG see ALT500
FUTRICAN see CDY325
FUVACILLIN see NGE500
FUXAL see SNN300
FUZIDIN see FQU000
F1V02 see EKG500
FW 50 see WCJ000
FW 293 see BIO750
FW 734 see DGI000
FW 925 see DFT800
FW 200 (mineral) see WCJ000
FWH 399 see TNV625
FWH 429 see TNV625
FX 703 see FMO150
FYDALIN see BNK000
FYDE see FMV000
FYFANON see MAK700
FYROL CEF see CGO500
FYROL FR 2 see FQU875
FYROL HB32 see TNC500
FYRQUEL 150 see TNP500
FYSIOQUENS see EEH575
FYTIC ACID see PIB250

G 0 see HDG000
G 1 see HIM000
G-11 see HCL000
G-52 see GAA100
G 130 see DTN775
G 301 see DCM750
G 338 see DER000
G 347 see DST200
G 475 see COY500
G 996 see CDS125
G 3063 see GAC000
G 3707 see DXY000
G 14744 see DWA500
G-2130A see DXY000
G 22150 see DLH630
G 22355 see DLH600, DLH630
G-23350 see ABF750
G 23992 see DER000
G 24,163 see PNH750
G-24480 see DCM750
G-24622 see DJX400
G-25804 see TBO500
G 27202 see HNI500
G 27365 see DJA200
G-29288 see MQH750
G 30027 see ARQ725
G 32883 see DCV200
G 33040 see DCV800
G 33182 see CLY600
G 34161 see BKL250
G 34360 see INR000
G 34586 see CDV000
G 35020 see DLS600
GA see EIF000, GEM000
GA 242 see CDF400
GA-10832 see CQG250
GA 56 (enzyme) see GGA800
GABA see PIM500
GABACET see NNE400
p-GABA HYDROCHLORIDE see GAD000
GABBROMICINA see APP500
GABBROMYCIN see APP500, NCF500
GABBROPAS see AMM250
GABBRORAL see APP500
GABBROROL see APP500
GABEXATE MESILATE see GAD400
GABEXATE MESYLATE see GAD400
GABOB see AKF375
GABOMADE see AKF375
GADEXYL see MQU750
GADOLINIA see GAP000
GADOLINIUM see GAF000
GADOLINIUM CHLORIDE see GAH000
GADOLINIUM CITRATE see GAJ000
GADOLINIUM(III) NITRATE (1583) see GAL000
GADOLINIUM(III) NITRATE, HEXAHYDRATE (1583586) see GAN000

GADOLINIUM OXIDE see GAP000
GADOLINIUM TRICHLORIDE see GAH000
GAFCOL EB see BPJ850
GAFCOTE see PKQ059
GAG ROOT see CCJ825
GAIMAR see RLU000
GAJAR, seed extract see GAP500
GALACTARIC ACID see GAR000
GALACTASOL see GLU000
GALACTASOL A see SLJ500
GALACTICOL see DDJ000
β-d-GALACTOPYRANOSIDE, (3-β)-SOLANID-5-EN-3-YL O-6-DEOXY-α-l-
    MANNOPYRANOSYL-(1-2)-O-(β-d-GLUCOPYRANOSYL-(1-3))-, HYDRO-
    CHLORIDE see SKS100
4-o-β-d-GALACTOPYRANOSYL-d-FRUCTOFURANOSE see LAR100
3-β-d-GALACTOPYRANOSYL-l-METHYL-l-NITROSOUREA see MNB000
GALACTOQUIN see GAX000
GALACTOSACCHARIC ACID see GAR000
d-GALACTOSAMINE HYDROCHLORIDE see GAT000
GALACTOSE see GAV000
d-GALACTOSE see GAV000
β-GALACTOSIDASE see GAV100
4-(β-d-GALACTOSIDO)-d-GLUCOSE see LAR000
GALACTURONIC ACID with α-(6-METHOXY-4-QUINOLYL)-5-VINYL-2-
    QUINUCLIDINEMETHANOL see GAX000
GALAN de DIA (CUBA) see DAC500
GALANGIN see GAZ000
GALAN de NOCHE (CUBA) see DAC500
GALANTHAMINE HYDROBROMIDE see GBA000
GALANTHUS NIVALIS see SED575
GALATONE see GBB500
4-o-β-d-GALATTOPIRANOSIL-d-FRUTTOFURANOSIO (ITALIAN) see LAR100
GALATUR see DPX200
GALBANUM OIL see GBC000
GALCTIN see PMH625
GALECRON see CJJ250
GALECRON SP see CJJ500
GALENA see LDZ000
GALFER see FBJ100
GALIPAN see BAV000
GALLACETOPHENONE see TKN250
GALLALDEHYDE-3,5-DIMETHYL ETHER see DOF600
GALLAMINE see PDD300
GALLIC ACID see GBE000
GALLIC ACID, PROPYL ESTER see PNM750
GALLIMYCIN see EDJ500
GALLITO (CUBA) see SBC550
GALLIUM see GBG000
GALLIUM ARSENIDE see GBK000
GALLIUM CHLORIDE see GBM000
GALLIUM (3+) CHLORIDE see GBM000
GALLIUM CITRATE see GBO000
GALLIUM COMPOUNDS see GBO500
GALLIUM LACTATE mixed with SODIUM LACTATE (1.6581 moles)
    see GBQ000
GALLIUM METAL, solid (DOT) see GBG000
GALLIUM METAL, liquid (DOT) see GBG000
GALLIUM MONOARSENIDE see GBK000
GALLIUM-NICKEL ALLOY see NDD500
GALLIUM NITRATE see GBS000
GALLIUM(III) NITRATE (1583) see GBS000
GALLIUM SULFATE see GBS100
GALLOCHROME see MCV000
GALLODESOXYCHOLIC ACID see CDL325
GALLOGAMA see BBQ500
GALLOTANNIC ACID see TAD750
GALLOTANNIN see TAD750
GALLOTOX see ABU500, PFO550
GALLOXON see DFH600
GALOFAK see BDY669
GALOPERIDOL see CLY500
GALOXANE see DFH600
GALOXOLIDE see GBU000
GALOZONE see CCL250
GALVANISONE see FLL000
GAMACID see BBQ500
GAMAPHEX see BBQ500
GAMAQUIL see PGA750
GAMAREX see PIM500
GAMASOL 90 see DUD800
GAMBIER see CCP800
GAMEFAR see RHZ000
GAMENE see BBQ500
GAMIBETAL see AKF375

GAMISO see BBQ500
GAMMA-COL see BBQ500
GAMMACORTEN see SOW000
GAMMAHEXA see BBQ500
GAMMAHEXANE see BBQ500
GAMMALIN see BBQ500
GAMMALON see PIM500
GAMMA OH see HJS500
GAMMAPHOS see AMD000
GAMMEXANE see BBP750
GAMMOPAZ see BBQ500
GAMONIL see IFZ900
GAMOPHENE see HCL000
GANCIDIN (unpurified) see GBU600
GANEAKE see ECU750
GANEX P 804 see PKQ250
GANGESOL see HII500
GANGLIOSTAT see HEA000
GANJA see CBD750
GANLION see MKU750
GANOCIDE see MLC250
GANOZAN see CHC500
GANPHEN see PMI750
GANSIL see CDP000
GANTANOL see SNK000
GANTAPRIN see TKX000
GANTRIM see TKX000
GANTRISINE see SNN500
GANU see CHA000
GARAMYCIN see GCO000, GCS000
GARANTOSE see BCE500
GARBANCILLO (CUBA) see GIW200
GARDCIDE see RAF100
GARDENAL SODIUM see SID000
GARDENTOX see DCM750
GARDEPANYL see EOK000
GARDIQUAT 1450 see AFP250
GARDONA see RAF100
GARDOPRIM see BQB000
GARGET see PJJ315
GARGON see TFQ275
GARI see CCO700
GARLIC see WBS850
GARLIC OIL see GBU800
GARLIC POWDER see GBU850
GARLON see TJE890
GARMIAN see BOV825
GARNITAN see DGD600
GAROX see BDS000
GARRATHION see TNP250
GARVOX see DQM600
GASHOUSE TANKAGE see SKZ000
GAS OIL see GBW000
GAS OILS (petroleum), hydrodesulfurized heavy vacuum see GBW010
GAS OILS (petroleum), light vacuum see GBW025
GASOLINE see GBY000
GASOLINE (100-130 octane) see GCA000
GASOLINE (115-145 octane) see GCC000
GASOLINE ENGINE EXHAUST CONDENSATE see GCE000
GASOLINE ENGINE EXHAUST ""TAR" see GCE000
GASOLINE, UNLEADED see GCE100
GASTER see FAB500
GASTOMAX see CDQ500
GASTRACID see PEK250
GASTRIDENE see BGC625
GASTRIDIN see FAB500
GASTRINIDE see TEH500
GASTRIPON see PEM750
GASTRODYN see GIC000
GASTROGRAFIN see AOO875
GASTROMET see TAB250
GASTRON see DJM800, XCJ000
GASTROPIDIL see CBF000
GASTROSEDAN see DJM800, XCJ000
GASTROTEST see PEK250
GASTROTOPIC see BGC625
GASTROZEPIN see GCE500
GATALONE see GBB500
GATINON see MHM500
GAULTHERIA OIL, ARTIFICIAL see MPI000
GB see IPX000
GB 94 see BMA625
GBH see BDD000
GBL see DWT600

GBS see SEG800
GBX see CDB770
GC 928 see CJR500
GC-1106 see HCL500
GC-2466 see MRV000
GC-2996 see CJY000
GC 3707 see SOY000
GC 4072 see CDS750
GC 6936 see ABX250
GC 7787 see HDA500
GC 8993 see CLU000
GC 3944-3-4 see PAX000
GCP 5126 see DED000
G-CURE see ADW200
GD see SKS500
Ge 132 see CCF125
GEA 6414 see CLK325
GEABOL see DAL300
GEARPHOS see MNH000, PAK000
GEBUTOX see BRE500
GECHLOREERDEDIFENYL (DUTCH) see PJM500
GEDEX see SMQ500
GEIGY 338 see DER000
GEIGY 2747 see PET250
GEIGY 12968 see DRU400
GEIGY 13005 see DSO000
GEIGY 19258 see DRL200
GEIGY 22870 see DQZ000
GEIGY 24480 see DCM750
GEIGY 27,692 see BJP000
GEIGY 30,027 see ARQ725
GEIGY 30,044 see BJP250
GEIGY 30494 see MNO750
GEIGY 32,293 see EGD000
GEIGY 32883 see DCV200
GEIGY-BLAU 536 see BGT250
GEIGY-444E see TBO500
GEIGY 444E see AKQ250
GEIGY G-23611 see DSK200
GEIGY G-27365 see DJA200
GEIGY G-28029 see PDC750
GEIGY G-29288 see MQH750
GEIGY G.S. 14254 see BQC250
GEIGY GS-19851 see IOS000
GEIGY HERBICIDE 444E see AKQ250
GEISSOSPERMINE see GCG300
GELACILLIN see BDY669
GELATIN see PCU360
GELATINE see PCU360
GELATIN DYNAMITE see DYG000
GELATIN-EPINEPHRINE see AES500
GELATIN FOAM see PCU360
GELATINS see PCU360
GELBER PHOSPHOR (GERMAN) see PHO740
GELBIN see CAP500
GELBIN YELLOW ULTRAMARINE see CAP750
GELBORANGE-S (GERMAN) see FAG150
GELCARIN see CCL250
GELCARIN HMR see CCL250
GELDANAMYCIN see GCI000
G-ELEVEN see HCL000
GELFOAM see PCU360
GEL II see SHF500
GELIOMYCIN see HAL000
GELOCATIL see HIM000
GELOSE see AEX250
GELOZONE see CCL250
GELSEMIN see GCK000
GELSEMINE see GCK000
GELSEMIUM SEMPERVIRENS see YAK100
GELSTAPH see SLJ000, SLJ050
GELTABS see VSZ100
GELUCYSTINE see CQK325
GELUTION see SHF500
GELVA CSV 16 see AAX250
GELVATOLS see PKP750
GEMCADIOL see TDO260
GEMEPROST see CDB775
GEMFIBROZIL see GCK300
GEMONIL see DJO800
GEMONIT see DJO800
GENACORT see CNS750
GENAZO RED KB SOLN see CLK225
GENDRIV 162 see GLU000

GENEP EPTC see EIN500
GENERAL CHEMICALS 1189 see KEA000
GENERAL CHEMICALS 3707 see SOY000
GENERAL CHEMICALS 8993 see CLU000
GENET ABSOLUTE see GCM000
GENETRON 11 see TIP500
GENETRON 12 see DFA600
GENETRON 13 see CLR250
GENETRON 21 see DFL000
GENETRON 22 see CFX500
GENETRON-23 see CBY750
GENETRON 100 see ELN500
GENETRON 101 see CFX250
GENETRON 112 see TBP050
GENETRON 113 see FOO000
GENETRON 114 see FOO509
GENETRON 115 see CJI500
GENETRON 316 see FOO509
GENETRON 1113 see CLQ750
GENETRON 133a see TJY175
GENETRON 142b see CFX250
GENETRON 1112A see DFA300
GENETRONE 1112A see DFA300
GENIPHENE see CDV100
GENIPIN see GCM300
GENISIS see ECU750
GENISTEIN see GCM350
GENISTEOL see GCM350
GENISTERIN see GCM350
GENITE see DFY400
GENITE 883 see CJT750
GENITHION see PAK000
GENITOL see DFY400
GENITOX see DAD200
GENITRON AC see ASM270
GENITRON AC 2 see ASM270
GENITRON AC 4 see ASM270
GENO-CRISTAUZ GREMY see TBF500
GENOGRIS see NNE400
GENOL see MGJ750
GENOPHYLLIN see TEP500
GENOPTIC see GCS000
GENOPTIC S.O.P. see GCS000
GENOTHERM see PKQ059
GENOXAL see CQC500, CQC650
GENOZYM see CMX700
GENPROPATHRIN see DAB825
GENTAMICIN see GCO000
GENTAMICIN C2b see MQS579
GENTAMICIN C(^{2}b) see MQS579
GENTAMICIN C COMPLEX see GCO200
GENTAMYCIN see GCO000
GENTAMYCIN-CREME (GERMAN) see GCO000
GENTAMYCIN SULFATE see GCS000
GENTIANAE SCABRAE RADIX (LATIN) see RSZ675
GENTIAN VIOLET see AOR500
GENTIMON see ALB000
GENTISATE see GCU000
GENTISIC ACID see GCU000
GENTRAN see DBD700
GENU see CCL250
GENUGEL see CCL250
GENUGEL CJ see CCL250
GENUGOL RLV see CCL250
GENUINE ACETATE CHROME ORANGE see LCS000
GENUINE ORANGE CHROME see LCS000
GENUINE PARIS GREEN see COF500
GENUVISCO J see CCL250
GENVIS see SLJ500
GEOFOS see DHH200
GEON see PJR000, PKQ059
GEON 135 see AAX175
GEON LATEX 151 see PKQ059
GEOPEN see CBO250
GEOTRICYN see MCH525
GERANIC ACID see GCW000
GERANIOL (FCC) see DTD000
GERANIOL ACETATE see DTD800
GERANIOL ALCOHOL see DTD000
GERANIOL CROTONATE see DTE000
GERANIOL EXTRA see DTD000
GERANIOL FORMATE see GCY000
GERANIOL TETRAHYDRIDE see DTE600
GERANIUM CRYSTALS see PFA850

GERANIUM LAKE N see FAG070
GERANIUM OIL see GDA000
GERANIUM OIL ALGERIAN TYPE see GDA000
GERANIUM OIL BOURBON see GDC000
GERANIUM OIL, EAST INDIAN TYPE see PAE000
GERANIUM OIL MOROCCAN see GDE000
GERANIUM OIL, TURKISH TYPE see PAE000
GERANIUM THUNBERGII Sieb. et Zucc., extract see GDE300
GERANONITRILE see GDM000
GERANOXY ACETALDEHYDE see GDM100
GERANYL ACETATE (FCC) see DTD800
GERANYL ACETOACETATE see AAY500
GERANYL ACETONE see GDE400
GERANYL ALCOHOL see DTD000
GERANYL BENZOATE see GDE800
GERANYL-2-BUTENOATE see DTE000
GERANYL BUTYRATE see GDE825
GERANYL CAPROATE see GDG000
GERANYL CROTONATE see DTE000
GERANYL ETHYL ETHER see GDG100
GERANYL FORMATE (FCC) see GCY000
GERANYL HEXANOATE see GDG000
GERANYL ISOBUTYRATE see GDI000
GERANYL ISOVALERATE see GDK000
GERANYL NITRILE see GDM000
GERANYL OXYACETALDEHYDE see GDM100
GERANYL PHENYLACETATE see GDM400
GERANYL PROPIONATE see GDM450
GERANYL TIGLATE see GDO000
GERASTOP see ARQ750
GERBITOX see BAV000
GERFIL see PMP500
GERHARDITE see CNN000
GERISON see SNM500
GERMALGENE see TIO750
GERMALL 115 see GDO800
GERMA-MEDICA see HCL000
GERMANATE(2-), BIS(2-CARBOXYLATOETHYL)TRIOXODI-, DIHYDROGEN
  (9CI) see CCF125
GERMAN CHAMOMILE OIL see CDH500
GERMANE (DOT) see GEI100
GERMANIA see GEC000
GERMANIC ACID see GEC000
GERMANIC OXIDE (crystalline) see GDS000
GERMANIN see BAT000
GERMANIUM see GDU000
GERMANIUM BROMIDE see GDW000
GERMANIUM CHLORIDE see GDY000
GERMANIUM COMPOUNDS see GEA000
GERMANIUM, (l-CYSTEINE)TETRAHYDROXY- see CQK100
GERMANIUM DIOXIDE see GEC000
GERMANIUM HYDRIDE see GEI100
GERMANIUM MONOHYDRIDE see GEG000
GERMANIUM OXIDE see GEC000
GERMANIUM OXIDE (GeO$_2$) see GEC000
GERMANIUM(II) SULFIDE see GEI000
GERMANIUM TETRABROMIDE see GDW000
GERMANIUM TETRACHLORIDE see GDY000
GERMANIUM TETRAHYDRIDE see GEI100
3,3'-(GERMANOIC ANHYDRIDE) DIPROPANOIC ACID see CCF125
GERMERIN (GERMAN) see VHF000
GERMICICLIN see MDO250
GERMIN see EBL500
GERMINE see EBL500
GERMINOL see BBA500
GERMISAN see PFO250
GERMITOL see BBA500
GERNEBCIN see TGI250
GEROBIT see DBA800
GERODYL see DWK400
GERONTINE see DCC400
GERONTINE TETRAHYDROCHLORIDE see GEK000
GEROSTOP see GEK200
GEROT-EPILAN see MKB250
GEROT-EPILAN-D see DKQ000
GEROVIT see DBA800
GEROVITAL see AIL750
GEROX see SLW500
GERTLEY BORATE see SFF000
GERVOT see MDT600
GESADURAL see BJP250
GESAFID see DAD200
GESAFRAM see MFL250
GESAGARD see BKL250

GESAKUR see PNH750
GESAMIL see PMN850
GESAPON see DAD200
GESAPRIM see ARQ725
GESARAN see BJP000, INQ000
GESAREX see DAD200
GESAROL see DAD200
GESATAMIN see EGD000
GESATOP see BJP000
GESFID see MQR750
GESOPRIM see ARQ725
GESTANIN see AGF750
GESTANOL see AGF750
GESTANON see AGF750
GESTANYN see AGF750
GESTATRON see DYF759
GESTEROL L.A. see HNT500
GESTID see MQR750
GESTONORONE CAPROATE see GEK510
GESTONORONE CAPRONATE see GEK510
GESTORAL see GEK500
GESTRIGONE see ENX575
GESTRINONE see ENX575
GESTRONOL CAPROATE see GEK510
GESTRONOL HEXANOATE see GEK510
GESTYL see SCA750
GETTYSOLVE-B see HEN000
GETTYSOLVE-C see HBC500
GEUM ELATUM (Royle) Hook. f., extract see GEK600
GEVILON see GCK300, SHL500
GF see DVR909
GFX-E see GEK875
GFX-ES see GEK880
G1V GARD DXN see ABC250
GHA 331 see AHC000
G 3063 HYDROCHLORIDE see GAC000
G-1 see AOO375
GIACOSIL HYDROCHLORIDE see LFK200
GIALLO CROMO (ITALIAN) see LCR000
GIANT MILKWOOD see COD675
GIARDIL see NGG500
GIARLAM see NGG500
GIATRICOL see MMN250
GIBBERELLIC ACID see GEM000
GIBBERELLIN see GEM000
GIBBREL see GEM000
GIB-SOL see GEM000
GIB-TABS see GEM000
GICHTEX see ZVJ000
GIE see DVR600
GIEGY GS-13798 see MPG250
GIFBLAAR see PLG000
GIFBLAAR POISON see FIC000
GIGANTIN see CMV000
GIHITAN see DCK759
GILEMAL see CEH700
GILOTHERM OM 2 see TBD000
GILSONITE see GEO000
GILURYTMAL see AFH250
GILUTENSIN see DWF200
GILVOCARCIN V see GEO200
GIMID see DYC800
GINANDRIN see AOO410
GINARSOL see ABX500
GINBEY (DOMINICAN REPUBLIC) see SLJ650
GINDARINE HYDROCHLORIDE see GEO600
GINDARIN HYDROCHLORIDE see GEO600
GINEFLAVIR see MMN250
GINGER OIL see GEQ000
GINGERONE see VFP100
GINGICAIN M see TBN000
GINGILLI OIL see SCB000
GINSENG see GEQ400
GINSENG, ROOT EXTRACT see GEQ425
GINSENG ROOT-NEUTRAL SAPONINS see GEQ425
GINSENGWURZEL, EXTRACT (GERMAN) see GEQ425
GINSENOSIDE RB1 see PAF450
GIQUEL see HKR500
GIRACID see PDC250
GIRL see CNE750
GIROSTAN see TFQ750
GITALIN see GES000
GITALOXIGENIN + DIGITALOSE (GERMAN) see VIZ200
GITALOXIGENIN-TRIDIGITOXOSID (GERMAN) see GES100

GITALOXIN see GES100
GITALOXIN-16-FORMATE see GES100
GITHAGENIN see GMG000
GITOFORMATE see FND100
GITOXIGENIN-3-o-MONODIGITALOSIDE see SMN275
GITOXIGENIN-TRIDIGITOXOSID (GERMAN) see GEU000
GITOXIGENIN TRIDIGITOXOXIDE-16-FORMATE see GES100
GITOXIN see GEU000
GITOXIN PENTAACETATE see GEW000
GITOXOGENIN-d-DIGITALOSID (GERMAN) see SMN275
G.L. 105 see PBH100
GL 2487 see DLO880
GLACIAL ACETIC ACID see AAT250
GLACIAL ACRYLIC ACID see ADS750
GLANDUBOLIN see EDV000
GLANDUCORPIN see PMH500
GLANIL see CMR100
GLARUBIN see GEW700
GLASS see FBQ000
GLASS FIBERS see FBQ000
GLAUCARUBIN see GEW700
GLAUCINE see TDI475
(+)-GLAUCINE see TDI475
d-GLAUCINE see TDI475
s-(+)-GLAUCINE see TDI475
GLAUCINE HYDROCHLORIDE see TDI500
dl-GLAUCINE PHOSPHATE see TDI600
dl-GLAUCINPHOSPHAT (GERMAN) see TDI600
GLAUMEBA see GEW700
GLAUPAX see AAI250
GLAURAMINE see IBB000
GLAUVENT see TDI475
GLAXORIDIN see TEY000
GLAZD PENTA see PAX250
GLAZIDIM see CCQ200
GLEEM see SHF500
GLENTONIN-RETARD see CCK125
GLIANIMON see FGU000, FLK100
GLIANIMON MITE see FLK100
GLIBENCLAMIDE see CEH700
GLIBORNURIDE see GFY100
GLIBUTIDE see BQL000
GLICLAZIDE see DBL700
GLICOL MONOCLORIDRINA (ITALIAN) see EIU800
GLICOSIL see CPR000
GLIKOCEL TA see SFO500
GLIMID see DYC800
GLIOTOXIN see ARO250
GLIPASOL see GEW750
GLIPORAL see BQL000
GLIRICIDIAL SEPIUM see RBZ000
GLISEMA see CKK000
GLISOLAMIDE see DBE885
GLISOXEPID see PMG000
GLISOXEPIDE see PMG000
GLOBENICOL see CDP250
GLOBOCICLINA see MDO250
GLOBOID see ADA725
GLOBULARIACITRIN see RSU000
GLOGAL see HNI500
GLOMAX see KBB600
GLONOIN see NGY000
GLONSEN see SHK800
GLORIOSA LILY see GEW800
GLORIOSA ROTHSCHILDIANA see GEW800
GLORIOSA SUPERBA see GEW800
GLOROUS see CDP250
GLORY LILY see GEW800
GLOSSO STERANDRYL see MPN500
GLOVER see LCF000
GLOXAZON see KFA100
GLOXAZONE see KFA100
GLUBORID see GFY100
GLUCAGON see GEW875
GLUCAZIDE see GBB500
GLUCID see BCE500
GLUCIDORAL see BSM000
GLUCINUM see BFO750
GLUCITOL see SKV200
d-GLUCITOL see SKV200
d-GLUCITOL, 1-DEOXY-1-(METHYLAMINO)-, 3,5-BIS(ACETYLAMINO)-2,4,6-
   TRIIODOBENZOATE (SALT) see AOO875
d-GLUCITOL, 1,1'-(SULFONYLBIS(4,1-PHENYLENEIMINO))BIS(1-DEOXY-1-
   SULFO-, DISODIUM SALT (9CI) see AOO800

GLUCOBASIN see ARQ500
GLUCOCHLORAL see GFA000
GLUCOCHLORALOSE see GFA000
α-d-GLUCOCHLORALOSE see GFA000
GLUCODIGIN see DKL800
GLUCO-FERRUM see FBK000
GLUCOFREN see BSM000
GLUCOGITOXIN see DAD650
GLUCOHEPTONIC ACID with ERYTHROMYCIN (1:1) see EDI500
GLUCOLIN see GFG000
GLUCONATE de CALCIUM (FRENCH) see CAS750
GLUCONATO di SODIO (ITALIAN) see SHK800
d-GLUCONIC ACID, CYCLIC ESTER with ANTIMONIC ACID (H8Sb2O9) (2:1),TRISODIUM SALT, NONAHYDRATE see AQH800
D-GLUCONIC ACID, CYCLIC ESTER with ANTIMONIC ACID (H8Sb2O9) (2:1),TRISODIUM SALT, NONAHYDRATE see AQI250
d-GLUCONIC ACID, MONOPOTASSIUM SALT (9CI) see PLG800
d-GLUCONIC ACID, 2,4:2',4'-O-(OXYDISTIBYLIDYNE)BIS-, Sb,Sb'-DIOXIDE, TRISODIUM SALT, NONAHYDRATE see AQH800
GLUCONIC ACID POTASSIUM SALT see PLG800
GLUCONIC ACID SODIUM SALT see SHK800
GLUCONSAN K see PLG800
GLUCOPHAGE see DQR600, DQR800
GLUCOPHAGE LA 6023 see DQR600
GLUCOPROSCILLARIDIN see GFC000
β-d-GLUCOPYRANOSIDE, 4-HYDROXYPHENYL- (9CI) see HIH100
4-(α-d-GLUCOPYRANOSIDO)-α-GLUCOPYRANOSE see MAO500
α-D-GLUCOPYRANOSIDURONIC ACID, (3-β,20-β)-20-CARBOXY-11-OXO-30-NOROLEAN-12-EN- 3-YL 2-O-β-D-GLUCOPYRANURONOSYL-, AMMONIATE see GIE100
β-D-GLUCOPYRANOSIDURONIC ACID, (METHYL-ONN-AZOXY)METHYL- see MGS700
10-GLUCOPYRANOSYL-1,8-DIHYDROXY-3-(HYDROXYMETHYL)-9(10H)-ANTHRACENONE see BAF825
α-d-GLUCOPYRANOSYL β-d-FRUCTOFURANOSIDE see SNH000
3-β-((α-d-GLUCOPYRANOSYL)OXY)-14-HYDROXY-19-OXO-BUFA-4,20,22-TRIENOLIDE see AGW625
2-(β-d-GLUCOPYRANOSYLOXY)ISOBUTYRONITRILE see GFC100
2-(β-d-GLUCOPYRANOSYLOXY)-2-METHYLPROPANENITRILE see GFC100
(d-GLUCOPYRANOSYLTHIO)GOLD see ART250
β-D-GLUCOPYRANURONIC ACID, 1-DEOXY-1-(9H-FLUOREN-2-YLHYDROXYAMINO)- see HLE750
GLUCOSE see GFG000
d-GLUCOSE see GFG000
d-GLUCOSE, anhydrous see GFG000
GLUCOSE AERODEHYDROGENASE see GFG100
GLUCOSE LIQUID see GFG000
GLUCOSE OXIDASE see GFG100
β-d-GLUCOSE OXIDASE see GFG100
GLUCOSE-RINGER'S SOLUTION (23.3%) see GFG200
GLUCOSE-RINGER'S SOLUTION (29.2%) see GFG205
GLUCOSIDE of ALLYL ISOTHIOCYANATE see AGH125
(α-d-GLUCOSIDO)-β-d-FRUCTOFURANOSIDE see SNH000
4-(α-d-GLUCOSIDO)-d-GLUCOSE see MAO500
GLUCOSTIBAMINE SODIUM see NCL000
GLUCOSTIMIDINE SODIUM see NCL000
GLUCOSULFONE see AOO800
GLUCOSULFONE SODIUM see AOO800
N-d-GLUCOSYL(2)-N'-NITROSOMETHYLHARNSTOFF (GERMAN) see SMD000
N-d-GLUCOSYL-(2)-N'-NITROSOMETHYLUREA see SMD000
β-d-GLUCOSYLOXYAZOXYMETHANE see COU000
(1-d-GLUCOSYLTHIO)GOLD see ART250
α-d-GLUCOTHIOPYRANOSE see GFK000
GLUCOXY see GFM000
GLUCURON see GFM000
GLUCURONE see GFM000
GLUCURONIC ACID, 1-DEOXY-1-(2-FLUORENYLHYDROXYAMINO)- see HLE750
GLUCURONIC ACID LACTONE see GFM000
d-GLUCURONIC ACID LACTONE see GFM000
GLUCURONIC ACID-γ-LACTONE see GFM000
d-GLUCURONIC ACID-γ-LACTONE see GFM000
N-GLUCURONIDE of N-HYDROXY-2-AMINOFLUORENE see HLE750
GLUCURONOLACTONE see GFM000
d-GLUCURONOLACTONE see GFM000
GLUCURONOSAN see GFM000
GLUDIASE see GFM200
GLUEOPHOGE see DQR600
GLU-P-I see AKS250
GLUKRESIN see GGS000
GLUMAL see AGX125
GLUMAMYCIN see AOB875
GLUMIN see GFO050
GLU-P-2 see DWW700

GLUPAN see TEH500
GLUPAX see AAI250
GLURONAZID see GBB500
GLURONAZIDE see GBB500
GLUSATE see GFO000
GLUSIDE see BCE500
GLUTACID see GFO000
GLUTACYL see MRL500
GLUTAMIC ACID see GFO000
l-GLUTAMIC ACID see GFO000
α-GLUTAMIC ACID see GFO000
GLUTAMIC ACID AMIDE see GFO050
GLUTAMIC ACID-5-AMIDE see GFO050
GLUTAMIC ACID, N-(p-(((2-AMINO-3,4,5,6,7,8-HEXAHYDRO-4-OXO-6-PTERIDINYL)METHYL)AMINO) BENZOYL)-, L- see TCR400
l-GLUTAMIC ACID, 5-(2-(4-CARBOXYPHENYL)HYDRAZIDE) see GFO100
GLUTAMIC ACID, N-(p-(((2,4-DIAMINO-6-PTERIDINYL)METHYL)METHYLAMINO)BENZOYL)-, DISODIUM SALT, L-(+)- see MDV600
GLUTAMIC ACID HYDROCHLORIDE see GFO025
l-GLUTAMIC ACID HYDROCHLORIDE see GFO025
α-GLUTAMIC ACID HYDROCHLORIDE see GFO025
GLUTAMIC ACID, IRON (2+) SALT (1:1) see FBM000
l-GLUTAMIC ACID, MAGNESIUM SALT (1:1), HYDROBROMIDE see MAG050
l-GLUTAMIC ACID, MONOPOTASSIUM SALT see MRK500
GLUTAMIC ACID, N-NITROSO-N-PTEROYL-, l- see NKG450
GLUTAMIC ACID, SODIUM SALT see MRL500
d-GLUTAMIENSUUR see GFO000
GLUTAMINE see GFO050
γ-GLUTAMINE see GFO050
GLUTAMINE, N-(1-CARBOXY-2-(METHYLENECYCLOPROPYL)ETHYL)- (7CI) see HOW100
l-GLUTAMINE (9CI, FCC) see GFO050
l-GLUTAMINE,N,N'-(SELENOBIS(THIO(1-(((CARBOXYMETHYL)AMINO)CARBONYL)-2,1-ET HANEDIYL)))BIS- see SBU500
GLUTAMINE, N,N'-((SELENODITHIO)BIS(1-((CARBOXYMETHYL)CARBAM-OYL)ETHYLENE))DI-, L- see SBU500
GLUTAMINIC ACID see GFO000
l-GLUTAMINIC ACID see GFO000
GLUTAMINIC ACID HYDROCHLORIDE see GFO025
l-GLUTAMINIC ACID HYDROCHLORIDE see GFO025
GLUTAMINOL see GFO000
GLUTAMMATO MONOSODICO (ITALIAN) see MRL500
N^2-(γ-l-(+)-GLUTAMYL)-4-CARBOXYPHENYLHYDRAZINE see GFO100
1-(l-α-GLUTAMYL)-2-ISOPROPYLHYDRAZINE see GFO200
GLUTANON see TEH500
GLUTARAL see GFQ000
GLUTARALDEHYD (CZECH) see GFQ000
GLUTARALDEHYDE see GFQ000
GLUTARDIALDEHYDE see GFQ000
GLUTARIC ACID see GFS000
GLUTARIC ACID DINITRILE see TLR500
GLUTARIC ANHYDRIDE see GFU000
GLUTARIC DIALDEHYDE see GFQ000
GLUTARIMIDE, 2-(p-AMINOPHENYL)-2-ETHYL- see AKC600
GLUTARIMIDE, 3-(1,3-DIOXO-2-METHYLINDAN-2-YL)- see DVS100
GLUTARIMIDE, 3-(1,3-DIOXO-2-PHENYLINDAN-2-YL)- see DVS300
GLUTARIMIDE, 3-(5,7-DIOXO-6-PHENYL-2,3,6,7-TETRAHYDRO-5H-CYCLOPENTA-p-DITHIIN-6-YL)- see DVS400
GLUTARODINITRILE see TLR500
GLUTARONITRILE see TLR500
GLUTARONITRILE, 2-BROMO-2-(BROMOMETHYL)- see DDM500
GLUTARYL DIAZIDE see GFU200
GLUTATHIMID see DYC800
GLUTATHIONE see GFW000
GLUTATHIONE (reduced) see GFW000
GLUTATIOL see GFW000
GLUTATIONE see GFW000
GLUTATON see GFO000
GLUTAVENE see MRL500
GLUTETHIMID see DYC800
GLUTETHIMIDE see DYC800
GLUTETIMIDE see DYC800
GLUTIDE see GFW000
GLUTINAL see GFW000
GLUTRIL see GFY100
GLYBENZCYCLAMIDE see CEH700
GLYBIGID see BRA625
GLYBURIDE see CEH700
GLYBUTAMIDE see BSM000
GLYBUTHIAZOL see GEW750
GLYBUTHIAZOLE see GEW750
GLYBUZOLE see GFM200
GLYCERIN see GGA000

GLYCERIN, anhydrous see GGA000
GLYCERIN, synthetic see GGA000
GLYCERINE see GGA000
GLYCERINE TRIACETATE see THM500
GLYCERINE TRIPROPIONATE see TMY000
GLYCERINFORMALE see DVR909
GLYCERIN GUAIACOLATE see RLU000
GLYCERIN-α-MONOCHLORHYDRIN see CDT750
GLYCERINMONOGUAIACOL ETHER see RLU000
GLYCERIN MONOSTEARATE see OAV000
GLYCERINTRINITRATE (CZECH) see NGY000
GLYCERITE see TAD750
GLYCERITOL see GGA000
GLYCEROL see GGA000
GLYCEROL-1-ACETATE see GGO000
GLYCEROLACETONE see DVR600
GLYCEROL-α-ALLYL ETHER see AGP500
GLYCEROL-1-p-AMINOBENZOATE see GGQ000
GLYCEROL CHLOROHYDRIN see CDT750
GLYCEROL-α-CHLOROHYDRIN see CDT750
GLYCEROL DIACETATE see DBF600
GLYCEROL-α,γ-DIBROMOHYDRINE see DDR800
GLYCEROL-α,β-DICHLOROHYDRIN see DGG600
GLYCEROL α,γ-DICHLOROHYDRIN see DGG400
sym-GLYCEROL DICHLOROHYDRIN see DGG400
GLYCEROL DIMETHYLKETAL see DVR600
GLYCEROL EPICHLORHYDRIN see EAZ500
GLYCEROL ESTER HYDROLASE see GGA800
GLYCEROL-α-ETHYL ETHER see EFF000
GLYCEROL FORMAL see DVR909
GLYCEROL GUAIACOLATE see RLU000
GLYCEROL-α-(o-METHOXYPHENYL)ETHER see RLU000
GLYCEROL MONOACETATE see GGO000
GLYCEROL-1-MONOACETATE see GGO000
GLYCEROL-α-MONOACETATE see GGO000
GLYCEROL-α-MONOCHLOROHYDRIN (DOT) see CDT750
GLYCEROL-α-MONOGUAIACOL ETHER see RLU000
GLYCEROL MONO(2-METHOXYPHENYL)ETHER see RLU000
GLYCEROL MONOSTEARATE see OAV000
GLYCEROL, NITRIC ACID TRIESTER see NGY000
GLYCEROL PHENYL ETHER DIACETATE see PFF300
GLYCEROL TRIACETATE see THM500
GLYCEROL TRIBROMOHYDRIN see GGG000
GLYCEROL TRIBUTYRATE see TIG750
GLYCEROL TRICAPRYLATE see TMO000
GLYCEROL TRICHLOROHYDRIN see TJB600
GLYCEROL (TRI(CHLOROMETHYL))ETHER see GGI000
GLYCEROL TRIHEXANOATE see GGK000
GLYCEROL TRIISOPENTANOATE see GGM000
GLYCEROL TRINITRATE see NGY000
GLYCEROL(TRINITRATE de) (FRENCH) see NGY000
GLYCEROLTRINTRAAT (DUTCH) see NGY000
GLYCEROL TRIOCTANOATE see TMO000
GLYCEROPLUMBONITRATE see LDH000
GLYCERYL ACETATE see GGO000
GLYCERYL-p-AMINOBENZOATE see GGQ000
β-GLYCERYL 1-p-CHLOROBENZYL-1H-INDAZOLE-3-CARBOXYLATE
    see GGR100
GLYCERYL-α-CHLOROHYDRIN see CDT750
GLYCERYL-1,3-DIACETATE see DBF600
GLYCERYL-α,γ-DIPHENYL ETHER see DVW600
GLYCERYL GUAIACOLATE see RLU000
GLYCERYLGUAIACOL CARBAMATE see GKK000
α-GLYCERYL GUAJACOLATE ETHER see RLU000
GLYCERYL GUAIACYL ETHER see RLU000
GLYCERYLGUAJACOL-CARBAMAT see GKK000
GLYCERYL MONOACETATE see GGO000
GLYCERYL MONOLAURATE see MRJ000
GLYCERYL MONOSTEARATE see OAV000
GLYCERYL NITRATE see NGY000
GLYCERYL-o-TOLYL ETHER see GGS000
GLYCERYL TRIACETATE see THM500
GLYCERYL TRIBENZOATE see GGU000
GLYCERYL TRIBROMOHYDRIN see GGG000
GLYCERYL TRICHLOROHYDRIN see TJB600
GLYCERYL TRINITRATE see NGY000
GLYCERYL TRINITRATE, solution see SLD000
GLYCERYL TRINITRATE, solution up to 1% in alcohol (DOT) see NGY000
GLYCERYL TRIOCTANOATE see TMO000
GLYCERYL TRIPROPIONATE see TMY000
GLYCIDAL see GGW000
GLYCIDALDEHYDE see GGW000
GLYCIDE see GGW500
GLYCIDOL see GGW500

GLYCIDOL OLEATE see ECJ000
GLYCIDOL STEARATE see SLK500
γ-GLYCIDOXYPROPYLTRIMETHOXYSILANE see ECH000
GLYCIDYL ACRYLATE see ECH500
GLYCIDYL ALCOHOL see GGW500
GLYCIDYLALDEHYDE see GGW000
GLYCIDYL BUTYL ETHER see BRK750
GLYCIDYLDIETHYLAMINE see GGW800
N-GLYCIDYL DIETHYL AMINE see GGW800
GLYCIDYL ESTER of DODECANOIC ACID see TJI500
GLYCIDYL ESTER of HEXANOIC ACID see GGY000
GLYCIDYL LAURATE see LBM000
GLYCIDYL METHACRYLATE see ECI000
GLYCIDYL-α-METHYL ACRYLATE see ECI000
GLYCIDYL METHYLPHENYL ETHER see TGZ100
GLYCIDYL OCTADECANOATE see SLK500
GLYCIDYL OCTADECENOATE see ECJ000
GLYCIDYL OLEATE see ECJ000
GLYCIDYL PHENYL ETHER see PFH000
N-GLYCIDYLPHTHALIMIDE see ECK000
GLYCIDYL PROPENATE see ECH500
GLYCIDYL STEARATE see SLK500
GLYCIDYLTRIMETHYLAMMONIUM CHLORIDE see GGY200
GLYCINE see GHA000
GLYCINE, N-(2,3-DIHYDRO-1H-INDEN-2-YL)-N-(N-(1-(ETHOXYCARBONYL)-3-
    PHENYLPROPYL)-l-A LANYL)-, MONOHYDROCHLORIDE, (S)-
    see DAM315
GLYCINE MUSTARD see GHE000
GLYCINE NITRILE see GHI000
GLYCINE NITROGEN MUSTARD see GHE000
GLYCINE, SODIUM SALT see GHG000
GLYCINOL see EEC600
GLYCINONITRILE see GHI000
GLYCINONITRILE HYDROCHLORIDE see GHI100
GLYCINONITRILE, MONOHYDROCHLORIDE see GHI100
GLYCIRENAN see VGP000
GLYCOALKALOID EXTRACT from POTATO BLOSSOMS see PLW550
GLYCOCOIL HYDROCHLORIDE see GHK000
GLYCODIAZINE SODIUM SALT see GHK200
GLYCODINE see TCY750
GLYCO-FLAVINE see DBX400
GLYCOHYDROCHLORIDE see GHK000
GLYCOL see EJC500
GLYCOL ALCOHOL see EJC500
GLYCOLANILIDE see GHK500
GLYCOL BIS(HYDROXYETHYL) ETHER see TJQ000
GLYCOL BROMIDE see EIY500
GLYCOL BROMOHYDRIN see BNI500
GLYCOL BUTYL ETHER see BPJ850
GLYCOL CARBONATE see GHM000
GLYCOL CHLOROHYDRIN see EIU800
GLYCOL CYANOHYDRIN see HGP000
GLYCOL DIACETATE see EJD759
GLYCOL DIBROMIDE see EIY500
GLYCOL DICHLORIDE see EIY600
GLYCOL, DIETHOXYTETRAETHYLENE see PBO250
GLYCOL DIFORMATE see EJF000
GLYCOL DIMETHACRYLATE see BKM250
GLYCOL DIMETHYL ETHER see DOE600
GLYCOLDINITRAAT (DUTCH) see EJG000
GLYCOL DINITRATE see EJG000
GLYCOL (DINITRATE DE) (FRENCH) see EJG000
GLYCOL ETHER see DJD600
GLYCOL ETHER de ACETATE see CBQ750
GLYCOL ETHER DB see DJF200
GLYCOL ETHER DB ACEATATE see BQP500
GLYCOL-ETHERDIAMINETETRAACETIC ACID see EIT000
GLYCOL ETHER EB see BPJ850
GLYCOL ETHER EB ACETATE see BPJ850
GLYCOL ETHER EE see EES350
GLYCOL ETHER EE ACETATE see EES400
GLYCOL ETHER EM see EJH500
GLYCOL ETHER EM ACETATE see EJJ500
GLYCOL ETHERS see GHN000
GLYCOL ETHYLENE ETHER see DVQ000
GLYCOL ETHYL ETHER see DJD600, EES350
GLYCOL FORMAL see DVR800
GLYCOLIC ACID see GHO000
GLYCOLIC ACID, ETHYL ESTER, METHYL PHTHALATE see MOD000
GLYCOLIC ACID PHENYL ETHER see PDR100
GLYCOLIC NITRILE see HIM500
GLYCOLIXIR see GHA000
GLYCOLLIC ACID PHENYL ETHER see PDR100
GLYCOL METHACRYLATE see EJH000

GLYCOLMETHYL ETHER see EJH500
GLYCOL MONOACETATE see EJI000
GLYCOL-MONOACETIN see EJI000
GLYCOL MONOBUTYL ETHER see BPJ850
GLYCOL MONOBUTYL ETHERACETATE see BPM000
GLYCOLMONOCHLOORHYDRINE (DUTCH) see EIU800
GLYCOL MONOCHLOROHYDRIN see EIU800
GLYCOL MONOETHYL ETHER see EES350
GLYCOL MONOETHYL ETHER ACETATE see EES400
GLYCOL MONOMETHACRYLATE see EJH000
GLYCOL MONOMETHYL ETHER see EJH500
GLYCOL MONOMETHYL ETHER ACETATE see EJJ500
GLYCOL MONOMETHYL ETHER ACETYLRICINOLEATE see MIF500
GLYCOL MONOMETHYL ETHER ACRYLATE see MIF750
GLYCOL MONOPHENYL ETHER see PER000
GLYCOL MONOSTEARATE see EJM500
GLYCOLONITRILE see HIM500
GLYCOLONITRILE ACETATE see COP750
(GLYCOLOYLOXY)TRIBUTYLSTANNANE see GHQ000
GLYCOL POLYETHYLENE MONOSTEARATE 200 see PJV500
GLYCOLPYRAMIDE see GHR609
GLYCOLS, POLYETHYLENE, (ALKYLIMINO)DIETHYLENE ETHER,
    MONOFATTY ACID ESTER see PKE550
GLYCOLS, POLYETHYLENE, DIMETHYL ETHER see DOM100
GLYCOLS, POLYETHYLENE, MONO(NONYLPHENYL) ETHER see NND500
GLYCOLS, POLYETHYLENE, MONO(1,1,3,3-TETRAMETHYLBUTYL)PHE-
    NYL) ETHER see GHS000
GLYCOLS, POLYETHYLENE MONO(TRIMETHYLNONYL) see GHU000
GLYCOLS, POLYETHYLENE POLYPROPYLENE, MONOBUTYL ETHER (non-
    ionic) see GHY000
GLYCOL STEARATE see EJM500
GLYCOL SULFATE see EJP000
GLYCOL SULFITE see COV750
GLYCOLYLUREA see HGO600
GLYCOMONOCHLORHYDRIN see EIU800
GLYCOMUL O see SKV100
GLYCOMUL S see SKV150
GLYCON S-70 see SLK000
GLYCON DP see SLK000
GLYCONIAZIDE see GBB500
GLYCONITRILE see HIM500
GLYCONORMAL see GHK200
GLYCON RO see OHU000
GLYCON TP see SLK000
GLYCON WO see OHU000
GLYCOPHEN see GIA000
GLYCOPHENE see GIA000
GLYCOPYRROLATE see GIC000
GLYCOPYRROLATE BROMIDE see GIC000
GLYCOPYRRONIUM BROMIDE see GIC000
GLYCOSPERSE L-20X see PKL000
GLYCOSPERSE O-20 see PKL100
GLYCOSPERSE TO-20 see TOE250
GLYCOSPERSE TS 20 see SKV195
GLYCO STEARIN see HKJ000
GLYCOTUSS see RLU000
GLYCOXALINEDICARBOXYLIC ACID see IAM000
GLYCURONE see GFM000
GLYCYCLAMIDE see CPR000
GLYCYL ALCOHOL see GGA000
2-GLYCYLAMINOFLUORENE see AKC000
α-GLYCYRRHETINIC ACID see GIE000
β-GLYCYRRHETINIC ACID see GIE050
GLYCYRRHETINIC ACID HYDROGEN SUCCINATE DISODIUM SALT
    see CBO500
18-β-GLYCYRRHETINIC ACID HYDROGEN SUCCINATE DISODIUM SALT
    see CBO500
GLYCYRRHIZA see LFN300
GLYCYRRHIZAE (LATIN) see LFN300
GLYCYRRHIZA EXTRACT see LFN300
GLYCYRRHIZIC ACID, AMMONIUM SALT see GIE100
GLYCYRRHIZINA see LFN300
GLYCYRRHIZINIC ACID see GIG000
GLYCYRRHIZINIC ACID DISODIUM SALT see DXD875
GLYECINE A see TFI500
GLYESTRIN see ECU750
GLYFERRO see FBD500
GLYFYLLIN see DNC000
GLYHEXYLAMIDE see AKQ250
GLYHEXYLAMINE ISODIANE see AKQ250
GLYKOKOLAN SODNY (CZECH) see SHT000
GLYKOLDINITRAT (GERMAN) see EJG000
GLYME see DOE600
GLYME-3 see TKL875

GLYME-23 see DOM100
GLYMIDINE SODIUM SALT see GHK200
GLYMOL see MQV750
GLYODEX 3722 see CBG000
GLYODIN see GII000
GLYODIN ACETATE see GII000
GLYOTOL see GGS000
GLYOXAL see GIK000
GLYOXAL, DIOXIME see EEA000
GLYOXALIN see IAL000
GLYOXALINE see IAL000
GLYOXALINE-5-ALANINE see HGE700
GLYOXIDE see GII000, GIO000
GLY-OXIDE see HIB500
GLYOXIDE DRY see GII000
GLYOXIME see EEA000
GLYOXYLALDEHYDE see GIK000
GLYOXYLIC ACID see GIQ000
GLYOXYLIC ACID, METHYL ESTER see MKI550
GLYPASOL see GEW750
GLYPED see THM500
GLYPHOSATE see PHA500
GLYPHOSINE see BLG250
GLYPHYLLIN see DNC000
GLYPHYLLINE see DNC000
GLYSANOL B see ART250
GLYSOLETTEN see EOK000
G-M-F see DVR200
GMI see DLS600
GMP DISODIUM SALT see GLS800
5'-GMP DISODIUM SALT see GLS800
GMP SODIUM SALT see GLS800
GM SULFATE see GCS000
GNOSCOPINE see NBP275
Gn-RH see LIU360
GnRH-A see GJI100
GNS see GEQ425
GO-80 see FBV100
GO 186 see SHL500
Go 919 see EOA500
GO-1261 see EIH000
GO 10213 see SAY950
GOE 1734 see DBQ125
GOE 687 (GERMAN) see MNG000
GOHSENOLS see PKP750
GOHSENYL E 50 Y see AAX250
D-GOITRIN see VQA100
(R)-GOITRIN see VQA100
GOLARSYL see ACN250
GOLD see GIS000
1721 GOLD see CNI000
GOLD(I) ACETYLIDE see GIT000
GOLDBALLS see FBS100
GOLD BOND see CCP250
GOLD BRONZE see CNI000
GOLD CHLORIDE see GIW176
GOLD(III) CHLORIDE see GIW176
GOLD COMPOUNDS see GIW179
GOLD(I) CYANIDE see GIW189
GOLD DUST see CNE750
GOLDEN ANTIMONY SULFIDE see AQF500
GOLDEN BROWN RK-FQ see CMP250
GOLDEN CEYLON CREEPER see PLW800
GOLDEN CHAIN see GIW195
GOLDEN DEWDROP see GIW200
GOLDEN HUNTER'S ROBE see PLW800
GOLDEN HURRICANE LILY see REK325
GOLDEN POTHOS see PLW800
GOLDEN RAIN see GIW300
GOLDEN SHOWER see GIW300
GOLDEN SPIDER LILY see REK325
GOLDEN YELLOW see DCZ000, DUX800
GOLD FLAKE see GIS000
GOLD(III) HYDROXIDE-AMMONIA see GIX300
GOLD LEAF see GIS000
GOLD NITRIDE AMMONIA see GIY000
GOLD(I) NITRIDE-AMMONIA see GIY300
GOLD(III) NITRIDE TRIHYDRATE see GIZ000
GOLD ORANGE see MND600
GOLD ORANGE MP see DOU600
GOLD POWDER see GIS000
GOLD SATINOBRE see LDS000
GOLD SODIUM THIOMALATE see GJC000
GOLD SODIUM THIOSULFATE see GJE000

GOLD SODIUM THIOSULFATE DIHYDRATE see GJG000
GOLD THIOGLUCOSE see ART250
GOLD TRICHLORIDE see GIW176
GOLDWEED see FBS100
GOLTIX see ALA500
GOMBARDOL see SNM500
GOMISIN A see SBE450
GONABION see CMG675
GONACRINE see DBX400, XAK000
GONADEX see CMG675
GONADOGRAPHON LUTEINIZING HORMONE see GJI075
GONADORELIN see LIU360
GONADORELIN DIACETATE see LIU380
GONADOTRAPHON FSH see SCA750
GONADOTROPIN-RELEASING FACTOR see LIU360
GONADOTROPIN RELEASING HORMONE see LIU360
GONADOTROPIN RELEASING HORMONE AGONIST see GJI100
GONADYL see SCA750
GONA-1,3,5(10)-TRIENE-3,16-α-17-β-TRIOL, 13-ETHYL- see HGI700
GONA-1,3,5(10)-TRIENE-3,16,17-TRIOL, 13-ETHYL-, (16-α-17-β)-(9CI) see HGI700
GONDAFON see GHK200
GONOCRIN see XAK000
GONOSAN see GJI250
GONTOCHIN see CLD000
GONTOCHIN PHOSPHATE see CLD250
GONYAULAX TOXIC DIHYDROCHLORIDE see SBA500
GOOD-RITE see PJR000
GOODRITE 1800X73 see SMR000
GOOD-RITE GP 264 see DVL700
GOOLS see MBU550
GOOSEBERRY TOMATO see JBS100
GOPHACIDE see BIM000
GORDOLOBO YERBA (MEXICO) see RBA400
GORDONA see RAF100
GORDON'S MECOMEC see CLO200
GORE-TEX see TAI250
GORMAN see SCA750
GOSLING see PAM780
GOSSYPLURE see GJK000
GOSSYPLURE H.F. see GJK000
GOSSYPOL see GJM000
(+)-GOSSYPOL see GJM025
(±)-GOSSYPOL see GJM030
racemic-GOSSYPOL see GJM030
GOSSYPOL ACETATE see GJM035, GJM259
GOSSYPOL ACETIC ACID see GJM259
GOTAMINE TARTRATE see EDC500
GOTA de SANGRE (CUBA) see PCU375
GOTHNION see ASH500
GOUGEROTIN see ARP000
GOUT STALK see CNR135
G 2 (OXIDE) see AHE250
GOYL see ABX500
GP-121 see AOO500
GP 130 see DTN775
GP 33679 see DPX400
GP 38383 see IBP309
GP 45840 see DEO600
GP-40-66:120 see HCD250
GPKh see HAR000
GR see GLS000
GR-23 see GFG200
GR-29 see GFG205
GR 2/234 see AFK875
GR 33040 see DCV800
GRAAFINA see EDP000
de GRAAFINA see EDP000
GRACET VIOLET 2R see DBP000
GRAFESTROL see DKA600
GRAHAM'S SALT see SII500
GRAIN ALCOHOL see EFU000
GRAINES D'EGLISE (GUADELOUPE) see RMK250
GRAINS de LIN PAYS (HAITI) see LED500
GRAIN SORGHUM HARVEST-AID see SFS000
GRAMAXIN see CCS250
GRAMEVIN see DGI400, DGI600
GRAMICIDIN see GJO000
GRAMICIDIN A see GJO025
GRAMIN see DYC000
GRAMINE see DYC000
GRAMINIC ACID see GJQ100
GRAMISAN see MEP250
GRAMOXONE METHYL SULFATE see PAJ250
GRAMOXONE S see PAI990

GRAMOZONE see PAJ000
GRAMPENIL see AIV500
GRANALINO (DOMINICAN REPUBLIC) see LED500
GRANATICIN see GJS000
GRANATICIN A see GJS000
GRANDAXIN see GJS200
GRANEX O see SFS000
GRANOSAN see CHC500
GRANOSAN M see EME100, EME500
GRANOX NM see HCC500
GRANOX PPM see CBG000
GRANULAR ZINC see ZBJ000
GRANULATED SUGAR see SNH000
GRANULIN see ACR300
GRANUREX see BRA250
GRANUTOX see PGS000
GRAPE BLUE A GEIGY see FAE100
GRAPEFRUIT OIL see GJU000
GRAPEFRUIT OIL, expressed see GJU000
GRAPEFRUIT OIL, coldpressed see GJU000
GRAPE SUGAR see GFG000
GRAPHITE (MAK) see CBT500
GRAPHITE, NATURAL (ACGIH) see CBT500
GRAPHITE, SYNTHETIC see CBT500
GRAPHOL see MGJ750
GRAPHTAL RED RL see CJD500
GRAPHTOL RED A-4RL see MMP100
GRAPHTOL RED 2GL see DVB800
GRAPHTOL YELLOW A-HG see DEU000
GRASAL BRILLIANT YELLOW see DOT300
GRASAL YELLOW see FAG130
GRASAN ORANGE 3R see XRA000
GRASAN ORANGE R see PEJ500
GRASCIDE see DGI000
GRASLAN see BSN000
GRASOL BLUE 2GS see TBG700
GRASOL VIOLET R see DBP000
GRASSLAND WEEDKILLER see BAV000
GRATIBAIN see OKS000
GRATUS STROPHANTHIN see OKS000
GRAVIDOX see PPK250
GRAVINOL see DYE600
GRAVISTAT see NNL500
GRAVOCAIN see DHK400
GRAVOCAIN HYDROCHLORIDE see DHK600
GRAVOL see DYE600
GRAY ACETATE see CAL750
GRAY AMBER see AHJ000
GRAYANOTOXANE-3,5,6,10,14,16-HEXOL 14 ACETATE see AOO375
GRAYANOTOXIN I see AOO375
GREAT BLUE LOBELIA see CCJ825
GREEN 5 see ADF000
1724 GREEN see FAG000
11091 GREEN see BLK000
11661 GREEN see CMJ900
GREEN CHLOROPHYL see CKN000
GREEN CHROME OXIDE see CMJ900
GREEN CHROMIC OXIDE see CMJ900
GREEN CINNABAR see CMJ900
GREEN CROSS COUCH GRASS KILLER see TII500
GREEN CROSS CRABGRASS KILLER see PLC250
GREEN CROSS WARBLE POWDER see RNZ000
GREEN-DAISEN M see DXI400
GREEN DRAGON see JAJ000
GREENHARTEN see HLY500
GREEN HELLEBORE see FAB100, VIZ000
GREEN HYDROQUINONE see QFJ000
GREEN LILY see GJU460
GREEN LOCUST see GJU475
GREEN MAMBA VENOM see GJU500
GREEN NICKEL OXIDE see NDF500
GREEN No. 2 see BLK000
GREEN No. 203 see FAF000
GREENOCKITE see CAJ750
GREEN OIL see APG500, COD750
GREEN ROUGE see CMJ900
GREEN SEAL-8 see ZKA000
GREEN VITRIOL see FBN100
GREEN VITROL see FBO000
GRELAN see AMK250
GREOSIN see GKE000
GRESFEED see GKE000
GREY ARSENIC see ARA750
GREY NICKER see CAK325

GRICIN see GKE000
GRIFFEX see ARQ725
GRIFFIN MANEX see MAS500
GRIFFITH'S ZINC WHITE see LHX000
GRIFOMIN see TEP500
GRIFULVIN see GKE000
GRILON see NOH000, PJY500
GRIPENIN see CBO250
GRIPPEX see TEH500
GRISACTIN see GKE000
GRISCOFULVIN see GKE000
GRISEFULINE see GKE000
GRISEIN see GJU800
GRISEMIN see CEX250
GRISEO see GKE000
(+)-GRISEOFULVIN see GKE000
GRISEOFULVIN-FORTE see GKE000
GRISEOFULVINUM see GKE000
GRISEOLUTEIN B see GJW000
GRISEOMYCIN see GJY000
GRISEORUBIN COMPLEX see GJY100
GRISEORUBIN I HYDROCHLORIDE see GKA000
GRISEOVIRIDIN see GKC000
GRISETIN see GKE000
GRISIN see CEX250, GJU800
GRISOFULVIN see GKE000
GRISOL see TCF250
GRISOVIN see GKE000
GRIS-PEG see GKE000
GROCEL see GEM000
GROCO see LGK000
GROCO 2 see OHU000
GROCO 4 see OHU000
GROCO 54 see SLK000
GROCO 5L see OHU000
GROCOLENE see GGA000
GROCOR 5500 see OAV000
GROCOR 6000 see OAV000
GROFAS see QSA000
GROUND HEMLOCK see YAK500
GROUNDNUT OIL see PAO000
GROUND RYANIA SPECISA(VAHL) STEMWOOD (ALKOLOID RYANODINE)
   see RSZ000
GROUNDSEL see RBA400
GROUND VOCLE SULPHUR see SOD500
GR 20263 PENTAHYDRATE see CCQ200
GRUNDIER ARBEZOL see PAX250
GRUNERITE see GKE900
GRYSIO see GKE000
GRYZBOL see DVF800
GRZYBOL see DVF800
GS-95 see TEZ000
GS 015 see BMB125
GS-3065 see DYE425
GS 6244 see FOI000
GS 13529 see BQB000
GS-13,798 see MPG250
GS 14259 see BQC500
GS 15254 see BQC250
GS-16068 see EPN500
GS 34360 see INR000
GSH see GFW000
G-STROPHANTHIN see OKS000
G 52 SULFATE see GAA120
GT see PCU360
GT41 see BOT250
GT-1012 see DNB000
GT 2041 see BOT250
GTB see GGU000
GTG see ART250
GTN see NGY000
GUABENXANE see GKG300
GUACALOTE AMARILLO (CUBA) see CAK325
GUACAMAYA (CUBA) see CAK325
GUACIS (MEXICO) see LED500
GUAIAC GUM see GLY100
GUAIACOL see GKI000
m-GUAIACOL see REF050
GUAIACOLGLICERINETERE see RLU000
GUAIACOL GLYCERYL ETHER see RLU000
GUAIACOL GLYCERYL ETHER CARBAMATE see GKK000
GUAIACURANE see RLU000
GUAIAC WOOD OIL see GKM000
GUAIACYL GLYCERYL ETHER see RLU000

GUAIACYLPROPANE see MFM750
GUAIA-1(5),7(11)-DIENE see GKO000
GUAIAMAR see RLU000
GUAIANESIN see RLU000
s-GUAIAZULENE see DSJ800
GUAICOL see GKI000
GUAIENE see GKO000
β-GUAIENE see GKO000
GUAIFENESIN see RLU000
GUAIMERCOL see GKO500
GUAIPHENESINE see RLU000
GUAJACOL-GLYCERINAETHER (GERMAN) see RLU000
GUAJACOL-α-GLYCERINETHER see RLU000
GUANABENZ see GKO750
GUANABENZ ACETATE see GKO750
GUANATOL HYDROCHLORIDE see CKB500
GUANAZODINE see GKO800
GUANAZODINE SULFATE MONOHYDRATE see CAY875
GUANAZOL see AJO500
GUANAZOLE see DCF200
GUANAZOLO see AJO500
GUANERAN see AKY250
GUANETHIDINE see GKQ000
GUANETHIDINE BISULFATE see GKS000
GUANETHIDINE MONOSULFATE see GKU000
GUANETHIDINE SULFATE see GKS000
GUANFACINE HYDROCHLORIDE see GKU300
GUANICAINE see PDN500
GUANICIL see AHO250
GUANIDINE see GKW000
GUANIDINE MONOHYDROCHLORIDE see GKY000
GUANIDINE MONONITRATE see GLA000
GUANIDINE NITRATE (DOT) see GLA000
GUANIDINE THIOCYANATE see TFF250
GUANIDINIUM CHLORIDE see GKY000
GUANIDINIUM DICHROMATE see GLB100
GUANIDINIUM NITRATE see GLB300
GUANIDINIUM PERCHLORATE see GLC000
GUANIDINIUM THIOCYANATE see TFF250
p-GUANIDINOBENZOIC ACID 4-ALLYL-2-METHOXYPHENYL ESTER
   see MDY300
p-GUANIDINOBENZOIC ACID 4-METHYL-2-OXO-2H-1-BENZOPYRAN-7-YL
   ESTER see GLC100
p-GUANIDINOBENZOIC ACID p-NITROPHENYL ESTER see NIQ500
γ-GUANIDINOBUTYRIC ACID AMIDE, NITROSATED see NJU000
1-(2-GUANIDINOETHYL)HEPTAMETHYLENIMINE see GKQ000
N-(2-GUANIDINO ETHYL)HEPTAMETHYLENIMINE SULFATE see GKU000
1-(2-GUANIDINOETHYL)OCTAHYDROAZOCINE see GKQ000
1-(2-GUANIDINOETHYL)OCTAHYDROAZOCINE SULFATE (2:1) see GKS000
γ-GUANIDINO-β-HYDROXYBUTYRIC ACID AMIDE, NITROSATED see
   HNA000
GUANIDINO-6-METHYL-1,4-BENZODIOXANE see GKG300
6-GUANIDINOMETHYL 1,4-BENZODIOXANE SULFATE see GKG300
GUANIDINO METHYL-6-BENZODIOXANNE-1,4 (FRENCH) see GKG300
GUANINE see GLI000
GUANINE DEOXYRIBOSIDE see DAR800
GUANINE HYDROCHLORIDE see GLK000
GUANINE-3-N-OXIDE see GLK100
GUANINE-7-N-OXIDE see GLM000
GUANINE-3-N-OXIDE HEMIHYDROCHLORIDE see GLO000
GUANINE RIBOSIDE see GLS000
GUANIOL see DTD000
GUANOCTINE see GLQ000
GUANOSINE see GLS000
GUANOTHIAZON see AHI875
GUANOXYFEN SULFATE see GLS700
GUANYL DISULFIDE DIHYDROCHLORIDE see DXN400
GUANYL HYDRAZINE see AKC750
GUANYLIC ACID SODIUM SALT see GLS800
1-GUANYL-4-NITROSAMINOGUANYLTETRAZENE see TEF500
GUANYL NITROSAMINO GUANYL TETRAZENE (DOT) see TEF500
GUANYL NITROSAMINO GUANYL TETRAZENE, containing, by weight, at
   least 30% water (DOT) see TEF500
N¹-GUANYLSULFANILAMIDE see AHO250
GUAR see GLU000
GUARANINE see CAK500
GUAR FLOUR see GLU000
GUAR GUM see GLU000
GUASTIL see EPD500
GUATEMALA LEMONGRASS OIL see LEH000
GUAVA see GLW000
GUAYIGA (DOMINICAN REPUBLIC) see CNH789
GUBERNAL see AGW000
GUDAKHU (INDIA) see SED400

GUESAPON see DAD200
GUESAROL see DAD200
GUIACOL-GLICERILETERE MONOCARBAMMATO
  see GKK000
GUICITRINA see AIV500
GUICITRINE see AIV500
GUIDAZIDE see GBB500
GUIGNER'S GREEN see CMJ900
GUINDA (PUERTO RICO) see APM875
GUINEA GREEN B see FAE950
GULF S-15126 see DED000
GULITOL see SKV200
l-GULITOL see SKV200
GULLIOSTIN see PCP250
GUM see PJR000
GUM ARABIC see AQQ500
GUM CAMPHOR see CBA750
GUM CARRAGEENAN see CCL250
GUM CHON 2 see CCL250
GUM CHROND see CCL250
GUM CYAMOPSIS see GLU000
GUM GHATTI see GLY000
GUM GUAIAC see GLY100
GUM GUAR see GLU000
GUM NAFKACRYSTAL see ROH900
GUM OPIUM see OJG000
GUM OVALINE see AQQ500
GUM SENEGAL see AQQ500
GUM STERCULIA see SLO500
GUM TARA see GMA000
GUM TRAGACANTH see THJ250
GUNACIN see GMC000
GUNCOTTON see CCU250
GUNPOWDER see ERF500, PLL750
GURJUN BALSAM see GME000
GURONSAN see GFM000
GUSATHION see ASH500
GUSATHION A see EKN000
GUSERVIN see GKE000
GUSTAFSON CAPTAN 30-DD
  see CBG000
GUTHION (DOT) see ASH500
GUTHION, liquid (DOT) see ASH500
GUTHION (ETHYL) see EKN000
GUTRON see MQT530
GUTTAGENA see PKQ059
GUTTALAX see SJJ175
β-GUTTIFERIN see CBA125
α-2-GUTTIFERIN see GME300
B''-GUTTIFERIN see CBA125
A''2-GUTTIFERIN see GME300
α-GUTTIFERIN (9CI) see GME300
GYKI 11679 see RFU800
GYKI 14166 see PEE100
GYNAESAN see EDU500
GYN-ANOVLAR see EEH520
GYNECLORINA see CDP000
GYNECORMONE see EDP000
GYNEFOLLIN see DAL600
GYNE-LOTRIMIN see MRX500
GYNE-MERFEN see MDH500
GYNERGEN see EDC500
GYNERGON see EDO000
GYNESTREL see EDO000
GYNFORMONE see EDP000
GYNOCHROME see MCV000
GYNOESTRYL see EDO000
GYNOFON see ABY900
GYNOKHELLAN see AHK750
GYNOLETT see DKB000
GYNONLAR 21 see EEH520
GYNO-PEVARYL see EAE000
GYNOPHARM see DKA600
GYNOPLIX see ABX500
GYNOREST see DYF759
GYNOVLAR see EQM500
GY-PHENE see CDV100
GYPSINE see LCK000
GYPSOGENIN see GMG000
GYPSOPHILASAPOGENIN see GMG000
GYPSOPHILASAPONIN see GMG000
GYPSUM see CAX500, CAX750
GYPSUM STONE see CAX750
GYROMITRIN see AAH000

GYRON see DAD200

H-28 see BSG250
H-33 see BPP250
H-34 see HAR000
H 46 see AHC000
H-69 see DTN200
H-88 see TKF250
H 95 see CKF750
H115 see CDY325
H 133 see DER800
H.17B see HHQ825
H 224 see BRA625
H241 see EQN259
H 321 see DST000
H-365 see ELL500
H-490 see ENG500
H 610 see EOM000
H 774 see BKO825
H-899 see AKG500
H 940 see ENG500
H 990 see AEX000
"H" see HBT500
H 1032 see MHJ500
H-1075 see HAA300
H-1286 see HAA310
H 1313 see DER800
H 1672 see PEF750
2903 H see CDS275
H3 111 see DBC510
H 3292 see DNN600
H 3452 see FBP100
H 4007 see MCH535
H4 099 see DTL200
H-4723 see CIR750
H 8717 see PMN250
96H60 see DFH600
H 56/28 see AGW250
H 59/64 see DUM100
H 93/26 see MQR144
H-1067 see BKO840
cis-H 102.09 see ZBA500
HA see HHQ800
HA 106 see TLG000
70H-2AAF see HIK500
HAAS see SMT500
HABA de SAN ANTONIO (PUERTO RICO) see CAK325
HABILLA (PUERTO RICO) see YAG000
HACHI-SUGAR see SGC000
H ACID see AKH000
HADACIDIN see FNO000
HADACIDINE see FNO000
HADACIN see FNO000
2,4-HADIYNYLENE CHLOROFORMATE see HAB600
HAEMATITE see HAO875
HAEMODAN see MGC350
HAEMOFORT see FBO000
HAEMOSTASIN see VGP000
HAFFKININE see ARQ250
HAFNIUM see HAC000
HAFNIUM (wet) see HAC500
HAFNIUM, wet with not less than 25% water (DOT) see HAC000
HAFNIUM CHLORIDE see HAC800
HAFNIUM CHLORIDE OXIDE see HAD500
HAFNIUM CHLORIDE OXIDE OCTAHYDRATE see HAD000
HAFNIUM DICYCLOPENTADIENE DICHLORIDE see HAE500
HAFNIUM METAL, dry (DOT) see HAC000
HAFNIUM METAL, wet (DOT) see HAC500, HAC000
HAFNIUM OXYCHLORIDE see HAD500
HAFNIUM OXYCHLORIDE OCTAHYDRATE see HAD000
HAFNIUM TETRACHLORIDE see HAC800
HAFNIUM(IV) TETRAHYDROBORATE see HAE000
HAFNOCENE DICHLORIDE see HAE500
HAIARI see RNZ000
HAIMASED see SIA500
HAIROXAL see PII100
HAIRY see HBT500
HAITIN see HON000
HALAMID see CDP000
HALANE see DFE200
HALARSOL see DFX400
HALAZONE see HAF000
HALBMOND see BAU750
HALCIDERM see HAF300

HALCIMAT see HAF300
HALCINONIDE see HAF300
HALCION see THS800
HALCORT see HAF300
HALDI RHIZOME EXTRACT see COF850
HALDOL see CLY500
HALDRONE see PAL600
HALF-CYSTEINE see CQK000
HALF-CYSTINE see CQK000
HALF-MUSTARD GAS see CGY750, CHC000
HALF-MYLERAN see EMF500
HALF SULFUR MUSTARD see CHC000
HALIDO see BAV250
HALIDOR see POD750
HALITE see SFT000
HALIZAN see TDW500
HALKAN see DYF200
HALLOYSITE see HAF375
HALLTEX see SEH000
HALOANISONE see HAF400
HALOANISONE COMPOSITUM see FDA880
HALOCARBON 11 see TIP500
HALOCARBON 14 see CBY250
HALOCARBON 23 see CBY750
HALOCARBON 112 see TBP050
HALOCARBON 113 see FOO000
HALOCARBON 114 see FOO509
HALOCARBON 115 see CJI500
HALOCARBON 112a see TBP000
HALOCARBON 152A see ELN500
HALOCARBON 1132A see VPP000
HALOCARBON C-138 see CPS000
HALOCARBON 13/UCON 13 see CLR250
HALOFANTRINE HYDROCHLORIDE
    see HAF500
HALOFANTRINO (SPANISH) see HAF500
HALOFLEX 208 see ADW200
HALOG see HAF300
HALOMICIN see HAF825
HALOMYCETIN see CDP250
HALON see DFA600
HALON 14 see CBY250
HALON 1001 see MHR200
HALON 1011 see CES650
HALON 1202 see DKG850
HALON 1211 see BNA250
HALON 1301 see TJY100
HALON 2001 see EGV400
HALONISON see HAF400
HALON TFEG 180 see TAI250
HALOPERIDOL see CLY500
HALOPERIDOL DECANOATE see HAG300
HALOPREDONE ACETATE see HAG325
HALOPROPANE see BOF750
HALOPYRAMINE see CKV625
HALOSTEN see CLY500
HALOTAN see HAG500
HALOTESTIN see AOO275
HALOTHANE see HAG500
HALOWAX see TBR000, TIT500
HALOXAZOLAM see HAG800
HALOXON see DFH600
HALSAN see HAG500
HALSO 99 see CLK100
HALTS see BAF250
HALVIC 223 see PKQ059
HALVISOL see HAH000
HAMAMELIS see WCB000
HAMIDOP see DTQ400
HAMILTON RED see CHP500
HAMOVANNID see HFG550
HAMP-ENE 100 see EIV000
HAMP-ENE 215 see EIV000
HAMP-ENE 220 see EIV000
HAMP-ENE ACID see EIX000
HAMP-ENE Na4 see EIV000
HAMP-EX ACID see DJG800
HAMP-OL ACID see HKS000
HAMPSHIRE see IBH000
HAMPSHIRE GLYCINE see GHA000
HAMPSHIRE NTA see SIP500
HAMYCIN see HAH800
HANANE see BJE750
HANCOCK YELLOW 10010 see DEU000

HANQ see HIX200
HANSACOR see DJS200
HANSAMID see NCR000
HANSA ORANGE RN see DVB800
HANSA RED B see MMP100
HANSOLAR see SNY500
HAPA-B SULFATE see SBE800
HAPLOS see EOK000
HAPPY DUST see CNE750
HAPTOCIL see CQE325, PPO000
HAQ see BKB300, BKB325
4HAQO see HIY500
HARMALOL HYDROCHLORIDE see HAI300
HARMAN see MPA050
HARMANE see MPA050
HARMAR see PNA500
HARMINE see HAI500
HARMONIN see MQU750
HARMONYL see RDF000
HARRICAL see CCK125
HARRY see HBT500
HARTOL see MQU750
HART'S HORN see MBU825
HARTSHORN PLANT see PAM780
HARVAMINE see WAK000
HARVATRATE see MGR500
HARVEN see SGD000
HARVEST-AID see SFS000
HAS (GERMAN) see HLB400
HASACH see CBD750
HASETHROL see PBC250
HASHISH see CBD750
HASTELLOY C see CNA750
HATCOL DOP see DVL700
HAURYMELLIN see DQR800
HAVERO-EXTRA see DAD200
HAVIDOTE see EIX000
HAVOC see TAC800
HAY ABSOLUTE see FMS000
HAYNES STELLITE 21 see CNA750
HAZODRIN see MRH209
HB see HEA000
HB[17] see HHQ825
HB-218 see AKM125
HB 419 see CEH700
m-HBA see HJI100
HBB see HCA500
HBBN see HJQ350
HBF 386 see AEA750, AEB000
HBK see HAA320
HBP see HHQ850
9,10-H2 B(e)P see DKU400
HC see CNS750
HC-3 see HAQ000
HC-064 see NGB700
292 HC see PIK400
336 HC see PIK375
HC 1281 see TIK500
1352 HC see DHY200
HC 2072 see NIM500
HC7901 see DUG500
8057HC see DSQ000
HCA see HCL500, HHQ800
HCB see HCC500
HCBD see HCD250
HC BLUE 1 see BKF250
HC BLUE No. 2 see HKN875
α-HCC see HJV000
HCCH see BBP750, BBQ500
HCCPD see HCE500
HCDD see HAJ500
HCE see EBW500
HCFU see CCK630
HC 20-511 FUMARATE see KGK200
HCG see CMG675
HCG-004 see CJP750
HCH see BBQ500
α-HCH see BBQ000
β-HCH see BBR000
Γ-HCH see BBQ500
HCL SALZ des p-AMINO-BENZOESAEURE-DIMETHYLAMINO-AETHYL-
    ESTER (GERMAN) see DPC200
HCl SALZ des p-AMINO-SALICYLSAEURE-DIAETHYLAMINOAETHYLESTER
    (GERMAN) see AMM750

HCl SALZ des p-AMINO-SALICYLSAEURE-DIMETHYLAMINOAETHYL-
ESTER (GERMAN) see AMN000
HCl SALZ DES p,N,N-
BUTYLAMINOSALICYLSAEUREDIAETHYLAMINOAETHYLESTER
(GERMAN) see BQH250
HCN see HHS000
HCN-CNCl MIXTURE see HHS500
HCP see ABD000, HCL000
HC RED No. 3 see ALO750
HCS 3260 see CDR750
HCT see HGO500
HCTZ see CFY000
H. CYANOCINCTUS VENOM see SBI900
HC YELLOW No. 4 see BKC000
HCZ see CFY000
HD 419 see CEH700
HD AMARANTH B see FAG020
7H-DB(c,g)C see DCY000
HDEHP see BJR750
H.D. EUTANOL see OBA000
HDMTX see MDV500
HDO see HEP500
HD OLEYL ALCOHOL 70/75 see OBA000
HD OLEYL ALCOHOL 80/85 see OBA000
HD OLEYL ALCOHOL 90/95 see OBA000
HD OLEYL ALCOHOL CG see OBA000
HD PONCEAU 4R see FMU080
(2-HDYROXYETHYL)TRIMETHYLAMMONIUM see CMF000
17-β-HDYROXY-17-α-METHYL-5-α-ANDROSTANO(2,3-c)FURAZAN see AOO300
HE 166 see DUS500
HEADACHE WEED see CMV390
HEALON see HGN600
HEART-OF-JESUS see CAL125
HEARTS see AOB250, BBK500
HEAT see HAJ700
HEAT PRE see MJG750
HEAVENLY BLUE see DJO000
#6 HEAVY FUEL OILS see HAJ750
HEAVY NAPHTHENIC DISTILLATE see MQV780
HEAVY NAPHTHENIC DISTILLATE SOLVENT EXTRACT see MQV857
HEAVY NAPHTHENIC DISTILLATES (PETROLEUM) see MQV780
HEAVY OIL see CMY825
HEAVY PARAFFINIC DISTILLATE see MQV785
HEAVY PARAFFINIC DISTILLATE, SOLVENT EXTRACT see MQV859
HEAVY WATER see HAK000
HEAVY WATER-d2 see HAK000
HEAZLEWOODITE see NDJ500
HEBABIONE HYDROCHLORIDE see PPK500
HEBANIL see CKP500
HEBARAL see EKT500
HEB-CORT see CNS750
HEBIN see FMT100
HEBUCOL see COW700
HECLOTOX see BBQ500
HECNU see CHB750
HECNU-MS see HKQ300
HEDAPUR M 52 see CIR250
HEDERA CANARIENSIS see AFK950
HEDERA HELIX see AFK950
HEDEX see HIM000
HEDGE PLANT see PMD550
HED-HEPARIN see HAQ550
HEDIONDA (PUERTO RICO) see CNG825
HEDIONDILLA (PUERTO RICO) see LED500
HEDOLIT see DUS700
HEDONAL see DGB000, MOU500
HEDONAL (The herbicide) see DAA800
HEDONAL DP see DGB000
HEDONAL MCPP see CIR500, CLO200
HEDP see HKS780
HEDTA see HKS000
HEDULIN see PFJ750
HEEDTA see HKS000
HEF-2 see JDA100
HEF-3 see JDA125
HE-HK-52 see HAA325
HEKBILIN see CDL325
HEKSAN (POLISH) see HEN000
HEKSOGEN (POLISH) see CPR800
HEKTALIN see VGP000
HELAKTYN BLACK DN see CMS227
HELANTHRENE YELLOW see DCZ000
HELENALIN see HAK300
HELFERGIN see AAE500, DPE000

HEL-FIRE see BRE500
HELFO DOPA see DNA200
HELIANTHINE see MND600
HELIANTHINE B see MND600
HELICHRYSUM OIL see HAK500
HELICOCERIN see ECE500
HELICON see ADA725
HELIC YELLOW GW see DEU000
HELIO FAST ORANGE RN see DVB800
HELIO FAST RED BN see MMP100
HELIOGEN see CDP000
HELIOMYCIN see HAL000
HELIOPAR see CLD000
HELIO RED TONER LCLL see CHP500
HELIOTRIDINE ESTER with LASIOCARPUM and ANGELIC ACID see LBG000
HELIOTRINE see HAL500
HELIOTRON see HAL500
HELIOTROPIN see PIW250
HELIOTROPIN ACETAL see PIZ499
HELIOTROPIUM SUPINUM L. see HAM000
HELIOTROPIUM TERNATUM see SAF500
HELIOTROPYL ACETATE see PIX000
HELIOTROPYL ACETONE see MJR250
HELIUM see HAM500
HELIUM, compressed (DOT) see HAM500
HELIUM, refrigerated liquid (DOT) see HAM500
HELIUM-OXYGEN (mixture) see HAN000
HELIUM-OXYGEN mixture (DOT) see HAN000
HELLEBORE see CMG700
HELLEBOREIN see HAN500
HELLEBORUS NIGER see CMG700
HELLEBRIGENIN-GLUCO-RHAMNOSID (GERMAN) see HAN600
HELLEBRIN see HAN600
HELLEGRIGENIN GLULCORHAMNOSIDE see HAN600
HELLIPIDYL see AMM250
HELMATAC see BQK000
HELMET FLOWER see MRE275
HELMETINA see PDP250
HELMIRANE see DFH600
HELMIRON see DFH600
HELMIRONE see DFH600
HELMOX see COH250
HELODERMA SUSPECTUM VENOM see HAN625
HELOTHION see SOU625
HELOXY WC68 see NCI300
HELVETICOSID (GERMAN) see HAN800
HELVETICOSIDE see HAN800
HELVETICOSIDE DIHYDRATE see HAO000
HEMACHATUS HAEMACHATES VENOM see HAO500
HEMACHATUS HAEMACHATUS VENOM see HAO500
HEMATIN-PROTEIN see CQM325
HEMATITE see HAO875
HEMATOIDIN see HAO900
HEMATOPORPHYRIN see PHV275
HEMATOPORPHYRIN MERCURY DISODIUM SALT see HAP000
HEMATOXYLIN see HAP500
HEMEL see HEJ500
HEMETOIDIN see HAO900
HEMICHOLINE see HAQ000
HEMICHOLINIUM-3 see HAQ000
HEMICHOLINIUM BROMIDE see HAQ000
HEMICHOLINIUM-3-BROMIDE see HAQ000
HEMICHOLINIUM DIBROMIDE see HAQ000
HEMICHOLINIUM-3-DIBROMIDE see HAQ000
HEMIMELLITENE see TLL500
HEMINEVRIN see CHD750
HEMISINE see VGP000
HEMITON see CMX760
HEMLOCK WATER DROPWORT see WAT315
HEMO-B-DOZE see VSZ000
HEMOCAPROL see AJD000
HEMOCOAGULASE see CMY325
HEMODAL see MMD500
HEMODESIS see PKQ250
HEMODEX see DBD700
HEMODEZ see PKQ250
HEMOFURAN see NGE500
HEMOMIN see VSZ000
HEMOPAR see AJD000
HEMOSTASIN see VGP000
HEMOSTYPTANON see EDU500
HEMOTON see FBJ100
HEMOTROPE see PEU000
HENBANE see HAQ100

HENDECANAL see UJJ000
HENDECANALDEHYDE see UJJ000
HENDECANE see UJS000
HENDECANOIC ACID see UKA000
HENDECANOIC ALCOHOL see UNA000
1-HENDECANOL see UNA000
2-HENDECANONE see UKS000
HENDECENAL see ULJ000
10-HENDECEN-1-YL ACETATE see UMS000
HENDECYL ALCOHOL see UNA000
n-HENDECYLENIC ALCOHOL see UNA000
10-HENEDECENOIC ACID see ULS000
HENKEL'S COMPOUND see TBC450
HENNOLETTEN see EOK000
HENU see HKW500
HEOD see DHB400
HEPACHOLINE see CMF750
HEPADIAL see DOK400
HEPAGON see VSZ000
HEPALIDINE see TEV000
HEPAR CALCIS see CAY000
HEPAREGENE see TEV000
HEPARIN see HAQ500
α-HEPARIN see HAQ500
HEPARINATE see HAQ500
HEPARINIC ACID see HAQ500
HEPARINOID see SEH450
HEPARIN SODIUM see HAQ550
HEPARIN SULFATE see HAQ500
HEPARLIPON see DXN800
HEPAR SULFUROUS see PLT250
HEPATHROM see HAQ550
HEPATION see MAO275
HEPAVIS see VSZ000
HEPCOVITE see VSZ000
HEPERAL see DXG600
HEPIN see AJD000
HEPINOID see SEH450
HEPORAL see AOO490
HEPT see TCF250
HEPTABARB see COY500
HEPTABARBITAL see COY500
HEPTABARBITONE see COY500
HEPTABARBUM see COY500
HEPTACAINE see PIO750
HEPTACHLOOR (DUTCH) see HAR000
1,4,5,6,7,8,8-HEPTACHLOOR-3a,4,7,7a-TETRAHYDRO-4,7-endo-METHANO-
INDEEN (DUTCH) see HAR000
HEPTACHLOR see HAR000
HEPTACHLOR (technical grade) see HAR500
HEPTACHLORE (FRENCH) see HAR000
HEPTACHLOR EPOXIDE (USDA) see EBW500
2,2,5-endo,6-exo,8,9,10-HEPTACHLOROBORNANE see THH575
3,4,5,6,7,8,8-HEPTACHLORODICYCLOPENTADIENE see HAR000
3,4,5,6,7,8,8a-HEPTACHLORODICYCLOPENTADIENE see HAR000
1,4,5,6,7,8,8-HEPTACHLORO-2,3-EPOXY-2,3,3a,4,7,7a-HEXAHYDRO-4,7-
METHANOINDENE see EBW500
1,4,5,6,7,8,8-HEPTACHLORO-2,3-EPOXY-3a,4,7,7a-TETRAHYDRO-4,7-
METHANOINDAN see EBW500
1,4,5,6,7,8,8-HEPTACHLORO-3a,4,5,6,7,7a-HEXAHYDRO-4,7-METHANO-1H-
INDENE see SCD500
2,3,4,5,6,7,7-HEPTACHLORO-1a,1b,5,5a,6,6a-HEXAHYDRO-2,5-METHANO-2H-
INDENO(1,2-b)OXIRENE see EBW500
1,4,5,6,7,8,8-HEPTACHLORO-3a,4,7,7a-TETRAHYDRO-4,7-EN-
DOMETHANOINDENE see HAR000
1,4,5,6,7,10,10-HEPTACHLORO-4,7,8,9,-TETRAHYDRO-4,7-EN-
DOMETHYLENEINDENE see HAR000
1,4,5,6,7,8,8a-HEPTACHLORO-3a,4,7,7a-TETRAHYDRO-4,7-METHANOINDANE
see HAR000
1(3a),4,5,6,7,8,8-HEPTACHLORO-3a(1),4,7,7a-TETRAHYDRO-4,7-
METHANOINDENE see HAR000
1,4,5,6,7,8,8-HEPTACHLORO-3a,4,7,7a-TETRAHYDRO-4,7-METHANOINDENE
see HAR000
1,4,5,6,7,8,8-HEPTACHLORO-3a,4,7,7a-TETRAHYDRO 4,7-METHANOINDENE
(technical grade) see HAR500
1,4,5,6,7,8,8-HEPTACHLORO-3a,4,7,7a-TETRAHYDRO-4,7-METHANOL-1H-
INDENE see HAR000
1,4,5,6,7,8,8-HEPTACHLORO-3a,4,7,7,7a-TETRAHYDRO-4,7-METHYLENE
INDENE see HAR000
1,4,5,6,7,8,8-HEPTACHLORO-3a,4,7,7,7a-TETRAHYDRO-4,7-endo-METHANO-
INDEN (GERMAN) see HAR000
HEPTACOSAFLUOROTRIBUTYLAMINE see HAS000
1-HEPTADECANECARBOXYLIC ACID see SLK000
HEPTADECANOIC ACID see HAS500

HEPTADECANOL (mixed primary isomers) see HAT000
2,8,10-HEPTADECATRIENE-4,6-DIYNE-1,14-DIOL see EAO500
8,10,12-HEPTADECATRIENE-4,6-DIYNE-1,14-DIOL, (E,E,E)-(-)- see CMN000
2-(8-HEPTADECENYL)-2-IMIDAZOLINE-1-ETHANOL see AHP500
n-HEPTADECOIC ACID see HAS500
HEPTADECYL CYANIDE see SLL500
2-HEPTADECYL-4,5-DIHYDRO-1H-IMIDAZOLYL MONOACETATE see GII000
2-HEPTADECYL GLYOXALIDINE see GIO000
2-HEPTADECYL GLYOXALIDINE ACETATE see GII000
2-HEPTADECYL-1-HYDROXYETHYLIMIDAZOLINE see HAT500
n-HEPTADECYLIC ACID see HAS500
2-HEPTADECYL-2-IMIDAZOLINE see GIO000
2-HEPTADECYL-2-IMIDAZOLINE ACETATE see GII000
2-HEPTADECYL-2-IMIDAZOLINE-1-ETHANOL see HAT500
HEPTADECYL-TRIMETHYLAMMONIUM METHYLSULFAT (CZECH)
see HAU500
HEPTADECYLTRIMETHYLAMMONIUM METHYLSULFATE see HAU500
2,4-HEPTADIENAL see HAV450
HEPTADIENAL-2,4 see HAV450
1,6-HEPTADIYNE see HAV500
HEPTADONE see MDO750
HEPTADON HYDROCHLORIDE see MDP750
HEPTADORM see COY500
HEPTAFLUORJODPROPAN see HAY300
HEPTAFLUORMASELNAN STRIBRNY (CZECH) see HAW000
HEPTAFLUOROBUTANOIC ACID, SILVER SALT see HAW000
2,2,3,3,4,4,4-HEPTAFLUOROBUTYRAMIDE see HAX000
HEPTAFLUOROBUTYRIC ACID see HAX500
HEPTAFLUOROBUTYRIC ACID, ETHYL ESTER see HAY000
HEPTAFLUOROBUTYRYL HYPOCHLORITE see HAY059
HEPTAFLUOROBUTYRYL HYPOFLUORITE see HAY100
HEPTAFLUOROBUTYRYL NITRATE see HAY200
HEPTAFLUOROIODOPROPANE see HAY300
HEPTAFLUOROISOBUTYLENE METHYL ETHER see HAY500
2-HEPTAFLUOROPROPYL-1,3,4-DIOXAZOLONE see HAY600
HEPTAFLUOROPROPYL HYPOFLUORITE see HAY650
HEPTAFLUR see CDF400
HEPTAGRAN see HAR000
2,3,3',4,4',5,7-HEPTAHYDROXYFLAVAN see HBA259
HEPTAKIS (DIMETHYLAMINO)TRIALUMINUM TRIBORON PENTAHY-
DRIDE see HBA500
Γ-HEPTALACTONE see HBA550
HEPTALDEHYDE see HBB500
HEPTALDEHYDE METHYLANTHRANILATE, Schiff's base see HBO700
HEPTAMAL see COY500
HEPTAMETHYLENEIMINE mixed with SODIUM NITRITE see SIT000
HEPTAMETHYLPHENYLCYCLOTETRASILOXANE see HBA600
HEPTAMINE see TNX750
HEPTAMINOL HYDROCHLORIDE see HBB000
HEPTAMUL see HAR000
HEPTAMYL HYDROCHLORIDE see HBB000
HEPTAN (POLISH) see HBC500
HEPTANAL see HBB500
HEPTANAL, CYCLIC (HYDROXYMETHYL)ETHYLENE ACETAL see HBC000
HEPTANAL-1,2-GLYCERYL ACETAL see HBC000
2-HEPTANAMINE see TNX750
2-HEPTANAMINE SULFATE (2:1) see AKD600
HEPTANDIOIC ACID see PIG000
HEPTANE see HBC500
n-HEPTANE see HBC500
1-HEPTANECARBOXYLIC ACID see OCY000
HEPTANEDICARBOXYLIC ACID see ASB750
1,7-HEPTANEDICARBOXYLIC ACID see ASB750
HEPTANEDINITRILE see HBD000
HEPTANEDIOIC ACID see PIG000
HEPTANE-1,7-DIOIC ACID see PIG000
1,7-HEPTANEDIOIC ACID see PIG000
HEPTANEN (DUTCH) see HBC500
1-HEPTANETHIOL see HBD500
1,1,1,3,5,5,5-HEPTANITROPENTANE see HBD650
HEPTANOIC ACID see HBE000
HEPTANOIC ACID, ISOBUTYL ESTER see IIS000
HEPTANOIC ACID, 2-METHYLPROPYL ESTER see IIS000
HEPTANOIC ACID, ester with TESTOSTERONE see TBF750
1-HEPTANOL see HBL500
2-HEPTANOL see HBE500
HEPTANOL-2 see HBE500
3-HEPTANOL see HBF000
n-HEPTANOL see HBL500
n-HEPTANOL-1 (FRENCH) see HBL500
3-HEPTANOL, 6-(DIMETHYLAMINO)-4,4-DIPHENYL-, ACETATE (ester), HY-
DROCHLORIDE, (3S,6S)-(-)- see ACQ690
HEPTANOL, FORMATE see HBO500
HEPTANOLIDE-1,4 see HBA550

HEPTANOLIDE-4,1 see HBA550
3-HEPTANOL-6-METHYL-3-PHENYL-1-(N-PIPERIDYL) HYDROCHLORIDE
　　see HBF500
HEPTANON see MDO750
HEPTAN-3-ON (DUTCH, GERMAN) see EHA600
2-HEPTANONE see MGN500
3-HEPTANONE see EHA600
HEPTAN-3-ONE see EHA600
4-HEPTANONE see DWT600
HEPTAN-4-ONE see DWT600
3-HEPTANONE, 6-(DIMETHYLAMINO)-4,4-DIPHENYL-, (±)- see MDO760
17-β-HEPTANOYLOXY-19-NOR-17-α-PREGNEN-20-YNONE see NNQ000
HEPTANYL ACETATE see HBL000
HEPTARINOID see SEH450
HEPTA SILVER NITRATE OCTAOXIDE see HBI500
HEPTA-1,3,5-TRIYNE see HBI725
2,4-HEPTDIENAL see HAV450
trans,trans-2,4-HEPTDIENAL see HAV450
HEPTEDRINE see TNX750
4-HEPTENAL see HBI800
cis-4-HEPTEN-1-AL see HBI800
2-HEPTENE see HBJ500
n-HEPTENE see HBJ000
3-HEPTENE (mixed isomers) see HBK350
1-HEPTENE-4,6-DIYNE see HBK450
2-HEPTENOIC ACID see HBK500
HEPTENOPHOS see HBK700
HEPTENYL ACROLEIN see DAE450
HEPTHLIC ACID see HBE000
HEPTOCOAGULASE see CMY325
n-HEPTOIC ACID see HBE000
HEPTYL ACETATE see HBL000
1-HEPTYL ACETATE see HBL000
n-HEPTYL ACETATE see HBL000
HEPTYL ALCOHOL see HBL500
2-HEPTYLAMINE see TNX750
3-HEPTYLAMINE see HBM500
2-HEPTYLAMINE SULFATE see AKD600
8-HEPTYLBENZ(a)ANTHRACENE see HBN000
5-n-HEPTYLBENZ(1:2)BENZANTHRACENE see HBN000
HEPTYL CARBINOL see OEI000
HEPTYL CELLOSOLVE see HBN250
α-HEPTYL CYCLOPENTANONE see HBN500
2-n-HEPTYL CYCLOPENTANONE see HBN500
HEPTYLDICHLORARSINE see HBN600
HEPTYLENE see HBK350
1-HEPTYLENE see HBJ000
HEPTYL ETHER see HBO000
HEPTYL FORMATE see HBO500
HEPTYL HYDRAZINE see HBO600
HEPTYL HYDRIDE see HBC500
n-HEPTYLIC ACID see HBE000
HEPTYLIDENE ALDEHYDE see NNA300
HEPTYLIDENE METHYL ANTHRANILATE see HBO700
HEPTYL MERCAPTAN see HBD500
n-HEPTYLMERCAPTAN see HBD500
n-HEPTYL METHANOATE see HBO500
HEPTYLMETHYLINITROSAMINE see HBP000
HEPTYL METHYL KETONE see NMY500
1-HEPTYL-1-NITROSOUREA see HBP250
n-HEPTYL NITROSUREA see HBP250
2-HEPTYLOXYCARBANILIC ACID-2-(1-PIPERIDINYL)ETHYL ESTER HYDRO-
　　CHLORIDE see PIO750
2-(HEPTYLOXY)ETHANOL see HBN250
2-(2-(HEPTYLOXY)ETHOXY)ETHANOL see HBP275
(2-(HEPTYLOXY)PHENYL)CARBAMIC ACID-2-(1-PIPERIDINYL)ETHYL
　　ESTER HYDROCHLORIDE see PIO750
N-(2-(HEPTYLOXYPHENYLCARBAMOYLOXY)ETHYL)PIPERIDINIUM CHLO-
　　RIDE see PIO750
HEPTYL PHTHALATE see HBP400
n-HEPTYL-Δ-VALEROLACTONE see HBP450
2-HEPTYN-1-OL see HBS500
HEPZIDE see ENV500
HEPZIDINE MALEATE see HBT000
HERBADOX see DRN200
HERB-ALL see MRL750
HERBAN see HDP500
HERBAN M see MRL750
HERBATIM see VFW009
HERBATOX see DXQ500
HERBAX TECHNICAL see DGI000
HERBAZIN see BJP000
HERBAZOLIN see BAV000
HERB BONNETT see PJJ300

HERB-CHRISTOPHER see BAF325
HERBE-A-BRINVILLIERS (HAITI) see PIH800
HERBE A PLOMB (HAITI) see LAU600
HERBE AUX GEAUX (CANADA) see CMV390
HERBESSER see DNU600
HERBEX see BJP000
HERBICIDE 82 see BNM000
HERBICIDE 273 see DXD000
HERBICIDE 326 see DGD600
HERBICIDE 976 see BMM650
HERBICIDE 6602 see MQR225
HERBICIDE C-2059 see DUK800
HERBICIDE M see CIR250
HERBICIDES, MONURON see CJX750
HERBICIDES, SILVEX see TIX500
HERBICIDE TOTAL see AMY050
HERBIDAL see DAA800
HERBITOX see PCT250
HERBIZOLE see AMY050
HERBOGIL see DRV200
HERBOXY see BJP000
HERCOFLAT 135 see PMP500
HERCOFLEX 260 see DVL700
HERCULES 3956 see CDV100
HERCULES 4580 see DIX000
HERCULES 5727 see COF250
HERCULES 7531 see HDP500
HERCULES 8717 see PMN250
HERCULES 9699 see MIB500
HERCULES 14503 see DBI099
HERCULES P6 see PBB750
HERCULES TOXAPHENE see CDV100
HERCULON see PMP500
HERKAL see DGP900
HERMAL see TFS350
HERMAT FEDK see ZHA000
HERMAT TMT see TFS350
HERMAT ZDM see BJK500
HERMAT Zn-MBT see BHA750
HERMESETAS see BCE500
HERMOPHENYL see MCU500
HEROIEN see HBT500
HEROIIN see HBT500
HEROIN see HBT500
HEROIN HYDROCHLORIDE see DBH400
HEROLAN see HBT500
HEROPON see DBA800
HERPESIL see DAS000
HERPIDU see DAS000
HERPLEX see DAS000
HERPLEX LIQUIFILM see DAS000
HERYL see TFS350
HERZO see POB500
HERZO PROSCILLAN see POB500
HES see HLB400
HESOFEN see EQL000
HESPANDER see HLB400
HESPANDER INJECTION see HLB400
HESPERIDIN see HBU000
HESPERIDIN METHYLCHALCONE see HBU400
HESPERIDOSIDE see HBU000
HESPERITIN-7-RHAMNOGLUCOSIDE see HBU000
HESTRIUM CHLORIDE see HEA500
HET see HCY000
2,3,3',4,4',5,7-HETAHYDROXYFLAVAN see HBA259
HETAMIDE ML see BKE500
HETAPHENONE see DHS200
HETEROAUXIN see ICN000
HETOLIN see MQH250
HETP see HCY000
HETRAZAN see DIW200
HEV-4 see CNA750
HEXA see BBP750
HEXAAMMINECHROMIUM(III) NITRATE see HBU500
HEXAAMMINECOBALT(III) CHLORATE see HBV000
HEXAAMMINECOBALT(III) CHLORITE see HBV500
HEXAAMMINECOBALT(III) HEXANITROCOBALTATE (3⁻) see HBW000
HEXAAMMINECOBALT(III) IODATE see HBW500
HEXAAMMINECOBALT(III) NITRATE see HBX000
HEXAAMMINECOBALT(III) PERCHLORATE see HBX500
HEXAAMMINECOBALT(III) PERMANGANATE see HBY000
HEXAAMMINETITANIUM(III) CHLORIDE see HBY500
HEXAAQUACHROMIUM CHLORIDE see CMK450
HEXAAQUACHROMIUM (III) CHLORIDE see CMK450

4,5,6,7,8,8-HEXACHLOR-$\Delta^{1,5}$-TETRAHYDRO-4,7-METHANOINDEN
  see HCN000
HEXACID 698 see HEU000
HEXACID 898 see OCY000
HEXACID 1095 see DAH400
HEXACID C-7 see HBE000
HEXACID C-9 see NMY000
HEXACOL BLACK PN see BMA000
HEXACOL BRILLIANT BLUE A see FAE000
HEXACOL CARMOISINE see HJF500
HEXACOL ERYTHROSINE BS see FAG040
HEXACOL OIL ORANGE SS see TGW000
HEXACOL ORANGE GG CRYSTALS see HGC000
HEXACOL PONCEAU 4R see FMU080
HEXACOL PONCEAU MX see FMU070
HEXACOL PONCEAU SX see FAG050
HEXACOL RHODAMINE B EXTRA see FAG070
HEXACOL TARTRAZINE see FAG140
HEXACOSE see HCS500
HEXACYANOFERRATE(3-) TRIPOTASSIUM see PLF250
HEXACYANOTRIS(3-DODECYL-1-METHYL-2-PHENYLBENZIMIIMIDAZOLIN-
  IUM FERRATE (3-) see TNH750
HEXACYCLEN TRISULFATE see HCN100
HEXACYCLONAS see SHL500
HEXACYCLONATE SODIUM see SHL500
HEXADECADROL see SOW000
HEXADECAFLUOROHEPTANE see PCH000
HEXADECAFLUORO-1-NONANOL see HCO000
2,2,3,3,4,4,5,5,6,6,7,7,8,8,9,9-HEXADECAFLUORONONANOL see HCO000
1-HEXADECANAMINE see HCO500
1-HEXADECANAMINE HYDROFLUORIDE (9CI) see CDF400
1-HEXADECANAMINIUM, N-HEXADECYL-,N,N-DIMETHYL-, CHLORIDE
  (9CI) see DRK200
HEXADECANOIC ACID see PAE250
HEXADECANOIC ACID, 2-ETHYLHEXYL ESTER (9CI) see OFE100
HEXADECANOIC ACID, ISOPROPYL ESTER see IQW000
HEXADECANOL see HCP000
1-HEXADECANOL see HCP000
HEXADECAN-1-OL see HCP000
n-HEXADECANOL see HCP000
1,16-HEXADECANOLACTONE see OKU000
HEXADECANOLIDE see OKU000
HEXADECENE EPOXIDE see EBX500
2-HEXADECEN-1-OL, 3,7,11,15-TETRAMETHYL-, (R-(R*,R*-(E)))-(9CI)
  see PIB600
n-HEXADECOIC ACID see PAE250
HEXADECYL ALCOHOL see HCP000
n-HEXADECYL ALCOHOL see HCP000
N-HEXADECYLAMINE see HCO500
HEXADECYLAMINE HYDROFLUORIDE see CDF400
HEXADECYL CYCLOPROPANECARBOXYLATE see HCP500
HEXADECYL 2-ETHYLHEXANOATE see HCP550
HEXADECYLIC ACID see PAE250
HEXADECYLMALEINAN DI-n-BUTYLCINICITY (CZECH) see DEI600
HEXADECYL NEODECANOATE see HCP600
1-HEXADECYL-PYRIDINIUM BROMIDE mixture with CHLORO(2-
  HYDROXYETHYL)MERCURY see DVJ500
HEXADECYLPYRIDINIUM CHLORIDE see CCX000
1-HEXADECYLPYRIDINIUM CHLORIDE see CCX000
n-HEXADECYLPYRIDINIUM CHLORIDE see CCX000
1-HEXADECYLPYRIDINIUM CHLORIDE MONOHYDRATE see CDF750
HEXADECYLTRICHLOROSILANE see HCQ000
HEXADECYLTRIETHYLAMMONIUM BROMIDE see CDF500
HEXADECYLTRIMETHYLAMMONIUM BROMIDE see HCQ500
(1-HEXADECYL)TRIMETHYLAMMONIUM BROMIDE see HCQ500
N-HEXADECYLTRIMETHYLAMMONIUM BROMIDE see HCQ500
N-HEXADECYL-N,N,N-TRIMETHYLAMMONIUM BROMIDE see HCQ500
HEXADECYLTRIMETHYLAMMONIUM PENTACHLOROPHENOL see TLN150
omega-h-HEXADEKAFLUORNONANOL-1 (GERMAN) see HCO000
HEXADENOL see HCS500
HEXA-2,4-DIENAL see SKT500
2,4-HEXADIENAL see SKT500
1,4-HEXADIENE see HCR000
1,5-HEXADIENE see HCR500
HEXA-1,5-DIENE see HCR500
1,3-HEXADIENE-5-YNE see HCS100
HEXADIENIC ACID see SKU000
HEXADIENOIC ACID see SKU000
2,4-HEXADIENOIC ACID see SKU000
trans-trans-2,4-HEXADIENOIC ACID see SKU000
2,4-HEXADIENOIC ACID POTASSIUM SALT see PLS750
2,4-HEXADIENOL see HCS500
2,4-HEXADIEN-1-OL see HCS500
2,4-HEXADIEN-1-OL ACETATE see AAU750

2,4-HEXADIENYL ACETATE see AAU750
2-(2,4-HEXADIENYLOXY)ETHANOL see HCT500
1,5-HEXADIEN-3-YNE see HCU500
4,5-HEXADIEN-2-YN-1-OL see HCV000
HEXADIMETHRINE BROMIDE see HCV500
HEXADIONA see DBB200
1,5-HEXADIYNE see HCV850
2,4-HEXADIYNE-1,6-DIOIC ACID see HCV875
1,5-HEXADIYNE-3-ONE see HCV880
2,4-HEXADIYNYLENE BISCHLOROFORMATE see HCW000
HEXADRIN see CQC650, EAT500
HEXADROL see SOW000
HEXAETHYLBENZENE see HCX000
HEXAETHYLDILEAD see TJS000
HEXAETHYLDISTANNOXANE see HCX050
1,1,1,3,3,3-HEXAETHYLDISTANNOXANE see HCX050
HEXAETHYLDISTANNTHIANE see HCX100
1,1,1,3,3,3-HEXAETHYLDISTANNTHIANE see HCX100
HEXAETHYL TETRAPHOSPHATE see HCY000
HEXAETHYL TETRAPHOSPHATE, liquid (DOT) see HCY000
HEXAETHYL TETRAPHOSPHATE and compressed gas mixture (DOT)
  see TEB500
HEXAETHYL TETRAPHOSPHATE, liquid, containing more than 25% hexaethyl
  tetraphosphate (DOT) see HCY000
HEXAETHYL TETRAPHOSPHATE MIXTURE, liquid (DOT) see TEB250
HEXAETHYL TETRAPHOSPHATE MIXTURE, DRY (DOT) see TEB000
HEXAETHYLTRIALUMINUM TRITHIOCYANATE see HCY500
HEXAFEN see HCL000
HEXAFERB see FAS000
HEXAFLUORENIUM DIBROMIDE see HEG000
HEXAFLUOROACETIC ANHYDRIDE see TJX000
HEXAFLUOROACETONE see HCZ000
HEXAFLUOROACETONE HYDRATE see HDA000
HEXAFLUOROACETONE SESQUIHYDRATE see HDE500
HEXAFLUORO ACETONE TRIHYDRATE see HDA500
HEXAFLUOROBENZENE see HDB000
1,1,1,4,5,5-HEXAFLUORO-2-CHLORO-2-BUTENE see CHK750
HEXAFLUORODICHLOROBUTENE see HDB500
HEXAFLUORODIETHYL ETHER see HDC000
HEXAFLUORO FERRATE (3-) TRIAMMONIUM SALT see ANI000
HEXAFLUOROGLUTARIC ACID DIETHYL ESTER see DJK100
HEXAFLUOROGLUTARONITRILE see HDC300
HEXAFLUOROGLUTARYL DIHYPOCHLORITE see HDC425
HEXAFLUOROISOBUTYRIC ACID METHYL ESTER see MKK750
HEXAFLUOROISOPROPANOL see HDC500
HEXAFLUOROISOPROPYLIDENEAMINE see HDD000
HEXAFLUOROISOPROPYLIDENEAMINOLITHIUM see HDD500
HEXAFLUOROKIESELSAIURE (GERMAN) see SCO500
HEXAFLUOROKIEZELZUUR (DUTCH) see SCO500
4,4,4,4',4',4'-HEXAFLUORO-N-NITROSODIBUTYLAMINE see NJN300
HEXAFLUOROPENTANEDIOIC ACID DIETHYL ESTER see DJK100
HEXAFLUOROPHOSPHORIC ACID see HDE000
1,1,1,3,3,3-HEXAFLUORO-2-PROPANOL see HDC500
HEXAFLUORO-2-PROPANONE HYDRATE see HDA000
HEXAFLUORO-2-PROPANONE SESQUIHYDRATE see HDE500
HEXAFLUOROPROPENE see HDF000
HEXAFLUOROPROPYLENE (DOT) see HDF000
HEXAFLUOROSILICATE(2-) DIHYDROGEN see SCO500
HEXAFLUOROSILICATE (2-1) LEAD(II) SALT DIHYDRATE see LDG000
HEXAFLUOROSILICATE (2-), NICKEL see NDD000
HEXAFLUOROSILICATE(2-) STRONTIUM see SMJ000
HEXAFLUORO-SILICATE(2-), THALLIUM see TEK250
HEXAFLUORO VANADATE (3-) TRIAMMONIUM SALT see ANI500
$\alpha,\alpha,\alpha,\alpha,\alpha,\alpha$-HEXAFLUORO-3,5-XYLIDINE see BLO250
1-($\alpha,\alpha,\alpha,\alpha',\alpha',\alpha'$-HEXAFLUORO-3,5-XYLYL)-4-METHYL-3-THIO-SEMICARBAZ-
  IDE see BLP325
HEXAFLUORURE de SOUFRE (FRENCH) see SOI000
HEXAFLUOSILICIC ACID see SCO500
HEXAFLURATE see PLH500
HEXAFLURONIUM BROMIDE see HEG000
HEXAFORM see HEI500
HEXAFUNGIN see HDF100
2,3,5,6,7,8-HEXAHYDRO-9-AMINO-1H-CYCLOPENTA(b)QUINOLINE HYDRO-
  CHLORIDE HYDRATE see AKD775
HEXAHYDROANILINE see CPF500
HEXAHYDROANILINE HYDROCHLORIDE see CPA775
1,2,3,7,8,9-HEXAHYDROANTHANTHRENE see HDF500
HEXAHYDROAZEPINE see HDG000
HEXAHYDRO-1H-AZEPINE see HDG000
HEXAHYDRO-2-AZEPINONE see CBF700
HEXAHYDRO-2H-AZEPIN-2-ONE see CBF700
HEXAHYDRO-2H-AZEPIN-2-ONE HOMOPOLYMER see PJY500
N-((HEXAHYDRO-1H-AZEPIN-1-YL)-AMINO)CARBONYL)-4-
  METHYLBENZENESULFONAMIDE see TGJ500

HEXAMETHYLENAMINE see HEI500
HEXAMETHYLEN-1,6-(N-DIMETHYLCARBODESOXYMETHYL)AMMONIUM DICHLORIDE see HEF200
HEXAMETHYLENE see CPB000
HEXAMETHYLENEAMINE see HEI500
HEXAMETHYLENEBIS((CARBOXYMETHYL)DIMETHYLAMMONIUM), DICHLORIDE, DIDODECYL ESTER see HEF200
1,1'-HEXAMETHYLENEBIS(5-(p-CHLOROPHENYL)BIGUANIDE see BIM250
1,1'-HEXAMETHYLENEBIS(5-(p-CHLOROPHENYL)BIGUANIDE DIACETATE see CDT125
1,1'-HEXAMETHYLENEBIS(5-(p-CHLOROPHENYL)BIGUANIDE DIGLUCON-ATE see CDT250
N,N'-HEXAMETHYLENEBIS(2,2-DICHLORO-N-ETHYLACETAMIDE) see HEF300
HEXAMETHYLENEBIS(DIMETHYL-9-FLUORENYLAMMONIUM BROMIDE) see HEG000
HEXAMETHYLENEBIS(FLUOREN-9-YLDIMETHYLAMMONIUM BROMIDE) see HEG000
HEXAMETHYLENE BIS(9-FLUORENYL DIMETHYLAMMONIUM)DIBRO-MIDE see HEG000
HEXAMETHYLENEBIS(TRIMETHYLAMMONIUM) BROMIDE see HEA000
HEXAMETHYLENE(BISTRIMETHYLAMMONIUM)CHLORIDE see HEA500
HEXAMETHYLENEBIS(TRIMETHYLAMMONIUM) DIBENZENESULFONATE see HEG100
HEXAMETHYLENEBIS(TRIMETHYLAMMONIUM IODIDE) see HEB000
1,6-HEXAMETHYLENEDIAMINE see HEO000
HEXAMETHYLENE DIAMINE, solid (DOT) see HEO000
HEXAMETHYLENE DIAMINE, solution (DOT) see HEO500
HEXAMETHYLENE DIISOCYANATE see DNJ800
HEXAMETHYLENE-1,6-DIISOCYANATE see DNJ800
HEXAMETHYLENEDIISOCYANATE (DOT) see DNJ800
1,6-HEXAMETHYLENE DIISOCYANATE (MAK) see DNJ800
HEXAMETHYLENE GLYCOL see HEP500
HEXAMETHYLENE IMINE (DOT) see HDG000
HEXAMETHYLENEIMINE-3,5-DINITROBENZOATE see ACI250
1-(2-HEXAMETHYLENEIMINOETHYL)-2-OXOCYCLOHEXANECARBOXYLIC ACID BENZYL ESTER HYDROCHLORIDE see HEI000
HEXAMETHYLENETETRAAMINE see HEI500
HEXAMETHYLENETETRAMINE see HEI500
HEXAMETHYLENE TETRAMINE TETRAIODIDE see HEI650
HEXAMETHYLENETETRAMMONIUM TETRAPEROXOCHROMATE(V) see HEJ000
HEXAMETHYLENETRIPEROXYDIAMINE see DCK700
HEXAMETHYLENIMINE see HDG000
2-(β-HEXAMETHYLENIMINOAETHYL)CYCLOHEXANON-2-CAR-BONSAUREBENZYLESTER-HYDROCHLORIDE (GERMAN) see HEI000
HEXAMETHYLENTETRAMIN (GERMAN) see HEI500
HEXAMETHYLERBIUM-HEXAMETHYLETHYLENEDIAMINE LITHIUM COM-PLEX see HEJ350
2,4,6,8,9,10-HEXAMETHYLHEXAAZA-1,3,5,7-TETRAPHOSPHA-ADAMANTANE see HEJ375
N,N,N,N',N',N'-HEXAMETHYL-1,6-HEXANEDIAMINIUM DIBROMIDE see HEA000
N,N,N,N',N',N'-HEXAMETHYL-1,6-HEXANEDIAMINIUM DICHLORIDE see HEA500
N,N,N,N',N',N'-HEXAMETHYL-1,6-HEXANEDIAMINIUM DIIODIDE see HEB000
2,3,3,4,4,5-HEXAMETHYL-2-HEXANETHIOL see DXT800
HEXAMETHYLMELAMINE see HEJ500
HEXAMETHYLOLMELAMIN (CZECH) see HDY000
HEXAMETHYLOLMELAMINE see HDY000
HEXAMETHYL PHOSPHORAMIDE see HEK000
HEXAMETHYLPHOSPHORIC ACID TRIAMIDE (MAK) see HEK000
HEXAMETHYLPHOSPHORIC TRIAMIDE see HEK000
N,N,N,N,N,N-HEXAMETHYLPHOSPHORIC TRIAMIDE see HEK000
HEXAMETHYLPHOSPHOROTRIAMIDE see HEK000
HEXAMETHYLPHOSPHOTRIAMIDE see HEK000
HEXAMETHYLRHENIUM see HEK550
HEXAMETHYL-p-ROSANILINE HYDROCHLORIDE see AOR500
HEXAMETHYLSILAZANE see HED500
N,N,N',N',N'',N''-HEXAMETHYL-1,3,5-TRIAZINE-2,4,6-TRIAMINE see HEJ500
2,2,4,4,6,6-HEXAMETHYLTRITHIANE see HEL000
HEXAMETHYL VIOLET see AOR500
HEXAMETON see HEA000
HEXAMETON CHLORIDE see HEA500
HEXAMIC ACID see CPQ625
HEXAMID see HEL500
HEXAMIDINE see DBB200
HEXAMIDINE (the antispasmodic) see DBB200
HEXAMINE (DOT) see HEI500
HEXAMITE see TCF250
HEXAMOL SLS see SIB600
HEXANAL see ERE000, HEM000
1-HEXANAL see HEM000

HEXANAMIDE see HEM500
1-HEXANAMINE see HFK000
HEXANAPHTHENE see CPB000
HEXANASTAB see ERE000
HEXANASTAB ORAL see ERD500
HEXANATE see TIX250
HEXANATE D see HCP600
HEXANATRIUMTETRAPOLYPHOSPHAT (GERMAN) see HEY500
n-HEXANE see HEN000
n-HEXANE see HEN000
HEXANE (DOT) see HEN000
1-HEXANECARBOXYLIC ACID see HBE000
HEXANEDIAMIDE (9CI) see AEN000
1,6-HEXANEDIAMINE see HEO000
1,6-HEXANEDIAMINE (solution) see HEO500
1,6-HEXANEDIAMINE, N,N'-DICARBOXYMETHYL-N,N-DIMETHYL-, DIMETHOCHLORIDE, DIDODECYL ESTER see HEF200
1,6-HEXANEDIAMINIUM, N,N'-BIS(2-(DODECYLOXY)-2-OXOETHYL)-N,N,N',N'-TETRAMETHYL-, DICHLORIDE see HEF200
1,6-HEXANEDIAMINIUM, N,N,N,N',N',N'-HEXAMETHYL-, DIBENZENESULFONATE (9CI) see HEG100
HEXANEDINITRILE see AER250
1,6-HEXANEDIOIC ACID see AEN250
HEXANEDIOIC ACID, BIS(2-BUTOXYETHYL) ESTER see BHJ750
HEXANEDIOIC ACID, BIS(2-ETHYLHEXYL) ESTER see AEO000
HEXANEDIOIC ACID, BIS(2-(HEXYLOXY)ETHYL)ESTER see AEQ000
HEXANEDIOIC ACID, BIS(3-METHYLBUTYL) ESTER see AEQ500
HEXANEDIOIC ACID, BIS(1-METHYLETHYL) ESTER see DNL800
HEXANEDIOIC ACID BIS(4-METHYL-7-OXABICYCLO(4.1.0)HEPT-3-YL)METHYL ESTER see AEN750
HEXANEDIOIC ACID-DIBUTYL ESTER see AEO750
HEXANEDIOIC ACID DIHYDRAZIDE see AEQ250
HEXANEDIOIC ACID DINITRILE see AER250
HEXANEDIOIC ACID, DIOCTYL ESTER see AEO000
HEXANEDIOIC ACID-DI-2-PROPENYL ESTER see AEO500
HEXANEDIOIC ACID ETHENYL METHYL ESTER see MQL000
HEXANEDIOIC ACID, compound with PIPERAZINE (1581) see HEP000
HEXANEDIOIC ACID, POLYMER with 1,4-BUTANEDIOL and 1,1'-METHYLENEBIS(4-ISOCYANATOBENZENE) see PKM250
HEXANEDIOIC ACID, POLYMER with 1,3-ETHANEDIOL and 1,1'-METHYLENEBIS(4-ISOCYANATOBENZENE) see PKL750
1,2-HEXANEDIOL see HFP875
1,6-HEXANEDIOL see HEP500
2,5-HEXANEDIOL see HEQ000
1,6-HEXANEDIOL DIISOCYANATE see DNJ800
2,5-HEXANEDIOL DIMETHYLSULFONATE see DSU000
2,3-HEXANEDIONE see HEQ200
2,5-HEXANEDIONE see HEQ500
2,5-HEXANEDIONE, 3,4-DIMETHYL- see DSF100
2,3-HEXANEDIONE, 5-METHYL- see MKL300
3,3'-(1,6-HEXANEDIYLBIS-((METHYLIMINO)CARBONYL)OXY)BIS(1-METHYLPYRIDINIUMDIBROMIDE) see DXG800
1,2,3,4,5,6-HEXANEHEXOL see HER000
6-HEXANELACTAM see CBF700
HEXA-NEMA see DFK600
HEXANEN (DUTCH) see HEN000
HEXANENITRILE see HER500
HEXANES (FCC) see HEN000
1-HEXANETHIOL see HES000
1,2,6-HEXANETRIOL see HES500
HEXANETRIOL-1,2,6 see HES500
HEXANE-1,2,6-TRIOL see HES500
HEXANHYDROPYRIDINE HYDROCHLORIDE see HET000
HEXANICIT see HFG550
HEXANICOTINOYL INOSITOL see HFG550
HEXANICOTOL see HFG550
HEXANITROBENZENE see HET350
HEXANITRODIFENYLAMINE (DUTCH) see HET500
HEXANITRODIPHENYLAMINE see HET500
HEXANITRODIPHENYLAMINE (FRENCH) see HET500
2,2',4,4',6,6'-HEXANITRODIPHENYLAMINE see HET500
2,4,6,2',4',6'-HEXANITRODIPHENYLAMINE see HET500
HEXANITRODIPHENYLSULFIDE see BLR750
HEXANITROETHANE see HET675
HEXANITROL see MAW250
HEXANOESTROL see DLB400
HEXANOIC ACID see HEU000
n-HEXANOIC ACID see HEU000
HEXANOIC ACID, BIS(2-ETHOXYETHYL) ESTER see BJO225
HEXANOIC ACID, BUTYL ESTER see BRK900
HEXANOIC ACID, 2-ETHYL-, HEXADECYL ESTER see HCP550
HEXANOIC ACID, HEXYL ESTER see HFQ500
HEXANOIC ACID, ISOBUTYL ESTER see IIT000
HEXANOIC ACID, 2-METHYLPROPYL ESTER see IIT000
HEXANOIC ACID, 3,5,5-TRIMETHYL-, ALLYL ESTER see AGU400

HEXANOIC ACID, VINYL ESTER (mixed isomers) see HEU500
HEXANOL see HFJ500
1-HEXANOL see HFJ500
n-HEXANOL (DOT) see HFJ500
sec-HEXANOL (DOT) see EGW000
6-HEXANOLACTONE see LAP000
tert-HEXANOL CARBAMATE see ENF000
tert-HEXANOL (9CI, DOT) see HFJ600
1,6-HEXANOLIDE see LAP000
HEXANON see CPC000
2-HEXANONE see HEV000
HEXANONE-2 see HEV000
3-HEXANONE see HEV500
HEXANONE ISOXIME see CBF700
3-HEXANONE, 2-METHYL- see MKL400
HEXANONISOXIM (GERMAN) see CBF700
1,4,7,10,13,16-HEXANOXACYCLOOCTADECANE see COD500
p-HEXANOYLANILINE see AJD250
1-HEXANOYLAZIRIDINE see HEW000
HEXANOYLETHYLENEIMINE see HEW000
17-α-HEXANOYLOXY-19-NOR-4-PREGNENE-3,20-DIONE see GEK510
17-α-HEXANOYLOXYPREGN-4-ENE-3,20-DIONE see HNT500
HEXAOCTYLDISTANNOXANE see HEW100
1,1,1,3,3,3-HEXAOCTYLDISTANNOXANE see HEW100
HEXAOCTYLDISTANNTHIANE see HEW150
1,1,1,3,3,3-HEXAOCTYLDISTANNTHIANE see HEW150
13,16,21,24-HEXAOXA-1,10-DIAZABICYCLO-(8,8,8)-HEXACOSANE see LFQ000
5,8,11,13,16,19-HEXAOXATRICOSANE (9CI) see BHK750
1,1,1,3,3,3-HEXAPHENYLDISTANNTHIANE see BLT250
HEXAPLAS M/1B see DNJ400
HEXAPLAS M/B see DEH200
HEXAPLAS M/O see ILR100
HEXAPROMIN see AJV500
1,1,1,3,3,3-HEXAPROPYLDISTANNOXANE see BLT300
HEXAPROPYLDISTANNTHIANE see HEW200
1,1,1,3,3,3-HEXAPROPYLDISTANNTHIANE see HEW200
HEXAPROPYMATE see POA250
HEXAPROPYNATE see POA250
HEXA-3-PYRIDINECARBOXYLATE-myo-INOSITOL (9CI) see HFG550
HEXAPYRIDINEIRON(II) TRIDECACARBONYL TETRAFERRATE(2⁻) see HEY000
HEXASODIUM TETRAPHOSPHATE see HEY500
HEXASODIUM TETRAPOLYPHOSPHATE see HEY500
HEXASTAT see HEJ500
HEXASUL see SOD500
HEXATHANE see EIR000
HEXATHIDE see HEB000
HEXATHIR see TFS350
HEXATOX see BBQ500
1,3,5-HEXATRIENE see HEZ000
1,3,5-HEXATRIYNE see HEZ375
HEXATRON see AJV500
HEXATYPE CARMINE B see EOJ500
HEXAUREA CHROMIC CHLORIDE see HEZ800
HEXAUREACHROMIUM(III) NITRATE see HFA000
HEXAUREAGALLIUM(III) PERCHLORATE see HFA225
HEXAVIBEX see PPK500
HEXAZANE see PIL500
HEXAZINONE see HFA300
HEXAZIR see BJK500
HEXEMAL see TDA500
HEXENAL see ERD500
2-HEXENAL see HFA500
HEX-2-ENAL see HFA500
HEX-2-EN-1-AL see HFA500
2-HEXENAL, (E)- see PNC500
trans-2-HEXENAL see PNC500
trans-2-HEXEN-1-AL see HFA525
HEXENAL (barbiturate) see ERD500
2-HEXEN-1-AL, 2-ISOPROPYL-5-METHYL- see IKM100
2-HEXEN-1-AL, 5-METHYL-2-(1-METHYLETHYL)- see IKM100
HEXENAL SODIUM see ERE000
HEXENE see HFB000
1-HEXENE see HFB000
2-HEXENE see HFB500
3-HEXENE DINITRILE see HFC000
HEXENE-OL see HCS500
trans-2-HEXENE OZONIDE see HFC500
4-HEXENE-1-YNE-3-OL see HFF000
4-HEXENE-1-YNE-3-ONE see HFF300
3-HEXENOIC ACID see HFD000
4-HEXENOIC ACID, 2-ACETYL-5-HYDROXY-3-OXO, Δ-LACTONE, SODIUM derivative see SGD000
HEXENOL see HCS500

2-HEXENOL see HFD500
β-γ-HEXENOL see HFE000
2-HEXEN-1-OL, (E)- see HFD500
cis-3-HEXENOL see HFE000
trans-2-HEXENOL see HFD500
cis-3-HEXEN-1-OL (FCC) see HFE000
trans-2-HEXEN-1-OL (FCC) see HFD500
2-HEXEN-1-OL ACETATE see HFE100
γ-HEXENOLACTONE see PAJ500
2-HEXEN-5,1-OLIDE see PAJ500
D″-HEXENOLLACTONE see PAJ500
4-HEXEN-1-OL, 5-METHYL-2-(1-METHYLETHENYL)-, ACETATE see LCA100
3-HEXEN-1-OL, PROPANOATE (Z)- see HFE650
HEX-2-ENYL ACETATE see HFE100
2-HEXENYL ACETATE see HFE100
2-HEXEN-1-YL-ACETATE see HFE100
(E)-2-HEXENYL ACETATE see HFE100
trans-2-HEXENYL ACETATE see HFE100
cis-3-HEXENYL BENZOATE see HFE500
3-HEXENYL ESTER, BENZOIC ACID (Z)- see HFE500
β,γ-HEXENYL ISOBUTANOATE see HFE520
cis-3-HEXENYL ISOBUTYRATE see HFE520
cis-3-HEXENYL ISOVALERATE (FCC) see ISZ000
cis-3-HEXENYL 2-METHYLBUTYRATE see HFE550
cis-HEXENYL OCYACETALDEHYDE see HFE600
cis-3-HEXENYL PHENYLACETATE see HFE625
β,γ-HEXENYL PROPANOATE see HFE650
cis-3-HEXENYL PROPIONATE see HFE650
cis-3-HEXENYL SALICYLATE see SAJ000
β,γ-cis-HEXENYL SALICYLATE see SAJ000
β,γ-HEXENYL α-TOLUATE see HFE625
4-HEXEN-1-YN-3-OL see HFF000
4-HEXEN-1-YN-3-ONE see HFF300
HEXERMIN see PPK500
HEXERMIN P see PII100
HEXESTROL see DLB400
meso-HEXESTROL see DLB400
HEXESTROL DIPHOSPHATE SODIUM see SHN150
HEXETHAL SODIUM see EKT500
HEXICIDE see BBQ500
HEXIDE see HCL000
HEXILMETHYLENAMINE see HEI500
HEXMETHYLPHOSPHORAMIDE see HEK000
HEXOBARBITAL see ERD500
HEXOBARBITAL Na see ERE000
HEXOBARBITAL SODIUM see ERE000
HEXOBARBITONE see ERD500
HEXOBARBITONE Na see ERE000
HEXOBARBITONE SODIUM see ERE000
HEXOBENDINE DIHYDROCHLORIDE see HFF500
HEXOBION see PPK500
HEXOCYCLIUM see HFG000
HEXOCYCLIUM METHYLSULFATE see HFG400
HEXOESTROL see DLB400
HEXOGEEN (DUTCH) see CPR800
HEXOGEN 5W see CPR800
HEXOGEN (explosive) see CPR800
n-HEXOIC ACID see HEU000
1,6-HEXOLACTAM see CBF700
HEXOLITE see CPR800
HEXOLITE, dry or containing, by weight, less than 15% water (DOT) see CPR800
HEXON (CZECH) see HFG500
HEXON CHLORIDE see HEA500
HEXONE see HFG500
HEXONE CHLORIDE see HEA500
HEXONIUM DIBROMIDE see HEA000
HEXONIUM DIIODIDE see HEB000
HEXOPAL see HFG550
HEXOPHENE see HCL000
HEXOPRENALINE DIHYDROCHLORIDE see HFG600
HEXOPRENALINE SULFATE see HFG650
HEXOSAN see HCL000
HEXOXYACETALDEHYDE DIMETHYLACETAL see HFG700
2-HEXOXYACETALDEHYDE DIMETHYLACETAL see HFG700
β-HEXOXYACETALDEHYDE DIMETHYLACETAL see HFG700
p-HEXOXYBENZOIC ACID-3-(2′-METHYLPIPERIDINO)PROPYL ESTER see HFH500
3-HEXOXY-1-(2′-CARBOXYPHENOXY)-PROPANOL-(2) SODIUM SALT see ERE200
n-HEXOXYETHOXYETHANOL see HFN000
HEXYCLAN see BBQ750
HEXYL (GERMAN, DUTCH) see HET500
HEXYL ACETATE see HFI500
1-HEXYL ACETATE see HFI500

sec-HEXYL ACETATE see HFJ000
n-HEXYL ACETATE (FCC) see HFI500
β-HEXYLACROLEIN see NNA300
HEXYL ACRYLATE see ADV000
N-HEXYL ACRYLATE see ADV000
HEXYL ALCOHOL see HFJ500
n-HEXYL ALCOHOL see HFJ500
sec-HEXYL ALCOHOL see EGW000
tert-HEXYL ALCOHOL see HFJ600
HEXYL ALCOHOL, ACETATE see HFI500
HEXYLAMINE see HFK000
N-HEXYLAMINE see HFK000
HEXYLAN see BBP750
8-HEXYL-BENZ(a)ANTHRACENE see HFL000
5-n-HEXYL-1,2-BENZANTHRACENE see HFL000
4-HEXYL-1,3-BENZENEDIOL see HFV500
HEXYL BENZOATE see HFL500
n-HEXYLBENZOATE see HFL500
HEXYL BROMIDE see HFM500
HEXYL BUTANOATE see HFM700
n-HEXYL BUTANOATE see HFM700
n-HEXYL n-BUTANOATE see HFM700
HEXYL-2-BUTENOATE see HFM600
n-HEXYL 2-BUTENOATE see HFM600
HEXYL BUTYRATE see HFM700
1-HEXYL BUTYRATE see HFM700
n-HEXYL BUTYRATE see HFM700
γ-N-HEXYL-γ-BUTYROLACTONE see HKA500
HEXYLCAINE HYDROCHLORIDE see COU250
HEXYL CAPROATE see HFQ500
1-HEXYLCARBAMOYL-5-FLUOROURACIL see CCK630
HEXYL CARBITOL see HFN000
n-HEXYL CARBITOL see HFN000
n-HEXYL CARBORANE see HFN500
N-HEXYL CELLOSOLVE see HFT500
2-HEXYL-4-CHLOROPHENOL see CHL500
HEXYL CINNAMALDEHYDE see HFO500
α-HEXYLCINNAMALDEHYDE (FCC) see HFO500
HEXYL CINNAMIC ALDEHYDE see HFO500
α-HEXYLCINNAMIC ALDEHYDE see HFO500
HEXYL CROTONATE see HFM600
2-n-HEXYL-2-CYCLOPENTEN-1-ONE see HFO700
2-HEXYLDECANOIC ACID see HFP500
2-HEXYLDECANSAURE (GERMAN) see HFP500
HEXYLDICARBADODECABORANE(12) see HFN500
HEXYLDICHLORARSINE see HFP600
5-HEXYLDIHYDRO-2(3H)-FURANONE see HKA500
4-HEXYL-1,3-DIHYDROXYBENZENE see HFV500
HEXYLENE see HFB000
HEXYLENE GLYCOL see HFP875
HEXYLENE GLYCOL DIACETATE see HFQ000
HEXYLENIC ALDEHYDE see HFA500
HEXYL ETHANOATE see HFI500
HEXYL ETHER see DKO800
N-HEXYL ETHER see DKO800
HEXYL FUMARATE see DKP000
n-HEXYL HEXANOATE see HFQ500
HEXYL HEXOATE see HFQ500
HEXYL ISOBUTANOATE see HFQ550
n-HEXYL ISOBUTANOATE see HFQ550
HEXYL ISOBUTYRATE see HFQ550
1-HEXYL ISOBUTYRATE see HFQ550
n-HEXYL ISOBUTYRATE see HFQ550
HEXYL MANDELATE see HFR000
HEXYL MERCAPTAN see HES000
HEXYLMERCURIC BROMIDE see HFR100
n-HEXYLMERCURIC BROMIDE see HFR100
HEXYL MERCURY BROMIDE see HFR100
n-HEXYL trans-2-METHYL-2-BUTENOATE see HFX000
HEXYL 2-METHYLBUTYRATE see HFR200
2-HEXYL-4-METHYL-1,3-DIOXOLANE see HFR500
1-HEXYL-1-NITROSOUREA see HFS500
2-(HEXYLOXY)BENZAMIDE see HFS759
o-HEXYLOXYBENZAMIDE see HFS759
2-n-HEXYLOXYBENZAMIDE see HFS759
2-(HEXYLOXY)ETHANOL see HFT500
2-((2-HEXYLOXY)ETHOXY)ETHANOL see HFN000
o-(3-(HEXYLOXY)-2-HYDROXYPROPOXY)BENZOIC ACID MONOSODIUM
    SALT see ERE200
4'-(HEXYLOXY)-3-PIPERIDINOPROPIOPHENONE HYDROCHLORIDE
    see HFU500
4-HEXYLOXY-β-(1-PIPERIDYL)PROPIOPHENONE HYDROCHLORIDE
    see HFU500
α-n-HEXYL-β-PHENYLACROLEIN see HFO500

2-(1-HEXYL-3-PIPERIDYL)ETHYL ESTER, BENZOIC ACID HYDROCHLO-
    RIDE see HFV000
HEXYL-2-PROPENOATE see ADV000
HEXYLRESORCIN (GERMAN) see HFV500
4-HEXYLRESORCINE see HFV500
HEXYLRESORCINOL see HFV500
4-HEXYLRESORCINOL see HFV500
p-HEXYLRESORCINOL see HFV500
4-n-HEXYLRESORCINOL see HFV500
3-HEXYL-7,8,9,10-TETRAHYDRO-6,6,9-TRIMETHYL-6H-DIBENZO(B,D)PYRAN-
    1-OL see HGK500
HEXYLTHIOCARBAM see EHT500
3-(5-(HEXYLTHIO)PENTYL)THIAZOLIDINE HYDROCHLORIDE see HFW500
n-HEXYL TIGLINATE see HFX000
HEXYL TILGLATE see HFX000
HEXYLTRICHLOROSILANE see HFX500
n-HEXYL VINYL SULFONE see HFY000
3-HEXYNE-2,5-DIOL see HFY500
HEXYNE-3-DIOL-2,5 see HFY500
HEXYNOL see HGA000
1-HEXYN-3-OL see HFZ000
HEYDEFLON see TAI250
HEYDEN 768 see TGC300
HF 264 see BRJ325
HF-1854 see CMY250
HF 1927 see DCW600, DCW800
HF-2333 see HOU059
HF3170 see DCS200
HFA see FIC000
H-35-F 87 (BVM) see DSQ000
HFCB see HDB500
HFE see HDC000
HF-2159 HYDROCHLORIDE see CIT000
HFIP see HDC500
81723 HFU see TET800
HG-203 see CHX250
Hg 532 see PGA750
H.G. BLENDING see SFT000
HGG-12 see HAA340
Hg-HEMATOPORPHYRIN-Na see HAP000
HGI see BBQ500
HH-197 see BOR350
H. HAEMACHATES VENOM see HAO500
HHDN see AFK250
H102/09 HYDROCHLORIDE see ZBA525
HI-6 see HAA345
HIADELON see PII100
HI-ALAZIN see TGB475
HI-A-VITA see VSK600
HIBANIL see CKP250, CKP500
HIBAWOOD OIL see HGA100
HIBERNA see DQA400
HIBERNAL see CKP250, CKP500
HIBERNON HYDROCHLORIDE see HGA500
HIBERNYL see TCY750
HIBESTROL see DKA600
HIBICON see BEG000
HIBISCOLIDE see OKW110
HIBISCUS MANIHOT Linn., extract see HGA550
HIBISCUS ROSA-SINENSIS, flower extract see HGA600
HIBITANE see BIM250
HIBITANE DIACETATE see CDT125
HIBROM see NAG400
HICAL-2 see HGB000
HI-CAL 3 see JDA125
HICHILLOS see KGK000
HICO CCC see CMF400
HIDA see LFO300
HIDACHROME ORANGE R see NEY000
HIDACIAN see COH250
HIDACIANN see COH250
HIDACID AMARANTH see FAG020
HIDACID AZO RUBINE see HJF500
HIDACID AZURE BLUE see FMU059
HIDACID BROMO ACID REGULAR see BNK700
HIDACID DIBROMO FLUORESCEIN see BNH500, BNK700
HIDACID FAST ORANGE G see HGC000
HIDACID FAST SCARLET 3R see FMU080
HIDACID FLUORESCEIN see FEV000
HIDACID METANIL YELLOW see MDM775
HIDACID SCARLET 2R see FMU070
HIDACID URANINE see FEW000
HIDACO BRILLIANT GREEN see BAY750
HIDACO MALACHITAE GREEN BASE see AFG500

HIDACO METHYLENE BLUE SALT FREE see BJI250
HIDACO OIL ORANGE see PEJ500
HIDACO OIL YELLOW see AIC250
HIDACO VICTORIA BLUE R see VKA600
HIDAN see DKQ000
HIDANTILO see DKQ000
HIDANTINA SENOSIAN see DKQ000
HIDANTINA VITORIA see DKQ000
HIDANTOMIN see DKQ000
HI-DERATOL see VSZ100
HIDRALAZIN see HGP495, HGP500
HIDRANIZIL see ILD000
HIDRIL see CFY000
HIDRIX see HOO500
HIDROCHLORTIAZID see CFY000
HIDRO-COLISONA see CNS750
HIDROESTRON see EDP000
HIDROMEDIN see DFP600
HIDRORONOL see CFY000
HIDROTIAZIDA see CFY000
N-(trans-4-HIDROXICICLOHEXIL)-(2-AMINO-3,5-DIBROMOBENCIL)AMINA
  (SPANISH) see AHJ250
HIDROXIFENAMATO see HNJ000
HIDROXITEOFILLINA see DNC000
HIDRULTA see ILD000
HI-DRY see TCE250
HI-ENTEROL see CHR500
HIERBA de SANTIAGO (MEXICO) see RBA400
HIESTRONE see EDV000
HI-FLASH NAPHTHAETHYLEN see NAI500
HIFOL see BIO750
HIGH BELIA see CCJ825
HIGILITE see AHC000
HIGOSAN see MEP250
HIGUERETA CIMARRONA see CNR135
HIGUERETA (CUBA, PUERTO RICO) see CCP000
HIGUERILLA (MEXICO) see CCP000
HIGUEROXYL DELABARRE see FBS000
HI-JEL see BAV750
HIKIZIMYCIN see APF000
HILDAN see EAQ750
HILDIT see DAD200
HILLS-OF-SNOW see HGP600
HILONG see CFZ000
HILTHION see MAK700
HILTHION 25WDP see MAK700
HILTONIL FAST BLUE B BASE see DCJ200
HILTONIL FAST ORANGE GR BASE see NEO000
HILTONIL FAST ORANGE R BASE see NEN500
HILTONIL FAST RED KB BASE see CLK225
HILTONIL FAST SCARLET 2G BASE see DEO400
HILTONIL FAST SCARLET G BASE see NMP500
HILTOSAL FAST BLUE B SALT see DCJ200
HILTOSAL FAST SCARLET 2G SALT see DEO400
HINDAMINE SCARLET GG see DEO400
HINDASOL BLUE B SALT see DCJ200
HINDASOL RED TR SALT see CLK235
HINOKITIOL see IRR000
HINOKITOL see IRR000
HINOSAN see EIM000
HINSALU see RQU300
HIOHEX CHLORIDE see HEA500
HIOXYL see HIB000
HIP see COF250
HIPERCILINA see BFD000
HIPHYLLIN see DNC000
HIPNAX see DLY000
HIPOFTALIN see HGP495, HGP500
HI-POINT 90 see MKA500
HIPPEASTRUM (VARIOUS SPECIES) see AHI635
HIPPOBROMA LONGIFLORA see SLJ650
HIPPODIN see HGB200
HIPPOMANE MANCINELLA see MAO875
HIPPOPHAIN see AJX500
HIPPURAN see HGB200
HIPPURIC ACID SODIUM SALT see SHN500
HIPPUROHYDROXAMIC ACID see BBB250
HIPPUZON see TEH500
HIPSAL see DLY000
HIPTAGENIC ACID see NIY500
HI-PYRIDOXIN see PII100
HIRATHIOL see IAD000
HIRSUTIC ACID N see HGB500
HI-SEL see SCH000

HISHIREX 502 see PKQ059
HISINDAMONE A see CDY000
HISMANAL see ARP675
HISPACID BRILLIANT SCARLET 3RF see FMU080
HISPACID FAST BLUE R see ADE750
HISPACID FAST ORANGE 2G see HGC000
HISPACID GREEN GB see FAE950
HISPACID ORANGE 1 see FAG010
HISPACID YELLOW MG see MDM775
HISPACROM ORANGE R see NEY000
HISPALIT FAST SCARLET RN see MMP100
HISPAMIN BLACK EF see AQP000
HISPAMIN BLUE 2B see CMO000
HISPAMIN BLUE 3BX see CMO250
HISPAMIN CONGO 4B see SGQ500
HISPAMIN SKY BLUE 3B see CMO500
HISPAMIN SKY BLUE 6B see CMN750
HISPAMIN VIOLET 3R see CMP000
HISPAVIC 229 see PKQ059
HISPERSE YELLOW G see AAQ250
HISPRIL see LJR000
HISPRIL HYDROCHLORIDE see DVW700
HISTABID see SPC500
HISTABUTYZINE DIHYDROCHLORIDE see BOM250
HISTACAP see WAK000
HISTADUR see TAI500
HISTADUR DURA-TABS see TAI500
HISTADYL see TEO250
HISTADYL HYDROCHLORIDE see DPJ400
HISTAFED see DPJ400
HISTAGLOBIN see HGC400
HISTALEN see TAI500
HISTALON see WAK000
HISTAMETHINE see HGC500
HISTAMETHIZINE see HGC500
HISTAMETIZINE see HGC500
HISTAMETIZYNE see HGC500
HISTAMINE see HGD000
HISTAMINE ACID PHOSPHATE see HGE000
HISTAMINE DICHLORIDE see HGD500
HISTAMINE DIHYDROCHLORIDE see HGD500
HISTAMINE DIPHOSPHATE see HGE000
HISTAMINE HYDROCHLORIDE see HGE500
HISTAMINE PHOSPHATE (1:2) see HGE000
HISTAMINOS see ARP675
HISTAN see WAK000
HISTANTIN see CFF500
HISTANTINE see CFF500
HISTANTINE DIHYDROCHLORIDE see CDR000
HISTAPAN see TAI500
HISTAPYRAN see WAK000
HISTARGAN see DQA400
HISTASAN see WAK000
HISTASPAN see CLD250
HISTATEX see DBM800
HISTAXIN see BBV500
HISTIDINE see HGE700
l-HISTIDINE see HGE700
l-HISTIDINE (FCC) see HGE700
l-HISTIDINE, N-β-ALANYL- see CCK665
HISTIDYL see DPJ400
HISTOCARB see CBJ000
HISTOSTAB see PDC000
HISTRYL see LJR000
HISTYN see LJR000
HISTYRENE S 6F see SMR000
HI-STYROL see SMQ500
HI-YIELD DESSICANT H-10 see ARB250
HIZAROCIN see CPE750
HJ 6 see HAA345
HK-141 see BAW500
HK 256 see LFJ000
H.K. FORMULA No. K. 7117 see FMU059
HL 267 see DGW600
HL-331 see ABU500
HL 2153 see DPD400
HL 2197 see CEP675
HL 2447 see DFV400
HL 8700 see PMI750
HL 8727 see DIG400
HL-8731 see MQF750
HLS 831 see BSQ000
HMB see HFR100
7-HMBA see BBH250
HMBD see HLX900, HLX925

HMD see PAN100
HMDA see HEO000
HMDI see DNJ800
HMDS see HED500
HMF see ORG000
HMG see HMC000
HMGA see HMC000
HMM see HEJ500
12-HM-7-MBA see HMF500
7-HM-12-MBA see HMF000
HMP see SHM500
HMPA see HEK000
HMPT see HEK000
HMT see HEI500
HMX (DOT) see CQH250
beta HMY see CQH250
HN1 see BID250
HN 1 see CGW000
HN2 see BIE250
HN$_2$ AMINE OXIDE see CFA500
HN1 CHLOROHYDRIN see EHK300
HNED CERVENAVA OSTANTHRENOVA 5 RF see CMU800
HNED KYPOVA 25 see CMU800
HN2.HCl see BIE500
HN1 HYDROCHLORIDE see EGU000
HN2 HYDROCHLORIDE see BIE500
HN1•HCl see EGU000
HN$_2$ OXIDE HYDROCHLORIDE see CFA750
HN$_2$ OXIDE MUSTARD see CFA500
HNT see HHD500
HNU see HKW500
HO 11513 see TMK000
3-HO-AAF see HIO875
HOCA see DXH250
HOCH see FMV000
HODAG GMS see OAV000
HODAG SMS see SKV150
HODAG SVO 9 see PKL100
HODOSTIN see DQY909
HODSON see DSK800
HOE 118 see PDW250
HOE 296 see BAR800
HOE 766 see LIU420
HOE 881 see FAL100
HOE 893 see PAP225
HOE 933 see CAZ125
HOE 984 see NMV725
HOE 2,671 see EAQ750
HOE 2747 see CKD500
HOE 2784 see BGB500
HOE 2810 see DGD600
HOE-2824 see ABX250
HOE 2872 see CLU000
HOE 2904 see ACE500
HOE 2982 see HBK700
HOE 766A see LIU420
HOE 837V see ECW550
HOE 893D see PAP225, PAP230
HOE 17411 see MHC750
HOE 33258 see MOD500
HOE 36801 see CGQ500
HOECHST 1082 see MDP000
HOECHST 10,820 see MDP750
12494 HOECHST see RDA375
33258 HOECHST see MOD500
HOECHST DYE 33258 see MOD500
HOECHST PA 190 see PJS750
HOE 2960 OJ see THT750
HOE 2982 OJ see HBK700
HOG see AOO500, PDC890
HOG APPLE see MBU800
HOG BUSH see CDM325
HOGGAR see BNK000
HOGGAR N see PGE775
HOG PHYSIC see CCJ825
HOG'S POTATO see DAE100
HOJA GRANDE (CUBA) see APM875
HOKKO-MYCIN see SLW500
HOKMATE see FAS000
HOLBAMATE see MQU750
HO LEAF OIL see HGF000
HOLIN see EDU500
HOLLICHEM HQ 3300 see BBA625
HOLLY see HGF100

HOLMIUM see HGF500
HOLMIUM CHLORIDE see HGG000
HOLMIUM CITRATE see HGG500
HOLMIUM NITRATE see HGH100
HOLMIUM(III) NITRATE, HEXAHYDRATE (1:3:6) see HGH000
HOLMIUM TRINITRATE see HGH100
HOLOCAINE see BJO500
HOLODORM see QAK000
HOLOPAN see SBH500
HOLOXAN see IMH000
HOMANDREN see MPN500
HOMANDREN (amps) see TBG000
HOMAPIN see MDL000
HOMATROMIDE see MDL000
HOMATROPINE METHYLBROMIDE see MDL000
HOMBITAN see TGG760
#2 HOME HEATING OILS see HGH200
HOMEOSTAN see ACL250
HOMIDIUM BROMIDE see DBV400
HOMIDIUM CHLORIDE see HGI000
HOMOANISIC ACID see MFE250
HOMOARGININE, nitrosated see HNA500
HOMOCATECHOL see DNE200
HOMOCHLORCYCLIZINE DIHYDROCHLORIDE see CJR809
HOMOCHLORCYCLIZINE HYDROCHLORIDE see HGI200
HOMOCHLOROCYCLIZINE DIHYDROCHLORIDE see CJR809
HOMOCHLOROCYCLIZINE HYDROCHLORIDE see HGI200
HOMOCODEINE see TCY750
18-HOMO-ESTRIOL see HGI700
HOMOFOLATE see HGI525
HOMOFOLIC ACID see HGI525
HOMOGUAIACOL see MEK250
HOMOHARRINGTONINE see HGI575
HOMOLLE'S DIGITALIN see DKN400
HOMOMENTHOL see TLO500
HOMOMYRETENOL see DTB800
18-HOMO-OESTRIOL see HGI700
HOMOOLAN see HIM000
HOMOPIPERAZINE see HGI900
HOMOPIPERIDINE see HDG000
HOMOPYROCATECHOL see DNE200
HOMOSALICYLIC ACID see CNX625
HOMOSTERONE see TBF500
3-HOMOTETRA HYDRO CANNIBINOL see HGK500
HOMOTRYPTAMINE HYDROCHLORIDE see AME750
HOMOVERATRONITRILE see VIK100
HOMOVERATRYLAMINE see DOE200
HON see AKG500
HONEY DIAZINE see PPP500
HONEYSUCKLE BUSH see HGK700
HONG KIEN see ABU500
HONVAN see DKA200
HONVAN TETRASODIUMTETRASODIUM SALT, (E)- see TEE300
HOOKER HRS-16 see DAE425
HOOKER HRS-1422 see DNS200
HOOKER HRS 1654 see DAE425
HOOKER No. 1 CHRYSOTILE ASBESTOS see ARM268
HOPA see CAT175
HOPANTENATE CALCIUM see CAT125
HOPANTENATE CALCIUM HEMIHDYRATE see CAT175
HOPCIDE see CKF000
HOPCIDE, NITROSATED (JAPANESE) see MMV250
HOPCIN see MOV000
HORACE VERNET'S BLUE see CNQ000
HORBADOX see DRN200
HORFEMINE see DKB000
HORMALE see MPN500
HORMATOX see DGB000
HORMEX ROOTING POWDER see ICP000
HORMIN see DFY800
HORMIT see SGH500
HORMOCEL-2CCC see CMF400
HORMODIN see ICP000
HORMOESTROL see DLB400
HORMOFEMIN see DAL600
HORMOFLAVEINE see PMH500
HORMOFOLLIN see EDV000
17-HORMOFORIN see AOO450
HORMOFORT see HNT500
HORMOGYNON see EDP000
HORMOLUTON see PMH500
HORMOMED see EDU500
HORMONIN see EDU500
HORMONISENE see CLO750

HORMOSLYR 64 see DAB000
HORMOSLYR 500T see TAH900
HORMOTESTON see TBG000
HORMOTUHO see CIR250
HORMOVARINE see EDV000
HORSE see HBT500
HORSE BLOB see MBU550
HORSE CHESTNUT see HGL575
HORSE-CYTOCHROME C see CQM325
HORSE GOLD see FBS100
HORSE HEAD A-410 see TGG760
HORSE HEART CYTOCHROME C see CQM325
HORSE NICKER see CAK325
HORSE POISON (JAMAICA) see SLJ650
HORTENSIA (CUBA) see HGP600
HORTFENICOL see CDP250
HORTOCRITT see EJQ000
HOSALON see DSK800
HOSDON GRANULE see DSK800
HOSTACAIN see BQA750
HOSTACAINE see BQA750
HOSTACAINE HYDROCHLORIDE see BQA750
HOSTACAIN HYDROCHLORIDE see BQA750
HOSTACORTIN see PLZ000, PMA000
HOSTACYCLIN see TBX000
HOSTADUR see PKF750
HOSTAFLEX VP 150 see AAX175
HOSTAFLON see TAI250
HOSTAGINAN see PEV750
HOSTALEN PP see PMP500
HOSTALIT see PKQ059
HOSTALIVAL see NMV725
HOSTAPHAN see PKF750
HOSTAPON T see SIY000
HOSTAQUICK see ABU500, HBK700
HOSTATHION see THT750
HOSTAVAT GOLDEN YELLOW see DCZ000
HOSTAVIK (RUSSIAN) see HBK700
HOSTYREN S see SMQ500
HOT PEPPER see PCB275
HOURBESE see DKE800
HOWFLEX GBP see DDY400, DJD700
HOX 1901 see EPR000
HP see PHV275
H.P. 34 see SAH000
HP 129 see FAO100
H.P. 165 see BPG750
H.P. 206 see PMY750
H.P. 209 see EEM000
H.P. 216 see HFS759
HP 1275 see PDV700
HP 1325 see HAA355
2M 4KHP see CIR500
HPA see EPI300
HPC see HNT500, OPK300
β-HPN see HGP000
HPOP see HNX500
HPP see ZVJ000
HPT see HEK000
HQ-275 see HAA360
HR 376 see CIR750
HR 756 see CCR950
HR 930 see FOL100
HRS-16 see DAE425
HRS 16A see DAE425
HRS 860 see CEP000
HRS 1276 see MQW500
HRS 1654 see DAE425
HRW 13 see SLJ500
HS see HGW500
HS 3 see HAA370
HS 4 see PPC000
HS 592 see FOS100
HS-119-1 see PEE750
HSP 2986 see SDZ000
HSR-902 see TGF075
5-HT see AJX500
5-HTA see AJX500
HT-400 E 1/8" see CNC250
H. TERNATUM see SAF500
HT-F 76 see SMQ500
HTH see HOV500
HTP see HCY000
5-HTP see HOO100

l-5-HTP see HOA600, HOO000
HT-2 TOXIN see THI250
HUBBUCK'S WHITE see ZKA000
HUELE de NOCHE (MEXICO) see DAC500
HUEVO de GATO (CUBA) see JBS100
HUILE d'ANILINE (FRENCH) see AOQ000
HUILE de CAMPHRE (FRENCH) see CBA750
HUILE de FUSEL (FRENCH) see FQT000
HUILE H50 see TFM250
HULS P 6500 see PMP500
HUMAN CHORIOGONADOTROPIN, DEGLYCOSYLATED see DAK325
HUMAN CHORIONIC GONADOTROPIN see CMG675
HUMAN IMMUNOGLOBULIN COG-78 see HGL800
HUMAN SPERM see HGM000
HUMATIN see APP500, NCF500
HUMEDIL see BCP650
HUMIDIN see HGM500
HUMIFEN WT 27G see DJL000
HUMULIN I see IDF325
HUMYCIN see NCF500
HUMYCIN SULFATE see APP500
HUNDRED PACE SNAKE VENOM see HGM600
HUNGARIAN CHAMOMILE OIL see CDH500
HUNGAZIN see ARQ725
HUNGAZIN DT see BJP000
HUNGAZIN PK see ARQ725
HUNGER WEED see FBS100
HUNTER'S ROBE see PLW800
HURA CREPITANS see SAT875
HURRICANE PLANT see SLE890
HUSEPT EXTRA see CLW000
HUSTODIL see RLU000
HUSTOSIL see RLU000
HVA 2 see BKL750
HVA-2 CURING AGENT see BKL750
HW 4 see CQH250
HW 920 see DXQ500
HWA 153 see HLF500
HX-868 see HGN000
HXR see IDE000
HY 951 see TJR000
HYACINTH see ZSS000
HYACINTH ABSOLUTE see HGN500
HYACINTHAL see COF000
HYACINTH BASE see BOF000
HYACINTHIN see BBL500
HYACINTH-OF-PERU see SLH200
HYADRINE see BBV500
HYADUR see DUD800
HYALURONIC ACID, SODIUM SALT see HGN600
HYAMINE see BEN000, BBU750
HYAMINE 10X see MHB500
HYAMINE 1622 see BEN000
HYAMINE 2389 see DYA600
HYAMINE 3500 see AFP250
HYASORB see BFD000
HYBAR X see UNJ800
HYBERNAL see CKP500
HYCANTHON see LIM000
HYCANTHONE see LIM000
HYCANTHONE MESYLATE see HGO500
HYCANTHONE METHANESULFONATE see HGO500
HYCANTHONE METHANESULPHONATE see HGO500
HYCANTHONE MONOMETHANESULPHONATE see HGO500
HYCAR see PJR000
HYCAR LX 407 see SMR000
HY-CHLOR see HOV500
HYCHOTINE see CJR909
HYCLORATE see ARQ750
HYCLORITE see SHU500
HYCORACE see HHR000
HYCORTOLE ACETATE see HHQ800
HYCOZID see ILD000
HYDAN see DFE200
HYDAN (antiseptic) see DFE200
HYDANTAL see DKQ000
HYDANTIN SODIUM see DNU000
HYDANTOIN see DKQ000, HGO600
HYDANTOIN, 5-ETHYL-5-PHENYL-, (-)- see EOL050
HYDANTOIN, 1-((5-(p-NITROPHENYL)FURFURYLIDENE)AMINO)- see DAB845
HYDANTOIN SODIUM see DNU000
HYDELTRA see PMA000
HYDELTRONE see PMA000
HYDERGIN see DLL400

HYDERGINE see DLL400
HYDNOCARPUS OIL see CDK750
HYDOUT see DXD000, EAR000
HYDOXIN see PPK250
HYDRACETIN see ACX750
HYDRACRYLIC ACID β-LACTONE see PMT100
HYDRACRYLONITRILE see HGP000
HYDRACRYLONITRILE ACRYLATE see ADT111
HYDRAL see CDO000
HYDRAL 705 see AHC000
HYDRALAZINE see HGP495
HYDRALAZINE CHLORIDE see HGP500
HYDRALAZINE HYDROCHLORIDE see HGP500
HYDRALAZINE MONOHYDROCHLORIDE see HGP500
HYDRAL de CHLORAL see CDO000
HYDRALIN see CPB750
HYDRALLAZINE see HGP495
HYDRALLAZINE HYDROCHLORIDE see HGP500
HYDRAM see EKO500
HYDRAMYCIN see HGP550
HYDRANGEA see HGP600
HYDRANGEA MACROPHYLLA see HGP600
HYDRAPHEN see PFN000
HYDRAPRESS see HGP500
HYDRARGAPHEN see PFN000
HYDRARGYRUM BIJODATUM (GERMAN) see MDD000
HYDRASTINE HYDROCHLORIDE see HGQ500
HYDRASTININE see HGR000
HYDRASTIS CANADENSIS L., ROOT EXTRACT see HGR500
HYDRATED ALUMINA see AHC250
HYDRATED LIME see CAT225
HYDRATROP ALDEHYDE see COF000
HYDRATROPIC ACETATE see PGB750
HYDRATROPIC ACID, p-ISOBUTYL-, SODIUM SALT see IAB100
HYDRATROPIC ALDEHYDE see COF000
HYDRATROPYL ACETATE see PGB750
HYDRAZID see ILD000
l-HYDRAZIDE CYSTEINE see CQK125
HYDRAZINE see HGS000
HYDRAZINE, anhydrous (DOT) see HGS000
HYDRAZINE, aqueous solution (DOT) see HGS000
HYDRAZINE BASE see HGS000
HYDRAZINE-BENZENE see PFI000
HYDRAZINE-BENZENE and BENZIDINE SULFATE see BBY300
HYDRAZINE BISBORANE see HGT500
HYDRAZINECARBOHYDRAZONOTHIOIC ACID see TFE250
HYDRAZINECARBOTHIOAMIDE see TFQ000
HYDRAZINECARBOTHIOAMIDE, 2,2'-(1-(1-ETHOXYETHYL)-1,2-
    ETHANEDIYLIDENE)BIS see KFA100
HYDRAZINECARBOXALDEHYDE see FNN000
HYDRAZINE CARBOXAMIDE see HGU000
HYDRAZINECARBOXAMIDE MONOHYDROCHLORIDE see SBW500
HYDRAZINECARBOXIMIDAMIDE see AKC750
HYDRAZINECARBOXYLIC ACID, ETHYL ESTER see EHG000
HYDRAZINE, 1-(2-(o-CHLOROPHENOXY)ETHYL)-, HYDROGEN SULFATE
    (1:1) see CJP500
HYDRAZINE, 1-(2-(o-CHLOROPHENOXY)ETHYL)-, SULFATE (1:1) see CJP500
1,2-HYDRAZINEDICARBOTHIOAMIDE see BLJ250
HYDRAZINE DIFLUORIDE see HGU100
HYDRAZINE, DIHYDROFLUORIDE see HGU100
HYDRAZINE, 1,1-DI-2-PROPENYL-(9CI) see DBK100
HYDRAZINE, HEPTYL- see HBO000
HYDRAZINE HYDRATE see HGU500
HYDRAZINE HYDROCHLORIDE see HGV000
HYDRAZINE HYDROGEN SULFATE see HGW500
HYDRAZINE, 1-(o-METHOXYPHENETHYL)-, SULFATE (1:1) see MFG200
HYDRAZINE, 1-(α-METHYLPHENETHYL)-2-PHENETHYL- see MNP400
HYDRAZINE, 1-(o-METHYLPHENETHYL)-, SULFATE (1:1) see MNU100
HYDRAZINE, 1-(p-METHYLPHENETHYL)-, SULFATE (1:1) see MNU150
HYDRAZINE, (1-METHYL-2-PHENOXYETHYL)-, (Z)-2-BUTENEDIOATE (1:1)
    (9CI) see PDV700
HYDRAZINE, 1-(2-(o-METHYLPHENOXY)ETHYL)-, HYDROGEN SULFATE
    (1:1) see MNR100
HYDRAZINE, (1-METHYL-2-PHENOXYETHYL)-, MALEATE see PDV700
HYDRAZINE, (2-(4-METHYLPHENYL)ETHYL)-, SULFATE (1:1) (9CI)
    see MNU150
HYDRAZINE MONOBORANE see HGV500
HYDRAZINE MONOHYDRATE see HGU500
HYDRAZINE MONOSULFATE see HGW500
HYDRAZINE, 1-(1-PHENOXY-2-PROPYL)-, MALEATE see PDV700
HYDRAZINE PROPANEMETHANE SULFONATE see HGW000
HYDRAZINE SULFATE (1581) see HGW500
HYDRAZINE SULPHATE see HGW500
HYDRAZINE, TRIFLUOROSTANNITE see HHA100

HYDRAZINE YELLOW see FAG140
HYDRAZINIUM CHLORATE see HGX000
HYDRAZINIUM CHLORITE see HGX500
HYDRAZINIUM DIPERCHLORATE see HGY000
HYDRAZINIUM HYDROGENSELENATE see HGY500
HYDRAZINIUM NITRATE see HGZ000
HYDRAZINIUM PERCHLORATE see HHA000
HYDRAZINIUM SULFATE see HGW500
HYDRAZINIUM TRIFLUOROSTANNITE see HHA100
2-HYDRAZINO-4-(4-AMINOPHENYL)THIAZOLE see HHB000
2-HYDRAZINO-4-(p-AMINOPHENYL)THIAZOLE see HHB000
HYDRAZINOBENZENE see PFI000
2-HYDRAZINOBENZOTHIAZOLE see HHB500
3-HYDRAZINO-6-(N,N-BIS-(2-HYDROXYETHYL)AMINO) PYRIDAZINE DIHY-
    DROCHLORIDE see BKB500
(S)-α-HYDRAZINO-3,4-DIHYDROXY-α-METHYL-BENZENEPROPANOIC ACID
    (9CI) see CBQ500
S(−)-α-HYDRAZINO-3,4-DIHYDROXY-α-METHYLHYDROCINNAMIC ACID
    MONOHYDRATE see CBQ529
(−)-l-α-HYDRAZINO-3,4-DIHYDROXY-α-METHYLHYDROCINNAMIC ACID
    MONOHYDRATE see CBQ529
2-HYDRAZINOETHANOL see HHC000
3-HYDRAZINO-6-((2-HYDROXYPROPYL)METHYLAMINO)PYRIDAZINE DIHY-
    DROCHLORIDE see HHD000
HYDRAZINO-α-METHYLDOPA see CBQ500
p-HYDRAZINONITROBENZENE see NIR000
2-HYDRAZINO-4-(5-NITRO-2-FURANYL)THIAZOLE see HHD500
2-HYDRAZINO-4-(5-NITRO-2-FURYL)THIAZOLE see HHD500
2-HYDRAZINO-4-(4-NITROPHENYL)THIAZOLE see HHE000
1-HYDRAZINO-2-PHENYLETHANE see PFC500
1-HYDRAZINO-2-PHENYLETHANE HYDROGEN SULPHATE see PFC750
2-HYDRAZINO-1-PHENYLPROPANE see PDN000
HYDRAZINOPHTHALAZINE see HGP495
1-HYDRAZINOPHTHALAZINE see HGP495
1-HYDRAZINOPHTHALAZINE ACETONE HYDRAZONE see HHE500
1-HYDRAZINOPHTHALAZINE HYDROCHLORIDE see HGP500
1-HYDRAZINOPHTHLAZINE MONOHYDROCHLORIDE see HGP500
4-HYDRAZINO-2-THIOURACIL see HHF500
HYDRAZOBENZEN (CZECH) see HHG000
HYDRAZOBENZENE see HHG000
HYDRAZODIBENZENE see HHG000
HYDRAZODICARBONSAEUREABIS(METHYLNITROSAMID) (GERMAN)
    see DRN400
HYDRAZODICARBOXYLIC ACID BIS(METHYLNITROSAMIDE) see DRN400
HYDRAZODIFORMIC ACID see DKJ600
HYDRAZOETHANE see DJL400
HYDRAZOIC ACID see HHG500
HYDRAZOMETHANE see DSF600, MKN000
HYDRAZONIUM SULFATE see HGW500
2,2'-HYDRAZONODIETHANOL see HHH000
HYDRAZYNA (POLISH) see HGS000
HYDREA see HOO500
HYDREL see HHH100
HYDRIDES see HHH500
HYDRIN-2 see HHQ800
HYDRINDANTIN, anhydrous see HHI000
1,2-HYDRINDENE see IBR000
HYDRINDONAPHTHENE see IBR000
HYDRIODIC ACID see HHI500
HYDRIODIC ACID, solution (DOT) see HHI500
HYDRIODIC ETHER see ELP500
HYDRIODIDE-ENTROL see CHR500
HYDRITE see KBB600
HYDRO-AQUIL see CFY000
HYDROAZODICARBOXYBIS(METHYLNITROSAMIDE) see DRN400
HYDROAZOETHANE see DJL400
HYDROBIS(2-METHYLPROPYL)ALUMINUM see DNI600
HYDROBROMIC ACID see HHJ000
HYDROBROMIC ACID, anhydrous (DOT) see HHJ000
HYDROBROMIC ACID MONOAMMONIATE see ANC250
HYDROCARBON GAS see HHJ500
HYDROCARBON GAS, compressed (DOT) see HHJ500
HYDROCARBON GAS, liquefied (DOT) see HHJ500
HYDROCARBON GAS, nonliquefied (DOT) see HHJ500
HYDROCELLULOSE see HHK000
HYDROCERIN see CMD750
HYDROCHINON (CZECH, POLISH) see HIH000
HYDROCHLORBENZETHYLAMINE DIMALEATE see HHK100
HYDROCHLORIC ACID see HHL000
HYDROCHLORIC ACID (mixture) see HHL500
HYDROCHLORIC ACID, anhydrous (DOT) see HHL000
HYDROCHLORIC ACID, solution, inhibited (DOT) see HHL000
HYDROCHLORIC ACID DICARBOXIDE see DEK000
HYDROCHLORIC ACID DIMETHYLAMINE see DOR600

HYDROCHLORIC ACID MIXTURE (DOT) see HHL500
HYDROCHLORIC ACID, mixed with NITRIC ACID (3581) see HHM000
HYDROCHLORIC ETHER see EHH000
HYDROCHLORIDE see HHL000
L-67 HYDROCHLORIDE see CMS250
1C50 HYDROCHLORIDE see EIJ000
d-8955 HYDROCHLORIDE see DXI800
191C49 HYDROCHLORIDE see DJP500
l-HYDROCHLORIDE ARGININE see AQW000
HYDROCHLORIDE of DI-n-BUTYLAMINOPROPYL-3-IODO-4-FLUOROBENZO-
    ATE see HHM500
HYDROCHLORIDE DIETHYLAMINE see DIS500
HYDROCHLORIDE OF 1-PHENYLCYCLOPENTANECARBOXYLIC ACID
    DIETHYLAMINOETHYL ESTER see PET250
HYDROCHLORID SALZ des p-N-n-BUTYLAMINO-BENZOESAURE-
    DIAETHYLAMINOAETHYLESTERS (GERMAN) see BQA000
HYDROCHLOROUS ACID, SODIUM SALT, PENTAHYDRATE see SHU525
HYDROCHLORTHIAZID see CFY000
HYDROCINNAMALDEHYDE see HHP000
HYDROCINNAMALDEHYDE, p-ISOPROPYL-α-METHYL-, DIETHYL ACETAL
    see COU510
HYDROCINNAMALDEHYDE, p-ISOPROPYL-α-METHYL-, DIMETHYL ACE-
    TAL see COU525
HYDROCINNAMIC ALCOHOL see HHP050
HYDROCINNAMIC ALDEHYDE see HHP000
HYDROCINNAMIQUE NITRILE (FRENCH) see HHP100
HYDROCINNAMONITRILE see HHP100
HYDROCINNAMYL ACETATE see HHP500
HYDROCINNAMYL ALCOHOL see HHP050
HYDROCINNAMYL FORMATE see HHQ000
HYDROCINNAMYL ISOBUTYRATE see HHQ500
HYDROCINNAMYL PROPIONATE see HHQ550
HYDROCODIN see DKW800
HYDROCODONE see OOI000
HYDROCODONE BITARTRATE see DKX050
HYDROCONQUININE see HIG500
11-β-HYDROCORTISONE see CNS750
Δ¹-HYDROCORTISONE see PMA000
HYDROCORTISONE-21-ACETATE see HHQ800
HYDROCORTISONE BUTYRATE see HHQ825
HYDROCORTISONE-17-BUTYRATE see HHQ825
HYDROCORTISONE-17-BUTYRATE-21-PROPIONATE see HHQ850
HYDROCORTISONE 17-α-BUYTRATE see HHQ825
HYDROCORTISONE BUYTRATE PROPIONATE see HHQ850
HYDROCORTISONE FREE ALCOHOL see CNS750
HYDROCORTISONE PHOSPHATE see HHQ875
HYDROCORTISONE-21-PHOSPHATE see HHQ875
21-HYDROCORTISONEPHOSPHORIC ACID see HHQ875
HYDROCORTISONE SODIUM SUCCINATE see HHR000
HYDROCORTISONE-21-SODIUM SUCCINATE see HHR000
HYDROCORTISYL see CNS750
HYDROCORTONE see CNS750
HYDROCOUMARIN see HHR500
HYDROCUPREINE ETHYL ESTER HYDROCHLORIDE see ELC500
HYDROCUPREINE ETHYL ETHER see HHR700
HYDROCYANIC ACID see HHS000
HYDROCYANIC ACID, liquefied (DOT) see HHS000
HYDROCYANIC ACID (unstabilized) see HHU000
HYDROCYANIC ACID, mixed with CYANOGEN CHLORIDE (3582 BY WT)
    see HHS500
HYDROCYANIC ACID, POTASSIUM SALT see PLC500
HYDROCYANIC ACID (PRUSSIC), unstabilized (DOT) see HHS000
HYDROCYANIC ACID, SALTS see HHT000
HYDROCYANIC ACID, SODIUM SALT see SGA500
HYDROCYANIC ACID, UNSTABILIZED (DOT) see HHU000
HYDROCYANIC ETHER see PMV750
HYDROCYCLIN see HOI000
HYDROCYCLINE see MCH525
HYDRODELTALONE see PMA000
HYDRODELTISONE see PMA000
HYDRODESULFURIZED HEAVY VACUUM GAS OIL see GBW010
HYDRODIISOBUTYLALUMINUM see DNI600
HYDRODIURETIC see CFY000
HYDRO-DIURIL see CFY000
17-β-HYDROESTR-4-EN-3-ONE see NNX400
HYDROFLUORIC ACID see HHU500
HYDROFLUORIC ACID, anhydrous (DOT) see HHU500
HYDROFLUORIC ACID, solution (DOT) see HHU500
HYDROFLUORIC ACID, SODIUM SALT (2:1) see SHQ500
HYDROFLUORIC ACID mixed with SULFURIC ACID see HHV000
HYDROFLUORIC and SULFURIC ACIDS, MIXTURE (DOT) see HHV000
HYDROFLUORIDE see HHU500
HYDROFLUORIDE-1927 WANDER see DCW800
HYDROFLUOSILICIC ACID see SCO500

HYDROFOL see PAE250
HYDROFOL ACID 1255 see LBL000
HYDROFOL ACID 1495 see MSA250
HYDROFOL ACID 1655 see SLK000
α-HYDROFORMAMINE CYANIDE see HHW000
HYDROFURAMIDE see FPS000
HYDROFURAN see TCR750
HYDROGEN see HHW500
HYDROGEN (DOT) see HHW500
HYDROGEN, compressed (DOT) see HHW500
HYDROGEN, refrigerated liquid (DOT) see HHW500
HYDROGEN ANTIMONIDE see SLQ000
HYDROGEN ARSENIDE see ARK250
HYDROGENATED COAL OIL FRACTION 1 see HHW509
HYDROGENATED COAL OIL FRACTION 3 see HHW519
HYDROGENATED COAL OIL FRACTION 4 see HHW529
HYDROGENATED COAL OIL FRACTION 7 see HHW539
HYDROGENATED COAL OIL FRACTION 9 see HHW549
HYDROGENATED TERPHENYLS see HHW800
HYDROGEN AZIDE see HHG500
HYDROGEN BROMIDE (OSHA ACGIH, MAK, DOT) see HHJ000
HYDROGEN CARBOXYLIC ACID see FNA000
HYDROGEN CHLORIDE see HHX000
HYDROGEN CHLORIDE (aerosol) see HHX500
HYDROGEN CHLORIDE, anhydrous (DOT) see HHL000
HYDROGEN CHLORIDE (OSHA, ACGIH, MAK, DOT) see HHL000
HYDROGEN CHLORIDE, refrigerated liquid (DOT) see HHL000
HYDROGEN, CRYOGENIC LIQUID (DOT) see HHY500
HYDROGEN CYANAMIDE see COH500
HYDROGEN CYANIDE, anhydrous, stabilized (DOT) see HHS000
HYDROGEN CYANIDE (OSHA, ACGIH) see HHS000
HYDROGEN CYANIDE-N-OXIDE see FOS050
HYDROGEN DIOXIDE see HIB000
HYDROGEN DISULFIDE see HHZ000
HYDROGENE SULFURE (FRENCH) see HIC500
(HYDROGEN(ETHYLENEDINITRILO)TETRAACETATO)IRON see HIA000
HYDROGEN FLUORIDE (OSHA, ACGIH, MAK, DOT) see HHU500
HYDROGEN HEXACHLOROIRIDATE (4+) see HIA500
HYDROGEN HEXACHLOROPLATINATE(4+) see CKO750
HYDROGEN HEXAFLUOROPHOSPHATE see HDE000
HYDROGEN HEXAFLUOROSILICATE see SCO500
HYDROGEN IODIDE see HHI500
HYDROGEN IODIDE, anhydrous (DOT) see HHI500
HYDROGEN IODIDE solution (DOT) see HHI500
HYDROGEN NITRATE see NED500
HYDROGEN OXALATE of AMITON see AMX825
HYDROGEN PEROXIDE see HIB000
HYDROGEN PEROXIDE, 30% see HIB010
HYDROGEN PEROXIDE, 8% to 20% see HIB005
HYDROGEN PEROXIDE, solution 30% see HIB010
HYDROGEN PEROXIDE, solution, 8% to 20% (DOT) see HIB005
HYDROGEN PEROXIDE, solution (over 52% peroxide) (DOT) see HIB000
HYDROGEN PEROXIDE, stabilized (over 60% peroxide) (DOT) see HIB000
HYDROGEN PEROXIDE CARBAMIDE see HIB500
HYDROGEN PEROXIDE, solution (8% to 40% PEROXIDE) (DOT) see HIB010
HYDROGEN PEROXIDE with UREA (1581) see HIB500
HYDROGEN PHOSPHIDE see PGY000
HYDROGEN SELENIDE see HIC000
HYDROGEN SELENIDE, anhydrous (DOT) see HIC000
21-(HYDROGEN SUCCINATE)CORTISOL, MONOSODIUM SALT see HHR000
HYDROGEN SULFIDE see HIC500
HYDROGEN SULFITE see HIC600
HYDROGEN SULFITE SODIUM see SFE000
HYDROGEN SULFURIC ACID see HIC500
HYDROGEN TRISULFIDE see HID000
HYDROGESTRONE see DYF759
α-HYDRO-omega-HYDROXY-POLY(OXY-1,2-ETHANEDIYL) see PJT500
HYDROL see DBI800
HYDROLIN see SHR500
HYDROLIT see FMW000
HYDROL SW see PKF500
HYDROMAGNESITE see MAC650
HYDROMEDIN see DFP600
HYDROMIREX see MRI750
HYDROMORPHONE see DLW600
HYDROMORPHONE HYDROCHLORIDE see DNU300
HYDRONITRIC ACID see HHG500
HYDRONOL see HID350
HYDRONSAN see GBB500
HYDROOT see SOI500
HYDROPERIT see HIB500
HYDROPEROXIDE see HIB000
HYDROPEROXIDE, ACETYL see PCL500
6-HYDROPEROXY-4-CHOLESTEN-3-ONE see HID500

6-β-HYDROPEROXYCHOLEST-4-EN-3-ONE see HID500
1-HYDROPEROXYCYCLOHEX-3-ENE see HIE000
1-HYDROPEROXY-3-CYCLOHEXENE see HIE000
1-((1-HYDROPEROXYCYCLOHEXYL)DIOXY)CYCLOHEXENOL with BIS(1-HYDROXYCYCLOHEXYL)PEROXIDE see CPC500
4-HYDROPEROXYCYCLOPHOSPHAMIDE see HIF000
HYDROPEROXYDE de BUTYLE TERTIAIRE (FRENCH) see BRM250
HYDROPEROXYDE de CUMENE (FRENCH) see IOB000
HYDROPEROXYDE de CUMYLE (FRENCH) see IOB000
10-β-HYDROPEROXY-17-α-ETHYNYL-4-ESTREN-17-β-OL-3-ONE see HIE525
6-β-HYDROPEROXY-Δ⁴CHOLESTEN-3-ONE see HID500
4-HYDROPEROXYIFOSFAMIDE see HIE550
N-(HYDROPEROXYMETHYL)-N-NITROSOPROPYLAMINE see HIE570
2-HYDROPEROXY-2-METHYLPROPANE see BRM250
HYDROPEROXY-N-NITROSODIBUTYLAMINE see HIE600
1-(HYDROPEROXY)-N-NITROSODIMETHYLAMINE see HIE700
4-HYDROPEROXYPHOSPHAMIDE see HIF000
2-HYDROPEROXYPROPANE see IPI000
1-HYDROPEROXY-1-VINYLCYCLOHEX-3-ENE see HIF575
4-HYDROPEROXY-4-VINYLCYCLOHEXENE see HIF575
HYDROPHENOL see CPB750
HYDROPHIS CYANOCINCTUS VENOM see SBI900
HYDROPHIS ELEGANS (AUSTRALIA) VENOM see HIG000
HYDROPRENE see EQD000
HYDROQUINIDINE see HIG500
HYDROQUINOL see HIH000
HYDROQUINONE see HIH000
m-HYDROQUINONE see REA000
o-HYDROQUINONE see CCP850
p-HYDROQUINONE see HIH000
α-HYDROQUINONE see HIH000
HYDROQUINONE, compounded with p-BENZOQUINONE see QFJ000
HYDROQUINONE BENZYL ETHER see AEY000
HYDROQUINONE CALCIUM SULFONATE see DMI300
HYDROQUINONECARBOXYLIC ACID see GCU000
HYDROQUINONE-β-d-GLUCOPYRANOSIDE see HIH100
HYDROQUINONE MONOBENZYL ETHER see AEY000
HYDROQUINONE MONOETHYL ETHER see EFA100
HYDROQUINONE MONOMETHYL ETHER see MFC700
HYDROQUINONE MUSTARD see BHB750
HYDROQUINONE, TRIMETHYL- see POG300
HYDRO-RAPID see CHJ750
HYDRORETROCORTIN see PMA000
HYDRORUBEANIC ACID see DXO200
HYDROSALURIC see CFY000
HYDROSARPAN see AFG750
HYDROSCINE HYDROBROMIDE see HOT500
HYDROSILICOFLUORIC ACID see SCO500
HYDROSORBIC ACID see HFD000
HYDROSULFITE ANION see HIC600
HYDROSULFITE AWC see FMW000
HYDROTHAL-47 see EAR000
HYDROTHIADENE see HII000
HYDROTHIDE see CFY000
HYDROTHOL see DXD000
HYDROTREATED HEAVY NAPHTHENIC DISTILLATE see MQV790
HYDROTREATED HEAVY NAPHTHENIC DISTILLATES (PETROLEUM) see MQV790
HYDROTREATED HEAVY PARAFFINIC DISTILLATE see MQV795
HYDROTREATED LIGHT NAPHTHENIC DISTILLATE see MQV800
HYDROTREATED LIGHT NAPHTHENIC DISTILLATES (PETROLEUM) see MQV800
HYDROTREATED LIGHT PARAFFINIC DISTILLATE see MQV805
HYDROTRICHLOROTHIAZIDE see HII500
HYDROTRICINE see TOG500
HYDROTROPALDEHYDE DIMETHYL ACETAL see HII600
HYDROTROPIC ALDEHYDE DIMETHYL ACETAL see HII600
HYDROXINE see CJR909
HYDROXON see HNT500
4-(4-HYDROXPHENYL)-2-BUTANONE see RBU000
HYDROXY No. 253 see NNC500
N-HYDROXY-AABP see ACD000
N-HYDROXY-AAF see HIP000
trans-4'-HYDROXY-AAS see HIK000
12-HYDROXYABIETIC ACID BIS(2-CHLOROETHYL)AMINE see HIJ000
12-HYDROXYABIETIC ACID BIS(2-CHLOROETHYL)AMINE SALT see HIJ000
N-HYDROXY-4-ACETAMIDOBIPHENYL see ACD000
N-4-(N-HYDROXYACETAMIDO)BIPHENYL see ACD000
N-HYDROXY-4-ACETAMIDODIPHENYL see ACD000
1-HYDROXY-2-ACETAMIDOFLUORENE see HIJ400
2-(N-HYDROXYACETAMIDO)FLUORENE see HIP000
N-HYDROXY-2-ACETAMIDOFLUORENE see HIP000
2-(N-HYDROXYACETAMIDO)NAPHTHALENE see NBD000
2-(N-HYDROXYACETAMIDO)PHENANTHRENE see PCZ000

HYDROXY-2-(β-(2'-ACETAMIDOPHENYL)ETHYLNAPHTHAMIDE see HIJ500
4-(N-HYDROXYACETAMIDO)STILBENE see SMT000
trans-N-HYDROXY-4-ACETAMIDOSTILBENE see SMT500
trans-4'-HYDROXY-4-ACETAMIDOSTILBENE see HIK000
HYDROXY(o-ACETAMINOBENZOATO)MERCURY SODIUM SALT see SJP500
7-HYDROXY-2-ACETAMINOFLUORENE see HIK500
3-HYDROXYACETANILIDE see HIL500
4-HYDROXYACETANILIDE see HIM000
m-HYDROXYACETANILIDE see HIL500
o-HYDROXYACETANILIDE see HIL000
p-HYDROXYACETANILIDE see HIM000
2'-HYDROXYACETANILIDE see HIL000
4'-HYDROXYACETANILIDE see HIM000
HYDROXYACETIC ACID see GHO000
HYDROXYACETIC ACID ETHYL ESTER see EKM500
HYDROXYACETIC ACID, MONOSODIUM SALT see SHT000
HYDROXYACETONE see ABC000
HYDROXYACETONITRILE see HIM500
2-HYDROXYACETONITRILE see HIM500
N-HYDROXY-p-ACETOPHENETIDIDE see HIN000
o-HYDROXYACETOPHENONE see HIN500
p-HYDROXYACETOPHENONE see HIO000
2'-HYDROXYACETOPHENONE see HIN500
4'-HYDROXYACETOPHENONE see HIO000
N-HYDROXY-4-ACETYLAMINOBIBENZYL see PDI250
N-HYDROXY-4-ACETYLAMINOBIPHENYL see ACD000
N-HYDROXY-2-ACETYLAMINOFLUORENE see HIP000
3-HYDROXY-N-ACETYL-2-AMINOFLUORENE see HIO875
N-HYDROXY-N-ACETYL-2-AMINOFLUORENE see HIP000
N-HYDROXY-2-ACETYLAMINOFLUORENE, COBALTOUS CHELATE see FDU875
N-HYDROXY-2-ACETYLAMINOFLUORENE, CUPRIC CHELATE see HIP500
N-HYDROXY-2-ACETYLAMINOFLUORENE, FERRIC CHELATE see HIQ000
N-HYDROXY-2-ACETYLAMINOFLUORENE-o-GLUCURONIDE see HIQ500
N-HYDROXY-2-ACETYLAMINOFLUORENE, MANGANOUS CHELATE see HIR000
N-HYDROXY-2-ACETYLAMINOFLUORENE, NICKELOUS CHELATE see HIR500
N-HYDROXY-2-ACETYLAMINOFLUORENE, POTASSIUM SALT see HIS000
N-HYDROXY-2-ACETYLAMINOFLUORENE, ZINC CHELATE see ZHJ000
N-HYDROXY-2-ACETYLAMINONAPHTHALENE see NBD000
N-HYDROXY-2-ACETYLAMINOPHENANTHRENE see PCZ000
N-HYDROXY-4-ACETYLAMINOSTILBENE see SMT000
trans-N-HYDROXY-4-ACETYLAMINOSTILBENE see SMT500
trans-N-HYDROXY-4-ACETYLAMINOSTILBENE CUPRIC CHELATE see SMU000
2-HYDROXYACLACINOMYCIN A see HIS300
3-(2-HYDROXY-1-ADAMANTYL)-N-METHYLPROPLYAMINE HYDROCHLORIDE see MGK750
N-HYDROXYADENINE see HIT000
6-HYDROXYADENOSINE see HIT100
N-HYDROXYADENOSINE see HIT100
6-N-HYDROXYADENOSINE see HIT100
N⁶-HYDROXYADENOSINE see HIT100
HYDROXYADIPALDEHYDE see HIT500
2-HYDROXYADIPALDEHYDE see HIT500
α-HYDROXYADIPALDEHYDE see HIT500
2-(2-HYDROXYETHOXY)AETHYLESTER der FLUTENAMINSAEURE (GERMAN) see HKK000
4-HYDROXYAFLATOXIN B1 see AEW000
17-HYDROXY-17-α-ALLYL-4-ESTRENE see AGF750
(−)-3-HYDROXY-N-ALLYLMORPHINAN see AGI000
l-3-HYDROXY-N-ALLYL MORPHINAN see AGI000
1-HYDROXY-4-AMINOANTHRAQUINONE see AKE250
2-HYDROXY-4-AMINOBENZOIC ACID see AMM250
β-HYDROXY-α-AMINOBUTYRIC ACID see AKF375
4-HYDROXY-3-AMINODIPHENYL HYDROCHLORIDE see AIW250
3-HYDROXY-4-AMINODIPHENYL SULPHATE see AKF250
5-HYDROXY-3-(β-AMINOETHYL)INDOLE see AJX500
N-HYDROXY-2-AMINOFLUORENE see HIU500
N-HYDROXY-3-AMINOFLUORENE see FEH000
N-HYDROXY-2-AMINOFLUORENE N-GLUCURONIDE see HLE750
2-HYDROXY-4-AMINO-5-FLUOROPYRIMIDINE see FHI000
3-HYDROXY-5-AMINOMETHYLISOXAZOLE see AKT750
3-HYDROXY-5-AMINOMETHYLISOXAZOLE-AGARIN see AKT750
4-(HYDROXYAMINO)-5-METHYLQUINOLINE-1-OXIDE see HIV000
4-(HYDROXYAMINO)-6-METHYLQUINOLINE-1-OXIDE see HIV500
4-(HYDROXYAMINO)-7-METHYLQUINOLINE-1-OXIDE see HIW000
4-(HYDROXYAMINO)-8-METHYLQUINOLINE-1-OXIDE see HIW500
N-HYDROXY-1-AMINONAPHTHALENE see HIX000
N-HYDROXY-2-AMINONAPHTHALENE see NBI500
2-HYDROXYAMINO-1,4-NAPHTHOQUINONE see HIX200
4-(HYDROXYAMINO)-6-NITROQUINOLINE-1-OXIDE see HIX500
4-(HYDROXYAMINO)-7-NITROQUINOLINE-1-OXIDE see HIY000

(8r)-3-α-HYDROXY-8-ISOPROPYL-1-α-H,5-α-H-TROPIUMBROMIDE-(±)-TROP-
ATE see IGG000
HYDROXYISOPROPYLMERCURY see IPW000
(2-(4-HYDROXY-2-ISOPROPYL-5-METHYLPHENOXY)ETHYL)DIMETHYLAM-
INE see DAD850
(2-(4-HYDROXY-2-ISOPROPYL-5-METHYLPHENOXY)ETHYL)METHYLAMINE
see DAD500
12-HYDROXY-13-ISOPROPYLPODOCARPA-7,13-DIEN-15-OIC ACID BIS(2-
CHLOROETHYL)AMINE SALT see HIJ000
HYDROXYISOXAZOLE see HLM000
3-HYDROXY-17-KETO-ESTRA-1,3,5-TRIENE see EDV000
4-HYDROXY-2-KETO-4-METHYLPENTANE see DBF750
3-HYDROXY-17-KETO-OESTRA-1,3,5-TRIENE see EDV000
HYDROXYKYNURENINE see HJD000
3-HYDROXYKYNURENINE see HJD000
3-HYDROXY-l-KYNURENINE see HJD500
l-3-HYDROXYKYNURENINE see HJD500
HYDROXYLAMINE see HLM500
HYDROXYLAMINE, N-ACETYL-N-(7-IODO-2-FLUORENYL)-O-MYRISTOYL-
see MSB100
HYDROXYLAMINE HYDROCHLORIDE see HLN000
HYDROXYLAMINE, N-METHYL-N-(4-QUINOLINYL)-, 1-OXIDE see HLX550
HYDROXYLAMINE NEUTRAL SULFATE see OLS000
HYDROXYLAMINE SULFATE see OLS000
HYDROXYLAMINE SULFATE (2:1) see OLS000
4-HYDROXYLAMINOBIPHENYL see BGI250
4-(HYDROXYLAMINO)-2-METHYL-QUINOLINE, 1-OXIDE see MKR000
6-HYDROXYLAMINOPURINE see HIT000
6-N-HYDROXYLAMINOPURINE see HIT000
6-HYDROXYLAMINOPURINE RIBOSIDE see HIT100
N⁶-HYDROXYLAMINOPURINE RIBOSIDE see HIT100
HYDROXYLAMMONIUM PHOSPHINATE see HLN500
HYDROXYLAMMONIUM SULFATE see OLS000
(2-HYDROXYLIMINIOBUTANE CHLORIDE) see BOV625
HYDROXYLIZARIC ACID see TKN750
13-HYDROXYLUPANINE-2-PYRROLE CARBOXYLIC ACID ESTER see CAZ125
HYDROXYLUREA see HOO500
N-HYDROXY-MAB see HLV000
1-HYDROXY-MA-144-N1 see RHA125
1-HYDROXY MA144 S1 see APV000
(E)-1-HYDROXY-MC-9,10-DIHYDRODIOL see DLT600
6-HYDROXY-2-MERCAPTOPYRIMIDINE see TFR250
o-(N-3-HYDROXYMERCURI-2-
HYDROXYETHOXYPROPYLCARBAMYL)PHENOXY-ACETIC ACID,
SODIUM SALT see HLO300
5-(3-HYDROXYMERCURI-2-METHOXYPROPYL) BARBITURIC ACID SODIUM
SALT see HLU000
o-((3-HYDROXYMERCURI-2-METHOXYPROPYL)CARBAM-
OYL)PHENOXYACETIC ACID MONOSODIUM SALT see SIH500
N-((3-(HYDROXYMERCURI)-2-METHOXYPROPYL)-CARBAMOYL)SUCCINA-
MIC ACID see MFC000
8-(γ-HYDROXYMERCURI-β-METHOXYPROPYL)-3-
COUMARINCARBOXYLICACID THEOPHYLLINE SODIUM see MCS000
8-(3-(HYDROXYMERCURI)-2-METHOXYPROPYL)-2-OXO-2H-1-BENZOPYRAN-
3-CARBOXYLIC ACID SODIUM SALT COMPOUND with THEOPHYLLINE
(1:1) see MCS000
N-(γ-HYDROXYMERCURI-β-METHOXYPROPYL)SALICYLAMIDE-o-ACETIC
ACID SODIUM SALT see SIH500
2-HYDROXYMERCURI-3-NITROBENZOIC ACID-1,2-CYCLIC ANHYDRIDE
see NHY500
HYDROXYMERCURI-o-NITROPHENOL see HLO400
o-(HYDROXYMERCURI)PHENOL see HLO500
HYDROXYMERCURIPROPANOLAMIDE of m-CARBOXYPHENOXYACETIC
ACID see HLP000
HYDROXYMERCURIPROPANOLAMIDE of o-CARBOXYPHENOXYACETIC
ACID see NCM800
HYDROXYMERCURIPROPANOLAMIDE of p-CARBOXYPHENOXYACETIC
ACID see HLP500
(5-(HYDROXYMERCURI)-2-THIENYL)MERCURY ACETATE see HLQ000
p-HYDROXYMETHAMPHETAMINE see FMS875
N-HYDROXY-METHANAMINE see MKQ875
HYDROXYMETHANESULFINIC ACID SODIUM SALT see FMW000
4'-HYDROXY-3'-METHOXYACETOPHENONE see HLQ500
4-HYDROXY-3-METHOXYALLYLBENZENE see EQR500
1-HYDROXY-2-METHOXY-4-ALLYLBENZENE see EQR500
4-HYDROXY-3-METHOXY-4-AMINOAZOBENZENE see HLR000
N-HYDROXY-3-METHOXY-4-AMINOAZOBENZENE see HLR000
3-HYDROXY-4'-METHOXY-4-AMINODIPHENYL see HLR500
3-HYDROXY-4'-METHOXY-4-AMINODIPHENYL HYDROCHLORIDE
see AKN250
2-HYDROXY-3-METHOXYBENZALDEHYDE see VFP000
3-HYDROXY-4-METHOXYBENZALDEHYDE see FNM000
4-HYDROXY-3-METHOXYBENZALDEHYDE see VFK000
1-HYDROXY-2-METHOXYBENZENE see GKI000

6-HYDROXY-7-METHOXY-5-BENZOFURANACRYLIC ACID Δ-LACTONE
see XDJ000
2-HYDROXY-3-METHOXYBENZOIC ACID see HJB500
3-HYDROXY-4-METHOXYBENZOIC ACID see HJC000
4-HYDROXY-3-METHOXYBENZOIC ACID see VFF000
2-HYDROXY-4-METHOXYBENZOPHENONE see MES000
o-(2-HYDROXY-4-METHOXYBENZOYL)BENZOIC ACID see HLS500
(α-HYDROXY-p-METHOXYBENZYL)-PHOSPHONIC ACID DIETHYL ESTER,
ESTER with BIS(2-CHLOROPROPYL) ANTIMONATE(III) see DHH600
4-HYDROXY-3-METHOXYCINNAMIC ACID see FBP200
4-HYDROXY-3-METHOXY-1-METHYLBENZENE see MEK250
3-HYDROXY-N-METHOXY-N-METHYL-cis-CROTONAMIDE, DIMETHYL
PHOSPHATE see DOL800
4-HYDROXY-3-METHOXY-5,6-METHYLENEDIOXY-APORPHIN see HLT000
6-HYDROXY-3-METHOXY-N-METHYL-4,5-EPOXYMORPHINAN see DKW800
5-HYDROXY-4-(METHOXYMETHYL)-6-METHYL-3-PYRIDINEMETHANOL
see MFN600
6-(4-HYDROXY-6-METHOXY-7-METHYL-3-OXO-5-PHTHALANYL)-4-METHYL-
4-HEXENOIC ACID, SODIUM SALT see SIN850
(E)-6-(4-HYDROXY-6-METHOXY-7-METHYL-3-OXO-5-PHTHALANYL)-4-
METHYL-4-HEXENOIC ACID see MRX000
2-HYDROXY-3-(o-METHOXYPHENOXY)PROPYL-1-CARBAMATE see GKK000
2-HYDROXY-3-(o-METHOXYPHENOXY)PROPYL NICOTINATE see HLT300
N-HYDROXY-2-METHOXY-4-(PHENYLAZO)BENZENAMINE see HLR000
4-(4-HYDROXY-3-METHOXYPHENYL)-2-BUTANONE see VFP100
1-(4-HYDROXY-3-METHOXYPHENYL)ETHANONE see HLQ500
(4-HYDROXY-3-METHOXYPHENYL)ETHYL METHYL KETONE see VFP100
5-HYDROXY-3-METHOXY-7-PHENYL-2,6-HEPTADIENOIC ACID γ-LACTONE
see GJI250
trans-N-((4-HYDROXY-3-METHOXYPHENYL)METHYL)-8-METHYL-6-NONEAM-
IDE see CBF750
N-((4-HYDROXY-3-METHOXYPHENYL)METHYL)-8-METHYL-6-NONENAM-
IDE see CBF750
(2-HYDROXY-4-METHOXYPHENYL)PHENYLMETHANONE see MES000
3-(4-HYDROXY-3-METHOXYPHENYL)PROPENOIC ACID see FBP200
1-(2-HYDROXY-3-METHOXY-3-PHENYLPROPYL)-4-(2-METHOXY-2-
PHENYLETHYL)PIPERAZINE DIHYDROCHLORIDE see MFG250
1-HYDROXY-2-METHOXY-4-PROPENYLBENZENE see IKQ000
4-HYDROXY-3-METHOXY-1-PROPENYLBENZENE see IKQ000
1-HYDROXY-2-METHOXY-4-PROP-2-ENYLBENZENE see EQR500
4-HYDROXY-3-METHOXYPROPYLBENZENE see MFM750
1-(2-HYDROXY-3-METHOXYPROPYL)-2-NITROIMIDAZOLE see NHH500
4-HYDROXY-8-METHOXYQUINALDIC ACID see HLT500
6-(HYDROXY(6-METHOXY-4-QUINOLINYL)METHYL)-1-ETHYL-3-VINYL-
QUINUCLIDINIUM, IODIDE see QIS000
6-(1-HYDROXY-1-(6-METHOXY-4-QUINOLINYL)METHYL-1-ETHYL-3-
VINYLQUINUCLIDINIUM, IODIDE see QIS000
4-HYDROXY-3-METHOXYTOLUENE see MEK250
HYDROXY(2-METHOXY-3-(2,4,6-TRIOXO-(1H,3H,5H)PYRIMID-5-YL)
PROPYL)MERCURY SODIUM SALT see HLU000
2-HYDROXY-3-METHYL-4-ACETYLTETRAHYDROFURANE see BMK290
N-(HYDROXYMETHYL)ACRYLAMIDE see HLU500
2-(HYDROXYMETHYL)ACRYLIC ACID, ETHYL ESTER see ELI500
2-HYDROXY-N-METHYL-1-ADAMANTANEPROPANAMINE HYDROCHLO-
RIDE see MGK750
17-β-HYDROXY-17-(2-METHYLALLYL)ESTR-4-EN-3-ONE see MDQ075
17-β-HYDROXY-17-α-(1-METHYLALLYL)ESTR-4-EN-3-ONE see MDQ100
17-β-HYDROXY-17-α-(2-METHYLALLYL)ESTR-4-EN-3-ONE see MDQ075
α-HYDROXY-β-METHYL AMINE PROPYLBENZENE see EAW000
N-HYDROXY-N-METHYL-4-AMINOAZOBENZENE see HLV000
(R)-4-(1-HYDROXY-2-(METHYLAMINO)ETHYL)-1,2-BENZENEDIOL (9CI)
see VGP000
(−)-α-HYDROXY-β-(METHYLAMINO)ETHYL-α-(3-HYDROXYBENZENE)
HYDROCHLORIDE see SPC500
3'-(1-HYDROXY-2-(METHYLAMINO)ETHYL)METHANESULFONANILIDE
METHANESULFONATE see AHL500
3'-(1-HYDROXY-2-(METHYLAMINO)ETHYL)METHANESULFONANILIDE
MONOMETHANESULFONATE SALT see AHL500
N-(3-(1-HYDROXY-2-(METHYLAMINO)ETHYL)PHENYL)-
METHANESULFONAMIDE MONOMETHANESULFONATE SALT
see AHL500
1-α-HYDROXY-β-METHYLAMINO-3-HYDROXY-1-ETHYLBENZENE see NCL500
4-HYDROXY-α-((METHYLAMINO)METHYL)BENZENEMETHANOL see HLV500
(R)-3-HYDROXY-α-((METHYLAMINO)METHYL)BENZENEMETHANOL
HYDROCHLORIDE see SPC500
(R)-3-HYDROXY-α-((METHYLAMINO)METHYL)BENZENEMETHANOOL
see NCL500
(−)-m-HYDROXY-α-(METHYLAMINOMETHYL)BENZYL ALCOHOL
see NCL500
p-HYDROXY-α-((METHYLAMINO)METHYL)BENZYL ALCOHOL see HLV500
1-m-HYDROXY-α-((METHYLAMINO)METHYL)-BENZYL ALCOHOL
see NCL500
1-m-HYDROXY-α-(METHYLAMINOMETHYL)BENZYL ALCOHOL HYDRO-
CHLORIDE see SPC500

2-HYDROXYMETHYL-2-NITROPROPANE-1,3-DIOL see HMJ500
2-(HYDROXYMETHYL)-2-NITRO-1,3-PROPANEDIOL see HMJ500
((HYDROXYMETHYL)NITROSAMINO)-BUTYRIC ACID METHYL ESTER, ACETATE see MEI250
N-(HYDROXYMETHYL)-N-NITROSO-β-ALANINE METHYL ESTER, ACETATE see MEH000
17-β-HYDROXY-17-METHYL-B-NORANDROST-4-EN-3-ONE see MNC150
2-HYDROXYMETHYLNORCAMPHANE see NNH000
17-β-HYDROXY-18-METHYL-19-NOR-17-α-PREGN-4-EN-20-YN-3-ONE see NNQ500, NNQ520
3-HYDROXY-4-METHYL-7-OXABICYCLO(4.1.0)HEPT-3-ENE-2,5-DIONE see TBF325
(1R-cis)-3-HYDROXY-4-METHYL-7-OXABICYCLO(4.1.0)HEPT-3-ENE-2,5-DIONE see TBF325
7-HYDROXY-4-METHYL-2-OXO-2H-1-BENZOPYRAN see MKP500
3-(HYDROXYMETHYL)-8-OXO-7-(2-(4-PYRIDYLTHIO)ACETAMIDO)-5-THIA-1-AZABICYCLO(4.2.0)OCT-2-ENE-2-CARBOXYLIC ACID, ACETATE (ESTER) see CCX500
3-(HYDROXYMETHYL)-8-OXO-7-(2-(4-PYRIDYLTHIO)ACETAMIDO)-5-THIA-1-AZABICYCLO(4,2,0)OCT-2-ENE-2-CARBOXYLIC ACID ACETATE(ESTER), MONOSODIUM SALT see HMK000
3-HYDROXY-3-METHYLPENTANEDIOIC ACID see HMC000
4-HYDROXY-4-METHYL-PENTAN-2-ON (GERMAN, DUTCH) see DBF750
4-HYDROXY-4-METHYLPENTANONE-2 see DBF750
4-HYDROXY-4-METHYL-2-PENTANONE see DBF750
4-HYDROXY-4-METHYL PENTAN-2-ONE see DBF750
2-HYDROXYMETHYLPHENOL see HMK100
o-(HYDROXYMETHYL)PHENOL see HMK100
4-HYDROXY-α-(1-((1-METHYL-2-PHENOXYETHYL)AMINO)ETHYL)-BENZENEMETHANOL HYDROCHLORIDE see VGA300
p-HYDROXY-N-(1-METHYL-2-PHENOXYETHYL)NOREPHEDRINE see VGF000
1-(3-HYDROXY-4-METHYLPHENYL)-2-AMINO-1-PROPANOL HYDROCHLORIDE see MKS000
HYDROXYMETHYLPHENYLARSINE OXIDE see HMK200
N-(4-((2-HYDROXY-5-METHYLPHENYL)AZO)PHENYL)ACETAMIDE see AAQ250
2-(2-HYDROXY-5-METHYLPHENYL)BENZOTRIAZOLE see HML500
p-HYDROXY-α-(1-((1-METHYL-3-PHENYLPROPYL)AMINO)ETHYL)BENZYL ALCOHOL HYDROCHLORIDE see DNU200
5-(1-HYDROXY-2-((1-METHYL-3-PHENYLPROPYL)AMINO)ETHYL)SALICYL-AMIDE HYDROCHLORIDE see HMM500
2-(HYDROXYMETHYL)-6-PHENYL-3(2H)-PYRIDAZINONE see HMN000
2-HYDROXYMETHYL-6-PHENYL-3-PYRIDAZONE see HMN000
3-HYDROXY-3-METHYL-1-PHENYLTRIAZENE see HMP000
3-HYDROXY-2-METHYL-5-((PHOSPHONOOXY)METHYL)-4-PYRIDINECARBOXALDEHYDE see PII100
HYDROXYMETHYLPHTHALIMIDE see HMP100
N-(HYDROXYMETHYL)PHTHALIMIDE see HMP100
17-HYDROXY-6-METHYLPREGNA-4,6-DIENE-3,20-DIONE ACETATE see VTF000
17-HYDROXY-6-METHYLPREGNA-4,6-DIENE-3,20-DIONE ACETATE mixed with 19-NOR-17-α-PREGNA-1,3,5(10)-TRIEN-2-YNE-3,17-DIOL see MCA500
17-HYDROXY-6-METHYLPREGN-4-ENE-3,20-DIONE ACETATE see HMQ000
17-HYDROXY-6-α-METHYLPREGN-4-ENE-3,20-DIONE ACETATE see MCA000
17-α-HYDROXY-6-α-METHYLPREGN-4-ENE-3,20-DIONE ACETATE see MCA000
17-α-HYDROXY-6-α-METHYLPROGESTERONE ACETATE see MCA000
1-HYDROXYMETHYLPROPANE see IIL000
HYDROXYMETHYL-PROPANEDIOIC ACID (9CI) see MPN250
N-(HYDROXYMETHYL)-2-PROPENAMIDE see HLU500
2-HYDROXY-2-METHYLPROPIONITRILE see MLC750
1-(HYDROXYMETHYL)PROPYLAMIDE of 1-METHYL-(+)-LYSERGIC ACID HYDROGEN MALEATE see MQP500
d-N,N'-(1-HYDROXYMETHYLPROPYL)ETHYLENEDINITROSAMINE see HMQ500
17-β-HYDROXY-6-α-METHYL-17-(1-PROPYNYL)ANDROST-4-EN-3-ONE see DRT200
(6-α,17-β)-17-HYDROXY-6-METHYL-17-(1-PROPYNYL)-ANDROST-4-EN-3-ONE see DRT200
3-HYDROXY-2-METHYL-4H-PYRAN-4-ONE see MAO350
2-(HYDROXYMETHYL)PYRIDINE see POR800
5-HYDROXY-6-METHYL-3,4-PYRIDINEDICARBINOL HYDROCHLORIDE see PPK500
5-HYDROXY-6-METHYL-3,4-PYRIDINEDIMETHANOL see PPK250
5-HYDROXY-6-METHYL-3,4-PYRIDINEDIMETHANOL HYDROCHLORIDE see PPK500
3-HYDROXY-1-METHYLPYRIDINIUM BROMIDE DIMETHYLCARBAMATE (ESTER) see MDL600
3-HYDROXY-1-METHYLPYRIDINIUM BROMIDE HEXAMETHYLENEBIS(METHYLCARBAMATE) see DXG800
3-HYDROXY-1-METHYL-PYRIDINIUM DIMETHYLCARBAMATE (ester) see PPI800
4-HYDROXY-2-METHYL-N-2-PYRIDINYL-2H-THIENO(2,3-e)-1,2-THIAZINE-3-CARBOXAMIDE 1,1-DIOXIDE see TAL485

4-HYDROXY-2-METHYL-N-(2-PYRIDYL)-2H-1,2-BENZOTHIAZIN-3-CAR-BOXYAMID-1,1-DIOXID (GERMAN) see FAJ100
4-HYDROXY-2-METHYL-N-(2-PYRIDYL)-2H-1,2-BENZOTHIAZINE-3-CAR-BOXAMIDE-1,1-DIOXIDE see FAJ100
3-HYDROXY-2-METHYL-4-PYRONE see MAO350
3-HYDROXY-2-METHYL-γ-PYRONE see MAO350
8-HYDROXY-1-METHYL QUINOLINIUM IODIDE, METHYLCARBAMATE see MID500
8-HYDROXY-1-METHYLQUINOLINIUM METHYLSULFATE DIMETHYLCARBAMATE see DQY400
4-HYDROXY-1-METHYL-2-QUINOLONE see HMR500
12-HYDROXY-4-METHYL-4,8-SECOSENECIONAN-8,11,16-TRIONE see DMX200
2-HYDROXYMETHYLTETRAHYDROPYRAN see MDS500
1-(HYDROXYMETHYL)-2,8,9-TRIOXA-5-AZA-1-SILABICYCLO(3.3.3) UNDECANEMETHACRYLATE see HMS500
3-α-HYDROXY-8-METHYL-1-αH,5-αH-TROPANIUM BROMIDE MANDELATE see MDL000
3-α-HYDROXY-8-METHYL-1-αH,5-αH-TROPANIUM BROMIDE 2-PRO-PYLVALERATE see LJS000
3-α-HYDROXY-8-METHYL-1-α,5-α-H-TROPANIUM BROMIDE (±)TROPATE see MGR250
3-α-HYDROXY-8-METHYL-1-αH,5-α-H-TROPANIUM NITRATE (±)-TROPATE (ESTER) see MGR500
3-HYDROXY-α-METHYL-l-TYROSINE see DNA800
3-HYDROXY-α-METHYL-l-TYROSINE-1-(2,2-DIMETHYL-1-OX-OPROPOXY)ETHYL ESTER PHOSPHATE HYDRATE see HMS875
4-HYDROXY-3-MORPHOLINOMETHYLBENZOIC ACID ETHYL ESTER see ELK000
HYDROXYMYCIN see NCF500
N-HYDROXY-N-MYRISTOYL-2-AMINOFLUORENE see HMU000
1-HYDROXYNAPHTHALENE see NAW500
2-HYDROXYNAPHTHALENE see NAX000
α-HYDROXYNAPHTHALENE see NAW500
β-HYDROXYNAPHTHALENE see NAX000
N-HYDROXY-2-NAPHTHALENECARBOXAMIDE see NAV000
5-HYDROXY-1,4-NAPHTHALENEDIONE see WAT000
2-HYDROXY-6-NAPHTHALENESULFONIC ACID see HMU500
6-HYDROXY-2-NAPHTHALENESULFONIC ACID see HMU500
N-HYDROXY-N-2-NAPHTHALENYLACETAMIDE see NBD000
4-((4-HYDROXY-1-NAPHTHALENYL)AZO)BENZENESULPHONIC ACID, MONOSODIUM SALT see FAG010
N-HYDROXYNAPHTHALIMIDE, DIETHYL PHOSPHATE see HMV000
N-HYDROXYNAPHTHALIMIDE-O,O-DIETHYL PHOSPHOROTHIOATE see NAQ500
3-HYDROXY-2-NAPHTHAMIDE see HMW500
2-HYDROXYNAPHTHENESULFONIC ACID see HMX000
2-HYDROXY-1-NAPHTHOIC ACID see HMX500
5-HYDROXY-1,4-NAPHTHOQUINONE see WAT000
N-HYDROXY-1-NAPHTHYLAMINE see HIX000
N-HYDROXY-2-NAPHTHYLAMINE see NBI500
1-HYDROXY-2-NAPHTHYLAMINE HYDROCHLORIDE see ALK250
2-HYDROXY-1-NAPHTHYLAMINE HYDROCHLORIDE see ALK000
p-((4-HYDROXY-1-NAPHTHYL)AZO)BENZENESULFONIC ACID, MONOSO-DIUM SALT see FAG010
p-((4-HYDROXY-1-NAPHTHYL)AZO)BENZENESULPHONIC ACID, MONOSO-DIUM SALT see FAG010
p-((4-HYDROXY-1-NAPHTHYL)AZO)BENZENESULPHONIC ACID, SODIUM SALT see FAG010
N-HYDROXYNAPHTHYLIMIDE DIETHYL PHOSPHATE see HMV000
3-HYDROXY-4-(1-NAPHTHYLOXY)BUTYRAMIDOXIME HYDROCHLORIDE see NAC500
2-HYDROXY-5-NITROANILINE see NEM500
4-HYDROXY-3-NITROANILINE see NEM480
2-HYDROXYNITROBENZENE see NIE500
4-HYDROXYNITROBENZENE see NIF000
4-HYDROXY-3-NITROBENZENEARSONIC ACID see HMY000
4-HYDROXY-3-NITROBENZENESULFONYL CHLORIDE see HMY050
N-HYDROXY-5-NITRO-2-FURANCARBOXIMIDAMIDE see NGD000
3-HYDROXY-2-(5-NITRO-2-FURYL)-2H-1,3-BENZOXAZIN-4(3H)-ONE see HMY100
2-HYDROXY-5-NITROMETANILIC ACID see HMY500
4-HYDROXY-3-NITROPHENYLARSONIC ACID see HMY000
2-HYDROXY-5-((4-NITROPHENYL)AZO)BENZOIC ACID see NEY000
4-HYDROXY-3-(1-(4-NITROPHENYL)-3-OXOBUTYL)-2H-1-BENZOPYRAN-2-ONE see ABF750
3-(HYDROXYNITROSOAMINO)-l-ALANINE (9CI) see AFH750
N-HYDROXY-N-NITROSO-BENZENAMINE, AMMONIUM SALT see ANO500
3-HYDROXYNITROSOCARBOFURAN see HMZ100
3-HYDROXY-4-(NITROSOCYANAMIDO)BUTYRAMIDE see HNA000
2-HYDROXY-6-(NITROSOCYANAMIDO)HEXANAMIDE see HNA500
2-HYDROXY-3-NITROSO-2,7-NAPHTHALENEDISULFONIC ACID DISODIUM SALT see HNB000
trans-4-HYDROXY-1-NITROSO-l-PROLINE see HNB500
3-HYDROXYNITROSOPYRROLIDINE see NLP700

8-HYDROXY-5-NITROSOQUINOLINE see NLR000
4-HYDROXYNONANOIC ACID, γ-LACTONE see CNF250
4-HYDROXYNONENAL see HNB600
4-HYDROXY-2-NONENAL see HNB600
HYDROXYNOREPHEDRINE see HNB875
m-HYDROXY NOREPHEDRINE see HNB875
1-m-HYDROXYNOREPHEDRINE see HNC000
3-α-HYDROXYNORETHYNODREL see NNU000
3-β-HYDROXYNORETHYNODREL see NNU500
17-HYDROXY-19-NOR-17-α-PREGNA-4,20-DIEN-3-ONE see NCI525
17-HYDROXY-19-NOR-17-α-PREGNA-4,9,11-TRIEN-20-YN-3-ONE see NNR125
17-HYDROXY-19-NORPREGN-4-ENE-3,20-DIONE HEXANOATE see GEK510
17-HYDROXY-19-NOR-4-PREGNENE-3,20-DIONE HEXANOATE see GEK510
17-HYDROXY-19-NORPREGN-4-EN-3-ONE see ENX600
17-HYDROXY-19-NOR-17-α-PREGN-4-EN-3-ONE see ENX600
17-β-HYDROXY-19-NOR-17-α-PREGN-4-EN-3-ONE see ENX600
17-β-HYDROXY-19-NORPREGN-4-EN-20-YN-3-ONE see NNP500
17-HYDROXY-19-NOR-17-α-PREGN-4-EN-20-YN-3-ONE see NNP500
(17-α)-17-HYDROXY-19-NORPREGN-4-EN-20-YN-3-ONE see NNP500
17-HYDROXY(17-α)-19-NORPREGN-5(10)-EN-20-YN-3-ONE see EEH550
17-HYDROXY-19-NOR-17-α-PREGN-5(10)-EN-20-YN-3-ONE see EEH550
(17-α)-17-HYDROXY-19-NORPREGN-5(10)-EN-20-YN-3-ONE see EEH550
17-HYDROXY-19-NOR-17-α-PREGN-4-EN-20-YN-3-ONE ACETATE see ABU000
17-β-HYDROXY-19-NOR-17-α-PREGN-4-EN-20-YN-3-ONE ACETATE see ABU000
17-α-HYDROXY-19-NORPROGESTERONE CAPROATE see GEK510
13-HYDROXY-9,11-OCTADECADIENOIC METHYL ESTER see MKR500
12-HYDROXY-cis-9-OCTADECENOIC ACID see RJP000
1-HYDROXYOCTANE see OEI000
5-HYDROXYOCTANOIC ACID LACTONE see OCE000
2-(3-HYDROXY-1-OCTENYL)-5-OXO-3-CYCLOPENTENE-1-HEPTANOIC ACID
  see POC250
7-(2-(3-HYDROXY-1-OCTENYL)-5-OXO-3-CYCLOPENTEN-1-YL)-5-HEPTENOIC
  ACID see MCA025
4-HYDROXY-5-OCTYL-2(5H)FURANONE see HND000
16-α-HYDROXYOESTRADIOL see EDU500
3-HYDROXY-OESTRA-1,3,5(10)-TRIEN-17-ONE see EDV000
3-HYDROXY-1,3,5(10)-OESTRATRIEN-17-ONE see EDV000
16-HYDROXY-11-OXAHEXADECANOIC ACID, omega-LACTONE see OKW100
16-HYDROXY-12 OXAHEXADECANOIC ACID, omega-LACTONE see OKW110
γ-HYDROXY-β-OXOBUTANE see ABB500
(11-β)-11-HYDROXY-17-(1-OXOBUTOXY)-21-(1-OXOPROPOXY)-PREGN-4-ENE-
  3,20-DIONE see HHQ850
1-HYDROXY-4-OXO-2,5-CYCLOHEXADIENE-1-SULFONIC ACID compound
  with DIETHYLAMINE see DIS600
5-β-HYDROXY-19-OXODIGITOXIGENIN see SMM500
6α-HYDROXY-3-OXO-11-EPIISOEUSANTONA-1,4-DIENIC ACID, Γ-LACTONE
  see SAU500
6α-HYDROXY-3-OXO-EUDESMA-1,4-DIEN-12-OIC ACID, Γ-LACTONE (11S)-
  (−)- see SAU500
4-HYDROXY-3-(3-OXO-1-FENYL-BUTYL) CUMARINE (DUTCH) see WAT200
5-HYDROXY-4-OXO-NORVALINE see AKG500
Δ-HYDROXY-γ-OXO-l-NORVALINE see AKG500
3-β-HYDROXY-11-OXOOLEAN-12-EN-30-OIC ACID see GIE000
3-β-HYDROXY-23-OXO-OLEAN-12-EN-28-OIC ACID see GMG000
3-β-HYDROXY-11-OXO-18α-OLEAN-12-EN-30-OIC ACID see GIEO5O
3-β-HYDROXY-11-OXO-OLEAN-12-EN-30-OIC ACID, HYDROGEN SUCCI-
  NATE see BGD000
3-β-HYDROXY-11-OXOOLEAN-12-EN-30-OIC ACID HYDROGEN SUCCINATE
  DISODIUM SALT see CBO500
4-HYDROXY-3-(3-OXO-1-PHENYLBUTYL)-2H-1-BENZOPYRAN-2-ONE POTASS-
  IUM SALT see WAT209
4-HYDROXY-3-(3-OXO-1-PHENYLBUTYL)-2H-1-BENZOPYRAN-2-ONE
  SODIUM SALT (9CI) see WAT220
4-HYDROXY-3-(3-OXO-1-PHENYL-BUTYL)-CUMARIN (GERMAN) see WAT200
HYDROXYOXOPHENYL IODANIUM PERCHLORATE see HND375
(HYDROXY)(OXO)(PHENYL)-LAMBDA³-IODANIUM PERCHLORATE
  see IFS000
3-β-HYDROXY-20-OXO-17-α-PREGN-5-ENE-16-β-CARBOXAMIDE ACETATE
  see HND800
(13E,15S)-15-HYDROXY-9-OXO-PROSTA-10,13-DIEN-1-OIC ACID (9CI)
  see POC250
15-HYDROXY-9-OXO-PROSTA-5,10-13-TRIEN-1-OIC ACID, (5Z,13E,15S)- (9CI)
  see MCA025
3-HYDROXY-2-OXOPURINE see HOB000
trans-6-(10-HYDROXY-6-OXO-1-UNDECENYL)-mu-LACTONE, RESORCYLIC
  ACID see ZAT000
6-(10-HYDROXY-6-OXO-trans-1-UNDECENYL)β-RESORCYCLIC ACID-N-LAC-
  TONE see ZAT000
2'-HYDROXYPELARGIDENOLON 1522 see MRN500
3-HYDROXY-2-PENTANEDIOIC ACID, DIMETHYL ESTER, DIMETHYL
  PHOSPHATE see SOY000
4-HYDROXYPENTANOIC ACID LACTONE see VAV000
4-HYDROXYPENT-3-ENOIC ACID LACTONE see AOO750, MKH250
4-HYDROXY-2-PENTENOIC ACID γ-LACTONE see MKH500

S-3-HYDROXY-4-PENTENONITRILE see COP400
1-HYDROXY-3-n-PENTYL-Δ⁸-TETRAHYDROCANNABINOL see HNE400
N-HYDROXYPHENACETIN see HIN000
HYDROXYPHENAMATE see HNJ000
N-HYDROXY-N-2-PHENANTHRENYL-ACETAMIDE (9CI) see PCZ000
N-HYDROXY-4'-PHENETHYLACETANILIDE see PDI250
β-HYDROXYPHENETHYL ALCOHOL-α-CARBAMATE see PFJ000
4-HYDROXYPHENETHYLAMINE see TOG250
p-HYDROXYPHENETHYLAMINE see TOG250
β-HYDROXYPHENETHYLAMINE see HNF000
p-HYDROXY-β-PHENETHYLAMINE see TOG250
2-(β-HYDROXYPHENETHYLAMINO)PYRIDINE see PGG350
2-(β-HYDROXY-β-PHENETHYLAMINO)-PYRIMIDINE HYDROCHLORIDE
  see FAR200
β-HYDROXYPHENETHYL CARBAMATE see PFJ000
(−)-2-(6-(β-HYDROXYPHENETHYL)-1-METHYL-2-PIPERIDYL)-
  ACETOPHENONE HYDROCHLORIDE see LHZ000
(m-HYDROXYPHENETHYL)TRIMETHYLAMMONIUM PICRATE see HNG000
N-HYDROXYPHENETIDINE see HNG100
N-HYDROXY-p-PHENETIDINE see HNG100
p-HYDROXYPHENETOLE see EFA100
2-HYDROXYPHENOL see CCP850
3-HYDROXYPHENOL see REA000
m-HYDROXYPHENOL see REA000
o-HYDROXYPHENOL see CCP850
p-HYDROXYPHENOL see HIH000
7-HYDROXY-3H-PHENOXAZIN-3-ONE-10-OXIDE see HNG500
1-HYDROXY-2-PHENOXYETHANE see PER000
(2-HYDROXYPHENOXY)PHENYLMERCURY (8CI) see PFO750
2-HYDROXY-N-PHENYLACETAMIDE see GHK500
N-(4-HYDROXYPHENYL)ACETAMIDE see HIM000
m-HYDROXYPHENYL ACETATE see RDZ900
α-HYDROXYPHENYLACETIC ACID see MAP000
HYDROXYPHENYLACETONITRILE see MAP250
2-HYDROXY-2-PHENYLACETOPHENONE see BCP250
α-HYDROXY-α-PHENYLACETOPHENONE see BCP250
p-HYDROXYPHENYLACRYLIC ACID see CNU825
β-(4-HYDROXYPHENYL)ACRYLIC ACID see CNU825
p-HYDROXYPHENYLLACTIC ACID see HNG700
l-β-(p-HYDROXYPHENYL)ALANINE see TOG300
α-(4-HYDROXYPHENYL)-β-AMINOETHANE see TOG250
1-(m-HYDROXYPHENYL)-2-AMINOETHANOL see AKT000
1-(p-HYDROXYPHENYL)-2-AMINOETHANOL see AKT250
1-(3'-HYDROXYPHENYL)-2-AMINOETHANOL see AKT000
l-1-(m-HYDROXYPHENYL)-2-AMINO-1-PROPANOL-d-HYDROGEN
  TARTRATE see HNC000
4-HYDROXYPHENYLARSENOUS ACID see HNG800
p-HYDROXYPHENYLARSONIC ACID see PDO250
(4-HYDROXYPHENYL)ARSONIC ACID polymer with FORMALDEHYDE
  see BCJ150
7-HYDROXY-8-(PHENYLAZO)-1,3-NAPHTHALENEDISULFONIC ACID,
  DISODIUM SALT see HGC000
7-HYDROXY-8-(PHENYLAZO)-1,3-NAPHTHALENEDISULPHONIC ACID,
  DISODIUM SALT see HGC000
6-((p-HYDROXYPHENYL)AZO)URACIL see HNH475
α-HYDROXY-α-PHENYLBENZENEACETIC ACID-2-(DIETHYLAMINO)ETHYL
  ESTER see DHU900
N-HYDROXYPHENYLBENZENESULFONAMIDE see HJH500
3-HYDROXYPHENYL BENZOATE see HNH500
2-(o-HYDROXYPHENYL)BENZOXAZOLE see HNI000
p-HYDROXYPHENYL BENZYL ETHER see AEY000
3-α-HYDROXY-8-(p-PHENYLBENZYL)-1-α-H,5-α-H-TROPANIUM BROMIDE,
  (±)-TROPATE see PEM750
1-(p-HYDROXYPHENYL)-3-BUTANONE see RBU000
4-(p-HYDROXYPHENYL)-2-BUTANONE (FCC) see RBU000
4-(p-HYDROXYPHENYL)-2-BUTANONE ACETATE see AAR500
p-HYDROXYPHENYLBUTAZONE see HNI500
1-(p-HYDROXYPHENYL)-2-BUTYLAMINOETHANOL see BQF250
2-HYDROXY-2-PHENYLBUTYL CARBAMATE see HNJ000
3-HYDROXY-2-PHENYLCINCHONINIC ACID see OPK300
4-(p-HYDROXYPHENYL)-2,5-CYCLOHEXADIENE-1-ONE SODIUM SALT
  see HNJ500
(3-HYDROXYPHENYL)DIETHYLMETHYLAMMONIUM BROMIDE see HNK000
(m-HYDROXYPHENYL)DIETHYLMETHYLAMMONIUM IODIDE,
  DIMETHYLCARBAMATE see HNK500
(m-HYDROXYPHENYL)DIETHYLMETHYLAMMONIUM METHOSULFATE
  METHYLCARBAMATE see HNK550
2-(p-HYDROXYPHENYL)-5,7-DIHYDROXYCHROMONE see CDH250
4-(4-(4-HYDROXYPHENYL)-2,3-DIISOCYANO-1,3-BUTADIENYL)-1,2-
  BENZENEDIOL see XCS680
(-)-(R)-1-(p-HYDROXYPHENYL)-2-((3,4-DIMETHOXYPHENETHYL)AMINO)
  ETHANOL see DAP850
3-HYDROXYPHENYLDIMETHYLETHYLAMMONIUM BROMIDE see TAL490
1-(4-HYDROXYPHENYL)-3,3-DIMETHYLTRIAZINE see DUI600

(4-HYDROXY-m-PHENYLENE)BIS(ACETATOMERCURY) see HNK575
m-HYDROXYPHENYLETHANOLAMINE see AKT000
p-HYDROXYPHENYLETHANOLAMINE see AKT250
m-HYDROXYPHENYLETHANOLETHYLAMINE see EGE500
1-(4-HYDROXYPHENYL)ETHANONE see HIO000
N-HYDROXY-N-(4-(2-PHENYLETHENYL)PHENYL)ACETAMIDE see SMT000
4-HYDROXY-PHENYLETHYLAMINE see TOG250
β-HYDROXYPHENYLETHYLAMINE see TOG250
2-(p-HYDROXYPHENYL)ETHYLAMINE see TOG250
1-(3'-HYDROXYPHENYL)-2-ETHYLAMINOETHANOL see EGE500
2-(β-(HYDROXYPHENYL)ETHYLAMINOMETHYL)TETRALONE HYDROCHLO-
    RIDE see HAJ700
2-(β-HYDROXY-β-PHENYL-ETHYL-AMINO)-PYRIMIDINE CHLORHYDRATE
    (FRENCH) see FAR200
2-HYDROXY-2-PHENYLETHYL CARBAMATE see PFJ000
β-HYDROXY-β-PHENYLETHYL DIMETHYLAMINE see DSH700
1-(3-HYDROXYPHENYL)-1-HYDROXY-2-AMINOETHANE see AKT000
1-(4-HYDROXYPHENYL)-1-HYDROXY-2-BUTYLAMINOETHANE see BQF250
(2-HYDROXYPHENYL)HYDROXYMERCURY see HLO500
7-HYDROXY-4-PHENYL-3-(4-HYDROXYPHENYL)COUMARIN see HNK600
1-(3-HYDROXYPHENYL)-2-ISOPROPYLAMINOETHANOL HYDROCHLORIDE
    see HLK500
β-(p-HYDROXYPHENYL)ISOPROPYLMETHYLAMINE see FMS875
(2R,4R)-2-(o-HYDROXYPHENYL)-3-(3-MERCAPTOPROPIONYL)-4-
    THIAZOLIDINECARBOX YLIC ACID see FAQ950
o-HYDROXYPHENYLMERCURIC CHLORIDE see CHW675
HYDROXYPHENYLMERCURY see PFN100
HYDROXYPHENYL-MERCURY compounded with
    NITRATOPHENYLMERCURY (1:1) see MDH500
p-HYDROXYPHENYLMETHYLAMINOETHANOL see HLV500
1-(4-HYDROXYPHENYL)-2-METHYLAMINOETHANOL see HLV500
1-1-(m-HYDROXYPHENYL)-2-METHYLAMINOETHANOL see NCL500
l-1-(m-HYDROXYPHENYL)-2-METHYL-AMINOETHANOL HYDROCHLORIDE
    see SPC500
(+ −)-1-(4-HYDROXYPHENYL)-2-METHYLAMINOETHANOL TARTRATE
    see SPD000
1-(p-HYDROXYPHENYL)-2-METHYLAMINOPROPANE see FMS875
α-(p-HYDROXYPHENYL)-β-METHYLAMINOPROPANE see FMS875
p-HYDROXYPHENYLMETHYLAMINOPROPANOL see HKH500
1-(4-HYDROXYPHENYL)-2-METHYLAMINOPROPANOL see HKH500
α-(p-HYDROXYPHENYL)-α-(2-METHYLCYCLOHEXYLIDENE)-p-CRESOL
    DIACETATE see BGQ325
1-(3-HYDROXYPHENYL)-N-METHYLETHANOLAMINE see NCL500
1-(4-HYDROXYPHENYL)-N-METHYLETHANOLAMINE see HLV500
o-HYDROXYPHENYL METHYL KETONE see HIN500
p-HYDROXYPHENYL METHYL KETONE see HIO000
1-(4-HYDROXYPHENYL)-2-(1-METHYL-2-PHENOXYETHYLAMINO)
    PROPANOL see VGF000
1-(p-HYDROXYPHENYL)-2-(1'-METHYL-2'-
    PHENOXY)ETHYLAMINOPROPANOL-1 HYDROCHLORIDE see VGA300
1-(p-HYDROXYPHENYL)-2-(1'-METHYL-2'-PHENOXYETHYLAMINO)
    PROPANOL-2-HYDROCHLORIDE see VGF000
α-(4-HYDROXYPHENYL)-β-METHYL-4-(PHENYLMETHYL)-1-
    PIPERIDINEETHANOL (9CI) see IAG600
1-p-HYDROXYPHENYL-2-(1'-METHYL-3'-PHENYLPROPYLAMINO)-1-
    PROPANOL HYDROCHLORIDE see DNU200
(±)-4-(2-((3-(4-HYDROXYPHENYL)-1-METHYLPROPYL)AMINO)ETHYL)-1,2-
    BENZENEDIOL HYDROCHLORIDE see DXS375
4-(2-((3-(p-HYDROXYPHENYL)-1-METHYLPROPYL)AMINO)ETHYL)PYROCAT-
    ECHOL HYDROCHLORIDE see DXS375
p-HYDROXYPHENYL-2-NAPHTHYLAMINE see NBF500
p-HYDROXYPHENYL-β-NAPHTHYLAMINE see NBF500
N-p-HYDROXYPHENYL-2-NAPHTHYLAMINE see NBF500
N-p-HYDROXYPHENYL-β-NAPHTHYLAMINE see NBF500
1-HYDROXY-1-PHENYLPENTANE see BQJ500
1-(p-HYDROXYPHENYL)-2-PHENYL-4-BUTYL-3,5-PYRAZOLIDINEDIONE
    see HNI500
1-p-HYDROXYPHENYL-2-PHENYL-3,5-DIOXO-4-N-BUTYLPYRAZOLIDINE
    see HNI500
1-(p-HYDROXYPHENYL)-2-(3'-PHENYLTHIOPROPYLAMINO)-1-PROPANOL
    HYDROCHLORIDE see HNL100
p-HYDROXYPHENYL-1-PROPANONE see ELL500
1-(4-HYDROXYPHENYL)-1-PROPANONE see ELL500
N-HYDROXY-3-PHENYL-2-PROPENAMIDE (9CI) see CMQ475
3-(4-HYDROXYPHENYL)-2-PROPENOIC ACID see CNU825
l-(1-HYDROXY-1-PHENYL-2-PROPYLAMINO)-1-(m-METHOXYPHENYL)-1-
    PROPANONE HYDROCHLORIDE see OQU100
3-((1-HYDROXY-1-PHENYL-2-PROPYL)METHYLAMINO)PROPIONITRILE
    see CCX600
p-HYDROXYPHENYLPYRUVIC ACID see HNL500
3-HYDROXY-2-PHENYL-4-QUINOLINECARBOXYLIC ACID see OPK300
HYDROXYPHENYL SALICYLAMIDE see DYE700
p-HYDROXYPHENYLSALICYLAMIDE see DYE700
N-(4-HYDROXYPHENYL)SALICYLAMIDE see DYE700

N-(p-HYDROXYPHENYL)SALICYLAMIDE see DYE700
4-(3-HYDROXY-3-PHENYL-3-(2-THIENYL)PROPYL)-4-
    METHYLMORPHOLINIUMIODIDE see HNM000
4-HYDROXY-α-(1-((3-(PHENYLTHIO)PROPYL)AMINO)ETHYL)-
    BENZENEMETHANOL HYDROCHLORIDE see HNL100
p-HYDROXY-α-(1-((3-(PHENYLTHIO)PROPYL)AMINO)ETHYL)BENZYL ALCO-
    HOL HYDROCHLORIDE see HNL100
2-((N-(m-HYDROXYPHENYL)-p-TOLUIDINO)METHYL)-2-IMIDAZOLINE
    see PDW400
3-HYDROXYPHENYLTRIMETHYLAMMONIUM BROMIDE
    DIMETHYLCARBAMIC ESTER see POD000
(m-HYDROXYPHENYL)TRIMETHYLAMMONIUM BROMIDE
    DIMETHYLCARBAMATE see POD000
(m-HYDROXYPHENYL)TRIMETHYLAMMONIUM CHLORIDE,
    METHYLCARBAMATE see HNN000
(m-HYDROXYPHENYL)TRIMETHYLAMMONIUM DIMETHYLCARBAMATE
    (ester) see NCL100
(m-HYDROXYPHENYL)TRIMETHYLAMMONIUM IODIDE see HNN500
(m-HYDROXYPHENYL)TRIMETHYLAMMONIUM IODIDE, BENZYLCARBA-
    MATE see HNO000
(m-HYDROXYPHENYL)TRIMETHYLAMMONIUM IODIDE, METHYLCARBA-
    MATE see HNO500
(p-HYDROXYPHENYL)TRIMETHYLAMMONIUM IODIDE, METHYLCARBA-
    MATE see HNP000
(m-HYDROXYPHENYL)TRIMETHYLAMMONIUM METHYLSULFATE
    BENZYLCARBAMATE see BED500
(m-HYDROXYPHENYL)TRIMETHYLAMMONIUM METHYLSULFATE, CARBA-
    MATE see HNP500
(m-HYDROXYPHENYL)TRIMETHYLAMMONIUM METHYLSULFATE
    DIETHYLCARBAMATE see HNQ000
(3-HYDROXYPHENYL)TRIMETHYLAMMONIUM METHYL SULFATE
    DIMETHYLCARBAMIC ESTER see DQY909
(m-HYDROXYPHENYL)TRIMETHYLAMMONIUM METHYL SULFATE
    DIMETHYLCARBAMATE see DQY909
(m-HYDROXYPHENYL)TRIMETHYLAMMONIUM METHYLSULFATE,
    ETHYLCARBAMATE see HNQ500
(m-HYDROXYPHENYL)TRIMETHYLAMMONIUM METHYLSULFATE,
    METHYLCARBAMATE see HNR000
(m-HYDROXYPHENYL)TRIMETHYLAMMONIUM METHYLSULFATE
    METHYLPHENYLCARBAMATE see HNR500
(m-HYDROXYPHENYL)TRIMETHYLAMMONIUM METHYLSULFATE, PEN-
    TAMETHYLENECARBAMATE see HNS000
(m-HYDROXYPHENYL)TRIMETHYLAMMONIUM METHYLSULFATE,
    PHENYLCARBAMATE see HNS500
3-HYDROXY-2-PICOLINE-4,5-DIMETHANOL see PPK250
3-(1-HYDROXY-2-PIPERIDINOETHYL)-5-PHENYLISOXAZOLE see PCJ325
3-(1-HYDROXY-2-PIPERIDINOETHYL)-5-PHENYLISOXAZOLE CITRATE
    see HNT075
3-(1-HYDROXY-2-PIPERIDINOETHYL)-5-PHENYLISOXAZOLE CITRATE (2:1)
    see PCJ325
2-HYDROXY-N-(3-(m-(PIPERIDINOMETHYL)PHENOXY)PROPYL)
    ACETAMIDE ACETATE (ester) HYDROCHLORIDE see HNT100
10-(3-(4-HYDROXYPIPERIDINO)PROPYL)PHENOTHIAZINE-2-CARBONITRILE
    see PIW000
HYDROXYPOLYETHOXYDODECANE see DXY000
17-HYDROXYPREGNA-4,6-DIENE-3,20-DIONE ACETATE see MCB375
17-HYDROXYPREGNA-4,6-DIENE-3,20-DIONE ACETATE (ESTER) see MCB375
3-HYDROXYPREGNANE-11,20-DIONE see AFK875
3-α-HYDROXY-5-α-PREGNANE-11,20-DIONE see AFK875
3-α-HYDROXY-5-α-PREGNANE-11,20-DIONE mixed with 3-α,21-DIHYDROXY-5-
    α-PREGNANE-11,20-DIONE 21-ACETATE (3:1) see AFK500
21-HYDROXYPREGNANE-3,20-DIONE SODIUM HEMISUCCINATE see VJZ000
21-HYDROXY-5-β-PREGNANE-3,20-DIONE SODIUM HEMISUCCINATE
    see VJZ000
21-HYDROXY-5-β-PREGNANE-3,20-DIONE, SODIUM SALT, HEMISUCCIN-
    ATE see VJZ000
3-β-HYDROXY-5-α-PREGNAN-20-ONE-20-ISONICOTINYLHYDRAZONE
    see AFT125
21-HYDROXYPREGN-4-ENE-3,20-DIONE see DAQ600
17-HYDROXYPREGN-4-ENE-3,20-DIONE ACETATE see PMG600
21-HYDROXYPREGN-4-ENE-3,20-DIONE-21-ACETATE see DAQ800
17-HYDROXYPREGN-4-ENE-3,20-DIONE HEXANOATE see HNT500
17α,21-HYDROXYPREGN-4-ENE-3,11,20-TRIONE see CNS800
17-HYDROXY-17-α-PREGN-4-EN-20-YN-3-ONE see GEK500
HYDROXYPROCAINE see DHO600
m-HYDROXYPROCAINE see DHO600
21-HYDROXYPROGESTERONE see DAQ600
17-HYDROXYPROGESTERONE ACETATE see PMG600
HYDROXYPROGESTERONE CAPROATE see HNT500
17-α-HYDROXYPROGESTERONE CAPROATE see HNT500
17-α-HYDROXY PROGESTERONE-N-CAPROATE see HNT500
17-α-HYDROXYPROGESTERONE HEXANOATE see HNT500
m-HYDROXYPROPADRINE see HNB875
1-HYDROXYPROPANE see PND000

2-HYDROXY-1,2,3-PROPANECARBOXYLIC ACID TERBIUM (3+) SALT (1:1) see TAM500
((2-HYDROXY-1,3-PROPANEDIYL)BIS(NITRILOBIS(METHYLENE)))TETRAKISPHOSPHONIC ACID see DYE550
3-HYDROXYPROPANENITRILE see HGP000
3-HYDROXY-1-PROPANESULFONIC ACID γ-SULTONE see PML400
3-HYDROXY-1-PROPANESULPHONIC ACID SULFONE see PML400
3-HYDROXY-1-PROPANESULPHONIC ACID SULTONE see PML400
2-HYDROXY-1,2,3-PROPANETRICARBOXYLIC ACID see CMS750
2-HYDROXY-1,2,3-PROPANETRICARBOXYLIC ACID COPPER(2+) SALT (1:2) (9CI) see CNK625
2-HYDROXY-1,2,3-PROPANETRICARBOXYLIC ACID EBRIUM(3+) salt (1:1) see ECY000
2-HYDROXY,1,2,3-PROPANETRICARBOXYLIC ACID MONOSODIUM SALT see MRL000
2-HYDROXY-1,2,3-PROPANETRICARBOXYLIC ACID, TRIETHYL ESTER see TJP750
2-HYDROXY-1,2,3-PROPANETRISCARBOXYLIC ACID CERIUM(3+) SALT (1:1) (9CI) see CCZ000
2-HYDROXYPROPANOIC ACID see LAG000
2-HYDROXYPROPANOIC ACID, BUTYL ESTER see BRR600
2-HYDROXYPROPANOIC ACID MONOSODIUM SALT see LAM000
1-HYDROXY-2-PROPANONE see ABC000
3-HYDROXY-2-PROPENAL SODIUM SALT see MAN700
3-HYDROXYPROPENE see AFV500
2-HYDROXYPROPIONIC ACID see LAG000
α-HYDROXYPROPIONIC ACID see LAG000
3-HYDROXYPROPIONIC ACID LACTONE see PMT100
2-HYDROXYPROPIONITRILE see LAQ000
3-HYDROXYPROPIONITRILE see HGP000
β-HYDROXYPROPIONITRILE see HGP000
N-HYDROXY-N-2-PROPIONYLAMINO FLUORENE see FEO000
1-5α-HYDROXYPROPIONYLAMINO-2,4,6-TRIIODOISOPHTHALIC ACID DI(1,3-DIHYDROXY-2-PROPYLAMIDE) see IFY000
HYDROXYPROPIOPHENONE see ELL500
4-HYDROXYPROPIOPHENONE see ELL500
p-HYDROXYPROPIOPHENONE see ELL500
2-(2-(2-HYDROXYPROPOXY)PROPOXY-1-PROPANOL see TMZ000
2-HYDROXYPROPYL ACRYLATE see HNT600
β-HYDROXYPROPYL ACRYLATE see HNT600
HYDROXY PROPYL ALGINATE see PNJ750
2-HYDROXYPROPYLAMINE see AMA500
3-HYDROXYPROPYLAMINE see PMM250
2'-(2-HYDROXY-3-(PROPYLAMINO)PROPOXY)-3-PHENYLPROPIOPHENONE HYDROCHLORIDE see PMJ525
(3-HYDROXYPROPYL)BENZENE see HHP050
α-HYDROXYPROPYLBENZENE see EGQ000
p-HYDROXYPROPYL BENZOATE see HNU500
3-HYDROXYPROPYL BROMIDE see BNY750
HYDROXYPROPYL CELLULOSE see HNV000
1-(3-HYDROXYPROPYL)-CNU see CHC250
1-(2-HYDROXYPROPYL)-3,7-DIMETHYLXANTHINE see HNY500, HNZ000
(2-HYDROXY-1,3-PROPYLENEDIAMINE-N,N,N¹,N'¹-TETRAMETHYLENEPHOSPHORIC ACID see DYE550
HYDROXYPROPYL ETHER of CELLULOSE see HNV000
HYDROXYPROPYL METHYLCELLULOSE see HNX000
2-HYDROXYPROPYLMETHYLNITROSAMINE see NKU500
3-((2-HYDROXYPROPYL)NITROSAMINO)-1,2-PROPANEDIOL see NOC400
1-((2-HYDROXYPROPYL)NITROSO)AMINO)ACETONE see HNX500
4-(N-(2-HYDROXYPROPYL)-N-NITROSOAMINO)-1-BUTANOL see HJR500
4-(N-(2-HYDROXYPROPYL)-N-NITROSOAMINO)-2-BUTANOL see HJR000
4-(N-(3-HYDROXYPROPYL)NITROSOAMINO)-1-BUTANOL see HJS000
1-((2-HYDROXYPROPYL)NITROSOAMINO)-2-PROPANONE see HNX500
1-(2-HYDROXYPROPYL)-1-NITROSOUREA see NKO400
1-(3-HYDROXYPROPYL)-1-NITROSOUREA see NKK000
2-HYDROXYPROPYL PHENYL ARSINIC ACID see HNX600
N-(3-HYDROXYPROPYL)-1,3-PROPANEDIAMINE see HNX800
β-HYDROXYPROPYLPROPYLNITROSAMINE see NLM500
(2-HYDROXYPROPYL)PROPYLNITROSOAMINE see NLM500
HYDROXYPROPYL STARCH see HNY000
1-(2-HYDROXYPROPYL)THEOBROMINE see HNY500
1-(3-HYDROXYPROPYL)THEOBROMINE see HNZ000, HNZ000
1-(β-HYDROXYPROPYL)THEOBROMINE see HNY500
HYDROXYPROPYLTHEOPHYLLINE see HOA000
β-HYDROXYPROPYLTHEOPHYLLINE see HOA000
7-(2-HYDROXYPROPYL))THEOPHYLLINE see HOA000
7-(β-HYDROXYPROPYL)THEOPHYLLINE see HOA000
(3-HYDROXYPROPYL)TRIMETHOXYSILANE METHACRYLATE see TLC250
(2-HYDROXYPROPYL)TRIMETHYLAMMONIUMCHLORIDE ACETATE see ACR000
(2-HYDROXYPROPYL)TRIMETHYLAMMONIUM CHLORIDE CARBAMATE see HOA500
17-β-HYDROXY-17-(1-PROPYNYL)ESTR-4-EN-3-ONE see MNC100
17-β-HYDROXY-17-α-(1-PROPYNYL)ESTR-4-EN-3-ONE see MNC100

4-(1-(2-HYDROXY)PROPYLOXY)BENZENEARSONIC ACID see PMQ000
N-HYDROXY-1H-PURIN-6-AMINE (9CI) see HIT000
3-HYDROXYPURIN-2(3H)-ONE see HOB000
4'-HYDROXYPYRAZOLOL(3,4-d)PYRIMIDINE see ZVJ000
4-HYDROXY-3,4-PYRAZOLOPYRIMIDINE see ZVJ000
4-HYDROXYPYRAZOLO(3,4-d)PYRIMIDINE see ZVJ000
4-HYDROXY-1H-PYRAZOLO(3,4-d)PYRIMIDINE see ZVJ000
4-HYDROXYPYRAZOLYL(3,4-d)PYRIMIDINE see ZVJ000
8-HYDROXYPYRENE-1,3,6-TRISULFONIC ACID SODIUM SALT see TNM000
8-HYDROXY-1,3,6-PYRENETRISULFONIC ACID TRISODIUM SALT see TNM000
6-HYDROXY-3(2H)-PYRIDAZINONE see DMC600
6-HYDROXY-3-(2H)-PYRIDAZINONE DIETHANOLAMINE see DHF200
1-HYDROXY-2-PYRIDINETHIONE see HOB500
1-HYDROXY-2-(1H)-PYRIDINETHIONE see HOB500
1-HYDROXY-2-(1H)-PYRIDINETHIONE SODIUM SALT see HOC000
(1-HYDROXY-2-PYRIDINETHIONE), SODIUM SALT, TECH see MCQ750
β-HYDROXY-N-(3-HYDROXY-4-PYRIDONE)-α-AMINOPROPIONIC ACID see HOC500
5-(α-HYDROXY-α-2-PYRIDYLBENZYL)-7-(α-2-PYRIDYLBENZYLIDENE)-5-NORBORNENE-2,3-DICARBOXIMIDE see NNF000
4-HYDROXY-2(1H)-PYRIMIDINETHIONE see TFR250
8-HYDROXYQUINALDIC ACID see HOE000
HYDROXYQUINOL see BBU250
4-HYDROXYQUINOLINAMINE-1-OXIDE MONOHYDROCHLORIDE see HIZ000
N-HYDROXY-4-QUINOLINAMINE-1-OXIDE MONOHYDROCHLORIDE see HIZ000
8-HYDROXYQUINOLINE see QPA000
8-HYDROXYQUINOLINE COPPER COMPLEX see BLC250
8-HYDROXYQUINOLINE SULFATE see QPS000
14-HYDROXY-3-β-(RHAMNOSYLOXY)BUFA-4,20,22-TRIENOLIDE see POB500
5-HYDROXY-1-β-d-RIBOFURANOSYL-1H-IMIDAZOLE-4-CARBOXAMIDE see BMM000
1'-HYDROXYSAFROLE see BCJ000
1'-HYDROXYSAFROLE-2',3'-OXIDE see HOE500
4'-HYDROXYSALICYLANILIDE see DYE700
p'-HYDROXYSALICYLANILIDE see DYE700
4-HYDROXYSALICYLIC ACID see HOE600
5-HYDROXYSALICYLIC ACID see GCU000
p-HYDROXYSALICYLIC ACID see HOE600
14-β-HYDROXY-3-β-SCILLOBIOSIDOBUFA-4,20,22-TRIENOLIDE see GFC000
3-HYDROXY-16,7-SECOESTRA-1,3,5(10)-TRIEN-17-OIC ACID see DYB000
12-HYDROXYSENECIONAN-11,16-DIONE see ARS500
(15E)-12-HYDROXY-SENECIONAN-11,16-DIONE (9CI) see IDG000
HYDROXYSENKIRKINE see HOF000
5-HYDROXY-1(N-SODIO-5-TETRAZOLYLAZO)TETRAZOLE see HOF500
3-HYDROXY-SPIRO(8-AZONIABICYCLO(3.2.1)OCTANE-8,1'-PYRROLIDINIUM CHLORIDE BENZILATE see BCA375
3-α-HYDROXY-SPIRO(1-α-H,5-α-H-NORTROPANE-8,1'-PYRROLIDINIUM) CHLORIDE BENZILATE see KEA300
12-HYDROXYSTEARIC ACID see HOG000
12-HYDROXYSTEARIC ACID, METHYL ESTER see HOG500
HYDROXYSTREPTOMYCIN see SLX500
4-HYDROXYSTYRENE see VQA200
p-HYDROXYSTYRENE see VQA200
(E)-4'-(p-HYDROXYSTYRYL) ACETANILIDE see HIK000
HYDROXYSUCCINIC ACID see MAN000
4'-HYDROXY-3-((4-SULFO-1-NAPHTHALENYL)AZO)-1-NAPHTHALENESULFONIC ACID, DISODIUM SALT see HJF500
3-HYDROXY-4-((4-SULFO-1-NAPHTHALENYL)AZO)-2,7-NAPHTHLENEDISULFONIC ACID, TRISODIUM SALT see FAG020
3-HYDROXY-4-((4-SULFO-1-NAPHTHYL)AZO)-2,7-NAPHTHALENEDISULFONIC ACID, TRISODIUM SALT see FAG020
1-HYDROXY-2-(3-SULFOPROPOXY)ANTHRAQUINONE SODIUM SALT see HOH000
4-HYDROXY-3-((5-SULFO-2,4-XYLYL)AZO)-1-NAPHTHALENESULFONIC ACID, DISODIUM SALT see FAG050
3-HYDROXY-4-((4-SULPHO-1-NAPHTHALENYL)AZO)-2,7-NAPHTHALENEDISULPHONIC ACID, TRISODIUM SALT see FAG020
3-HYDROXY-4-((4-SULPHO-1-NAPHTHYL)AZO)-2,7-NAPHTHALENEDISULPHONIC ACID, TRISODIUM SALT see FAG020
4-HYDROXY-3-((5-SULPHO-2,4-XYLYL)AZO)-1-NAPHTHALENESULPHONIC ACID, DISODIUM SALT see FAG050
5-HYDROXYTETRACYCLINE see HOH500
5-HYDROXYTETRACYCLINE HYDROCHLORIDE see HOI000
N-HYDROXY-N-TETRADECANOYL-2-AMINOFLUORENE see HMU000
4-HYDROXY-3-(1,2,3,4-TETRAHYDRO-1-NAFTYL)-4-CUMARINE (DUTCH) see EAT600
4-HYDROXY-3-(1,2,3,4-TETRAHYDRO-1-NAPHTHALENYL)-2H-1-BENZOPYRAN-2-ONE (9CI) see EAT600
4-HYDROXY-3-(1,2,3,4-TETRAHYDRO-1-NAPHTHYL)CUMARIN see EAT600
6-HYDROXY-2,5,7,8-TETRAMETHYLCHROMAN-2-CARBOXYLIC ACID see TNR625
7-HYDROXY-3,4-TETRAMETHYLENECOUMARIN-O,O-DIETHYL THIOPHOSPHATE see DXO000

14-HYDROXY-6-β-THEBAINOL 4-METHYL ETHER see ORE000
7-HYDROXYTHEOPHYLLIN (GERMAN) see HOJ000
7-HYDROXYTHEOPHYLLINE see HOJ000
9-(2-HYDROXYTHEOXYMETHYL)GUANINE see AEC700
α-HYDROXY-THYMIDINE (9CI) see HMB550
HYDROXYTOLUENE see BDX500
4-HYDROXYTOLUENE see CNX250
m-HYDROXYTOLUENE see CNW750
o-HYDROXYTOLUENE see CNX000
p-HYDROXYTOLUENE see CNX250
α-HYDROXYTOLUENE see BDX500
α-HYDROXY-α-TOLUIC ACID see MAP000
1-HYDROXY-4-(p-TOLUIDINO)ANTHRAQUINONE see HOK000
HYDROXYTOLUOLE (GERMAN) see CNW500
2-(m-HYDROXY-N-p-TOLYLANILINOMETHYL)-2-IMIDAZOLINE see PDW400
4'-((6-HYDROXY-m-TOLYL)AZO)ACETANILIDE see AAQ250
2-HYDROXY-3-o-TOLYLOXYPROPYL-1-CARBAMATE see CBK500
(3-HYDROXY-p-TOLYL)TRIMETHYLAMMONIUM CHLO-
    RIDE,METHYLCARBAMATE see HOL000
4-HYDROXY-3H-o-TRIAZOLO(4,5-d)PYRIMIDINE-5,7(4H-6H)-DIONE
    see HJE575
(HYDROXY)TRIBUTYLSTANNANE see TID500
HYDROXYTRIBUTYLSTANNANE-4,4-DIMETHYLOCTANOATE see TIF250
HYDROXYTRIBUTYLSTANNANE, SULFATE (2:1) see TIF600
β-HYDROXYTRICARBALLYLIC ACID see CMS750
1-HYDROXY-2,2,2-TRICHLOROETHYL DIETHYL PHOSPHONITE see EPY600
1-HYDROXY-2,2,2-TRICHLOROETHYLPHOSPHONIC ACID DIMETHYL
    ESTER see TIQ250
2'-HYDROXY-2,4,4'-TRICHLORO-PHENYLETHER see TIQ000
2-HYDROXYTRIETHYLAMINE see DHO500
(S)-N-(3-HYDROXY-9,10,11-TRIMETHOXY-5H-DIBENZO(a,c)CYCLOHEPTEN-5-
    YL)-ACETAMIDE (9CI) see ACG250
17-β-HYDROXY-4,4,17-α-TRIMETHYL-ANDROST-5-ENE(2,3-d)ISOXAZOLE
    see HOM259
3-HYDROXY-N,N,N-TRIMETHYLBENZENAMINIUM IODIDE see HNN500
3-HYDROXY-N,N,N-TRIMETHYLBENZENEETHANAMINIUM PICRATE
    see HNG000
2-HYDROXY-N,N,N-TRIMETHYLETHANAMINIUM see CMF000
2-HYDROXY-N,N,N-TRIMETHYLETHANAMINIUM SALT with 3,7-DIHYDRO-
    1,3-DIMETHYLPURINE-2,6-DIONE see CMF500
2-HYDROXY-N,N,N-TRIMETHYLETHANAMINIUM SALT with 2-
    HYDROXYBENZOIC ACID (1:1) see CMG000
3-HYDROXY-2,2,4-TRIMETHYL-3-PENTENOIC ACID, β-LACTONE see IPL000
3-HYDROXY-4-((2,4,5-TRIMETHYLPHENYL)AZO)-2,7-NAPH-
    THALENEDISULFONIC ACID, DISODIUM SALT see FAG018
3-HYDROXY-4-((2,4,5-TRIMETHYLPHENYL)AZO)-2,7-NAPH-
    THALENEDISULPHONIC ACID, DISODIUM SALT see FAG018
3-HYDROXY-N,N,5-TRIMETHYLPYRAZOLE-1-CARBOXAMIDE
    DIMETHYLCARBAMATE (ESTER) see DQZ000
HYDROXYTRIMETHYLSTANNANE see TMI250
2-HYDROXY-1,3,5-TRINITROBENZENE see PID000
3-HYDROXY-2,4,6-TRINITROPHENOL see SMP500
HYDROXYTRIPHENYLSTANNANE see HON000
HYDROXYTRIPHENYLTIN see HON000
3-HYDROXYTROPOLONE see HON500
5-HYDROXYTRYPTAMINE see AJX500
5-HYDROXYTRYPTAMINE CREATININE SULFATE see AJX750
5-HYDROXYTRYPTAMINE CREATININE SULFATE MONOHYDRATE
    see AJX750
HYDROXYTRYPTOPHAN see HOO100
5-HYDROXYTRYPTOPHAN see HOA575, HON800, HOO100
5-HYDROXY-l-TRYPTOPHAN see HOA600, HOO000
dl-HYDROXYTRYPTOPHAN see HOA575, HON800
l-HYDROXYTRYPTOPHAN see HOA600, HOO000
(±)-5-HYDROXYTRYPTOPHAN see HOA575, HON800
dl-5-HYDROXYTRYPTOPHAN see HOA575
5-HYDROXYTRYPTOPHANE see HOO100
3-HYDROXYTYRAMINE HYDROCHLORIDE see DYC600
m-HYDROXYTYRAMINE HYDROCHLORIDE see DYC600
3-HYDROXY-l-TYROSINE see DNA200
l-o-HYDROXYTYROSINE see DNA200
3-HYDROXY-l-TYROSINE HYDROCHLORIDE see DYC200
5-HYDROXYUNDECANOIC ACID LACTONE see UKJ000
HYDROXYUREA see HOO500
N-HYDROXYUREA see HOO500
N-HYDROXYURETHAN see HKQ025
N-HYDROXYURETHANE see HKQ025
3-HYDROXYURIC ACID see HOO875
4-HYDROXYVALERIC ACID LACTONE see VAV000
17-HYDROXY-17-α-VINYL-4-ESTREN-3-ONE see NCI525
HYMINAL see QAK000
HYMORPHAN see DLW600, DNU300
HYONIC PE-250 see PKF500
HYOSAN see MJM500

HYOSCIN-N-BUTYLBROMID (GERMAN) see SBG500
HYOSCIN-N-BUTYL BROMIDE see SBG500
HYOSCINE see SBG000
(−)-HYOSCINE see SBG000
HYOSCINE BROMIDE see HOT500
HYOSCINE BUTOBROMIDE see SBG500
HYOSCINE BUTYL BROMIDE see SBG500
HYOSCINE-N-BUTYL BROMIDE see SBG500
HYOSCINE F HYDROBROMIDE see HOT500
HYOSCINE HYDROBROMIDE see HOT500
(−)-HYOSCINE HYDROBROMIDE see HOT500
l-HYOSCINE HYDROBROMIDE see HOT500
HYOSCINE METHYL BROMIDE see SBH500
(−)-HYOSCYAMINE see HOU000
HYOSCYAMINE see HOU000
l-HYOSCYAMINE see HOU000
dl-HYOSCYAMINE see ARR000
HYOSCYAMINE METHYLBROMIDE see MGR250
dl-HYOSCYAMINE METHYLNITRATE see MGR500
HYOSCYAMUS NIGER see HAQ100
HYOSCYINE HYDROBROMIDE see HOT500
HYOSOL see SBG000
dl-HYOSYAMINE METHYLNITRATE see MGR500
HYOZID see ILD000
HYPAQUE see SEN500
HYPAQUE 60 see AOO875
HYPAQUE 13.4 see AOO875
HYPAQUE CYSTO see AOO875
HYPAQUE M 30 see AOO875
HYPAQUE MEGLUMINE see AOO875
HYPAQUE SODIUM see SEN500
HYPCOL see QAK000
HYPERAN see HFS759
HYPERAZIN see HGP500
HYPERBUTAL see BPF500
HYPERNEPHRIN see VGP000
HYPEROL see HIB500
HYPERPAX see DNA800, MJE780
HYPERSIN see BFW250
HYPERSTAT see DCQ700
HYPERTENAIN see MAW250
HYPERTENSIN see AOO900
HYPERTONALUM see DCQ700
HY-PHI 1055 see OHU000
HY-PHI 1088 see OHU000
HY-PHI 1199 see SLK000
HY-PHI 2066 see OHU000
HY-PHI 2088 see OHU000
HY-PHI 2102 see OHU000
HYPHYLLINE see DNC000
HYPNODIN see HOU059
HYPNOGEN see EOK000
HYPNOGENE see BAG000
HYPNOMIDATE see HOU100
HYPNON see DLV000
HYPNONE see ABH000
HYPNOREX see LGZ000
HYPNORM see HAF400
HYPNOSTAN see PBT500
HYPNO-TABLINETTEN see EOK000
HYPO see SKI000, SKI500
HYPOCHLORITE, solution (DOT) see PLK000
HYPOCHLORITES see HOU500
HYPOCHLOROUS ACID see HOV000
HYPOCHLOROUS ACID, CALCIUM SALT see HOV500
HYPOCHLOROUS ACID, CALCIUM SALT (dry mixture) see HOW000
HYPODERMACID see TIQ250
HYPOGLYCIN see MJP500
HYPOGLYCIN A see MJP500
HYPOGLYCIN B see HOW100
HYPOGLYCINE A see MJP500
HYPOGLYCINE B see HOW100
HYPONITROUS ACID see HOW500
HYPONITROUS ACID ANHYDRIDE see NGU000
α-HYPOPHAMINE see ORU500
HYPOPHENON see ELL500
HYPOPHOSPHOROUS ACID see PGY250
HYPOPHTHALIN see HGP495, HGP500
HYPOPHYSEAL GROWTH HORMONE see PJA250
HYPOPRESOL see OJD300
HYPORENIN see VGP000
HYPOS see HGP500
HYPOTHIAZIDE see CFY000
HYPOTROL see SBM500, SBN000

HYPOXANTHINE see DMC000
HYPOXANTHINE NUCLEOSIDE see IDE000
HYPOXANTHINE RIBONUCLEOSIDE see IDE000
HYPOXANTHINE RIBOSIDE see IDE000
HYPOXANTHINE-d-RIBOSIDE see IDE000
HYPOXANTHOSINE see IDE000
HYPROVAL-PA see HNT500
HYPTOR BASE see QAK000
HYRE see RIK000
HYSCO see HOT500
HYSCYLENE P see PDX000
HYSONE-A see HHQ800
HYSSOP OIL see HOX000
HYSTRENE 80 see SLK000
HYSTRENE 8016 see PAE250
HYSTRENE 9014 see MSA250
HYSTRENE 9512 see LBL000
HYTAKEROL see DME300
HYTONE LOTION see CNS750
HYTOX see MIA250
HYTRIN see TEF700
HYVAR see BMM650, BNM000
HYVAREX see BMM650
HYVAR X see BMM650
HYVAR X BROMACIL see BMM650
HYVAR X WEED KILLER see BMM650
HYVERMECTIN see ITD875
HYZYD see ILD000

I 677 see OLK200
I-1431 see CQH000
I 337A see CDP250
IA see IDZ000, IHN200
IA-4 see CFP750
IAA see ICN000
IAB see AOC500
I ACID see AKI000
IAMBOLEN see PKF750
IA-4 N-OXIDE see CFQ000
IA-PRAM see DLH630
IARACTAN see TAF675
IBA see ICP000
IBATRAN see ACE000
IBD see CCK125
IBD 78 see MGL500
IBDU see IIV000
IBENZMETHYZINE see PME250
IBENZMETHYZINE HYDROCHLORIDE see PME500
IBENZMETHYZIN HYDROCHLORIDE see PME500
IBERIN see MPN100
IBIFUR see FPI000
IBINOLO see TAL475
IBIODORM see SKS700
IBIODRAL see BBV500
IBIOFURAL see NGE500
IBIOSUC see SGC000
IBIOTON see TAI500
IBIOTYZIL see BCA000
IBIOZEDRINE see AOB250
IBN see IJD000
IBOGAMINE-18-CARBOXYLIC ACID, METHYL ESTER, HYDROCHLORIDE
  see CNS200
IBOGAMINE-18-CARBOXYLIC ACID, METHYL ESTER,
  MONOHYDROCHLORIDE (9CI) see CNS200
IBOMAL see QCS000
IBOTENIC ACID see AKG250
IBOTENSAURE (GERMAN) see AKG250
IBP see BKS750, DIU800
IBUFEN see IIU000
IBUFENAC see IJG000
IBUNAC see IJG000
IBUPROCIN see IIU000
IBUPROFEN see IIU000
IBUPROFEN GUAIACOL ESTER see IJJ000
IBUPROFEN PICONOL see IAB000
IBUPROFEN SODIUM see IAB100
IBYLCAINE HYDROCHLORIDE see IAC000
IBZ see PME500
I-C 26 see DWC100
IC 6002 see PIW000
ICCSH see LIU300
ICG see CCK000
ICHDEN see IAD000
ICHTAMMON see IAD000

ICHTHADONE see IAD000
ICHTHALUM see IAD000
ICHTHAMMOL see IAD000
ICHTHAMMONIUM see IAD000
ICHTHIUM see IAD000
ICHTHOSAN see IAD000
ICHTHOSAURAN see IAD000
ICHTHOSULFOL see IAD000
ICHTHYMALL see IAD000
ICHTHYNAT see IAD000
ICHTHYOL see IAD000
ICHTHYOPON see IAD000
ICHTHYSALLE see IAD000
ICI 350 see AJH129
ICI 543 see PMP500
ICI 3435 see SNI500
ICI 28257 see ARQ750
ICI 32525 see SNL800
ICI-32865 see TJQ333
I.C.I. 33,828 see MLJ500
ICI 38174 see INT000
ICI 42464 see CHT500
ICI 43823 see BPI300
ICI 45520 see ICB000, ICC000
ICI 46,474 see NOA600, TAD175
ICI 46638 see DMZ000
ICI 48213 see FBP100
ICI 50172 see ECX100
ICI 51426 see CQH625
ICI 54,450 see CKK250
ICI-58834 see VKA875
ICI 58,834 see VKF000
ICI 59118 see RCA375
ICI 66,082 see TAL475
ICI 74205 see EPC200
ICI 79,939 see IAD100
ICI 80996 see CMX880
ICI 81008 see ECW550
ICI 118587 see XAH000
ICI 123215 see LIU420
ICI 151291 see BJW600
racemic-ICI 74205 see EPC200
racemic-ICI 79,939 see IAD100
racemic-ICI 80,996 see CMX880
racemic-ICI 81,008 see ECW550
ICI 47,699 CITRATE see DVY875
ICI 156834 DISODIUM see CCS371
ICIG 770 see EAI850
ICIG 772 see MEL775
I.C.I.G. 1105 see RFU600
ICIG 1109 see CGV250
I.C.I.G. 1163 see ROF200
I.C.I.G. 1325 see BIF625
I.C.I. HYDROCHLORIDE see INT000
I.C.I. LTD. COMPOUND NUMBER 80996 see CMX880
ICI 54856 METHYL ESTER see MIO975
ICIPEN see PDT750
ICI-PP 557 see AHJ750
ICN-1229 see RJA500
ICORAL B see HNB875
ICR 10 see QDS000
ICR-125 see QCS875
ICR 170 see ADJ875
ICR 180 see CGS750
ICR-191 see MEJ250
ICR 217 see CGX750
ICR 220 see BHZ000
ICR-25A see CLD500
ICR 290 see CGX500
ICR 292 see EHJ000
ICR 311 see EHI500
ICR 340 see IAE000
ICR 342 see CFA250
ICR 355 see BPI500
ICR 368 see CGY000
ICR 372 see CGS500
ICR 377 see EHJ500
ICR 394 see CHH000
ICR 395 see CGR500
ICR 433 see CIN500
ICR 449 see ADQ500
ICR-450 see BIJ750
ICR 451 see CIG250
ICR 486 see CFB500

ICR-48b see DFH000
ICR 498 see BNO750
ICR 502 see BNR000
ICR 506 see BNO500
ICR DIHYDROCHLORIDE see MEJ250
ICRF-159 see PIK250
ICRF 159 see RCA375
ICR 170-OH see CHY750
ICSH see LIU300
ICTALIS SIMPLE see DKQ000
ID 540 see FDB100
ID-622 see IAG300
ID-1229 see FGQ000
IDA see IBH000
IDALENE see MOV500
IDANTOIL see DNU000
IDANTOIN see DKQ000
IDANTOINAL see DNU000
IDARAC see TKG000
ID 480 DIHYDROCHLORIDE see DAB800
IDEXUR see DAS000
IDMT see IOW500
IDOCYL NOVUM see SJO000
IDOMETHINE see IDA000
IDONOR see PJB500
IDOXENE see DAS000
IDOXURIDIN see DAS000
IDOXURIDINE see DAS000
IDPN see BIQ500
IDRAGIN see ADA725
IDRALAZINA (ITALIAN) see HGP495
IDRAZIDE DELL'ACIDO ISONICOTINICO see ILD000
IDRAZIL see ILD000
IDRAZINA IDRATA (ITALIAN) see HGU500
IDRAZINA SOLFATO (ITALIAN) see HGW500
2-IDRAZINO-6-(N,N-BIS(2-IDROSSIETIL)-AMINO)-PIRIDAZINACLORIDRATO
    (ITALIAN) see BKB500
3-IDRAZINO-6-(N-(2-IDROSSIPROPIL)METILAMINO)PIRIDAZINA
    DICLORIDRATO (ITALIAN) see HHD000
IDRIANOL see SPC500
IDROBUTAZINA see HNI500
IDROCHINONE (ITALIAN) see HIH000
IDROESTRIL see DKA600
IDROGENO SOLFORATO (ITALIAN) see HIC500
IDROGESTENE see HNT500
IDROPEROSSIDO di CUMENE (ITALIAN) see IOB000
IDROPEROSSIDO di CUMOLO (ITALIAN) see IOB000
IDROSSIDO DI STAGNO TRIFENILE (ITALIAN) see HON000
1',1-(2-IDROSSIETIL)4-(3-(2-CLORO-10-FENOTIAZIL)PROPILPIPERAZINA
    (ITALIAN) see CJM250
(+ −)-1-(4-IDROSSIFENIL)-2-METILAMINOETANOLO TARTRATO (ITALIAN)
    see SPD000
4-IDROSSI-4-METIL-PENTAN-2-ONE (ITALIAN) see DBF750
d-N,N'-(1-IDROSSIMETIL PROPIL)-ETILENDINITROSAMINA (ITALIAN)
    see HMQ500
4-IDROSSI-3-(3-OXO-)-FENIL-BUTIL)-CUMARINE (ITALIAN) see WAT200
l-5α-IDROSSIPROPIONILAMINO-2,4,6-TRIIODOISOFTAL-DI(1,3-DIIDROSSI-2-
    PROPILAMIDE) see IFY000
4-IDROSSI-3-(1,2,3,4-TETRAIDRO-1-NAFTIL)CUMARINA (ITALIAN)
    see EAT600
IDROSSIZINA see CJR909
IDROTIADENE see HII000
IDROTIAZIDE see CFY000
IDRYL see FDF000
IDU see DAS000
IDUCHER see DAS000
IDULEA see DAS000
IDULIAN see DLV800
IDUOCULOS see DAS000
IDUR see DAS000
IDURIDIN see DAS000
IEM-1-15 see CEX250
IEM 455 see CJN750
IERGIGAN see DQA400
IEROIN see HBT500
IF (fumigant) see TBV750
IFC see CBM000
IFC-45 see CGG500
IFENEC see EAE000
IFENPRODIL see IAG600
IFENPRODIL TARTRATE see IAG625
IFENPRODIL TARTRATE (2:1) see IAG625
IFENPRODIL l-(+)-TARTRATE see IAG625
IFIBRIUM see LFK000

IFOSFAMID see IMH000
IFOSFAMIDE see IMH000
IF ROM 203 see IAG700
IGE see IPD000
IGELITE F see PKQ059
IGEPAL CA see GHS000
IGEPAL CA-63 see PKF500
IGEPAL CO-630 see PKF000
IGEPAL GAS see IAH000
IGEPON T-33 see SIY000
IGEPON T-43 see SIY000
IGEPON T 51 see SIY000
IGEPON T-71 see SIY000
IGEPON T-73 see SIY000
IGEPON T-77 see SIY000
IGEPON TE see SIY000
IGIG 929 see HKH000
IGNAZIN see AFH250
IGNOTINE see CCK665
IGROSIN see SIH500
IGROTON see CLY600
II-C-2 see DOQ400
IIH see ILE000
IKACLOMIN see CMX700
IKADA RHODAMINE B see FAG070
IKTEROSAN see PGG000
IKURIN see ANU650
IL 6001 see DLH200
ILBION see FNF000
ILETIN see IDF300
ILETIN U 40 see LEK000
ILEX AQUIFOLIUM see HGF100
ILEX OPACA see HGF100
ILEX VOMITORIA see HGF100
ILIADIN see AEX000
ILIDAR see AGD500, ARZ000
ILIDAR BASE see ARZ000
ILIDAR PHOSPHATE see AGD500
ILITIA see VSZ450
ILIXATHIN see RSU000
ILLOXOL see DHB400
ILLUDINE S see LIO600
ILLUDIN S see LIO600
IL-6302 MESYLATE see FMU039
ILOPAN see PAG200
ILOTYCIN see EDH500
ILOTYCIN HYDROCHLORIDE see EDJ000
IM see DLH600
IMADYL see CCK800
IMAGON see CLD000
IMAKOL see AFL750
IMAVATE see DLH630
IMAVEROL see FPB875
IMAZALIL see FPB875
IMBARAL see SOU550
IMBRILON see IDA000
IMC 3950 see SAZ000
IMD 760 see ARX800
IMESONAL see SBM500, SBN000
IMET 3106 see IAK100
IMET 3393 see CQM750
IMETRO see IBP200
IMFERON see IGS000
IMI 115 see TGF250
IMIDA-LAB see TEH500
IMIDALIN see BBW750
IMIDALINE HYDROCHLORIDE see BBJ750
IMIDAMINE see PDC000
IMIDAN see PHX250
IMIDAN (PEYTA) see TEH500
IMIDAZOL see IAL000
IMIDAZOLE see IAL000
α-β-IMIDAZOLECARBOXYLIC ACID see IAM000
4,5-IMIDAZOLEDICARBOXYLATE see IAM000
IMIDAZOLE-4,5-DICARBOXYLIC ACID see IAM000
1H-IMIDAZOLE-4-ETHANAMINE see HGD000
1H-IMIDAZOLE-4-ETHANAMINE PHOSPHATE (1:2) see HGE000
IMIDAZOLE-4-ETHYLAMINE see HGD000
4-IMIDAZOLEETHYLAMINE see HGD000
5-IMIDAZOLEETHYLAMINE see HGD000
IMIDAZOLE-2-HYDROXYBENZOATE see IAM100
IMIDAZOLE MUSTARD see IAN000
IMIDAZOLEPYRAZOLE see IAN100
IMIDAZOLE with SALICYLIC ACID see IAM100

IMIDAZOLE-2-THIOL see IAO000
IMIDAZOL-2-HYDROXYBENZOAT (GERMAN) see IAM100
2,4-IMIDAZOLIDINEDIONE (9CI) see HGO600
2,4-IMIDAZOLIDINEDIONE, 5-ETHYL-5-PHENYL-, (R)- (9CI) see EOL050
2,4-IMIDAZOLIDINEDIONE, 1-(((5-(4-NITROPHENYL)-2-FURANYL)METHY-
  LENE)AMINO)-, SODIUM SALT see DAB840
4,5-IMIDAZOLIDINEDITHIONE see IAP000
2-IMIDAZOLIDINETHIONE see IAQ000
2-IMIDAZOLIDINETHIONE mixed with SODIUM NITRITE see IAR000
2-IMIDAZOLIDINONE see IAS000
2-IMIDAZOLIDINONE, 1-(1-METHYL-5-NITRO-1H-IMIDAZOL-2-YL)-3-
  (METHYLSULFONYL)- see SAY950
2-IMIDAZOLIDONE see IAS000
IMIDAZOLINE see IAT000
2-IMIDAZOLINE see IAT000
2-IMIDAZOLINE, 2-(2,6-DICHLOROANILINO)- see DGB500
IMIDAZOLINE-2,4-DITHIONE see IAT100
N-(2-IMIDAZOLINE-2-YL)-N-(4-INDANYL)AMINE MONOHYDROCHLORIDE
  see IBR200
N-(2-IMIDAZOLINE-2-YL)-N-(4-INDANYL)AMIN-MONOHYDROCHLORID
  (GERMAN) see IBR200
1H-IMIDAZOLIUM, 4,5-DIHYDRO-1-(CARBOXYMETHYL)-1-(2-
  HYDROXYETHYL)-2-UNDECYL-, HYDROGEN SULFATE (salt), MONOSO-
  DIUM SALT see AOC275
2-(4-IMIDAZOLYL)ETHYLAMINE see HGD000
2-IMIDAZOL-4-YL-ETHYLAMINE see HGD000
β-IMIDAZOLYL-4-ETHYLAMINE see HGD000
1H-IMIDAZO(2,1-f)PURINE-2,4(3H,6H)-DIONE,7,8-DIHYDRO-8-ALLYL-1,3-
  DIMETHYL-(8CI) see KHU100
4-IMIDAZO(1,2-a)PYRIDIN-2-YL-α-METHYLBENZENEACETIC ACID
  see IAY000
2-(4-(IMIDAZO(1,2-a)PYRIDIN-2-YL)PHENYL)PROPIONIC ACID see IAY000
2-(p-(2-IMIDAZO(1,2-a)PYRIDYL)PHENYL)PROPIONIC ACID see IAY000
7H-IMIDAZO(4,5-D)PYRIMIDINE see POJ250
IMIDAZO(5,1-d)-1,2,3,5-TETRAZINE-8-CARBOXAMIDE, 3-(2-CHLOROETHYL)-
  3,4-DIHYDRO-4-OXO- see MQY110
IMIDAZO(2,1-β)THIAZOLE MONOHYDROCHLORIDE see LFA020
IMIDENE see TEH500
IMIDIN see NAH500
IMIDOBENZYLE see DLH600, DLH630
4,4'-(IMIDOCARBONYL)BIS(N,N-DIMETHYLAMINE) MONOHYDROCHLOR-
  IDE see IBA000
4,4'-(IMIDOCARBONYL)BIS(N,N-DIMETHYLANILINE) see IBB000
IMIDODICARBONIC DIHYDRAZIDE (9 CI) see IBC000
IMIDODICARBOXYLIC ACID, DIHYDRAZIDE see IBC000
3,3'-IMIDODI-1-PROPANOL, DIMETHANESULFONATE (ester), HYDROCHLO-
  RIDE see YCJ000
IMIDOL see DLH630
IMIDOLE see PPS250
IMILANYLE see DLH630
IMINAZOLE see IAL000
IMINOALDEHYDE see FNL000
IMINO-1,1'-BIANTHRAQUINONE see IBI000
IMINOBIS(ACETIC ACID) see IBH000
1,1'-IMINOBIS(4-AMINO-9,10-ANTHRACENEDIONE) see IBD000
1,1'-IMINOBIS(4-AMINOANTHRAQUINONE) see IBD000
1,1'-IMINOBIS-9,10-ANTHRACENEDIONE see IBI000
1,1'-IMINOBIS(4-BENZAMIDOANTHRAQUINONE) see IBJ000
4,5'-IMINOBIS(4-BENZAMIDOANTHRAQUINONE) see IBE000
2,2'-IMINOBISETHANOL see DHF000
2,2'-IMINOBISETHYLAMINE see DJG600
IMINOBISFORMALDEHYDE see FNL000
5,6'-IMINOBIS(1-HYDROXY-2-NAPHTHALENESULFONIC ACID) see IBF000
(IMINOBIS(OCTAMETHYLENE))-DIGUANIDINE SULFATE see GLQ000
3,3'-IMINOBISPROPANENITRILE see BIQ500
3,3'-IMINOBIS-1-PROPANOL DIMETHANESULFONATE (ester),
  4-METHYLBENZENESULFONATE (salt) see IBQ100
IMINOBIS(PROPYLAMINE) see AIX250
3,3'-IMINOBIS(PROPYLAMINE) see AIX250
4-IMINO-2,5-CYCLOHEXADIEN-1-ONE see BDD500
4,4'-((4-IMINO-2,5-CYCLOHEXADIEN-1-YLIDENE)METHYLENE)DIANILINE
  MONOHYDROCHLORIDE-o-TOLUIDINE see RMK020
IMINODIACETIC ACID see IBH000
2,2'-IMINODIACETIC ACID see IBH000
1,1'-IMINODIANTHRAQUINONE see IBI000
N,N'-(IMINODI-4,1-ANTHRAQUINONYLENE)BISBENZAMIDE see IBJ000
IMINODIBENZYL see DKY800
IMINODIETHANOIC ACID see IBH000
2,2'-IMINODIETHANOL see DHF000
2,2'-IMINODI-ETHANOL with 1,2-DIHYDRO-3,6-PYRIDAZINEDIONE (1:1)
  see DHF200
2,2'-IMINODI-N-NITROSOETHANOL see NKM000
IMINODIOCTAN SODNY (CZECH) see DXE200
1,1'-IMINODI-2-PROPANOL see DNL600

β,β-IMINODIPROPIONITRILE see BIQ500
3,3'-IMINODIPROPIONITRILE see BIQ500
IMINO-β,β'-DIPROPIONITRILE see BIQ500
β,β'-IMINODIPROPIONITRILE see BIQ500
IMINODIPROPYL DIMETHANESULFONATE 4-TOLUENESULPHONATE
  see IBQ100
5-IMINO-1,2,4-DITHIAZOLIDINE-3-THIONE see IBL000
2-IMINOHEXAFLUOROPROPANE see HDD000
(2-((IMINO(METHYLAMINO)METHYL)THIO)ETHYL)TRIETHYLAMMONIUM
  BROMIDE HYDROBROMIDE see MRU775
(2-IMINO-5-PHENYL-4-OXAZOLIDINONATO(2-))DIAQUOMAGNESIUM
  see PAP000
2-IMINO-5-PHENYL-4-OXAZOLIDINONE see IBM000
2-IMINO-5-PHENYL-4-OXAZOLIDINONE MAGNESIUM CHELATE see PAP000
IMINOPHOSPHATE see DXN600
2(3H)-IMINO-9-β-d-RIBOFURANOSYL-9H-PURIN-6(1H)-ONE see GLS000
IMINOUREA see GKW000
IMIPHOS see BGY000
IMIPRAMINA (ITALIAN) see DLH600, DLH630
IMIPRAMINE see DLH600, DLH630
IMIPRAMINEDEMETHYL HYDROCHLORIDE see DLS600
IMIPRAMINE HYDROCHLORIDE see DLH630
IMIPRAMINE MONOHYDROCHLORIDE see DLH630
IMIPRAMINE-N-OXIDE HYDROCHLORIDE see IBP000
IMIPRIN see DLH600, DLH630
IMIZIN see DLH600
IMIZINUM see DLH600
IMMENOCTAL see SBM500, SBN000
IMMENOX see SBM500
IMMETROPAN see IBP200
IMMORTELLE see HAK500
IMODIUM see LIH000
IMOL S 140 see TNP500
IMOTRYL see BBW500
IMOVANCE see ZUA450
IMOVANE see ZUA450
IMP see IDE200
5'-IMP see IDE200
o-IMPC see PMY300
IMP DISODIUM SALT see DXE500
5'-IMP DISODIUM SALT see DXE500
IMPERATOR see RLF350
IMPERATORIN see IHR300
IMPERIAL GREEN see COF500
IMPERON FIXER T see TND250
IMPERVOTAR see CMY800
IMPF see IPX000
IMP HYDROCHLORIDE see DLH630
IMPIRAMINE-N-OXIDE see IBP309
IMPOSIL see IGS000
IMPRAMINE see DLH600
IMPROMEN see BNU725
IMPROMIDINE HYDROCHLORIDE see IBQ075
IMPROMIDINE TRIHYDROCHLORIDE see IBQ075
IMPROSULFAN-p-TOLUENESULFONATE see IBQ100
IMPROSULFAN TOSILATE see IBQ100
IMPROSULFAN TOSYLATE see IBQ100
IMPROVED WILT PRUF see PKQ059
IMPRUVOL see BFW750
IMP SODIUM SALT see DXE500
IMPY see IAN100
IMS see IPY000
IMUGAN see CDP750
IMURAN see ASB250
IMUREK see ASB250
IMUREL see ASB250
IMUTEX see IAL000
IMVITE I.G.B.A. see BAV750
IMWITOR 191 see OAV000
IMWITOR 900K see OAV000
IN-117 see HEG000
IN 399 see MKW000
IN 511 see PGG350, PGG355
IN 836 see FAR200
INACID see IDA000
INACILIN see AOD000
INACTIVE LIMONENE see MCC250
INAKOR see ARQ725
INALONE O see AFJ625
INALONE R see AFJ625
INAMIL see AIF000
INAMYCIN see NOB000, SMB000
INAPPIN see DYF200
INAPSIN see DYF200

INBESTAN see FOS100
INBUTON see BSM000
INCASAN see IBQ300
INCAZANE see IBQ300
INCIDOL see BDS000
2-(4-(3-INCOLYLMETHYL)-1-PIPERAZINYL)-QUINOLINE DIMALEATE
    see ICZ100
INCORTIN see CNS800, CNS825
INCRECEL see CMF400
INDACRINONE see IBQ400
INDAFLEX see IBV100
INDALCA AG see GLU000
INDALONE see BRT000
INDAMOL see IBV100
INDAN see IBR000
INDANAL see CFH825, CMV500
INDANAZOLIN (GERMAN) see IBR200
INDANAZOLINE see IBR200
INDANAZOLINE HYDROCHLORIDE see IBR200
1,3-INDANDIONE see IBS000
INDANE see IBR000
5-INDANOL see IBU000
INDAN-5-OL see IBU000
2-INDANONE see IBV000
INDANTHRENE GOLDEN YELLOW see DCZ000
INDANTHRENE NAVY BLUE G see CMU500
INDANTHRENE RED BROWN 5RF see CMU800
INDANTHRENE REDDISH BROWN 5RF see CMU800
INDANTHREN GREY M see CMU475
INDANTHREN GREY MG see CMU475
INDANTHREN NAVY BLUE TRR see CMU750
INDANTHREN RED BROWN 5RF see CMU800
1,2,3-INDANTRIONE-2-HYDRATE see DMV200
1,2,3-INDANTRIONE MONOHYDRATE see DMV200
2-INDANYLAMINE HYDROCHLORIDE see AKL000
INDAPAMIDE see IBV100
INDAR see BPU000
1H-INDAZOLE-3-CARBOXYLIC ACID, 1-(p-CHLOROBENZYL)-, 1,3-DIHY-
    DROXY-2-PROPYL ESTER see GGR100
2-(p-(2H-INDAZOL-2-YL)PHENYL)PROPIONIC ACID see IBW100
INDECAINIDE HYDROCHLORIDE see IBW400
INDELOXAZINE HYDROCHLORIDE see IBW500
INDEMA see PFJ750
INDENE see IBX000
1H-INDENE-1,3(2H)-DIONE see IBS000
INDENE TRIPROPYLAMINE see IBY000
INDENOLOL HYDROCHLORIDE see IBY600, ICA000
13H-INDENO(1,2-1)PHENANTHRENE see DCX000
INDENO(1,2,3-cd)PYRENE see IBZ000
INDENO(1,2,3-cd)PYRENE-1,2-OXIDE see DLK750
INDENO(1,2,3-cd)PYREN-8-OL see IBZ100
4-(1H-INDEN-1-YLIDENEMETHYL)-N,N-DIMETHYLBENZENAMINE
    see DOT600
4-(1-H-INDEN-1-YLIDENEMETHYL)-N-METHYL-N-NITROSOBENZENAMINE
    see MMR750
1-(4(or 7)-INDENYLOXY)-3-(ISOPROPYLAMINO)-2-PROPANOL HYDROCHLO-
    RIDE see IBY600
(±)-1-(7-INDENYLOXY)-3-ISOPROPYLAMINOPROPAN-2-OL HYDROCHLO-
    RIDE see ICA000
2-(7-INDENYLOXYMETHYL)MORPHOLINE HYDROCHLORIDE see IBW500
2-((1H-INDEN-7-YLOXY)METHYL)MORPHOLINE HYDROCHLORIDE
    see IBW500
INDEPENDENCE RED see MMP100
INDERAL see ICB000
INDERAL HYDROCHLORIDE see ICC000
INDEREX see ICC000
INDEROL see ICC000
INDI see ICD100
INDIAN APPLE see MBU800
INDIAN BERRY see PIE500
INDIAN BLACK DRINK see HGF100
INDIAN CANNABIS see CBD750
INDIAN COBRA VENOM see ICC700
INDIAN GUM see AQQ500, GLY000
INDIAN HEMP see CBD750
INDIAN JACK-IN-THE-PULPIT see JAJ000
INDIAN LABURNUM see GIW300
INDIAN LAUREL see MBU780
INDIAN LICORICE SEED see AAD000
INDIAN LILAC see CDM325
INDIAN PINK see CCJ825, PIH800
INDIAN POKE see FAB100, VIZ000
INDIAN POLK see PJJ315
INDIAN RED see IHD000, LCS000

INDIAN SAVIN TREE (JAMAICA) see CAK325
INDIAN TOBACCO see CCJ825
INDIAN WALNUT see TOA275
INDIA RUBBER see ROH900
INDIA RUBBER VINE see ROU450
INDICAN (POTASSIUM SALT) see ICD000
INDICINE-N-OXIDE see ICD100
INDIGENOUS PEANUT OIL see PAO000
INDIGO BLUE 2B see CMO000
INDIGO CARMINE see FAE100
INDIGO CARMINE (BIOLOGICAL STAIN) see FAE100
INDIGO CARMINE DISODIUM SALT see FAE100
INDIGO EXTRACT see FAE100
INDIGO-KARMIN (GERMAN) see FAE100
5,5'-INDIGOTIN DISULFONIC ACID see FAE100
INDIGOTINE see FAE100
INDIGOTINE DISODIUM SALT see FAE100
INDIGO YELLOW see ICE000
INDION see PFJ750
INDISAN see IKA000
INDISULFAT (GERMAN) see ICJ000
INDIUM see ICF000
INDIUM ACETYLACETONATE see ICG000
INDIUM CHLORIDE see ICK000
INDIUM CITRATE see ICH000
INDIUM NITRATE see ICI000
INDIUM SULFATE see ICJ000
INDIUM TRICHLORIDE see ICK000
INDOBLOC see ICC000
INDOCID see IDA000
INDOCYANINE GREEN see ICL000
INDOCYBIN see PHU500
INDOKLON see HDC000
INDOL (GERMAN) see ICM000
3-INDOLACETONITRILE see ICW000
INDOLACIN see CMP950
β-INDOLAETHYLAMIN-CHLORHYDRAT (GERMAN) see AJX250
α-INDOLAETHYLAMIN SALZSAEURE (GERMAN) see LIU100
INDOLAPRIL HYDROCHLORIDE see ICL500
INDOLE see ICM000
3-INDOLEACETIC ACID see ICN000
β-INDOLEACETIC ACID see ICN000
β-INDOLE-3-ACETIC ACID see ICN000
1H-INDOLE-3-ACETIC ACID see ICN000
INDOLEACETONITRILE see ICW000
INDOLE-3-ACETONITRILE see ICW000
1H-INDOLE-3-ACETONITRILE see ICW000
INDOLE-3-ACRYLIC ACID see ICO000
INDOLE-3-ALANINE see TNX000
INDOLE-3-(2-AMINOBUTYL) ACETATE see AJB250
INDOLE, 3-(2-AMINOETHYL)-5-METHOXY- see MFS400
1H-INDOLE-3-BUTANOIC ACID see ICP000
INDOLE BUTYRIC see ICP000
INDOLE BUTYRIC ACID ICP000
3-INDOLEBUTYRIC ACID see ICP000
β-INDOLEBUTYRIC ACID see ICP000
Γ-(INDOLE-3)-BUTYRIC ACID see ICP000
INDOLE-3-CARBINOL see ICP100
INDOLE-3-CARBOXALDEHYDE, 2-(m-AMINOPHENYL)-, 4-(m-TOLYL)-3-
    THIOSEMICARBAZONE see ALW900
INDOLE-2,3-DIONE see ICR000
1H-INDOLE-3-ETHANAMINE see AJX000
INDOLE ETHANOL see ICS000
INDOLE-3-ETHYLAMINE HYDROCHLORIDE see AJX250
α-INDOLEETHYLAMINE HYDROCHLORIDE see LIU100
β-INDOLE-ETHYLAMINE HYDROCHLORIDE see AJX250
INDOLE-3-METHANOL see ICP100
1H-INDOLE-3-METHANOL (9CI) see ICP100
INDOLENE see ICS100
INDOLE-3-PROPYLAMINE HYDROCHLORIDE see AME750
INDOL-3-ETHYLAMINE see AJX000
INDOLIN see BBW500
2,3-INDOLINEDIONE see ICR000
INDOL-N-METHYLHARMINE HYDROCHLORIDE see ICU100
INDOL-3-OL, HYDROGEN SULFATE (ESTER), POTASSIUM SALT see ICD000
INDOL-3-OL, POTASSIUM SULFATE see ICD000
INDOLYACETIC ACID see ICN000
Δ-INDOLYBUTYLAMINE HYDROCHLORIDE see AJB500
INDOLYL-3-ACETIC ACID see ICN000
3-INDOLYLACETIC ACID see ICN000
β-INDOLYLACETIC ACID see ICN000
α-INDOL-3-YL-ACETIC ACID see ICN000
INDOLYLACETONITRILE see ICW000
3-INDOLYLACETONITRILE see ICW000

3-INDOLYLACRYLIC ACID see ICO000
1-β-3-INDOLYLALANINE see TNX000
ω-3-INDOLYLAMYLAMINE ADIPINATE. see ALR500
INDOLYL-3-BUTYRIC ACID see ICP000
4-(INDOLYL)BUTYRIC ACID see ICP000
3-INDOLYL-Γ-BUTYRIC ACID see ICP000
4-(INDOL-3-YL)BUTYRIC ACID see ICP000
4-(3-INDOLYL)BUTYRIC ACID see ICP000
Γ-(3-INDOLYL)BUTYRIC ACID see ICP000
Γ-(INDOL-3-YL)BUTYRIC ACID see ICP000
3-INDOLYLCARBINOL see ICP100
3-INDOLYLETHANOL see ICS000
2-(3-INDOLYL)ETHYLAMINE see AJX000
β-3-INDOLYLETHYLAMINE HYDROCHLORIDE see AJX250
N-(2-(3-INDOLYL)ETHYL)-NICOTINAMIDE see NDW525
N-(2-INDOL-3-YLETHYL)NICOTINAMIDE see NDW525
(±)-1-(3-INDOLYL)-2-METHYLAMINOETHANOL see ICY000
1-(INDOLYL-3)-2-METHYLAMINOETHANOL-1 RACEMATE see ICY000
1-(3′-INDOLYLMETHYL)-4-(2″-QUINOLYL)PIPERAZINE DIMALEATE
    see ICZ100
1-(4-INDOLYLOXY)-3-ISOPROPYLAMINO)-2-PROPANOL see VSA000
1-(1H-INDOL-4-YLOXY)-3-((1-METHYLETHYL)AMINO)-2-PROPANOL
    see VSA000
INDOL-3-YL POTASSIUM SULFATE see ICD000
3-(1-H-INDOL-3-YL)-2-PROPENOIC ACID see ICO000
Γ-3-INDOLYLPROPYLAMINE HYDROCHLORIDE see AME750
INDOL-3-YL SULFATE, POTASSIUM SALT see ICD000
INDOMECOL see IDA000
INDOMED see IDA000
INDOMETHACIN see IDA000
INDOMETHAZINE see IDA000
INDOMETICINA (SPANISH) see IDA000
INDON see PFJ750
INDONAPHTHENE see IBX000
INDOPAN see AME500
INDOPROFEN see IDA400
INDOPTIC see IDA000
INDO-RECTOLMIN see IDA000
INDORM see IDA500
INDO-TABLINEN see IDA000
INDOXAMIC ACID see OLM300
INDUSTRENE 105 see OHU000
INDUSTRENE 205 see OHU000
INDUSTRENE 206 see OHU000
INDUSTRENE 4516 see PAE250
INDUSTRENE 5016 see SLK000
INERTEEN see PJL750
INETOL see INS000, INT000
INEXIT see BBQ500
INF 1837 see TKH750
INF 3355 see XQS000
INF 4668 see DGM875
INFAMIL see HNI500
INFERNO see DJA400
INFILTRINA see DUD800
INFLAMEN see BMO000
INFLATINE see LHY000
INFLAZON see IDA000
INFRON see VSZ100
INFUSORIAL EARTH see DCJ800
INGALAN see DFA400
INGALAN (RUSSIAN) see DFA400
INGENANE HEXADECANOATE see IDB000
INHALAN see DFA400
INH-G see GBB500
INHG-SODIUM see IDB100
INHIBINE see HIB000
INHIBISOL see MIH275
INHISTON see TMK000
'INIA (HAWAII) see CDM325
INICARDIO see DJS200
INIPROL see PAF550
INITIATING EXPLOSIVE DIAZODINITROPHENOL (DOT) see DUR800
INITIATING EXPLOSIVE FULMINATE of MERCURY (DOT) see MDC250
INITIATING EXPLOSIVE IMINOBISPROPYLAMINE (DOT) see AIX250
INITIATING EXPLOSIVE LEAD AZIDE, DEXTRINATED TYPE ONLY (DOT)
    see LCM000
INITIATING EXPLOSIVE LEAD MONONITRORESORCINATE (DOT)
    see LDP000
INITIATING EXPLOSIVE LEAD STYPHNATE (DOT) see LEE000
INITIATING EXPLOSIVE LEAD TRINITRORESORCINATE (DOT)
    see LEE000
INITIATING EXPLOSIVE NITRO MANNITE (DOT) see MAW250
INITIATING EXPLOSIVE NITROSOGUANIDINE (DOT) see NKH000

INITIATING EXPLOSIVE PENTAERYTHRITE TETRANITRATE (DOT)
    see PBC250
INITIATING EXPLOSIVE-TETRAZENE (DOT) see TEF500
INJERTO (TEXAS, MEXICO) see MQW525
INKASAN see IBQ300
INK BERRY see PJJ315
INK ORANGE JSN see HGC000
INNOVAN see DYF200
INNOVAR see DYF200
INNOXALON see EID000
INO see IDE000
INOCOR see AOD375
INOKOSTERONE see IDD000
INOLIN see IDD100
INOPHYLLINE see TEP500
INOPSIN see DYF200
INOSIE see IDE000
INOSINE see IDE000
β-INOSINE see IDE000
INOSINE-5′-MONOPHOSPHATE see IDE200
INOSINE-5′-MONOPHOSPHATE DISODIUM see DXE500
INOSINE-5′-MONOPHOSPHORIC ACID see IDE200
INOSINE-5′-PHOSPHATE see IDE200
INOSINIC ACID see IDE200
5′-INOSINIC ACID, DISODIUM SALT, mixed with DISODIUM-5′-GUANYLATE
    (1:1) see RJF400
5′-INOSINIC ACID, HOMOPOLYMER complex with 5′-CYTIDYLIC ACID
    HOMOPOLYMER (1:1) see PJY750
INOSIN-5′-MONOPHOSPHATE DISODIUM see DXE500
INOSITHEXAPHOSPHORSAURE (GERMAN) see PIB250
INOSITOL HEXANICOTINATE see HFG550
m-INOSITOL HEXANICOTINATE see HFG550
myo-INOSITOL HEXANICOTINATE see HFG550
meso-INOSITOL HEXANICOTINATE see HFG550
INOSITOL HEXAPHOSPHATE see PIB250
INOSITOL NIACINATE see HFG550
INOSITOL NICOTINATE see HFG550
INOSTRAL see CNX825
INOTREX see DXS375
INOVAL see DYF200
INOVITAN PP see NCR000
INPC see PMS825
INPROQUONE see DCN000
INSANE ROOT see HAQ100
INSARIOTOXIN see FQS000
INSECTICIDE 1,179 see MDU600
INSECTICIDE-NEMATICIDE 1410 see DSP600
INSECTICIDE No. 497 see DHB400
INSECTICIDE No. 4049 see MAK700
INSECTOPHENE see EAQ750
INSECT POWDER see POO250
INSECT REPELLENT 448 see PES000
6-12-INSECT REPELLENT see EKV000
INSIDON see DCV800
INSIDON DIHYDROCHLORIDE see IDF000
INSOLUBLE SACCHARINE see BCE500
INSOMNOL see EQL000
INSOM-RAPIDO see CAV000
IN-SONE see PLZ000
INSPIR see ACH000
INSULAMIN see BOM750, BQL000
INSULAMINA see DNA200
INSULAR see IDF300
INSULATARD see IDF325
INSULIN see IDF300
INSULIN INJECTION see IDF300
INSULIN LENTE see LEK000
INSULIN NOVO LENTE see LEK000
INSULIN PROTAMINE ZINC see IDF325
INSULIN RETARD R1 see IDF325
INSULIN ZINC COMPLEX see LEK000
INSULIN ZINC PROTAMINATE see IDF325
INSULIN ZINC PROTAMINE see IDF325
INSULIN ZINC SUSPENSION see LEK000
INSULTON see MKB250
INSULYL see IDF300
INSULYL-RETARD see IDF325
INSUMIN see DAB800
INTAL see CNX825
INTALBUT see BRF500
INTALPRAM see DLH600, DLH630
INTEBAN SP see IDA000
INTEGERRIMINE see IDG000
INTEGRIN see ECW600

INTENKORDIN see CBR500
INTENSAIN see CBR500
INTENSAIN HYDROCHLORIDE see CBR500
INTENSE BLUE see FAE100
INTERCAIN see BQA010
INTERCHEM ACETATE BLUE B see MGG250
INTERCHEM ACETATE DEVELOPED BLACK see DPO200
INTERCHEM ACETATE VIOLET R see DBP000
INTERCHEM DIRECT BLACK Z see AQP000
INTERKELLIN see AHK750
INTERNATIONAL ORANGE 2221 see LCS000
INTEROX see HIB010
INTERSTITIAL CELL STIMULATING HORMONE see LIU300
INTERTULLE FUCIDIN see SHK000
INTEXAN LB-50 see AFP250
INTEXAN SB-85 see DTC600
INTEXSAN CPC see CCX000
INTEXSAN LQ75 see LBW000
INTOCOSTRIN see TOA000
INTOCOSTRINE see COF750
INTOLEX see BEQ625
INTRABILIX see EQC000
INTRABLIX see BGB315
INTRACID FAST ORANGE G see HGC000
INTRACID PURE BLUE L see FAE000
INTRACORT see HHR000
INTRACRON BLUE 3G see IDH000
INTRADERM TYROTHRICIN see TOG500
INTRADEX see DBD700
INTRAMYCETIN see CDP250
INTRANEFRIN see VGP000
INTRASPERSE YELLOW GBA EXTRA see AAQ250
INTRASPORIN see TEY000
INTRASTIGMINA see NCL100
INTRATHION see PHI500
INTRATION see PHI500
INTRAVAL see PBT250
INTRAVAL SODIUM see PBT500
INTROMENE see HII500
INTROPIN see DYC600
INVENOL see BSM000
INVERSAL see AHI875
INVERSINE see VIZ400
INVERSINE HYDROCHLORIDE see MQR500
INVERTON 245 see TAA100
INVISI-GARD see PMY300
IOB 82 see AFQ575
IOCARMATE MEGLUMINE see IDJ500
IOCARMIC ACID see TDQ230
IOCARMIC ACID DI-N-METHYLGLUCAMINE SALT see IDJ500
IODAIRAL see HGB200
IODAMIDE see AAI750
IODAMIDE 380 see IDJ600
IODAMIDE MEGLUMINE see IDJ600
IODAMIDE METHYLGLUCAMINE see IDJ600
IODATES see IDJ700
IODE (FRENCH) see IDM000
IODEIKON see TDE750
IODENTEROL see CHR500
IODIC ACID see IDK000
IODIC ACID, SODIUM SALT see SHV500
IODIC ACIODIC ACID, POTASSIUM SALT see PLK250
IODIDES see IDL000
IODIMIDE see DNE600
IODINE see IDM000
IODINE AZIDE see IDN000
IODINE(I) AZIDE see IDN000
IODINE BROMIDE see IDN200
IODINE CHLORIDE see IDS000
IODINE CRYSTALS see IDM000
IODINE CYANIDE see COP000
IODINE DIOXIDE TRIFLUORIDE see IDP000
IODINE DIOXYGEN TRIFLUORIDE see IDP000
IODINE HEPTAFLUORIDE see IDQ000
IODINE ISOCYANATE see IDR000
IODINE MONOCHLORIDE see IDS000
IODINE(V) OXIDE see IDS300
IODINE PENTAFLUORIDE see IDT000
IODINE(III) PERCHLORATE see IDU000
IODINE PHOSPHIDE see PHA000
IODINE-POTASSIUM IODIDE see PLW285
IODINE and PROPYLTHIOURACIL see PNX100
IODINE SUBLIMED see IDM000
IODINE TRIACETATE see IDV000

IODIO (ITALIAN) see IDM000
IODIPAMIDE MEGLUMINE see BGB315
IODIPAMIDE MEGLUMINE SALT see BGB315
IODIPAMIDE METHYLGLUCAMINE SALT see BGB315
IODOACETAMIDE see AOC500, IDW000
2-IODOACETAMIDE see IDW000
α-IODOACETAMIDE see IDW000
4-IODOACETANILIDE see IDY000
p-IODOACETANILIDE see IDY000
4'-IODOACETANILIDE see IDY000
IODOACETATE see IDZ000
IODOACETATE SODIUM SALT see SIN000
IODOACETIC ACID see IDZ000
IODOACETIC ACID ETHYL ESTER see ELQ000
IODOACETIC ACID SODIUM SALT see SIN000
(IODOACETOXY)TRIBUTYLSTANNANE see TID750
(IODOACETOXY)TRIPROPYLSTANNANE see TNB500
IODOACETYLENE see IDZ400
IODOALPHIONIC ACID see PDM750
3-IODOANILINE see IEB000
4-IODOANILINE see IEC000
m-IODOANILINE see IEB000
p-IODOANILINE see IEC000
IODOAZIDE see IDN000
4-IODOBENZENAMINE see IEC000
IODOBENZENE see IEC500
4-IODOBENZENEDIAZONIUM-2-CARBOXYLATE see IED000
o-IODOBENZOIC ACID see IEE000
o-IODOBENZOIC ACID SODIUM SALT see IEE050
p-IODOBENZOIC ACID SODIUM SALT see IEE100
o-IODOBENZOIC ACID TRIBUTYLSTANNYL ESTER see TIE000
p-IODOBENZOIC ACID TRIBUTYLSTANNYL ESTER see TIE250
N-(2-IODOBENZOYL)-GLYCIN MONOSODIUM SALT (9CI) see HGB200
N-p-IODOBENZOYL-N',N',N'-DIETHYLENETRIAMIDE of PHOSPHORIC
     ACID see BGY140
(4-IODOBENZOYLOXY)TRIBUTYLSTANNANE see TIE250
(o-IODOBENZOYLOXY)TRIPROPYLSTANNANE see IEF000
IODOBIL see PDM750
1-IODO-1,3-BUTADIYNE see IEG000
1-IODOBUTANE see BRQ250
2-IODOBUTANE see IEH000
IODOCHLORHYDROXYQUINOL see CHR500
IODOCHLORHYDROXYQUINOLINE see CHR500
7-IODO-5-CHLORO-8-HYDROXYQUINOLINE see CHR500
7-IODO-5-CHLOROXINE see CHR500
5-IODODEOXYURIDINE see DAS000
5-IODO-2'-DEOXYURIDINE see DAS000
IODODIMETHYLARSINE see IEI000
4-IODO-3,5-DIMETHYLISOXAZOLE see IEI600
2-IODO-3,5-DINITROBIPHENYL see IEJ000
IODOENTEROL see CHR500
IODOETHANE see ELP500
IODOETHANOL see IEL000
2-IODOETHANOL see IEL000
2-(2-IODOETHYL)-1,3-DIOXOLANE see IEL700
(2-IODOETHYL)TRIMETHYLAMMONIUM IODIDE see IEM300
3-IODO-2-FAA see IEO000
IODOFENOPHOS see IEN000
N-(3-IODO-2-FLUORENYL)ACETAMIDE see IEO000
IODOFORM see IEP000
IODOGNOST see TDE750
IODOHEPTAFLUOROPROPANE see HAY300
IODOHIPPURA see HGB200
IODOHIPPURATE SODIUM see HGB200
o-IODOHIPPURATE SODIUM see HGB200
4-(3-IODO-4-HYDROXYPHENOXY)-3,5-DIIODOPHENYLALANINE see LGK050
7-IODO-8-HYDROYQUINOLINE-5-SULFONIC ACID see IEP200
IODOMETANO (ITALIAN) see MKW200
IODOMETHANE see MKW200
IODOMETHANESULFONIC ACID SODIUM SALT see SHX000
IODOMETHYLMAGNESIUM see MLE250
7-IODOMETHYL-12-METHYLBENZ(a)ANTHRACENE see IER000
1-IODO-2-METHYLPROPANE see IIV509
2-IODO-2-METHYLPROPANE see TLU000
IODOMETHYLTRIMETHYLARSONIUM IODIDE see IET000
IODOMETHYLZINC see MQP250
IODOPACT see AAN000
IODOPANIC ACID see IFY100
IODOPANOIC ACID see IFY100
IODOPAQUE see AAN000
1-IODO-3-PENTEN-1-YNE see IEU000
IODOPHENE see IEU100
IODOPHENE SODIUM see TDE750
2-IODOPHENOL see IEV000

4-IODOPHENOL see IEW000
o-IODOPHENOL see IEV000
p-IODOPHENOL see IEW000
1-IODO-3-PHENYL-2-PROPYNE see IEX000
IODOPHOS see IEN000
IODOPHTHALEIN SODIUM see TDE750
1-IODOPROPANE see PNO750
2-IODOPROPANE see IPS000
3-IODOPROPIONIC ACID see IEY000
(IODOPROPIONYLOXY)TRIBUTYLSTANNANE see TIE500
3-IODOPROPYNE see IEZ800
3-IODO-2-PROPYNYL-2,4,5-TRICHLOROPHENYL ETHER see IFA000
IODOPYRACET see DNG400
IODORAYORAL see TDE750
IODOSOBENZENE see IFC000
IODOSOBENZENE DIACETATE see IFD000
IODOSYLBENZENE see IFE000
4-IODOSYLTOLUENE see IFE875
2-IODOSYLVINYL CHLORIDE see IFE879
3-IODOTETRAHYDROTHIOPHENE-1,1-DIOXIDE see IFG000
5-IODO-2-THIOURACIL, SODIUM SALT see SHX500
4-IODOTOLUENE see IFK509
IODO(p-TOLYL)MERCURY see IFL000
IODOTRIBUTYLSTANNANE see IFM000
IODOTRIMETHYLSTANNANE see IFN000
IODOTRIMETHYLTIN see IFN000
IODOTRIPHENYLSTANNANE see IFO000
IODOTRIPROPYLSTANNANE see TNB250
IODO-UNDECINIC ACID see IFO700
11-IODO-10-UNDECINIC ACID see IFO700
11-IODO-10-UNDECYNOIC ACID see IFO700
5-IODOURACIL see IFP000
5-IODOURACIL DEOXYRIBOSIDE see DAS000
IODOXAMATE MEGLUMINE see IFP800
IODOXAMIC ACID MEGLUMINE SALT see IFP800
IODTETRAGNOST see TDE750
IODURE de MERCURE (FRENCH) see MDC750
IODURE de METHYLE (FRENCH) see MKW200
l'IODURE de α-THIENYL-1 PHENYL-1 N-METHYL MORPHOLINIUM-3 PRO-
PANOL-1 (FRENCH) see HNM000
IODURIL see SHW000
IODURON B see DNG400
4-IODYLANISOLE see IFQ775
IODYLBENZENE see IFR000
IODYLBENZENE PERCHLORATE see IFS000
4-IODYL TOLUENE see IFS350
2-IODYLVINYL CHLORIDE see IFS385
IOGLUCOMIDE see IFS400
IOMESAN see DFV400
IOMEX see IFT100
IOMEZAN see DFV400
6,3-IONENE BROMIDE see IFT300
IONET S-80 see SKV100
IONET MO-400 see PJY100
IONET S 60 see SKV150
IONOL see BFW750
IONOL (antioxidant) see BFW750
IONONE see IFV000
α-IONONE see IFW000
β-IONONE see IFX000
IONOX 100 see IFX200
IONOX 100 ANTIOXIDANT see IFX200
IOP see AES000
IOPAMIDOL see IFY000
IOPAMIRON see IFY000
IOPANOIC ACID see IFY100
IOPEZITE see PKX250
IOPHENDYLATE see ELQ500
IOPHENOXIC ACID see IFZ800
IOPODATE SODIUM see SKM000
IOPRAMINE HYDROCHLORIDE see IFZ900
IOPRONIC ACID see IGA000
IOPYRACIL see DNG400
IOSALIDE see JDS200
IOTALAMATE de METHYLGLUCAMINE (FRENCH) see IGC000
IOTHALAMATE MEGLUMINE see IGC000
IOTHALAMATE METHYLGLUCAMINE see IGC000
IOTHALAMATE METHYLGLUCAMINE SALT see IGC000
IOTHALAMIC ACID see IGD000
IOTOX see HKB500
IOTROXATE MEGLUMINE see IGD075
IOTROXATE METHYLGLUCMINE SALT see IGD075
IOTROXIC ACID see IGD100
IOTROXINSAEURE (GERMAN) see IGD100

IOXAGLIC ACID see IGD200
IOXYNIL see HKB500
IOXYNIL, LITHIUM SALT see DNF400
IOXYNIL OCTANOATE see DNG200
IP-10 see IGE100
IP-82 see IIU000
IPA see DMV600, IMJ000
o-IPA see INA200
IPABUTONA see HNI500
IPAMIX see IBV100
IPANER see DAA800
IPC COMBINED with ENDOTHAL (1:1) see EAR500
IPD see YCJ000
IP-1,2-DIOL see DLE400
IPECAC SYRUP see IGF000
IPECACUANHA see IGF000
IPE-TOBACCO WOOD see HLY500
IPHOSPHAMIDE see IMH000
IPN see ILE000, PHX550
IPNO see IBP309
IPNOFIL see QAK000
IPO 8 see TBW100
IPODATE SODIUM see SKM000
IPOGLICONE see BSQ000
IPOLINA see HGP500
1,4-IPOMEADIOL see IGF200
IPOMEANIN see IGF300
IPOMEANINE see IGF300
IPOMEANOL see FQL100, IGF325
4-IPOMEANOL see FQL100
IPORES see RCA200
IPOTENSIVO see MBW750
IPPC see CBM000
IPRAL see ELX000
IPRAL SODIUM see NBU000
IPRATROPIUMBROMID (GERMAN) see IGG000
IPRATROPIUM BROMIDE see IGG000
IPRAZID see ILE000
IPRINDOLE see DPX200
IPROCLOZIDE see CJN250
IPRODIONE see GIA000
IPROGEN see DLH630
IPROHEPTINE HYDROCHLORIDE see IGG300
IPRONIAZID see ILE000
IPRONIAZID DIHYDROCHLORIDE see IGG600
IPRONIAZID HYDROCHLORIDE see IGG600
IPRONIAZID PHOSPHATE see IGG700
IPRONID see ILE000
IPRONIDAZOLE (USDA) see IGH000
IPRONIN see ILE000
IPROPLATIN see IGG775
IPROPRAN see IGH000
IPROVERATRIL see IRV000
IPROVERATRIL HYDROCHLORIDE see VHA450
IPROX see CMY250
IPSILON see AJD000
IPSOFLAME see BRF500
IPSOTIAN see MQU750
IPT (GERMAN) see INP100
I.P.Z. see IDF325
IQ 1 see IRV300
IQ DIHYDROCHLORIDE see AKT620
IRADICAV see SHF500
IRAGEN RED L-U see FAG070
IRALDEINE see IFV000
IRAMIL see DLH600, DLH630
IRC 453 see CIN750
IRC-50 ARVIN see VGU700
IRCON see FBJ100
IREHDIAMINE A see IGH700
IRENAL see ELX000
IRENE see DLS600
IRETIN see AQR000
IRG see AKY250
IRGACHROME ORANGE OS see LCS000
IRGALITE BRONZE RED CL see BNH500, BNK700
IRGALITE FAST RED 2GL see DVB800
IRGALITE RED CBN see CHP500
IRGALITE RED PRR see CJD500
IRGALITE SCARLET RB see MMP100
IRGALITE YELLOW BO see DEU000
IRGALON see EIV000
IRGAPYRIN see IGI000
IRGAPYRINE see IGI000

IRGASAN see TIQ000
IRGASAN DP300 see TIQ000
IRIDIL see HNI500
IRIDIUM see IGJ000
IRIDIUM CHLORIDE see IGJ300
IRIDIUM(IV) CHLORIDE see IGJ499
IRIDIUM MURIATE see IGJ300
IRIDIUM TETRACHLORIDE see IGJ499
IRIDOCIN see EPQ000
IRIDOZIN see EPQ000
IRIFAN see TDA500
IRISH GUM see CCL250
IRISH MOSS EXTRACT see CCL250
IRISH MOSS GELOSE see CCL250
IRISOL BASE see HOK000
IRISONE see IFV000
IRISONE ACETATE see CNS825
IRIUM see SIB600
IRMIN see DLH600
IROCAINE see AIT250
IROINI see HBT500
IRO-JEX see IGS000
IROMIN see FBK000
IRON see IGK800
IRON(2+) see FBN000
IRON(2+) ACETATE see FBH000
IRON(II) ACETATE see FBH000
IRON ACETYLACETONATE see IGL000
IRON ALLOY, BASE, Fe,P (FERROPHOSPHORUS) see FBF000
IRON ARSENATE (DOT) see IGM000
IRON(II) ARSENATE (3:2) see IGM000
IRON(III) ARSENATE (1:1) see IGN000
IRON(III)-o-ARSENITE PENTAHYDRATE see IGO000
IRON(II) ASCORBATE see FBH050
IRONATE see FBO000
IRON, (N,N-BIS(CARBOXYMETHYL)GLYCINATO(3-)-N,O,O',O'')-, (T-4)-(9CI) see IHC100
IRON BIS(CYCLOPENTADIENE) see FBC000
IRON BLUE see IGY000
IRON(II) BROMIDE see IGP000
IRON(III) BROMIDE see IGQ000
IRON CARBIDE see IGQ750
IRON CARBOHYDRATE COMPLEX see IGU000
IRON(II) CARBONATE see FBH100
IRON CARBONYL see IHG500, NMV740
IRON, CARBONYL (FCC) see IGK800
IRON CHLORIDE see FAU000
IRON(III) CHLORIDE see FAU000
IRON(II) CHLORIDE (1:2) see FBI000
IRON CHLORIDE, solid (DOT) see FAU000
IRON(III) CHLORIDE (solution) see FAV000
IRON(3+) CHLORIDE HEXAHYDRATE see FAW000
IRON(III), CHLORIDE HEXAHYDRATE see FAW000
IRON CHLORIDE TETRAHYDRATE see FBJ000
IRON(2+) CHLORIDE TETRAHYDRATE see FBJ000
IRON (II) CHLORIDE TETRAHYDRATE see FBJ000
IRON CHOLINE CITRATE COMPLEX see FBC100
IRON CHROMITE see CMI500
IRON COMPOUNDS see IGR499
IRON DEXTRAN see IGS000
IRON-DEXTRAN COMPLEX see IGS000
IRON DEXTRAN GLYCEROL GLYCOSIDE see IGT000
IRON DEXTRAN INJECTION see IGS000
IRON-DEXTRIN COMPLEX see IGU000
IRON DEXTRIN INJECTION see IGU000
IRON DIACETATE see FBH000
IRON DICHLORIDE see FBI000
IRON DICHLORIDE TETRAHYDRATE see FBJ000
IRON DICYCLOPENTADIENYL see FBC000
IRON DIMETHYLDITHIOCARBAMATE see FAS000
IRON DISULFIDE see IGV000
IRON DUST see IGW000
α-IRONE see IGW500
IRON(II) EDTA COMPLEX see IGX000
IRON(III)-EDTA SODIUM see IGX875
IRON, ELECTROLYTIC see IGK800
IRON, ELEMENTAL see IGK800
IRON (Fe 2+) see FBN000
IRON complex with N-FLUOREN-2-YL ACETOHYDROXAMIC ACID see HIQ000
IRON FLUORIDE see FAX000
IRON FUMARATE see FBJ100
IRON GLUCONATE see FBK000
IRON(III) HEXACYANOFERRATE(4−) see IGY000
IRON HYDROGENATED DEXTRAN see IGS000

IRON(+3) HYDROXIDE COMPLEX with NITRILO-TRI-PROPIONIC ACID see FAY000
IRON(III)HYDROXIDE-POLYMALTOSE see IHA000
IRON(II,III) OXIDE see IHC550
IRON(II) ION see FBN000
IRON(2+) LACTATE see LAL000
IRON(II) MALEATE see IHB675
IRON METHANEARSONATE see IHB680
IRON MONOSULFATE see FBN100
IRON NICKEL SULFIDE see NDE500
IRON NICKEL ZINC OXIDE see IHB800
IRON(III) NITRATE, NONAHYDRATE (1:3:9) see IHC000
IRON NITRILOTRIACETATE see IHC100
IRON-NITRILOTRIACETATE CHELATE see IHC100
IRON(3+) NTA see IHC100
IRON ORE see HAO875
IRONORM INJECTION see IGS000
IRON OXIDE see IHD000
IRON(II) OXIDE see IHC500
IRON(III) OXIDE see IHD000
IRON OXIDE, CHROMIUM OXIDE, and NICKEL OXIDE FUME see IHE000
IRON OXIDE FUME see IHF000
IRON OXIDE RED see IHD000
IRON OXIDE, SACCHARATED see IHG000
IRON PENTACARBONYL see IHG500
IRON PERSULFATE see FBA000
IRON-POLYSACCHARIDE COMPLEX see IHH000
IRON-POLY(SORBITOL-GLUCONIC ACID) COMPLEX see IHH300
IRON PROTOCHLORIDE see FBI000
IRON PROTOSULFATE see FBN100
IRON PYRITES see IGV000
IRON, REDUCED (FCC) see IGK800
IRON SACCHARATE see IHG000
IRON SESQUICHLORIDE, solid (DOT) see FAU000
IRON SESQUIOXIDE see IHD000
IRON SESQUIOXIDE HYDRATED see LFW000
IRON SESQUISULFATE see FBA000
IRON-SILICON see IHJ000
IRON-SILICON ALLOY see FBG000
IRON SODIUM GLUCONATE see IHK000
IRON SORBITEX see IHL000
IRON-SORBITOL see IHK100
IRON SORBITOL CITRATE see IHL000
IRON-SORBITOL-CITRIC ACID see IHL000
IRON SUGAR see IHG000
IRON SULFATE (2:3) see FBA000
IRON(III) SULFATE see FBA000
IRON(II) SULFATE (1:1) see FBN100
IRON(II) SULFATE (1:1), HEPTAHYDRATE see FBO000
IRON SULFIDE see IGV000
IRON(II) SULFIDE see IHN000
IRON(III) SULFIDE see IHN050
IRON TERSULFATE see FBA000
IRON TRICHLORIDE see FAU000
IRON TRICHLORIDE HEXAHYDRATE see FAW000
IRON TRIFLUORIDE see FAX000
IRON VITRIOL see FBN100
IRON VITROL see FBO000
IROQUINE see CLD000
IROSPAN see FBN100
IROSUL see FBN100, FBO000
IROX (GADOR) see FBK000
IRRADIATED ERGOSTA-5,7,22-TRIEN-3-β-OL see VSZ100
IRTRAN 1 see MAF500
IRTRAN 3 see CAS000
1205 I.S. see MHD300
1530 I.S. see MRR125
1665 I.S. see BKS825
2341 I.S. see BIQ500
2723 I.S. see PFY100
ISACONITINE see PIC250
ISADRINE see IMR000
ISADRINE-HYDROCHLORIDE see IMR000
dl-ISADRINE HYDROCHLORIDE see IQS500
ISALIZINA see BET000
ISAMID see DAB875
ISAMIN see WAK000
ISAROL see AIF000
ISATEX see ACD500
ISATIC ACID LACTAM see ICR000
ISATIDINE see RFU000
ISATIN see ICR000
ISATINIC ACID ANHYDRIDE see ICR000
ISATOIC ACID ANHYDRIDE see IHN200

ISATOIC ANHYDRIDE see IHN200
ISAZOPHOS see PHK000
ISCEON 22 see CFX500
ISCEON 113 see FOO000
ISCEON 122 see DFA600
ISCEON 131 see TIP500
ISCHELIUM see DLL400
ISCOBROME see MHR200
ISCOBROME D see EIY500
ISCOVESCO see DKA600
ISDIN see CCK125
I-SEDRIN see EAW000
ISEPAMICIN SULFATE see SBE800
ISETHION see GFW000
ISETHIONIC ACID see HKI500
ISF 2001 see TNP275
ISF 2469 see CAK275
ISF 2508 see CFJ000
ISICAINA see DHK400
ISICAINE HYDROCHLORIDE see DHK600
ISIDRINA see ILD000
ISINDONE see IDA400
ISKIA-C see ARN125
ISLACTID see AES650
ISLANDITOXIN see COW750
ISMAZIDE see ILD000
ISMELIN see GKQ000, GKS000
ISMELIN SULFATE see GKS000
ISMICETINA see CDP250
ISMIPUR see POK000
5-ISMN see ISC500
ISMOTIC see HID350
ISOACETOPHORONE see IMF400
ISOADRENALINE HYDROCHLORIDE see AMB000
ISOAMIDONE II see IKZ000
ISOAMINILE CITRATE see PCK639
ISOAMINILE CYCLAMATE see IHO200, PCK669
ISOAMYCIN see BBK000
ISOAMYGDALIN see IHO700
ISOAMYL ACETATE see IHO850
ISOAMYL ALCOHOL see IHP000, IHP010
ISOAMYL ALKOHOL (CZECH) see IHP000
ISO-AMYLALKOHOL (GERMAN) see IHP000
ISOAMYL AMINOFORMATE see IHQ000
ISOAMYL BENZOATE see IHP100
ISOAMYL BENZYL ETHER see BES500
ISOAMYL BROMIDE see BNP250
ISOAMYL BUTYRATE see IHP400
ISOAMYL CAPROATE see IHU100
ISOAMYL CAPRYLATE see IHP500
ISOAMYL CARBAMATE see IHQ000
ISOAMYL CINNAMATE see AOG600
ISOAMYL CYANIDE see MNJ000
ISOAMYLDICHLORARSINE see IHQ100
ISOAMYLDICHLOROARSINE see IHQ100
ISOAMYL N-(β-DIETHYLAMINOETHYL)-α-AMINOPHENYLACETATE
    see NOC000
ISOAMYL α-(N-(β-DIETHYLAMINOETHYL))-AMINOPHENYLACETATE
    see NOC000
ISOAMYL 5,6-DIHYDRO-7,8-DIMETHYL-4,5-DIOXO-4H-PYRANO(3,2-c)QUINO-
    LINE-2-CARBOXYLATE see IHR200
β-ISOAMYLENE see AOI750
ISOAMYLENE ALCOHOL see PBL000
β-ISOAMYLENE OXIDE see TLY175
8-ISOAMYLENOXYPSORALEN see IHR300
4-(β-ISOAMYLENYL)-1,2-DIPHENYL-3,5-PYRAZOLIDINEDIONE see PEW000
ISOAMYL ETHANOATE see IHO850
ISOAMYLETHYLBARBITURIC ACID see AMX750
5-ISOAMYL-5-ETHYLBARBITURIC ACID see AMX750
5-ISOAMYL-5-ETHYLBARBITURIC ACID, SODIUM DERIVATIVE see AON750
ISOAMYL FORMATE see IHS000
ISOAMYL GERANATE see IHS100
1-ISOAMYL GLYCEROL ETHER see IHT000
ISOAMYL HEXANOATE see IHU100
ISOAMYLHYDRIDE see EIK000
ISOAMYL HYDROCUPREINE DIHYDROCHLORIDE see EQQ500
ISOAMYL o-HYDROXYBENZOATE see IME000
ISOAMYL ISOVALERATE (FCC) see ITB000
ISOAMYL METHANOATE see IHS000
ISOAMYL METHYL KETONE see MKW450
ISOAMYLNITRILE see ITD000
ISOAMYL NITRITE see IMB000
ISOAMYL OCTANOATE see IHP500
ISOAMYLOL see IHP000

ISOAMYL 3-PENTYL PROPENATE see AOG600
ISOAMYL PHENYLACETATE see IHV000
ISOAMYL PHENYLAMINOACETATE HYDROCHLORIDE see PCV750
ISOAMYL PHENYLETHYL ETHER see IHV050
ISOAMYL POTASSIUM XANTHATE see PLK580
ISOAMYL PROPIONATE see AON350, ILW000
ISOAMYL SALICYLATE (FCC) see IME000
1-ISOAMYL THEOBROMINE see IHX200
ISOANETHOLE see AFW750
trans-ISOASARONE see IHX400
ISOBAC 20 see HCL000, MRM000
ISOBAMATE see IPU000
ISOBARB see NBU000
ISOBENZAMIC ACID see AMU750
ISOBENZAN see OAN000
1,3-ISOBENZOFURANDIONE see PHW750
1(3H)-ISOBENZOFURANONE, 3-BUTYLIDENE-(9CI) see BRQ100
1(3H)-ISOBENZOFURANONE, 3-(3-FURANYL)-3a,4,5,6-TETRAHYDRO-3a,
    7-DIMETHYL-, (3R-cis)-(9CI) see FOM200
ISO-BID see CCK125
ISOBIDE see HID350
ISO-BNU see IJF000
ISOBORNEOL see IHY000
dl-ISOBORNEOL see IHY000
ISOBORNEOL THIOCYANATOACETATE see IHZ000
ISOBORNYL ACETATE see IHX600
ISOBORNYL ALCOHOL see IHY000
ISOBORNYL THIOCYANATOACETATE see IHZ000
ISOBORNYL THIOCYANOACETATE see IHZ000
ISOBROMYL see BNP750
ISOBUTANAL see IJS000
ISOBUTANDIOL-2-AMINE see ALB000
ISOBUTANE see MOR750
ISOBUTANOL (DOT) see IIL000
ISOBUTANOLAMINE see IIA000
ISOBUTANOL-2-AMINE see IIA000
ISOBUTAZINA see HNI500
ISOBUTENAL see MGA250
ISOBUTENE see IIC000
ISOBUTENYL CHLORIDE see CIU750
ISOBUTENYL METHYL KETONE see MDJ750
ISOBUTIL see HNI500
ISOBUTINYL-N-(3-CHLORPHENYL)-CARBAMAT (GERMAN) see CEX250
2-ISOBUTOXYETHANOL see IIP000
ISOBUTOXY-2-ETHOXY-2-ETHANOL see DJF800
2-(2-ISOBUTOXYETHOXY)ETHANOL see DJF800
(2-ISOBUTOXYETHYL)CARBAMATE see IIE000
1-ISOBUTOXY-2-PROPANOL see IIG000
2-ISOBUTOXY TETRAHYDROPYRAN see III000
ISOBUTYL ACETATE see IIJ000
N-ISOBUTYL-N-(ACETOXYMETHYL)NITROSAMINE see ABR125
ISOBUTYL ACRYLATE see IIK000
ISOBUTYL ACRYLATE, inhibited (DOT) see IIK000
ISOBUTYL ADIPATE see DNH125
ISOBUTYL ALCOHOL see IIL000
ISOBUTYLALDEHYDE see IJS000
ISOBUTYL ALDEHYDE (DOT) see IJS000
ISOBUTYLALKOHOL (CZECH) see IIL000
ISO-BUTYLALLYLBARBITURIC ACID see AGI750
ISOBUTYLALLYLBARTURIC ACID see AGI750
ISOBUTYLAMINE see IIM000
2-ISOBUTYLAMINOETHANOL HYDROCHLORIDE ACID SALT, p-AMINO-
    BENZOIC ACID ESTER see IAC000
2-(ISOBUTYLAMINO)ETHYL-p-AMINOBENZOATE HYDROCHLORIDE
    see IAC000
2-(ISOBUTYLAMINO)-2-METHYL-1-PROPANOL BENZOATE (ESTER), HYDRO-
    CHLORIDE see IIM100
2-(ISOBUTYLAMINO)-2-METHYL-1-PROPANOL BENZOATE HYDROCHLO-
    RIDE see IIM100
3-(ISOBUTYLAMINO)PROPYL)-2-(3,4-METHYLENEDIOXYPHENYL)-4-
    THIAZOLIDINONE HYDROCHLORIDE see WBS855
ISOBUTYLBENZENE see IIN000
ISOBUTYLBIS(2-CHLOROETHYL)AMINE HYDROCHLORIDE see BIE750
ISOBUTYLBIS(β-CHLOROETHYL)AMINE HYDROCHLORIDE see BIE750
ISOBUTYL BROMIDE see BNR750
ISOBUTYL BUTANOATE see BSW500
3-ISOBUTYL-6-sec-BUTYL-2-HYDROXYPYRAZINE-1-OXIDE
    see ARO000
ISOBUTYL BUTYRATE (FCC) see BSW500
ISOBUTYL CAPROATE see IIT000
ISOBUTYLCARBINOL see IHP000
(2-(p-(5-(ISOBUTYLCARBOZMOYL)-2-OCTYLOXYBENZAMIDO)BENZAMIDO)
    ETHYL)-TRIETHYLAMMONIUM IODIDE see IIO750
ISOBUTYL CELLOSOLVE see IIP000

ISOBUTYL CHLORIDE see CIU500
ISOBUTYL CINNAMATE see IIQ000
2-((1-ISOBUTYL-3,5-DIMETHYLHEXYL)OXY)ETHANOL see EJL000
1-ISOBUTYL-3,4-DIPHENYLPYRAZOLE-5-ACETIC ACID SODIUM SALT
  see DWD400
ISOBUTYLDIUREA see IIV000
ISOBUTYLENE (DOT) see IIC000
ISOBUTYLENE CHLORIDE see IIQ200
ISOBUTYLENEDIUREA see IIV000
ISOBUTYL FORMATE see IIR000
ISOBUTYL-2-FURANPROPIONATE see IIR100
ISOBUTYL FURYLPROPIONATE see IIR100
ISOBUTYL HEPTANOATE see IIS000
ISOBUTYL HEPTYLATE see IIS000
ISOBUTYL HEXANOATE see IIT000
4-ISOBUTYLHYDRATROPIC ACID see IIU000
p-ISOBUTYLHYDRATROPIC ACID see IIU000
p-ISOBUTYLHYDRATROPIC ACID SODIUM SALT see IAB100
ISOBUTYL-o-HYDROXYBENZOATE see IJN000
1,1'-ISOBUTYLIDENEBISUREA see IIV000
ISOBUTYLIDENEDIUREA see IIV000
ISOBUTYL IODIDE see IIV509
ISOBUTYL ISOBUTYRATE see IIW000
ISOBUTYLISOBUTYRATE (DOT) see IIW000
ISOBUTYL ISOPENTANOATE see ITA000
ISOBUTYL ISOVALERATE see ITA000
ISOBUTYL KETONE see DNI800
ISOBUTYL MERCAPTAN see IIX000
ISOBUTYL METHACRYLATE see IIY000
ISOBUTYL-α-METHACRYLATE see IIY000
2-(ISOBUTYL-3-METHYLBUTOXY)ETHANOL see IJA000
2-(2-(1-ISOBUTYL-3-METHYLBUTOXY)ETHOXY)ETHANOL see IJB000
ISOBUTYL METHYL CARBINOL see MKW600
ISOBUTYL-METHYLKETON (CZECH) see HFG500
ISOBUTYL METHYL KETONE see HFG500
ISOBUTYLMETHYLMETHANOL see MKW600
5-ISOBUTYL-5-(METHYLTHIOMETHYL)BARBITURIC ACID SODIUM SALT
  see SHY000
ISOBUTYL(2-NAPHTHYL)ETHER see NBJ000
ISOBUTYL NITRITE see IJD000
1-ISOBUTYL-3-NITRO-1-NITROSOGUANIDINE see IJE000
N-ISOBUTYL-N'-NITRO-N-NITROSOGUANIDINE see IJE000
2-ISOBUTYL-3-NITROSOTHIAZOLIDINE see NKN500
1-ISO-BUTYL-1-NITROSOUREA see IJF000
N-ISOBUTYL-N-NITROSOUREA see IJF000
ISOBUTYL ORTHOVANADATE see VDK000
ISOBUTYL PHENYLACETATE see IJF400
4-ISOBUTYLPHENYLACETIC ACID see IJG000
(p-ISOBUTYLPHENYL)ACETIC ACID see IJG000
2-(4-ISOBUTYLPHENYL)BUTYRIC ACID see IJH000
2-(4-ISOBUTYLPHENYL)PROPANOIC ACID see IIU000
2-(p-ISOBUTYLPHENYL)PROPIONIC ACID see IIU000
α-(4-ISOBUTYLPHENYL)PROPIONIC ACID see IIU000
α-p-ISOBUTYLPHENYLPROPIONIC ACID see IIU000
2-(p-ISOBUTYLPHENYL)PROPIONIC ACID-o-METHOXYPHENYL ESTER
  see IJJ000
2-(p-ISOBUTYLPHENYL)PROPIONIC ACID-2-PYRIDYLMETHYL ESTER
  see IAB000
ISOBUTYL POTASSIUM XANTHATE see PLK600
ISOBUTYL PROPENOATE see IIK000
ISOBUTYL-2-PROPENOATE see IIK000
2-ISOBUTYLQUINOLINE see IJM000
α-ISOBUTYLQUINOLINE see IJM000
ISOBUTYL SALICYLATE see IJN000
ISOBUTYLSULFHYDRATE see IJO000
p-ISOBUTYL-α-TOLUIC ACID see IJG000
ISOBUTYLTRIMETHYLETHANE see TLY500
ISOBUTYL VINYL ETHER see IJQ000
p-ISOBUTYOXYBENZOIC ACID-3-(2'-METHYLPIPERIDINO)PROPYL ESTER
  see IJR000
ISOBUTY PROPIONATE (DOT) see PMV250
ISOBUTYRALDEHYD (CZECH) see IJS000
ISOBUTYRALDEHYDE see IJS000
ISOBUTYRALDEHYDE, OXIME see IJT000
ISOBUTYRIC ACID see IJU000
ISOBUTYRIC ACID, BENZYL ESTER see IJV000
ISOBUTYRIC ACID, CINNAMYL ESTER see CMR750
ISOBUTYRIC ACID, ETHYL ESTER see ELS000
ISOBUTYRIC ACID, HEXYL ESTER see HFQ550
ISOBUTYRIC ACID, ISOBUTYL ESTER see IIW000
ISOBUTYRIC ACID, 1-ISOPROPYL-2,2-DIMETHYLTRIMETHYLENE ESTER
  see TLZ000
ISOBUTYRIC ACID, METHYL ESTER see MKX000
ISOBUTYRIC ACID-3-PHENYLPROPYL ESTER see HHQ500

ISOBUTYRIC ACID, p-TOLYL ESTER see THA250
ISOBUTYRIC ALDEHYDE see IJS000
ISOBUTYRIC ANHYDRIDE see IJW000
ISOBUTYRONITRILE see IJX000
3-ISOBUTYRYL-2-ISOPROPYLPYRAZOLO(1,5-a)PYRIDINE see KDA100
ISOCAINE see IJZ000
ISOCAINE-ASID see AIT250
ISOCAINE BASE see PIV750
ISOCAINE-HEISLER see AIT250
ISOCAMPHANYL CYCLOHEXANOL (mixed isomers) see IKA000
ISOCAMPHOL see IHY000
ISOCAMPHYL CYCLOHEXANOL (mixed isomers) see IKA000
ISOCAPROALDEHYDE see MQJ500
ISOCAPRONITRILE see MNJ000
ISOCARAMIDINE SULFATE see IKB000
ISOCARB see PMY300
ISOCARBONAZID see IKC000
ISOCARBOSSAZIDE see IKC000
ISOCARBOXAZID see IKC000
ISOCARBOXAZIDE see IKC000
ISOCARBOXYZID see IKC000
ISOCHINOL see DNX400
ISOCHLOORTHION (DUTCH) see NFT000
ISOCHRYSENE see TMS000
ISOCID see ILD000
ISOCIL see BNM000
ISOCILLIN see PDT750
cis-ISOCITRATO-(1,2-DIAMINO-CYCLOHEXAN)-PLATIN(II) (GERMAN)
  see PGQ275
ISO-CORNOX see CIR500
ISOCOTIN see ILD000
ISOCROTYL CHLORIDE see IKE000
ISOCTENE see IKF000
ISOCUMENE see IKG000
ISOCYANATE de METHYLE (FRENCH) see MKX250
ISOCYANATES see IKG349
ISOCYANATOETHANE see ELS500
2-ISOCYANATOETHYL METHACRYLATE see IKG700
β-ISOCYANATOETHYL METHACRYLATE see IKG700
ISO-CYANATOMETHANE see MKX250
3-ISOCYANATOMETHYL-3,5,5-TRIMETHYLCYCLOHEXYLISOCYANATE
  see IMG000
4-ISOCYANATO-N-(4-NITROPHENYLBENZENAMINE (9CI) see IKI000
1-ISOCYANATOPROPANE see PNP000
ISOCYANIC ACID ALLYL ESTER see AGJ000
ISOCYANIC ACID, BENZ(a)ANTHRACEN-7-YL ESTER see BBJ250
ISOCYANIC ACID, BUTYL ESTER see BRQ500
ISOCYANIC ACID-2-CHLOROETHYL ESTER see IKH000
ISOCYANIC ACID-6-CHLOROHEXYL ESTER see CHL250
ISOCYANIC ACID-m-CHLOROPHENYL ESTER see CKA750
ISOCYANIC ACID-p-CHLOROPHENYL ESTER see CKB000
ISOCYANIC ACID, CYCLOHEXYL ESTER see CPN500
ISOCYANIC ACID-3,4-DICHLOROPHENYL ESTER see IKH099
ISOCYANIC ACID, DIESTER with 1,6-HEXANEDIOL see DNJ800
ISOCYANIC ACID, 3,3'-DIMETHYL-4,4'-BIPHENYLENE ESTER see DQS000
ISOCYANIC ACID, ESTER with DI-o-TOLUENEMETHANE see MJN750
ISOCYANIC ACID, ETHYL ESTER see ELS500
ISOCYANIC ACID, HEXAHYDRO-4,7-METHANOINDAN-1,8-YLENE ESTER
  see TJG500
ISOCYANIC ACID, HEXAMETHYLENE ESTER see DNJ800
ISOCYANIC ACID, HEXAMETHYLENE ESTER, POLYMER with 1,4-
  BUTANEDIOL see PKO750
ISOCYANIC ACID, METHYLENEDI-p-PHENYLENE ESTER, POLYMER with
  1,4-BUTANEDIOL see PKP000
ISOCYANIC ACID, METHYL ESTER see MKX250
ISOCYANIC ACID, METHYLPHENYLENE ESTER see TGM740, TGM750
ISOCYANIC ACID, 4-METHYL-m-PHENYLENE ESTER see TGM750
ISOCYANIC ACID-1,5-NAPHTHYLENE ESTER see NAM500
ISOCYANIC ACID, OCTADECYL ESTER see OBG000
ISOCYANIC ACID, PHENYL ESTER see PFK250
ISOCYANIC ACID, PROPYL ESTER see PNP000
ISOCYANIC ACID, (m-TRIFLUOROMETHYLPHENYL) ESTER see TKJ250
ISOCYANIDE see COI500
ISOCYANIDES see IKH339
ISOCYANOAMIDE see IKH669
ISOCYANOETHANE see ELT000
4-ISOCYANO-4'-NITRODIPHENYLAMINE see IKI000
ISOCYANURIC ACID see THS000
ISOCYANURIC CHLORIDE see TIQ750
ISOCYCLEX see IOE000
ISOCYCLOCITRAL see IKJ000
ISODECANOL see IKK000
ISODECYL ACRYLATE see IKL000
ISODECYL ALCOHOL see IKK000

ISODECYL ALCOHOL ACRYLATE see IKL000
ISODECYL METHACRYLATE see IKM000
ISODECYL PROPENOATE see IKL000
ISODEMETON see DAP200
ISODIAZOMETHANE see IKH669
ISODIENESTROL see DAL600
ISODIHYDROLAVANDULOL see IQH000
ISODIHYDRO LAVANDULYL ACETATE see AAV250
ISODIHYDROLAVANDULYL ALDEHYDE see IKM100
ISODILAN see IJG000
ISODIN see CFZ000
ISODINE see PKE250
ISODIPHENYLBENZENE see TBC620
ISODIPRENE see CCK500
ISODONAZOLE NITRATE see IKN200
ISODORMID see IQX000
ISODRIN see IKO000
ISODRINE see FMS875
ISODRINUM see FMS875
ISODUR see IIV000
ISODURENE see TDM500
ISOENDOXAN see IMH000
d-ISOEPHEDRINE see POH000
ISOESTRAGOLE see PMQ750
ISOETHADIONE see PAH500
ISOEUGENOL see IKQ000
ISOEUGENOL ACETATE see AAX750
ISOEUGENOL AMYL ETHER see AOK000
1,3,4-ISOEUGENOL METHYL ETHER see IKR000
ISOEUGENYL ACETATE (FCC) see AAX750
ISOEUGENYL METHYL ETHER see IKR000
ISOFEDROL see EAW000, EAY500
ISOFENPHOS see IMF300
ISOFLAV see DBN400
ISOFLAV BASE see DBN600
ISOFLUOROPHATE see IRF000
ISOFLURANE see IKS400
ISOFLUROPHATE see IRF000
ISOFORON see IMF400
ISOFORONE (ITALIAN) see IMF400
ISOFOSFAMIDE see IMH000
ISOFTALODINITRIL (CZECH) see PHX550
ISOGAMMA ACID see AKI000
ISOGLAUCON see CMX760
ISOHARRINGTONINE see IKS500
ISOHEXANE see DQT400, IKS600
ISOHEXANOIC ACID (mixed isomers) see IKT000
ISOHEXYL ALCOHOL see AOK750
ISOHOL see INJ000
ISOHOMOGENOL see IKR000
ISOINDOLE-1,3-DIONE see PHX000
1,3-ISOINDOLEDIONE see PHX000
1H-ISOINDOLE-1,3(2H)-DIONE,2-(2,6-DIOXO-3-PIPERIDINYL)-, (R)- see TEH510
1H-ISOINDOLE-1,3(2H)-DIONE, 2-(HYDROXYMETHYL)- see HMP100
1H-ISOINDOLE-5-SULFONAMIDE, 6-CHLORO-2,3-DIHYDRO-1,3-DIOXO-
    see CGM500
ISOINDOLINE, 2-(2-DIMETHYLAMINOETHYL)-4,5,6,7-
    TETRACHLORODIMETHO CHLORIDE see CDY000
1,3-ISOINDOLINEDIONE see PHX000
5-ISOINDOLINESULFONAMIDE, 6-CHLORO-1,3-DIOXO- see CGM500
ISOINOKOSTERONE see HKG500
ISOJASMONE see IKV000
ISO-K see BDU500
ISOKET see CCK125
ISOL see HFP875
ISOLAN see DSK200
ISOLANE (FRENCH) see DSK200
ISOLANID see LAU000
ISOLANIDE see LAU000
ISOLASALOCID A see IKW000
ISOL BENZIDINE YELLOW G see DEU000
ISOLEUCINE see IKX000
l-ISOLEUCINE (FCC) see IKX000
ISOL FAST RED 2G see DVB800
ISOL FAST RED HB see MMP100
ISOL FAST RED R see CJD500
ISOL LAKE RED LCS 12527 see CHP500
ISOLYN see ILD000
ISOMACK see CCK125
ISOMALIC ACID see MPN250
ISOMEBUMAL see EJY000
ISOMELIN see GKS000
ISOMENTHONE see IKY000
ISOMEPROBAMATE see IPU000

β-ISOMER see BBR000
ISOMERIC CHLORTHION see NFT000
ISOMETASYSTOX see DAP400
ISOMETASYSTOX SULFONE see DAP600
ISOMETHADONE see IKZ000
ISOMETHEPTENE see ILK000
ISOMETHEPTENE HYDROCHLORIDE see ODY000
ISOMETHYLSYSTOX see DAP400
ISOMETHYLSYSTOX SULFONE see DAP600
ISOMETHYLSYSTOX SULFOXIDE see DAP000
ISOMIN see TEH500
ISOMYN see BBK000
ISOMYST see IQN000
ISONAL see AFT000, ENB500
ISONAL (ROUSSEL) see ENB500
ISONAPHTHOIC ACID see NAV500
ISONAPHTHOL see NAX000
ISONATE see MJP400
ISONEX see ILD000
ISONIACID see ILD000
ISONIAZIDE see ILD000
ISONIAZID SODIUM METHANESULFONATE see SIJ000
ISONICAZIDE see ILD000
ISONICO see ILD000
ISONICOTAN see ILD000
ISONICOTINALDEHYDE THIOSEMICARBAZONE see ILB000
ISONICOTINAMIDE PENTAAMMINE RUTHENIUM(II) PERCHLORATE
    see ILB150
ISONICOTINHYDRAZID see ILD000
ISONICOTINIC ACID see ILC000
ISONICOTINIC ACID, ETHYL ESTER see ELU000
ISONICOTINIC ACID HYDRAZIDE see ILD000
ISONICOTINIC ACID-2-ISOPROPYLHYDRAZIDE see ILE000
ISONICOTINIC ACID, 2-ISOPROPYLHYDRAZIDE, PHOSPHATE see IGG700
ISONICOTINIC ACID, SODIUM SALT see ILF000
ISONICOTINIC ACID, VANILLYLIDENEHYDRAZIDE see VEZ925
N-ISONICOTINOYL-N'-GLUCURONID-HYDRAZIN-NATRIUMSALZ
    DIHYDRAT (GERMAN) see IDB100
N-ISONICOTINOYL-N'-GLUCURONSAEURE-Γ-LACTON-HYDRAZON
    (GERMAN) see GBB500
ISONICOTINOYL HYDRAZIDE see ILD000
2-(2-ISONICOTINOYLHYDRAZINO)-d-GLUCOPYRANURONIC ACID SODIUM
    SALT DIHYDRATE see IDB100
4-(ISONICOTINOYLHYDRAZONE)PIMELIC ACID see ILG000
1-ISONICOTINOYL-2-ISOPROPYLHYDRAZINE see ILE000
N-ISONICOTINOYL-N'(β-N-BENZYLCARBOXAMIDOETHYL)HYDRAZINE
    see BET000
ISONICOTINSAEUREHYDRAZID see ILD000
ISONICOTINYL HYDRAZIDE see ILD000
1-ISONICOTINYL-2-ISOPROPYLHYDRAZINE see ILE000
ISONIDE see ILD000
ISONIKAZID see ILD000
ISONIN see ILD000
ISONIPECAINE see DAM600
ISONIPECAINE HYDROCHLORIDE see DAM700
ISONIPECOTIC ACID, 1-(2-HYDROXY-3-PHENOXYPROPYL)-4-PHENYL-,
    ETHYL ESTER, HYDROCHLORIDE see EEQ000
ISONIRIT see ILD000
ISONITOX see MIW250
ISONITROPROPANE see NIY000
ISONITROSOACETOPHENONE see ILH000
5-ISONITROSOBARBITURIC ACID see AFU000
β-ISONITROSOPROPANE see ABF000
ISONITROSOPROPIOPHENONE see ILI000
ISONIXIN see XLS300
ISONIXINE see XLS300
ISONIZIDE see ILD000
ISONONYL ACETATE see TLT500
ISONONYL ALCOHOL see ILJ000
ISONORENE see DMV600
ISONYL see ILK000
ISOOCTANE (DOT) see TLY500
ISOOCTANOL see ILL000
ISOOCTYL ALCOHOL see ILL000
ISOOCTYL ALCOHOL (2,4-DICHLOROPHENOXY)ACETATE see ILO000
2-ISOOCTYL AMINE see ILM000
ISOOCTYL-2,4-DICHLOROPHENOXYACETATE see ILO000
ISOOCTYL ESTER, MERCAPTOACETATE ACID see ILR000
ISOOCTYL MERCAPTOACETATE see ILR000
ISOOCTYL PHTHALATE see ILR100
p-ISOOCTYLPOLYOXYETHYLENEPHENOL FORMALDEHYDE POLYMER
    see TDN750
ISOOCTYL THIOGLYCOLATE see ILR000
ISOPAL see IQW000

ISOPARATHION see DJT000
ISOPELLETIERINE see PAO500
ISOPENTALDEHYDE see MHX500
ISOPENTANE see EIK000
ISOPENTANOIC ACID (DOT) see ISU000
ISOPENTANOIC ACID, PHENYLMETHYL ESTER see ISW000
ISOPENTANOL see IHP000
ISOPENT-2-ENYL ACETATE see DOQ350
4-(2-ISOPENTENYL)-1,2-DIPHENYL-3,5-PYRAZOLIDINEDIONE see PEW000
5-(1-ISOPENTENYL)-5-ISOPROPYLBARBITURIC ACID see ILT000
5-(2-ISOPENTENYL)-5-ISOPROPYL-1-METHYLBARBITURIC ACID see ILU000
8-ISOPENTENYLOXYPSORALENE see IHR300
ISOPENTYL ACETATE see IHO850
ISOPENTYL ALCHOL, ESTER with N-(2-(DIETHYLAMINO)ETHYL)-2-
    PHENYLGLYCINE see NOC000
ISOPENTYL ALCOHOL see IHP000
ISOPENTYL ALCOHOL ACETATE see IHO850
ISOPENTYL ALCOHOL, FORMATE see IHS000
ISOPENTYL ALCOHOL NITRITE see IMB000
ISOPENTYL ALCOHOL, PROPIONATE see ILW000
ISOPENTYL BENZOATE see IHP100
ISOPENTYL BROMIDE see BNP250
ISOPENTYL FORMATE see IHS000
ISOPENTYL-2-HYDROXYPHENYL METHANOATE see IME000
ISOPENTYL ISOVALERATE see ITB000
ISOPENTYL METHYL KETONE see MKW450
ISOPENTYL NITRITE see IMB000
3-((ISOPENTYL)NITROSOAMINO)-2-BUTANONE see MHW350
ISOPENTYL OCTANOATE see IHP500
2-(ISOPENTYLOXY)-4-METHYL-1-PENTANOL see IJA000
ISOPENTYLPHENYLACETATE see IHV000
ISOPENTYL-2-PHENYLGLYCINATE HYDROCHLORIDE see PCV750
ISOPENTYL PROPIONATE see ILW000
ISOPENTYL SALICYLATE see IME000
1-ISOPENTYL-THEOBROMINE see IHX200
ISOPERTHIOCYANIC ACID see IBL000
ISOPESTOX see PHF750
ISOPHANE INSULIN see IDF325
ISOPHANE INSULIN INJECTION see IDF325
ISOPHANE INSULIN SUSPENSION see IDF325
ISOPHEN see CBW000, DBA800, MDT600
ISOPHEN (pesticide) see CBW000
ISOPHENERGAN see DQA400
ISOPHENICOL see CDP250
ISOPHENPHOS see IMF300
ISOPHORONE see IMF400
ISOPHORONE DIAMINE DIISOCYANATE see IMG000
ISOPHORONE DIISOCYANATE see IMG000
ISOPHOSPHAMIDE see IMH000
ISOPHRIN see NCL500
ISOPHRINE see SPC500
ISOPHRIN HYDROCHLORIDE see SPC500
ISOPHTALDEHYDES (FRENCH) see IMI000
ISOPHTHALALDEHYDE see IMI000
ISOPHTHALIC ACID see IMJ000
ISOPHTHALIC ACID CHLORIDE see IMO000
ISOPHTHALIC ACID, DIALLYL ESTER see IMK000
ISOPHTHALIC ACID DICHLORIDE see IMO000
ISOPHTHALIC ACID, DIMETHYL ESTER see IML000
ISOPHTHALODINITRILE see PHX550
ISOPHTHALONITRILE see PHX550
3,3'-(ISOPHTHALOYLBIS(IMINO-p-PHENYLENECARBONYLIMINO))BIS
    (1-ETHYLPYRIDINIUM, DI-p-TOLUENESULFONATE see IMM000
ISOPHTHALOYL CHLORIDE see IMO000
ISOPHTHALOYL DICHLORIDE see IMO000
ISOPHTHALYL DICHLORIDE see IMO000
ISOPIRINA see INM000
ISOPLAIT see VGA300
ISO PPC see PMS825
ISOPRAL see IMQ000
ISOPREGNENONE see DYF759
ISOPRENALINE see DMV600
ISOPRENALINE CHLORIDE see IMR000
ISOPRENALINE HYDROCHLORIDE see IMR000
(±)-ISOPRENALINE HYDROCHLORIDE see IQS500
dl-ISOPRENALINE HYDROCHLORIDE see IQS500
racemic ISOPRENALINE HYDROCHLORIDE see IQS500
(±)-ISOPRENALINE SULFATE see IRU000
dl-ISOPRENALINE SULFATE see IRU000
ISOPRENE see IMS000, MNH750
ISOPRENE, INHIBITED (DOT) see IMS000
ISOPRENYL CHALCONE see IMS300
ISOPROCARB see MIA250
ISOPROCIL (FRENCH) see BNM000

ISOPROMEDOL see IMT000
ISOPROMETHAZINE see DQA400
ISOPROPAMIDE see DAB875
ISOPROPAMIDE IODIDE see DAB875
ISOPROPANETHIOL see IMU000
ISOPROPANOL (DOT) see INJ000
ISOPROPANOLAMINE see AMA500
ISOPROPANOLAMINES see IMV000
ISOPROPANOLAMINES, mixed see IMV000
ISOPROPENE CYANIDE see MGA750
ISOPROPENIL-BENZOLO (ITALIAN) see MPK250
ISOPROPENYL ACETATE see MQK750
ISOPROPENYL-BENZEEN (DUTCH) see MPK250
ISOPROPENYLBENZENE see MPK250
ISOPROPENYL-BENZOL (GERMAN) see MPK250
ISOPROPENYL CARBINOL see IMW000
4-ISOPROPENYL-1-CYCLOHEXENE-1-CARBOXALDEHYDE see DKX100
4-ISOPROPENYL-1-CYCLOHEXENE-1-CARBOXALDEHYDE, ANTI-OXIME
    see PCJ000
4-ISOPROPENYL-CYCLOHEX-1-ENE-1-METHANOL see PCI550
2'-ISOPROPENYL-2-(2-METHOXYETHYLAMINO)PROPRIONANILIDE
    OXALATE see IMX000
(+)-4-ISOPROPENYL-1-METHYLCYCLOHEXENE see LFU000
ISOPROPENYLNITRILE see MGA750
ISOPROPHYL METHYLPHOSPHONOFLUORIDATE see IPX000
ISOPROPILAMINA (ITALIAN) see INK000
2-ISOPROPILAMINO-4-METILAMINO-6-METILTIO-1,3,5-TRIAZINA (ITALIAN)
    see INR000
ISOPROPILBENZENE (ITALIAN) see COE750
1-ISOPROPILBIGUANIDE CLORIDRATO (ITALIAN) see IOF100
(2-ISOPROPILCROTONIL)UREA (ITALIAN) see PMS825
ISOPROPILE (ACETATO di) (ITALIAN) see INE100
ISOPROPIL-N-FENIL-CARBAMMATO (ITALIAN) see CBM000
ISOPROPILIDRAZIDE dell'ac. p-CLORO-FENOSSIACETICO (ITALIAN)
    see CJN250
(1-ISOPROPIL-3-METIL-1H-PIRAZOL-5-IL)-N,N-DIMETIL-CARBAMMATO
    (ITALIAN) see DSK200
o-ISOPROPOXYANILINE see INA200
(3)-O-2-ISOPROPOXY-CARBONYL-1-METHYLVINYL-O-METHYL
    ETHYLPHOSPHORAMIDOTHIOATE see MKA000
11-ISOPROPOXY-15,16-DIHYDRO-17-CYCLOPENTA(a)PHENANTHREN-17-
    ONE see INA400
4-ISOPROPOXYDIPHENYLAMINE see HKF000
p-ISOPROPOXYDIPHENYLAMINE see HKF000
2-ISOPROPOXYETHANOL see INA500
(2-ISOPROPOXYETHYL)CARBAMATE see IND000
N-(ISOPROPOXY-3-HYDROXYMERCURIPROPYL)BARBITAL see INE000
ISOPROPOXYMETHYLPHORYL, FLUORIDE see IPX000
2-ISOPROPOXYNITROBENZENE see NIR550
o-ISOPROPOXYNITROBENZENE see NIR550
N-(4-ISOPROPOXYPHENYL)ANILINE see HKF000
o-ISOPROPOXYPHENYL METHYLCARBAMATE see PMY300
2-ISOPROPOXYPHENYL-N-METHYLCARBAMATE see PMY300
o-ISOPROPOXYPHENYL-N-METHYLCARBAMATE see PMY300
2-ISOPROPOXYPHENYL N-METHYLCARBAMATE, nitrosated see PMY310
o-ISOPROPOXYPHENYL METHYLNITROSOCARBAMATE see PMY310
o-ISOPROPOXYPHENYL N-METHYL-N-NITROSOCARBAMATE see PMY310
2-ISOPROPOXYPROPANE see IOZ750
ISOPROPYDRIN see DMV600
ISOPROPYLACETAAT (DUTCH) see INE100
ISOPROPYLACETAT (GERMAN) see INE100
ISOPROPYL ACETATE see INE100
ISOPROPYL (ACETATE d') (FRENCH) see INE100
ISOPROPYLACETIC ACID see ISU000
ISOPROPYL ACETIC ACID, BENZYL ESTER see ISW000
ISOPROPYLACETONE see HFG500
N-ISOPROPYL-N-(ACETOXYMETHYL)NITROSAMINE see ING300
ISOPROPYL-N-ACETYL-N-PHENYLCARBAMATE see ING400
ISOPROPYL ACID PHOSPHATE solid see PHE500
2-ISOPROPYL ACRYLALDEHYDE OXIME see ING509
ISOPROPYL ACRYLAMIDE see INH000
N-ISOPROPYLACRYLAMIDE see INH000
ISOPROPYL ADIPATE see DNL800
ISOPROPYLADRENALINE see DMV600
ISOPROPYL ALCOHOL see INJ000
ISO-PROPYLALKOHOL (GERMAN) see INJ000
ISOPROPYLALLYLAZETYLKARBAMID (GERMAN) see IQX000
ISOPROPYLALLYLBARBITURIC ACID see AFT000
ISOPROPYLAMINE see INK000
ISOPROPYLAMINE HYDROCHLORIDE see INL000
4-(2-ISOPROPYLAMINE-1-HYDROXYETHYL)METHANESULFOANILIDE
    HYDROCHLORIDE see CCK250
1-ISOPROPYLAMINE-3-(1-NAPHTHYLOXY)-2-PROPANOL see ICB000
ISOPROPYLAMINOANTIPYRINE see INM000

4-(ISOPROPYLAMINO)ANTIPYRINE see INM000
ISOPROPYLAMINO-4-AZIDO-6-METHYLTHIO-1,3,5-TRIAZIN (GERMAN)
  see ASG250
1-ISOPROPYLAMINO-3-(2-CYCLOHEXYLPHENOXY)-2-PROPANOL HYDRO-
  CHLORIDE see ERE100
4-ISOPROPYLAMINO-2,3-DIMETHYL-1-PHENYL-3-PYRAZOLIN-5-ONE
  see INM000
4-ISOPROPYLAMINODIPHENYLAMINE see PFL000
ISOPROPYLAMINOETHANOL see INN400
2-ISOPROPYLAMINOETHANOL see INN400
N-ISOPROPYLAMINOETHANOL see INN400
ISOPROPYLAMINO ETHANOL HYDROCHLORIDE see INN500
2-(ISOPROPYLAMINO)ETHANOL HYDROCHLORIDE see INN500
ISOPROPYLAMINO-O-ETHYL-(4-METHYLMERCAPTO-3-
  METHYLPHENYL)PHOSPHATE see FAK000
4-(2-ISOPROPYLAMINO-1-HYDROXYAETHYL)METHANESULFONALID
  HYDROCHLORID (GERMAN) see CCK250
1-(ISOPROPYLAMINO)-2-HYDROXY-3-(o-(ALLYLOXY)PHENOXY)PROPANE
  see CNR500
ISOPROPYLAMINOHYDROXYETHYLMETHANESULFONALIDE HYDRO-
  CHLORIDE see CCK250
1-(1-(2-(3-ISOPROPYLAMINO-2-HYDROXYPROPOXY)-3,6-
  DICHLOROPHENYL)VINYL)-1H-IMIDAZOLE HCl see DFM875
4-(3-ISOPROPYLAMINO)-2-HYDROXYPROPOXY)-3-METHOXY-BENZOIC
  ACID METHYL ESTER see VFP200
7-(ISOPROPYLAMINOISOPROPYL)THEOPHYLLINE HYDROCHLORIDE
  see INP100
(±)-1-(ISOPROPYLAMINO)-3-(p-(2-METHOXYETHYL)PHENOXY)-2-PRO-
  PANOL HEMI-I-TARTRATE see MQR150
1-ISOPROPYLAMINO-3-(p-(2-METHOXYETHYL)PHENOXY)-2-PROPANOL
  TARTRATE see SBV500
1-(ISOPROPYLAMINO)-3-(p-(2-METHOXYETHYL)PHENOXY)-2-PROPANOL
  TARTRATE (2:1) see MQR150
2-ISOPROPYLAMINO-4-(3-METHOXYPROPYLAMINO)-6-METHYLTHIOs-
  TRIAZINE see INQ000
2-ISOPROPYLAMINO-4-(3-METHOXYPROPYLAMINO)-6-METHYLTHIO-1,3,5-
  TRIAZIN (GERMAN) see INQ000
4-ISOPROPYLAMINO-6-(3'-METHOXYPROPYLAMINO)-2-METHYTHIO-1,3,5-
  TRIAZIN (GERMAN) see INQ000
2-ISOPROPYLAMINO-4-METHYLAMINO-6-METHYLMERCAPTO-s-TRIAZINE
  see INR000
2-ISOPROPYLAMINO-4-METHYLAMINO-6-METHYLTHIO-1,3,5-TRIAZINE
  see INR000
α-(ISOPROPYLAMINOMETHYL)-3,4-DIHYDROXYBENZYL ALCOHOL
  HYDROCHLORIDE see IMR000
ISOPROPYLAMINOMETHYL-3,4-DIHYDROXYPHENYL CARBINOL
  see DMV600
7-(2-ISOPROPYLAMINO-2-METHYLETHYL)THEOPHYLLINE HYDROCHLO-
  RIDE see INP100
α-((ISOPROPYLAMINO)METHYL)-2-NAPHTHALENEMETHANOL see INS000
α-((ISOPROPYLAMINO)METHYL)NAPHTHALENEMETHANOL, HYDROCHLO-
  RIDE see INT000
α-(ISOPROPYLAMINOMETHYL)-4-NITROBENZYL ALCOHOL see INU000
2-((ISOPROPYLAMINO)METHYL)-7-NITRO-1,2,3,4-TETRAHYDRO-6-
  QUINOLINEMETHANOL see OLT000
α-(ISOPROPYLAMINOMETHYL)PROTOCATECHUYL ALCOHOL see DMV600
1-ISOPROPYLAMINO-3-(1-NAPHTHOXY)-PROPAN-2-OL HYDROCHLORIDE
  see ICC000
1-(ISOPROPYLAMINO)-3-(α-NAPHTHOXY)-2-PROPANOL HYDROCHLORIDE
  see ICC000
2-ISOPROPYLAMINO-1-(NAPHTH-2-YL)ETHANOL see INS000
2-ISOPROPYLAMINO-1-(2-NAPHTHYL)ETHANOL see INS000
2-ISOPROPYLAMINO-1-(2-NAPHTHYL)ETHANOL HYDROCHLORIDE
  see INT000
1-ISOPROPYLAMINO-3-(1-NAPHTHYLOXY)-2-PROPANOL see ICB000
1-(ISOPROPYLAMINO)-3-(1-NAPHTHYLOXY)PROPAN-2-OL HYDROCHLO-
  RIDE see ICC000
(±)-1-(ISOPROPYLAMINO)-3-(1-NAPHTHYLOXY)-2-PROPANOL HYDRO-
  CHLORIDE see PNB800
1-(ISOPROPYLAMINO)-3-(1-NAPTHYLOXY)-2-PROPANOL HYDROCHLORIDE
  see ICC000
ISOPROPYLAMINOPHENAZON see INM000
ISOPROPYLAMINOPHENAZONE see INM000
(±)-1-(ISOPROPYLAMINO)-3-(o-PHENOXYPHENOXY)-2-PROPANOL HYDRO-
  CHLORIDE see INU200
4-ISOPROPYLAMINO-1-PHENYL-2,3-DIMETHYL-3-PYRAZOLIN-5-ONE
  see INM000
9-((3-(ISOPROPYLAMINO)PROPYL)AMINO)-1-NITROACRIDINE DIHYDRO-
  CHLORIDE see INW000
7-(2-(ISOPROPYLAMINO)PROPYL)-THEOPHYLLINE MONOHYDROCHLOR-
  IDE see INP100
2-ISOPROPYL ANILINE see INX000
N-ISOPROPYLANILINE see INX000
o-ISOPROPYLANILINE see INX000

ISOPROPYLANTIPYRIN see INY000
ISOPROPYLANTIPYRINE see INY000
4-ISOPROPYLANTIPYRINE see INY000
ISOPROPYLANTIPYRINE with ETHENZAMIDE and CAFFEINE MONOHY-
  DRATE see INY100
ISOPROPYLARTERENOL see DMV600
ISOPROPYLARTERENOL HYDROCHLORIDE see IMR000
4-ISOPROPYLBENZALDEHYDE see COE500
p-ISOPROPYLBENZALDEHYDE see COE500
8-ISOPROPYLBENZ(a)ANTHRACENE see INZ000
9-ISOPROPYLBENZ(a)ANTHRACENE see IOA000
5-ISOPROPYL-1:2-BENZANTHRACENE see INZ000
6-ISOPROPYL-1:2-BENZANTHRACENE see IOA000
ISOPROPYLBENZEEN (DUTCH) see COE750
ISOPROPYL BENZENE see COE750
p-ISOPROPYLBENZENECARBOXALDEHYDE see COE500
ISOPROPYLBENZENE HYDROPEROXIDE see IOB000
ISOPROPYLBENZENE PEROXIDE see DGR600
ISOPROPYL BENZOATE see IOD000
p-ISOPROPYLBENZOIC ACID-2-(DIETHYLAMINO)ETHYL ESTER HYDRO-
  CHLORIDE see IOI600
ISOPROPYLBENZOL see COE750
ISOPROPYL-BENZOL (GERMAN) see COE750
p-ISOPROPYLBENZONITRILE see IOD050
3-ISOPROPYL-2,1,3-BENZOTHIADIAZINON-(4)-2,2-DIOXID (GERMAN)
  see MJY500
3-ISOPROPYL-1H-2,1,3-BENZOTHIADIAZIN-4(3H)-ONE-2,2-DIOXIDE
  see MJY500
N-ISOPROPYL BENZOTHIAZOLE SULFONAMIDE see IOE000
2-ISOPROPYL-3:4-BENZPHENANTHRENE see IOF000
p-ISOPROPYLBENZYL ALCOHOL see CQI250
1-ISOPROPYLBIGUANIDE HYDROCHLORIDE see IOF100
N-ISOPROPYLBIGUANIDE HYDROCHLORIDE see IOF100
ISOPROPYLBIPHENYL see IOF200
ISOPROPYL-BIS(β-CHLOROETHYL)AMINE see IOF300
ISOPROPYL BIS(β-CHLOROETHYL)AMINE HYDROCHLORIDE see IPG000
ISOPROPYL BORATE see IOI000
5-ISOPROPYL-5-BROMALLYLBARBITURIC ACID see QCS000
5-ISOPROPYL-5-(2-BROMOALLYL)BARBITUATE see QCS000
3-ISOPROPYL-5-BROMO-6-METHYLURACIL see BNM000
ISOPROPYLCAINE HYDROCHLORIDE see IOI600
ISOPROPYL CARBAMATE see IOJ000
ISOPROPYLCARBAMIC ACID, ESTER with 2-(HYDROXYMETHYL)-2-
  METHYLPENTYL CARBAMATE see IPU000
ISOPROPYLCARBAMIC ACID, ETHYL ESTER see IPA000
2-(p-ISOPROPYL CARBAMOYL BENZYL)-1-METHYLHYDRAZINE see PME250
1-(p-ISOPROPYLCARBAMOYLBENZYL)-2-METHYLHYDRAZINE HYDRO-
  CHLORIDE see PME500
2-(p-(ISOPROPYLCARBAMOYL)BENZYL)-1-METHYLHYDRAZINE HYDRO-
  CHLORIDE see PME500
1-ISOPROPYL CARBAMOYL-3-(3,5-DICHLOROPHENYL)-HYDANTOIN
  see GIA000
ISOPROPYL CARBANILATE see CBM000
ISOPROPYL CARBANILIC ACID ESTER see CBM000
ISOPROPYLCARBINOL see IIL000
4-ISOPROPYLCATECHOL see IOK000
ISOPROPYL CELLOSOLVE see INA500
ISOPROPYL CHLORIDE see CKQ000
N-ISOPROPYL-2-CHLOROACETANILIDE see CHS500
N-ISOPROPYL-α-CHLOROACETANILIDE see CHS500
ISOPROPYL-3-CHLOROCARBANILATE see CKC000
ISOPROPYL-m-CHLOROCARBANILATE see CKC000
ISOPROPYL CHLOROCARBONATE see IOL000
ISOPROPYL CHLOROFORMATE see IOL000
ISOPROPYL CHLOROMETHANOATE see IOL000
ISOPROPYL-3-CHLOROPHENYLCARBAMATE see CKC000
ISOPROPYL-N-(3-CHLOROPHENYL)CARBAMATE see CKC000
o-ISOPROPYL-N-(3-CHLOROPHENYL)CARBAMATE see CKC000
ISOPROPYL-N-(3-CHLORPHENYL)-CARBAMAT (GERMAN) see CKC000
3-ISOPROPYLCHOLANTHRENE see ION000
20-ISOPROPYLCHOLANTHRENE see ION000
ISOPROPYL CINNAMATE see IOO000
ISOPROPYL CRESOL see TFX810
ISOPROPYL-o-CRESOL see CCM000
6-ISOPROPYL-m-CRESOL see TFX810
2-ISOPROPYLCROTONYLUREA see PMS825
ISOPROPYL CYANIDE see IJX000
4-ISOPROPYLCYCLOHEXANOL see IOO300
p-ISOPROPYLCYCLOHEXANOL see IOO300
ISOPROPYL-2,4-D ESTER see IOY000
14-ISOPROPYLDIBENZ(a,j)ACRIDINE see IOR000
10-ISOPROPYL-3,4,5,6-DIBENZACRIDINE (FRENCH) see IOR000
ISOPROPYL-4,4'-DIBROMOBENZILATE see IOS000
ISOPROPYL-4,4'-DICHLOROBENZILATE see PNH750

ISOPROPYL DIETHYLDITHIOPHOSPHORYLACETAMIDE see IOT000
N-ISOPROPYL-β-DIHYDROXYPHENYL-β-HYDROXYETHYLAMINE
  see DMV600
4'-ISOPROPYL-4-DIMETHYLAMINOAZOBENZENE see IOT875
α-ISOPROPYL-α-(2-DIMETHYLAMINOETHYL)-1-NAPHTHACETAMIDE
  see IOU000
α-ISOPROPYL-α-(2-(DIMETHYLAMINO)ETHYL)-1-NAPHTHALENEACETA-
  MIDE see IOU000
α-ISOPROPYL-α-(2-DIMETHYLAMINOETHYL)-1-NAPHTHYLACETAMIDE
  see IOU000
m-ISOPROPYL-p-DIMETHYL-AMINO-PHENOL-DIMETHYL-URETHANE
  METHIODIDE see IOU200
α-(ISOPROPYL)-α-(β-DIMETHYLAMINOPROPYL)PHENYLACETONITRIL
  CITRATE see PCK639
α-(ISOPROPYL)-α-(β-DIMETHYLAMINOPROPYL)PHENYLACETONITRIL
  CYCLAMATE see PCK669
α-(ISOPROPYL)-α-(β-DIMETHYLAMINOPROPYL)PHENYLACETONITRILE
  CYCLAMATE see IHO200
(N-(2-ISOPROPYL-4-DIMETHYLCARBAMOYLOXY)PHENYL)
  TRIMETHYLAMMONIUM IODIDE see HJZ000
ISOPROPYL DIMETHYL CARBINOL see AOK750
ISOPROPYL-O,O-DIMETHYLDITHIOPHOSPHORYL-1-PHENYLACETATE
  see IOV000
4-ISOPROPYL-2,3-DIMETHYL-1-PHENYL-3-PYRAZOLIN-5-ONE see INY000
N'-ISOPROPYL-N,N-DIMETHYL-1,3-PROPANE-DIAMINE see IOW000
3-ISOPROPYL-5,7-DIMORPHOLINOMETHYLTROPOLONE DIHYDROCHLO-
  RIDE see IOW500
ISOPROPYL-2,4-DINITRO-6-SEC-BUTYLPHENYL CARBONATE see CBW000
4-ISOPROPYL-2,6-DINITROPHENOL see IOX000
ISOPROPYLDIPHENYL see IOF200
ISOPROPYLENE BROMIDE see BOA250
ISOPROPYL ESTER MERCAPTOPHENYLACETIC ACID S-ESTER with O-O-
  DIMETHYL PHOSPHORODITHIOATE see IOV000
ISOPROPYL ETHANE FLUOROPHOSPHONATE see ELX100
N-ISOPROPYLETHANOLAMINE see INN400
ISOPROPYL ETHER see IOZ750
ISOPROPYL 3-(((ETHYLAMINO)METHOXYPHOSPHINOTHIOYL)OXY)
  CROTONIC ACID see IOZ800
ISOPROPYL ETHYL URETHAN see IPA000
ISOPROPYL-N-FENYL-CARBAMAAT (DUTCH) see CBM000
N-ISOPROPYL-N'-FENYL-p-FENYLENDIAMIN (CZECH) see PFL000
ISOPROPYL FLUOPHOSPHATE see IRF000
ISOPROPYL FORMATE see IPC000
ISOPROPYLFORMIC ACID see IJU000
ISOPROPYL GLYCIDYL ETHER see IPD000
ISOPROPYL GLYCOL see INA500
ISOPROPYL HEXADECANOATE see IQW000
ISOPROPYL-n-HEXADECANOATE see IQW000
N'-ISOPROPYL HYDRAZINOMETHANESULFONIC ACID, SODIUM SALT
  see HGW000
ISOPROPYL HYDROPEROXIDE see IPI000
ISOPROPYL HYPOCHLORITE see IPI350
ISOPROPYLIDENEACETONE see MDJ750
4,4'-ISOPROPYLIDENE-BIS(2-tert-BUTYLPHENOL) see IPJ000
4,4'-ISOPROPYLIDENEBIS(2,6-DICHLOROPHENOL) see TBO750
4,4'-ISOPROPYLIDENEBISPHENOL see BLD500
p,p'-ISOPROPYLIDENEBISPHENOL see BLD500
4,4'-ISOPROPYLIDENEDI-o-CRESOL see IPK000
4-ISOPROPYLIDENE-3,3-DIMETHYL-2-OXETANONE see IPL000
2,3-ISOPROPYLIDENEDIOXYPHENYL METHYLCARBAMATE see DQM600
p,p'-ISOPROPYLIDENEDIPHENOL see BLD500
4,4'-ISOPROPYLIDENEDIPHENOL DIGLYCIDYL ETHER see BLD750
4,4'-ISOPROPYLIDENEDIPHENOL DIMER with 1-CHLORO-2,3-EPOXYPRO-
  PANE see IPM000
4,4'-ISOPROPYLIDENEDIPHENOL, MONOMER with 1-CHLORO-2,3-
  EPOXYPROPANE see IPN000
4,4'-ISOPROPYLIDENE DIPHENOL POLYMER with 1-CHLORO-2,3-
  EPOXYPROPANE see ECL000
4,4'-ISOPROPYLIDENEDIPHENOL, POLYMER with 1-CHLORO-2,3-
  EPOXYPROPANE see IPO000
4,4'-ISOPROPYLIDENEDIPHENOL, TETRAMER with 1-CHLORO-2,3-
  EPOXYPROPANE see IPP000
ISOPROPYLIDENE GLYCEROL see DVR600
1,2-o-ISOPROPYLIDENE GLYCEROL see DVR600
2-(2-ISOPROPYLIDENEHYDRAZONO)-4-(5-NITRO-2-FURYL)-THIAZOLE
  see NGN000
1-ISOPROPYLIDENE-4-METHYL-2-CYCLOHEXANONE see MCF500
13-ISOPROPYLIDENEPODOCARP-8(14)-EN-15-OIC ACID see NBU800
ISOPROPYLIDINE AZASTREPTONIGRIN see IPR000
ISOPROPYL IODIDE see IPS000
ISOPROPYL ISOBUTYL ARSINIC ACID see IPS100
ISOPROPYL ISOCYANIDE DICHLORIDE see IPS400
N-ISOPROPYL ISONICOTINHYDRAZIDE see ILE000
ISOPROPYL LANOLATE see IPS500

ISOPROPYL MANDELATE see IPT000
ISOPROPYL MEPROBAMATE see IPU000
ISOPROPYL MERCAPTAN (DOT) see IMU000
N-ISOPROPYL-2-MERCAPTOACETAMIDE-S-ESTER with O,O-DIETHYL PHOS-
  PHORODITHIOATE see IOT000
ISOPROPYL MERCAPTOPHENYLACETATE-O,O-DIMETHYL PHOS-
  PHORODITHIOATE see IOV000
ISOPROPYLMERCURY HYDROXIDE see IPW000
ISOPROPYL MESYLATE see IPY000
ISOPROPYL METHANEFLUOROPHOSPHONATE see IPX000
N-ISOPROPYL-β-(4-METHANESULFONAMIDOPHENYL)ETHANOLAMINE
  HYDROCHLORIDE see CCK250
ISOPROPYLMETHANESULFONATE see IPY000
ISOPROPYL METHANE SULPHONATE see IPY000
2'-ISOPROPYL-2-(2-METHOXYETHYLAMINO)-PROPIONANILIDE HYDRO-
  CHLORIDE see IPZ000
3-ISOPROPYL-2-(p-METHOXYPHENYL)-3H-NAPHTH(1,2-d)IMIDAZOLE
  see THG700
ISOPROPYL(2E,4E)-11-METHOXY-3,7,11-TRIMETHYL-2,4-DODECADIENOATE
  see KAJ000
8-((ISOPROPYLMETHYLAMINO)METHYL)QUINOLINE see IPZ500
N-ISOPROPYL-α-(2-METHYLAZO)-p-TOLUAMIDE see IQB000
4-ISOPROPYL-1-METHYLBENZENE see CQI000
ISOPROPYL 2-METHYL-2-BUTENOATE see IRN100
2-ISOPROPYL-6-(1-METHYLBUTYL)PHENOL see IQD000
ISOPROPYL α-METHYL CROTONATE see IRN100
1-ISOPROPYL-4-METHYLCYCLOHANE HYDROPEROXIDE see IQE000
4-ISOPROPYL-1-METHYL-1,5-CYCLOHEXADIENE see MCC000
5-ISOPROPYL-2-METHYL-1,3-CYCLOHEXADIENE see MCC000
5-ISOPROPYL-2-METHYL-2,5-CYCLOHEXADIENE-1,4-DIONE see IQF000
2-ISOPROPYL-5-METHYL-CYCLOHEXANOL see MCF750
4-ISOPROPYL-1-METHYLCYCLOHEXAN-3-OL see MCG000
l-2-ISOPROPYL-5-METHYL-CYCLOHEXAN-1-OL ACETATE see MCG750
2-ISOPROPYL-5-METHYL-CYCLOHEXANONE see IKY000
2-ISOPROPYL-5-METHYL-CYCLOHEXAN-1-ONE, racemic see MCE250
1-ISOPROPYL-2-METHYL ETHYLENE see MNK500
1-ISOPROPYL-7-METHYL-4-(p-FLUOROPHENYL)-2(1H)-QUINAZOLINONE
  see THH350
ISOPROPYL METHYLFLUOROPHOSPHATE see IPX000
N-ISOPROPYL-6-METHYL-2-HEPTYLAMINE HYDROCHLORIDE see IGG300
2-ISOPROPYL-5-METHYL-2-HEXEN-1-AL see IKM100
2-ISOPROPYL-5-METHYL-2-HEXEN-1-OL see IQH000
2-ISOPROPYL-5-METHYL-2-HEXEN-1-YL ACETATE see AAV250
N-ISOPROPYL-p-(2-METHYLHYDRAZINOMETHYL)
  BENZAMIDEHYDROCHLORIDE see PME500
N-ISOPROPYL-α-(2-METHYLHYDRAZINO)-p-TOLUAMIDE see PME250
N-ISOPROPYL-α-(2-METHYLHYDRAZINO)-p-TOLUAMIDE HYDROCHLO-
  RIDE see PME500
p-ISOPROPYL-α-METHYLHYDROCINNAMIC ALDEHYDE see COU500
2-ISOPROPYL-4-METHYL-6-HYDROXYPYRIMIDINE see IQL000
ISOPROPYL METHYL KETONE see MLA750
2-ISOPROPYL-1-METHYL-5-NITROIMIDAZOLE see IGH000
2-ISOPROPYL-4-METHYLPHENOL see IQJ000
2-ISOPROPYL-5-METHYLPHENOL see TFX810
5-ISOPROPYL-2-METHYLPHENOL see CCM000
p-ISOPROPYL-α-METHYLPHENYLPROPYL ALDEHYDE see COU500
ISOPROPYL-7-METHYL-4-PHENYL-2(1H)-QUINAZOLINONE see POB300
ISOPROPYL METHYLPHOSPHONOFLUORIDATE see IPX000
O-ISOPROPYL METHYLPHOSPHONOFLUORIDATE see IPX000
ISOPROPYL-METHYL-PHOSPHORYL FLUORIDE see IPX000
ISOPROPYL-2-(1-METHYL-N-PROPYL)-4,6-DINITROPHENYL CARBONATE
  see CBW000
N-ISOPROPYL-2-METHYL-2-PROPYL-1,3-PROPANEDIOL DICARBAMATE
  see IPU000
(1-ISOPROPYL-3-METHYL-1H-PYRAZOL-5-YL)-N,N-DIMETHYLCARBAMAAT
  (DUTCH) see DSK200
(1-ISOPROPYL-3-METHYL-1H-PYRAZOL-5-YL)-N,N-DIMETHYL-CARBAMAT
  (GERMAN) see DSK200
ISOPROPYLMETHYLPYRAZOLYL DIMETHYLCARBAMATE see DSK200
1-ISOPROPYL-3-METHYL-5-PYRAZOLYL DIMETHYLCARBAMATE
  see DSK200
1-ISOPROPYL-3-METHYLPYRAZOLYL-(5)-DIMETHYLCARBAMATE
  see DSK200
2-ISOPROPYL-4-METHYL-6-PYRIMIDOL see IQL000
O-2-ISOPROPYL-4-METHYLPYRIMIDYL-O,O-DIETHYL PHOSPHOROTHIO-
  ATE see DCM750
ISOPROPYLMETHYLPYRIMIDYL DIETHYL THIOPHOSPHATE see DCM750
ISOPROPYLMORPHOLINE see IQM000
ISOPROPYL MYRISTATE see IQN000
ISOPROPYL NITRATE see IQP000
ISOPROPYL NITRITE see IQQ000
N-ISOPROPYL-5-NITRO-2-FURANACRYLAMIDE see FPO000
N-ISOPROPYL-3-(5-NITRO-2-FURYL)-ACRYLAMIDE see FPO000
ISOPROPYL o-NITROPHENYL ETHER see NIR550

2-ISOPROPYL-3-NITROSOTHIAZOLIDINE see IQS000
1-ISOPROPYL-1-NITROSOUREA see NKO425
ISOPROPYL NORADRENALINE see DMV600
1-ISOPROPYLNORADRENALINE see DMV600
N-ISOPROPYLNORADRENALINE see DMV600
dl-ISOPROPYLNORADRENALINE HYDROCHLORIDE see IQS500
dl-N-ISOPROPYLNORADRENALINE HYDROCHLORIDE see IQS500
8-ISOPROPYLNORATROPINE METHOBROMIDE see IGG000
N-ISOPROPYLNORATROPINIUM BROMOMETHYLATE see IGG000
ISOPROPYLNOREPINEPHRINE-HYDROCHLORIDE see IMR000
dl-ISOPROPYLNOREPINEPHRINE HYDROCHLORIDE see IQS500
ISOPROPYLOCTADECYLAMINE see IQT000
N-ISOPROPYLOCTADECYLAMINE see IQT000
ISOPROPYL OILS see IQU000
ISOPROPYL PALMITATE see IQW000
(2-ISOPROPYL-4-PENTENOYL)UREA see IQX000
ISOPROPYL PERCARBONATE see DNR400
ISOPROPYL PERCARBONATE, stabilized (DOT) see DNR400
ISOPROPYL PERCARBONATE, unstabilized (DOT) see DNR400
ISOPROPYL PEROXYDICARBONATE see DNR400
ISOPROPYL PEROXYDICARBONATE, technically pure (DOT) see DNR400
ISOPROPYLPHENAZONE see INY000
4-ISOPROPYLPHENOL see IQZ000
p-ISOPROPYLPHENOL see IQZ000
ISOPROPYLPHENOL METHYLCARBAMATE see MIA250
m-ISOPROPYLPHENOL-N-METHYLCARBAMATE see COF250
4-ISOPROPYL PHENYLACETALDEHYDE see IRA000
ISOPROPYL-N-PHENYLCARBAMAT (GERMAN) see PMS825
ISOPROPYL-N-PHENYL-CARBAMAT (GERMAN) see CBM000
ISOPROPYL PHENYLCARBAMATE see CBM000
ISOPROPYL-N-PHENYLCARBAMATE see CBM000
o-ISOPROPYL-N-PHENYL CARBAMATE see CBM000
p-ISOPROPYLPHENYLETHYL ALCOHOL see IRC000
3-ISOPROPYLPHENYL METHYLCARBAMATE see COF250
m-ISOPROPYLPHENYL METHYLCARBAMATE see COF250
2-ISOPROPYL-PHENYL-N-METHYLCARBAMATE see MIA250
m-ISOPROPYLPHENYL-N-METHYLCARBAMATE see COF250
o-ISOPROPYLPHENYL-N-METHYLCARBAMATE see MIA250
2-ISOPROPYLPHENYL N-METHYLCARBAMATE, nitrosated
    see MMW250
N-ISOPROPYL-N'-PHENYL-p-PHENYLENEDIAMINE see PFL000
ISOPROPYL-N-PHENYLURETHAN (GERMAN) see CBM000
ISOPROPYL PHOSPHONATE see DNQ600
ISOPROPYLPHOSPHORAMIDOTHIOIC ACID-O-2,4-DICHLOROPHENYL-O-
    METHYL ESTER see DGD800
ISOPROPYL-PHOSPHORAMIDOTHIOIC ACID O-ETHYL O-(2-
    ISOPROPOXYCARBONYLPHENYL) ESTER see IMF300
N-ISOPROPYLPHOSPHORAMIDOTHIOIC ACID-O-(2,4,5-
    TRICHLOROPHENYL)ESTER see DYD200
ISOPROPYL PHOSPHORIC ACID see PHE500
ISOPROPYL PHOSPHOROFLUORIDATE see IRF000
N-ISOPROPYLPHTHALIMIDE see IRG000
13-ISOPROPYLPODOCARPA-7,13-DIEN-15-OIC ACID see AAC500
13-ISOPROPYLPODOCARPA-8,11,13-TRIEN-15-OIC ACID see DAK400
ISOPROPYL PYROPHOSPHATE see TDF750
N-ISOPROPYLPYRROLIDINONE see IRG100
1-ISOPROPYL-2-PYRROLIDINONE see IRG100
ISOPROPYL QUINOLINE see IRL000
6-ISOPROPYL QUINOLINE see IRL000
p-ISOPROPYL QUINOLINE see IRL000
ISOPROPYL-S see IOF300
ISOPROPYL SALICYLATE O-ESTER with O-ETHYLISOPROPYL-
    PHOSPHORAMIDOTHIOATE see IMF300
ISOPROPYL-S HYDROCHLORIDE see IPG000
ISOPROPYL SULFATE see DNO900
N-ISOPROPYL TEREPHTHALAMIC ACID see IRN000
S-2-ISOPROPYLTHIOETHYL-O,O-DIMETHYL PHOSPHORODITHIOATE
    see DSK800
ISOPROPYLTHIOL see IMU000
(ISOPROPYLTHIO)-METHANETHIOL-S-ESTER with O,O-DIETHYL PHOS-
    PHORODITHIOATE see DJN600
erythro-p-(ISOPROPYLTHIO)-α-(1-(OCTYLAMINO)ETHYL)BENZYL ALCOHOL
    see SOU600
erythro-1-(4-ISOPROPYLTHIOPHENYL)-2-n-OCTYLAMINOPROPANOL see
    SOU600
ISOPROPYL TIGLATE see IRN100
p-ISOPROPYLTOLUENE see CQI000
ISOPROPYL-2-(4-TRIAZOLYL)-5-BENZIMIDAZOLECARBAMATE see CBA100
4-ISOPROPYL-2-(α,α,α-TRIFLUORO-m-TOLYL)MORPHOLINE see IRP000
N-ISOPROPYL-4-(3,4,5-TRIMETHOXYCINNAMOYL)-1-PIPERAZINEACETA-
    MIDE MALEATE see IRQ000
4-ISOPROPYL-2,6,7-TRIOXA-1-ARSABICYCLO(2.2.2)OCTANE see IRQ100
β-ISOPROPYLTROPOLON see IRR000
4-ISOPROPYLTROPOLONE see IRR000

ISOPROPYLUREA and SODIUM NITRITE see SIS675
ISOPROPYL VINYL ETHER see IRS000
ISOPROPYLXANTHIC ACID, SODIUM SALT see SIA000
ISOPROTERENOL see DMV600
l-ISOPROTERENOL see DMV600
racemic ISOPROTERENOL HDYROCHLORIDE see IQS500
ISOPROTERENOL HYDROCHLORIDE see IMR000
(±)-ISOPROTERENOL HYDROCHLORIDE see IQS500
dl-ISOPROTERENOL HYDROCHLORIDE see IQS500
dl(±)-ISOPROTERENOL HYDROCHLORIDE see IQS500
(±)-ISOPROTERENOL SULFATE see IRU000
dl-ISOPROTERENOL SULFATE see IRU000
ISOPROTERNOL MONOHYDROCHLORIDE see IMR000
ISOPSORALIN see FQC000
ISOPTIN see IRV000, VHA450
ISOPTO CARBACHOL see CBH250
ISOPTO-CARPINE see PIF250
ISOPTO CETAMIDE see SNP500
ISOPTO FENICOL see CDP250
ISOPTO-HYDROCORTISONE see HHQ800
ISOPTO HYOSCINE see SBG000
ISOPULEGOL (FCC) see MCE750
ISO-PUREN see CCK125
ISOPURINE see POJ250
ISOPYRIN see INM000
ISOPYRINE see INM000, INY000
ISOQUINALDEHYDE THIOSEMICARBAZONE see IRV300
ISOQUINAZEPON see IRW000
ISOQUINOLINE see IRX000
2-(1-ISOQUINOLINYLMETHYLENE)-HYDRAZINECARBOTHIOAMIDE (9CI)
    see IRV300
ISORBID see CCK125
ISORDIL see CCK125
ISORDIL TEMBIDS see CCK125
ISOREN see CLY600
ISORENIN see DMV600
ISORETINENE a see VSK975
ISOSAFROEUGENOL see IRY000
ISOSAFROLE see IRZ000
ISOSAFROLE, OCTYL SULFOXIDE see ISA000
ISOSAFROLE-n-OCTYLSULFOXIDE see ISA000
ISOSCOPIL see HOT500
ISOSORBIDE see HID350
(+)-d-ISOSORBIDE see HID350
ISOSORBIDE DINITRATE see CCK125
ISOSORBIDE 5-MONONITRATE see ISC500
ISOSORBIDE 5-NITRATE see ISC500
ISOSTEARYL NEOPENTANOATE see ISC550
ISOSTENASE see CCK125
ISOSUMITHION see MKC250
ISO SYSTOX SULFOXIDE see ISD000
ISOTACTIC POLYPROPYLENE see PMP500
ISOTAZIN see DIR000
ISOTEBEZID see ILD000
ISOTENIC ACID see AKG250
ISOTENSE see RCA200
ISOTHAN see LBW000
ISOTHAZINE see DIR000
ISOTHIAZINE see DIR000
ISOTHIN see EPQ000
ISOTHIOATE see DSK800
ISOTHIOCYANATE d'ALLYLE (FRENCH) see AGJ250
ISOTHIOCYANATE de METHYLE (FRENCH) see ISE000
1-ISOTHIOCYANATE-NAPHTHALENE see ISN000
ISOTHIOCYANATOBENZENE see ISQ000
(2-ISOTHIOCYANATOETHYL)BENZENE see ISP000
ISOTHIOCYANATOMETHANE see ISE000
1-ISOTHIOCYANATONAPHTHALENE see ISN000
4-ISOTHIOCYANATO-4'-NITRODIPHENYLAMINE see AOA050
4-ISOTHIOCYANATO-N-(4-NITROPHENYL)-BENZENAMINE (9CI) see AOA050
3-ISOTHIOCYANATO-1-PROPENE see AGJ250
(ISOTHIOCYANATO)TRIMETHYLSTANNE see ISF000
ISOTHIOCYANATOTRIMETHYLTIN see ISF000
(ISOTHIOCYANATO)TRIPROPYLSTANNANE see TNB750
ISOTHIOCYANIC ACID-m-ACETAMIDOPHENYL ESTER see ISG000
ISOTHIOCYANIC ACID BENZYL ESTER see BEU250
ISOTHIOCYANIC ACID,-p-CHLOROBENZYL ESTER see CEQ750
ISOTHIOCYANIC ACID, p-CHLOROPHENYL ESTER see ISH000
ISOTHIOCYANIC ACID-p-CYANOPHENYL ESTER see ISI000
ISOTHIOCYANIC ACID, CYCLOHEXYL ESTER see ISJ000
ISOTHIOCYANIC ACID, ETHYLENE ESTER see ISK000
ISOTHIOCYANIC ACID-m-FLUOROPHENYL ESTER see ISL000
ISOTHIOCYANIC ACID-p-FLUOROPHENYL ESTER see ISM000
ISOTHIOCYANIC ACID, METHYL ESTER see ISE000

ISOTHIOCYANIC ACID, 3-(METHYLSULFINYL)PROPYL ESTER see MPN100
ISOTHIOCYANIC ACID-1-NAPHTHYL ESTER see ISN000
ISOTHIOCYANIC ACID-m-NITROPHENYL ESTER see ISO000
ISOTHIOCYANIC ACID, PHENETHYL ESTER see ISP000
ISOTHIOCYANIC ACID-p-PHENYLENE ESTER see PFA500
ISOTHIOCYANIC ACID, PHENYL ESTER see ISQ000
4-ISOTHIOCYANO-4'-NITRO DIPHENYLAMINE see AOA050
ISOTHIONIC ACID see HKI500
ISOTHIOUREA see ISR000
ISOTHIOURONIUM CHLORIDE, BENZYL see BEU500
ISOTHIPENDYL HYDROCHLORIDE see AEG625
ISOTHYMOL see CCM000
ISOTIAMIDA see EPQ000
ISOTIOCIANATO di METILE (ITALIAN) see ISE000
ISOTONIL see DRM000
ISOTOX see BBQ500
ISOTRATE see CCK125
ISOTRETINOIN see VSK955
ISOTRICYCLOQUINAZOLINE NITRATE see TBI750
ISOTRON 11 see TIP500
ISOTRON 12 see DFA600
ISOTRON 22 see CFX500
ISOUREA see USS000
ISOVAL see BNP750
ISOVALERAL see MHX500
ISOVALERALDEHYDE see MHX500
8-ISOVALERATE see FQS000
ISOVALERIANIC AICD see ISU000
ISOVALERIC ACID see ISU000
ISOVALERIC ACID, ALLYL ESTER see ISV000
ISOVALERIC ACID, BENZYL ESTER see ISW000
ISOVALERIC ACID, BUTYL ESTER see ISX000
ISOVALERIC ACID, (4,7-DIMETHYL-1,6-OCTADIEN-3-YL) ESTER
    see LGC000
(E)-ISOVALERIC ACID-3,7-DIMETHYL-2,6-OCTADIENYL ESTER see GDK000
ISOVALERIC ACID-8-ESTER with 3-FORMAMIDO-N-(7-HEXYL-8-HYDROXY-
    4,9-DIMETHYL-2,6-DIOXO-1,5-DIOXONAN-3-YL)SALICYLAMIDE
    see DUO350
ISOVALERIC ACID, ETHYL ESTER see ISY000
(Z)-ISOVALERIC ACID-3-HEXENYL see ISZ000
ISOVALERIC ACID, ISOBUTYL ESTER see ITA000
ISOVALERIC ACID, ISOPENTYL ESTER see ITB000
ISOVALERIC ACID, METHYL ESTER see ITC000
ISOVALERIC ALDEHYDE see MHX500
ISOVALERONE see DNI800
ISOVALERONITRILE see ITD000
ISOVANILLIC ACID see HJC000
ISOVANILLIN see FNM000
ISOVANILLINE see FNM000
ISOXAL see PCJ325
ISOXAMIN see SNN500
ISOXANTHINE see XCA000
ISOXATHION see DJV600
ISOXAZOL see HOM259
ISOXSUPRINE see VGF000
ISOXSUPRINE HYDROCHLORIDE see VGA300
ISOXSUPRIN HYDROCHLORIDE see VGA300
ISOZIDE see ILD000
ISPENORAL see PDT750
ISPHAMYCIN see CMB000
ISRAVIN see DBX400
ISTIN see DMH400
ISTONYL see DRM000
ISUPREL see DMV600, IMR000
ISUPREL HYDROCHLORIDE see IMR000
ISUPREN see DMV600
ISZILIN see IDF300
IT 40 see SMQ500
IT 931 see DLK200
IT 3456 see CDT000
DL-111-IT see EOL600
ITA-104 see PJB500
ITA 312 see LII400
ITACHIGARDEN see HLM000
ITALCHIN see CFU750
ITALIAN ARUM see ITD050
ITAMID see PJY500
ITAMIDONE see DOT000
ITAMO REAL (PUERTO RICO) see SDZ475
ITAMYCIN see HAL000
ITCH WEED see FAB100
ITF 182 see IAM100
ITF 611 see TKF699
ITF 1016 see ITD100

ITINEROL see HGC500
ITIOCIDE see EPQ000
ITOBARBITAL see AGI750
ITOPAZ see EEH600
ITROP see IGG000
ITURAN see NGE000
I.U. 7 see CAL075
IUDR see DAS000
5-IUDR see DAS000
IVALON see FMV000
IVAUGAN see CFY000
IVE see IJQ000
IVERMECTIN see ITD875
IVERSAL see AHI875
IVERTOL see AHI875
IVIRON see IHG000
IVORAN see DAD200
IVORIT see EJM500
IVORY see ITE000
IVOSIT see ACE500
IVY see AFK950
IVY ARUM see PLW800
IVY BUSH see MRU359
IXODEX see DAD200
IXOTEN see TNT500
IXPER 25M see MAH750
IYLOMYCIN see ITF000
IYOMYCIN see ITG000
IYOMYCIN B1 see ITH000
IZ 914 see DYC875
IZADRIN see IMR000
IZOACRIDINA see AHS500
IZOFORON (POLISH) see IMF400
o-IZOPROPOKSYANILINA (POLISH) see INA200
o-IZOPROPOKSYNITROBENZEN (CZECH) see NIR550
IZOPROPYLOWY ETER (POLISH) see IOZ750
IZOPTIN see VHA450
IZOPTIN HYDROCHLORIDE see VHA450
IZOSYSTOX (CZECH) see DAP200
IZSAB see LEK000

J3 see BPP750
J 400 see PMP500
JACINTO de PERU (CUBA) see SLH200
JACK-IN-THE-PULPIT see JAJ000
JACK WILSON CHLORO 51 (oil) see CKC000
JACOBINE see JAK000, SBX500
JACODINE HYDROCHLORIDE see SBX525
JACUTIN see BBQ500
JADE GREEN BASE see JAT000
JA-FA IPP see IQW000
JAFFNA TOBACCO see BFW135
JAGUAR GUM A-20-D see GLU000
JAGUAR No. 124 see GLU000
JAGUAR PLUS see GLU000
JAIKIN see SFF000
JALAN see EKO500
JAMAICA GINGER EXTRACT see JBA000
JAMAICAN WALNUT see TOA275
JAMBERRY see JBS100
JAMESTOWN WEED see SLV500
JANIMINE see DLH630
JANUPAP see HIM000
JANUS GREEN B see DHM500
JANUS GREEN V see DHM500
JAPAN AGAR see AEX250
JAPAN CAMPHOR see CBA750
JAPANESE AUCUBA see JBS050
JAPANESE BEAD TREE see CDM325
JAPANESE CAMPHOR see CBB250
JAPANESE CAMPHOR OIL see CBB500
JAPANESE LANTERN PLANT see JBS100
JAPANESE LAUREL see JBS050
JAPANESE MEDLAR see LIH200
JAPANESE, OIL of CAMPHOR see CBB500
JAPANESE PLUM see LIH200
JAPANESE POINTSETTIA see SDZ475
JAPAN ISINGLASS see AEX250
JAPAN LACQUER see JCA000
JAPAN OIL TREE see TOA275
JAPANOL VIOLET J see CMP000
JASAD see ZBJ000
JASMINALDEHYDE see AOG500
JASMIN de NUIT (HAITI) see DAC500

JASMOLIN I or II see POO250
JATRONEURAL see TKE500, TKK250
JATROPHA CATHARTICA see CNR135
JATROPHA CURCAS see CNR135
JATROPHA GOSSYPIIFOLIA see CNR135
JATROPHA INTEGERRIMA see CNR135
JATROPHA MACRORHIZA see CNR135
JATROPHA MULTIFIDA see CNR135
JATROPHA PODAGRICA see CNR135
JATROPUR see UVJ450
JAUNE AB see FAG130
JAUNE de BEURRE (FRENCH) see DOT300
JAUNE OB see FAG135
JAVA AMARANTH see FAG020
JAVA CHROME YELLOW 3R see NEY000
JAVA METANIL YELLOW G see MDM775
JAVA ORANGE 2G see HGC000
JAVA ORANGE I see FAG010
JAVA PONCEAU 2R see FMU070
JAVA RUBINE N see HJF500
JAVA SCARLET 3R see FMU080
JAVA UNICHROME YELLOW 3R see NEY000
JAVILLO (DOMINICAN REPUBLIC, PUERTO RICO) see SAT875
JAYANTI, extract see SCB100
JAYFLEX DTDP see DXQ200
JAYSOL see EFU000
JAYSOL S see EFU000
J.B. 305 see EOY000
JB-323 see PJA000
JB 329 see DXP800
JB 336 see MON250
JB 340 see CBF000
JB 516 see PDN000, PDN250
JB 8181 see DLS600
177 J.D. see AJD000
JECTOFER see IHK100
JEFFERSOL DB see DJF200
JEFFERSOL EB see BPJ850
JEFFERSOL EE see EES350
JEFFERSOL EM see EJH500
JEFFOX see PJT000, PJT200, PKI500
JEFFOX OL 2700 see MKS250
JELLIN see SPD500
JENACAINE see AIL750
JEN-DIRIL see CFY000
JERUSALEM OAK see SAF000
JERVINE see JCS000
JERVINE-3-ACETATE see JDA000
JERVINE, 11-DEOXO-12-β,13-α-DIHYDRO-11-β-HYDROXY- see DAQ002
JESTRYL see CBH250
JESUIT'S BALSAM see CNH792
JET BEAD see JDA075
JET FUEL HEF-2 see JDA100
JET FUEL HEF-3 see JDA125
JET FUEL JP-4 see JDA135
JET FUELS see JDJ000
JETRIUM see AFJ400
JETRIUM R see AFJ400
JEW BUSH see SDZ475
JEWELER'S ROUGE see IHD000
JF 5705F see RLF350
JICAMA de AQUA (CUBA) see YAG000
JICAMILLA (MEXICO, TEXAS) see CNR135
JICAMO (MEXICO) see YAG000
JIFFY GROW see ICP000
JILKON HYDROBROMIDE see GBA000
JIMBAY BEAN (BAHAMAS) see LED500
JIMSON WEED see SLV500
JISC 3108 see AGX000
JISC 3110 see AGX000
JL 130 see COW700
J-LIBERTY see MDQ250
JM 8 see CCC075
JM-28 see IGG775
JMC 45498 see DAF300
JN-21 see CQH000
JODAIROL see HGB200
JODAMID (GERMAN) see AAI750
JODCYAN see COP000
JODDEOXIURIDIN see DAS000
JODFENPHOS see IEN000
o-JODHIPPURSAEURE NATRIUM (GERMAN) see HGB200
JODID SODNY see SHW000
JOD-METHAN (GERMAN) see MKW200

JODOBIL see PDM750
JODOMIRON see AAI750
JODOPAX see AAN000
JODPHOSPHONIUM (GERMAN) see PHA000
JOD (GERMAN, POLISH) see IDM000
JOGEN see JDJ100
JOHNKOLOR see TKQ250
JOLIPEPTIN see JDS000
JOLT see EIN000
JOMYBEL see JDS200
JONIT see PFA500
JONNIX see SNQ500
JONQUIL see DAB700
JON-TROL see DXE600
JOOD (DUTCH) see IDM000
JOODMETHAAN (DUTCH) see MKW200
JOPAGNOST see IFY100
JORCHEM 400 ML see PJT000
JOSAMINA see JDS200
JOSAMYCIN see JDS200
JOSCINE see SBG500
JOTA (GERMAN) see IGD000
JOTALAMSAEURE (GERMAN) see IGD000
JOY POWDER see HBT500
JP-10 see TLR675
JP 61 see AJU625
JP 428 HYDROCHLORIDE see DAI200
J SOFT C 4 see DTC600
J-SUL see SNN500
JUCA see CCO680
JUDEAN PITCH see ARO500
JUDOLOR see FQJ100
JULIN'S CARBON CHLORIDE see HCC500
JULODIN see CKL250
JUMBEE BEADS (VIRGIN ISLANDS) see RMK250
JUMBLE BEAD see AAD000
JUMEX see DAZ125
JUMP-AND-GO (BAHAMAS) see LED500
JUNIPER BERRY OIL see JEA000
JUNIPERIC ACID LACTONE see OKU000
JUNIPER TAR see JEJ000
JUNIPERUS COMMUNIS auct. non. Linn., extract excluding roots see JEJ100
JUNIPERUS COMMUNIS Linn. var. SAXATILIS Pallas, extract excluding roots see JEJ100
JURIMER AC 10P see ADW200
JUSONIN see SFC500
JUSQUIAME (CANADA) see HAQ100
JUSTAMIL see AIE750
JUVASON see PLZ000
JUVASTIGMIN see NCL100
JUVENIMICIN A3 see RMF000
JUVOCAINE see AIT250
JZF see BLE500

K-9 see MEC250
06K see AJI250
K 17 see TEH500
K 52 see OHU000
6FK see HCZ000
K2₂₀ see VTA650
K-300 see CQH000
K-315 see THJ750
K315 see KAH000
K 351 see NEA100
K-373 see DAP700
K 55E see SMR000
K 653 see HGW000
K 694 see CDR550
K-708 see AAE625
K 1875 see NCM700
K 1900 see NHH000
K-1902 see PPW000
K 2680 see AHP125
K3917 see CFY750
K 4277 see IDA400
K 6451 see CJT750
K-10033 see BPP250
K 22023 see DGD800
K-30052 see EGC500
K62-105 see LEN000
K25 (polymer) see PKQ250
K 31 (pharmaceutical) see CME675
KABAT see KAJ000
KABI 1774 see CMS241

KABIKINASE see SLW450
KADMIUM (GERMAN) see CAD000
KADMIUMCHLORID (GERMAN) see CAE250
KADMIUMSTEARAT (GERMAN) see OAT000
KADMU TLENEK (POLISH) see CAH500
KADOX-25 see ZKA000
KAERGONA see MMD500
KAFAR COPPER see CNI000
KAFFIR LILLY see KAJ100
KAFIL SUPER see RLF350
KAFOCIN see CCR890
KAISER CHEMICALS 11 see FOO000
KAISER CHEMICALS 12 see DFA600
KAKEN see CPE750
KAKO BLUE B SALT see DCJ200
KAKO RED TR BASE see CLK220
KAKO SCARLET GG SALT see DEO400
KALEX see EIV000
KALGAN see PFC750
K'-ALGILINE see SEH000
KALINITE see AHF200
KALITABS see PLA500
KALIUMARSENIT (GERMAN) see PKV500
KALIUM-BETA see PLG800
KALIUMCARBONAT (GERMAN) see PLA000
KALIUMCHLORAAT (DUTCH) see PLA250
KALIUMCHLORAT (GERMAN) see PLA250
KALIUMCYANAT (GERMAN) see PLC250
KALIUM-CYANID (GERMAN) see PLC500
KALIUMDICHROMAT (GERMAN) see PKX250
KALIUMHYDROXID (GERMAN) see PLJ500
KALIUMHYDROXYDE (DUTCH) see PLJ500
KALIUMNITRAT (GERMAN) see PLL500
KALIUMPERMANGANAAT (DUTCH) see PLP000
KALIUMPERMANGANAT (GERMAN) see CAV250, PLP000
KALLIDIN see BML500
KALLIKREIN-TRYPSIN INACTIVATOR see PAF550
KALLOCRYL K see PKB500
KALLODENT CLEAR see PKB500
KALMETTUMSOMNIFERUM see GFA000
KALMIA ANGUSTIFOLIA see MRU359
KALMIA LATIFOLIA see MRU359
KALMIA MICROPHYLLA see MRU359
KALMOCAPS see LFK000, MDQ250
KALMUS OEL (GERMAN) see OGK000
KALO see EAI600
KALOMEL (GERMAN) see MCW000
KALZIUMARSENIAT (GERMAN) see ARB750
KALZIUMZYKLAMATE (GERMAN) see CAR000
KAM 1000 see SLK000
KAM 2000 see SLK000
KAM 3000 see SLK000
KAMALIN see KAJ500
KAMANI (HAWAII) see MBU780
KAMAVER see CDP250
KAMBAMINE RED TR see CLK220
KAMBAMINE SCARLET GG BASE see DEO400
KAMFOCHLOR see CDV100
KAMILLENOEL (GERMAN) see DRV000
KAMPFER (GERMAN) see CBA750
KAMPOSAN see CDS125
KAMPSTOFF ""LOST"" see BIH250
KAMYCIN see KAV000
KAMYNEX see KAV000
KANABRISTOL see KAV000
KANACEDIN see KAV000
KANAMICINA (ITALIAN) see KAL000
KANAMYCIN see KAL000
KANAMYCIN A see KAL000
KANAMYCIN A SULFATE see KAV000
KANAMYCIN B see BAU270
KANAMYCIN B SULFATE see KBA100
KANAMYCIN B, SULFATE (1:1) (SALT) see KBA100
KANAMYCIN MONOSULFATE see KAV000
KANAMYCIN SULFATE see KAM000, KAV000
KANAMYCIN SULFATE (1581) SALT see KAV000
KANAMYTREX see KAL000, KAV000
KANAQUA see KAV000
KANASIG see KAV000
KANATROL see KAV000
KANDISET see BCE500
KANECHLOR see PJL750
KANECHLOR 300 see PJO500
KANECHLOR 400 see PJO750

KANECHLOR 500 see PJP000, PJP250
KANEKALON see ADY250
KANEKROL 500 see PAV600
KANENDOMYCIN see BAU270
KANENDOMYCIN SULFATE see KBA100
KANEPAR see TIY500
KANESCIN see KAV000
KANNASYN see KAM000, KAV000
KANO see KAV000
KANOCHOL see DYE700
KANONE see MMD500
KANTEC see MAO275
KAN-TO-KA (JAPANESE) see CNH250
KANTREX see KAL000, KAV000
KANTREXIL see KAV000
KANTREX SULFATE see KAM000
KANTROX see KAV000
KANZO (JAPANESE) see LFN300
KAO 264 see KBB000
KAOCHLOR see PLA500
KAOLIN see KBB600
KAON see PLG800
KAON-Cl see PLA500
KAON ELIXIR see PLG800
KAOPAOUS see KBB600
KAOPHILLS-2 see KBB600
KAPPAXAN see MMD500
e-KAPROLAKTAM (CZECH) see CBF700
KAPROLIT see PJY500
KAPROLON see PJY500
KAPROMIN see PJY500
KAPRON see NOH000, PJY500
KAPRONAN DI-N-BUTYLCINICITY (CZECH) see BJX750
KAPRONAN DICYKLOHEXYLAMINU (CZECH) see DGU400
KAPRYLAN DI-N-BUTYLCINICITY (CZECH) see BLB250
KAPTAN see CBG000
KARAMATE see DXI400
KARAYA GUM see KBK000
KARBAM BLACK see FAS000
KARBAM WHITE see BJK500
KARBARYL (POLISH) see CBM750
KARBATION see VFU000, VFW009
KARBOFOS see MAK700
KARBOKROMEN (RUSSIAN) see CBR500
5-KARBOKSIMETIL-3-p-TOLIL-TIAZOLIDIN-2,4-DION-2-
ACETOFENONHIDRAZON (CZECH) see CCH800
KARBORAFIN see CDI000
1-(4'-KARBOXYLAMIDOFENYL)-3,3-DIMETHYLTRIAZENU (CZECH)
see CCC325
1-(4'-KARBOXYLAMIDOPHENYL)-3,3-DIMETHYLTRIAZEN (GERMAN)
see CCC325
KARBROMAL see BNK000
KARBUTILATE see DUM800
KARCON see MMD500
KARDIAMID see DJS200
KARDIN see TJL250
KARDONYL see DJS200
KAREON see MMD500
KARIDIUM see SHF500
KARIGEL see SHF500
KARION see SKV200
KARI-RINSE see SHF500
KARLAN see RMA500
KARMESIN see HJF500
KARMEX see DXQ500
KARMEX DIURON HERBICIDE see DXQ500
KARMEX DW see DXQ500
KARMEX MONURON HERBICIDE see CJX750
KARMEX W. MONURON HERBICIDE see CJX750
KARMINOMYCIN see KBU000
KARMINOMYCIN HYDROCHLORIDE see KCA000
KARNOZZN see CCK665
KARO TARTRAZINE see FAG140
KARPHOS see DJV600
KARSAN see FMV000
KARSULPHAN see SNR000
KARTRYL see BNK000
KARWINSKIA HUMBOLDTIANA see BOM125
KASAL see SEM305
KASEBON see CBW000
KASH, LEAF EXTRACT see KCA100
KASIMID see BBW750
KASSIA OEL (GERMAN) see CCO750
KASTONE see HIB010

KASUGAMYCIN see AEB500
KASUGAMYCIN HYDROCHLORIDE see KCK000
KASUGAMYCIN MONOHYDROCHLORIDE see KCK000
KASUGAMYCIN PHOSPHATE see KCU000
KASUGAMYCIN SULFATE see KDA000
KASUMIN see KCK000
KAT 256 see CMW500
KATAGRIPPE see EEM000
KATALYSIN see AKK750
KATAMINE AB see AFP250, DTC600
KATAPYRIN see AFW500
KATCHUNG OIL see PAO000
KATEXOL 300 see KDA025
KATHA see CCP800
KATHON BIOCIDE see KDA035
KATHON LP PRESERVATIVE see OFE000
KATHON MW 886 BIOCIDE see KDA035
KATHON SP 70 see OFE000
KATHRO see CMD750
KATIV-G see MMD500
KATIV N see VTA000
KATLEX see CHJ750
KATONIL see CHX250
KATORIN see PLG800
KATOVIT HYDROCHLORIDE see PNS000
KATRON see PDN000, BCA000
KATRONIAZID see PDN000
KAURIT S see DTG700
KAUTSCHIN see MCC250
KAVAIN see GJI250
(+)-KAVAIN see GJI250
KAWAIN see GJI250
KAYAFUME see MHR200
KAYAKU ACID BRILLIANT SCARLET 3R see FMU080
KAYAKU AMARANTH see FAG020
KAYAKU BLUE B BASE see DCJ200
KAYAKU CONGO RED see SGQ500
KAYAKU DIRECT see CMO000
KAYAKU DIRECT DEEP BLACK EX see AQP000
KAYAKU DIRECT SKY BLUE 6B see CMN750
KAYAKU SCARLET G BASE see NMP500
KAYAKU SCARLET GG BASE see DEO400
KAYALON FAST BLUE BR see TBG700
KAYALON FAST BLUE FN see MGG250
KAYAZINON see DCM750
KAYAZOL see DCM750
KAY CIEL see PLA500
KAYDOL see MQV750
KAYEXALATE see SJK375
KAYKLOT see MMD500
KAYLITE see PKQ059
KAYQUINONE see MMD500
KAYTRATE see PBC250
KAYTWO see VTA650
KAYVISYN see AKX500
KAZOE see SFA000
KB-16 see MIH250
KB-53 see CCP875
KB 95 see BCP650
KB 227 see RGP450
KB-227 see TCY260
KB 227 see THJ825
KB-509 see FMR100
KB-944 see DIU500
KB (POLYMER) see SMQ500
K-BRITE see SHR500
KBT-1585 see LEJ500
KC-400 see PJO750
KC-404 see KDA100
KC-500 see PJP000
KC 9147 see MQA000
KCA CHROME ORANGE R see NEY000
KCA FOODCOL AMARANTH A see FAG020
KCA METHYL ORANGE see MND600
KD-136 see HAG300
KDM see BAU270
KE see CNS800
KEBILIS see CDL325
KEBUZONE see KGK000
KEDACILLIN see SNV000
KEDAVON see TEH500
KEESTAR see SLJ500
KEFENID see BDU500
KEFGLYCIN see CCR890

KEFLEX see ALV000
KEFLIN see SFQ500
KEFLODIN see TEY000
KEFORAL see ALV000
KEFZOL see CCS250
KEIMSTOP see CBL750
KELACID see AFL000
KELAMERAZINE see ALF250
KELCO GEL LV see SEH000
KELCOLOID see PNJ750
KELCOSOL see SEH000
KELENE see EHH000
KEL-F see KDK000
KELFIZIN see MFN500
KELGIN see SEH000
KELGUM see SEH000
KELICORIN see AHK750
KELINCOR see AHK750
KELOFORM see EFX000
KELSET see SEH000
KELSIZE see SEH000
KELTANE see BIO750
KELTEX see SEH000
KELTHANE (DOT) see BIO750
p,p'-KELTHANE see BIO750
KELTHANE DUST BASE see BIO750
KELTHANETHANOL see BIO750
KELTONE see SEH000
KEMADRINE see CPQ250
KEMAMIDE S see OAR000
KEMAMINE P 989 see OHM700
KEMAMINE S 190 see KDU100
KEMATE see DEV800
KEMDAZIN see MHC750
KEMESTER 105 see OHW000
KEMESTER 115 see OHW000
KEMESTER 205 see OHW000
KEMESTER 213 see OHW000
KEMI see ICC000
KEMICETINE see CDP250
KEMICETINE SUCCINATE see CDP725
KEMIKAL see CAT225
KEMITHAL see TES500
KEMODRIN see DBA800
KEMOLATE see PHX250
KEMOVIRAN see MKW250
KEMPORE see ASM270
KEMPORE 125 see ASM270
KEMPORE R 125 see ASM270
KENACHROME BLUE 2R see HJF500
KENACHROME ORANGE see NEY000
KENACORT see AQX250
KENACORT-A see AQX500
KENACORT DIACETATE SYRUP see AQX750
KENALOG see AQX500
KENAPON see DGI400
KENDALL'S COMPOUND B see CNS625
KENDALL'S COMPOUND E see CNS800
KENDALL'S COMPOUND F see CNS750
KENGSHENGMYCIN see AEB500
KEOBUTANE-JADE see KGK000
KEPHRINE see MGC350
KEPHTON see VTA000
KEPINOL see TKX000
KEPMPLEX 100 see EIV000
KEPONE see KEA000
KEPTAN see KEA300
KER 710 see EAL100
KERALYT see SAI000
KERAPHEN see TDE750
KERASALICYL see SJO000
KERB see DTT600
KERECID see DAS000
KERLONE see KEA350
KERNECHTROT see AJQ250
KEROCAINE see AIT250
KEROPUR see BAV000
KEROSAL see SJO000
KEROSENE see KEK000
KESELAN see CLY500
KESSCO 40 see OAV000
KESSCOCIDE see ISN000
KESSCOFLEX see BHK000
KESSCOFLEX MCP see DOF400

KESSCOFLEX TRA see THM500
KESSCO ISOPROPYL see IQW000
KESSCOMIR see IQN000
KESSOBAMATE see MQU750
KESSODANTEN see DKQ000
KESSODRATE see CDO000
KESTREL (Pesticide) see AHJ750
KESTRIN see ECU750
KESTRONE see EDV000
KETAJECT see CKD750
KETALAR see CKD750
KETALGIN see MDO750
KETALGIN HYDROCHLORIDE see MDP750
KETAMAN see HKR500
KETAMINE see CKD750, KEK200
KETAMINE HYDROCHLORIDE see CKD750
KETANEST see CKD750
KETANRIFT see ZVJ000
KETASET see CKD750
KETASON see KGK000
KETAVET see CKD750
KETAZONE see KGK000
KETENE see KEU000
KETENE DIMER see KFA000
KETHOXAL-BIS-THIOSEMICARBAZIDE see KFA100
3-(3-KETO-7-α-ACETYLTHIO-17-β-HYDROXY-4-ANDROSTEN-17-α-YL)PROPIONIC ACID LACTONE see AFJ500
4-KETOAMYLTRIMETHYLAMMONIUM IODIDE see TLY250
KETOBEMIDONE HYDROCHLORIDE see KFK000
4-KETOBENZOTRIAZINE see BDH000
KETOBUN-A see ZVJ000
β-KETOBUTYRANILIDE see AAY000
KETOCAINE HYDROCHLORIDE see DNN000
KETOCHOL see DAL000
KETOCHOLANIC ACID see DAL000
7-KETOCHOLESTEROL see ONO000
KETOCONAZOLE see KFK100
KETOCYCLOPENTANE see CPW500
KETODESTRIN see EDV000
2-KETO-3-ETHOXY-BUTYRALDEHYDE-BIS(THIOSEMICARBAZONE) see KFA100
KETO-ETHYLENE see KEU000
KETOGAZE see MGC350
3-KETO-l-GULOFURANOLACTONE see ARN000
KETOHEXAMETHYLENE see CPC000
2-KETOHEXAMETHYLENIMINE see CBF700
KETOHYDROXY-ESTRATRIENE see EDV000
KETOHYDROXYESTRIN see EDV000
KETOHYDROXYESTRIN BENZOATE see EDV500
KETOHYDROXYOESTRIN see EDV000
2,3-KETOINDOLINE see ICR000
KETOLAR see CKD750
KETOLE see ICM000
KETOLIN-H see CJU250
γ-KETO-β-METHOXY-Δ-METHYLENE-Δ^α-HEXENOIC ACID see PAP750
15-KETO-20-METHYLCHOLANTHRENE see MIM250
KETONE, 5-AMINO-1,3-DIMETHYLPYRAZOL-4-YL o-FLUOROPHENYL see AJR400
KETONE-METHYL-5-OXO-5H-(1)BENZOPYRANO(2,3-b)PYRIDYL see ACU125
KETONE METHYL PHENYL see ABH000
KETONE PROPANE see ABC750
KETONES see KGA000
4-KETONIRIDAZOLE see KGA100
KETOPENTAMETHYLENE see CPW500
KETOPHENYLBUTAZONE see KGK000
KETOPROFEN see BDU500
KETOPROFEN SODIUM see KGK100
KETOPRON see BDU500
β-KETOPROPANE see ABC750
1-KETOPROPIONALDEHYDE see PQC000
2-KETOPROPIONALDEHYDE see PQC000
α-KETOPROPIONALDEHYDE see PQC000
2-KETOPYRROLIDINE-1-YLACETAMIDE see NNE400
2-KETO-4-QUINAZOLINONE see QEJ800
4-KETOSTEARIC ACID see KGK150
l-3-KETOTHREOHEXURONIC ACID LACTONE see ARN000
KETOTIFEN FUMARATE see KGK200
8-KETOTRICYCLO(5.2.1.0^{2,6})DECANE see OPC000
2-KETO-1,7,7-TRIMETHYLNORCAMPHANE see CBA750
4-KETOVALERIC ACID see LFH000
γ-KETOVALERIC ACID see LFH000
KEUTEN see DEE600
KEVADON see TEH500
KEY-TUSSCAPINE see NOA000

KEY-TUSSCAPINE HYDROCHLORIDE see NOA500
KF-868 see KGK300
KF-1820 see DOQ400
K-FLEBO see PKV600
K-FLEX DP see DWS800
K-GRAN see PLA000
KH 360 see TGG760
KHAINI (INDIA) see SED400
KHAROPHEN see ABX500
KHAT LEAF EXTRACT see KGK350
KHE 0145 see MIA250
KHIMCOCCID see RLK890
KHIMCOECID see RLK890
KHIMKOKTSID see RLK890
KHIMKOKTSIDE see RLK890
KHINGAMIN see CLD250
KHINOTILIN see KGK400
KHLADON 113 see FOO000
KHLORAKON see BEG000
KHLORIDIN see TGD000
KHLORTRIANIZEN see CLO750
KHLOTAZOL see CMB675
KHOMECIN see ZJS300
KHOMEZIN see ZJS300
KHP 2 see AHE250
K-IAO see PLG800
KIATRIUM see DCK759
KID KILL see MRU359
KIDNEY BEAN TREE see WCA450
KIDOLINE see VGP000
KIEFERNADEL OEL (GERMAN) see PIH500
KIESELGUHR see DCJ800
KIESELSAURE (GERMAN) see SCL000
KIEZELFLUORWATERSTOFZUUR (DUTCH) see SCO500
K III see DUS700
KIKUTHRIN see PMN700
KILACAR see PMF550
KILDIP see DGB000
KILEX 3 see BSQ750
KILL-ALL see SEY500
KILLAX see TCF250
KILL COW see PJJ300
KILLEEN see CCL250
KILL KANTZ see AQN635
KILMITE 40 see TCF250
KILOSEB see BRE500
KILPROP see CIR500
KILRAT see ZLS000
KILSEM see CIR250
KILVAL see MJG500
KIMAVOXYL see CBK500
KINADION see VTA000
KINAVOSYL see GGS000
KINEKS see AKO500
KINEX see AKO500
KINGCUP see MBU550
KING'S GREEN see COF500
KING'S YELLOW see ARI000, LCR000
KINNIKINNIK see CCJ825
KINOPRENE see POB000
KINOTOMIN see FOS100
KIPCA see MMD500
KIRESUTO B see EIX500
KIRESUTO NTB see DXF000
KIRKSTIGMINE BROMIDE see POD000
KIRKSTIGMINE METHYL SULFATE see DQY909
α-KIRONDRIN see GEW700
β-KIRONDRIN see GEW700
KIR RICHTER see PAF550
KIRTICOPPER see CNK500
KITASAMYCIN see SLC000
KITASAMYCIN A3 see JDS200
KITASAMYCIN TARTRATE see LEX000
KITAZIN see DIU800
KITAZIN P see BKS750
KITON CRIMSON 2R see HJF500
KITON FAST ORANGE G see HGC000
KITON FAST YELLOW A see SGP500
KITON ORANGE MNO see MDM775
KITON PONCEAU R see FMU070
KITON PURE BLUE L see FMU059
KITON RUBINE S see FAG020
KITON SCARLET 4R see FMU080
KITON YELLOW MS see MDM775

KITON YELLOW T see FAG140
KIVATIN see HKR500
KIWAM (INDIA) see SED400
KIWI LUSTR 277 see BGJ250
K. IXINA see CAC500
KL-001 see BMN750
KL 255 see BQB250
(−)-KL 255 see BQB250
KL 373 see BGD500
KLAVI KORDAL see NGY000
KLEBCIL see KAV000
KLEER-LOT see AMY050
KLEESALZ (GERMAN) see OLE000
KLEGECELL see PKQ059
KLERAT see TAC800
KLIMORAL see EDU500
KLINE see BBK250
KLINGTITE see NAK500
KLINIT see XPJ000
KLINOSORB see CMV850
KLION see MMN250
KLOBEN see BRA250
KLOBEN NEBURON see BRA250
KLOFIRAN see ARQ750
K-LOR see PLA500
KLORALFENAZON see SKS700
KLORAMIN see BIE500, CDP000
KLORAMINE-T see CDP000
KLOREX see SFS000
KLORINOL see TIX000
KLOROKIN see CLD000
KLORPROMAN see CKP500
KLORPROMEX see CKP500
KLORT see MQU750
KLOT see AJP250
KLOTRIX see PLA500
KLOTTONE see MMD500
KLT 40 see PKF750
KLUCEL see HNV000
KM see KAL000
4K-2M see CIR250
KM-208 see BAC175
KM-1146 see KGU100
KM 2210 see BFV325
KM (the antibiotic) see KAL000
KMTS 212 see SFO500
KNEE PINE OIL see PIH400
KNIGHT'S SPUR see LBF000
KNITTEX ASL see DTG700
KNITTEX LE see DTG000
KNOCKMATE see FAS000
KNOLL H75 see FMS875
KO 7 see EAL100
KO 08 see SCR400
KO 1173 see MQR775
KOA-HAOLE (HAWAII) see LED500
KOAXIN see MMD500
KOBALT CHLORID (GERMAN) see CNB599
KOBALT-EDTA (GERMAN) see DGQ400
KOBALT HISTIDIN (GERMAN) see BJY000
KOBALT (GERMAN, POLISH) see CNA250
KOBAN see EFK000
KOBU see PAX000
KOBUTOL see PAX000
KOCHINEAL RED A FOR FOOD see FMU080
KOCIDE see CNM500, SOD500
KO 1366-CL see BON400
KODAFLEX see TIG750
KODAFLEX DBS see DEH600
KODAFLEX DOA see AEO000
KODAFLEX DOP see DVL700
KODAFLEX TRIACETIN see THM500
KODAK SILVER HALIDE SOLVENT HS-103 see BGT500
KODOCYTOCHALASIN-1 see PAM775
KOE 1366 CHLORIDE see BON400
KOE 1173 HYDROCHLORIDE see MQR775
KOFFEIN (GERMAN) see CAK500
KOHLENDIOXYD (GERMAN) see CBU250
KOHLENDISULFID (SCHWEFELKOHLENSTOFF) (GERMAN) see CBV500
KOHLENMONOXID (GERMAN) see CBW750
KOHLENOXYD (GERMAN) see CBW750
KOHLENSAURE (GERMAN) see CBU250
KOJIC ACID see HLH500
KOKAIN see CNE750

KOKAN see CNE750
KOKAYEEN see CNE750
KOKOTINE see BBQ500
KOLALES HALOMTANO (GUAM) see RMK250
KOLCHAMIN see MIW500
KOLI (HAWAII) see CCP000
KOLKLOT see MMD500
KOLLIDON see PKQ250
KOLOFOG see SOD500
KOLOSPRAY see SOD500
KOLPHOS see PAK000
KOLPON see EDV000
KOLTON see PIZ250
KOLTONAL see PIZ250
KOMBE-STROPHANTHIN see SMN000
KOMBETIN see SMN000
KOMPLXON see EIV000
KONAKION see VTA000
KONDREMUL see MQV750
KONESSIN DIHYDROBROMIDE see DOX000
KONESTA see TII250
KONLAX see DJL000
KONTRAST-U see SHX000
KOOLMONOXYDE (DUTCH) see CBW750
KOOLSTOFDISULFIDE (ZWAVELKOOLSTOF) (DUTCH) see CBV500
KOOLSTOFOXYCHLORIDE (DUTCH) see PGX000
KOPFUME see EIY500
KOP KARB see CNJ750
KOP MITE see DER000
KOPOLYMER BUTADIEN STYRENOVY (CZECH) see SMR000
KOPROSTERIN (GERMAN) see DKW000
KOPSOL see DAD200
KOP-THIODAN see EAQ750
KOP-THION see MAK700
KORAD see PKB500
KORAX see CJE000
KORBUTONE see AFJ625
KORDIAMIN see DJS200
KOREON see NBW000
KORGLYKON see CNH780
KORIUM see MJM500
KORLAN see RMA500
KORLANE see RMA500
KORMOGRIZEIN see GJU800
KORODIL see CCK125
KOROSEAL see PKQ059
KORO-SULF see SNN500
KOROTRIN see CMG675
KORUM see HIM000
KORUND see EAL100
KOSATE see DJL000
KOSTIL see ADY500
K-OTHRIN see DAF300
KOTION see DSQ400
KOTOL see BBQ750
KOTORAN see KHK000
KP 2 see PAX000
KP 140 see BPK250
KPB see KGK000
KPE see KHK100
K PHENETHICILLIN see PDD350
K-PIN see PIB900
K-PRENDE-DOME see PLA500
K PREPARATION see BJU000
06K-QUINONE see AJI250
KR 492 see MKK500
KRAMERIA IXINA see CAC500
KRAMERIA TRIANDRA see RGA000
KRASTEN 1.4 see SMQ500
KRATEDYN see EAW000
KREBON see BQL000
KRECALVIN see DGP900, PHC750
KREGASAN see TFS350
KRENITE see ANG750
KRENITE (OBS.) see DUS700, DUU600
KRESAMONE see DUS700
KRESIDIN see MGO750
m-KRESOL see CNW750
p-KRESOL see CNX250
o-KRESOL (GERMAN) see CNX000
KRESOLE (GERMAN) see CNW500
KRESOLEN (DUTCH) see CNW500
o-KRESOL-GLYCERINAETHER (GERMAN) see GGS000
KRESONIT E see DUT800

KRESOXYPROPANDIOL see GGS000
KREZAMON see DUT800
KREZIDINE see MGO750
KREZOL (POLISH) see CNW500
KREZONE see CIR250
KREZONIT E see DUT800
KREZONITE see DUU600
KREZOTOL 50 see DUS700
KRINOCORTS see DAQ800
KRIPLEX see DEO600
KRIPTIN see WAK000
KRISOLAMINE see DBP000
KRISTALLOSE see SJN700
KRMD 58 see MAE000
KRO 1 see SMR000
KROKYDOLITH (GERMAN) see ARM275
KROMAD see KHU000
KROMFAX SOLVENT see TFI500
KROMON GREEN B see CLK235
KROMON HELIO FAST RED see MMP100
KROMON RED R see CJD500
KROMON YELLOW MTB see DEU000
KRONISOL see BHK000
KRONITEX see TNP500
KRONITEX KP-140 see BPK250
KRONITEX TOF see TNI250
KRONOS TITANIUM DIOXIDE see TGG760
KROTENAL see DXH250
KROTILINE see DAA800
KROTONALDEHYD (CZECH) see COB250
KROVAR II see BMM650
KRUMKIL see ABF500
KRYOGENIN see CBL000
KRYOLITH (GERMAN) see SHF000
KRYPTOCUR see LIU360
KRYSID see AQN635
KRZEWOTOKS see BSQ750
KS 1675 see DBA600
K-STROPHANTHIDIN see SMM500
K-STROPHANTHIN-α see CQH750
K-STROPHANTHIN-β see SMN002
KSYLEN (POLISH) see XGS000
KT 136 see KHU100
K-THROMBYL see MMD500
KTS (PHARMACEUTICAL) see KFA100
KU 5-3 see EAL100
KUBACRON see HII500
KUBARSOL see ABX500
KU 13-032-C see DFL200
KUE 13032c see DFL200
KUEMMEL OIL (GERMAN) see CBG500
KUKUI (HAWAII, GUAM) see TOA275
KUMADER see WAT200
KUMIAI see MIB750
KUMORAN see BJZ000
KUMULUS see SOD500
KUPAOA (HAWAII) see DAC500
KUPFERCARBONAT (GERMAN) see CNJ750
KUPFEROXYCHLORID (GERMAN) see CNK559
KUPFEROXYDUL (GERMAN) see CNO000
KUPFERRON (CZECH) see ANO500
KUPFERSULFAT-PENTAHYDRAT (GERMAN) see CNP500
KUPFERVITRIOL (GERMAN) see CNP500
KUPPERSULFAT (GERMAN) see CNP250
KUPRABLAU see CNM500
KUPRATSIN see EIR000
KUPROTSIN see ZJS300
KURAN see TIX500
KURARE OM 100 see AAX250
KURCHICINE see KHU136
KURDUMANA, root extract see CNH750
KURON see TIX500, TIX750
KUROSAL see TIX500
KUSA-TOHRU see SFS000
KUSATOL see SFS000
KUSNARIN see EID000
KUTROL see UVJ475
K-VITAN see MMD500
KW-066 see PNX750
KW 110 see AGX125
KW-125 see AES750
06K-50W see AJI250
KW-1062 see MQS579
KW-1070 see FOK000

KW-1100 see BAC325
KW-3149 see FDD080
KW-4354 see OMG000
KW-5338 see DYB875
KWAS BENZYDYNODWUKAROKSYLOWY (POLISH) see BFX250
KWAS METANIOWY (POLISH) see FNA000
KWD 2019 see TAN100, TAN250
KWELL see BBQ500
KWELLS see HOT500
KWIETAL see QCS000
KWIK (DUTCH) see MCW250
KWIK-KIL see SMN500
KWIKSAN see ABU500
KWIT see EEH600
KW-2-LE-T see LFA020
KYAMEPROMAZINE MALEATE see COS899
KYANACETHYDRAZID see COH250
KYANID SODNY (CZECH) see SGA500
KYANID STRIBRNY (CZECH) see SDP000
KYANITE see AHF500
2-KYANMETHYLBENZIMIDAZOL (CZECH) see BCC000
KYANOSTRIBRNAN DRASELNY (CZECH) see PLS250
KYANURCHLORID (CZECH) see TJD750
KYLAR see DQD400
KYNEX see AKO500
KYOCRISTINE see LEZ000
KYONATE see PLV750
KYPCHLOR see CDR750
KYPFOS see MAK700
KYPMAN 80 see MAS500
KYPTHION see PAK000
KYPZIN see EIR000
KYSELINA ADIPOVA (CZECH) see AEN250
KYSELINA AKRYLOVA see ADS750
KYSELINA AMIDOSULFONOVA (CZECH) see SNK500
KYSELINA-4-AMINOANISOL-3-SULFONOVA see AIA500
KYSELINA 4-AMINOAZOBENZEN-3,4'-DISULFONOVA (CZECH) see AJS500
KYSELINA 1-AMINO-2-ETHOXYNAFTALEN-6-SULFONOVA (CZECH)
    see AJU500
KYSELINA-3-AMINO-4-METHOXYBENZOOVA (CZECH) see AIA250
KYSELINA 1-AMINO-8-NAFTOL-3,6-DISULFONOVA (CZECH) see AKH000
KYSELINA 1-AMINO-8-NAFTOL-4-SULFONOVA (CZECH) see AKH750
KYSELINA 2-AMINO-5-NAFTOL-7-SULFONOVA (CZECH) see AKI000
KYSELINA-p-AMINOSALICYLOVA (CZECH) see AMM250
KYSELINA ANILIN-2,5-DISULFONOVA (CZECH) see AIE000
KYSELINA ANILIN-3-SULFONOVA (CZECH) see SNO000
KYSELINA BENZIDIN-2,2'-DISULFONOVA (CZECH) see BBX500
KYSELINA BENZOOVA (CZECH) see BCL750
KYSELINA C (CZECH) see ALH250
KYSELINA CEROMSALICYLOVA (CZECH) see BHA000
KYSELINA 2-CHLOR-6-AMINOFENOL-4-SULFONOVA (CZECH) see AJH500
KYSELINA o-CHLORBENZOOVA (CZECH) see CEL250
KYSELINA-S-(8-CHLORMETHYL-1-NAFTYL)THIOGLYKOLOVA (CZECH)
    see CIQ000
KYSELINA-2-CHLORO-4-NITROBENZOOVA (CZECH) see CJC250
KYSELINA 4-CHLORO-3-NITROBENZOOVA (CZECH) see CJC500
KYSELINA 2-CHLOR-4-TOLUIDIN-5-SULFONOVA (CZECH) see AJJ250
KYSELINA CITRONOVA (CZECH) see CMS750
KYSELINA CLEVE (CZECH) see ALI250
KYSELINA-2,4-DIAMINOBENZENSULFONOVA (CZECH) see PFA250
KYSELINA DICHLORISOKYANUROVA (CZECH) see DGN200
KYSELINA 2,5-DICHLOR-4-(3'-METHYL-5'-PYRAZOLON-1'-
    YL)BENZENSULFONOVA (CZECH) see DFQ200
KYSELINA 3,6-DICHLORPIKOLINOVA see DGJ100
KYSELINA O,O-DIETHYLDITHIOFOSFORECNA (CZECH) see PHG500
KYSELINA DI-I (CZECH) see IBF000
KYSELINA 3,5-DIKARBOXYBENZENSULFONOVA (CZECH) see SNU500
KYSELINA O,O-DIMETHYLDITHIOFOSFORCNA (CZECH) see PHH500
KYSELINA-2,4-DINITROBENZENSULFONOVA (CZECH) see DUR400
KYSELINA-4,4'-DINITROSTILBEN-2,2'-DISULFONOVA (CZECH) see DVF600
KYSELINA 3,6-ENDOMETHYLEN-3,4,5,6,7,7-HEXACHLOR-Δ⁴-
    TETRAHYDROFTALOVA (CZECH) see CDS000
KYSELINA ETHOXY-CLEVE-1,6 (CZECH) see AJU500
KYSELINA-1,3-FENYLENDIAMIN-4-SULFONOVA (CZECH) see PFA250
KYSELINA FUMAROVA (CZECH) see FOU000
KYSELINA H (CZECH) see AKH000
KYSELINA HET (CZECH) see CDS000
KYSELINA ISOFTALOVA (CZECH) see IMJ000
KYSELINA JABLECNA see MAN000
KYSELINA KOCHOVA (CZECH) see ALI750
KYSELINA KYANUROVA (CZECH) see THS000
KYSELINA METANILOVA (CZECH) see SNO000
KYSELINA 2-METHYLAMINO-5-NAFTOL-7-SULFONOVA (CZECH)
    see HLX000

KYSELINA N-METHYLANTHRANILOVA (CZECH) see MGQ000
KYSELINA METHYLARSONOVA see MGQ530
KYSELINA-N-METHYL-1 (CZECH) see HLX000
KYSELINA MLECNA (CZECH) see LAG000
KYSELINA MUKOCHLOROVA see MRU900
KYSELINA-2-NAFTOL-1-SULFONOVA (CZECH) see HMX000
KYSELINA 2-NAFTYLAMIN-1,5-DISULFONOVA (CZECH) see ALH000
KYSELINA 2-NAFTYLAMIN-4,8-DISULFONOVA (CZECH) see ALH250
KYSELINA-1-NAFTYLAMIN-6-SULFONOVA (CZECH) see ALI250
KYSELINA-2-NAFTYLAMIN-1-SULFONOVA (CZECH) see ALH750
KYSELINA 2-NAFTYLAMIN-3,6,8-TRISULFONOVA (CZECH) see ALI750
KYSELINA-4-NITRO-2-AMINOFENOL-6-SULFONOVA (CZECH) see HMY500
KYSELINA 6-NITRO-2-AMINOFENOL-4-SULFONOVA (CZECH) see AKI250
KYSELINA 4-NITRO-4'-AMINOSTILBEN-2,2'-DISULFONOVA (CZECH)
　　see ALP750
KYSELINA NITROBENZEN-m-SULFONOVA (CZECH) see NFB500
KYSELINA-p-NITROBENZOOVA (CZECH) see CCI250
KYSELINA-4-NITROTOLUEN-2-SULFONOVA (CZECH) see MMH250
KYSELINA STAVELOVA (CZECH) see OLA000
KYSELINA SULFAMINOVA (CZECH) see SNK500
KYSELINA 1-SULFOMETHYL-2-NAFTYLAMIN-6-SULFONOVA (CZECH)
　　see AMP000
KYSELINA SULFO-TOBIAOVA (CZECH) see ALH000
KYSELINA TERFTALOVA (CZECH) see TAN750
KYSELINA-β,β'-THIODIPROPIONOVA (CZECH) see BHM000
KYSELINA TOBIASOVA (CZECH) see ALH750
KYSELINA p-TOLUENESULFONOVA (CZECH) see TGO000
KYSELINA 2-TOLUIDIN-4-SULFONOVA (CZECH) see AMT000
KYSELINA-3-TOLUIDIN-6-SULFONOVA (CZECH) see AMT250
KYSELINA-4-TOLUIDIN-3-SULFONOVA (CZECH) see AKQ000
KYSELINA o-TOSYL-H (CZECH) see AKH500
KYSELINA TRICHLOISOKYANUROVA (CZECH) see TIQ750
KYSELINA WOLFRAMOVA (CZECH) see TOD000
KYSLICNIK DI-n-AMYLCINICITY (CZECH) see DVV000
KYSLICNIK DI-n-BUTYLCINICITY (CZECH) see DEF400
KYSLICNIK DIISOAMYLCINICITY (CZECH) see DNL400
KYSLICNIK DIISOBUTYLCINICITY (CZECH) see DNJ000
KYSLICNIK DIISOPROPYLCINICITY (CZECH) see DNR200
KYSLICNIK DI-N-PROPYLCINICITY (CZECH) see DWV000
KYSLICNIK TRI-N-BUTYLCINICITY (CZECH) see BLL750
KYURINETT see YCJ200
KZ 3M see SCQ000
KZ 5M see SCQ000
KZ 7M see SCQ000
K-ZINC see ZKA000

84L see DIW000
L-99 see CDG250
L-105 see CCS635
L-310 see LGK000
L343 see IOT000
L-395 see DSP400
L-561 see DRR400
L. 1633 see DEE600
L 1718 see DYE700
L 1811 see DJT400
L-2103 see POA250
L2214 see DDP200
L. 3428 see AJK750
L-5103 see RKP000
L 6150 see BKB500
L 8580 see MNB250
L-01748 see DJT800
L-10492 see MFJ105
L 10499 see PEU650
L-10503 see MFF650
L 11204 see EFC259
L 11373 see PGN840
L 11752 see MFG275
L 12236 see CKA630
L 12717 see CKL325
L 14105 see BGO325
L-36352 see DUV600
L 586.153 see NNX300
LA see DXY000
LA 1 see DLY000
LA 1221 see PEU000
LA 6023 see DQR600
LAAM see ACQ666
LAAM HYDROCHLORIDE see ACQ690
LA'AU-'AILA (HAWAII) see CCP000
LABA see LGK100
LABAZ see AJK750
LABAZENE see PNX750

LABDANOL see IIQ000
LABDANUM OIL see LAC000
LABETALOL HYDROCHLORIDE see HMM500
LABICAN see MDQ250
LABILITE see MAP300
LABITON see DHF600
LABOPAL see DKQ000
LABOR-NR 2683 see DHR800
LABURNUM see GIW195
LABURNUM ANAGYROIDES see GIW195
LABYRIN see CMR100
LAC-43 see BOO000
LACHESIN see ELF500
LACHESINE CHLORIDE see ELF500
LAC LSP-1 see LIJ000
LACOLIN see LAM000
LACQREN 550 see SMQ500
LACQUER DILUENT see ROU000
LACQUER ORANGE V see TGW000
LACQUER ORANGE VG see PEJ500
LACQUER ORANGE VR see XRA000
LACQUER RED V3B see EOJ500
LACQUERS see LAD000
LACQUERS, NITROCELLULOSE see LAE000
LACRETIN see FOS100
LACRIMIN see OPI300
LACTASE see GAV100
LACTATE DEHYDROGENASE X see LAO300
LACTATE d'ETHYLE (FRENCH) see LAJ000
LACTIC ACID see LAG000
dl-LACTIC ACID see LAG000
racemic LACTIC ACID see LAG000
LACTIC ACID, ANTIMONY SALT see AQE250
LACTIC ACID, BERYLLIUM SALT see LAH000
LACTIC ACID, BUTYL ESTER see BRR600
LACTIC ACID, BUTYL ESTER, BUTYRATE see BQP000
LACTIC ACID, CADMIUM SALT see CAG750
LACTIC ACID, ETHYL ESTER see LAJ000
LACTIC ACID, IRON(2+) SALT (2581) see LAL000
LACTIC ACID, LEAD(2+) SALT (2:1) see LDL000
LACTIC ACID LITHIUM SALT see LHL000
LACTIC ACID, MONOSODIUM SALT see LAM000
LACTIC ACID, NEODYMIUM SALT see LAN000
LACTIC ACID SODIUM SALT see LAM000
LACTIC ACID, SODIUM ZIRCONIUM SALT (4584581) see LAO000
LACTIC ACID, TRIS(2-HYDROXYETHYL)(PHENYLMERCURI)AMMONIUM
　　derivative see TNI500
LACTIC ACID, ion(1−), TRIS(2-HYDROXYETHYL)PHENYLMERCURIO)
　　AMMONIUM see TNI500
LACTIC ACID, ZIRCONIUM SALT (3:1) see ZRJ000
LACTIC ACID, ZIRCONIUM SALT (4:1) see ZRS000
LACTIC DEHYDROGENASE X see LAO300
LACTIN see LAR000
LACTOBACILLUS LACTIS DORNER FACTOR see VSZ000
LACTOBARYT see BAP000
LACTOBIOSE see LAR000
LACTOCAINE see AIT250
LACTOFLAVIN see RIK000
LACTOFLAVINE see RIK000
LACTOGEN see PMH625
LACTOGENIC HORMONE see PMH625
ϵ-LACTONE HEXANOIC ACID see LAP000
LACTONITRILE see LAQ000
LACTOSE see LAR000
d-LACTOSE see LAR000
LACTOSOMATOTROPIC HORMONE see PMH625
LACTULOSE see LAR100
LACUMIN see MOQ250
LADAKAMYCIN see ARY000
LADIE'S THIMBLES see FOM100
LADOGAL see DAB830
LADY LAUREL see LAR500
LAE-32 see LJI000
LAETRILE see LAS000
LAEVORAL see LFI000
LAEVOSAN see LFI000
LAEVOXIN see LEQ300
LAEVULIC ACID see LFH000
LAEVULINIC ACID see LFH000
LAGOSIN see FPC000
LAGRIMAS de MARIA see CAL125
LAIDLOMYCIN see DAK000
LAI (HAITI) see WBS850
LAKANA, MIKINOLIA-HIHIU (HAWAII) see LAU600

LAKE BLUE B BASE see DCJ200
LAKE PONCEAU see FMU070
LAKE RED 4R see MMP100
LAKE RED C see CHP500
LAKE RED KB BASE see CLK225
LAKE RED 2GL see DVB800
LAKE SCARLET G BASE see NMP500
LAKE SCARLET GG BASE see DEO400
LAKE YELLOW see FAG140
LALOI (HAITI) see AGV875
LAMAR see FMB000
LAMBAST see CFW750
LAMBDAMYCIN see CDK250
LAMBETH see PMP500
LAMB KILL see MRU359
LAMBRATEN see AJT250
LAMBRIL see BBW750
LAMBROL see FDB200, FIP999
LAMDIOL see EDO000
LAMIDON see IIU000
LAMITEX see SEH000
LAMIUM ALBUM LINN., EXTRACT see LAS500
LAMORYL see GKE000
LAMP BLACK see CBT750
LAMPIT see NGG000
LAMPTEROL see LIO600
LAMURAN see AFG750
LANACORT see HHQ800
LANADIGENIN see DKN300
LANADIN see TIO750
LANALENE L see IPS500
LANALENE P see IPS500
LANALENE S see IPS500
LANASYN GREEN BL see CMM400
LANATOSID A (GERMAN) see LAT000
LANATOSID B (GERMAN) see LAT500
LANATOSID C (GERMAN) see LAU000
LANATOSIDE A see LAT000
LANATOSIDE B see LAT500
LANATOSIDE C see LAU000
LANATOSIDES see LAU400
LANATOXIN see DKL800
LANAZINE see DBA800
LANCOL see OBA000
LANDALGINE see AFL000
LANDAMYCINE see RIP000
LANDISAN see MEO750
LANDOCAINE see BQA010
LAND PLASTER see CAX750
LANDRAX see CQM325
LANDRIN see TMC750, TMD000
LANDRIN, NITROSO DERIVATIVE see NLY500
LANDRUMA see NDX500
LANESTA see CDV700
LANESTA L see IPS500
LANESTA P.S. see IPS500
LANETTE WAX-S see SIB600
LANEX see DUK800
LANGFORD see KBB600
LANGORAN see CCK125
LANI-ALI'I (HAWAII) see AFQ625
LANIAZID see ILD000
LANICOR see DKN400
LANIMERCK see DBH200
LANIRAPID see MJD300
LANITOP see MJD300
LANNAGOL LF see PJY100
LANNATE see MDU600
LANOL see CMD750
LANOPHYLLIN see TEP000
LANOSTABIL see LAU400
LANOXIN see DKN400
LANSTAN see CJE000
LANTANA see LAU600
LANTANA CAMARA see LAU600
LANTHANACETAT (GERMAN) see LAW000
LANTHANUM see LAV000
LANTHANUM ACETATE see LAW000
LANTHANUM AMMONIUM NITRATE see ANL500
LANTHANUM CHLORIDE see LAX000
LANTHANUM DIHYDRIDE see LAY499
LANTHANUM EDETATE see LAZ000
LANTHANUM HYDRIDE see LBC000
LANTHANUM NITRATE see LBA000

LANTHANUM SULFATE see LBB000
LANTHANUM(III) SULFATE (2583) see LBB000
LANTHANUM TRIACETATE see LAW000
LANTHANUM TRIHYDRIDE see LBC000
LANTOSIDE see LAU400
LANVIS see AMH250
LAPACHIC ACID see HLY500
LAPACHOL see HLY500
LAPACHOL WOOD see HLY500
LAPAQUIN see CLD000
LAPEMIS HARDWICKII VENOM see SBI910
LAPPACONITINE see LBD000
LARAHA see LBE000
LARD FACTOR see VSK600
LARGACTIL see CKP250
LARGACTIL MONOHYDROCHLORIDE see CKP500
LARGACTILOTHIAZINE see CKP250
LARGACTYL see CKP250
LARGAKTYL see CKP500
LARGON HYDROCHLORIDE see PMT500
LARIXIC ACID see MAO350
LARIXIN see ALV000
LARIXINIC ACID see MAO350
LARKSPUR see LBF000
LAROCAINE see DNY000
LARODON see INY000
LARODOPA see DNA200
LAROXIL see EAH500
LAROXYL see EAH500
LARTEN see MQU750
LARVACIDE see CKN500
LAS see AFO500, CMV325
LASALOCID see LBF500
LASERDIL see CCK125
LASEX see CHJ750
LASIOCARPINE see LBG000
LASIX see CHJ750
LAS, MAGNESIUM SALT see LGF800
LAS-Mg see LGF800
LAS-Na see LGF825
LASODEX see TEP500
LASSO see CFX000
LAS, SODIUM SALT see LGF825
LATAMOXEF SODIUM see LBH200
LATEX see PJR000
LATEXOL RED J see CJD500
LATEXOL SCARLET R see CHP500
LATHOSTEROL see CMD000
LATHYRUS ODORATUS, SEEDS see SOZ000
LATIBON see DII200
LATICATOXIN see LBI000
LATICAUDA SEMIFASCIATA VENOM see LBI000
LATRODECTUS M. MACTANS VENOM see BLW500
LATSCHENKIEFEROL see PIH400
LATTULOSIO (ITALIAN) see LAR100
LATUSATE see AFY500
LAUDICON see DLW600
LAUDOCAINE see BQA010
LAUDRAN DI-n-BUTYLCINICITY (CZECH) see DDV600
LAUGHING GAS see NGU000
LAURAMIDE DEA see BKE500
LAUREL CAMPHOR see CBA750
LAUREL LEAF OIL see BAT500, LBK000
LAURELWOOD see MBU780
LAURETH see DXY000
LAURIC ACID see LBL000
LAURIC ACID, CADMIUM SALT (2:1) see CAG775
LAURIC ACID, DIBUTYLSTANNYLENE derivative see DDV600
LAURIC ACID, DIBUTYLSTANNYLENE SALT see DDV600
LAURIC ACID DIETHANOLAMIDE see BKE500
LAURIC ACID-2,3-EPOXYPROPYL ESTER see LBM000
LAURIC ACID ESTER with 2-HYDROXYETHYL THIOCYANATE see LBO000
LAURIC ACID, SODIUM SALT see LBN000
LAURIC ACID, 2-THIOCYANATOETHYL ESTER see LBO000
LAURIC ALCOHOL see DXV600
LAURIC DIETHANOLAMIDE see BKE500
LAURIER ROSE (HAITI) see OHM875
LAURINE see CMS850
LAURINIC ALCOHOL see DXV600
LAURODIN see AJS750
LAUROLINIUM ACETATE see AJS750
LAUROLITSINE see LBO100
LAUROMACROGOL 400 see DXY000
LAURONITRILE see DXT400

LAUROSCHOLTZINE see TKX700
LAUROSTEARIC ACID see LBL000
LAUROTETANIN see LBO200
LAUROTETANINE see LBO200
(+)-LAUROTETANINE see LBO200
LAUROX see LBR000
1-LAUROYLAZIRIDINE see LBQ000
LAUROYL DIETHANOLAMIDE see BKE500
LAUROYLETHYLENEIMINE see LBQ000
(LAUROYLOXY)TRIBUTYLSTANNANE see TIE750
LAUROYL PEROXIDE see LBR000
LAUROYL PEROXIDE, TECHNICALLY PURE (DOT) see LBR000
LAURYDOL see LBR000
LAURYL 24 see DXV600
LAURYL ACETATE see DXV400
LAURYL ALCOHOL (FCC) see DXV600
LAURYL ALCOHOL CONDENSED with 4 MOLES ETHYLENE OXIDE
  see LBS000
LAURYL ALCOHOL CONDENSED with 23 MOLES ETHYLENE OXIDE
  see LBU000
LAURYL ALCOHOL EO (4) see LBS000
LAURYL ALCOHOL EO (7) see LBT000
LAURYL ALCOHOL EO (23) see LBU000
LAURYL ALCOHOL, ETHOXYLATED see DXY000
n-LAURYL ALCOHOL, PRIMARY see DXV600
LAURYL ALDEHYDE (FCC) see DXT000
LAURYLAMINE see DXW000
LAURYL AMMONIUM SULFATE see SOM500
LAURYL DIETHANOLAMIDE see BKE500
LAURYLDIETHYLENETRIAMINE see LBV000
LAURYLDIMETHYLAMINE see DRR800
N-LAURYLDIMETHYLAMINE see DRR800
LAURYLDIMETHYLAMINE OXIDE see DRS200
LAURYLESTER KYSELINY DUSICNE (CZECH) see NEE000
LAURYLESTER KYSELINYMETHAKRYLOVE (CZECH) see DXY200
LAURYL GALLATE see DXX200
LAURYLGUANIDINE ACETATE see DXX400
LAURYLISOQUINOLINIUM BROMIDE see LBW000
LAURYL MERCAPTAN see LBX000
m-LAURYL MERCAPTAN see LBX000
LAURYL METHACRYLATE see DXY200
LAURYLNITRAT (CZECH) see NEE000
LAURYL NITRATE see NEE000
LAURYL POLYETHYLENE GLYCOL ETHER see DXY000
LAURYL RHODANATE see DYA200
LAURYL SODIUM SULFATE see SIB600
LAURYL SULFATE see MRH250
LAURYL SULFATE AMMONIUM SALT see SOM500
LAURYL SULFATE, SODIUM SALT see SIB600
LAURYL SULFURIC ACID see MRH250
LAURYL THIOCYANATE see DYA200
LAUSIT see IDA000
LAUTHSCHES VIOLETT (GERMAN) see AKK750
LAUXTOL see PAX250
LAUXTOL A see PAX250
LAV see CMY800
LAVANDIN OIL see LCA000
LAVANDULYL ACETATE see LCA100
LAVATAR see CMY800
LAVENDEL OEL (GERMAN) see LCD000
LAVENDER ABSOLUTE see LCC000
LAVENDER OIL see LCD000
LAVENDER OIL, SPIKE see SLB500
LAVSAN see PKF750
LAWN-KEEP see DAA800
LAWSONITE see PKF750
LAXAGEN see ACD500
LAXAGETTEN see ACD500
LAXANORM see DMH400
LAXANTHREEN see DMH400
LAXESIN see ELF500
LAXIDOGOL see SJJ175
LAXINATE see DJL000
LAXIPUR see DMH400
LAXIPURIN see DMH400
LAXOBERAL see SJJ175
LAXOBERON see SJJ175
LA XVII see BMN750
LAYOR CARANG see AEX250
LAZETA see CMR100
LAZO see CFX000
LB-46 see VSA000
LB 502 see CHJ750
(L)-BC-2605 see CQF079

LBF DISULFIDE see PAG150
LBI see LEF200
LB-ROT 1 see FAG040
LC 44 see FMO129
LC-80 see CCK660
LCR see LEY000
LD-813 see LCE000
L.D. 3055 see IBP200
LDA see BKE500
LDE see BKE500
LDH-X see LAO300
LD NORGESTREL (FRENCH) see NNQ500
LD RUBBER RED 16913 see CHP500
Le-100 see EIF000
29060 LE see VLA000
LEA-COV see SHF500
LEAD see LCF000
LEAD ACETATE see LCG000, LCV000
LEAD (2+) ACETATE see LCV000
LEAD(II) ACETATE see LCV000
LEAD(IV) ACETATE AZIDE see LCG500
LEAD ACETATE, BASIC see LCH000
LEAD ACETATE BROMATE see LCI000
LEAD ACETATE-LEAD BROMITE see LCI600
LEAD ACETATE TRIHYDRATE see LCJ000
LEAD ACETATE(II), TRIHYDRATE see LCJ000
LEAD ACID ARSENATE see LCK000
LEAD ARSENATE see ARC750, LCK000
LEAD ARSENATE, solid (DOT) see LCK000
LEAD ARSENATE (standard) see LCK000
LEAD(II) ARSENITE see LCL000
LEAD ARSENITE, solid (DOT) see LCL000
LEAD(II) AZIDE see LCM000
LEAD(IV) AZIDE see LCN000
LEAD AZIDE, DRY (DOT) see LCM000
LEAD BOTTOMS see LDY000
LEAD BROMATE see LCO000
LEAD BROWN see LCX000
LEAD CARBONATE see LCP000
LEAD(2+) CARBONATE see LCP000
LEAD CHLORIDE see LCQ000
LEAD (2+) CHLORIDE see LCQ000
LEAD (II) CHLORIDE see LCQ000
LEAD(II) CHLORITE see LCQ300
LEAD CHROMATE see LCR000
LEAD CHROMATE(VI) see LCR000
LEAD CHROMATE, BASIC see LCS000
LEAD CHROMATE OXIDE (MAK) see LCS000
LEAD CHROMATE, RED see LCS000
LEAD CHROMATE, SULPHATE and MOLYBDATE see LDM000
LEAD COMPOUNDS see LCT000
LEAD(II) CYANIDE see LCU000
LEAD CYANIDE (DOT) see LCU000
LEAD DIACETATE see LCV000
LEAD DIACETATE TRIHYDRATE see LCJ000
LEAD DIBASIC ACETATE see LCV000
LEAD DICHLORIDE see LCQ000
LEAD DIFLUORIDE see LDF000
LEAD DIMETHYLDITHOCARBAMATE see LCW000
LEAD DINITRATE see LDO000
LEAD DIOXIDE see LCX000
LEAD DIPHENYL ACID PROPIONATE see LCZ000
LEAD DIPHENYL NITRATE see LDA000
LEAD DIPICRATE see LDA500
LEAD DISODIUM EDTA see LDB000
LEAD DISODIUM ETHYLENEDINITRILOTETRACETATE see LDB000
LEAD DROSS (DOT) see LDC000, LDY000
LEAD DROSS (containing 3% or more free acid) (DOT) see LDC000
LEAD(II) EDTA COMPLEX see LDD000
LEAD FLAKE see LCF000
LEAD FLUOBORATE see LDE000
LEAD(II) FLUORIDE see LDF000
LEAD FLUORIDE (DOT) see LDF000
LEAD(II) FLUOROSILICATE see LDG000
LEAD GLYCERONITRATE see LDH000
LEAD HYPONITRITE see LDI000
LEAD HYPOPHOSPHITE see LDJ000
LEAD IMIDE see LDK000
LEAD LACTATE see LDL000
LEAD-MOLYBDENUM CHROMATE see LDM000
LEAD MONONITRORESORCINATE (DRY) (DOT) see LDP000
LEAD MONOSUBACETATE see LCH000
LEAD MONOXIDE see LDN000
LEAD NAPHTHENATE see NAS500

LEAD NITRATE see LDO000
LEAD (2+) NITRATE see LDO000
LEAD(II) NITRATE see LDO000
LEAD(II) NITRATE (1:2) see LDO000
LEAD NITRORESORCINATE see LDP000
LEAD(II) OLEATE (1:2) see LDQ000
LEAD ORTHOPHOSPHATE see LDU000
LEAD ORTHOPLUMBATE see LDS000
LEAD OXIDE see LDN000
LEAD(II) OXIDE see LDN000
LEAD(IV) OXIDE see LCX000
LEAD OXIDE BROWN see LCX000
LEAD OXIDE RED see LDS000
LEAD OXIDE YELLOW see LDN000
LEAD(II) PERCHLORATE see LDS499
LEAD(II) PERCHLORATE, HEXAHYDRATE (1:2:6) see LDT000
LEAD PEROXIDE (DOT) see LCX000
LEAD PHOSPHATE see LDU000
LEAD (2+) PHOSPHATE see LDU000
LEAD PHOSPHATE (3:2) see LDU000
LEAD(II) PHOSPHATE (3:2) see LDU000
LEAD(II) PHOSPHINATE see LDJ000
LEAD POTASSIUM THIOCYANATE see LDV000
LEAD PROTOXIDE see LDN000
LEAD S2 see LCF000
LEAD SCRAP (DOT) see LDC000
LEAD SILICATE see LDW000
LEAD STEARATE see LDX000
LEAD STYPHNATE (DRY) (DOT) see LEE000
LEAD SUBACETATE see LCH000
LEAD(II) SULFATE (1:1) see LDY000
LEAD SULFATE, solid, containing more than 3% free acid (DOT) see LDY000
LEAD SULFIDE see LDZ000
LEAD SUPEROXIDE see LCX000
LEAD(II) TARTRATE (1:1) see LEA000
LEAD TETRACETATE see LEB000
LEAD TETRACHLORIDE see LEC000
LEAD TETRAOXIDE see LDS000
LEAD(II) THIOCYANATE see LEC500
LEAD TITANATE see LED000
LEAD TREE see LED500
LEAD TRINITRORESORCINATE see LEE000
LEAD TRINITRORESORCINATE (DOT) see LEE000
LEAD 2,4,6-TRINITRORESORCINOXIDE see LEE000
LEAD(II) TRINITROSOBENZENE-1,3,5-TRIOXIDE see LEF000
LEAD(II) TRINITROSOPHLOROGLUCINOLATE see LEF000
LEAD TRIPROPYL see TNA500
LEAF ALCOHOL see HFE000
LEAF ALDEHYDE see HFA500
LEAF DROP see SFS500
LEAF GREEN see CMJ900
LEALGIN COMPOSITUM see CLY500
LEANDIN see COH250
LEATHER BLUE G see ADE500
LEATHER BUSH see LEF100
LEATHER FLOWER see CMV390
LEATHER GREEN B see FAE950
LEATHER GREEN SF see FAF000
LEATHER ORANGE HR see PEK000
LEATHER PURE BLUE HB see BJI250
LEATHERWOOD see LEF100
LEAVER WOOD see LEF100
LEBAYCID see FAQ900
LEBON 15 HYDROCHLORIDE see DYA850
LE CAPTANE (FRENCH) see CBG000
LECASOL see FOS100
LECENINE see HOC500
LECITHIN-BOUND IODINE see LEF200
LECITHIN IODIDE see LEF200
LECTOPAM see BMN750
LEDAKRIN see LEF300, NFW500
LEDERCILLIN VK see PDT750
LEDERFEN see BGL250
LEDERKYN see AKO500
LEDERLE AA223 see AFI625
LEDERMYCIN see MIJ500
LEDERMYCIN HYDROCHLORIDE see DAI485
LE DINITROCRESOL-4,6 (FRENCH) see DUS700
LEDON 11 see TIP500
LEDON 12 see DFA600
LEDON 114 see FOO509
LEDOSTEN see DJT400
LEFEBAR see EOK000
LEGUMEX see CLN750

LEGUMEX D see DGA000
LEGUMEX DB see CIR250
LEGUMEX EXTRA see BAV000
LEGURAME see CBL500
LEHYDAN see DKQ000
LEINOLEIC ACID see LGG000
LEIOPLEGIL see LEF400
LEIOPYRROLE see LEF400
LEIPZIG YELLOW see LCR000
LEIVASOM see TIQ250
LEKAMIN see TNF500
LEMAC 1000 see AAX250
LEMBROL see DCK759
LEMMATOXIN see LEF800
LEMOFLUR see SHF500
LEMON AJAX see AFG625
LEMON CHROME see BAK250
LEMONENE see BGE000
LEMONGRAS OEL (GERMAN) see LEG000
LEMONGRASS OIL EAST INDIAN see LEG000
LEMONGRASS OIL WEST INDIAN see LEH000
LEMON OIL see LEI000
LEMON OIL, desert type, coldpressed see LEI025
LEMON OIL, distilled see LEI030
LEMON OIL, COLDPRESSED (FCC) see LEI000
LEMON OIL, EXPRESSED see LEI000
LEMONOL see DTD000
LEMON PETITGRAIN OIL see LEJ000
LEMON YELLOW see BAK250, LCR000
LEMORAN see DYF000
LENAMPICILLIN HYDROCHLORIDE see LEJ500
LENAMYCIN see ACJ250
LENDINE see BBQ500
LENDORM see LEJ600
LENDORMIN see LEJ600
LENETRAN see MFD500
LENETRANAT see MFD500
LENETRAN TAB see MFD500
LENGUNA de VACA (CUBA, PUERTO RICO) see APM875
LENOREMYCIN see LEJ700
LENOTAN see BAV350
LENTAC see AKO500
LENTE see LEK000
LENTE INSULIN see LEK000
LENTIN see CBH250
LENTINAN see LEK100
LENTINE (FRENCH) see CBH250
LENTIZOL see EAI000
LENTOCILLIN see BFC750
LENTOPENIL see BFC750
LENTOTRAN see MDQ250
LENTOX see BBQ500
LEO 72a see BHO250
LEO 1727 see NNX600
LEO 640 HYDROCHLORIDE see IFZ900
LEOMYPEN see BFC750
LEOPENTAL see PBT500
LEOSTESIN see DHK400
LEOSTESIN HYDROCHLORIDE see DHK600
LEOSTIGMINE BROMIDE see POD000
LEOSTIGMINE METHYL SULFATE see DQY909
LEPARGYLIC ACID see ASB750
LEPASEN see SEP000
LEPENIL see MQU750
LEPETOWN see MQU750
LEPHEBAR see EOK000
LEPIDINE see EOK000
4-LEPIDINE see LEL000
LEPIDINE-1-OXIDE see LEM000
LEPIDINE-N-OXIDE see LEM000
LEPIMIDIN see DBB200
LEPINAL see EOK000
LEPITOIN see DKQ000, DNU000
LEPITOIN SODIUM see DNU000
LEPONEX see CMY250
LEPOTEX see CMY250
LEPSIN see DKQ000
LEPSIRAL see DBB200
LEPTAMIN see DJS200
LEPTANAL see DYF200, PDW750
LEPTODACTYLINE PICRATE see HNG000
LEPTOFEN see DYF200
LEPTON see DJT400
LEPTOPHOS see LEN000
LEPTRYL see LEO000

LERBEK see CMX850
LERCIGAN see DQA400
LERENOX see BDD000
LERGIGAN see DQA400, PMI750
LERGINE see CPQ250
LERGINE CHLORIDE see EAI875
LERGITIN see BEM500
LERGOTRILE MESYLATE see LEP000
LERTUS see BDU500
LESAN see DOU600
LESCOPINE BROMIDE see SBH500
LESSER CELANDINE see FBS100
LESSER HEMLOCK see FMU200
LESTEMP see HIM000
LETHALAIRE G-52 see TCF250
LETHALAIRE G-54 see PAK000
LETHALAIRE G-57 see SOD100
LETHALAIRE G-58 see CJT750
LETHALAIRE G-59 see OCM000
LETHANE see BPL250
LETHANE 60 see LBO000
LETHANE 384 see BPL250
LETHANE (special) see LEQ000
LETHANE 384 REGULAR see BPL250
LETHELMIN see PDP250
LETHIDROME see AFT500
LETHOX see TNP250
LETHURIN see TIO750
LETIDRONE see AFT500
LETTER see LEQ300
LETUSIN see PNA250
LETYL see MQU750
LEUCAENA LEUCOCEPHAIA see LED500
LEUCAENINE see HOC500
LEUCAENOL see HOC500
LEUCARSONE see CBJ000
LEUCENOL see HOC500
LEUCETHANE see UVA000
LEUCIDIL see BCA000
LEUCIN (GERMAN) see LES000
LEUCINE see LES000
l-LEUCINE see LES000
dl-LEUCINE see LER000
epsilon-LEUCINE see AJD000
LEUCINE, N-CARBOXY-, N-BENZYL 1-VINYL ESTER see CBR215
l-LEUCINE, N-((PHENYLMETHOXY)CARBONYL)-, ETHENYL ESTER
    see CBR215
LEUCINOCAINE see LET000
LEUCINOCAINE MESYLATE see LEU000
LEUCINOCAINE METHANESULFONATE see LEU000
LEUCO-4 see AEH000
LEUCO-1,4-DIAMINOANTHRAQUINONE see DBO600
LEUCOGEN see ARN800
LEUCOHARMINE see HAI500
LEUCOL see QMJ000
LEUCOLINE see IRX000, QMJ000
LEUCOMYCIN see SLC000
LEUCOMYCIN A3 see JDS200
LEUCOMYCIN A6 see LEV025
LEUCOMYCIN B see LEW000
LEUCOMYCIN TARTRATE see LEX000
LEUCOPARAFUCHSIN see THP000
LEUCOPARAFUCHSINE see THP000
LEUCOPIN see CMV000
LEUCOSOL GOLDEN YELLOW see DCZ000
LEUCOSULFAN see BOT250
1,4,5,8-LEUCOTETRAOXYANTHRAQUINONE see TDD250
LEUCOTHANE see UVA000
LEUCOTHOE (VARIOUS SPECIES) see DYA875
LEUCOVYL PA 1302 see AAX175
LEUKAEMOMYCIN C see DAC000
LEUKAEMOMYCIN D see DAC300
LEUKERAN see CDO500, POK000
LEUKERSAN see CDO500
LEUKICHTHOL see IAD000
LEUKO-1,4-DIAMINOANTHRACHINON (CZECH) see DBO600
LEUKOL see QMJ000
LEUKOMYAN see CDP250
LEUKOMYCIN A6 see LEV025
LEUKORAN see CDO500
LEUNA M see CIR250
LEUPEPTIN Ac-LL see LEX400
LEUPURIN see POK000
LEUROCRISTINE see LEY000

LEUROCRISTINE SULFATE (1581) see LEZ000
LEVADONE see MDP250
(−)-LEVALLORPHAN see AGI000
LEVALLORPHAN TARTRATE see LIH300
l-LEVALLORPHAN TARTRATE see LIH300
LEVALLORPHINE TARTRATE see LIH300
LEVAMISOLE see LFA000, LFA020
LEVAMISOLE HYDROCHLORIDE see LFA020
LEVANIL see EHP000
LEVANOX GREEN GA see CMJ900
LEVANOX RED 130A see IHD000
LEVANOX WHITE RKB see TGG760
LEVANXENE see CFY750
LEVANXOL see CFY750
LEVARGIN see AQW000
LEVARTERENOL see NNO500
LEVARTERENOL BITARTRATE see NNO699
LEVATROM see ARQ750
LEVAXIN see LEQ300
LEVEDRINE see BBK750
LEV HYDROCHLORIDE see LFA020
LEVIL see EHP000
LEVIUM see DCK759
LEVOARTERENOL see NNO500
LEVOGLUTAMID see GFO050
LEVOGLUTAMIDE see GFO050
LEVOMEPATE HYDROCHLORIDE see LFC000
LEVOMEPROMAZINE see MCI500
LEVOMETHADONE see MDO775
LEVOMETHORPHAN HYDROBROMIDE see LFD200
LEVOMYCETIN see CDP250
LEVOMYCETIN HEMISUCCINATE see CDP725
LEVOMYCETIN SUCCINATE see CDP725
LEVOMYCIN see EAD500
LEVOMYSOL HYDROCHLORIDE see LFA020
LEVONORADRENALINE see NNO500
LEVONOREPINEPHRINE see NNO500
LEVONORGESTREL see NNQ525
LEVOPHACETOPERANE HYDROCHLORIDE see LFO000
LEVOPHACETOPERAN HYDROCHLORIDE see LFO000
LEVOPHED see NNO500
LEVOPHED HYDROCHLORIDE see NNP000
LEVOPROMAZINE see MCI500
LEVORENIN see VGP000
LEVORIN see LFF000
LEVOROXINE see LEQ300
LEVORPHAN see LFG000
LEVORPHANOL see LFG000
LEVORPHANOL TARTRATE see DYF000
LEVORPHAN TARTRATE see DYF000
LEVOSAN see LFG100
LEVOTHROID see LEQ300
LEVOTHYL see MDO775, MDP250
LEVOTHYROX see LEQ300
LEVOTHYROXINE see TFZ275
LEVOTHYROXINE SODIUM see LEQ300, LEQ300
LEVOTOMIN see MCI500
LEVOXADROL HYDROCHLORIDE see LFG100
LEVUGEN see LFI000
LEVULIC ACID see LFH000
LEVULINIC ACID see LFH000
LEVULINIC ACID, ETHYL ESTER see EFS600
LEVULINIC ACID, TRIPHENYLSTANNYL ESTER see TMW250
LEVULOSE see LFI000
LEWISITE see CLV000
LEWISITE (ARSENIC COMPOUND) see CLV000
LEWISITE II see BIQ250
LEWISITE I OXIDE see DEW000
LEWIS-RED DEVIL LYE see SHS000
LEXATOL see SPC500
LEXIBIOTICO see ALV000
LEXOMIL see BMN750
LEXONE see MQR275
LEXOTAN see BMN750
LEXOTANIL see BMN750
LEY-CORNOX see BAV000
LEYMIN see BAV000
LEYSPRAY see CIR250
LEYTOSAN see ABU500, PFP500
LF 62 see POA250
LFA 2043 see GIA000
LFP 83 see FJT100
LG 61 see DJR700
LG 50043 see TKX250

L.G. 11,457 HYDROCHLORIDE see DHS200
L-GRUEN No. 1 (GERMAN) see CKN000
LGYCOSPERSE S-20 see PKL030
LH see ILE000, LIU300
LH 3012 see ZMA000
LH ANTISERUM see LIU350
LHAS see LIU350
L'HEMISULFATE de GUANIDINO METHYL-6-BENZODIOXANNE-1,4
    (FRENCH) see GKG300
LH RELEASING FACTOR see LIU360
LH-RELEASING HORMONE see LIU360
LH-RF see LIU360
LHRH see LIU360, LIU360
LH-RH see LIU380
LHRH DIACETATE see LIU380
LHRH DIACETATE TETRAHYDRATE see LIU400
LH-RH/FSH-RH see LIU360
LH 30/Z see ZMA000
LIANE BON GARCON (HAITI) see CMV390
LIATRIS see DAJ800
LIATRIX OLEORESIN see DAJ800
LIBAVIUS FUMING SPIRIT see TGC250
LIBERETAS see DCK759
LIBEXIN see LFJ000
LIBIOLAN see MQU750
LIBRATAR see CDQ500
LIBRAX see LFK000
LIBRININ see LFK000
LIBRITABS see LFK000
LIBRIUM see LFK000, MDQ250
LIBRIUM HYDROCHLORIDE see MDQ250
LICABILE HYDROCHLORIDE see LFK200
LICARAN HYDROCHLORIDE see LFK200
LICAREOL ACETATE see LFY100
LICHENIC ACID see FOU000
LICHENIFORMIN A see LFL500
LICHENOL see IRL000
LICHTGRUEN (GERMAN) see FAF000
LICIDRIL see DPE000
LICORICE see LFN300
LICORICE COMPONENT FM 100 see LFN000
LICORICE EXTRACT see LFN300
LICORICE ROOT see LFN300
LICORICE ROOT EXTRACT see LFN300
LICORICE VINE see RMK250
LIDA-MANTLE see DHK400
LIDAMYCIN CREME see NCG000
LIDENAL see BBQ500
LIDEPRAN HYDROCHLORIDE see LFO000
LIDEX see FDD150
LIDOCAINE see DHK400
LIDOCAINE HYDROCHLORIDE see DHK600
LIDOFENIN see LFO300
LIDOL see DAM600, DAM700
LIDOTHESIN HYDROCHLORIDE see DHK600
LIENOMYCIN see LFP000
LIFEAMPIL see AIV500, AOD125
LIFENE see MNZ000
LIFRIL see FMB000
LIGAND 222 see LFQ000
LIGHT CAMPHOR OIL see CBB500
LIGHT FAST YELLOW ES see SGP500
LIGHT GAS OIL see GBW025
LIGHT GREEN FCF YELLOWISH see FAF000
LIGHT GREEN LAKE see FAF000
LIGHT GREEN N see AFG500
LIGHTHOUSE CHROME BLUE 2R see HJF500
LIGHTHOUSE CHROME ORANGE see NEY000
LIGHT NAPHTHENIC DISTILLATE see MQV810
LIGHT NAPHTHENIC DISTILLATE, SOLVENT EXTRACT see MQV860
LIGHT NAPHTHENIC DISTILLATES (PETROLEUM) see MQV810
LIGHT OIL of CAMPHOR see CBB500
LIGHT ORANGE CHROME see LCS000
LIGHT ORANGE R see DVB800
LIGHT PARAFFINIC DISTILLATE see MQV815
LIGHT PARAFFINIC DISTILLATE, SOLVENT EXTRACT see MQV862
LIGHT RED see IHD000
LIGHT SF YELLOWISH (BIOLOGICAL STAIN) see FAF000
LIGHT SPAR see CAX750
LIGHT YELLOW JB see DEU000
LIGNASAN see EME100
LIGNASAN FUNGICIDE see BJT250
LIGNASAN-X see BJT250
LIGNOCAINE see DHK400

LIGNOCAINE HYDROCHLORIDE see DHK600
LIGNYL ACETATE see DTC000
LIGROIN see PCT250
LIGUSTRUM VULGARE see PMD550
LIHOCIN see CMF400
LIKUDEN see GKE000
LILACILLIN see SNV000
LILAILA see CDM325
LILAS (HAITI, DOMINICAN REPUBLIC) see CDM325
LILAS de NUIT (HAITI) see DAC500
LILIA-O-KE-AWAWA (HAWAII) see LFT700
LILIAL see LFT000
LILIAL-METHYLANTHRANILATE, Schiff's base see LFT100
LILLY 1516 see DQA400
LILLY 01516 see DQA400
LILLY 22113 see AGL875
LILLY 22451 see MDU500
LILLY 22641 see RDF000
LILLY 34,314 see DRP800
LILLY 36,352 see DUV600
LILLY 37231 see LEZ000
LILLY 38253 see SFQ500
LILLY 39435 see CCR890
LILLY 40602 see TEY000
LILLY-68169 see MRX000
LILLY 69323 see FAP100
LILLY 99094 see VGU750
LILLY 99638 see PAG075
LILLY COMPOUND LY133314 see TMP175
LILLY 99638 HYDRATE see CCR850
LILLY-OF-THE-VALLEY see LFT700
LILYAL see LFT000
LILYL ALDEHYDE see CMS850
LILY OF THE FIELD see PAM780
LILY-OF-THE-VALLEY BUSH see FBP520
LIMARSOL MALAGRIDE see ABX500
LIMAS see LGZ000
LIMBIAL see CFZ000
LIME see CAU500
LIME ACETATE see CAL750
LIME, BURNED see CAU500
LIME CHLORIDE see HOV500
LIMED ROSIN see CAW500
LIME-NITROGEN (DOT) see CAQ250
LIME OIL see OGO000
LIME OIL, distilled (FCC) see OGO000
LIME PYROLIGNITE see CAL750
LIMESTONE (FCC) see CAO000
LIME, UNSLAKED (DOT) see CAU500
LIME WATER see CAT225
LIMIT see DKE800
LIMONENE see MCC250
1-LIMONENE see MCC500
d-LIMONENE see LFU000
(+)-R-LIMONENE see LFU000
d-(+)-LIMONENE see LFU000
dl-LIMONENE see MCC250
(−)-LIMONENE (FCC) see MCC500
LIMONENE DIOXIDE see LFV000
LIMONENE OXIDE see CAL000
LIMONITE see LFW000
LIMONIUM NASHII see MBU750
LIN 1418 see EPD100
LINADRYL HYDROCHLORIDE see LFW300
LINALOL see LFX000
LINALOL ACETATE see LFY100
LINALOOL see LFX000
p-LINALOOL see LFY000
LINALOOL ACETATE see LFY100
LINALOOL ISOBUTYRATE see LGB000
LINALOOL OXIDE see LFY500
LINALYL ACETATE see LFY100
LINALYL ALCOHOL see LFX000
LINALYL-o-AMINOBENZOATE see APJ000
LINALYL ANTHRANILATE see APJ000
LINALYL BENZOATE see LFZ000
LINALYL CINNAMATE see LGA000
LINALYL FORMATE see LGA050
LINALYL ISOBUTYRATE see LGB000
LINALYL ISOVALERATE see LGC000
LINALYL PROPIONATE see LGC100
LINAMARIN see GFC100
LINAMIN see FQJ100
LINAMPHETA see AOB250, TEP500

LINARIS see TKX000
LINARODIN see MDW750
LINASEN see EOK000
LINAXAR see PFJ000
LINCOCIN see LGC200, LGD000, LGE000
LINCOCIN HYDROCHLORIDE HYDRATE see LGC200
LINCOLCINA see LGD000
LINCOLNENSIN see LGD000
LINCOLN RED 1002 see CJD500
LINCOMYCIN see LGD000
LINCOMYCINE (FRENCH) see LGD000
LINCOMYCIN HYDROCHLORIDE see LGE000
LINCOMYCIN, HYDROCHLORIDE, HEMIHYDRATE see LGC200
LINCOMYCIN, HYDROCHLORIDE HYDRATE see LGC200
LINCOMYCIN HYDROCHLORIDE MONOHYDRATE see LGC200
LINCTUSSAL see DEE600
LINDAGRAIN see BBQ500
LINDAN see DGP900
LINDANE see BBP750
α-LINDANE see BBQ000
β-LINDANE see BBR000
Δ-LINDANE see BFW500
LINDANE (ACGIH, DOT, USDA) see BBQ500
LINDATOX see EEM000
LINDEROL see NCQ820
LIN-DIBENZANTHRACENE see PAV000
LINDOL see TNP500
LINEAR ALKYLBENZENE SULFONATE see AFO500
LINEAR ALKYLBENZENESULFONATE, MAGNESIUM SALT see LGF800
LINEAR ALKYLBENZENESULFONATE, SODIUM SALT see LGF825
LINEAR ALKYLBENZENE SULPHONATE see AFO500
LINE RIDER see TAA100
LINESTRENOL see NNV000
LINEVOL 7-9 see LGF875
LINEVOL 79 see LGF875
LINEX 4L see DGD600
LINFADOL see AFJ400
LINFOLIZIN see CDO500
LINFOLYSIN see CDO500
LINGEL see BRF500
LINGRAINE see EDC500
LINGRAN see EDC500
LINGUSORBS see PMH500
LIN-NAPHTHOANTHRACENE see PAV000
LINODIL see HFG550
LINOLEIC ACID see LGG000
9,12-LINOLEIC ACID see LGG000
LINOLEIC ACID (oxidized) see LGH000
LINOLEIC ACID mixed with OLEIC ACID see LGI000
LINOLEIC DIETHANOLAMIDE see BKF500
LINOLENATE HYDROPEROXIDE see PCN000
LINOLENIC ACID see OAX100
LINOLENIC HYDROPEROXIDE see PCN000
(LINOLEOYLOXY)TRIBUTYLSTANNANE see LGJ000
LINORMONE see CIR250
LINOROX see DGD600
LINSEED OIL see LGK000
LINTON see CLY500
LINTOX see BBQ500
LINUREX see DGD600
LINURON see DGD600
LINURON (herbicide) see DGD600
LIOMYCIN see HGP550
LION'S BEARD see PAM780
LION'S MOUTH see FOM100
LIORESAL see BAC275
LIOTHYRONIN see LGK050
LIOTHYRONINE see LGK050
l-LIOTHYRONINE see LGK050
LIOXIN see VFK000
LIPAL 30W see PJY100
LIPAL 4LA see DXY000
LIPAL 400-DL see PJY100
LIPAL 20-OA see PJW500
LIPAMID see ARQ750
LIPAMIDE see DXN709
α-LIPAMIDE see DXN709
4-(dl-α-LIPAMIDO)BUTYRIC ACID see LGK100
LIPAMONE see DHB500
LIPAN see DUS700
LIPARITE see CAS000
LIPARON see DPA000
LIPASE see GGA800
LIPASE AP6 see LGK150

LIPAVIL see ARQ750
LIPAVLON see ARQ750
LIPAZIN see GGA800
LIPENAN see LGK200
LIPHADIONE see CJJ000
LIPIDE 500 see ARQ750
LIPIDSENKER see ARQ750
LIPOACIN see DXN709
LIPOAMID see DXN709
LIPOAMIDE see DXN709
α-LIPOAMIDE see DXN709
LIPOCLIN see CMV700
LIPOCOL L-4 see DXY000
LIPOCOL O-n20 see PJW500
LIPOCTON see DXN709
LIPO-DIAZINE see PPP500
LIPO EGMS see EJM500
LIPOFACTON see ARQ750
LIPOGLUTAREN see HMC000
LIPO GMS 410 see OAV000
LIPO GMS 450 see OAV000
LIPO GMS 600 see OAV000
LIPO-HEPIN see HAQ500
LIPOIC ACID see DXN800
α-LIPOIC ACID see DXN800
α-LIPOIC ACID AMIDE see DXN709
LIPOICIN see DXN709
LIPO-LEVAZINE see PPP500
LIPO-LUTIN see PMH500
LIPOMID see ARQ750
LIPONEURINA see DXO300
α-LIPONIC ACID see DXN800
LIPONORM see ARQ750
α-LIPONSAEURE (GERMAN) see DXN800
LIPOPILL see DTJ400
LIPOPOLYSACCHARIDE see PPQ600
LIPOPOLYSACCHARIDE, from B. ABORTUS Bang. see LGK350
LIPOPOLYSACCHARIDE, ESCHERICHIA COLI see LGK300
LIPOREDUCT see ARQ750
LIPORIL see ARQ750
LIPOSAN see DXN800
LIPOSID see ARQ750
LIPOSORB O-20 see SKV100
LIPOSORB S-20 see SKV150
LIPOSORB O see SKV100
LIPOSORB O-20 see PKL100
LIPOSORB S see SKV150
LIPOSORB S-20 see PKL030
LIPOTHION see DXN800
LIPOTRIL see CMF750
LIPOZYME see DXN709
LIPRIN see ARQ750
LIPRINAL see ARQ750
LIPTAN see IIU000
LIPUR see GCK300
LIQUACILLIN see BDY669
LIQUADIAZINE see PPP500
LIQUAEMIN see HAQ500
LIQUAEMIN SODIUM see HAQ550
LIQUAGESIC see HIM000
LIQUAMYCIN see TBX000
LIQUAMYCIN INJECTABLE see HOI000
LIQUA-TOX see WAT200
LIQUEFIED CARBON DIOXIDE see LGL000
LIQUEFIED PETROLEUM GAS see LGM000
LIQUEFIED PETROLEUM GAS (DOT) see IIC000
LIQUEMIN see HAQ500, HAQ550
LIQUIBARINE see BAP000
LIQUIDAMBAR STYRACIFLUA see SOY500
LIQUID BRIGHT PLATINUM see PJD500
LIQUID CAMPHOR see CBB500
LIQUID CARBONIC GAS see LGL000
LIQUID DERRIS see RNZ000
LIQUID DETERGENTS containing AES see DBB450
LIQUID ETHYENE see EIO000
LIQUIDOW see CAO750
LIQUID PITCH OIL see CMY825
LIQUID ROSIN see TAC000
LIQUIGEL see AHC000
LIQUIMETH see MDT750
LIQUIPHENE see ABU500
LIQUIPRIN see SAH000
LIQUIPRON see LGM200
LIQUI-SAN see MLH000

LIQUI-STIK see NAK500
LIQUITAL see EOK000
LIQUOPHYLLINE see TEP000
LIQUORICE see GIG000
LIRANOL see DQA600
LIRANOX see CIR500
LIRCAPYL see PFB350
LIRIO see AHI635
LIRIO (SPANISH) see BAR325, SLB250
LIRIO CALA (SPANISH) see CAY800
LIROBETAREX see CJX750
LIRO CIPC see CKC000
LIROHEX see TCF250
LIROMAT see OPK250
LIROMATIN see ABX250
LIRONOX see BQZ000
LIROPON see DGI400
LIROPREM see PAX250
LIROSTANOL see ABX250
LIROTAN see EIR000
LIROTHION see PAK000
LISACORT see PLZ000
LISERGAN see ABH500
LISERGAN HYDROCHLORIDE see DPI000
LISKONUM see LGZ000
LISOLIPIN see SKJ300
LISSAMINE AMARANTH AC see FAG020
LISSAMINE FAST YELLOW AE see SGP500
LISSAMINE LAKE GREEN SF see FAF000
LISSENPHAN see GGS000
LISSOLAMINE see HCQ500
LISTENON see HLC500
LISTICA see HNJ000
LISULFEN see AKO500
LISURIDE HYDROGEN MALEATE see LJE500
LITAC see ADY500
LITALER see HOO500
LITEX CA see SMR000
LITHAMIDE see LGT000
LITHANE see LGZ000
LITHARGE see LDN000
LITHARGE YELLOW L-28 see LDN000
LITHIC ACID see UVA400
LITHICARB see LGZ000
LITHIDRONE see AFT500
LITHINATE see LGZ000
LITHIUM see LGO000
LITHIUM ACETYLIDE see LGP875
LITHIUM ACETYLIDE COMPLEXED with ETHYLENEDIAMINE see LGQ000
LITHIUM ACETYLIDE-ETHYLENEDIAMINE COMPLEX see LGQ000
LITHIUM ALANATE see LHS000
LITHIUM ALUMINOHYDRIDE see LHS000
LITHIUM ALUMINUM HYDRIDE (DOT) see LHS000
LITHIUM ALUMINUM HYDRIDE, ETHEREAL (DOT) see LHS000
LITHIUM ALUMINUM TETRAHYDRIDE see LHS000
LITHIUM ALUMINUM TRI-tert-BUTOXYHYDRIDE see LGS000
LITHIUM AMIDE see LGT000
LITHIUM AMIDE, POWDERED (DOT) see LGT000
LITHIUM ANTIMONIOTHIOMALATE see LGU000
LITHIUM ANTIMONY THIOMALATE see LGU000
LITHIUM ANTIMONY THIOMALATE NONAHYDRATE see AQE500
LITHIUM AZIDE see LGV000
LITHIUM BENZENEHEXOXIDE see LGV700
LITHIUM BENZOATE see LGW000
LITHIUM BIS(TRIMETHYLSILYL)AMIDE see LGX000
LITHIUM BOROHYDRIDE (DOT) see LHT000
LITHIUM BROMIDE see LGY000
LITHIUM CARBONATE see LGZ000
LITHIUM CARBONATE (2581) see LGZ000
LITHIUM CARMINE see LHA000
LITHIUM CHLORIDE see LHB000
LITHIUM CHLOROACETYLIDE see LHC000
LITHIUM CHLOROETHYNIDE see LHC000
LITHIUM CHROMATE see LHD000
LITHIUM CHROMATE(VI) see LHD099
LITHIUM COMPOUNDS see LHE000
LITHIUM DIAZOMETHANIDE see LHE450
LITHIUM DIETHYL AMIDE see LHE475
LITHIUM-2,2-DIMETHYLTRIMETHYLSILYL HYDRAZIDE see LHE525
LITHIUM FERRO SILICON see LHK000
LITHIUM FLUORIDE see LHF000
LITHIUM FLUORURE (FRENCH) see LHF000
LITHIUM-1-HEPTYNIDE see LHF625
LITHIUM HEXAAZIDOCUPRATE(4⁻) see LHG000

LITHIUM HYDRIDE see LHH000
LITHIUM HYDRIDE (fused solid form) see LHI000
LITHIUM HYDRIDE in fused solid form (DOT) see LHI000
LITHIUM HYPOCHLORITE see LHJ000
LITHIUM HYPOCHLORITE COMPOUND, dry, containing more than 39% available chlorine (DOT) see LHJ000
LITHIUM IRON SILICON see LHK000
LITHIUM LACTATE see LHL000
LITHIUM METAL (DOT) see LGO000
LITHIUM METAL, IN CARTRIDGES (DOT) see LGO000
LITHIUM NITRIDE see LHM000
LITHIUM-4-NITROTHIOPHENOXIDE see LHM750
LITHIUM PENTAMETHYLTITANATE-BIS(2,2'-BIPYRIDINE) see LHM850
LITHIUM PERCHLORATE see LHM875
LITHIUM PERCHLORATE TRIHYDRATE see LHN000
LITHIUM PEROXIDE see LHO000
LITHIUMRHODANID see TFF500
LITHIUM SILICON see LHP000
LITHIUM SODIUM NITROXYLATE see LHQ000
LITHIUM SULFATE (2:1) see LHR000
LITHIUM SULFOCYANATE see TFF500
LITHIUM SULPHATE see LHR000
LITHIUM TETRAAZIDOALUMINATE see LHR650
LITHIUM TETRAAZIDOBORATE see LHR675
LITHIUM TETRADEUTEROALUMINATE see LHR700
LITHIUM TETRAHYDROALUMINATE see LHS000
LITHIUM TETRAHYDROBORATE see LHT000
LITHIUM TETRAMETHYLBORATE see LHT400
LITHIUM TETRAMETHYL CHROMATE(II) see LHT425
LITHIUM TETRAZIDO ALUMINATE see LHU000
LITHIUM THIOCYANATE see TFF500
LITHIUM-TIN ALLOY see LHV000
LITHIUM TRIETHYLSILYL AMIDE see LHV500
LITHOBID see LGZ000
LITHOCHOLIC ACID see LHW000
LITHOGRAPHIC STONE see CAO000
LITHOL see IAD000
LITHOL FAST SCARLET RN see MMP100
LITHONATE see LGZ000
LITHOPONE see LHX000
LITHOSOL ORANGE R BASE see NMP500
LITHOSPERMUM RUDERALE, root extract see LHX300
LITHOTABS see LGZ000
LITICON see DOQ400
LITMOMYCIN see GJS000
LITSEA CUBEBA OIL see LHX325
LITSOEINE see LBO200
LIVAZONE see FNF000
LIVIATIN see DYE425
LIVICLINA see CCS250
LIVIDOMYCIN see LHX350
LIVIDOMYCIN B see DAS600
LIVIDOMYCIN SULFATE see LHX360
LIVODYMYCIN see LHX350
LIVONAL see EGQ000
LIXIL see BON325
L.J. 48 see MBX800
L.J. 206 see CBR675
LJ 206 see CCH125
LL 1530 see NAC500
LL 1656 see BOM600
LLD FACTOR see VSZ000
LLONCEFAL see TEY000
LM-16 see CMS125
LM 91 see CJJ000
LM-209 see MCJ400
LM 280 see IDJ500
LM 433 see GKG300
LM-637 see BMN000
LM-2717 see CIR750
LM 2910 see LHX498
LM 2911 see LHX495
LM 2916 see LHX500
LM 2917 see LHX510
LM 2918 see LHX515
LM 2930 see LHX600
LM 22070 see DWD400
LM SEED PROTECTANT see MLH000
LOBAK see CKF500
LOBAMINE see MDT740
LO-BAX see HOV500
LOBELIA, INDIAN TOBACCO see LHY000
LOBELIA INFLATA see CCJ825
LOBELIA NICOTIANIFOLIA Roth ex R. & S., extract see LHX700

LOBELIA SIPHILITICA see CCJ825
LOBELINE see LHY000
LOBELINE HYDROCHLORIDE see LHZ000
(−)-LOBELINE HYDROCHLORIDE see LHZ000
LOBELIN HYDROCHLORIDE see LHZ000
LOBELLOA CARDINALIS see CCJ825
LOBENZARIT DISODIUM see LHZ600
LOBENZARIT SODIUM see LHZ600
LOBETRIN see ARQ750
LOBNICO see LHY000
LOBULANTINA see SCA750
LOBUTEROL see BQE250
LOCALYN see SPD500
LOCOID see HHQ825
LOCOWEED see LHZ700
LOCRON EXTRA see AHA000
LOCUNDIEZIDE see RFZ000
LOCUNDIOSIDE see RFZ000
LOCUST BEAN GUM see LIA000
LOCUTURINE see MPA050
LODESTONE YELLOW YB-57 see DEU000
LODIBON see DII200
LODITAC see OMY700
LODOPIN see ZUJ000
LODOSIN see CBQ500
LODOSYN see CBQ500
LOFEPRAMINE see DLH630
LOFEPRAMINE HYDROCHLORIDE see IFZ900
LOFETENSIN see LIA400
LOFETENSIN HYDROCHLORIDE see LIA400
LOFEXIDINE HYDROCHLORIDE see LIA400
LOFTYL see BOM600
LOGGERHEAD WEED (BARBADOS) see PIH800
LOISOL see TIQ250
LOKARIN see DPE200
LOKUNDJOSID (GERMAN) see RFZ000
LOKUNDJOSIDE see RFZ000
LOMAPECT see LFJ000
LOMBRICERA (PUERTO RICO) see PIH800
LOMBRICERO (CUBA) see APM875
LOMBRISTOP see TEX000
LO MICRON TALC 1 see TAB750
LO MICRON TALC, BC 1621 see TAB750
LO MICRON TALC USP, BC 2755 see TAB750
LOMIDIN see DBL800
LOMIDINE see DBL800
LOMIDINE ISOETHIONATE see DBL800
LOMOTIL see LIB000
LOMUDAL see CNX825
LOMUDAS see CNX825
LOMUPREN see DMV600
LOMUSTINE see CGV250
LOMYCIN see GJY000
LON 41 see DTM600
LON 798 see GKU300
LON-954 see PFV000
LONACOL see EIR000
LONAMIN see DTJ400
LONARID see HIM000
LONAVAR see AOO125
LONDOMIN see RCA200
LONDOMYCIN see MDO250
LONDON PURPLE see LIC000
LONDON PURPLE, solid (DOT) see LIC000
LONETHYL see LID000
LONETIL see LID000
LONGACILIAN see BFC750
LONGANOCT see BPF500
LONGATIN see NOA000
LONGATIN HYDROCHLORIDE see NOA500
LONGICIL see BFC750
LONGIFENE see BOM250, HGC500
β-LONGILOBINE see RFP000
LONGIN see AKO500
LONG PEPPER see PCB275
LONGUM see MFN500
LONICERA PERICLYMENUM see HGK700
LONICERA TATARICA see HGK700
LONICERA XYLOSTEUM see HGK700
LONIDAMINE see DEL200
LONIN 3 see BHO500
LONITEN see DCB000
LONOCOL M see MAS500
LONOMYCIN see LIF000

LONOMYCIN, MONOSODIUM SALT see LIG000
LONOMYCIN, SODIUM SALT see LIG000
LONTREL see DGJ100
LONTREL 3 see DGJ100
LONZA G see PKQ059
LOOPLURE INHIBITOR see DXU800
LO/OVRAL see NNL500
LOPEMID see LIH000
LOPEMIN see LIH000
LOPERAMIDE HYDROCHLORIDE see LIH000
LOPERYL see LIH000
LOPHINE see TMS750
LOPID see GCK300
LOPIRIN see MCO750
LOPRAMINE HYDROCHLORIDE see IFZ900
LOPREMONE see TNX400
LOPRESOR see SBV500
LOPRESS see HGP500
LOPRESSOR see SBV500
LOPURIN see ZVJ000
LOQUAT see LIH200
LORAMET see MLD100
LORAX see CFC250
LORAZEPAM see CFC250
LORDS-AND-LADIES see ITD050
LORETIN see IEP200
LOREX see DGD600
LORFAN see AGI000, LIH300
LORFAN TARTRATE see LIH300
LORIDIN see TEY000
LORINAL see CDO000
LORMETAZEPAM see MLD100
LORMIN see CBF250
LORO see DYA200
LOROL see DXV600
LOROL 20 see OEI000
LOROL 22 see DAI600
LOROL 24 see HCP000
LOROL 28 see OAX000
LOROL THIOCYANATE see DYA200
LOROMISIN see CDP250
LOROTHIDOL see TFD250
LOROX see BNM000, DGD600
LOROXIDE see BDS000
LOROX LINURON WEED KILLER see DGD600
LORPHEN see TAI500
LORPOX see BAR800
LORSBAN see CMA100
LORSILAN see CFC250
LOSANTIN see HOV500
N-LOST see BIE500
N-LOST (GERMAN) see BIE250
S-LOST see BIH250
LOSUNGSMITTEL APV see CBR000
LOTRIFEN see CKL325
LOTRIMIN see MRX500
(−)-LOTUCAINE HYDROCHLORIDE see TGJ625
LOTURINE see MPA050
LOTUSATE see AFY500
LOUISIANA LOBELIA see CCJ825
LOVAGE see PMD550
LOVAGE OIL see LII000
LOVE BEAD see RMK250
LO-VEL see SCH000
LOVISCOL see CBR675
LOVOSA see SFO500
LOVOZAL see DGA200
LOW BELIA see CCJ825
LOWETRATE see PBC250
LOWPSTRON see CHJ750
LOXACOR see LIA400
LOXACOR HYDROCHLORIDE see LIA400
LOXANOL 95 see OBA000
LOXANOL K see HCP000
LOXANOL M see OBA000
LOXAPINE see DCS200
LOXON see DFH600
LOXOPROFEN SODIUM DIHYDRATE see LII300
LOXURAN see DIW200
LOXYNIL (GERMAN) see HKB500
LOZILUREA see LII400
LOZOL see IBV100
LPG see BFC750, LGM000
L.P.G. (OSHA, ACGIH) see LGM000

LPG ETHYL MERCAPTAN 1010 see EMB100
LPS see EDK700, PPQ600
LPT see PKB500
LR-529 see CFO750
LRF see LIU360
LRH see LIU360
LS 121 see NAE100
L-S 519 see GCE500
LS 1727 see NNX600
L.S. 3394 see BLL750
LS 4442 see CLU000
LSD see DJO000
d-LSD see DJO000
LSD-25 see DJO000
LS 519 DIHYDROCHLORIDE see GCE500
LSD-25-PYRROLIDATE see LJM000
LSD TARTRATE see LJG000, LJM600
(+)-LSD TARTRATE see LJM600
d-LSD TARTRATE see LJM600
LSD 25 TARTRATE see LJM600
LSM-775 see LJJ000
LSP 1 see LIJ000
LTAN see DMH400
1,4,5,8-LTOA (RUSSIAN) see TDD250
LU 1631 see MQS100
LUBALIX see CMY125
LUBERGAL see AGQ875, EOK000
LUBRICATING OIL see LIK000
LUBRICATING OIL (mainly mineral) see LIK000
LUBRICATING OIL, CYLINDER see LIK000
LUBRICATING OIL, MOTORS see LIK000
LUBRICATING OIL, SPINDLE see LIK000
LUBRICATING OIL, TURBINE see LIK000
LUBROL 12A9 see DXY000
LUCALOX see AHE250
LUCAMIDE see EEM000
LUCANTHON see DHU000
LUCANTHONE see DHU000
LUCANTHONE HYDROCHLORIDE see SBE500
LUCANTHONE METABOLITE see LIM000
LUCANTHONE MONOHYDROCHLORIDE see SBE500
LUCEL see CMA500
LUCEL (polysaccharide) see SFO500
LUCEL ADA see ASM270
LUCENSOMYCIN see LIN000
LUCIDIN ETHYL ETHER see DMT200
LUCIDOL see BDS000
LUCIDRIL see AAE500
LUCIDRYL see DPE000
LUCIDRYL HYDROCHLORIDE see AAE500
LUCIJET see LIN400
LUCITE see PKB500
LUCKNOMYCIN see LIN600
LUCKY NUT see YAK350
LUCOFEN see ARW750
LUCOFENE see ARW750
LUCOFLEX see PKQ059
LUCORTEUM ORAL see GEK500
LUCORTEUM SOL see PMH500
LUCOVYL PE see PKQ059
LUDIGOL F,60 see NFC500
LUDIOMIL see LIN800, MAW850
LUDOX see SCH000
LUEH 6 see BGS250
LUETOCRIN DEPOT see HNT500
LUFYLLIN see DNC000
LUGOL'S IODINE see PLW285
LUGOL'S SOLUTION see PLW285
LUH6 see BGS250
LUH(6) see BGS250
LUH6-CHLORIDE see BGS250
LUH(6)-Cl2 see BGS250
LUH6-DINITRAT (GERMAN) see OPY000
LULAMIN see TEH500, TEO250
LULIBERIN see LIU360
LULLAMIN see DPJ400, TEO250
LUMBANG (GUAM) see TOA275
LUMBRICAL see PIJ000
LUMEN see EOK000
LUMESETTES see EOK000
LUMICREASE BLUE 4GL see CMN750
LUMILAR 100 see PKF750
LUMINAL see EOK000
LUMINAL SODIUM see SID000

LUMIRROR see PKF750
LUMNITZERA RACEMOSA Willd., extract excluding roots see LIN850
LUMOFRIDETTEN see EOK000
LUNACOL see ZJS300
LUNAMIN see POA250
LUNAMYCIN see LIO600
LUNAR CAUSTIC see SDS000
LUNARINE HYDROCHLORIDE see LIP000
LUNETORON see BON325
LUNIPAK see DAB800
LUNIS see FDD085
LUPANIN see LIQ800
LUPANINE see LIQ800
(+)-LUPANINE see LIQ800
d-LUPANINE see LIQ800
LUPAREEN see PMP500
LUPERCO see BDS000, DGR600
LUPERCO CST see BIX750
LUPEROX see DGR600
LUPEROX 500R see DGR600
LUPEROX 500T see DGR600
LUPEROX FL see BDS000
LUPERSOL see MKA500
LUPERSOL 70 see BSC250
LUPIN see LIT000
LUPINIDINE see SKX500
LUPINUS see LIT000
LUPINUS ANGUSTIFOLIUS, seed alkaloid mixture see LIT000
β-LUPULIC ACID see LIU000
LUPULON see LIU000
LUPULONE see LIU000
LURAFIX BLUE FFR see MGG250
LURAMIN see EOK000
LURAN see ADY500
LURAZOL BLACK BA see AQP000
LURGO see DSP400
LURIDE see SHF500
LUSIL see SNM500
LUSTRAN see ADY500
LUSTREX see SMQ500
LUTALYSE see POC750
LUTAMIN see LIU100
LUTATE see HNT500
LUTEAL HORMONE see PMH500
LUTEINIZING GONADOTROPIC HORMONE see LIU300
LUTEINIZING HORMONE see LIU300
LUTEINIZING HORMONE ANTISERUM see LIU350
LUTEINIZING HORMONE, GONADOGRAPHON RELEASING HORMONE see GJI075
LUTEINIZING HORMONE-RELEASING FACTOR see LIU360
LUTEINIZING HORMONE-RELEASING HORMONE see LIU360
LUTEINIZING HORMONE-RELEASING HORMONE, DIACETATE (SALT) see LIU380
LUTEINIZING HORMONE-RELEASING HORMONE, DIACETATE, TETRAHYDRATE see LIU400
LUTEINIZING HORMONE-RELEASING HORMONE (PIG), 6-(O-(1,1-DIMETHYLETHYL)-d-SERINE)-9-(N-ETHYL-l-PROLINAMIDE)-10-DEGLYCINAMIDE- see LIU420
LUTEINIZING HORMONE-RELEASING HORMONE, (d-SER(TBU)^6-EA^{10})- see LIU420
LUTEOANTINE see FMT100
LUTEOCRIN see HNT500
LUTEOHORMONE see PMH500
LUTEOSAN see PMH500
LUTEOSKYRIN see LIV000
(−)-LUTEOSKYRIN see LIV000
LUTEOSTIMULIN see LIU360
LUTEOTROPHIN see PMH625
LUTEOTROPIC HORMONE see PMH625
LUTEOTROPIC HORMONE LTH see PMH625
LUTEOTROPIN see PMH625
LUTEOZIMAN see LIU300
LUTESTRAL (FRENCH) see CNV750
LUTETIA FAST ORANGE 3R see CJD500
LUTETIA FAST ORANGE R see DVB800
LUTETIA FAST SCARLET RF see MMP100
LUTETIA RED CLN see CHP500
LUTETIUM CHLORIDE see LIW000
LUTETIUM CITRATE see LIX000
LUTETIUM(III) NITRATE (1583) see LIY000
LUTEX see PMH500
2,5-LUTIDINE see LJA000
3,4-LUTIDINE see LJB000
LUTIDON ORAL see GEK500

LUTINYL see CBF250
LUTOCYCLIN see PMH500
LUTOCYLOL see GEK500
LUTOFAN see PKQ059
LUTOGAN see DRT200
LUTO-METRODIOL see EQJ500
LUTOPRON see HNT500
LUTOSAN see DRT200
LUTOSOL see INJ000
LUTRELEF see LIU360
LUTROL see EAE675, EIM000, PJT000
LUTROL-9 see EJC500
LUTROMONE see PMH500
LUTROPIN see LIU300
LUVATRENE see MNN250
LUVISKOL see PKQ250
LUXISTELM see PHI500
LUXOMNIN see AGQ875
LUXON see DFH600
LVM see LHX350
LW 3170 see DCS200
LX 14-0 see CQH250
LXON see DFH600
LY 61017 see BGE125
LY 127935 see LBH200
LY133314 see TMP175
LY150720 see PIB700
LYCANOL see GHK200
LYCEDAN see AOA125
LYCOBETAINE see LJB800
LYCOID DR see GLU000
LYCOPERSICIN see THG250
LYCOPERSIN see LJB700
LYCORANIUM, 1,2,3,3a,6,7,12b,12c-OCTADEHYDRO-2-HYDROXY- see LJB800
LYCORCIDINOL see LJC000
LYCOREMINE HYDROBROMIDE see GBA000
LYCORICIDIN-A see LJC000
LYCORICIDINOL see LJC000
LYCORIS AFRICANA see REK325
LYCORIS RADIATA see REK325
LYCORIS SQUAMIGERA see REK325
LYCURIM see LJD500
LYDOL see DAM700
LYE see PLJ500
LYE (DOT) see SHS000
LYE, solution see SHS500
LYGODIUM FLEXUOSUM (LINN.) SWARTZ., EXTRACT see LJD600
LYGOMME CDS see CCL250
LYKURIM see LJD500
LYMECYCLINE see MRV250
LYMPHCHIN see AOQ875
LYMPHOCIN see AOQ875
LYMPHOQUIN see AOQ875
LYMPHOSCAN see AQL500
LYNAMINE see AHK750
LYNDIOL see LJE000
LYNENOL see NNV000
LYNESTRENOL see NNV000
LYNESTRENOL mixed with MESTRANOL see LJE000
LYNESTROL see EEH550
LYNESTROL mixed with MESTRANOL see LJE000
LYNOESTRENOL see NNV000
LYNOESTRENOL mixed with ETHINYLOESTRADIOL see EEH575
LYNOESTRENOL mixed with MESTRANOL see LJE000
LYOBEX see NOA000
LYOBEX HYDROCHLORIDE see NOA500
LYONIATOXIN see LJE100
LYONIOL A see LJE100
LYOPECT see CNG250
LYOPHRIN see AES000, VGP000
LYOVAC COSMEGEN see AEB000
LYP 97 see LBR000
LYPOARAN see DXN709
LYSALGO see XQS000
LYSANXIA see DAP700
LYSENYL see LJE500
LYSENYL BIMALEATE see LJE500
LYSENYL HYDROGEN MALEATE see LJE500
LYSERGAMID see DJO000
LYSERGAURE DIETHYLAMID see DJO000
d-LYSERGIC ACID DIETHYLAMIDE see DJO000, LJF000
LYSERGIC ACID DIETHYLAMIDE-25 see DJO000
LYSERGIC ACID DIETHYLAMIDE, 1-ISOMER see LJF000
LYSERGIC ACID DIETHYLAMIDE TARTRATE see LJG000

d-LYSERGIC ACID DIETHYLAMIDE TARTRATE see LJG000
d-LYSERGIC ACID DIMETHYLAMIDE see LJH000
LYSERGIC ACID ETHYLAMIDE see LJI000
d-LYSERGIC ACID-1-HYDROXYMETHYLETHYLAMIDE see LJL000
d-LYSERGIC ACID MONOETHYLAMIDE see LJI000
LYSERGIC ACID MORPHOLIDE see LJJ000
d-LYSERGIC ACID MORPHOLIDE see LJJ000
d-LYSERGIC ACID MORPHOLIDE, TARTARIC ACID SALT see LJK000
LYSERGIC ACID PROPANOLAMIDE see LJL000
d-LYSERGIC ACID-l,2-PROPANOLAMIDE see LJL000
LYSERGIC ACID PYROLIDATE see LJM000
d-LYSERGIC ACID PYRROLIDIDE see LJM000
LYSERGIDE see DJO000
LYSERGIDE TARTRATE see LJM600
LYSERGSAUEREDIAETHYLAMID see DJO000
LYS (HAITI) see SLB250
LYSINE see LJM700
l-(+)-LYSINE see LJM700
l-LYSINE (9CI) see LJM700
l-LYSINE ACETATE see LJM800
dl-LYSINE ACETYLSALICYLATE see ARP125
dl-LYSINE ACETYLSALICYLIC ACID SALT see ARP125
LYSINE ACID see LJM700
d-LYSINE HYDROCHLORIDE see LJN000
l-LYSINE HYDROCHLORIDE see LJO000
l-LYSINE, MONOACETATE see LJM800
dl-LYSINE MONO(2-(ACETYLOXY)BENZOATE see ARP125
LYSINE MONOHYDROCHLORIDE see LJO000
l-LYSINE MONOHYDROCHLORIDE see LJO000
N-LYSINOMETHYLTETRACYCLINE see MRV250
LYSIVANE see DIR000
LYSOCELLIN see LJP000
LYSOCOCCINE see SNM500
LYSOFON see AGW750
LYSOFORM see FMV000
LYSOL see LJP500
LYSOSTAPHIN see LJQ000
LYSSIPOLL see LJR000
LYSTENON see HLC500
LYSTHENONE see HLC500
LYSURON see ZVJ000
LYTECA SYRUP see HIM000
LYTHIDATHION see DRU400
LYTISPASM see LJS000
LYTRON 5202 see SMR000

M-2 see CAR875
M-14 see RKZ000
M-30 see DAE695
M-32 see DAE700
M 40 see CIR250
M 73 see DFV600
M-74 see DXH325
M 81 see PHI500
M-101 see TFY000
M-108 see TAL575
M 140 see CDR750
M 176 see DUD800
M 259 see AAC000
M 410 see CDR750
M 551 see PFB350
M 1703 see IOV000
M 2060 see FIP999
M 3180 see CQH650
M-4209 see CBT250
M-4212 see MAB050
M 4888 see CKB500
M 5055 see CDV100
5512-M see PDC000
M5943 see DGD100
M 6/42 see RSU450
M-9500 see TND500
M-12210 see MAB055
M 3/158 see DAP600
M73101 see EAN700
M-6029 see BOO630
M-4365A2 see RMF000
M 4212 (pesticide) (9CI) see MAB050
MA see DAL300, MNQ500
4-MA see DJP700
MA-110 see EHP000
MA-117 see EHO700
MA-1214 see DXV600
MA 1291 see QWJ500

MA 1337 see CMX840
MAA see AFI850, MGQ530
MA-144A2 see TAH675
1-MA-40EAA (RUSSIAN) see MGG250
MAAC see HFJ000
MAA SODIUM SALT see DXE600
MAB see MNR500
MA144 B2 see TAH650
MABERTIN see CFY750
MABLIN see BOT250
MABUTEROL see MAB300
MACASIROOL see CHJ750
MACBAL see DTN200
MACBECIN II see MAB400
MACE (lachrymator) see CEA750
MACE OIL see OGQ100
MACHETE see CFW750
MACHETE (herbicide) see CFW750
MACHETTE see CFW750
MACH-NIC see NDN000
MACKREAZID see COH250
MACLEYINE see FOW000
MACODIN see MDP750
MACQUER'S SALT see ARD250
MACRABIN see VSZ000
MACROCYCLON see TDN750
MACRODANTIN see NGE000
MACRODIOL see EDO000
MACROGOL 1000 see PJT250
MACROGOL 4000 see PJT750
MACROGOL 400 BPC see EJC500
MACROGOL OLEATE 600 see PJY100
MACROL see EDO000
MACROMOMYCIN see MAB750
MACRONDRAY see DAA800
MACROPAQUE see BAP000
MACROSE see DBD700
MACULOTOXIN see FOQ000
MAD see AOO475
MADAGASCAR LEMONGRASS OIL see LEH000
MADAME FATE (JAMAICA) see SLJ650
MADAR see CGA500
MADECASSOL see ARN500
MADEIRA IVY see AFK950
MADHURIN see SJN700
MADIOL see AOO475, MQU750
MADRESELVA see HGK700
MADRESELVA (MEXICO) see YAK100
MADRIBON see SNN300
MADRIGID see SNN300
MADRINE see DBA800
MADRIQID see SNN300
MADROXIN see SNN300
MADROXINE see SNN300
MAFENIDE ACETATE see MAC000
MAFU see DGP900
MAG see MRF000
MAGBOND see BAV750
MAGENTA see MAC250
MAGENTA BASE see MAC500
MAGIC LILY see REK325
MAGIC METHYL see MKG250
MAGK'S T.B.1 CONTEBEN see FNF000
MAGMASTER see MAC650
MAGNACAT see MAP750
MAGNACIDE H see ADR000
MAGNAMYCIN see CBT250
MAGNAMYCIN A see CBT250
MAGNAMYCIN HYDROCHLORIDE see MAC600
MAGNESIA see MAH500
MAGNESIA ALBA see MAC650
MAGNESIA MAGMA see MAG750
MAGNESIA USTA see MAH500
MAGNESIA WHITE see CAX750
MAGNESIO (ITALIAN) see MAC750
MAGNESITE see MAC650
MAGNESIUM see MAC750
MAGNESIUM ACETATE see MAD000
MAGNESIUM ALUMINUM PHOSPHIDE (DOT) see AHD250
MAGNESIUM ARSENATE see ARD000
MAGNESIUM ARSENATE PHOSPHOR see ARD000
MAGNESIUM AUREOLATE see MAD050
MAGNESIUM BIS(2,3-DIBROMOPROPYL)PHOSPHATE see MAD100
MAGNESIUM BORIDE see MAD250

MAGNESIUM BORINGS see MAC750
MAGNESIUM CARBONATE see MAC650
MAGNESIUM(II) CARBONATE (1581) see MAC650
MAGNESIUM CARBONATE, PRECIPITATED see MAC650
MAGNESIUM CHLORATE see MAE000
MAGNESIUM CHLORIDE see MAE250
MAGNESIUM CHLORIDE HEXAHYDRATE see MAE500
MAGNESIUM CLIPPINGS (DOT) see MAC750
MAGNESIUM COMPOUNDS see MAE750
MAGNESIUM DIACETATE see MAD000
MAGNESIUM DICHLORATE see MAE000
MAGNESIUM DROSS (HOT) see MAF000
MAGNESIUM DROSS, WET (DOT) see MAF000
MAGNESIUM FLUORIDE see MAF500
MAGNESIUM FLUORURE (FRENCH) see MAF500
MAGNESIUM FLUOSILICATE see MAG250
MAGNESIUMFOSFIDE (DUTCH) see MAI000
MAGNESIUM FUMARATE see MAF750
MAGNESIUM GLUCONATE see MAG000
MAGNESIUM GLUTAMATE HYDROBROMIDE see MAG050
MAGNESIUM GOLD PURPLE see GIS000
MAGNESIUM GRANULES COATED, particle size not less than 149 microns (DOT) see MAC750
MAGNESIUM HEXAFLUOROSILICATE see MAG250
MAGNESIUM HYDRATE see MAG750
MAGNESIUM HYDRIDE see MAG500
MAGNESIUM HYDROXIDE see MAG750
MAGNESIUM LINEAR ALKYLBENZENE SULFONATE see LGF800
MAGNESIUM METAL (DOT) see MAC750
MAGNESIUM NITRATE (DOT) see MAH000
MAGNESIUM(II) NITRATE (1:2) see MAH000
MAGNESIUM(II) NITRATE (1:2), HEXAHYDRATE see MAH250
MAGNESIUM OXIDE see MAH500
MAGNESIUM-5-OXO-2-PYRROLIDINECARBOXYLATE see PAN775
MAGNESIUM PELLETS see MAC750
MAGNESIUM PEMOLINE see PAP000
MAGNESIUM PERCHLORATE see PCE000
MAGNESIUM PEROXIDE see MAH750
MAGNESIUM PEROXIDE, solid (DOT) see MAH750
MAGNESIUM PHOSPHATE, DIBASIC see MAH775
MAGNESIUM PHOSPHATE, TRIBASIC see MAH780
MAGNESIUM PHOSPHIDE see MAI000
MAGNESIUM PHTHALOCYANINE see MAI250
MAGNESIUM POLYOXYETHYLENE ALKYL ETHER SULFATE see MAI500
MAGNESIUM POTASSIUM ASPARTATE see MAI600
MAGNESIUM POTASSIUM-l-ASPARTATE see MAI600
MAGNESIUM POWDER (DOT) see MAC750
MAGNESIUM RIBBONS see MAC750
MAGNESIUM SALT of AUREOLIC ACID see MAD050
MAGNESIUM SCALPINGS (DOT) see MAC750
MAGNESIUM SCRAP (DOT) see MAC750
MAGNESIUM SHAVINGS (DOT) see MAC750
MAGNESIUM SHEET see MAC750
MAGNESIUM SILICATE HYDRATE see MAJ000
MAGNESIUM SILICOFLUORIDE see MAG250
MAGNESIUM SULFATE (1:1) see MAJ250
MAGNESIUM SULFATE and CALOMEL (8:5) see CAZ000
MAGNESIUM SULFATE adduct of 2,2-DITHIO-BIS-PYRIDINE 1-OXIDE see OIU850
MAGNESIUM SULFATE HEPTAHYDRATE see MAJ500
MAGNESIUM SULPHATE see MAJ250
MAGNESIUM TETRAHYDROALUMINATE see MAJ775
MAGNESIUM THIOSULFATE see MAK000
MAGNESIUM THIOSULFATE HEXAHYDRATE see MAK250
MAGNESIUM TURNINGS (DOT) see MAC750
MAGNETIC 70,90 and 95 see SOD500
MAGNETITE see IHC550
MAGNEZU TLENEK (POLISH) see MAH500
MAGNOFENYL see OPK300
MAGNOLIONE see OOO100
MAGNOPHENYL see OPK300
MAGNOSULF see MAK000
MAGRACROM ORANGE see NEY000
MAGRON see MAE000
MAGSALYL see SJO000
MAHOGANY see MAK300
MAIKOHIS see FOS100
MAINTAIN A see CDT000
MAINTAIN CF125 see CDT000
MAIORAD see TGF175
MAIPEDOPA see DNA200
MAIS BOUILLI (HAITI) see GIW200
MAITANSINE see MBU820
MAITENIN see TGD125

MAIZENA see SLJ500
MAJEPTYL see TFM100
MAJOL PLX see SFO500
MAJSOLIN see DBB200
MAKAHALA (HAWAII) see DAC500
MAKAROL see DKA600
MAKI see BMN000
MALACHITE see CNJ750
MALACHITE GREEN CARBINOL see MAK500
MALACHITE GREEN G see BAY750
MALACHITE GREEN OXALATE see MAK600
MALACID see TGD000
MALACIDE see MAK700
MALAFOR see MAK700
MALAGRAN see MAK700
MALAKILL see MAK700
MALAMAR see MAK700
MALAMAR 50 see MAK700
MALA MUJER (MEXICO) see CNR135
MALANGA CARA de CHIVO (CUBA) see EAI600
MALANGA (CUBA) see XCS800
MALANGA DEUX PALLES (HAITI) see EAI600
MALANGA ISLENA (CUBA) see EAI600
MALANGA de JARDIN (CUBA) see EAI600
MALANGA TREPADORA (CUBA) see PLW800
MALAOXON see OPK250
MALAOXONE see OPK250
MALAPHELE see MAK700
MALAPHOS see MAK700
MALAQUIN see CLD000
MALAREN see CLD000
MALAREX see CLD000
MALARICIDA see CFU750
MALARIDINE see PPQ650
MALASOL see MAK700
MALASPRAY see MAK700
MALATHION see MAK700
MALATHION-O-ANALOG see OPK250
MALATHION ULV CONCENTRATE see MAK700
MALATHIOZOO see MAK700
MALATHON see MAK700
MALATHYL LV CONCENTRATE & ULV CONCENTRATE see MAK700
MALATION (POLISH) see MAK700
MALATOL see MAK700
MALATOX see MAK700
MALAYAN CAMPHOR see BMD000
MALCOTRAN see MDL000
MALDISON see MAK700
MALDOCIL see TEN750
MALEALDEHYDIC ACID, DICHLORO-4-OXO-, (Z)-(9CI) see MRU900
MALEATE ACIDE de l'ACETYL-3-DIMETHYLAMINO-3-PROPYL-10-PHENO-
    THIAZINE (FRENCH) see AAF750
MALEATE de 2-((p-CHLOROPHENYL)-2-(PYRIDYL)HYDROXY METHYL)IM-
    IDAZOLINE (FRENCH) see SBC800
MALEATE de CINEPAZIDE (FRENCH) see VGK000
MALEATE de CINPROPAZIDE see IRQ000
MALEIC ACID see MAK900
MALEIC ACID ANHYDRIDE (MAK) see MAM000
MALEIC ACID, DIALLYL ESTER see DBK200
MALEIC ACID, DIBUTLY ESTER see DED600
MALEIC ACID DI(1,3-DIMETHYLBUTYL) ESTER see DKP400
MALEIC ACID, DIETHYL ESTER see DJO200
MALEIC ACID DIHEXYL ESTER see DKP400
MALEIC ACID, DIPENTYL ESTER see MAL000
MALEIC ACID-N-ETHYLIMIDE see MAL250
MALEIC ACID HYDRAZIDE see DMC600
MALEIC ACID, METHYL ERGONOVINE see MJV750
MALEIC ACID, MONO(HYDROXYETHOXYETHYL) ESTER see MAL500
MALEIC ACID, MONO(2-HYDROXYPROPYL) ESTER see MAL750
MALEIC ANHYDRIDE see MAM000
MALEIC ANHYDRIDE adduct of BUTADIENE see TDB000
MALEIC ANHYDRIDE OZONIDE see MAM250
MALEIC HYDRAZIDE see DMC600
MALEIC HYDRAZIDE DIETHANOLAMINE SALT see DHF200
MALEIMIDE see MAM750
MALEINAN SODNY (CZECH) see SIE000
MALEINIC ACID see MAK900
MALEINIMIDE see MAM750
MALENIC ACID see MAK900
N,N-MALEOYLHYDRAZINE see DMC600
MALESTRONE see MPN500
MALESTRONE (AMPS) see TBF500
MALGESIC see BRF500
MALIASIN see CPO500

MALIC ACID see MAN000
MALIC ACID, SODIUM SALT see MAN250
MALIL see AFS500
MALILUM see AFS500
MALIPUR see CBG000
MALIVAN see DNV000
MALIX see EAQ750
MALLERMIN-F see FOS100
MALLOFEEN see PDC250
MALLOPHENE see PDC250, PEK250
MALLOROL see MOO250
MALLOTOXIN see KAJ500
MALMED see MAK700
MALOCID see TGD000
MALOCIDE see TGD000
MALOGEN see MPN500
MALOGEN L.A.200 see TBF750
MALON, combustion products see ADX750
MALONAL see BAG000
MALONALDEHYDE see PMK000
MALONALDEHYDE DIETHYL ACETAL see MAN750
MALONALDEHYDE, ION(1-), SODIUM see MAN700
MALONALDEHYDE SODIUM SALT see MAN700
MALONALDEHYDE TETRAETHYL DIACETAL see MAN750
MALONAMIDE see MAO000
MALONAMIDE NITRILE see COJ250
MALONAMONITRILE see COJ250
MALONDIALDEHYDE see PMK000
MALONDIAMIDE see MAO000
MALONIC ACID DIAMIDE see MAO000
MALONIC ACID, DIETHYL ESTER see EMA500
MALONIC ACID DIMETHYL ESTER see DSM200
MALONIC ACID, (ETHOXYMETHYLENE)-, DIETHYL ESTER see EEV200
MALONIC ACID ETHYL ESTER NITRILE see EHP500
MALONIC ACID, THALLIUM SALT (1:2) see TEM399
MALONIC ALDEHYDE see PMK000
MALONIC DIALDEHYDE see PMK000
MALONIC DINITRILE see MAO250
MALONIC ESTER see EMA500
MALONIC MONONITRILE see COJ500
MALONITRILE HYDRAZIDE see COH250
MALONOBEN see DED000
MALONODIALDEHYDE see PMK000
MALONONITRILE see MAO250
MALONONITRILE HYDRAZIDE see COH250
MALONYLDIALDEHYDE see PMK000
MALONYLDIAMIDE see MAO000
MALOPRIM see SOA500, TGD000
MALORAN see CES750
MALOTILATE see MAO275
MALPHOS see MAK700
MALTA RED X 2284 see NAY000
MALTOBIOSE see MAO500
MALTOL see MAO350
MALTOSE see MAO500
d-MALTOSE see MAO500
MALTOX see MAK700
MALTOX MLT see MAK700
MALT SUGAR see MAO500
α-MALT SUGAR see MAO500
MALUS (VARIOUS SPECIES) see AQP875
MALVAVISCUS CONZATTI Greenm., flower extract see ADE125
MALVIDIN CHLORIDE see MAO600
MALVIDOL see MAO750
MALYSOL see MKA250
MAM see HMG000
MAM AC see MGS750
MAM ACETATE see MGS750
MAMALLET-A see DOT000
MAMAN LAMAN (HAITI) see JBS100
MAMBNA see MHW350
MA144 MI see MAB250
MAMIVERT see SLC300
MAMMEX see NGE500
MAMMOTROPIN see PMH625
MAMN see AAW000
MAM-TOX see TCY275
MA144 N2 see RHA125
MANADRIN see EAW000
MANAM see MAS500
MANCENILLIER (HAITI) see MAO875
MANCHESTER YELLOW see DUX800
MANCHINEEL see MAO875
MANCOFOL see DXI400

MANCOZEB see DXI400
MANDELAMIDE, N-2-PYRIDYL-, HYDROCHLORIDE see PPM725
MANDELIC ACID see MAP000
racemic MANDELIC ACID see MAP000
MANDELIC ACID, ETHYL ESTER see EMB000
MANDELIC ACID NITRILE see MAP250
d(−)-MANDELONITRILE-β-d-GENTIOBIOSIDE see AOD500
d,l-MANDELONITRILE-β-d-GLUCOSIDO-6-β-GLUCOSIDE see IHO700
d-MANDELONITRILE-β-d-GLUCOSIDO-6-β-d-GLUCOSIDE see AOD500
l-MANDELONITRILE-β-β-GLUCURONIC ACID see LAS000
MANDELSAEUREAETHYLESTER (GERMAN) see EMB000
MANDOKEF see CCX300
MANDRIN see EAW000
MANEB see MAS500
MANEB, stabilized against self-heating (DOT) see MAS500
MANEB, with not less than 60% maneb (DOT) see MAS500
MANEBE (FRENCH) see MAS500
MANEB-METHYLTHIOPHANATE mixture see MAP300
MANEB-ZINC see DXI400
MANEB plus ZINC ACETATE (50:1) see EIQ500
MANEB-ZINEB-KOMPLEX (GERMAN) see DXI400
MANETOL see MAP600
MANEXIN see MAW250
MANGAANBIOXYDE (DUTCH) see MAS000
MANGAANDIOXYDE (DUTCH) see MAS000
MANGAAN(II)-(N,N'-ETHYLEEN-BIS(DITHIOCARBAMAAT)) (DUTCH) see MAS500
MANGAN (POLISH) see MAP750
MANGAN(II)-(N,N'-AETHYLEN-BIS(DITHIOCARBAMATE)) (GERMAN) see MAS500
MANGANATE(3-), HEXACYANO-, TRIPOTASSIUM see TMX600
MANGANATE(3-), HEXAKIS(CYANO-C)-, TRIPOTASSIUM, (OC-6-11)- see TMX600
MANGANATE, TETRACHLORO-, BIS(3,4,5-TRIMETHOXYPHENETHYLAMMONIUM) see BKM100
MANGANDIOXID (GERMAN) see MAS000
MANGANESE see MAP750
MANGANESE ACETATE see MAQ000
MANGANESE(2+) ACETATE see MAQ000
MANGANESE(II) ACETATE see MAQ000
MANGANESE ACETATE TETRAHYDRATE see MAQ250
MANGANESE(II) ACETATE TETRAHYDRATE see MAQ250
MANGANESE ACETYLACETONATE see MAQ500
MANGANESE, (4-AMINOBUTANOATO-N,O)(N-(2,4-DIHYDROXY-3,3-DIMETHYL-1-OXOBUTYL)-β-ALANINATO)- see MAQ600
MANGANESE γ-AMINOBUTYRATOPANTOTHENATE see MAQ600
MANGANESE BINOXIDE see MAS000
MANGANESE (BIOSSIDO di) (ITALIAN) see MAS000
MANGANESE (BIOXYD de) (FRENCH) see MAS000
MANGANESE(II) BIS(ACETYLIDE) see MAQ780
MANGANESE, BIS(l-HISTIDINATO-N,O)-, TETRAHYDRATE see BJX800
MANGANESE, BIS(l-HISTIDINATO)-, TETRAHYDRATE see BJX800
MANGANESE, BIS(5-SULFO-8-QUINOLINOLATO)- see BKJ275
MANGANESE BLACK see MAS000
MANGANESE(II) CHLORIDE (1582) see MAR000
MANGANESE(II) CHLORIDE TETRAHYDRATE see MAR250
MANGANESE COMPOUNDS see MAR500
MANGANESE CYCLOPENTADIENYL TRICARBONYL see CPV000
MANGANESE DIACETATE see MAQ000
MANGANESE DIACETATE TETRAHYDRATE see MAQ250
MANGANESE DICHLORIDE see MAR000
MANGANESE DIMETHYL DITHIOCARBAMATE see MAR750
MANGANESE (DIOSSIDO di) (ITALIAN) see MAS000
MANGANESE DIOXIDE see MAS000
MANGANESE (DIOXYDE de) (FRENCH) see MAS000
MANGANESE EDTA COMPLEX see MAS250
MANGANESE, ((1,2-ETHANEDIYLBIS(CARBAMODITHIOATO))(2-))-, mixed with DIMETHYL (1,2-PHENYLENEBIS(IMINOCARBONOTHIOYL))BIS(CARBAMATE) see MAP300
MANGANESE ETHYLENE-1,2-BISDITHIOCARBAMATE see MAS500
MANGANESE(II) ETHYLENEBIS(DITHIOCARBAMATE) see MAS500
MANGANESE(II) ETHYLENE DI(DITHIOCARBAMATE) see MAS500
MANGANESE complex with N-FLUOREN-2-YL ACETOHYDROXAMIC ACID see HIR000
MANGANESE FLUORIDE see MAS750
MANGANESE(II) FLUORIDE see MAS750
MANGANESE FLUORURE (FRENCH) see MAS750
MANGANESE GREEN see MAT250
MANGANESE MANGANATE see MAT500
MANGANESE MONOXIDE see MAT250
MANGANESE OXIDE see MAS000, MAU800
MANGANESE(II) OXIDE see MAT250
MANGANESE(IV) OXIDE see MAS000
MANGANESE(III) OXIDE see MAT500

MANGANESE(VII) OXIDE see MAT750
MANGANESE(II) PERCHLORATE see MAT899
MANGANESE PERCHLORATE HEXAHYDRATE see MAU000
MANGANESE PEROXIDE see MAS000
MANGANESE SISQUIOXIDE see MAT500
MANGANESE(II) SULFATE (1581) see MAU250
MANGANESE(II) SULFATE TETRAHYDRATE (1581584) see MAU750
MANGANESE(II) SULFIDE see MAV000
MANGANESE(II) TELLURIDE see MAV250
MANGANESE, TETRACARBONYL(TRIFLUOROMETHYLTHIO)-, dimer see TBN200
MANGANESE(II) TETRAHYDROALUMINATE see MAV500
MANGANESE TETROXIDE see MAU800
MANGANESE TRICARBONYL METHYLCYCLOPENTADIENYL see MAV750
MANGANESE TRIFLUORIDE see MAW000
MANGANESE TRIOXIDE see MAT500
MANGANESE ZINC BERYLLIUM SILICATE see BFS750
MANGANIC OXIDE see MAT500
MANGAN NITRIDOVANY (CZECH) see MAP750
MANGANOMANGANIC OXIDE see MAU800
MANGANOUS ACETATE see MAQ000
MANGANOUS ACETATE TETRAHYDRATE see MAQ250
MANGANOUS ACETYLACETONATE see MAQ500
MANGANOUS CHLORIDE see MAR000
MANGANOUS CHLORIDE TETRAHYDRATE see MAR250
MANGANOUS DIMETHYLDITHIOCARBAMATE see MAR750
MANGANOUS OXIDE see MAT250
MANGANOUS SULFATE see MAU250
MANGAN-ZINK-AETHYLENDIAMIN-BIS-DITHIO-CARBAMAT (GERMAN) see DXI400
MANGENESE SUPEROXIDE see MAS000
MAN-GRO see MAU250
MANICOLE see MAW250
MANIHOT ESCULENTA see CCO680
MANIHOT UTILISSIMA see CCO700
MANINIL see CEH700
MANIOC see CCO680
MANIOKA see CCO680
MANIOL see CCR875
MANITE see MAW250
MANNA SUGAR see HER000
MANNITE see HER000
MANNIT-LOST (GERMAN) see MAW500
MANNIT-MUSTARD (GERMAN) see MAW500
MANNITOL see HER000
d-MANNITOL see HER000
d-MANNITOL BUSULFAN see BKM500
MANNITOL HEXANITRATE see MAW250
d-MANNITOL HEXANITRATE see MAW250
MANNITOL HEXANITRATE, containing, by weight, at least 40% water (DOT) see MAW250
MANNITOL MUSTARD DIHYDROCHLORIDE see MAW750
MANNITOL MYLERAN see BKM500
MANNITOL NITROGEN MUSTARD see MAW500
MANNOGRANOL see BKM500
MANNOMUSTINE see MAW500
MANNOMUSTINE DIHYDROCHLORIDE see MAW750
MANNOPYRANOSIDE, STROPHANTHIDIN-3 6-DEOXY-, α-l- see CNH780
MANNOSULFAN see MAW800
MANOSEB see DXI400
MANOXAL OT see DJL000
MANSIL see OLT000
MAN'S MOTHERWORT see CCP000
MANTA see KAJ000
MANTADAN see AED250
MANTHELINE see DJM800, XCJ000
MANUCOL see SEH000
MANUCOL DM see SEH000
MANUFACTURED IRON OXIDES see IHD000
MANUTEX see SEH000
MANZANA (SPANISH) see AQP875
MANZANILLO (CUBA, PUERTO RICO, DOMINICAN REPUBLIC) see MAO875
MANZATE see MAS500
MANZATE 200 see DXI400
MANZATE MANEB FUNGICIDE see MAS500
MANZEB see DXI400
MANZIN 80 see DXI400
MAOA see QAK000
MAOH see MKW600
MAOLATE see CJQ250
MAO-REM see PFC750
4M2AP see ALC250
MAPHARSAL see ARL000

MAPHARSEN see ARL000
MAPHARSIDE see ARL000
MAPHTHYPRAMIDE see IOU000
MAPLE AMARANTH see FAG020
MAPLE BRILLIANT BLUE FCF see FMU059
MAPLE ERYTHROSINE see FAG040
MAPLE INDIGO CARMINE see FAE100
MAPLE LACTONE see HMB500
MAPLE PONCEAU 3R see FAG018
MAPLE TARTRAZOL YELLOW see FAG140
MAPO see TNK250
MAPOSOL see VFU000, VFW009
MAPP see MFX600
MAPROFIX 563 see SIB600
MAPROFIX TLS 65 see SON000
MAPROFIX WAC-LA see SIB600
MAPROTILINE see LIN800
MAPROTILINE HYDROCHLORIDE see MAW850
MAPTC see CMP900
MAQBARL see DTN200
MAQBARL, nitrosated (JAPANESE) see MMW750
MARALATE see MEI450
MARANHIST see WAK000
MARANTA see SLJ500
MARANYL F 114 see PJY500
MARAPLAN see IKC000
MARASMIC ACID see MAW875
MARASMIC ACID METHYL ESTER see MLE600
MARAVILLA (MEXICO) see CAK325
MARAZINE see EAN600
MAR BATE see MQU750
MARBLE see CAO000
MARBON 9200 see SMR000
MARBORAN see MKW250
MARCAIN see BOO000
MARCELLOMYCIN see MAX000
MARCOPHANE see BGY000
MARETIN see HMV000
MAREVAN (SODIUM SALT) see WAT220
MAREX see HGC500
MAREZINE see EAN600
MAREZINE HYDROCHLORIDE see MAX275
MARFOTOKS see DJI000
MARGARIC ACID see HAS500
MARGONIL see MQU750
MARGONOVINE see LJL000
MARGOSA OIL see NBS300
MARIA GRANDE (PUERTO RICO) see MBU780
MARICAINE see DHK400
MARIHUANA see CBD750
MARIJUANA see CBD750
MARIJUANA, SMOKE RESIDUE see CBD760
MARINDININ see DLR000
MARINEURINA see DXO300
MARISAN see CMR100
MARISILAN see AIV500
MARITUS YELLOW see DUX800
MARJORAM OIL, SPANISH see MBU500
MARKOFANE see BGY000
MARKS 4-CPA see CJN000
MARLATE see MEI450
MARLEX 9400 see PMP500
MARLIPAL 1217 see DXY000
MARLOPHEN 820 see PKF500
MARMELOSIN see IHR300
MARMER see DXQ500
MAROMERA (CUBA) see RBZ400
MAROXOL-50 see DUZ000
MARPLAN see IKC000
MARPLON see IKC000
MARRONNIER (CANADA) see HGL575
MARSALID see ILE000
MARS BROWN see IHD000
MARSH GAS see MDQ750
MARSH MARIGOLD see MBU550
MARSH ROSEMARY EXTRACT see MBU750
MARSILID see ILE000
MARSIN see MNV750
MARS RED see IHD000
MARSTHINE see FOS100
MARTRATE-45 see PBC250
MARUCOTOL see AAE500
MARUNGUEY (PUERTO RICO) see CNH789
MARVEX see DGP900

MARVINAL see PKQ059
MARYGIN-M see DAB875
MARZIN see DXI400
MARZINE see EAN600
MARZULENE S see MBU775
MAS see MGQ750
MA144 S2 see APV000
MASCHITT see CFY000
MASDIOL see AOO475
MASELNAN DI-n-BUTYLCINICITY (CZECH) see DDX600
MASENATE see TBG000
MASENONE see MPN500
MASEPTOL see HJL500
MASHERI (INDIA) see SED400
MASLETINE see FOS100
MASOTEN see TIQ250
MASSICOT see LDN000
MASSICOTITE see LDN000
MASTESTONA see MPN500
MASTIPHEN see CDP250
MASTWOOD see MBU780
MATACIL see DOR400
MATANOL see AMK250
MATAVEN see ARW000
MATCH see MLJ500
MATENON see MQS225
MATO AZUL (PUERTO RICO) see CAK325
MATO de PLAYA (PUERTO RICO) see CAK325
MATRICARIA CAMPHOR see CBA750
MATRIGON see DGJ100
MATROMYCIN see OHO200
MATSUTAKE ALCOHOL (JAPANESE) see ODW000
MATTING ACID (DOT) see SOI500
MATULANE see PME250, PME500
MAURYLENE see PMP500
MAXATASE see BAC000
MAXIBALIN see EJT600
MAXIBOLIN see EJT600
MAXIDEX see SOW000
MAXIFEN see AOD000
MAXIPEN see PDD350
MAXITATE see MAW250
MAXOLON see AJH000
MAXULVET see SNN300
MA 144 Y see ADG425
MAY APPLE see MBU800
MAY & BAKER S-4084 see COQ399
MAY BLOB see MBU550
MAYCOR see CCK125
MAYFLOWER see LFT700
MAYSANINE see MBU820
MAYT see MBU820
MAYTANSINE see MBU820
MAYTANSINOL ISOVALERATE see APE529
MAYTENIN see TGD125
2-MAYTHIC ACID see NAV500
MAY THORN see MBU825
MAYVAT GOLDEN YELLOW see DCZ000
MAZATICOL HYDROCHLORIDE see MBV100
MAZEPTYL see TFM100
MAZINDOL see MBV250
MAZOTEN see TIQ250
MB see MHR200
4HMB see HLY400
M & B 800 see DBL800
M + B 695 see PPO000
M + B 760 see TEX250
MB 2878 see EAK500
MB 9057 see SNQ500
M&B 3046 see CLO000
M & B 4620 see TDS250
M&B 8873 see HKB500
MB 10064 see DDP000
M&B 11,461 see DNG200
M & B 39565 see MQY110
M&B 17,803A see AAE125
MBA see BIE250
7-MBA see MGW750
MB 2050A see PBT000
7-MBA-3,4-DIHYDRODIOL see MBV500
MBA HYDROCHLORIDE see BIE500
MBAO see CFA500
MBAO HYDROCHLORIDE see CFA750
MBBA see CQK600

MBC see BAV575, MHC750
MBCP see LEN000
M & B 782 DIHYDROCHLORIDE see DBM400
MBDZ see MHL000
MBH see PME500
MBK see HEV000
MBMA see MJQ250
MBNA see MHW500
MBOCA see MJM200
MBOT see MJO250
MBP see MRF525
MB PYRETHROID see MBV700, MNG525
MBR 6168 see DRR000
MBR 8251 see TKF750
MBR 12325 see DUK000
MBR 3092-42 see MBV710
MBT see BDF000
MBTH see MHJ250
MBTS see BDE750
MBTS RUBBER ACCELERATOR see BDE750
MBX see MHR200
6-MC see MIP750
2M-4C see CIR250
MC 338 see NIW500
MC 474 see DJI000
MC 1053 see CBW000
MC 1108 see DVJ400
MC 1415 see PMB250
MC 1478 see NIW500
MC 1488 see BRU750
MC 1945 see DVI800
MC 2188 see CDY299
MC 2303 see CBK500, GGS000
MC 2420 see MLJ000
MC 3199 see BCP685
MC6897 see DQM600
MCA see CEA000
3-MCA see MIJ750
MCA-600 see BDG250
MCA-11,12-EPOXIDE see ECA500
MCA-11,12-OXIDE see ECA500
M. CARINICAUDUS DUMERILII VENOM see MQT000
MCB see CEJ125
4-(MCB) see CLN750
MCC see DEV600
MC DEFOLIANT see MAE000
(E)-MC 11,12-DIHYDRODIOL see DML800
M.C. DUMERILLI VENOM see MQT000
MCE see EIF000
MCEABN see MEG750
MCEAMN see MEH000
MCF see MIG000
MCI-C54875 see DXY000
MCIT see CEX275
MCM-JR-4584 see FLK100
MCN-485 see AJF500
MCN 1025 see NNF000
MCN 2559 see TGJ850
MCN-3113 see XGA725
MCN 2783-21-98 see ZUA300
MCNAMEE see KBB600
MCNEIL 481 see DQU200
MCN-JR-2498 see TKK500
MCN-JR-3345 see FHG000
McN-JR 4263 see PDW750
MCN-JR-4584 see FGU000
MCN-JR-4749 see DYF200
McN-JR-6238 see PIH000
MCN-JR-8299 see TDX750
McN-JR-16,341 see PAP250
MCN-JR-4263-49 see PDW750
M1 (COPPER) see CNI000
M2 (COPPER) see CNI000
MCP see CIR250
MCP 875 see AFJ400
MCPA see CIR250
MCPA DIETHANOLAMINE SALT see MIH750
MCPA-ETHYL see EMR000
MCPAMN see MEI250
MCPA SODIUM SALT see SIL500
MCPB see CLN750, CLO000
MCPB-ETHYL see EHL500
MCP-BUTYRIC see CLN750
4-(MCPD) see CLO000

MCPEE see EMR000
MCPP see CIR500
2-MCPP see CIR500
MCPP-D-4 see CIR500
MCPP 2,4-D see CIR500
MCPP-K-4 see CIR500
MCPP POTASSIUM SALT see CLO200
MCR see MAB750
MCT see CPV000, TMO000
MCZ NITRATE see MQS560
10-MD see MEK700
l-(α-MD) see DNA800
MD 141 see DIS600
516 MD see CMR100
MD 2028 see HAF400
2028 MD see HAF400
5579 MD see DAB875
MD 67350 see VGK000
68111 M.D. see IRQ000
MD 780515 see MEW800
MDA see MJQ000, MJQ775
MDAB see DUH600
3'-MDAB see DUH600
MDBA see MEL500
MDBCP see MBV720
MDI see MJP400
bu-MDI see BRC500
pr-MDI see DPY600
M 141 DIHYDROCHLORIDE PENTAHYDRATE see SKY500
MDI-PC see DGE200
MDL-035 see THG700
MDL-899 see MBV735
MDS see DBD750, DXL400
2-ME see MCN250
ME 277 see PAM000
ME-1700 see BIM500
ME 3625 see DFD000
MEA see AJT250, EEC600, MCN750
MEAD JOHNSON 1999 see CCK250
MEADOW BRIGHT see MBU550
MEADOW GARLIC see WBS850
MEADOW GREEN see COF500
MEADOW ROSE LEEK see WBS850
MEADOW SAFFRON see CNX800
MEARLMAID see GLI000
MEASURIN see ADA725
MEAVERIN see CBR250
MEB see MAS500
MEB 6447 see CJO250
MEBACID see ALF250
MEBALLYMAL see SBM500
MEBALLYMAL SODIUM see SBN000
MEBANAZINE OXALATE see MBV750
MEBANAZINE SULPHATE see MHO000
MEBANE SODIUM SALT see MBV775
MEBARAL see ENB500
MEBENDAZOLE (USDA) see MHL000
MEBEREL see ENB500
MEBHYDROLIN NAPADISYLATE see MBW100
MEBICAR see MBW250
MEBICAR-A see MBW500
MEBICHLORAMINE see BIE500
MEBR see MHR200
ME4 BROMINAL see DDP000
MEBRON see MCH550
MEBROPHENHYDRAMINE see BMN250
MEBROPHENHYDRAMINE HYDROCHLORIDE see BMN250
MEBRYL see BMN250
MEBUBARBITAL see NBT500
MEBUBARBITAL SODIUM see NBU000
MEBUMAL NATRIUM see NBU000
MEBUMAL SODIUM see NBU000
MEBUTAMATE see MBW750
MEBUTINA see MBW750
MEC see MBW775
MECADOX see FOI000
MECALMIN see CCR875
MECAMILAMINA (ITALIAN) see VIZ400
MECAMINE see VIZ400
MECAMINE HYDROCHLORIDE see MQR500
MECAMYLAMINE see VIZ400
MECAMYLAMINE HYDROCHLORIDE see MQR500
MECARBAM see DJI000
MECARBENIL see MBW780

MECAROL see EQL000
MECARPHON see MLJ000
MECB see DJG000
ME-CCNU see CHD250
Me-CCNU see CHD500
MECHLORETHAMINE see BIE250
MECHLORETHAMINE HYDROCHLORIDE see BIE500
MECHLORETHAMINE OXIDE see CFA500
MECHLORETHAMINE OXIDE HYDROCHLORIDE see CFA750
MECHOLYL see ACR000
MECHOTHANE see HOA500
MECICLIN see DAI485
MECINARONE see MBX000
MECLASTINE HYDROGEN FUMARATE see FOS100
MECLIZINE see HGC500
MECLIZINE DIHYDROCHLORIDE see MBX250
MECLIZINE HYDROCHLORIDE see MBX500
MECLOFENAMATE SODIUM see SIF425
MECLOFENAMIC ACID see DGM875
MECLOFENOXANE see DPE000
MECLOFENOXATE HYDROCHLORIDE see AAE500
MECLOMEN see SIF425
MECLOPHENAMIC ACID see DGM875
MECLOPHENOXATE see DPE000
MECLOZINE see HGC500
MECLOZINE HYDROCHLORIDE see MBX500
MECOBALAMIN see VSZ050
MECODIN see MDO750
MECODRIN see BBK000
MECOMEC see CIR500
MECOPEOP see CIR500
MECOPER see CIR500
MECOPEX see CIR500, CLO200
MECOPROP see CIR500
MECOPROP POTASSIUM SALT see CLO200
MECOTURF see CIR500
MECPROP see CIR500
MECRAMINE see AJT250
MECRLAT see MIQ075
MECRYL see CFU750
MECS see EJH500
MeCsAc see EJJ500
MECYSTEINE HYDROCHLORIDE see MBX800
2-MeDAB see TLE750
MEDADDY BUSH see HGK700
MEDAMYCIN see TBX250
MEDAPAN see COY500
MEDARON see NGG500
MEDARSED see BPF000
MEDAZEPAM see CGA000
MEDAZEPAM HYDROCHLORIDE see MBY000
MEDAZEPOL see CGA000
MEDEMYCIN see MBY150
MEDFALAN see SAX200
MEDIAMID see DJS200
MEDIAMYCETINE see CDP250
MEDIBEN see MEL500
MEDICAINE see BQA010
MEDI-CALGON see SHM500
MEDICEL see AKO500
MEDICON see DBE200
MEDIDRYL see BBV500
MEDIFENAC see AGN000
MEDIFLAVIN see DBX400
MEDIFURAN see FPI000
MEDIGOXIN see MJD300
MEDIHALER-EPI see AES000, VGP000
MEDIHALER-TETRACAINE see BQA010
MEDILLA see MKP500
MEDINAL see BAG250
MEDINOTERB ACETATE see BRU750
MEDIQUIL see IPU000
MEDIREX see IJG000
MEDITRENE see IEP200
MEDIUM BLUE EMBL see ADE750
MEDLEXIN see ALV000
MEDOMET see DNA800, MJE780
MEDOMIN see COY500
MEDOMINE see COY500
α-MEDOPA see MJE780
MEDOPAQUE see HGB200
MEDOPREN see DNA800, MJE780
2,4-MEDP see MBY500
4,4-MEDP see MBZ000

MEDPHALAN see SAX200
MEDROGESTERONE see MBZ100
MEDROGESTONE see MBZ100
MEDROGLUTARIC ACID see HMC000
MEDROL see MOR500
MEDROL ACETATE see DAZ117
MEDROL DOSEPAK see MOR500
MEDRONE see MOR500
MEDROXYPROGESTERON see MBZ150
MEDROXYPROGESTERONE see MBZ150
MEDROXYPROGESTERONE ACETATE see MCA000
MEDROXYPROGESTERONE ACETATE and ETHINYLESTRADIOL see POF275
MEDULLIN see MCA025
MEE see EDP000
MEETHOBALM see BQA010
MEFEDINA see DAM700
MEFENAL see SNJ350
MEFENAMIC ACID see XQS000
MEFENOXALONA see MFD500
MEFENOXALONE see MFD500
MEFENSINA see GGS000
MEFLUIDIDE see DUK000
MEFOXIN see CCS510
MEFOXITIN see CCS510
MEFRUSID see MCA100
MEFRUSIDE see MCA100
MEFURINA see AHK750
M.E.G. see EJC500
MEGABA see AJC625
MEGABION see VSZ000
MEGABION (JAPANESE) see AOO475
MEGACE see VTF000
MEGACILLIN ORAL see PDT750
MEGACILLIN SUSPENSION see BFC750
MEGACILLIN TABLETS see BFD000
MEGACORT see DAE525
MEGADIURIL see CFY000
MEGALOMICIN A see MCA250
MEGALOMYCIN-A see MCA250
MEGALOVEL see VSZ000
MEGAMYCINE see MDO250
MEGAPHEN see CKP250, CKP500
MEGASEDAN see CGA000
MEGATOX see FFF000
MEGEPTIL see TFM100
MEGESTROL ACETATE see VTF000
MEGESTROL ACETATE + ETHINYLOESTRADIOL see MCA500
MEGESTROL ACETATE 4 MG., ETHINYLOESTRADIOL 50 μg see MCA500
MEGESTRYL ACETATE see VTF000
MEGIMIDE see MKA250
6-ME-GLU-P-2 see AKS250
MEGLUMINE AMIDOTRIZOATE see AOO875
MEGLUMINE CONRAY see IGC000
MEGLUMINE DIATRIZOATE see AOO875
MEGLUMINE IOCARMATE see IDJ500
MEGLUMINE IODAMIDE see IDJ600
MEGLUMINE IODIPAMIDE see BGB315
MEGLUMINE IODOXAMATE see IFP800
MEGLUMINE IOTHALAMATE see IGC000
MEGLUMINE IOTROXATE see IGD075
MEGLUMINE ISOTHALAMATE see IGC000
MEGLUMINE SODIUM IODAMIDE see MCA775
MEGLUTOL see HMC000
MEGUAN see DQR800
MEHP see MRI100
4-ME-I see MKU000
MEICELIN see CCS365
MEISEI SCARLET GG SALT see DEO400
MEISEI TERYL DIAZO BLACK CR see DPO200
MEISEI TERYL DIAZO BLUE HR see DCJ200
MEITO MY 30 see GGA800
MEK see MKA400
MEKAMINE see VIZ400
MEKAMIN HYDROCHLORIDE see MQR500
MEK-OXIME see EMU500
MEKP see MKA500
MEK PEROXIDE see MKA500
MELABON see ABG750
MELADININ see XDJ000
MELADININE see XDJ000
MELAMINE see MCB000
MELAMINKYANURAT (CZECH) see THO750
MELAMIN-N,N',N''-TRIMETHYLSULFONSAURES NATRIUM (GERMAN) see THS500

MELAN BLACK see BMA000
MELANEX see AKQ250
MELANILINE see DWC600
MELANOL LP20T see SON000
MELANOMYCIN see MCB100
MELANOSPORIN see MCB250
MELARSONYL POTASSIUM SALT see PLK800
MELATONIN see MCB350
MELATONINE see MCB350
MELBEX see MRX000
MELBIN see DQR600
MELDIAN see CKK000
MELDONE see CNU750
MELENGESTRO ACETATE see MCB375
MELENGESTROL ACETATE see MCB380
MELERIL see MOO250
MELETIN see QCA000
MELEX see MQR760
MELIA AZEDARACH see CDM325
MELIFORM see PKF750
MELILOTAL see MFW250
MELILOTIN see HHR500
MELILOTOL see HHR500
MELIN see RSU000
MELINEX see PKF750
MELINITE see PID000
MELIPAN see MCB500
MELIPAX see CDV100
MELIPRAMIN see DLH600, DLH630
MELIPRAMINE see DLH600, DLH630
MELIPRAMINE HYDROCHLORIDE see DLH630
MELIPRAMIN HYDROCHLORIDE see DLH630
MELITASE see CKK000
MELITOXIN see BJZ000
MELITRACEN see AEG875
MELITRACENE see AEG875
MELITRACENE HYDROCHLORIDE see TDL000
MELITRACEN HYDROCHLORIDE see TDL000
MELITTIN see MCB525
MELITTIN-I see MCB525
MELLARIL see MOO250
MELLARIL HYDROCHLORIDE see MOO500
MELLERETTE see MOO250
MELLERETTEN see MOO250
MELLERIL see MOO250
MELLINESE see CKK000
MELLITE 131 see DEH650
MELOCHIA TOMENTOSA see BAR500
MELOCOTON (SPANISH) see AQP890
MELOGEL see SLJ500
MELONEX see AKQ250
MELOXINE see XDJ000
MELPHALAN see PED750
MELPHALAN (RUSSIAN) see BHV000
MELPHALAN HYDROCHLORIDE see BHV250
MELPREX see DXX400
MELSEDIN BASE see QAK000
MELSMON see MCB535
MELSOMIN see QAK000
MELTROL see PDF250
MELUNA see SLJ500
MEL W see PLK800
MELYSIN see MCB550
MEMA see MEO750, MLF250
MEMC see MEP250
MEMCOZINE see SNN300
ME-MDA see MJO250
MEMINE see TCY750
MEMMI see MLF500
Me$_2$NMOR see DTA000
MEMORY ROOT see JAJ000
MEMPA see HEK000
MENADIOL see MMC250
MENADION see MMD500
MENADIONE see MMD500
MENADION-NATRIUM-BISULFIT TRIHYDRAT (GERMAN) see MMD750
MENAGEN see EDV000
MENAPHTHON see MMD500
MENAPHTHONE see MMD500
MENAQUINONE-4 see VTA650
MENAQUINONE K$_4$ see VTA650
MENATENSINA see RCA200
MENATETRENONE see VTA650
MENAZON see ASD000

dd-MENCS see MLC000
MENDEL see MQU750
MENDIAXON see MKP500
MENDON see CDQ250
MENDRIN see EAT500
MENEST see ECU750
MENETYL see EJR500
MENFORMON see EDV000
MENHYDRINATE see DYE600
MENICHLOPHOLAN see DFD000
MENIDRABOL see NNX400
MENIPHOS see MQR750
MENISPERMUM CANADENSE see MRN100
MENITE see MQR750
MENOCIL see ALX250
MENOGAROL see MCB600
MENOGEN see ECU750
MENOMYCIN see MRA250
MENONASAL see TBN000
MENOQUENS see MCA500
MENOSTILBEEN see DKA600
MENOTAB see ECU750
MENOTROL see ECU750
MENOTROPHIN see FMT100
MENOTROPINS see FMT100
MENSISO see APY500
MENTA-BAL see ENB500
MENTHA ARVENSIS, OIL see MCB625
MENTHA ARVENSIS OIL, PARTIALLY DEMENTHOLIZED (FCC) see MCB625
p-MENTHA-1,8-DIEN-7-AL see DKX100
o-1,4-MENTHADIENE see MCB700
p-MENTHA-1,3-DIENE see MLA250
p-MENTHA-1,4-DIENE see MCB750
p-MENTHA-1,5-DIENE see MCC000
p-MENTHA-1,8-DIENE see LFU000, MCC250
1,8(9)-p-MENTHADIENE see MCC250
d-p-MENTHA-1,8-DIENE see LFU000
(S)-(−)-p-MENTHA-1,8-DIENE see MCC500
p-MENTHA-1,8-DIEN-7-OL see PCI550
l-p-MENTHA-6,8-DIEN-2-OL see MKY250
p-MENTHA-6,8-DIEN-2-OL, PROPIONATE see MCD000
p-MENTHA-6,8-DIEN-2-ONE see MCD250
6,8(9)-p-MENTHADIEN-2-ONE see MCD250
(R)-(−)-p-MENTHA-6,8-DIEN-2-ONE see CCM120
1-6,8(9)-p-MENTHADIEN-2-ONE see CCM120
d-p-MENTHA-6,8,(9)-DIEN-2-ONE see CCM100
1-p-MENTHA-6(8,9)-DIEN-2-YL ACETATE see CCM750
(−)-trans-2-p-MENTHA-1,8-DIEN-3-YL-5-PENTYLRESORCINOL see CBD599
1-p-MENTHA-6,8(9)-DIEN-2-YL PROPIONATE see MCD000
MENTHANE DIAMINE see MCD750
p-MENTHANE-1,8-DIAMINE see MCD750
p-MENTHANE HYDROPEROXIDE see MCE000
p-MENTHANE-8-HYDROPEROXIDE see MCE000
p-MENTHANE HYDROPEROXIDE, TECHNICALLY PURE (DOT) see IQE000
p-MENTHAN-3-OL see MCF750
dl-3-p-MENTHANOL see MCG000
p-MENTHAN-8-OL ACETATE see DME400
l-p-MENTHAN-3-ONE see MCG275
(Z)-p-MENTHAN-3-ONE see IKY000
p-MENTHAN-3-ONE racemic see MCE250
O-1-MENTHENE see MCE275
p-MENTH-3-ENE see MCE500
1-p-MENTHEN-4-OL see TBD825
8-p-MENTHEN-2-OL see DKV150
p-MENTH-1-EN-8-OL see TBD500
p-MENTH-8-EN-3-OL see MCE750
8(9)-p-MENTHEN-3-OL see MCE750
p-MENTH-1-EN-8-OL (8CI) see TBD750
p-MENTH-8-EN-2-OL, ACETATE see DKV160
p-MENTH-1-EN-8-OL, FORMATE (mixed isomers) see TBE500
8-p-MENTHEN-2-ONE see DKV175
p-MENTH-1-EN-3-ONE see MCF250
p-MENTH-8-EN-2-ONE see DKV175
4(8)-p-MENTHEN-3-ONE see MCF500
p-MENTH-4(8)-EN-3-ONE see MCF500
d-p-MENTH-4(8)-EN-3-ONE see MCF500
p-MENTH-8-EN-2-YL ACETATE see DKV160
MENTHENYL KETONE see MCF525
1-(p-MENTHEN-6-YL)-1-PROPANONE see MCF525
MENTHEN-1-YL-8 PROPIONATE see TBE600
MENTHOL see MCF750
1-MENTHOL see MCF750
l-MENTHOL see MCG250
3-p-MENTHOL see MCG000

dl-MENTHOL see MCG000
MENTHOL racemic see MCG000
MENTHOL racemique (FRENCH) see MCG000
MENTHOL, ACETATE (8CI) see MCG500
MENTHOL ACETOACETATE see MCG850
MENTHONE see MCG275
p-MENTHONE see MCG275
l-MENTHONE (FCC) see MCG275
trans-MENTHONE see MCG275
MENTHONE, racemic see MCE250
MENTHYL ACETATE see MCG500
(−)-MENTHYL ACETATE see MCG750
dl-MENTHYL ACETATE see MCG500
l-p-MENTH-3-YL ACETATE see MCG750
l-p-MENTH-3-YL ACETATE see MCG750
l-MENTHYL ACETATE (FCC) see MCG750
MENTHYL ACETATE racemic see MCG500
MENTHYL ACETOACETATE see MCG850
(-)-MENTHYL ACETOACETATE see MCG850
(−)-MENTHYL ALCOHOL see MCG250
p-MENTH-3-YL ESTER-dl-ACETIC ACID see MCG500
MEOBAL see XTJ000
MEOBAL, NITROSATED (JAPANESE) see XTS000
MEONAL see BPF500
MEONINE see MDT740
MEOTHRIN see DAB825
MEP see EOS000
MEP (Pesticide) see DSQ000
MEPACRINE see ARQ250
MEPACRINE DIHYDROCHLORIDE see CFU750
MEPACRINE HYDROCHLORIDE see CFU750
MEPADIN see DAM700
MEPAMTIN see MQU750
ME-PARATHION see MNH000
MEPARFYNOL CARBAMATE see MNM500
MEPATON see MNH000
MEPAVLON see MQU750
MEPAZIN see MOQ250
MEPAZINE BASE see MOQ250
MEPEDYL see PIZ250
MEPENICYCLINE see MCH525
MEPENTAMATE see MNM500
MEPENTAMATO see EQL000, MNM500
MEPENTIL see EQL000
MEPENZOLATE see CBF000
MEPENZOLATE BROMIDE see CBF000
MEPERIDIDE see FLV000
MEPERIDINE see DAM600
MEPERIDINE HYDROCHLORIDE see DAM700
9-ME-PGA see BMM125
MEPHABUTAZONE see BRF500
MEPHACYCLIN see TBX250
MEPHADRYL see BBV500
MEPHANAC see CIR250
MEPHATE see GGS000
MEPHEDAN see GGS000
MEPHEDINE see DAM700
MEPHELOR see GGS000
MEPHENAMINE HYDROCHLORIDE see OJW000
MEPHENAMIN HYDROCHLORIDE see OJW000
MEPHENAMINIC ACID see XQS000
MEPHENESIN CARBAMATE see CBK500
MEPHENON see MDP750
MEPHENOXALONE see MFD500
MEPHENSIN see GGS000
MEPHENYTOIN see MKB250
MEPHEXAMIDE see DIB600, MCH250
MEPHOBARBITAL see ENB500
MEPHOBARBITONE see ENB500
MEPHOSAL see GGS000
MEPHOSFOLAN see DHH400
MEPHSON see GGS000
MEPHYTAL see ENB500
MEPHYTON see VTA000
MEPIBEN see LJR000
MEPICYCLINE PENICILLINATE see MCH525
MEPIOSINE see MQU750
MEPIPRAZOLE DIHYDROCHLORIDE see MCH535
MEPIPRAZOLE HYDROCHLORIDE see MCH535
MEPIRESERPATE HYDROCHLORIDE see MDW100
MEPIRIZOL see MCH550
MEPISERATE HYDROCHLORIDE see MDW100
MEPITIOSTANE see MCH600
MEPIVACAINE see SBB000

dl-MEPiVACAINE see SBB000
MEPIVACAINE HYDROCHLORIDE see CBR250
dl-MEPIVACAINE HYDROCHLORIDE see CBR250
MEPIVASTESIN see CBR250
2'-MePO₄' see DSS200
MEPOSED see MQU750
MEPRANIL see MQU750
MEPRIN see MCI375
MEPRIN (detoxicant) see MCI375
MEPRO see CIR500
MEPROBAM see MQU750
MEPROBAMAT (GERMAN) see MQU750
MEPROBAMATE see MQU750
MEPROBAMATO (ITALIAN) see MQU750
MEPROCOMPREN see MQU750
MEPROCON CMC see MQU750
MEPRODIL see MQU750
MEPRODINE (GERMAN) see NOE550
MEPROFEN see BDU500
MEPROLEAF see MQU750
MEPROMAZINE see MCI500
MEPROSAN see MQU750
MEPROSCILLARIN see MCI750
MEPROTABS see MQU750
MEPROZINE see MQU750
MEPRYLCAINE HYDROCHLORIDE see OJI600
MEPTIN see PME600
MEPTOX see MNH000
MEPTRAN see MQU750
MEPYRAMIN (GERMAN) see WAK000
MEPYRAMINE HYDROCHLORIDE see MCJ250
MEPYRAMINE MALEATE see DBM800
MEPYRAMINE 7-THEOPHYLLINE ACETATE see MCJ300
MEPYRAPONE see MCJ370
MEPYREN see WAK000
MEQUIN see QAK000
MEQUINOL see MFC700
MEQUITAZINE see MCJ400
MER 25 see DHS000
MER 29 see TMP500
MER-41 see CMX700
MERABITOL see AFH250
MERACTINOMYCIN see AEB000
MERADAN see AJU625
MERADANE see AJU625
MERAKLON see PMP500
MERALEIN DISODIUM see SIF500
MERALEN see TKH750
MERALLURIDE see MFC000, TEQ000
MERANTINE BLUE EG see FAE000
MERANTINE GREEN SF see FAF000
MERATRAN see DWK400
MERBAPHEN see CCG500
MERBENTUL see CLO750
MERBROMIN see MCV000
MERCALEUKIN see POK000
MERCAMINE see AJT250
MERCAMINE DISULFIDE see MCN500
MERCAPTAMINE see AJT250
MERCAPTAN AMYLIQUE (FRENCH) see PBM000
MERCAPTAN METHYLIQUE (FRENCH) see MLE650
MERCAPTAN METHYLIQUE PERCHLORE (FRENCH) see PCF300
MERCAPTANS see MCJ500
MERCAPTAZOLE see MCO500
2-MERCAPTOACETANILIDE see MCK000
α-MERCAPTOACETANILIDE see MCK000
MERCAPTOACETATE see TFJ100
(MERCAPTOACETATO(2-)-O,S)METHYL-MERCURATE(1-), SODIUM
    see SIM000
MERCAPTOACETIC ACID see TFJ100
2-MERCAPTOACETIC ACID see TFJ100
α-MERCAPTOACETIC ACID see TFJ100
MERCAPTOACETIC ACID, DIESTER with DITHIO-p-URE-
    IDOBENZENEARSONOUS ACID see CBI250
MERCAPTOACETIC ACID ETHYL ESTER see EMB200
MERCAPTOACETIC ACID-2-ETHYLHEXYL ESTER see EKW300
MERCAPTOACETIC ACID METHYL ESTER see MLE750
MERCAPTOACETIC ACID, SODIUM-BISMUTH SALT see BKX750
MERCAPTOACETIC ACID SODIUM SALT see SKH500
MERCAPTOACETONITRILE see MCK300
4-(MERCAPTOACETYL)MORPHOLINE O,O-DIMETHYL PHOSPHORODITHIO-
    ATE see MRU250
β-MERCAPTOALANINE see CQK000
6-MERCAPTO-2-AMINOPURINE see AMH250

2-MERCAPTO-5-AMINO-1,3,4-THIADIAZOLE see AKM000
4-MERCAPTOANILINE see AIF750
o-MERCAPTOANILINE see AIF500
p-MERCAPTOANILINE see AIF750
2-MERCAPTOBARBITURIC ACID see MCK500
7-MERCAPTOBENZ(a)ANTHRACENE see BBH750
2-MERCAPTOBENZIMIDAZOLE see BCC500
MERCAPTOBENZIMIDAZOLE ZINC SALT see ZIS000
2-MERCAPTOBENZIMIDAZOLE ZINC SALT (2:1) see ZIS000
o-MERCAPTOBENZOIC ACID see MCK750
o-MERCAPTOBENZOIC ACID, DIESTER with DITHIO-p-URE-
   IDOBENZENEARSONOUS ACID see TFD750
MERCAPTOBENZOIMIDAZOLE see BCC500
2-MERCAPTOBENZOIMIDAZOLE see BCC500
MERCAPTOBENZOTHIAZOLE see BDF000
2-MERCAPTOBENZOTHIAZOLE see BDF000
2-MERCAPTOBENZOTHIAZOLEDISULFIDE see BDE750
2-MERCAPTOBENZOTHIAZOLE SODIUM DERIVATIVE see SIG500
2-MERCAPTOBENZOTHIAZOLE SODIUM SALT see SIG500
2-MERCAPTOBENZOTHIAZOLE ZINC SALT see BHA750
2-MERCAPTOBENZOTHIAZYLDISULFIDE see BDE750
(MERCAPTOBUTANEDIOATO(1-))GOLD DISODIUM SALT see GJC000
(2-MERCAPTOCARBAMOYL)DI-ACETANILIDE see MCL500
MERCAPTODIACETIC ACID see MCM750
MERCAPTO DI-ACETIC ACID, ETHYLENE ESTER see MCN000
MERCAPTODIMETHUR (DOT) see DST000
1-MERCAPTODODECANE see LBX000
2-MERCAPTOETHANESULFONIC ACID MONOSODIUM SALT see MDK875
MERCAPTOETHANOL see MCN250
2-MERCAPTOETHANOL see MCN250
β-MERCAPTOETHANOL see MCN250
(2-MERCAPTOETHYL)AMINE see AJT250
β-MERCAPTOETHYLAMINE see AJT250
β-MERCAPTOETHYLAMINE DISULFIDE see MCN500
2-MERCAPTOETHYLAMINE HYDROCHLORIDE see MCN750, MCN750
β-MERCAPTOETHYLAMINE HYDROCHLORIDE see MCN750
2-MERCAPTOETHYLAMINE (OXIDIZED) see MCN500
N-(2-MERCAPTOETHYLBENZENESULFONAMIDE-S-(O,O-DIISOPROPYL
   PHOSPHORODITHIOATE) see DNO800
(2-MERCAPTOETHYL)CARBAMIC ACID, ETHYL ESTER, S-ESTER with O,O-
   DIMETHYL PHOSPHORODITHIOATE see EMC000
N-(2-MERCAPTOETHYL)DIMETHYLAMINE HYDROCHLORIDE see DOY600
4-MERCAPTOETHYLMORPHOLINE see MCO000
2-MERCAPTOETHYL TRIMETHOXY SILANE see MCO250
(2-MERCAPTOETHYL)TRIMETHYLAMMONIUM S-ESTER with O,O'-
   DIETHYLPHOSPHOROTHIOATE see PGY600
(2-MERCAPTOETHYL)TRIMETHYLAMMONIUM IODIDE S-ESTER with O,O-
   DIETHYL PHOSPHOROTHIOATE see TLF500
MERCAPTOFOS (RUSSIAN) see DAO500
1-MERCAPTOGLYCEROL see MRM750
6-MERCAPTOGUANINE see AMH250
6-MERCAPTOGUANOSINE see TFJ500
2-MERCAPTO-4-HYDROXY-6-METHYLPYRIMIDINE see MPW500
2-MERCAPTO-4-HYDROXY-6-N-PROPYLPYRIMIDINE see PNX000
2-MERCAPTO-4-HYDROXYPYRIMIDINE see TFR250
2-MERCAPTOIMIDAZOLE see IAO000
2-MERCAPTOIMIDAZOLINE see IAQ000
MERCAPTOMERIN SODIUM see TFK270
(MERCAPTOMETHYL)BENZENE see TGO750
3-(MERCAPTOMETHYL)-1,2,3-BENZOTRIAZIN-4(3H)-ONE-O,O-DIMETHYL
   PHOSPHORODITHIOATE see ASH500
3-(MERCAPTOMETHYL)-1,2,3-BENZOTRIAZIN-4(3H)-ONE-O,O-DIMETHYL
   PHOSPHORODITHIOATE-S-ESTER see ASH500
(trans)-2-MERCAPTOMETHYLCYCLOBUTYLAMINE HYDROCHLORIDE
   see AKL750
2-MERCAPTO-1-METHYLIMIDAZOLE see MCO500
N-(2-MERCAPTO-2-METHYL-1-OXOPROPYL)-l-CYSTEINE see MCO775
N-(2-MERCAPTO-2-METHYL-1-OXOPROPYL)-l-CYSTEINE SODIUM SALT
   see SAY875
1-(3-MERCAPTO-2-METHYL-1-OXOPROPYL)-l-PROLINE see MCO750
1-(d-3-MERCAPTO-2-METHYL-1-OXOPROPYL)-l-PROLINE (S,S) see MCO750
N-(MERCAPTOMETHYL)PHTHALIMIDE S-(O,O-DIMETHYL PHOS-
   PHORODITHIOATE) see PHX250
N-(2-MERCAPTO-2-METHYLPROPANOYL)-l-CYSTEINE see MCO775
N-(2-MERCAPTO-2-METHYLPROPANOYL)-l-CYSTEINE SODIUM SALT
   see SAY875
1-((2S)-3-MERCAPTO-2-METHYLPROPIONYL)-l-PROLINE see MCO750
2-MERCAPTO-6-METHYLPYRIMID-4-ONE see MPW500
2-MERCAPTO-6-METHYL-4-PYRIMIDONE see MPW500
MERCAPTOMETHYLTRIETHOXYSILANE see MCP000
2-MERCAPTO-N-(2-METHYOXYETHYL)-ACETAMIDE S-ESTER with O,O-
   DIMETHYL PHOSPHORODITHIOATE see AHO750
2-MERCAPTONAPHTHALENE see NAP500
2-MERCAPTO-6-NITROBENZOTHIAZOLE see MCP250

5-MERCAPTO-1-PHENYLTETRAZOLE see PGJ750
5-MERCAPTO-3-PHENYL-2H-1,3,4-THIADIAZOLE-2-THIONE see MCP500
MERCAPTOPHOS see DAO600, FAQ900
2-MERCAPTOPROPANE see IMU000
1-MERCAPTO-2,3-PROPANEDIOL see MRM750
3-MERCAPTO-1,2-PROPANEDIOL see MRM750
α-MERCAPTOPROPANOIC ACID see TFK250
3-MERCAPTOPROPANOL see PML500
2-MERCAPTOPROPIONIC ACID see TFK250
3-MERCAPTOPROPIONIC ACID see MCQ000
α-MERCAPTOPROPIONIC ACID see TFK250
MERCAPTOPROPIONIC ACID, DIBUTYLTIN SALT see DEJ200
MERCAPTOPROPIONYLGLYCINE see MCI375
(2-MERCAPTOPROPIONYL)GLYCINE see MCI375
α-MERCAPTOPROPIONYLGLYCINE see MCI375
N-(2-MERCAPTOPROPIONYL)GLYCINE see MCI375
2-MERCAPTO-6-PROPYL-4-PYRIMIDONE see PNX000
2-MERCAPTO-6-PROPYLPYRIMID-4-ONE see PNX000
3-MERCAPTOPROPYLTRIMETHOXYSILANE see TLC000
γ-MERCAPTOPROPYLTRIMETHOXYSILANE see TLC000
6-MERCAPTOPURIN see POK000
MERCAPTOPURIN (GERMAN) see POK000
6-MERCAPTOPURINE see POK000
6-MERCAPTOPURINE MONOHYDRATE see MCQ100
MERCAPTOPURINE-3-N-OXIDE see MCQ250
6-MERCAPTOPURINE 3-N-OXIDE see MCQ250
6-MERCAPTOPURINE 3-N-OXIDE MONOHYDRATE see OMY800
MERCAPTOPURINE RIBONUCLEOSIDE see MCQ500
6-MERCAPTOPURINE RIBOSIDE see MCQ500
2-MERCAPTOPYRIDINE-N-OXIDE SODIUM SALT see MCQ750
2-MERCAPTO-4-PYRIMIDINOL see TFR250
2-MERCAPTO-4-PYRIMIDONE see TFR250
2-MERCAPTOPYRIMID-4-ONE see TFR250
MERCAPTOSUCCINIC ACID see MCR000
MERCAPTOSUCCINIC ACID ANTIMONATE(III) HEXALITHIUM SALT
   see LGU000
MERCAPTOSUCCINIC ACID-S-ANTIMONY DERIVATIVE LITHIUM SALT
   see LGU000
MERCAPTOSUCCINIC ACID DIETHYL ESTER see MAK700
MERCAPTOSUCCINIC ACID, GOLD SODIUM SALT see GJC000
MERCAPTOSUCCINIC ACID, THIOANTHIMONATE(III), DILITHIUM SALT
   see LGU000
p-MERCAPTO SULFADIAZINE see MCR250
7-MERCAPTO-1,3,4,6-TETRAZAINDENE see POK000
2-MERCAPTOTHIAZOLINE see TFS250
MERCAPTOTHION see MAK700
MERCAPTOTION (SPANISH) see MAK700
o-MERCAPTOTOLUENE see TGP000
p-MERCAPTOTOLUENE see TGP250
α-MERCAPTOTOLUENE see TGO750
3-MERCAPTO-1H-1,2,4-TRIAZOLE see THT000
d-MERCAPTOVALINE see MCR750
d,3-MERCAPTOVALINE see MCR750
dl-α-MERCAPTOVALINE see PAP500
3-MERCAPTO-dl-VALINE (9CI) see PAP500
MERCAPTURIC ACID see ACH000
(R)-MERCAPTURIC ACID see ACH000
MERCAPURIN see POK000
MERCARDAN see SIG000
MERCAZOLYL see MCO500
MERCHLORATE see MEP250
MERCHLORETHANAMINE see BIE500
MERCKOGEN 6000 see AAX250
MERCLORAN see CHX250
MERCOL 25 see DXW200
MERCORAL see CHX250
MERCUFENOL CHLORIDE see CHW675
MERCUHYDRIN see TEQ000
MERCUMATILIN SODIUM see MCS000
MERCUPURIN see MCV750
MERCURAM see TFS350
MERCURAMIDE see SIH500
MERCURAN see MEO750
MERCURANINE see MCV000
MERCURATE(1-), ACETATOPHENYL-, AMMONIUM SALT see PFO550
MERCURE (FRENCH) see MCW750
MERCURHYDRIN SODIUM see SIG000
MERCURIACETATE see MCS750
MERCURIALIN see MGC250
2,2'-MERCURIBIS(6-ACETOXYMERCURI-4-NITRO)ANILINE see MCS250
MERCURIBIS(DIETHYL(2,2-DIMETHYL-4-
   DITHIOCARBOXYAMINO)BUTYLAMMONIUM DICHLORIDE see MCS500
MERCURIBIS-o-NITROPHENOL see MCS600
MERCURIC ACETATE see MCS750

MERCURIC AMMONIUM CHLORIDE, solid see MCW500
MERCURIC ARSENATE see MDF350
MERCURIC BASIC SULFATE see MDG000
MERCURIC BENZOATE see MCX500
MERCURIC BENZOATE, solid (DOT) see MCX500
MERCURIC BROMIDE see MCY000
MERCURIC BROMIDE, solid (DOT) see MCY000
MERCURIC CHLORIDE (DOT) see MCY475
MERCURIC CHLORIDE, AMMONIATED see MCW500
MERCURIC CHLORIDE-1,4-OXATHIANE see OMA000
MERCURIC CYANIDE, solid (DOT) see MDA250
MERCURIC DIACETATE see MCS750
MERCURIC-8,8-DICAFFEINE see MCT000
MERCURIC DINAPHTHYLMETHANE DISULPHONATE see MCT250
MERCURIC IODIDE see MDD000
MERCURIC IODIDE, solid (DOT) see MDD000
MERCURIC IODIDE, solution (DOT) see MDD000
MERCURIC IODIDE, RED see MDD000
MERCURIC LACTATE see MDD500
MERCURIC NITRATE see MDF000
MERCURIC OLEATE, solid (DOT) see MDF250
MERCURIC OXIDE see MCT500
MERCURIC OXIDE, solid (DOT) see MCT500
MERCURIC OXIDE, RED see MCT500
MERCURIC OXIDE, YELLOW see MCT500
MERCURIC OXYCYANIDE see MDA500
MERCURIC OXYCYANIDE, solid (desensitized) (DOT) see MDA500
MERCURIC PEROXYBENZOATE see MCT750
MERCURIC POTASSIUM CYANIDE (DOT) see PLU500
MERCURIC POTASSIUM CYANIDE, solid (DOT) see PLU500
MERCURIC POTASSIUM IODIDE see NCP500
MERCURIC POTASSIUM IODIDE, solid (DOT) see NCP500
MERCURIC SALICYLATE see MCU000
MERCURIC SALICYLATE, solid (DOT) see MCU000
MERCURIC SUBSULFATE, solid (DOT) see MDG000
MERCURIC SULFATE, solid (DOT) see MDG500
MERCURIC SULFOCYANATE see MCU250
MERCURIC SULFOCYANIDE see MCU250
MERCURIC SULFOCYANTE, solid (DOT) see MCU250
MERCURIC THIOCYANATE see MCU250
MERCURIC THIOCYANATE, solid (DOT) see MCU250
MERCURIDIACETALDEHYDE see BJW800
N,N′-MERCURIDIANILINE see DCJ000
3,3′-MERCURIDI-2-PROPYN-1-OL see BKJ250
MERCURIDISALICYLIC ACID, DISODIUM SALT see SJR000
MERCURI-HEMATOPORPHYRIN DISODIUM SALT see HAP000
MERCURIO (ITALIAN) see MCW250
MERCURIPHENOLDISULFONATE SODIUM see MCU500
MERCURIPHENYL ACETATE see ABU500
MERCURIPHENYL CHLORIDE see PFM500
MERCURIPHENYL NITRATE see MCU750
MERCURISALICYLIC ACID see MCU000
MERCURITAL see SIH500
MERCUROCHLORIDE (DUTCH) see MCW000
MERCUROCHROME see MCV000
MERCUROCHROME-220 SOLUBLE see MCV000
MERCUROCOL see MCV000
MERCUROL see MCV250
MERCUROME see MCV000
MERCUROPHAGE see MCV000
MERCUROPHEN see MCV500
MERCUROPHYLLINE see MCV750
MERCUROTHIOLATE see MDI000
MERCUROUS ACETATE see MDE250
MERCUROUS ACETATE, solid (DOT) see MDE250
MERCUROUS AZIDE (DOT) see MCX000
MERCUROUS BROMIDE, solid (DOT) see MCX750
MERCUROUS CHLORIDE see MCW000
MERCUROUS GLUCONATE see MDC500
MERCUROUS GLUCONATE, solid (DOT) see MDC500
MERCUROUS IODIDE see MDC750
MERCUROUS IODIDE, solid (DOT) see MDC750
MERCUROUS NITRATE, solid (DOT) see MDE750
MERCUROUS OXIDE, BLACK, solid (DOT) see MDF750
MERCUROUS SULFATE, solid (DOT) see MDG250
MERCURY see MCW250
MERCURY ACETATE see MCS750, MDE250
MERCURY(2+) ACETATE see MCS750
MERCURY(II) ACETATE see MCS750
MERCURY, (ACETATO-O)PHENYL-, AMMONIATE see PFO550
MERCURY ACETYLIDE see MCW349
MERCURY(II) ACETYLIDE see MCW350
MERCURY ACETYLIDE (DOT) see MCW349
MERCURY AMIDE CHLORIDE see MCW500

MERCURY AMINE CHLORIDE see MCW500
MERCURY AMMONIATED see MCW500
MERCURY AZIDE see MCX000
MERCURY(I) AZIDE see MCX000
MERCURY(II) AZIDE see MCX250
MERCURY(II) BENZOATE see MCX500
MERCURY BICHLORIDE see MCY475
MERCURY BINIODIDE see MDD000
MERCURY, BIS(ACETATO)(mu-(3′,6′-DIHYDROXY-2′,7′-FLUORANDIYL))DI- see FEV100
MERCURY BIS(CHLOROACETYLIDE) see MCX600
MERCURY, BIS(4-HYDROXY-3-NITROPHENYL)- see MCS600
MERCURY BISULFATE see MDG500
MERCURY(I) BROMATE see MCX700
MERCURY(I) BROMIDE (1581) see MCX750
MERCURY(II) BROMIDE (1582) see MCY000
MERCURY(II) BROMIDE COMPLEX with TRIS(2-ETHYLHEXYL) PHOSPHITE see MCY250
MERCURY, BROMOHEXYL see HFR100
MERCURY, BROMO(2-HYDROXYETHYL)-, compound with AMMONIA (1:0.8 moles) see BNL275
MERCURY, (3-(α-CARBOXY-o-ANISAMIDO)-2-HYDROXYPROPYL)HYDROXY- see NCM800
MERCURY, (3-(α-CARBOXYMETHOXYPROPYL)HYDROXY MONOSODIUM SALT, COMPOUNDED with THEOPHYLLINE (1:1) see MDH750
MERCURY(I) CHLORIDE see MCW000
MERCURY(II) CHLORIDE see MCY475
MERCURY(II) CHLORIDE COMPLEX with TRIS(2-ETHYLHEXYL) PHOSPHITE see MCY500
MERCURY(I) CHLORITE see MCY750
MERCURY(II) CHLORITE see MCY755
MERCURY (E)-CHLORO(2-(3-BROMOPROPIONAMIDO)CYCLOHEXYL) see CET000
MERCURY COMPOUNDS, INORGANIC see MCZ000
MERCURY COMPOUNDS, ORGANIC see MDA000
MERCURY(I) CYANAMIDE see MDA100
MERCURY(II) CYANATE see MDA150
MERCURY(II) CYANIDE see MDA250
MERCURY CYANIDE OXIDE see MDA500
MERCURY DIACETATE see MCS750
MERCURY-O,O-DI-n-BUTYL PHOSPHORODITHIOATE see MDA750
MERCURY DICHLORITE see MCY755
MERCURY, (DIHYDROGEN PHOSPHATO)METHYL- see MLF520
MERCURY, DIMETHYL see DSM450
MERCURY(II) aci-DINITROMETHANIDE see MDA800
MERCURY DITHIOCYANATE see MCU250
MERCURY(II) EDTA COMPLEX see MDB250
MERCURY(II) FLUOROACETATE see MDB500
MERCURY(II) FORMOHYDROXAMATE see MDB775
MERCURY FULMINATE (DOT) see MDC000
MERCURY FULMINATE (wet) see MDC250
MERCURY(II) FULMINATE (dry) see MDC000
MERCURY(I) GLUCONATE see MDC500
MERCURY, HYDROXY(4-HYDROXY-3-NITROPHENYL)- see HLO400
MERCURY, HYDROXYPHENYL- see PFN100
MERCURY, (4-HYDROXY-m-PHENYLENE)BIS(ACETATO- see HNK575
MERCURY(I) IODIDE see MDC750
MERCURY(II) IODIDE see MDD000
MERCURY(II) IODIDE (solution) see MDD250
MERCURYL ACETATE see MCS750
MERCURY(2+) LACTATE see MDD500
MERCURY, METALLIC (DOT) see MCW250
MERCURY, (METHANETHIOLATO)METHYL- see MLX300
MERCURY METHYLCHLORIDE see MDD750
MERCURY METHYLMERCURY SULFIDE see MLX300
MERCURY, METHYL(METHYLTHIO)- see MLX300
MERCURY(II) METHYLNITROLATE see MDE000
MERCURY, (2-METHYL-5-NITROPHENOLATO(2-)-C^6,O^1)-(9CI) see NHK900
MERCURY, METHYL(PHOSPHATO(1-)-O)-(9CI) see MLF520
MERCURY MONOACETATE see MDE250
MERCURY MONOCHLORIDE see MCW000
MERCURY-2-NAPHTHALENEDIAZONIUM TRICHLORIDE see MDE500
MERCURY NITRATE see MDF000
MERCURY(I) NITRATE (1581) see MDE750
MERCURY(II) NITRATE (1582) see MDF000
MERCURY NITRIDE see TKW000
MERCURY(II) 5-NITROTETRAZOLIDE see MDF100
MERCURY NUCLEATE, solid (DOT) see MCV250
MERCURY OLEATE see MDF250
MERCURY(II) ORTHOARSENATE see MDF350
MERCURY(II) OXALATE see MDF500
MERCURY(I) OXIDE see MDF750
MERCURY(II) OXIDE see MCT500
MERCURY OXIDE SULFATE see MDG000

MERCURY OXYCYANIDE see MDA500
MERCURY(II) PERCHLORATE see MDG200
MERCURY PERCHLORIDE see MCY475
MERCURY PERNITRATE see MDF000
MERCURY(II) PEROXYBENZOATE see MCT750
MERCURY PERSULFATE see MDG500
MERCURY(II) POTASSIUM IODIDE see NCP500
MERCURY PROTOCHLORIDE see MCW000
MERCURY PROTOIODIDE see MDC750
MERCURY SALICYLATE see MCU000
MERCURY and SODIUM PHENOLSULFONATE see MCU500
MERCURY SUBSALICYLATE see MCU000
MERCURY(I) SULFATE see MDG250
MERCURY(II) SULFATE (1581) see MDG500
MERCURY, SULFATOBIS(METHYL- see BKS810
MERCURY(II) SULFIDE see MDG750
MERCURY TETRAVANADATE see MDH000
MERCURY(II) THIOCYANATE see MCU250
MERCURY THIOCYANATE (DOT) see MCU250
MERCURY(I) THIONITROSYLATE see MDH250
MERCURY, (((p-TOLYL)SULFAMOYL)IMINO)BIS(METHYL- see MLH100
MERCURY ZINC CHROMATE COMPLEX see ZJA000
MERCUSAL see SIH500
MERCUTAL see PFO750
MERCUZANTHIN see MCV750
MEREPRINE see PGE775
MEREX see KEA000
MERFALAN see BHT750
MERFAMIN see MDI000
MERFAZIN see PFM500
MERFEN see PFP250
MERFEN-STYLI see MDH500
MERGE see MRL750
MERIAN see AIF000
MERIDIL see MNQ000
MERILID see CHX250
MERINAX see POA250
MERIT see WBJ700
MERITAL see NMV725
MERITIN see DDT300
MERIZONE see BRF500
MERKAPTOBENZIMIDAZOL (CZECH) see BCC500
2-MERKAPTOBENZOTIAZOL (POLISH) see BDF000
5-MERKAPTO-3-FENYL-1,3,4-THIADIAZOL-2-THION DRASELNY (CZECH) see MCP500
2-MERKAPTOIMIDAZOLIN (CZECH) see IAQ000
MERKAZIN see BKL250
MERMETH see SNJ000
MERN see POK000
MERODICEIN see SIF500
MERONIDAL see MMN250
MEROPENIN see PDT500
MEROPHAN see BHT250
o-MEROPHAN see BHT250
MERPAN see CBG000
MERPHALAN see BHT750
o-MERPHALAN see BHT750
MERPHALAN HYDROCHLORIDE see BHV000
MERPHENYL NITRATE see MCU750, MDH500
MERPHOS see TIG250
MERPHYLLIN see HAP000
MERPHYRIN see HAP000
MERPOL see EJN500
MERRILLITE see ZBJ000
MERSALIN see SIH500
MERSALYL see SIH500
MERSALYL THEOPHYLLINE see MDH750
MERSALYL with THEOPHYLLINE see SAQ000
MERSOLITE 1 see PFN100
MERSOLITE 2 see PFM500
MERSOLITE 7 see MCU750
MERTEC see TEX000
MERTESTATE see TBF500
MERTHIOLATE see MDI000
MERTHIOLATE SALT see MDI000
MERTHIOLATE SODIUM see MDI000
MERTIONIN see MDT740
MERTORGAN see MDI000
MERURAN see MDI200
MERVAMINE see DJL000
MERVAN see AGN000
MERVAN ETHANOLAMINE SALT see MDI225
MERXIN see CCS510
MERZONIN SODIUM see MDI000

MES see MJW250
MESA see TGE300
MESACONATE see MDI250
MESACONIC ACID see MDI250
MESAMATE see MRL750
MESAMATE CONCENTRATE see MRL750
MESANOLON see MJE760
MESANTOIN see MKB250
MESATON see NCL500
MESCAL BEAN see NBR800
MESCALINE see MDI500
MESCALINE ACID SULFATE see MDJ000
MESCALINE HYDROCHLORIDE see MDI750
MESCALINE SULFATE see MDJ000
MESCOPIL see SBH500
MESECLAZONE see CIL500
MESENTOL see ENG500
MESEREIN see MDJ250
MESIDICAINE HYDROCHLORIDE see DHL800
MESIDIN (CZECH) see TLG500
MESIDINE see TLG500
MESIDINE HYDROCHLORIDE see TLH000
MESITYLAMINE see TLG500
MESITYLAMINE HYDROCHLORIDE see TLH000
MESITYLENE see TLM050
MESITYL ESTER of 1-PIPERIDINEACETIC ACID HYDROCHLORIDE see FAC175
MESITYLOXID (GERMAN) see MDJ750
MESITYL OXIDE see MDJ750
MESITYL OXIDE (1-PHTHALAZINYL)HYDRAZONE see DQU400
MESITYLOXYDE (DUTCH) see MDJ750
2-MESITYLOXYDIISOPROPYLAMINE HYDROCHLORIDE see MDK000
N-(2-MESITYLOXYETHYL)-N-METHYL-2-(2-METHYLPIPERIDINO)
  ACETAMIDE HYDROCHLORIDE see MDK250
N-(2-MESITYLOXYETHYL)-N-METHYL-2-(MORPHOLINE)ACETAMIDE HYDROCHLORIDE see MRT250
(1-MESITYLOXY-2-PROPYL)-N-METHYLCARBAMIC ACID-2-
  (DIETHYLAMINO)ETHYL ESTER, HYDROCHLORIDE see MDK500
N-(1-MESITYLOXY-2-PROPYL)-N-METHYL-2-(2-METHYLPIPERIDINO)ACET-
  AMIDE HYDROCHLORIDE see MDK750
MESNA see MDK875
MESNUM see MDK875
MESOCAINE HYDROCHLORIDE see DHL800
MESOFOLIN see MDV000
MESOKAIN HYDROCHLORIDE see DHL800
MESOMILE see MDU600
MESONEX see HFG550
MESOPIN see MDL000
MESORANIL see ASG250
MESORIDAZINE see MON750
MESOTAL see HFG550
MESOXALONITRILE see MDL250
MESOXALYLCARBAMIDE see AFT750
MESOXALYLCARBAMIDE MONOHYDRATE see MDL500
MESOXALYLUREA see AFT750
MESOXALYLUREA MONOHYDRATE see MDL500
MESPAFIN see HGP550
MESTALONE see MJE760
MESTANOLONE see MJE760
MESTENEDIOL see AOO475
MESTERONE see MPN500
MESTINON see MDL600
MESTRANOL see MKB750
MESTRANOL mixed with ANAGESTONE ACETATE (1:10) see AOO000
MESTRANOL mixed with CHLORMADINONE ACETATE see CNV750
MESTRANOL mixed with 6-CHLORO-6-DEHYDRO-17-α-
  ACETOXYPROGESTERONE see CNV750
MESTRANOL mixed with CHLOROETHYNYL NORGESTREL (1:20) see CHI750
MESTRANOL mixed with ETHYNERONE (1:20) see EQJ000
MESTRANOL mixed with ETHYNODIOL see EQK100
MESTRANOL mixed with ETHYNODIOL DIACETATE see EQK010
MESTRANOL mixed with LYNESTRENOL see LJE000
MESTRANOL mixed with LYNESTROL see LJE000
MESTRANOL mixed with NORETHINDRONE see MDL750
MESTRANOL mixed with NORETHISTERONE see MDL750
MESTRANOL mixed with NORETHYNODREL see EAP000
MESTRANOL mixed with NORGESTREL see NNR000
MESTRENOL see MKB750
MESTRENOL mixed with 6-CHLORO-6-DEHYDRO-17-α-
  ACETOXYPROGESTERONE see CNV750
MESULFA see ALF250
MESUPRINE HYDROCHLORIDE see MDM000
MESURAL see LFK000
MESUROL see DST000

MESUXIMIDE see MLP800
MESYLITH see CLD000
META see TDW500
METAARSENIC ACID see ARB000
METABARBITAL see DJO800
META BLACK see AQP000
METABOLITE C see SOA500
METABOLITE I see HNI500
METACAINE see EFX500
METACARDIOL see AKT000
METACE see CLO750
METACEN see IDA000
METACETALDEHYDE see TDW500
METACETONE see DJN750
METACETONIC ACID see PMU750
METACHLOR see CFX000
METACHROME ORANGE R see NEY000
METACIDE see MNH000
METACIL see MPW500
METACIN see ORQ000
METACLOPROMIDE see AJH000
METACORTANDRACIN see PLZ000
METACORTANDRALONE see PMA000
METACRATE see MIB750
METADEE see VSZ100
METADELPHENE see DKC800
METADIAZINE see PPO750
METADOMUS see MDO250
METAFOS see MNH000, SII500
METAFUME see MHR200
METAHEXAMIDE see AKQ250
METAHYDRIN see HII500
METAHYDROXYPROCAINE see DHO600
METAISOSEPTOX see DAP400
METAISOSYSTOX see DAP400
METAISOSYSTOX-SOLFON 20 315 see DAP600
METAISOSYSTOXSULFOXIDE see DAP000
METAKRYLAN METYLU (POLISH) see MLH750
METALCAPTASE see MCR750, PAP550
METALDEHYD (GERMAN) see TDW500
METALDEHYDE (DOT) see TDW500
METALDEIDE (ITALIAN) see TDW500
METALKAMATE see BTA250
METALLIBURE see MLJ500
METALLIC ARSENIC see ARA750
METALLIC OSMIUM see OKE000
METALUTIN see MDM350
METAM see VFW009
METAMFETAMINA see DBA800
METAMID see PJY500
METAMIDOFOS ESTRELLA see DTQ400
METAMIN see TJL250
METAMINE see TJL250
METAMITON see ALA500
METAMITRON (GERMAN) see ALA500
METAMPHETAMIN see DBA800
METAMPHETAMINE HYDROCHLORIDE see MDT600
METAM-SODIUM (DUTCH, FRENCH, GERMAN, ITALIAN) see VFU000
METANA ALUMINUM PASTE see AGX000
METANABOL see DAL300
METANDIENON see DAL300
METANDIENONE see DAL300
METANDIENONUM see DAL300
METANDIOL see AOO475
METANDREN see MPN500
METANDRIOL see AOO475
METANDROSTENOLON see DAL300
METANDROSTENOLONE see DAL300
METANEPHRIN see VGP000
METANFETAMINA see DBA800
METANICOTINE see MDM750
METANILAN SODNY (CZECH) see AIF250
METANILE YELLOW O see MDM775
METANILIC ACID see SNO000
METANIL YELLOW see MDM775
METANIL YELLOW 1955 see MDM775
METANIL YELLOW C see MDM775
METANIL YELLOW E see MDM775
METANIL YELLOW EXTRA see MDM775
METANIL YELLOW F see MDM775
METANIL YELLOW G see MDM775
METANIL YELLOW GRIESBACH see MDM775
METANIL YELLOW K see MDM775
METANIL YELLOW KRSU see MDM775

METANIL YELLOW M3X see MDM775
METANIL YELLOW O see MDM775
METANIL YELLOW PL see MDM775
METANIL YELLOW S see MDM775
METANIL YELLOW SUPRA P see MDM775
METANIL YELLOW VS see MDM775
METANIL YELLOW WS see MDM775
METANIL YELLOW Y see MDM775
METANIL YELLOW YK see MDM775
METANIN see CPQ250
METANITE see MGR500
METANOLO (ITALIAN) see MGB150
METANTYL see DJM800, XCJ000
METAOKSEDRIN see SPC500
METAOXEDRIN see NCL500, SPC500
METAOXEDRINUM see SPC500
METAPHEN see NHK900
METAPHENYLALANINE MUSTARD see BHU500
METAPHENYLENEDIAMINE see PEY000
METAPHOR see MNH000
METAPHOS see MNH000
METAPHOSPHORIC ACID, CALCIUM SODIUM SALT see CAX260
METAPHYLLIN see TEP500
METAPHYLLINE see TEP500
METAPLEXAN see MCJ400
METAPLEX NO see PKB500
METAPREL see MDM800
METAPROTERENOL see DMV800
METAPROTERENOL SULFATE see MDM800
METAQUALON see QAK000
METAQUEST A see EIX000
METAQUEST B see EIX500
METAQUEST C see EIV000
METARADRINE see HNB875
METARAMINOL see HNB875
(-)-METARAMINOL see HNB875
1-METARAMINOL see HNB875
METARAMINOL BITARTRATE see HNC000
METARAMINOL TARTRATE (1:1) see HNC000
METARSENOBILLON see SNR000
METARTRIL see IDA000
METASILICIC ACID see SCL000
METASOL see MLH000
METASOL P-6 see PFO000
METASOL TK-100 see TEX000
METASON see TDW500
METASQUALENE see TMP500
METASTENOL see DAL300
METASYMPATOL see NCL500
METASYNEPHRINE see NCL500
METASYSTEMOX see DAP000
METASYSTOX see MIW100
METASYSTOX FORTE see DAP400
METASYSTOX-R see DAP000
METASYSTOX-S see DSK600
METATHIONE see DSQ000
METATHION, S-METHYL ISOMER see MKC250
METATION see DSQ000
META TOLUYLENE DIAMINE see TGL750
METATOLYLENEDIAMINE DIHYDROCHLORIDE see DCE000
METATSIN see ORQ000
METATYL see MGJ750
d,l-METATYROSINE see TOG275
METAUPON see OHU000
METAUPON PASTE see SIY000
METAXALONE see XVS000
METAXAN see DJM800, XCJ000
METAXITE see ARM268
METAXON see CIR250
METAXONE see SIL500
METAZALONE see XVS000
METAZIN see SNJ000
METAZOL see MLH000
METAZOLO see MCO500
METAZOLONE see XVS000
METEBANYL see ORE000
METELILACHLOR see MQQ450
METENDIOL see AOO475
METENIX see ZAK300
METENOLONE ACETATE see PMC700
METEPA see TNK250
METERAZIN MALEATE see PMF250
METERFER see FBJ100
METERFOLIC see FBJ100

METET see MDN100
METFENOSSIDIOLO see RLU000
METFORMIN see DQR600
METFORMIN HYDROCHLORIDE see DQR800
"METH" see MDQ500
METHAANTHIOL (DUTCH) see MLE650
METHABOL see PAN100
METHACETONE see DJN750
METHACHLOR see CFX000
METHACHLORPHENPROP see CFC750
METHACHOLINE CHLORIDE see ACR000
METHACHOLINIUM CHLORIDE see ACR000
METHACIDE see TGK750
METHACIN see ORQ000
METHACON see DPJ400
METHACRALDEHYDE (DOT) see MGA250
METHACROLEIN see MGA250
METHACROLEIN DIMER see DLI600
METHACRYLALDEHYDE (DOT) see MGA250
METHACRYLALDEHYDE DIMER see DLI600
METHACRYLATE de BUTYLE (FRENCH) see MHU750
METHACRYLATE de METHYLE (FRENCH) see MLH750
METHACRYL CHLORIDE see MDN899
METHACRYLIC ACID see MDN250
METHACRYLIC ACID ALLYL ESTER see AGK500
METHACRYLIC ACID AMIDE see MDN500
METHACRYLIC ACID ANHYDRIDE see MDN699
METHACRYLIC ACID CHLORIDE see MDN899
METHACRYLIC ACID-3,4-DICHLOROANILIDE see DFO800
METHACRYLIC ACID DODECYL ESTER see DXY200
METHACRYLIC ACID, INHIBITED (DOT) see MDN250
METHACRYLIC ACID, ISOBUTYL ESTER see IIY000
METHACRYLIC ACID, ISODECYL ESTER see IKM000
METHACRYLIC ACID LAURYL ESTER see DXY200
METHACRYLIC ACID, METHYL ESTER (MAK) see MLH750
METHACRYLIC ACID METHYL ESTER POLYMERS see PKB500
METHACRYLIC ALDEHYDE see MGA250
METHACRYLIC AMIDE see MDN500
METHACRYLIC ANHYDRIDE see MDN699
METHACRYLIC CHLORIDE see MDN899
γ-METHACRYLOXYPROPYLTRIMETHOXYSILANE see TLC250
METHACRYLOYL ANHYDRIDE see MDN699
METHACRYLOYL CHLORIDE see MDN899
α-METHACRYLOYL CHLORIDE see MDN899
METHACRYLOYLOXYETHYL ISOCYANATE see IKG700
METHACRYLSAEUREBUTYLESTER (GERMAN) see MHU750
METHACRYLSAEUREMETHYL ESTER (GERMAN) see MLH750
METHACRYLYL CHLORIDE see MDN899
METHACYCLINE HYDROCHLORIDE see MDO250
METHACYCLINE MONOHYDROCHLORIDE see MDO250
METHADON see MDO760
METHADONE see MDO750, MDO760
(−)-METHADONE see MDO775
(+)-METHADONE see DBE100
d-METHADONE see DBE100
l-METHADONE see MDO775
(±)-METHADONE see MDO760
6s-METHADONE see DBE100
dl-METHADONE see MDO760
l-(+)-METHADONE see DBE100
s-(+)-METHADONE see DBE100
METHADONE HYDROCHLORIDE see MDP000
d-METHADONE HYDROCHLORIDE see MDP500
l-METHADONE HYDROCHLORIDE see MDP250
(±)-METHADONE HYDROCHLORIDE see MDP750
dl-METHADONE HYDROCHLORIDE see MDP750
racemic METHADONE HYDROCHLORIDE see MDP750
METHADRENE see MJV000
METHAFORM see ABD000
METHAFRONE see AHK500
METHAFURILEN FUMARATE see MDP800
METHAFURYLENE FUMARATE see MDP800
METHAGON see FDD075
METHAHEXAMIDE see AKQ250
METHAKRYLOXYMETHYLSILATRAN (CZECH) see HMS500
METHALLIBURE see MLJ500
METHALLYL ALCOHOL (DOT) see IMW000
METHALLYL CHLORIDE see CIU750
α-METHALLYL CHLORIDE see CIU750
1-METHALLYL-3-METHYL-6-AMINOTETRAHYDROPYRIMIDINEDIONE
   see AKL625
METHALLYL-19-NORTESTOSTERONE see MDQ075
17-α-(1-METHALLYL)-19-NORTESTOSTERONE see MDQ100
METHALUTIN see MDM350

METHAMBUCAINE HYDROCHLORIDE see AJA500
METHAMBUTOXYCAINE HYDROCHLORIDE see AJA500
METHAMIDOPHOS see DTQ400
METHAMIN see HEI500
METHAMINOACETOCATECHOL see MGC350
METHAMINODIAZEPINE HYDROCHLORIDE see MDQ250
METHAMINODIAZEPOXIDE see LFK000
METHAMINODIAZEPOXIDE HYDROCHLORIDE see MDQ250
METHAMPHETAMINE see DBB000
(+)-METHAMPHETAMINE CHLORIDE see MDT600
METHAMPHETAMINE HYDROCHLORIDE see DBA800, MDT600
(+)-METHAMPHETAMINE HYDROCHLORIDE see MDT600
d-METHAMPHETAMINE HYDROCHLORIDE see MDT600
l-METHAMPHETAMINE HYDROCHLORIDE see MDQ500
dl-METHAMPHETAMINE HYDROCHLORIDE see DAR100
METHAMPHETAMINIUM CHLORIDE see MDT600
METHAM SODIUM see VFU000, VFW009
METHANABOL see AOO475
METHANAL see FMV000
METHANAMIDE see FMY000
METHANAMINE (9CI) see MGC250
METHANAMINIUM NITRATE see MGN150
METHANDIENONE see DAL300
METHANDIOL see AOO475
METHANDRIOL see AOO475
METHANDROLAN see AOO475
METHANDROLONE see DAL300
METHANDROSTENOLONE see DAL300
METHANE see MDQ750
METHANE, compressed (DOT) see MDQ750
METHANE, refrigerated liquid (DOT) see MDQ750
METHANEARSONIC ACID see MGQ530
METHANEARSONIC ACID DIMERCURY SALT see DNW000
METHANEARSONIC ACID, IRON SALT see IHB680
METHANE BASE see MJN000
METHANEBIS(N,N'-(5-UREIDO-2,4-DIKETOTETRAHYDROIMIDAZOLE)-N,N-
   DIMETHYLOL) see GDO800
METHANE BORONIC ANHYDRIDE-PYRIDINE COMPLEX see MDQ800
METHANECARBONITRILE see ABE500
METHANECARBOTHIOLIC ACID see TFA500
METHANECARBOXAMIDE see AAI000
METHANECARBOXYLIC ACID see AAT250
METHANEDICARBOXYLIC ACID see CCC750
METHANEDICARBOXYLIC ACID, DIETHYL ESTER see EMA500
METHANE DICHLORIDE see MJP450
METHANEDISULFONIC ACID, CALCIUM SALT (1581) see CAT700
METHANEDITHIOL-S,S-DIESTER with O,O-DIETHYL ESTER PHOS-
   PHORODITHIOIC ACID see EEH600
METHANEPEROXOIC ACID see PCM500
METHANE, PHENYL- see TGK750
METHANESULFONAMIDE, N-(4-(9-ACRIDINYLAMINO)-3-
   METHOXYPHENYL)-, MONO(2-HYDROXYPROPANOATE)
   see AOD425
METHANESULFON-m-ANISIDIDE, 4'-(9-ACRIDINYLAMINO)-, compounded
   with LACTIC ACID see AOD425
METHANESULFONIC ACID see MDR250
METHANESULFONIC ACID CHLOROETHYL ESTER see CHC750
METHANESULFONIC ACID, COMPOUND with 2-(DIETHYLAMINO)-4-
   METHYLPENTYL PAMINOBENZOATE (1:1) see LEU000
METHANESULFONIC ACID, METHYLENE ESTER see MJQ500
METHANESULFONIC ACID-1-METHYLETHYL ESTER see IPY000
METHANESULFONIC ACID, 2-PROPENYL ESTER (9CI) see AGK750
METHANESULFONIC ACID, PROPYL ESTER see PNQ000
METHANESULFONIC ACID, SILVER SALT see SDR500
METHANESULFONIC ACID, SILVER(1+) SALT see SDR500
METHANESULFONIC ACID TETRAMETHYLENE ESTER see BOT250
METHANESULFONYL FLUORIDE see MDR750
1-(3-(2-(METHANESULFONYL)PHENOTHIAZIN-10-YL)PROPYL-4-
   PIPERIDINECARBOXAMIDE see MPM750
METHANESULPHONIC ACID ETHYL ESTER see EMF500
METHANESULPHONIC ACID METHYL ESTER see MLH500
METHANESULPHONYL FLUORIDE see MDR750
METHANETELLUROL see MDR775
METHANE TETRACHLORIDE see CBY000
METHANE TETRAMETHYLOL see PBB750
1,1',1'',1'''-METHANETHETRAYLTETRAKISBENZENE see TEA750
METHANETHIOL see MLE650
(METHANETHIOLATO)METHYLMERCURY see MLX300
METHANE TRICHLORIDE see CHJ500
METHANIDE see DJM800, XCJ000
1H-3a,7-METHANOAZULENE, OCTAHYDRO-6-METHOXY-3,6,8,8-
   TETRAMETHYL-,(3R-(3-α-3a-β, 6-α-7-β,8aα)- see CCR525
1H-3-α-7-METHANOAZULEN-6-OL, OCTAHYDRO-3,6,8,8-TETRAMETHYL-,
   FORMATE, (3R-(3-α-3a-β,6-α-7-β,8aα)- see CCR524

2,6-METHANO-3-BENZAZOCIN-8-OL-3-ALLYL-6-ETHYL-1,2,3,4,5,6-HEXAHYDRO-11-METHYL see AGG250
4,7-METHANOBENZOTRITHIOLE, HEXAHYDRO- see TNO300
METHANOIC ACID see FNA000
4,7-METHANOINDAN, 2,2,4,5,6,7,8,8-OCTACHLORO-3a,4,7,7a-TETRAHYDRO- see CDR575
4,7-METHANOINDENE, 4,5,6,7,8,8-HEXACHLORO-3a,4,7,7a-TETRAHYDRO- see HCN000
4,7-METHANO-1H-INDENE, 2,2,4,5,6,7,8,8-OCTACHLORO-2,3,3a,4,7,7a-HEXAHYDRO- (9CI) see CDR575
METHANOL see MGB150
METHANOLACETONITRILE see HGP000
N-METHANOLACRYLAMIDE see HLU500
METHANOL, (METHYL-ONN-AZOXY)-, BENZOATE (ester) (9CI) see MGS925
METHANOL, SODIUM SALT see SIK450
2-METHANOL TETRAHYDROPYRAN see MDS500
METHANOMETHENE see MCE500
4,7-METHANO-2,3,8-METHENOCYCLOPENT(a)INDENE, DODECAHYDRO-, stereoisomer see DLJ500
METHANONE, (5-AMINO-1,3-DIMETHYL-1H-PYRAZOL-4-YL)(2-FLUOROPHENYL)- see AJR400
6,9-METHANO-8H-PYRIDO(1',2':1,2)AZEPINO(4,5-b)INDOLE-6(6aH)-CARBOXYLIC ACID, 7,8,9,10,12,13- HEXAHYDRO-6-ETHYL-13a-HYDROXY-, METHYL ESTER, HYDROCHLORIDE see CNS200
4,7-METHANOSELENOPHENE, OCTAHYDRO-2,2,3,3-TETRAFLUORO- see TCI100
o-(1,4-METHANO-1,2,3,4-TETRAHYDRO-6-NAPHTHYL)-N-METHYL-N-(m-TOLYL)-THIOCARBAMATE see MQA000
METHANTHELINE BROMIDE see DJM800, XCJ000
METHANTHELINIUM BROMIDE see XCJ000
METHANTHINE BROMIDE see XCJ000
METHANTHIOL (GERMAN) see MLE650
METHAPHENILENE HYDROCHLORIDE see DCJ850
METHAPHOXIDE see TNK250
METHAPYRAPONE see MCJ370
METHAPYRILENE see DPJ200, TEO250
METHAPYRILENE HYDROCHLORIDE see DPJ400
METHAPYRILENE HYDROCHLORIDE (L.A.) see DPJ400
METHAPYRILENE HYDROCHLORIDE (S.A.) see DPJ400
METHAPYRILENE mixed with SODIUM NITRITE (1:2) see MDT000
METHAQUALONE see QAK000
METHAQUALONE HYDROCHLORIDE see MDT250
METHAQUALONEINONE see QAK000
METHAR see DXE600
METHARBITAL see DJO800
METHARBITONE see DJO800
METHARBUTAL see DJO800
METHARSINAT see DXE600
METHASAN see BJK500
β-METHASONE-17-VALERATE see VCA000
METHASQUIN see DBY500
METHAZATE see BJK500
METHAZINE see IDA000
METHAZOLE see BGD250
METHAZONIC ACID see NEJ600
METHCAINE HYDROCHLORIDE see IJZ000
METHDILAZINE see MPE250
METHDILAZINE HYDROCHLORIDE see MDT500
METHEDRINE see DBA800, DBB000, MDT600
METHEDRINE HYDROCHLORIDE see DBA800, MDT600
METHEGRIN see MJV750
METHELINA see DJM800, XCJ000
METHENAMIC ACID see XQS000
METHENAMINE see HEI500
METHENOLONE ACETATE see PMC700
N,N'-METHENYL-o-PHENYLENEDIAMINE see BCB750
METHENYL TRIBROMIDE see BNL000
METHENYL TRICHLORIDE see CHJ500
METHERGINE see PAM000
METHETOIN see ENC500
METHEXAMIDE see AKQ250
METHEXENYL see ERD500
METHEXENYL SODIUM see ERE000
METHIACIL see MPW500
METHIAMAZOLE see MCO500
METHIAZIC ACID see MNQ500
METHIAZINIC ACID see MNQ500
METHIDATHION see DSO000
METHILANIN see MDT740
METHIOCARB see DST000
METHIOCIL see MPW500
METHIODAL SODIUM see SHX000
METHIODIDE of N-BENZYLURETHANE of 3-DIMETHYLAMINOPHENOL see HNO000

METHIODIDE of N-METHYLURETHANE of 3-DIETHYLAMINOPHENOL see MID250
METHIODIDE of N-METHYLURETHANE of 3-DIMETHYLAMINOPHENOL see HNO500
METHIODIDE of N-METHYLURETHANE of 4-DIMETHYLAMINOPHENOL see HNP000
METHIONINE see MDT750
d-METHIONINE see MDT730
l-METHIONINE see MDT750
l-(-)-METHIONINE see MDT750
(±)-METHIONINE see MDT740
dl-METHIONINE see MDT740
l-METHIONINE, N-(N-(3-(p-FLUOROPHENYL)-l-ANANYL)-3-(m-(BIS(2-CHLOROETHYL)AMINO)PHENYL)-l-ALANYL)-, ETHYL ESTER, HYDROCHLORIDE see POI550
METHIONINE SULFOXIDE see ALF600
dl-METHIONINE SULFOXIDE see ALF600
METHIONINE SULFOXIMINE see MDU100
dl-METHIONINE-dl-SULFOXIMINE see MDU100
METHIOPLEGIUM see TKW500
METHISAZONE see MKW250
METHIUM CHLORIDE see HEA500
METHIXENE HYDROCHLORIDE see THL500
METHOCEL HG see HNX000
METHOCHLOPRAMIDE see AJH000
METHOCHLORIDE of N-METHYLURETHANE of 3-DIMETHYLAMINOPHENOL see HNN000
METHOCILLIN-S see SLJ000
METHOCROTOPHOS see DOL800
METHODICHLOROPHEN see MQR100
METHOFADIN see MDU300, MDU300
METHOFAZINE see MDU300
METHOFLURANE see DFA400
METHOGAS see MHR200
METHOHEXITAL SODIUM see MDU500
METHOHEXITONE SODIUM see MDU500
METHOIDAL SODIUM see SHX000
METHOIN see MKB250
METHOLENE 2218 see MJW000
METHOMYL see MDU600
METHOPHENAZATE ACID FUMARATE see MDU750
METHOPHENAZINE DIFUMARATE see MDU750
METHOPHOLINE see MDV000
METHOPHYLLINE see TEP500
METHOPIRAPONE see MCJ370
METHOPLAIN see DNA800, MJE780
METHOPRENE see KAJ000
METHOPROMAZINE MALEATE see MFK750
METHOPROPTRYNE see INQ000
METHOPTERIN see MDV500
METHOPYRAPONE see MCJ370
METHOPYRININE see MCJ370
METHOPYRONE see MCJ370
METHOQUINE see CFU750
METHORATE HYDROBROMIDE see DBE200
d-METHORPHAN see DBE150
Δ-METHORPHAN see DBE150
d-METHORPHAN HYDROBROMIDE see DBE200
METHORPHINAN see MKR250
METHORPHINAN HYDROBROMIDE see MDV250
METHOSCOPYLAMINE BROMIDE see SBH500
METHOSERPEDINE see MEK700
METHOSERPIDINE see MEK700
METHOSTAN see AOO475
METHOTEXTRATE see MDV500
METHOTREXATE see MDV500
METHOTREXATE DISODIUM SALT see MDV600
METHOTREXATE SODIUM see MDV750
METHOTRIMEPRAZINE see MCI500
METHOXA-DOME see XDJ000
METHOXADONE see MFD500
METHOXAMINE HYDROCHLORIDE see MDW000
METHOXANE see DFA400
METHOXCIDE see MEI450
METHOXERPATE HYDROCHLORIDE see MDW100
METHOXIPHENADRIN HYDROCHLORIDE see OJY000
METHOXO see MEI450
METHOXOLONE see XVS000
METHOXONE see CIR250, CIR500, SIL500
METHOXONE M see CLO200
METHOXSALEN see XDJ000
METHOXYACETALDEHYDE see MDW250
α-METHOXYACETALDEHYDE see MDW250
1-METHOXY-2-ACETAMIDOFLUORENE see MER000

2-METHOXYACETANILIDE see AAR000
3-METHOXYACETANILIDE see AAQ750
4-METHOXYACETANILIDE see AAR250
m-METHOXYACETANILIDE see AAQ750
o-METHOXYACETANILIDE see AAR000
p-METHOXYACETANILIDE see AAR250
2'-METHOXYACETANILIDE see AAR000
3'-METHOXYACETANILIDE see AAQ750
4'-METHOXYACETANILIDE see AAR250
METHOXYACETIC ACID see MDW275
2-METHOXYACETIC ACID see MDW275
2-METHOXYACETOACETANILIDE see ABA500
o-METHOXYACETOACETANILIDE see ABA500
2'-METHOXYACETOACETANILIDE see ABA500
p-METHOXYACETOPHENONE see MDW750
4'-METHOXYACETOPHENONE see MDW750
METHOXYACETYL CHLORIDE see MDW780
METHOXYACETYLENE see MDX000
(METHOXYACETYL)METHYLCARBAMIC ACID-o-ISOPROPOXYPHENYL
    ESTER see MDX250
5-METHOXY-N-ACETYLTRYPTAMINE see MCB350
4'-(3-METHOXY-9-ACRIDINYLAMINO)METHANESULFONALIDE see MDY000
4'-(1-METHOXY-9-ACRIDINYLAMINO)METHANESULFONANILIDE
    see MDX500
4'-(4-METHOXY-9-ACRIDINYLAMINO)METHANESULFONANILIDE
    see MDY250
N-(4-((3-METHOXY-9-ACRIDINYL)AMINO)PHENYL)-METHANESULFONAM-
    IDE see MDY000
2-METHOXY-AETHANOL (GERMAN) see EJH500
2-METHOXYAETHYLACETAT (GERMAN) see EJJ500
1-METHOXY-AETHYL-AETHYLNITROSAMIN (GERMAN) see MEO500
1-METHOXY-AETHYL-METHYLNITROSAMIN (GERMAN) see MEP500
METHOXYAETHYLQUECKSILBERCHLORID (GERMAN) see MEP250
METHOXYAETHYLQUECKSILBERSILIKAT (GERMAN) see MEP000
p-METHOXYALLYLBENZENE see AFW750
2-METHOXY-4-ALLYLPHENOL see EQR500
2'-METHOXY-4'-ALLYLPHENYL 4-GUANIDINOBENZOATE see MDY300
METHOXYAMINE see MKQ880
METHOXYAMINE HYDROCHLORIDE see MNG500
2-METHOXY-1-AMINOBENZENE see AOV900
4-METHOXY-2-AMINOBENZOTHIAZOLE see AKM750
6-METHOXY-2-AMINOBENZOTHIAZOLE see MDY750
2-METHOXY-4-AMINO-5-CHLORO-N-β-DIETHYLAMINOETHYL)BENZAMIDE
    DIHYDROCHLORIDE MONOHYDRATE see MQQ300
2-METHOXY-3-AMINODIBENZOFURAN see MDZ000
3-METHOXY-4-AMINODIPHENYL see MEA000
3-METHOXY-2-AMINODIPHENYLENE OXIDE see AKN500
2-METHOXY-4-AMINO-5-HYDROXYMETHYLPYRIMIDINE see AKO750
6-METHOXY-8-(4-AMINO-1-METHYLBUTYLAMINO)QUINOLINE see PMC300
1-METHOXY-2-AMINONAPHTHALENE see MFA250
4-METHOXY-4-AMINO-2-PENTANOL see MEA250
3-METHOXY-4-AMINOSTILBENE see MFP500
4-METHOXYAMPHETAMINE HYDROCHLORIDE see MEA500
dl,4-METHOXYAMPHETAMINE HYDROCHLORIDE see MEA500
4-METHOXYANILINE see AOW000
o-METHOXYANILINE see AOV900
p-METHOXYANILINE see AOW000
2-METHOXYANILINE HYDROCHLORIDE see AOX250
2-METHOXYANILINIUM NITRATE see MEA600
METHOXYAZOXYMETHANOLACETATE see MEA750
2-METHOXYBENZALDEHYDE see AOT525
4-METHOXYBENZALDEHYDE see AOT530
6-METHOXYBENZALDEHYDE see AOT525
o-METHOXYBENZALDEHYDE see AOT525
p-METHOXYBENZALDEHYDE (FCC) see AOT530
2-METHOXYBENZAMIDE see AOT750
o-METHOXYBENZAMIDE see AOT750
2-(p-METHOXYBENZAMIDO)ACETOHYDROXAMIC ACID see MEA800
5-METHOXY-BENZ(a)ANTHRACENE see MEB000
7-METHOXY-BENZ(a)ANTHRACENE see MEB500
8-METHOXY-BENZ(a)ANTHRACENE see MEB250
3-METHOXY-1,2-BENZANTHRACENE see MEB000
5-METHOXY-1,2-BENZANTHRACENE see MEB250
10-METHOXY-1,2-BENZANTHRACENE see MEB500
5-METHOXY-1,2-BENZ(a)ANTHRACENE see MEB250
3-METHOXY-7H-BENZ(de)ANTHRACEN-7-ONE see MEB750
3-METHOXYBENZANTHRONE see MEB750
4-METHOXYBENZENAMINE see AOW000
2-METHOXY-BENZENAMINE (9CI) see AOV900
METHOXYBENZENE see AOX750
4-METHOXYBENZENEACETIC ACID see MFE250
p-METHOXYBENZENEACETONITRILE see MFF000
4-METHOXYBENZENEAMINE see AOW000
2-METHOXYBENZENECARBOXALDEHYDE see AOT525

4-METHOXY-1,3-BENZENEDIAMINE see DBO000
4-METHOXY-1,2-BENZENEDIAMINE (9CI) see MFG000
4-METHOXY-1,3-BENZENEDIAMINE SULFATE see DBO400
4-METHOXY-1,3-BENZENEDIAMINE SULPHATE see DBO400
4-METHOXYBENZENEMETHANOL see MED500
2-METHOXY-4H-1,2,3-BENZODIOXAPHOSPHORINE-2-SULFIDE see MEC250
o-METHOXYBENZOHYDRAZIDE see AOV250
2-METHOXYBENZOIC ACID see MPI000
3-METHOXYBENZOIC ACID see AOU500
m-METHOXYBENZOIC ACID see AOU500
o-METHOXYBENZOIC ACID see MPI000
2-METHOXYBENZOIC ACID HYDRAZIDE see AOV250
4-METHOXYBENZOIC ACID HYDRAZIDE see AOV500
o-METHOXYBENZOIC ACID HYDRAZIDE see AOV250
p-METHOXYBENZOIC ACID HYDRAZIDE see AOV500
p-METHOXYBENZO HYDRAZIDE see AOV500
6-METHOXYBENZO(a)PYRENE see MEC500
METHOXYBENZOYL CHLORIDE see AOY250
2-METHOXYBENZOYL HYDRAZIDE see AOV250
4-METHOXYBENZOYL HYDRAZIDE see AOV500
o-METHOXYBENZOYLHYDRAZIDE see AOV250
2-METHOXYBENZOYLHYDRAZINE see AOV250
4-METHOXYBENZOYLHYDRAZINE see AOV500
(p-METHOXYBENZOYL)HYDRAZINE see AOV500
N-(4-METHOXY)BENZOYLOXYPIPERIDINE see MED000
8-METHOXY-3,4-BENZPYRENE see MED250
p-METHOXYBENZYL ACETATE see AOY400
4-METHOXYBENZYL ALCOHOL see MED500
p-METHOXYBENZYL ALCOHOL see MED500
p-METHOXYBENZYL BUTYRATE see MED750
p-METHOXYBENZYL CHLORIDE see MEE000
p-METHOXYBENZYL CYANIDE see MFF000
N-(p-METHOXYBENZYL)-N',N'-DIMETHYL-N-2-PYRIDYLETHYLENEDIAM-
    INE see WAK000
N-p-METHOXYBENZYL-N',N'-DIMETHYL-N-α-PYRIDYLETHYLENEDIAMINE
    see WAK000
N-p-METHOXYBENZYL-N'-N'-DIMETHYL-N-α-PYRIDYLETHYLENEDIAMINE
    MALEATE see DBM800
N-p-METHOXYBENZYL-N',N'-DIMETHYL-N-2-PYRIMIDINYLETHYLENE
    DIAMINE HYDROCHLORIDE see RDU000
p-METHOXYBENZYL FORMATE see MFE250
3-(p-METHOXYBENZYL)-4-(p-METHOXYPHENYL)-3-HEXENE see CNH525
4-METHOXYBENZYL METHYL KETONE see AOV875
p-METHOXYBENZYL METHYL KETONE see AOV875
p-METHOXYBENZYL PHENYLACETATE see APE000
2-METHOXY BIPHENYL see PEG000
4-METHOXYBIPHENYL see PEG250
p-METHOXYBIPHENYL see PEG250
3-METHOXYBIPHENYLAMINE see MEA000
1-((4-METHOXY(1,1'-BIPHENYL)-3-YL)METHYL)PYRROLIDINE
    see MEF400
2-METHOXY-4,6-BIS(ETHYLAMINO)-s-TRIAZINE see BJP250
2-METHOXY-4,6-BIS(ISOPROPYLAMINO)-s-TRIAZINE see MFL250
2-METHOXY-4,6-BIS(ISOPROPYLAMINO)-1,3,5-TRIAZINE see MFL250
4-METHOXY-α-(BIS(4-METHOXYPHENYL)METHYLENE)-N,N-
    DIMETHYLBENZENEETHANAMINE HCl see TNJ750
3-METHOXY BUTANOIC ACID see MEF500
4-((1-METHOXYBUTOXY)(5-VINYL-2-QUINUCLIDINYL)METHYL)-6-
    QUINOLINOL DIHYDROCHLORIDE see EQQ500
3-METHOXYBUTYL ACETATE see MHV750
2-METHOXY-4-sec-BUTYLAMINO-6-AETHYLAMINO-s-TRIAZIN (GERMAN)
    see BQC250
2-METHOXY-4-tert-BUTYLAMINO-6-AETHYLAMINO-s-TRIAZIN (GERMAN)
    see BQC500
4-METHOXY-2-tert-BUTYLPHENOL see BRN000
3-METHOXY BUTYRALDEHYDE see MEF750
3-METHOXYBUTYRIC ACID see MEF500
4'-METHOXYCARBONYL-N-ACETOXY-N-METHYL-4-AMINOAZOBENZENE
    see MEG000
2-(METHOXY-CARBONYLAMINO)-BENZIMIDAZOL (GERMAN) see MHC750
2-(METHOXYCARBONYLAMINO)-BENZIMIDAZOLE see MHC750
3-METHOXYCARBONYLAMINOPHENYL-N-3-METHYLPHENYLCARBA-
    MATE see MEG250
2-(METHOXYCARBONYL)ANILINE see APJ250
4'-METHOXYCARBONYL-N-BENZOYLOXY-N-METHYL-4-
    AMINOAZOBENZENE see MEG500
METHOXYCARBONYL CHLORIDE see MIG000
N-(2-METHOXYCARBONYLETHYL)-N-(1-ACETOXYBUTYL)NITROSAMINE
    see MEG750
N-(2-METHOXYCARBONYLETHYL)-N-(ACETOXYMETHYL)NITROSAMINE
    see MEH000
METHOXYCARBONYLETHYLENE see MGA500
4'-METHOXYCARBONYL-N-HYDROXY-N-METHYL-4-AMINOAZOBENZENE
    see MEH250

2-METHOXYETHYL VINYL ETHER see VPZ000
3-METHOXY-17-α-ETHYNOESTRADIOL see MKB750
3-METHOXYETHYNYLESTRADIOL see MKB750
3-METHOXY-17-α-ETHYNYLESTRADIOL see MKB750
3-METHOXY-17-α-ETHYNYL-1,3,5(10)-ESTRATRIEN-17-β-OL see MKB750
3-METHOXYETHYNYLOESTRADIOL see MKB750
3-METHOXY-17-ETHYNYLOESTRADIOL-17β see MKB750
3-METHOXY-17-α-ETHYNYL-1,3,5(10)-OESTRATRIEN-17-β-OL see MKB750
1-METHOXY-2-FAA see MER000
7-METHOXY-2-FAA see MER250
1-p-METHOXYFENYL-3,3-DIMETHYLTRIAZEN (CZECH) see DSN600
METHOXYFLUORAN see DFA400
METHOXYFLUORANE see DFA400
1-METHOXY-2-FLUORENAMINE HYDROCHLORIDE see MEQ750
1-METHOXYFLUOREN-2-AMINE HYDROCHLORIDE see MEQ750
7-METHOXY-N-2-FLUORENYLACETAMIDE see MER250
N-(1-METHOXYFLUOREN-2-YL)ACETAMIDE see MER000
N-(1-METHOXY-2-FLUORENYL)ACETAMIDE see MER000
N-(7-METHOXY-2-FLUORENYL)ACETAMIDE see MER250
N-(7-METHOXYFLUOREN-2-YL)ACETAMIDE see MER250
METHOXYFLURANE see DFA400
8-METHOXY-(FURANO-3′.2′:6.7-COUMARIN) see XDJ000
9-METHOXY-7H-FURO(3,2-g)BENZOPYRAN-7-ONE see XDJ000
4-METHOXY-7H-FURO(3,2-g)(1)BENZOPYRAN-7-ONE see MFN275
8-METHOXY-2′,3′,6,7-FUROCOUMARIN see XDJ000
8-METHOXY-4′,5′,6,7-FUROCOUMARIN see XDJ000
6-METHOXYHARMAN see HAI500
METHOXYHYDRASTINE see NOA000
METHOXYHYDRASTINE HYDROCHLORIDE see NOA500
3-METHOXY-4-HYDROXYBENZALDEHYDE see VFK000
3-METHOXY-4-HYDROXYBENZOIC ACID see VFF000
3-METHOXY-4-HYDROXYBENZOIC ACID DIETHYLAMIDE see DKE200
3-METHOXY-4-HYDROXYBENZOIC ACID, ETHYL ESTER see EQE500
4-METHOXY-2-HYDROXYBENZOPHENONE see MES000
3-METHOXY-4-HYDROXY-BENZYLACETONE see VFP100
β-METHOXY-β′-HYDROXYDIETHYL ETHER see DJG000
METHOXYHYDROXYETHANE see EJH500
N-(2-METHOXY-3-HYDROXYMERCURIPROPYL)BARBITAL see MES250
METHOXYHYDROXYMERCURIPROPYLSUCCINYLUREA see MFC000
l-3-METHOXY-ω-(1-HYDROXY-1-PHENYLISOPROPYLAMINO)PRO-
   PIOPHENONE HYDROCHLORIDE see OQU000
2-METHOXY-10-(3-(4-HYDROXYPIPERIDINO)-2-METHYLPROPYL)PHENOTHI-
   AZINE see LEO000
3-METHOXY-10-(3-(4-HYDROXYPIPERIDYL)-2-METHYLPROPYL)PHENOTHI-
   AZINE see LEO000
8-METHOXY-4-HYDROXYQUINOLINE-2-CARBOXYLIC ACID see HLT500
3-METHOXY-4-HYDROXYTOLUENE see MEK250
1-METHOXY IMIDAZOLE-N-OXIDE see MES300
3-METHOXY-5,4′-IMINOBIS(1-BENZAMIDO-ANTHRAQUINONE) see MES500
METHOXYINDOLEACETIC ACID see MES850
5-METHOXYINDOLEACETIC ACID see MES850
5-METHOXYINDOLE-3-ACETIC ACID see MES850
5-METHOXYINDOLE-3-ETHANOL see MFT300
5-METHOXY-1H-INDOLE-3-ETHANOL (9CI) see MFT300
2-METHOXY-4-ISOPROPYLAMINO-6-ETHYLAMINO-s-TRIAZINE see EGD000
METHOXYLAMINE HYDROCHLORIDE see MNG500
METHOXYLENE see DPJ400
1-(2-METHOXY-4-METHOXYCARBONYL-1-PHENOXY)-3-ISOPROPYLAMINO-2-
   PROPANOL see VFP200
2-(METHOXYMETHOXY)ETHANOL see MET000
4-METHOXY-2-(5-METHOXY-3-METHYL-1H-PYRAZOL-1-YL)-6-
   METHYLPYRIMIDINE see MCH550
2-METHOXY-3-METHYL-4-ACETYLTETRAHYDROFURAN see MHR100
METHOXYMETHYLACRYLAMIDE see MEX300
N-(METHOXYMETHYL)ACRYLAMIDE see MEX300
METHOXYMETHYL-AETHYLNITROSAMINE (GERMAN) see MEV750
3-METHOXYMETHYLAMINOAZOBENZENE see MNS000
(E)-(3-(METHOXYMETHYLAMINO)-1-METHYL-3-OXO-1-PROPE-
   NYL)DIMETHYL PHOSPHATE see DOL800
2-METHOXY-5-METHYLANILINE see MGO750
4-METHOXY-2-METHYLANILINE see MGO500
3-METHOXY-7-METHYLBENZ(c)ACRIDINE see MET875
5-METHOXY-7-METHYL-BENZ(a)ANTHRACENE see MEU000
8-METHOXY-7-METHYL-BENZ(a)ANTHRACENE see MEU250
7-METHOXY-12-METHYLBENZ(a)ANTHRACENE see MEU500
3-METHOXY-10-METHYL-1,2-BENZANTHRACENE see MEU000
5-METHOXY-10-METHYL-1,2-BENZANTHRACENE see MEU250
4-METHOXY-2-METHYLBENZENAMINE see MGO500
2-METHOXY-5-METHYL-BENZENAMINE (9CI) see MGO750
S-(((METHOXYMETHYL-CARBAMOYL)METHYL) O,O-
   DIMETHYLPHOSPHORODITHIOATE see FNE000
S-(N-METHOXYMETHYLCARBAMOYLMETHYL) DIMETHYL PHOS-
   PHOROTHIOLOTHIONONATE see FNE000
3-METHOXY-17-METHYL-15H-CYCLOPENTAPHENANTHRENE see MEU750

3-METHOXY-17-METHYL 15H-CYCLOPENTA(a)PHENANTHRENE see MEU750
11-METHOXY-17-METHYL-15H-CYCLOPENTA(a)PHENANTHRENE
   see MEV000
1-METHOXY-1-METHYL-3-(3,4-DICHLOROPHENYL)UREA see DGD600
6-METHOXY-11-METHYL-15,16-DIHYDRO-17H-CYCLOPENTA(a)
   PHENANTHREN-17-ONE see MEV250
19-METHOXY-1,2-(METHYLENEDIOXY)-6a-α-APORPHIN-11-OL see HLT000
(2-(2-METHOXY METHYL ETHOXY)METHYL ETHOXY)PROPANOL
   see MEV500
METHOXYMETHYL ETHYL NITROSAMINE see MEV750
4-METHOXYMETHYL-5-HYDROXY-6-METHYL-3-PYRIDINEMETHANOL
   HYDROCHLORIDE see MEY000
2-METHOXY-10-(2-METHYL-3-(4-HYDROXYPIPERIDINO)PROPYL)PHENOTHI-
   AZINE see LEO000
7-METHOXYMETHYL-12-METHYLBENZ(a)ANTHRACENE see MEW000
METHOXYMETHYL-METHYLNITROSAMIN (GERMAN) see MEW250
METHOXYMETHYL METHYLNITROSAMINE see MEW250
8-METHOXY-1-METHYL-4-(p-((p-((1-METHYLPYRIDINIUM-4-YL)AMINO)PHE-
   NYL)CARBAMOYL)ANILINO)QUINOLINIUM), DIBROMIDE see MEW500
6-METHOXY-1-METHYL-4-(p-((p-((1-METHYLPYRIDINIUM-4-YL)AMINO)
   PHENYL)CARBAMOYL)ANILINO)QUINOLINIUM), DI-p-TOLUENESULFON-
   ATE see MEW750
4-METHOXY-5-METHYL-6-(7-METHYL-8-(TETRAHYDRO-3,4-DIHYDROXY-
   2,4,5-TRIMETHYL-2-FURANYL)-1,3,5,7-OCTATETRAENYL)-2H-PYRAN-2-
   ONE see CMS500
3-METHOXY-17-METHYLMORPHINAN HYDROBROMIDE see LFD200
d-3-METHOXY-N-METHYLMORPHINAN HYDROBROMIDE see DBE200
(±)-3-METHOXY-17-METHYLMORPHINAN HYDROBROMIDE see RAF300
3-METHOXY-17-METHYL-9-α,13-α,14-α-MORPHINAN HYDROBROMIDE
   see DBE200
l-(-)-6-METHOXY-α-METHYL-2-NAPHTHALENEACETIC ACID SODIUM SALT
   see NBO550
α-(METHOXYMETHYL)-2-NITROIMIDAZOLE-1-ETHANOL see NHH500
α-(METHOXYMETHYL)-2-NITRO-1H-IMIDAZOLE-1-ETHANOL see NHH500
α-(METHOXYMETHYL)-2-NITRO-1H-IMIDAZOLE-1-ETHANOL (9CI)
   see NHH500
7-METHOXY-1-METHYL-2-NITRONAPHTHO(2,1-b)FURAN see MEW775
N-METHOXYMETHYL-N-NITROSO-sec-BUTYLAMINE see BRU000
N-METHOXY-N-METHYLNONANAMIDE see EQS500
3-METHOXY-5-METHYL-4-OXO-2,5-HEXADIENOIC ACID see PAP750
3-((4-(5-(METHOXYMETHYL)-2-OXO-3-OX-
   AZOLIDINYL)PHENOXY)METHYL)BENZONITRILE see MEW800
N-(7-METHOXY-3-METHYL-4-OXO-2-PHENYL-4H-CHROMEN-8-YL)METHYL-
   N,N-DIMETHYLAMINE see DNV000
5-METHOXY-2-METHYL-1-(1-OXO-3-PHENYL-2-PROPENYL)-1H-INDOLE-3-
   ACETIC ACID (9CI) see CMP950
4-METHOXY-4-METHYL-2-PENTANONE see MEX250
4-METHOXY-4-METHYLPENTAN-2-ONE (DOT) see MEX250
dl-p-METHOXY-α-METHYL-PHENETHYLAMINE HYDROCHLORIDE
   see MEA500
2-METHOXY-4-METHYLPHENOL see MEK250
4-β-METHOXY-1-METHYL-4-α-PHENYL-3-α,5-α-PROPANOPIPERIDINE
   HYDROGEN CITRATE see ARX150
N-(METHOXYMETHYL)-2-PROPENAMIDE see MEX300
2-METHOXY-3(5)-METHYLPYRAZINE see MEX350
2-(3-METHOXY-5-METHYLPYRAZOL-2-YL)-4-METHOXY-6-METHYLPYRIMID-
   INE see MCH550
7-METHOXY-1-METHYL-9H-PYRIDO(3,4-b)INDOLE see HAI500
4-METHOXYMETHYLPYRIDOXINE HYDROCHLORIDE see MEY000
4-METHOXYMETHYLPYRIDOXOL HYDROCHLORIDE see MEY000
1-(4-METHOXY-6-METHYL-2-PYRIMIDINYL)-3-METHYL-5-
   METHOXYPYRAZOLE see MCH550
p-METHOXY-β-METHYLSTYRENE see PMQ750
N-(4-(METHOXYMETHYL)-1-(2-(2-THIENYL)ETHYL)-4-PIPERIDINYL)-N-
   PHENYLPROPIONAMIDE CITRATE see SNH150
N-(4-(METHOXYMETHYL)-1-(2-(2-THIENYL)ETHYL)-4-PIPERIDYL)
   PROPIONANILIDE CITRATE see SNH150
2-(METHOXY(METHYLTHIO)PHOSPHINYLIMINO)-3-ETHYL-5-METHYL-1,3-
   OXAZOLIDINE see MOA750
6-METHOXY-3-METHYL-1,7,8-TRIHYDROXYANTHRAQUINONE see MEY750
3-METHOXY-4-MONOMETHYLAMINOAZOBENZENE see MNS000
METHOXYN see DBA800
METHOXYNAL see AAE500
4-(6-METHOXY-2-NAPHTHALENYL)-2-BUTANONE see MFA300
3-(4-METHOXY-1-NAPHTHOYL)PROPIONIC ACID see MEZ300
β-(1-METHOXY-4-NAPHTHOYL)-PROPIONSAEURE (GERMAN) see MEZ300
1-METHOXY-2-NAPHTHYLAMINE see MFA000
1-METHOXY-2-NAPHTHYLAMINE HYDROCHLORIDE see MFA250
4-(6-METHOXY-2-NAPHTHYL)-2-BUTANONE see MFA300
(+)-2-(METHOXY-2-NAPHTHYL)-PROPIONIC ACID see MFA500
(+)-2-(METHOXY-2-NAPHTHYL)-PROPIONSAEURE (GERMAN) see MFA500
4-METHOXY-2-NITROANILIN (CZECH) see MFB000
2-METHOXY-5-NITROANILINE see NEQ500
4-METHOXY-2-NITROANILINE see MFB000

3-METHOXY-4-NITROAZOBENZENE see MFB250
2-METHOXY-5-NITROBENZENAMINE see NEQ500
2-METHOXYNITROBENZENE see NER000
4-METHOXYNITROBENZENE see NER500
p-METHOXYNITROBENZENE see NER500
1-METHOXY-2-NITROBENZENE see NER000
1-METHOXY-4-NITROBENZENE see NER500
5-METHOXY-2-NITROBENZOFURAN see MFB350
3-METHOXY-2-(2-NITRO-1-IMIDAZOLYL)-1-PROPANOL see DBA700
7-METHOXY-2-NITRONAPHTHO(2,1-b)FURAN see MFB400
8-METHOXY-2-NITRONAPHTHO(2,1-b)FURAN see MFB410
8-METHOXY-6-NITROPHENANTHOL-(3,4-d)-1,3-DIOXOLE-5-CARBOXYLIC
    ACID see AQY250
8-METHOXY-6-NITROPHENANTHRO(3,4-d)-1,3-DIOXOLE-5-CARBOXYLIC
    ACID SODIUM SALT see AQY125
(3-METHOXY-4-NITROPHENYL)PHENYLDIAZENE see MFB250
N-METHOXY-N-NITROSOMETHYLAMINE see DSZ000
1-METHOXY-N-NITROSO-N-PROPYLPROPYLAMINE see DWU800
3-METHOXY-17-α-19-NORPREGNA-K,3,5(10)-TRIEN-20-YN-17-OL see MKB750
11-β-METHOXY-19-NOR-17-α-PREGNA-1,3,5(10)-TRIEN-20-YNE-3,17-DIOL
    see MRU600
3-METHOXY-19-NOR-17-α-PREGNA-1,3,5(10)-TRIEN-10-YN-17-OL see MKB750
(17-α)-3-METHOXY-19-NORPREGN-1,3,5(10)-TRIEN-20-YN-17-OL see MKB750
d-threo-METHOXY-3-(1-OCTENYL-ONN-AZOXY)-2-BUTANOL see EAG000
3-METHOXYOESTRADIOL see MFB775
METHOXYOXIMERCURIPROPYLSUCCINYL UREA see MFC000
S-5-METHOXY-4-OXOPYRAN-2-YLMETHYL DIMETHYL PHOSPHOROTHIO-
    ATE see EAS000
S-((5-METHOXY-2-OXO-1,3,4-THIADIAZOL-3(2H)-YL)METHYL)-O,O-
    DIMETHYL PHOSPHORODITHIOATE see DSO000
2-METHOXY-1-(PENTYLOXY)-4-(1-PROPENYL)-BENZENE see AOK000
METHOXYPHENAMINE HYDROCHLORIDE see OJY000
METHOXYPHENAMINIUM CHLORIDE see OJY000
8-METHOXYPHENANTHRO(3,4-d)-1,3-DIOXOLE-5-CARBOXYLIC ACID
    see AQX825
8-METHOXYPHENANTHRO(3,4-d)-1,3-DIOXOLE-5-CARBOXYLIC ACID
    METHYL ESTER see MGQ525
p-METHOXYPHENETHYLAMINE see MFC500
1-(p-METHOXYPHENETHYL)HYDRAZINE HYDROGEN SULFATE see MFC600
1-(p-METHOXYPHENETHYL)-HYDRAZINE SULFATE (1:1) see MFC600
4-(β-METHOXYPHENETHYL)-α-PHENYL-1-PIPERAZINEPROPANOL DIHY-
    DROCHLORIDE see ECU600
2-METHOXYPHENOL see GKI000
3-METHOXYPHENOL see REF050
4-METHOXYPHENOL see MFC700
m-METHOXYPHENOL see REF050
o-METHOXYPHENOL see GKI000
p-METHOXYPHENOL see MFC700
METHOXYPHENOTHIAZINE see MCI500
1-(3-(2-METHOXYPHENOTHIAZIN-10-YL)-2-METHYLPROPYL)-4-
    PIPERIDINOL see LEO000
2-(p-METHOXYPHENOXY)-N-(2-(DIETHYLAMINO)ETHYL)ACETAMIDE
    see DIB600, MCH250
METHOXYPHENOXYDIOL see RLU000
N-(2-o-METHOXYPHENOXYETHYL)-6,7-DIMETHOXY-3,4-DIHYDRO-2-(1H)-
    ISOQUINOLINE CARBOXAMIDINE HBr see SBA875
(2-(p-METHOXYPHENOXY)ETHYL)HYDRAZINE HYDROCHLORIDE
    see MFD250
3-(2-METHOXYPHENOXY)-1-GLYCERYL CARBAMATE see GKK000
3-(o-METHOXYPHENOXY-2-HYDROXYPROPYL CARBAMATE see GKK000
3-(o-METHOXYPHENOXY)-2-HYDROXYPROPYL-1-NICOTINATE see HLT300
(3R-trans)-3-((4-METHOXYPHENOXY)METHYL)-1-METHYL-4-
    PHENYLPIPERIDINE HYDROCHLORIDE see FAJ200
5-(o-METHOXYPHENOXYMETHYL)-2-OXAZOLIDINONE see MFD500
5-(o-METHOXYPHENOXYMETHYL)-2-OXAZOLIDONE see MFD500
3-o-METHOXYPHENOXYPROPANE 1:2-DIOL see RLU000
3-(o-METHOXYPHENOXY)-1,2-PROPANEDIOL see RLU000
3-(o-METHOXYPHENOXY)-1,2-PROPANEDIOL-1-CARBAMATE see GKK000
4-METHOXYPHENYLACETIC ACID see MFE250
p-METHOXYPHENYLACETIC ACID see MFE250
p-METHOXYPHENYLACETONE see AOV875
4-METHOXYPHENYLACETONITRILE see MFF000
p-METHOXYPHENYLACETONITRILE see MFF000
2-METHOXY-2-PHENYLACETOPHENONE see MHE000
β-(o-METHOXYPHENYL)ACROLEIN see MEJ750
o-METHOXYPHENYLAMINE see AOV900
p-METHOXYPHENYLAMINE see AOW000
N-(p-METHOXYPHENYL)-1-AZIRIDINECARBOXAMIDE see MFF250
2-METHOXY-4-PHENYLAZOANILINE see MFF500
p-((METHOXYPHENYL)AZO)ANILINE see MFF550
4-((p-METHOXYPHENYL)AZO)-o-ANISIDINE see DNY400
N-(2-METHOXY-4-(PHENYLAZO)PHENYL)HYDROXYLAMINE see HLR000
1-(2-(p-(6-METHOXY-2-PHENYLBENZOFURAN-3-YL)PHENOXY)ETHYL)
    PYRROLIDINE HYDROCHLORIDE see PGF100

1-(2-(p-(6-METHOXY-2-PHENYL-3-BENZOFURANYL)PHENOXY)ETHYL)
    PYRROLIDINE HYDROCHLORIDE see PGF100
1-(2-(4-(6-METHOXY-2-PHENYLBENZO(b)THIEN-3-YL)PHENOXY)ETHYL)
    PYRROLIDINE HYDROCHLORIDE see MFF575
4-p-METHOXYPHENYL-2-BUTANONE see MFF580
p-METHOXYPHENYL-N-CARBAMOYLAZIRIDINE see MFF250
(E)-o-METHOXY-α-PHENYL-CINNAMIC ACID see MFG600
α-p-METHOXYPHENYL-α-DI-n-BUTYLAMINOACETAMIDE see DDT300
3-p-METHOXYPHENYL-5-DIETHYLAMINOETHYL-1,2,4-OXADIAZOLE
    see CPN750
2-(p-(6-METHOXY-2-PHENYL-3,4-DIHYDRO-1-NAPHTHYL)PHENOXY)TRIETH-
    YLAMINE HYDROCHLORIDE see MFF625
2-(3-METHOXYPHENYL)-5,6-DIHYDRO-s-TRIAZOLO(5,1-a)ISOQUINOLINE
    see MFF650
2-(m-METHOXYPHENYL)-5,6-DIHYDRO-s-TRIAZOLO(5,1-a)ISOQUINOLINE
    see MFF650
1-(m-METHOXYPHENYL)-2-DIMETHYLAMINOMETHYLCYCLOHEXAN-1-OL
    HYDROCHLORIDE see THJ750
trans-1-(m-METHOXYPHENYL)-2-DIMETHYLAMINOMETHYLCYCLOHEXAN-1-
    OL HYDROCHLORIDE see THJ750
1-(p-METHOXYPHENYL)-3,3-DIMETHYLTRIAZENE see DSN600
2-(p-METHOXYPHENYL)-3,3-DIPHENYLACRYLONITRILE see MFF750
α-(p-METHOXYPHENYL)-β,β-DIPHENYLACRYLONITRILE see MFF750
5-(p-METHOXYPHENYL)-1,2-DITHIOCYCLOPENTEN-3-THIONE see AOO490
5-(p-METHOXYPHENYL)-3H-1,2-DITHIOLE-3-THIONE see AOO490
5-(4-METHOXYPHENYL)-3H-1,2-DITHIOLE-3-THIONE (9CI) see AOO490
4-METHOXY-m-PHENYLENEDIAMINE see DBO000, MFG000
p-METHOXY-m-PHENYLENEDIAMINE see DBO000
4-METHOXY-m-PHENYLENEDIAMINE SULFATE see DBO400
4-METHOXY-m-PHENYLENEDIAMINE SULPHATE see DBO400
p-METHOXY-m-PHENYLENEDIAMINE SULPHATE see DBO400
p-METHOXYPHENYLETHYLAMINE see MFC500
o-METHOXY-β-PHENYLETHYLHYDRAZINE DIHYDROGEN SULFATE
    see MFG200
p-METHOXY-β-PHENYLETHYLHYDRAZINE DIHYDROGEN SULFATE
    see MFC600
1-(2-METHOXY-2-PHENYL)ETHYL-4-(2-HYDROXY-3-METHOXY-3-
    PHENYL)PROPYLPIPERAZINE DIHYDROCHLORIDE see MFG250
1-(2-METHOXY-2-PHENYLETHYL)-4-(3-HYDROXY-3-PHENYLPROPYL)
    PIPERAZINE DIHYDROCHLORIDE see ECU600
4-(2-METHOXY-2-PHENYLETHYL)-α-PHENYL-1-PIPERAZINEPROPANOL
    (9CI) see EQY600
o-METHOXYPHENYL GLYCERYL ETHER see RLU000
2-(p-(6-METHOXY-2-PHENYL-3-INDENYL)PHENOXY)TRIETHYLAMINE
    HYDROCHLORIDE see MFG260
2-(p-(6-METHOXY-2-PHENYLINDEN-3-YL)PHENOXY)TRIETHYLAMINE HY-
    DROCHLORIDE see MFG260
β-(o-METHOXYPHENYL)ISO-p-TROPYLMETHYLAMINEHYDROCHLORIDE
    see OJY000
2-(3-METHOXYPHENYL)-8-METHOXY-5H-s-TRIAZOLO(5,1-a)ISOINDOLE
    see MFG275
α-(2-METHOXYPHENYL)-β-METHYLAMINOPROPANEHYDROCHLORIDE
    see OJY000
9-β-METHOXY-9-α-PHENYL-3-METHYL-3-AZABICYCLO(3.3.1)NONANE CI-
    TRATE see ARX150
2-(4-METHOXYPHENYL)-3-(1-METHYLETHYL)-3H-NAPHTH(1,2-d)IMIDAZ
    OLE see THG700
4-METHOXYPHENYL METHYL KETONE see MDW750
p-METHOXYPHENYL METHYL KETONE see MDW750
1-(p-METHOXYPHENYL)-3-METHYL-3-NITROSOUREA see MFG400
1-(4-METHOXYPHENYL)-3-METHYL TRIAZENE see MFG510
1-(2-(p-(α-(p-METHOXYPHENYL)-β-NITROSTYRYL)PHENOXY)ETHYL)
    PYRROLIDINE CITRATE (1:1) see NHP500
1-(2-(p-(α-(p-METHOXYPHENYL)-β-NITROSTYRYL)PHENOXY)ETHYL)
    PYRROLIDINE MONOCITRATE see NHP500
2-(p-(p-METHOXY-α-PHENYLPHENETHYL)PHENOXY)TRIETHYLAMINE
    see MFG525
2-(p-(p-METHOXY-α-PHENYLPHENETHYL)PHENOXY)TRIETHYLAMINE
    HYDROCHLORIDE see MFG530
trans-3-(o-METHOXYPHENYL)-2-PHENYLACRYLIC ACID see MFG600
2-(p-(2-(p-METHOXYPHENYL)-1-PHENYL-1-BUTENYL)PHENOXY)TRIETHYLA-
    MINE CITRATE see HAA310
2-(3-o-METHOXYPHENYLPIPERAZINO)-PROPYL)-3-METHYL-7-
    METHOXYCHROMONE see MFH000
4-(4-(o-METHOXYPHENYL)-1-PIPERAZINYL)-p-FLUOROBUTYROPHENONE
    see HAF400
7-(4-(3-METHOXYPHENYL)-1-PIPERAZINYL)-4-NITROBENZOFURAZAN-1-
    OXIDE see MFH750
6-(3-(4-(o-METHOXYPHENYL)-1-PIPERAZINYL)PROPYLAMINO)-1,3-
    DIMETHYLURACIL see USJ000
1-(p-METHOXYPHENYL)-2-PROPANONE see AOV875
3-(2-METHOXYPHENYL)-2-PROPENAL see MEJ750
1-(p-METHOXYPHENYL)PROPENE see PMQ750
(Z)-3-(2-METHOXYPHENYL)-2-PROPENOIC ACID (9CI) see MEJ775

2-(3-METHOXYPHENYL)PYRAZOLO(5,1-a)ISOQUINOLINE see MFH800
2-(m-METHOXYPHENYL)-PYRAZOLO(5,1-a)ISOQUINOLINE see MFH800
(2-(p-METHOXYPHENYLTHIO)ETHYL)HYDRAZINE MALEATE see MFJ000
5-(m-METHOXYPHENYL-3-(o-TOLYL)-s-TRIAZOLE see MFJ100
2-(3-METHOXYPHENYL)-5H-s-TRIAZOLO(5,1-a)ISOINDOLE see MFJ105
2-(m-METHOXYPHENYL)-s-TRIAZOLO(5,1-a)ISOQUINOLINE see MFJ110
5-(p-METHOXYPHENYL)TRITHIONE see AOO490
4-METHOXY-6-(β-PHENYLVINYL)-5,6-DIHYDRO-α-PYRONE see GJI250
5-(m-METHOXYPHENYL)-3-(2,4-XYLYL)-s-TRIAZOLE see MFJ115
METHOXY POLYETHYLENE GLYCOL 350 see MFJ750
METHOXY POLYETHYLENE GLYCOL 550 see MFK000
METHOXY POLYETHYLENE GLYCOL 750 see MFK250
α-omega-METHOXYPOLY(ETHYLENE OXIDE) see DOM100
2-METHOXYPROMAZINE see MFK500
METHOXYPROMAZINE MALEATE see MFK750
METHOXYPROPANEDIOL see RLU000
1-METHOXY-2-PROPANOL see PNL250
3-METHOXY-1-PROPANOL see MFL000
METHOXYPROPAZINE see MFL250
4-METHOXYPROPENYLBENZENE see PMQ750
1-METHOXY-4-PROPENYLBENZENE see PMQ750
1-METHOXY-4-(2-PROPENYL)BENZENE see AFW750
(E)-1-METHOXY-4-(1-PROPENYL)BENZENE see PMR250
2-METHOXY-4-PROPENYLPHENOL see IKQ000
2-METHOXY-4-PROP-2-ENYLPHENOL see EQR500
2-METHOXY-4-(2-PROPENYL)PHENOL see EQR500
2-METHOXY-4-PROPENYLPHENYL ACETATE see AAX750
3-METHOXYPROPIONITRILE see MFL750
2-(2-(2-METHOXYPROPOXY)PROPOXY PROPANOL see TNA000
3-METHOXYPROPYLAMINE see MFM000
2-(2-METHOXYPROPYLAMINO)-2',6'-ACETOXYLIDIDE HYDROCHLORIDE
  see DSN800
1-METHOXY-4-PROPYLBENZENE see PNE250
2-METHOXY-4-PROPYLPHENOL see MFM750
1-METHOXYPROPYLPROPYLNITROSAMIN (GERMAN) see DWU800
1-METHOXYPROPYLPROPYLNITROSAMINE see DWU800
9-(2-METHOXY-4-(PROPYLSULFONAMIDO)ANILINO)-4-
  ACRIDINECARBOXAMIDE HYDROCHLORIDE see MFN000
(2-METHOXYPROPYL)UREA, MERCURY COMPLEX see CHX250
3-METHOXYPROPYNE see MFN250
5-METHOXY PSORALEN see MFN275
8-METHOXYPSORALEN see XDJ000
9-METHOXYPSORALEN see XDJ000
2-METHOXYPYRAZINE see MFN285
3-METHOXYPYRAZINE SULFANILAMIDE see MFN500
N1-(3-METHOXY-2-PYRAZINYL)SULFANILAMIDE see MFN500
N1-(6-METHOXY-3-PYRIDAZINYL)SULFANILAMIDE see AKO500
4-METHOXYPYRIDOXINE see MFN600
METHOXYPYRIMAL SODIUM see MFO000
N1-(5-METHOXY-2-PYRIMIDINYL)SULFANILAMIDE, SODIUM SALT
  see MFO000
4-METHOXY-7H-PYRIMIDO(4,5-b)(1,4)THIAZIN-6-AMINE
  MONOHYDROCHLORIDE see THG600
S-((5-METHOXY-4H-PYRON-2-YL)-METHYL)-O,O-DIMETHYL-
  MONOTHIOFOSFAAT (DUTCH) see EAS000
S-((5-METHOXY-4H-PYRON-2-YL)-METHYL)-O,O-DIMETHYL-
  MONOTHIOPHOSPHAT(GERMAN) see EAS000
S-(5-METHOXY-4-PYRON-2-YLMETHYL) DIMETHYL PHOSPHOROTHIOL-
  ATE see EAS000
5-METHOXY-3-(2-PYRROLIDINOETHYL)INDOLE see MFO250
METHOXY-5-PYRROLIDINO-2'-ETHYL-3-INDOLE see MFO250
3-METHOXY-4-PYRROLIDINYLMETHYLDIBENZOFURAN see MFO500
6-METHOXY-3-(p-(2-(N-PYRROLIDYL)ETHOXY)PHENYL)-2-
  PHENYLBENZO(b)FURAN HYDROCHLORIDE see PGF100
6-METHOXY-3-(p-2-(1-PYRROLIDYL)ETHOXY)PHENYL)-2-
  PHENYLBENZO(b)THIOPHENE HYDROCHLORIDE see MFF575
6-METHOXYQUINOLINE see MFP000
N4-(6-METHOXY-8-QUINOLINYL)-1,4-PENTANEDIAMINE PHOSPHATE (1:2)
  (9CI) see PMC310
1-(6-METHOXY-4-QUINOLYL)-3-(3-VINYL-4-PIPERIDYL)-1-PROPANONE OXA-
  LATE see QFJ300
α-(6-METHOXY-4-QUINOLYL)-5-VINYL-2-QUINUCLIDINEMETHANOL
  see QFS000
α-(6-METHOXY-4-QUINOLYL)-5-VINYL-2-QUINUCLIDINEMETHANOL
  GALACTURONATE (salt) see GAX000
α-(6-METHOXY-4-QUINOYL)-5-VINYL-2-QUINCLIDINEMETHANOL see QHJ000
3-METHOXYSALICYLALDEHYDE see VFP000
3-METHOXYSALICYLIC ACID see HJB500
3-METHOXY-16,17-SECOESTRA-1,3,5(10),6,8-PENTAEN-17-OIC ACID see BIS750
METHOXY SIMAZINE see BJP250
3-METHOXY-4-STILBENAMINE see MFP500
5-METHOXYSULFADIAZINE SODIUM see MFO000
6-METHOXY-3-SULFANILAMIDOPYRIDAZINE see AKO500
3-METHOXY-2-SULFAPYRAZINE see MFN500

4-METHOXY-2-SULFOANILINE see AIA500
6-METHOXY-1,2,3,4-TETRAHYDRO-β-CARBOLINE HYDROCHLORIDE
  see MFT250
6-METHOXY-1,2,3,4-TETRAHYDRO-9H-PYRIDO(3,4-B)INDOLE HYDROCHLO-
  RIDE see MFT250
5-METHOXY-1,2,3,4-THIATRIAZOLE see MFQ250
4-METHOXYTOLUENE see MGP000
p-METHOXYTOLUENE see MGP000
4-METHOXY-m-TOLUIDINE see MFQ500, MGO750
2-METHOXY-3,5,6-TRICHLOROBENZOIC ACID see TIK000
2-METHOXYTRICYCLOQUINAZOLINE see MFQ750
3-METHOXYTRICYCLOQUINAZOLINE see MFR000
METHOXY TRIETHYLENE GLYCOL VINYL ETHER see MFR250
1-METHOXY-2,2,2-TRIFLUOROETHANOL see MFR500
METHOXYTRIGLYCOL see TJQ750
METHOXYTRIGLYCOL ACETATE see AAV500
METHOXYTRIMEPRAZINE see MCI500
(E,E)-11-METHOXY-3,7,11-TRIMETHYL-2,4-DODECANDIENOATE see KAJ000
(all-E)-9-(4-METHOXY-2,3,6-TRIMETHYLPHENYL)-3,7-DIMETHYL-2,4,6,8-NON-
  ATETRAENOIC ACID see REP400
1-METHOXY-3,4,5-TRIMETHYL PYRAZOLE-N-OXIDE see MFR775
METHOXYTRYPTAMINE see MFS400
5-METHOXYTRYPTAMINE see MFS400
6-METHOXYTRYPTAMINE see MFS500
5-METHOXYTRYPTAMINE HYDROCHLORIDE see MFT000
6-METHOXYTRYPTOLINE HYDROCHLORIDE see MFT250
5-METHOXYTRYPTOPHOL see MFT300
N-METHOXYURETHANE see MEO000
p-METHOXY-α-VINYLBENZYL ALCOHOL see HKI000
p-METHOXY-α-VINYLBENZYL ALCOHOL ACETATE (ester) see ABN725
1-METHOXY-2-(VINYLOXY)ETHANE see VPZ000
6-METHOXY-α-(5-VINYL-2-QUINUCLIDINYL)-4-QUINOLINEMETHANOL
  see QFS000
METHRAZONE see PEW000
METHSCOPOLAMINE BROMIDE see SBH500
METHSUXIMIDE see MLP800
METHURAL see DTG700
METHURIN (RUSSIAN) see DTG700
METHVTIOLO (ITALIAN) see MLE650
12-METHYBENZ(a)ANTHRACENE-7-METHANOL see HMF000
METHYBOL see MJE760
METHYCOBAL see VSZ050
METHYL ABIETATE see MFT500
METHYLACETAAT (DUTCH) see MFW100
METHYLACETALDEHYDE see PMT750
METHYLACETAMIDE see MFT750
N-METHYLACETAMIDE see MFT750
2-(2-(3-(N-METHYLACETAMIDO)-2,4,6-TRIIODOPHENOXY)ETHOXY)ACETIC
  ACID SODIUM SALT see MFU000
2-(2-(3-(N-METHYLACETAMIDO)-2,4,6-TRIIODOPHENOXY)ETHOXY)
  BUTYRIC ACID SODIUM SALT see MFU250
2-(2-(3-(N-METHYLACETAMIDO)-2,4,6-TRIIODOPHENOXY)ETHOXY)-2-
  PHENYLACETIC ACID SODIUM SALT see MFU500
2-METHYLACETANILIDE see ABJ000
3-METHYLACETANILIDE see ABI750
4-METHYLACETANILIDE see ABJ250
m-METHYLACETANILIDE see ABI750
N-METHYLACETANILIDE see MFW000
o-METHYLACETANILIDE see ABJ000
p-METHYLACETANILIDE see ABJ250
2'-METHYLACETANILIDE see ABJ000
3'-METHYLACETANILIDE see ABI750
4'-METHYLACETANILIDE see ABJ250
METHYLACETAPHOS see DRB600
METHYLACETAT (GERMAN) see MFW100
METHYL ACETATE see MFW100
METHYL ACETIC ACID see PMU750
METHYLACETIC ANHYDRIDE see PMV500
2'-METHYLACETOACETANILIDE see ABA000
METHYLACETOACETATE see MFX250
METHYL ACETONE (DOT) see MKA400
p-METHYL ACETOPHENONE see MFW250
4'-METHYL ACETOPHENONE see MFW250
METHYL ACETOPHOS see DRB600
METHYLACETOPYRONONE see MFW500
METHYL ACETOXON see DRB600
N-METHYL-N-(α-ACETOXYBENZYL)NITROSAMINE see NKP000
METHYL-β-ACETOXYETHYL-β-CHLOROETHYLAMINE see MFW750
METHYLACETOXYMALONONITRILE see MFX000
METHYL(ACETOXYMETHYL)NITROSAMINE see AAW000
METHYL-12-ACETOXY-9-OCTADECENOATE see MFX750
METHYL-12-ACETOXYOLEATE see MFX750
6-METHYL-17-α-ACETOXYPREGNA-4,6-DIENE-3,20-DIONE see VTF000
6-α-METHYL-17-α-ACETOXYPREGN-4-ENE-3,20-DIONE see MCA000

2-METHYLAMINO METHYL BENZOATE see MGQ250
1-METHYLAMINOMETHYLDIBENZO(b,c)BICYCLO(2,2,2)OCTADIENE HYDRO-
    CHLORIDE see BCH750
6-METHYLAMINO-2-METHYLHEPTENE see ILK000
METHYLAMINO-METHYLHEPTENE HYDROCHLORIDE see ODY000
METHYLAMINOMETHYL(4-HYDROXYPHENYL)CARBINOL see HLV500
2-METHYLAMINO-4-METHYLTHIO-6-ISOPROPYLAMINO-1,3,5-TRIAZINE
    see INR000
2-METHYL-4-AMINO-1-NAPHTHOL see AKX500
o-METHYL-2-AMINO-1-NAPHTHOL HYDROCHLORIDE see MFA250
4-METHYLAMINO-N-NITROSOAZOBENZENE see MMY250
1-METHYLAMINO-4-OXYETHYLAMINOANTHRAQUINONE (RUSSIAN)
    see MGG250
3-(α-METHYLAMINOPHENETHYL)PHENOL HYDROCHLORIDE see MGJ600
N-(α-METHYLAMINOPHENETHYL)PHENOL HYDROCHLORIDE see MGJ600
p-METHYLAMINOPHENOLSULFATE see MGJ750
METHYL-p-AMINOPHENOL SULFATE see MGJ750
p-(METHYLAMINO)PHENOL SULFATE (2:1) (SALT) see MGJ750
6-METHYL-2-(p-AMINO PHENYL) see ALX000
METHYLAMINOPHENYLDIMETHYLPYRAZOLONE METHANESULFONATE
    SODIUM see AMK500
1-2-METHYLAMINO-1-PHENYLPROPANOL see EAW000
d-psi-2-METHYLAMINO-1-PHENYL-1-PROPANOL see POH000
METHYL ((4-AMINOPHENYL)SULFONYL)CARBAMATE see SNQ500
3-METHYL-4-AMINO-6-PHENYL-1,2,4-TRIAZIN(4H)-ON (GERMAN) see ALA500
1-(3-METHYLAMINOPROPYL)-2-ADAMANTANOL HYDROCHLORIDE
    see MGK750
1-(3-METHYLAMINOPROPYL)DIBENZO(b,e)BICYCLO(2.2.2)OCTADIENE
    HYDROCHLORIDE see MAW850
5-(3-METHYLAMINOPROPYL)-5H-DIBENZO(a,d)CYCLOHEPTENE see DDA600
5-(3-METHYLAMINOPROPYL)-5H-DIBENZO(a,d)CYCLOHEPTENE HYDRO-
    CHLORIDE see POF250
9-(Γ-METHYLAMINOPROPYL)-9,10-DIHYDRO-9,10-ETHANOANTHRACENE
    HYDROCHLORIDE see MAW850
5-(3-(METHYLAMINO)PROPYLIDENE)DIBENZO(a,e)CYCLOHEPTA(1,5)DIENE
    see NNY000
4-(3'-METHYLAMINOPROPYLIDENE)-9,10-DIHYDRO-4H-
    BENZO(4,5)CYCLOHEPTA(1,2-b)THIOPHEN see MGL500
5-(3-METHYLAMINOPROPYLIDENE)-10,11-DIHYDRO-5H-
    DIBENZO(a,d)CYCLOHEPTENE see NNY000
METHYLAMINOPROPYLIMINODIBENZYL see DSI709
N-(Γ-METHYLAMINOPROPYL)IMINODIBENZYL HYDROCHLORIDE
    see DLS600
p-(2-METHYLAMINOPROPYL)PHENOL see FMS875
METHYLAMINOPTERIN see MDV500
4-METHYL-2-AMINOPYRIDINE see ALC250
METHYL-4-AMINO-2-PYRIDINE see ALC250
1-METHYL-3-AMINO-5H-PYRIDO(4,3-b)INDOLE see ALD500
N-METHYL-4-AMINO-1,2,5-SELENADIAZOLE-3-CARBOXAMIDE see MGL600
2-METHYL-4-AMINOSTILBENE see MPJ250
3-METHYL-4-AMINOSTILBENE see MPJ500
3-β-METHYLAMINO-2,2,3-TRIMETHYLBICYCLO(2.2.1)HEPTANE see VIZ400
2-METHYLAMINO-2,3,3-TRIMETHYLNORBORANE see VIZ400
2-METHYLAMINO-2,3,3-TRIMETHYLNORBORNANE see VIZ400
METHYLAMMONIUM CHLORITE see MGN000
METHYLAMMONIUM NITRATE see MGN150
METHYLAMMONIUM PERCHLORATE see MGN250
METHYLAMPHETAMINE see DBB000
N-METHYLAMPHETAMINE see DBB000
METHYLAMPHETAMINE HYDROCHLORIDE see DBA800, MDT600
d-METHYLAMPHETAMINE HYDROCHLORIDE see MDT600
N-METHYLAMPHETAMINE HYDROCHLORIDE see MDT600
METHYLAMYL ACETATE see HFJ000
METHYL AMYL ACETATE (DOT) see HFJ000
METHYL AMYL ALCOHOL see MKW600
METHYLAMYL ALCOHOL see AOK750
METHYL AMYL CARBINOL see HBE500
METHYL-AMYL-CETONE (FRENCH) see MGN500
METHYL n-AMYL KETONE see MGN500
METHYL AMYL KETONE (DOT) see MGN500
METHYLAMYLNITROSAMIN (GERMAN) see AOL000
METHYLAMYLNITROSAMINE see AOL000
METHYL-N-AMYLNITROSAMINE see AOL000
17-METHYL-5-α-ANDROSTANO(2,3-c)(1,2,5)OXADIAZOL-17-β-OL see AOO300
17-α-METHYL-5-α-ANDROSTANO(2,3-c)(1,2,5)OXADIAZOL-17-β-OL see AOO300
METHYLANDROSTENDIOL see AOO475
METHYLANILINE see MGN750
2-METHYLANILINE see TGQ750
3-METHYLANILINE see TGQ500
4-METHYLANILINE see TGR000
m-METHYLANILINE see TGQ500
o-METHYLANILINE see TGQ750
p-METHYLANILINE see TGR000
N-METHYL ANILINE (MAK) see MGN750

2-METHYLANILINE HYDROCHLORIDE see TGS500
4-METHYLANILINE HYDROCHLORIDE see TGS750
o-METHYLANILINE HYDROCHLORIDE see TGS500
METHYLANILINE and SODIUM NITRITE (1.2:1) see MGO000
N-METHYLANILINE mixed with SODIUM NITRITE (1:35) see MGO250
2-(N-METHYLANILINO)ETHANOL see MKQ250
N-METHYLANILIN UND NATRIUMNITRIT (GERMAN) see MGO250
METHYL-p-ANISATE see AOV750
2-METHYL-p-ANISIDINE see MGO500
5-METHYL-o-ANISIDINE see MGO750
p-METHYL ANISOLE see MGP000
2-α-METHYL-A-NOR-17-α-PREGN-20-YNE-2-β,17-β-DIOL see MNC125
METHYLANTALON see MJE760
6-METHYLANTHANTHRENE see MGP250
2-METHYLANTHRACENE see MGP500
9-METHYLANTHRACENE see MGP750
3-METHYL-1,8,9-ANTHRACENETRIOL see CML750
3-METHYLANTHRALIN see CML750
METHYL ANTHRANILATE (FCC) see APJ250
N-METHYLANTHRANILIC ACID see MGQ000
N-METHYLANTHRANILIC ACID, METHYL ESTER see MGQ250
2-METHYL-1-ANTHRAQUINONYLAMINE see AKP750
METHYLANTIFEBRIN see MFW000
METHYL APHOXIDE see TNK250
METHYL ARISTOLATE see MGQ525
METHYLARSENIC ACID see MGQ530
METHYLARSENIC ACID, SODIUM SALT see MRL750
METHYLARSENIC DIMETHYL DITHIOCARBAMATE see USJ075
METHYLARSENIC SULFIDE see MGQ750
METHYL ARSINE-BIS(DIMETHYLDITHIOCARBAMATE) see USJ075
METHYLARSINE DICHLORIDE see DFP200
METHYLARSINE DIIODIDE see MGQ775
METHYLARSINE SULFIDE see MGQ750
METHYLARSINIC ACID see MGQ530
METHYLARSINIC SULFIDE see MGQ750
METHYLARSINIC SULPHIDE see MGQ750
METHYLARSONIC ACID see MGQ530
METHYLARSONOUS DICHLORIDE see DFP200
METHYLARTERENOL see VGP000
METHYL ASPARTYLPHENYLALANATE see ARN825
1-METHYL N-l-α-ASPARTYL-l-PHENYLALANINE see ARN825
METHYLATROPINE BROMIDE see MGR250
METHYL ATROPINE NITRATE see MGR500
N-METHYLATROPINE NITRATE see MGR500
METHYLATROPINIUM BROMIDE see MGR250
8-METHYLATROPINIUM BROMIDE see MGR250
8-METHYLATROPINIUM NITRATE see MGR500
N-METHYLATROPINIUM NITRATE see MGR500
8-METHYL-8-AZABICYCLO(3.2.1)OCTAN-3-OL BENZOATE (ester), HYDRO-
    CHLORIDE, exo- see TNS200
2-METHYLAZACYCLOPROPANE see PNL400
3-METHYL-3-AZAPENTANE-1,5-BIS(ETHYLDIMETHYLAMMONIUM) BRO-
    MIDE see MKU750
METHYL AZIDE see MGR750
METHYL-2-AZIDOBENZOATE see MGR800
METHYLAZINPHOS see ASH500
2-METHYLAZIRIDINE see PNL400
N-METHYL-1-AZIRIDINECARBOXAMIDE see EJN400
METHYL-AZOXY-BUTANE see MGS500
METHYLAZOXYMETHANOL see HMG000
METHYLAZOXYMETHANOL ACETATE see MGS750
METHYLAZOXYMETHANOL GLUCOSIDE see COU000
METHYLAZOXYMETHANOL-β-d-GLUCOSIDE see COU000
METHYLAZOXYMETHANOL-β-D-GLUCOSIDURONIC ACID see MGS700
METHYL AZOXYMETHYL ACETATE see MGS750
METHYLAZOXYMETHYL BENZOATE see MGS925
METHYLAZOXYOCTANE see MGT000
METHYL-AZULENO(5,6,7-c,d)PHENALENE see MGT250
METHYL-B₁₂ see VSZ050
METHYLBARBITAL see DJO800
1-METHYLBARBITAL see DJO800
N-METHYLBARBITAL see DJO800
METHYLBEN see HJL500
3-METHYLBENZ(j)ACEANTHRYLENE see DAL200
6-METHYL-11H-BENZ(bc)ACEANTHRYLENE see MLN500
3-METHYLBENZ(e)ACEPHENANTHRYLENE see MGT400
7-METHYLBENZ(e)ACEPHENANTHRYLENE see MGT410
8-METHYLBENZ(e)ACEPHENANTHRYLENE see MGT415
12-METHYLBENZ(e)ACEPHENANTHRYLENE see MGT420
7-METHYLBENZ(c)ACRIDINE see MGT500
12-METHYLBENZ(a)ACRIDINE see MGU500
9-METHYL-1,2-BENZACRIDINE see MGU500
9-METHYL-3,4-BENZACRIDINE see MGT500
10-METHYL-5,6-BENZACRIDINE see MGU500

10-METHYL-7,8-BENZACRIDINE (FRENCH) see MGT500
9-METHYLBENZ(c)ACRIDINE-7-CARBOXALDEHYDE see FNR000
11-METHYLBENZ(c)ACRIDINE-7-CARBOXALDEHYDE see FNS000
7-METHYLBENZ(c)ACRIDINE 3,4-DIHYDRODIOL see MGU550
N-METHYLBENZAMIDE see MHA250
10-METHYL-1,2-BENZANTHRACEN (GERMAN) see MGW750
1-METHYLBENZ(a)ANTHRACENE see MGU750
2-METHYLBENZ(a)ANTHRACENE see MGV000
3-METHYLBENZ(a)ANTHRACENE see MGV250
4-METHYLBENZ(a)ANTHRACENE see MGV500
5-METHYLBENZ(a)ANTHRACENE see MGV750
6-METHYLBENZ(a)ANTHRACENE see MGW000
7-METHYLBENZ(a)ANTHRACENE see MGW750
8-METHYLBENZ(a)ANTHRACENE see MGW250
9-METHYLBENZ(a)ANTHRACENE see MGW500
10-METHYLBENZ(a)ANTHRACENE see MGX000
11-METHYLBENZ(a)ANTHRACENE see MGX250
12-METHYLBENZ(a)ANTHRACENE see MGX500
3-METHYL-1,2-BENZANTHRACENE see MGV750
4-METHYL-1,2-BENZANTHRACENE see MGW000
5-METHYL-1,2-BENZANTHRACENE see MGW250
6-METHYL-1,2-BENZANTHRACENE see MGW500
7-METHYL-1,2-BENZANTHRACENE see MGX000
8-METHYL-1:2-BENZANTHRACENE see MGX250
9-METHYL-1,2-BENZANTHRACENE see MGX500
1'-METHYL-1,2-BENZANTHRACENE see MGU750
10-METHYL-1,2-BENZANTHRACENE see MGW750
2'-METHYL-1,2-BENZANTHRACENE see MGV000
4'-METHYL-1:2-BENZANTHRACENE see MGV500
10-METHYL-1,2-BENZANTHRACENE-5-CARBONAMIDE see MGX750
7-METHYLBENZ(a)ANTHRACENE-8-CARBONITRILE see MGY000
7-METHYLBENZ(a)ANTHRACENE-10-CARBONITRILE see MGY250
12-METHYLBENZ(a)ANTHRACENE-7-CARBOXALDEHYDE see FNT000
7-METHYLBENZ(a)ANTHRACENE-12-CARBOXALDEHYDE see MGY500
12-METHYL BENZ(A)ANTHRACENE-7-ETHANOL see HKU000
7-METHYLBENZ(a)ANTHRACENE-12-METHANOL see HMF500
12-METHYLBENZ(a)ANTHRACENE-7-METHANOL ACETATE (ESTER)
   see ABR250
12-METHYLBENZ(a)ANTHRACENE-7-METHANOL BENZOATE (ESTER)
   see BDQ250
7-METHYLBENZ(a)ANTHRACENE-5,6-OXIDE see MGZ000
7-METHYLBENZ(a)ANTHRACEN-8-YL CARBAMIDE see MGX750
S-(12-METHYL-7-BENZ(a)ANTHRYLMETHYL)HOMOCYSTEINE see MHA000
N-METHYLBENZAZIMIDE, DIMETHYLDITHIOPHOSPHORIC ACID ESTER
   see ASH500
METHYLBENZEDRIN see DBA800
METHYL-1H-BENZEMEDAZOL-2-YLCARBAMATE see MHC750
N-METHYLBENZENAMIDE see MHA250
2-METHYLBENZENAMINE see TGQ750
3-METHYLBENZENAMINE see TGQ500
4-METHYLBENZENAMINE see TGR000
m-METHYLBENZENAMINE see TGQ500
N-METHYLBENZENAMINE see MGN750
o-METHYLBENZENAMINE see TGQ750
p-METHYLBENZENAMINE see TGR000
2-METHYLBENZENAMINE HYDROCHLORIDE see TGR500, TGS500
4-METHYLBENZENAMINE HYDROCHLORIDE see TGS750
o-METHYLBENZENAMINE HYDROCHLORIDE see TGS500
5-METHYL-1,3-BENZENDIOL see MPH500
METHYLBENZENE see TGK750
METHYL BENZENEACETATE see MHA500
2-METHYLBENZENECARBONITRILE see TGT500
METHYL BENZENECARBOXYLATE see MHA750
ar-METHYLBENZENEDIAMINE see TGL500
2-METHYL-1,2-BENZENEDIAMINE see TGM100
2-METHYL-1,4-BENZENEDIAMINE see TGM000, TGM400
4-METHYL-1,2-BENZENEDIAMINE see TGM250
4-METHYL-1,3-BENZENEDIAMINE see TGL750
2-METHYL-1,4-BENZENEDIAMINE DIHYDROCHLORIDE see DCE200
2-METHYL-1,4-BENZENEDIAMINE SULFATE see DCE600
METHYL BENZENEDIAZOATE see MHB000
3-METHYL-1,2-BENZENEDIOL see DNE000
4-METHYL-1,2-BENZENEDIOL see DNE200
(S)-α-METHYL-BENZENEETHANAMINE SULFATE (2:1) see BBK500
α-METHYLBENZENEETHANEAMINE see BBK000
4-METHYL-BENZENEMETHANOL see MHB250
2-METHYLBENZENESULFONAMIDE see TGN250
4-METHYLBENZENESULFONAMIDE see TGN500
o-METHYLBENZENESULFONAMIDE see TGN250
p-METHYLBENZENESULFONAMIDE see TGN500
4-METHYLBENZENESULFONIC ACID see TGO000
p-METHYLBENZENESULFONIC ACID see TGO000
4-METHYL-BENZENESULFONIC ACID BUTYL ESTER (9CI) see BSP750
4-METHYL-BENZENESULFONIC ACID, SODIUM SALT see SKK000

N-(4-METHYLBENZENESULFONYL)-N-(3-AZABICYCLO(3.3.0)OCT-3-
   YL)UREA see DBL700
p-METHYL BENZENE SULFONYL CHLORIDE see TGO250
N-(4-METHYLBENZENESULFONYL)-N-CYCLOHEXYLUREA see CPR000
4-METHYL BENZENE SYLFONYL CHLORIDE see TGO250
2-METHYLBENZENETHIOL see TGP000
4-METHYLBENZENETHIOL see TGP250
o-METHYLBENZENETHIOL see TGP000
p-METHYLBENZENETHIOL see TGP250
METHYLBENZETHONIUM CHLORIDE see MHB500
METHYL BENZETHONIUM CHLORIDE MONOHYDRATE see MHB500
10-METHYL-7H-BENZIMIDAZOL(2,1-a)BENZ(de)ISOQUINOLIN-7-ONE
   see MHC000
METHYL-2-BENZIMIDAZOLE see MHC250
2-METHYLBENZIMIDAZOLE see MHC250
5-METHYLBENZIMIDAZOLE see MHC500
METHYL 2-BENZIMIDAZOLECARBAMATE see MHC750
METHYL 2-BENZIMIDAZOLE CARBAMATE and SODIUM NITRITE
   see CBN375
METHYL BENZIMIDAZOLE-2-YL CARBAMATE see MHC750
7-METHYLBENZO(c)ACRIDINE see MGT500
METHYL BENZOATE (FCC) see MHA750
6-METHYL-3,4-BENZOCARBAZOLE see MHD000
9-METHYL-1582-BENZOCARBAZOLE see MHD250
10-METHYL-7H-BENZO(c)CARBAZOLE see MHD000
N-(2-METHYLBENZODIOXAN)-N-ETHYL-β-ALANINAMIDE see MHD300
N-(2-METHYL-1,4-BENZODIOXAN)-N-METHYL-β-ALANINAMIDE
   see MHD300
4-(2-(2-METHYL-1,3-BENZODIOXOL-2-YL)ETHYL)PIPERAZIN-1-YL-2-ETHA-
   NOL DIHYDROCHLORIDE see MHD750
3-METHYLBENZO(b)FLUORANTHENE see MGT400
7-METHYLBENZO(b)FLUORANTHENE see MGT410
8-METHYLBENZO(b)FLUORANTHENE see MGT415
12-METHYLBENZO(b)FLUORANTHENE see MGT420
2-METHYLBENZOIC ACID see TGQ000
3-METHYLBENZOIC ACID see TGP750
m-METHYLBENZOIC ACID see TGP750
o-METHYLBENZOIC ACID see TGQ000
4-METHYLBENZOIC ACID METHYL ESTER see MNR250
METHYL BENZOIN see MHE000
METHYLBENZOL see TGK750
2-METHYLBENZONITRILE see TGT500
4-METHYLBENZONITRILE see TGT750
o-METHYLBENZONITRILE see TGT500
p-METHYLBENZONITRILE see TGT750
5-METHYLBENZO(rat)PENTAPHENE see MHE250
8-METHYLBENZO(rst)PENTAPHENE-5-CARBOXALDEHYDE see FNU000
METHYL-1,12-BENZOPERYLENE see MHE500
7-METHYLBENZO(a)PHENALENO(1,9-hi)ACRIDINE see MHE750
7-METHYLBENZO(h)PHENALENO(1,9-bc)ACRIDINE see MHF000
4-METHYLBENZO(c)PHENANTHRENE see MHL750
5-METHYLBENZO(c)PHENANTHRENE see MHF250
6-METHYLBENZO(c)PHENANTHRENE see MHF500
4-METHYL-p-BENZOPHENONE see MHF750
6-METHYL-2H-1-BENZOPYRAN-2-ONE see MIP750
7-METHYL-2H-1-BENZOPYRAN-2-ONE (9CI) see MIP775
α-METHYL-5H-(1)-BENZOPYRANO(2,3-b)PYRIDINE-7-ACETIC ACID
   see PLX400
1-METHYLBENZO(a)PYRENE see MHG250
2-METHYLBENZO(a)PYRENE see MHG500
3-METHYLBENZO(a)PYRENE see MHM250
4-METHYLBENZO(a)PYRENE see MHG750
5-METHYLBENZO(a)PYRENE see MHH200
6-METHYLBENZO(a)PYRENE see MHM000
7-METHYLBENZO(a)PYRENE see MHH000
10-METHYLBENZO(a)PYRENE see MHH500
11-METHYLBENZO(a)PYRENE see MHH750
12-METHYLBENZO(a)PYRENE see MHI000
4'-METHYLBENZO(a)PYRENE see MHH000
5-METHYL-3,4-BENZOPYRENE see MHM000
6-METHYLBENZOPYRONE see MIP750
6-METHYL-1,2-BENZOPYRONE see MIP750
METHYL-p-BENZOQUINONE see MHI250
METHYL-1,4-BENZOQUINONE see MHI250
2-METHYL-p-BENZOQUINONE see MHI250
2-METHYLBENZOQUINONE-1,4 see MHI250
5-METHYL-2,1,3-BENZOSELENADIAZOLE see MHI300
2-METHYLBENZOSELENAZOLE see MHI500
METHYLBENZOTHIADIAZINE CARBAMATE see MHI600
2-METHYLBENZOTHIAZOLE see MHI750
3-METHYL-BENZOTHIAZOLIUM SALT with 4-METHYLBENZENESULFONIC
   ACID (1:1) see MHJ000
3-METHYLBENZOTHIAZOLIUM-p-TOLUENE SULFONATE see MHJ000
3-METHYL-2-BENZOTHIAZOLONE HYDRAZONE see MHJ250

2-((2-METHYLBENZO(b)THIEN-3-YL)METHYL)-2-IMIDAZOLINE HYDRO-
CHLORIDE see MHJ500
7-METHYL-6H-(1)BENZOTHIOPYRANO(4,3-b)QUINOLINE see MHJ750
5-METHYLBENZOTRIAZOLE see MHK250
METHYL-1H-BENZOTRIAZOLE see MHK000
5-METHYL-1,2,3-BENZOTRIAZOLE see MHK250
2-METHYLBENZOXAZOL (CZECH) see MHK500
2-METHYLBENZOXAZOLE see MHK500
METHYL-5-BENZOYL BENZIMIDAZOLE-2-CARBAMATE see MHL000
1-METHYL-3,4-BENZPHENANTHRENE see MHF500
2-METHYL-3,4-BENZPHENANTHRENE see MHF250
6-METHYL-3584-BENZPHENANTHRENE see MHL250
7-METHYL-3,4-BENZPHENANTHRENE see MHL500
8-METHYL-3584-BENZPHENANTHRENE see MHL750
5-METHYL-3,4-BENZPYRENE see MHM000
6-METHYL-3,4-BENZPYRENE see MHH750
8-METHYL-3,4-BENZPYRENE see MHM250
9-METHYL-3,4-BENZPYRENE see MHG500
4'-METHYL-3:4-BENZPYRENE see MHH000
2-METHYLBENZSELENAZOL (CZECH) see MHI500
1-METHYL-3-(2-BENZTHIAZOLYL)UREA see MHM500
N-METHYL-N'-BENZYHYDRYLPIPERAZINE see EAN600
4-METHYLBENZYL ALCOHOL see MHB250
p-METHYLBENZYLALCOHOL see MHB250
α-METHYLBENZYL ALCOHOL (FCC) see PDE000
α-METHYLBENZYL ALCOHOL, PROPIONATE see PFR000
p-METHYLBENZYLALKOHOL (GERMAN) see MHB250
α-METHYLBENZYLAMINE see ALW250
METHYLBENZYLAMINE mixed with SODIUM NITRITE (2:3) see MHN250
N-METHYLBENZYLAMINE mixed with SODIUM NITRITE (1:1)
see MHN000
N-METHYLBENZYLAMINE mixed with SODIUM NITRITE (2:3) see MHN250
((α-METHYLBENZYL)AMINO)ETHANOL see HKU500
2-METHYL-4-BENZYLAMINO-PYRROLO-(2,3d)PYRIMIDINE see RLZ000
(α-METHYLBENZYL)-1-BENZOYL-2-HYDRAZINE see NCQ100
METHYLBENZYLCARBINYL ACETATE see ABU800
4-METHYLBENZYL CHLORIDE see MHN300
1-METHYLBENZYL-3-(DIMETHOXYPHOSPHINYLOXO) ISOCROTONATE
see COD000
α-METHYL BENZYL-3-(DIMETHOXY-PHOSPHINYLOXY)-cis-CROTONATE
see COD000
α-METHYL BENZYL ETHER see MHN500
1-METHYL-2-BENZYLHYDRAZINE see MHN750
(α-METHYLBENZYL)HYDRAZINE SULFATE see MHO000
α-METHYLBENZYL-3-HYDROXY-CROTONATE DIMETHYL PHOSPHATE
see COD000
1-(α-METHYLBENZYLIDENE)THIOSEMICARBIZIDE see ABH250
(R)-(+)-1-(α-METHYLBENZYL)IMIDAZOLE-5-CARBOXYLIC ACID ETHYL
ESTER see HOU100
1-(α-METHYLBENZYL)-1H-IMIDAZOLE-5-CARBOXYLIC ACID ETHYL ESTER
SULFATE see HOU100
1-(α-METHYLBENZYL)IMIDAZOLE-5-CARBOXYLIC ACID METHYL ESTER,
HYDROCHLORIDE see MQQ750
N-METHYL-N-BENZYLNITROSAMINE see MHP250
METHYL-BENZYL-NITROSOAMIN (GERMAN) see MHP250
α-METHYLBENZYL PROPANOATE see PFR000
N-METHYL-N-BENZYLPROPYNYLAMINE see MOS250
METHYLBICYCLO(2.2.1)HEPTENE-2,3-DICARBOXYLIC ANHYDRIDE ISO-
MERS see NAC000
1-METHYLBIGUANIDE HYDROCHLORIDE see MHP400
2-METHYL-2,2'-BIOXIRANE (9CI) see DHD200
α-METHYL-(1,1'-BIPHENYL)-4-ACETIC ACID 3-SULFOPROPYL ESTER
SODIUM SALT see PFR350
4'-METHYLBIPHENYLAMINE see MGF750
METHYLBIS(3-AMINOPROPYL)AMINE see BGU750
N-METHYL-BIS-CHLORAETHYLAMIN (GERMAN) see BIE250
Γ(1-METHYL-5-BIS(β-CHLORAETHYL)AMINOBENZIMIDAZOLYL)
BUTTERSAEUREHYDROCHLORID(GERMAN) see CQM750
METHYL-BIS-(β-CHLORAETHYL)-AMIN-N-OXYD-HYDROCHLORID
(GERMAN) see CFA750
N-METHYL-BIS-β-CHLORETHYLAMINE HYDROCHLORIDE see BIE500
Γ'-(1-METHYL-5-BIS(β-CHLOROAETHYL) AMINOBENZIMIDAZOYL)
BUTTERSAUERHYDROCHLORID(GERMAN) see CQM750
METHYLBIS(β-CHLOROETHYL)AMINE see BIE250
N-METHYL-BIS(β-CHLOROETHYL)AMINE see BIE250
METHYLBIS(2-CHLOROETHYL)AMINE HYDROCHLORIDE see BIE500
METHYLBIS(β-CHLOROETHYL)AMINE HYDROCHLORIDE see BIE500
N-METHYLBIS(2-CHLOROETHYL)AMINE HYDROCHLORIDE
see BIE500
N-METHYL-BIS(2-CHLOROETHYL)AMINE (MAK) see BIE250
METHYL-BIS(2-CHLOROETHYL)AMINE OXIDE see CFA500
METHYLBIS(β-CHLOROETHYL)AMINE N-OXIDE see CFA500
METHYLBIS(β-CHLOROETHYL)AMINE-N-OXIDE HYDROCHLORIDE
see CFA750

N-METHYLBIS(2-CHLOROETHYL)AMINE-N-OXIDE HYDROCHLORIDE
see CFA750
4-METHYL-5-(BIS(β-CHLOROETHYL)AMINO)URACIL see DYC700
6-METHYL-5-BIS(2-CHLOROETHYL)AMINO)URACIL see DYC700
METHYL-BIS(2-CHLOROETHYL-MERCAPTOETHYL)AMINE HYDROCHLO-
RIDE see MHQ500
METHYLBIS(β-CHLOROETHYLTHIOETHYL)AMINE HYDROCHLORIDE
see MHQ500
METHYL BIS(β-CYANOETHYL)AMINE see MHQ750
7-METHYLBISDEHYDRODOISYNOLIC ACID see BIT000
METHYLBIS(DIMETHYLDITHIOCARBAMOYLTHIO)ARSINE see USJ075
2-METHYL-4,5-BIS(HYDROXYMETHYL)-3-HYDROXYPYRIDINE see PPK250
7-METHYL-2,3:9,10-BIS(METHYLENEDIOXY)-7,13a-SECOBERBIN-13a-ONE
see FOW000
N-METHYL-N,N-BIS(3-METHYLSULFONYLOXYPROPYL)AMINE 4,4'-
BIPHENYLDISULFONATE see MHQ775
METHYLBISMUTH OXIDE see MHR000
β-METHYLBIVINYL see IMS000
17-α-METHYL-B-NORTESTOSTERONE see MNC150
METHYL BORATE see TLN000
METHYL BOTRYODIPLODIN see MHR100
1-METHYL-BP see MHG250
5-METHYL-BP see MHH200
METHYLBROMID (GERMAN) see MHR200
METHYL BROMIDE see MHR200
dl-METHYLBROMIDE see MDL000
METHYL BROMIDE and ETHYLENE DIBROMIDE MIXTURE, liquid (DOT)
see BNM750
METHYL BROMOACETATE see MHR250
METHYL α-BROMOACETATE see MHR250
2-METHYL-4-BROMOANILINE see MHR500
METHYL-4-BROMOBENZENEDIAZOATE see MHR750
O-METHYL-O-(4-BROMO-2,5-DICHLOROPHENYL)PHENYL THIOPHOSPHON-
ATE see LEN000
METHYL-(BROMOMERCURI)FORMATE see MHS250
1-METHYL-7-BROMOMETHYLBENZ(a)ANTHRACENE see BNQ750
1-METHYL-3-(p-BROMOPHENYL)-1-NITROSOUREA see BNX125
1-METHYL-3-(p-BROMOPHENYL)UREA see MHS375
1-METHYL-3-(p-BROMOPHENYL)UREA mixed with SODIUM NITRITE
see SIQ675
1-METHYL-3-(p-BROMPHENYL)HARNSTOFF see MHS375
1-METHYL-3-(p-BROMPHENYL)-1-NITROSOHARNSTOFF (GERMAN)
see BNX125
N-METHYLBUTABARBITAL see BRJ250
2-METHYLBUTADIENE see IMS000
2-METHYL-1,3-BUTADIENE (DOT) see IMS000
2-METHYLBUTANAL see MJX500
3-METHYLBUTANAL see MHX500
α-METHYLBUTANAL see MJX500
2-METHYL-1-BUTANAL see MJX500
2-METHYLBUTANE see EIK000
2-METHYL-1,4-BUTANEDIOL see MHS500
α-(3-METHYL-BUTANE)-α-(2-PIPERIDYLETHYL) BENZYL ALCOHOL HYDRO-
CHLORIDE see HBF500
2-METHYLBUTANE SECONDARY MONONITRILE see ITD000
METHYL n-BUTANOATE see MHY000
3-METHYLBUTANOIC ACID see ISU000
3-METHYLBUTANOIC ACID, BUTYL ESTER see ISX000
3-METHYLBUTANOIC ACID, ETHYL ESTER see ISY000
2-METHYLBUTANOIC ACID, n-HEXYL ESTER see HFR200
3-METHYLBUTANOIC ACID, METHYL ESTER see ITC000
3-METHYLBUTANOIC ACID-2-METHYLPROPYL ESTER see ITA000
3-METHYLBUTANOIC ACID, PHENYLETHYL ESTER see ISW000
3-METHYL-BUTANOIC ACID 2-PHENYLETHYL ESTER see PDF775
3-METHYLBUTANOIC ACID, 2-PROPENYL ESTER see ISV000
2-METHYLBUTANOL see MHS750
3-METHYL BUTANOL see IHP000
2-METHYL BUTANOL-1 see MHS750
2-METHYL BUTANOL-2 see PBV000
2-METHYL-2-BUTANOL see PBV000
2-METHYL-4-BUTANOL see IHP000
3-METHYLBUTAN-1-OL see IHP000
3-METHYLBUTAN-3-OL see PBV000
3-METHYL-1-BUTANOL (CZECH) see IHP000
3-METHYL-1-BUTANOL CARBAMATE see IHQ000
3-METHYLBUTANOL NITRITE see IMB000
3-METHYL-2-BUTANONE see MLA750
3-METHYL BUTAN 2-ONE (DOT) see MLA750
2-METHYL-1-BUTENE see MHT000
2-METHYL-2-BUTENE see AOI750
3-METHYL-1-BUTENE see MHT250
2-METHYL-2-BUTENEDIOIC ACID see CMS320
cis-METHYLBUTENEDIOIC ACID see CMS320
(Z)-2-METHYL-2-BUTENEDIOIC ACID (9CI) see CMS320

trans-2-METHYL-2-BUTENEDIOIC ACID see MDI250
cis-2-METHYL-2-BUTENEDIOIC ACID, DIMETHYL ESTER see DRF200
METHYL trans-2-BUTENOATE see COB825
3-METHYL-2-BUTENOIC ACID see MHT500
trans-2-METHYL-2-BUTENOIC ACID see TGA700
2-METHYLBUTENOIC ACID-7-((2,3-DIHYDROXY-2-(1-METHYLETHYL)-1-OX-
OBUTOXY)METHYL-2,3,5,7a-TETRAHYDRO-1H-PYRROLIZIN-1-YL ESTER
see SPB500
3-METHYL-2-BUTENOIC ACID ETHYL ESTER see MIP800
2-METHYL-2-BUTENOIC ACID, HEXYL ESTER see HFX000
METHYLBUTENOL see MHU100
2-METHYL-3-BUTEN-2-OL see MHU100
3-METHYL-1-BUTEN-3-OL see MHU100
3-METHYL-2-BUTEN-1-OL see MHU110
3-METHYL-BUTEN-(1)-OL-(3) (GERMAN) see MHU100
3-METHYL-3-BUTEN-2-ON (GERMAN) see MKY500
2-METHYL-1-BUTEN-3-ONE see MKY500
3-METHYL-2-BUTENYL ACETATE see DOQ350
3-METHYL-2-BUTENYL BENZOATE see MHU150
4-(3-METHYL-2-BUTENYL)-1,2-DIPHENYL-3,5-PYRAZOLIDINEDIONE
see PEW000
3-(3-METHYL-2-BUTENYL)-1,2,3,4,5,6-HEXAHYDRO-6,11-DIMETHYL-2,6-
METHANO-3-BENZAZOCIN-8-OL see DOQ400
9-((3-METHYL-2-BUTENYL)OXY)-7H-FURO(3,2-g)(1)BENZOPYRAN-7-ONE
see IHR300
5-(1-METHYL-1-BUTENYL)-5-PROPYLBARBITURIC ACID see MHU200
3-METHYL-2-BUTENYL SALICYLATE see PMB600
2-METHYL-1-BUTEN-3-YNE see MHU250
3-METHYL-3-BUTEN-1-YNYLTRIETHYLLEAD see MHU500
3-METHYL-BUTIN-(1)-OL-(3) (GERMAN) see MHX250
(2-(3-METHYLBUTOXY)ETHYL)BENZENE see IHV050
3'-METHYLBUTTERGELB (GERMAN) see DUH600
1-METHYLBUTYL ACETATE see AOD735
3-METHYLBUTYL ACETATE see IHO850
3-METHYL-1-BUTYL ACETATE see IHO850
2-METHYL-BUTYLACRYLAAT (DUTCH) see MHU750
2-METHYL-BUTYLACRYLAT (GERMAN) see MHU750
2-METHYL BUTYLACRYLATE see MHU750
METHYLBUTYLAMINE see MHV000
N-(METHYL) BUTYL AMINE see MHV000
N-METHYL-n-BUTYLAMINE see MHV000
3-METHYLBUTYL α-AMINOBENZENEACETATE HYDROCHLORIDE (±)-
see PCV750
p-METHYL-tert-BUTYLBENZENE see BSP500
1-METHYL-4-tert-BUTYLBENZENE see BSP500
METHYL-5-BUTYL-2-BENZIMIDAZOLECARBAMATE see BQK000
1-(3-METHYL)BUTYL BENZOATE see IHP100
3-METHYLBUTYL BROMIDE see BNP250
METHYL-1-(BUTYLCARBAMOYL)-2-BENZIMIDAZOLYLCARBAMATE
see BAV575
METHYL-1,3-BUTYLENE GLYCOL ACETATE see MHV750
1-METHYLBUTYL ESTER CARBAMIC ACID see MOU500
3-METHYLBUTYL ETHANOATE see IHO850
METHYL tert-BUTYL ETHER see MHV859
3-METHYLBUTYL FORMATE see IHS000
METHYLBUTYL HYDRAZINE see MHW000
1-METHYLBUTYL-HYDRAZINE DIHYDROCHLORIDE see MHW250
α-METHYL-p-(tert-BUTYL)HYDROCINNAMALDEHYDE see LFT000
3-METHYLBUTYL 2-HYDROXYBENZOATE see IME000
METHYL-tert-BUTYL KETONE see DQU000
METHYL n-BUTYL KETONE (ACGIH) see HEV000
N-3-METHYLBUTYL-N-1-METHYL ACETONYLNITROSAMINE see MHW350
METHYL-N-(p-tert-BUTYL-α-METHYLHYDROCINNAMYLIDENE) AN-
THRANILATE see LFT100
3-METHYLBUTYL NITRITE see IMB000
METHYL-BUTYL-NITROSAMIN (GERMAN) see MHW500
METHYLBUTYLNITROSAMINE see MHW500
METHYL-N-BUTYLNITROSAMINE see MHW500
METHYL-tert-BUTYLNITROSAMINE see MHW750
m-(1-METHYLBUTYL)PHENYL METHYLCARBAMATE see AON250
α-METHYL-β-(p-tert-BUTYLPHENYL)PROPIONALDEHYDE see LFT000
2-METHYL-2-sec-BUTYL-1,3-PROPANEDIOL DICARBAMATE see MBW750
5-(1-METHYLBUTYL)-5-(2-PROPENYL)-2,4,6(1H,3H,5H)-PYRIMIDINETRIONE
MONOSODIUM SALT see SBN000
5-(1-METHYLBUTYL)-5-(2-PROPENYL)-2,4,6(1H,3H,5H)-PYRIMIDINITRIONE
see SBM500
5-METHYL-3-BUTYLTETRAHYDROPYRAN-4-YL ACETATE see MHX000
2-METHYL-3-BUTYN-2-OL see MHX250
METHYL-3-BUTYN-2-OL CARBAMATE see CBM875
2-METHYLBUTYRALDEHYDE see MJX500
3-METHYLBUTYRALDEHYDE see MHX500
α-METHYLBUTYRALDEHYDE see MJX500
METHYL BUTYRATE see MHY000
METHYL-n-BUTYRATE see MHY000

3-METHYLBUTYRIC ACID see ISU000
β-METHYLBUTYRIC ACID see ISU000
3-METHYLBUTYRIC ACID, ALLYL ESTER see ISV000
(E)-3-METHYLBUTYRIC ACID-3,7-DIMETHYL-2,6-OCTADIENYL ESTER
see GDK000
3-METHYLBUTYRIC ACID, ETHYL ESTER see ISY000
4-METHYL-γ-BUTYROLACTONE see VAV000
Γ-METHYL-γ-BUTYROLACTONE see VAV000
2-METHYLBUTYRONITRILE see ITD000
3-METHYLBUTYRONITRILE see ITD000
8-(3-METHYLBUTYRYLOXY)-DIACETOXYSCIRPENOL see FQS000
METHYL CADMIUM AZIDE see MHY550
METHYL-CALMINAL see ENB500
METHYLCAMPHENOATE see MHY600
1-METHYLCAPROLACTAM see EGN500
N-METHYLCAPROLACTAM see EGN500
N-METHYL-ε-CAPROLACTAM see MHY750
METHYL CARBAMATE see MHZ000
N-METHYLCARBAMATE de 4-DIMETHYLAMINO-3-METHYL PHENYLE
(FRENCH) see DOR400
METHYLCARBAMATE-1-NAPHTHALENOL see CBM750
METHYLCARBAMATE-1-NAPHTHOL see CBM750
N-METHYLCARBAMATE de 1-NAPHTYLE (FRENCH) see CBM750
METHYLCARBAMIC ACID-(4-ALLYLDIMETHYLAMMONIO)-m-TOLYL
ESTER, IODIDE see TGH685
METHYLCARBAMIC ACID-3-sec-BUTYL-6-CHLOROPHENYL ESTER
see MOU750
METHYLCARBAMIC ACID-2-CHLORO-5-(1-METHYLPROPYL)PHENYL
ESTER see MOU750
METHYLCARBAMIC ACID-2-CHLORO-5-tert-PENTYLPHENYL ESTER
see MIA000
METHYLCARBAMIC ACID-o-CUMENYL ESTER see MIA250
METHYLCARBAMIC ACID-m-CYM-5-YL ESTER see CQI500
METHYLCARBAMIC ACID-2,4-DICHLORO-5-ETHYL-m-TOLYL ESTER
see MIA500
N-METHYL-CARBAMIC ACID-3-DIETHYLAMINOPHENYL ESTER,
DIMETHYLSULFATE see HNK550
N-METHYLCARBAMIC ACID-3-DIETHYLAMINOPHENYL ESTER, METHIOD-
IDE see MID250
N-METHYLCARBAMIC ACID-3-DIETHYLAMINOPHENYL ESTER,
METHOCHLORIDE see TGH665
N-METHYL-CARBAMIC ACID-3-DIETHYLAMINOPHENYL ESTER,
METHOSULFATE see HNK550
METHYLCARBAMIC ACID-m-(DIETHYLMETHYLAMMONIO)PHENYL
ESTER, CHLORIDE see TGH665
N-METHYLCARBAMIC ACID-3-(DIETHYLMETHYLAMMONIO)PHENYL
ESTER, IODIDE see MID250
METHYL-CARBAMIC ACID-m-(DIETHYLMETHYLAMMONIO)PHENYL
ESTER, METHOSULFATE see HNK550
METHYL CARBAMIC ACID 2,3-DIHYDRO-2,2-DIMETHYL-7-
BENZOFURANYL ESTER see CBS275
N-METHYLCARBAMIC ACID, m-(α-DIMETHYLAMINOETHYL)PHENYL
ESTER, HYDROCHLORIDE see MIC250
N-METHYLCARBAMIC ACID-3-DIMETHYLAMINOPHENYL ESTER, ALLYL
BROMIDE see TGH680
N-METHYL-CARBAMIC ACID-3-DIMETHYLAMINOPHENYL ESTER, HYDRO-
CHLORIDE see MID000
N-METHYL CARBAMIC ACID, 4-DIMETHYLAMINOPHENYL ESTER
METHIODIDE see HNP000
N-METHYLCARBAMIC ACID-3-DIMETHYLAMINOPHENYL ESTER
METHOCHLORIDE see HNN000
N-METHYLCARBAMIC ACID-3-DIMETHYLAMINOPHENYL ESTER,
METHOSULFATE see HNR000
N-METHYLCARBAMIC ACID-3-DIMETHYLAMINOPHENYL ESTER, PROPYL
BROMIDE see TGH675
METHYLCARBAMIC ACID-5-DIMETHYLAMINO-2,4-XYLYL ESTER, HYDRO-
CHLORIDE see MIA775
N-METHYL-CARBAMIC ACID 2,4-DIMETHYL-5-DIMETHYLAMINOPHENYL
ESTER, HYDROCHLORIDE see MIA775
N-METHYLCARBAMIC ACID, (3,4-DIMETHYLPHENYL)ESTER
see XTJ000
METHYL-CARBAMIC ACID, ESTER with ESEROLINE see PIA500
METHYLCARBAMIC ACID, ETHYL ESTER see EMQ500
N-METHYLCARBAMIC ACID-3-ISOPROPYL-4-DIMETHYLAMINOPHENYL
ESTER, METHOCHLORIDE see HJY500
METHYLCARBAMIC ACID-2,3-(ISOPROPYLIDENEDIOXY)PHENYL ESTER
see DQM600
METHYLCARBAMIC ACID-m-(1-METHYL)BUTYL)PHENYL ESTER mixed
with CARBAMIC ACID, METHYL-m-(1-ETHYLPROPYL)PHENYL ESTER
(3:1) see BTA250
N-METHYLCARBAMIC ACID-3-METHYL-4-DIMETHYLAMINOPHENYL
ESTER, ALLYLIODIDE see TGH685
N-METHYLCARBAMIC ACID-2-METHYL-5-DIMETHYLAMINOPHENYL
ESTER, METHOSULFATE see TGH650

N-METHYLCARBAMIC ACID 2-METHYL-5-DIMETHYLAMINOPHENYL ESTER, METHOSULFURIC ACID see TGH655

N-METHYLCARBAMIC ACID-4-METHYL-3-DIMETHYLAMINOPHENYL ESTER, METHOSULFATE see TGH660

METHYLCARBAMIC ACID-4-METHYLTHIO-m-CUMENYL ESTER see MIB000

METHYLCARBAMIC ACID-4-METHYLTHIO-m-TOLYL ESTER see MIB250

METHYL CARBAMIC ACID-4-(METHYLTHIO)-3,5-XYLYL ESTER see DST000

METHYLCARBAMIC ACID-1-NAPHTHYL ESTER see CBM750

METHYLCARBAMIC ACID PHENYL ESTER see PFS350

METHYLCARBAMIC ACID-o-(2-PROPYNYLOXY)PHENYL ESTER see MIB500

METHYLCARBAMIC ACID-2,6-PYRIDINEDIYLDIMETHYLENE ESTER see PPH050

N-METHYL CARBAMIC ACID-8-QUINOLINYL ESTER, METHIODIDE see MID500

METHYLCARBAMIC ACID-m-TOLYL ESTER see MIB750

METHYLCARBAMIC ACID, (4-TRIMETHYLAMMONIO)-m-CUMENYL ESTER, CHLORIDE see HJY500

METHYLCARBAMIC ACID, (m-(TRIMETHYLAMMONIO)PHENYL) ESTER CHLORIDE see HNN000

METHYLCARBAMIC ACID, (m-(TRIMETHYLAMMONIO)PHENYL)ESTER, IODIDE see HNO500

N-METHYLCARBAMIC ACID-p-(TRIMETHYLAMMONIO)PHENYL ESTER IODIDE see HNP000

N-METHYLCARBAMIC ACID-(3-(TRIMETHYLAMMONIO)PHENYL) ESTER, METHYLSULFATE see HNR000

METHYLCARBAMIC ACID 5-(TRIMETHYLAMMONIO)-o-TOLYL ESTER, BISULFATE see TGH655

METHYLCARBAMIC ACID-5-(TRIMETHYLAMMONIO)-o-TOLYL ESTER, CHLORIDE see HOL000

METHYLCARBAMIC ACID-3-(TRIMETHYLAMMONIO)-p-TOLYL ESTER, METHYLSULFATE see TGH660

METHYLCARBAMIC ACID-5-(TRIMETHYLAMMONIO)-o-TOLYL ESTER, METHYLSULFATE see TGH650

METHYLCARBAMIC ACID, TRIMETHYLPHENYL ESTER see TMC750

METHYLCARBAMIC ACID-3,4-XYLYL ESTER see XTJ000

METHYLCARBAMIC ESTER of α-3-HYDROXYPHENYLETHYLDIMETHYLAMINE HYDROCHLORIDE see MIC250

METHYLCARBAMIC ESTER of 3-OXYPHENYLDIMETHYLAMINE HYDROCHLORIDE see MID000

METHYLCARBAMIC ESTER of OXYPHENYLMETHYLDIETHYLAMMONIUM IODIDE see MID250

METHYLCARBAMIC ESTER of p-OXYPHENYLTRIMETHYLAMMONIUM IODIDE see HNP000

METHYLCARBAMIC ESTER of 3-OXYPHENYLTRIMETHYLAMMONIUM METHYLSULFATE see HNR000

METHYLCARBAMIC ESTER of 8-OXYQUINOLINE METHIODIDE see MID500

METHYLCARBAMOYLETHYL ACRYLATE see MID750

2-METHYL-4-CARBAMOYL-5-HYDROXYIMIDAZOLE see MID800

S-METHYLCARBAMOYLMETHYL-O,O-DIMETHYL PHOSPHORODITHIOATE see DSP400

o-METHYLCARBAMOYL-2-METHYLPROPENEALDOXIME see MBW780

(4-METHYLCARBAMOYLOXY-o-CUMENYL)TRIMETHYLAMMONIUM CHLORIDE see HJY500

(3-(N-METHYLCARBAMOYLOXY)PHENYL)ALLYLDIMETHYLAMMONIUM BROMIDE see TGH680

(3-(N-METHYLCARBAMOYLOXY)PHENYL)DIETHYLMETHYLAMMONIUM CHLORIDE see TGH665

(3-(N-METHYLCARBAMOYLOXY)PHENYL)DIETHYLMETHYL-AMMONIUM IODIDE see MID250

(3-(N-METHYLCARBAMOYLOXY)PHENYL)DIMETHYLPROPYLAMMONIUM BROMIDE see TGH675

((p-METHYLCARBAMOYLOXY)PHENYL)TRIETHYLAMMONIUM IODIDE see HNP000

(3-(METHYLCARBAMOYLOXY)PHENYL)TRIMETHYLAMMONIUM CHLORIDE see HNN000

(3-(METHYLCARBAMOYLOXY)PHENYL)TRIMETHYLAMMONIUM IODIDE see HNO500

(4-(N-METHYLCARBAMOYLOXY)PHENYL)TRIMETHYLAMMONIUM IODIDE see HNP000

(3-(N-METHYLCARBAMOYLOXY)PHENYL)TRIMETHYLAMMONIUM METHYLSULFATE see HNR000

(3-(N-METHYLCARBAMOYLOXY)PHENYL)TRIMETHYL-ARSONIUM IODIDE see MID900

METHYL N-(CARBAMOYLOXY)THIOACETIMIDATE see MPS250

(4-METHYLCARBAMOYLOXY-o-TOLYL)ALLYLDIMETHYLAMMONIUM IODIDE see TGH685

(3-(METHYLCARBAMOYLOXY)-p-TOLYL)TRIMETHYLAMMONIUM BISULFATE see TGH655

(3-(METHYLCARBAMOYLOXY)-p-TOLYL)TRIMETHYLAMMONIUM METHYLSULFATE see TGH650

o-METHYLCARBANILIC ACID-N-ETHYL-3-PIPERDINYL ESTER see MIE250

1-METHYL-1-CARBETHOXY-4-PHENYL HEXAMETHYLENIMINE CITRATE see MIE600

1-METHYL-4-CARBETHOXY-4-PHENYLPIPERIDINE HYDROCHLORIDE see DAM700

METHYL CARBETHOXYSYRINGOYL RESERPATE see RCA200

METHYLCARBINOL see EFU000

METHYL CARBITOL see DJG000

METHYL CARBITOL ACETATE see MIE750

2-METHYL-β-CARBOLINE see MPA050

3-METHYL-β-CARBOLINE see MPA050

METHYL-4-CARBOMETHOXY BENZOATE see DUE000

METHYL CARBONATE see MIF000

METHYLCARBONYL FLUORIDE see ACM000

METHYLCARBOPHENOTHION see MQH750

17-β-(1-METHYL-3-CARBOXYPROPYL)ETHIOCHOLAN-3-α-OL see LHW000

17-β-(1-METHYL-3-CARBOXYPROPYL)ETIOCHOLANE-3-α,7-β-DIOL see DMJ200

17-β-(1-METHYL-3-CARBOXYPROPYL)-ETIOCHOLANE-3-α,12-α-DIOL see DAQ400

N-METHYL-N-(3-CARBOXYPROPYL)NITROSAMINE see MIF250

METHYLCATECHOL see GKI000

3-METHYLCATECHOL see DNE000

4-METHYLCATECHOL see DNE200

METHYL-CCNU see CHD250

trans-METHYL-CCNU see CHD250

METHYL CEDRYL ETHER see CCR525

METHYL CELLOSOLVE (OSHA, DOT) see EJH500

METHYL CELLOSOLVE ACETATE (OSHA, DOT) see EJJ500

METHYL CELLOSOLVE ACETYLRICINOLEATE see MIF500

METHYL CELLOSOLVE ACRYLATE see MIF750

METHYL CELLOSOLYE ACETAAT (DUTCH) see EJJ500

β-METHYLCHALCONE see MPL000

METHYL CHAVICOL see AFW750

METHYL CHEMOSEPT see HJL500

6-METHYL-CHINOXALIN-2,3-DITHIOL-CYCLO-CARBONAT (GERMAN) see ORU000

METHYLCHLOORFORMIAT (DUTCH) see MIG000

METHYL-2-CHLORAETHYLNITROSAMIN (GERMAN) see CIQ500

METHYLCHLORID (GERMAN) see MIF765

METHYL CHLORIDE see MIF765

METHYL CHLORIDE-METHYLENE CHLORIDE MIXTURE (DOT) see CHX750

p-METHYLCHLOROACETANILIDE see CEB500

4-METHYL-α-CHLOROACETANILIDE see CEB500

METHYL CHLOROACETATE see MIF775

METHYL CHLOROACETATE (DOT) see MIF775

METHYL-2-CHLOROACRYLATE see MIF800

METHYL-α-CHLOROACRYLATE see MIF800

2-METHYL-4-CHLOROANILINE see CLK220

2-METHYL-4-CHLOROANILINE HYDROCHLORIDE see CLK235

4-METHYL-3-CHLOROANILINE HYDROCHLORIDE see CLK230

2-(2'-METHYL-3'-CHLORO)ANILINONICOTINIC ACID see CMX770

2-METHYLCHLOROBENZENE see CLK100

1-METHYL-2-CHLOROBENZENE see CLK100

N-METHYL-N'-(4-CHLOROBENZHYDRYL)PIPERAZINE DIHYDROCHLORIDE see CDR000

o-METHYL-o-2-CHLORO-4-tert-BUTYLPHENYL-N-METHYLAMIDOPHOSPHATE see COD850

METHYL CHLOROCARBONATE see MIG000

METHYL-2-CHLORO-3-(4-CHLOROPHENYL)PROPIONATE see CFC750

2-METHYL-5-CHLORO-2-(N,N-DIETHYLAMINOETHOXYETHYL)-1,3-BENZODIOXOLE see CFO750

METHYL-β-CHLOROETHYLAMINE HYDROCHLORIDE see MIG250

N'-METHYL-N'-β-CHLOROETHYLBENZALDEHYDE HYDRAZONE see MIG500

N'-METHYL-N'-β-CHLOROETHYL-(p-DIMETHYLAMINO)-BENZALDEHYDE HYDRAZONE see MIG750

1-METHYL-1-(β-CHLOROETHYL)ETHYLENIMONIUM see MIG800

METHYL-β-CHLOROETHYL-ETHYLENIMONIUM PICRYLSULFONATE see MIG850

1-METHYL-1-(β-CHLOROETHYL)ETHYLENIMONIUM PICRYLSULFONATE see MIG850

METHYL-β-CHLOROETHYL-β-HYDROXYETHYLAMINE HYDROCHLORIDE see MIH000

METHYL(2-CHLOROETHYL)NITROSAMINE see CIQ500

METHYL-N-(β-CHLOROETHYL)-N-NITROSOCARBAMATE see MIH250

4-METHYL-5-(β-CHLOROETHYL)THIAZOLE see CHD750

5-METHYL-3-(2-CHLORO-6-FLUOROPHENYL)-4-ISOXAZOLYLPENICILLIN SODIUM see FDA100

METHYL CHLOROFORM see MIH275

METHYL CHLOROFORMATE (DOT) see MIG000

METHYL-2-CHLORO-9-HYDROXYFLUORENE-9-CARBOXYLATE see CDT000

1-METHYL-5-CHLOROINDOLINE METHYLBROMIDE see MIH300

METHYLCHLOROMETHYL ETHER see CIO250

METHYLCHLOROMETHYL ETHER (DOT) see CIO250

METHYL CHLOROMETHYL ETHER, anhydrous (DOT) see CIO250

4-METHYL-6-(((2-CHLORO-4-NITRO)PHENYL)AZO)-m-ANISIDINE see MIH500

3-METHYL-4-CHLOROPHENOL see CFE250

4,4'-METHYLENEBIS(3-HYDROXY-2-NAPHTHOIC ACID) with TRIS-(p-AMINOPHENYL)METHYLIUM SALT (1:2) see TNC725
4,4'-METHYLENEBIS(3-HYDROXY-2-NAPHTHOIC ACID) with TRIS-(p-AMINOPHENYL)CARBONIUM SALT (1:2) see TNC750
4,4'-METHYLENEBIS(3-HYDROXY-2-NAPHTHOIC ACID) with TRIS-(p-AMINOPHENYL)METHYLIUM SALT (1:2) see TNC750
4,4'-METHYLENEBIS(3-HYDROXY-2-NAPTHOIC ACID ESTER with 2-(2-(4-(p-CHLORO-α-PHENYLBENZYL)-1-PIPERAZINYL)-ETHOXY)ETHANOL see HOR500
METHYLENEBIS(4-ISOCYANATOBENZENE) see MJP400
1,1-METHYLENEBIS(4-ISOCYANATOBENZENE) see MJP400
5,5'-METHYLENEBIS(2-ISOCYANATO)TOLUENE see MJN750
METHYLENE BIS(METHANESULFONATE) see MJQ500
4,4'-METHYLENEBIS(2-METHYLANILINE) see MJO250
4,4'-METHYLENEBIS(N-METHYLANILINE) see MJO000
METHYLENE-BIS(METHYL ANTHRANILATE) see MJQ250
4,4'-METHYLENEBIS(2-METHYLBENZENAMINE) see MJO250
2,2''-METHYLENEBIS(4-METHYL-6-tert-BUTYLPHENOL) see MJO500
2,2'-METHYLENE-BIS(4-METHYL-6-NONYLPHENOL) see MJO750
N,N''-METHYLENEBIS(N'-(1-(HYDROXYMETHYL)-2,5-DIOXO-4-IM-IDAZOLIDINYL)UREA) see GDO800
METHYLENE BIS(NITRAMINE) see MJO775
2,2'-METHYLENEBIS(6-NONYL)-p-CRESOL see MJO750
METHYLENE-BIS-ORTHOCHLOROANILINE see MJM200
2,2'-METHYLENEBIS OXIRANE (9CI) see DHE000
1,1'-(METHYLENEBIS(OXY))BIS(4-CHLORO)BENZENE see NCM700
1,1'-(METHYLENEBIS(OXY)BIS(2-CHLOROETHANE) see BID750
4,4'-METHYLENEBISPHENOL see BKI250
METHYLENEBIS(4-PHENYLENE ISOCYANATE) see MJP400
METHYLENEBIS(p-PHENYLENE ISOCYANATE) see MJP400
METHYLENE BISPHENYL ISOCYANATE see MJP400
METHYLENEBIS(4-PHENYL ISOCYANATE) see MJP400
METHYLENEBIS(p-PHENYL ISOCYANATE) see MJP400
4,4'-METHYLENEBIS(PHENYL ISOCYANATE) see MJP400
p,p'-METHYLENEBIS(PHENYL ISOCYANATE) see MJP400
2,2'-METHYLENEBIS(3,4,6-TRICHLOROPHENOL) see HCL000
METHYLENE BLUE see BJI250
METHYLENE BLUE (medicinal) see BJI250
METHYLENE BLUE A see BJI250
METHYLENE BLUE BB see BJI250
METHYLENE BLUE BB ZINC FREE see BJI250
METHYLENE BLUE CHLORIDE see BJI250
METHYLENE BLUE CHLORIDE (biological stain) see BJI250
METHYLENE BLUE D see BJI250
METHYLENE BLUE I (medicinal) see BJI250
METHYLENE BLUE NF (medicinal) see BJI250
METHYLENE BLUE POLYCHROME see BJI250
METHYLENE BLUE USP (medicinal) see BJI250
METHYLENE BLUE USP XII (medicinal) see BJI250
METHYLENE BROMIDE see DDP800
(4-(2-METHYLENEBUTYRYL)-2,3-DICHLOROPHENOXY)ACETIC ACID see DFP600
METHYLENEBUTYRYL PHENOXYACETIC ACID see DFP600
METHYLENE CHLORIDE see MJP450
METHYLENE CHLOROBROMIDE see CES650
1,2-α-METHYLENE-6-CHLORO-Δ⁶-17-α-HYDROXYPROGESTERONE ACETATE see CQJ500
1,2-α-METHYLENE-6-CHLORO-PREGNA-4,6-DIENE-3,20-DIONE 17-α-ACE-TATE see CQJ500
1,2-α-METHYLENE-6-CHLORO-Δ⁻⁴,⁶-PREGNADIENE-17-α-OL-3,20-DIONE 17-α-ACETATE see CQJ500
-α-METHYLENE-6-CHLORO-Δ⁻⁴,⁶-PREGNADIENE-17-α-OL-3,20-DIONE ACE-TATE see CQJ500
4,5-METHYLENECHRYSENE see CPU000
METHYLENE CYANIDE see MAO250
2-METHYLENE-CYCLOPENTAENE-3-ONO-4,5-EPOXY-4,5-DIMETHYL-1-CARBOXYLIC ACID see MJU500
2-METHYLENECYCLOPROPANEALANINE see MJP500
2-METHYLENECYCLOPROPANYLALANINE see MJP500
β-(METHYLENECYCLOPROPYL)ALANINE see MJP500
N,N'-METHYLENEDIACRYLAMIDE see MJL500
METHYLENEDIANILINE see MJQ000
2,4'-METHYLENEDIANILINE see MJP750
4,4'-METHYLENEDIANILINE see MJQ000
p,p'-METHYLENEDIANILINE see MJQ000
4,4'-METHYLENEDIANILINE DIHYDROCHLORIDE see MJQ100
METHYLENEDIANTHRANILIC ACID DIMETHYL ESTER see MJQ250
1',9-METHYLENE-1,2;5,6-DIBENZANTHRACENE see DCR600
METHYLENE DIBROMIDE see DDP800
METHYLENE DICHLORIDE see MJP450
3,4-METHYLENE-DIHYDROXYBENZALDEHYDE see PIW250
METHYLENE DIIODIDE see DNF800
METHYLENE DIISOCYANATE see DNK100
METHYLENEDILITHIUM see MJQ325

METHYLENE DIMETHANESULFONATE see MJQ500
METHYLENE DIMETHYL ETHER see MGA850
4,4'-METHYLENEDIMORPHOLINE see MJQ750
METHYLENEDINAPHTHALENESULFONIC ACID BISPHENYLMERCURI SALT see PFN000
3,3'-METHYLENEDI-2-NAPHTHALENESULFONIC ACID, MERCURY SALT see MCT250
1,5-METHYLENE-3,7-DINITROSO-1,3,5,7-TETRAAZACYCLOOCTAINE see DVF400
1,5-METHYLENE-3,7-DINITROSO-1,3,5,7-TETRAAZACYCLOOCTANE see DVF400
3,4-METHYLENEDIOXY-ALLYBENZENE see SAD000
1,2-METHYLENEDIOXY-4-ALLYLBENZENE see SAD000
METHYLENEDIOXYAMPHETAMINE see MJQ775
3,4-METHYLENEDIOXY-AMPHETAMINE see MJQ775
3,4-METHYLENEDIOXYBENZALDEHYDE see PIW250
METHYLENEDIOXYBENZENE see MJR000
1,2-METHYLENEDIOXYBENZENE see MJR000
3,4-METHYLENEDIOXYBENZYL ACETATE see PIX000
3,4-METHYLENEDIOXYBENZYL ACETONE see MJR250
2-(4-(3,4-METHYLENEDIOXYBENZYL)PIPERAZINO)PYRIMIDINE see TNR485
1-(3,4-METHYLENEDIOXYBENZYL)-4-(2-PYRIMIDYL)PIPERAZINE see TNR485
1,2-(METHYLENEDIOXY)-4-(3-BROMO-1-PROPENYL)BENZENE see BOA750
3,4-METHYLENEDIOXY-N,α-DIMETHYL-β-PHENYLETHYLAMINE HYDRO-CHLORIDE see MJR500
3,4-METHYLENEDIOXY-α-ETHYL-β-PHENYLETHYLAMINE see MJR750
1,2-METHYLENEDIOXY-4-(1-HYDROXYALLYL)BENZENE see BCJ000
3,4-METHYLENEDIOXY-α-METHYL-β-PHENYLETHYLAMINE HYDROCHLO-RIDE see MJS750
1,2-(METHYLENEDIOXY)-4-(2-(OCTYLSULFINYL)PROPYL)BENZENE see ISA000
2-(3,4-(METHYLENEDIOXY)PHENOXY)-1-((3,4-(METHYLENEDIOXY)PHENOXY)METHYL)ETHYLAMINE see MJS250
1-(5-(3,4-(METHYLENEDIOXY)PHENOXY)-4-((3,4-(METHYLENEDIOXY)PHENOXY)METHYL)PENTYLPIPERIDINE CITRATE see MJS500
1-(3,4-METHYLENEDIOXYPHENYL)-2-AMINOPROPANE see MJS750
4-(3,4-METHYLENEDIOXYPHENYL)-2-BUTANONE see MJR250
3,4-METHYLENEDIOXY-β-PHENYLETHYLAMINE HYDROCHLORIDE see MJT000
1,2-METHYLENEDIOXY-4-PROPENYLBENZENE see IRZ000
3,4-METHYLENEDIOXY-1-PROPENYL BENZENE see IRZ000
1,2-(METHYLENEDIOXY)-4-PROPYLBENZENE see DMD600
(3,4-METHYLENEDIOXY-6-PROPYLBENZYL)(BUTYL)DIETHYLENE GLYCOL ETHER see PIX250
3,4-METHYLENEDIOXY-6-PROPYLBENZYL-n-BUTYL DIETHYLENEGLYCOL ETHER see PIX250
4,4'-METHYLENE DIPHENOL see BKI250
4,4'-METHYLENEDIPHENYL DIISOCYANATE see MJP400
METHYLENEDI-p-PHENYLENE DIISOCYANATE see MJP400
METHYLENEDI-p-PHENYLENE ISOCYANATE see MJP400
4,4'-METHYLENEDIPHENYLENE ISOCYANATE see MJP400
METHYLENE DI(PHENYLENE ISOCYANATE) (DOT) see MJP400
4,4'-METHYLENEDIPHENYL ISOCYANATE see MJP400
METHYLENE DITHIOCYANATE see MJT500
4,4'-METHYLENE DI-o-TOLUIDINE see MJO250
METHYLENE DIURETHAN see MJT750
METHYLENE ETHER of OXYHYDROQUINONE see MJU000
N-METHYLENE GLYCINONITRILE see HHW000
METHYLENE GLYCOL see FMV000
METHYLENE GREEN see MJU250
METHYLENE IODIDE see DNF800
METHYLENEMAGNESIUM see MJU350
(1S-cis)-5-METHYLENE-6-(1-METHYLETHENYL)-2-CYCLOHEXEN-1-OL see CCL109
3-METHYLENE-7-METHYL-1,6-OCTADIENE see MRZ150
3-METHYLENE-7-METHYLOCTAN-7-YL ACETATE see DLX100
3-METHYLENE-7-METHYL-1-OCTEN-7-OL see MLO250
3-METHYLENE-7-METHYL-1-OCTEN-7-YL ACETATE see AAW500
4-METHYLENE-2-OXETANONE see KFA000
METHYLENE OXIDE see FMV000
2-METHYLENE-3-OXO-CYCLOPENTANECARBOXYLIC ACID see SAX500
2-METHYLENE-3-OXO-1-CYCLOPENTANECARBOXYLIC ACID compounded with NICOTINIC HYDRAZIDE see SAY000
2-METHYLENE-3-OXOCYCLOPENTANECARBOXYLIC ACID, SODIUM SALT see SJS500
METHYLENESUCCINYLOXYBIS(TRIBUTYLSTANNANE) see BLL500
S,S'-METHYLENE O,O,O',O'-TETRAETHYL PHOSPHORODITHIOATE see EEH600
2,5-endo-METHYLENE-Δ³-TETRAHYDROBENZYL ACRYLATE see BFY250
8-METHYLENE-4,11,11-(TRIMETHYL)BICYCLO(7.2.0)UNDEC-4-ENE see CCN000
METHYLENIMINOACETONITRILE see HHW000
METHYLENIUM CERULEUM see BJI250

N-(1-METHYLETHYL)-N'-(1-NITRO-9-ACRIDINYL)-1,3-PROPANEDIAMINE DIHYDROCHLORIDE see INW000
METHYLETHYLNITROSAMINE see MKB000
N,N-METHYLETHYLNITROSAMINE see MKB000
((1-METHYLETHYL)NITROSOAMINO)-METHANOL ACETATE (ESTER) (9CI) see ING300
N-(1-METHYLETHYL)-N-NITROSO-UREA (9CI) see NKO425
1-METHYL-4-ETHYLOCTYLAMINE see EMY000
METHYLETHYLOLAMINE see MGG000
4-(1-METHYLETHYL)PHENOL see IQZ000
1-METHYL-5-ETHYL-5-PHENYLBARBITURIC ACID see ENB500
2-METHYL-3-ETHYL-4-PHENYL-4-CYCLOHEXENE CARBOXYLIC ACID see FAO200
2-METHYL-3-ETHYL-4-PHENYL-Δ^4-CYCLOHEXENECARBOXYLIC ACID see FAO200
2-METHYL-3-ETHYL-4-PHENYL-Δ^4-CYCLOHEXENE CARBOXYLIC ACID SODIUM SALT see MBV775
METHYLETHYL PHENYLETHYL CARBINYL ACETATE see PFR300
3-METHYL-5-ETHYL-5-PHENYLHYDANTOIN see MKB250
2-(1-METHYLETHYL)PHENYL METHYLCARBAMATE see MIA250
(1-METHYLETHYL)PHOSPHORAMIDOTHIOIC ACID O-(2,4-DICHLOROPHENYL)-O-METHYL ESTER see DGD800
N-(1-METHYLETHYL)-2-PROPANAMINE see DNM200
2-METHYL-5-ETHYLPYRIDINE see EOS000
6-METHYL-3-ETHYLPYRIDINE see EOS000
METHYL ETHYL PYRIDINE (DOT) see EOS000
2-METHYL-5-ETHYLPYRIDINE (DOT) see EOS000
N-METHYL-N-ETHYL-4-(4'-(PYRIDYL-1'OXIDE)AZO)ANILINE see MKB500
3-METHYL-3-ETHYLPYRROLIDINE-2,5-DIONE see ENG500
γ-METHYL-γ-ETHYL-SUCCINIMIDE see ENG500
METHYLETHYLTHIOFOS see ENI175
(R*,S*)-4-((1-METHYLETHYL)THIO)-α-(1-OC-TYLAMINO)ETHYL)BENZENEMETHANOL see SOU600
METHYLETHYLTHIOPHOS see ENI175
O-METHYL-O-ETHYL-O-2,4,5-TRICHLOROPHENYL THIOPHOSPHATE see TIR250
4-(1-METHYLETHYL)-2-(3-(TRIFLUOROMETHYL)PHENYL)MORPHOLINE see IRP000
1-(METHYLETHYL)UREA and SODIUM NITRITE see SIS675
3-METHYLETHYNYLESTRADIOL see MKB750
17-α-METHYLETHYNYL-19-NORTESTOSTERONE see MNC100
18-METHYL-17-α-ETHYNYL-19-NORTESTOSTERONE see NNQ500, NNQ520
3-METHYLETHYNYLOESTRADIOL see MKB750
METHYL EUGENOL (FCC) see AGE250
METHYL EUGENOL GLYCOL see MKC000
S-METHYL FENITROOXON see MKC250
S-METHYL FENITROTHION see MKC250
2-METHYLFLUORANTHENE see MKC500
3-METHYLFLUORANTHENE see MKC750
METHYL FLUORIDE (DOT) see FJK000
METHYL FLUOROACETATE see MKD000
7-METHYL-9-FLUOROBENZ(c)ACRIDINE see MKD250
7-METHYL-11-FLUOROBENZ(c)ACRIDINE see MKD500
10-METHYL-3-FLUORO-5,6-BENZACRIDINE see MKD750
7-METHYL-2-FLUOROBENZ(a)ANTHRACENE see FJN000
METHYL N-(5-(p-FLUOROBENZOYL)-2-BENZIMIDAZOLYL)CARBAMATE see FDA887
METHYL-4-FLUOROBUTYRATE see MKE000
METHYL-γ-FLUOROBUTYRATE see MKE000
METHYL-ω-FLUOROBUTYRATE see MKE000
METHYL-γ-FLUOROCROTONATE see MKE250
16-α-METHYL-9-α-FLUORO-1-DEHYDROCORTISOL see SOW000
16-α-METHYL-9-α-FLUORO-Δ¹-HYDROCORTISONE see SOW000
METHYL-γ-FLUORO-β-HYDROXYBUTYRATE see MKE750
17-α-METHYL-9-α-FLUORO-11-β-HYDROXYTESTERONE see AOO275
METHYL-γ-FLUORO-β-HYDROXYTHIOLBUTYRATE see MKF000
METHYLFLUOROPHOSPHORIC ACID, ISOPROPYL ESTER see IPX000
METHYLFLUORO-PHOSPHORYLCHOLINE see MKF250
16-α-METHYL-9-α-FLUOROPREDNISOLONE see SOW000
16-α-METHYL-6-α-FLUORO-PREDNISOLONE-21-ACETATE see PAL600
16-α-METHYL-9-α-FLUORO-1,4-PREGNADIENE-11-β,17,21-TRIOL-3,20-DIONE see SOW000
METHYL FLUOROSULFATE see MKG250
METHYL FLUOROSULFONATE see MKG250
16-α-METHYL-9-α-FLUORO-11-β,17-α,21-TRIHYDROXYPREGNA-1,4-DIENE-3,20-DIONE see SOW000
METHYLFLUORPHOSPHORSAEURECHOLINESTER (GERMAN) see MKF250
METHYLFLUORPHOSPHORSAEUREISOPROPYLESTER (GERMAN) see IPX000
METHYLFLUORPHOSPHORSAEUREPINAKOLYLESTER (GERMAN) see SKS500
METHYL FLUORSULFONATE see MKG250
METHYLFLURETHER see EAT900
9-METHYLFOLIC ACID see HGI525
x-METHYLFOLIC ACID see MKG275

METHYLFORMAMIDE see MKG500
N-METHYLFORMAMIDE see MKG500
METHYL FORMATE see MKG750
METHYLFORMIAAT (DUTCH) see MKG750
METHYLFORMIAT (GERMAN) see MKG750
N-METHYL-N-FORMLYHYDRAZINE see FNW000
METHYL FORMYL see DOO600
N-METHYL-N-FORMYL HYDRAZONE of ACETALDEHYDE see AAH000
1-METHYL-2-FORMYLPYRIDINIUM CHLORIDE OXIME see FNZ000
METHYL FOSFERNO see MNH000
METHYLFUMARIC ACID see MDI250
METHYLFURAN see MKH000
2-METHYLFURAN see MKH000
METHYL 2-FURANCARBOXYLATE see MKH600
3-METHYL-2,5-FURANDIONE see CMS322
5-METHYL-2(3H)-FURANONE see MKH250
5-METHYL-2(5H)-FURANONE see MKH500
METHYL FUROATE see MKH600
METHYL 2-FUROATE see MKH600
5-(5-METHYL-2-FURYL)-1-PHENYL-3-(2(PIPERIDINO)ETHYL)-1H-PYRAZLO-INE HYDROCHLORIDE see MKH750
METHYL GAG see MKI000
METHYL GALLATE see MKI100
6'-N-METHYLGENTAMICIN C(^{1}a) see MQS579
METHYL GLUCAMINE BILIGRAFIN see BGB315
METHYLGLUCAMINE DIATRIXOATE see AOO875
METHYLGLUCAMINE-3,5-DIIODO-4-PYRIDONE-N-ACETATE see DNG400
METHYL GLUCAMINE IODIPAMIDE see BGB315
METHYLGLUCAMINE IOTALAMATE see IGC000
METHYLGLUCAMINE IOTHALAMATE see IGC000
N-METHYLGLUCAMINE and SULFALENE see SNJ400
N-METHYLGLUCAMINE and SULFAMONOMETOXINE see SNL830
3-METHYL GLUTARALDEHYDE see MKI250
METHYL GLYCOL see EJH500, PML000
METHYL GLYCOL ACETATE see EJJ500
METHYL GLYCOL MONOACETATE see EJJ500
METHYLGLYKOL (GERMAN) see EJH500
METHYLGLYKOLACETAT (GERMAN) see EJJ500
METHYLGLYOXAL see PQC000
METHYL GLYOXYLATE see MKI550
4-METHYLGUAIACOL see MEK250
p-METHYLGUAIACOL see MEK250
METHYLGUANIDIN (GERMAN) see MKI750
METHYLGUANIDINE see MKI750
METHYLGUANIDINE mixed with SODIUM NITRITE (1:1) see MKJ000
METHYL GUTHION see ASH500
METHYLGUVACINE see AQT625
N-METHYLGUVACINE see AQT625
METHYLHARNSTOFF and NATRIUMNITRIT (GERMAN) see MQJ250
11-METHYL-11H-BENZO(a)CARBAZOLE see MHD250
METHYL HEPTANETHIOL see MKJ250
3-METHYL-5-HEPTANONE see EGI750
5-METHYL-3-HEPTANONE see EGI750
METHYLHEPTENOL see MKJ750
6-METHYL-6-HEPTEN-2-OL see MKJ750
METHYL HEPTENONE see MKK000
6-METHYL-5-HEPTEN-2-ONE see MKK000
METHYL HEPTINE CARBONATE see MND275
2-METHYL-2-HEPTYLAMINE see ILM000
6-METHYL-2-HEPTYLAMINE see ILM000
(6-(1-METHYL-HEPTYL)-2,4-DINITRO-FENYL)-CROTONAAT (DUTCH) see AQT500
(6-(1-METHYL-HEPTYL)-2,3-DINITRO-PHENYL)-CROTONAT (GERMAN) see AQT500
2-(1-METHYLHEPTYL)-4,6-DINITROPHENYL CROTONATE see AQT500
6-METHYL-2-HEPTYLHYDRAZINE see MKK250
6-METHYL-2-HEPTYL-ISOPROPYLIDENHYDRAZIN (GERMAN) see MKK500
6-METHYL-2-HEPTYLISOPROPYLIDENHYDRAZINE see MKK500
METHYL HEPTYL KETONE see NMY500
METHYLHEPTYLNITROSAMIN (GERMAN) see HBP000
METHYLHESPERIDIN see MKK600
METHYLHEXABARBITAL see ERD500
METHYLHEXABITAL see ERD500
3-β-14-METHYLHEXADECANOATE-CHOLEST-5-EN-3-OL see CCJ500
2-METHYL-1,5-HEXADIENE-3-YNE see MJL000
METHYL HEXAFLUOROISOBUTYRATE see MKK750
N-METHYLHEXAHYDRO-2-PICOLINIC ACID, 2,6-DIMETHYLANILIDE see SBB000
2-METHYLHEXANE see MKL250
5-METHYL-2,3-HEXANEDIONE see MKL300
2-METHYL-3-HEXANONE see MKL400
2-METHYL-5-HEXANONE see MKW450
5-METHYL-2-HEXANONE see MKW450
5-METHYL-3-HEXEN-2-ONE see MKM250

METHYL ISOBUTYL KETOXIME see MKW750
METHYL ISOBUTYRATE see MKX000
N-METHYL-2-ISOCAMPHANAMINE see VIZ400
METHYLISOCYANAAT (DUTCH) see MKX250
METHYL ISOCYANAT (GERMAN) see MKX250
METHYL ISOCYANATE see MKX250
METHYL ISOCYANATE, solutions (DOT) see MKX250
METHYL ISOCYANIDE see MKX500
METHYL ISOCYANOACETATE see MKX575
METHYL ISOEUGENOL (FCC) see IKR000
METHYLISOMIN see DBA800
METHYLISOMYN see MDT600
METHYL ISONITRILE see MKX500
METHYLISOOCTENYLAMINE see ILK000
METHYL ISOPENTANOATE see ITC000
1-METHYL-4-ISOPROPENYLCYCLOHEXAN-3-OL see MCE750
1-METHYL-4-ISOPROPENYL-1-CYCLOHEXENE see MCC250
1-METHYL-4-ISOPROPENYL-6-CYCLOHEXEN-2-OL see MKY250
1-1-METHYL-4-ISOPROPENYL-6-CYCLOHEXEN-2-ONE see CCM120
d-1-METHYL-4-ISOPROPENYL-6-CYCLOHEXEN-2-ONE see CCM100
Δ-1-METHYL-4-ISOPROPENYL-6-CYCLOHEXEN-2-ONE see MCD250
METHYL ISOPROPENYL KETONE see MKY500
METHYL ISOPROPENYL KETONE INHIBITED (DOT) see MKY500
N-METHYL-2-ISOPROPOXYPHENYLCARBAMATE see PMY300
p-METHYLISOPROPYL BENZENE see CQI000
1-METHYL-4-ISOPROPYLBENZENE see CQI000
1-METHYL-2-p-(ISOPROPYLCARBAMOYL)BENZOHYDRAZINE HYDROCHLO-
RIDE see PME500
1-METHYL-2-(-ISOPROPYLCARBAMOYL)BENZYL)HYDRAZINE see PME250
1-METHYL-2-(p-ISOPROPYLCARBAMOYLBENZYL)HYDRAZINE HYDRO-
CHLORIDE see PME500
1-METHYL-4-ISOPROPYLCYCLOHEXADIENE-1,3 see MLA250
1-METHYL-4-ISOPROPYL-1,3-CYCLOHEXADIENE see MLA250
1-METHYL-4-ISOPROPYLCYCLOHEXADIENE-1,4 see MCB750
2-METHYL-5-ISOPROPYL-1,3-CYCLOHEXADIENE see MCC000
6-METHYL-3-ISOPROPYLCYCLOHEXANOL see DKV150
1-METHYL-4-ISOPROPYL-1-CYCLOHEXEN-3-ONE see MCF250
α-METHYL-p-ISOPROPYLHYDROCINNAMALDEHYDE see COU500
α-METHYL-p-ISOPROPYL HYDROCINNAMIC ALDEHYDE DIETHYL ACE-
TAL see COU510
α-METHYL-p-ISOPROPYLHYDROCINNAMIC ALDEHYDE DIMETHYL ACE-
TAL see COU525
1-METHYL-4-ISOPROPYLIDENE-3-CYCLOHEXANONE see MCF500
METHYL ISOPROPYL KETONE see MLA750
2-METHYL-5-ISOPROPYLPHENOL see CCM000
5-METHYL-2-ISOPROPYL-1-PHENOL see TFX810
N-METHYL-3-ISOPROPYLPHENYL CARBAMATE see COF250
N-METHYL-m-ISOPROPYLPHENYL CARBAMATE see COF250
(3-METHYL-5-ISOPROPYLPHENYL)-N-METHYLCARBAMAT (GERMAN)
see CQI500
3-METHYL-5-ISOPROPYLPHENYL-N-METHYLCARBAMATE see CQI500
2-METHYL-3-(p-ISOPROPYLPHENYL)PROPIONALDEHYDE see COU500
5-METHYL-2-ISOPROPYL-3-PYRAZOLYL DIMETHYLCARBAMATE
see DSK200
METHYL(5-ISOPROPYL-N-(p-TOLYL)-o-TOLUENESULFONAMIDO)MERCURY
see MLB000
METHYL ISOSYSTOX see DAP400
3-METHYL-5-ISOTHIAZOLAMINE HYDROCHLORIDE see MLB600
3-METHYL-5-ISOTHIAZOLAMINE MONOHYDROCHLORIDE see MLB600
METHYLISOTHIOCYANAAT (DUTCH) see ISE000
METHYL-ISOTHIOCYANAT (GERMAN) see ISE000
d-d-METHYLISOTHIOCYANATE see MLC000
METHYL ISOTHIOCYANATE (DOT) see ISE000
METHYL ISOVALERATE see ITC000
METHYLISOVALERATE (DOT) see ITC000
3-METHYL-5-ISOXAZOLECARBOXYLIC ACID 2-BENZYLHYDRAZIDE
see BEW000
5-METHYL-3-ISOXAZOLECARBOXYLIC ACID-2-BENZYLHYDRAZIDE
see IKC000
3-METHYL-4,5-ISOXAZOLEDIONE-4-((2-CHLOROPHENYL)HYDRAZONE)
see MLC250
N'-(5-METHYL-3-ISOXAZOLE)SULFANILAMIDE see SNK000
5-METHYL-3-ISOXAZOLOL see HLM000
5-METHYL-3(2H)-ISOXAZOLONE see HLM000
N'-(5-METHYL-3-ISOXAZOLYL)SULFANILAMIDE see SNK000
N'-(5-METHYLISOXAZOL-3-YL)SULPHANILAMIDE see SNK000
N1-(5-METHYL-3-ISOXAZOLYL)SULPHANILAMIDE see SNK000
METHYLIUM, TRIS(4-AMINOPHENYL)-, 4,4'-METHYLENEBIS(3-HYDROXY-
NAPHTHOATE) (2:1) see TNC725
N-METHYLJERVINE see MLC500
METHYLJODID (GERMAN) see MKW200
METHYLJODIDE (DUTCH) see MKW200
METHYL KETONE see ABC750
2-METHYLLACTONITRILE see MLC750

N-METHYLLAUROTETANINE see TKX700
METHYLLCISTOX see DAO800
METHYL LEDATE see LCW000
METHYLLITHIUM see MLD000
METHYL-LOMUSTINE see CHD250
METHYLLORAZEPAM see MLD100
N-METHYLLORAZEPAM see MLD100
N-METHYL-LOST see BIE250
1-METHYL-LUMILYSERGOL-8-(5-BROMONICOTINATE)-10-METHYL ETHER
see NDM000
METHYLLYSERGIC ACID BUTANOLAMIDE see MLD250
1-METHYLLYSERGIC ACID BUTANOLAMIDE see MLD250
1-METHYLLYSERGIC ACID ETHYLAMIDE see MLD500
d-1-METHYL LYSERGIC ACID MONOETHYLAMIDE see MLD500
METHYLMAGNESIUM BROMIDE (ethyl ether solution) see MLE000
METHYL MAGNESIUM BROMIDE in ETHYL ETHER (DOT) see MLE000
METHYLMAGNESIUM IODIDE see MLE250
METHYL MALEATE see DSL800
METHYLMALEIC ACID see CMS320
METHYLMALEIC ACID, DIMETHYL ESTER see DRF200
METHYLMALEIC ANHYDRIDE see CMS322
2-METHYLMALEIC ANHYDRIDE see CMS322
3-METHYLMALEIC ANHYDRIDE see CMS322
α-METHYLMALEIC ANHYDRIDE see CMS322
METHYL MALONATE see DSM200
METHYL MARASMATE see MLE600
METHYLMERCAPTAAN (DUTCH) see MLE650
METHYL MERCAPTAN see MLE650
METHYLMERCAPTOACETATE see MLE750
METHYL-2-MERCAPTOACETATE see MLE750
2-METHYLMERCAPTO-4,6-BIS(ISOPROPYLAMINO)-s-TRIAZINE see BKL250
4-METHYLMERCAPTO-3,5-DIMETHYLPHENYL N-METHYLCARBAMATE
see DST000
2-METHYLMERCAPTO-4-ETHYLAMINO-6-ISOPROPYLAMINO-s-TRIAZINE
see MPT500
METHYL-MERCAPTOFOS TEOLOVY see DAP400
1-METHYL-2-MERCAPTOIMIDAZOLE see MCO500
1-METHYL-2-MERCAPTO-5-IMIDAZOLE CARBOXYLIC ACID see MLF000
2-METHYLMERCAPTO-4-ISOPROPYLAMINO-6-ETHYLAMINO-s-TRIAZINE see
MPT500
METHYLMERCAPTO-4-ISOPROPYLAMINO-6-METHYLAMINO-s-TRIAZINE
see INR000
4-METHYLMERCAPTO-3-METHYLPHENYL DIMETHYL THIOPHOSPHATE
see FAQ900
2-METHYLMERCAPTO-10-(2-N-METHYL-2-PIPERIDYL)ETHYL)PHENOTHI-
AZINE see MOO250
2-METHYLMERCAPTO-10-(2-(N-METHYL-2-PIPERIDYL)ETHYLPHENOTHIAZ-
INE HYDROCHLORIDE see MOO500
METHYL-MERCAPTOPHOS see MIW100
METHYLMERCAPTOPHOS see DAO800
d-2-METHYL-3-MERCAPTOPROPANOYL-l-PROLINE see MCO750
6-METHYLMERCAPTOPURINE RIBONUCLEOSIDE see MPU000
6-METHYLMERCAPTOPURINE RIBOSIDE see MPU000
1-METHYL-5-MERCAPTO-1,2,3,4-TETRAZOLE see MPQ250
4-METHYLMERCAPTO-3,5-XYLYL METHYLCARBAMATE see DST000
N-METHYLMERCURI-BIS-p-TOLUENSULFONAMID (CZECH) see BLK250
METHYLMERCURIC CHLORIDE see MDD750
METHYLMERCURIC CYANOGUANIDINE see MLF250
METHYLMERCURIC DICYANDIAMIDE see MLF250
METHYLMERCURICHLORENDIMIDE see MLF500
METHYLMERCURIC HYDROXIDE see MLG000
METHYLMERCURIC PHOSPHATE see MLF520
METHYLMERCURIC SULFATE see BKS810
S-(METHYLMERCURIC)THIOGLYCOLIC ACID SODIUM SALT see SIM000
N-(METHYLMERCURI)-1,4,5,6,7,7-HEXACHLOROBICYCLO(2.2.1)HEPT-5-ENE-
2,3-DICARBOXIMIDE see MLF500
8-(METHYLMERCURIOXY)QUINOLINE see MLH000
METHYLMERCURIPENTACHLORFENOLAT (CZECH) see MLG250
N-METHYLMERCURI-1,2,3,6-TETRAHYDRO-3,6-ENDOMETHANO-3,4,5,6,7,7-
HEXACHLOROPHTHALIMIDE see MLF500
N-METHYLMERCURI-1,2,3,6-TETRAHYDRO-3,6-METHANO-3,4,5,6,7,7-
HEXACHLOROPHTHALIMIDE see MLF500
METHYLMERCURY see MLF550
METHYL-MERCURY(1+) (9CI) see MLF550
METHYLMERCURY(II) CATION see MLF550
METHYLMERCURY CHLORIDE see MDD750
METHYLMERCURY DICYANDIAMIDE see MLF250
METHYLMERCURY DIMERCAPTOPROPANOL see MLF750
METHYLMERCURY HYDROXIDE see MLG000
METHYLMERCURY β-HYDROXYQUINOLATE see MLH000
METHYLMERCURY 8-HYDROXYQUINOLINATE see MLH000
METHYLMERCURY ION see MLF550
METHYLMERCURY ION(1+) see MLF550
METHYLMERCURY OXINATE see MLH000

N-METHYL-2-((o-METHYL-α-PHENYLBENZYL)OXY)ETHYLAMINE HYDRO-
CHLORIDE see TGJ250
1-METHYL-4-(o-METHYL-α-PHENYLBENZYL)PIPERAZINE DIHYDROCHLO-
RIDE see MIQ725
METHYL (3-METHYLPHENYL)CARBAMOTHIOIC ACID, o-2-NAPH-
THALENYL ESTER see TGB475
O-METHYL-S-(4-METHYLPHENYL) ETHYLPHOSPHONODITHIOATE
see EOO000
N-METHYL-N′-(p-METHYLPHENYL)-N-NITROSOUREA see MNA650
N-METHYL-α-((2-METHYLPHENYL)PHENYLMETHOXY)ETHANAMINE HY-
DROCHLORIDE see TGJ250
1-METHYL-4-((2-METHYLPHENYL)PHENYLMETHYL)PIPERAZINE DIHYDRO-
CHLORIDE (9CI) see MIQ725
2-METHYL-3-(2-METHYLPHENYL)-4-QUINAZOLINONE see QAK000
2-METHYL-3-(2-METHYLPHENYL)-4(3H)-QUINAZOLINONE see QAK000
N-METHYL-α-METHYL-α-PHENYLSUCCINIMIDE see MLP800
N-METHYL-α-α-METHYLPHENYLSUCCINIMIDE see MLP800
3-METHYL-1-(4-METHYLPHENYL)TRIAZENE see MQB250
6-METHYL-α-(4-METHYL-1-PIPERAZINYLCARBONYL)ERGOLINE-8β-PRO-
PIONITRILE see MLR400
2-METHYL-11-(4-METHYL-1-PIPERAZINYL)-DIBENZO(b,f)(1,4)THIAZEPINE
see MQQ000
2-METHYL-2-(2-(4-METHYL-1-PIPERAZINYL)ETHYL)-1,3-BENZODIOXOLE HY-
DROCHLORIDE see MLL500
5-METHYL-3-(4-METHYL-1-PIPERAZINYL)-5H-PYRIDAZINO(3,4-
b)(1,4)BENZOXAZINE HYDROCHLORIDE see ASC250
5-METHYL-4-(4-METHYL-1-PIPERAZINYL)THIENO(2,3-d)PYRIMIDINE HY-
DROCHLORIDE see MLR500
N-METHYL-2-(2-METHYLPIPERIDINO)-N-(2-PHENOXYETHYL)ACETAMIDE
HYDROCHLORIDE see MLS250
N-METHYL-2-(2-METHYLPIPERIDINO)-N-(2-(o-TOLYLOXY)ETHYL)-
ACETAMIDE HYDROCHLORIDE see MLS750
METHYL-2-METHYL-2-PROPENOATE see MLH750
α-METHYL-4-(2-METHYLPROPYL)BENZENEACETIC ACID see IIU000
α-METHYL-4-(2-METHYLPROPYL)BENZENEACETIC ACID-2-
PYRIDINYLMETHYL ESTER see IAB000
2-METHYL-N-(2-METHYLPROPYL)-2-PROPANAMINE see DNH400
1-METHYL-4-(p-(p-((1-METHYLPYRIDINIUM-4-YL)AMINO)BENZAMIDO)
ANILINO)QUINOLINIUM), DI-p-TOLUENESULFONATE see MLT250
1-METHYL-4-((4-(3-((4-((1-METHYLPYRIDINIUM-4-YL)AMINO)PHE-
NYL)AMINO)3-OXO-1-PROPENYL)PHENYL)AMINO)QUINOLINIUM),
DIBROMIDE see MLT500
1-METHYL-4-(p-((p-((1-METHYLPYRIDINIUM-4-YL)AMINO)PHENYL)CARBAM-
OYL)ANILINO)QUINOLINIUM), DIBROMIDE see MLU000
1-METHYL-4-(p-((p-((1-METHYLPYRIDINIUM-4-YL)AMINO)PHENYL)CARBAM-
OYL)ANILINO)-7-NITROQUINOLINIUM), DI-p-TOLUENE SULFONATE
see MLT750
1-METHYL-4-(p-(p-((1-METHYLPYRIDINIUM-4-YL)AMINO)STYRYL)
ANILINO)QUINOLIUM DIBROMIDE see MLU250
1-METHYL-4-(p-((p-(1-METHYLPYRIDINIUM-4-YL)PHENYL)CARBAMOYL)
ANILINO)QUINOLINIUM), DI-p-TOLUENE SULFONATE see MLU750
1-METHYL-6-(p-(p-((1-METHYLQUINOLINIUM-6-YL)CARBAM-
OYL)BENZAMIDO)BENZAMIDO)QUINOLINIUM), DI-p-TOLUENE SULFO-
NATE see MLW250
1-METHYL-6-((p-(p-((1-METHYLQUINOLINIUM-6-YL)CARBAM-
OYL)BENZAMIDO)PHENYL)CARBAMOYL)QUINOLINIUM), DI-p-TOLU-
ENE SULFONATE see MLW500
2-METHYL-2-(METHYLSULFONYL)PROPANAL-o-((METHYLAMINO)
CARBONYL)OXIME see AFK000
2-METHYL-2-(METHYLSULFONYL)PROPIONALDEHYDE-o-
(METHYLCARBAMOYL)OXIME see AFK000
METHYL ((METHYLTHIO)ACETYL)CARBAMIC ACID-o-
ISOPROPOXYPHENYL ESTER see MLX000
3-METHYL-2-(METHYLTHIO)-BENZOTHIAZOLIUM-p-TOLUENESULFONATE
see MLX250
METHYL(METHYLTHIO)MERCURY see MLX300
3-METHYL-4-METHYLTHIOPHENOL see MLX750
2-METHYL-2-(METHYLTHIO)PROPANAL-O-((METHYLAMINO)
CARBONYL)OXIME see CBM500
2-METHYL-2-(METHYLTHIO)PROPANAL-o-((METHYLNITROSOAMINO)
CARBONYL)OXIME see NJJ500
2-METHYL-2-(METHYLTHIO)PROPIONALDEHYDE-O-
(METHYLCARBAMOYL)OXIME see CBM500
2-METHYL-2-(METHYLTHIO)PROPIONALDEHYDE-o-
((METHYLNITROSO)CARBAMOYL) OXIME see NJJ500
2-METHYL-2-(METHYLTHIO)PROPIONALDEHYDE OXIME see CBM500
2-METHYL-2-METHYLTHIO-PROPIONALDEHYD-O-(N-METHYL-CARBAM-
OYL)-OXIM (GERMAN) see CBM500
METHYL METIRAM see MLX850
METHYLMITOMYCIN see MLY000
N-METHYLMITOMYCIN C see MLY000
METHYL MONOBROMOACETATE see MHR250
METHYL MONOCHLORACETATE see MIF775
METHYL MONOCHLOROACETATE see MIF775

3-METHYL-4-MONOMETHYLAMINOAZOBENZENE see MLY250
3′-METHYL-4-MONOMETHYLAMINOAZOBENZENE see MPY000
N-METHYLMONOTHIOSUCCINIMIDE see MLY500
(+−)-17-METHYLMORPHINAN-3-OL HYDROBROMIDE see HMH500
(−)-17-METHYLMORPHINAN-3-OL TARTRATE DIHYDRATE see MLZ000
(+)-17-METHYLMORPHINAN-3-OL TARTRATE HYDRATE see MMA000
METHYLMORPHINE see CNF500
N-METHYL MORPHINE CHLORIDE see MRP000
N-METHYLMORPHINIUM CHLORIDE see MRP000
4-METHYLMORPHOLINE see MMA250
N-METHYL MORPHOLINE see MMA250
METHYLMORPHOLINE (DOT) see MMA250
α-METHYL-4-MORPHOLINEACETIC ACID-2,6-XYLYL ESTER HYDROCHLO-
RIDE see MMA525
1-METHYL-MORPHOLINO-3-PHTHALIMIDO-GLUTARIMIDE see MRU080
6-METHYL-MP-RIBOSIDE see MPU000
METHYL MUSTARD OIL see ISE000
METHYLNAFTALEN see MMB500
5-METHYLNAFTOYLBENZIMIDAZOL (CZECH) see MHC000
N-METHYL-1-NAFTYL-CARBAMAAT (DUTCH) see CBM750
METHYL NAMATE see SGM500
2-METHYL-1,4-NAPHTHALENEDIONE see MMD500
METHYLNAPHTHALENE see MMB500
1-METHYLNAPHTHALENE see MMB750
2-METHYLNAPHTHALENE see MMC000
α-METHYLNAPHTHALENE see MMB750
β-METHYLNAPHTHALENE see MMC000
2-METHYL-1,4-NAPHTHALENEDIOL see MMC250
2-METHYL-1,4-NAPHTHALENEDIONE see MMD500
2-METHYL-1,4-NAPHTHOCHINON (GERMAN) see MMD500
2-METHYL-1,4-NAPHTHOCHINON-NATRIUM-BISULFIT TRIHYDRAT
(GERMAN) see MMD750
5-METHYLNAPHTHO(1,2,3,4-def)CHRYSENE see MMD000
6-METHYLNAPHTHO(1,2,3,4-def)CHRYSENE see MMD250
METHYLNAPHTHOHYDROQUINONE see MMC250
2-METHYL-1,4-NAPHTHOHYDROQUINONE see MMC250
2-METHYL-1,4-NAPHTHOQUINOL see MMC250
2-METHYL-1,4-NAPHTHOQUINONE see MMD500
3-METHYL-1,4-NAPHTHOQUINONE see MMD500
2-METHYL-1,4-NAPHTHOQUINONE, SODIUM BISULFITE, TRIHYDRATE
see MMD750
3-METHYL-2-NAPHTHYLAMINE see MME500
N-METHYL-1-NAPHTHYLAMINE see MME250
3-METHYL-2-NAPHTHYLAMINE HYDROCHLORIDE see MME750
N-METHYL-1-NAPHTHYL-CARBAMAT (GERMAN) see CBM750
N-METHYL NAPHTHYLCARBAMATE see MME800
N-METHYL-1-NAPHTHYL CARBAMATE see CBM750
N-METHYL-α-NAPHTHYLCARBAMATE see CBM750
N-METHYL-N-(1-NAPHTHYL)FLUOROACETAMIDE see MME809
METHYL 1-NAPHTHYL KETONE see ABC475
METHYL-2-NAPHTHYL KETONE see ABC500
α-METHYL NAPHTHYL KETONE see ABC475
METHYL α-NAPHTHYL KETONE see ABC475
β-METHYL NAPHTHYL KETONE see ABC500
METHYL-β-NAPHTHYL KETONE (FCC) see ABC500
N-METHYL-N-(1-NAPHTHYL)MONOFLUOROACETAMIDE see MME809
N-METHYL-α-NAPHTHYLURETHAN see CBM750
METHYL NICOTINATE see NDV000
METHYL NIRAN see MNH000
METHYLNITRAMINE see NHN500
METHYL NITRATE see MMF500
2-METHYL-1-NITRATODIMERCURIO-2-NITRATOMERCURIO PROPANE
see MMF600
1-METHYLNITRAZEPAM see DLV000
METHYL NITRITE see MMF750
1-METHYL-4-NITRO-5-(2′-AMINO-6′-PURINYL)MERCAPTOIMIDAZIDE
see AKY250
4-METHYL-3-NITROANILINE see NMP000
6-METHYL-3-NITROANILINE see NMP500
N-METHYL-4-NITROANILINE see MMF800
2-METHYL-1-NITRO-9,10-ANTHRACENEDIONE see MMG000
2-METHYL-1-NITROANTHRAQUINONE see MMG000
2-METHYLNITROBENZENE see NMO525
3-METHYLNITROBENZENE see NMO500
4-METHYLNITROBENZENE see NMO550
m-METHYLNITROBENZENE see NMO500
o-METHYLNITROBENZENE see NMO525
p-METHYLNITROBENZENE see NMO550
2-METHYL-5-NITRO-BENZENEAMINE see NMP500
9-METHYL-ω-(p-NITROBENZENEAZO)-3,4-BENZACRIDINE see NIK500
METHYL-2-NITROBENZENE DIAZOATE see MMH000
2-METHYL-5-NITROBENZENESULFONIC ACID see MMH250
4-METHYL-3-NITROBENZENE SULFONIC ACID see MMH400
2-METHYL-5-NITROBENZENESULFONYL CHLORIDE see MMH500

1-METHYL-2-NITROBENZIMIDAZOLE see MMH740
METHYL-p-NITROBENZOATE see MMI000
3-METHYL-2-NITROBENZOYL CHLORIDE see MMI250
3-METHYL-4-NITRO-1-BUTEN-3-YL ACETATE see MMI640
3-METHYL-4-NITRO-2-BUTEN-1-YL ACETATE see MMI650
1-METHYL-7-NITRO-5-(2-FLUOROPHENYL)-3H-1,4-BENZODIAZEPIN-2(1H)-
  ONE see FDD100
4-METHYL-1-((5-NITROFURFURYLIDENE)AMINO-2-IMIDAZOLIDINONE
  see MMJ000
3-METHYL-4-(5'-NITROFURYLIDENE-AMINO)-TETRAHYDRO-4H-1,4-THIA-
  ZINE-1,1-DIOXIDE see NGG000
5-METHYL-3-(5-NITRO-2-FURYL)ISOXAZOLE see MMJ950
N-METHYL-3-(5-NITRO-2-FURYL)-N-NITROSO-1H-1,2,4-TRIAZOL-5-AMINE
  see MMU000
5-METHYL-3-(5-NITRO-2-FURYL)PYRAZOLE see MMJ960
2-METHYL-4-(5-NITRO-2-FURYL)THIAZOLE see MMJ975
N-(1-METHYL-3-(5-NITRO-2-FURYL)-s-TRIAZOL-5-YL-ACETAMIDE see
  MMK000
N-(1-METHYL-3-(5-NITRO-2-FURYL)-1H-1,2,4-TRIAZOL-5-YL)ACETAMIDE
  see MMK000
N-METHYL-N-NITROGLYCINE see NJH500
1-METHYL-3-NITROGUANIDINE see MML250
N-METHYL-N'-NITROGUANIDINE see MML250
1-METHYL-3-NITROGUANIDINE mixed with SODIUM NITRITE (1:1)
  see MML500
1-METHYL-3-NITROGUANIDINIUM NITRATE see MML550
1-METHYL-3-NITROGUANIDINIUM PERCHLORATE see MML575
1-METHYL-2-NITROIMIDAZOLE see MML750
1-METHYL-4-NITROIMIDAZOLE see MMM000
2-METHYL-5-NITROIMIDAZOLE see MMM500
4-METHYL-5-NITROIMIDAZOLE see MMM750
1-METHYL-4-NITRO-1H-IMIDAZOLE see MMM000
2-METHYL-5-NITRO-1H-IMIDAZOLE see MMM500
4-METHYL-5-NITRO-1H-IMIDAZOLE see MMM750
2-METHYL-5-NITROIMIDAZOLE-1-ETHANOL see MMN250
1-METHYL-5-NITROIMIDAZOLE-2-METHANOL see MMN500
1-METHYL-5-NITROIMIDAZOLE-2-METHANOL CARBAMATE (ESTER)
  see MMN750
1-METHYL-5-NITRO-1H-IMIDAZOLE-2-METHANOL CARBAMATE ESTER
  see MMN750
4-((E)-2-(1-METHYL-5-NITRO-1H-IMIDAZOL-2-YL)-AETHENYL)-2-
  PYRIMIDINAMIN (GERMAN) see TJF000
4-((E)-2-(1-METHYL-5-NITRO-1H-IMIDAZOL-2-YL)-ETHENYL)-2-
  PYRIMIDINAMINE see TJF000
METHYLNITROIMIDAZOLYLMERCAPTOPURINE see ASB250
6-(1'-METHYL-4'-NITRO-5'-IMIDAZOLYL)-MERCAPTOPURINE see ASB250
1-(1-METHYL-5-NITRO-1H-IMIDAZOL-2-YL)-3-(METHYLSULFONYL)-2-
  IMIDAZOLIDINO NE see SAY950
6-(METHYL-p-NITRO-5-IMIDAZOLYL)-THIOPURINE see ASB250
6-((1-METHYL-4-NITROIMIDAZOL-5-YL)THIO)PURINE see ASB250
6-(1-METHYL-4-NITROIMIDAZOL-5-YLTHIO)PURINE see ASB250
6-(1-METHYL-p-NITRO-5-IMIDAZOLYL)-THIOPURINE see ASB250
6-((1-METHYL-4-NITRO-1H-IMIDAZOL-5-YL)THIO)-1H-PURINE see ASB250
METHYLNITROLIC ACID see NHY250
3-METHYL-4-NITRO-1-(p-NITROPHENYL)-2-PYRAZOLIN-5-ONE see PIE000
METHYLNITRONITROSOGUANIDINE see MMP000
1-METHYL-3-NITRO-1-NITROSOGUANIDINE see MMP000
N-METHYL-N'-NITRO-N-NITROSOGUANIDINE see MMP000
N-METHYL-N'-NITRO-N-NITROSOGUANIDINE, not exceeding 25 grams in one
  outside packaging (DOT) see MMP000
5-METHYL-2-NITRO-7-OXA-8-MERCURABICYCLO(4.2.0)OCTA-1,3,5-TRIENE
  see NHK900
4-METHYL-2-NITROPHENOL see NFU500
1-((4-METHYL-2-NITROPHENYL)AZO)-2-NAPHTHALENOL see MMP100
1-METHYL-7-NITRO-5-PHENYL-1,3-DIHYDRO-2H-1,4-BENZODIAZEPIN-2-ONE
  see DLV000
METHYLNITROPHOS see DSQ000
2-METHYL-2-NITROPROPANE see NFQ500
2-METHYL-2-NITROPROPANE-1,3-DIOL see NHO500
N-(2-METHYL-2-NITROPROPYL)-p-NITROSOANILINE see NHK800
N-(2-METHYL-2-NITROPROPYL)-4-NITROSOBENZAMINE see NHK800
2-METHYL-4-NITROPYRIDINE-1-OXIDE see MMP500
3-METHYL-4-NITROPYRIDINE-1-OXIDE see MMP750
2-METHYL-4-NITROQUINOLINE-1-OXIDE see MMQ250
3-METHYL-4-NITROQUINOLINE-1-OXIDE see MMQ500
5-METHYL-4-NITROQUINOLINE-1-OXIDE see MMQ750
6-METHYL-4-NITROQUINOLINE-1-OXIDE see MMR000
7-METHYL-4-NITROQUINOLINE-1-OXIDE see MMR250
8-METHYL-4-NITROQUINOLINE-1-OXIDE see MMR500
1-(4-N-METHYL-N-NITROSAMINOBENZYLIDENE)INDENE see MMR750
1-(METHYLNITROSAMINO)-2-BUTANONE see MMR800
4-(METHYLNITROSAMINO)-2-BUTANONE see MMR810
1-N-METHYL-N-NITROSAMINO-1-DEOXY-d-GLUCITOLE see DAS400
1-(METHYLNITROSAMINO)ETHYL ACETATE see ABR500

METHYLNITROSAMINOMETHYL-d3 ESTER ACETIC ACID see MMS000
2-METHYLNITROSAMINO-2-METHYLPENTANON(4) (GERMAN) see MMX750
3-(METHYLNITROSAMINO)-1,2-PROPANEDIOL see NMV450
3-METHYLNITROSAMINOPROPIONITRILE see MMS200
6-(METHYLNITROSAMINO)PURINE see MMT250
2-(METHYLNITROSAMINO)PYRIDINE see NKQ000
γ-(METHYLNITROSAMINO)-3-PYRIDINEBUTYRALDEHYDE see MMS250
4-(N-METHYL-N-NITROSAMINO)-4-(3-PYRIDYL)BUTANAL see MMS250
4-(N-METHYL-N-NITROSAMINO)-1-(3-PYRIDYL)-1-BUTANONE see MMS500
4-(4-N-METHYL-N-NITROSAMINOSTYRYL)QUINOLINE see MMS750
METHYLNITROSOACETAMID (GERMAN) see MMT000
METHYLNITROSOACETAMIDE see MMT000
N-METHYL-N-NITROSOACETAMIDE see MMT000
1-METHYL-1-NITROSOACETYLUREA see ACR400
N-METHYL-N-NITROSO-N'-ACETYLUREA see ACR400
N-METHYL-N-NITROSOADENINE see MMT250
N⁶-(METHYLNITROSO)ADENOSINE see MMT300
N-METHYL-N-NITROSOALLYLAMINE see MMT500
2-(N-METHYL-N-NITROSO)AMINOACETONITRILE see MMT750
α-METHYLNITROSOAMINOBENZYL ALCOHOL ACETATE (ESTER)
  see NKP000
4-(METHYLNITROSOAMINO)BUTYRIC ACID see MIF250
1-(N-METHYL-N-NITROSOAMINO)-1-DEOXY-d-GLUCITOL see DAS400
1-N-METHYL-N-NITROSOAMINO-1-DESOXY-d-GLUCIT (GERMAN) see DAS400
2-(METHYLNITROSOAMINO)ETHANOL see NKU350
α-(1-(N-METHYL-N-NITROSOAMINO)ETHYL)BENZYL ALCOHOL see NKC000
5-(N-METHYL-N-NITROSO)AMINO-3-(5-NITRO-2-FURYL)-s-TRIAZOLE
  see MMU000
2-(N-METHYL-N-NITROSOAMINO)-1-PHENYL-1-PROPANOL see NKC000
1-(METHYLNITROSOAMINO)-2-PROPANOL see NKU500
1-(METHYLNITROSOAMINO)2-PROPANONE see NKV000
N-METHYL-N-NITROSO-2-AMINOPYRIDINE see NKQ000
Γ-(METHYLNITROSOAMINO)-3-PYRIDINEBUTANAL see MMS250
4-(N-METHYL-N-NITROSOAMINO)-4-(3-PYRIDYL)-1-BUTANONE see MMS500
N-METHYL-N-NITROSOANILINE see MMU250
N-METHYL-N-NITROSOBENZAMIDE see MMU500
N-METHYL-N-NITROSOBENZENAMINE see MMU250
o-METHYLNITROSOBENZENE see NLW500
1-METHYL-2-NITROSO-BENZENE (9CI) see NLW500
2-METHYL-N-NITROSO-BENZIMIDAZOLE CARBAMATE and SODIUM NI-
  TRITE (1:1) see SIQ700
N-METHYL-N-NITROSO-N'-(2-BENZOTHIAZOLYL)-HARNSTOFF (GERMAN)
  see NKR000
N-METHYL-N-NITROSO-N'-(2-BENZOTHIAZOLYL)-UREA see NKR000
N-METHYL-N-NITROSOBENZYLAMINE see MHP250
1-METHYL-1-NITROSOBIURET see MMV000
N-METHYL-N-NITROSOBIURET see MMV000
1-METHYL-1-NITROSO-3-(p-BROMOPHENYL)UREA see BNX125
N-METHYL-N-NITROSOBUTYLAMINE see MHW500
METHYLNITROSOCARBAMIC ACID-o-CHLOROPHENYL ESTER see MMV250
METHYLNITROSOCARBAMIC ACID-2,3-DIHYDRO-2,3-DIMETHYL-7-
  BENZOFURANYL ESTER see NJQ500
METHYLNITROSO-CARBAMIC ACID-2,3-DIHYDRO-2,2-DIMETHYL-3-
  HYDROXY-7-BENZOFURANYL ESTER see HMZ100
METHYLNITROSOCARBAMIC ACID-3,4-DIMETHYLPHENYL ESTER
  see DTN400
METHYLNITROSOCARBAMIC ACID-o-(1,3-DIOXOLAN-2-YL)PHENYL ESTER
  see MMV500
N-METHYL-N-NITROSOCARBAMIC ACID, ETHYL ESTER see MMX250
METHYLNITROSOCARBAMIC ACID-α-(ETHYLTHIO)-o-TOLYL ESTER
  see MMV750
METHYLNITROSOCARBAMIC ACID o-ISOPROPOXYPHENYL ESTER
  see PMY310
METHYLNITROSOCARBAMIC ACID-o-ISOPROPYLPHENYL ESTER
  see MMW250
METHYLNITROSOCARBAMIC ACID-3-METHYLPHENYL ESTER see MNV500
METHYL-NITROSOCARBAMIC ACID-1-NAPHTHYL ESTER see NBJ500
N-METHYL-N-NITROSOCARBAMIC ACID-m-3-PENTYLPHENYL ESTER
  see PBX750
N-METHYL-N-NITROSOCARBAMIC ACID, PHENYL ESTER see NKV500
N-METHYL-N-NITROSOCARBAMIC ACID, TRIMETHYLPHENYL ESTER
  see NLY500
METHYLNITROSOCARBAMIC ACID 3,4-XYLYL ESTER see XTS000
METHYLNITROSOCARBAMIC ACID-3,5-XYLYL ESTER see MMW750
N⁵-(METHYLNITROSOCARBAMOYL)-l-ORNITHINE see MQY325
N⁵-(N-METHYL-N-NITROSOCARBAMOYL)-l-ORNITHINE see MQY325
Nᐃ-(N-METHYL-N-NITROSOCARBAMOYL)-l-ORNITHINE see MQY325
N-((METHYLNITROSOCARBAMOYL)OXY)-2-METHYLTHIOACETIMIDIC
  ACID see NKX000
N-METHYL-N-NITROSO-N'-CARBAMOYLUREA see MMV000
1-METHYL-1-NITROSO-3-(p-CHLOROPHENYL)UREA see MMW775
METHYLNITROSOCYANAMIDE see MMX000
N-METHYL-N-NITROSOCYCLOHEXYLAMINE see NKT500
N-METHYL-N-NITROSODECYLAMINE see MMX200

1-METHYL-N-NITROSODIETHYLAMINE see ELX500
N-METHYL-N-NITROSO-ETHAMINE see MKB000
N-METHYL-N-NITROSO-ETHENYLAMINE see NKY000
N-METHYL-N-NITROSOETHYLAMINE see MKB000
N-METHYL-N-NITROSOETHYLCARBAMATE see MMX250
N-METHYL-N-NITROSO-β-d-GLUCOSAMINE see MMX500
N-METHYL-N-NITROSO-β-d-GLUCOSYLAMIN (GERMAN) see MMX500
N-METHYL-N-NITROSO-β-d-GLUCOSYLAMINE see MMX500
N-METHYL-N-NITROSOGLYCINE see NLR500
METHYLNITROSOGUANIDINE see MMP000
METHYLNITROSO-HARNSTOFF (GERMAN) see MNA750
N-METHYL-NITROSO-HARNSTOFF (GERMAN) see MNA750
N-METHYL-N-NITROSOHEPTYLAMINE see HBP000
N-METHYL-N-NITROSO-1-HEXANAMINE see NKU400
N-METHYL-N-NITROSOHEXYLAMINE see NKU400
N-METHYL-N-NITROSOLAURYLAMINE see NKU000
N-METHYL-N-NITROSOMETHANAMINE see NKA600
4-METHYL-4-N-NITROSOMETHYLAMINO)-2-PENTANONE see MMX750
2-METHYL-4-NITROSOMORPHOLINE see NKU550
N-METHYL-N-NITROSONITROGUANIDIN (GERMAN) see MMP000
1-METHYL-1-NITROSO-3-NITROGUANIDINE see MMP000
N-METHYL-N-NITROSO-N'-NITROGUANIDINE see MMP000
N-METHYL-N-NITROSOOCTANAMIDE see MMY000
N-METHYL-N-NITROSOOCTYLAMINE see NKU590
2-METHYL-3-NITROSO-1,3-OXAZOLIDINE see NKU600
5-METHYL-3-NITROSO-1,3-OXAZOLIDINE see NKU875
N-METHYL-N-NITROSO-4-OXO-4-(3-PYRIDYL)BUTYL AMINE see MMS500
N-METHYL-N-NITROSOPENTYLAMINE see AOL000
N-METHYL-N-NITROSOPHENETHYLAMINE see MNU250
N-METHYL-N-NITROSO-4-(PHENYLAZO)ANILINE see MMY250
N-METHYL-N-NITROSO-1-PHENYLETHYLAMINE see NKW000
METHYL-(4-NITROSOPHENYL)NITROSAMINE see MJG750
1-METHYL-1-NITROSO-3-PHENYLUREA see MMY500
N-METHYL-N-NITROSO-N'-PHENYLUREA see MMY500
1-METHYL-4-NITROSOPIPERAZINE see NKW500
N'-METHYL-N-NITROSOPIPERAZINE see NKW500
2-METHYLNITROSOPIPERIDINE see NLI000
3-METHYLNITROSOPIPERIDINE see MMY750
4-METHYLNITROSOPIPERIDINE see MMZ000
3-METHYL-1-NITROSO-4-PIPERIDONE see MMZ800
N-METHYL-N-NITROSO-1-PROPANAMINE see MNA000
N-METHYL-N-NITROSO-2-PROPEN-1-AMINE see MMT500
METHYLNITROSO-PROPIONAMIDE see MNA250
N-METHYL-N-NITROSOPROPIONAMIDE see MNA250
METHYL-NITROSOPROPIONSAEUREAMID (GERMAN) see MNA250
METHYLNITROSOPROPIONYLAMIDE see MNA250
N-METHYL-N-NITROSO-1H-PURIN-6-AMINE see MMT250
N⁶-METHYL-N⁶-NITROSO-1H-PURIN-6-AMINE see MMT250
N-METHYL-N-NITROSO-4-(2-(4-QUINOLINYL)ETHENYL)BENZENAMINE
    see MMS750
N⁶-METHYL-N⁶-NITROSO-9b-d-RIBOFURANOSYL-9H-PURIN-6-AMINE
    see MMT300
2-METHYL-N-NITROSOTHIAZOLIDINE see MNA500
METHYLNITROSO-p-TOLUENESULFONAMIDE see THE500
1-METHYL-1-NITROSO-3-(p-TOLYL)UREA see MNA650
N-METHYL-N-NITROSOUNDECYLAMINE see NKX500
METHYLNITROSOUREA see MNA750
1-METHYL-1-NITROSOUREA see MNA750
N-METHYL-N-NITROSOUREA see MNA750
METHYLNITROSOUREE (FRENCH) see MNA750
d-1-(3-METHYL-3-NITROSOUREIDO)-1-DEOXYGALACTOPYRANOSE
    see MNB000
METHYLNITROSOURETHAN (GERMAN) see MMX250
METHYLNITROSOURETHANE see MMX250
N-METHYL-N-NITROSO-URETHANE see MMX250
N-METHYL-N-NITROSOVINYLAMINE see NKY000
4-METHYL-5-(5-NITRO-2-4H-1,2,4-TRIAZOL-3-AMINE see AKY000
2-METHYL-N-(4-NITRO-3-(TRIFLUOROMETHYL)PHENYL)PROPANAMIDE
    (9CI) see FMR050
1-METHYL-2-NITRO-5-VINYL-1H-IMIDAZOLE see MNB250
o-METHYL NOGALAROL see MNB300
7-o-METHYL NOGALAROL see MNB300
7-o-METHYLNOGAROL see MCB600
3-METHYL-2(3)-NONENENITRILE see MNB500
METHYL-2-NONENOATE see MNB750
METHYL NONYLENATE see MNB750
METHYL NONYL KETONE see UKS000
METHYL-n-NONYL KETONE see UKS000
1-METHYL-7-(NONYLOXY)-9H-PYRIDO(3,4-b)INDOLE HYDROCHLORIDE
    see NNC100
METHYL-2-NONYNOATE see MNC000
α-METHYLNORADRENALINE HYDROCHLORIDE see AMB000
N-METHYLNORAPORMORPHINE HYDROCHLORIDE see AQP500
N-METHYL-NORDOCEINE see CNF500

N-METHYLNOREPHEDRINE see EAW000
21-METHYLNORETHISTERONE see MNC100
7(R)-o-METHYLNORGAROL see MCB600
1-METHYLNORHARMAN see MPA050
11-β-METHYL-19-NOR-17-α-PREGNA-1,3,5(10)-TRIEN-20-YNE-3,17-DIOL
    see MJW875
METHYLNORTESTOSTERONE see MDM350
17-METHYL-19-NORTESTOSTERONE see MDM350
17-α-METHYL-19-NORTESTOSTERONE see MDM350
4-METHYLNORVALINE see LES000
METHYL OCTADECANOATE see MJW000
METHYL-9-OCTADECENOATE see OHW000
METHYL (Z)-9-OCTADECENOATE see OHW000
METHYL cis-9-OCTADECENOATE see OHW000
2-METHYLOCTANAL see MNC175
α-METHYLOCTANAL see MNC175
2-METHYLOCTANE see MNC250
3-METHYLOCTANE see MNC500
4-METHYLOCTANE see MNC750
METHYL OCTANOATE and METHYL DECANOATE see MND000
3-METHYL-3-OCTANOL see MND050
3-METHYLOCTAN-3-OL see MND050
7-METHYL-1-OCTANOL see ILJ000
3-METHYL-1-OCTEN-OL see MND100
METHYLOCTENYLAMINE see ILK000
METHYL 2-OCTINATE see MND275
METHYL OCTINE CARBONATE see MNC000
METHYL OCTYL ACETALDEHYDE see MIW000
METHYLOCTYLDIAZENE 1-OXIDE see MGT000
METHYL OCTYNE CARBONATE see MNC000
METHYL 2-OCTYNOATE see MND275
METHYLOESTRENOLONE see MDM350
METHYLOL see MGB150
N-METHYLOLACRYLAMIDE see HLU500
N'-METHYLOL-o-CHLORTETRACYCLINE see MND500
METHYL OLEATE see OHW000
N-METHYL-N-OLEOYLTAURINE SODIUM SALT see SIY000
3-METHYLOLPENTANE see EGW000
2-METHYLOLPHENOL see HMK100
o-METHYLOLPHENOL see HMK100
N-METHYLOLPHTHALIMIDE see HMP100
METHYLOLPROPANE see BPW500
METHYL-4-OMBELLIFERONE SODEE (FRENCH) see HMB000
(METHYL-ONN-AZOXY)METHANOL see HMG000
(METHYL-ONN-AZOXY)METHANOL, ACETATE (ester) see MGS750
(METHYL-ONN-AZOXY)METHYL-β-d-GLUCOPYRANOSIDE
    see COU000
(METHYL-ONN-AZOXY)METHYL-β-D-GLUCOPYRANOSIDURONIC ACID
    see MGS700
METHYL ORANGE see MND600
METHYL ORANGE B see MND600
METHYL ORTHOFORMATE see TLX600
METHYL ORTHOSILICATE see MPI750
9-METHYL-3-OXA-9-AZATRICYCLO(3.3.1.0²,⁴)NONAN-7-OL,TROPATE (ester)
    see SBG000
4-METHYL-7-OXABICYCLO(4.1.0)HEPTANE-3-CARBOXYLIC ACID, ALLYL
    ESTER see AGF500
2-(3-(5-METHYL-1,3,4-OXADIAZOL-2-YL)-3,3-DIPHENYLPROPYL)-2-
    AZABICYCLO(2.2.2)OCTANE see DWF700
METHYLOXAZEPAM see CFY750
N-METHYLOXAZEPAM see CFY750
3-METHYL-2-OXAZOLIDONE see MND750
4-METHYL-2-OXETANONE see BSX000
METHYL OXIRANE see PNL600
METHYLOXIRANE, POLYMER with OXIRANE, MONOBUTYL ESTER
    see MNE250
(2R-cis)-(3-METHYLOXIRANYL)PHOSPHONIC ACID DISODIUM SALT
    see DXF600
METHYL OXITOL see EJH500
METHYL-3-OXOBUTYRATE see MFX250
(5-METHYL-2-OXO-1,3-DIOXOLEN-4-YL)METHYL-d-α-
    AMINOBENZYLPENICILLINATE HYDROCHLORIDE see LEJ500
6-METHYL-2-OXO-1,3-DITHIOLO(4,5-b)QUINOXALINE see ORU000
METHYL OXOETHANOATE see MKI550
METHYL-12-OXO-trans-10-OCTADECENOATE see MNF250
N-(2-(2-METHYL-4-OXOPENTYL))ACRYLAMIDE see DTH200
3-METHYL-4-OXO-2-PHENYL-4H-1-BENZOPYRAN-8-CARBOXYLATE 1-
    PIPERIDINEETHANOL HYDROCHLORIDE see FCB100
3-METHYL-4-OXO-5-PIPERIDINO-Δ²,α-THIAZOLIDINEACETIC ACID ETHYL
    ESTER see MNG000
(3-METHYL-4-OXO-5-PIPERIDINO-2-THIAZOLIDINYLIDENE)ACETIC ACID
    ETHYL ESTER see MNG000
Z-(3-METHYL-4-OXO-5-PIPERIDINO-THIAZOLIDIN-2-YLIDEN)-
    ESSIGSAEUREAETHYLESTER (GERMAN) see MNG000

(3-METHYL-4-OXO-5-(1-PIPERIDINYL)-2-THIAZOLIDINYLIDENE)ACETIC
ACID ETHYL ESTER see MNG000
11-METHYL-1-OXO-1,2,3,4-TETRAHYDROCHRYSENE see MNG250
METHYL p-OXYBENZOATE see HJL500
METHYLOXYLAMMONIUM CHLORIDE see MNG500
4-(METHYLOXYMETHYL)BENZYL CHRYSANTHEMUMMONOCARBOXYL-
ATE see MBV700
4-(METHYLOXYMETHYL)BENZYL CHRYSANTHEMUM MONOCARBOXYL-
ATE see MNG525
4-METHYL-5-OXYMETHYLURACIL see HMH300
1-(4-METHYLOXYPHENYL)-3,3-DIMETHYLTRIAZINE see DSN600
2-METHYL-3-OXY-γ-PYRONE see MAO350
METHYLPARABEN (FCC) see HJL500
METHYLPARAFYNOL CARBAMATE see MNM500
METHYL PARAHYDROXYBENZOATE see HJL500
METHYL PARAOXON see PHD500
METHYL PARASEPT see HJL500
METHYL PARATHION see MNH000
METHYL PARATHION (dry) see MNH010
METHYL PARATHION MIXTURE, liquid see MNH020
N-METHYLPAREDRINE see FMS875
METHYL PCT see DTQ600
METHYL PENTACHLOROPHENATE see MNH250
METHYL(PENTACHLOROPHENOXY)MERCURY see MLG250
METHYL PENTACHLOROPHENYL ESTER see MNH250
METHYLPENTADIENE see MNH500
2-METHYL-1,3-PENTADIENE see MNH750
4-METHYL-1,3-PENTADIENE see MNI000
2-METHYLPENTANE see IKS600
3-METHYLPENTANE see MNI500
3-METHYL PENTANEDIAL see MKI250
2-METHYL-2,4-PENTANEDIAMINE see MNI525
2-METHYL PENTANE-2,4-DIOL see HFP875
2-METHYL-2,4-PENTANEDIOL see HFP875
3-METHYL-1,5-PENTANEDIOL see MNI750
2-METHYL-2,4-PENTANEDIOL ESTER with BORIC ACID (H3BO3) (1:2) cyclic
BIS(1,1,3-TRIMETHYLMETHYLENE) ESTER see TKM250
4-METHYLPENTANENITRILE see MNJ000
2-METHYLPENTANOIC ACID see MQJ750
2-METHYLPENTANOL-1 see AOK750
2-METHYL-4-PENTANOL see MKW600
3-METHYL-3-PENTANOL see MNJ100
4-METHYLPENTANOL-2 see MKW600
4-METHYL-2-PENTANOL (MAK) see MKW600
3-METHYL-PENTANOL-(3) (GERMAN) see MNJ100
4-METHYL-2-PENTANOL, ACETATE see HFJ000
3-METHYL-3-PENTANOL CARBAMATE see ENF000
2-METHYL-2-PENTANOL-4-ONE see DBF750
4-METHYL-2-PENTANON (CZECH) see HFG500
4-METHYL-PENTAN-2-ON (DUTCH, GERMAN) see HFG500
2-METHYL-4-PENTANONE see HFG500
4-METHYL-2-PENTANONE (FCC) see HFG500
3-METHYL-2-n-PENTANYL-2-CYCLOPENTEN-1-ONE see DLQ600
2-METHYL-2-PENTENAL see MNJ750
2-METHYL-2-PENTEN-1-AL see MNJ750
2-METHYLPENTENE see MNK000
2-METHYL-1-PENTENE see MNK000
2-METHYL-PENTENE-1 see MNK000
4-METHYL-1-PENTENE see MNK250
4-METHYL-2-PENTENE see MNK500
trans-4-METHYL-2-PENTENE see MNK750
2-METHYL-2-PENTENE-1-AL see MNJ750
4-METHYL-3-PENTENE-2-ONE see MDJ750
3-METHYL-1-PENTEN-3-OL see MNL250
4-METHYL-2-PENTEN-4-OL see MNL500
3-METHYL-PENTEN-(1)-OL-(3) (GERMAN) see MNL250
4-METHYL-3-PENTEN-2-ON (DUTCH, GERMAN) see MDJ750
2-METHYL-2-PENTEN-4-ONE see MDJ750
4-METHYL-3-PENTEN-2-ONE see MDJ750
4-METHYL-3-PENTEN-2-ONE (1-PHTHALAZINYL)HYDRAZONE see DQU400
3-METHYL-2-PENTEN-4-YN-1-OL see MNL775
3-METHYLPENTIN-3-OL see EQL000
3-METHYL-PENTIN-(1)-OL-(3) (GERMAN) see MNM500
4-METHYL-2-PENTYL ACETATE see HFJ000
2-METHYLPENTYL CARBITOL see DJG200
2-METHYLPENTYL CELLOSOLVE see EJK000
METHYL N-(β-PENTYLCINNAMYLIDENE)ANTHRANILATE see AOH100
3-METHYL-2-PENTYL-2-CYCLOPENTEN-1-ONE see DLQ600
2-METHYL-2,3-PENTYLENE OXIDE see ECC600
METHYL PENTYL KETONE see MGN500
METHYL-N-PENTYLNITROSAMINE see AOL000
2-((2-METHYLPENTYL)OXY)ETHANOL see EJK000
2-(2-((2-METHYLPENTYL)OXY)ETHOXY)ETHANOL see DJG200
3-METHYLPENT-1-YN-3-OL see EQL000

3-METHYL-1-PENTYN-3-OL see EQL000
METHYLPENTYNOL CARBAMATE see MNM500
3-METHYL-1-PENTYN-3-OL CARBAMATE see MNM500
METHYL PERCHLORATE see MNM750
METHYL PERFLUOROMETHACRYLATE see MNN000
METHYLPERIDIDE see FLV000
METHYLPERIDOL see MNN250
METHYLPERIDOL HYDROCHLORIDE see MNN250
METHYLPERONE HYDROCHLORIDE see FKI000
9-METHYL-PGA see BMM125
15(s)-15-METHYL-PGE2 see MOV800
METHYL-PGF2-α see CCC100
15(S)-15-METHYL PGF2-α see CCC100
15-METHYL-PGF2-α-METHYL ESTER see MLO300
15(s)15-METHYL-PGF2-α-METHYL ESTER see MLO300
15(2)15-METHYL PGF2-α TROMETHAMINE SALT see CCC110
dl-N-METHYLPHEDRINE HYDROCHLORIDE see MNN350
1-METHYLPHENANTHRENE see MNN500
2-METHYLPHENANTHRO(2,1-d)THIAZOLE see MNO250
5-METHYLPHENAZINE METHYLSULFATE see MNO500
N-METHYLPHENAZIUM METHOSULFATE see MNO500
5-METHYLPHENAZIUM METHYL SULFATE see MNO500
METHYLPHENAZONIUM METHOSULFATE see MRW000
N-METHYLPHENAZONIUM METHOSULFATE see MNO500
5-N-METHYLPHENAZONIUM METHOSULFATE see MNO500
N-METHYLPHENAZONIUM METHOSULPHATE see MNO500
METHYL PHENCAPTON see MNO750
2-METHYLPHENELZINE see MNO775
α-METHYLPHENETHYLAMINE see AOA250
(±)-α-METHYLPHENETHYLAMINE see BBK000
dl-α-METHYLPHENETHYLAMINE see BBK000
α-METHYL-PHENETHYLAMINE compounded with AMBERLITE XE-69
see AOB000
α-METHYLPHENETHYLAMINE, d-FORM see AOA500
dl-α-METHYL-PHENETHYLAMINE HYDROCHLORIDE see AOA750
dl-α-METHYL-PHENETHYLAMINE PHOSPHATE see AOB500
α-METHYLPHENETHYLAMINE PHOSPHATE, dl-MIXTURE see AOB500
d-α-METHYLPHENETHYLAMINE SULFATE see BBK500
(±)-α-METHYLPHENETHYLAMINE SULFATE see AOB250
dl-α-METHYLPHENETHYLAMINE SULFATE see BBK250
2-(α-METHYLPHENETHYL)AMINOETHANOL see PFI750
2-METHYL-2-(2-(PHENETHYLAMINO)ETHYL)-1,3-BENZODIOXOLE HYDRO-
CHLORIDE see MNP250
4-(2-(α-METHYLPHENETHYLAMINO)ETHYL)-3-PHENYL-1,2,4-OXADIAZOLE
MONOHYDROCHLORIDE see WBA600
7-(2-((α-METHYLPHENETHYL)AMINO)ETHYL)THEOPHYLLINE HYDRO-
CHLORIDE see CBF825
7-(2-(1-METHYL-2-PHENETHYLAMINO)ETHYL)THEOPHYLLINE HYDRO-
CHLORIDE see CBF825
METHYL PHENETHYL ETHER see PFD325
METHYL 2-PHENETHYL ETHER see PFD325
(o-METHYLPHENETHYL)HYDRAZINE see MNO775
(α-METHYLPHENETHYL)HYDRAZINE see PDN000
METHYL PHENETHYL KETONE see PDF800
1-(α-METHYLPHENETHYL)-2-(5-METHYL-3-ISOXAZOLYLCARBONYL)
HYDRAZINE see MNP300
1-(α-METHYLPHENETHYL)-2-PHENETHYLHYDRAZINE see MNP400
3-(α-METHYLPHENETHYL)SYDONE IMINE MONOHYDROCHLORIDE
see SPA000
2-(p-METHYLPHENETHYL)-3-THIOSEMICARBAZIDE see MNP500
METHYLPHENIDAN see MNQ000
METHYL PHENIDATE see MNQ000
METHYLPHENIDATE HYDROCHLORIDE see RLK000
METHYL PHENIDYL ACETATE see MNQ000
METHYLPHENIDYLACETATE HYDROCHLORIDE see RLK000
2-METHYLPHENISOPROPYLAMINE SULFATE see MNQ250
4-METHYLPHENISOPROPYLAMINE SULFATE see AQQ000
METHYLPHENOBARBITAL see ENB500
1-METHYLPHENOBARBITAL see ENB500
N-METHYLPHENOBARBITAL see ENB500
METHYLPHENOBARBITONE see ENB500
2-METHYLPHENOL see CNX000
3-METHYLPHENOL see CNW750
4-METHYLPHENOL see CNX250
m-METHYLPHENOL see CNW750
o-METHYLPHENOL see CNX000
p-METHYLPHENOL see CNX250
N-METHYLPHENOLBARBITOL see ENB500
4-METHYLPHENOL METHYL ETHER see MGP000
10-METHYLPHENOTHIAZINE-2-ACETIC ACID see MNQ500
N-METHYL-3-PHENOTHIAZINYLACETIC ACID see MNQ500
(10-METHYL-2-PHENOTHIAZINYL)ACETIC ACID see MNQ500
2-(2-(4-(2-METHYL-3-PHENOTHIAZIN-10-YLPROPYL)-1-
PIPERAZINYL)ETHOXY)ETHANOL see MKQ000

METHYL-3-PHENYLPROPENOATE see MIO500
2-(p-METHYLPHENYL)PROPIONALDEHYDE see THD750
1-METHYL-4-PHENYL-4-PROPIONOXYPIPERIDINE HYDROCHLORIDE
   see MNW775
1-METHYL-4-PHENYL-4-PROPIONOXYPIPERIDINE N-OXIDE HYDROCHLO-
   RIDE see MNW780
1-(4-METHYLPHENYL)-2-PROPYLAMINE SULFATE see AQQ000
N-(2-METHYLPHENYL)-2-(PROPYLAMINO)-PROPANAMIDE
   MONOHYDROCHLORIDE see CMS250
2-METHYL-1-PHENYL-2-PROPYL HYDROPEROXIDE see MNX000
N-METHYL-N-(3-PHENYLPROPYL)-2-(PYRROLIDINYL)ACETAMIDE HYDRO-
   CHLORIDE see MNX260
N-METHYL-N-(1-PHENYL-2-PROPYL)-2-(PYRROLIDINYL)ACETAMIDE
   HYDROCHLORIDE see MNX250
1-METHYL-1-PHENYL-2-PROPYNYL CYCLOHEXANECARBAMATE
   see MNX850
1-METHYL-1-PHENYL-2-PROPYNYL-N-CYCLOHEXYLCARBAMATE
   see MNX850
3-METHYL-1-PHENYL-1H-PYRAZOLE-5-AMINE see PFQ350
3-METHYL-1-PHENYL-PYRAZOLIN-5-ONE see NNT000
3-METHYL-1-PHENYL-5-PYRAZOLONE see NNT000
3-METHYL-1-PHENYL PYRAZOL-5-YL DIMETHYL CARBAMATE see PPQ625
3-METHYL-1-PHENYL-5-PYRAZOLYL DIMETHYL CARBAMATE see PPQ625
trans-1-(4'-METHYLPHENYL)-1-(2'-PYRIDYL)-3-PYRROLIDINOPROP-1-N E HY-
   DROCHLORIDE see TMX775
1-METHYL-3-PHENYLPYRROLIDIN-2,5-DIONE see MNZ000
1-METHYL-3-PHENYL-2,5-PYRROLIDINEDIONE see MNZ000
5-METHYL-1-PHENYL-2-(PYRROLIDINYL)IMIDAZOLE see MNY750
METHYL-5 PHENYL-1 (PYRROLIDINYL-1)-2 IMIDAZOLE (FRENCH)
   see MNY750
(E)-2-(1-(4-METHYLPHENYL)-3-(1-PYRROLIDINYL)-1-PROPENYL)-PYRIDINE
   MONOHYDROCHLORIDE see TMX775
4-METHYLPHENYL SALICYLATE see THD850
METHYLPHENYLSUCCINIMIDE see MNZ000
N-METHYL-2-PHENYL-SUCCINIMIDE see MNZ000
N-METHYL-α-PHENYLSUCCINIMIDE see MNZ000
METHYL 5-(PHENYLSULFINYL)-2-BENZIMIDAZOLECARBAMATE
   see OMY500
METHYL (5-PHENYLSULFINYL)-1H-BENZIMIDAZOL-2-YL CARBAMATE
   see OMY500
3-METHYL-1-PHENYL-3-(2-SULFOETHYL)TRIAZENE SODIUM SALT
   see PEJ250
p-METHYLPHENYLSULFONIC ACID see TGO000
5-METHYL-1-PHENYL-3,4,5,6-TETRAHYDRO-1H-2,5-BENZOXAZOCINE
   HYDROCHLORIDE see NBS500
2-METHYL-9-PHENYL-2,3,4,9-TETRAHYDRO-1H-INDENO(2,1-c)PYRIDINE
   TARTRATE see PDD000
3-METHYL-2-PHENYLTETRAHYDRO-2H-1,4-OXAZINE HYDROCHLORIDE
   see MNV750
2-METHYL-9-PHENYL-2,3,4,9-TETRAHYDRO-1-PYRIDINDENE HYDROCHLO-
   RIDE see TEQ700
2-METHYL-9-PHENYL-2,3,4,9-TETRAHYDRO-1-PYRIDINDENE TARTRATE
   see PDD000
N-METHYL-N'-(4'-PHENYL-THIAZOLYL(2'))-HARNSTOFF (GERMAN)
   see MOA000
1-METHYL-3-(4-PHENYL-2-THIAZOLYL)UREA see MOA000
N-METHYL-N'-(4-PHENYL-2-THIAZOLYL)UREA see MOA000
N-METHYL-N'-(4'-PHENYL-THIAZOLYL(2'))-UREA see MOA000
1-METHYL-4-(PHENYLTHIO)PYRIDINIUM IODIDE see MOA250
N-METHYL-N'-PHENYL THIOUREA see MOA500
1-(1-METHYL-2-((α-PHENYL-o-TOLYL)OXY)ETHYL)PIPERIDINE see MOA600
3-METHYL-1-PHENYLTRIAZENE see MOA725
4-METHYL-5-PHENYL-2-TRIFLUOROMETHOXAZOLIDINE see MOA750
3-METHYLPHENYLUREA see THF750
4-METHYLPHENYLUREA see THG000
3-METHYL-2-PHENYLVALERIC ACID-2-DIETHYLAMINOETHYL ESTER
   METHYL BROMIDE see VBK000
3-METHYL-2-PHENYLVALERIC ACID DIETHYL(2-
   HYDROXYETHYL)METHYLAMMONIUM BROMIDE ESTER see VBK000
2-(((3-METHYL-2-PHENYLVALERYL)OXY)-N,N-DIETHYL-N-
   METHYLETHANAMINIUM BROMIDE see VBK000
METHYL PHOSPHATE see DLO800, TMD250
o-METHYLPHOSPHATE see DLO800
METHYL PHOSPHINE see MOB000
METHYL PHOSPHITE see TMD500
METHYLPHOSPHODITHIOIC ACID-S-(((p-CHLOROPHENYL)THIO)METHYL)-
   O-METHYL ESTER see MOB250
METHYLPHOSPHONIC ACID, (2-(BIS(1-METHYLETHYL)AMINO)ETHYL)
   ETHYL ESTER see MOB275
METHYLPHOSPHONIC ACID DIMETHYL ESTER see DSR400
METHYL PHOSPHONIC DICHLORIDE see MOB399
METHYLPHOSPHONIC DIFLUORIDE see MJD275
METHYLPHOSPHONODITHIOIC ACID O-METHYL ESTER, S-ESTER with 2-
   MERCAPTO-N-METHYLACETAMIDE see MOB500

METHYLPHOSPHONOFLUORIDIC ACID, 3,3-DIMETHYL-2-BUTYL ESTER
   see SKS500
METHYLPHOSPHONOFLUORIDIC ACID ISOPROPYL ESTER see IPX000
METHYLPHOSPHONOFLUORIDIC ACID-1-METHYLETHYL ESTER see IPX000
METHYLPHOSPHONOFLUORIDIC ACID 1,2,2-TRIMETHYLPROPYL ESTER
   see SKS500
METHYLPHOSPHONOTHIOIC ACID-S-(2-
   (BIS(METHYLETHYL)AMINO)-ETHYL ESTER see EIG000
METHYLPHOSPHONOTHIOIC ACID-O,S-DIETHYL ESTER see DJR700
METHYLPHOSPHONOTHIOIC ACID-O-ETHYL O-(p-(METHYLTHIO)
   PHENYL)ESTER see MOB599
METHYLPHOSPHONOTHIOIC ACID-O-ETHYL O-(4-(METHYLTHIO)
   PHENYL)ESTER (9CI) see MOB599
METHYLPHOSPHONOTHIOIC ACID-O-(4-NITROPHENYL)-O-PHENYL
   ESTER see MOB699
METHYLPHOSPHONOTHIOIC ACID-O-(p-NITROPHENYL)-O-PHENYL
   ESTER see MOB699
METHYLPHOSPHONOTHIOIC ACID-O-PHENYL ESTER, O-ESTER with p-
   HYDROXYBENZONITRILE see MOB750
METHYL PHOSPHONOTHIOIC DICHLORIDE anhydrous see MOC000
METHYLPHOSPHONOUS DICHLORIDE see MOC250
METHYLPHOSPHONYLDIFLUORIDE see MJD275
METHYLPHOSPHORAMIDIC-2-CHLORO-4-(1,1-DIMETHYLPROPYL)PHENYL
   METHYL ESTER see DYE200
METHYL-PHOSPHORAMIDOTHIOIC ACID o-(tert-BUTYL-2-
   CHLOROPHENYL)ESTER see BQU500
N-METHYL-PHOSPHORAMIDOTHIOIC ACID-O-ISOPROPYL ESTER
   see DYD800
METHYLPHOSPHOROTHIOATE see DUG500
METHYL PHTHALATE see DTR200
METHYL-4-PHTHALIMIDO-dl-GLUTARAMATE see MOC500
N-METHYL-2-PHTHALIMIDOGLUTARIMIDE see MOC750
METHYL PHTHALYL ETHYL GLYCOLATE see MOD000
2-METHYL-3-PHYTHYL-1,4-NAPHTHOCHINON (GERMAN) see VTA000
METHYL PINACOLYLOXY PHOSPHORYLFLUORIDE see SKS500
METHYL PINACOLYL PHOSPHONOFLUORIDATE see SKS500
N-METHYL-2-PIPECOLIC ACID, 2,6-DIMETHYLANILIDE see SBB000
N-METHYL-2-PIPECOLIC ACID, 2,6-XYLIDIDE see SBB000
(±)-1-METHYL-2',6'-PIPECOLOXYLIDIDE see SBB000
dl-1-METHYL-2',6'-PIPECOLOXYLIDIDE HYDROCHLORID see CBR250
1-METHYL-2',6'-PIPECOLOXYLIDIDE HYDROCHLORIDE see CBR250
1-METHYLPIPERAZINE see MOD250
N-METHYLPIPERAZINE see MOD250
4-METHYL-1-PIPERAZINEACETIC, ACID ((5-NITRO-2-FURANYL)METHY-
   LENE)HYDRAZIDE, ACETATE see NDY360
4-METHYL-1-PIPERAZINEACETIC ACID (5-
   NITROFURFURYLIDENE)HYDRAZIDE see NDY350
4-METHYL-1-PIPERAZINEACETIC, ACID (5-
   NITROFURFURYLIDENE)HYDRAZIDE ACETATE see NDY360
4-METHYL-1-PIPERAZINEACETIC, ACID (5-
   NITROFURFURYLIDENE)HYDRAZIDE DIHYDROCHLORIDE see NDY370
4-METHYL-1-PIPERAZINEACETIC, ACID (5-
   NITROFURFURYLIDENE)HYDRAZIDE HYDROCHLORIDE see NDY380
4-METHYL-1-PIPERAZINEACETIC ACID, (5-
   NITROFURFURYLIDENE)HYDRAZIDE MALEATE (1:1) see NDY390
4-METHYL-1-PIPERAZINEACETIC ACID 2,6-XYLYL ESTER DIHYDROCHLO-
   RIDE see FAC060
4-METHYL-PIPERAZINO-ACETOHYDRAZONE-5-NITROFURFUROL see
   NDY350
10-(Γ-(N'-METHYLPIPERAZINO)PROPYL)-2-
   TRIFLUOROMETHYLPHENOTHIOZINE see TKE500
(4-METHYLPIPERAZINYLACETYL)HYDRAZONE-5-NITRO-2-FURALDEHYDE
   see NDY350
N-METHYL-PIPERAZINYL-N'-AETHYL-PHENOTHIAZIN (GERMAN)
   see MOE250
p-(5-(5-(4-METHYL-1-PIPERAZINYL)-2-BENZIMIDAZOLYL)-2-
   BENZIMIDAZOLYL)-PHENOL TRIHYDROCHLORIDE see MOD500
4-(5-(4-METHYL-1-PIPERAZINYL)(2,5'-BI-1H-BENZIMIDAZOL)-2'-YL)-PHENOL
   TRIHYDROCHLORIDE see MOD500
4-(4-METHYL-1-PIPERAZINYLCARBONYL)-1-PHENYL-2-PYRROLIDINONE
   HYDROCHLORIDE see MOD750
6-(4-METHYL-1-PIPERAZINYL)-11H-DIBENZ(b,e)AZEPINE see HOU059
10-(2-(4-METHYL-1-PIPERAZINYL)ETHYL)PHENOTHIAZINE see MOE250
3-(4-METHYLPIPERAZINYLIMINOMETHYL)-RIFAMYCIN SV see RKP000
8-(4-METHYLPIPERAZINYLIMINOMETHYL) RIFAMYCIN SV see RKP000
8-(((4-METHYL-1-PIPERAZINYL)IMINO)METHYL)RIFAMYCIN SV see RKP000
2-(4-METHYL-1-PIPERAZINYL)-10-METHYL-3,4-
   DIAZAPHENOXAZINDIHYDROCHLORID (GERMAN) see ASC250
6-(4-METHYL-1-PIPERAZINYL)MORPHANTHRIDINE see HOU059
N-(Γ-(4'-METHYLPIPERAZINYL-1')PROPYL)-3-CHLOROPHENOTHIAZINE
   see PMF500
N-METHYLPIPERAZINYL-N'-PROPYL-PHENOTHIAZIN (GERMAN)
   see PCK500
N-(3-(4-METHYL-1-PIPERAZINYL)PROPYL)PHENOTHIAZINE see PCK500

10-(3-(4-METHYL-1-PIPERAZINYL)PROPYL)-10H-PHENOTHIAZINE (9CI) see PCK500

1-(10-(3-(4-METHYL-1-PIPERAZINYL)PROPYL)PHENOTHIAZIN-2-YL)-1-BUTA-NONE DIMALEATE see BSZ000

10-(3-(4-METHYL-1-PIPERAZINYL)PROPYL)-2-(TRIFLUOROMETHYL) PHENO-THIAZINE see TKE500

10-(3-(4-METHYL-1-PIPERAZINYL)PROPYL)-2-TRIFLUOROMETHYLPHENOTHIAZINE DIHYDROCHLORIDE see TKK250

4-(4-METHYL-1-PIPERAZINYL)-5,6,7,8-TETRAHYDRO-(1)-BENZOTHIENO(2,3-d) PYRIMIDINE HYDROCHLORIDE see MOF750

4-(4-METHYL-1-PIPERAZINYL)THIENO(2,3-d) PYRIMIDINE HYDROCHLO-RIDE see MOG000

2-(4-METHYL-1-PIPERAZINYL)-11-(p-TOLYL)-10,11-DIHYDROPYRIDAZINO(3,4-b)(1,4)BENZOXAZEPINE see MOG250

2-METHYLPIPERIDINE see MOG750

3-METHYLPIPERIDINE see MOH000

4-METHYLPIPERIDINE see MOH250

N-METHYLPIPERIDINE see MOG500

1-METHYLPIPERIDINE (DOT) see MOG500

1-METHYL-PIPERIDINE-4-CARBONSAURE-O,O-XYLIDID HYDROCHLORID (GERMAN) see MQO250

(1-METHYL-dl-PIPERIDINE-2-CARBOXYLIC ACID)-2,6-DIMETHYLANILIDE HYDROCHLORIDE see CBR250

γ-(4-METHYLPIPERIDINE)-p-FLUOROBUTYROPHENONE HYDROCHLORIDE see FKI000

2-METHYL-1-PIPERIDINEPROPANOL BENZOATE HYDROCHLORIDE see IJZ000

β-4-METHYLPIPERIDINOETHYL BENZOATE HYDROCHLORIDE see MOI250

2-METHYL-2-(2-PIPERIDINOETHYL)-1,3-BENZODIOXOLE HYDROCHLORIDE see MOI500

N-(1-METHYL-2-PIPERIDINOETHYL)-N-2-PYRIDYLPROPIONAMIDE FUMAR-ATE see PMX250

(±)-6-((2-METHYLPIPERIDINO)METHYL)-5-INDANOL MALEATE see PJI600

β-METHYL-4-PIPERIDINOPHENETHYLAMINE DIHYDROCHLORIDE see MOJ500

2-METHYL-2-(4-PIPERIDINOPHENYL)ETHYLAMINE DIHYDROCHLORIDE see MOJ500

2-METHYL-1-PIPERIDINOPROPANOL, BENZOATE see PIV750

METHYL-4-(3-PIPERIDINOPROPIONYLAMINO)SALICYLATE, METHIODIDE see MOK000

γ-3-METHYLPIPERIDINOPROPYL-p-AMINOBENZOATE HYDROCHLORIDE see MOK500

(2-METHYLPIPERIDINO)PROPYL BENZOATE see PIV750

3-(2-METHYLPIPERIDINO)PROPYL BENZOATE HYDROCHLORIDE see IJZ000

Γ-(2-METHYLPIPERIDINO)PROPYL BENZOATE HYDROCHLORIDE see IJZ000

dl-(2-METHYLPIPERIDINO)PROPYL BENZOATE HYDROCHLORIDE see IJZ000

3-(2-METHYLPIPERIDINO)PROPYL-3,4-DICHLOROBENZOATE see MOL250

Γ-(2-METHYLPIPERIDINO)PROPYL-3,4-DICHLOROBENZOATE see MOL250

2-METHYL-3-PIPERIDINOPYRAZINE MONOSULFATE see MOM750

2-METHYL-3-PIPERIDINOPYRAZINE SULFATE see MOM750

2-METHYL-3-PIPERIDINO-1-p-TOLYLPROPAN-1-ONE see TGK200

2-METHYL-3-PIPERIDINO-1-p-TOLYLPROPAN-1-ONE HYDROCHLORIDE see MRW125

N-(1-METHYL-2-(1-PIPERIDINYL)ETHYL)-N-2-PYRIDINYLPROPANAMIDE-(E)-2-BUTENEDIOATE (1:1) see PMX250

1-METHYL-4-PIPERIDYL-p-AMINOBENZOATE HYDROCHLORIDE see MON000

N-METHYLPIPERIDYL-(4)-BENZHYDRYLAETHER SALZSAUREN SALZE (GERMAN) see LJR000

N-METHYL-3-PIPERIDYL BENZILATE see MON250

N-METHYL-3-PIPERIDYL BENZILATE METHOBROMIDE see CBF000

1-METHYL-4-PIPERIDYL BENZOATE HYDROCHLORIDE see MON500

N-METHYL-3-PIPERIDYL-α-CYCLOHEXYL MANDELATE see MOQ500

N-METHYL-3-PIPERIDYLDIPHENYLGLYCOLATE METHOBROMIDE see CBF000

9-(4'-(N-METHYLPIPERIDYLENE)THIOXANTHENE MALEATE see MOP000

1-METHYL-3-PIPERIDYL ESTER METHOBROMIDE BENZILIC ACID see CBF000

10-(2-(1-METHYL-2-PIPERIDYL)ETHYL)-2-METHYLSULFINYL PHENOTHI-AZINE see MON750

10-(2-(1-METHYL-2-PIPERIDYL)ETHYL)-2-(METHYLTHIO)PHENOTHIAZINE see MOO250

10-(2-(1-METHYL-2-PIPERIDYL)ETHYL)-2-METHYLTHIOPHENOTHIAZINE HYDROCHLORIDE see MOO500

(1-METHYL-4-PIPERIDYLIDENE)-9-ANTHROL-9,10-DIHYDRO-10-HYDRO-CHLORIDE see WAJ000

1-METHYL-3-PIPERIDYLIDENEDI(2-THIENYL)METHANE see BLV000

9-(1-METHYL-4-PIPERIDYLIDENE)THIOXANTHENE see MOO750

9-(N-METHYL-PIPERIDYLIDEN-4)THIOXANE MALEATE see MOP000

9-(1-METHYL-PIPERIDYL-(2)-METHYL)-CARBAZOL (GERMAN) see MOP500

9-(METHYL-2-PIPERIDYL)METHYLCARBAZOLE see MOP500

(N-METHYL-3-PIPERIDYL)METHYLPHENOTHIAZINE see MOQ250

1-(1-METHYL-2-PIPERIDYL)METHYLPHENOTHIAZINE see MOQ000

10-(1-METHYLPIPERIDYL-3-METHYL)PHENOTHIAZINE see MOQ250

10-(1-METHYL-3-PIPERIDYL)METHYL PHENOTHIAZINE see MOQ250

9-(1-METHYL-3-PIPERIDYLMETHYL)THIAXANTHENE HYDROCHLORIDE see THL500

9-((N-METHYL-3-PIPERIDYL)METHYL)-THIOXANTHENHYDROCHLORID (GERMAN) see THL500

1-METHYL-3-PIPERIDYL-α-PHENYLCYCLOHEXANEGLYCOLATE see MOQ500

d-1-METHYL-3-PIPERIDYL-dl-α-PHENYLCYCLOHEXANEGLYCOLATE HYDROCHLORIDE see PMS800

γ-(2-METHYLPIPERIDYL)PROPYL BENZOATE see PIV750

(+-)-γ-(2-METHYLPIPERIDYL)PROPYL BENZOATE HYDROCHLORIDE see IJZ000

METHYL PIRIMIPHOS see DIN800

METHYL POTASSIUM see MOR250

METHYLPREDNISOLONE see MOR500

6-α-METHYLPREDNISOLONE see MOR500

METHYLPREDNISOLONE ACETATE see DAZ117

6-METHYLPREDNISOLONE ACETATE see DAZ117

METHYLPREDNISOLONE 21-ACETATE see DAZ117

6-α-METHYLPREDNISOLONE ACETATE see DAZ117

METHYLPREDNISOLONE SODIUM SUCCINATE see USJ100

6-α-METHYLPREDNISOLONE SODIUM SUCCINATE see USJ100

16-β-METHYL-1,4-PREGNADIENE-9-α-FLUORO-11-β,17-α,21-TRIOL- 3,20-DIONE see BFV750

6-METHYL-Δ^{4,6}-PREGNADIEN-17-α-OL-3,20-DIONE ACETATE see VTF000

6-α-METHYL-4-PREGNENE-3,20-DION-17-α-OL ACETATE see MCA000

METHYLPROMAZINE see AFL500

METHYLPROPAMINE see DBA800

2-METHYLPROPANAL see IJS000

2-METHYL-1-PROPANAL see IJS000

2-METHYL-1-PROPANAL OXIME see IJT000

2-METHYLPROPANE see MOR750

2-METHYLPROPANETHIOL see IIX000

2-METHYL-2-PROPANETHIOL see MOS000

METHYL PROPANOATE see MOT000

2-METHYLPROPANOIC ACID see IJU000

2-METHYL-PROPANOIC ACID-3-PHENYL-2-PROPENYL ESTER see CMR750

2-METHYL PROPANOL see IIL000

2-METHYL-1-PROPANOL see IIL000

2-METHYLPROPAN-1-OL see IIL000

2-METHYL-2-PROPANOL see BPX000

N-METHYL-N-PROPARGYLBENZYLAMINE see MOS250

N-METHYL-N-PROPARGYL-3-(2,4-DICHLOROPHENOXY)PROPYLAMINE HYDROCHLORIDE see CMY000

METHYL PROPARGYL ETHER see MFN250

2-METHYLPROPENAL (CZECH) see MGA250

2-METHYLPROPENAMIDE see MDN500

METHYL PROPENATE see MGA500

2-METHYLPROPENE see IIC000

2-METHYLPROPENENITRILE see MGA750

2-METHYL-1-PROPENE-1-ONE see DSL289

METHYL PROPENOATE see MGA500

METHYL-2-PROPENOATE see MGA500

2-METHYLPROPENOIC ACID see MDN250

2-METHYL-2-PROPENOIC ACID ANHYDRIDE (9CI) see MDN699

2-METHYLPROPENOIC ACID CHLORIDE see MDN899

2-METHYL-2-PROPENOIC ACID 2-(DIMETHYLAMINO)-1-((DIMETHYLAMINO)METHYL)ETHYL ESTER (9CI) see BJI125

2-METHYL-2-PROPENOIC ACID, ETHYL ESTER see EMF000

2-METHYL-2-PROPENOIC ACID METHYL ESTER see MLH750

2-METHYL-2-PROPENOIC ACID METHYL ESTER HOMOPOLYMER see PKB500

2-METHYL-2-PROPENOIC ACID METHYL ESTER, POLYMER with TRIBUTYL(92-METHYL-1-OXO-2-PROPENYL)OXY)STANNANE see OIY000

2-METHYL-2-PROPENOIC ACID METHYL ESTER, POLYMER with TRIBUTYL(92-METHYL-1-OXO-2-PROPENYL)OXY)STANNANE and ((2-METHYL-1-OXO-2-PROPENYL)OXY)TRIPOPYLSTANNANE see OIW000

2-METHYL-2-PROPENOIC ACID-2-METHYLPROPYL ESTER see IIY000

2-METHYL-2-PROPENOIC ACID-3-(TRIMETHOXYSILYL)PROPYL ESTER see TLC250

1-METHYL PROPENOL see MQL250

2-METHYL-2-PROPEN-1-OL see IMW000

2-METHYL-2-PROPENOYL CHLORIDE see MDN899

2-METHYLPROPENYL CHLORIDE see MDN899

METHYL PROPENYL KETONE see PBR500

METHYL PROPIOLATE see MQS875

2-METHYLPROPIONALDEHYDE see IJS000

METHYL PROPIONATE see MOT000

2-METHYLPROPIONIC ACID see IJU000

α-METHYLPROPIONIC ACID see IJU000

2-METHYLPROPIONIC ACID, ETHYL ESTER see ELS000

2-METHYLPROPIONITRILE see IJX000

METHYL 5-n-PROPOXY-2-BENZIMIDAZOLE CARBAMATE see OMY700

1-(2-METHYLPROPOXY)-2-PROPANOL (9CI) see IIG000
2-METHYLPROPYL ACETATE see IIJ000
2-METHYL-1-PROPYL ACETATE see IIJ000
METHYLPROPYLACETIC ACID see MQJ750
2-METHYLPROPYL ALCOHOL see IIL000
1-METHYLPROPYLAMINE see BPY000
(2-METHYLPROPYL)-p-AMINOBENZOATE see MOT750
2-METHYL-2-(PROPYLAMINO)-1-PROPANOL BENZOATE (ester), HYDRO-
    CHLORIDE see OJI600
2-METHYL-2-(PROPYLAMINO)-1-PROPANOL BENZOATE HYDROCHLORIDE
    see OJI600
METHYLPROPYLARSINIC ACID see MOT800
METHYL PROPYLATE see MOT000
12-METHYL-7-PROPYLBENZ(a)ANTHRACENE see MOU250
4-(2-METHYLPROPYL)BENZENEACETIC ACID see IJG000
2-METHYLPROPYL BUTYRATE see BSW500
METHYL PROPYL CARBINOL see PBM750
METHYLPROPYLCARBINOL CARBAMATE see MOU500
METHYL-PROPYL-CETONE (FRENCH) see PBN250
3-(1-METHYLPROPYL)-6-CHLOROPHENYL METHYLCARBAMATE
    see MOU750
METHYL PROPYL DIKETONE see HEQ200
6-(1-METHYL-PROPYL)-2,4-DINITROFENOL (DUTCH) see BRE500
(6-(1-METHYL-PROPYL)-2,4-DINITRO-FENYL)-3,3-DIMETHYL ACRYLAAT
    (DUTCH) see BGB500
2-(1-METHYLPROPYL)-4,6-DINITROPHENOL see BRE500
2-(1-METHYL-N-PROPYL) 4,6-DINITROPHENOL AMMONIUM SALT
    see BPG250
2-(1-METHYL-N-PROPYL)-4,6-DINITROPHENOL TRIETHANOLAMINE SALT
    see BRE750
2-(1-METHYLPROPYL)-4,6-DINITROPHENYL ACETATE see ACE500
2-(1-METHYLPROPYL)-4,6-DINITROPHENYL-β,β-DIMETHACRYLATE
    see BGB500
(6-(1-METHYL-PROPYL)-2,4-DINITRO-PHENYL)-3,3-DIMETHYL ACRYLAT
    (GERMAN) see BGB500
2-(1-METHYL-2-PROPYL)-4,6-DINITROPHENYL ISOPROPYLCARBONATE
    see CBW000
β-METHYLPROPYL ETHANOATE see IIJ000
2-METHYL-2-PROPYLETHANOL see AOK750
2-METHYLPROPYL HEXANOATE see IIT000
N,N''-(2-METHYLPROPYLIDENE)BISUREA (9CI) see IIV000
2-METHYLPROPYL ISOBUTYRATE see IIW000
2-METHYLPROPYL ISOVALERATE see ITA000
METHYL-n-PROPYL KETONE see PBN250
METHYL PROPYL KETONE (ACGIH, DOT) see PBN250
2-METHYLPROPYL METHACRYLATE see IIY000
2-METHYLPROPYL-3-METHYLBUTYRATE see ITA000
METHYL-N-PROPYLNITROSAMINE see MNA000
METHYLPROPYLNITROSOAMINE see MNA000
N-(2-METHYLPROPYL)-N-NITROSOUREA see IJF000
8-METHYL-3-(2-PROPYLPENTANOYLOXY)TROPINIUM BROMIDE see LJS000
2-(1-METHYLPROPYL)PHENYL METHYLCARBAMATE see MOV000
m-(1-METHYLPROPYL)PHENYLMETHYLCARBAMATE see BSG250
2-METHYL-2-PROPYL-1,3-PROPANEDIOL BUTYLCARBAMATE CARBA-
    MATE see MOV500
2-METHYL-2-PROPYL-1,3-PROPANEDIOL CARBAMATE ISOPROPYLCARBA-
    MATE see IPU000
2-METHYL-2-N-PROPYL-1,3-PROPANEDIOL DICARBAMATE see MQU750
2-METHYLPROPYLPROPANOIC ACID-2-METHYLPROPYL ESTER (9CI)
    see IIW000
5-(1-METHYLPROPYL)-5-(2-PROPENYL)-2,4,6(1H,3H,5H)-PYRIMIDINETRIONE
    (9CI) see AFY500
5-(2-METHYLPROPYL)-5-(2-PROPENYL)-2,4,6(1H,3H,5H)-PYRIMIDINETRIONE
    (9CI) see AGI750
2-METHYLPROPYL PROPIONATE see PMV250
6-METHYL-2-PROPYL-4-PYRIMIDINYL DIMETHYL CARBAMATE see PNQ250
METHYL 5-(PROPYLTHIO)-2-BENZIMIDAZOLECARBAMATE see VAD000
2-METHYL-2-PROPYLTRIMETHYLENE BUTYLCARBAMATE CARBAMATE
    see MOV500
2-METHYL-2-PROPYLTRIMETHYLENE CARBAMATE see MQU750
3-METHYL-5-PROPYL-1,2,4-TRIOXOLANE see HFC500
METHYL PROPYNOATE see MOS875
N-METHYL-N-2-PROPYNYLBENZYLAMINE see MOS250
N-METHYL-N-(2-PROPYNYL)BENZYLAMINE HYDROCHLORIDE see BEX500
1-METHYL-2-PROPYNYL-m-CHLOROCARBANILATE see CEX250
1-METHYL-2-PROPYNYL-m-CHLOROPHENYLCARBAMATE see CEX250
1-METHYLPROPYNYL 3-CHLOROPHENYLCARMATE see CEX250
1-METHYLPROPYNYL ESTER of 3-CHLOROPHENYLCARBAMIC ACID
    see CEX250
2-METHYL-5-(2-PROPYNYL)-3-FURYLMETHYL-cis-trans-CHRYSANTHEMATE
    see PMN700
6-α-METHYL-17-(1-PROPYNYL)TESTOSTERONE see DRT200
6-α-METHYL-17-α-PROPYNYLTESTOSTERONE see DRT200
15(R)-METHYLPROSTAGLANDIN E2 see AQT575

15(R)-15-METHYLPROSTAGLANDIN E2 see AQT575
15(s)-15-METHYL-PROSTAGLANDIN E2 see MOV800
15-METHYLPROSTAGLANDIN F2-α see CCC100
15(S)-METHYLPROSTAGLANDIN F2-α see CCC100
15(S)-15-METHYLPROSTAGLANDIN F2-α see CCC100
(15S)-15-METHYLPROSTAGLANDIN F2-α see CCC100
15-METHYL-PROSTAGLANDIN-F2-α-METHYL ESTER see MLO300
15(s)15-METHYL-PROSTAGLANDIN-F2-α-METHYL ESTER see MLO300
15(S)15-METHYL PROSTAGLANDIN F2-α TROMETHAMINE see CCC110
N-METHYL-4-PROTOADAMANTANEAMINE HYDROCHLORIDE see MOW000
N-METHYL-4-PROTOADAMANTANEMETHANAMINE MALEATE see
    MOW250
METHYL PROTOANEMONIN see MOW500
METHYLPROTOCATECHUALDEHYDE see VFK000
9-METHYL PTEROYLGLUTAMIC ACID see BMM125
x-METHYLPTEROYLGLUTAMIC ACID see MKG275
2-METHYLPYRAZINE see MOW750
4-METHYLPYRAZOLE see MOX000
METHYLPYRAZOLYL DIETHYLPHOSPHATE see MOX250
3-METHYLPYRAZOLYL-5-DIETHYLPHOSPHATE see MOX250
5-METHYL-1H-PYRAZOL-3-YL DIMETHYLCARBAMATE see DQZ000
1-METHYLPYRENE see MOX875
3-METHYLPYRENE see MOX875
2-METHYLPYRIDINE see MOY000
3-METHYLPYRIDINE see PIB920
4-METHYLPYRIDINE see MOY250
α-METHYLPYRIDINE see MOY000
N-METHYLPYRIDINE-2-ALDOXIME IODIDE see POS750
2-METHYLPYRIDINE-4-AZO-p-DIMETHYLANILINE see MOY500
N-METHYL-2-PYRIDINEETHANAMINE DIMETHANESULFONATE see BFV350
2-METHYLPYRIDINE-1-OXIDE-4-AZO-p-DIMETHYLANILINE see DSS200
3-METHYLPYRIDINE-1-OXIDE-4-AZO-p-DIMETHYL-ANILINE see MOY750
1-METHYL-2-PYRIDINIUM ALDOXIME CHLORIDE see FNZ000
N-METHYLPYRIDINIUM-2-ALDOXIME IODIDE see POS750
1-METHYLPYRIDINIUM-2-ALDOXIME METHANESULFONATE see PLX250
N-METHYLPYRIDINIUM-2-ALDOXIME METHANESULPHONATE see PLX250
METHYL PYRIDINIUM CHLORIDE see MOY875
1-METHYLPYRIDINIUM CHLORIDE see MOY875
N-METHYLPYRIDINIUM CHLORIDE see MOY875
N-METHYLPYRIDINIUM CHLORIDE-2-ALDOXIME see FNZ000
1-METHYLPYRIDINIUM IODIDE see MOZ000
N-METHYLPYRIDINIUM METHANE SULFONATE-2-ALDOXIME see PLX250
1-METHYL-9H-PYRIDO(3,4-b)INDOLE see MPA050
α⁴-O-METHYLPYRIDOXOL see MFN600
α-((4-METHYL-2-PYRIDYLAMINO)METHYL)BENZYL ALCOHOL HYDRO-
    CHLORIDE see MPA100
9-(2-(2-METHYLPYRIDYL-5)ETHYL)-3,6-DIMETHYL-1,2,3,4-TETRAHYDRO-
    Γ-CARBOLINE 2HCl see DNU875
METHYL PYRIDYL KETONE see ABI000
METHYL-3-PYRIDYL KETONE see ABI000
METHYL-β-PYRIDYL KETONE see ABI000
3-(6-(5-METHYL-2-PYRIDYLOXY)HEXYL)THIAZOLIDINE DIHYDROCHLO-
    RIDE see MPA250
1-METHYL-2-(3-PYRIDYL)PYRROLE see NDX300
1-METHYL-2-(3-PYRIDYL)PYRROLIDINE see NDN000
1-1-METHYL-2-(3-PYRIDYL)-PYRROLIDINE SULFATE see NDR500
METHYLPYRIMAL see ALF250
6-METHYL-2,4(1H,3H)-PYRIMIDINEDIONE see MQI750
N¹-(4-METHYL-2-PYRIMIDINYL)SULFANILAMIDE see ALF250
N¹-(4-METHYL-2-PYRIMIDINYL)SULFANILAMIDE SODIUM SALT see SJW475
3-METHYLPYROCATECHOL see DNE000
4-METHYLPYROCATECHOL see DNE200
p-METHYLPYROCATECHOL see DNE200
2-METHYL PYROMECONIC ACID see MAO350
METHYL PYROMUCATE see MKH600
METHYL PYRONIN see PPQ750
METHYL PYROPHOSPHATE see TDV000
1-METHYLPYRROLE see MPB000
1-METHYLPYRROLE-2,3-DIMETHANOL see MPB175
1-METHYLPYRROLIDINE see MPB250
β-METHYL-1-PYRROLIDINEPROPIONANILIDE see MPB500
N-(N'-METHYLPYRROLIDINIUMMETHYL)-2,2-DIPHENYL SULFOXIMIDE
    BROMIDE see HAA325
N-METHYLPYRROLIDINONE see MPF200
1-METHYL-2-PYRROLIDINONE see MPF200
1-METHYL-5-PYRROLIDINONE see MPF200
N-METHYL-2-PYRROLIDINONE see MPF200
1-(2-METHYL-5-PYRROLIDINO-2,4-PENTADIENYLIDENE)PYRROLIDINIUM
    PERCHLORATE see MPC250
3-(N-METHYLPYRROLIDINO)PYRIDINE see NDN000
6-METHYL-α-(1-PYRROLIDINYLCARBONYL)ERGOLINE-8β-PROPIONITRILE
    see MPC300
1-(1-METHYL-2-PYRROLIDINYLIDENE)-3-(2,6-XYLYL)UREA see XGA725
3-(1-METHYL-2-PYRROLIDINYL)INDOLE see MPD000

3-(1-METHYL-3-PYRROLIDINYL)INDOLE see MPD250
10-((1-METHYL-3-PYRROLIDINYL)METHYL)-PHENOTHIAZINE see MPE250
10-((1-METHYL-3-PYRROLIDINYL)METHYL)PHENOTHIAZINE, HYDROCHLO-
  RIDE see MDT500
α-1-(1-METHYL-3-PYRROLIDINYL)-1-PHENYLPROPYL PROPIONATE FUMAR-
  ATE see ENC600
3-(1-METHYL-2-PYRROLIDINYL)PYRIDINE see NDN000
(S)-3-(1-METHYL-2-PYRROLIDINYL)PYRIDINE (9CI) see NDN000
(S)-3-(1-METHYL-2-PYRROLIDINYL-PYRIDINE (R-(R,R))-2,3-
  DIHYDROXYBUTANEDIOATE (1:2) see NDS500
(S)-3-(1-METHYL-2-PYRROLIDINYL)PYRIDINE SULFATE (2:1) see NDR500
METHYLPYRROLIDONE see MPF200
N-METHYLPYRROLIDONE see MPF200, MPF500
1-METHYL-2-PYRROLIDONE see MPF200
N-METHYL-2-PYRROLIDONE see MPF200
1-METHYL-2-(PYRROLID-2-YL)METHYL PHENYLCYCLOHEXYL GLYCOL-
  LATE METHOBROMIDE see IBP200
1-METHYL-3-PYRROLIDYL-α-PHENYLCYCLOPENTANEGLYCOLATE HYDRO-
  CHLORIDE see AEY400
(−)-3-(1-METHYL-2-PYRROLIDYL)PYRIDINE see NDN000
l-3-(1-METHYL-2-PYRROLIDYL)PYRIDINE see NDN000
1-3-(1-METHYL-2-PYRROLIDYL)PYRIDINE SULFATE see NDR500
3-(1-METHYL-2-PYRROLYL)PYRIDINE see NDX300
3-(1-METHYL-1H-PYRROL-2-YL)-PYRIDINE (9CI) see NDX300
METHYL-QUECKSILBER-TOLUOLSULFAMID see MLH100
3-METHYL-4-QUINAZOLINONE see MPF750
2-METHYL-4(3H)-QUINAZOLINONE see MPF500
METHYLQUINAZOLONE HYDROCHLORIDE see MDT250
2-METHYLQUINOLINE see QEJ000
4-METHYLQUINOLINE see LEL000
6-METHYLQUINOLINE see MPF800
p-METHYLQUINOLINE see MPF800
Γ-METHYLQUINOLINE see LEL000
4-METHYLQUINOLINE OXIDE see LEM000
1-METHYLQUINOLINIUM CHLORIDE see MPG000
8-(METHYLQUINOLYL)-N-METHYL CARBAMATE see MPG250
2-METHYL-1,4-QUINONE see MHI250
6-METHYL-2,3-QUINOXALINE DITHIOCARBONATE see ORU000
6-METHYL-2,3-QUINOXALINEDITHIOL CYCLIC CARBONATE see ORU000
6-METHYL-2,3-QUINOXALINEDITHIOL CYCLIC DITHIOCARBONATE
  see ORU000
6-METHYL-QUINOXALINE-2,3-DITHIOLCYCLOCARBONATE see ORU000
METHYL RED see CCE500
METHYL RESERPATE see MPH300
METHYL RESERPATE-4-ETHOXYCARBONYL-3,5-DIMETHOXYBENZOIC
  ACID ESTER see RCA200
METHYL RESERPAT ESTER of SYRINGIG ACID ETHYL CARBONATE
  see RCA200
METHYL RESERPATE 3,4,5-TRIMETHOXYBENZOIC ACID see RDK000
METHYL RESERPATE 3,4,5-TRIMETHOXYBENZOIC ACID ESTER
  see RDK000
METHYL RESERPATE 3,4,5-TRIMETHOXYCINNAMIC ACID ESTER
  see TLN500
METHYL RESERPINOLATE see MPH300
5-METHYLRESORCINOL see MPH500
5-METHYLRESORCINOL ORCINOL see MPH500
METHYLRHODANID (GERMAN) see MPT000
3-METHYLRHODANINE see MPH750
6-METHYL-9-RIBOFURANOSYLPURINE-6-THIOL see MPU000
METHYLROSANILINE see TJK000
METHYLROSANILINE CHLORIDE see AOR500
METHYL SALICYLATE see MPI000
3-METHYLSALICYLIC ACID see CNX625
METHYLSCOPOLAMINE BROMIDE see SBH500
METHYLSCOPOLAMINE HYDROBROMIDE see SBH500
N-METHYLSCOPOLAMINE METHOSULFATE see DAB750
METHYLSCOPOLAMINE METHYL SULFATE see DAB750
N-METHYLSCOPOLAMINE METHYL SULFATE see DAB750
METHYL SCOPOLAMINE NITRATE see SBH000
N-METHYLSCOPOLAMMONIUM BROMIDE see SBH500
METHYLSCOPOLAMMONIUM METHYLSULFATE see DAB750
METHYL SELENAC see SBQ000
METHYL SELENIDE see DUB200
METHYL SELENIUM see DUB200
METHYLSELENO-2-BENZOIC ACID see MPI200
o-(METHYLSELENO)BENZOIC ACID see MPI200, MPI205
(METHYLSELENO)TRIS(DIMETHYLAMINO)PHOSPHONIUM IODIDE
  see MPI225
METHYLSENFOEL (GERMAN) see ISE000
METHYLSERGIDE BIMALEATE see MQP500
1-METHYLSILACYCLOPENTA-2,4-DIENE see MPI600
METHYLSILANE see MPI625
METHYL SILICATE see MPI750
METHYL SILICONE see PJR000

METHYLSILVER see MPI800
METHYL SODIUM see MPJ000
METHYL STEARATE see MJW000
METHYL STIBINE see MPJ100
2-METHYL-4-STILBENAMINE see MPJ250
3-METHYL-4-STILBENAMINE see MPJ500
METHYL STREPONIGRIN see SMA500
β-METHYLSTREPTOZOTOCIN see MPJ800
α-METHYLSTYREEN (DUTCH) see MPK250
METHYL STYRENE see VQK650
α-METHYL STYRENE see MPK250
METHYL STYRENE (mixed isomers) see MPK500
4-METHYLSTYRENE OXIDE see MPK750
p-METHYLSTYRENE OXIDE see MPK750
α-METHYL-STYROL (GERMAN) see MPK250
METHYL STYRYLPHENYL KETONE see MPL000
6-METHYL-7-SULFAMIDO-THIOCHROMAN-1,1-DIOXIDE see MQQ050
5-METHYL-3-SULFANILAMIDOISOXAZOLE see SNK000
5-METHYL-2-SULFANILAMIDO-1,3,4-THIADIAZOLE see MPQ750
METHYL SULFANILYL CARBAMATE see SNQ500
METHYL SULFATE (DOT) see DUD100
METHYLSULFAZIN see ALF250
METHYL SULFIDE (DOT) see TFP000
METHYL SULFIDE compound with BORANE (1581) see MPL250
2-((METHYLSULFINYL)ACETYL)PYRIDINE see MPL500
METHYLSULFINYL ETHYLTHIAMINE DISULFIDE see MPL600
METHYLSULFINYLMETHANE see DUD800
METHYL SULFOCYANATE see MPT000
METHYLSULFONAL see BJT750
9-(p-(METHYLSULFONAMIDO)ANILINO)-3-ACRIDINE CARBAMIC ACID
  METHYL ESTER see MPL750
N-(9-(p-(METHYLSULFONAMIDO)ANILINO)ACRIDIN-3-YL)ACETAMIDE
  METHANESULFONATE see MPM000
METHYLSULFONIC ACID, ETHYL ESTER see EMF500
N-(9-((4-(((METHYLSULFONYL)AMINO)PHENYL)AMINO)-3-ACRIDINYL)
  ACETAMIDE METHANESULFONATE see MPM000
2-METHYLSULFONYL-10-(3-(4-CARBAMOYLPIPERIDINO)PROPYL)PHENO-
  THIAZINE see MQR000
2-METHYLSULFONYL-10-(3-(4'-CARBAMOYLPIPERIDINO)PROPYL)-PHENO-
  THIAZINE see MPM750
METHYLSULFONYL CHLORAMPHENICOL see MPN000
2-METHYLSULFONYL-o-(N-METHYL-CARBAMOYL)-BUTANON-(3)-
  OXIM (GERMAN) see SOB500
1-METHYLSULFONYL-3-(1-METHYL-5-NITRO-2-IMIDAZOLYL)-2-
  IMIDAZOLIDINONE see SAY950
1,4-(METHYLSULFONYLOXYETHYLAMINO)-1,4-DIDEOXY-
  ERYTHRIOLDIMETHYLSULFONATE see LJD500
((METHYLSULFONYL)OXY)TRIBUTYLSTANNANE see TIF000
((METHYLSULFONYL)OXY)TRIPHENYLSTANNANE see TMW500
1-(3-(2-(METHYLSULFONYL)PHENOTHIAZIN-10-YL)PROPYL)ISONIPECOTAM-
  IDE see MQR000
1-(3-(2-METHYLSULFONYL)PHENOTHIAZIN-10-YL)PROPYL)-4-PIPERIDINE
  CARBOXAMIDE see MQR000
1-(3-(2-(METHYLSULFONYL)-10H-PHENOTHIAZIN-10-YL)PROPYL)-4-PIPERDI-
  INE CARBOXAMIDE see MQR000
2-(4-(METHYLSULFONYL)PHENYL)-IMIDAZO(1,2-a)PYRIDINE see ZUA200
METHYL SULFOXIDE see DUD800
5-METHYL-3-SULPHANIL-AMIDOISOXAZOLE see SNK000
METHYL SULPHIDE see TFP000
3-METHYLSULPHINYLPROPYLISOTHIOCYANATE see MPN100
METHYLSULPHONAL see BJT750
METHYL SYSTOX see MIW100
METHYLSYSTOX see DAO800
METHYL TARTRONIC ACID see MPN250
o-(METHYLTELLURO)BENZOIC ACID see MPN275
17-METHYLTESTOSTERON see MPN500
METHYLTESTOSTERONE see MPN500
17-METHYLTESTOSTERONE see MPN500
17-α-METHYLTESTOSTERONE see MPN500
Δ'-17-METHYLTESTOSTERONE see DAL300
Δ(¹)-17-α-METHYLTESTOSTERONE see DAL300
2-METHYL-3,4,5,6-TETRABROMOPHENOL see TBJ500
4-o-METHYL-12-o-TETRADECANOYLPHORBOL-13-ACETATE see MPN600
2-METHYL-3,3,4,5-TETRAFLUORO-2-BUTANOL see MPN750
10-METHYL-1,2-TETRAHYDRO-1,2:5,6-BENZACRIDINE see MPO400
10-METHYL-1,2-TETRAHYDRO-1,2:7,8-BENZACRIDINE see MPO390
2-METHYL-1,2,3,6-TETRAHYDROBENZALDEHYDE see MPO000
6-METHYL-1,2,3,4-TETRAHYDROBENZ(a)ANTHRACENE see MPO250
4-METHYL-1',2',3',4'-TETRAHYDRO-1,2-BENZANTHRACENE see MPO250
7-METHYL-1,2,3,4-TETRAHYDRODIBENZ(c,h)ACRIDINE see MPO390
14-METHYL-8,9,10,11-TETRAHYDRODIBENZ(a,h)ACRIDINE see MPO400
6-METHYL-2,3,5,7-TETRAHYDRO-6H-p-DITHIINO-(2,3-C)PYRROLE-5,7-DIONE
  see MJJ250
METHYLTETRAHYDROFURAN see MPO500

2-METHYLTETRAHYDROFURAN (CZECH) see MPO500
METHYL TETRAHYDRO-5-HYDROXY-4-METHYL-3-FURYL KETONE see BMK290
N-METHYL-1,2,3,4-TETRAHYDROISOQUINOLINE HYDROCHLORIDE see MPO750
METHYL-1,2,5,6-TETRAHYDRO-1-METHYLNICOTINATE see AQT750
METHYL-1,2,5,6-TETRAHYDRO-1-METHYLNICOTINATE, HYDROBROMIDE see AQU000
2-METHYL-2-(4-(1,2,3,4-TETRAHYDRO-1-NAPHTHALENYL)PHENOXY)PROPANOIC ACID see MCB500
2-METHYL-2-(4-(1,2,3,4-TETRAHYDRO-1-NAPHTHYL)PHENOXY)PROPANOIC ACID see MCB500
2-METHYL-2-(p-(1,2,3,4-TETRAHYDRO-1-NAPHTHYL)PHENOXY)PROPIONIC ACID see MCB500
α-METHYL-α-(p-1,2,3,4-TETRAHYDRONAPHTH-1-YLPHENOXY)PROPIONIC ACID see MCB500
N-METHYL-Δ-TETRAHYDRONICOTINIC ACID METHYL ESTER see AQT750
METHYLTETRAHYDROPHTHALIC ANHYDRIDE see MPP000
N-METHYL-1,2,3,6-TETRAHYDROPHTHALIMIDE see MIS250
N-METHYL-3,4,5,6-TETRAHYDROPHTHALIMIDE see MPP250
N-METHYLTETRAHYDROPYRIDINE-β-CARBOXYLIC ACID METHYL ESTER see AQT750
N-METHYLTETRAHYDROPYRROLE see MPB250
N-METHYL-TETRAHYDROTHIAMIDINTHIONE ACETIC ACID see MPP750
2-METHYL-3-(3,7,11,15-TETRAMETHYL-2,6,10,14-HEXADECATETRAENYL)-1,4-NAPHTHOQUINONE see VTA650
2-METHYL-3-(3,7,11,15-TETRAMETHYL-2-HEXADECENYL)-1,4-NAPHTHALENEDIONE see VTA000
2-METHYL-3-trans-TETRAMETHYL-1,4-NAPHTHQUINONE see VTA650
N-METHYL-N,2,4,6-TETRANITROANILINE see TEG250
1-METHYL-1H-TETRAZOLE-5-THIOL see MPQ250
α-METHYL TETRONIC ACID see MPQ500
METHYL 5-(2-THENOYL)-2-BENZIMIDAZOLECARBAMATE see OJD100
METHYLTHEOBROMIDE see CAK500
1-METHYLTHEOBROMINE see CAK500
7-METHYLTHEOPHYLLINE see CAK500
N¹-(5-METHYL-1,3,4-THIADIAZOL-2-YL)-SULFANILAMIDE see MPQ750
N-METHYL-3-THIA-2-METHYL-VALERAMID DER O,O-DIMETHYLTHIOLPHOSPHORSAEURE (GERMAN) see MJG500
2-METHYLTHIAZOLIDINE see MPR000
METHYL-2-THIAZOLIDINE (FRENCH) see MPR000
5-METHYL-2-(6-(3-THIAZOLIDINYL)HEXYLOXY)PYRIDINE DIHYDROCHLORIDE see MPA250
N-(3-METHYL-2-THIAZOLIDINYLIDENE)NICOTINAMIDE see MPR250
α-METHYL-4-(2-THIENYLCARBONYL)BENZENEACETIC ACID see TEN750
METHYL (5-(2-THIENYLCARBONYL)-1H-BENZIMIDAZOLE-2-YL)CARBAMATE see OJD100
2-METHYLTHIIRANE see PNL750
METHYLTHIOACETALDEHYDE-o-(CARBAMOYL)OXIME see MPS250
2-METHYLTHIO-ACETALDEHYD-O-(METHYLCARBAMOYL)-OXIM (GERMAN) see MDU600
METHYL(THIOACETAMIDO) MERCURY see MPS300
1-Γ-METHYLTHIO-α-AMINOBUTYRIC ACID see MDT750
(METHYLTHIO)BENZENE see TFC250
2-METHYLTHIOBENZIMIDAZOLE see MPS500
2-METHYLTHIO-4,6-BIS(ISOPROPYLAMINO)-s-TRIAZINE see BKL250
6-METHYLTHIOCHROMAN-7-SULFONAMIDE 1,1-DIOXIDE see MQQ050
4-(METHYLTHIO)-m-CRESOL see MLX750
4-(METHYLTHIO)-m-CRESOL-O-ESTER with O-ETHYL METHYLPHOSPHORAMIDOTHIOATE see ENI500
METHYL THIOCYANATE see MPT000
4-METHYLTHIO-3,5-DIMETHYLPHENYL METHYLCARBAMATE see DST000
17-β-(METHYLTHIO)ESTRA-1,3,5(10)-TRIEN-3-OL see MPT100
2-(METHYLTHIO)-ETHANETHIOL-O,O-DIMETHYL PHOSPHOROTHIOATE see MIW250
2-(METHYLTHIO)-ETHANETHIOL-S-ESTER with O,O-DIMETHYL PHOSPHOROTHIOATE see MIW250
2-(METHYLTHIO)ETHANOL ACRYLATE see MPT250
2-METHYLTHIOETHYL ACRYLATE see MPT250
2-METHYLTHIO-4-ETHYLAMINO-6-tert-BUTYLAMINO-s-TRIAZINE see BQC750
2-(2-METHYLTHIOETHYLAMINO)ETHYLGUANIDINE SULFATE see MPT300
2-METHYLTHIO-4-ETHYLAMINO-6-ISOPROPYLAMINO-s-TRIAZINE see MPT500
7-(2-METHYLTHIO)ETHYL)-THEOPHYLLINE see MDN100
7-(β-METHYLTHIOETHYL)THEOPHYLLINE see MDN100
METHYLTHIOGLYCOLATE see MLE750
METHYLTHIOINOSINE see MPU000
6-METHYLTHIOINOSINE see MPU000
2-METHYLTHIO-4-ISOPROPYLAMINO-6-METHYLAMINO-s-TRIAZINE see INR000
METHYLTHIOKYANAT see MPT000
METHYLTHIOMETHANE see TFP000
3-(METHYLTHIO)-o-((METHYLAMINO)CARBONYL)OXIME-2-BUTANONE see MPU250

2-(METHYLTHIO)-4-(METHYLAMINO)-6-(ISOPROPYLAMINO)-s-TRIAZINE see INR000
2-METHYLTHIO-o-(N-METHYLCARBAMOYL)-BUTANONOXIM-3 (GERMAN) see MPU250
8-β-((METHYLTHIO)METHYL)-6-PROPYLERGOLINE METHANESULFONATE see MPU500
METHYLTHIONINE CHLORIDE see BJI250
METHYLTHIONIUM CHLORIDE see BJI250
METHYL THIOPHANATE see PEX500
METHYLTHIOPHANATE-MANEB mixture see MAP300
2-METHYLTHIOPHENE see MPV000
3-METHYLTHIOPHENE see MPV250
2-METHYLTHIOPHENOL see TGP000
4-METHYLTHIOPHENOL see TGP250
o-METHYLTHIOPHENOL see TGP000
p-METHYLTHIOPHENOL see TGP250
4-METHYLTHIOPHENYLDIMETHYL PHOSPHATE see PHD250
METHYLTHIOPHOS see MNH000
2-METHYLTHIO-PROPIONALDEHYD-O-(METHYLCARBAMOYL)-OXIM (GERMAN) see MDU600
METHYL THIOPSEUDOUREA SULFATE see MPV750
2-METHYL-2-THIOPSEUDOUREA SULFATE see MPV750
2-METHYL-2-THIOPSEUDOUREA SULFATE (1581) SODIUM SALT see MPV800
6-(METHYLTHIO)PURINE RIBONUCLEOSIDE see MPU000
6-METHYLTHIOPURINE RIBOSIDE see MPU000
6-METHYL-2-THIO-2,4-(1H3H)PYRIMIDINEDIONE see MPW500
METHYLTHIOURACIL see MPW500
6-METHYLTHIOURACIL see MPW500
4-METHYL-2-THIOURACIL see MPW500
6-METHYL-2-THIOURACIL see MPW500
METHYL THIOUREA see MPW600
1-METHYLTHIOUREA see MPW600
1-METHYL-3-(THIOXANTHEN-9-YLMETHYL)-1-PIPERIDINE HYDROCHLORIDE see THL500
METHYLTHIOXOARSINE see MGQ750
5-METHYL-6-THIOXOTETRAHYDRO-3-THIADIAZINEACETIC ACID see MPP750
4-(METHYLTHIO)-3,5-XYLENOL-O-ESTER with O,O-DIETHYL PHOSPHOROTHIOATE see DJR800
4-(METHYLTHIO)-3,5-XYLENOL METHYLCARBAMATE see DST000
4-(METHYLTHIO)-3,5-XYLYL METHYLCARBAMATE see DST000
METHYL THIRAM see TFS350
METHYL THIURAMDISULFIDE see TFS350
METHYLTIN TRICHLORIDE see MQC750
METHYL-4-TOLUATE see MPX850
METHYL-p-TOLUATE see MPX850
METHYL-α-TOLUATE see MHA500
METHYL TOLUENE see XGS000
o-METHYLTOLUENE see XHJ000
p-METHYLTOLUENE see XHS000
METHYL TOLUENE-4-SULFONATE see MLL250
METHYL-p-TOLUENESULFONATE see MLL250
α-METHYL-α-TOLUIC ALDEHYDE see COF000
2-METHYL-p-TOLUIDINE see XMS000
4-METHYL-o-TOLUIDINE see XMS000
5-METHYL-o-TOLUIDINE see XNA000
6-METHYL-m-TOLUIDINE see XNA000
2-METHYL-p-TOLUIDINE HYDROCHLORIDE see XOJ000
4-METHYL-o-TOLUIDINE HYDROCHLORIDE see XOJ000
5-METHYL-o-TOLUIDINE HYDROCHLORIDE see XOS000
6-METHYL-m-TOLUIDINE HYDROCHLORIDE see XOS000
1-METHYL-5-(p-TOLUOYL)-2-PYRROLEACETIC ACID see SKJ340
1-METHYL-5-p-TOLUOYL-PYRROLE-2-ACETIC ACID see TGJ850
2-METHYL-3-o-TOLYL-6-AMINO-CHINAZOLINON-4 (GERMAN) see AKM125
N-METHYL-p-(m-TOLYLAZO)ANILINE see MPY000
N-METHYL-p-(o-TOLYLAZO)ANILINE see MPY250
N-METHYL-p-(p-TOLYLAZO)ANILINE see MPY500
2-METHYL-4-((o-TOLYL)AZOANILINE HYDROCHLORIDE see MPY750
2'-METHYL-4'-(o-TOLYLAZO)OXANILIC ACID see OLG000
N-METHYL-2-(α-(2-TOLYLBENZYL)OXY)ETHYLAMINE HYDROCHLORIDE see TGJ250
8-METHYL-3-(α-(o-TOLYL)BENZYLOXY)TROPANIUM IODIDE see MPZ000
1-METHYL-4-(α-o-TOLYLBENZYL)PIPERAZINE DIHYDROCHLORIDE see MIQ725
N-METHYL-N-(m-TOLYL)CARBAMOTHIOIC ACID-(1,2,3,4-TETRAHYDRO-1,4-METHANONAPHTHALEN-6-YL) ESTER see MQA000
METHYL 3-(m-TOLYLCARBAMOYLOXY)PHENYLCARBAMATE see MEG250
METHYL-p-TOLYLCARBINOL see TGZ000
2-METHYL-3-o-TOLYL-4(3H)-CHINAZOLINON (GERMAN) see QAK000
2-METHYL-3-o-TOLYL-4(3H)-CHINAZOLONE see QAK000
2-METHYL-3-TOLYLCHINAZOLON-4 HYDROCHLORIDE (GERMAN) see MDT250
(2-METHYL-3-(o-TOLYL)-3,4-DIHYDRO-4-(QUINAZOLINONE) see QAK000, QAK000

METHYL-p-TOLYL ETHER see MGP000
METHYL-p-TOLYL KETONE see MFW250
2-METHYL-3-TOLYL-4-OXYBENSDIAZINE see QAK000
2-METHYL-3-o-TOLYL-4(3H)-QUINAZOLINONE see QAK000
2-METHYL-3-o-TOLYL-4(3H)-QUINAZOLINONE HYDROCHLORIDE
   see MDT250
2-METHYL-3-(2-TOLYL)QUINAZOL-4-ONE see QAK000
2-METHYL-3-o-TOLYL-4-QUINAZOLONE see QAK000
2-METHYL-3-(o-TOLYL)-4-QUINAZOLONE HYDROCHLORIDE see MDT250
2-METHYL-3-o-TOLYL-6-SULFAMYL-7-CHLORO-1,2,3,4-TETRAHYDRO-4-
   QUINAZOLINONE see ZAK300
3-METHYL-5-(p-TOLYLSULFONYL)-1,2,4-THIADIAZOLE see MQB100
3-METHYL-1-(p-TOLYL)-TRIAZENE see MQB250
METHYL TOSYLATE see MLL250
METHYL-p-TOSYLATE see MLL250
4-o-METHYL-TPA see MPN600
METHYLTRETAMINE see TNK000
METHYLTRIACETOXYSILANE see MQB500
5-(3-METHYL-1-TRIAZENO)IMIDAZOLE-4-CARBOXAMIDE see MQB750
5-METHYL-1,2,4-TRIAZOLE(3,4-b)BENZOTHIAZOLE see MQC000
5-METHYL-s-TRIAZOLO(3,4-b)BENZOTHIAZOLE see MQC000
2'-METHYL-1,2:4,5:8,9-TRIBENZOPYRENE see MJA750
METHYLTRICAPRYLYLAMMONIUMCHLORIDE see MQH000
METHYL TRICHLORIDE see CHJ500
METHYL TRICHLOROACETATE see MQC150
N-METHYL-2,4,5-TRICHLOROBENZENESULFONAMIDE see MQC250
METHYLTRICHLOROMETHANE see MIH275
METHYLTRICHLOROSILANE see MQC500
METHYLTRICHLOROSTANNANE see MQC750
METHYLTRICHLOROTIN see MQC750
METHYL-TRICHLORSILAN (CZECH) see MQC500
α-METHYLTRICYCLO(3.3.1.1(3,7))DECANE-1-METHANAMINE HYDROCHLO-
   RIDE see AJU625
1-METHYLTRICYCLOQUINAZOLINE see MQD000
3-METHYLTRICYCLOQUINAZOLINE see MQD250
4-METHYLTRICYCLOQUINAZOLINE see MQD500
METHYLTRIETHOXYSILANE see MQD750
METHYL TRIFLUORIDE see CBY750
5-METHYL-2-TRIFLUOROMETHYLOXAZOLIDINE see MQE000
8-METHYL-3-(α-(α,α,α-TRIFLUORO-o-TOLYL)BENZYLOXY)TROPANIUM IO-
   DIDE see MQF000
METHYL-3,3,3-TRIFLUORO-2-(TRIFLUOROMETHYL)PROPIONATE
   see MKK750
METHYL TRIFLUOROVINYL ETHER see MQF200
6-METHYL-1,3,8-TRIHYDROXYANTHRAQUINONE see MQF250
METHYL 3,4,5-TRIHYDROXYBENZOATE see MKI100
1-METHYL-3,7,9-TRIHYDROXY-6H-DIBENZO(b,d)PYRAN-6-ONE see AGW476
METHYL-18-o-(3,4,5-TRIMETHOXYCINNAMOYL RESERPATE
   see TLN500
α-METHYL-3,4,5-TRIMETHOXYPHENETHYLAMINE HYDROCHLORIDE
   see TKX500
METHYLTRIMETHOXYSILANE see MQF500
(2-METHYL-2-(2-(TRIMETHYLAMMONIO)ETHOXY)ETHYL)
   DIETHYLMETHYL AMMONIUM DIIODIDE see MQF750
METHYLTRIMETHYLENE GLYCOL see BOS500
4-METHYL TRIMETHYLENE SULFITE see MQG500
N-METHYL-N-TRIMETHYLSILYLTRIFLUOROACETAMIDE see MQG750
1-METHYL-2,4,5-TRINITROBENZENE see TMN400
3-METHYL-2,4,6-TRINITROPHENOL see TML500
METHYLTRIOCTYLAMMONIUM CHLORIDE see MQH000
4-METHYL-2,6,7-TRIOXA-1-ARSABICYCLO(2.2.2)OCTANE see MQH100
3-METHYL-1,2,4-TRIOXOLANE see PMP250
1-METHYL-4-(3,3,3-TRIS(p-CHLOROPHENYL)PROPIONYLPIPERAZINE
   MONOHYDROCHLORIDE see MQH250
METHYLTRIS(2-ETHYLHEXYLOXYCARBONYLMETHYLTHIO)STANNANE
   see MQH500
METHYL TRITHION see MQH750
8-METHYLTROPINIUM BROMIDE 2-PROPYLVALERATE see LJS000
4-METHYL-TROPOLONE see MQI000
α-METHYLTRYPTAMINE see AME500
l-5-METHYLTRYPTOPHAN see MQI250
METHYL TUADS see TFS350
α-METHYL-m-TYRAMINE see AMF500
(±)-α-METHYL-p-TYROSINE METHYL ESTER HYDROCHLORIDE see MQI500
dl-α-METHYL-p-TYROSINE METHYL ESTER HYDROCHLORIDE see MQI500
4-METHYLUMBELLIFERON (CZECH) see MKP500
4-METHYLUMBELLIFERONE see MKP500
β-METHYLUMBELLIFERONE see MKP500
4-METHYLUMBELLIFERONE-O,O-DIETHYL THIOPHOSPHATE see PKT000
4-METHYLUMBELLIFERONE-4-GUANIDINOBENZOATE see GLC100
METHYL-4-UMBELLIFERONE SODIUM see HMB000
METHYL 9-UNDECENOATE see ULS400
METHYL 10-UNDECENOATE see ULS400
METHYL UNDECYLENATE see ULS400

4-METHYLURACIL see MPW500, MQI750
5-METHYLURACIL see TFX800
6-METHYLURACIL see MQI750
METHYLUREA see MQJ000
1-METHYLUREA see MQJ000
N-METHYLUREA see MQJ000
METHYL UREA and SODIUM NITRITE see MQJ250
METHYLURETHAN see MHZ000
N-METHYL URETHAN see EMQ500
METHYLURETHANE see MHZ000
N-METHYLURETHANE of HYDROCHLORIDE of 2-DIMETHYLAMINO-p-CRE-
   SOL see DPL900
N-METHYLURETHANE of HYDROCHLORIDE of 3-
   DIMETHYLAMINOPHENOL see MID000
N-METHYLURETHANE HYDROCHLORIDE of 6-DIMETHYLAMINO-o-4-
   XYLENOL see MIA775
4-METHYL VALERALDEHYDE see MQJ500
2-METHYLVALERIC ACID see MQJ750
α-METHYLVALERIC ACID see MQJ750
2-METHYL-1,5-VALERODINITRILE see MQK000
4-METHYLVALERONITRILE see MNJ000
METHYLVANILLIN see VHK000
4-o-METHYLVANILLIN see VHK000
trans-8-METHYL-N-VANILLYL-6-NONEAMIDE see CBF750
N-METHYL-N-VINYLACETAMIDE see MQK500
METHYLVINYL ACETATE see MQK750
METHYL VINYL ADIPATE see MQL000
α-METHYLVINYL BROMIDE see BOA250
METHYL VINYL CARBINOL see MQL250
METHYL-VINYL-CETONE (FRENCH) see BOY500
METHYL VINYL ETHER see MQL750
METHYL VINYL ETHER (inhibited) see MQM000
METHYLVINYLKETON (GERMAN) see BOY500
METHYL VINYL KETONE see BOY500, MQM100
METHYLVINYLNITROSAMIN (GERMAN) see NKY000
METHYLVINYLNITROSAMINE see NKY000
2-METHYL-5-VINYLPYRIDINE see MQM500
METHYL VINYL SULFONE see MQM750
2-METHYL-5-VINYL TETRAZOLE see MQM775
METHYL VIOLET see MQN025
METHYL VIOLET 2B see MQN025
METHYL VIOLET 6B see MQN000
METHYL VIOLET BB see MQN025
METHYL VIOLET CARBINOL see MQN250
METHYLVIOLET CARBINOL BASE see MQN250
METHYL VIOLET FN see MQN025
METHYL VIOLET N see MQN025
METHYL-VIOLETT (GERMAN) see MQN025
METHYLVIOLOGEN see PAJ000
METHYL VIOLOGEN (2+) see PAI990
3-METHYLXANTHINE see MQN500
1-METHYL-N-(2,6-XYLYL)ISONIPECOTAMIDE HYDROCHLORIDE
   see MQO250
1-METHYL-2-(2,6-XYLYLOXY)-ETHYLAMINE HYDROCHLORIDE see MQR775
N-METHYL-N-(1-(3,5-XYLYLOXY)-2-PROPYL)CARBAMIC ACID-2-
   DIETHYLAMINO)ETHYL ESTER, HYDROCHLORIDE see MQO500
N-METHYL-N-(1-(3,5-XYLYLOXY)-2-PROPYL)CARBAMIC ACID-2-(2-
   METHYLPIPERIDINO)ETHYL ESTER HYDROCHLORIDE see MQO750
N-METHYL-N-(1-(3,5-XYLYLOXY)-2-PROPYL)CARBAMIC ACID-2-
   (PYRROLIDINYL)ETHYL ESTER, HCl see MQP000
N-METHYL-N'-2,4-XYLYL-N-(N-2,4-XYLYLFORMIMIDOYL)FORMAMIDINE
   see MJL250
METHYL YELLOW see DOT300
METHYL ZIMATE see BJK500
METHYLZINC IODIDE see MQP250
METHYL ZINEB see BJK500, ZMA000
METHYL ZIRAM see BJK500
METHYPHENYLMETHANOL see PDE000
METHYPROLON see DNW400
METHYPRYLON see DNW400
METHYRIMOL see BRD000
METHYSERGID see MLD250
METHYSERGIDE see MLD250
METHYSERGIDE DIMALEATE see MQP500
METIAPINE see MQQ000
METIAZIC ACID see MNQ500
METIAZINIC ACID see MNQ500
METICORTELANE see PMA100
METICORTELONE see PMA000
METICORTELONE ACETATE see SOV100
METICORTELONE SOLUBLE see PMA100
METICRAN see MQQ050
METICRANE see MQQ050
METI-DERM see PMA000

METIDIONE see AOO475
METIFEX see EDW500
METIFONATE see TIQ250
METIGUANIDE see DQR800
METILACRILATO (ITALIAN) see MGA500
METILAMIL ALCOHOL (ITALIAN) see MKW600
METILAMINE (ITALIAN) see MGC250
2-METIL-6-AMINO-EPTANO (ITALIAN) see ILM000
1-METILBIGUANIDE CLORIDRATO (ITALIAN) see MHP400
3-METIL-BUTANOLO (ITALIAN) see IHP000
METIL CELLOSOLVE (ITALIAN) see EJH500
2-METILCICLOESANONE (ITALIAN) see MIR500
METILCLOROFORMIATO (ITALIAN) see MIG000
METILCLORPINDOL see CMX850
N-METIL-N-(β-DICICLOESILAMINOETIL)PIPERIDINIO BROMURO (ITAL-
    IAN) see MJC775
METILDIGOXIN see MJD300, MJD500
METILDIGOXINA (SPANISH) see MJD300
2′-METIL-3′-DIMETILAMINO-PROPIL-5-IMINODIBENZILE (ITALIAN)
    see DLH200
METILDIOLO see AOO475
N-METIL-DITIOCARBAMMATO di SODIO (ITALIAN) see VFU000
METILE (ACETATO di) (ITALIAN) see MFW100
METILENBIOTIC see MDO250
O,O-METILEN-BIS(4-CLOROFENOLO) (ITALIAN) see MJM500
3,3′-METILEN-BIS(4-IDROSSI-CUMARINA) (ITALIAN) see BJZ000
4,4-METILENE-BIS-o-CLOROANILINA (ITALIAN) see MJM200
(6-(1-METIL-EPITL)-2,4-DINITRO-FENIL)-CROTONATO (ITALIAN) see AQT500
METILESTER del ACIDO BISDEHIDROISYNOLICO (SPANISH) see BIT000
7-METILETER del ACIDO BISDEHIDRODOISYNOLICO see BIT030
METILETILCHETONE (ITALIAN) see MKA400
METIL (FORMIATO di) (ITALIAN) see MKG750
METILISOBUTILCHETONE (ITALIAN) see HFG500
METIL ISOCIANATO (ITALIAN) see MKX250
METILMERCAPTANO (ITALIAN) see MLE650
METILMERCAPTOFOSOKSID see DAP000
METIL METACRILATO (ITALIAN) see MLH750
α-METIL-β-(2-METILENE-4,5-DIIDROIMIDAZOLIL)BENZOTIOFANE
    CLORIDRATO (ITALIAN) see MHJ500
N-METIL-1-NAFTIL-CARBAMMATO (ITALIAN) see CBM750
N-(4-(β-(5-METILOSSAZOL-3-CARBOSSAMIDO)-ETIL)-BENZENESOLFONIL)-N-
    CICLOESIL-UREA see DBE885
METILPARATION (HUNGARIAN) see MNH000
4-METILPENTAN-2-OLO (ITALIAN) see MKW600
4-METILPENTAN-2-ONE (ITALIAN) see HFG500
4-METIL-3-PENTEN-2-ONE (ITALIAN) see MDJ750
1-(N-METIL-PIPERIDIL-4′)-3-FENIL-4-BENZIL-PIRAZOLONE-5 (ITALIAN)
    see BCP650
3-Γ-(α-METIL-1-PIPERIDINO)PROPIL-5,5-DIFENILTIOIDANTOINA
    CLORIDRATO (ITALIAN) see DWF875
(6-(1-METIL-PROPIL)-2,4-DINITRO-FENIL)-3,3-DIMETIL-ACRILATO (ITAL-
    IAN) see BGB500
6-(1-METIL-PROPIL)-2,4-DINITRO-FENOLO (ITALIAN) see BRE500
α-METIL-STIROLO (ITALIAN) see MPK250
2-METIL-2-TIOMETIL-PROPIONALDEID-O-(N-METIL-CARBAMOIL)-OSSIMA
    (ITALIAN) see CBM500
6-METIL-TIOURACILE (ITALIAN) see MPW500
METILTRIAZOTION see ASH500
METINDOL see IDA000
METIONE see MDT740
d-METIONIEN (AUSTRALIAN) see MDT730
METIPREGNONE see MCA000
METIPRILONE see DNW400
METIRAM see MQQ250
METISAZONUM see MKW250
METIZOL see MCO500
METIZOLIN see BAV000
METIZOLINE HYDROCHLORIDE see MHJ500
METMERCAPTURON see DST000
METOBROMURON see PAM785
METOCHLOPRAMIDE see AJH000
METOCLOL see AJH000
METOCLOPRAMIDE DIHYDROCHLORIDE MONOHYDRATE see MQQ300
METOCRYST see AOO475
METOFANE see DFA400
METOFOLINE see MDV000
2-METOKSY-4-ALLILOFENOL (POLISH) see EQR500
METOKSYCHLOR (POLISH) see MEI450
METOKSYETYLOWY ALKOHOL (POLISH) see EJH500
METOLACHLOR see MQQ450
METOLAZONE see ZAK300
METOLQUIZOLONE see QAK000
METOMIDATE see MQQ500
METOMIDATE HYDROCHLORIDE see MQQ750

METOMIL (ITALIAN) see MDU600
METON see HEA500
METOPIMAZINE see MQR000
METOPIRON see MCJ370, MJJ000
METOPIRONE see MCJ370
METOPIRONE DITARTRATE see MJJ000
METOPRINE see MQR100
METOPROLOL see MQR144
(±)-METOPROLOL see MQR144
METOPROLOL HEMITARTRATE see MQR150
METOPROLOL TARTRATE see MQR150, SBV500
METOPYRONE see MCJ370
METOQUINE see CFU750
METOSERPATE HYDROCHLORIDE see MDW100, MQR200
(2-METOSSICARBONIL-1-METIL-VINIL)-DIMETIL-FOSFATO (ITALIAN)
    see MQR750
2-METOSSIETANOLO (ITALIAN) see EJH500
2-METOSSIETILACETATO(ITALIAN) see EJJ500
S-((5-METOSSI-4H-PIRON-2-IL)-METIL)-O,O-DIMETIL-MONOTIOFOSFATO
    (ITALIAN) see EAS000
METOSSIPROPANDIOLO see RLU000
METOSYN see FDD150
METOTHYRINE see MCO500
METOX see CEP000, MEI450
METOXADONE see MFD500
METOXAL see SNK000
METOXFLURAN see DFA400
METOXIDON see SNN300
METOXIFLURAN see DFA400
METOXON see CKC000
METOXURON see MQR225
METOXYDE see HJL500
Me-TPA see MPN600
4-o-METPA see MQR250
METRACTYL see MQU750
METRAMAC see DJA400
METRAMAK see DJA400
METRAMINOL BITARTRATE see HNC000
(−)-METRAMINOL (+)-BITARTRATE see HNC000
l-METRAMINOL BITARTRATE see HNC000
d-(−)-METRAMINOL BITARTRATE see HNC000
METRANIL see PBC250
METRASPRAY see BQA010
METRIBEN see TIK000
METRIBUZIN see MQR275
METRIFONATE see TIQ250
METRIPHONATE see TIQ250
METRISONE see MOR500
METRIZAMIDE see MQR300
METRIZOATE see MQR350
METRIZOIC ACID see MQR350
METRODIOL see EQJ500
METRODIOL DIACETATE see EQJ500
METROGEN RED FORMER KB SOLN see CLK225
METROGESTONE see MBZ100
METRON see MNH000
METRONE see MPN500
METRONIDAZ see MMN250
METRONIDAZOL see MMN250
METRONIDAZOLO see MMN250
METROPINE see MGR500
METRORAT see DBE200
METRO TALC 4604 see TAB750
METRO TALC 4608 see TAB750
METRO TALC 4609 see TAB750
METROXEDRINE see SPC500
METSO 20 see SJU000
METSO BEADS 2048 see SJU000
METSO BEADS, DRYMET see SJU000
METSO PENTABEAD 20 see SJU000
MET-SPAR see CAS000
METSUCCIMIDE see MLP800
METURAL see DTG700
METYCAINE see PIV750
METYLAL (POLISH) see MGA850
METYLENU CHLOREK (POLISH) see MJP450
METYLESTER KYSELINY SALICYLOVE (CZECH) see MPI000
METYLFENEMAL see ENB500
METYLOAMINA (POLISH) see MGC250
METYLOCYKLOHEKSAN (POLISH) see MIQ740
METYLOCYKLOHEKSANOL (POLISH) see MIQ745
METYLOCYKLOHEKSANON (POLISH) see MIR250
METYLOETYLOKETON (POLISH) see MKA400
METYLOHYDRAZYNA (POLISH) see MKN000

METYLOIZOBUTYLOKETON (POLISH) see HFG500
1-METYLO-2-MERKAPTOIMIDAZOLEM (POLISH) see MCO500
N-METYLO-N'-NITRO-N-NITROZOGOUANIDYNY (POLISH) see MMP000
METYLOPARATION (POLISH) see MNH000
METYLOPROPYLOKETON (POLISH) see PBN250
7-(β-METYLOTIOETYLO)-TEOFILINA (POLISH) see MDN100
METYLOWY ALKOHOL (POLISH) see MGB150
METYLPARATION (CZECH) see MNH000
METYLU BROMEK (POLISH) see MHR200
METYLU CHLOREK (POLISH) see MIF765
METYLU JODEK (POLISH) see MKW200
METYNA see ENB500
METYPRAPONE BITARTRATE see MJJ000
METYRAPON see MCJ370
METYRAPONE see MCJ370
METYRAPONE DITARTRATE see MJJ000
MEV see MNB250
MEVASINE see VIZ400
MEVASIN HYDROCHLORIDE see MQR500
MEVINFOS (DUTCH) see MQR750
MEVINPHOS see MQR750
MEXACARBATE (DOT) see DOS000
MEXAMINE HYDROCHLORIDE see MFT000
MEXAZOLAM see MQR760
MEXENE see BJK500
MEXEPHENAMIDE see DIB600, MCH250
MEXICA FLAME LEAF see EQX000
MEXICAN BREADFRUIT see SLE890
MEXICAN FLAME TREE see EQX000
MEXICAN FLOWER PLANT see EQX000
MEXICAN HUSK TOMATO see JBS100
MEXICO WEED see CCP000
MEXIDE see RNZ000
MEXIDEX see SOW000
MEXILETINE HYDROCHLORIDE see MQR775
MEXITIL see MQR775
MEXOCINE see DAI485, MIJ500
MEXOLAMINE see CPN750
MEXYL see ABX500
MEYPRALGIN R/LV see SEH000
MEZARONIL see ASG250
MEZATON see NCL500, SPC500
R(−)-MEZATON see NCL500
MEZCALINE see MDI500
MEZCALINE SULFATE see MDJ000
MEZCLINE see MDI500
MEZENE see BJK500
MEZEPAN see CGA000
MEZEREIN see MDJ250
MEZEREUM see LAR500
MEZIDINE see TLG500
MEZINEB see ZMA000
MEZINIUM METHYL SULFATE see MQS100
MEZLOCILLIN see MQS200
MEZOLIN see IDA000
MEZOTOX see DFT800
MEZURON see ASG250
MF-344 see EFK000
MF 565 see MAP300
MF 598 see MAP300
MFA see FIC000, MKD000, TLX175
M.F. FULVIUS VENOM see CNR150
MFH see FNW000
MFI see IPX000
MFI-PC see FDA100
MFM see TLX175
MFNA see MME809
M. FULVIUS FULVIUS VENOM see CNR150
h-MG see CGY750
MG 46 see LGK200
M.G. 8823 see ERE100
MG 8926 see CPP000
MG 18037 see BQX000
MG 18370 see DTJ400
MG 18415. see AJK000
MG 18512 see CPP750
MG 18570 see DTJ400
M.G. 18590 see TEY600
M.G.18755 see IBW100
MGA see MCB380
MGA 100 (STEROID) see MCB380
M.G. 8948-2HCl see BJW000
M.G. 18001-3HCl see BJV750
M7-GIFTKOERNER see TEM000

MGK 11 see BHJ500
MGK-264 see OES000
MGK 326 see EAU500
MGK DIETHYLTOLUAMIDE see DKC800
MGK DOG AND CAT REPELLENT see UKS000
MGK REPELLENT 11 see BHJ500
MGK REPELLENT-326 see EAU500
MGK REPELLENT 874 see OGA000
MGK REPELLENT 1,207 see CKU750
MH see MKK600
M.H. see HAP000
M-2H see BPF825
MH-30 see DHF200
MH-532 see PGA750
2M-4KH see CIR250
MHOROMER see EJH000
MHP see NKU500
683 M HYDROCHLORIDE see OOE100
3-MI see MKV750
MI 85 see AQN750
N-6-MI see NKI000
217 MI see TLF500
MIA see IDZ000
MIADONE see MDP750
MIAK see MKW450
MIALEX see CKK250
MIANESINA see GGS000
MIANG TEA LEAF EXTRACT see MQS215
MIANSERIN see MQS220
MIANSERINE see MQS220
MIANSERINE HYDROCHLORIDE see BMA625
MIANSERIN HYDROCHLORIDE see BMA625
MIANSERYNA see MQS220
MIARSENOL see NCJ500
MIAZOLE see IAL000
MIBC see MKW600
MIBK see HFG500
MIBOLERON see MQS225
MIBOLERONE see MQS225
MIBP see MRI775
MIBT see MKW250
MIC see ISE000, MKW600
3-MIC see MKW600
MICA see MQS250
MICA SILICATE see MQS250
MICASIN see CDS500, CKL500
MICHLER'S BASE see MJN000
MICHLER'S HYDRIDE see MJN000
p,p'-MICHLER'S HYDROL see TDO750
MICHLER'S KETONE see MQS500
p,p'-MICHLER'S KETONE see MQS500
MICHLER'S METHANE see MJN000
MICHROME No. 226 see AJQ250
MICIDE see EIR000
MICOCHLORINE see CDP250
MICOFENOLICO ACIDO (SPANISH) see MRX000
MICOFUME see DSB200
MICOFUR see NGC000
MICOL see HCQ500
MICONAZOLE see MQS550
MICONAZOLE NITRATE see MQS560
MICREST see DKA600
MICROCETINA see CDP250
MICRO-CHECK 12 see CBG000
MICRO-CHEK 11 see OFE000
MICRO-CHEK SKANE see OFE000
MICROCID see GFG100
MICROCILLIN see CBO250
MICRO DDT 75 see DAD200
MICRODIOL see EDO000
MICRO DRY see AHA000
MICROEST see DKA600
MICROFLOTOX see SOD500
MICROGRIT WCA see AHE250
MICROGYNON see NNL500
MICROIODIDE see PLW285
MICRO-LEX GREEN 5B see BLK000
MICROLYSIN see CKN500
MICROMICIN see MQS579
MICROMYCIN see MQS579
MICRONASE see CEH700
MICRONOMICIN SULFATE see MQS600
MICRONOMYCIN SULFATE see MQS600
MICROPENIN see MNV250

MICROSETILE BLUE EB see TBG700
MICROSETILE BLUE FF see MGG250
MICROSETILE DIAZO BLACK G see DPO200
MICROSETILE ORANGE RA see AKP750
MICROSETILE VIOLET 3R see DBP000
MICROSETILE YELLOW GR see AAQ250
MICROTAN PIRAZOLO see AIF000
MICROTEX LAKE RED CR see CHP500
MICROTHENE see PJS750
MICROTIN see SFQ500
MICROTRIM see TKX000
MICROZUL see CJJ000
MICRURUS ALLENI YATESI VENOM see MQS750
MICRURUS CARINICAUDUS DUMERILII VENOM
    see MQT000
MICRURUS FULVIUS FULVIUS VENOM see CNR150
MICRURUS FULVIUS VENOM see MQT100
MICRURUS MIPARTIUS HERTWIGI VENOM see MQT250
MICRURUS NIGROCINCTUS VENOM see MQT500
MICTINE see AFW500
MIDARINE see HLC500
MIDAZOLAM see MQT525
MIDECAMYCIN see LEV025, MBY150
MIDECAMYCIN A₁ see MBY150
MIDELID see BGC625
MIDETON FAST RED VIOLET R see DBP000
MI 85 DI see ASA000
MIDICEL see AKO500
MIDIKEL see AKO500
MIDODRINE see MQT530
(±)-MIDODRINE HYDROCHLORIDE see MQT530
MIDONE see DBB200
MIDOXIN see HGP550
MIDRONAL see CMR100
MIEDZ (POLISH) see CNM000
MIELUCIN see BOT250
MIEOBROMOL see DDP600
MIERENZUUR (DUTCH) see FNA000
MIFEPRISTONE see HKB700
MIFUROL see CCK630
MI-GEE see DNF800
MIGHTY 150 see NAJ500
MIGLYOL 810 NEUTRAL OIL see CBF710
MIGLYOL 812 NEUTRAL OIL see CBF710
MIGRISTENE see FMU039
MIH see PME250
MIH HYDROCHLORIDE see PME500
MIK see HFG500
MIKACION BRILLIANT RED 5BS see PMF540
MIKAMETAN see IDA000
MIKAMYCIN see VRF000
MIKAMYCIN B see VRA700
MIKAMYCIN IA see VRA700
MIKEDIMIDE see MKA250
MIKELAN see MQT550
MIKETHRENE GOLD YELLOW see DCZ000
MIKETHRENE GREY K see IBJ000
MIKETHRENE GREY M see CMU475
MIKETHRENE GREY MG see CMU475
MIKETHRENE MARINE BLUE G see CMU500
MIKETHRENE RED BROWN 5RF see CMU800
MIKETON BRILLIANT BLUE B see MGG250
MIKETON FAST BLUE see TBG700
MIKETORIN see EAI000
MIKIPALAOA (HAWAII) see CNG825
MIL see MQU000
MILAXEN see HEG000
MILBAM see BJK500
MILBAN see BJK500
MIL-B-4394-B see CES650
MILBEDOCE see VSZ000
MILBEMYCIN D see MQT600
MILBEX see CKL500
MILBEX mixed with PHOSALON (2:1) see MQT750
MILBEX mixed with PHOSALONE (2:1)/ see MQT750
MILBOL see BIO750
MILBOL 49 see BBQ500
MILCHSAURE (GERMAN) see LAG000
MIL-COL see MLC250
MILCURB see BRD000, BRI750
MILCURB SUPER see BRI750
MILD ACONITATE see ADH500
MILD ACONITINE see ADH500
MILDIOMYCIN see MQU000

MILDMEN see LFK000, MDQ250
MILD MERCURY CHLORIDE see MCW000
MIL-DU-RID see BGJ750
MILEPSIN see DBB200
MILESTROL see DKA600
MILEZIN see MCI500
MILFARON see CDP750
MILGO see BRI750
MILGO E see BRI750
MILID see BGC625
MILIDE see BGC625
MILK ACID see LAG000
MILK of MAGNESIA see MAG750
MILK SUGAR see LAR000
MILK WHITE see LDY000
MIL L 7808 see MQU250
MIL L 17535 see MQU500
MILLED CASSAVA POWDER see CCO700
MILLER NU SET see TIX500
MILLER P.C. WEEDKILLER see PLC250
MILLICORTEN see SOW000
MILLIPHYLLINE see CNR125
MILLON'S BASE ANHYDRIDE see DNW200
MILLOPHYLLINE see CNR125
MILMER see BLC250
MILOCEP see MQQ450
MILOGARD see PMN850
MILONTIN see MNZ000
MILOXACIN see MQU525
MILPREM see ECU750, MQU750
MILPREX see DXX400
MILSAR see CKL500
MILSTEM see BRI750
MILSTEM SEED DRESSING see BRI750
MILTANN see MQU750
MILTON see SHU500
MILTOWN see MQU750
MILTOX see EIR000, ZJS300
MILTOX SPECIAL see EIR000
MIMEDRAN see PIK625
MIMOSA ABSOLUTE see MQV000
MIMOSA TANNIN see MQV250
MIMOSINE see HOC500
MINACIDE see CQI500
MINAPHIL see TEP500
MINCARD see AFW500
MINEPENTATE see DPA800
MINERAL DUSTS see MQV500
MINERAL FIRE RED 5GS see MRC000
MINERAL GREEN see COF500
MINERAL NAPHTHA see BBL250
MINERAL OIL see MQV750
MINERAL OIL, PETROLEUM CONDENSATES, VACUUM TOWER
    see MQV755
MINERAL OIL, PETROLEUM DISTILLATES, ACID-TREATED HEAVY NAPH-
    THENIC see MQV760
MINERAL OIL, PETROLEUM DISTILLATES, ACID-TREATED HEAVY PAR-
    AFFINIC see MQV765
MINERAL OIL, PETROLEUM DISTILLATES, ACID-TREATED LIGHT NAPH-
    THENIC see MQV770
MINERAL OIL, PETROLEUM DISTILLATES, ACID-TREATED LIGHT PAR-
    AFFINIC see MQV775
MINERAL OIL, PETROLEUM DISTILLATES, HEAVY NAPHTHENIC
    see MQV780
MINERAL OIL, PETROLEUM DISTILLATES, HEAVY PARAFFINIC
    see MQV785
MINERAL OIL, PETROLEUM DISTILLATES, HYDROTREATED HEAVY
    NAPHTHENIC see MQV790
MINERAL OIL, PETROLEUM DISTILLATES, HYDROTREATED HEAVY PAR-
    AFFINIC see MQV795
MINERAL OIL, PETROLEUM DISTILLATES, HYDROTREATED LIGHT
    NAPHTHENIC see MQV800
MINERAL OIL, PETROLEUM DISTILLATES, HYDROTREATED LIGHT PAR-
    AFFINIC see MQV805
MINERAL OIL, PETROLEUM DISTILLATES, LIGHT NAPHTHENIC
    see MQV810
MINERAL OIL, PETROLEUM DISTILLATES, LIGHT PARAFFINIC
    see MQV815
MINERAL OIL, PETROLEUM DISTILLATES, SOLVENT-DEWAXED HEAVY
    NAPHTHENIC see MQV820
MINERAL OIL, PETROLEUM DISTILLATES, SOLVENT-DEWAXED HEAVY
    PARAFFINIC see MQV825
MINERAL OIL, PETROLEUM DISTILLATES, SOLVENT-DEWAXED LIGHT
    NAPHTHENIC see MQV835

MINERAL OIL, PETROLEUM DISTILLATES, SOLVENT-DEWAXED LIGHT PARAFFINIC see MQV840
MINERAL OIL, PETROLEUM DISTILLATES, SOLVENT-REFINED HEAVY NAPHTHENIC see MQV845
MINERAL OIL, PETROLEUM DISTILLATES, SOLVENT-REFINED HEAVY PARAFFINIC see MQV850
MINERAL OIL, PETROLEUM DISTILLATES, SOLVENT-REFINED LIGHT NAPHTHENIC see MQV852
MINERAL OIL, PETROLEUM DISTILLATES, SOLVENT-REFINED LIGHT PARAFFINIC see MQV855
MINERAL OIL, PETROLEUM EXTRACTS, HEAVY NAPHTHENIC DISTIL-LATE SOLVENT see MQV857
MINERAL OIL, PETROLEUM EXTRACTS, HEAVY PARAFFINIC DISTIL-LATE SOLVENT see MQV859
MINERAL OIL, PETROLEUM EXTRACTS, LIGHT NAPHTHENIC DISTIL-LATE SOLVENT see MQV860
MINERAL OIL, PETROLEUM EXTRACTS, LIGHT PARAFFINIC DISTIL-LATE SOLVENT see MQV862
MINERAL OIL, PETROLEUM EXTRACTS, RESIDUAL OIL SOLVENT see MQV863
MINERAL OIL, PETROLEUM NAPHTHENIC OILS, CATALYTIC DEWAXED HEAVY see MQV865
MINERAL OIL, PETROLEUM NAPHTHENIC OILS, CATALYTIC DEWAXED LIGHT see MQV867
MINERAL OIL, PETROLEUM PARAFFIN OILS, CATALYTIC DEWAXED HEAVY see MQV868
MINERAL OIL, PETROLEUM PARAFFIN OILS, CATALYTIC DEWAXED LIGHT see MQV870
MINERAL OIL, PETROLEUM RESIDUAL OILS, ACID-TREATED see MQV872
MINERAL OIL, SLAB OIL see MQV875
MINERAL OIL, WHITE (FCC) see MQV750
MINERAL ORANGE see LDS000
MINERAL PITCH see ARO500
MINERAL RED see LDS000
MINERAL SPIRITS see PCT250
MINERAL THINNER see PCT250
MINERAL TURPENTINE see PCT250
MINERAL WHITE see CAX750
MINETOIN see DKQ000, DNU000
MINGIT see DFP600
MINIDRIL see NNL500
MINIHIST see DBM800, WAK000
MINILYN see EEH575
MINIMYCIN see OMK000
MINING REAGENT see MQW000
MINING REAGENT, liquid (containing 20% or more cresylic acid) (DOT) see MQW000
MINIPLANOR see ZVJ000
MINIUM see LDS000
MINIUM NON-SETTING RL-95 see LDS000
MINOALEUIATIN see TLP750
MINOCIN see MQW100
MINOCYCLIN see MQW250
MINOCYCLINE see MQW250
MINOCYCLINE CHLORIDE see MQW100
MINOCYCLINE HYDROCHLORIDE see MQW100
MINOPHAGEN A see AQW000
MINORAN see MEK700
MINORLAR see EEH520
MINOSSIDILE (ITALIAN) see DCB000
MINOVLAR see EEH520
MINOXIDIL see DCB000
MINOZINAN see MCI500
MINPROG see POC350
MINTACO see NIM500
MINTACOL see NIM500
MINTAL see NBU000
MINTEZOL see TEX000
MINTUSSIN see MGR250
MINURIC see DDP200
MINUS see DKE800, SEH000
MINUSIN see NNW500
MINZIL see CLH750
MINZOLUM see TEX000
MIOARTRINA see IPU000
MIOBLOCK see PAF625
MIOCURIN see RLU000
MIODAR see PGG350
MIOFILIN see TEP500
MIOLISODAL see IPU000
MIOLISODOL see IPU000
MIONAL see EAV700
MIORATRINA see IPU000
MIORELAX see RLU000

MIORIL see IPU000
MIORILAX see CKF500
MIORIODOL see IPU000
MIO-SED see CKF500
MIOSTAT see CBH250
MIOTICOL see DNR309
MIOTINE see MIC250
MIOTISAL see NIM500
MIOTISAL A see NIM500
MIOTOLON see AOO300
MIPAFOX (DOT) see PHF750
MIPAX see DTR200
MIPC see MIA250
MIPCIN see MIA250
MI-PILO OPHTH SOL see PIF250
MIPK see MLA750
MIPSIN see MIA250
MIPSIN, nitrosated (JAPANESE) see MMW250
MIRACIL D see DHU000, SBE500
MIRACIL D HYDROCHLORIDE see SBE500
MIRACLE see DAA800
MIRACOL see SBE500
MIRADOL see EPD500
MIRAL see PDN000
MIRAMEL see EQL000
MIRAMID WM 55 see PJY500
MIRANOL C2M-SF CONC see AOC250
MIRANOL MHT see AOC275
MIRAPRONT see DTJ400
MIRBANE OIL see NEX000
MIRBANIL see EPD500
MIRCOL see MCJ400
MIRCOSULFON see PPP500
MIREX see MQW500
MIRLON see NOH000
MIROISTONIL see DRM000
MIROMORFALIL see NAG500
MIROPROFEN see IAY000
MIROSERINA see CQH000
MIROTIN see MNZ000
MIRREX MCFD 1025 see PKQ059
MISASIN see CKL500
MISCLERON see ARQ750
MISHERI (INDIA) see SED400
MISHRI (INDIA) see SED400
MISODINE see DBB200
MISOLYNE see DBB200
MISONIDAZOLE see NHH500
MISOPROSTOL see MJE775
MISPICKEL see ARJ750
MISTABRON see MDK875
MISTABRONCO see MDK875
MISTLETOE (AMERICAN) see MQW525
MISTRON 2SC see TAB750
MISTRON FROST P see TAB750
MISTRON RCS see TAB750
MISTRON STAR see TAB750
MISTRON SUPER FROST see TAB750
MISTRON VAPOR see TAB750
MISTURA C see CBH250
MISULBAN see BOT250
MISULVAN see EPD500
MISURAN see MSC100
MIT see ISE000
MITABAN see MJL250
MITAC see MJL250
MITACIL see DOR400
MITANOLINE see PMC250
MIT-C see AHK500
MITC see ISE000
MITENON see MMD500
MITEXAN see MDK875
MITHRACIN see MQW750
MITHRAMYCIN see MQW750
MITHRAMYCIN A see MQW750
MITHRAMYCIN, MAGNESIUM SALT see MAD050
MITICIDE K-101 see CJT750
MITIGAN see BIO750
MITION see CKM000
MITIS GREEN see COF500
MITOBRONITOL see DDP600
MITO-C see AHK500
MITOCHROMIN see MQX000
MITOCIN-C see AHK500

MITOCROMIN see MQX000
MITOLAC see DDJ000
MITOLACTOL see DDJ000
MITOMALCIN see MQX250
MITOMEN see CFA500, CFA750
MITOMIN see CFA500
MITOMYCIN see AHK500
MITOMYCIN A see MQX500
MITOMYCIN B see MQX750
MITOMYCIN-C see AHK500
MITOMYCINUM see AHK500
MITONAFIDE see MQX775
MITOSTAN see BOT250
MITOTANE see CDN000
MITOX see CEP000
MITOXAN see CQC500, CQC650
MITOXANA see IMH000
MITOXANTHRONE see MQY090
MITOXANTRONE see MQY090
MITOXANTRONE HYDROCHLORIDE see MQY100
MITOXINE see BIE500
MITOZOLOMIDE see MQY110
MITRAMYCIN see MQW750
MITRONAL see CMR100
MITSUI ALIZARINE B see DMG800
MITSUI AURAMINE O see IBA000
MITSUI BLUE B BASE see DCJ200
MITSUI BRILLIANT GREEN G see BAY750
MITSUI CHROME ORANGE A see NEY000
MITSUI CONGO RED see SGQ500
MITSUI CROME ORANGE AN see NEY000
MITSUI DIRECT BLACK EX see AQP000
MITSUI DIRECT BLUE 2BN see CMO000
MITSUI METANIL YELLOW see MDM775
MITSUI METHYLENE BLUE see BJI250
MITSUI RED TR BASE see CLK220
MITSUI RHODAMINE BX see FAG070
MITSUI SCARLET G BASE see NMP500
MITUSI SCARLET GG BASE see DEO400
MIVIZON see FNF000
MIX SODIUM see MDV750
MIXTURE of p-METHENOLS see TBD500
MIZODIN see DBB200
MIZOLIN see DBB200
MIZORIBINE see BMM000
MJ 505 see PGG350, PGG355
MJ 1992 see SKW000, SKW500
MJ 1998 see HLX500
MJ 1999 see CCK250
MJ 5022 see MPE250
MJ 5190 see AHL500
MJ 10061 see DDP200
MJ 4309-1 see OPK000
MJF 9325 see IMH000
MJF-12264 see FMB000
MJ 1999 HYDROCHLORIDE see CCK250
MK₄ see VTA650
MK-33 see GEW700
MK 56 see POL500
MK 125 see SOW000
MK 141 see PCI500
MK 142 see MQY125
MK 184 see TAF675
MK-188 see RBF100
MK 196 see IBQ400
MK 231 see SOU550
MK 240 see DDA600
MK-351 see MJE780
MK 351 see DNA800
MK 360 see TEX000
MK-366 see BAB625
MK 421 see EAO100
MK 486 see CBQ500
MK-595 see DFP600
MK 647 see DK1600
MK 665 see CHP750
MK-872 see HMS875
MK 905 see CBA100
MK 933 see ITD875
MK 950 see TGB185
MK-955 see PHA550
(±)-MK 196 see IBQ400
MK.B51 see MJE780
MK. B51 see DNA800

MK-142 DIMETHANESULFONATE see MQY125
2M-4KH SODIUM SALT see SIL500
MK 421 MALEATE see EAO100
ML 97 see FAB400, PGX300
ML 33F see SKV100
ML 55F see SKV100
ML 1024 see TEQ500
MLA-74 see MLD500
MLO-5277 (ORGANO-SILICATE) see MQY250
MLT see MAK700
MM see BKM500
MM 4462 see AOP250
MM 14151 see CMV250
MMA see MGQ250
3M MBR 6168 see DRR000
MMC see AHK500, MDD750
MMD see MLF250
MME see MFC700, MLH750
MMH see MKN000
4-MMPD see DBO000
4-MMPD SULPHATE see DBO400
MMS see MLH500
MMT see MAV750
MMTP see MLX750
MN-1695 see MQY300
MNA see MMU250, NEN500
6-MNA see MMT250
MNB see MMU500
MNBK see HEV000
MNC see MMX000
MNCO see MQY325
MNE see NDM000
MNFA see MME809
MNG see MMP000
MNNG see MMP000
MNPN see MMS200
MNQ see MMD500
MNT see MDQ075, MNA500, NMO500
MNU see MMX250, MNA750
MNU and CYCLOPHOSPHAMIDE (2:1) see CQC600
MO see NIW500
MO 338 see NIW500
MO 709 see DRM000
MOB see MES000
M-2-OB see MMR800
M-3-OB see MMR810
MOBAM see BDG250
MOBAM PHENOL see BDG250
MOBENOL see BSQ000
MOBILAN see IDA000
MOBILAT see MQY350
MOBILAWN see DFK600
MOBIL DBHP see DEG800
MOBIL MC-A-600 see BDG250
MOBIL V-C 9-104 see EIN000
MOBUTAZON see MQY400
MOBUZON see MQY400
MOCA see MJM200
MOCAP see EIN000
MOCK AZALEA see DBA450
MOCTYNOL see GGS000
MODALINE SULFATE see MOM750
MODANE see DMH400
MODANE SOFT see DJL000
MODECCIN see MRA000
MODECCIN TOXIN see MRA000
MODERAMIN see MFD500
MODERIL see TLN500
MODIFIER 113-63 see MRA100
MODIRAX see POA250
MODITEN see FMP000, TJW500
MODITEN ENANTHATE see PMI250
MODITEN-RETARD see PMI250
MODOCOLL 1200 see SFO500
MODR ALIZARINOVA CISTA B (CZECH) see AIY750
MODR BRILANTNI ALIZARINOVA BRL (CZECH) see DKR400
MODR BRILANTNI OSTAZINOVA S-R (CZECH) see DGN400
MODRENAL see EBY600
MODR FRALOSTANOVA 3G (CZECH) see DNE400
MODR METHYLENOVA (CZECH) see BJI250
MODR MIDLONOVA STALA ER (CZECH) see TMA750
MODR NAMORNICKA OSTANTHRENOVA G (CZECH) see CMU500
MODR NAMORNICKA OSTANTHRENOVA RA (CZECH) see CMU750
MODR OSTACETOVA LG (CZECH) see BND250

MODR OSTACETOVA LR see BNC800
MODR OSTACETOVA P3R (CZECH) see MGG250
MODR OSTACETOVA SE-LB (CZECH) see DBT000
MOENOMYCIN see MRA250
MOENOMYCIN A see MRA250
MOF see DFA400
MOFEBUTAZONE see MQY400
MOGADAN see DLY000
MOGALAROL see MRA275
MOGETON GRANULE see AJI250
MOHEPTAN see MDP750
MOHICAN RED A-8008 see CHP500
MOLANTIN P see MRA300
MOLASSES ALCOHOL see EFU000
MOLATOC see DJL000
MOLCER see DJL000
MOLDAMIN see BFC750
MOLDCIDIN B see FPC000
MOLDEX see HJL500
MOLECULAR CHLORINE see CDV750
MOLECULAR SIEVE 13X with 14.6% DI-n-BUTYLAMINE see MRA750
MOLE DEATH see SMN500
MOLEVAC see PQC500
MOLINATE see EKO500
MOLINDONE HYDROCHLORIDE see MRB250
MOLINILLO (PUERTO RICO) see SAT875
MOLIPAXIN see THK880, CKJ000
MOL-IRON see FBO000
MOLIVATE see CMW400
MOLLAN O see DVL700
MOLLINOX see QAK000
MOLLUSCICIDE BAYER 73 see DFV600
MOLMATE see EKO500
MOLOFAC see DJL000
MOLOL see MQV750
MOLSIDOLAT see MRN275
MOLSIDOMINE see MRN275
MOLTEN ADIPIC ACID see AEN250
MOLURAME see BJK500
MOLYBDATE see MRC250
MOLYBDATE ORANGE see MRC000
MOLYBDATE RED see MRC000
MOLYBDENITE see MRD750
MOLYBDEN RED see MRC000
MOLYBDENUM see MRC250
MOLYBDENUM AZIDE PENTACHLORIDE see MRC500
MOLYBDENUM AZIDE TRIBROMIDE see MRC600
MOLYBDENUM BORIDE see MRC650
MOLYBDENUM-COBALT-CHROMIUM ALLOY see VSK000
MOLYBDENUM COMPOUNDS see MRC750
MOLYBDENUM DIAZIDE TETRACHLORIDE see MRD000
MOLYBDENUM DIOXIDE see MRD250
MOLYBDENUM-LEAD CHROMATE see LDM000
MOLYBDENUM ORANGE see LDM000
MOLYBDENUM OXIDE see MRD250
MOLYBDENUM(VI) OXIDE see MRD250, MRE000
MOLYBDENUM PENTACHLORIDE see MRD500
MOLYBDENUM RED see MRC000
MOLYBDENUM(IV) SULFIDE see MRD750
MOLYBDENUM, TRICARBONYL(1,3,5-CYCLOHEPTATRIENE)- see COY100
MOLYBDENUM TRIOXIDE see MRE000
MOLYBDIC ACID DIAMMONIUM SALT see ANM750
MOLYBDIC ACID, DISODIUM SALT see DXE800
MOLYBDIC ANHYDRIDE see MRE000
MOLYBDIC SULFIDE see MRD750
MOLYBDIC TRIOXIDE see MRE000
MOLYKOTE see MRD750
MOLYKOTE 522 see TAI250
MOMENTOL see TKX000
MOMENTUM see HIM000
MOMETINE see CJL500
MOMORDICA BALSAMINA see FPD100
MOMORDICA CHARANTIA see FPD100
MOMORDIQUE A FUEILLES de VIGNE see FPD100
MON 0573 see PHA500
MONACETYLFERROCENE see ABA750
MONACRIN see AHS500, AHS750
MONACRIN HYDROCHLORIDE see AHS750
MONAGYL see MMN250
MONALIDE see CGL250
MONAM see VFW009
MONAMID 150-LW see BKE500
MONAQUEST see DJG800
MONARCH see ZVJ000

MONARGAN see ABX500
MONASIRUP see PMQ750
MONATE see MRL750
MONAWET MD 70E see DJL000
MONAZAN see MQY400
MONDUR P see PFK250
MONDUR-TD see TGM740, TGM750
MONDUR-TD-80 see TGM740, TGM750
MONDUR TDS see TGM750
MONELAN see MRE225
MONELGIN see OAV000
MONENSIC ACID see MRE225
MONENSIN (USDA) see MRE225
MONENSIN A see MRE225
MONENSIN, MONOSODIUM SALT (9CI) see MRE230
MONENSIN SODIUM see MRE230
MONENSIN SODIUM SALT see MRE230
MONEX see BJL600
MONGARE see MJK750
MONHYDRIN see PMJ500
MONILIFORMIN see MRE250
MONITAN see PKL100
MONITOR see DTQ400
MONKEY PISTOL see SAT875
MONKEY'S DINNER BELL see SAT875
MONKIL WP see MGQ750
MONKSHOOD see MRE275
MONOACETIN see GGO000
1-MONOACETIN see GGO000
α-MONOACETIN see GGO000
MONOACETYL GLYCERINE see GGO000
MONOACETYLHYDRAZINE see ACM750
MONOACETYL 4-HYDROXYAMINOQUINOLINE see QQA000
MONOACETYLISOSPIRAMYCIN see MRE500
MONOACETYLSPIRAMYCIN see MRE750
MONOAETHANOLAMIN (GERMAN) see EEC600
MONOALLYLAMINE see AFW000
MONOALLYLUREA see AGV000
MONOALUMINUM PHOSPHATE see PHB500
MONOAMMONIUM CARBONATE see ANB250
MONOAMMONIUM GLUTAMATE see MRF000
MONOAMMONIUM l-GLUTAMATE see MRF000
MONOAMMONIUM GLYCYRRHIZINATE see GIE100
MONOAMMONIUM SULFAMATE see ANU650
MONOAMMONIUM SULFIDE see ANJ750
MONOAZO see MDM775
MONOBASIC racemic AMPHETAMINE PHOSPHATE see AOB500
MONOBASIC CHROMIUM SULFATE see NBW000
MONOBASIC CHROMIUM SULPHATE see NBW000
MONOBASIC LEAD ACETATE see LCH000
MONOBASIC dl-α-METHYLPHENETHYLAMINE PHOSPHATE see AOB500
MONOBENZALPENTAERYTHRITOL see MRF200
MONOBENZONE see AEY000
MONOBENZYL-p-AMINOPHENOL HYDROCHLORIDE see MRF250
MONOBENZYL ETHER HYDROQUINONE see AEY000
MONOBENZYL HYDROQUINONE see AEY000
MONOBOR-CHLORATE see SFS500
MONOBROMESSIGSAEURE (GERMAN) see BMR750
MONOBROMOACETALDEHYDE see BMR000
MONOBROMOACETIC ACID see BMR750
MONOBROMOACETONE see BNZ000
MONOBROMOBENZENE see PEO500
MONOBROMODIFLUOROPHOSPHINE SULFIDE see TFO500
MONOBROMOETHANE see EGV400
MONOBROMOGLYCEROL see MRF275
MONOBROMOISOVALERYLUREA see BNP750
2-MONOBROMOISOVALERYLUREA see BNP750
MONOBROMOMETHANE see MHR200
MONOBUTYL see MQY400
MONO-n-BUTYLAMINE see BPX750
MONOBUTYL DIPHENYL SODIUM MONOSULFONATE see AQU750
MONOBUTYL GLYCOL ETHER see BPJ850
MONO-tert-BUTYLHYDROQUINONE see BRM500
MONOBUTYL PHOSPHITE see MRF500
MONOBUTYL PHTHALATE see MRF525
MONO-n-BUTYL PHTHALATE see MRF525
MONOBUTYLTIN TRICHLORIDE see BSO750
MONOBUTYLTIN TRILAURATE see BSO750
MONOCAINE HYDROCHLORIDE see IAC000
MONOCALCIUM ARSENITE see CAM500
MONOCALCIUM PHOSPHATE see CAW110
MONOCARBETHOXYHYDRAZINE see EHG000
MONOCHLOORAZIJNZUUR (DUTCH) see CEA000
MONOCHLOORBENZEEN (DUTCH) see CEJ125

MONOCHLORACETIC ACID see CEA000
MONOCHLORACETONE see CDN200
MONOCHLORAMIDE see CDO750
MONOCHLORAMINE see CDO750
MONOCHLORBENZENE see CEJ125
MONOCHLORBENZOL (GERMAN) see CEJ125
MONOCHLORESSIGSAEURE (GERMAN) see CEA000
MONOCHLORETHANE see EHH000
MONOCHLORHYDRIN see CDT750
MONOCHLORHYDRINE du GLYCOL (FRENCH) see EIU800
MONOCHLORIMIPRAMINE see CDU750
MONOCHLOROACETALDEHYDE see CDY500
MONOCHLOROACETIC ACID see CEA000
MONOCHLOROACETIC ACID METHYL ESTER see MIF775
MONOCHLOROACETONE see CDN200
MONOCHLOROACETONE, inhibited (DOT) see CDN200
MONOCHLOROACETONE, stabilized (DOT) see CDN200
MONOCHLOROACETONE, unstabilized (DOT) see CDN200
MONOCHLOROACETONITRILE see CDN500
MONOCHLOROACETYL CHLORIDE see CEC250
MONOCHLOROAMINE see CDO750
MONOCHLOROAMMONIA see CDO750
α-MONOCHLOROANTHRAQUINONE see CEI000
MONOCHLOROBENZENE see CEJ125
MONOCHLORODIFLUOROMETHANE see CFX500
MONOCHLORODIMETHYL ETHER (MAK) see CIO250
MONOCHLORO DIPHENYL OXIDE see MRG000
MONOCHLOROETHANOIC ACID see CEA000
2-MONOCHLOROETHANOL see EIU800
MONOCHLOROETHENE see VNP000
MONOCHLOROETHYLENE (DOT) see VNP000
MONOCHLOROETHYLENE OXIDE see CGX000
MONOCHLOROHYDRIN see CDT750
α-MONOCHLOROHYDRIN see CDT750
MONOCHLOROISOTHYMOL see CEX275
MONOCHLOROMETHANE see MIF765
MONOCHLOROMETHYL CYANIDE see CDN500
MONO-CHLORO-MONO-BROMO-METHANE see CES650
MONOCHLOROMONOFLUOROMETHANE see CHI900
MONOCHLOROPENTAFLUOROETHANE (DOT) see CJI500
MONOCHLOROPHENYLETHER see MRG000
p-MONOCHLOROPHENYL PHENYL SULFONE see CKI625
MONOCHLOROPOLYOXYETHYLENE see PJU250
β-MONOCHLOROPROPIONIC ACID see CKS500
MONOCHLOROSULFURIC ACID see CLG500
MONOCHLOROTETRAFLUOROETHANE (DOT) see CLH000
MONOCHLOROTRIFLUOROETHYLENE see CLQ750
MONOCHLOROTRIFLUOROMETHANE (DOT) see CLR250
MONOCHORIA VAGINALIS (Burm. f.) Presl., extract see MRG100
MONOCHROME ORANGE R see NEY000
MONOCHROME YELLOW 3R see NEY000
MONOCHROMIUM OXIDE) see CMK000
MONOCHROMIUM TRIOXIDE see CMK000
MONOCIL 40 see MRH209
"MONOCITE" METHACRYLATE MONOMER see MLH750
MONOCLAIR see CCK125
MONOCLOROBENZENE (ITALIAN) see CEJ125
MONOCOBALT OXIDE see CND125
MONOCOPPER MONOSULFIDE see CNQ000
MONOCORTIN see PAL600
MONOCRATILIN see MRH000
MONOCRESYL DIPHENYL PHOSPHATE see TGY750
MONOCRON see MRH209
MONOCROTALINE see MRH000
MONOCROTALINE, 3,8-DIDEHYDRO- see DAL350
MONOCROTOPHOS see ASN000, MRH209
MONOCYANOACETIC ACID see COJ500
MONOCYCLOHEXYLTIN ACID see MRH212
MONODEHYDROSORBITOL MONOOLEATE see SKV100
MONODEMETHYLIMIPRAMINE see DSI709
MONO-DIGITOXID (GERMAN) see DKL875
MONODION see VTA000
MONODODECYL ESTER SULFURIC ACID see MRH250
MONODORM see BPF500
MONODRAL see PBS000
MONODRAL BROMIDE see PBS000
MONOESTER with 4-BUTYL-4-(HYDROXYMETHYL)-1,2-DIPHENYL-SUC-
    CINIC ACID 3,5-PYRAZOLIDINEDIONE see SOX875
MONOETHANOLAMINE see EEC600
MONOETHANOLETHYLENEDIAMINE see AJW000
N-MONOETHYLAMIDE of O,O-DIMETHYLDITHIOPHOSPHORYLACETIC
    ACID see DNX600
MONOETHYLAMINE (DOT) see EFU400
MONOETHYLAMINE, anhydrous (DOT) see EFU400

2-N-MONOETHYLAMINOETHANOL see EGA500
MONOETHYLDICHLOROTHIOPHOSPHATE see MRI000
MONOETHYLENE GLYCOL see EJC500
MONOETHYLENE GLYCOL DIMETHYL ETHER see DOE600
MONOETHYL ETHER of DIETHYLENE GLYCOL see CBR000
MONOETHYLHEXYL PHTHALATE see MRI100
MONO(2-ETHYLHEXYL)PHTHALATE see MRI100
MONO(2-ETHYLHEXYL)SULFATE SODIUM SALT see TAV750
MONOETHYLISOSELENOURONIUMBROMIDE-HYDROBROMIDE see AJY000
MONOETHYLPHENYLTRIAZENE see MRI250
MONOETHYLTIN TRICHLORIDE see EPS000
MONOFLUORAZIJNZUUR (DUTCH) see FIC000
MONOFLUORESSIGSAURE (GERMAN) see FIC000
MONOFLUORESSIGSAURES NATRIUM (GERMAN) see SHG500
MONOFLUORETHANOL see FID000
MONOFLUOROACETAMIDE see FFF000
MONOFLUOROACETATE see FIC000
MONOFLUOROACETIC ACID see FIC000
MONOFLUOROETHANE see FIB000
MONOFLUOROETHANOL see FID000
MONOFLUOROETHYLENE see VPA000
MONOFLUOROPHOSPHORIC ACID, anhydrous see PHJ250
MONOFLUOROTRICHLOROMETHANE see TIP500
MONOFURACIN see NGE500
MONOGERMANE see GEI100
MONOGLYCEROL-p-AMINOBENZOATE see GGQ000
MONOGLYCIDYL ETHER of N-PHENYLDIETHANOLAMINE see MRI500
MONO-GLYCOCOARD see DKL800
MONOGLYME see DOE600
MONO-N-HEXYLAMINE see HFK000
8-MONOHYDRO MIREX see MRI750
MONOHYDROXYBENZENE see PDN750
MONO(HYDROXYETHYL) ESTER MALEIC ACID see EJG500
MONOHYDROXY-MERCURI-DI-IODORESORCIN-SULPHONPHTHALEIN
    DISODIUM see SIF500
MONOHYDROXYMETHANE see MGB150
MONOIODOACETAMIDE see IDW000
MONOIODOACETATE see IDZ000
MONOIODOACETIC ACID see IDZ000
MONOIODOMETHANESULFONIC ACID, SODIUM SALT see SHX000
MONOIODURO di METILE (ITALIAN) see MKW200
MONOISOBUTYLAMINE see IIM000
MONO-ISO-BUTYL PHTHALATE see MRI775
MONO-ISO-PROPANOLAMINE see AMA500
N-MONOISOPROPYLAMIDE of O,O-DIETHYLDITHIOPHOSPHORYLACETIC
    ACID see IOT000
MONOISOPROPYLAMINE see INK000
MONOISOPROPYLAMINE HYDROCHLORIDE see INL000
MONOISOPROPYLAMINOETHANOL see INN400
4-MONOISOPROPYLAMINO-1-PHENYL-2,3-DIMETHYL-5-PYRAZOLONE
    see INM000
MONOISOPROPYLBIPHENYL see IOF200
MONOISOPROPYL ETHER of ETHYLENE GLYCOL see INA500
MONO-KAY see VTA000
1-MONOLAURIN see MRJ000
MONOLAURYL DIMETHYLAMINE see DRR800
MONOLINURON see CKD500
MONOLITE FAST ORANGE R see DVB800
MONOLITE FAST RED G see CJD500
MONOLITE FAST SCARLET CA see MMP100
MONOLITE YELLOW GT see DEU000
MONOLITHIUM ACETYLIDE-AMMONIA see MRJ125
MONOMER FA see FPQ900
MONOMER MG-1 see EJH000
MONOMETHYLACETAMIDE see MFT750
N-MONOMETHYLAMIDE of O,O-DIMETHYLDITHIOPHOSPHORYLACETIC
    ACID see DSP400
MONOMETHYLAMINE see MGC250
MONOMETHYL-AMINOAETHANOL (GERMAN) see MGG000
4-MONOMETHYLAMINOAZOBENZENE see MNR500
p-MONOMETHYLAMINOAZOBENZENE see MNR500
MONOMETHYLAMINOETHANOL see MGG000
N-MONOMETHYLAMINOETHANOL see MGG000
2-MONOMETHYLAMINOFLUORENE see FEI500
N-MONOMETHYL-2-AMINOFLUORENE see FEI500
N-MONOMETHYLANILINE see MGN750
MONOMETHYL ANILINE (OSHA) see MGN750
MONOMETHYLARSINIC ACID see MGQ530
10-MONOMETHYLBENZO(a)PYRENE see MHH500
MONOMETHYL DIHYDROGEN PHOSPHATE see DLO800
MONOMETHYL ETHER of ETHYLENE GLYCOL see EJH500
MONO METHYL ETHER HYDROQUINONE see MFC700
MONOMETHYLFORMAMIDE see MKG500
MONOMETHYLFOSFIT (CZECH) see PGZ950

MONOMETHYL GUANIDIN (GERMAN) see MKI750
MONOMETHYLGUANIDINE see MKI750
MONOMETHYL HYDRAZINE see MKN000
MONOMETHYLHYDRAZINE NITRATE see MRJ250
MONOMETHYLMALEIC ANHYDRIDE see CMS322
MONOMETHYL MERCURY CHLORIDE see MDD750
MONOMETHYLOLACRYLAMIDE see HLU500
MONOMETHYLTIN TRICHLORIDE see MQC750
MONOMYCIN see MRJ600
MONOMYCIN A see NCF500
MONONITROCHLOROBENZENE see CJA950
MONONITROSOCIMETIDINE see MRJ700
MONONITROSOPIPERAZINE see MRJ750
N-MONONITROSOPYRIDINOL CARBAMATE see POR250
MONO-N-OCTYLTIN TRICHLORIDE see OGG000
MONO-N-OCTYL-ZINN-TRICHLORID (GERMAN) see OGG000
MONOPEN see BFD000
MONOPENTEK see PBB750
MONOPEROXY SUCCINIC ACID see MRK000
MONOPHEN see PFC750
MONOPHENOL see PDN750
MONOPHENYLBUTAZONE see MQY400
MONOPHENYLHEPTAMETHYLCYCLOTETRASILOXANE see HBA600
MONOPHENYLUREA see PGP250
MONOPHOR see AOB500
MONOPHOS see AOB500
MONOPHYLLINE see HOA000
MONOPLEX DBS see DEH600
MONOPLEX DOA see AEO000
MONOPLEX DOS see BJS250
MONOPOTASSIUM ARSENATE see ARD250
MONOPOTASSIUM DIHYDROGEN ARSENATE see ARD250
MONOPOTASSIUM aci-1-DINITROETHANE see MRK250
MONOPOTASSIUM GLUTAMATE see MRK500
MONOPOTASSIUM l-GLUTAMATE (FCC) see MRK500
MONOPOTASSIUM PHOSPHATE see PLQ405
MONOPOTASSIUM SALT of ACETYLENEDICARBOXYLIC ACID see ACJ500
MONOPOTASSIUM SULFATE see PKX750
MONOPRIM see TKZ000
MONO-N-PROPYLAMINE see PND250
MONOPROPYLENE GLYCOL see PML000
MONOPROPYL ETHER of ETHYLENE GLYCOL see PNG750
MONOPYRROLE see PPS250
MONORHEUMETTEN see MQY400
MONOSAN see DAA800
MONOSILANE see SDH575
MONOSODIOGLUTAMMATO (ITALIAN) see MRL500
MONOSODIUM ACETYLIDE see MRK609
MONOSODIUM ACID METHANEARSONATE see MRL750
MONOSODIUM ACID METHARSONATE see MRL750
MONOSODIUM-β-AMINOETHYL THIOPHOSPHATE see AKB500
MONOSODIUM ARSENATE see ARD600
MONOSODIUM ASCORBATE see ARN125
MONOSODIUM BARBITURATE see MRK750
MONOSODIUM CARBONATE see SFC500
MONOSODIUM CEFAZOLIN see CCS250
MONOSODIUM CEROXITIN see CCS510
MONOSODIUM CITRATE see MRL000
MONOSODIUM DIHYDROGEN CITRATE see MRL000
MONOSODIUM DIHYDROGEN PHOSPHATE see SJH100
MONOSODIUM-5-ETHYL-5-(1-METHYLBUTYL) THIOBARBITURATE
   see PBT500
MONOSODIUM FERRIC EDTA see EJA379
MONOSODIUM FLUCLOXACILLIN see FDA100
MONOSODIUM GLUCONATE see SHK800
MONOSODIUM GLUTAMATE see MRL500
α-MONOSODIUM GLUTAMATE see MRL500
MONOSODIUM-l-GLUTAMATE (FCC) see MRL500
MONOSODIUM GLYCOLATE see SHT000
MONOSODIUM METHANEARSONATE see MRL750
MONOSODIUM METHANEARSONIC ACID see MRL750
MONOSODIUM METHYLARSONATE see MRL750
MONOSODIUM NOVOBIOCIN see NOB000
MONOSODIUM PHOSPHATE see SJH100
MONOSODIUM SALT of 2,2'-METHYLENE BIS(3,4,6-TRICHLOROPHENOL)
   see MRM000
MONOSODIUM-2-SULFANILAMIDOPYRIMIDINE see MRM250
MONOSODIUM-2-SULFANILAMIDOTHIAZOLE see TEX500
MONOSODIUM THYROXINE see LEQ300
MONOSODIUM URATE see SKO575
MONOSORB XP-4 see SJH100
MONOSPAN see SBG500
MONOSTEARIN see OAV000
MONOSTEOL see SLL000

MONOSULFUR DICHLORIDE see SOG500
MONOTARD see LEK000
MONOTEN see PFC750
MONOTHIOETHYLENEGLYCOL see MCN250
MONOTHIOGLYCEROL see MRM750
α-MONOTHIOGLYCEROL see MRM750
MONOTHIOSUCCINIMIDE see MRN000
MONO-THIURAD see BJL600
MONOTHIURAM see BJL600
MONOTRICHLOR-AETHYLIDEN-α-GLUCOSE (GERMAN) see GFA000
MONOVAR see NNQ500
MONOVERIN see SDZ000
MONOVINYL PHOSPHATE see VQA400
MONOXONE see SFU500
β-MONOXYNAPHTHALENE see NAX000
MONSANTO CP-16226 see OAL000
MONSANTO CP-19203 see MOB500
MONSANTO CP-19699 see CON300
MONSANTO CP-40294 see MOB699
MONSANTO CP-40507 see MOB750
MONSANTO CP-43858 see TIP750
MONSANTO CP 47114 see DSQ000
MONSANTO CP-48985 see CFC500
MONSANTO CP-49674 see DOP200
MONSANTO CP 51969 see BNL250
MONSTERA DELICIOSA see SLE890
MONTAN 80 see SKV100
MONTANE 60 see SKV150
MONTANOA TOMENTOSA, leaf extract, crude see ZTS600
MONTANOA TOMENTOSA, leaf extract, semi-purified see ZTS625
MONTANOX 80 see PKL100
MONTAR see PJL750
MONTECATINI L-561 see DRR400
MONTHYBASE see EJM500
MONTHYLE see EJM500
MONTMORILLONITE see BAV750
MONTREL see COD850
MONTROSE PROPANIL see DGI000
MONUREX see CJX750
MONURON see CJX750
MONURON-TCA see CJY000
MONUROX see CJX750
MONURUON see CJX750
MONUURON see CJX750
MONZAOMYCIN see TAB300
MONZET see USJ075
MOON see GGA000
MOONSEED see MRN100
MOOSEWOOD see LEF100
MOP see NKV000
8-MOP see XDJ000
MOPA see MFE250
MOPARI see DGP900
MOPAZIN see MFK500
MOPAZINE see MFK500
MOPERONE CHLORHYDRATE see MNN250
MOPERONE HYDROCHLORIDE see MNN250
MOPLEN see PMP500
MOPOLM see MRD750
MOQUIZONE see MRN250
MOQUIZONE HYDROCHLORIDE see PCJ350
MORAMIDE see AFJ400
MORBOCID see FMV000
MORBUSAN see ENB500
MORDANT YELLOW 3R see SIU000
MOREPEN see AOD125
MORESTAN see ORU000
MORESTANE see ORU000
MORESTIN see EDV600
MORFAMQUAT see BJK750
MORFAX see BDF750
MORFINA (ITALIAN) see MRO500
MORFOTHION (DUTCH) see MRU250
MORFOXONE see BJK750
MORIAL see MRN275
MORIN see MRN500
MORINGA OLEIFERA Lamk., extract excluding roots see MRN550
MORIPERAN see AJH000
MORISYLYTE CITRATE see MRN600
MORNIDINE see CJL500
MOROCIDE see BGB500
MORONAL see NOH500
MOROSAN see DCK759
MORPHACETIN see HBT500

MORPHACTIN see CDT000
MORPHANQUAT DICHLORIDE see BJK750
MORPHERIDINE see MRN675
MORPHERIDINE DIHYDROCHLORIDE see MRN675
MORPHIA see MRO500
MORPHINA see MRO500
MORPHINAN-3,6-DIOL, 7,8-DIDEHYDRO-4,5-EPOXY-17-METHYL- (5-α-6-α)-, MONOHYDRATE see MRP100
MORPHINAN, 3-METHOXY-17-PHENETHYL-, TARTRATE, (−)- see MRN700
MORPHINAN-6-ONE, 3,14-DIHYDROXY-4,5-α-EPOXY-17-METHYL-, HYDRO-CHLORIDE see ORG100
MORPHINAN-6-ONE, 4,5-EPOXY-3,14-DIHYDROXY-17-METHYL-, HYDRO-CHLORIDE, (5-α)- (9CI) see ORG100
MORPHINAN-6-ONE, 4,5-α-EPOXY-3,14-DIHYDROXY-17-(2-PROPENYL)- see NAG550
MORPHINAN-6-ONE, 4,5-α-EPOXY-3-METHOXY-17-METHYL-, TARTRATE (1:1) see DKX050
MORPHINAN, 6,7,8,14-TETRADEHYDRO-4,5-α-EPOXY-3,6-DIMETHOXY-17-METHYL-, HYDROCHLORIDE see TEN100
(−)-MORPHINE see MRO500
MORPHINE see MRO500
MORPHINE CHLORHYDRATE see MRO750
MORPHINE CHLORIDE see MRO750
MORPHINE DIACETATE see HBT500
MORPHINE HYDROCHLORIDE see MRO750
MORPHINE METHOCHLORIDE see MRP000
MORPHINE METHYLCHLORIDE see MRP000
MORPHINE-3-METHYL ETHER see CNF500
MORPHINE MONOHYDRATE see MRP100
MORPHINE MONOMETHYL ETHER see CNF500
MORPHINE SULFATE see MRP250
MORPHINE SULPHATE see MRP250
MORPHINISM see MRO500
MORPHINUM see MRO500
MORPHIUM see MRO500
MORPHOCYCLINE see MRP500
MORPHOLINE see MRP750
MORPHOLINE, AQUEOUS MIXTURE (DOT) see MRP750
MORPHOLINEBORANE see MRQ250
MORPHOLINE, compounded with BORANE (1581) see MRQ250
MORPHOLINE, 4-BUTYL- see BRV100
4-MORPHOLINECARBOXAMIDE, N-(2-((2-HYDROXY-3-(4-HYDROXYPHENOXY)PROPYL)AMINO)ETHYL)-, (E)-2-BUTENEDIOATE (2:1) (salt) see XAH000
MORPHOLINE, 4-DECANOYL- see CBF725
MORPHOLINE, 2,6-DIMETHYL-4-NITROSO-, (Z)- see NKA695
MORPHOLINE, 2,6-DIMETHYL-N-NITROSO-, (cis)- see NKA695
MORPHOLINE DISULFIDE see BKU500
4-MORPHOLINEETHANAMINE see AKA750
MORPHOLINE ETHANOL see MRQ500
4-MORPHOLINEETHANOL, HYDROCHLORIDE see HKW300
MORPHOLINE HYDROCHLORIDE see MRQ600
4-MORPHOLINENONYLIC ACID see MRQ750
MORPHOLINE, 4-(1-OXODECYL)-(9CI) see CBF725
MORPHOLINE SALICYLATE see SAI100
MORPHOLINE, componded with SALICYLIC ACID (1:1) see SAI100
MORPHOLINE and SODIUM NITRITE (1:1) see SIN675
4-MORPHOLINE SULFENYL CHLORIDE see MRR075
MORPHOLINIUM, HEXACHLOROSTANNATE(2-) (2:1) see DUO500
MORPHOLINIUM, (3-INDOLYLMETHYLENE)-, HEXACHLOROSTANNATE (2-) (2:1) see BKJ500
MORPHOLINIUM PERCHLORATE see MRR100
1-MORPHOLINOACETYL-3-PHENYL-2,3-DIHYDRO-4(1H)-QUINAZOLINONE HYDROCHLORIDE see PCJ350
N-MORPHOLINO-β-(2-AMINOMETHYLBENZODIOXAN)-PROPIONAMIDE see MRR125
MORPHOLINO-2-BENZOTHIAZOLYL DISULFIDE see BDF750
MORPHOLINOCARBONYLACETONITRILE see MRR750
MORPHOLINO-CNU see MRR775
MORPHOLINODAUNOMYCIN see MRR850
3′-MORPHOLINO-3′-DEAMINODAUNORUBICIN see MRT100
MORPHOLINODISULFIDE see BKU500
2-(MORPHOLINODITHIO)BENZOTHIAZOLE see BDF750
β-MORPHOLINOETHYL BENZHYDRYL ETHER HYDROCHLORIDE see LFW300
β-MORPHOLINOETHYLMORPHINE see TCY750
3-O-(2-MORPHOLINOETHYL)MORPHINE see TCY750
O³-(2-MORPHOLINOETHYL)MORPHINE see TCY750
N-2-MORPHOLINOETHYL-5-NITROIMIDAZOLE see NHH000
MORPHOLINOETHYL NORPETHIDINE DIHYDROCHLORIDE see MRN675
1-(2-MORPHOLINOETHYL)-4-PHENYLISONIPECOTIC ACID ETHYL ESTER DIHYDROCHLORIDE see MRN675
N-(1-(MORPHOLINOMETHYL)-2,6-DIOXO-3-PIPERIDYL)PHTHALIMIDE see MRU080

2-(MORPHOLINO)-N-METHYL-N-(2-MESITYLOXYETHYL)ACETAMIDE HYDROCHLORIDE see MRT250
1-5-(MORPHOLINOMETHYL)-3-((5-NITROFURFURYLIDENE)AMINO)-2-OX-AZOLIDINONEHYDROCHLORIDE see FPI150
5-MORPHOLINOMETHYL-3-(5-NITROFURFURYLIDINE)AMINO-2-OX-AZOLIDINONE HYDROCHLORIDE see FPI100
5-MORPHOLINOMETHYL-3-(5-NITRO-2-FURFURYLIDINE-AMINO)-2-OX-AZOLIDINONE see FPI000
1-MORPHOLINOMETHYLTHALIDOMIDE see MRU080
4-MORPHOLINO-2-(5-NITRO-2-THIENYL)QUINAZOLINE see MRU000
N-MORPHOLINO NONANAMIDE see MRQ750
MORPHOLINOPHOSPHONIC ACID DIMETHYL ESTER see DST800
1-(6-MORPHOLINO-3-PYRIDAZINYL)-2-(1-(tert-BUTOXYCARBONYL)-2-PROPYLIDENE)HYDRAZINE see RFU800
3-MORPHOLINOSYDNONE IMINE HYDROCHLORIDE see MRU075
MORPHOLINO-THALIDOMIDE see MRU080
2-(MORPHOLINOTHIO)BENZOTHIAZOLE see BDG000
MORPHOLIN SALICYLAT see SAI100
N-MORPHOLINYL-2-BENZOTHIAZOLYL DISULFIDE see BDF750
4-MORPHOLINYL-2-BENZOTHIAZYL DISULFIDE see BDF750
MORPHOLINYLETHYLMORPHINE see TCY750
3-(2-(4-MORPHOLINYL)ETHYL)MORPHINE see TCY750
1-(2-N-MORPHOLINYLETHYL)-5-NITROIMIDAZOLE see NHH000
1-(2-MORPHOLINYL)ETHYL)-4-PHENYL-4-PIPERIDINECARBOXYLIC ACID ETHYL ESTER DIHYDROCHLORIDE see MRN675
MORPHOLINYLMERCAPTOBENZOTHIAZOLE see BDG000
4-MORPHOLINYLPHOSPHONIC ACID DIMETHYL ESTER see DST800
3-((6-(4-MORPHOLINYL)-3-PYRIDAZINYL)HYDRAZONO)BUTANOIC ACID 1,1-DIMETHYL ETHYL ESTER see RFU800
4-(2-MORPHOLINYL)PYROCATECHOL see MRU100
2-(4-MORPHOLINYLTHIO)BENZOTHIAZOLE see BDG000
3-MORPHOLYLAETHYLMORPHIN (GERMAN) see TCY750
N-MORPHOLYLCYSTEAMIN (GERMAN) see MCO000
MORPHOTHION see MRU250
MORROCID see BGB500
MORSODREN see MLF250
MORSYDOMINE see MRN275
MORTON EP-316 see CQI500
MORTON EP332 see DSO200
MORTON EP 333 see CJJ500
MORTON SOIL DRENCH see MLF250
MORTON WP-161E see ISE000
MORTOPAL see TCF250
MORYL see CBH250
MOS-708 see BDG250
MOSANON see SEH000
MOSCARDA see MAK700
MOSCHUS KETONE see MIT625
MOSE see MRU255
MOSS GREEN see COF500
MOSTEN see PMP500
MOSYLAN see DNG400
5-MOT see MFS400
MOTAZOMIN see MRN275
MOTH BALLS (DOT) see NAJ500
MOTHER-IN-LAW PLANT see CAL125
MOTHER-IN-LAW'S TONGUE PLANT see DHB309
MOTH FLAKES see NAJ500
MOTIAX see FAB500
MOTILIUM see DYB875
MOTILYN see PAG200
MOTIORANGE R see PEJ500
MOTIROT G see XRA000
MOTOLON see QAK000
MOTOR BENZOL see BBL250
MOTOR FUEL (DOT) see GBY000
MOTOR SPIRIT (DOT) see GBY000
MOTOX see CDV100
MOTRIN see IIU000
MOTTENHEXE see HCI000
MOUNTAIN GREEN see COF500
MOUNTAIN LAUREL see MRU359
MOUNTAIN TOBACCO see AQY500
MOUS-CON see ZLS000
MOUSE-NOTS see SMN500
MOUSE PAK see WAT200
MOUSE-RID see SMN500
MOUSE-TOX see SMN500
MOVINYL 100 see PKQ059
MOVINYL 114 see AAX250
MOXADIL see AOA095
MOXALACTAM DISODIUM see LBH200
MOXAM see LBH200
MOXESTROL see MRU600

MOXIE see MEI450
MOXISYLYTE HYDROCHLORIDE see TFY000
MOXNIDAZOLE see MRU750
MOXONE see DAA800
MOZAMBIN see QAK000
MP see POK000
8-MP see XDJ000
MP 655 see CHE750
MP 1023 see AAN000
MP 12-50 see TAB750
MP 25-38 see TAB750
MP 45-26 see TAB750
MP 1 (refractory) see EAL100
MPA see DAZ117
3-MPA see MFM000
3MPA see MCQ000
M-PARATHION see MNH000
3-MPC see MNM500
M.P. CHLORCAPS T.D. see TAI500
MPCM see XTJ000
MPE see MPU500
MPG see MRK500
15-M3-PGF2-α see CCC100
M 4212 (PHARMACEUTICAL) see MQX775
MPI-PC see DSQ800
MPI-PENICILLIN see DSQ800
MPK see PBN250
MPK 90 see PKQ250
MPN see MNA000, NKV100
MPNU see MMY500
MPP see FAQ900
1-MPPN see DWU800
MPS see USJ100
MPT see MPX850
MR see MPH300
MR 56 see BOO630
MRA-CN see COP765
MRAVENCAN DI-n-BUTYLCINICITY (CZECH) see DDZ000
MRAVENCAN TRIBENZYLCINICITY (CZECH) see FOE000
MRAVENCAN VAPENATY (CZECH) see CAS250
MRC 910 see GIA000
MRD 108 see DWK400
MRL-37 see MFG525
MRL 41 see CMX700
MRL-37 HYDROCHLORIDE see MFG530
MROWCZAN ETYLU (POLISH) see EKL000
3-MS see CNX625
MS 33 see SKV150
MS 53 see SNK000
MS 33F see SKV150
MS. 752 see SBE500
MS 1053 see DJI000
MS-1112 see BFV760
MS 1143 see DJI000
MS-4101 see FMR075
MS-5075 see FBP850
MS-ANTIGEN 40 see MRU755
MS-BENZANTHRONE see BBI250
MSC-102824 see AQN750
MSF see MDR750
MSG see MRL500
MSMA see MRL750
MSMED see ECU750
MSTFA see MQG750
MSZYCOL see BBQ500
MT see HOA000
MT-45 see MRU757
MT-141 see CCS365
MT 14-411 see CMY125
MTB see HNY500
MTB 51 see DJM800, XCJ000
MTBHQ see BRM500
MTD see DTQ400, TGL750
MTDQ see MRU760
MTEAI see MRU775
MTIC see MQB750
MTIQ see MPO750
MTMC see MIB750
M.T. MUCORETTES see MPN500
MTQ see QAK000
MTU see MPW500
MTX see MDV500
MTX DISODIUM see MDV600
MUCAESTHIN see BQA010

MUCAINE see DTL200
MUCALAN see IHO200
MUCIC ACID see GAR000
MUCICLAR see CBR675
MUCIDRIL see DPE000
MUCIDRINA see VGP000
MUCINOL see AOO490
MUCITUX see ECU550
MUCOCHLORIC ACID see MRU900
MUCOCHLORIC ANHYDRIDE see MRV000
MUCOCIS see CBR675
MUCODYNE see CBR675
MUCOFLUID see MDK875
MUCOLASE see CBR675
MUCOLEX see CBR675
MUCOLYSIN see MCI375
MUCOLYTICUM see ACH000
MUCOLYTICUM LAPPE see ACH000
MUCOMYCIN see MRV250
MUCOMYST see ACH000
MUCONOMYCIN A see MRV500
MUCOPOLYSACCHARIDE, POLYSULFURIC ACID ESTER see MRV525
MUCOPOLYSACCHARIDIPOLY SCHWEFELSAEUREESTER (GERMAN)
  see MRV525
MUCOPRONT see CBR675
MUCORAMA see PMJ500
MUCOSOLVAN see AHJ500
MUCOSOLVIN see ACH000
MUCOXIN see DTL200
MUCUNA MONOSPERMA DC. ex Wight (extract excluding roots) see MRV600
MUDAR see COD675
MUGB see GLC100
MUGUET (CANADA) see LFT700
MUIRAMID see AAI250
MULDAMINE see MRV750
MUL F 66 see PKL750
MULHOUSE WHITE see LDY000
MULSIFEROL see VSZ100
MULTAMAT see DQM600
MULTERGAN see MRW000
MULTERGAN METHYL SULFATE see MRW000
MULTEZIN see MRW000
MULTICHLOR see CDP000
MULTIFUGE CITRATE see PIJ500
MULTIN see HIM000
MULTIPROP see CDT000
MUNDISAL see CMG000
MU OIL TREE see TOA275
MURACIL see MPW500
MURATOX see DJI000
MURCIL see MDQ250
MUREL see VBK000
MUREX see GFA000
MURFOS see PAK000
MURFOTOX see DJI000
MURFULVIN see GKE000
MURIATE of PLATINUM see PJE000
MURIATIC ACID (DOT) see HHL000
MURIATIC ETHER see EHH000
MURIOL see CJJ000
MURITAN see DEQ000
MUROTOX see DJI000
MURPHOTOX see DJI000
MURUTOX see DJI000
MURVESCO see CJR500
MUSARIL see CFG750
MUSCALM see MRW125
MUSCARIN see MRW250
MUSCARINE see MRW250
dl-MUSCARINE see MRW250
MUSCIMOL see AKT750
MUSCLE ADENYLIC ACID see AOA125
MUSCONE see MIT625
MUSCULAMINE see DCC400
MUSCULAMINE TETRAHYDROCHLORIDE see GEK000
MUSCULARON see IHH000
MUSETTAMYCIN see APV000
MUSHROOM AMNITA RUBESCENS TOXIN see AHI500
MUSHROOMS see MRW269
MUSK see MRW250
MUSK 36A see ACL750
MUSK AMBRETTE see BRU500
MUSKARIN see MRW250
MUSKEL see CKF500

MUSKEL-TRANCOPAL see CKF500
MUSKONE see MIT625
MUSK R 1 see OKW100
MUSK-T see EJQ500
MUSK XYLDL see TML750
MUSK XYLENE see TML750
MUSQUASH POISON see WAT325
MUSQUASH ROOT see WAT325
MUSSEL POISON DIHYDROCHLORIDE see SBA500
MUSTARD CHLOROHYDRIN see CHC000
MUSTARD GAS see BIH250
MUSTARD GAS SULFONE see BIH500
MUSTARD HD see BIH250
MUSTARD OIL see AGJ250
MUSTARD SULFONE see BIH500
MUSTARD VAPOR see BIH250
MUSTARGEN see BIE250, BIE500
MUSTARGEN HYDROCHLORIDE see BIE500
MUSTINE see BIE250
MUSTINE HYDROCHLOR see BIE500
MUSTINE HYDROCHLORIDE see BIE500
MUSTRON see CFA750
MUSUET SYNTHETIC see CMS850
MUSUETTINE PRINCIPLE see CMS850
MUTABASE see DCQ700
MUTAGEN see BIE250
MUTAMYCIN see AHK500
MUTAMYCIN 6 see EBC000
MUTAMYCIN (MITOMYCIN for INJECTION) see AHK500
MUTHESA see DTL200
MUTHMANN'S LIQUID see ACK250
MUTOXIN see DAD200
MUZOLIMINE see EAE675
MV 119A see DLK200
MVEEG (RUSSIAN) see EJL500
MVNA see NKY000
2M-4X see SIL500
MX 5517-02 see SMQ500
MXDA see XHS800
MY/68 see FAF000
MY 33-7 see TGJ625
MY 41-6 see AJC000
MY-5116 see IHR200
MYACINE see NCD550
MYACYNE see NCE000
MYALEX see CKK250
MYAMBUTOL see EDW875
MYANIL see GGS000
MYARSENOL see SNR000
MYASUL see AKO500
MYBASAN see ILD000
MY-B-DEN see AOA125
MYBORIN see MRW275
MYCAIFRADIN SULFATE see NCG000
MYCARDOL see PBC250
MYCELAX see MRX500
MYCELEX see MRX500
MYCHEL see CDP250
MYCIFRADIN see NCE000
MYCIFRADIN-N see NCG000
MYCIGIENT see NCG000
ωMYCIN see TBX000
MYCINAMICIN 1 see MRW500
MYCINAMICINS II see MRW750
MYCINOL see CDP250
MYCIVIN see LGC200
MYCOBACIDIN see CMP885
1-MYCOBACIDIN see CCI500
MYCOBACTYL see GBB500
MYCOBUTOL see EDW875
MYCOCURAN see GGS000
MYCOFARM see BFD250
MYCOHEPTIN see MRW800
MYCOHEPTYNE see MRW800
MYCOIN see CMV000
MYCOLUTEIN see DTU200
MYCOPHENOLIC ACID see MRX000
MYCOPHYT see PIF750
MYCO-POLYCID see CDY325
MYCOSPOR see BGA825
MYCOSPORIN see MRX500
MYCOSTATIN see NOH500
MYCOSTATIN 20 see NOH500
MYCOTICIN see MRY000

MYCOTICIN (1581) see MRY000
MYCOTOXIN F2 see ZAT000
MYCOZOL see TEX000
MYCRONIL see BJK500
MYDECAMYCIN see MBY150
MYDETON see TGK200
MYDFRIN see SPC500
MYDOCALM see MRW125, TGK200
MYDRIAL see AOA250
MYDRIASIN see MGR250
MYDRIATIN see NNM000
MYDRIATINE see NNN000, PMJ500
MYEBROL see DDP600
MYELOBROMOL see DDP600
MYELOLEUKON see BOT250
MYELOTRAST see SHX000, TDQ230
MYELOTRAST DI-N-METHYLGLUCAMINE SALT see IDJ500
MYGAL see BET000
MYKOSTIN see VSZ100
MYLAR see PKF750
MYLAXEN see HEG000
MYLEPSIN see DBB200
MYLEPSINUM see DBB200
MYLERAN see BOT250
MYLIS see DAL040
MYLOFANOL see PGG000
MYLON (CZECH) see DSB200
MYLONE see DSB200
MYLONE 85 see DSB200
MYLOSAR see ARY000
MYLOSUL see AKO500
MYNOSEDIN see IIU000
MYOCAINE see RLU000
MYOCHOLINE see HOA500
MYOCHRYSINE see GJC000
MYOCOL see AEH750
MYOCON see NGY000
MYOCORD see TAL475
MYOCRISIN see GJC000
MYODETENSINE see GGS000
MYODIGIN see DKL800
MYOFER 100 see IGS000
MYOFLEXINE see CDQ750
MYOHEMATIN see CQM325
MYO-INOSISTOL HEXAKISPHOSPHATE see PIB250
MYO-INOSITOL HEXAPHOSPHATE see PIB250
MYOLASTAN see CFG750
MYOLAX see GGS000
MYOLYSEEN see PIL550
MYOMYCIN B see MRY100
MYOMYCIN SULFATE see MRY250
MYOPAN see GGS000
MYOPLEGINE see HLC500
MYOPONE see WBJ700
MYOPORUM LAETUM see MRY600
MYORDIL see TNJ750
MYORELAX see RLU000
MYOREXON see CCK125
MYOSALVARSAN see SNR000
MYOSCAINE see RLU000
MYOSEROL see GGS000
MYOSPAZ see PFJ000
MYOSTHENINE see VGP000
MYOSTIBIN see AQH800, AQI250
MYOSTON see AOA125
MYOTOLON see AOO300
MYOTRATE ""10" see PBC250
MYOTRIPHOS see ARQ500
MYOXANE see GGS000
MYPROZINE see PIF750
MYRABOLAM TANNIN see MRZ100
MYRAFORM see PKQ059
MYRCENE see MRZ150
MYRCENOL see MLO250
MYRCENYL ACETATE see AAW500
MYRCIA OIL see BAT500
MYRICA CERIFERA see WBA000
MYRICIA OIL see BAT500
MYRINGACAINE DROPS see CHW675
MYRISIIC ALCOHOL see TBY250
MYRISTAN DI-n-BUTYLCINICITY (CZECH) see BLH309
MYRISTICA see NOG000
MYRISTIC ACID see MSA250
MYRISTIC ALDEHYDE see TBX500

MYRISTICA OIL see NOG500
MYRISTICIN see MSA500
9-MYRISTOYL-1,7,8-ANTHRACENETRIOL see MSA750
10-MYRISTOYL-1,8,9-ANTHRACENETRIOL see MSA750
1-MYRISTOYLAZIRIDINE see MSB000
MYRISTOYLETHYLENEIMINE see MSB000
N-MYRISTOYLOXY-AAF see ACS000
N-MYRISTOYLOXY-AAIF see MSB100
N-MYRISTOYLOXY-N-ACETYL-2-AMINOFLUORENE see ACS000
N-MYRISTOYLOXY-N-ACETYL-2-AMINO-7-IODOFLUORENE see MSB100
N-MYRISTOYLOXY-N-MYRISTOYL-2-AMINOFLUORENE see MSB250
MYRISTYL ALCOHOL (mixed isomers) see TBY500
MYRISTYL-Γ-PICOLINIUM CHLORIDE see MSB500
MYRISTYL STEARATE see TCB100
MYRISTYL SULFATE, SODIUM SALT see SIO000
MYRITICALORIN see RSU000
MYRITICOALORIN see RSU000
MYRITOL 318 see CBF710
MYROBALANS TANNIN see MSB750
MYRTAN TANNIN see MSC000
MYRTLE OIL see OGU000
MYSEDON see DBB200
MYSOLINE see DBB200
MYSONE see HHQ800
MYSORITE see ARM262
MYSTECLIN-F see AOC500
MYSTER GRASS see DAE100
MYSTERIA see CNX800
MYSTOX WFA see BGJ750
MYSURAN see MSC100
MYSURAN CHLORIDE see MSC100
MYTELASE see MSC100
MYTELASE CHLORIDE see MSC100
MYTOMYCIN see AHK500
MYTRATE see VGP000
MYVAK see VSK900
MYVAX see VSK900
MYVIZONE see FNF000
MYVPACK see VSK600
MYXOVIROMYCIN see AHN625

N-9 see NNB300
N 68 see BRS000
N-399 see PEM750
N 521 see DSB200
N-553 see MRW125
N 642 see PIR100
N 714 see TAF675
N 715 see THL500
N-746 see CLX550
N-1544A see PFC750
N 2038 see ZJS300
N 2404 see CJD650
N 2790 see FMU045
N 3051 see BSG000
N 4328 see EOO000
N 4548 see MOB250
N 7001 see AEG875
N 7009 see FMO129, FMO150
NA see EID000, NBE500
NA-22 see IAQ000
NA-53 see BCP000
NA 97 see PAF625
NA-101 see TDX000
NA-872 see AHJ250, AHJ500
NAA 800 see NAK500
Na-AESCINAT see EDM000
NAAM see NAK000
NaATP see AEM250
NAB see NJK150
NAB 365 see VHA350
NABAC see HCL000
NABADIAL see AOO475
NABAM see DXD200
NABAME (FRENCH) see DXD200
NAB 365Cl see VHA350
NABOLIN see MPN500
NABUMETONE see MFA300
NAC see ACH000
NACARAT A EXPORT see HJF500
NACCANOL NR see DXW200
NACCONATE 1OO see TGM750
NACCONATE-100 see TGM740
NACCONATE 300 see MJP400

NACCONATE 400 see BBP000
NACCONATE H 12 see MJM600
NACELAN BLUE G see TBG700
NACELAN BLUE KLT see MGG250
NACELAN FAST YELLOW CG see AAQ250
NACELAN VIOLET 4R see DBP000
NACIMYCIN see RKK000
NACLEX see BDE250
NACM-CELLULOSE SALT see SFO500
NACRICLASINE see LJC000
NAC-TB see ACH000
NACYCLYL see EDR000
NAD see NAK000
NADEINE see DKW800
NA-DESOXYCHOLAT (GERMAN) see SGE000
NADIC METHYL ANHYDRIDE see NAC000
NADISAL see SJO000
NADISAN see BSM000
NADIZAN see BSM000
NADOLOL see CNR675
NADONE see CPC000
NADOXOLOL HYDROCHLORIDE see NAC500
NADOZONE see BRF500
NADROTHYRON D see SKJ300
NAEPAINE see PBV750
NAFCILLIN SODIUM SALT see SGS500
NAFEEN see SHF500
NAFENOIC ACID see MCB500
NAFENOPIN see MCB500
NAFIVERINE DIHYDROCHLORIDE see NAD000
NAFKA see ROH900
NAFKA CRYSTAL GUM see ROH900
NAFKA KRISTALGOM see ROH900
NAFOXIDINE see NAD500
NAFOXIDINE HYDROCHLORIDE see NAD750
NaFPAK see SHF500
NAFRINE see AEX000
Na FRINSE see SHF500
NAFRONYL see NAE000
NAFRONYL OXALATE see NAE100
NAFTALEN (POLISH) see NAJ500
NAFTALIN-BUTIL-SOLFONATO (ITALIAN) see NBS700
NAFTALOFOS see HMV000
NAFTIDROFURYL see NAE000
NAFTIDROFURYL OXALATE see NAE100
1-NAFTILAMINA (SPANISH) see NBE000
β-NAFTILAMINA (ITALIAN) see NBE500
1-NAFTIL-TIOUREA (ITALIAN) see AQN635
NAFTIPRAMIDE see IOU000
NAFTIZIN see NAH500
2-NAFTOL (DUTCH) see NAX000
β-NAFTOL (DUTCH) see NAX000
2-NAFTOLO (ITALIAN) see NAX000
β-NAFTOLO (ITALIAN) see NAX000
NAFTOPEN see SGS500
α-NAFTYLAMIN (CZECH) see NBE000
β-NAFTYLAMIN (CZECH) see NBE500
2-NAFTYLAMIN-5,7-DISULFONAN SODNY (CZECH) see ALH500
1-NAFTYLAMINE (DUTCH) see NBE000
2-NAFTYLAMINE (DUTCH) see NBE500
α-NAFTYL-N-METHYLKARBAMAT (CZECH) see CBM750
β-NAFTYLOAMINA (POLISH) see NBE500
1-NAFTYLTHIOUREUM (DUTCH) see AQN635
NAFTYPRAMIDE see IOU000
NAFUSAKU see NAK500
NAGARMOTHA OIL see NAE505
NAGARMUSTA OIL see NAE505
NAGARSE see BAC000
NAGRAVON see VSZ000
NAH see NCQ900
NAH 80 see SHO500
NAHP see NJJ950
NA 65 HYDROCHLORIDE see PIV500
Na III HYDROCHLORIDE see DWF200
NAJA FLAVA VENOM see NAE510
NAJA HAJE ANNULIFERA VENOM see NAE512
NAJA HAJE VENOM see NAE515
NAJA MELANOLEUCA VENOM see NAE875
NAJA NAJA ATRA VENOM see NAF000
NAJA NAJA KAOUTHIA VENOM see NAF200
NAJA NAJA NAJA VENOM see ICC700
NAJA NAJA SIAMENSIS VENOM see NAF250
NAJA NIGRICOLLIS VENOM see NAG000
NAJA NIVEA VENOM see NAG200

NAPHTHALOXIMIDE-O,O-DIETHYL PHOSPHOROTHIOATE see NAQ500
NAPHTHALOXIMIDODIETHYL THIOPHOSPHATE see NAQ500
α-NAPHTHALTHIOHARNSTOFF (GERMAN) see AQN635
NAPHTHANE see DAE800
NAPHTHANIL BLUE B BASE see DCJ200
NAPHTHANIL SCARLET 2G BASE see DEO400
NAPHTHANIL SCARLET G BASE see NMP500
NAPHTHANTHRACENE see BBC250
NAPHTHANTHRONE see BBI250
Δ5,7,9-NAPHTHANTRIENE see TCX500
NAPHTHA PETROLEUM (DOT) see NAI500
1,2 NAPHTHAQUINONE see NBA000
NAPHTHA SAFETY SOLVENT see SLU500
NAPHTHA, SOLVENT (DOT) see NAI500
NAPHTHENATE de COBALT (FRENCH) see NAR500
NAPHTHENE see NAJ500
NAPHTHENIC ACID see NAR000
NAPHTHENIC ACID, COBALT SALT see NAR500
NAPHTHENIC ACID, COPPER SALT see NAS000
NAPHTHENIC ACID, LEAD SALT see NAS500
NAPHTHENIC ACID, PHENYLMERCURY SALT see PFP000
NAPHTHENIC ACID, ZINC SALT see NAT000
NAPHTHENIC OILS (PETROLEUM), CATALYTIC DEWAXED HEAVY(9CI)
    see MQV865
NAPHTHENIC OILS (PETROLEUM), CATALYTIC DEWAXED LIGHT (9CI)
    see MQV867
NAPHTHIOMATE T see TGB475
NAPHTHIONIC ACID see ALI000
1,4-NAPHTHIONIC ACID see ALI000
NAPHTHIPRAMIDE see IOU000
NAPHTHISEN see NAH550
NAPHTHIZEN see NAH550
NAPHTHIZINE see NAH500
NAPHTHOCAINE HYDROCHLORIDE see NAH800
NAPHTHOCARD YELLOW O see FAG140
NAPHTHO(1,2,3,4-def)CHRYSENE see NAT500
α-β-NAPHTHO-2,3-DIPHENYL-TRIAZOLIUM CHLORID (GERMAN)
    see DWH875
NAPHTHO(1,8-gh:4,5-g'h')DIQUINOLINE see NAU000
NAPHTHO(1,8-gh:5,4-g'h')DIQUINOLINE see NAU500
β-NAPHTHOFLAVONE see NAU525
2-NAPHTHOHYDROXAMIC ACID see NAV000
2-NAPHTHOHYDROXAMIC ACID-o-ACETATE ESTER see ACS250
2-NAPHTHOHYDROXAMIC ACID-o-PROPIONATE ESTER see NBC500
2-NAPHTHOHYDROXIMIC ACID see NAV000
2-NAPHTHOIC ACID see NAV500
β-NAPHTHOIC ACID see NAV500
2-NAPHTHOIC ACID, 4,4'-METHYLENEBIS(3-HYDROXY-, compounded with
    (E)-1,4,5,6-TETRAHYDRO-1-METHYL-2-(2-(2-THIENYL)VINYL)PYRIMIDINE
    (1:1) see POK575
NAPHTHOL see NAW000
1-NAPHTHOL see NAW500
2-NAPHTHOL see NAX000
α-NAPHTHOL see NAW500
β-NAPHTHOL see NAX000
NAPHTHOL B see NAX000
2-NAPHTHOL-3-CARBOXYLIC ACID see HMX500
2-NAPHTHOL, DECAHYDRO-, ACETATE see DAF100
2-NAPHTHOL, DECAHYDRO-, FORMATE see DAF150
2-NAPHTHOL-3,6-DISULFONIC ACID SODIUM SALT see ROF300
2-NAPHTHOL ETHYL ETHER see EEY500
β-NAPHTHOL ETHYL ETHER see EEY500
NAPHTHOL GREEN B see NAX500
1-NAPHTHOL-N-METHYLCARBAMATE see CBM750
NAPHTHOL ORANGE see FAG010
α-NAPHTHOL ORANGE see FAG010
NAPHTHOL RED B see FAG020, NAY000
NAPHTHOL RED B 20-7575 see NAY000
NAPHTHOL RED DEEP 10459 see NAY000
NAPHTHOL RED D TONER 35-6001 see NAY000
2-NAPHTHOL-6-SULFONIC ACID see HMU500
β-NAPHTHOL-6-SULFONIC ACID see HMU500
β-NAPHTHOLSULFONIC ACID S see HMU500
1-NAPHTHOL, 1,2,3,4-TETRAHYDRO- see TDI350
NAPHTHOL YELLOW see DUX800
1H-NAPHTHO(2,1-b)PYRAN-1-ONE, 3-PHENYL- see NAU525
NAPHTHOPYRIN see NAY500
β-NAPHTHOQUINALDINE see BDB750
α-NAPHTHOQUINOLINE see BDC000
NAPHTHO(2,3-f)QUINOLINE see NAZ000
α-NAPHTHOQUINONE see NBA500
β-NAPHTHOQUINONE see NBA000
1,2-NAPHTHOQUINONE see NBA000
1,4-NAPHTHOQUINONE see NBA500

β-NAPHTHOQUINONE-4-SULFONATE SODIUM SALT see DLK000
NAPHTHORESORCINOL see NAN000
NAPHTHOSOL FAST RED KB BASE see CLK225
NAPHTHO(1,2-e)THIANAPHTHENO(3,2-b)PYRIDINE see BCF500
NAPHTHO(2,1-e)THIANAPHTHENO(3,2-b)PYRIDINE see BCF750
α-NAPHTHOTHIOUREA see AQN635
2H-NAPHTHO(1,2-d)TRIAZOLE, 2-(4-STYRYL-3-SULFOPHENYL)-7-SULFO-,
    DISODIUMSALT see DXG025
NAPHTHO(1,2-d)TRIAZOLE-7-SULFONIC ACID,2-(4-(2-PHENYLETHENYL)-3-
    SULFOPHENYL)-, DISODIUM see DXG025
4-(2H-NAPHTHO(1,2-d)TRIAZOL-2-YL)-2-STILBENESULFONIC ACID SODIUM
    SALT see TGE155
1-(2-NAPHTHOYL)-AZIRIDINE see NBC000
β-NAPHTHOYLETHYLENEIMINE see NBC000
2-NAPHTHOYLHYDROXAMIC ACID see NAV000
N-(2-NAPHTHOYL)-o-PROPIONYLHYDROXYLAMINE see NBC500
1-NAPHTHYLACETAMIDE see NAK000
α-NAPHTHYLACETAMIDE see NAK000
α-NAPHTHYLACETIC see NAK500
NAPHTHYLACETIC ACID see NAK500
1-NAPHTHYLACETIC ACID see NAK500
α-NAPHTHYLACETIC ACID see NAK500
N-2-NAPHTHYLACETOHYDROXAMIC ACID see NBD000
α-NAPHTHYL ACETONITRILE see NBD500
α-(1-NAPHTHYL)ACETONITRILE see NBD500
β-NAPHTHYL ALCOHOL see NAX000
1-NAPHTHYLAMIN (GERMAN) see NBE000
2-NAPHTHYLAMIN (GERMAN) see NBE500
β-NAPHTHYLAMIN (GERMAN) see NBE500
1-NAPHTHYLAMINE see NBE000
2-NAPHTHYLAMINE see NBE500
6-NAPHTHYLAMINE see NBE500
α-NAPHTHYLAMINE see NBE000
β-NAPHTHYLAMINE see NBE500
NAPHTHYLAMINE BLUE see CMO250
β-NAPHTHYLAMINEDISULFONIC ACID see ALH250
2-NAPHTHYLAMINE-4,8-DISULFONIC ACID see ALH250
2-NAPHTHYLAMINE-6,8-DISULFONIC ACID see NBE850
β-NAPHTHYLAMINE-4,8-DISULFONIC ACID see ALH250
2-NAPHTHYLAMINE-1-d-GLUCOSIDURONIC ACID see ALL000
1-NAPHTHYLAMINE HYDROCHLORIDE see NBF000
α-NAPHTHYLAMINE HYDROCHLORIDE see NBF000
NAPHTHYLAMINE MUSTARD see BIF250
2-NAPHTHYLAMINE MUSTARD see NBE500
1-NAPHTHYLAMINE-4-SULFONIC ACID see ALI000
1-NAPHTHYLAMINE-6-SULFONIC ACID see ALI250
2-NAPHTHYLAMINE-1-SULFONIC ACID see ALH750
2-NAPHTHYLAMINE-8-SULFONIC ACID see ALI300
5-NAPHTHYLAMINE-2-SULFONIC ACID see ALI250
α-NAPHTHYLAMINE-p-SULFONIC ACID see ALI000
p-(2-NAPHTHYLAMINO)PHENOL see NBF500
N-(1-NAPHTHYL)ANILINE see PFT250
N-(2-NAPHTHYL)ANILINE see PFT500
4-(1-NAPHTHYLAZO)-2-NAPHTHYLAMINE see NBG000
4-(1-NAPHTHYLAZO)-m-PHENYLENEDIAMINE see NBG500
5-(β-NAPHTHYLAZO)-2,4,6-TRIAMINOPYRIMIDINE see NBO500
2-NAPHTHYLBIS(2-CHLOROETHYL)AMINE see BIF250
β-NAPHTHYL-BIS-(β-CHLOROETHYL)AMINE see BIF250
1-NAPHTHYL-N-BUTYL-N-NITROSOCARBAMATE see BRY750
1-NAPHTHYL CHLOROCARBONATE see NBH200
1-NAPHTHYL CHLOROFORMATE see NBH200
α-NAPHTHYL CHLOROFORMATE see NBH200
β-NAPHTHYL-DI-(2-CHLOROETHYL)AMINE see BIF250
o-2-NAPHTHYL m,N-DIMETHYLTHIOCARBANILATE see TGB475
α-NAPHTHYLENEACETIC ACID see NAK500
1,2-(1,8-NAPHTHYLENE)BENZENE see FDF000
1,5-NAPHTHYLENEDIAMINE see NAM000
1,5-NAPHTHYLENE DISULFONIC ACID see AQY400
NAPHTHYLENE YELLOW see DUX800
α-NAPHTHYLESSIGSAEURE (GERMAN) see NAK500
N-(1-NAPHTHYL)ETHYLENEDIAMINE DIHYDROCHLORIDE see NBH500
1-NAPHTHYL-N-ETHYL-N-NITROSOCARBAMATE see NBI000
2-NAPHTHYLHYDROXAMIC ACID see NAV000
β-NAPHTHYL HYDROXIDE see NAX000
1-NAPHTHYLHYDROXYLAMINE see HIX000
2-NAPHTHYLHYDROXYLAMINE see NBI500
N-1-NAPHTHYLHYDROXYLAMINE see HIX000
β-NAPHTHYL ISOBUTYL ETHER see NBJ000
(2-NAPHTHYL)-1-ISOPROPYLAMINOETHANOL see INS000
NAPHTHYLISOPROTERENOL see INS000
NAPHTHYLISOPROTERENOL HYDROCHLORIDE see INT000
1-NAPHTHYL ISOTHIOCYANATE see ISN000
α-NAPHTHYL ISOTHIOCYANATE see ISN000
2-NAPHTHYL MERCAPTAN see NAP500

β-NAPHTHYL MERCAPTAN see NAP500
1-NAPHTHYL METHYLCARBAMATE see CBM750
1-NAPHTHYL-N-METHYLCARBAMATE see CBM750
α-NAPHTHYL N-METHYLCARBAMATE see CBM750
α-NAPHTHYLMETHYL IMIDAZOLINE see NAH500
2-(α-NAPHTHYLMETHYL)-IMIDAZOLINE see NAH500
2-(1-NAPHTHYLMETHYL)-2-IMIDAZOLINE see NAH500
2-(1-NAPHTHYLMETHYL)IMIDAZOLINE HYDROCHLORIDE see NCW000
2-(1-NAPHTHYLMETHYL)-2-IMIDAZOLINE HYDROCHLORIDE see NCW000
1-NAPHTHYL METHYL KETONE see ABC475
2-NAPHTHYL METHYL KETONE see ABC500
α-NAPHTHYL METHYL KETONE see ABC475
β-NAPHTHYL METHYL KETONE see ABC500
1-NAPHTHYL METHYLNITROSOCARBAMATE see NBJ500
1-NAPHTHYL-N-METHYL-N-NITROSOCARBAMATE see NBJ500
2-NAPHTHYL N-METHYL-N-(3-TOLYL)THIONOCARBAMATE see TGB475
(2-(α-NAPHTHYLOXY)ETHYL)HYDRAZINE MALEATE see NBK000
4-(α-NAPHTHYLOXY)-3-HYDROXYBUTYRAMIDOXIME HYDROCHLORIDE
  see NAC500
4-α-NAPHTHYLOXY-3-HYDROXY-BUTYRAMIDOXIM-HYDROCHLORID
  (GERMAN) see NAC500
1-(1-NAPHTHYLOXY)-2-HYDROXY-3-ISOPROPYLAMINOPROPANE HYDRO-
  CHLORIDE see ICC000
4,4'-(3-(1-NAPHTHYL)-1,5-PENTAMETHYLENEDIMORPHOLINE see DUO600
2-NAPHTHYLPHENYLAMINE see PFT500
β-NAPHTHYLPHENYLAMINE see PFT500
2-NAPHTHYL-p-PHENYLENEDIAMINE see NBL000
α-NAPHTHYLPHTHALAMIC ACID SODIUM SALT see SIO500
N-1-NAPHTHYLPHTHALAMIC ACID SODIUM SALT see SIO500
1-NAPHTHYL-N-PROPYL-N-NITROSOCARBAMATE see NBM500
β-NAPHTHYLSULFONIC ACID see NAP000
3-(1-NAPHTHYL)-2-TETRAHYDROFURFURYLPROPIONIC ACID 2-
  (DIETHYLAMINO)ETHYL ESTER see NAE000
1-NAPHTHYL THIOACETAMIDE see NBN000
α-NAPHTHYLTHIOCARBAMIDE see AQN635
1-NAPHTHYL-THIOHARNSTOFF (GERMAN) see AQN635
2-NAPHTHYL THIOL see NAP500
α-NAPHTHYLTHIOUREA see AQN635
1-(1-NAPHTHYL)-2-THIOUREA see AQN635
N-(1-NAPHTHYL)-2-THIOUREA see AQN635
1-NAPHTHYL THIOUREA (MAK) see AQN635
α-NAPHTHYLTHIOUREA (DOT) see AQN635
1-NAPHTHYL-THIOUREE (FRENCH) see AQN635
5-β-NAPHTHYL-2:4:6-TRIAMINOAZOPYRIMIDINE see NBO500
N-1-NAPHTHYL-N,N',N'-TRIETHYLETHYLENEDIAMINE see NBO515
4-(1-NAPHTHYLVINYL)-PYRIDINE HYDROCHLORIDE see NBO525
4-(2-(1-NAPHTHYL)VINYL)PYRIDINE HYDROCHLORIDE see NBO525
NAPHTOELAN FAST SCARLET G SALT see NMP500
NAPHTOELAN MITSUI SCARLET GG SALT see DEO400
NAPHTOELAN ORANGE R BASE see NEN500
NAPHTOELAN RED GG BASE see NEO500
2-NAPHTOL (FRENCH) see NAX000
β-NAPHTOL (GERMAN) see NAX000
NAPHTOL AS-KG see TGR000
NAPHTOL AS-KGLL see TGR000
2-NAPHTOL-6-SULFOSAURE (GERMAN) see HMU500
NAPHTO(1,2-c-d-e)NAPHTACENE (FRENCH) see BCP750
NAPHTOX see AQN635
NAPHTYZIN see NAH550
NAPHYL-1-ESSIGSAEURE (GERMAN) see NAK500
NAPOTON see LFK000, MDQ250
NAPRINOL see HIM000
NAPROSINE see MFA500
NAPROSYN see MFA500
NAPROXEN SODIUM see NBO550
NAPRUX see MFA500
NAPTHOLROT S see FAG020
NAQUA see HII500
NARAMYCIN see CPE750
NARCAN see NAH000
NARCEOL see MNR250
NARCICLASINE see LJC000
NARCISO (CUBA, MEXICO) see DAB700
NARCISSUS ABSOLUTE see NBP000
NARCISSUS POETICUS see DAB700
NARCISSUS PSEUDONARCISSUS see DAB700
NARCOGEN see TIO750
NARCOLAN see ARW250, THV000
NARCOLO see AFJ400
NARCOMPREN see NOA000
NARCOMPREN HYDROCHLORIDE see NOA500
NARCOSAN see ERD500
NARCOSAN SOLUBLE see ERE000

NARCOSINE see NOA000
NARCOSINE HYDROCHLORIDE see NOA500
NARCOSINE, HYDROCHLORIDE (8CI) see NOA500
NARCOTANE see HAG500
NARCOTANN NE-SPOFA (RUSSIAN) see HAG500
NARCOTILE see EHH000
NARCOTINE see NBP275, NOA000
1-α-NARCOTINE see NOA000
1-α-NARCOTINE HYDROCHLORIDE see NOA500
NARCOTINE N-OXIDE HYDROCHLORIDE see NBP300
NARCOTUSSIN see NOA000
NARCOZEP see FDD100
NARCYLEN see ACI750
NARDELZINE see PFC750
NARDIL see PFC500, PFC750
NAREA see HDP500
NARGOLINE see NDM000
NARIGIX see EID000
NARITHERACIN see VGZ000
NARKOLAN see ARW250, THV000
NARKOSOID see TIO750
NARLENE see BQU500
NARPHEN see PMD325
NARSIS see CGA000
NASALIDE see FDD085
NASDOL see TBG000
NASIVINE see ORA100
NASMIL see CNX825
NASOL see EAW000
NASS (IRAN) see SED400
NASTENON see PAN100
NASTYN see EHP000
NASWAR (PAKISTAN and AFGHANISTAN) see SED400
NATA see TII500
NATACYN see PIF750
NATAMYCIN see PIF750
NATASOL FAST ORANGE GR SALT see NEO000
NATASOL FAST RED TR SALT see CLK235
NATERETIN see BEQ625
NATHULANE see PME500
NAT-333 HYDROCHLORIDE see DAI200
NATICARDINA see GAX000
NATIL see DNU100
NATIONAL 120-1207 see AAX250
NATIVE CALCIUM SULFATE see CAX750
NATRASCORB see ARN125
NATRASCORB INJECTABLE see ARN000
NATREEN see BCE500, SGC000
NATRI-C see ARN125
NATRILIX see IBV100
NATRINAL see BAG250
NATRIN HERBICIDE see SKL000
NATRIONEX see AAI250
NATRI-PAS see SEP000
NATRIPHENE see BGJ750
NATRIUM see SEE500
NATRIUMACETAT (GERMAN) see SEG500
NATRIUMALUMINUMFLUORID (GERMAN) see SHF000
NATRIUMANTIMONYLTARTRAT (GERMAN) see AQI750
NATRIUMARSENIT (GERMAN) see ARJ500
NATRIUMAZID (GERMAN) see SFA000
NATRIUMBARBITALS (GERMAN) see BAG250
NATRIUMBICHROMAAT (DUTCH) see SGI000
NATRIUMCHLORAAT (DUTCH) see SFS000
NATRIUMCHLORAT (GERMAN) see SFS000
NATRIUMCHLORID (GERMAN) see SFT000
NATRIUM CITRICUM (GERMAN) see DXC400
NATRIUMDEHYDROCHOLAT (GERMAN) see SGD500
NATRIUM-2,4-DICHLORPHENOXYATHYLSULFAT (GERMAN) see CNW000
NATRIUMDICHROMAAT (DUTCH) see SGI000
NATRIUMDICHROMAT (GERMAN) see SGI000
NATRIUM-3-α,12-α-DIHYDROXYCHOLANAT (GERMAN) see SGE000
NATRIUMFLUORACETAAT (DUTCH) see SHG500
NATRIUMFLUORACETAT (GERMAN) see SHG500
NATRIUM FLUORIDE see SHF500
NATRIUMGLUTAMINAT (GERMAN) see MRL500
NATRIUMHEXAFLUOROALUMINATE (GERMAN) see SHF000
NATRIUMHYDROXID (GERMAN) see SHS000
NATRIUMHYDROXYDE (DUTCH) see SHS000
NATRIUMHYPOPHOSPHIT (GERMAN) see SHV000
NATRIUMJODAT (GERMAN) see SHV500
NATRIUMJODID (GERMAN) see SHW000
NATRIUMMALAT (GERMAN) see MAN250
NATRIUMMAZIDE (DUTCH) see SFA000

NATRIUM-N-METHYL-DITHIOCARBAMAAT (DUTCH) see VFU000
NATRIUMMETHYLDITHIOCARBAMAT (GERMAN) see VFW009
NATRIUM-N-METHYL-DITHIOCARBAMAT (GERMAN) see VFU000
NATRIUMMOLYBDAT (GERMAN) see DXE800
NATRIUMNICOTINAT (GERMAN) see NDW500
NATRIUMNITRAT UND N-METHYLANILIN (GERMAN) see MGO250
NATRIUM NITRIT (GERMAN) see SIQ500
NATRIUMOXALAT (GERMAN) see SIY500
NATRIUMPERCHLORAAT (DUTCH) see PCE750
NATRIUMPERCHLORAT (GERMAN) see PCE750
NATRIUMPHOSPHAT (GERMAN) see SJH090
NATRIUMPROPIONAT (GERMAN) see SJL500
NATRIUMPYROPHOSPHAT see TEE500
NATRIUMRHODANID (GERMAN) see SIA500
NATRIUMSALZ DER 2,2-DICHLORPROPIONSAURE see DGI600
NATRIUMSELENIAT (GERMAN) see DXG000
NATRIUMSELENIT (GERMAN) see SJT500
NATRIUMSILICOFLUORID (GERMAN) see DXE000
NATRIUMSUFAT (GERMAN) see SJY000
NATRIUMSULFID (GERMAN) see SJZ000
NATRIUMTARTRAT (GERMAN) see SKB000
NATRIUMTRICHLOORACETAAT (DUTCH) see TII500
NATRIUMTRICHLORACETAT (GERMAN) see TII500
NATRIUMTRIPOLYPHOSPHAT (GERMAN) see SKN000
NATRIUMZYKLAMATE (GERMAN) see SGC000
NATRIURAN see CLY600
NATRIURETIC HORMONE (bovine pineal) see AQW125
NATULAN see PME250, PME500
NATULANAR see PME500
NATULAN HYDROCHLORIDE see PME500
NATURAL CALCIUM CARBONATE see CAO000
NATURAL GASOLINE (DOT) see GBY000
NATURAL IRON OXIDES see IHD000
NATURAL LEAD SULFIDE see LDZ000
NATURAL RED OXIDE see IHD000
NATURAL RUBBER see ROH900
NATURAL WINTERGREEN OIL see MPI000
NATURETIN see BEQ625
NATURINE see BEQ625
NATYL see PCP250
NAUCAINE see AIT250
NAUGARD TJB see DWI000
NAUGARD TKB see NKB500
NAUGATUCK D-014 see SOP000
NAUGATUCK DET see DKC800
NAUGA WHITE see MJO750
NAULI "GUM" see PMQ750
NAUROCTIL see MCI500
NAUSEN see BBV500
NAUSIDOL see CJL500
NAUTAZINE see EAN600
NAVADYL see NOC000
NAVANE see NBP500
NAVARON see NBP500
NAVICALM see HGC500
NAVIDREX see CPR750
NAVIDRIX see CPR750
NAVRON see FFF000
NAXAMIDE see IMH000
NAXEN see MFA500
NAXOFEM see NHH000
NAXOGIN see NHH000
NAXOL see CPB750
NAYPER B and BO see BDS000
NB2B see CMO000
NBBA see NJO150
NBD-C 1 see NFS550
NBD-CHLORIDE see NFS550
NBHA see HJQ350
N.B. MECOPROP see CIR500
NBN see BRV500
NBS 706 see SMQ500
NBT see DVF800
NC5 see HER500
NC12 see TJH750
NC 26 see BHO250
NC11F see FIA000
NC 150 see PDC250, PEK250
NC-262 see DSP400
N 714C see TAF675
NC-2962 see DRU400
NC 3363 see DGO400
NC 5016 see DGA200
NCI see COP000

NCI 1136 see DBA175
NCI 143-418 see RKA000
NCI 145-604 see RKK000
NCI-C00044 see AFK250
NCI-C00055 see AJM000
NCI-C00066 see ASH500
NCI-C00077 see CBG000
NCI-C00099 see CDR750
NCI-C00102 see TBQ750
NCI-C00113 see DGP900
NCI-C00124 see DHB400
NCI-C00135 see DSP400
NCI-C00157 see EAT500
NCI-C00168 see TBW100
NCI-C00180 see HAR000
NCI-C00191 see KEA000
NCI-C00204 see BBQ500
NCI-C00215 see MAK700
NCI-C00226 see PAK000
NCI-C00237 see PIB900
NCI-C00259 see CDV100
NCI-C00260 see HON000
NCI-C00395 see DVQ709
NCI-C00408 see DER000
NCI-C00419 see PAX000
NCI-C00420 see DFT800
NCI-C00431 see DFV600
NCI-C00442 see DUV600
NCI-C00453 see CDO250
NCI-C00464 see DAD200
NCI-C00475 see BIM500
NCI-C00486 see BIO750
NCI-C00497 see MEI450
NCI-C00500 see DDL800
NCI-C00511 see EIY600
NCI-C00522 see EIY500
NCI-C00533 see CKN500
NCI-C00544 see DOS000
NCI-C00555 see BIM750
NCI-C00566 see EAQ750
NCI-C00588 see FAB400
NCI-C00599 see PHV250
NCI-C00920 see EJC500
NCI-C01445 see NEI000
NCI-C01478 see LBG000
NCI-C01514 see AES750
NCI-C01536 see ADR750
NCI-C01547 see YCJ000
NCI-C01558 see CME250
NCI-C01569 see ARY000
NCI-C01570 see EDR500
NCI-C01592 see BOT250
NCI-C01605 see EAN500
NCI-C01616 see IAN000
NCI-C01627 see PIK250
NCI-C01638 see IMH000
NCI-C01649 see TFQ750
NCI-C01661 see DDG800
NCI-C01672 see PDC250
NCI-C01683 see TGD000
NCI-C01694 see EPQ000
NCI-C01707 see TFI000
NCI-C01718 see SOA500
NCI-C01729 see TNX000
NCI-C01730 see API500
NCI-C01741 see PDF000
NCI-C01785 see POL500
NCI-C01810 see PME500
NCI-C01821 see DSY600
NCI-C01832 see DCE600, TGM400
NCI-C01843 see NMP500
NCI-C01854 see HHG000
NCI-C01865 see DVH000
NCI-C01876 see AIB000
NCI-C01887 see AJT750
NCI-C01901 see AKP750
NCI-C01912 see NFD500
NCI-C01923 see MMG000
NCI-C01934 see NEQ500
NCI-C01945 see NES500
NCI-C01956 see NHQ000
NCI-C01967 see NEJ500
NCI-C01978 see NEL000
NCI-C01989 see DBO400

NCI-C09007 see ARM275
NCI-C50011 see BCP250
NCI-C50033 see SBT000
NCI-C50044 see BII250
NCI-C50055 see DUE000
NCI-C50077 see PMO500
NCI-C50088 see EJN500
NCI-C50099 see PNL600
NCI-C50102 see MJP450
NCI-C50124 see PDN750
NCI-C50135 see EIU800
NCI-C50146 see OPM000
NCI-C50157 see RDK000
NCI-C50168 see CNU000
NCI-C50191 see SIB600
NCI-C50204 see TAV750
NCI-C50226 see ZAT000
NCI-C50259 see HEJ500
NCI-C50260 see DBR400
NCI-C50282 see BGE325
NCI-C50317 see DCE400
NCI-C50351 see BGJ250
NCI-C50362 see HER000
NCI-C50384 see EFT000
NCI-C50395 see GLU000
NCI-C50419 see LIA000
NCI-C50442 see BJK500
NCI-C50453 see EQR500
NCI-C50464 see AGJ250
NCI-C50475 see AEX250
NCI-C50533 see TGM750
NCI-C50544 see WCB000
NCI-C50602 see BOP500
NCI-C50613 see AMW000
NCI-C50635 see BLD500
NCI-C50646 see CBF700
NCI-C50657 see DBL200
NCI-C50668 see MJP400
NCI-C50680 see MLH750
NCI-C50715 see MCB000
NCI-C50737 see CQK600
NCI-C50748 see AQQ500
NCI-C52459 see TBQ000
NCI-C52733 see DVL700
NCI-C52904 see NAJ500
NCI-C53634 see HCA500
NCI-C53781 see AAQ250
NCI-C53792 see CHP500
NCI-C53838 see HGC000
NCI-C53849 see HJF500
NCI-C53894 see PAW500
NCI-C53929 see PEJ500
NCI-C54262 see VPK000
NCI-C54364 see GMA000
NCI-C54375 see BEC500
NCI-C54386 see AEO000
NCI-C54546 see SBW000
NCI-C54557 see AQP000
NCI-C54568 see CMO750
NCI-C54579 see CMO000
NCI-C54604 see MJQ000, MJQ100
NCI-C54626 see MRC000
NCI-C54660 see DHF200
NCI-C54706 see FEW000
NCI-C54717 see ISV000
NCI-C54728 see DTD800
NCI-C54739 see RMK020
NCI-C54740 see DST800
NCI-C54751 see TNI250
NCI-C54773 see DSG600
NCI-C54795 see DHE800
NCI-C54808 see ARN000
NCI-C54819 see IKE000
NCI-C54820 see CIU750
NCI-C54831 see TIQ250
NCI-C54842 see PMK000
NCI-C54853 see EES350
NCI-C54886 see CEJ125
NCI-C54897 see HKN875
NCI-C54900 see TBG700
NCI-C54911 see SGP500
NCI-C54922 see ALO750
NCI-C54933 see PAX250
NCI-C54944 see DEP600

NCI-C54955 see DEP800
NCI-C54966 see REF000
NCI-C54977 see EBR000
NCI-C54988 see TCF000
NCI-C54999 see CPD750
NCI-C55005 see CPC000
NCI-C55050 see TDI000
NCI-C55061 see TDH750
NCI-C55072 see CDS000
NCI-C55107 see CEA750
NCI-C55118 see CEQ600
NCI-C55129 see DRS200
NCI-C55130 see BNL000
NCI-C55141 see PNJ400
NCI-C55152 see AQF000
NCI-C55163 see CCP250
NCI-C55174 see DHF000
NCI-C55185 see GEO000
NCI-C55196 see NGE000
NCI-C55209 see OLA000
NCI-C55210 see RNZ000
NCI-C55221 see SHF500
NCI-C55232 see XGS000
NCI-C55243 see BND500
NCI-C55254 see CFK500
NCI-C55265 see TAI500
NCI-C55276 see BBL250
NCI-C55287 see PAU500
NCI-C55298 see QPA000
NCI-C55301 see POP250
NCI-C55323 see BKE500
NCI-C55345 see DFX800
NCI-C55367 see BPX000
NCI-C55378 see PAX250
NCI-C55425 see GFQ000
NCI-C55436 see DDS000
NCI-C55447 see MKA500
NCI-C55458 see AJL500
NCI-C55470 see WCJ000
NCI-C55481 see EGV400
NCI-C55492 see PEO500
NCI-C55505 see SEM000
NCI-C55516 see DDQ400
NCI-C55527 see BOX750
NCI-C55538 see EBX500
NCI-C55549 see GGW500
NCI-C55550 see TEO250
NCI-C55561 see TBX250
NCI-C55572 see LFU000
NCI-C55583 see NKM000
NCI-C55594 see MHZ000
NCI-C55607 see HCE500
NCI-C55618 see IMF400
NCI-C55630 see IMR000
NCI-C55641 see SPC500
NCI-C55652 see EAY500
NCI-C55663 see AES500
NCI-C55674 see EDJ500
NCI-C55685 see PDE000
NCI-C55696 see SNC000
NCI-C55709 see CDP250
NCI-C55710 see AOB250
NCI-C55721 see DNA800
NCI-C55743 see PBC250
NCI-C55765 see DKQ000
NCI-C55776 see PJD000
NCI-C55787 see HFV500
NCI-C55798 see PDO750
NCI-C55801 see HIM000
NCI-C55812 see MIP750
NCI-C55823 see GEM000
NCI-C55834 see HIH000
NCI-C55845 see QQS200
NCI-C55856 see CCP850
NCI-C55867 see MCD250
NCI-C55878 see BOV000
NCI-C55889 see HAP500
NCI-C55890 see HHR500
NCI-C55903 see XDJ000
NCI-C55925 see CFY000
NCI-C55936 see CHJ750
NCI-C55947 see TDY250
NCI-C55958 see NEM500
NCI-C55969 see AOR500

NCI C55970 see ALO000
NCI-C55981 see ASM270
NCI-C55992 see NIF000
NCI-C56019 see BKC000
NCI-C56031 see DFI100
NCI C56042 see UVJ450
NCI-C56064 see NGE500
NCI-C56075 see BAU750
NCI-C56086 see AOD125
NCI-C56097 see DWW000
NCI-C56100 see BFC750
NCI-C56111 see CMP969
NCI-C56122 see RGW000
NCI-C56133 see BAY500
NCI-C56144 see IDA000
NCI-C56155 see TMN490
NCI-C56166 see BCK250
NCI-C56177 see FPQ875
NCI-C56188 see XNJ000
NCI-C56199 see EID000
NCI-C56202 see FPK000
NCI-C56213 see ABC250
NCI-C56224 see FPU000
NCI-C56235 see IPU000
NCI-C56246 see QFS000
NCI-C56257 see HKR500
NCI-C56279 see COB260
NCI-C56280 see MNQ000
NCI-C56291 see BSU250
NCI-C56304 see PFL000
NCI-C56326 see AAG250
NCI-C56337 see BJL000
NCI-C56348 see DTC800
NCI-C56359 see PAH250
NCI-C56360 see DBB200
NCI-C56382 see BIE500
NCI-C56393 see EGP500
NCI-C56406 see VQK650
NCI-C56417 see BMC000
NCI-C56428 see DQF800
NCI-C56439 see IPD000
NCI-C56440 see HCZ000
NCI-C56451 see PKL500
NCI-C56462 see MRH000
NCI-C56473 see HOH500
NCI-C56484 see OGQ100
NCI-C56508 see HMY000
NCI-C56519 see BDF000
NCI-C56520 see MNH250
NCI-C56531 see BRF500
NCI-C56553 see BRV500
NCI-C56564 see CBF750
NCI-C56575 see CAL000
NCI-C56586 see CHP250
NCI-C56597 see SNH000
NCI-C56600 see SNJ000
NCI-C56611 see TBO780
NCI-C56633 see TKV000
NCI-C56644 see HOO100
NCI-C56655 see PAX250
NCI-C56666 see AGH150
NCI-C56762 see DSR400
NCI-C60015 see COG000
NCI-C60026 see AHP000
NCI-C60048 see DJX000
NCI-C60060 see MKQ880
NCI-C60066 see MKQ875
NCI-C60071 see MRL750
NCI-C60082 see NEX000
NCI-C60093 see PFJ250
NCI-C60102 see QCJ000
NCI-C60106 see QCA000
NCI-C60117 see TAJ000
NCI-C60128 see CGO500
NCI-C60139 see VOA000
NCI-C60173 see MCY475
NCI-C60184 see PAD500
NCI-C60195 see PEC500
NCI-C60208 see VPP000
NCI-C60219 see PGX000
NCI-C60220 see TJB600
NCI-C60231 see CEA000
NCI-C60242 see TIK500
NCI-C60286 see PKL100

NCI-C60297 see CNV000
NCI-C60311 see CNA250
NCI-C60322 see DTG000
NCI-C60333 see HLU500
NCI-C60344 see NDK500
NCI-C60366 see MMP100
NCI-C60377 see NAY000
NCI-C60388 see NER000
NCI-C60399 see MCW250
NCI-C60402 see EEA500
NCI-C60413 see DER000
NCI-C60537 see NMO550
NCI-C60559 see CHY250
NCI-C60560 see TCR750
NCI-C60571 see HEN000
NCI-C60582 see PKQ250, PKQ500, PKQ750, PKR000, PKR250, PKR750, PKS000
NCI-C60606 see TAG750
NCI-C60628 see TFD250
NCI-C60639 see DYE600
NCI-C60640 see DPJ200
NCI-C60651 see WAK000
NCI-C60662 see TMP750
NCI-C60673 see DQA400
NCI C60684 see DYE500
NCI-C60695 see TMJ750
NCI-C60708 see NCD500
NCI-C60719 see BEM500
NCI-C60720 see MPE250
NCI-C60742 see DMN400
NCI-C60753 see DUP600
NCI-C60786 see NEO500
NCI-C60797 see PKQ059
NCI-C60811 see ZSJ000
NCI-C60822 see ABE500
NCI-C60866 see BRR900
NCI-C60899 see DIX200
NCI-C60902 see TLP500
NCI-C60913 see DSB000
NCI-C60924 see DWC600
NCI-C60935 see LBX000
NCI-C60946 see AFW750
NCI-C60957 see MES000
NCI-C60968 see IJS000
NCI-C60979 see IKQ000
NCI-C60980 see BCC500
NCI-C61018 see NMW500
NCI-C61029 see PMT750
NCI-C61041 see TNP500
NCI-C61052 see IJD000
NCI-C61074 see BAK000
NCI-C61109 see CMN750
NCI-C61143 see MAU250
NCI-C61165 see SBX525
NCI-C61176 see ARA500
NCI-C61187 see TIV750
NCI-C61201 see CJU250
NCI-C61289 see CMO250
NCI-C61290 see CMO500
NCI-C61405 see HEO000
NCI-C61494 see BEN000
NCI-C60253A see ARM262
NCI-C61223A see ARM268
NCI-CO1752 see CKK000
NCI-CO1763 see BSQ000
NCI-CO2131 see BSS250
NCI-CO4035 see CFK000
NCI-CO5210 see CKP500
NCI-CO03247 see ABB000
NCR CE EE DOV7 see DDM200
NCR DR DCN see BGF250
NCS see NBV500
NCS-14210 see BHV000
NCS-27381 see BHU500
NCS 110435 see CDB800
NDBA see BRY500
NDC 0082-4155 see DAC200
NDC 002-1452-01 see VKZ000
NDD see CGA500
NDEA see NJW500
NDELA see NKM000
NDFDA see PCG725
NDGA see NBR000
NDHU see NJY000

N¹-3,4-DICHLOROPHENYL-N⁵-ISOPROPYLDIGUANIDE HYDROCHLORIDE
  see DGD100
2-(N'-(2-(DIETHYLAMINO)ETHYL)ACETAMIDO)-2,6'-ACETOXYLIDIDE
  HYDROCHLORIDE see DHS600
4-(N,4-DIMETHYL-l-ALLOISOLEUCINE)-8-(N,4-DIMETHYL-l-ALLOISOLEUC-
  INE)-QUINOMYCIN A see QQS075
N(¹)-(3,4-DIMETHYL-5-ISOXAZOLYL)SULFANILAMIDE LITHIUM SALT
  see DSL000
N(¹)-(3,4-DIMETHYL-5-ISOXAZOLYL)SULFANILAMIDE SODIUM SALT
  see DSL200
N,N'DIMETHYL-3-METHYL-4-(4'-(2'-METHYLPYRIDYL-1'OXIDE)AZO)
  ANILINE see DQD800
(E)-N,N,-DIMETHYL-4-(2-PHENYLETHENYL)BENZENAMINE see DUC000
α,N-DIMETHYL-3-TRIFLUOROMETHYLPHENETHYLAMINE see DUJ800
α,N-DIMETHYL-m-TRIFLUOROMETHYLPHENETHYLAMINE see DUJ800
NDMA see DSY600, NKA600
NDMI see NJK500
NDPA see DWI000, NKB700
NDPEA see NBR100
NDPhA see DWI000
NDR A-3682 see FKW000
NDRC-143 see AHJ750
NDR-5998A HYDROCHLORIDE see DAI200
NEA see NKD000
NEAMIN see NCE500
NEANTINE see DJX000
NEARGAL see LJR000
NEASINA see SNJ000
NEATHALIDE see INS000
NEAT OIL of SWEET ORANGE see OGY000
NEAUFATIN see TEH500
NEAZINE see PPP500
NEBCIN see TGI500
NEBERK see FMB000
NEBRAMYCIN see NBR500
NEBRAMYCIN FACTOR 5 see BAU270
NEBRAMYCIN FACTOR 6 see TGI250
NEBRAMYCIN V see BAU270
NEBRAMYCIN X see NCE500
NEBRASKA FERN see PJJ300
NEBULARIN(E) see RJF000
NEBULARINE see RJF000
NEBUREA see BRA250
NEBUREX see BRA250
NEBURON see BRA250
NECARBOXYLIC ACID see AFR250
NECATORINA see CBY000
NECATORINE see CBY000
NECCANOL SW see DXW200
NECKLACEPOD SOPHORA see NBR800
NECKLACEWEED see BAF325
NECTADON see NOA000
NECTADON HYDROCHLORIDE see NOA500
NEDCIDOL see DCM750
NEEDLE ANTIMONY see AQL500
NEEM OIL see NBS300
NEFCO see NGE500
NEFOPAM see FAL000
NEFOPAM HYDROCHLORIDE see NBS500
NEFRECIL see PDC250
NEFRIX see CFY000
NEFTIN see NGG500
NEFURTHIAZOLE see NDY500
NEFUSAN see DSB200
NEGALIP see ARQ750
NEGAMICIN see NCE500
NEGRAM see EID000
NEGUVON see TIQ250
NEGUVON A see TIQ250
NEIRINE see VQR300
NEKAL see NBS700
NEKAL WT-27 see DJL000
NEKLACID FAST ORANGE 2G see HGC000
NEKLACID ORANGE 1 see FAG010
NEKLACID RED 3R see FMU080
NEKLACID RED A see FAG020
NEKLACID RED RR see FMU070
NEKLACID RUBINE W see HJF500
NEKTROHAN see ZVJ000
NELBON see DLY000
NELIPRAMIN see DLH600
NELKEN OEL (GERMAN) see CMY075
NELLITE see PEV500
NELPON see TJK100

NEMA see MKB000, PCF275
NEMABROM see DDL800
NEMACIDE see DFK600
NEMACTIL see PIW000
NEMACUR see FAK000
NEMAFENE see DGG000
NEMAFOS see EPC500
NEMAFUME see DDL800
NEMAGON see DDL800
NEMAGONE see DDL800
NEMAGON SOIL FUMIGANT see DDL800
NEMALITE see NBT000
NEMALITH (GERMAN) see NBT000
NEMANAX see DDL800
NEMAPAN see TEX000
NEMAPAZ see DDL800
NEMAPHOS see EPC500
NEMASET see DDL800
NEM-A-TAK see DHH200
NEMATOCIDE see EPC500, DDL800
NEMATOLYT see PAG500
NEMATOX see DDL800
NEMAZENE see PDP250
NEMAZINE see PDP250
NEMAZON see DDL800
NEMBUSEN see GGS000
NEMBUTAL see NBT500
NEMBUTAL CALCIUM see CAV000
NEMBUTAL SODIUM see NBU000
NEMEROL see DAM600
NEMICIDE see LFA020
NEMISPOR see DXI400
NENDRIN see EAT500
NENESIN see BNK000
NEO see TEH500
NEOABIETIC ACID see NBU800
NEOANTERGAN see WAK000
NEOANTERGAN HYDROCHLORIDE see MCJ250
NEOANTERGAN MALEATE see DBM800
NEOANTERGAN PHOSPHATE see NBV000
NEOANTIMOSAN see AQH500
NEOARSOLUIN see NCJ500
NEOARSPHENAMINE see NCJ500
NEOASYCODILE see DXE600
NEO-ATOMID see ARQ750
NEO-AVAGAL see SBH500
NEOBAN see CEW500
NEOBAR see BAP000
NEOBEE M-5 see CBF710
NEOBEE O see CBF710
NEO-BENODINE see TGJ475
NEOBIOTIC see NCG000
NEOBOR see SFF000
NEOBRETTIN see NCD550
NEOBRIDAL see WAK000
NEOCAINE see AIL750, AIT250
NEO-CALMA see CJR909
NEOCARCINOSTATIN see NBV500
NEOCARZINOSTATIN see NBV500
NEOCARZINOSTATIN K see NBV500
NEOCHIN see CLD000
NEOCHROMIUM see NBW000
NEOCID see DAD200
NEOCIDOL see DCM750
NEOCLYM see FBP100
NEOCOCCYL see SNM500
NEO-CODEMA see CFY000
NEOCOMPENSAN see PKQ250
NEO-COROVAS see PBC250
NEOCROSEDIN see EHP000
NEOCRYL A-1038 see ADW200
NEO-CULTOL see MQV750
NEOCYCLINE see TBX000
NEOCYCLOHEXIMIDE see CPE750
NEODALIT see DCW800
NEODECADRON see BFW325
NEODECANOIC ACID see NBW500
NEODECANOIC ACID, HEXADECYL ESTER see HCP600
NEODELPREGNIN see MCA500
NEO-DEMA see CLH750
NEO-DEVOMIT see EAN600
NEODICOUMARIN see BKA000
NEODICOUMAROL see BKA000
NEODICUMARINUM see BKA000

NEODORM (NEW) see NBT500
NEODRENAL see DMV600
NEODRINE see DBA800
NEODROL see DME500
NEODURABOLIN see EJT600
NEODYMACETAT (GERMAN) see NBX500
NEODYMIUM see NBX000
NEODYMIUM ACETATE see NBX500
NEODYMIUM CHLORIDE see NBY000
NEODYMIUM LACTATE see LAN000
NEODYMIUM(III) NITRATE (1:3) see NCB000
NEODYMIUM(III) NITRATE, HEXAHYDRATE (1:3:6) see NCB500
NEODYMIUM OXIDE see NCC000
NEODYMIUM TRIACETATE see NBX500
NEODYMLACTAT (GERMAN) see LAN000
NEO-EPININE see DMV600
NEO-ERGOTIN see EDC500
NEOESERINE BROMIDE see POD000
NEOESERINE METHYL SULFATE see DQY909
NEO-ESTRONE see ECU750
NEO-FARMADOL see HNI500
NEO-FAT 8 see OCY000
NEO-FAT 10 see DAH400
NEO-FAT 12 see LBL000
NEO-FAT 18-61 see SLK000
NEO-FAT 90-04 see OHU000
NEO-FAT 92-04 see OHU000
NEO-FAT 18-S see SLK000
NEOFEMERGEN see LJL000
NEOFEN see HNI500
NEO-FERRUM see IHG000
NEOFIX FP see CNH125
NEOFLUMEN see CFY000
NEOFOLLIN see EDS100
NEO-FULCIN see GKE000
NEOGEL see CBO500
NEO GERM-I-TOL see AFP250
NEO-GILURYTMAL see DNB000
NEOGLAUCIT see IRF000
NEOGYNON see NNL500
NEOHETRAMINE see NCD500
NEOHETRAMINE HYDROCHLORIDE see RDU000
NEOHEXANE (DOT) see DQT200
NEO-HIBERNEX see DQA600
NEO-HOMBREOL see TBG000
NEO-HOMBREOL-M see MPN500
NEOHYDRAZID see COH250
NEOHYDRIN see CHX250
NEOISCOTIN see SIJ000
NEO-ISTAFENE see HGC500
NEOL see DTG400
NEOLEXINA see ALV000
NEOLIN see BFC750
NEOLOID see CCP250
NEOLUTIN see HNT500
NEO-MANTLE CREME see NCG000
NEOMCIN see NCE000
NEOMETANTYL see HKR500
NEOMETHIODAL see DNG400
NEOMIX see NCD550, NCG000
NEOMYCIN see NCE000
NEOMYCIN A see NCE500
NEOMYCIN B see NCF000
NEOMYCIN B SULFATE (SALT) see NCD550
NEOMYCIN C see NCF300
NEOMYCIN E see NCF500
NEOMYCINE SULFATE see NCG000
NEOMYCIN SULFATE see NCG000
NEOMYCIN SULPHATE see NCG000
NEOMYSON G HYDROCHLORIDE see UVA150
NEON see NCG500
NEON, compressed (DOT) see NCG500
NEON, refrigerated liquid (DOT) see NCG500
NEO-NACLEX see BEQ625
NEONAL see BPF500
NEO-NAVIGAN see DYE600
NEONICOTINE see AON875
NEONIKOTIN see AON875
NEO-NILOREX see DKE800
NEONOL see NCG700
NEONOL 2B1317-12 see NCG725
NEONOL AF-14 see NCG715
NEO-OESTRANOL I see DKA600
NEO-OESTRANOL II see DKB000

NEO-ORMONAL see AOO275
NEOOXEDRINE see SPC500
NEO-OXYPATE see PQC500
NEOPANTANOYL CHLORIDE see DTS400
NEOPAX see SJJ175
NEOPELLIS see TFD250
NEOPENTANE see NCH000
NEOPENTANETETRAYL NITRATE see PBC250
1,1',1'',1'''-(NEOPENTANE TETRAYLTETRAOXY)TETRAKIS(2,2,2-
    TRICHLOROETHANOL) see NCH500
NEOPENTANOIC ACID see PJA500
NEOPENTYLENE GLYCOL see DTG400
NEOPENTYL GLYCOL see DTG400
NEOPENTYL GLYCOL DIACRYLATE see DUL200
NEOPENTYL GLYCOL DIGLYCIDYL ETHER see NCI300
NEOPEPULSAN see HKR500
NEOPHARMEDRINE see DBA800
NEOPHEDAN see PEC250
NEOPHENAL see PEC250
NEOPHRYN see SPC500
NEOPHYILINE see TEP500
NEOPHYLLIN see DNC000
NEOPHYLLINE see DNC000
NEOPHYLLIN M see DNC000
NEOPLATIN see PJD000
NEOPRENE see NCI500
NEOPROGESTIN see NCI525
NEOPROMA see MFK500
NEOPROSERINE see NCI550
NEOPROTOVERATRIN see NCI600
NEOPROTOVERATRINE see NCI600
NEOPSICAINE HYDROCHLORIDE see NCJ000
NEORESERPAN see RCA200
NEORESTAMIN see TAI500
NEORON see IOS000
NEO-RONTYL see BEQ625
NEOSABENYL see CJU250
NEO-SALICYL see SJO000
NEOSALVARSAN see NCJ500
NEOSAR see CQC650
NEO-SCABICIDOL see BBQ500
NEOSEDYN see TEH500
NEOSEPT see HCL000
NEOSERINE BROMIDE see POD000
NEOSETILE BLUE EB see TBG700
NEOS-HIDANTOINA see DKQ000
NEO-SINEFRINA see SPC500
NEO-SKIODAN see DNG400
NEOSOLANIOL see NCK000
NEOSONE D see MFA250
NEOSPIRAMYCIN see NCK500
NEOSTAM see NCL000
NEOSTAM STIBAMINE GLUCOSIDE see NCL000
NEOSTEN see AGN000
NEOSTENE see AOO475
NEOSTENOVASAN see DNC000
NEOSTERON see AOO475
NEOSTIBOSAN see SLP600
NEOSTIGMETH see DQY909
NEOSTIGMINE see NCL100
NEOSTIGMINE BROMIDE see POD000
NEOSTIGMINE METHOSULFATE see DQY909
NEOSTIGMINE METHYL BROMIDE see POD000
NEOSTIGMINE METHYL SULFATE see DQY909
NEOSTIGMINE MONOMETHYLSULFATE see DQY909
NEOSTON see AGN000
NEOSTREPAL see SNN300
NEOSTREPSAN see TEX250
NEOSULF see NCD550
NEOSUPRANOL HYDROCHLORIDE see NCL300
NEO-SUPRIMAL see HGC500
NEO-SUPRIMEL see HGC500
NEOSYDYN see TEH500
NEOSYMPATOL see SPC500
NEOSYNEPHRINE see NCL500, SPC500
NEOSYNEPHRINE HYDROCHLORIDE see SPC500
NEOSYNESINE see SPC500
NEO-TENEBRYL see DNG400
NEO-TESTIS see TBF500
NEOTHESIN see PIV750
NEOTHESIN HYDROCHLORIDE see IJZ000
NEOTHRAMYCIN see NCM275
NEOTHRAMYCIN A see TCP000
NEOTHYLLINE see DNC000

NEOTIBIL see FNF000
NEOTILINA see DNC000
NEOTIZIDE see SIJ000
NEO-TIZIDE SODIUM SALT see SIJ000
NEOTONZIL see MCI500
NEOTOPSIN see PEX500
NEO-TRAN see MQU750
NEOTRAN see NCM700
NEO-TRIC see MMN250
NEO-VASOPHYLINE see DNC000
NEOVITAMIN A ACID see VSK955
NEOVLETTA see NNL500
NEOXAZOL see SNN500
NEOXIN see ILD000
NEOZEPAM see DLY000
NEOZINE see MCI500
NEO-ZINE see MNV750
NEOZONE A see PFT250
NEOZONE D see PFT500
NEPHELOR see GGS000
NEPHENTINE see MQU750
NEPHIS see EIY500
NEPHOCARP see TNP250
NEPHRAMIDE see AAI250
NEPHRIDINE see VGP000
NEPRESOL see OJD300
NEPRESOLIN see OJD300
NEPRESSOL see OJD300
NEPTAL see NCM800
NEPTUNE BLUE BRA CONCENTRATION see FMU059
NEPTUNIUM see NCN500
NERA see ZLS200
NERACID see CBG000
NERA EMULZE (CZECH) see ZLS200
NERAL see DTC800
NERIINE DIHYDRBROMIDE see DOX000
NERIODIN see DEO600
NERIOL see OHQ000
NERIOLIN see OHQ000
NERIOSTENE see OHQ000
NERISONA see DKF130
NERISONE see DKF130
NERIUM OLEANDER see OHM875
NERKOL see DGP900
NEROBIL see DYF450
NEROBIOLIL see DYF450
NEROBOL see DAL300
NEROBOLETTES see DAL300
NEROL (FCC) see DTD200
NEROLI BIGARADE OIL, TUNISIAN see NCO000
NEROLIDYL ACETATE see NCN800
NEROLIN see EEY500
NEROLINE see EEY500
NEROLIN II see EEY500
NEROLIN NEW see EEY500
NEROLI OIL, ARTIFICAL see APJ250
NEROLI OIL, TUNISIAN see NCO000
NERONE see MCF525
NERPRUN (CANADA) see MBU825
NERUSIL see NCQ100
NERVACTON see BCA000
NERVANAID B ACID see EIX000
NERVANAID B LIQUID see EIV000
NERVANAID B see EIV000
NERVATIL see BCA000
NERVILON see DXO300
NERVOSETON see BAG250
NERVOTON see DOZ000
NERYL FORMATE see FNC000
NERYL ISOVALERIANATE see NCO500
NERYL-β-METHYL BUTYRATE see NCO500
NERYL PROPIONATE see NCP000
NESDONAL see PBT250
NESDONAL SODIUM see PBT500
NESOL see MCC250
NESONTIL see CFZ000
NESPOR see MAS500
NESSLER REAGENT see NCP500
NESTON see MDT740
NESTYN see EHP000
NETAGRONE 600 see DAA800
NETAL see BNL250
NETALID see INS000
NETAZOL see CIR250

NETH see INS000
NETHALIDE see INS000
NETHALIDE HYDROCHLORIDE see INT000
NETHAMINE HYDROCHLORIDE see EJR500
NETILLIN see NCP550
NETILMICIN see SBD000
NETILMICIN SULFATE see NCP550
NETOCYD see DJT800
NETROMYCIN see NCP550
NETROPSIN see NCP875
NETROSPIN HYDROCHLORIDE see NCQ000
NETSUSARIN see DOT000
NEU see ENV000, NKE500
NEUCHLONIC see DLY000
NEUCOCCIN see FMU080
NEUFIL see DNC000
NEULACTIL see PIW000
NEULEPTIL see PIW000
NEUMOLISINA see ELC500
NEUPERM GFN see DTG000
NEURACEN see BEG000
NEURACTIL see MCI500
NEURAKTIL see BCA000
NEURALEX see NCQ100
NEURAZINE see CKP500
NEURIDINE see DCC400
NEURIDINE TETRAHYDROCHLORIDE see GEK000
NEURIN see VQR300
NEURINE see VQR300
NEURIPLEGE see CLY750
NEUROBARB see EOK000
NEUROBENZIL see BCA000
NEUROCAINE see CNE750
NEUROCIL see MCI500
NEURODYN see TEH500
NEUROLEPTONE see BCA000
NEURONIKA see ADA725
NEUROPROCIN see EHP000
NEUROSEDIN see TEH500
NEUROSTAIN see NCQ500
NEUROTONE see RLU000
NEUROZINA see CJR909
NEURVITA see DXO300
NEURYL see SHL500
NEUSTAB see FNF000
NEUT see SFC500
NEUTHION see GFW000
NEUTRAFIL see DNC000
NEUTRAFILLINA see DNC000
NEUTRAL ACRIFLAVINE see DBX400, XAK000
NEUTRAL AMMONIUM FLUORIDE see ANH250
NEUTRAL BERBERINE SULFATE see BFN625
NEUTRAL POTASSIUM CHROMATE see PLB250
NEUTRAL PROFLAVINE SULPHATE see DBN400
NEUTRAL RED see AJQ250
NEUTRAL RED CHLORIDE see AJQ250
NEUTRAL RED W see AJQ250
NEUTRAL SAPONINS of PANAX GINSENG ROOT see GEQ425
NEUTRAL SODIUM CHROMATE see DXC200
NEUTRAL VERDIGRIS see CNI250
NEUTRAPHYLLIN see DNC000
NEUTRAPHYLLINE see DNC000
NEUTRAZYME see SIB600
NEUTROFLAVINE see XAK000
NEUTRONYX 600 see PKF000
NEUTRONYX 605 see PKF500
NEUTRONYX 622 see GHS000
NEUTRORMONE see AOO475
NEUTROSEL NAVY BN see DCJ200
NEUTROSEL RED TRVA see CLK235
NEUTROSTERON see AOO475
NEUTROXANTINA see DNC000
NEUVITAN see GEK200
NEUWIED GREEN see COF500
NEVANAID-B POWDER see TNL250
NEVAX see DJL000
NEVIGRAMON see EID000
NEVIN see ILD000, SNL850
NEVIN (mixture) see SNL850
NEVRACTEN see CDQ250
NEVRITON see DXO300
NEVRODYN see TEH500
NEW BLITANE see ZJS300
NEW COCCIN see FMU080

NEWCOL 60 see SKV150
NEW GREEN see COF500
NEW IMPROVED CERESAN see BJT250
NEW IMPROVED GRANOSAN see BJT250
NEWLASE see NCQ600
NEW MILSTEM see BRI750
NEWPHRINE see SPC500
NEW PINK BLUISH GEIGY see FAG040
NEWPOL LB3000 see BRP250
NEW PONCEAU 4R see FMU070
NEW VICTORIA GREEN EXTRA I see AFG500
NEXAGAN see EGV500
NEXION see BNL250
NEXION 40 see BNL250
NEXIT see BBQ500
NEXOVAL see CKC000
NF see NGE500
NF 6 see TGI250
β-NF see NAU525
NF 44 see PEX500
NF 64 see NGB700
NF 246 see NDY000
NF 260 see FPI000
NF-269 see FPI100
NF 323 see NDY400
NF 1010 see EAE500
NF 35 (fungicide) see DJV000
5-NFA see NGI000
NFHAA see FNO000
NF-902 HYDROCHLORIDE see FPI100
NFIP see NGI800
NFN see DLU800
NG see NGY000
NG-180 see NGG500
NGAI CAMPHOR see NCQ820
NH 188 see RDU000
NHA see NBI500
NHC see HFN500
NH-LOST see BHN750
NHMI see OBY000
NHU see HKQ025
NHYD see NKJ000
Ni 270 see NCW500
Ni 4303T see NCW500
NIA 249 see AFR250
NIA 2,995 see DEV600
NIA 4556 see DFO800
NIA 5273 see PIX250
NIA 5462 see EAQ750
NIA 5488 see CKM000
NIA-5767 see EAS000
NIA 5996 see DER800
NIA 9044 see BGB500
NIA 9102 see MQQ250
NIA-9241 see BDJ250
NIA 10242 see CBS275
NIA 11092 see DUM800
NIA 16,388 see PNV750
NIA 17170 see BEP500
NIA-18739 see BEP750
NIA 2995J see DEV600
NIA 33297 see AHJ750
NIACEVIT see NCR000
NIACIDE see FAS000
NIACIN see NCQ900
NIACINAMIDE see NCR000
NIAGARA 1,137 see BIT250
NIAGARA 1240 see EEH600
NIAGARA 4512 see SKQ400
NIAGARA 4556 see DFO800
NIAGARA 5006 see DER800
NIAGARA 5,462 see EAQ750
NIAGARA 5767 see EAS000
NIAGARA 5943 see AIX000
NIAGARA 5,996 see DER800
NIAGARA 9044 see BGB500
NIAGARA 9241 see BDJ250
NIAGARA BLUE see CMO250
NIAGARA BLUE 2B see CMO000
NIAGARAMITE see SOP500
NIAGARA P.A. DUST see NDN000
NIAGARA SKY BLUE see CMO500
NIAGARA-STIK see NAK500
NIAGARATHAL see DXD000

NIAGARATRAN see CJT750
NIAGARIL see BEQ625
NIAGRA 10242 see CBS275
NIALAMIDE see BET000
NIALATE see EEH600
NIALK see TIO750
NIAMID see BET000
NIAMIDAL see BET000
NIAMIDE see NCR000
NIAMINE see DJS200
NIAPROOF 4 see SIO000
NIA PROOF 08 see TAV750
NIAQUITIL see BET000
NIAX CATALYST AL see BJH750
NIAX CATALYST ESN see NCS000
NIAX FLAME RETARDANT 3 CF see CGO500
NIAX ISOCYANATE TDI see DNK200, TGM740
NIAX POLYOL LG-168 see NCT000
NIAX POLYOL LHT-42 see NCT500
NIAX TDI see TGM750, TGM800
NIAX TDI-P see TGM750
NIAX TRIOL 6000 see NCV500
NIAZOL see NCW000
NIBAL see PMC700
NIBREN WAX see TIT500
NIBROL see TEH500
NIBUFIN see NIM000
NICACID see NCQ900
NICAMETATE CITRATE see EQQ100
NICAMETATE DIHYDROGEN CITRATE see EQQ100
NICAMIDE see DJS200, NCR000
NICAMIN see NCQ900
NICAMINA see NCR000
NICAMINDON see NCR000
NICANGIN see NCQ900
NICARDIPINE HYDROCHLORIDE see PCG550
NICASIR see NCR000
NICAZIDE see ILD000
NICELATE see EID000
N. I. CERESAN see EME100
NICERGOLIN (GERMAN) see NDM000
NICERGOLINE see NDM000
NICERITROL see NCW300
NICETAMIDE see DJS200
NICETHAMIDE see DJS200
NICHEL TETRACARBONILE (ITALIAN) see NCZ000
NICHOLIN see CMF350
NICIZINA see ILD000
NICKEL see NCW500
NICKEL 270 see NCW500
NICKEL (ITALIAN) see NCW500
NICKEL(II) ACETATE (1582) see NCX000
NICKEL ACETATE TETRAHYDRATE see NCX500
NICKEL ALLOY, Ni,Be see NCY000
NICKEL AMMONIUM SULFATE see NCY050
NICKEL ANTIMONIDE see NCY100
NICKEL ARSENIDE (As$_2$-Ni$_5$) see NDJ399
NICKEL ARSENIDE (As$_8$-Ni$_{11}$) see NDJ400
NICKEL ARSENIDE SULFIDE see NCY125
NICKEL(2+), BIS(N,N′-BIS(2-AMINOETHYL)-1,2-ETHANEDIAMINE-N,N,N^N)-,
  (T-4)-TETRAOXOTUNGSTATE(2-) (1:1) see BLN100
NICKEL BISCYCLOPENTADIENE see NDA500
NICKEL, BIS(TRIETHYLENETETRAMINE)TUNGSTATO- see BLN100
NICKEL BISTRIPHENYLPHOSPHINE DITHIOCYANATE see BLS500
NICKEL BLACK see NDE010
NICKEL BOROFLUORIDE see NDC000
NICKEL(II) CARBONATE (1:1) see NCY500
NICKEL CARBONYL see NCZ000
NICKEL CARBONYLE (FRENCH) see NCZ000
NICKEL CHLORIDE (DOT) see NDH000
NICKEL(II) CHLORIDE (1:2) see NDH000
NICKEL(II) CHLORIDE HEXAHYDRATE (1:2:6) see NDA000
NICKEL COMPLEX with N-FLUOREN-2-YL ACETOHYDROXAMIC ACID
  see HIR500
NICKEL, COMPOUND with pi-CYCLOPENTADIENYL (1:2) see NDA500
NICKEL COMPOUNDS see NDB000
NICKEL CYANIDE (DOT) see NDB500
NICKEL CYANIDE (solid) see NDB500
NICKEL DIBUTYLDITHIOCARBAMATE see BIW750
NICKEL DIFLUORIDE see NDC500
NICKEL DINITRITE see NDG550
NICKEL DISULFIDE see NDB875
NICKEL (DUST) see NCW500
NICKEL(II) EDTA COMPLEX see EJA500

NICOTINIC ACID, BUTYL ESTER see NDT500
NICOTINIC ACID-7,8-DIDEHYDRO-4,5-α-EPOXY-3-METHOXY-17-
    METHYLMORPHINAN-6-α-YL ESTER see CNG250
NICOTINIC ACID DIETHYLAMIDE see DJS200
NICOTINIC ACID, 3-(2,6-DIMETHYLPIPERIDINO)PROPYL ESTER, HYDRO-
    CHLORIDE see NDU000
NICOTINIC ACID, ESTER with CODEINE see CNG250
NICOTINIC ACID, HYDRAZIDE see NDU500
NICOTINIC ACID-2-HYDROXY-3-(o-METHOXYPHENOXY)PROPYL ESTER
    see HLT300
NICOTINIC ACID, METHYL ESTER see NDV000
NICOTINIC ACID, 3-(2-METHYLPIPERIDINO)PROPYL ESTER, HYDROCHLO-
    RIDE see NDW000
NICOTINIC ACID, SODIUM SALT see NDW500
NICOTINIC ACID-1,2,5,6-TETRAHYDRO-1-METHYL-, METHYL ESTER,
    HYDROCHLORIDE see AQU250
NICOTINIC AMIDE see NCR000
NICOTINIPCA see NCQ900
NICOTINOYL HYDRAZINE see NCQ900, NDU500
N-NICOTINOYLTRYPTAMIDE see NDW525
NICOTINSAURE (GERMAN) see NCQ900
NICOTINSAUREAMID (GERMAN) see NCR000
NICOTINYLHYDRAZIDE see NDU500
NICOTION see EPQ000
NICOTOL see NCR000
NICOTYLAMIDE see NCR000
NICOTYRINE see NDX300
β-NICOTYRINE see NDX300
3,2'-NICOTYRINE see NDX300
NICOULINE see RNZ000
NICOUMALONE see ABF750
NICOVASAN see NCQ900
NICOVASEN see NCQ900
NICOVEL see NCQ900, NCR000
NICOVIT see NCR000
NICOVITOL see NCR000
NICOZIDE see ILD000
NICOZYMIN see NCR000
NICYL see NCQ900
NIDA see MMN250
NIDANTIN see OOG000
NIDRAFUR see NGB700
NIDRAFUR P see NGB700
NIDRAN see ALF500
NIDRANE see BEG000
NIDRAZID see ILD000
NIDROXYZONE see FPE100
NIEA see HKW475
NIERALINE see VGP000
NIESYMETRYCZNA DWU METYLOHYDRAZYNA (POLISH) see DSF400
NIFEDIN see AEC750
NIFEDIPINE see AEC750
NIFELAT see AEC750
NIFENALOL see INU000
NIFLAN see PLX400
NIFLEX see SAV000
NIFLUMIC ACID see NDX500
NIFLURIL see NDX500
NIFOS T see TCF250
NIFTHOLIDE see FMR050
NIFTOLIDE see FMR050
NIFULIDONE see NGG500
NIFURADENE see NDY000
NIFURAN see NGG500
NIFURATRONE see HKW450
NIFURDAZIL see EAE500
NIFURHYDRAZONUM see NGB700
NIFUROXAZID see DGQ500
NIFUROXAZIDE see DGQ500
NIFUROXIME see NGC000
NIFURPIPONE see NDY350
NIFURPIPONE ACETATE see NDY360
NIFURPIPONE DIHYDROCHLORIDE see NDY370
NIFURPIPONE HYDROCHLORIDE see NDY380
NIFURPIPONE MALEATE (1581) see NDY390
NIFURPIRINOL see NDY400
NIFURPRAZINE HYDROCHLORIDE see ALN250
NIFURTHIAZOLE see NDY500
NIFURTIMOX see NGG000
NIFUZON see NGE500
NIGERICIN SODIUM SALT see SIO788
NIGHT BLOOMING JESSAMINE see DAC500
NIGHTCAPS see PAM780
NIGHTSAGE (JAMAICA) see YAK300

NIGHTSHADE see DAD880
NIGLYCON see NGY000
NIGRIN see SMA000
NIH 204 see TKZ000
NIH 3127 see SBE500
NIH 7574 see BBU625
NIH 7672 see MDV000
NIH 7958 see DOQ400
NIH 7981 see COV500
NIH 7519 HYDROBROMIDE see PMD325
NIH-4185 HYDROCHLORIDE see DJP500
NIH-5145 HYDROCHLORIDE see EIJ000
NIH 7440 HYDROCHLORIDE see AGV890
NIH-LH-B 9 see LIU300
NIHONTHRENE GOLDEN YELLOW see DCZ000
NIHONTHRENE GREY M see CMU475
NIHONTHRENE NAVY BLUE G see CMU500
NIHYDRAZON see NGB700
NIHYDRAZONE see NGB700
NIKAFLOC D 1000 see CNH125
NIKARDIN see DJS200
NIKA-TEMP see PKQ059
NIKAVINYL SG 700 see PKQ059
NIKETAMID see DJS200
NIKETHAROL see DJS200
NIKETHYL see DJS200
NIKETILAMID see DJS200
NIKION see BEQ625
NIKKELTETRACARBONYL (DUTCH) see NCZ000
NIKKOL BL see DXY000
NIKKOL MYO 2 see PJY100
NIKKOL MYO 10 see PJY100
NIKKOL OTP 70 see DJL000
NIKKOL SO 10 see SKV100
NIKKOL SO-15 see SKV100
NIKKOL SO-30 see SKV100
NIKKOL SS 30 see SKV150
NIKKOL TO see PKL100
NIKORIN see DJS200
NIKO-TAMIN see NCR000
NIKOTIN (GERMAN) see NDN000
NIKOTINSAEUREAMID (GERMAN) see NCR000
NIKOTYNA (POLISH) see NDN000
NILACID see ABX500
NILE BLUE A see AJN250
NILERGEX HYDROCHLORIDE see AEG625
NILEVA see ENX600
NILEVAR see ENX600
NILHISTIN see DCJ850
NILODIN see DHU000, SBE500
NILOX see BPF000
NILOX PBNA see PFT500
NILSTAT see NOH500
NILUDIPINE see NDY600
NILVADIPINE see NDY650
NILVERM see TDX750
NILVEROM see TDX750
NIMBLE WEED see PAM780
NIMCO CHOLESTEROL BASE H see CMD750
NIMERGOLINE see NDM000
NIMETAZEPAM see DLV000
NIMITEX see TAL250
NIMITOX see TAL250
NIMODIPINE see NDY700
NIM OIL see NBS300
NIMORAZOLE see NHH000
NIMORAZOLO see NHH000
NIMOTOP see NDY700
NIMROD see BRJ000
NIMUSTINE see NDY800
NIMUSTINE HYDROCHLORIDE see ALF500
NINCALUICOLFLASTINE see VKZ000
NINHYDRIN see DMV200
NINHYDRIN HYDRATE see DMV200
NINOL 4821 see BKE500
NINOL AA62 see BKE500
NINOL AA-62 EXTRA see BKE500, LBL000
NINOPTERIN see BMM125
NIOBATE, OCTAPOTASSIUM see PLL250
NIOBE OIL see MHA750
NIOBIUM see NDZ000
NIOBIUM CHLORIDE see NEA000
NIOBIUM PENTACHLORIDE see NEA000
NIOBIUM POTASSIUM FLUORIDE see PLN500

NIOBIUM POTASSIUM OXIDE see PLL250
NIOCINAMIDE see NCR000
NIOFORM see CHR500
NIOGEN ES 160 see PJY100
NIOI-PEPA (HAWAII) see PCB275
NIOMIL see DQM600
NIONATE see FBK000
NIONG see NGY000
NIOPAM see IFY000
NIOZYMIN see NCR000
NIP see DFT800
NIPA 49 see PNM750
NIPABUYL see BSC000
NIPAGALLIN LA see DXX200
NIPAGALLIN P see PNM750
NIPAGIN see HJL500
NIPAGIN A see HJL000
NIPAM see INH000
NIPANTIOX 1-F see BQI000
NIPAR S-20 see NIY000
NIPAR S-20 SOLVENT see NIY000
NIPAR S-30 SOLVENT see NIY000
NIPASOL see HNU500
NIPAXON see NOA000
NIPAZIN A see HJL000
NIPELLEN see NCQ900
NIPEON A 21 see PKQ059
NIPERYT see PBC250
NIPERYTH see PBC250
NIPHANOID see BQA010
NIPLEN see ILD000
NIPODAL see PMF500
NIPOL 407 see SMR000
NIPOL 576 see PKQ059
NIPPAS see SEP000
NIPPON BLUE BB see CMO000
NIPPON DEEP BLACK see AQP000
NIPPON ORANGE X-881 see DVB800
NIPRADILOL see NEA100
NIPRIDE see SIU500
NIPRIDE DIHYDRATE see SIW500
NIPSAN see DCM750
NIQUETAMIDA see DJS200
NIRAN see CDR750, PAK000
NIRATIC HYDROCHLORIDE see LFA020
NIRATIC-PURON HYDROCHLORIDE see LFA020
NIRIDAZOLE see NML000
NIRIT see DVF800
NIRVANOL see EOL000
(-)-NIRVANOL see EOL050
l-NIRVANOL see EOL050
(R)-NIRVANOL see EOL050
NIRVOTIN see PGE000
Ni 0901-S see NCW500
NISENTIL see NEA500, NEB000
NISENTIL HYDROCHLORIDE see NEB000
NISETAMIDE see DJS200
NISIDANA see DCV800
NISOLONE see SOV100
NISONE see PLZ100
NISOTIN see EPQ000
NISPERO DEL JAPON (CUBA, PUERTO RICO) see LIH200
NISSAN CATION M2-100 see TCA500
NISSAN CATION S2-100 see DTC600
NISSAN DIAPION S see SIY000
NISSAN DIAPON T see SIY000
NISSAN NONION SP 60 see SKV150
NISSOCAINE see AIL750
NISSOL EC see MME809
NISY see DRB400
NITARSONE see NIJ500
NITEBAN see ILD000
NITER see PLL500
NITHIAZID see ENV500
NITHIAZIDE see ENV500
NITHIOCYAMINE see AOA050
NITICID see CHS500
NITICOLIN see CMF350
NITOBANIL see FMB000
NITOFEN see DFT800
NITOL see BIE500
NITOL "TAKEDA" see BIE500
NITOMAN see TBJ275
NITRADOR see DUS700

NITRADOS see DLY000
NITRAFEN see DFT800
NITRALAMINE HYDROCHLORIDE see NEC000
NITRALDONE see FPI000
NITRAMIDE see NEG000
NITRAMIN see ALQ000
NITRAMINE see ALQ000
NITRAMINOACETIC ACID see NEC500, NGY700
NITRAN see DUV600
4-NITRANILINE see NEO500
m-NITRANILINE see NEN500
o-NITRANILINE see NEO000
p-NITRANILINE see NEO500
NITRANOL see TJL250
NITRAPHEN see DFT800
NITRAPYRIN (ACGIH) see CLP750
NITRATE d'AMYLE (FRENCH) see AOL250
NITRATE d'ARGENT (FRENCH) see SDS000
NITRATE de BARYUM (FRENCH) see BAN250
NITRATE MERCUREUX (FRENCH) see MDE750
NITRATE MERCURIQUE (FRENCH) see MDF000
NITRATE PHENYL MERCURIQUE (FRENCH) see MDH500
NITRATE de PLOMB (FRENCH) see LDO000
NITRATE de PROPYLE NORMAL (FRENCH) see PNQ500
NITRATES see NED000
NITRATE de SODIUM (FRENCH) see SIO900
NITRATE de STRONTIUM (FRENCH) see SMK000
NITRATE de ZINC (FRENCH) see ZJJ000
NITRATINE see SIO900
NITRATION BENZENE see BBL250
NITRAZEPAM see DLY000
NITRAZOL CF EXTRA see NEO500
NITRE see PLL500
NITRE CAKE see SEG800
NITRENDIPINE see EMR600
NITRENPAX see DLY000
NITRETAMIN see TJL250
NITRETAMIN PHOSPHATE see TJL250
NITRIC ACID see NED500
NITRIC ACID, over 40% (DOT) see NED500
NITRIC ACID, ALUMINUM(3+) SALT see AHD750
NITRIC ACID, ALUMINUM SALT, NONAHYDRATE (8CI,9CI) see AHD900
NITRIC ACID, ALUMIUM SALT see AHD750
NITRIC ACID, AMMONIUM LANTHANUM SALT see ANL500
NITRIC ACID, AMMONIUM SALT see ANN000
NITRIC ACID, ANHYDRIDE with PEROXYACETIC ACID see PCL750
NITRIC ACID, BARIUM SALT see BAN250
NITRIC ACID, BERYLLIUM SALT see BFT000
NITRIC ACID, BISMUTH(3+) SALT see BKW250
NITRIC ACID, CADMIUM SALT see CAH000
NITRIC ACID, CADMIUM SALT, TETRAHYDRATE see CAH250
NITRIC ACID, CALCIUM SALT, TETRAHYDRATE see CAU250
NITRIC ACID, CERIUM(3+) SALT, HEXAHYDRATE see CDB250
NITRIC ACID, CERIUM(3+) SALT (8CI, 9CI) see CDB000
NITRIC ACID, CESIUM SALT see CDE250
NITRIC ACID, CHROMIUM (3+) SALT see CMJ600
NITRIC ACID, COBALT (2+) SALT see CNC500
NITRIC ACID COPPER(2+) SALT TRIHYDRATE see CNN000
NITRIC ACID, DODECYL ESTER see NEE000
NITRIC ACID DYSPROSIUM(3+) SALT HEXAHYDRATE see DYH000
NITRIC ACID, ERBIUM (3⁻) SALT see ECY500
NITRIC ACID, ERBIUM (3⁻) SALT, HEXAHYDRATE see ECZ000
NITRIC ACID, ETHYL ESTER see ENM500
NITRIC ACID, EUROPIUM(3+) SALT, HEXAHYDRATE see ERC000
NITRIC ACID, FUMING (DOT) see NEE500
NITRIC ACID, GADOLINIUM(3+) SALT see GAL000
NITRIC ACID, GALLIUM(3+) SALT see GBS000
NITRIC ACID, HOLMIUM(3⁻) SALT, HEXAHYDRATE see HGH000
NITRIC ACID, IRON (3⁻) SALT, NONAHYDRATE see IHC000
NITRIC ACID, ISOPROPYL ESTER see IQP000
NITRIC ACID, LANTHANUM AMMONIUM SALT see ANL500
NITRIC ACID, LEAD (2+) SALT see LDO000
NITRIC ACID, LUTETIUM(3+) SALT see LIY000
NITRIC ACID, MAGNESIUM SALT (2:1) see MAH000
NITRIC ACID, MERCURY(I) SALT see MDE750
NITRIC ACID, MERCURY(II) SALT see MDF000
NITRIC ACID METHYL ESTER see MMF500
NITRIC ACID, NEODYMIUM SALT see NCB000
NITRIC ACID, NEODYMIUM (3+) SALT, HEXAHYDRATE see NCB500
NITRIC ACID, NICKEL(II) SALT see NDG000
NITRIC ACID, NICKEL(2+) SALT, HEXAHYDRATE see NDG500
NITRIC ACID, PHENYLMERCURY SALT see MCU750
NITRIC ACID, POTASSIUM SALT see PLL500
NITRIC ACID, PRASEODYMIUM(3+) SALT see PLY250

NITRIC ACID, PROPYL ESTER see PNQ500
NITRIC ACID (RED FUMING) see NEE500
NITRIC ACID, RED FUMING (DOT) see NEE500
NITRIC ACID, SAMARIUM(3+) SALT, HEXAHYDRATE see SAT000
NITRIC ACID, SILVER(1+) SALT see SDS000
NITRIC ACID, SODIUM SALT see SIO900
NITRIC ACID, STRONTIUM SALT see SMK000
NITRIC ACID, THALLIUM(1+) SALT see TEK750
NITRIC ACID, THORIUM(4+) SALT see TFT500
NITRIC ACID, THULIUM(3+) SALT, HEXAHYDRATE see TFX250
NITRIC ACID TRIESTER OF GLYCEROL see NGY000
NITRIC ACID (WHITE FUMING) see NEF000
NITRIC ACID, YTTERBIUM(3+) SALT see YDS800
NITRIC ACID, YTTERBIUM(3+) SALT, HEXAHYDRATE see YEA000
NITRIC ACID, YTTRIUM(3+) SALT see YFJ000
NITRIC ACID, YTTRIUM(3+)SALT, HEXAHYDRATE see YFS000
NITRIC ACID, ZINC SALT see ZJJ000
NITRIC ACID, ZINC SALT, HEXAHYDRATE see NEF500
NITRIC AMIDE see NEG000
NITRIC ETHER (DOT) see ENM500
NITRIC OXIDE see NEG100
NITRIC OXIDE and NITROGEN TETROXIDE MIXTURES (DOT) see NGT500
NITRIDAZOLE see NML000
NITRIDES see NEH000
NITRILE ACRILICO (ITALIAN) see ADX500
NITRILE ACRYLIQUE (FRENCH) see ADX500
NITRILE ADIPICO (ITALIAN) see AER250
NITRILES see NEH500
NITRILE TRICHLORACETIQUE (FRENCH) see TII750
NITRIL KISELINY DIETHYLAMINOOCTOVE (CZECH) see DHJ600
NITRIL KYSELINY-o-CHLORBENZOOVE (CZECH) see CEM000
NITRIL KYSELINY p-CHLORBENZOOVE (CZECH) see CEM250
NITRIL KYSELINY ISOFTALOVE (CZECH) see PHX550
NITRIL KYSELINY MALONOVE (CZECH) see MAO250
NITRIL KYSELINY MANDLOVE (CZECH) see MAP250
NITRIL KYSELINY STEAROVE (CZECH) see SLL500
NITRIL KYSELINY TEREFTALOVE (CZECH) see BBP250
NITRIL KYSELINY β,β'-THIODIPROPIONOVE (CZECH) see DGS600
NITRIL KYSELINY m-TOLUYLOVE (CZECH) see TGT250
NITRIL KYSELINY p-TOLUYLOVE (CZECH) see TGT750
NITRIL MASTNE KYSELINY S (CZECH) see AFQ000
NITRILOACETIC ACID TRISODIUM SALT MONOHYDRATE see NEI000
NITRILOACETONITRILE see COO000
NITRILOMALONAMIDE see COJ250
NITRILOTRIACETIC ACID see AMT500
NITRILOTRIACETIC ACID, DISODIUM SALT see DXF000
NITRILOTRIACETIC ACID SODIUM SALT see SEP500
NITRILOTRIACETIC ACID, TRISODIUM SALT see SIP500
NITRILOTRIACETIC ACID TRISODIUM SALT MONOHYDRATE see NEI000
NITRILO-2,2',2''-TRIETHANOL see TKP500
2,2',2''-NITRILOTRIETHANOL see TKP500
(2,2',2''-NITRILOTRIETHANOL)PHENYLMERCURY(1+) LACTATE (salt) see TNI500
2,2',2''-NITRILOTRIETHANOL TRINITRATE PHOSPHATE see TJL250
NITRILOTRIMETHYLPHOSPHONIC ACID ZINC complex TRISODIUM TETRA-HYDRATE see ZJJ200
1,1',1''-NITRILOTRI-2-PROPANOL see NEI500
NITRILOTRISILANE see TNJ250
NITRIMIDAZINE see NHH000
NITRINE-TDC see NGY000
N-NITRISO-N-(2,3-DIHYDROXYPROPYL)-N-(2-HYDROXYETHYL)AMINE see NBR100
NITRITE see DVF800
NITRITES see NEJ000
NITRITE de SODIUM (FRENCH) see SIQ500
NITRITO see NGR500
5-NITROACENAPHTHENE see NEJ500
5-NITROACENAPHTHYLENE see NEJ500
5-NITROACENAPTHENE see NEJ500
2-NITROACETALDEHYDE OXIME see NEJ600
2-NITRO-4-ACETAMINOFENETOL (CZECH) see NEL000
4-NITROACETANILIDE see NEK000
p-NITROACETANILIDE see NEK000
4'-NITROACETANILIDE see NEK000
3-NITRO-p-ACETOPHENETIDE see NEL000
3-NITRO-p-ACETOPHENETIDIDE see NEL000
5-NITRO-p-ACETOPHENETIDIDE see NEL000
3'-NITRO-p-ACETOPHENETIDIN see NEL000
m-NITROACETOPHENONE see NEL500
o-NITROACETOPHENONE see NEL450
2'-NITROACETOPHENONE see NEL450
3'-NITROACETOPHENONE see NEL500
NITRO ACID 100 percent see HMY000
NITRO ACID SULFITE see NMJ000

2,2'-((2-(8-NITRO-9-ACRIDINYLAMINO)ETHYL)DIETHANOL HYDROCHLO-RIDE see NFW350
4'-(3-NITRO-9-ACRIDINYLAMINO)METHANESULFONANILIDE see NEM000
N-(4-((3-NITRO-9-ACRIDINYL)AMINO)PHENYL)METHANESULFONAMIDE see NEM000
N'-(1-NITRO-9-ACRIDINYL)-N,N,1-TRIMETHYL-1,2-ETHANEDIAMINE HYDROCHLORIDE see NFW435
NITROALKANES see NEM300
m-NITROAMINOBENZENE see NEN500
4-NITRO-2-AMINOFENOL (CZECH) see NEM500
p-NITROAMINOFENOL (POLISH) see NEM500
4'-NITRO-4-AMINO-3-HYDROXYDIPHENYL HYDROCHLORIDE see AKI500
4'-NITRO-4-AMINO-3-HYDROXYDIPHENYL HYDROGEN CHLORIDE see AKI500
2-NITRO-4-AMINOPHENOL see NEM480
o-NITRO-p-AMINOPHENOL see NEM480
p-NITRO-o-AMINOPHENOL see NEM500
2-(N-NITROAMINO)PYRIDINE-N-OXIDE see NEM600
5-N-NITROAMINOTETRAZOLE see NEN000
5-NITRO-2-AMINOTHIAZOLE see ALQ000
2-NITRO-4-AMINOTOLUENE see NMP500
4-NITRO-2-AMINOTOLUENE (MAK) see NMP500
4-NITROAMINO-1,2,4-TRIAZOLE see NEN300
p-NITROANILINA (POLISH) see NEO500
3-NITROANILINE see NEN500
m-NITROANILINE see NEN500
o-NITROANILINE see NEO000
p-NITROANILINE see NEO500
4-NITROANILINE (MAK) see NEO500
4-NITROANILINE, 2,6-DICHLORO- see RDP300
p-NITROANILINE MERCURY(II) derivative see NEP000
4-NITROANILINE-2-SULFONIC ACID see NEP500
4-NITROANILINIUM PERCHLORATE see NEP600
(p-(p-NITROANILINO)PHENYL)ISOCYANIC ACID see IKI000
4-NITRO-o-ANISIDINE see NEQ000
5-NITRO-o-ANISIDINE see NEQ500
p-NITROANISOL see NER500
2-NITROANISOLE see NER000
4-NITROANISOLE see NER500
o-NITROANISOLE see NER000
p-NITROANISOLE see NER500
2-NITROANTHRACENE see NES000
1-NITRO-9,10-ANTHRACENEDIONE see NET000
1-NITROANTHRACHINON (CZECH) see NET000
4-NITROANTHRANILIC ACID see NES500
1-NITROANTHRAQUINONE see NET000
α-NITROANTHRAQUINONE see NET000
3-NITRO-3-AZAPENTANE-1,5-DIISOCYANATE see NHI500
4-NITROAZOBENZENE see NET500
p-NITROAZOBENZENE see NET500
2-NITROBENZALDEHYDE see NEU500
3-NITROBENZALDEHYDE see NEV000
4-NITROBENZALDEHYDE see NEV500
m-NITROBENZALDEHYDE see NEV000
o-NITROBENZALDEHYDE see NEU500
p-NITROBENZALDEHYDE see NEV500
2-(p-NITROBENZAMIDO)ACETOHYDROXAMIC ACID see NEW550
NITROBENZEEN (DUTCH) see NEX000
NITROBENZEN (POLISH) see NEX000
3-NITROBENZENAMINE see NEN500
4-NITROBENZENAMINE see NEO500
NITROBENZENE see NEX000
NITROBENZENE, liquid (DOT) see NEX000
2-NITROBENZENEACETIC ACID see NII500
4-NITROBENZENEACETONITRILE see NIJ000
p-NITROBENZENEACETONITRILE see NIJ000
4-NITROBENZENEARSONIC ACID see NIJ500
o-NITROBENZENEARSONIC ACID see NEX500
p-NITROBENZENEAZOSALICYLIC ACID see NEY000
m-NITROBENZENEBORONIC ACID see NEY500
2-NITRO-1,4-BENZENEDIAMINE see ALL750
4-NITRO-1,2-BENZENEDIAMINE see ALL500
4-NITROBENZENEDIAZONIUM AZIDE see NEZ000
3-NITROBENZENEDIAZONIUM CHLORIDE see NEZ100
4-NITROBENZENEDIAZONIUM NITRATE see NFA000
3-NITROBENZENEDIAZONIUM PERCHLORATE see NFA500
2-NITROBENZENEDIAZONIUM SALTS see NFA600
4-NITROBENZENEDIAZONIUM SALTS see NFA625
p-NITROBENZENESULFONAMIDE see NFB000
3-NITROBENZENESULFONIC ACID see NFB500
m-NITROBENZENESULFONIC ACID see NFB500
m-NITROBENZENESULFONIC ACID, SODIUM SALT see NFC500
NITROBENZEN-m-SULFONAN SODNY (CZECH) see NFC500
2-NITROBENZIMIDAZOLE see NFC700

6-NITRO-BENZIMIDAZOLE see NFD500
5-NITRO-1H-BENZIMIDAZOLE see NFD500
2-NITRO-1H-BENZIMIDAZOLE (9CI) see NFC700
2-NITROBENZOFURAN see NFD700
2,2'-((7-NITRO-4-BENZOFURAZANYL)IMINO)BISETHANOL OXIDE
   see NFF000
2-NITROBENZOIC ACID see NFG500
4-NITROBENZOIC ACID see CCI250
m-NITROBENZOIC ACID see NFG000
o-NITROBENZOIC ACID see NFG500
p-NITROBENZOIC ACID see CCI250
4-NITROBENZOIC ACID CHLORIDE see NFK100
p-NITROBENZOIC ACID CHLORIDE see NFK100
p-NITROBENZOIC ACID HYDRAZIDE see NFH000
p-NITROBENZOIC ACID, PIPERIDINO ESTER see NFL100
NITROBENZOL (DOT) see NEX000
NITROBENZOL, liquid (DOT) see NEX000
3-NITROBENZONITRILE see NFH500
m-NITROBENZONITRILE see NFH500
o-NITROBENZONITRILE see NFI000
1-NITROBENZO(a)PYRENE see NFI100
5-NITROBENZOTRIAZOL (DOT) see NFJ000
5-NITROBENZOTRIAZOLE see NFJ000
5-NITRO-1H-BENZOTRIAZOLE see NFJ000
6-NITRO-1H-BENZOTRIAZOLE see NFJ000
3-NITROBENZOTRIFLUORIDE see NFJ500
m-NITROBENZOTRIFLUORIDE (DOT) see NFJ500
7-NITRO-3H-2,1-BENZOXAMERCUROL-3-ONE see NHY500
2-NITROBENZOYL CHLORIDE see NFL000
4-NITROBENZOYL CHLORIDE see NFK100
m-NITROBENZOYL CHLORIDE see NFK500
o-NITROBENZOYL CHLORIDE see NFL000
p-NITROBENZOYL CHLORIDE see NFK100
3-NITROBENZOYL NITRATE see NFL100
N-(4-NITRO)BENZOYLOXYPIPERIDINE see NFL100
3-NITROBENZ(a)PYRENE see NFM000
6-NITROBENZ(a)PYRENE see NFM500
2-NITROBENZYL ALCOHOL see NFM550
4-NITROBENZYL ALCOHOL see NFM560
2-NITROBENZYL BROMIDE see NFM700
4-NITROBENZYL BROMIDE see NFN000
p-NITROBENZYL BROMIDE see NFN000
2-NITROBENZYL CHLORIDE see NFN500
o-NITROBENZYL CHLORIDE see NFN500
p-NITROBENZYL CHLORIDE see NFN400
4-NITRO-BENZYL-CYANID (GERMAN) see NIJ000
4-NITROBENZYL CYANIDE see NIJ000
p-NITROBENZYLCYANIDE see NIJ000
2-((m-NITROBENZYLIDENE)AMINO)ETHANOL N-OXIDE see HKW345
2-((p-NITROBENZYLIDENE)AMINO)ETHANOL N-OXIDE see HKW350
m-NITROBENZYLIDENE-1,2-BENZ-9-METHYLACRIDINE see NMF000
m-NITROBENZYLIDENE-3,4-BENZ-9-METHYLACRIDINE see NMD500
o-NITROBENZYLIDENE-1,2-BENZ-9-METHYLACRIDINE see NMF500
o-NITROBENZYLIDENE-3,4-BENZ-9-METHYLACRIDINE see NME000
p-NITROBENZYLIDENE-1,2-BENZ-9-METHYLACRIDINE see NMG000
p-NITROBENZYLIDENE-3,4-BENZ-9-METHYLACRIDINE see NME500
1-(p-NITROBENZYL)-2-NITROIMIDAZOLE see NFO700
4-NITROBIPHENYL see NFQ000
o-NITROBIPHENYL see NFP500
4'-NITRO-4-BIPHENYLAMINE see ALM000
4-NITROBIPHENYL ESTER see NIU000
2-NITRO-1,1-BIS(p-CHLOROPHENYL)PROPANE see BIN500
NITROBROMOFORM see NMQ000
tert-NITROBUTANE see NFQ500
1-NITRO-3-BUTENE see NFR000
2-NITRO-2-BUTENE see NFR500
m-NITROCARBAMIDE see NMQ500
NITROCARBOL see NHM500
NITRO CARBO NITRATE see NFS500
NITROCARDIOL see TJL250
NITROCELLULOSE see CCU250
4-NITROCHINOLIN N-OXID (SWEDISH) see NJF000
p-NITROCHLOORBENZEEN (DUTCH) see NFS525
NITROCHLOR see DFT800
4-NITRO-2-CHLOROANILINE see CJA175
1-NITRO-5-CHLOROANTHRAQUINONE see CJA250
NITROCHLOROBENZENE see CJA950
m-NITROCHLOROBENZENE see CJB250
o-NITROCHLOROBENZENE see CJB750
p-NITROCHLOROBENZENE see NFS525
m-NITROCHLOROBENZENE, solid (DOT) see CJB250
p-NITROCHLOROBENZENE solid (DOT) see NFS525
o-NITROCHLOROBENZENE, liquid (DOT) see CJB750
4-NITRO-7-CHLOROBENZOFURAZAN see NFS550

p-NITROCHLOROBENZOL (GERMAN) see NFS525
3-NITRO-4-CHLOROBENZOTRIFLUORIDE see NFS700
NITROCHLOROFORM see CKN500
p-NITRO-m-CHLOROPHENYL DIMETHYL THIONOPHOSPHATE see MIJ250
p-NITRO-o-CHLOROPHENYL DIMETHYL THIONOPHOSPHATE see NFT000
3-NITRO-4-CHLORO-α,α,α-TRIFLUOROTOLUENE see NFS700
6-NITROCHRYSENE see NFT400
p-NITROCINNAMALDEHYDE see NFT425
p-NITROCLOROBENZENE (ITALIAN) see NFS525
NITRO COMPOUNDS see NFT459
NITRO COMPOUNDS of AROMATIC HYDROCARBONS see NFT500
NITROCOTTON see CCU250
2-NITRO-p-CRESOL see NFU500
4-NITRO-m-CRESOL see NFV000
4-NITRO-m-CRESOL DIMETHYL PHOSPHATE see PHD750
NITROCYCLOHEXANE see NFV500
2-NITRO-1,4-DIAMINOBENZENE see ALL750
4-NITRO-1,2-DIAMINOBENZENE see ALL500
NITRO-p-DICHLOROBENZENE see DFT400
4'-NITRO-2,4-DICHLORODIPHENYL ETHER see DFT800
NITRODIETHANOLAMINE DINITRATE see NFW000
N-NITRODIETHANOLAMINE DINITRATE see NFW000
N-NITRODIETHYLAMINE see DJS500
1-NITRO-9-(DIETHYLAMINOETHYLAMINE)-ACRIDINE DIHYDROCHLORIDE
   see NFW100
1-NITRO-9-(DIETHYLAMINOPROPYLAMINE)-ACRIDINE DIHYDROCHLO-
   RIDE see NFW200
4-NITRODIFENYLAMIN (CZECH) see NFY000
4-NITRODIFENYLAMIN-2-SULFONAN SODYN (CZECH) see AOS500
4-NITRODIFENYLETHER (CZECH) see NIU000
1-NITRO-9-(2-DIHYDROXYETHYLAMINO-ETHYLAMINO)-ACRIDINE HYDRO-
   CHLORIDE see NFW350
N-NITRODIMETHYLAMINE see DSV200
1-NITRO-9-(DIMETHYLAMINE)-ACRIDINE DIHYDROCHLORIDE see NFW400
1-NITRO-9-(DIMETHYLAMINO)-ACRIDINE HYDROCHLORIDE see NFW425
2'-NITRO-4-DIMETHYLAMINOAZOBENZENE see DSW800
3'-NITRO-4-DIMETHYLAMINOAZOBENZENE see DSW600
1-NITRO-10-(DIMETHYLAMINOETHYL)-9-ACRIDONE HYDROCHLORIDE
   see NFW430
5-NITRO-2-(2-DIMETHYLAMINOETHYL)BENZO(de)ISOQUINOLINE-1,3-DIONE
   HYDROCHLORIDE see MQX775
1-NITRO-9-((2-DIMETHYLAMINO)-1-METHYLETHYLAMINO)-ACRIDINE
   DIHYDROCHLORIDE see NFW435
1-NITRO-9-(5-DIMETHYLAMINOPENTYLAMINO)-ACRIDINE DIHYDROCHLO-
   RIDE see NFW450
1-NITRO-14-(DIMETHYLAMINOPROPYL)-ACRIDONE HYDROCHLORIDE
   see NFW460
1-NITRO-10-(3-DIMETHYLAMINOPROPYL)-ACRIDONE HYDROCHLORIDE
   see NFW460
1-NITRO 10-(3-DIMETHYLAMINOPROPYL)-ACRIDON HYDROCHLORIDE
   see NFW460
1-NITRO-9-(3-DIMETHYLAMINOPROPYLAMINE)ACRIDINE-N^{10}-OXIDE
   DIHYDROCHLORIDE see CAB125
1-NITRO-9-(3-DIMETHYLAMINOPROPYLAMINE)-ACRIDINE-N-OXIDE
   DIHYDROCHLORIDE see NFW470
1-NITRO-9-(3'-DIMETHYLAMINOPROPYLAMINO)-ACRIDINE see NFW500
1-NITRO-9-(3-DIMETHYLAMINOPROPYLAMINO)-ACRIDINE DIHYDROCHLO-
   RIDE see LEF300
NITRODIMETHYLBENZENE see NMS000
1-NITRO-3-(2,4-DINITROPHENYL)UREA see NFX500
2-NITRODIPHENYL see NFP500
o-NITRODIPHENYL see NFP500
p-NITRODIPHENYL see NFQ000
4-NITRODIPHENYLAMINE see NFY000
p-NITRODIPHENYLAMINE see NFY000
2-NITRODIPHENYL ETHER see NIT500
4-NITRODIPHENYL ETHER see NIU000
p-NITRODIPHENYL ETHER see NIU000
2-(p-(2-NITRO-1,2-DIPHENYLVINYL)PHENOXY)TRIETHYLAMINE CITRATE
   see EAF100
N-NITRO-DMA see DSV200
4-NITRODRACYLIC ACID see CCI250
NITRODURAN see TJL250
dl-1-(2-NITRO-3-EMTHYLPHENOXY)-3-tert-BUTYLAMINO-PROPAN-2-OL
   see BQF750
NITROETAN (POLISH) see NFY500
NITROETHANE see NFY500
2-NITROETHANOL see NFY550
4-NITRO-2-ETHYLQUINOLINE-N-OXIDE see NGA500
NITROFAN see DUS700
NITRO FAST GREEN GB see BLK000
NITROFEN see DFT800
NITROFENE (FRENCH) see DFT800
4-NITROFENOL (DUTCH) see NIF000

1-p-NITROFENYL-3,3-DIMETHYLTRIAZEN (CZECH) see DSX400
3-NITROFLUORANTHENE see NGA700
4-NITROFLUORANTHENE see NGA700
2-NITROFLUORENE see NGB000
3-NITROFLUORENONE see NGB500
3-NITRO-9-FLUORENONE see NGB500
3-NITRO-9H-FLUOREN-9-ONE see NGB500
4-NITROFLUOROBENZENE see FKL000
p-NITROFLUOROBENZENE see FKL000
NITROFORM see TMM500
5-NITRO-2-FURALDEHYDE see NGE775
5-NITRO-2-FURALDEHYDE ACETYLHYDRAZONE see NGB700
5-NITRO-2-FURALDEHYDE-2-(2-HYDROXYETHYL)SEMICARBAZONE
see FPE100
5-NITRO-2-FURALDEHYDE OXIME see NGC000
5-NITROFURALDEHYDE SEMICARBAZIDE see NGE500
6-NITROFURALDEHYDE SEMICARBAZIDE see NGE500
5-NITRO-2-FURALDEHYDE SEMICARBAZONE see NGE500
5-NITRO-2-FURALDEHYDE THIOSEMICARBAZONE see NGC400
5-NITRO-2-FURALDOXIME see NGC000
5-NITRO-2-FURAMIDOXIME see NGD000
NITROFURAN see NGD400
2-NITROFURAN see NGD400
5-NITROFURAN see NGD400
5-NITRO-2-FURANACRYLAMIDE see NGL500
5-NITRO-2-FURANACRYLIC ACID see NGI000
5-NITROFURAN-2-ALDEHYDE SEMICARBAZONE see NGE500
5-NITRO-5-FURANCARBOXALDEHYDE (9CI) see NGE775
5-NITRO-2-FURANCARBOXALDEHYDE SEMICARBAZONE see NGE500
5-NITROFURANCARBOXYLIC ACID see NGH500
5-NITRO-2-FURANCARBOXYLIC ACID (9CI) see NGH500
5-NITRO-2-FURANMETHANOL see NGD600
NITROFURANTOIN see NGE000
6-(2-(5-NITRO-2-FURANYL)ETHENYL)-2-PYRIDINEMETHANOL (9CI)
see NDY400
1-(((5-NITRO-2-FURANYL)METHYLENE)AMINO)-2-IMIDAZOLIDINONE
see NDY000
3-(((5-NITRO-2-FURANYL)METHYLENE)AMINO)-2-OXAZOLIDINONE
see NGG500
1-((5-NITROFURANYL-2)METHYLENEAMINO)TETRAHYDROPYRIMIDONE-2-
ONE see FPO100
((5-NITRO-2-FURANYL)METHYLENE)HYDRAZIDEACETIC ACID (9CI)
see NGB700
2((5-NITRO-2-FURANYL)METHYLENE)HYDRAZINECARBOXAMIDE
see NGE500
3-(5-NITRO-2-FURANYL)-2-PROPENOIC ACID see NGI000
5-(5-NITRO-2-FURANYL)-1,3,4-THIADIAZOL-2-AMINE see NGI500
N-(4-(5-NITRO-2-FURANYL)-2-THIAZOLYL)ACETAMIDE see AAL750
2-(4-(5-NITRO-2-FURANYL)-2-THIAZOLYL)-HYDRAZINECARBOXALDEHYDE
see NDY500
NITROFURATE see NGH500
NITROFURAZOLIDONE see NGG500
NITROFURAZOLIDONUM see NGG500
NITROFURAZONE see NGE500
NITROFURFURAL see NGE775
5-NITROFURFURAL see NGE775
5-NITROFURFURALACETYLHYDRAZONE see NGB700
3-(5'-NITROFURFURALAMINO)-2-OXAZOLIDONE see NGG500
5-NITROFURFURALDEHYDE see NGE775
5-NITROFURFURAL SEMICARBAZONE see NGE500
5-NITROFURFURYL ALCOHOL see NGD600
2-(5-NITRO-2-FURFURYLIDENE)AMINOETHANOL N-OXIDE see HKW450
1-((5-NITROFURFURYLIDENE)AMINO)HYDANTOIN see NGE000
N-(5-NITROFURFURYLIDENE)-1-AMINOHYDANTOIN see NGE000
N-(5-NITRO-2-FURFURYLIDENE)-1-AMINOHYDANTOIN see NGE000
N-(5-NITRO-2-FURFURYLIDENEAMINO)-2-IMIDAZOAIDINONE see NDY000
1-((5-NITROFURFURYLIDENE)AMINO)-2-IMIDAZOLIDINONE see NDY000
N-(5-NITRO-2-FURFURYLIDENE)-1-AMINO-2-IMIDAZOLIDONE see NDY000
1-((5-NITROFURFURYLIDENE)AMINO)-2-METHYLTETRAHYDRO-1,4-THIA
ZINE-4,4-DIOXIDE see NGG000
4-((5-NITROFURFURYLIDENE)AMINO)-3-METHYLTHIOMORPHOLINE-1,1-DI
OXIDE see NGG000
N-(5-NITRO-2-FURFURYLIDENE)-3-AMINOOXAZOLIDINE-2-ONE see NGG500
3-((5-NITROFURFURYLIDENE)AMINO)-2-OXAZOLIDONE see NGG500
N-(5-NITRO-2-FURFURYLIDENE)-3-AMINO-2-OXAZOLIDINONE see NGG500
(5-NITRO-2-FURFURYLIDENEAMINO)UREA see NGE500
(NITRO-5' FURFURYLIDENE-2') HYDROXY-4 BENZHYDRAZIDE (FRENCH)
see DGQ500
N-(6-(5-NITROFURFURYLIDENEMETHYL)-1,2,4-TRIAZIN-3-YL)IM
INODIMETHANOL see BKH500
NITROFURMETHONE see FPI000
NITROFURMETON see FPI000
5-NITROFUROIC ACID see NGH500
NITROFUROXIME see NGC000

NITROFUROXON see NGG500
NITROFURYLACRYLAMIDE see NGI000
5-NITRO-2-FURYLACRYLAMIDE see NGL500
3-(5-NITRO-2-FURYL)ACRYLAMIDE see NGL500
5-NITRO-2-FURYL ACRYLIC ACID see NGI000
3-(5-NITRO-2-FURYL)ACRYLIC ACID see NGI000
(1-(5-NITRO-2-FURYL)-2-(6-AMINO-3-PYRIDAZYL)-ETHYLENE HYDROCHLO-
RIDE see ALN250
2-(5-NITRO-2-FURYL)-5-AMINO-1,3,4-THIADIAZOLE see NGI500
5-(5-NITRO-2-FURYL)-2-AMINO-1,3,4-THIADIAZOLE see NGI500
3-((5-NITROFURYLIDENE)AMINO)-2-OXAZOLIDONE see NGG500
3-(5-NITRO-2-FURYL)-IMIDAZO(1,2-a)PYRIDINE see NGI800
(5-NITRO-2-FURYL) METHYL KETONE see ACT250
((3-NITRO-2-FURYL)-1-(2-(5-NITRO-2-FURYL)VINYL)
ALLYIDENE)AMINO)GUANIDINE see PAF500
N-(3-(5-NITRO-2-FURYL)-6H-1,2,4-OXADIAZINYL)ACETAMIDE see AAL500
5-(5-NITRO-2-FURYL)-1,3,4-OXADIAZOLE-2-OL see NGK000
N-((3-(5-NITRO-2-FURYL)-1,2,4-OXADIAZOL-5-YL)METHYL)ACETAMIDE
see NGK500
3-(5-NITRO-2-FURYL)-2-PHENYLACRYLAMIDE see NGL000
3-(5-NITRO-2-FURYL)-2-PHENYL-2-PROPENAMIDE see NGL000
3-(5-NITRO-2-FURYL)-2-PROPENAMIDE see NGL500
N-(5-(5-NITRO-2-FURYL)-1,3,4-THIADIAZOL-2-YL)ACETAMIDE see FQJ000
4-(5-NITRO-2-FURYL)THIAZOLE see NGM400
N-(4-(5-NITRO-2-FURYL)-2-THIAZOLYL)ACETAMIDE see AAL750
N-(4-(5-NITRO-2-FURYL)THIAZOL-2-YL)ACETAMIDE see AAL750
N-(4-(5-NITRO-2-FURYL)-2-THIAZOLYL)FORMAMID (GERMAN) see NGM500
N-(4-(5-NITRO-2-FURYL)-2-THIAZOLYL)FORMAMIDE see NGM500
(4-(5-NITRO-2-FURYL)THIAZOL-2-YL)HYDRAZONOACETONE see NGN000
N-(4-(5-NITRO-2-FURYL)-2-THIAZOLYL)-2,2,2-TRIFLUOROACETAMIDE
see NGN500
N,N'-(6-(5-NITRO-2-FURYL)-s-TRIAZINE-2,4-DIYL)BISACETAMIDE see DBF400
3-(5-NITRO-2-FURYL)-1H-1,2,4-TRIAZOL-5-AMINE see ALM750
N-(3-(5-NITRO-2-FURYL)-s-TRIAZOL-5-YL)-N-NITROSOETHYLAMINE
see ENS000
6-(5-NITRO-2-FURYLVINYL)-3-(DIHYDROXYDIMETHYLAMINO)-1,2,4-
TRIAZENE see BKH500
6-(2-(5-NITRO-2-FURYL)VINYL-2-PYRIDINE-METHANOL see NDY400
N-(4-(2-(5-NITRO-2-FURYL)VINYL)2-THIAZOLYL)FORMAMIDE see FNG000
((6-(2-(5-NITRO-2-FURYL)VINYL)-as-TRIAZIN-3-YL)IMINO)DIMETHANOL
see BKH500
N-(6-(2-(5-NITRO-2-FURYL)VINYL)-1,2,4-TRIAZIN-3-YL)IMINODIMETHANOL
see BKH500
NITROGEN see NGP500
NITROGEN, compressed (DOT) see NGP500
NITROGEN (cryogenic liquid) see NGP510
NITROGEN, refrigerated liquid (DOT) see NGP500
NITROGEN CHLORIDE see NGQ500
NITROGEN CHLORIDE DIFLUORIDE see NGR000
NITROGEN DI-DIOXIDE see NGV000
NITROGEN DIOXIDE see NGR500
NITROGEN DIOXIDE (liquid) see NGS000
NITROGEN FLUORIDE see NGW000
NITROGEN FLUORIDE OXIDE see NGS500
NITROGEN GAS see NGP500
NITROGEN GLUCOSIDE of SODIUM-p-AMINOPHENYLSTIBONATE
see NCL000
NITROGEN HALF MUSTARD see CGW000
NITROGEN IODIDE see IDN000, NGW500
NITROGEN LIME see CAQ250
NITROGEN MONOXIDE see NEG100
NITROGEN MONOXIDE, mixed with NITROGEN TETROXIDE see NGT500
NITROGEN MUSTARD see BIE250
NITROGEN MUSTARD HYDROCHLORIDE see BIE500
NITROGEN MUSTARD OXIDE see CFA500, CFA750
NITROGEN MUSTARD-N-OXIDE see CFA500, CFA750
NITROGEN MUSTARD-N-OXIDE HYDROCHLORIDE see CFA750
NITROGEN OXIDE see NGU000
NITROGEN OXIDES mixed with OZONE (47%:53%) see ORY000
NITROGEN OXYCHLORIDE see NMH000
NITROGEN OXYFLUORIDE see NMH500
NITROGEN PEROXIDE, liquid (DOT) see NGR500, NGV000
NITROGEN TETROXIDE see NGU500
NITROGEN TETROXIDE (liquid) see NGV000
NITROGEN TETROXIDE, liquid (DOT) see NGU500
NITROGEN TETROXIDE-NITRIC OXIDE MIXTURE (DOT) see NGT500
NITROGEN TRIBROMIDE HEXAAMMONIATE see NGV500
NITROGEN TRIFLUORIDE see NGW000
NITROGEN TRIIODIDE see NGW500
NITROGEN TRIIODIDE-AMMONIA see NGX000
NITROGEN TRIIODIDE-SILVER AMIDE see NGX500
NITROGLICERINA (ITALIAN) see NGY000
NITROGLICERYNA (POLISH) see NGY000
N-NITROGLICIN see NGY700

NITROGLYCERIN see NGY000
NITROGLYCERIN, liquid, desensitized (DOT) see NGY000
NITROGLYCERIN, liquid, not desensitized (DOT) see NGY000
NITROGLYCERINE see NGY000
NITROGLYCERIN mixed with ETHYLENE GLYCOL DINITRATE (1581)
    see NGY500
NITROGLYCEROL see NGY000
N-NITROGLYCINE see NGY700
NITROGLYCOL see EJG000
NITROGLYKOL (CZECH) see EJG000
NITROGLYN see NGY000
NITROGRANULOGEN see BIE500
NITROGRANULOGEN HYDROCHLORIDE see BIE500
4-NITROGUAIACOL see NHA000
NITROGUANIDINE see NHA500
1-NITROGUANIDINE see NHA500
2-NITROGUANIDINE see NHA500
α-NITROGUANIDINE see NHA500
NITROGUANIDINE, containing less than 20% water (DOT) see NHA500
NITROGUANIDINE DRY (DOT) see NHA500
NITROGUANIL see NIJ400
2-NITRO-2-HEPTENE see NHB000
3-NITRO-2-HEPTENE see NHB500
2-NITRO-2-HEXENE see NHD000
3-NITRO-3-HEXENE see NHE000
NITROHYDRENE see NHE500
NITROHYDROCHLORIC ACID (DOT) see HHM000
NITROHYDROCHLORIC ACID, diluted (DOT) see HHM000
3-NITRO-4-HYDROXYBENZENEARSONIC ACID see HMY000
2-NITRO-1-HYDROXYBENZENE-4-ARSONIC ACID see HMY000
6-NITRO-4-HYDROXYLAMINOQUINOLINE-1-OXIDE see HIX500
7-NITRO-4-HYDROXYLAMINOQUINOLINE-1-OXIDE see HIY000
4-NITRO-5-HYDROXYMERCURIORTHOCRESOL see NHK900
2-NITRO-2-(HYDROXYMETHYL)-1,3-PROPANEDIOL see HMJ500
3-NITRO-4-HYDROXYPHENYLARSENOUS ACID see NHE600
3-NITRO-4-HYDROXYPHENYLARSONIC ACID see HMY000
5-NITRO-8-HYDROXYQUINOLINE see NHF500
2-NITROIMIDAZOLE see NHG000
2-NITRO-1H-IMIDAZOLE see NHG000
1-NITRO-2-IMIDAZOLIDONE see NKL000
N-NITRO-2-IMIDAZOLIDONE see NKL000
4-(2-(5-NITROIMIDAZOL-1-YL)ETHYL)MORPHOLINE see NHH000
1-(2-NITROIMIDAZOL-1-YL)-3-METHOXYPROPAN-2-OL see NHH500
1-(2-NITRO-1-IMIDAZOLYL)-3-METHOXY-2-PROPANOL see NHH500
3-(2-NITROIMIDAZOL-1-YL)-1,2-PROPANEDIOL see NHI000
3-(2-NITRO-1H-IMIDAZOL-1-YL)-1,2-PROPANEDIOL see NHI000
2,2'-(NITROIMINO)BISETHANOL, DINITRATE (ESTER) (9CI) see NFW000
2,2'-NITROIMINOBIS(ETHYLNITRATE) see NFW000
2,2'-NITROIMINODIETHANOL NITRATE see NFW000
NITROIMINODIETHYLENEDIISOCYANIC ACID see NHI500
2,2'-(NITROIMINO)ETHANOL DINITRATE see NFW000
4-NITROINDANE see NHJ000
5-NITROINDANE see NHJ009
7-NITROINDAZOLE see NHK000
7-NITRO-1H-INDAZOLE see NHK000
5-NITROINDOLE see NHK500
5-NITRO-1H-INDOLE see NHK500
NITROISOPROPANE see NIY000
1-NITRO-9-(3-ISOPROPYLAMINOPROPYLAMINE)-ACRIDINE DIHYDROCHLO-
    RIDE see INW000
1-NITRO-9-(3-ISOPROPYLAMINOPROPYLAMINO)-ACRIDINE DIHYDRO-
    CHLORIDE see INW000
NITRO KLEENUP see DUZ000
NITROL see NGY000, NHK800
NITROLIME see CAQ250
NITROLINGUAL see NGY000
NITROLOWE see NGY000
NITROL (PROMOTER) see NHK800
NITROMANNITE (DOT) see MAW250
NITROMANNITE (DRY) (DOT) see MAW250
NITROMANNITOL see MAW250
NITROMERSOL see NHK900
NITROMERSOL SOLUTION see NHK900
NITROMESITYLENE see NHM000
NITROMETAN (POLISH) see NHM500
N-NITROMETHANAMINE see NHN500
NITROMETHANE see NHM500
3-NITRO-6-METHOXYANILINE see NEQ500
5-NITRO-2-METHOXYANILINE see NEQ500
2-NITRO-5-METHOXYBENZOFURAN see MFB350
2-NITRO-7-METHOXYNAPHTHO(2,1-b)FURAN see MFB400
2-NITRO-8-METHOXYNAPHTHOL(2,1-b)-FURAN see MFB410
N-NITROMETHYLAMINE see NHN500
3-NITRO-4-METHYLANILINE see NMP000

1-NITRO-2-METHYLANTHRAQUINONE see MMG000
2-NITRO-3-METHYL-5-CHLOROBENZOFURAN see NHN550
((α-NITROMETHYL)-o-CHLOROBENZYLTHIO)ETHYLAMINE HYDROCHLO-
    RIDE see NEC000
3-(5-NITRO-1-METHYL-2-IMIDAZOLYL)-METHYLENE-AMINO-5-
    MORPHOLINO-METHYL-2-OXAZOLIDONE HCl see MRU750
2-NITRO-2-METHYL-1,3-PROPANEDIOL see NHO500
2-NITRO-2-METHYL-1-PROPANOL see NHP000
NITROMIDINE see TJF000
NITROMIFENE CITRATE see NHP500
NITROMIM see CFA750
NITROMIN see CFA500
NITROMIN HYDROCHLORIDE see CFA750
NITROMIN IDO see ALQ000
NITROMURIATIC ACID (DOT) see HHM000
1-NITRONAPHTHALENE see NHQ000
2-NITRONAPHTHALENE see NHQ500
α-NITRONAPHTHALENE see NHQ000
β-NITRONAPHTHALENE see NHQ500
5-NITRONAPHTHALENE ETHYLENE see NEJ500
2-NITRONAPHTHO(2,1-b)FURAN see NHQ950
3-NITRO-2-NAPHTHYLAMINE see NHR500
NITRONET see NGY000
NITRONG see NGY000
2-NITRO-1-(p-NITROBENZYL)IMIDAZOLE see NFO700
N-NITRO-N-(3-(5-NITRO-2-FURYL)-s-TRIAZOL-5-YL)ETHYLAMINE see ENN500
2-NITRO-1-(4-NITROPHENOXY)-4-(TRIFLUOROMETHYL)BENZENE see NIX000
N'-NITRO-N-NITROSO-N-METHYLGUANIDINE see MMP000
3-NITRO-1-NITROSO-1-PENTYLGUANIDINE see NLC500
3-NITRO-1-NITROSO-1-PROPYLGUANIDINE see NHS000
NITRONIUM TETRAFLUOROBORATE(1−) see NHS500
NITRON LAVSAN see PKF750
2-NITRO-2-NONENE see NHT000
3-NITRO-3-NONENE see NHU000
5-NITRO-4-NONENE see NHV500
NITRON (POLYESTER) see PKF750
2-NITRO-2-OCTENE see NHW000
3-NITRO-2-OCTENE see NHW500
3-NITRO-3-OCTENE see NHX000
1-NITRO-1-OXIMINOETHANE see NHY100
NITROOXIMINOMETHANE see NHY250
7-NITRO-3-OXO-3H-2,1-BENZOXAMERCUROLE see NHY500
5-NITRO-N-(2-OXO-3-OXAZOLIDINYL)-2-FURANMETHANIMINE see NGG500
NITROPENTA see PBC250
NITROPENTAERYTHRITE see PBC250
NITROPENTAERYTHRITOL see PBC250
1-NITROPENTANE see AOL500
2-NITRO-2-PENTENE see NHZ000
3-NITRO-2-PENTENE see NIA000
3-NITROPERCHLORYLBENZENE see NIA700
p-NITROPEROXYBENZOIC ACID see NIB000
NITROPHEN see DFT800
2-NITROPHENANTHRENE see NIC000
NITROPHENE see DFT800
p-NITROPHENETOL (GERMAN) see NID000
p-NITROPHENETOLE see NID000
2-NITROPHENOL see NIE500
3-NITROPHENOL see NIE000
4-NITROPHENOL see NIF000
o-NITROPHENOL see NIE500
m-NITROPHENOL (DOT) see NIE000
p-NITROPHENOL (DOT) see NIF000
p-NITROPHENOL ACETATE see ABS750
3-(α-(p-NITROPHENOL)-β-ACETYLETHYL)-4-HYDROXYCOUMARIN
    see ABF750
NITROPHENOLARSONIC ACID see HMY000
P-NITROPHENOL, ESTER with DIETHYL PHOSPHATE see NIM500
p-NITROPHENOL, O-ESTER with O,O-DIETHYLPHOSPHOROTHIOATE
    see PAK000
p-NITROPHENOL, MERCURY(II) SALT see NIG500
p-NITROPHENOL TIN(IV) SALT see NIH000
p-NITROPHENOL ZINC SALT see NIH500
(4-NITROPHENOXY)ACETIC ACID see NII000
p-NITROPHENOXYACETIC ACID see NII000
1-NITRO-4-PHENOXYBENZENE (9CI) see NIU000
p-NITROPHENOXYTRIBUTYLTIN see NII200
N-(4-NITROPHENYL)ACETAMIDE see NEK000
4-NITROPHENYL ACETATE see ABS750
p-NITROPHENYL ACETATE see ABS750
(2-NITROPHENYL)ACETIC ACID see NII500
o-NITROPHENYLACETIC ACID see NII500
2-(o-NITROPHENYL)ACETIC ACID see NII500
4-NITROPHENYLACETONITRILE see NIJ000
p-NITROPHENYLACETONITRILE see NIJ000

2-NITROPHENYLACETYL CHLORIDE see NIJ200
NITROPHENYL ACETYLENE see NIJ300
3-(α-p-NITROPHENYL-β-ACETYLETHYL)-4-HYDROXYCOUMARIN see ABF750
3-(α-(4'-NITROPHENYL)-β-ACETYLETHYL)-4-HYDROXYCOUMARIN
 see ABF750
1-p-NITROPHENYL-3-AMIDINOUREA HYDROCHLORIDE see NIJ400
m-NITROPHENYLAMINE see NEN500
p-NITROPHENYLAMINE see NEO500
4-NITROPHENYLARSONIC ACID see NIJ500
p-NITROPHENYLARSONIC ACID see NIJ500
4-((p-NITROPHENYL)AZO)DIPHENYLAMINE see NIK000
7-((p-NITROPHENYLAZO)METHYLBENZ(c)ACRIDINE see NIK500
5-(p-NITROPHENYL)AZO)SALICYLIC ACID see NEY000
4-NITRO-N-PHENYLBENZENAMINE see NFY000
1-(p-NITROPHENYL)BIGUANIDE HYDROCHLORIDE see NIL500
N-(2-NITROPHENYL)-1,3-DIAMINOETHANE see NIL625
p-NITROPHENYLDIBUTYLPHOSPHINATE see NIM000
p-NITROPHENYLDI-N-BUTYLPHOSPHINATE see NIM000
d-(−)-threo-1-p-NITROPHENYL-2-DICHLORACETAMIDO-1,3-PROPANEDIOL
 see CDP250
d-threo-1-(p-NITROPHENYL)-2-(DICHLOROACETYLAMINO)-1,3-
 PROPANEDIOL see CDP250
p-NITROPHENYL DIETHYLPHOSPHATE see NIM500
7-NITRO-5-PHENYL-2,3-DIHYDRO-1H-1,4-BENZODIAZEPIN-2-ONE see DLY000
4-(2'-NITROPHENYL)-2,6-DIMETHYL-3,5-DICARBOMETHOXY-1,4-
 DIHYDROPYRIDINE see AEC750
p-NITROPHENYLDIMETHYLTHIONOPHOSPHATE see MNH000
1-(p-NITROPHENYL-3,3-DIMETHYL-TRIAZEN (GERMAN) see DSX400
1-(4-NITROPHENYL)-3,3-DIMETHYLTRIAZENE see DSX400
1-(p-NITROPHENYL)-3,3-DIMETHYLTRIAZENE see DSX400
NITRO-p-PHENYLENEDIAMINE see ALL750
2-NITRO-p-PHENYLENEDIAMINE see ALL750
4-NITRO-o-PHENYLENE-DIAMINE see ALL500
p-NITRO-o-PHENYLENEDIAMINE see ALL500
2-NITRO-1,4-PHENYLENEDIAMINE see ALL750
4-NITRO-1,2-PHENYLENEDIAMINE see ALL500
o-NITRO-p-PHENYLENEDIAMINE (MAK) see ALL750
p-NITROPHENYL ESTER of DIETHYLPHOSPHINIC ACID see DJW200
1-(2-NITROPHENYL)ETHANONE see NEL450
1-(3-NITROPHENYL)ETHANONE see NEL500
p-NITROPHENYL ETHYLBUTYLPHOSPHONATE see NIN000
p-NITROPHENYL ETHYL PENTYLPHOSPHONATE see ENQ500
1-((5-(p-NITROPHENYL)FURFURYLIDENE)AMINO)HYDANTOIN SODIUM
 see DAB840
4-NITROPHENYL 4-GUANIDINOBENZOATE see NIQ500
p-NITROPHENYL p-GUANIDINO-BENZOATE see NIQ500
p-NITROPHENYL-p'-GUANIDINOBENZOATE see NIQ500
2-NITROPHENYLHYDRAZINE see NIQ800
4-NITROPHENYLHYDRAZINE see NIR000
(o-NITROPHENYL)HYDRAZINE see NIQ800
p-NITROPHENYLHYDRAZINE see NIR000
N-(4-NITROPHENYL)-IMIDODICARBONIMIDIC DIAMIDE
 MCNOHYDROCHLORIDE see NIL500
o-NITROPHENYL ISOPROPYL ETHER see NIR550
3-NITROPHENYL ISOTHIOCYANATE see ISO000
m-NITROPHENYL ISOTHIOCYANATE see ISO000
o-NITROPHENYLMERCURY ACETATE see NIS000
o-NITROPHENYLMERCURY (ACETATO) see NIS000
o-NITROPHENYL METHYL ETHER see NER000
p-NITROPHENYL-2-NITRO-4-(TRIFLUOROMETHYL) PHENYL ETHER
 see NIX000
2-NITROPHENYL PHENYL ETHER see NIT500
4-NITROPHENYL PHENYL ETHER see NIU000
p-NITROPHENYL PHENYLETHER see NIU000
O-(4-NITROPHENYL) O-PHENYLMETHYL PHOSPHONOTHIOATE
 see MOB699
4-NITRO-5-(4-PHENYL-1-PIPERAZINYL)BENZOFURAZAN OXIDE see NIV000
N-(4-NITROPHENYL)-N'-(3-PYRIDINYLMETHYL)UREA see PPP750
2-NITROPHENYL SULFONYL DIAZOMETHANE see NIW300
4-(4-NITROPHENYL)THIAZOLE see NIW400
p-NITROPHENYL-2,4,6-TRICHLOROPHENYL ETHER see NIW500
p-NITROPHENYL-α,α,α-TRIFLUORO-2-NITRO-p-TOLYL ETHER see NIX000
NITROPHOS see DSQ000
NITROPONE C see BRE500
NITROPORE see ASM270
NITROPORE OBSH see OPE000
1,1'-(2-NITROPORPYLIDENE)BIS(4-CHLOROBENZENE) see BIN500
1-NITROPROPANE see NIX500
2-NITROPROPANE see NIY000
β-NITROPROPANE see NIY000
2-NITROPROPENE see NIY200
3-NITROPROPIONIC ACID see NIY500
β-NITROPROPIONIC ACID see NIY500
NITROPRUSSIATE de SODIUM see SIW500

NITROPRUSSIDNATRIUM see SIW500
NITROPRUSSIDNATRIUM (GERMAN) see SIU500
1-NITROPYRENE see NJA000
3-NITROPYRENE see NJA000
4-NITROPYRENE see NJA100
4-NITROPYRIDINE-1-OXIDE see NJA500
4-NITROPYRIDINE-N-OXIDE see NJA500
5-NITROPYROMUCATE see NGH500
4-NITROQUINALDINE-N-OXIDE see MMQ250
2-NITROQUINOLINE see NJB500
5-NITROQUINOLINE see NJC000
6-NITROQUINOLINE see NJC500
8-NITROQUINOLINE see NJD500
4-NITROQUINOLINE-6-CARBOXYLIC ACID-1-OXIDE see CCG000
4-NITRO-6-QUINOLINECARBOXYLIC ACID-1-OXIDE see CCG000
4-NITROQUINOLINE-1-OXIDE see NJF000
4-NITROQUINOLINE-N-OXIDE see NJF000
5-NITRO-8-QUINOLINOL see NHF500
NITROSAMINES see NJH000
N-NITROSAMINO DIACETONITRIL (GERMAN) see NKL500
N-NITROSARCOSINE see NJH500
NITROSATED COAL DUST EXTRACT see NJH750
N-NITROSAZETIDINE see NJL000
NITROSIMINODIACETONITRILE see NKL500
N-NITROSOACETANILIDE see NJI700
N-NITROSO-N-(1-ACETOXYMETHYL)BUTYLAMINE see BRX500
N-NITROSO-N-(ACETOXYMETHYL)-N-ISOBUTYLAMINE see ABR125
N-NITROSO-N-(ACETOXY)METHYL-N-METHYLAMINE see AAW000
N-NITROSO-N-(1-ACETOXYMETHYL)PROPYL AMINE see PNR250
NITROSO-N-(1-ACETOXYMETHYL)TRIDEUTEROMETHYLAMINE see MMS000
1-NITROSO-4-ACETYL-3,5-DIMETHYLPIPERAZINE see NJI850
N-NITROSOAETHYLAETHANOLAMIN (GERMAN) see ELG500
NITROSOAETHYLDIMETHYLHARNSTOFF see DRV600
NITROSOALDICARB see NJJ500
N-NITROSOALLYL-2,3-DIHYDROXYPROPYLAMINE see NJY500
N-NITROSOALLYLETHANOLAMINE see NJJ875
NITROSO-ALLYL-2-HYDROXYPROPYLAMINE see NJJ950
N-NITROSOALLYL-2-HYDROXYPROPYLAMINE see NJJ950
N-NITROSO-N-ALLYL-N-(2-HYDROXYPROPYL)AMINE see NJJ950
N-NITROSOALLYLMETHYLAMINE see MMT500
N-NITROSOALLYL-2-OXOPROPYLAMINE see AGM125
NITROSOALLYLUREA see NJK000
N-NITROSOAMINODIETHANOL see NKM000
4-(NITROSOAMINO-N-METHYL)-1-(3-PYRIDYL)-1-BUTANONE see MMS500
1-NITROSOANABASINE see NJK150
N-NITROSOANABASINE see NJK150
N'-NITROSOANABASINE see NJK150
N-(p-NITROSOANILINOMETHYL)-2-NITROPROPANE see NHK800
N-NITROSOAZACYCLOHEPTANE see NKI000
N-NITROSOAZACYCLONONANE see OCA000
N-NITROSOAZACYCLOOCTANE see OBY000
1-NITROSOAZACYCLOTRIDECANE see NJK500
NITROSO-AZETIDIN (GERMAN) see NJL000
NITROSOAZETIDINE see NJL000
1-NITROSOAZETIDINE see NJL000
N-NITROSOAZETIDINE see NJL000
NITROSO-BAYGON see PMY310
1-NITROSO-4-BENZOYL-3,5-DIMETHYLPIPERAZINE see NJL850
N-NITROSO-4-BENZOYL-3,5-DIMETHYLPIPERAZINE see NJL850
N-NITROSOBENZTHIAZURON see NKR000
N-NITROSOBENZYLMETHYLAMINE see MHP250
NITROSOBENZYLUREA see NJM000
4-NITROSOBIPHENYL see NJM400
4-NITROSO-1,1'-BIPHENYL see NJM400
N-NITROSOBIS(ACETOXYETHYL)AMINE see NKM500
N-NITROSOBIS(2-ACETOXYPROPYL)AMINE see NJM500
NITROSOBIS(2-CHLOROETHYL)AMINE see BIF500
NITROSOBIS(2-CHLOROPROPYL)AMINE see DFW000
N-NITROSOBIS(2-ETHOXYETHYL)AMINE see BJO250
N-NITROSOBIS(2-HYDROXYETHYL)AMINE see NKM000
N-NITROSOBIS(2-HYDROXYPROPYL)AMINE see DNB200
N-NITROSOBIS(2-METHOXYETHYL)AMINE see BKO000
N-NITROSOBIS(2-OXOBUTYL)AMINE see NKL300
N-NITROSOBIS(2-OXOPROPYL)AMINE see NJN000
N-NITROSO-BIS-(4,4,4-TRIFLUORO-n-BUTYL)AMINE see NJN300
NITROSOBROMOETHYLUREA see NJN500
N-NITROSOBUTANAMINE see NJO000
N-NITROSOBUTYLAMINE see NJO000
N-NITROSO-1-BUTYLAMINO-2-PROPANONE see BRY000
N-NITROSO-N-BUTYLBUTYRAMIDE see NJO150
N-NITROSO-N-(BUTYL-N-BUTYROLACTONE)AMINE see NJO200
N-NITROSO-N-BUTYL-N-(3-CARBOXYPROPYL)AMINE see BQQ250
N-NITROSO-N-BUTYLETHYLAMINE see EHC000
N-NITROSO-tert-BUTYLETHYLAMINE see NKD500

N-NITROSO-n-BUTYL-(4-HYDROXYBUTYL)AMINE see HJQ350
N-NITROSO-N-BUTYLMETHYLAMINE see MHW500
N-NITROSO-N-BUTYLPENTYLAMINE see BRY250
N-NITROSO-N-BUTYL-N-PENTYLAMINE see BRY250
N-NITROSO-4-tert-BUTYLPIPERIDINE see BRZ200, NJO300
N-NITROSO-n-BUTYLTHIAZOLIDINE see BSA000
N-NITROSOBUTYLUREA see BSA250
NITROSO-sec-BUTYLUREA see NJO500
N-NITROSO-N-BUTYROXY-BUTYLAMINE see NJO150
N-NITROSO-N-(1-BUTYROXYMETHYL)METHYL AMINE see BSX500
NITROSO-BUX-TEN see NJP000, PBX750
N-NITROSO-N'-CARBAETHOXYPIPERAZIN (GERMAN) see NJQ000
N-NITROSOCARBANILIC ACID ISOPROPYLESTER see NJP500
N-NITROSOCARBARYL see NBJ500
N-NITROSO-N'-CARBETHOXYPIPERAZINE see NJQ000
N-NITROSO-N-(3-CARBOETHOXYPROPIONYL)BUTYLAMINE see EHC800
NITROSOCARBOFURAN see NJQ500
N-NITROSO-N-(3-CARBOXYPROPYL)ETHYLAMINE see NKE000
NITROSOCHLOROETHYLDIETHYLUREA see NJR000
NITROSOCHLOROETHYLDIMETHYLUREA see NJR500
1-NITROSO-1-(2-CHLOROETHYL)-3,3-DIMETHYLUREA see NJR500
N-NITROSO-2-CHLOROETHYLUREA see CHE750
1-NITROSO-1-(2-CHLOROETHYL)UREA see CHE750
N-NITROSO-3-CHLOROPIPERIDINE see CJG375
N-NITROSO-4-CHLOROPIPERIDINE see CJG500
NITROSOCIMETIDINE see NJS300
N-NITROSOCIMETIDINE see NJS300
NITROSO-l-CITRULLINE see NJT000
NITROSO-dl-CITRULLINE see NJS500
NITROSO COMPOUNDS see NJT550
N-NITROSO COMPOUNDS see NJT550
4-(NITROSOCYANAMIDO)BUTYRAMIDE see NJU000
4-(NITROSOCYANAMIDO)-3-HYDROXYBUTYRAMIDE see HNA000
5-(NITROSOCYANAMIDO)-2-HYDROXYVALERAMIDE see NJU500
NITROSOCYCLOHEXYLUREA see NJV000
1'-NITROSO-1'-DEMETHYLNICOTINE see NLD500
N-NITROSODIACETONITRILE see NKL500
N-NITROSODIAETHANOLAMIN (GERMAN) see NKM000
N-NITROSODIAETHYLAMINE (GERMAN) see NJW500
N-NITROSODIALLYL AMINE see NJV500
N-NITROSO-3,4-DIBROMOPIPERIDINE see DDQ800
N-NITROSODIBUTYLAMINE see BRY500
NITROSODI-sec-BUTYLAMINE see NJW000
N-NITROSO-DI-sec-BUTYLAMINE see NJW000
N-NITROSODI-n-BUTYLAMINE (MAK) see BRY500
N-NITROSO-2,2'-DICHLORODIETHYLAMINE see BIF500
N-NITROSO-3,4-DICHLOROPIPERIDINE see DFW200
N-NITROSODI(CYANOMETHYL)AMINE see NKL500
N-NITROSODIETHANOLAMINE (MAK) see NKM000
NITROSODIETHYLAMINE see NJW500
N-NITROSODIETHYLAMINE see NJW500
NITROSO-1,1-DIETHYL-3-METHYLUREA see DJP600
N-NITROSODIFENYLAMIN (CZECH) see DWI000
p-NITROSODIFENYLAMIN (CZECH) see NKB500
N-NITROSO-3,6-DIHYDROOXAZIN-1,2 (GERMAN) see NJX000
N-NITROSO-3,6-DIHYDRO-1,2-OXAZINE see NJX000
1-NITROSO-5,6-DIHYDROTHYMINE see NJX500
1-NITROSO-5,6-DIHYDROURACIL see NJY000
N-NITROSO-2,3-DIHYDROXYPROPYLALLYLAMINE see NJY500
N-NITROSO-N,N-DI(2-HYDROXYPROPYL)AMINE see DNB200
N-NITROSODIHYDROXYPROPYLETHANOLAMINE see NBR100
N-NITROSO-2,3-DIHYDROXYPROPYL-2-HYDROXYETHYLAMINE see NBR100
N-NITROSO-2,3-DIHYDROXYPROPYL-2-HYDROXYPROPYLAMINE
  see NOC400
NITROSO-DIHYDROXYPROPYLOXOPROPYLAMINE see NJY550
N-NITROSODIHYDROXYPROPYL-2-OXOPROPYLAMINE see NJY550
NITROSODIISOBUTYLAMINE see DRQ200
N-NITROSODIISOBUTYLAMINE see DRQ200
N-NITROSODI-ISO-BUTYLAMINE see DRQ200
N-NITROSODIISOPROPYLAMINE see NKA000
NITROSODIMETHOATE see NKA500
NITROSODIMETHYLAMINE see NKA600
N-NITROSODIMETHYLAMINE see NKA600
4-NITROSODIMETHYLANILINE see DSY600
p-NITROSO-N,N-DIMETHYLANILINE see DSY600
p-NITROSODIMETHYLANILINE (DOT) see DSY600
N-NITROSO-2,2'-DIMETHYLDI-n-PROPYLAMINE see DRQ200
NITROSO-1,1-DIMETHYL-3-ETHYLUREA see DRV600
NITROSO-2,6-DIMETHYLMORPHOLINE see DTA000
N-NITROSO-2,6-DIMETHYLMORPHOLINE see DTA000
cis-NITROSO-2,6-DIMETHYLMORPHOLINE see NKA695
cis-N-NITROSO-2,6-DIMETHYLMORPHOLINE see NKA695
trans-NITROSO-2,6-DIMETHYLMORPHOLINE see NKA700
trans-N-NITROSO-2,6-DIMETHYLMORPHOLINE see NKA700

NITROSO-3,5-DIMETHYLPIPERAZINE see NKA850
1-NITROSO-3,5-DIMETHYLPIPERAZINE see NKA850
N-NITROSO-3,5-DIMETHYLPIPERAZINE see NKA850
N-NITROSO-2,6-DIMETHYLPIPERIDINE see DTA400
N-NITROSO-3,5-DIMETHYLPIPERIDINE see DTA600
NITROSO-3,5-DIMETHYLPIPERIDINE cis-isomer see DTA690
NITROSO-3,5-DIMETHYLPIPERIDINE trans-isomer see DTA700
NITROSODIMETHYLUREA see DTB200
N-NITROSODIMETHYLUREA see DTB200
NITROSODIOCTYLAMINE see NKB000
N-NITROSO-DI-N-OCTYLAMINE see NKB000
NITROSO DIOXACARB see MMV500
N-NITROSO-N,N-DI(2-OXYPROPYL)AMINE see NJN000
N-NITROSODIPENTYLAMINE see DCH600
N-NITROSODI-n-PENTYLAMINE see DCH600
NITROSODIPHENYLAMINE see DWI000
4-NITROSODIPHENYLAMINE see NKB500
N-NITROSODIPHENYLAMINE see DWI000
p-NITROSODIPHENYLAMINE see NKB500
N-NITROSODIPROPYLAMINE see NKB700
N-NITROSODI-N-PROPYLAMINE see NKB700
N-NITROSO-N-DIPROPYLAMINE see NKB700
N-NITROSODI-i-PROPYLAMINE (MAK) see NKA000
N-NITROSODODECAMETHYLENEIMINE see NJK500
NITROSODODECAMETHYLENIMINE see NJK500
N-NITROSODODECAMETHYLENIMINE see NJK500
N-NITROSOEPHEDRINE see NKC000
N-NITROSO-3,4-EPOXYPIPERIDINE see NKC300
NITROSOETHANECARBAMONITRILE see ENT500
N-NITROSOETHANOLISOPROPANOLAMINE see HKW475
NITROSOETHIOFENCARB see MMV750
NITROSOETHOXYETHYLAMINE see NKC500
NITROSOETHYLANILINE see NKD000
N-NITROSO-N-ETHYL ANILINE see NKD000
N-NITROSO-N-ETHYLBENZYLAMIN (GERMAN) see ENS500
N-NITROSO-N-ETHYLBENZYLAMINE see ENS500
N-NITROSO-N-ETHYL BIURET see ENT000
N-NITROSOETHYL-N-BUTYLAMINE see EHC000
N-NITROSOETHYL-tert-BUTYLAMINE see NKD500
N-NITROSO-ETHYL(3-CARBOXYPROPYL)AMINE see NKE000
NITROSOETHYLDIMETHYLUREA see DRV600
1-NITROSO-1-ETHYL-3,3-DIMETHYLUREA see DRV600
N-NITROSOETHYLENETHIOUREA see NKK500
N-NITROSOETHYLETHANOLAMINE see ELG500
N-NITROSOETHYL-2-HYDROXYETHYLAMINE see ELG500
N-NITROSO-N-ETHYL-N-(2-HYDROXYETHYL)AMINE see ELG500
N-NITROSOETHYLISOPROPYLAMINE see ELX500
N-NITROSOETHYLMETHYLAMINE see MKB000
N-NITROSOETHYLPHENYLAMINE (MAK) see NKD000
N-NITROSO-2-ETHYLTHIAZOLIDINE see ENU500
NITROSOETHYLUREA see ENV000
NITROSOETHYLURETHAN see NKE500
N-NITROSO-N-ETHYLURETHAN see NKE500
N-NITROSOETHYLVINYLAMINE see NKF000
N-NITROSO-N-ETHYLVINYLAMINE see NKF000
N-NITROSOFENYLHYDROXYLAMIN AMONNY (CZECH) see ANO500
2-NITROSOFLUORENE see NKF500
NITROSOFLUOROETHYLUREA see NKG000
NITROSOFOLIC ACID see NKG450, NLP000
NITROSOGLYPHOSATE see NKG500
NITROSOGUANIDIN (GERMAN) see NKH000
NITROSOGUANIDINE see NKH000
N-NITROSOGUANIDINE see NKH000
N-NITROSOGUVACINE see NKH500
NITROSOGUVACOLINE see NKH500
N-NITROSOGUVACOLINE see NKH500
NITROSOHEPTAMETHYLENEIMINE see OBY000
N-NITROSOHEPTAMETHYLENEIMINE see OBY000
NITROSO-HEPTAMETHYLENIMIN (GERMAN) see OBY000
N-NITROSO-4,4,4',4',4'-HEXAFLUORODIBUTYLAMINE see NJN300
N-NITROSOHEXAHYDROAZEPINE see NKI000
N-NITROSOHEXAMETHYLENEIMINE see NKI000
NITROSOHEXAMETHYLENIMINE see NKI000
NITROSO-n-HEXYLMETHYLAMINE see NKU400
NITROSO-N-HEXYLUREA see HFS500
NITROSO HYDANTOIC ACID see NKI500
1-NITROSOHYDANTOIN see NKJ000
1-NITROSO-1-HYDROXYETHYL-3-CHLOROETHYLUREA see NKJ050
1-NITROSO-1-(2-HYDROXYETHYL)-3-(2-CHLOROETHYL)UREA see NKJ050
NITROSO-2-HYDROXYETHYLUREA see HKW500
N-NITROSOHYDROXYETHYLUREA see HKW500
1-NITROSO-1-(2-HYDROXYETHYL)UREA see HKW500
N-NITROSOHYDROXYLAMINE see HOW500
N-NITROSO-3-HYDROXYPIPERIDINE see NLK000

N-NITROSOHYDROXYPROLINE see HNB500
1-NITROSO-1-HYDROXYPROPYL-3-CHLOROETHYLUREA see NKJ100
1-NITROSO-1-(2-HYDROXYPROPYL)-3-(2-CHLOROETHYL)UREA see NKJ100
N-NITROSO-(2-HYDROXYPROPYL)-(2-HYDROXYETHYL)AMINE see HKW475
N-NITROSO(2-HYDROXYPROPYL)(2-OXOPROPYL)AMINE see HNX500
N-NITROSO-2-HYDROXY-N-PROPYL-N-PROPYLAMINE see NLM500
NITROSO-3-HYDROXYPROPYLUREA see NKK000
NITROSO-2-HYDROXY-N-PROPYLUREA see NKO400
NITROSO-3-HYDROXY-N-PROPYLUREA see NKK000
NITROSO-2-HYDROXY-N-PROPYLUREA see NKO400
N-NITROSO-3-HYDROXYPYRROLIDINE see NLP700
N-NITROSOIMIDAZOLIDINETHIONE see NKK500
1-NITROSOIMIDAZOLIDINONE see NKL000
1-NITROSO-2-IMIDAZOLIDINONE see NKL000
N-NITROSO-IMIDAZOLIDON (GERMAN) see NKL000
N-NITROSOIMIDAZOLIDONE see NKL000
2,2'-(NITROSOIMINO)BISACETONITRILE see NKL500
1,1'-(NITROSOIMINO)BIS-2-BUTANONE see NKL300
2,2'-(NITROSOIMINO)BISETHANOL see NKM000
(NITROSOIMINO)DIACETONE see NJN000
2,2'-(N-NITROSOIMINO)DIACETONITRILE see NKL500
NITROSOIMINO DIETHANOL see NKM000
N-NITROSO-2,2'-IMINODIETHANOLDIACETATE see NKM500
1,1'-NITROSOIMINODI-2-PROPANOL see DNB200
1-NITROSO-1,1'-IMINODI-2-PROPANOL see DNB200
1,1-(N-NITROSOIMINO)DI-2-PROPANOL, DIACETATE see NJM500
N-NITROSOINDOLIN (GERMAN) see NKN000
1-NITROSOINDOLINE see NKN000
N-NITROSOINDOLINE see NKN000
N-NITROSOISOBUTYLTHIAZOLIDINE see NKN500
N-NITROSO-ISO-BUTYLUREA see IJF000
N-NITROSOISONIPECTOIC ACID see NKO000
NITROSOISOPROPANOL-ETHANOLAMINE see HKW475
NITROSOISOPROPANOLUREA see NKO400
NITROSOISOPROPYLUREA see NKO425
NITROSO-LANDRIN see NLY500
NITROSO LINURON see NKO500
NITROSO-METHOMYL see NKX000
N-NITROSO-2-METHOXY-2,6-DIMETHYLMORPHOLINE see NKO600
1-NITROSOMETHOXYETHYLUREA see NKO900
NITROSO-2-METHOXYETHYLUREA see NKO900
N-NITROSOMETHOXYMETHYLAMINE see DSZ000
N-NITROSO-N-METHOXYMETHYLMETHYLAMINE see MEW250
N-NITROSO-N-METHYLACETAMIDE see MMT000
N-NITROSO-N-METHYL-N-α-ACETOXYBENZYLAMINE see NKP000
N-NITROSO-N-METHYL-N-ACETOXYMETHYLAMINE see AAW000
o-(N-NITROSO-N-METHYL-β-ALANYL)-l-SERINE see NKP500
NITROSOMETHYLALLYLAMINE see MMT500
N-NITROSOMETHYLALLYLAMINE see MMT500
N-NITROSOMETHYLAMINACETONITRIL (GERMAN) see MMT750
N-NITROSOMETHYLAMINOACETONITRILE see MMT750
N-NITROSO-4-METHYLAMINOAZOBENZENE see MMY250
N-NITROSO-4-METHYLAMINOAZOBENZOL (GERMAN) see MMY250
2-NITROSOMETHYLAMINOPYRIDINE see NKQ000
N-NITROSO-N-METHYL-2-AMINOPYRIDINE see NKQ000
4-(N-NITROSO-N-METHYLAMINO)-4-(3-PYRIDYL)BUTANAL see MMS250
4-(N-NITROSO-N-METHYLAMINO)-1-(3-PYRIDYL)-1-BUTANONE see MMS500
3-(N-NITROSOMETHYLAMINO)SULFOLAN (GERMAN) see NKQ500
N-NITROSOMETHYLAMINOSULFOLANE see NKQ500
N-NITROSO-N-METHYL-N-AMYLAMINE see AOL000
NITROSOMETHYLANILINE see MMU250
N-NITROSO-N-METHYLANILINE see MMU250
N-NITROSO-N-METHYL-N'-(2-BENZOTHIAZOLYL)UREA see NKR000
N-NITROSOMETHYLBENZYLAMINE see MHP250
N-NITROSO-N-(2-METHYLBENZYL)-METHYLAMIN (GERMAN) see NKR500
N-NITROSO-N-(3-METHYLBENZYL)-METHYLAMIN (GERMAN) see NKS000
N-NITROSO-N-(4-METHYLBENZYL)-METHYLAMIN (GERMAN) see NKS500
N-NITROSO-N-(2-METHYLBENZYL)METHYLAMINE see NKR500
N-NITROSO-N-(3-METHYLBENZYL)METHYLAMINE see NKS000
N-NITROSO-N-(4-METHYLBENZYL)METHYLAMINE see NKS500
NITROSOMETHYLBIS(CHLOROETHYL)UREA see NKT000
1-NITROSO-1-METHYL-3,3-BIS-(2-CHLOROETHYL)UREA see NKT000
N-NITROSO-N-METHYLBIURET see MMV000
NITROSOMETHYL-d³-n-BUTYLAMINE see NKT105
N-NITROSOMETHYL-N-BUTYLAMINE see MHW500
N-NITROSO-N-METHYL-N-n-BUTYL-1-d²-AMINE see NKT100
N-NITROSO-N-METHYLCAPRYLAMIDE see MMY000
N-NITROSOMETHYLCARBAMIDE see MNA750
N-NITROSOMETHYL-3-CARBOXYPROPYLAMINE see MIF250
N-NITROSOMETHYL-2-CHLOROETHYLAMINE see CIQ500
N-NITROSOMETHYLCYCLOHEXYLAMINE see NKT500
N-NITROSO-N-METHYLCYCLOHEXYLAMINE see NKT500
NITROSOMETHYL-n-DECYLAMINE see MMX200
NITROSOMETHYLDIAETHYLHARNSTOFF see DJP600

NITROSOMETHYLDIETHYLUREA see DJP600
1-NITROSO-1-METHYL-3,3-DIETHYLUREA see DJP600
N-NITROSOMETHYL(2,3-DIHYDROXYPROPYL)AMINE see NMV450
N-NITROSO-N-METHYL-N-DODECYLAMIN (GERMAN) see NKU000
NITROSOMETHYL-n-DODECYLAMINE see NKU000
N-NITROSOMETHYL-N-METHYL-N-DODECYLAMINE see NKU000
N-NITROSOMETHYLETHANOLAMINE see NKU350
N-NITROSOMETHYLETHYLAMINE (MAK) see MKB000
N-NITROSO-N-METHYL-4-FLUOROANILINE see FKF800
N-NITROSOMETHYLGLYCINE see NLR500
N-NITROSO-N-METHYL-HARNSTOFF (GERMAN) see MNA750
N-NITROSO-N-METHYLHEPTYLAMINE see HBP000
NITROSOMETHYL-n-HEXYLAMINE see NKU400
N-NITROSO-N-METHYL-(4-HYDROXYBUTYL)AMINE see MKP000
N-NITROSOMETHYL-(2-HYDROXYETHYL)AMINE see NKU350
N-NITROSOMETHYL-2-HYDROXYPROPYLAMINE see NKU500
N-NITROSOMETHYLMETHOXYAMINE see DSZ000
N-NITROSO-N-METHYL-o-METHYLHYDROXYLAMIN (GERMAN) see DSZ000
N-NITROSO-N-METHYL-o-METHYL-HYDROXYLAMINE see DSZ000
NITROSO-2-METHYLMORPHOLINE see NKU550
N-NITROSO-2-METHYLMORPHOLINE see NKU550
NITROSOMETHYLNEOPENTYLAMINE see NKU570
N-NITROSO-N-METHYLNITROGUANIDINE see MMP000
N-NITROSO-N-METHYL-4-NITROSO-ANILINE see MJG750
NITROSOMETHYL-n-NONYLAMINE see NKU580
NITROSO-N-METHYL-n-NONYLAMINE see NKU580
N-NITROSOMETHYL-n-NONYLAMINE see NKU580
NITROSOMETHYL-n-OCTYLAMINE see NKU590
NITROSO-N-METHYL-n-OCTYLAMINE see NKU590
N-NITROSOMETHYL-n-OCTYLAMINE see NKU590
NITROSO-2-METHYL-1,3-OXAZOLIDINE see NKU600
NITROSO-5-METHYL-1,3-OXAZOLIDINE see NKU875
N-NITROSO-2-METHYL-1,3-OXAZOLIDINE see NKU600
N-NITROSO-5-METHYL-1,3-OXAZOLIDINE see NKU875
NITROSO-5-METHYLOXAZOLIDONE see NKU875
N-NITROSOMETHYL(2-OXOBUTYL)AMINE see MMR800
N-NITROSOMETHYL(3-OXOBUTYL)AMINE see MMR810
N-NITROSOMETHYL-2-OXOPROPYLAMINE see NKV000
NITROSOMETHYL-N-PENTYLAMINE see AOL000
N-NITROSOMETHYLPENTYLNITROSAMINE see NKV100
N-NITROSOMETHYLPHENYLAMINE (MAK) see MMU250
NITROSOMETHYLPHENYLCARBAMATE see NKV500
N-NITROSO-N-METHYLPHENYLCARBAMATE see NKV500
N-NITROSO-N-METHYL-(1-PHENYL)-ETHYLAMIN (GERMAN) see NKW000
N-NITROSO-N-METHYL-2-PHENYLETHYLAMINE see MNU250
N-NITROSO-N-METHYL-1-(1-PHENYL)ETHYLAMINE see NKW000
NITROSOMETHYLPHENYLUREA see MMY500
N-NITROSO-N'-METHYLPIPERAZIN (GERMAN) see NKW500
1-NITROSO-4-METHYLPIPERAZINE see NKW500
N-NITROSO-N'-METHYLPIPERAZINE see NKW500
R(−)-N-NITROSO-2-METHYL-PIPERIDIN (GERMAN) see NLI500
s(+)-N-NITROSO-2-METHYL-PIPERIDIN (GERMAN) see NLJ000
R(−)-N-NITROSO-2-METHYLPIPERIDINE see NLI500
s(+)-N-NITROSO-2-METHYLPIPERIDINE see NLJ000
NITROSO-3-METHYL-4-PIPERIDONE see MMZ800
NITROSOMETHYLPROPYLAMINE see MNA000
NITROSOMETHYL-N-PROPYLAMINE see MNA000
N-NITROSO-N-METHYL-n-TETRADECYLAMINE see NKW800
NITROSO-2-METHYLTHIOPROPIONALDEHYDE-o-METHYL CARBAMOYL-
  OXIME see NKX000
N-NITROSO-N-METHYL-4-TOLYLSULFONAMIDE see THE500
NITROSOMETHYL-2-TRIFLUOROETHYLAMINE see NKX300
NITROSOMETHYLUNDECYLAMINE see NKX500
NITROSOMETHYLUREA see MNA750
NITROSOMETHYL UREA see MOA500
1-NITROSO-1-METHYLUREA see MNA750
N-NITROSO-N-METHYLUREA see MNA750
NITROSOMETHYLURETHAN (GERMAN) see MMX250
NITROSOMETHYLURETHANE see MMX250
N-NITROSO-N-METHYLURETHANE see MMX250
N-NITROSOMETHYLVINYLAMINE see NKY000
NITROSOMETOXURON see NKY500
N-NITROSOMORPHOLIN (GERMAN) see NKZ000
NITROSOMORPHOLINE see NKZ000
4-NITROSOMORPHOLINE see NKZ000
N-NITROSOMORPHOLINE (MAK) see NKZ000
NITROSO-MPMC see DTN400
NITROSO-MTMC see MNV500
NITROSO-NAC see NBJ500
1-NITROSONAPHTHALENE see NLA000
2-NITROSONAPHTHALENE see NLA500
NITROSO-β-NAPHTHOL see NLB000
1-NITROSO-2-NAPHTHOL see NLB000
2-NITROSO-1-NAPHTHOL see NLB500

α-NITROSO-β-NAPHTHOL see NLB000
1-NITROSO-2-NITROAMINO-2-IMIDAZOLINE see NLB700
1-NITROSO-3-NITRO-1-BUTYLGUANIDINE see NLC000
N-NITROSO-N'-NITRO-N-BUTYLGUANIDINE see NLC000
1-NITROSO-3-NITRO-1-PENTYLGUANIDINE see NLC500
3-NITROSO-1-NITRO-1-PROPYLGUANIDINE see NLD000
N-NITROSO-N'-NITRO-N-PROPYLGUANIDINE see NHS000
N-NITROSO-N-(NITROSOMETHYL)PENTYLAMINE see NKV100
NITROSONIUM BISULFITE see NMJ000
N'-NITROSONORNICOTINE see NLD500
N'-NITROSONORNICOTINE-1-N-OXIDE see NLD525
N-NITROSOOCTAMETHYLENEIMINE see OCA000
1-NITROSO-1-OCTYLUREA see NLD800
N-NITROSO-O,N-DIAETHYLHYDROXYLAMIN (GERMAN) see NKC500
N-NITROSO-O,N-DIETHYLHYDROXYLAMINE see NKC500
3-NITROSO-7-OXA-3-AZABICYCLO(4.1.0)HEPTANE see NKC300
N-NITROSOOXAZOLIDIN (GERMAN) see NLE000
3-NITROSOOXAZOLIDINE see NLE000
N-NITROSOOXAZOLIDINE see NLE000
N-NITROSO-1,3-OXAZOLIDINE see NLE000
NITROSOOXAZOLIDONE see NLE000
N-NITROSO-N-(2-OXOBUTYL)BUTYLAMINE see BSB500
N-NITROSO-N-(3-OXOBUTYL)BUTYLAMINE see BSB750
N-NITROSO(2-OXOBUTYL)(2-OXOPROPYL)AMINE see NLE400
N-NITROSO-(2-OXOPROPYL)-N-BUTYLAMINE see BRY000
N-NITROSO-2-OXOPROPYL-2,3-DIHYDROXYPROPYLAMINE see NJY550
N-NITROSO-2-OXO-N-PROPYL-N-PROPYLAMINE see ORS000
N-NITROSO-N-PENTYLCARBAMIC ACID-ETHYL ESTER see AOL750
N-NITROSO-N-PENTYL-(4-HYDROXYBUTYL)AMINE see NLE500
1-NITROSO-1-PENTYLUREA see PBX500
N-NITROSOPERHYDROAZEPINE see NKI000
N-NITROSOPHENACETIN see NLE550
2-NITROSOPHENANTHRENE see NLF000
1-NITROSO-1-PHENETHYLUREA see NLG000
NITROSOPHENOL see NLF200
2-NITROSOPHENOL see NLF300
4-NITROSOPHENOL see NLF200
p-NITROSOPHENOL see NLF200
4-NITROSO-N-PHENYLANILINE see NKB500
N-NITROSO-N-PHENYLANILINE see DWI000
p-NITROSO-N-PHENYLANILINE see NKB500
4-NITROSO-N-PHENYLBENZENAMINE see NKB500
N-NITROSOPHENYLBENZYLAMINE see BFE750
NITROSOPHENYLETHYLUREA see NLG000
N-NITROSO-N-(2-PHENYLETHYL)UREA see NLG000
N-NITROSOPHENYLHYDROXYLAMIN AMMONIUM SALZ (GERMAN)
　　see ANO500
N-NITROSOPHENYLHYDROXYLAMINE AMMONIUM SALT see ANO500
α-(NITROSO(PHENYLMETHYL)AMINO)-BENZENEMETHANOL ACETATE
　　(ester) see ABL875
N-NITROSO-N-(PHENYLMETHYL)UREA see NJM000
NITROSO-4-PHENYLPIPERIDINE see PFT600
1-NITROSO-4-PHENYLPIPERIDINE see PFT600
N-NITROSO-4-PHENYLPIPERIDINE see PFT600
NITROSOPHENYLUREA see NLG000
N-NITROSO-N-PHENYLUREA see NLG500
N-NITROSO-N-(PHOSPHONOMETHYL)GLYCINE see NKG500
N-NITROSO-4-PICOLYLETHYLAMINE see NLH000
1-NITROSOPIPECOLIC ACID see NLH500
N-NITROSOPIPECOLIC ACID see NLH500
1-NITROSO-2-PIPECOLINE see NLI000
1-NITROSO-3-PIPECOLINE see MMY750
1-NITROSO-4-PIPECOLINE see MMZ000
R(-)-N-NITROSO-α-PIPECOLINE see NLI500
S(+)-N-NITROSO-α-PIPECOLINE see NLJ000
1-NITROSOPIPERAZINE see MRJ750
N-NITROSOPIPERAZINE see MRJ750
NITROSOPIPERIDIN (GERMAN) see NLJ500
N-NITROSO-PIPERIDIN (GERMAN) see NLJ500
1-NITROSOPIPERIDINE see NLJ500
N-NITROSOPIPERIDINE see NLJ500
N-NITROSO-Δ²-PIPERIDINE see NLU480
N-NITROSO-Δ³-PIPERIDINE see NLU500
1-NITROSO-2-PIPERIDINECARBOXYLIC ACID see NLH500
1-NITROSO-4-PIPERIDINECARBOXYLIC ACID see NKO000
NITROSO-3-PIPERIDINOL see NLK000
NITROSO-4-PIPERIDINOL see NLK500
N-NITROSO-3-PIPERIDINOL see NLK000
NITROSO-4-PIPERIDINONE see NLL000
N-NITROSO-4-PIPERIDINONE see NLL000
NITROSO-4-PIPERIDONE see NLL000
1-NITROSO-4-PIPERIDONE see NLL000
1-NITROSO-l-PROLINE see NLL500
N-NITROSO-l-PROLINE see NLL500

N-NITROSO-N-2-PROPENYLUREA see NJK000
NITROSOPROPHAM see NJP500
NITROSOPROPOXUR see PMY310
N-NITROSO-N-PROPYLACETAMIDE see NLM000
N-NITROSO-N(N-PROPYL)ACETAMIDE see NLM000
4-(N-NITROSOPROPYLAMINO)BUTYRIC ACID see PNG500
1-(NITROSOPROPYLAMINO)-2-PROPANOL see NLM500
1-(NITROSOPROPYLAMINO)-2-PROPANONE see ORS000
N-NITROSO-N-PROPYL-1-BUTANAMINE see PNF750
N-NITROSO-N-PROPYLBUTYLAMINE see PNF750
N-NITROSO-N-PROPYLCARBAMIC ACID ETHYL ESTER see PNR500
N-NITROSO-N-PROPYLCARBAMIC ACID-1-NAPHTHYL ESTER see NBM500
N-NITROSO-N-PROPYL-(4-HYDROXYBUTYL)AMINE see NLN000
N-NITROSO-N-PROPYLPROPANAMINE see NKB700
N-NITROSO-N-PROPYL-1-PROPANAMINE see NKB700
N-NITROSO-N-PROPYLPROPIONAMIDE see NLN500
N-NITROSO-N-PROPYL-PROPRIONAMID (GERMAN) see NLN500
N-NITROSOPROPYLTHIAZOLIDINE see IQS000
3-NITROSO-2-PROPYLTHIAZOLIDINE see NLO000
N-NITROSO-2-N-PROPYLTHIAZOLIDINE see NLO000
NITROSOPROPYLUREA see NLO500
NITROSO-N-PROPYLUREA see NLO500
N-NITROSO-N-PROPYLUREA see NLO500
N-NITROSO-N-PTEROYL-l-GLUTAMIC ACID see NLP000
1-NITROSOPYRENE see NLP375
N-NITROSO-2-(3'-PYRIDYL)PIPERIDINE see NJK150
1-NITROSO-2-(3-PYRIDYL)PYRROLIDINE see NLD500
N-NITROSOPYRROLIDIN (GERMAN) see NLP500
NITROSOPYRROLIDINE see NLP480
1-NITROSOPYRROLIDINE see NLP500
N-NITROSOPYRROLIDINE see NLP500
N-NITROSO-2-PYRROLIDINE see NLP600
1-NITROSO-3-PYRROLIDINOL see NLP700
1-NITROSO-2-PYRROLIDINONE see NLP600
3-(1-NITROSO-2-PYRROLIDINYL)PYRIDINE see NLD500
(s)-3-(1-NITROSO-2-PYRROLIDINYL)PYRIDINE see NLD500
NITROSO-3-PYRROLIN (GERMAN) see NLQ000
N-NITROSO-3-PYRROLINE see NLQ000
4-NITROSOQUINOLINE-1-OXIDE see NLQ500
5-NITROSO-8-QUINOLINOL see NLR000
NITROSORBID see CCK125
NITROSORBIDE see CCK125
NITROSORBON see CCK125
N-NITROSOSARCOSINE see NLR500
N-NITROSOSARCOSINE, ETHYL ESTER see NLS000
o-(N-NITROSOSARCOSYL)-l-SERINE see NLS500
NITROSO SARKOSIN (GERMAN) see NLR500
NITROSO-SARKOSIN-AETHYLESTER see NLS000
4-NITROSO-trans-STILBENE see NLT000
NITROSOSWEP see DFV800
1-NITROSO-1,2,5,6-TETRAHYDRONICOTINIC ACID METHYL ESTER
　　see NKH500
N-NITROSOTETRAHYDRO-1,2-OXAZIN see NLT500
N-NITROSO-TETRAHYDROOXAZIN-1,2 (GERMAN) see NLT500
N-NITROSOTETRAHYDROOXAZIN-1,3 (GERMAN) see NLU000
3-NITROSO-TETRAHYDRO-1,3-OXAZINE see NLU000
N-NITROSO-TETRAHYDRO-1,2-OXAZINE see NLT500
N-NITROSO-TETRAHYDRO-1,3-OXAZINE see NLU000
2-NITROSOTETRAHYDRO-2H-1,2-OXAZINE see NLT500
1-NITROSO-1,2,3,4-TETRAHYDROPYRIDINE see NLU480
N-NITROSO-1,2,3,4-TETRAHYDROPYRIDINE see NLU480
N-NITROSO-1,2,3,6-TETRAHYDROPYRIDINE see NLU500
N-NITROSO-Δ²-TETRAHYDROPYRIDINE see NLU480
N-NITROSO-Δ³-TETRAHYDROPYRIDINE see NLU500
1-NITROSO-TETRAHYDRO-2(1H)-PYRIMIDINONE see PNL500
1-NITROSO-2,2,6,6-TETRAMETHYLPIPERIDINE see TDS000
3-NITROSOTHIAZOLIDINE see NLV500
N-NITROSOTHIAZOLIDINE see NLV500
4-NITROSOTHIOMORPHOLINE see NLW000
N-NITROSOTHIOMORPHOLINE see NLW000
2-NITROSOTOLUENE see NLW500
o-NITROSOTOLUENE see NLW500
NITROSOTRIAETHYLHARNSTOFF (GERMAN) see NLX500
NITROSOTRIDECYLUREA see NLX000
1-NITROSO-1-TRIDECYLUREA see NLX000
NITROSOTRIETHYLUREA see NLX500
N-NITROSO-2,2,2-TRIFLUORODIETHYLAMINE see NLX700
N-NITROSO-2,2,2-TRIFLUOROETHYL-ETHYLAMINE see NLX700
2-(NITROSO(3-(TRIFLUOROMETHYL)PHENYL)AMINO)-BENZOIC ACID
　　see TKF699
N-NITROSO-N-(α,α,α-TRIFLUORO-m-TOLYL)ANTHRANILIC ACID see TKF699
NITROSOTRIHYDROXY-DIPROPYLAMINE see NOC400
1-NITROSO-2,2,4-TRIMETHYL-1,2,-DIHYDRO-QUINOLINE (POLYMER)
　　see COG250

NMDHP see NMV450
NMEA see MKB000
NMH see MNA750
NMHP see NKU500
NMNA see NKU570
NMO see CFA500
NMOP see NKV000
NMOR see NKZ000
NMP see MPF200
NMU see MNA750
NMUM see MMX250
NMUT see MMX250
NMVA see NKY000
NNA see MMS250
N. NAJA NAJA VENOM see ICC700
NNDG see DQR600
1000 NN FERRITE see IHB800
N. NIGRICOLLIS VENOM see NAG000
NNK see MMS500
NNN see NLD500
NNN-1-N-OXIDE see NLD525
N-N-PIP see NLJ500
N-N-PYR see NLP500
No. 1249 see CKN000
No. 1403 see CKN000
No. 48-80 see PMB800
No. 75810 see CKN000
NOBECUTAN see TFS350
NOBEDON see HIM000
NOBEDORM see QAK000
NOBFELON see IIU000
NOBFEN see IIU000
NOBGEN see IIU000
NOBILEN see TFR250
NOBITOCIN S see ORU500
NOBLEN see PMP500
NOBRIUM see CGA000
NO BUNT LIQUID see HCC500
NOCA see SLL000
NOCARDICIN A see APT750
NOCARDICIN COMPLEX see NMV480
NOCBIN see DXH250
NOCODAZOLE see OJD100
No. 49 CONCENTRATED BENZIDINE YELLOW see DEU000
No. 3 CONC. SCARLET see CHP500
NOCRAC NS 6 see MJO500
NOCTAL see QCS000
NOCTAMID see MLD100
NOCTAN see DNW400
NOCTAZEPAM see CFZ000
NOCTEC see CDO000
NOCTENAL see QCS000
NOCTILENE see QAK000
NOCTIVANE SODIUM see ERE000
NOCTOSEDIV see TEH500
NOCTOSOM see FMQ000
NOCTOVANE see ERD500
NODAPTON see GIC000
NO-DHU see NJY000
NO-DOZ see CAK500
NOE (FRENCH) see CBA100
NO-ETU see NKK500
NOFLAMOL see PJL750
No. 1 FORTHFAST RED R see CJD500
No. 2 FORTHFAST SCARLET see MMP100
NOGALAMYCIN see NMV500
NOGAL de LA INDIA (CUBA) see TOA275
NOGALOMYCIN see NMV500
NOGAMYCIN see NMV600
NOGEDAL see DPH600
NOGEST see MCA000, POF275
NOGOS see DGP900
NOGRAM see EID000
NOHDABZ see HKB000
NOHFAA see HIP000
NOHO-MALIE (HAWAII) see YAK350
NOIGEN 160 see DXY000
NOIR BRILLANT BN (FRENCH) see BMA000
NOISETTE des GRANDS-FONDS (GUADELOUPE) see TOA275
NOIX de BANCOUL (GUADELOUPE) see TOA275
NOIX des MOLLUQUES (GUADELOUPE) see TOA275
NOKHEL see AHP375
NOLEPTAN see FMU000

NOLTRAN see CMA250
NOLUDAR see DNW400
NOLVADEX see TAD175
NOMATE PBW see GJK000
NOMERSAN see TFS350
NOMETIC see DWK200
NOMETINE see CJL500
No. 907 METRO TALC see TAB750
NOMIFENSIN see NMV700, NMV725
NOMIFENSINE see NMV700
NOMIFENSINE HYDROGEN MALEATE see NMV725
NOMIFENSINE MALEATE see NMV725
NOMIFENSIN HYDROGEN MALEATE see NMV725
NONABROMOBIPHENYL see NMV735
NONACARBONYL DIIRON see NMV740
NONACHLAZINE see NMV750
NONADECAFLUORODECANOIC ACID see PCG725
NONADECAFLUORO-n-DECANOIC ACID see PCG725
2,6-NONADIENAL see NMV760
trans-2,cis-6-NONADIENAL see NMV760
trans,cis-2,6-NONADIENAL see NMV760, NMV760
NONADIENOL see NMV780
trans,cis-2,6-NONADIENOL see NMV780
2-trans-6-cis-NONADIEN-1-OL see NMV780
1,1,2,2,3,3,4,4,4-NONAFLUORO-N,N-BIS(NONAFLUOROBUTYL)-1-BUTANAM-
    INE see HAS000
Γ-NONALACTONE (FCC) see CNF250
1-NONALDEHYDE see NMW500
NONALOL see NNB500
1,4-NONALOLIDE see CNF250
1-NONANAL see NMW500
NONANE see NMX000
1-NONANECARBOXYLIC ACID see DAH400
4-NONANECARBOXYLIC ACID see NMX500
NONANEDIOIC ACID see ASB750
NONANEDIOTIC ACID, DIHEXYL ESTER see ASC000
NONANOIC ACID see NMY000
NONANOIC ACID, ETHYL ESTER see ENW000
NONANOIC ACID OXYDI-3,1-PROPANEDIYL ESTER (9CI) see DWT000
NONANOIC ACID OXYDIPROPYLENE ESTER see DWT000
NONANOIC ACID, TRIBUTYLSTANNYL ESTER see TIF500
1-NONANOL see NNB500
NONAN-1-OL see NNB500
2-NONANONE see NMY500
NONAN-2-ONE see NMY500
5-NONANONE see NMZ000
NONAN-5-ONE see NMZ000
1-NONANOYLAZIRIDINE see NNA000
NONANOYLETHYLENEIMINE see NNA000
4-NONANOYLMORPHOLINE see MRQ750
(NONANOYLOXY)TRIBUTYLSTANNANE see TIF500
2-NONENAL see NNA300
2-NONEN-1-AL see NNA300
6-NONENAL, (Z)- see NNA325
(Z)-6-NONENAL see NNA325
cis-6-NONENAL see NNA325
cis-6-NONEN-1-AL see NNA325
trans-2-NONENAL (FCC) see NNA300
2-NONENENITRILE, 3-METHYL- see MNB500
2-NONENOIC ACID, METHYL ESTER see MNB750
2-NONEN-4,6,8-TRIYN-1-AL see NNB000
α-NONENYL ALDEHYDE see NNA300
NONEX 25 see PJY100
NONEX 30 see PJY100
NONEX 52 see PJY100
NONEX 64 see PJY100
NONEX 411 see HKJ000
NONFLAMIN see TGE165
NONIDET P40 see GHS000
NONION 06 see PJY100
NONION HS 206 see GHS000
NONION O2 see PJY100
NONION O4 see PJY100
NONION OP80R see SKV100
NONION SP 60 see SKV150
NONION SP 60R see SKV150
NONISOL 200 see PJY100
NONISOLD see CNH125
n-NONOIC ACID see NMY000
NONOX CL see NBL000
NONOX D see PFT500
NONOX DPPD see BLE500
NONOX TBC see BFW750
NONOXYNOL see NND500

NONOXYNOL-9 see NNB300
NONOX ZA see PFL000
NONYL ACETATE see NNB400
NONYL ALCOHOL see NNB500
n-NONYL ALCOHOL see NNB500
sec-NONYL ALCOHOL see DNH800
1-NONYL ALDEHYDE see NMW500
1-NONYLAMINE, N-METHYL-N-NITROSO- see NKU580
NONYLCARBINOL see DAI600
o-NONYL-HARMOL HYDROCHLORIDE see NNC100
n-NONYLIC ACID see NMY000
NONYL METHYL KETONE see UKS000
NONYL PHENOL (mixed isomers) see NNC500
NONYLPHENOL, POLYOXYETHYLENE ETHER see NND500
NONYLPHENOXYPOLY(ETHYLENEOXY)ETHANOL see NNB300
NONYLPHENOXYPOLYETHYLENEOXY ETHANOL-IODINE COMPLEX
    see NND000
NONYL PHENYL POLYETHYLENE GLYCOL see NND500
NONYL PHENYL POLYETHYLENE GLYCOL ETHER see NND500
NONYLTRICHLOROSILANE see NNE000
2-NONYNAL DIMETHYLACETAL see NNE100
2-NONYNAL, DIMETHYL ACETAL see NNE100
2-NONYN-1-AL DIMETHYLACETAL see NNE100
2-NONYNOIC ACID, METHYL ESTER see MNC000
No. 156 ORANGE CHROME see LCS000
NOOTRON see NNE400
NOOTROPIL see NNE400
NOOTROPYL see NNE400
NOPALCOL 1-0 see PJY100
NOPALCOL 4-0 see PJY250
NOPALCOL 6-0 see PJY100
NOPALCOL 6-L see PJY000
NOPALMATE see PLH500
NOPCOCIDE see TBQ750
NOPCO 2272-R see NNE500
4-NOPD see ALL500
NO-PEST see DGP900
NO-PEST STRIP see DGP900
NOPIL see TKX000
NOPINEN see POH750
NOPINENE see POH750
NO-PIP see NLJ500
NOPOL see DTB800
NOPOL ACETATE see DTC000
NOPOL (TERPENE) see DTB800
NO-PRESS see MBW750
NO-Pro see NLL500
NOPTIL see EOK000
NOPYL ACETATE see DTC000
NO-PYR see NLP500
NORACYCLINE see LJE000
(−)-NORADREC see NNO500
NORADRENALIN see NNO500
NORADRENALINA (ITALIAN) see NNO500
NORADRENALINE see NNO500
(−)-NORADRENALINE see NNO500
d-(−)-NORADRENALINE see NNO500
l-NORADRENALINE see NNO500
(−)-NORADRENALINE ACID TARTRATE see NNO699
(−)-NORADRENALINE BITARTRATE see NNO699
l-NORADRENALINE BITARTRATE see NNO699
(−)-NORADRENALINE BITARTRATE MONOHYDRATE see NNO699
NORADRENALINE HYDROCHLORIDE see NNP000
(±)-NORADRENALINE HYDROCHLORIDE see NNP050
dl-NORADRENALINE HYDROCHLORIDE see NNP050
NORADRENALINE HYDROGEN TARTRATE see NNO699
(−)-NORADRENALINE TARTRATE see NNO699
l-NORADRENALINE TARTRATE see NNO699
NORADRENALINE TARTRATE (1:1) see NNO699
NOR-ADRENALIN HYDROCHLORIDE see NNP000
NORADRENLINE see NNO500
NORAL INK GRADE ALUMINUM see AGX000
NOR-AM EP 332 see DSO200
NOR-AM EP 333 see CJJ500
NORAMITRIPTYLINE see NNY000
NORAM O see OHM700
NORANAT see IBV100
NORANDROLONE PHENYLPROPIONATE see DYF450
NORANDROSTENOLON see NNX400
NORANDROSTENOLONE see NNX400
19-NORANDROSTENOLONE see NNX400
NORANDROSTENOLONE DECANOATE see NNE550
19-NORANDROSTENOLONE DECANOATE see NNE550
NORANDROSTENOLONE PHENYLPROPIONATE see DYF450

NORARTRINAL see NNO500
NORBILAN see TGZ000
NORBOLDINE see LBO100
(+)-NORBOLDINE see LBO100
NORBOLETHONE see NNE600
NORBORAL see BSM000
NORBORMIDE see NNF000
NORBORNADIENE see NNG000
2,5-NORBORNADIENE see NNG000
endo-2-NORBORNANECARBOXYLIC ACID ETHYL ESTER see NNG500
2-NORBORNANEMETHANOL see NNH000
2-NORBORNANONE see NNK000
2-NORBORNENE see NNH500
trans-5-NORBORNENE-2,3-DICARBONYL CHLORIDE see NNI000
cis-5-NORBORNENE-2,3-DICARBOXYLIC ACID DIMETHYL ESTER
    see DRB400
5-NORBORNENE-2-METHANOL ACRYLATE see BFY250
5-NORBORNENE-2-METHYLOLACRYLATE see BFY250
α-5-NORBORNEN-2-YL-α-PHENYL-PIPERIDINE PROPANOL HYDROCHLO-
    RIDE see BGD750
NORBORNYLENE see NNH500
NORBUTORPHANOL TARTRATE see NNJ600
NORCAIN see EFX000
NORCAMPHANE see EOM000
2-NORCAMPHANE METHANOL see NNH000
NORCAMPHENE see NNH500
NORCAMPHOR see NNK000
NORCHLORCYCLIZINE see NNK500
NORCHLORCYCLIZINE HYDROCHLORIDE see NNL000
NORCOZINE see CKP500
20-NORCROTALANAN-11,15-DIONE, 3,8-DIDEHYDRO-14,19-DIHYDRO-12,13-
    DIHYDROXY-, (13-α-14-α)- see DAL350
NORCURON see VGU075
NORDEN see AKT250
NORDETTE see NNL500
NORDHAUSEN ACID (DOT) see SOI500
NORDIALEX see DBL700
NORDIAZEPAM see CGA500
NORDICOL see EDO000
NORDICORT see HHR000
NORDIHYDROGUAIARETIC ACID see NBR000
NORDIHYDROGUAIRARETIC ACID see NBR000
NORDIMAPRIT see NNL400
NORDIOL see NNL500
NORDIOL-28 see NNL500
NORDOPAN see BIA250
NOREA see HDP500
NORENOL see AKT000
NOREPHEDRANE see AOB250, BBK000
(−)-NOREPHEDRINE see NNM000
dl-NOREPHEDRINE see NNM500
NOREPHEDRINE HYDROCHLORIDE see NNN000
(−)-NOREPHEDRINE HYDROCHLORIDE see NNN500, NNO000
d-NOREPHEDRINE HYDROCHLORIDE see NNO000
l-NOREPHEDRINE HYDROCHLORIDE see NNN500
dl-NOREPHEDRINE HYDROCHLORIDE see PMJ500
NOREPINEPHRINE see NNO500
(−)-NOREPINEPHRINE see NNO500
l-NOREPINEPHRINE see NNO500
(−)-NOREPINEPHRINE BITARTRATE see NNO699
l-NOREPINEPHRINE BITARTRATE see NNO699
NOREPINEPHRINE HYDROCHLORIDE see NNP000
(±)-NOREPINEPHRINE HYDROCHLORIDE see NNP050
dl-NOREPINEPHRINE HYDROCHLORIDE see NNP050
(−)-NOREPINEPHRINE TARTRATE see NNO699
NOREPIRENAMINE see NNO500
NORES see HDP500
NORETHANDROL see EAP000
NORETHANDROLONE see ENX600
NORETHINDRONE-17-ACETATE see ABU000
NORETHINDRONE ACETATE 3-CYCLOPENTYL ENOL ETHER see QFA275
NORETHINDRONE ACETATE and ETHINYLESTRADIOL see EEH520
NORETHINDRONE mixed with ETHYNYLESTRADIOL see EQM500
NORETHINDRONE mixed with MESTRANOL see MDL750
NORETHINODREL see EEH550
19-NOR-ETHINYL-4,5-TESTOSTERONE see NNP500
19-NOR-ETHINYL-5,10-TESTOSTERONE see EEH550
NORETHINYNODREL see EEH550
19-NORETHISTERONE see NNP500
19-NORETHISTERONE ACETATE see ABU000
NORETHISTERONE ACETATE mixed with ETHINYL OESTRADIOL
    see EEH520
NORETHISTERONE ENANTHATE see NNQ000
NORETHISTERONE mixed with ETHINYL OESTRADIOL (60:1) see EQM500

NORETHISTERONE mixed with MESTRANOL see MDL750
NORETHISTERONE OENANTHATE see NNQ000
19-NORETHYLTESTOSTERONE see ENX600
19-NOR-17-α-ETHYLTESTOSTERONE see ENX600
NORETHYNODRAL see EEH550
NORETHYNODREL see EEH550
19-NORETHYNODREL see EEH550
NORETHYNODREL and ETHINYLESTRADIOL-3-METHYL ETHER (50:1)
    see EAP000
NORETHYNODREL mixed with MESTRANOL see EAP000
19-NOR-17-α-ETHYNYLANDROSTEN-17-β-OL-3-ONE see NNP500
19-NOR-17-α-ETHYNYL-17-β-HYDROXY-4-ANDROSTEN-3-ONE see NNP500
19-NOR-17-α-ETHYNYLTESTOSTERONE see NNP500
19-NORETHYNYLTESTOSTERONE ACETATE see ABU000
NORETHYSTERONE ACETATE see ABU000
NOREX see CJQ000
NORFENEFRINE HYDROCHLORIDE see NNT100
NORFLOXACIN see BAB625
NORFORMS see ABU500
NORGAMEM see TEV000
NORGESTREL see NNQ500
d(−)-NORGESTREL see NNQ500
d-NORGESTREL see NNQ500
d(−)-NORGESTREL see NNQ520
d-NORGESTREL see NNQ520
l-NORGESTREL see NNQ525
α-NORGESTREL see NNQ500
(±)-NORGESTREL see NNQ500
dl-NORGESTREL see NNQ500
NORGESTREL mixed with ETHINYL ESTRADIOL see NNL500
dl-NORGESTREL mixed with ETHINYLESTRADIOL see NNL500
NORGESTREL mixed with MESTRANOL see NNR000
NORGESTRIENONE see NNR125
NORGINE see AFL000
NORGLAUCIN see NNR200
NORGLAUCINE see NNR200
d-N-NORGLAUCINE see NNR200
NORHARMAN see NNR300
NOR-HN2 see BHO250
NOR-HN2 HYDROCHLORIDE see BHO250
NORHOMOEPINEPHRINE HYDROCHLORIDE see AMB000
NORIDIL see UVJ450
NORIDYL see UVJ450
NORIGEST see NNQ000
NORILGAN-S see SNN500
NORIMIPRAMINE see DSI709
NORIMYCIN V see CDP250
NORINYL-1 see MDL750
NORISODRINE see DMV600
NORISODRINE HYDROCHLORIDE see IMR000
NORIZALPININ see GAZ000
NORKEL see AHK750
NORKOOL see EJC500
NORLESTRIN see EEH520
epsilon-NORLEUCINE see AJD000
NOR-LOST HYDROCHLORID (GERMAN) see BHO250
NORLUTATE see ABU000
NORLUTIN see NNP500
NORLUTINE ACETATE see ABU000
NORLUTIN ENANTHATE see NNQ000
NORMABRAIN see NNE400
NORMALIP see ARQ750
NORMAL LEAD ACETATE see LCV000
NORMAL LEAD ORTHOPHOSPHATE see LDU000
NORMASTIGMIN see DQY909, NCL100
NORMAT see ARQ750
NORMERSAN see TFS350
NORMETANDRONE see MDM350
NORMETHANDROLONE see MDM350
NORMETHANDRONE see MDM350
NORMETHYL EX4442 see POF250
19-NOR-17-α-METHYLTESTOSTERONE see MDM350
NORMETOL see AKT000
NORMI-NOX see QAK000
NORMISON see CFY750
NORMOCYTIN see VSZ000
NORMOLIPOL see ARQ750
NORMONSON see CHG000
NORMORESCINA see TLN500
NORMORPHINONE, N-ALLYL-7,8-DIHYDRO-14-HYDROXY-, (-)- see NAG550
NORMOSAN see CHG000
NORMOSON see CHG000
NORMUSCONE see CPU250
NORNICOTIN see NNR500

NORNICOTINE see NNR500
NORNICOTINE (+)-HYDROCHLORIDE see NNS500
d-NORNICOTINE HYDROCHLORIDE see NNS500
NORNICOTYRINE see PQB500
NOR-NITROGEN MUSTARD see BHN750
NORNITROGEN MUSTARD HYDROCHLORIDE see BHO250
NOROCAINE see AIL750
NORODIN see DBA800
NORODIN HYDROCHLORIDE see MDT600
NOROX BZP-250 see BDS000
NOROXIN see BAB625
19-NOR-P see NNT500
NORPACE see RSZ600
NORPHEN see AKT250
NORPHENAZONE see NNT000
NORPHENYLEPHRINE see AKT000
(±)-NORPHENYLEPHRINE HYDROCHLORIDE see NNT100
NORPLANT see NNQ525
NORPRAMIN see DLS600
NORPRAZEPAM see CGA500
19-NOR-17-α-PREGNA-1,3,5(10)-TRIEN-2-YNE-3,17-DIOL see EEH500
(17-α)-19-NORPREGNA-1,3,5(10)-TRIEN-20-YNE-3,17,DIOL see EEH500
19-NORPREGN-4-ENE-3,20-DIONE see NNT500
19-NOR-17-α-PREGN-5(10)-EN-20-YNE-3-α,17-DIOL see NNU000
19-NOR-17-α-PREGN-5(10)-EN-20-YNE-3-β, 17-DIOL see NNU500
(3-β,17-α)-19-NORPREGN-4-EN-20-YNE-3,17-DIOL DIACETATE see EQJ500
19-NOR-17-α-PREGN-4-EN-20-YNE-3-β,17-DIOL DIACETATE mixed with 3-
    METHOXY-19-NOR-17-α-PREGNA-1,3,5(10)-TRIEN-20-YN-17-OL see EQK010
10-NOR-17-α-PREGN-4-EN-20-YNE-3-β,17-DIOL mixed with 3-METHOXY-17-α-19-
    NORPREGNA-1-3-5(10)-TRIEN-20-YN-17-OL see EQK100
(17-α)-19-NORPREGN-4-EN-20-YN-17-OL see NNV000
19-NOR-17-α-PREGN-4-EN-20-YN-17-OL see NNV000
19-NOR-17-α-PREGN-4-EN-20-YN-3-ONE, 17-HYDROXY-, mixed with 19-NOR-17-
    α-PREGNA-1,3,5(10)-TRIEN-2-YNE-3,17-DIOL (60:1) see EQM500
NOR-PRESS 25 see HGP500
NOR-PROGESTELEA see NCI525
19-NORPROGESTERONE see NNT500
(−)-NORPSEUDOEPHEDRINE see NNV500
(+)-NORPSEUDOEPHEDRINE HYDROCHLORIDE see NNW500
d-NORPSEUDOEPHEDRINE HYDROCHLORIDE see NNW500
1-NOR-PSI-EPHEDRIN (GERMAN) see NNV500
19-NORSPIROXENONE see NNX300
NORSTEARANTHRENE see TCJ500
NORSULFASOL see TEX250
NORSULFAZOLE see TEX250
NORSYMPATHOL see AKT250
NORSYNEPHRINE see AKT000, AKT250
NORTEC see CDO000
NORTESTONATE see NNX400
NORTESTOSTERONE see NNX400
19-NORTESTOSTERONE see NNX400
(+)-19-NORTESTOSTERONE see NNX400
17-α-NORTESTOSTERONE see ENX600
19-NORTESTOSTERONE-17-N-(2-CHLOROETHYL)-N-NITROSOCARBAMATE
    see NNX600
NORTESTOSTERONE DECANOATE see NNE550
19-NORTESTOSTERONE DECANOATE see NNE550
19-NORTESTOSTERONE HOMOFARNESATE see NNX650
19-NORTESTOSTERONE PHENYLPROPIONATE see DYF450
NORTHERN COPPERHEAD VENOM see NNX700
NORTIMIL see DLS600
NORTRIPTYLINE see NNY000
NORURON see HDP500
NORVAL see BMA625, DJL000
NORVALAMINE see BPX750
NORVEDAN see CKI750
NORVINISTERONE see NCI525
NORVINYL see PKQ059
NORVINYL P 6 see AAX175
19-NOR-17-α-VINYLTESTOSTERONE see NCI525
NORZETAM see NNE400
NO SCALD see DVX800
NOSCAPAL see NOA000
NOSCAPALIN see NOA000
NOSCAPINE see NBP275, NOA000
NOSCAPINE HYDROCHLORIDE see NOA500
NOSCOSED see CLD250
NOSIM see CCK125
NOSOPHENE SODIUM see TDE750
NOSPAN see MOV500
NOSPASM see IPU000
NOSTAL see EHP000, QCS000
NOSTEL see CHG000
NOSTIN see EHP000

NOSTRAL see QCS000
NOSYDRAST see DNG400
NOTANDRON see AOO475
NOTANDRON-DEPOT see AOO475
NOTARAL see BFD000, CFZ000
NOTATIN see GFG100
NOTECHIS SCUTATUS VENOM see ARV550
NOTENQUIL see ABH500
NOTENSIL see AAF750, ABH500
NOTESIL see ABH500
NOTEZINE see DIW000
NOTOMYCIN A1 see CNV500
NOURALGINE see SEH000
NOURITHION see PAK000
NOVACRYSIN see GJG000
NOVADELOX see BDS000
NOVADEX see NOA600
NOVADOX see HGP550
NOVADRAL see AKT000, NNT100
NOVAFED see POH000, POH250
NOVAKOL see FNF000
NOVALICHIN see NOA700
NOVALLYL see AFS500
NOVAMIDON see DOT000
NOVAMIN see APT000, DYE600, PMF500
NOVAMINE see DYE600
NOVAMONT 2030 see PMP500
NOVANTOINA see DKQ000, DNU000
NOVA-PHENO see EOK000
NOVARSAN see NCJ500
NOVARSENOBENZOL see NCJ500
NOVARSENOBILLON see NCJ500
NOVASMASOL see MDM800
NOVASUROL see CCG500
NOVATHION see DSQ000
NOVATONE see MDW750
NOVATRIN see MDL000
NOVATROPINE see MDL000
NOVAZOLE see CBA100
NOVECYL see SAH000
NOVEDRIN HYDROCHLORIDE see EJR500
NOVEGE see PBK000
NOVEMBICHIN see NOB700
NOVEMBIKHIN see NOB700
NOVERIL see DCW800
NOVERYL see DCW800
NOVESIN see OPI300
NOVESINE see OPI300
NOVIBEN see CBA100
NOVICET see VLF000
NOVICODIN see DKW800
NOVID see ADA725
NOVIDIUM CHLORIDE see HGI000
NOVIDORM see THS800
NOVIGAM see BBQ500
NOVIZIR see EIR000
NOVOBIOCIN see SMB000
NOVOBIOCIN MONOSODIUM see NOB000
NOVOBIOCIN, MONOSODIUM SALT see NOB000
NOVOBIOCIN, SODIUM derivative see NOB000
NOVOCAINAMIDE see AJN500
NOVOCAIN-CHLORHYDRAT (GERMAN) see AIT250
NOVOCAINE see AIL750
NOVOCAINE AMIDE see AJN500
NOVOCAINE HYDROCHLORIDE see AIT250
NOVOCAIN HYDROCHLORID (GERMAN) see AIT250
NOVOCAMID see AJN500
NOVOCEBRIN HYDROCHLORIDE see DXI800
NOVOCHLOROCAP see CDP250
NOVOCILLIN see BFD250
NOVOCOLIN see DAL000
NOVOCONESTRON see ECU750
NOVODIL see DNU100
NOVODIPHENYL see DNU000
NOVODOLAN see TKG000
NOVODRIN see DMV600
NOVOEMBICHIN see NOB700
NOVOHEPARIN see HAQ500
NOVOHETRAMIN see RDU000
NOVOL see OBA000
NOVOLEN see PMP500
NOVOL POE 20 see PJW500
NOVOMAZINA see CKP250
NOVOMYCETIN see CDP250

NOVON see PBK000
NOVON 712 see PKQ059
NOVONAL see DJU200
NOVONIDAZOL see MMN250
NOVOPHENICOL see CDP250
NOVOPHENYL see BRF500
NOVOPHONE see SOA500
NOVO-R see SMB000
NOVOSAXAZOLE see SNN500
NOVOSCABIN see BCM000
NOVOSED see MDQ250
NOVOSERIN see CQH000
NOVOSERPINA see RCA200
NOVOSIR N see EIR000
NOVOSPASMIN see NOC000
NOVOX see BSC500
NOVYDRINE see BBK000
NOXABEN see DGE200
NOXAL see DXH250
NOXFISH see RNZ000
N-OXIDE de PHENAZINE (FRENCH) see PDB750
NOXIPTILINE HYDROCHLORIDE see DPH600
NOXIPTILIN HYDROCHLORID (GERMAN) see DPH600
NOXIPTYLINE HYDROCHLORIDE see DPH600
NOXODYN see TEH500
NOXOKRATIN see EQL000
NOXYRON see DYC800
1-NP see NIX500
NP 2 see NCW500
2-NP see ALL750, NIY000
NP-9 see NNB300
NP 212 see CJR909
NP 246 see CJN750
NPA see DNB000
NPAB see PNC925
NPD see TED500
NPG see DTG400
NPGB see NIQ500
NPH 83 see MJG500
NPH-1091 see BDJ250
NPH ILETIN see IDF325
NPH INSULIN see IDF325
NPH 50 INSULIN see IDF325
NPIP see NLJ500
NPOH see DVD200
NPP see DYF450
2-NPPD see ALL750
NPRO see NLL500
NPU see NLO500
NPYR see NLP500
4-NQO see NJF000
NR.C 2294 see DNA800
NRDC 104 see BEP500
NRDC 107 see BEP750
NRDC 119 see RDZ875
NRDC 149 see RLF350
NRDC 160 see RLF350
NRDC 161 see DAF300
NRDC 166 see RLF350
NRL-18-B see TNI000
NS 11 see DTG000
NSC-185 see CPE750
NSC 339 see DVF200
NSC 423 see DAA800
NSC 729 see PIB925
NSC 739 see AMG750
NSC-740 see MDV500
NSC 741 see HHQ800
NSC-742 see ASA500
NSC 743 see POJ500
NSC 744 see CKV500
NSC 746 see UVA000
NSC-749 see AJO500
NSC-750 see BOT250
NSC 751 see EEI000
NSC-752 see AMH250
NSC 753 see POJ250
NSC 755 see POK000
NSC 756 see THT350
NSC 757 see CNG830
NSC 759 see BCB750
NSC 762 see BIE250, BIE500
NSC-763 see DUD800
NSC 1026 see AJK250

NSC-1390 see ZVJ000
NSC 1393 see POM600
NSC 1532 see DUZ000
NSC 1895 see DCF200
NSC 2066 see TEP000
NSC 2083 see AMM250
NSC-2100 see NGE500
NSC-2101 see HMY000
NSC 2107 see NGE000
NSC 2752 see FOU000
NSC 3051 see MKG500
NSC 3053 see AEB000
NSC-3055 see AEI000
NSC 3058 see BDH250
NSC 3060 see PKV500
NSC 3061 see TGD000
NSC 3063 see DAY825
NSC 3069 see CDP250
NSC-3070 see DKA600
NSC 3072 see SFA000
NSC 3073 see FMT000
NSC-3088 see CDO500
NSC 3094 see DTP000
NSC 3095 see IBC000
NSC 3096 see MIW500
NSC 3138 see DOO800
NSC-3424 see QDJ000
NSC 3425 see THR750
NSC 4170 see MMD500
NSC 4320 see SMN002
NSC 4729 see TES000
NSC 4730 see EGI000
NSC 4911 see MCQ500
NSC 5088 see SAL500
NSC 5159 see CDK250
NSC 5340 see AHL000
NSC 5354 see PEY250
NSC 5356 see DSB000
NSC-5366 see NBP275
NSC 5366 see NOA000
NSC-6091 see SOA500
NSC-6396 see TFQ750
NSC-6470 see NDY000
NSC-6738 see DGP900
NSC 7365 see DCQ400
NSC-7571 see AHS750
NSC-7760 see POS750
NSC-7778 see MES000
NSC-8806 see BHV250, PED750
NSC 8819 see ADR000
NSC 9166 see TBG000
NSC 9169 see HOI000
NSC-9324 see AFS500
NSC 9369 see MMP000
NSC 9659 see ILD000
NSC-9698 see MAW750
NSC-9701 see MPN500
NSC-9704 see PMH500
NSC 9706 see TND500
NSC 9717 see TND250
NSC-9895 see EDO000
NSC 10023 see PLZ000
NSC-10107 see CFA500, CFA750
NSC-10108 see CLO750
NSC 10483 see CNS750
NSC 10815 see AMX750
NSC-10873 see BHN750, BHO250
NSC 11595 see PNW800
NSC-11905 see HLY500
NSC-12165 see AOO275
NSC-12169 see EDU500
NSC 13002 see DOS800
NSC-13252 see CMB000
NSC 13875 see HEJ500
NSC 14083 see SLW500
NSC-14210 see BHT750
NSC 14574 see SBE500
NSC 14575 see EMC500
NSC 15193 see DAR600
NSC-15432 see EEH550
NSC 15747 see BFL125
NSC-15780 see AOD500
NSC-16498 see BIA000
NSC-16895 see LGZ000

NSC-17118 see CLD500
NSC-17262 see BDC750
NSC-17591 see TBF750
NSC-17592 see HNT500
NSC 17661 see GHE000
NSC 17663 see BHN500
NSC-17777 see PGG355
NSC 18016 see CHC750
NSC-18268 see AEA750
NSC 18321 see BHB750
NSC-18429 see AOQ875
NSC-18439 see BIC600
NSC-19477 see HEG000
NSC 19488 see BRS500
NSC-19494 see MQR100
NSC-19893 see FMM000
NSC-19962 see CBO625
NSC-19987 see MOR500
NSC-20264 see AOA125
NSC 21206 see ALL250
NSC 21402 see AKM000
NSC 22420 see DCP400
NSC 23519 see DCQ600
NSC-23890 see DSU000
NSC-23892 see BCC250
NSC 23909 see MNA750
NSC 24559 see MQW750
NSC 24818 see PJJ225
NSC-24970 see BMZ000
NSC-25154 see BHJ250
NSC 25855 see TEV000
NSC 25958 see NLB700
NSC-26154 see AJD000
NSC-26198 see PAN100
NSC 26271 see CQC500, CQC650
NSC 26805 see EMF500
NSC-26806 see VMA000
NSC 26812 see AQO000
NSC 26980 see AHK500
NSC-27640 see DAR400
NSC 28693 see MRH000
NSC-29215 see TND000
NSC 29421 see BDG325
NSC-29422 see TFJ500
NSC-29630 see DFO000
NSC-29863 see GKS000
NSC-30152 see PMO250
NSC-30211 see TNF500
NSC 30622 see SBX525
NSC 31083 see AEA109
NSC 31712 see TCQ500
NSC 32065 see HOO500
NSC 32074 see RJA000
NSC-32606 see BOP750
NSC 32743 see ABN000
NSC 32946 see MKI000
NSC-33531 see DMH600
NSC-33669 see EAL500, EAN000
NSC-34372 see DFH000
NSC-34462 see BIA250
NSC 34533 see GKE000
NSC-34652 see MKB250
NSC-35051 see BHU750, SAX200
NSC 35770 see CMX700
NSC-36354 see DBA175
NSC 37095 see EHV500
NSC 37448 see PDT250
NSC-37538 see BKM500
NSC-37725 see NNV000
NSC 38270 see OIU499
NSC 38721 see CDN000
NSC-38887 see AKY250
NSC-39069 see TFU500
NSC-39084 see ASB250
NSC 39147 see SMC500
NSC-39470 see BFV750
NSC 39661 see DAS000
NSC-39690 see DIG400
NSC 40774 see MPU000
NSC-40902 see AOO500
NSC 42076 see BER500
NSC-42722 see DAL300
NSC 44185 see EAM500
NSC 44690 see DIS200

NSC 45383 see SMA000
NSC-45384 see SMA500
NSC-45388 see DAB600
NSC 45403 see ENV000
NSC-45463 see DOY600
NSC-45624 see SKI500
NSC-46015 see TCW500
NSC 47547 see CHE750
NSC 47842 see VKZ000
NSC-48841 see BHP250
NSC 49842 see VLA000
NSC-50256 see MLH500
NSC-50364 see MMN250
NSC-50413 see AQY250
NSC-50982 see BHW500
NSC 51143 see IAN100
NSC-52695 see CQN000
NSC-52760 see DAK200
NSC-52947 see PAB500
NSC 53396 see AEA250
NSC 56408 see TNY500
NSC-56410 see MLY000
NSC 56654 see ASO510
NSC-57199 see BHT250
NSC-58404 see DCO800
NSC-58514 see CMK650
NSC-58775 see DLY000
NSC 59729 see SKX000
NSC-60195 see MQG500
NSC 60380 see CMA250
NSC-62209 see BIF250
NSC 62579 see DJB800
NSC 62580 see DRO200
NSC-62709 see IPR000
NSC-62939 see TKX250
NSC 63346 see DQD200
NSC-63701 see VGZ000
NSC 63878 see AQQ750, AQR000
NSC 64375 see BDW000
NSC-64393 see SLC000
NSC-65104 see SBZ000
NSC-65346 see SAU000
NSC 65426 see PML250
NSC-66847 see TEH500
NSC-67239 see THM750
NSC-67574 see LEY000
NSC 67574 see LEZ000
NSC 68075 see TEH250
NSC-68626 see ADA000
NSC 69187 see MEL775
NSC-69536 see MLJ500
NSC-69811 see MKW250
NSC 69856 see NBV500
NSC-69945 see PHA750
NSC-70731 see LGD000
NSC-70762 see TGJ500
NSC-70845 see NMV500
NSC-71045 see HKQ025
NSC-71261 see TFJ250
NSC-71423 see VTF000
NSC-71795 see EAI850
NSC-71936 see CMQ725
NSC 71964 see PCU425
NSC 73438 see NKL000
NSC 74437 see DHE100
NSC 75520 see TKH325
NSC-76098 see ARW750
NSC-76411 see OIU000
NSC-77213 see PME250, PME500
NSC 77471 see MQX000
NSC 77517 see MKN750
NSC-77518 see DCK759
NSC 78409 see BIH325
NSC-78559 see DAB800
NSC 78572 see HIZ000
NSC 79019 see PQC000
NSC-79037 see CGV250
NSC-79389 see ARQ750
NSC-80087 see DOT600
NSC-81430 see CQJ500
NSC-82116 see KFA100
NSC-82151 see DAC000, DAC200
NSC-82174 see EID000
NSC-82196 see IAN000

NSC-82261 see GCS000
NSC-82699 see TKH750
NSC 83265 see TNR475
NSC-83629 see HKQ025
NSC 83653 see AED250
NSC-84423 see HGI000
NSC 84954 see CFH500
NSC 84963 see AMN300
NSC 85598 see SMD000
NSC-85998 see SMD000
NSC-86078 see SMM500
NSC 87419 see CPN500
NSC 87974 see CPL750, FIJ000
NSC 88104 see CFH750
NSC-88106 see CHE500
NSC 89936 see JAK000
NSC-89945 see DMX200
NSC-91523 see ICB000, ICC000
NSC 91726 see CGV000
NSC-92338 see CBF250
NSC-93150 see AQR500
NSC 93169 see MGL600
NSC-94100 see DDP600
NSC 94600 see CBB870
NSC-95441 see CHD250
NSC-95466 see CGW250
NSC 100880 see CBB870, CBB875
NSC-101983 see MMR750
NSC-101984 see MMS750
NSC 102627 see YCJ000
NSC-102816 see ARY000
NSC 103548 see CHF000
NSC 104469 see CME250
NSC-104800 see DDJ000
NSC-104801 see CQK600, SIK000
NSC 105023 see RQF350
NSC-105388 see FPC000
NSC-106563 see BFW250
NSC-106568 see TKZ000
NSC-106959 see CBQ625
NSC-106962 see SKM000
NSC-107429 see COV500
NSC-107430 see DOQ400
NSC-107434 see SKO500
NSC-107654 see CFC000
NSC-108034 see HNJ000
NSC-108160 see AEG750
NSC-109229 see ARN800
NSC 109723 see TNT500
NSC-109724 see IMH000
NSC-110326 see CNH800
NSC-110364 see OOG000
NSC-110430 see CBR500
NSC-110431 see MIV500
NSC-110432 see DFA400
NSC-111071 see CBO250
NSC 111180 see ACH000
NSC 112259 see EDR500
NSC-113233 see MQX250
NSC 113619 see AAE500
NSC 113926 see RKP000
NSC-114649 see FCB100
NSC-114650 see DNV200
NSC 114900 see DLH630
NSC-114901 see DLS600
NSC-115944 see EAT900
NSC-119875 see PJD000
NSC 119876 see TBO776
NSC-120949 see PJY750
NSC 122023 see VBZ000
NSC 122053 see PCV775
NSC-122402 see LJD500
NSC-122758 see VSK950
NSC-122819 see EQP000
NSC 122870 see DBY500
NSC-123127 see AES750
NSC-125973 see TAH775
NSC 126771 see DFN500
NSC-129185 see MRX000
NSC-129943 see PIK250
NSC-130044 see TEZ000
NSC 132313 see DCI600
NSC-132319 see ICD100
NSC-133099 see RKA000

NSC-133129 see DCI800
NSC-134434 see LIM000
NSC-135758 see PIK075
NSC 137679 see SBU700
NSC 138780 see FQS000
NSC 139105 see THR500
NSC-140117 see IBQ100
NSC-141537 see AOP250
NSC 141540 see EAV500
NSC 141549 see ADL500, ADL750
NSC 141633 see HGI575
NSC 141634 see IKS500
NSC-143019 see MJL750
NSC 143504 see CNH625
NSC-143969 see TGB000
NSC-145668 see COW900
NSC-145669 see DLX300
NSC-148958 see FMB000
NSC-150014 see HGW500
NSC-153858 see MBU820
NSC-154020 see TJE870
NSC 154890 see CNR250
NSC-164011 see ROZ000
NSC-165563 see BOL500
NSC 166100 see POC000
NSC-172112 see SLD900
NSC 176319 see QOJ250
NSC-177023 see LFA020
NSC 178248 see CLX000
NSC-180024 see CCK625
NSC 181815 see HIF000
NSC 182986 see ASK875
NSC 190935 see CJJ250
NSC 190945 see DOP200
NSC 190947 see CFC500
NSC 190948 see MDX250
NSC 190949 see MLX000
NSC 190955 see DTP800
NSC 190956 see DTP600
NSC 190978 see DRR400
NSC 190981 see DVS000
NSC 190986 see DJY200
NSC 190987 see DTQ400
NSC 190997 see MPG250
NSC 190998 see IEN000
NSC 191000 see DRP600
NSC 191001 see MPS250
NSC 191025 see DGA200
NSC 192965 see SLD800
NSC 194814 see DAD040
NSC 195022 see BEP500
NSC 195058 see PHE250
NSC 195087 see IOS000
NSC 195102 see CJJ500
NSC 195106 see FAK000
NSC 195154 see DOL800
NSC 195164 see CLJ875
NSC 217697 see MRU760
NSC-218321 see PBT100
NSC-224131 see PGY750
NSC 226080 see RBK000
NSC 238159 see OJD100
NSC-239717 see CHB000
NSC 239724 see CHA500
NSC-241240 see CCC075
NSC-245382 see ALF500
NSC-246131 see TJX350
NSC 247516 see PCC000
NSC 249992 see ADL750
NSC 251222 see IHO700
NSC-253272 see CBG075
NSC 253947 see CHA750
NSC-256439 see DAN000
NSC 259968 see BMK325
NSC 261036 see NHI000
NSC-261036 see DBA700
NSC-261037 see NHH500
NSC 262666 see BIC325
NSC-267703 see QBS000
NSC-269148 see MCB600
NSC 271674 see CCJ350
NSC 271675 see DEU115
NSC 279836 see MQY090
NSC 281278 see SNS100

NSC-286193 see RJF500
NSC 287513 see BKB300, BKB325
NSC 294895 see CHB750
NSC-294896 see CHF250
NSC-296934 see TBC450
NSC-296961 see AMD000
NSC 298223 see APT375
NSC 299195 see DBE875
NSC 301739 see MQY100
NSC 311056 see SLE875
NSC-322921 see MOD500
NSC-325319 see DHA300
NSC 330500 see MAB400
NSC 337766 see BGW325
NSC-339004 see CMA600
NSC 353451 see MQY110
NSC 403169 see ADR750
NSC-405124 see TIQ750
NSC-407347 see MQB750
NSC 409425 see DRN400
NSC-409962 see BIF750
NSC 521778 see FNO000
NSC-525334 see NDY500
NSC-525816 see ADC250
NSC-526062 see DVP400
NSC 526417 see EAD500
NSC 527017 see AOC500
NSC-528986 see AIV500
NSC 616586 see PGQ500
NSC-1461711 see AFQ575
NSC A15920 see AEA250
NSC B-2992 see MQX25MDV250
NSC D 254157 see CHA000, CLX000
NSC-280594 HYDRATE see TJE875
NSC-762 HYDROCHLORIDE see BIE500
NSC 5366 HYDROCHLORIDE see NOA500
NSD 1055 see BMM600
NSD 256927 see IGG775
NT see NLV500
NTA see SIP500
NTA-194 see CJH750
NTA SODIUM HYDRATE see NEI000
NTD 2 see BHL750
NTG see NGY000
NTL see SBD000
NTL SULFATE see NCP550
NTM see DTR200
NTN-8629 see DGC800
NTOI see NML000
N-TOIN see NGE000
NTPA see NOC400
NTPP see DYF450
19NTPP see DYF450
NU-1196 see NEA500, NEB000
NU-1932 see NOE550
NU 2017 see ABV250
NU 2206 see MKR250
NU-2221 see DDF200
NU 2222 see TKW500
NU-BAIT II see MDU600
NUCES NUCISTAE see NOG000
NUCHAR 722 see CDI000
NUCIDOL see DCM750
NUCIFERIN see NOE500
NUCIFERINE see NOE500
(−)-NUCIFERINE see NOE500
1-NUCIFERINE see NOE500
NUCLEAR FAST RED see AJQ250
NUCLEOCARDYL see AEH750
NUCLON ARGENTINIAN see FNF000
NUCTALON see CKL250
NUDRIN see MDU600
NUEZ de la INDIA (PUERTO RICO) see TOA275
NUEZ NOGAL (PUERTO RICO) see TOA275
NUFENOXOLE see DWF700
NUFLUOR see SHF500
NUGESTORAL see GEK500
NU-1326 HYDROBROMIDE see NOE525
NU 1604 HYDROCHLORIDE see TEQ700
NUISANCE DUSTS and AEROSOLS see NOF000
NUJOL see MQV750
NULANS see CFY500
NU-LAWN WEEDER see DDP000
NULLAPON B see EIV000

NULLAPON BF-78 see EIV000
NULLAPON BF ACID see EIX000
NULLAPON BFC CONC see EIV000
NULOGYL see NHH000
NULSA see BGC625
NUMAL see AFT000
NU MAN see MPN500
NU-MANESE see MAT250
NUMOQUIN see HHR700
NUMOQUIN HYDROCHLORIDE see ELC500
NUMORPHAN HYDROCHLORIDE see ORG100
NUOPLAZ see DXQ200
NUOPLAZ DOP see DVL700
NUPERCAINAL see DDT200
NUPERCAINE see DDT200
NUPERCAINE HYDROCHLORIDE see NOF500
NUPRIN see AIE750
NUREDAL see BET000
NURELLE see TIV750
NUSYN-NOXFISH see PIX250
NUTINAL see BCA000
NUTMEG see NOG000
NUTMEG OIL see NOG500
NUTMEG OIL, EAST INDIAN see NOG500
NU-TONE see NAK500
NUTRASWEET see ARN825
NUTRIFOS STP see SJH200
NUTROP see SBH500
NUTROSE see SFQ000
NUVA see DGP900
NUVACRON see MRH209
NUVANOL see DSQ000
NUVANOL N see IEN000
NUVAPEN see AIV500
NUVELBI V.C.A. see DXO300
NUX MOSCHATA see NOG000
NUX-VOMICA TREE see NOG800
NVP see NBO525
NYACOL see SCH000
NYACOL 830 see SCH000
NYACOL 1430 see SCH000
NYAD 10 see WCJ000
NYAD 325 see WCJ000
NYA G see WCJ000
NYANTHRENE GOLDEN YELLOW see DCZ000
NYAZIN see BET000
NYCOLINE see PJT200
NYCOR 200 see WCJ000
NYCOR 300 see WCJ000
NYCOTON see CDO000
NYCTAL see BNK000
NYDRAN see BEG000
NYDRANE see BEG000
NYDRAZID see ILD000
NY IV-34-1 see TEH250
NYLIDRIN HYDROCHLORIDE see DNU200
NYLMERATE see ABU500
NYLOMINE ACID RED P4B see HJF500
NYLOMINE ACID YELLOW B-RD see SGP500
NYLON see NOH000
NYLON-6 see PJY500
NYLOQUINONE BLUE 2J see TBG700
NYLOQUINONE ORANGE JR see AKP750
NYLOQUINONE PURE BLUE see MGG250
NYLOQUINONE VIOLET R see DBP000
NYLOSAN GREEN F-BL see CMM400
NYMCEL S see SFO500
NYSCAPS see WAK000
NYSTAN see NOH500
NYSTATIN see NOH500
NYSTATINE see NOH500
NYSTAVESCENT see NOH500
NYTAL see TAB750

O 250 see SKV100
O-2857 see BDE500
OAAT see AIC250
OAP see HED500
OBB see OAH000
OBDZ see OMY700
OBELINE PICRATE see ANS500
OBEPAR see DKE800
OBESIN see PNN300
OBESITABS see AOB500

OBIDOXIME CHLORIDE see BGS250
OBIDOXIME DICHLORIDE see BGS250
OBIDOXIME HYDROCHLORIDE see BGS250
OBLEVIL see EQL000
OBLITEROL see EMT500
m-OBLIVON see EQL000
OBLIVON C see MNM500
OBLIVON CARBAMATE see MNM500
OBOB see NLE400
OBPA see OMY850
OBRACINE see TGI500
OBRAMYCIN see TGI250
OBSTON see DJL000
OC 1085 see MQT550
OCDD see OAJ000
OCENOL see OBA000
OCEOL see OBA000
OCHRATOXIN see OAD090
OCHRATOXIN A see CHP250
OCHRATOXIN A SODIUM SALT see OAD100
OCHRE see IHD000
OCI 56 see SGG500
OCIMENE see DTF200
OCIMENOL see DTD400
OCIMUM BASILICUM OIL see BAR250
OCIMUM SANCTUM LINN., LEAF EXTRACT see OAD200
OCNA see CJA200
OCOTEA CYMBARUM OIL see OAF000
OCPA see CEH750
OCPNA see CJA175
OCRITEN see DAQ800
OCTABROMOBIPHENYL see OAH000
ar,ar,ar,ar,ar',ar',ar',ar' OCTABROMO-1,1'-BIPHENYL see OAH000
OCTABROMODIPHENYL see OAH000
OCTACAINE HYDROCHLORIDE see MLO750
OCTACARBONYLDICOBALT see CNB500
1,2,4,5,6,7,8,8-OCTACHLOOR-3a,4,7,7a-TETRAHYDRO-4,7-endo-METHANO-
  INDAAN (DUTCH) see CDR750
OCTACHLOR see CDR750
OCTACHLOROCAMPHENE see CDV100
OCTACHLORODIBENZODIOXIN see OAJ000
OCTACHLORODIBENZO-p-DIOXIN see OAJ000
OCTACHLORODIBENZO(b,e)(1,4)DIOXIN see OAJ000
1,2,3,4,6,7,8,9-OCTACHLORODIBENZODIOXIN see OAJ000
OCTACHLORODIHYDRODICYCLOPENTADIENE see CDR750
OCTACHLORODIPROPYLETHER see OAL000
1,2,4,5,6,7,8,8-OCTACHLORO-2,3,3a,4,7,7a-HEXAHYDRO-4,7-
  METHANOINDENE see CDR750
1,2,4,5,6,7,8,8-OCTACHLORO-2,3,3a,4,7,7a-HEXAHYDRO-4,7-METHANO-1H-
  INDENE see CDR750
OCTACHLORO-HEXAHYDRO-METHANOISOBENZOFURAN see OAN000
1,3,4,5,6,8,8-OCTACHLORO-1,3,3a,4,7,7a-HEXAHYDRO-4,7-METHANO
  ISOBENZOFURAN see OAN000
1,2,4,5,6,7,8,8-OCTACHLORO-3a,4,7,7a-HEXAHYDRO-4,7-METHYLENE
  INDANE see CDR750
OCTACHLORO-4,7-METHANOHYDROINDANE see CDR750
OCTACHLORO-4,7-METHANOTETRAHYDROINDANE see CDR750
1,2,4,5,6,7,8,8-OCTACHLORO-4,7-METHANO-3a,4,7,7a-TETRAHYDROINDANE
  see CDR750
1,3,4,5,6,7,10,10-OCTACHLORO-4,7-endo-METHYLENE-4,7,8,9-
  TETRAHYDROPHTHALAN see OAN000
OCTACHLORONAPHTHALENE see OAP000
1,3,4,5,6,7,8,8-OCTACHLORO-2-OXA-3a,4,7,7a-TETRAHYDRO-4,7-
  METHANOINDENE see OAN000
1,2,4,5,6,7,8,8-OCTACHLORO-3a,4,7,7a-TETRAHYDRO-4,7-METHANOINDAN
  see CDR750
2,2,4,5,6,7,8,8-OCTACHLORO-3a,4,7,7a-TETRAHYDRO-4,7-METHANOINDAN
  see CDR575
1,2,4,5,6,7,8,8-OCTACHLORO-3a,4,7,7a-TETRAHYDRO-4,7-METHANOINDANE
  see CDR750
1,2,4,5,6,7,10,10-OCTACHLORO-4,7,8,9-TETRAHYDRO-4,7-METHYLENE-
  INDANE see CDR750
1,2,4,5,6,7,8,8-OCTACHLOR-3a,4,7,7a-TETRAHYDRO-4,7-endo-METHANO-
  INDAN (GERMAN) see CDR750
OCTACIDE 264 see OES000
9,12-OCTADECADIENOIC ACID see LGG000
cis-9,cis-12-OCTADECADIENOIC ACID see LGG000
cis,cis-9,12-OCTADECADIENOIC ACID see LGG000
OCTADECANENITRILE see SLL500
OCTADECANNAMIDE see OAR000
OCTADECANOIC ACID see SLK000
OCTADECANOIC ACID, BARIUM CADMIUM SALT (4:1:1) (9CI) see BAI800
OCTADECANOIC ACID, BARIUM SALT (9CI) see BAO825
OCTADECANOIC ACID, CADMIUM SALT see OAT000

OCTADECANOIC ACID, METHYL ESTER see MJW000
OCTADECANOIC ACID, MONOESTER with 1,2-PROPANEDIOL see SLL000
OCTADECANOIC ACID, MONOESTER with 1,2,3-PROPANETRIOL
   see OAV000
OCTADECANOIC ACID, SODIUM SALT see SJV500
OCTADECANOIC ACID, TETRADECYL ESTER (9CI) see TCB100
OCTADECANOIC ACID, ZINC SALT see ZMS000
OCTADECANOL see OAX000
1-OCTADECANOL see OAX000
n-OCTADECANOL see OAX000
OCTADECANONITRILE see SLL500
5,7,11,13-OCTADECATETRAYNE-1,18-DIOL see OAX050
9,12,15-OCTADECATRIENEPEROXOIC ACID (Z,Z,Z) (9CI) see PCN000
9,12,15-OCTADECATRIENOIC ACID see OAX100
9,10-OCTADECENOIC ACID see OHU000
cis-OCTADEC-9-ENOIC ACID see OHU000, OHU000
trans-OCTADEC-9-ENOIC ACID see EAF500
trans-9-OCTADECENOIC ACID see EAF500
cis-Δ⁹-OCTADECENOIC ACID see OHU000
trans-Δ⁹-OCTADECENOIC ACID see EAF500
(Z)-9-OCTADECENOIC ACID BUTYL ESTER see BSB000
(Z)-9-OCTADECENOIC ACID METHYL ESTER see OHW000
(Z)-9-OCTADECENOIC ACID, TIN (2+) SALT see OAZ000
(Z)-9-OCTADECENOIC ACID mixed with (Z,Z)-9,12-OCTADECADIENOIC
   ACID see LGI000
(Z)-9-OCTADECEN-1-OL see OBA000
cis-9-OCTADECEN-1-OL see OBA000
cis-9-OCTADECENYLAMINE see OHM700
(Z)-α-9-OCTADECENYL-omega-HYDROXYPOLY(OCY-1,2-ETHANEDIYL)
   see OIG000
(Z)-α-9-OCTADECENYL-omega-HYDROXYPOLY(OXY-1,2-ETHANEDIYL)
   see OIG040
α-9-OCTADECENYL-omega-HYDROXYPOLY(OXY-1,2-ETHANEDIYL, (Z)
   see PJW500
α-9-OCTADECENYL-ω-HYDROXYPOLY(OXY(METHYL-1,2-ETHANEDIYL))
   see UHA000
OCTA DECYL ALCOHOL see OAX000
n-OCTADECYL ALCOHOL see OAX000
OCTADECYLAMINE see OBC000
N-OCTADECYLAMINE see OBC000
OCTADECYLAMINE HYDROCHLORIDE see OBE000
N-OCTADECYL-N-BENZYL-N,N-DIMETHYLAMMONIUMCHLORIDE
   see DTC600
OCTADECYLDIMETHYLBENZYLAMMONIUM CHLORIDE see DTC600
OCTADECYL ISOCYANATE see OBG000
OCTADECYLTRICHLOROSILANE see OBI000
OCTADECYLTRIMETHYLAMMONIUM CHLORIDE see TLW500
1,2,3,3a,6,7,12b,12c-OCTADEHYDRO-2-HYDROXYLYCORANIUM
   see LJB800
1,7-OCTADIENE see OBK000
1,6-OCTADIENE, 3,7-DIMETHYL- see CMT050
2,6-OCTADIENE, 1-ETHOXY-3,7-DIMETHYL- see GDG100
2,6-OCTADIENOIC ACID, 3,7-DIMETHYL-, ISOPENTYL ESTER, (E)-
   see IHS100
OCTAFLUOROADIPAMIDE see OBK100
OCTAFLUOROADIPONITRILE see PCG600
OCTAFLUOROAMYL ALCOHOL see OBS800
OCTAFLUOROBUTENE-2 see OBO000
OCTAFLUORO-sec-BUTENE see OBM000
OCTAFLUOROBUT-2-ENE (DOT) see OBO000
1,1,1,2,3,4,4,4-OCTAFLUORO-2-BUTENE see OBO000
OCTAFLUOROCYCLOBUTANE (DOT) see CPS000
OCTAFLUOROISOBUTENE see OBM000
OCTAFLUOROISOBUTYLENE see OBM000
OCTAFLUOROPENTANOL see OBS800
OCTAFLUORO-1-PENTANOL see OBS800
2,2,3,3,4,4,5,5-OCTAFLUORO-1-PENTANOL see OBU000
OCTAFLUOROPENTYL ALCOHOL see OBS800
OCTAFLUOROTOLUENE see PCH500
OCTAHYDROAZOCINE see SMV500
OCTAHYDROAZOCINE HYDROCHLORIDE mixed with SODIUM NITRITE
   (1:1) see SIT000
(2-(OCTAHYDRO-1-AZOCINYL)ETHYL)GUANIDINE SULFATE see GKS000
2-(OCTAHYDRO-1-AZOCINYL)ETHYL GUANIDINE SULPHATE
   see GKU000
((OCTAHYDRO-2-AZOCINYL)METHYL)GUANIDINE see GKO800
((OCTAHYDRO-2-AZOCINYL)METHYL)GUANIDINE SULFATE HYDRATE
   see CAY875
OCTAHYDRO-1-BENZOPYRAN-2-ONE see OBV200
OCTAHYDRO-2H-1-BENZOPYRAN-2-ONE (9CI) see OBV200
OCTAHYDRO-2H-BISOXIRENO(a,f)INDENE see BFY750
OCTAHYDROCOUMARIN see OBV200
OCTAHYDRODIBENZ(a,h)ANTHRACENE see OBW000
OCTAHYDRO-1:2:5:6-DIBENZANTHRACENE see OBW000

6,7,9,10,17,18,20,21-OCTAHYDRODIBENZO(b,k)(1,4,7,10,13,16)HEXAOXY-
   CYCLOOCTADECIN see COD575
(1S,cis)1,2,3,4,5,6,7,8-OCTAHYDRO-1,4-DIMETHYL-7-(1-
   METHYLETHYLIDENE)-AZULENE see GKO000
1,2,3,4,5,6,7,8-OCTAHYDRO-1,4-DIMETHYL-7-(1-
   METHYLETHYLIDENE)AZULENEMONOEP OXIDE see EBU100
(2S-(1(4*(4*)),2-α,3-α-β,7-α-β))-OCTAHYDRO-1-(2-((1-(ETHOXYCARBONYL)-3-
   PHENYLPROPYL)AMINO)-1-OXYPROPYL)-1H-INDOLE-2-CARBOXYLIC
   ACID MONOHYDROCHLORIDE see ICL500
1,2,3,4,6,7,12A-OCTAHYDRO-2-(1-(p-FLUOROPHENYL)-1-OXO-4-BUTYL)-
   PYRAZINO(2,1:6,1)PYRIDO(3,4-B)INDOLE see CCW500
2,5,5a,6,9,10,10a,1a-OCTAHYDRO-4-HYDROXYMETHYL-1,1,7,9-
   TETRAMETHYL-5,5a-6-TRIHYDROXY-1H-2,8a-
   METHANOCYCLOPENTA(a)CYCLOPROPA(e)CYCLODECEN-11-ONE-5-
   HEXADECANOATE see IDB000
1,2,3,3a,4,5,6,8a-OCTAHYDRO-2-ISOPROPYLIDENE-6-AZULENOL-4,8-
   DIMETHYL ACETATE see AAW750
OCTAHYDRO-1,6-METHANONAPHTHALEN-1(2H)-AMINE see AKD875
1,2,3,4,6,7,7a,11c-OCTAHYDRO-9-METHOXY-2-METHYL-BENZOFURO(4,3,2-
   efg)(2)BENZAZOCIN-6-OL HBr see GBA000
1,2,6,7,8,9,10,12b-OCTAHYDRO-3-METHYLBENZ(j)ACEANTHRYLENE
   see HDR500
OCTAHYDRO-1-NITROSOAZOCINE see OBY000
OCTAHYDRO-1-NITROSO-1H-AZONINE see OCA000
OCTAHYDRO-3,6,8,8-TETRAMETHYL-1H-3a,7-METHANOAZULEN-6-OL ACE-
   TATE see CCR250
OCTAHYDRO-3,6,9-TRIMETHYL-3,12-EPOXY-12H-PYRANO(4,3-j)-1,2-
   BENZODIOXEPIN-10(3H)-ONE see ARL375
OCTA-KLOR see CDR750
Γ-OCTALACTONE see OCE000
OCTALENE see AFK250
OCTALOX see DHB400
OCTAMETHYL-DIFOSFORZUUR-TETRAMIDE (DUTCH) see OCM000
OCTAMETHYLDIPHOSPHORAMIDE see OCM000
OCTAMETHYL-DIPHOSPHORSAEURE-TETRAMID (GERMAN)
   see OCM000
5,5'-(OCTAMETHYLENEBIS(CARBONYLIMINO))BIS(N-METHYL-2,4,6-
   TRIIODOISOPHTHALAMIC ACID) see OCI000
N,N'-OCTAMETHYLENEBIS(2,2-DICHLOROACTAMIDE) see BIX250
OCTAMETHYLPYROPHOSPHORAMIDE see OCM000
OCTAMETHYL PYROPHOSPHORTETRAMIDE see OCM000
OCTAMETHYL TETRAMIDO PYROPHOSPHATE see OCM000
1-OCTANAL see OCO000
OCTANALDEHYDE see OCO000
OCTANAL, 2-METHYL- see MNC175
1-OCTANAMINE see OEK000
OCTAN AMYLU (POLISH) see AOD725
OCTAN ANTIMONITY (CZECH) see AQJ750
OCTAN BARNATY (CZECH) see BAH500
OCTAN n-BUTYLU (POLISH) see BPU750
OCTAN CYKLOHEXYLAMINU (CZECH) see CPG000
OCTANE see OCU000
1-OCTANECARBOXYLIC ACID see NMY000
1,8-OCTANEDIAMINE, DIHYDROCHLORIDE see OCW000, OCW000
1,8-OCTANEDICARBOXYLIC ACID see SBJ500
OCTANE-1-NNO-AZOXYMETHANE see MGT000
OCTAN ETOKSYETYLU (POLISH) see EES400
OCTAN ETYLU (POLISH) see EFR000
OCTAN FENYLRTUTNATY (CZECH) see ABU500
OCTANIL see ILK000
OCTAN KOBALTNATY (CZECH) see CNA500
OCTAN MANGANATY (CZECH) see MAQ000
OCTAN MEDNATY (CZECH) see CNI250
OCTAN METYLU (POLISH) see MFW100
OCTANOIC ACID see OCY000
OCTANOIC ACID ALLYL ESTER see AGM500
OCTANOIC ACID, CADMIUM SALT (2:1) see CAD750
OCTANOIC ACID, 2,2-DIMETHYL-, HEXADECYL ESTER see HCP600
OCTANOIC ACID, ETHYL ESTER see ENY000
OCTANOIC ACID, ISOPENTYL ESTER see IHP500
OCTANOIC ACID, METHYL ESTER mixed with METHYL DECANOATE
   see MND000
OCTANOIC ACID, POTASSIUM SALT see PLN250
OCTANOIC ACID, 1,2,3-PROPANETRIYL ESTER see TMO000
OCTANOIC ACID-2-PROPENYL ESTER see AGM500
OCTANOIC ACID, p-TOLYL ESTER see THB000
OCTANOIC ACID TRIGLYCERIDE see TMO000
OCTANOIC ACID, ZINC SALT (2:1) see ZEJ000
OCTANOIC/DECANOIC ACID TRIGLYCERIDE see CBF710
OCTANOL see OEI000
3-OCTANOL see OCY100, OCY100
n-OCTANOL see OEI000
1-OCTANOL (FCC) see OEI000
OCTANOL, mixed isomers see ODE000

1-OCTANOL ACETATE see OEG000
OCTANOLIC ACID (mixed isomers) see ODE200
OCTANOLIDE-1,4 see OCE000
3-OCTANOL, 3-METHYL- see MND050
2-OCTANONE see ODG000
3-OCTANONE see ODI000
2-OCTANOYL-1,2,3,4-TETRAHYDROISOQUINOLINE see ODO000
OCTAN PROPYLU (POLISH) see PNC250
OCTAN WINYLU (POLISH) see VLU250
n-OCTANYL ACETATE see OEG000
OCTAN ZINECNATY (CZECH) see ZCA000
OCTATENSIN see GKQ000
OCTATENZINE see GKQ000
1,3,7-OCTATRIEN-5-YNE see ODQ300
OCTATROPINE METHYLBROMIDE see LJS000
2-OCTENE, 8-ETHOXY-2,6-DIMETHYL- see EES100
1-OCTEN-3-OL see ODW000
7-OCTEN-2-OL, 2,6-DIMETHYL-8-(1H-INDOL-1-YL)- see ICS100
1-OCTEN-3-OL, 3-METHYL- see MND100
OCTHILINONE see OFE000
OCTIC ACID see OCY000
OCTILIN see OEI000
OCTIN see ILK000
OCTIN HYDROCHLORIDE see ODY000
OCTINUM see ILK000
OCTOCLOTHEPIN see ODY100
OCTOCLOTHEPINE see ODY100
OCTODRINE see ILM000
OCTOGEN see CQH250
n-OCTOIC ACID see OCY000
OCTOIL see DVL700
OCTOIL S see BJS250
OCTON see ILK000
OCTOPAMINE see AKT250
m-OCTOPAMINE see AKT000
dl-m-OCTOPAMINE HYDROCHLORIDE see NNT100
OCTOTIAMINE see GEK200
OCTOWY ALDEHYD (POLISH) see AAG250
OCTOWY BEZWODNIK (POLISH) see AAX500
OCTOWY KWAS (POLISH) see AAT250
OCTOXINOL see PKF500
OCTOXYNOL see PKF500
OCTOXYNOL 3 see PKF500
OCTOXYNOL 9 see PKF500
OCTYL ACETATE see OEE000, OEG000
1-OCTYL ACETATE see OEG000
3-OCTYL ACETATE see OEG100
n-OCTYL ACETATE see OEG000
β-OCTYL ACROLEIN see DXU280
OCTYL ACRYLATE see ADU250
OCTYL ADIPATE see AEO000
OCTYL ALCOHOL see OEI000
OCTYL ALCOHOL ACETATE see OEG000
OCTYL ALCOHOL, NORMAL-PRIMARY see OEI000
n-OCTYL ALDEHYDE see OCO000
N-OCTYLAMINE see OEK000
1-OCTYLAMINE, N-METHYL-N-NITROSO- see NKU590
N-n-OCTYLATROPINE BROMIDE see OEM000
N-n-OCTYL-ATROPINIUMBROMID (GERMAN) see OEM000
8-OCTYLATROPINIUM BROMIDE see OEM000
N-OCTYLATROPINIUM BROMIDE see OEM000
N-n-OCTYLATROPINIUM BROMIDE see OEM000
p-OCTYLBENZOIC ACID see OEQ000
4-n-OCTYLBENZOIC ACID see OEQ000
N-OCTYL BICYCLOHEPTENE DICARBOXIMIDE see OES000
N-OCTYLBICYCLO-(2.2.1)-5-HEPTENE-2,3-DICARBOXIMIDE see OES000
n-OCTYL BROMIDE see BNU000
OCTYL CARBINOL see NNB500
OCTYL DECYL PHTHALATE see OEU000
N-OCTYL-N-DECYL PHTHALATE see OEU000
OCTYL-DIMETHYL-p-AMINOBENZOIC ACID see AOI500
OCTYL-DIMETHYL-BENZYLAMMONIUM CHLORIDE see OEW000
OCTYLENE EPOXIDE see ECE000
OCTYLENE GLYCOL see EKV000
OCTYL EPOXYTALLATE see FAB920
sec-OCTYL EPOXYTALLATE see FAB900
OCTYL ETHER see OEY000
OCTYL FORMATE see OEY100
n-OCTYL FORMATE see OEY100
OCTYL GALLATE see OFA000
N-OCTYL-o-HYDROXYBENZOATE see SAJ500
Γ-OCTYL-β-HYDROXY-Δ^{α},β-BUTENOLID (GERMAN) see HND000
n-OCTYLIC ACID see OCY000
n-OCTYLISOSAFROLE SULFOXIDE see ISA000

2-OCTYL-4-ISOTHIAZOLIN-3-ONE see OFE000
2-OCTYL-3(2H)-ISOTHIAZOLONE see OFE000
tert-OCTYL MERCAPTAN see MKJ250
n-OCTYL NITROSUREA see NLD800
OCTYL-OCTADECYL DIMETHYL ETHYLBENZYL AMMONIUM CHLORIDES
    see BBA500
cis-3-OCTYL-OXIRANEOCTANOIC ACID see ECD500
4'-(OCTYLOXY)-3-PIPERIDINOPROPIOPHENONE HYDROCHLORIDE
    see OFG000
4-OCTYLOXY-β-(1-PIPERIDYL)PROPIOPHENONE HYDROCHLORIDE
    see OFG000
OCTYL PALMITATE see OFE100
OCTYLPEROXIDE see OFI000
OCTYL PHENOL see OFK000
p-tert-OCTYLPHENOL see TDN500
OCTYL PHENOL CONDENSED with 12-13 MOLES ETHYLENE OXIDE
    see PKF500
OCTYLPHENOL EO (3) see OFM000
OCTYLPHENOL EO (10) see OFQ000
OCTYL PHENOL EO (16) see OFS000
OCTYL PHENOL EO (20) see OFU000
OCTYL PHENOL condensed with 3 MOLES ETHYLENE OXIDE see OFM000
OCTYL PHENOL condensed with 5 MOLES ETHYLENE OXIDE see OFO000
OCTYL PHENOL condensed with 16 MOLES ETHYLENE OXIDE see OFS000
OCTYL PHENOL condensed with 20 MOLES ETHYLENE OXIDE see OFU000
OCTYL PHENOL condensed with 8-10 MOLES ETHYLENE OXIDE see OFQ000
p-tert-OCTYLPHENOXYETHOXYETHYLDIMETHYLBENZYL AMMONIUM
    CHLORIDE see BEN000
OCTYLPHENOXYPOLY(ETHOXYETHANOL) see GHS000
tert-OCTYLPHENOXYPOLY(ETHOXYETHANOL) see GHS000
p-tert-OCTYLPHENOXYPOLYETHOXYETHANOL see PKF500
OCTYLPHENOXYPOLY(ETHYLENEOXY)ETHANOL see GHS000
tert-OCTYLPHENOXY POLY(OXYETHYLENE)ETHANOL see GHS000
OCTYL PHTHALATE see DVL600, DVL700
n-OCTYL PHTHALATE see DVL600
4'-OCTYL-3-PIPERIDINOPROPIOPHENONE HYDROCHLORIDE see PIY000
N-OCTYLPYRROLIDINONE see OFU100
1-OCTYL-2-PYRROLIDINONE see OFU100
N-OCTYLPYRROLIDONE see OFU100
OCTYL SALICYLATE see SAJ500
OCTYL SEBACATE see BJS250
2-(OCTYLTHIO)ETHANOL see OGA000
OCTYLTRICHLOROSILANE see OGE000
OCTYLTRICHLOROSTANNANE see OGG000
OCTYLTRIS(2-ETHYLHEXYLOXYCARBONYLMETHYLTHIO)STANNANE
    see OGI000
OCTYNECARBOXYLIC ACID, METHYL ESTER see MNC000
OCUSOL see SPC500
OCU-VINC see VLF000
ODA see SJN700
ODANON see AER666
ODB see DEP600, EDP000
ODCB see DEP600
ODISTON see DCK000
ODODIBORANE see OGI100
OE3 see EDU500
OE 7 see EDJ500
OEDROPHANIUM see TAL490
OEKOLP see DKA600
OENANTHALDEHYDE see HBB500
OENANTHE CROCATA see WAT315
OENANTHIC ACID see HBE000
OENANTHOL see HBB500
OENANTHOTOXIN see EAO500
OENANTHYLIC ACID see HBE000
OENETHYL HYDROCHLORIDE see NCL300
OESTERGON see EDO000
OESTRADIOL see EDO000
d-OESTRADIOL see EDO000
α-OESTRADIOL see EDO000
β-OESTRADIOL see EDO000
OESTRADIOL-17-α see EDO500
OESTRADIOL-17-β see EDO000
cis-OESTRADIOL see EDO000
3,17-β-OESTRADIOL see EDO000
d-3,17-β-OESTRADIOL see EDO000
OESTRADIOL BENZOATE see EDP000
OESTRADIOL-3-BENZOATE see EDP000
β-OESTRADIOL BENZOATE see EDP000
β-OESTRADIOL-3-BENZOATE see EDP000
17-β-OESTRADIOL-3-BENZOATE see EDP000
OESTRADIOL DIPROPIONATE see EDR000
β-OESTRADIOL DIPROPIONATE see EDR000
17-β-OESTRADIOL DIPROPIONATE see EDR000

OESTRADIOL-3,17-DIPROPIONATE see EDR000
3,17-β-OESTRADIOL DIPROPIONATE see EDR000
OESTRADIOL 3-METHYL ETHER see MFB775
OESTRADIOL MONOBENZOATE see EDP000
OESTRADIOL MUSTARD see EDR500
OESTRADIOL PHOSPHATE POLYMER see EDS000
OESTRADIOL POLYESTER with PHOSPHORIC ACID see EDS000
OESTRADIOL R see EDO000
OESTRAFORM (BDH) see EDP000
OESTRASID see DAL600
OESTRA-1,3,5(10)-TRIENE-3,17-β-DIOL see EDO000
17-β-OESTRA-1,3,5(10)-TRIENE-3,17-DIOL see EDO000
1,3,5(10)-OESTRATRIENE-3,17-β-DIOL 3-BENZOATE see EDP000
OESTRA-1,3,5(10)-TRIENE-3,16-α,17-β-TRIOL see EDU500
1,3,5-OESTRATRIENE-3-β-3,16-α,17-β-TRIOL see EDU500
(16-α,17-β)-OESTRA-1,3,5(10)-TRIENE-3,16,17-TRIOL see EDU500
1,3,5-OESTRATRIEN-3-OL-17-ONE see EDV000
1,3,5(10)-OESTRATRIEN-3-OL-17-ONE see EDV000
Δ-1,3,5-OESTRATRIEN-3-β-OL-17-ONE see EDV000
OESTRATRIOL see EDU500
OESTRENOLON see NNX400
OESTRILIN see ECU750
OESTRIN see EDV000
OESTRIOL see EDU500
16-α,17-β-OESTRIOL see EDU500
3,16-α,17-β-OESTRIOL see EDU500
OESTRODIENE see DAL600
OESTRODIENOL see DAL600
OESTRO-FEMINAL see ECU750
OESTROFORM see EDV000
OESTROGENINE see DKA600
OESTROGLANDOL see EDO000
OESTROGYNAEDRON see DKB000
OESTROGYNAL see EDO000
OESTROL VETAG see DKA600
OESTROMENIN see DKA600
OESTROMENSIL see DKA600
OESTROMENSYL see DKA600
OESTROMIENIN see DKA600
OESTROMON see DKA600
OESTRONBENZOAT (GERMAN) see EDV500
OESTRONE see EDV000
OESTRONE-3-SULPHATE SODIUM SALT see EDV600
OESTROPAK MORNING see ECU750
OESTROPEROS see EDV000
OESTRORAL see DAL600
OF see PMI250
OFDZ see OMY500
OFF see DKC800
OFFITRIL see HNI500
OFF-SHOOT-O see MND000
OFHC Cu see CNI000
OFLOXACIN see OGI300
OFNACK see POP000
OFNA-PERL SALT RRA see CLK235
OFTALENT see CDP250
OFTALFRINE see SPC500
OFTANOL see IMF300
OFUNACK see POP000
OG 1 see SEH000
OGEEN 515 see OAV000
OGEEN GRB see OAV000
OGEEN M see OAV000
OGINEX see FBP100
OGOSTAL see CBF675
OGYLINE see NNR125
3-OHAA see AKE750
'OHAI-ALI'I (HAWAII) see CAK325
'OHAI (HAWAII) see SBC550
'OHAI-KE'OKE'O (HAWAII) see SBC550
'OHAI-'ULA'ULA (HAWAII) see SBC550
OHB see SAH000
OH-BBN see HJQ350
1-α-OH-CC see HJV000
cis-4-OH-CCNU see CHA500
trans-2-OH-CCNU see CHA750
trans-4-OH CCNU see CHB000
4-OH-CP see HKA200
1-α-OH-D³ see HJV000
4-OH-E2 see HKH850
4-OH-ESTRADIOL see HKH850
17-β-OH-ESTRADIOL see EDO000
OHIO 347 see EAT900
7-OHM-MBA see HMF000

7-OHM-12-MBA see HMF000
17-β-OH-OESTRADIOL see EDO000
OHRIC see DGF000
1-α-OH VITAMIN D3 see HJV000
OIL of ABIES ALBA see AAC250
OIL de ACORUS CALAMUS (SPANISH) see OGL020
OIL of ACORUS CALAMUS Linn. see OGL020
OIL of AMERICAN WORMSEED see CDL500
OIL of ANISE see AOU250
OIL of ANISEED see PMQ750
OIL of ARBOR VITAE see CCQ500
OIL of ARGEMONE mixed with OIL of MUSTARD see OGS000
OIL, ARTEMISIA see ARL250
OIL of BASIL see BAR250
OIL of BAY see BAT500
OIL of BERGAMOT, coldpressed see BFO000
OIL of BERGAMOT, rectified see BFO000
OIL, BITTER ALMOND see BLV500
OIL BLUE see CNQ000
OIL of CALAMUS, GERMAN see OGK000
OIL of CALAMUS, SPANISH see OGL020
OIL of CAMPHOR RECTIFIED see CBB500
OIL CAMPHOR SASSAFRASSY see CBB500
OIL of CAMPHOR WHITE see CBB500
OIL of CARAWAY see CBG500
OIL of CARDAMON see CCJ625
OIL of CASSIA see CCO750
OIL CEDAR see CCR000
OIL of CEDAR LEAF see CCQ500
OIL of CHENOPODIUM see CDL500
OIL of CHINESE CINNAMON see CCO750
OIL of CINNAMON see CCO750
OIL of CINNAMON, CEYLON see CCO750
OIL of CITRONELLA see CMT000
OIL of CLOVE see CMY075
OIL of CORIANDER see CNR735
OIL-DRI see AHF500
OIL of EUCALYPTUS see EQQ000
OIL of FENNEL see FAP000
OIL of FUR see AAC250
OIL GARLIC see AGS250
OIL GAS see OGM000
OIL of GERANIUM see GDA000
OIL GERANIUM REUNION see GDC000
OIL of GRAPEFRUIT see GJU000
OIL GREEN see CMJ900
OIL of HARTSHORN see BMA750
OIL, HIBAWOOD see HGA100
OIL of JUNIPER BERRY see JEA000
OIL of LABDANUM see LAC000
OIL of LAUREL LEAF see LBK000
OIL of LAVANDIN, ABRIAL TYPE see LCA000
OIL of LAVENDER see LCD000
OIL of LEMON see LEI000
OIL of LEMON, desert type, coldpressed see LEI025
OIL of LEMON, distilled see LEI030
OIL of LEMONGRASS, EAST INDIAN see LEG000
OIL of LEMONGRASS, WEST INDIAN see LEH000
OIL of LIME, distilled see OGO000
OIL of MACE see OGQ100
OIL of MARJORAM, SPANISH see MBU500
OIL of MIRBANE (DOT) see NEX000
OIL MIST, MINERAL (OSHA, ACGIH) see MQV750
OIL of MOUNTAIN PINE see PIH400
OIL of MUSTARD, artificial see AGJ250
OIL of MUSTARD, EXPRESSED mixed with OIL of ARGEMONE see OGS000
OIL of MYRBANE see NEX000
OIL of MYRCIA see BAT500
OIL of MYRISTICA see NOG500
OIL of MYRTLE see OGU000
OIL of NIOBE see MHA750
OIL of NUTMEG see NOG500
OIL of NUTMEG, expressed see OGQ100
OIL of ONION see OJD200
OIL of ORANGE see OGY000
OIL ORANGE see PEJ500
OIL ORANGE 2R see XRA000
OIL ORANGE KB see XRA000
OIL ORANGE N EXTRA see XRA000
OIL ORANGE O'PEL see TGW000
OIL ORANGE R see XRA000
OIL ORANGE SS see TGW000
OIL ORANGE X see XRA000
OIL ORANGE XO see XRA000

OIL of ORIGANUM see OJO000
OIL of PALMA CHRISTI see CCP250
OIL of PALMAROSA see PAE000
OIL of PARSLEY see PAL750
OIL of PELARGONIUM see GDA000
OIL of PIMENTA LEAF see PIG730
OIL of PINE see PIH750
OIL PINK see CMS242
OIL RED see OHA000
OIL RED GRO see XRA000
OIL RED O see XRA000
OIL RED RO see XRA000
OIL RED XO see XRA000, FAG080
OIL of ROSE BULGARIAN see RNF000
OIL ROSE GERANIUM see GDC000
OIL of ROSE GERANIUM see GDA000
OIL ROSE GERANIUM ALGERIAN see GDA000
OIL of ROSE MOROCCAN see RNK000
OIL ROSE TURKISH see OHC000
OIL of RUE see OHE000
OIL of SANDALWOOD, EAST INDIAN see OHG000
OILS, ANGELICA ROOT see AOO760
OIL of SASSAFRAS see OHI000
OIL of SASSAFRAS BRAZILIAN see OAF000
OIL SCARLET see OHA000, XRA000
OIL SCARLET 6G see XRA000
OIL SCARLET 371 see XRA000
OIL SCARLET APYO see XRA000
OIL SCARLET BL see XRA000
OIL SCARLET L see XRA000
OIL SCARLET YS see XRA000
OILS, CEDAR LEAF see CCQ500
OILS, CINNAMON see CCO750
OILS, CINTONELLA see CMT000
OILS, CLOVE see CMY075
OILS, CLOVE LEAF see CMY100
OILS, CORIANDER see CNR735
OILS, CUMIN see COF325
OIL of SHADDOCK see GJU000
OIL-SHALE PYROLYSE LAC LSP-1 see LIJ000
OIL of SILVER FIR see AAC250
OIL of SILVER PINE see AAC250
OILS, LIME see OGO000
OILS, NIM see NBS300
OIL SOLUBLE ANILINE YELLOW see PEI000
OILS, PALM see PAE500
OIL of SPEARMINT see SKY000
OIL of SPIKE LAVENDER see SLB500
OILS, SWEET BIRCH see SOY100
OIL of SWEET FLAG see OGK000
OIL of SWEET ORANGE see OGY000
OILS, WHEAT GERM see WBJ700
OIL THUJA see CCQ500
OIL of THUJA see CCQ500
OIL of THYME see TFX500
OIL of TURPENTINE see TOD750
OIL of TURPENTINE, rectified see TOD750
OIL of VETIVER see VJU000
OIL VIOLET see EOJ500
OIL VIOLET IRS see HOK000
OIL VIOLET R see DBP000
OIL of VITRIOL (DOT) see SOI500
OIL of WHITE CEDAR see CCQ500
OIL of WINTERGREEN see MPI000
OIL YELLOW see AIC250, DOT300
OIL YELLOW 21 see AIC250
OIL YELLOW 2R see AIC250
OIL YELLOW 2635 see OHI875
OIL YELLOW 2681 see AIC250
OIL YELLOW A see AIC250, FAG130
OIL YELLOW AAB see PEI000
OIL YELLOW AT see AIC250
OIL YELLOW C see AIC250
OIL YELLOW DE see OHI875
OIL YELLOW DEA see OHI875
OIL YELLOW E190 see OHI875
OIL YELLOW ENC see OHI875
OIL YELLOW GA see OHI875
OIL YELLOW HA see OHK000
OIL YELLOW I see AIC250
OIL YELLOW NB see OHI875
OIL YELLOW OB see FAG135
OIL YELLOW OPS see OHK000
OIL YELLOW T see AIC250

OJO de CANGREJO (PUERTO RICO) see RMK250
OK 7 see PJY100
OK 622 see PAJ000
OKASA-MASCUL see TBG000
OKO see DGP900
OK PRE-GEL see SLJ500
OKSAZIL see MSC100
OKTADECYLAMIN (CZECH) see OBC000
omega-H-OKTAFLUORPENTANOL-1 (GERMAN) see OBS800
OKTAN (POLISH) see OCU000
OKTANEN (DUTCH) see OCU000
OKTATERR see CDR750
OKTOGEN see CQH250
p-terc.OKTYLFENOL (CZECH) see TDN500
OKULTIN M see CIR250
OKY-046 SODIUM see SHV100
OL 27-400 see CQH100
OLAMINE see EEC600
OLCADIL see CMY125
OLD 01 see ADW200
OLDHAMITE see CAY000
OLDREN see ZAK300
OLEAL ORANGE R see PEJ500
OLEAL ORANGE SS see TGW000
OLEAL YELLOW 2G see DOT300
OLEAL YELLOW RE see OHK000
OLEAMINE see OHM700
'OLEANA (HAWAII) see OHM875
OLEANDER see OHM875
OLEANDOMYCIN HYDROCHLORIDE see OHO000
OLEANDOMYCIN MONOHYDROCHLORIDE see OHO000
OLEANDOMYCIN PHOSPHATE see OHO200
OLEANDRIN see OHQ000
OLEANDRINE see OHQ000
OLEANOGLYCOTOXIN A see OHQ300
OLEANOGLYCOTOXIN B see LEF800
OLEATE of MERCURY see MDF250
OLEFIANT GAS see EIO000
OLEFINS see OHS000
α-OLEFIN SULFONATE see AFN500
α-OLEFIN SULPHONATE see AFN500
OLEIC ACID see OHU000
OLEIC ACID GLYCIDYL ESTER see ECJ000
OLEIC ACID LEAD SALT see LDQ000
OLEIC ACID, LEAD(2+) SALT (2:1) see LDQ000
OLEIC ACID mixed with LINOLEIC ACID see LGI000
OLEIC ACID, 2-METHOXYETHYL ESTER see MEP750
cis-OLEIC ACID, METHYL ESTER see OHW000
OLEIC ACID POLY(OXYETHYLENE) ESTER see PJY100
OLEIC ACID, POTASSIUM SALT see OHY000
OLEIC ACID, SODIUM SALT see OIA000
OLEIC ACID, ZINC SALT see ZJS000
OLEINAMINE see OHM700
OLEOAKARITHION see TNP250
OLEOCUIVRE see CNO000
OLEOFAC see IOT000
OLEOFOS 20 see PAK000
OLEOGESAPRIM see ARQ725
OLEOL see OBA000
OLEOMYCETIN see CDP250
OLEO NORDOX see CNO000
OLEOPARAPHENE see PAK000
OLEOPARATHION see PAK000
OLEOPHOSPHOTHION see MAK700
OLEORESIN TUMERIC see TOD625
OLEOSUMIFENE see DSQ000
OLEOVITAMIN A see VSK600
OLEOVITAMIN D see VSZ100
OLEOVITAMIN D3 see CMC750
OLEOVOFOTOX see MNH000
OLEOX 5 see PJY100
1-OLEOYLAZIRIDINE see OIC000
OLEOYLETHYLENEIMINE see OIC000
OLEOYLMETHYLTAURINE SODIUM SALT see SIY000
OLEPAL I see PJY100
OLEPAL III see PJY100
OLEPTAN see FMU000
OLETAC 100 see PMP500
OLETETRIN see TBX000
OLEUM (DOT) see SOI520
OLEUM ABIETIS see PIH750
OLEUM SINAPIS VOLATILE see AGJ250
OLEUM TIGLII see COC250
OLEYL ALCOHOL see OBA000

OLEYL ALCOHOL EO (2) see OIG000
OLEYL ALCOHOL EO (10) see OIG040
OLEYL ALCOHOL EO (20) see PJW500
OLEYL ALCOHOL condensed with 2 moles ETHYLENE OXIDE see OIG000
OLEYL ALCOHOL condensed with 10 moles ETHYLENE OXIDE see OIG040
OLEYL ALCOHOL condensed with 20 MOLES ETHYLENE OXIDE see PJW500
OLEYLAMIN (GERMAN) see OHM700
OLEYL AMINE see OHM700
OLEYLAMINE-HF see OII200
OLEYLAMINE HYDROFLUORIDE see OII200
OLEYLAMINHYDROFLUORID (GERMAN) see OII200
OLEYLPOLYOXETHYLENE GLYCOL ETHER see OIK000
OLIBANUM GUM see OIM000
OLIBANUM OIL see OIM025
OLIGOMYCIN A, mixed with OLIGOMYCIN B see OIS000
OLIGOMYCIN B, mixed with OLIGOMYCIN A see OIS000
OLIMPEN see MCH525
OLIN see DXN300
'OLINANA (HAWAII) see OHM875
OLIN MATHIESON 2,424 see EFK000
OLIN MO. 2174 see AQO000
OLIO DI CROTON (ITALIAN) see COC250
OLITENSOL see BBW750
OLITREF see DUV600
OLIVE AR, ANTHRIMIDE see IBJ000
OLIVE OIL see OIQ000
OLIVE R BASE see IBJ000
OLIVOMITSIN see OIS000, OIU499
OLIVOMYCIN see OIS000, OIU499
OLIVOMYCIN A see OIU000, OIU499
OLIVOMYCIN D see OIU499
OLIVOMYCINE see OIU499
OLIVOMYCIN 1 see OIU000
OLIVOMYCIN, SODIUM SALT see CMK750
'OLIWA (HAWAII) see OHM875
OLOSED see MNM500
OLOTHORB see PKL100
OLOW (POLISH) see LCF000
OLPISAN see PAX000
OLVASAN see BIM250
OM 518 see MFD500
OM-1455 see LIN400
OM-1463 see TIP750
OM-1563 see ZMJ200
OM 2424 see EFK000
OMACIDE 24 see HOC000
OMADINE see HOB500
OMADINE MDS see OIU850
OMADINE ZINC see ZMJ000
OMAHA see LCF000
OMAHA & GRANT see LCF000
OMAINE see MIW500
OMAIT see SOP000
OMAL see TIW000
OMCA see TJW500
OMCHLOR see DFE200
OMEGA 127 see MKP500
OMEGA CHROME BLUE FB see HJF500
7-OMEN see MCB600
OMETHOAT see DNX800
OMETHOATE see DNX800
OMF 59 see PHA750
OM-HYDANTOINE see DKQ000
OM-HYDANTOINE SODIUM see DNU000
OMIFIN see CMX700
OMITE see SOP000
OMNIBON see SNN300
OMNI-PASSIN see DJT800
OMNIPEN see AIV500
OMNIPEN-N see SEQ000
OMNIZOLE see TEX000
OMNYL see QAK000
OMP-1 see OIW000
OMP 2 see OIY000
OMP-4 see OJA000
OMP-5 see OJC000
OMPA see OCM000
OMPACIDE see OCM000
OMPATOX see OCM000
OMPAX see OCM000
OMPERAN see EPD500
OMS 2 see FAQ900
OMS 14 see DGP900
OMS-15 see COF250

OMS-33 see PMY300
OMS 43 see DSQ000
OMS-47 see DOS000
OMS-93 see DST000
OMS 115 see DGD800
OMS 174 see CGI500
OMS 570 see EAQ750
OMS-597 see TMD000
OMS-658 see BNL250
OMS-659 see EGV500
OMS-708 see BDG250
OMS-771 see CBM500
OMS-0971 see CMA100
OMS 1075 see DRR400
OMS-1092 see IOV000
OMS-1155 see CMA250
OMS-1206 see BEP500
OMS-1211 see IEN000
OMS 1325 see FAB400
OMS 1328 see CDS750
OMS 1342 see CLJ875
OMS 1804 see CJV250
OMSAT see TKX000
OMT see SIY000
OMTAN see OAN000
ONAMYCIN POWDER V see NCE000
ONAONA-IAPANA (HAWAII) see DAC500
ONCB see CJB750
ONCO-CARBIDE see HOO500, TGN250
ONCODAZOLE see OJD100
ONCOSTATIN see AEA500
ONCOSTATIN K see AEB000
ONCOTEPA see TFQ750
ONCOTIOTEPA see TFQ750
ONCOVEDEX see TND000
ONCOVIN see LEY000, LEZ000
ONDENA see DAC200
ONDOGYNE see FBP100
ONDONID see FBP100
ONE-IRON see FBJ100
ONEX see CJT750
ONGROVIL S 165 see PKQ059
ONION see WBS850
ONION OIL see OJD200
ONION TREE see WBS850
ONKOTIN see DBD700
ONO 802 see CDB775
ON OXIDE (CO) see CBW750
ONQUININ see PAF550
ONT see NMO525
ONTOSEIN see OJM400
ONTRACK see MFL250
ONTRACK 8E see MQQ450
ONYX see SCI500, SCJ500
ONYX BTC (ONYX OIL & CHEM CO) see AFP250
ONYXIDE 3300 see BBA625
ONYXIDE (ONYX OIL & CHEM CO) see AFN750
ONYXOL 345 see BKE500
OOS see TAB750
OP 1062 see GHS000
OPACIN see TDE750
OPALON see PKQ059
OPALON 400 see AAX175
OPARENOL see DNG400
OPC 1085 see MQT550
OPC 2009 see PME600
OPC-8212 see DLF700
OPC-13013 see CMP825
OPE-3 see OFM000
OPE 30 see PKF500
OPERIDINE see DAM700
OPERTIL see ECW600
OPHIOBOLIN see CNF109
OPHIOBOLIN A see CNF109
OPHIOPHAGUS HANNAH VENOM see ARV125
OPHTALMOKALIKAN see KAV000
OPHTHALAMIN see VSK600
OPHTHALMADINE see DAS000
OPHTHAZIN see OJD300
OPHTHOCILLIN see DVU000
OPIAN see NBP275, NOA000
OPIAN HYDROCHLORIDE see NOA500
OPIANINE see NOA000
OPIANINE HYDROCHLORIDE see NOA500

OPILON HYDROCHLORIDE see TFY000
OPINSUL see AKO500
OPIPRAMOL see DCV800
OPIPRAMOL DIHYDROCHLORIDE see IDF000
OPIUM see OJG000
OPIUM, PYROLYZATE see OJG509
OPLOSSINGEN (DUTCH) see FMV000
OPOBALSAM see BAF000
OPOSIM see ICC000
OPP see BGJ250
OPP-Na see BGJ750
OPRAMIDOL see DCV800
OPREN see OJI750
OPSB see BRP250
OP-SULFA 30 see SNP500
OPTAL see PND000
OPTALIDON see AGI750
OPTEF see CNS750
OP-THAL-ZIN see ZNA000
OPTHEL-S see SNP500
OPTHOCHLOR see CDP250
OPTIMINE see DLV800
OPTIMYCIN see MDO250
OPTINOXAN see QAK000
OPTIPEN see PDD350
OPTIPHYLLIN see TEP000
OPTISULIN LONG see IDF300
OPTOCHIN see HHR700
OPTOCHIN HYDROCHLORIDE see ELC500
OPTOQUINE see HHR700
OPTOQUINHYDROCHLORIDE see ELC500
8-OQ see QPA000
OR 1191 see FAB400, PGX300
OR-1549 see AJP125
OR-1550 see AJL875
OR-1556 see AJN375
OR-1578 see AJO625
ORABET see BSQ000
ORABILEX see EQC000
ORABOLIN see EJT600
ORACAINE HYDROCHLORIDE see OJI600
ORACEF see ALV000
ORACET SAPPHIRE BLUE G see TBG700
ORACET VIOLET 2R see DBP000
ORACILLIN see PDT500
α-ORACILLIN see PDD350
ORACIL-VK see PDT750
ORACON see DNX500
ORACONAL see MCA500
ORADEXON see SOW000
ORADEXON PHOSPHATE see BFW325
ORADIAN see CDR550, CKK000
ORADIL see CLY600
ORAFLEX see OJI750
ORAFURAN see NGE000
ORAGEST see MCA000
ORAGESTON see AGF750
ORAGRAFIN-SODIUM see SKM000
ORALCID see ABX500
ORALIN see BSQ000
ORALITH RED see CJD500
ORALITH RED 2GL see DVB800
ORALITH RED P4R see MMP100
ORALOPEN see PDD350
ORALSONE see HHR000
ORALSTERONE see AOO275
ORA-LUTIN see GEK500
ORAMID see SAH000
ORANGE 3 see MND600
ORANGE #10 see HGC000
1333 ORANGE see FAG010
ORANGE A l'HUILE see PEJ500
ORANGE BASE CIBA II see NEO000
ORANGE BASE IRGA I see NEN500
ORANGE CHROME see LCS000
ORANGE CRYSTALS see ABC500
ORANGE G (biological stain) see HGC000
ORANGE G (indicator) see HGC000
ORANGE GC BASE see CEH675
ORANGE G DYE see HGC000
ORANGE I see FAG010
ORANGE III see MND600
ORANGE INSOLUBLE OLG see PEJ500, XRA000
ORANGE INSOLUBLE RR see XRA000

ORANGE LEAD see LDS000
ORANGE NITRATE CHROME see LCS000
ORANGE No. 203 see DVB800
ORANGE OIL see OGY000
ORANGE OIL, coldpressed (FCC) see OGY000
ORANGE OIL KB see XRA000
ORANGE PEL see PEJ500
ORANGE PIGMENT X see DVB800
ORANGE RESENOLE No. 3 see PEJ500
ORANGE 3R SOLUBLE IN GREASE see TGW000
ORANGES see BBK500
ORANGE SOLUBLE A l'HUILE see PEJ500
ORANIL see BSM000
ORANIXON see GGS000
ORANYL see BSM000
ORANZ BRILANTNI OSTAZINOVA S-2R (CZECH) see DGO000
ORANZ G (POLISH) see HGC000
ORANZ ZASADITA 2 (CZECH) see DBP999
ORAP see PIH000
ORAPEN see PDT750
ORARSAN see ABX500
ORASECRON see NNL500
ORASONE see PLZ000
ORASPOR see CCS530
ORASTHIN see ORU500
ORASULIN see BSM000
ORATESTIN see AOO275
ORA-TESTRYL see AOO275
ORATRAST see BAP000
ORATREN see PDT500
ORATROL see DEQ200
ORAVIRON see MPN500
ORAVUE see IGA000
ORBENIN SODIUM see SLJ050
ORBENIN SODIUM HYDRATE see SLJ000
ORBICIN see DCQ800, PAG050
ORBINAMON see NBP500
ORBON see OAV000
ORCANON see MPW500
ORCHARD BRAND ZIRAM see BJK500
ORCHIOL see TBG000
ORCHISTIN see TBG000
ORCIN see MPH500
ORCINOL see MPH500
ORCIPRENALINE see DMV800
ORCIPRENALINE SULFATE see MDM800
ORDIMEL see ABB000
ORDINARY AZOXYBENZENE see ASO750
ORDINARY LACTIC ACID see LAG000
ORDRAM see EKO500
OREGON HOLLY see HGF100
OREMET see TGF250
ORESOL see RLU000
ORESON see RLU000
ORESTOL see DKB000
ORETIC see CFY000
ORETON see TBG000
ORETON-F see TBF500
ORETON-M see MPN500
ORETON METHYL see MPN500
ORETON PROPIONATE see TBG000
OREZAN see BSQ000
ORF 2166 see CBO625
ORF 3858 see FAO200
ORF 4563 see MBV775
ORF 5513 see BJG150
ORF 8511 see DWF300
ORF 13811 see OJK300
ORG-483 see EJT600
ORG 2969 see DBA750
ORG 485-50 see NNV000
ORGA-414 see AMY050
ORGABOLIN see EJT600
ORGABORAL see EJT600
ORGAMETIL see NNV000
ORGAMETRIL see NNV000
ORGAMETROL see NNV000
ORGAMIDE see PJY500
ORGANEX see CAK500
ORGANIC GLASS E 2 see PKB500
ORGANIL 644 see MAP300
ORGANOL BORDEAUX B see EOJ500
ORGANOL BROWN 2R see NBG500
ORGANOL FAST GREEN J see BLK000

ORGANOL ORANGE see PEJ500
ORGANOL ORANGE 2R see TGW000
ORGANOL SCARLET see OHA000
ORGANOL YELLOW see PEI000
ORGANOL YELLOW 25 see AIC250
ORGANOL YELLOW ADM see DOT300
ORGANOMETALS see OJM000
ORGANON see AGF750
ORGANON'S DOCA ACETATE see DAQ800
ORGASEPTINE see SNM500
ORGASTERON see MDM350
ORGASTYPTIN see EDU500
ORG GB 94 see BMA625
ORG NA 97 see PAF625
ORG NC 45 see VGU075
ORGOTEIN see OJM400
ORGOTEINS see OJM400
ORICUR see CHX250
ORIDIN see OJV500
ORIENTAL BERRY see PIE500
ORIENT BASIC MAGENTA see MAC250
ORIENT OIL ORANGE PS see PEJ500
ORIENT OIL YELLOW GG see DOT300
ORIENTOMYCIN see CQH000
ORIGANUM OIL see OJO000
ORIMETEN see AKC600
ORIMON see PDP250
ORINASE see BSQ000
ORINAZ see BSQ000
ORION BLUE 3B see CMO250
ORIPAVINE, 6,14-endo-ETHYLENETETRAHYDRO-7-(1-HYDROXY-1-
  METHYLBUTYL)-(7CI) see EQO450
ORISUL see AIF000
ORISULF see AIF000
ORLON, combustion products see ADX750
ORLUTATE see ABU000
ORMETEIN see OJM400
ORNAMENTAL WEED see AJM000
ORNID see BMV750
ORNIDAZOLE see OJS000
dl-ORNITHINE, 2-(DIFLUOROMETHYL)-, MONOHYDROCHLORIDE
  see EAE775
l-ORNITHINE HYDROCHLORIDE see OJU000
l-ORNITHINE MONOHYDROCHLORIDE see OJU000
ORNITROL see ARX800
OROBETINA see DXO300
OROLEVOL see MQU750
ORONOL see ART250
OROPUR see OJV500
OROSOMYCIN see TFC500
OROTIC ACID see OJV500
OROTIC ACID mixed with CHOLESTEROL and CHOLIC ACID (2:2:1)
  see OJV525
OROTONIN see OJV500
OROTSAURE (GERMAN) see OJV500
OROTURIC see OJV500
OROTYL see OJV500
OROVERMOL see TDX750
OROXIN see ALV000
OROXINE see LEQ300
ORPHENADINE see MJH900
ORPHENADRIN see MJH900
ORPHENADRINE see MJH900
ORPHENADRINE CITRATE see DPH000
ORPHENADRINE HYDROCHLORIDE see OJW000
ORPHENOL see BGJ750
ORPIMENT see ARI000
ORQUISTERONE see TBF500
ORSILE see BEQ625
ORSIN see PEY500
ORTAL SODIUM see EKT500
ORTEDRINE see BBK000
ORTHAMINE see PEY250
ORTHENE see DOP600
ORTHENE-755 see DOP600
ORTHESIN see EFX000
ORTHO 4355 see NAG400
ORTHO-5353 see BTA250
ORTHO-5655 see MOU750
ORTHO 5865 see CBF800
ORTHO 9006 see DTQ400
ORTHO 12420 see DOP600
l'ORTHO, p-AMINONITROPHENOL, SEL SODIQUE (FRENCH) see ALO500
ORTHOARSENIC ACID see ARB250

ORTHOARSENIC ACID HEMIHYDRATE see ARC500
ORTHOBENZOIC ACID, cyclic 7,8,10a-ESTER with 5,6-EPOXY-
  4,5,6,6a,7,8,9,10,10a,10b-DECAHYDRO-3a,4,7,8,10a-PENTAHYDROXY-5-
  (HYDROXYMETHYL)-8-ISOPROPENYL-2,10-DIMETHYLBENZ(e)AZULEN-
  3(3aH)-ONE see DAB850
ORTHOBORIC ACID see BMC000
ORTHOCAINE see AKF000
ORTHO C-1 DEFOLIANT & WEED KILLER see SFS000
ORTHOCHLOROPARANISIDINE see CEH750
ORTHOCHROME ORANGE R see NEY000
ORTHOCIDE see CBG000
ORTHOCRESOL see CNX000
ORTHODERM see AKF000
ORTHODIBROM see NAG400
ORTHODIBROMO see NAG400
ORTHODICHLOROBENZENE see DEP600
ORTHODICHLOROBENZOL see DEP600
ORTHO N-4 DUST see NDN000
ORTHO N-5 DUST see NDN000
ORTHO EARWIG BAIT see DXE000
ORTHOFORM see AKF000
ORTHOFORMIC ACID, ETHYL ESTER see ENY500
ORTHOFORMIC ACID, TRIETHYL ESTER see ENY500
ORTHOFORMIC ACID, TRIMETHYL ESTER see TLX600
ORTHO GRASS KILLER see CBM000
ORTHOHYDROXYBENZOIC ACID see SAI000
ORTHOHYDROXYDIPHENYL see BGJ250
ORTHOIODIN see HGB200
ORTHO-KLOR see CDR750
ORTHO L10 DUST see LCK000
ORTHO L40 DUST see LCK000
ORTHO-LM APPLE SPRAY see MLH000
ORTHO LM CONCENTRATE see MLH000
ORTHO LM SEED PROTECTANT see MLH000
ORTHO MALATHION see MAK700
ORTHO MC see MAE000
ORTHO-MITE see SOP500
ORTHOMRAVENCAN ETHYLNATY (CZECH) see ENY500
ORTHOMRAVENCAN METHYLNATY (CZECH) see TLX600
ORTHONAL see QAK000
ORTHONITROANILINE (DOT) see NEO000
ORTHO-NOVUM see MDL750
ORTHO P-G BAIT see COF500
ORTHOPHALTAN see TIT250
ORTHOPHENANTHROLINE see PCY250
ORTHOPHENYLALANINE MUSTARD see BHT250
ORTHOPHENYLPHENOL see BGJ250
ORTHOPHOS see PAK000
ORTHO PHOSPHATE DEFOLIANT see BSH250
ORTHOPHOSPHORIC ACID see PHB250
ORTHOSAN MB see DTC600
ORTHOSIL see SJU000
ORTHOSILICATE see SCN500
ORTHOTELLURIC ACID see TAI750
ORTHOTOLUIC ACID see TGQ000
ORTHO-TOLUOL-SULFONAMID (GERMAN) see TGN250
ORTHOTRAN see CJT750
ORTHOVANILLINE see VFP000
ORTHO WEEVIL BAIT see DXE000
ORTHOXENOL see BGJ250
ORTHOXINE HYDROCHLORIDE see OJY000
ORTIN see TJL250
ORTISPORINA see ALV000
ORTIZON see HIB500
ORTODRINEX HYDROCHLORIDE see OJY000
ORTOL see GGS000
ORTONAL see QAK000
ORTRAN see DOP600
ORTRIL see DOP600
ORTUDUR see PKQ059
ORUDIS see BDU500
ORUVAIL see BDU500
ORVAGIL see MMN250
ORVINYLCARBINOL see AFV500
ORVUS WA PASTE see SIB600
ORYZAEMATE see PMD800
ORYZANIN see TES750
ORYZANINE see TES750
OS 1836 see CLV375
OS 1897 see DDL800
OS 2046 see MQR750
OS-3966-A see RMK200
OSACYL see AMM250
OSAGE ORANGE see MRN500

OSAGE ORANGE CRYSTALS see MRN500
OSAGE ORANGE EXTRACT see MRN500
OSALMID see DYE700
OSALMIDE see DYE700
OSARSAL see ABX500
OSARSOLE see ABX500
OSBAC see MOV000
OSCINE see SBG000
OSCOPHEN see ARP250
O. SCUTELLATUS (AUSTRALIA) VENOM see ARV500
OSDMP see DJR700
OSIREN see AFJ500
OSMIC ACID see OKK000
OSMITROL see HER000
OSMIUM see OKE000
OSMIUM HEXAFLUORIDE see OKG000
OSMIUM(IV) OXIDE see OKI000
OSMIUM(VIII) OXIDE see OKK000
OSMIUM TETROXIDE see OKK000
OSMOFERRIN see IHK000
OSMOSOL EXTRA see PND000
OSNERVAN see CPQ250
OSOCIDE see CBG000
OSPEN see PDT500
OSPENEFF see PDT750
OS-40 (PHOSPHATE ESTER) see OKK500
OSSALIN see SHF500
OSSIAMINA see CFA750
OSSIAN see OOG000
OSSICHLORIN see CFA750
OSSIDO di MESITILE (ITALIAN) see MDJ750
OSSIN see SHF500
OSTAMER see CDV625
OSTANTHREN GREY M see CMU475
OSTANTHREN REDDISH BROWN 5RF see CMU800
OSTAZIN BLACK H-N see CMS227
OSTAZIN BRILLIANT RED S 5B see PMF540
OSTELIN see VSZ100
OSTENOL see TLG000
OSTENSIN see TLG000
OSTREOGRYCIN see VRF000
OSTREOGRYCIN B see VRA700
OSVARSAN see ABX500
OSYRITRIN see RSU000
OSYROL see AFJ500
OT see HLC000
OTACRIL see DFP600
OTAHEITE WALNUT (VIRGIN ISLANDS) see TOA275
OTAMOL see OKO200
OTB see ONI000
OTBE (FRENCH) see BLL750
OTC see HOH500
OTERBEN see BSQ000
OTETRYN see HOI000
OTIFURIL see FP1000
OTILONIUM BROMIDE see OKO400
OTOBIOTIC see NCG000
OTODYNE see EQQ500
OTOFURAN see NGE500
OTOKALIXIN see KAV000
OTOPHEN see CDP250
OTRACID see CJT750
OTRIVINE HYDROCHLORIDE see OKO500
OTRIVIN HYDROCHLORIDE see OKO500
OTS see TGN250
OTS 11 see DVM800
OTTAFACT see CFE250
OTTANI (ITALIAN) see OCU000
OTTASEPT see CLW000
OTTASEPT EXTRA see CLW000
OTTILONIO BROMURO (ITALIAN) see OKO400
1,2,4,5,6,7,8,8-OTTOCHLORO-3A,4,7,7A-TETRAIDRO-4,7-endo-METANO-
INDANO (ITALIAN) see CDR750
OTTOMETIL-PIROFOSFORAMMIDE (ITALIAN) see OCM000
OTTO of ROSE see RNF000
OTTO ROSE TURKISH see OHC000
OUABAGENIN-I-RHAMNOSID (GERMAN) see OKS000
OUABAGENIN-I-RHAMNOSIDE see OKS000
OUABAIN see OKS000
OUABAINE see OKS000
OUBAIN see OKS000
OUDENONE see OKS100
OUDENONE SODIUM SALT see OKS150
OURARI see COF750

OUTFLANK see AHJ750
OUTFLANK-STOCKADE see AHJ750
OUTFOX see CQI750
OV₁) see HNK600
OVABAN see VTF000
OVADOFOS see DSQ000
OVADZIAK see BBQ500
OVAHORMON see EDO000
OVAHORMON BENZOATE see EDP000
OVAMIN 30 see DAP810
OVANON see LJE000
OVARELIN see LIU360
OVARIOSTAT (FRENCH) see LJE000
OVASTEROL see EDO000
OVASTEROL-B see EDP000
OVASTEVOL see EDO000
OVATRAN see CJT750
OVEST see ECU750
OVESTERIN see EDU500
OVESTIN see EDU500
OVESTINON see EDU500
OVESTRION see EDU500
OVETTEN see MNM500
OVEX see CJT750, EDP000, EDV000
OVIDON see NNL500
OVIFOLLIN see EDV000
OVIN see DNX500
OVINE PITUIRARY INTERSTITIAL CELL STIMULATING HORMONE see
LIU300
OVISOT see ABO000, CMF250
OVITELMIN see MHL000
OVOCHLOR see CJT750
OVOCICLINA see EDO000
OVOCYCLIN see EDO000
OVOCYCLIN BENZOATE see EDP000
OVOCYCLIN DIPROPIONATE see EDR000
OVOCYCLINE see EDO000
OVOCYCLIN M see EDP000
OVOCYCLIN-MB see EDP000
OVOCYCLIN-P see EDR000
OVOCYLIN see EDO000
OVOSTAT see EEH575
OVOSTAT 1375 see EEH575
OVOSTAT E see EEH575
OVOTOX see CJT750
OVOTRAN see CJT750
OVRAL see NNL500
OVRAL 21 see NNL500
OVRAL 28 see NNL500
OVRAN see NNL500
OVRANETT see NNL500
O-V STATIN see NOH500
OVULEN see EQK100
OVULEN 30 see DAP810
OVULEN 50 see EQJ500
OWISPOL GF see SMQ500
OX see CFZ000
OXAALZUUR (DUTCH) see OLA000
10H-9-OXAANTHRACENE see XAT000
5-OXA-1-AZABICYCLO(4.2.0)OCT-2-ENE-2-CARBOXYLIC ACID, 7-(2-((DIF-
LUOROMETHYL)THIO)ACETAMIDO)-3-(((1-(2-HYDROXYETHYL)-1H-
TETRAZOL-5-YL)THIO)METHYL)-7-METHOXY-8-OXO-, SODIUM SALT,
(6R,7R)- see FCN100
1-OXA-4-AZACYCLOHEXANE see MRP750
1-OXA-3-AZAINDENE see BDI500
OXABEL see MNV250
7-OXABICYCLO(4.1.0)HEPTANE see CPD000
7-OXABICYCLO(2.2.1)HEPTANE-2,3-DICARBOXYLIC ACID see EAR000
OXACILLIN see DSQ800
OXACILLIN-AMPICILLIN MIXTURE see AOC875
OXACILLIN SODIUM SALT see MNV250
OXACYCLOHEPTADECAN-2-ONE see OKU000
OXACYCLOPENTADIENE see FPK000
OXACYCLOPENTANE see TCR750
OXACYCLOPROPANE see EJN500
4H-1,3,5-OXADIAZINE-4-THIONE, TETRAHYDRO-3,5-DIMETHYL- see TCQ275
(1,2,3)OXADIAZOLO(5,4-d)PYRIMIDIN-5(4H)-ONE see DCQ600
8-OXA-3,5-DITHIA-4-STANNAHEPTADECANOIC ACID, 4,4-DIBUTYL-7-OXO-,
NONYLESTER (9CI) see DEH650
OXAF see BHA750
OXAFLOZANE see IRP000
OXAFLOZANO (SPANISH) see IRP000
OXAFLUMAZINE DISUCCINATE see OKW000
OXAFLUMINE see OKW000

OXAFURADENE see NDY000
3-OXA-1-HEPTANOL see BPJ850
11-OXAHEXADECANOLIDE see OKW100
12-OXAHEXADECANOLIDE see OKW110
OXAINE see DTL200
8-OXA-9-KETOTRICYCLO(5.3.1.0^{2,6})UNDECANE see OOK000
OXAL see GIK000
OXALALDEHYDE see GIK000
OXALAMIDE see OLO000
OXALATES see OKY000
OXALDIN see OGI300
OXALIC ACID see OLA000
OXALIC ACID DIAMIDE see OLO000
OXALIC ACID, DIETHYL ESTER see DJT200
OXALIC ACID DINITRILE see COO000
OXALIC ACID, DISODIUM SALT see SIY500
OXALIC ACID, MONOPOTASSIUM SALT see OLE000
OXALIC ACID, TIN(2+) SALT (1:1) (8CI) see TGE250
OXALID see HNI500
OXALONITRILE see COO000
OXALSAEURE (GERMAN) see OLA000
(OXALYBIS(IMINOETHYLENE)BIS((o-CHLOROBENZYL)DIETHYLAM-
    MONIUM) DICHLORIDE see MSC100
2,2'-OXALYDIFURAN OXIME see FQB000
OXALYL-o-AMINOAZOTOLUENE see OLG000
4'-OXALYLAMINO-2,3'-DIMETHYLAZOBENZENE see OLG000
OXALYL CHLORIDE see OLI000
OXALYL CYANIDE see COO000
OXALYL FLUORIDE see OLK000
OXALYSINE see OLK200
l-4-OXALYSINE see OLK200
7-OXA-8-MERCURABICYCLO(4.2.0)OCTA-1,3,5-TRIENE, 5-METHYL-2-NITRO-
    see NHK900
1,4-OXAMERCURANE see OLM000
OXAMETACIN see OLM300
OXAMETACINE see OLM300
OXAMETHACIN see OLM300
d-OXAMICINA (ITALIAN) see CQH000
OXAMID (CZECH) see OLO000
OXAMIDE see OLO000
OXAMIMIDIC ACID see OLO000
OXAMIZIL see MSC100
OXAMMONIUM see HLM500
OXAMMONIUM HYDROCHLORIDE see HLN000
OXAMMONIUM SULFATE see OLS000
OXAMNIQUINE see OLT000
d-OXAMYCIN see CQH000
OXAMYL see DSP600
OXAN 600 see ARQ750
OXANAL YELLOW T see FAG140
OXANAMIDE see OLU000
OXANDROLONE see AOO125
OXANE see EJN500
OXANILIC ACID see OLW000
OXANTHRENE see DDA800
3-OXAPENTANE-1,5-DIOL see DJD600
3-OXA-1,5-PENTANEDIOL see DJD600
OXAPHENAMID see DYE700
OXAPHENAMIDE see DYE700
OXAPIUM IODIDE see OLW400
OXAPRIM see TKX000
OXAPRO see OLW600
OXAPROPANIUM IODIDE see FMX000
OXAPROZIN see OLW600
OXARSANILIC ACID see PDO250
OXASULFA see AIE750
6-OXA-3-THIABICYCLO(3.1.0)HEXANE-3,3-DIOXIDE see ECP000
OXATHIANE see OLY000
1,4-OXATHIANE see OLY000
p-OXATHIANE-4,4-DIOXIDE see DVR000
1,4-OXATHIANE compound with MERCURIC CHLORIDE
    see OMA000
1,3,2-OXATHIASTANNOLANE, 2,2-DIBUTYL- see DEF150
1,2-OXATHIETANE-2,2-DIOXIDE see OMC000
1,2-OXATHIOLANE-2,2-DIOXIDE see PML400
OXATIMIDE see OMG000
OXATOMIDA see OMG000
OXATOMIDE see OMG000
5-OXATRICYCLO(8.2.0.0^{4,6})DODECANE, 4,12,12-TRIMETHYL-9-METHYL-
    ENE-, (1R,4R,6R,10S)- see CCN100
3-OXATRICYCLO(3.2.1.0^{2,4})OCTANE-6-CARBOXYLIC ACID, ETHYL ESTER
    see ENZ000
2-H-1,3,2-OXAZAPHOSPHORINANE see CQC650
OXAZEPAM see CFZ000

OXAZIL see MSC100
OXAZIMEDRINE see PMA750
OXAZINOMYCIN see OMK000
OXAZOCILLIN see DSQ800
OXAZOLAM see OMK300
OXAZOLAZEPAM see OMK300
OXAZOLIDIN see HNI500
OXAZOLIDINE A see DTG750
2-OXAZOLIDINETHIONE, 5-ETHENYL-, (R)-(9CI)
    see VQA100
2-OXAZOLIDINETHIONE, 5-VINYL-, (R)- see VQA100
OXAZOLIDIN-GEIGY see HNI500
OXAZOLIDINONE see MFD500
2-OXAZOLIDINONE see OMM000
OXAZOLIDONE see OMM000
OXAZYL see MSC100
OX BILE EXTRACT see SFW000
OXCORD see AEC750
OXEDRINE see HLV500
m-OXEDRINE see NCL500
(−)-m-OXEDRINE see NCL500
m-OXEDRINE see SPC500
p-OXEDRINE see HLV500
OXEDRINE TARTRATE see SPD000
OXELADIN see DHQ200
OXELADIN CITRATE see OMS400
OXELADINE CITRATE see OMS400
OXENDOLONE see ELF100
2-OXEPANONE (8CI, 9CI) see LAP000
OXEPINAC see OMU000
OXEPINACO (SPANISH) see OMU000
OXEPIN-3,6-ENDOPEROXIDE see EBQ550
OXETACAINE see DTL200
OXETANE see OMW000
OXETHACAINA (ITALIAN) see DTL200
OXETHAZINE see DTL200
OXFENDAZOLE see OMY500
OXIAMIN see IDE000
OXIARSOLAN see ARL000
OXIBENDAZOLE see OMY700
OXIBUTININA HYDROCHLORIDE see OPK000
OXIBUTOL see HNI500
OXIDASE GLUCOSE see GFG100
OXIDATION BASE 22 see ALL750
OXIDATION BASE 25 see NEM480
OXIDATION BASE 12A see DBO400
OXIDATION BASE 10A see PEY650
OXIDE of CHROMIUM see CMJ900
N^{10}-OXIDE-1-NITRO-9-(3-DIMETHYLAMINOPROPYLAMINO)-DIHYDRO-
    CHLORIDE ACRIDINE see CAB125
3-N-OXIDE PURIN-6-THIOL MONOHYDRATE see OMY800
1-OXIDE-4-QUINOLINAMINE see AML500
10,10'-OXIDIPHENOXARSINE see OMY850
OXIDIZED l-CYSTEINE see CQK325
OXIDIZED LINOLEATE see LGH000
OXIDIZED LINOLEIC ACID see LGH000
OXIDOETHANE see EJN500
α,β-OXIDOETHANE see EJN500
1,8-OXIDO-p-MENTHANE see CAL000
OXIDOPAMINE see HKF875
OXI-FENIBUTOL see HNI500
OXIFENON see ORQ000
OXIFENYLBUTAZON see HNI500
OXIFLAVIL see ELH600
OXIKON see DLX400
OXILAPINE see DCS200
OXILORPHAN see CQF079
α-OXIME BENZOIN see BCP500
OXIME COPPER see BLC250
OXIMES see OMY899
OXIMETHOLONUM see PAN100
OXIMETOLONA see PAN100
2-OXIMIMINOBUTANE see EMU500
1-OXINDENE see BCK250
OXINE see QPA000
OXINE COPPER see BLC250
OXINE CUIVRE see BLC250
OXINE SULFATE see QPS000
OXINOFEN see OPK300
2-OXI-PROPYL-PROPYLNITROSAMIN (GERMAN)
    see ORS000
OXIRAAN (DUTCH) see EJN500
OXIRANE see EJN500
OXIRANE-CARBOXALDEHYDE see GGW000

OXIRANE CARBOXALDEHYDE OXIME see ECG100
OXIRANECARBOXYLIC ACID, 3-(((3-METHYL-1-(((3-METHYLBUTYL)
AMINO)CARBONYL)BUTYL)AMINO) CARBONYL)-, ETHYL ESTER,
(2S-(2-α-3-β(R*))- see OMY925
OXIRANE, 2-CHLORO-3-METHYL-, cis-(9CI) see CKS099
OXIRANE, 2-CHLORO-3-METHYL-, trans-(9CI) see CKS100
OXIRANE, 2-(3,5-DICHLOROPHENYL)-2-(2,2,2-TRICHLOROETHYL)- see TJK100
OXIRANEMETHANOL see GGW500
OXIRANEMETHANOL NITRATE see ECI600
OXIRANE ((METHYLPHENOXY)METHYL) (9CI) see TGZ100
OXIRANE, 2,2'-(2,5,8,11-TETRAOXADODECANE-1,12-DIYL)BIS- (9CI) see TJQ333
7-OXIRANYLBENZ(a)ANTHRACENE see ONC000
6-OXIRANYLBENZO(a)PYRENE see ONE000
OXIRANYLMETHANOL see GGW500
OXIRANYLMETHYL ESTER of OCTADECANOIC ACID see SLK500
OXIRANYLMETHYL ESTER of 9-OCTADECENOIC ACID see ECJ000
2-(OXIRANYLMETHYL)-1H-ISOINDOLE-1,3(2H)-DIONE see ECK000
N-(OXIRANYLMETHYL)-N-PHENYL-OXIRANEMETHANAMINE (9CI)
see DKM120
3-OXIRANYL-7-OXABICYCLO(4.1.0)HEPTENE see VOA000
1-OXIRANYLPYRENE see ONG000
OXISURAN see MPL500
OXITETRACYCLIN see HOH500
OXITOL see EES350
OXITOSONA-50 see PAN100
OXITROPIUM BROMIDE see ONI000
OXLOPAR see HOI000
OXO see TAB750
3'-(3-OXO-7-α-ACETYLTHIO-17-β-HYDROXYANDROST-4-EN-17-β-YL)
PROPIONIC ACID LACTONE see AFJ500
7-OXOBENZ(de)ANTHRACENE see BBI250
5-OXO-5H-BENZO(E)ISOCHROMENO(4,3-b)INDOLE see DLY800
2-OXO-1,2-BENZOPYRAN see CNV000
Γ-OXO(1,1-BIPHENYL)-4-BUTANOIC ACID see BGL250
OXOBIS(SULFATO(2-)-O)-ZIRCONATE(2-), DISODIUM (9CI) see ZTS100
OXOBOI see OOG000
2-OXOBORNANE see CBA750
3-OXO-BUTANOIC ACID BUTYL ESTER see BPV250
3-OXOBUTANOIC ACID ETHYL ESTER see EFS000
3-OXOBUTANOIC ACID METHYL ESTER see MFX250
3-OXO-BUTANOIC ACID 2-((1-OXO-2-PROPENYL)OXY)ETHYL ESTER (9CI)
see AAY750
1-(1-OXOBUTYL)AZIRIDINE see BSY000
4-(3-OXOBUTYL)-1,2-DIPHENYL-3,5-PYRAZOLIDINEDIONE see KGK000
Γ-OXO-α-BUTYLENE see BOY500
p-(3-OXOBUTYL)PHENYL ACETATE see AAR500
trans-pi-OXOCAMPHOR see VSF400
19-OXO-CARDOGENEN-(20α22)-TRIOL(3-β,5,14) (GERMAN) see SMM500
4-OXO-2-(β-CHLOROETHYL)-2,3-DIHYDROBENZO-1,3-OXAZINE see CDS250
7-OXOCHOLESTEROL see ONO000
2-OXOCHROMAN see HHR500
18-OXOCORTICOSTERONE see AFJ875
N-(2-OXO-3,5,7-CYCLOHEPTATRIEN-1-YL)AMINOOXOACETIC ACID ETHYL
ESTER see ONQ000
((4-OXO-2,5-CYCLOHEXADIEN-1-YLIDENE)AMINO)GUANIDINE
THIOSEMICARBAZONE see AHI875
N-(6-OXO-6H-DIBENZO(b,d)PYRAN-1-YL)ACETAMIDE see BCH250
5-OXO-5,13-DIHYDROBENZO(E)(2)BENZOPYRANO(4,3-b)INDOLE see DLY800
OXODIOCTYLSTANNANE see DVL400
OXODIPEROXODIPYRIDINECHROMIUM(VI) see ONU000
OXODIPEROXOPYRIDINE CHROMIUM-N-OXIDE see ONU100
α-OXODIPHENYLMETHANE see BCS250
OXODIPHENYLSTANNANE, POLYMER see DWO600
OXODISILANE see ONW000
OXODOLIN see CLY600
9-OXO-2-FLUORENYLACETAMIDE see ABY250
N-(9-OXO-2-FLUORENYL)ACETAMIDE see ABY250
3-OXO-l-GULOFURANOLACTONE see ARN000
4-OXOHEPTANEDIOIC ACID, ISONICOTINOYL HYDRAZONE see ILG000
(17-β)17-((1-OXOHEPTYL)OXY)-ANDROST-4-EN-3-ONE see TBF750
2-OXOHEXAMETHYLENIMINE see CBF700
1-(5-OXOHEXYL)-3,7-DIMETHYLXANTHINE see PBU100
17-((1-OXOHEXYL)OXY)PREGN-4-ENE-3,20-DIONE see HNT500
1-(5-OXOHEXYL)THEOBROMINE see PBU100
(E)-7-OXO-3-β-HYDROXY-14-α-METHYL-8-β-PODOCARPANE-Δ13-α-ACETIC
ACID-2-(DIMETHYLAMINO)ETHYL ESTER HYDROCHLORIDE see CCO675
4-OXO-4H-IMIDAZO(4,5-D)-v-TRIAZINE see ARY500
2-(3-OXO-1-INDANYLIDENE)-1,3-INDANDIONE see ONY000
2-OXO-3-ISOBUTYL-9,10-DIMETHOXY-1,3,4,6,7,11-β-HEXAHYDRO-2H-
BENZOQUIN OLIZINE see TBJ275
8-OXO-8H-ISOCHROMENO(4',3':4,5)PYRROLO(2,3-f)QUINOLINE see OOA000
p-(1-OXO-2-ISOINDOLINYL)-HYDRATROPIC ACID see IDA400
2-(p-(1-OXO-2-ISOINDOLINYL)PHENYL)-PROPIONIC ACID see IDA400
α-(4-(1-OXO-2-ISO-INDOLINYL)-PHENYL)-PROPIONIC ACID see IDA400

OXOLAMINA CLORIDRATO (ITALIAN) see OOE100
OXOLAMINE see OOC000
OXOLAMINE CITRATE see OOE000
OXOLAMINE HYDROCHLORIDE see OOE100
OXOLANE see TCR750
OXOLE see FPK000
OXOLINIC ACID see OOG000
OXOMEMAZINE see AFL750
OXOMETHANE see FMV000
6-OXO-3-METHOXY-N-METHYL-4,5-EPOXYMORPHINAN see OOI000
1-OXO-2-(p-((α-METHYL)CARBOXYMETHYL)PHENYL)ISOINDOLINE
see IDA400
3-(1-OXO-4,7-NONADIENYL)OXIRANECARBOXAMIDE see ECE500
(2R-(2-α,3-α(4E,7E)))-3-(1-OXO-4,7-NONADIENYL)OXIRANECARBOXAMIDE
see ECE500
5-OXONONANE see NMZ000
(E)-12-OXO-10-OCTADECENOIC ACID, METHYL ESTER see MNF250
12-OXO-trans-10-OCTADECENOIC ACID, METHYL ESTER see MNF250
α-1-(OXOOCTADECYL)-omega-HYDROXYPOLY(OXY-1,2-ETHANEDIYL)
see PJW750
9-OXO-8-OXATRICYCLO(5.3.1.0^{2,6})UNDECANE see OOK000
4-OXOPENTANOIC ACID see LFH000
OXOPHENARSINE see OOK100
OXOPHENARSINE HYDROCHLORIDE see ARL000
((4-OXO-2-PHENYL-4H-1-BENZOPYRAN-7-YL)OXY)ACETIC ACID ETHYL
ESTER see ELH600
(12-β(E,E))-12-((1-OXO-5-PHENYL-2,4-PENTADIENYL)OXY)-DAPHNETOXIN
see MDJ250
(17-β)-17-(1-OXO-3-PHENYLPROPOXY)-ESTR-4-EN-3-ONE (9CI) see DYF450
N-(2-OXO-3-PIPERIDYL)PHTHALIMIDE see OOM300
5-OXO-l-PROLYL-l-HISTIDYL-l-PROLINAMIDE TARTRATE see POE050
2-OXOPROPANAL see PQC000
OXOPROPANEDINITRILE see OOM400
OXOPROPANEDINITRILE (CARBONYL DICYANIDE) see MDL250
2-OXOPROPANOIC ACID see PQC100
17-(1-OXOPROPOXY)-(17-β)-ANDROST-4-EN-3-ONE see TBG000
p-(2-OXOPROPOXY)BENZENEARSONIC ACID see OOO000
3-(2-OXOPROPYL)-2-PENTYLCYCLOPENTANONE see OOO100
1-OXOPROPYLPROPYLNITROSAMIN (GERMAN) see NLN500
1-OXOPROPYLPROPYLNITROSAMINE see NLN500
2-OXO-PROPYL-PROPYLNITROSAMINE see ORS000
(2-OXOPROPYL)PROPYLNITROSOAMINE see ORS000
2-OXO-PYRROLIDINE ACETAMIDE see NNE400
2-OXO-1-PYRROLIDINEACETAMIDE see NNE400
5-OXO-2-PYRROLIDINECARBOXYLIC ACID MAGNESIUM SALT (2:1)
see PAN775
2'-OXOPYRROLIDINO-1-PYRROLIDINO-4-BUTYNE see OOY000
2-OXOPYRROLIDIN-1-YLACETAMIDE see NNE400
OXOSILANE see OOS000
2-OXOSPARTEINE see LIQ800
OXOSULFATOVANADIUM PENTAHYDRATE see VEZ100
OXOSUMITHION see PHD750
1-4-OXO-2-THIAZOLIDINEHEXANOIC ACID see CCI500
4-OXO-2-THIONOTHIAZOLIDINE see RGZ550
OXOTREMORIN see OOY000
OXOTREMORINE see OOY000
8-OXOTRICYCLO(5.2.1.0^{2,6})DECANE see OPC000
2,2'-(2-OXO-TRIMETHYLENE)BIS(1,1-DIMETHYLPYRROLIDINIUM
DIBENZENESUFLONATE see COG500
2,2'-(2-OXOTRIMETHYLENE)BIS(1,1-DIMETHYLPYRROLIDINIUM) DIIODIDE
see COG750
(17-β)-17-((1-OXO-4,8,12-TRIMETHYL-3,7,11-TRIDECATRIENYL)OXY)ESTR-4-
EN-3-ONE see NNX650
6-OXOUNDECANE see ULA000
4-OXOVALERIC ACID see LFH000
β-(Γ-OXOVALEROYL)FURAN see IGF300
(4-OXOVALERYLOXY)TRIPHENYLSTANNANE see TMW250
22-OXOVINCALEUKOBLASTINE see LEY000
OXPENTIFYLLINE see PBU100
OXPRENOLOL see AGW000, CNR500
OXPRENOLOL HYDROCHLORIDE see THK750
OXSORALEN see XDJ000
OXTRIMETHYLLINE see CMF500
OXTRIPHYLLINE see CMF500
OXUCIDE see PIJ500
OXURASIN see HEP000
OXY-5 see BDS000
OXY-10 see BDS000
p-OXYACETOPHENONE see HIO000
β-OXYAETHYL-MORPHOLIN (GERMAN) see MRQ500
OXYAETHYLTHEOPHYLLIN (GERMAN) see HLC000
OXYAMINE see CFA750
3-OXYANTHRANILIC ACID see AKE750
p-OXYBENZALDEHYDE see FOF000

OXYBENZENE see PDN750
p-OXYBENZOESAEUREAETHYLESTER (GERMAN) see HJL000
p-OXYBENZOESAURE (GERMAN) see SAI500
p-OXYBENZOESAUREMETHYLESTER (GERMAN) see HJL500
p-OXYBENZOESAUREPROPYLESTER (GERMAN) see HNU500
OXYBENZONE see MES000
OXYBENZOPYRIDINE see QPA000
OXYBIS(4-AMINOBENZENE) see OPM000
4,4'-OXYBISANILINE see OPM000
p,p'-OXYBIS(ANILINE) see OPM000
4,4'-OXYBISBENZENAMINE see OPM000
p,p'-OXYBISBENZENE DISULFONYLHYDRAZIDE see OPE000
OXYBISBENZENESULFONIC ACID DIHYDRAZIDE see OPE000
OXYBIS(BENZENESULFONYL HYDRAZIDE) see OPE000
p,p'-OXYBIS(BENZENESULFONYL) HYDRAZINE see OPE000
1,1'-OXYBIS(BUTANE) see BRH750
4,4'-OXYBIS(2-CHLOROANILINE) see BGT000
4,4'-OXYBIS(2-CHLORO-BENZENAMINE) see BGT000
1,1'-OXYBIS(1-CHLOROETHANE) see BID000
1,1'-OXYBIS(2-CHLORO)ETHANE see DFJ050
1,1'-OXYBIS(2-(2-CHLOROETHYL)THIOETHANE see DFK200
OXYBIS(CHLOROMETHANE) see BIK000
OXYBIS(DIBUTYL(2,4,5-TRICHLOROPHENOXY)TIN) see OPE100
5,5'-OXYBIS(3,4-DICHLORO-2(5H)-FURANONE see MRV000
4,4'-OXYBIS(2,3-DICHLORO-4-HYDROXYCROTONIC ACID-DI-Γ-LACTONE
     see MRV000
OXYBIS(N,N-DIMETHYLACETAMIDETRIPHENYLSTIBONIUM) DIPERCHLOR-
     ATE see OPG000
1,1'-OXYBISETHANE see EJU000
1,1'-(OXYBIS(2,1-ETHANEDIYLOXY))BISBUTANE see DDW200
(OXYBIS(2,1-ETHANEDIYLOXY))BIS-PROPANAMINE see EBV100
3,3'-(OXYBIS(2,1-ETHANEDIYLOXY))BIS-1-PROPANAMINE see DJD800
2,2'-OXYBISETHANOL see DJD600
1,1'-OXYBISETHENE see VOP000
2,2'-OXYBIS(ETHYLENEOXY))DIETHANOL see TCE250
1,1'-OXYBIS(2-ETHYLHEXANE) see DJK600
1,1'-OXYBISHEXANE see DKO800
OXYBISMETHANE see MJW500
1,1'-(OXYBIS(METHYLENE))BIS(4-(1,1-DIMETHYLETHYL)-PYRIDINIUM,
     DICHLORIDE (9CI) see SAB800
1,1'-(OXYBIS(METHYLENE))BIS(4-(HYDROXYIMINO)METHYL)PYRIDINIUM
     DICHLORIDE see BGS250
2,2'-OXYBIS-6-OXABICYCLO-(3.1.0)HEXANE see BJN250
2,2'-OXYBIS(6-OXABICYCLO(3.1.0)HEXANE) mixed with 2,2-BIS(p-(2,3-
     EPOXYPROPOXY)PHENYL)PROPANE see OPI200
1,1'-OXYBIS(2,3,4,5,6-PENTABROMOBENZENE (9CI) see PAU500
1,1-OXYBIS PENTANE see PBX000
10-10' OXYBISPHENOXYARSINE see OMY850
1,1'-OXYBISPROPANE see PNM000
3,3'-OXYBIS(1-PROPENE) see DBK000
OXYBIS(TRIBUTYLTIN) see BLL750
OXYBIS(TRIMETHYLSILANE) see HEE000
OXYBUPROCAINE HYDROCHLORIDE see OPI300
OXYBUTANAL see AAH750
β-OXYBUTTERSAEURE-p-PHENETIDID see HJS850
OXYBUTYNIN CHLORIDE see OPK000
OXYBUTYRIC ALDEHYDE see AAH750
OXYCAINE see DHO600
OXYCARBON SULFIDE see CCC000
OXYCARBOPHOS see OPK250
OXYCARBOXIN see DLV200
OXYCARBOXINE see DLV200
OXY CHEK 114 see MJO500
OXYCHINOLIN see QPA000
o-OXYCHINOLIN (GERMAN) see QPA000
OXYCHLORURE CHROMIQUE (FRENCH) see CML125
OXYCHOLIN see DAL000
OXYCIL see SFS000
OXYCINCHOPHEN see OPK300
OXYCLIPINE see MOQ500
OXYCLOZANID see DMZ000
OXYCLOZANIDE see DMZ000
OXYCODEINONE see PCG500
OXYCODONE HYDROCHLORIDE see DLX400
OXYCODON HYDROCHLORIDE see DLX400
OXYCON see DLX400
OXY DBCP see DDL800
OXYDE d'ALLYLE et de GLYCIDYLE (FRENCH) see AGH150
OXYDE de BARYUM (FRENCH) see BAO000
OXYDE de CALCIUM (FRENCH) see CAU500
OXYDE de CARBONE (FRENCH) see CBW750
OXYDE de CHLORETHYLE (FRENCH) see DFJ050
OXYDE d'ETHYLE (FRENCH) see EJU000
OXYDE de MERCURE (FRENCH) see MCT500

OXYDE de MESITYLE (FRENCH) see MDJ750
OXYDEMETONMETHYL see DAP000
OXYDEMETON-METILE (ITALIAN) see DAP000
OXYDE NITRIQUE (FRENCH) see NEG100
OXYDEPROFOS see DSK600
OXYDE de PROPYLENE (FRENCH) see PNL600
OXYDE de TRIBUTYLETAIN see BLL750
OXYDIACETIC ACID, DISODIUM SALT see SIZ000
OXYDIANILINE see OPM000
4,4'-OXYDIANILINE see OPM000
p,p'-OXYDIANILINE see OPM000
OXYDIAZEPAM see CFY750
OXYDIAZOL see BGD250
2,2'-OXYDIETHANOL see DJD600
N-OXYDIETHYL-2-BENZOTHIAZOLSULFENAMID (CZECH) see BDF750
OXYDIETHYLENE ACRYLATE see ADT250
N-(OXYDIETHYLENE)BENZOTHIAZOLE-2-SULFENAMIDE see BDG000
OXYDIETHYLENE BIS(CHLOROFORMATE) see OPO000
OXYDIETHYLENE CHLOROFORMATE see OPO000
OXYDIETHYLENE DIACRYLATE see ADT250
3,3'-(2,2'-OXYDIETHYLENEDIOXYBISACETAMIDO)BIS(2,4,6-
     TRIIODOBENZOIC ACID) see IGD100
(N,N'-OXYDIETHYLENEDIOXYDIETHYLENE)BIS(TRIETHYLAMMONIUM
     IODIDE) see OPQ000
OXYDIFORMIC ACID DIETHYL ESTER see DIZ100
β-(4-OXY-3,5-DIJOD-PHENYL)-α-PHENYL-PROPIONSAEURE (GERMAN)
     see PDM750
1,1'-OXYDIMETHYLENE BIS(4-tert-BUTYLPYRIDINIUM CHLORIDE)
     see SAB800
3,3'-(OXYDIMETHYLENEBIS(CARBONYLIMINO)BIS(2,4,6-TRIIODOBENZOIC
     ACID DISODIUM SALT) see BGB350
1,1'-(OXYDIMETHYLENE)BIS(4-FORMYLPYRIDINIUM)DICHLORIDE
     DIOXIME see BGS250
1,1'-(OXYDIMETHYLENE)BIS(4-FORMYLPYRIDINIUM) DINITRATE
     DIOXIME see OPY000
1,1'-(OXYDIMETHYLENE)BIS(4-FORMYLPYRIDINIUM) DIOXIME DICHLO-
     RIDE see BGS250
OXYDIMETHYLQUINAZINE see AQN000
4,4'-OXYDIPHENOL see OQI000
p,p'-OXYDIPHENOL see OQI000
p-OXYDIPHENYLAMINE see AOT000
4,4'-OXYDIPHENYLAMINE see OPM000
OXYDI-p-PHENYLENEDIAMINE see OPM000
1,1'-OXYDI-2-PROPANOL see OQM000
3,3'-OXYDI-1-PROPANOL DIBENZOATE see DWS800
OXYDIPROPANOL PHOSPHITE (3:1) see OQO000
3,3'-OXYDIPROPIONITRILE see OQQ000
β,β'-OXYDIPROPIONITRILE see OQQ000
OXYDISULFOTON see OQS000
N-OXYD-LOST see CFA500, CFA750
N-OXYD-MUSTARD see CFA500
OXYDOL see HIB000
OXYDRENE see DBA800
OXYEPHEDRINE see HKH500
OXYETHYLATEED TERTIARY OCTYL-PHENOL-FORMALDEHYDE POLY-
     MER see TDN750
OXYETHYLENATED DODECYL ALCOHOL see DXY000
OXYETHYLIDENEDIPHOSPHONIC ACID see HKS780
1-(β-OXYETHYL)-2-METHYL-5-NITROIMIDAZOLE see MMN250
OXYETHYLTHEOPHYLLINE see HLC000
OXYFED see DBA800
1-OXYFEDRIN see OQU000
OXYFENAMATE see HNJ000
OXYFENON see ORQ000
OXYFLAVIL see ELH600
OXYFUME see EJN500
OXYFUME 12 see EJN500
OXYFUME 20 see EJO000
OXYFUME 30 see EJO000
OXYFURADENE see NDY000
β-OXY-GABA see AKF375
OXYGEN see OQW000
OXYGEN, compressed (DOT) see OQW000
OXYGEN, refrigerated liquid (DOT) see OQW000
OXYGEN DIFLUORIDE see ORA000
OXYGEN FLUORIDE see ORA000
OXYGERON see VLF000
OXYHYDROCHINON (GERMAN) see BBU250
OXYHYDROQUINONE see BBU250
OXYJECT 100 see HOI000
OXYKODAL see DLX400
OXYKON see DLX400
OXYLAN see DKQ000
OXYLITE see BDS000

OXY MBC see SFS500
OXYMEMAZINE see AFL750
OXYMETAZOLINE see ORA100
OXYMETAZOLINE CHLORIDE see AEX000
OXYMETAZOLINE HYDROCHLORIDE see AEX000
OXYMETEBANOL see ORE000
OXYMETHALONE see PAN100
OXYMETHAZOLINE see ORA100
OXYMETHEBANOL see ORE000
OXYMETHENOLONE see PAN100
OXYMETHOLONE see PAN100
OXY-2-METHOXY-3-BENZALDEHYDE (FRENCH) see VFP000
OXY-3-METHOXY-4 BENZALDEHYDE (FRENCH) see FNM000
OXYMETHUREA see DTG700
OXYMETHYLENE see FMV000
5-OXYMETHYLFURFUROLE see ORG000
OXYMETHYLPHTHALIMIDE see HMP100
OXYMETOZOLINE see ORA100
OXYMORPHINONE HYDROCHLORIDE see ORG100
OXYMORPHONE HYDROCHLORIDE see ORG100
OXYMURIATE OF POTASH see PLA250
OXYMYCIN see CQH000
OXYMYKOIN see HOH500
β-OXYNAPHTHOIC ACID see HMX500
OXY-NH2 see CFA500
OXYPAAT see HEP000
OXYPANAMINE see ORI300
OXYPARATHION see NIM500
OXYPERTIN see ECW600
OXYPERTINE see ECW600
OXYPHENALON see RBU000
OXYPHENAMATE see HNJ000
OXYPHENBUTAZONE see HNI500
OXYPHENIC ACID see CCP850
OXYPHENON see ORQ000
OXYPHENONIUM see ORQ000
OXYPHENONIUM BROMIDE see ORQ000
OXYPHENYLBUTAZONE see HNI500
1-(p-OXYPHENYL)-2-METHYLAMINOPROPAN (GERMAN) see FMS875
α-(p-OXYPHENYL)-β-METHYL AMINOPROPANE see FMS875
3-OXYPHENYL TRIMETHYLAMMONIUM IODIDE see HNN500
OXYPHIONFOS see DSK600
OXYPHYLLINE see HLC000
OXYPHYLLINE (AMIDO) see HLC000
OXYPROCAIN see DHO600
OXYPROCAINE see DHO600
p-OXYPROPIOPHENONE see ELL500
2-OXY-1,3-PROPYLENEDIAMINE-N,N,N,N'-TETRAMETHYLENEPHOS-
  PHONIC ACID see DYE550
β-OXYPROPYLPROPYLNITROSAMINE see ORS000
Γ-OXYPROPYLTHEOBROMIN (GERMAN) see HNZ000
Γ-(Γ-OXYPROPYL)-THEOBROMIN (GERMAN) see HNZ000
β-OXYPROPYLTHEOBROMINE see HNY500
β-OXYPROPYLTHEOPHYLLIN see HOA000
OXYPROPYLTHEOPHYLLINE see HOA000
OXYPSORALEN see XDJ000
OXYPYRRONIUM see IBP200
OXYPYRRONIUM BROMIDE see IBP200
OXYQUINOLINE see QPA000
8-OXYQUINOLINE see QPA000
OXYQUINOLINE SULFATE see QPS000
OXYQUINOLINOLEATE de CUIVRE (FRENCH) see BLC250
5-OXYRESORCINOL see PGR000
OXYRITIN see RSU000
OXYSTIN see ORU500
OXYSULFATOVANADIUM see VEZ000
OXYTERRACINE see HOH500
OXYTETRACYCLINE see HOH500
OXYTETRACYCLINE AMPHOTERIC see HOH500
OXYTETRACYCLINE HYDROCHLORIDE see HOI000
OXYTHANE see NCM700
OXYTHEONYL see HLC000
OXYTHIAMIN see ORS200
OXYTHIAMINE see ORS200
OXYTHIOQUINOX see ORU000
OXYTOCIC see EDB500
OXYTOCIN see ORU500
OXYTOL ACETATE see EES400
m-OXYTOLUENE see CNW750
o-OXYTOLUENE see CNX000
p-OXYTOLUENE see CNX250
OXYTRIL see HKB500
OXYTRIL M see DDP000
OXYTROPIUM BROMIDE see ONI000

OXYURANUS SCUTELLATUS (AUSTRALIA) VENOM see ARV500
OXYUREA see HOO500, TGN250
OXY WASH see BDS000
OXYZINE see PIJ500
OXYZIN (TABL.) see HEP000
OZAGREL see SHV100
OZIDE see ZKA000
OZLO see ZKA000
OZOCERITE see ORU900
OZOKERITE see ORU900
OZON (POLISH) see ORW000
OZONE see ORW000
OZONE mixed with NITROGEN OXIDES (53%:47%) see ORY000
OZONIDES see ORY499

P07 see PKO500
P 12 see MNV250
P-25 see SLJ000
P-40 see DXG000
P 48 see CMV400
P-50 see AIV500
P 55 see VJZ000
134 P see MLK800
P-165 see ASA500
P 237 see BTA000
P 241 see PEO750
P 253 see LJR000
P 271 see PAB000
P 284 see DWF000
P 301 see HNJ000
P-314 see DHR000
P-329 see DPU800
P 391 see MOQ250
P-398 see ACX500
501 P see AOO800
P 527 see MOE250
P 652 see PDU250
P 887 see ILE000
P 892 see MOQ000
P 1011 see DGE200
P-1108 see DVJ400
P 1133 see BET000
P 1134 see COS500
P 1142 see PDN000
P 1297 see BEQ500
P 1393 see BDE250
P 1488 see BRU750
P1496 see RBF100
P 1531 see PFC750
P-2292 see ADA000
P 2647 see BCL250
P-5048 see DRT200
P-5307 see PDY870
P-7138 see NDY400
P-267 see BPR500
P 71-0129 see FAO100
P 286 (contrast medium) see IGD200
PA see PAP750, PEG500
PA 93 see SMB000
PA 94 see CQH000
PA 144 see MQW750
3N4HPA see HMY000
PA 6 (polymer) see PJY500
PAA see BBL500
PAA-25 see ADW200
PAA-701 DIHYDROCHLORIDE see BFX125
PA'AILA (HAWAII) see CCP000
PAB see SEO500
PA 114B see VRA700
PABA see AIH600
PABAVJT see SEO500
PABESTROL see DKA600, DKB000
PABIALGIN see IGI000
PABRACORT see HHQ800
PABS see SNM500
PACAMINE HYDROCHLORIDE see NCL300
PACATAL see MOQ250
PACATAL BASE see MOQ250
PACEMO see HIM000
PACETYN see EHP000
PACHYCARPINE see PAB250
6-β,7-α,9-α,11-α-PACHYCARPINE see SKX500
PACHYRHIZUS EROSUS see YAG000
PACIENCIA see DAB700

PACIENX see CFZ000
PACIFAN see NBU000
PACINOL see TJW500
PACITANE see BBV000
PACITRAN see DCK759, MQR200
PACTAMYCIN see PAB500
PADAN see BHL750
PADARYL see BQW000
PADISAL see MRW000
PADOPHENE see PDP250
PADRIN see PAB750
PAE see INN500
PAEONIA MOUTAN see PAC000
PAEONIFLORIN see PAC000
PAEONOL see PAC250
PAEONY ROOT see PAC000
PAFE see PKE550
PAGANO-COR see DHS200
PAGITANE HYDROCHLORIDE see CQH500
PAGODA TREE see NBR800
PAIDAZOLO see AIF000
PA'INA (HAWAII) see JBS100
PAIN de COULEUVRE (CANADA) see BAF325
PAINTERS' NAPHTHA see PCT250
PAISAJE (PUERTO RICO) see PGQ285
PAISLEY POLYMER see PMP500
PAKA (HAWAII) see TGI100
PAKHTARAN see DUK800
PAL see PEC750
PALA see PGY750
PALACOS see PKB500
PALACRIN see CFU750
PALAFER see FBJ100
PALANILCARRIER A see PAC500
PALANTHRENE GOLDEN YELLOW see DCZ000
PALANTHRENE NAVY BLUE G see CMU500
PALANTHRENE NAVY BLUE RB see CMU750
PALAPENT see NBU000
PALASH SEED EXTRACT see BOV800
PALATINOL A see DJX000
PALATINOL AH see DVL700
PALATINOL BB see BEC500
PALATINOL C see DEH200
PALATINOL IC see DNJ400
PALATINOL M see DTR200
PALATONE see MAO350
PALAVALE see EAE000
PALE ORANGE CHROME see LCS000
PALESTROL see DKA600
PALETA de PINTOR (PUERTO RICO) see CAL125
PALFADONNA see AFJ400
PALFIUM see AFJ400
PALINUM see TDA500
PALIUROSIDE see RSU000
PALLADIUM see PAD250
PALLADIUM CHLORIDE see PAD500
PALLADIUM(2+) CHLORIDE see PAD500
PALLADIUM CHLORIDE, DIHYDRIDE see PAD750
PALLADOUS CHLORIDE see PAD500
PALLETHRINE see AFR250
PALLICID see ABX500
PALMA CHRISTI (HAITI) see CCP000
PALMAROSA OIL see PAE000
PALMATINE HYDROXIDE see PAE100
PALMATINIUM HYDROXIDE see PAE100
PALM BUTTER see PAE500
PALMITA see PGA750
PALMITA de JARDIN (PUERTO RICO) see CNH789
PALMITIC ACID see PAE250
PALMITIC ACID, 2-ETHYLHEXYL ESTER see OFE100
PALMITYL ALCOHOL see HCP000
PALMITYLAMINE see HCO500
PALMO-ARA-C see AQS875
PALM OIL see PAE500
PALO GUACO (CUBA) see CDH125
PALOHEX see HFG550
PALO de NUEZ (PUERTO RICO) see TOA275
PALOPAUSE see ECU750
PALOSEIN see OJM400
PAL-P see PII100
PALTET see TBX250
PALUDRINE see CKB250
PALUDRINE DIHYDROCHLORIDE see PAE600
PALUDRINE HYDROCHLORIDE see CKB500

PALUSIL HYDROCHLORIDE see CKB500
PALYGORSCITE see PAE750
PALYGORSKIT (GERMAN) see PAE750
PALYTHOATOXIN see PAE875
PALYTOXIN see PAF000
PAM see ALW500
l-PAM see PED750
PAM (CZECH) see POS750
PAMA see PGV250
PAMACYL see AMM250
PAMAQUIN see RHZ000
PAMAZONE see CJR909
2-PAM CHLORIDE see FNZ000
PAMINE see SBH500
PAMINE BROMIDE see SBH500
2-PAM IODIDE see POS750
PAMISAN see ABU500
PAMISYL see AMM250
PAMISYL SODIUM see SEP000
2-PAM METHANESULFONATE see PLX250
PAMN see PNR250
PAMOIC ACID see PAF100
PAMOLYN see OHU000
PAMOSOL 2 FORTE see EIR000
PAMOVIN see PQC500
PAM-PERCHLORAT (GERMAN) see FOA000
PAN see BFW120, PCL750
PANA see PFT250
PANACAINE see PDU250
PANACEF see CCR850
PANACELAN see POC500
PANACID see PAF250
PANACIDE see MJM500
PANACUR see FAL100
PANADOL see HIM000
PANADON see PAG200
PANALDINE see TGA525
PANAMIDIN DIHYDROCHLORIDE see DBM400
PANAX see GEQ400
PANAX GINSENG, ROOT EXTRACT see GEQ425
PANAX SAPONIN E see PAF450
PANAZON see DKK100
PANAZONE see PAF500
PANCAL see CAU750
PANCALMA see MQU750
PANCID see SNN500
PANCIL see OFE000
PANCODINE see DLX400
PANCORAL see BQJ500
PANCREATIC BASIC TRYPSIN INHIBITOR see PAF550
PANCREATIC TRYPSIN INHIBITOR see PAF550
PANCREATIC TRYPSIN INHIBITOR (KUNITZ) see PAF550
PANCREATIN see PAF600
PANCREATOPEPTIDASE E see EAG875
PANCRIDINE see DBN400
PANCURONIUM BROMIDE see PAF625
PANCURONIUM BROMIDE HYDRATE see PAF630
PANCURONIUM DIBROMIDE HYDRATE see PAF630
PANDEX see PKM250
PANDIGAL see LAU400
PANDRINOX see MLF250
PANDUROL see BHD250
PANESTIN see TBG000
PANETS see HIM000
PANEX see HIM000
PANFLAVIN see DBX400, XAK000
PANFORMIN see BQL000
PANFURAN-S see BKH500
PANGUL see TEH500
PANHIBIN see SKS600
PANIMYCIN see PAG050
PANINI-'AWA'AWA (HAWAII) see AGV875
PANITHAL see SAH000
PANLANAT see LAU400
PANMYCIN see TBX000
PANMYCIN HYDROCHLORIDE see TBX250
PANO-DRENCH 4 see MLF250
PANOFEN see HIM000
PANOGEN see MEO750, MLF250
PANOGEN TURF FUNGICIDE see MLF250
PANOGEN TURF SPRAY see MLF250
PANOLID see DWE800
PANORAL see CCR850, PAG075
PANORAL HYDRATE see CCR850

PANORAM 75 see TFS350
PANORAM D-31 see DHB400
PANOSINE see MMD500
PANOSPRAY 30 see MLF250
PANOXYL see BDS000
PANPARNIT see PET250
PANRONE see FNF000
PANSOIL see EFK000
PANTALGINE see DAM700
PANTAS see HKR500
PANTASOTE R 873 see PKQ059
PANTELMIN see MHL000
PANTETHINE see PAG150
d-PANTETHINE see PAG150
PANTETINA see PAG150
PANTHELINE see HKR500
PANTHENOL see PAG200
d-PANTHENOL see PAG200
d(+)-PANTHENOL (FCC) see PAG200
PANTHER CREEK BENTONITE see BAV750
PANTHESIN see LEU000
PANTHESINE see LEU000
PANTHION see PAK000
PANTHODERM see PAG200
PANTHOJECT see CAU750
PANTHOLIC-L see IDE000
PANTHOLIN see CAU750
PANTOBROMINO see HNY500
PANTOCAINE see BQA010
PANTOCAINE HYDROCHLORIDE see TBN000
PANTOCID see HAF000
PANTOCRIN see PAG225
PANTOGAM see CAT125
PANTOL see PAG200
PANTOLAX see HLC500
PANTOMICINA see EDH500
PANTOMIN see PAG150
PANTONSILETTEN see DBX400
PANTOPRIM see TKX000
PANTOSEDIV see TEH500
PANTOSIN see PAG150
PANTOTHENATE CALCIUM see CAU750
PANTOTHENATE de ZINC (FRENCH) see ZKS000
PANTOTHENIC ACID, CALCIUM SALT see CAU750
(+)-PANTOTHENIC ACID, CALCIUM SALT see CAU750
PANTOTHENIC ACID, ZINC SALT see ZKS000
PANTOTHENOL see PAG200
d-PANTOTHENOL see PAG200
PANTOTHENYL ALCOHOL see PAG200
d-PANTOTHENYL ALCOHOL see PAG200
d(+)-PANTOTHENYL ALCOHOL see PAG200
PANTOVERNIL see CDP250
PANTOZOL 1 see COD000
PAN-TRANQUIL see MQU750
PANURIN see CFY000
PANWARFIN see WAT220
PAP see ALT250, DRR400, PDC250
PAP-1 see AGX000
PAPAIN see PAG500
PAPANERINE see PAH000
PAPANERIN-HCL (GERMAN) see PAH250
PAPAO-APAKA (GUAM) see EAI600
PAPAO-ATOLONG (GUAM) see EAI600
PAPAVARINE CHLORHYDRATE see PAH250
PAPAVERINA (ITALIAN) see PAH000
PAPAVERIN CARBOXYLIC ACID, SODIUM SALT see PAG750
PAPAVERINE see PAH000
PAPAVERINE CHLOROHYDRATE see PAH250
PAPAVERINE HYDROCHLORIDE see PAH250
PAPAVERINE MONOHYDROCHLORIDE see PAH250
PAPAYOTIN see PAG500
PAPER BLACK BA see AQP000
PAPER RED HRR see FMU070
PAP H see PAH250
PAPOOSE ROOT see BMA150
PAPP see AMC000
PAPTHION see DRR400
PARA see PEY500
PARAAMINODIPHENYL see AJS100
PARABAR 441 see BFW750
PARABEN see HJL500, HNU500
PARABROMODYLAMINE MALEATE see BNE750
PARACAIN see AIT250
PARACETALDEHYDE see PAI250

PARACETAMOLE see HIM000
PARACETAMOLO (ITALIAN) see HIM000
PARACETANOL see HIM000
PARACETOPHENETIDIN see ABG750
PARACHLORAMINE see HGC500
PARACHLOROCIDUM see DAD200
PARACIDE see DEP800
PARACODIN see DKW800
PARACODINE see DKW800
PARACORT see PLZ000
PARACORTOL see PMA000
PARACOTOL see PMA000
PARACRESYL ACETATE see MNR250
PARACRESYL ISOBUTYRATE see THA250
PARA CRYSTALS see DEP800
d-PARACURARINE CHLORIDE see TOA000
PARACYMENE see CQI000
PARACYMOL see CQI000
PARADERIL see RNZ000
PARADI see DEP800
PARADICHLORBENZOL (GERMAN) see DEP800
PARADICHLOROBENZENE see DEP800
PARADICHLOROBENZOL see DEP800
PARA-DIEN see DAL600
PARADIONE see PAH500
PARADISE TREE see CDM325
(0)-PARADOL see VFP100
PARADONE GOLDEN YELLOW see DCZ000
PARADONE GREY M see CMU475
PARADONE GREY MG see CMU475
PARADONE NAVY BLUE G see CMU500
PARADONE OLIVE GREEN B see ALT000
PARADORMALENE see TEO250
PARADOW see DEP800
PARADUST see PAK000
PARAESIN see BQH250
PARAFFIN see PAH750
PARAFFIN HYDROCARBONS see PAH770
PARAFFIN OILS (PETROLEUM), CATALYTIC DEWAXED HEAVY (9CI)
  see MQV868
PARAFFIN OILS (PETROLEUM), CATALYTIC DEWAXED LIGHT (9CI)
  see MQV870
PARAFFIN WAX see PAH750
PARAFFIN WAXES and HYDROCARBON WAXES, CHLORINATED
  (C12, 60% CHLORINE) see PAH800
PARAFFIN WAXES and HYDROCARBON WAXES, CHLORINATED
  (C23, 43% CHLORINE) see PAH810
PARAFFIN WAX FUME (ACGIH) see PAH750
PARAFLEX see CDQ750
PARAFLU see TKH750
PARAFORM see FMV000
PARAFORMALDEHYDE see PAI000
PARAFORSN see PAI000
PARAFUCHSIN (GERMAN) see RMK020
PARAGLAS see PKB500
PARAGLYINE see MOS250
PARAHEXYL see HGK500
PARAHYDROXYBENZALDEHYDE see FOF000
PARAISO (MEXICO) see CDM325
PARAL see PAI250
PARALCTIN see PMH625
PARALDEHYD (GERMAN) see PAI250
PARALDEHYDE see PAI250
PARALDEIDE (ITALIAN) see PAI250
PARALERGIN see ARP675
PARALEST see BBV000
PARALKAN see BBA500
PARALYTIC SHELLFISH POISON DIHYDROCHLORIDE see SBA500
PARA-MAGENTA see RMK020
PARAMAL see DBM800
PARAMANDELIC ACID see MAP000
PARAMAR see PAK000
PARAMELACONITE see CNO250
PARAMEL DC see CNH125
PARAMENTHANE HYDROPEROXIDE (DOT) see IQE000
PARAMETADIONE see PAH500
PARAMETHADIONE see PAH500
PARAMETHASONE 21-ACETATE see PAL600
PARAMETHAZONE ACETATE see PAL600
PARAMETHYL PHENOL see CNX250
PARAMIBE see DDS600
PARAMICINA see APP500
PARAMID see AKO500
PARAMIDIN see BQW825

PARAMIDINE see BQW825
PARAMID SUPRA see AKO500
PARAMINE BLACK B see AQP000
PARAMINE BLUE 2B see CMO000
PARAMINE BLUE 3B see CMO250
PARAMINE FAST VIOLET N see CMP000
PARAMINOPROPIOPHENONE see AMC000
PARAMINYL see WAK000
PARAMINYL MALEATE see DBM800
PARAMORFAN see DNU310
PARAMORPHAN see DLW600
PARAMORPHINE see TEN000
PARAMOTH see DEP800
PARAMYCIN see AMM250
PARANAPHTHALENE see APG500
PARANEPHRIN see VGP000
PARANIT ETHANE DISULFONATE see CBG250
PARANITROANILINE, solid (DOT) see NEO500
PARANITROFENOL (DUTCH) see NIF000
PARANITROFENOLO (ITALIAN) see NIF000
PARANITROPHENOL (FRENCH, GERMAN) see NIF000
PARANITROSODIMETHYLANILIDE see DSY600
PARANOL see ALT250
PARANUGGETS see DEP800
PARAOXON see NIM500
PARAOXONE see NIM500
PARAOXON-METHYL see PHD500
PARAPAN see HIM000
PARA-PAS see AMM250
PARAPEST M-50 see MNH000
PARAPHENOLAZO ANILINE see PEI000
PARAPHENYLEN-DIAMINE see PEY500
PARAPHOS see PAK000
PARAPLEX P 543 see PKB500
PARAPLEX RG-2 (60%) see PAI750
PARAQUAT see PAI990
PARAQUAT BIS(METHYL SULFATE) see PAJ250
PARAQUAT CHLORIDE see PAJ000
PARAQUAT DICATION see PAI990
PARAQUAT DICHLORIDE see PAJ000
PARAQUAT DIHYDRIDE see PAI995
PARAQUAT DIMETHYL SULFATE see PAJ250
PARAQUAT DIMETHYL SULPHATE see PAJ250
PARAROSANILINE see RMK020
PARAROSANILINE CHLORIDE see RMK020
PARAROSANILINE HYDROCHLORIDE see RMK020
PARASAL see AMM250
PARASALICIL see AMM250
PARASALINDON see AMM250
PARASAN see BCA000
PARASCORBIC ACID see PAJ500
PARASEPT see BSC000, HJL500, HNU500
PARASOL see CNM500
PARASORBIC ACID see PAJ500
(+)-PARASORBINSAEURE (GERMAN) see PAJ500
PARASPAN see SBH500
PARASPEN see HIM000
PARASTARIN see EJM500
PARASULFONDICHLORAMIDO BENZOIC ACID see HAF000
PARASYMPATOL see HLV500
PARATAF see MNH000
PARATHENE see PAK000
PARATHESIN see EFX000
PARATHIAZINE see PAJ750
PARATHION see PAK000
PARATHION and compressed gas mixture (DOT) see PAK230
PARATHION, liquid (DOT) see PAK000
PARATHION (mixture, dry) see PAK250
PARATHION-ETHYL see PAK000
PARATHION METHYL see MNH000
PARATHION-METILE (ITALIAN) see MNH000
PARATHION MIXTURE, liquid (DOT) see PAK260
PARATHION S see DJS800
PARATOX see MNH000
PARAWET see PAK000
PARAXANTHINE see PAK300
PARAXENOL see BGJ500
PARAXIN see CDP250
PARAXIN SUCCINATE see CDP725
PARAZENE see DEP800
PARAZINE see PIJ500
PARAZONE see FNF000
PARBENDAZOLE see BQK000
PARBOCYL-REV see SJO000

PARCIDOL see DIR000
PARCLAY see KBB600
PARCLOID see PKQ059
PARDA see DNA200
PARDIDOL see DIR000
PARDISOL see DIR000
PARDROYD see PAN100
PAREDRINOL see FMS875
PARENTERAL see CJR909
PARENTRACIN see BAC250
PARENZYME see TNW000
PARENZYMOL see TNW000
PAREPHYLLIN see CNR125
PAREST see QAK000
PAR ESTRO see ECU750
PARFENAC see BPP750
PARFENAL see BPP750
PARFEZINE see DIR000
PARFURAN see NGE000
PARGITAN see BBV000
PARGLYAMINE see MOS250
PARGONYL see NCF500
PARGYLINE see MOS250
PARGYLINE HYDROCHLORIDE see BEX500
PARICINA see APP500
PARIDINE RED LCL see CHP500
PARIDOL see HJL500
PARIS GREEN see COF500
PARIS RED see LDS000
PARIS VIOLET R see MQN025
PARIS YELLOW see LCR000
PARITOL see SEH450
PARKE DAVIS CI-628 see NHP500
PARKEMED see XQS000
PARKIBLEU see CMO250
PARKIN see DIR000
PARKINSAN see BBV000
PARKIPAN see CMO250
PARKISOL see DIR000
PARKOPAN see BBV000, PAL500
PARKOPHYLLIN see TEP000
PARKOSED see BNK000
PARKOTAL see EOK000
PAR KS-12 see PMC250
PARLEF see TKH750
PARLIF see TKH750
PARLODEL see BNB325
PARMAL see WAK000
PARMATHASONE ACETATE see PAL600
PARMAVERT see NNE100
PARMETOL see CFE250
PARMIDIN see PPH050
PARMIDINE see PPH050
PARMIDINE R see PPH050
PARMINAL see QAK000
PARMOL see HIM000
PARMONE see NAK500
PARNATE see PET500, PET750
PAROL see CFE250, MQV750
PAROLEINE see MQV750
PAROMOMYCIN see NCF500
PAROMOMYCIN SULFATE see APP500
PAROXAN see NIM500
PAROXON see ELL500
PAROXYL see ABX500
PAROXYPROPIONE see ELL500
PARPANIT see PET250
PARPHEZEIN see DIR000
PARPON see BCA000
PARRAFIN OIL see MQV750
PARROT GREEN see COF500
PARSAL see AJC000
PARSALMIDE see AJC000
PARSIDOL see DIR000
PARSITAN see DIR000
PARSLEY APIOL see AGE500
PARSLEY CAMPHOR see AGE500
PARSLEY HERB OIL (FCC) see PAL750
PARSLEY OIL see PAL750
PARSLEY SEED OIL (FCC) see PAL750
PARTEL see DJT800
PARTERGIN see PAM000
PARTEROL see DME300
PARTHENICIN see PAM175

PARTHENIN see PAM175
PARTOCON see ORU500
PARTREX see TBX250
PARTUSISTEN see FAQ100
PARVOLEX see ACH000
PARZATE see DXD200, EIR000
PARZONE see DKW800
PAS see AMM250
PASA see AMM250
PASADE see SEP000
PASALIN see PAM500
PASALON see AMM250
PASALON-RAKEET see SEP000
PASARA see AMM250
PAS-C see AMM250
PASCO see ZBJ000, ZKA000
PASCORBIC see AMM250
PASEM see AMM250
PASEPTOL see HNU500
PASEXON 100T see SHK800
PASILLA (PUERTO RICO) see CDM325
PASK see AMM250
PASMED see AMM250
PASNAL see SEP000
PASNODIA see AMM250
PASOLAC see AMM250
PASOTOMIN see PMF250
PASPALIN see PAM775
PASPALIN P I see PAM775
PASQUE FLOWER see PAM780
PASSIFLORAE INCARNATAE EXTRACTUM see PAM782
PASSIFLORA EXTRACT see PAM782
PASSIFLORA INCARNATA, EXTRACT see PAM782
PASSIFLORIN see MPA050
PASSIONFLOWER EXTRACT see PAM782
PASSODICO see SEP000
PAT see PKE600
PATAP see BHL750
PATENTBLAU V (GERMAN) see ADE500
PATENT BLUE AE see FMU059
PATENT GREEN see COF500
PATHOCIDIN see AJO500
PATHOCIDINE see AJO500
PATHOCIL see DGE200
PATHOCLON see DXN709
PATHOMYCIN see APY500
PATORAN see PAM785
PATRICIN see VRF000
PATRINOSIDE see PAM789
PATRINOSIDE-AGLYCONE see PAM800
PATROVINE see DHX800, THK000
PATTINA V 82 see PKQ059
PATTONEX see PAM785
PATULIN see CMV000
PAUCIMYCIN see NCF500
PAUSITAL see CKE750
P. AUSTRALIS VENOM see ARV000
PAVATRIN see CBS250
PAVATRINE see FDK000
PAVATRINEAT see CBS250
PAVISOID see PAN100
PAVULON see PAF625, PAF630
PAXAREL see ACE000
PAXATE see DCK759
PAXILON see BGD250
PAXISTIL see CJR909
PAXISYN see DLY000
PAXITAL see MOQ250
PAXYL see TAF675
PAY-OFF see DRN200
PAYZE see BLW750
PAYZONE see DKK100, PAF500
PAZITAL see CGA000
PB see PIX250, POQ250
PB-106 see PIX800
P-4657-B see NBP500
PBA, DIMETHYLAMINE SALT see PJQ000
PBAN-560 see PAN250
PBAN-560 (degassed) see PAN500
PBB see FBU000, FBU509, PJL335
PB 89 CHLORIDE see FMU000
PBDZ see BQK000
PB-89 HYDROCHLORIDE see FMU000
PBNA see PFT500

PBQI see BDD500
PB-1,6-QUINONE see BCU500
PBS see SID000
PBX(AF) 108 see CPR800
PBZ see TMP750
PC 1 see BQJ500
PC 222 see CKD250
PC 603 see CJN250
PC-1421 see PII500
PCA see MCR750, PEE750
PC ALCOHOL DF see PAN750
PCB see PME250
PCB2 see HCD500
PCB (DOT, USDA) see PJL750
PCB HYDROCHLORIDE see PME500
PCBS see CJR500
PCB's see PJM500, PJN000
PCC see CDV100, PIN225
P.C. 80 CRABGRASS KILLER see PLC250
PCDF see PJP100
PCEO see TBQ275
PCHO see PAI250
PCI see CJR500
PCL see HCE500
PCM see PCF300
pCMA see CIF250
PCMC see CFE250
PCMH see PAN775
PCMX see CLW000
PCNB see PAX000
PCNU see CGW250
PCP see PAX250
PCP (anesthetic) see PDC890
PCPA see CJN000, CJR125
PCPBS see CJR500
PCPCBS see CJT750
PCP HYDROCHLORIDE see AOO500
PCPI see CKB000
PCTP see PAY500
PD 93 see PAF250
PD 71627 see AJR400
PD 109394 see AAI100
2-N-p-PDA see ALL750
P.D.A.B. see DOT300
p-PDA HCl see PEY650
PDB see DEP800
PDC see DGG800
PDCB see DEP800
PDD see PGT250
PDD 60401 see CJV250
PDH see POK300
p-PD HCl see PEY650
PDMT see DTP000
m-PDN see PHX550
p-PDN see BBP250
PDP see PDC250
PDQ see CLN750
PDT see DTP000
PDTA-Sb see AQI500
PDU see DTP400
PE see PBB750
PE-043 see PAN800
PEACE PILL see AOO500
PEACH see AQP890
PEACH ALDEHYDE see UJA800
PEACHES see AOB250
PEACOCK BLUE X-1756 see FMU059
PEA FLOWER LOCUST see GJU475
PEANUT OIL see PAO000
PEARL ASH see PLA000, PLA250
PEARLPUSS see CCL250
PEARL STEARIC see SLK000
PEARLY GATES see DJO000
PEAR OIL see AOD725, IHO850
PEARSALL see AGY750
PEB1 see DAD200
PEBC see PNF500
PEBULATE see PNF500
PECAN SHELL POWDER see PAO000
PECATAL see MOQ250
PECAZINE see MOQ250
PECNON see KGK000
PECTA-DIAZINE, suspension see PPP500
PECTALGINE see SEH000

PECTAMOL see DHQ200
PECTITE see MBX800
PECTOLIN see TCY750
PECTOX see CBR675
PED see LBT000
PEDIAFLOR see SHF500
PEDIDENT see SHF500
PEDILANTHUS TITHYMALOIDES see SDZ475
PEDINEX (FRENCH) see CPK500
PEDIPEN see PDT750
PEDRACZAK see BBQ500
PEDRIC see HIM000
PEERACHROME YELLOW R see NEY000
PEERAMINE BLACK E see AQP000
PEERAMINE CONGO RED see SGQ500
PEERLESS see KBB600
PEG see PJT000
PEG 200 see PJT200
PEG 300 see PJT225
PEG 400 see PJT230
PEG 600 see PJT240
PEG 1000 see PJT250
PEG 1500 see PJT500
PEG 4000 see PJT750
PEG 6000 see PJU000
PEGANONE see EOL100
PEG 1000MO see PJY100
PEG-9 NONYL PHENYL ETHER see PKF000
PEG 200MO see PJY100
PEG 600MO see PJY100
PEGOANONE see EOL100
PEG-9 OCTYL PHENYL ETHER see PKF500
PEG-6 OLEATE see PJY100
PEG-20 OLEATE see PJY100
PEG-32 OLEATE see SEQ000
PEG-20 OLEYL ETHER see PJW500
PEGOSPERSE 400MO see PJY100
PEGOTERATE see PKF750
PEHA see PBD000
PEHANORM see TEM500
PELADOW see CAO750
PELAGOL 3GA see ALT000
PELAGOL CD see PEY650
PELAGOL D see PEY500
PELAGOL DA see DBO000
PELAGOL DR see PEY500
PELAGOL EG see ALT500
PELAGOL GREY see DBO400
PELAGOL GREY C see CCP850
PELAGOL GREY CD see PEY650
PELAGOL GREY D see PEY500
PELAGOL GREY GG see ALT000
PELAGOL GREY J see TGL750
PELAGOL GREY L see DBO000
PELAGOL GREY P BASE see ALT250
PELAGOL GREY RS see REA000
PELAGOL L see DBO000
PELAGOL P BASE see ALT250
PELA PUERCO (DOMINICAN REPUBLIC) see DHB309
PELARGIC ACID see NMY000
PELARGIDENOLON see ICE000
PELARGIDENON 1449 see CDH250
PELARGOL see DTE600
PELARGON (RUSSIAN) see NMY000
PELARGONIC ACID see NMY000
PELARGONIC ALCOHOL see NNB500
PELARGONIC ALDEHYDE see NMW500
PELARGONIC MORPHOLIDE see MRQ750
PELARGONIUM OIL see GDA000
PELAZID see ILD000
PELENTAN see BKA000
PELIDORM see BNK000
PELLAGRAMIN see NCQ900
PELLAGRA PREVENTIVE FACTOR see NCQ900
PELLAGRIN see NCQ900
PELLCAFS see BBK500
PELLCAP see BBK500
PELLCAPS see BBK500
PELLETIERINE see PAO500
(R)-(−)-PELLETIERINE see PAO500
PELLIDOL see ACR300
PELLIDOLE see ACR300
PELLON 2506 see PMP500
PELLUGEL see CCL250

PELMIN see NCR000
PELMINE see NCR000
PELONIN see NCQ900
PELONIN AMIDE see NCR000
PELSON see DLY000
PELT-44 see PEX500
PELTAR see MAP300
PELTOL D see PEY500
PELT SOL see DJV000
PELUCES see CLY500
PELVIRAN see DNG400
PEMAL see ENG500
PEMALIN see ENG500
PEMOLINE MAGNESIUM see PAP000
PEMOLINE MAGNESIUM CHELATE see PAP000
PEMOLIN and MAGNESIUM HYDROXIDE see PAP000
PEMPIDINA TARTRATO (ITALIAN) see PAP110
PEMPIDINE HYDROCHLORIDE see PAP100
PEMPIDINE TARTRATE see PAP110
PEN 200 see PDD350
PEN A see AOD125
PEN-A-BRASIVE see BFD250
PENADUR see BFC750
PENADUR L-A see BFC750
PENAGEN see PDT750
PENALEV see BFD000
d-PENAMINE see MCR750
PEN A/N see SEQ000
PENAR see DRS000
PENATIN see CMV000, GFG100
PENBAR see NBU000
PENBRISTOL see AIV500
PENBRITIN see AIV500
PENBRITIN PAEDIATRIC see AIV500
PENBRITIN-S see SEQ000
PENBRITIN SYRUP see AIV500
PENBROCK see AIV500
PENBUTOLOL see PAP225
(−)-PENBUTOLOL see PAP225
PENBUTOLOL SULFATE see PAP230
PENCARD see PBC250
PENCHLOROL see PAX250
PENCIL GREEN SF see FAF000
PENCILLIC ACID see PAP750
PENCOGEL see CCL250
PENCOMPREN see PDT750
PENDEPON see BFC750
PENDEROL see PMB250
PEN-DI-BEN see BFC750
PENDIMETHALIN see DRN200
PENDIOMID see MKU750
PENDIOMIDE BROMIDE see MKU750
PENDIOMIDE DIBROMIDE see MKU750
PENDITAN see BFC750
PENDURAN see BFC750
PENETECK see MQV750
PENETRACYNE see MCH525
PENFLURIDOL see PAP250
PENFORD 260 see HLB400
PENFORD 280 see HLB400
PENFORD 290 see HLB400
PENFORD GUM 380 see SLJ500
PENFORD P 208 see HLB400
PENGITOXIN see GEW000
PENGLOBE see BAB250
PENIALMEN see SEQ000
PENICIDIN see CMV000
PENICILLAMIN see MCR750
(S)-PENICILLAMIN see MCR750
PENICILLAMINE see MCR750
d-PENICILLAMINE see MCR750
dl-PENICILLAMINE see PAP500
PENICILLAMINE HYDROCHLORIDE see PAP550
d-PENICILLAMINE HYDROCHLORIDE see PAP550
PENICILLANIC ACID DIOXIDE SODIUM SALT see PAP600
PENICILLANIC ACID 1,1-DIOXIDE SODIUM SALT see PAP600
PENICILLANIC ACID SULFONE SODIUM SALT see PAP600
PENICILLIC ACID see PAP750
PENICILLIN see PAQ000
PENICILLIN P-12 see DSQ800, MNV250
PENICILLIN, compounded with 9-AMINOACRIDINE see AHT000
PENICILLIN AT see AGK250
PENICILLIN BT see BRS250
PENICILLIN compounded with CHOLINE CHLORIDE see PAQ060

PENICILLIN-CHOLINESTER CHLORID (GERMAN) see PAQ060
PENICILLIN G see BDY669
PENICILLIN G BENETHAMINE see PAQ100
PENICILLIN G, compounded with N,N'-DIBENZYLETHYLENEDIAMINE (2:1)
   see BFC750
PENICILLIN G EPHEDRINE SALT see PAQ120
PENICILLIN-G, MONOSODIUM SALT see BFD250
PENICILLIN G POTASSIUM see BFD000
PENICILLIN G POTASSIUM SALT see BFD000
PENICILLIN G SALT of N,N'-DIBENZYLETHYLENEDIAMINE see BFC750
PENICILLIN G, SODIUM see BFD250
PENICILLIN G, SODIUM SALT see BFD250
PENICILLIN O see AGK250
PENICILLIN PHENOXYMETHYL see PDT500
PENICILLIN POTASSIUM PHENOXYMETHYL see PDT750
PENICILLIN V see PDT500
PENICILLIN V POTASSIUM see PDT750
PENICILLIN V POTASSIUM SALT see PDT750
PENICILLIUM ROQUEFORTI TOXIN see PAQ875
PENICIN see PCU500
PENICLINE see AIV500
PENIDURAL see BFC750
PENIDURE see BFC750
PENILARYN see BFD250
PENILENTE see BFC750
PENILTETRA see MCH525
PENIMEPICYCLINE see MCH525
PENISEM see BFD000
PENITE see SEY500
PENITRACIN see BAC250
PENITREM A see PAR250
PENIZILLIN (GERMAN) see PAQ000
PENNAC see SGF500
PENNAC CBS see CPI250
PENNAC CRA see IAQ000
PENNAC MBT POWDER see BDF000
PENNAC MS see BJL600
PENNAC TBBS see BQK750
PENNAC ZT see BHA750
PENNAMINE see DAA800
PENNCAP-M see MNH000
PENNFLOAT M see LBX000
PENNFLOAT S see LBX000
PENNSALT TD-72 see DJI000
PENN SALT TD-183 see TBV750
PENNSALT TD 5032 see HEE500
PENNWALT C-4852 see DSQ000
PENNWHITE see SHF500
PENNYROYAL OIL see PAR500
PENNZONE B see DEI000
PENNZONE E see DKC400
PENOCTONIUM BROMIDE see PAR600
PEN-ORAL see PDT500
PENOTRANE see PFN000
PENOXALINE see DRN200
PENPHENE see CDV100, TBV750
PENRECO see MQV750
PENSIG see PDD350
PEN-SINT see DGE200
PENSTAPHOCID see MNV250
PENSYN see AOD125
PENTA see PAX250
PENTAACETYLGITOXIN see GEW000
PENTA-o-ACETYLGITOXIN see GEW000
PENTAAMMINEAQUACOBALT(III) CHLORATE see PAR750
PENTAAMMINEPYRAZINERUTHENIUM(II) PERCHLORATE see PAR799
PENTAAMMINEPYRIDINERUTHENIUM(II) PERCHLORATE see PAS829
PENTAAMMINETHIOCYANATOCOBALT(III) PERCHLORATE see PAS859
PENTAAMMINETHIOCYANATORUTHENIUM(II) PERCHLORATE see PAS879
PENTAAZAACENAPHTHYLENE-5-PHOSPHATE ESTER MONOHYDRATE
   see TJE875
PENTAAZACENTOPTHYLENE see TJE870
1,4,7,10,13-PENTAAZATRIDECANE see TCE500
1,3,3,5,5-PENTAAZIRIDINO-1-THIA-2,4,6-TRIAZA-3,5-DIPHOSPHORINE-1-
   OXIDE see SED700
PENTABARBITAL SODIUM see NBU000
PENTABARBITONE see NBT500
PENTABORANE(9) see PAT750
PENTABORANE(11) see PAT799
PENTABORANE(9)DIAMMONIATE see DCF725
PENTABROMOPHENOL see PAU250
PENTABROMOPHENYL ETHER see PAU500
PENTABROMO PHOSPHORANE see PHR250
PENTABROMO PHOSPHORUS see PHR250

PENTAC see DAE425
PENTACAINE see PPW000
PENTACAINE HYDROCHLORIDE see PPW000
PENTACARBONYLIRON see IHG500
PENTACENE see PAV000
PENT-ACETATE see AOD725
PENTACHLOORETHAAN (DUTCH) see PAW500
PENTACHLOORFENOL (DUTCH) see PAX250
PENTACHLORAETHAN (GERMAN) see PAW500
PENTACHLORETHANE (FRENCH) see PAW500
PENTACHLORIN see DAD200
PENTACHLORNITROBENZOL (GERMAN) see PAX000
PENTACHLOROACETONE see PAV225
PENTACHLOROACETOPHENONE see PAV250
2',3',4',5',6'-PENTACHLOROACETOPHENONE see PAV250
PENTACHLOROANISOLE see MNH250
2,3,4,5,6-PENTACHLOROANISOLE see MNH250
PENTACHLOROANTIMONY see AQD000
PENTACHLOROBENZENE see PAV500
PENTACHLORO-BENZENETHIOL see PAY500
PENTACHLOROBIPHENYL see PAV600
PENTACHLOROBUTANE see PAV750
PENTACHLORO-3-BUTENOIC ACID see PAV775
2,2,3,4,4-PENTACHLORO-3-BUTENOIC ACID see PAV775
1,2,3,7,8-PENTACHLORODIBENZO-p-DIOXIN see PAW000
2,3,4,7,8-PENTACHLORODIBENZOFURAN see PAW100
3,3',5,5',6-PENTACHLORO-2,2'-DIHYDROXYBENZANILIDE see DMZ000
3,5,6,3',5'-PENTACHLORO-2,2'-DIHYDROXYBENZANILIDE see DMZ000
PENTACHLORODIPHENYL see PAV600
PENTACHLORO DIPHENYL OXIDE see PAW250
PENTACHLOROETHANE see PAW500
PENTACHLOROFENOL see PAX250
2',2'',4',4'',5-PENTACHLORO-4-HYDROXY-ISOPHTHALANILIDE see PAW600
3,3',5,5',6-PENTACHLORO-2'-HYDROXYSALICYLANILIDE see DMZ000
PENTACHLOROMETHOXYBENZENE see MNH250
PENTACHLORONAPHTHALENE see PAW750
PENTACHLORONITROBENZENE see PAX000
PENTACHLOROPHENATE see PAX250
PENTACHLOROPHENATE SODIUM see SJA000
PENTACHLOROPHENOL see PAX250
2,3,4,5,6-PENTACHLOROPHENOL see PAX250
PENTACHLOROPHENOL (GERMAN) see PAX250
PENTACHLOROPHENOL, DOWICIDE EC-7 see PAX250
PENTACHLOROPHENOL, DP-2 see PAX250
PENTACHLOROPHENOL, SODIUM SALT see SJA000
PENTACHLOROPHENOL, SODIUM derivative mixed with
   TETRACHLOROPHENOL SODIUM derivative (4581) see PAX750
PENTACHLOROPHENOL, TECHNICAL see PAX250
PENTACHLOROPHENOXY SODIUM see SJA000
PENTACHLOROPHENYL CHLORIDE see HCC500
PENTACHLOROPHENYL METHYL ETHER see MNH250
1,1,2,2,3-PENTACHLOROPROPANE see PAY000
1,1,1,3,3-PENTACHLOROPROPANONE see PAV225
1,1,1,3,3-PENTACHLORO-2-PROPANONE see PAV225
1,1,2,3,3-PENTACHLOROPROPENE see PAY200
1,1,2,3,3-PENTACHLORO-1-PROPENE see PAY200
1,1,2,3,3-PENTACHLOROPROPYLENE see PAY200
2,3,4,5,6-PENTACHLOROPYRIDINE see PAY250
PENTACHLOROTHIOFENOLAT ZINECNATY (CZECH) see BLC500
PENTACHLOROTHIOPHENOL see PAY500
PENTACHLORTHIOFENOL (CZECH) see PAY500
PENTACHLORURE d'ANTIMOINE (FRENCH) see AQD000
PENTACIN see CAY500
PENTACINE see CAY500
PENTACLOROETANO (ITALIAN) see PAW500
PENTACLOROFENOLO (ITALIAN) see PAX250
PENTACON see PAX250
PENTAC WP see DAE425
PENTACYANONITROSYLFERRATE BARIUM see PAY600
PENTACYANONITROSYLFERRATE COBALT see PAY610
PENTACYANONITROSYLFERRATE ZINC see ZJJ400
1-(2,2,3,3,4,4,5,5,6,6,7,7,8,8,8-PENTADECAFLUOROOCTYL)PYRIDINIUM
   TRIFLUOROMETHANESULFONATE see MBV710
1-PENTADECANAMINE see PBA000
N-PENTADECANE see PAY750
1-PENTADECANECARBOXYLIC ACID see PAE250
7-PENTADECANECARBOXYLIC ACID see HFP500
PENTADECANOIC ACID see PAZ000
PENTADECYLIC ACID see PAZ000
PENTADECYLAMINE see PBA000
1-PENTADECYLAMINE see PBA000
n-PENTADECYLAMINE see PBA000
(E)-1,3-PENTADIENE see PBA250
trans-1,3-PENTADIENE see PBA250

2,4-PENTANEDIAMINE, 2-METHYL- see MNI525
1,5-PENTANEDICARBOXYLIC ACID see PIG000
PENTANEDINITRILE see TLR500
PENTANEDINITRILE, 2-BROMO-2-(BROMOMETHYL)- see DDM500
PENTANEDIOIC ACID see GFS000
1,5-PENTANEDIOIC ACID see GFS000
1,5-PENTANEDIOL see PBK750
PENTANE-1,5-DIOL see PBK750
2,4-PENTANEDIOL see PBL000
PENTANEDIOL-2,4 see PBL000
(2,4-PENTANEDIONATO-O,O′)PHENYLMERCURY see PBL250
PENTANEDIONE see ABX750
1,5-PENTANEDIONE see GFQ000
2,3-PENTANEDIONE see PBL350
2,4-PENTANEDIONE (FCC) see ABX750
2,4-PENTANEDIONE, NICKEL(II) DERIVATIVE see PBL500
2,4-PENTANEDIONE, PHENYLMERCURIC SALT see PBL250
2,4-PENTANEDIONE, ZIRCONIUM COMPLEX see PBL750
4,4′-(1,5-PENTANEDIYLBIS(OXY))BIS-BENZENECARBOXIMIDAMIDE, (9CI) see DBM000
PENTANEN (DUTCH) see PBK250
1,2,3,4,5-PENTANEPENTOL see RIF000
1-PENTANETHIOL see PBM000
PENTANI (ITALIAN) see PBK250
PENTANITROANILINE see PBM100
PENTANOCHLOR see SKQ400
PENTANOIC ACID see VAQ000
n-PENTANOIC ACID see VAQ000
tert-PENTANOIC ACID see PJA500
PENTANOIC ACID, 2,2-DIPHENYL, 2-(N,N-DIETHYLAMINO)ETHYL ESTER, HYDROCHLORIDE see PBM500
PENTANOIC ACID, 4-OXO-, ETHYL ESTER (9CI) see EFS600
PENTANOL-1 see AOE000
PENTAN-1-OL see AOE000
2-PENTANOL see PBM750
PENTANOL-2 see PBM750
3-PENTANOL see IHP010
PENTANOL-3 see IHP010
PENTAN-3-OL see IHP010
N-PENTANOL see AOE000
tert-PENTANOL see PBV000
1-PENTANOL ACETATE see AOD725
2-PENTANOL, ACETATE see AOD735
2-PENTANOL CARBAMATE see MOU500
4-PENTANOLIDE see VAV000
3-PENTANOL, 3-METHYL-1-PHENYL- see PFR200
2-PENTANONE see PBN250
PENTANONE-3 see DJN750
3-PENTANONE see DJN750
N-PENTAN-4-ONE-N,N,N-TRIMETHYLAMMONIUM IODIDE see TLY250
PENTANTIN see DAM700
PENTANYL see PDW750
1,4,7,10,13-PENTAOXACYLOPENTADECANE see PBO000
3,6,9,12,15-PENTAOXAHEPTADECANE see PBO250
2,5,8,11,14-PENTAOXAPENTADECANE see PBO500
PENTAPHEN see AON000
PENTAPHENATE see SJA000
PENTAPHENE HYDROCHLORIDE see PET250
PENTAPOTASSIUM TRIPHOSPHATE see PLW400
PENTAPYRROLIDIUM BITARTRATE see PBT000
PENTASILVER DIAMIDOPHOSPHATE see PBO800
PENTASILVER DIIMIDOTRIPHOSPHATE see PBP000
PENTASILVER ORTHODIAMIDOPHOSPHATE see PBP100
PENTASODIUM COLISTINMETHANESULFONATE see SFY500
PENTASODIUM TRIPHOSPHATE see SKN000
PENTASOL see AOE000, PAX250
PENTASULFURE de PHOSPHORE (FRENCH) see PHS000
PENTAZOCINE see DOQ400
PENTAZOCINE HYDROCHLORIDE see PBP300
PENTECH see DAD200
PENTEK see PBB750
2-PENTENAL see PBP500
4-PENTENAL see PBP750
2-PENTENAL, 4,5-EPOXY- see ECE550
1-PENTENE see PBQ000
2-PENTENE see PBQ250
trans-2-PENTENE OZONIDE see PBQ300
4-PENTENOIC ACID see PBQ750
4-PENTEN-1-OL see PBR000
1-PENTEN-3-ONE see PBR250
3-PENTEN-2-ONE see PBR500
1-PENTEN-3-ONE, 1-(2,6,6-TRIMETHYL-2-CYCLOHEXEN-1-YL)-2-METHYL- see COW780
4-PENTENONITRILE, 3-HYDROXY-, S- see COP400

2-PENTEN-4-YN-3-OL see PBR750
PENTESTAN-80 see PBC250
PENTETATE TRISODIUM CALCIUM see CAY500
PENTETRATE UNICELLES see PBC250
PENTHAMIL see CAY500, DJG800
PENTHAZINE see PDP250
PENTHIENATE BROMIDE see PBS000
PENTHIOBARBITAL see PBT250
PENTHIOBARBITAL SODIUM see PBT500
PENTHRANE see DFA400
PENTICORT see COW825
PENTID see BFD000
PENTIDS see BFD000
PENTIFORMIC ACID see HEU000
PENTILEN see CFU750
PENTILIUM see PBT000
PENTIN C see MNM500
PENTINIMID see ENG500
PENTISOMICIN see EBC000
PENTISOMICINA (SPANISH) see EBC000
PENTOBARBITAL see NBT500, PBS250
PENTOBARBITAL CALCIUM see CAV000
R(+)-PENTOBARBITAL SODIUM see PBS500
PENTOBARBITONE SODIUM see NBU000
PENTOBARBITURATE see NBT500
PENTOBARBITURIC ACID see NBT500
PENTOFRAN see DSI709
PENTOFURYL see DGQ500
PENTOKSIL see HMH300
PENTOLE see CPU500
PENTOLINIUM BITARTRATE see PBT000
PENTOLINIUM DITARTRATE see PBT000
PENTOLINIUM TARTRATE see PBT000
PENTOMID see MKU750
PENTONAL see NBU000
PENTOSTAM see AQH800, AQI250
PENTOSTATIN see PBT100
PENTOTHAL see PBT250
PENTOTHAL SODIUM see PBT500
PENTOTHIOBARBITAL see PBT250
PENTOXIFYLLIN see PBU100
PENTOXIFYLLINE see PBU100
PENTOXIL see HMH300
PENTOXIPHYLLIUM see PBU100
PENTOXYL see HMH300
1-PENTOXY-2-METHOXY-4-PROPENYLBENZENE see AOK000
PENTOXYPHYLLINE see PBU100
PENTOYL see EAN700
PENTRAN see DFA400
PENTRANE see DFA400
PENTRATE see PBC250
PENTREX see AIV500
PENTREXL see AIV500
PENTRIOL see PBC250
PENTRYATE 80 see PBC250
PENTYDORM see EQL000
PENTYL see NBU000
PENTYL ACETATE see AOD725
1-PENTYL ACETATE see AOD725
2-PENTYL ACETATE see AOD735
n-PENTYL ACETATE see AOD725
PENTYL ALCOHOL see AOE000
sec-PENTYL ALCOHOL see PBM750
tert-PENTYL ALCOHOL see PBV000
PENTYLAMINE (mixed isomers) see PBV500
2,N-PENTYLAMINOETHYL-p-AMINOBENZOATE see PBV750
8-PENTYLBENZ(a)ANTHRACENE see AOE750
tert-PENTYLBENZENE see AOF000
p-PENTYLBENZOIC ACID see PBW000
4-n-PENTYLBENZOIC ACID see PBW000
PENTYLBIPHENYL see AOF500
PENTYL BUTYRATE see AOG000
PENTYLCANNABICHROMENE see PBW400
PENTYLCARBINOL see HFJ500
3-PENTYLCARBINOL see EGW000
sec-PENTYLCARBINOL see EGW000
PENTYL CHLORIDE see PBW500
α-PENTYLCINNAMALDEHYDE see AOG500
α-PENTYL CINNAMYL ACETATE see AOG750
PENTYLCYCLOHEXANOL ACETATE see AOI000
4-tert-PENTYLCYCLOHEXANONE see AOH750
PENTYLCYCLOPENTANONEPROPANONE see OOO100
PENTYLDICHLOROARSINE see AOI200
1,5-PENTYLENE GLYCOL see PBK750

PENTYLENETETRAZOL see PBI500
PENTYL ESTER PHOSPHORIC ACID see PBW750
PENTYL ETHER see PBX000
PENTYL FORMATE see AOJ500
m-PENTYL FORMATE see AOJ500
PENTYLFORMIC ACID see HEU000
n-PENTYLHYDRAZINE HYDROCHLORIDE see PBX250
tert-PENTYL HYDROPEROXIDE see PBX325
N-PENTYL-N-(4-HYDROXYBUTYL)NITROSAMINE see NLE500
2-PENTYLIDENECYCLOHEXANONE see PBX350
PENTYLIDENE GYROMITRIN see PBJ875
PENTYL MERCAPTAN see PBM000
2-PENTYL-3-METHYL-2-CYCLOPENTEN-1-ONE see DLQ600
PENTYL NITRITE see AOL500
1-PENTYL-3-NITRO-1-NITROSOGUANIDINE see NLC500
4-(PENTYLNITROSAMINO)-1-BUTANOL see NLE500
n-PENTYLNITROSOUREA see PBX500
m-PENTYLOXYCARBANILIC ACID, trans-2-(1-
    PYRROLIDINYL)CYCLOHEXYL ESTER HYDROCHLORIDE see PPW000
(3-(PENTYLOXY)PHENYL)CARBAMIC ACID, 2-(1-
    PYRROLIDINYL)CYCLOHEXYL ESTER, HCl, (E)- see PPW000
PENTYL PENTYLAMINE see DCH200
p-PENTYLPHENOL see AOM250
o-(sec-PENTYL) PHENOL see AOM500
p-(sec-PENTYL) PHENOL see AOM750
p-tert-PENTYLPHENOL see AON000
m-(3-PENTYL)PHENYL-N-METHYL-N-NITROSOCARBAMATE see PBX750
PENTYLTRICHLOROSILANE see PBY750
PENTYLTRIETHOXYSILANE see PBZ000
PENTYLTRIMETHYLAMMONIUM IODIDE see TMA500
1-PENTYL-6,6,9-TRIMETHYL-6a,7,10,10a-TETRAHYDRO-6H-
    DIBENZO(b,d)PYRAN-3-OL see HNE400
3-PENTYL-6,6,9-TRIMETHYL-6a,7,8,10a-TETRAHYDRO-6H-
    DIBENZO(b,d)PYRAN-1-OL see TCM250
1-PENTYNE see PCA250
2-PENTYNE see PCA300
m-PENTYNOL see EQL000
PENTYREST see EQL000
PEN V see PDT500
PEN-VEE see PDT500
PEN-VEE-K see PDT750
PEN-VEE-K POWDER see PDT750
PENVIKAL see PDT750
PEN-V-K POWDER see PDT750
PENWAR see PAX250
PENZYLPENICILLIN SODIUM SALT see BFD250
PEONIA (CUBA) see RMK250
PEP see BLY770, EDS000, PCC000, PJR750
PEPCID see FAB500
PEPCIDINE see FAB500
PEPDUL see FAB500
PEPLEOMYCIN see BLY770
PEPLEOMYCIN SULFATE see PCB000
PEPLOMYCIN see BLY770
PEPPER BUSH see DYA875
PEPPERMINT CAMPHOR see MCF750
PEPPERMINT OIL see PCB250
PEPPER PLANTS see PCB275
PEPPER-ROOT see FAB100
PEPPER TREE see PCB300
PEPPER TURNIP see JAJ000
PEPSTATIN see PCB750
PEPSTATIN A see PCB750
PEPTICHEMIO see PCC000
PER-ABRODIL see DNG400
PERACETIC ACID (MAK) see PCL500
PERACETIC ACID, solution (DOT) see PCL500
PERACON see PCK639
PERAGAL ST see PKQ250
PERAGIT see BBV000
PERANDREN see TBF500, TBG000
PERATOX see PAX250
PERAWIN see PCF275
PERAZIL see CDR000, CDR500
PERAZIL DIHYDROCHLORIDE see CDR000
PERAZIN see CJM250
PERAZINE see PCK500
PERAZINE MALEATE see PCC475
PERBENZOATE de BUTYLE TERTIAIRE (FRENCH) see BSC500
PERBENZOIC ACID see PCM000
PERBROMYL FLUORIDE see PCC750
PERBUTYL H see BRM250
PERCAIN see NOF500
PERCAINE HYDROCHLORIDE see NOF500

PERCAPYL see CHX250
PERCARBAMIDE see HIB500
PERCELINE OIL see HCP550
PERCHLOORETHYLEEN, PER (DUTCH) see PCF275
PERCHLOR see PCF275
PERCHLORAETHYLEN, PER (GERMAN) see PCF275
PERCHLORATE ACID, LEAD SALT, HEXAHYDRATE see LDT000
PERCHLORATE de MAGNESIUM (FRENCH) see PCE000
PERCHLORATES see PCD000
PERCHLORATE de SODIUM (FRENCH) see PCE750
PERCHLORETHYLENE see PCF275
PERCHLORETHYLENE, PER (FRENCH) see PCF275
PERCHLORIC ACID see PCD250
PERCHLORIC ACID, AMMONIUM SALT see PCD500
PERCHLORIC ACID, BARIUM SALT•3H₂O see PCD750
PERCHLORIC ACID, COPPER(II) SALT, DIHYDRATE (8CI, 9CI) see CNO500
PERCHLORIC ACID, MAGNESIUM SALT see PCE000
PERCHLORIC ACID, MANGANESE(2+) SALT, compounded with 3 mols. of
    OCTAMETHYLPYROPHOSPHORAMIDE see TNK400
PERCHLORIC ACID, NICKEL(II) SALT compounded with OCTAMETHYL
    PYROPHOSPHORAMIDE see PCE250
PERCHLORIC ACID, SODIUM SALT see PCE750
PERCHLORIDE of MERCURY see MCY475
PERCHLORMETHYLMERKAPTAN (CZECH) see PCF300
PERCHLOROBENZENE see HCC500
PERCHLOROBUTADIENE see HCD250
PERCHLORO-2-CYCLOBUTENE-1-ONE see PCF250
PERCHLORODIHOMOCUBANE see MQW500
PERCHLOROETHANE see HCI000
PERCHLOROETHYLENE see PCF275
PERCHLOROMETHANE see CBY000
PERCHLOROMETHYL MERCAPTAN see PCF300
PERCHLORON see HOV500
PERCHLOROPENTACYCLODECANE see MQW500
PERCHLOROPENTACYCLO(5.2.1.0²,⁶.0³,⁹.0⁵,⁸)DECANE see MQW500
PERCHLOROPYRIMIDINE see TBU250
PERCHLOROTHIOPHENE see TBV750
PERCHLORURE d'ANTIMOINE (FRENCH) see AQD000
PERCHLORURE de FER see FAU000
PERCHLORYLBENZENE see PCF500
PERCHLORYL FLUORIDE see PCF750
PERCHLORYL HYPOFLUORITE see FFD000
PERCHLORYL PERCHLORATE see PCF775
1-PERCHLORYLPIPERIDINE see PCG000
PERCHROMATES see PCG250
PERCIN see ILD000
PERCIPITATED CALCIUM PHOSPHATE see CAW120
PERCLENE see PCF275
PERCLOROETILENE (ITALIAN) see PCF275
PERCLUSON see CMW750
PERCLUSONE see CMW750
PERCOBARB see ABG750, PCG500
PERCOCCIDE see ALF250
PERCODAN see ABG750, PCG500
PERCODAN HYDROCHLORIDE see DLX400
PERCOLATE see PHX250
PERCORAL see DJS200
PERCORTEN see DAQ800
PERCOSOLVE see PCF275
PERCOTOL see DAQ800
PERCUTACRINE see PMH500
PERCUTACRINE ANDROGENIQUE see TBF500
PERCUTATRINE OESTROGENIQUE ISCOVESCO see DKA600
PERCUTINA see SPD500
PERDIPINE see PCG550
PEREBRAL see DNU100
PEREGRINA (PUERTO RICO, CUBA, US) see CNR135
PEREMESIN see HGC500
PEREQUIL see MQU750
PERFECTA see MQV750
PERFECTHION see DSP400
PERFENAZINA (ITALIAN) see CJM250
PERFLUIDONE see TKF750
PERFLURIDE see FEX875
PERFLUOROACETIC ACID see TKA250
PERFLUOROACETIC ANHYDRIDE see TJX000
PERFLUOROACETYL CHLORIDE see TJX500
PERFLUOROADIPIC ACID DINITRILE see PCG600
PERFLUOROADIPINIC ACID DINITRILE see PCG600
PERFLUOROADIPONITRILE see PCG600
PERFLUOROAMMONIUM OCTANOATE see ANP625
PERFLUOROBUT-2-ENE see OBO000
PERFLUORO-2-BUTENE (DOT) see OBO000
PERFLUORO-tert-BUTYL PEROXYHYPOFLUORITE see PCG650

PERFLUOROCYCLOBUTANE see CPS000
PERFLUORODECANOIC ACID see PCG725
PERFLUORO-N-DECANOIC ACID see PCG725
PERFLUORODIBUTYLETHER see PCG755
PERFLUORO-n-DIBUTYL ETHER see PCG755
PERFLUOROETHENE see TCH500
PERFLUORO ETHER see PCG760
PERFLUOROETHYLENE see TCH500
PERFLUOROFORMAMIDINE see PCG775
PERFLUOROGLUTARIC ACID DINITRILE see HDC300
PERFLUOROGLUTARONITRILE see HDC300
PERFLUOROHEPTANE see PCH000
PERFLUORO-n-HEPTANE see PCH000
PERFLUOROHEXYL IODIDE see PCH100
PERFLUORO HYDRAZINE see TCI000
PERFLUOROISOBUTYLENE (ACGIH) see OBM000
PERFLUOROMETHANE see CBY250
PERFLUOROMETHOXYPROPIONIC ACID METHYL ESTER see PCH275
PERFLUOROMETHYLCYCLOHEXANE see PCH290
PERFLUORO-tert-NITROSOBUTANE see PCH300
PERFLUOROPROPENE see HDF000
PERFLUOROPROPIONIC ACID, METHYL ESTER see PCH350
PERFLUOROPROPYLENE see HDF000
PERFLUOROSUCCINIC ACID see TCJ000
PERFLUOROTOLUENE see PCH500
PERFMID see BSN000
PERGACID VIOLET 2B see FAG120
PERGANTENE see SHF500
PERGITRAL see PBC250
PERGLOTTAL see NGY000
PER-GLYCERIN see LAM000
PERGOLIDE MESYLATE see MPU500
PERHEXILINE MALEATE see PCH800
PERHYDRIT see HIB500
PERHYDROAZEPINE see HDG000
2-PERHYDROAZEPINONE see CBF700
PERHYDROGERANIOL see DTE600
PERHYDROL see HIB000
PERHYDROL-UREA see HIB500
PERHYDRONAPHTHALENE see DAE800
PERIACTIN see PCI500
PERIACTIN HYDROCHLORIDE see PCI250
PERIACTINOL see PCI500
PERICHLOR see NCH500
PERICHLORAL see NCH500
PERICHTHOL see IAD000
PERICIAZINE see PIW000
PERICLOR see NCH500
PERICYAZINE see PIW000
PERIDAMOL see PCP250
PERIDEX-LA see PBC250
PERI-DINAPHTHALENE see PCQ250
PERIFUNAL see PJB500
PERILENE see PCQ250
PERILLA ALCOHOL see PCI550
PERILLA ALDEHYDE see DKX100
PERILLA FRUTESCENS (Linn.) Britt., extract see PCI600
PERILLA KETONE see PCI750
PERILLAL see DKX100
PERILLALDEHYDE see DKX100
1-PERILLALDEHYDE-α-ANTIOXIME see PCJ000
PERILLA OCIMOIDES Linn., extract see PCI600
PERILLARTINE see PCJ000
PERILLA SUGAR see PCJ000
PERILLOL see PCI550
PERILLYL ALCOHOL see PCI550
PERILLYL ALDEHYDE see DKX100
PERIMETAZINE see LEO000
PERIMETHAZINE see LEO000
O-PERIODIC ACID see PCJ250
PERIODIN see PLO500
PERIPHERINE see BBW750
PERIPHERMIN see ACR300
PERIPHETOL see BOV825
PERISOLOL see NDL800
PERISOXAL CITRATE see PCJ325
PERISTIL see PCJ350
PERISTON see PKQ250
PERITRATE see PBC250
PERITYL see PBC250
PERK see PCF275
PERKE see BBK500
PERKLONE see PCF275
PERLAPINE see HOU059

PERLATAN see EDV000
PERLITE see PCJ400
PERLITON BLUE B see TBG700
PERLITON BLUE FFR see MGG250
PERLITON ORANGE 3R see AKP750
PERLITON VIOLET 3R see DBP000
PERLON see NOH000
PERLUTEX see MCA000
PERM-A-CHLOR see TIO750
PERMACIDE see PAX250
PERMAFRESH 183 see DTG000
PERMAFRESH 477 see DTG700
PERMAGARD see PAX250
PERMA KLEER see DJG800
PERMA KLEER 50 ACID see EIX000
PERMA KLEER 50 CRYSTALS see EIV000
PERMA KLEER 50 CRYSTALS DISODIUM SALT see EIX500
PERMA KLEER TETRA CP see EIV000
PERMA KLEER 50, TRISODIUM SALT see TNL250
PERMANENT ORANGE see DVB800
PERMANENT RED 4R see MMP100
PERMANENT RED TONER R see CJD500
PERMANENT WHITE see BAP000, ZKA000
PERMANENT YELLOW see BAK250
PERMANENT YELLOW GHG see DEU000
PERMANGANATE of POTASH (DOT) see PLP000
PERMANGANATE de POTASSIUM (FRENCH) see PLP000
PERMANGANATES see PCJ500
PERMANGANATE de SODIUM (FRENCH) see SJC000
PERMANGANIC ACID AMMONIUM SALT see PCJ750
PERMANGANIC ACID, BARIUM SALT see PCK000
PERMANGANIC ACID, SODIUM SALT see SJC000
PERMANSA RED see CJD500
PERMAPEN see BFC750
PERMASAN see PAX250
PERMASEAL see PJH500
PERMATONE ORANGE see DVB800
PERMATON RED XL 20-7015 see CJD500
PERMATOX DP-2 see PAX250
PERMATOX PENTA see PAX250
PERMETHRIN (USDA) see AHJ750
PERMETRINA (PORTUGUESE) see AHJ750
PERMETRIN (HUNGARIAN) see AHJ750
PERMICORT see CNS750
PERMINAL FC-P see CNH125
PERMITAL see DYE600
PERMITE see PAX250
PERMITIL see TJW500
PERMITIL HYDROCHLORIDE see FMP000
PERMONID see DKX600
PERNAEMON see VSZ000
PERNAEVIT see VSZ000
PERNAZINE see PCK500
PERNETTYA (various species) see PCK600
PERNIPURON see VSZ000
PERNOCTON see BOR000
PERNOSTON see BOR000
PERNOX see CLY500
PERNSATOR-WIRKSTOFF see PNN300
PEROCAN CITRATE see PCK639
PEROCAN CYCLAMATE see PCK669
PEROLYSEN see PBJ000
PERONE see HIB000
PERONE 30 see HIB010
PERONE 35 see HIB010
PERONE 50 see HIB010
PERONIAS (PUERTO RICO) see RMK250
PEROPAL see THT500
PEROPYRENE see DDC400
PEROSIN see EIR000
PEROSSIDO di BENZOILE(ITALIAN) see BDS000
PEROSSIDO di BUTILE TERZIARIO (ITALIAN) see BSC750
PEROSSIDO di IDROGENO (ITALIAN) see HIB000
PEROXAN see HIB000
PEROXIDE see HIB000
PEROXIDE, (PHENYLENEBIS(1-METHYLETHYLIDENE))BIS(1,1-DIMETHYLETHYL)- see BHL100
PEROXIDE, (PHENYLENEDIISOPROPYLIDENE)BIS(tert-BUTYL- see BHL100
PEROXIDES, INORGANIC see PCL000
PEROXIDES, ORGANIC see PCL250
1,4-PEROXIDO-p-MENTHENE-2 see ARM500
PEROXOMONOPHOSPHORIC ACID see PCN500
PEROXOMONOSULFURIC ACID see PCN750
PEROXYACETIC ACID see PCL500

PEROXYACETIC ACID, maximum concentration 43% in acetic acid (DOT) see PCL500
PEROXYACETYL NITRATE see PCL750
PEROXYACETYL PERCHLORATE see PCL775
PEROXYBENZOIC ACID see PCM000
4-PEROXY-CPA see HIF000
PEROXYDE de BARYUM (FRENCH) see BAO250
PEROXYDE de BENZOYLE (FRENCH) see BDS000
PEROXYDE de BUTYLE TERTIAIRE (FRENCH) see BSC750
PEROXYDE d'HYDROGENE (FRENCH) see HIB000
PEROXYDE de LAUROYLE (FRENCH) see LBR000
PEROXYDE de PLOMB (FRENCH) see LCX000
PEROXYDICARBONATE D'ISOPROPYLE (FRENCH) see DNR400
PEROXYDICARBONIC ACID, BIS(1-METHYLETHYL) ESTER see DNR400
PEROXYDICARBONIC ACID DIPROPYL ESTER see DWV400
PEROXYDISULFIRIC ANHYDRIDE see DXH600
PEROXYDISULFURIC ACID DIPOTASSIUM SALT see DWQ000
PEROXYDISULFURYL DIFLUORIDE see PCM250
PEROXYFORMIC ACID see PCM500
PEROXYFUROIC ACID see PCM550
PEROXYHEXANOIC ACID see PCM750
PEROXYLINOLEIC ACID, SODIUM SALT see SIC250
PEROXYLINOLENIC ACID see PCN000
PEROXYMALEIC ACID, O,O-(tert-BUTYL) ESTER see PCN250
PEROXYMONOPHOSPHORIC ACID see PCN500
PEROXYMONOSULFURIC ACID see PCN750
PEROXYNITRIC ACID see PCO000
PEROXYPROPIONIC ACID see PCO100
PEROXYPROPIONYL NITRATE see PCO150
PEROXYPROPIONYL PERCHORATE see PCO175
PEROXYSULFURIC ACID, POTASSIUM SALT see PLP750
PEROXYTRIFLUOROACETIC ACID see PCO250
PERPHENAZIN see CJM250
PERPHENAZINE see CJM250
PERPHENAZINE DIHYDROCHLORIDE see PCO500
PERPHENAZINE HYDROCHLORIDE see PCO750
PERPHENAZINE MALEATE see PCO850
PER-RADIOGRAPHOL see DNG400
PERSADOX see BDS000
PERSAMINE see DLH630
PERSANTIN see PCP250
PERSANTINAT see ARQ750
PERSANTINE see PCP250
PERSEC see PCF275
PERSIAN BERRY see MBU825
PERSIAN BERRY LAKE see MQF250
PERSIAN LILAC see CDM325
PERSIAN RED see LCS000
PERSIA-PERAZOL see DEP800
PERSIMMON see PCP500
PERSISTEN see DJT400
PERSISTOL see TND500
PERSPEX see PKB500
PERSULFATE d'AMMONIUM (FRENCH) see ANR000
PERSULFATE de SODIUM (FRENCH) see SJE000
PERSULFEN see SNN300
PERTESTIS see TBF600
PERTHANE see DJC000
PERTHIOXANTHATE, TRICHLOROMETHYL ALLYL see TIR750
PERTHIOXANTHATE, TRICHLOROMETHYL METHYL see TIS500
PERTOFRAM see DLH630
PERTOFRAN see DLS600, DSI709
PERTOFRANE see DLS600, DSI709
PERTOXIL see CMW500
PERU BALSAM see PCP750
PERU BALSAM OIL see PCQ000
PERUSCABIN see BCM000
PERUVIAN BALSAM see BAE750
PERUVIAN JACINTH see SLH200
PERUVIAN MASTIC TREE see PCB300
PERUVOSID see EAQ050
PERUVOSIDE see EAQ050
PERVAGAL see HKR500
PERVAL see VLF000
PERVERTIN see DBB000
PERVINCAMINE see VLF000
PERVITIN see DBA800, MDT600
PERVONE see VLF000
PERYCIT see NCW300
PERYLENE see PCQ250
PESTAN see DJI000
PESTMASTER see EIY500
PESTMASTER EDB-85 see EIY500
PESTMASTER (OBS.) see MHR200

PESTON XV see PHF750
PESTOX see OCM000
PESTOX 3 see OCM000
PESTOX 14 see BJE750
PESTOX 15 see PHF750
PESTOX 101 see NIM500
PESTOX III see OCM000
PESTOX IV see BJE750
PESTOX PLUS see PAK000
PESTOX XIV see BJE750
PESTOX XV see PHF750
PETA see PBC750
PETANTIN HYDROCHLORIDE see DAM700
PETASITENINE see PCQ750
PETASITES JAPONICUS MAXIM see PCR000
PETEHA see PNW750
PETE-PETE (HAITI) see RBZ400
PETERPHYLLIN see TEP500
PETERSILIENSAMEN OEL (GERMAN) see PAL750
PETHIDINE CHLORIDE see DAM700
PETHIDINETER see DAM600
PETHIDOINE see DAM600
PETHION see PAK000
PETIDIN see DAM700
PETIDION see TLP750
PETIDON see TLP750
PETILEP see TLP750
PETINIMID see ENG500
PETINUTIN see MLP800
PETIT PRECHEUR (CANADA) see JAJ000
PETNAMYCETIN see CDP250
PETNIDAN see ENG500
PETRICHLORAL see NCH500
PETRISUL see AKO500
PETROGALAR see MQV750
PETROHOL see INJ000
PETROL (DOT) see GBY000
PETROLATUM, liquid see MQV750
PETROLEUM see PCR250
PETROLEUM ASPHALT see PCR500
PETROLEUM BENZIN see NAI500
PETROLEUM COKE (calcined) see PCR750
PETROLEUM COKE (uncalcined) see PCS000
PETROLEUM COKE CALCINED (DOT) see PCR750
PETROLEUM COKE UNCALCINED (DOT) see PCS000
PETROLEUM CRUDE see PCR250
PETROLEUM DISTILLATE see PCS250
PETROLEUM DISTILLATES, CLAY-TREATED HEAVY NAPHTHENIC see PCS260
PETROLEUM DISTILLATES, CLAY-TREATED LIGHT NAPHTHENIC see PCS270
PETROLEUM DISTILLATES, HYDROTREATED HEAVY NAPHTHENIC see MQV790
PETROLEUM DISTILLATES (NAPHTHA) see NAI500
PETROLEUM DISTILLATES, SOLVENT-DEWAXED HEAVY PARAFFINIC see MQV825
PETROLEUM ETHER (DOT) see NAI500
PETROLEUM GAS, LIQUEFIED see LGM000
PETROLEUM NAPHTHA (DOT) see NAI500
PETROLEUM PITCH see ARO500
PETROLEUM ROOFING TAR see PCR500
PETROLEUM 60 SOLVENT see PCS750
PETROLEUM 70 SOLVENT see PCT000
PETROLEUM SPIRIT (DOT) see NAI500
PETROLEUM SPIRITS see PCT250
PETROLEUM 50 THINNER see PCT500
PETROL ORANGE Y see PEJ500
PETROL YELLOW WT see DOT300
PETROSULPHO see IAD000
PETUNIDOL see PCU000
PETZINOL see TIO750
PEVARYL see EAE000
PEVIKON D 61 see PKQ059
PEVITON see NCQ900
PEXID see PCH800
PEYRONE'S CHLORIDE see PJD000
PF-1 see DSA800
PF-3 see IRF000
PF-26 see DWK700
PF 38 see FQU875
PF-82 see DWK900
PF 1593 see BON325
PFDA see PCG725
PFEFFERMINZ OEL (GERMAN) see PCB250

PFETFFER'S SUBSTANCE see AMK250
PFH see PNM650
PFIB see OBM000
PFIKLOR see PLA500
PFIZER 1393 see BDE250
PFIZER-E see EDJ500
PFIZERPEN see BFD000
PFIZERPEN A see AIV500
PFIZERPEN VK see PDT750
PFPA see FLF000
PFT see CML835
PG see PML250
PG 12 see PML000
PG 501 see KBB000
PGA see AHC000
PGA^1 see POC250
PGA2 see MCA025
PGA^2 see MCA025
(15S)-PGA^2 see MCA075
5,6-cis-PGA^2 see MCA025
PGABA see PEE500
PGD2 see POC275
PGDN see PNL000
PGE see PFH000
PGE-1 see POC350
PGE2 see DVJ200
PGE2 SODIUM SALT see POC360
PGF 1-α see POC400
PGF2-α see POC500
(±)-PGF2-α see POC525
dl-PGF2-α see POC525
15M-PGF2-α see MLO300
PGF2-α racemic mixture see POC525
PGF2 METHYL ESTER see DVJ100
PGF2-α METHYL ESTER see DVJ100
PGF2-α THAM see POC750
PGF2-α TRIS SALT see POC750
PGF2-α TROMETHAMINE see POC750
Ph. 458 see SCA525
PH 1882 see BNV500
PH 60-40 see CJV250
PHA see PIB575
PHACETUR see PEC250
PHALDRONE see CDO000
PHALLOIDIN see PCU350
PHALLOIDINE see PCU350
PHALTAN see TIT250
PHANAMIPHOS see FAK000
PHANANTIN see DKQ000
PHANODORM see TDA500
PHANODORN see TDA500
PHANQUINONE see PCY300
PHANQUINONUM see PCY300
PHANQUONE see PCY300
PHARGAN see DQA400
PHARLON see EDS100
PHARMAGEL A see PCU360
PHARMAGEL AdB see PCU360
PHARMAGEL B see PCU360
PHARMANTHRENE GOLDEN YELLOW see DCZ000
PHARMAZOID RED KB see CLK225
PHARMEDRINE see AOB250
PHAROS 100.1 see SMR000
PHASEOLUNATIN see GFC100
PHASOLON see BDJ250
PH BC see BQJ500
PHBN see NLN000
PHC see PMY300
M-PHDM see BKL750
PHEASANT'S EYE see PCU375
PHEBUZIN see BRF500
PHELIPAEA CALOTROPIDIS Walp., extract see CMS248
α-PHELLANDRENE (FCC) see MCC000
PHELLOBERIN A see PCU390
PHEM see PCU400
PHEMERIDE see BEN000
PHEMERNITE see MDH500
PHEMEROL CHLORIDE see BEN000
PHEMEROL CHLORIDE MONOHYDRATE see BBU750
PHEMETONE see ENB500
PHEMITHYN see BEN000
PHEMITON see ENB500
PHEMITONE see ENB500
PHENACAINE see BJO500

PHENACALUM see PEC250
PHENACEMIDE see PEC250
PHENACEREUM see PEC250
PHENACETALDEHYDE DIMETHYL ACETAL see PDX000
p-PHENACETIN see ABG750
PHENACETUR see PEC250
PHENACETYLCARBAMIDE see PEC250
PHENACETYLUREA see PEC250
PHENACHLOR see TIW000
PHENACID see PCU425
PHENACIDE see CDV100
PHENACITE see SCN500
PHENACTYL see CKP250
PHENACYLAMINE see AHR250
PHENACYL-6-AMINOPENICILLINATE see PCU500
PHENACYL CHLORIDE see CEA750
PHENACYLIDENE CHLORIDE see DEN200
PHENACYLPIVALATE see PCV350
PHENADONE see MDO750
PHENADONE HYDROCHLORIDE see MDP750
PHENADOR-X see BGE000
PHENAEMAL see EOK000
PHENAGLYCODOL see CKE750
PHENAKITE see SCN500
PHENALCO see MCU750
PHENALENO(1,9-gh)QUINOLINE see PCV500
PHENALGENE see AAQ500
PHENALLYMAL see AGQ875
PHENALLYMALUM see AGQ875
PHENALZINE see PFC750
PHENALZINE DIHYDROGEN SULFATE see PFC750
PHENALZINE HYDROGEN SULPHATE see PFC750
PHENAMACIDE HYDROCHLORIDE see PCV750
PHENAMIDE see FAJ150, PCV775
PHENAMINE see AOB250
PHENAMINE BLUE BB see CMO000
PHENAMINE SKY BLUE A see CMO500
PHENAMIZOLE HYDROCHLORIDE see DCA600
PHENANTHRA-ACENAPHTHENE see PCW000
9,10-PHENANTHRAQUINONE see PCX250
PHENANTHREN (GERMAN) see PCW250
PHENANTHRENE see PCW250
PHENANTHRENE-1,2-DIHYDRODIOL see DMA000
PHENANTHRENE-3,4-DIHYDRODIOL see PCW500
9,10-PHENANTHRENEDIONE see PCX250
PHENANTHRENE-9,10-EPOXIDE see PCX000
9,10-PHENANTHRENE OXIDE see PCX000
PHENANTHRENEQUINONE see PCX250
9,10-PHENANTHRENEQUINONE see PCX250
PHENANTHRENETETRAHYDRO-3,4-EPOXIDE see ECR500
4,7-PHENANTHROLENE-5,6-QUINONE see PCY300
o-PHENANTHROLINE see PCY250
β-PHENANTHROLINE see PCY250
1,10-PHENANTHROLINE see PCY250
1,10-o-PHENANTHROLINE see PCY250
4,7-PHENANTHROLINE-5,6-DIONE see PCY300
PHENANTHRO(2,1-d)THIAZOLE see PCY400
2-PHENANTHRYLACETAMIDE see AAM250
9-PHENANTHRYLACETAMIDE see PCY750
N-2-PHENANTHRYLACETAMIDE see AAM250
N-(2-PHENANTHRYL)ACETAMIDE see AAM250
N-3-PHENANTHRYLACETAMIDE see PCY500
N-9-PHENANTHRYLACETAMIDE see PCY750
2-PHENANTHRYLACETHYDROXAMIC ACID see PCZ000
N-(2-PHENANTHRYL)ACETOHYDROXAMIC ACETATE see ABK250
N-2-PHENANTHRYLACETOHYDROXAMIC ACID see PCZ000
2-PHENANTHRYLAMINE see PDA250
3-PHENANTHRYLAMINE see PDA250
9-PHENANTHRYLAMINE see PDA750
PHENANTOIN see MKB250
PHENANTRIN see PCW250
PHENAROL see CKF500
PHENARONE see PEC250
PHENARSAZINE CHLORIDE see PDB000
PHENARSAZINE OXIDE see PDB250
PHENARSEN see OOK100
PHENARSENAMINE see SAP500
PHENASAL see DFV400
10-PHENASAZINETHIOL, S-ESTER with O,O-DIISOOCTYLPHOS-
     PHORODITHIOATE see PDB300
S-(10-PHENASAZINYL)-O,O-DIISOOCTYLPHOSPHORODITHIOATE see PDB300
PHENATHYL see CKP250
PHENATOINE see DKQ000
PHENATOX see CDV100

PHENAZEPAM see PDB350
PHENAZIN see PDB750
PHENAZINE see DKE800, PDB500
PHENAZINE ETHOSULFATE see EOE000
PHENAZINE-9-OXIDE see PDB750
PHENAZINE-N-OXIDE see PDB750
PHENAZINE-5N-OXIDE see PDB750
PHENAZIN OXIDE see PDB750
PHENAZIN-5-OXIDE see PDB750
PHENAZITE see SCN500
PHENAZO see PDC250
PHENAZOCINE HYDROBROMIDE see PMD325
PHENAZODINE see PDC250, PEK250
PHENAZOLINE see PDC000
PHENAZONE (pharmaceutical) see AQN000
PHENAZOPYRIDINE see PEK250
PHENAZOPYRIDINE HYDROCHLORIDE see PDC250
PHENAZOPYRIDINIUM CHLORIDE see PDC250
PHENBENZAMINE see BEM500
PHENBENZAMINE HYDROCHLORIDE see PEN000
PHENBUTAZOL see BRF500
PHENCAPTON see PDC750
PHENCARBAMID see PDC850
PHENCARBAMIDE see PDC850
PHENCARBAMIDE HYDROCHLORIDE see PDC875
PHENCARBAMID HYDROCHLORIDE (GERMAN) see PDC875
PHENCAROL see DWM400
PHENCEN see PMI750
PHENCYCLIDINE see PDC890
PHENCYCLIDINE HYDROCHLORIDE see AOO500
PHENDAL see DRR400
PHENDIMETRAZINE BITARTRATE see DKE800
PHENDIMETRAZINE HYDROCHLORIDE see DTN800
PHENDIMETRAZINE TARTRATE see PDD000
PHENE see BBL250
PHENEDRINE see AOB250, BBK000
PHENEENE GERMICIDAL SOLUTION and TINCTURE see AFP250
PHENEGIC see PDP250
PHENELZIN see PFC750
PHENELZINE see PFC500
PHENELZINE ACID SULFATE see PFC750
PHENELZINE BISULPHATE see PFC750
PHENELZINE SULFATE see PFC750
PHENEMALUM see SID500
(v-PHENENYLTRIS(OXYETHYLENE))TRIS(TRIETHYLAMMONIUM IODIDE)
     see PDD300
PHENERGAN see DQA400
PHENERGAN HYDROCHLORIDE see PMI750
PHENESTERINE see CME250
PHENESTRIN see CME250
PHENETAMINE HYDROCHLORIDE see LFK200
PHENETHAMINE HYDROCHLORIDE see LFK200
PHENETHANOL see PDD750
PHENETHECILLIN POTASSIUM see PDD350
PHENETHECILLIN POTASSIUM SALT see PDD350
PHENETHICILLIN K see PDD350
PHENETHICILLIN K SALT see PDD350
PHENETHIDINE see PDD500
β-PHENETHYBIGUANIDE see PDF000
2-PHENETHYL ACETATE see PFB250
β-PHENETHYL ACETATE see PFB250
PHENETHYL ALCOHOL see PDD750
2-PHENETHYL ALCOHOL see PDD750
α-PHENETHYL ALCOHOL see PDE000
β-PHENETHYL ALCOHOL see PDD750
PHENETHYL ALCOHOL, BENZOATE see PFB750
PHENETHYL ALCOHOL, FORMATE see PFC250
β-PHENETHYLAMINE see PDE250
β-PHENETHYLAMINE HYDROCHLORIDE see PDE500
PHENETHYLAMINE, α-METHYL-, SULFATE (2:1) see BBK250
β-PHENETHYL-o-AMINOBENZOATE see APJ500
PHENETHYL ANTHRANILATE see APJ500
PHENETHYL BENZOATE see PFB750
1-PHENETHYLBIGUANIDE see PDF000
β-PHENETHYLBIGUANIDE see PDF000
PHENETHYLBIGUANIDE HYDROCHLORIDE see PDF250
1-PHENETHYLBIGUANIDE HYDROCHLORIDE see PDF250
N'-β-PHENETHYLBIGUANIDE HYDROCHLORIDE see PDF250
2-PHENETHYL BUTANOATE see PFB800
β-PHENETHYL N-BUTANOATE see PFB800
PHENETHYL BUTYRATE see PFB800
PHENETHYLCARBAMID (GERMAN) see EFE000
PHENETHYL CHLORACETATE see PDF500
PHENETHYL CINNAMATE see BEE250

β-PHENETHYL CINNAMATE see BEE250
PHENETHYL CYANIDE see HHP100
PHENETHYLDIGUANIDE see PDF000
PHENETHYLENE see SMQ000
PHENETHYLENE OXIDE see EBR000
PHENETHYL ESTER HYDRACRYLIC ACID see PFC100
PHENETHYL ESTER ISOVALERIC ACID see PDF775
N'-β-PHENETHYLFORMAMIDINYLLIMINOUREA see PDF000
PHENETHYL FORMATE see PFC250
PHENETHYLHYDRAZINE see PFC500
PHENETHYLHYDRAZINE SULFATE (1:1) see PFC750
1-(1-PHENETHYL)-IMIDAZOLE-5-CARBOXYLIC ACID, METHYL ESTER,
     HYDROCHLORIDE see MQQ750
PHENETHYL ISOBUTYRATE see PDF750
β-PHENETHYL ISOTHIOCYANATE see ISP000
PHENETHYL ISOVALERATE see PDF775
2-PHENETHYL 2-METHYLBUTYRATE see PDF790
PHENETHYLMETHYLETHYLCARBINOL see PFR200
PHENETHYL METHYL KETONE see PDF800
PHENETHYL-8-OXA-1-DIAZA-3,8-SPIRO(4,5)DECANONE-2-HYDROCHLRO-
     IDE see DAI200
PHENETHYL PHENYLACETATE see PDI000
N-(p-PHENETHYL)PHENYLACETOHYDROXAMIC ACID see PDI250
α-(p-PHENETHYLPHENYL)-1-IMIDAZOLEETHANOL MONOHYDROCHLOR-
     IDE see DAP880
1-PHENETHYLPIPERIDINE see PDI500
N-(1-PHENETHYL-4-PIPERIDI-NYL)PROPIONANILIDE DIHYDROGEN
     CITRATE see PDW750
1-PHENETHYL-4-PIPERIDYL-p-AMINOBENZOATE HYDROCHLORIDE
     see PDJ000
1-PHENETHYL-4-PIPERIDYL BENZOATE HYDROCHLORIDE see PDJ250
N-(1-PHENETHYL-4-PIPERIDYL)PROPIONANILIDE CITRATE see PDW750
N-(1-PHENETHYL-4-PIPERIDYL)PROPIONANILIDE DIHYDROGEN CITRATE
     see PDW750
PHENETHYL PROPIONATE see PDK000
1-PHENETHYL-4-N-PROPIONYLANILINOPIPERIDINE see PDW500
N-PHENETHYL-4-(N-PROPIONYLANILINO)PIPERIDINE see PDW500
PHENETHYL SALICYLATE see PDK200
1-PHENETHYLSEMICARBAZIDE see PDK300
2-PHENETHYL-3-THIOSEMICARBAZIDE see PDK500
PHENETHYL TIGLATE see PFD250
PHENETHYLUREA see PDK750
PHENETICILLIN POTASSIUM see PDD350
p-PHENETIDINANTIMONYLTARTRAT (GERMAN) see PDL500
p-PHENETIDINE (DOT) see PDD500
m-PHENETIDINE ANTIMONYL TARTRATE see PDL000
o-PHENETIDINE ANTIMONYL TARTRATE see PDK900
p-PHENETIDINE ANTIMONYL TARTRATE see PDL500
PHENETIDINE HYDROCHLORIDE see PDL750
p-PHENETOLCARBAMID (GERMAN) see EFE000
p-PHENETOLCARBAMIDE see EFE000
PHENETOLE see PDM000
p-PHENETOLECARBAMIDE see EFE000
PHENETURIDE see PFB350
p-PHENETYLUREA see EFE000
PHENFLUORAMINE HYDROCHLORIDE see PDM250
PHENFORMINE see PDF000
PHENGLYKODOL see CKE750
PHENHYDREN see PFJ750
PHENIBUT HYDROCHLORIDE see GAD000
PHENIC ACID see PDN750
PHENICARB see PEC250
PHENICOL see EGQ000
PHENIDONE see PDM500
PHENIDYLATE see MNQ000
PHENIGAM see PEE500
PHENIGAMA HYDROCHLORIDE see GAD000
PHENIGAM HYDROCHLORIDE see GAD000
PHENINDAMINE HYDROCHLORIDE see TEQ700
PHENINDAMINE HYDROGEN TARTRATE see PDD000
PHENINDIONE see PFJ750
PHENIODOL see PDM750
PHENIPRAZINE see PDN000
PHENIPRAZINE HYDROCHLORIDE see PDN250
PHENIRAMINE MALEATE see TMK000
PHENISATIN see ACD500
PHENISOBROMOLATE see IOS000
(±)-PHENISOPROPYLAMINE SULFATE see AOB250
PHENISTAN see SPC500
PHENITOL see MCU750
PHENITROTHION see DSQ000
PHENIZIDOLE see PEL250
PHENIZINE see PDN000
PHENLINE see PFC750

PHENMAD see ABU500
PHENMEDIPHAM see MEG250
PHENMERZYL NITRATE see MCU750
PHENMETHYL TRIMETHYLAMMONIUM IODIDE see BFM750
PHENMETRAZIN see PMA750
PHENMETRAZINE see PMA750
PHENMETRAZINE HYDROCHLORIDE see MNV750
PHENOBAL see EOK000
PHENOBAL SODIUM see SID000
PHENOBARBITAL see EOK000
PHENOBARBITAL ELIXIR see SID000
PHENOBARBITAL Na see SID000
PHENOBARBITAL SODIUM see SID000
PHENOBARBITAL SODIUM SALT see SID000
PHENOBARBITOL and DIPHENYLHDANTOIN see DWD000
PHENOBARBITONE see EOK000
PHENOBARBITONE and PHENOBARBITONE see DWD000
PHENOBARBITONE SODIUM see SID000
PHENOBARBITONE SODIUM SALT see SID000
PHENOBARBITURIC ACID see EOK000
PHENO BLACK EP see AQP000
PHENO BLUE 2B see CMO000
PHENOBOLIN see DYF450
PHENOCAINE see BQH250
PHENOCHLOR see PJL750
PHENOCLOR DP6 see PJN250
PHENODIANISYL see PDN500
PHENODIANISYL HYDROCHLORIDE see PDN500
PHENODIOXIN see DDA800
PHENODODECINIUM BROMIDE see DXX000
PHENODYNE see PFC750
PHENOFORMINE HYDROCHLORIDE see PDF250
PHENOHEP see HCI000
PHENOL see PDN750
PHENOL (liquid) see PDO000
PHENOL, molten (DOT) see PDN750
PHENOL ACETATE see PDY750
PHENOL ALCOHOL see PDN750
PHENOL, 4-ALLYL-2-METHOXY-, FORMATE (ester) see EQS100
PHENOL, 2-AMINO-4-ARSENOSO- see OOK100
PHENOL, 2-AMINO-4-ARSENOSO-, SODIUM SALT see ARJ900
PHENOL, 5-((AMINOOXY)METHYL)-2-BROMO-(9CI) see BMM600
PHENOL, p-ARSENOSO- see HNG800
PHENOL, 4-ARSENOSO-2-NITRO- see NHE600
PHENOL-p-ARSONIC ACID see PDO250
PHENOL, 2,4-BIS(ACETOXYMERCURI)- see HNK575
PHENOL, 6-t-BUTYL-3-(2-IMIDAZOLIN-2-YLMETHYL)-2,4-DIMETHYL-
    see ORA100
PHENOL, 4-(3-CARBAZOLYLAMINO)- see CBN100
PHENOLCARBINOL see BDX500
PHENOL, 2-(((2-CHLOROETHYL)AMINO)METHYL)-4-NITRO- see CGQ280
PHENOL, 4,4'-(1,2-DIETHYLETHYLENE)BIS(2-AMINO- see DJI250
PHENOL, 4,4'-(1,2-DIETHYLIDENE-1,2-ETHANEDIYL)BIS-, (E,E)-(9CI)
    see DHB550
PHENOLE (GERMAN) see PDN750
PHENOL, 4-ETHENYL- (9CI) see VQA200
PHENOL, 4-ETHOXY-(9CI) see EFA100
PHENOL-GLYCIDAETHER (GERMAN) see PFH000
PHENOL GLYCIDYL ETHER (MAK) see PFH000
PHENOLPHTHALEIN see PDO750
PHENOL SODIUM SALT see SJF000
PHENOL TRINITRATE see PID000
PHENOLURIC see EOK000
PHENOL, p-VINYL- see VQA200
PHENOMERCURIC ACETATE see ABU500
PHENOMET see EOK000
PHENONYL see EOK000
PHENOPENICILLIN see PDT500
PHENO-m-PENICILLIN see PDD350
PHENOPLASTE ORGANOL RED B see SBC500
PHENOPROMIN see BBK500
PHENOPROPAZINE see DIR000
PHENOPROZINE see DIR000
PHENOPYRIDINE see QPA000
PHENOPYRINE see BRF500
PHENOQUIN see PGG000
PHENOSAN see PDP250
PHENOSANE see PEE750
PHENOSELENAZIN-5-IUM, 3-AMINO-7-(DIMETHYLAMINO)-2-METHYL-,
    CHLORIDE see SBU950
PHENOSMOLIN see PDP100
PHENOSUCCIMIDE see MNZ000
PHENOTAN see ACE500, BRE500
PHENOTEROL HYDROBROMIDE see FAQ100

PHENOTHIAZINE see PDP250
PHENOTHIAZINE-10-CARBODITHIOIC ACID-2-(DIETHYLAMINO)ETHYL
    ESTER see PDP500
10-PHENOTHIAZINECARBOXYLIC ACID β-DIETHYLAMINOETHYL ESTER
    HYDROCHLORIDE see THK600
PHENOTHIAZINE, 2-CHLORO-10-(3-(1-METHYL-4-PIPERAZINYL)PROPYL)-,
    ETHANEDISULFONATE see PME700
PHENOTHIAZINE, 10-(3-(DIMETHYLAMINO)-2-METHYLPROPYL)-3-ME-
    THOXY- see PDP600
PHENOTHIAZINE HYDROCHLORIDE see CKP500
10H-PHENOTHIAZINE, 10-(3-(4-METHYL-1-PIPERAZINYL)PROPYL)- (Z)-2-
    BUTENEDIOATE see PCC475
10H-PHENOTHIAZINE-10-PROPANAMINE, 3-METHOXY-N,N,β-TRIMETHYL-
    see PDP600
3-PHENOTHIAZINESULFONAMIDE, N,N-DIMETHYL- see DTK300
10H-PHENOTHIAZINE-3-SULFONAMIDE, N,N-DIMETHYL- see DTK300
PHENOTHIAZINYL-10-DITHIACARBOXYLATE de DIETHYLAMINOETHYLE
    (FRENCH) see PDP500
N-(β-(10-PHENOTHIAZINYL)PROPYL)TRIMETHYLAMMONIUM METHYL
    SULFATE see MRW000
PHENOTHIOXIN see PDQ750
PHENOTOLAMINE see PDW400
PHENOTURIC see EOK000
PHENOVERM see PDP250
PHENO VIOLET N see CMP000
PHENOVIS see PDP250
PHENOX see DAA800
PHENOXAKSINE OXIDE see OMY850
PHENOXALIN see DRN200
PHENOXATHINE see PDQ750
PHENOXATHRIN see PDQ750
PHENOXAZOLE see IBM000
PHENOXENE HYDROCHLORIDE see CIS000
PHENOXETHOL see PER000
PHENOXETOL see PER000
PHENOXTHIN see PDQ750
PHENOXUR see PDP250
PHENOXYACETALDEHYDE see PDR000
6-PHENOXYACETAMIDOPENICILLANIC ACID see PDT500
PHENOXYACETIC ACID see PDR100
PHENOXYACETYLENE see PDR250
PHENOXYAETHYLPENICILLIN K-SALZ (GERMAN) see PDD350
4-PHENOXYANILINE see PDR500
p-PHENOXYANILINE see PDR500
PHENOXYBENZAMIDE HYDROCHLORIDE see DDG800
PHENOXYBENZAMINE see PDT250
PHENOXYBENZENE see PFA850
3-PHENOXYBENZYL (±)-3-(2,2-DICHLOROVINYL)-2,2-
    DIMETHYLCYCLOPROPANECARBOXYLATE see AHJ750
(4-PHENOXYBUTYL)HYDRAZINE MALEATE see PDS250
3-PHENOXY-1,2-EPOXYPROPANE see PFH000
PHENOXYETHANOIC ACID see PDR100
PHENOXYETHANOL see PER000
2-PHENOXYETHANOL see PER000
2-PHENOXY-ETHANOL, ACRYLATE see PER250
2-PHENOXYETHANOL PROPIONATE see EJK500
2-(2-PHENOXYETHOXY)ETHANOL see PEQ750
PHENOXYETHYL ALCOHOL see PER000
β-PHENOXYETHYLDIMETHYLDODECYLAMMONIUM BROMIDE see DXX000
(2-PHENOXYETHYL)GUANIDINE see PDS500
(2-PHENOXYETHYL)HYDRAZINE HYDROCHLORIDE see PDS750
PHENOXYETHYL ISOBUTYRATE see PDS900
α-PHENOXYETHYLPENICILLIN see PDD350
α-PHENOXYETHYLPENICILLIN POTASSIUM see PDD350
α-PHENOXYETHYLPENICILLIN POTASSIUM SALT see PDD350
PHENOXYETHYL PROPIONATE see EJK500
N-PHENOXYISOPROPYL-N-BENZYL-β-CHLOROETHYLAMINE see PDT250
N-2-PHENOXYISOPROPYL-N-BENZYL-CHLOROETHYLAMINE HYDROCHLO-
    RIDE see DDG800
N-PHENOXYISOPROPYL-N-BENZYL-β-CHLOROETHYLAMINE HYDROCHLO-
    RIDE see DDG800
PHENOXYLENE see SIL500
PHENOXYLENE SUPER see CIR250
PHENOXYMETHYLENEPENICILLINIC ACID see PDT500
PHENOXYMETHYLPENICILLIN see PDT500
d-α-PHENOXYMETHYLPENICILLINATE K SALT see PDT750
PHENOXYMETHYLPENICILLIN POTASSIUM see PDT750
N-(4-(PHENOXY-METHYL)-Γ-PHENYL-PROPYL)-MORPHOLIN (GERMAN)
    see PDU250
N-(Γ-(5-PHENOXYMETHYL-PHENYL)-PROPYL)-MORPHOLIN (GERMAN)
    see PDU250
4-(3-(p-PHENOXYMETHYLPHENYL)PROPYL)MORPHOLINE see PDU250
4-(3-(4-(PHENOXYMETHYL)PHENYL)PROPYL)-MORPHOLINE, (9CI)
    see PDU250

((1-PHENOXYMETHYL)PROPYL)HYDRAZINE HYDROCHLORIDE see PDU500
PHENOXYMETHYL-6-TETRAHYDROXAZINE-1,3-THIONE-2 see PDU750
6-PHENOXYMETHYL-3,4,5,6-TETRAHYDROXAZINE-1,3 THIONE-2 see
    PDU750
(5-PHENOXYPENTYL)HYDRAZINE MALEATE see PDV000
m-PHENOXYPHENOL see PDV250
(3-PHENOXYPHENYL)METHYL-3-(2,2-DICHLORETHENYL)-2,2-
    DIMETHYLCYCLOPROPANECARBOXYLATE see AHJ750
dl-2-(3-PHENOXYPHENYL)-PROPIONIC ACID CALCIUM SALT, DIHYDRATE see
    FAP100
d,l-2-(3-PHENOXYPHENYL)PROPIONIC ACID SODIUM SALT see FAQ000
3-PHENOXY-1,2-PROPANEDIOL DIACETATE see PFF300
PHENOXYPROPAZINE MALEATE see PDV700
PHENOXYPROPENE OXIDE see PFH000
2-(PHENOXY-2-PROPYLAMINO)-1-(p-HYDROXYPHENYL)-1-PROPANOL HY-
    DROCHLORIDE see VGF000
PHENOXYPROPYLENE OXIDE see PFH000
(3-PHENOXYPROPYL)GUANIDINE SULFATE see GLS700
(3-PHENOXYPROPYL)HYDRAZINE MALEATE see PDV750
4-PHENOXY-3-(PYRROLIDINYL)-5-SULFAMOYLBENZOIC ACID see PDW250
16-PHENOXY-17,18,19,20 TETRANOR PROSTAGLANDIN E$_2$ METHYL SUL-
    FONYLAMIDE see SOU650
16-PHENOXY-omega-17,18,19,20-TETRANOR PROSTAGLANDIN E2
    METHYLSULFONYLAMIDE see SOU650
PHENOXYTOL see PER000
4-(3-(α-PHENOXY-p-TOLYL)PROPYL)-MORPHOLINE see PDU250
PHENOXYTRIETHYLSTANNANE see TJV250
PHENSEDYL see DQA400
PHENTALAMINE see PDW400
PHENTANYL see PDW500
PHENTANYL CITRATE see PDW750
PHENTERMINE see DTJ400
PHENTHIAZINE see PDP250
PHENTHIURAM see FAQ930
PHENTHOATE see DRR400
PHENTIN ACETATE see ABX250
PHENTINOACETATE see ABX250
PHENTOLAMINE see PDW400
PHENTOLAMINE MESILATE see PDW950
PHENTOLAMINE MESYLATE see PDW950
PHENTOLAMINE METHANESULFONATE see PDW950
PHENTOLAMINE METHANESULPHONATE see PDW950
PHENTYRIN see BHY500
PHENURIDE see PFB350
PHENURON see PEC250
PHENUTAL see PEC250
PHENVALERATE see FAR100
PHENYBUT HYDROCHLORIDE see GAD000
PHENYCHOLON see EGQ000
PHENYGAM HYDROCHLORIDE see GAD000
PHENYL-Γ see PEE500
PHENYLACETALDEHYDE DIMETHYL ACETAL see PDX000
PHENYLACETALENEGLYCOL ACETAL see BEN250
PHENYLACETALDEHYDE (FCC) see BBL500
PHENYLACETALDEHYDE GLYCERYL ACETAL see PDX250
2-PHENYLACETAMIDE see PDX750
N-PHENYLACETAMIDE see AAQ500, PDX500
α-PHENYLACETAMIDE see PDX750
PHENYLACETAMIDOPENICILLANIC ACID see BDY669
p-PHENYLACETANILIDE see PDY500
2'-PHENYLACETANILIDE see PDY000
3'-PHENYLACETANILIDE see PDY250
4'-PHENYLACETANILIDE see PDY500
PHENYL ACETATE see PDY750
N'-PHENYLACETHYDRAZIDE see ACX750
PHENYLACETIC ACID see PDY850
omega-PHENYLACETIC ACID see PDY850
PHENYLACETIC ACID ALLYL ESTER see PMS500
PHENYLACETIC ACID AMIDE see PDX750
PHENYLACETIC ACID 2-BENZYLHYDRAZIDE see PDY870
PHENYL ACETIC ACID DIETHYLAMINOETHOXYETHANOL ESTER
    CITRATE see BOR350
PHENYLACETIC ACID, ETHYL ESTER see EOH000
PHENYLACETIC ACID, p-METHOXYBENZYL ESTER see APE000
PHENYLACETIC ACID, METHYL ESTER see MHA500
PHENYLACETIC ACID, PHENETHYL ESTER see PDI000
PHENYLACETIC ALDEHYDE see BBL500
N-PHENYLACETOACETAMIDE see AAY000
PHENYLACETO-ACETONITRILE see PEA500
α-PHENYLACETOACETONITRILE see PEA500
PHENYLACETONITRILE see PEA750
2-PHENYLACETONITRILE see PEA750
PHENYLACETONITRILE, liquid (DOT) see PEA750
2-PHENYLACETOPHENONE see PEB000

4'-PHENYL-o-ACETOTOLUIDE see PEB250
(PHENYLACETOXY)TRIETHYL PLUMBANE see TJT250
PHENYL-β-ACETYLAMINE see PDX750
PHENYLACETYLENE see PEB750
3-(α-PHENYL-β-ACETYLETHYL)-4-HYDROXYCOUMARIN see WAT200
3-(1'-PHENYL-2'-ACETYLETHYL)-4-HYDROXYCOUMARIN see WAT200
(PHENYL-1 ACETYL-2 ETHYL)-3-HYDROXY-4 COUMARINE (FRENCH) see
    WAT200
PHENYLACETYLGLYCINE DIMETHYLAMIDE see PEB775
(PHENYLACETYL)UREA see PEC250
α-PHENYLACETYLUREA see PEC250
PHENYLACETYLUREE (FRENCH) see PEC250
PHENYLACROLEIN see CMP969
3-PHENYLACROLEIN see CMP969
PHENYLACRYLIC ACID see CMP975
3-PHENYLACRYLIC ACID see CMP975
tert-β-PHENYLACRYLIC ACID see CMP975
3-PHENYLACRYLOPHENONE see CDH000
β-PHENYLACRYLOPHENONE see CDH000
β-PHENYLAETHYLAMINE see PDE250
1-PHENYL-2-AETHYLAMINO-PROPAN (GERMAN) see EGI500
PHENYLAETHYLBERNSTEINSAEURE-AETHYL-DIAETHYL-AMINOAETHYL-
    DI-ESTER (GERMAN) see SBC700
1-PHENYLAETHYLBIGUANID HYDROCHLORID (GERMAN) see PDF250
PHENYLAETHYLCARBINOL (GERMAN) see EGQ000
5,5-PHENYL-AETHYL-3-(β-DIAETHYLAMINO-AETHYL)-2,4,6-TRIOXO-
    HEXAHYDROPYRIMIDIN-HCl (GERMAN) see HEL500
PHENYLAETHYL-HYDRAZIN see PFC750
PHENYLAETHYLMALONSAEURE-AETHYL-DIAETHYLAMINOAETHYL-DI-
    ESTER (GERMAN) see PCU400
PHENYLAETHYLMALONSAEURE-AETHYLESTER-
    DIAETHYLAMINOAETHYL-AMID (GERMAN) see FAJ150
PHENYLAETHYLSENFOEL (GERMAN) see ISP000
PHENYLALANINE see PEC750
3-PHENYLALANINE see PEC750
d-PHENYLALANINE see PEC500
l-PHENYLALANINE see PEC750
(S)-PHENYLALANINE see PEC750
PHENYL-α-ALANINE see PEC750
β-PHENYLALANINE see PEC750
d-β-PHENYLALANINE see PEC500
l-β-PHENYLALANINE see PEC750
dl-PHENYLALANINE (FCC) see PEC500
PHENYLALANINE MUSTARD see BHU750
d-PHENYLALANINE MUSTARD see BHU750, SAX200
l-PHENYLALANINE MUSTARD see PED750
o-PHENYLALANINE MUSTARD see BHT250
dl-PHENYLALANINE MUSTARD see BHT750
(±)-o-PHENYLALANINE MUSTARD see BHT250
l-PHENYLALANINE MUSTARD HYDROCHLORIDE see BHV250
dl-PHENYLALANINE MUSTARD HYDROCHLORIDE see BHV000
PHENYLALANINE NITROGEN MUSTARD see PED750
l-PHENYLALANINE, N-((PHENYLMETHOXY)CARBONYL)-, ETHENYL
    ESTER see CBR235
d,l-PHENYLALANINE, PYROLYZATE see ALY000
PHENYLALANIN-LOST (GERMAN) see BHT750
(S)-d-PHENYLALANYL-N-(4-((AMINOIMINOMETHYL)AMINO)-1-
    FORMYLBUTYL)-l-PROLINAMIDE SULFATE (1:1) see PEE100
d-PHENYLALANYL-l-PROLYL-l-ARGININE ALDEHYDE SULFATE (1:1)
    see PEE100
Γ-PHENYLALLYL ACETATE see CMQ730
3-PHENYLALLYL ALCOHOL see CMQ740
Γ-PHENYLALLYL ALCOHOL see CMQ740
5-PHENYL-5-ALLYLBARBITURIC ACID see AGQ875
PHENYLAMINE see AOQ000
PHENYLAMINE HYDROCHLORIDE see BBL000
PHENYLAMINOACETIC ACID ISOAMYL ESTER HYDROCHLORIDE
    see PCV750
N-PHENYL-p-AMINOANILINE see PFU500
1-PHENYL-2-AMINO-ATHAN (GERMAN) see PDE250
2-(PHENYLAMINO)BENZOIC ACID see PEG500
4-(PHENYLAMINO)BUTANE see BQH850
PHENYL-α-AMINO-n-BUTYRAMIDE-p-ARSONIC ACID see CBK750
β-PHENYL-Γ-AMINOBUTYRIC ACID see PEE500
N-PHENYLAMINOCARBONYL)AZIRIDINE see PEH250
17-β-PHENYLAMINOCARBONYLOXYOESTRA-1,3,5(10)-TRIENE-3-METHYL
    ETHER see PEE600
1-PHENYL-4-AMINO-5-CHLOROPYRIDAZON-(6) (GERMAN) see PEE750
1-PHENYL-4-AMINO-5-CHLORO-6-PYRIDAZONE see PEE750
1-PHENYL-4-AMINO-5-CHLOROPYRIDAZONE-6 see PEE750
1-PHENYL-4-AMINO-5-CHLORPYRIDAZ-6-ONE see PEE750
trans-2-PHENYL-1-AMINOCYCLOPROPANE see PET750
2,2'-(PHENYLAMINO)DIETHANOL see BKD500
4-PHENYLAMINODIPHENYLAMINE see BLE500

m-PHENYLENEBIS(1-METHYLETHYLENE)BIS(TRIMETHYLAMMONIUM BRO-
MIDE) see PEX350
p-PHENYLENEBIS(1-METHYLETHYLENE)BIS(TRIMETHYLAMMONIUM BRO-
MIDE) see PEX355
(m-PHENYLENEBIS(1-METHYLETHYLENE)BIS(TRIMETHYLAMMONIUM)
DIBROMIDE see PEX350
(p-PHENYLENEBIS(1-METHYLETHYLENE)BIS(TRIMETHYLAMMONIUM)
DIBROMIDE see PEX355
1,1'-(p-PHENYLENEBIS(OXYETHYLENE))DIHYDRAZINE DIHYDROCHLO-
RIDE see HAA355
2,2'-(1,3-PHENYLENEBIS(OXYMETHYLENE))BISOXIRANE see REF000
1,1'-(m-PHENYLENE)BIS-1H-PYROLE-2,5-DIONE (9CI) see BKL750
4,4'-o-PHENYLENEBIS(3-THIOALLOPHANIC ACID)DIMETHYL ESTER
see PEX500
PHENYLENE-m,m'-BIS(2-TRIMETHYLAMMONIUMPROPYL) DIBROMIDE
see PEX350
PHENYLENE-p,p'-BIS(2-TRIMETHYLAMMONIUMPROPYL) DIBROMIDE
see PEX355
1,1'-(p-PHENYLENEBIS(VINYLENE-p-PHENYLENE))BIS(PYRIDINIUM)
DI-p-TOLUENESULFONATE see PEX750
m-PHENYLENEDIACETATE see REA100
m-PHENYLENEDIAMINE see PEY000
o-PHENYLENEDIAMINE see PEY250
p-PHENYLENEDIAMINE see PEY500
1,3-PHENYLENEDIAMINE see PEY000
1,4-PHENYLENEDIAMINE see PEY500
m-PHENYLENEDIAMINE (DOT) see PEY000
1,2-PHENYLENEDIAMINE (DOT) see PEY250
p-PHENYLENEDIAMINE, N,N-DIETHYL-, SULFATE (1:1) see DJV250
o-PHENYLENEDIAMINE, DIHYDROCHLORIDE see PEY600
p-PHENYLENEDIAMINE DIHYDROCHLORIDE see PEY650
1,3-PHENYLENEDIAMINE DIHYDROCHLORIDE see PEY750
1,4-PHENYLENEDIAMINE DIHYDROCHLORIDE see PEY650
m-PHENYLENEDIAMINE HYDROCHLORIDE see PEY750
p-PHENYLENEDIAMINE HYDROCHLORIDE see PEY650
PHENYLENEDIAMINE, META, solid (DOT) see PEY000
PHENYLENEDIAMINE, PARA, solid (DOT) see PEY500
m-PHENYLENEDIAMINESULFONIC ACID see PFA250
m-PHENYLENEDIAMINE-4-SULFONIC ACID see PFA250
1,3-PHENYLENEDIAMINE-4-SULFONIC ACID see PFA250
p-PHENYLENEDIAZIDE see DCL125
p-PHENYLENEDICARBONYL DICHLORIDE see TAV250
PHENYLENE-m,m'-DI-(2-DIMETHYLBENZYLAMMONIUMPROPYL) DICHLO-
RIDE see PEX300
PHENYLENE-m,m'-DI(2-DIMETHYLETHYLAMMONIUMPROPYL)DIBROMIDE
see PEX325
PHENYLENE-p,p'-DI(2-DIMETHYLETHYLAMMONIUMPROPYL) DIBROMIDE
see PEX330
m-PHENYLENE DIISOCYANATE see BBP000
(PHENYLENEDIISOPROPYLIDENE)BIS(tert-BUTYLPEROXIDE) see BHL100
PHENYLENE-1,4-DIISOTHIOCYANATE see PFA500
1,4-PHENYLENEDIISOTHIOCYANIC ACID see PFA500
N,N'-(m-PHENYLENEDIMALEIMIDE) see BKL750
N,N'-(p-PHENYLENEDIMETHYLENE)BIS(2,2-DICHLORO-N-ETHYLACETAM-
IDE) see PFA600
o-PHENYLENEDIOL see CCP850
m-PHENYLENE ISOCYANATE see BBP000
2,3-PHENYLENEPYRENE see IBZ000
2,3-o-PHENYLENEPYRENE see IBZ000
1,10-(o-PHENYLENE)PYRENE see IBZ000
1,10-(1,2-PHENYLENE)PYRENE see IBZ000
PHENYLENE THIOCYANATE see PFA500
o-PHENYLENETHIOUREA see BCC500
PHENYLEPHRINE see NCL500
(−)-PHENYLEPHRINE see NCL500
R(−)-PHENYLEPHRINE see NCL500
PHENYLEPHRINE HYDROCHLORIDE see SPC500
d-(−)-PHENYLEPHRINE HYDROCHLORIDE see SPC500
1-PHENYL-1,2-EPOXYETHANE see EBR000
PHENYL-2,3-EPOXYPROPYL ETHER see PFH000
PHENYLETHANAL see BBL500
PHENYLETHANE see EGP500
PHENYLETHANOIC ACID BUTYL ESTER see BQJ350
1-PHENYLETHANOL see PDE000
2-PHENYLETHANOL see PDD750
β-PHENYLETHANOL see PDD750
PHENYL ETHANOLAMINE see AOR750
N-PHENYLETHANOLAMINE see AOR750
PHENYLETHANOL AMINE HYDROCHLORIDE see PFA750
1-PHENYLETHANONE see ABH000
PHENYLETHENE see SMQ000
4-(2-PHENYLETHENYL)BENZENAMINE see SLQ900
4-(2-PHENYLETHENYL)BENZENAMINE,(E) see AMO000
N-(4-(2-PHENYLETHENYL)PHENYL)ACETAMIDE see SMR500

1-((2-PHENYLETHENYL)PHENYL)ETHANONE see MPL000
2-(4-(2-PHENYLETHENYL)-3-SULFOPHENYL)-2H-NAPHTHO(1,2-d)TRIAZOLE
SODIUM SALT see TGE155
PHENYL ETHER see PFA850
PHENYL ETHER-BIPHENYL MIXTURE see PFA860
1-PHENYL ETHER-2,3-DIACETATE GLYCEROL see PFF300
PHENYL ETHER DICHLORO see DFE800
PHENYL ETHER, HEXACHLORO derivative (8CI) see CDV175
PHENYL ETHER MONO-CHLORO see MRG000
PHENYL ETHER PENTACHLORO see PAW250
PHENYL ETHER TETRACHLORO see TBP250
cis-2-(1-PHENYLETHOXY)CARBONYL-1-METHYLVINYL DIMETHYLPHOS-
PHATE see COD000
1-(2-PHENYL-2-ETHOXYETHYL)-4-(2-BENZYLOXYPROPYL)PIPERAZINE
see PFB000
1-(2-PHENYL-2-ETHOXY)ETHYL-4-(2-BENZYLOXY)PROPYLPIPERAZINE
DIHYDROCHLORIDE see ECU550
2-PHENYLETHYL ACETATE see PFB250
α-PHENYL ETHYL ACETATE see MNT075
β-PHENYLETHYL ACETATE see PFB250
PHENYLETHYLACETYLUREA see PFB350
PHENYLETHYLACETYLUREE (FRENCH) see PFB350
2-PHENYLETHYL ALCOHOL see PDD750
β-PHENYLETHYL ALCOHOL see PDD750
PHENYLETHYLAMINE see PDE250
1-PHENYLETHYLAMINE see ALW250
2-PHENYLETHYLAMINE see PDE250
α-PHENYLETHYLAMINE see ALW250
ω-PHENYLETHYLAMINE see PDE250
2-PHENYLETHYLAMINE HYDROCHLORIDE see PDE500
β-PHENYLETHYLAMINE HYDROCHLORIDE see PDE500
2-PHENYLETHYL-o-AMINOBENZOATE see APJ500
2-PHENYLETHYLAMINOETHANOL see PFB500
1-PHENYL-2-ETHYLAMINOPROPANE see EGI500
α-PHENYL-β-ETHYLAMINOPROPANE see EGI500
2-PHENYLETHYL ANTHRANILATE see APJ500
PHENYLETHYLBARBITURATE see EOK000
PHENYL-ETHYL-BARBITURIC ACID see EOK000
5-PHENYL-5-ETHYLBARBITURIC ACID see EOK000
PHENYLETHYLBARBITURIC ACID, SODIUM SALT see SID000
PHENYLETHYL BENZOATE see PFB750
2-PHENYLETHYL BENZOATE see PFB750
β-PHENYLETHYL BENZOATE see PFB750
PHENYLETHYL BUTYRATE see PFB800
2-PHENYLETHYL BUTYRATE see PFB800
PHENYLETHYL CARBAMATE see CBL750
N-PHENYL-1-(ETHYLCARBAMOYL-1)-ETHYLCARBAMATE, D ISOMER
see CBL500
PHENYLETHYL CINNAMATE see BEE250
β-PHENYLETHYL CINNAMATE see BEE250
2-PHENYLETHYL CYANIDE see HHP100
3-PHENYL-3-ETHYL-2,6-DIKETOPIPERIDINE see DYC800
PHENYLETHYL DIMETHYL CARBINOL see BEC250
3-PHENYL-3-ETHYL-2,6-DIOXOPIPERIDINE see DYC800
PHENYLETHYLENE see SMQ000
PHENYLETHYLENE OXIDE see EBR000
β-PHENYLETHYL ESTER HYDRACRYLIC ACID see PFC100
2-PHENYLETHYL FORMATE see PFC250
β-PHENYLETHYL FORMATE see PFC250
2-PHENYL-2-ETHYLGLUTARIC ACID IMIDE see DYC800
α-PHENYL-α-ETHYLGLUTARIC ACID IMIDE see DYC800
α-PHENYL-α-ETHYLGLUTARIMIDE see DYC800
5-PHENYL-5-ETHYL-HEXAHYDROPYRIMIDINE-4,6-DIONE see DBB200
2-(1-PHENYLETHYL)HYDRAZIDE BENZOIC ACID see NCQ100
2-PHENYLETHYLHYDRAZINE see PFC500
β-PHENYLETHYLHYDRAZINE see PFC500
β-PHENYLETHYLHYDRAZINE DIHYDROGEN SULFATE see PFC750
2-PHENYLETHYLHYDRAZINE DIHYDROGEN SULPHATE see PFC750
β-PHENYLETHYLHYDRAZINE HYDROGEN SULPHATE see PFC750
α-PHENYLETHYLHYDRAZINE OXALATE see MBV750
β-PHENYLETHYLHYDRAZINE SULFATE see PFC750
PHENYLETHYLHYDRAZINE SULPHATE see PFC750
2-PHENYLETHYL HYDROPEROXIDE see PFC775
1-(1-PHENYLETHYL)-1H-IMIDAZOLE-5-CARBOXYLIC ACID ETHYL ESTER
see HOU100
PHENYLETHYL ISOAMYL ETHER see IHV050
PHENYLETHYL ISOBUTYRATE see PDF750
2-PHENYLETHYL ISOBUTYRATE see PDF750
β-PHENYLETHYL ISOBUTYRATE see PDF750
PHENYLETHYL ISOTHIOCYANATE see ISP000
2-PHENYLETHYL ISOTHIOCYANATE see ISP000
β-PHENYLETHYL ISOTHIOCYANATE see ISP000
PHENYLETHYL ISOVALERATE see PDF775
β-PHENYLETHYL ISOVALERATE see PDF775

PHENYL ETHYL KETONE see EOL500
PHENYLETHYLMALONYLUREA see EOK000
5-PHENYL-5-ETHYL-3-METHYLBARBITURIC ACID see ENB500
PHENYLETHYL-α-METHYLBUTENOATE see PFD250
2-PHENYLETHYL-3-METHYLBUTIRATE see PDF775
PHENYLETHYL METHYL ETHER see PFD325
β-PHENYLETHYL METHYL ETHER see PFD325
PHENYLETHYL METHYL ETHYL CARBINOL see PFR200
PHENYLETHYL METHYLETHYLCARBINYL ACETATE see PFR300
PHENYLETHYLMETHYLHYDANTOIN see MKB250
β-PHENYLETHYL METHYL KETONE see PDF800
2-PHENYLETHYL-2-METHYLPROPIONATE see PDF750
PHENYLETHYL MUSTARD OIL see ISP000
8-(2-PHENYLETHYL)-1-OXA-3,8-DIAZASPIRO(4.5)DECAN-2-ONE HYDRO-
    CHLORIDE see DAI200
2-PHENYLETHYL PHENYLACETATE see PDI000
β-PHENYLETHYL PHENYLACETATE see PDI000
N-(β-(4-(β-PHENYLETHYL)PHENYL)-β-HYDROXYETHYL)IMIDAZOLE
    HYDROCHLORIDE see DAP880
β-2-PHENYLETHYLPIPERIDINOETHYL BENZOATE HYDROCHLORIDE
    see PFD500
Γ-2-PHENYLETHYLPIPERIDINOPROPYL BENZOATE HYDROCHLORIDE
    see PFD750
1-PHENYLETHYL PROPIONATE see PFR000
2-PHENYLETHYL PROPIONATE see PDK000
PHENYLETHYL TIGLATE see PFD250
2-PHENYLETHYL-α-TOLUATE see PDI000
(2-PHENYLETHYL)TRICHLOROSILANE see PFE500
PHENYLETHYNLCARBINOL CARBAMATE see PGE000
PHENYLETTEN see EOK000
N-PHENYL-2-FLUORENAMINE see PFE900
N-PHENYL ISOPROPYL CARBAMATE see PFE900
N-PHENYL-2-FLUORENYLHYDROXYLAMINE see PFF000
N-PHENYL-N-9H-FLUOREN-2-YLHYDROXYLAMINE see PFF000
PHENYL FLUORIDE see FGA000
PHENYLFLUOROFORM see BDH500
PHENYL FORMAMIDE see FNJ000
N-PHENYLFORMAMIDE see FNJ000
PHENYLFORMIC ACID see BCL750
PHENYLGAMMA HYDROCHLORIDE see GAD000
PHENYLGLYCERYL ETHER DIACETATE see PFF300
dl-2-PHENYLGLYCINEDECYLESTERHYDROCHLORIDE see PFF500
dl-2-PHENYLGLYCINEHEPTYLESTERHYDROCHLORIDE see PFF750
dl-2-PHENYLGLYCINEHEXYLESTERHYDROCHLORIDE see PFG000
dl-2-PHENYLGLYCINENONYLESTERHYDROCHLORIDE see PFG250
dl-2-PHENYLGLYCINEOCTYLESTERHYDROCHLORIDE see PFG500
dl-2-PHENYLGLYCINEPENTYLESTERHYDROCHLORIDE see PFG750
dl-2-PHENYLGLYCINISOAMYLESTERHYDROCHLORID (GERMAN) see
    PCV750
PHENYLGLYCOLIC ACID see MAP000
o-PHENYLGLYCOLIC ACID see PDR100
PHENYLGLYCOLONITRILE see MAP250
PHENYL GLYCYDYL ETHER see PFH000
PHENYLGLYOXAL see PFH250
PHENYLGLYOXYLONITRILE OXIME-O,O-DIETHYL PHOSPHOROTHIOATE
    see BAT750
PHENYLGOLD see PFH275
2-PHENYLHYDRACRYLIC ACID-3-α-TROPANYL ESTER see ARR000
PHENYL HYDRATE see PDN750
2-PHENYLHYDRAZIDEBENZENECARBOXIMIDIC ACID
    MONOHYDROCHLORIDE see PEK675
2-PHENYLHYDRAZIDE, CARBAMIC ACID see CBL000
PHENYLHYDRAZIN (GERMAN) see PFI000
PHENYLHYDRAZINE see PFI000
1-PHENYLHYDRAZINE CARBOXAMIDE see CBL000
2-PHENYLHYDRAZINECARBOXAMIDE see CBL000
PHENYLHYDRAZINE HYDROCHLORIDE see PFI250
PHENYLHYDRAZINE MONOHYDROCHLORIDE see PFI250
PHENYLHYDRAZINE-p-SULFONIC ACID see PFI500
PHENYLHYDRAZIN HYDROCHLORID (GERMAN) see PFI250
PHENYLHYDRAZINIUM CHLORIDE see PFI250
1-PHENYL-2-HYDRAZINOPROPANE see PDN000
PHENYL HYDRIDE see BBL250
PHENYLHYDROQUINONE see BGG250
PHENYLHYDROXAMIC ACID see BCL500
PHENYL HYDROXIDE see PDN750
PHENYLHYDROXYACETIC ACID see MAP000
o-(PHENYLHYDROXYARSINO)BENZOIC ACID see PFI600
1-PHENYL-2-(β-HYDROXYETHYL)AMINOPROPANE see PFI750
2-PHENYL-2-HYDROXYETHYL CARBAMATE see PFJ000
2-PHENYL-2-HYDROXYETHYL, m-CHLOROPHENYL ARSINIC ACID
    see CKA575
N-PHENYLHYDROXYLAMINE see PFJ250
β-PHENYLHYDROXYLAMINE see PFJ250

PHENYLHYDROXYLAMINIUM CHLORIDE see PFJ275
PHENYL HYDROXYMERCURY see PFN100
1-PHENYL-1-HYDROXYPENTANE see BQJ500
1-PHENYL-2-(p-HYDROXYPHENYL)-3,5-DIOXO-4-BUTYLPYRAZOLIDINE
    see HNI500
PHENYLIC ACID see PDN750
PHENYLIC ALCOHOL see PDN750
PHENYL-IDIUM see PDC250
PHENYL-IDIUM 200 see PDC250
N-PHENYL-IMIDODICARBONIMIDIC DIAMIDE MONOHYDROCHLORIDE
    see PEO000
2,2'-(PHENYLIMINO)DIETHANOL see BKD500
5-PHENYL-2-IMINO-4-OXAZOLIDINONE see IBM000
5-PHENYL-2-IMINO-4-OXOOXAZOLIDINE see IBM000
2-PHENYLINDAN-1,3-DIONE see PFJ750
2-PHENYL-1,3-INDANDIONE see PFJ750
PHENYLINDIONE see PFJ750
PHENYLIODINE(III) CHROMATE see PFJ775
PHENYLIODINE(III) NITRATE see PFJ780
9-PHENYL-9-IODOFLUORENE see PFK000
PHENYL ISOCYANATE see PFK250
PHENYL ISOHYDANTOIN see IBM000
2-PHENYLISOPROPANOL see DTN100
β-PHENYLISOPROPYLAMIN (GERMAN) see AOA250
(PHENYLISOPROPYL)AMINE see AOA250
β-PHENYLISOPROPYLAMINE see AOA250
dl-β-PHENYLISOPROPYLAMINE HYDROCHLORIDE see AOA750
β-PHENYL ISOPROPYLAMINE SULFATE see AOB250
d-β-PHENYLISOPROPYLAMINE SULFATE see BBK500
7-(PHENYL-ISOPROPYL-AMINO-AETHYL)-THEOPHYLLIN-HYDROCHLORID
    (GERMAN) see CBF825
N-PHENYL ISOPROPYL CARBAMATE see CBM000
PHENYLISOPROPYLHYDRAZINE see PDN000
β-PHENYLISOPROPYLHYDRAZINE see PDN000
PHENYLISOPROPYLHYDRAZINE HYDROCHLORIDE see PDN250
(−)-PHENYLISOPROPYLMETHYLPROPYNYLAMINE see DAZ125
(+)-PHENYLISOPROPYLMETHYLPROPYNYLAMINE HYDROCHLORIDE
    see DAZ120
(±)-PHENYLISOPROPYLMETHYLPROPYNYLAMINE HYDROCHLORIDE
    see DAZ118
N-PHENYL-N'-ISOPROPYL-p-PHENYLENEDIAMINE see PFL000
3-(β-PHENYLISOPROPYL)-SIDNONIMINE HYDROCHLORIDE see SPA000
PHENYL ISOTHIOCYANATE see ISQ000
α-(5-PHENYL-3-ISOXAZOLYL)-1-PIPERIDINEETHANOL CITRATE (2:1)
    see PCJ325
PHENYLIUM see PDR100
PHENYL KETONE see BCS250
PHENYLLIN see PFJ750
PHENYLLITHIUM see PFL500
PHENYLMAGNESIUM BROMIDE see PFL600
N-PHENYLMALEIMIDE see PFL750
α-PHENYLMANDELIC ACID-N,N-DIMETHYLPIPERIDINIUM-2-METHYL
    ESTER METHYLSULFATE see CDG250
PHENYL MERCAPTAN see PFL850
1-PHENYL-5-MERCAPTOTETRAZOLE see PGJ750
1-PHENYL-5-MERCAPTO-1,2,3,4-TETRAZOLE see PGJ750
PHENYLMERCURIACETATE see ABU500
PHENYL MERCURIC ACETATE see ABU500
PHENYLMERCURIC BROMIDE see PFM250
PHENYL MERCURIC CHLORIDE see PFM500
PHENYLMERCURIC DINAPHTHYLMETHANEDISULFONATE see PFN000
PHENYL MERCURIC FIXTAN see PFN000
PHENYLMERCURIC HYDROXIDE see PFN100
PHENYLMERCURIC HYDROXIDE (DOT) see PFN100
PHENYL MERCURIC-8-HYDROXYQUINOLINATE see PFN250
PHENYLMERCURIC 3,3'-METHYLENEBIS(2-NAPHTHALENESULFONATE)
    see PFN000
PHENYLMERCURIC NITRATE see MCU750
PHENYLMERCURIC NITRATE, BASIC see MDH500
PHENYLMERCURIC TRIETHANOLAMMONIUM LACTATE see TNI500
PHENYLMERCURIC UREA see PFP500
N-(PHENYLMERCURI)-1,4,5,6,7,7-HEXACHLOROBICYCLO(2.2.1)HEPTENE-5-
    DICARBOXIMIDE see PFN500
PHENYLMERCURILAURYL THIOETHER see PFN750
8-PHENYLMERCURIOXY-QUIN-OLINE see PFN250
PHENYLMERCURI PROPIONATE see PFO000
PHENYLMERCURIPYROCATECHIN see PFO250
PHENYLMERCURITRIETHANOLAMMONIUM LACTATE see TNI500
PHENYLMERCURIUREA see PFP500
PHENYLMERCURY ACETATE see ABU500
PHENYLMERCURY ACETATE 95% plus ETHYLMERCURY CHLORIDE 5%
    see PFO500
PHENYL MERCURY AMMONIUM ACETATE see PFO550
PHENYLMERCURY BROMIDE see PFM250

1-PHENYL-3-(β-PIPERIDINO-AETHYL)-5-(α'-METHYL-α-)-FURYL-PYRAZOLIN-HCL (GERMAN) see MKH750
α-PHENYL-α-(2-PIPERIDINOETHYL)-β-ETHYLBUTYRIC ACID NITRILE HYDROCHLORIDE see EQZ000
α-PHENYL-α-(2-PIPERIDINOETHYL)-β-ETHYLBUTYRONITRILE HYDROCHLORIDE see EQZ000
2-PHENYL-6-PIPERIDINOHEXYNOPHENONE see PFY750
1-PHENYL-3-PIPERIDINOPROPAN-1-ONE HYDROCHLORIDE see PIV500
1-PHENYL-1-(2-PIPERIDYL)-1-ACETOXYMETHANE HYDROCHLORIDE see LFO000
PHENYL-(2-PIPERIDYL)METHYL ACETATE HYDROCHLORIDE see LFO000
1-PHENYL-4-(4-PIPERONYL-1-PIPERAZINYLCARBONYL)-2-PYRROLIDINONE HYDROCHLORIDE see PGA250
2-PHENYLPROPANAL see COF000
3-PHENYLPROPANAL see HHP000
3-PHENYL-1-PROPANAL see HHP000
1-PHENYLPROPANE see IKG000
2-PHENYLPROPANE see COE750
1-PHENYL-1,2-PROPANEDIONE see PGA500
3-PHENYLPROPANENITRILE see HHP100
1-PHENYLPROPANOL see EGQ000
3-PHENYLPROPANOL see HHP050
Γ-PHENYLPROPANOL see HHP050
1-PHENYL-1-PROPANOL see EGQ000
3-PHENYL-1-PROPANOL (FCC) see HHP050
1-PHENYL-2-PROPANOL ACETATE see ABU800
3-PHENYL-1-PROPANOL ACETATE see HHP500
PHENYLPROPANOLAMINE see NNN000
PHENYLPROPANOLAMINE HYDROCHLORIDE see NNN000, PMJ500
3-PHENYL-1-PROPANOL CARBAMATE see PGA750
1-PHENYL-1-PROPANONE see EOL500
3-PHENYLPROPENAL see CMP969
3-PHENYL-2-PROPENAL see CMP969
2-PHENYLPROPENE see MPK250
β-PHENYLPROPENE see MPK250
3-PHENYLPROPENOIC ACID see CMP975
3-PHENYL-2-PROPENOIC ACID see CMP975
3-PHENYL-2-PROPENOIC ACID-1,5,DIMETHYL-1-VINYL-4-HEXEN-1-YL ESTER see LGA000
3-PHENYL-2-PROPENOIC ACID-1-ETHENYL-1,5-DIMETHYL-4-HEXENYL ESTER see LGA000
3-PHENYL-2-PROPENOIC ACID METHYL ESTER (9CI) see MIO500
3-PHENYL-2-PROPENOIC ACID, 2-METHYLPROPYL ESTER see IIQ000
3-PHENYL-2-PROPENOIC ACID PHENYLMETHYL ESTER (9CI) see BEG750
3-PHENYL-2-PROPEN-1-OL see CMQ740
3-PHENYL-2-PROPEN-1-YL ACETATE see CMQ730
3-PHENYL-2-PROPENYLANTHRANILATE see API750
3-PHENYL-2-PROPEN-1-YL ANTHRANILATE see API750
PHENYLPROPENYL n-BUTYRATE see CMQ800
3-PHENYL-2-PROPEN-1-YL FORMATE see CMR500
3-PHENYL-2-PROPENYL PROPIONATE see CMR850
3-PHENYL-2-PROPEN-1-YL PROPIONATE see CMR850
5-PHENYL-5-(2-PROPENYL)-2,4,6(1H,3H,5H)-PYRIMIDINETRIONE (9CI) see AGQ875
α-PHENYLPROPIONALDEHYDE see COF000
β-PHENYLPROPIONALDEHYDE see HHP000
2-PHENYLPROPIONALDEHYDE (FCC) see COF000
3-PHENYLPROPIONALDEHYDE (FCC) see HHP000
2-PHENYLPROPIONALDEHYDE DIMETHYL ACETAL see HII600
PHENYLPROPIONITRILE see HHP100
3-PHENYLPROPIONITRILE see HHP100
β-PHENYLPROPIONITRILE see HHP100
3-PHENYLPROPIONYL AZIDE see PGB500
17-β-PHENYLPROPIONYLOXY-4-ESTREN-3-ONE see DYF450
PHENYL(PROPIONYLOXY)MERCURY see PFO000
PHENYLPROPYL ACETATE see HHP500
2-PHENYLPROPYL ACETATE see PGB750
3-PHENYL-1-PROPYL ACETATE see HHP500
3-PHENYLPROPYL ACETATE (FCC) see HHP500
PHENYLPROPYL ALCOHOL see HHP050
1-PHENYLPROPYL ALCOHOL see EGQ000
3-PHENYLPROPYL ALCOHOL see HHP050
Γ-PHENYLPROPYL ALCOHOL see HHP050
3-PHENYLPROPYL ALDEHYDE see HHP000
α-PHENYL-α-PROPYLBENZENEACETIC ACID-2-(DIETHYLAMINO)ETHYL ESTER HYDROCHLORIDE see PBM500
α-PHENYL-α-PROPYL-BENZENEACETIC ACID-S-(DIETHYLAMINO)ETHYL ESTER see DIG400
Γ-PHENYLPROPYLCARBAMAT (GERMAN) see PGA750
Γ-PHENYLPROPYL CARBAMATE see PGA750
2-PHENYLPROPYLENE see MPK250
β-PHENYLPROPYLENE see MPK250
PHENYLPROPYL FORMATE see HHQ000
3-PHENYL-1-PROPYL FORMATE see HHQ000

α-(β-PHENYL-PROPYL)-β-HYDROXY-Δ^α,β-BUTENOLID (GERMAN) see PGC750
3-(3-PHENYLPROPYL)-4-HYDROXY-2(5H)FURANONE see PGC750
3-PHENYLPROPYL ISOBUTYRATE see HHQ500
O-PHENYL-S-PROPYL METHYL PHOSPHONODITHIOATE see PGD000
PHENYLPROPYL PROPIONATE see HHQ550
3-PHENYLPROPYL PROPIONATE see HHQ550
β-PHENYLPROPYL PROPIONATE see HHQ550
1-PHENYL-2-PROPYNYL CARBAMATE see PGE000
PHENYLPSEUDOHYDANTOIN see IBM000
6-PHENYL-2,4,7-PTERIDINETRIAMINE see UVJ450
1-PHENYL-3-PYRAZOLIDINONE see PDM500
1-PHENYL-3-PYRAZOLIDONE see PDM500
1-PHENYL-4-PYRAZOLIN-3-ONE see PGE250
2-PHENYL-8H-PYRAZOLO(5,1-a)ISOINDOLE see PGE300
2-PHENYL-PYRAZOLO(5,1-a)ISOQUINOLINE see PGE350
p-(3-PHENYL-1-PYRAZOLYL)BENZENESULFONIC ACID see PGE500
N'-(1-PHENYLPYRAZOL-5-YL)SULFANILAMIDE see AIF000
N(1)-(1-PHENYLPYRAZOL-5-YL)-SULFANILAMIDE (8CI) see AIF000
4-PHENYLPYRIDINE see PGE750
p-PHENYLPYRIDINE see PGE750
PHENYL(2-PYRIDYL)(β-N,N-DIMETHYLAMINOMETHYL) METHANE MALEATE see TMK000
1-PHENYL-1-(2-PYRIDYL)-3-DIMETHYLAMINOPROPANE see TMJ750
1-PHENYL-1-(2-PYRIDYL)-3-DIMETHYLAMINOPROPANE HYDROCHLORIDE see PGF000
1-PHENYL-1-(2-PYRIDYL)-3-DIMETHYLAMINOPROPANE MALEATE see TMK000
3-PHENYL-3-(2-PYRIDYL)-N,N-DIMETHYLPROPYLAMINE see TMJ750
1-(2-(PHENYL(2-PYRIDYLMETHYL)AMINO)ETHYL)PIPERIDINE HYDROCHLORIDE see CNE375
PHENYL-2-PYRIDYLMETHYL-β-N,N-DIMETHYLAMINOETHYL ETHER see DYE500
PHENYL-2-PYRIDYLMETHYL-β-N,N-DIMETHYLAMINOETHYL ETHER SUCCINATE see PGE775
N-PHENYL-N-(2-PYRIDYLMETHYL)-2-PIPERIDINOETHYLAMINE HYDROCHLORIDE see CNE375
1-PHENYL-3-(p-2-PYRIDYLSULFAMOYLANILINO)-1,3-PROPANEDISULFONIC ACID DISODIUM SALT see DXF400
2-PHENYL-3-p-(β-PYRROLIDINOETHOXY)PHENYL-6-METHOXY-BENZOFURAN HYDROCHLORIDE see PGF100
2-PHENYL-3-p-(β-PYRROLIDINOETHOXY)PHENYL-(2581-b)NAPHTHOFURAN see PGF125
2-PHENYL-3-p-(β-PYRROLIDINOETHOXY)-PHENYL-NAPHTHO(21-b)FURAN see PGF125
4-PHENYL-2-PYRROLIDINONE see PGF800
PHENYL-2-PYRROLIDINOPENTANE HYDROCHLORIDE see PNS000
1-PHENYL-2-N-PYRROLIDINOPENTANE HYDROCHLORIDE see PNS000
2-PHENYL-1-(p-2-(1-PYRROLIDINYL)ETHOXY)PHENYL)NAPHTHO(2,1-b)FURAN see PGF125
2-PHENYL-N-(2-(1-PYRROLIDINYL)ETHYL)GLYCINE ISOPENTYL ESTER see CBA375
PHENYLPYRROLIDONE see PGF800
4-PHENYLPYRROLIDONE-2 see PGF800
PHENYLQUECKSILBERACETAT (GERMAN) see ABU500
PHENYLQUECKSILBERBRENZKATECHIN (GERMAN) see PFO750
PHENYLQUECKSILBERCHLORID (GERMAN) see PFM500
2-PHENYLQUINOLINE-4-CARBOXYLIC ACID see PGG000
2-PHENYL-4-QUINOLINECARBOXYLIC ACID see PGG000
2-PHENYL-4-QUINOLONECARBOXYLIC AICD SODIUM SALT see SJH000
PHENYLQUINONE see PEL750
PHENYLRAMIDOL see PGG350
PHENYLRAMIDOL HYDROCHLORIDE see PGG355
3-PHENYL-RHODANIN (GERMAN) see PGG500
3-PHENYLRHODANINE see PGG500
PHENYL SALICYLATE see PGG750
3-PHENYLSALICYLIC ACID see PGH000
PHENYL SELENIDE see PGH250
PHENYLSELENONIC ACID see PGH500
PHENYLSEMICARBAZIDE see CBL000
1-PHENYLSEMICARBAZIDE see CBL000
PHENYLSENFOEL (GERMAN) see ISQ000
PHENYLSILATRANE see PGH750
PHENYLSILICON TRICHLORIDE see TJA750
PHENYLSILVER see PGI000
PHENYL SODIUM see PGI100
PHENYL STYRYL KETONE see CDH000
1-PHENYL-5-SULFANILAMIDOPYRAZOLE see AIF000
PHENYL SULFIDE see PGI500
5-(PHENYLSULFINYL)-2-BENZIMIDAZOLECARBAMIC ACID METHYL ESTER see OMY500
(5-(PHENYLSULFINYL)-1H-BENZIMIDAZOL-2-YL)CARBAMIC ACID METHYL ESTER see OMY500

5-PHENYLSULFINYL-2-CARBOMETHOXYAMINOBENZIMIDAZOLE
   see OMY500
PHENYL SULFONE see PGI750
PHENYLSULFONIC ACID see BBS250
2-(PHENYLSULFONYLAMINO)-1,3,4-THIADIAZOLE-5-SULFONAMIDE
   see PGJ000
2-(PHENYLSULFONYL)ETHANOL see BBT000
N-(4-PHENYLSULFONYL-o-TOLYL)-1,1,1-TRIFLUOROMETHANESULFONAM-
   IDE see TKF750
4-(PHENYLSULFOXYETHYL)-1,2-DIPHENYL-3,5-PYRAZOLIDINEDIONE
   see DWM000
PHENYLSUXIMIDE see MNZ000
5'-PHENYL-m-TERPHENYL see TMR000
dl-6-PHENYL-2,3,5,6-TETRAHYDROIMIDAZOLE(2,1-b)THIAZOLE-HYDRO-
   CHLORIDE see TDX750
6-PHENYL-2,3,5,6-TETRAHYDROIMIDAZO(2,1-b)THIAZOLE see LFA000
5-PHENYLTETRAZOLE see PGJ500
1-PHENYLTETRAZOLE-5-THIOL see PGJ750
3-PHENYL-1-TETRAZOLYL-1-TETRAZENE see PGK000
PHENYLTHALLIUM DIAZIDE see PGK100
N-PHENYL-N'-1,2,3-THIADIAZOL-5-YL-UREA see TEX600
5-PHENYL-2,4-THIAZOLEDIAMINE MONOHYDROCHLORIDE (9CI)
   see DCA600
N-PHENYLTHIOACETAMIDE see TFA250
1-(PHENYLTHIO)ANTHRAQUINONE see PGL000
N-PHENYLTHIOBENZAMIDE see PGL250
(5-(PHENYLTHIO)-2-BENZIMIDAZOLECARBAMIC ACID, METHYL ESTER
   see FAL100
(4-PHENYLTHIOBUTYL)HYDRAZINE MALEATE see PGL500
PHENYLTHIOCARBAMIDE see PGN250
4-(PHENYLTHIOETHYL)-1,2-DIPHENYL-3,5-PYRAZOLIDINE-DIONE
   see DWJ400
((1-PHENYLTHIOMETHYL)ETHYL)HYDRAZINE MALEATE see PGM250
PHENYLTHIOPHOSPHONATE de O-ETHYLE et O-4-NITROPHENYLE
   (FRENCH) see EBD700
PHENYLTHIOPHOSPHONIC DIAZIDE see PGM500
1-PHENYLTHIOSEMICARBAZIDE see PGM750
4-PHENYL-3-THIOSEMICARBAZIDE see PGN000
1-PHENYLTHIOUREA see PGN250
N-PHENYLTHIOUREA see PGN250
1-PHENYL-2-THIOUREA see PGN250
PHENYLTOLOXAMINE DIHYDROGEN CITRATE see DTO600
PHENYLTOLOXAMINE HYDROCHLORIDE see DTO800
(N-PHENYL-p-TOLUENESULFONAMIDO)ETHYLMERCURY see EME500
PHENYL-p-TOLYL KETONE see MHF750
(2-(PHENYL-o-TOLYLMETHOXY)ETHYL)METHYLAMINE HYDROCHLORIDE
   see TGJ250
PHENYL-(o-TOLYLMETHYL) METHYLAMINOETHYL ETHER HYDROCHLO-
   RIDE see TGJ250
5-PHENYL-3-(o-TOLYL)-s-TRIAZOLE see PGN400
6-PHENYL-2,4,7-TRIAMINOPTERIDINE see UVJ450
2-PHENYL-5H-1,2,4-TRIAZOLO(5,1-a)ISOINDOLE see PGN825
2-PHENYL-s-TRIAZOLO(5,1-a)ISOQUINOLINE see PGN840
2-PHENYL-(1,2,4)TRIAZOLO(5,1-a)ISOQUINOLINE (9CI) see PGN840
1-PHENYL-1,2,4-TRIAZOLYL-3-(O,O-DIETHYLTHIONOPHOSPHATE)
   see THT750
PHENYLTRICHLOROMETHANE see BFL250
PHENYL TRICHLOROSILANE (DOT) see TJA750
PHENYLTRIETHOXYSILANE see PGO000
PHENYLTRIFLUOROSILANE see PGO500
PHENYLTRIMETHOXYSILANE see PGO750
PHENYL TRIMETHYL AMMONIUM BROMIDE see TMB000
PHENYLTRIMETHYLAMMONIUM CHLORIDE see TMB250
PHENYL TRIMETHYL AMMONIUM HYDROXIDE see TMB500
PHENYLTRIMETHYLAMMONIUM IODIDE see TMB750
4-PHENYL-1,2,5-TRIMETHYL-4-PIPERIDINOL PROPIONATE see IMT000
PHENYL-2,8,9-TRIOXA-5-AZA-1-SILABICYCLO(3.3.3) UNDECANE see PGH750
1-PHENYLUREA see PGP250
PHENYL UREA-p-DI(CARBOXYMETHYL) THIOARSENITE see CBI250
PHENYLURETHAN see CBL750
PHENYLURETHAN(E) see CBL750
N-PHENYLURETHANE see CBL750
α-PHENYL-VALERATE du DIETHYLAMINO-ETHANOL CHLORHYDRATE
   (FRENCH) see PGP500
2-PHENYLVALERIC ACID-2-(DIETHYLAMINO)ETHYL ESTER HYDROCHLO-
   RIDE see PGP500
PHENYLVANADIUM(V) DICHLORIDE OXIDE see PGP600
PHENYLVINYL KETONE see PMQ250
PHENYL XYLYL KETONE see PGP750
1-PHENYL-1-(3,4-XYLYL)-2-PROPYNYL N-CYCLOHEXYLCARBAMATE
   see PGQ000
PHENYRAL see AGQ875, EOK000
PHENYRAMIDOL see PGG350
PHENYRAMIDOL HYDROCHLORIDE see PGG355

PHENYRIT see PEC250
PHENYTOIN and PHENOBARBITONE see DWD000
PHENYTOIN SODIUM see DNU000
PHENZIDOLE see PEL250
PHERMERNITE see MCU750
PHETADEX see BBK500
PHETIDINE see DAM600
PHETYLUREUM see PEC250
PHGABA HYDROCHLORIDE see GAD000
PHIC see PGQ275
PHILIPS 2133 see CPW250
PHILIPS 2605 see BIK750
PHILIPS-DUPHAR PH 60-40 see CJV250
PHILIPS-DUPHAR V-101 see CKL750
PHILLIPS 1863 see AMJ250
PHILLIPS 1908 see AMA000
PHILLIPS 2038 see AAM750
PHILLIPS R-11 see BHJ500
PHILODENDRON (various species) see PGQ285
PHILODORM see TDA500
PHILOPON see DBA800, MDT600
PHILOSOPHER'S WOOL see ZKA000
PHILOSTIGMIN BROMIDE see POD000
PHILOSTIGMIN METHYL SULFATE see DQY909
PHISODANV see HCL000
PHISOHEX see HCL000
PHIX see ABU500
PHIXIA see CMS850
PHLEOCIDIN see PGQ350
PHLEOMYCIN see PGQ500
PHLEOMYCIN COMPLEX see PGQ750
PHLEOMYCIN D2 see BLY760
PHLOROGLUCIN see PGR000
PHLOROGLUCINOL see PGR000
PHLOROGLUCINOL TRIMETHYL ETHER see TKY250
PHLOROL see PGR250
PHLORONE see XQJ000
PHLOXIN see CMM000
PHLOXINE see CMM000
PHLOXINE B see ADG250
PHLOXINE K see CMM000
PHLOXINE RED 20-7600 see BNK700
PHLOXINE TONER B see BNH500
PHLOXINRHODAMINE see PGR750
PHLOX RED TONER X-1354 see BNH500
α-PHNEYLTHIOUREA see PGN250
PHOB see EOK000
PHOBEX see BCA000
PHOLATE see AQO000
PHOLCODIN see TCY750
PHOLCODINE see TCY750
PHOLEDRINE see FMS875
PHOLETONE see FMS875
PHOMIN see CQM125
PHOMOPSIN see PGR775
PHONURIT see AAI250
PHORADENDRON RUBRUM see MQW525
PHORADENDRON SEROTINUM see MQW525
PHORADENDRON TOMENTOSUM see MQW525
PHORAT (GERMAN) see PGS000
PHORATE see PGS000
PHORATE-10G see PGS000
PHORBOL see PGS250
PHORBOL ACETATE, CAPRATE see ACZ000
PHORBOL ACETATE, LAURATE see PGS500, PGU250
PHORBOL ACETATE, MYRISTATE see PGV000
PHORBOL-12-o-BUTYROYL-13-DODECANOATE see BSX750
PHORBOL CAPRATE, (+)-(S)-2-METHYLBUTYRATE see PGU500
PHORBOL CAPRATE, TIGLATE see PGU750
PHORBOL-12,13-DIACETATE see PGS750
PHORBOL-12,13-DIBENZOATE see PGT000
PHORBOL-12,13-DIDECANOATE see PGT250
PHORBOL-12,13-DIHEXA(Δ-2,4)-DIENOATE see PGT500
PHORBOL-12,13-DIHEXANOATE see PGT750
PHORBOL LAURATE, (+)-S-2-METHYLBUTYRATE see PGU000
PHORBOL MONOACETATE MONOLAURATE see PGS500, PGU250
PHORBOL MONOACETATE MONOMYRISTATE see PGV000
PHORBOL MONODECANOATE (S)-(+)-MONO(2-METHYLBUTYRATE)
   see PGU500
(E)-PHORBOL MONODECANOATE MONO(2-METHYLCROTONATE)
   see PGU750
PHORBOL MONOLAURATE MONO(S)-(+)-2-METHYLBUTYRATE see PGU000
PHORBOL MYRISTATE ACETATE see PGV000
PHORBOL-9-MYRISTATE-9a-ACETATE-3-ALDEHYDE see PGV250

PHORBOLOL ACETATE MYRISTATE see PGV500
PHORBOLOL MYRISTATE ACETATE see PGV500
PHORBOL-12-o-TIGLYL-13-BUTYRATE see PGV750
PHORBOL-12-o-TIGLYL-13-DODECANOATE see PGW000
PHORBYOL see CCP250
PHORDENE see DFY800
PHORON (GERMAN) see PGW250
PHORONE see PGW250
PHORTOX see TAA100
PHOSADEN see AOA125
PHOSALON see BDJ250
PHOSALONE see BDJ250
PHOSALON mixed with MILBEX (1:2) see MQT750
PHOSAN-PLUS see PGW500
PHOSAZETIM see BIM000
PHOSCHLOR R50 see TIQ250
PHOSCOLIC ACID see PGW600
PHOSDRIN (OSHA) see MQR750
PHOSFENE see MQR750
PHOSFLEX 179-C see TMO600
PHOSFLEX T-BEP see BPK250
PHOS-FLUR see SHF500
PHOSFOLAN see DXN600, PGW750
PHOSFON D see THY500
PHOSGEN (GERMAN) see PGX000
PHOSGENE see PGX000
PHOSHOROTHIOIC ACID-S-2-(ETHYLAMINO)-2-OXOETHYL)-O,O-
   DIMETHYL ESTER see DNX600
PHOSKIL see PAK000
PHOSMET see PHX250
PHOSPHACOL see NIM500
PHOSPHADEN see AOA125
PHOSPHALUGEL see PHB500
PHOSPHAM see PGX250
PHOSPHAMID see DSP400
PHOSPHAMIDON see FAB400
(E)-PHOSPHAMIDON see PGX300
trans-PHOSPHAMIDON see PGX300
PHOSPHATE 100 see EAS000
PHOSPHATE de O,O-DIETHYLE et de O-2-CHLORO-1-(2,4-
   DICHLOROPHENYL) VINYLE (FRENCH) see CDS750
PHOSPHATE de DIETHYLE et de 3-METHYL-5-PYRAZOLYLE (FRENCH)
   see MOX250
PHOSPHATE de O,O-DIMETHLE et de O-(1,2-DIBROMO-2,2-
   DICHLORETHYLE) (FRENCH) see NAG400
PHOSPHATE de DIMETHYLE et de (2-CHLORO-2-DIETHYLCARBAMOYL-1-
   METHYL-VINYLE) see FAB400
PHOSPHATE de DIMETHYLE et de 2,2-DICHLOROVINYL (FRENCH)
   see DGP900
PHOSPHATE de DIMETHYLE et de 2-DIMETHYLCARBAMOYL-1-METHYL
   VINYLE (FRENCH) see DGQ875
PHOSPHATE de DIMETHYLE et de 2-METHOXYCARBONYL-1
   METHYLVINYLE (FRENCH) see MQR750
PHOSPHATE de DIMETHYLE et de 2-METHYLCARBAMOYL-1-METHYL
   VINYLE (FRENCH) see MRH209
PHOSPHATE ROCK see FDH000
PHOSPHATES see PGX500
PHOSPHATE, SODIUM HEXAMETA see SHM500
PHOSPHATE de TRICRESYLE (FRENCH) see TNP500
PHOSPHEMOL see PAK000
PHOSPHENE (FRENCH) see MQR750
PHOSPHENOL see PAK000
PHOSPHENTASIDE see AOA125
PHOSPHENYL CHLORIDE see DGE400
PHOSPHESTROL see DKA200
PHOSPHIDES see PGX750
PHOSPHINE see PGY000
PHOSPHINE OXIDE, BIS(1-AZIRIDINYL)METHYLAMINO- see MGE100
PHOSPHINE OXIDE, TRIS(p-DIMETHYLAMINOPHENYL)-, compd. with
   STANNIC CHLORIDE (2:1) see BLT775
PHOSPHINE SELENIDE, TRIMETHYL- see TMD400
PHOSPHINE SELENIDE, TRIPIPERIDINO- see TMX350
PHOSPHINE SULFIDE, TRIBUTYL- see TIA450
PHOSPHINE, TRIBUTYL-, SULFIDE see TIA450
PHOSPHINIC ACID see PGY250
PHOSPHINIC AMIDE, P,P-BIS(1-AZIRIDINYL)-N-ETHYL- see BGX775
PHOSPHINIC AMIDE, p,p-BIS(1-AZIRIDINYL)-N-ISOPROPYL- see BGX850
PHOSPHINIC AMIDE, P,P-BIS(1-AZIRIDINYL)-N-METHYL- see MGE100
PHOSPHINIC AMIDE, p,p-BIS(1-AZIRIDINYL)-N-(1-METHYLETHYL)-(9CI)
   see BGX850
PHOSPHINIC AMIDE, p,p-BIS(1-AZIRIDINYL)-N-PROPYL- see BGY500
2,2'-PHOSPHINICOBIS(2-HYDROXY-PROPANOIC) ACID (9CI) see PGW600
2,2'-PHOSPHINICODILACTIC ACID see PGW600
PHOSPHINOETHANE. see EON000

1,1',1''-PHOSPHINOTHIOYLIDYNETRISAZIRIDINE see TFQ750
1,1',1''-PHOSPHINYLIDYNETRISAZIRIDINE see TND250
1,1',1''-PHOSPHINYLTRIS(2-METHYL)AZRIDINE see TNK250
o-PHOSPHOETHANOLAMINE see EED000
PHOSPHOLINE see PGY600
PHOSPHOLINE (pharmaceutical) see TLF500
PHOSPHOLINE IODIDE see TLF500
PHOSPHOMYCIN see PHA550
N-PHOSPHONACETYL-l-ASPARTATE see PGY750
PHOSPHONACETYL-l-ASPARTIC ACID see PGY750
PHOSPHON D see THY500
PHOSPHONE D see THY500
PHOSPHONIC ACID see PGZ899
PHOSPHONIC ACID, BIS(1-METHYLETHYL) ESTER see DNQ600
PHOSPHONIC ACID, DIMETHYL ESTER see DSG600
PHOSPHONIC ACID, 1-HYDROXY-1,1-ETHANEDIYL ESTER see HKS780
PHOSPHONIC ACID, (1-HYDROXYETHYLIDENE)BIS- see HKS780
PHOSPHONIC ACID, MONOMETHYL see PGZ950
PHOSPHONIUM IODIDE see PHA000
PHOSPHONIUM, (METHYLSELENO)TRIS(DIMETHYLAMINO)-, IODIDE
   see MPI225
PHOSPHONIUM PERCHLORATE see PHA250
PHOSPHONOACETIC ACID, TRIETHYL ESTER see EIC000
9-(5-o-PHOSPHONO-β-d-ARABINOFURANOSYL)-9H-PURIN-6-AMINE
   see AQQ905
N-(PHOSPHONOMETHYL)GLYCINE see PHA500
PHOSPHONOMYCIN see PHA550
PHOSPHONOMYCIN DISODIUM SALT see DXF600
PHOSPHONOMYCIN SODIUM see DXF600
21-(PHOSPHONOOXY)PREGNA-1,4-DIENE-3,20-DIONE DISODIUM SALT
   see CCS675
PHOSPHONOSELENOIC ACID, ETHYL-, Se-(2-(DIETHYLAMINO)ETHYL)
   O-ETHYL ESTER see SBU900
2-PHOSPHONOXYBENZOESAEURE see PHA575
PHOSPHONOXYBENZOIC ACID see PHA575
2-PHOSPHONOXYBENZOIC ACID see PHA575
PHOSPHOPYRIDOXAL see PII100
PHOSPHOPYRON see EAS000
PHOSPHOPYRONE see EAS000
PHOSPHORAMIDE MUSTARD CYCLOHEXYLAMINE SALT see PHA750
PHOSPHORE BLANC (FRENCH) see PHO740
PHOSPHORE(PENTACHLORURE de) (FRENCH) see PHR500
PHOSPHORE(TRICHLORURE de) (FRENCH) see PHT275
PHOSPHORIC ACID see PHB250
PHOSPHORIC ACID, ALUMINUM SALT (1581), (solution) see PHB500
PHOSPHORIC ACID, BERYLLIUM SALT (1:1) see BFS000
PHOSPHORIC ACID, 2-CHLORO-1-(2,4,5-TRICHLOROPHENYL)ETHENYL
   DIMETHYL ESTER see TBW100
(Z)-PHOSPHORIC ACID-2-CHLORO-1-(2,4,5-TRICHLOROPHENYL)ETHENYL
   DIMETHYL ESTER see RAF100
PHOSPHORIC ACID CHROMIUM (III) SALT see CMK300
PHOSPHORIC ACID, CHROMIUM(3+) SALT (1:1) see CMK300
PHOSPHORIC ACID, DIBUTYL PHENYL ESTER see DEG600
PHOSPHORIC ACID-2,2-DICHLOROETHENYL DIMETHYL ESTER
   see DGP900
PHOSPHORIC ACID-2,2-DICHLOROVINYL METHYL ESTER, CALCIUM
   SALT mixed with 2,2-DICHLOROVINYL PHOSPHORIC ACID CALCIUM
   SALT see PHC750
PHOSPHORIC ACID, DIETHYL ESTER, with 3-CHLORO-7-HYDROXY-4-
   METHYLCOUMARIN see CIK750
PHOSPHORIC ACID, DIETHYL ESTER, with ETHYL 3-HYDROXYCROTON-
   ATE see CBR750
PHOSPHORIC ACID, DIETHYL ESTER-N-NAPHTHALIMIDE derivative
   see HMV000
PHOSPHORIC ACID, DIETHYL ESTER, NAPHTHALIMIDO derivative
   see HMV000
PHOSPHORIC ACID-DIETHYL-(3-METHYL-5-PYRAZOLYL) ESTER
   see MOX250
PHOSPHORIC ACID DIETHYL 4-NITROPHENYL ESTER see NIM500
PHOSPHORIC ACID, DIMETHYL ESTER with p-CHLOROBENZYL-3-
   HYDROXYCROTONATE see CEQ500
PHOSPHORIC ACID, DIMETHYL ESTER, ESTER with DIMETHYL 3-
   HYDROXYGLUTACONATE see SOY000
PHOSPHORIC ACID, DIMETHYL ESTER, ESTER with cis-3-HYDROXY-N-
   METHYLCROTONAMIDE see MRH209
PHOSPHORIC ACID, DIMETHYL ESTER, ester with N-
   HYDROXYNAPHTHALIMIDE see DOL400
PHOSPHORIC ACID DIMETHYL-p-(METHYLTHIO)PHENYL ESTER
   see PHD250
PHOSPHORIC ACID, DIMETHYL-p-NITROPHENYL ESTER see PHD500
PHOSPHORIC ACID, DIMETHYL-4-NITROPHENYL ESTER (9CI) see PHD500
PHOSPHORIC ACID DIMETHYL-4-NITRO-m-TOLYL ESTER see PHD750
PHOSPHORIC ACID, DIMETHYL-3,5,6-TRICHLORO-2-PYRIDYL ESTER
   see PHE250

PHOSPHOROUS THIOCHLORIDE see TFO000
PHOSPHOROUS TRICHLORIDE SULFIDE see TFO000
PHOSPHOROUS TRIFLUORIDE see PHQ500
PHOSPHORPENTACHLORID (GERMAN) see PHR500
PHOSPHORSAEURELOESUNGEN (GERMAN) see PHB250
PHOSPHORTRICHLORID (GERMAN) see PHT275
PHOSPHORUS (red) see PHO500
PHOSPHORUS (white) see PHO740
PHOSPHORUS (white in water) see PHP000
PHOSPHORUS (yellow) see PHO740
PHOSPHORUS, AMORPHOUS, RED (DOT) see PHO500
PHOSPHORUS AZIDE DIFLUORIDE see PHP250
PHOSPHORUS AZIDE DIFLUORIDE-BORANE see PHP500
PHOSPHORUS CHLORIDE see PHT275
PHOSPHORUS COMPOUNDS, INORGANIC see PHQ000
PHOSPHORUS CYANIDE see PHQ250
PHOSPHORUS FLUORIDE see PHQ500
PHOSPHORUS HEPTASULFIDE see PHQ750
PHOSPHORUS(V) OXIDE see PHS250
PHOSPHORUS OXYCHLORIDE see PHQ800
PHOSPHORUS OXYFLUORIDE see PHR000
PHOSPHORUS OXYTRICHLORIDE see PHQ800
PHOSPHORUS PENTABROMIDE see PHR250
PHOSPHORUS PENTACHLORIDE see PHR500
PHOSPHORUS PENTAFLUORIDE see PHR750
PHOSPHORUS PENTASULFIDE see PHS000
PHOSPHORUS PENTOXIDE see PHS250
PHOSPHORUS PERCHLORIDE see PHR500
PHOSPHORUS PERSULFIDE see PHS000
PHOSPHORUS SESQUISULFIDE see PHS500
PHOSPHORUS(III) SULFIDE(IV) see PHS500
PHOSPHORUS SULFOBROMIDE see TFN750
PHOSPHORUS THIOCYANATE see PHS750
PHOSPHORUS TRIAZIDE see PHT000
PHOSPHORUS TRIBROMIDE see PHT250
PHOSPHORUS TRICHLORIDE see PHT275
PHOSPHORUS TRICYANIDE see PHQ250
PHOSPHORUS TRIHYDRIDE see PGY000
PHOSPHORUS TRIOXIDE see PHT500
PHOSPHORUS TRISULFIDE see PHT750
PHOSPHORWASSERSTOFF (GERMAN) see PGY000
PHOSPHORYL BROMIDE see PHU000
PHOSPHORYL CHLORIDE see PHQ800
PHOSPHORYLETHANOLAMINE see EED000
PHOSPHORYL FLUORIDE see PHR000
PHOSPHORYL HEXAMETHYLTRIAMIDE see HEK000
O-PHOSPHORYL-4-HYDROXY-N,N-DIMETHYLTRYPTAMINE see PHU500
PHOSPHORYL TRIBROMIDE see PHU000
PHOSPHOSTIGMINE see PAK000
PHOSPHOTEX see TEE500
PHOSPHOTHION see MAK700
PHOSPHOTOX E see EEH600
PHOSPHOTUNGSTIC ACID see PHU750
PHOSPHURE de MAGNESIUM (FRENCH) see MAI000
PHOSPHURE de POTASSIUM (FRENCH) see PLQ500
PHOSPHURES d'ALUMIUM (FRENCH) see AHE750
PHOSPHURE de SODIUM (FRENCH) see SJI500
PHOSPHURE de STRONTIUM (FRENCH) see SML000
PHOSPHURE de ZINC (FRENCH) see ZLS000
PHOSTEX see BIT250
PHOSVEL see LEN000
PHOSVIN see ZLS000
PHOSVIT see DGP900
PHOTOBILINE see TDE750
PHOTODIELDRIN see PHV250
PHOTODYN see PHV275
PHOTOL see MGJ750
PHOTOMIREX see MRI750
PHOTRIN see FOM000
PHOTRINE see FOM000
PHOXIME see BAT750
PHOXIN see BAT750
PHOZALON see BDJ250
PHP see ELL500
PHPH see BGE000
PHPS see DYE700
Ph-QA 33 see INU200
PHRENOLAN see MDU750
PHRILON see NOH000
PHT see TMB750
PHTALALDEHYDES (FRENCH) see PHV500
PHTALOPHOS see HMV000
PHTHALALDEHYDE see PHV500
p-PHTHALALDEHYDE see TAN500

PHTHALAMODINE see CLY600
PHTHALAMUDINE see CLY600
1,3-PHTHALANDIONE see PHW750
PHTHALAZINE, 1,4-DIHYDRAZINO- OJD300
1,4-PHTHALAZINEDIONE, 2,3-DIHYDRO-, DIHYDRAZONE (9CI) see OJD300
PHTHALAZINOL see PHV725
PHTHALAZINOL (PHOSPHODIESTERASE INHIBITOR) see PHV725
PHTHALAZINONE see PHV750
1(2H)PHTHALAZINONE see PHV750
1(2H)-PHTHALAZINONE (1,3-DIMETHYL-2-BUTENYLIDENE)HYDRAZONE
   see DQU400
1(2H)-PHTHALAZINONE HYDRAZONE see HGP495
1(2H)-PHTHALAZINONE HYDRAZONE HYDROCHLORIDE see HGP500
3-(1-PHTHALAZINYL)CARBAZIC ACID ETHYL ESTER see CBS000
PHTHALAZOL see PHY750
PHTHALAZONE see PHV750
PHTHALIC ACID see PHW250
PHTHALIC ACID ANHYDRIDE see PHW750
PHTHALIC ACID BIS(2-METHOXYETHYL) ESTER see DOF400
PHTHALIC ACID, DECYL HEXYL ESTER see PHW500
PHTHALIC ACID, DIALLYL ESTER see DBL200
o-PHTHALIC ACID, DIALLYL ESTER see DBL200
PHTHALIC ACID, DIETHYL ESTER see DJX000
PHTHALIC ACID, DIGLYCIDYL ESTER see DKM600
PHTHALIC ACID, DIHEPTYL ESTER see HBP400
PHTHALIC ACID DIHEXYL ESTER see DKP600
PHTHALIC ACID DINITRILE see PHY000
PHTHALIC ACID DIOCTYL ESTER see DVL700
PHTHALIC ACID, DIPENTYL ESTER see AON300
PHTHALIC ACID, DIPROPYL ESTER see DWV500
PHTHALIC ACID, DITRIDECYL ESTER see DXQ200
PHTHALIC ACID METHYL ESTER see DTR200
PHTHALIC ANHYDRIDE see PHW750
m-PHTHALIC DICHLORIDE see IMO000
o-PHTHALIC IMIDE see PHX000
PHTHALIDE, 3-BUTYLIDENE- see BRQ100
PHTHALIDE, 3-(3-FURYL)-3a,4,5,6-TETRAHYDRO-3a,7-DIMETHYL-
   see FOM200
PHTHALIMIDE see PHX000
PHTHALIMIDE, N-(HYDROXYMETHYL)- see HMP100
PHTHALIMIDIMIDE see DNE400
PHTHALIMIDO-O,O-DIMETHYL PHOSPHORODITHIOATE see PHX250
2-PHTHALIMIDOGLUTARIC ACID ANHYDRIDE see PHX100
2-PHTHALIMIDOGLUTARIMIDE see TEH500
3-PHTHALIMIDOGLUTARIMIDE see TEH500
α-PHTHALIMIDOGLUTARIMIDE see TEH500
α-(N-PHTHALIMIDO)GLUTARIMIDE see TEH500
PHTHALIMIDOMETHYL ALCOHOL see HMP100
PHTHALIMIDOMETHYL-O,O-DIMETHYL PHOSPHORODITHIOATE
   see PHX250
(PHTHALOCYANINATO(2–))MAGNESIUM see MAI250
PHTHALOCYANINE BLUE 01206 see DNE400
PHTHALODINITRILE see PHY000
m-PHTHALODINITRILE see PHX550
o-PHTHALODINITRILE see PHY000
p-PHTHALODINITRILE see BBP250
PHTHALOGEN see DNE400
PHTHALOL see DJX000
PHTHALONITRILE see PHY000
PHTHALOPHOS see PHX250
N-PHTHALOYL-l-ASPARTIC ACID see PHY250
m-PHTHALOYL CHLORIDE see IMO000
p-PHTHALOYL CHLORIDE see TAV250
PHTHALOYL DIAZIDE see PHY275
p-PHTHALOYL DICHLORIDE see TAV250
N-PHTHALOYLGLUTAMIMIDE see TEH500
PHTHALOYL PEROXIDE see PHY500
PHTHALOYLSULFATHIAZOLE see PHY750
PHTHALSAEUREANHYDRID (GERMAN) see PHW750
PHTHALSAEUREDIAETHYLESTER (GERMAN) see DJX000
PHTHALSAEUREDIMETHYLESTER (GERMAN) see DTR200
PHTHALTAN see TIT250
2-(N^4)-PHTHALYAMINOBENZENESULFONAMIDE)THIAZOLE see PHY750
2-(N^4-PHTHALYAMINOBENZENESULFONAMIDO)THIAZOLE see PHY750
2-(p-PHTHALYLAMINOBENZENESULFAMIDO)THIAZOLE see PHY750
N-PHTHALYL-dl-ASPARTIMIDE see PIA000
N-PHTHALYLGLUTAMIC ACID IMIDE see TEH500
N-PHTHALYL-GLUTAMINSAEURE-IMID (GERMAN) see TEH500
4-PHTHALYLGLUTARAMIC ACID see PIA250
α-N-PHTHALYLGLUTARAMIDE see TEH500
N-PHTHALYLISOGLUTAMINE see PIA250
PHTHALYLNORSULFAZOLE see PHY750
2-(N^4-PHTHALYLSULFANILAMIDO)THIAZOLE see PHY750
2-(p-N-PHTHALYLSULFANILYL)AMINOTHIAZOLE see PHY750

PHTHALYLSULFATHIAZOLE see PHY750
PHTHALYLSULFONAZOLE see PHY750
PHTHALYLSULPHATHIAZOLE see PHY750
PHTHIOCOL see HMI000
1(2H)-PHTHLAZINONE, HYDRAZONE, MONOHYDROCHLORIDE see HGP500
1-PHTHLAZINYLHYDRAZONE ACETONE see HHE500
PHTIVAZID see VEZ925
PHTIVAZIDE see VEZ925
PHYBAN see MRL750
PHYGON see DFT000
PHYGON PASTE see DFT000
PHYGON SEED PROTECTANT see DFT000
PHYGON XL see DFT000
PHYLCARDIN see TEP500
PHYLLINDON see TEP500
PHYLLOCHINON (GERMAN) see VTA000
PHYLLOCORMIN N see HLC000
PHYLLOQUINONE see VTA000
α-PHYLLOQUINONE see VTA000
trans-PHYLLOQUINONE see VTA000
PHYMONE see NAK500
PHYOL see PJA250
PHYONE see PJA250
PHYSALIA PHYSALIS TOXIN see PIA375
PHYSALIN-X see PIA400
PHYSALIS (VARIOUS SPECIES) see JBS100
PHYSEPTONE see MDO750
PHYSEPTONE HYDROCHLORIDE see MDP750
PHYSEX see CMG675
PHYSIC NUT see CNR135
PHYSIOMYCINE see MDO250
PHYSOSTIGMINE see PIA500
PHYSOSTIGMINE SALICYLATE (1581) see PIA750
PHYSOSTIGMINE SO4 see PIB000
PHYSOSTIGMINE SULFATE see PIB000
PHYSOSTIGMINE SULFATE (2:1) see PIB000
PHYSOSTOL see PIA500
PHYSOSTOL SALICYLATE see PIA750
PHYTAR see HKC000
PHYTAR 560 see HKC500
PHYTIC ACID see PIB250
PHYTOGERMINE see VSZ450
PHYTOHAEMAGGLUTININ see PIB575
PHYTOHEMAGGLUTININ see PIB575
PHYTOHEMAGLUTININS see PIB575
PHYTOL see PIB600
trans-PHYTOL see PIB600
PHYTOLACCA AMERICANA see PJJ315
PHYTOLACCA DODECANDRA, extract see EAQ100
PHYTOMELIN see RSU000
PHYTOMENADIONE see VTA000
PHYTOMYCIN see SLY500
PHYTONADIONE see VTA000
PHYTOSOL see EPY000
PIACCAMIDE see DAB875
PIANADALIN see BNK000
PIAPONON see PMH500
PIBECARB see PCV350
PIC-CLOR see CKN500
PICCOLASTIC see SMQ500
PICENADOL see PIB700
PICENE see PIB750
PICEOL see HIO000
PICFUME see CKN500
PICHUCO (MEXICO) see ROU450
PICLORAM see PIB900
2-PICOLINAMINE see ALB750
2-PICOLINE see MOY000
3-PICOLINE see PIB920
4-PICOLINE see MOY250
α-PICOLINE see MOY000
β-PICOLINE see PIB920
Γ-PICOLINE see MOY250
m-PICOLINE (DOT) see PIB920
o-PICOLINE (DOT) see MOY000
p-PICOLINE (DOT) see MOY250
PICOLINE-2-ALDEHYDE THIOSEMICARBAZONE see PIB925
PICOLINIC ACID see PIB930
α-PICOLINIC ACID see ILC000
PICOLINIC ACID, 5-BUTYL-, CALCIUM SALT, HYDRATE see FQR100
PICOLINIC ACID, 3,6-DICHLORO- see DGJ100
2-PICOLYIDENEBIS(p-PHENYL SODIUM SULFATE) see SJJ175
α-PICOLYL ALCOHOL see POR800
2-PICOLYLAMINE see ALB750

4-PICOLYLAMINE see ALC250
2-PICOLYL CHLORIDE HYDROCHLORIDE see PIC000
4,4'-(2-PICOLYLIDENE)BIS(PHENYLSULFURIC ACID) DISODIUM SALT
   see SJJ175
N-(2-PICOLYL)-N-PHENYL-N-(2-PIPERIDINOETHYL)AMINE HYDROCHLO-
   RIDE see CNE375
N-(2-PICOLYL)-N-PHENYL-N-(2-PIPERIDINOETHYL)AMINE TRIPALMITATE
   see PIC100
PICOPERIDAMINE HYDROCHLORIDE see CNE375
PICOPERINE HYDROCHLORIDE see CNE375
PICOPERINE TRIPALMITATE see PIC100
PICOSULFATE SODIUM see SJJ175
PICOSULFOL see SJJ175
PICRACONITINE see PIC250
PICRAMIC ACID see DUP400
PICRAMIC ACID, SODIUM SALT see PIC500
PICRAMIC ACID, ZIRCONIUM SALT (WET) see PIC750
PICRATE of AMMONIA (DOT) see ANS500
PICRATE, DRY (DOT) see PID500
PICRATE ION see PID500
PICRATES see PIC899
PICRIC ACID see PID000
PICRIC ACID (dry) see PID250
PICRIC ACID, ion(1⁻) (dry) see PID500
PICRIC ACID (wet) see PID750
PICRIC ACID, WET (DOT) see PID750
PICRIC ACID, AMMONIUM SALT see ANS500
PICRIDE see CKN500
PICRITE (THE EXPLOSIVE) see NHA500
PICROLONIC ACID see PIE000
PICRONITRIC ACID see PID000
PICROTIN, compounded with PICROTOXININ (1:1) see PIE500
PICROTOXIN see PIE500
PICROTOXINE see PIE500
PICROTOXININ see PIE510
PICROTOXININE see PIE510
PICRYL AZIDE see PIE525
PICRYLMETHYLNITRAMINE see TEG250
PICRYLNITROMETHYLAMINE see TEG250
2-PICRYL-5-NITROTETRAZOLE see PIE550
PICRYL SULFIDE see BLR750
PID see DVV600, PFJ750
PIECIOCHLOREK FOSFORU (POLISH) see PHR500
PIED PIPER MOUSE SEED see SMN500
PIELIK see DAA800
PIELIK E see SGH500
PIE PLANT see RHZ600
PIERIS FLORIBUNDA see FBP520
PIERIS JAPONICA see FBP520
PIETIL see OOG000
PIFARNINE METHANESULFONATE see PIE750
PIFAZINE METHANESULFONATE see PIE750
PIGEON-BERRY see PJJ315
PIGEON BERRY see GIW200, ROA300
PIGMENT FAST ORANGE see DVB800
PIGMENT GREEN 15 see LCR000
PIGMENT PONCEAU R see FMU070
PIGMENT RED 4 see CJD500
PIGMENT RED 23 see NAY000
PIGMENT RED BH see NAY000
PIGMENT RED CD see CHP500
PIGMENT RUBY see MMP100
PIGMENT SCARLET (RUSSIAN) see MMP100
PIGMENT YELLOW 33 see CAP750
PIGMENT YELLOW GT see DEU000
PIGMEX see AEY000
PIGTAIL PLANT see APM875
PIG-WRACK see CCL250
PIH see PDN000
PIK-OFF see EEA000
PIKRINEZUUR (DUTCH) see PID000
PIKRINSAEURE (GERMAN) see PID000
PIKRYNOWY KWAS (POLISH) see PID000
PILEWORT see FBS100
PILIOPHEN see TDE750
PILLARDRIN see MRH209
PILLARON see DTQ400
PILLARZO see CFX000
PILLS (INDIA) see SED400
PILOCARPINE see PIF000
PILOCARPINE HYDROCHLORIDE see PIF250
PILOCARPINE MONOHYDROCHLORIDE see PIF250
PILOCARPINE MONONITRATE see PIF500
PILOCARPINE MURIATE see PIF250

PILOCARPINE NITRATE see PIF500
PILOCARPOL see PIF000
PILOCEL see PIF250
PILOMIOTIN see PIF250
PILORAL see FOS100
PILOT 447 see MKP500
PILOT HD-90 see DXW200
PILOT SF-40 see DXW200
PILOVISC see PIF250
PILPOPHEN see DQA400
PIMACOL-SOL see NAK500
PIMAFUCIN see PIF750
PIMARICIN see PIF750
PIMELIC ACID see PIG000
PIMELIC ACID DINITRILE see HBD000
PIMELIC KETONE see CPC000
PIMELONITRILE see HBD000
4,4'-(PIMELOYLBIS(IMINO-p-PHENYLENEIMINO))BIS(1-ETHYLPYRIDIN-
   IUM) DIPERCHLORATE see PIG250
4,4'-(PIMELOYLBIS(IMINO-p-PHENYLENEIMINO))BIS(1-METHYLPYRIDIN-
   IUM) DIBROMIDE see PIG500
PIMENTA BERRIES OIL see PIG740
PIMENTA LEAF OIL see PIG730
PIMENTA OIL see PIG740
PIMENT BOUC (HAITI) see PCB275
PIMENT (HAITI) see PCB275
PIMENTO OIL see PIG740
PIMEPROFEN see IAB000
PIMETON see BJP250
PIMIENTA del BRAZIL (PUERTO RICO) see PCB300
PIMIENTO de AMERICA (CUBA) see PCB300
PIMM see PFN500
PIMOZIDE see PIH000
PIN see EBD700
PINACIDIL see COS500
PINACOL see TDR000
PINACOLIN see DQU000
PINACOLINE see DQU000
PINACOLONE see DQU000
PINACOLOXYMETHYLPHOSPHORYL FLUORIDE see SKS500
PINACOLYL METHYLFLUOROPHOSPHONATE see SKS500
PINACOLYL METHYLPHOSPHONOFLUORIDATE see SKS500
PINACOLYL METHYLPHOSPHONOFLUORIDE see SKS500
PINACOLYLOXY METHYLPHOSPHORYL FLUORIDE see SKS500
PINAKOLIN (GERMAN) see DQU000
PINAKON see HFP875
PINANG see BFW000
PINANONA (MEXICO, PUERTO RICO) see SLE890
PINAZEPAM see PIH100
PINDIONE see PFJ750
PINDOLOL see VSA000
PINDON (DUTCH) see PIH175
PINDONE see PIH175
PINEAPPLE SHRUB see CCK675
2-PINENE see PIH250
2(10)-PINENE see POH750
α-PINENE (FCC) see PIH250
β-PINENE (FCC) see POH750
PINE NEEDLE OIL see FBV000
PINE NEEDLE OIL, DWARF see PIH400
PINE NEEDLE OIL, SCOTCH see PIH500
2-PINENE-10-METHYL ACETATE see DTC000
(1R,5R)-(+)-2-PINEN-4-ONE see VIP000
PINE OIL see PIH750
PINERORO see CCR875
PINGYANGMYCIN (CHINESE) see BLY500
PINHOLE AK 2 see ASM270
PINKROOT see PIH800
PINK WEED see PIH800
PINON (PUERTO RICO, DOMINICAN REPUBLIC) see CNR135
PIN-TEGA see PIJ500
PINUS MONTANA OIL see PIH400
PINUS PUMILIO OIL see PIH400
PIO see PII150
PIODEL see PII100
PIOMY see PII150
PIOMYCIN see PII150
PIOPEN see CBO250
PIPADOX see HEP000
PIPAMAZINE see CJL500
PIPAMPERONE see FHG000
PIPAMPERONE DICHLORHYDRATE see PII200
PIPAMPERONE DIHYDROCHLORIDE see PII200
PIPAMPERONE HYDROCHLORIDE see PII200

PIPANEPERONE see FHG000
PIPANOL see BBV000
PIPA de TURCO (CUBA) see GEW800
α-PIPECOLIN see MOG750
16β-PIPECOLINIO-2β-PIPERIDINO-5α-ANDROSTAN-3α,17β-DIOL BROMIDE
   DIACETATE see VGU075
dl-6-(N-α-PIPECOLINOMETHYL)-5-HYDROXY-INDANE MALEATE see PJI600
PIPECURIUM BROMIDE see PII250
PIPECURONIUM BROMIDE see PII250
PIPEDAC see PIZ000
PIPEMIDIC ACID see PIZ000
PIPEMIDIC ACID TRIHYDRATE see PII350
PIPENZOLATE BROMIDE see PJA000
PIPENZOLATE METHYLBROMIDE see PJA000
PIPERACETAZINE see PII500
PIPERACILLIN SODIUM see SJJ200
PIPERADROL HYDROCHLORIDE see PII750
PIPERALIN see MOL250
PIPERAMIC ACID see PIZ000
PIPERAZIDINE see PIJ000
PIPERAZIN (GERMAN) see PIJ000
PIPERAZINE see PIJ000
PIPERAZINE, anhydrous see PIJ000
PIPERAZINE ADIPATE see HEP000
PIPERAZINE, 1-(4-AMINO-6,7-DIMETHOXY-2-QUINAZOLINYL)-4-
   ((TETRAHYDRO-2-FURANYL)CARBONYL)-, MONOHYDROCHLORIDE,
   DIHYDRATE see TEF700
PIPERAZINE, 1-BENZOYL-2,6-DIMETHYL-4-NITROSO-(9CI) see NJL850
1-PIPERAZINECARBOXYLIC ACID, 4-METHYL-, 6-(5-CHLORO-2-
   PYRIDINYL)-6,7-DIHYDRO-7-OXO-5H-PYRROLO(3,4-b)PYRAZIN-5-YL
   ESTER see ZUA450
PIPERAZINE, 1-(8-CHLORO-10,11-DIHYDRODIBENZO(b,f)THIEPIN-10-YL)-4-
   METHYL- see ODY100
PIPERAZINE, 1-CINNAMYL-4-(DIPHENYLMETHYL)- see CMR100
PIPERAZINE CITRATE (3582) see PIJ500
PIPERAZINE CITRATE TELRA see PIJ500
PIPERAZINE DIANTIMONY TARTRATE see PIJ600
1,4-PIPERAZINEDIETHANOL see PIJ750
PIPERAZINE DIHYDROCHLORIDE see PIK000
PIPERAZINE, 1-(3,4-DIMETHOXYBENZOYL)-4-(1,2,3,4-TETRAHYDRO-2-OXO-
   6-QUINOLINYL)- see DLF700
PIPERAZINEDIONE see PIK075
PIPERAZINEDIONE 593A see PIK075
2,6-PIPERAZINEDIONE-4,4'-PROPYLENE DIOXOPIPERAZINE see PIK250
PIPERAZINE, 1-(DIPHENYLMETHYL)-4-(3-PHENYL-2-PROPENYL)-(9CI)
   see CMR100
1,4-PIPERAZINEDIYLBIS(BIS(1-AZIRIDINYL)PHOSPHINE OXIDE see BJC250
1,1'-(1,4-PIPERAZINEDIYLDIETHYLENE)BIS(1-ETHYLPIPERIDINIUM IO-
   DIDE) see PIK375
(1,4-PIPERAZINEDIYLDIETHYLENE)BIS(TRIETHYLAMMONIUM IODIDE)
   see PIK400
1-PIPERAZINEETHANOL see HKY500
PIPERAZINE HEXAHYDRATE see PIK500
PIPERAZINE HYDROCHLORIDE see PIK000
PIPERAZINE and SODIUM NITRITE (4581) see PIJ250
PIPERAZINE SULTOSILATE see PIK625
PIPERAZINIUM, 4-(β-CYCLOHEXYL-β-HYDROXYPHENETHYL)-1,1-
   DIMETHYL-, METHYL SULFATE see HFG400
2-(1-PIPERAZINYL)ETHANOL see HKY500
2-(1-PIPERAZINYL)-QUINOLINE (Z)-2-BUTENEDIOATE (1:1) (9CI) see QWJ500
2-(1-PIPERAZINYL)-QUINOLINE MALEATE (1:1) see QWJ500
PIPERIDILATE HYDROCHLORIDE see EOY000
PIPERIDIN (GERMAN) see PIL500
PIPERIDINE see PIL500
PIPERIDINE, 4-tert-BUTYL-1-NITROSO see NJO300
2-PIPERIDINECARBOXAMIDE,1-BUTYL-N-(2,6-
   DIMETHYLPHENYL)MONOHYDROCHLORIDE MONOHYDRATE
   see BOO000
PIPERIDINE, 1-(CYCLOHEXYLCARBONYL)-3-METHYL- see CPI350
PIPERIDINE, 2,6-DIMETHYL-1-((2-METHYLCYCLOHEXYL)CARBONYL)-
   see DSP650
PIPERIDINE, 3,5-DIMETHYL-1-NITROSO-, (E)- see DTA700
PIPERIDINE, 3,5-DIMETHYL-1-NITROSO-, (Z)- see DTA690
2,6-PIPERIDINEDIONE, 3-(4-AMINOPHENYL)-3-ETHYL- see AKC600
3-PIPERIDINE-1,1-DIPHENYL-PROPANOL-(1) METHANESULPHONATE
   see PIL550
2-PIPERIDINEETHANOL-m-AMINOBENZOATE (ester) HYDROCHLORIDE
   see AIQ880
2-PIPERIDINEETHANOL-o-AMINOBENZOATE (ester) HYDROCHLORIDE
   see AIQ885
2-PIPERIDINEETHANOL-p-AMINOBENZOATE (ester) HYDROCHLORIDE
   see AIQ890
1-PIPERIDINEETHANOL BENZILATE HYDROCHLORIDE see PIM000
2-PIPERIDINEETHANOL CARBANILATE (ester) HYDROCHLORIDE see PIU800

PIRACAPS see TBX250
PIRACETAM see NNE400
PIRAFLOGIN see HNI500
PIRAMIDON see DOT000
PIRARREUMOL "B" see BRF500
PIRAZETAM see NNE400
PIRAZINON see DJX400
PIRAZOXON (ITALIAN) see MOX250
PIRBUTEROL DIHYDROCHLORIDE see POM800
PIRECIN see POF500
PIREF see DIZ100
PIRENZEPINE HYDROCHLORIDE see GCE500
PIRETANIDE see PDW250
PIRETRINA 1 (PORTUGUESE) see POO050
PIREVAN see PJA120
PIREXYL see MOA600
PIREXYL PHOSPHATE see PJA130
PIRIA'S ACID see ALI000
PIRIBEDIL see TNR485
PIRIBENZIL see TMP750
PIRIBENZIL METHYL SULFATE see CDG250
PIRID see PDC250, PEK250
PIRIDACIL see PDC250
7-(2-(PIRIDIL)-METILAMMINO-ETIL)-TEOFILLINO NICOTINATO (ITALIAN)
    see PPN000
PIRIDINA (ITALIAN) see POP250
PIRIDINOL CARBAMATO (SPANISH) see PPH050
PIRIDISIR see PPP500
PIRIDOL see DOT000
PIRIDOLAN see PJA140
PIRIDOLO see AKO500
PIRIDOSAL see DAM600, DAM700
PIRIDROL see DWK400
PIRIDROL HYDROCHLORIDE see PII750
PIRIEX see TAI500
PIRIMAL-M see ALF250
PIRIMECIDAN see TGD000
PIRIMETAMINA (SPANISH) see TGD000
PIRIMICARB see DOX600
PIRIMIFOSETHYL see DIN600
PIRIMIFOS-METHYL see DIN800
PIRIMIPHOS-ETHYL see DIN600
PIRIMOR see DOX600
PIRINITRAMIDE see PJA140
PIRINIXIL see CLW500
PIRISTIN see POO750
PIRITON see TAI500
PIRITRAMIDE see PJA140
PIRMAZIN see SNJ000
PIRMENOL HYDROCHLORIDE see PJA170
PIROAN see PCP250
PIROCRID see POE100
PIROD see UNJ800
PIRODAL see PAF250
PIROFOS see SOD100
PIROHEPTINE see PJA190
PIROHEPTINE HYDROCHLORIDE see TMK150
PIROMIDIC ACID see PAF250
PIROMIDINA see DOT000
PIROSOLVINA see EEM000
PIROXICAM see FAJ100
PIRPROFEN see PJA220
PIRROLIDINOMETIL-TETRACICLINA (ITALIAN) see PPY250
PIRROXIL see NNE400
PIRVINIUM PAMOATE see PQC500
PIRYDYNA (POLISH) see POP250
PISCAROL see IAD000
PISCIDEIN see HJO500
PISCIDIC ACID see HJO500
PISCIOL see IAD000
PISS-A-BED (JAMAICA) see CNG825
PITAYINE see QFS000
PITC see ISQ000
PITCH see CMZ100
PITCH APPLE see BAE325
PITCH, COAL TAR see CMZ100
PITMAL see TLP750
PITOCIN see ORU500
PITON S see ORU500
PITREX see TGB475
PITTCHLOR see HOV500
PITTCIDE see HOV500
PITTCLOR see HOV500
PITTSBURGH PX-138 see DVL700

PITUITARY GLAND ADRENO CORTICO-TROPIC HORMONE see AES650
PITUITARY GROWTH HORMONE see PJA250
PITUITARY LACTOGENIC HORMONE see PMH625
PITUITARY LEUTEINIZING HORMONE see LIU300
PIVACIN see PIH175
PIVADORM see BNP750
PIVADORN see BNP750
PIVAL see PIH175
PIVALDION (ITALIAN) see PIH175
PIVALDIONE (FRENCH) see PIH175
PIVALIC ACID see PJA500
PIVALIC ACID CHLORIDE see DTS400
PIVALIC ACID LACTONE see DTH000
PIVALOLACTONE see DTH000
PIVALOLYL CHLORIDE see DTS400
PIVALONITRILE see PJA750
PIVALOYL AZIDE see PJB000
PIVALOYL CHLORIDE see DTS400
2-PIVALOYL-INDAAN-1,3-DION (DUTCH) see PIH175
2-PIVALOYL-INDAN-1,3-DION (GERMAN) see PIH175
2-PIVALOYL-1,3-INDANDIONE see PIH175
2-PIVALOYLINDANE-1,3-DIONE see PIH175
PIVALOYLOXYMETHYL d-α-AMINOBENZYLPENICILLINATE HYDRO-
    CHLORIDE see AOD000
PIVALYL CHLORIDE see DTS400
2-PIVALYL-1,3-INDANDIONE see PIH175
PIVALYL VALONE see PIH175
PIVALYN see PIH175
PIVAMPICILLIN HYDROCHLORIDE see AOD000
PIVATIL see AOD000
PIVMECILLINAM HYDROCHLORIDE see MCB550
PIXALBOL see CMY800
PIX CARBONIS see CMY800
PJ185 see LIH000
PK 10169 see HAQ550
PKhNB see PAX000
PK-MERZ see TJG250
PLACIDAL see EQL000, MNM500
PLACIDAS see MNM500
PLACIDEX see MFD500
PLACIDIL see CHG000
PLACIDOL see CJR909
PLACIDOL E see DJX000
PLACIDON see MQU750
PLACIDYL see CHG000
PLAFIBRIDA (SPANISH) see PJB500
PLAFIBRIDE see PJB500
PLANADALIN see BNK000
PLANOCAINE see AIT250
PLANOCHROME see MCV000
PLANOFIX see NAK500
PLANOMIDE see TEX250
PLANOMYCIN see FBP300
PLANOTOX see DAA800
PLANT DITHIO AEROSOL see SOD100
PLANTDRIN see MRH209
PLANTFUME 103 SMOKE GENERATOR see SOD100
PLANTGARD see DAA800
PLANTIFOG 160M see MAS500
PLANTOMYCIN see SLY500
PLANT PIN see SOB500
PLANT PROTEASE CONCENTRATE see BMO000
PLANT PROTECTION PP511 see DIN800
PLANTULIN see PMN850
PLANTVAX see DLV200
PLANT WAX see DLV200
PLANUM see CFY750
PLAQUENIL see PJB750
PLASDONE see PKQ250
PLASIL see AJH000
PLASKON 201 see PJY500
PLASMA COAGULASE see CMY325
PLASMASTERIL see HLB400
PLASMOCHIN see RHZ000
PLASMOCIDE see RHZ000
PLASMOCOAGULASE see CMY325
PLASMOQUINE see RHZ000
PLASTANOX 2246 see MJO500
PLASTANOX 425 ANTIOXIDANT see MJN250
PLASTER of PARIS see CAX500
PLASTIBEST 20 see ARM268
PLASTICIZER BDP see BQX250
PLASTICIZER G-316 see PJC000
PLASTICIZER GPE see PJC500

PLASTICIZER 4GO see PJC250
PLASTICIZER Z-88 see PJC750
PLASTOLEIN 9058 see BJQ500
PLASTOLEIN 9214 see FAB920
PLASTOLEIN 9058 DOZ see BJQ500
PLASTOMOLL DOA see AEO000
PLASTORESIN ORANGE F4A see PEJ500
PLATENOMYCIN B1 see MBY150
PLATENOMYCIN-B3 see LEV025
PLATIBLASTIN see PJD000
cis-PLATIN see PJD000
PLATIN (GERMAN) see PJD500
PLATINATE(2-), HEXACHLORO-, DIHYDROGEN, HEXAHYDRATE
    see DLO400
PLATINATE(2-), NITROTRICHLORO-, DIPOTASSIUM see PLN050
PLATINATE(2-), TETRACHLORO-, DIAMMONIUM see ANV800
PLATINATE(2-), TETRAKIS(THIOCYANATO)-, DIPOTASSIUM see PLU590
PLATINATE(1-), TRICHLOROETHYLENE-, DIPOTASSIUM see PLW200
PLATINATE(2-), TRICHLORO(NITRITO-N-), DIPOTASSIUM (SP-4-2)-
    see PLN050
PLATINEX see PJD000
PLATINIC AMMONIUM CHLORIDE see ANF250
PLATINIC CHLORIDE see CKO750
PLATINIC POTASSIUM CHLORIDE see PLR000
PLATINIC SODIUM CHLORIDE see SJJ500
PLATINOL see PJD000
PLATINOL AH see DVL700
PLATINOL DOP see DVL700
PLATINOUS CHLORIDE see PJE000
cis-PLATINOUS DIAMMINE DICHLORIDE see PJD000
PLATINOUS POTASSIUM CHLORIDE see PJD250
PLATINUM see PJD500
PLATINUM(II) AMMINE TRICHLOROPOTASSIUM see PJD750
PLATINUM (II), BIS(METHYL SELENIDE)DICHLORO-, cis- DEU115
PLATINUM (II), BIS(METHYL SELENIDE)SULFATO-, HYDRATE see SNS100
PLATINUM BLACK see PJD500
PLATINUM CHLORIDE see PJE000
PLATINUM(IV) CHLORIDE see PJE250
PLATINUM COMPOUNDS see PJE500
PLATINUM(II) (CYCLOHEXANE-1,2-DIAMMINE)ISOCITRATO-, (Z)-
    see PGQ275
cis-PLATINUM(II) DIAMINEDICHLORIDE see PJD000
trans-PLATINUM(II)DIAMMINEDICHLORIDE see DEX000
cis-PLATINUMDIAMMINE TETRACHLORIDE see TBO776
cis-PLATINUM(IV) DIAMMINOTETRACHLORIDE see TBO776
PLATINUM DIARSENIDE see PJE750
PLATINUM, DICHLOROBIS(METHYL SELENIDE)-, cis- see DEU115
PLATINUM, DICHLOROBIS(SELENOBIS(METHANE))-(SP-4-2) see DEU115
PLATINUM(II) DINITRODIAMMINE see DCF800
PLATINUM ETHYLENEDIAMINE DICHLORIDE see DFJ000
PLATINUM FULMINATE see PJF000
PLATINUM (2-), NITROTRICHLORO-, DIPOTASSIUM see PLN050
PLATINUM SPONGE see PJD500
PLATINUM SULFATE see PJF500
PLATINUM(II) SULFATE see PJF500
PLATINUM SULFATE TETRAHYDRATE see PJF750
PLATINUM(II) SULFATE TETRAHYDRATE see PJF750
PLATINUM TETRACHLORIDE see PJE250
cis-PLATINUM-2-THYMINE see PJG000
PLATINUM THYMINE BLUE see PJG000
PLATIPHILLIN HYDROCHLORIDE see PJG150
PLATIPHYLLIN HYDROCHLORIDE see PJG150
PLAVOLEX see DBD700
PLAXIDOL see CJR909
PLE 1053 see ASO375
PLECYAMIN see VSZ000
PLEGANGIN see VIZ400
PLEGATIL see MQF750
PLEGECYL see ABH500
PLEGICIN see ABH500
PLEGINE see DKE800, PDD000
PLEGOMAZIN see CKP250, CKP500
PLENASTRIL see PAN100
PLENOLIN see DLO875
PLENUR see LGZ000
PLEOCIDE see ABY900
PLESSY'S GREEN (HEMIHEPTAHYDRATE) see CMK300
PLESTROVIS see QFA250
PLEXIGLAS see PKB500
PLEXIGUM M 920 see PKB500
PLH see LIU300
PLICTRAN see CQH650
PLIDAN see DCK759
PLIMASINE see MNQ000

PLIOFILM see PJH500
PLIOFLEX see SMR000
PLIOLITE S5 see SMR000
PLIOVAC AO see AAX175
PLIOVIC see PKQ059
PLISULFAN see AIF000
PLIVA see BIE500
PLIVAPHEN see ABH500
PLLETIA see PMI750
PLOMB FLUORURE (FRENCH) see LDF000
PLOSTENE see TKH750
PLOYMANNURONIC ACID see AFL000
PLP see PII100
PLTU see MJZ000
PLUCHEA LANCEOLATA (DC.) Cl., extract excluding roots see PJH550
PLUCKER see NAK500
PLUM see AQP890
PLUMBAGIN see PJH610
PLUMBAGO ZEYLANICA Linn., root extract see PJH615
PLUMBOPLUMBIC OXIDE see LDS000
PLUMBOUS ACETATE see LCJ000, LCV000
PLUMBOUS CHLORIDE see LCQ000
PLUMBOUS CHROMATE see LCR000
PLUMBOUS FLUORIDE see LDF000
PLUMBOUS OXIDE see LDN000
PLUMBOUS PHOSPHATE see LDU000
PLUMBOUS SULFIDE see LDZ000
PLURACOL E see PJT200
PLURACOL P-410 see PJT000
PLURAFAC RA 43 see DXY000
PLURONIC L-81 see PJH630
PLURYL see BEQ625
PLURYLE see BEQ625
PLUSURIL see BEQ625
PLUTONIUM see PJH750
PLUTONIUM BISMUTHIDE see PJH775
PLUTONIUM COMPOUNDS see PJI000
PLUTONIUM(III) HYDRIDE see PJI250
PLUTONIUM NITRATE (solution) see PJI500
PLYCTRAN see CQH650
PLYMOUTH IPP see IQW000
PM245 see BEM500
PM 334 see MNZ000
P.M. 388 see PBH100
PM 396 see MLP800
PM 671 see ENG500
P.M. 434,526 see DBE885
PMA see ABU500, PGV000
PMAC see ABU500
PMACETATE see ABU500
PMAL see ABU500
PMAS see ABU500
PMB see ECU750
PMC see PFM500
2M-4CP see CIR500
PMCG see PJI575
PMCG HYDROCHLORIDE see PJI575
PMDT see PBG500
P.M.F. see PFN000
PMFP see SKS500
PMH see PAP000
(±)-PMHI MALEATE see PJI600
PMMA see PKB500
PM(1)MM(3) see HBA600
PMP see PHX250, TGK200
PMP SODIUM GLUCONATE see SHK800
PMS see MRW000, PNQ000, SCA750
PMS 1.5 see SCR400
PMS 300 see SCR400
PMS 154A see SCR400
PMS 200A see SCR400
PMSG see SCA750
PMS No. 1 see PJW750
PMT see MOA725, PFS500
PM-TC see PMI000
2M-4XP see CIR325
PN see CFC000
PN6 see AQO000
PNA see NEO500
19583RP-Na see KGK100
PNB see NFQ000
PNBAS see NEY000
PNCB see NFS525
PND see COS500

PNEUMOREL see DAI200
PNNG see NLD000
PNOT see NMP500
PNS 25 see SCR400
PNT see NMO550
PNU see NLO500
PO-20 see PMD800
POCAN BUSH see PJJ315
PO-DIMETHOATE see DNX800
PODOPHYLLIN see PJJ000
PODOPHYLLINIC ACID LACTONE see PJJ225
PODOPHYLLOTOXIN see PJJ225
PODOPHYLLOTOXIN-o-BENXYLIDENE-β-d-GLUCOPYRANOSIDE see BER500
PODOPHYLLOTOXIN-BENZILIDEN-GLUCOSID (GERMAN) see BER500
PODOPHYLLUM see PJJ000
PODOPHYLLUM PELTATUM see MBU800
PODOPHYLLUM RESIN see PJJ000
POHA (HAWAII) see JBS100
POINSETTIA see EQX000
POINSETTIA PULCHERRIMA GRAH see EQX000
POINT TWO see SHF500
POIS COCHON (HAITI) see YAG000
POIS MANIOC (HAITI) see YAG000
POISON BULB see SLB250
POISON BUSH see BOO700
POISON de COULEUVRE (CANADA) see BAF325
POISON HEMLOCK see PJJ300
POISON PARSLEY see PJJ300
POISON ROOT see PJJ300
POISON SEGO see DAE100
POISON TOBACCO see HAQ100
POISON TREE see BOO700
POIS PUANTE (HAITI) see CNG825
POIS VALLIERE (HAITI) see SBC550
POKE see PJJ315
POKEBERRY see PJJ315
POKEWEED see PJJ315
POLAAX see OAX000
POLACARITOX see CKM000
POLACTINE G YELLOW see CMS230
POLAKTYN YELLOW G see CMS230
POLAMIDON see MDP250
POLAMIDONE see MDO750
POLARAMIN see PJJ325
POLARAMINE MALEATE see PJJ325
POLARIS see BLG250
POLARONIL (GERMAN) see TAI500
POLCILLIN see PJJ350
POLCOMINAL see EOK000
POLECAT WEED see SDZ450
POLEON see EID000
POLFOSCHLOR see TIQ250
POLICAPRAN see PJY500
POLICAR MZ see DXI400
POLICAR S see DXI400
POLIDOCANOL see DXY000
POLIFEN see TAI250
POLIFLOGIL see HNI500
POLIFUNGIN see PJJ500
POLIGOSTYRENE see SMQ500
POLIKARBATSIN (RUSSIAN) see MQQ250
POLINALIN see DOT000
POLISEPTIL see TEX250
POLISIN see BKL250
POLITEF see TAI250
POLIURON see BEQ625
POLIVAL see TEX000
POLIVINIT see PKQ059
POLKWEED see PJJ315
POLLACID see HCQ500
POLONIUM see PJJ750
POLONIUM CARBONYL see PJK000
POLOPIRYNA see ADA725
POLOXAL RED 2B see HJF500
POLSTIGMINE see DQY909
POLY see SKN000
POLYAC see PJR500
POLYACRYLATE see ADW200
POLY(ACRYLIC ACID) see ADW200
POLYACRYLONITRILE see ADX750
POLYACRYLONITRILE, COMBUSTION PRODUCTS see PJK250
POLYAETHYLENGLYKOLE 200 (GERMAN) see PJT200
POLYAETHYLENGLYKOLE 300 (GERMAN) see PJT225
POLYAETHYLENGLYKOLE 400 (GERMAN) see PJT230

POLYAETHYLENGLYKOLE 600 (GERMAN) see PJT240
POLYAETHYLENGLYKOLE 1000 (GERMAN) see PJT250
POLYAETHYLENGLYKOLE 1500 (GERMAN) see PJT500
POLYAETHYLENGLYKOLE 4000 (GERMAN) see PJT750
POLYAETHYLENGLYKOLE 6000 (GERMAN) see PJU000
POLYAMID (GERMAN) see NOH000
POLYAMIDE 6 see PJY500
POLYAMIDE-6 (combustion products) see PJK750
POLYAMINE D see PJL000
POLYAMINE H SPECIAL see PJL100
POLY-p-AMINOBENZALDEHYD (CZECH) see DXS000
POLY(epsilon-AMINOCAPROIC ACID) see PJY500
POLYBENZARSOL see BCJ150
POLYBIS(2-ETHYLBUTYL)SILOXANE see EHC900
POLYBOR see DXF200, SFF000
POLYBOR 3 see DXF200
POLY[BORANE(1)] see PJL325
POLYBOR-CHLORATE see SFS500
POLYBREME see HCV500
POLYBROMINATED BIPHENYL see FBU509, HCA500
POLYBROMINATED BIPHENYL (FF-1) see FBU509
POLYBROMINATED BIPHENYLS see FBU000, PJL335
POLYBROMINATED SALICYLANILIDE see THW750
POLYBROMOETHYLENE see PKQ000
cis-POLY(BUTADIENE) see PJL350
POLY(1,3-BUTADIENE PEROXIDE) see PJL375
POLYBUTADIENE-POLYSTYRENE COPOLYMER see SMR000
POLYCAPROAMIDE see PJY500
POLY(epsilon-CAPROAMIDE) see PJY500
POLYCAPROLACTAM see PJY500
POLY(epsilon-CAPROLACTAM) see PJY500
POLYCARBACIN see MQQ250
POLYCARBACINE see MQQ250
POLYCARBAMATE see BJK000
POLYCARBAZIN see MQQ250
POLYCARBAZINE see MQQ250
POLY(CARBON MONOFLUORIDE) see PJL600
POLYCAT 8 see DRF709
POLYCHLORCAMPHENE see CDV100
POLYCHLORINATED BIPHENYL see PJL750
POLYCHLORINATED BIPHENYL (AROCLOR 1016) see PJL800
POLYCHLORINATED BIPHENYL (AROCLOR 1221) see PJM000
POLYCHLORINATED BIPHENYL (AROCLOR 1232) see PJM250
POLYCHLORINATED BIPHENYL (AROCLOR 1242) see PJM500
POLYCHLORINATED BIPHENYL (AROCLOR 1248) see PJM750
POLYCHLORINATED BIPHENYL (AROCLOR 1254) see PJN000
POLYCHLORINATED BIPHENYL (AROCLOR 1260) see PJN250
POLYCHLORINATED BIPHENYL (AROCLOR 1262) see PJN500
POLYCHLORINATED BIPHENYL (AROCLOR 1268) see PJN750
POLYCHLORINATED BIPHENYL (AROCLOR 2565) see PJO000
POLYCHLORINATED BIPHENYL (AROCLOR 4465) see PJO250
POLYCHLORINATED BIPHENYL (KANECHLOR 300) see PJO500
POLYCHLORINATED BIPHENYL (KANECHLOR 400) see PJO750
POLYCHLORINATED BIPHENYL (KANECHLOR 500) see PJP000
POLYCHLORINATED CAMPHENES see CDV100
POLYCHLORINATED DIBENZOFURANS see PJP100
POLYCHLORINATED TERPHENYL see PJP250
POLYCHLORINATED TRIPHENYL (AROCLOR 5442) see PJP750
POLYCHLOROBENZOIC ACID, DIMETHYLAMINE SALTS see PJQ000
POLYCHLOROBIPHENYL see PJL750
POLYCHLOROCAMPHENE see CDV100
POLYCHLORODICYCLOPENTADIENE see BAF250
POLYCHLORODICYCLOPENTADIENE ISOMERS see BAF250
POLY(CHLOROETHYLENE) see PKQ059
POLYCHLOROPINENE see PJQ250
POLYCHOM see ZJS300
POLYCHOME see ZJS300
POLYCIDAL see MFN500
POLYCILLIN see AIV500, AOD125
POLYCILLIN-N see SEQ000
POLYCIZER 962-BPA see DXQ200
POLYCIZER DBP see DEH200
POLYCIZER DBS see DEH600
POLYCLAR L see PKQ250
POLYCLENE see DGB000
POLYCO 2410 see SMR000
POLY C POLY I see PJY750
POLYCRON see BNA750
POLYCTYIDYLIC-POLYINOSINIC ACID see PJY750
POLYCYCLIC MUSK see ACL750
POLYCYCLINE see TBX000
POLYCYCLINE HYDROCHLORIDE see TBX250
POLYCYCLOPENTADIENYLTITANIUM DICHLORIDE see PJQ275
POLYDAZOL see PJQ350

POLYDIBROMOSILANE see PJQ500
POLYDIBROMOSILYLENE see PJQ500
POLY(1,2-DIHYDRO-2,2,4-TRIMETHYLQUINOLINE) see PJQ750
POLY(DIMERCURYIMMONIUM ACETYLIDE) see PJQ775
POLY(DIMERCURYIMMONIUM BROMATE) see PJQ780
POLY(DIMERCURYIMMONIUM HYDOXIDE) see DNW200
POLY((DIMETHYLIMINIO)HEXAMETHYLENE(DIMETHYLIMINO)
  TRIMETHYLENE DIBROMIDE) see IFT300
POLYDIMETHYL SILOXANE see PJR000
POLYDIMETHYLSILOXANE see DTR850
POLYDIMETHYLSILOXANE RUBBER see PJR250
POLY-p-DINITROSOBENZENE see PJR500
POLYESTRADIOL PHOSPHATE see PJR750
POLY(ESTRADIOL PHOSPHATE) see EDS000
POLYETHER DIAMINE L-1000 see PJS000
POLYETHYLENE see PJS750
POLYETHYLENE AS see PJS750
POLYETHYLENE 600 DIBENZOATE see PKE750
POLYETHYLENE GLYCOL see PJT000
POLYETHYLENE GLYCOL 200 see PJT200
POLYETHYLENE GLYCOL 300 see PJT225
POLYETHYLENE GLYCOL 400 see PJT230
POLYETHYLENE GLYCOL 600 see PJT240
POLYETHYLENE GLYCOL 1000 see PJT250
POLYETHYLENE GLYCOL 1500 see PJT500
POLYETHYLENE GLYCOL 4000 see PJT750
POLYETHYLENE GLYCOL 6000 see PJU000
POLYETHYLENE GLYCOL CHLORIDE 210 see PJU250
POLYETHYLENE GLYCOL DIBENZOATE see PKE750
POLYETHYLENE GLYCOL 220 DIBENZOATE see PKE750
POLYETHYLENE GLYCOL 200 DI(2-ETHYLHEXOATE) see FCD512
POLYETHYLENE GLYCOL DIMETHYL ETHER see DOM100
POLYETHYLENE GLYCOL DISTEARATE see PJU500
POLYETHYLENE GLYCOL 300 DISTEARATE see PJU500
POLYETHYLENE GLYCOL 400 (DI) STEARATE see PJU500
POLYETHYLENE GLYCOL 600 (DI) STEARATE see PJU500
POLYETHYLENE GLYCOL DISTEARATE 1000 see PJU750
POLYETHYLENE GLYCOL DODECYL ETHER see DXY000
POLYETHYLENE GLYCOL E 600 see PJV000
POLYETHYLENE GLYCOL MONOETHER with p-tert-OCTYLPHENYL
  see PKF500
POLYETHYLENE GLYCOL MONOLEYL ETHER see OIG040
POLYETHYLENE GLYCOL MONO(OCTYLPHENYL) ETHER see GHS000
POLYETHYLENE GLYCOL MONO(4-OCTYLPHENYL) ETHER see PKF500
POLYETHYLENE GLYCOL MONO(4-tert-OCTYLPHENYL) ETHER see PKF500
POLYETHYLENE GLYCOL MONO(p-tert-OCTYLPHENYL) ETHER see PKF500
POLYETHYLENE GLYCOL MONOOLEATE see PJY100
POLYETHYLENE GLYCOL MONOSTEARATE see PJV250
POLYETHYLENE GLYCOL MONOSTEARATE 200 see PJV500
POLYETHYLENE GLYCOL MONOSTEARATE 400 see PJV750
POLYETHYLENE GLYCOL MONOSTEARATE 1000 see PJW000
POLYETHYLENE GLYCOL MONOSTEARATE 6000 see PJW250
POLYETHYLENE GLYCOL MONO(p-(1,1,3,3-TETRAMETHYLBUTYL)
  PHENYL) ETHER see PKF500
POLYETHYLENE GLYCOL 450 NONYL PHENYL ETHER see PKF000
POLYETHYLENE GLYCOL OCTYLPHENOL ETHER see PKF500
POLYETHYLENE GLYCOL OCTYLPHENYL ETHER see GHS000
POLYETHYLENE GLYCOL p-OCTYLPHENYL ETHER see PKF500
POLYETHYLENE GLYCOL 450 OCTYL PHENYL ETHER see PKF500
POLYETHYLENE GLYCOL p-tert-OCTYLPHENYL ETHER see PKF500
POLYETHYLENE GLYCOL OLEATE see PJY100
POLYETHYLENE GLYCOL 1000 OLEYL ETHER see PJW500
POLYETHYLENEGLYCOLS MONOSTEARATE see PJW750
POLYETHYLENE GLYCOL p-1,1,3,3,-TETRAMETHYLBUTYLPHENYL
  ETHER see PKF500
POLYETHYLENEIMIN (CZECH) see PJX000
POLYETHYLENE IMINE see PJX000
POLY(ETHYLENE OXIDE) see PJT000
POLY(ETHYLENE OXIDE) DODECYL ETHER see DXY000
POLYETHYLENE OXIDE MONOOLEATE see PJY100
POLY(ETHYLENE OXIDE)OCTYLPHENYL ETHER see GHS000
POLY(ETHYLENE OXIDE) OLEATE see PJY100
POLYETHYLENE-POLYPROPYLENE GLYCOLS PLURONIC L-81 see PJH630
POLYETHYLENE TEREPHTHALATE see PKF750
POLYETHYLENE TEREPHTHALATE FILM see PKF750
POLY(ETHYLENE TETRAFLUORIDE) see TAI250
POLYETHYLENE Y-141-A see PJX750
POLYETHYLENIMINE (10,000) see PJX800
POLYETHYLENIMINE (20,000) see PJX825
POLYETHYLENIMINE (35,000) see PJX835
POLYETHYLENIMINE (40,000) see PJX845
POLY(ETHYLIDENE PEROXIDE) see PJX850
POLYFENE see TAI250
POLYFER see IGS000

POLYFIBRON 120 see SFO500
POLYFLON see TAI250
POLYFOAM PLASTIC SPONGE see PKL500
POLYFOAM SPONGE see PKL500
POLYFUNGIN see PJJ500
POLY-G see PJT200
POLY G 400 see PJT230
POLY-GIRON see TEH500
POLYGLUCIN see DBD700
POLYGLYCOL 1000 see PJT250
POLYGLYCOL 4000 see PJT750
POLYGLYCOLAMINE H-163 see AMC250
POLYGLYCOLDIAMINE H 221 see EBV100
POLYGLYCOL DISTEARATE see PJU500
POLYGLYCOL E see PJT200
POLYGLYCOL E1000 see PJT250
POLYGLYCOL E-4000 see PJT750
POLYGLYCOL E-4000 USP see PJT750
POLYGLYCOL LAURATE see PJY000
POLYGLYCOL MONOOLEATE see PJY100
POLYGLYCOL OLEATE see PJY250
POLYGON see SKN000
POLYGONUM HYDROPIPER L., dry powdered whole plant see WAT350
POLYGRIPAN see TEH500
POLY-G SERIES see PJT000
POLY I:C see PJY750
POLY(IMINOCARBONYLPENTAMETHYLENE) see PJY500
POLY(IMINO(1-OXO-1,6-HEXANEDIYL)) see PJY500
POLYINOSINATE:POLYCYTIDYLATE see PJY750
POLYINOSINIC58POLYCYTIDYLIC ACID COPOLYMER see PJY750
POLY(GERMANIUM MONOHYDRIDE) see GEG000
POLYMARCIN see MQQ250
POLYMARCINE see MQQ250
POLYMARSIN see MQQ250
POLYMARZIN see MQQ250
POLYMARZINE see MQQ250
POLYMAT see MQQ250
POLYMERIC DIALDEHYDE see PKA000
POLYMERS of EPICHLOROHYDRIN and 2,2-BIS(4-HYDROXY PHENYL)
  PIPERAZINE see ECM500
POLYMERS, WATER INSOLUBLE see PKA850
POLYMERS, WATER SOLUBLE see PKA860
POLY(2-METHOXY-5(2-(METHYLAMINO)ETHYL)-m-PHENYLENE-
  METHYLENE) see PMB800
POLY-(METHYLBIS(THIOCYANATO)ARSINE) see MJK750
POLY(METHYLENEMAGNESIUM) see PKB000
POLYMETHYLMETHACRYLATE see PKB500
POLYMINE D see BEN000
POLYMONE see DGB000
POLYMO RED FON see MMP100
POLYMYCIN see PKB775
POLYMYXIN see PKC000
POLYMYXIN A see PKC250
POLYMYXIN B see PKC500
POLYMYXIN B1 see PKC550
POLYMYXIN B SULFATE see PKC750
POLYMYXIN D1 see PKD050
POLYMYXIN E see PKD250
POLYOESTRADIOL PHOSPHATE see EDS000
POLYOX see PJT000
POLYOXIETHYLENE (6) ALKYL (13) ETHER see TJJ250
POLYOXIN AL see PKE100
POLYOXIN B see PKE100
POLY(1-(2-OXO-1-PYRROLIDINYL)ETHYLENE) see PKQ250
POLY(1-(2-OXO-1-PYRROLIDINYL)ETHYLENE)IODINE COMPLEX see PKE250
POLY(OXY(DIMETHYLSILYLENE)) see PJR000
POLY(OXY-1,2-ETHANEDIYL), $\alpha-\alpha',\alpha'',\alpha'''$-(1,2-ETHANEDIYLBIS(NITRILODI-
  2,1-ETHANEDIYL))TETRAKIS(omega-HYDROXY- see EIV750
POLY(OXY-1,2-ETHANEDIYL), α-METHYL-omega-METHOXY-(9CI)
  see DOM100
POLY(OXY-1,2-ETHANEDIYLOXYCARBONYL-1,4-PHENYLENECARBONYL)
  see PKF750
POLYOXYETHYLATED VEGETABLE OIL see PKE500
POLYOXYETHYLENE (75) see PJT750
POLYOXYETHYLENE 1500 see PJT500
POLYOXYETHYLENE ALKYLAMINE MONO-FATTY ACID ESTER see PKE550
POLYOXYETHYLENE-ALKYL CITRIC DIESTER-TRIETHANOLAMINE
  see PKE600
POLYOXYETHYLENE (7) ALKYL (14) ETHER see TBY750
POLYOXYETHYLENE-sec-ALKYL ETHER CITRIC DIESTER
  TRIETHANOLAMINE see PKE600
POLYOXYETHYLENE DIBENZOATE see PKE750
POLYOXYETHYLENE DIMETHYL ETHER see DOM100
POLYOXYETHYLENE DODECANOL see LBT000

POLYOXYETHYLENE LAURIC ALCOHOL see DXY000
POLYOXYETHYLENE LAURYL ETHER see DXY000
POLYOXYETHYLENE MONO(OCTYLPHENYL) ETHER see PKF500
POLYOXYETHYLENE MONOOCTYLPHENYL ETHER see GHS000
POLY(OXYETHYLENE) MONOOLEATE see PJY100
POLYOXYETHYLENE MONOSTEARATE see PJW750
POLYOXYETHYLENE-8-MONOSTEARATE see PJV250
POLYOXYETHYLENE NONYLPHENOL see NND500
POLYOXYETHYLENE (9) NONYL PHENYL ETHER see PKF000
POLY(OXYETHYLENE)OCTYLPHENOL ETHER see GHS000
POLYOXYETHYLENE (9) OCTYLPHENYL ETHER see PKF500
POLYOXYETHYLENE (13) OCTYLPHENYL ETHER see PKF500
POLY(OXYETHYLENE)-p-tert-OCTYLPHENYL ETHER see PKF500
POLY(OXYETHYLENE) OLEATE see PJY100
POLY(OXYETHYLENE) OLEIC ACID ESTER see PJY100
POLY(OXYETHYLENE) (20) OLEYL ETHER see PJW500
POLY(OXYETHYLENEOXYTEREPHTHALOYL) see PKF750
POLYOXYETHYLENE SORBITAN MONOLAURATE see PKG000
POLYOXYETHYLENE (20) SORBITAN MONOLAURATE see PKL000
POLYOXYETHYLENE SORBITAN MONOOLEATE see PKL100
POLYOXYETHYLENE SORBITAN MONOPALMITATE see PKG500
POLYOXYETHYLENE SORBITAN MONOSTEARATE see PKL030
POLYOXYETHYLENE 20 SORBITAN MONOSTEARATE see PKL030
POLYOXYETHYLENE SORBITAN OLEATE see PKL100
POLYOXYETHYLENE (20) SORBITAN TRIOLEATE see TOE250
POLYOXYETHYLENE(8)STEARATE see PJV250
POLYOXYL 10 OLEYL ETHER see OIG040
POLYOXYMETHYLENE see TMP000
POLYOXYMETHYLENE GLYCOLS see FMV000
POLY(OXYPROPYLENE) BUTYL ETHER see BRP250
POLYOXYPROPYLENE MONOBUTYL ETHER see BRP250
POLYPEPTIN see CMS225
POLY(PEROXYISOBUTYROLACTONE) see PKH260
POLYPHENOL FRACTION OF BETEL NUT see BFW010
POLY p-PHENYLENE TEREPTHALAMIDE ARAMID FIBER see PKH850
POLYPHLOGIN see PGG000
POLYPHOS see SHM500
POLYPODINE A see HKG500
POLYPRO 1014 see PMP500
POLYPROPENE see PMP500
POLY(2-PROPYL-m-DIOXANE-4,6-DIYLENE) see PKI000
POLYPROPYLENE see PMP500
POLYPROPYLENE, combustion products see PKI250
POLYPROPYLENE GLYCOL see PKI500
POLYPROPYLENE GLYCOL 425 see PKI550
POLYPROPYLENE GLYCOL 750 see PKI750
POLYPROPYLENE GLYCOL 1200 see PKJ250
POLYPROPYLENE GLYCOL 2025 see PKJ500
POLYPROPYLENE GLYCOL 4025 see PKK000
POLYPROPYLENE GLYCOL METHYL ETHER see MKS250
POLYPROPYLENE GLYCOL MONOBUTYL ETHER see BRP250
POLYPROPYLENE GLYCOL 400, MONOBUTYL ETHER see PKK500
POLYPROPYLENEGLYCOL 800, MONOBUTYL ETHER see PKK750
POLYPROPYLENE GLYCOL MONOMETHYLETHER see MKS250
POLYPROPYLENGLYKOL (CZECH) see PKI500
POLYRAM see MQQ250
POLYRAM 80 see MQQ250
POLYRAM COMBI see MQQ250
POLYRAM M see MAS500
POLYRAM 80WP see MQQ250
POLYRAM ULTRA see TFS350
POLYRAM Z see EIR000
POLYSILICONE see PJR250
POLYSILYLENE see PKK775
POLY(SODIUM p-STYRENESULFONATE) see SJK375
POLY-SOLV see CBR000
POLY-SOLV DB see DJF200
POLY-SOLV DM see DJG000
POLY-SOLV EB see BPJ850
POLY-SOLV EE see EES350
POLY-SOLV EE ACETATE see EES400
POLY-SOLV EM see EJH500
POLY-SOLVE MPM see PNL250
POLY-SOLV TB see TKL750
POLY-SOLV TE see EFL000
POLY-SOLV TM see TJQ750
POLYSORBAN 80 see PKL100
POLYSORBATE 20 see PKL000
POLYSORBATE 60 see PKL030
POLYSORBATE 65 see SKV195
POLYSORBATE 80 see PKL100
POLYSORBATE 80, U.S.P. see PKL100
POLYSTICHOCITRIN see FBY000
POLYSTROL D see SMQ500

POLYSTYRENE see SMQ500
POLYSTYRENE-ACRYLONITRILE see ADY500
POLYSTYRENE BEADS (DOT) see SMQ500
POLYSTYRENE LATEX see SMQ500
POLYSTYROL see SMQ500
POLYTAC see PMP500
POLYTAR BATH see CMY800
POLYTEF see TAI250
POLYTETRAFLUOROETHENE see TAI250
POLYTETRAFLUOROETHYLENE see TAI250
POLY(N,N,N',N'-TETRAMETHYL-N-TRIMETHYLENEHEXAMETHYLENE-
    DIAMMONIUM DIBROMIDE) see HCV500
POLYTEX 973 see ADW200
POLYTHERM see PKQ059
POLYTHIAZIDE see PKL250
POLYTOX see DGB000
POLYURETHANE ESTER FOAM see PKL500
POLYURETHANE ETHER FOAM see PKL500
POLYURETHANE FOAM see PKL500
POLYURETHANE SPONGE see PKL500
POLYURETHANE Y-195 see PKL750
POLYURETHANE Y-217 see PKM000
POLYURETHANE Y-218 see PKM250
POLYURETHANE Y-221 see PKM500
POLYURETHANE Y-222 see PKM750
POLYURETHANE Y-223 see PKN000
POLYURETHANE Y-224 see PKN250
POLYURETHANE Y-225 see PKN500
POLYURETHANE Y-226 see PKN750
POLYURETHANE Y-227 see PKO000
POLYURETHANE Y-238 see CDV625
POLYURETHANE Y-290 see PKO500
POLYURETHANE Y-299 see PKO750
POLYURETHANE Y-302 see PKP000
POLYURETHANE Y-304 see PKP250
POLYVIDONE see PKQ250
POLYVINYL ACETATE (FCC) see AAX250
POLYVINYL ACETATE CHLORIDE see PKP500
POLYVINYL ALCOHOL see PKP750
POLY(VINYL ALCOHOL) see PKP750
POLYVINYLBROMIDE see PKQ000
POLYVINYLBUTYRAL (CZECH) see PKI000
POLY(n-VINYLBUTYROLACTAM) see PKQ250
POLYVINYLCHLORID (GERMAN) see PKQ059
POLYVINYL CHLORIDE see PKQ059
POLYVINYLCHLORIDE ACETATE see PKP500
POLYVINYL CHLORIDE-POLYVINYL ACETATE see AAX175
POLYVINYLIDENE FLUORIDE (PYROLYSIS) see DKH600
POLY(VINYLPYRIDINE 1-OXIDE) see PKQ100
POLY(VINYLPYRIDINE N-OXIDE) see PKQ100
POLY(1-VINYL-2-PYRROLIDINONE) HOMOPOLYMER
    see PKQ250
POLY(1-VINYL-2-PYRROLIDINONE) Hueper's polymer No. 1 see PKQ500
POLY(1-VINYL-2-PYRROLIDINONE) Hueper's polymer No. 2 see PKQ750
POLY(1-VINYL-2-PYRROLIDINONE) Hueper's polymer No. 3 see PKR000
POLY(1-VINYL-2-PYRROLIDINONE) Hueper's polymer No. 4 see PKR250
POLY(1-VINYL-2-PYRROLIDINONE) Hueper's polymer No. 5 see PKR500
POLY(1-VINYL-2-PYRROLIDINONE) Hueper's polymer No. 6 see PKR750
POLY(1-VINYL-2-PYRROLIDINONE) Hueper's polymer No. 7 see PKS000
POLYVINYLPYRROLIDONE see PKQ250
POLYVINYL SULFATE, POTASSIUM SALT see PKS250
POLYWAX 1000 see PJS750
POLY-ZOLE AZDN see ASL750
POMARSOL see TFS350
POMARSOL Z FORTE see BJK500
POMASOL see TFS350
POMME EPINEUSE (FRENCH) see SLV500
POMMIER (FRENCH) see AQP875
PONALAR see XQS000
PONALID see DWE800
PONALIDE see DWE800
PONCEAU 3R see FAG018
PONCEAU 4R see FMU080
PONCEAU 4R ALUMINUM LAKE see FMU080
PONCEAU BNA see FMU070
PONCEAU INSOLUBLE OLG see XRA000
PONCEAU R (BIOLOGICAL STAIN) see FMU070
PONCEAU SX see FAG050
PONCEAU XYLIDINE (BIOLOGICAL STAIN) see FMU070
PONCYL see GKE000
PONDERAL see PDM250
PONDERAX see PDM250
PONDIMIN see PDM250
PONDINIL see PKS500

PONDOCIL see AOD000
PONDOCILLIN see AOD000
PONECIL see AIV500
PONSTAN see XQS000
PONSTEL see XQS000
PONSTIL see XQS000
PONSTYL see XQS000
PONTACHROME YELLOW 3RN see NEY000
PONTACYL FAST BLUE R see ADE750
PONTACYL GREEN BL see FAE950
PONTACYL RUBINE R see HJF500
PONTACYL SCARLET RR see FMU080
PONTAL see XQS000
PONTALITE see PKB500
PONTAMINE BLACK E see AQP000
PONTAMINE BLUE BB see CMO000
PONTAMINE BLUE 3BX see CMO250
PONTAMINE DEVELOPER TN see TGL750
PONTAMINE SKY BLUE see CMN750
PONTAMINE VIOLET N see CMP000
PONTOCAINE see BQA010
PONTOCAINE HYDROCHLORIDE see TBN000
POOA see PJY100
POOGIPHALAM, nut extract see BFW000
POOR MAN'S TREACLE see WBS850
POP see ELL500
POP-DOCK see FOM100
POPO-HAU (HAWAII) see HGP600
POPROLIN see PMP500
POPULAGE see MBU550
PORAMINE MALEATE see PJJ325
PORCINE-TRYPSIN see PKS600
PORFIROMYCIN see MLY000
PORFIROMYCINE see MLY000
PORK TRYPSIN see PKS600
POROFOR 57 see ASL750
POROFOR 505 see ASM270
POROFOR ADC/R see ASM270
POROFOR CHKHC-18 see DVF400
POROFOR ChKhZ 21 see ASM270
POROFOR ChKhZ 21R see ASM270
POROPHOR B see DVF400
PORPHYROMYCIN see MLY000
PORTAMYCIN see SLW475
PORTLAND CEMENT see PKS750
PORTLAND CEMENT SILICATE see PKS750
PORTLAND STONE see CAO000
PORTUGUESE MAN-OF-WAR TOXIN see PIA375
POSEDRAN see BEG000
POSEDRINE see BEG000
POSSUM WOOD see SAT875
POSTAFEN see HGC500
POSTERIOR PITUITARY EXTRACT see ORU500
POSTINOR see NNQ500, NNQ520
POST LOCUST see GJU475
PO-SYSTOX see DAP200
POTABLAN see CGL250
POTALIUM see PLG800
POTASAN see PKT000
POTASH see PLA000
POTASH CHLORATE (DOT) see PLA250
POTASORAL see PLG800
POTASSA see PLJ500
POTASSE CAUSTIQUE (FRENCH) see PLJ500
POTASSIO (CHLORATO di) (ITALIAN) see PLA250
POTASSIO (IDROSSIDO di) (ITALIAN) see PLJ500
POTASSIO (PERMANGANATO di) (ITALIAN) see PLP000
POTASSIUM see PKT250
POTASSIUM (liquid alloy) see PKT500
POTASSIUM, metal liquid alloy (DOT) see PKT500
POTASSIUM ACETATE see PKT750
POTASSIUM ACETYLENE-1,2-DIOXIDE see PKT775
POTASSIUM ACETYLIDE see PKU000
POTASSIUM ACID ARSENATE see ARD250
POTASSIUM ACID FLUORIDE see PKU250
POTASSIUM ACID FLUORIDE (solution) see PKU500
POTASSIUM ACID SULFATE see PKX750
POTASSIUM ACID TARTRATE see PKU600
POTASSIUM ALUM see AHF100, AHF200
POTASSIUM ALUM DODECAHYDRATE see AHF200
POTASSIUM AMALGAM see PKU750
POTASSIUM AMIDE see PKV000
POTASSIUM, AMMINETRICHLOROPLATINATE (1⁻) see PJD750
POTASSIUM ANTIMONYL TARTRATE see AQG250

POTASSIUM ANTIMONYL-d-TARTRATE see AQG250, AQG500
POTASSIUM ANTIMONYL-l-TARTRATE see AQH000
POTASSIUM ANTIMONYL-d,l-TARTRATE see AQG750
POTASSIUM ANTIMONYL-meso-TARTRATE see AQH250
POTASSIUM ANTIMONY TARTRATE see AQG250
POTASSIUM ARSENATE see ARD250
POTASSIUM ARSENITE see PKV500
POTASSIUM ARSENITE solution see FOM050
POTASSIUM ASPARTATE see PKV600
POTASSIUM-l-ASPARTATE see PKV600
POTASSIUM AZAOROTATE see AFQ750
POTASSIUM AZIDE see PKW000
POTASSIUM AZIDODISULFATE see PKW250
POTASSIUM AZIDOSULFATE see PKW500
POTASSIUM BENZENEHEXOXIDE see PKW550
POTASSIUM BENZENESULFONYLPEROXYSULFATE see PKW750
POTASSIUM BENZOATE see PKW760
POTASSIUM-O-O-BENZOYLMONOPEROXOSULFATE see PKW775
POTASSIUM BENZOYLPEROXYSULFATE see PKX000
POTASSIUM BENZYLPENICILLIN see BFD000
POTASSIUM BENZYLPENICILLINATE see BFD000
POTASSIUM BENZYLPENICILLIN G see BFD000
POTASSIUM BICARBONATE see PKX150
POTASSIUM BICHROMATE see PKX250
POTASSIUM BIFLUORIDE see PKU250
POTASSIUM BIFLUORIDE, solution (DOT) see PKU500
POTASSIUM BIPHOSPHATE see PLQ405
POTASSIUM BIS(2-HYDROXYETHYL)DITHIOCARBAMATE see PKX500
POTASSIUM BIS(PROPYNYL)PALLADATE see PKX639
POTASSIUM BIS(PROPYNYL)PLATINATE see PKX700
POTASSIUM BISULFATE see PKX750
POTASSIUM BISULPHATE see PKX750
POTASSIUM BITARTRATE see PKU600
POTASSIUM BOROFLUORIDE see PKY000
POTASSIUM BOROHYDRATE see PKY250
POTASSIUM BOROHYDRIDE (DOT) see PKY250
POTASSIUM BROMATE see PKY300
POTASSIUM BROMIDE see PKY500
POTASSIUM-tert-BUTOXIDE see PKY750
POTASSIUM BUTYLXANTHATE see PKY850
POTASSIUM-o-BUTYL XANTHATE see PKY850
POTASSIUM BUTYLXANTHOGENATE see PKY850
POTASSIUM CANRENOATE see PKZ000
POTASSIUM CAPRYLATE see PLN250
POTASSIUM CARBONATE (2581) see PLA000
POTASSIUM CARBONYL see PKW550
POTASSIUM CHLORATE see PLA250
POTASSIUM CHLORATE (DOT) see PLA250
POTASSIUM (CHLORATE de) (FRENCH) see PLA250
POTASSIUM CHLORIDE see PLA500
POTASSIUM CHLORITE see PLA525
POTASSIUM 7-CHLORO-2,3-DIHYDRO-2-OXO-5-PHENYL-1H-1,4-BENZODIAZEPINE-3-CARBOXYLATE KOH see CDQ250
POTASSIUM CHLOROPALLADATE see PLA750
POTASSIUM CHLOROPLATINATE see PLR000
POTASSIUM CHLOROPLATINITE see PJD250
POTASSIUM CHROMATE(VI) see PLB250
POTASSIUM CHROMIC SULFATE see PLB500
POTASSIUM CHROMIC SULPHATE see PLB500
POTASSIUM CHROMIUM ALUM see CMG850, PLB500
POTASSIUM CITRATE see PLB750
POTASSIUM CITRATE TRI(HYDROGEN PEROXIDATE) see PLB759
POTASSIUM CLAVULANATE see PLB775
POTASSIUM CLAVULANATE mixed with AMOXICILLIN (1:2) see ARS125
POTASSIUM COLUMBATE see PLL250
POTASSIUM COMPOUNDS see PLC100
POTASSIUM COPPER(I) CYANIDE see PLC175
POTASSIUM CUPROCYANIDE see PLC175
POTASSIUM CYANATE see PLC250
POTASSIUM CYANIDE see PLC500
POTASSIUM CYANIDE (solid) see PLC750
POTASSIUM CYANIDE, solution (DOT) see PLC500
POTASSIUM CYANIDE-POTASSIUM NITRITE see PLC775
POTASSIUM CYANONICKELATE HYDRATE see TBW250
POTASSIUM CYCLOHEXANEHEXONE-1,3,5-TRIOXIMATE see PLC780
POTASSIUM CYCLOPENTADIENIDE see PLC800
POTASSIUM DICHLOROISOCYANURATE see PLD000
POTASSIUM DICHLORO-s-TRIAZINETRIONE see PLD000
POTASSIUM DICHROMATE(VI) see PKX250
POTASSIUM DICHROMATE, ZINC CHROMATE and ZINC HYDROXIDE (1:3:1) see ZFJ150
POTASSIUM DICYANOCUPRATE(1-) see PLC175
POTASSIUM DIETHYNYLPALLADATE(2⁻) see PLD100
POTASSIUM DIETHYNYLPLATINATE(2⁻) see PLD150

POTASSIUM DIFLUOROPHOSPHATE see PLD250
POTASSIUM DIHYDROGEN ARSENATE see ARD250
POTASSIUM DIHYDROGEN PHOSPHATE see PLQ405
POTASSIUM DIMETHYL DITHIOCARBAMATE see PLD500
POTASSIUM-2,5-DINITROCYCLOPENTANONIDE see PLD550
POTASSIUM DINITROMETHANIDE see PLD575
POTASSIUM DINITROOXALATOPLATINATE(2⁻) see PLD600
POTASSIUM aci-1,1-DINITROPROPANE see PLD700
POTASSIUM-1,1-DINITROPROPANIDE see PLD700
POTASSIUM DINITROSOSULFITE see PLD710
POTASSIUM-3,5-DINITRO-2(1-TETRAZENYL)PHENOLATE see PLD730
POTASSIUM DIOXIDE see PLE260
POTASSIUM DIPEROXY ORTHOVANADATE see PLE500
POTASSIUM DISULPHATOCHROMATE(III) see PLB500
POTASSIUM DODECANOATE see PLK750
POTASSIUM ETHOXIDE see PLE575
POTASSIUM O-ETHYL DITHIOCARBONATE see PLF000
POTASSIUM ETHYLMERCURIC THIOGLYCOLLATE see PLE750
POTASSIUM ETHYLXANTHATE see PLF000
POTASSIUM ETHYL XANTHOGENATE see PLF000
POTASSIUM FERRICYANATE see PLF250
POTASSIUM FERRICYANIDE see PLF250
POTASSIUM FERROCYANATE see TEC500
POTASSIUM FERROCYANIDE see PLH100, TEC500
POTASSIUM FLUOBORATE see PKY000
POTASSIUM N-FLUOREN-2-YL ACETOHYDROXAMATE see HIS000
POTASSIUM FLUORIDE see PLF500
POTASSIUM FLUORIDE, solution (DOT) see PLF500
POTASSIUM FLUORIDE, DIHYDRATE see PLF750
POTASSIUM FLUOROACETATE see PLG000
POTASSIUM FLUOROBORATE see PKY000
POTASSIUM FLUORURE (FRENCH) see PLF500
POTASSIUM FLUOSILICATE see PLH750
POTASSIUM FLUOTANTALATE see PLH000
POTASSIUM FLUOZIRCONATE see PLG500
POTASSIUM FORMATE see PLG750
POTASSIUM GLUCONATE see PLG800
POTASSIUM d-GLUCONATE see PLG800
POTASSIUM GLUTAMATE see MRK500
POTASSIUM GLUTAMINATE see MRK500
POTASSIUM GRAPHITE see PLG825
POTASSIUM HEPTAFLUOROTANTALATE see PLH000
POTASSIUM HEXACHLOROPLATINATE(IV) see PLR000
POTASSIUM HEXACYANOFERRATE(II) see PLH100
POTASSIUM HEXACYANOFERRATE see TEC500
POTASSIUM HEXACYANOFERRATE(II) see TEC500
POTASSIUM HEXACYANOFERRATE(III) see PLF250
POTASSIUM HEXAETHYNYLCOBALTATE see PLH250
POTASSIUM HEXAFLUOROARSENATE see PLH500
POTASSIUM HEXAFLUOROSILICATE see PLH750
POTASSIUM HEXAFLUOROTITANATE see PLI000
POTASSIUM HEXAFLUORSTANNATE see PLI250
POTASSIUM HEXAHYDRATE ALUMINATE see PLI500
POTASSIUM HEXANITROCOBALTATE see PLI750
POTASSIUM HEXAOXYXENONATE(4-) XENON TRIOXIDE see PLJ000
POTASSIUM HYDRATE (DOT) see PLJ500
POTASSIUM HYDRATE (solution) see PLJ750
POTASSIUM HYDRIDE see PLJ250
POTASSIUM HYDROGEN ARSENATE see ARD250
POTASSIUM HYDROGEN FLUORIDE see PKU250
POTASSIUM HYDROGEN FLUORIDE, solution (DOT) see PKU500
POTASSIUM HYDROGEN OXALATE see OLE000
POTASSIUM HYDROGEN SULFATE, solid (DOT) see PKX750
POTASSIUM HYDROXIDE see PLJ500
POTASSIUM HYDROXIDE, dry, solid, flake, bead, or granular (DOT) see PLJ500
POTASSIUM HYDROXIDE, liquid or solution (DOT) see PLJ500
POTASSIUM HYDROXIDE (solution) see PLJ750
POTASSIUM-4-HYDROXYAMINE-5,7-DINITRO-4,5-DIHYDROBENZO-FURAZANIDE-3-OXIDE see PLJ775
POTASSIUM (HYDROXYDE de) (FRENCH) see PLJ500
POTASSIUM-4-HYDROXY-5,7-DINITRO-4,5-DIHYDROBENZOFURAZANIDE see PLJ780
POTASSIUM HYDROXYFLUORONIOBATE see PLN500
POTASSIUM-3-(17-β-HYDROXY-3-OXOANDROSTA-4,6-DIEN-17-YL)PROPINO-ATE see PKZ000
POTASSIUM-17-HYDROXY-3-OXO-17-α-PREGNA-4,6-DIENE-21-CARBOXYL-ATE see PKZ000
POTASSIUM HYPERCHLORIDE see PLO500
POTASSIUM HYPOBORATE see PLJ790
POTASSIUM HYPOCHLORITE (solution) see PLK000
POTASSIUM HYPOPHOSPHITE see PLQ750
POTASSIUM HYPOSULFITE see PLW000
POTASSIUM INDOL-3-YL SULFATE see ICD000
POTASSIUM IODATE see PLK250

POTASSIUM IODIDE see PLK500
POTASSIUM IODOHYDRAGYRATE see NCP500
POTASSIUM ISOAMYL XANTHATE see PLK580
POTASSIUM ISOAMYL XANTHOGENATE see PLK580
POTASSIUM ISOBUTYL XANTHATE see PLK600
POTASSIUM-o-ISOBUTYL XANTHATE see PLK600
POTASSIUM ISOBUTYL XANTHOGENATE see PLK600
POTASSIUM ISOCYANATE see PLC250
POTASSIUM ISOPENTYL XANTHATE see PLK580
POTASSIUM ISOTHIOCYANATE see PLV750
POTASSIUM KURROL'S SALT see PLR125
POTASSIUM LAURATE see PLK750
POTASSIUM MANGANOCYANIDE see TMX600
POTASSIUM MELARSONYL see PLK800
POTASSIUM MERCURIC IODIDE see NCP500
POTASSIUM METAARSENITE see PKV500
POTASSIUM METABISULFITE (DOT, FCC) see PLR250
POTASSIUM, METAL (DOT) see PKT250
POTASSIUM METAPHOSPHATE see PLR125
POTASSIUM METAVANADATE see PLK900, PLK900
POTASSIUM METHANEDIZOATE see PLK825
POTASSIUM METHOXIDE see PLK850
POTASSIUM-4-METHOXY-1-aci-NITRO-3,5-DINITRO-2,5-CYCLOHEXADIENONE see PLK860
POTASSIUM METHYLAMIDE see PLL000
POTASSIUM METHYLDIAZENE OXIDE see PLK825
POTASSIUM-4-METHYLFURAZAN-5-CARBOXYLATE-2-OXIDE see PLL100
POTASSIUM METHYLPHENOXYMETHYLPENICILLIN see PDD350
POTASSIUM MONOCHLORIDE see PLA500
POTASSIUM MONOPERSULFATE see PLP750
POTASSIUM MONOSULFIDE see PLT250
POTASSIUM NIOBATE see PLL250
POTASSIUM NITRATE see PLL500
POTASSIUM NITRATE mixed with CHARCOAL and SULFUR (15:3:2) see PLL750
POTASSIUM NITRATE mixed with SODIUM NITRITE see PLM000
POTASSIUM NITRATE mixed (fused) with SODIUM NITRITE (DOT) see PLM000
POTASSIUM NITRIDE see PLM250
POTASSIUM NITRITE (1581) see PLM500
POTASSIUM NITRITE (DOT) see PLM500
POTASSIUM-4-NITROBENZENEAZOSULFONATE see PLM550
POTASSIUM-6-aci-NITRO-2,4-DINITRO-2,4-CYCLOHEXADIENIMINIDE see PLM575
POTASSIUM 1-NITROETHANE-1-OXIMATE see MRK250
POTASSIUM-1-NITROETHOXIDE see PLM650
POTASSIUM-4-NITROPHENOXIDE see PLM700
POTASSIUM NITROSODISULFATE see PLM750
POTASSIUM N-NITROSOHYDROXYLAMINE-N-SULFONATE see PLD710
POTASSIUM NITROSOOSMATE(1⁻) see PLN000
POTASSIUM NITROTRICHLOROPLATINATE see PLN050
POTASSIUM OCTACYANODICOBALTATE see PLN100
POTASSIUM cis-9-OCTADECENOIC ACID see OHY000
POTASSIUM OCTANOATE see PLN250
POTASSIUM OCTATITANATE see PLW150
POTASSIUM OLEATE see OHY000
POTASSIUM-3-(3-OXO-17-β-HYDROXY-4,6-ANDROSTADIEN-17-α-YL)PRO-PANOATE see PKZ000
POTASSIUM OXONATE see AFQ750
POTASSIUM OXOTETRAFLUORONIOBATE(V) see PLN500
POTASSIUM OXYMURIATE see PLA250
POTASSIUM PALLADIUM CHLORIDE see PLN750
POTASSIUM PALLADOUS CHLORIDE see PLN750
POTASSIUM PALMITATE see PLO000
POTASSIUM PENICILLIN G see BFD000
POTASSIUM PENICILLIN V SALT see PDT750
POTASSIUM PENTACYANODIPEROXOCHROMATE(5⁻) see PLO150
POTASSIUM PENTAPEROXODICHROMATE see PLO250
POTASSIUM PENTAPEROXYDICHROMATE see PLO250
POTASSIUM PENTYL THIARSAPHENYLMELAMINE see PLK800
POTASSIUM PERCHLORATE see PLO500
POTASSIUM PERIODATE see PLO750
POTASSIUM PERMANGANATE see PLP000
POTASSIUM (PERMANGANATE de) (FRENCH) see PLP000
POTASSIUM PEROXIDE see PLP250
POTASSIUM PEROXOFERRATE(2-) see PLP500
POTASSIUM PEROXYDISULFATE see DWQ000
POTASSIUM PEROXYDISULPHATE see DWQ000
POTASSIUM PEROXYFERRATE see PLP500
POTASSIUM PEROXYSULFATE see PLP750
POTASSIUM PERRHENATE see PLQ000
POTASSIUM PERSULFATE (DOT) see DWQ000
POTASSIUM PHENETHICILLIN see PDD350
POTASSIUM (1-PHENOXYETHYL)PENICILLIN see PDD350

POTASSIUM-α-PHENOXYETHYL PENICILLIN see PDD350
POTASSIUM PHENOXYMETHYLPENICILLIN see PDT750
POTASSIUM-6-(α-PHENOXYPROPIONAMIDO)PENICILLANATE see PDD350
POTASSIUM PHENYL DINITROMETHANIDE see PLQ275
POTASSIUM PHOSPHATE, DIBASIC see PLQ400
POTASSIUM PHOSPHATE, MONOBASIC see PLQ405
POTASSIUM PHOSPHATE, TRIBASIC see PLQ410
POTASSIUM PHOSPHIDE see PLQ500
POTASSIUM PHOSPHINATE see PLQ750
POTASSIUM PICRATE see PLQ775
POTASSIUM PLATINIC CHLORIDE see PLR000
POTASSIUM PLATINOCHLORIDE see PJD250
POTASSIUM POLYMETAPHOSPHATE see PLR125
POTASSIUM-O-PROPIONOHYDROXAMATE see PLR175
POTASSIUM PYROPHOSPHATE see PLR200
POTASSIUM PYROSULFITE see PLR250
POTASSIUM RHENATE see PLR500
POTASSIUM RHODANATE see PLV750
POTASSIUM RHODANIDE see PLV750
POTASSIUM SALT of BENZYLPENICILLIN see BFD000
POTASSIUM SALT OF POLYVINYL SULFATE see PKS250
POTASSIUM SALT of SORREL see OLE000
POTASSIUM SELENATE see PLR750
POTASSIUM SELENOCYANATE see PLS000
POTASSIUM SILICOFLUORIDE (DOT) see PLH750
POTASSIUM SILVER CYANIDE see PLS250
POTASSIUM SODIUM ALLOY see PLS500
POTASSIUM SORBATE see PLS750
POTASSIUM STANNATE TRIHYDRATE see PLS760
POTASSIUM SULFATE (2581) see PLT000
POTASSIUM SULFIDE (2581) see PLT250
POTASSIUM SULFITE see PLT500
POTASSIUM SULFOCYANATE see PLV750
POTASSIUM SULFURDIIMIDE see PLT750
POTASSIUM SULFUR DIIMIDE see PLT750
POTASSIUM TELLURITE see PLU000
POTASSIUM TETRABROMOPLATINATE see PLU250
POTASSIUM TETRACHLOROPALLADATE see PLN750
POTASSIUM TETRACHLOROPLATINATE(II) see PJD250
POTASSIUM TETRACYANOMERCURATE(II) see PLU500
POTASSIUM TETRACYANONICKELATE see NDI000
POTASSIUM TETRACYANONICKELATE(II) see NDI000
POTASSIUM TETRACYANOTITANATE(IV) see PLU550
POTASSIUM TETRAETHYNYL NICKELATE(2⁻) see PLU575
POTASSIUM TETRAETHYNYL NICKELATE(4⁻) see PLU580
POTASSIUM TETRAIODOMERCURATE(II) see NCP500
POTASSIUM TETRAKISTHIOCYANATOPLATINATE see PLU590
POTASSIUM-1,1,2,2-TETRANITROETHANDIIDE see PLU600
POTASSIUM TETRAPEROXYCHROMATE see PLU750
POTASSIUM TETRAPEROXYMOLYBDATE see PLV000
POTASSIUM TETRAPEROXYTUNGSTATE see PLV250
POTASSIUM-1-TETRAZOLACETATE see PLV275
POTASSIUM THALLIUM(I)AMIDE AMMONIATE see PLV500
POTASSIUM THIOCYANATE see PLV750
POTASSIUM THIOCYANIDE see PLV750
POTASSIUM THIOSULFATE see PLW000
POTASSIUM TITANIUM OXIDE see PLW150
POTASSIUM TRIAZIDOCOBALTATE see PLI750
POTASSIUM-s-TRIAZINE-2,4-DIONE-6-CARBOXYLATE see AFQ750
POTASSIUM, TRICHLOROAMMINEPLATINATE (2) see PJD750
POTASSIUM TRICHLOROETHYLENEPLATINATE see PLW200
POTASSIUM TRICYANODIPEROXOCHROMATE (3⁻) see PLW275
POTASSIUM TRIIODIDE see PLW285
POTASSIUM TRINITROMETHANIDE see PLW300
POTASSIUM-2,4,6-TRINITROPHENOXIDE see PLQ775
POTASSIUM TRIPHOSPHATE see PLW400
POTASSIUM TRIPOLYPHOSPHATE see PLW400
POTASSIUM VANADIUM TRIOXIDE see PLK900
POTASSIUM WARFARIN see WAT209
POTASSIUM XANTHATE see PLF000
POTASSIUM XANTHOGENATE see PLF000
POTASSIUM XANTHOGENATE BUTYL ETHER see PKY850
POTASSIUM ZINC CHROMATE see CMK400, PLW500
POTASSIUM ZINC CHROMATE HYDROXIDE see PLW500
POTASSURIL see PLG800
POTATO ALCOHOL see EFU000
POTATO BLOSSOMS, GLYCOALKALOID EXTRACT see PLW550
POTATO, GREEN PARTS see PLW750
POTAVESCENT see PLA500
POTCRATE see PLA250
POTENTIATED ACID GLUTARALDEHYDE see GFQ000
POTHOS see PLW800
POTOMAC RED see CHP500
POUNCE see AHJ750

POVAN see PQC500
POVANYL see PQC500
POVIDONE-IODINE see PKE250
POVIDONE (USP XIX) see PKQ250
POWDERED METALS see PLX000
POWDER GREEN see COF500
POWDER and ROOT see RNZ000
POX see PHS250, PHS250
POYAMIN see VSZ000
PP-12 see HMN000
PP010 see MHI600
PP 062 see DOX600
PP149 see BRI750
PP175 see ASD000
PP211 see DIN600
PP383 see RLF350
PP511 see DIN800
PP 557 see AHJ750
PP 581 see TAC800
PP 675 see BRD000
PP 745 see BJK750
PP781 see MLC250
PP-1466 see RSZ375
PPA see NNM000
PPBP see BCP650
PPD see PEY500
PPE201 see PKO500
PP FACTOR see NCQ900
PP-FACTOR see NCR000
P.P. FACTOR-PELLAGRA PREVENTIVE FACTOR see NCQ900
PPG-14 BUTYL ETHER see BRP250
PPG-16 BUTYL ETHER see BRP250
PPG-33 BUTYL ETHER see BRP250
P. PHYSALIS TOXIN see PIA375
P. PORPHYRIACUS (AUSTRALIA) VENOM see ARV250
PPTC see PNI750
PPZEIDAN see DAD200
PQ see PCY300
PQD see DVR200
PRACARBAMIN see UVA000
PRACARBAMINE see UVA000
PRACTALOL see ECX100
PRACTOLOL see ECX100
PRADUPEN see BDY669
PRAECIRHEUMIN see BRF500
PRAEDYN see CMG675
PRAENITRON see TJL250
PRAENITRONA see TJL250
PRAEQUINE see RHZ000
PRAGMAZONE see CKJ000, THK880
PRAIRIE CROCUS see PAM780
PRAIRIE HEN FLOWER see PAM780
PRAIRIE RATTLESNAKE VENOM see PLX100
PRAJMALINE see PNC875
PRAJMALINE BITARTRATE see DNB000
PRAJMALINE HYDROGEN TARTRATE see DNB000
PRAJMALIUM see PNC875
PRAKTOLOLU (POLISH) see ECX100
PRALIDOXIME CHLORIDE see FNZ000
PRALIDOXIME IODIDE see POS750
PRALIDOXIME MESYLATE see PLX250
PRALIDOXIME METHANESULFONATE see PLX250
PRALIDOXIME METHIODIDE see POS750
PRALIDOXIME METHOSULFATE see PLX250
PRALUMIN see TDA500
PRAMIL see SBN000
PRAMINDOLE see DPX200
PRAMITOL see MFL250
PRAMIVERINE HYDROCHLORIDE see SDZ000
PRAMIVERIN HYDROCHLORIDE see SDZ000
PRAMUSCIMOL see AKG250
PRANDIOL see PCP250
PRANONE see GEK500
PRANONE and DIETHYLSTILBESTROL see EEI050
PRANONE and STILBESTROL see EEI050
PRANOPROFEN see PLX400
PRANTAL see DAP800
PRANTAL METHYLSULFATE see DAP800
PRAPAR see MKU750
PRAPARAT 5968 see HGP500
PRAPARAT 9295 see MKU750
PRASEODYMIUM see PLX500
PRASEODYMIUM CHLORIDE see PLX750
PRASEODYMIUM(III) NITRATE (1:3) see PLY250

PRASTERONE see AOO450
PRASTERONE SODIUM SULFATE see DAL040
PRASTERONE SODIUM SULFATE DIHYDRATE see DAL030
PRAXILENE see NAE100
PRAXIS see IDA400
PRAXITEN see CFZ000
PRAYER BEAD see AAD000
PRAYER BEADS see RMK250
PRAZEPAM see DAP700
PRAZEPINE see DLH600
PRAZIL see CKP250
PRAZINIL see CCK780
PRAZIQUANTEL see BGB400
PRAZOSIN see AJP000
PRAZOSIN HYDROCHLORIDE see FPP100
PRD EXPERIMENTAL NEMATOCIDE see DGL200
PREAN see MBW750
PRECEPTIN see PKF500
PRECIPITATED BARIUM SULPHATE see BAP000
PRECIPITATED CALCIUM SULFATE see CAX750
PRECIPITATED SILICA see SCL000
PRECIPITATED SULFUR see SOD500
PRECIPITE BLANC see MCW000
PRECOCENE 2 see AEX850
PRECOCENE II see AEX850
PRECORT see PLZ000
PRECORTANCYL see PMA000
PRECORTISYL see PMA000
PREDALON-S see SCA750
PREDENT see SHF500
PREDIACORTINE see SOV100
PREDICORT see SOV100
PREDNE-DOME see PMA000
PREDNELAN see PMA000
PREDNELAN-N see SOV100
PREDNICEN-M see PLZ000
PREDNIDOREN see SOV100
PREDNILONGA see PLZ000
PREDNIS see PMA000
PREDNI-SEDIV see TEH500
PREDNISOLONBISUCCINAT see PLY300
PREDNISOLONE see PMA000
PREDNISOLONE ACETATE see SOV100
PREDNISOLONE 21-ACETATE see SOV100
PREDNISOLONE BISUCCINATE see PLY300
PREDNISOLONE HEMISUCCINATE see PLY300
PREDNISOLONE 21-HEMISUCCINATE see PLY300
PREDNISOLONE SODIUM HEMISUCCINATE see PMA100
PREDNISOLONE SODIUM SUCCINATE see PMA100
PREDNISOLONE 21-SODIUM SUCCINATE see PMA100
PREDNISOLONE SUCCINATE see PLY300
PREDNISOLONE 21-SUCCINATE see PLY300
PREDNISOLONE 21-SUCCINATE SODIUM see PMA100
PREDNISOLONE VALERATE ACETATE see AAF625
PREDNISOLONE-17-VALERATE-21-ACETATE see AAF625
PREDNISOLUT see PLY300
PREDNISON see PLZ000
PREDNISONE see PLZ000
PREDNISONE ACETATE see PLZ100
PREDNISONE 21-ACETATE see PLZ100
PREDNIZON see PLZ000
PREDONIN see PMA000
PREDONINE see PMA000
PREDONINE INJECTION see SOV100
PREDONINE SOLUBLE see PMA100
PREEGLONE see DWX800
PREFAR see DNO800
PREFAXIL see BBW750
PREFEMIN see CHJ750
PREFERID see BOM520
PREFLAN see BSN000
PREFMID see BSN000
PREFORAN see NIX000
PREFRIN see SPC500
PREGARD see CQG250
PREGICIL see AAF750
PREGLANDIN see CDB775
1,4-PREGNADIEN-17-α-21-DIOL-3,11,20-TRIONE-21-ACETATE
     see PLZ100
1,4-PREGNADIENE-17-α,21-DIOL-3,11,20-TRIONE see PLZ000
9-β,10-α-PREGNA-4,6-DIENE-3,20-DIONE see DYF759
(9-β,10-α)-PREGNA-4,6-DIENE-3,20-DIONE (9CI) see DYF759
PREGNA-1,4-DIENE-3,20-DIONE, 21-(ACETYLOXY)-11,17-DIHYDROXY-, (11-β)-
     see SOV100

PREGNA-1,4-DIENE-3,20-DIONE, 21-(3-CARBOXY-1-OXOPROPOXY)-11,17-DI-
     HYDROXY-, (11-β)- (9CI) see PLY300
PREGNA-1,4-DIENE-3,20-DIONE, 7-CHLORO-11-HYDROXY-16-METHYL-17,21-
     BIS(1-OXOPROPOXY)-, (7-α-11-β,16-α)- see AFI980
PREGNA-4,6-DIENE-3,20-DIONE, 6-CHLORO-17-HYDROXY-1-α,2-α-METHY-
     LENE-, ACETATE see CQJ500
PREGNA-1,4-DIENE-3,20-DIONE,9-FLUORO-11,16,17,21-TETRAHYDROXY-,(11-
     β,16-α) see AQX250
PREGNA-1,4-DIENE-3,20-DIONE, 9-FLUORO-11-β,17,21-TRIHYDROXY-16-β-
     METHYL-, 21-ACETATE and 9-FLUORO-11-β,17,21-TRIHYDROXY-16-β-
     METHYL-PREGNA-1,4-DIENE-3,20-DIO NE 21-(DIHYDROGEN PHOS-
     PHATE) see BFV755
9-β,10-α-PREGNA-4,6-DIENE-3,20-DIONE and 17-α-HYDROXYPREGN-4-ENE-
     3,20-DIONE (95810) see PMA250
PREGNA-1,4-DIENE-3,20-DIONE, 11-β,17,21-TRIHYDROXY-, 21-(HYDROGEN
     SUCCINATE) see PLY300
PREGNA-1,4-DIENE-3,20-DIONE, 11-β,17,21-TRIHYDROXY-, 21-(HYDROGEN
     SUCCINATE), MONOSODIUM SALT see PMA100
1,4-PREGNADIENE-3,20-DIONE-11-β,17-α,21-TRIOL see PMA000
1,4-PREGNADIENE-11-β,17-α,21-TRIOL-3,20-DIONE see PMA000
PREGNA-1,4-DIENE-3,11,20-TRIONE, 21-(ACETYLOXY)-17-HYDROXY- (9CI)
     see PLZ100
PREGNA-1,4-DIENE-3,11,20-TRIONE, 17,21-DIHYDROXY-, 21-ACETATE
     see PLZ100
17-α-2,4-PREGNADIEN-20-YNO(2,3-d)ISOXAZOL-17-OL see DAB830
17-α-PREGNA-2,4-DIEN-20-YNO(2,3-d)ISOXAZOL-17-OL see DAB830
5-α-17-α-PREGNA-2-EN-20-YN-17-OL, ACETATE see PMA450
PREGNANT MARE SERUM GONADOTROPIN see SCA750
PREGN-4-EN-17-α,21-DIOL-3,11,20-TRIONE see CNS800
3,20-PREGNENE-4 see PMH500
17-α-PREGN-4-ENE-21-CARBOXYLIC ACID, 1-HYDROXY-7-α-MERCAPTO-3-
     OXO-α-LACTONE see AFJ500
PREGN-5-ENE-3-β,20-α-DIAMINE see IGH700
4-PREGNENE-17α,21-DIOL-3,11,20-17,21-DIHYDROXY- see CNS800
4-PREGNENE-11-α,21-DIOL-3,20-DIONE see CNS625
Δ⁴-PREGNENE-17α,21-DIOL-3,11,20-TRIONE see CNS800
4-PREGNENE-17,α,21-DIOL-3,11,20-TRIONE 21-ACETATE see CNS825
PREGNENEDIONE see PMH500
PREGNENE-3,20-DIONE see PMH500
PREGN-4-ENE-3,20-DIONE see PMH500
4-PREGNENE-3,20-DIONE see PMH500
Δ⁴-PREGNENE-3,20-DIONE see PMH500
4-PREGNENE-3,20-DIONE-21-OL ACETATE see DAQ800
4-PREGNENE-11-β,17-α,21-TRIOL 3,20-DIONE see CNS750
PREGNENINOLONE see GEK500
4-PREGNEN-21-OL-3,20-DIONE see DAQ600
PREGN-4-EN-20-ONE, 3-β,17-DIHYDROXY-, 17-ACETATE, and 3-METHOXY-
     19-NOR-17-α-PREGNA-1,3,5(10)-TRIEN-20-YN-17-OL (20:1) see ABV600
17-α-PREGN-5-EN-20-YNE-3-β,17-DIOL-3-(3-CYCLOHEXYLPROPIONATE)
     see EDY600
17-α-PREGN-4-EN-20-YNO(2,3-d)ISOXAZOL-17-OL see DAB830
PREGN-2-EN-20-YN-17-OL, ACETATE, (5-α-17α— see PMA450
17-α-PREGN-4-EN-20-YN-3-ONE, 17-HYDROXY-, and trans-α-α′-DIETHYL-4,4′-
     STILBENEDIOL see EEI050
PREGNIN see GEK500
PREGNYL see CMG675
PRELIS see SBV500
PRELUDIN see PMA750
PRELUDIN HYDROCHLORIDE see MNV750
PREMALIN see CKD500, DGD600
PREMALOX see CBM000
PREMARIN see ECU750, PMB000
PREMAZINE see BJP000
PREMERGE see BRE500
PREMERGE 3 see BRE500
PREMERGE PLUS see BQI000
PREMGARD see BEP500
PREMOBOLAN see PMC700
PREMODRIN see DBA800
PRENAZONE see PEW000
PRENDEROL see PMB250
PRENDIOL see PMB250
PRENEMA see SOV100
PRENIMON see TND000
PRENITRON see TJL250
PRENOL see MHU110
PRENORMINE see TAL475
PRENOXDIAZINE HYDROCHLORIDE see LFJ000
PRENT see AAE100
PRENTOX see PIX250, RNZ000
PRENYL ACETATE see DOQ350
PRENYL ALCOHOL see MHU110
PRENYLAMINE see PEV750
PRENYLAMINE LACTATE see DWL200

PRENYL BENZOATE see MHU150
α-PRENYL-α-(2-DIMETHYLAMINOETHYL)-1-NAPHTHYLACETAMIDE
  see PMB500
4-PRENYL-1,2-DIPHENYL-3,5-PYRAZOLIDINEDIONE see PEW000
PRENYL SALICYLATE see PMB600
PREP see SFV250
PREPALIN see VSK600
PRE-PAR see RLK700
PREPARATION 84 see TCQ260
PREPARATION 125 see DFT800
PREPARATION 48-80 see PMB800
PREPARATION AF see HEI500
PREPARATION C (the Russian Drug) see HEF200
PREPARATION HE 166 see DUS500
PREPARATION S see HEF200
PREQUINE see RHZ000
PREROIDE see MJE760
PRESAMINE see DLH630
PRE-SAN see DNO800
PRE-SATE see ARW750
PRESATE HYDROCHLORIDE see ARW750
PRESCRIN see FAQ950
PRESDATE see HMM500
PRESERVAL M see HJL500
PRESERVAL P see HNU500
PRESERV-O-SOTE see CMY825
PRESFERSUL see FBO000
PRESINOL see DNA800, MJE780
PRESOLISIN see DNA800, MJE780
PRESOMEN see ECU750
PRESOXIN see ORU500
PRESPERSION, 75 UREA see USS000
PRESSITAN see FMS875
PRESSOMIN HYDROCHLORIDE see MDW000
PRESSONEX see HNB875, HNC000
PRESSONEX BITARTRATE see HNC000
PRESSOROL see HNC000
PRESTOCICLINA see MCH525
PRESUREN see VJZ000
PRETILACHLOR see PMB850
PRETILACHLORE see PMB850
PRETONINE see HOA575
PREVANGOR see PBC250
PREVENCILINA P see SLJ050
PREVENOL see CKC000
PREVENOL 56 see CKC000
PREVENTOL see CKC000, MJM500
PREVENTOL 1 see SKK500
PREVENTOL 56 see CKC000
PREVENTOL CMK see CFE250
PREVENTOL I see TIV750
PREVENTOL O EXTRA see BGJ250
PREVENTOL-ON see BGJ750
PREVEX see PNI250
PREVICUR see EIH500
PREVICUR-N see PNI250
PREVOCEL #12 see PMC100
PREWEED see CKC000
PREZA see HBT500
PREZERVIT see DSB200
PR-F 36 Cl see CID825
PRIADEL see LGZ000
PRIAMIDE see DAB875
PRIARIE SMOKE see PAM780
PRIATIN see SCA750
PRIAZIMIDE see DAB875
PRIDE of CHINA see CDM325
PRIDE of INDIA see CDM325
PRIDINOL see PMC250
PRIDINOL HYDROCHLORIDE see PMC250
PRIDINOL MESILATE see PMC275
PRIESTS PENTLE see JAJ000
PRILEPSIN see DBB200
PRILOCAINE HYDROCHLORIDE see CMS250
PRILTOX see PAX250
PRIM see PMD550
PRIMACAINE see AJA500
PRIMACAINE HYDROCHLORIDE see AJA500
PRIMACHIN (GERMAN) see AKR250
PRIMACIONE see DBB200
PRIMACLONE see DBB200
PRIMACOL see NAK500
PRIMACONE see DBB200

PRIMAGRAM see MQQ450
PRIMAKTON see DBB200
PRIMAL see AHI875
PRIMALAN see MCJ400
PRIMAMYCIN see HAH800
PRIMANTRON see SCA750
PRIMAQUINE see PMC300
PRIMAQUINE DIPHOSPHATE see PMC310
PRIMAQUINE PHOSPHATE see PMC310
PRIMARY AMYL ACETATE see AOD725
PRIMARY AMYL ALCOHOL see AOE000
PRIMARY DECYL ALCOHOL see DAI600
PRIMARY ISOBUTYL IODIDE see IIV509
PRIMARY OCTYL ALCOHOL see OEI000
PRIMARY SODIUM PHOSPHATE see SJH100
PRIMATENE MIST see VGP000
PRIMATOL see ARQ725, EGD000
PRIMATOL-M80 see BQB000
PRIMATOL P see PMN850
PRIMATOL Q see BKL250
PRIMATOL S see BJP000
PRIMAZE see ARQ725
PRIMAZIN see SNJ000
PRIMBACTAM see ARX875
PRIMENE JM-T see PMC400
PRIMEXTRA see MQQ450
PRIMICID see DIN600
PRIMIDOLOL HYDROCHLORIDE see PMC600
PRIMIDON see DBB200
PRIMIDONE see DBB200
PRIMIN see DSK200
PRIMINE see PMI750
PRIMOBOLAN see PMC700
PRIMOBOLONE see PMC700
PRIMOCARCIN see PMC750
PRIMOCORT see DAQ800
PRIMOCORTAN see DAQ800
PRIMODOS see EEH520
PRIMOFOL see EDO000
PRIMOGONYL see CMG675
PRIMOGYN B see EDP000
PRIMOGYN BOLEOSUM see EDP000
PRIMOGYN I see EDP000
PRIMOL 335 see MQV750
PRIMOLUT C see GEK500
PRIMOLUT DEPOT see HNT500
PRIMONABOL see PMC700
PRIMOSTAT see GEK510
PRIMOTEC see DIN600
PRIMOTEST see TBF500
PRIMOVLAR see NNL500
PRIMPERAN see AJH000
PRIMROSE YELLOW see ZFJ100
PRIMUN see FDD100
PRINADOL HYDROBROMIDE see PMD325
PRINALGIN see AGN000
PRINCILLIN see AOD125
PRINCIPAL BILE PIGMENT see HAO900
PRINCIPEN see AIV500
PRINCIPEN/N see SEQ000
PRINICID see DIN600
PRINODOLOL see VSA000
PRINTEL'S see SMQ500
PRIODAX see PDM750
PRIODERM see MAK700
PRIOSPEN see PDD350
PRISCOL see BBJ750, BBW750
PRISCOLINE see BBW750
PRISCOLINE HYDROCHLORIDE see BBJ750
PRISILIDENE HYDROCHLORIDE see NEB000
PRISILIDINE see NEA500
PRISMANE see PMD350
PRIST see EJH500
PRISTACIN see CCX000
PRISTANE see PMD500
PRISTIMERIN see PMD525
PRISTINAMYCIN see VRF000
PRISTINAMYCIN IA see VRA700
PRIVENAL see ERE000
PRIVET see PMD550
PRIVINE see NAH500, NAH550
PRIVINE HYDROCHLORIDE see NCW000
PRIVINE NITRATE see NAH550
PRIZOLE HYDROCHLORIDE see NCW000

PRM-TC see PPY250
PRO-ACTIDIL see TMX775
PROADIFEN see DIG400
PROADIFEN HYDROCHLORIDE see PBM500
PROASMA HYDROCHLORIDE see OJY000
PROAZAIMINE see DQA400
PROAZAMINE see DQA400
PRO-BAN M see TEH500
PRO-BANTHINE see HKR500
PROBARBITAL see ELX000
PROBARBITONE see ELX000
PROBE see BGD250
PROBECID see DWW000
PROBEDRYL see BBV500
PROBEN see DWW000
PROBENAZOLE see PMD800
PROBENECID ACID see DWW000
PROBENECID SODIUM SALT see DWW200
PROBENEMID see DWW000
PROBESE-P see PMA750
PROBESE-P HYDROCHLORIDE see MNV750
PROBILIN see EOA500
PROBON see PMD825
PROBONAL see PMD825
PROCAINAMIDE see AJN500
PROCAINAMIDE HYDROCHLORIDE see PME000
PROCAINAMIDE SULFATE see AJN750
PROCAINE see AIL750
PROCAINE AMIDE see AJN500
PROCAINE AMIDE HYDROCHLORIDE see PME000
PROCAINE AMIDE SULFATE see AJN750
PROCAINE, BASE see AIL750
PROCAINE FLUOBORATE see TCH000
PROCAINE HYDROCHLORIDE see AIT250
PROCALM see BCA000
PROCALMIDOL see MQU750
PROCAMIDE see AJN500
PROCARBAZIN (GERMAN) see PME500
PROCARBAZINE see PME250
PROCARBAZINE HYDROCHLORIDE see PME500
PROCARDIA see AEC750
PROCARDIN see POB500
PROCARDINE see DJS200
PROCASIL see PNX000
PROCATEROL HYDROCHLORIDE see PME600
PROCHLOROPERAZINE see PMF500
PROCHLOROPROAZINE HYDROGEN MALEATE see PMF250
PROCHLORPEMAZINE see PMF500
PROCHLORPERAZINE see PMF500
PROCHLORPERAZINE BIMALEATE see PMF250
PROCHLORPERAZINE DIMALEATE see PMF250
PROCHLORPERAZINE EDISYLATE see PME700
PROCHLORPERAZINE ETHANE DISULFONATE see PME700
PROCHLORPERAZINE HYDROGEN MALEATE see PMF250
PROCHLORPERAZINE MALEATE see PMF250
PROCHLORPERIZINE MALEATE see PMF250
PROCHLORPROMAZINE see PMF500
PROCHOLON see DAL000
PROCIDLIDINA see CPQ250
PROCILAN see POB500
PROCINOLOL HYDROCHLORIDE see PMF525
(±)-PROCINOLOL HYDROCHLORIDE see PMF535
PROCION BLACK H-N see CMS227
PROCION BRILLIANT RED M 5B see PMF540
PROCION BRILLIANT RED MX 5B see PMF540
PROCION BRILLIANT RED 5BS see PMF540
PROCION RED MX 5B see PMF540
PROCION YELLOW MX 4R see RCZ000
PROCIT see DQA400
PROCLONOL see PMF550
PROCOL OA-20 see PJW500
PRO-COR see HNY500
PROCORMAN see DJS200
PROCTIN see SEH000
PROCTODON see DVV500
PROCURAN see DAF800
PROCYAZINE see PMF600
PROCYCLIDINE see CPQ250
PROCYKLIDIN see CPQ250
PROCYMIDONE see PMF750
PROCYTOX see CQC500, CQC650
PRODALUMNOL see SEY500
PRODALUMNOL DOUBLE see SEY500
PRODAN see DXE000

PRODARAM see BJK500
PRODECTINE see PPH050
PRODEL see DYF759
PRODHYBASE ETHYL see EJM500
PRODHYPHORE B see PJY100
PRO-DIABAN see PMG000
PRODICTAZIN see DIR000
PRODIERAZINE see DIR000
PRODILIDINE see DTO200
α-PRODINE see NEA500
α-PRODINE HYDROCHLORIDE see NEB000
PRODIXAMON see HKR500
PRO-DORM see QAK000
PRODOX 133 see IQZ000
PRODOX 146 see DEG000
PRODOX 156 see DCI000
PRODOX 340 see BST000
PRODOX ACETATE see PMG600
PRODOXAN see GEK500
PRODOX 146A-85X see DEG000
PRODOXOL see OOG000
PRODROMINE see TCY750
PRODROXAN see GEK500
PRODUCER GAS see PMG750
PRODUCT 308 see HCP000
PRODUCT 5022 see MPE250
PRODUCT 5190 see AHL500
PRODUCT No. 161 see SIB600
PRO-DUOSTERONE see NNL500
PRODUXAN see GEK500
PROEPTATRIENE (ITALIAN) see PMH600
PROFALVINE SULPHATE see DBN400
PROFAM see CBM000
PROFAMINA see BBK000
PROFARMIL see TEH500
PROFAX see PMP500
PROFECUNDIN see VSZ450
PROFEMIN see CHJ750
PROFENAMINA (ITALIAN) see DIR000
PROFENAMINUM see DIR000
PROFENID see BDU500
PROFENOFOS see BNA750
PROFENONE see ELL500
PROFERRIN see IHG000
PROFETAMINE see AOB500
PROFETAMINE PHOSPHATE see AOB500
PROFIROMYCIN see MLY000
PROFLAVIN see DBN600
PROFLAVINE see DBN600, PMH100
PROFLAVINE HEMISULPHATE see PMH100
PROFLAVINE HYDROCHLORIDE see PMH250
PROFLAVINE MONOHYDROCHLORIDE see PMH250
PROFLAVINE MONOHYDROCHLORIDE HEMIHYDRATE see DBN200
PROFLAVINE (SULFATE) see DBN400
PROFLAVIN SULFATE see DBN400
PROFLURALIN see CQG250
PROFOLIOL see DBN600, EDO000
PROFORMIPHEN see DBN600
PROFUME A see CKN500
PROFUME (OBS.) see MHR200
PROFUNDOL see AFY500, DBN600
PROFURA see DBN600
PROGALLIN LA see DXX200
PROGALLIN P see PNM750
PROGARMED see DBN600
PRO-GASTRON see HKR500
PROGEKAN see PMH500
PRO-GEN see DBN600
PROGESIC see DBN600, FAP100
PROGESTAB see GEK500
PROGESTEROL see PMH500
PROGESTERONE see PMH500
β-PROGESTERONE see PMH500
PROGESTERONE CAPROATE see HNT500
PROGESTERONE mixed with ESTRADIOL BENZOATE (14:1 moles) see EDQ000
PROGESTERONE mixed with ESTRA-1,3,5(10)-TRIENE-3,17-β-DIOL-3-BENZO-ATE (14:1 moles) see EDQ000
PROGESTERONE RETARD PHARLON see HNT500
PROGESTERONUM see PMH500
PROGESTIN see PMH500
PROGESTIN P see GEK500
PROGESTOLETS see GEK500
PROGESTONE see PMH500
PROGESTORAL see GEK500

PRO-GIBB see GEM000
PROGLICEM see DCQ700
PROGLUMETACINA (SPANISH) see POF550
PROGLUMETACIN MALEATE see POF550
PROGLUMIDE see BGC625
PROGLUMIDE SODIUM see PMH575
PROGUANIL see CKB250
PROGUANIL HYDROCHLORIDE see CKB500
PROGYNON see EDO000, EDS100
PROGYNON B see EDP000
PROGYNON BENZOATE see EDP000
PROGYNON-DEPOT see EDS100
PROGYNON-DH see EDO000
PROGYNON-DP see EDR000
PROGYNOVA see EDS100
PROHEPTADIENE see EAH500
PROHEPTADIEN MONOHYDROCHLORIDE see EAI000
PROHEPTATRIENE see PMH600
PROHEPTATRIENE HYDROCHLORIDE see DPX800
PROHEPTATRIEN MONOHYDROCHLORIDE see DPX800
PROKARBOL see DUS700
PROKAYVIT see MMD500
PROLACTIN see PMH625
PROLAN see BIN500
PROLAN B see FMT100
PROLAN (CSC) see BIN500
PROLATE see PHX250
PROLAX see GGS000
PROLIDON see PMH500
PROLINOMETHYLTETRACYCLINE see PMI000
PROLINTANE HYDROCHLORIDE see PNS000
PROLIXAN see AQN750, ASA000
PROLIXIN see FMP000
PROLIXINE see TJW500
PROLIXIN ENANTHATE see PMI250
PROLONGAL see IGS000
PROLONGINE see DWW000
PROLOPRIM see TKZ000
PROLOXIN see GGS000
PROLUTOL see GEK500
PROLUTON C see GEK500
PROLUTON DEPOT see HNT500
PROMACID see CKP500
PROMACTIL see CKP250
PROMAMIDE see DTT600
PROMANIDE see AOO800
PROMANTINE see PMI750
PROMAPAR see CKP500
PROMAQUID see FMU039
PROMAR see DVV600
PROMARIT see ECU750
PROMASSOL see AHI875
PROMAZIL see CKP250
PROMAZINAMIDE see DQA400
PROMAZINE see DQA600
PROMAZINE HYDROCHLORIDE see PMI500
PROMECARB see CQI500
PROMEDOL see IMT000
PROMERAN see CHX250
PROMETASIN see DQA400
PROMETAZIN see DQA400
PROMETHAZINE N-(2'-DIMETHYLAMINO-2'-METHYLETHYL)PHENOTHI-
   AZINE HYDROCHLORIDE see PMI750
PROMETHAZINE HYDROCHLORIDE see PMI750
PROMETHIAZIN (GERMAN) see PMI750
PROMETHIAZINE see DQA400
PROMETHIN see FMS875
PROMETHIUM see PMJ000
PROMETIN see FMS875
PROMETON see MFL250
PROMETREX see BKL250
PROMETRIN see BKL250
PROMETRYN see BKL250
PROMETRYNE (USDA) see BKL250
PROMEZATHINE see DQA400
PROMIBEN see DLH600, DLH630
PROMIDE (parasympatholytic) see BGC625
PROMIDIONE see GIA000
PROMIN see AOO800
PROMINAL see ENB500
PROMIN SODIUM see AOO800
PROMIT see DBD700
PROMONTA see MOQ000
PROMOTESTON see TBF500

PROMOTIL see PNS000
PROMOTIN see AOO800
PROMPT INSULIN ZINC SUSPENSION see LEK000
PROMPTONAL see EOK000
PROMUL 5080 see HKJ000
PROMURIT see DEQ000
PROMURITE see DEQ000
PRONABOL see ENX600
PRONAMIDE see DTT600
PRONASE see PMJ100
PRONDOL see DPX200
PRONE see GEK500
PRONESTYL see AJN500
PRONESTYL HYDROCHLORIDE see PME000
PRONETALOL see INS000
PRONETHALOL see INS000, INT000
PRONETHALOL HYDROCHLORIDE see INT000
PRONEURIN see DXO300
PRONTALBIN see SNM500
PRONTODIN see DUO400
PRONTOSIL 1 see SNM500
PROPACHLOR see CHS500
PROPACHLORE see CHS500
PROPACIL see PNX000
PROPADERM see AFJ625
PROPADRINE see NNM000
PROPADRINE HYDROCHLORIDE see NNN000, PMJ500
PROPAL see IQW000, CIR500
PROPALDON see QCS000
PROPALGYL see DPE200
PROPALLYLONAL see QCS000
PROPAMIDINE DIHYDROCHLORIDE see DBM400
PROPAMINE D see TDQ750
PROPANAL, 2-CHLORO-(9CI) see CKP700
PROPANALOL see ICB000
2-PROPANAMINE see INK000
1-PROPANAMINE, 3-DIBENZ(b,e)OXEPIN-11(6H)-YLIDENE-N,N-DIMETHYL-,
   HYDROCHLORIDE see AEG750
1-PROPANAMINE, 3-(10,11-DIHYDRO-5H-DIBENZO(A,D)CYCLOHEPTEN-5-
   YLIDENE)-N,N-DI-METHYL-N-OXIDE see AMY000
1-PROPANAMINIUM, 3-CARBOXY-2-HYDROXY-N,N,N-TRIMETHYL-,
   CHLORIDE, (R)- (9CI) see CCK660
1,3-PROPAN-BIS-(4-HYDROXYIMINOMETHYL-PYRIDINIUM-(1))-DIBROMIDS
   (GERMAN) see TLQ500
1-PROPANECARBOXYLIC ACID see BSW000
PROPANE, 1-CHLORO-1,2-EPOXY-, (E)- see CKS100
PROPANE, 1-CHLORO-1,2-EPOXY-, (Z)- see CKS099
PROPANEDIAL, ION(1-), SODIUM (9CI) see MAN700
1,3-PROPANEDICARBOXYLIC ACID see GFS000
PROPANE DIETHYL SULFONE see ABD500
PROPANE-1,3-DIMETHANESULFONATE see TLR250
PROPANEDINITRILE see MAO250
PROPANEDINITRILE((2-CHLOROPHENYL)METHYLENE) see CEQ600
PROPANEDIOIC ACID see CCC750
PROPANEDIOIC ACID, DIETHYL ESTER see EMA500
PROPANEDIOIC ACID DIMETHYL ESTER (9CI) see DSM200
PROPANEDIOIC ACID, DITHALLIUM SALT see TEM399
1,2-PROPANEDIOL-1-ACRYLATE see HNT600
1,2-PROPANEDIOL, 3-(6-AMINO-9H-PURIN-9-YL)-, (S)- see AMH800
1,3-PROPANEDIOL BIS(α-(p-CHLOROPHENOXY)ISOBUTYRATE)
   see SDY500
1,3-PROPANEDIOL BIS(2-(4-CHLOROPHENOXY)-2-METHYLPROPIONATE)
   see SDY500
1,2-PROPANEDIOL CARBONATE see CBW500
1,2-PROPANEDIOL, 3-CHLORO-, 1-BENZOATE see CKQ500
1,2-PROPANEDIOL, 3-CHLORO-, DIACETATE see CIL900
1,2-PROPANEDIOL CYCLIC CARBONATE see CBW500
1,1-PROPANEDIOL, 2,3-DICHLORO-, DIACETATE see DBF875
1,2-PROPANEDIOL, MONOSTEARATE see SLL000
1,2-PROPANEDIOL, 3-(NITROSO-2-PROPENYLAMINO)- see NJY500
1,1'-(1,3-PROPANEDIYL)BIS(4-(HYDROXYIMINO)METHYLPYRIDINIUM,
   DIBROMIDE see TLQ500
1,1'-(1,3-PROPANEDIYL)BIS((4-HYDROXYIMINO)METHYL)-PYRIDINIUM
   DICHLORIDE see TLQ750
4,4'-(1,3-PROPANEDIYLBIS(OXY))BIS-BENZENECARBOXIMIDAMIDE,
   DIHYDROCHLORIDE see DBM400
1,2-PROPANEDIYL CARBONATE see CBW500
PROPANE, HEPTAFLUOROIODO- see HAY300
PROPANE, 1-ISOTHIOCYANATO-3-(METHYLSULFINYL)-(9CI) see MPN100
PROPANE, 2-METHOXY-2-METHYL (9CI) see MHV859
PROPANENITRILE, 2-(β-d-GLUCOPYRANOSYLOXY)-2-METHYL- see GFC100
PROPANENITRILE, 3-(METHYLNITROSOAMINO)- see MMS200
PROPANENITRILE, TRICHLORO- see TJC800
2-PROPANETHIOL see IMU000

1,2,3-PROPANETRICARBOXYLIC ACID, 2-HYDROXY-, TIN(2+) SALT (2:3) (9CI) see TGC285

1,2,3-PROPANETRICARBOXYLIC ACID, 2-HYDROXY-, ZINC SALT (2:3) (9CI) see ZFJ250

1,2,3-PROPANETRIOL see GGA000

1,2,3-PROPANETRIOL MONOACETATE see GGO000

1,2,3-PROPANETRIOL TRIACETATE see THM500

1,2,3-PROPANETRIOL, TRINITRATE see NGY000

1,2,3-PROPANETRIYL NITRATE see NGY000

PROPANEX see DGI000

PROPANID see DGI000

PROPANIDE see DGI000

PROPANIL see DGI000

PROPANOIC ACID BUTYLESTER (9CI) see BSJ500

PROPANOIC ACID, 2-((4'-CHLORO(1,1'-BIPHENYL)-4-YL)OXY)-2-METHYL-, METHYL ESTER (9CI) see MIO975

PROPANOIC ACID, 2-(4-CHLORO-2-METHYLPHENOXY)-, POTASSIUM SALT (9CI) see CLO200

PROPANOIC ACID, 2-CHLORO-, SODIUM SALT see CKT100

PROPANOIC ACID, 2-(4-(2,2-DICHLOROCYCLOPROPYL)PHENOXY)-2-METHYL- see CMS210

PROPANOIC ACID, 2,2-DIMETHYL-, ISOOCTADECYL ESTER see ISC550

PROPANOIC ACID, ETHENYL ESTER see VQK000

PROPANOIC ACID, METHYL ESTER see MOT000

PROPANOIC ACID, 2-METHYL-, 3-HEXENYL ESTER, (Z)- see HFE520

PROPANOIC ACID, 2-METHYL-, HEXYL ESTER (9CI) see HFQ550

PROPANOIC ACID, SODIUM SALT see SJL500

PROPAN-2-OL see INJ000

2-PROPANOL see INJ000

i-PROPANOL (GERMAN) see INJ000

2-PROPANOL, 1-AMINO-3-CHLORO-, (-)- see AJI520

2-PROPANOL, 1-AMINO-3-CHLORO-, (±)- see AJI530

2-PROPANOL, 1,1'-(2-BUTYNYLENEDIOXY)BIS(3-CHLORO- see BST900

2-PROPANOL, 1-CHLORO-3-ISOPROPOXY- see CHS250

2-PROPANOL, 1-(4-(2-(CYCLOPROPYLMETHOXY)ETHYL)PHENOXY)-3-((1-METHYLETHYL)AMINO)-, HYDROCHLORIDE see KEA350

1-PROPANOL, 3-(DIETHYLAMINO)-2,2-DIMETHYL-, TROPATE, PHOSPHATE see AOD250

PROPANOL NITRITE see IQQ000

1-PROPANOL, 3-PHENYL-, PROPIONATE see HHQ550

2-PROPANOL, 1-(2-((3,3,5-TRIMETHYLCYCLOHEXYL)OXY)PROPOXY)-(9CI) see TLO600

PROPANONE see ABC750

2-PROPANONE see ABC750

2-PROPANONE, 1-(4-(CHLOROMETHYL)-1,3-DIOXOLAN-2-YL)- see CIL850

1-PROPANONE, 1-p-MENTH-6-EN-2-YL- see MCF525

2-PROPANONE OXIME see ABF000

PROPANOSEDYL see RLU000

PROPANTEL see HKR500

PROPANTHELINE BROMIDE see HKR500

PROPAPHEN see CKP500

PROPAPHENIN see CKP250

PROPAPHENIN HYDROCHLORIDE see CKP500

2-PROPARGILOSSI-5-AMINO-N-(n-BUTIL)-BENZAMIDE (ITALIAN) see AJC000

PROPARGITE (DOT) see SOP000

PROPARGYL CHLORIDE see CKV275

PROPARSAMIDE see SJG000

PROPASA see AMM250

PROPASOL SOLVENT B see BPS250

PROPASTE 6708 see TAV750

PROPASTE T see SON000

PROPAVAN see IDA500

PROPAX see CFZ000

PROPELLANT 12 see DFA600

PROPELLANT 22 see CFX500

PROPELLANT 114 see FOO509

PROPELLANT C318 see CPS000

2-PROPENAL see ADR000

PROP-2-EN-1-AL see ADR000

PROPENAL (CZECH) see ADR000

PROPENAMIDE see ADS250

2-PROPENAMIDE see ADS250

2-PROPENAMIDE, N-(METHOXYMETHYL)- see MEX300

2-PROPENAMIDE, N-METHYL- (9CI) see MGA300

2-PROPENAMINE see AFW000

2-PROPEN-1-AMINE see AFW000

PROPENE ACID see ADS750

trans-1-PROPENE-1,2-DICARBOXYLIC ACID see MDI250

PROPENENITRILE see ADX500

2-PROPENENITRILE see ADX500

2-PROPENENITRILE HOMOPOLYMER (9CI) see ADX750

2-PROPENENITRILE, POLYMER with CHLOROETHENE see ADY250

2-PROPENENITRILE POLYMER with ETHENYLBENZENE see ADY500

2-PROPENE-1-THIOL see AGJ500

1-PROPENE-1,2,3-TRICARBOXYLIC ACID see ADH000

PROPENOIC ACID see ADS750

2-PROPENOIC ACID (9CI) see ADS750

2-PROPENOIC ACID BICYCLO(2,2,1)HEPT-5-EN-2-YLMETHYL ESTER see BFY250

2-PROPENOIC ACID 2-BUTOXYETHYL ESTER see BPK500

2-PROPENOIC ACID-2-CHLOROETHYL ESTER see ADT000

2-PROPENOIC ACID-2-CYANOETHYL ESTER see ADT111

2-PROPENOIC ACID-2,2-DIMETHYL-1,3-PROPANEDIYL ESTER see DUL200

2-PROPENOIC ACID, 1,2-ETHANEDIYLBIS(OXY-2,1-ETHANEDIYL) ESTER (9CI) see TJQ100

2-PROPENOIC ACID-1,2-ETHANEDIYL ESTER see EIP000

2-PROPENOIC ACID-2-ETHOXYETHYL ESTER see ADT500

2-PROPENOIC ACID-2-ETHYLBUTYL ESTER see EGZ000

2-PROPENOIC ACID, ETHYL ESTER (MAK) see EFT000

2-PROPENOIC ACID-2-ETHYLHEXYL ESTER see ADU250

2-PROPENOIC ACID, HEXEL ESTER see ADV000

2-PROPENOIC ACID HOMOPOLYMER see ADW200

2-PROPENOIC ACID, HOMOPOLYMER, ZINC SALT see ADW250

2-PROPENOIC ACID-2-HYDROXYETHYL ESTER (9CI) see ADV250

2-PROPENOIC ACID-2-HYDROXYPROPYL ESTER see HNT600

2-PROPENOIC ACID ISODECYL ESTER (9CI) see IKL000

2-PROPENOIC ACID-2-METHOXYETHYL ESTER see MEM250

PROPENOIC ACID METHYL ESTER see MGA500

2-PROPENOIC ACID METHYL ESTER see MGA500

2-PROPENOIC ACID, 2-METHYL-, 2-((3a,4,5,6,7,7a-HEXAHYDRO-4,7-METHANO-1H-INDEN-5-YL)OXY) ETHYL ESTER see DGW450

2-PROPENOIC ACID-1-METHYL-13-PROPANEDIYL ESTER see BRG500

2-PROPENOIC ACID-2-METHYLPROPYL ESTER see IIK000

2-PROPENOIC ACID-2-(METHYLTHIO)ETHYL ESTER see MPT250

2-PROPENOIC ACID OXIRANYLMETHYL ESTER see ECH500

2-PROPENOIC ACID, OXYBIS(2,1-ETHANEDIYLOXY-2,1-ETHANEDIYL)ESTER see ADT050

2-PROPENOIC ACID, OXYBIS(METHYL-2,1-ETHANEDIYL) ESTER see DWS650

2-PROPENOIC ACID TRIDECYL ESTER see ADX000

PROPENOL see AFV500

PROPEN-1-OL-3 see AFV500

1-PROPEN-3-OL see AFV500

2-PROPEN-1-OL see AFV500

2-PROPEN-1-ONE see ADR000

2-PROPENOYL CHLORIDE see ADZ000

2-PROPENYLACRYLIC ACID see SKU000

PROPENYL ALCOHOL see AFV500

2-PROPENYL ALCOHOL see AFV500

5-(1-PROPENYL)-1,3-BENZODIOXOLE see IRZ000

5-(2-PROPENYL)-1,3-BENZODIOXOLE see SAD000

1-PROPENYL BROMIDE see BOA000

4-PROPENYLCATECHOL METHYLENE ETHER see IRZ000

2-PROPENYL CHLORIDE see AGB250

PROPENYL CINNAMATE see AGC000

17-(2-PROPENYL)ESTR-4-EN-17-OL see AGF750

PROPENYL ETHER see DBK000

PROPENYLGUAETHOL (FCC) see IRY000

4-PROPENYLGUAIACOL see IKQ000

2-PROPENYL HEPTANOATE see AGH250

2-PROPENYL-N-HEXANOATE see AGA500

2-PROPENYL ISOTHIOCYANATE see AGJ250

2-PROPENYL ISOVALERATE see ISV000

3-PROPENYL METHANOATE see AGH000

4-(PROPENYL)-2-METHOXYPHENYL FORMATE see EQS100

2-PROPENYL 3-METHYLBUTANOATE see ISV000

4-PROPENYL-1,2-METHYLENEDIOXYBENZENE see IRZ000

(2-PROPENYLOXY)BENZENE see AGR000

((2-PROPENYLOXY)METHYL)OXIRANE see AGH150

N-2-PROPENYL-2-PROPEN-1-AMINE see DBI600

(2-PROPENYL)THIOUREA see AGT500

2-PROPENYL 3,5,5-TRIMETHYLHEXANOATE see AGU400

N-2-PROPENYLUREA see AGV000

4-PROPENYL VERATROLE see IKR000

PROPERIDOL see DYF200

PROPETAMPHOS see MKA000

PROPHAM see CBM000

PROPHENAL see AGQ875

PROPHENATIN see DEO600

PROPHENPYRIDAMINE MALEATE see TMK000

PROPHOS see EIN000

PROPICOL see DNR309

PROPILDAZINA (ITALIAN) see HHD000

PROPILENTIOUREA see MJZ000

PROPINAN DI-n-BUTYLCINICITY (CZECH) see DEB400

PROPINE see MFX590

PROPINEB see ZMA000

PROPINEBE see ZMA000

5-(PROPYLTHIO)-2-CARBOMETHOXYAMINOBENZIMIDAZOLE see VAD000
PROPYL THIOPYROPHOSPHATE see TED500
1-PROPYLUREA and SODIUM NITRITE see SIT500
n-PROPYLUREA and SODIUM NITRITE see SIT500
2-PROPYLVALERIC ACID CALCIUM SALT (2:1) see CAY675
PROPYNE see MFX590
PROPYNE mixed with PROPADIENE see MFX600
2-(2-PROPYNYLOXY)PHENYL METHYLCARBAMATE see MIB500
PROPYPERONE see FLN000
PROPYPHENAZONE see INY000
N-3'-a-PROPYPHENAZONYL-2-ACETOXYBENZAMIDE see PNR800
PROPYZAMIDE see DTT600
PROQUANIL see MQU750
PRORALONE-MOP see XDJ000
PRORESIDOR see BER500
PROREX see DQA400
PROSCOMIDE see SBH500
PROSEPTINE see SNM500
PROSEPTOL see SNM500
PROSERIN see DQY909
PROSERINE METHYL SULFATE see DQY909
PROSEROUT see AAE500
PROSERYL see DPE000
PRO-SONIL see TDA500
PROSTAGLANDIN A2 see MCA025
(+)-PROSTAGLANDIN A^2 see MCA025
PROSTAGLANDIN E2 see DVJ200
(−)-PROSTAGLANDIN E2 see DVJ200
(15S)-PROSTAGLANDIN E2 see DVJ200
PROSTAGLANDIN F2-α METHYL ESTER see DVJ100
PROSTAPHILIN see MNV250
PROSTAPHILIN A see SLJ050
PROSTAPHLIN-A see SLJ000
PROSTAPHLYN see DSQ800
PROSTEARIN see SLL000
PROST-13-EN-1-OIC ACID, 11,16-DIHYDROXY-16-METHYL-9-OXO-, METHYL
     ESTER, (11-α-13E)-(±)- see MJE775
PROSTETIN see ELF100
PROSTIGMIN see NCL100
PROSTIGMINE see NCL100
PROSTIGMINE METHYLSULFATE see DQY909
PROSTIN see CCC100
PROSTIN E2 see DVJ200
PROSTOGLANDIN E$_2$-METHYLHESPERIDIN COMPLEX see KHK100
PROSTRUMYL see MPW500
PROSULTHIAMINE see DXO300
PROSULTIAMINE see DXO300
PROSYKLIDIN see CPQ250
PROTABEN P see HNU500
PROTABOL see TFK300
PROTACELL 8 see SEH000
PROTACHEM GMS see OAV000
PROTACTYL see DQA600
PROTAMINE ZINC INSULIN see IDF325
PROTAMINE ZINC INSULIN INJECTION see IDF325
PROTAMINE ZINC INSULIN SUSPENSION see IDF325
PROTANAL see SEH000
PROTANDREN see AOO475
PROTATEK see SEH000
PROTAZINE see DQA400
PROTECT see NAQ000
PROTECTON see IBQ100
PROTECTONA see DKA600
PROTEINA see DME500
PROTEINASE, ASPERFILLUS ALKALINE see SBI860
PROTEINASE, ASPERGILLUS TERRICOLA NEUTRAL see TBF350
PROTERGURIDE see DJX300
PROTERNOL see DMV600
PROTESINE DMU see DTG700
PROTEX (POLYMER) see AAX250
PROTHAZIN see DQA400
PROTHAZIN METHOSULFATE see MRW000
PROTHEOBROMINE see HNY500
PROTHEOPHYLLINE see DNC000
PROTHIADEN see DYC875
PROTHIADENE HYDROCHLORIDE see DPY200
PROTHIADEN HYDROCHLORIDE see DPY200
PROTHIADEN SPOFA see DYC875
PROTHIL see MBZ100
PROTHIOCARB see EIH500
PROTHIOPHOS see DGC800
PROTHIPENDYL see DYB600
PROTHIPENDYL HYDROCHLORIDE see DYB800
PROTHOATE see IOT000

PROTHROMADIN see WAT200
PROTHROMBIN see WAT220
PROTIOAMPHETAMINE see AOA250
PROTIRELIN see TNX400
PROTIVAR see AOO125
PROTOAT (HUNGARIAN) see IOT000
PROTOBOLIN see DAL300
PROTOCATECHUALDEHYDE DIMETHYL ETHER see VHK000
PROTOCATECHUIC ACID, 3-METHYL ESTER see VFF000
PROTOCATECHUIC ALDEHYDE DIMETHYL ETHER see VHK000
PROTOCATECHUIC ALDEHYDE ETHYL ETHER see EQF000
PROTOCHLORURE D'IODE (FRENCH) see IDS000
PROTOCOL C see DTG000
PROTOMIN see AOO800
PROTONA see DME500
PROTOPAM CHLORIDE see FNZ000
PROTOPET see MQV750
PROTOPHENICOL see CDP500
PROTOPINE see FOW000
PROTOPORPHYRIN DISODIUM see DXF700
PROTOPORPHYRIN SODIUM see DXF700
PROTOPORPHYRIN SODIUM SALT see DXF700
PROTOPYRIN see EEM000
PROTOVERATRINE B see NCI600
PROTOX TYPE 166 see ZKA000
PROTRIPTYLINE see DDA600
PROTRYPTYLINE see DDA600
PROVAMYCIN see SLC000
PROVASAN see EQQ100
PROVENTIL see BQF500
PROVIGAN see DQA400
PROVITAMIN D see CMD750
PROVITAMIN D$_3$ see DAK600
PROVITAR see AOO125
PROWL see DRN200
PROXAGESIC see DAB879
PROXANSODIUM see SIA000
PROX DW see DTG000
PROXEN see MFA500
PROXIPHYLLINE see HOA000
PROXOL see TIQ250
PROXYPHYLLINE see HOA000
PROZIL see CKP250
PROZIME 10 see SBI860
PROZIN see CKP250
PRS 640 see BML500
PRUMYCIN see AFI500
PRUNETOL see GCM350
PRUNOLIDE see CNF250
PRUNUS (VARIOUS SPECIES) see AQP890
PRURALGAN see DNX400
PRURALGIN see DNX400
PRUSSIAN BROWN see IHD000
PRUSSIC ACID (DOT) see HHS000
PRUSSIC ACID, unstabilized see HHU000, HHS000
PRUSSITE see COO000
PRYNACHLOR see CDS275
PRYSOLINE see DBB200
PS see CKN500
PS 1 see AHE250
PS 2383 see TKX250
PSC CO-OP WEEVIL BAIT see DXE000
PSEUDECHIS AUSTRALIS VENOM see ARV000
PSEUDECHIS PORPHYRIACUS (AUSTRALIA) VENOM see ARV250
PSEUDECHIS PORPHYRIACUS VENOM see ARV250
PSEUDOBUTYLBENZENE see BQJ250
PSEUDO-BUTYLENE see BOW500
PSEUDOCUMENE see TLL750
PSEUDOCUMIDINE see TLG750
PSEUDOCUMIDINE HYDROCHLORIDE see TLG750
PSEUDOCUMOL see TLL750
PSEUDOCYANURIC ACID see THS000
PSEUDODIGITOXIN see GEU000
PSEUDOHEXYL ALCOHOL see EGW000
PSEUDONAJA TEXTILIS (AUSTRALIA) VENOM see ARU500
PSEUDONAJA TEXTILIS VENOM see TEG650
PSEUDOTHEOPHYLLINE see TEP000
PSEUDO-THIOHYDANTOIN see IAP000
PSEUDOTHIOUREA see ISR000
PSEUDOTHYMINE see MQI750
PSEUDOTROPINE BENZOATE HYDROCHLORIDE see TNS200
PSEUDOUREA see USS000
PSEUDOXANTHINE see XCA000
PSICAINE-NEU HYDROCHLORIDE see NCJ000

PSICAIN-NEW HYDROCHLORIDE see NCJ000
PSICHIAL see MDQ250
PSICODISTEN see BET000
PSICOPAX see CFC250, CFZ000
PSICOPERIDOL-R see TKK500
PSICOPLEGIL see MNM500
PSICOSAN see LFK000, MDQ250
PSICOSEDINA see MNM500
PSICOSTERONE see AOO450
PSICRONIZER see NMV725
PSIDIUM GUAJAVA see GLW000
PSILOCINE see HKE000
PSILOTSIN see HKE000
PSIQUIM see CGA000
PSL see LEN000
PSORADERM see MFN275
PSORALEN see FQD000
PSYCHAMINE A 66 HYDROCHLORIDE see MNV750
PSYCHEDRINE see BBK000
PSYCHEDRINUM see AOB250
PSYCHEDRYNA see AOB250
PSYCHODRINE see BBK500
PSYCHOLIQUID see TEH500
PSYCHOPERIDOL see TKK500
PSYCHO-SOMA see MAG050
PSYCHOTABLETS see TEH500
PSYCHOZINE see CKP500
PSYCOPERIDOL HYDROCHLORIDE see TKK750
Pt-09 see TBO776
PT 155 see DAD040
PTAB see TNI500
PTAP see AON000
PTEGLU see FMT000
PTERIDIUM AQUILINUM see BML000
PTERIDIUM AQUILINUM TANNIN see BML250
PTERIS AQUALINA see BML000
PTEROFEN see UVJ450
PTEROPHENE see UVJ450
PTEROYLGLUTAMIC ACID see FMT000
PTEROYL-l-GLUTAMIC ACID see FMT000
PTEROYLMONOGLUTAMIC ACID see FMT000
PTEROYL-l-MONOGLUTAMIC ACID see FMT000
P. TEXTILIS (AUSTRALIA) VENOM see ARU500
PTFE see TAI250
PTG see EQP000
PTMS see AKQ000
PTMSA see AKQ000
PTO see HOB500
PU 603 see CJN250
PUA-HOKU (HAWAII) see SLJ650
PUA KALAUNU (HAWAII) see COD675
PUDDING-PIPE TREE see GIW300
PUD (HERBICIDE) see DTP400
PUKE WEED see CCJ825
PUKIAWE-LEI (HAWAII) see RMK250
PULANOMYCIN see FBP300
PULARIN see HAQ550
PULEGONE see MCF500
d-PULEGONE see MCF500
PULMICORT see BOM520
PULMOCLASE see CBR675
PULSAN see ICA000
PULSOTYL see FMS875
PURADIN see NGG500
PURALIN see TFS350
PURASAN-SC-10 see ABU500
PURATIZED see TNI500
PURATIZED AGRICULTURAL SPRAY see TNI500
PURATIZEDAT AGRICULTURAL SPRAY see TNI500
PURATIZED B-2 see MDD500
PURATIZED N5E see TNI500
PURATRONIC CHROMIUM CHLORIDE see CMJ250
PURATRONIC CHROMIUM TRIOXIDE see CMK000
PURATURF see TNI500
PURATURF 10 see ABU500
PURE EOSINE YY see BNH500
PURE LEMON CHROME L3GS see LCR000
PURE ORANGE CHROME M see LCS000
PURE QUARTZ see SCI500, SCJ500
PUREX see SFT000
PURE ZINC CHROME see ZFJ100
PURGING BUCKTHORN see MBU825
PURGING FISTULA see GIW300
PURIFIED CHARCOAL see CBT500

PURIFIED OXGALL see SFW000
PURIFILIN see DNC000
1H-PURIN-2-AMINE see AMH000
1H-PURIN-6-AMINE see AEH000
9H-PURIN-6-AMINE, 9-(PHENYLMETHYL)- (9CI) see BDX100
1H-PURINE-2-AMINE, 6-((1-METHYL-4-NITRO-1H-IMIDAZOL-5-YL)-THIO)-
   see AKY250
PURINE-2,6-DIOL see XCA000
9H-PURINE-2,6-DIOL see XCA000
2,6(1,3)-PURINEDION see XCA000
PURINE-2,6-(1H,3H)-DIONE see XCA000
PURINE RIBONUCLEOSIDE see RJF000
9-PURINE RIBONUCLEOSIDE see RJF000
PURINE RIBOSIDE see RJF000
PURINE-6-THIOL, MONOHYDRATE see MCQ100
6H-PURINE-6-THIONE, 2-AMINO-1,7-DIHYDRO- (9CI) see AMH250
1H-PURINE-2,6,8(3H)-TRIONE, 7,9-DIHYDRO- (9CI) see UVA400
1H-PURINE-2,6,8(3H)-TRIONE, 7,9-DIHYDRO-, MONOSODIUM SALT (9CI)
   see SKO575
9H-PURIN-6-OL see DMC000
6(1H)-PURINONE see DMC000
PURIN-6(3H)-ONE see DMC000
PURIVEL see MQR225
PUROCYCLINA see TBX000
PURODIGIN see DKL800
PUROMYCIN see AEI000
PUROPHYLLIN see HOA000
PUROSTROPHAN see OKS000
PURPLE 4R see FAG050
PURPLE ALLAMANDA see ROU450
PURPLE RED see FMU080
PURPUREA B see DAD650
PURPUREA GLYCOSIDE B see DAD650
PURPUREAGLYKOSID B (GERMAN) see DAD650
PURPURID see DKL800
PURPURIN see TKN750
PURPURINE see TKN750
PURPUROCATECHOL see TNV550
PURPUROGALLIN see TDD500
PURTALC USP see TAB750
PUTRESCIN see BOS000
PUTRESCINE see BOS000
PVA see AAF625
P.V. CARBACHOL see CBH250
PVC CORDO see AAX175
PX 104 see DEH200
PX-138 see DVL600
PX-238 see AEO000
PX 404 see DEH600
PX 438 see BJS250
PX-806 see FAB920
PXO see OMY850
PYCAZIDE see ILD000
PYDRIN see FAR100
PYDT see DJY000
PYELOKON-FR see AAN000
PYELOSIL see DNG400
PYKNOLEPSINUM see ENG500
PYLAPRON see ICC000
PYLOSTROPIN see MGR500
PYLUMBRIN see DNG400
PYMAFED see DBM800, WAK000
PYNACOLYL METHYLFLUOROPHOSPHONATE see SKS500
PYNAMIN see AFR250
PYNAMIN-FORTE see AFR250
PYNDT see DTV200
PYNOSECT see BEP500
PYOPEN see CBO250
PYOPENE see CBO250
PYOSTACINE see VRF000
PYQUITON see BGB400
PYRA see WAK000
PYRABITAL see AMK250
PYRACETAM see NNE400
PYRACETON see DNG400
PYRACRYMYCIN-1 see COV125
PYRADEX see DBI200
PYRADONE see DOT000
PYRAHEXYL see HGK500
PYRALCID see ALF250
PYRALIN see CCU250
PYRAMAL see WAK000
PYRA MALEATE see DBM800
PYRAMEM see NNE400

PYRAMIDON see DOT000
PYRAMIDONE see DOT000
PYRAMON see AMK250
PYRAN ALDEHYDE see ADR500
PYRANILAMINE MALEATE see DBM800
PYRANINYL see DBM800
PYRANISAMINE see WAK000
PYRANISAMINE MALEATE see DBM800
2H-PYRAN-2-ONE, 6-HEPTYLTETRAHYDRO- see HBP450
4H-PYRAN-4-ONE, TETRAHYDRO-2,6-DIMETHYL- see TCQ350
4H-PYRANO(3,2-c)QUINOLINE-2-CARBOXYLIC ACID, 5,6-DIHYDRO-7,8-
    DIMETHYL-4,5-DIOXO-, ISOPENTYL ESTER see IHR200
PYRANTEL TARTRATE see TCW750
PYRANTON see DBF750
PYRATHYN see DPJ400, TEO250
PYRAZINOL-O-ESTER with O,O-DIETHYL PHOSPHOROTHIOATE see EPC500
PYRAZOLANTHRONE see APK000
PYRAZOL BLUE 3B see CMO250
PYRAZOLEANTHRONE see APK000
PYRAZOL FAST BRILLIANT BLUE VP see CMN750
PYRAZOLIDIN see BRF500
1,9-PYRAZOLOANTHRONE see APK000
PYRAZOLO(2,3-a)IMIDAZOLIDINE see IAN100
4H-PYRAZOLO(3,4-d)PYRIMIDIN-4-ONE see ZVJ000
PYRAZOLO(1,5-a)QUINOLINE, 2-(m-(BENZYLOXY)PHENYL)- see BFC450
2H-PYRAZOLO(4,3-c)QUINOLINE, 3,3a,4,5-TETRAHYDRO-2-ACETYL-5-((4-
    METHYLPHENYL)SULFONYL)-3-PHENYL-, cis- see ACY700
1-PYRENYLOXIRANE see ONG000
PYREQUAN TARTRATE see TCW750
PYRESIN see AFR250
PYRESYN see AFR250
PYRETHERM see BEP500
PYRETHIA see DQA400
PYRETHIAZINE see DQA400
PYRIBENZAMINE see TMP750
PYRICAROYL see DJS200
PYRICIDIN see ILD000
PYRIDAMAL-100 see TAI500
3,3'-(3,6-PYRIDAZINEDIYLBIS(OXY)BIS(N,N-DIETHYL-N-METHYL-1-PRO-
    PANAMINIUM DIIODIDE (9CI) see BJA825
(3,6-PYRIDAZINEDIYLBIS(OXY-
    TRIMETHYLENE))BIS(DIETHYLMETHYLAMMONIUM IODIDE) see BJA825
4-PYRIDINAMIDE HYDROCHLORIDE see AMJ000
3-PYRIDINAMINE see AMI250
4-PYRIDINAMINE see AMI500
α-PYRIDINAMINE see AMI000
3-PYRIDINAMINE HYDROCHLORIDE see AMI750
PYRIDIN-4-CARBONSAEURE-(DEXAMETHASON-21')-ESTER (GERMAN)
    see DBC510
PYRIDIN-2,5-DICARBONSAEURE-DI-N-PROPYLESTER (GERMAN) see EAU500
2-PYRIDINEALDOXIME CHLORIDE see FNZ000
PYRIDINE-2-ALDOXIME METHOCHLORIDE see FNZ000
2-PYRIDINE ALDOXIME METHYL CHLORIDE see FNZ000
PYRIDINE-3-CARBONIC ACID see NCQ900
PYRIDINE-4-CARBOXALDEHYDE THIOSEMICARBAZONE see ILB000
3-PYRIDINECARBOXAMIDE, 1,2-DIHYDRO-2-OXO-N-(2,6-XYLYL)- see XLS300
3-PYRIDINECARBOXAMIDE, N-(2,6-DIMETHYLPHENYL)-1,2-DIHYDRO-2-
    OXO-(9CI) see XLS300
3-PYRIDINECARBOXAMIDE, N-(2-(1H-INDOL-3-YL)ETHYL)-(9CI) see NDW525
PYRIDINE-3-CARBOXYDIETHYLAMIDE see DJS200
PYRIDINE-3-CARBOXYLIC ACID see NCQ900
3-PYRIDINECARBOXYLIC ACID see NCQ900
4-PYRIDINECARBOXYLIC ACID see ILC000
PYRIDINE-β-CARBOXYLIC ACID see NCQ900
4-PYRIDINECARBOXYLIC ACID-2-ACETYLHYDRAZIDE see ACO750
2-PYRIDINECARBOXYLIC ACID-2-ACETYLHYDRAZIDE (9CI) see ADA000
PYRIDINE-3-CARBOXYLIC ACID AMIDE see NCR000
3-PYRIDINECARBOXYLIC ACID AMIDE see NCR000
PYRIDINE-4-CARBOXYLIC ACID-(DEXAMETHASONE-21') ESTER see DBC510
2-PYRIDINECARBOXYLIC ACID, 3,6-DICHLORO-(9CI) see DGJ100
PYRIDINE-3-CARBOXYLIC ACID DIETHYLAMIDE see DJS200
3-PYRIDINECARBOXYLIC ACID, 3-(2,6-DIMETHYLPIPERIDINO)PROPYL
    ESTER, HYDROCHLORIDE see NDU000
4-PYRIDINECARBOXYLIC ACID, ETHYL ESTER see ELU000
4-PYRIDINECARBOXYLIC ACID, HYDRAZIDE see ILD000
3-PYRIDINECARBOXYLIC ACID, (2-HYDROXY-1,3-
    CYCLOHEXANEDIYLIDENE)TETRAKIS(METHYLENE) ESTER
    see CME675
3-PYRIDINECARBOXYLIC ACID-2-HYDROXY-3-(2-METHOXYPHENOXY)
    PROPYL ESTER see HLT300
4-PYRIDINECARBOXYLIC ACID, ((4-HYDROXY-3-METHOXYPHENYL)
    METHYLENE)HYDRAZIDE see VEZ925
3-PYRIDINECARBOXYLIC ACID, 3-(2-METHYLPIPERIDINO)PROPYL ESTER,
    HYDROCHLORIDE see NDW000

4-PYRIDINECARBOXYLIC ACID 2-(3-OXO-3-((PHENYLMETHYL)AMINO)
    PROPYL)HYDRAZIDE see BET000
3-PYRIDINECARBOXYLIC ACID-1,2,5,6-TETRAHYDRO-1-METHYL ESTER,
    HYDROCHLORIDE see AQU250
PYRIDINE-CARBOXYLIQUE-3 (FRENCH) see NCQ900
3,5-PYRIDINEDICARBOXYLIC ACID, 2-CYANO-1,4-DIHYDRO-6-METHYL-4-(3-
    NITROPHENYL)-, 3-METHYL-5-(1-METHYLETHYL) ESTER see NDY650
3,5-PYRIDINEDICARBOXYLIC ACID, 1,4-DIHYDRO-2,6-DIMETHYL-4-(3-
    NITROPHENYL)-, ETHYL METHYL ESTER see EMR600
2,5-PYRIDINEDICARBOXYLIC ACID, DIPROPYL ESTER see EAU500
PYRIDINE, 2-((2-(DIMETHYLAMINO)ETHYL)(SELENOPHENE-2-
    YLMETHYL)AMINO)- see DPI750
2-PYRIDINEMETHYLAMINE see ALB750
PYRIDINE, 3-(1-NITROSO-2-PIPERIDINYL)-, (S)-(9CI) see NJK150
PYRIDINE PERCHROMATE see ONU000
PYRIDINE, 1,2,3,6-TETRAHYDRO-1-((6-METHYL-3-CYCLOHEXEN-1-YL)
    CARBONYL)- see TCV375
PYRIDINE, 1,2,3,6-TETRAHYDRO-1-((2-METHYLCYCLOHEXYL)CARBONYL)-
    see TCV400
PYRIDINE, 3-(TETRAHYDRO-1-METHYLPYRROL-2-YL) see NDN000
2-PYRIDINETHIOL-1-OXIDE SODIUM SALT see MCQ750
2-PYRIDINETHIOL-1-OXIDE, ZINC SALT see ZMJ000
PYRIDINIUM ALDOXIME METHOCHLORIDE see FNZ000
PYRIDINIUM, 2-FORMYL-4'-METHYL-1,1'-(OXYDIMETHYLENE)DI-, DICHLO-
    RIDE, OXIME see MBY500
4,4'-(2-PYRIDINYLMETHYLENE)BISPHENOL BIS(HYDROGEN SULFATE)
    (ESTER) DISODIUM SALT see SJJ175
PYRIDO(3',2':5,6)CHRYSENE see BCQ000
PYRIDO(3',2':3,4)FLUORANTHENE see FDP000
5H-PYRIDO(4,3-b)INDOL-3-AMINE, 1-METHYL-1, MONOACETATE
    see ALE750
9H-PYRIDO(3,4-b)INDOLE see NNR300
1H-PYRIDO(4,3-b)INDOLE, 2,3,4,5-TETRAHYDRO-2,8-DIMETHYL-5-(2-(6-
    METHYL-3-PYRIDYL)ETHYL)- see TCQ260
PYRIDOSTIGMINE BROMIDE see MDL600
PYRIDOXAL HYDROCHLORIDE see VSU000
PYRIDOXAMIN-5-THIOACETAT HYDROBROMID (GERMAN) see ADD000
PYRIDOXIN-5'-DISULFID DIHYDROCHLORID HYDRAT (GERMAN)
    see BMB000
3-PYRIDOYL HYDRAZINE see NDU500
N-3-PYRIDOYLTRYPTAMINE see NDW525
PYRIDROL see DWK400
PYRIDROLE see DWK400
3-PYRIDYLAMINE see AMI250
4-PYRIDYLAMINE see AMI500
α-PYRIDYLAMINE see AMI000
N-(2-PYRIDYL)BENZYLAMINE HYDROCHLORIDE see BEA000
N¹-α-PYRIDYL-N¹-BENZYL-N,N-DIMETHYL ETHYLENEDIAMINE
    MONOHYDROCHLORIDE see POO750
1-(PYRIDYL-3-)-3,3-DIAETHYL-TRIAZEN (GERMAN) see DJY000
m-PYRIDYL-DIETHYL-TRIAZENE see DJY000
1-PYRIDYL-3,3-DIETHYLTRIAZENE see DJY000
1-(PYRIDYL-3)-3,3-DIETHYLTRIAZENE see DJY000
1-(3-PYRIDYL)-3,3-DIETHYLTRIAZENE see DJY000
N-α-PYRIDYL-N-p-METHOXYBENZYL-N,N'-DIMETHYLETHYLENEDIAMINE
    PHOSPHATE see NBV000
(2-PYRIDYLMETHYL)AMINE see ALB750
1-(2-(N-(2-PYRIDYLMETHYL)ANILINO)ETHYL)PIPERIDINE HYDROCHLO-
    RIDE see CNE375
4,4'-(2-PYRIDYLMETHYLENE)DIPHENOLBIS(HYDROGEN SULFATE)
    (ESTER) DISODIUM SALT see SJJ175
2-PYRIDYLMETHYL-2-(p-(2-METHYLPROPYL)PHENYL)PROPIONATE
    see IAB000
N-(2-PYRIDYLMETHYL)-N-PHENYL-N-2-(PIPERIDINOETHYL)AMINE HYDRO-
    CHLORIDE see CNE375
β-PYRIDYL-α-N-METHYLPYRROLIDINE see NDN000
1-(PYRIDYL-3-N-OXID)-3,3-DIMETHYL-TRIAZEN (GERMAN) see DTV200
1-(PYRIDYL-3-N-OXIDE)-3,3-DIMETHYLTRIAZENE see DTV200
2-(3-PYRIDYL)-PIPERIDINE see AON875
2-(3'-PYRIDYL) PIPERIDINE see AON875
(−)-2-(3'-PYRIDYL)PIPERIDINE see AON875
2-(3-PYRIDYL)PYRROLIDINE see NNR500
N¹-2-PYRIDYLSULFANILAMIDE see PPO000
N¹-2-PYRIDYLSULFANILAMIDE SODIUM SALT see PPO250
N-(α-PYRIDYL)-N-(α-THENYL)-N',N'-DIMETHYLETHYLENEDIAMINE
    see TEO250
N-(2-PYRIDYL)-N-(β-THENYL)-N',N'-DIMETHYLETHYLENEDIAMINE
    see DPJ200
N-(2-PYRIDYL)-N-(2-THIENYL)-N,N-DIMETHYL-ETHYLENEDIAMINE
    HYDROCHLORIDE see DPJ400
PYRILAMINE see WAK000
PYRILAMINE HYDROCHLORIDE see MCJ250
PYRILAMINE MALEATE see DBM800
PYRIMAL M see ALF250

PYRIMETHAMINE see MQR100
PYRIMIDINE-DEOXYRIBOSE N1-2'-FURANIDYL-5-FLUOROURACIL see
  FMB000
2,4-PYRIMIDINEDIOL see UNJ800
2,4-PYRIMIDINEDIONE see UNJ800
2,4(1H,3H)-PYRIMIDINEDIONE (9CI) see UNJ800
2,4(1H,3H)-PYRIMIDINEDIONE, 5-(HYDROXYMETHYL)-6-METHYL- (9CI) see
  HMH300
PYRIMIDINE PHOSPHATE see DIN800
2,4,5,6(1H,3H)-PYRIMIDINETETRONE see AFT750
2,4,5,6(1H,3H)-PYRIMIDINETETRONE HYDRATE see MDL500
2,4,5,6(1H,3H)-PYRIMIDINETETRONE 5-OXIME see AFU000
2,4,6(1H,3H,5H)-PYRIMIDINETRIONE, 1-BENZOYL-5-ETHYL-5-PHENYL-(9CI)
  see BDS300
2,4,6(1H,3H,5H)-PYRIMIDINETRIONE, 5-(2-BROMO-2-PROPENYL)-5-(1-
  METHYLPROPYL)-(9CI) see BOR000
2,4,6(1H,3H,5H)-PYRIMIDINETRIONE, 5,5-DIETHYL-, MONOSODIUM SALT
  (9CI) see BAG250
2,4,6(1H,3H,5H)-PYRIMIDINETRIONE, 5-ETHYL-5-(1-METHYLBUTYL)-, CAL-
  CIUM SALT (9CI) see CAV000
2,4,6(1H,3H,5H)-PYRIMIDINETRIONE, MONOSODIUM SALT (9CI) see
  MRK750
2(1H)-PYRIMIDINONE-4-AMINO-1-(4-AMINO-6-0-(3-AMINO-3-DEOXYβ-d-
  GLUCOPYRANOSYL)-4-DEOXY-d-GLYCERO-d-GALACTOβ-d-GLUCO-UN-
  DECAPYRANOSYL) see APF000
4(1H)-PYRIMIDINONE, 2-AMINO-5-BROMO-6-PHENYL- see AIY850
2,4,5,6-PYRIMIDINTETRON (CZECH) see AFT750
α-((2-PYRIMIDINYLAMINO)METHYL)BENZYL ALCOHOL HYDROCHLORIDE
  see FAR200
N^1-2-PYRIMIDINYL-SULFANILAMIDE see PPP500
N^1-2-PYRIMIDINYLSULFANILAMIDE MONOSODIUM SALT
  see MRM250
PYRIMIDONE MEDI-PETS see DBB200
1-(2-PYRIMIDYL)-4-PIPERONYLPIPERAZINE see TNR485
1-(2''-PYRIMIDYL)-4-PIPERONYL PIPERAZINE see TNR485
PYRIMIPHOS METHYL see DIN800
PYRIMOR see DOX600
PYRINAMINE BASE see TMP750
PYRINAZINE see HIM000
PYRINEX see CMA100
PYRINISTAB see TEO250
PYRINISTOL see TEO250
PYRISEPT see CCX000
PYRISTAN see SPC500
PYRITHEN see CHY250
PYRITHIONE see HOB500
PYRITHIONE ZINC see ZMJ000
PYRITHIOXIN see BMB000
PYRKAPPL see EPD500
PYRLEUGAN see DLH630
PYROACE see CFC000
PYROACETIC ACID see ABC750
PYROACETIC ETHER see ABC750
PYROBENZOL see BBL250
PYROBENZOLE see BBL250
PYROCARBONATE d'ETHYLE (FRENCH) see DIZ100
PYROCARBONIC ACID, DIETHYL ESTER see DIZ100
PYROCATECHIN see CCP850
PYROCATECHINIC ACID see CCP850
PYROCATECHOL see CCP850
PYROCATECHOL DIMETHYL ETHER see DOA200
PYROCATECHUIC ACID see CCP850
PYROCHOL see DAQ400
PYROD see UNJ800
PYRODINE see ACX750
PYRODONE see OES000
PYROGALLOL DIMETHYLETHER see DOJ200
PYROGALLOL-1,3-DIMETHYL ETHER see DOJ200
PYROGASTRONE see CBO500
PYROGENS see TGJ850
l-PYROGLUTAMYL-l-HISTIDYL-l-PROLINEAMIDE see TNX400
PYROGUAIAC ACID see GKI000
PYROKOHLENSAEURE DIAETHYL ESTER (GERMAN) see DIZ100
M-PYROL see MPF200
PYROLLNITRIN see CFC000
PYROLUSITE BROWN see MAS000
PYROMUCIC ACID see FQF000
PYROMUCIC ACID METHYL ESTER see MKH600
PYROMUCIC ALDEHYDE see FPQ875
PYRONALROT R see XRA000
PYRONIN B see DIS200
PYROPENTYLENE see CPU500
PYROPHOSPHATE see TEE500
PYROPHOSPHATE de TETRAETHYLE (FRENCH) see TCF250

PYROPHOSPHORIC ACID OCTAMETHYLTETRAAMIDE see OCM000
PYROPHOSPHORIC ACID, TETRAETHYL ESTER (dry mixture)
  see TCF270
PYROPHOSPHORIC ACID, TETRAETHYL ESTER (liquid mixture)
  see TCF280
PYROPHOSPHORIC ACID, TETRAMETHYL ESTER see TDV000
PYROPHOSPHORODITHIOIC ACID, TETRAETHYL ESTER
  see SOD100
PYROPHOSPHORODITHIOIC ACID-O,O,O,O-TETRAETHYL ESTER see
  SOD100
PYROPHOSPHORYLTETRAKISDIMETHYLAMIDE see OCM000
PYROSET FLAME RETARDANT TKP see TDH500
PYROSET TKO see TDI000
PYROSET TKP see TDH500
PYROSTIB see AQH500
PYROSULPHURIC ACID see SOI520
PYROTARTARIC ACID NITRILE see TLR500
PYROTONE RED TONER RA-5520 see CJD500
PYROTROPBLAU see CMO250
PYROXYLIC SPIRIT see MGB150
PYROXYLIN see CCU250
PYROXYLIN PLASTICS (DOT) see CCU250
PYROXYLIN PLASTIC SCRAP (DOT) see CCU250
PYRROCYCLINE N see SPE000
PYRROLAMIDOL see AFJ400
PYRROLAMIDOLUM see AFJ400
(2S-(2-α,7-β,7A-β,14-β,14a-α))-1H-PYRROLE-2-CARBOXYLIC ACID-
  DODECAHYDRO-11-OXO-7,14-METHANO-2H,6H-DIPYRIDO(1,2α:1',2'-
  e)(1,5)DIAZOCIN-2-YL) ESTER see CAZ125
PYRROLE-2,5-DIONE see MAM750
1,2-PYRROLIDINECARBOXYLIC ACID, 1-BENZYL-2-VINYL ESTER see
  CBR247
1,2-PYRROLIDINEDICARBOXYLIC ACID-1-BENZYL 2-(1,2-DIBROMOETHYL)
  ESTER see CBR245
2,5-PYRROLIDINEDIONE see SND000
PYRROLIDINEETHANOL, 3,4-DIMETHYL- see HKR550
PYRROLIDINE, 1-(p-(p-ETHYLPHENYL)AZO)PHENYL- see EPC950
PYRROLIDINE, 1-((4-METHOXY(1,1'-BIPHENYL)-3-YL)METHYL)-
  see MEF400
PYRROLIDINE, NITROSO- see NLP480
4-(2,1-PYRROLIDINOETHOXY)QUINOLINE HYDROCHLORIDE
  see QUJ300
3-PYRROLIDINO-N-METHYL-4-METHOXYBIPHENYL see MEF400
2-PYRROLIDINONEACETAMIDE see NNE400
2-PYRROLIDINONE, 1-CYCLOHEXYL- see CPQ275
2-PYRROLIDINONE, 1-DODECYL- see DXY750
2-PYRROLIDINONE, 1-ETHYL- see EPC700
2-PYRROLIDINONE, 1-ETHYL-4-(2-MORPHOLINOETHYL)-3,3-DIPHENYL- see
  ENL100
2-PYRROLIDINONE, 1-ETHYL-4-(2-(4-MORPHOLINYL)ETHYL)-3,3-DIPHENYL-
  (9CI) see ENL100
2-PYRROLIDINONE, 1-ISOPROPYL- see IRG100
2-PYRROLIDINONE, 1-(1-METHYLETHYL)-(9CI) see IRG100
2-PYRROLIDINONE, 1-OCTYL- see OFU100
2-PYRROLIDINONE, 1-TALLOW ALKYL derivs. see TAC200
1-(4-(1-PYRROLIDINYL)-2-BUTYNYL)-2-PYRROLIDINONE see OOY000
1-((1-PYRROLIDINYLCARBONYL)METHYL)-4-(3,4,5-
  TRIMETHOXYCINNAMOYL)PIPERAZINE MALEATE see VGK000
2-(PYRROLIDINYL)ETHYL-2-CHLORO-6-METHYLCARBANILATE HYDRO-
  CHLORIDE see CIK500
2-(2-(PYRROLIDINYL)ETHYL)-5-NITRO-1H-BENZ(de)ISOQUINOLINE-1,3(2H)-
  DIONE see MAB055
3-(2-PYRROLIDINYL)PYRIDINE see NNR500
3-(2-PYRROLIDINYL)-(s)-PYRIDINE see NNR500
trans-2-(3-(1-PYRROLIDINYL)-1-p-TOLYLPROPENYL)PYRIDINE
  MONOHYDROCHLORIDE see TMX775
4-(1-PYRROLIDINYL)-1-(2,4,6-TRIMETHOXYPHENYL)-1-BUTANONE HYDRO-
  CHLORIDE see BOM600
2-PYRROLIDONEACETAMIDE see NNE400
1-PYRROLINE-2-ACRYLAMIDE see COV125
3-PYRROLINE-2,5-DIONE see MAM750
PYRROLNITRIN see CFC000
(E)-1H-PYRROLO(2,1-C)(1,4)BENZODIAZEPINE-2-ACRYLAMIDE, 5,10,11,11a-
  TETRAHYDRO-9,11-DIHYDROXY-8-METHYL-5-OXO see API000
5H-PYRROLO(2,2-c)(1,4)BENZODIAZEPIN-5-ONE, 2-ETHYLIDENE-
  1,2,3,10,11,11A-HEXAHYDRO-8-HYDROXY-7,11-DIMETHOXY-
  see THG500
PYRROLYLENE see BOP500
PYRRO(b)MONAZOLE see IAL000
PYSOCOCCINE see SNM500
PY-TETRAHYDROSERPENTINE see AFG750
PZ 177 see CJW500
PZ 222 see CKD250
PZ 1511 see CCK780

Q-137 see DJC000
QA 71 see PNW800
QB-1 see CMX920
QCB see PAV500
QDO see DVR200
QGH see BDD000
QIDAMP see AIV500
QIDMYCIN see EDJ500
QIDPEN G see BFD000
QIDPEN VK see PDT750
QIDTET see TBX250
QIKRON see BIN000
QINGHAOSU (CHINESE) see ARL375
QING HAU SAU (CHINESE) see ARL375
QM 657 see DGW450
QM-1143 see MOF750
QM-1148 see MLR500
QM-1149 see MOG000
QM-6008 see BAV625
QO THFA see TCT000
QPB see PJA000
QSAH 7 see PKQ059
QUAALUDE see QAK000
QUADRACYCLINE see TBX250
QUADRATIC ACID see DMJ600
QUADROL see QAT000
QUADROSILAN see CMS241
QUAMAQUIL see PGA750
QUAMONIUM see HCQ500
QUANTALAN see CME400
QUANTRIL see BCL250
QUANTROVANIL see EQF000
QUANTRYL see BCL250
QUARTZ see SCJ500
QUARTZ GLASS see SCK600
QUATERNARIO CPC see CCX000
QUATERNARY AMMONIUM COMPOUNDS, ALKYLBENZYLDIMETHYL,
   CHLORIDES see AFP250
QUATERNIUM-12 see DGX200
QUATERNIUM 15 see CEG550
QUATERNOL 1 see DTC600
QUATRACHLOR see BEN000
QUATRESIN see MSB500
QUATREX see TBX250
QUAZO PURO (ITALIAN) see SCJ500
QUEBRACHIN see YBJ000
QUEBRACHINE see YBJ000
QUEBRACHO TANNIN see QBJ000
QUECKSILBER (GERMAN) see MCW250
QUECKSILBER CHLORID (GERMAN) see MCY475
QUECKSILBER(I)-CHLORID (GERMAN) see MCW000
QUECKSILBER CHLORUER (GERMAN) see MCW000
QUECKSILBEROXID (GERMAN) see MCT500, MDF750
QUELAMYCIN see QBS000
QUELETOX see FAQ900
QUELICIN see CMG250, HLC500
QUELICIN CHLORIDE see HLC500
QUELLADA see BBQ500
QUEMICETINA see CDP250
QUEN see CFZ000
QUERCETIN see QCA000
QUERCETIN, 3-(6-DEOXY-α-l-MANNOPYRANOSIDE) see QCJ000
QUERCETIN-3-(6-o-(6-DEOXY-β-l-MANNOPYRANOSYL-β-d-GLUCOPYRANOS-
   IDE) see RSU000
QUERCETIN DIHYDRATE see QCA175
QUERCETINE see QCA000
QUERCETIN RHAMNOGLUCOSIDE see RSU000
QUERCETIN-3-RHAMNOGLUCOSE see RSU000
QUERCETIN-3-(6-o-α-l-RHAMNOPYRANOSYL-β-d-GLUCOPYRANOSIDE)
   see RSU000
QUERCETIN-3-l-RHAMNOSIDE see QCJ000
QUERCETIN-3-RUTINOSIDE see RSU000
QUERCETOL see QCA000
QUERCITIN see QCA000
QUERCITRIN see QCJ000
QUERCUS AEGILOPS L. TANNIN see VCK000
QUERCUS FALCATA PAGODAEFOLIA see CDL750
QUERTINE see QCA000
QUERYL see CNR125
QUESTEX 4 see EIV000
QUESTEX 4H see EIX000
QUESTIOMYCIN A see QCJ275
QUESTRAN see CME400
QUETIMID see TEH500

QUETINIL see DCK759
QUIACTIN see OLU000
QUIADON see MCH535
QUIATRIL see DCK759
QUICK see CJJ000
QUICKLIME (DOT) see CAU500
QUICKSAN see ABU500
QUICKSET EXTRA see MKA500
QUICK SILVER see MCW250
QUIDE see PII500
QUIETAL see QCS000
QUIETALUM see QCS000
QUIETIDON see MQU750
QUIETOPLEX see TEH500
QUI-LEA see QFA250
QUILIBREX see CFZ000
QUILONUM RETARD see LGZ000
QUIMAR see CML850
QUIMOTRASE see CML850
QUINACETOPHENONE see DMG600
QUINACHLOR see CLD000
QUINACRINE see ARQ250
QUINACRINE DIHYDROCHLORIDE see CFU750
QUINACRINE ETHYL M/2 see QCS875
QUINACRINE ETHYL MUSTARD see DFH000
QUINACRINE HYDROCHLORIDE see CFU750
QUINACRINE MUSTARD see QDJ000, QDS000
QUINACRINE MUSTARD DIHYDROCHLORIDE see QDS000
QUINAGAMINE see CLD000
QUINAGLUTE see QDS225
QUINALBARBITAL see SBM500
QUINALBARBITONE see SBM500
QUINALBARBITONE SODIUM see SBN000
QUINALDIC ACID see QEA000
QUINALDINE see QEJ000
QUINALDINIC ACID see QEA000
QUINALIZARIN see TDD000
QUINALPHOS see DJY200
QUINALSPAN see SBN000
QUINAMBICIDE see CHR500
QUINAPYRAMINE see AQN625
QUINAPYRAMINE METHYL SUFLATE see AQN625
QUINAZINE see QRJ000
QUINAZOLINEDIONE see QEJ800
QUINAZOLINE-2,4-DIONE see QEJ800, QEJ800
(1H,3H)QUINAZOLINE DIONE-2,4 see QEJ800
2,4(1H,3H)-QUINAZOLINEDIONE see QEJ800
4-QUINAZOLINONE see QFA000
4(3H)-QUINAZOLINONE see QFA000
10-(3-QUINCULIDINYLMETHYL)PHENOTHIAZINE see MCJ400
QUINDOXIN see QSA000
QUINERCYL see CLD000
QUINESTROL see QFA250
QUINEX see PFN250
QUINEXIN see SAN600
QUINGESTANOL ACETATE see QFA275
QUINGFENGMYCIN see ARP000
QUINHYDRONE see QFJ000
QUINICARDINE see QFS000, QHA000
QUINICINE OXALATE see QFJ300
QUINIDATE see QHA000
QUINIDEX see QFS000, QHA000
QUINIDINE see QFS000
(+)-QUINIDINE see QFS000
QUINIDINE GLUCONATE see QDS225
QUINIDINE-d-GLUCONATE (salt) see QDS225
QUINIDINE MONO-d-GLUCONATE (salt) see QDS225
QUINIDINE POLYGALACTURONATE see GAX000
QUINIDINE SULFATE (2581) (salt) see QHA000
QUINILON see CLD000
QUININE see QHJ000, QHJ000
β-QUININE see QFS000
QUININE BIMURIATE see QIJ000
QUININE BISULFATE see QMA000
QUININE CHLORIDE see QJS000
QUININE DIHYDROCHLORIDE see QIJ000, QIJ000
QUININE ETHIODIDE see QIS000
QUININE FORMATE see QIS300
QUININE, FORMATE (SALT) see QIS300
QUININE HYDROBROMIDE see QJJ100
QUININE HYDROCHLORIDE see QJS000
QUININE HYDROGEN SULFATE see QMA000
QUININE MONOHYDROCHLORIDE see QJS000
QUININE MURIATE see QJS000

QUININE SULFATE see QMA000
QUINIPHEN see IEP200
QUINITEX see QHA000
QUINIZARIN see DMH000
QUINIZARINE GREEN BASE see BLK000
QUINOFEN see PGG000
QUINOFORM see QIS300
8-QUINOL see QPA000
β-QUINOL see HIH000
QUINOL DIMETHYL ETHER see DOA400
4-QUINOLINAMINE, N-HYDROXY-N-METHYL-, 1-OXIDE see HLX550
QUINOLINE see QMJ000
QUINOLINE-2-ALDEHYDECHLOROMETHYLATE-(p-(BISβ-CHLOROETHYL)-AMINO)ANIL) see IAK100
QUINOLINE-6-AZO-p-DIMETHYLANILINE see DPR000
6-QUINOLINE CARBONYL AZIDE see QMS000
QUINOLINE-2-CARBOXYLIC ACID see QEA000, QEA000
2,4-QUINOLINEDIOL see QNA000
2-QUINOLINEETHANOL see QNJ000
5-QUINOLINESULFONIC ACID, 8,8'-((HYDROXYSTIBYLENE)BIS(OXY))BIS(7-FORMYL-, DISODIUM SALT see AQE320
2-QUINOLINE THIOACETAMIDE HYDROCHLORIDE see QOJ000
QUINOLINIC ACID IMIDE see POQ750
QUINOLINIUM, 6-AMINO-1-METHYL-4-((4-((((1-METHYLPYRIDINIUM-4-YL)AMINO)PHENYL)CARBAMOYL)PHENYL)AMINO)-, DIBROMIDE see QOJ250
QUINOLINIUM DIBROMIDE see QOJ250
2-QUINOLINOL see CCC250
8-QUINOLINOL see QPA000
8-(QUINOLINOLATO)METHYL MERCURY see MLH000
(8-QUINOLINOLATO)TRIBUTYLSTANNANE see TIB000
8-QUINOLINOL HYDROGEN SULFATE (2:1) see QPS000
8-QUINOLINOLIUM-4',7'-DIBROMO-3'-HYDROXY-2'-NAPHTHOATE see QPJ000
8-QUINOLINOL, MERCURY COMPLEX see MLH000
8-QUINOLINOL SULFATE see QPS000
8-QUINOLINOL SULFATE (2581) (SALT) see QPS000
QUINOLOR COMPOUND see BDS000
N-(4-QUINOLYL)ACETOHYDROXAMIC ACID see QQA000
N-(4-QUINOLYL)HYDROXYLAMINE-1'-OXIDE see HIY500
N-(8-QUINOLYLMETHYL)-N-METHYL-2-PROPYLAMINE see IPZ500
α-4-QUINOLYL-5-VINYL-2-QUINUCLIDINEMETHANOL see CMP925
QUINOMETHIONATE see ORU000
QUINOMYCIN A see EAD500
QUINOMYCIN C see QQS075
QUINONDO see BLC250
QUINONE see QQS200
o-QUINONE see BDC250
p-QUINONE see QQS200
QUINONE DIOXIME see DVR200
p-QUINONE DIOXIME see DVR200
QUINONE MONOXIME see NLF200
QUINONE OXIME see NLF200
p-QUINONE OXIME see DVR200
QUINONE OXIME BENZOYLHYDRAZONE see BDD000
QUINON I see BGW750
p-QUINONIMINE see BDD500
QUINOPHEN see PGG000
QUINOPHENOL see QPA000
QUINORA see QHA000
QUINOSCAN see CLD000
QUINOSEPTYL see AKO500
QUINOTILINE see KGK400
QUINOXALINE see QRJ000
2,3-QUINOXALINEDIOL see QRS000
QUINOXALINE DIOXIDE see QSA000
QUINOXALINE DI-N-OXIDE see QSA000
QUINOXALINE 1,4-DIOXIDE see QSA000
QUINOXALINE-1,4-DI-N-OXIDE see QSA000
2,3-QUINOXALINEDITHIOL see QSJ000
2,3-QUINOXALINETHIOL see QSJ000
(2,3-QUINOXALINYLDITHIO)DIMETHYLTIN see QSJ800
(2,3-QUINOXALINYLDITHIO)DIPHENYL STANNANE see QTJ000
(2,3-QUINOXALINYLDITHIO)DIPHENYLTIN see QTJ000
(2-QUINOXALINYLMETHYLENE)-HYDRAZINECARBOXYLIC ACID METHYL ESTER-N,N'-DIOXIDE see FOI000
N-(2-QUINOXALINYL)SULFANILAMIDE see QTS000
N¹-2-QUINOXALINYLSULFANILAMIDE see QTS000
N'-2-QUINOXALYLSULFANILAMIDE see QTS000
QUINOXYL see IEP200
QUINPYRROLIDINE see QUJ300
QUINSORB 010 see DMI600
QUINTAR see DFT000
QUINTAR 540F see DFT000

QUINTESS-N see NCG000
QUINTOCENE see PAX200
QUINTOMYCIN C see NCF500
QUINTOMYCIN D see DAS600
QUINTOX see DHB400
QUINTOZEN see PAX000
QUINTOZENE see PAX000
QUINTRATE see PBC250
3-QUINUCLIDINOL ACETATE see AAE250
3-QUINUCLIDINOL ACETATE (ESTER) HYDROCHLORIDE see QUS000
3-QUINUCLIDINOL BENZILATE see QVA000
3-QUINUCLIDINOL, DIPHENYLACETATE (ESTER), HYDROGEN SULFATE (2:1) DIHYDRATE see QVJ000
QUINUCLIDYL BENZILATE see QWJ000
QUINUCLIDYL-3-DIPHENYLCARBINOL HYDROCHLORIDE see DWM400
QUINURONIUM SULFATE see PJA120
QUIPAZINE MALEATE see QWJ500
QUIPENYL see RHZ000
QUIRVIL see PKQ059
QUOLAC EX-UB see SIB600
QUOTANE see DNX400
QUOTANE HYDROCHLORIDE see DNX400
QYSA see PKQ059
QZ 2 see QAK000

R 2 see BCM250
R 5 see DDT300
R 10 see CBY000
R-11 see BHJ500
R 13 see CLR250
R 14 see CBY250, PKL750
R 21 see BGR750
R 23 see CBY750
R 31 see CHI900
R-47 see TNF500
R48 see BIF250
R50 see DAD200
R-52 see MAW800
R 54 see DDP600
R 79 see DAB875
88-R see SOP500
R 113 see FOO000
R 114 see FOO509
R161 see FIB000
R-242 see CKI625
R-246 see TND500
R 261 see DBA200
R-326 see EAU500
R 400 see NCF500
R 516 see CMR100
R-548 see TIH800
R 661 see BTA325
R 694 see MEK700
R-874 see OGA000
R 875 see AFJ400
R 951 see CBP325
R 1067 see CPN750
R-1303 see TNP250
R-1492 see MQH750
R 1504 see PHX250
R 1513 see EKN000
R 1575 see CMR100
R-1607 see PNI750
R-1608 see EIN500
R 1625 see CLY500
R 1658 see MNN250
R 1892 see FLL000
R-1910 see EID500
R 1929 see FLU000
R 2010 see NNR125
R 2028 see HAF400
R-2061 see PNF500
R 2063 see EHT500
R 2113 see DBA875
R 2167 see HAF400
R 2170 see DAP000
R 2267 see AGW675
R-2323 see ENX575
R-2498 see TKK500, TKK750
R 2858 see MRU600
R 3345 see FHG000
R 3365 see PJA140
R-3588 see CNR125
R 40B1 see MHR200

R 4263 see PDW500, PDW750
R 4318 see TKG000
R-4461 see DNO800
R-4572 see EKO500
R 4584 see FGU000, FLK100
R 4749 see DYF200
R-4845 see BCE825
R 4929 see BBU800
R-5,158 see DJA400
R 5240 see PDW750
R-5461 see BJD000
R 6238 see PIH000
R6597 see NHQ950
R 6700 see OAN000
R6998 see MFB410
R7000 see MFB400
R 7158 see FLJ000
R 7372 see MEW775
R 8284 see PMF550
R 8299 see TDX750
R 9298 see CLS250
R 9985 see MDV500
R 11333 see BNU725
R-12,564 see LFA020
R 12 (DOT) see DFA600
R 14,827 see EAE000
R 14889 see MQS560
R 14950 see FDD080
R 15,175 see ASD000
R 16341 see PAP250
R 16470 see DBE000
R 16659 see HOU100
R 17635 see MHL000
R 17899 see FDA887
R 17934 see OJD100
R 18134 see MQS550
R 22 (DOT) see CFX500
R 23979 see FPB875
R 25061 see TEN750
R-25788 see DBJ600
R-28627 see DNT200
R 31520 see CFC100
R 35443 see OMG000
R 38486 see HKB700
R 51619 see CMS237
R 151885 see BJW600
R 14 (refrigerant) see CBY250
R 20 (refrigerant) see CHJ500
R 31 (refrigerant) see CHI900
RA 8 see PCP250
β-RA see VSK950
13-RA see VSK955
R 133a see TJY175
RA 1226 see ALC250
RABANO (PUERTO RICO) see DHB309
RABBIT ANTISERUM TO OVINE LUTEINIZING HORMONE see LIU350
RABBIT-BELLS see RBZ400
RABBIT FLOWER see FOM100
RABON see RAF100
RABOND see RAF100
RACCOON BERRY see MBU800
RACEMETHORPHAN HYDROBROMIDE see RAF300
RACEMOMYCIN A see RAG300
RACEMORPHAN see MKR250
RACEMORPHAN HYDROBROMIDE see MDV250
RACEPHEDRINE HYDROCHLORIDE see EAX500
RACEPHEN see AOB250, AOB500
RACUMIN see EAT600
RACUSAN see DSP400
RADAPON see DGI400, DGI600
RADAZIN see ARQ725
RADDLE see IHD000
RAD-E CATE 16 see HKC500
RAD-E-CATE 25 see HKC000
RADEDORM see DLY000
RADEPUR see LFK000
RADIASURF 7125 see SKV000
RADIASURF 7155 see SKV100
RADIATION see RAQ000
RADIATION, IONIZING see RAQ010
RADICALISIN see RRA000
RADIOCIN see SPD500
RADIOGRAPHOL see SHX000
RADIOL GERMICIDAL SOLUTION see EKN500

RADIOSTOL see VSZ100
RADIOTETRANE see TDE750
RADIUM see RAV000
RADIUM F see PJJ750
RADIZINE see ARQ725
RADOCON see BJP000
RADOKOR see BJP000
RADON see RBA000
RADONIL see SNK000
RADONIN see SNN300
RADONNA see CHJ750
RADOSAN see MEO750
RADOX see CFK000
RADOXONE TL see AMY050
RADSTERIN see VSZ100
RAFEX see DUS700
RAFLUOR see SHF500
RAGADAN see HBK700
RAGUAR (GUAM) see TOA275
RAGWORT see RBA400
RAIN LILY see RBA500
RAISIN de COULEUVRE (CANADA) see MRN100
RAKUTO AMARANTH see FAG020
RALABOL see RBF100
RALGRO see RBF100
RALONE see RBF100
RAM-326 see ORE000
RAMETIN see HMV000
RAMIK see DVV600
RAMIZOL see AMY050
RAMOR see TEI000
R/AMP see RKP000
RAMPART see PGS000
RAMROD see CHS500
RAM'S CLAWS see FBS100
RAMUCIDE see CJJ000
RAMYCIN see FQU000
RANAC see CJJ000
RANDOLECTIL see BSZ000
RANDONOS see CMG675
RANDOX see CFK000
RANEY ALLOY see NCW500
RANEY COPPER see CNI000
RANEY NICKEL see NCW500
RANGOON YELLOW see DEU000
RANIDIL see RBF400
RANITIDINE HYDROCHLORIDE see RBF400
RANITOL see AFG750
RANKOTEX see CIR500
RANTOX T see CFK000
RANTUDIL see AAE625
RANUNCULUS ACRIS see FBS100
RANUNCULUS BULBOSUS see FBS100
RANUNCULUS SCELERATUS see FBS100
RAPACODIN see DKW800
RAPAMYCIN see RBK000
RAPHATOX see DUS700
RAPHETAMINE see BBK000
RAPHETAMINE PHOSPHATE see AOB500
RAPHONE see CIR250
RAPID see DOX600
RAPIDOSEPT see DET000
RAPISOL see DJL000
RAPYNOGEN see ICC000
RARE EARTHS see RBP000
RAS-26 see NIJ500
RASCHIT see CFE250
RASEN-ANICON see CFE250
RASIKAL see SFS000
RASPBERIN see SAH000
RASPBERRY KETONE see RBU000
RASPBERRY RED for JELLIES see FAG020
RASTINON see BSQ000
RATAFIN see ABF500
RATAK see TAC800
RATAK PLUS see TAC800
RAT-A-WAY see ABF500, WAT200
RATBANE 1080 see SHG500
RAT-B-GON see WAT200
RAT-O-CIDE RAT BAIT see RCF000
RAT-GARD see WAT200
RATICATE see NNF000
RATIMUS see BMN000
RATINDAN 1 see DVV600

RAT & MICE BAIT see WAT200
RAT-NIP see PHO740
RATO see TMO000
RATOMET see CJJ000
RATON see RBZ000
RATOX see TEL750
RAT'S END see RCF000
RATS-NO-MORE see WAT200
RATSUL SOLUBLE see WAT220
RATTENGIFTKONSERVE see TEM000
RATTEX see CNV000
RATTLE BOX see RBZ400
RATTLESNAKE WEED see FAB100
RATTLEWEED (JAMAICA) see RBZ400
RATTRACK see AQN635
RAUBASINE see AFG750
RAUBASNE HYDROCHLORIDE see AFH000
RAUCUMIN 57 see EAT600
RAUGALLINE see AFH250
RAUMALINA see AFG750
RAUMANON see ILD000
RAUNORINE see RDF000
RAUNOVA see RCA200
RAUPYROL see TLN500
RAURESCINE see TLN500
RAUSERPIN see RDK000
RAUTRAX see RCA275
RAUWOLEAF see RDK000
RAUWOLFIN see AFH250
RAUWOLFINE see AFH250
RAUWOLSCINE HYDROCHLORIDE see YCA000
RAVIAC see CJJ000
RAVINYL see PKQ059
RAVONA see CAV000
RAVONAL see PBT500
RAWETIN see HMV000
RAW SHALE OIL see COD750
RAYBAR see BAP000
RAY-GLUCIRON see FBK000
RAYOX see TGG760
RAZIOSULFA see AIF000
RAZOL DOCK KILLER see CIR250
RAZOXANE see RCA375
RAZOXIN see PIK250, RCA375
RB see RJF400
R-242-B see CKI625
R-451-B see ERD000
RB 1489 see PIZ000
RB 1509 see CGV250
RBA 777 see CLW000
R-BASE see CBN100
R 74 BASE see LJD500
1-RBO-12-A see PNA200
RC 146 see CNG250
R-C 318 see CPS000
RC 5629 see DRR800
RC 61-91 see IAG600
RC 61-96 see BEU800
RC 72-01 see RCA435
RC 72-02 see RCA450
R.C. 27-109 see DGQ500
RC 30-109 see DGQ500
RCA WASTE NUMBER P105 see SFA000
RCA WASTE NUMBER U203 see SAD000
RCA WASTE NUMBER U205 see SBR000
RC 172DBM see AHE250
RC COMONOMER DBM see DED600
RC COMONOMER DOF see DVK600
RC COMONOMER DOM see BJR000
R-CHLORODIPHENYL SULPHONE see CKI625
R 30 730 CITRATE SALT see SNH150
RC PLASTICIZER DOP see DVL700
RCRA WASTE NUMBER 7066 see DDL800
RCRA WASTE NUMBER P001 see WAT200
RCRA WASTE NUMBER P002 see ADD250
RCRA WASTE NUMBER P003 see ADR000
RCRA WASTE NUMBER P004 see AFK250
RCRA WASTE NUMBER P005 see AFV500
RCRA WASTE NUMBER P006 see AHE750
RCRA WASTE NUMBER P007 see AKT750
RCRA WASTE NUMBER P008 see AMI500, POO050
RCRA WASTE NUMBER P010 see ARB250
RCRA WASTE NUMBER P011 see ARH500
RCRA WASTE NUMBER P012 see ARI750

RCRA WASTE NUMBER P013 see BAK750
RCRA WASTE NUMBER P014 see PFL850
RCRA WASTE NUMBER P015 see BFO750
RCRA WASTE NUMBER P016 see BIK000
RCRA WASTE NUMBER P017 see BNZ000
RCRA WASTE NUMBER P018 see BOL750
RCRA WASTE NUMBER P020 see BRE500
RCRA WASTE NUMBER P021 see CAQ500
RCRA WASTE NUMBER P022 see CBV500
RCRA WASTE NUMBER P023 see CDY500
RCRA WASTE NUMBER P024 see CEH680
RCRA WASTE NUMBER P026 see CKL000
RCRA WASTE NUMBER P027 see CKT250
RCRA WASTE NUMBER P028 see BEE375
RCRA WASTE NUMBER P029 see CNL000
RCRA WASTE NUMBER P030 see COI500
RCRA WASTE NUMBER P031 see COO000
RCRA WASTE NUMBER P033 see COO750
RCRA WASTE NUMBER P034 see CPK500
RCRA WASTE NUMBER P036 see DGB600
RCRA WASTE NUMBER P037 see DHB400
RCRA WASTE NUMBER P039 see DXH325
RCRA WASTE NUMBER P040 see DJX400
RCRA WASTE NUMBER P042 see VGP000
RCRA WASTE NUMBER P043 see IRF000
RCRA WASTE NUMBER P044 see DSP400
RCRA WASTE NUMBER P045 see DAB400
RCRA WASTE NUMBER P046 see DTJ400
RCRA WASTE NUMBER P047 see DUS700
RCRA WASTE NUMBER P048 see DUZ000
RCRA WASTE NUMBER P049 see DXL800
RCRA WASTE NUMBER P050 see EAQ750
RCRA WASTE NUMBER P051 see EAT500
RCRA WASTE NUMBER P054 see EJM900
RCRA WASTE NUMBER P056 see FEZ000
RCRA WASTE NUMBER P057 see FFF000
RCRA WASTE NUMBER P058 see SHG500
RCRA WASTE NUMBER P059 see HAR000
RCRA WASTE NUMBER P060 see IKO000
RCRA WASTE NUMBER P062 see HCY000, TEB000
RCRA WASTE NUMBER P063 see HHS000
RCRA WASTE NUMBER P064 see MKX250
RCRA WASTE NUMBER P065 see MDC000, MDC250
RCRA WASTE NUMBER P066 see MDU600
RCRA WASTE NUMBER P067 see PNL400
RCRA WASTE NUMBER P068 see MKN000
RCRA WASTE NUMBER P069 see MLC750
RCRA WASTE NUMBER P070 see CBM500
RCRA WASTE NUMBER P071 see MNH000
RCRA WASTE NUMBER P072 see AQN635
RCRA WASTE NUMBER P073 see NCZ000
RCRA WASTE NUMBER P074 see NDB500
RCRA WASTE NUMBER P075 see NDN000
RCRA WASTE NUMBER P076 see NEG100
RCRA WASTE NUMBER P077 see NEO500
RCRA WASTE NUMBER P078 see NGR500
RCRA WASTE NUMBER P081 see NGY000, SLD000
RCRA WASTE NUMBER P082 see NKA600
RCRA WASTE NUMBER P084 see NKY000
RCRA WASTE NUMBER P085 see OCM000
RCRA WASTE NUMBER P087 see OKK000
RCRA WASTE NUMBER P088 see DXD000, EAR000
RCRA WASTE NUMBER P089 see PAK000
RCRA WASTE NUMBER P092 see ABU500
RCRA WASTE NUMBER P093 see PGN250
RCRA WASTE NUMBER P094 see PGS000
RCRA WASTE NUMBER P095 see PGX000
RCRA WASTE NUMBER P096 see PGY000
RCRA WASTE NUMBER P097 see FAB600
RCRA WASTE NUMBER P098 see PLC500
RCRA WASTE NUMBER P099 see PLS250
RCRA WASTE NUMBER P101 see PMV750
RCRA WASTE NUMBER P102 see PMN450
RCRA WASTE NUMBER P103 see SBV000
RCRA WASTE NUMBER P104 see SDP000
RCRA WASTE NUMBER P106 see SGA500
RCRA WASTE NUMBER P108 see SMN500
RCRA WASTE NUMBER P109 see SOD100
RCRA WASTE NUMBER P110 see TCF000
RCRA WASTE NUMBER P111 see TCF250
RCRA WASTE NUMBER P112 see TDY250
RCRA WASTE NUMBER P113 see TEL050
RCRA WASTE NUMBER P114 see TEL500
RCRA WASTE NUMBER P115 see TEM000

RCRA WASTE NUMBER P116 see TFQ000
RCRA WASTE NUMBER P118 see PCF300
RCRA WASTE NUMBER P119 see ANY250
RCRA WASTE NUMBER P120 see VDU000
RCRA WASTE NUMBER P121 see ZGA000
RCRA WASTE NUMBER P122 see ZLS000
RCRA WASTE NUMBER P123 see CDV100
RCRA WASTE NUMBER P401 see NIM500
RCRA WASTE NUMBER P404 see EPC500
RCRA WASTE NUMBER U001 see AAG250
RCRA WASTE NUMBER U002 see ABC750
RCRA WASTE NUMBER U003 see ABE500
RCRA WASTE NUMBER U005 see FDR000
RCRA WASTE NUMBER U006 see ACF750
RCRA WASTE NUMBER U007 see ADS250
RCRA WASTE NUMBER U008 see ADS750
RCRA WASTE NUMBER U009 see ADX500
RCRA WASTE NUMBER U010 see AHK500
RCRA WASTE NUMBER U011 see AMY050
RCRA WASTE NUMBER U014 see IBB000
RCRA WASTE NUMBER U015 see ASA500
RCRA WASTE NUMBER U016 see BAW750
RCRA WASTE NUMBER U017 see BAY300
RCRA WASTE NUMBER U018 see BBC250
RCRA WASTE NUMBER U019 see BBL250
RCRA WASTE NUMBER U020 see BBS750
RCRA WASTE NUMBER U021 see BBX000
RCRA WASTE NUMBER U022 see BCS750
RCRA WASTE NUMBER U023 see BFL250
RCRA WASTE NUMBER U024 see BID750
RCRA WASTE NUMBER U025 see DFJ050
RCRA WASTE NUMBER U026 see BIF250
RCRA WASTE NUMBER U027 see BII250
RCRA WASTE NUMBER U028 see DVL700
RCRA WASTE NUMBER U029 see MHR200
RCRA WASTE NUMBER U031 see BPW500
RCRA WASTE NUMBER U032 see CAP500
RCRA WASTE NUMBER U033 see CCA500
RCRA WASTE NUMBER U035 see CDO500
RCRA WASTE NUMBER U036 see CDR750
RCRA WASTE NUMBER U037 see CEJ125
RCRA WASTE NUMBER U038 see DER000
RCRA WASTE NUMBER U039 see CFE250
RCRA WASTE NUMBER U041 see EAZ500
RCRA WASTE NUMBER U042 see CHI250
RCRA WASTE NUMBER U043 see VNP000
RCRA WASTE NUMBER U044 see CHJ500
RCRA WASTE NUMBER U045 see MIF765
RCRA WASTE NUMBER U046 see CIO250
RCRA WASTE NUMBER U047 see CJA000
RCRA WASTE NUMBER U048 see CJK250
RCRA WASTE NUMBER U049 see CLK235
RCRA WASTE NUMBER U050 see CML810
RCRA WASTE NUMBER U051 see BAT850, CMY825
RCRA WASTE NUMBER U052 see CNW500, CNW750, CNX000, CNX250
RCRA WASTE NUMBER U053 see COB250, COB260, POD000
RCRA WASTE NUMBER U055 see COE750
RCRA WASTE NUMBER U056 see CPB000
RCRA WASTE NUMBER U057 see CPC000
RCRA WASTE NUMBER U058 see CQC650
RCRA WASTE NUMBER U059 see DAC000
RCRA WASTE NUMBER U060 see BIM500
RCRA WASTE NUMBER U061 see DAD200
RCRA WASTE NUMBER U062 see DBI200
RCRA WASTE NUMBER U063 see DCT400
RCRA WASTE NUMBER U064 see BCQ500
RCRA WASTE NUMBER U067 see EIY500
RCRA WASTE NUMBER U068 see DDP800
RCRA WASTE NUMBER U069 see DEH200
RCRA WASTE NUMBER U070 see DEP600, DEP800
RCRA WASTE NUMBER U071 see DEP800
RCRA WASTE NUMBER U072 see DEP800
RCRA WASTE NUMBER U073 see DEQ600
RCRA WASTE NUMBER U074 see DEV000
RCRA WASTE NUMBER U075 see DFA600
RCRA WASTE NUMBER U076 see DFF809
RCRA WASTE NUMBER U077 see EIY600
RCRA WASTE NUMBER U078 see VPK000
RCRA WASTE NUMBER U079 see ACK000
RCRA WASTE NUMBER U080 see MJP450
RCRA WASTE NUMBER U081 see DFX800
RCRA WASTE NUMBER U082 see DFY000
RCRA WASTE NUMBER U083 see DGG800, PNJ400

RCRA WASTE NUMBER U084 see DGG950
RCRA WASTE NUMBER U085 see BGA750
RCRA WASTE NUMBER U086 see DJL400
RCRA WASTE NUMBER U088 see DJX000
RCRA WASTE NUMBER U089 see DKA600
RCRA WASTE NUMBER U090 see DMD000
RCRA WASTE NUMBER U091 see DCJ200
RCRA WASTE NUMBER U092 see DOQ800
RCRA WASTE NUMBER U093 see DOT300
RCRA WASTE NUMBER U094 see DQJ200
RCRA WASTE NUMBER U095 see TGJ750
RCRA WASTE NUMBER U096 see IOB000
RCRA WASTE NUMBER U097 see DQY950
RCRA WASTE NUMBER U098 see DSF400
RCRA WASTE NUMBER U099 see DSF600
RCRA WASTE NUMBER U101 see XKJ500
RCRA WASTE NUMBER U102 see DTR200
RCRA WASTE NUMBER U103 see DUD100
RCRA WASTE NUMBER U105 see DVH000
RCRA WASTE NUMBER U106 see DVH400
RCRA WASTE NUMBER U107 see DVL600
RCRA WASTE NUMBER U108 see DVQ000
RCRA WASTE NUMBER U109 see HHG000
RCRA WASTE NUMBER U110 see DWR000
RCRA WASTE NUMBER U111 see NKB700
RCRA WASTE NUMBER U112 see EFR000
RCRA WASTE NUMBER U113 see EFT000
RCRA WASTE NUMBER U115 see EJN500
RCRA WASTE NUMBER U116 see IAQ000
RCRA WASTE NUMBER U117 see EJU000
RCRA WASTE NUMBER U118 see EMF000
RCRA WASTE NUMBER U119 see EMF500
RCRA WASTE NUMBER U120 see FDF000
RCRA WASTE NUMBER U121 see TIP500
RCRA WASTE NUMBER U122 see FMV000
RCRA WASTE NUMBER U123 see FNA000
RCRA WASTE NUMBER U124 see FPK000
RCRA WASTE NUMBER U125 see FPQ875
RCRA WASTE NUMBER U126 see GGW000
RCRA WASTE NUMBER U127 see HCC500
RCRA WASTE NUMBER U128 see HCD250
RCRA WASTE NUMBER U129 see BBQ500
RCRA WASTE NUMBER U130 see HCE500
RCRA WASTE NUMBER U131 see HCI000
RCRA WASTE NUMBER U132 see HCL000
RCRA WASTE NUMBER U133 see HGS000
RCRA WASTE NUMBER U134 see HHU000
RCRA WASTE NUMBER U135 see HIC500
RCRA WASTE NUMBER U136 see HKC000
RCRA WASTE NUMBER U137 see IBZ000
RCRA WASTE NUMBER U138 see MKW200
RCRA WASTE NUMBER U139 see IGS000
RCRA WASTE NUMBER U140 see IIL000
RCRA WASTE NUMBER U141 see IRZ000
RCRA WASTE NUMBER U142 see KEA000
RCRA WASTE NUMBER U143 see LBG000
RCRA WASTE NUMBER U144 see LCV000
RCRA WASTE NUMBER U146 see LCH000
RCRA WASTE NUMBER U147 see MAM000
RCRA WASTE NUMBER U149 see MAO250
RCRA WASTE NUMBER U150 see PED750
RCRA WASTE NUMBER U151 see MCW250
RCRA WASTE NUMBER U152 see MGA750
RCRA WASTE NUMBER U153 see MLE650
RCRA WASTE NUMBER U154 see MGB150
RCRA WASTE NUMBER U155 see TEO250
RCRA WASTE NUMBER U156 see MIG000
RCRA WASTE NUMBER U157 see MIJ750
RCRA WASTE NUMBER U158 see MJM200
RCRA WASTE NUMBER U159 see MKA400
RCRA WASTE NUMBER U160 see MKA500
RCRA WASTE NUMBER U161 see HFG500
RCRA WASTE NUMBER U162 see MLH750
RCRA WASTE NUMBER U163 see MMP000
RCRA WASTE NUMBER U164 see MPW500
RCRA WASTE NUMBER U165 see NAJ500
RCRA WASTE NUMBER U166 see NBA500
RCRA WASTE NUMBER U167 see NBE000
RCRA WASTE NUMBER U168 see NBE500
RCRA WASTE NUMBER U169 see NEX000
RCRA WASTE NUMBER U170 see NIF000
RCRA WASTE NUMBER U171 see NIY000
RCRA WASTE NUMBER U172 see BRY500
RCRA WASTE NUMBER U173 see NKM000

RCRA WASTE NUMBER U174 see NJW500
RCRA WASTE NUMBER U176 see ENV000
RCRA WASTE NUMBER U177 see MNA750
RCRA WASTE NUMBER U178 see MMX250
RCRA WASTE NUMBER U179 see NLJ500
RCRA WASTE NUMBER U180 see NLP500
RCRA WASTE NUMBER U181 see NMP500
RCRA WASTE NUMBER U182 see PAI250
RCRA WASTE NUMBER U183 see PAV500
RCRA WASTE NUMBER U184 see PAW500
RCRA WASTE NUMBER U185 see PAX000
RCRA WASTE NUMBER U187 see ABG750
RCRA WASTE NUMBER U188 see PDN750
RCRA WASTE NUMBER U189 see PHS000
RCRA WASTE NUMBER U190 see PHW750
RCRA WASTE NUMBER U191 see MOY000
RCRA WASTE NUMBER U192 see DTT600
RCRA WASTE NUMBER U193 see PML400
RCRA WASTE NUMBER U194 see PND250
RCRA WASTE NUMBER U196 see POP250
RCRA WASTE NUMBER U197 see QQS200
RCRA WASTE NUMBER U201 see REA000
RCRA WASTE NUMBER U202 see BCE500
RCRA WASTE NUMBER U204 see SBO000, SBQ500
RCRA WASTE NUMBER U206 see SMD000
RCRA WASTE NUMBER U207 see TBN750
RCRA WASTE NUMBER U208 see TBQ000
RCRA WASTE NUMBER U209 see TBQ100
RCRA WASTE NUMBER U210 see PCF275
RCRA WASTE NUMBER U211 see CBY000
RCRA WASTE NUMBER U212 see TBT000
RCRA WASTE NUMBER U213 see TCR750
RCRA WASTE NUMBER U214 see TEI250
RCRA WASTE NUMBER U215 see TEJ000
RCRA WASTE NUMBER U216 see TEJ250
RCRA WASTE NUMBER U217 see TEK750
RCRA WASTE NUMBER U218 see TFA000
RCRA WASTE NUMBER U219 see ISR000
RCRA WASTE NUMBER U220 see TGK750
RCRA WASTE NUMBER U221 see TGL750
RCRA WASTE NUMBER U222 see TGS500
RCRA WASTE NUMBER U223 see TGM740, TGM750
RCRA WASTE NUMBER U225 see BNL000
RCRA WASTE NUMBER U226 see MIH275
RCRA WASTE NUMBER U227 see TIN000
RCRA WASTE NUMBER U228 see TIO750
RCRA WASTE NUMBER U230 see TIV750
RCRA WASTE NUMBER U231 see TIW000
RCRA WASTE NUMBER U232 see TAA100
RCRA WASTE NUMBER U233 see TIX500
RCRA WASTE NUMBER U234 see TMK500, TMK750
RCRA WASTE NUMBER U235 see TNC500
RCRA WASTE NUMBER U236 see CMO250
RCRA WASTE NUMBER U237 see BIA250
RCRA WASTE NUMBER U238 see UVA000
RCRA WASTE NUMBER U239 see XGS000
RCRA WASTE NUMBER U240 see DAA800
RCRA WASTE NUMBER U242 see PAX250
RCRA WASTE NUMBER U244 see TFS350
RCRA WASTE NUMBER U246 see COO500
RCRA WASTE NUMBER U247 see MEI450
RCTA see RJA500
RD 292 see CKG000
RD 406 see DGB000
R.D. 1403 see EPD500
RD 1572 see DBV400
RD 2195 see CEP000
R.D. 2786 see WAK000
RD 4593 see CIR500
RD-6584 see RDP300
RD7693 see BAV000
RD 11654 see IJG000
R.D. 13621 see IIU000
RDGE see REF000
RDX see CPR800
RE-4355 see NAG400
RE 5454 see MIA000
RE 5655 see MOU750
RE 1-0185 see ELH600
RE 12420 see DOP600
REACID see COH250
REACTHIN see AES650
REACTIVE BLACK 8 see CMS227
REACTIVE BLUE 19 see BMM500

REACTIVE BRILLIANT RED 5SKH see PMF540
REACTROL see CMV400
READPRET KPN see DTG000
REALGAR see ARF000
REANIMIL see DNV000
REAZID see COH250
REAZIDE see COH250
REBELATE see DSP400
REBEMID see BCM250
REBRAMIN see VSZ000
REBUGEN see IIU000
REC 1-0060 see AJE350
REC 1-0185 see ELH600
REC 7-0040 see FCB100
REC 7-0518 see DNN000
REC 7-0591 see AJO750
REC 15-1533 see DAP880
REC 7/0267 see DNV000, DNV200
RE 5305 (CALIFORNIA CHEMICAL) see BSG250
RECANESCIN see RDF000
RECHETON see KGK000
RECINNAMINE see TLN500
RECITENSINA see TLN500
RECOFNAN see CMF350
RECOGNAN see CMF350
RECOLIP see ARQ750
RECOLITE FAST RED RL see MMP100
RECOLITE RED LAKE C see CHP500
RECOLITE YELLOW GB see DEU000
RECONIN see FOS100
RECONOX see PDP250
RECORDIL see ELH600
RECREIN see DPA000
RECTALAD-AMINOPHYLLINE see TEP500
RECTHORMONE OESTRADIOL see EDP000
RECTHORMONE TESTOSTERONE see TBG000
RECTODELT see PLZ000
RECTULES see CDO000
RED #14 see HJF500
1306 RED see FAG050
1427 RED see FAG040
1671 RED see FAG040
1695 RED see FMU070
1860 RED see CHP500
11411 RED see FAG070
11445 RED see BNH500
11554 RED see IHD000
11959 RED see HJF500
11969 RED see ADG250
12094 RED see CJD500
12101 RED see FAG050
RED 4B ACID see AKQ000
REDAMINA see VSZ000
REDAX see DWI000
RED B see XRA000
RED BASE CIBA IX see CLK220, CLK235
RED BASE IRGA IX see CLK220, CLK235
RED BASE NTR see CLK220
RED BEAD VINE see RMK250
RED BEAN see NBR800
REDBIRD FLOWER OR CACTUS see SDZ475
RED CEDARWOOD OIL see CCR000
RED COPPER OXIDE see CNO000
REDDON see TAA100
REDDOX see TAA100
RED DYE No. 2 see FAG020
REDERGIN see DLL400
RED FUMING NITRIC ACID see NEE500
RED 2G BASE see NEO500
RED HOTS see NBR800
REDIFAL see SNN300
REDI-FLOW see BAP000
RED INK PLANT see PJJ315
RED IRON ORE see HAO875
RED IRON OXIDE see IHD000
REDISOL see VSZ000
RED KB BASE see CLK225
RED LEAD see LDS000
RED LEAD CHROMATE see LCS000
RED LEAD OXIDE see LDS000
RED LOBELIA see CCJ825
RED MERCURIC IODIDE see MDD000
RED MOROCCO see PCU375
RED No. 1 see FAG050

RED No. 2 see FAG020
RED No. 4 see FAG050
RED No. 5 see XRA000
RED NO 213 see FAG070
RED No. 228 see CJD500
RED OCHRE see IHD000
RED OIL see OHU000, TOD500
RED OXIDE of MERCURY see MCT500
RED PEPPER see PCB275
RED PRECIPITATE see MCT500
RED R see FMU070
RED ROOT see AHJ875
RED SALT CIBA IX see CLK235
RED SALT IRGA IX see CLK235
RED SCARLET see CHP500
RED-SEAL-9 see ZKA000
REDSKIN see AGJ250
RED SPIDER LILY see REK325
RED SQUILL see RCF000
RED TR BASE see CLK220
RED TRS SALT see CLK235
REDUCED-d-PENICILLAMINE see MCR750
REDUCTO see DKE800
REDUCTONE see SHR500
REDUFORM see NNW500
REDUL see GHK200
REDUTON see RLU000
RED WEED see PJJ315
REE see TFF100
REED AMINE 400 see DFY800
REED LV 2,4-D see ILO000
REED LV 400 2,4-D see ILO000
REED LV 600 2,4-D see ILO000
REELON see PAF250
REFINED SOLVENT NAPHTHA see PCT250
REFORMIN see DJS200
REFOSPOREN see RCK000
REFRACTORY CERAMIC FIBERS see RCK725
REFRIGERANT 12 see DFA600
REFRIGERANT 22 see CFX500
REFRIGERANT 112 see TBP050
REFRIGERANT 113 see FOO000
REFRIGERANT 112a see TBP000
REFUGAL see CMW700
REFUSAL see DXH250
REGARDIN see ARQ750
REGELAN see ARQ750
REGENERATED CELLULOSE see HHK000
REGENON see DIP600
REGIM 8 see TKQ250
REGIN 8 see TKQ250
REGION see SNP500
REGITIN see PDW400
REGITINE see PDW400
REGITINE MESYLATE see PDW950
REGITINE METHANESULFONATE see PDW950
REGITIN METHANESULPHONATE see PDW950
REGITIPE see PDW400
REGLISSE (QUADELOUPE and HAITI) see RMK250
REGLON see DWX800
REGLONE see DWX800
REGONAL see MDL600
REGONOL see GLU000
REGONON HYDROCHLORIDE see DIP600
REGULIN see BFW250, TBJ275
REGULTON see MQS100
REGUTOL see DJL000
REHIBIN see FBP100
REHORMIN see DJS200
REICHSTEIN'S F see MIW500
REICHSTEIN'S SUBSTANCE FA see CNS800
REICHSTEIN'S SUBSTANCE H see CNS625
REICHSTEIN'S SUBSTANCE M see CNS750
REICHSTEIN X see AFJ875
REIN GUARIN see GLU000
REISE-ENGLETTEN see DYE600
REKAWAN see PLA500
RELA see IPU000
RELACT see DLY000
RELAMINAL see DCK759
RELAN BETA see BEQ625
RELANE see DVP400
RELANIUM see DCK759
RELASOM see IPU000

RELAX see IPU000
RELAXAN see PDD300
RELAXANT see GGS000
RELAXAR see GGS000
RELAXYL-G see RLU000
RELBAPIRIDINA see PPO000
RELDAN see CMA250
RELEFACT LH-RH see LIU360
RELIBERAN see MDQ250
RELICOR see DHS200
RELICOR HYDROCHLORIDE see DHS200
RELIVERAN see AJH000
RELON P see PJY500
RELUTIN see HNT500
REMADERM YELLOW HPR see MDM775
REMALAN BRILLIANT BLUE R see BMM500
REMANTADIN see AJU625
REMASAN CHLOROBLE M see MAS500
REMAZIN see HNI500
REMAZOL BLACK B see RCU000
REMAZOL BRILLIANT BLUE R see BMM500
REMAZOL YELLOW G see RCZ000
REMEFLIN see DNV000, DNV200
REMESTAN see CFY750
REMESTYP see MGC350
REMICYCLIN see TBX250
REMID see ZVJ000
REMIN see CKE750
REMOL TRF see BGJ250
REMONOL see RDZ900
REMSED see PMI750
REMYLINE Ac see SLJ500
REMZYME PL 600 see GGA800
RENAFUR see NDY000
RENAGLADIN see VGP000
RENAL AC see ALT250
RENAL EG see ALT500
RENALEPTINE see VGP000
RENALINA see VGP000
RENAL MD see TGL750
RENAL PF see PEY500
RENAL SLA see DBO400
RENAL SO see CEG625
RENAMYCIN see ACJ250
RENARCOL see ARW250, GGS000
RENARDIN see DMX200
RENARDINE see DMX200
RENBORIN see DCK759
RENGASIL see PJA220
RENO-M-DIP see AOO875
RENOFORM see VGP000
RENOGRAFFIN M-76 see AOO875
RENOGRAFIN see AOO875
RENOLBLAU 3B see CMO250
RENO M see AOO875
RENO M 60 see AOO875
RENON see CLY600
RENONCULE (CANADA) see FBS100
RENOSTYPRICIN see VGP000
RENOSTYPTIN see VGP000
RENOSULFAN see SNN500
REN O-SAL see HMY000
RENSTAMIN see DBM800
RENTIAPRIL see FAQ950
RENTOVET see TAF675
RENTYLIN see PBU100
RENUMBRAL see HGB200
RENURIX see AOO875
REOMAX see DFP600
REOMOL D 79P see DVL700
REOMOL DOA see AEO000
REOMOL DOP see DVL700
REOMUCIL see CBR675
REORGANIN see RLU000
REOXYL see HFF500
REP see BCM250
REPAIRSIN see PKB500
REPARIL see EDK875
β-REPARIL see EDL500
REPARIL SODIUM SALT see EDM000
REPEL see DKC800
REPELLENT 612 see EKV000
REPELTIN see AFL500
REPHOXITIN see CCS500

REPICIN see BEQ625
REPOCAL see CAV000
REPOISE see MFD500
REPOISE MALEATE see BSZ000
REPOSO-TMD see TBF750
REPPER-DET see DKC800
REPRISCAL see SHL500
REPRODAL see AQH500
REPROMIX see MCA000
REPROTEROL HYDROCHLORIDE see DNA600
REPTILASE see RDA350
REPTILASE R see RDA350
REPTILASE S see CMY325
REPUDIN-SPECIAL see DKC800
REPULSON see PAF550
REQUTOL see DJL000
RERANIL see TBO500
RESACETOPHENONE see DMG400
β-RESACETOPHENONE see DMG400
RESACTIN A see SMC500
RESANTIN see RDA375
RESARIT 4000 see PKB500
RESCALOID see TLN500
RESCAMIN see TLN500
RESCIDAN see TLN500
RESCIN see TLN500
RESCINNAMINE see TLN500
RESCINPAL see TLN500
RESCINSAN see TLN500
RESCITEN see TLN500
RESCUE SQUAD see SHF500
RESERPATE de METHYLE (FRENCH) see MPH300
RESERPIC ACID-METHYL ESTER, ESTER with 4-HYDROXY-3,5-
    DIMETHOXYBENZOIC ACID ETHYL CARBONATE see RCA200
RESERPIDINE see RDF000
RESERPINE see RDK000
RESERPINENE see TLN500
RESERPINE PHOSPHATE see RDK050
RESIBUFOGENIN see BOM650
RESIDUAL OIL SOLVENT EXTRACT see MQV863
RESIDUAL OILS (PETROLEUM), ACID-TREATED (9CI) see MQV872
RESIL see RLU000
RESIN (solution) see RDP000
RESIN, solution, in flammable liquid (DOT) see RDP000
RESIN, solution (resin compound, liquid) (DOT) see RDP000
RESINOL BROWN RRN see NBG500
RESINOL ORANGE R see PEJ500
RESINOL YELLOW GR see DOT300
RESIN SCARLET 2R see XRA000
RESIN, SMA 1440-H see SEA500
RESIN T see CNH125
RESIN TOLU see BAF000
RESIPAL see TLN500
RESIREN VIOLET TR see DBP000
RESISAN see RDP300
RESISTAB see RDU000
RESISTAMINE see POO750, TMP750
RESISTOMYCIN see HAL000
RESISTOMYCIN (BAYER) see KAV000
RESISTOPHEN see MNV250
RESITAN see VBK000
RESKINNAMIN see TLN500
RESLOOM M 75 see HDY000
RESMETHRIN see BEP500
(+)-cis-RESMETHRIN see RDZ875
(+)-trans-RESMETHRIN see BEP750
d-trans-RESMETHRIN see BEP750
RESMETRINA (PORTUGUESE) see BEP500
RESMIT see CGA000
RESOACETOPHENONE see DMG400
RESOBANTIN see DJM800, XCJ000
RESOCHIN see CLD000, CLD250
RESOCHIN DIPHOSPHATE see CLD250
RESOFORM ORANGE G see PEJ500
RESOFORM ORANGE R see XRA000
RESOFORM YELLOW GGA see DOT300
RESOIDAN see CCK125
RESOQUINA see CLD000
RESOQUINE see CLD000, CLD250
RESORCIN see REA000
RESORCIN ACETATE see RDZ900
RESORCINE see REA000
RESORCINE BROWN J see XMA000
RESORCINE BROWN R see XMA000

RESORCIN MONOACETATE see RDZ900
RESORCINOL see REA000
RESORCINOL BIS(2,3-EPOXYPROPYL)ETHER see REF000
RESORCINOL, DIACETATE see REA100
RESORCINOL DIGLYCIDYL ETHER see REF000
β-RESORCINOLIC ACID see HOE600
RESORCINOL METHYL ETHER see REF050
RESORCINOL, MONOACETATE see RDZ900
RESORCINOL, MONOBENZOATE see HNH500
RESORCINOL MONOMETHYL ETHER see REF050
RESORCINOLPHTHALEIN see FEV000
RESORCINOL PHTHALEIN SODIUM see FEW000
RESORCINYL DIGLYCIDYL ETHER see REF000
RESORCITATE see RDZ900
β-RESORCYLIC ACID see HOE600
RESOTROPIN see HEI500
RESOXOL see SNN500
RESPAIRE see ACH000
RESPENYL see RLU000
RESPIFRAL see DMV600
RESPILENE see MFG250
RESPIRIDE see DAI200
RESPRAMIN see AJD000
RESTAMIN see BBV500
RESTAMINE see BBV500
RESTAS see FMR100
RESTENIL see MQU750
REST-ON see TEO250
RESTORIL see CFY750
RESTOVAR see LJE000
RESTROL see DAL600
RESTROPIN see SBH500
RESTRYL see TEO250
RESULFON see AHO250
RESURRECTION LILY see REK325
RETABOLIL see NNE550
RETACEL see CMF400
RETALON see DAL600
RETALON-ORAL see DHB500
RETAMA see YAK350
RETAMID see AKO500
RETAR-B₁ see FQJ100
RETARCYL see SAI100
RETARDER AK see PHW750
RETARDER BA see BCL750
RETARDER ESEN see PHW750
RETARDER J see DWI000
RETARDER PD see PHW750
RETARDER W see SAI000
RETARDEX see BCL750
RETARPEN see BFC750
RETASULFIN see AKO500
RETENS see HGP550
RETENSIN see PDD300
RETICULIN see SLX500
RETIN-A see VSK950
9-cis-RETINAL see VSK975
9-cis-RETINALDEHYDE see VSK975
RETINOIC ACID see VSK950
β-RETINOIC ACID see VSK950
13-cis-RETINOIC ACID see VSK955
all-trans-RETINOIC ACID see VSK950
RETINOIC ACID, M, 7,8-DIDEHYDRO- see DHA200
RETINOIC ACID, 7,8-DIHYDRO- see DMD100
RETINOIC ACID, SODIUM SALT see SJN000
RETINOID ETRETIN see REP400
RETINOL see VSK600
all-trans RETINOL see VSK600
RETINOL ACETATE see VSK900
RETINOL PALMITATE see VSP000
RETINYL ACETATE see VSK900
all-trans-RETINYL ACETATE see VSK900
all-trans-RETINYLIDENE METHYL NITRONE see REZ200
RETINYL PALMITATE see VSP000
RETOZIDE see ILD000
RETRO-6-DEHYDROPROGESTERONE see DYF759
RETRONE see DYF759
cis-RETRONECIC ACID ESTER of RETRONECINE see RFP000
cis-RETRONECIC ACID ESTER of RETRONECINE-N-OXIDE see RFU000
RETRONECINE HYDROCHLORIDE see RFK000
Δ⁶-RETROPROGESTERONE see DYF759
RETRORSINE see RFP000
RETRORSINE-N-OXIDE see RFU000
RETROVITAMIN A see VSK600

REUBLONIL see BCP650
REUDO see BRF500
REUFENAC see AGN000
REULATT S.S. see GFM000
REUMACHLOR see CLD000
REUMACIDE see IDA000
REUMALON see OPK300
REUMAQUIN see CLD000
REUMARTRIL see OPK300
REUMASYL see BRF500
REUMATOX see MQY400
REUMAZOL see BRF500
REUMOFENE see IDA400
REUMOFIL see SOU550
REUMOX see HNI500
REUPOLAR see BRF500
REVAC see DJL000
REVELOX-WIRKSTOFF see RCA200
REVENGE see DGI400
REVERIN see PPY250
REVERTINA see VIZ400
REVIDEX see BBK500
REVIENTA CABALLOS (CUBA) see SLJ650
REVONAL see QAK000
REWOMID DLMS see BKE500
REWOPOL HV-9 see PKF000
REWOPOL NLS 30 see SIB600
REWOPOL TLS 40 see SON000
REXALL 413S see PMP500
REXAN see CKF500
REXENE see EJA379, PMP500
REXENE 106 see ADY500
REXOCAINE see BQA010
REXOLITE 1422 see SMQ500
REX REGULANS see GGS000
REZIFILM see TFS350
REZIPAS see AMM250
RF 46-790 see THH350
RFCNU see RFU600
RFNA see NEE500
RG 235 see AMV375
R-GENE see AQW000
RGH-1106 see PII250
RGH 2957 see YGA700
RGH-2958 see PEE100
RGH-4405 see EGM100
RGH-5526 see RFU800
Rh-110 see RHF150
RH-124 see BPU000
RH 315 see DTT600
RH 893 see OFE000
RH-8218 see EPC175
RHABARBERONE see DMU600
RHABDOPHIS TIGRINUS TIGRINUS VENOM see RFU875
RHAMNOLUTEIN see ICE000
RHAMNOLUTIN see ICE000
3-β(α-l-RHAMNOPYRANOSIDE)-5,11-α,14-β-TRIHYDROXY-5-β-CARD(20,22)EN-OLIDE see RFZ000
RHAMNOSIDE, STROPHANTHIDIN-3, α-l- see CNH780
3-β-RHAMNOSIDO-14-β-HYDROXY-Δ4,20,22-BUFATRIENOLIDE see POB500
RHAMNUS CALIFORNICA see MBU825
RHAMNUS CATHARTICA see MBU825
RHAMNUS FRANGULA see MBU825
RHATHANI see RGA000
RHEMATAN see PGG000
RHENIC ACID, POTASSIUM SALT see PLR500
RHENIUM see RGF000
RHENIUM(VII) SODIUM OXIDE see SJD500
RHENIUM(VII) SULFIDE see RGK000
RHENIUM TRICHLORIDE see RGP000
RHENOCURE CA see DWN800
RHEONINE B see FAG070
RHEOPYRINE see IGI000
RHEOSMIN see RBU000
RHEUMATOL see CCD750
RHEUM EMODIN see MQF250
RHEUMIN TABLETTEN see ADA725
RHEUMON see HKK000
RHEUMON GEL see HKK000
RHEUMOX see AQN750
RHEUM RHABARBARUM see RHZ600
RHINALAR see FDD085
RHINALL see SPC500
RHINANTIN see NCW000

RHINASPRAY see RGP450
RHINATHIOL see CBR675
RHINAZINE see NAH500
RHINE BERRY see MBU825
RHINOCORT see BOM520
RHINOGUTT see RGP450
RHINOPERD see NCW000
RHINOSPRAY see RGP450
RHIZOCTOL see MGQ750
RHIZOPIN see ICN000
RHODACRYST see VSZ000
RHODALLIN see AGT500
RHODALLINE see AGT500
RHODAMINE see FAG070
RHODAMINE 6G (biological stain) see RGW000
RHODAMINE 6GEX ETHYL ESTER see RGW000
RHODAMINE 6G EXTRA BASE see RGW000
RHODAMINE S (RUSSIAN) see FAG070
RHODAMINE WT see RGZ100
RHODANDINITROBENZOL see DVF800
RHODANIC ACID see RGZ550
RHODANID see ANW750
RHODANIDE see ANW750, PLV750
RHODANIN (CZECH) see RGZ550
RHODANINE see RGZ550
RHODANINIC ACID see RGZ550
RHODIA see DAA800
RHODIA-6200 see DJA400
RHODIACHLOR see HAR000
RHODIACID see BJK500
RHODIACIDE see EEH600
RHODIA RP 11974 see BDJ250
RHODIASOL see PAK000
RHODIATOX see PAK000
RHODIATROX see PAK000
RHODINE see ADA725
RHODINOL see CMT250
RHODINOL (FCC) see DTF400
RHODINOL ACETATE see RHA000
RHODINYL ACETATE see RHA000
RHODINYL BUTYRATE see DTF800
RHODIRUBIN A see RHA125
RHODIRUBIN B see RHA150
RHODIRUBIN C see TAH675
RHODIUM see RHF000
RHODIUM(II) ACETATE see RHF150
RHODIUM(II) BUTYRATE see RHK250
RHODIUM CHLORIDE see RHK000, RHP000
RHODIUM(III) CHLORIDE (1583) see RHK000
RHODIUM CHLORIDE, TRIHYDRATE see RHP000
RHODIUM, (1,5-CYCLOOCTADIENE)(2,4-PENTANEDIONATO)- see CPR840
RHODIUM, ((1,2,5,6-eta)-1,5-CYCLOOCTADIENE)(2,4-PENTANEDIONATO-O,O')- see CPR840
RHODIUM DIACETATE see RHF150
RHODIUM DIBUTYRATE see RHK250
RHODIUM(II) PROPIONATE see RHK850
RHODIUM TRICHLORIDE see RHK000
RHODIUM TRICHLORIDE TRIHYDRATE see RHP000
RHODOCIDE see EEH600
RHODODENDRON see RHU500
RHODOL see MGJ750
RHODOLNE see SMQ500
RHODOPAS 6000 see AAX175
RHODOPAS M see AAX250
RHODOQUINE see RHZ000
RHODORA (CANADA) see RHU500
RHODOTOXIN see AOO375
RHODOTYPOS SCANDENS see JDA075
RHODULINE ORANGE see BJF000
RHODULINE ORANGE NO see BAQ250
RHOMBIC see ELC500
RHOMENE see CIR250
RHOMEX see PIJ500
RHONOX see CIR250
RHOPLEX AC-33 (ROHM and HAAS) see EMF000
RHOPLEX B 85 see PKB500
RHOTHANE see BIM500
RHOTHANE D-3 see BIM500
RHUBARB see RHZ600
RHUBARBE (CANADA) see RHZ600
RHUS COPALLINA see SCF000
R 18553 HYDROCHLORIDE see LIH000
RHYTHMONORM see PMJ525
RHYTMATON see AFH250

RHYUNO OIL see SAD000
RIANIL see CLO750
RIBALL see ZVJ000
RIBAVIRIN see RJA500
RIBBON CACTUS see SDZ475
RIBIPCA see RIK000
RIBITOL see RIF000
RIBO-AZAURACIL see RJA000
RIBO-AZURACIL see RJA000
RIBODERM see RIK000
RIBOFLAVIN see RIK000
RIBOFLAVINE see RIK000
RIBOFLAVINEQUINONE see RIK000
RIBOFLAVINE SULFATE see RIP000
RIBOFURANOSIDE, 9H-PURINE-6-THIOL-9 see MCQ500
9-β-d-RIBOFURANOSIDOADENINE see AEH750
1-β-d-RIBOFURANOSYLCYTOSINE see CQM500
9-β-d-RIBOFURANOSYLGUANINE see GLS000
2-β-d-RIBOFURANOSYLMALEIMIDE see RIU000
5-β-d-RIBOFURANOSYL-1,3-OXAZINE-2,4-DIONE see OMK000
5-β-d-RIBOFURANOSYL-2H-1,3-OXAZINE-2,4(3H)-DIONE see OMK000
9-(β-d-RIBOFURANOSYL)PURINE see RJF000
9-(β-d-RIBOFURANOSYL)-9H-PURINE see RJF000
9-β-d-RIBOFURANOSYL-9H-PURINE-6-THIOL see TFJ825
3-β-d-RIBOFURANOSYL-1H-PYRROLE-2,5-DIONE see RIU000
7-β-d-RIBOFURANOSYL-7H-PYRROLO(2,3-D)PYRIMIDINE-4-AMINE
    see TNY500
7-β-d-RIBOFURANOSYL-7H-PYRROLO(2,3-d)PYRIMIDIN-4-OL see DAE200
2-β-d-RIBOFURANOSYLTHIAZOLE-4-CARBOXAMIDE see RJF500
2-β-d-RIBOFURANOSYL-4-THIAZOLECARBOXAMIDE see RJF500
2-β-d-RIBOFURANOSYL-as-TRIAZINE-3,5(2H,4H)-DIONE see RJA000
2-β-d-RIBOFURANOSYL-1,2,4-TRIAZINE-3,5(2H,4H)-DIONE see RJA000
2-β-d-RIBOFURANOSYL-as-TRIAZINE-3,5(2H,4H)-DIONE 2′,3′,5′-TRIACETATE
    see THM750
1-β-d-RIBOFURANOSYL-1,2,4-TRIAZOLE-3-CARBOXAMIDE see RJA500
1-β-d-RIBOFURANOSYLURACIL see UVJ000
RIBOMYCINE see RIP000
RIBONOSINE see IDE000
RIBOSTAMIN see RIP000
RIBOSTAMYCIN see XQJ650
RIBOSTAMYCIN SULFATE see RIP000
β-d-RIBOSYL-6-METHYLTHIOPURINE see MPU000
RIBOSYLPURINE see RJF000
RIBOSYLTHIOGUANINE see TFJ500
RIBOSYL-6-THIOPURINE see MCQ500
RIBOTIDE see RJF400
RIBOXAMIDE see RJF500
RICE STARCH see SLJ500
RICHAMIDE 6310 see BKE500
RICHONATE 1850 see DXW200
RICHONOL C see SIB600
RICHONOL T see SON000
RICID see DIU800
RICIFON see TIQ250
RICIN see RJK000
RICIN (HAITI) see CCP000
RICINIC ACID see RJP000
RICINOLEIC ACID see RJP000
RICINOLEIC ACID, BARIUM SALT see RJU000
RICINOLEIC ACID, SODIUM SALT see SJN500
RICINOLEIC ACID TRIESTER with GLYCEROL 12-ETHER with
    TRIDECAETHYLENE GLYCOL see EAN800
RICINOLIC ACID see RJP000
RICINO (PUERTO RICO) see CCP000
RICIN-TOXIN CON A see CNH625
RICINUS COMMUNIS see CCP000
RICINUS OIL see CCP250
RICIRUS OIL see CCP250
RICKAMICIN see SDY750
RICKAMICIN SULFATE see APY500
RICKETON see CMC750
RICON 100 see SMR000
RICORTEX see CNS825
RICYCLINE see TBX250
RIDAURA see ARS150
RIDAZOLE see MMN750
RIDDELLINE see RJZ000
RIDZOL P see MMN750
RIFA see RKP000
RIFADINE see RKP000
RIFAGEN see RKP000
RIFALDAZINE see RKP000
RIFALDIN see RKP000
RIFAMATE see RKP000

RIFAMICINE SV see RKZ000
RIFAMIDE see RKA000
RIFAMPICIN see RKP000
RIFAMPICINE (FRENCH) see RKP000
RIFAMPICIN M/14 see RKA000
RIFAMPICINUM see RKP000
RIFAMPIN see RKP000
RIFAMYCIN see RKK000, RKZ000
RIFAMYCIN AMP see RKP000
RIFAMYCIN B see RKK000
RIFAMYCIN B N,N-DIETHYLAMIDE see RKA000
RIFAMYCIN DIETHYLAMIDE see RKA000
RIFAMYCIN M14 see RKA000
RIFAMYCIN O see RKP400
RIFAMYCIN S see RKU000
RIFAMYCIN SV see RKZ000
RIFAPRODIN see RKP000
RIFATHYROIN see TNX400
RIFINAH see RKP000
RIFIT see PMB850
RIFLE POWDER (DOT) see PLL750
RIFLOC RETARD see CCK125
RIFOBAC see RKP000
RIFOCIN see RKZ000
RIFOCINA M see RKA000
RIFOCYN see RKZ000
RIFOLDIN see RKP000
RIFOMIDE see RKA000
RIFOMYCIN B see RKK000
RIFOMYCIN B DIETHYLAMIDE see RKA000
RIFOMYCIN O see RKP400
RIFOMYCIN S see RKU000
RIFOMYCIN SV see RKZ000
RIFORAL see RKP000
RIGEDAL see CCK125
RIGENICID see EPQ000
RIGETAMIN see EDC500
RIGEVIDON see NNL500
RIGIDIL see BBV500
RIGIDYL see BBV500
RIKAVARIN see AJV500
RIKEMAL S 250 see SKV150
RIKER 548 see TIH800
RIKER 595 see BSZ000
RIKER 601 see TND000
RIKER 52G see PAF550
RILANSYL see CKF500
RILAQUIL see CKF500
RILASSOL see CKF500
RILAX see CKF500
RILLASOL see CKF500
RIMACTAN see RKP000
RIMACTAZID see RKP000
RIMADYL see CCK800
RIMANTADINE HYDROCHLORIDE see AJU625
RIMAON see EDW500
RIMAZOLIUM METHYL SULFATE see PMD825
RIMICID see ILD000
RIMIDIN see FAK100
RIMITSID see ILD000
RIMSO-50 see DUD800
RINATIOL see CBR675
RINAZIN see NAH550
RINDEX see MDO250
RINEPTIL see TNX750
RINGER'S GLUCOSE SOLUTION see GFG200
RINO-CLENIL see AFJ625
RIOL see FMB000
RIOMITSIN see HOH500
RIOMYCIN see FMR500
RIP-15830 see DNG200
RIPAZEPAM see POL475
RIPCORD see RLF350
RIPENTHOL see DXD000
RIPERCOL see TDX750
RIPERCOL-L see LFA020
RIPEREOL see TDX750
RIPOSON see EQL000
RIRILIM see BBW500
RIRIPEN see BBW500
RISE see CKA000
RISELECT see DGI000
RISELF see MFD500
RISPASULF see PDC850

RISTAT see HMY000
RISTOGEN see TKH750
RITALIN see MNQ000, RLK000
RITALINE see MNQ000
RITALIN HYDROCHLORIDE see RLK000
RITCHER WORKS see MNQ000
RITMENAL see DKQ000
RITMODAN see DNN600
RITMOS see AFH250
RITODRINE HYDROCHLORIDE see RLK700
RITOSEPT see HCL000
RITROSULFAN see LJD500
RITSIFON see TIQ250
RITUSSIN see RLU000
RIVADORM see NBT500
RIVADORN see NBU000
RIVANOL see EDW500
RIVINA HUMILIS see ROA300
RIVINOL see EDW500
RIVIVOL see ILE000
RIVOMYCIN see CDP250
RIVOTRIL see CMW000
RIZABEN see RLK800
RJ 5 see DLJ500
RL-50 see MRL500
RMI 9918 see TAI450
RMI 11002 see FDN000
RMI 12,936 see HLX600
RMI 71782 see DBE835, DKH875
RMI9,384A see DLS600
RMI 10,482A see MHJ500
RMI-14042A see LIA400
RMI 10024DA see BJA500
RMI 10874DA see BJG500
RMI 11002 DA see FDN000
RMI 11513 DA see XBA000
RMI 11567 DA see DDB400
RMI 11877 DA see DDE000
RO 1-5130 see MDL600
RO 1-5431 see MDV250, MKR250
RO 1-5470 see RAF300
RO 1-6463 see DNW400
RO 1-6794 see DBE825
RO-1-7700 see AGI000
RO 1-7788 see LFD200
Ro 1-7977 see DNV800
RO-1-9213 see CQH000
RO 1-9569 see TBJ275
RO 2-1160 see CCH250
RO 2-2222 see TKW500
RO 2-2980 see HNK000
RO 2-3208 see FDK000
RO 2-3245 see ARX750
RO 2-3248 see AGD500
RO 2-3308 see QWJ000
RO 2-3599 see DLP000
RO 2-3742 see AGE000
RO 2-3951 see ARX770
RO 2-4572 see ILE000
RO 2-5803 see BHR750
Ro 2-7113 see AGV890
RO 2-9757 see FMM000
RO 2-9915 see FHI000
RO 2-9945 see TCQ500
RO 4-0403 see TAF675
RO 4-1385 see GFO200
RO 4-1778 see MDV000
RO 4-2130 see SNK000
RO 4-3476 see SNL800
RO-4-3780 see VSK955
RO 4-3816 see DBK400
RODANCA see AIE500
RO 4-4393 see AIE500
RO 4-4602 see SCA400
RO 4-5360 see DLY000
RO 4-6316 see DNA200
RO 4-6467 see PME250, PME500
RO-4-6861 see CIF250
RO 4-8180 see CMW000
RO 4-9253 see BMN750
RO 5-0360 see DAR400
RO 5-0690 see MDQ250
RO 5-0831 see IKC000
RO 5-1226 see MNP300
RO 5-2180 see CGA500

RO 5-3059 see DLY000
RO 5-3307 see DAI475
RO 5-3350 see BMN750
RO 5-3438 see FDB100
RO 5-4023 see CMW000
RO 5-4200 see FDD100
RO 5-5345 see CFY750
RO 5-5516 see MLD100
RO 5-6789 see CFZ000
RO 5-6901 see DAB800
RO 5-9963 see NHI000
RO 6-4563 see GFY100
RO 7-0207 see OJS000
Ro 7-0582 see NHH500
RO 7-1554 see IGH000
RO 7-2340 see BMA650
Ro 04-7683 see INY100
RO 10-1670 see REP400
RO 10-6338 see BON325
Ro 10-9359 see EMJ500
RO 11-1781 see TGA275
RO 12-0068 see TAL485
RO 13-6298 see AQZ400
RO 13-7410 see AQZ200
RO 13-8320 see AQZ300
Ro 215535 see DMJ400
RO 21-9738 see DYE415
RO 1-5431/7 see DYF000
RO 1-5470/5 see DBE200
RO 1-5470/6 see LFD200
RO-5-0810/1 see TJE880
RO 5-3307/1 see IKB000
Ro-5-6901/3 see FMQ000
RO-6-4563/8 see GFY100
RO 8-6270/9 see TNX400
RO 1-9569/12 see TBJ275
ROACH SALT see SHF500
ROAD ASPHALT see PCR500
ROAD ASPHALT (DOT) see ARO500, ARO750
ROAD TAR (DOT) see ARO500
ROAD TAR, liquid (DOT) see ARO750
RO-AMPEN see AIV500, AOD125
ROBAMATE see MQU750
ROBANUL see GIC000
ROBAVERON see RLK875
ROBENIDINE see RLK890
ROBICILLIN VK see PDT750
ROBIGRAM see ARQ750
ROBIMYCIN see EDH500
ROBINETIN see RLP000
ROBINIA PSEUDOACACIA see GJU475
ROBINUL see GIC000
ROBIOCINA see SMB000
ROBISELLIN see ILD000
ROBITET see TBX000
ROBITUSSIN see RLU000
ROBORAL see PAN100
ROC-101 see RLU550
ROCALTROL see DMJ400
ROCCAL see BBA500
ROCHE 5-9000 see API125
ROCHIPEL see DIR000
RO-CILLIN see PDD350
ROCILLIN-VK see PDT750
ROCIPEL see DIR000
ROCK CANDY see SNH000
ROCK OIL see PCR250
ROCK SALT see SFT000
ROCORNAL see DIO200
RO-CYCLINE see TBX250
RODALON see AFP250, BBA500
RODANCA see PLV750
RODANIN S-62 (CZECH) see IAQ000
RODATOX 60 see DVF800
RODENTIN see EAT600
RO-DETH see WAT200
RO-DEX see SMN500
RODEX see FFF000
RODILONE see SNY500
RODINAL see ALT250
RODINE see RCF000
RODINOL see CMT250
RODINOLONE see AQX250
RODIPAL see DIR000

RODOCID see EEH600
RODY see DAB825
ROE 101 see DBB200
ROERIDORM see CHG000
ROGERSINE see TKX700
ROGITINE see PDW400
ROGODIAL see DRR400, DSP400
ROGOR see DSP400
ROGUE see DGI000
ROHM & HAAS RH-218 see EPC175
ROHYDRA see BAU750
RO-HYDRAZIDE see CFY000
ROHYPNOL see FDD100
ROIDENIN see IIU000
ROIPNOL see FDD100
ROKACET see PJY100
ROKACET O 7 see PJY100
ROKANOL L see DXY000
RO-KO see RNZ000
ROKON see BDF000
ROL see RAF100
ROLAMID CD see BKE500
ROLAZINE see HGP500
ROLAZOTE see PAJ750, POL000
ROLICTON see AKL625
ROLITETRACYCLINE see PPY250, SPE000
ROLL-FRUCT see CDS125
ROLODIN see RLZ000
ROLODINE see RLZ000
ROM 203 see IAG700
ROMACRYL see PKB500
ROMANTRENE GOLDEN YELLOW see DCZ000
ROMANTRENE GREY K see IBJ000
ROMANTRENE NAVY BLUE FG see CMU500
ROMAN VITRIOL see CNP250, CNP500
ROMERGAN see DQA400
ROMETIN see CHR500
ROMEZIN see ALF250
ROMILAR see DBE200
ROMILAR HYDROBROMIDE see DBE200
ROMOSOL see ART250
ROMPARKIN see BBV000
ROMPHENIL see CDP250
ROMPUN see DMW000
ROMULGIN O see PKL100
RONDAR see CFZ000
RONDOMYCIN see MDO250
RO-NEET see EHT500
RONGALIT see FMW000
RONGALITE C see FMW000
RONIDAZOLE see MMN750
RONILAN see RMA000
RONIN see PPO000
RONIT see EHT500
RONNEL see RMA500
RONONE see RNZ000
RONTON see ENG500
RONTYL see TKG750
ROOTONE see ICP000, NAK000, NAK500
ROPE BARK see LEF100
ROP 500 F see GIA000
ROPTAZOL see NGG500
ROQUESSINE DIHYDROBROMIDE see DOX000
ROQUINE see CLD000
RORASUL see ASB250, PPN750
RORER 148 see QAK000
ROSACETOL see TIT000
ROSA FRANCESA (CUBA) see OHM875
ROSA LAUREL (MEXICO) see RHU500
ROSAMICIN see RMF000
ROSANIL see DGI000
p-ROSANILINE see RMK000
ROSANILINE BASE see MAC500
ROSANILINE CHLORIDE see MAC250
p-ROSANILINE HCL see RMK020
ROSANILINE HYDROCHLORIDE see MAC250
p-ROSANILINE HYDROCHLORIDE see RMK020
ROSANILINIUM CHLORIDE see MAC250
ROSANOMYCIN A see RMK200
ROSARAMICIN see RMF000
ROSARY PEA see RMK250
ROSCOPENIN see PDT750
ROSCOSULF see SNN300
ROSE ABSOLUTE FRENCH see RMP000

ROSE BAY see OHM875
ROSEBAY see RHU500
ROSE BENGAL SODIUM see RMP175
ROSE CRYSTALS see TIT000
ROSE ETHER see PER000
ROSE-FLOWERED JATROPHA see CNR135
ROSE GERANIUM OIL ALGERIAN see GDA000
ROSE LAUREL (MEXICO) see OHM875
ROSE de MAI ABSOLUTE see RMP000
ROSEMARIE OIL see RMU000
ROSEMARY OIL see RMU000
ROSEMIDE see CHJ750
ROSEN OEL (GERMAN) see RNA000
ROSENSTHIEL see MAT250
ROSE OIL see RNA000
ROSE OIL BULGARIAN see RNF000
ROSE OIL MOROCCAN see RNK000
ROSE OXIDE LEVO see RNU000
ROSE QUARTZ see SCI500, SCJ500
ROSES see AOB250
ROSETONE see NAK000
ROSMARIN OIL (GERMAN) see RMU000
ROSPAN see PNH750
ROSPIN see PNH750
RO-SULFIRAM see DXH250
ROTATE see DQM600
ROTAX see BDF000
ROT B see XRA000
ROTEFIVE see RNZ000
ROTEFOUR see RNZ000
ROTENONA (SPANISH) see RNZ000
ROTENONE see RNZ000
ROTERSEPT see BIM250
ROTESAR see BOV825
ROTESSENOL see RNZ000
ROT GG FETTLOESLICH see XRA000
ROTHANE see BIM500
ROTOCIDE see RNZ000
ROTOX see MHR200
ROTTLERIN see KAJ500
ROUGE see IHD000
ROUGE CERASINE see OHA000
ROUGE de COCHENILLE A see FMU080
ROUGE PLANT see ROA300
ROUGH & READY MOUSE MIX see WAT200
ROUGH & READY RAT BAIT & RAT PASTE see RCF000
ROUGOXIN see DKN400
ROUQUALONE see QAK000
ROUTRAX see TKG750
ROVAMICINA see SLC000
ROVRAL see GIA000
ROWACHOL see ROA400
ROWATIN see ROA425
ROWMATE see DET400, DET600
ROXARSONE (USDA) see HMY000
ROXBURY WAXWORK see AHJ875
ROXICAM see FAJ100
ROXIMYCIN see HGP550
ROXION U.A. see DSP400
ROXOSUL TABLETS see SNN500
ROYAL BLUE see DJO000
ROYAL MBTS see BDE750
ROYALTAC see DAI600, ODE000
ROYAL TMTD see TFS350
ROZOL see CJJ000
ROZTOZOL see CKM000
RP-093 see CMY325
866 R.P. see CFU750
2259 R.P. see GEW750
RP 2259 see GEW750
2325 RP see DRV850
2339 RP see BEM500
2512 R.P. see DBL800
R.P. 2512 see DBL800
R.P. 2591 see DFX400
RP 2616 see PPP500
RP 2632 see ALF250
2643-RP see ALF250
RP 2786 see WAK000
RP 2929 see TFH500
2987 R.P. see DII200
RP 2990 see TEX250
RP 3203 see DNG400
3277 R.P. see PMI750

3277 RP see DQA400
RP 3356 see DIR000
3359 RP see CKB500
RP 3377 see CLD000
3389 R.P. see DQA400
RP 3554 see MRW000
RP 3602 see GGS000
RP 3697 see PDD300
3718 RP see CGX625
3735 R.P. see SBE500
RP 3799 see DIW000
4182 R.P. see DQA400
RP 4207 see FNF000
RP 4632 see MFK500
4753 R.P. see TGD000
RP 4909 see CLY750
5171 RP see PBM500
RP5171 see DIG400
5278 R.P. see SMA000
5337 R.P. see SLC000
6140 RP see PMF500
6710 RP see CIO500
6847 R.P. see AFL750
6909 RP see PIW000
7162 RP see DLH200
7204 RP see COS899
RP7293 see VRF000
RP 7522 see AKO500
7843 R.P. see TFM100
RP 7843 see TFM100
RP 7891 see GFM200
RP 8167 see EEH600
RP 8228 see LFO000
8595 R.P. see DSV800
RP 8599 see FMU039
RP 8823 see MMN250
RP 8908 see PIW000
RP 9159 see LEO000
9159 RP see LEO000
9260 RP see DSS600
9778 R.P. see PNW750
RP-9921 see PAF550
9965 RP see MPM750
RP 9965 see MQR000
RP 10192 see MIJ500
10257 R.P. see TND000
11,561 RP see CBL500
RP 13057 see DAC000
13,057 R.P. see DAC000
13245 R. P. see PMM000
RP 16,091 see MNQ500
17190RP see POE100
RP 18,429 see AJV500
19583 RP see BDU500
22050 R.P. see ROZ000
RP 22410 see PMG000
RP 26019 see GIA000
27267 R.P. see ZUA450
RPCNU see ROF200
R-PENTINE see CPU500
R-874 PHILLIPS see OGA000
2339 R.P. HYDROCHLORIDE see PEN000
4560 RP HYDROCHLORIDE see CKP500
RP 13057 HYDROCHLORIDE see DAC200
RP 22,050 HYDROCHLORIDE see ROZ000
2786 R.P. MALEATE see DBM800
RP 8599 MESYLATE see FMU039
19583 RP SODIUM see KGK100
RS 141 see CJJ250
RS 1280 see CBF250
RS-2177 see FDD075
RS-2196 see MBW775
RS-2290 see EDU100
RS 2874 see BIS750
R-3422-S see EMC000
RS-8858 see OMY500
RS 44872 see SNH480
RS-79216 see PCG550
R SALT see ROF300
RS-1401 AT see SPD500
R105 SODIUM see RMP175
RTEC (POLISH) see MCW250
R-010-TK see BIX250
RU 486 see HKB700

RU 2010 see NNR125
RU 2323 see ENX575
RU 2858 see MRU600
RU-4723 see CIR750
RU 486-6 see HKB700
RU-11484 see BEP750
RU 15060 see SOX400
RU 15750 see TKG000
RU 22974 see DAF300
RU 24756 see CCR950
RU 38486 see HKB700
RU 43715 see POB300
RUBATONE see BRF500
RUBBER see ROH900
RUBBER (DOT) see ROH900
"522" RUBBER ACCELERATOR see PIY500
RUBBER BUFFINGS (DOT) see ROK000
RUBBER HYDROCHLORIDE see PJH500
RUBBER HYDROCHLORIDE POLYMER see PJH500
RUBBER, NATURAL see ROH900
RUBBER RED R EXTRA see CJD500
RUBBER SCRAP see ROK000
RUBBER SHODDY (DOT) see ROK000
RUBBER SOLVENT see ROU000
RUBBER VINE see ROU450
RUBEANE see DXO200
RUBEANIC ACID see DXO200
RUBENS BROWN see MAT500
RUBERON see EME100
RUBESCENCE RED MT-21 see NAY000
RUBESCENSLYSIN see AHI500
RUBESOL see VSZ000
RUBIAZOL A see SNM500
RUBIDAZON see ROU800
RUBIDAZONE see ROU800
RUBIDAZONE MONOHYDROCHLORIDE see ROZ000
RUBIDIUM see RPA000
RUBIDIUM ACETYLIDE see RPB100
RUBIDIUM CHLORIDE see RPF000
RUBIDIUM DICHROMATE see RPK000
RUBIDIUM FLUORIDE see RPP000
RUBIDIUM HYDRIDE see RPU000
RUBIDIUM HYDROXIDE see RPZ000
RUBIDIUM HYDROXIDE, solid and solution (DOT) see RPZ000
RUBIDIUM IODIDE see RQA000
RUBIDIUM METAL (DOT) see RPA000
RUBIDIUM METAL, IN CARTRIDGES (DOT) see RPA000
RUBIDIUM NITRIDE see RQF000
RUBIDOMYCIN see DAC000, DAC200
RUBIDOMYCINE see DAC000
RUBIDOMYCIN HYDROCHLORIDE see DAC200
RUBIFLAVIN see RQF350
RUBIGAN see FAK100
RUBIGEN see TBJ275
RUBIGO see IHD000
RUBIJERVINE see SKR000
RUBINATE 44 see MJP400
RUBINATE TDI see TGM740
RUBINATE TDI 80/20 see TGM740, TGM750
RUBITOX see BDJ250
RUBOMYCIN see RQK000
RUBOMYCIN C see DAC000, DAC200
RUBOMYCIN C 1 see DAC000
RUBRAMIN see VSZ000
RUBRATOXIN B see RQP000
RUBRIPCA see VSZ000
RUBROCITOL see VSZ000
RUBRUM SCARLATINUM see SBC500
RUBUS ELLIPTICUS Smith, extract excluding roots see RQU300
RUCAINA see DHK400
RUCAINA HYDROCHLORIDE see DHK600
RUCOFLEX PLASTICIZER DOA see AEO000
RUCON B 20 see PKQ059
RUDBECKIA BICOLOR nutt., extract see RQU650
RUDOTEL see CGA000
RUELENE see COD850
RUELENE DRENCH see COD850
RUELENE 25E see COD850
RUFOCHROMOMYCIN see SMA000
RUFOCROMOMYCINE see SMA000
RUGULOSIN see RRA000
(+)-RUGULOSIN see RRA000
RUKSEAM see DAD200
RULENE see COD850

RUMAPAX see HNI500
RUMENSIN see MRE230
RUMESTROL 1 see DKA600
RUMESTROL 2 see DKA600
RUMETAN see ZLS000
RUN see PDN000
RUNA RH20 see TGG760
RUNCATEX see CIR500
RUSSELL'S VIPER VENOM see VQZ635
RUSSIAN COMFREY LEAVES see RRK000
RUSSIAN COMFREY ROOTS see RRP000
RUTA GRAVEOLENS, extract see RRP675
RUTERGAN HYDROCHLORIDE see DPI000
RUTGERS 612 see EKV000
RUTHENIUM see RRU000
RUTHENIUM CHLORIDE see RRZ000
RUTHENIUM CHLORIDE HYDROXIDE see RSA000
RUTHENIUM COMPOUNDS see RSF000
RUTHENIUMHYDROXIDE TRICHLORIDE see RSA000
RUTHENIUM HYDROXYCHLORIDE see RSA000
RUTHENIUM OXIDE see RSF875
RUTHENIUM(VIII) OXIDE see RSK000
RUTHENIUM SALT of TETRAMETHYLPHENANTHRENE see RSP000
RUTHENIUM TETRAOXIDE see RSK000
RUTHENIUM TRICHLORIDE see RRZ000
RUTHENIUM TRICHLORIDE HYDROXIDE see RSA000
RUTILE see TGG760
RUTIN see RSU000
RUTINIC ACID see RSU000
RUTOSIDE see RSU000
RUVAZONE see RSU450
RVK see DXO200
RX 6029M see TAL325
RYANEXEL see RSZ000
RYANIA see RSZ000
RYANIA POWDER see RSZ000
RYANIA SPECIOSA see RSZ000
RYANICIDE see RSZ000
RYANODINE see RSZ000
RYEMYCIN-B2 see TAH675
RYLUX BSP (CZECH) see BHN250
RYLUX PRS (CZECH) see BKO750
RYNACROM see CNX825
RYODIPINE see RSZ375
RYOMYCIN see HOH500
RYTHMODAN see RSZ600
RYTHMOL see BHR750
RYTHRITOL see PBC250
RYTMONORM see PMJ525
RYUTAN (JAPANESE) see RSZ675

S-6 see MIH300
S 13 see BHD250
S 22 see SAK500
S-33 see CIF250
S 46 see PFB350
S51 see BBV500
S-62 see ARW750
S 77 see DRC600
S 84 see BNF750
S115 see NCQ900
S 140 see DAM700
S 142 see PNB250
S 151 see EJM500, SAA000
S 154 see BPR500
S 187 see DPI700
S 190 see KDU100
S 202 see DHK600
S 222 see BKB250
S 276 see DXH325
S410 see DSK600
S 421 see OAL000
S-578 see DYF700
S 596 see BQE000
S 62-2 see ARW750
S 650 see BQH250
S 767 see FAQ800
S-805 see DCS200
S-847 see CEW500
S 852 see DBL700
S 940 see HMV000
S-1000 see DXS375
S 1006 see PJY100
S-1046 see XTJ000

S 112A see DSQ000
S 1132 see PJY100
S 1520 see IBV100
S 1530 see DLV000
S 1544 see PFC750
S 1688 see AMJ500
S 1702 see DBL700
S 1752 see FAQ900
S 1942 see BNL250
S 2225 see EGV500
S 2940 see DRR400
S 2957 see CLJ875
S-3151 see AHJ750
S 32-06 see DAB825
S 3308 see DGC100
S 4004 see YCJ200
S 4084 see COQ399
S 4087 see CON300
S 5602 see FAR100
S 5660 see DSQ000
6059S see LBH200
S-6115 see CQI750
6315-S see FCN100
S 640P see APT250
S 6437 see ALV000
S 6900 see DRR200
S-6,999 see NNF000
S 7131 see PMF750
S 8527 see CMV700
S-9115 see CQI750
S-9700 see DAB880
S 10165 see DGI000
10275-S see EBD500
10364S see MCH600
S-15126 see DED000
31252-S see HNT075, PCJ325
S-3440 see BFV765
450191-S see AHP875
710674-S see CMW550
711389-S see DFM875
S 73 4118 see PDW250
S 65 (polymer) see PKQ059
S6 (pharmaceutical) see MIH300
SA see SAI000
SA 79 see PMJ525
SA96 see MCO775
S 99 A see SLD800
SA 111 see SNJ000
SA 217 see CBI000
Sa 267 see DGW600
SA 446 see FAQ950
SA 504 see TGB160
SA 546 see OMY850
SA 42-548 see MBV250
SAATBEIZFUNGIZID (GERMAN)
     see HCC500
SABACIDE see VHZ000
SABADILLA see VHZ000
SABADININE see EBL000
SABANE DUST see VHZ000
SABARI see HGC500
SABILA (CUBA, PUERTO RICO) see AGV875
SABLIER (HAITI) see SAT875
SACABUCHE (PUERTO RICO) see JBS100
SACARINA see BCE500
SACCAHARIMIDE see BCE500
SACCHARATED FERRIC OXIDE see IHG000
SACCHARATED IRON see IHG000
SACCHARIN see SJN700
SACCHARINA see BCE500
SACCHARIN ACID see BCE500
SACCHARIN AMMONIUM see ANT500
SACCHARINATE AMMONIUM see ANT500
SACCHARIN CALCIUM see CAM750
SACCHARINE see BCE500
SACCHARINE SOLUBLE see SJN700
SACCHARINNATRIUM see SJN700
SACCHARINOL see BCE500
SACCHARINOSE see BCE500
SACCHARIN, SODIUM see SJN700
SACCHARIN, SODIUM SALT see SJN700
SACCHARIN SOLUBLE see SJN700
SACCHAROIDUM NATRICUM see SJN700
SACCHAROL see BCE500

SACCHAROLACTIC ACID see GAR000
SACCHAROSE see SNH000
SACCHARUM see SNH000
SACCHARUM LACTIN see LAR000
SACERIL see DKQ000, DNU000
SACERNO see MKB250
SACHSISCHBLAU see FAE100
SACROMYCIN see AHL000
SAD see EDA500
SAD-128 see SAB800
SADAMIN see XCS000
SADH see DQD400
SA 97 DIHYDROCHLORIDE see CJR809
SADOFOS see MAK700
SADOPHOS see MAK700
SADOPLON see TFS350
SADOREUM see IDA000
SAEURE FLUORIDE (GERMAN) see FEZ000
SAFE-N-DRI see AHF500
SAFFAN see AFK500
SAFFLOWER OIL see SAC000
SAFFLOWER OIL (UNHYDROGENATED) (FCC) see SAC000
SAFFRON YELLOW see DUX800
SAFRASIN see SAC875
SAFROL see SAD000
SAFROLE see SAD000
SAFROLE MF see SAD000
SAFROTIN see MKA000, SAD100
SAFSAN see DXE000
SAGAMICIN see MQS579
SAGAMICIN (OBS.) see MQS579
SAGATAL see NBU000
SAGE OIL see SAE500, SAE550
SAGE OIL, DALMATIAN TYPE see SAE500
SAGE OIL, SPANISH TYPE see SAE550
SAGO CYCAS see CNH789
SAGRADO see SAF000
SAH see SAG000
SAH 22 see SEM500
SaH 43-715 see POB300
SaH 46-790 see THH350
SAICLATE see DNU100
SAIKOSIDE, CRUDE see SAF300
SAIPHOS see ASD000
SAISAN see MLC250
SAKOLYSIN (GERMAN) see BHT750
SAKURAI No. 864 see YCJ000
SAL see HMK100
SALACETIN see ADA725
SALACHLOR see SHJ000
SALAMID see SAH000
SALAMIDE see SAH000
SAL AMMONIA see ANE500
SAL AMMONIAC see ANE500
SALAZOLON see AQN250
SALAZOPYRIN see PPN750
SALAZOSULFAPYRIDINE see PPN750
SALBEI OEL (GERMAN) see SAE500, SAE550
SALBUTAMOL see BQF500
SALBUTAMOL HEMISULFATE see SAF400
SALBUTAMOL SULFATE see SAF400
SALCETOGEN see ADA725
SALCOMIN see BLH250
SALCOMINE POWDER see BLH250
SAL ENIXUM see PKX750
SALETIN see ADA725
SALI see SAF500
SALICILAMIDE (ITALIAN) see SAH000
SALICIM see SAH000
SALICOL see CMG000
SALICRESIN FLUID see CHW675
SALICYLADEHYDE see SAG000
SALICYLAL see SAG000
SALICYL ALCOHOL see HMK100
SALICYLALDEHYDE see SAG000
SALICYLALDEHYDE ETHYLENEDIIMINE COBALT see BLH250
SALICYLALDEHYDE METHYL ETHER see AOT525
SALICYLALDEHYDE OXIME see SAG500
SALICYLALDOXIME see SAG500
SALICYLAMID see SAH000
SALICYLAMIDE see SAH000
SALICYLANILIDE see SAH500
SALICYLAZOSULFAPYRIDINE see PPN750
SALICYL-DIAETHYL (GERMAN) see BQH250

SALICYLDIETHYLAMIDE see DJY400
SALICYL-DIMETHYL (GERMAN) see BQH500
SALICYLDIMETHYLAMIDE see DUA800
SALICYLHYDROXAMIC ACID see SAL500
SALICYLIC ACID see SAI000
m-SALICYLIC ACID see HJI100
p-SALICYLIC ACID see SAI500
SALICYLIC ACID, 4-AMINO-, 2-(DIETHYLAMINO)ETHYL ESTER, HYDRO-
    CHLORIDE see AMM750
SALICYLIC ACID CHOLINE SALT see CMG000
SALICYLIC ACID, DIHYDROGEN PHOSPHATE see PHA575
SALICYLIC ACID-2-ETHYLHEXYL ESTER see ELB000
SALICYLIC ACID, FLUOROACETATE see FFT000
SALICYLIC ACID-3-HEXEN-1-YL ESTER see SAJ000
SALICYLIC ACID, ISOBUTYL ESTER see IJN000
SALICYLIC ACID, ISOPENTYL ESTER see IME000
SALICYLIC ACID ISOPROPYL ESTER O-ESTER with O-ETHYL
    ISOPROPYLPHOSPHORAMIDOTHIOATE see IMF300
SALICYLIC ACID, METHYL ESTER see MPI000
SALICYLIC ACID, METHYL ESTER, ESTER with p-GUANIDINOBENZOIC
    ACID see CBT175
SALICYLIC ACID, compounded with MORPHOLINE (1:1) see SAI100
SALICYLIC ACID OCTYL ESTER see SAJ500
SALICYLIC ACID PENTYL ESTER see SAK000
SALICYLIC ACID-3-PHENYLPROPYL ESTER see SAK500
SALICYLIC ACID with PHYSOSTIGMINE (1:1) see PIA750
SALICYLIC ACID, SODIUM SALT see SJO000
SALICYLIC ACID, p-TOLYL ESTER see THD850
SALICYLIC ALDEHYDE see SAG000
SALICYLIC ETHER see SAL000
SALICYLIC ETHYL ESTER see SAL000
SALICYLOHYDROXAMIC ACID see SAL500
SALICYLOHYDROXIMIC ACID see SAL500
SALICYLOYLGLYCINE see SAN200
SALICYLOYLOXYTRIBUTYLSTANNANE see SAM000
SALICYLOYLSALICYLIC ACID see SAN000
SALICYL PHOSPHATE see PHA575
o-SALICYLSALICYLIC ACID see SAN000
SALICYLSULFONIC ACID see SOC500
SALICYLURIC ACID see SAN200
SALIGENIN see HMK100
SALIGENOL see HMK100
SALIMED see CPR750
SALIMID see CPR750
SALINE see SFT000
SALIPHENAZON see AQN250
SALIPRAN see SAN600
SALIPUR see SAH000
SALIPYRAZOLAN see AQN250
SALIPYRINE see AQN250
SALISAN see CLH750
SALISOD see SJO000
SALITHION see MEC250
SALITHION-SUMITOMO see MEC250
SALIX see CHJ750
SALIZELL see SAH000
SAL de MERCK see CNF000
SALMIDOCHOL see DYE700
SALMINE SULFATE see SAO200
SALMINE SULFATE (1:1) see SAO200
SALMONELLA ENTERITIDIS ENDOTOXIN see SAO250
SALMONELLA TYPHI ENDOTOXIN see EAS100
SALMONELLA TYPHOSA ENDOTOXIN see EAS100
SALOL see PGG750
SALP see SEM300
SALPETERSAURE (GERMAN) see NED500
SALPETERZUUROPLOSSINGEN (DUTCH) see NED500
SALPIX see AAN000
SALRIN see SAH000
SALSOCAIN see SAO475
SALSONIN see SJO000
SALT see SFT000
SALT BATHS (NITRATE OR NITRITE) see SAO500
SALT CAKE see SJY000
SALT OF TARTER see PLA250
SALTPETER see PLL500
SALT of SATURN see LCV000
SALUFER see DXE000
SALUNIL see CLH750
SALURAL see BEQ625
SALURES see BEQ625
SALURETIL see CLH750
SALURETIN see CLY600
SALURIC see CLH750

SALURIN see HII500, SIH500
SALVACARD see DJS200
SALVACORIN see DJS200
SALVADERA (CUBA) see SAT875
SALVARSAN see SAP500
SALVIS see SEP000
SALVO see DAA800, HKC000
SALVO LIQUID see BCL750
SALVO POWDER see BCL750
SALYMID see SAH000
SALYRGAN see SIH500
SALYRGAN THEOPHYLLINE see MDH750, SAQ000
SALYZORON see BBW500
SALZBURG VITRIOL see CNP500
SAM see DUA800, SAH000
SAMARIUM see SAQ500
SAMARIUMACETAT (GERMAN) see SAR000
SAMARIUM ACETATE see SAR000
SAMARIUM(III) CHLORIDE see SAR500
SAMARIUM CITRATE see SAS000
SAMARIUM NITRAT (GERMAN) see SAT000
SAMARIUM NITRATE see SAT200
SAMARIUM(III) NITRATE, HEXAHYDRATE (1583586) see SAT000
SAMARIUM TRINITRATE see SAT200
SAMBUCUS CAERULEA see EAI100
SAMBUCUS CANADENSIS see EAI100
SAMBUCUS MELANOCARPA see EAI100
SAMBUCUS MEXICANA see EAI100
SAMBUCUS RACEMOSA see EAI100
SAMECIN see HGP550
SAMID see SAH000
SAMURON see INR000
SAN see ALQ625
SAN 155 see TFH750
SAN 230 see PHI500
SAN 1551 see TFH750
SANAI, seed extract see CNX827
SANASEED see SMN500
SANASPASMINA see CBA375
SANASTHYMYL see AFJ625
SANATRICHOM see MMN250
SANCAP see EPN500
SANCHINOSIDE E1 see PAF450
SANCLOMYCINE see TBX000
SAN-CYAN see COI250
SANCYCLAN see DNU100
SAND see SCI500, SCJ500
SAND ACID see SCO500
SANDALWOOD OIL, EAST INDIAN see OHG000
SANDAMYCIN see AEA750
SANDBOX TREE see SAT875
SAND CORN see DAE100
SANDESIN see AEH750
SANDIMMUN see CQH100
SANDIMMUNE see CQH100
SANDIX see LDS000
SANDOCRYL BLUE BRL see BJI250
SANDOLIN see DUS700
SANDOPEL BLACK EX see AQP000
SANDOPTAL see AGI750
SANDORMIN see TEH500
SANDOSCILL see POB500
SANDOTHRENE PRINTING YELLOW see DCZ000
SANDOZ 6538 see DJY200
SANDOZ 43-715 see POB300
SANDOZ 52139 see MKA000
SANDRIX see DAB750
SANEDRINE see EAW000
SANEGYT see CAY875
SAN-EI BRILLIANT SCARLET 3R see FMU080
SANEPIL see DKQ000
SANFURAN see NGE500
SANG gamma see BBQ500
SANGIVAMYCIN see SAU000
SANGUICILLIN see AOD000
SANGUIRITRIN see SAU350
SANGUIRITRINE see SAU350
SANGUIRYTHRINE see SAU350
SAN 155 1 see TFH750
SAN 244 1 see DRR200
SAN 6538 1 see DJY200
SAN 6913 1 see DRR200
SAN 7107 1 see DRR200
SAN 52 139 1 see MKA000

SANICLOR 30 see PAX000
SANIPIROL see SEP000
SANIPRIOL-4 see AMM250
SANITIZED SPG see ABU500
SANLOSE SN 20A see SFO500
SANMARTON see FAR100
SANMORIN OT 70 see DJL000
SANOCHOLEN see DAL000
SANOCHRYSINE see GJG000
SANOCIDE see HCC500
SANOCRISIN see FBP100
SANODIAZINE see PPP500
SANODIN see CBO500
SANOFLAVIN see DBN400
SANOHIDRAZINA see ILD000
SANOMA see IPU000
SANOQUIN see CLD000, CLD250
SANORIN see NAH500
SANORIN-SPOFA see NCW000
SANPRENE LQX 31 see PKP000
SANQUINON see DFT000
SANREX see ADY500
SANSDOLOR see GGS000
SANSEL ORANGE G see PEJ500
SANSERT see MQP500
SANSPOR see CBF800
α-SANTALOL (FCC) see OHG000
SANTALOZONE see DAF150
SANTALYL ACETATE see SAU400
SANTAR see MCT500
SANTAVY'S SUBSTANCE F see MIW500
SANTHEOSE see TEO500
SANTICIZER 160 see BEC500
SANTICIZER 711 see DXQ400
SANTICIZER 141 (MONSANTO) see DWB800
SANTICIZIER B-16 see BQP750
SANTOBANE see DAD200
SANTOBRITE see PAX250, SJA000
SANTOCEL see SCH000
SANTOCHLOR see DEP800
SANTOCURE see CPI250
SANTOCURE MOR see BDG000
SANTOFLEX 17 see BJT500
SANTOFLEX 36 see PFL000
SANTOFLEX 77 see BJL000
SANTOFLEX 134 see SAU475
SANTOFLEX A see SAV000
SANTOFLEX AW see SAV000
SANTOFLEX IC see PEY500
SANTOMERSE 3 see DXW200
SANTONIN see SAU500
α-SANTONIN see SAU500
l-α-SANTONIN see SAU500
SANTONINIC ANHYDRIDE see SAU500
SANTONOX see TFC600
SANTOPHEN see CJU250, PAX250
SANTOPHEN 20 see PAX250
SANTOPHEN I GERMICIDE see CJU250
SANTOQUIN see SAV000
SANTOQUINE see SAV000
SANTOQUIN EMULSION see TMJ000
SANTOQUIN MIXTURE 6 see TMJ000
SANTOTHERM see PJL750
SANTOUAR A see DCH400
SANTOVAR A see DCH400
SANTOWAX see TBC750
SANTOWAX M see TBC620
SANTOWHITE POWDER see BRP750
SANTOX see EBD700
SANULCIN see DAB875
SANVEX see BHL750
SANWAPHYLLIN see HOA000
SANYO BENZIDINE YELLOW-B see DEU000
SANYO CARMINE L2B see DMG800
SANYO FAST BLUE SALT B see DCJ200
SANYO FAST RED 10B see NAY000
SANYO FAST RED SALT TR see CLK235
SANYO FAST RED TR BASE see CLK220
SANYO FAST SCARLET GG BASE see DEO400
SANYO LAKE RED C see CHP500
SANYO SCARLET PURE see MMP100
SAOLAN see DSK200
SAPECRON see CDS750
SAPEP see AMD000

SAPHICOL see ASD000
SAPHIZON see ASD000
SAPHIZON-DP see ASD000
SAPHOS see ASD000
SAPHRAZINE see SAC875
SAPILENT see DLH200
SAPONA RED LAKE RL-6280 see NAY000
SAPONIN see SAV500
SAPONIN-GYPSOPHILA see GMG000
SAPPILAN see CJT750
SAPPIRAN see CJT750
SAPRECON C see DRP600
SAPWOOD of LINDEN see SAW300
SARAN see SAX000
SARCELL TEL see SFO500
SARCLEX see DGD600
SARCOCLORIN see BHT750
l-SARCOLYSIN see PED750
dl-SARCOLYSIN see BHT750
p-l-SARCOLYSIN see PED750
o-dl-SARCOLYSIN see BHT250
d-SARCOLYSINE see SAX200
dl-SARCOLYSINE see BHT750
m-l-SARCOLYSINE see SAX210
l-SARCOLYSINE HYDROCHLORIDE see BHV250
dl-SARCOLYSINE HYDROCHLORIDE see BHV000
SARCOLYSIN HYDROCHLORIDE see BHV000
SARCOMYCIN see SAX500
SARCOSET GM see DTG000
SARGOTT see YAG000
SARIFA, seed extract see SAX275
SARIN see IPX000
SARIN II see IPX000
SARKOKLORIN see BHV000
SARKOMYCIN see SAX500
SARKOMYCIN B*, SODIUM SALT see SJS500
SARKOMYCIN compounded with NICOTINIC HYDRAZIDE see SAY000
SARKOMYCIN, SODIUM SALT see SJS500
SARODORMIN see DYC800
SAROLEX see DCM750
SAROMET see DCK759
SAROTEN see EAI000
SAROTENE see EAI000
SARPAN see AFG750
SARPIFAN HP 1 see AAX175
SARTOMER SR 206 see BKM250
SARTOMER SR 351 see TLX175
SARTOSONA see ALV000
SARYUURIN see DYE700
S.A.S.-500 see PPN750
SAS 538 see CFY500
SAS 643 see DKV400
SASAPYRIN see SAN000
SA 96 SODIUM SALT see SAY875
SASP see PPN750
SASSAFRAS see SAY900
SASSAFRAS ALBIDUM see SAY900
SASTRIDEX see TKH750
SATANOLON see CCR875
SATECID see CHS500
SATIAGEL GS 350 see CCL250, CCL350
SATIAGUM 3 see CCL250
SATIAGUM STANDARD see CCL250
SATILAN PLZ see CNH125
SATILAN RL 2 see CNH125
SATINITE see CAX750
SATIN SPAR see CAX750
SATOL see OBA000
SATOX 20WSC see TIQ250
SATRANIDAZOLE see SAY950
SATURN see SAZ000
SATURN BROWN LBR see CMO750
SATURNO see SAZ000
SATURN RED see LDS000
SAUCO BLANCO (CUBA) see EAI100
SAUCO (MEXICO, PUERTO RICO) see EAI100
SAURE DES PHYTINS (GERMAN) see PIB250
SAUROL see IAD000
SAUTERALGYL see DAM700
SAUTERZID see ILD000
SAVAC see DNG400
SAVENTRINE see DMV600, IMR000
SAVORY OIL (summer variety) see SBA000
SAX see SAI000

SAXIN see BCE500, SJN700
SAXITOXIN (8CI) see SBA600
SAXITOXIN DIHYDROCHLORIDE see SBA500
SAXITOXIN HYDRATE see SBA600
SAXITOXIN HYDROCHLORIDE see SBA500
SAXOL see MQV750
SAXOSOZINE see SNN500
SA 50 Y see TGB160
SAYFOR see ASD000
SAYFOS see ASD000
SAYPHOS see ASD000
SAYTEX 102 see PAU500
SAYTEX 102E see PAU500
SAZZIO see AFL000
SB-8 see HJZ000
SB-23 see DQY909
SB 502 see BOO650
SB 5833 see CFY250
S.B.A. see BPW750
SBa 0108E see BAK000
SBO see DJL000
SBP-1382 see BEP500
SBP-1390 see BEP750
SBP-1513 see AHJ750
S.B. PENICK 1382 see BEP500
SBS see SMR000
SBTPC see SOU675
SC10 see DCD000
SC-110 see ABU500
SC. 121 see PAW600
SC-1950 see DRK600
SC-2292 see MEK350
SC 2538 see DIR000
SC-2644 see DOB300
SC-2657 see DOA875
SC-2798 see DOB325
SC 2910 see DJM800, XCJ000
SC-3171 see HKR500
SC-3296 see BKP300
SC-3402 see BKI750
S.C. 3497 see AFW500
SC 4641 see NCI525
SC-4642 see EEH550
SC 4722 see ONO000
SC 5914 see ENX600
SC 7031 see DNN600
SC 8016 see CJL500
SC-8117 see MDQ100
SC 9022 see MDQ075
SC 9387 see CJL500
SC 9420 see AFJ500
SC 9794 see PII500
SC/3123 see SBA875
SC 10295 see MMN250
SC10363 see VTF000
SC 11927 see CCP750
SC 12937 see ARX800
SC 13957 see RSZ600
SC 15090 see TEO500
SC 15983 see AFJ500
SC 27166 see DWF700
SC 28762 see VRP200
SC-29333 see MJE775
SC-11800M see SBA625
SCABANCA see BCM000
SCALDIP see DVX800
SCANBUTAZONE see BRF500
SCANDICAIN see CBR250, SBB000
SCANDICAINE see SBB000
SCANDICANE see SBB000
SCANDISIL see SNN300
SCANDIUM see SBB500
SCANDIUM CHLORIDE see SBC000
SCANDIUM (3+) CHLORIDE see SBC000
SCAPUREN see SBE500
SCARCLEX see DGD600
SCARLET BASE CIBA I see DEO400
SCARLET BASE CIBA II see NMP500
SCARLET LOBELIA see CCJ825
SCARLET PIGMENT RN see MMP100
SCARLET R see FMU070
SCARLET RED see SBC500
SCARLET TR BASE see CLK200
SCARLET WISTERIA TREE see SBC550

SCATOLE see MKV750
SCE 129 see CCS550
SC 11800 EE see DAP810
SCE 1365 HYDROCHLORIDE see CCS300
SCF see TEE225
SCH see FMR050
SCH 286 see SOU650
SCH 412 see DJO800
Sch 1000 see IGG000
Sch 4831 see BFV750
SCH 5472 see BBW000
Sch 5705 see PCU400
Sch 5706 see FAJ150
Sch 5712 see BRJ125
SCH 6673 see ABG000
SCH 7307 see IPU000
SCH 9384 see AEX000
SCH 9724 see GCS000
SCH 10015 see HIE525
SCH 10304 see CMX770
SCH 10649 see DLV800
SCH 11527 see TEY000
SCH 12650 see SBC800
SCH 13475 see SDY750
SCH 13521 see FMR050
Sch 14947 see RMF000
SCH-17726 see GAA100
SCH 20569 see NCP550, SBD000
SCH 22219 see AFI980
SCH 15719w see HMM500
SCH 18020W see AFJ625
SCHAARDINGER α-DEXTRIN see COW925
SCHAEFFER'S-β-NAPHTHOLSULFONIC ACID see HMU500
SCHARLACH B see PEJ500
SCH 5802 B see SBC700
SCH CHLORIDE see HLC500
SCHEELES GREEN see CNN500
SCHEELE'S MINERAL see CNN500
SCHERCEMOL 85 see ISC550
SCHERING 34615 see CQI500
SCHERING-35830 see CGL250
SCHERING 36056 see DSO200
SCHERING 36103 see FNE500
SCHERING 36268 see CJJ250, CJJ500
SCHERING 38107 see EEO500
SCHERING-38584 see MEG250
SCHERISOLON see PMA000
SCHEROSON see CNS800, CNS825
SCHEROSON F see CNS750
SCH 31846 HYDROCHLORIDE see ICL500
SCHINOPSIS LORENTZII TANNIN see QBJ000
SCHINUS MOLLE see PCB300
SCHINUS MOLLE OIL see SBE000
SCHINUS TEREBINTHIFOLIUS see PCB300
SCHISANDRIN see SBE400
SCHISANDROL B see SBE450
SCHISTOSOMICIDE see SBE500
SCHIZANDRIN see SBE400
SCHIZANDROL B see SBE400, SBE450
SCHLAFEN see CAV000
SCHLEIMSAURE see GAR000
SCHOENOCAULON DRUMMONDII see GJU460
SCHOENOCAULON OFFICIANLIS see GJU460
SCHOENOCAULON TEXANUM see GJU460
SCHOENOPLECTUS ARTICULATUS (Linn.) Palla, extract see SBF550
SCHRADAN see OCM000
SCHRADANE (FRENCH) see OCM000
SCH 21420 SULFATE see SBE800
SCHULTENITE see LCK000
SCHULTZ No. 39 see HGC000
SCHULTZ No. 95 see FMU070
SCHULTZ No. 770 see FMU059
SCHULTZ No. 853 see PPQ750
SCHULTZ No. 1038 see BJI250
SCHULTZ No. 1041 see AJP250
SCHULTZ Nr. 208 (GERMAN) see HJF500
SCHULTZ Nr. 826 (GERMAN) see ADE500
SCHULTZ Nr. 836 (GERMAN) see ADF000
SCHULTZ Nr. 1309 (GERMAN) see FAE100
SCHULTZ-TAB No. 779 (GERMAN) see RMK020
SCHUTTGELB see MQF250
L-SCHWARZ 1 see BMA000
SCHWEFELDIOXYD (GERMAN) see SOH500
SCHWEFELKOHLENSTOFF (GERMAN) see CBV500

SCHWEFEL-LOST see BIH250
SCHWEFELSAEURELOESUNGEN (GERMAN) see SOI500
SCHWEFELWASSERSTOFF (GERMAN) see HIC500
SCHWEFLIGE SAURE (GERMAN) see SOO500
SCHWEINFURTERGRUN see COF500
SCHWEINFURT GREEN see COF500
SCILLACRIST see POB500
SCILLA "DIDIER" see POB500
SCILLAGLYKOSID A (GERMAN) see GFC000
SCILLAREN A see GFC000
SCILLAREN-3,6-DEOXY-4-o-β-d-GLUCOPYRANOSYL-α-l-MANNOPYRANOS-
IDE see GFC000
α-l-SCILLARENIN-3-RHAMNOPYRANOSIDE see POB500
SCILLARENIN-RHAMNOSE (GERMAN) see POB500
SCILLAREN & RHAMNOSE & GLUCOSE (GERMAN) see GFC000
SCILLA (VARIOUS SPECIES) see SLH200
SCILLIROSID see SBF500
SCILLIROSIDE see SBF500
(3-β,6-β)SCILLIROSIDE see SBF500
SCILLIROSIDE GLYCOSIDE see RCF000
SCILLIROSIDIN + GLUCOSE (GERMAN) see SBF500
3-β-SCILLOBIOSIDO-14-β-HYDROXY-Δ-4,20,22-BUFATRIENOLID (GERMAN)
see GFC000
SCINNAMINA see TLN500
SCINTILLAR see XHS000
SCIRPUS ARTICULATUS Linn., extract see SBF550
SCLAVENTEROL see NGG500
SCOBRO see SBG500
SCOBRON see SBG500
SCOBUTIL see SBG500
SCOBUTYL see SBG500
SCOKE see PJJ315
SCOLINE see HLC500
SCOLINE CHLORIDE see HLC500
SCON 5300 see PKQ059
SCONATEX see AAX175, VPK000
SCOPAMIN see HOT500
SCOPARON see DRS800
SCOPARONE see DRS800
SCOPINE TROPATE see SBG000
SCOPOLAMINE see SBG000
(−)-SCOPOLAMINE see SBG000
SCOPOLAMINE BROMIDE see HOT500
(−)-SCOPOLAMINE BROMIDE see HOT500
SCOPOLAMINE BROMOBUTYLATE see SBG500
SCOPOLAMINE BUTOBROMIDE see SBG500
SCOPOLAMINE BUTYL BROMIDE see SBG500
SCOPOLAMINE-N-BUTYL BROMIDE see SBG500
SCOPOLAMINE HYDROBROMIDE see HOT500
(−)-SCOPOLAMINE HYDROBROMIDE see HOT500
SCOPOLAMINE METHOBROMIDE see SBH500
SCOPOLAMINE METHYLBROMIDE see SBH500
(−)-SCOPOLAMINE METHYL BROMIDE see SBH500
SCOPOLAMINE METHYL NITRATE see SBH000
SCOPOLAMINE compounded with METHYL NITRATE (1581) see SBH000
SCOPOLAMINIUM BROMIDE see HOT500
SCOPOLAMMONIUM BROMIDE see HOT500
SCOPOLAN see SBG500
SCOPOS see HOT500
SCORBAMIC ACID see AJL125
l-SCORBAMIC ACID see AJL125
SCORBAMINIC ACID see AJL125
SCOT see HBT500
SCOTCH ATTORNEY see BAE325
SCOTCH PAR see PKF750
SCOTCH PINE NEEDLE OIL see PIH500
SCOTCIL see BFD000
SC 7031 PHOSPHATE see RSZ600
SCROBIN see ARQ750
SCTZ see CHD750
SCURENALINE see VGP000
SCUROCAINE see AIL750, AIT250
SCYAN see SIA500
SD 25 see DHY200
40 SD see IMF300
SD-345 see ADR250
SD 354 see SMR000
SD 709 see DRM000
SD-1750 see DGP900
SD 1836 see CLV375
SD 1897 see DDL800
SD 3450 see HCI475
SD 3562 see DGQ875
SD 4,239 see CEQ500

SD 4294 see COD000
SD 4402 see OAN000
SD 5532 see CDR750
SD 7438 see BES250
SD 7727 see DGD400
SD 7961 see DGM600
SD 8447 see RAF100
SD 8530 see TMD000
SD 9098 see DIX600
SD-10576 see SCD500
SD 14114 see BLU000
SD 15418 see BLW750
SD 171-02 see MQQ050
SD 210-32 see PIR000
SD 210-37 see PIQ750
SD 270-31 see OKW000
SD 30,053 see EGS000
SD 417-06 see DAB825
SD 43775 see FAR100
SD ALCOHOL 23-HYDROGEN see EFU000
S DC 200 see SCR400
SDD see DXH300
SDDC see SGM500
SDDS see AOT125
SDEH see DJL400
SD 2124-01 HYDROCHLORIDE see PMF525
(±)-SD 2124-01 HYDROCHLORIDE see PMF535
SDIC see SGG500
(3aS-cis)-3a,12a-DIHYDRO-4,6-DIHYDROXY-ANTHRA(2,3-b)FURO(3,2-d)FURAN-
5,10-DIONE (9CI) see DAZ100
SDM see SNN300
SDMH see DSF600
SDM No. 5 see MAW250
SDM No. 23 see PBC250
SDMO see AIE750, SNN300
SDmP see SNI500
SDPH see DNU000
SDT 1041 see FNF000
S. DYSENTERIAE TOXIN see SCD750
SE-1520 see IBV100
SE 1702 see DBL700
SEA ANEMONE TOXIN II see ARS000
SEA ANEMONE VENOM see SBI800
SEA COAL see CMY635
SEACYL VIOLET R see DBP000
SEA DAFFODIL see BAR325
SEAKEM CARRAGEENIN see CCL250
SEA-LEGS see HGC500
SEA ONION see SLH200
SEAPROSE S see SBI860
SEARLE 703 see DNN600
SEARLE 27166 see DWF700
SEARS HEAVY DUTY LAUNDRY DETERGENT see SBI875
SEA SALT see SFT000
SEA SNAKE VENOM, AIPYSURUS LAEVIS see SBI880
SEA SNAKE VENOM, ASTROTIA STOKESII see SBI890
SEA SNAKE VENOM, HYDROPHIS CYANOCINCTUS see SBI900
SEA SNAKE VENOM, LAPEMIS HARDWICKII see SBI910
SEA SNAKE VENOM, MICROCEPHALOPHIS GRACILIS see SBI929
SEATREM see CCL250
SEAWATER MAGNESIA see MAH500
SEAZINA see SNJ000
SEBACIC ACID see SBJ500
SEBACIC ACID, DIBUTYL ESTER see DEH600
SEBACIC ACID, DIETHYL ESTER see DJY600
SEBACIC ACID, DIHYDRAZIDE see SBK000
SEBACIL see BAT750
SEBACONITRILE see SBK500
SEBACOYLDIOXYBIS(TRIBUTYLSTANNANE) see BLL000
SEBAR see SBN000
SEBATROL see FMR050
SEBERCIM see BAB625
SEBIZON see SNP500
SECACORNIN see LJL000
SECAGYN see EDC500
SECALONIC ACID D see EDA500
SECBUBARBITAL see BPF000
SECBUBARBITAL SODIUM see BPF250
SECBUTABARBITAL see BPF000
SECBUTOBARBITONE see BPF000
SECHVITAN see PII100
SECLAR see BEG000
SECO 8 see SBN000
SECOBARBITAL see SBM500

SECOBARBITAL SODIUM see SBN000
R(+)-SECOBARBITAL SODIUM see SBL500
SECOBARBITONE see SBM500, SBN000
SECOBARBITONE SODIUM see SBN000
21,22-SECOCAMPTOTHECIN-21-OIC ACID LACTONE see CBB870
(5Z,7E)-9,10-SECOCHESTA-5.7.10(19)-TRIENE-1-α,3-β,25-TRIOL see DMJ400
9,10-SECOCHOLESTA-5,7,10(19)-TRIENE-1-α,3-β-DIOL see HJV000
(1-α,3-β,5Z,7E)-9,10-SECOCHOLESTA-5,7,10(19)-TRIENE-1,3,25-TRIOL
see DMJ400
9,10-SECOCHOLESTA-5,7,10(19)-TRIEN-3-β-OL see CMC750
9,10,SECOERGOSTA-5,7,10(19),22-TETRAEN-3-β-OL see VSZ100
(E-β,5E,7E,10-α,22E)-9,10-SECOERGOSTA-5,7,22-TRIEN-3-OL (9CI) see DME300
16,17-SECOESTRA-1,3,5(10),6,8-PENTAEN-17-OIC ACID, 3-METHOXY-
see BIT030
16,17-SECO-13-α-ESTRA-1,3,5,6,7,9-PENTAEN-17-OIC ACID, METHYL ESTER
see BIT000
SECOMETRIN see LJL000
SECONAL see SBM500
SECONAL SODIUM see SBN000
SECONDARY AMMONIUM ARSENATE see DCG800
SECONDARY AMMONIUM PHOSPHATE see ANR500
SECONDARY ZINC PHOSPHITE see ZLS200
SECONESINZ see GGS000
SECOPAL OP 20 see GHS000
SECOVERINE see SBN300
SECOVERINE HYDROCHLORIDE see SBN300
SECROSTERON see DRT200
SECROVIN see DNX500
SECT HYDROCHLORIDE see EEQ500
SECTRAL see AAE125
SECUPAN see EDC500
SECURINAN-11-ONE (9CI) see SBN350
SECURININ see SBN350
(-)-SECURININE see SBN350
SECURININE see SBN350
SECURITY see LCK000
SEDABAMATE see MQU750
SEDAFORM see ABD000
SEDALANDE see HAF400
SEDALIS SEDI-LAB see TEH500
SEDALON see AMK250
SEDAMYL see ACE000
SEDANTOINAL see MKB250
SEDAPRAN see DAP700
SEDAPSIN see CKE750
SEDAREX see EHP000
SEDA-TABLINEN see EOK000
SEDAVIC see HAF400
SEDESTRAN see DKA600
SEDETINE see OAV000
SEDETOL see EJM500
SEDEVAL see BAG000
SEDICAT see EOK000
SEDIMIDE see TEH500
SEDIN see TEH500
SEDIPAM see DCK759
SEDIRANIDO see DBH200
SEDISPERIL see TEH500
SEDIZORIN see EOK000
SEDMYNOL see ACE000
SEDOFEN see EOK000
SEDOMETIL see DNA800, MJE780
SEDONAL see EOK000
SEDONEURAL see SFG500
SEDOPHEN see EOK000
SEDOR see SKS700
SEDORMID see IQX000
SED OSTANTHRENOVA M see CMU475
SEDOVAL see TEH500
SEDRENA see BBV000
SEDRESAN see MEP250
SED-TEMS see MDL000
SEDTRAN see ACE000
SEDURAL see PDC250, PEK250
SEDUTAIN see SBN000
SEDUXEN see DCK759
SEED of FIDDLENECK see TAG250
SEEDRIN see AFK250
SEEDTOX see ABU500
SEEDVAX see AKR500
SEEKAY WAX see TIT500
SEFACIN see TEY000
SEFONA see IEP200
SEFRIL see SBN440

SEFRIL HYDRATE see SBN450
SEGNALE LIGHT ORANGE RNG see DVB800
SEGNALE LIGHT RED B see MMP100
SEGNALE LIGHT RED PRG see CJD500
SEGNALE LIGHT RUBINE RG see NAY000
SEGNALE LIGHT YELLOW 2GR see DEU000
SEGNALE RED LC see CHP500
SEGONTIN see PEV750
S. EGRELTRI ATUNUN (TURKISH) see BML000
SEGUREX see BON325
SEGURIL see CHJ750
SELACRYN see TGA600
SELECRON see BNA750
SELECTIVE see BPG250
SELECTOMYCIN see SLC000
SELECTROL see SBN475
SELEKTIN see BKL250
SELEKTON B 2 see EIX500
SELEN (POLISH) see SBO500
1,2,5-SELENADIAZOLE-3-CARBOXAMIDE, 4-AMINO- see AMN300
1,2,5-SELENADIAZOLE-3-CARBOXAMIDE, 4-AMINO-N-METHYL- see MGL600
SELENIC ACID see SBN500
SELENIC ACID, liquid (DOT) see SBN500
SELENIC ACID, DIPOTASSIUM SALT see PLR750
SELENIC ACID, DISODIUM SALT, DECAHYDRATE see SBN505
SELENIDE, DIALLYL- see DBL300
SELENINIC ACID, PROPYL- see PNV760
β-SELENINOPROPIONIC ACID see SBN510
SELENINYL BIS(DIMETHYLAMIDE) see SBN525
SELENINYL BROMIDE see SBN550
SELENINYL CHLORIDE see SBT500
SELENIOUS ACID see SBO000
SELENIOUS ACID, DISODIUM SALT see SJT500
SELENIOUS ACID DISODIUM SALT PENTAHYDRATE see SJT600
SELENIOUS ANHYDRIDE see SBQ500
SELENIUM see SBO500
SELENIUM (colloidal) see SBP000
SELENIUM ALLOY see SBO500
SELENIUM BASE see SBO500
SELENIUM CHLORIDE see SBS500
SELENIUM(IV) CHLORIDE (1:4) see SBU000
SELENIUM CHLORIDE OXIDE see SBT500
SELENIUM COMPOUNDS see SBP500
SELENIUM CYSTINE see DWY800, SBP600
SELENIUM, DICHLOROBIS(2-CHLOROCYCLOHEXYL)- see DEU100
SELENIUM, DICHLOROBIS(2-ETHOXYCYCLOHEXYL)- see DEU125
SELENIUM DIETHYLDITHIOCARBAMATE see DJD400, SBP900
SELENIUM DIMETHYLDITHIOCARBAMATE see SBQ000
SELENIUM DIOXIDE see SBO000, SBQ500
SELENIUM(IV) DIOXIDE (1582) see SBQ500
SELENIUM DIOXIDE mixed with ARSENIC TRIOXIDE (1:1) see ARJ000
SELENIUM(IV) DISULFIDE (1582) see SBR000
SELENIUM DISULFIDE (2.5%) SHAMPOO see SBR500
SELENIUM(IV) DISULFIDE SHAMPOO (2.5%) see SBR500
SELENIUM DISULPHIDE (DOT) see SBR000
SELENIUM DUST see SBO500
SELENIUM ELEMENTAL see SBO500
SELENIUM FLUORIDE see SBS000
SELENIUM HEXAFLUORIDE see SBS000
SELENIUM HOMOPOLYMER see SBO500
SELENIUM HYDRIDE see HIC000
SELENIUM METAL POWDER, NON-PYROPHORIC (DOT) see SBO500
SELENIUM MONOCHLORIDE see SBS500
SELENIUM MONOSULFIDE see SBT000
SELENIUM OXIDE see SBQ500
SELENIUM OXYCHLORIDE see SBT500
SELENIUM SULFIDE see SBR000, SBT000
SELENIUM SULPHIDE see SBT000
SELENIUM TETRACHLORIDE see SBU000
SELENOASPIRINE see SBU100
2,2'-SELENOBIS(BENZOIC ACID) see SBU150
o,o'-SELENOBIS(BENZOIC ACID) see SBU150
SELENOCYANIC ACID, POTASSIUM SALT see PLS000
SELENOCYSTINE see DWY800, SBP600, SBU200
d,l-SELENOCYSTINE see SBU200
SELENO-dl-CYSTINE see SBU200
4,4'-(SELENODI-2,1-ETHANEDIYL)BISMORPHOLINE DIHYDROCHLORIDE see MRU255
SELENODIGLUTATHIONE see SBU500
2-SELENOETHYLGUANIDINE see SBU600
6-SELENOGUANOSINE see SBU700
6-SELENO-GUANOSINE (9CI) see SBU700
SELENOHOMOCYSTINE see SBU710
SELENOMERCAPTOAETHYLGUANIDIN (GERMAN) see SBU600

SELENOMETHIONINE see SBU725
SELENONIUM, TRIMETHYL- see TME500
SELENONIUM, TRIMETHYL-, CHLORIDE see TME600
SELENOPHOS see SBU900
SELENOSULFURIC ACID, 2-AMINOETHYL ESTER see AJS950
SELENO-TOLUIDINE BLUE see SBU950
SELENOUREA see SBV000
SELENSULFID (GERMAN) see SBT000
SELEPHOS see PAK000
SELES BETA see TAL475
SELEXID see MCB550
SELEXOL see DOM100
SELFER see IDE000
SELF ROCK MOSS see CCL250
SELINON see DUS700
SELLAITE see MAF500
SELOKEN see SBV500
SELSUN see SBW000
SELSUN BLUE see SBR000
SEL-TOX SSO2 and SS-20 see DXG000
SEM (cytostatic) see TND500
SEMAP see PAP250
SEMBRINA see DNA800, MJE780
SEMDOXAN see CQC500, CQC650
SE-MEG see SBU600
SEMERON see INR000
SEMESAN see CHO750
SEMICARBAZIDE see HGU000
SEMICARBAZIDE, 2-(o-CHLOROPHENETHYL)-3-THIO- see CJK100
SEMICARBAZIDE HYDROCHLORIDE see SBW500
SEMICARBAZIDE, 2-(p-METHYLPHENETHYL)-3-THIO- see MNP500
SEMICARBAZIDE, 1-PHENETHYL- see PDK300
SEMICARBAZIDE, 2-PHENETHYL-3-THIO- see PDK500
SEMICILLIN see AIV500
SEMIKON see DPJ400, TEO250
SEMIKON HYDROCHLORIDE see DPJ400
SEMILLA de CULEBRA (MEXICO) see RMK250
SEMINOLE BEAD see RMK250
SEMINOLE BREAD see CNH789
SEMOPEN see PDD350
SEMOXYDRINE see DBA800
SEMPERVIVUM (JAMAICA) see AGV875
SEMUSTINE see CHD250
SENAQUIN see CLD000
SENCEPHALIN see ALV000
SENCOR see MQR275
SENCORAL see MQR275
SENCORER see MQR275
SENCOREX see MQR275
SENDOXAN see CQC500
SENDRAN see PMY300
SENDUXAN see CQC500, CQC650
SENECA OIL see PCR250
SENECIA JACOBAEA see RBA400
SENECIA LONGILOBUS see RBA400
SENECIO CANNABIFOLIUS, leaves and stalks see SBW950
SENECIOIC ACID see MHT500
SENECIO LONGILOBUS see SBX000
SENECIONANIUM, 12-(ACETYLOXY)-14,15,20,21-TETRADEHYDRO-15,20-DIHYDRO-8-HYDROXY-4-METHYL-11,16-DIOXO-, (8-xi,12-β,14-Z)- see CMV950
SENECIO NEMORENSIS FUCHSII, alkaloidal extract see SBX200
SENECIONINE see ARS500
SENECIO VULGARIS see RBA400
SENECIPHYLLIN see SBX500
SENECIPHYLLINE see SBX500
SENECIPHYLLIN HYDROCHLORIDE see SBX525
SENEGAL GUM see AQQ500
SENEGENIN see SBY000
SENEGIN see SBY000
SENF OEL (GERMAN) see AGJ250
SENFOL (GERMAN) see ISK000
SENIRAMIN see RCA200
SENKIRKIN see DMX200
SENKIRKINE see DMX200
SENTIPHENE see CJU250
SENTONIL see PDW500
SENTRY see HOV500
SENTRY GRAIN PRESERVER see PMU750
SEOTAL see SBN000
SEPAN see CMR100
SEPAZON see CMY125
SEPERIDOL see CLS250
SEPEROL see CLS250

SEPPIC MMD see CIR250
SEPSINOL see FPI000
SEPTACIDIN see SBZ000
SEPTACIL see ALF250
SEPTAMIDE ALBUM see SNM500
SEPTAMYCIN see SCA000
SEPTICOL see CDP250
SEPTINAL see SNM500
SEPTIPULMON see PPO000
SEPTISOL see HCL000
SEPTOCHOL see DAQ400
SEPTOFEN see HCL000
SEPTOPLEX see SNM500
SEPTOS see HJL500
SEPTOTAN see PFN000
SEPTRA see SNK000, TKX000
SEPTRAN see SNK000, TKX000
SEPTRIM see TKX000
SEPTRIN see TKX000
SEPTURAL see PAF250
SEPYRON see DNU100
SEQ 100 see EIX000
SEQUAMYCIN see SLC000
SEQUESTRENE 30A see EIV000
SEQUESTRENE AA see EIX000
SEQUESTRENE Na3 see TNL250
SEQUESTRENE Na 4 see EIV000
SEQUESTRENE NaFe IRON CHELATE see EJA379
SEQUESTRENE SODIUM 2 see EIX500
SEQUESTRENE ST see EIV000
SEQUESTRENE TRISODIUM see TNL250
SEQUESTRENE TRISODIUM SALT see TNL250
SEQUESTRIC ACID see EIX000
SEQUESTROL see EIX000
SEQUILAR see NNL500
SEQUOSTAT see NNL500
SERAGON see SCA750
SERAGONIN see SCA750
SERAX see CFZ000
(d-SER(BU^6))-LH-RH-(1-9)NONAPEPTIDE-ETHYLAMIDE see LIU420
SEREEN see HOT500
SERENACE see CLY500
SERENACK see DCK759
SERENAL see CFZ000, OMK300
SERENID see CFZ000
SERENID-D see CFZ000
SERENIL see CHG000
SERENIUM see CGA000
SERENSIL see CHG000
SEREN VITA see MDQ250
SERENZIN see DCK759
SEREPAX see CFZ000
SERESTA see CFZ000
SERGOSIN see SHX000
SERIAL see MCA500
SERIBAK see MRM000
SERICOSOL-N see DVR909
SERIEL see GJS200
SERIL see MQU750
l-SERINE DIAZOACETATE see ASA500
l-SERINE DIAZOACETATE (ester) see ASA500
l-SERINE, ESTER with N-METHYL-N-NITROSOGLYCINE see NLS500
dl-SERINE 2-(2,3,4-TRIHYDROXYBENZYL)HYDRAZINE HYDROCHLORIDE
    see SCA400
dl-SERINE 2-((2,3,4-TRIHYDROXYPHENYL)METHYL)HYDRAZIDE
    MONOHYDROCHLORIDE (9CI) see SCA400
SERINGINA see RCA200
SERINYL BLUE 2G see TBG700
SERINYL HOSIERY BLUE see MGG250
SERISOL BRILLIANT BLUE BG see MGG250
SERISOL BRILLIANT VIOLET 2R see DBP000
SERISOL ORANGE YL see AKP750
SERISTAN BLACK B see AQP000
SERITOX 50 see DGB000
SERMION see NDM000
SERNAS see CLY500
SERNEL see CLY500
SERNEVIN see EPD500
SERNYL see AOO500
SERNYLAN see AOO500
SERNYL HYDROCHLORIDE see AOO500
SERODEN see FNF000
SEROFINEX see ARQ750
SEROGAN see SCA750

SEROMYCIN see CQH000
SEROTIN CREATININE SULFATE see AJX750
SEROTINEX see ARQ750
SEROTONIN see AJX500
SEROTONIN BENZYL ANALOG see BEM750
SEROTONIN CREATININE SULFATE MONOHYDRATE see AJX750
SEROTROPIN see SCA750
SERPAGON see RCA200
SERPASIL see RDK000
SERPASIL APRESOLINE see RDK000
SERPASIL APRESOLINE No. 2 see HGP500
SERPAX see CFZ000
SERPENT see YAK350
SERPENTINE see ARM250, ARM268, SCA475
SERPENTINE CHRYSOTILE see ARM268
SERPENTINE (alkaloid) HYDROXIDE, inner salt see SCA475
SERPENTINE HYDROXIDE, inner salt see SCA475
SERPENTINIC ACID ISOBUTYL ESTER see SCA525
SERPENTINTARTRAT (GERMAN) see SCA550
SERRAL see DKA600
SERRATIA EXTRACELLULAR PROTEINASE see SCA625
SERRATIA MARCESCENS ENDOTOXIN see SCA600
SERRATIA PISCVATORUM POLYSACCHARIDE see SCA610
SERRATIO PEPTIDASE see SCA625
SERTAN see DBB200
(d-SER(TBU)^6-EA^{10})-LHRH see LIU420
(d-SER(TBU)^6-EA^{10})-LUTEINIZING HORMONE-RELEASING HORMONE
    see LIU420
d-SER(TBU^6)-LH-RH-(1-9)-NONAPEPTIDE ETHYLAMIDE see LIU420
SERTINON see EPQ000
SERTOFRAN see DSI709
SERUM GONADOTROPHIN see SCA750
SERUM GONADOTROPIC HORMONE see SCA750
SERUM GONADOTROPIN see SCA750
N-SERVE NITROGEN STABILIZER see CLP750
SERVISONE see PLZ000
SES see CNW000
SESAGARD see BKL250
SESAME OIL see SCB000
SESAMOL see MJU000
SESBANIA SESBAN (L.) Merr. var. BICOLOR W. & A., extract excluding roots
    see SCB100
SESBANIA (VARIOUS SPECIES) see SBC550
SESDEN see TGB160
SESONE (ACGIH) see CNW000
SESO VEGETAL (CUBA, PUERTO RICO) see ADG400
SESQUIMUSTARD see SCB500
SESQUIMUSTARD Q see SCB500
SESQUISULFURE de PHOSPHORE (FRENCH) see PHS500
SET see MDI000
SETACYL BLUE BN see MGG250
SETACYL DIAZO NAVY R see DCJ200
SETACYL VIOLET R see DBP000
SETHYL see MDL000
SETILE VIOLET 3R see DBP000
SETONIL see DCK759
SETRETE see PFO550
SETTIMA see DAP700
S-SEVEN see SCC000
SEVENAL see EOK000
SEVEN BARK see HGP600
SEVICAINE see AIT250
SEVIN see CBM750
SEVINOL see TJW500
SEXADIEN see DAL600
SEXADIENO see FBP100
SEXOCRETIN see DKA600
SEXOVAR see FBP100
SEXOVID see FBP100
SEXTONE see CPC000
SEXTONE B see MIQ740
SEXTRA see SCB000
SF 60 see MAK700
SF-337 see AJO500
SF 733 see XQJ650
SF 837 see MBY150
SF-6505 see HLM000
SF6539 see HDC000
S 7481F1 see CQH100
SF 733 ANTIBIOTIC SULFATE see RIP000
SFERICASE see SCC550
S6F HISTYRENE RESIN see SMR000
SG-67 see SCH000
SGD-SCHA 1059 see BGC500

S. GRISEUS PROTEASE see PMJ100
S. GRISEUS PROTEINASE see PMJ100
SH 100 see OLW400
SH 261 see EGQ000
SH 393 see NNQ000
SH 514 see SKM000
SH 567 see PMC700
SH 582 see GEK510
SH 714 see CQJ500
SH 717 see GHK200
SH 742 see FDA925
SH 771 see AMV375
SH 850 see NNQ500
SH 926 see AAI750
SH 567a see PMC700
SH 30858 see FAJ150
SH 66752 see PNI250
SH 70850 see NNQ500
SH 71121 see NNL500
SH 80582 see GEK510
SHA see SAL500
SHADOCOL see TDE750
SHAKE-SHAKE see RBZ400
SHALE OIL (DOT) see COD750
SHAMMAH (SAUDI ARABIA) see SED400
SHAMROX see CIR250
SHARSTOP 204 see SGM500
SHB 286 see SOU650
SH 213AB see IGD100
SHB 261AB see NNL500
SHB 264AB see NNL500
SHED-A-LEAF see SFS000
SHED-A-LEAF ""L" see SFS000
SHEEP DIP see ARE250
SHEEP LAUREL see MRU359
SHELL 40 see BQZ000
SHELL 300 see SMQ500
SHELL 345 see ADR250
SHELL 4072 see CDS750
SHELL 4402 see OAN000
SHELL 5520 see PMP500
SHELL ATRAZINE HERBICIDE see ARQ725
SHELL GOLD see GIS000
SHELL MIBK see HFG500
SHELL OS 1836 see CLV375
SHELLOYNE H see DLJ500
SHELL SD 345 see ADR250
SHELL SD-3450 see HCI475
SHELL SD-3562 see DGQ875
SHELL SD 4,239 see CEQ500
SHELL SD 4294 see COD000
SHELL SD-5532 see CDR750
SHELL SD 7,438 see BES250
SHELL SD 7,727 see DGD400
SHELL SD-8530 see TMD000
SHELL SD-9098 see DIX600
SHELL SD 9129 see MRH209
SHELL SD-10576 see SCD500
SHELL SD-14114 see BLU000
SHELL SILVER see SDI500
SHELLSOL 140 see NMX000
SHELL UNDRAUTTED A see AFV500
SHELL WL 1650 see OAN000
SHENG BAI XIN (CHINESE) see CMV375
SHIGATOX see AHO250
SHIGA TOXIN see SCD750
SHIGELLA DYSENTERIAE TOXIN see SCD750
SHIGRODIN see BRF500
SHIKIMATE see SCE000
SHIKIMIC ACID see SCE000
SHIKIMOLE see SAD000
SHIKISO AMARANTH see FAG020
SHIKISO DIRECT SKY BLUE 5B see CMO500
SHIKISO DIRECT SKY BLUE 6B see CMN750
SHIKISO METANIL YELLOW see MDM775
SHIKOMOL see SAD000
SHINGLE PLANT see SLE890
SHINING SUMAC see SCF000
SHINKOLITE see PKB500
SHIN-NAITO S see TEH500
SHINNIBROL see TEH500
SHINNIPPON FAST RED GG BASE see NEO500
SHIONOGI 6059S see LBH200
SHIRLAN EXTRA see SAH500

SHM see SLD900
SHMP see SHM500
SHOALLOMER see PMP500
SHOCK-FEROL see VSZ100
SHOWDOMYCIN see RIU000
SHOXIN see NNF000
cis-SHP see SCF025
trans(−)-SHP see SCF050
trans(+)-SHP see SCF075
SHRUBBY BITTERSWEET see AHJ875
SHRUB VERBENA see LAU600
SHULTZ No. 737 see FAG140
SI see LCF000
SI-6711 see DJV600
SIACARB see SAZ000
SIARKI CHLOREK (POLISH) see SON510
SIARKI DWUTLENEK (POLISH) see SOH500
SIARKOWODOR (POLISH) see HIC500
SIBEPHYLLIN see DNC000
SIBEPHYLLINE see DNC000
SIBIROMYCIN see SCF500
SIBOL see DKA600
SICILIAN CERISE TONER A-7127 see FAG070
SICOCLOR see FAM000
SICOL 150 see DVL700
SICOL 160 see BEC500
SICOL 250 see AEO000
SICO LAKE RED 2L see CHP500
SICRON see PKQ059
SIDDIQUI see AFH250
SIDNOFEN see SPA000
SIDVAX see AKR500
501 SIEGFRIED see AOO800
SIEGLE RED BB see MMP100
SIENNA see IHD000
SIERRA C-400 see TAB750
SIGACALM see CFZ000
SIGAPRIN see TKX000
SIGETIN see SPA650
SIGMAFON see MBW750
SIGMAMYCIN see TBX000
SIGMART see NDL800
SIGNAL ORANGE ORANGE Y-17 see DVB800
SIGNOPAM see CFY750
SIGOPHYL see HOA000
SIGURAN see HGC500
SILAK M 10 see SCR400
SILANE see SDH575
SILANE, (3-CYANOPROPYL)DIETHOXY(METHYL)- see COR500
SILANE, (3-CYANOPROPYL)TRIETHOXY- see COS800
SILANE, PHENYLTRIFLUORO- see PGO500
2,2',2'',2'''-SILANETETRAYLTETRAKISETHANOL see TDH100
SILANE, TRIETHOXY(3-CYANOPROPYL)- see COS800
SILANE, VINYL TRICHLORO 1-150 see TIN750
SILANE Y-4086 see EBO000
SILANE-Y-4087 see ECH000
SILANOL, DIPHENYLMETHYL- see DWH550
SILANTIN see DKQ000
SILASTIC see PJR250
SILATRANE see TMO750
SILBER (GERMAN) see SDI500
SILBERNITRAT see SDS000
SILBESAN see CLD000
SILICA (crystalline) see SCI500
SILICA AEROGEL see SCH000, SCI000
SILICA, AMORPHOUS see SCH000
SILICA, AMORPHOUS FUMED see SCH000
SILICA, AMORPHOUS FUSED see SCK600
SILICA, AMORPHOUS HYDRATED see SCI000
SILICA, CRYSTALLINE-CRISTOBALITE see SCJ000
SILICA, CRYSTALLINE-QUARTZ see SCJ500
SILICA, CRYSTALLINE-TRIDYMITE see SCK000
SILICA FLOUR see SCI500, SCK500
SILICA FLOUR (powdered crystalline silica) see SCJ500
SILICA, FUSED see SCK600
SILICA GEL see SCI000, SCL000
SILICA, GEL and AMORPHOUS-PRECIPITATED see SCL000
SILICANE see SDH575
SILICATE D'ETHYLE (FRENCH) see EPF550
SILICATES see SCM500
SILICATE SOAPSTONE see SCN000
SILICA, VITREOUS see SCK600
SILICA XEROGEL see SCI000
SILICIC ACID see SCI000, SCL000

SILICIC ACID ALUMINUM SALT see AHF500
SILICIC ACID, BERYLLIUM SALT see SCN500
SILICIC ACID, 2-ETHYLBUTYL ESTER see EHC900
SILICIC ACID TETRAETHYL ESTER see EPF550
SILICIC ACID, ZIRCONIUM(4+) SALT (1:1) see ZSS000
SILICI ANHYDRIDE see SCH000, SCJ500
SILICI-CHLOROFORME (FRENCH) see TJD500
SILICIO(TETRACLORURO di) see SCQ500
SILICIUMCHLOROFORM (GERMAN) see TJD500
SILICIUMTETRACHLORID (GERMAN) see SCQ500
SILICIUMTETRACHLORIDE (DUTCH) see SCQ500
SILICIUM(TETRACHLORURE de) (FRENCH) see SCQ500
SILICOBROMOFORM see THX000
SILICOCHLOROFORM see TJD500
SILICOETHANE see DXA000
SILICOFLUORIC ACID see SCO500
SILICON see SCP000
SILICON BROMIDE see SCP500
SILICON CARBIDE see SCQ000
SILICON CHLORIDE see SCQ500
SILICONCHROME (exothermic) see SCR000
SILICON DIBROMIDE SULFIDE see SCR100
SILICON DIOXIDE see SCI500, SCK600
SILICON DIOXIDE (FCC) see SCH000
SILICONE 360 see SCR400
SILICONE 1-174 see TLC250
SILICONE A-186 see EBO000
SILICONE A-187 see ECH000
SILICONE A-189 see TLC000
SILICONE A-1120 see TLC500
SILICONE DC 200 see SCR400
SILICONE DC 360 see SCR400
SILICONE DC 360 FLUID see SCR400
SILICONE RELEASE L 45 see SCR400
SILICONE RUBBER see PJR250
SILICONES see SDC000, SDF000
SILICONE Y-6607 see SDF000
SILICON FLUORIDE see SDF650
SILICON MONOCARBIDE see SCQ000
SILICON ORGANIC LACQUER MODIFICATOR 113-63 see MRA100
SILICON OXIDE see SDH000
SILICON PHENYL TRICHLORIDE see TJA750
SILICON SODIUM FLUORIDE see DXE000
SILICON TETRAAZIDE see SDH500
SILICON TETRACHLORIDE (DOT) see SCQ500
SILICON TETRAFLUORIDE (DOT) see SDF650
SILICON TETRAHYDRIDE see SDH575
SILICON TRIETHANOLAMIN see SDH670
SILIKILL see SCH000
SILIKON ANTIFOAM FD 62 see SCR400
SILK see SDI000
SILK FAST GREEN B see CMM400
SILMURIN see RCF000, SBF500
SILOGOMMA RED RLL see MMP100
SILOMAT see CMW500
SILON see NOH000
SILONIST see CMW500
SILOPOL ORANGE R see DVB800
SILOPOL RED G see CJD500
SILOSAN see DIN800
SILOSOL RED RN see MMP100
SILOTON RED RLL see MMP100
SILOTON YELLOW GTX see DEU000
SILOTRAS ORANGE TR see PEJ500
SILOTRAS YELLOW T2G see DOT300
SILOXANES see SDC000, SDF000
SILOXANES and SILICONES, DI Me see SCR400
SILUBIN see BQL000
SILUNDUM see SCQ000
SILVADENE see PPP500, SNI425
SILVAN (CZECH) see MKH000
SILVER see SDI500
SILVER (colloidal) see SDI750
SILVER ACETYLIDE see SDJ000
SILVER ACETYLIDE-SILVER NITRATE see SDJ025
SILVER AMIDE see SDJ500
SILVER 5-AMINOTETRAZOLIDE see SDK000
SILVER AMMONIUM COMPOUNDS see SDK500
SILVER AMMONIUM LACTATE see SDL000
SILVER AMMONIUM NITRATE see SDL500
SILVER AMMONIUM SULFATE see SDM000
SILVER ARSENITE see SDM100
SILVER ATOM see SDI500
SILVER AZIDE see SDM500

SILVER 2-AZIDO-4,6-DINITROPHENOXIDE see SDM525
SILVER AZIDODITHIOFORMATE see SDM550
SILVER BENZO-1,2,3-TRIAZOLE-1-OXIDE see SDM575
SILVER BETA-STANNATE see DXA800
SILVER BOROFLUORIDE see SDN000
SILVER BUSH see NBR800
SILVER BUTEN-3-YNIDE see SDN100
SILVER CHAIN see GJU475
SILVER CHLORATE see SDN399
SILVER CHLORITE, dry see SDN500
SILVER CHLOROACETYLIDE see SDN525
SILVER COMPOUNDS see SDO500
SILVER CUP see CDH125
SILVER CYANATE see SDO525
SILVER CYANIDE see SDP000
SILVER 3-CYANO-1-PHENYLTRIAZEN-3-IDE see SDP025
SILVER CYCLOPROPYLACETYLIDE see SDP100
SILVER DIFLUORIDE see SDQ500
SILVER DINITRITODIOXYSULFATE see SDP500
SILVER DINITROACETAMIDE see SDP550
SILVER 3,5-DINITROANTHRANILATE see SDP600
SILVER FIR NEEDLE OIL see AAC250
SILVER FIR OIL see AAC250
SILVER(I) FLUORIDE see SDQ000
SILVER(II) FLUORIDE see SDQ500
SILVER FLUOROBORATE see SDN000
SILVER FULMINATE, dry see SDR000
SILVER HALIDE SOLVENT (HS103) see BGT750
SILVER 1,3,5-HEXATRIENIDE see SDR150
SILVER 3-HYDROXYPROPYNIDE see SDR175
SILVER MALONATE see SDR350
SILVER MATT POWDER see TGB250
SILVER METHANESULFONATE see SDR500
SILVER 3-METHYLISOXAZOLIN-4,5-DIONE-4-OXIMATE see SDR400
SILVER METHYLSULFONATE see SDR500
SILVER MONOACETYLIDE see SDR759
SILVER(1+) NITRATE see SDS000
SILVER NITRATE (DOT) see SDS000
SILVER(I) NITRATE (1581) see SDS000
SILVER NITRIDE see SDS500
SILVER NITRIDOOSMITE see SDT000
SILVER 4-NITROPHENOXIDE see SDT300
SILVER NITROPRUSSIDE see SDT500
SILVER OSMATE see SDT750
SILVER OXALATE, dry see SDU000
SILVER (1+) OXIDE see SDU500
SILVER PERCHLORATE see SDV000
SILVER PERCHLORYL AMIDE see SDV500
SILVER N-PERCHLORYL BENZYLAMIDE see SDV600
SILVER PEROXIDE see SDV700
SILVER PEROXYCHROMATE see SDW000
SILVER PHENOXIDE see SDW100
SILVER PINE OIL see AAC250
SILVER POTASSIUM CYANIDE see PLS250
SILVER(II) SALT see SDN500
SILVER SULFADIAZINE see SNI425
SILVER SULPHADIAZINE see SNI425
SILVER TETRAFLUOROBORATE see SDN000
SILVER TETRAZOLIDE see SDW500
SILVER TRICHLOROMETHANEPHOSPHONATE see SDW600
SILVER TRIFLUORO METHYL ACETYLIDE see SDX000
SILVER TRIFLUOROPROPYNIDE see SDX000
SILVER TRINITROMETHANIDE see SDX200
SILVEX (USDA) see TIX500
SILVIC ACID see AAC500
SILVI-RHAP see TIX500
SILVISAR see HKC500
SILVISAR 510 see HKC000
SILVISAR 550 see MRL750
SILYL BROMIDE see BOE750
SILYL PEROXIDE Y-5712 see SDX300
SILYMARIN see SDX625
SILYMARIN HYDROGEN BUTANEDIOATE SODIUM SALT see SDX630
SILYMARIN SODIUM HEMISUCCINATE see SDX630
SIM see SNK000
SIMAN see CGA000
SIMANEX see BJP000
SIMARUBACEAE see GEW700
SIMATIN(E) see ENG500
SIMAZIN see BJP000
SIMAZINE 80W see BJP000
SIMAZINE (USDA) see BJP000
SIMAZOL see AMY050
SIMEON see POB500

SIMESKELLINA see AHK750
SIMETON see BJP250
SIMETONE see BJP250
SIMFIBRATE see SDY500
SIMPADREN see SPD000
SIMPALON see HLV500
SIMPAMINA-D see BBK500
SIMPATEDRIN see BBK000
SIMPATOBLOCK see HEA000
SIMPATOL see HLV500
SIMPLA see SGG500
SIMULOL 330 M see DXY000
SIN-10 see MRN275
SINAFID M-48 see MNH000
SINALAR see SPD500
SINALGICO see NBS500
SINALOST see TNF500
SINAN see GGS000
SINAXAR see PFJ000
SINBAR see BQT750
SINCICLAN see DKB000
SINCODEEN see BOR350
SINCODEX see BOR350
SINCODIN see BOR350
SINCODIX see BOR350
SINCORTEX see DAQ800
SINDERESIN see CJN250
SINDESVEL see QAK000
SINDIATIL see BQL000
SINDRENINA see VGP000
SINECOD see BOR350
SINEFLUTTER see GAX000
SINEFRINA TARTRATO (ITALIAN) see SPD000
SINEQUAN see AEG750
SINESALIN see BEQ625
SINESTROL see DLB400
SINFIBRATE see SDY500
SINFORIL see CKE750
SINGOSERP see RCA200
SINIFIBRATE see SDY500
SINIGRIN see AGH125
SINITUHO see PAX250
SINKLE BIBLE (JAMAICA) see AGV875
SINKUMAR see ABF750
SINNAMIN see AQN750
SINOACTINOMYCIN see AEA625
SINOFLUROL see FMB000
SINOGAN see MCI500
SINOMENI CAULIS et RHIZOMA (LATIN) see SDY600
SINOMENINE A BIS(METHYL IODIDE) see TDX835
SINOMENINE A DIMETHIODIDE see TDX835
SINOMENIUM ACUTUM, crude extract see SDY600
SINOMIN see SNK000
SINORATOX see DSP400
SINOX see DUS700, DUU600
SINOX GENERAL see BRE500
SINOX W see BPG250
SINTECORT see PAL600
SINTESPASMIL see NOC000
SINTESTROL see DKA600
SINTHROME see ABF750
SINTOFENE see OPK300
SINTOMICETINA see CDP250
SINTOSIAN see DYF200
SINTOTIAMINA see DXO300
SINUFED see POH250
SINURON see DGD600
SINUTAB see ABG750
SIOCARBAZONE see FNF000
SIONIT see SKV200
SIONON see SKV200
SIPCAVIT see PEX500
SIPERIN see RLF350
SIPEX BOS see TAV750
SIPEX OP see SIB600
SIPLARIL see FMO129
SIPLAROL see FMO129
SIPOL L8 see OEI000
SIPOL L10 see DAI600
SIPOL L12 see DXV600
SIPOL O see OBA000
SIPOL S see OAX000
SIPOMER DAM see DBK200
SIPONIC L see DXY000

SIPONIC Y-501 see PJW500
SIPON LT see SON000
SIPONOL S see OAX000
SIPON WD see SIB600
SIPPR-113 see SDY675
SIPTOX I see MAK700
SIRAGAN see CLD000
SIRAN N,N-DIETHYL-p-FENYLENDIAMINU see DJV250
SIRAN HYDRAZINU (CZECH) see HGW500
SIRBIOCINA see SMB000
SIRINGAL see RCA200
SIRINGINA see RCA200
SIRINGONE see RCA200
SIRISERPIN see RCA200
SIRLEDI see NHH000
SIRLENE see PML000
SIRMATE see DET400, DET600
SIRNIK AMONNY see ANJ750
SIRNIK FOSFORECNY (CZECH) see PHS000
SIRNIK TRIBENZYLCINICTY (CZECH) see BLK750
SIROKAL see PLG800
SIROMYCIN see VGZ000
SIROTOL see RLU000
SIRUP see GFG000
SISEPTIN see SDY750
SISOLLINE see SDY750
SISOMICIN see SDY750
SISOMICIN HYDROCHLORIDE see SDY755
SISOMICIN SULFATE see APY500
SISOMIN see APY500
SISTALGIN see SDZ000
SISTAN see VFU000
SISTOMETRENOL see LJE000
SISTRURUS MILARIUS BARBOURI VENOM see SDZ300
SITFAST see FBS100
SIX HUNDRED SIX see SAP500
SIXTY-THREE SPECIAL E.C. INSECTICIDE see MNH000, PAK000
SK 1 see NCW300
SK 65 see DAB879
SK 74 see CMW700
SK-100 see TNF500
SK 555 see BHO250
SK-598 see CFA750
SK1133 see TND500
SK 1150 see AJO500
SK-3818 see TND250
SK 6048 see CKV500
SK 6882 see TFQ750
SK-15673 see PED750
SK 18615 see TFJ500
SK-19849 see BIA250
SK 20501 see CQC650
SK 22591 see HOO500
SK 27702 see BIF750
SK 29836 see DBY500
SK 331 A see XCS000
SK-AMITRIPTYLINE see EAI000
SK-AMPICILLIN see AIV500
SKANE M8 see OFE000
SK-Apap see HIM000
SKATOL see MKV750
SKATOLE see MKV750
ω-SKATOLE CARBOXYLIC ACID see ICN000
SK-CHLOROTHIAZIDE see CLH750
SK-CHORAL HYDRATE see CDO000
SK-DEXAMETHASONE see SOW000
SK-DIGOXIN see DKN400
SK-DIPHENHYDRAMINE see BAU750
SKDN see WBS675
SKEDULE see CBF250
SKEKhG see EAZ500
SKELAXIN see XVS000
SKELLY-SOLVE-F see NAI500
SKELLY-SOLVE-L see ROU000
SKELLY-SOLVE S see PCT250
SKEROLIP see ARQ750
SK-ERYTHROMYCIN see EDJ500
SK-ESTROGENS see ECU750
SKF 51 see ILM000
SKF 250 see DBA175
SKF 385 see PET750
SKF 478 see DWK200
SKF 1045 see MGD500
SKF 1498 see DQA400

SKF 1717 see AOO490
SKF-183A see ALW000
SKF 2170 see AOO425
SKF 2538 see DIR000
SKF-2601 see CKP250
SKF 4740 see DAB875
SKF 5019 see TKK250
SKF 5116 see MCI500
SKF 5137 see AFJ400
SKF 525A see PBM500
SKF 5654 see CNS650
SKF 5883 see TFM100
SKF 6539 see HDC000
SKF 688A see DDG800
SKF 7261 see FKW000
SKF 7690 see MNC150
SKF 7988 see VRA700, VRF000
SKF 8542 see UVJ450
SK+F 1340 see DRX400
SKF 12141 see BGK250
SKF 20,716 see PIW000
SKF 2208K see CDF400
SKF 24260 see FOO875
SKF 28175 see DMF800
SKF 29044 see BQK000
SKF 30310 see OMY700
SKF 41588 see CCS250
SKF 60771 see APT250
SKF 62698 see TGA600
SKF 62979 see VAD000
SKF-69,634 see CMX860
SKF 83088 see CCS350
SKF-88373 see EBE100
SKF 91487 see NNL400
SKF 92334 see TAB250
SK&F 14287 see DAS000
SK&F 36914 see CLQ500
SK&F 39162 see ARS150
SK&F 92676 see IBQ075
SKF 16805A see BQB250
SKF-525-A see DIG400
SKF 70643-A see DPM200
SKF 16214-A2 see ADI750
SKF-501 HYDROCHLORIDE see FEE100
SK&F No. 478-A see DWK200
SKF No. 769-J(^{2}) see CBG250
SKF 83088 SODIUM see CCS360
SKI 21739 see BHV000
SKI 24464 see MNA750
SKI 27013 see SAU000
SKI 28404 see CHE750
SKINO #1 see BSU500
SKINO #2 see EMU500
SKIODAN see SHX000
SKLEROMEX see ARQ750
SKLEROMEXE see ARQ750
SKLERO-TABLINEN see ARQ750
SKLERO-TABULS see ARQ750
SK-LYGEN see MDQ250
SK-106N see NGY000
SK-NIACIN see NCQ900
SKOLIN see HLC500
SK-PENICILLIN G see BFD000
SK-PENICILLIN VK see PDT750
SK-PHENOBARBITAL see EOK000
SK-PRAMINE see DLH630
SK-PRAMINE HYDROCHLORIDE see DLH630
SK-PREDNISONE see PLZ000
SKS 85 see SMR000
SK-TETRACYCLINE see TBX000, TBX250
SK-TOLBUTAMIDE see BSQ000
SK-TRIAMCINOLONE see AQX250
SKUNK CABBAGE see FAB100, SDZ450
SKY BLUE 5B see CMO500
SKY FLOWER see GIW200
SL 31 see MNO775
SL-90 see RLU000
SL 501 see CMW700
SL-512 see CQF125
1000SL see HKS780
SL-6057 see CMB125
SL-75212 see KEA350
SLAB OIL (9CI) see MQV875
SLAKED LIME see CAT225

SL-D.212 see KEA350
SLEEPAN see TEH500
SLEEPING NIGHTSHADE see DAD880
SLEEPWELL see TEO250
S-426-S (LEPETIT) see EAD500
SLIMICIDE see ADR000
SLIPPER FLOWER see SDZ475
SLIPPER PLANT see SDZ475
SLIPRO see TEH500
SLOE see AQP890
S-LON see PKQ059
SLO-PHYLLIN see TEP000
SLOSUL see AKO500
SLOVASOL A see PJY100
SLOW-FE see FBN100
SLOW-K see PLA500
SLS see SIB600
SLUDGE ACID see SEA000
SLUG-TOX see TDW500
S 75M see SFO500
SM-307 see DLG000
SMA see SFU500
SMA 1440-H RESIN see SEA500
SMALL JACK-IN-THE-PULPIT see JAJ000
S. MARCESCENS LIPOPOLYSACCHARIDE see SEA400
SMCA see SFU500
SMDC see VFU000, VFW009
SMEDOLIN see AHP125
SMEESANA see AQN635
SMIDAN see PHX250
S. MILARIUS BARBOURI VENOM see SDZ300
SMOG see SEB000
SMOKE BROWN G see TKN750
SMOKE CONDENSATE, cigarette see SEC000
SMOKELESS POWDER see SED000
SMOKELESS TOBACCO see SED400
SMOP see AKO500
SMP see AKO500
S MUSTARD see BIH250
SMUT-GO see HCC500
SN see SMA000
SN 20 see ADY500
SN 46 see CPK500
S.N. 112 see PPP500
S. N. 166 see AOO800
SN 186 see ADI750
SN 203 see CMW400
SN 390 see CFU750
SN 407 see CCS625
SN-4395 see PQC500
SN 6718 see CLD000
SN 7618 see CLD000
SN 12,837 see CKB500
SN 13,272 see PMC300
SN 35830 see CGL250
SN 36056 see DSO200
SN 36268 see CJJ250
SN 38107 see EEO500
SN-41703 see EIH500
SN 49537 see TEX600
SNAKEBERRY see BAF325
SNAKE FLOWER see VQZ675
SNAKEROOT OIL, CANADIAN see SED500
SNAKE VENOM BITIS ARIETANS see BLV075
SNAKE VENOM BITIS GABONICA see BLV080
SNAKE WEED see PJJ300
SNAPPING HAZEL see WCB000
SNATOWHITE CRYSTALS see TFC600
SN 6771 DIHYDROCHLORIDE see BFX125
SNEEZING GAS see CGN000
SNIECIOTOX see HCC500
SNIP see DQZ000
SNIP FLY see DQZ000
SNIP FLY BANDS see DQZ000
SNOMELT see CAO750
SNOW ALGIN H see SEH000
SNOWBERRY see SED550
SNOWDROP see SED575
SNOWGOOSE see TAB750
SNOW TEX see AHF500, KBB600
SNOW WHITE see ZKA000
SNP see PAK000
SN 401 PENTAHYDRATE see CCQ200
SNUFF see SED400

SO see LCF000
SOAP see CNF175
SOAP PLANT see DAE100
SOAPSTONE see SCN000
SOAP YELLOW F see FEV000
SOAz see SED700
SOBELIN see CMV675, CMV690
SOBENATE see SFB000
SOBIC ALDEHYDE see SKT500
SOBIODOPA see DNA200
SOBITAL see DJL000
SOBRIL see CFZ000
SOC see CIL775
SOCLIDAN see EQQ100
SODA ALUM see AHG500
SODA ASH see SFO000
SODA CHLORATE (DOT) see SFS000
SODA LIME (solid) see SEE000
SODA LYE see SHS000, SHS500
SODAMIDE see SEN000
SODA MINT see SFC500
SODANIT see SEY500
SODA NITER see SIO900
SODANTON see DKQ000, DNU000
SODA PHOSPHATE see SJH090
SODAR see DXE600
SODASCORBATE see ARN125
SODESTRIN-H see ECU750
SODIO (CLORATO di) (ITALIAN) see SFS000
SODIO (DICROMATO di) (ITALIAN) see SGI000
SODIO, FLUORACETATO di (ITALIAN) see SHG500
SODIO(IDROSSIDO di) (ITALIAN) see SHS000
3-SODIO-5-(5'-NITRO-2'-FURFURYLIDENAMINO)IMIDAZOLIDIN-2,4-DIONE
   see SEE250
SODIOPAS see SEP000
SODIO (PERCLORATO DI) (ITALIAN) see PCE750
SODIO(TRICLOROACETATO di) (ITALIAN) see TII500
SODITAL see NBU000
SODIUM see SEE500
1719 SODIUM see ASO510
SODIUM (dispersions) see SEF500
SODIUM (liquid alloy) see SEF600
SODIUM, metal liquid alloy (DOT) see SEF600
SODIUM-4-(2-ACETAMIDOETHYLDITHIO)BUTANESULFINATE see AAK250
SODIUM-3-ACETAMIDO-2,4,6-TRIIODOBENZOATE see AAN000
SODIUM ACETARSONE see SEG000
SODIUM ACETATE see SEG500
SODIUM ACETATE, anhydrous (FCC) see SEG500
SODIUM ACETATE MONOHYDRATE see SEG650
SODIUM ACETAZOLAMIDE see AAS750
SODIUM ACETRIZOATE see AAN000
SODIUM p-ACETYLAMINOPHENYLANTIMONATE see SLP500
SODIUM-3-ACETYLAMINO-2,4,6-TRIIODOBENZOATE see AAN000
SODIUM ACETYLARSANILATE see AQZ900, ARA000
SODIUM ACETYLIDE see SEG700
SODIUM ACID ARSENATE see ARC000
SODIUM ACID ARSENATE, HEPTAHYDRATE see ARC250
SODIUM ACID CARBONATE see SFC500
SODIUM ACID FLUORIDE see SHQ500
SODIUM ACID HEPARIN see HAQ550
SODIUM ACID METHANEARSONATE see MRL750
SODIUM ACID PHOSPHATE see SJH100
SODIUM ACID PYROPHOSPHATE (FCC) see DXF800
SODIUM ACID SULFATE see SEG800
SODIUM ACID SULFATE (solid) see SEG800
SODIUM ACID SULFATE, solution (DOT) see SEG800
SODIUM ACID SULFITE see SFE000
SODIUM ACYCLOVIR see AEC725
SODIUM ADENOSINE-5'-MONOPHOSPHATE see AEM750
SODIUM ADENOSINE TRIPHOSPHATE see AEM250
SODIUM ADENOSINE-5'-TRIPHOSPHATE see AEM250
SODIUM AESCINATE see EDM000
SODIUM ALBAMYCIN see NOB000
SODIUM ALGINATE see SEH000
SODIUM ALGINATE SULFATE see SEH450
SODIUM ALLYLBENZYL THIOBARBITURATE see AFX500
SODIUM-5-ALLYL-5-(1-(BUTYLTHIO)ETHYL) BARBITURATE see AGA250
SODIUM-5-ALLYL-5-ISOPROPYLBARBITURATE see BOQ750
SODIUM-5-ALLYL-5-(1-METHYLBUTYL)BARBITURATE see SBN000
SODIUM-5-ALLYL-5-(1-METHYLBUTYL)-2-THIOBARBITURATE see SOX500
SODIUM-dl-5-ALLYL-1-METHYL-5-(1-METHYL-2-PENTYNYL)BARBITURATE
   see MDU500
SODIUM ALUMINATE, solid (DOT) see AHG000
SODIUM ALUMINOFLUORIDE see SHF000

SODIUM ALUMINOSILICATE see SEM000
SODIUM ALUMINUM FLUORIDE see SHF000
SODIUM ALUMINUM HYDRIDE (DOT) see SEM500
SODIUM ALUMINUM OXIDE see AHG000
SODIUM ALUMINUM PHOSPHATE, ACIDIC see SEM300
SODIUM ALUMINUM PHOSPHATE, BASIC see SEM305
SODIUM ALUMINUM SULFATE see AHG500
SODIUM ALUMINUM TETRAHYDRIDE see SEM500
SODIUM AMAZOLENE see ADE750
SODIUM A-dl-1-METHYL-5-ALLYL-5-(1-METHYL-2-PENTYNYL)BARBITU-
   RATE see MDU500
SODIUM AMIDE see SEN000
SODIUM AMIDOTRIZOATE see SEN500
SODIUM AMINARSONATE see ARA500
SODIUM-p-AMINOBENZENEARSONATE see ARA500
SODIUM-p-AMINOBENZENESTIBONATE GLUCOSIDE see NCL000
SODIUM-4-AMINOBENZOATE see SEO500
SODIUM p-AMINOBENZOATE see SEO500
SODIUM (2-AMINO-3-BENZOYLPHENYL)ACETATE MONOHYDRATE
   see AHK625
SODIUM d-(−)-α-AMINOBENZYLPENICILLIN see SEQ000
SODIUM AMINOPHENOL ARSONATE see ARA500
SODIUM-p-AMINOPHENYLARSONATE see ARA500
SODIUM AMINOSALICYLATE see SEP000
SODIUM p-AMINOSALICYLATE see SEP000
SODIUM p-AMINOSALICYLIC ACID see SEP000
SODIUM-7-(2-(2-AMINO-4-THIAZOLYL)-2-METHOXYIMINOACETAMIDO)
   CEPHALOSPORANATE see CCR950
SODIUM AMINOTRIACETATE see SEP500
SODIUM AMPICILLIN see SEQ000
SODIUM AMYLOBARBITONE see AON750
SODIUM ANAZOLENE see ADE750
SODIUM ANDROST-5-EN-17-ONE-3-β-YL SULFATE DIHYDRATE see DAL030
SODIUM ANILARSONATE see ARA500
SODIUM-ANILINE ARSONATE see ARA500
SODIUM-2-ANTHRACHINONESULPHONATE see SER000
SODIUM ANTHRAQUINONE-1,5-DISULFONATE see DLJ700
SODIUM ANTHRAQUINONE-1-SULFONATE see DLJ800
SODIUM-2-ANTHRAQUINONESULFONATE see SER000
SODIUM-β-ANTHRAQUINONESULFONATE see SER000
SODIUM-9,10-ANTHRAQUINONE-2-SULFONATE see SER000
SODIUM ANTIMONATE see AQB250
SODIUM ANTIMONY see AQB250
SODIUM ANTIMONY BIS(PYROCATECHOL-2,4-DISULFONATE) see AQH500
SODIUM ANTIMONY (III) BIS-PYROCATECHOL-3,5-DISULFONATE
   HEPTAHYDRATE see AQH500
SODIUM ANTIMONY(III)-3-CATECHOL THIOSALICYLATE see SEU000
SODIUM ANTIMONY-2,3-meso-DIMERCAPTOSUCCINATE see AQD750
SODIUM ANTIMONY ERYTHRITOL see SER500
SODIUM ANTIMONY GLUCONATE see AQI000
SODIUM ANTIMONY(V) GLUCONATE see AQI250
SODIUM ANTIMONY(III) GLUCONATE see AQI000
SODIUM ANTIMONYL ADONITOL see SES000
SODIUM ANTIMONYL-d-ARABITOL see SES500
SODIUM ANTIMONYL BISCATECHOL see SET000
SODIUM ANTIMONYL tert-BUTYL CATECHOL see SET500
SODIUM ANTIMONYL CATECHOL THIOSALICYLATE see SEU000
SODIUM ANTIMONYL CITRATE see SEU500
SODIUM ANTIMONYL DIMETHYLCYSTEINE TARTRATE see AQH750
SODIUM ANTIMONYL-d-FUNCITOL see SEV000
SODIUM ANTIMONYL GLUCO-GULOHEPTITOL see SEV500
SODIUM ANTIMONYL GLYCEROL see SEW000
SODIUM ANTIMONYL-d-MANNITOL see SEW500
SODIUM ANTIMONYL-2,5-METHYLENE-d-MANNITOL see SEX000
SODIUM ANTIMONYL-2,4-METHYLENE-d-SORBITOL see SEX500
SODIUM ANTIMONYL TARTRATE see AQI750
SODIUM ANTIMONYL XYLITOL see SEY000
SODIUM ANTIMONY TARTRATE see AQI750
SODIUM ANTIMOSAN see AQH500
SODIUM ARSANILATE see ARA500
SODIUM-p-ARSANILATE see ARA500
SODIUM ARSENATE see ARC000, ARD500, ARD600
SODIUM ARSENATE (DOT) see ARD750
SODIUM ARSENATE DIBASIC, anhydrous see ARC000
SODIUM ARSENATE, DIBASIC, HEPTAHYDRATE see ARC250
SODIUM ARSENATE HEPTAHYDRATE see ARC250
SODIUM ARSENITE see SEY500
SODIUM ARSENITE (liquid) see SEZ000
SODIUM ARSENITE, liquid (solution) (DOT) see SEY500
SODIUM ARSENITE, solid (DOT) see SEY500
SODIUM ARSONILATE see ARA500
SODIUM-l-ASCORBATE see ARN125
SODIUM ASCORBATE (FCC) see ARN125
SODIUM ASPARTATE see SEZ350

SODIUM l-ASPARTATE see SEZ350, SEZ355
SODIUM ASPIRIN see ADA750
SODIUM ATP see AEM250
SODIUM AUROTHIOMALATE see GJC000
SODIUM AUROTHIOSULPHATE DIHYDRATE see GJG000
SODIUM AZAPROPAZONE see ASA250
SODIUM AZIDE see SFA000
SODIUM-5-AZIDOTETRAZOLIDE see SFA100
SODIUM AZO-α-NAPHTHOLSULFANILATE see FAG010
SODIUM AZO-α-NAPHTHOLSULPHANILATE see FAG010
SODIUM AZOTOMYCIN see ASO510
SODIUM, AZOTURE de (FRENCH) see SFA000
SODIUM, AZOTURO di (ITALIAN) see SFA000
SODIUM BARBITAL see BAG250
SODIUM BARBITONE see BAG250
SODIUM BARBITURATE see MRK750
SODIUM BENZENE HEXOIDE see SFA600
SODIUM-1,2 BENZISOTHIAZOLIN-3-ONE-1,1-DIOXIDE see SJN700
SODIUM BENZOATE see SFB000
SODIUM BENZOATE and CAFFEINE see CAK800
SODIUM BENZOIC ACID see SFB000
SODIUM o-BENZOSULFIMIDE see SJN700
SODIUM BENZOSULPHIMIDE see SJN700
SODIUM-2-BENZOSULPHIMIDE see SJN700
SODIUM-o-BENZOSULPHIMIDE see SJN700
SODIUM-2-BENZOTHIAZOLYLSULFIDE see SFB100
SODIUM BENZYLPENICILLIN see BFD250
SODIUM BENZYLPENICILLINATE see BFD250
SODIUM BENZYLPENICILLIN G see BFD250
SODIUM BERYLLIUM MALATE see SFB500
SODIUM BERYLLIUM TARTRATE see SFC000
SODIUM BIBORATE see SFF000
SODIUM BIBORATE DECAHYDRATE see SFF000
SODIUM BICARBONATE see SFC500
SODIUM BICHROMATE see SGI000
SODIUM BIFLUORIDE (VAN) see SHQ500
SODIUM BINOTAL see SEQ000
SODIUM BIPHOSPHATE see SJH100
SODIUM BIPHOSPHATE anhydrous see SJH100
SODIUM BIS(2-ETHYLHEXYL) SULFOSUCCINATE see DJL000
SODIUM BISMUTHATE see SFD000
SODIUM BISMUTH THIOGLYCOLATE see BKX750
SODIUM BISMUTH THIOGLYCOLLATE see BKX750
SODIUM-3,5-BIS(aci-NITRO)CYCLOHEXENE-4,6-DIIMINIDE see SFD300
SODIUM BISPROPYLACETATE see PNX750
SODIUM BISULFATE, fused see SEG800
SODIUM BISULFATE, solid (DOT, FCC) see SEG800
SODIUM BISULFATE, solution (DOT) see SEG800
SODIUM BISULFIDE see SHR000, SFE000
SODIUM BISULFITE see SFE000
SODIUM BISULFITE (1581) see SFE000
SODIUM BISULFITE, solid (DOT) see SFE000
SODIUM BISULFITE, solution (DOT) see SFE000
SODIUM BORATE see SFE500
SODIUM BORATE anhydrous see SFE500
SODIUM BORATE DECAHYDRATE see SFF000
SODIUM BOROHYDRIDE see SFF500
SODIUM BROMATE see SFG000
SODIUM BROMEBRATE see SIK000
SODIUM BROMIDE see SFG500
SODIUM BROMOACETYLIDE see SFG600
SODIUM-5-(2-BROMOALLYL)-5-sec-BUTYLBARBITURATE see BOR250
SODIUM BUTABARBITAL see BPF250
SODIUM BUTANOATE see SFN600
SODIUM BUTAZOLIDINE see BOV750
SODIUM-5-sec-BUTYL-5-ETHYLBARBITURATE see BPF250
SODIUM BUTYLMERCURIC THIOGLYCOLLATE see SFJ500
SODIUM-5-(1-(BUTYLTHIO)ETHYL)-5-ETHYLBARBITURATE see SFJ875
SODIUM-3-BUTYRAMIDO-α-ETHYL-2,4,6-TRIIODOCINNAMATE see EQC000
SODIUM-3-BUTYRAMIDO-α-ETHYL-2,4,6-TRIIODOHYDROCINNAMATE
   see SKO500
SODIUM BUTYRATE see SFN600
SODIUM n-BUTYRATE see SFN600
SODIUM CACODYLATE (DOT) see HKC500
SODIUM CALCIUM ALUMINOSILICATE, HYDRATED see SFN700
SODIUM CARBENICILLIN see CBO250
SODIUM CARBOLATE see SJF000
SODIUM CARBONATE (2581) see SFO000
SODIUM CARBOXYMETHYL CELLULOSE see SFO500
SODIUM CARRAGHEENATE see SFP000
SODIUM CARRIOMYCIN see SFP500
SODIUM CASEINATE see SFQ000
SODIUM CEFACETRIL see SGB500
SODIUM CEFAMANDOLE see CCR925

SODIUM CEFAPIRIN see HMK000
SODIUM CEFAZOLIN see CCS250
SODIUM CEFUROXIME see SFQ300
SODIUM CELLULOSE GLYCOLATE see SFO500
SODIUM CEPHACETRILE see SGB500
SODIUM CEPHALOTHIN see SFQ500
SODIUM CEPHALOTIN see SFQ500
SODIUM CEPHAPIRIN see HMK000
SODIUM CEPHAZOLIN see CCS250
SODIUM CEZ see CCS250
SODIUM CHAULMOOGRATE see SFR000
SODIUM CHENODEOXYCHOLATE see CDL375
SODIUM CHENODESOXYCHOLATE see CDL375
SODIUM CHLORAMBUCIL see CDO625
SODIUM CHLORAMINE T see CDP000
SODIUM CHLORAMPHENICOL SUCCINATE see CDP500
SODIUM CHLORATE see SFS000
SODIUM (CHLORATE de) (FRENCH) see SFS000
SODIUM CHLORATE, aqueous solution (DOT) see SFS000
SODIUM CHLORATE BORATE see SFS500
SODIUM CHLORIDE see SFT000
SODIUM CHLORITE see SFT500
SODIUM CHLORITE (solution) see SFU000
SODIUM CHLOROACETATE see SFU500
SODIUM-4-CHLOROACETOPHENONE OXIMATE see SFU600
SODIUM CHLOROACETYLIDE see SFV000
SODIUM cis-3-CHLOROACRYLATE see SFV250
SODIUM cis-β-CHLOROACRYLATE see SFV250
SODIUM N-CHLOROBENZENESULFONAMIDE see SFV275
SODIUM-5-(4-CHLOROBENZOYL)-1,4-DIMETHYL-1H-PYRROLE-2-ACETATE
   DIHYDRATE see ZUA300
SODIUM CHLOROETHYNIDE see SFV000
SODIUM-4-CHLORO-2-METHYL PHENOXIDE see SFV300
SODIUM (4-CHLORO-2-METHYLPHENOXY)ACETATE see SIL500
SODIUM-2-CHLORO-6-PHENYL PHENATE see SFV500
SODIUM CHLOROPLATINATE see SJJ500
SODIUM-2-CHLOROPROPIONATE see CKT100
SODIUM N-CHLORO-4-TOLUENE SULFONAMIDE see SFV550
SODIUM CHOLATE see SFW000
SODIUM CHOLIC ACID see SFW000
SODIUM CHONDROITIN POLYSULFATE see SFW300
SODIUM CHONDROITIN SULFATE see SFW300
SODIUM CHROMATE see SGI000
SODIUM CHROMATE (VI) see DXC200
SODIUM CHROMATE (DOT) see DXC200
SODIUM CHROMATE DECAHYDRATE see SFW500
SODIUM CINCHOPHEN see SJH000
SODIUM CINNAMATE see SFX000
SODIUM CITRATE see MRL000
SODIUM CITRATE (FCC) see DXC400
SODIUM CITRATE, anhydrous see TNL000
SODIUM CITRATE, POTASSIUM CITRATE, CITRIC ACID (2:2:1)
   see SFX725
SODIUM CLOXACILLIN see SLJ050
SODIUM CLOXACILLIN MONOHYDRATE see SLJ000
SODIUM CMC see SFO500
SODIUM CM-CELLULOSE see SFO500
SODIUM COBALTINITRITE see SFX750
SODIUM COCOMETHYLAMINOETHYL-2-SULFONATE see SFY000
SODIUM COCO METHYL TAURIDE see SFY000
SODIUM COLISTIMETHATE see SFY500
SODIUM COLISTINEMETHANESULFONATE see SFY500
SODIUM COLISTIN METHANESULFONATE see SFY500
SODIUM COMPOUNDS see SFZ000
SODIUM COUMADIN see WAT220
SODIUM-4-CRESOLATE see SIM100
SODIUM p-CRESYLATE see SIM100
SODIUM CROMOGLYCATE see CNX825
SODIUM CROMOLYN see CNX825
SODIUM CUMENEAZO-β-NAPHTHOL DISULPHONATE see FAG018
SODIUM CUPROCYANIDE (DOT) see SFZ100
SODIUM CUPROCYANIDE, solution (DOT) see SFZ100
SODIUM CYANIDE see SGA500
SODIUM CYANIDE, solid and solution (DOT) see SGA500
SODIUM CYANIDE (solution) see SGB000
SODIUM-7-(2-CYANOACETAMIDO)CEPHALOSPORANIC ACID see SGB500
SODIUM CYCLAMATE see SGC000
SODIUM CYCLOHEXANESULFAMATE see SGC000
SODIUM CYCLOHEXANESULPHAMATE see SGC000
SODIUM-5-(1-CYCLOHEXEN-1-YL)-1,5-DIMETHYLBARBITURATE see ERE000
SODIUM CYCLOHEXYL AMIDOSULPHATE see SGC000
SODIUM CYCLOHEXYL SULFAMATE see SGC000
SODIUM CYCLOHEXYL SULFAMIDATE see SGC000
SODIUM CYCLOHEXYL SULPHAMATE see SGC000

SODIUM-2,4-D see SGH500
SODIUM DALAPON see DGI600
SODIUM DBDT see SGF500
SODIUM DECYLBENZENESULFONAMIDE see DAJ000
SODIUM DECYLBENZENESULFONATE see DAJ000
SODIUM DECYL SULFATE see SOK000
SODIUM DEDT see SGJ000
SODIUM DEHYDROACETATE (FCC) see SGD000
SODIUM DEHYDROACETIC ACID see SGD000
SODIUM DEHYDROCHOLATE see SGD500
SODIUM DEHYDROEPIANDROSTERONE SULFATE see DAL040
SODIUM DELVINAL see VKP000
SODIUM DEOXYCHOLATE see SGE000
SODIUM DEOXYCHOLIC ACID see SGE000
SODIUM DESOXYCHOLATE see SGE000
SODIUM DEXAMETHASONE PHOSPHATE see DAE525
SODIUM DEXTROTHYROXINE see SKJ300
SODIUM-3,5-DIACETAMIDO-2,4,6-TRIIODOBENZOATE see SEN500
SODIUM DIACETYLDIAMINETRIIODOBENZOATE see SEN500
SODIUM DIATRIZOATE see SEN500
SODIUM-1,2585,6-DIBENZANTHRACENE-9,10-endo-α,β-SUCCINATE see SGF000
SODIUM DIBUTYLDITHIOCARBAMATE see SGF500
SODIUM DIBUTYLNAPHTHALENE SULFATE see NBS700
SODIUM DIBUTYLNAPHTHALENESULFONATE see NBS700
SODIUM-2,6-DI-tert-BUTYLNAPHTHALENESULFONATE see DEE600
SODIUM DIBUTYLNAPHTHYLSULFONATE see NBS700
SODIUM-N⁶,2'-o-DIBYTYRYLADENOSINE 3',5'-CYCLIC PHOSPHATE see DEJ300
SODIUM DICHLORISOCYANURATE see SGG500
SODIUM DICHLOROACETATE see SGG000
SODIUM (o-(2,6-DICHLOROANILINO)PHENYL)ACETATE see DEO600
SODIUM DICHLOROCYANURATE see SGG500
SODIUM DICHLOROISOCYANURATE see SGG500
SODIUM-2-((2,6-DICHLORO-3-METHYLPHENYL)AMINO)BENZOATE see SIF425
SODIUM-2,4-DICHLOROPHENOXYACETATE see SGH500
SODIUM-2-(2,4-DICHLOROPHENOXY)ETHYL SULFATE see CNW000
SODIUM-2,4-DICHLOROPHENOXYETHYL SULPHATE see CNW000
SODIUM (o-((2,6-DICHLOROPHENYL)AMINO)PHENYL)ACETATE see DEO600
SODIUM-2,4-DICHLOROPHENYL CELLOSOLVE SULFATE see CNW000
SODIUM-2,2-DICHLOROPROPIONATE see DGI600
SODIUM-α,α-DICHLOROPROPIONATE see DGI600
SODIUM-1,3-DICHLORO-1,3,5-TRIAZINE-2,4-DIONE-6-OXIDE see SGG500
1-SODIUM-3,5-DICHLORO-s-TRIAZINE-2,4,6-TRIONE see SGG500
1-SODIUM-3,5-DICHLORO-1,3,5-TRIAZINE-2,4,6-TRIONE see SGG500
SODIUM DICHLORO-s-TRIAZINETRIONE, dry, containing more than 39% available chlorine (DOT) see SGG500
SODIUM DICHROMATE see SGI000
SODIUM DICHROMATE(VI) see SGI000
SODIUM DICHROMATE de (FRENCH) see SGI000
SODIUM DICHROMATE DIHYDRATE see SGI500
SODIUM DICLOXACILLIN see DGE200
SODIUM DICLOXACILLIN MONOHYDRATE see DGE200
SODIUM DIETHYLBARBITURATE see BAG250
SODIUM-5,5-DIETHYLBARBITURATE see BAG250
SODIUM DIETHYLDITHIOCARBAMATE see SGJ000
SODIUM N,N-DIETHYLDITHIOCARBAMATE see SGJ000
SODIUM DIETHYLDITHIOCARBAMATE TRIHYDRATE see SGJ500
SODIUM DI-(2-ETHYLHEXYL) SULFOSUCCINATE see DJL000
SODIUM DIFORMYLNITROMETHANIDE HYDRATE see SGK600
SODIUM DIHYDROBIS(2-METHOXYETHOXY)ALUMINATE see SGK800
SODIUM DIHYDROGEN ARSENATE see ARD600
SODIUM DIHYDROGEN CITRATE see MRL000
SODIUM DIHYDROGEN ORTHOARSENATE see ARD600
SODIUM DIHYDROGEN PHOSPHATE (1582581) see SJH100
SODIUM DIHYDROGENPHOSPHIDE see SGM000
SODIUM-m-DIISOPROPYLBENZOL (Na-m) DIHYDROPEROXIDE (RUSSIAN) see DNN840
SODIUM-p-DIISOPROPYLBENZOL (Na-p) DIHYDROPEROXIDE (RUSSIAN) see DNN850
SODIUM-2,3-DIMERCAPTOPROPANE-1-SULFONATE see DNU860
SODIUM-4,4-DIMETHOXY-1-aci-NITRO-3,5-DINITRO-2,5-CYCLOHEXADIENE see SGM100
SODIUM-4-(DIMETHYLAMINO)BENZENEDIAZOSULFONATE see DOU600
SODIUM-p-(DIMETHYLAMINO)BENZENEDIAZOSULFONATE see DOU600
SODIUM-4-(DIMETHYLAMINO)BENZENEDIAZOSULPHONATE see DOU600
SODIUM-p-(DIMETHYLAMINO)BENZENEDIAZOSULPHONATE see DOU600
SODIUM-3-(DIMETHYLAMINOMETHYLENEAMINO)-2,4,6-TRIIODOHYDROCINNAMATE see SKM000
SODIUM-(4-(DIMETHYLAMINO)PHENYL)DIAZENESULFONATE see DOU600
SODIUM DIMETHYLARSINATE see HKC500
SODIUM DIMETHYLARSINIC ACID TRIHYDRATE see HKC550
SODIUM DIMETHYLARSONATE see HKC500
SODIUM N,N-DIMETHYLDITHIOCARBAMATE see SGM500

SODIUM-4-(2,4-DINITROANILINO)DIPHENYLAMINE-2-SULFONATE see SGP500
SODIUM-4,6-DINITRO-o-CRESOXIDE see DUU600
SODIUM DINITROMETHANIDE see SGP600
SODIUM-5-DINITROMETHYLTETRAZOLIDE see SGQ000
SODIUM-2,4-DINITROPHENOL see DVA800
SODIUM-2,4-DINITROPHENOLATE see DVA800
SODIUM-2,4-DINITROPHENOXIDE see SGQ100
SODIUM DIOCTYL SULFOSUCCINATE see DJL000
SODIUM DIOCTYL SULPHOSUCCINATE see DJL000
SODIUM DIOXIDE see SJC500
SODIUM-1,1-DIOXOPENICILLANATE see PAP600
SODIUM DIPHENYL-4,4'-BIS-AZO-2''-8''-AMINO-1''-NAPHTHOL-3'',6'' DISULPHONATE see CMO000
SODIUM DIPHENYLDIAZO-BIS(α-NAPHTHYLAMINESULFONATE) see SGQ500
SODIUM DIPHENYLHYDANTOIN see DNU000
SODIUM DIPHENYL HYDANTOINATE see DNU000
SODIUM-5,5-DIPHENYLHYDANTOINATE see DNU000
SODIUM-5,5-DIPHENYL-2,4-IMIDAZOLIDINEDIONE see DNU000
SODIUM DIPROPYLACETATE see PNX750
SODIUM-α,α-DIPROPYLACETATE see PNX750
SODIUM DISULFIDE see SGR500
SODIUM-2,3-DITHIOLPROPANESULFONATE see DNU860
SODIUM DITHIONITE (DOT) see SHR500
SODIUM DITOLYLDIAZOBIS-8-AMINO-1-NAPHTHOL-3,6-DISULFONATE see CMO250
SODIUM DITOLYLDIAZOBIS-8-AMINO-1-NAPHTHOL-3,6-DISULPHONATE see CMO250
SODIUM DNP see DVA800
SODIUM DODECANOATE see LBN000
SODIUM DODECYLBENZENESULFONATE (DOT) see DXW200
SODIUM DODECYLBENZENESULFONATE, dry see DXW200
SODIUM DODECYL SULFATE see SIB600
SODIUM EDETATE see EIV000
SODIUM EDTA see EIV000
SODIUM EOSINATE see BNH500
SODIUM EQUILIN 3-MONOSULFATE see ECW520
SODIUM EQUILIN SULFATE see ECW520
SODIUM ESTRONE SULFATE see EDV600
SODIUM ESTRONE-3-SULFATE see EDV600
SODIUM ETASULFATE see TAV750
SODIUM ETHAMINAL see NBU000
SODIUM ETHANEPEROXOATE see SJB000
SODIUM ETHASULFATE see TAV750
SODIUM ETHIDRONATE see DXD400
SODIUM ETHOXIDE see SGR800
SODIUM ETHOXYACETYLIDE see SGS000
SODIUM-6-(2-ETHOXY-1-NAPHTHAMIDO)PENICILLANATE see SGS500
SODIUM ETHYDRONATE see DXD400
SODIUM ETHYLBARBITAL see BAG250
SODIUM ETHYL-N-BUTYL BARBITURATE see BPF750
SODIUM-5-ETHYL-5-sec-BUTYLBARBITURATE see BPF250
SODIUM ETHYLENEDIAMINETETRAACETATE see EIV000
SODIUM ETHYLENEDIAMINETETRAACETIC ACID see EIV000
SODIUM-4-ETHYL-1-(3-ETHYLPENTYL)-1-OCTYL SULFATE see DKD400
SODIUM(2-ETHYLHEXYL)ALCOHOL SULFATE see TAV750
SODIUM-5-ETHYL-5-HEXYLBARBITURATE see EKT500
SODIUM-2-ETHYLHEXYL SULFATE see TAV750
SODIUM-2-ETHYLHEXYLSULFOSUCCINATE see DJL000
SODIUM ETHYLISOAMYLBARBITURATE see AON750
SODIUM ETHYLMERCURIC THIOSALICYLATE see MDI000
SODIUM-o-(ETHYLMERCURITHIO)BENZOATE see MDI000
SODIUM ETHYLMERCURITHIOSALICYLATE see MDI000
SODIUM-5-ETHYL-5-(1-METHYL-1-BUTENYL) BARBITURATE see VKP000
SODIUM-5-ETHYL-5-(1-METHYLBUTYL)BARBITURATE see NBU000
SODIUM-5-ETHYL-5-(1-METHYLBUTYL)-2-THIOBARBITURATE see PBT500
SODIUM-5-ETHYL-5-(1-METHYLPROPYL)BARBITURATE see BPF250
SODIUM-7-METHYL-2-METHYL-4-UNDECANOL SULFATE see EMT500
SODIUM-7-ETHYL-2-METHYLUNDECYL-4-SULFATE see EMT500
SODIUM-5-ETHYL-5-PHENYLBARBITURATE see SID000
SODIUM ETHYLXANTHATE see SHE500
SODIUM ETHYLXANTHOGENATE see SHE500
SODIUM ETHYNIDE see SEG500
SODIUM ETIDRONATE see DXD400
SODIUM EVIPAL see ERE000
SODIUM EVIPAN see ERE000
SODIUM FEREDETATE see EJA379
SODIUM FERRIC EDTA see EJA379
SODIUM FERRIC PYROPHOSPHATE see SHE700
SODIUM FLUCLOXACILLIN see FDA100
SODIUM FLUOACETATE see SHG500
SODIUM FLUOACETIC ACID see SHG500
SODIUM FLUOALUMINATE see SHF000

SODIUM FLUORACETATE de (FRENCH) see SHG500
SODIUM FLUORESCEIN see FEW000
SODIUM FLUORESCEINATE see FEW000
SODIUM FLUORIDE see SHF500
SODIUM FLUORIDE, solid and solution (DOT) see SHF500
SODIUM FLUORIDE (solution) see SHG000
SODIUM FLUORIDE(Na(HF$_2$)) see SHQ500
SODIUM FLUOROACETATE see SHG500
SODIUM γ-FLUORO-β-HYDROXYBUTYRATE see SHI000
SODIUM FLUOROPHOSPHATE (Na$_2$PO$_3$F) see DXD600
SODIUM FLUOROSILICATE see DXE000
SODIUM FLUORURE (FRENCH) see SHF500
SODIUM FLUOSILICATE see DXE000
SODIUM FORMALDEHYDE BISULFITE see SHI500
SODIUM FORMALDEHYDE SULFOXYLATE see FMW000
SODIUM FORMALDEHYDE SULFOXYLATE of 3-AMINO-4-
  HYDROXYPHENYLARSONIC ACID see SHI625
SODIUM FORMATE see SHJ000
SODIUM FOSFOMYCIN see DXF600
SODIUM FOSFOMYCIN HYDRATE see FOL200
SODIUM FULMINATE see SHJ500
SODIUM FUMARATE see DXD800
SODIUM FUSIDATE see SHK000
SODIUM FUSIDIN see SHK000
SODIUM GERMANIDE see SHK500
SODIUM GLUCONATE see SHK800
SODIUM d-GLUCONATE see SHK800
SODIUM GLUCOSULFONE see AOO800
SODIUM GLUTAMATE see MRL500
SODIUM l-GLUTAMATE see MRL500
l(+) SODIUM GLUTAMATE see MRL500
SODIUM GMP see GLS800
SODIUM GUANOSINE-5'-MONOPHOSPHATE see GLS800
SODIUM GUANYLATE see GLS800
SODIUM-5'-GUANYLATE see GLS800
SODIUM HEPARIN see HAQ550
SODIUM HEPARINATE see HAQ550
SODIUM HEXABARBITAL see ERE000
SODIUM HEXACHLOROPLATINATE HEXAHYDRATE see HCL300
SODIUM HEXACYCLONATE see SHL500
SODIUM HEXAFLUOROALUMINATE see SHF000
SODIUM HEXAFLUOROARSENATE see SHM000
SODIUM HEXAFLUOROSILICATE see DXE000
SODIUM HEXAFLUOSILICATE see DXE000
SODIUM HEXAMETAPHOSPHATE see SHM500, SII500
SODIUM HEXANITROCOBALTATE see SFX750
SODIUM HEXAVANADATE see SHN000
SODIUM HEXESTROL DIPHOSPHATE see SHN150
SODIUM HEXETHAL see EKT500
SODIUM HEXOBARBITONE see ERE000
SODIUM-N-HEXYLETHYL BARBITURATE see EKT500
SODIUM HEXYLETHYL THIOBARBITURATE see SHN275
SODIUM HIPPURATE see SHN500
SODIUM HYALURONATE see HGN600
SODIUM HYDRATE (DOT) see SHS000
SODIUM HYDRATE, solution see SHS500
SODIUM HYDRAZIDE see SHO000
SODIUM HYDRIDE see SHO500
SODIUM HYDROCORTISONE SUCCINATE see HHR000
SODIUM HYDROCORTISONE-21-SUCCINATE see HHR000
SODIUM HYDROFLUORIDE see SHF500
SODIUM HYDROGEN-S-(2-AMINOETHYL)PHOSPHOROTHIOATE see AKB500
SODIUM HYDROGEN-S-(2-AMINOETHYL)PHOSPHOROTHIOIC ACID see
  AKB500
SODIUM HYDROGEN CARBONATE see SFC500
SODIUM HYDROGEN S-((N-CYCLOOCTYLMETHYLAMIDINO)METHYL)
  PHOSPHOROTHIOATE HYDRATE (4584585) see SHQ000
SODIUM HYDROGEN DIFLUORIDE see SHQ500
SODIUM HYDROGEN FLUORIDE see SHQ500
SODIUM HYDROGEN PHOSPHATE see SJH090
SODIUM HYDROGEN SULFATE, solid (DOT) see SEG800
SODIUM HYDROGEN SULFATE, solution (DOT) see SEG800
SODIUM HYDROGEN SULFIDE see SHR000
SODIUM HYDROGEN SULFITE see SFE000
SODIUM HYDROGEN SULFITE, solid (DOT) see SFE000
SODIUM HYDROGEN SULFITE, solution (DOT) see SFE000
SODIUM HYDROGEN TRILACTATOZIRCONYLATE see ZTA000
SODIUM HYDROSULFIDE see SHR000
SODIUM HYDROSULFIDE, solution (DOT) see SHR000
SODIUM HYDROSULFITE (DOT) see SHR500
SODIUM HYDROSULPHIDE, solid (DOT) see SHR000
SODIUM HYDROSULPHIDE, with less than 25% water of crystallization (DOT)
  see SHR000
SODIUM HYDROSULPHITE see SHR500

SODIUM HYDROXIDE see SHS000
SODIUM HYDROXIDE, bead (DOT) see SHS000
SODIUM HYDROXIDE, dry (DOT) see SHS000
SODIUM HYDROXIDE, flake (DOT) see SHS000
SODIUM HYDROXIDE, granular (DOT) see SHS000
SODIUM HYDROXIDE (liquid) see SHS500
SODIUM HYDROXIDE, solid (DOT) see SHS000
SODIUM HYDROXIDE, solution (FCC) see SHS500
SODIUM HYDROXYACETATE see SHT000
SODIUM α-HYDROXYACETATE see SHT000
SODIUM-o-HYDROXYBENZOATE see SJO000
SODIUM-4-HYDROXYBUTYRATE see HJS500
SODIUM-γ-HYDROXYBUTYRATE see HJS500
SODIUM(HYDROXYDE de) (FRENCH) see SHS000
SODIUM-2-HYDROXYDIPHENYL see BGJ750
SODIUM-1-HYDROXYETHANESULFONATE see AAH500
SODIUM-2-HYDROXYETHOXIDE see SHT100
SODIUM o-HYDROXYMERCURIBENZOATE see SHT500
SODIUM p-HYDROXYMERCURIBENZOATE see SHU000
SODIUM o-((3-(HYDROXYMERCURI)-2-METHOXYPROPYL)CARBAM-
  OYL)PHENOXY ACETATE see SIH500
SODIUM-o-((3-HYDROXYMERCURI-2-METHOXYPROPYL)CAR-
  BAMYL)PHENOXYACETATE and THEOPHYLLINE see MDH750
SODIUM-3-HYDROXYMERCURIO-2,6-DINITRO-4-aci-NITRO-2,5-
  CYCLOHEXADIENONIDE see SHU175
SODIUM-2-HYDROXYMERCURIO-4-aci-NITRO-2,5-CYCLOHEXADIENONIDE
  see SHU250
SODIUM-2-HYDROXYMERCURIO-6-NITRO-4-aci-NITRO-2,5-
  CYCLOHEXADIENONIDE see SHU275
SODIUM HYDROXYMETHANESULFINATE see FMW000
SODIUM-1-(HYDROXYMETHYL)CYCLOHEXANEACETATE see SHL500
SODIUM-3-β-HYDROXY-11-OXO-12-OLEANEN-30-OATE SODIUM SUCCI-
  NATE see CBO500
SODIUM-5-(5'-HYDROXYTETRAZOL-3'-YLAZO)TETRAZOLIDE see SHU300
SODIUM-5(5'-HYDROXYTETRAZOL-3'-YLAZO)TETRAZOLIDE see HOF500
SODIUM HYPOBORATE see SHU475
SODIUM HYPOCHLORITE see SHU500
SODIUM HYPOCHLORITE PENTAHYDRATE see SHU525
SODIUM HYPOPHOSPHITE see SHV000
SODIUM HYPOSULFITE see SKI000, SKI500
SODIUM IBUPROFEN see IAB500
SODIUM (E)-3-(p-(1H-IMIDAZOL-1-YLMETHYL)PHENYL)-2-PROPENOATE
  see SHV100
SODIUM-5,5'-INDIGOTIDISULFONATE see FAE100
SODIUM INOSINATE see DXE500
SODIUM-5'-INOSINATE see DXE500
SODIUM IODATE see SHV500
SODIUM IODIDE see SHW000
SODIUM IODINE see SHW000
SODIUM-o-IODOBENZOATE see IEE050
SODIUM-p-IODOBENZOATE see IEE100
SODIUM IODOHIPPURATE see HGB200
SODIUM-2-IODOHIPPURATE see IEE050
SODIUM o-IODOHIPPURATE see HGB200
SODIUM IODOMETHANESULFONATE see SHX000
SODIUM-5-IODO-2-THIOURACIL see SHX500
SODIUM IOPODATE see SKM000
SODIUM IPODATE see SKM000
SODIUM IRON EDTA see EJA379
SODIUM IRON PYROPHOSPHATE see SHE700
SODIUM ISOAMYLETHYL BARBITURATE see AON750
SODIUM-5-ISOBUTYL-5-(METHYLTHIOMETHYL)BARBITURATE see SHY000
SODIUM ISOCYANATE see COI250
SODIUM ISONICOTINYL HYDRAZINE METHANSULFONATE see SIJ000
SODIUM ISOPROPOXIDE see SHY300
SODIUM ISOPROPYLETHYL THIOBARBITURATE see SHY500
SODIUM ISOPROPYLXANTHATE see SIA000
SODIUM ISOTHIOCYANATE see SIA500
SODIUM KETOPROFEN see KGK100
SODIUM LACTATE see LAM000
SODIUM LAURATE see LBN000
SODIUM LAURYLBENZENESULFONATE see DXW200
SODIUM LAURYL ETHER SULFATE see SIB500
SODIUM LAURYL ETHOXYSULPHATE see DYA000
SODIUM-N-LAURYL SARCOSINE see DXZ000
SODIUM LAURYL SULFATE see SIB600
SODIUM LAURYL TRIOXYETHYLENE SULFATE see SIC000
SODIUM LEVOTHYROXINE see LEQ300
SODIUM LINOLEATE HYDROPEROXIDE see SIC250
SODIUM LINOLEIC ACID HYDROPEROXIDE see SIC250
SODIUM LITHOCHOLATE see SIC500
SODIUM LUMINAL see SID000
SODIUM MALEATE see SIE000
SODIUM MALONATE see SIE500

SODIUM MALONDIALDEHYDE see MAN700
SODIUM MALONYLUREA see BAG250
SODIUM MANNITOL ANTIMONATE see SIF000
SODIUM MCPA see SIL500
SODIUM MECLOFENAMATE see SIF425
SODIUM MECLOPHENATE see SIF425
SODIUM MERALEIN see SIF500
SODIUM MERALLURIDE see SIG000
SODIUM MERCAPTAN see SHR000
SODIUM MERCAPTIDE see SHR000
SODIUM MERCAPTOACETATE see SKH500
SODIUM-2-MERCAPTOBENZOTHIAZOLE see SIG500
SODIUM-2-MERCAPTOETHANESULFONATE see MDK875
SODIUM MERCAPTOMERIN see TFK270
SODIUM MERSALYL see SIH500
SODIUM MERTHIOLATE see MDI000
SODIUM METAARSENATE see ARD500
SODIUM METAARSENITE see SEY500
SODIUM METABISULFITE see SII000
SODIUM METABOSULPHITE see SII000
SODIUM METAL (DOT) see SEE500
SODIUM, METAL DISPERSION IN ORGANIC SOLVENT see SEF500
SODIUM METAPERIODATE see SJB500
SODIUM METAPHOSPHATE see SII500, TKP750
SODIUM METASILICATE see SJU000
SODIUM METASILICATE, anhydrous see SJU000
SODIUM METAVANADATE see SKP000
SODIUM METHANALSULFOXYLATE see FMW000
SODIUM METHANEARSONATE see DXE600, MRL750
SODIUM METHANESULFONATE see SIJ000
4-SODIUM METHANESULFONATE METHYLAMINE-ANTIPYRINE see
   AMK500
SODIUM METHARSONATE see DXE600
SODIUM METHIODAL see SHX000
SODIUM METHOHEXITAL see MDU500
SODIUM METHOHEXITONE see MDU500
SODIUM METHOTREXATE see MDV600
SODIUM METHOXIDE see SIK450
SODIUM METHOXYACETYLIDE see SIJ600
SODIUM β-4-METHOXYBENZOYL-β-BROMOACRYLATE see SIK000
SODIUM METHYLAMINOANTIPYRINE METHANESULFONATE see AMK500
SODIUM-4-METHYLAMINO-1,5-DIMETHYL-2-PHENYL-3-PYRAZOLONE 4-
   METHANESULFONATE see AMK500
SODIUM METHYLARSONATE see DXE600
SODIUM METHYLATE see SIK450
SODIUM METHYLATE (alcohol mixture) see SIK500
SODIUM METHYLATE, DRY (DOT) see SIK450
SODIUM-4-METHYLBENZENESULFINATE see SKJ500
SODIUM-p-METHYLBENZENESULFONATE see SKK000
SODIUM (2-METHYL-4-CHLOROPHENOXY)ACETATE see SIL500
SODIUM-N-METHYL CYCLOHEXENYLMETHYBARBITURATE see ERE000
SODIUM METHYLDITHIOCARBAMATE see VFU000
SODIUM N-METHYLDITHIOCARBAMATE see VFU000
SODIUM N-METHYLDITHIOCARBAMATE DIHYDRATE see VFW009
SODIUM-2-METHYL-7-ETHYLUNDECANOL-4-SULFATE see EMT500
SODIUM-2-METHYL-7-ETHYLUNDECYL SULFATE-4 see EMT500
SODIUM METHYLHEXABITAL see ERE000
SODIUM METHYLHEXABITOL see ERE000
SODIUM-3-METHYLISOXAZOLIN-4,5-DIONE-4-OXIMATE see SIL600
SODIUM METHYLMERCURIC THIOGLYCOLLATE see SIM000
SODIUM-2-(N-METHYLOLEAMIDO)ETHANE-1-SULFONATE see SIY000
SODIUM METHYL OLEOYL TAURATE see SIY000
SODIUM N-METHYL-N-OLEOYLTAURATE see SIY000
SODIUM-4-METHYLPHENOXIDE see SIM100
SODIUM-1-METHYL-5-SEMICARBAZONO-6-OXO-2,3,5,6-
   TETRAHYDROINDOLE-3-SULFONATE TRIHYDRATE see AER666
SODIUM-1-METHYL-5-p-TOLUOYLPYRROLE-2-ACETATE DIHYDRATE
   see SKJ350
SODIUM MOLYBDATE see DXE800
SODIUM MOLYBDATE(VI) see DXE800
SODIUM MOLYBDATE DIHYDRATE see DXE875
SODIUM MONENSIN see MRE230
SODIUM MONOCHLORACETATE see SFU500
SODIUM MONODODECYL SULFATE see SIB600
SODIUM MONOFLUORIDE see SHF500
SODIUM MONOFLUOROACETATE see SHG500
SODIUM MONOHYDROGEN ARSENATE see ARD500
SODIUM MONOHYDROGEN PHOSPHATE (2:1:1) see SJH090
SODIUM MONOIODIDE see SHW000
SODIUM MONOIODOACETATE see SIN000
SODIUM MONOIODOMETHANESULFONATE see SHX000
SODIUM MONOSULFIDE see SJY500
SODIUM MONOXIDE see SIN500
SODIUM MONOXIDE, solid (DOT) see SIN500

SODIUM MORPHOLINECARBODITHIOATE see SIN650
SODIUM MORPHOLINEDITHIOCARBAMATE see SIN650
SODIUM MORPHOLINE and NITRITE (1581) see SIN675
SODIUM MORPHOLINODITHIOCARBAMATE see SIN650
SODIUM MYCOPHENOLATE see SIN850
SODIUM MYRISTYL SULFATE see SIO000
SODIUM NAFCILLIN see SGS500
SODIUM NAPHTHIONATE see ALI500
SODIUM-β-NAPHTHOQUINONE-4-SULFONATE see DLK000
SODIUM-1,2-NAPHTHOQUINONE-4-SULFONATE see DLK000
SODIUM-α-NAPHTHYLAMINE-6-SULPHONATE see ALI500
SODIUM N-1-NAPHTHYLPHTHALAMATE see SIO500
SODIUM N-1-NAPHTHYLPHTHALAMIC ACID see SIO500
SODIUM NEMBUTAL see NBU000
SODIUM-22 NEOPRENE ACCELERATOR see IAQ000
SODIUM NICOTINATE see NDW500
SODIUM NIGERICIN see SIO788
SODIUM NITRATE (DOT) see SIO900
SODIUM(I) NITRATE (1581) see SIO900
SODIUM NITRIDE see SIP000
SODIUM NITRILOACETATE see SEP500
SODIUM NITRILOTRIACETATE see SEP500, SIP500
SODIUM NITRITE see SIQ500
SODIUM NITRITE mixed with AMINOPYRINE (1:1) see DOT200
SODIUM NITRITE and BENLATE see BAV500
SODIUM NITRITE mixed with BIS(DIETHYLTHIOCARBAMOYL)DISULFIDE
   see SIS200
SODIUM NITRITE mixed with 1-(p-BROMOPHENYL)-3-METHYLUREA
   see SIQ675
SODIUM NITRITE and CARBENDAZIM (1:5) see CBN375
SODIUM NITRITE and CARBENDAZIME (1:1) see SIQ700
SODIUM NITRITE mixed with CHLORDIAZEPOXIDE (1:1) see SIS000
SODIUM NITRITE mixed with CIMETIDINE (1:4) see CMP875
SODIUM NITRITE and l-CITRULLINE (1:2) see SIS100
SODIUM NITRITE mixed with 4-(DIMETHYLAMINO)ANTIPYRINE (1:1)
   see DOT200
SODIUM NITRITE mixed with DIMETHYLDODECYLAMINE (8:7) see SIS150
SODIUM NITRITE and DIMETHYLUREA see DUM400
SODIUM NITRITE mixed with DISULFIRAM see SIS200
SODIUM NITRITE with DODECYL GUANIDINE ACETATE (5:3) see DXX600
SODIUM NITRITE mixed with ETHAMBUTOL (1:1) see SIS500
SODIUM NITRITE mixed with ETHYLENETHIOUREA see IAR000
SODIUM NITRITE and ETHYL UREA (1:1) see EQD900
SODIUM NITRITE and ETHYLUREA (1:2) see EQE000
SODIUM NITRITE mixed with HEPTAMETHYLENEIMINE see SIT000
SODIUM NITRITE and ISOPROPYLUREA see SIS675
SODIUM NITRITE mixed with METHAPYRILENE (2:1) see MDT000
SODIUM NITRITE mixed with N-METHYLADENOSINE (4:1) see SIS650
SODIUM NITRITE and METHYLANILINE (1:1.2) see MGO000
SODIUM NITRITE mixed with N-METHYLANILINE (35:1) see MGO250
SODIUM NITRITE and METHYL-2-BENZIMIDAZOLE CARBAMATE see
   CBN375
SODIUM NITRITE mixed with METHYLBENZYLAMINE (3:2) see MHN250
SODIUM NITRITE mixed with N-METHYLBENZYLAMINE (1:1) see MHN000
SODIUM NITRITE mixed with 1-METHYL-3-(p-BROMOPHENYL)UREA see
   SIQ675
SODIUM NITRITE and 1-(METHYLETHYL)UREA see SIS675
SODIUM NITRITE mixed with METHYLGUANIDINE (1:1) see MKJ000
SODIUM NITRITE mixed with 1-METHYL-3-NITROGUANIDINE (1:1) see
   MML500
SODIUM NITRITE and 2-METHYL-N-NITROSO-BENZIMIDAZOLE CARBA-
   MATE (1:1) see SIQ700
SODIUM NITRITE and 1-METHYL-1-NITROSO-3-PHENYLUREA see SIS700
SODIUM NITRITE and METHYL UREA see MQJ250
SODIUM NITRITE mixed with OCTAHYDROAZOCINE HYDROCHLORIDE
   (1:1) see SIT000
SODIUM NITRITE and PIPERAZINE (1:4) see PIJ250
SODIUM NITRITE and 1-PROPYLUREA see SIT500
SODIUM NITRITE and n-PROPYLUREA see SIT500
SODIUM NITRITE, mixed with SODIUM NITRATE and POTASSIUM NI-
   TRATE see SIT750
SODIUM NITRITE and TRIFORINE see SIT800
SODIUM p-NITROBENZENEAZOSALICYLATE see SIU000
SODIUM NITROCOBALTATE (III) see SFX750
SODIUM NITROFERRICYANIDE see SIU500
SODIUM NITROMALONALDEHYDE see SIV000
SODIUM aci-NITROMETHANE see SIV500
SODIUM aci-NITROMETHANIDE see SIV500
SODIUM-4-NITROPHENOXIDE see SIV600
SODIUM NITROPRUSSATE see SIU500
SODIUM NITROPRUSSIATE see SIW000
SODIUM NITROPRUSSIDE see SIU500
SODIUM NITROPRUSSIDE DIHYDRATE see SIW500
SODIUM NITROSOPENTACYANOFERRATE (3) see SIW000

SODIUM-4-NITROSOPHENOXIDE see SIW550
SODIUM NITROSYLPENTACYANOFERRATE see SIU500
SODIUM NITROSYLPENTACYANOFERRATE(III) see SIU500
SODIUM NITROSYLPENTACYANOFERRATE(III) DIHYDRATE see SIW500
SODIUM-5-NITROTETRAZOLIDE see SIW600
SODIUM-2-NITROTHIOPHENOXIDE see SIW625
SODIUM NITROXYLATE see SIX000
SODIUM NONYL SULFATE see SIX500
SODIUM NORAMIDOPYRINE METHANESULFONATE see AMK500
SODIUM NORSULFAZOLE see TEX500
SODIUM NOVOBIOCIN see NOB000
SODIUM OCTADECANOATE see SJV500
SODIUM OCTAHYDROTRIBORATE see SIX550
SODIUM OLEATE see OIA000
SODIUM N-OLEOYL-N-METHYLATAURINE see SIY000
SODIUM N-OLEOYL-N-METHYLTAURATE see SIY000
SODIUM OLEYLMETHYLTAURIDE see SIY000
SODIUM OMADINE see HOC000
SODIU-4-MORPHOLINECARBODITHIOATE see SIN650
SODIUM ORTHOARSENATE see ARD750
SODIUM ORTHOARSENITE see ARJ500
SODIUM ORTHOVANADATE see SIY250
SODIUM OXACILLIN see MNV250
SODIUM OXALATE see SIY500
SODIUM OXIDE see SIN500
SODIUM OXIDE (Na2-O2) see SJC500
SODIUM OXYBATE see HJS500
SODIUM OXYDIACETATE see SIZ000
SODIUM P-50 see SEQ000
SODIUM PANTOTHENATE see SIZ050
SODIUM PARATOLUENE SULPHONATE see SKK000
SODIUM PATENT BLUE V see ADE500
SODIUM PCP see SJA000
SODIUM PENICILLANATE 1,1-DIOXIDE see PAP600
SODIUM PENICILLIN see BFD250
SODIUM PENICILLIN G see BFD250
SODIUM PENICILLIN II see BFD250
SODIUM-PENT see NBU000
SODIUM PENTABARBITAL see NBU000
SODIUM PENTABARBITONE see NBU000
SODIUM PENTACARBONYL RHENATE see SIZ100
SODIUM PENTACHLOROPHENATE see SJA000
SODIUM PENTACHLOROPHENATE (DOT) see SJA000
SODIUM PENTACHLOROPHENOL see SJA000
SODIUM PENTACHLOROPHENOLATE see SJA000
SODIUM PENTACHLOROPHENOXIDE see SJA000
SODIUM PENTAFLUOROSTANNITE see SJA500
SODIUM-β,β-PENTAMETHYLENE-Γ-HYDROXYBUTYRATE see SHL500
SODIUM PENTHIOBARBITAL see PBT500
SODIUM PENTOBARBITAL see NBU000
SODIUM (R)(+)-PENTOBARBITAL see PBS500
SODIUM PENTOBARBITONE see NBU000
SODIUM PENTOBARBITURATE see NBU000
SODIUM PENTOTHAL see PBT500
SODIUM PENTOTHIOBARBITAL see PBT500
SODIUM PERACETATE see SJB000
SODIUM PERBORATE see SJD000
SODIUM PERBORATE TETRAHYDRATE see SJB350
SODIUM PERCHLORATE see PCE750
SODIUM PERCHLORATE (DOT) see PCE750
SODIUM PERIODATE see SJB500
SODIUM PERMANGANATE see SJC000
SODIUM PEROXIDE see SJC500
SODIUM PEROXOBORATE see SJD000
SODIUM PEROXYACETATE see SJB000
SODIUM PEROXYBORATE see SJD000
SODIUM PEROXYDISULFATE see SJE000
SODIUM PERRHENATE see SJD500
SODIUM PERSULFATE see SJE000
SODIUM PHENATE see SJF000
SODIUM PHENOBARBITAL see SID000
SODIUM PHENOBARBITONE see SID000
SODIUM PHENOLATE, solid (DOT) see SJF000
SODIUM PHENOXIDE see SJF000
SODIUM-α-PHENOXYCARBONYLBENZYLPENICILLIN see CBO000
SODIUM PHENYLACETYLIDE see SJF500
SODIUM PHENYL-β-AMINOPROPIONAMIDE-p-ARSONATE see SJG000
SODIUM PHENYLBUTAZONE see BOV750
SODIUM-2-PHENYLCINCHONINATE see SJH000
SODIUM-1-PHENYL-2,3-DIMETHYL-4-METHYLAMINOPYRAZOLON-N-METHANESULFONATE see AMK500
SODIUM-1-PHENYL-2,3-DIMETHYL-5-PYRAZOLONE-4-METHYLAMINO METHANESULFONATE see AMK500

SODIUM PHENYLDIMETHYLPYRAZOLONMETHYLAMINOMETHANE SULFONATE see AMK500
SODIUM PHENYLETHYLBARBITURATE see SID000
SODIUM PHENYLETHYLMALONYLUREA see SID000
SODIUM-N-PHENYLGLYCINAMIDE-p-ARSONATE see CBJ750
SODIUM-2-PHENYLPHENATE see BGJ750
SODIUM-o-PHENYLPHENATE see BGJ750
SODIUM-o-PHENYLPHENOLATE see BGJ750
SODIUM-o-PHENYLPHENOXIDE see BGJ750
SODIUM PHOSPHATE see SJH200, TKP750
SODIUM PHOSPHATE, anhydrous see SJH200
SODIUM PHOSPHATE, DIBASIC see SJH090
SODIUM PHOSPHATE, MONOBASIC see SJH100
SODIUM PHOSPHATE, TRIBASIC see SJH200
SODIUM PHOSPHIDE see SJI500
SODIUM PHOSPHINATE see SHV000
SODIUM PHOSPHOROFLUORIDATE see DXD600
SODIUM PHOSPHOROFLURIDATE see DXD600
SODIUM PHOSPHOROTHIOATE see TNM750
SODIUM PHOSPHOTUNGSTATE see SJJ000
SODIUM PICOSULFATE see SJJ175
SODIUM PICRAMATE, WET (DOT) see PIC500
SODIUM PICRATE see SJJ190
SODIUM PIPERACILLIN see SJJ200
SODIUM PLATINIC CHLORIDE see SJJ500
SODIUM POLYACRYLATE see SJK000
SODIUM POLYALUMINATE see AHG000
SODIUM POLYANHYDROMANNURONIC ACID SULFATE see SEH450
SODIUM POLYANTIMONATE see AQB250
SODIUM POLYMANNURONATE see SEH000
SODIUM POLYOXYETHYLENE ALKYL ETHER SULFATE see SJK200
SODIUM POLYPHOSPHATES, GLASSY see SII500
SODIUM POLYSTYRENE SULFONATE see SJK375
SODIUM POTASSIUM ALLOY, liquid and solid (DOT) see PLS500
SODIUM POTASSIUM BISMUTH TARTRATE (SOLUBLE) see BKX250
SODIUM PRASTERONE SULFATE see SJK400
SODIUM PRASTERONE SULFATE DIHYDRATE see SJK410
SODIUM p-2-PROPANOL-OXY-PHENYLARSONATE see SJK475
SODIUM PROPIONATE see SJL500
SODIUM-2-PROPYLPENTANOATE see PNX750
SODIUM-2-PROPYLVALERATE see PNX750
SODIUM PYRIDINETHIONE see HOC000
SODIUM PYRITHIONE see MCQ750
SODIUM PYRITHIONE SQ 3277 see HOC000
SODIUM PYROBORATE see SFF000
SODIUM PYROBORATE DECAHYDRATE see SFF000
SODIUM PYROPHOSPHATE see DXF800
SODIUM PYROPHOSPHATE (FCC) see TEE500
SODIUM PYROSULFATE see SEG800
SODIUM PYROSULFITE see SII000
SODIUM PYROVANADATE see SJM500
SODIUM QUINALBARBITONE see SBN000
SODIUM RETINOATE see SJN000
SODIUM RHODANATE see SIA500
SODIUM RHODANIDE see SIA500
SODIUM RICINOLEATE see SJN500
SODIUM RIFOMYCIN SV see SJN650
SODIUM SACCHARIDE see SJN700
SODIUM SACCHARIN see SJN700
SODIUM SACCHARINATE see SJN700
SODIUM SACCHARINE see SJN700
SODIUM SALICYLATE see SJO000
SODIUM SALICYL-(Γ-HYDROXYMERCURI-β-METHOXYPROPYL)AMIDE-o-ACETATE see SIH500
SODIUM SALICYLIC ACID see SJO000
SODIUM SALT of 1-AMINO-2-NAPHTHOL-3,6-DISULPHONIC ACID see AKH250
SODIUM SALT of CACODYLIC ACID see HKC500
SODIUM SALT of CARBOXYMETHYLCELLULOSE see SFO500
SODIUM SALT-m-CRESOL see SJP000
SODIUM SALT of DICHLORO-s-TRIAZINETRIONE see SGG500
SODIUM SALT of N,N-DIETHYLDITHIOCARBAMIC ACID see SGJ000
SODIUM SALT of 3-(3-DIMETHYLAMINOMETHYLENEAMINO-2,4,6-TRIIODOPHENYL) PROPIONIC ACID see SKM000
SODIUM SALT of 4,6-DINITRO-o-CRESOL see DUU600
SODIUM SALT of ETHYLENEDIAMINETETRAACETIC ACID see EIV000
SODIUM SALT of HYDROXY-o-CARBOXY-PHENYL-FLUORONE see FEW000
SODIUM SALT of HYDROXYMERCURIACETYLAMINOBENZOIC ACID see SJP500
SODIUM SALT of HYDROXYMERCURIPROPANOL PHENYLACETIC ACID see SJQ000
SODIUM SALT of HYDROXYMERCURIPROPANOL PHENYL MALONIC ACID see SJQ500
SODIUM SALT of ISONICOTINIC ACID see ILF000

SODIUM SALT of MERCURI-BIS SALICYLIC ACID see SJR000
SODIUM SALT of MERCURY DITHIOCARBAMATE PIPERAZINE ACETIC ACID see SJR500
SODIUM SALT of β-4-METHOXYBENZOYL-β-BROMOACRYLIC ACID see SIK000
SODIUM SALT of 4-MORPHOLINECARBODITHIOIC ACID see SIN650
SODIUM SALT of PHENYLBUTAZONE see BOV750
SODIUM SALT of SULFONATED NAPHTHALENEFORMALDEHYDE CONDENSATE see BLX000
SODIUM SALT of 1,1,3,3-TETRACYANOPROPENE see TAI050
SODIUM SALT of 2,4,5-TRICHLOROPHENOL see SKK500
SODIUM SARKOMYCIN see SJS500
SODIUM SECOBARBITAL see SBN000
SODIUM SECONAL see SBN000
SODIUM SELENATE see DXG000
SODIUM SELENIDE see SJT000
SODIUM SELENITE see SJT500
SODIUM SELENITE PENTAHYDRATE see SJT600
SODIUM SESQUICARBONATE see SJT750
SODIUM SILICATE see SJU000
SODIUM SILICIDE see SJU500
SODIUM SILICOALUMINATE see SEM000
SODIUM SILICOFLUORIDE (DOT) see DXE000
SODIUM SORBATE see SJV000
SODIUM SOTRADECOL see EMT500, SIO000
SODIUM STANNOUS TARTRATE see TGF000
SODIUM STEARATE see SJV500
SODIUM STIBANILATE GLUCOSIDE see NCL000
SODIUM STIBINIVANADATE see SJW000
SODIUM STIBOGLUCONATE see AQH800, AQI250
SODIUM-2-(4-STYRYL-3-SULFOPHENYL)-2H-NAPHTHO-(1,2-d)-TRIAZOLE see TGE155
SODIUM SUCARYL see SGC000
SODIUM SULBACTAM see PAP600
SODIUM SULFACETAMIDE see SNQ000
SODIUM SULFADIAZINE see MRM250
SODIUM SULFAETHYLTHIADIAZOLE see SJW300
SODIUM SULFAMERAZINE see SJW475
SODIUM SULFAMETAZINE see SJW500
SODIUM SULFAMETHAZINE see SJW500
SODIUM SULFAMETHIAZINE see SJW500
SODIUM SULFAMEZATHINE see SJW500
SODIUM-2-SULFANILAMIDOTHIAZOLE see TEX500
SODIUM SULFANILYLACETAMIDE see SNQ000
SODIUM N-SULFANILYLACETAMIDE see SNQ000
SODIUM SULFAPYMONOHYDRATE see PPO250
SODIUM SULFAPYRIDINE see PPO250
SODIUM SULFAPYRIMIDINE see MRM250
SODIUM SULFATE (2581) see SJY000
SODIUM SULFATE anhydrous see SJY000
SODIUM SULFATHIAZOLE see TEX500
SODIUM SULFHYDRATE see SHR000
SODIUM SULFIDE (anhydrous) see SJY500
SODIUM SULFITE (2581) see SJZ000
SODIUM SULFITE, anhydrous see SJZ000
SODIUM SULFOCYANATE see SIA500
SODIUM SULFOCYANIDE see SIA500
SODIUM SULFODI-(2-ETHYLHEXYL)SULFOSUCCINATE see DJL000
SODIUM SULFOXYLATE see SHR500
SODIUM SULFOXYLATE FORMALDEHYDE see FMW000
SODIUM SULHYDRATE see SFE000
SODIUM SULPHAMERAZINE see SJW475
SODIUM SULPHATE see SJY000
SODIUM SULPHIDE see SJY500
SODIUM SULPHITE see SJZ000
SODIUM SYNTARPEN see SLJ050
SODIUM d-T4 see SKJ300
SODIUM TARTRATE see SKB000
SODIUM l-(+)-TARTRATE see BLC000
SODIUM TARTRATE (FCC) see BLC000
SODIUM TAUROCHOLATE see SKB500
SODIUM TCA, solution see TII250
SODIUM TCA INHIBITED see TII500
SODIUM TELLURATE see SKC000
SODIUM TELLURATE(IV) see SKC500
SODIUM TELLURATE, DIHYDRATE see SKC100
SODIUM TELLURATE VI see SKC000
SODIUM TELLURITE see SKC500
SODIUM TETRABORATE see SFF000
SODIUM TETRABORATE DECAHYDRATE see SFF000
SODIUM TETRACYANATOPALLADATE(II) see SKD600
SODIUM TETRADECYL SULFATE see SIO000
SODIUM TETRAFLUOROBORATE see SKE000
SODIUM TETRAHYDROALUMINATE(1−) see SEM500

(T-4) SODIUM, TETRAHYDROALUMINATE(1−) (9CI) see SEM500
SODIUM TETRAHYDROBORATE(1-) see SFF500
SODIUM TETRAHYDROGALLATE see SKE500
SODIUM TETRAPEROXYCHROMATE see SKF000
SODIUM TETRAPEROXYMOLYBDATE see SKF500
SODIUM TETRAPOLYPHOSPHATE see SII500
SODIUM TETRAVANADATE see SKG500
SODIUM THEOPHYLLINE see SKH000
SODIUM THIAMYLAL see SOX500
SODIUM THIOCYANATE see SIA500
SODIUM THIOCYANIDE see SIA500
SODIUM THIOGLYCOLATE see SKH500
SODIUM THIOGLYCOLLATE see SKH500
SODIUM THIOPENTAL see PBT500
SODIUM THIOPENTOBARBITAL see PBT500
SODIUM THIOPENTONE see PBT500
SODIUM THIOPHOSPHATE see TNM750
SODIUM THIOSULFATE see SKI000
SODIUM THIOSULFATE, anhydrous see SKI000
SODIUM THIOSULFATE, PENTAHYDRATE see SKI500
SODIUM THYROXIN see LEQ300
SODIUM THYROXINATE see LEQ300
SODIUM THYROXINE see LEQ300
SODIUM d-THYROXINE see SKJ300
SODIUM l-THYROXINE see LEQ300
SODIUM TOLMETIN see SKJ340
SODIUM TOLMETIN DIHYDRATE see SKJ350
SODIUM-4-TOLUENESULFINATE see SKJ500
SODIUM-p-TOLUENESULFINATE see SKJ500
SODIUM-p-TOLUENE SULFINIC ACID see SKJ500
SODIUM-p-TOLUENESULFONATE see SKK000
SODIUM p-TOLUENESULFONYLCHLORAMIDE see CDP000
SODIUM-p-TOLYL SULFONATE see SKK000
SODIUM TOSYLATE see SKK000
SODIUM TOSYLCHLORAMIDE see CDP000
SODIUM TRIAZIDOAURATE see SKK100
SODIUM (TRICHLORACETATE de) (FRENCH) see TII500
SODIUM TRICHLOROACETATE see TII500
SODIUM-2,4,5-TRICHLOROPHENATE see SKK500
SODIUM-2,4,5-TRICHLOROPHENOXYETHYL SULFATE see SKL000
SODIUM TRIFLUOROSTANNITE see SKL500
SODIUM TRIIODOHYDROCINNAMATE see SKM000
SODIUM-3,7,12-TRIKETOCHOLANATE see SGD500
SODIUM TRIMETAPHOSPHATE see SKM500, TKP750
SODIUM-2,4,6-TRINITROPHENOXIDE see SJJ190
SODIUM-3,7,12-TRIOXO-5-β-CHOLANATE see SGD500
SODIUM TRIPHOSPHATE see SKN000
SODIUM TRIPOLYPHOSPHATE see SKN000
SODIUM TUNGSTATE see SKN500
SODIUM TUNGSTATE, DIHYDRATE see SKO000
SODIUM TUNGSTOPHOSPHATE see SJJ000
SODIUM TYROPANOATE see SKO500
SODIUM URATE see SKO575
SODIUM VALPROATE see PNX750
SODIUM VANADATE see SIY250, SKP000
SODIUM VANADITE see SKQ000
SODIUM VANADIUM OXIDE see SIY250
SODIUM VERONAL see BAG250
SODIUM VERSENATE see EIX500
SODIUM VINBARBITAL see VKP000
SODIUM WARFARIN see WAT220
SODIUM XANTHATE see SHE500
SODIUM XANTHOGENATE see SHE500
SODIUM ZIRCONIUM LACTATE see LAO000
SODIUM ZIRCONIUM OXIDE SULFATE see ZTS100
SODIUM ZIRCONYL SULPHATE see ZTS100
SODIUM ZOMEPIRAC see ZUA300
SODIURETIC see BEQ625
SODIZOLE see SNN500
SODNA SUL KYSELINY cis-β-4-METHOXYBENZOYL-β-BROMAKRYLOVE see CQK600
SODOTHIOL see SKI500
SOFALCONE see IMS300
SO-FLO see SHF500
SOFRAMYCIN see NCF000
SOFRIL see SOD500
SOFTENIL see TEH500
SOFTENON see TEH500
SOFTIL see DJL000
SOFTRAN see BOM250
SOG-4 see CJN250
SOHNHOFEN STONE see CAO000
SOILBROM-40 see EIY500
SOILBROM-85 see EIY500

SOILFUME see EIY500
SOIL FUNGICIDE 1823 see CJA100
SOILSIN see EME100
SOIL STABILIZER 661 see SMR000
SOL see PKB500
SOLABAR see BAO300
SOLACEN see MOV500
SOLACIL see POF500
SOLACIN see MOV500
SOLACTHYL see AES650
SOLACTOL see LAJ000
SOLADREN see VGP000
SOLAMINE see BEN000
SOLAN see SKQ400
SOLANCARPIDINE see POJ000
SOLANDRA (VARIOUS SPECIES) see CDH125
(22r,25s)-5-α-SOLANIDANINE-3-β-OL see SKQ510
(22s,25s)-5-α-SOLANIDAN-3-β-OL see SKQ510
(22s,25r)-5-α-SOLANIDAN-3-β-OL see SKQ500
SOLANID-5-ENE-3-β, 12-α-DIOL see SKR000
(22s,25r)-SOLANID-5-EN-3-β-OL see SKR500
SOLANIDINE-S see POJ000
SOLANIN see SKR875
α-SOLANIN see SKS000
SOLANINE see SKS000
α-SOLANINE see SKS000
SOLANINE HYDROCHLORIDE see SKS100
SOLANIN HYDROCHLORIDE see SKS100
SOLANTAL see CJH750
SOLANTHRENE BRILLIANT YELLOW see DCZ000
SOLANTIN see DKQ000
SOLANTOIN see DNU000
SOLANTYL see DNU000
SOLANUM SODOMEUM, extract see AQP800
SOLANUM TUBEROSUM L see PLW750
SOLAR 40 see DXW200
SOLAR BROWN PL see CMO750
SOLAR LIGHT ORANGE GX see HGC000
SOLAR PURE YELLOW 8G see CMM750
SOLAR RUBINE see HJF500
SOLARSON see CEI250
SOLAR VIOLET 5BN see FAG120
SOLAR WINTER BAN see PML000
SOLASKIL see LFA020
SOLASOD-5-EN-3-β-OL see POJ000
SOLASODINE see POJ000
SOLASON see CEI250
SOLAXIN see CDQ750
SOLBAR see BAO300, BAP000
SOLBASE see PJT200
SOLBROL A see HJL000
SOLBROL B see BSC000
SOLBROL M see HJL500
SOLBUTAMOL see BQF500
SOLCAIN see DHK400
SOLCOSERYL see ADZ125, SKS299
SOLDEP see TIQ250
SOLDESAM see DAE525
SOLDIER'S BUTTONS see MBU550
SOLDIER'S CAP see MRE275
SOLESTRIL see POB500
SOLESTRO see EDP000
SOLEVAR see ENX600
2-SOLFAMONYL-4,4'-DIAMINOPHENYLSULFONE see AOT125
SOLFARIN see WAT200
SOLFO BLACK B see DUZ000
SOLFO BLACK BB see DUZ000
SOLFO BLACK G see DUZ000
SOLFO BLACK SB see DUZ000
SOLFO BLACK 2B SUPRA see DUZ000
SOLFOCRISOL see GJG000
SOLFONE see AOO800
SOLFURO di CARBONIO (ITALIAN) see CBV500
SOLGANAL see ART250
SOLGANAL B see ART250
SOLGOL see CNR675
SOLID GREEN CRYSTALS O see AFG500
SOLID GREEN FCF see FAG000
SOLIMIDIN see ZUA200
SOLIPHYLLINE see CMF500
SOLIWAX see DJL000
SOLKETAL see DVR600
SOLLICULIN see EDV000
SOLMETHINE see MJP450

SOLNET see PMB850
SOLOCALM see FAJ100
SOLOCHROME BLUE FB see HJF500
SOLOCHROME ORANGE M see NEY000
SOLOMON ISLAND IVY see PLW800
SOLOMON'S LILY see ITD050
SOLON see IMS300
SOLOSIN see TEP000
SOLOZONE see SJC500
SOL PHENOBARBITAL see SID000
SOL PHENOBARBITONE see SID000
SOLPRENE 300 see SMR000
SOLPRINA see CLD000
SOLPYRON see ADA725
SOL SODOWA KWASU LAURYLOBENZENOSULFONOWEGO (POLISH)
    see DXW200
SOLSOL NEEDLES see SIB600
SOLUBLE BARBITAL see BAG250
SOLUBLE FLUORESCEIN see FEW000
SOLUBLE GLUSIDE see SJN700
SOLUBLE GUN COTTON see CCU250
SOLUBLE INDIGO see FAE100
SOLUBLE IODOPHTHALEIN see TDE750
SOLUBLE PENTOBARBITAL see NBU000
SOLUBLE PHENOBARBITAL see SID000
SOLUBLE PHENOBARBITONE see SID000
SOLUBLE PHENYTOIN see DNU000
SOLUBLE SACCHARIN see SJN700
SOLUBLE SULFACETAMIDE see SNQ000
SOLUBLE SULFADIAZINE see MRM250
SOLUBLE SULFAMERAZINE see SJW475
SOLUBLE SULFAPYRIDINE see PPO250
SOLUBLE SULFATHIAZOLE see TEX500
SOLUBLE TARTRO-BISMUTHATE see BKX250
SOLUBLE THIOPENTONE see PBT500
SOLUBLE VANDYKE BROWN see MAT500
SOLUBOND 0-869 see RDP000
SOLUBOND 3520 see RDP000
SOLU-CORTEF see HHR000
SOLU-DECADRON see DAE525
SOLU-DECORTIN see PMA100
SOLUFILIN see DNC000
SOLUFILINA see CNR125, DIH600
SOLUFYLLIN see DNC000
SOLUGLACIT see NIM500
SOLU-GLYC see HHR000
SOLUMEDINE see SJW475
SOLU-MEDROL see USJ100
SOLU-MEDRONE see USJ100
SOLUPHYLINE see CNR125
SOLUPHYLLINE see HLC000
SOLUPYRIDINE see DXF400
SOLUROL see SOX875
SOLUSEDIV 2% see HAF400
SOLUSOL-75% see DJL000
SOLUSOL-100% see DJL000
SOLUSTIBOSAN see AQH800, AQI250
SOLUSTIN see AQH800, AQI250
SOLUSURMIN see AQH800, AQI250
SOLUTION CONCENTREE T271 see AMY050
SOLUTRAST see IFY000
SOLVANOL see DJX000
SOLVANOM see DTR200
SOLVARONE see DTR200
SOLVAT 14 see BFL000
SOLVENT 111 see MIH275
SOLVENT-DEWAXED HEAVY NAPHTHENIC DISTILLATE see MQV820
SOLVENT-DEWAXED HEAVY PARAFFINIC DISTILLATE see MQV825
SOLVENT-DEWAXED LIGHT NAPHTHENIC DISTILLATE see MQV835
SOLVENT-DEWAXED LIGHT PARAFFINIC DISTILLATE see MQV840
SOLVENT ETHER see EJU000
SOLVENT NAPHTHA see PCT250
SOLVENT ORANGE 15 see BJF000
SOLVENT RED 1 see CMS242
SOLVENT RED 19 see EOJ500
SOLVENT RED 72 see DDO200
SOLVENT REFINED COAL-II, HEAVY DISTILLATE see SLH300
SOLVENT REFINED COAL MATERIALS, HEAVY DISTILLATE see SLH300
SOLVENT-REFINED HEAVY NAPHTHENIC DISTILLATE see MQV845
SOLVENT-REFINED HEAVY PARAFFINIC DISTILLATE see MQV850
SOLVENT-REFINED LIGHT NAPHTHENIC DISTILLATE see MQV852
SOLVENT-REFINED LIGHT PARAFFINIC DISTILLATE see MQV855
SOLVENT YELLOW 1 see PEI000
SOLVENT YELLOW 14 see PEJ500

SOLVIC see PKQ059
SOLVIC 523KC see AAX175
SOLVIREX see DXH325
SOLVOSOL see CBR000
SOLYACORD see DJS200
SOLYUSURMIN see AQH800, AQI250
SOMA see IPU000
SOMACTON see PJA250
SOMADRIL see IPU000
SOMALGIT see IPU000
SOMALIA ORANGE A2R see XRA000
SOMALIA ORANGE I see PEJ500
SOMALIA ORANGE 2R see XRA000
SOMALIA RED III see OHA000
SOMALIA YELLOW A see DOT300
SOMALIA YELLOW R see AIC250
SOMAN see SKS500
SOMANIL see IPU000
SOMAR see DXE600
SOMATOSTATIN see SKS600
SOMATOSTATIN 14 see SKS600
SOMATOTROPIC HORMONE see PJA250
SOMATOTROPIN see PJA250
SOMAZINA see CMF350
SOMBEROL see QAK000
SOMBREVIN see PMM000
SOMBUCAPS see ERD500
SOMBULEX see ERD500
SOMBUTOL see EOK000
SOMELIN see HAG800
SOMILAN see ENE500
SOMINAT see SKS700
SOMIO see GFA000
SOMIPRONT see DUD800
SOMLAN see DAB800
SOMNAFAC see QAK000
SOMNALERT see ERD500
SOMNASED see DLY000
SOMNESIN see EQL000
SOMNEVRIN see CHD750
SOMNIBEL see DLY000
SOMNICAPS see DPJ400
SOMNI SED see CDO000
SOMNITE see DLY000
SOMNOLETTEN see EOK000
SOMNOMED see QAK000
SOMNOPENTYL see NBU000
SOMNOS see CDO000
SOMNOSAN see EOK000
SOMNUROL see BNP750
SOMONAL see EOK000
SOMONIL see DSO000
SOMOPHYLLIN see TEP500
SOMOPHYLLIN O see TEP500
SOMSANIT see HJS500
SONACIDE see GFQ000
SONACON see DCK759
SONAFORM see TDA500
SONAL see QAK000
SONAPAX see MOO250
SONAZINE see CKP500
SONBUTAL see BOR000
SONEBON see DLY000
SONERILE see BPF500
SONERYL see BPF500
SONG-SAM, ROOT EXTRACT see GEQ425
SONISTAN see NBU000
SONNOLIN see DLY000
SONTEC see CDO000
SONTOBARBITAL NABITONE see NBU000
SOOT see SKS750
SOPAQUIN see CLD000
SOPENTAL see NBU000
SOPHIA see MDL750
SOPHIAMIN see MDQ250
SOPHOCARPINE HYDROBROMIDE see SKS800
SOPHORA SECUNDIFLORA see NBR800
SOPHORA TOMENTOSA see NBR800
SOPHORETIN see QCA000
SOPHORICOL see GCM350
SOPHORIN see RSU000
SOPHORINE see CQL500
SOPP see BGJ750
SOPRABEL see LCK000

SOPRACOL see MLC250
SOPRACOL 781 see MLC250
SOPRATHION see EEH600, PAK000
SOPRINAL see BAG250
SOPRINTIN see ABH500
SOPROCIDE see BBQ750
SOPRONTIN see AAF750, ABH500
SOPROTIN see ABH500
SORBALDEHYDE see SKT500
SORBANGIL see CCK125
SORBA-SPRAY Mn see MAU250
SORBESTER P 17 see SKV100
SORBIC ACID see SKU000
SORBIC ACID, COPPER SALT see CNP000
SORBIC ACID, POTASSIUM SALT see PLS750
SORBIC ACID, SODIUM SALT see SJV000
SORBIC ALCOHOL see HCS500
SORBIC OIL see PAJ500
SORBICOLAN see SKV200
SORBID see CCK125
SORBIDE NITRATE see CCK125
SORBIDILAT see CCK125
SORBIDINITRATE see CCK125
SORBIMACROGOL OLEATE see PKL100
SORBIMACROGOL TRIOLEATE 300 see TOE250
SORBIMACROGOL TRISTEARATE 300 see SKV195
SORBINIC ALCOHOL see HCS500
SORBISLO see CCK125
SORBISTAT see SKU000
SORBISTAT-K see PLS750
SORBISTAT-POTASSIUM see PLS750
SORBITAL O 20 see PKL100
SORBITAN C see SKV150
SORBITAN MONODODECANOATE see SKV000
SORBITAN MONOLAURATE see SKV000
SORBITAN, MONOLAURATE POLYOXYETHYLENE derivative see PKG000
SORBITAN MONOOCTADECANOATE see SKV150
SORBITAN, MONOOCTADECANOATE, POLY(OXY-1,2-ETHANEDIYL) DERIV-
  ATIVES see PKL030
SORBITAN MONOOLEATE see SKV100
SORBITAN MONOOLEIC ACID ESTER see SKV100
SORBITAN MONOSTEARATE see SKV150
SORBITAN O see SKV100
SORBITAN OLEATE see SKV100
SORBITAN STEARATE see SKV150
SORBITAN, TRIOCTADECANOATE, POLY(OXY-1,2-ETHANEDIYL) derivs.
  (9CI) see SKV195
SORBITAN, TRISTEARATE, POLYOXYETHYLENE derivatives. see SKV195
SORBITE see SKV200
SORBITOL see SKV200
d-SORBITOL see SKV200
SORBITRATE see CCK125
SORBO see SKV200
SORBO-CALCIAN see CAL750
SORBO-CALCION see CAL750
SORBOL see SKV200
SORBONIT see CCK125
SORBON S 60 see SKV150
SORBOSTYL see SKV200
SORBYL ACETATE see AAU750
SORBYL ALCOHOL see HCS500
SORCI see FPD100
SORDENAC see CLX250
SORDINOL see CLX250
SOREFLON 604 see TAI250
SORETHYTAN (20) MONOOLEATE see PKL100
SORGEN 40 see SKV100
SORGEN 50 see SKV150
SORGHUM GUM see SLJ500
SORGOA see TGB475
SORGOPRIM see BQB000
SORIDERMAL see MNQ500
SORIPAL see MNQ500
SORLATE see PKL100
SORQUAD see CCK125
SORQUAT see CCK125
SORREL SALT see OLE000
SORROSIE (HAITI) see FPD100
SORSAKA see SKV500
l-SORSAKA, LEAF and STEM EXTRACT see SKV500
SORVILANDE see SKV200
SOSIGON see DOQ400
SOSPITAN see PPH050
SOTACOR see CCK250

SOTALEX see CCK250
SOTALOL see CCK250
SOTALOL HYDROCHLORIDE see CCK250
SOTERENOL see SKW000
SOTERENOL HYDROCHLORIDE see SKW500
SOTIPOX see TIQ250
SOTRADECOL see EMT500
SOTYL see NBU000
SOUCI d'EAU (CANADA) see MBU550
SOUDAN I see PEJ500
SOUDAN II see XRA000
SOUFRAMINE see PDP250
SOUP see NGY000
SOUTHERN BAYBERRY see WBA000
SOUTHERN BENTONITE see BAV750
SOUTHERN COPPERHEAD VENOM see SKW775
SOVCAIN see NOF500
SOVCAINE see DDT200
SOVCAINE HYDROCHLORIDE see NOF500
SOVERIN see QAK000
SOVIET TECHNICAL HERBICIDE 2M-4C see CIR250
SOVIOL see AAX250
SOVKAIN see NOF500
SOVOCAINE HYDROCHLORIDE see NOF500
SOVOL see PJL750
SOWBUG & CUTWORM BAIT see COF500
SOXINAL PZ see BJK500
SOXINOL PZ see BJK500
SOXISOL see SNN500
SOXYSYMPAMINE see MDT600
SOYEX see NIX000
SP see AIF000
75 SP see DOP600
SP 104 see TCM250
S.P. 147 see LEU000
SP 725 see PEF500
SPA see DWA600
SPACTIN see BTA325
SPAMOL see CPQ250
SPAN 20 see SKV000
SPAN 55 see SKV150
SPAN 60 see SKV150
SPAN 80 see SKV100
SPANBOLET see SNJ000
SPANISH BROOM see GCM000
SPANISH CARNATION see CAK325
SPANISH FLY see CBE250
SPANISH MARJORAM OIL see MBU500
SPANISH THYME OIL see TFX750
SPANON see CJJ250
SPANONE see CJJ250
SPANTOL see PGA750
SPANTRAN see MQU750
SPARIC see BRE500
SPARICON see SBG500
SPARINE see DQA600
SPARINE HYDROCHLORIDE see PMI500
SPARSOMYCIN see SKX000
SPARTAKON see TDX750
SPARTASE see MAI600
SPARTEINE see SKX500
(−)-SPARTEINE see SKX500
l-SPARTEINE see SKX500
d-SPARTEINE see PAB250
SPARTEINE, d-ISOMER see PAB250
SPARTEINE SULFATE see SKX750
SPARTOCIN see SKX750
SPARTOLOXYN see GGS000
SPASMAMIDE see FAJ150
SPASMEDAL see DAM700
SPASMENTRAL see BBU800
SPASMEX see KEA300
SPASMEXAN HYDROCHLORIDE see LFK200
SPASMIONE see DNU100
SPASMO 3 see KEA300
SPASMOCAN see NOC000
SPASMOCYCLON see DNU100
SPASMOCYCLONE see DNU100
SPASMODOLIN see DAM700
SPASMOLYSIN see HOA000
SPASMOLYTIN see DHX800
SPASMOLYTON see THK000
SPASMOPHEN see ORQ000
SPASURET see FCB100

SPATHE FLOWER see SKX775
SPATHIPHYLLUM (VARIOUS SPECIES) see SKX775
SPATONIN see DIW000
SPAVIT see VSZ450
SP 60 (CHLOROCARBON) see PKQ059
SPD PHOSPHATE see AMD500
SPEARMINT OIL see SKY000
SPEARWORT see FBS100
SPECIAL BLUE X 2137 see EOJ500
SPECIAL TERMITE FLUID see DEP600
SPECIFEN see EID000
SPECILLINE G see BDY669
SPECTAZOLE see EAE000
SPECTINOMYCIN DIHYDROCHLORIDE see SLI325
SPECTINOMYCIN, DIHYDROCHLORIDE, PENTAHYDRATE see SKY500
SPECTINOMYCIN HYDROCHLORIDE see SLI325
SPECTOGARD see SKY500
SPECTRACIDE see DCM750
SPECTRAR see INJ000
SPECTRA-SORB UV 9 see MES000
SPECTROBID see BAB250
SPECTROLENE BLUE B see DCJ200
SPECTROLENE RED KB see CLK225
SPECTROLENE SCARLET 2G see DEO400
SPECULAR IRON see IHD000
SPEED see DBA800
"SPEED" see MDQ500
SPENCER 401 see PJY500
SPENCER S-6538 see DJY200
SPENCER S-6900 see DRR200
S(−)-PENTOBARBITAL SODIUM see PBS750
SPENT OXIDE see SKZ000
SPENT SULFURIC ACID (DOT) see SOI500
SPERGON I see TBO500
SPERGON TECHNICAL see TBO500
SPERLOX-S see SOD500
SPERLOX-Z see EIR000
SPERMIDINE see SLA000
SPERMIDINE PHOSPHATE see AMD500
SPERMINE see DCC400
SPERMINE, PURISS see DCC400
SPERMINE TETRAHYDROCHLORIDE see GEK000
SPERM SPECIFIC ISOZYME of LACTATE DEHYDROGENASE see LAO300
SPERSADOX see DAE525
SPERSUL see SOD500
SPERSUL THIOVIT see SOD500
SP 60 ESTER see AAX250
SP G see BER500
SPG 827 see BER500
SPH see HGN600
SPHEROIDINE see FOQ000
SPHEROMYCIN see SMB000
SPHYGMOGENIN see VGP000
SPICEBUSH see CCK675
SPICLOMAZINE HYDROCHLORIDE see SLB000
SPICY JATROPHA see CNR135
SPIDER LILY see BAR325, REK325, SLB250
SPIGELIA ANTHELMIA see PIH800
SPIGELIA MARILANDICA see PIH800
SPIKE see BSN000
SPIKE LAVENDER OIL see SLB500
SPINOCAINE see AIL750
SPIPERONE see SLE500
SPIRAMIN see AJV500
SPIRAMYCIN see SLC000
SPIRAMYCIN ADIPATE see SLC300
SPIRAMYCIN, HEXANEDIOATE see SLC300
SPIRAMYCINS see SLC000
SPIRESIS see AFJ500
SPIRIDON see AFJ500
SPIRIT of GLONOIN see SLD000
SPIRIT of GLYCERYL TRINITRATE see SLD000
SPIRIT of HARTSHORN see AMY500
SPIRIT of NITER (sweet) see SLD500
SPIRIT ORANGE see PEJ500
SPIRITS of NITROGLYCERIN (DOT) see SLD000
SPIRITS of SALT (DOT) see HHL000
SPIRITS of TURPENTINE see TOD750
SPIRITS of WINE see EFU000
SPIRIT of TRINITROGLYCERIN see SLD000
SPIRIT of TURPENTINE see TOD750
SPIRIT YELLOW I see PEJ500
SPIRO-32 see SLD800
SPIROBROMIN see SLD600

SPIROCID see ABX500
SPIROCTANIE see AFJ500
SPIRO(CYCLOHEXANE-1,5'-HYDANTOIN) see DVO600
SPIRO(17H-CYCLOPENTA(a)PHENANTHRENE-17,2(5'H)FURAN), PREGN-4-ENE-21-CARBOXYLIC ACID DERIV. see AFJ500
SPIRO(17H-CYCLOPENTA(a)PHENAUTHRENE-17,2-(3'H)-FURAN) see AFJ500
SPIROFULVIN see GKE000
SPIROGERMANIUM DIHYDROCHLORIDE see SLD800
SPIROGERMANIUM HYDROCHLORIDE see SLD800
SPIROHYDANTOIN MUSTARD see SLD900
SPIRO(ISOBENZOFURAN-1(3H),9-(9H)XANTHENE-3-ONE, 3',6'-DIHYDROXY-DISODIUM SALT see FEW000
SPIRO(ISOBENZOFURAN-1(3H),9-(9H)XANTHEN)-3-ONE, 4',5'-DIBROMO-3',6'-DIHYDROXY- (9CI) see DDO200
SPIROLACTONE see AFJ500
SPIROLAKTON see AFJ500
SPIROLANG see AFJ500
SPIROMUSTINE see SLD900
SPIRO(NAPHTHALENE-2(1H),2'-OXIRANE)-3'-CARBOXALDEHYDE, 3,5,6,7,8,8a-HEXAHYDRO-7-ACETOXY-5,6-EPOXY-3,8,8a-TRIMETHYL-3-OXO- see POF800
SPIRONE see AFJ500
SPIRONOLACTONE see AFJ500
SPIRONOLACTONE A see AFJ500
SPIROPENT see VHA350
SPIROPERIDOL see SLE500
SPIROPITAN see SLE500
SPIROPLATIN see SLE875
SPIROZID see ABX500
SPIRT see EFU000
SPIZOFURONE see SLE880
SPLIT LEAF PHILODENDRON see SLE890
SPLOTIN see EPD500
SPOFADAZINE see AKO500
SPOFADRIZINE see PPP500
SPONGIOFORT see PCU360
SPONGOADENOSINE see AEH100
SPONGOCYTIDINE HYDROCHLORIDE see AQR000
SPONTOX see TAA100
SPOONWOOD IVY see MRU359
SPORAMIN see SBG500
SPORARICIN A see SLF500
SPORILINE see TGB475
SPOR-KIL see ABU500
SPOROSTATIN see GKE000
SPOTRETE see TFS350
SPOTTED ALDER see WCB000
SPOTTED COWBANE see WAT325
SPOTTED HEMLOCK see PJJ300
SPOTTED PARSLEY see PJJ300
SPOTTON see FAQ900
SPP see AIF000
SPRAY-DERMIS see NGE500
SPRAY-FORAL see NGE500
SPRAY-HORMITE see SGH500
SPRAYSET MEKP see MKA500
SPRAY-TROL BRANCH RODEN-TROL see WAT200
SPRENGEL EXPLOSIVES see SLG500
SPRING-BAK see DXD200
SPRITZ-HORMIN/2,4-D see DAA800
SPRITZ-HORMIT see SGH500
SPROUT NIP see CKC000
SPROUT-NIP EC see CKC000
SPROUT-OFF see DAI800, ODE000
SPUD-NIC see CKC000
SPUD-NIE see CKC000
SPURGE see BRE500
SPURGE LAUREL see LAR500
SPURGE OLIVE see LAR500
SPUTOLYSIN see DXO800
SQ 1089 see HOO500
SQ 1156 see GGS000
SQ 1489 see TFS350
SQ 2113 see HOB500
SQ 2303 see CBK500
SQ 2321 see FNF000
SQ 3277 see HOC000
SQ 8388 see AQO000
SQ 9453 see DUD800
SQ 10269 see CBQ625
SQ 10,643 see CMP900
SQ 11725 see CNR675
SQ 13396 see IFY000
SQ 14055 see TET800

SQ 14,225 see MCO750
SQ 15,659 see PPY250
SQ-15761 see SKM000
SQ 16,114 see PMI250
SQ 16360 see SHK000
SQ 16423 see MNV250
SQ 16496 see PMC700
SQ 16603 see FQU000
SQ 16819 see HKQ025
SQ-18,566 see HAF300
SQ 21065 see POJ500
SQ 21611 see BDX100
SQ 21977 see MPU000
SQ-21983 see IGA000
SQ 22451 see AMH000
SQ 22947 see DAR000, TET800
SQ 26,776 see ARX875
SQUALIDIN see IDG000
SQUALIDINE see IDG000
SQUARIC ACID see DMJ600
SQUAW ROOT see BMA150
SQUIBB see HNT500
SQUIBB 4918 see FMP000
SQUILL see RCF000, SLH200
SQUIRREL FOOD see DAE100
SR 73 see DFV600
SR-202 see CMU875
SR 206 see BKM250
SR 247 see DUL200
SR 351 see TLX175
SR406 see CBG000
SR 1354 see NHH500
SR 720-22 see ZAK300
SRA 5172 see DTQ400
SRA 7312 see DJY200
SRA 7847 see EIM000
SRA 12869 see IMF300
SRC-7 see DWQ850
SRC-15 see BHJ625
SRC-16 see BLD325
SRC-226 see DSH800
SRC-II, HEAVY DISTILLATE see SLH300
SRC-II HEAVY DISTILLATE see CMY625
SR E2 see SMF000
SRG 95213 see DCQ700
SRI 702 see BFL125
SRI 753 see BRS500
SRI 759 see TFJ500
SRI 859 see MNA750
SRI 1354 see NHH500
SRI 1666 see DRN400
SRI 1720 see BIF750
SRI 1869 see NKL000
SRI 2200 see CGV250
SRI 2489 see IAN000
SRI 2619 see CPL750, FIJ000
SRI 2638 see CHE500
SRI 2656 see CFH750
SRIF see SKS600
SS-094 see SKS299
SS 2074 see ORU000
S.T. 37 see HFV500
ST 155 see CBF250
ST-155 see CMX760
ST 1085 see MQT530
ST-1191 see EQO000
ST-1435 see HMB650
1-ST-2121 see SAA040
ST 720 (FRENCH) see CGB000
ST A 307 see TFK300
STABICILLIN see PDT500
STABILAN see CMF400
STABILENE see BKA000, BRP250
STABILENE FLY REPELLENT see BRP250
STABILISATOR C see DWN800
STABILIZATOR AR see PFT500
STABILIZER D-22 see DDV600
STABILLIN VK SYRUP 125 see PDT750
STABILLIN VK SYRUP 62.5 see PDT750
STABINOL see CKK000
STABLE RED KB BASE see CLK225
STADOL see BPG325
H-STADUR see CLD250
STAFAC see VRA700, VRF000

STAFAST see NAK500
STA-FAST see TIX500
STAFF TREE see AHJ875
STAFLEX DBM see DED600
STAFLEX DBP see DEH200
STAFLEX DBS see DEH600
STAFLEX DOP see DVL700
STAFLEX DOX see BJQ500
STA-FRESH 615 see SGM500
STAGGER WEED see LBF000
STAGNO (TETRACLORURO di) (ITALIAN) see TGC250
STALFEX DTDP see DXQ200
STALFLEX DOS see BJS250
STALLIMYCIN see SLI300
STALLIMYCIN HYDROCHLORIDE see DXG600
STAM see DGI000
STAM M-4 see DGI000
STAM F 34 see DGI000
STAMINE see WAK000
STAM LV 10 see DGI000
STAMPEDE see DGI000
STAMPEDE 3E see DGI000
STAMPEN see DGE200
STAM SUPERNOX see DGI000
STANAPROL see DME500
STANDACOL CARMOISINE see HJF500
STANDACOL ORANGE G see HGC000
STANDAK see AFK000
STANDAMIDD LD see BKE500
STANDAMUL DIPA see DNL800
STANDAMUL O20 see PJW500
STANDAPOL 112 CONC see SIB600
STANDAPOL TLS 40 see SON000
STANDARD LEAD ARSENATE see LCK000
STANEPHRIN see SPC500
STANGEN see WAK000
STANGEN MALEATE see DBM800
STAN-GUARD 156 see BKO250
STANILO see SKY500, SLI325
STAN-MAG MAGNESIUM CARBONATE see MAC650
STANNANE, ((p-ACETAMIDOBENZOYL)OXY)TRIPHENYL- see TMV800
STANNANE, ACETOXYTRICYCLOHEXYL- see ABW600
STANNANE, ACETOXYTRIOCTYL- see ABX150
STANNANE, ACETOXYTRIPENTYL- see ABX175
STANNANE, (ACETYLENECARBONYLOXY)TRIPHENYL- see TMW600
STANNANE, BROMOTRIETHYL-, compd. with 2-PIPECOLINE (1:1) see TJU850
STANNANE, BUTYLDIETHYLIODO- see BRA550
STANNANE, (BUTYLTHIO)TRIOCTYL- see BSO200
STANNANE, (BUTYLTHIO)TRIPROPYL- see TMY850
STANNANE, CYANATOTRIMETHYL- see TMI100
STANNANE, (CYANOACETOXY)TRIPHENYL- see TMV825
STANNANE, CYCLOHEXYLHYDROXYOXO- see MRH212
STANNANE, DICYCLOHEXYLOXO- see DGV900
STANNANE, DIMETHYL(2,3-QUINOXALINYLDITHIO)- see QSJ800
STANNANE, ((DIMETHYLTHIOCARBAMOYL)THIO)TRIBUTYL- see TID150
STANNANE, (ETHYLTHIO)TRIOCTYL- see EPR200
STANNANE, ETHYNYLENEBIS(CARBONYLOXY)BIS(TRIPHENYL- see BLS900
STANNANE, FLUOROTRIS(p-CHLOROPHENYL)- see TNG050
STANNANE, (ISOPROPYLSUCCINYLOXY)TRIBUTYL- see TIE600
STANNANE, (p-NITROPHENOXY)TRIBUTYL- see NII200
STANNANE, OXYBIS(DIBUTYL(2,4,5-TRICHLOROPHENOXY)- see OPE100
STANNANE, TRIBUTYLSALICYLOYLOXY see SAM000
STANNANE, TRIISOPROPYL(UNDECANOYLOXY)- see TKT850
STANNIC BROMIDE see TGB750
STANNIC CHLORIDE, anhydrous (DOT) see TGC250
STANNIC CITRIC ACID see SLI335
STANNIC DICHLORIDE DIIODIDE see TGC280
STANNIC IODIDE see TGD750
STANNIC PHOSPHIDE (DOT) see TGE500
STANNOCHLOR see TGC275
STANNOPLUS see DAE600
STANNORAM see DAE600
STANNOUS CHLORIDE (FCC) see TGC000
STANNOUS CHLORIDE, solid (DOT) see TGC000
STANNOUS CHLORIDE DIHYDRATE see TGC275
STANNOUS CITRIC ACID see TGC285
STANNOUS DIBUTYLDITRIFLUOROACETATE see BLN750
STANNOUS FLUORIDE see TGD100
STANNOUS IODIDE see TGD500
STANNOUS OLEATE see OAZ000
STANNOUS OXALATE see TGE000
STANNOUS POTASSIUM TARTRATE see TGE750
STANNOXYL see TGE300
STANNPLOUS see DAE600

STANOLONE see DME500
STANOMYCETIN see CDP250
STANOZOLOL see AOO400
STANSIN see SNN500
ST. ANTHONY'S TURNIP see FBS100
STANZAMINE see POO750
STAPENOR see DSQ800, MNV250
STAPHCILLIN V see MNV250
STAPHOBRISTOL-250 see SLJ000
STAPHYBIOTIC see SLJ000
STAPHYBIOTIC I see SLJ050
STAPHYLEX see CHJ000
STAPHYLOCOAGULASE see CMY325
STAPHYLOCOCCAL PHAGE LYSATE see SLJ100
STAPHYLOMYCIN see VRA700, VRF000
STAPHYLOMYCINE see VRA700
STAPYOCINE see VRF000
STAR see GGA000
STARAMIC 747 see SLJ500
STAR ANISE OIL see AOU250
STAR CACTUS (HAWAII) see AGV875
STARCH see SLJ500
α-STARCH see SLJ500
STARCH (OSHA) see SLJ500
STARCH, CORN see SLJ500
STARCH DUST see SLJ500
STARCH GUM see DBD800
STARCH HYDROXYETHYL ETHER see HLB400
STARCHWORT see JAJ000
STAR DUST see CNE750
STARFOL GMS 450 see OAV000
STARFOL GMS 600 see OAV000
STARFOL GMS 900 see OAV000
STARFOL IPP see IQW000
STAR HYACINTH see SLH200
STARIFEN see EOK000
STARLICIDE see CLK230
STAR-OF-BETHLEHEM see SLJ650
STARSOL No. 1 see AQQ500
STARVE-ACRE see FBS100
STA-RX 1500 see SLJ500
ST52-ASTA see DKA200
STATHION see PAK000
STATOBEX see DKE800
STATOCIN see CCK550
STATOMIN see WAK000
STATOMIN MALEATE see DBM800
STAUFFER N-2404 see CJD650
STAUFFER N-3049 see EPY000
STAUFFER N-3051 see BSG000
STAUFFER N-4328 see EOO000
STAUFFER N-4548 see MOB250
STAUFFER ASP-51 see TED500
STAUFFER B-10094 see CON300
STAUFFER B-11163 see DTQ800
STAUFFER CAPTAN see CBG000
STAUFFER MV242 see MLG250
STAUFFER MV-119A see DLK200
STAUFFER N 521 see DSB200
STAUFFER N 2790 see FMU045
STAUFFER R-1,303 see TNP250
STAUFFER R-1492 see MQH750
STAUFFER R 1504 see PHX250
STAUFFER R-1571 see PHL750
STAUFFER R-1910 see EID500
STAUFFER R-2061 see PNF500
STAUFFER R-3413 see DTV400
STAUFFER R-4,572 see EKO500
STAUFFER R-25788 see DBJ600
STAUFFER R-3442-S see EMC000
STAURODERM see FMQ000
STAVELAN CINATY (CZECH) see TGE250
STAVINOR 40 see BAO825
STAY-FLO see SHF500
STAZEPIN see DCV200
STB see SKE000
ST. BENNET'S HERB see PJJ300
STCA see TII500
ST 1396 CLORHIDRATO (SPANISH) see SBN475
ST-1512 DIHYDROCHLORIDE see HFG600
STEADFAST see CCP000
STEAPSIN see GGA800
STEARALKONIUM CHLORIDE see DTC600
STEARAMIDE see OAR000

STEAREX BEADS see SLK000
STEARIC ACID see SLK000
STEARIC ACID, ALUMINIUM SALT see AHA250
STEARIC ACID, BARIUM CADMIUM SALT (4:1:1) see BAI800
STEARIC ACID, BARIUM SALT see BAO825
STEARIC ACID, CADMIUM SALT see OAT000
STEARIC ACID with CYCLOHEXYLAMINE (1:1) see CPH500
STEARIC ACID-2,3-EPOXYPROPYL ESTER see SLK500
STEARIC ACID, LEAD SALT see LDX000
STEARIC ACID, MONOESTER with ETHYLENE GLYCOL see EJM500
STEARIC ACID, MONOESTER with GLYCEROL see OAV000
STEARIC ACID, MONOESTER with 1,2-PROPANEDIOL see SLL000
STEARIC ACID, SODIUM SALT see SJV500
STEARIC ACID, TETRADECYL ESTER see TCB100
STEARIC ACID, ZINC SALT see ZMS000
STEARIC MONOGLYCERIDE see OAV000
STEARIX ORANGE see PEJ500
STEAROL see OAX000
Γ-STEAROLACTONE see SLL400
STEARONITRILE see SLL500
STEAROPHANIC ACID see SLK000
1-STEAROYLAZIRIDINE see SLM000
STEAROYL ETHYLENEIMINE see SLM000
STEAR YELLOW JB see DOT300
STEARYL ALCOHOL see OAX000
STEARYL ALCOHOL EO (10) see SLM500
STEARYL ALCOHOL EO (20) see SLN000
STEARYL ALCOHOL condensed with 10 moles ETHYLENE OXIDE see SLM500
STEARYL ALCOHOL condensed with 20 moles ETHYLENE OXIDE see SLN000
STEARYLAMINE see OBC000
STEARYLDIMETHYLBENZYLAMMONIUM CHLORIDE see DTC600
STEARYLTRIMETHYLAMMONIUM CHLORIDE see TLW500
STEAWHITE see TAB750
STEBAC see DTC600
STECKAPFUL (GERMAN) see SLV500
STECLIN see TBX000
STECLIN HYDROCHLORIDE see TBX250
STEDIRIL see NNL500
STEDIRIL D see NNL500
STEINAMID DL 203 S see BKE500
STEINAPOL TLS 40 see SON000
STEINBUHL YELLOW see BAK250, CAP750
STELADONE see CDS750
STELAZINE see TKE500, TKK250
STELAZINE DIHYDROCHLORIDE see TKK250
STELLAMINE see DJS200
STELLARID see POB500
STELLAZINE see TKE500
STELLITE see VSK000
STELLON PINK see PKB500
STEMAX see PAL600
STEMETIL see PMF500
STEMETIL DIMALEATE see PMF250
STENANDIOL see AOO410
STENDOMYCIN A see SLN509
STENDOMYCIN SALICLATE see SLO000
STENEDIOL see AOO475
STENIBELL see AOO475
STENOLON see DAL300, MPN500
STENOLONE see DAL300
STENOSINE see DXE600
STENOSTERONE see AOO475
STENOVASAN see TEP500
STENTAL EXTENTABS see EOK000
STEPAN D-70 see IQW000
STEPANOL WAQ see SIB600
STEPANOL WAT see SON000
STERAFFINE see OAX000
STERAL see HCL000
STERAMIDE see SNP500
STERANDRYL see TBG000
STERANE see PMA000
STERAQ see DAQ800
STERASKIN see HCL000
STERAZINE see PPP500
STERCORIN see DKW000
STERCULIA FOETIDA OIL see SLO450
STERCULIA GUM see SLO500
STERIDO see BIM250
STERIGMATOCYSTIN see SLP000
STERILE TETRACAINE HYDROCHLORIDE see TBN000
STERILIZING GAS ETHYLENE OXIDE 100% see EJN500
STERINOLU (POLISH) see BEO000
STERISEAL LIQUID #40 see SGM500

STERLING see SFT000
STERLING WAQ-COSMETIC see SIB600
STERLING WAT see SON000
STEROGYL see VSZ100
STEROIDS see SLP199
STEROLAMIDE see TKP500
STEROLONE see PMA500
STERONYL see MPN500
STESIL see CKE750
STESOLID see DCK759
STEVIA REBAUDIANA Bertoni, extract see SLP350
ST 1396 HYDROCHLORID (GERMAN) see SBN475
ST 1396 HYDROCHLORIDE see SBN475
STIBACETIN see SLP500
STIBAMINE GLUCOSIDE see NCL000
STIBANATE see AQH800, AQI250
STIBANILIC ACID see SLP600
STIBANOSE see AQH800, AQI250
STIBATIN see AQH800, AQI250
STIBENYL see SLP500
STIBIC ANHYDRIDE see AQF750
2,2',2''-(STIBILIDYNETRIS(THIO))TRIS-BUTANEDIOIC ACID HEXALITHIUM
    SALT see LGU000
STIBILIUM see DKA600
STIBINE see SLQ000
STIBINE OXIDE, TRIPHENYL- see TMQ550
STIBINE SULFIDE, TRIPHENYL- see TMQ600
STIBINE, TRICHLORO- see AQC500
STIBINE, TRIS(DODECYLTHIO)- see TNH850
STIBINOL see AQH800, AQI250
STIBIUM see AQB750
STIBNAL see AQI750
STIBOCAPTATE see AQD750
((2-STIBONOPHENYL)THIO)ACETIC ACID see CCH250
((2-STIBONOPHENYL)THIO)ACETIC ACID DIETHANOLAMINE SALT
    see SLQ500
STIBOPHEN see AQH500
STIBOSAMIN see SLP600
STIBSOL see SEU000
STIBUNAL see AQI750
STICKMONOXYD (GERMAN) see NEG100
STICKSTOFFDIOXID (GERMAN) see NGR500
STICKSTOFFLOST see BIE500
STICKSTOFFWASSERSTOFFSAEURE (GERMAN) see HHG500
STIEDEX see DBA875
STIGMANOL BROMIDE see POD000
STIGMANOL METHYL SULFATE see DQY909
5-α-STIGMASTANE-3-β,5,6-β-TRIOL 3-MONOBENZOATE see SLQ625
STIGMATELLIN see SLQ650
STIGMOSAN BROMIDE see POD000
STIGMOSAN METHYL SULFATE see DQY909
STIK see NAK500
STIKSTOFDIOXYDE (DUTCH) see NGR500
STIL see DKA600
STILALGIN see GGS000
STILBAMIDINE DIHYDROCHLORIDE see SLS500
STILBEN (GERMAN) see SLR000
4-STILBENAMINE see SLQ900
4-N-STILBENAMINE see SLQ900
trans-4-N-STILBENAMINE see AMO000
STILBENE see SLR000
STILBENE 3 see TGE150
trans-4-STILBENE see AMO000
STILBENE, 4-(BENZYLOXY)-2'-FLUORO-α-PHENYL- see BFC400
STILBENE, α'-BROMO-α-PHENYL- see BOK500
α-STILBENECARBONITRILE see DVX600
4,4'-STILBENEDIAMINE see SLR500
4,4'-STILBENEDICARBOXAMIDINE see SLS000
4,4'-STILBENEDICARBOXAMIDINE, DIHYDROCHLORIDE see SLS500
STILBENE, α-α'-DIETHYL-4,4'-DIMETHYL-, (E)- see DRK500
2,2'-STILBENEDISULFONIC ACID, 4,4'-BIS((4-ANILINO-6-((2-
    HYDROXYETHYL)METHYLAMINO)-s-TRIAZIN- 2-YL)AMINO)-, DISOD-
    IUM SALT see TGE100
2,2'-STILBENEDISULFONIC ACID, 4,4'-BIS(4-PHENYL-1,2,3-TRIAZOL-2-YL),
    DIPOTASSIUM SALT see BLG100
2,2'-STILBENEDISULFONIC ACID, 4,4'-BIS(4-PHENYL-2H-1,2,3-TRIAZOL-2-
    YL)-, DIPOTASSIUM SALT see BLG100
2-STILBENESULFONIC ACID, 4-(7-SULFO-2H-NAPHTHO(1,2-d)TRIAZOL-2-YL)-,
    DISODIUM SALT see DXG025
4-STILBENYL-N,N-DIETHYLAMINE see DKA000
STILBENYL-N,N-DIMETHYLAMINE see DUB800
STILBESTROL see DKA600
STILBESTROL DIETHYL DIPROPIONATE see DKB000
STILBESTROL DIMETHYL ETHER see DJB200

STILBESTROL DIPHOSPHATE see DKA200
STILBESTROL DIPROPIONATE see DKB000
STILBESTROL and PRANONE see EEI050
STILBESTROL PROPIONATE see DKB000
STILBESTRONATE see DKB000
STILBESTRONE see DKA600
STILBETIN see DKA600
STILBIOCINA see SMB000
STILBOEFRAL see DKA600
STILBOESTROFORM see DKA600
STILBOESTROL see DKA600
STILBOESTROL DIPROPIONATE see DKB000
STILBOFAX see DKB000
STILBOFOLLIN see DKA600
STILBOL see DKA600
STILBRON see SBG500
STILKAP see DKA600
STILNY see CGA500
STILON see PJY500
STILPHOSTROL see DKA200
STIL-ROL see DKA600
STILRONATE see DKB000
STIMAMIZOL HYDROCHLORIDE see LFA020
STIMATONE see FMS875
STIMINOL see DJS200
STIMULAN see AOB250
STIMULEX see DBA800, DBB000
STIMULEXIN see SLU000
STIMULIN see DJS200
STIMULINA see GFO050
STINERVAL see PFC500, PFC750
STINK DAMP see HIC500
STINKING NIGHTSHADE see HAQ100
STINKING WEED see CNG825
STINKING WILLIE see RBA400
STIPINE see SEH000
STIPOLAC see TDE750
STIPTANON see EDU500
STIRAMATO see PFJ000
STIROFOS see RAF100
STIROLO (ITALIAN) see SMQ000
STIROPHOS see RAF100
ST. JOHN'S BREAD see LIA000
STOCKADE see RLF350
STODDARD SOLVENT see PCT250, SLU500
STOIKON see BCA000
STOMACAIN see DTL200
STOMOLD B see BGJ750
STOMP see DRN200
STONE RED see IHD000
STOPAETHYL see DXH250
STOP-DROP see NAK500
STOPETHYL see DXH250
STOPETYL see DXH250
STOPGERME-S see CKC000
STOP-MOLD see SFV500
STOP-SCALD see SAV000
STOPSPOT see PFM500
STOPTON ALBUM see SNM500
STOVAINE see AOM000
STOVARSAL see ABX500
STOVARSOL see ABX500
STOVARSOLAN see ABX500
STOXIL see DAS000
STPHCILLIN A BANYU see DGE200
STPP see SKN000
STR see SMD000
STRABOLENE see DYF450
STRADERM see FDD150
STRAIGHT-CHAIN ALKYL BENZENE SULFONATE see LGF825
STRAIGHT-RUN KEROSENE see KEK000
STRAMONA (ITALIAN) see SLV500
STRAMONIUM see SLV500
STRATHION see PAK000
STRAWBERRY ALDEHYDE see ENC000
STRAWBERRY BUSH see CCK675
STRAWBERRY RED A GEIGY see FMU080
STRAWBERRY TOMATO see JBS100
STRAW OIL see LIK000
STRAZINE see ARQ725
STREL see DGI000
STREPAMIDE see SNM500
STREPCEN see SLW500
STREPCIN see SLY500

STREP-GRAN see SLY500
STREPSULFAT see SLY500
STREPTAGOL see SNM500
d-STREPTAMINE, O-3-AMINO-3-DEOXY-α-d-GLUCOPYRANOSYL-(1-6)-O-(2,6-
    DIAMINO-2,3,4,6-TETRADEOXY-α-d-erythro-HEXOPYRANOSYL-(1-4))-N-(4-
    AMINO-2-HYDROXY-1-OXOBUTYL)-2-DEOXY-, (S)-, SULFATE (salt) see
    HAA320
STREPTASE see SLW450
STREPTOCLASE see SNM500
STREPTOCOCCAL FIBRINOLYSIN see SLW450
STREPTOGRAMIN see VRF000
STREPTOGRAMIN B see VRA700
STREPTOKINASE see SLW450
STREPTOKINASE (ENZYME-ACTIVATING) see SLW450
STREPTOL see SNM500
STREPTOLYDIGIN see SLW475
STREPTOMICINA (ITALIAN) see SLW500
STREPTOMYCES GRISEUS PROTEASE see PMJ100
STREPTOMYCES GRISEUS PROTEINASE see PMJ100
STREPTOMYCES PEUCETIUS see DAC000
STREPTOMYCIN see SLW500
STREPTOMYCIN A see SLW500
STREPTOMYCIN C see SLX500
STREPTOMYCIN CALCIUM CHLORIDE see SLY000
STREPTOMYCIN compounded with CALCIUM CHLORIDE (1:1) see SLY000
STREPTOMYCIN and DIHYDROSTREPTOMYCIN see SLY200
STREPTOMYCINE see SLW500
STREPTOMYCIN SESQUISULFATE see SLY500
STREPTOMYCIN SULFATE see SLY500
STREPTOMYCIN, SULFATE (1:3) SALT see SLZ000
STREPTOMYCIN SULPHATE see SLZ000
STREPTOMYCIN SULPHATE B.P. see SLY500
STREPTOMYCINUM see SLW500
STREPTOMYZIN (GERMAN) see SLW500
STREPTONIGRAN see SMA000
STREPTONIGRIN see SMA000
STREPTONIGRIN METHYL ESTER see SMA500
STREPTONIVICIN see SMB000
STREPTOREX see SLY500
STREPTOSIL see SNM500
STREPTOSILTHIAZOLE see TEX250
STREPTOTHRICIN F see RAG300
STREPTOTHRICIN VI see RAG300
STREPTOVARICIN C see SMB850
STREPTOVIRUDIN see SMC375
STREPTOVITACIN A see SMC500
STREPTOVITACIN E 73 see ABN000
STREPTOZOCIN see SMD000
STREPTOZONE see SNM500
STREPTOZOTICIN see SMD000
STREPTROCIDE see SNM500
STREPVET see SLY500
STRESNIL see FLU000
STRESSON see BON400
STREUNEX see BBQ500
STRIADYNE see ARQ500
STRICNINA (ITALIAN) see SMN500
STRICYLON see NCW000
STRIPED ALDER see WCB000
STROBANE see MIH275, PJQ250, TBC500
STROBANE-T-90 see CDV100
STRONCYLATE see SML500
STRONTIUM see SMD500
STRONTIUM ACETATE see SME000
STRONTIUM ACETYLIDE see SME100
STRONTIUM ARSENITE see SME500
STRONTIUM ARSENITE, solid (DOT) see SME500
STRONTIUM BROMIDE see SMF000
STRONTIUM CHLORATE see SMF500
STRONTIUM CHLORATE (wet) see SMG000
STRONTIUM CHLORATE, WET (DOT) see SMF500
STRONTIUM CHLORIDE see SMG500
STRONTIUM CHLORIDE, HEXAHYDRATE see SMH525
STRONTIUM CHROMATE (1:1) see SMH000
STRONTIUM CHROMATE (VI) see SMH000
STRONTIUM CHROMATE 12170 see SMH000
STRONTIUM COMPOUNDS see SMH500
STRONTIUM DICHLORIDE HEXAHYDRATE see SMH525
STRONTIUM FLUOBORATE see SMI000
STRONTIUM FLUORIDE see SMI500
STRONTIUM FLUOSILICATE see SMJ000
STRONTIUM IODIDE see SMJ500
STRONTIUM MONOSULFIDE see SMM000
STRONTIUM(II) NITRATE (1:2) see SMK000

STRONTIUM PEROXIDE see SMK500
STRONTIUM PHOSPHIDE see SML000
STRONTIUM SALICYLATE see SML500
STRONTIUM SULFIDE see SMM000
STRONTIUM SULPHIDE see SMM000
STRONTIUM YELLOW see SMH000
STROPHANTHIDIN see SMM500
STROPHANTHIDIN-d-CYMAROSID (GERMAN) see CQH750
STROPHANTHIDIN-3-(6-DEOXY-α-l-MANNOPYRANOSIDE) see CNH780
STROPHANTHIDIN-β-d-DIGITOXOSID (GERMAN) see HAN800
STROPHANTHIDINE see SMM500
k-STROPHANTHIDINE see SMM500
STROPHANTHIDIN-GLUCOCYMAROSID (GERMAN) see SMN002
STROPHANTHIDIN-α-l-RHAMNOSIDE see CNH780
α-l-STROPHANTHIDOL-3,6-DEOXY-MANNOPYRANOSIDE see CNH785
STROPHANTHIN see SMN000
k-STROPHANTHIN see SMN000
β-k-STROPHANTHIN see SMN002
STROPHANTHIN G see OKS000
STROPHANTHIN K see SMN002
STROPHANTHIN K (crystalline) see SMN002
β-STROPHANTHOBIOSIDE, STROPHANTHIDIN-3 see SMN002
STROPHANTHUS GRATUS Franch., leaf and stem bark extract see SMN100
STROPHOPERM see OKS000
STROSPESID (GERMAN) see SMN275
STROSPESIDE see SMN275
STROSPESIDE-16-FORMATE see VIZ200
STROSPESIDE TRIFORMATE see TKL175
STRUMACIL see MPW500
STRUMAZOLE see MCO500
STRYCHININE SULFATE see SMP000
STRYCHNIDIN-10-ONE see SMN500
STRYCHNIDIN-10-ONE, SULFATE (2:1) see SMP000
STRYCHNIN (GERMAN) see SMN500
STRYCHNINE see NOG800, SMN500
STRYCHNINE, solid and liquid (DOT) see SMN500
STRYCHNINE MONONITRATE see SMO000
STRYCHNINE NITRATE see SMO000
STRYCHNINE SALT (solid) see SMO500
STRYCHNINE SULFATE (2581) see SMP000
STRYCHNOS see SMN500
STRYCHNOS NUX-VOMICA see NOG800
STRYCIN see SLZ000
STRYPHNON see MGC350
STRYPHNONE see MGC350
STRYPTIRENAL see VGP000
STRZ see SMD000
STS see EMT500, SIO000
STS 153 see PEE600
STS 557 see SMP400
ST 5066/S (GERMAN) see AMU750
ST-1512 SULFATE see HFG650
STUDAFLUOR see SHF500
STUGERON see CMR100
STUPENONE see DLX400
STUTGERON see CMR100
STUTGIN see CMR100
STX DIHYDROCHLORIDE see SBA500
STYLOMYCIN see AEI000, POK250
STYPHNATE of LEAD (DOT) see LEE000
STYPHNIC ACID see SMP500
STYPHNONE see MGC350
STYPNON see MGC350
STYPTIC WEED see CNG825
STYRAFOIL see SMQ500
STYRAGEL see SMQ500
STYRALLYL ALCOHOL see PDE000
STYRALLYL PROPIONATE see PFR000
STYRALYL ALCOHOL see PDE000
STYRAMATE see PFJ000
STYREEN (DUTCH) see SMQ000
STYREN (CZECH) see SMQ000
STYREN-ACRYLONITRILEPOLYMER see ADY500
STYRENE see SMQ000
STYRENE-ACRYLONITRILE COPOLYMER see ADY500
STYRENE-BUTADIENE COPOLYMER see SMR000
STYRENE-1,3-BUTADIENE COPOLYMER see SMR000
STYRENE-BUTADIENE POLYMER see SMR000
STYRENE EPOXIDE see EBR000
STYRENE MONOMER (ACGIH) see SMQ000
STYRENE MONOMER, inhibited (DOT) see SMQ000
STYRENE OXIDE see EBR000
STYRENE-7,8-OXIDE see EBR000
STYRENE POLYMER see SMQ500

STYRENE POLYMER with 1,3-BUTADIENE see SMR000
STYRENE POLYMERS see SMQ500
STYROFOAM see SMQ500
STYROL (GERMAN) see SMQ000
STYROLE see SMQ000
STYROLENE see SMQ000
STYROLUX see SMQ500
STYROLYL PROPIONATE see PFR000
STYRON see SMQ000, SMQ500
STYRONE see CMQ740
STYROPOR see SMQ000
STYRYL 430 see AMO250
trans-4'-STYRYLACETANILIDE see SMR500
ar-STYRYLACETOPHENONE see MPL000
p-STYRYLANILINE see SLQ900
6-STYRYL-BENZO(a)PYRENE see SMS000
5-STYRYL-3,4-BENZOPYRENE see SMS000
STYRYL CARBINOL see CMQ740
STYRYL METHYL KETONE see SMS500
STYRYL OXIDE see EBR000
N-(p-STYRYLPHENYL)ACETOHYDROXAMIC ACETATE see ABW500
N-(p-STYRYLPHENYL)ACETOHYDROXAMIC ACID see SMT000
trans-N-(p-STYRYLPHENYL)ACETOHYDROXAMIC ACID see SMT500
N-(p-STYRYLPHENYL)ACETOHYDROXAMIC ACID ACETATE see ABW500
trans-N-(p-STYRYLPHENYL)ACETOHYDROXAMIC ACID, COPPER(2+) COM-
PLEX see SMU000
(E)-N-(p-STYRYLPHENYL)HYDROXYLAMINE see HJA000
trans-N-(4-STYRYLPHENYL)HYDROXYLAMINE see HJA000
trans-N-(p-STYRYLPHENYL)HYDROXYLAMINE see HJA000
STZ see SMD000
SU-88 see IMS300
SU 3088 see CDY000
SU 3118 see RCA200
SU-4885 see MCJ370
SU 5864 see GKQ000
SU-5864 see GKS000
SU 5879 see CFY000
SU 8341 see CPR750
SU 9064 see MDW100, MQR200
SU-10568 see DRC600
SU 11927 see CCP750
SU-13320 see TCN250
SU-13437 see MCB500
SU 21524 see PJA220
SUANOVIL see SLC300
SUAVITIL see BCA000
SUBACETATE LEAD see LCH000
SUBAMYCIN see TBX250
SUBARI see HGC500
SUBCHLORIDE of MERCURY see MCW000
SUBERANE see COX500
SUBERONE see SMV000
SUBERONE OXIME see SMV500
SUBERONISOXIM (GERMAN) see SMV500
4,4'-(SUBEROYLBIS(IMINO-p-PHENYLENEIMINO))BIS(1-ETHYLPYRIDINIUM)
DIBROMIDE see SMW000
4,4'-(SUBEROYLBIS(IMINO-p-PHENYLENEIMINO))BIS(1-METHYLPYRIDIN-
IUM)DIBROMIDE see PPD500
SUBICARD see PBC250
SUBITEX see BRE500
SUBITOL see IAD000
SUBLIMAT (CZECH) see MCY475
SUBLIMAZE see PDW750
SUBLIMAZE CITRATE see PDW750
SUBLIMED SULFUR see SOD500
SUBLINGULA see HAQ500
SUBSTANCE DS see AJX500
SUBSTANCE H 20 see MFG600
SUBSTANCE H 36 see MEJ775
SUBSTANCE II see AFG750
SUBSTANZ DS see AJX500
SUBSTANZ NR. 2135 see CKE250
SUBSTANZ NR. 1602 (GERMAN) see CJR959
SUBSTANZ NR. 1766 (GERMAN) see CIS000
SUBSTANZ NR. 1925 (GERMAN) see CKE000
SUBSTANZ NR. 1934 (GERMAN) see DSQ600
SUBTIGEN see SMW400
SUBTILIN see SMW500
SUBTILISIN CARLSBURG see BAC000
SUBTILISIN (9CI) see BAC000
SUBTILISIN NOVO see BAC000
SUBTILISINS (ACGIH) see BAB750
SUBTILISINS BPN see BAB750
SUBTILOPEPTIDASE A see BAC000

SUBTILOPEPTIDASE B see BAC000
SUBTILOPEPTIDASE BPN' see BAC000
SUBTILOPEPTIDASE C see BAC000
SUBTOSAN see PKQ250
SUCARYL see CPQ625
SUCARYL ACID see CPQ625
SUCARYL CALCIUM see CAR000
SUCARYL SODIUM see SGC000
SUCCARIL see SGC000, SJN700
SUCCICURAN see HLC500
SUCCIMAL see ENG500
SUCCIMER see DNV800
SUCCIMITIN see ENG500
SUCCINALDEHYDE DISODIUM BISULFITE see SMX000
SUCCINATE SODIQUE de 21-HYDROXYPREGNANDIONE (FRENCH) see VJZ000
SUCCINATO ACIDO DI 1-p-CLOROFENILPENTILE (ITALIAN) see CKI000
SUCCINATO de CLORANFENICOL (SPANISH) see CDP725
SUCCINBROMIMIDE see BOF500
SUCCINCHLORIMIDE see SND500
SUCCINIBROMIMIDE see BOF500
SUCCINIC ACID see SMY000
SUCCINIC ACID ANHYDRIDE see SNC000
SUCCINIC ACID BIS(β-DIMETHYLAMINOETHYL) ESTER BISMETHIODIDE see BJI000
SUCCINIC ACID BIS(β-DIMETHYLAMINOETHYL) ESTER, DIHYDROCHLORIDE see HLC500
SUCCINIC ACID BIS(β-DIMETHYLAMINOETHYL)ESTER DIMETHOCHLORIDE see HLC500
SUCCINIC ACID, BIS(2-(HEXYLOXY)ETHYL) ESTER see SMZ000
SUCCINIC ACID-α-BUTYL-p-CHLOROBENZYL ESTER see CKI000
SUCCINIC ACID, CADMIUM SALT (1:1) see CAI750
SUCCINIC ACID, DIBUTYL ESTER see SNA500
SUCCINIC ACID DI-N-BUTYL ESTER see SNA500
SUCCINIC ACID DICHLORIDE see SNG000
SUCCINIC ACID DIESTER with CHOLINE see CMG250
SUCCINIC ACID DIESTER with CHOLINE CHLORIDE see HLC500
SUCCINIC ACID, DIESTER with CHOLINE IODIDE see BJI000
SUCCINIC ACID, DIETHYL ESTER see SNB000
SUCCINIC ACID, DI-2-HEXYLOXYETHYL ESTER see SMZ000
SUCCINIC ACID-2,2-DIMETHYLHYDRAZIDE see DQD400
SUCCINIC ACID DINITRILE see SNE000
SUCCINIC ACID DIPROPYL ESTER see DWV800
SUCCINIC ACID, (4-ETHOXY-1-NAPTHYLCARBONYLMETHYL) ESTER see SNB500
SUCCINIC ACID, O-ISOPROPYL-O'-TRIBUTYLSTANNYL ESTER see TIE600
SUCCINIC ACID, MERCAPTO-, DIETHYL ESTER, S-ESTER with O,O-DIMETHYLPHOSPHOROTHIOATE see OPK250
SUCCINIC ACID MONOESTER with 7-CHLORO-1,3-DIHYDRO-3-HYDROXY-5-PHENYL-2H-1,4-BENZODIAZEPIN-2-ONE see CFY500
SUCCINIC ACID-α-MONOESTER with d-threo-(−)-2,2-DICHLORO-N-(β-HYDROXY-α-(HYDROXYMETHYL)-p-NITROPHENETHYL)ACETAMIDE see CDP725
SUCCINIC ACID MONOESTER with N-(2-ETHYLHEXYL)-3-HYDROXYBUTYRAMIDE see BPF825
SUCCINIC ACID PEROXIDE (DOT) see SNC500
SUCCINIC ANHYDRIDE see SNC000
SUCCINIC CHLORIDE see SNG000
SUCCINIC-1,1-DIMETHYL HYDRAZIDE see DQD400
SUCCINIC DINITRILE see SNE000
SUCCINIC IMIDE see SND000
SUCCINIC PEROXIDE see SNC500
SUCCINIMIDE see SND000
SUCCINIMIDE, N,2-DIMETHYL-2-PHENYL- see MLP800
SUCCINOCHLORIMIDE see SND500
SUCCINOCHOLINE see CMG250
SUCCINODINITRILE see SNE000
SUCCINONITRILE see SNE000
4'-SUCCINOYLAMINO-2,3'-DIMETHYLAZOBENZENE see SNF500
SUCCINOYL CHLORIDE see SNG000
SUCCINOYLCHOLINE see CMG250
SUCCINOYLCHOLINE CHLORIDE see HLC500
SUCCINOYL DIAZIDE see SNF000
4'-SUCCINYLAMINO-2,3'-DIMETHYLAZOBENZOL see SNF500
SUCCINYL-ASTA see HLC500
SUCCINYLBISCHOLINE see CMG250
SUCCINYL BISCHOLINE CHLORIDE see HLC500
SUCCINYLBISCHOLINE DICHLORIDE see HLC500
SUCCINYL CHLORIDE see SNG000
SUCCINYLCHOLINE CHLORIDE see HLC500
SUCCINYLCHOLINE DICHLORIDE see HLC500
SUCCINYLCHOLINE HYDROCHLORIDE see HLC500
SUCCINYL DICHLORIDE see SNG000
SUCCINYLDICHOLINE see CMG250

SUCCINYLDICHOLINE CHLORIDE see HLC500
SUCCINYLDICHOLINE IODIDE see BJI000
o,o-SUCCINYLDICHOLINE IODIDE see BJI000
SUCCINYLFORTE see HLC500
SUCCINYLNITRILE see SNG500
SUCCINYL OXIDE see SNC000
SUCCINYL PEROXIDE see SNC500
SUCCITIMAL see MNZ000
SUCKER PLUCKER see ODE000
SUCOSTRIN see HLC500
SUCOSTRIN CHLORIDE see HLC500
SUCRA see SJN700
SUCRAPHEN see SPC500
SUCRE EDULCOR see BCE500
SUCRETS see HFV500
SUCRETTE see BCE500
SUCROFER see IHG000
SUCROL see EFE000
SUCROSA see SGC000
SUCROSE see SNH000
SUCROSE, POLYALLYL ETHER, POLYMER with ACRYLIC ACID see ADW000
SUDAFED see POH000, POH250
SUDAN AX see XRA000
SUDAN BROWN RR see NBG000
SUDAN GREEN 4B see BLK000
SUDAN III see OHA000
SUDAN ORANGE see XRA000
SUDAN ORANGE R see PEJ500
SUDAN ORANGE RPA see XRA000
SUDAN ORANGE RRA see XRA000
SUDAN RED see XRA000
SUDAN RED 7B see EOJ500
SUDANROT 7B see EOJ500
SUDAN SCARLET 6G see XRA000
SUDAN X see XRA000
SUDAN YELLOW see DOT300
SUDAN YELLOW GGN see OHI875
SUDAN YELLOW R see PEI000
SUDAN YELLOW RRA see AIC250
SUDINE see SNN300
SU 4885 DITARTRATE see MJJ000
SUESSETTE see SGC000
SUESSTOFF see EFE000
SUESTAMIN see SGC000
SUFENTA see SNH150
SUFENTANIL CITRATE see SNH150
SUFENTANYL see SNH150
SUFENTANYL CITRATE see SNH150
SUFFIX see EGS000
SUFFIX 25 see EGS000
SUFRALEM see AOO490
SUGAI CHRYSOIDINE see PEK000
SUGAI CONGO RED see SGQ500
SUGAI FAST SCARLET G BASE see NMP500
SUGAI TARTRAZINE see FAG140
SUGAR see SNH000
SUGAR-BOWLS see CMV390
SUGARIN see SGC000
SUGAR of LEAD see LCV000
SUGARON see SGC000
SUGRACILLIN see BFD000
SU 8629 HYDROCHLORIDE see AKL000
SU 8842 HYDROCHLORIDE see MQR200
SUICALM see FLU000
SUISYNCHRON see MLJ500
SUKHTEH see SNH450
SULADYNE see PDC250
SULAMYD see SNP500
SUL ANILINOVA (CZECH) see BBL000
SULBENICILLIN see SNV000
SULBENICILLIN DISODIUM see SNV000
SULCOLON see PPN750
SULCONAZOLE NITRATE see SNH480
SULDIXINE see SNN300
SULEMA (RUSSIAN) see MCY475
SULF-10 see SNP500
SULFAAFENAZOLO (ITALIAN) see AIF000
SULFABENZAMIDE see SNH800
SULFABENZID see SNH800
SULFABENZIDE see SNH800
SULFABENZOYLAMIDE see SNH800
SULFABENZPYRAZINE see QTS000
SULFABID see AIF000

SULFABUTIN see AIE750
SULFACARBAMIDE see SNQ550
4-SULFACARBAMIDE see SNQ550
SULFACET see SNP500
SULFACETAMIDE see SNP500
SULFACETIMIDE see SNP500
SULFACID BRILLIANT GREEN 1B see FAE950
SULFACID LIGHT ORANGE J see HGC000
SULFACOMBIN see SNI000
SULFACTIN see BAD750
SULFACYL see SNP500
SULFACYL SODIUM SALT see SNQ000
SULFADENE see BDF000
SULFADIAMINE see SNY500
SULFADIAZINE see PPP500
SULFADIAZINE SILVER see SNI425
SULFADIAZINE SILVER SALT see SNI425
SULFADIMERAZINE see SNJ000
SULFADIMETHOXIN see SNN300
SULFADIMETHOXINE see SNN300
SULFADIMETHOXYDIAZINE see SNN300
SULFADIMETHOXYPYRIMIDINE see SNI500
SULFADIMETHYLDIAZINE see SNJ000
SULFADIMETHYLISOXAZOLE see SNN500
SULFADIMETHYLOXAZOLE see AIE750
SULFADIMETHYLPYRIMIDINE see SNJ000
4-SULFA-2,6-DIMETHYLPYRIMIDINE see SNJ350
SULFADIMETINE see SNJ000, SNJ350
SULFADIMETOSSINA (ITALIAN) see SNN300
SULFADIMETOXIN see SNN300
SULFADIMEZINE see SNJ000
SULFADIMIDINE see SNJ000
SULFADIMIDINE SODIUM see SJW500
SULFADINE see SNJ000
SULFADOXINE see AIE500
SULFADSIMESINE see SNJ000
SULFAETHYLTHIADIAZOLE SODIUM see SJW300
SULFAFURAZOL see SNN500
SULFAGAN see SNN500
SULFAGUANIDINE see AHO250
SULFA-ISODIMERAZINE see SNJ000
SULFAISODIMERAZINE see SNJ350
SULFAISODIMIDINE see SNJ000, SNJ350
SULFALENE see MFN500
SULFALENE N-METHYLGLUCAMINE SALT see SNJ400
SULFALEX see AKO500
SULFALLATE see CDO250
SULFAMATE see ANU650
SULFAMERADINE see ALF250
SULFAMERAZIN see ALF250
SULFAMERAZINE SODIUM see SJW475
SULFAMETER SODIUM see MFO000
SULFAMETHALAZOLE see SNK000
SULFAMETHAZINE SODIUM see SJW500
SULFAMETHIAZINE see SNJ000
SULFAMETHIN see SNJ000, SNJ350
SULFAMETHIZOLE see MPQ750
SULFAMETHOMIDINE see MDU300
SULFAMETHOPYRAZINE see MFN500
SULFAMETHOXAZOL see SNK000
SULFAMETHOXAZOLE see SNK000
SULFAMETHOXAZOL-TRIMETHOPRIM see TKX000
SULFAMETHOXYDIAZINE SODIUM see MFO000
SULFAMETHOXYPYRAZINE see MFN500
3-SULFA-6-METHOXYPYRIDAZINE see AKO500
SULFA-5-METHOXYPYRIMIDINE SODIUM SALT see MFO000
SULFAMETHYLDIAZINE see ALF250
SULFAMETHYLISOXAZOLE see SNK000
SULFAMETHYLTHIADIAZOLE see MPQ750
SULFAMETOMIDINE see MDU300
SULFAMETOPYRAZINE see MFN500
SULFAMETORINE SODIUM see MFO000
SULFAMETOSSIPIRIDAZINA (ITALIAN) see MFN500
SULFAMETOXIPIRIDAZINE see AKO500
SULFAMETOXYPYRIDAZIN (GERMAN) see MFN500
SULFAMEZATHINE see SNJ000
SULFAMIC ACID see SNK500
SULFAMIC ACID, MONOAMMONIUM SALT see ANU650
SULFAMIC ACID, TETRAMETHYLENE ESTER see BOU100
SULFAMIDIC ACID see SNK500
p-SULFAMIDOANILINE see SNM500
SULFAMIDYL see SNM500
SULFAMINSAURE (GERMAN) see ANU650
SULFAMONOMETHOXIN see SNL800

SULFAMONOMETHOXINE see SNL800
SULFAMONOMETOXINE N-METHYLGLUCAMINE SALT see SNL830
SULFAMOXOLE see AIE750
SULFAMOXOLE and TRIMETHOPRIM see SNL850
SULFAMOXOLE-TRIMETHOPRIM mixture see SNL850
SULFAMOXOLUM see AIE750
5'-SULFAMOYL-2-CHLOROADENOSINE see CEF125
3-SULFAMOYLMETHYL-1,2-BENZISOXAZOLE see BCE750
N-(5-SULFAMOYL-1,3,4-THIADIAZOL-2-YL)ACETAMIDE see AAI250
SULFAMUL see TEX250
N-SULFAMYLBENZAMIDE see SNH800
SULFAMYLON ACETATE see MAC000
SULFAN see SOR500
SULFANA see SNM500
SULFANALONE see SNM500
SULFANIL see SNM500
SULFANILACETAMIDE see SNP500
SULFANILAMIDE see SNM500
3-SULFANILAMIDE-6-METHOXYPYRIDAZINE see AKO500
SULFANILAMIDE, N^1-(5-tert-BUTYL-1,3,4-THIADIAZOL-2-YL)- see GEW750
SULFANILAMIDE, N^1-(5-ETHYL-1,3,4-THIADIAZOL-2-YL)-, MONOSODIUM SALT see SJW300
4-SULFANILAMIDO-3,6-DIMETHOXYPYRIDAZINE see DON700
4-SULFANILAMIDO-5,6-DIMETHOXYPYRIMIDINE see AIE500
6-SULFANILAMIDO-2,4-DIMETHOXYPYRIMIDINE see SNN300
5-SULFANILAMIDO-3,4-DIMETHYL-ISOXAZOLE see SNN500
2-SULFANILAMIDO-4,6-DIMETHYLPYRIMIDINE see SNJ000
4-SULFANILAMIDO-2,6-DIMETHYLPYRIMIDINE see SNJ350
6-SULFANILAMIDO-2,4-DIMETHYLPYRIMIDINE see SNJ350
2-SULFANILAMIDO-3-METHOXYPYRAZINE see MFN500
3-SULFANILAMIDO-6-METHOXYPYRIDAZINE see AKO500
6-SULFANILAMIDO-3-METHOXYPYRIDAZINE see AKO500
2-SULFANILAMIDO-5-METHOXYPYRIMIDINE SODIUM SALT see MFO000
3-SULFANILAMIDO-5-METHYLISOXAZOLE see SNK000
2-SULFANILAMIDO-5-METHYL-1,3,4-THIADIAZOLE see MPQ750
5-SULFANILAMIDO-1-PHENYLPYRAZOLE see AIF000
SULFANILAMIDOPYRIMIDINE see PPP500
2-SULFANILAMIDOPYRIMIDINE SODIUM SALT see MRM250
2-SULFANILAMIDOQUINOXALINE see QTS000
2-SULFANILAMIDOTHIAZOLE see TEX250
2-SULFANILAMIDOTHIAZOLE SODIUM SALT see TEX500
SULFANILCARBAMID see SNQ550
N-SULFANILCARBAMIDE see SNQ550
SULFANILGUANIDINE see AHO250
m-SULFANILIC ACID see SNO000
N-SULFANILYLACETAMIDE see SNP500
N-SULFANILYLACETAMIDE SODIUM see SNQ000
N-SULFANILYLACETAMIDE, SODIUM SALT see SNQ000
2-SULFANILYL AMINOPYRIDINE see PPO000
2-SULFANILYLAMINOPYRIMIDINE see PPP500
2-(SULFANILYLAMINO)THIAZOLE see TEX250
N-SULFANILYLBENZAMIDE see SNH800
SULFANILYLCARBAMIC ACID, METHYL ESTER see SNQ500
SULFANILYLGUANIDINE see AHO250
N^1-SULFANILYL-N^2-BUTYLCARBAMIDE see BSM000
N^1-SULFANILYL-N^2-BUTYLUREA see BSM000
N-SULFANILYL-N'BUTYLUREE (FRENCH) see BSM000
SULFANILYLUREA see SNQ550
SULFANO see AIE750
SULFANOL NP 1 see SNQ700
SULFANYLHARNSTOFF see SNQ550
SULFANYLUREE see SNQ550
SULFAPHENAZOLE see AIF000
SULFAPHENAZON see AIF000
SULFAPHENYLPIPAZOL see AIF000
SULFAPHENYLPYRAZOLE see AIF000
SULFAPOL see DXW200
SULFAPOLU (POLISH) see DXW200
SULFAPYRAZINEMETHOXINE see MFN500
SULFAPYRAZINEMETHOXYNE see MFN500
SULFAPYRIDAZINE see AKO500
SULFAPYRIDINE see PPO000
2-SULFAPYRIDINE see PPO000
SULFAPYRIDINE NEUTRAL SOLUBLE see DXF400
SULFAPYRIMIDIN (GERMAN) see PPP500
2-SULFAPYRIMIDINE see PPP500
SULFAQUINOXALINE see QTS000
SULFARLEM see AOO490
SULFARSENOBENZENE see SNR000
SULFARSENOL see SNR000
SULFARSPHENAMINE see SNR000
SULFARSPHENAMINE BISMUTH see BKV250
SULFASALAZINE see PPN750
SULFASAN see BKU500

SULFASAN R POWDER see BKU500
SULFASOL see SNN300
SULFASOMIDINE see SNJ350
SULFASOXAZOLE see SNN500
SULFASTOP see SNN300
SULFATE d'ATROPINE (FRENCH) see ARR500
SULFATE de CUIVRE (FRENCH) see CNP250
SULFATED AMYLOPECTIN see AOM150
SULFATED CASTOR OIL see TOD500
SULFATE DIMETHYLIQUE (FRENCH) see DUD100
SULFATE ESTER of N-HYDROXY-N-2-FLUORENYL ACETAMIDE see FDV000
SULFATE MERCURIQUE (FRENCH) see MDG500
SULFATE de METHYLE (FRENCH) see DUD100
SULFATE de NICOTINE (FRENCH) see NDR500
SULFATE de PLOMB (FRENCH) see LDY000
SULFATES see SNS000
SULFATE de ZINC (FRENCH) see ZNA000
SULFATHALIDINE see PHY750
SULFATHIAZOL see TEX250
SULFATHIAZOLE (USDA) see TEX250
2-SULFATHIAZOLE SODIUM see TEX500
SULFATOBIS(DIMETHYLSELENIDE)PLATINUM(II) HYDRATE see SNS100
cis-SULFATO-1,2-DIAMINOCYCLOHEXANEPLATINUM(II) see SCF025
trans(−)-SULFATO-1,2-DIAMINOCYCLOHEXANEPLATINUM(II) see SCF050
trans(+)-SULFATO-1,2-DIAMINOCYCLOHEXANEPLATINUM(II) see SCF075
SULFAUREA see SNQ550
SULFAVIGOR see AIE750
SULFAZOLE see SNN500
SULFDURAZIN see AKO500
SULFENAMIDE M see BDG500
SULFENAMIDE TS see CPI250
SULFENAX MOB (CZECH) see BDF750
SULFENAZIN see TFM100
SULFENONE see CKI625
SULFENTANIL see SNH150
SULFERROUS see FBN100, FBO000
SULFIDAL see SOD500
SULFIDES see SNT000
SULFIDINE see PPO000
SULFIMEL DOS see DJL000
SULFINPYRAZINE see DWM000
SULFINYLBIS(METHANE) see DUD800
SULFINYL BROMIDE see SNT200
SULFINYL CHLORIDE see TFL000
SULFINYL CYANAMIDE see SNT300
SULFISIN see SNN500
SULFISOMEZOLE see SNK000
SULFISOMIDIN see SNJ000, SNJ350
SULFISOMIDINE see SNJ000, SNJ350
SULFISOXAZOLE see SNN500
SULFITE LYE see HIC600
SULFITES see SNT500
SULFIZOLE see SNN500
SULFMETHOXIPIRIDAZINE see AKO500
SULFMIDIL see AIE750
SULFOACETIC ACID see SNU000
2-SULFOANTHRAQUINONE SODIUM SALT see SER000
5-SULFOBENZEN-1,3-DICARBOXYLIC ACID see SNU500
SULFOBENZIDE see PGI750
3-SULFOBENZIDINE see BBY250
o-SULFOBENZIMIDE see BCE500
o-SULFOBENZOIC ACID IMIDE see BCE500
SULFOBENZYLPENICILLIN see SNV000
α-SULFOBENZYLPENICILLIN DISODIUM see SNV000
SULFOCARBANILIDE see DWN800
SULFOCIDINE see SNM500
SULFOCILLIN see SNV000
SULFODIAMINE see SNY500
SULFODIMESIN see SNJ000
SULFODIMEZINE see SNJ000
SULFODOR (CZECH) see EPH000
SULFOETHANOIC ACID see SNU000
SULFOGAL see AOO490
SULFOGENOL see IAD000
SULFO GREEN J see FAF000
SULFOGUANIDINE see AHO250
SULFOGUENIL see AHO250
SULFOLANE see SNW500
SULFOL-3-ENE see DMF000
3-SULFOLENE see DMF000
β-SULFOLENE see DMF000
4-SULFOMETANILIC ACID see AIE000
SULFOMETURON METHYL see SNW550
SULFONA see SOA500

SULFONAL see ABD500
SULFONAL NP 1 see SNQ700
SULFONAMIDE see SNM500
4-SULFONAMIDE-4'-DIMETHYLAMINOAZOBENZENE see SNW800
SULFONAMIDE P see SNM500
4-SULFONAMIDO-3'-METHYL-4'-AMINOAZOBENZENE see SNX000
2-SULFONAMIDOTHIAZOLE see TEX250
SULFONA P see AOO800
SULFO-3-NAPHTHALENEFURANE see SNX250
2-(4-SULFO-1-NAPHTHYLAZO)-1-NAPHTHOL-4-SULFONIC ACID, DISODIUM SALT see HJF500
SULFONATED CASTOR OIL see TOD500
SULFONATES see SNY000
o-SULFONBENZOIC ACID IMIDE SODIUM SALT see SJN700
p-SULFONDICHLORAMIDOBENZOIC ACID see HAF000
SULFONE, p-BROMOPHENYL CHLOROMETHYL see BNV850
SULFONE-2,4,4',5-TETRACHLORODIPHENYL see CKM000
SULFONETHYLMETHANE see BJT750
SULFONIC ACID, MONOCHLORIDE see CLG500
SULFONIMIDE see CBF800
SULFONINE ACID BLUE R see ADE750
SULFONMETHANE see ABD500
N-SULFONOXY-AAF see FDV000
N-SULFONOXY-N-ACETYL-2-AMINOFLUORENE see FDV000
4,4'-SULFONYLBISACETANILIDE see SNY500
p,p'-SULFONYLBISACETANILIDE see SNY500
4',4'''-SULFONYLBIS(ACETANILIDE) see SNY500
1,1'-SULFONYLBIS(4-AMINOBENZENE) see SOA500
3,3'-SULFONYLBIS(ANILINE) see SOA000
4,4'-SULFONYLBISANILINE see SOA500
p,p-SULFONYLBISBENZAMINE see SOA500
4,4'-SULFONYLBISBENZAMINE see SOA500
p,p-SULFONYLBISBENZENAMINE see SOA500
1,1'-(SULFONYLBIS(4,1-PHENYLENEIMINO))BIS(1-DEOXY-1-SULFO-d-GLUCITOL) DISODIUM SALT see AOO800
4,4'-SULFONYLBIS(4-PHENYLENEOXY)DIANILINE see SNZ000
SULFONYL CHLORIDE see SOT000
SULFONYL CHLORIDE FLUORIDE see SOT500
3,3'-SULFONYLDIANILINE see SOA000
4,4'-SULFONYLDIANILINE see SOA500
p,p'-SULFONYLDIANILINE see SOA500
p,p'-SULFONYLDIANILINE-N,N-DI-d-GLUCOSE SODIUM BISULFITE see AOO800
p,p'-SULFONYLDIANILINE N,N'-DIGLUCOSIDE DISODIUM DISULFONALTE see AOO800
4,4'-SULFONYLDIPHENOL see SOB000
O,O'-(SULFONYLDI-1,4-PHENYLENE) O,O,O',O'-TETRAMETHYL DIPHOSPHOROTHIOATE see PHN250
3-(SULFONYL)-o-((METHYLAMINO)CARBONYL)OXIME-2-BUTANONE see SOB500
N-(SULFONYL-p-METHYLBENZENE)-N-N-BUTYLUREA see BSQ000
(3-β)-3-(SULFOOXY)-ANDROST-5-EN-17-ONE SODIUM SALT (9CI) see SJK400
4-p-SULFOPHENYLAZO-1-NAPHTHOL MONOSODIUM SALT see FAG010
1-(p-SULFOPHENYL)-3-PHENYL-PYRAZOL (GERMAN) see PGE500
SULFOPLAN see SNN300
SULFOPON WA 1 see SIB600
SULFORON see SOD500
SULFORTHOMIDINE see AIE500
SULFOSALICYLIC ACID see SOC500
5-SULFOSALICYLIC ACID see SOC500
SULFOSFAMIDE see SOD000
SULFOSUCCINIC ACID DIHEXYL ESTER, SODIUM SALT see DKP800
SULFOTEP see SOD100
SULFOTEPP see SOD100
SULFOTEX WALA see SIB600
SULFOTHIORINE see SKI500
5-SULFO-TOBIAS ACID see ALH000
SULFOTRIM see TKX000
SULFOTRIMIN see TKX000
SULFOTRINAPHTHYLENOFURAN, SODIUM SALT see SOD200
SULFOX-CIDE see ISA000
SULFOXIDE see ISA000
SULFOXYL see BDS000, ISA000
2-(6-SULFO-2,4-XYLYLAZO)-1-NAPHTHOL-4-SULFONIC ACID, DISODIUM SALT see FAG050
SULFOXYPHENYLPYRAZOLIDINE see DWM000
SULFOZONA see AKO500
SULFRALEM see AOO490
SULFRAMIN 85 see DXW200
SULFRAMIN 40 FLAKES see DXW200
SULFRAMIN 40 GRANULAR see DXW200
SULFRAMIN 1238 SLURRY see DXW200
SULFTECH see SJZ000
SULFUNE see AIE750

SULFUNO see AIE750
SULFUR see SOD500
SULFUR BROMIDE see SOE000
SULFUR CHLORIDE see SOG500, SON510
SULFUR CHLORIDE(DI) (DOT) see SON510
SULFUR CHLORIDE FLUORINE see SOF000
SULFUR CHLORIDE (MONO) (DOT) see SOG500
SULFUR CHLORIDE OXIDE see TFL000
SULFUR CHLORIDE PENTAFLUORIDE see SOF000
SULFUR CHLOROPENTAFLUORIDE see SOF000
SULFUR COMPOUNDS see SOF500
SULFUR DECAFLUORIDE see SOQ450
SULFUR DICHLORIDE see SOG500
SULFUR DIFLUORIDE MONOXIDE see TFL250
SULFUR DIFLUORIDE OXIDE see TFL250
SULFUR DINITRIDE see SOH000
SULFUR DIOXIDE see SOH500
SULFUR DIOXIDE, solution see SOO500
SULFURE de 4-CHLOROBENZYLE et de 4-CHLOROPHENYLE (FRENCH)
   see CEP000
SULFURE de METHYLE (FRENCH) see TFP000
SULFURETED HYDROGEN see HIC500
SULFUR FLOWER (DOT) see SOD500
SULFUR FLUORIDE see SOI000
SULFUR FLUORIDE OXIDE see BLD000
SULFUR HALF-MUSTARD see CHC000
SULFUR HEXAFLUORIDE see SOI000
SULFUR HYDRIDE see HIC500
SULFURIC ACID see SOI500
SULFURIC ACID, aromatic see SOI510
SULFURIC ACID, fuming see SOI520
SULFURIC ACID (mist) see SOI530
SULFURIC ACID, sec-ALKYL ESTER, SODIUM SALT see SOJ000
SULFURIC ACID, ALUMINUM POTASSIUM SALT (2:1:1), DODECAHY-
   DRATE see AHF200
SULFURIC ACID, ALUMINUM SALT (3:2) see AHG750
SULFURIC ACID, AMMONIUM NICKEL(2+) SALT (2:2:1) see NCY050
SULFURIC ACID, BARIUM SALT (1:1) see BAP000
SULFURIC ACID, BERYLLIUM SALT (1:1) see BFU250
SULFURIC ACID, BERYLLIUM SALT (1:1), TETRAHYDRATE see BFU500
SULFURIC ACID, BIS(METHYLMERCURY) SALT see BKS810
SULFURIC ACID, CADMIUM(2+) SALT see CAJ000
SULFURIC ACID, CADMIUM SALT, HYDRATE see CAJ250
SULFURIC ACID, CADMIUM SALT, TETRAHYDRATE see CAJ500
SULFURIC ACID, CALCIUM(2+) SALT, DIHYDRATE see CAX750
SULFURIC ACID, CERIUM SALT (2:1) see CDB400
SULFURIC ACID, CHROMIUM (3+) POTASSIUM SALT (2:1:1) see PLB500
SULFURIC ACID, CHROMIUM(3+)POTASSIUM SALT(2:1:1), DODECAHY-
   DRATE see CMG850
SULFURIC ACID, CHROMIUM SALT, BASIC see NBW000
SULFURIC ACID, CHROMIUM(3+) SALT (3:2), PENTADECAHYDRATE
   see CMK425
SULFURIC ACID, COBALT(2+) SALT (1:1) see CNE125
SULFURIC ACID, COPPER(2+) SALT (1:1) see CNP250
SULFURIC ACID, COPPER(2+) SALT, PENTAHYDRATE see CNP500
SULFURIC ACID, CYCLIC ETHYLENE ESTER see EJP000
SULFURIC ACID, DECYL ESTER, SODIUM SALT see SOK000
SULFURIC ACID, DIAMMONIUM SALT see ANU750
SULFURIC ACID, DICESIUM SALT see CDE500
SULFURIC ACID, DILITHIUM SALT see LHR000
SULFURIC ACID, DIMETHYL ESTER see DUD100
SULFURIC ACID, DIPOTASSIUM SALT see PLT000
SULFURIC ACID, DISODIUM SALT see SJY000
SULFURIC ACID, DITHALLIUM(1+) SALT (8CI, 9CI) see TEM000
SULFURIC ACID, DODECYL ESTER, TRIETHANOLAMINE SALT
   see SON000
SULFURIC ACID, GALLIUM SALT (3:2) see GBS100
SULFURIC ACID, INDIUM SALT see ICJ000
SULFURIC ACID, IRON(2⁻) SALT (1:1) see FBN100
SULFURIC ACID, IRON (3⁻) SALT (3:2) see FBA000
SULFURIC ACID LANTHANUM(3+) SALT (3:2) see LBB000
SULFURIC ACID, LAURYL ESTER, AMMONIUM SALT see SOM500
SULFURIC ACID, LEAD (2+) SALT (1:1) see LDY000
SULFURIC ACID, LITHIUM SALT (1:2) see LHR000
SULFURIC ACID, MAGNESIUM SALT (1:1), compounded with 2,2'-
   DITHIOBIS(PYRIDINE) 1,1'-OXIDE see OIU850
SULFURIC ACID, MAGNESIUM SALT (1:1) HEPTAHYDRATE see MAJ500
SULFURIC ACID, MANGANESE(2+) SALT see MAU250
SULFURIC ACID, MANGANESE (2+) SALT (1:1) TETRAHYDRATE
   see MAU750
SULFURIC ACID, MERCURY(2+) SALT (1:1) see MDG500
SULFURIC ACID, MONOANHYDRIDE with NITROUS ACID see NMJ000
SULFURIC ACID, MONODECYL ESTER, SODIUM SALT see SOK000
SULFURIC ACID, MONODODECYL ESTER, AMMONIUM SALT see SOM500

SULFURIC ACID, MONODODECYL ESTER, compounded with 2,2',2''-
   NITRILOTRIS(ETHANOL) see SON000
SULFURIC ACID, MONODODECYL ESTER, compounded with 2,2',2''-
   NITRILOTRIETHANOL (1581) see SON000
SULFURIC ACID, MONODODECYL ESTER, SODIUM SALT see SIB600
SULFURIC ACID, MONO(2-(2-(2-(DODECYLOXY)ETHOXY)ETHOXY)ETHYL)
   ESTER, SODIUM SALT see SIC000
SULFURIC ACID, MONO(2-ETHYLHEXYL)ESTER see ELB400
SULFURIC ACID, MONO(2-ETHYLHEXYL)ESTER, SODIUM SALT (8CI)
   see TAV750
SULFURIC ACID, MONONONYL ESTER, SODIUM SALT see SIX500
SULFURIC ACID, MONOPOTASSIUM SALT see PKX750
SULFURIC ACID, MONOSODIUM SALT see SEG800
SULFURIC ACID, MONOTETRADECYL ESTER, SODIUM SALT see SIO000
SULFURIC ACID, MYRISTYL ESTER, SODIUM SALT see SIO000
SULFURIC ACID, NICKEL(2+)SALT see NDK500
SULFURIC ACID, NICKEL(2+) SALT (1:1) see NDK500
SULFURIC ACID, NICKEL(2+) SALT, HEXAHYDRATE see NDL000
SULFURIC ACID, THALLIUM SALT see TEL750
SULFURIC ACID, THALLIUM(2+) SALT see TEM250
SULFURIC ACID, THALLIUM(1+) SALT (1:2) see TEM000
SULFURIC ACID, VANADIUM SALT see VEA100
SULFURIC ACID, ZINC SALT (1:1) see ZNA000
SULFURIC ACID, ZINC SALT (1:1), HEPTAHYDRATE see ZNJ000
SULFURIC ACID, ZIRCONIUM(4+) SALT (2:1) see ZTJ000
SULFURIC AND HYDROFLUORIC ACIDS, MIXTURE (DOT) see HHV000
SULFURIC ANHYDRIDE (DOT) see SOR500
SULFURIC CHLOROHYDRIN see CLG500
SULFURIC OXIDE see SOR500
SULFURIC OXYCHLORIDE see SOT000
SULFURIC OXYFLUORIDE see SOU500
SULFUR MONOBROMIDE see SOE000
SULFUR MONOCHLORIDE see SON510
SULFUR MUSTARD see BIH250
SULFUR MUSTARD GAS see BIH250
SULFURMYCIN A see SON520
SULFURMYCIN B see SON525
SULFUR NITRIDE see SOO000
SULFUROUS ACID see SOO500
SULFUROUS ACID ANHYDRIDE see SOH500
SULFUROUS ACID, 2-(p-tert-BUTYLPHENOXY)CYCLOHEXYL-2-PROPYNYL
   ESTER see SOP000
SULFUROUS ACID, 2-(p-tert-BUTYLPHENOXY)-1-METHYLETHYL-2-
   CHLOROETHYL ESTER see SOP500
SULFUROUS ACID, DIBENZYL ESTER see BFK000
SULFUROUS ACID, DIPOTASSIUM SALT see PLT500
SULFUROUS ACID, cyclic ester with 1,4,5,6,7,7-HEXACHLORO-5-
   NORBORNENE-2,3-DIMETHANOL see EAQ750
SULFUROUS ACID, MONOSODIUM SALT see SFE000
SULFUROUS ACID, SODIUM SALT (1:2) see SJZ000
SULFUROUS ANHYDRIDE see SOH500
SULFUROUS DICHLORIDE see TFL000
SULFUROUS OXIDE see SOH500
SULFUROUS OXYCHLORIDE see TFL000
SULFUROUS OXYFLUORIDE see TFL250
SULFUR OXIDE see SOH500
SULFUR OXIDE (N-FLUOROSULPHONYL)IMIDE see SOQ000
SULFUR PENTAFLUORIDE see SOQ450
SULFUR PHOSPHIDE see PHS000
SULFUR SELENIDE see SBT000
SULFUR SUBCHLORIDE see SON510
SULFUR TETRACHLORIDE see SOQ500
SULFUR TETRAFLUORIDE see SOR000
SULFUR THIOCYANATE see SOI200
SULFUR TRIOXIDE see SOR500
SULFUR TRIOXIDE (stabilized) see SOS000
SULFUR TRIOXIDE, STABILIZED (DOT) see SOR500
SULFURYL AZIDE CHLORIDE see SOS500
SULFURYL CHLORIDE see SOT000
SULFURYL CHLORIDE FLUORIDE see SOT500
SULFURYL CHLOROFLUORIDE see SOT500
SULFURYL DIAZIDE see SOU000
SULFURYL FLUORIDE see SOU500
SULFURYL FLUOROCHLORIDE see SOT500
SULGIN see AHO250
SULINDAC see SOU550
SULINOL see SOU550
SULKOL see SOD500
SULMET see SJW500, SNJ000
SULOCTIDIL see SOU600
SULOCTIDYL see SOU600
SULOCTON see SOU600
SULODYNE see PDC250
SULOUREA see ISR000

SULPELIN see SNV000
SULPHABUTIN see BOT250
SULPHACETAMIDE see SNP500
SULPHADIAZINE see PPP500
SULPHADIMETHOXINE see SNN300
SULPHADIMETHYLISOXAZOLE see SNN500
SULPHADIMETHYLPYRIMIDINE see SNJ000
SULPHADIMIDINE see SNJ000
SULPHADIONE see SOA500
SULPHAFURAZ see SNN500
SULPHAMERAZINE see ALF250
SULPHAMETHALAZOLE see SNK000
SULPHAMETHOXAZOL see SNK000
SULPHAMETHOXAZOLE see SNK000
SULPHAMETHOXYPYRIDAZINE see AKO500
SULPHAMETHYLISOXAZOLE see SNK000
SULPHAMIC ACID (DOT) see SNK500
SULPHAMOPRINE see SNI500
SULPHANILAMIDE see SNM500
5-SULPHANILAMIDO-3,4-DIMETHYL-ISOXAZOLE see SNN500
3-SULPHANILAMIDO-5-METHYLISOXAZOLE see SNK000
SULPHASALAZINE see PPN750
SULPHASIL see SNP500
SULPHASOMIDINE see SNJ350
SULPHATHIAZOLE see TEX250
SULPHAUREA see SNQ550
SULPHEIMIDE see CBF800
SULPHENAZOLE see AIF000
SULPHENONE see CKI625
SULPHISOMEZOLE see SNK000
SULPHISOXAZOL see SNN500
2-SULPHOBENZOIC IMIDE see BCE500
SULPHOBENZOIC IMIDE CALCIUM SALT see CAM750
SULPHOBENZOIC IMIDE, SODIUM SALT see SJN700
SULPHOCARBONIC ANHYDRIDE see CBV500
SULPHOFURAZOLE see SNN500
SULPHON ACID BLUE R see ADE750
SULPHON ACID BLUE RA see ADE750
SULPHONAL NP 1 see SNQ700
1-(4-SULPHO-1-NAPHTHYLAZO)-2-NAPHTHOL-3,6-DISULPHONIC ACID,
    TRISODIUM SALT see FAG020
SULPHON-MERE see SOA500
1,1'-SULPHONYLBIS(4-AMINOBENZENE) see SOA500
p,p-SULPHONYLBISBENZAMINE see SOA500
4,4'-SULPHONYLBISBENZAMINE see SOA500
p,p-SULPHONYLBISBENZENAMINE see SOA500
4,4'-SULPHONYLBISBENZENAMINE see SOA500
SULPHONYLDIANILINE see SOA500
p,p-SULPHONYLDIANILINE see SOA500
SULPHORMETHOXINE see AIE500
SULPHOS see PAK000
SULPHOXIDE see ISA000
SULPHUR (DOT) see SOD500
SULPHUR, lump or power (DOT) see SOD500
SULPHUR, molten (DOT) see SOD500
SULPHUR DIOXIDE, LIQUEFIED (DOT) see SOH500
SULPHURIC ACID see SOI500
SULPHURIC ACID, CADMIUM SALT (1:1) see CAJ000
SULPHUR MUSTARD GAS see BIH250
SULPIRID see EPD500
SULPIRIDE see EPD500
SULPRIM see TKX000
SULPROFOS see SOU625
SULPROSTONE see SOU650
SULPYRID see EPD500
SULSOL see SOD500
SULTAMICILLIN TOSILATE see SOU675
SULTANOL see BQF500
SULTIRENE see AKO500
SULTOPRIDE see EPD100
SULTOPRIDE HYDROCHLORIDE see SOU725
SULTOSILATO de PIPERACINA (SPANISH) see PIK625
SULTOSILIC ACID, PIPERAZINE SALT see PIK625
SULTROPRIDE CHLORHYDRATE see SOU725
SULXIN see SNN300
SULZOL see TEX250
SUM 3170 see DCS200
SUMAC TANNIN see SOU750
SUMADIL see DVW700
SUMAPEN VK see PDT750
SUMATRA CAMPHOR see BMD000
SUMEDINE see ALF250
SUMETROLIM see TKX000
SUMICIDIN see FAR100

SUMIFLY see FAR100
SUMILEX see PMF750
SUMILIT BBM see BRP750
SUMILIT EXA 13 see PKQ059
SUMILIT PCX see AAX175
SUMINE 2015 see DQP800
SUMIOXON see PHD750
SUMIPLEX LG see PKB500
SUMIPOWER see FAR100
SUMISCLEX see PMF750
SUMISET D see CNH125
SUMITEX FSK see DTG000
SUMITEX NS see DTG000
SUMITHIAN see DSQ000
SUMITHION S-ISOMER see MKC250
SUMITOL see BQC250
SUMITOL 80W see BQC250
SUMITOMO LIGHT GREEN SF YELLOWISH see FAF000
SUMITOMO PX 11 see PKQ059
SUMITOMO S 4084 see COQ399
SUMITOX see MAK700
SUMMETRIN see PAG500
SUMPOCAINE see SPB800
SUNAPTIC ACID B see NAR000
SUNAPTIC ACID C see NAR000
SUNCHOLIN see CMF350
SUNCIDE see PMY300
SUNCIDE, nitrosated see PMY310
SUNFRAL see FMB000
SUNITOMO S 4084 see COQ399
SUNLIGHT see SOV000
SUNRABIN see EAU075
SUNSET YELLOW FCF see FAG150
SUPACAL see CNG835
SUPARI (INDIA) see AQT650
SUPARI, nut extract see BFW000
SUPERACRYL AE see PKB500
SUPER AMIDE L-9A see BKE500
SUPERANABOLON see DYF450
SUPER-CAID see BMN000
SUPERCEL 3000 see USS000
SUPERCICLIN see PPY250
SUPER COBALT see CNA250
SUPERCOL see LIA000
SUPERCOL U POWDER see GLU000
SUPERCORTIL see PLZ000
SUPERCORTYL see SOV100
SUPER COSAN see SOD500
SUPER CRAB-E-RAD-CALAR see CAM000
SUPER DAL-E-RAD see CAM000
SUPER DAL-E-RAD-CALAR see CAM000
SUPER-DENT see SHF500
SUPER D WEEDONE see DAA800, TAA100
SUPERELGETOL see DUT800
SUPERFLAKE ANHYDROUS see CAO750
SUPERFLOC see PKF750
SUPER HARTOLAN see CMD750
SUPERIAN YELLOW R see SGP500
SUPERINONE see TDN750
SUPERIUONE (FRENCH) see TDN750
SUPERLYSOFORM see FMV000
SUPERNOX see DGI000
SUPEROL see GGA000
SUPEROL RED C RT-265 see CHP500
SUPERORMONE CONCENTRE see DAA800
SUPEROX see BDS000
SUPEROXOL see HIB000
SUPERPHOSPHATE see SOV500
SUPERPREDNOL see SOW000
SUPER PRODAN see DXE000
SUPER RODIATOX see PAK000
SUPER-ROZOL see BMN000
SUPERSEPTIL see SNJ000
SUPERTAH see CMY800
SUPICAINE AMIDE SULFATE see AJN750
SUPININ see SOW500
SUPININE see SOW500
SUPLEXEDIL see BPM750
SUPONA see CDS750
SUPONE see CDS750
SUP'OPERATS see BMN000
SUPOTRAN see CKF500
SUPRA see IHD000
SUPRACAPSULIN see VGP000

SUPRACET BRILLIANT BLUE BG see MGG250
SUPRACET BRILLIANT BLUE 2GN see TBG700
SUPRACET BRILLIANT VIOLET 3R see DBP000
SUPRACET DIAZO BLACK A see DPO200
SUPRACET ORANGE R see AKP750
SUPRACHOL see SGD500
SUPRACID see SNQ000
SUPRADIN see VGP000
SUPRAMIKE see BAP000
SUPRAMYCIN see TBX250
SUPRANEPHRANE see VGP000
SUPRANEPHRINE see VGP000
SUPRANEPHRIN SOLUTION see AES500
SUPRANOL see VGP000
SUPRARENIN see AES000, VGP000
SUPRARENIN HYDROCHLORIDE see AES500
SUPRASEC see LIH000
SUPRASTIN see CKV625
SUPRATONIN see MQS100
SUPREFACT see LIU420
SUPREL see VGP000
SUPREMAL see DYE600
SUPREME DENSE see TAB750
SUP'R FLO see DXQ500
SUP'R FLO FERBAM FLOWABLE see FAS000
SUPRIFEN see HKH500
SUPRIFENE see HKH500
SUPRIFEN PSB HYDROCHLORIDE see DNU200
SUPRILENT see VGA300
SUPRIMAL see HGC500
SUPRIN see TKX000
SUPRISTOL see SNL850
SUPROFEN see TEN750
SUPROTAN see CKF500
SURAMETHINIUM see HLC500
SURAMINE see BAT000
SURAUTO see IDW000
SURCHLOR see SHU500
SURCOPUR see DGI000
SUREAU (CANADA, HAITI) see EAI100
SURECIDE see CON300
SUREM see DLY000, PEU000
SURENINE see VGP000
SURE-SET see CJN000
SURESTRINE see BIT000
SURESTRYL see BIT000, MRU600
SURFACTANT WK see DXY000
SURGAM see SOX400
SURGEX see BET000
SURGICAL SIMPLEX see PKB500
SURGI-CEN see HCL000
SURHEME see PEU000
SURIKA see TKH750
SURINAM GREENHEART WOOD see HLY500
SURIRENE see AKO500
SURITAL see AGL375, SOX500
SURITAL SODIUM see SOX500
SURITAL SODIUM (derivative) see SOX500
SURITAL SODIUM SALT see SOX500
SURMONTIL see DLH200
SURMONTIL MALEATE see SOX550
SUROFENE see HCL000
SURPLIX see DLH600, DLH630
SURPRACIDE see DSO000
SURPUR see DGI000
SURRECTAN see AJY250
SURSUM see CJN250
SURSUMID see EPD500
SU SEGURO CARPIDOR see DUV600
SUSPEN see PDT750
SUSPENDOL see ZVJ000
SUSPHRINE see VGP000
SUSTANE see BFW750, BQI000, BRM500
SUSTANE 1-F see BQI000
SUSTANONE see TBF500
SUSVIN see MRH209
SUTAN see EID500
SUTICIDE see HCQ500
SUTOPROFEN see TEN750
SUVREN see BRS000
SUXAMETHONIUM CHLORIDE see HLC500
SUXAMETHONIUM see CMG250
SUXAMETHONIUM CHLORIDE see HLC500
SUXAMETHONIUM DICHLORIDE see HLC500

SUXAMETHONIUM IODIDE see BJI000
SUXCERT see HLC500
SUXEMETHONIUM see CMG250
SUXETHONIUM CHLORIDE see HLC500
SUXIBUZONE see SOX875
SUXIL see SNE000
SUXILEP see ENG500
SUXIMAL see ENG500
SUXIN see ENG500
SUXINUTIN see ENG500
SUXINYL see HLC500
SUZORITE MICA see MQS250
SUZU see ABX250
SUZU H see HON000
SV-052 see SMW400
SV-1522 see ABH500
SVC see ABX500
SVO 9 see PKL100
SW-751 see POM275
SWALLOW WORT see CCS650
SWAMP HELLEBORE see FAB100
SWAMP WOOD see LEF100
SWAT see SOY000
SWEBATE see TAL250
SWEDISH GREEN see CNN500, COF500
SWEENEY'S ANT-GO see ARD750
SWEEP see TBQ750
SWEETA see SJN700
SWEET BELLS see DYA875
SWEET BETTIE see CCK675
SWEET BIRCH OIL see MPI000, SOY100
SWEET DIPEPTIDE see ARN825
SWEET GUM see SOY500
SWEET MYRTLE see WBA000
SWEET ORANGE OIL see OGY000
SWEET PEA SEEDS see SOZ000
SWEET POTATO PLANT see CCO680
SWEET SHRUB see CCK675
SWEP see DEV600
SWIETENIA MAHAGONI see MAK300
SWINE PROSTATE EXTRACT see RLK875
SWISS BLUE see BJI250
SWISS CHEESE PLANT see SLE890
SWITCH IVY see DYA875
SWP (ANTIOXIDANT) see BRP750
SYBIROMYCIN see SCF500
SYCOTROL see PIM000
SYD 230 see CGB250
SYDNONE IMINE, 3-(1-METHYL-2-PHENYLETHYL)-, MONOHYDROCHLOR-
    IDE see SPA000
SYDNOPHENE see SPA000
SYDNOPHEN HYDROCHLORIDE see SPA000
SYEP see SPA500
SYGETHIN see SPA650
SYGETIN see SPA650
SYKOSE see BCE500, SJN700
SYLANTOIC see DKQ000, DNU000
SYLGARD 184 CURING AGENT see SPB000
SYLLIT see DXX400
SYLODEX see ARM268
SYMBIO see SNN300
SYMCLOSEN see TIQ750
SYMCLOSENE see TIQ750
SYMETRA see DKE800
SYMMETREL see AED250, TJG250
SYMPAMINA-D see BBK500
SYMPAMINE see BBK000
SYMPATEDRINE see BBK000
SYMPATEKTOMAN see TCC000
SYMPATHIN E see NNO500
SYMPATHIN I see VGP000
SYMPATHOL see HLV500, SPD000
m-SYMPATHOL see NCL500, SPC500
SYMPATHOLYTIN see DCR000, DCR200
SYMPATOL see SPD000
m-SYMPATOL see NCL500, SPC500
SYMPATOL TARTRATE see SPD000
SYMPHORICARPOS (various species) see SED550
SYMPHYTINE see SPB500
SYMPHYTUM OFFICINALE L see RRK000, RRP000
SYMPLOCARPUS FOETIDUS see SDZ450
SYMPOCAINE HYDROCHLORIDE see SPB800
SYMPROPAMIN see FMS875
SYMULER FAST YELLOW GF see DEU000

SYMULER LAKE RED C see CHP500
SYMULEX MAGENTA F see FAG070
SYMULEX PINK F see FAG070
SYMULON ACID BRILLIANT SCARLET 3R see FMU080
SYMULON METANIL YELLOW see MDM775
SYMULON SCARLET G BASE see NMP500
SYMULON SCARLET 2G SALT see DEO400
SYNADRIN see PEV750
SYNALAR see SPD500
SYNALATE see FDD150
SYNALGOS see DQA400
SYNAMOL see SPD500
SYNANCEJA HORRIDA Linn. VENOM see SPB875
SYNANDONE see SPD500
SYNANDRETS see MPN500
SYNANDROL see TBG000
SYNANDROL F see TBF500
SYNANDRONE see SPD500
SYNANDROTABS see MPN500
SYNANTHIC see OMY500
SYNAPAUSE see EDU500
SYNAPEN see PDD350
SYNAPHORIN see CMG675
SYNASAL see SPC500
SYNASTERON see PAN100
SYNATE see SBN000
SYNCAINE see AIT250
SYNCAL see BCE500
SYNCHROMATE ORANGE AOR see NEY000
SYNCILLIN see PDD350
SYNCL see ALV000
SYNCORDAN see FMS875
SYNCORT see DAQ800
SYNCORTA see DAQ800
SYNCORTYL see DAQ800
SYNCURARINE see PDD300
SYNCURINE see DAF600
SYNDIOL see EDO000
SYNDIOTACTIC POLYPROPYLENE see PMP500
SYNDROX see MDQ500, MDT600
SYNELAUDINE see DAM700
SYNEPHRIN see HLV500
m-SYNEPHRINE see NCL500, SPC500
p-SYNEPHRINE see HLV500
dl-SYNEPHRINE see SPD000
dl-p-SYNEPHRINE see SPD000
m-SYNEPHRINE HYDROCHLORIDE see SPC500
SYNEPHRINE TARTRATE see SPD000
(+ −)-SYNEPHRINE TARTRATE see SPD000
SYNERGID R see DWW000
SYNERGIST 264 see OES000
SYNERONE see TBG000
SYNERPENIN see PDD350
SYNESTRIN see DKA600, DKB000
SYNESTROL see DAL600, DLB400
SYNETHENATE see SPC500
SYNFEROL AH EXTRA see SPD100
SYNGACILLIN see AJJ875
SYNGESTERONE see PMH500
SYNGESTROTABS see GEK500
SYNGUM D 46D see GLU000
SYNGYNON see HNT500
SYNHEXYL see HGK500
SYNISTAMIN see TAI500
SYNKAMIN see AKX500
SYNKAMIN BASE see AKX500
SYNKAY see MMD500
SYNKLOR see CDR750
SYNMIOL see DAS000
SYNOESTRON see DKB000
SYNOPEN see CKV625
SYNOPEN R see CKV625
SYNOTODECIN see PPY250
SYNOTOL L-60 see BKE500
SYNOVEX S see PMH500
SYNOX 5LT see MJO500
SYNOX TBC see BSK000
SYNPEN see CKV625
SYNPENIN see AIV500
SYNPERONIC OP see GHS000
SYNPITAN see ORU500
SYNPOL 1500 see SMR000
SYNPREN-FISH see PIX250
SYNSAC see SPD500

SYNSTIGMIN BROMIDE see POD000
SYNSTIGMINE see DER600
SYNTAR see CMY800
SYNTARIS see FDD085
SYNTARPEN see DGE200
SYNTASE 62 see MES000
SYNTASE 100 see DMI600
SYNTEDRIL see BBV500
SYNTEFIX see CNH125
SYNTES 12A see EIV000
SYNTESTRIN see DKB000
SYNTESTRINE see DKB000
SYNTETREX see PPY250
SYNTETRIN see PPY250
SYNTETRIN NITRATE see SPE000
SYNTEXAN see DUD800
SYNTHAMIDE 5 see SPE500
SYNTHARSOL see ACN250
SYNTHECILLIN see PDD350
SYNTHECILLINE see PDD350
SYNTHENATE see HLV500
SYNTHETIC 3956 see CDV100
SYNTHETIC AMORPHOUS SILICA see SCH000
SYNTHETIC BRADYKININ see BML500
SYNTHETIC EUGENOL see EQR500
SYNTHETIC GLYCERIN see GGA000
SYNTHETIC IRON OXIDE see IHD000
SYNTHETIC LH-RH see LIU360
SYNTHETIC MUSTARD OIL see AGJ250
SYNTHETIC OXYTOCIN see ORU500
SYNTHETIC PYRETHRINS see AFR250
SYNTHETIC TRF see TNX400
SYNTHETIC TRH see TNX400
SYNTHETIC TSH-RELEASING FACTOR see TNX400
SYNTHETIC TSH-RELEASING HORMONE see TNX400
SYNTHETIC WINTERGREEN OIL see MPI000
SYNTHILA see DJB200
SYNTHOESTRIN see DKA600
SYNTHOFOLIN see DKA600
SYNTHOMYCINE see CDP250
SYNTHOPHYLLINE see DNC000
SYNTHOSTIGMINE BROMIDE see POD000
SYNTHOSTIGMINE METHYL SULFATE see DQY909
SYNTHOVO see DLB400
SYNTHRIN see BEP500
SYNTHROID see LEQ300
SYNTHROID SODIUM see LEQ300
SYNTOCIN see ORU500
SYNTOCINON see ORU500
SYNTOCINONE see ORU500
SYNTODRIL see BBV500
SYNTOFOLIN see DKA600
SYNTOLUTAN see PMH500
SYNTOMETRINE see LJL000
SYNTOPHEROL see VSZ450
SYNTOSTIGMIN see DER600
SYNTOSTIGMIN (tablet) see POD000
SYNTOSTIGMIN BROMIDE see POD000
SYNTOSTIGMINE BROMIDE see POD000
SYNTROGENE see DLB400
SYNTRON B see EIV000
SYNTROPAN see AOD250
SYPHOS see ASD000
SYRAP see CMG000
SYRAPRIM see TKZ000
SYRIAN BEAD TREE see CDM325
SYRINGALDEHYDE see DOF600
SYRINGEALDEHYDE see DOF600
SYRINGIC ACID ETHYL CARBONATE ESTER with METHYL RESERPATE
   see RCA200
SYRINGIC ALDEHYDE see DOF600
SYRINGOL see DOJ200
SYRINGOPINE see RCA200
SYRINGYLALDEHYDE see DOF600
SYROSINGOPIN see RCA200
SYROSINGOPINE see RCA200
SYRUP of IPECAC, U.S.P. see IGF000
SYS 67ME see SIL500
SYS 67MPROP see CLO200
SYSTAM see OCM000
SYSTAMEX see OMY500
SYSTEMOX see DAO600
SYSTODIN see QHA000
SYSTOGENE see TOG250

SYSTOPHOS see OCM000
SYSTOX see DAO600
SYSTOX SULFONE see SPF000
SYSTRAL see CIS000
SYTAM see OCM000
SYTASOL see CBW000
SYTLOMYCIN AMINONUCLEOSIDE see ALQ625
SYTOBEX see VSZ000
SYTON FAST RED 2G see DVB800
SYTON FAST RED R see CJD500
SYTON FAST SCARLET RB see MMP100
SYTRON see EJA379
SYZYGIUM JAMBOS (Linn.) Alston, extract excluding roots see SPF200
SZAZ see MNB000
SZESCIOMETYLENODWUIZOCYJANIAN (POLISH) see DNJ800
SZKLARNIAK see DGP900

T-2 see TIK500
T3 see LGK050
α-T see MIH275
β-T see TIN000
T 72 see PNX000
L-T3 see LGK050
l-T4 see TFZ275
T 100 see TGM740
T-113 see BPG000
T-125 see TBX000
T-144 see IPX000
2,4,5-T see TAA100
864T see IBQ100
T-1035 see DSA800
T-1036 see DJJ400
T-1088 see HNP000
T-1123 see TGH665
T-1125 see HNO000
T-1152 see HNO500
T 1220 see SJJ200
T-1551 see CCS369
T-1690 see HNN000
T-1703 see IRF000
T-1768 see DPL900
T-1770 see MIA775
T 1824 see BGT250
T-1835 see DEC200
T-1843 see MIC250
T-1982 see TAA400
T-2002 see BJE750
T-2104 see EIF000
T-2106 see IPX000
T-2588 see TAA420
79T61 see SBE500
T(sub 3) see LGK050
T-42082 see SFP500
T4 (hormone) see TFZ275
T 1 (Catalyst) see DBF800
TA 1 see MRN675
TA 12 see TAN750
TA 064 see DAP850
TAA see TFA000
TA-AZUR see THM750
TABAC (FRENCH) see TGI100
TABAC du DIABLE (CANADA) see SDZ450
TABACHIN (MEXICO) see CAK325
TABACO (SPANISH) see TGI100
TABALGIN see HIM000
TABASCO HOT PEPPER SAUCE see TAA875
TABASCO PEPPER see PCB275
TABILIN see BFD000
TABLE SALT see SFT000
TABLOID see AMH250
TABUN see EIF000
TAC-28 see VIK150
TAC 121 see TGG250
TAC 131 see TGG250
TACARYL see MDT500, MPE250
TACAZYL see MPE250
TACE see CLO750
TACE-FN see CLO750
TACHIGAREN see HLM000
TACHIONIN see HII500
TACHMALIN see AFH250
TACITIN see BCH750
TACOSAL see DKQ000, DNU000
TACP see TNC725

TACRINE see TCJ075
TACRYL see MPE250
TACTARAN see TAF675
TACUMIL see TKX000
TAD see TEP500
TAFASAN see MEP250
TAFAZINE see BJP000
TAFIL see XAJ000
TAG see ABU500
TAG-39 see ECU750
TAGAMET see TAB250
TAGAT see SEH000
TAGATHEN see CHY250
TAGETES OIL see TAB275
TAG FUNGICIDE see ABU500
TAGN see THN800
TAHMABON see DTQ400
TAI-284 see CFH825
TAI 284 see CMV500
(±)-TAI 284 see CFH825
dl-TAI 284 see CFH825
TAIFEN see ZMA000
TAIGUIC ACID see HLY500
TAIGU WOOD see HLY500
TAIL FLOWER see APM875
TAJMALIN see AFH250
TAK see MAK700
TAKACIDIN see TAB300
TAKAMINA see VGP000
TAKANARUMIN see ZVJ000
TAKAOKA AMARANTH see FAG020
TAKAOKA BRILLIANT SCARLET 3R see FMU080
TAKAOKA METANIL YELLOW see MDM775
TAKAOKA RHODAMINE B see FAG070
TAKEDO 1969-4-9 see GGA800
TAKILON see PKQ059
TAKTIC see MJL250
TAKYCOR see AFH250
TALADREN see DFP600
TALAN see CBW000
TALARGAN see TEH500
TALATROL see TEM500
TALBOT see LCK000
TALBUTAL see AFY500
TALC see TAB750
TALC, containing asbestos fibers see TAB775
TALCORD see AHJ750
TALCUM see TAB750
TALIBLASTIN see TEH250
TALIBLASTINE see TEH250
TALIMOL see TEH500
TALISOMYCIN see TAC500
TALISOMYCIN B see TAC750
TALISOMYCIN S^{10b} see TAB785
TALL OIL see TAC000
TALLOL see TAC000
TALLOW BENZYL DIMETHYLAMMONIUM CHLORIDE see DTC600
N-TALLOWPYRROLIDINONE see TAC200
TALLYSOMYCIN A see TAC500
TALLYSOMYCIN B see TAC750
TALLYSOMYCIN S^{10b} see TAB785
TALMON see MAO350
TALODEX see FAQ900
TALON see TAC800
TALON RODENTICIDE see TAC800
TALPHENO see EOK000
TALUCARD see POB500
TALUSIN see POB500
TALWAN see DOQ400
TALWIN see DOQ400
TAMARIZ see SAZ000
TAMARON see DTQ400
TAMAS see BBW500
TAMBALISA (CUBA) see NBR800
TAMCHA see AJV500
TAMETIN see TAB250
TAMILAN see PAP000
TAMOXIFEN see NOA600
TAMOXIFEN CITRATE see TAD175
TA-33MP see TAN750
TAMPOVAGAN STILBOESTROL see DKA600
TAMPULES see CDP000
TANAFOL see CKF500
TANAGER RED X-761 see CJD500

TANAKAN see CLD000, CLD250
TANAMICIN see HGP550
TANASUL see MDU300
TANDACOTE see HNI500
TANDALGESIC see HNI500
TANDEARIL see HNI500
TANDEM see TJK100
TANDERAL see HNI500
TANDEX see DUM800
TANDIX see IBV100
TANGANTANGAN OIL see CCP250
TANGELO OIL see TAD250
TANGERINE OIL see TAD500
TANGERINE OIL, COLDPRESSED (FCC) see TAD500
TANGERINE OIL, EXPRESSESED (FCC) see TAD500
TANICAINE see BJO500
TANIDIL see DQA400
TANNEX see IDA000
TANNIC ACID see TAD750
TANNIN see TAD750
TANNIN from ACORN see ADI625
TANNIN from BETEL NUT see BFW050
TANNIN from BRACKEN FERN see BML250
TANNIN from CHERRY BARK OAK see CDL750
TANNIN from CHESTNUT see CDM250
TANNIN-FREE FRACTION of BRACKEN FERN see TAE250
TANNIN from LIMONIUM NASHII see MBU750
TANNIN from MARSH ROSEMARY see MBU750
TANNIN from MIMOSA see MQV250
TANNIN from MYRABOLAM see MRZ100
TANNIN from MYROBALANS see MSB750
TANNIN from MYRTAN see MSC000
TANNIN from PERSIMMON see PCP500
TANNIN from QUEBRACHO see QBJ000
TANNIN from SUMAC see SOU750
TANNIN from SWEET GUM see SOY500
TANNIN from VALONEA see VCK000
TANNIN from WAX MYRTLE see WBA000
TANONE see DRR400
TANRUTIN see RSU000
TANSTON see XQS000
TANSY OIL see TAE500
TANSY RAGWORT see RBA400
TANTALIC ACID ANHYDRIDE see TAF500
TANTALIUM PENTAFLUORIDE see TAF250
TANTALUM see TAE750
TANTALUM-181 see TAE750
TANTALUM CHLORIDE see TAF000
TANTALUM FLUORIDE see TAF250
TANTALUM OXIDE see TAF500
TANTALUM(V) OXIDE see TAF500
TANTALUM PENTACHLORIDE see TAF000
TANTALUM PENTAOXIDE see TAF500
TANTALUM PENTOXIDE see TAF500
TANTALUM POTASSIUM FLUORIDE see PLH000
TANTAN (PUERTO RICO) see LED500
TANTARONE see MJE760
TANTUM see BBW500
TAOMYCIN see HOH500
TAOMYXIN see HOH500
TAORYL see CBG250
TAP see MPN000
TAP 85 see BBQ500
TAPAR see HIM000
TAPAZOLE see MCO500
TAPHAZINE see BJP000
TAPIOCA see CCO680, DBD800
TAPIOCA STARCH see SLJ500
TAPIOCA STARCH HYDROXYETHYL ETHER see HLB400
TAPON see SLJ500
TAP 9VP see DGP900
TAR see CMY800
TAR, from tobacco see CMP800
TAR, liquid (DOT) see CMY800
TARACTAN see TAF675
TARA GUM see GMA000
TARAPACAITE see PLB250
TARAPON K 12 see SIB600
TARARACO see AHI635
TARARACO BLANCO (CUBA) see BAR325
TARARACO DOBLE (CUBA) see AHI635
TARASAN see TAF675
TAR CAMPHOR see NAJ500
TAR, COAL see CMY800

TARDAMID see AIE750
TARDAMIDE see AIE750
TARDEX 100 see PAU500
TARDIGAL see DKL800
TARDOCILLIN see BFC750
TARFLEN see TAI250
TARGET MSMA see MRL750
TARICHATOXIN see FOQ000
TARIMYL see SOA500
TARIVID see OGI300
TARLON XB see PJY500
TARNAMID T see PJY500
TARO see EAI600
TAROCTYL see CKP500
TARODYL see GIC000
TARODYN see GIC000
TAR OIL see CMY825
TARO VINE see PLW800
TARPAN see CMW250
TARRAGON see AFW750
TARRAGON OIL see TAF700
TARTAGO (PUERTO RICO) see CNR135
TARTAN see PHK250
TARTAR EMETIC see AQG250
TARTARIC ACID see TAF750
l-TARTARIC ACID, AMMONIUM SALT see DCH000
l-TARTARIC ACID, ANTIMONY POTASSIUM SALT see AQH000
dl-TARTARIC ACID, ANTIMONY POTASSIUM SALT see AQG750
TARTARIC ACID, DIAMMONIUM SALT see DCH000
TARTARIC ACID, DIBENZOATE see DDE300
meso-TARTARIC ACID, ION(2−) see TAF775
TARTARIC ACID, MONOSODIUM SALT see SKB000
TARTARIZED ANTIMONY see AQG250
TARTAR YELLOW see FAG140
meso-TARTARTE see TAF775
TARTRATE ANTIMONIO-POTASSIQUE (FRENCH) see AQG250
TARTRATED ANTIMONY see AQG250
TARTRATE de NICOTINE (FRENCH) see NDS500
TARTRAZINE see FAG140
TARTRAZOL YELLOW see FAG140
TARWEED see TAG250
TARZOL see DGA200
TASK see DGP900
TASK TABS see DGP900
TASMIN see AOO800
TAT-1 see DIK000
TATBA see AQY375
TAT CHLOR 4 see CDR750
TATD see DXH250, GEK200
TATERPEX see CKC000
TATHIONE see GFW000
TAT-3 HYDROCHLORIDE see CNE375
TATTOO see DQM600
TAT-3 TRIPALMITATE see PIC100
TATURIL see UVJ450
TAURE(o)DON see GJC000
TAURINE see TAG750
TAURINOPHENETIDINE HYDROCHLORIDE see TAG875
TAUROCHOLATE see TAH250
TAUROCHOLIC ACID see TAH250
TAUROMYCETIN see TAH500
TAUROMYCETIN-III see TAH650
TAUROMYCETIN-IV see TAH675
TAUTUBA (PUERTO RICO) see CNR135
TAVEGIL see FOS100
TAVEGYL see FOS100
TA-VERM see PIJ500
TAVOR see CFC250
TAVROMYCETIN III see TAH650
TAVROMYCETIN-IV see TAH675
TAXIFOLIN see DMD000
TAXIFOLIOL see DMD000
TAXILAN see PCK500
TAXIN (GERMAN) see TAH750
TAXINE see TAH750
TAXOL see TAH775
TAXUS BACCATA LINN., LEAF EXTRACT see KCA100
TAXUS (VARIOUS SPECIES) see YAK500
TAYO BAMBOU (HAITI) see EAI600
TAYSSATAO see MEP250
TAZEPAM see CFZ000
TAZICEF see CCQ200
TAZIDIME see CCQ200
TAZIN see AER666

TAZONE see BRF500
TB see CMO250
TB 2 see TFD000
2,3,6-TBA (herbicide) see TIK500
TBB see THX500
TBBA see BQK500
TB 1 (BAYER) see FNF000
TBDZ see TEX000
TBE see ACK250
TBEP see BPK250
TBF-43 see TKP850
TBHP-70 see BRM250
TBHQ (FCC) see BRM500
TBOT see BLL750
TBP see TFD250, TIA250
TBT see BSP500
TBTO see BLL750
2,4,5-T BUTOXYETHANOL ESTER see TAH900
2,4,5-T BUTOXYETHYL ESTER see TAH900
2,4,5-TC see TIX500
TCA see TII250, TII500
T-4CA see TEV000
TCAB see TBN500
TCAOB see TBN550
TCA-PE see AMU000
TCA-PR see AMU500
T-250 CAPSULES see TBX250
TCA SODIUM see TII500
TCAT see TIJ500
TCB see TBO700
2,3,6-TCB see TIK500
2,3,6-TCBA see TIK500
TCBC see TIL250
TCBN see TBS000
TCC see TIQ000
TCC mixed with TFC (2:1) see TIL526
TCDBD see TAI000
TCDD see TAI000
2,3,7,8-TCDD see TAI000
TCE see TBQ100
1,1,1-TCE see MIH275
TCEO see ECT600
TCH see TFE250
TC HYDROCHLORIDE see TBX250
TCIN see TBQ750
TCM see CHJ500
TCMTB see BOO635
TCNA see TBR250
TCNB see TBR750
TCNP SODIUM SALT see TAI050
TCPA see TIY500
TCPE see TIX000
m-TCPN see TBQ750
o-TCPN see TBT200
TCPO see TJB750
TCPP see FQU875
2,4,5-TCPPA see TIX500
TCSA see TBV000
TCT see TFZ000
TCTH see CQH650
TCTP see TBV750
TCV-3B see EGM100
TD-183 see TBV750
TD-758 see TBC200
TD 1771 see PEX500
TD-5032 see HEE500
TDA see TGL750
TDBP (CZECH) see TNC500
TDCPP see FQU875
TDE see TJQ333
o,p-TDE see CDN000
TDE (DOT) see BIM500
o,p'-TDE see CDN000
p,p'-TDE see BIM500
TDI see TGM740
2,4-TDI see TGM750
2,6-TDI see TGM800
TDI-80 see TGM740, TGM750
TDI 80-20 see TGM740
TDI (OSHA) see TGM750
T-82 DIFUMARATE see MDU750
TDOT see TCQ275
TDPA see BHM000
TDS see CHK825

2,5-TDS see TGM400
TE see TBF750
TE 114 see HNM000
TEA see TCB725, TCC000, TJN750
TEAB see TCC000
TEABERRY OIL see MPI000
TEAC see TCC250
TEA CHLORIDE see TCC250
TEAEI see TAI100
TEA LAURYL SULFATE see SON000
T.E.A.S. see SDH670
TEB see TDG500
TEBALON see FNF000
TEBE see HNY500
TEBECID see ILD000
TEBECURE see FNF000
TEBEFORM see PNW750
TEBEMAR see FNF000
TEBERUS see EPQ000
TEBESONE I see FNF000
TEBETHIONE see FNF000
TEBEXIN see ILD000
TEBEZON see FNF000
TEBRAZID see POL500
TEBULAN see BSN000
TEBUTHIURON see BSN000
TEC see TJP750
TECACIN see HGP550
TECH DDT see DAD200
TECHNETIUM TC 99M SULFUR COLLOID see SOD500
TECHNICAL BHC see BBQ750
90 TECHNICAL GLYCERINE see GGA000
TECHNICAL HCH see BBQ750
TECHNOPOR see PKQ059
TECH PET F see MQV750
TECNAZEN (GERMAN) see TBR750
TECNAZENE see TBR750
TECODIN see DLX400
TECODINE see DLX400
TECOFLEX HR see PKN000
TECOMIN see HLY500
TECQUINOL see HIH000
TECRAMINE see TKH750
TECSOL see EFU000
TECTILON ORANGE 3GT see SGP500
TECTO see TEX000
TECZA see TJR000
TEDION see CKM000
TEDION V-18 see CKM000
TEDP (OSHA, MAK) see SOD100
TEEBACONIN see ILD000
TEF see TND250
TEFAMIN see TEP000, TEP500
TEFILAN see DNC000
TEFILIN see TBX250
TEFLON see TAI250
TEFLON (various) see TAI250
TEFSIEL C see FMB000
TEG see TJQ000
TEGAFUR see FMB000
TEGAFUR mixture with URACIL (1:4) see UNJ810
TEGDH see TCE375
TEGENCIA HYDROCHLORIDE see DAI200
TEGESTER ISOPALM see IQW000
TEGIN see OAV000
TEGIN 503 see OAV000
TEGIN 515 see OAV000
TEGIN P see SLL000
TEGO 51 see DYA850
TEGOLAN see CMD750
TEGO-OLEIC 130 see OHU000
TEGOPEN see SLJ000, SLJ050, AOC500
TEGOSEPT B see BSC000
TEGOSEPT E see HJL000
TEGOSEPT M see HJL500
TEGOSEPT P see HNU500
TEGO-STEARATE see EJM500
TEGOSTEARIC 254 see SLK000
TEGRETAL see DCV200
TEGRETOL see DCV200
TEIB see TND000
TEICHOMYCIN A2 see TAI400
TEKKAM see NAK500
TEKODIN see DLX400

TEKRESOL see CNW500
TEKWAISA see MNH000
TEL see TCF000
TELAGAN see TEH500
TELARGAN see TEH500
TELARGEAN see TEH500
TELDANE see TAI450
TELDRIN see CLD250, TAI500
TELEBRIX 38 see TAI600
TELEBRIX 300 see MQR300
TELEFOS see IOT000
TELEMID see MDU300
TELEMIN-SOFT see TAI725
TELEPAQUE see IFY100
TELEPATHINE see HAI500
TELEPRIN see TKX000
TELESMIN see DCV200
TELETRAST see IFY100
TELGIN-G see FOS100
TELIDAL see HNI500
TELINE see TBX250
TELIPEX see TBG000
TELLOY see TAJ000
TELLUR (POLISH) see TAJ000
TELLURANE-1,1-DIOXIDE see TAI735
TELLURATE see TAI750
TELLURIC ACID see TAI750
TELLURIC(VI) ACID see TAI750
TELLURIC ACID, AMMONIUM SALT see ANV750
TELLURIC ACID, DISODIUM SALT, PENTAHYDRATE see TAI800
TELLURIC CHLORIDE see TAJ250
TELLURIUM see TAJ000
TELLURIUM (dust or fume) see TAJ010
TELLURIUM CHLORIDE see TAJ250
TELLURIUM COMPOUNDS see TAJ500
TELLURIUM DIETHYLDITHIOCARBAMATE see EPJ000
TELLURIUM DIOXIDE see TAJ750
TELLURIUM HEXAFLUORIDE see TAK250
TELLURIUM HYDROXIDE see TAI750
TELLURIUM NITRIDE see TAK500
TELLURIUM OXIDE see TAJ750
TELLURIUM TETRABROMIDE see TAK600
TELLURIUM TETRACHLORIDE see TAJ250
TELLUROUS ACID, DISODIUM SALT see SKC500
TELMICID see DJT800
TELMID see DJT800
TELMIDE see DJT800
TELMIN see MHL000
TELOCIDIN B see TAK750
TELODRIN see OAN000
TELODRON see CLD250
TELOMYCIN see TAK800
TELON see BCP650
TELONE see DGG000, DGG950
TELONE II SOIL FUMIGANT see DGG950
TELON FAST BLACK E see AQP000
TELOTREX see TBX250
TELTOZAN see CAL750
TELVAR see CJX750, DXQ500
TELVAR DIURON WEED KILLER see DXQ500
TELVAR MONURON WEEDKILLER see CJX750
TEMARIL see TAL000
TEMAZEPAM see CFY750
TEMECHINE see TAL275
TEMED see TDQ750
TEMEFOS see TAL250
TEMENTIL see PMF500
TEMEPHOS see TAL250
TEMEQUINE see TAL275
TEMESTA see CFC250
TEMETEX see DKF130
TEMGESIC see TAL325
TEM-HISTINE see DPJ400
TEMIC see CBM500
TEMIK see CBM500
TEMIK G10 see CBM500
TEMLO see HIM000
TEMOPHOS see TAL250
TEMPANAL see HIM000
TEMPARIN see BJZ000
TEMPLIN OIL see AAC250
TEMPODEX see BBK500
TEMPRA see HIM000
TEMUR see TDX250

TEMUS see BMN000
TEN see TJO000
TENAC see DGP900
TENALIN see TEO250
TENAMENE see BJL000
TENAMENE 2 see DEG200
TENAMENE 31 see BJT500
TENAMINE see TLN500
TENATHAN see BFW250
TENCILAN see CDQ250
TENDEARIL see HNI500
TENDIMETHALIN see DRN200
TENDOR see IKB000
TENDUST see NDN000
TENEBRIMYCIN see NBR500, TAL350
TENEMYCIN see NBR500, TAL350
TENESDOL see CCR875
TENFIDIL see DPJ200
TENIATHANE see MJM500
TENICID see PDM750
TENIPOSIDE see EQP000
TENITE 423 see PMP500
TENITE 800 see PJS750
TENNECETIN see PIF750
TENNECO 1742 see PKQ059
TENN-PLAS see BCL750
TENNUS 0565 see AAX175
TENORAN see CJQ000
TENORMIN see TAL475
TENOSIN-WIRKSTOFF see TOG250
TENOX BHA see BQI000
TENOX BHT see BFW750
TENOX HQ see HIH000
TENOXICAM see TAL485
TENOX PG see PNM750
TENOX P GRAIN PRESERVATIVE see PMU750
TENOX TBHQ see BRM500
TENSIBAR see DIH000
TENSIFEN see HNJ000
TENSILON see EAE600, TAL490
TENSILON BROMIDE see TAL490
TENSILON CHLORIDE see EAE600
TENSINASE D see DWF200
TENSINYL see MDQ250
TENSIVAL see TEH500
TENSOL 7 see PKB500
TENSOPAM see DCK759
TENSYL see AFG750
TENTON see MFK500
TENTONE see MFK500
TENTONE MALEATE see MFK750
TENTRATE-20 see PBC250
TENUATE see DIP600
TENUATE HYDROCHLORIDE see DIP600
TENUAZONIC ACID see VTA750
l-TENUAZONIC ACID see VTA750
TENULIN see TAL550
TENURID see DXH250
TENUTEX see DXH250
TEOBOND SURGICAL BONE CEMENT see PKB500
TEOBROMIN see TEO500
TEODRAMIN see DYE600
TEOFILCOLINA see CMF500
2-(7'-TEOFILLINMETIL)-1,3-DIOSSOLANO (ITALIAN) see TEQ175
TEOFYLLAMIN see TEP000
TEOKOLIN see CMF500
TEONANACATL see PHU500
TEONICON see PPN000
TEOPROLOL see TAL560
TEOS see EPF550
TEP see TCF280, TJT750
TEPA see TND250
TEPANIL see DIP600
TEPERINE see DLH630
TEPIDONE see SGF500
TEPIDONE RUBBER ACCELERATOR see SGF500
TEPILTA see DTL200
TEPOGEN see SLJ000
TEPP see TCF250
TERABOL see MHR200
TERALEN see AFL500
TERALIN see DPJ400
TERALLETHRIN see TAL575
TERALUTIL see HNT500

TERAMETHYL THIURAM DISULFIDE see TFS350
TERAMYCIN HYDROCHLORIDE see HOI000
TERANOL see ECX100
TERAPINYL see FMS875
TERAZOSIN see FPP100
TERBACIL see BQT750
TERBENOL see NOA000
TERBENOL HYDROCHLORIDE see NOA500
TERBENZENE see TBD000
TERBIUM see TAL750
TERBIUM CHLORIDE see TAM000
TERBIUM CITRATE see TAM500
TERBIUM OXIDE see TAN000
(−)-TERBUCLOMINE see PAP230
TERBUFOS see BSO000
TERBUMETON see BQC500
TERBUTALIN see TAN100
TERBUTALINE see TAN100
TERBUTALINE SULFATE see TAN250
TERBUTALINE SULPHATE see TAN250
TERBUTHYLAZINE see BQB000
TERCININ see CMV000
TEREBENTHINE (FRENCH) see TOD750
TEREFTALODINITRIL (CZECH) see BBP250
TEREPHTAHLIC ACID-ETHYLENE GLYCOL POLYESTER see PKF750
TEREPHTALDEHYDE see TAN500
TEREPHTALDEHYDES (FRENCH) see TAN500
TEREPHTHALALDEHYDE see TAN500
TEREPHTHALALDEHYDONITRILE see COK250
TEREPHTHALIC ACID see TAN750
TEREPHTHALIC ACID CHLORIDE see TAV250
TEREPHTHALIC ACID DICHLORIDE see TAV250
TEREPHTHALIC ACID ISOPROPYLAMIDE see IRN000
TEREPHTHALIC ACID METHYL ESTER see DUE000
TEREPHTHALIC ALDEHYDE see TAN500
TEREPHTHALIC DICHLORIDE see TAV250
TEREPHTHALONITRILE see BBP250
5,5′-(TEREPHTHALOYLBIS(IMINO-p-PHENYLENE))BIS(2,4-DIAMINO-1-ETHYLPYRIMIDINIUM)-DI-p-TOLUENESULFONATE see TAO750
5,5′-(TEREPHTHALOYLBIS(IMINO-p-PHENYLENE))BIS(2,4-DIAMINO-1-METHYLPYRIMIDINIUM)-DI-p-TOLUENESULFONATE see TAP000
5,5′-(TEREPHTHALOYLBIS(IMINO-p-PHENYLENE))BIS(2,4-DIAMINO-1-PRO-PYLPYRIMIDINIUM)-DI-p-TOLUENESULFONATE see TAP250
3,3′-(TEREPHTHALOYLBIS(IMINO-p-PHENYLENE))BIS(1-PROPYLPYRIDIN-IUM)-DI-p-TOLUENESULFONATE see TAR000
4,4′-(TEREPHTHALOYLBIS(IMINO-p-PHENYLENE))BIS(1-PROPYLPYRIDIN-IUM)-DI-p-TOLUENESULFONATE see TAR250
TEREPHTHALOYL DICHLORIDE see TAV250
3,3′-(TEREPHTHALOYLDIIMINOBIS(p-PHENYLENE))BIS(1-PROPYLPYRIDIN-IUM)-DI-p-TOLUENESULFONATE see TAR000
TERETON see MFW100
TERFAN see PKF750
TERFENADINE see TAI450
TERFLUZINE see TKE500, TKK250
TERFLUZINE DIHYDROCHLORIDE see TKK250
TERGAL see PKF750
TERGEMIST see TAV750
TERGIMIST see TAV750
TERGITOL see EMT500
TERGITOL 4 see SIO000
TERGITOL 7 see DKD400
TERGITOL 08 see TAV750
TERGITOL 12-P-9 see TAX500
TERGITOL 15-S-5 see TAY250
TERGITOL 15-S-20 see TBA500
TERGITOL 15-S-9 (nonionic) see TAZ000
TERGITOL ANIONIC 4 see EMT500
TERGITOL ANIONIC 08 see TAV750
TERGITOL ANIONIC P-28 see TBA750
TERGITOL MIN-FOAM 1X see TBA800
TERGITOL NONIONIC XD see GHY000
TERGITOL NP-14 see TAW250
TERGITOL NP-27 see TAW500
TERGITOL NP-35 (nonionic) see TAX000
TERGITOL NP-40 (nonionic) see TAX250
TERGITOL NPX see NND500
TERGITOL PENETRANT 4 see EMT500
TERGITOL PENETRANT 7 see TBB750
TERGITOL TMN-6 see GHU000
TERGITOL TMN-10 see TBB775
TERGITOL TP-9 (NONIONIC) see PKF000
TERGITOL XD (nonionic) see GHY000
TERGURID see DLR100
TERGURIDE see DLR100

TERGURIDE HYDROGEN MALEATE see DLR150
TERIAM see UVJ450
TERIDAX see IFZ800
TERIDIN see UVJ450
TERININ see CMV000
TERIT see BAR800
TERMIL see TBQ750
TERMINALIA CHEBULA RETZ TANNING see MSB750
TERMITKIL see DEP600
TERM-I-TROL see PAX250
TERMOSOLIDO RED LCG see CHP500
TERODILINE CHLORIDE see TBC200
TERODILINE HYDROCHLORIDE see TBC200
TEROFENAMATE see DGN000
TEROLUT see DYF759
TEROM see PKF750
TEROXIRONE see TBC450
Δ6,8-(9)-TERPADIENONE-2 see MCD250
TERPENE POLYCHLORINATES see TBC500
TERPENTIN OEL (GERMAN) see PIH750, TOD750
TERPHAN see PKF750
m-TERPHENYL see TBC620
o-TERPHENYL see TBC640
p-TERPHENYL see TBC750
1,3-TERPHENYL see TBC620
TERPHENYLS see TBD000
p-TERPHENYL-4-YLACETAMIDE see TBD250
α-TERPINENE (FCC) see MLA250
Γ-TERPINENE (FCC) see MCB750
TERPINENOL-4 see TBD825
TERPINEOL see TBD500
4-TERPINEOL see TBD825
α-TERPINEOL see TBD750
α-TERPINEOL (FCC) see TBD500
α-TERPINEOL ACETATE see TBE250
TERPINEOLS see TBD500
TERPINEOL SCHLECHTHIN see TBD750
TERPINOLENE see TBE000
TERPINYL ACETATE see TBE250
TERPINYL FORMATE see TBE500
TERPINYL PROPIONATE see TBE600
TERPINYL THIOCYANOACETATE see IHZ000
Δ-1,8-TERPODIENE see MCC250
TERRA ALBA see CAX750
TERRACHLOR see PAX000
TERRACHLOR-SUPER X see EFK000
TERRACOAT see EFK000
TERRACOTTA 2RN see NEY000
TERRA COTTA RRN see NEY000
TERRACUR P see FAQ800
TERRAFLO see EFK000
TERRAFUN see PAX000
TERRAFUNGINE see HOH500
TERRAMITSIN see HOH500
TERRAMYCIN see HOH500
TERRAMYCIN SODIUM see TBF000
TERRA-SYSTAM see BJE750
TERRA-SYTAM see BJE750
TERRASYTUM see BJE750
TERRAZOLE see EFK000
TERREIC ACID see TBF325
TERR-O-GAS 100 see MHR200
TERRILITIN see TBF350
TERRILYTIN see TBF350
TERSAN see TFS350
TERSAN 1991 see BAV575
TERSAN-SP see CJA100
TERSASEPTIC see HCL000
TERSAVIN see PAQ120
TERTRACID FAST BLUE SR see ADE750
TERTRACID LIGHT ORANGE G see HGC000
TERTRACID LIGHT YELLOW 2R see SGP500
TERTRACID ORANGE I see FAG010
TERTRACID PONCEAU 2R see FMU070
TERTRACID RED A see FAG020
TERTRACID RED CA see HJF500
TERTRACID YELLOW M see MDM775
TERTRAL D see PEY500
TERTRAL EG see ALT500
TERTRAL ERN see NAW500
TERTRAL P BASE see ALT250
TERTROCHROME BLUE FB see HJF500
TERTROCHROME YELLOW 3R see NEY000
TERTRODIRECT BLACK E see AQP000

TERTRODIRECT BLUE 2B see CMO000
TERTRODIRECT BLUE F see CMQ500
TERTRODIRECT BLUE FF see CMN750
TERTRODIRECT RED C see SGQ500
TERTRODIRECT VIOLET N see CMP000
TERTROGRAS ORANGE SV see PEJ500
TERTROPHENE BRILLIANT GREEN G see BAY750
TERTROPHENE BROWN CG see PEK000
TERTROPIGMENT ORANGE LRN see DVB800
TERTROPIGMENT RED HAB see MMP100
TERTROSULPHUR BLACK PB see DUZ000
TERTROSULPHUR PBR see DUZ000
TERULAN KP 2540 see ADY500
2,4,5-TES see SKL000
TESCOL see EJC500
TESERENE see DAL600
TESLEN see TBF500
TESNOL see ICC000
TESOPREL see BNU725
TESPAMINE see TFQ750
TESTAFORM see TBG000
TESTANDRONE see TBF500
TESTATE see TBF750
TESTAVOL see VSK600
TESTEX see TBG000
TESTHORMONE see MPN500
TESTICULOSTERONE see TBF500
TESTOBASE see TBF500
TESTODET see TBG000
TESTODIOL see AOO475
TESTODRIN see TBG000
TESTODRIN PROLONGATUM see TBF600
TESTOGEN see TBG000
TESTONIQUE see TBG000
TESTOPROPON see TBF500
TESTORA see MPN500
TESTORAL see AOO275
TESTORMOL see TBG000
TESTOSTEROID see TBF500
TESTOSTERONE see TBF500
trans-TESTOSTERONE see TBF500
TESTOSTERONE, CYCLOPENTANEPROPIONATE see TBF600
TESTOSTERONE CYCLOPENTYLPROPIONATE see TBF600
TESTOSTERONE CYPIONATE see TBF600
TESTOSTERONE 17-β-CYPIONATE see TBF600
TESTOSTERONE ENANTHATE see TBF750
TESTOSTERONE ENANTHATE and DEPO-MEDROXYPROGESTERONE
    ACETATE see DAM325
TESTOSTERONE ENANTHATE and DEPO-PROVERA see DAM325
TESTOSTERONE ETHANATE see TBF750
TESTOSTERONE HEPTANOATE see TBF750
TESTOSTERONE HEPTOATE see TBF750
TESTOSTERONE HEPTYLATE see TBF750
TESTOSTERONE HYDRATE see TBF500
TESTOSTERONE OENANTHATE see TBF750
TESTOSTERONE PROPIONATE see TBG000
TESTOSTERONE-17-PROPIONATE see TBG000
TESTOSTERONE-17-β-PROPIONATE see TBG000
TESTOSTERON PROPIONATE see TBG000
TESTOSTOSTERONE see TBF500
TESTOSTROVAL see TBF750
TESTOVIRON see MPN500, TBG000
TESTOVIRON SCHERING see TBF500
TESTOVIRON T see TBF500
TESTOXYL see TBG000
TESTRED see MPN500
TESTREX see TBG000
TESTRONE see TBF500
TESTRYL see TBF500
TESULOID see SOD500
TETA see TJR000
TETAINE see BAC175
TETAMON IODIDE see TCC750
TETD see DXH250
TETIDIS see DXH250
TETIOTHALEIN SODIUM see TDE750
TETLEN see PCF275
2,3,4,6-TETRA-o-ACETYL-1-THIO-β-d-GLUCOPYRANOSATO-S-
    (TRIETHYLPHOSPHINE)GOLD see ARS150
TETRAACRYLONITRILECOPPER(I) PERCHLORATE see TBG250
TETRAACRYLONITRILECOPPER(II) PERCHLORATE see TBG500
O,O,O',O'-TETRAAETHYL-BIS(DITHIOPHOSPHAT) (GERMAN) see EEH600
O,O,O,O-TETRAAETHYL-DIPHOSPHAT, BIS(O,O-DIAETHYLPHOSPHORSAE-
    URE-ANHYDRID (GERMAN) see TCF250

O,O,O,O-TETRAAETHYL-DITHIONOPYROPHOSPHAT (GERMAN) see SOD100
TETRAAMINE-2,3-BUTANEDIIMINE RUTHENIUM(III) PERCHLORATE see
    TBG600
TETRAAMINEDITHIOCYANATO COBALT(III) PERCHLORATE see TBG625
1,4,5,8-TETRAAMINO-9,10-ANTHRACENEDIONE see TBG700
1,4,5,8-TETRAAMINOANTHRAQUINONE see TBG700
3,3',4,4'-TETRAAMINOBIPHENYL see BGK500
3,3',4,4'-TETRAAMINOBIPHENYL TETRAHYDROCHLORIDE see BGK750
TETRA-(4-AMINOPHENYL)ARSONIUM CHLORIDE see TBG710
TETRAAMMINECOPPER(II) AZIDE see TBG750
TETRAAMMINECOPPER(II) NITRATE see TBH000
TETRAAMMINECOPPER(II) NITRITE see TBH250
TETRAAMMINEDICHLOROPLATINUM(II) see TBI100
TETRAAMMINEHYDROXYNITRATOPLATINUM(IV) NITRATE see TBH500
TETRAAMMINELITHIUM DIHYDROGENPHOSPHIDE see TBH750
TETRAAMMINEPALLADIUM(II) NITRATE see TBI000
TETRAAMMINEPLATINNIUM(II) DICHLORIDE (SP-4-1)- (9CI) see TBI100
TETRAAMMINEPLATINUM DICHLORIDE see TBI100
TETRAAMMINEPLATINUM(2+) DICHLORIDE see TBI100
TETRAAMMINEPLATINUM(II) DICHLORIDE see TBI100
TETRAAMMINEPLATINUM(2+) DICHLORIDE (SP-4-1)- see TBI100
TETRA-N-AMYLAMMONIUM BROMIDE see TEA250
TETRAAMYLSTANNANE see TBI600
1,3,5,7-TETRAAZAADAMANTANE see HEI500
1,4,7,10-TETRAAZADECANE see TJR000
9,10,14c-15-TETRAAZANAPHTHO(1,2,-3-fg)NAPHTHACENE NITRATE
    see TBI750
3,6,9,12-TETRAAZATETRADECANE-1,14-DIAMINE see PBD000
4a,8a,9a,10a-TETRAAZA-2,3,6,7-TETRAOXAPERHYDROANTHRACENE
    see TBI775
1,4,6,9-TETRAAZA-TRICYCLO-(4.4.1.1^{9,4}) DODECAN (GERMAN) see TBJ000
1,3,6,8-TETRAAZATRICYCLO(4.4.1.1^{3,8})DODECANE see TBJ000
TETRAAZIDO-p-BENZOQUINONE see TBJ250
TETRABAKAT see TBX250
TETRA-BASE see MJN000
TETRABENAZINE see TBJ275
TETRABENZAINE see TBJ275
TETRABENZINE see TBJ275
TETRABLET see TBX250
TETRABON see TBX000
TETRABORANE(10) see TBJ300
TETRA(BORON NITRIDE)FLUOROSULFATE see TBJ400
TETRABORON TETRACHLORIDE see TBJ475
1,1,2,2-TETRABROMAETHAN (GERMAN) see ACK250
TETRABROMIDE METHANE see CBX750
TETRABROMOACETYLENE ACK250
3,3',5,5'-TETRABROMO-2,2'-BIPHENYLDIOL see DAZ110
2,4,5,7-TETRABROMO-9-o-CARBOXYPHENYL-6-HYDROXY-3-ISOXANTHONE,
    DISODIUM SALT see BNH500
3,4,5,6-TETRABROMO-o-CRESOL see TBJ500
2,4,5,7-TETRABROMO-12,15-DICHLOROFLUORESCEIN, DIPOTASSIUM SALT
    see CMM000
2',4',5',7'-TETRABROMO-4,7-DICHLORO-FLUORESCEIN DIPOTASSIUM SALT
    see CMM000
1,1,2,2-TETRABROMOETANO (ITALIAN) see ACK250
S-TETRABROMOETHANE see ACK250
1,1,2,2-TETRABROMOETHANE see ACK250
2,4,5,7-TETRABROMO-3,6-FLUORANDIOL see BMO250, BNH500
TETRABROMOFLUORESCEIN see BMO250, BNH500
2',4',5',7'-TETRABROMOFLUORESCEIN see BMO250
2',4',5',7'-TETRABROMOFLUORESCEIN DISODIUM SALT see BNH500
TETRABROMOFLUORESCEIN S see BNH500
TETRABROMOFLUORESCEIN SOLUBLE see BNH500
2-(2,4,5,7-TETRABROMO-6-HYDROXY-3-OXO-3H-XANTHENE-9-YL)BENZOIC
    ACID, DISODIUM SALT see BNH500
TETRABROMOMETHANE see CBX750
TETRABROMO-PLATINUM(2-), DIPOTASSIUM see PLU250
TETRABROMOSILANE see SCP500
TETRABROMOSILICANE see SCP500
TETRABROMO-p-XYLEN (CZECH) see TBK250
TETRABROMO-p-XYLENE see TBK250
α,α,α',α'-TETRABROMO-m-XYLENE see TBK000
1,1,2,2-TETRABROOMETHAAN (DUTCH) see ACK250
TETRABUTYLAMMONIUM BROMIDE see TBK500
TETRA-N-BUTYLAMMONIUM BROMIDE see TBK500
TETRABUTYLAMMONIUM HYDROXIDE see TBK750
TETRABUTYLAMMONIUM IODIDE see TBL000
TETRA-N-BUTYLAMMONIUMJODID (CZECH) see TBL000
TETRABUTYLAMMONIUM NITRATE see TBL250
TETRA-n-BUTYLCIN (CZECH) see TBM250
TETRABUTYL DICHLOROSTANNOXANE see TBL500
TETRA-N-BUTYLPHOSPHONIUM BROMIDE see TBL750
TETRA-N-BUTYLPHOSPHONIUM CHLORIDE see TBM000
TETRABUTYLSTANNANE see TBM250

TETRABUTYLTHIURAM DISULPHIDE see TBM750
TETRABUTYLTIN see TBM250
TETRABUTYLTITANATE (CZECH) see BSP250
TETRABUTYLUREA see TBM850
1,1,3,3-TETRABUTYLUREA see TBM850
TETRACAINE see BQA010
TETRACAINE HYDROCHLORIDE see TBN000
TETRACAP see PCF275
TETRACAPS see TBX250
TETRACARBON MONOFLUORIDE see TBN100
TETRACARBONYLMOLYBDENUM DICHLORIDE see TBN150
TETRACARBONYL(TRIFLUOROMETHYLTHIO)MANGANESE dimer
see TBN200
TETRACEMATE DISODIUM see EIX500
TETRACEMIN see EIV000
TETRACENE see NAI000, TEF500
TETRACENE EXPLOSIVE see TEF500
2,4,4′,5-TETRACHLOOR-DIFENYL-SULFON (DUTCH) see CKM000
1,1,2,2-TETRACHLOORETHAAN (DUTCH) see TBQ100
TETRACHLOORETHEEN (DUTCH) see PCF275
TETRACHLOORKOOLSTOF (DUTCH) see CBY000
TETRACHLOORMETAAN see CBY000
1,1,2,2-TETRACHLORAETHAN (GERMAN) see TBQ100
TETRACHLORAETHEN (GERMAN) see PCF275
N-(1,1,2,2-TETRACHLORAETHYLTHIO)CYCLOHEX-4-EN-1,4-
DIACARBOXIMID (GERMAN) see CBF800
N-(1,1,2,2-TETRACHLORAETHYLTHIO)TETRAHYDROPHTHALAMID
(GERMAN) see CBF800
TETRACHLORDIAN (CZECH) see TBO750
3,4,6,R′-TETRACHLOR-DIPHENYLSULFID (GERMAN) see CKL750
2,4,4′,5-TETRACHLOR-DIPHENYL-SULFON (GERMAN) see CKM000
TETRACHLORETHANE see TBQ100
1,1,2,2-TETRACHLORETHANE (FRENCH) see TBQ100
TETRACHLORKOHLENSTOFF, TETRA (GERMAN) see CBY000
TETRACHLORMETHAN (GERMAN) see CBY000
2,3,5,6-TETRACHLOR-3-NITROBENZOL (GERMAN) see TBR750
TETRACHLOROACETONE see TBN250
1,1,3,3-TETRACHLOROACETONE see TBN300
N,N,N′,N′-TETRACHLOROADIPAMIDE see TBN400
3,3′,4,4′-TETRACHLOROAZOBENZENE see TBN500
3,4,3′,4′-TETRACHLOROAZOBENZENE see TBN500
3,3′,4,4′-TETRACHLOROAZOXYBENZENE see TBN550
3,4,3′,4′-TETRACHLOROAZOXYBENZENE see TBN550
1,2,3,4-TETRACHLOROBENZENE see TBN740
1,2,3,5-TETRACHLOROBENZENE see TBO700
1,2,4,5-TETRACHLOROBENZENE see TBN750
2,3,5,6-TETRACHLORO-1,4-BENZENEDICARBOXYLIC ACID, DIMETHYL
ESTER see TBV250
3,4,5,6-TETRACHLORO-1,2-BENZENEDIOL see TBU500
TETRACHLOROBENZIDINE see TBO000
2,2′,5,5′-TETRACHLOROBENZIDINE see TBO000
3,3′,6,6′-TETRACHLOROBENZIDINE see TBO000
TETRACHLOROBENZOQUINONE see TBO500
TETRACHLORO-p-BENZOQUINONE see TBO500
TETRACHLORO-1,4-BENZOQUINONE see TBO500
2,3,5,6-TETRACHLORO-p-BENZOQUINONE see TBO500
2,3,5,6-TETRACHLORO-1,4-BENZOQUINONE see TBO500
3,3′,4,4′-TETRACHLOROBIPHENYL see TBO700
3,4,3′,4′-TETRACHLOROBIPHENYL see TBO700
2,2′,5,5′-TETRACHLORO-(1,1′-BIPHENYL)-4,4′-DIAMINE, (9CI)
see TBO000
2,2′,6,6′-TETRACHLOROBISPHENOL A see TBO750
1,1,2,3-TETRACHLORO-1,3-BUTADIENE see TBO760
1,1,4,4-TETRACHLOROBUTATRIENE see TBO765
TETRACHLOROCARBON see CBY000
9-(3′,4′,5′,6′-TETRACHLORO-o-CARBOXYPHENYL)-6-HYDROXY-2,4,5,7-
TETRAIODO-3-ISOXANTHONE●2Na see RMP175
TETRACHLOROCATECHOL see TBU500
2,4,5,6-TETRACHLORO-3-CYANOBENZONITRILE see TBQ750
2,3,4,5-TETRACHLORO-2-CYCLOBUTEN-1-ONE see PCF250
2,3,5,6-TETRACHLORO-2,5-CYCLOHEXADIENE-1,4-DIONE see TBO500
2,2′,5,5′-TETRACHLORO-4,4′-DIAMINODIPHENYL see TBO000
cis-TETRACHLORODIAMMINE PLATINUM(IV) see TBO776
trans-TETRACHLORODIAMMINE PLATINUM(IV) see TBO777
TETRACHLORODIAZOCYCLOPENTADIENE see TBO778
2,3,7,8-TETRACHLORODIBENZO(b,e)(1,4)DIOXAN see TAI000
1,2,3,8-TETRACHLORODIBENZO-p-DIOXIN see CDV125
2,3,6,7-TETRACHLORODIBENZO-p-DIOXIN see TAI000
2,3,7,8-TETRACHLORODIBENZO-p-DIOXIN see TAI000
2,3,7,8-TETRACHLORODIBENZO-1,4-DIOXIN see TAI000
2,3,7,8-TETRACHLORODIBENZOFURAN see TBO780
1,1,1,2-TETRACHLORO-2,2-DIFLUOROETHANE see TBP000
1,1,2,2-TETRACHLORO-1,2-DIFLUOROETHANE see TBP050
3,3,4,5-TETRACHLORO-3,6-DIHYDRO-1,2-DIOXIN see TBP075

4,5,6,7-TETRACHLORO-2-(2-DIMETHYLAMINOETHYL)-ISOINDOLINE
DIMETHOCHLORIDE see CDY000
TETRACHLORODIPHENYLETHANE see BIM500
TETRACHLORODIPHENYL OXIDE see TBP250
2,4,4′,5-TETRACHLORODIPHENYL SULFIDE see CKL750
2,4,5,4′-TETRACHLORODIPHENYL SULFIDE see CKL750
2,4,4′,5-TETRACHLORODIPHENYL SULFONE see CKM000
2,4,5,4′-TETRACHLORODIPHENYLSULPHONE see CKM000
TETRACHLORODIPHOSPHANE see TBP500
TETRACHLOROEPOXYETHANE see TBQ275
TETRACHLOROETHANE see TBP750
sym-TETRACHLOROETHANE see TBQ100
1,1,1,2-TETRACHLOROETHANE see TBQ000
1,1,2,2-TETRACHLOROETHANE see TBQ100
TETRACHLOROETHENE see PCF275
TETRACHLOROETHYLENE (DOT) see PCF275
1,1,2,2-TETRACHLOROETHYLENE see PCF275
TETRACHLOROETHYLENE CARBONATE see TBQ255
TETRACHLOROETHYLENE OXIDE see TBQ275
N-1,1,2,2-TETRACHLOROETHYLMERCAPTO-4-CYCLOHEXENE-1,2-CAR-
BOXIMIDE see CBF800
N-((1,1,2,2-TETRACHLOROETHYL)SULFENYL)-cis-4-CYCLOHEXENE-1,2-
DICARBOXIMIDE see CBF800
N-(1,1,2,2-TETRACHLOROETHYLTHIO)-4-CYCLOHEXENE-1,2-DICARBOXIM-
IDE see CBF800
2,3,4,5-TETRACHLOROHEXATRIENE see TBQ300
TETRACHLOROHYDROQUINONE see TBQ500
TETRACHLOROISOPHTHALONITRILE see TBQ750
TETRACHLOROMETHANE see CBY000
1,2,4,5-TETRACHLORO-3-METHOXY-6-NITROBENZENE (9CI) see TBR250
TETRACHLORONAPHTHALENE see TBR000
TETRACHLORONITROANISOLE see TBR250
2,3,5,6-TETRACHLORO-4-NITROANISOLE see TBR250
2,3,4,5-TETRACHLORONITROBENZENE see TBS000
2,3,4,6-TETRACHLORONITROBENZENE see TBR500
2,3,5,6-TETRACHLORONITROBENZENE see TBR750
1,2,3,4-TETRACHLORO-5-NITROBENZENE see TBS000
1,2,3,5-TETRACHLORO-4-NITROBENZENE see TBR500
1,2,4,5-TETRACHLORO-3-NITROBENZENE see TBR750
TETRACHLOROPALLADATE(2-) DIPOTASSIUM see PLN750
TETRACHLOROPHENOL see TBS250
2,3,4,5-TETRACHLOROPHENOL see TBS500
2,3,4,6-TETRACHLOROPHENOL see TBT000
2,3,5,6-TETRACHLOROPHENOL see TBS750
2,4,5,6-TETRACHLOROPHENOL see TBT000
TETRACHLOROPHENYL ETHER see TBP250
o-TETRACHLOROPHTHALODINITRILE see TBT200
TETRACHLOROPHTHALONITRILE see TBT200
m-TETRACHLOROPHTHALONITRILE see TBQ750
3,4,5,6-TETRACHLOROPHTHALONITRILE see TBT200
1,1,1,3-TETRACHLOROPROPANE see TBT250
1,1,3,3-TETRACHLOROPROPANONE see TBN300
1,1,3,3-TETRACHLORO-2-PROPANONE see TBN300
1,1,2,3-TETRACHLOROPROPENE see TBT500
2,3,4,5-TETRACHLOROPYRIDINE see TBT750
2,3,5,6-TETRACHLOROPYRIDINE see TBU000
TETRACHLOROPYRIMIDINE see TBU250
2,4,5,6-TETRACHLOROPYRIMIDINE see TBU250
TETRACHLOROPYROCATECHOL see TBU500
TETRACHLOROPYROCATECHOL HYDRATE see TBU750
TETRACHLOROQUINONE see TBO500
TETRACHLORO-p-QUINONE see TBO500
5,6,7,8-TETRACHLOROQUINOXALINE see CMA500
3,3′,4′,5-TETRACHLOROSALICYLANILIDE see TBV000
TETRACHLOROSILANE see SCQ500
TETRACHLOROTELLURIUM see TAJ250
TETRACHLOROTEREPHTHALIC ACID DIMETHYL ESTER see TBV250
4,5,6,7-TETRACHLORO-2′,4′,5′,7′-TETRAIODOFLUORESCEIN DISODIUM
SALT see RMP175
TETRACHLOROTHIOFENE see TBV750
TETRACHLOROTHIOPHENE see TBV750, TBV750
2,3,4,5-TETRACHLOROTHIOPHENE see TBV750
TETRACHLOROTHORIUM see TFT000
4,5,6,7-TETRACHLORO-2-(TRIFLUOROMETHYL)BENZIMIDAZOLE
see TBW000
TETRACHLOROTRIFLUOROMETHYLPHOSPHORANE see TBW025
5-endo,6-exo-2,2,5,6-TETRACHLORO-1,7,7-TRIS(CHLOROMETHYL)-
BICYCLO(2.2.1) HEPTANE see THH575
2,3,5,6-TETRACHLORPHTHALSAURE-DIMETHYLESTER (GERMAN)
see TBV250
TETRACHLORPYROKATECHIN (CZECH) see TBU750
TETRACHLORURE d'ACETYLENE (FRENCH) see TBQ100
TETRACHLORURE de CARBONE (FRENCH) see CBY000
TETRACHLORURE de SILICIUM (FRENCH) see SCQ500

TETRACHLORURE de TITANE (FRENCH) see TGH350
TETRACHLORVINPHOS see TBW100
TETRACICLINA CLORIDRATO (ITALIAN) see TBX250
TETRACICLINA-l-METILENLISINA (ITALIAN) see MRV250
2,4,4',5-TETRACLORO-DIFENIL-SOLFONE (ITALIAN) see CKM000
1,1,2,2-TETRACLOROETANO (ITALIAN) see TBQ100
TETRACLOROETENE (ITALIAN) see PCF275
TETRACLOROMETANO (ITALIAN) see CBY000
TETRACLORURO di CARBONIO (ITALIAN) see CBY000
TETRACOMPREN see TBX250
TETRACOSAHYDRO DIBENZ(b,n)(1,4,7,10,13,16,19,22)OC-
    TAOXACYCLOTETRACOSIN see DGT300
(1,1,3,3-TETRACYANOALLYL) SODIUM see TAI050
TETRACYANOETHYLENE see EEE500
N,N,N',N'-TETRACYANOMETHYLAETHYLENEDIAMIN (GERMAN) see EJB000
TETRACYANONICKELATE(2-) DIPOTASSIUM, HYDRATE see TBW250
TETRACYANOOCTAETHYLTETRAGOLD see TBW500
TETRACYANOQUINODIMETHAN see TBW750
TETRACYCLINE see TBX000
TETRACYCLINE CHLORIDE see TBX250
TETRACYCLINE HYDROCHLORIDE see TBX250
TETRACYCLINE I see TBX000
TETRACYCLINE-l-METHYLENE LYSINE see MRV250
TETRACYDIN see ABG750
TETRACYN see TBX000
TETRA-D see TBX250
1-TETRADECANAL see TBX500
TETRADECANE see TBX750
TETRADECANOIC ACID see MSA250
TETRADECANOIC ACID, ISOPROPYL see IQN000
TETRADECANOL, mixed isomers see TBY500
1-TETRADECANOL see TBY250
N-TETRADECANOL-1 see TBY250
TETRADECANOL condensed with 7 moles ETHYLENE OXIDE see TBY750
2-TETRADECANOL, 1-(2-HYDROXYETHYLAMINO)-, and 1-(2-
    HYDROXYETHYL)-2-DODECANOL (1:1) see HKS400
TETRADECANOYLETHYLENEIMINE see MSB000
12-TETRADECANOYLPHORBOL-13-ACETATE see PGV000
N-TETRADECANOYL-N-TETRADECANOYLOXY-2-AMINOFLUORENE
    see MSB250
12-o-TETRADECA-2-cis-4-trans,6,8-TETRAENOYLPHORBOL-13-ACETATE
    see TCA250
TETRADECIN see TBX000
n-TETRADECOIC ACID see MSA250
TETRADECYL ALCOHOL see TBY250, TBY500
N-TETRADECYL ALCOHOL see TBY250
1-TETRADECYL ALDEHYDE see TBX500
TETRADECYL DIMETHYL BENZYLAMMONIUM CHLORIDE see TCA500
TETRADECYLHEPTAETHOYLATE see TCA750
TETRADECYL OCTADECANOATE see TCB100
TETRADECYL PHOSPHONIC ACID see TCB000
TETRADECYL SODIUM SULFATE see SIO000
TETRADECYL STEARATE see TCB100
TETRADECYL SULFATE, SODIUM SALT see SIO000
1,2,6,7-TETRADEHYDRO-14,17-DIHYDRO-3-METHOXY-16(15H)-OX- AER-
    YTHRINAN-15-ONE, HYDROCHLORIDE see EDH000
7,8,13,13A-TETRADEHYDRO-9,10-DIMETHOXY-2,3-
    (METHYLENEDIOXY)BERBINIUM SULFATE TRIHYDRATE see BFN750
TETRADEHYDRODOISYNOLIC ACID METHYL ETHER see BIT000
6,7,8,14-TETRADEHYDRO-4,5-α-EPOXY-3,6-DIMETHOXY-17-
    METHYLMORPHINAN see TEN000
6,7,8,14-TETRADEHYDRO-4,5-α-EPOXY-3,6-DIMETHOXY-17-
    METHYLMORPHINAN HYDROCHLORIDE see TEN100
4,5,6,6a-TETRADEHYDRO-9-HYDROXY-1,2,10-TRIMETHOXYNORAPORPHIN-
    7-ONE see ARQ325
(3-β)-1,2,6,7-TETRADEHYDRO-3-METHOXYERYTHRINAN-16-OL HYDRO-
    CHLORIDE see EDE500
(3-β)-1,2,6,7-TETRADEHYDRO-3-METHOXY-15,16-
    (METHYLENEBIS(OXY))ERYTHRINAN, HYDROBROMIDE see EDF000
7,8,13,13a-TETRADEHYDRO-2,3,9,10-TETRAMETHOXYBERBINIUM HYDROX-
    IDE see PAE100
12-o-TETRADEKANOYLPHORBOL-13-ACETAT (GERMAN) see PGV000
2,3,4,6-TETRADEOXY-4-((5,6,11,12,13,14,15,16,16A,17,17A,18,21,22-
    TETRADECAHYDRO-2,12,14,16-TETRAHYDROXY-10-METHOXY-
    3,6,11,13,15,20-HEXAMETHYL-5,18,21,26-TETRAOXO-4,6-EPOXY-1,23-
    METHANOBANZO(d)CYCLOPROP(n)(1,9)OXAAZACYCLOTETRACOSIN-25(10
    H)-YLIDENE)AMINO)-l-erythro-HEXOPYRANOSE, 12-ACETATE see THD300
TETRADICHLONE see CKM000
TETRADIFON see CKM000
TETRADIN see DXH250
TETRADINE see DXH250
TETRADIOXIN see TAI000
TETRADIPHON see CKM000
TETRADOX see HGP550

TETRAETHANOL AMMONIUM HYDROXIDE see TCB500
TETRAETHOXY PROPANE see MAN750
1,1,3,3-TETRAETHOXYPROPANE see MAN750
TETRAETHOXYSILANE see EPF550
TETRAETHYLAMMONIUM see TCB725
TETRAETHYLAMMONIUM BOROHYDRIDE see TCB750
TETRAETHYL AMMONIUM BROMIDE see TCC000
TETRAETHYL AMMONIUM CHLORIDE see TCC250
TETRAETHYLAMMONIUM HYDROXIDE see TCC500
TETRAETHYLAMMONIUM IODIDE see TCC750
TETRAETHYLAMMONIUM PERCHLORATE see TCD000
TETRA(2-ETHYLBUTOXY) SILANE see TCD250
TETRA(2-ETHYLBUTYL) ORTHOSILICATE see TCD250
TETRAETHYLDIAMINO-o-CARBOXY-PHENYL-XANTHENYL CHLORIDE
    see FAG070
TETRAETHYLDIARSANE see TCD500
O,O,O,O-TETRAETHYL-DIFOSFAAT (DUTCH) see TCF250
O,O,O,O-TETRAETHYL-DITHIO-DIFOSFAAT (DUTCH) see SOD100
TETRAETHYL DITHIONOPYROPHOSPHATE see SOD100
TETRAETHYL DITHIOPYROPHOSPHATE see SOD100
O,O,O,O-TETRAETHYL DITHIOPYROPHOSPHATE see SOD100
TETRAETHYL DITHIO PYROPHOSPHATE, liquid (DOT) see SOD100
TETRAETHYLENE GLYCOL see TCE250
TETRAETHYLENE GLYCOL DIACRYLATE see ADT050
TETRAETHYLENE GLYCOL, DIBUTYL ETHER see TCE350
TETRAETHYLENE GLYCOL DIETHYL ETHER see PBO250
TETRAETHYLENE GLYCOL DI(2-ETHYLHEXOATE) see FCD512
TETRAETHYLENE GLYCOL-DI-n-HEPTANOATE see TCE375
TETRAETHYLENE GLYCOL DIMETHYL ETHER see PBO500
TETRAETHYLENEIMIDEPIPERAZINE-N,N-DIPHOSPHORIC ACID see BJC250
2,3,5,6-TETRA-ETHYLENEIMINO-1,4-BENZOQUINONE see TDG500
TETRAETHYLENEPENTAMINE see TCE500
TETRAETHYL GERMANE see TCE750
TETRAETHYL GERMANIUM see TCE750
TETRA(2-ETHYLHEXOXY)SILANE see EKQ500
TETRA(2-ETHYLHEXYL)ORTHOSILICATE see EKQ500
TETRA(2-ETHYLHEXYL)SILICATE see EKQ500
TETRAETHYL LEAD see TCF000
O,O,O',O'-TETRAETHYL S,S'-METHYLENEBISPHOSPHORDITHIOATE
    see EEH600
O,O,O',O'-TETRAETHYL-S,S'-METHYLENEBISPHOSPHORODITHIOATE
    see EEH600
TETRAETHYL S,S'-METHYLENE BIS(PHOSPHOROTHIOLOTHIONATE)
    see EEH600
O,O,O',O'-TETRAETHYL S,S'-METHYLENE DI(PHOSPHORODITHIOATE)
    see EEH600
TETRAETHYL ORTHOSILICATE see EPF550
TETRAETHYL ORTHOSILICATE (DOT) see EPF550
TETRAETHYLPLUMBANE see TCF000
TETRAETHYL PYROFOSFAAT (BELGIAN) see TCF250
TETRAETHYL PYROPHOSPHATE see TCF250
TETRAETHYLPYROPHOSPHATE and compressed gas mixtures see TCF260
TETRAETHYL PYROPHOSPHATE, liquid (DOT) see TCF250
TETRAETHYLPYROPHOSPHATE MIXTURE (dry) see TCF270
TETRAETHYL PYROPHOSPHATE MIXTURE (liquid) see TCF280
TETRAETHYLRHODAMINE see FAG070
TETRAETHYL SILICATE see EPF550
TETRAETHYL SILICATE (DOT) see EPF550
TETRAETHYLSTANNANE see TCF750
4,4,8,8-TETRAETHYL-3,3a,4,8-TETRAHYDRO-3a,4a,4-DIAZABORA-5-
    INDACENE see MRW275
TETRAETHYLTHIOPEROXYDICARBONIC DIAMIDE see DXH250
TETRAETHYLTHIRAM DISULPHIDE see DXH250
TETRAETHYLTHIURAM see DXH250
TETRAETHYLTHIURAM DISULFIDE see DXH250
TETRAETHYLTHIURAM DISULPHIDE see DXH250
N,N,N',N'-TETRAETHYLTHIURAM DISULPHIDE see DXH250
TETRAETHYL TIN see TCF750
TETRAETHYNYLGERMANIUM see TCG000
TETRAETHYNYLTIN see TCG250
TETRAETIL see DXH250
TETRAETILAMMONIO (ITALIAN) see TCB725
O,O,O-TETRAETIL-DITIO-PIROFOSFATO (ITALIAN) see SOD100
O,O,O,O-TETRAETIL-PIROFOSFATO (ITALIAN) see TCF250
TETRAFIDON see CKM000
TETRAFINOL see CBY000
TETRAFLUORETHYLENE see TCH500
TETRAFLUOROAMMONIUM PERBROMATE see TCG450
TETRAFLUORO AMMONIUM TETRAFLUOROBORATE see TCG500
3,3',5,5'-TETRAFLUOROBENZIDINE see TCG750
3,3',5,5'-TETRAFLUORO(1,1'-BIPHENYL)-4,4'-DIAMINE see TCG750
TETRAFLUOROBORATE(1-) compound with p-AMINOBENZOIC ACID 2-
    (DIETHYLAMINO)ETHYL ESTER see TCH000
TETRAFLUORO BORATE(1-) LEAD (2+) see LDE000

TETRAMETHYLENIMINE see PPS500
TETRAMETHYL ETHYLENE DIAMINE see TDQ750
N,N,N',N'-TETRAMETHYLETHYLENEDIAMINE see TDQ750
TETRAMETHYLETHYLENE GLYCOL see TDR000
TETRAMETHYLETHYLFORMYLTETRALIN see FNK200
3',6',10,11b-TETRAMETHYL-1'a,8,11,11a,11b-HEXADECAHYDRO-SPIRO(9H-
    BENZO(a)FLUORENE-9,2'(3'H)-FURO(3,2-b)PYRIDIN)-3(1H)-ONE see DAQ125
3,7,11,15-TETRAMETHYL-2-HEXADECEN-1-OL see PIB600
N,N,N',N'-TETRAMETHYLHEXAMETHYLENEDIAMINE-1,3-DIBROMOPROP-
    ANE copolymer see HCV500
TETRAMETHYLHYDRAZINE HYDROCHLORIDE see TDR250
TETRAMETHYL LEAD see TDR500
N,N,N'N'-TETRAMETHYLMETHANEDIAMINE see TDR750
TETRAMETHYL METHYLENE DIAMINE (DOT) see TDR750
2,2,6,6-TETRAMETHYLNITROSOPIPERIDINE see TDS000
N,2,3,3-TETRAMETHYL-2-NORBORNAMINE see VIZ400
N,2,3,3-TETRAMETHYL-2-NORBORNANAMINE see TDS250
N,2,3,3-TETRAMETHYL-2-NORBORNANAMINE HYDROCHLORIDE
    see MQR500
N,2,3,3-TETRAMETHYL-2-NORCAMPHANAMINE see VIZ400
1,(5,5,7,7-TETRAMETHYL-2-OCTANYL)-2-METHYL-5-ETHYLPYRIDINIUM
    CHLORIDE see TDS500
p-1',1',4',4'-TETRAMETHYLOKTYLBENZENSULFONAN SODNY (CZECH)
    see DXW200
TETRAMETHYLOLMETHANE see PBB750
2,6,10,14-TETRAMETHYLPENTADECANE see PMD500
1582583584-TETRAMETHYLPHENANTHRENE see TDS750
N,N,N-α-TETRAMETHYL-10H-PHENOTHIAZINE-10-ETHANAMINIUM
    METHYL SULFATE see MRW000
2-(2,3,5,6-TETRAMETHYLPHENOXY)PROPIONIC AICD see BHM000
TETRAMETHYL-p-PHENYLENEDIAMINE see BJF500
N,N,N',N'-TETRAMETHYL-o-PHENYLENEDIAMINE see TDT000
N,N,N',N'-TETRAMETHYL-p-PHENYLENEDIAMINE see BJF500
N,N,N',N'-TETRAMETHYL-p-PHENYLENEDIAMINE DIHYDROCHLORIDE
    see TDT250
(2,3,5,6-TETRAMETHYLPHENYL)MERCURY ACETATE see AAS500
TETRAMETHYLPHOSPHONIUM IODIDE see TDT500
TETRAMETHYLPHOSPHORODIAMIDIC FLUORIDE see BJE750
N,N,N,N-TETRAMETHYLPHOSPHORODIAMIDIC FLUORIDE see BJE750
2,2,6,6-TETRAMETHYLPIPERIDINE see TDT750
2,2,6,6-TETRAMETHYL-4-PIPERIDONE OXIME see TDU000
TETRAMETHYLPLATINUM see TDU250
TETRAMETHYLPLUMBANE see TDR500
N,N,N',N'-TETRAMETHYL-1,3-PROPANEDIAMINE see TDU500
N,N,N',N'-TETRAMETHYL PROPENE-1,3-DIAMINE see TDU750
N,N,N',N'-TETRAMETHYL-1,3-PROPENYLDIAMINE see TDU750
TETRAMETHYLPYRAZINE see TDV725
2,3,5,6-TETRAMETHYL PYRAZINE (FCC) see TDV725
N,N,2,3-TETRAMETHYL-4-(4'-(PYRIDYL-1'-OXIDE)AZO)ANILINE see DQF200
N,N,2,5-TETRAMETHYL-4-(4'-(PYRIDYL-1'-OXIDE)AZO)ANILINE see DQF400
N,N,2,6-TETRAMETHYL-4-(4'-(PYRIDYL-1'-OXIDE)AZO)ANILINE see DQF600
TETRAMETHYL PYRONIN see PPQ750
TETRAMETHYLPYROPHOSPHATE see TDV000
2,2,5,5-TETRAMETHYLPYRROLIDINE see TDV250
2,2,6,6-TETRAMETHYLQUINUCLIDINE HYDROBROMIDE see TAL275
TETRAMETHYLSILANE see TDV500
TETRAMETHYLSILICATE see MPI750
TETRAMETHYLSTANNANE see TDV750
TETRAMETHYLSTIBONIUM IODIDE see TDW000
TETRAMETHYLSUCCINONITRILE see TDW250
TETRA-o-METHYL-SULPHONYL-d-MANNITOL see MAW800
1,1,4,4-TETRAMETHYL-1,2,3,4-TETRAHYDRO-6-ETHYLNAPHTHALENE
    see TCR250
3,3,6,6-TETRAMETHYL-1,2,4,5-TETRAOXANE see TDW275
TETRAMETHYL-2-TETRAZENE see TDW300
2,4,6,8-TETRAMETHYL-1,3,5,7-TETROXOCANE see TDW500
TETRAMETHYLTHIOCARBAMOYLDISULPHIDE see TFS350
TETRAMETHYL-O,O'-THIODI-p-PHENYLENE PHOSPHOROTHIOATE
    see TAL250
O,O,O'O'-TETRAMETHYL-O,O'-THIODI-p-PHENYLENE PHOSPHOROTHIO-
    ATE see TAL250
TETRAMETHYLTHIONINE CHLORIDE see BJI250
TETRAMETHYLTHIORAMDISULFIDE (DUTCH) see TFS350
TETRAMETHYLTHIOUREA see TDX000
1,1,3,3-TETRAMETHYLTHIOUREA see TDX000
TETRAMETHYL-THIRAM DISULFID (GERMAN) see TFS350
TETRAMETHYLTHIURAM BISULFIDE see TFS350
TETRAMETHYLTHIURAM DISULFIDE see TFS350
N,N,N',N'-TETRAMETHYLTHIURAM DISULFIDE see TFS350
TETRAMETHYLTHIURAM DISULFIDE mixed with FERRIC
    NITROSODIMETHYLDITHIOCARBAMATE see FAZ000
N,N-TETRAMETHYLTHIURAM DISULPHIDE see TFS350
TETRAMETHYLTHIURAMMONIUM SULFIDE see BJL600
TETRAMETHYLTHIURAM MONOSULFIDE see BJL600

TETRAMETHYLTHIURAMONOSULFIDE see BJL600
TETRAMETHYLTHIURAM SULFIDE see BJL600
TETRAMETHYL THIURANE DISULFIDE see TFS350
TETRAMETHYL TIN see TDV750
(−)-2,2,5,5-TETRAMETHYL-α-((o-TOLYLOXY)METHYL)-1-
    PYRROLIDINEETHANOL HYDROCHLORIDE see TGJ625
TETRAMETHYLTRITHIO CARBAMIC ANHYDRIDE see BJL600
N,N',o,o-TETRAMETHYL-(+)-TUBOCURINE see DUL800
TETRAMETHYLUREA see TDX250
1,1,3,3-TETRAMETHYLUREA see TDX250
TETRAMETHYLUREE (FRENCH) see TDX250
N,N,N',N'-TETRAMETIL-FOSFORODIAMMIDO-FLUORURO (ITALIAN)
    see BJE750
TETRAMIN see VMA000
TETRAMINE see HOI000, TDX500
TETRAMINE FAST BROWN BRS see CMO750
TETRAMINE PLATINUM(II) CHLORIDE see ANV800
1,4,5,8-TETRAMINOANTHRAQUINONE see TBG700
TETRAMISOLE HYDROCHLORIDE see TDX750
l-TETRAMISOLE HYDROCHLORIDE see LFA020
dl-TETRAMISOLE HYDROCHLORIDE see TDX750
TETRAMISOL HYDROCHLORIDE see TDX750
TETRAM MONOOXALATE see AMX825
TETRAMON see TCB725
TETRAN see HOH500
TETRANAP see TCX500
TETRANATRIUMPYROPHOSPHAT (GERMAN) see TEE500
TETRANDRINE see TDX830
(+)-TETRANDRINE see TDX830
d-TETRANDRINE see TDX830
TETRANDRINE DIMETHIODIDE see TDX835
TETRANGOMYCIN see TDX840
TETRAN HYDROCHLORIDE see HOI000
TETRANICOTINIC ACID-2-HYDROXYCYCLOHEXA-1,1,3,3-TETRAMETHYL
    ESTER see CME675
TETRANICOTYLFRUCTOFURANOSE see TDX860
TETRANICOTYLFRUCTOSE see TDX860
TETRANITRANILINE (FRENCH) see TDY000
TETRANITROANILINE see TDY000
2,3,4,6-TETRANITROANILINE see TDY000
N,2,3,5-TETRANITROANILINE see TDY050
N,2,4,6-TETRANITROANILINE see TDY075
TETRANITROMETHANE see TDY250
N,2,4,5-TETRANITRO-N-METHYLANILINE see TEG250
1,3,6,8-TETRANITRO NAPHTHALENE see TDY500
TETRANITROPENTAERYTHRITE see PBC250
1,3,5,7-TETRANITROPERHYDRO-1,3,5,7-TETRAZOCINE see CQH250
2,3,4,6-TETRANITROPHENOL see TDY600
trans-1,4,5,8-TETRANITRO-1,4,5,8-TETRAAZADECAHYDRONAPHTHALENE
    see TDY775
TETRA OLIVE N2G see APG500
2,5,7,8-TETRAOXA[4.2.0]BICYCLOOCTANE see DVQ759
1,4,7,10-TETRAOXACYCLODODECANE see COD475
2,5,8,11-TETRAOXADODECANE see TKL875
2,2'-(2,5,8,11-TETRAOXA-1,2-DODECANEDIYL)BISOXIRANE see TJQ333
2,2'-(2,5,8,11-TETRAOXA-1,12-DODECANE DIYL)BISOXIRANE see TJQ333
2,4,10,12-TETRAOXA-6,16,17,18-TETRAAZA-3,11-DISTIBATRICYCLO
    (11.3.1.1^{5,9})OCTADECA-1(17),5,7,9(18),13,15-HEXAENE, 3,11-DIHYDROXY-
    see AQE305
2,4,10,12-TETRAOXA-6,16,17,18-TETRAAZA-3,11-DISTIBATRICYCLO
    (11.3.1.1^{5,9})OCTADECA-1(17),5,7,9(18),13,15-HEXAENE-8,14-
    DIMETHANOL, 3,11-DIHYDROXY- see AQE300
(±)-(3,5,3',5'-TETRAOXO)-1,2-DIPIPERAZINOPROPANE see PIK250
2,4,5,6-TETRAOXOHEXAHYDROPYRIMIDINE see AFT750
2,4,5,6-TETRAOXOHEXAHYDROPYRIMIDINE HYDRATE see MDL500
TETRAPENTYLAMMONIUM BROMIDE see TEA250
TETRA-N-PENTYLAMMONIUM BROMIDE see TEA250
TETRAPENTYLSTANNANE see TBI600
TETRAPHAN see TEF775
TETRAPHENE see BBC250
TETRAPHENYLARSENIUM CHLORIDE see TEA300
TETRAPHENYLARSONIUM CHLORIDE see TEA300
1,1,4,4-TETRAPHENYL-1,4-BUTANEDIOL see TEA500
TETRAPHENYL METHANE see TEA750
TETRAPHOSPHATE HEXAETHYLIQUE (FRENCH) see HCY000
TETRAPHOSPHOR (GERMAN) see PHO740
TETRAPHOSPHORIC ACID, HEXAETHYL ESTER, mixed with compressed gas
    see TEB500
TETRAPHOSPHORIC ACID, HEXAETHYL ESTER (dry mixture) see TEB000
TETRAPHOSPHORIC ACID, HEXAETHYL ESTER (liquid mixture) see TEB250
TETRAPHOSPHORUS IODIDE see TEB750
TETRAPHOSPHORUS TRISELENIDE see TEC000
TETRAPHOSPHORUS TRISULFIDE see PHS500

TETRAPOM see TFS350
TETRAPOTASSIUM ETIDRONATE see TEC250
TETRAPOTASSIUM FERROCYANIDE see TEC500
TETRAPOTASSIUM HEXACYANOFERRATE see TEC500
TETRAPOTASSIUM HEXACYANOFERRATE(4-) see TEC500
TETRAPOTASSIUM HEXACYANOFERRATE(II) see TEC500
TETRAPOTASSIUM PYROPHOSPHATE see PLR200
TETRAPROPYLAMMONIUM BROMIDE see TEC750
TETRAPROPYLAMMONIUM HYDROXIDE see TED000
TETRA-N-PROPYLAMMONIUM HYDROXIDE see TED000
TETRAPROPYLAMMONIUM IODIDE see TED250
TETRA-N-PROPYLAMMONIUM IODIDE see TED250
TETRAPROPYLAMMONIUM OXIDE see TED000
TETRA-n-PROPYL DITHIONOPYROPHOSPHATE see TED500
TETRA-n-PROPYL DITHIOPYROPHOSPHATE see TED500
O,O,O,O-TETRAPROPYL DITHIOPYROPHOSPHATE see TED500
TETRAPROPYLENE see PMP750
TETRAPROPYL LEAD see TED750
TETRASELENIUM TETRANITRIDE see TEE000
TETRASILVER DIIMIDODIOXOSULFATE see SDP500
TETRASILVER DIIMIDOTRIPHOSPHATE see TEE100
TETRASILVER ORTHODIAMIDOPHOSPHATE see TEE125
TETRASIPTON see TFS350
TETRASODIUM BIS(CITRATE(3-)FERRATE(4-)) see TEE225
TETRASODIUM BISCITRATO FERRATE see TEE225
TETRASODIUM DIETHYLSTILBESTROL PHOSPHATE see TEE300
TETRASODIUM DIPHOSPHATE see TEE500
TETRASODIUM EDTA see EIV000
TETRASODIUM ETHYLENEDIAMINETETRAACETATE see EIV000
TETRASODIUM ETHYLENEDIAMINETETRACETATE see EIV000
TETRASODIUM (ETHYLENEDINITRILO)TETRAACETATE see EIV000
TETRASODIUM ETIDRONATE see TEE250
TETRASODIUM FOSFESTROL see TEE300
TETRASODIUM PYROPHOSPHATE see TEE500
TETRASODIUM PYROPHOSPHATE, ANHYDROUS see TEE500
TETRASODIUM SALT of EDTA see EIV000
TETRASODIUM SALT of ETHYLENEDIAMINETETRACETICACID see EIV000
TETRASOL see CBY000
TETRASTIGMINE see TCF250
TETRASUL see CKL750
TETRASULE see PBC250
TETRASULFIDE, BIS(ETHOXYTHIOCARBONYL) see BJU250
TETRASULFUR DINITRIDE see TEE750
TETRASULFUR TETRANITRIDE see SOO000
TETRA SYTAM see BJE750
TETRATELLURIUM TETRANITRIDE see TNO250
TETRATHIURAM DISULFIDE see TFS350
TETRATHIURAM DISULPHIDE see TFS350
TETRAVEC see PCF275
TETRAVERINE see TBX000
TETRAVINYLLEAD see TEF250
TETRAVOS see DGP900
TETRA-WEDEL see TBX250
7,8,9,10-TETRAZABICYCLO(5.3.0)-8,10-DECADIENE see PBI500
1,2,3,3A-TETRAZACYCLOHEPTA-8A,2-CYCLOPENTADIENE see PBI500
TETRAZENE see TEF500
TETRAZEPAM see CFG750
1,2,4,5-TETRAZINE see TEF600
TETRAZOBENZENE-β-NAPHTHOL see OHA000
TETRAZO DEEP BLACK G see AQP000
TETRAZOL see TEF650
1H-TETRAZOL-5-AMINE see AMR000
TETRAZOLE-5-DIAZONIUM CHLORIDE see TEF675
TETRAZOSIN HYDROCHLORIDE DIHYDRATE see TEF700
TETRINE see EIV000
TETRINE ACID see EIX000
TETROCARCIN A see TEF725
TETRODIRECT BLACK EFD see AQP000
TETRODONTOXIN see FOQ000
TETRODOTOXIN see FOQ000
TETRODOXIN see FOQ000
TETROFAN see TEF775
TETROGUER see PCF275
TETROLE see FPK000
TETRON see TCF250, TCF280
TETRON-100 see TCF250
TETRONE A see TEF750
TETROPHAN see TEF775
TETROPHENE GREEN M see AFG500
TETROPHINE see TEF775
TETROPIL see PCF275
TETROSAN see AFP750
TETROSIN OE see BGJ250
TETROSOL see TBX250

TETRYL see TEG250
2,4,6-TETRYL see TEG250
TETRYLAMMONIUM see TCB725
TETRYLAMMONIUM BROMIDE see TCC000
TETRYL FORMATE see IIR000
TETTERWORT see CCS650
TETURAM see DXH250
TETURAMIN see DXH250
TEVCOCIN see CDP250
TEVCODYNE see BRF500
TEXACO LEAD APPRECIATOR see BPV100
TEXANOL see TEG500
TEXAN RED TONER D see CHP500
TEXAPON T-42 see SON000
TEXAPON ZHC see SIB600
TEXAS MOUNTAIN LAUREL see NBR800
TEXAS SARSAPARILLA see MRN100
TEXAS UMBRELLA TREE see CDM325
TEXIN 192A see PKO500
TEXIN 445D see PKM250
TEXTILE see SFT500
TEXTILE RED WD-263 see NAY000
TEXTILON (OBS.) see TEG650
TEXTILOTOXIN see TEG650
TEX WET 1001 see DJL000
TF see GJS200
TFA see TJX900
2,3,4-TFA see TJX900
TFC mixed with TCC (1:2) see TIL526
TFDU see TKH325
TFE see TKA350
TFF see TNT500
T. FLAVOVIRIDIS VENOM see TKW100
T-FLUORIDE see SHF500
TFPHM see TCI500
TFT THILO see TKH325
TG see AMH250
TGA-EXTRACT see PLW550
T-GAS see EJN500
β-TGDR see TFJ250
T-GELB BZW, GRUN 1 see QCA000
TGR see TFJ500
TGS see TFJ500
TH-152 see MDM800
TH-218 see EPC175
TH/100 see TEP500
1064 TH see DUO400
TH 1314 see EPQ000
1321 TH see PNW750
TH-1395 see GFM200
2180 TH see PMM000
TH 346-1 see DRR400
TH 6040 see CJV250
Th 1165a see FAQ100
THACAPZOL see MCO500
THAIMYCIN B see TEG750
THAIMYCIN C see TEH000
THALAMONAL see DYF200
THALIBLASTINE see TEH250
THALICARPIN see TEH250
THALICARPINE see TEH250
(+)-THALICARPINE see TEH250
THALIDOMIDE see TEH500
(-)-THALIDOMIDE see TEH520
(+)-THALIDOMIDE see TEH510
(+,-)-THALIDOMIDE see DVT200
s-(-)-THALIDOMIDE see TEH520
(±)-THALIDOMIDE see TEH520
R-(+)-THALIDOMIDE see TEH510
THALIN see TEH500
THALINETTE see TEH500
THALLIC OXIDE see TEL050
THALLINE see TCU700
THALLIUM see TEI000
THALLIUM ACETATE see TEI250
THALLIUM(I) ACETATE see TEI250
THALLIUM(1+) ACETATE see TEI250
THALLIUM(I) AZIDE see TEI500
THALLIUM(I) AZIDODITHIOCARBONATE see TEI600
THALLIUM BROMATE see TEI625
THALLIUM BROMIDE see TEI750
THALLIUM(I) CARBONATE (2581) see TEJ000
THALLIUM CHLORATE see TEJ100
THALLIUM CHLORIDE see TEJ250

THALLIUM(1+) CHLORIDE see TEJ250
THALLIUM COMPOUNDS see TEJ500
THALLIUM(I) DITHIOCARBONAZIDATE see TEI600
THALLIUM(I) FLUOBORATE see TEJ750
THALLIUM(I) FLUORIDE see TEK000
THALLIUM(I) FLUOSILICATE see TEK250
THALLIUM FULMINATE see TEK300
THALLIUM IODIDE see TEK500
THALLIUM(I) IODIDE see TEK500
THALLIUM(1+) IODIDE see TEK500
THALLIUM(I) IODOACETYLIDE see TEK525
THALLIUM MALONATE see TEM399
THALLIUM MONOACETATE see TEI250
THALLIUM MONOCHLORIDE see TEJ250
THALLIUM MONOFLUORIDE see TEK000
THALLIUM MONOIODIDE see TEK500
THALLIUM MONONITRATE see TEK750
THALLIUM MONOSELENIDE see TEL500
THALLIUM NITRATE see TEK750
THALLIUM(I) NITRIDE see TEL000
THALLIUM OXIDE see TEL050
THALLIUM (3+) OXIDE see TEL050
THALLIUM(III) OXIDE see TEL050
THALLIUM PEROXIDE see TEL050
THALLIUM(I) PEROXODIBORATE see TEL100
THALLIUM aci-PHENYLNITROMETHANIDE see TEL150
THALLIUM SALT (solid) see TEL250
THALLIUM SELENIDE see TEL500
THALLIUM SESQUIOXIDE see TEL050
THALLIUM SULFATE see TEL750
THALLIUM(I) SULFATE (2581) see TEM000
THALLIUM(II) SULFATE (1581) see TEM250
THALLIUM SULFATE, solid (DOT) see TEL750
THALLOUS ACETATE see TEI250
THALLOUS CARBONATE see TEJ000
THALLOUS CHLORIDE see TEJ250
THALLOUS FLUORIDE see TEK000
THALLOUS IODIDE see TEK500
THALLOUS MALONATE see TEM399
THALLOUS NITRATE see TEK750
THALLOUS SULFATE see TEM000
THAM see POC750, TEM500
THAM-E see TEM500
THAM SET see TEM500
THANISOL see IHZ000
THANITE see IHZ000
THBP see BCS325, TKO250
THC see TCM250
Δ^1-THC see TCM250
Δ^6-THC see TCM000
Δ^8-THC see TCM000
Δ^9-THC see TCM250
TH-DMBA see TCP600
THE-7 see HKG500
THEACITIN see TEQ250
THEAL see DNC000
THEAL AMPULES see DNC000
THEAL TABL. see TEP000
THEAN see HOA000
THEBAINE see TEN000
THEBAINE HYDROCHLORIDE see TEN100
(-)-THEBAINE HYDROCHLORIDE see TEN100
THEBAIN HYDROCHLORID see TEN100
THECODIN see DLX400
THECODINE see DLX400
THEELIN see EDV000
THEELOL see EDU500
THEFANIL see DPJ200
THEFYLAN see DNC000
THEIN see CAK500
THEINE see CAK500
THEKODIN see DLX400
THELESTRIN see EDV000
THELYKININ see EDV000
THEMALON HYDROCHLORIDE see DJP500
THENALTON see PAG200
THENARDITE see SJY000
THENARDOL see HIB500
THENFADIL see DPJ200
THENFADIL HYDROCHLORIDE see TEO000
THENOBARBITAL see EOK000
N-(2-THENOYL)GLYCINOHYDROXAMIC ACID see TEN725
p-(2-THENOYL)HYDRATROPIC ACID see TEN750
(±)-2-(p-(2-THENOYL)PHENYL)PROPIONIC ACID see TEN750

2-THENYLAMINE, 5-CHLORO-N-(2-(DIMETHYLAMINO)ETHYL)-N-2-
    PYRIDYL- see CHY250
THENYLDIAMINE see DPJ200
THENYLDIAMINE CHLORIDE see TEO000
THENYLDIAMINE HYDROCHLORIDE see TEO000
THENYLENE see DPJ400, TEO250
THENYLENE HYDROCHLORIDE see DPJ400
THENYLPYRAMINE see TEO250
THENYLPYRAMINE HYDROCHLORIDE see DPJ400
THEOBROMINE see TEO500
THEOBROMINE SODIUM SALICYLATE see TEO750
THEOCIN see TEP000
THEOCIN SOLUBLE see TEQ250
THEOCOR see HNY500
THEODEN see HOA000
THEODROX see TEP500
THEOESBERIVEN see TEO800
THEOFIBRATE see TEQ500
THEOFOL see TEP000
THEOGEN see ECU750
THEOGRAD see TEP000
THEOLAIR see TEP000
THEOLAMINE see TEP500
THEOLIX see TEP000
THEOMIN see TEP500
THEOMINAL see EOK000
THEON see HOA000
THEOPHILCHOLINE see TEH500
THEOPHORIN TARTRATE see PDD000
THEOPHYL-225 see TEP000
THEOPHYLDINE see TEP500
THEOPHYLLAMINE see TEP500
THEOPHYLLAMINIUM see TEP500
THEOPHYLLIN see TEP000
THEOPHYLLIN AETHYLENDIAMIN (GERMAN) see TEP500
THEOPHYLLINE see TEP000
THEOPHYLLINE, anhydrous see TEP000
THEOPHYLLINE CHOLINATE see CMF500
7-THEOPHYLLINEETHANOL see HLC000
THEOPHYLLINE ETHYLENEDIAMINE see TEP500
THEOPHYLLINE, compound with ETHYLENEDIAMINE (2581) see TEP500
8-THEOPHYLLINE MERCURIC ACETATE see TEP750
THEOPHYLLINE METHOXYOXIMERCURIPROPYL SUCCINYLUREA
    see TEQ000
2-(7'-THEOPHYLLINEMETHYL)-1,3-DIOXOLANE see TEQ175
THEOPHYLLINE SALT of CHOLINE see CMF500
THEOPHYLLINE SODIUM ACETATE see TEQ250
THEOPHYLLINE, compound with SODIUM ACETATE (1:1) see TEQ250
THEOPHYLLIN ETHYLENEDIAMINE see TEP500
THEOPHYLLINUM NATRIUM ACETICUM (GERMAN) see TEQ250
1-(THEOPHYLLIN-7-YL)AETHYL-2-(2-(p-CHLORPHENOXY)-2-
    METHYLPROPIONAT (GERMAN) see TEQ500
1-(THEOPHYLLIN-7-YL)ETHYL-2-(2-(p-CHLOROPHENOXY)-2-
    METHYLPROPIONATE see TEQ500
THEOSALVOSE see TEO500
THEOSTENE see TEO500
THEOXYLLINE see CMF500
THEPHORIN HYDROCHLORIDE see TEQ700
THEPHORIN TARTRATE see PDD000
THEPHYLDINE see TEP500
THERACANZAN see SNN300
THERADERM see BDS000
THERADIAZINE see PPP500
THERA-FLUR-N see SHF500
THERALEPTIQUE see DUO400
THERAMINE see HGD000
THERAPAV see PAH250
THERAPOL see SNM500
THERAPTIQUE see DUO400
THERAZONE see BRF500
THERMACURE see MKA500
THERMALOX see BFT250
THERM CHEK 820 see DDV600
THERMINOL FR-1 see PJL750
"THERMIT" see TER000
THERMOASE PC-10 see BAC000
THERMOLITE 831 see BKK750
THERMOPLASTIC 125 see SMR000
THERUHISTIN HYDROCHLORIDE see AEG625
THESAL see TEO500
THESODATE see TEO500
THETAMID see ENG500
THEVETIA PERUVIANA see YAK350
THEVETIGENIN see DMJ000

dl-β-2-THIENYLALANINE see TEY250
2-THIENYLALDEHYDE see TFM500
5-(2-THIENYLCARBONYL)-2-BENZIMIDAZOLECARBAMIC ACID METHYL
   ESTER see OJD100
(5-(2-THIENYLCARBONYL)-1H-BENZIMIDAZOL-2-YL)-CARBAMIC ACID
   METHYL ESTER see OJD100
2-THIENYLCARBOXALDEHYDE see TFM500
THIENYLIC ACID see TGA600
4-(2-THIENYLKETO)-2,3-DICHLOROPHENOXYACETIC ACID see TGA600
2-(2-THIENYL)MORPHOLINE HYDROCHLORIDE see TEY600
1-α-THIENYL-1-PHENYL-3-N-METHYLMORPHOLINIUM-1-PROPANOL IO-
   DIDE see HNM000
2-THIEPANONE see TEY750
THIEPAN-2-ONE see TEY750
THIERGAN see DQA400
THIETHYLPERAZINE DIMALEATE see TEZ000
THIETHYLPERAZINE MALEATE see TEZ000
THIFOR see EAQ750
THIIRANE see EJP500
THILAVEN see IAD000
THILLATE see TFS350
THILOPEMAL see ENG500
THILOPHENYL see DKQ000
THILOPHENYT see DNU000
THIMBLES see FOM100
THIMBLEWEED see PAM780
THIMECIL see MPW500
THIMER see TFS350
THIMEROSALATE see MDI000
THIMEROSOL see MDI000
THIMET see PGS000
THIMUL see EAQ750
50 THINNER see PCT500
THIOACETAMIDE see TFA000
THIOACETANILIDE see TFA250
THIOACETARAMIDE see TFA350
THIOACETAZONE see FNF000
THIOACETIC ACID see TFA500
THIOALKOFEN BM see TFD000
THIOALKOFEN BM 4 see TFC600
THIOALLATE see CDO250
THIOALLYL ETHER see AGS250
THIOAMI see DXN709
THIOAMIDE see EPQ000
THIOANILINE see TFI000
4,4'-THIOANILINE see TFI000
THIOANISOLE see TFC250
THIOANTIMONIC(III) ACID, TRIESTER with MERCAPTO SUCCINIC ACID
   DILITHIUM SALT, NONAHYDRATE see AQE500
THIOARSENITE see TFA350
(THIOARSENOSO)METHANE see MGQ750
THIOARSMINE see SNR000
THIOAURIN see TFC500
2-THIOBARBITURIC ACID see MCK500
THIOBEL see BHL750
THIOBENCARB see SAZ000
THIOBENZAMIDE see BBM250
THIOBENZANILIDE see PGL250
THIOBENZYL ALCOHOL see TGO750
2-THIO-2-BENZYL-PSEUDOUREA HYDROCHLORIDE see BEU500
4,4'-THIOBIS(ANILINE) see TFI000
4,4'-THIOBISBENZENAMINE see TFI000
1,1'-THIOBIS(BENZENE) see PGI500
4,4'-THIOBIS(6-tert-BUTYL-m-CRESOL) see TFC600
4,4'-THIOBIS(6-tert-BUTYL-o-CRESOL) see TFD000
4,4'-THIOBIS(2-tert-BUTYL-5-METHYLPHENOL) see TFC600
4,4'-THIOBIS(6-tert-BUTYL-3-METHYLPHENOL) see TFC600
1,1'-THIOBIS(2-CHLOROETHANE) see BIH250
2,2'-THIOBIS(4-CHLORO-6-METHYLPHENOL) see DMN000
2,2'-THIOBIS(4,6-DICHLOROPHENOL) see TFD250
4,4'-THIOBIS(2-(1,1-DIMETHYLETHYL)-6-METHYLPHENOL see TFD000
1,1'-THIOBIS(N,N-DIMETHYLTHIO)FORMAMIDE see BJL600
THIOBIS(DODECYL PROPIONATE) see TFD500
1,1'-THIOBISETHANE see EPH000
4,4'-THIOBIS(3-METHYL-6-tert-BUTYLPHENOL) see TFC600
1,1'-THIOBIS(2-METHYL-4-HYDROXY-5-tert-BUTYLBENZENE) see TFC600
THIOBISMOL see BKX750, BKY500
3,3-THIOBIS(1-PROPENE) see AGS250
THIOBIS(TRIBENZYL-TIN (8CI) see BLK750
THIOBORIC ACID, ESTER with 2,2'-
   ((DIBUTYLSTANNYLENE)DIOXY)DIETHANETHIOL (2:3) see TNG250
epsilon-THIOCAPROLACTONE see TEY750
THIOCARB see SGJ000
THIOCARBAMATE see ISR000

THIOCARBAMIC ACID-S,S-(2-(DIMETHYLAMINO)TRIMETHYLENE)ESTER
   HYDROCHLORIDE see BHL750
THIOCARBAMIDE see ISR000
THIOCARBAMISIN see TFD750
THIOCARBAMIZINE see TFD750
THIOCARBAMYLHYDRAZINE see TFQ000
THIOCARBANIL see ISQ000
THIOCARBANILIDE see DWN800
THIOCARBARSONE see CBI250
THIOCARBAZIDE see TFE250
THIOCARBAZIL see FNF000
THIOCARBOHYDRAZIDE see TFE250
THIOCARBONIC DICHLORIDE see TFN500
THIOCARBONIC DIHYDRAZIDE see TFE250
THIOCARBONOHYDRAZIDE see TFE250
THIOCARBONYL AZIDE THIOCYANATE see TFE275
THIOCARBONYL CHLORIDE (DOT) see TFN500
THIOCARBONYL DICHLORIDE see TFN500
THIOCHROMAN-4-ONE, OXIME see TEJ000
THIOCHRYSINE see GJG000
THIOCOLCHICOSIDE see TFE325
10-THIOCOLCHICOSIDE see TFE325
4-THIOCRESOL see TGP250
o-THIOCRESOL see TGP000
p-THIOCRESOL see TGP250
THIOCRON see AHO750
THIOCTACID see DXN800
THIOCTAMID see DXN709
THIOCTAMIDE see DXN709
THIOCTIC ACID see DXN800
6-THIOCTIC ACID see DXN800
6,8-THIOCTIC ACID see DXN800
THIOCTIC ACID AMIDE see DXN709
THIOCTIDASE see DXN800
THIOCTSAN see DXN800
THIOCYANATES see TFE500
THIOCYANATE SODIUM see SIA500
THIOCYANATOACETIC ACID CYCLOHEXYL ESTER see TFF000
THIOCYANATOACETIC ACID ETHYL ESTER see TFF100
THIOCYANATOACETIC ACID ISOBORNYL ESTER see IHZ000
1-THIOCYANATODODECANE see DYA200
THIOCYANATOETHANE see EPP000
α-THIOCYANATOTOLUENE see BFL000
THIOCYANIC ACID, ALLYL ESTER see AGS750
THIOCYANIC ACID, AMYL ESTER see AON500
THIOCYANIC ACID, 2-(BENZOTHIAZOLYLTHIO)METHYL ESTER
   see BOO635
THIOCYANIC ACID, CALCIUM SALT (2:1) see CAY250
THIOCYANIC ACID-p-DIMETHYLAMINOPHENYL ESTER see TFH500
THIOCYANIC ACID, DODECYL ESTER see DYA200
THIOCYANIC ACID, ETHYL ESTER see EPP000
THIOCYANIC ACID compounded with GUANIDINE (1581) see TFF250
THIOCYANIC ACID, 2-HYDROXYETHYL ESTER, LAURATE see LBO000
THIOCYANIC ACID, LITHIUM SALT (1581) see TFF500
THIOCYANIC ACID, MERCURY(2⁻) SALT see MCU250
THIOCYANIC ACID, TRIMETHYLSTANNYL ESTER see TMI750
THIOCYANIC ACID, TRIPHENYLSTANNYL ESTER see TMV750
1-THIOCYANOBUTANE see BSN500
p-THIOCYANODIMETHYLANILINE see TFH500
4-THIOCYANO-N,N-DIMETHYLANILINE see TFH500
THIOCYANO-ESSIGSAEURE-AETHYL-ESTER (GERMAN) see TFF100
2-THIOCYANOETHYL COCONATE see LBO000
2-THIOCYANOETHYL DODECANOATE see LBO000
2-THIOCYANOETHYL LAURATE see LBO000
β-THIOCYANOETHYL LAURATE see LBO000
THIOCYANOGEN see TFH600
2-(THIOCYANOMETHYLTHIO)BENZOTHIAZOLE, 60% see BOO635
THIOCYCLAM (ETHANEDIOATE 1:1) see TFH750
THIOCYCLAM HYDROGEN OXALATE see TFH750
THIOCYMETIN see MPN000
THIOCYNAMINE see AGT500
THIODAN see EAQ750
THIODELONE see MCH600
THIODEMETON see DAO500, DXH325
THIODEMETRON see DXH325
6-THIODEOXYGUANOSINE see TFJ500
THIODERON see MCH600
p,p-THIODIANILINE see TFI000
4,4'-THIODIANILINE see TFI000
2,2'-THIODIETHANOL see TFI500
THIODIETHYLENE GLYCOL see TFI500
THIODIFENYLAMINE (DUTCH) see PDP250
THIODIGLYCOL see TFI500
β-THIODIGLYCOL see TFI500

THIODIGLYCOLIC ACID see MCM750
2,2'-THIODIGLYCOLIC ACID see MCM750
β,β'-THIODIGLYCOLIC ACID see MCM750
THIODIGLYCOLLIC ACID see MCM750
2-THIO-3,5-DIMETHYLTETRAHYDRO-1,3,5-THIADIAZINE see DSB200
4,4'-THIODIPHENOL see TFJ000
THIODIPHENYLAMIN (GERMAN) see PDP250
THIODIPHENYLAMINE see PDP250
O,O'-(THIODI-4,1-PHENYLENE)BIS(O,O-DIMETHYL PHOSPHOROTHIOATE)
   see TAL250
THIODI-p-PHENYLENEDIAMINE see TFI000
O,O'-(THIODI-p-PHENYLENE)-O,O,O',O'-TETRAMETHYL BIS(PHOS-
   PHOROTHIOATE) see TAL250
THIODIPROPIONIC ACID see BHM000
3,3'-THIODIPROPIONIC ACID see BHM000
β,β'-THIODIPROPIONIC ACID see BHM000
β,β'-THIODIPROPIONITRILE see DGS600
THIODOW see EIR000
THIODRIL see CBR675
THIODROL see EBD500
THIOETHANOL see EMB100
2-THIOETHANOL see MCN250
THIOETHANOLAMINE see AJT250
THIOETHYL ALCOHOL see EMB100
(THIOETHYLENE)BIS(BIS(2-HYDROXYETHYL)SULFONIUM DICHLORIDE
   see BHD000
THIOETHYL ETHER see EPH000
THIOFACO M-50 see EEC600
THIOFACO T-35 see TKP500
THIOFANATE see DJV000
THIOFANOX see DAB400
THIOFIDE see BDE750
THIOFOR see EAQ750
THIOFORM (CZECH) see TLS500
THIOFOSGEN (CZECH) see TFN500
THIOFOZIL see TFQ750
THIOFURAM see TFM250
THIOFURAN see TFM250
THIOFURFURAN see TFM250
(1-THIO-d-GLUCOPYRANOSATO)GOLD see ART250
5-THIO-α-d-GLUCOPYRANOSE see GFK000
1-THIO-GLUCOPYRANOSE, MONOGOLD(1+) SALT see ART250
1-THIO-β-d-GLUCOPYRANOSE 1-(N-(SULFOOXY-3-BUTENIMIDATE),
   MONOPOTASSIUM see AGH125
5-THIO-d-GLUCOSE see GFK000
THIOGLUCOSE d'OR (FRENCH) see ART250
THIOGLYCERIN see MRM750
1-THIOGLYCEROL see MRM750
α-THIOGLYCEROL see MRM750
THIOGLYCOL (DOT) see MCN250
THIOGLYCOLANILIDE see MCK000
THIOGLYCOLATESODIUM see SKH500
THIOGLYCOLIC ACID see TFJ100
2-THIOGLYCOLIC ACID see TFJ100
THIOGLYCOLIC ACID ANILIDE see MCK000
THIOGLYCOLIC ACID ETHYL ESTER see EMB200
THIOGLYCOLIC ACID-2-ETHYLHEXYL ESTER see EKW300
THIOGLYCOLIC ACID METHYL ESTER see MLE750
THIOGLYCOLIC ACID, AMMONIUM SALT see ANM500
THIOGLYCOLLIC ACID, SODIUM SALT see SKH500
THIOGLYKOLSAEURE-AETHYLESTER (GERMAN) see EMB200
THIOGLYKOLSAEURE-2-AETHYLHEXYL ESTER (GERMAN) see EKW300
THIOGLYKOLSAEURE-METHYLESTER (GERMAN) see MLE750
THIOGUANINE see AMH250
6-THIOGUANINE see AMH250
β-THIOGUANINE DEOXYRIBOSIDE see TFJ250
6-THIOGUANINE RIBONUCLEOSIDE see TFJ500
THIOGUANINE RIBOSIDE see TFJ500
THIOGUANOSINE see TFJ500
6-THIOGUANOSINE see TFJ500
THIOHEXITAL see AGL875
2-THIOHYDANTOIN see TFJ750
2-THIO-4-HYDRAZINOURACIL see HHF500
THIOHYPOXANTHINE see POK000
THIOINOSINE see TFJ825
THIOISONICOTINAMIDE see TFK000
THIOKARBONYLCHLORID (CZECH) see TFN500
2-THIO-4-KETOIHIAZOLIDINE see RGZ550
THIOKOL NVT see ROH900
THIOKTSAFURE (GERMAN) see DXN800
THIOLA see MCI375
THIOLACETIC ACID see TFA500
2-THIOLACTIC ACID see TFK250
THIOLDEMETON see DAP200

2-THIOL-DIHYDROGLYOXALINE see IAQ000
THIOLE see TFM250
THIOLIN see IAD000
THIOLMECAPTOPHOS see DAO500
THIOL SYSTOX see DAP200
THIOLUTIN see ABI250
THIOLUX see SOD500
THIOMALIC ACID see MCR000
2-THIO-MALIC ACID see MCR000
THIOMEBUMAL see PBT250
THIOMEBUMAL SODIUM see PBT500
THIOMECIL see MPW500
THIOMERIN SODIUM see TFK270
THIOMERSALATE see MDI000
THIOMESTERONE see TFK300
THIOMESTRONE see TFK300
THIOMETAN see DSK600
2-THIO-6-METHYL-1,3-PYRIMIDIN-4-ONE see MPW500
6-THIO-4-METHYLURACIL see MPW500
THIOMETON see PHI500
THIOMICID see FNF000
THIOMIDIL see MPW500
THIOMONOGLYCOL see MCN250
THIOMUL see EAQ750
THIOMYLAL SODIUM see SOX500
THIO-β-NAPHTHOL see NAP500
β-THIONAPHTHOL see NAP500
THIONAZIN see EPC500
THIONEMBUTAL see PBT500
THIONEX see BJL600, EAQ750
THIONEX RUBBER ACCELERATOR see BJL600
THIONICID see FNF000
THIONIDEN see EPQ000
THIONIN see AKK750
THIONINE see AKK750
THIONOACETIC ACID see TFA500
THIONOBENZENEPHOSPHONIC ACID ETHYL-p-NITROPHENYL ESTER
   see EBD700
THIONODEMETON SULFONE see SPF000
THIONOSINE see MCQ500
THIONTAN see DHF600
THIONYLAN see TEO250
THIONYL BROMIDE see SNT200
THIONYL CHLORIDE see TFL000
THIONYL DICHLORIDE see TFL000
THIONYL DIFLUORIDE see TFL250
THIONYL FLUORIDE see TFL250
THIOOCTANOIC ACID see DXN800
THIOORATIN see TET250
2-THIO-4-OXO-6-METHYL-1,3-PYRIMIDINE see MPW500
2-THIO-4-OXO-6-PROPYL-1,3-PYRIMIDINE see PNX000
2-THIO-6-OXYPURINE see TFL500
2-THIO-6-OXYPYRIMIDINE see TFR250
THIOPARAMIZONE see FNF000
THIOPENTAL see PBT250
THIOPENTAL SODIUM see PBT500
THIOPENTAL SODIUM SALT see PBT500
THIOPENTEX see CPK000
THIOPENTOBARBITAL see PBT250
THIOPENTONE see PBT250
THIOPENTONE SODIUM see PBT500
THIOPERAZINE see TFM100
THIOPEROXYDICARBONIC ACID DIETHYL ESTER see BJU000
THIOPEROXYDICARBONIC ACID DIMETHYL ESTER see DUN600
THIOPEROXYDICARBONIC DIAMIDE, N,N'-DIETHYL-N,N'-DIPHENYL-
   see DJC200
THIOPEROXYDICARBONIC DIAMIDE, TETRAMETHYL-, mixture with (1-α-2-
   α-3-β,4-α- 5-α-6-β)-1,2,3,4,5,6-HEXACHLOROCYCLOHEXANE and
   TRICHLOROPHENOL COPPER(2+) SALT see FAQ930
THIOPHAL see TIT250
THIOPHANAT (GERMAN) see DJV000
THIOPHANATE ETHYL see DJV000
THIOPHANATE METHYL-MANEB mixture see MAP300
THIOPHANE DIOXIDE see SNW500
THIOPHEN see TFM250
THIOPHENE see TFM250
7-(THIOPHENE-2-ACETAMIDO)CEPHALOSPORANIC ACID see CCX250
7-(THIOPHENE-2-ACETAMIDO)CEPHALOSPORANIC ACID SODIUM SALT
   see SFQ500
7-(THIOPHENE-2-ACETAMIDO)-3-(1-PYRIDYLMETHYL)-3-CEPHEM-4-CAR-
   BOXYLIC ACID BETAINE see TEY000
2-THIOPHENEALANINE see TEY250
2-THIOPHENEALDEHYDE see TFM500
2-THIOPHENECARBOXALDEHYDE see TFM500

α-THIOPHENECARBOXALDEHYDE see TFM500
(THIOPHENE-2,5-DIYL)BIS(HYDROXYMERCURY)MONOACETATE
   see HLQ000
THIOPHENICOL see MPN000
THIOPHENICOL GLYCINATE HYDROCHLORIDE see UVA150
THIOPHENIT see MNH000
THIOPHENITE see DJV000
THIOPHENOL (DOT) see PFL850
1-THIOPHENYLANTHRAQUINONE see PGL000
THIOPHOS see PAK000
THIOPHOSGENE see TFN500
THIOPHOSPHAMIDE see TFQ750
THIOPHOSPHATE de S-N-(1-CYANO-1-METHYLETHYL)CAR-
   BAMOYLMETHYLE et de O,O-DIETHYLE (FRENCH) see PHK250
THIOPHOSPHATE de O-2,4-DICHLOROPHENYLE et de O,O-DIETHYLE
   (FRENCH) see DFK600
THIOPHOSPHATE de O,O-DIETHYLE et de O-(3-CHLORO-4-METHYL-7-
   COUMARINYLE) (FRENCH) see CNU750
THIOPHOSPHATE de O,O-DIETHYLE et de O-(2,5-DICHLORO-4-BROMO)
   PHENYLE (FRENCH) see EGV500
THIOPHOSPHATE de O,O-DIETHYLE et de S-(2-ETHYLTHIO-ETHYLE)
   (FRENCH) see DAP200
THIOPHOSPHATE de O,O-DIETHYLE et de o-2-ISOPROPYL-4-METHYL-6-
   PYRIMIDYLE (FRENCH) see DCM750
THIOPHOSPHATE de O,O-DIETHYLE et de S-(N-METHYLCARBAMOYL)
   METHYLE (FRENCH) see DNX800
THIOPHOSPHATE de O,O-DIETHYLE et de O-(4-METHYL-7-COUMARINYLE)
   (FRENCH) see PKT000
THIOPHOSPHATE de O,O-DIETHYLE et de O-(4-NITROPHENYLE) (FRENCH)
   see PAK000
THIOPHOSPHATE de O,O-DIMETHYLE et de O-4-BROMO-2,5-
   DICHLOROPHENYLE (FRENCH) see BNL250
THIOPHOSPHATE de O,O-DIMETHYLE et de O-3-CHLORO-4-
   NITROPHENYLE (FRENCH) see MIJ250
THIOPHOSPHATE de O,O-DIMETHYLE et de O-4-CHLORO-3-
   NITROPHENYLE (FRENCH) see NFT000
THIOPHOSPHATE de O,O-DIMETHYLE ET DE S-2-(ISOPROPYLSULFINYL)-
   ETHYLE see DSK600
THIOPHOSPHATE de O,O-DIMETHYLE et de S-2-ETHYLSULFINYLETHYLE
   (FRENCH) see DAP000
THIOPHOSPHATE de O,O-DIMETHYLE et de O-2-ETHYLTHIO-ETHYLE
   (FRENCH) see DAO800
THIOPHOSPHATE de O,O-DIMETHYLE et de S-2-ETHYLTHIOETHYLE
   (FRENCH) see DAP400
THIOPHOSPHATE de O,O-DIMETHYLE et de S-((5-METHOXY-4-PYRONYL)-
   METHYLE) (FRENCH) see EAS000
THIOPHOSPHATE de O,O-DIMETHYLE et de O-(3-METHYL-4-
   METHYLTHIOPHENYLE) (FRENCH) see FAQ900
THIOPHOSPHATE de O,O-DIMETHYLE et de O-(3-METHYL-4-
   NITROPHENYLE) (FRENCH) see DSQ000
THIOPHOSPHATE de O,O-DIMETHYLE et de O-(4-NITROPHENYLE)
   (FRENCH) see MNH000
THIOPHOSPHATE de O,O-DIMETHYLE et de O-(2,4,5-
   TRICHLOROPHENYLE) (FRENCH) see RMA500
THIOPHOSPHATE de O,S-DIMETHYL et de O-(3-METHYL-4-
   NITROPHENYLE) (FRENCH) see MKC250
THIOPHOSPHORIC ACID, TRISODIUM SALT see TNM750
THIOPHOSPHORIC ANHYDRIDE see PHS000
THIOPHOSPHOROUS TRIBROMIDE see TFN750
THIOPHOSPHORSAEURE-O,S-DIMETHYLESTERAMID (GERMAN)
   see DTQ400
THIOPHOSPHORYL BROMIDE see TFN750
THIOPHOSPHORYL CHLORIDE see TFO000
THIOPHOSPHORYL DIBROMIDEMONOFLUORIDE see TFO250
THIOPHOSPHORYL DIFLUORIDEMONOBROMIDE see TFO500
THIOPHOSPHORYL FLUORIDE see TFO750
THIOPHOSPHORYL MONOBROMODIFLUORIDE see TFO500
THIOPHOSPHORYL TRICHLORIDE see TFO000
THIOPHYLLINE CHOLINATE see CMF500
THIOPHYLLINE with CHOLLINE see CMF500
THIOPROLINE see TEV000
THIOPRONIN see MCI375
THIOPRONINE see MCI375
2-THIOPROPANE see TFP000
THIOPROPAZATE DIHYDROCHLORIDE see TFP250
THIOPROPAZATE HYDROCHLORIDE see TFP250
THIOPROPERAZIN see TFM100
THIOPROPERAZINE see TFM100
(THIOPROPERAZINE)-2-DIMETHYLSULFAMIDO-(10-(3-1-
   METHYLPIPERAZINYL-4)PROPYL)-PHENOTHIAZINE see TFM100
2-THIO-6-PROPYL-1,3-PYRIMIDIN-4-ONE see PNX000
6-THIO-4-PROPYLURACIL see PNX000
β-THIOPSEUDOUREA see ISR000
6-THIOPURINE RIBONUCLEOSIDE see MCQ500

6-THIOPURINE RIBOSIDE see MCQ500
2-THIO-1,3-PYRIMIDIN-4-ONE see TFR250
THIORIDAZIEN THIOMETHYL SULFOXIDE see MON750
THIORIDAZIN see MOO250
THIORIDAZINE see MOO250
THIORIDAZINE HYDROCHLORIDE see MOO500
THIORYL see MPW500
THIOSALICYLIC ACID see MCK750
THIOSAN see DXH250, TFS350
THIOSCABIN see DXH250
THIOSECONAL see AGL375
THIOSEMICARBARZONE see FNF000
THIOSEMICARBAZIDE see TFQ000
3-THIOSEMICARBAZIDE see TFQ000
THIOSEMICARBAZONE ACETONE see TFQ250
THIOSEMICARBAZONE (PHARMACEUTICAL) see FNF000
THIOSEPTAL see PPO000
THIOSERINE see CQK000
THIOSINAMIN see AGT500
THIOSINAMINE see AGT500
THIOSOL see MCI375
THIOSTOP N see SGM500
THIOSTREPTON see TFQ275
THIOSUCCINIMIDE see MRN000
THIOSULFAN see EAQ750
THIOSULFAN TIONEL see EAQ750
THIOSULFATES see TFQ500
THIOSULFIL-A FORTE see PDC250
THIOSULFURIC ACID, DIPOTASSIUM SALT see PLW000
THIOSULFURIC ACID, DISODIUM SALT, PENTAHYDRATE see SKI500
THIOSULFURIC ACID, GOLD(1+) SODIUM SALT (2:1:3) see GJE000
THIOSULFURIC ACID, GOLD(1+) SODIUM SALT(2:1:3), DIHYDRATE
   see GJG000
THIOSULFURIC ACID, S-(2-((4-(p-TOLYLOXY)BUTYL)AMINO)ETHYL) ESTER
   see THB250
THIOSULFUROUS DICHLORIDE see SON510
THIOTEBESIN see FNF000
THIO-TEP see TFQ750
THIOTEPP see SOD100
THIOTETROLE see TFM250
THIOTEX see TFS350
THIOTHAL see PBT250
THIOTHAL SODIUM see PBT500
2-THIOTHIAZOLIDONE see TFS250
THIOTHIXENE see NBP500
THIOTHIXENE DIHYDROCHLORIDE see TFQ600
THIOTHIXINE see NBP500
THIOTHYMIN see MPW500
THIOTHYRON see MPW500
6-THIOTIC ACID see DXN800
6,8-THIOTIC ACID see DXN800
3-THIOTOLENE see MPV250
THIOTOMIN see DXN709
THIOTOX see TFS350
5-THIO-1H-v-TRIAZOLO(4,5-d)PYRIMIDINE-5,7(4H,6H)-DIONE see MLY000
THIOTRIETHYLENEPHOSPHORAMIDE see TFQ750
THIOTRITHIAZYL NITRATE see TFR000
THIOURACIL see TFR250
2-THIOURACIL see TFR250
6-THIOURACIL see TFR250
2-THIOUREA see ISR000
THIOUREA (DOT) see ISR000
THIOVANIC ACID see TFJ100
THIOVANOL see MRM750
THIOVIT see SOD500
THIOXAMYL see DSP600
p-THIOXANE see OLY000
3-(THIOXANTHEN-9-YL)METHYL-1-PIPECOLINE, HYDROCHLORIDE
   see THL500
THIOXIDIL see GGS000
THIOXIDRENE see TCZ000
6-THIOXOPURINE see POK000
2-THIOXO-4-THIAZOLIDINONE see RGZ550
4-(THIOXO(3,4,5-TRIMETHOXYPHENYL)METHYL)MORPHOLINE see TNP275
THIOZAMIDE see TEX250
THIOZIN see IAD000, ZJS300
2-THIOZOLIDINETHIONE see TFS250
THIPENTAL SODIUM see PBT500
THIPHEN see DHY400
THIPHENAMIL HYDROCHLORIDE see DHY400
THIRAM see TFS350
THIRAMAD see TFS350
THIRAME (FRENCH) see TFS350

THIRASAN see TFS350
THIRERANIDE see DXH250
THISTROL see CLN750, CLO000
THIULIX see TFS350
THIURAD see TFS350
THIURAGYL see PNX000
THIURAM see TFS350
THIURAM DISULFIDE see TFS500
THIURAM E see DXH250
THIURAM EF see DJC200
THIURAMIN see TFS350
THIURAMYL see TFS350
THIURANIDE see DXH250
THIURETIC see CFY000
THIURYL see MPW500
THIXOKON see AAN000
THIZONE see FNF000
THLARETIC see CFY000
THN see TCY250
THOMAPYRIN see ARP250
THOMAS BALSAM see BAF000
THOMBRAN see CKJ000, THK880
THOMIZINE see THG600
THOMPSON-HAYWARD TH6040 see CJV250
THOMPSON'S WOOD FIX see PAX250
THONZYLAMINE HYDROCHLORIDE see RDU000
THONZYLAMINIUM CHLORIDE see RDU000
THORAZINE see CKP250, CKP500
THORAZINE HYDROCHLORIDE see CKP500
THORIA see TFT750
THORIDAZINE HYDROCHLORIDE see MOO500
THORIUM see TFS750
THORIUM-232 see TFS750
THORIUM CHLORIDE see TFT000
THORIUM DICARBIDE see TFT100
THORIUM DIOXIDE see TFT750
THORIUM HYDRIDE see TFT250
THORIUM METAL, PYROPHORIC (DOT) see TFS750
THORIUM (4+) NITRATE see TFT500
THORIUM(IV) NITRATE see TFT500
THORIUM OXIDE see TFT750
THORIUM OXIDE SULFIDE see TFU000
THORIUM TETRACHLORIDE see TFT000
THORIUM TETRANITRATE see TFT500
THORN APPLE see SLV500
THOROTRAST see TFT750
THORTRAST see TFT750
THPA see TDB000
THPC see TDH750
THPS see TDI000
THQ see MKO250
THREADLEAF GROUNDSEL see RBA400
THREAMINE see AMA500
THREE ELEPHANT see BMC000
THREE-LEAVED INDIAN TURNIP see JAJ000
l-THREITOL-1,4-BISMETHANESULFONATE see TFU500
d-THREO-α-BENZYL-N-ETHYLTETRAHYDROFURFURYLAMINE see ZVA000
THREONINE see TFU750
l-THREONINE see TFU750
THRETHYLENE see TIO750
THROATWORT see FOM100
THROMBIN COAGULASE see CMY325
THROMBOCID see SEH450
THROMBOCYTIN see AJX500
THROMBOLIQUINE see HAQ500
THROMBOTONIN see AJX500
THS-101 see DBC500
THS-201 see HAG325
THU see ISR000
(1S,4R,5R)-(−)-3-THUJANONE see TFW000
THUJA OIL see CCQ500
β-THUJAPLICIN see IRR000
Γ-THUJAPLICIN see TFV750
β-THUJAPLICINE see IRR000
THUJON see TFW000
THUJONE see TFW000
(−)-THUJONE see TFW000
l-THUJONE see TFW000
α-THUJONE see TFW000
THULIUM see TFW250
THULIUM CHLORIDE see TFW500
THULIUM EDETATE see TFX000
THULIUM(III) NITRATE, HEXAHYDRATE (1583586) see TFX250
THULOL see EDU500

THX see TFZ275
THYALONE see BAG250
THYCAPSOL see MCO500
THYLAKENTRIN see FMT100
THYLATE see TFS350
THYLFAR M-50 see MNH000
THYLOGEN see WAK000
THYLOGEN MALEATE see DBM800
THYLOQUINONE see MMD500
THYME CAMPHOR see TFX810
THYME OIL see TFX500
THYME OIL RED see TFX750
THYMEOL see AEG875
THYMIAN OEL (GERMAN) see TFX500
THYMIC ACID see TFX810
THYMIDIN see TFX790
THYMIDINE see TFX790
THYMINE see TFX800
THYMINEDEOXYRIBOSIDE see TFX790
THYMINE-2-DEOXYRIBOSIDE see TFX790
THYMIN (PURINE BASE) see TFX800
THYM OIL see TFX500
THYMOL see TFX810
m-THYMOL see TFX810
o-THYMOL see CCM000
THYMOQUINONE see IQF000
THYMOXAMINE HYDROCHLORIDE see TFY000
THYMUS KOTSCHYANUS, OIL EXTRACT see TFY100
THYNESTRON see EDV000
THYNON see DLK200
THYRADIN see TFZ100
THYREOIDEUM see TFZ275
THYREONORM see MPW500
THYREOSTAT see MPW500
THYREOSTAT II see PNX000
THYRIL see MPW500
THYROCALCITONIN see TFZ000
THYROID see TFZ100
THYROID RELEASING HORMONE see TNX400
THYROLIBERIN see TNX400
THYROTROPIC-RELEASING FACTOR see TNX400
THYROTROPIC RELEASING HORMONE see TNX400
THYROTROPIN-RELEASING FACTOR see TNX400
THYROTROPIN-RELEASING HORMONE see TNX400
THYROXEVAN see LEQ300
THYROXIN see TFZ275
l-THYROXIN see TFZ275
THYROXINE see TFZ275
(−)-THYROXINE see TFZ275
THYROXINE see TFZ300
l-THYROXINE see TFZ275
l-THYROXINE MONOSODIUM SALT see LEQ300
d-THYROXINE SODIUM see SKJ300
THYROXINE SODIUM SALT see LEQ300
d-THYROXINE SODIUM SALT see SKJ300
l-THYROXINE SODIUM SALT see LEQ300
TI-8 see TCA250
TI-1258 see BHL750
TIACETAZON see FNF000
TIADIPONE see BAV625
TIAMIPRINE see AKY250
TIAMIZID see CIP500
TIAMIZIDE see CIP500
TIAMULIN see TET800
TIAMULINA (ITALIAN) see TET800
TIAMUTIN see DAR000
TIAPAMIL see TGA275
TIAPRIDE HYDROCHLORIDE see TGA375
TIAPROFENIC ACID see SOX400
TIARAMIDE HYDROCHLORIDE see CJH750
TIAZOFURIN see RJF500
TIAZON see DSB200
TIB see TKQ250
TIBA see TKQ250
2,3,5-TIBA see TKQ250
TIBAMATO see MOV500
TIBAZIDE see ILD000
TIBENZATE see BFK750
TIBERAL see OJS000
TIBEXIN see LFJ000
TIBEY (PUERTO RICO) see SLJ650
TIBICUR see FNF000
TIBINIDE see ILD000
TIBIONE see FNF000

TIBIVIS see ILD000
TIBIZAN see FNF000
TIBRIC ACID see CGJ250
TIBUTOL see TGA500
TIC see TKJ250
TICARCILLIN SODIUM see TGA520
TICKLE WEED see FAB100
TICLID see TGA525
TICLOBRAN see ARQ750
TICLODIX see TGA525
TICLODONE see TGA525
TICLOPIDINE HYDROCHLORIDE see TGA525
TIC MUSTARD see IAN000
TICOLIN see DXN709
TICREX see TGA600
TICRYNAFEN see TGA600
TIDEMOL see BQL000
TIEMONIUM IODIDE see HNM000
TIEMOZYL see HNM000
TIEMPE see TKZ000
TIENILIC ACID see TGA600
2-(2-TIENYL)MORFOLINA CLORIDRATE (ITALIAN) see TEY600
TIEZENE see EIR000
TIFEMOXONE see PDU750
TIFEN see DHY400
TIFERRON see DXH300
TIFOMYCINE see CDP250
TIGAN see TKW750
TIGAN HYDROCHLORIDE see TKW750
TIGER ORANGE see CJD500
TIGLIC ACID see TGA700
TIGLIC ACID, GERANIOL ESTER see GDO000
TIGLINIC ACID see TGA700
12-o-TIGLYL-PHORBOL-13-BUTYRATE see PGV750
12-o-TIGLYL-PHORBOL-13-DODECANOATE see PGW000
7-TIGLYLRETRONECINE VIRIDIFLORATE see SPB500
7-TIGLYL-9-VIRIDIFLORYLRETRONECINE see SPB500
TIGUVON see FAQ900
TIKLID see TGA525
TIKOFURAN see NGG500
TIL see FHI000
TILCAREX see PAX000
TILCOTIL see TAL485
TILDIEM see DNU600
TILDIN see BNK000
TILIDATE see EIH000
TILIDINE HYDROCHLORIDE see EIH000
TIL4K8 see TOC250
TILLAM (RUSSIAN) see PNF500
TILLAM-6-E see PNF500
TILLANTOX see BDD000
TILLRAM see DXH250
TILORONE HYDROCHLORIDE see TGB000
TIMACOR see TGB185
TIMAZIN see FMM000
TIMBER RATTLESNAKE VENOM see TGB150
TIMEPIDIUM BROMIDE see TGB160
TIMET see PGS000
TIMIPERONE see TGB175
TIMOLET see DLH600, DLH630
TIMOLOL MALEATE see TGB185
l-TIMOLOL MALEATE see TGB185
TIMONIL see DCV200
TIMOPED see TGB475
TIMOPTOL see TGB185
TIMOSIN see MDQ250
TIN see TGB250
TIN (α) see TGB250
TINACTIN see TGB475
TINADERM see TGB475
TINAJA (PUERTO RICO) see CNR135
TIN AZIDE TRICHLORIDE see TGB500
TIN BIFLUORIDE see TGD100
TIN(IV) BROMIDE (1584) see TGB750
TIN(II) CHLORIDE (1582) see TGC000
TIN(IV) CHLORIDE (1584) see TGC250
TIN CHLORIDE, fuming (DOT) see TGC250
TIN(II) CHLORIDE DIHYDRATE (1582582) see TGC275
TIN CHLORIDE IODIDE see TGC280
TIN(2+) CITRATE see TGC285
TIN CITRATE (7CI) see TGC285
TIN, COMPLEX with PYROCATECHOL see TGC300
TIN COMPOUNDS see TGC500
TINDAL see ABG000

TIN DIBUTYL DILAURATE see DDV600
TIN DIBUTYLDITRIFLUOROACETATE see BLN750
TIN DIBUTYL MERCAPTIDE see DEI200
TIN DICHLORIDE see TGC000
TIN DIFLUORIDE see TGD100
TINDURIN see TGD000
TINESTAN see ABX250
TINESTAN 60 WP see ABX250
TIN FLAKE see TGB250
TIN FLUORIDE see TGD100
TINGENIN A see TGD125
TINGENON see TGD125
TINGENONE see TGD125
TINIC see NCQ900
TINIDAZOLE see TGD250
TIN(II) IODIDE see TGD500
TIN(IV) IODIDE (1584) see TGD750
TINMATE see CLU000
TINNING GLUX (DOT) see ZFA000
TIN(II) NITRATE OXIDE see TGE000
TINOFEDRINE HYDROCHLORIDE see DXI800
TINON GOLDEN YELLOW see DCZ000
TINOPAL CBS see TGE150
TINOPAL CBS-X see TGE150
TINOPAL 5BM see TGE100
TINOPAL RBS see TGE155
TINOPAL RBS 200 see TGE155
TINORIDINE HYDROCHLORIDE see TGE165
TINOSTAT see DDV600
TINOX see MIW250
TIN OXALATE see TGE250
TIN(2+) OXALATE see TGE250
TIN(II) OXALATE see TGE250
TIN OXIDE see TGE300
TIN P see HML500
TIN PERBROMIDE see TGB750
TIN PERCHLORIDE (DOT) see TGC250
TIN (IV) PHOSPHIDE see TGE500
TIN POTASSIUM TARTRATE see TGE750
TIN POWDER see TGB250
TIN PROTOCHLORIDE see TGC000
TIN-SAN see TIC500
TINSET see OMG000
TIN SODIUM TARTRATE see TGF000
TIN TETRABROMIDE see TGB750
TIN TETRACHLORIDE, anhydrous (DOT) see TGC250
TINTETRACHLORIDE (DUTCH) see TGC250
TIN TETRAIODIDE see TGD750
TINTORANE see WAT220
TIN TRIAETHYLZINNACETAT (GERMAN) see ABW750
TIN TRIPHENYL ACETATE see ABX250
TINUVIN 144 see PMK800
TINUVIN 1130 see TGF025
TINUVIN P see HML500
TIOACETAZON see FNF000
TIOALKOFEN BM see TFD000
TIOBENZAMIDE (ITALIAN) see BBM250
TIOCARONE see FNF000
TIOCONAZOLE see TGF050
TIOCTACID see DXN800
TIOCTAN see DXN709
TIOCTIDASI see DXN800
TIOCTIDASI ACETATE REPLACING FACTOR see DXN800
TIODIFENILAMINA (ITALIAN) see PDP250
TIOFANATE ETILE (ITALIAN) see DJV000
TIOFINE see TGG760
TIOFOS see PAK000
TIOFOSFAMID see TFQ750
TIOFOZIL see TFQ750
TIOGUANIN see AMH250
TIOGUANINE see AMH250
TIOINOSINE see MCQ500
TIOMERACIL see MPW500
TIOMESTERONE see TFK300
TIONAL see BJT750
TIOPENTALE (ITALIAN) see PBT250
TIOPENTAL SODIUM see PBT500
TIOPRONIN see MCI375
TIOPROPEN see AOO490
TIORALE M see MPW500
TIORIDAZIN see MOO500
TIOTIAMINE see DXO300
TIOTIRON see MPW500
TIOTIXENE see NBP500

TOLESTAN see CMY125
TOLFENAMIC ACID see CLK325
TOLFERAIN see FBJ100
TOLHEXAMIDE see CPR000
o-TOLIDIN see TGJ750
2-TOLIDIN (GERMAN) see TGJ750
2-TOLIDINA (ITALIAN) see TGJ750
TOLIDINE see TGJ750
2-TOLIDINE see TGJ750
o-TOLIDINE see TGJ750
3,3'-TOLIDINE see TGJ750
o,o'-TOLIDINE see TGJ750
TOLIFER see FBJ100
TOLIMAN see CMR100
TOLINASE see TGJ500
TOLIT see TMN490, TMN500
TOLITE see TMN490, TMN500
TOLL see MNH000
TOLMETIN see TGJ850
TOLMETINE see TGJ850
TOLMETIN SODIUM see SKJ340
TOLMETIN SODIUM DIHYDRATE see SKJ350
TOLNAFTATE see TGB475
TOLNAPHTHATE see TGB475
TOLNIDAMIDE see TGJ875
TOLOFREN see GGS000
TOLOMOL see AIV500
TOLONIDINE NITRATE see TGJ885
TOLONIUM CHLORIDE see AJP250
TOLOPELON see TGB175
(−)-1-(o-TOLOSSI-3-(2,2,5,5-TETRAMETIL-PIRROLIDIN-1-IL)-PROPAN-2-OLO CLORIDRATO (ITALIAN) see TGJ625
3-o-TOLOXY-2-HYDROXYPROPYL-1-CARBAMATE see CBK500
3-o-TOLOXY-1,2-PROPANEDIOL see GGS000
3-o-TOLOXY-1,2-PROPANEDIOL-1-CARBAMIC ACID ESTER see CBK500
TOLPAL see BBJ750
TOLPERISONE see TGK200
TOLPERISONE HYDROCHLORIDE see MRW125
TOLPRONINE HYDROCHLORIDE see TGK225
TOLSANIL see TGB475
TOLSERAM see CBK500
TOLSEROL see GGS000
TOLSERON see RLU000
TOLU see BAF000
TOLUALDEHYDE see TGK250
α-TOLUALDEHYDE see BBL500
TOLUALDEHYDE GLYCERYL ACETAL see TGK500
α-TOLUAMIDE see PDX750
TOLU BALSAM GUM see BAF000
TOLU BALSAM TINCTURE see BAF000
TOLUBUTEROL HYDROCHLORIDE see BQE250
TOLUEEN (DUTCH) see TGK750
TOLUEEN-DIISOCYANAAT see TGM750
TOLUEN (CZECH) see TGK750
TOLUEN-DISOCIANATO see TGM750
TOLUENE see TGK750
TOLUENE, p-ARSENOSO- see TGV100
TOLUENE-2-AZONAPHTHOL-2 see TGW000
o-TOLUENE-1-AZO-2-NAPHTHYLAMINE see FAG135
o-TOLUENEAZO-o-TOLUENEAZOβ-NAPHTHOL see SBC500
o-TOLUENEAZO-o-TOLUENE-β-NAPHTHOL see SBC500
o-TOLUENEAZO-o-TOLUIDINE see AIC250
TOLUENECARBOXALDEHYDE see TGK250
p-TOLUENE CHLOROMETHYLAMIDE see CEB500
TOLUENE, 2-CHLORO-4-NITRO- see CJG800
TOLUENEDIAMINE see TGL500
m-TOLUENEDIAMINE see TGL750
p-TOLUENEDIAMINE see TGM000
TOLUENE-2,4-DIAMINE see TGL750
2,4-TOLUENEDIAMINE see TGL750
TOLUENE-2,5-DIAMINE see TGM000
TOLUENE-2,6-DIAMINE see TGM100
TOLUENE-3,4-DIAMINE see TGM250
p-TOLUENEDIAMINE DIHYDROCHLORIDE see DCE200
p-TOLUENEDIAMINE SULFATE see DCE600, TGM400
2,5-TOLUENEDIAMINE SULFATE see DCE600
TOLUENE-2,5-DIAMINE, SULFATE (1:1) (8CI) see DCE600
p-TOLUENEDIAMINE SULPHATE see DCE600
TOLUENE-2,5-DIAMINE SULPHATE see DCE600
2-TOLUENEDIAZONIUM BROMIDE see TGM425
2-TOLUENEDIAZONIUM PERCHLORATE see TGM525
o-TOLUENEDIAZONIUM PERCHLORATE see TGM525
2-TOLUENEDIAZONIUM SALTS see TGM550
3-TOLUENEDIAZONIUM SALTS see TGM575

4-TOLUENEDIAZONIUM SALTS see TGM580
4-TOLUENEDIAZONIUM TRIIODIDE see TGM590
TOLUENE DIISOCYANATE see TGM740, TGM750
TOLUENE-1,3-DIISOCYANATE see TGM740
TOLUENE-2,4-DIISOCYANATE see TGM750
2,4-TOLUENEDIISOCYANATE see TGM750
TOLUENE-2,6-DIISOCYANATE see TGM800
2,6-TOLUENE DIISOCYANATE see TGM800
2,3-TOLUENEDIOL see DNE000
2,5-TOLUENEDIOL see MKO250
TOLUENE-3,4-DIOL see DNE200
TOLUENE-3,4-DITHIOL see TGN000
TOLUENE-α,α-DITHIOL BIS(O,O-DIMETHYL PHOSPHORODITHIOATE) see BES250
TOLUENE HEXAHYDRIDE see MIQ740
p-TOLUENENITRILE see TGT750
4-TOLUENESULFANAMIDE see TGN500
p-TOLUENESULFANAMIDE see TGN500
4-TOLUENESULFINIC ACID SODIUM SALT see SKK000
4-TOLUENESULFINYL AZIDE see TGN100
TOLUENE-2-SULFONAMIDE see TGN250
TOLUENE-4-SULFONAMIDE see TGN500
o-TOLUENESULFONAMIDE see TGN250
p-TOLUENESULFONAMIDE see TGN500
p-TOLUENESULFONAMIDE, N-(α-(CHLOROACETYL)PHENETHYL)-, (-)- see THH450
TOLUENESULFONIC ACID see TGO000
4-TOLUENESULFONIC ACID see TGO000
p-TOLUENESULFONIC ACID see TGO000
TOLUENESULFONIC ACID (liquid) see TGO100
p-TOLUENE SULFONIC ACID CHLORIDE see TGO250
p-TOLUENESULFONIC ACID HYDRAZIDE see THE250
o-TOLUENESULFONYLAMIDE see TGN000
p-TOLUENESULFONYLAMIDE see TGN500
1-p-TOLUENESULFONYL-3-BUTYLUREA see BSQ000
p-TOLUENE SULFONYL CHLORIDE see TGO250
p-TOLUENESULFONYL FLUORIDE see TGO500
N-(p-TOLUENESULFONYL)-N'-HEXAMETHYLENIMINOUREA see TGJ500
4-TOLUENESULFONYL HYDRAZIDE see TGO550
o-4-TOLUENE SULFONYL HYDROXYLAMINE see TGO575
TOLUENE-p-SULFONYLMETHYLNITROSAMIDE see THE500
TOLUENE-p-SULPHONAMIDE see TGN500
p-TOLUENESULPHONIC ACID see TGO000
2-TOLUENETHIOL see TGP000
4-TOLUENETHIOL see TGP250
o-TOLUENETHIOL see TGP000
p-TOLUENETHIOL see TGP250
α-TOLUENETHIOL see TGO750
TOLUENE TRICHLORIDE see BFL250
o-TOLUENO-AZO-β-NAPHTHOL see TGW000
α-TOLUENOL see BDX500
ar-TOLUENOL see CNW500
o-TOLUENSULFAMID (CZECH) see TGN000
p-TOLUHYDROQUINOL see MKO250
m-TOLUIC ACID see TGP750
o-TOLUIC ACID see TGQ000
p-TOLUIC ACID see TGQ250
α-TOLUIC ACID see PDY850
m-TOLUIC ACID DIETHYLAMIDE see DKC800
α-TOLUIC ACID, ETHYL ESTER see EOH000
α-TOLUIC ALDEHYDE see BBL500
o-TOLUIC NITRILE see TGT500
p-TOLUIC NITRILE see TGT750
TOLUIDENE BLUE O CHLORIDE see AJP250
m-TOLUIDIN (CZECH) see TGQ500
o-TOLUIDIN (CZECH) see TGQ750
p-TOLUIDIN (CZECH) see TGR000
2-TOLUIDINE see TGQ750
3-TOLUIDINE see TGQ500
4-TOLUIDINE see TGR000
m-TOLUIDINE see TGQ500
o-TOLUIDINE see TGQ750
p-TOLUIDINE see TGR000
m-TOLUIDINE ANTIMONYL TARTRATE see TGR250
o-TOLUIDINE ANTIMONYL TARTRATE see TGR500
p-TOLUIDINE ANTIMONYL TARTRATE see TGR750
TOLUIDINE BLUE see AJP250, TGS000
TOLUIDINE BLUE O see AJP250
2-TOLUIDINE HYDROCHLORIDE see TGS500
m-TOLUIDINE HYDROCHLORIDE see TGS250
o-TOLUIDINE HYDROCHLORIDE see TGS500
p-TOLUIDINE HYDROCHLORIDE see TGS750
TOLUIDINE RED see MMP100
p-TOLUIDINE-m-SULFONIC ACID see AKQ000

TOLYLSULFONYLBUTYLUREA see BSQ000
3-(p-TOLYL-4-SULFONYL)-1-BUTYLUREA see BSQ000
1-(p-TOLYLSULFONYL)-3-CYCLOHEXYLUREA see CPR000
4-(p-TOLYLSULFONYL)-1,1-HEXAMETHYLENESEMICARBAZIDE see TGJ500
p-TOLYLSULFONYLHYDRAZINE see THE250
p-TOLYLSULFONYL-METHYL-NITROSAMID (GERMAN) see THE500
p-TOLYLSULFONYLMETHYLNITROSAMIDE see THE500
p-TOLYLSULFONYLMETHYLNITROSAMINE see THE500
p-TOLYLTHIOL see TGP250
o-TOLYL THIOUREA see THF250
1-o-TOLYL-2-THIOUREA see THF250
N-(4'-o-TOLYL-o-TOLYLAZOSUCCINAMIC ACID see SNF500
3-TOLYLUREA see THF750
m-TOLYLUREA see THF750
p-TOLYLUREA see THG000
p-TOLYMERCURY IODIDE see IFL000
TOLYN see RLU000
m-TOLYNITRILE see TGT250
TOLYNOL see GGS000
TOLYSPAZ see GGS000
p-TOLYUREA see THG000
TOMANOL-WIRKSTOFF see INM000
TOMATES (MEXICO) see JBS100
TOMATHREL see CDS125
TOMATIDINE GLYCOSIDE see THG250
TOMATIN see THG250
TOMATINE see THG250
α-TOMATINE see THG250
TOMATO FIX CONCENTRATE see CJN000
TOMATO HOLD see CJN000
TOMATOTONE see CJN000
TOMAYMYCIN see THG500
TOMBRAN see CKJ000, THK880
TOMIL see DIR000
TOMIZINE see THG600
TOMOBIL see TGZ000
TOMOXIPROL see THG700
TOMOXIPROLE see THG700
TONARIL see TMP750
TONARSEN see DXE600
TONCARINE see MIP750
TONEDRIN see DBA800
TONEDRON see MDT600
TONER LAKE RED C see CHP500
TONEXOL see TBN000
TONITE see CDN200
TONKA ABSOLUTE see THG750
TONKA BEAN CAMPHOR see CNV000
TONOCARD see DJS200, TGI699
TONOCOR see DJS200
TONOFTAL see TGB475
TONOGEN see VGP000
TONOLYSIN see OJD300
TONOLYT ISOPROPYL MEPROBAMATE see IPU000
TONOX see MJQ000
TO NTU see BMR750
TONY RED see OHA000
TOOSENDANIN see CML825
TOOT POISON see TOE175
TOPANE see BGJ750
TOPANEL see COB260
TOPANOL see BFW750
TOPAZ see THH000
TOPAZONE see NGG500
TOPCAINE see EFX000
TOPCIDE see ANB000
TOPE-TOPES (DOMINICAN REPUBLIC) see JBS100
TOPEX see BDS000
TOP FLAKE see SFT000
TOP FORM WORMER see TEX000
TOPHOL see PGG000
TOPICAIN see DTL200
TOPICHLOR 20 see CDR750
TOPICLOR see CDR750
TOPICLOR 20 see CDR750
TOPICON see HAG325
TOPICORTE see DBA875
TOPICYCLINE see TBX250
TOPIDERM see DBA875
TOPISOLON see DBA875
TOPITOX see CJJ000
TOPITRACIN see BAC250
TOPOKAIN see AIT250
TOPOREX 855-51 see SMQ500

TOPSIN see DJV000
TOPSIN WP METHYL see PEX500
TOPSYM see DUD800, FDD150
TOPSYN see FDD150
TOPSYNE see FDD150
TOPUSYN see INR000
TOPZOL see RCF000
TORAK see DBI099
TORATE see BPG325
TORAZINA see CKP250, CKP500
TORBUTROL see BPG325
TORDON see PIB900
TORDON 10K see PIB900
TORDON 22K see PIB900
TORDON 101 MIXTURE see PIB900
TORECAN see TEZ000
TORECAN BIMALEATE see TEZ000
TORECAN DIMALEATE see TEZ000
TORECAN MALEATE see TEZ000
TORELLE see PHE250
TORENTAL see PBU100
TORESTEN see TEZ000
TORINAL see QAK000
TORMONA see BSQ750, TAA100
TORMOSYL see THH350
TORQUE see BLU000
TORSITE see BGJ250
TORULOX see GGS000
TORYN see CBG250
TOSIC ACID see TGO000
5-TOSILOXI 2-HYDROXIBENCENO SULFONATO de PIPERACINA (SPANISH)
    see PIK625
TOSTRIN see TBG000
TOSYL see CDO000
TOSYLAMIDE see TGN500
p-TOSYLAMIDE see TGN500
l-1-TOSYLAMIDO-2-PHENYLETHYL CHLOROMETHYL KETONE see THH450
TOSYLCHLORAMIDE SODIUM see CDP000
TOSYL CHLORIDE see TGO250
TOSYL FLUORIDE see TGO500
TOSYLLYSINE CHLOROMETHYL KETONE see THH500
TOSYL-l-LYSINE CHLOROMETHYL KETONE see THH500
TOSYLLYSYL CHLOROMETHYL KETONE see THH500
α-TOSYL-l-LYSYLCHLOROMETHYL KETONE see THH500
N-α-TOSYL-l-LYSYL-CHLOROMETHYLKETONE see THH500
N-TOSYL-l-PHENYLALANINE CHLOROMETHYL KETONE see THH550
TOSYL-l-PHENYLALANYLCHLOROMETHYL KETONE see THH550
TOTACEF see CCS250
TOTACILLIN see AIV500
TOTALCICLINA see AIV500
TOTAPEN see AIV500
TOTAZINA see PKD250
TOTOCAINE HYDROCHLORIDE see AIT750
TOTOMYCIN see TBX250
TOTP see TMO600
TOTRIL see DNG200, HKB500
2,5-TOULENEDIAMINE SULFATE see TGM400
TOWK see RGP450
TOX 47 see PAK000
TOXADUST see CDV100
TOXAFEEN (DUTCH) see CDV100
TOXAKIL see CDV100
TOXALBUMIN see AAD000
TOXAPHEN (GERMAN) see CDV100
TOXAPHENE see CDV100
TOXAPHENE TOXICANT A see THH560
TOXAPHENE TOXICANT B see THH575
TOXICHLOR see CDR750
TOXIFERINE DICHLORIDE see THI000
TOXIFREN see FBY000
TOXILIC ACID see MAK900
TOXILIC ANHYDRIDE see MAM000
TOXIN, ADENIA DIGITATA (MODECCA DIGITATA), MODECCIN
    see MRA000
TOXIN, ANABAENA FLOS-AQUAE NRC-44-1 see AON825
TOXIN, BACTERIUM CORYNE-BACTERIUM DIPHTHERIAE, DIPTHERIA
    see DWP300
TOXIN HT 2 see THI250
TOXIN, JELLYFISH, CHIRONEX FLECKERI see CDM700
TOXIN, JELLYFISH, SCYPHOZOAN, CYANEA CAPILLATA see COI125
TOXIN, NEMATOCYST, PHYSALIA PHYSALIS see PIA375
TOXIN, PENICILLIUM ROQUEFORTI see PAQ875
TOXIN (PENICILLIUM ROQUEFORTII) see POF800
TOXIN PR see POF800

TOXIN, SHIGELLA DYSENTERIAE see SCD750
TOXIN T2 see FQS000
TOXIN T-2 TRIOL see DAD600
TOXOBIDIN see BGS250
TOXOFACTOR see THI425
TOXOGONIN see BGS250
TOXOGONIN DICHLORIDE see BGS250
TOXOGONINE see BGS250
TOXON 63 see CDV100
TOXYLON POMIFERUM see MRN500
TOXYNON see SJP500
TOXYPHEN see CDV100
2,2'-(o-TOYLYIMINO)DIETHANOL see DMT800
TOYO ACID PHLOXINE see CMM000
TOYO AMARANTH see FAG020
TOYOCAMYCIN NUCLEOSIDE see VGZ000
TOYO EOSINE G see BNH500
TOYOMYCIN see CMK650
TOYO OIL ORANGE see PEJ500
TOYO OIL YELLOW G see DOT300
TOYO ORIENTAL OIL BLUE G see BLK000
TP see TBG000
TP-21 see MOO250, MOO500
2,4,5-TP see TIX500
TP540 see DTQ800
TP 90B see BHK750
TPA see PGV000
TPA-3-β-OL see PGV500
TPD see DXO300
TP DIOL-EPOXIDE-1 see DMT000
TP DIOL EPOXIDE-2 see DMS800
2,4,5-T PGBEE see THJ100
TPG HYDROCHLORIDE see UVA150
TP H4-1,2-DIOL see DND000
TP H4-1,2-EPOXIDE see ECT000
TPN (pesticide) see TBQ750
T 45 (POLYGLYCOL) see GHS000
TPP see TMT750
2,4,5-T PROPYLENE GLYCOL BUTYL ETHER ESTER see THJ100
TPTA see ABX250
TPTC see CLU000
TPTH see HON000
TPU 10M see PKM250
TPU 2T see PKO500
TPZA see ABX250
TR 201 see SMR000
TR5379M see COW675
TRABEST see FOS100
TRACHOSEPT see DBX400
TRADENAL see POB500
TRAFARBIOT see AOD125
TRAGACANTH see THJ250
TRAGACANTH GUM see THJ250
TRAGAYA see SEH000
TRAKIPEAL see MDQ250
TRAL see HFG000
TRALGON see HIM000
TRAMACIN see AQX500
TRAMADOL see THJ500
TRAMADOL (2) see THJ755
TRAMADOL HYDROCHLORIDE see THJ750
TRAMAL see THJ500, THJ750
TRAMAZOLINE HYDROCHLORIDE see THJ825
TRAMAZOLINE HYDROCHLORIDE MONOHYDRATE see RGP450
TRAMETAN see TFS350
TRAMISOL see LFA020
TRAMISOLE see LFA020
TRANCALGYL see EEM000
TRANCOLON see CBF000
TRANCOPAL see CKF500
TRANCYLPROMINE SULFATE see PET500
TRANDATE see HMM500
TRANEX see AJV500
TRANEXAMIC ACID see AJV500
TRANHEXAMIC ACID see AJV500
TRANID see CFF250
TRANILAST see RLK800
TRANILCYPROMINE see PET750
TRANIMUL see DCK759
TRANITE D-LAY see PBC250
TRANK see AOO500
TRAN-Q see CJR909
TRANQUIL see PGA750
TRANQUILAN see MQU750, TAF675

TRANQUILAX see CGA000
TRANQUILLIN see BCA000
TRANQUINIL see RDF000
TRANQUIRIT see DCK759
TRANQUO-BUSCOPAN-WIRKSTOFF see CFZ000
TRANSAMINE see DAA800, PET750, TAA100
TRANSAMINE SULFATE see PET500
TRANSAMLON see AJV500
TRANSANNON see ECU750
TRANSBRONCHIN see CBR675
TRANSCYCLINE see PPY250
TRANSENE see CDQ250
TRANSENTINE see DHX800, THK000
TRANSERGAN see THK600
TRANSETILE BLUE P-FER see MGG250
TRANSETILE VIOLET P 3R see DBP000
TRANSILIUM see CDQ250
TRANSIT see CHJ750
TRANSPARENT BRONZE SCARLET see CHP500
TRANSPLANTONE see NAK500
TRANSPOISE see MFD500
TRANSVAALIN see GFC000
TRANXENE see CDQ250
TRANXILEN see CDQ250
TRANXILENE see CDQ250
TRANXILIUM see CDQ250
TRANYLCYPRAMINE see PET750
TRANYLCYPRAMINE SULFATE see PET500
TRANYLCYPROMINE see PET750
TRANYLCYPROMINE SULFATE see PET500
TRANYLCYPROMINE SULPHATE see PET500
TRANZER see EHP000
TRANZETIL see DHX800
TRANZINE see CKP500
TRAPANAL see PBT500
TRAPANAL SODIUM see PBT500
TRAPEX see ISE000, VFU000
TRAPEXIDE see ISE000
TRAPIDIL see DIO200
TRAPYMIN see DIO200
TRAQUILAN see TAF675
TRAQUIZINE see CJR909
TRASAN see CIR250
TRASENTIN see DHX800
TRASENTINE see DHX800
TRASICOR see THK750
TRASULPHANE see IAD000
TRASYLOL see PAF550
TRATUL see TAB250
TRAUMANASE see BMO000
TRAUMASEPT see PKE250
TRAUSABUM see AEG875
TRAUSABUN see AEG875, TDL000
TRAVAD see BAP000
TRAVELERS JOY see CMV390
TRAVELIN see DYE600
TRAVELMIN see DYE600
TRAVELON see HGC500
TRAVENON see SKJ300
TRAVEX see SFS000
TRAWOTOX see CDO000
TRAXANOX SODIUM PENTAHYDRATE see THK850
TRAZENTYNA see DHX800
TRAZENTYNA (POLISH) see THK000
TRAZININ see PAF550
TRAZODON see THK875
TRAZODONE see THK875
TRAZODONE HYDROCHLORIDE see CKJ000, THK880
TRAZOLINONE see EQO000
TRECALMO see CKA000
TRECATOR see EPQ000
TREDIONE see TLP750
TREDUM see CKG000
TREESAIL see GJU475
TREFANOCIDE see DUV600
TREFICON see DUV600
TREFLAM see DUV600
TREFLAN see DUV600
TREFLANOCIDE ELANCOLAN see DUV600
TRE-HOLD see NAK500
TRELMAR see MQU750
TREMARIL HYDROCHLORIDE see THL500
TREMARIL WONDER see THL500
TREMBLEX see DBE000

TREMERAD see CGB250
TREMIN see BBV000
TREMOLITE ASBESTOS see ARM280
TREMORINE see DWX600
TREMORTIN A see PAR250
TRENIMON see TND000
TRENTADIL see THL750
TRENTADIL HYDROCHLORIDE see BEO750, THL750
TRENTAL see PBU100
TREOMICETINA see CDP250
TREOSULFAN see TFU500
TREPENOL WA see SIB600
TREPIBUTONE see CNG835
TREPIDAN see DAP700
TREPIDONE see MFD500
TREPOL (FRENCH) see BKX250
TRESANIL see TNP275
TRESCATYL see EPQ000
TRESCAZIDE see EPQ000
TRESITOPE see LGK050
TRESOCHIN see CLD000
TRESPAPHAN see PMP500
TREST see THL500
TRESTEN see TEZ000
TRESULFAN see TFU500
TRETAMINE see TND500
TRETINOIN see VSK950
TREVINTIX see PNW750
TRF see TNX400
TRH-T see POE050
TRIABARB see EOK000
TRI-ABRODIL see AAN000
TRIACETALDEHYDE (FRENCH) see PAI250
TRIACETIN (FCC) see THM500
TRIACETINASE see GGA800
TRIACETONITRILE TUNGSTEN TRICARBONYL see THM250
1,8,9-TRIACETOXYANTHRACENE see APH500
4-β,8-α,15-TRIACETOXY-3-α-HYDROXY-12,13-EPOXY-TRICHOTHEC-9-ENE
   see ACS500
TRIACETYL-6-AZAURIDINE see THM750
2',3',5'-TRIACETYL-6-AZAURIDINE see THM750
2',3',5'-TRI-o-ACETYL-6-AZAURIDINE see THM750
TRIACETYLDIPHENOLISATIN see ACD500
TRIACETYL GLYCERIN see THM500
2-(2',3',5'-TRIACETYL-β-d-RIBOFURANOSYL)-as-TRIAZINE-3,5-(2H,4H)-DIONE
   see THM750
TRIACYLGLYCEROL HYDROLASE see GGA800
TRIACYLGLYCEROL LIPASE see GGA800
TRIAD see TIO750
TRIADENYL see ARQ500
TRIADIMEFON see CJO250
TRIADIMENOL see MEP250
TRIAETHANOLAMIN-NG see TKP500
TRIAETHYLAMIN (GERMAN) see TJO000
TRIAETHYLENMELAMIN (GERMAN) see TND500
TRIAETHYLENPHOSPHORSAEUREAMID (GERMAN) see TND250
TRIAETHYLZINNSULFAT (GERMAN) see BLN500
TRIALLAT (GERMAN) see DNS600
TRIALLATE see DNS600
TRIALLYLAMINE see THN000
TRIALLYL BORATE see THN250
TRIALLYL CYANAURATE see THN500
TRIALLYL PHOSPHATE see THN750
TRIAMCINCOLONE ACETONIDE see AQX500
TRIAMCINOLONE see AQX250
TRIAMCINOLONE ACETONIDE see AQX500
TRIAMCINOLONE-16,17-ACETONIDE see AQX500
TRIAMCINOLONE DIACETATE see AQX750
TRIAMCINOLONE HEXACETONIDE see AQY375
TRIAMELIN see TND500
TRIAMIFOS (GERMAN, DUTCH, ITALIAN) see AIX000
2,4,7-TRIAMINO-6-FENILPTERIDINA (ITALIAN) see UVJ450
TRIAMINOGUANIDINE NITRATE see THN800
TRIAMINOGUANIDINIUM PERCHLORATE see THO250
2,4,6-TRIAMINOPHENOL TRIHYDROCHLORIDE see THO500
TRIAMINOPHENYL PHOSPHATE see THO550
2,4,7-TRIAMINO-6-PHENYLPTERIDINE see UVJ450
2,4,6-TRIAMINO-s-TRIAZINE see MCB000
2,4,6-TRIAMINO-s-TRIAZINE compounded with s-TRIAZINE-TRIOL
   see THO750
1,3,5-TRIAMINOTRINITROBENZENE see THO775
TRIAMINOTRIPHENYLMETHANE see THP000
4,4',4''-TRIAMINOTRIPHENYLMETHANE see THP000
p,p',p''-TRIAMINOTRIPHENYLMETHANE see THP000

4,4'4''-TRIAMINOTRIPHENYLMETHAN-HYDROCHLORID (GERMAN)
   see RMK020
TRIAMIPHOS see AIX000
TRIAMMINEDIPEROXOCHROMIUM(IV) see THP250
TRIAMMINE GOLDTRIHYDROXIDE see THP500
TRIAMMINENITRATOPLANTINUM(II) NITRATE see THP750
cis-TRIAMMINETRICHLORORUTHENIUM(III) see THQ000
TRIAMMINETRINITROCOBALT(III) see THQ250
TRIAMMONIUM ALUMINUM HEXAFLUORIDE see THQ500
TRIAMMONIUM HEXAFLUOROALUMINATE see THQ500
TRIAMMONIUM TRIS-(ETHANEDIOATO(2-)-O,O')FERRATE(3-1) see ANG925
TRIAMPHOS see AIX000
TRIAMPUR see UVJ450
TRIAMTEREN see UVJ450
TRIAMTERENE see UVJ450
TRIAMTERIL see UVJ450
TRIAMTERIL COMPLEX see UVJ450
TRI-n-AMYL BORATE see TMQ000
TRIANATE see TJL250
TRIANGLE see CNP500
TRI-p-ANISYLCHLOROETHYLENE see CLO750
TRIANOL DIRECT BLUE 3B see CMO250
TRIANON see PPO000
1,4-TRIANTHRIMID (CZECH) see APL750
1,5-TRIANTHRIMID (CZECH) see APM000
TRIANTINE LIGHT BROWN 3RN see FMU059
TRIANTOIN see MKB250
TRIASOL see TIO750
TRIASYN see DEV800
TRIATOMIC OXYGEN see ORW000
TRIATOX see MJL250
3,5,7-TRIAZA-1-AZONIAADAMANTANE, 1-(3-CHLOROALLYL)-, CHLORIDE
   see CEG550
1,2,3-TRIAZAINDENE see BDH250
3,5,7-TRIAZAINDOLE see POJ250
3,6,9-TRIAZATETRACYCLO[6.1.0.0^{2,4}.0^{5,7}] NONANE see THQ600
TRIAZENE see THQ750
1-TRIAZENE see THQ750
p,p'-TRIAZENYLENEDIBENZENESULFONAMIDE see THQ900
TRIAZICHON (GERMAN) see TND000
TRIAZIDOBORANE see BMG325
TRIAZIDOMETHYLIUM HEXACHLOROANTIMONATE see THR100
2,4,6-TRIAZIDO-1,3,5-TRIAZINE see THR250
TRIAZIN see DEV800
TRIAZINATE see THR500
1,3,5-TRIAZINE see THR525
TRIAZINE (pesticide) see DEV800
TRIAZINE A 1294 see ARQ725
1,3,5-TRIAZINE-2,4-DIAMINE, 6-CHLORO-N,N,N',N'-TETRAETHYL- (9CI)
   see AKQ250
s-TRIAZINE, 1,2-DIHYDRO-1-(p-CHLOROPHENYL)-4,6-DIAMINO-2,2-
   DIMETHYL- see COX400
s-TRIAZINE-3,5(2H,4H)-DIONE see THR750
s-TRIAZINE TRICHLORIDE see TJD750
1,3,5-TRIAZINE-2,4,6-TRIMERCAPTAN see THS250
sym-TRIAZINETRIOL see THS000
s-TRIAZINE-2,4,6-TRIOL see THS000
s-2,4,6-TRIAZINETRIOL see THS000
s-TRIAZINE-2,4,6-TRIOL, TRI(2,3-EPOXYPROPYL) ESTER see TKL250
s-TRIAZINE-2,4,6(1H,3H,5H)-TRIONE see THS000
2,4,6-TRIAZINETRITHIOL see THS250
s-TRIAZINE-2,4,6-TRITHIOL see THS250
1,3,5-TRIAZINE-2,4,6(1H,3H,5H)-TRITHIONE see THS250
(1,3,5-TRIAZINE-2,4,6-TRIYLTRINITRILO)HEXAKIS METHANOL see HDY000
(s-TRIAZINE-2,4,6-TRIYLTRINITRILO)HEXAMETHANOL see HDY000
1,1',1''-s-TRIAZINE-2,4,6-TRIYLTRISAZIRIDINE see TND500
s-TRIAZIN-2,4,6-TRIYLTRIAMINOMETHANESULFONIC ACID TRISODIUM
   SALT see THS500
(s-TRIAZIN-2,4,6-TRIYLTRIAMINO)TRISMETHANESULFONIC ACID TRI-
   SODIUM SALT see THS500
TRIAZIQUINONE see TND000
TRIAZIQUONE see TND000
TRIAZIRIDINOPHOSPHINE OXIDE see TND250
2,3,5-TRI-(1-AZIRIDINYL)-p-BENZOQUINONE see TND000
TRI(AZIRIDINYL)PHOSPHINE OXIDE see TND250
TRI-1-AZIRIDINYL)PHOSPHINE OXIDE see TND250
TRIAZIRIDINYLPHOSPHINE SULFIDE see TFQ750
TRIAZIRIDINYL TRIAZINE see TND500
TRIAZOFOSZ (HUNGARIAN) see THT750
TRIAZOIC ACID see HHG500
TRIAZOLAM see THS800
TRIAZOLAMINE see AMY050
1H-1,2,4-TRIAZOL-3-AMINE see AMY050
1,2,3-TRIAZOLE see THS850

s-TRIAZOLE, 5-(m-(ALLYLOXY)PHENYL)-3-(o-TOLYL)- see AGP400
1H-1,2,4-TRIAZOLE, 3,5-BIS(3,4-DIMETHOXYPHENYL)- see BJE700
s-TRIAZOLE, 3,5-BIS(o-TOLYL)- see BLK500
s-TRIAZOLE, 3-(o-BUTYLPHENYL)-5-(m-METHOXYPHENYL)- see BSG100
s-TRIAZOLE, 3-(o-BUTYLPHENYL)-5-PHENYL- see BSH075
s-TRIAZOLE, 5-(o-CHLOROPHENYL)-3-(o-TOLYL)- see CKL100
s-TRIAZOLE, 3,5-DIPHENYL- see DWO950
1H-1,2,4-TRIAZOLE, 3,5-DIPHENYL- see DWO950
1H-1,2,4-TRIAZOLE-1-ETHANOL, α-α-BIS(4-FLUOROPHENYL)- see BJW600
1H-1,2,4-TRIAZOLE-1-ETHANOL, β-((2,4-DICHLOROPHENYL)METHYLENE)-α-
    (1,1-DIMETHYLETHYL)-, (E)- see DGC100
1H-1,2,4-TRIAZOLE-3-THIOL see THT000
TRIAZOLOGUANINE see AJO500
1,2,4-TRIAZOLO[4,3-a]PYRIDINE-SILVER NITRATE see THT250
s-TRIAZOLO(4,3-a)PYRIDIN-3(2H)-ONE, 2-(3-(4-(m-CHLOROPHENYL)-1-
    PIPERAZINYL)PROPYL)-, MONOHYDROCHLORIDE see THK880
v-TRIAZOLO(4,5-d)PYRIMIDINE-5,7-DIOL see THT350
1H-v-TRIAZOLO(4,5-d)PYRIMIDINE-5,7(4H,6H)-DIONE see THT350
(1H-1,2,4-TRIAZOLYL)TRICYCLOHEXYLSTANNANE see THT500
1H-1,2,4-TRIAZOL-1-YL)TRIPHENYLSTANNANE see TMX000
(1H-1,2,4-TRIAZOLYL-1-YL)TRICYCLOHEXYLSTANNANE see THT500
TRIAZOPHOS see THT750
TRIAZOTION (RUSSIAN) see EKN000
TRIAZURE see THM750
TRIB see SNK000
TRIBAC see TIK500
TRI-BAN see PIH175
TRIBASIC SODIUM PHOSPHATE see SJH200
TRIBENZOIN see GGU000
TRIBENZO(a,e,i)PYRENE see DDC200
(1,2,4,5,7,8)TRIBENZOPYRENE see DDC200
(1,2,4,5,8,9)TRIBENZOPYRENE see DDC200
1,2:4,5:8,9-TRIBENZOPYRENE see DDC200
TRIBENZYLARSINE see THU000
TRIBENZYLCHLOROSTANNANE see CLP000
TRIBENZYLSULFONIUM IODIDE MERCURIC IODIDE see THU250
TRIBENZYLSULFONIUM IODIDE, compounded with MERCURY IODIDE
    (1:1) see THU250
TRIBENZYLTIN CHLORIDE see CLP000
TRIBENZYLTIN FORMATE see FOE000
TRIBONATE see DVC200
TRIBORON PENTAFLUORIDE see THU275
TRIBROMAMINE HEXAAMMONIATE see NGV500
TRIBROMETHANOL see ARW250, THV000
TRIBROMMETHAAN (DUTCH) see BNL000
TRIBROMMETHAN (GERMAN) see BNL000
TRIBROMOALDEHYDE HYDRATE see THU500
TRIBROMOALUMINUM see AGX750
2,4,6-TRIBROMOANILINE see THU750
sym-TRIBROMOANILINE see THU750
TRIBROMOARSINE see ARF250
TRIBROMOETHANOL see THV000
2,2,2-TRIBROMOETHANOL see ARW250
TRIBROMOETHYL ALCOHOL see ARW250, THV000
2,2,2-TRIBROMOETHYL ALCOHOL see ARW250
1,3,7-TRIBROMO-2-FLUORENAMINE see THV250
1,3,7-TRIBROMOFLUOREN-2-AMINE see THV250
2,4,5-TRIBROMOIMIDAZOLE see THV450
2,4,5-TRIBROMOIMIDAZOLE CADMIUM SALT (2581) see THV500
TRIBROMOMETAN (ITALIAN) see BNL000
TRIBROMOMETHANE see BNL000
TRIBROMONITROMETHANE see NMQ000
TRIBROMOPHENOL see THV750
2,4,6-TRIBROMOPHENOL see THV750
1-(2,4,6-TRIBROMOPHENYL)-3,3-DIMETHYLTRIAZENE see DUI800
TRIBROMOPHOSPHINE see PHT250
1,2,3-TRIBROMOPROPANE see GGG000
sym-TRIBROMOPROPANE see GGG000
3,4′,5-TRIBROMOSALICYLANILIDE see THW750
TRIBROMOSILANE see THX000
TRIBROMO STIBINE see AQK000
TRIBROMOTRIMETHYLDIALUMINUM see MGC225
TRIBUFON see BPG000
TRIBURON see TJE880
TRIBURON CHLORIDE see TJE880
TRIBUTILFOSFATO (ITALIAN) see TIA250
TRIBUTON see DAA800, TAA100
TRIBUTOXYBORANE see THX750
TRI-n-BUTOXYBORANE see THX750
TRI(2-BUTOXYETHANOL PHOSPHATE) see BPK250
TRIBUTOXYETHYL PHOSPHATE see BPK250
TRI(2-BUTOXYETHYL) PHOSPHATE see BPK250
TRIBUTYLAMINE see THX250
TRI-n-BUTYLAMINE see THX250

TRI-n-BUTYL BORANE see THX500
TRIBUTYL BORATE see THX750
TRI-n-BUTYL BORATE see THX750
TRI-sec-BUTYL BORATE see THY000
TRIBUTYLBORINE see THX500
TRIBUTYL CELLOSOLVE PHOSPHATE see BPK250
TRIBUTYLCHLOROSTANNANE see CLP500
TRIBUTYL(2,4-DICHLOROBENZYL)PHOSPHONIUM CHLORIDE see THY500
TRI-n-BUTYL-2,4-DICHLOROPHENOXYTIN see DGB400
TRIBUTYLE (PHOSPHATE de) (FRENCH) see TIA250
TRIBUTYL((2-ETHYLHEXANOYL)OXY)STANNANE see TID250
TRIBUTYL(2-ETHYL-1-OXOHEXYL)OXY)STANNANE see TID250
TRIBUTYLFOSFAAT (DUTCH) see TIA250
TRIBUTYLFOSFINSULFID see TIA450
TRIBUTYL(GLYCOLOYLOXY)STANNANE see GHQ000
TRIBUTYL(GLYCOLOYLOXY)TIN see GHQ000
TRIBUTYLHYDROXYSTANNANE see TID500
TRIBUTYLISOCYANATOSTANNANE see THY750
TRIBUTYLLEAD ACETATE see THY850
TRIBUTYL(METHACRYLOXY)STANNANE see THZ000
TRIBUTYL(METHACRYLOYLOXY)STANNANE see THZ000
TRIBUTYL(METHACRYLOYLOXY)-STANNANE POLYMER with METHYL
    METHACRYLATE (8CI) see OIY000
TRIBUTYL((2-METHYL-1-OXO-2-PROPENYL)OXY)STANNANE see THZ000
TRIBUTYL(NEODECANOYLOXY)STANNANE see TIF250
TRIBUTYL(OLEOYLOXY)STANNANE see TIA000
TRIBUTYL(OLEOYLOXY)TIN see TIA000
TRIBUTYL((1-OXODODECYL)OXY)STANNANE (9CI) see TIE750
(TRIBUTYL)PEROXIDE see BSC750
TRIBUTYL-o-PHENYLPHENOXYTIN see BGK000
TRIBUTYLPHOSPHAT (GERMAN) see TIA250
TRIBUTYL PHOSPHATE see TIA250
TRI-n-BUTYL PHOSPHATE see TIA250
TRIBUTYL-PHOSPHINE compounded with NICKELCHLORIDE (2:1) see BLS250
TRIBUTYLPHOSPHINE SULFIDE see TIA450
TRIBUTYL PHOSPHITE see TIA750
S,S,S-TRIBUTYL PHOSPHOROTRITHIOATE see BSH250
TRIBUTYL PHOSPHOROTRITHIOITE see TIG250
S,S,S-TRIBUTYL PHOSPHOROTRITHIOITE see TIG250
TRI-n-BUTYLPLUMBYL ACETATE see THY850
TRIBUTYL(8-QUINOLINOLATO)TIN see TIB000
TRIBUTYLSTANNANECARBONITRILE see TIB250
TRIBUTYLSTANNANE FLUORIDE see FME000
TRI-n-BUTYLSTANNANE HYDRIDE see TIB500
TRI-n-BUTYL-STANNANE OXIDE see BLL750
TRIBUTYLSTANNIC HYDRIDE see TIB500
TRIBUTYLSTANNYL ISOCYANATE see THY750
TRIBUTYLSTANNYL METHACRYLATE see THZ000
TRIBUTYLTIN-p-ACETAMIDOBENZOATE see TIB750
TRIBUTYLTIN ACETATE see TIC000
TRIBUTYLTIN BENZOATE see BDR750
TRI-n-BUTYLTIN BROMIDE see TIC250
TRI-n-BUTYLTIN CHLORIDE see CLP500
TRIBUTYLTIN CHLORIDE COMPLEX see TIC500
TRIBUTYLTIN CHLOROACETATE see TIC750
TRIBUTYLTIN-γ-CHLOROBUTYRATE see TID000
TRIBUTYLTIN CYANATE see COI000
TRI-n-BUTYLTIN CYANIDE see TIB250
TRIBUTYLTIN CYCLOHEXANECARBOXYLATE see TID100
TRIBUTYLTIN-S,S'-DIBUTYLDITHIOCARBAMATE see DEB600
TRIBUTYLTIN DIMETHYLDITHIOCARBAMATE see TID150
TRIBUTYLTIN DODECANOATE see TIE750
TRIBUTYLTIN-2-ETHYLHEXANOATE see TID250
TRIBUTYLTIN FLUORIDE see FME000
TRIBUTYLTIN HYDRIDE see TIB500
TRI-n-BUTYLTIN HYDRIDE see TIB500
TRIBUTYLTIN HYDROXIDE see TID500
TRI-N-BUTYL TIN IODIDE see IFM000
TRIBUTYLTIN IODOACETATE see TID750
TRIBUTYLTIN-o-IODOBENZOATE see TIE000
TRIBUTYLTIN-p-IODOBENZOATE see TIE250
TRIBUTYLTIN-β-IODOPROPIONATE see TIE500
TRIBUTYLTIN ISOCYANATE see THY750
TRI-n-BUTYLTIN ISOCYANATE see THY750
TRIBUTYLTIN ISOOCTYLTHIOACETATE see TDO000
TRIBUTYLTIN ISOPROPYLSUCCINATE see TIE600
TRIBUTYLTIN ISOTHIOCYANATE see THY750
TRIBUTYLTIN LAURATE see TIE750
TRIBUTYL TIN LINOLEATE see LGJ000
TRIBUTYLTIN METHACRYLATE see THZ000
TRI-n-BUTYLTIN METHANESULFONATE see TIF000
TRIBUTYLTIN MONOLAURATE see TIE750
TRIBUTYLTIN NEODECANOATE see TIF250
TRIBUTYLTIN NONANOATE see TIF500

TRI-N-BUTYLTIN OLEATE see TIA000
TRIBUTYLTIN OXIDE see BLL750
TRIBUTYLTIN-o-PHENYLPHENOXIDE see BGK000
TRIBUTYLTIN SALICYLATE see SAM000
TRI-N-BUTYLTIN SALICYLATE see SAM000
TRIBUTYLTIN SULFATE see TIF600
TRIBUTYLTIN SULFIDE see HCA700
TRIBUTYLTIN-α-(2,4,5-TRICHLOROPHENOXY)PROPIONATE see TIF750
TRI-n-BUTYLTIN UNDECYLATE see TIG500
TRIBUTYLTIN UNDECYLENATE see TIG500
TRIBUTYL(2,4,5-TRICHLOROPHENOXY)STANNANE see TIG000
TRIBUTYL(2,4,5-TRICHLOROPHENOXY)TIN see TIG000
S,S,S-TRIBUTYL TRITHIOPHOSPHATE see BSH250
S,S,S-TRIBUTYL TRITHIOPHOSPHITE see TIG250
TRIBUTYL(UNDECANOYLOXY)STANNANE see TIG500
TRI-n-BUTYL-ZINN-ACETAT (GERMAN) see TIC000
TRI-N-BUTYL-ZINN BENZOATE (GERMAN) see BDR750
TRI-n-BUTYLZINN-CHLORID (GERMAN) see CLP500
TRI-n-BUTYLZINN-LAURAT (GERMAN) see TIE750
TRI-n-BUTYL-ZINN OLEAT (GERMAN) see TIA000
TRI-n-BUTYL-ZINN SALICYLAT (GERMAN) see SAM000
TRI-n-BUTYL-ZINN UNDECYLAT (GERMAN) see TIG500
TRIBUTYRASE see GGA800
TRIBUTYRIN see TIG750
TRIBUTYRINASE see GGA800
TRIBUTYRIN ESTERASE see GGA800
TRIBUTYROIN see TIG750
TRICADMIUM DINITRIDE see TIH000
TRICAINE METHANE SULFONATE see EFX500
TRICALCIUMARSENAT (GERMAN) see ARB750
TRICALCIUM ARSENATE see ARB750
TRICALCIUM DINITRIDE see TIH000
TRICALCIUM DIPHOSPHIDE see CAW250
TRICALCIUM PHOSPHATE see CAW120
TRICALCIUM SILICATE see TIH600
TRICAMBA see TIK000
TRICAPRONIN see GGK000
TRICAPRYLIC GLYCERIDE see TMO000
TRICAPRYLIN see TMO000
TRICAPRYLMETHYLAMMONIUM CHLORIDE see MQH000
TRICAPRYLYLAMINE see DVL000
TRICAPRYLYLMETHYLAMMONIUM CHLORIDE see MQH000
TRICARBALLYLIC ACID-β-ACETOXYTRIBUTYL ESTER see ADD750
TRICARBAMIX Z see BJK500
TRICESIUM NITRIDE see TIH750
TRICESIUM TRICHLORIDE see CDD000
TRICESIUM TRIFLUORIDE see CDD500
TRICETAMIDE see TIH800
TRICETO 3-7-12 CHOLANATE de Na (FRENCH) see SGD500
TRICHAZOL see MMN250
TRICHLOORAZIJNZUUR (DUTCH) see TII250
1,1,1-TRICHLOOR-2,2-BIS(4-CHLOOR FENYL)-ETHAAN (DUTCH) see DAD200
2,2,2-TRICHLOOR-1,1-BIS(4-CHLOOR FENYL)-ETHANOL (DUTCH) see BIO750
1,1,1-TRICHLOORETHAAN (DUTCH) see MIH275
TRICHLOORETHEEN (DUTCH) see TIO750
TRICHLOORETHYLEEN, TRI (DUTCH) see TIO750
(2,4,5-TRICHLOOR-FENOXY)-AZIJNZUUR (DUTCH) see TAA100
2-(2,4,5-TRICHLOOR-FENOXY)-PROPIONZUUR (DUTCH) see TIX500
O-(2,4,5-TRICHLOOR-FENYL)-O,O-DIMETHYL-MONOTHIOFOSFAAT
   (DUTCH) see RMA500
TRICHLOORFON (DUTCH) see TIQ250
TRICHLOORMETHAAN (DUTCH) see CHJ500
TRICHLOORMETHYLBENZEEN (DUTCH) see BFL250
TRICHLOORNITROMETHAAN (DUTCH) see CKN500
TRICHLOORSILAAN (DUTCH) see TJD500
TRICHLORACETALDEHYD-HYDRAT (GERMAN) see CDO000
TRICHLOR-ACETONITRIL (GERMAN) see TII750
TRICHLORACRYLYL CHLORIDE (GERMAN) see TIJ500
1,1,1-TRICHLORAETHAN (GERMAN) see MIH275
TRICHLORAETHEN (GERMAN) see TIO750
TRICHLORAETHYLEN, TRI (GERMAN) see TIO750
2,3,3-TRICHLORALLYL-N,N-(DIISOPROPYL)-THIOCARBAMAT (GERMAN)
   see DNS600
TRICHLORAMINE see NGQ500
TRICHLORAN see TIO750
2,3,6-TRICHLORBENZOESAEURE (GERMAN) see TIK500
1,1,1-TRICHLOR-2,2-BIS(4-CHLOR-PHENYL)-AETHAN (GERMAN) see DAD200
1,1,1-TRICHLOR-2,2-BIS(4-CHLORPHENYL)-AETHANOL (GERMAN)
   see BIO750
2,2,2-TRICHLOR-1,1-BIS(4-CHLOR-PHENYL)-AETHANOL (GERMAN)
   see BIO750
1,1,1-TRICHLOR-2,2-BIS(4-METHOXY-PHENYL)-AETHAN (GERMAN)
   see MEI450
TRICHLORESSIGSAEURE (GERMAN) see TII250

TRICHLORESSIGSAURES NATRIUM (GERMAN) see TII500
1,1,2-TRICHLORETHANE see TIN000
TRICHLORETHANOL see TIN500
TRICHLORETHENE (FRENCH) see TIO750
TRICHLORETHYLENE, TRI (FRENCH) see TIO750
TRICHLORETHYL PHOSPHATE see CGO500
2,4,6-TRICHLORFENOL (CZECH) see TIW000
2,3,5-TRICHLORFENOXYOCTAN 1-FENYL-2,2-DIMETHYL-4,6-DIAMINO-1,2-
   DIHYDRO-1,3,5-TRIAZINU (CZECH) see DMF400
TRICHLORFENSON (OBS.) see CJT750
2,4,6-TRICHLORFENYLESTER KYSELINY OCTOVE (CZECH) see TIY250
TRICHLORFON (USDA) see TIQ250
TRICHLORINATED ISOCYANURIC ACID see TIQ750
TRICHLORINE NITRIDE see NGQ500
TRICHLORMETAFOS-3 see TIR250
TRICHLORMETAZID see HII500
TRICHLORMETHAN (CZECH) see CHJ500
TRICHLORMETHIAZIDE see HII500
TRICHLORMETHINE see TNF250, TNF500
TRICHLORMETHINIUM CHLORIDE see TNF500
TRICHLORMETHYLBENZOL (GERMAN) see BFL250
TRICHLORMETHYLESTER KYSELINY DICHLORMETHANTHIOSULFONOVE
   (CZECH) see DFS600
N-(TRICHLOR-METHYLTHIO)-PHTHALAMID (GERMAN) see TIT250
N-(TRICHLOR-METHYLTHIO)-PHTHALIMID (GERMAN) see CBG000
TRICHLORNITROMETHAN (GERMAN) see CKN500
TRICHLOROACETALDEHYDE HYDRATE see CDO000
TRICHLOROACETALDEHYDE MONOETHYLACETAL see TIO000
TRICHLOROACETALDEHYDE MONOHYDRATE see CDO000
TRICHLOROACETALDEHYDE OXIME see TIH825
TRICHLOROACETAMIDE see TII000
2,2,2-TRICHLOROACETAMIDE see TII000
α,α,α-TRICHLOROACETAMIDE see TII000
TRICHLOROACETIC ACID see TII250
TRICHLOROACETIC ACID, solid (DOT) see TII250
TRICHLOROACETIC ACID, solution (DOT) see TII250
TRICHLOROACETIC ACID CHLORIDE see TIJ150
TRICHLOROACETIC ACID compounded with N'-(4-CHLOROPHENYL)-N,N-
   DIMETHYLUREA (1:1) see CJY000
TRICHLOROACETIC ACID SODIUM SALT see TII500
TRICHLOROACETIC ACID-2-(2,4,5-TRICHLOROPHENOXY)ETHYL ESTER
   see TIX250
TRICHLOROACETIC ACID TRIPROPYLSTANNYL ESTER see TNC000
TRICHLOROACETOCHLORIDE see TIJ150
1,1,3-TRICHLOROACETONE see TII550
α,α',α'-TRICHLOROACETONE see TII550
TRICHLOROACETONITRILE see TII750
(TRICHLOROACETOXY)TRIPROPYLSTANNANE see TNC000
10-TRICHLOROACETYL-1,2-BENZANTHRACENE see TIJ000
TRICHLOROACETYL CHLORIDE see TIJ150
TRICHLOROACETYL FLUORIDE see TIJ175
2,3,3-TRICHLOROACROLEIN see TIJ250
TRICHLOROACRYLOYL CHLORIDE see TIJ500
2,3,3-TRICHLOROACRYLOYL CHLORIDE see TIJ500
TRICHLOROACRYLYL CHLORIDE see TIJ500
2,3,3-TRICHLOROALLYL DIISOPROPYLTHIOCARBAMATE see DNS600
S-2,3,3-TRICHLOROALLYL-N,N-DIISOPROPYLTHIOCARBAMATE see DNS600
TRICHLOROALLYLSILANE see AGU250
TRICHLOROALUMINUM see AGY750
3,5,6-TRICHLORO-4-AMINOPICOLINIC ACID see PIB900
2,4,6-TRICHLOROANILINE see TIJ750
sym-TRICHLOROANILINE see TIJ750
3,5,6-TRICHLORO-o-ANISIC ACID see TIK000
TRICHLOROARSINE see ARF500
2,4,6-TRICHLOROBENZENAMINE see TIJ750
1,2,4-TRICHLOROBENZENE see TIK250
unsym-TRICHLOROBENZENE see TIK250
1,2,4-TRICHLOROBENZENE, liquid (DOT) see TIK250
2,3,6-TRICHLOROBENZENEACETIC ACID see TIY500
2,4,5-TRICHLOROBENZENEDIAZO p-CHLOROPHENYL SULFIDE see CDS500
TRICHLOROBENZOIC ACID see TIK500
2,3,6-TRICHLOROBENZOIC ACID see TIK500
2,3,6-TRICHLOROBENZOIC ACID DIMETHYLAMINE SALT see DOR800
N,N-TRICHLORO-p-BENZOQUINONE IMINE see CHR000
TRICHLOROBENZYL CHLORIDE see TIL250
1,1,1-TRICHLORO-2,2-BIS(p-ANISYL)ETHANE see MEI450
TRICHLOROBIS (4-CHLOROPHENYL) ETHANE see DAD200
2,2,2-TRICHLORO-1,1-BIS(4-CHLOROPHENYL)-ETHANOL (FRENCH) see
   BIO750
2,2,2-TRICHLORO-1,1-BIS(4-CLORO-FENIL)-ETANOLO (ITALIAN) see BIO750
1,1,1-TRICHLORO-2,2-BIS(p-FLUOROPHENYL) ETHANE see FHJ000
1,1,1-TRICHLORO-2,2-BIS(p-HYDROXYPHENYL)ETHANE see BKI500
1,1,1-TRICHLORO-2,2-BIS(p-METHOXYPHENOL)ETHANOL see MEI450
1,1,1-TRICHLORO-2,2-BIS(p-METHOXYPHENYL)ETHANE see MEI450

2,4,5-TRICHLOROPHENOL, SODIUM SALT see SKK500
2,4,5-TRICHLOROPHENOXYACETIC ACID see TAA100
(2,4,5-TRICHLOROPHENOXY)ACETIC ACID-2-BUTOXYETHYL ESTER
   see TAH900
2,4,5-TRICHLOROPHENOXYACETIC ACID, BUTYL ESTER see BSQ750
2,4,5-TRICHLOROPHENOXY, ACETIC ACID, ISOOCTYL ESTER see TGF200
(2,4,5,-TRICHLOROPHENOXY)ACETIC ACID, ISOPROPYL ESTER see TGF210
(2,4,5,-TRICHLOROPHENOXY)ACETIC ACID-1-METHYL ESTER (9CI)
   see TGF210
2,4,5-TRICHLOROPHENOXYACETIC ACID, PROPYLENE GLYCOL BUTYL
   ETHER ESTERS see THJ100
4-(2,4,5-TRICHLOROPHENOXY)BUTYRIC ACID see TIW750
2-(2,4,5-TRICHLOROPHENOXY)ETHANOL see TIX000
2,4,5-TRICHLOROPHENOXYETHYL-α,α-DICHLOROPROPIONATE see PBK000
2-(2,4,5-TRICHLOROPHENOXY)ETHYL-2,2-DICHLOROPROPIONATE
   see PBK000
2,4,5-TRICHLOROPHENOXYETHYL-α,α,α-TRICHLOROACETATE see TIX250
2-(2,4,5-TRICHLOROPHENOXY)PROPIONIC ACID see TIX500
2,4,5-TRICHLOROPHENOXY-α-PROPIONIC ACID see TIX500
α-(2,4,5-TRICHLOROPHENOXY)PROPIONIC ACID see TIX500
2-(2,4,5-TRICHLOROPHENOXY)PROPIONIC ACID PROPYLENE GLYCOL
   BUTYL ETHER ESTER see TIX750
2-(2,4,5-TRICHLOROPHENOXY)PROPIONIC ACID TRIBUTYLSTANNYL
   ESTER see TIF750
(2,4,5-TRICHLOROPHENOXY)SODIUM see SKK500
2,4,6-TRICHLOROPHENYL ACETATE see TIY250
2,3,6-TRICHLOROPHENYLACETIC ACID see TIY500
2,4,5-TRICHLOROPHENYLAZO-4'-CHLOROPHENYL-SULFIDE see CDS500
TRI-o-CHLOROPHENYL BORATE see TIY750
2,4,6-TRICHLORO-PHENYLDIMETHYLTRIAZENE see TJA000
1-(2,4,6-TRICHLOROPHENYL)-3,3-DIMETHYLTRIAZENE see TJA000
2,4,5-TRICHLOROPHENYL-γ-IODOPROPARGIL ETHER see IFA000
2,4,5-TRICHLOROPHENYL IODOPROPARGYL ETHER see IFA000
TRICHLOROPHENYLMETHANE see BFL250
1-(2,4,6-TRICHLOROPHENYL)-3-p-NITROANILINO-2-PYRAZOLIN-5-ONE
   see TJA500
2,4,6-TRICHLOROPHENYL-4-NITROPHENYL ETHER see NIW500
TRICHLOROPHENYLSILANE see TJA750
TRICHLOROPHON see TIQ250
TRICHLOROPHOSPHINE SULFIDE see TFO000
2,4,6-TRICHLORO-PMDT see TJA000
1,1,1-TRICHLOROPROPANE see TJB000
1,1,2-TRICHLOROPROPANE see TJB250
1,2,2-TRICHLOROPROPANE see TJB500
1,2,3-TRICHLOROPROPANE see TJB600
TRICHLOROPROPANE OXIDE see TJC250
1,2,3-TRICHLOROPROPANE-2,3-OXIDE see TJB750
1,1,1-TRICHLORO-2-PROPANOL see IMQ000
1,1,3-TRICHLORO-2-PROPANONE see TII550
2,3,3-TRICHLOROPROPENAL see TIJ250
2,3,3-TRICHLORO-2-PROPENAL see TIJ250
1,2,3-TRICHLOROPROPENE see TJC000
TRICHLOROPROPENE OXIDE see TJC250
1,1,1-TRICHLOROPROPENE OXIDE see TJC250
3,3,3-TRICHLOROPROPENE OXIDE see TJC250
1,1,1-TRICHLOROPROPENE-2,3-OXIDE see TJC250
2,3,3-TRICHLORO-2-PROPEN-1-OL see TJC500
2,3,4-TRICHLORO-2-PROPENOYL CHLORIDE see TIJ500
2,2,3-TRICHLOROPROPIONALDEHYDE see TJC750
TRICHLOROPROPIONITRILE see TJC800
TRICHLOROPROPYLENE see TJB750
1,1,1-TRICHLOROPROPYLENE OXIDE see TJC250
3,3,3-TRICHLOROPROPYLENE OXIDE see TJC250
TRICHLOROPROPYLSILANE see PNX250
3,5,6-TRICHLORO-2-PYRIDINOL-O-ESTER with O,O-DIETHYL PHOS-
   PHOROTHIOATE see CMA100
3,5,6-TRICHLORO-2-PYRIDYLOXYACETIC ACID see TJE890
TRICHLOROSILANE see TJD500
TRICHLOROSTIBINE see AQC500
2,2,2-TRICHLORO-1-(2-THIAZOLYLAMINO)ETHANOL see CMB675
α,α,α-TRICHLOROTOLUENE see BFL250
ω,ω,ω-TRICHLOROTOLUENE see BFL250
TRICHLORO-s-TRIAZINE see TJD750
1,3,5-TRICHLOROTRIAZINE see TJD750
2,4,6-TRICHLOROTRIAZINE see TJD750
sym-TRICHLOROTRIAZINE see TJD750
2,4,6-TRICHLORO-s-TRIAZINE see TJD750
2,4,6-TRICHLORO-1,3,5-TRIAZINE see TJD750
TRICHLORO-s-TRIAZINETRIONE see TIQ750
1,3,5-TRICHLORO-1,3,5-TRIAZINETRIONE see TIQ750
TRICHLORO-s-TRIAZINE-2,4,6(1H,3H,5H)-TRIONE see TIQ750
2,2',2"-TRICHLOROTRIETHYLAMINE see TNF250
2,2',2"-TRICHLOROTRIETHYLAMINE HYDROCHLORIDE see TNF500
1,3,5-TRICHLORO-2,4,6-TRIFLUOROBORAZINE see TJE050
TRICHLOROTRIFLUOROETHANE see FOO000
1,1,2-TRICHLORO-1,2,2-TRIFLUOROETHANE (OSHA, ACGIH, MAK)
   see FOO000
TRICHLOROTRIMETHYLDIALUMINUM see MGC230
1,3,5-TRICHLORO-2,4,6-TRIOXOHEXAHYDRO-s-TRIAZINE see TIQ750
TRICHLORO TRITANIUM see TGG250
TRICHLORO(VINYL)SILANE see TIN750
TRICHLOROVINYL SILICANE see TIN750
TRICHLORPHENE see TIQ250
(2,4,5-TRICHLOR-PHENOXY)-ESSIGSAEURE (GERMAN) see TAA100
2-(2,4,5-TRICHLOR-PHENOXY)-PROPIONSAEURE (GERMAN) see TIX500
O-(2,4,5-TRICHLOR-PHENYL)-O,O-DIMETHYL-MONOTHIOPHOSPHAT
   (GERMAN) see RMA500
2,3,6-TRICHLORPHENYLESSIGSAEURE (GERMAN) see TIY500
TRICHLORPHON see TIQ250
TRICHLORPHON FN see TIQ250
TRICHLORSILAN (GERMAN) see TJD500
TRICHLOR-TRIAETHYLAMIN-HYDROCHLORID (GERMAN) see TNF500
TRICHLORURE d'ANTIMOINE see AQC500
TRICHLORURE d'ARSENIC (FRENCH) see ARF500
sym-TRICHLOTRIAZIN (CZECH) see TJD750
TRICHLOURACETONITRIL (DUTCH) see TII750
TRICHROCHROMOGENIC FACTOR see AIH600
TRICHOCIDE see MMN250
TRICHOFURON see NGG500
TRICHOMOL see MMN250
TRICHOMONACID ""PHARMACHIM" see MMN250
TRICHOPOL see MMN250
TRICHORAD see ABY900
TRICHORAL see ABY900
TRICHOSANTHIN see TJF350
TRICHOTHECIN see TJE750
TRICICLIDINA see CPQ250
TRICILOID see CPQ250
TRICIONE see TLP750
TRICIRIBINE see TJE870
TRICIRIBINE PHOSPHATE HYDRATE see TJE875
TRICLABENDAZOLE see CFL200
TRI-CLENE see TIO750
TRICLOBISONIUM CHLORIDE see TJE880
TRICLOPYR see TJE890
TRI-CLOR see CKN500
TRICLORDIURIDE see HII500
TRICLORETENE (ITALIAN) see TIO750
TRICLORMETIAZIDE (ITALIAN) see HII500
1,1,1-TRICLORO-2,2-BIS(4-CLORO-FENIL)-ETANO (ITALIAN) see DAD200
1,1,1-TRICLOROETANO (ITALIAN) see MIH275
TRICLOROETILENE (ITALIAN) see TIO750
O-(2,4,5-TRICLORO-FENIL)-O,O-DIMETIL-MONOTIOFOSFATO (ITALIAN)
   see RMA500
TRICLOROMETANO (ITALIAN) see CHJ500
TRICLOROMETILBENZENE (ITALIAN) see BFL250
TRICLORO-NITRO-METANO (ITALIAN) see CKN500
TRICLOROSILANO (ITALIAN) see TJD500
TRICLOROTOLUENE (ITALIAN) see BFL250
TRICLOSAN see TIQ000
TRICLOSE see TJF000
2,4,6-TRICI-PDMT see TJA000
TRICOFURON see NGG500
TRICOLOID see CPQ250
TRICOLOID CHLORIDE see EAI875
TRICOM see MMN250
TRICON BW see EIX000
TRI-CORNOX SPECIAL see BAV000
TRICORYL see TJL250
12-TRICOSANONE see TJF250
TRICOSANTHIN see TJF350
TRICOWAS B see MMN250
TRICRESILFOSFATI (ITALIAN) see TNP500
TRI-o-CRESYL BORATE see TJF500
TRICRESYLFOSFATEN (DUTCH) see TNP500
TRICRESYL PHOSPHATE see TMO600, TNP500
TRI-o-CRESYL PHOSPHATE see TMO600
TRICRESYLPHOSPHATE, with more than 3% ortho isomer (DOT) see TNP500
TRI-o-CRESYL PHOSPHITE see TJF750
TRI-p-CRESYL PHOSPHITE see TJG000
TRICTAL see TAF675
TRICURAN see PDD300
TRICYANIC ACID see THS000
TRICYANOGEN CHLORIDE see TJD750
TRICYCLAMOL see CPQ250
TRICYCLAMOL CHLORIDE see EAI875
TRICYCLAMOL METHOCHLORIDE see EAI875
TRICYCLAMOL SULFATE see TJG225

TRICYCLAZOLE see MQC000
TRICYCLAZONE see MQC000
TRICYCLIC ANTIDEPRESSANTS see TJG239
TRICYCLO(3.3.1.1^{3,7})DECAN-1-AMINE see TJG250
TRICYCLO(3.3.1.1.$^{(3,7)}$)DECAN-1-AMINE, HYDROCHLORIDE (9CI) see AED250
exo-TRICYCLO(5.2.1.0^{2,6})DECANE see TLR675
TRICYCLODECANE(5.2.1.0^{2,6})-3,10-DIISOCYANATE see TJG500
TRICYCLODECEN-4-YL-8-ACETATE see DLY400
1,1'-(TRICYCLO(3.3.2.2^{3,7})DEC-1-YLPHOSPHINYLIDENE)BIS-AZIRIDINE see BJP325
TRI(2-CYCLOHEXYLCYCLOHEXYL)BORATE see TJG750
TRICYCLOHEXYLHYDROXYSTANNANE see CQH650
TRICYCLOHEXYLHYDROXYTIN see CQH650
1-(TRICYCLOHEXYLSTANNYL)-1H-1,2,4-TRIAZOLE see THT500
TRICYCLOHEXYLTIN HYDROXIDE see CQH650
TRICYCLOHEXYLZINNHYDROXID (GERMAN) see CQH650
TRICYCLOQUINAZOLINE see TJH250
TRIDECANE see TJH500
n-TRIDECANE see TJH500
1-TRIDECANECARBOXYLIC ACID see MSA250
TRIDECANEDIOIC ACID, CYCLIC ETHYLENE ESTER see EJQ500
TRIDECANENITRILE see TJH750
TRIDECANITRILE (mixed isomers) see TJI000
TRIDECANOIC ACID see TJI250
TRIDECANOIC ACID-2,3-EPOXYPROPYL ESTER see TJI500
TRIDECANOL see TJI750
1-TRIDECANOL see TJI750
n-TRIDECANOL see TJI750
TRIDECANOL condensed with 6 moles ETHYLENE OXIDE see TJJ250
1-TRIDECANOL PHTHALATE see DXQ200
n-TRIDECOIC ACID see TJI250
TRIDECYL ACRYLATE see ADX000
TRIDECYL ALCOHOL see TJI750
n-TRIDECYL ALCOHOL see TJI750
N-TRIDECYL-2,6-DIMETHYLMORPHOLIN (GERMAN) see DUJ400
4-TRIDECYL-2,6-DIMETHYLMORPHOLINE see DUJ400
N-TRIDECYL-2,6-DIMETHYLMORPHOLINE see DUJ400
TRIDECYLIC ACID see TJI250
TRIDEMORF see TJJ500
TRIDEMORPH see TJJ500
TRIDESTRIN see EDU500
TRIDEUTEROMETHYL ACETOXYMETHYLNITROSAMINE see MMS000
TRIDEX see DUN600
TRIDEZIBARBITUR see EOK000
1,2,3-TRI(β-DIETHYLAMINOETHOXY)BENZENE TRIETHIODIDE see PDD300
TRI(β-DIETHYLAMINOETHOXY)-1,2,3-BENZENE TRI-IODOETHYLATE see PDD300
TRI-DIGITOXOSIDE (GERMAN) see DKL800
TRI(DIISOBUTYLCARBINYL) BORATE see TJJ750
TRIDILONA see TLP750
2,4,6-TRI(DIMETHYLAMINOMETHYL)PHENOL see TNH000
TRI(p-DIMETHYLAMINOPHENYL)METHANOL see TJK000
TRI(DIMETHYLAMINO)PHOSPHINEOXIDE see HEK000
TRIDIMITE (FRENCH) see SCK000
TRIDIONE see TLP750
TRIDIPAM see TFS350
TRIDIPHANE see TJK100
TRI-n-DODECYL BORATE see TJK250
TRI-(3-DODECYL-1-METHYL-2-PHENYLBENZIMIDAZOLIUM) FERRICYANIDE see TNH750
TRIDONE see TLP750
TRIDYMITE see SCI500, SCK000
TRIELINA (ITALIAN) see TIO750
TRIEN see TJR000
TRI-ENDOTHAL see EAR000, DXD000
TRIENTINE see TJR000
1,2,4,5,9,10-TRIEPOXYDECANE see TJK500
TRI-ERVONUM see MCA500
TRIESIFENIDILE see BBV000
TRIESTE FLOWERS see POO250
TRIETAZINE see TJL500
TRI-ETHANE see MIH275
N,N',N''-TRI-1,2-ETHANEDIYL PHOSPHORIC TRIMIDE see TND250
N,N',N''-TRI-1,2-ETHANEDIYLPHOSPHOROTHIOIC TRIAMIDE see TFQ750
N,N',N''-TRI-1,2-ETHANEDIYLTHIOPHOSPHORAMIDE see TFQ750
TRIETHANOLAMIN see TKP500
TRIETHANOLAMINE see TKP500
TRIETHANOLAMINE BORATE see TJK750
TRIETHANOLAMINE DODECYL SULFATE see SON000
TRIETHANOLAMINE LAURYL SULFATE see SON000
TRIETHANOLAMINE SILICIEE (FRENCH) see SDH670
TRIETHANOLAMINE TRINITRATE BIPHOSPHATE see TJL250
TRIETHANOLAMINE TRINITRATE DIPHOSPHATE see TJL250

TRIETHANOMELAMINE see TND500
TRIETHAZINE see TJL500
TRIETHOXY(3-AMINOPROPYL)SILANE see TJN000
3-(2,4,5-TRIETHOXYBENZOYL)PROPIONIC ACID see CNG835
TRIETHOXYDIALUMINUM TRIBROMIDE see TJL775
1,1,3-TRIETHOXYHEXANE see TJM000
TRIETHOXY-3-KYANPROPYLSILAN see COS800
TRIETHOXYMETHANE see ENY500
TRIETHOXYMETHYLSILANE see MQD750
2,4,5-TRIETHOXY-Γ-OXOBENZENEBUTANOIC ACID see CNG835
TRIETHOXYPHENYLSILANE see PGO000
1,3,3-TRIETHOXYPROPANE see TJM250
1,3,3-TRIETHOXYPROPENE see TJM500
1,3,3-TRIETHOXY-1-PROPENE see TJM500
TRIETHOXYSILANE see TJM750
4-(TRIETHOXYSILYL)BUTYLAMINE see AJC250
3-(TRIETHOXYSILYL)-1-PROPANAMINE see TJN000
3-(TRIETHOXYSILYL)PROPYLAMINE see TJN000
TRIETHOXYVINYLSILANE see TJN250
TRIETHOXYVINYLSILICANE see TJN250
TRIETHYL ACETYLCITRATE see ADD750
TRIETHYLALUMINUM see TJN750
TRIETHYLAMINE see TJO000
s-(2-TRIETHYLAMINOETHYL)ISOTHIURONIUM BROMIDE HYDROBROMIDE see TAI100
S-(2-TRIETHYLAMINOETHYL)-Γ-METHYLISOTHIURONIUM BROMIDE HYDROBROMIDE see MRU775
2-(2-TRIETHYLAMINOETHYLTHIO)-Δ$^{(2)}$-IMIDAZOLINE BROMIDE HYDROBROMIDE see EDW300
TRIETHYL AMMONIUM NITRATE see TJO100
TRIETHYLANTIMONY see TJO250
TRIETHYLARSINE see TJO500
TRIETHYLBENZENE see TJO750
TRIETHYL-BENZENE (mixed isomers) see TJO750
TRIETHYLBISMUTH see TJP000
TRIETHYLBORANE see TJP250
TRIETHYL BORATE see TJP500
TRIETHYLBORINE see TJP250
TRIETHYLCHLOROPLUMBANE see TJS250
TRIETHYLCHLOROSTANNANE see TJV000
TRIETHYLCHLOROTIN see TJV000
TRIETHYL CITRATE see TJP750
TRIETHYL DIALUMINUM TRICHLORIDE see TJP775
TRIETHYLDIBORANE see TJP780
TRIETHYLENEDIAMINE see DCK400
TRIETHYLENE GLYCOL see TJQ000
TRIETHYLENE GLYCOL-n-BUTYL ETHER see TKL750
TRIETHYLENE GLYCOL, DIACETATE see EJB500
TRIETHYLENE GLYCOL DIACRYLATE see TJQ100
TRIETHYLENE GLYCOL DICHLORIDE see TKL500
TRIETHYLENEGLYCOL DIETHYL BUTYRATE see TJQ250
TRIETHYLENE GLYCOL DI(2-ETHYL-BUTYRATE) see TJQ250
TRIETHYLENE GLYCOL DI(2-ETHYLHEXOATE) see FCD560
TRIETHYLENE GLYCOL DIGLYCIDYL ETHER see TJQ333
TRIETHYLENE GLYCOL DIMETHYL ETHER see TKL875
TRIETHYLENE GLYCOL, DINITRATE see TJQ500
TRIETHYLENE GLYCOL ETHYL ETHER see EFL000
TRIETHYLENE GLYCOL MONOBUTYL ETHER see TKL750
TRIETHYLENE GLYCOL MONOETHYL ETHER see EFL000
TRIETHYLENE GLYCOLMONOMETHYL ETHER see TJQ750
2,3,5-TRIETHYLENEIMINO-1,4-BENZOQUINONE see TND000
TRI(ETHYLENEIMINO)THIOPHOSPHORAMIDE see TFQ750
2,4,6-TRIETHYLENEIMINO-s-TRIAZINE see TND500
2,4,6-TRI(ETHYLENEIMINO)-1,3,5-TRIAZINE see TND500
TRIETHYLENEMELAMINE see TND500
N,N',N''-TRIETHYLENEPHOSPHOROTHIOIC TRIAMIDE see TFQ750
TRIETHYLENEPHOSPHOROTRIAMIDE see TND250
TRIETHYLENETETRAMINE see TJR000
N,N',N''-TRIETHYLENETHIOPHOSPHAMIDE see TFQ750
N,N',N''-TRIETHYLENETHIOPHOSPHORAMIDE see TFQ750
TRIETHYLENETHIOPHOSPHOROTRIAMIDE see TFQ750
TRIETHYLENIMINOBENZOQUINONE see TND000
2,4,6-TRIETHYLENIMINO-s-TRIAZINE see TND500
2,4,6-TRIETHYLENIMINO-1,3,5-TRIAZINE see TND500
O,O,O-TRIETHYLESTER KYSELINY THIOFOSFORECNE (CZECH) see TJU000
N,N,N-TRIETHYL-ETHANAMINIUM (9CI) see TCB725
N,N,N-TRIETHYLETHANAMINIUM CHLORIDE see TCC250
TRIETHYLGALLIUM see TJR250
TRIETHYLHEXADECYLAMMONIUM BROMIDE see CDF500
TRI(2-ETHYLHEXYL) BORATE see TJR500
TRIETHYLHEXYL PHOSPHATE see TNI250
TRI(2-ETHYLHEXYL)PHOSPHATE see TNI250
TRIETHYL(2-HYDROXYETHYL)-AMMONIUM BROMIDE DICYCLOPENTYLACETATE see DGW600

TRIETHYL(2-HYDROXYETHYL)AMMONIUM-p-TOLUENESULFONATE 3,4,5-TRIMETHOXYBENZOATE see TNV625
TRIETHYLHYDROXY-STANNANE SULFATE (2:1) (8CI) see BLN500
TRIETHYLHYDROXYTIN SULFATE see BLN500
N,N,N-TRIETHYL-2-((IMINO(METHYLAMINO)METHYL)THIO)-ETHANAMINIUM BROMIDE, MONOHYDROBROMIDE see MRU775
TRIETHYL INDIUM see TJR750
TRIETHYL LEAD see TJS000
TRIETHYL LEAD CHLORIDE see TJS250
TRIETHYL LEAD FLUOROACETATE see TJS500
TRIETHYL LEAD FUROATE see TJS750
TRIETHYL LEAD OLEATE see TJT000
TRIETHYL LEAD PHENYL ACETATE see TJT250
TRIETHYL LEAD PHOSPHATE see TJT500
1,1,3-TRIETHYL-3-NITROSOUREA see NLX500
TRIETHYLOLAMINE see TKP500
TRIETHYL ORTHOFORMATE see ENY500
TRIETHYL PHOSPHATE see TJT750
TRIETHYL PHOSPHINE see TJT775
TRIETHYLPHOSPHINEAUROUS CHLORIDE see CLQ500
TRIETHYLPHOSPHINE GOLD see ARS150
TRIETHYL PHOSPHINE GOLD NITRATE see TJT780
TRIETHYL PHOSPHITE see TJT800
TRIETHYL PHOSPHONOACETATE see EIC000
O,S,S-TRIETHYL PHOSPHORODITHIOATE see TJT900
TRIETHYL PHOSPHOROTHIOATE see TJU000
O,O,O-TRIETHYL PHOSPHOROTHIOATE see TJU000
O,O,S-TRIETHYL PHOSPHOROTHIOATE see TJU800
TRIETHYLPLUMBYL ACETATE see TJU150
TRIETHYLPROPYL GERMANE see TJU250
TRIETHYLSILYL PERCHLORATE see TJU500
TRIETHYLSTANNIUM BROMIDE see BOI750
TRIETHYLSTANNYL CHLORIDE see TJV000
TRIETHYLSULFONIUM IODIDE BIS(MERCURIC IODIDE) addition compound see TJU600
TRIETHYLSULFONIUM IODIDE MERCURIC IODIDE addition compound see TJU750
TRIETHYLSULFONIUM, IODIDE compounded with MERCURY IODIDE (1:1) see TJU750
TRIETHYL-SULFONIUM, IODIDE, compound with MERCURY IODIDE (1:2) see TJU600
TRIETHYLTHIOFOSFAT (CZECH) see TJU000
O,O,S-TRIETHYL THIOPHOSPHATE see TJU800
TRIETHYLTIN ACETATE see ABW750
TRIETHYL TIN BROMIDE see BOI750
TRIETHYLTIN BROMIDE-2-PIPECOLINE see TJU850
TRIETHYLTIN CHLORIDE see TJV000
TRIETHYLTIN HYDROPEROXIDE see TJV100
TRIETHYLTIN PHENOXIDE see TJV250
TRIETHYLTIN SULPHATE see BLN500
3,6,9-TRIETHYL-3,6,9-TRIAZAUNDECANE see TJV500
TRIETHYL(TRIFLUOROACETOXY) STANNANE see TJX250
2,4,6-TRIETHYL-1,3,5-TRIISOPROPYLBORAZINE see TKU000
TRIETHYL-2-(3,4,5-TRIMETHOXYBENZOYLOXY)ETHYLAMMONIUM-p-TOLUENESULFATE see TNV625
TRIETHYL-2-(3,4,5-TRIMETHOXYBENZOYLOXY)ETHYLAMMONIUM TOSYLATE see TNV625
TRIETHYNYL ALUMINUM see TJV775
TRIETHYNYL ANTIMONY see TJV785
TRIETHYNYLARSINE see TJW000
TRIETHYNYLPHOSPHINE see TJW250
TRIETILAMINA (ITALIAN) see TJO000
TRIEXIFENIDILA see BBV000
TRIFARON see CJM250
TRI-FEN see TIY500
TRIFENOXYFOSFIN (CZECH) see TMU250
TRIFENSON see CJR500
TRIFENYLFOSFIT (CZECH) see TMU250
TRIFENYLTINACETAAT (DUTCH) see ABX250
TRIFENYL-TINHYDROXYDE (DUTCH) see HON000
TRIFERRIC ADRIAMYCIN see QBS000
TRIFERRIC DOXORUBICIN see QBS000
TRIFLIC ACID see TKB310
TRIFLORPERAZINE DIHYDROCHLORIDE see TKK250
TRIFLUMEN see HII500
TRIFLUMETHAZINE see TJW500
TRIFLUOMETHYLTHIAZIDE see TKG750
TRIFLUOPERAZINA (ITALIAN) see TKE500
TRIFLUOPERAZINE see TKE500
TRIFLUOPERAZINE DIMALEATE see TJW600
TRIFLUOPERAZINE HYDROCHLORIDE see TKK250
TRIFLUORACETIC ACID see TKA250
TRIFLUORALIN (USDA) see DUV600

3-(5-TRIFLUORMETHYLPHENYL)-,1-DIMETHYLHARNSTOFF (GERMAN) see DUK800
TRIFLUOROACETALDEHYDE HYDRATE see TJZ000
2-(2,2,2-TRIFLUOROACETAMIDO)-4-(5-NITRO-2-FURYL)THIAZOLE see NGN500
TRIFLUOROACETIC ACID (DOT) see TKA250
TRIFLUOROACETIC ACID ANHYDRIDE see TJX000
TRIFLUOROACETIC ACID TRIETHYLSTANNYL ESTER see TJX250
TRIFLUOROACETIC ANHYDRIDE see TJX000
TRIFLUOROACETYLADRIAMYCIN-14-VALERATE see TJX350
N-TRIFLUOROACETYLADRIAMYCIN-14-VALERATE see TJX350
2-TRIFLUOROACETYLAMINOFLUORENE see FER000
2-TRIFLUOROACETYLAMINOFLUOREN-9-ONE see TKH000
TRIFLUOROACETYL ANHYDRIDE see TJX000
TRIFLUOROACETYL AZIDE see TJX375
TRIFLUOROACETYL CHLORIDE see TJX500
2-TRIFLUOROACETYL-1,3,4-DIOXAZALONE see TJX600
O-TRIFLUOROACETYL-S-FLUOROFORMYL THIOPEROXIDE see TJX625
TRIFLUOROACETYL HYPOCHLORITE see TJX650
TRIFLUOROACETYL HYPOFLUORITE see TJX750
TRIFLUOROACETYLIMINOIODOBENZENE see TJX775
TRIFLUOROACETYL NITRITE see TJX780
TRIFLUOROACETYL TRIFLUOROMETHANE SULFONATE see TJX800
TRIFLUOROACRYLOYL FLUORIDE see TJX825
TRIFLUOROAMINE OXIDE see NGS500
2,3,4-TRIFLUOROANILINE see TJX900
TRIFLUOROANTIMONY see AQE000
TRIFLUOROARSINE see ARI250
1,1,1-TRIFLUORO-2-BROMO-2-CHLOROETHANE see HAG500
1,1,2-TRIFLUORO-1-BROMO-2-CHLOROETHANE see TJY000
TRIFLUOROBROMOETHYLENE see BOJ000
TRIFLUOROBROMOMETHANE see TJY100
1,1,1-TRIFLUORO-2-CHLORO-2-BROMOETHANE see HAG500
2,2,2-TRIFLUORO-1-CHLORO-1-BROMOETHANE see HAG500
2,2,2-TRIFLUOROCHLOROETHANE see TJY175
1,1,1-TRIFLUORO-2-CHLOROETHANE see TJY175
TRIFLUOROCHLOROETHYLENE (DOT) see CLQ750
1,1,2-TRIFLUORO-2-CHLOROETHYLENE see CLQ750
TRIFLUOROCHLOROMETHANE (DOT) see CLR250
1,1,1-TRIFLUORO-3-CHLOROPROPANE see TJY200
α,α,α-TRIFLUORO-4-CHLOROTOLUENE see CEM825
α,α,α-TRIFLUORO-m-CRESOL see TKE750
5-TRIFLUORO-2'-DEOXYTHYMIDINE see TKH325
2,2,2-TRIFLUORODIAZOETHANE see TJY275
1,1,1-TRIFLUORO-2,2-DICHLOROETHANE see TJY500
1,1,2-TRIFLUORO-1,2-DICHLOROETHANE see TJY750
2',4',6'-TRIFLUORO-4-DIMETHYLAMINOAZOBENZENE see DUK200
α,α,α-TRIFLUORO-2,6-DINITRO-N,N-DIPROPYL-p-TOLUIDINE see DUV600
2,2,2-TRIFLUORO-1,1-ETHANEDIOL see TJZ000
TRIFLUOROETHANOIC ACID see TKA250
2,2,2-TRIFLUOROETHANOL see TKA350
(2,2,2-TRIFLUOROETHOXY)ETHENE see TKB250
2,2,2-TRIFLUOROETHYLAMINE see TKA500
TRIFLUOROETHYLAMINE HYDROCHLORIDE see TKA750
1,1,1-TRIFLUOROETHYL CHLORIDE see TJY175
2,2,2-TRIFLUOROETHYL VINYL ETHER see TKB250
2,2,2-TRIFLUORO-N-(FLUOREN-2-YL)ACETAMIDE see FER000
N,N,N'-TRIFLUOROHEXANAMIDINE see TKB275
TRIFLUOROMETHANE (DOT) see CBY750
TRIFLUOROMETHANESULFENYL CHLORIDE see TKB300
TRIFLUOROMETHANE SULFONIC ACID see TKB310
1,3,3-TRIFLUORO-2-METHOXYCYCLOPROPENE see TKB325
4-TRIFLUOROMETHYL ALLYLOXY-2 N-(β-DIETHYLAMINO-ETHYL)BENZAMIDE (FRENCH) see FDA875
3-(TRIFLUOROMETHYL)ANILINE see AID500
m-(TRIFLUOROMETHYL)ANILINE see AID500
p-TRIFLUOROMETHYLANILINE see TKB750
2-(3-(TRIFLUOROMETHYL)ANILINO)NICOTINIC ACID see NDX500
3-(TRIFLUOROMETHYL)BENZENAMINE see AID500
(TRIFLUOROMETHYL)BENZENE see BDH500
3-(TRIFLUOROMETHYL)BENZENEACETONITRILE see TKF000
4-TRIFLUOROMETHYL-6H-BENZO(e)(1)BENZOTHIOPYRANO(4,3-b)INDOLE see TKC000
m-TRIFLUOROMETHYL BENZOIC ACID, THALLIUM SALT see TKH500
6-(TRIFLUOROMETHYL)-1,2,4-BENZO-THIADIAZINE-7-SULFONAMIDE-1,1-DIOXIDE see TKG750
6-(TRIFLUOROMETHYL)-1,4,2-BENZOTHIADIAZINE-7-SULFONAMIDO-1,1-DIOXIDE see TKG750
6-TRIFLUOROMETHYL-3-BENZYL-7-SULFAMYL-3,4-DIHYDRO-1,2,4-BENZOTHIADIAZINE-1,1-DIOXIDE see BEQ625
3-(TRIFLUOROMETHYL)BROMOBENZENE see BOJ500
m-(TRIFLUOROMETHYL)BROMOBENZENE see BOJ500
o-(TRIFLUOROMETHYL)BROMOBENZENE see BOJ750
TRIFLUOROMETHYL CHLORIDE see CLR250

TRIHEXYLENE GLYCOL BIBORATE see TKM250
TRIHEXYLPHENIDYL see PAL500
TRIHEXYLPHENIDYL HYDROCHLORIDE see BBV000
TRI-n-HEXYLPHOSPHINE OXIDE see TKM500
TRIHEXYLTIN ACETATE see ABX000
TRI-N-HEXYLZINNACETAT (GERMAN) see ABX000
TRIHISTAN see CDR000
TRIHYDRATED ALUMINA see AHC000
TRIHYDRAZINECOBALT(II) NITRATE see TKM750
TRIHYDRAZINENICKEL(II) NITRATE see TKN000
TRIHYDROGENPYROPHOSPHATE(ESTER)THIAMINE see TET750
2,3,4-TRIHYDROXYACETOPHENONE see TKN250
3,7,15-TRIHYDROXY-4-ACETOXY-8-OXO-12,13-EPOXY-Δ^9-TRICHOTHECENE
  see FQR000
1,8,9-TRIHYDROXYANTHRACENE see APH250
1,2,3-TRIHYDROXY-9,10,ANTHRACENEDIONE see TKN500
1,2,4-TRIHYDROXY-9,10-ANTHRACENEDIONE see TKN750
1,2,4-TRIHYDROXYANTHRACHINON (CZECH) see TKN750
1,2,3-TRIHYDROXYANTHRAQUINONE see TKN500
1,2,4-TRIHYDROXYANTHRAQUINONE see TKN750
1,2,3-TRIHYDROXYBENZEN (CZECH) see PPQ500
s-TRIHYDROXYBENZENE see PGR000
1,2,3-TRIHYDROXYBENZENE see PPQ500
1,2,4-TRIHYDROXYBENZENE see BBU250
1,3,5-TRIHYDROXYBENZENE see PGR000
sym-TRIHYDROXYBENZENE see PGR000
3,4,5-TRIHYDROXYBENZENE-1-PROPYLCARBOXYLATE see PNM750
2,4,6-TRIHYDROXYBENZOIC ACID see TKO000
3,4,5-TRIHYDROXYBENZOIC ACID see GBE000
3,4,5-TRIHYDROXYBENZOIC ACID, n-PROPYL ESTER see PNM750
2-(2,3,4-TRIHYDROXYBENZYL)HYDRAZIDE SERINE MONOHYDROCHLOR-
  IDE, dl- see SCA400
3-β,14,16-β-TRIHYDROXY-5-β-BUFA-20,22-DIENOLIDE see BOM655
3-β,14,16-β-TRIHYDROXY-5-β-BUFA-20,22-DIENOLIDE-16-ACETATE
  see BON000
2,4,5-TRIHYDROXYBUTYROPHENONE see TKO250
2',4',5'-TRIHYDROXYBUTYROPHENONE see TKO250
3-β,12-β,14-TRIHYDROXY-CARD-20(22)-ENOLIDE see DKN300
3-β,12-β,14-TRIHYDROXY-5-β-CARD-20(22)-ENOLIDE-3,12-DIFORMATE
  see FNK075
3β,12β,14β-TRIHYDROXY-5-β-CARD-20(22)-ENOLIDE-3-(4'')-o-METHYL-
  TRIDIGITOXOSIDE) see MJD300
3-β,12-β,14-β-TRIHYDROXY-5-β-CARD-20(22)-ENOLIDE-3-(R'''-O-
  METHYLTRIDIGITOXOSIDE), ACETONE (2:1) see MJD500
TRIHYDROXY-3,7,12-CHOLANATE de Na (FRENCH) see SFW000
3-α,7-α,12-α-TRIHYDROXY-5-β-CHOLAN-24-OIC ACID see CME750
3,7,12-TRIHYDROXY-CHOLAN-24-OIC ACID (3-α,5-β,7-α,12-α)
  see CME750
(3-α,5-β,7-α,12-α)3,7,12-TRIHYDROXY-CHOLAN-24-OIC ACID, MONOSODIUM
  SALT see SFW000
3-α,7-α,12-α-TRIHYDROXYCHOLANSAEURE (GERMAN) see CME750
TRIHYDROXYCYANIDINE see THS000
1,3,5-TRIHYDROXYCYCLOHEXATRIENE see PGR000
3,4,5-TRIHYDROXY-1-CYCLOHEXENE-1-CARBOXYLIC ACID see SCE000
Z-(-)-4,6,8-TRIHYDROXY-3a,12a-DIHYDROANTHRA(2,3-b)FURO(3,2-d)FURAN-
  5,10-DIONE see TKO500
1,9,10-TRIHYDROXY-9,10-DIHYDRO-3-METHYLCHOLANTHRENE see TKO750
3,16-α,17-β-TRIHYDROXYESTRA-1,3,5(10)-TRIENE see EDU500
3,16-α,17-β-TRIHYDROXY-Δ-1,3,5-ESTRATRIENE see EDU500
TRIHYDROXYESTRIN see EDU500
TRI(HYDROXYETHYL)AMINE see TKP500
3,5,7-TRIHYDROXYFLAVONE see GAZ000
4',5,7-TRIHYDROXYFLAVONE see CDH250
5,7,4'-TRIHYDROXYFLAVONOL see ICE000
9,10,16-TRIHYDROXYHEXADECANOIC ACID see TKP000
dl-ertho-9,10,16-TRIHYDROXYHEXADECANOIC ACID see TKP000
3,5,7-TRIHYDROXY-2-(4-HYDROXY-3,5-DIMETHOXYHPENYL)-BENZOPYRYL-
  IUM ACID ANION see MAO750
4',5,7-TRIHYDROXYISOFLAVONE see GCM350
5,7,4'-TRIHYDROXYISOFLAVONE see GCM350
1,8,9-TRIHYDROXY-3-METHYLANTHRACENE see CML750
1,3,8-TRIHYDROXY-6-METHYL-9,10-ANTHRACENEDIONE see MQF250
1,3,8-TRIHYDROXY-6-METHYLANTHRAQUINONE see MQF250
3-α,4-β,15-TRIHYDROXY-8,α-(3-METHYLBUTYRYLOXY)-12,13-
  EPOXYTRICHOTHEC-9-ENE see DAD600
8,12,18-TRIHYDROXY-4-METHYL-11,16-DIOXOSENECIONANIUM see HOF000
N,N',N''-TRIHYDROXYMETHYLMELAMINE see TLW750
2,4a,7-TRIHYDROXY-1-METHYL-8-METHYLENEGIBB-3-ENE-1,10-CARBOX-
  YLIC ACID 1-4-LACTONE see GEM000
TRIHYDROXYMETHYLNITROMETHANE see HMJ500
11-β,17,21-TRIHYDROXY-6-α-METHYLPREGNA-1,4-DIENE-3,20-DIONE
  see MOR500
11-β,17-α,21-TRIHYDROXY-6-α-METHYL-1,4-PREGNADIENE-3,20-DIONE
  see MOR500

11-β,17,21-TRIHYDROXY-6-α-METHYL-PREGNA-1,4-DIENE-3,20-DIONE 21-AC-
  ETATE see DAZ117
11-β,17,21-TRIHYDROXY-6-α-METHYLPREGNA-1,4-DIENE-3,20-DIONE, 21-(HY-
  DROGEN SUCCINATE), MONOSODIUM SALT see USJ100
11-β,17,21-TRIHYDROXY-6-α-METHYL-1,4-PREGNADIENE-3,20-DIONE 21-(SO-
  DIUM SUCCINATE) see USJ100
1,1,1-(TRIHYDROXYMETHYL)PROPANE TRIESTER ACRYLIC ACID see
  TLX175
(5Z,9-α,11-α,13E,15S)-9,11,15-TRIHYDROXY-15-METHYLPROSTA-5,13-DIEN-1-
  OIC ACID see CCC100
9,11,15-TRIHYDROXY-15-METHYL-PROSTA-5,13-DIEN-1-OIC ACID, (5Z,9-α,11-
  α,13E,15S)-compd. with 2-AMINO-2-(HYDROXYMETHYL)-1,3-PRO-
  PANEDIOL (1:1) see CCC110
Δ-20:22-3,14,21-TRIHYDROXYNORCHOLENIC ACID LACTONE see DMJ000
3,16-α,17-β-TRIHYDROXYOESTRA-1,3,5(10)-TRIENE see EDU500
3,16-α,17-β-TRIHYDROXY-Δ-1,3,5-OESTRATRIENE see EDU500
TRIHYDROXYOESTRIN see EDU500
3,5,14-TRIHYDROXY-19-OXOCARD-20(22)-ENOLIDE see SMM500
3-β,5,14-TRIHYDROXY-19-OXO-5-β-CARD-20(22)-ENOLIDE see SMM500
2-((3-α,7-α,12-α-TRIHYDROXY-24-OXO-5-β-CHOLAN-24-
  YL)AMINO)ETHANESULFONIC ACID see TAH250
8,9,15-TRIHYDROXYPENTADECANE-1-CARBOXYLIC ACID see TKP000
2,4,5-TRIHYDROXYPHENETHYLAMINE see HKF875
3,4,5-TRIHYDROXYPHENETHYLAMINE HYDROCHLORIDE see HKG000
7-(3-((β,3,5-TRIHYDROXYPHENETHYL)AMINO)PROPYL)THEOPHYLLINE
  MONOHYDROCHLORIDE see DNA600
3,5,7-TRIHYDROXY-2-PHENYL-4H-BENZOPYRAN-4-ONE see GAZ000
11-β,17,21-TRIHYDROXYPREGNA-1,4-DIENE-3,20-DIONE see PMA000
11-β,17-α,21-TRIHYDROXYPREGNA-1,4-DIENE-3,20-DIONE see PMA000
11-β,17-α,21-TRIHYDROXY-1,4-PREGNADIENE-3,20-DIONE see PMA000
11-β,17-α,21-TRIHYDROXYPREGNA-1,4-DIENE-3,20-DIONE, 21-ACETATE see
  SOV100
11-β,17-α,21-TRIHYDROXY-1,4-PREGNADIENE-3,20-DIONE-21-ACETATE-17-
  VALERATE see AAF625
11-β,17,21-TRIHYDROXYPREGN-4-ENE-3,20-DIONE see CNS750
11-β,17-α,21-TRIHYDROXY-4-PREGNENE-3,20-DIONE see CNS750
11-β,17,21-TRIHYDROXY-PREGN-4-ENE-3,20-DIONE 21-ACETATE see HHQ800
11-β,17,21-TRIHYDROXY-PREGN-4-ENE-3,30-DIONE 17-BUTYRATE, 21-PROPI-
  ONATE see HHQ850
TRIHYDROXYPROPANE see GGA000
1,2,3-TRIHYDROXYPROPANE see GGA000
9,11,15-TRIHYDROXYPROSTA-5,13-DIEN-1-OIC ACID see POC500
(5Z,9-α,11-α,13E,15S)-9,11,15-TRIHYDROXYPROSTA-5,13-DIEN-1-OIC ACID
  see POC500
(5Z,9-α,11-α,13E,15S)-(±)-8,11,15-TRIHYDROXY-PROSTA-5,13-DIEN-1-OIC
  ACID (9CI) see POC525
(5Z,9-α,11-α,13E,15S)-9,11,15-TRIHYDROXY-PROSTA-5,13-DIEN-1-OIC ACID,
  METHYL ESTER see DVJ100
(9-α,11-α,13E,15S)-11,15-TRIHYDROXY-PROST-13-EN-1-OIC ACID (9CI)
  see POC400
2,6,8-TRIHYDROXYPURINE see UVA400
3,7,15-TRIHYDROXYSCIRP-4-ACETOXY-9-EN-8-ONE see FQR000
2,4,6-TRIHYDROXY-1,3,5-TRIAZINE see THS000
TRIHYDROXYTRIETHYLAMINE see TKP500
2,2',2''-TRIHYDROXYTRIETHYLAMINE see TKP500
3,4,7-TRIHYDROXY-2-(3,4,5-TRIHYDROXYPHENYL)BENZOPYRYLIUM, ACID
  ANION see DAM400
2,4,6-TRIHYDROXY-1,3,5,2,4,6-TRIOXATRIPHOSPHORINANE TRISODIUM
  SALT see TKP750
3,5,3'-TRIIODO-4'-ACETYLTHYROFORMIC ACID see TKP850
TRIIODOARSINE see ARG750
3,4,5-TRIIODOBENZENEDIAZONIUM NITRATE see TKQ000
TRIIODOBENZOIC ACID see TKQ250
2,3,5-TRIIODOBENZOIC ACID see TKQ250
α-(2,4,6-TRIIODO-3-BUTYRYLAMINOBENZYLIDENE)BUTYRIC ACID SO-
  DIUM SALT see EQC000
TRIIODOETHIONIC ACID see IFZ800
TRIIODOETHYLATE de GALLAMINE (FRENCH) see PDD300
TRIIODOETHYLATE of TRI(DIETHYLAMINOETHYLOXY)-1,2,3-BENZENE
  see PDD300
α-(2,4,6-TRIIODO-3-HYDROXYBENZYL)BUTYRIC ACID see IFZ800
TRIIODOMETHANE see IEP000
N,N'-(2,4,6-TRIIODO-5-((METHYLAMINO)CARBONYL)-1,3-PHENYLENE)BIS-d-
  GLUCONAMIDE see IFS400
2,4,6-TRIIODOPHENOL see TKR000
TRIIODO(6',7',10,11-TETRAMETHOXYEMETAN)BISMUTH see EAM500
TRIIODOTHYRONINE see LGK050
TRIIODO-L-THYRONINE see LGK050
l-TRIIODOTHYRONINE see LGK050
3,3',5-TRIIODOTHYRONINE see LGK050
3,5,3'-TRIIODOTHYRONINE see LGK050
TRIIODRAST see AAN000
TRIIODYL see AAN000
TRIIOTRAST see AAN000

TRIMETHYLCOLCHICINSAEUREMETHYL ESTER (GERMAN) see TLO000
3,3,5-TRIMETHYLCYCLOHEXANECARBOXALDEHYDE see TLO250
3,3,5-TRIMETHYLCYCLOHEXANOL see TLO500
3,5,5-TRIMETHYLCYCLOHEXANOL see TLO500
3,3,5-TRIMETHYL-1-CYCLOHEXANOL see TLO500
3,3,5-TRIMETHYLCYCLOHEXANOL-α-PHENYL-α-HYDROXYACETATE
   see DNU100
α,α,4-TRIMETHYL-3-CYCLOHEXENE-1-METHANOL see TBD750
1,1,3-TRIMETHYL-3-CYCLOHEXENE-5-ONE see IMF400
3,5,5-TRIMETHYL-2-CYCLOHEXENE-1-ONE see IMF400
3,5,5-TRIMETHYL-2-CYCLOHEXEN-1-ON (GERMAN, DUTCH) see IMF400
4-(2,6,6-TRIMETHYL-2-CYCLOHEXEN-1-YL)-2-BUTANONE see DLP800
4-(2,6,6-TRIMETHYL-1-CYCLOHEXEN-1-YL)-3-BUTEN-2-ONE see IFX000
4-(2,6,6-TRIMETHYL-2-CYCLOHEXEN-1-YL)-3-BUTEN-2-ONE see IFW000
1-(2,6,6-TRIMETHYL-2-CYCLOHEXEN-1-YL)-1,6-HEPTADIEN-3-ONE see
   AGI500
3,5,5-TRIMETHYLCYCLOHEXYL AMYGDALATE see DNU100
3,3,5-TRIMETHYLCYCLOHEXYL DIPROPYLENE GLYCOL see TLO600
3,3,5-TRIMETHYLCYCLOHEXYL MANDELATE see DNU100
11,12-17-TRIMETHYL-15H-CYCLOPENTA(a)PHENANTHRENE see TLP000
3,8,13-TRIMETHYLCYCLOQUINAZOLINE see TLP250
TRIMETHYLDIBORANE see TLP275
11,12,17-TRIMETHYL-16,17-DIHYDRO-15H-CYCLOPENTA(a)PHENANTHRENE
   see DMF600
2,2,4-TRIMETHYL-1,2-DIHYDROQUINOLINE see TLP500
TRIMETHYLDIHYDROQUINOLINE POLYMER see PJQ750
2,2,4-TRIMETHYL-1,2-DIHYDROQUINOLINE POLYMER see PJQ750
3,3,5-TRIMETHYL-2,4-DIKETOOXAZOLIDINE see TLP750
TRIMETHYL(2-(2,6-DIMETHYLPHENOXY)PROPYL)AMMONIUM CHLORIDE
   MONOHYDRATE see TLQ000
N,N,N-TRIMETHYL-1,3-DIOXOLANE-4-METHANAMINIUM IODIDE see
   FMX000
1,3,7-TRIMETHYL-2,6-DIOXOPURINE see CAK500
1-(1,3,7-TRIMETHYL-2,6-DIOXOPURIN-8-YL)-4-(2-PHENYL-1-METHYL)ETHYL-
   4-METHYL-ETHYLENEDIAMINE, HYDROCHLORIDE see FAM000
3,7,11-TRIMETHYL-2,4-DODECADIENOIC ACID 2-PROPYNYL ESTER
   see POB000
3,7,11-TRIMETHYL-2,6,10-DODECATRIEN-1-OL see FAB800
3,7,11-TRIMETHYL-1,6,10-DODECATRIEN-3-YL ACETATE see NCN800
TRIMETHYLEENTRINITRAMINE (DUTCH) see CPR800
1,3-TRIMETHYLEN-BIS-(4-HYDROXIMINOFORMYLPYRIDINIUM)-DIBROMID
   (GERMAN) see TLQ500
N,N-TRIMETHYLEN-BIS-(PYRIDINIUM-4-ALDOXIM)-DIBROMID (GERMAN)
   see TLQ500
TRIMETHYLENE see CQD750
5:10-TRIMETHYLENE-1:2-BENZANTHRACENE see DKT400
1,1'-TRIMETHYLENEBIS(4-FORMYLPYRIDINIUM BROMIDE)DIOXIME
   see TLQ500
1,1'-TRIMETHYLENEBIS(4-FORMYLPYRIDINIUM CHLORIDE) DIOXIME
   see TLQ750
1,1'-TRIMETHYLENEBIS(4-FORMYLPYRIDINIUM) DIOXIME DICHLORIDE
   see TLQ750
1,1'-TRIMETHYLENEBIS(4-(HYDROXYIMINOMETHYL)PYRIDINIUM CHLO-
   RIDE) see TLQ750
N,N-TRIMETHYLENE BIS(PYRIDINIUM-4-ALDOXIME)DICHLORIDE
   see TLQ750
2,2'-TRIMETHYLENE-BIS-(2-THIOPSEUDOUREA), DIHYDROBROMIDE
   see PMK750
TRIMETHYLENE BROMIDE see TLR000
TRIMETHYLENE BROMIDE CHLORIDE see BNA825
TRIMETHYLENE CHLOROBROMIDE see BNA825
1:12-TRIMETHYLENECHRYSENE see DKT400
TRIMETHYLENEDIAMINE see PMK500
TRIMETHYLENE DIBROMIDE see TLR000
TRIMETHYLENE DICHLORIDE see DGF800
TRIMETHYLENEDIMETHANESULFONATE see TLR250
TRIMETHYLENE DIMETHANESULPHONATE see TLR250
1,3-TRIMETHYLENEDINITRILE see TLR500
4,4'-(TRIMETHYLENEDIOXY)DIBENZAMIDINE DIHYDROCHLORIDE
   see DBM400
TRIMETHYLENEETHYLENEDIAMINE see HGI900
TRIMETHYLENEFUROXAN see CPW325
TRIMETHYLENE GLYCOL see PML250
exo-TRIMETHYLENENORBORNANE see TLR675
exo-5,6-TRIMETHYLENENORBORNANE see TLR675
TRIMETHYLENE OXIDE see OMW000
TRIMETHYLENE SULFATE see TLR750
N,N'-TRIMETHYLENETHIOUREA see TLS000
TRIMETHYLENETRINITRAMINE see CPR800
sym-TRIMETHYLENETRINITRAMINE see CPR800
TRIMETHYLENE TRISULFIDE see TLS500
TRIMETHYLENOXID (GERMAN) see OMW000
TRIMETHYLENTRISULFID (CZECH) see TLS500
N,N,N-TRIMETHYLETHENAMINIUM HYDROXIDE see VQR300

2,2,4-TRIMETHYL-6-ETHOXY-1,2-DIHYDROQUINOLINE see SAV000
TRIMETHYLETHYLENE see AOI750
TRIMETHYLETHYLENE OXIDE see TLY175
N,N,N-TRIMETHYL-2-FURANMETHANAMINIUM IODIDE see FPY000
TRIMETHYLFURFURYLAMMONIUM IODIDE see FPY000
TRIMETHYLGALLIUM see TLT000
TRIMETHYLGERMYL PHOSPHINE see TLT100
TRIMETHYL GLYCOL see PML000
TRIMETHYLHARNSTOFF (GERMAN) see TMJ250
N,N,N-TRIMETHYL-1-HEXADECANAMINIUM BROMIDE see HCQ500
TRIMETHYLHEXADECYLAMMONIUM BROMIDE see HCQ500
3,5,5-TRIMETHYLHEXANOIC ACID ALLYL ESTER see AGU400
N-1,5-TRIMETHYL-4-HEXENYLAMINE see ILK000
3,5,5-TRIMETHYLHEXYL ACETATE see TLT500
3,5,5-TRIMETHYLHEXYL ACETIC ACID see TLT500
TRIMETHYLHYDRAZINE HYDROCHLORIDE see TLT750
TRIMETHYLHYDROQUINONE see POG300
2,3,5-TRIMETHYLHYDROQUINONE see POG300
TRIMETHYLHYDROXYLAMMONIUM PERCHLORATE see TLE500
TRIMETHYL INDIUM see TLT775
TRIMETHYLIODOMETHANE see TLU000
TRI(METHYLISOBUTYLCARBINYL) BORATE see BMC750
TRIMETHYL-ε-LACTONE (mixed isomers) see TLY000
TRIMETHYL LEAD CHLORIDE see TLU175
N^2,N^4,N^6-TRIMETHYLMELAMINE see TNJ825
TRIMETHYL((2-METHYL-1,3-DIOXOLAN-4-YL)METHYL)AMMONIUM IO-
   DIDE (8CI) see MJH250
4,11,11-TRIMETHYL-8-METHYLENE-5-OXATRICYCLO(8.2.0.0(4,6))DODECANE
   see CCN100
TRIMETHYL(1-METHYL-2-PHENOTHIAZIN-10-YLETHYL)AMMONIUM
   METHYL SULFATE see MRW000
TRIMETHYL(1-METHYL-2-(10-PHENOTHIAZINYL)ETHYL)AMMONIUM
   METHYL SULFATE see MRW000
N,N,N'-TRIMETHYL-N'-(5-METHYL-3-PHENYL-1H-INDOL-1-YL)-1,2-
   ETHANEDIAMINE HYDROCHLORIDE see MNU750
N,N,2-TRIMETHYL-4-(4'-(3'-METHYLPYRIDYL-1'-OXIDE)AZO)ANILINE
   see DQE100
N,N,3-TRIMETHYL-4-(4'-(3'-METHYLPYRIDYL-1'-OXIDE)AZO)ANILINE
   see DQE200
1,3,3-TRIMETHYL-6'-NITROINDOLINE-2-SPIRO-2'-BENZOPYRAN see TLU500
TRIMETHYLNITROSOHARNSTOFF (GERMAN) see TLU750
N,2,2-TRIMETHYL-N-NITROSO-1-PROPYLAMINE see NKU570
N-TRIMETHYL-N-NITROSOUREA see TLU750
1,1,3-TRIMETHYL-3-NITROSOUREA see TLU750
2,6,8-TRIMETHYLNONANOL-4 see TLV000
2,6,8-TRIMETHYL-4-NONANOL see TLV000
TRIMETHYL NONANONE see TLV250
2,6,8-TRIMETHYLNONYL VINYL ETHER see VQU000
(−)-endo-1,3,3-TRIMETHYL-2-NORBORNANOL see TLW000
1,3,3-TRIMETHYL-2-NORBORNANOL ACETATE see FAO000
1,3,3-TRIMETHYL-2-NORBORNANONE see TLW250
1,3,3-TRIMETHYL-2-NORBORNANYL ACETATE see FAO000
1,3,3-TRIMETHYL-2-NORCAMPHANONE see TLW250
1,7,7-TRIMETHYLNORCAMPHOR see CBA750
3,7,7-TRIMETHYL-3-NORCARENE see CCK500
4,7,7-TRIMETHYL-3-NORCARENE see CCK500
TRIMETHYLOCTADECYLAMMONIUM CHLORIDE see TLW500
TRIMETHYLOLAMINOMETHANE see TEM500
TRIMETHYLOLMELAMINE see TLW750
TRIMETHYLOLNITROMETHANE see HMJ500
TRIMETHYLOLPROPANE DIALLYL ETHER see TLX000
TRIMETHYLOLPROPANE TRIACRYLATE see TLX175
TRIMETHYLOLPROPANE TRIMETHACRYLATE see TLX250
TRIMETHYLOLPROPANE TRIMETHANCRYLATE see TLX250
TRIMETHYLOPROPANE MONOALLYL ETHER see TLX100
TRIMETHYL ORTHOFORMATE see TLX600
1,3,3-TRIMETHYL-2-OXABICYCLO(2.2.2)OCTANE see CAL000
TRIMETHYLOXACYCLOPROPANE see TLY175
3,5,5-TRIMETHYL-2,4-OXAZOLIDINEDIONE see TLP750
TRIMETHYL-2-OXEPANONE (mixed isomers) see TLY000
TRIMETHYLOXIRANE see TLY175
2,2,3-TRIMETHYLOXIRANE see TLY175
N,N,N-TRIMETHYL-4-OXO-1-PENTANAMINIUM IODIDE see TLY250
TRIMETHYL(4-OXYPENTHYL)AMMONIUM IODIDE see TLY250
N,N,N-TRIMETHYL-1-PENTANAMINIUM IODIDE see TMA500
2,2,4-TRIMETHYLPENTANE see TLY500
2,2,4-TRIMETHYL-1,3-PENTANEDIOL see TLY750
TRI(2-METHYL-2,4-PENTANEDIOL)BIBORATE see TKM250
2,2,4-TRIMETHYL-1,3-PENTANEDIOL DIISOBUTYRATE see TLZ000
2,2,4-TRIMETHYLPENTANEDIOL-1,3-DIISOBUTYRATE see TLZ000
2,2,4-TRIMETHYL-1,3-PENTANEDIOL MONOISOBUTYRATE see TEG500
2,4,4-TRIMETHYL PENTENE see TMA250
2,2,4-TRIMETHYL-1-PENTENE see IKF000
TRIMETHYLPENTYLAMMONIUM IODIDE see TMA500

6,6,9-TRIMETHYL-3-PENTYL-6H-DIBENZO(b,d)PYRAN-1-OL see CBD625
6,6,9-TRIMETHYL-3-PENTYL-7,8,9,10-TETRAHYDRO-6H-
 DIBENZO(B,D)PYRAN-1-OL see TCM250
N(8),N(8),3-TRIMETHYL-2,8-PHENAZINEDIAMINE MONOHYDROCHLORIDE
 see AJQ250
N,N,α-TRIMETHYL-10H-PHENOTHIAZINE-10-ETHANAMINE
 MONOHYDROCHLORIDE see PMI750
N,N,β-TRIMETHYL-10H-PHENOTHIAZINE-10-PROPANAMINE see AFL500
N-(2,4,6-TRIMETHYLPHENYL)ACETAMIDE see TLD250
1-(2,4,6-TRIMETHYLPHENYLAMINO)ANTHRAQUINONE see TMA750
TRIMETHYLPHENYLAMMONIUM BROMIDE see TMB000
TRIMETHYLPHENYL AMMONIUM CHLORIDE see TMB250
TRIMETHYLPHENYLAMMONIUM HYDROXIDE see TMB500
TRIMETHYLPHENYLAMMONIUM IODIDE see TMB750
N-sym-TRIMETHYLPHENYLDIETHYLAMINOACETAMIDE HYDROCHLO-
 RIDE see DHL800
N,N,N'-TRIMETHYL-N'-(3-PHENYL-1H-INDOL-1-YL)-1,2-ETHANEDIAMINE
 MONOHYDROCHLORIDE see BGC500
TRIMETHYLPHENYLMETHANE see BQJ250
TRIMETHYLPHENYL METHYLCARBAMATE see TMC750
3,4,5-TRIMETHYLPHENYL METHYLCARBAMATE see TMD000
TRI 2-METHYLPHENYL PHOSPHATE see TMO600
1,2,5-TRIMETHYL-4-PHENYL-4-PIPERIDINOL, PROPIONATE (ESTER)
 see IMT000
1,2,5-TRIMETHYL-4-PHENYL-4-PROPIONOXYPIPERIDINE see IMT000
TRIMETHYL PHOSPHATE see TMD250
O,O,O-TRIMETHYL PHOSPHATE see TMD250
TRIMETHYLPHOSPHINE see TMD275
TRIMETHYLPHOSPHINE SELENIDE see TMD400
TRIMETHYL PHOSPHITE see TMD500
O,S,S-TRIMETHYL PHOSPHORODITHIOATE see TMD625
TRIMETHYL PHOSPHOROTHIOATE see TMH525
O,O,S-TRIMETHYL PHOSPHOROTHIOATE see DUG500
S,S,S-TRIMETHYL PHOSPHOROTRITHIOATE see TMD699
2,2,6-TRIMETHYL-4-PIPERIDINOL BENZOATE (ESTER) see EQP500
TRIMETHYLPLATINUM HYDROXIDE see TME250
1,2,2-TRIMETHYLPROPYL METHYLPHOSPHONOFLUORIDATE see SKS500
TRIMETHYLPYRAZINE see TME270
2,3,5-TRIMETHYLPYRAZINE see TME270
2,4,6-TRIMETHYLPYRILIUM PERCHLORATE see TME275
TRIMETHYLSELENONIUM see TME500
TRIMETHYLSELENONIUM CHLORIDE see TME600
TRIMETHYLSELENONIUM ION see TME500
N-TRIMETHYLSILYLACETAMIDE see TME750
N-(TRIMETHYLSILYL)ACETIMIDIC ACID, TRIMETHYLSILYL ESTER
 see TMF000
TRIMETHYLSILYL AZIDE see TMF100
TRIMETHYL SILYL-1-CHLORO-7-DIHYDRO-1,3-PHENYL-5,2H-BENZODIAZEP-
 INE-1,4-ONE-2 (FRENCH) see CGB000
N-(TRIMETHYLSILYL)ETHANIMIDIC ACID, TRIMETHYLSILYL ESTER
 see TMF000
TRIMETHYL SILYL HYDROPEROXIDE see TMF125
N-(TRIMETHYLSILYL)-IMIDAZOL see TMF250
(TRIMETHYLSILYL)IMIDAZOLE see TMF250
1-(TRIMETHYLSILYL)IMIDAZOLE see TMF250
N-(TRIMETHYLSILYL)IMIDAZOLE see TMF250
1-(TRIMETHYLSILYL)-1H-IMIDAZOLE see TMF250
TRIMETHYLSILYLMETHANOL see TMF500
N-TRIMETHYLSILYLMETHYL-N-NITROSOUREA see TMF600
TRIMETHYLSILYL PERCHLORATE see TMF625
TRIMETHYLSTANNANE SULPHATE see TMI500
TRIMETHYLSTANNYL CHLORIDE see CLT000
TRIMETHYLSTANNYL IODINE see IFN000
TRIMETHYLSTEARYLAMMONIUM CHLORIDE see TLW500
TRINETHYL STIBINE see AQL000
N,N,2'-TRIMETHYL-4-STILBENAMINE see TMF750
N,N,3'-TRIMETHYL-4-STILBENAMINE see TMG000
N,N,4'-TRIMETHYL-4-STILBENAMINE see TMG250
TRIMETHYLSTILBINE see TLH100
TRIMETHYLSULFONIUM IODIDE see TMG500
TRIMETHYLSULFONIUM, IODIDE, COMPOUND with MERCURY IODIDE
 (1:1) see TMG750
TRIMETHYLSULFONIUM IODIDE MERCURIC IODIDE addition compound
 see TMG750
TRIMETHYLSULFOXONIUM BROMIDE see TMG775
TRIMETHYL(TETRAHYDRO-4-HYDROXY-5-METHYLFURFURYL)
 AMMONIUM see MRW250
TRIMETHYLTHALLIUM see TMH250
TRIMETHYL THIOPHOSPHATE see TMH525
TRIMETHYLTHIOUREA see TMH750
1,1,3-TRIMETHYL-2-THIOUREA see TMH750
N,N,N'-TRIMETHYLTHIOUREA see TMH750
TRIMETHYLTIN ACETATE see TMI000
TRIMETHYLTIN CHLORIDE see CLT000

TRIMETHYLTIN CYANATE see TMI100
TRIMETHYLTIN HYDROXIDE see TMI250
TRIMETHYLTIN IODIDE see IFN000
TRIMETHYLTIN ISOTHIOCYANATE see ISF000
TRIMETHYLTIN SULPHATE see TMI500
TRIMETHYLTIN THIOCYANATE see TMI750
5,7,8-TRIMETHYLTOCOL see VSZ450
6,6',6''-TRIMETHYLTRIHEPTYLAMINE see TKS500
2,2'-((1,1,3-TRIMETHYLTRIMETHYLENE)DIOXY)BIS(4,4,6-TRIMETHYL)1,3,2-
 DIOXABORINANE see TKM250
α,α,α'-TRIMETHYLTRIMETHYLENE GLYCOL see HFP875
1,1,1-TRIMETHYL-N-(TRIMETHYLSILYL)SILANAMINE see HED500
2,4,6-TRIMETHYL-1,3,5-TRIOXAAN (DUTCH) see PAI250
2,4,6-TRIMETHYL-s-TRIOXANE see PAI250
2,4,6-TRIMETHYL-1,3,5-TRIOXANE see PAI250
s-TRIMETHYLTRIOXYMETHYLENE see PAI250
1,3,5-TRIMETHYL-2,4,6-TRIS(3,5-DI-tert-BUTYL-4-HYDROXYBENZYL)
 BENZENE see TMJ000
TRIMETHYLUREA see TMJ250
1,1,3-TRIMETHYLUREA see TMJ250
TRIMETHYL VINYL AMMONIUM HYDROXIDE see VQR300
1,3,7-TRIMETHYLXANTHINE see CAK500
TRIMETHYL(2-(2,6-XYLYLOXY)ETHYL)AMMONIUM BROMIDE see XSS900
2-(TRIMETIL-ACETIL)-INDAN-1,3-DIONE (ITALIAN) see PIH175
3,5,5-TRIMETIL-2-CICLOESEN-1-ONE (ITALIAN) see IMF400
2,4,6-TRIMETIL-1,3,5-TRIOSSANO (ITALIAN) see PAI250
TRIMETIN see TLP750
TRIMETION see DSP400
TRIMETOGINE see TKX250
TRIMETON see PGF000, TMJ750
TRIMETON MALEATE see TMK000
TRIMETOPRIM-SULFA see SNK000
TRIMETOQUINOL see IDD100
1-TRIMETOQUINOL see IDD100
TRIMETOQUINOL HYDROCHLORIDE see IDD100
(+)-TRIMETOQUINOL HYDROCHLORIDE see TMJ800
(±)-TRIMETOQUINOL HYDROCHLORIDE see TKX125
dl-TRIMETOQUINOL HYDROCHLORIDE see TKX125
R-(+)-TRIMETOQUINOL HYDROCHLORIDE see TMJ800
TRIMETOSE see TMK000
TRIMETOZIN see TKX250
TRIMETOZINA see TKX250
TRIMETOZINE see TKX250
TRIMFORTE see TKX000
TRI-MINO see MQW100
TRI-MINOCYCLINE see MQW100
TRIMIPRAMINE see DLH200
TRIMIPRAMINE ACID MALEATE see SOX550
TRIMIPRAMINE MALEATE see SOX550
TRIMIPROMINE HYDROCHLORIDE see TAL000
TRIMITAN see TNF500
TriML see TLU175
TRIMOL see TMK150
TRIMON see AQH500
TRIMOPAN see TKZ000
TRIMOSULFA see TKX000
TRIMPEX see TKZ000
TRIMSTAT see DKE800
TRIMTABS see DKE800
TRIMUSTINE see TNF500
TRIMUSTINE HYDROCHLORIDE see TNF500
TRIMYSTEN see MRX500
TRINAGLE see CNP250
TRINATRIUMPHOSPHAT (GERMAN) see SJH200
TRINEX see TIQ250
TRINITRIN see NGY000
TRINITROACETONITRILE see TMK250
TRINITROBENZEEN (DUTCH) see TMK500
TRINITROBENZENE see TMK500
1,3,5-TRINITROBENZENE see TMK500
TRINITROBENZENE, dry (DOT) see TMK500
TRINITROBENZENE (wet) see TMK750
2,3,5-TRINITROBENZENEDIAZONIUM-4-OXIDE see TMK775
2,4,6-TRINITROBENZENE-1,3-DIOL see SMP500
2,4,6-TRINITRO-1,3-BENZENEDIOL see SMP500
2,4,6-TRINITROBENZENE-1,3,5-TRIOL see TMM775
TRINITROBENZENE, WET, at least 10% water, over 15 ounces in one outside
 packaging (DOT) see TMK750
TRINITROBENZENE, WET, containing at least 10% water (DOT) see TMK750
TRINITROBENZENE, WET, containing less than 30% water (DOT) see TMK750
2,4,6-TRINITROBENZOIC ACID see TML000
TRINITROBENZOIC ACID (dry) see TML000
TRINITROBENZOIC ACID (wet) see TML250
TRINITROBENZOL (GERMAN) see TMK500

TRIS(N-(α,α,α-TRIFLUORO-m-TOLYL)ANTHRANILATO)ALUMINUM see AHA875
TRIS(3,3,5-TRIMETHYLCYCLOHEXYL)ARSINE see TNO000
TRIS-(TRIMETHYLSILYL)FOSFAT (CZECH) see PHE750
TRISULFIDE, BIS(ETHOXYTHIOCARBONYL)- see DKE400
TRISULFON CONGO RED see SGQ500
TRISULFON VIOLET N see CMP000
TRISULFURATED PHOSPHORUS see PHS500
TRISUSTAN see TJL250
TRITELLURIUM TETRANITRIDE see TNO250
TRITEREN see UVJ450
TRITERPENE SAPONINS mixture from AESCULUS HIPPOCASTONUM
    see TNO275
TRITHENE see CLQ750
TRITHEOM see ABY900
1,3,5-TRITHIACYCLOHEXANE see TLS500
sym-TRITHIAN (CZECH) see TLS500
1,3,5-TRITHIANE see TLS500
3,4,5-TRITHIATRICYCLO(5.2.1.0^{2,6})DECANE see TNO300
TRITHIO see AOO490
TRITHIOANETHOLE see AOO490
TRITHIOBIS(TRICHLOROMETHANE) see BLM750
TRITHIOCYANURIC ACID see THS250
TRITHIOFORMALDEHYDE see TLS500
TRITHIO-(p-METHOXYPHENYL)PROPENE see AOO490
TRITHION see TNP250
TRITHION MITICIDE see TNP250
TRITHIOZINE see TNP275
TRITIOZINA (ITALIAN) see TNP275
TRITIOZINE see TNP275
TRITISAN see PAX000
TRITOFTOROL see EIR000
TRITOL see TMN490, TMN500
TRITOLYL PHOSPHATE see TNP500
TRI-2-TOLYL PHOSPHATE see TMO600
TRI-o-TOLYL PHOSPHATE see TMO600
TRITON N-100 see NND500
TRITON A-20 see TDN750
TRITON GR-5 see DJL000
TRITON K-60 see AFP250
TRITON WR 1339 see TDN750
TRITON X 15 see GHS000
TRITON X 35 see PKF500
TRITON X-40 see DTC600
TRITON X 45 see PKF500
TRITON X100 see OFQ000
TRITON X 100 see PKF500
TRITON X 102 see PKF500
TRITON X 165 see PKF500
TRITON X 305 see PKF500
TRITON X 405 see PKF500
TRITON X 705 see PKF500
TRITOX see TII750
TRITROL see CLN750
TRITTICO see CKJ000, THK880
TRITYL CHLORIDE see CLT500
S-TRITYLCYSTEINE see TNR475
S-TRITYL-l-CYSTEINE see TNR475
3-TRITYLTHIO-l-ALANINE see TNR475
3-(TRITYLTHIO)-l-ALANINE see TNR475
TRITYLTHIOL see TMT500
TRIUMBREN see AAN000
TRIUROL see AAN000
TRIUROPAN see AAN000
TRIVALENT SODIUM ANTIMONYL GLUCONATE see AQI000
TRIVASTAL see TNR485
TRIVASTAN see TNR485
TRIVAZOL see MMN250
TRI-VC 13 see DFK600
TRIVINYLANTIMONY see TNR490
TRIVINYLBISMUTH see TNR500
TRIVINYLTIN CHLORIDE see CLU500
TRIVITAN see CMC750
TRIZILIN see DFT800
TRIZIMAN see DXI400
TRIZIMAN D see DXI400
TROBICIN see SKY500
TROCHIN see CLD000
TROCINAT see DHY400
TROCINATE see DHY400
TROCINATE HYDROCHLORIDE see DHY400
TROCLOSENE see DGN200
TROCOSONE see ECU750
TRODAX see HLJ500
TROFORMONE see AOO475

TROFOSFAMID see TNT500
TROFURIT see CHJ750
TROGAMID T see NOH000
TROGUM see SLJ500
TROJCHLOREK FOSFORU (POLISH) see PHT275
TROJCHLOROBENZEN (POLISH) see TIK250
TROJCHLOROETAN(1,1,2) (POLISH) see TIN000
TROJCHLOROWODOREK 4-ACETYLO-4-(3-CHLOROFENYLO)-1-(3-(4-
    METYLOPIPERAZYNO)-PROPYLO)-PIPERYDYNY see ACG125
TROJKREZYLU FOSFORAN (POLISH) see TMO600
TROJNITROTOLUEN (POLISH) see TMN490
TROLAMINE see TKP500
TROLEN see RMA500
TROLENE see RMA500
TROLITUL see SMQ500
TROLMINE see TJL250
TROLNITRATE PHOSPHATE see TJL250
TROLOVOL see MCR750
TROLOX see TNR625
TROLOX C see TNR625
TROMASEDAN see BEA825
TROMASIN see PAG500
TROMBARIN see BKA000
TROMBAVAR see SIO000
TROMBIL see BKA000
TROMBOLYSAN see BKA000
TROMBOSAN see BJZ000
TROMBOVAR see EMT500, SIO000
TROMEDONE see TLP750
TROMETAMOL see TEM500
TROMETE see SJH200
TROMETHAMINE see TEM500
TROMETHAMINE PROSTAGLANDIN F2-α see POC750
TROMETHANE see TEM500
TROMETHANMIN see TEM500
TROMEXAN see BKA000
TROMEXAN ETHYL ACETATE see BKA000
TRONA see BMG400, SFO000, SJY000
TRONAMANG see MAP750
TRONOX see TGG760
TROPACAINE HYDROCHLORIDE see TNS200
TROPACOCAINE HYDROCHLORIDE see TNS200
TROPAEOLIN see MND600
TROPAEOLIN 1 see FAG010
TROPAEOLIN D see DOU600
TROPAEOLIN G see MDM775
TROPAKOKAIN HYDROCHLORID (GERMAN) see TNS200
2-β-TROPANECARBOXYLIC ACID, 3-β-HYDROXY-, METHYL ESTER, BEN-
    ZOATE (ESTER) see CNE750
3-TROPANYLBENZOATE-2-CARBOXYLIC ACID METHYL ESTER see CNE750
dl-TROPANYL-2-HYDROXY-1-PHENYLPROPIONATE see ARR000
dl-TROPANYL-2-HYDROXY-1-PHENYLPROPIONATE SULFATE see ARR500
3-α-TROPANYL (−)-2-METHYL-2-PHENYLHYDRACRYLATE HYDROCHLO-
    RIDE see LFC000
dl-TROPASAEUREESTER DES 3-DIAETHYLAMINO-2,2-DIMETHYL-1-PRO-
    PANOL PHOSPHAT (GERMAN) see AOD250
TROPASAEUREESTER DES 3-TRIAETHYLAMMONIUM-2,2-DIMETHYL-1-PRO-
    PANOLBROMID (GERMAN) see DSI200
(−+)-TROPATE-3-α-HYDROXY-8-OCTYL-1-α-H,5-α-H-TROPANIUM see OEM000
TROPAX see OPK000
TROPEOLIN see BFL000
TROPHICARD see MAI600
TROPHICARDYL see IDE000
TROPHOSPHAMID see TNT500
TROPHOSPHAMIDE see TNT500
TROPIC ACID, ESTER with SCOPINE see SBG000
TROPIC ACID, ESTER with TROPINE see ARR000
(−)-TROPIC ACID ESTER with TROPINE see HOU000
TROPIC ACID, 9-METHYL-3-OXA-9-AZATRICYCLO(3.3.1.0^{2,4})NON-7-YL
    ESTER see SBG000
TROPIC ACID, 3-α-TROPANYL ESTER see AQO250
TROPIC ACID-3-α-TROPANYL ESTER see ARR000
TROPIDECHIS CARINATUS VENOM see ARV375
TROPILIDENE see COY000
TROPIN see MGR250
TROPINE, ATROPATE (ESTER) see AQO250
TROPINE BENZOHYDRYL ETHER METHANESULFONATE see TNU000
TROPINE-4-CHLOROBENZHYDRYL ETHER HYDROCHLORIDE see CMB125
TROPINE (−)-α-METHYLTROPATE HYDROCHLORIDE see LFC000
TROPINE TROPATE see ARR000
TROPINIUM METHOBROMIDE MANDELATE see MDL000
TROPINTRAN see ARR500
TROPISTON see PGP500
TROPITAL see PIZ499

TROPOCOCAIN HYDROCHLORIDE see TNS200
TROPOLONE see TNV550
TROPOTOX see CLN750, CLO000
TROPYLIUM PERCHLORATE see TNV575
(±)-TROPYL TROPATE see ARR000
dl-TROPYLTROPATE see ARR000
TROSINONE see GEK500
TROSPIUM CHLORIDE see KEA300
TROTOX see CLN750
TROTYL see TMN490
TROTYL OIL see TMN490, TMN500
TROVIDUR see PKQ059, VNP000
TROVITHERN HTL see PKQ059
TROXIDONE see TLP750
TROXILAN see AFJ400
TROXONIUM TOSILATE see TNV625
TROXONIUM TOSYLATE see TNV625
TROYKYD ANTI-SKIN B see EMU500
TROYKYD ANTI-SKIN BTO see BSU500
TROYSAN 142 see DSB200
TROYSAN ANTI-MILDEW O see TIT250
TRP-P-2 see ALD500
TRP-P-1 (ACETATE) see AJR500
TRP-P-2(ACETATE) see ALE750
D-TRP-LH-RH see GJI100
TRP-P-1 see TNX275
TRU see PQC500
TRUBAN see EFK000
TRUCIDOR see MJG500
TRUE AMMONIUM SULFIDE see ANJ750
TRUFLEX DOA see AEO000
TRUFLEX DOP see DVL700
TRUFLEX DOX see BJQ500
TRUFLEX DTDP see DXQ200
TRUMPET PLANT see CDH125
TRUSONO see EQL000, MNM500
TRUXAL see TAF675
TRUXALETTEN see TAF675
TRUXIL see TAF675
TRYBEN see TIK500
TRYCOL HCS see PJW500
TRYCOL LAL SERIES see DXY000
TRYCOL NP-1 see NND500
TRYDET OS SERIES see PJY100
TRYDET SA SERIES see PJV250
TRYOPANOATE SODIUM see SKO500
TRYPAFLAVIN see XAK000
TRYPAFLAVINE see DBX400
TRYPANBLAU (GERMAN) see CMO250
TRYPAN BLUE see CMO250
TRYPAN BLUE SODIUM SALT see CMO250
TRYPARSAMIDE see CBJ750
TRYPSIN see TNW000
TRYPSIN INHIBITOR see PAF550
TRYPSIN INHIBITOR, PANCREATIC BASIC see PAF550
TRYPSIN-KALLIKREIN INHIBITOR (KUNITZ) see PAF550
TRYPTAMIDE see NDW525
TRYPTAMINE see AJX000
TRYPTAMINE HYDROCHLORIDE see AJX250
TRYPTAR see TNW000
TRYPTAZINE DIHYDROCHLORIDE see TKK250
TRYPTIZOL see EAH500, EAI000
TRYPTIZOL HYDROCHLORIDE see EAI000
(−)-TRYPTOPHAN see TNX000
d-TRYPTOPHAN see TNW250
dl-TRYPTOPHAN see TNW500
l-TRYPTOPHAN (FCC) see TNX000
l-TRYPTOPHAN, pyrolyzate see TNW950
dl-TRYPTOPHAN, pyrolyzate 1 see TNX275
TRYPTOPHANE see TNX000
d-TRYPTOPHANE see TNW250
l-TRYPTOPHANE see TNX000
l-TRYPTOPHAN, 5-HYDROXY-, (9CI) see HOA600
TRYPTOPHAN P1 see TNX275
TRYPTOPHAN P2 see ALD500
9-l-TRYPTOPHAN-TRYCODINE A HYDROCHLORIDE (9CI) see TOF825
TRYPURE see TNW000
TRYSBEN 200 see TIK500
TS 160 see TNF250, TNF500
TS 219 see NIM500
TS 1801 see AHA875
TSAA 291 see ELF100
TSAA-328 see ELF110
TSC see TFQ000

TSD see MCR250
TSELATOX see TNX375
TSERENOX see BDD000
TSH-RELEASING FACTOR see TNX400
TSH-RELEASING HORMONE see TNX400
TSH-RF see TNX400
TSIAZID see COH250
TSIDIAL see DRR400
TSIKLAMID see ABB000
TSIKLODOL see BBV000
TSIKLOMETIAZID see CPR750
TSIKLOMITSIN see TBX000
TSIM see TMF250
TSIMAT see BJK500
TSINEB (RUSSIAN) see EIR000
TSIPROMAT (RUSSIAN) see ZMA000
TSIRAM (RUSSIAN) see BJK500
TSITREX see DXX400
TSIZP 34 see ISR000
TSLT see TNX375
d-T4 SODIUM see SKJ300
TSP see SJH200
TSPA see TFQ750
TSPP see TEE500
TST see EIV000
T-STUFF see HIB000
TSUDOHMIN see DEO600
TSUMACIDE see MIB750
TSUSHIMYCIN see TNX650
TTD see DXH250, TFS350
TTFB see TBW000
TTFD see FQJ100
T (²)-TRICHOTHECENE see FQS000
TTS see DXH250
TTT see HDV500
TTX see FOQ000
TU see TFR250
2-TU see TFR250
TUADS see TFS350
TUAMINE see TNX750
TUAMINE SULFATE see AKD600
TUAMINOHEPTANE see TNX750
TUAMINOHEPTANE SULFATE see AKD600
TUAREG see TLN500
TUATUA (PUERTO RICO) see CNR135
TUAZOLE see QAK000
TUAZOLONE see QAK000
TUBADIL see TOA000
TUBARINE see TOA000
TUBA ROOT see DBA000
TUBATOXIN see RNZ000
TUBAZIDE see ILD000
TUBERACTINOMYCIN B see VQZ000
TUBERACTINOMYCIN-N SULFATE see TNY250, VQZ100
TUBERACTIN SULFATE see TNY250, VQZ100
TUBERCID see ILD000
TUBERCIDIN see TNY500
TUBEREX see PNW750
TUBERGAL see AGQ875
TUBERIT see CBM000
TUBERITE see CBM000
TUBERMIN see EPQ000
TUBEROID see EPQ000
TUBEROSON see EPQ000
TUBERSAN see SEP000
TUBEX see AES650
TUBICON see ILD000
TUBIGAL see FNF000
TUBIN see FNF000
TUBOCIN see RKP000
TUBOCURARIN see TNY750
TUBOCURARINE see TNY750
(+)-TUBOCURARINE see TNY750
d-TUBOCURARINE see TNY750
TUBOCURARINE CHLORIDE see TOA000
(+)-TUBOCURARINE CHLORIDE see TOA000
d-TUBOCURARINE CHLORIDE see TOA000
TUBOCURARINE, CHLORIDE, HYDROCHLORIDE, (+)- (8CI) see TOA000
d-TUBOCURARINE CHLORIDE PENTAHYDRATE see TNZ000
d-TUBOCURARINE DICHLORIDE see TOA000
(+)-TUBOCURARINE DICHLORIDE PENTAHYDRATE see TNZ000
TUBOCURARINE DIMETHYL ETHER IODIDE see DUM000
TUBOCURARINE DIMETHYL ETHER IODIDE see DUM000
TUBOCURARINE HYDROCHLORIDE see TOA000

(+)-TUBOCURARINE HYDROCHLORIDE see TOA000
d-TUBOCURARINE HYDROCHLORIDE see TOA000
d-TUBOCURARINE IODIDE DIMETHYL ETHER see DUM000
TUBOPHAN see DWW000
TUBOTIN see ABX250, HON000
TU CILLIN see BFD000
TUEX see TFS350
TUFF-LITE see PMP500
TUFT ROOT see DHB309
TUGON see TIQ250
TUGON FLIEGENKUGEL see PMY300
TUGON FLY BAIT see TIQ250
TUGON STABLE SPRAY see TIQ250
TULABASE FAST GARNET GB see AIC250
TULABASE FAST GARNET GBC see AIC250
TULABASE FAST RED TR see CLK220
TULA (PUERTO RICO) see COD675
TULISAN see TFS350
TULLIDORA see BOM125
TULUYLENDIISOCYANAT see TGM750
TULYL see RLU000
TUMBLEAF see SFS000
TUMENOL see IAD000
TUMESCAL OPE see BGJ250
TUMEX see QPA000
TUNG NUT see TOA275
TUNG NUT MEALS see TOA500
TUNG NUT OIL see TOA510
TUNG OIL TREE see TOA275
TUNGSTEN see TOA750
TUNGSTEN AZIDE PENTABROMIDE see TOB000
TUNGSTEN AZIDE PENTACHLORIDE see TOB250
TUNGSTEN BLUE see TOC750
TUNGSTEN CARBIDE see TOB500
TUNGSTEN CARBIDE, mixed with COBALT (92%:8%) see TOC000
TUNGSTEN CARBIDE, mixed with COBALT (85%:15%) see TOB750
TUNGSTEN CARBIDE, mixed with COBALT and TITANIUM (78%:14%:8%)
  see TOC250
TUNGSTEN COMPOUNDS see TOC500
TUNGSTEN FLUORIDE see TOC550
TUNGSTEN HEXAFLUORIDE see TOC550
TUNGSTEN-NICKEL CATALYST DUST see TOC600
TUNGSTEN OXIDE see TOC750
TUNGSTEN TRIOXIDE see TOC750
TUNGSTIC ACID see TOD000
TUNGSTIC ACID, DISODIUM SALT see SKN500
TUNGSTIC ACID, SODIUM SALT, DIHYDRATE see SKO000
TUNGSTIC ANHYDRIDE see TOC750
TUNGSTIC OXIDE see TOC750
TUNGSTOPHOSPHORIC ACID (8CI) see PHU750
TUNGSTOPHOSPHORIC ACID, SODIUM SALT see SJJ000
TUNIC see BGD250
TUPHETAMINE see BBK500
TUR see CMF400
TURBINAIRE see DAE525
TURBSVIL see BQT750
TURBULETHYLAZIN (GERMAN) see BQB000
TURCAM see DQM600
3336 TURF FUNGICIDE see DJV000
TURGEX see HCL000
TURIMYCIN A3 see LEV025
TURIMYCIN A5 see JDS200
TURIMYCIN P3 see MBY150
TURINAL see AGF750
TURISYNCHRON see MLJ500
TURKEY RED see DMG800
TURKEY-RED OIL see TOD500
TURKISH ROSE OIL see OHC000
TURMERIC see TOD625
TURMERIC rhizome extract see COF850
TURMERIC OIL see COG000
TURMERIC OLEORESIN see COG000
TURPENTINE see TOD750
TURPENTINE, steam distilled see TOD750
TURPENTINE OIL, rectifier see TOD750
TURPETH MINERAL see MDG000
TURPINAL SL see HKS780
TUS-1 see XAJ000
TUSILAN see DBE200
TUSSADE see DBE200
TUSSAL see MDP750
TUSSAPAP see HIM000
TUSSAPHED see POH250
TUSSCAPINE see NOA000

TUSSCAPINE HYDROCHLORIDE see NOA500
TUSSILAGO FARFARA L see CNH250
TUTANE see BPY000
TUTIN see TOE175
TUTINE see TOE175
TUTOCAINE see TOE150
TUTOCAINE HYDROCHLORIDE see AIT750
TUTOFUSIN TRIS see TEM500
TUTTOMYCIN see NCD550
TUTU see TOE175
TUZET see USJ075
TV 485 see HKK000
TV 1322 see AAE625
TVX 485 see HKK000
TWEEN 20 see PKG000
TWEEN 40 see PKG500
TWEEN 60 see PKL030
TWEEN 65 see SKV195
TWEEN 80 see PKL100
TWEEN 85 see TOE250
TWEENASE see GGA800
TWEEN ESTERASE see GGA800
TWEEN HYDROLASE see GGA800
TWIN LIGHT RAT AWAY see WAT200
TX 100 see PKF500
TX 380 see PMA450
TYBAMATE see MOV500
TYBATRAN see MOV500
TYCLAROSOL see EIV000
TYDEX see BBK500
TYGON see AAX175
TYLAN see TOE600
TYLENOL see HIM000
TYLON see TOE600
TYLOROL LT 50 see SON000
TYLOSE 666 see SFO500
TYLOSIN see TOE600
TYLOSIN HYDROCHLORIDE see TOE750
TYLOSTERONE see DKA600
TYLOXAPOL see TDN750
TYLOXYPAL see TDN750
TYOX A see BHM000
TYPOGEN CARMINE see EOJ500
TYRAMINE see TOG250
p-TYRAMINE see TOG250
TYRAMINE, 3-DIAZO-, HYDROCHLORIDE see DCQ575
TYRAMINE MONOCHLORIDE see TOF750
TYRANTON see DBF750
TYRIL see ADY500
TYRIMIDE see DAB875
TYRION YELLOW see DCZ000
TYROCIDIN B see TOF825
TYROCIDINE A, HYDROCHLORIDE see GJQ100
TYROCIDINE B, HYDROCHLORIDE see TOF825
TYROSAMINE see TOG250
TYROSINE see TOG300
l-TYROSINE see TOG300
p-TYROSINE see TOG300
l-p-TYROSINE see TOG300
dl-m-TYROSINE see TOG275
m-TYROSINE, dl- see TOG275
l-TYROSINE, O-(4-HYDROXY-3-IODOPHENYL)-3,5-DIIODO- (9CI) see LGK050
l-TYROSINE, 3-HYDROXY-α-METHYL-, SESQUIHYDRATE see MJE780
TYROTHRICIN see TOG500
TYVID see ILD000
TYZANOL HYDROCHLORIDE see VRZ000
TYZINE see VRZ000
TYZINE HYDROCHLORIDE see VRZ000
TZT see THR500
TZU-0460 see HNT100

U-14 see POC750
U 46 see CIR500, DAA800, TAA100
U46 see DGB000
U-197 see MQR100
U-0045 see BKP200
U-0172 see DPO100
U-0290 see UAG000
U-1149 see FOU000
U-1247 see BOQ500
U 1363 see DVV600
U-1434 see EEI000
U-1804 see UAG025
U-2069 see RDP300

U 2134 see MGC350
U-2363 see DTS625, UAG050
U-2397 see UAG075
U-3818 see CKK000
U-3886 see SFA000
U-4224 see DSB000
U 4513 see DRP800
U-4527 see CPE750
U-4568 see EAQ500
U-4748 see POK000
U-4761 see AHL000
U-4783 see ACJ500
U-4858 see TNW000
U 4905 see HHR000
U-5043 see DAA800
U-5227 see AQN635
U-5438 see FJF100
U 5446 see PBM500
U 5533 see PMG600
U-5897 see CDT750
U-5954 see DOO800
U 5963 see FHH100
U-5965 see TBX250
U 6020 see PLZ000
U 6040 see AOO275
U-6062 see CDP250
U-6233 see BDH250
U 6324 see PGN250
U-6421 see NGE500
U-6591 see NOB000
U-6596 see FKF100
U-6658 see AOB875
U-6780 see CMB000
U 6987 see BSM000
U-7118 see FBP300
U-7257 see CDK250
U 7524 see ALQ650
U-7726 see KFA100
U-7743 see CHW675
U 8210 see DAZ117
U-8344 see BIA250
U 8771 see EHP000
U-8774 see ALL250
U 8840 see MBZ150
U-8953 see FMM000
U 9088 see USJ100
U-9361 see SMC500
U-9889 see SMD000
U-10149 see LGD000
U-10293 see DFE700
U-10,858 see DCB000
U-10974 see FDD075
U 10997 see MQS225
U 11100 see NAD750
U-11,634 see TKJ750
U-12062 see DVJ200
U 12241 see CMS232
U 12927 see CGI500
U-14583 see POC500
U-14743 see MLY000
U 15030 see DKC400
U-15167 see NMV500
U-15,646 see PBH150
U-15800 see PAB500
U-17004 see CGI500, DET600
U 17556 see MIA500
U-17835 see TGJ500
U 18496 see ARY000
U-19183 see SKX000
U 19571 see AGT500
U-19,646 see CJQ250
U-21,251 see CMV675
U-24544 see PPI775
U-24973 see AEG875
U 25,352 see CHS250
U 25,354 see DGG400
U 26452 see CEH700
U 27,151 see BST900
U 27,574 see CKQ500
U-28,508 see CMV690
U-29135 see GCI000
U-29,409 see DLX300
U 31889 see XAJ000
U-33,030 see THS800

U 48160 see QQS075
U-52047 see MCB600
U-54461 see AIY850
U 11100A see NAD750
U-22,304A see LFG100
U-22394A see HDQ500
U-22,559A see DVP400
U 32921E see CCC100
U 631963 see CCS525
U 1 (polymer) see PKQ059
U-11555A see MFG260
U 12062A see POC360
U-24973A see TDL000
UBATOL U 2001 see SMQ500
UBIDECARENONE see UAH000
UBIQUINONE 10 see UAH000
UBIQUINONE 50 see UAH000
UBRETID see DXG800
UBRITIL see DXG800
UC 8305 see CGM400
UC 8,454 see TCY275
UC 9880 see CQI500
UC 10854 see COF250
UC 19786 see CBW000
UC 20047 see CFF250
UC 20,299 see SFV250
UC-21149 see CBM500
UC-21865 see AFK000
UC 22,463 see DET400, DET600, DWT200
UC-25074 see FNE500
UC 26089 see CFF250
UC 20,047A see CFF250
UCANE ALKYLATE 12 see UAK000
UCAR 17 see EJC500
UCAR 130 see AAX250
UCAR AMYL PHENOL 4T see AON000
UCAR BUTYLPHENOL 4-T see BSE500
UCAR TRIOL HG-170 see UBA000
UCB 170 see HGC500
UCB 492 see CJR909
UCB 1402 see BBV750
U.C.B. 1474 see CDQ500
UCB 1549 see DPA800
UCB 2493 see CJL409
UCB 3412 see MKQ000
UCB 4208 see COL250
UCB 4268 see HDS200
UCB 4445 see BOM250
U.CB 4492 see CJR909
UCB 6215 see NNE400
UCB 6249 see DIF600
UCB 1545 HYDROCHLORIDE see LFK200
UCC 974 see DSB200
UCET TEXTILE FINISH 11-74 (OBS.) see VOA000
UC LIQUID G see SCR400
U-COMPOUND see UVA000
UCON 12 see DFA600
UCON 113 see FOO000
UCON 114 see FOO509
UCON FLUID AP-1 see UBJ000
UCON FLUID 50-HB 260 see MNE250
UCON FLUOROCARBON 113 see FOO000
UCON FLUOROCARBON 122 see TIM000
UCON 12/HALOCARBON 12 see DFA600
UCON 22/HALOCARBON 22 see CFX500
UCON 113/HALOCARBON 113 see FOO000
UCON 500/HALOCARBON 500 see DFB400
UCON 50-HB-55 see UBS000
UCON 50-HB-100 see UCA000
UCON 50-HB-260 see UCJ000
UCON 50-HB-400 see UDA000
UCON 50-HB-660 see UDJ000
UCON 50-HB-2000 see UDS000
UCON 50-HB-3520 see UEA000
UCON 50-HB-5100 see UEJ000
UCON 50-HB-280-X see UFA000
UCON LB-250 see BRP250
UCON LB 1145 see BRP250
UCON LB-1715 see MKS250
UCON LB 1800X see BRP250
UCON LO-500 see UHA000
UCON LUBRICANT DLB-62-E see UIA000
UCON LUBRICANT DLB-140-E see UIJ000
UCON LUBRICANT DLB-200-E see UIS000

UCON REFRIGERANT 11 see TIP500
UDANTOL HYDROCHLORIDE see AEG625
UDCA see DMJ200
UDMH (DOT) see DSF400
UDOLAC see SOA500
U46 DP-FLUID see DGB000
UF 1 see DMI600
UFT see UNJ810
UG 767 see PAR600
UGUROL see AJV500
UK-738 see DWE800
UK 4261 see OLT000
UK 4271 see OLT000
UKOPEN see AOD125
U 46 KV-ESTER see CIR500
U 46 KV-FLUID see CIR500
ULACORT see PMA000
ULATKAMBAL ROOT EXTRACT see UIS300
ULCEDINE see TAB250
ULCERFEN see TEH500
ULCIMET see TAB250
ULCINE see DJM800, XCJ000
ULCOLIND see CLY500
ULCOMET see TAB250
ULCUDEXTER see DJM800, XCJ000
ULCUS-TABLINEN see CBO500
ULCUTIN see BGC625
SYMULER EOSIN TONER see BNH500
SYMULER FAST SCARLET 4R see MMP100
ULEXINE see CQL500
ULIOLIND see CLY500
ULMENIDE see CDL325
ULO see CMW700
ULONE see CMW700
ULSTRON see PMP500
ULTANDREN see AOO275
ULTANDRENE see AOO275
ULTRABIL see BGB315
ULTRABION see AIV500
ULTRA BRILLIANT BLUE P see DSY600
ULTRABRON see AIV500
ULTRACIDE see DSO000
ULTRACILLIN see AJJ875
ULTRACORTEN see PLZ000
ULTRACORTENE-H see PMA000
ULTRADINE see PKE250
ULTRAFUR see FPI000
ULTRALAN ORAL see FDA925
ULTRALENTE INSULIN see LEK000
ULTRA LENTE ISZILIN see LEK000
ULTRAMARINE BLUE see UJA200
ULTRAMARINE GREEN see CMJ900
ULTRAMARINE YELLOW see BAK250
ULTRAMID BMK see PJY500
ULTRAN see CKE750
ULTRAPAL CHLORIDE see HLC500
ULTRA SULFATE SL-1 see SIB600
ULTRAWET K see DXW200
ULTRON see PKQ059
ULUP see FMM000
ULV see DAO600
ULVAIR see MRH209
UM-792 see CQF099
UMBELLATINE SULFATE TRIHYDRATE see BFN750
UMBRADIL see DNG400
UMBRATHOR see TFT750
UMBRELLA LEAF see MBU800
UMBRIUM see DCK759
U46 MCPB see CLN750
U 27 METHANESULFONATE see PIE750
U 46 M-FLUID see CIR250
UML 491 see MLD250
UNADS see BJL600
UNAMIDE J-56 see BKE500
UNAMYCIN-B see VGZ000
1,2,3,4,5,5,6,7,9,10,10-UNDECACHLOROPENTACYCLO(5.3.0.O^{2,6}.0^{3,9}.0^{4,8})
    DECANE see MRI750
5,9-UNDECADIEN-2-ONE, 6,10-DIMETHYL- see GDE400
γ-UNDECALACTONE see UJA800
Δ-UNDECALACTONE see UKJ000
UNDECANAL see UJJ000
1-UNDECANAL see UJJ000
n-UNDECANAL see UJJ000
UNDECANALDEHYDE see UJJ000

UNDECANE see UJS000
n-UNDECANE see UJS000
1-UNDECANECARBOXYLIC ACID see LBL000
1,1'-UNDECANEDICARBOXYLIC ACID ESTER with ETHYLENE GLYCOL
    see EJQ500
UNDECANOIC ACID see UKA000
UNDECANOIC ACID, TRIBUTYLSTANNYL ESTER see TIG500
UNDECANOIC ACID, TRIISOPROPYLSTANNYL ESTER see TKT850
n-UNDECANOL see UNA000
UNDECANOLIDE-1,5 see UKJ000
2-UNDECANONE see UKS000
6-UNDECANONE see ULA000
10-UNDECENAL see ULJ000
1-UNDECEN-10-AL see ULJ000
10-UNDECENAL DIGERANYL ACETAL see UNA100
1-UNDECENE, 11,11-BIS((3,7-DIMETHYL-2,6-OCTADIENYL)OXY)- see UNA100
10-UNDECENOIC ACID see ULS000
10-UNDECENOIC ACID, BUTYL ESTER see BSS100
9-UNDECENOIC ACID, METHYL ESTER see ULS400
10-UNDECENOL see UMA000
1-UNDECEN-11-OL see UMA000
10-UNDECENOYL CHLORIDE see UMJ000
UNDECENYL ACETATE see UMS000
10-UNDECENYL ACETATE see UMS000
ω-UNDECENYL ALCOHOL see UMA000
n-UNDECOIC ACID see UKA000
UNDECYL ALCOHOL see UNA000
UNDECYL ALDEHYDE see UJJ000
N-UNDECYL ALDEHYDE see UJJ000
UNDECYLENALDEHYDE see ULJ000
10-UNDECYLENEALDEHYDE see ULJ000
UNDECYLENIC ACID see ULS000
9-UNDECYLENIC ACID see ULS000
UNDECYL-10-ENIC ACID see ULS000
10-UNDECYLENIC ACID see ULS000
ω-UNDECYLENIC ACID CHLORIDE see UMJ000
UNDECYLENIC ALCOHOL see UMA000
UNDECYLENIC ALDEHYDE see ULJ000
UNDECYLENIC ALDEHYDE DIGERANYL ACETAL see UNA100
10-UNDECYLENOYL CHLORIDE see UMJ000
UNDECYLIC ACID see UKA000
UNDECYLIC ALDEHYDE see UJJ000
UNDEN see EDV000, PMY300
UNFINISHED LUBRICATING OIL see COD750
UNGEREMINE see LJB800
UNIBARYT see BAP000
UNIBOLDINA see DNZ100
UNICELLES see BPF000
UNICEL-ND see DVF400
UNICEL NDX see DVF400
UNICHEM see PKQ059
UNICIN see TBX250
UNICOCYDE see ILD000
UNICROP CIPC see CKC000
UNICROP DNBP see BRE500
UNICROP MANEB see MAS500
UNIDERM WGO see WBJ700
UNIDIGIN see DKL800
UNIDOCAN see DAQ800
UNIDRON see DXQ500
UNIFLEX DOS see BJS250
UNIFOAM AZ see ASM270
UNIFORM AZ see ASM270
UNIFUME see EIY500
UNIFUR see FPI000
UNI-GUAR see GLU000
UNILAX see ACD500
UNIMATE GMS see OAV000
UNIMATE IPP see IQW000
UNIMOLL BB see BEC500
UNIMYCETIN see CDP250
UNIMYCIN see TBX250
UNION BLACK EM see AQP000
UNION CARBIDE 1-174 see TLC250
UNION CARBIDE 1-189 see TLC000
UNION CARBIDE 19786 see CBW000
UNION CARBIDE 20299 see SFV250
UNION CARBIDE A-15 see EQA000
UNION CARBIDE A-150 see TIN750
UNION CARBIDE A-162 see MQD750
UNION CARBIDE A-163 see MQF500
UNION CARBIDE A-186 see EBO000
UNION CARBIDE A-187 see ECH000
UNION CARBIDE LIQUID G see SCR400

UNION CARBIDE UC-8305 see CGM400
UNION CARBIDE UC-8454 see TCY275
UNION CARBIDE UC-9880 see CQI500
UNION CARBIDE UC-10,854 see COF250
UNION CARBIDE UC 20047 see CFF250
UNIPAQUE see AOO875
UNIPEN see SGS500
UNIPINE see PIH750
UNIPON see DGI400, DGI600
UNIPROFEN see OJI750
UNIROYAL see DFT000
UNIROYAL D014 see SOP000
UNIROYAL F849 see AKR500
UNISEDIL see DCK759
UNISOL 4-O see PJY100
UNISOL RH see SFO500
UNISOM see PGE775
UNISOMNIA see DLY000
UNISTRADIOL see EDP000
UNISULF see SNN500
UNITANE O-110 see TGG760
UNITENE see MCC250
UNITENSEN see CKP500
UNITERTRACID LIGHT ORANGE G see HGC000
UNITERTRACID YELLOW TE see FAG140
UNITESTON see TBG000
UNITHIOL see DNU860
UNITIOL see DNU860
UNITOL see DNU860
UNITOX see CDS750
UNIVERM see CBY000
UNIVOL U 316S see MSA250
UNJECOL 50 see OBA000
UNJECOL 70 see OBA000
UNJECOL 90 see OBA000
UNJECOL 110 see OBA000
UNLEADED GASOLINE see GCE100
UNLEADED MOTOR GASOLINE see GCE100
UNON P see TAI250
UNOSPASTON see DGW600
UNOX 201 see ECB000
UNOXAT EPOXIDE 269 see LFV000
UNOX EPOXIDE 201 see ECB000
UNOX EPOXIDE 206 see VOA000
UNYSH A see FAQ930
UOP 88 see BJT500
UP 1E see SMR000
UP 83 see NDX500
U 46DP see DAA800
UPIOL see BNP750
UPJOHN U-12,927 see CGI500
UPJOHN U-18120 see MDX250
UPJOHN U-22023 see MLX000
UPJOHN U-32714 see CON300
UPJOHN U-36059 see MJL250
UR 606 see CEH700
UR 1522 see PHA575
URACIL see UNJ800
6-URACILCARBOXYLIC ACID see OJV500
URACIL mixture with FT (4:1) see UNJ810
URACIL, 5-(HYDROXYMETHYL)-6-METHYL- see HMH300
URACILLOST see BIA250
URACILMOSTAZA see BIA250
URACIL MUSTARD see BIA250
URACIL, 6-PROPYL-2-THIO-, and IODINE see PNX100
URACIL RIBOSIDE see UVJ000
URACIL mixture with TEGAFUR (4581) see UNJ810
URACIL mixture with 1-(2-TETRAHYDROFURYL)-5-FLUOROURACIL (4:1)
    see UNJ810
URACTONE see AFJ500
URACTYL see SNQ550
URADAL see BNK000
URAGAN see BMM650, SIH500
URAGON see BMM650
URALGIN see EID000
URALYT-U see SFX725
URAMID see SNQ550
URAMINE T 80 see HLU500
URAMUSTIN see BIA250
URAMUSTINE see BIA250
URAMYCIN B see VGZ000
URANIN see FEW000
URANINE A EXTRA see FEW000

URANINE USP XII see FEW000
URANINE YELLOW see FEW000
URANIUM see UNS000
URANIUM ACETATE see UPS000
URANIUM AZIDE PENTACHLORIDE see UOA000
URANIUM, BIS(ACETATO)DIOXO-, DIHYDRATE see UQT700
URANIUM, BIS(ACETO-O)DIOXO-, DIHYDRATE (9CI) see UQT700
URANIUM, BIS(NITRATO-O,O')DIOXO-, (OC-6-11)- see URA100
URANIUM CARBIDE see UOB100
URANIUM(IV) CHLORIDE see UQJ000
URANIUM DICARBIDE see UOC200
URANIUM FLUORIDE (fissile) see UOJ000
URANIUM FLUORIDE (low specific activity) see UOS000
URANIUM FLUORIDE OXIDE see UQA000
URANIUM HEXAFLUORIDE, FISSILE (containing more than 1% U-235) (DOT)
    see UOJ000
URANIUM HEXAFLUORIDE, LOW SPECIFIC ACTIVITY (containing 0.7% or
    less U-235) (DOT) see UOS000
URANIUM(III) HYDRIDE see UPA000
URANIUM METAL, PYROPHORIC (DOT) see UNS000
URANIUM(IV) OXIDE see UPJ000
URANIUM OXYACETATE see UPS000
URANIUM OXYFLUORIDE see UQA000
URANIUM TETRACHLORIDE see UQJ000
URANIUM(IV) TETRAHYDROBORATE see UQT300
URANIUM(III) TETRAHYDROBORATE see UQS000
URANYL ACETATE see UPS000
URANYL ACETATE DIHYDRATE see UQT700
URANYL CHLORIDE see URA000
URANYL FLUORIDE see UQA000
URANYL NITRATE see URA100
URANYL NITRATE (solid) see URA200
URANYL NITRATE HEXAHYDRATE see URS000
URANYL NITRATE HEXAHYDRATE, solution (DOT) see URS000
URAPIDIL see USJ000
URARI see COF750
URAZIUM see PDC250
URBACID see USJ075
URBACIDE see USJ075
URBANYL see CIR750
URBASON see MOR500
URBASON CRYSTAL SUSPENSION see DAZ117
URBASONE see MOR500
URBASON SOLUBLE see USJ100
URBASULF see MGQ750
URBAZID see USJ075
URBIL see MQU750
URBOL see ZVJ000
UREA see USS000
UREA, 1-AMIDINO-3-(p-NITROPHENYL)-, MONOHYDROCHLORIDE see NIJ400
UREA ANTIMONYL TARTRATE see UTA000
UREA, N-(4-BROMOPHENYL)-N'-METHYL-(9CI) see MHS375
UREA, 1-((p-CHLOROPHENYL)SULFONYL)-3-CYCLOHEXYL- see CDR550
UREA, 1-((p-CHLOROPHENYL)SULFONYL)-3-ISOPROPYL- see CDY100
UREA, N,N-DIETHYL-N'-((8-α)-6-METHYLERGOLIN-8-YL)- see DLR100
UREA, N,N-DIETHYL-N'-((8-α)-6-METHYLERGOLIN-8-YL)-, (Z)-2-
    BUTENEDIOATE (1:1) see DLR150
UREA, N,N-DIETHYL-N'-((8-α)-6-PROPYLERGOLIN-8-YL)- see DJX300
UREA, N,N-DIETHYL-N'-((8-α)-6-PROPYLERGOLIN-8-YL)-, (Z)-2-BUTENEDIO-
    ATE (1:1) see DJX350
UREA DIOXIDE see HIB500
UREA HYDROGEN PEROXIDE (DOT) see HIB500
UREA HYDROGEN PEROXIDE SALT see HIB500
UREA HYDROPEROXIDE see HIB500
UREA, N-(2-METHOXYETHYL)-N-NITROSO- see NKO900
UREA, 1-METHYL-2-THIO- see MPW600
UREA MONONITRATE (8CI, 9CI) see UTJ000
UREA NITRATE see UTJ000
UREA NITRATE (wet) see UTJ000
UREA PERCHLORATE see UTU400
UREA PEROXIDE (DOT) see HIB500
UREAPHIL see USS000
UREA, SULFANILYL- see SNQ550
UREA, TETRABUTYL- see TBM850
UREA, 1,1,3,3-TETRABUTYL- see TBM850
URECHITES LUTEA see YAK300
URECHOLINE see HOA500
URECHOLINE CHLORIDE see HOA500
UREGIT see DFP600
p-UREIDOBENZENEARSONIC ACID see CBJ000
(p-UREIDOBENZENEARSYLENEDITHIO)DI-o-BENZOIC ACID see TFD750
4-UREIDO-1-PHENYLARSONIC ACID see CBJ000
(p-UREIDOPHENYLARSYLENEDITHIO)DIACETIC ACID see CBI250
(p-UREIDOPHENYLARSYLENEDITHIO)DI-o-BENZOIC ACID see TFD750

URENIL see SNQ550
UREOL P see DTG700
UREOPHIL see USS000
URESE see BDE250
URETAN ETYLOWY (POLISH) see UVA000
URETHAN see UVA000
URETHANE see UVA000
URETHYLANE see MHZ000
UREVERT see USS000
UREX see CHJ750
6,6'-UREYLENEBIS(1,1'-DIMETHYLQUINOLINIUM) SULFATE see PJA120
6,6'-UREYLENEBIS(1-METHYLQUINOLINIUM)BIS(METHOSULFATE) see PJA120
URFAMICIN HYDROCHLORIDE see UVA150
URGENEA MARITIMA see RCF000
URGILAN see POB500
URI see SPC500
URIBEN see EID000
URIC ACID see UVA400
URIC ACID, MONOSODIUM SALT see SKO575
URICEMIL see ZVJ000
URICOSID see DWW000
URICOVAC see DDP200
URIDINAL see PDC250, PEK250
URIDINE see UVJ000
β-URIDINE see UVJ000
URIDION see UVJ400
URINARY HEBIN see FMT100
URINARY INDICAN see ICD000
URINEX see CLH750
URIODONE see DNG400
URIPLEX see PDC250
URISOXIN see SNN500
URISPAS see FCB100
URITAS see ZVJ000
URITONE see HEI500
URITRATE see OOG000
URITRISIN see SNN500
URIZEPT see NGE000
URLEA see BEQ625
URNER'S LIQUID see DEL000
URO-ALVAR see OOG000
UROANTHELONE see UVJ475
UROBENYL see ZVJ000
UROBIOTIC-250 see PDC250
UROCALUM see UVJ425
UROCAUDAL see UVJ450
UROCONTRAST see HGB200
URODIAZIN see CFY000
URODINE see PDC250, PEK250
URODIXIN see EID000
UROENTERONE see UVJ475
UROFEEN see PDC250
UROGAN see SNN500
UROGASTRON see UVJ475
UROGASTRONE see UVJ475
UROGRAFIN ACID see DCK000
UROGRANOIC ACID see DCK000
UROKINASE see UVS500
UROKINASE (ENZYME-ACTIVATING) see UVS500
UROKON SODIUM see AAN000
UROMAN see EID000
UROMIDE see PDC250
UROMIRO see AAI750
UROMIRO 380 see MCA775
UROMIRON see AAI750
UROMITEXAN see MDK875
UROMUCAESTHIN see BQA010
UROMYCINE see GCO000
URONAL see BAG000
URONEG see EID000
URONIUM NITRATE see UTJ000
URONIUM PERCHLORATE see UTU400
UROPHENYL see PDC250
UROPYRIDIN see PDC250
UROPYRINE see PDC250
UROSEMIDE see CHJ750
URO-SEPTRA see TKX000
UROSIN see ZVJ000
UROSULFAN see SNQ550
UROSULFANE see SNQ550
UROSULFON see SNP500
UROSULFONE see SNP500
UROTRAST see DCK000
UROTRATE see OOG000

UROTROPIN see HEI500
UROTROPINE see HEI500
UROVISON see SEN500
UROVIST see AOO875
UROX 379 see CJY000
UROX B WATER SOLUBLE CONCENTRATE WEED KILLER see BMM650
UROX HX GRANULAR WEED KILLER see BMM650
UROXIN see AFU250
UROXOL see OOG000
URSACOL see DMJ200
URSIN see HIH100
URSO see DMJ200
URSOCHOL see DMJ200
URSODEOXYCHOL see DMJ200
URSODEOXYCHOLIC ACID see DMJ200
URSODESOXYCHOLIC ACID see DMJ200
URSOFALK see DMJ200
URSOFERRAN see IGS000
URSOL BROWN O see CEG600
URSOL BROWN RR see ALL750
URSOL D see PEY500
URSOL EG see ALT500
URSOL ERN see NAW500
URSOL OLIVE 6G see CFK125
URSOL P see ALT250
URSOL P BASE see ALT250
URSOL SLA see DBO400
URSOLVAN see DMJ200
URSOL YELLOW BROWN A see ALO000
URTOSAL see SAH000
US 2 see CNH125
USACERT BLUE No. 1 see FAE000
USACERT BLUE No.2 see FAE100
USACERT FD & C RED No. 4 see FAG050
USACERT RED No. 1 see FAG018
USACERT RED No. 3 see FAG040
USACERT YELLOW No. 5 see FAG140
USAF D-1 see IDW000
USAF D-3 see CHU500
USAF D-5 see BSO500
USAF D-9 see CBM000
USAF M-2 see MCR000
USAF M-4 see ISQ000
USAF M-5 see ALI000
USAF M-6 see FOF000
USAF M-7 see CEI500
USAF P-2 see BJK500
USAF P-7 see DXQ500
USAF P-8 see CJX750
USAF S-1 see GLK000
USAF A-233. see AIR250
USAF A-3701 see TEV000
USAF A-3803 see PIK500
USAF A-4600 see MAO250
USAF A-6598 see DXP200
USAF A-8354 see TES250
USAF A-8564 see BIQ500
USAF A-8565 see HIM500
USAF A-8798 see CKT250
USAF A-9230 see EEW000
USAF A-9442 see SNE000
USAF A-9789 see DVX600
USAF A-11074 see DXN400
USAF A -15972 see DGX000
USAF A-19120 see BEX500
USAF AB-315 see DXJ800
USAF AM-1 see DJI400
USAF AM-3 see EMU500
USAF AM-4 see MKW750
USAF AM-5 see AAH250
USAF AM-6 see BSU500
USAF AM-7 see MKW500
USAF AM-8 see IJT000
USAF AN-7 see MFC700
USAF AN-8 see TDK000
USAF AN-9 see DOA400
USAF AN-11 see CAM000
USAF B-7 see ADC750
USAF B-15 see TFC600
USAF B-17 see BKU500
USAF B-19 see DWC600
USAF B-21 see DCH400
USAF B-22 see TFD250
USAF B-24 see SAV000

USAF B-30 see TFS350
USAF B-31 see TEF750
USAF B-32 see BJL600
USAF B-33 see BDE750, DXH250
USAF B-35 see SGF500
USAF B-40 see MRM750
USAF B-44 see BLJ250, DXL800
USAF B-45 see DCF000
USAF B-51 see PAY500
USAF B-58 see FPM000
USAF B-59 see TGN000
USAF B-74 see DTG600
USAF B-100 see DJY800
USAF B-121 see MAL250
USAF BE-25 see TFJ750
USAF BE-4-5 see IAP000
USAF BE-0405 see ADC750
USAF BO-1 see TKM250
USAF BO-2 see BBM000
USAF BV-8 see CJF500
USAF C-1 see TEV000
USAF CB-2 see ILD000
USAF CB-7 see BAC250
USAF CB-10 see AEH750
USAF CB-11 see GLS000
USAF CB-13 see FMT000
USAF CB-17 see XCA000
USAF CB-18 see AEH000
USAF CB-19 see NCG000
USAF CB-20 see TET300
USAF CB-21 see TFA000
USAF CB-22 see NBE500
USAF CB-26 see THT350
USAF CB-27 see RDK000
USAF CB-29 see ICW000
USAF CB-30 see THR750
USAF CB-34 see CQJ750
USAF CB-35 see TFJ100
USAF CB-36 see MCM750
USAF CB-37 see MRM750
USAF CB-96 see HOO100
USAF CF-2 see QCJ000
USAF CF-3 see HBU000
USAF CF-5 see RSU000
USAF CS-1 see DPM400
USAF CS-4 see AKS750
USAF CS-6 see NNM000
USAF CY-2 see CAQ250
USAF CY-4 see NAP500
USAF CY-5 see BDE750
USAF CY-6 see MJN250
USAF CY-7 see BDG000
USAF CY-9 see MES000
USAF CY-10 see NBA500
USAF CY-14 see DCF000
USAF CZ-1 see LFH000
USAF D-4 see POD750
USAF DM-1 see IAP000
USAF DO-1 see CEB250
USAF DO-4 see DES000
USAF DO-5 see HHW000
USAF DO-6 see BMY500
USAF DO-11 see ELB000
USAF DO-12 see HHR500
USAF DO-14 see HKI500
USAF DO-17 see PDQ750
USAF DO-20 see BQW000
USAF DO-21 see EKR500
USAF DO-22 see HKY500
USAF DO-23 see AGR000
USAF DO-28 see DMI600
USAF DO-29 see CDY850
USAF DO-30 see BGG500
USAF DO-32 see TCC000
USAF DO-36 see DVF200
USAF DO-37 see HKM500
USAF DO-38 see DVR000
USAF DO-41 see TNC500
USAF DO-42 see DDS000
USAF DO-43 see THU750
USAF DO-44 see ABD250, EMU500
USAF DO-45 see AAG000
USAF DO-46 see AKB000
USAF DO-47 see AGO000

USAF DO-50 see MGG000
USAF DO-51 see BOB500
USAF DO-52 see MKU250
USAF DO-54 see MHF750
USAF DO-55 see IBH000
USAF DO-59 see PGH000
USAF DO-61 see BHJ000
USAF DO-62 see TBQ500
USAF DO-63 see DWY200
USAF DO-65 see HCE000
USAF DO-68 see DGK200
USAF E-1 see PEW250
USAF E-2 see MCM750
USAF E-4 see DHO400
USAF EA-1 see NGG500
USAF EA-2 see NGE000
USAF EA-3 see FPE100
USAF EA-4 see NGE500
USAF EA-5 see NGC000
USAF EA-14 see NGC400
USAF EE-3 see MCN750
USAF EK see TKO250
USAF EK-3 see AAQ500
USAF EK-206 see PEY750
USAF EK-218 see QMJ000
USAF EK-245 see DWN800
USAF EK-338 see DOT300
USAF EK-356 see HIH000
USAF EK-394 see PEY500
USAF EK-442 see BBL000
USAF EK-488 see ABE500
USAF EK-496 see ABH000
USAF EK-497 see ISR000
USAF EK-510 see TGP250
USAF EK-534 see CBM250
USAF EK-572 see IEE000
USAF EK-600 see CBN000
USAF EK-631 see AHR750
USAF EK-660 see MCK500
USAF EK-678 see PEY600
USAF EK-695 see CCC750
USAF EK-704 see ASL250
USAF EK-705 see TFF250
USAF EK-743 see DVY000
USAF EK-749 see GKY000
USAF EK-794 see QPA000
USAF EK-906 see EMU500
USAF EK-982 see MHK500
USAF EK-1047 see DJC400
USAF EK-1231 see MPV750
USAF EK-1235 see EOL500
USAF EK-1239 see AAY000
USAF EK-1270 see DWC600
USAF EK-1275 see TFQ000
USAF EK-1375 see PEI000
USAF EK-1509 see TGO750
USAF EK-1569 see PGN250
USAF EK-1597 see EEC600
USAF EK-1651 see DXP600
USAF EK-1719 see TFA000
USAF EK-1803 see DKC400
USAF EK-1853 see MHI750
USAF EK-1860 see TFM250
USAF EK-1902 see TFA250
USAF EK-1995 see COH500
USAF EK-2010 see PGL250
USAF EK-2070 see EMB200
USAF EK-2089 see TFS350
USAF EK-2122 see HBD500
USAF EK-2124 see BEU500
USAF EK-2138 see DEI000
USAF EK-2219 see BGJ250
USAF EK-2596 see SGJ000
USAF EK-2635 see DJD000
USAF EK-2676 see TGP000
USAF EK-2784 see CEM500
USAF EK-3092 see DWN200
USAF EK-3110 see DWN400
USAF EK-3302 see ELL500
USAF EK-3941 see AIS500
USAF EK-3967 see HHB500
USAF EK-3991 see MCP250
USAF EK-4037 see AIG000
USAF EK-4196 see MCN250

USAF EK-4376 see AIF500
USAF EK-4394 see DXO200
USAF EK-4628 see HES000
USAF EK-4733 see IAL000
USAF EK-4812 see BDE500
USAF EK-4890 see ADD250
USAF EK-5017 see BDI500
USAF EK-5185 see MMD500
USAF EK-5199 see SKH500
USAF EK-5296 see CKT500
USAF EK-5426 see PGM750
USAF EK-5429 see BDJ000
USAF EK-5432 see BDE750
USAF EK-5496 see TEV500
USAF EK-6232 see QRS000
USAF EK-6279 see SIN000
USAF EK-6454 see MPW500
USAF EK-6540 see BCC500
USAF EK-6561 see ALQ000
USAF EK-6583 see MCK000
USAF EK-6754 see HNI000
USAF EK-6775 see DWI200
USAF EK-7087 see DWO000
USAF EK-7094 see QRJ000
USAF EK-7119 see MLE750
USAF EK-7162 see ENM000
USAF EK-7317 see QSJ000
USAF EK-7372 see TFE250
USAF EK-7423 see DTM000
USAF EK-8413 see AJT500
USAF EK-P-433 see ANW750
USAF EK-P-583 see FOU000
USAF EK-P-737 see TFA500
USAF EK-T-434 see SIA500
USAF EK-P-5430 see ACQ250
USAF EK-P-5501 see AMS250
USAF EK-P-5976 see AQN635
USAF EK-P-6255 see BJL600
USAF EK-P-6281 see BLJ250, DXL800
USAF EK-P-6297 see MCR000
USAF EK-T-2805 see MCK750
USAF EK-T-6645 see BKU500
USAF EL-23 see PAP550
USAF EL-30 see MCO500
USAF EL-42 see EEH000
USAF EL-44 see CKE750
USAF EL-45 see PGM750
USAF EL-52 see MFC500
USAF EL-54 see BEQ500
USAF EL-57 see IAO000
USAF EL-62 see IAQ000
USAF EL-76 see PDE500
USAF EL-78 see DQQ000
USAF EL-82 see BDY000
USAF EL-97 see MLF000
USAF EL-101 see BQP250
USAF EL-108 see MNM500
USAF FA-4 see TES250
USAF FA-5 see BCP500
USAF FO-1 see BQP250
USAF GE-1 see MNV750
USAF GE-12 see PCP250
USAF GE-13 see DWM000
USAF GE-14 see HNI500
USAF GY-1 see DXP400
USAF GY-2 see BLE500
USAF GY-3 see BDF000
USAF GY-5 see BIX000
USAF GY-7 see BHA750
USAF H-1 see QSA000
USAF HA-2 see RGZ550
USAF HA-5 see DGS600
USAF HC-1 see SBJ500
USAF IN-399 see MKW000
USAF KE-3 see MEQ000
USAF KE-5 see IQW000
USAF KE-7 see OAV000
USAF KE-8 see HKJ000
USAF KE-9 see PJW000
USAF KE-11 see EJM500
USAF KE-12 see PJV500
USAF KE-13 see SLL000
USAF KE-14 see PJW250
USAF KE-20 see HIN500

USAF KF-1 see COK250
USAF KF-2 see MCK750
USAF KF-3 see PES750
USAF KF-4 see BCK750
USAF KF-5 see CDN500
USAF KF-10 see EEW000
USAF KF-11 see CEQ600
USAF KF-13 see DVX200
USAF KF-14 see COJ250
USAF KF-15 see HIO000
USAF KF-17 see COJ500
USAF KF-18 see COH250
USAF KF-21 see PEA750
USAF KF-22 see MIQ000
USAF KF-25 see EHP500
USAF KF-26 see MAN750
USAF LO-3 see MCR250
USAF MA-1 see NEL500
USAF MA-2 see BDH000
USAF MA-3 see CEK000
USAF MA-4 see AID500
USAF MA-5 see NFJ500
USAF MA-10 see CDQ750
USAF MA-12 see AJF500
USAF MA-16 see BDH500
USAF MA-17 see MLX750
USAF ME-1 see BAD750
USAF MK-3 see BHM250
USAF MK-6 see DXO200
USAF MK-43 see BJJ000
USAF MO-2 see ANM500
USAF NB-1 see CEG750
USAF ND-09 see PHY000
USAF ND-54 see DMI600
USAF ND-59 see DMH400
USAF P-5 see TFS350
USAF P-220 see QQS200
USAF PD-20 see DOT800
USAF PD-25 see AKM000
USAF PD-58 see EEU500
USAF PE-1 see PEA500
USAF Q-1 see FPW000
USAF Q-2 see TCS500
USAF RH-1 see MDN500
USAF RH-3 see DPG600
USAF RH-4 see MCD750
USAF RH-6 see OFK000
USAF RH-7 see HGP000
USAF RH-8 see MLC750
USAF SC-2 see GKE000
USAF SN-9 see TEX250
USAF SO-1 see EGL500
USAF SO-2 see EPK000
USAF ST-40 see MGA750
USAF SZ-1 see BNM250
USAF SZ-3 see MOO500
USAF SZ-B see MOO500
USAF T-1 see CKJ750
USAF T-2 see MPH750
USAF TH-3 see THS250
USAF TH-9 see DXM600
USAF UCTL-7 see AES639
USAF UCTL-8 see MJV750
USAF UCTL-958 see BHY750
USAF UCTL-974 see BII500
USAF UCTL-1766 see MCL500
USAF UCTL-1791 see DOA400
USAF UCTL-1856 see AMC000
USAF VI-6 see PDF250
USAF WI-1 see MRN000
USAF WI-3 see MLY500
USAF XF-21 see BCC500
USAF XR-10 see DBL800
USAF XR-19 see PFL850
USAF XR-22 see AMY050
USAF XR-27 see AIS500
USAF XR-29 see BDF000
USAF XR-30 see AIS550
USAF XR-31 see AJY250
USAF XR-32 see AJY500
USAF XR-35 see MCK750
USAF XR-41 see CJX750
USAF XR-42 see DXQ500
USAN-DIAZIQUONE see ASK875

USB-3153 see DWS200
USB-3584 see CNE500
U.S. BLENDED LIGHT TOBACCO CIGARETTE REFINED TAR see CMP800
USNEIN see UWJ000
USNIACIN see UWJ000
USNIC ACID see UWJ000
(+)-USNIC ACID see UWJ100
d-USNIC ACID see UWJ100
USNIC ACID, (R)-(8CI) see UWJ100
USNINIC ACID see UWJ000
(+)-USNINIC ACID see UWJ100
d-USNINIC ACID see UWJ100
USNINSAEURE (GERMAN) see UWJ000
U.S.P. MENTHOL see MCG250
USP SODIUM CHLORIDE see SFT000
USP XIII STEARYL ALCOHOL see OAX000
USR 604 see DFT000
U.S. RUBBER 604 see DFT000
U.S. RUBBER D-014 see SOP000
USTINEX see CIR250
U 46T see TGF200
UTEDRIN see ORU500
UTERACON see ORU500
UTERAMINE see TOG250
UTIBID see OOG000
UTICILLIN see CBO000
UTICILLIN VK see PDT750
UTIL see EQL000
UTOPAR see RLK700
UTOSTAN see PDC250
UV ABSORBER-1 see HML500
UV ABSORBER-2 see DGX000
UVALERAL see BNP750
UVASOL see HIH100
UV CHEK AM 104 see BIW750
UVINOL D-50 see BCS325
UVINUL 400 see DMI600
UVINUL D-50 see BCS325
UVINUL M 40 see MES000
U46KW see BSQ750
UZONE see BRF500

V 7 see TKX250
V-18 see CKM000
V 252 see CGD000
V 255 see CKN750
V 315 see CIT750
V 316 see CFU500
V 317 see DIK800
V331 see DIM400
V 340 see DIS000
V 343 see DHJ400
V 346 see DIK600
V 374 see DTR800
V 375 see CGI750
V 377 see DNM600
V 17004 see XTJ000
VA see VFF000
VA 0112 see AAX250
VABEN see CFZ000
VABROCID see NGE500
VAC see VLU250
VACATE see CIR250
VACUUM RESIDUUM see MQV755
VADEBEX see NOA000
VADEBEX HYDROCHLORIDE see NOA500
VADILEX see IAG625
VADOSILAN see VGA300
VADROCID see NGE500
VAFLOL see VSK600
VAGAMIN see DJM800, XCJ000
VAGANTIN see DJM800, XCJ000
VAGD see AAX175
VAGESTROL see DKA600
VAGILEN see MMN250
VAGILIA see SNN500
VAGIMID see MMN250
VAGISEPT see ABX500
VAGOFLOR see ABX500
VAGOPHEMANIL see DAP800
VAGOPHEMANIL METHYL SULFATE see DAP800
VAGOSIN see CPQ250
VAGOSTIGMIN see DER600
VAGOSTIGMINE see NCL100

VAGOSTIGMINE BROMIDE see POD000
VAGOSTIGMINE METHYL SULFATE see DQY909
VAL 13081 see VCK100
VALACIDIN see SMA000
VALADOL see HIM000
VALAMINA see EEH000
VALAMINE see IIM000
VALAMINETTEN see EEH000
VALAN see MRW000
VALANAS see MFD500
VALBAZEN see VAD000
VALBIL see BPP250
VALCOR see CPP000
VALDRENE see BAU750
VAL-DROP see SFS000
VALECOR see PEV750
VALEO see DCK759
VALERAL see VAG000
n-VALERALDEHYDE see VAG000
VALERAMIDE, 2-(2-ACETAMIDO-4-METHYLVALERAMIDO)-N-(1-FORMYL-4-GUANIDINOBUTYL)-4-METHYL-(S)- see AAL300
VALERAMIDE-OM see DOY400
VALERAN DI-n-BUTYLCINICITY (CZECH) see DEA600
VALERIANIC ACID see VAQ000
VALERIANIC ALDEHYDE see VAG000
VALERIC ACID see VAQ000
n-VALERIC ACID see VAQ000
VALERIC ACID ALDEHYDE see VAG000
VALERIC ALDEHYDE see VAG000
4-VALEROLACTONE see VAV000
Γ-VALEROLACTONE (FCC) see VAV000
VALERONE see DNI800
Δ-VALEROSULTONE see BOU250
VALERYLALDEHYDE see VAG000
VALERYL CHLORIDE see VBA000
VALETAN see DEO600
VALETHAMATE see VBK000
VALETHAMATE BROMIDE see VBK000
VALEXONE see BAT750
VALFLON see TAI250
VALFOR see AHF500
VALGESIC see HIM000
VALGIS see TEH500
VALGRAINE see TEH500
VALINE see VBP000
d-VALINE see VBU000
l-VALINE (FCC) see VBP000
VALINE ALDEHYE see IJS000
VALINOMYCIN see VBZ000
VALIOIL see HNI500
VALISONE see VCA000
VALITRAN see DCK759
VALIUM see DCK759
VALKACIT CA see DWN800
VALLADAN see BCA000
VALLENE see MBW750
VALLERGINE see DQA400
VALMAGEN see TAB250
VALMETHAMIDE see ENJ500
VALMID see EEH000
VALMIDATE see EEH000
VALOID see EAN600
VALONEA TANNIN see VCK000
VALONTIN see CAY675
VALOPRIDE see VCK100
VALORON HYDROCHLORIDE see EIH000
VALPIN see LJS000
VALPROATE see PNR750
VALPROATE SODIUM see PNX750
VALPROIC ACID see PNR750
VALPROIC ACID CALCIUM SALT see CAY675
VALPROIC ACID HEMI-CALCIUM SALT see CAY675
VALPROIC ACID SODIUM SALT see PNX750
VALPROMIDE see PNX600
VALSYN see FPI000
VALYL GRAMICIDIN A see GJO025
VALZIN see EFE000
VAMIDOATE see MJG500
VAMIDOTHION see MJG500
V. AMMODYTES VENOM see VQZ425
VANADIC ACID, AMMONIUM SALT see ANY250
VANADIC ACID, MONOSODIUM SALT see SKP000
VANADIC ACID, POTASSIUM SALT see PLK900
VANADIC(II) ACID, TRISODIUM SALT see SIY250

VANADIC ANHYDRIDE see VDU000
VANADIC OXIDE see VEA000
VANADIO, PENTOSSIDO di (ITALIAN) see VDU000
VANADIOUS(4+) ACID, DISODIUM SALT see SKQ000
VANADIUM see VCP000
VANADIUM AZIDE TETRACHLORIDE see VCU000
VANADIUM BROMIDE see VEK000
VANADIUM CHLORIDE see VEF000
VANADIUM(III) CHLORIDE see VEP000
VANADIUM CHLORIDE OXIDE see DFW800
VANADIUM COMPOUNDS see VCZ000
VANADIUM DICHLORIDE see VDA000
VANADIUM DICHLORIDE OXIDE see DFW800
VANADIUM DUST and FUME (ACGIH) see VDU000, VDZ000
VANADIUM ORE see VDF000
VANADIUM OXIDE see VEA000
VANADIUM(V) OXIDE see VDU000
VANADIUM OXIDE TRIISOBUTOXIDE see VDK000
VANADIUM OXYCHLORIDE see DFW800
VANADIUM OXYDICHLORIDE see DFW800
VANADIUM OXYTRICHLORIDE see VDP000
VANADIUM PENTAOXIDE see VDU000
VANADIUMPENTOXID (GERMAN) see VDU000
VANADIUM PENTOXIDE (dust) see VDU000
VANADIUM PENTOXIDE (fume) see VDZ000
VANADIUM PENTOXIDE, non-fused form (DOT) see VDU000
VANADIUMPENTOXYDE (DUTCH) see VDU000
VANADIUM, PENTOXYDE de (FRENCH) see VDU000
VANADIUM SESQUIOXIDE see VEA000
VANADIUM SULFATE see VEA100
VANADIUM SULPHATE see VEA100
VANADIUM TETRACHLORIDE see VEF000
VANADIUM TRIBROMIDE see VEK000
VANADIUM TRIBROMIDE OXIDE see VEK100
VANADIUM TRICHLORIDE see VEP000
VANADIUM TRICHLORIDE OXIDE see VDP000
VANADIUM TRIOXIDE see VEA000
VANADYL AZIDE DICHLORIDE see VEU000
VANADYL SULFATE see VEZ000
VANADYL SULFATE PENTAHYDRATE see VEZ100
VANADYL TRICHLORIDE see VDP000
VANAY see THM500
VANCENASE see AFJ625
VANCERIL see AFJ625
VANCIDA TM-95 see TFS350
VANCIDE see MAS500, VEZ925
VANCIDE 89 see CBG000
VANCIDE BL see TFD250
VANCIDE FE95 see FAS000
VANCIDE KS see HON000
VANCIDE MZ-96 see BJK500
VANCIDE P see ZMJ000
VANCIDE PA see BLG500
VANCIDE PA DISPERSION see BLG500
VANCIDE PB see DVI600
VANCIDE TH see HDW000
VANCIDE TM see TFS350
VANCOCIN see VFA000
VANCOCINE HYDROCHLORIDE see VFA050
VANCOCIN HYDROCHLORIDE see VFA050
VANCOMYCIN see VFA000
VANDEX see SBO500
VAN DYK 264 see OES000
VANDYKE BROWN see MAT500
VANGARD K see CBG000
VANGUARD GF see FAZ000
VANGUARD N see BIW750
VANICID see VEZ925
VANICIDE see CBG000
VANILLA see VFK000
VANILLABERON see VEZ925
VANILLAL see EQF000
VANILLALDEHYDE see VFK000
VANILLA PLANT see DAJ800
VANILLA TINCTURE see VFA200
VANILLIC ACID see VFF000
o-VANILLIC ACID see HJB500
p-VANILLIC ACID see VFF000
VANILLIC ACID DIETHYLAMIDE see DKE200
VANILLIC ACID-N,N-DIETHYLAMIDE see DKE200
VANILLIC ALDEHYDE see VFK000
VANILLIN see VFK000
o-VANILLIN see VFP000
p-VANILLIN see VFK000

VANILLIN METHYL ETHER see VHK000
VANILLINSAEURE-DIAETHYLAMID (GERMAN) see DKE200
VANILLYL ACETONE see VFP100
VANILOL see VFP200
VANIROM see EQF000
VANIZIDE see VEZ925
VANLUBE PCX see BFW750
VANNAX NS see BQK750
VANOBID see LFF000
VANOXIDE see BDS000
VANQUIN see PQC500
VANSIL see OLT000
VANSIL W 10 see WCJ000
VANSIL W 20 see WCJ000
VANSIL W 30 see WCJ000
VANTYL see PGG000
VANZOATE see BCM000
VAPAM see VFU000, VFW009
VAPIN see LJS000
VAPO-ISO see IMR000
VAPONA see DGP900
VAPONEFRIN see VGP000
VAPO-N-ISO see DMV600
VAPONITE see DGP900
VAPOPHOS see PAK000
VAPOROLE see IMB000
VAPOROOTER see VFW009
VAPORPAC see ILM000
VAPOTONE see TCF250
VAPTONE see TCF280
VARAMID ML 1 see BKE500
VARDHAK see NAK500
VARFINE see WAT220
VARICOL see EMT500
VARIEGATED PHILODENDRON see PLW800
VARIOFORM II see USS000
VARISOFT SDC see DTC600
VARITOL see FMS875
VARITON see DAP800
VARITOX see TII250, TII500
VARNISH MAKERS' AND PAINTERS' NAPHTHA see PCT250
VARNISH MAKERS' NAPHTHA see PCT250
VARNOLINE see SLU500
VAROX see DRJ800
VAROX DCP-R see DGR600
VAROX DCP-T see DGR600
VARSOL see PCT250
VASAL see PAH250
VASALGIN see ELX000
VASAPRIL see PPH050
VASAZOL see DJS200
VASCARDIN see CCK125
VASCOPIL see HNY500
VASCUALS see VSZ450
VASCULAT see BOV825
VASCULIT see BOV825
VASCUNICOL see BOV825
VASE FLOWER see CMV390
VASE VINE see CMV390
VASICIN (GERMAN) see VGA000
VASICINE see VGA000
VASICINONE HYDROCHLORIDE see VGA025
VASIMID see BBW750
VASIODONE see DNG400
VASITOL see PBC250
VASKULAT see BOV825
VASOBRIX 32 see VGA100
VASOCIL see PPH050
VASOCON see NCW000
VASOCONSTRICTINE see VGP000
VASOCONSTRICTOR see VGP000
VASODIATOL see PBC250
VASODIL see BBW750
VASODILAN see VGA300, VGF000
VASODILATAN see BBW750
VASODILATATEUR 2249F see FMX000
VASODILIAN see VGF000
VASODISTAL see VGK000
VASODRINE see VGP000
VASOFILINA see TEP500
VASOKELLINA see AHK750
VASOLAN see IRV000, VHA450
VASOMED see TJL250
VASOPERIF see CBH250

VASOPLEX see VGA300
VASORBATE see CCK125
VASORELAX see POD750
VASOROME see AOO125
VASOTON see VGP000
VASOTONIN see VGP000
VASOTRAN see VGA300
VASOTRATE see CCK125
VASO-80 UNICELIES see PBC250
VASOVERIN see PPH050
VASOXINE see MDW000
VASOXINE HYDROCHLORIDE see MDW000
VASOXYL HYDROCHLORIDE see MDW000
VASTAREL see YCJ200
VASTCILLIN see AJJ875
VATERITE see CAO000
VAT GOLDEN YELLOW see DCZ000
VAT GRAY S see CMU475
VAT GREY S see CMU475
VATRACIN see AJJ875
VATRAN see DCK759
VATROLITE see SHR500
VATSOL OT see DJL000
VAZADRINE see ILD000
VAZO 64 see ASL750
VAZOFIRIN see PBU100
V-BRITE see SHR500
VC see VNP000
V-C 3-670 see DWU200
V-C 3-764 see PGD000
V-C CHEMICAL V-C 9-104 see EIN000
V-CIL see PDT500
V-CIL-K see PDT750
V-CILLIN see PDT500
V-CILLIN K see MNV250, PDT750
VCM see VNP000
VC13 NEMACIDE see DFK600
VCR see LEY000
VCR SULFATE see LEZ000
VCS 438 see BGD250
VDC see VPK000
VDF see VPP000
VDM see VFU000
VD 1827 p-TOLUENESULFONATE see SOU675
VEATCHINE HYDROCHLORIDE see VGU000
VEBECILLIN see PDT500
VECTAL see ARQ725
VECTAL SC see ARQ725
VECTREN see CIP500
VECURONIUM BROMIDE see VGU075
VEDERON see ILD000
VEDRIL see PKB500
VEE GEE GELATIN see PCU360
VEETIDS see PDT750
VEGABEN see AJM000
VEGADEX see CDO250
VEGADEX SUPER see CDO250
VEGANTINE see DHX800, THK000
VEGASERPINA see RCA200
VEGETABLE GUM see DBD800
VEGETABLE-HUMMING-BIRD see SBC550
VEGETABLE OIL see VGU200
VEGETABLE OIL 1400 see CBF710
VEGETABLE OIL MIST (OSHA) see VGU200
VEGETABLE PEPSIN see PAG500
VEGETABLE (SOYBEAN) OIL, brominated see BMO825
VEGETOX see BHL750
VEGFRU see PGS000
VEGFRU FOSMITE see EEH600
VEGFRU MALATOX see MAK700
VEGOLYSEN see HEA000
VEGOLYSIN see HEA000
VEHAM-SANDOZ see EQP000
VEHEM see EQP000
VEJIGA de PERRO (CUBA) see JBS100
VEL 3973 see DUK000
VEL 4283 see MKA000
VEL 4284 see DRR200
VELACICLINE see PPY250
VELACYCLINE see PPY250
VELARDON see PAG500
VELBAN see VLA000
VELBE see VLA000
VELDOPA see DNA200

VELFLON see TAI250
VELLORITA (CUBA) see CNX800
VELMOL see DJL000
VELO de NOVIA (MEXICO) see GIW200
VELOSEF see SBN440
VELOSEF HYDRATE see SBN450
VELPAR see HFA300
VELPAR WEED KILLER see HFA300
VELSICOL see TIK000
VELSICOL 104 see HAR050
VELSICOL 506 see LEN000
VELSICOL 1068 see CDR750
VELSICOL COMPOUND C see TIK000
VELSICOL COMPOUND "R" see MEL500
VELSICOL 53-CS-17 see EBW500
VELSICOL 58-CS-11 see MEL500
VELSICOL FCS-303 see BND750
VELSICOL VCS 506 see LEN000
VENA see BBV500
VENACIL see VGU700
VENCIPON see EAW000
VENDACID LIGHT ORANGE 2G see HGC000
VENDESINE SULFATE see VGU750
VENDEX see BLU000
VENETIAN RED see IHD000
VENETLIN see BQF500
VENGICIDE see VGZ000
VENOM, AUSTRALIAN ELAPIDAE SNAKE, ACANTHOPHIS ANTARCTICUS
    see ARU875
VENOM, AUSTRALIAN ELAPIDAE SNAKE, AUSTRELAPS SUPERBA
    see ARU750
VENOM, AUSTRALIAN ELAPIDAE SNAKE, OXYURANUS SCUTELLATUS
    see ARV500
VENOM, AUSTRALIAN ELAPIDAE SNAKE, PSEUDECHIS PORPHYRIACUS
    see ARV250
VENOM, AUSTRALIAN SNAKE, AUSTRELAPS SUPERBA see ARV625
VENOM, AUSTRALIAN SNAKE, NOTECHIS SCUTATUS see ARV550
VENOM, AUSTRALIAN SNAKE, OPHIOPHAGUS HANNAH see ARV125
VENOM, AUSTRALIAN SNAKE, PSEUDECHIS see ARV000
VENOM, AUSTRALIAN SNAKE, PSEUDONAJA TEXTILIS see TEG650
VENOM, AUSTRALIAN SNAKE, TROPIDECHIS CARINATUS see ARV375
VENOM, COSTA RICAN SNAKE, BOTHROPS ASPER see BMI000
VENOM, COSTA RICAN SNAKE, BOTHROPS ATROX see BMI125
VENOM, COSTA RICAN SNAKE, BOTHROPS GODMANI see BMI500
VENOM, COSTA RICAN SNAKE, BOTHROPS LATERALIS see BMI750
VENOM, COSTA RICAN SNAKE, BOTHROPS NASUTUS see BMJ000
VENOM, COSTA RICAN SNAKE, BOTHROPS NIGROVIRIDIS NIGROVIRIDIS
    see BMJ250
VENOM, COSTA RICAN SNAKE, BOTHROPS NUMMIFER MEXICANUS
    see BMJ500
VENOM, COSTA RICAN SNAKE, BOTHROPS OPHRYOMEGA see BMJ750
VENOM, COSTA RICAN SNAKE, BOTHROPS PICADOI see BMK000
VENOM, COSTA RICAN SNAKE, BOTHROPS SCHLEGLII see BMK250
VENOM, COSTA RICAN SNAKE, CROTALUS (CROTALUS) DURISSUS
    DURISSUS see CNX835
VENOM, COSTA RICAN SNAKE, MICRURUS ALLENI YATESI see MQS750
VENOM, COSTA RICAN SNAKE, MICRURUS MIPARTITUS HERTWIGI
    see MQT250
VENOM, COSTA RICAN SNAKE, MICRURUS NIGROCINCTUS see MQT500
VENOM, EGYPTIAN COBRA, NAJA HAJE ANNULIFERA see NAE512
VENOM, FORMOSAN BANDED KRAIT, BUNGARUS MULTICINCTUS
    see BON370
VENOM, INDIAN COBRA, NAJA NAJA NAJA see ICC700
VENOM, LIZARD, HELODERMA SUSPECTUM see HAN625
VENOM, MIDGET FADED RATTLESNAKE, CROTALUS VIRIDIS CON-
    COLOR see CNY750
VENOM, MIDGET FADED RATTLESNAKE, CROTALUS VIRIDIS HELLERI
    see COA000
VENOM, MIDGET FADED RATTLESNAKE, CROTALUS VIRIDIS LUTOSUS
    see COA250
VENOM, MIDGET FADED RATTLESNAKE, CROTALUS VIRIDIS OR-
    EGANUS see COA500
VENOM, MIDGET FADED RATTLESNAKE, CROTALUS VIRIDIS VIRIDIS
    see COA750
VENOM, MIDGET FADED RATTLESNAKE, CROTALUS VIRIDUS
    CERBERUS see CNZ000
VENOM, ORIENTAL HORNET, VESPA ORIENTALIS see VJP900
VENOM, SCORPION, ANDROCTONUS AMOREUXI see AOO250
VENOM, SCORPION, ANDROCTONUS AUSTRALIS HECTOR see AOO265
VENOM, SCORPION, CENTRUROIDES SUFFUSUS SUFFUSUS see CCW925
VENOM, SEA ANEMONE, AIPTASIA PALLIDA see SBI800
VENOM, SEA SNAKE, AIPYSURUS LAEVIS see SBI880
VENOM, SEA SNAKE, AIPYSURUS LAEVIS (AUSTRALIA) see AFG000
VENOM, SEA SNAKE, ASTROTIA STOKESII see SBI890

VENOM, SEA SNAKE, ENHYDRINA SCHISTOSA see EAU000
VENOM, SEA SNAKE, HYDROPHIS CYANOCINCTUS see SBI900
VENOM, SEA SNAKE, HYDROPHIS ELEGANS see HIG000
VENOM, SEA SNAKE, LAPEMIS HARDWICKII see SBI910
VENOM, SEA SNAKE, LATICAUDA SEMIFASCIATA see LBI000
VENOM, SEA SNAKE, MICROCEPHALOPHIS GRACILIS see SBI929
VENOM, SEA SNAKE, NAJA NAJA ATRA see NAF000
VENOM, SNAKE, AGKISTRODON ACUTUS see HGM600
VENOM, SNAKE, AGKISTRODON CONTORTRIX see AEY125
VENOM, SNAKE, AGKISTRODON CONTORTRIX CONTORTRIX see SKW775
VENOM, SNAKE, AGKISTRODON CONTORTRIX MOKASEN see NNX700
VENOM, SNAKE, AGKISTRODON PISCIVORUS see AEY130
VENOM, SNAKE, AGKISTRODON PISCIVORUS PISCIVORUS see EAB200
VENOM, SNAKE, AGKISTRODON RHODOSTOMA see AEY135
VENOM, SNAKE, ANCISTRODON PISCIVORUS see AOO135
VENOM, SNAKE, BOTHROPS COLOMBIENSIS see BMI250
VENOM, SNAKE, BUNGARUS CAERULEUS see BON365
VENOM, SNAKE, BUNGARUS FASCIATUS see BON367
VENOM, SNAKE, CERASTES CERASTES see CCX620
VENOM, SNAKE, CROTALUS ADAMANTEUS see EAB225
VENOM, SNAKE, CROTALUS ATROX see WBJ600
VENOM, SNAKE, CROTALUS CERASTES see CNX830
VENOM, SNAKE, CROTALUS DURISSUS TERRIFICUS see CNY000
VENOM, SNAKE, CROTALUS HORRIDUS see CNY300
VENOM, SNAKE, CROTALUS HORRIDUS HORRIDUS see TGB150
VENOM, SNAKE, CROTALUS RUBER RUBER see CNY325
VENOM, SNAKE, CROTALUS SCUTULATUS SCUTULATUS see CNY350,
   CNY375
VENOM, SNAKE, CROTALUS SCUTULATUS SCUTULATUS (CALIFORNIA)
   see CNY339
VENOM, SNAKE, CROTALUS VIRIDIS VIRIDIS see PLX100
VENOM, SNAKE, DENDROASPIS ANGUSTICEPS see DAP815
VENOM, SNAKE, DENDROASPIS JAMESONI see DAP820
VENOM, SNAKE, DENDROASPIS VIRIDIS see GJU500
VENOM, SNAKE, DISPHOLIDUS TYPHUS see DXG150
VENOM, SNAKE, ECHIS CARINATUS see EAD600
VENOM, SNAKE, ECHIS COLORATUS see EAD650
VENOM, SNAKE, HEMACHATUS HAEMCHATES see HAO500
VENOM, SNAKE, NAJA FLAVA see NAE510
VENOM, SNAKE, NAJA HAJE see NAE515
VENOM, SNAKE, NAJA MELANOLEUCA see NAE875
VENOM, SNAKE, NAJA NAJA KAOUTHIA see NAF200
VENOM, SNAKE, NAJA NIGRICOLLIS see NAG000
VENOM, SNAKE, NICRURUS FULVIUS see MQT100
VENOM, SNAKE, RHABDOPHIS TIGRINUS TIGRINUS see RFU875
VENOM, SNAKE, SISTRURUS MILARIUS BARBOURI see SDZ300
VENOM, SNAKE, TRIMERESURUS FLAVOVIRIDIS see TKW100
VENOM, SNAKE, VIPERA AMMODYTES see VQZ425
VENOM, SNAKE, VIPERA ASPIS see VQZ475
VENOM, SNAKE, VIPERA BORNMULLERI see VQZ500
VENOM, SNAKE, VIPERA LATIFII see VQZ525
VENOM, SNAKE, VIPERA LEBETINA see VQZ550
VENOM, SNAKE, VIPERA PALESTINAE see VQZ575
VENOM, SNAKE, VIPERA RUSSELLII see VQZ635
VENOM, SNAKE, VIPERA RUSSELLII FORMOSENSIS see VQZ625
VENOM, SNAKE, VIPERA XANTHINA PALAESTINAE see VQZ650
VENOM, SNAKE, WALTERINNESIA AEGYPTIA see WAT100
VENOM, SOUTH AFRICAN CAPE COBRA, NAJA NIVEA see NAG200
VENOM, STONEFISH, SYNANCEJA HORRIDA Linn. see SPB875
VENOM, THAILAND COBRA, NAJA NAJA SIAMENSIS see NAF250
VENOPEX see TAB250
VENOPIRIN see VHA275
VENTILAT see ONI000
VENTIPULMIN see VHA350
VENTOLIN see BQF500
VENTOX see ADX500
VENTRAMINE see DJS200
VENTUROL see DXX400
VENZONATE see BCM000
VEON 245 see TAA100
VEPEN see PDT750
VEPESID see EAV500
VER A see MRV500
VERACILLIN see DGE200
VERACTIL see MCI500
VERAMID see AMK250
VERAMIN ED 4 see EIV750
VERAMIN ED 40 see EIV750
VERAMON see AMK250
VERANTEROL see PPH050
VERANTIN see TKN750
VERAPAMIL see IRV000
VERAPAMIL HYDROCHLORIDE see VHA450
VERAPRET DH see DTG000

VERATENSINE see VHF000
VERATRALDEHYDE see VHK000
VERATRAMINE see VHP500
VERATRIC ALDEHYDE see VHK000
VERATRIDINE see VHU000, VHZ000
VERATRIN see NCI600
VERATRIN (GERMAN) see VHZ000
VERATRINE see CDG000, NCI600, VHZ000
VERATRINE (amorphous) see VHU000
VERATRINE (crystallized) see CDG000
VERATROL see DOA200
VERATROLE see DOA200
VERATROLE METHYL ETHER see AGE250
N-(o-VERATROYL)GLYCINOHYDROXAMIC ACID see VIA875
3-VERATROYLVERACEVINE see VHU000
VERATRUM ALBUM see FAB100
VERATRUM CALIFORNICUM see FAB100
VERATRUM VIRIDE see FAB100, VIZ000
VERATRUM VIRIDE ALKALOIDS EXTRACT see VIZ000
VERATRYL ALDEHYDE see VHK000
VERATRYLAMINE see VIK050
VERATRYL CHLORID (GERMAN) see BKM750
VERATRYL CHLORIDE see BKM750
VERATRYL CYANIDE see VIK100
VERATRYLHYDRAZINE see VIK150
1-VERATRYLPIPERAZINE DIHYDROCHLORIDE see VIK200
VERAX see BBW500
VERAZINC see ZNA000
VERBENA HYBRIDA Cornol. & Rpl., extract see VIK400
d-VERBENONE see VIP000
VERBINDUNG S 557 HCl see AMM750
VERCIDON see DJT800
VERCYTE see BHJ250
VERDANTIOL see LFT100
VERDICAN see DGP900
VERDIPOR see DGP900
VERDONE see CIR250
VERDYL ACETATE see DLY400
VEREDEN see MNM500
VERESENE DISODIUM SALT see EIX500
VERGEMASTER see DAA800
VERGFRU FORATOX see PGS000
VERILOID see VIZ000
VERINA see DNU200
3427 VERI PUR PINK see ADG250
VERITAIN see FMS875
VERITOL see FMS875
VERMAGO see PIJ500
VERMICIDE BAYER 2349 see TIQ250
VERMICOMPREN (TABL.) see HEP000
VERMICULIN see VIZ100
VERMICULINE see VIZ100
VERMILASS see HEP000
VERMINUM see BQK000
VERMIRAX see MHL000
VERMISOL see PIK500
VERMITIN see DFV400, PDP250
VERMIZYM see PAG500
VERMOESTRICID see CBY000
VERMOX see MHL000
VERNAM see PNI750
VERNAMYCIN see VRF000
VERNAMYCIN BA see VRA700
VERNINE see GLS000
VERNOLATE see PNI750
VERODIGEN see GES000
VERODOXIN see VIZ200
VEROL see MGJ750
VEROLETTIN see BAG000
VERONAL see BAG000
VERONAL SODIUM see BAG250
VERONA YELLOW X-1791 see DEU000
VERON P 130/1 see PKQ059
VEROPHEN see DQA600
VEROS 030 see VIZ250
VEROSPIRON see AFJ500
VEROSPIRONE see AFJ500
VERROL see VSZ450
VERRUCARIN A see MRV500
VERSAL FAST RED R see CJD500
VERSALIDE see ACL750
VERSAL ORANGE RNL see DVB800
VERSAL SCARLET PRNL see MMP100
VERSAMINE see VIZ400

VERSAR DSMA LQ see DXE600
VERSATIC 9-11 see VIZ500
VERSATIC 9-11 ACID see VIZ500
VERSATIC ACID 911 see VIZ500
VERSENE 9 see TNL250
VERSENE 100 see EIV000
VERSENE ACID see EIX000
VERSENE CA see CAR800
VERSENE NTA ACID see AMT500
VERSENE POWDER see EIV000
VERSENE SODIUM 2 see EIX500
VERSENOL see HKS000
VERSENOL 120 see HKS000
VERSICOLORIN A see TKO500
VERSICOLORIN B see VJP800
VERSIDYNE see MDV000
VERSNELLER NL 63/10 see DQF800
VERSOMNAL see EOK000
VERSOTRANE see PFN000
VERSTRAN see DAP700
VERSULIN see CDH250
VERSUS see BAV325
TRANS-VERT see MRL750
VERTAC see BQI000, DGI000
VERTAC 90% see CDV100
VERTAC DINITRO WEED KILLER see BRE500
VERTAC GENERAL WEED KILLER see BRE500
VERTAC METHYL PARATHION TECHNISCH 80% see MNH000
VERTAC SELECTIVE WEED KILLER see BRE500
VERTAC TOXAPHENE 90 see CDV100
VERTAVIS see VIZ000
VERTENEX see BQW500
VERTHION see DSQ000
VERTISAL see MMN250
VERTOFIX COEUR see ACF250
VERTOLAN see SNJ000
VERTON 2T see TAA100
VERTON D see DAA800
VESADIN see SJW500
VESAKONTUHO MCPA see CIR250
VESALIUM see CLY500
VESAMIN see AAN000
VESPA ORIENTALIS VENOM see VJP900
VESPARAZ-WIRKSTOFF see CJR909
VESPERAL see BAG000
VESPRIN see TKL000
VESTIN see PDC250
VESTINOL AH see DVL700
VESTINOL OA see AEO000
VESTOLIT B 7021 see PKQ059
VESTROL see TIO750
VESTYRON see SMQ500
VESTYRON HI see SMR000
VETACALM see TAF675
VETAFLAVIN see DBX400
VETALAR see CKD750
VETALOG see AQX500
VETA-MERAZINE see ALF250
VETAMOX see AAI250
VETARCILLIN see BFC750
VETARSENOBILLON see NCJ500
VETBUTAL see NBU000
VETERINARY NITROFURAZONE see NGE500
VETICILLIN see BFD250
VETICOL see CDP250
VETIDREX see CFY000
VETIOL see MAK700
VETIVER ACETATE see AAW750
VETIVEROL ACETATE see AAW750
VETIVERT ACETATE see AAW750
VETIVERT OIL see VJU000
VETIVERYL ACETATE see AAW750
VETKALM see DYF200
VETOL see MAO350
VETQUAMYCIN-324 see TBX250
VETRANQUIL see ABH500
VETRAZIN see VIK150
VETRAZINE see VIK150
VETREN see HAQ500
VETSIN see MRL500
VETSTREP see SLY500
VFR 3801 see PKF750
VIADRIL see VJZ000
VIALIDON see XQS000

VI-ALPHA see VSK600
VIANSIN see MDQ250
VIARESPAN HYDROCHLORIDE see DAI200
VIAROX see AFJ625
VIBALT see VSZ000
VIBAZINE see BOM250, HGC500
VIBISONE see VSZ000
VIBRADOX see HGP550
VIBRAMYCIN see DYE425
VIBRAMYCIN HYCLATE see HGP550
VIBRA-TABS see HGP550
VI-CAD see CAE250
VICCILLIN see AIV500
VICCILLIN S see AIV500
VICILAN see VKF000
VICILLIN see AIV500
VICIN see BFC750
VICKNITE see PLL500
VICTAN see EKF600
VICTORIA BLUE R see VKA600
VICTORIA BLUE RS see VKA600
VICTORIA GREEN see AFG500
VICTORIA LAKE BLUE R see VKA600
VICTORIA ORANGE see DUT600
VICTORIA RUBINE O see FAG020
VICTORIA SCARLET 3R see FMU080
VICTORIA YELLOW see DUT600
VICTOR TSPP see TEE500
VIDANGA DRIED BERRY EXTRACT see VKA650
VIDARABIN see AQQ900
VIDARABINE see AEH100, AQQ900
VIDDEN D see DGG000, DGG950
VIDEOBIL see IGA000
VIDEOPHEL see TDE750
VIDLON see PJY500
VIDON 638 see DAA800
VIDOPEN see AOD125
VIDR-2GD see VKA675
VIENNA GREEN see COF500
VIGANTOL see VSZ100
VIGORSAN see CMC750
VIKANE see SOU500
VIKANE FUMIGANT see SOU500
VIKROL RQ see AFP250
VILESCON see PNS000
VILLESCON see PNS000
VILLIAUMITE see SHF000, SHF500
VILOXAZIN see VKA875
VILOXAZINE see VKA875
VILOXAZINE HYDROCHLORIDE see VKF000
VINAC B 7 see AAX250
VINACETIN A see VQZ000
VINADINE see OMY850
VINAMAR see EQF500
VINBARBITAL SODIUM see VKP000
VINBLASTIN see VKZ000
VINBLASTINE see VKZ000
VINBLASTINE SULFATE see VLA000
VINCADAR see VLF000
VINCAFOLINA see VLF000
VINCAFOR see VLF000
VINCAGIL see VLF000
VINCAIN see AFG750
VINCALEUCOBLASTIN see VKZ000
VINCALEUKOBLASTINE see VKZ000
VINCALEUKOBLASTINE SULFATE see VLA000
VINCALEUKOBLASTINE SULFATE (1:1) (SALT) see VLA000
VINCAMIDOL see VLF000
VINCAMIN COMPOSITUM see ARQ750
VINCAMINE see VLF000
(+)-VINCAMINE see VLF000
VINCA MINOR L., TOTAL ALKALOIDS see VLF300
VINCAPAN see VLF000
VINCAPRONT see VLF000
VINCASAUNIER see VLF000
VINCEINE see AFG750
VINCES see AKO500
VINCIMAX see VLF000
VINCLOZOLIN (GERMAN) see RMA000
VINCOBLASTINE see VKZ000
VINCRISTINE see LEY000
VINCRISTINE SULFATE ONCORIN see LEZ000
VINCRISTINSULFAT (GERMAN) see LEZ000
VINCRISUL see LEZ000

VINCRYSTINE see LEY000
VINDESINA SULFATO (SPANISH) see VGU750
VINEGAR ACID see AAT250
VINEGAR NAPHTHA see EFR000
VINEGAR SALTS see CAL750
VINESTHENE see VOP000
VINESTHESIN see VOP000
VINETHEN see VOP000
VINETHENE see VOP000
VINETHER see VOP000
VINICIZER 80 see DVL700
VINICIZER 85 see DVL600
VI-NICOTYL see NCR000
VI-NICTYL see NCR000
VINIDYL see VOP000
VINIKA KR 600 see PKQ059
VINIKULON see PKQ059
VINILE (ACETATO di) (ITALIAN) see VLU250
VINILE (BROMURO di) (ITALIAN) see VMP000
VINILE (CLORURO di) (ITALIAN) see VNP000
VINIPLAST see PKQ059
VINIPLEN P 73 see PKQ059
VINISIL see PKQ250
VINKEIL 100 see EIX000
VINKRISTIN see LEY000
4-VINLYCYCLOHEXENE DIOXIDE see VOA000
VINNOL E 75 see PKQ059
VINNOL H 10/60 see AAX175
VINODREL RETARD see VLF000
VINOFLEX see PKQ059
VINOFLEX MO 400* see IJQ000
VINOTHIAM see TES750
VINPOCETINE see EGM100
VINSTOP see SGM500
VINTHIONINE see VLU200
l-VINTHIONINE see VLU210
VINYDAN see VOP000
VINYLACETAAT (DUTCH) see VLU250
VINYLACETAT (GERMAN) see VLU250
VINYL ACETATE see VLU250
VINYL ACETATE HOMOPOLYMER see AAX250
VINYL ACETATE OZONIDE see VLU310
VINYL ACETATE POLYMER see AAX250
VINYL ACETATE RESIN see AAX250
VINYL ACETATE-VINYL CHLORIDE COPOLYMER see AAX175
VINYL ACETATE-VINYL CHLORIDE POLYMER see AAX175
VINYLACETONITRILE see BOX500
VINYL ACETYLENE see BPE109
α-VINYL AE see VMA000
VINYL ALCOHOL POLYMER see PKP750
VINYLAMINE see VLU400
VINYL A MONOMER see VLU250
VINYL AZIDE see VLY300
1-VINYL AZIRIDINE see VLZ000
α-VINYL-1-AZIRIDINEETHANOL see VMA000
7-VINYLBENZ(a)ANTHRACENE see VMF000
VINYLBENZEN (CZECH) see SMQ000
VINYLBENZENE see SMQ000
VINYLBENZENE POLYMER see SMQ500
VINYL BENZOATE see VMK000
VINYLBENZOL see SMQ000
VINYL(β-BIS(β-CHLOROETHYL)AMINO)ETHYL SULFONE see BHP500
VINYLBROMID (GERMAN) see VMP000
VINYL BROMIDE see VMP000
VINYL BROMIDE, inhibited (DOT) see VMP000
VINYL 2-BUTENOATE see VNU000
VINYL-2-(BUTOXYETHYL) ETHER see VMU000
VINYL BUTYL ETHER see VMZ000
VINYL-n-BUTYL ETHER see VMZ000
VINYL 2-(BUTYLMERCAPTOETHYL) ETHER see VNA000
VINYL BUTYRATE see VNF000
VINYL BUTYRATE, INHIBITED (DOT) see VNF000
VINYLBUTYROLACTAM see EEG000
N-VINYLBUTYROLACTAM POLYMER see PKQ250
VINYL CARBAMATE see VNK000
VINYLCARBINOL see AFV000
VINYL CARBINYL BUTYRATE see AFY250
VINYL CARBINYL CINNAMATE see AGC000
VINYLCHLON 4000LL see PKQ059
VINYLCHLORID (GERMAN) see VNP000
VINYL CHLORIDE see VNP000
VINYL CHLORIDE ACETATE COPOLYMER see PKP500
VINYL CHLORIDE HOMOPOLYMER see PKQ059
VINYL CHLORIDE MONOMER see VNP000

VINYL CHLORIDE POLYMER see PKQ059
VINYL CHLORIDE VINYL ACETATE COPOLYMER see PKP500
VINYL CHLORIDE-VINYL ACETATE POLYMER see AAX175
VINYL-2-CHLOROETHYL ETHER see CHI250
VINYL-β-CHLOROETHYL ETHER see CHI250
VINYL C MONOMER see VNP000
VINYL CROTONATE see VNU000
VINYL CYANIDE see ADX500
VINYLCYCLOHEXANE MONOXIDE see VNZ000
1-VINYLCYCLOHEXENE-3 see CPD750
1-VINYLCYCLOHEX-3-ENE see CPD750
4-VINYLCYCLOHEXENE-1 see CPD750
4-VINYL-1-CYCLOHEXENE see CPD750
VINYL CYCLOHEXENE DIEPOXIDE see VOA000
4-VINYLCYCLOHEXENE DIEPOXIDE see VOA000
4-VINYL-1-CYCLOHEXENE DIEPOXIDE see VOA000
4-VINYL-1,2-CYCLOHEXENE DIEPOXIDE see VOA000
VINYL CYCLOHEXENE DIOXIDE see VOA000
1-VINYL-3-CYCLOHEXENE DIOXIDE see VOA000
4-VINYL-1-CYCLOHEXENE DIOXIDE (MAK) see VOA000
4-VINYLCYCLOHEXENE-1,2-EPOXIDE see VNZ000
VINYLCYCLOHEXENE MONOXIDE see VNZ000
4-VINYLCYCLOHEXENE MONOXIDE see VNZ000
5-VINYL-DEOXYURIDINE see VOA550
VINYLDIAZOMETHANE see DCQ550
VINYL DIHYDROGEN PHOSPHATE see VQA400
VINYL-2-(N,N-DIMETHYLAMINO)ETHYL ETHER see VOF000
VINYLE (ACETATE de) (FRENCH) see VLU250
VINYLE (BROMURE de) (FRENCH) see VMP000
VINYLE(CHLORURE de) (FRENCH) see VNP000
VINYLENE CARBONATE see VOK000
4,4'-VINYLENEDIANILINE see SLR500
4,4'-VINYLENEDIBENZAMIDINE, DIHYDROCHLORIDE see SLS500
N,N'-VINYLENEFORMAMIDINE see IAL000
1-VINYL-3,4-EPOXYCYCLOHEXANE see VNZ000
VINYL ETHER see VOP000
VINYLETHYLENE see BOP500
N-VINYLETHYLENEIMINE see VLZ000
VINYL ETHYL ETHER see EQF500
VINYL ETHYL ETHER, inhibited (DOT) see EQF500
VINYL-2-ETHYLHEXOATE see VOU000
VINYL-2-ETHYLHEXYL ETHER see ELB500
VINYLETHYLNITROSAMIN (GERMAN) see NKF000
VINYLETHYLNITROSAMINE see NKF000
VINYL FLUORIDE see VPA000
VINYL FORMATE see VPF000
VINYLFORMIC ACID see ADS750
VINYLGLYCOLONITRILE see HJQ000
S-VINYL-l-HOMOCYSTEINE see VLU210
S-VINYL-dl-HOMOCYSTEINE see VLU200
VINYLIDENE CHLORIDE see VPK000
VINYLIDENE CHLORIDE (II) see VPK000
VINYLIDENE DICHLORIDE see VPK000
VINYLIDENE FLUORIDE see VPP000
VINYLIDINE CHLORIDE see VPK000
VINYL ISOBUTYL ETHER (DOT) see IJQ000
VINYL ISOBUTYL ETHER, inhibited (DOT) see IJQ000
VINYLITE VYDR 21 see AAX175
VINYLLITHIUM see VPU000
VINYL-2-METHOXYETHYL ETHER see VPZ000
N-VINYLMETHYLACETAMIDE see MQK500
N-VINYL-N-METHYLACETAMIDE see MQK500
VINYL METHYLADIPATE see MQL500
VINYL METHYL ETHER (DOT) see MQL750
VINYL METHYL ETHER, INHIBITED (DOT) see MQM000
VINYL METHYL KETONE see BOY500
VINYLNORBORNENE see VQA000
2-VINLNORBORNENE see VQA000
5-VINYLNORBORNENE see VQA000
5-VINYL-2-NORBORNENE see VQA000
17-VINYL-19-NORTESTOSTERONE see NCI525
17-α-VINYL-19-NORTESTOSTERONE see NCI525
VINYLOFOS see DGP900
VINYLOPHOS see DGP900
3-VINYL-7-OXABICYCLO(4.1.0)HEPTANE see VNZ000
R-5-VINYL-2-OXAZOLIDINETHIONE see VQA100
VINYLOXYETHANOL see EJL500
2-(VINYLOXY)ETHANOL see EJL500
VINYLPHATE see CDS750
4-VINYLPHENOL see VQA200
p-VINYLPHENOL see VQA200
VINYL PHOSPHATE see VQA400
VINYLPHOSPHONIC ACID BIS(2-CHLOROETHYL) ESTER see VQF000
5-VINYL-2-PICOLINE see MQM500

α-VINYLPIPERONYL ALCOHOL see BCJ000
VINYL PRODUCTS R 3612 see SMQ500
VINYL PRODUCTS R 10688 see AAX250
VINYL PROPIONATE see VQK000
1-VINYLPYRENE see EEF000
3-VINYLPYRENE see EEF000
4-VINYLPYRENE see EEF500
VINYL PYRIDINE see VQK550
N-VINYLPYRROLIDINONE see EEG000
1-VINYL-2-PYRROLIDINONE see EEG000
N-VINYL-2-PYRROLIDINONE see EEG000
1-VINYL-2-PYRROLIDINONE POLYMER, compounded with IODINE see PKE250
VINYLPYRROLIDONE see EEG000
N-VINYLPYRROLIDONE see EEG000
1-VINYL-2-PYRROLIDONE see EEG000
N-VINYL-2-PYRROLIDONE see EEG000
N-VINYLPYRROLIDONE POLYMER see PKQ250
α-(5-VINYL-2-QUINOLYL)-2-QUINUCLIDINEMETHANOL see CMP925
VINYLSILICON TRICHLORIDE see TIN750
m-VINYLSTYRENE see DXQ745
VINYL SULFONE see DXR200
VINYL TOLUENE see VQK650, VQK650
VINYLTOLUENE (mixed isomers) (OSHA) see MPK500
VINYL TOLUENES (mixed isomers), inhibited (DOT) see VQK650
VINYL TRICHLORIDE see TIN000
VINYL TRICHLOROSILANE (DOT) see TIN750
VINYL TRICHLOROSILANE, INHIBITED (DOT) see TIN750
VINYLTRIETHOXYSILANE see TJN250
VINYL TRIMETHOXY SILANE see TLD000
VINYLTRIMETHYLAMMONIUM HYDROXIDE see VQR300
VINYL-2,6,8-TRIMETHYLNONYL ETHER see VQU000
VINYLTRIS(METHOXYETHOXY)SILANE see TNJ500
VINYLTRIS(2-METHOXYETHOXY)SILANE see TNJ500
VINYLTRIS(β-METHOXYETHOXY)SILANE see TNJ500
VINYON N see ADY250
VINYZENE see OMY850
VINYZENE (pesticide) see OMY850
VINYZENE bp 5 see OMY850
VINYZENE bp 5-2 see OMY850
VINYZENE SB 1 see OMY850
VIOACTANE see VQZ000
VIOCIN see VQZ000
VIOFORM see CHR500
VIOFORM N.N.R. see CHR500
VIOFURAGYN see NGG500
VIOLACETIN see VQU450
VIOLANTHRONE see DCU800
VIOLAQUERCITRIN see RSU000
VIOLET 3 see XGS000
11092 VIOLET see HOK000
VIOLET BNP see VQU500
VIOLET LEAF ALCOHOL see NMV780
VIOLET LEAF ALDEHYDE see NMV760
VIOLET POWDER H 2503 see MQN025
VIOLURIC ACID see AFU000
VIOMICINAE (ITALIAN) see VQZ000
VIOMYCIN see VQZ000
VIOMYCIN, 1-(threo-4-HYDROXY-L-3,6-DIAMINOHEXANOIC ACID)-6-(L-2-(2-AMINO-1,4,5,6-TETRAHYDRO- 4-PYRIMIDINYL)GLYCINE)-, (R)-, SESQUISULFATE see VQZ100
VIOMYCIN SULFATE SALT (2:3) see TNY250
VIOMYCIN SULFATE SALT (2:3) see VQZ100
VIOPSICOL see MDQ250, LFK000
VIOSTEROL see VSZ100
VIOZENE see RMA500
VI-PAR see CIR500
VIPERA AMMODYTES VENOM see VQZ425
VIPERA BERUS VENOM see VQZ475
VIPERA BORNMULLERI VENOM see VQZ500
VIPERA LATIFII VENOM see VQZ525
VIPERA LEBETINA VENOM see VQZ550
VIPERA PALESTINAE VENOM see VQZ575
VIPERA RUSSELLII FORMOSENSIS VENOM see VQZ625
VIPERA RUSSELLII VENOM see VQZ635
VIPERA XANTHINA PALAESTINAE VENOM see VQZ650
VIPERINE (CANADA) see VQZ675
VIPER'S BUGLOSS see VQZ675
VI-PEX see CIR500, CLO200
VIPICIL see AJJ875
VIPRYNIUM EMBONATE see PQC500
VIRA-A see AEH100
VIRACTIN see VRA000
VIRAMID see RJA500
VIRAZOLE see RJA500

VIRCHEM see SHR500
VIRGIMYCIN see VRA700, VRF000
VIRGINIA-CAROLINA 3-670 see DWU200
VIRGINIA CAROLINA VC 9-104 see EIN000
VIRGINIAMYCIN see VRF000
VIRGIN'S BOWER see CMV390
VIRIDICATUMTOXIN see VRP000
VIRIDINE see PDX000
VIRIDITOXIN see VRP200
VIROFRAL see AED250
VIROPHTA see TKH325
VIROPTIC see TKH325
VIRORMONE see TBF500
VIROSIN see AQM250
VIROSTERONE see TBF500
VIRTEX CC see SHR500
VIRTEX D see SHR500
VIRTEX L see SHR500
VIRTEX RD see SHR500
VIRUBRA see VSZ000
VIRUSTOMYCIN A see VRP775
VIRUZONA see MKW250
VISADRON see NCL500, SPC500
VISCARIN see CCL250
VISCARIN 402 see CAO250
VISCERALGIN see HNM000
VISCERALGINE see HNM000
VISCLAIR see MBX800
VISCOL 350P see PMP500
VISCOTOXIN see VRU000
VISCUM ALBUM see ERA100
VISINE see VRZ000
VISINE HYDROCHLORIDE see VRZ000
VISKEN see VSA000
VISKING CELLOPHANE see CCT250
VISKO-RHAP see DGB000
VISKO-RHAP DRIFT HERBICIDES see DAA800
VISKO RHAP LOW VOLATILE ESTER see TAA100
VISNAGALIN see AHK750
VISOTRAST see AAN000
VISTABAMATE see MQU750
VISTAMYCIN see RIP000, XQJ650
VISTAR see DUK000
VISTAR HERBICIDE see DUK000
VISTARIL HYDROCHLORIDE see VSF000
VISUBUTINA see HNI500
VITACAMPHER see VSF400
VITACIN see ARN000
VITAHEXIN P see PII100
VITALLIUM see CNA750, VSK000
VITALOID see VQR300
VITAMIN A see VSK600
VITAMIN A1 see VSK600
VITAMIN A ACETATE see VSK900
trans-VITAMIN A ACETATE see VSK900
VITAMIN A ACID see VSK950
13-cis-VITAMIN A ACID see VSK955
VITAMIN A1 ALCOHOL see VSK600
all-trans-VITAMIN A ALCOHOL see VSK600
VITAMIN A ALCOHOL ACETATE see VSK900
9-cis-VITAMIN A ALDEHYDE see VSK975
VITAMIN AB see HAQ500
VITAMIN A PALMITATE see VSP000
VITAMIN B1 see TES750
VITAMIN B1 see TET300
VITAMIN B2 see RIK000
VITAMIN B3 see NCR000
VITAMIN B4 see AEH000
VITAMIN B-5 see CAU750
VITAMIN B6 see PPK250
VITAMIN B7 see VSU100
VITAMIN B12 (FCC) see VSZ000
VITAMIN Bc see FMT000
VITAMIN B12 COMPLEX see VSZ000
VITAMIN B HYDROCHLORIDE see TET300
VITAMIN B6-HYDROCHLORIDE see PPK500
VITAMIN B6 HYDROCHLORIDE see VSU000
VITAMIN B12, METHYL see VSZ050
VITAMIN B1 MONONITRATE see TET500
VITAMIN B1 NITRATE see TET500
VITAMIN B1 PROPYL DISULFIDE see DXO300
VITAMIN C see ARN000, ARN125
VITAMIN C SODIUM see ARN125
VITAMIN D2 see VSZ100

VITAMIN D3 see CMC750
VITAMIN D^3 see HJV000
VITAMIN E see VSZ450
VITAMIN G see RIK000
VITAMIN H see AIH600, VSU100
VITAMIN K see VSZ500
VITAMIN K1 see VTA000
VITAMIN K3 see MMD500
VITAMIN K5 see AKX500
VITAMIN K2$_{20}$ see VTA650
VITAMIN K3-NATRIUM-BISULFIT TRIHYDRAT (GERMAN) see MMD750
VITAMIN K2(O) see MMD500
VITAMIN L see API500
VITAMIN M see FMT000
VITAMIN MK 4 see VTA650
VITAMIN P see RSU000
VITAMIN PP see NCR000
VITAMISIN see ARN000
VITANEURON see TES750
VITAPLEX E see VSZ450
VITAPLEX N see NCQ900
VITARUBIN see VSZ000
VITA-RUBRA see VSZ000
VITASCORBOL see ARN000
VITAVAX see CCC500
VITAVEL-A see VSK600
VITAVEL-D see VSZ100
VITAVEX see DLV200
VITAYONON see VSZ450
VITAZECHS see PII100
VITEOLIN see VSZ450
VITESTROL see DLB400
VITICOSTERONE see HKG500
VITINC DAN-DEE-3 see CMC750
VITON see BBQ500
VITPEX see VSK600
VITRAL see VSZ000
VITRAN see TIO750
VITREOUS QUARTZ see SCK600
VITREX see PAK000
VITRIOL BROWN OIL see SOI500
VITRIOL, OIL of (DOT) see SOI500
VITRIOL RED see IHD000
VITRUM AB see HAQ500
VITUF see PKF750
VI-TWEL see VSZ000
VIVACTIL see DDA600, POF250
VIVAL see DCK759
VIVALAN see VKF000
VIVICIL see FMR500
VIVOTOXIN see VTA750
VK-738 see DWE800
VLB see VKZ000
VLB MONOSULFATE see VLA000
VLVF see AAX175
VM-26 see EQP000
VMCC see AAX175
VMI 10-3 see AMI500
VM&P NAPHTHA see PCT250
VM & P NAPHTHA (ACGIH) see NAI500
VOFATOX see MNH000
VOGALENE see MQR000
VOGAN see VSK600
VOGAN-NEU see VSK600
VOGEL'S IRON RED see IHD000
VOLAMIN see EEH000
VOLATILE OIL of MUSTARD see AGJ250
VOLATON see BAT750
VOLCLAY see BAV750
VOLCLAY BENTONITE BC see BAV750
VOLDYS see MCA500
VOLFARTOL see TIQ250
VOLFAZOL see COD000
VOLID see TAC800
VOLIDAN see MCA500, VTF000
VOLPO 20 see PJW500
VOLTALEF 10 see KDK000
VOLTAREN see DEO600
VOLTAROL see DEO600
VOLUNTAL see TIO500
VOMEX A see DYE600
VOMISSELS see HGC500
VOMITOXIN see VTF500
VONDACEL BLACK N see AQP000

VONDACEL BLUE 2B see CMO000
VONDACEL BLUE FF see CMN750
VONDACEL BLUE HH see CMO500
VONDACEL RED CL see SGQ500
VONDACID FAST YELLOW AE see SGP500
VONDACID METANIL YELLOW G see MDM775
VONDACID TARTRAZINE see FAG140
VONDAMOL FAST BLUE R see ADE750
VONDCAPTAN see CBG000
VONDODINE see DXX400
VONDURON see DXQ500
VONEDRINE see DBA800
VONTIL see TFM100
VONTROL see DWK200
VOPCOLENE 27 see OHU000
VORANIL see DRC600
VOREN see DBC510
V. ORIENTALIS VENOM see VJP900
VORLEX see ISE000, MLC000
VORONITE see FQK000
VOROX see AMY050
VORTEX see ISE000
VOTEXIT see TIQ250
VP16 see EBB600
VP 1940 see ABX250
VP 16213 see EAV500
VPM see VFU000, VFW009
V-PYROL see EEG000
VROKON see AAN000
VTI 1 see AOT000
VUAGT-I-4 see TFS350
VUCINE see EDW500
6605 VUFB see MIQ400
VUFB 6638 see DLR150
VULCACEL B-40 see DVF400
VULCACEL BN see DVF400
VULCACID D see DWC600
VULCACURE see BIX000, BJC000, SGF500
VULCAFIX SCARLET R see CHP500
VULCAFOR BSM see BDG000
VULCAFOR FAST YELLOW GTA see DEU000
VULCAFOR ORANGE R see CJD500
VULCAFOR SCARLET A see MMP100
VULCAFOR TMTD see TFS350
VULCALENT A see DWI000
VULCAN RED LC see CHP500
VULCAN RED R see CJD500
VULCATARD see DWI000
VULCOL FAST RED L see CHP500
VUL-CUP see BHL100
VUL-CUP 40KE see BHL100
VUL-CUP R see BHL100
VULCUREN 2 see BDF750
VULKACIT 4010 (CZECH) see PET000
VULKACIT D/C see DWC600
VULKACIT DM see BDE750
VULKACIT DM/MGC see BDE750
VULKACIT DOTG/C see DXP200
VULKACIT LDA see BJC000
VULKACIT LDB/C see BIX000
VULKACIT MTIC see TFS350
VULKACIT NPV/C2 see IAQ000
VULKACIT P EXTRA N see ZHA000
VULKACIT THIURAM see TFS350
VULKACIT THIURAM/C see TFS350
VULKACIT THIURAM MS/C see BJL600
VULKACIT ZM see BHA750
VULKALENT A (CZECH) see DWI000
VULKANOX 4020 see PEY500
VULKANOX BKF see MJO500
VULKANOX PAN see PFT250
VULKASIL see SCH000
VULKAZIT see DWC600
VULKLOR see TBO500
VULNOPOL NM see SGM500
VULTROL see DWI000
VULVAN see TBG000
VUMON see EQP000
VX see EIG000
VYAC see VLU250
VYDATE see DSP600
VYDATE L INSECTICIDE/NEMATICIDE see DSP600
VYDATE L OXAMYL INSECTICIDE/NEMATOCIDE see DSP600
VYDYNE see NOH000

VYGEN 85 see PKQ059
VYNAMON ORANGE CR see LCS000
VYNW
see AAX175

W 37 see BRA625
W 45 see ALC250
W 101 see PMP500
W 483 see DIR000
W 491 see CCP500
W 713 see MOV500
W-801 see CMY250
W 1544 see PFC500
W 1655 see PDC250, PEK250
W 1803 see PGJ000
W 1929 see SFY500
2317-W see DVF800
W-2395 see CIL500
W 2426 see ARW750
W-2429 see DLQ400
W 2451 see MJE250
W 3566 see QFA250
W 3699 see EOA500
W 4565 see OOG000
W 5219 see BGC625
W5769 see CFC750
W 6309 see DKJ300
W 7320 see AGN000
W 7618 see CLD000
W 5759A see EIH000
W 2900 A see MNG000
WACHOLDERBEER OEL (GERMAN) see JEA000
WACKER S 14/10 see BJE750
W. AEGYPTIA VENOM see WAT100
WA 335 HYDROCHLORIDE see WAJ000
WAIT'S GREEN MOUNTAIN ANTIHISTAMINE see WAK000
WAKE ROBIN see JAJ000
WALKO-NESIN see GGS000
WALNUT EXTRACT see WAT000
WALNUT STAIN see MAT500
WALTERINNESIA AEGYPTIA VENOM see WAT100
WAMPOCAP see NCQ900
WANADU PIECIOTLENEK (POLISH) see VDU000
WANSAR see CCR875
WAPNIOWY TLENEK (POLISH) see CAU500
WARAN see WAT220
WARBEX see FAB600
WARCO F 71 see CNH125
WARCOUMIN see WAT220
WARDAMATE see MQU750
WARDUZIDE see CLH750
WARECURE C see IAQ000
WARFARIN see WAT200
WARFARINE (FRENCH) see WAT200
WARFARIN K see WAT209
WARFARIN POTASSIUM see WAT209
WARFARIN SODIUM see WAT220
WARFILONE see WAT220
WARKEELATE ACID see EIX000
WARKEELATE PS-43 see EIV000
WASH OIL see CMY825
WASSERINA see CQH000
WASSERSTOFFPEROXID (GERMAN) see HIB000
WATAPANA SHIMARON see AAD750
WATER see WAT259
WATER-d2 (9CI) see HAK000
WATER ARUM see WAT300
WATER CROWFOOT see FBS100
WATER DRAGON see WAT300
WATER DROPWART see WAT315
WATER GLASS see SJU000
WATER GOGGLES see MBU550
WATER²-H2 see HAK000
WATER HEMLOCK see WAT325
WATER LILY see DAE100
WATER-PEPPER HERB see WAT350
WATER QUENCH PYROLYSIS FUEL OIL see FOP100
WATERSTOFPEROXYDE (DUTCH) see HIB000
WATERWEED see PIH800
WATTLE GUM see AQQ500
WAXAKOL ORANGE GL see PEJ500
WAXAKOL VERMILION L see XRA000
WAXAKOL YELLOW NL see AIC250
WAXBERRY see SED550

WAX MYRTLE see WBA000
WAXOLINE GREEN see BLK000
WAXOLINE ORANGE A see BJF000
WAXOLINE PURPLE A see HOK000
WAXOLINE RED A see MAC500
WAXOLINE YELLOW AD see DOT300
WAXOLINE YELLOW ED see OHI875
WAXOLINE YELLOW I see PEJ500
WAXOLINE YELLOW O see IBB000
WAXSOL see DJL000
WAYNECOMYCIN see LGC200
WAYNE RED X-2486 see CHP500
W3207B see MOM750
WBA 8119 see TAC800
WD 67/2 see WBA600
WE 941 see LEJ600
WE 941-BS see LEJ600
WEATHERBEE MUSTARD see BHB750
WEATHER PLANT see RMK250
WEC 50 see TIQ250
WECKAMINE see BBK000
WECOLINE 1295 see LBL000
WECOLINE OO see OHU000
WEDDING BELLS see DJO000
WEECON see COI250
WEED 108 see MRL750
WEED-AG-BAR see DAA800
WEEDANOL CYANOL see PLC250
WEEDAR see TAA100
WEEDAR-64 see DAA800
WEEDAR ADS see AMY050
WEEDAR MCPA CONCENTRATE see CIR250
WEEDAZIN see AMY050
WEEDAZOL see AMY050
WEEDAZOL TL see ANW750
WEEDBEADS see SJA000
WEED-B-GON see DAA800, TIX500
WEED BROOM see DXE600
WEED DRENCH see AFV500
WEED-E-RAD see DXE600, MRL750
WEEDEX A see ARQ725
WEEDEX GRANULAT see AMY050
WEEDEZ WONDER BAR see DAA800
WEED-HOE see DXE600, MRL750
WEEDMASTER GRASS KILLER see TII500
WEEDOCLOR see AMY050
WEEDONE see PAX250, TAA100
WEEDONE 40 see EHY600
WEEDONE 128 see IOY000
WEEDONE 170 see DGB000
WEEDONE CONCENTRATE 48 see EHY600
WEEDONE CRAB GRASS KILLER see PLC250
WEEDONE DP see DGB000
WEEDONE LV4 see DAA800
WEEDONE MCPA ESTER see CIR250
WEED-RHAP see CIR250
WEED TOX see DAA800
WEEDTRINE-D see DWX800
WEEDTRINE-II see ILO000
WEEDTROL see DAA800
WEEVILTOX see CBV500
WEG 147 see CJP500
WEG 148 see MNR100
WEGANTYNA see DHX800
WEGANTYNA (POLISH) see THK000
WEGLA DWUSIARCZEK (POLISH) see CBV500
WEGLA TLENEK (POLISH) see CBW750
WEHYDRYL see BAU750
WEIFACODINE see TCY750
WEISS PHOSPHOR (GERMAN) see PHO740
WELDING FUMES see WBJ000
WELFURIN see NGE000
WELLBATRIN see WBJ500
WELLCOME-248U see AEC700
WELLCOME PREPARATION 47-83 see EAN600
WELLCOME U3B see AMH250
WELLCOPRIM see TKZ000
WELLDORM see SKS700
WELVIC G 2/5 see PKQ059
WEMCOL see IOF200
WEPSIN see AIX000
WEPSYN see AIX000
WEPSYN 155 see AIX000
WESCOZONE see BRF500

WESPURIL see MJM500
WESTERN DIAMONDBACK RATTLESNAKE VENOM see WBJ600
WEST INDIAN LEMONGRASS OIL see LEH000
WEST INDIAN LILAC see CDM325
WEST INDIAN PINKROOT see PIH800
WESTOCAINE see AIT250
WESTRON see TBQ100
WESTROSOL see TIO750
WETAID SR see DJL000
WET-TONE B see MSB500
WEX 1242 see PMP500
WF-104 see OKK500
WFNA see NEF000
W-GUM see SLJ500
WH 5668 see PMM000
WH 7286 see DMW000
WH 7508 see PIW000
WHEAT GERM OIL see WBJ700
WHEAT HUSK OIL see WBJ700
WHEY FACTOR see OJV500
WHISKEY see WBS000
1700 WHITE see TGG760
WHITE ACID (DOT) see ANG250
WHITE ANTHURIUM see SKX775
WHITE ARSENIC see ARI750
WHITE ASBESTOS see ARM268
WHITE CAMPHOR OIL see CBB500
WHITE CAUSTIC see SHS000
WHITE CAUSTIC, solution see SHS500
WHITE CEDAR see CDM325
WHITE CEDAR OIL see CCQ500
WHITE COPPERAS see ZNA000
WHITE CRYSTAL see SFT000
WHITE FUMING NITRIC ACID see NEF000
WHITE HELLEBORE see FAB100
WHITE HONEY FLOWER see GJU475
WHITE KERRIA see JDA075
WHITE LEAD see LCP000
WHITE LOCUST see GJU475
WHITE MERCURY PRECIPITATED see MCW500
WHITE MINERAL OIL see MQV750, MQV875
WHITE OIL of CAMPHOR see CBB500
WHITE OSTER see DYA875
WHITE PHOSPHORUS see PHO740
WHITE POPINAC see LED500
WHITE PRECIPITATE see MCW500
WHITE SEAL-7 see ZKA000
WHITE SPIRIT see WBS675
WHITE SPIRIT, DILUTINE 5 see WBS700
WHITE SPIRITS see PCT250, SLU500
WHITE STREPTOCIDE see SNM500
WHITE STUFF see HBT500
WHITE TAR see NAJ500
WHITE VITRIOL see ZNA000, ZNJ000
WHORTLEBERRY RED see FAG020
W-53 HYDROCHLORIDE see DPJ400
WHYO TREE see GJU475
WICKENOL 111 see IQW000
WICKENOL 116 see DNL800
WICKENOL 155 see OFE100
WICKENOL 158 see AEO000
WICKENOL 324 see AHA000
WICKERBY BUSH see LEF100
WICKUP see LEF100
WICKY see MRU359
WICOPY see LEF100
WIDLON see PJY500
WIE OBEN see DJY200
WIJS' CHLORIDE see IDS000
WILD ALLAMANDA (FLORIDA) see YAK300
WILD BALSAM APPLE see FPD100
WILD CALLA see WAT300
WILD COFFEE see CNG825
WILD CROCUS see PAM780
WILD GARLIC see WBS850
WILD GINGER OIL see SED500
WILD JALAP see MBU800
WILD LEMON see MBU800
WILD LICORICE see RMK250
WILD MAMEE see BAE325
WILD NIGHTSHADE see YAK300
WILD ONION see DAE100, WBS850
WILD TAMARIND (HAWAII, PUERTO RICO) see LED500
WILD TOBACCO see CCJ825

WILD UNCTION (BAHAMAS) see YAK300
WILD YAM BEAN see YAG000
WILELAIKI (HAWAII) see PCB300
WILKINITE see BAV750
WILLBUTAMIDE see BSQ000
WILLESTROL see DKB000
WILLNESTROL see DAL600
WILLOSETTEN see SJN700
WILPO see DTJ400
WILT PRUF see PKQ059
WILTZ-65 see NBW500
WIN 244 see CLD000
WIN 833 see HLK000
WIN 1258 see PJB750
WIN 1593 see CPM750
WIN 1701 see DOS800
WIN 2022 see AMM750
WIN 2661 see WBS855
WIN 2663 see WBS860
WIN-2848 see DPJ200, TEO000
WIN 3074 see PPM550
WIN 3706 see SPB800
WIN 4369 see PBS000
WIN 4981 see BJA825
WIN 5095 see AFY500
WIN 5162 see DMV600
WIN 5501 see AKT500
WIN 5512 see AKT250
WIN 5606 see BCA000
WIN 6703 see MGJ600
WIN 8077 see MSC100
WIN 11450 see SAN600
WIN-1161-3 see DWC100
WIN 12267 see DEM000
WIN 13,099 see PFA600
WIN 14833 see AOO400
WIN 17,416 see HEF300
WIN 17625 see HOM259
WIN 17757 see DAB830
WIN 18,320 see EID000
WIN 18441 see BIX250
WIN 18,446 see BIX250
WIN 20228 see DOQ400
WIN 20740 see COV500
WIN 22005 see UVS500
WIN 24450 see EBY600
WIN 24933 see LIM000
WIN 32729 see EBH400, EBY100
WIN 32784 see BLV125
WIN 35833 see CMS210
WIN 39103 see MQR300
WIN 40680 see AOD375
WIN-5063-2 see MPN000
WIN 5522-2 see CPB500
WIN 8308-3 see SEN500
WIN 8851-2 see SKO500
WIN 18501-2 see ECW600
WINACET D see AAX250
WINCORAM see AOD375
WINDFLOWER see PAM780
WINDOWLEAF see SLE890
WINDOW PLANT see SLE890
WINE see WCA000
WINE ETHER see ENW000
WINE PALM see FBW100
WINE PLANT see RHZ600
WING STOP B see SGM500
WIN-2299 HYDROCHLORIDE see DHW400
WIN 2848 HYDROCHLORIDE SALT see DPJ400
WINIDUR see PKQ059
WIN-KINASE see UVS500
WINOBANIN see DAB830
WINSTROL see AOO400
WINSTROL V see AOO400
WINTER BLOOM see WCB000
WINTER CHERRY see JBS100
WINTER FERN see PJJ300
WINTERGREEN OIL (FCC) see MPI000
WINTERGREEN OIL, SYNTHETIC see MPI000
WINTERMIN see CKP250
WINTERSTEINER'S COMPOUND F see CNS800
WINTERSWEET see BOO700
WINTERWASH see DUS700
WINTOMYLON see EID000

WINYLU CHLOREK (POLISH) see VNP000
WIRKSTOFF 37289 see EPY000
WIRNESIN see POB500
WISTARIA see WCA450
WISTERIA see WCA450
WISTERIA FLORIBUNDA see WCA450
WISTERIA SINENSIS see WCA450
WITAMINA PP see NCR000
WITAMOL 320 see AEO000
WITCHES' THIMBLES see FOM100
WITCH HAZEL see WCB000
WITCIZER 300 see DEH200
WITCIZER 312 see DVL700
WITCO 31 see PJY100
WITCONATE see AFO250
WITCONOL MS see OAV000
WITCONOL MST see OAV000
WITEPSOL E-75 see WCB100
WITTOX C see NAS000
WL 19 see PDK500
WL 20 see NNL500, PDK300
WL 23 see MNP400
WL 28 see CJK000
WL 29 see MFG200
WL 31 see MNU100
WL 32 see MNU150
WL 33 see NNL500
WL 34 see CJK100
WL 39 see MNP500
WL 1650 see OAN000
WL-5792 see DGM600
WL 17,731 see EGS000
WL 18236 see MDU600
WL 19805 see BLW750
WL 417-06 see DAB825
WL 43467 see RLF350
WL 43479 see AHJ750
WL 43775 see FAR100
W-2946M see DNA600
WN 12 see ISE000
WNM see DTG000
WNYESTRON see EDV000
WOCHEM No. 320 see OHU000
WODE WHISTLE see PJJ300
WOFACAIN A see BQH250
WOFATOS see MNH000
WOFATOX see MNH000
WOFOTOX see MNH000
WOJTAB see PLZ000
WOLFEN see CNL500
WOLFRAM see TOA750
WOLFRAMITE see TOC750
WOLFSBANE see AQY500, MRE275
WOLLASTOKUP see WCJ000
WOLLASTONITE see WCJ000
WONDER BULB see CNX800
WONDER TREE see CCP000
WONUK see ARQ725
WOOD ALCOHOL (DOT) see MGB150
WOODBINE HONEYSUCKLE see HGK700
WOOD ETHER see MJW500
WOOD LAUREL see MRU359
WOOD NAPHTHA see MGB150
WOOD OIL see GME000
WOOD SPIRIT see MGB150
WOOD VINE see YAK100
WOOL BLUE RL see ADE750
WOOL BORDEAUX 6RK see FAG020
WOOL BRILLIANT GREEN SF see FAF000
WOOL FAST BLUE R see ADE750
WOOL GREEN S (BIOLOGICAL STAIN) see ADF000
WOOLLEY'S ANTISEROTONIN see BEM750
WOOL MORDANT see CMK425
WOOL ORANGE 2G see HGC000
WOOL RED see FAG020
WOOL VIOLET see FAG120
WOOL YELLOW see FAG140
WOORALI see COF750
WOORARI see COF750
WORM-AGEN see HFV500
WORM-CHEK see LFA020
WORM GRASS see PIH800
WORM GUARD see BQK000
WORMWOOD PLANT see SAF000

WORT-WEED see CCS650
WOTEXIT see TIQ250
WOURARA see COF750
WP 155 see AIX000
W-PYRIDINYLMETHANOL see POR800
WR 141 see GJS000
WR 448 see SOA500
WR 638 see AKB500
WR 2721 see AMD000
WR 2978 see TGD000
WR 3121 see THB250
WR 5473 see COX400
WR 9838 see KFA100
WR 09792 see FMH000
WR 14,997 see AJK250
WR 81844 see WCJ750
WR-171,669 see HAF500
W 45 RASCHIG see ALC250
WRIGHTINE DIHYDROBROMIDE see DOX000
WS 102 see NDU500
WS 2434 see MJH900
WSQ 1 see MDR250
W-13 STABILIZER see SLJ500
W-T SASP ORAL see PPN750
WURM-THIONAL see PDP250
WURSTER'S BLUE see BJF500
WURSTER'S REAGENT see BJF500
WUWEIZI ALCOHOL A see SBE400
WUWEIZI ALCOHOL B see SBE450
WUWEIZICHUN A see SBE400
WUWEIZICHUN B see SBE450
WV 365 see TFY000
WV 761 see BMW000
W VII/117 see FQK000
WY 509 see DQA400
WY 554 see DAM700
WY 806 see DTL200
WY 1094 see DQA600
WY-1172 see ABH500
WY-1395 see TLG000
WY 1485 see CHD750
WY 2149 see CMB125
WY 2917 see CFY750
WY-3263 see DPX200
WY 3277 see SGS500
WY-3478 see HJS500
WY-3498 see CFZ000
WY 3707 see NNQ500
WY 3917 see CFY750
WY 4036 see CFC250
WY 4508 see AJJ875
WY-5090 see HGI700
WY-5103 see AIV500
WY 5244 see CKA500
WY 5256 see AKN750
WY-8678 see GKO750
WY-14,643 see CLW250
WY-21743 see OLW600
WY 8678 ACETATE see GKO750
WYACORT see MOR500
WYCILLINA see BFC750
WY-E 104 see NNL500
WYMESONE see BFW325
WY-4355 mixed with MESTRANOL (20:1) see CHI750
WYPAX see CFC250
WYSEALS see MQU750
WYVITAL see AJJ875
WZ 884
see PMJ525

X 34 see IBC000
X 79 see POJ500
X 119 see DTP000
X 134 see TES000
X 146 see TFQ275
X 149 see BOT250
X 201 see AJK250
X-340 see HAL000
XA 2 see CFA750
X-AB see PKQ059
X-ALL LIQUID see AMY050
XAMAMINA see DYE600
XAMOTEROLFUMARATE see XAH000
XAN see XCA000

XANAX see XAJ000
XANTELINE see DJM800, XCJ000
XANTHACRIDINE see XAK000
XANTHACRIDINUM see DBX400
XANTHAHYDROGEN see IBL000
XANTHANOL see XBJ000
XANTHAURINE see QCA000
XANTHENE see XAT000
9H-XANTHENE see XAT000
XANTHENE-9-CARBOXYLIC ACID, ESTER with DIETHYL(2-
   HYDROXYETHYL)METHYLAMMONIUM BROMIDE see XCJ000
XANTHENE-9-CARBOXYLIC ACID ESTER with DIETHYL(2-
   HYDROXYETHYL) METHYLAMMONIUM BROMIDE see DJM800
XANTHENE-9-CARBOXYLIC ACID, ESTER with (2-
   HYDROXYETHYL)DIISOPROPYLMETHYLAMMONIUM BROMIDE
   see HKR500
1,1'-(9H-XANTHENE-2,7-DIYL)BIS(2-(DIMETHYLAMINO)ETHANONE DIHY-
   DROCHLORIDE SESQUIHYDRATE see XBA000
XANTHEN-9-OL see XBJ000
9-XANTHENONE see XBS000
XANTHIC OXIDE see XCA000
XANTHINE see XCA000
XANTHINE BROMIDE see XCJ000
XANTHINE-3-N-OXIDE see HOP000
XANTHINE-7-N-OXIDE see HOP259
XANTHINE-x-N-OXIDE see HOP000
XANTHINOL NIACINATE see XCS000
XANTHINOL NICOTINATE see XCS000
XANTHIUM CANADENSE, KERNAL EXTRACT see XCS400
XANTHOCILLIN see DVU000
XANTHOCILLIN Y 1 see XCS680
XANTHOCILLIN Y 2 see XCS700
XANTHOPUCCINE see TCJ800
XANTHOSOMA (various species) see XCS800
XANTHOTOXIN see XDJ000
XANTHURENIC ACID see DNC200
XANTHURENIC ACID-8-METHYL ETHER see HLT500
XANTHYDROL see XBJ000
XANTURAT see ZVJ000
XANTYRID see DVU000
XARIL see ABG750
X465A SODIUM SALT see CDK500
XAVIN see XCS000
XAXA see ADA725
XB 2615 see TBC450
XB 2793 see DKM130
1-X-5-BAA (RUSSIAN) see BDK750
XENENE see BGE000
o-XENOL see BGJ250
XENON see XDS000
XENON, refrigerated liquid (DOT) see XDS000
XENON(II) FLUORIDE METHANESULFONATE see XEA000
XENON(II) FLUORIDE TRIFLUOROACETATE see XEJ000
XENON(II) FLUORIDE TRIFLUOROMETHANESULFONATE see XEJ100
XENON HEXAFLUORIDE see XEJ300
XENON(II) PERCHLORATE see XES000
XENON TETRAFLUORIDE see XFA000
XENON(II) TRIFLUOROACETATE see XFJ000
XENON TRIOXIDE see XFS000
XENYLAMIN (CZECH) see AJS100
XENYLAMINE see AJS100
1-XENYL-3,3-DIMETHYLTRIAZIN (CZECH) see BGL500
XENYTROPIUM BROMIDE see PEM750
XERAC see BDS000
XERAL see GGS000
XERENE see MFD500
XEROSIN see XFS600
XF-408 see XGA000
XF-13-563 see SCR400
m-XHQ see DSG700
XIBENOL HYDROCHLORIDE see XGA500
XILENOLI (ITALIAN) see XKA000
XILIDINE (ITALIAN) see XMA000
XILOBAM see XGA725
XILOCAINA (ITALIAN) see DHK400
XILOLI (ITALIAN) see XGS000
XIPAMID see CLF325
XIPAMIDE see CLF325
XITIX see ARN000
XK-62-2 see MQS579
XK-70-1 see FOK000
XL 7 see TFD250
XL-50 see PDP250
XL-90 see RLU000

XL ALL INSECTICIDE see NDN000
3,5-XMC see DTN200
XN see XCS000
XOE 2982 (RUSSIAN) see HBK700
XOLAMIN see FOS100
XORU-OX see CJY000
XP-470 see CJI000
X-539-R see PJY100
XRM 3972 see DGJ100
XS-89 see SMM500
XUMBRADIL see DNG400
XX 212 see DAD050
XYCAINE HYDROCHLORIDE see DHK600
XYCIANE see DHK400
XYDE see BGC625
XYDURIL see ARQ750
XYLAMIDE see BGC625
XYLAMIDE (gastroprotective agent) see BGC625
XYLAZINE (USDA) see DMW000
XYLENE see XGS000
m-XYLENE see XHA000
o-XYLENE see XHJ000
p-XYLENE see XHS000
1,2-XYLENE see XHJ000
1,3-XYLENE see XHA000
1,4-XYLENE see XHS000
XYLENE ACID MILLING GREEN BL see CMM400
XYLENE BLACK F see BMA000
XYLENE BLUE VS see ADE500
XYLENE BLUE VSG see FMU059
m-XYLENE, 5-CHLORO- see CLV500
m-XYLENE-α,α'-DIAMINE see XHS800
m-XYLENE-α,α'-DIISOCYANATE see XIJ000
XYLENE FAST ORANGE G see HGC000
XYLENE FAST YELLOW ES see SGP500
XYLENE FAST YELLOW GT see FAG140
XYLENEN (DUTCH) see XGS000
XYLENE RED see ADG125
XYLENESULFONIC ACID see XJA000
m-XYLENESULFONIC ACID see XJJ000
2,4-XYLENESULFONIC ACID see XJJ000
m-XYLENE-4-SULFONIC ACID see XJJ000
XYLENOL see XKA000
m-XYLENOL see XKJ500
2,3-XYLENOL see XKJ000
2,4-XYLENOL see XKJ500
2,5-XYLENOL see XKS000
2,6-XYLENOL see XLA000
3,4-XYLENOL see XLJ000
3,5-XYLENOL see XLS000
1,2,5-XYLENOL see XKS000
1,3,4-XYLENOL see XLJ000
1,3,5-XYLENOL see XLS000
m-XYLENOL (DOT) see XKJ500
o-XYLENOL (DOT) see XKJ000
p-XYLENOL (DOT) see XKS000
XYLENOLEN (DUTCH) see XKA000
3,5-XYLENOL METHYLCARBAMATE see DTN200
3,5-XYLENYL-N-METHYLCARBAMATE see DTN200
XYLESTESIN see DHK400
XYLESTESIN HYDROCHLORIDE see DHK600
2,4-XYLIDENE (MAK) see XMS000
2,6-XYLIDIDE of 2-PYRIDONE-3-CARBOXYLIC ACID see XLS300
2,6-XYLIDID der 2-PYRIDON-3-CARBOXYLSAURE
   see XLS300
XYLIDINE see XMA000
o-XYLIDINE see XNJ000
2,3-XYLIDINE see XMJ000
2,4-XYLIDINE see XMS000
2,5-XYLIDINE see XNA000
2,6-XYLIDINE see XNJ000
3,4-XYLIDINE see XNS000
3,5-XYLIDINE see XOA000
m-4-XYLIDINE see XMS000
m-XYLIDINE (DOT) see XMS000
o-XYLIDINE (DOT) see XMJ000
p-XYLIDINE (DOT) see XNA000
m-XYLIDINE HYDROCHLORIDE see XOJ000
p-XYLIDINE HYDROCHLORIDE see XOS000
2,4-XYLIDINE HYDROCHLORIDE see XOJ000
2,5-XYLIDINE HYDROCHLORIDE see XOS000
XYLIDINEN (DUTCH) see XMA000
XYLIDINE PONCEAU see FMU070
XYLITE (SUGAR) see XPJ000

XYLITOL see XPJ000
l-XYLOASCORBIC ACID see ARN000
XYLOCAIN see DHK400
XYLOCAINE HYDROCHLORIDE see DHK600
XYLOCARD see DHK600
XYLOCHOLINE see XSS900
XYLOCHOLINE BROMIDE see XSS900
XYLOCITIN see DHK400
XYLOCITIN HYDROCHLORIDE see DHK600
m-XYLOHYDROQUINONE see DSG700
2,6-XYLOHYDROQUINONE see DSG700
XYLOIDIN see CCU250
XYLOL (DOT) see XGS000
m-XYLOL (DOT) see XHA000
o-XYLOL (DOT) see XHJ000
p-XYLOL (DOT) see XHS000
XYLOLE (GERMAN) see XGS000
5-(3,5-XYLOLOXYMETHYL)OXAZOLIDIN-2-ONE see XVS000
XYLOMETAZOLINE HYDROCHLORIDE see OKO500
XYLONEURAL see DHK600
2,6-XYLOQUINOL see DSG700
p-XYLOQUINONE see XQJ000
XYLOSTATIN see XQJ650
XYLOTOX see DHK400
XYLOTOX HYDROCHLORIDE see DHK600
2,3-XYLYLAMINE see XMJ000
2,6-XYLYLAMINE see XNJ000
3,4-XYLYLAMINE see XNS000
3,5-XYLYLAMINE see XOA000
N-(2,3-XYLYL)-2-AMINOBENZOIC ACID see XQS000
N-(2,3-XYLYL)ANTHRANILIC ACID see XQS000
1-XYLYLAZO-2-NAPHTHOL see FAG080, XRA000
1-(o-XYLYLAZO)-2-NAPHTHOL see XRA000
1-(2,4-XYLYLAZO)-2-NAPHTHOL see XRA000
1-(2,5-XYLYLAZO)-2-NAPHTHOL see FAG080
1-XYLYLAZO-2-NAPHTHOL-3,6-DISULFONIC ACID, DISODIUM SALT
    see FMU070
1-XYLYLAZO-2-NAPHTHOL-3,6-DISULPHONIC ACID, DISODIUM SALT
    see FMU070
1-(2,4-XYLYLAZO)-2-NAPHTHOL-3,6-DISULPHONIC ACID, DISODIUM SALT
    see FMU070
XYLYL BROMIDE see XRS000
(((2,6-XYLYLCARBAMOYL)METHYL)IMINO)DI-ACETIC ACID see LFO300
m-XYLYLENDIAMIN (CZECH) see XHS800
p-XYLYLENDIAMINE (CZECH) see PEX250
XYLYLENDIISOKYANAT (CZECH) see XIJ000
1,1'-(p-XYLYLENE)BIS(3-(1-AZIRIDINYL)UREA) see XSJ000
p-XYLYLENEBIS(TRIPHENYLPHOSPHONIUM CHLORIDE) see XSS000
XYLYLENE CHLORIDE see XSS250
XYLYLENE DICHLORIDE see XSS250
m-XYLYLENE DICHLORIDE see DGP200
o-XYLYLENE DICHLORIDE see DGP400
s,6-XYLYL ESTER of 1-PIPERIDINEACETIC ACID HYDROCHLORIDE
    see XSS375
2,6-XYLYL ETHER BROMIDE see XSS900
3,4-XYLYL METHYLCARBAMATE see XTJ000
3,5-XYLYL-N-METHYLCARBAMATE see DTN200
3,4-XYLYL-N-METHYLCARBAMATE, nitrosated see XTS000
3,5-XYLYL-N-METHYLCARBAMATE, NITROSATED see MMW750
p-XYLYLMETHYLCARBINAMINE SULFATE see AQQ000
(2-(3,5-XYLYLOXY)ETHYL)HYDRAZINE HYDROCHLORIDE see XVA000
(2-(3,4-XYLYLOXY)ETHYL)HYDRAZINE MALEATE see XVJ000
5-((3,5-XYLYLOXY)METHYL)-2-OXAZOLIDINONE see XVS000
2-(2,6-XYLYLOXY)TRIETHYLAMINE HYDROCHLORIDE see XWS000
XYLZIN see DMW000
XYMOSTANOL see DKW000

Y 2 see CBM000
Y 3 see CKC000
Y 195 see PKL750
Y 218 see PKM250
Y 221 see PKM500
Y-223 see PKN000
Y-238 see CDV625
Y 299 see PKO750
Y 302 see PKP000
Y-4153 see DCV400
Y 5350 see CCK550
Y 6047 see CKA000
Y-7131 see EQN600
Y-8004 see PLX400
Y-9000 see ACU125
Y-9213 see IAY000
YADALAN see CLH750

YAGEINE see HAI500
YAJEINE see HAI500
YALAN see EKO500
YALTOX see CBS275
YAMAFUL see CCK630
YAMAMOTO METHYLENE BLUE B see BJI250
YAM BEAN see YAG000
YANOCK see FFF000
YARMOR see PIH750
YARMOR PINE OIL see PIH750
YASOKNOCK see SHG500
YATREN see IEP200
YATROCIN see NGE500
YATROZIDE see ILE000
YAUPON see HGF100
YAUTIA MALANGA (PUERTO RICO) see EAI600
YAUTIA (PUERTO RICO) see XCS800
YAZUMYCIN A see RAG300
dl-YB2 see ICA000
YC 93 see PCG550
YEAST (active) see YAK000
YEDRA (CUBA) see AFK950
YEH-YAN-KU see ARW000
YELLON see IEP200
YELLOW see DCZ000
11363 YELLOW see MDM775
11712 YELLOW see FEV000
11824 YELLOW see FEW000
12417 YELLOW see FEW000
YELLOW AB see FAG130
YELLOW ALLAMANDA see AFQ625
YELLOW CROSS LIQUID see BIH250
YELLOW CUPROCIDE see CNO000
YELLOW FALSE JESSAMINE see YAK100
YELLOW FERRIC OXIDE see IHD000
YELLOW GINSENG see BMA150
YELLOW GOWAN see FBS100
YELLOW G SOLUBLE in GREASE see DOT300
YELLOW HENBANE see JBS100
YELLOW JESSAMINE see YAK100
YELLOW LAKE 69 see FAG140
YELLOW LEAD OCHER see LDN000
YELLOW LOCUST see GJU475
YELLOW MERCURIC OXIDE see MCT500
YELLOW MERCURY IODIDE see MDC750
YELLOW NIGHTSHADE see YAK300
YELLOW No. 2 see FAG130
YELLOW OB see FAG135
YELLOW OLEANDER see YAK350
YELLOW OXIDE of IRON see IHD000
YELLOW OXIDE of MERCURY see MCT500
YELLOW PARILLA (CANADA) see MRN100
1903 YELLOW PINK see BNH500
YELLOW POLAKTIN G see CMS230
YELLOW PRECIPITATE see MCT500
YELLOW PYOCTANINE see IBB000
YELLOW SAGE see LAU600
YELLOW SARSAPARILLA see MRN100
YELLOW ULTRAMARINE see CAP500
YELLOW Z see AAQ250
YERBA HEDIONDA (CUBA) see CNG825
YERBA de PORDIOSEROS (CUBA) see CMV390
YERMONIL see EEH575
YESDOL see CCR875
YESPAZINE see TJW500
YESQUIN (HAITI) see FPD100
YETRIUM see AFJ400
YEW see YAK500
Y-GUTTIFERIN see GME300
Y-3642-HCl see AJU000, TGE165
Y-3642 HYDROCHLORIDE see TGE165
YL-704 A3 see JDS200
YLANG YLANG OIL see YAT000
YL 704 B₁ see MBY150
YL 704 B3 see LEV025
YM-08054 see IBW500
YM-08310 see AMD000
YM-09229 see FAP100
YM 09330 see CCS371, CCS373
YM-09538 see AOA075
YM-11170 see FAB500
YM-08054-1 see IBW500
YOCLO see ARQ750
YODOCHROME METANIL YELLOW see MDM775

YODOCHROME ORANGE A see NEY000
YODOMIN see TEH500
YOHIMBAN-16-CARBOXYLIC ACID DERIVATIVE of BENZ(G)INDOLO(2,3-A)QUINOLIZINE see RDK000
3-β,20-α-YOHIMBAN-16-β-CARBOXYLIC ACID, 18-β-HYDROXY-10,17-α-DIMETHOXY-, METHYL ESTER, 3,4,5-TRIMETHOXYBENZOATE (ester) see MEK700
3-β,20-α-YOHIMBAN-16-β-CARBOXYLIC ACID, 18-β-HYDROXY-11,17-α-DIMETHOXY-METHYL ESTER, 3,4,5-TRIMETHOXYBENZOATE (ESTER), TARTRATE see SCA550
YOHIMBIC ACID METHYL ESTER see YBJ000
YOHIMBINE see YBJ000
Δ-YOHIMBINE see AFG750
YOHIMBINE HYDROCHLORIDE see YBS000
Γ-YOHIMBINE HYDROCHLORIDE see AFH000
YOHIMBINE MONOHYDROCHLORIDE see YBS000
α-YOHIMBIN HYDROCHLORIDE see YCA000
YOMESAN see DFV400
YOSHI 864 see YCJ000
YOSHINOX S see TFC600
YOSIMILON see YCJ200
YPERITE see BIH250
YPERITE SULFONE see BIH500
Y-5712 SILYL PEROXIDE see SDX300
YTTERBIUM see YDA000
YTTERBIUM CHLORIDE see YDJ000
YTTERBIUM NITRATE see YDS800
YTTERBIUM(III) NITRATE, HEXAHYDRATE (1583586) see YEA000
YTTERBIUM TRICHLORIDE see YDJ000
YTTRIA see YGA000
YTTRIUM see YEJ000
YTTRIUM-89 see YEJ000
YTTRIUM CHLORIDE see YES000
YTTRIUM CITRATE see YFA000
YTTRIUM EDETATE complex see YFA100
YTTRIUMNITRAT (GERMAN) see YFS000
YTTRIUM(III) NITRATE (1583) see YFJ000
YTTRIUM(III) NITRATE HEXAHYDRATE (1583586) see YFS000
YTTRIUM OXIDE see YGA000
YTTRIUM TRICHLORIDE see YES000
YU 7802 see BDN125
YuB-25 see NBO525
YUCA see CCO680
YUCA BRAVA see CCO680
YUGILLA (CUBA) see CNH789
YUGOVINYL see PKQ059
YULAN see EKO500
YUNIHOMU AZ see ASM270
YUROTIN A see ERC600
YUTAC see YGA700
YUTOPAR see RLK700

Z-4 see DOX100
Z-78 see EIR000
Z 102 see BEQ500
Z-134 see BRE255
Z-905 see PIH100
Z 4828 see TNT500
Z 4942 see IMH000
Z 7557 see ARP625
ZABILA (MEXICO, DOMINICAN REPUBLIC) see AGV875
ZACLON DISCOIDS see HHS000
ZACTANE CITRATE see MIE600
ZACTIRIN COMPOUND see ABG750
ZACTRAN see DOS000
ZADINE see DLV800
ZADITEN see KGK200
ZADONAL see EOK000
Z-(3-AETHYL-4-OXO-5-PIPERIDINO-THIAZOLIDIN-2-YLIDEN)-ESSIGSAURE-AETHYLESTER (GERMAN) see EOA500
ZAFFRE see CND125
ZAGREB see BIE500
ZAHARINA see BCE500
ZAHLREICHE BEZEICHNUNGEN (GERMAN) see DUS700
ZAMBESIL see CLY600
ZAMI 905 see PIH100
ZAMI 1305 see BQF750
dl-ZAMI 1305 see BQF750
ZAMIA DEBILIS see ZAK000
ZAMIA PUMILA see CNH789
ZAMINE see BBK500
ZANIL see DMZ000
ZANILOX see DMZ000

ZANOSAR see SMD000
ZANTAC see RBF400
ZANTEDESCHIA AETHIOPICA see CAY800
ZARAONDAN see ENG500
ZARCILLA (PUERTO RICO) see LED500
ZARDA (INDIA) see SED400
ZARDANCHU see RQU300
ZARDEX see HCP500
ZARODAN see ENG500
ZARONDAN-SAFT see ENG500
ZARONTIN see ENG500
ZAROXOLYN see ZAK300
ZARTALIN see ENG500
ZASSOL see COI250
ZAVILA (PUERTO RICO) see AGV875
ZEAPUR see BJP000
ZEARALANOL see RBF100
ZEARALENONE see ZAT000
(−)-ZEARALENONE see ZAT000
(s)-ZEARALENONE see ZAT000
(10s)-ZEARALENONE see ZAT000
trans-ZEARALENONE see ZAT000
ZEARANOL see RBF100
ZEAZIN see ARQ725
ZEAZINE see ARQ725
ZEBENIDE see EIR000
ZEBTOX see EIR000
ZECTANE see DOS000
ZECTRAN see DOS000
ZEFRAN, combustion products see ADX750
ZEIDANE see DAD200
ZELAN see CIR250
ZELAZA TLENKI (POLISH) see IHF000
ZELEN ALIZARINOVA BRILANTNI G-EXTRA (CZECH) see APL500
ZELEN KYSELA 40 see CMM400
ZELEN MIDLONOVA BLS (CZECH) see APM250
ZELEN OSTANTHRENOVA BRILANTNI FFB (CZECH) see JAT000
ZELIO see TEL750
ZENADRID (VETERINARY) see PLZ000
ZENALOSYN see PAN100
ZENITE see BHA750
ZENITE SPECIAL see BHA750
ZENTAL see VAD000
ZENTINIC see PAG200
ZENTRONAL see DKQ000
ZENTROPIL see DKQ000, DNU000
ZEOTIN see MBX800
ZEPC see EOK550
ZEPELIN see PEW000
ZEPHIRAN CHLORIDE see AFP250, BBA500
ZEPHIROL see BBA500
ZEPHROL see EAW000
ZEPHYRANTHES ATAMASCO see RBA500
ZEPHYR LILY see RBA500
ZERANOL (USDA) see RBF100
ZERDANE see DAD200
ZERLATE see BJK500
ZERTELL see CMA250
ZEST T see VLU250
ZEST see MRL500
ZESTE see ECU750
ZET see ZBA000
ZETAR see CMY800
ZETAR EMULSION see ZBA000
ZETAX see BHA750
ZETIFEX ZN see TNC500
ZEXTRAN see DOS000
Z-GLY see CBR125
ZHENGGUANGMYCIN A2 (CHINESE) see BLY250
ZHENGGUANGMYCIN A5 (CHINESE) see BLY500
ZIARNIK see ABU500
ZIAVETINE see BQL000
ZIDAN see EIR000
ZIDE see CFY000
ZIGADENUS (VARIOUS SPECIES) see DAE100
ZIKHOM see ZJS300
ZILDASAC see BAV325
ZIMALLOY see CNA750
ZIMANAT see DXI400
ZIMANEB see DXI400
ZIMATE see BJK500, EIR000
ZIMATE METHYL see BJK500
ZIMCO see VFK000
ZIMELIDINE see ZBA500

(Z)-ZIMELIDINE see ZBA500
cis-ZIMELIDINE see ZBA500
ZIMELIDINE DIHYDROCHLORIDE see ZBA525
ZIMELIDINE HYDROCHLORIDE see ZBA525
ZIMET 3106 see IAK100
ZIMMAN-DITHANE see DXI400
ZIMTALDEHYDE see CMP969
ZIMTSAEURE (GERMAN) see CMP975
ZINACEF see CCS600
ZINADON see ILD000
ZINC see ZBJ000
ZINC ACETATE see ZBS000
ZINC ACETATE, DIHYDRATE see ZCA000
ZINC ACETATE plus MANEB (1:50) see EIQ500
ZINC ACETOACETONATE see ZCJ000
ZINC ALLYL DITHIO CARBAMATE see ZCS000
ZINC AMMONIUM NITRITE see ZDA000
ZINC ARSENATE see ZDJ000
ZINC ARSENATE, solid (DOT) see ZDJ000
ZINC ARSENATE, BASIC see ZDJ000
ZINC-m-ARSENITE see ZDS000
ZINC ARSENITE, solid (DOT) see ZDS000
ZINC BENZIMIDAZOLE-2-THIOLATE see ZIS000
ZINC-2-BENZOTHIAZOLETHIOLATE see BHA750
ZINC BENZOTHIAZOLYL MERCAPTIDE see BHA750
ZINC BENZOTHIAZOL-2-YLTHIOLATE see BHA750
ZINC BENZOTHIAZYL-2-MERCAPTIDE see BHA750
ZINC BERYLLIUM SILICATE see BFV250
ZINC-BIBUTYLDITHIOCARBAMATE see BIX000
ZINC BIS(1H-BENZIMIDAZOLE-2-THIOLATE) see ZIS000
ZINC BIS(DIMETHYLDITHIOCARBAMATE) see BJK500
ZINC BIS(DIMETHYLDITHIOCARBAMATE)CYCLOHEXYLAMINE COM-
    PLEX see ZEA000
ZINC BIS(DIMETHYLDITHIOCARBAMOYL)DISULPHIDE see BJK500
ZINC, BIS(d-GLUCONATO-O^1),O^2)- (9CI) see ZIA750
ZINC CAPRYLATE see ZEJ000
ZINC CARBONATE (1:1) see ZEJ050
ZINC CHLORATE see ZES000
ZINC CHLORIDE see ZFA000
ZINC CHLORIDE, anhydrous (DOT) see ZFA000
ZINC CHLORIDE, solid (DOT) see ZFA000
ZINC CHLORIDE (solution) see ZFJ000, ZFA000
ZINC CHLORIDE, solution (DOT) see ZFJ000
ZINC (CHLORURE de) (FRENCH) see ZFA000
ZINC CHROMATE see ZFJ100
ZINC CHROMATE HYDROXIDE see CMK500
ZINC CHROMATE(VI) HYDROXIDE see CMK500, ZFJ100
ZINC CHROMATE, POTASSIUM DICHROMATE and ZINC HYDROXIDE
    (3:1:1) see ZFJ150
ZINC CHROMATE with ZINC HYDROXIDE and CHROMIUM OXIDE (9:1)
    see ZFJ120
ZINC CHROME see PLW500
ZINC CHROME YELLOW see ZFJ100
ZINC CHROMIUM OXIDE see ZFJ100
ZINC CITRATE see ZFJ250
ZINC COMPOUNDS see ZFS000
ZINC-COPPER CHROMATE COMPLEX see CNQ750
ZINC CYANIDE see ZGA000
ZINC DIACETATE see ZBS000
ZINC DIACETATE, DIHYDRATE see ZCA000
ZINC-DIBUTYLDITHIOCARBAMATE see BIX000
ZINC-N,N-DIBUTYLDITHIOCARBAMATE see BIX000
ZINC DICHLORIDE see ZFA000
ZINC DICYANIDE see ZGA000
ZINC DIETHYLDITHIOCARBAMATE see BJC000
ZINC-N,N-DIETHYLDITHIOCARBAMATE see BJC000
ZINC DIHYDRAZIDE see ZGJ000
ZINC DIMETHYLDITHIOCARBAMATE see BJK500
ZINC N,N-DIMETHYLDITHIOCARBAMATE see BJK500
ZINC, DIMETHYLDITHIOCARBAMATE CYCLOHEXYLAMINE COMPLEX
    see ZEA000
ZINC DISTERATE see ZMS000
ZINC DITHIOPHOSPHATE see ZGS000
ZINC DUST see ZBJ000
ZINC(II) EDTA COMPLEX see ZGS100
ZINC ETHIDE see DKE600
ZINC ETHOXIDE see ZGW100
ZINC ETHYL (DOT) see DKE600
ZINC ETHYLENEBISDITHIOCARBAMATE see EIR000
ZINC ETHYLENE-1,2-BISDITHIOCARBAMATE see EIR000
ZINC ETHYLPHENYLDITHIOCARBAMATE see ZHA000
ZINC ETHYLPHENYLTHIOCARBAMATE see ZHA000
ZINC-N-FLUOREN-2-YLACETOHYDROXAMATE see ZHJ000
ZINC FLUORIDE see ZHS000

ZINC FLUORURE (FRENCH) see ZHS000
ZINC FLUOSILICATE see ZIA000
ZINC GLUCONATE see ZIA750
ZINC HEXAFLUOROSILICATE see ZIA000
ZINC HYDRIDE see ZIJ000
ZINC HYDROXYCHROMATE see CMK500, ZFJ100
ZINC INSULIN see LEK000
ZINCITE see ZKA000
ZINC MANGANESE BERYLLIUM SILICATE see BFS750
ZINCMATE see BJK500
ZINC MERCAPTOBENZIMIDAZOLATE see ZIS000
ZINC MERCAPTOBENZIMIDAZOLE see ZIS000
ZINC MERCAPTOBENZOTHIAZOLATE see BHA750
ZINC-2-MERCAPTOBENZOTHIAZOLE see BHA750
ZINC MERCAPTOBENZOTHIAZOLE SALT see BHA750
ZINC MERCURY CHROMATE COMPLEX see ZJA000
ZINC METAARSENITE see ZDS000
ZINC METHARSENITE see ZDS000
ZINC METIRAM see MQQ250
ZINC MURIATE, solution (DOT) see ZFA000, ZFJ000
ZINC NAPHTHENATE see NAT000
ZINC NITRATE see ZJJ000
ZINC(II) NITRATE, HEXAHYDRATE (1:2:6) see NEF500
ZINC NITRILOTRIMETHYLPHOSPHONIC ACID TRISODIUM TETRAHY-
    DRATE see ZJJ200
ZINC NITROSYLPENTACYANOFERRATE see ZJJ400
ZINCO (CLORURO di) (ITALIAN) see ZFA000
ZINC OCTADECANOATE see ZMS000
ZINCO(FOSFURO di) (ITALIAN) see ZLS000
ZINCOID see ZKA000
ZINC OLEATE (1:2) see ZJS000
ZINC OMADINE see ZMJ000
ZINCOP see ZJS300
ZINC OXIDE see ZKA000
ZINC OXIDE (ointment) see ZKJ000
ZINC OXIDE FUME (MAK) see ZKA000
ZINC PANTOTHENATE see ZKS000
ZINC PENTACYANONITROSYLFERRATE(2−) see ZJJ400
ZINC 2,4-PENTANEDIONATE see ZCJ000
ZINC PERMANGANATE see ZLA000
ZINC PEROXIDE see ZLJ000
ZINC PHOSPHIDE see ZLS000
ZINC PHOSPHITE see ZLS200
ZINC(PHOSPHURE de) (FRENCH) see ZLS000
ZINC POLYACRYLATE see ADW250
ZINCPOLYANEMINE see ZMJ000
ZINC POLYCARBOXYLATE see ADW250
ZINC POTASSIUM CHROMATE see CMK400
ZINC POWDER see ZBJ000
ZINC, POWDER or DUST, non-pyrophoric (DOT) see ZBJ000
ZINC, POWDER or DUST, pyrophoric (DOT) see ZBJ000
ZINC (N,N'-PROPYLENE-1,2-BIS(DITHIOCARBAMATE)) see ZMA000
ZINC PROTAMINE INSULIN see IDF325
ZINC PT see ZMJ000
ZINC PYRIDINE-2-THIOL-1-OXIDE see ZMJ000
ZINC PYRIDINETHIONE see ZMJ000
ZINC PYRION see ZMJ000
ZINC PYRITHIONE see ZMJ000
ZINC STEARATE see ZMS000
ZINC SULFATE see ZNA000, ZNJ000
ZINC SULFATE (1:1) HEPTAHYDRATE see ZNJ000
ZINC SULFATE HEPTAHYDRATE (1:1:7) see ZNJ000
ZINC SULPHATE see ZNA000
ZINC SUPEROXIDE see ZLJ000
ZINC TETRAOXYCHROMATE 76A see ZFJ100
ZINC-TOX see ZLS000
ZINC TRIFLUOROSTANNITE, HEPTAHYDRATE see ZNS000
ZINC TRISODIUM DIETHYLENETRIAMINEPENTAACETATE see TNM850
ZINC TRISODIUM DTPA see TNM850
ZINC UVERSOL see NAT000
ZINC VITRIOL see ZNA000, ZNJ000
ZINC WHITE see ZKA000
ZINC YELLOW see PLW500, CMK500, ZFJ100, ZFJ120
ZINEB see EIR000
ZINEB-COPPER OXYCHLORIDE mixture see ZJS300
ZINEB-ETHYLENE THIURAM DISULFIDE ADDUCT see MQQ250
ZINGERONE see VFP100
ZINGIBER ROSEUM (Roxb.) Rosc., extract see ZNS200
ZINK-(N,N'-AETHYLEN-BIS(DITHIOCARBAMAT)) (GERMAN) see EIR000
ZINK-BIS(N,N-DIMETHYL-DITHIOCARBAMAAT) (DUTCH) see BJK500
ZINK-BIS(N,N-DIMETHYL-DITHIOCARBAMAT) (GERMAN) see BJK500
ZINKCARBAMATE see BJK500
ZINKCHLORID (GERMAN) see ZFA000
ZINKCHLORIDE (DUTCH) see ZFA000

ZINK-(N,N-DIMETHYL-DITHIOCARBAMAT) (GERMAN) see BJK500
ZINKFOSFIDE (DUTCH) see ZLS000
ZINKOSITE see ZNA000
ZINKPHOSPHID (GERMAN) see ZLS000
ZINK-(N,N'-PROPYLEN-1,2-BIS(DITHIOCARBAMAT)) (GERMAN) see ZMA000
ZINN (GERMAN) see TGB250
ZINNTETRACHLORID (GERMAN) see TGC250
ZINOCHLOR see DEV800
ZINOPHOS see EPC500
ZINOSAN see EIR000
ZINOSTATIN see NBV500
ZIPAN see DCK759
ZIPEPROL see MFG250
ZIPEPROL DIHYDROCHLORIDE see MFG250
ZIPROMAT see ZMA000
ZIRAM see BJK500
ZIRAM CYCLOHEXYLAMINE COMPLEX see ZEA000
ZIRAMVIS see BJK500
ZIRASAN see BJK500
ZIRBERK see BJK500
ZIRCAT see ZOA000
ZIRCON see ZSS000
ZIRCONATE, BARIUM (1:1) see BAP750
ZIRCONATE(2-), OXODISULFATO-, DISODIUM (8CI) see ZTS100
ZIRCONIUM see ZOA000
ZIRCONIUM (wet) see ZOS000
ZIRCONIUM CHLORIDE see ZPA000
ZIRCONIUM(II) CHLORIDE see ZQJ000
ZIRCONIUM(IV) CHLORIDE (1:4) see ZPA000
ZIRCONIUM CHLORIDE HYDROXIDE see ZPJ000
ZIRCONIUM CHLORIDE OXIDE OCTAHYDRATE see ZPS000
ZIRCONIUM CHLOROHYDRATE see ZPJ000
ZIRCONIUM COMPOUNDS see ZQA000
ZIRCONIUM DIBROMIDE see ZQB100
ZIRCONIUM DICARBIDE see ZQC200
ZIRCONIUM DICHLORIDE see ZQJ000
ZIRCONIUM, DICHLORO-DI-pi-CYCLOPENTADIENYL- see ZTK400
ZIRCONIUM FLUORIDE see ZQS000
ZIRCONIUM GLUCONATE see ZQS100
ZIRCONIUM HYDRIDE see ZRA000
ZIRCONIUM HYDROXYCHLORIDE see ZPJ000
ZIRCONIUM(IV) LACTATE see ZRS000
ZIRCONIUM(III) LACTATE (1:3) see ZRJ000
ZIRCONIUM METAL see ZOA000
ZIRCONIUM METAL, dry (DOT) see ZOA000
ZIRCONIUM METAL, wet (DOT) see ZOS000
ZIRCONIUM NITRATE see ZSA000
ZIRCONIUM OXYCHLORIDE see ZSJ000
ZIRCONIUM PICRAMATE, WET (DOT) see PIC750
ZIRCONIUM POTASSIUM FLUORIDE see PLG500
ZIRCONIUM SHAVINGS see ZOA000
ZIRCONIUM SHEETS (DOT) see ZOA000
ZIRCONIUM(IV) SILICATE (1:1) see ZSS000
ZIRCONIUM SODIUM LACTATE see LAO000, ZTA000
ZIRCONIUM(IV) SULFATE (1:2) see ZTJ000
ZIRCONIUM TETRACHLORIDE (DOT) see ZPA000
ZIRCONIUM TETRACHLORIDE, solid (DOT) see ZPA000
ZIRCONIUM TETRAFLUORIDE see ZQS000
ZIRCONIUM(IV) TETRAHYDROBORATE see ZTK300
ZIRCONIUM TURNINGS see ZOA000
ZIRCONOCENE, DICHLORIDE see ZTK400
ZIRCONYL ACETATE see ZTS000
ZIRCONYL CHLORIDE see ZSJ000
ZIRCONYL CHLORIDE OCTAHYDRATE see ZPS000
ZIRCONYL HYDROXYCHLORIDE see ZPJ000
ZIRCONYL NITRATE see BLA000
ZIRCONYL SODIUM SULPHATE see ZTS100
ZIRCONYL SULFATE see ZTJ000
ZIREX 90 see BJK500
ZIRIDE see BJK500
ZIRPON see MQU750
ZIRTHANE see BJK500
ZITEX H 662-124 see TAI250
ZITHIOL see MAK700
ZITOSTOP see MAW800
ZITOX see BJK500
ZITOXIL see MFG250
ZITRONELL OEL (GERMAN) see CMT000
ZITRONEN OEL (GERMAN) see LEI000
ZK 26173 see DPB200
ZK 57671 see SOU650
ZK-Nr.26173 see DPB200
ZLUT MASELNA (CZECH) see DOT300
ZMA see ZDS000

ZMBT see BHA750
Z-2-(METHYL(1-OXO-9-OCTADECENYL)AMINO)-ETHANESULFONIC ACID, SODIUM SALT see SIY000
Z-METHYLPROPYL ACRYLATE see IIK000
ZN 6 see SHK000
ZnMB see BHA750
ZN 6-NA see SHK000
ZOALENE see DUP300
ZOAMIX see DUP300
ZOAPATLE, crude leaf extract see ZTS600
ZOAPATLE, semi-purified leaf extract see ZTS625
ZOBA 4R see DUP400
ZOBA 3GA see ALT000
ZOBA BLACK D see PEY500
ZOBA BROWN P BASE see ALT250
ZOBA BROWN RR see ALL750
ZOBA EG see ALT500
ZOBA ERN see NAW500
ZOBA GKE see TGL750
ZOBAR see PJQ000, TIK500
ZOBA SLE see DBO400
ZOGEN DEVELOPER H see TGL750
ZOLAMIDE SOLUTION see AMR750
ZOLAMINE HYDROCHLORIDE see ZUA000
ZOLAPHEN see BRF500
ZOLCIENI POLACTYNOWEJ G (POLISH) see CMS230
ZOLICEF see CCS250
ZOLIDINUM see BRF500
ZOLIMIDIN see ZUA200
ZOLIMIDINE see ZUA200
ZOLIRIDINE see ZUA200
ZOLON see BDJ250
ZOLONE see BDJ250
ZOLONE PM see BDJ250
ZOMAX see ZUA300
ZOMEPIRAC see ZUA300
ZOMEPIRAC SODIUM see ZUA300
ZOMEPIRAC SODIUM SALT see ZUA300
ZONAZIDE see ILD000
ZONDEL see NNT100
ZONGNON (HAITI) see WBS850
ZONIFUR see NGB700
ZOOFURIN see NGE000
ZOOLON see BDJ250
ZOPICLONE see ZUA450
ZORANE see BRF500
ZORIFLAVIN see DBX400
ZOROXIN see BAB625
ZORUBICIN see ROU800
ZORUBICIN HYDROCHLORIDE see ROZ000
ZOTEPINE see ZUJ000
ZOTHELONE see PJA120
ZOTIL see ABF750
ZOTOX see ARB250, ARH500
ZOTOX CRAB GRASS KILLER see ARB250
ZOVIRAX see AEC700
ZOVIRAX SODIUM see AEC725
ZOXAMIN see AJF500
ZOXAZOLAMINE see AJF500
ZOXINE see AJF500
ZP see ZLS000
Z 3 (PESTICIDE) see PLF000
ZR 512 see EQD000
ZR-777 see POB000
ZR-856 see HCP500
Z10-TR see CFZ000
ZUTRACIN see BAC250
ZWAVELWATERSTOF (DUTCH) see HIC500
ZWAVELZUUROPLOSSINGEN (DUTCH) see SOI500
ZWITSALAX see DMH400
ZYBAN see PEX500
ZYGADENUS (VARIOUS SPECIES) see DAE100
ZYGOMYCIN A1 see NCF500
ZYGOSPORIN A see ZUS000
N-(ZYKLOHEXYL)-SYDNONIMIN HYDROCHLORID (GERMAN) see CPQ650
ZYLOFURAMINE see ZVA000
ZYLOPRIM see ZVJ000
ZYLORIC see ZVJ000
ZYMOFREN see PAF550
ZYTEL 211 see PJY500
ZYTRON see DGD800
ZZL-0814 see EBU500
ZZL-0816 see EBV000
ZZL-0854 see EBV500

SECTION 3
References

In this section is listed the general toxicological bibliography of nearly 2200 entries keyed to an alpha numeric system.

AAAHAN Australian Journal of Experimental Agriculture and Animal Husbandry. (Commonwealth Scientific and Industrial Research Organization, POB 89, E. Melbourne, Vic 3002, Australia) V.1-1961-

AAATAP Annali dell'Accademia di Agricoltura di Torino Academia de Agricoltura, Via Andrea Doria 10, Turin, Italy) V.119- 1976/77-

AABEAJ Acta Anaesthesiologica Belgica. (Publications Acta Medica Belgica, rue des Champs-Elysees 43, 1050 Brussels 5, Belgium) V.1-1950-

AABIAV Annals of Applied Biology. (Biochemical Society, P.O. Box 32, Commerce Way, Whitehall Industrial Estate, Colchester CO2 8HP, Essex, England) V.1- 1914-

AACHAX Antimicrobial Agents and Chemotherapy. (Ann Arbor, MI) 1961-70. For publisher information, see AMACCQ

AACRAT Anesthesia and Analgesia; Current Research. (International Anesthesia Research Society, 3645 Warrensville Center Rd., Cleveland, OH 44122) V.36- 1957-

AAGAAW Antimicrobial Agents Annual. (New York, NY) 1960-60. For Publisher information, see AMACCQ

AAGEAA Arkhiv Anatomii, Gistologii i Embriologii. Archives of Anatomy, Histology and Embryology. (v/o Mezhdunarodnaya Kniga, Kuznetskii Most 18, Moscow G-200, U.S.S.R.) V.1- 1916-

AAHEAF Archives d'Anatomie, d'Histologie et d'Embryologie. (Editions Alsatia, 10 rue Bartholdi, 68001 Colmar, France) V.1- 1922-

AAJNDL AJNR, American Journal of Neuroradiology. (Williams and Wilkins, Dept. 5J, P.O. Box 1496, Baltimore, MD 21203) V.1-1980-

AAJRDX AJR, American Journal of Roentgenology. (Charles C. Thomas Publisher, 301-327 E. Lawrence Ave., Springfield, IL 62717) V.126- 1976-

AAMLAR Atti della Accademia Medica Lombarda. (The Academy, c/o Prof. W. Montorsi, Festa del Perdono 3, Milan, Italy) V.20-29, 1931-40; V.15- 1960-

AAMMAU Archives d'Anatomie Microscopique et de Morphologie Experimentale. (Masson Publishing USA, Inc. 14 E. 60th St., New York, NY 10022) V.36- 1947-

AANEAB Acta Anaesthesiologica Scandinavica. (Munksgaard, 35 Noerre Soegade, DK-1370, Copenhagen K, Denmark) V.1- 1957-

AANFAE Annales de l'Anesthesiologie Francaise. (Doin Editeurs, 8 Place de l'Odeon, F-75006 Paris, France) V.1- 1960-

AANIBO Acta Anaesthesiologica Italica. (Tipografia la Garangola, Via Montana 4, 35100 Padua, Italy) V.23- 1972-

AANLAW Atti della Accademia Nazionale dei Lincei, Rendiconti della Classe di Scienze Fisiche, Matematiche e Naturali. (Academia Nazionale dei Lincei, Ufficio Pubblicazioni, Via della Lungara, 10, I-00165 Rome, Italy) V.1- 1946-

AAOPAF AMA Archives of Ophthalmology. (Chicago, IL) V.44, No.4-63, 1950-60. For publisher information, see AROPAW

AAREAV Anesthesie, Analgesie, Reanimation. (Masson et Cie, Editeurs, 120 Blvd. Saint Germain, P-75280, Paris 06, France) V.14-1957-

AASXAP Acta Anaesthesiologica Scandinavica, Supplementum. (Munksgaard 35 Noerre Soegade, DK-1370, Copenhagen K, Denmark) No.1- 1959-

ABABAC Annales de Biologie Animale, Biochimie, Biophysique. (Versailles, France) V.1-19, 1961-79.

ABAHAU Acta Biologica Academiae Scientiarum Hungaricae. (Kultura, P.O. Box 149, H-1389 Budapest, Hungary) V.1-1950-

ABANAE Antibiotics Annual. (New York, NY) 1953-60. For publisher information, see AMACCQ

ABBIA4 Archives of Biochemistry and Biophysics. (Academic Press, 111 5th Ave., New York, NY 10003) V.31- 1951-

ABBPAP Acta Biochimica et Biophysica. (Kultura, PO Box 149, Budapest 62, Hungary) V.1- 1966-

ABCHA6 Agricultural and Biological Chemistry. (Maruzen Co. Ltd., P.O.Box 5050 Tokyo International, Tokyo 100-31, Japan) V.25-1961-

ABEMAV Annals of Biochemistry and Experimental Medicine. (Calcutta, India) V.1-23, 1941-63. For publisher information, see IJEBA6

ABHUE6 Acta Biologica Hungarica. (Kultura, POB 149, H-1389 Budapest, Hungary) V.34- 1983-

ABHYAE Abstracts on Hygiene. (Bureau of Hygiene and Tropical Diseases, Keppel St., London WC1E 7HT, England) V.1- 1926-

ABILAE Archives de Biologie. (Vaillant-Carmanne, SA rue Fond Saint-Servais, 14 B.P.22, B-4000 Liege, Belgium) V.1- 1880-

ABMGAJ Acta Biologica et Medica Germanica. (Berlin, Ger. Dem. Rep.) V.1-41, 1958-82. For publisher information, see BBIADT

ABMHAM Archives Belges de Medicine Social, Hygiene, Medicine du Trevail et Medicine Legale. (Publications Acta Medica Belgica, rue des Champs- Elysees 43, 1050 Brussels 5, Belgium) V.4- 1946-

ABMPAC Advances in Biological and Medical Physics. (Academic Press, 111 5th Ave., New York, NY 10003) V.1- 1948-

ABPAAG Acta Biologiae Experimentalis (Warsaw). (Warsaw, Poland) V.1-13, 1928-39; V.14-29, 1947-69.

ABPYBL Advances in Biochemical Psychopharmacology. (Raven Press, 1140 Avenue of the Americas, New York, NY 10036) V.1-1969

ABSYB2 Alfred Benzon Symposium. (Academic Press, 111 5th Ave., New York, NY 10003) V.1- 1968-

ABTTAP Arquivos de Biologia Tecnologia. (Instituto de Biologia e Pesquisas Tecnologicas, Caixa Postal, 939 Curitiba, Parana, Brazil) V.1-11, 1946-56; V.12- 1966-

ACAEAS Acta Anaesthesiologica. (Padua, Italy) V.1-22, 1950-71. For publisher information, see AANIBO

ACATA5 Acta Anatomica. (S. Karger AG, Arnold-Boecklin-St 25, CH-4000 Basel 11, Switzerland) V.1- 1945-

ACCBAT Acta Clinica Belgica. (Association des Societes Scientifiques Medicales Belges, rue des Champs-Elysees, 43, 1050 Brussels 5, Belgium) V.1- 1946-

ACEDAB Acta Endocrinologica, Supplementum. (Periodica, Skolegade 12E, 2500 Copenhagen-Valby, Denmark) No.1- 1948-

ACENA7 Acta Endocrinologica. (Periodica, Skolegade 12E, 2500 Copenhagen-Valby, Denmark) V.1- 1948-

ACGHD2 Annals of the American Conference of Governmental Industrial Hygienists. (American Conference of Governmental Industrial Hygienists, Inc., 6500 Glenway Ave., Bldg. D-5, Cincinnati, Ohio, 54211) V.1- 1981-

ACHAAH Acta Haematologica. (S. Karger AG, Arnold-Boecklin-St 25, CH-4000 Basel 11, Switzerland) V.1- 1948-

ACHCBO Acta Histochemica et Cytochemica. (Japan Society of Histochemistry and Cytochemistry, Dept. Anatomy, Faculty of Medicine, Kyoto University, Konoe-cho, Yoshida, Sakyo-ku, Kyoto, 606, Japan) V.1- 1968-

ACHSA3 Acta Chirurgica Scandinavica. (Almqvist and Wiksell, P.O. Box 62, 26 Gamla Brogatan, S-101 20 Stockholm, Sweden) V.52- 1919-

ACHTA6 Antibiotica et Chemotherapia. (Basel, Switzerland) V.1-16, 1954-70. For publisher information, see ANBCB3

ACIEAY Angewandte Chemie, International Edition in English. (Verlag Chemie GmbH, Postfach 1260/1280, D6940, Weinheim, Germany) V.1- 1962-

ACJOAZ American Chemical Journal. (Baltimore, MD) V.1-50, 1879-1913. For publisher information, see JACSAT

ACLRBL Annals of Clinical Research. (Finnish Medical Society, Duodecim Runeberginkatu 47A, SF-00260 Helsinki 26, Finland) V.1- 1969-

ACLSCP Annals of Clinical Laboratory Science. (1833 Delancey Pl., Philadelphia, PA 19103) V.1- 1971-

ACMDBI Acta Medica. (Instituto Politecnico Nacional, Escvele Superior de Medicina, Prolongacion de Diaz Miron y Plan de San Luis, Mexico) V.1- 1965-

ACNSAX Activitas Nervosa Superior. (Avicenum, Malostranske nam. 28, Prague 1, Czechoslovakia) V.1- 1959-

ACPAAN Acta Paediatrica. (Stockholm, Sweden) V.1-53, 1921-64. For publisher information, see APSVAM

ACPADQ Acta Pathologica, Microbiologica et Immunologica Scandinavica, Section A: Pathology. (Munksgaard, 35 Noerre Soegade, DK-1370, Copenhagen K, Denmark) V.90- 1982-

ACPMAP Actualites Pharmacologiques. (Paris, France) No.1-33, 1949-81. For publisher information, see JNPHAG

ACRAAX Acta Radiologica. (Stockholm, Sweden) V.1-58, 1921-62. For publisher information, see ACRDA8

ACRDA8 Acta Radiologica: Diagnosis. (Box 7449, S-10391 Stockholm, Sweden) NS.V.1- 1963-

ACRSAJ Advances in Cancer Research. (Academic Press, 111 5th Ave., New York, NY 10003) V.1- 1953-

ACRSDM Alcoholism: Clinical and Experimental Research. (Grune and Stratton, Inc., 111 5th Ave., New York, NY 10003) V.1- 1977-

ACSAA4 Acta Chemica Scandinavica. (Munksgaard, 35 Noerre Soegade, DK-1370, Copenhagen K, Denmark) V.1-27, 1947-73.

ACTRAQ Acta Tropica. (Schwabe and Co., Steintorstr., 13, CH-4010 Basel, Switzerland) V.1 1944-

ACTTDZ Acta Therapeutica. (Pl. de Bastogne 8, B13, 1080 Brussels, Belgium) V. 1- 1975-.

ACVEAV Acta Cientifica Venezolana. (Apartado del Este 5843, Caracas, Venezuela) V.1- 1950-

ACVIA9 Acta Vitaminologica. (Milan, Italy) V.1-20, 1947-66. For publisher information, see AVEZA6

ACVTA8 Acta Veterinaria. (Jugoslovenska Knjiga, Terazije 27/II, Belgrade, Yugoslavia) V.1- 1951-

ACYTAN Acta Cytologica. (Williams and Wilkins Co., 428 E. Preston St., Baltimore, MD 21202) V.1- 1957-

ADBBBW Advances in Behavioral Biology. (Plenum Publishing Corp., 233 Spring Spring St., New York, NY 10013) V.1- 1971-

ADCHAK Archives of Disease in Childhood. (British Medical Journal, 1172 Commonwealth Avenue, Boston, MA 02134) V.1- 1926-

ADCMAZ Advances in Chemotherapy. (New York, NY) V.1-3, 1964-68. For publisher information, see AVPCAQ

ADCSAJ Advances in Chemistry Series. (American Chemical Society, 1155 16th St., N.W., Washington, DC 20036) No.1- 1950-

ADENAE Advances in Enzymology and Related Subjects of Biochemistry. (New York, NY) V.1-28, 1941-66. For publisher information, see AERAAD

ADIRDF Advances in Inflammation Research. (Raven Press, 1140 Avenue of the Americas, New York, NY 10036) V.1- 1979-

ADMFAU Archiv fuer Dermatologische Forschung. (Secaucus, NJ) V.240-252, No.4, 1971-75. For publisher information, see ADREDL

ADRCAC Annali Italiani di Dermatologia Clinica e Sperimentale. (Clinica Dermosifilopatica Universita degli Studi, Policlinica Monteluce, 06100 Perugia, Italy) V.16- 1962-

ADREDL Archives of Dermatological Research. (Springer-Verlag New York, Inc., Service Center, 44 Hartz Way, Secaucus, NJ 07094) V.253- 1975-

ADRPBI Advances in Reproductive Physiology. (Academic Press, 111 Fifth Ave., New York, NY 10003) V.1-6, 1966-73, Discontinued

ADSRDV Advances in Shock Research. (Alan R. Liss, Inc., 150 Fifth Ave., New York, NY 10011) V.1- 1979-

ADSYAF Archives of Dermatology and Syphilology. (Chicago, IL) V.1-62, 1920-60. For publisher information, see ARDEAC

ADTEAS Advances in Teratology. (Academic Press, 111 5th Ave., New York, NY 10003) V.1-5, 1966-72, Discontinued

ADVEA4 Acta Dermato-Venereologica. (Almqvist and Wiksell International, Box 62, S-101 20 Stockholm, Sweden) V.1- 1920-

ADVED7 Annales de Dermatologie et de Venereologie. (Masson Publishing USA Inc., 133 E. 58th St., New York, NY 10022) V.104- 1977-

ADVPA3 Advances in Pharmacology. (New York, NY) V.1-6, 1962-68. For publisher information, see AVPCAQ

ADVPB4 Advances in Planned Parenthood. (Excerpta Medica, Inc., POB 3085 Princeton, New Jersey 08540) V.1- 1967-

ADWMAX Abhandlungen der Deutschen Akademie der Wissenschaften zu Berlin, Klasse fuer Medizin. (Akademie-Verlag GmbH, Leipziger Str. 3-4, DDR-108 Berlin, German Democratic Republic) 1950-

AECTCV Archives of Environmental Contamination and Toxicology. (Springer-Verlag New York, Inc., Service Center, 44 Hartz Way, Secaucus, NJ 07094) V.1- 1973-

AEEDDS Adverse Effects of Environmental Chemicals and Psychotropic Drugs. (Elsevier North Holland, Inc., 52 Vanderbilt Ave., New York, NY 10017) V.1- 1973-

AEEXAH Acta Embryologiae Experimentalis. (Via Archirafi 18, 90123 Palermo, Italy) No. 1- 1969-

AEFTAA Acta Europaea Fertilitatis. (Piccin Medical Books, Via Brunacci, 12, 35100 Padua, Italy) V.1 1969-

AEHA** U. S. Army, Environmental Hygiene Agency Reports. (Edgewood Arsenal, MD 21010)

AEHLAU Archives of Environmental Health. (Heldreff Publications, 4000 Albemarle St., N.W., Washington, D.C. 20016) V.1- 1960-

AEMBAP Advances in Experimental Medicine and Biology. (Plenum Publishing Corp., 233 Spring St., New York, NY 10013) V.1- 1967-

AEMED3 Annals of Emergency Medicine. (American College of Emergency Physicians, 1125 Executive Circle, Irving, TX 75038)

AEMIDF Applied and Environmental Microbiology. (American Society for Microbiology, 1913 I St., N.W., Washington, DC 20006) V.31- 1976-

AEMMBP Annales d'Embryologie et de Morphogenese. (Paris, France) V.1-7, 1968-74. Discontinued

AENBAX Zeitschrift fuer Naturforschung, Teil B: Anorganische Chemie, Organische Chemie, Biochemie, Biophysik, Biologie.

AEPPAE Naunyn-Schmiedeberg's Archiv fuer Experimentelle Pathologie und Pharmakologie. (Berlin, Germany) V.110-253, 1925-66. For publisher information, see NSAPCC

AERAAD Advances in Enzymology and Related Areas of Molecular Biology. (Wiley, 605 3rd Ave., New York, NY 10016) V.29- 1967-

AESAAI Annals of the Entomological Society of America. (Entomological Society of America, 4603 Calvert Road, College Park, MD 20740) V.1- 1908-

AETODY Advances in Modern Environmental Toxicology. (Senate Press, Inc., P.O. Box 252, Princeton Junction, NJ 08550) V.1- 1980-

AEXPBL Archiv fuer Experimentelle Pathologie und Pharmakologie. (Leipzig, Germany) V.1-109, 1873-1925. For publisher information, see NSAPCC

AEZRA2 Advances in Enzyme Regulation. (Pergamon Press, Headington Hill Hall, Oxford OX3 OBW, England) V.1- 1963-

AFCPDR Annual Symposium on Fundamental Cancer Research, Proceedings. (University of Texas System Cancer Center, M.D. Anderson Hospital and Tumor Institute, Houston, TX 77030) V.30- 1978-

AFDOAQ Association of Food and Drug Officials of the United States, Quarterly Bulletin. (Editorial Committee of the Association, P.O. Box 20306, Denver, CO 80220) V.1- 1937-

AFECAT Annales des Falsifications et de l'Expertise Chimique. (Paris, France) V.53, No.613-No.779, 1960-79.

AFPBAD Acta Facultatis Pharmaceutica Brunensis Bratislavensis. (Brno, Czech.) V.1-3, 1958-60. For publisher information, see AFPCAG

AFPCAG Acta Facultatis Pharmaceuticae, Universitatis Comenianae. (Ustredna Kniznica Farmaceuticka Fakulta Univerzity Komenskeho, Kolinciakova 8, 886 34 Bratislava, Czechoslovakia) V.14- 1967-

AFPEAM Archives Francaises de Pediatrie. (Doin Editeurs, 8 Place de l'Odeon, F-75006 Paris, France) V.1- 1942-

AFPYAE American Family Physician. (American Academy of Family Physicians, 1740 W. 92nd Street, Kansas City, MO 64114) V.2-17, 1962-1969: New series: V.2- 1970-

AFQSAZ Anais de Farmacia e Quimica de Sao Paulo. (Sociedade de Farmacia e Quimica de Sao Paulo, Avenida Brigadeiro Luis Antonio, 393, 7.o, Sao Paulo, Brazil) V.6- 1953-

AFREAW Advances in Food Research. (Academic Press, 111 5th Ave., New York, NY 10003) V.1- 1948-

AFRNDS Acta Facultatis Rerum Naturalium Universitatis Comenianae, Formatio et Protectio Naturae. (Slovenske Pedagogicke Nakladatel'stvo Sasinkova 5, 89112 Bratislava, Czechoslovakia) No.1- 1976-

AFSPA2 Archivo di Farmacologia Sperimentale e Scienze Affini. (Rome, Italy) V.1-82, 1902-1954. Discontinued.

AFTOD7 Archivos de Farmacologia y Toxicologia. (Universidad Complutense, Facultad de Medicina, Departamento Coordinado de Farmacologia, Ciudad Universitaria, Madrid 3, Spain) V.1- 1975-

AGACBH Agents and Actions, A Swiss Journal of Pharmacology. (Birkhaeuser Verlag, P.O. Box 34, Elisabethenst 19, CH-4010, Basel, Switzerland) V.1- 1969/70-

AGDJAI Journal of the Academy of General Dentistry. (Academy of General Denistry, 211 E. Chicago Ave., Suite 1200, Chicago, IL 60611) V.1- 1952(?)-

AGEFAB Archiv fuer Gefluegelkunde. (Verlag Eugen Ulmer, Postfach 700561, D-7000 Stuttgart 70, Fed. Rep Ger.) V.1- 1927-

AGGHAR Archiv fuer Gewerbepathologie und Gewerbehygiene. (Berlin, Germany) V.1-18, 1930-61. For publisher information, see IAEHDW

AGMGAK Acta Geneticae Medicae et Gemellologiae. (Cappelli Editore, Via Marsili 9, I-40124 Bologna, Italy) V.1- 1952-

AGNKA3 Archiv fuer Genetik. (Orell Fuessli Graphische Betriebe AG, Postfach A8036, CH-8022 Zurich, Switzerland) V.45- 1972-

AGNYDE Aging. (Raven Press, 1140 Avenue of the Americas, New York, NY 10036) V.1- 1975-

AGPSA3 AMA Archives of General Psychiatry. (Chicago, IL) V.1-2, 1959-60. For publisher information, see ARGPAQ

AGSOA6 Agressologie. Revue Internationale de Physio-Biologie et de Pharmacologie Appliquees aux Effets de l'Agression. (Masson et Cie, Editeurs, 120 Blvd. Saint-Germain, P-75280, Paris 06, France) V.1- 1960-

AGTQAH Annales de Genetique. (Expansion Scientifique Francaise, 5 rue Saint- Benoit, F-75278, Paris 06, France) V.1- 1958-

AHBAAM Archiv fuer Hygiene und Bakteriologie. (Munich, Germany) V.101-154, 1929-71. For publisher information, see ZHPMAT

AHEMA5 Anatomia, Histologia, Embryologia. (Verlag Paul Parey, Linderstr. 44-47, D-1000 Berlin 61, Fed. Rep. Ger.) V.2- 1973-

AHISA9 Acta Histochemica. Zeitschrift fuer Histologische Topochemie. (VEB Gustav Fischer Verlag, Postfach 176, Villengang 2, DDR-69, Jena, Germany) V.1- 1954-

AHJOA2 American Heart Journal. (C.V. Mosby Co., 11830 Westline Industrial Dr., St. Louis, MO 63141) V.1- 1925-

AHRAAY Arhiv za Higijenu Rada. Archives of Industrial Hygiene. (Zagreb, Yugoslavia) V.1-6, 1950-55. For publisher information, see AHRTAN

AHRTAN Arhiv za Higijenu Rada i Toksikologiju. Archives of Industrial Hygiene and Toxicology. (Jugoslovenska Knjiga, P.O. Box 36, Terazije 27, 11001, Belgrade, Yugoslavia) V.7- 1956-

AHYGAJ Archiv fuer Hygiene. (Munich, Germany) V.1-100, 1883-1928. For publisher information, see ZHPMAT

AICCA6 Acta Unio Internationalis Contra Cancrum. (Louvain, Belgium) V.1-20, 1936-64. For publisher information, see IJCNAW

AICMA2 Archivos del Instituto de Cardiologia de Mexico. (Instituto de Cardiologia de Mexico, Juan Badiano No.1, Mexico City 22, Mexico) V.14- 1944-

AIDZAC Aichi Ika Daigaku Igakkai Zasshi. Journal of the Aichi Medical Univ. Assoc. (Aichi Ika Daigaku, Yazako, Nagakutemachi, Aichi-gun, Aichi- Ken 480-11, Japan) V.1- 1973-

AIHAAP American Industrial Hygiene Association Journal. (AIHA, 475 Wolf Ledges Pkwy., Akron, OH 44311) V.19- 1958-

AIHAM* Annual Meeting of American Industrial Hygiene Association. For publisher information, see AIHAAP

AIHOAX Archives of Industrial Hygiene and Occupational Medicine. (Chicago, IL) V.1-2, No.3, 1950. For publisher information, see AEHLAU

AIHQA5 American Industrial Hygiene Association Quarterly. (Baltimore, MD) V.7-18, 1946-57. For publisher information, see AIHAAP

AIMDAP Archives of Internal Medicine. (American Medical Association, 535 N. Dearborn St., Chicago, IL 60610) V.1- 1908-

AIMEAS Annals of Internal Medicine. (American College of Physicians, 4200 Pine St., Philadelphia, PA 19104) V.1- 1927-

AIMJA9 Ain Shams Medical Journal (Ain Shams Clinical and Scientific Society, c/o Ain Shams University Hospital, Abbassia, Cairo, Egypt) V.1- 1950-

AIPAAV Annales de l'Institut Pasteur. (Paris, France) V.1-123, 1887-1972. For publisher information, see ANMBCM

AIPBAY Archives Internationales de Physiologie et de Biochimie. (Madame Duchateau-Bosson, Secretaire de la Redaction, 17, Pl. Delcour, Liege 4020, Belgium) V.63- 1955-

AIPHAI Archives Internationales de Physiologie. (Liege, Belgium) V.1-62, 1904-54. For publisher information, see AIPBAY

AIPSAH Archives de l'Institut Pasteur d'Algerie. (Bibliotheque, Institut Pasteur d'Algerie, rue du Docteur-Laveran, Algiers, Algeria) V.1- 1923-

AIPTAK Archives Internationales de Pharmacodynamie et de Therapie. (Editeurs, Institut Heymans de Pharmacologie, De Pintelaan 135, B-9000 Ghent, Belgium) V.4- 1898-

AIPUAN Archivio Italiano di Patologia e Clinica dei Tumori. (Istituto di Farmacologia della Universita, Via Vanvitelli 32, 20129 Milan, Italy) V.1-15, 1957-72, Discontinued

AISFAR Archivio Italiano di Scienze Farmacologiche. (Modena, Italy) 1932-65. Discontinued.

AISMAE Archivio Italiano di Scienze Mediche Tropical e di Parassitologia. Italian Archives of the Science of Tropical Medicine and Parasitology. (Rome, Italy) V.1-54, 1920-73. Suspended.

AISSAW Annali dell'Istituto Superiore di Sanita. (Istituto Poligrafico

dello Stato, Libreria dello Stato, Piazza Verdi, 10 Rome, Italy) V.1-
1965-

AITDAQ Archiwum Immunologii i Terapii Doswiadczalnej. (Warsaw, Poland) V.1-9, 1953-61. For publisher information, see AITEAT

AITEAT Archivum Immunologiae et Therapiae Experimentalis. (Ars
Polona-RUCH, P.O. Box 1001, P-00 068 Warsaw, 1, Poland) V.10-
1962-

AITRD3 Annual Report of the Inhalation Toxicology Research Institute. (NTIS, 5285 Port Royal Rd., Springfield, VA 22161)
1972/73-

AIVMBU Archivos de Investigacion Medica. (POB 12,631, Col. Nar-
varte, Deleg. Benito Juarez, 03020 Mexico City, DF, Mex.) V.1-
1970-

AJANA2 American Journal of Anatomy. (Alan R. Liss Inc., 150 5th
Ave., New York, NY 10011) V.1- 1901-

AJBSAM Australian Journal of Biological Sciences. (Commonwealth
Scientific and Industrial Research Organization, POB 89, E. Melbourne, Vic 3002, Australia) V.6- 1953-

AJCAA7 American Journal of Cancer. (New York, NY) V.15-40,
1931-40. For publisher information, see CNREA8

AJCDAG American Journal of Cardiology. (Technical Publishing
Co., 666 5th Ave., New York, NY 10103) V.1- 1958-

AJCHAS Australian Journal of Chemistry. (Commonwealth Scientific and Industrial Research Organization, 314 Albert St., P.O. Box
89, E. Melbourne, Victoria, Australia) V.6- 1953-

AJCNAC American Journal of Clinical Nutrition. (American Society
for Clinical Nutrition, Inc., 9650 Rockville Pike, Bethesda, MD
20814) V.2- 1954-

AJCPAI American Journal of Clinical Pathology. (J.B. Lippincott
Co., (Keystone Industrial Park, Scanton, PA 18512) V.1- 1931-

AJDCAI American Journal of Diseases of Children. (American Medical Assoc., 535 N. Dearborn St., Chicago, IL 60610) V.1-80(3),
1911-50; V.100- 1960-

AJDCBJ AMA American Journal of Diseases of Children. (Chicago,
IL) V.80,No.4-V.90, 1950-55. For publisher information, see
AJDCAI

AJDDAL American Journal of Digestive Diseases. (Plenum Publishing Corp., 233 Spring St., New York, NY 10013) V.5-22, 1938-55.,
V.1-23, 1956-78

AJDEBP Australasian Journal of Dermatology. (Australasian College of Dermatologists, 271 Bridge Rd., Glebe, NSW 2037, Australia) V.9- 1967-

AJDKA8 Azabu Juika Daigaku Kenkyu Hokoku. Bulletin of the
Azabu Veterinary College. (Azabu Juika Daigaku, 1-17-71,
Fuchinobe, Sagamihara, Kanagawa-Ken 229, Japan) No.1-31, 1954-
76; V.1-1976-

AJDNAH American Journal of Digestive Diseases and Nutrition.
(New York, NY) V.1-4, 1934-38. For publisher information, see
AJDDAL

AJDRAT Australian Journal of Dermatology. (Sydney, Australia)
V.1-8, 1951-66. For publisher information, see AJDEBP

AJEBAK Australian Journal of Experimental Biology and Medical
Science. (Univ. of Adelaide Registrar, Adelaide, S.A. 5000, Australia) V.1- 1924-

AJEPAS American Journal of Epidemiology. (Johns Hopkins Univ.,
550 N. Broadway, Suite 201, Baltimore, MD 21205) V.81- 1965-

AJGAAR American Journal of Gastroenterology. (American College
of Gastroenterology, Inc., 299 Broadway, New York, NY 10007)
V.21-1954-

AJHEAA American Journal of Public Health. (American Public
Health Assoc., Inc., 1015 15th St., N.W., Washington, D.C. 20005)
V.2-17, 1912-27; V.61- 1971-

AJHGAG American Journal of Human Genetics. (University of Chicago Press, 5801 S. Ellis Ave., Chicago, IL 60637) V.1- 1949-

AJHPA9 American Journal of Hospital Pharmacy. (American Society of Hospital Pharmacists, 4630 Montgomery Ave., Washington,
DC 20014) V.15- 1958-

AJHYA2 American Journal of Hygiene. (Baltimore, MD) V.1-80,
1921-64. For publisher information, see AJEPAS

AJIMD8 American Journal of Industrial Medicine. (Alan R. Liss,
Inc., 150 Fifth Ave., New York, NY 10011) V.1- 1980-

AJINO* Ajinomoto Co., Inc. (9 W. 57th St., Suite 4625, New York,
NY 10019)

AJKDDP American Journal of Kidney Diseases. (Grune & Stratton,
Inc., Journal Subscription Department, POB 6280, Duluth, MN
55806) V.1- 1981-

AJMEAZ American Journal of Medicine. (Yorke Medical Group, 666
5th Ave., New York, NY 10103) V.1- 1946-

AJMSA9 American Journal of the Medical Sciences. (Charles B.
Slack, Inc., 6900 Grove Rd., Thorofare, NJ 08086) V.1- 1841-

AJNED9 American Journal of Nephrology. (S. Karger Pub., Inc., 79
Fifth Ave., New Uork, NY 10003) V.1- 1981-

AJOGAH American Journal of Obstetrics and Gynecology. (C.V.
Mosby Co., 11830 Westline Industrial Dr., St. Louis, MO 63141)
V.1- 1920-

AJOMAZ Alabama Journal of Medical Sciences. (University of Alabama Medical Center, Univ. Station, Birmingham, AL 35294) V.1-
1964-

AJOPAA American Journal of Ophthalmology. (Ophthalmic Publishing Co., 435 N. Michigan Ave., Chicago, Il 60611) V.1- 1918-

AJPAA4 American Journal of Pathology. (Harper and Row, Medical
Dept., 2350 Virginia Ave., Hagerstown, MD 21740) V.1- 1925-

AJPEAG American Journal of Public Health and the Nation's
Health. (New York, NY) V.18-60, 1928-60. For publisher information, see AJHEAA

AJPHAP American Journal of Physiology. (American Physiological
Society, 9650 Rockville Pike, Bethesda, MD 20814) V.1- 1898-

AJPPAF American Journal of Physiological Optics. (Southbridge,
MA) V.1- 1920-

AJPRAL American Journal of Pharmacy and the Sciences Support-
ing Public Health. (Philadelphia College of Pharmacy and Science,
43rd St., Woodland Ave. and Kingsessing Mall, Philadelphia, PA
19104) V.109-150, 1937-78

AJPSAO American Journal of Psychiatry. (American Psychiatric Association, Publications Services Div., 1700 18th St., N.W., Washington, DC 20009) V.78- 1921-

AJRRAV American Journal of Roentgenology, Radium Therapy and
Nuclear Medicine. (Springfield, IL) V.67-125, 1952-75. For publisher information, see AAJRDX

AJSGA3 American Journal of Syphilis, Gonorrhea and Venereal Diseases. (St. Louis, MO) V.20-38, 1936-54. Discontinued

AJSNAO American Journal of Syphilis and Neurology. (St. Louis,
MO) V.1-19, 1917-35. For publisher information, see AJSGA3

AJTHAB American Journal of Tropical Medicine and Hygiene.
(Allen Press, 1041 New Hampshire St., Lawrence, KS 66044) V.1-
1952-

AJTMAQ American Journal of Tropical Medicine. (Baltimore, MD)
V.1-31, 1921-51. For publisher information, see AJTHAB

AJVRAH American Journal of Veterinary Research. (American Veterinary Medical Association, 930 N. Meacham Road, Schaumburg,
IL 60196) V.1- 1940-

AKBNAE Arkhiv Biologicheskikh Nauk. Archives of Biological Sciences. (Moscow, U.S.S.R.) V.1-64, 1892-1941. Discontinued

AKEDAX Archiv fuer Klinische und Experimentelle Dermatologie.
(Berlin, Germany) V.201-239, 1955-71. For publisher information,
see ADMFAU

AKGIAO Akushcherstvo i Ginekologiya. (v/o Mezhdunarodnaya
Kniga, Kuznetskii Most 18, Moscow G-200, USSR) No. 1- 1936-

AKMEA8 Archiv fuer Klinische Medizin. (Berlin, Germany) V.212-
216, 1966-69. For publisher information, see EJCIB8

ALACBI Aldrichimica Acta. (Aldrich Chemical Co., Inc., 940 W. St.
Paul Ave., Milwaukee, WI 53233) V.1- 1968-

ALBRW* Albright and Wilson Inc., P.O. Box 26229, Richmond, VA
23260-6229

ALEPA8 Acta Leprol. (Publisher unknown) V.1- 1920(?)-

ALJMAO Antonie van Leeuwenhoek. Journal of Microbiology and Serology. (H. Veenman en Zonen BV, PO Box 7, 6700 AA Wageninen, Netherlands) V.1- 1934-

ALLVAR Alimentation et la Vie. (Societe Scientifique d'Hygiene Alimentaire, 16, rue de l'Estrapade, 75005 Paris, France) V.39- 1951-

AMACCQ Antimicrobial Agents and Chemotherapy. (American Society for Microbiology, 1913 I St., N.W., Washington, DC 20006) V.1- 1972-

AMADBS Archives Francaises des Maladies de l'Appareil Digestif. (Paris, France) V.55-65, 1966-76, For publisher information, see GCBIDC

AMAHA5 Acta Microbiologica Academiae Scientiarum Hungaricae. (Akademiai Kiado, P.O. Box 24, Budapest 502, Hungary) V.1- 1954-

AMASA4 Acta Medica Academiae Scientiarum Hungaricae. (Kultura, POB 149 H-1389, Budapest, Hungary) V.1- 1950-

AMBBBR Acta Microbiologica Polonica, Series B: Microbiologia Applicata. (Warsaw, Poland) V.1-7, No. 4, 1969-75, For publisher information, see AMPOAX

AMBED4 AMI-Berichte. (Berlin, Fed. Rep. Ger.) No. 1-2, 1978-82

AMBNAS Acta Medica et Biologica. (Niigata University School of Medicine, 1 Asahi-machi-dori, Niigata 951, Japan) V.1- 1953-

AMBOCX Ambio. A Journal of the Human Environment, Research and Management. (Pergamon Press, Inc., Maxwell House, Fairview Park, Elmsford, NY 10523) V.1- 1972-

AMBPBZ Acta Pathologica et Microbiologica Scandinavica, Section A: Pathology. (Copenhagen K, Denmark) V.78-89, 1970-81. For publisher information, see ACPADQ

AMBUCH Acta Medica Bulgarica. (Durzhavno Izdatelstvo Meditsina i Fizkultura, Pl. Slaveikov 11, Sofia, Bulgaria) V.1- 1973-

AMCTAH Antibiotic Medicine and Clinical Therapy. (New York, NY) V.3-8, 1956-61. For publisher information, see CLMEA3

AMCYC* Toxicological Information on Cyanamid Insecticides. (American Cyanamid Co., Agricultural Division, Princeton, NJ 08540)

AMDCA5 AMA Journal of Diseases of Children. (Chicago, IL) V.91-99, 1956-60. For publisher information, see AJDCAI

AMDEAB AMA Archives of Dermatology. (Chicago, IL) V.71-81, 1955-60. For publisher information, see ARDEAC

AMDNA4 Annali di Medicina Navale. (Lungo Tevere delle Navi, 00100 Rome, Italy) V.66- 1961-

AMESES Abstracts of the Annual Meeting, Environmental Mutagen Society. (Environmental Mutagen Society, 1340 Old Chain Bridge Road, Suite 300, McLean, Va 22101.

AMEBA7 Annales Medicinae Experimentalis et Biologiae Fenniae. (Mikonkatu 8, Helsinki, Finland) V.25- 1947-

AMICCW Archives of Microbiology. (Springer-Verlag New York, Inc., Service Center, 44 Hartz Way, Secaucus, NJ 07094) V.95- 1974-

AMIGB9 Acta Microbiologica Polonica, Series A: Microbiologica Generalis. (Warsaw, Poland) V.1-7, No.4, 1969-75, For publisher information, see AMPOAX

AMIHAB AMA Archives of Industrial Health. (Chicago, IL) V.11-21, 1955-60. For publisher information, see AEHLAU

AMIHBC AMA Archives of Industrial Hygiene and Occupational Medicine. (Chicago, IL) V.2-10, 1950-54. For publisher information, see AEHLAU

AMILAN Annales de Medecine Legale. Criminologie, Police Scientifique et Toxicologie. (Paris, France) V.1, 1920-51; V.39-47, 1958-67. For publisher information, see MLDCAS

AMIUAG Acta Medica Iugoslavica. (Akademija Zbora Lijecnika Hrvatske, Subiceva 9/1, Zagreb, Yugoslavia) V.1- 1947-

AMJPA6 American Journal of Pharmacy (1835-1936). (Philadelphia, PA) V.1-108, 1835-1936. For publisher information, see AJPRAL

AMLSAP Acta Medicinae Legalis et Socialis. (Liege, Belgium) V.1-21, 1948-68(?). Discontinued.

AMLTAS Annales de Medecine Legale et de Criminologie. (Paris, France) V. 31-8, 1951-8. For publisher information, see MLDCAS

AMNIB6 Acta Manilana, Series A: Natural and Applied Sciences. (University of Santo Tomas Research Center, Manila, D-403, Philippines) V.4- 1968-

AMNTA4 American Naturalist. (University of Chicago Press, 5801 S. Ellis Ave., Chicago, IL 60637) V.1- 1867-

AMNUA7 AMA Archives of Neurology and Psychiatry. (Chicago, IL) V.64-81, 1950-59. For publisher information, see ARNEAS

AMOKAG Acta Medicia Okayama. (Okayama University Medical School, 2-5-1 Shikata-cho, Okayama 700, Japan) V.8- 1952-

AMONDS Applied Methods in Oncology. (Elsevier North Holland, Inc., 52 Vanderbilt Ave., New York, NY 10017) V.1- 1978-

AMPIAF Acta Medica Philippina. (547 Herran St., Manila D-406, Philippines) V.1-19, 1939-63; Series 2: V.1- 1964-

AMPLAO AMA Archives of Pathology. (American Medical Association, 535 N. Dearborn St., Chicago, IL 60610) V.50,No.4-V.69, 1950-60.

AMPMAR Archives des Maladies Professionnelles de Medecine du Travail et de Securite Sociale. (Masson et Cie, Editeurs, 120 Blvd. Saint-Germain, P-75280, Paris 06, France) V.7- 1946-

AMPOAX Acta Microbiologica Polonica. (Ars Polona-RUCH, P.O. Box 1001 P-00 068 Warsaw, 1, Poland) V.1-17, No. 4, 1952-68; V.25 1976-

AMPYAT Annales Medico-Psychologiques. (Masson et Cie, Editeurs, 120 Blvd. Saint-Germain, P-75280, Paris 06, France) V.80- 1922-

AMRL** Aerospace Medical Research Laboratory Report. (Aerospace Technical Div., Air Force Systems Command, Wright-Patterson Air Force Base, OH 45433)

AMSGAO AMA Archives of Surgery. (Chicago, IL) V.61-80, 1950-60. For publisher information, see ARSUAX

AMSHAR Acta Morphologica Academiae Scientiarum Hungaricae. (Akademiai Kiado, P.O. Box 24, H-1389 Budapest 502, Hungary) V.1- 1951-

AMSSAQ Acta Medica Scandinavica, Supplement. (Almqvist and Wiksell, P.O. Box 62, 26 Gamla Brogatan, S-101, 20 Stockholm, Sweden) No.1- 1921-

AMSVAZ Acta Medica Scandinavica. (Almqvist and Wiksell, P.O. Box 62, 26 Gamla Brogatan, S-101, 20 Stockholm, Sweden) V.52- 1919-

AMTODM Advances in Modern Toxicology. (John Wiley and Sons, 605 3rd Ave., New York, NY 10016) V.1, No.1-2, 1976-79. Discontinued

AMTUA3 Acta Medica Turcica. (Dr. Ayhan Okcuoglu, Cocuk Hastalikari Klinig i, c/o Ankara University Tip Facultesi, PK 48, Cebeci, Ankara, Turkey) V.1- 1964-

AMUK** Acta Medica University Kyoto. (Kyoto, Japan)

AMZOAF American Zoologist. (American Society of Zoologists, Box 2739, California Lutheran College, Thousand Oaks, California 91360) V.1- 1961-

ANAEA3 Annals of Allergy. (American College of Allergists, Box 20671, Bloomington, MN 55420) V.1- 1943-

ANAIAF Annales de la Nutrition et de l'Alimentation. (Centre National de la Recherche Scientifique, 15 Quai Anatole-France, F-75700, Paris, France) V.1- 1941-

ANANAU Anatomischer Anzeiger. (VEB Gustav Fischer Verlag, Postfach 176, DDR-69, Jena, E. Germany) V.1 1886-

ANASAB Anaesthesia. (Blackwell Scientific, Osney Mead, Oxford OX2 OEL, England) V.1- 1946-

ANATAE Anaesthesist. (Springer-Verlag, Heidelberger Pl. 3, D-1000 Berlin 33, Federal Republic of Germany) V.1- 1952-

ANBCA2 Analytical Biochemistry. (Academic Press, 111 5th Ave., New York, NY 10003) V.1- 1960-

ANBCB3 Antibiotics and Chemotherapy. (S. Karger, AG, Arnold-Boecklin-St 25, Postfach CH-4009 Basel, Switzerland) V.17- 1971-

ANCHAM Analytical Chemistry. (American Chemical Society, 1155 16th St., N.W., Washington, DC 20036) V.19- 1947-

ANDRDQ Andrologia. (Grosse Verlag, Kurfuerstendamm 152, D-1000 Berlin 31, Federal Republic of Germany) V.6- 1974-

ANENAG Annales d'Endocrinologie. (Masson et Cie, Editeurs, 120 Blvd. Saint-Germain, P-75280, Paris 06, France) V.1- 1939-

ANESAV Anesthesiology. (J.B. Lippincott Co., Keystone Industrial Park, Scranton, PA 18512) V.1- 1940-

ANGIAB Angiology. (Williams and Wilkins Co., 428 E. Preston St., Baltimore, MD 21202) V.1- 1950-

ANIFAC Annales de Chirurgie Infantile. (Paris, France) V.1- 1960(?)-

ANJOAA Astronomical Journal. (American Institute of Physics, 335 E. 45th St., New York, NY 10017) V.1- 1849-

ANMBCM Annales de Microbiologie (Paris). (Masson et Cie, Editeurs, 120 Blvd. Saint-Germain, P-75280, Paris 06, France) V.124- 1973-

ANOBAU Bulletin de l'Association des Anatomistes. (U.E.R. Sciences Medicales A., Boite Postale 1080, 54019 Nancy Cedex, France)

ANOPB5 Annals of Ophthalmology. (American Society of Contemporary Ophthalmology, 211 E. Chicago Ave., Suite 1044, Chicago, IL 60611) V.1- 1969-

ANPBAZ Acta Neurologia et Psychiatrica Belgica. (Brussels, Belgium) V.48-69, 1948-69. For publisher information, see ANUBBR

ANPIAM Arquivos de Neuro-Psiquiatria. (Caixa Postal 30657, 01000 Sao Paulo, Brazil) V.1- 1943-

ANPSAI Archives of Neurology and Psychiatry. (Chicago, IL) V.1-64, 1919-50. For publisher information, see ARNEAS

ANPTAL Acta Neuropathologica. (Springer-Verlag New York, Inc., Service Center, 44 Hartz Way, Secaucus, NJ 07094) V.1- 1961-

ANREAK Anatomical Record. (Alan R. Liss, Inc., 150 5th Ave., New York, NY 10011) V.1- 1906/08-

ANRSAS Acta Neurologica Scandinavica. (Munksgaard, 35 Noerre Soegade, DK-1370, Copenhagen K, Denmark) V.37- 1961-

ANSUA5 Annals of Surgery. (J.B. Lippincott Co., Keystone Industrial Park, Scranton, PA 18512) V.1- 1885-

ANTBAL Antibiotiki. (Moscow, USSR) V.1-29, 1956-84. For publisher information, see AMBIEH

ANTCAO Antibiotics and Chemotherapy. (Washington, DC) V.1-12, 1951-62. For publisher information, see CLMEA3

ANTRD4 Anticancer Research. (Anticancer Research, 5 Argyropoulou St., Kato Patissia, Athens 907, Greece) V.1- 1981-

ANUBBR Acta Neurologica Belgica. (Association des Societiés Scientifiques, Medicales Belges, rue des Champs-Elysees 43, B-1050 Brussels, Belgium) V.1- 1900-

ANYAA9 Annals of the New York Academy of Sciences. (The Academy, Exec. Director, 2 E. 63rd St., New York, NY 10021) V.1- 1877-

ANZJA7 Australian and New Zealand Journal of Surgery. (Blackwell Scientific Publications, College of Surgeons' Gardens, 99 Barry St., Carlton 3053, Australia) V.1- 1931-

ANZJB8 Australian and New Zealand Journal of Medicine. (Modern Medicine of Australia Pty., Ltd., 100 Pacific Highway, North Sydney, 2060, Australia) V.1- 1971-

AOBIAR Archives of Oral Biology. (Pergamon Press, Headington Hill Hall, Oxford OX3 OBW, England) V.1- 1959-

AOGLAR Acta Obstetrica et Gynaecologica Japonica, English Edition. (Tokyo, Japan) V.16-23, 1969-76. For publisher information, See NISFAY

AOGMAU Annali di Ostetricia, Ginecologia, Medicina Perinatale. (Via Commenda, 12, 20122 Milan, Italy) V.93- 1972-

AOGNAX Archivio di Ostetricia e Ginecologia. (C.C. Postale N 6/19773, Naples, Italy) V.1- 1937-

AOGSAE Acta Obstetricia et Gynecologica Scandinavica. (Kvinnokliniken, Lasarettet, Lund, Sweden) V.1- 1921-

AOHYA3 Annals of Occupational Hygiene. (Pergamon Press, Headington Hill Hall, Oxford OX3 OBW, England) V.1- 1958-

AOISDR Annual Report of Osaka City Institute of Public Health and Environmental Sciences. (Osaka-shiritsu Kankyo Kagaku Kenkyusho, 8-34 Tojo-cho, Tennoji-ku, Osaka 543, Japan) No.43-1980-

AORLCG Archives of Oto-Rhino-Laryngology. (Springer-Verlag New York, Inc., Service Center, 44 Hartz Way, Secaucus, NJ 07094) V.206- 1973-

AOSAAK Acta Orthopaedica Scandinavica. (Munksgaard, 35 Noerre Soegade, DK-1370, Copenhagen K, Denmark) V.1- 1930-

AOUNAZ Archiv fuer Orthopaedische und Unfall-Chirurgie. (Munich, Fed. Rep. Germany) V.1-90 1903-77. See AOTSDE

APACAB Acta Physiologica Academiae Scientiarum Hungaricae. (Akademiai Kiado, P.O. Box 24, H-1389 Budapest 502, Hungary) V.1- 1950-

APAVAY Virchows Archiv fuer Pathologische, Anatomie und Physiologie, und fuer Klinische Medizin. (Berlin, Germany) V.1-343, 1847-1967. For publisher information, see VAAPB7

APBDAJ Archiv der Pharmazie und Berichte der Deutschen Pharmazeutischen Gesellschaft. (Weinheim, Germany) V.262-304, 1924-71. For publisher information, see ARPMAS

APBOAI Acta Facultatis Pharmaceuticae Bohemoslovenicae. (Bratislava, Czech.) V.4-13, 1961-67. For publisher information, see AFPCAG

APCRAW Advances in Pest Control Research. (New York) V.1-8, 1957-68. discontinued

APDCDT Advances in Tumour Prevention, Detection and Characterization. (Elsevier North Holland, Inc., 52 Vanderbilt Ave., New York, NY 10017) V.1- 1974-

APEPA2 Naunyn-Schmiedebergs Archiv fuer Pharmakologie und Experimentelle Pathologie. (Berlin, Germany) V.254-263, 1966-69. For publisher information, see NSAPCC

APFRAD Annales Pharmaceutiques Francaises. (Masson et Cie, Editeurs, 120 Blvd. Saint-Germain, P-75280, Paris 06, France) V.1- 1943-

APHGAO Acta Pharmaceutica Hungarica. (Kultura, POB 149, H-1389 Butapest, Hungary) 1953- Adopts V.24 in 1955.

APHGBP Acta Pathologica (Belgrade). (Belgrade Univerzitet, Institute de Pathologie, Belgrade, Yugoslavia) V.2/3- 1938-

APHNAB Acta Paediatrica Academiae Scientiarum Hungaricae. (Akademiai Kiada, P.O. Box 24, H-1389 Budapest 502, Hungary) V.1- 1960-

APHRDQ Archives of Pharmacal Research. (Pharmaceutical Society of Korea, College of Pharmacy, Seoul National Univ., San 56-1, Sinrim-Dong, Kwanak-ku, Seoul 151, S. Korea) V.1- 1978-

APINAG Acta Pharmaceutica Internationalia. (Copenhagen, Denmark) V.1-2, 1950-53. Discontinued.

APJAAG Acta Pathologica Japonica. (Nippon Byori Gakkai, 7-3-1, Hongo, Bunkyo-Ku, Tokyo 113, Japan) V.1- 1951-

APJUA8 Acta Pharmaceutica Jugoslavica. (Jugoslovenska Knjiga, P.O. Box 36, Terazije 27, YU-11001 Belgrade, Yugoslavia) V.1-1951-

APLAAQ Acta Paediatrica Latina. (Editrice Arti Grafiche Emiliane, postale 10110427, 42100 Reggio Emilia, Italy) V.1- 1948-

APLMAS Archives of Pathology and Laboratory Medicine. (American Medical Association, 535 N. Dearbon St., Chicago, Il. 60610) V.1-5, No. 2, 1926-28, V.100- 1976-

APMBAY Applied Microbiology. (Washington, DC) V.1-30, 1953-75. For publisher information, see AEMIDF

APMIAL Acta Pathologica et Microbiologica Scandinavica. (Copenhagen, Denmark) V.1-77, 1924-69. For publisher information, see AMBPBZ

APMIBM Acta Pathologica et Microbiologica Scandinavica, Section B: Microbiology and Immunology. (Copenhagen, Denmark) V.78B-82B, 1970-74. For publisher information, see APSCD2

APMUAN Acta Pathologica et Microbiologica Scandinavica, Supplementum. (Munksgaard, 35 Noerre Soegade, DK-1370 Copenhagen K, Denmark) No.1- 1926-

APPBAF Annales de Physiologie et de Physicochimie Biologique. (Paris, France) V.1-16, 1925-40. Discontinued.

APPBDI Acta Physiologica et Pharmacologica Bulgarica. (Izdatelstvo na Bulgarskata Akademiya na Naukite, St. Geo Milev ul 36, Sofia 13, Bulgaria) V.1- 1974-

APPHAX Acta Poloniae Pharmaceutica. (Ars Polona-RUCH, P.O. Box 1001, P-00 068 Warsaw, 1, Poland) V.1- 1937-

APPNAH Acta Physiologica et Pharmacologica Neerlandica. (Amsterdam, Netherlands) V.1-15, 1950-69. For publisher information, see EJPHAZ

APPYAG Annual Review of Phytopathology. (Annual Reviews, Inc., 4139 El Camino Way, Palo Alto, CA 94306) V.1- 1963-

APRCAS American Perfumer and Cosmetics. (Oak Park, IL) V.77-86, 1962-71. For publisher information, see CSPEAX

APSCAX Acta Physiologica Scandinavica. (Karolinska Institutet, S-10401 Stockholm, Sweden) V.1- 1940-

APSCD2 Acta Pathologica et Microbiologica Scandinavica, Section C: Immunology. (Munksgaard, 35 Noerre Soegade, DK-1370, Copenhagen CK, Denmark) V.83C- 1975-

APSQA7 Acta Paediatrica Scandinavica, Supplement. (Almqvist and Wiksell, POB 159, 26 Gamla Brogatan, S-101 22 Stockholm, Sweden) No.158- 1965-

APSVAM Acta Paediatrica Scandinavica. (Almqvist and Wiksell, P.O. Box 62, 26 Gamla Brogatan, S-101, 20 Stockholm, Sweden) V.54- 1965-

APSXAS Acta Pharmaceutica Suecdca. (Apotekarsocieteten, Wallingatan 26, Box 1136, S-111, 81 Stockholm, Sweden) V.1- 1964-

APTOA6 Acta Pharmacologica et Toxicologica. (Munksgaard, 35 Noerre Soegade, DK-1370, Copenhagen K, Denmark) V.1- 1945-

APTOD9 Abstracts of Papers, Society of Toxicology. Annual Meetings. (Academic Press, 111 5th Ave., New York, NY 10003)

APTRDI Advances in Prostaglandin and Thromboxane Research. (Raven Press 1140 Ave. of the America, New York, NY 10036) V.1- 1976-

APTSAI Acta Pharmacologica et Toxicologica, Supplementum. (Munksgaard, 35 Noerre Soegade, DK-1370, Copenhagen K, Denmark) No.1- 1947-

APTUAO Archives de L'Institut Pasteur de Tunis. (The Institute, 13 Pl Pasteur, Tunis, Tunisia) V.1- 1906-

APYAAN Archiv fuer Anatomie und Physiologie, Physiologische Abteilung. (Berlin, Germany) V.1-43, 1877-1919. For publisher information, see PFLABK

APYPAY Acta Physiologica Polonica. (Panstwowy Zaklad Wydawnictw Lekarskich, ul. Dluga 38-40, P-00 238 Warsaw, Poland) V.1- 1950-

AQMOAC Air Quality Monographs. (American Petroleum Institute, 2101 L St., N.W., Washington, D.C. 20037) No. 69-1- 1969-

ARANDR Archives of Andrology. (Elsevier North Holland, Inc., 52 Vanderbilt Ave., New York, NY 10017) V.1- 1978-

ARBIAE Archives of Biochemistry. (New York, NY) V.1-30, 1942-51. For publisher information, see ABBIA4

ARCGDG Archives of Gynecology. (Springer-Verlag New York, Inc., Service Center, 44 Hartz Way, Secaucus, NJ 07094) V.226- 1978-

ARCVBP Annales des Recherches Veterinaires. (Institut National de la Recherche Agronomique, Service des Publ., route de Saint-Cyr, 78000 Versailles, France) V.1- 1970-

ARDEAC Archives of Dermatology. (American Medical Association, 535 N. Dearborn St., Chicago, IL 60610) V.82- 1960-

ARDIAO Annals of the Rheumatic Diseases. (BMA or British Medical Journal, 1172 Commonwealth Ave., Boston, Mass. 02134) V.1- 1939-

ARDSAK Archiv fuer Dermatologic und Syphilis. (Berlin, Germany) V.1-5, 1869-73; V.21-200, 1889-1955. For publisher information, See AKEDAX

ARDSBL American Review of Respiratory Disease. (American Lung Association, 1740 Broadway, New York, NY 10019) V.80- 1959-

AREAD8 Anesteziologiya i Reanimatologiya. (v/o Mezhdunarodnaya Kniga, Kuznetskii Most 18, Moscow G-200, USSR) No.1- 1977-

ARENAA Annual Review of Entomology. (Annual Reviews, Inc., 4139 El Camino Way, Palo Alto, CA 94306) V.1- 1956-

ARGEAR Archiv fuer Geschwulstforschung. (VEB Verlag Volk und Gesundheit Neue Gruenstr. 18, DDR-102 Berlin, German Democratic Republic) V.1- 1949-

ARGPAQ Archives of General Psychiatry. (American Medical Association, 535 N. Dearborn St., Chicago, IL 60610) V.3- 1960-

ARGSAZ Arbeit und Gesundheit. (Georg Thieme Verlag, Postfach 732, Herdweg 63, 7000 Stuttgart, Federal Republic of Germany) V.1- 1926-

ARGYAJ Archiv fuer Gynaekologie. (Munich, Germany) V.1-225, 1870-1978. For publisher information, see ARCGDG

ARHEAW Arthritis and Rheumatism. (Arthritis Foundation, 3400 Peachtree Road, N.E., Atlanta, Ga 30326) V.1- 1958-

ARINAU Annual Report of the Research Institute of Environmental Medicine, Nagoya University. (Nagoya, Japan) V.1-25, 1951-80.

ARKIAP Archiv fuer Kinderheilkunde. (Stuttgart, Fed. Rep Ger.) V.1-183, 1880-1971.

ARMCAH Annual Review of Medicine. (Annual Reviews, Inc., 4139 El Camino Way, Palo Alto, CA 94306) V.1- 1950-

ARMCBI Annual Reports in Medicinal Chemistry. (Academic Press, 111 5th Ave., New York, NY 10003) 1965-

ARMIAZ Annual Review of Microbiology. (Annual Reviews, Inc., 4139 El Camino Way, Palo Alto, CA 94306) V.1- 1947-

ARMKA7 Archiv fuer Mikrobiologie. (Springer (Berlin)) V.1-13, 1930-43; V.14-94, 1948-73. For publisher information, see AMICCW

ARNEAS Archives of Neurology. (American Medical Association, 535 N. Dearborn St., Chicago, IL 60610) V.3- 1960-

AROPAW Archives of Ophthalmology. (American Medical Assn., 535 N. Dearborn St., Chicago, IL 60610) V.1-44, No. 3, 1929-50; V.64- 1960-

AROTAA Archives of Otolaryngology. (American Medical Association, 535 N. Dearborn St., Chicago, IL 60610) V.1-52, 1925-50; V.72- 1960-

ARPAAQ Archives of Pathology. (American Medical Assn., 535 N. Dearborn St., Chicago, IL 60610) V.5, no.3-V.50, no.3, 1928-50; V.70-99, 1960-75.

ARPMAS Archiv der Pharmazie. (Verlag Chemie GmbH, Postfach 1260/1280 D-6940 Weinheim, Federal Republic of Germany) V.51-261, 1835-1923; V.305- 1972-

ARPTAF Arkhiv Patologii. Archives of Pathology. (v/o Mezhdunarodnaya Kniga, Kuznetskii Most 18, Moscow G-200 U.S.S.R.) V.1- 1959-

ARPTDI Annual Review of Pharmacology and Toxicology. (Annual Reviews, Inc., 4139 El Camino Way, Palo Alto, CA 94306) V.16- 1976-

ARSIM* Agricultural Research Service, USDA Information Memorandum. (Beltsville, MD 20705)

ARSMA9 American Review of Soviet Medicine. (New York, NY) V.1-5, 1943-48. Discontinued.

ARSUAX Archives of Surgery. (American Medical Association, 535 N. Dearborn St., Chicago, IL 60610) V.1-61, 1920-50; V.81- 1960-

ARTODN Archives of Toxicology. (Springer-Verlag, Heidelberger, Pl. 3, D-1 Berlin 33, Germany) V.32- 1974-

ARTUA4 American Review of Tuberculosis. (New York, NY) V.1-70, 1917-54. For publisher information, see ARDSBL

ARVPAX Annual Review of Pharmacology. (Palo Alto, CA) V.1-15, 1961-75. For publisher information, see ARPTDI

ARZFAN Aerztliche Forschung. (Munich, Fed. Rep. Ger.) V.1-26, 1947-72. Discontinued.

ARZNAD Arzneimittel-Forschung. (Edition Cantor Verlag fuer Medizin und Naturwissenschaften KG, D-7960 Aulendorf, Germany) V.1- 1951-

ARZWA6 Aerztliche Wochenschrift. (Berlin, Germany) V.1-15, 1946-60. For publisher information, see INTEAG

ASBDD9 Advances in the Study of Birth Defects. (University Park Press, 233 E. Redwood St., Baltimore, MD 21202) V.1- 1979-

ASBIAL Archivio di Science Biologiche. (Cappelli Editore, Via Marsili 9, I-40124 Bologna, Italy) V.1- 1919-

ASBMAX Annales de la Societe Belge de Medicine Tropicale. (J. Goemaere, Publisher, Grensstraat 21, B-1030 Brussels, Belgium) V.1- 1920-

ASBUAN Archives des Sciences Biologiques. (Leningrad, U.S.S.R.) V.1-22, 1892-1922. Discontinued.

ASCHAN Anzeiger fuer Schaedlingskunde. (Berlin, Germany) V.1-41, 1925-68. For publisher information, see ASPUCR

ASHRDD Advances in Sex Hormone Research. (Baltimore, MD) V.1-3, 1975-77. Discontinued.

ASLBAG Atti della Societa Lombarda di Scienze Mediche e Biologiche. (Milan, Italy) V.1-19, 1912-30; V.1-14, 1945-59. For publisher information, see AAMLAR

ASMACK Abstracts of the Annual Meeting of the American Society for Microbiology. (American Society for Microbiology, 1913 I Street, NW Washington, DC 20006)

ASMUAA Acta Scholae Medicinalis Universitatis in Kyoto. (Kyoto, Japan) V.27-40, 1947-70. Discontinued

ASPHAK Archives des Sciences Physiologique. (Paris, France) V.1-28, 1947-74. Discontinued.

ASPUCR Anzeiger fuer Schaedlingskunde, Pflanzen-und Umweltschutz. (Paul Parey, Linderst 44-47, D-1000 Berlin 61, Federal Republic of Germany) V.46-47, 1973-74.

ASTTA8 ASTM Special Technical Publication. (American Society for Testing Materials, 1916 Race St., Philadelphia. PA 19103) No.1-1911-

ASUPAZ Acta Societatis Medicorum Upsaliensis. (Uppsala, Sweden) V.55-76, 1950-71. For publisher information, see UJMSAP

ASVMAV Abstracts of Soviet Medicine. (Amsterdam, Netherlands) V.5, No.1-12, 1961. Discontinued.

ASXBA8 Archives of Sexual Behavior. (Plenum Publishing Corp., 233 Spring St., New York, NY 10013) V.1- 1971-

ATAREK AAMI Technology Assessment Report. (Association for the Advancement of Medical Instrumentation, 1901 N. Ft. Myer Dr., Suite 602, Arlington, VA 22209) No.1-81- 1981-

ATENBP Atmospheric Environment. Air Pollution, Industrial Aerodynamics, Micrometerology, Aerosols. (Pergamon Press, Headington Hill Hall, Oxford OX3 OBW, England) V.1- 1967-

ATHBA3 Acta Radiologica, Therapy, Physics, Biology. (P.O. Box 7449, S-103 91 Stockholm, Sweden) V.1-16, No.6, 1963-77

ATHSBL Atherosclerosis (Shannon, Ireland). (Elsevier/North-Holland Scientific Publishers Ltd., POB 85, Limerick, Ireland) V.11-1970-

ATMPA2 Annals of Tropical Medicine and Parasitology. (Academic Press, 24-28 Oval Rd., London NW1 7DX, England) V.1- 1907-

ATORAI A.P.I. Toxicological Review. (American Petroleum Institute, 2101 L St., N.W., Washington D.C. 20037)

ATPNAB Ateneo Parmense, Acta Naturalia. (Ospedale Maggiore, Via Gramsci 14, 43100 Parma, Italy) V.1- 1965-

ATSUDG Archives of Toxicology, Supplement. (Springer-Verlag, Heidelberger Pl. 3, D-1000 Berlin 33, Fed. Rep. Ger.) No. 1- 1978-

ATXKA8 Archiv fuer Toxikologie. (Berlin, Germany) V.15-31, 1954-74. For publisher information, see ARTODN

AUAAB7 Acta Universitatis Agriculturae, Facultas Agronomica (Brno) (Ustredni Knihovna Vysoke Skoly Zemedelske, Zemedelska 1, 662-65 Brno, Czechoslovakia) V.15- 1967-

AUCMBJ Acta Universitatis Carolinae, Medica, Monographia. (Univerzita Karlova, Ovocny Trh.3, CS-116-36 Prague 1, Czechoslovakia) No.1- 1954-

AUHPAI Australian Journal of Hospital Pharmacy. (Mr. B.R. Miller, Secretary, P.O. Box 125, Heidelberg, Victoria 3084, Australia) V.1- 1971-

AUMKAS Annales Universitatis Mariae Curie-Sklodowska, Section D. (Uniw Marii Curie-Sklodowskiej, Biuro Wydawn) V.1- 1946-

AUODDK Acta Universitatis Ouluensis, Series D: Medica. (Oulu University Library, Box 186, SF-90101 Oulu 10, Finland) Number 1-1972-

AUPJB7 Australian Paediatric Journal. (Royal Childrens Hospital, Parkville, Victoria 3052, Australia) V.1- 1965-

AUVJA2 Australian Veterinary Journal. (Australian Veterinary Association, Executive Director, 134-136 Hampden Road, Artarmon, N.S.W. 2064, Australia) V.2- 1927-

AVBIB9 Advances in the Biosciences. (Pergamon Press Ltd., Headington Hill Hall, Oxford OX3 0BW, England) V.1- 1969-

AVBNAN Arhiv Bioloskih Nauka. (Jugoslovenska Knjigu, P.O. Box 36, Terazije 27, 11001, Belgrade, Yugoslavia)

AVCPAY Advances in Clinical Pharmacology. (Urban and Schwarzenberg, Pettenkoferstr. 18, D-8000 Munich 2, Federal Republic of Germany) V.1 1969-

AVERAG American Veterinary Review. (Chicago, IL) V.1-47, 1877-1915. For publisher information, see JAVMA4

AVEZA6 Acta Vitaminologica et Enzymologica. (Gruppo Lepetit SpA, Via R. Lepetit N. 8, 20124 Milan, Italy) V.21- 1967-

AVIRA2 Acta Virologica; English edition. (Academic Press, Inc., Ltd., 24-28 Oval Rd., London NW1 7DX, England) V.1- 1957-

AVPCAQ Advances in Pharmacology and Chemotherapy. (Academic Press, 111 5th Ave., New York, NY 10003) V.7- 1969-

AVSCA7 Acta Veterinaria Scandinavica. (Danske Dyrlargeforening, Alhambravej 15, DK-1826, Copenhagen V, Denmark) V.1- 1959-

AVSCB8 Advances in Veterinary Science and Comparative Medicine. (Academic Press, 111 5th Ave., New York, NY 10003)

AVSUAR Acta Dermato-Venereologica, Supplementum. (Almqvist and Wiksell Periodical Co., P.O. Box 62, 26 Gamla Brogatan, S-101 20 Stockholm, Sweden) No.1- 1929-

AWLRAO Australian Wildlife Research. (Commonwealth Scientific and Industrial Research Organization, POB 89, E. Melbourne, Vic. 3002, Australia) V.1- 1974-

AXVMAW Archiv fuer Experimentelle Veterinaermedizin. (S. Hirzel Verlag, Postfach 506, DDR-701 Leipzig, German Democratic Republic) V.6- 1952-

AZMZA6 Azerbaidzhanskii Meditsinskii Zhurnal. Azerbaidzhan Medical Journal. (v/o Mezhdunarodnaya Kniga, Kuznetskii Most 18, Moscow G-200, U.S.S.R.) 1928-41; 1955-

AZOGBS Australian and New Zealand Journal of Obstetrics and Gynaecology. (Australian and New Zealand Journal of Obstetrics and Gynaecology, 8 Latrobe St., Melbourne 3000, Australia) V.1- 1961-

AZOOAH Annales de Zootechnie. (Institut National de la Recherche Agronomique, Service des Publ., Rt. de St.-Cyr, F-78000 Versailles, France) V.10- 1961-

BurLW# Personal Communication from Mr. L.W. Burnette, Material Safety Dept., GAF Corp., 1361 Alps Rd., Wayne, NJ 07470, to Dr. A. Friedman, Tracor Jitco, Inc., November 2, 1978

BACCAT Bulletin of the Academy of Sciences of the U.S.S.R., Division of Chemical Science (English Translation). (Plenum Publishing Corp., 233 Spring St., New York, NY 10013) 1952-

BAFEAG Bulletin de l'Association Francaise pour l'Etude du Cancer. (Paris, France) V.1-52, 1908-65. For publisher information, see BUCABS

BANMAC Bulletin de l'Academie Nationale de Medicine. (Masson et Cie, Editeurs, 120 Blvd. Saint-Germain, P-75280, Paris 06, France) V.1- 1836-

BANRDU Banbury Report. (Cold Spring Harbor Laboratory, POB 100, Cold Spring Harbor, NY 11724) V.1- 1979-

BAPBAN Bulletin de l'Academie Polonaise des Sciences, Series des Sciences Biologiques. (Ars Polona-RUCH, P.O. Box 1001 P-00 068 Warsaw 1, Poland) V.5- 1957-

BAREA8 Bacteriological Reviews. (American Society of Microbiology, 1913 I St., N.W., Washington, D.C. 20006) V.1-41, 1937-77

BATTL* Reports produced for the National Institute for Occupational Safety and Health by Battelle Pacific Northwest Laboratories, Richland, WA 99352

BAXXDU British UK Patent Application. (U.S. Patent Office, Science Library, 2021 Jefferson Davis Highway, Arlington, VA 22202)

BBACAQ Biochimica et Biophysica Acta. (Elsevier Publishing Co., POB 211, Amsterdam C, Netherlands) V.1- 1947-

BBIADT Biomedica Biochimica Acta. (Akademie-Verlag GmbH, Postfach 1233, DDR-1086 Berlin, Ger. Dem. Rep.) V.42- 1983-

BBKCA8 Bruns' Beitraege zur Klinische Chirurgie. (Springer-Verlag, Heidelberger Pl. 3, D-1000 Berlin 33, Federal Republic of Germany) V.1-221, No.8, 1883-1974

BBMS** "Medicaments du Systeme Nerveux Vegetalif," Bovet, D., and F. Bovet-Nitti, New York, S. Karger, 1948

BBRCA9 Biochemical and Biophysical Research Communications. (Academic Press Inc., 111 5th Ave., New York, NY 10003) V.1- 1959-

BCFAAI Bollettino Chimico Farmaceutico. (Societa Editoriale Farmaceutica, Via Ausonio 12, 20123 Milan, Italy) V.33- 1894-

BCPCA6 Biochemical Pharmacology. (Pergammon Press Inc., Maxwell House, Fairview Park, Elmsford, NY 10523) V.1- 1974-

BCPHBM British Journal of Clinical Pharmacology. (Macmillan Journals, Houndmills Estate, Basingstoke, Hants RG21 2XS, England) V.1- 1974-

BCSTA4 Bulletin of the Calcutta School of Tropical Medicine. (Calcutta School of Tropical Medicine, Chittaranjan Ave., Calcutta 12, India) V.1- 1953-

BCSTB5 Biochemical Society Transactions. (Biochemical Society, P.O. Box 32, Commerce Way, Whitehall Rd., Industrial Estate, Colchester CO2 8HP, Essex, England) V.1- 1973-

BCSYDM Bristol-Myers Cancer Symposia. (Academic Press, 111 Fifth Ave., New York, NY 10003) V.1- 1979-

BCTKAG Bromatologia i Chemia Toksykologiczna. (Ars Polona-RUCH, P.O. Box 1001, P-00 068 Warsaw, 1, Poland) V.4- 1971-

BCTRD6 Breast Cancer Research and Treatment. (Kluwer Academic Publishers Group, Distribution Center, POB 322, 3300 AH Dordrecht, Neth.) V.1- 1981-

BCZVDE Biologizace a Chemizace Zivocisne Vyroby-Veterinaria. Biological and Chemical Factors in Animal Production-Veterinary Science. (Statni Zemedelske Nakladatelstvi, Vaclavske nam. 47, 11311 Prague, Czechoslovakia) V.14- 1978-

BDBU** Zur Pharmakologie Ungesattigter Alkohole, Harald Bock Dissertation. (Pharmakologischen Institut der Universitat Breslau, Germany, 1930)

BDCBAD Berichte der Deutschen Chemischen Gesellschaft. Abteilung B: Abhandlungen. (Heidelberg, Germany) V.62-77, 1929-45. For publisher information, see CHBEAM

BDCGAS Berichte der Deutschen Chemischen Gesellschaft. (Leipzig/Berlin) V.1-61, 1868-1928. For publisher information, see CHBEAM

BDHU** Nachprufung der Toxicitat von Novocain und Tutocain bei Chlorali- sierten Tieren, August Barke Dissertation. (Pharmakologischen Institut der Tierarztlichen Hochschule zu Hannover, Germany, 1936)

BDKS** Studien uber die Pharmakologie des Pinakolins und einiger seiner Derivate mit besonderer Berucksichtigung der Kreislaufwirkung, Lennart Bang Dissertation. (Pharmakologischen Abteilung des Karolinischen Instituts, Stockholm, Sweden, 1934)

BDVU** Uber die Pharmakologische Wirkung des 1,3-Dioxy-2-Nicotinsaureamid- Tetrazols. Eine Experimentelle Studie zur Kenntnis der Wirkung neuer Tetrazolderivate, Rudolf Barwanietz Dissertation. (Institut fuer Veterinar-Pharmakolgie der Universitat Berlin, Germany, 1937)

BEBMAE Byulleten' Eksperimental'noi Biologii i Meditsiny. Bulletin of Experimental Biology and Medicine. (v/o Mezhdunarodnaya Kniga, Kuznetskii Most 18, Moscow G-200, U.S.S.R.) V.1- 1936-

BECCAN British Empire Cancer Campaign Annual Report. (Cancer Research Campaign, 2 Carlton House Terrace, London SW1Y 5AR, England) V.1- 1924-

BECTA6 Bulletin of Environmental Contamination and Toxicology.

(Springer- Verlag New York, Inc., Service Center, 44 Hartz Way, Secaucus, NJ 07094) V.1- 1966-

BEGMA5 Beitraege zur Gerichtlichen Medizin. (Verlag Franz Deuticke, Helferstorferstr 4, A-1010 Vienna, Austria) V.1- 1911-

BEMTAM Berliner und Muenchener Tieraerztliche Wochenschrift. (Verlag Paul Parey, Lindenstr. 44-47, D-1000 Berlin 61, Fed. Rep. Germany) 1938-43; No.1- 1946-

BENPBG Behavioral Neuropsychiatry. (Behavioral Neuropsychiatry Medical Publ., Inc., 61 E. 86th St., New York, NY 10028) V.1- 1969-

BEPRDY Bulletin Europeen de Physiopathologie Respiratoire. (Pergamon Press, Inc., Journals Dept., Maxwell House, Fairview Park, Elmsford, NY 10523) V.12- 1976-

BEREA2 Bulletin of Entomological Research. (Commonwealth Agricultural Bureaux, Central Sales Branch, Farnham House, Farnham Royal, Slough SL2 3BN, England) V.1- 1910-

BERUAG Berufs-Dermatosen. (Editio Cantor KG, Postfach 1310, D-79640 Aulendorf/Wuertt, Federal Republic of Germany) V.2-25, No.6, 1953-77

BESAAT Bulletin of the Entomological Society of America. (The Society, 4603 Calvert Rd., College Park, MD 20740) V.1- 1955-

BEXBAN Bulletin of Experimental Biology and Medicine. Translation of BEBMAE. (Plenum Publishing Corp., 233 Spring St., New York, NY 10013) V.41- 1956-

BEXBBO Biochemistry and Experimental Biology. (Piccin Medical Books, Via Brunacci, 12, 35100 Padua, Italy) V.10- 1971/72-

BFPHA8 Bulletin of the Faculty of Pharmacy, Cairo University. (Faculty of Pharmacy, Cairo University, Cairo, Egypt) V.1- 1961/62-

BFSGAK Bulletin de la Federation des Societes de Gynecologie et d'Obstetrique de Langue Francaise. (Paris, France) V.1-23, 1949-71. For publisher information, see JGOBAC

BHBLAW Behavioral Biology. (Academic Press, Inc., 111 Fifth Ave., New York, NY 10033) V.7-24, 1972-78

BHJUAV British Heart Journal. (British Medical Journal, 1172 Commonwealth Ave., Boston, MA 02134) V.1- 1939-

BIALAY Biochimica Applicata. (Parma, Italy) V.1-19, 1954-72(?). Discontinued.

BIANA6 Bibliotheca Anatomica. (S. Karger AG, Postfach CH-4009, Basel, Switzerland) No.1- 1961-

BIATDR Bulletin of the International Association of Forensic Toxicologists.

BIBIAU Biotechnology and Bioengineering. (John Wiley and Sons, 605 3rd Ave., New York, NY 10016) V.4- 1962-

BIBUBX Biological Bulletin (Woods Hole, Mass.). (Biological Bulletin, Marine Biological Laboratories, Woods Hole, MA 02543) V.1- 1898-

BIBUDZ Biology Bulletin of the Academy of Sciences of the USSR. English translation of IANBAM. (Plenum Publishing Corp., 227 W. 17th St., New York, NY 10011) V.1- 1974-

BICHAW Biochemistry. (American Chemical Society Publications, 1155 16th St., N.W., Washington, DC 20036) V.1- 1962-

BICHBX Bioinorganic Chemistry. (Elsevier North Holland, Inc., 52 Vanderbilt Ave., New York, NY 10017) V.1-9, 1971-78

BICMBE Biochimie. (Masson et Cie, Editeurs, 120 Blvd. Saint-Germain, P-75280, Paris 06, France) V.53- 1971-

BIHAA2 Bibliotheca Haematologica. (S. Karger AG, Arnold-Boecklin-St 25, CH-4000, Basel 11, Switzerland) No.1- 1955-

BIJOAK Biochemical Journal. (Biochemical Society, P.O. Box 32, Commerce Way, Whitehall Rd., Industrial Estate, Colchester CO2 8HP, Essex, England) V.1- 1906-

BIMADU Biomaterials. (Quadrant Subscription Services Ltd., Oakfield House, Perrymount Rd., Haywards Heath, W. Sussex, RH16 3DH, U.K.) V.1- 1980-

BIMDA2 Biochemical Medicine. (Academic Press, 111 5th Ave., New York, NY 10003) V.1- 1967-

BIMDB3 Biomedicine. (Masson et Cie, Editeurs, 120 Blvd. Saint-Germain, P-75280, Paris 06, France) V.18- 1973-

BIMEA5 Biologie Medicale. (Paris) V.1-60, 1903-71; new series 1972-V.4,1975

BINEAA Biologia Neonatorum. (Basel, Switzerland) V.1-14, No.5/6, 1956- 1969, For publisher information, see BNEOBV

BINKBT Biologicheskie Nauki. (v/o Mezhdunarodnaya Kniga, Kuznetskii Most 18, Moscow G-200, USSR) No.3- 1965-

BIOFX* BIOFAX Industrial Bio-Test Laboratories, Inc., Data Sheets. (1810 Frontage Rd., Northbrook, IL 60062)

BIOGAL Biologico. (Instituto Biologica, Av. Rodriques Alves, 1252, C.P. 4185, Sao Paulo, Brazil) V.1- 1935-

BIOHAO Biokhimiia. (v/o Mezhdunarodnaya Kniga, Kuznetskii Most 18, Moscow G-200, U.S.S.R.) V.1- 1936-

BIOJAU Biophysical Journal. (Rockefeller Univ. Press, 1230 York Ave., New York, NY 10021) V.1- 1960-

BIORAK Biochemistry. Translation of BIOHAO. (Plenum Publishing Corp., 233 Spring St., New York, NY 10013) V.21- 1956-

BIPAA8 Bulletin de l'Institut Pasteur. (Masson et Cie, Editeurs, 120 Blvd. Saint-Germain, P-75280, Paris 06, France) V.1- 1903-

BIPCBF Biological Psychiatry. (Plenum Publishing Corp., 233 Spring St., New York, NY 10013) V.1- 1969-

BIPMAA Biopolymers. (John Wiley and Sons, 605 3rd Ave., New York, NY 10158) V.1- 1963-

BIREBV Biology of Reproduction. (Society for the Study of Reproduction, 309 West Clark Street, Champaign, IL 61820) V.1- 1969-

BIRSB5 Biology of Reproduction, Supplement. (Champaign, Ill.) For publisher information, see BIREBV

BIRUAA Biologische Rundschau. (VEB Gustav Fischer Verlag, Postfach 176, Villengang 2, DDR-69 Jena, German Democratic Republic) V.1- 1963-

BISNAS BioScience. (American Instutite of Biological Sciences, 1401 Wilson Blvd., Arlington, VA 22209) V.14- 1964-

BIZEA2 Biochemische Zeitschrift. (Berlin, Germany) V.1-346, 1906-67. For publisher information, see EJBCAI

BIZNAT Biologisches Zentralblatt. (VEB Georg Thieme, Hainst 17/19, Postfach 946, 701 Leipzig, E. Germany) V.1- 1881-

BJANAD British Journal of Anesthesia. (Scientific and Medical Division, Macmillan Publishers Ltd., Houndmills, Basingstoke, Hampshire RG21 2XS, England) V.1- 1923-

BJCAAI British Journal of Cancer. (H.K. Lewis and Co., 136 Gower St., London WC1E 6BS, England) V.1- 1947-

BJCPAT British Journal of Clinical Practice. (Harvey and Blythe Ltd., Lloyd's Bank Chambers, 216 Church Rd., Hove, Sussex BN3 2DJ, England) V.10, No.10- 1956-

BJCSB5 British Journal of Cancer, Supplement. (H.K. Lewis and Co., Ltd., 136 Gower St., London WC1E 6BS, England) No.1-1973-

BJDEAZ British Journal of Dermatology. (Blackwell Scientific Publications, Osney Mead, Oxford OX2 OEL, England) V.63- 1951-

BJEPA5 British Journal of Experimental Pathology. (H.K. Lewis and Co., 136 Gower St., London WC1E 6BS, England) V.1- 1920-

BJHEAL British Journal Haematology. (Blackwell Scientific Pub. Ltd., POB 88, Oxford, UK) V.1- 1955-

BJHMAB British Journal of Hospital Medicine. (Hospital Medicine Publications Ltd. Theba House, 49/50 Hatton garden, London EC1N 8XS, England) V.1- 1968-

BJIMAG British Journal of Industrial Medicine. (British Medical Journal, 1172 Commonwealth Ave., Boston, MA 02134) V.1- 1944-

BJLSAF Botanical Journal of the Linnean Society. (Academic Press, 111 5th Ave., New York, NY 10003) V.62- 1969-

BJNUAV British Journal of Nutrition. (Cambridge University Press, The Edinburgh Bldg., Shaftesburg Rd., Cambridge CB2 2RU, England) V.1- 1947-

BJOGAS British Journal of Obstetrics and Gynaecology. (British Journal of Obstetrics and Gynaecology, 27 Sussex Place, Regent's Park, London NW1 4RG, England) V.82- 1975-

BJOPAL British Journal of Ophthalmology. (British Medical Journal, 1172 Commonwealth Ave., Boston, MA 02134) V.1- 1917-

BJPCAL British Journal of Pharmacology and Chemotherapy. (London, England) V.1-33, 1946-68. For publisher information, see BJPCBM

BJPCBM British Journal of Pharmacology. (MacMillan Journals Ltd., Houndmills Estate, Basingstroke, Hampshire RG21 2XS, England) V.34- 1968-

BJPVAA British Journal of Preventive and Social Medicine. (British Medical Assoc., BMA House, Tavistock Sq., London WC1H 9JR, England) V.7- 1953-

BJPYAJ British Journal of Psychiatry. (Headley Brothers, Ashford TN24 8HH, Kent, England) V.109- 1963-

BJRAAP British Journal of Radiology. (British Institute of Radiology, 36 Portland Place, London W1N 3DG, England) V.1- 1928-

BJRHDF British Journal of Rheumatology. (Bailliere Tinball, 33 The Ave., Eastborne BN21 3UN, UK) V.1- 1928-

BJSUAM British Journal of Surgery. (John Wright and Sons Ltd., 42-44 Triangle West, Bristol BS8 1EX, England) V.1- 1913-

BJURAN British Journal of Urology. (Williams and Wilkens Co., 428 E. Preston St., Baltimore, MD 21202) V.1- 1929-

BKNJA5 Biken Journal. (Research Institute for Microbial Diseases, Osaka Univ., Yamada-Kami, Suita, Osaka, Japan) V.1- 1958-

BKWOAV Berliner Klinische Wochenschrift. (Berlin, Germany) V.1-58, 1864-1921. For publisher information, see KLWOAZ

BKZHAP Biokhimichnii Zhurnal. (Kiev, U.S.S.R.) V.10-17, 1937-41. For publisher information, see UBZHAZ

BLASR* Blaser and Co., Ltd. Reports, (CH-3415 Hasle-Ruegsau/ Switzerland)

BLFSBY Basic Life Sciences. (Plenum Publishing Corp., 227 W. 17th St., New York, NY 10011) V.1- 1973-

BLLIAX Bratislavske Lekarske Listy. (PNS-Ustredna Expedicia Tlace, Gottwaldovo Namestie 48/7, CS-884 19 Bratislava, Czechoslovakia) V.1- 1921-

BLOAAO Biologia (Bratislava). (PNS-Ustredna Expedicia Tlace, Gottwaldovo Namestie 48/7, CS-884 19 Bratislava, Czechoslovakia) V.24- 1969-

BLOOAW Blood. (Grune and Stratton, 111 5th Ave., New York, NY 10003) V.1- 1946-

BLUTA9 Blut. (Springer-Verlag New York, Inc., Service Center, 44 Hartz Way, Secaucus, NJ 07094) V.1- 1955-

BMAOA3 Biologicheskii Zhurnal. Biological Journal. (Moscow, U.S.S.R.) V.1-7, No.6, 1932-38. For publisher information, see ZOBIAU

BMBUAQ British Medical Bulletin. (Churchill Livingstone, Robert Stevenson House, 1-3 Baxter's Place, Leith Walk, Edinburgh, EH1 3AF, UK) V.1- 1943-

BMJOAE British Medical Journal. (British Medical Association, BMA House, Travistock Square, London WC1H 9JR, England) V.1- 1857-

BMRII* "U. S. Bureau of Mines Report of Investigation No. 2979," Patty, F.A., and W.P. Yant, 1929

BNEOBV Biology of the Neonate. (S. Karger AG, Postfach, CH-4009 Basel, Switzerland) V.15- 1970-

BNUNA5 Bulletin on Narcotics. (United Nations, Sales Section, Rm 1059, New York, NY 10017) V.1- 1949-

BNYMAM Bulletin of the New York Academy of Medicine. (2 East 103rd Street, New York, NY 10029) V.1 1925-

BOCKAE Bochu Kagaku. Scientific Pest Control. (Kyoto, Japan) V.1-42, 1937-77. Discontinued.

BOOKA7 "Biological Effects of Air Pollutants," 2nd ed., A.C. Stern, ed., New York, Academic Press, 1968

BPAAAG Beitraege zur Pathologischen Anatomie und Allgemeinen Pathologie. (Stuttgart, Germany) V.3-109, No.2, 1888-1944; V.109, No.3-V.140, No.4, 1947-70, For publisher information, see BTPGAZ

BPBSDA Bulletin of the Philippine Biochemical Society. (Philippine Biochemical Society, POB 593, Manila, Philippines) V.1- 1978-

BPCT** Progress Report from the Environmental Protection Agency to NIOSH on Benzenepolycarboxylates, 1974

BPJLAQ Bangladesh Pharmaceutical Journal. (Bangladesh Pharmaceutical Society, Department of Pharmacy, University of Dacca, Dacca 2, Bangladesh) V.1- 1972-

BPNSBY Bulletin of the Psychonomic Society. (Psychonomic Society, 1108 W. 34th St., Austin, TX 78705) V.1- 1973-

BPOSA4 British Poultry Science. (Longman Group Ltd., Journals Division, 43-45 Annandale St., Edinburgh EH47 4AT, Scotland) V.1- 1960-

BPYKAU Biophysik. (Berlin, Germany) V.1-10, 1963-73.

BRAIAK Brain; Journal of Neurology. (Oxford Univ. Press, Walton St., Oxford OX2 6DP, England) V.1- 1878-

BRCAB7 Basic Research in Cardiology. (Dr. Dietrich Steinkopff Verlag, Postfach 11-10-08, D-6000 Darmstadt 11, Fed. Rep. Ger) V.68- 1973-

BRGOAY Bulletin de la Societa Royale Belge de Gynecologie et d'Obstetrique.

(Brussels, Belgium) V.1-39, 1925-69.

BRMEAY Bruxelles-Medical. Bruxelles-Medical, Boite 1, 1180 Brussels, Belgium) V.1- 1921-

BRPTDT Bioscience Reports. (Biochemical Society Book Depot, POB 32, Commerce Way, Colchester, Essex CO2 8HP, England) V.1- 1981-

BRREAP Brain Research. (Elsevier Scientific Publishing Co., P.O. Box 211, Amsterdam, Netherlands) V.1- 1966-

BRXXAA British Patent Document. (U.S. Patent Office, Science Library, 2021 Jefferson Davis Highway, Arlington, VA 22202)

BSAMA5 Bulletin der Schweizerische Akademie der Medizinischen Wissenschaften. (Schwabe und Co., Steintorstr. 13, 4000 Basel 10, Switzerland) V.1- 1944-

BSBGAQ Berichte der Schweizerischen Botanischen Gesellschaft. (KRYPTO F. Flueck-Wirth, CH-9053 Teufen, Switzerland) V.1- 1891-

BSBSAS Boletin de la Sociedad de Biologia de Santiago de Chile. (Santiago, Chile) V.1-12, 1943-55. Discontinued.

BSCFAS Bulletin de la Societe Chimique de France. (Masson et Cie, Editeurs, 120 Blvd. Saint-Germain, P-75280, Paris 06, France) V.1- 1864-

BSCIA3 Bulletin de la Societe de Chimie Biologique. (Paris, France) V.1-52, 1914-70. For publisher information, see BICMBE

BSECBU Biochemical Systematics and Ecology. (Pergamon Press Inc., Maxwell House, Fairview Park, Elmsford, NY 10523) V.2- 1974-

BSFDA3 Bulletin de la Societe Francaise de Dermatologie et de Syphiligraphie. (Paris, France) V.1-83, 1890-1976. Discontinued.

BSIBAC Bolletino della Societe Italiana di Biologia Sperimentale. (Casa Editrice Libraria V. Idelson, Via Alcide De Gasperi, 55, 80133 Naples, Italy) V.2- 1927-

BSOFAK Bulletin des Societes d'Ophthalmologie de France. (Diffusion Litteraire et Scientifique, 11 rue Moliere (ler etage), 13 Marseilles, France) 1949-

BSPBAD Bulletin de la Societe de Pharmacie de Bordeaux. (Societe de Pharmacie de Bordeaux, Faculte de Medecine et de Pharmacie, 91, rue Leyteire, 33000 Bordeaux, France) V.89- 1951-

BSPHAV Bulletin des Sciences Pharmacologiques. (Paris, France) V.1-49, 1899-1942. For publisher information, see APFRAD

BSPII* SPI Bulletin. (Society of the Plastics Industry, 250 Park Ave., New York, NY 10017)

BSPMAC Bulletin de la Societe de Pharmacie de Marseille. (La Societe, Faculte de Medecine et de Pharmacie, Blvd. Jean-Moulin, Marseilles 503-67, France) V.1- 1952-

BSPPDO Bulletin of the Society of Pharmacological and Environmental Pathologists. (Toxicologic Pathology, E.I. du Pont de Nemours Co., Inc., Elkton Rd., Newark, DE 19711) V.1-6 No.2, 1972-78

BSRSA6 Bulletin de la Society Royale des Sciences de Liege. (Societe

Royale des Sciences de Liege, Universite de Liege, 15, Ave des Tilleuls, B-4000 Liege, Belgium) V.1- 1932-

BSVMA8 Bulletin de la Societe des Sciences Veterinaires et de Medecine Compare de Lyon. (Societe des Sciences Veterinaires et de Medecine Comparee de Lyon, Ecole National Veterinaire de Lyon, Route de Sain-Bel, Marcy-l'Etoile, 69260 Charbonnieres-les-Bains, France) V.1- 1898-

BTDCAV Bulletin of Tokyo Dental College. (Tokyo Dental College, Misaki-cho, Chiyoda-Ku, Tokyo 101, Japan) V.1- 1960-

BTERDG Biological Trace Element Research. (The Humana Press, Inc., Crescent Manor, P.O. Box 2148, Clifton NJ 07015) V.1- 1979-

BTHDAK Birth Defects, Original Article Series. (National Foundation-March of Dimes, 1275 Mamaroneck Ave., White Plains, NY 10605) V.1- 1965-

BTMNA7 Bitamin. (Nippon Bitamin Gakkai, 4 Ushinomiya-cho, Yoshida, Sakyo-Ku, Kyoto 606, Japan) V.1- 1948-

BTPGAZ Beitraege zur Pathologie. (Gustav Fischer Verlag, Postfach 72-0143 D-7000 Stuttgart 70, Germany) V.141- 1970-

BTSRAF Biochimica e Terapia Sperimentale. (Milan, Italy) V.1-30, 1909-43. For publisher information, see NIVAAY

BUCABS Bulletin du Cancer. (Masson et Cie, Editeurs, 120 Blvd. Saint- Germain, P-75280, Paris 06, France) V.53- 1966-

BUCKL* Buckman Laboratories, Inc. (1256 N. McLean, Memphis, TN 38108)

BUFRAP Bulletin de l'Institut International du Froid. (l'Institut, 177 Blvd. Molesherbes, 75017 Paris 17, France) V.1- 1920-

BUMMAB Bulletin of the University of Miami School of Medicine and Jackson Memorial Hospital. (Coral Gables, FL) V.1-20, 1939-66. Discontinued.

BUMOAH Bulletin Mensuel de l'Office International d'Hygiene Publique. (New York, NY) V.1-38, 1909-46. For publisher information, see BWHOA6

BUYRAI Bulletin of Parenteral Drug Association. (The Association, Western Saving Fund Bldg., Broad and Chestnut Sts., Philadelphia, PA 19107) V.1- 1946-

BVIPA7 Bulletin of the Veterinary Institute in Pulawy. (Instytut Weterynaryjny, Aleja Partyzantow 55, Pulawy, Poland) V.1- 1957-

BVJOA9 British Veterinary Journal. (Bailliere Tindall, 35 Red Lion Sq., London WClR 4SG, England) V.105- 1949-

BWHOA6 Bulletin of the World Health Organization. (WHO, 1211 Geneva 27, Switzerland) V.1- 1947-

BYYADW Byoin Yakugaku. Hospital Pharmacology. (Yakuji Nippon Sha, 1-11 Izumi-cho, Kanda, Chiyoda-ku, Tokyo 101, Japan) V.1- 1975-

BZARAZ Biologicheskii Zhurnal Armenii. Biological Journal of Armenia. (v/o Mezhdunarodnaya Kniga, Kuznetskii Most 18, Moscow G-200, USSR) V.19- 1966-

ChaEB# Personal Communication from Dr. E.B. Chappel, Abbott Labs., N. Chicago, Illinois, to I. Pigman, Tracor Jitco, Inc., January 19, 1978

ChrHE# Personal Communication to Editor, nToxic Substances List, from Dr. H.E. Christensen, Bethesda, MD, October 18, 1973

CroHP# Personal Communication from H.P. Crocker, Ultimate Holding Co.-Reckitt and Colman Ltd., London, England, to R.L. Tatken, NIOSH, Cincinnati, OH, September 2, 1980

CAANAT Comptes Rendus de l'Association des Anatomistes. (Paris, France) V.1- 1916(?)-

CAES** Reports supported by the Pennwalt Corp. and the Center for Air Environment Studies at the Pennsylvania State University.

CAJOBA Canadian Journal of Ophthalmology. (Canadian Ophthalmological Society, Box 8844, Ottawa, Ont K1G 3G2, Canada) V.1- 1966-

CAJPBD Proceedings of the Congenital Anomalies Research Association of Japan. (Kyoto, Japan) No.1- 1961-

CALEDQ Cancer Letters. (Elsevier Publishing, P.O. Box 211, Amsterdam C, Netherlands) V.1- 1975-

CAMCAM Ca - A Cancer Journal for Clinicians. (American Cancer Society, 777 3rd Ave., New York, NY 10017) V.1- 1950-

CAMEAS California Medicine. (San Francisco, CA) V.65-119, 1946-73.

CANCAR Cancer. (J. B. Lippincott Co., E. Washington Sq., Philadelphia, PA 19105) V.1- 1948-

CANJAE Canadian Anaesthetist's Society Journal. (178 St. George St., Toronto 5, Ont M5R 2M7, Canada) V.1- 1954-

CANRA4 Cancer, Revue Internationale d'Etudes Cancerologiques. (Brussels, Belgium) V.1-13, 1923-37. Discontinued.

CAREBK Caries Research. (S. Karger AG, Postfach, CH-4009, Basel, Switzerland) V.1- 1967-

CARYAB Caryologia. (Caryologia, Via Lamarmora 4, 50121 Florence, Italy) V.1- 1948-

CAXXA4 Canadian Patents. (U.S. Patent Office, Science Library, 2021 Jefferson Davis Highway, Arlington, VA 22202)

CBCCT* "Summary Tables of Biological Tests," National Research Council Chemical-Biological Coordination Center. (National Academy of Science Library, 2101 Constitution Ave., N.W., Washington, DC 20418)

CBINA8 Chemico-Biological Interactions. (Elsevier Publishing, P.O. Box 211, Amsterdam C, Netherlands) V.1- 1969-

CBPBB8 Comparative Biochemistry and Physiology, B: Comparative Biochemistry. (Pergamon Press, Headington Hill Hall, Oxford 0X3 0BW, England) V.38- 1971-

CBPCBB Comparative Biochemistry and Physiology, C: Comparative Pharmacology. (Oxford, England) V.50-73, 1975-82.

CBRPDS Cell Biology International Reports. (Academic Press Inc. Ltd. 24-28 Oval Rd., London NW1 7DX, England) V.1- 1977-

CBTIAE Contributions from Boyce Thompson Institute. (Yonkers, NY) V.1-24, 1925-71. Discontinued.

CBTOE2 Cell Biology and Toxicology. (Princeton Scientific Publishers, Inc., 301 N. Harrison St., CN 5279, Princeton, NJ 08540) V.1- 1984-

CCCCAK Collection of Czechoslovak Chemical Communications. (Academic Press, 24-28 Oval Rd., London NW1 7DX, England) V.1- 1929-

CCHCDE Zhonghua Jiehe He Huxixi Jibing Zazhi. Chinese Journal of Tuberculosis and Respiratory Diseases. (China International Book Trading Corp., POB 2820, Beijing, Peop. Rep. China) V.1- 1978-

CCLCDY Zhonghua Zhongliu Zazhi. Chinese Journal of Oncology. (Guozi Sudian, Beijing, Peop. Rep. China) V.1- 1978-

CCPHDZ Cancer Chemotherapy and Pharmacology. (Springer-Verlag, Heidelberger, Pl. 3, D-1 Berlin 33, Germany) V.1- 1978-

CCPNAG Canadian Cancer Society, British Columbia Division, Provincial News.

CCPTAY Contraception. (Geron-X, Publishers, P.O. Box 1108, Los Altos, CA 94022) V.1- 1970-

CCROBU Cancer Chemotherapy Reports, Part 1. (Washington, DC) V.52, no.6-V.59, 1968-75. For publisher information, see CTRRDO

CCSUBJ Cancer Chemotherapy Reports, Part 2. (Washington, DC) V.1-5, 1964-75. For publisher information, see CTRRDO

CCSUDL Carcinogenesis - A Comprehensive Survey (Raven Press, 1140 Ave. of the Americas, New York, NY 10036) V.1- 1976-

CCTRDH Cancer Clinical Trials. (Masson Publishing USA, Inc., 14 E. 60th St., New York, NY 10022) V.1- 1978-

CCUPAD Ciencia e Cultura. Science and Culture. (Sociedade Brasileira para o Progresso da Ciencia, Caixa Postal 1 1008, 01000 Sao Paulo, Brazil) V.1- 1949

CCYPBY Cancer Chemotherapy Reports, Part 3. (Washington, DC) V.1-6, 1968-75. For publisher information, see CTRRDO

CDESDK Contraceptive Delivery Systems. (Kluwer Academic Publishers Group Distribution Centre, POB 322, 3300 AH Dordrecht, Netherlands) V.1- 1980-

CDGU** Uber die Pharmakologische Wirkung eines dem Pentamethyltetrazol (Cardiazol) nahestehenden stickstoffhaltigen Kampherab-Kommlings, Hans Hermann Czygan Dissertation. (Hessischen Ludwigs-Universitat zu Giessen, Germany, 1934)

CDPRD4 Cancer Detection and Prevention. (Marcel Dekker, Inc., POB 11305, Church St. Station, New York, NY 10249) V.1- 1979-

CECED9 Commission of the European Communities, Report EUR. (Office for Official Publications, POB 1003, Luxembourg 1, Luxembourg) 1967-

CEDEDE Clinical and Experimental Dermatology. (Blackwell Scientific Publications Ltd., Osney Mead, Oxford OX2 0EL, England) V.1- 1976-

CEFYAD Ceskoslovenska Fysiologie. (Academia, Vodickova 40, Prague 1, Czechoslovakia) V.1- 1952-

CEKNA5 Chiba-ken Eisei Kenkyusho Nenpo. (666-2 Nitona-cho, Chiva 280, Japan)

CEHYAN Ceskoslovenska Hygiena. Czechoslovak Hygiena. (Ve Smeckach 30, Prague 1, Czech). V.1- 1956-

CELLB5 Cell (Cambridge, Mass.). (Massachusetts Institute of Technology Press, 28 Carleton St., Cambridge, MA 02142) V.1- 1974-

CENEAR Chemical and Engineering News. (American Chemical Society, 1155 16th St., N.W., Washington, DC 20036) V.20- 1942-

CESTAT Ceskoslovenska Stomatologie. (PNS-Ustredni Expedice Tisku, Jindriska 14, Prague 1, Czechoslovakia) V.36- 1936-

CEXIAL Clinical and Experimental Immunology. (Blackwell Scientific Publications Ltd., Osney Mead, Oxford OX2 0EL, England) V.1- 1966-

CFRGBR Code of Federal Regulations. (U.S. Government Printing Office, Supt. of Doc., Washington, DC 20402)

CGCGBR Cytogenetics and Cell Genetics. (S. Karger AG, Arnold-Boecklin Str. 25, CH-4011 Basel, Switzerland) V.12- 1973-

CGCYDF Cancer Genetics and Cytogenetics. (Elsevier North Holland, Inc., 52 Vanderbilt Ave., New York, NY 10017) V.1- 1979-

CHABA8 Chemical Abstracts. (Chemical Abstracts Service, Box 3012, Columbus, OH 43210) V.1- 1907-

CHBEAM Chemische Berichte. (Verlag Chemie GmbH, Postfach 129/149, Pallelallee 3, D-6940, Weinheim/Bergst, Germany) V.80-1947-

CHDDAT Comptes Rendus Hebdomadaires des Seances de l'Academie des Sciences, Serie D. (Centrale des Revues Dunod-Gauthier-Villars, 24-26 Blvd. de l'Hopital, 75005 Paris, France) V.262- 1966-

CHETBF Chest: The Journal of Circulation, Respiration and Related Systems. (American College of Chest Physicians, 911 Busse Hwy., Park Ridge, IL 60068) V.57- 1970-

CHFCA2 Zhonghua Fuchanke Zazhi (Beijing). Chinese Journal of Obstetrics and Gynecology. (Guozi Shudian, Beijing, Peop. Rep. China) V.1, 1953-66; 1978-

CHHTAT Zhonghua Yixue Zazhi. Chinese Medical Journal. (Guozi Shudian, Beijing, Peop. Rep. China) V.1- 1915-

CHIMAD Chimia. (Sauerlaender AG, 5001 Aarau, Switzerland) V.1-1947-

CHINAG Chemistry and Industry. (Society of Chemical Industry, 14 Belgrave Sq., London SW1X 8PS, England) V.1-21, 1923-43; No.1-1944-

CHIP** Chemical Hazard Information Profile. Draft Report. (U.S. Environmental Protection Agency, Office of Toxic Substances, 401 M St., S.W., Washington, DC 20460)

CHMBAY Chemistry in Britain. (Chemical Society, Publications, Sales Office, Blackhorse Rd., Letchworth SG6 1HN, Herts, England) V.1- 1965-

CHMEBA China's Medicine. (Guozi Shudian, Peking, China) 1966-68. Discontinued.

CHMTBL Chemical Technology. (American Chemical Society, 1155 16th St., N.W., Washington, DC 20036) V.1- 1917-

CHPUA4 Chemicky Prumysl. Chemical Industry. (ARTIA, Ve Smeckach 30, 111-27 Prague 1, Czechoslovakia) V.1- 1951-

CHREAY Chemical Reviews. (American Chemical Society, 1155 16th St., N.W., Washington, DC 20036) V.1- 1924-

CHROAU Chromosoma. (Springer-Verlag, Heidelberger Platz 3, D-1000 Berlin 33, Federal Republic of Germany) V.1- 1939-

CHRTBC Chromosomes Today. (Elsevier Scientific Publishing Co., P.O. Box 211 Amsterdam, The Netherlands) V.1- 1966-

CHRYAQ Chemistry. (American Chemical Society Publications, 1155 16th St., N.W., Washington, DC 20036) V.18- 1944-

CHTHBK Chemotherapy. (S. Karger AG, Arnold-Boecklin-St 25, CH-4000, Basel 11, Switzerland) V.13- 1968-

CHTPBA Chimica Therapeutica. (Editions DIMEO, Arcueil, France) V.1-8, 1965-73.

CHWEAP Chemisch-Weekblad. (Koninklijke Nederlandse Chemische Vereniging, Afdeling Tijdschriften, Brunierstraat 1, The Hague, Netherlands) V.1- 1903-

CHWKA9 Chemical Week. (McGraw-Hill, Inc., Distribution Center, Princeton Rd., Hightstown, NJ 08520) V.68- 1951-

CHYCDW Zhonghua Yufangyixue Zazhi. Chinese Journal of Preventive Medicine. (42 Tung Szu Hsi Ta Chieh, Beijing, Peop. Rep. China) Beginning history not known.

CIFAA3 Circular Farmaceutica. (Colegio Oficial de Farmaceuticos, Via Layetana, 94, Barcelona, Spain) V.1- 1943-

CIGET* Ciba-Geigy Toxicology Data/Indexes, 1977. (Ciba-Geigy Corp., Ardsley, NY 10502)

CIGZAF Chiba Igakkai Zasshi. Journal of the Chiba Medical Society. (Chiba, Japan) V.1-49, 1923-73. For publisher information, see CIZAAZ

CIHPDR Zhongguo Yixue Kexueyuan Xuebao. Journal of the Chinese Academy of Medicine. (China Book Trading Corp., POB 2820, Beijing, Peop. Rep. China) V.1- 1979-

CIIT** Chemical Industry Institute of Toxicology, Docket Reports. (POB 12137, Research Triangle Park, NC 27709)

CIRCAZ Circulation. (American Heart Assoc., 7320 Greenville Ave., Dallas, TX 75231) V.1- 1950-

CIRUAL Circulation Research. (American Heart Association, Publishing Director, 7320 Greenville Ave., Dallas, TX 75231) V.1- 1953-

CISCB7 CIS, Chromosome Information Service. (Maruzen Co. Ltd., POB 5050, Tokyo International, Tokyo 100-31, Japan) No.1- 1961-

CIWYAO Carnegie Institute of Washington, Year Book. (Carnegie Institution of Washington, 1530 P St., N.W. Washington, DC 20005) V.1- 1902-

CIYADX Jiangsu Yiyao. Chiang-su Medicine. (Chiang-su Jen Min Ch'u Pan She, 13 Hu Nan Lu, Nanking, Peop. Rep. China) V.1- 1975(?)-

CIYPDA Sichuan Yixueyuan Xuebao. Acta Akademiae Medicinae Sichuan. (Guozi Sudian, Beijing, Peop. Rep. China) V.1- 1970-

CIZAAZ Chiba Igaku Zasshi. Chiba Medical Journal. (Chiba Igakkai, Inohana 1-8-8, Chiba 280, Japan) V.50- 1974-

CJBBDU Canadian Journal of Biochemistry and Cell Biology. (National Research Council of Canada, Ottawa, Canada ON K1A OR6) V.61- 1983-

CJBIAE Canadian Journal of Biochemistry. (Ontario, Canada) V.42-60, 1964-82. For publisher information, see CJBBDU

CJBPAZ Canadian Journal of Biochemistry and Physiology. (Ottawa, Canada) V.32-41, 1954-63. For publisher information, see CJBIAE

CJCHAG Canadian Journal of Chemistry. (National Research Council of Canada, Ottawa K1A OR6, Ontario, Canada) V.29- 1951-

CJCMAV Canadian Journal of Comparative Medicine. (360 Bronson Ave., Ottawa K1R 6J3, Ontario, Canada) V.1-3, 1937-39; V.32-1968-

CJMIAZ Canadian Journal of Microbiology. (National Research Council of Canada, Administration, Ottawa, Canada K1A OR6) V.1- 1954-

CJNSA2 Canadian Journal of Neurological Sciences. (Editor, 1516-233 Kennedy St., Winnipeg, Manitoba R3C 3J5, Canada) V.1- 1974-

CJPEA4 Canadian Journal of Public Health. (Canadian Public Health Assoc., 1335 Carling Avenue, Suite 210, Ottawa, Ontario K1Z 8N8, Canada) V.20- 1929-

CJPPA3 Canadian Journal of Physiology and Pharmacology. (National Research Council of Canada, Ottawa K1A OR6, Ontario, Canada) V.42- 1964-

CJPSDF Canadian Journal of Psychiatry. (Keith Health Care Communications, 289 Rutherford Road S., Suite 11, Brampton, ON L6W 3R9, Canada)

CJSUAX Canadian Journal of Surgery. (Canadian Medical Association, P.O. Box 8650, Ottawa, Ontario K1G OG8, Canada) V.1-1957-

CJVRE9 Canadian Journal of Veterinary Research. (339 Booth St. Ottawa, K1R 7K1, Canada) V.50- 1986-

CKFRAY Ceskoslovenska Farmacie. (PNS-Ustredni Expedice Tisku, Jindriska 14, Prague 1, Czechoslovakia) V.1- 1952-

CLANA4 Clinical Anesthesia. (F.A. Davis Co., 1915 Arch St., Philadelphia, PA 19103) No.1- 1963-

CLBIAS Clinical Biochemistry. (Canadian Society of Clinical Chemists, 151 Slater St., Ottawa K1P 5H3, Ontario, Canada) V.1- 1967-

CLCEAL Casopis Lekaru Ceskych. Journal of Czech Physicians. (ARTIA, Ve Smeckach 30, Prague 1, Czechoslovakia) V.1- 1862-

CLCHAU Clinical Chemistry. (American Assoc. of Clinical Chemists, 1725 K St., N.W., Washington, DC 20006) V.1- 1955-

CLDFAT Cell Differentiation. (Elsevier/North-Holland Scientific Publishers Ltd., POB 85, Limerick, Ireland) V.1- 1972-

CLDND* Compilation of LD50 Values of New Drugs. (J.R. MacDougal, Dept. of National Health and Welfare, Food and Drug Divisions, 35 John St., Ottawa, Ontario, Canada)

CLECAP Clinical Endocrinology. (Blackwell Scientific Publications, Osney Mead, Oxford OX2 OEL, England) V.1- 1972-

CLMEA3 Clinical Medicine. (Clinical Medicine Publications, 444 Frontage Rd., Northfield, IL 60093) V.69- 1962-

CLNHBI Clinical Nephrology. (Dustri-Verlag Dr. Karl Feistle, Postfach 49, D-8024 Munich-Deisenhofen, Fed. Rep. Ger.) V.1-1973-

CLONEA Clinics in Oncology. (W.B. Saunders Co., W. Washington Sq., Philadelphia, PA 19105) V.1- 1982-

CLPJAX Cleft Palate Journal. (American Cleft Palate Assoc., c/o B.J. McWilliams, ed., Univ. of Pittsburgh, 331 Salk Hall, Pittsburgh, PA 15261) V.1- 1964-

CLPTAT Clinical Pharmacology and Therapeutics. (C.V. Mosby Co., 11830 Westline Industrial Dr., St. Louis, MO 63141) V.1- 1960-

CLREAS Clinical Research. (American Federation for Clinical Research, 6900 Grove Rd., Thorotare, NJ 08086) V.6- 1958-

CLTEA4 Clinica Terapeutica (Rome). (Societa Edtrice Universo, Via G.B. Morgangni, 1 I-00161 Rome, Italy) V.1- 1955-

CMAJAX Canadian Medical Association Journal. (CMA House, Box 8650, Ottawa K1G OG8, Ontario, Canada) V.1- 1911-

CMBID4 Cellular and Molecular Biology. (Pergamon Press Ltd., Headington Hill Hall, Oxford OX3 0BW, England) v.22- 1977-

CMDT** Christensen, Herbert E., Thesis, National Institute for Occupational Safety and Health, Rockville, MD 20852

CMEP** "Clinical Memoranda on Economic Poisons," U. S. Dept. HEW, Public Health Service, Communicable Disease Center, Atlanta, GA, 1956

CMJOAP Chinese Medical Journal. (Peking, China) V.46-85, 1932-66. For publisher information, see CHMEBA

CMJODS Chinese Medical Journal (Beijing, English Edition). New Series. (Guozi Shudian, Beijing, Peop. Rep. China) V.1- 1975- (Adopted vol. no. 92 in 1979)

CMJRAY Calcutta Medical Journal. (Calcutta Medical Club, CMC House, 91-B Chittaranjan Ave., Calcutta, India) V.1- 1906-

CMLMDW Collection de Medecine Legale et de Toxicologie Medicale. (Masson et Cie, Editeurs, 120 Blvd. Saint-Germain, F-75280 Paris 06, France) No.53- 1970-

CMLTAG Chemistry Letters. (Japan Publications Trading Co., Ltd., POB 5030, Tokyo International, Tokyo 101-31, Japan) No.1- 1972-

CMMUAO Chemical Mutagens. Principles and Methods for Their Detection (Plenum Publishing Corp., 233 Spring St., New York, NY 10013) V.1- 1971-

CMROCX Current Medical Research and Opinion. (Clayton-Wray Publications, 27 Sloane Sq., London SW1W 8AB, England) V.1- 1973-

CMSHAF Chemosphere. (Pergamon Press, Headington Hill Hall, Oxford OX3 OEW, England) V.1- 1971-

CMTRAG Chemotherapia. (Basel, Switzerland) V.1-12, 1960-67. For publisher information, see CHTHBK

CMTVAS Cahiers de Medecine du Travail. (Association Professionnelle Belge des Medecins du Travail, rue du Bultia, 6280 Gerpinnes, Belgium) V.1- 1963-

CNCRA6 Cancer Chemotherapy Reports. (Bethesda, MD) V.1-52, 1959-68. For publisher information, see CCROBU

CNDIBJ Cahiers de Notes Documentaires. (Institut National de Recherche et de Securite, 30 rue Olivier-Noyer, 75680 Paris Cedex 14, France) Note 1- 1955-

CNDQA8 Cahiers de Nutrition et de Dietetique. (Presses Universitaires de France, 108 Blvd. Saint-Germain, 75 Paris 6, France) V.1- 1966-

CNJGA8 Canadian Journal of Genetics and Cytology. (Genetics Society of Canada, 151 Slater St., Suite 907, Ottawa, Ont. K1P 5H4, Canada) V.1- 1959-

CNJMAQ Canadian Journal of Comparative Medicine and Veterinary Science. (Gardenvale, Quebec, Canada) V.4-32, 1940-68. For publisher information, see CJCMAV

CNJPAZ Canadian Journal of Pharmaceutical Sciences. (Canadian Pharmaceutical Association, Inc., 175 College St., Toronto, Ontario M5T 1P8, Canada) V.1- 1966-

CNREA8 Cancer Research. (Waverly Press, Inc., 428 E. Preston St. Baltimore, MD 21202) V.1- 1941-

CNRMAW Canadian Journal of Research, Section E, Medical Sciences. (Ottawa, Canada) V.22-28, 1944-50. For publisher information, see CJBIAE

CNVJA9 Canadian Veterinary Journal. (Canadian Veterinary Journal, 360 Bronson Ave., Ottawa, Ontario K1R 6J3, Canada) V.1- 1960-

CODEDG Contact Dermatitis. Environmental and Occupational Dermatitis. (Munksgaard, 35 Norre Sogade, DK 1370 Copenhagen K, Denmark) V.1- 1975-

COGYAK Clinical Obstetrics and Gynecology. (Harper and Row, Medical Dept., 2350 Virginia Ave., Hagerstown, MD 21740) V.1- 1958-

COINAV Colloques Internationaux du Centre National de la Recherche Scientifique. (Centre National de la Recherche Scientifique, 15, Quai Anatole-France, F-75700 Paris, France) V.1- 1946-

CONEAT Confinia Neurologica. (S. Karger AG, Arnold-Boecklin-St 25, CH-4000, Basel 11, Switzerland) V.1- 1938-

COPYAV Comprehensive Psychiatry. (Grune and Stratton, Inc., 111 5th Ave., New York, NY 10003) V.1- 1960-

COREAF Comptes Rendus Hebdomadaires des Seances de l'Academie des Sciences. (Paris, France) V.1-261, 1835-1965. For publisher information, see CHDDAT

CORPAE Clinical Orthopaedics. (Philadelphia, PA) No.1-25, 1953-62. For publisher information, see CORTBR

CORTBR Clinical Orthopaedics and Related Research. (J. B. Lippincott Co., E. Washington Sq., Philadelphia, PA 19105) No.26- 1963-

COTODO Journal of Fire and Flammability/Combustion Toxicology Supplement. (Technomic Publishing Co., 265 Post Rd. W., Westport, CT 06880) V.1-2, 1974-75. For Publisher information, see JCTODH

COTPAO Clinica Ortopedica. (Clinica Ortopedica dell'Universita Via Giustiniani, 35100 Padua, Italy) V.1- 1949-

COVEAZ Cornell Veterinarian. (Cornell University, New York State College of Veterinary Medicine, Ithaca, NY 14853) V.1- 1911-

CPAJAK Canadian Psychiatric Association Journal. (Suite 103, 225 Lisgar St., Ottawa K2P OC6, Ontario, Canada) V.1- 1956-

CPBTAL Chemical and Pharmaceutical Bulletin. (Pharmaceutical Society of Japan, 12-15-501, Shibuya 2-chome, Shibuya-ku, Tokyo, 150, Japan) V.6- 1958-

CPCHAO Clinical Proceedings of the Children's Hospital of the District of Columbia. (Washington, DC) V.1-27, 1971. For publisher information, see CPNMAQ

CPEDAM Clinical Pediatrics. (J. B. Lippincott Co., E. Washington Sq., Philadelphia, PA 19105) V.1- 1962-

CPGPAY Comparative and General Pharmacology. (New York, NY) V.1-5, 1970-74. For publisher information, see GEPHDP

CPHADV Clinical Pharamacy. (American Soc. of Hospital Pharmacists, 4630 Montgomery Ave., Bethesda, MD 20814) V.1- 1982-

CPHPA5 Jiepou Xuebao. Journal of Anatomy. (Guozi Shudian, Beijing, Peop. Rep. China) V.1-9, 1953-66; V.10- 1979-

CPJOAC Canadian Pharmaceutical Journal. (175 College St., Toronto M5T 1P8, Ontario, Canada) V.1- 1868-

CPNMAQ Clinical Proceedings, Children's Hospital National Medical Center. (2125 13th St., N.W., Washington, DC 20009) V.28- 1972-

CRAAA7 Current Researches in Anesthesia and Analgesia. (Cleveland, OH) V.1-35, 1922-56. For publisher information, see AACRAT

CRBCAI Critical Reviews in Biochemistry. (CRC Press, Inc., 2000 N.W. 24th Street, Boca Raton, Florida 33431) V.1- 1974-

CRDLP* U. S. Army Chemical Research and Development Laboratory, Special Publication. (Edgewood Arsenal, MD 21010)

CRDLR* U. S. Army Chemical Research and Development Laboratory, Technical Report. (Edgewood Arsenal, MD 21010)

CRNGDP Carcinogenesis. (Information Retrieval, 1911 Jefferson Davis Highway, Arlington, VA 22202) V.1- 1980-

CRSBAW Comptes Rendus des Seances de la Societe de Biologie et de Ses Filiales. (Masson et Cie, Editeurs, 120 Blvd. Saint-Germain, P-75280, Paris 06, France) V.1- 1849-

CRSHAG Circulatory Shock. (Alan R. Liss Inc., 150 5th Ave., New York, NY 10011) V.1- 1974-

CRSUBM Cancer Research Supplement (Williams and Wilkins Company, 428 E. Preston St., Baltimore, MD 21202) No.1-4, 1953-56

CRTXB2 CRC Critical Reviews in Toxicology. (CRC Press, Inc., 2000 N.W. 24th St., Boca Raton, FL 33431) V.1- 1971-

CSFUDY Cell Structure and Function. (Japan Society for Cell Biology, Nakanishi Printing Co., Shimotachiuri-ogawa, Kamigyo-Ku, Kyoto 602, Japan) V.1- 1975-

CSHCAL Cold Spring Harbor Conferences on Cell Proliferation. (Cold Spring Harbor Laboratory, POB 100, Cold Spring Harbor, NY 11724) V.1- 1974-

CSHSAZ Cold Spring Harbor Symposia on Quantitative Biology. (Cold Spring Harbor Laboratory of Quantitative Biology, Cold Spring Harbor, NY 11724) V.1- 1933-

CSLNX* U. S. Army Armament Research and Development Command, Chemical Systems Laboratory, NIOSH Exchange Chemicals. (Aberdeen Proving Ground, MD 21010)

CSPEAX Cosmetics and Perfumery. (Allured Publishing Corp., Box 318, Wheaton, IL 60187) V.88- 1973-

CTCEA9 Current Therapeutic Research, Clinical and Experimental. (Therapeutic Research Press, P.O. Box 514, Tenafly, NJ 07670) V.1- 1959-

CTKIAR Cell and Tissue Kinetics. (Blackwell Scientific Publications Ltd., Osney Mead, Oxford OX2 0EL, England) V.1- 1968-

CTOIDG Cosmetics and Toiletries. (Allured Publishing Corp., P.O. Box 318, Wheaton, IL 60187) V.91- 1976-

CTOXAO Clinical Toxicology. (New York, NY) V.1-18, 1968-81. For publisher information, see JTCTDW

CTPHBG Current Topics in Pathology. (Springer-Verlag, Heidelberger Pl. 3, D-1 Berlin 33, Germany) V.51- 1970-

CTRRDO Cancer Treatment Reports. (U. S. Government Printing Office, Supt. of Doc., Washington, DC 20402) V.60- 1976-

CTSRCS Cell and Tissue Research. (Springer-Verlag New York, Inc., Service Center, 44 Hartz Way, Secaucus, NJ 07094) V.148- 1974-

CTYAD8 Zhongcaoyao. Chinese Herbal Medicine. (Tsa Chih Pien Chi Pu, Hu-nan Yao Kung Yeh Yen Chiu So, Shao-Yang, Hu-nan, Peop. Rep. China) V.11- 1980-

CUGED5 Current Genetics. (Springer-Verlag, Heidelberger, Pl. 3, D-1 Berlin 33, Germany) V.1- 1979(?)-

CUMIDD Current Microbiology. (Springer-Verlag New York, Inc., Service Center, 44 Hartz Way, Secaucus, NJ 07094) V.1- 1978-

CUSCAM Current Science. (Current Science Assoc., Mgr., Raman Research Institute, Bangalore 6, India) V.1- 1932-

CUTIBC CUTIS; Cutaneous Medicine for the Practitioner. (Technical Pub., 875 Third Ave., New York, NY 10022) 1965-

CVREAU Cardiovascular Research. (British Medical Journal, 1172 Commonwealth Ave., Boston, MA 02134) V.1- 1967-

CWLTM* Chemical Warfare Laboratories Technical Memorandum. (U. S. Army Chemical Center, Edgewood Arsenal, MD 21010)

CYGEDX Cytology and Genetics. English Translation of Tsitologiya i Genetika. (Allerton Press, Inc., 150 Fifth Ave., New York, NY 10011) V.8- 1974-

CYLPDN Zhongguo Yaoli Xuebao. Acta Pharmacologica Sinica. (Shanghai K'o Hsueh Chi Shu Ch'u Pan She, 450 Shui Chin Erh Lu, Shanghai 200020, Peop Rep. China) V.1- 1980-

CYTBAI Cytobios. (The Faculty Press, 88 Regent St., Cambridge, England) V.1- 1969-

CYTGAX Cytogenetics. (Albert J. Phiebig Inc., POB 352, White Plains, NY 10602) V.1-11, 1962-72

CYTOAN Cytologia. (Maruzen Co. Ltd., P.O. Box 5050, Tokyo International, Tokyo 100-31, Japan) V.1- 1929-

CYTPDT Zhongyao Tongbao. Bulletin of Chinese Materia Medica. (Chinese International Book Trading Corp., POB 2820, Beijing, Peop. Rep. China) V.1- 1955- Suspended 1960-80?

CYTZAM Cytobiologie. (Stuttgart 1, Fed. Rep. Ger.) V.1-18, 1969-79. For publisher information, see EJCBDN

CZXXA9 Czechoslovakian Patent Document. (Commissioner of Patents and Trademarks, Washington, DC 20231)

DABBBA Dissertation Abstracts International, B: The Sciences and Engineering. (University Microfilms, A Xerox Co., 300 N. Zeeb Rd., Ann Arbor, MI 48106) V.30- 1969-

DABSAQ Dissertation Abstracts, B: The Sciences and Engineering. (University Microfilms, A Xerox Co., 300 N. Zeeb Rd., Ann Arbor, MI 48106) V.27-29, 1966-69.

DADEDV Drug and Alcohol Dependence. (Elsevier Sequoia SA, POB 851, CH-1001 Lausanne, Switzerland) V.1- 1975-

DAKMAJ Deutsches Archiv fuer Klinische Medizin. (Munich, Germany) V.1-211, 1865-1965. For publisher information, see EJCIB8

DANKAS Doklady Akademii Nauk S.S.S.R. (v/o Mezhdunarodnaya Kniga, Kuznetskii Most 18, Moscow G-200, U.S.S.R.) V.1- 1933-

DANND6 Dopovidi Akademii Nauk Ukrains'koi RSR, Seriya B: Geologichini, Khimichni ta Biologichni Nauki. (v/o Mezhdunarodnaya Kniga, 121200 Moscow, USSR) 1976-

DANTAL Doklady Akademii Nauk Tadzhikskoi SSR. Proceedings of the Academy of Sciences of the Tadzhik SSR. (v/o Mezhdunarodnaya Kniga, Kuznetskii Most 18, Moscow G-200, U.S.S.R.) V.1- 1958-

DAPOAG Deutsche Apotheker, Die Aktuelle Zeitschrift Fuer Pharmazeutische Berufe. (DM28, Postfach 350, 637 Oberursel, Germany) V.1- 1949-

DAZEA2 Deutsche Apotheker-Zeitung. (Deutscher Apotheker-Verlag, Postfach 40, D-7000 Stuttgart 1, Fed. Rep. Ger.) V.49- 1934-

DAZRA7 Doklady Akademii Nauk Azerbaidzhnskoi SSR. Proceedings of the Academy of Sciences of the Azerbaidzhan SSR. (v/o Mezhdunarodnaya Kniga, Kuznetskii Most 18, Moscow G-200, USSR) V.1- 1945-

DBABEF Doga Bilim Dergisi, Seri A2: Biyoloji. Natural Science Journal, Series A2. (Turkiye Bilimsel ve Teknik Arastirma Kurumu, Ataturk Bul. No. 221, Kavaklidere, Ankara, Turkey) V.8- 1984-

DBANAD Doklady Bolgarskoi Akademii Nauk. (Hemus, Blvd. Russki 6, Sofia, Bulgaria) V.1- 1948-

DBIOAM Doklady Biochemistry. English of Doklady Akademii Nauk SSSR. (Plenum Publishing Corp., 233 Spring St., New York, NY 11013) V.112- 1957-

DBLRAC Doklady Akademii Nauk BSSR. Proceedings of the Academy of Sciences of the Belorussian SSR. (v/o Mezhdunarodnaya Kniga, Kuznetskii Most 18, Moscow G-200, USSR) V.1- 1957-

DBRRDB Developmental Brain Research. (Elsevier North-Holland Biomedical Press, Amsterdam, Netherlands) V.1- 1981-

DBTEAD Diabete. (Le Raincy, France) V.1-22, 1953-1974

DBTGAJ Diabetologia. (Springer-Verlag New York, Inc., Service Center, 44 Hartz Way, Secaucus, NJ 07094) V.1- 1965-

DCINAQ Drug and Cosmetic Industry. (Drug Markets, 101 W. 31st St., New York, NY 10001) V.30- 1932-

DCTODJ Drug and Chemical Toxicology. (Marcel Dekker, POB 11305, Church St. Station, New York, NY 10249) V.1- 1977/78-

DCUNAI "Drugs in Current Use and New Drugs," Modell, W., ed., New York, Springer Verlag, 1976.

DDEVD6 Drug Development and Evaluation. (Gustav Fischer Verlag, Postfach 720143, D-7000 Stuttgart 70, Fed. Rep. Ger.) V.1- 1977-

DDIPD8 Drug Development and Industrial Pharmacy. (Marcel Dekker, POB 11305, Church St. Station, New York, NY 10249) V.3- 1977-

DDREDK Drug Development Research. (Alan R. Liss, Inc., 150 5th Ave., New York, NY 10011) V.1- 1981-

DDSCDJ Digestive Diseases and Sciences. (Plenum Publishing Corp., 233 Spring St., New York, NY 10013) V.24- 1979-

DEBIAO Developmental Biology. (Academic Press, Inc., 111 5th Ave., New York, NY 10003) V.1- 1959-

DEBIDR Developments in Biochemistry. (Elsevier North Holland, Inc., 52 Vanderbilt Ave., New York, NY 10017) V.1- 1978-

DECRDP Drugs under Experimental and Clinical Research. (J.R. Prous Pub., Apartado de Correos 1179, Barcelona, Spain) V.1- 1977-

DEGEA3 Deutsche Gesundheitswesen. (VEB Verlag Volk und Gesundheit, Neue Gruenstr 18, 102 Berlin, E. Germany) V.1- 1946-

DENED7 Developmental Neuroscience (Basel). (S. Karger AG, Arnold- Boecklin Str. 25, CH-4011 Basel, Switzerland) V.1- 1978-

DEPBA5 Developmental Psychobiology. (John Wiley and Sons Ltd., Baffins Lane, Chichester, Sussex P01 1UD, England) V.1- 1968-

DERAAC Dermatologica. (Albert J. Phiebig, Inc., P.O. Box 352, White Plains, NY 10602) V.79- 1939-

DFSCDX Developments in Food Science. (Elsevier Science Pub. Co., Inc. 52 Vanderbilt Ave., New York, NY 10017) V.1- 1978-

DGDFA5 Development, Growth and Differentiation. (Maruzen Co. Ltd., P.O. Box 5050, Tokyo International, Tokyo 100-31, Japan) V.11- 1969-

DHEFDK HEW Publication (FDA. United States). (Washington, DC) 19??-1979(?). For publisher information, see HPFSDS

DIAEAZ Diabetes. (American Diabetes Assoc., 600 5th Ave., New York, NY 10020) V.1- 1952-

DICHAK Diseases of the Chest. (Chicago, IL) V.1-56, 1935-69. For publisher information, see CHETBF

DICPBB Drug Intelligence and Clinical Pharmacy. (Drug Intelligence and Clinical Pharmacy, Inc., University of Cincinnati, Cincinnati, OH 45267) V.3- 1969-

DICRAG Diseases of the Colon and Rectum. (Harper and Row, Publishers, 10 E. 53rd St., New York, NY 10022) V.1- 1958-

DIGEBW Digestion. (S. Karger AG, Arnold-Boecklin Street 25, CH-4011 Basel, Switzerland) V.1- 1968-

DIMCAL Developments in Industrial Microbiology. (Society for Industrial Microbiology, 1401 Wilson Blvd., Arlington, VA 22209) V.1- 1960-

DIPHAH Dissertationes Pharmaceuticae. (Warsaw, Poland) V.1-17, 1949-65. For publisher information, see PJPPAA

DKBSAS Doklady Biological Sciences (English Translation). (Plenum Publishing Corp., 233 Spring St., New York, NY 10013) V.112-1957-

DMBUAE Danish Medical Bulletin. (Ugeskrift for Laeger, Domus Medica, 2100 Copenhagen, Denmark) V.1- 1954-

DMCNAW Developmental Medicine and Child Neurology. (J. B. Lippincott Co., E. Washington Sq., Philadelphia, PA 19105) V.1-1958-

DMCSAD Developmental Medicine and Child Neurology, Supplement. (William Heinemann Medical Books Ltd., 23 Bedford Sq., London WC1B 3HT, England) V.5- 1962-

DMDSAI Drug Metabolism and Disposition. (Williams and Wilkins Co., 428 E. Preston St., Baltimore, MD 21202) V.1- 1973-

DMTRAR Drug Metabolism Reviews. (Marcel Dekker, POB 11305, Church St. Station, New York, NY 10249) V.1- 1972-

DMWOAX Deutsche Medizinische Wochenschrift. (Georg Thieme Verlag, Herdweg 63, Postfach 732, 7000 Stuttgart 1, Federal Republic of Germany) V.1- 1875-

DNEUD5 Developments in Neuroscience (Amsterdam). (Elsevier North Holland, Inc., 52 Vanderbilt Ave., New York, NY 10017) V.1- 1977-

DNSSAW Diseases of the Nervous System, Supplement. (Irvington, NJ) For publisher information, see DNSYAG

DNSYAG Diseases of the Nervous System. (Physicians Postgraduate Press, Box 38293, Memphis, TN 38138) V.1-38, 1940-77.

DNXUDW Dongbei Nongxueyuan Xuebao. Journal of Northeast Agricultural College. (Dongbei Nongxueyuan Xuebao Beinjibu, Majia Huayuan, Xiangfangqu Harbin, Peop. Rep. China) History not known.

DOEAAH Down to Earth. A Review of Agricultural Chemical Progress. (Dow Chemical U.S.A., 1703 S. Saginaw Rd., Midland, MI 48640) V.1- 1945-

DOELB* Reports prepared for the Office of Environmental Research of the Assistant Secretary for the Environment, U.S. Department of Energy, Under Contract Number EY-76-C-04-1013 by the Lovelace Biomedical and Environmental Research Institute

DOESD6 DOE Symposium Series. (NTIS, 5285 Port Royal Rd., Springfield, VA 22161) No.45- 1978-

DOWCC* Dow Chemical Company Reports. (Dow Chemical U.S.A., Health and Environment Research, Toxicology Research Lab., Midland, MI 48640)

DPHFAK Dissertationes Pharmaceuticae et Pharmacologicae. (Warsaw, Poland) V.18-24, 1966-72. For publisher information, see PJPPAA

DPIRDU Dangerous Properties of Industrial Materials Report. (Van Nostrand Reinhold Co., Inc., 135 West 50th St., New York, NY 10020) V.1- 1981-

DPTHDL Developmental Pharmacology and Therapeutics. (S. Karger AG, Postfach CH-4009 Basel, Switzerland) V.1- 1980-

DRFUD4 Drugs of the Future. (J.R. Prous, S.A. International Publishers, Apartado de Correos 1641, Barcelona, Spain) V.1- 1975/76-

DRISAA Drosophila Information Service (Cold Spring Harbor Laboratory, POB 100, Cold Spring Harbor, NY 11724) No.1- 1934-

DRSTAT Drug Standards. (Washington, DC) V.19-28, 1951-60. For publisher information, see JPMSAE

DRUGAY Drugs. International Journal of Current Therapeutics and Applied Pharmacology Reviews. (ADIS Press Ltd., 18/F., Tung Sun Commercial Centre, 194-200 Lockhart Road, Wanchai, Hong Kong)

DSMJAA Delaware Medical Journal. (Medical Society of Delaware, 1925 Lovering Ave., Wilmington, DE 19806) V.32- 1960-

DTESD7 Developments in Toxicology and Environmental Science. (Elsevier, Scientific Publishing Co., POB 211, 1000 AE Amsterdam, Netherlands) V.1- 1977-

DTLVS* "Documentation of Threshold Limit Values for Substances in Workroom Air." For publisher information, see 85INA8

DTLWS* "Documentation of the Threshold Limit Values for Substances in Workroom Air," Supplements. For publisher information, see 85INA8

DTTIAF Deutsche Tieraerztliche Wochenschrift. (Verlag M. und H. Schaper, Postfach 260669, 3 Hanover 26, Germany) V.1- 1893-

DUPON* E. I. Dupont de Nemours and Company, Technical Sheet. (1007 Market St., Wilmington, DE 19898)

DUSA** "Dispensatory of the United States of America," Osol, A. and G.E. Farrar, 24th ed., Philadelphia, J.B. Lippincott, 1947

DVBSA3 Developments in Biological Standardization. (S. Karger AG, Postfach CH-4009 Basel, Switzerland) V.23- 1974-

DZGGAK Deutsche Zeitschrift fuer die Gesamte Gerichtliche Medizin. (Heidelberg, Germany) V.1-66, 1922-69. For publisher information, see ZRMDAN

DZMKAS Deutsche Zahn-, Mund- und Kieferheilkunde. (Leipzig, German Democratic Republic) V.1-10, 1934-43; V.11-62, 1948-74.

DZZEA7 Deutsche Zahnaerztliche Zeitschrift. (Carl Hanser Verlag, Postfach 860420, D-8000 Munich 86, Federal Republic of Germany) V.1- 1946-

EbeAG# Personal Communication to NIOSH from A.G. Ebert, International Glutamate Technical Committee, 85 Walnut St., Watertown, MA 02172

EAGRDS Experimental Aging Research. (Beech Hill Publishing Co., P.O. Box 136, Southwest Harbor, ME 04679) V.1- 1975-

EAHB** Environmental and Health Criteria for Benzidine. Meeting of DHEW Committee to Coordinate Toxicology and Related Programs, February 15, 1974 at Sterling Forest, NY

EAMJAV East African Medical Journal (P.O. Box 41632, Nairobi, Kenya) V.9- 1932-

EAPHA6 Eastern Pharmacist. (Eastern Pharmacist, 507, Ashok Bhawan, 93, Neru Place, New Delhi 110019, India) V.1- 1958-

EATR** U. S. Army, Edgewood Arsenal Technical Report. (Aberdeen Proving Ground, MD 21010)

ECBUDQ Ecological Bulletins. (Swedish National Science Research Council, Stockholm) Number 19- 1975-

ECEBDI Experimental Cell Biology. (Phiebig Inc., POB 352, White Plains, NY 10602) V.44- 1976-

ECIET* European Chemical Industry Ecology and Toxicology Centre. Joint Assessment of Commodity Chemicals. (Avenue Louise 250, B-63, B-1050 Brussels, Belgium)

ECJPAE Endocrinologia Japonica. (Japan Publications Trading Co., 1255 Howard St., San Francisco, CA 94103) V.1- 1954-

ECMAAI Economie et Medecine Animales. (Librairie des Facultes de Medecine et de Pharmacie, 174 Blvd. Saint-Germain, 75280 Paris CEDEX 06, France) V.1- 1960-

ECNEAZ Electroencephalography and Clinical Neurophysiology. (Elsevier Scientific Publishing Co., POB 211, 1000 AE Amsterdam, Netherlands) V.1- 1949-

ECREAL Experimental Cell Research. (Academic Press, 111 5th. Ave., New York, NY 10003) V.1- 1950-

EDRCAM Endocrine Research Communcations. (Marcel Dekker, POB 11305, Church St. Station, New York, NY 10249) V.1- 1974-

EDWU** Beitrag zur Toxikologie Technischer Weichmachungsmittel, Heinrich Eller Dissertation. (Pharmakologischen Institut der Universitat Wurzburg, Germany, 1937)

EESADV Ecotoxicology and Environmental Safety. (Academic Press, 111 5th Ave., New York, NY 10003) V.1- 1977-

EGESAQ Egeszsegtudomany. (Kultura, POB 149, H-1389 Budapest, Hungary) V.1- 1957-

EGJBAY Egyptian Journal of Botany. (National Information and Documentation Centre, Al-Tahrir St., Awgaf P.O. Dokki, Cairo, Egypt) V.1- 1958-

EISOAU Eiyo to Shokuryo. Food and Nutrition. (Nippon Eiyo, Shokuryo Gakkai, C/o Nippon Gakkai Jimu Senta 2-4-16 Yayoi, Bunkyo-ku, Tokyo 113, Japan) V.10- 1957-

EJBCAI European Journal of Biochemistry. (Springer-Verlag, Heidelberger, Pl. 3, D-1 Berlin 33, Germany) V.1- 1967-

EJBLAB Egyptian Journal of Bilharziasis. (National Information Documentation Center, Tahrir St., Dokki, Cairo, Egypt) V.1- 1974-

EJCAAH European Journal of Cancer. (Pergamon Press, Headington Hill Hall, Oxford OX3 OEW, England) V.1- 1965-

EJCBDN European Journal of Cell Biology. (Wissenschaftliche Verlagsgesellschaft mbH, Postfach 40, D-7000, Stuttgart l, Fed. Rep. Ger.) V.19- 1979-

EJCIB8 European Journal of Clinical Investigation. (Springer-Verlag, Heidelberger, Pl. 3, D-1 Berlin 33, Germany) V.1- 1970-

EJCODS European Journal of Cancer and Clinical Oncology. (Pergamon Press Ltd., Headington Hill Hall, Oxford OX3 0BW, England) V.17, No.7- 1981-

EJCPAS European Journal of Clinical Pharmacology. (Springer-Verlag, Heidelberger Pl. 3, D-1 Berlin 33, Germany) V.3- 1970-

EJGCA9 Egyptian Journal of Genetics and Cytology. (Egyptian Society of Genetics, c/o Alexandria University, Faculty of Agriculture, Dept. of Genetics, Alexandria, Egypt) V.1- 1972-

EJMBA2 Egyptian Journal of Microbiology. (National Information and Documentation Centre, A1-Tahrir St., Awqaf P.O. Dokki, Cairo, Egypt) V.7- 1972-

EJMCA5 European Journal of Medicinal Chemistry. Chimie Therapeutique. (Center National de la Recherche Scientifique, 3 rue J.B. Clement, F-92290 Chatenay-Malabry, France) V.9- 1974-

EJNMD9 European Journal of Nuclear Medicine. (Springer-Verlog New York, Inc., Service Center, 44 Hartz Way, Secaucus, NJ 07094) V.1- 1975-

EJPEDT European Journal of Pediatrics. (Springer-Verlag New York, Inc., Service Center, 44 Hart Way, Secaucus, New Jersey 07094) V.121- 1975-

EJPHAZ European Journal of Pharmacology. (North-Holland Publishing, POB 211, 1000AE Amsterdam, Netherlands) V.1- 1967-

EJRDD2 European Journal of Respiratory Diseases. (Munksgaard, 35 Norre Sogade, DK 1370 Copenhagen K, Denmark) V.61- 1980-

EJTXAZ European Journal of Toxicology and Environmental Hygiene. (Paris, France) v.7-9, 1974-76. For publisher information, see TOERD9

EKFMA7 Eksperimental'naya i Klinicheskaya Farmakoterapiya. (Akademiya Nauk Latviiskoi SSR, Inst., Organicheskogo Sinteza, ul. Aizkraukles 21, Riga, U.S.S.R.) No.1- 1970-

EKMMA8 Eksperimentalna Meditsina i Morfologiya. (Hemus, Blvd. Russki 6, Sofia, Bulgaria) V.1- 1962-

EKSODD Eksperimental'naya Onkologiya. Experimental Oncology. (v/o Mezhdunarodnaya Kniga, 121200 Moscow, USSR) V.1- 1979-

EMERY* Emery Industries, Inc., Data Sheets. (4900 Este Ave., Cincinnati, OH 45232)

EMGMAQ Emergency Medicine. (Fischer Medical Publications, 280 Madison Ave., New York, NY 10006) V.1- 1969-

EMJBA8 E. Merck's Jahresberichte. (Darmstadt, Germany) V.?-V.69, 1908-56. For publisher information, see WBEMAZ

EMMUEG Environmental and Molecular Mutagenesis. (Alan R. Liss, Inc., 4 E. 11th St., New York, NY 10003) V.10- 1987-

EMPSAL Experimental and Molecular Pathology, Supplement. (Academic Press, 111 5th Ave., New York, NY 10003) No.1- 1963-

EMSPAR Electron Microscopy Society of America, Proceedings of Annual Meeting: (Claitor's Publishing Division, POB 239, Baton Rouge, LA 70821) V.35- 1977-

EMSUA8 Experimental Medicine and Surgery. (Brooklyn Medical Press, 600 Lafayette Ave., Brooklyn, NY 11216) V.1- 1943-

ENDKAC Endokrinologie. (Leipzig, Ger. Dem. Rep.) V.1-80, 1928-82. For publisher information, see EXCEDS

ENDOAO Endocrinology. (Williams and Wilkins Co., Dept. 260, P.O. Box 1496, Baltimore, MD 21203) V.1- 1917-

ENMUDM Environmental Mutagenesis. (Alan. R. Liss, Inc., 150 Fifth Ave., New York, NY 10011) V.1- 1979-

ENPBBC Environmental Physiology and Biochemistry. (Copenhagen, Denmark) V.2-5, No. 6, 1972-5, Discontinued.

ENVIDV Environmental Internationl. (Pergamon Press, Ltd., Headington Hill Hall, Oxford OX3 0BW, England) V.1- 1978-

ENVPAF Environmental Pollution. (Applied Science Publishers, Ltd., 510 Ripple Rd., Barking, Essex 1911 OSA, England) V.1- 1970-

ENVRAL Environmental Research. (Academic Press, 111 5th Ave., New York, NY 10003) V.1- 1967-

ENZYAS Enzymologica. (Dr. W. Junk bv Publishers, POB 13713, 2501 ES The Hague, Netherlands) V.1-43, 1936-72.

EOGRAL European Journal of Obstetrics, Gynecology and Reproductive Biology. (Elsevier Science Pub. BV, POB 211,1000 AE Amsterdam, Netherlands) V3- 1973-

EPASR* United States Environmental Protection Agency, Office of Pesticides and Toxic Substances. (U.S. Environmental Protection Agency, 401 M St., S.W. Washington, DC 20460) History Unknown

EPILAK Epilepsia. (Elsevier Publishing Co., P.O. Box 211, Amsterdam C, Netherlands) V.1- 1959-

EPXXDW European Patent Application. (U.S. Patent Office, Science Library, 2021 Jefferson Davis Highway, Arlington, VA 22202)

EQSFAP Environmental Quality and Safety. (Academic Press, 111 5th Ave., New York, NY 10003) V.1- 1972-

EQSSDX Environmental Quality and Safety, Supplement. (Academic Press, 111 5th Ave., New York, NY 10003) V.1- 1975-

ERNFA7 Ernaehrungsforschung. (Akademie-Verlag GmbH, Leipziger St. 3-4, 108 Berlin, E. Germany) V.1- 1956-

ESDBAK Eisei Dobutsu. Sanitary Zoology. (Japan Society of Sanitary Zoology, c/o Institute of Medical Science, Unversity of Tokyo, 4-6- 1 Shiroganedai, Minato-ku, Tokyo 108, Japan) V.1- 1950-

ESKGA2 Eisei Kagaku. (Nippon Yakugakkai, 2-12-15 Shibuya, Shibuya-Ku, Tokyo 150, Japan) V.1- 1953-

ESKHA5 Eisei Shikenjo Hokoku. Bulletin of the National Hygiene Sciences. (Kokuritsu Eisei Shikenjo, 18-1 Kamiyoga 1 chome, Setagaya-ku, Tokyo, Japan) V.1- 1886-

ESTHAG Environmental Science and Technology. (American Chemical Society, 1155 16th St., N.W., Washington, DC 20036) V.1- 1967-

ETATAW Eesti N.S.V. Teaduste Akadeemia Toimetised, Biologia. (v/o Mezhdunarodnaya Kniga, Kuznetskii Most 18, Moscow G-200, U.S.S.R.) V.5- 1956-

ETEAAT Entomologia Experimentalis et Applicata. (Nederlandse Entomologische Vereniging, Pl. Middenlaan 64, 1018 DH Amsterdam, Netherlands) V.1- 1958-

ETOCDK Environmental Toxicology and Chemistry. (Pergamon Press Inc., Maxwell House, Fairview Park, Elmsford, NY 10523) V.1- 1982-

ETOXAC Essays in Toxicology. (Academic Press, 111 5th Ave., New York, NY 10003) V.1- 1969-

EUSRBM European Surgical Research. (S. Karger AG, Arnold-Boecklin Str. 25, CH-4011, Basel, Switzerland) V.1- 1969-

EUURAV European Urology. (S. Karger Publishers, Inc., 150 Fifth Ave., Suite 1105, New York, NY 10011) V.1- 1975-

EVETBX Environmental Entomology. (Entomological Society of America, 4603 Calvert Rd., College Park, MD 20740) V.1- 1972-

EVHPAZ EHP, Environmental Health Perspectives. Subseries of DHEW Publications. (U.S. Government Printing Office, Superintendent of Documents, Washington, DC 20402) No.1- 1972-

EVPHBI Environmental Physiology. (Copenhagen, Denmark) V.1 (No.1-4), 1971, For publisher information, see ENPBBC

EVSRBT Environmental Science Research. (Plenum Publishing Corp., 233 Spring St., New York, NY 10013) V.1- 1972-

EVSSAV Environmental Space Science. English Translation of Kosmicheskaya Biologiya Meditsina. 1967-70.

EXCEDS Experimental and Clinical Edocrinology. (Johann Ambrosius Barth Verlag, Postfach 109, DDR-7010 Leipzig, Ger. Dem. Rep.) V.81- 1983-

EXERA6 Experimental Eye Research. (Academic Press, Inc. 1 E. 1st., Duluth, MN 55802) V.1- 1961-

EXHMA6 Experimental Hematology. (Munksgaard, 35 Norre Sogade, DK 1370 Copenhagen K, Denmark) V.1- 1973-

EXMDA4 International Congress Series - Excerpta Medica. (Elsevier North Holland, Inc., 52 Vanderbilt Ave., New York, NY 10017) No.1- 1952-

EXMPA6 Experimental and Molecular Pathology. (Academic Press, 111 5th Ave., New York, NY 10003) V.1- 1962-

EXNEAC Experimental Neurology. (Academic Press, 111 5th Ave., New York, NY 10003) V.1- 1959-

EXPAAA Experimental Parasitology. (Academic Press, 111 5th Ave., New York, NY 10003) V.1- 1951-

EXPADD Experimental Pathology. (Elsevier Scientific Publishers Ireland Ltd., POB 85, Limerick, Ireland) V.19- 1981-

EXPEAM Experientia. (Birkhaeuser Verlag, P.O. Box 34, Elisabethenst 19, CH-4010, Basel, Switzerland) V.1- 1945-

EXPTAX Experimentelle Pathologie. (Jena, Ger. Dem. Rep.) V.1-18, 1967-80. For publisher information, see EXPADD

FerJJ# Personal Communication to NIOSH from J.J. Ferry, Area Manager, Industrial and Environmental Hygiene, General Electric Company, 1 River Rd., Schenectady, NY 12345

FAATDF Fundamental and Applied Toxicology (Official Journal of the Society of Toxicology, 475 Wolf Ledges Parkway, Akron, OH 44311) V.1- 1981-

FACOEB Food Additives and Contaminants. (Taylor and Francis Inc., 242 Cherry St., Philadelphia, PA 19106) V.1- 1984-

FAONAU Food and Agriculture Organization of United Nations, Report Series. (FAO-United Nations, Room 101, 1776 F Street, NW, Washington, DC 20437)

FAPOA4 Farmacja Polska. Polish Pharmacy. (RUCH, ul Wronia 23, Warsaw 1, Poland) V.1- 1945-

FARMA8 Farmacognosia. (Anales del Instituto Jose Celestino Mutis de Farmacognosia, Madrid, Spain) V.1-27, 1942-62. Discontinued.

FARVAZ Farmakologiya Alkaloidov. (Tashkent, U.S.S.R.) No.1- 1962-

FATOAO Farmakologiya i Toksikologiya (Moscow). (v/o Mezhdunarodnaya Kniga, Kuznetskii Most 18, Moscow G-200, U.S.S.R.) V.2- 1939-

FATOBP Farmakologiya i Toksikologiya (Kiev). Pharmacology and Toxicology. (Kievskii Nauchno-Issledovatel'skii Institut Farmakologii i Toksikologii, Kiev, U.S.S.R.) No.1- 1964-

FAVUAI Fiziologicheski Aktivnye Veshchestva. Physiologically Active Substances. (Akademiya Nauk Ukrainskoi S.S.R., Kiev, U.S.S.R.) No.1- 1966-

FAZMAE Fortschritte der Arzneimittelforschung. (Birkhauser Verlag, P.O. Box 34, Elisabethenst 19, CH-4010, Basel, Switzerland) V.1- 1959-

FCHNA9 Food Chemical News. (Food Chemical News, Inc., 1101 Pennsylvania Ave., S.E., Washington, DC 20003) V.1- 1959-

FCTOD7 Food and Chemical Toxicology. (Pergamon Press, Headington Hill Hall, Oxford OX3 OBW, England) V.20- 1982-

FCTXAV Food and Cosmetics Toxicology. (Pergamon Press Ltd., Maxwell House, Fairview Park Elmsford, NY 10523) V.1-19, 1963-81.

FDHU** Uber die lokalanasthetischen Eigenschaften des salzsauren Salzes des Benzoyl-d-ekgonin-n-propylesters, Gerda Felgner Dissertation. (Pharmakologischen Institut der Martin-Luther-Universitat Halle- Wittenburg, Germany, 1933)

FDMU** Zur Pharmakologie des Octins, Adolf Fiegenbaum Dissertation. (Pharmakologischen Institut der Westfalischen Wilhelms-Universitat, Munster, Germany, 1935)

FDRLI* Food and Drug Research Labs., Papers. (Waverly, NY 14892)

FDWU** Uber die Wirkung Verschiedener Gifte Auf Vogel, Ludwig Forchheimer Dissertation. (Pharmakologischen Institut der Universitat Wurzburg, Germany, 1931)

FEBLAL FEBS Letters. (Elsevier Scientific Publishing Co., POB 211, 1000 AE Amsterdam, Netherlands) V.1- 1968-

FEPRA7 Federation Proceedings, Federation of American Societies for Experimental Biology. (9650 Rockville Pike, Bethesda, MD 20014) V.1- 1942-

FEREAC Federal Register. (U. S. Government Printing Office, Sup. of Doc., Washington, DC 20402) V.1- 1936-

FESTAS Fertility and Sterility. (American Fertility Society, 1608 13th Ave. S., Birmingham, AL 35205) V.1- 1950-

FGIGDO Fujita Gakuen Igakkaishi. Journal of the Fujita Gakuen Medical Society. (Nagoya Hoken Eisei Daigaku Fujita Gakuen Igakkai, 1-98 Dengakuga, Kubo, Kutsukake-Cho, Toyoake, Aichi-Ken 470-11, Japan) V. 1- 1977-

FGREDT Frontiers of Gastrointestinal Research. (S. Karger AG, Arnold- Boecklin Str. 25, CH-4011 Basel, Switzerland) V.1- 1975-

FHCYAI Folia Histochemica et Cytochemica. (Ars Polona-RUCH, P.O. Box 1001, P-00 068 Warsaw, 1, Poland) V.1- 1963-

FKGCAR Fortschritte der Kiefer und Gesichts-Chirurgie. (Georg Thieme Verlag, Stuttgart, Germany) V.1- 1955-

FKIZA4 Fukuoka Igaku Zasshi. (Fukuoka Igakkai, c/o Kyushu Daigaku Igakubu, Tatekasu, Fukuoka-shi, Fukuoka, Japan) V.33- 1940-

FLABAZ Fluoride Abstracts. (Kettering Lab., University of Cincinnati, College of Medicine, Dept. of Envir. Health, Cincinnati, OH 45219) No.1- 1955/58-

FLCRAP Fluorine Chemistry Reviews. (Marcel Dekker Inc., 305 E. 45th St., New York, NY 10017) V.1- 1967-

FLNWAR Forschungsberichte des Landes Nordrhein-Westfalen. (Westdeutscher Verlag GmbH, Postfach 300 620, 5090 Leverkusen 3, Fed. Rep. Ger.) 1960-

FLUOA4 Fluoride. (International Society for Fluoride Research, P.O. Box 692, Warren, MI 48090) V.3- 1970-

FMCHA2 Farm Chemicals Handbook. (Meister Publishing, 37841 Euclid Ave., Willoughy, OH 44094)

FMDZAR Fortschritte der Medizin. (Dr. Schwappach und Co., Wessobrunner Str 4, 8035 Gauting vor Muenchen, Germany) V.1- 1883-

FMLED7 FEMS Microbiology Letters. (Elsevier Scientific Publishing Co., POB 211, Amsterdam, Neth.) V.1- 1977-

FMNUAS Farmacia Nueva. (Eugenia de Montijo, 68 Madrid 25, Spain) V.1- 1934-

FMORAO Folia Morphologica (Prague). (Plenum Publishing Corp. 233 Spring St., New York, NY 10013) V.13- 1965-

FMTYA2 Farmatsiya (Sofia). (Hemus, Blvd. Russki 6, Sofia, Bulgaria) V.1- 1951-

FMXXAJ French Medicament Patent Document. (U.S. Patent Office, Science Library, 2021 Jefferson Davis Highway, Arlington, VA 22202) Discontinued.

FNSCA6 Forensic Science. (Elsevier Sequoia SA, P.O. Box 851, CH 1001, Lausanne 1, Switzerland) V.1- 1972-

FOBGA8 Folia Biologica (Krakow). (Ars Polona-RUCH, POB 154, Warsaw 1, Poland) V.1- 1953-

FOBLAN Folia Biologica (Prague). (Academic Press, 111 Fifth Ave., New York, NY 10003) V.1- 1955-

FOCIAJ Folia Clinica Internacional. (Editorial Glarma, Folgarolas, 15 Barcelona, Spain) V.1- 1951-

FOHEAW Folia Haematologica (Leipzig). (Buchexport, Postfach 160, Leninstr. 16, DDR-701, Leipzig, Ger. Dem. Rep.) V.69- 1949-

FOMAAB Food Manufacture. (Bayard, S. Co., 20 Vesey St., New York, NY 10007) V.1- 1927-

FOMDAK Folia Medica. (Via Raffaele de Caesare, 31 Naples, Italy) V.1- 1915-

FOMIAZ Folia Microbiologica (Prague). (Academia, Vodickova 40, CS-112 29 Prague 1, (Czechoslavakia) V.4- 1959-

FOMOAJ Folia Morphologica (Warsaw). (Panstwowy Zaklad Wydawnictw Lekarskich, ul. Druga 38-40, P-00 238 Warsaw, Poland) V.1- 1929-

FOREAE Food Research. (Champaign, IL) V.1-25, 1936-60. For publisher information, see JFDSAZ

FOTEAO Food Technology. (Institute of Food Technolgists, 221 N. La Salbe St., Chicago, IL 60601) V.1- 1947-

FPNJAG Folia-Psychiatrica et Neurologica Japonica. (Folia Publishing Society, Todai YMCA Bldg., 1-20-6 Mukogaoka, Bunkyo-Ku, Tokyo 113, Japan) 1947-

FRBGAT Frontiers of Biology. (Elsevier North Holland Inc., 52 Vanderbilt Ave., New York, NY 10017) V.1- 1966-

FRCHD6 JFF/Fire Retardant Chemistry. (Technomic Publishing Co., Inc., 265 W. State St., Westport, CN 06880) V.1- 1974-

FRMBAZ Farmacia. (Rompresfilatelia, P.O. Box 2001, Calea Grivitei 64-66, Bucharest, Romania) V.1- 1953-

FRMTAL Farmatsiya (Moscow). Pharmacy. (v/o Mezhdunarodnaya Kniga, 121200 Moscow, USSR) V.16- 1967-

FRPPAO Farmaco, Edizione Pratica. (Casella Postale 114, 27100 Pavia, Italy) V.8- 1953-

FRPSAX Farmaco, Edizione Scientifica. (Casella Postale 114, 27100 Pavia, Italy) V.8- 1953-

FRXXBL French Demande Patent Document. (Commissioner of Patents and Trademarks, Washington, DC 20231)

FRZKAP Farmatsevtichnii Zhurnal (Kiev). (v/o Mezhdunarodnaya Kniga, Kuznetskii Most 18, Moscow G-200, USSR) V.3- 1930-

FSASAX Fette, Seifen, Anstrichmittel. (Industrieverlag von Hernhaussen KG, Roedingsmarkt 24, 2 Hamburg 11, Fed. Rep. Ger.) V.55- 1953-

FSDZD4 Fukuoka Shika Daigaku Gakkai Zasshi. Journal of Fukuoka Dental College. (700 Ta, O-aza, Sawara-ku, Fukuoka, 814-01, Japan) V.1- 1974-

FSIZAQ Fukushima Igaku Zasshi. (c/o Fukushima-Kenritsu Ika Daigaku Toshokan, 5175 Sugizuma-cho, Fukushima, Japan) V.1- 1951-

FSSJAS Forensic Science Society Journal. (The Society, 107 Fenchurch St., London EC3M 5JB, England) V.1- 1960-

FSTEAI Farmaco (Pavia); Scienza e Tecnica. (Pavia, Italy) V.1-7, 1946-52. For publisher information, see FRPSAX

FTNZAO Food Technology in New Zealand. (R. Hill and Son, Ideal House, Eden St., Newmarket, Auckland, New Zealand) V.1- 1966-

FTSEAK Fette und Seifen. (Hamburg, Fed, Rep. Ger.) V.43-54, 1936-52. For publisher information, see FSASAX

FUJAD7 Fragrance Journal. (Fureguransu Janaru Sha, c/o Okumura Bldg., 2-7-4 Iidabashi, Chiyoda-Ku, Tokyo 102, Japan) 1974(?)-

FZPAAZ Frankfurter Zeitschrift fuer Pathologie. (Munich, Germany) V.1-77, 1907-67. For publisher information, see VAAZA2

GANMAX Gann Monograph. (Tokyo, Japan) No.1-10, 1966-71. For publisher information, see GMCRDC

GANNA2 Gann. Japanese Journal of Cancer Research. (Tokyo, Japan) V.1-75, 1907-84. For publisher information, see JJCREP

GANRAE Gan No Rinsho. Cancer Clinics. (Ishiyaku Publishers, Inc., 7-10 Honkomagome 1-chome, Bunkyo-ku, Tokyo, Japan) V.1- 1954-

GASTAB Gastroenterology. (Elsevier North Holland, Inc., 52 Vanderbilt Avenue, New York, NY 10017) V.1- 1943-

GCBIDC Gastroenterologie Clinique et Biologique. (Masson et Cie, Editeurs, 120 Blvd. Saint-Germain, P-75280, Paris 06, France) V.1- 1977-

GCENA5 General and Comparative Endocrinology. (Academic Press, 111 Fifth Ave., New York, NY 10003) V.1- 1961-

GCTB** Givaudan Corporation, Corporate Communications. (100 Delawanna Ave., Clifton, NJ 07014)

GDIKAN Gifu Daigaku Igakubu Kiyo. Papers of the Gifu University School of Medicine. (Gifu Daigaku Igakubu, 40 Tsukasa-cho, Gifu, Japan) V.15- 1967-

GDNU** "Klinische Untersuchungen Mit Einem Neuen Schlafmitten Eldoral", Marianne Gielen Dissertation. (Martin Luther Universitat, Halle- Wittenberg, Germany, 1935)

GEFRA2 Geburtshilfe und Frauenheilkunde. (Intercontinental Medical Books Corp., 381 Park Ave., S., New York, NY 10016) V.1- 1939-

GEIGAI Gendai Igaku. Current Medicine. (Aichi-ken Ishikai, 1-10 Minami Hisaya-cho, Naka-ku, Nagoya, Japan) V.1- 1950-

GEIRDK Gendai Iryo. Modern Medical Care. (Gendai Iryosha, Kandabashi Bldg., 1-21 Nishiki-cho, Kanada Chiyoda-ku, Tokyo 101, Japan) V.1- 1969-

GENEA3 Genetica (The Hague). (Dr. W. Junk bv Publishers, POB 13713, 2501 ES The Hague, Netherlands) V.1- 1919-

GENRA8 Genetical Research. (Cambridge University Press, P.O. Box 92, Bentley House, 200 Euston Rd., London NW1 2DB, England) V.1- 1960-

GENTAE Genetics. (P.O. Drawer U, University Station, Austin, TX 78712) V.1- 1916-

GEPHDP General Pharmacology. (Pergamon Press Inc., Maxwell House Fairview Park, Elmsford, NY 10523) V.1- 1970-

GERNDJ Gerontology. (S. Karger AG, Postfach CH-4009, Basel, Switzerland) V.22- 1976-

GESKAC Genetika i Selektsiya. Genetics and Plant Breeding. (Izdatelstvo na Naukite, ul. Akad. G. Bonchev, 1113 Sofia, Bulgaria) V.1- 1968-

GETRE8 Gematologiya i Transfuziologiya. Hematology and Transfusion Science. (v/o Mezhdunarodnaya Kniga, Kuznetskii Most 18, Moscow G-200, USSR) V.28- 1983-

GEXXA8 German (East) Patent Document. (U.S. Patent Office, Science Library, 2021 Jefferson Davis Highway, Arlington, VA 22202)

GHREAA Guy's Hospital Reports. (London) V.1-123, 1836-1974. Discontinued

GICTAL Giornale Italiano di Chemioterapia. Italian Journal of Chemotherapy. (Edizioni Minerva Medica, Casella Postale 491, I-10126 Turin, Italy) V.1-5, 1954-58; V.5- 1962-

GIFT** Gigiena i Fiziologiya, Truda, Pro. Toksikologiya Klinika. Hygiene and Physiology of Labor, Industrial Toxicology and Clinical Aspects. (USSR)

GISAAA Gigiena i Sanitariya. (English Translation is HYSAAV). (v/o Mezhdunarodnaya Kniga, Kuznetskii Most 18, Moscow G-200, U.S.S.R.) V.1- 1936-

GJMSA7 Gunma Journal of Medical Sciences. (Gunma Daigaku Igakubu, 3-39-22, Showa-Machi, Maebashi, Gunma 371, Japan) V.1- 1952-

GJSCDQ Georgia Journal of Science. (Univ. of Georgia, Institute of Natural Resources, Athens, GA 30602) V.35- 1977-

GMCRDC Gann Monograph on Cancer Research. (Japan Scientific Societies Press, Hongo 6-2-10, Bunkyo-ku, Tokyo 113, Japan) No. 11- 1971-

GMITAB Gazzetta Medica Italiana. (Minerva Medica, Casella Postale 491, Turin, Italy) V.95- 1936-

GMJOAZ Glasgow Medical Journal. (Glasgow, Scotland) 1828-1955. For publisher information, see SMDJAK

GMMIAW Giornale di Medicina Militare. Journal of Military Medicine. (Giornale di Medicina Militare, Via S. Stefano Rotondo, 4, Rome, Italy) V.56- 1908-

GNAMAP Gigiena Naselennykh Mest. Hygiene in Populated Places. (Kievskii Nauchno - Issledovatel'skii Institut Obshchei i Kommunol'noi Gigieny, Kiev, U.S.S.R.) V.7- 1967-

GNKAA5 Genetika. (see SOGEBZ for English Translation) (v/o "Mexhdunarodnaya Kniga," Kuznetskii Most 18, Moscow G-200, USSR) No.1- 1965-

GNLEAM Genetics Lectures. (Corvallis, Oregon) V.1-5, 1969-77. Discontinued.

GNRIDX Gendai no Rinsho. (Tokyo, Japan) V.1-10, 1967-76(?)

GNTKDF Genetika (Zemun, Yugoslavia). (Unija Bioloskih Naucnih Drustava Jugoslavije, Postanski fah 127, Zemun, Yugoslavia) V.1- 1969-

GOBIDS Gynecologic and Obstetric Investigation. (S. Karger AG, Postfach CH-4009, Basel, Switzerland) V.9- 1978-

GRCSB* Goodrich Chemical Co., Service Bulletin (B. F. Goodrich Chemical Group, 6100 Oak Tree Blvd., Cleveland, OH 44131)

GRMDAI German Medicine. (Stuttgart, Germany) V.1-3, 1971-73. Discontinued.

GRMMAB German Medical Monthly. (Stuttgart, Germany) 1956-70. (English Translation of DMWOAX) For publisher information, see GRMDAI

GROWAH Growth. (Southern Bio-Research Institute, Florida Southern College, Lakeland, FL 33802) V.1- 1937-

GSAMAQ Geological Society of America, Memoir. (The Society, Colorado Bldg., P.O. Box 1719, Boulder, CO 80302) No.1- 1934-

GSLNAG Gaslini(Genoa). Rivista de Pediatria e di Specialita Pediatriche. (Istituto "Giannina Gaslini", Via Cinque Maggio, 39, 16148 Genoa, Italy) V.1- 1969-?

GTKRDX Gan to Kagaku Ryoho. Cancer and Chemotherapy. (Gan to Kagaku Ryohosha, Yaesu Bldg., 1-8-9 Yaesu, Chuo-Ku, Tokyo 103, Japan) V.1- 1974-

GTNIAI Geneeskundig Tijdschrift voor Nederlandsch-Indie. Medical Journal for the Netherlands Indies. (Djakarta, Indonesia) V.1-82, 1852-1942. Discontinued.

GTPPAF Gigiena Truda i Professional'naya Patologiya v Estonskoi SSR. Labor Hygiene and Occupational Pathology in the Estonian SSR. (Institut Eksperimental'noi i Klinicheskoi Meditsiny Ministerstva Zdravookhraneniya Estonskoi SSR, Tallinn, USSR) V.8- 1972-

GTPZAB Gigiena Truda i Professional'nye Zabolevaniia. Labor Hygiene and Occupational Diseases. (v/o Mezhdunarodnaya Kniga, Kuznetskii Most 18, Moscow G-200, U.S.S.R.) V.1- 1957-

GUCHAZ "Guide to the Chemicals Used in Crop Protection," Information Canada, 171 Slater St., Ottawa, Ontario, Canada (Note that although the Registry cites data from the 1968 edition, this reference has been superseded by the 1973 edition.)

GUTTAK Gut. (British Medical Journal, 1172 Commonwealth Ave., Boston, MA 02134) V.1- 1960-

GVOGAV Galvano-Organo (Champs-Elysees, 75008 Paris, France) V.43- 1974-

GWXXAW German Patent Document. (U.S. Patent Office, Science Library, 2021 Jefferson Davis Highway, Arlington, VA 22202)

GWXXBX German Offenlegungsschrift Patent Document. (U.S. Patent Office, Science Library, 2021 Jefferson Davis Highway, Arlington, VA 22202)

GYNIAJ Gynecologic Investigation. (Basel, Switzerland) V.1-8, 1970-77. For publisher information, see GOBIDS

GYNOA3 Gynecologic Oncology. (Academic Press, 111 Fifth Ave., New York, NY 10003) V.1- 1972-

GYOGAI Gyogyszereszet. Pharmacy. (Kultura, POB 149, H-1389 Budapest, Hungary) V.1- 1957-

HarPN# Personal Communication to Henry Lau, Tracor Jitco, from Paul N. Harris, M.D., Eli Lily Co., Greenfield, IN 46140

HunNJ# Personal Communication from Nancy J. Hunt, E.I. DuPont de Nemours and Co., 1007 Market St., Wilmington, DE 19898, to Henry Lau, Tracor Jitco, Inc., May 10, 1977

HAEMAX Haematologica. (Il Pensiero Scientifico, Via Panama 48, I-00198, Rome, Italy) V.1- 1920-

HAEMBY Haematologia. (Akademiai Kiado, POB 24, H-1389, Budapest 502, Hungary) V.1- 1967-

HAONDL Hematological Oncology. (John Wiley and Sons Ltd., Baffins Lane, Chichester, Sussex, PO19 1UD, U.K.) V.1- 1983-

HAREA6 Harefuah Medicine. (Israel Medical Association, 39 Shaul Hamlech Blvd., Tel Aviv-Jaffa, Israel) V.1- 1924

HAZL** Hazelton Laboratories, Reports. (Falls Church, VA)

HBAMAK "Abdernalden's Handbuch der Biologischen Arbeitsmethoden." (Leipzig, Germany)

HBTXAC "Handbook of Toxicology, Volumes I-V," Philadelphia, W.B. Saunders, 1956-1959

HCACAV Helvetica Chimica Acta. (Verlag Helvetica Chimica Acta, Postfach, CH-4002 Basel, Switzerland) V.1- 1918-

HCKHDV Huanjing Kexue. Environmental Sciences. (China International Book Trading Corp., POB 2820, Beijing, Peop. Rep. China) 1976- (Adopted vol. no. with Vol. 1 in 1980)

HCMYAL Histochemistry. (Springer-Verlag New York, Inc., Service Center, 44 Hartz Way, Secaucus, NJ 07094) V.38- 1974-

HDIZAB Hiroshima Daigaku Zasshi, Medical Journal of the Hiroshima University (Hiroshima Daigaku Igakubu Saikingaku Kyoshitsu, Kasumi-cho, Hiroshima, Japan) V.10- 1962-

HDKU** Die Krampfmischung der Lokalanaesthetica, ihre Beeinflussung durch Mineralsalze und Adrenalin, Gerhard Hoppe Dissertation. (Pharmakologischen Institut der Albertus-Universitat zu Konigsberg Pr., Germany, 1933)

HDSKEK Hiroshima Daigaku Sogo Kagakubu Kiyo, 4: Kiso, Kankyo Kagaku Kenkyu. Bulletin of the Faculty of Integrated Arts and Sciences, Hiroshima University, Series 4: Fundamental and Environmental Sciences. (1-1-89 Higashisenda-machi, Naka-ku, Hiroshima, 730, Japan) V.6- 1980-

HDTU** Pharmakologische Prufung von Analgetika, Gunter Herrlen Dissertation. (Pharmakologischen Institut der Universitat Tubingen, Germany, 1933)

HDTYAT Heredity. (Oliver and Boyd, Tweeddale Court, 14 High St., Edinburgh EH1 1YL, Scotland) V.1- 1947-

HDWU** Beitrage zur Toxikologie des Athylenoxyds und der Glykole, Arnold Hofbauer Dissertation. (Pharmakologischen Institut der Universitat Wurzburg, Germany, 1933)

HEADAE Headache. (American Assoc. for the Study of Headache, Rm 537, 621 S. New Ballas Rd., St. Louis, MO 63141) V.1- 1961-

HEGAD4 Hepatogastroenterology. (Georg Thieme Verlag, Postfach 732, Herdweg 63, D-7000 Stuttgart 1, Fed. Rep. Ger.) V.1- 1980-

HEPHD2 Handbook of Experimental Pharmacology. (Springer-Verlag, Heidelberger Pl. 3, D-1000 Berlin 33, Federal Republic of Germany) V.50- 1978-

HERBU* Hercules Incorporated, Hercules Bulletin. (Hercules Incorporated, Pine and Paper Chemicals Department, Wilmington, DE)

HEREAY Hereditas. (J.L. Toernqvist Book Dealers, S-26122 Landskrona, Sweden) V.1- 1947-

HFHBAG Hartford Hospital Bulletin. (Hartford Hospital, Hartford, CT 06115) V.1- 1938/39-

HGANAO Haigan. Lung Cancer. (Nippon Haigan Gakkai, c/o Chiba Daigaku Igakubu Haigan Kenkyusho, 1-8-1 Inohona, Chiba 280, Japan)

HHBLAU Harper Hospital, Bulletin. (Detroit, Michigan) V.1-26, No.5, 1890-1968, Discontinued.

HIFUAG Hifu. Skin. (Nihon Hifuka Gakkai Osaka Chihokai, c/o Osaka Daigaku Hifuka Kyoshitsu, 3-1, Dojima Hamadori, Fukushima-ku, Osaka 553, Japan) V.1- 1959-

HIKYAJ Hinyokika Kiyo. (Acta Urologica Japonica). (Kyoto University Hospital, Department of Urology) V.1- 1955-

HINAAU Hindustan Antibiotics Bulletin. (Hindustan Antibiotics, Pimpri, India) V.1- 1958-

HINEL* Hine Laboratories Reports. (San Francisco, CA)

HIRGAY Hiroshima Igaku. Journal of Hiroshima Medical Association. (Hiroshima Ishi Kaikan, 8-2 Naka-machi, Hiroshima, Japan) No.1- 1948-

HIRIA6 Hirosaki Igaku. (Hirosaki Daigaku Igakubu, Hirosaki, Japan) V.1- 1950-

HIUN** Department of Cancer Research, Research Institute for Nuclear Medicine and Biology, Hiroshima University, Kasumi 1-2-3, Hiroshima 734, Japan

HKXUDL Huanjing Kexue Xuebao. Environmental Sciences Journal. (Guoji Shudian, POB 399, Beijing, Peop. Rep. China) V.1- 1981-

HLSCAE Health Laboratory Science. (American Public Health Assoc., 1015 18th St., N.W., Washington, DC 20036) V.1- 1964-

HLTPAO Health Physics. (Pergamon Press, Maxwell House, Fairview Park, Elmsford, NY 10523) V.1- 1958-

HMMRA2 Hormone and Metabolic Research. (Georg Thieme Verlag, Postfach 732, Herdweg 63, D-7000 Stuttgart 1, Federal Republic of Germany) V.1- 1969-

HOBEAO Hormones and Behavior. (Academic Press, 111 5th Ave., New York, NY 10003) V.1- 1969-

HOCC** Hooker Chemical Corporation Reports. (Niagara Falls, NY)

HOEKAN Hokkaidoritsu Eisei Kenkyushoho. (Hokkaidoritsu Eisei Kenkyusho, Nishi-12-chome, Kita-19-jo, Kita-ku, Sapporo, Japan) No.1- 1951-

HOIZAK Hokkaido Igaku Zasshi. Hokkaido Journal of Medical Science. (Hokkaido Daigaku Igakubu, Nishi-5-chome, Kita-12-jo, Sapporo, Japan) V.1- 1923-

HOKBAQ Hoken Butsuri. Journal of the Japan Health Physics Society. (Nihon Hoken Butsuri Kyogikai, Sakamoto Bldg., Kitaaoyama 2-12-4, Minato-ku, Tokyo, Japan) V.1- 1966-

HPAOAM Heating, Piping and Air Conditioning. (25 Sullivan St., Westwood, NJ 07675) V.1- 1929-

HPCQA4 Human Pathology. (W. B. Saunders Co., W. Washington Sq., Philadelphia, PA 19105) V.1- 1970-

HPFSDS HHS Publication (FDA. United States). (U.S. Government Printing Office, Superintendent of Documents, Washington, DC 20402) 1980-

HPPAAL Helvetica Physiologia et Pharmacologica Acta. (Basel, Switzerland) V.1-26, 1943-68. Discontinued.

HRMRA3 Hormone Research. (S. Karger AG, Postfach CH-4009, Basel, Switzerland) V.4- 1978-

HSZPAZ Hoppe-Seyler's Zeitschrift fuer Physiologische Chemie. (Walter de Gruyter and Co., Genthiner Street 13, D-1000, Berlin 30, Federal Republic of Germany) V.21- 1895/96-

HUDJAN Hiroshima Daigaku Shigaku Zasshi. Hiroshima University Dentistry Journal. (Hiroshima Daiguka Shigakkai, 2-3, 1-chrome, Kasumi, Hiroshima 734, Japan) V.1- 1969-

HUGEDQ Human Genetics. (Springer-Verlag, Neuenheimer Landst 28-30, D-6900 Heidelberg 1, Germany) V.31- 1976-

HUHEAS Human Heredity. (S. Karger AG, Postfach, CH-4009, Basel, Switzerland)

V.19- 1969-

HUMAA7 Humangenetik. (Heidelberg, Germany) V.1-30, 1964-1975. For publisher information, see HUGEDQ

HURC** Huntingdon Research Center Reports. (Box 527, Brooklandville, MD 21022)

HUTODJ Human Toxicology. (Macmillan Press Ltd., Houndmills, Bassingstoke, Hants., RG21 2XS, UK) V.1- 1981-

HWMJAE Hawaii Medical Journal. (Hawaii Medical Association, 320 Ward Ave., Honolulu, Hawaii 96814) V.1- 1941-

HXPHAU Handbuch der Experimentellen Pharmakologie. (Berlin, Germany) V.1-49, 1935-78. For publisher information, see HEPHD2

HYDKAK Hoshi Yakka Daigaku Kiyo. Proceedings of the Hoshi College of Pharmacy. (2-chome, Ebara, Shinagawa-ku, Tokyo, Japan) No.1- 1951-

HYDRDA Hydrometallurgy. (Elsevier Science Publishers B.V., POB 211, 1000 AE Amsterdam, Netherlands) V.1- 1975-

HYHPDO Hunan Yixueyuan Xuebao. Journal of Hunan Medical College. (China International Book Trading Corp., POB 2820, Beijing, Peop. Rep. China) Beginning history unknown.

HYSAAV Hygiene and Sanitation: English Translation of Gigiena Sanitariya. (Springfield, VA) V.29-36, 1964-71. Discontinued.

IAAAAM International Archives of Allergy and Applied Immunology. (S. Karger, Postfach CH-4009, Basel Switzerland) V.1- 1950-

IAANBS Internationales Archiv fuer Arbeitsmedizin. (Berlin, Germany) V.26-34, 1970-74. For publisher information, see IAEHDW

IAEC** Interagency Collaborative Group on Environmental Carcinogenesis, National Cancer Institute, Memorandum, June 17, 1974

IAEHDW International Archives of Occupational and Environmental Health. Springer-Verlag, Heidelberger, Platz 3, D-1000 Berlin 33, Federal Republic of Germany) V.35- 1975-

IANBAM Izvestiya Akademii Nauk S.S.S.R., Seriya Biologi-cheskaya. (v/o Mezhduharodnaya Kniga, Kuznetskii Most 18, Moscow G-200, U.S.S.R.) No.1- 1936-

IAPUDO IARC Publications. (World Health Organization, CH-1211 Geneva 27, Switzerland) No.27- 1979-

IAPWAR International Journal of Air and Water Pollution. (London, England) V.4, No.1-4, 1961. For publisher information, see ATENBP

IARC** IARC Monographs on the Evaluation of Carcinogenic Risk of Chemicals to Man. (World Health Organization, Internation Agency for Research on Cancer, Lyon, France) (Single copies can be ordered from WHO Publications Centre U.S.A., 49 Sheridan Avenue, Albany, NY 12210)

IARCCD IARC Scientific Publications. (Geneva Switzerland) V.1-No.26, 1971-78, For publisher information, see IAPUDO

IASKA6 Izvestiya Akademii Nauk S.S.S.R., Seriya Khimicheskaya. (v/o Mezhdunarodnaya Kniga, Kuznetskii Most 18, Moscow G-200, U.S.S.R.) No.1- 1936-

IBSCDQ Information Bulletin on the Survey of Chemicals Being Tested for Carcinogenicity. (World Health Organization, International Agency for Research on Cancer, Lyon, France) (Single copies can be ordered from WHO Publications Centre U.S.A., 49 Sheridan Avenue, Albany, NY 12210) No. 1 1973-

ICHAA3 Inorganica Chimica Acta. (Elsevier Sequoia SA, POB 851, CH-1001 Lausanne 1, Switzerland) V.1- 1967-

ICHUDW Yichuan. Heredity. (China International Book Trading Corp., POB 2820, Beijing, Peop. Rep. China) V.1- 1979-

ICMED9 Intensive Care Medicine. (Springer-Verlag New York, Inc., Service Center, 44 Hart Way, Secaucus, New Jersey 07094) V.3-1977-

IDJOAS International Dental Journal. (A. Sijthoff, 37 Wagenstr, The Hague, Netherlands) V.1- 1950-

IDPYAK Industrial Pharmacology. (Futura Publishing Co., 295 Main St., Mount Kisco, NY 10549) V.1- 1974-

IDZAAW Idengaku Zasshi. Japanese Journal of Genetics. (Genetics Society of Japan, Nippon Iden Gakkai, Tanida 111, Mishima-shi, Shizuoka 411, Japan) V.1- 1921-

IECHAD Industrial and Engineering Chemistry. (Washington, DC) V.15-62, 1923-70. For publisher information, see CHMTBL

IGAYAY Igaku No Ayumi. Progress in Medicine. (Ishiyaku Shuppan K.K., 7-10, Honkomagone 1 chome, Bunkyo-ku, Tokyo, Japan) V.1- 1946-

IGIBA5 Igiena. (Editura Medicala, St 13 Decembrie 14, Bucharest, Romania) V.5- 1956-

IGKEAO Igaku Kenkyu. (Daido Gakkan Shuppanbu, Kyushu Daigaku Igakubu, Hoigaku Kyoshitsu, Fukuoka 812, Japan) V.1- 1927-

IGMPAX Igiene Moderna. Modern Hygiene. (Amministrazione de l'Igiene Moderna, Tipografia "La Commerciale," Piazza Pontida, 11, Fidenza, Parma, Italy) V.1- 1908-

IGSBAL Igaku to Seibutsugaku. Medicine and Biology. (1-11-4 Higashi-Kanda, Chiyoda, Tokyo 101, Japan) V.1- 1942-

IGSBDO Izvestiya Akademii Nauk Gruzinskoi SSR, Seriya Biologicheskaya. Proceedings of the Academy of Sciences of the Georgian SSR, Biological Series. (v/o Mezhdunarodnaya Kniga, Kuznetskii Most 18, Moscow G-200, USSR) V.1- 1975-

IHFCAY Industrial Hygiene Foundation of America, Chemical and Toxicological Series, Bulletin. (5231 Centre Ave., Pittsburgh, PA 15232) 1947-69.

IHFMAU Industrial Hygiene Foundation of America, Medical Series, Bulletin. (Pittsburgh, PA) No.1-17, 1937-70. Discontinued.

IHYDAU Industrial Hygiene Digest. (Industrial Health Foundation, 5231 Centre Ave., Pittsburgh, PA 15232) V.1- 1937-

IIFBA4 Izvestiya na Instituta po Fiziologiya, Bulgarska Akademiya na Naukite. (Izdatelstvo na Bulgarskata Akademiya na Naukite, St. Geo. Milev ul 36, Sofia 13, Bulgaria) V.4- 1960-

IIZAAX Iwate Igaku Zasshi. Journal of the Iwate Medical Association. (Iwate Igakkai, c/o Iwate Ika Daigaku, Uchimaru, Morioka, Japan) V.1- 1947-

IJANDP International Journal of Andrology. (Scriptor Publisher ApS, 15 Gasvaerksvej, DK-1656 Copenhagen V, Denmark) V.1- 1978-

IJBBBQ Indian Journal of Biochemistry and Biophysics. (Council of Scientific and Industrial Research, Publication and Information Director, Hillside Rd., New Delhi 110012, India) V.8- 1971-

IJBOBV International Journal of Biochemistry. (Pergamon Press Ltd., Headington Hill Hall, Oxford 0X3 0BW, England) V.1- 1970-

IJCAAR Indian Journal of Cancer. (Dr. D.J. Jussawalla, Hospital Ave, Parel, Bombay 12, India) V.1- 1963-

IJCBDX International Journal of Clinical Pharmacology and Biopharmacy. (Urban and Schwarzenberg, Pettenkoferst 18, D-8000 Munich 15, Germany) V.11- 1975-

IJCCE3 International Journal of Cell Cloning. (AlphaMed Press, Inc., 3525 Southern Blvd., Dayton, OH 45429) V.1- 1983-

IJCNAW International Journal of Cancer. (International Union Against Cancer, 3 rue du Conseil-General, 1205 Geneva, Switzerland) V.1- 1966-

IJCPB5 International Journal of Clinical Pharmacology, Therapy and Toxicology. (Munich, Germany) V.6-10, 1972-74. For publisher information, see IJCBDX

IJCREE International Journal of Crude Drug Research. (Swets and Zeitlinger B. V., POB 825, 2160 SZ Lisse, Netherlands) V.20- 1982-

IJDEBB International Journal of Dermatology. (J. B. Lippincott Co., E. Washington Sq., Philadelphia, PA 19105) V.9- 1970-

IJEBA6 Indian Journal of Experimental Biology. V.1- 1963- For publisher information, see IJBBBQ

IJEVAW International Journal of Environmental Studies. (Gordon and Breach Science Publishers Ltd., 41-42 William IV St., London WC2N 4DE, England) V.1- 1970-

IJGOAL International Journal of Gynaecology and Obstetrics. (International Journal of Gynaecology and Obstetrics, POB 12037, Baltimore, NC 27709) V.7- 1969-

IJIMDS International Journal of Immunopharmacology. (Pergamon Press Ltd., Headington Hill Hall, Oxford OX3 0BW, England) V.1- 1979-

IJLAA4 Indian Journal Animal Science. (Indian Council of Agricultural Research, Krishi Bhavan, New Delhi 110 001, India) V.39- 1969-

IJLEAG International Journal of Leprosy. (Box 39088, Washington, DC 20016) V.1- 1933-

IJMDAI Israel Journal of Medical Sciences. (P.O. Box 2296, Jerusalem, Israel) V.1- 1965-

IJMRAQ Indian Journal of Medical Research. (Indian Council of Medical Research, P.O. Box 4508, New Delhi 110016, India) V.1- 1913-

IJMSAT Irish Journal of Medical Science. (Royal Academy of Medicine, 6 Kildare St., Dublin, Ireland) bV.1- 1968-

IJNDAN Indian Journal of Nutrition and Dietetics. (Sri Avinashilingam Home Science College, Coimbatore 641011, India) V.7- 1970-

IJNEAQ International Journal of Neuropharmacology. (Oxford, England/New York, NY) V.1-8, 1962-69. For publisher information, see NEPHBW

IJNMCI International Journal of Nuclear Medicine and Biology. (Pergamon Press Inc., Maxwell House, Fairview Park, Elmsford, NY 10523) V.1- 1973-

IJNUB7 International Journal of Neuroscience. (Gordon and Breach Science Publication S.A., 50 W. 23rd St., New York, New York 10010) V.1- 1970-

IJOCAP Indian Journal of Chemistry. (Council of Scientific and Industrial Research, Publication and Information Director, Hillside Rd., New Delhi 110012, India) V.1-13, 1963-75.

IJPAAO Indian Journal of Pharmacy. (Bombay, India) V.1-40, No.1, 1939-78 For publisher information, see IJSIDW

IJPBAR Indian Journal of Pathology and Bacteriology. (Dr. S.G. Deodhare, Medical College, Miraj Maharashtra, India) V.1- 1958-

IJPPAZ Indian Journal of Physiology and Pharmacology. (Indian Journal of Physiology and Pharmacology, All India Institute of Medical Sciences, Department of Physiology, Ansari Nagar, New Delhi 110016, India) V.1- 1957-

IJRBA3 International Journal of Radiation Biology and Related Studies in Physics, Chemistry and Medicine. (Taylor and Francis Ltd., 4 John Street, London WC1N 2ET, UK) V.1- 1959-

IJSBDB Indian Journal of Chemistry. Section B: Organic Chemistry, Including Medicinal Chemistry. (Council of Scientific and Industrial Research, Publication and Information Director, Hillside Rd., New Delhi 110012, India) V.14- 1976-

IJSIDW Indian Journal of Pharmaceutical Sciences. (Indian Journal of Pharmaceutical Sciences, Kalina, Santa Cruz (East), Bombay 400 029, India) V.40, no.2- 1978-

IJSPDJ International Journal of Andrology, Supplement. (Scriptor Publisher ApS, 15 Gasvaerksvej, DK-1656 Copenhagen V, Denmark) No.1- 1978-

IJTEDP International Journal on Tissue Reactions. Cell and Tissue Injuries, Experimental and Clinical Aspects. (Biosciences Ediprint, Rue Winkelried 8, 1211 Geneva 1, Switzerland) V.1- 1979-

IJVNAP International Journal for Vitamin and Nutrition Research. (Verlag Hans Huber, Laenggasstr. 76, CH-3000 Bern 9, Switzerland) V.41- 1971-

IJZOAE Israel Journal of Zoology. (Weizman Science Press of Israel, P.O. Box 801, Jerusalem, Israel) V.12- 1963-

IMEMDT IARC Monographs on the Evaluation of Carcinogenic Risk of Chemicals to Man. (World Health Organization, Internation Agency for Research on Cancer, Lyon, France) (Single copies can be ordered from WHO Publications Centre U.S.A., 49 Sheridan Avenue, Albany, NY 12210)

IMGAAY Indian Medical Gazette. (Indian Medical Gazette, Block F, 105-C, New Alipore, Calcutta 700053, India) v.1-90, 1866-1955; new series V.1- 1961-

IMLCAV Immunological Communications. (Marcel Dekker, POB 11305, Church St. Station, New York, NY 10249) V.1- 1972-

IMMUAM Immunology. (Blackwell Scientific Publications, Osney Mead, Oxford OX2 OEL, England) V.1- 1958-

IMNGBK Immunogenetics (New York). (Springer-Verlag New York, Inc., Service Center, 44 Hartz Way, Secaucus, NJ 07094) V.1- 1974-

IMSCE2 IRCS Medical Science. (Lancaster, UK) V.12-14 1984-86.

IMSUAI Industrial Medicine and Surgery. (Chicago, IL/Miami, FL) V.18-42, 1949-73. For publisher information see IOHSA5

INDRBA Indian Drugs. (Indian Drugs Manufacturers' Associations c/o Dr. A. Patani, 102B, Poonam Chambers, Dr. A. B. Rd., Worli Bobmay 400 018, India) V.1- 1963-

INFIBR Infection and Immunity. (American Society for Microbiology, 1913 I St., N.W., Washington, DC 20006) V.1- 1970-

INHEAO Industrial Health. (2051 Kizukisumiyoshi-cho, Nakahara-ku, Kawasaki, Japan) V.1- 1963-

INJFA3 International Journal of Fertility. (Allen Press, 1041 New Hampshire, St., Lawrence, KS66044) V.1- 1955-

INJHA9 Indian Journal of Heredity. (Genetic Assoc. of India, Indian Veterinary Research Institute, Izatnagar, India) V.1- 1969-

INJMAO Indian Journal of Medical Sciences. (Backbay View, Mama Parmanand Marg, Bombay 400004, India) V.1- 1947-

INJPD2 Indian Journal of Pharmacology. (Indian Pharmacological Society, Jawaharlal Institute of Postgraduate Medical Education and Research, Pharmacology Dept., Pondicherry 605006, India) V.1- 1968(?)-

INMEAF Industrial Medicine. (Chicago, IL) V.1-18, 1932-49. For publisher information, see IOHSA5

INNDDK Investigational New Drugs. The Journal of New Anticancer Agents. (Kluwer Boston Inc., 190 Old Derby St., Hingham, MA 02043) V.1- 1983-

INOPAO Investigative Ophthalmology. (C.V. Mosby Co., 11830 Westline Industrial Dr., St. Louis, MO 63141) V.1- 1962-

INPDAR Indian Pediatrics. (46-B Garcha Rd., Calcutta, India) V.1- 1964-

INPHB6 International Pharmacopsychiatry. (S. Karger AG, Postfach CH-4009, Basel, Switzerland) V.1- 1968-

INSSDM INSERM Symposium. (Elsevier North Holland, Inc., 52 Vanderbilt Ave., New York, NY 10017) No.1- 1975-

INTEAG Internist. (Springer-Verlag, Heidelberger Pl. 3, D-1 Berlin 33, Germany) V.1- 1960-

INTSAO International Surgery. (International College of Surgeons, 1516 N. Lake Shore Dr., Chicago, IL 60610) V.45- 1966-

INURAQ Investigative Urology. (Williams and Wilkins Co., 428 E. Preston St., Baltimore, MD 21202) V.1- 1963-

INVRAV Investigative Radiology. (J.B. Lippincott Co., E. Washington Sq., Philadelphia, PA 19105) V.1- 1966-

IOBPD3 International Journal of Radiation Oncology, Biology and Physics. Pergamon Press Ltd., Headington Hill Hall, Oxford OX3 0BW, England) V.1- 1975-

IOHSA5 International Journal of Occupational Health and Safety. (Medical Publications, Inc., 3625 Woodhead Dr., Northbrook, IL 60093) V.43- 1974-

IOVSDA Investigative Ophthalmology and Visual Science. (C.V. Mosby Co., 11830 Westline Industrial Dr., St. Louis, MO 63146) V.16- 1977-

IPCLBZ International Pest Control. (McDonald Publications, 268 High St., Uxbridge UB8 1UA, Middx., England) V.5- 1962

IPPABX Iugoslavica Physiologica et Pharmacologica Acta. (Unija Bioloskih Naucnik Drustava Jugoslavije, Postanski fah 127, Belgrade-Zemun, Yugoslavia) V.1- 1965-

IPRAA8 Indian Practitioner. (231 Dr. Dadabhoy Naoroji Rd., Bombay 1, India) V.1- 1947/48-

IPSTB3 International Polymer Science and Technology. (Rapra Technology Ltd., Shawbury, Shrewsbury, Shropshire SY4 4NR, UK)

IRGGAJ Internationales Archiv fuer Gewerbepathologie und Gewerbehygiene. (Heidelberg, Germany) V.19-25, 1962-69. For publisher information, see IAEHDW

IRIPDP I.T.R.I. Publication. (Tin Research Institute, 483 W. 6th Ave., Columbus, OH 43201) No.530- 1976-

IRLCDZ IRCS Medical Science: Library Compendium. (MTP Press Ltd., St. Leonards House, St. Leonards Gate, Lancaster LA1 3BR, England) V.3-11, 1975-83.

IRMEA9 International Record of Medicine. (New York, NY) V.170-174, 1957-61. For publisher information, see CLMEA3

IRNEAE International Review of Neurobiology. (Academic Press, 111 5th Ave., New York, NY 10003) V.1- 1959-

IRXPAT International Review of Experimental Pathology. (Academic Press, 24-28 Oval Rd., London NW1 7DX, England) V.1- 1962-

ISBNBN Izvestiya Sibirskogo Otdeleniya Akademii Nauk S.S.S.R., Seriya Biologicheskikh Nauk. (v/o Mezhdunarodnaya Kniga, Kuznetskii Most 18, Moscow G-200, U.S.S.R.) No.1- 1969-

ISGEAK Issledovaniya po Genetike (Leningrad). (Leningradskii Gosudarstvennyi Universitet, Leningrad, USSR) V.1- 1961-

ISMJAV Israel Medical Journal. (Jerusalem, Israel) V.17-23, 1958-64. For publisher information, see IJMDAI

ISYAM* "Amphetamines and Related Compounds," E. Costa, ed., Proceedings of the Mario Negri Institute for Pharmacological Research, Milan, Italy, New York, Raven Press, 1970

ITCSAF In Vitro. (Tissue Culture Association, 12111 Parklawn Dr., Rockville, MD 20852) V.1- 1965-

ITFCBH Interface. (Society for Environmental Geochemistry and Health, West Virginia Univ., Morgantown, WV 26506) V.1- 1972-

ITMZBJ Intensivmedizin. (Dr. Dietrich Steinkopff Verlag, Postfach 11-10-08, D-6100 Darmstadt 11, Fed. Rep. Germany) V.1- 1964-

ITOBAO Izvestiya Akademii Nauk Tadzhikskoi SSR, Otdelenie Biologicheskikh Nauk. Proceedings of the Academy of Sciences of the Tadzhik SSR, Department of Biological Sciences. (v/o Mezhdunarodnaya Kniga, Kuznetskii Most 18, Moscow G-200, USSR) No.1- 1962-

IUSMDJ ICN-UCLA Symposia on Molecular and Cellular Biology. (Academic Press, 111 5th Ave., New York, NY 10003) V.1- 1974-

IVEBBN Beiheft zur Internationalen Zeitschrift fuer Vitamin-und Ernaehrungsforschung. (Verlag Hans Huber Laenggastr. 76 CH-3000 Bern 9, Switzerland) No.12- 1972-

IVEJAC Indian Veterinary Journal. (Indian Veterinary Journal, Dr. R. Krishnamurti, G.M.V.C., 10 Avenue Rd., Madras 600 034, India) V.1- 1924-

IVIVE4 In Vivo. (POB 51359, Kiffisia, GR-145 10, Greece) V.1- 1987-

IVRYAK Intervirology. (S. Karger AG, Postfach CH-4009, Basel, Switzerland) V.1- 1973-

IYKEDH Iyakuhin Kenkyu. (Nippon Koteisho Kyokai, 12-15, 2-chome, Shibuya-Ku, Tokyo 150, Japan) V.1- 1970-

IZKPAK Internationale Zeitschrift fuer Klinische Pharmakologie, Therapie und Toxikologie. (Munich, Fed. Rep. Ger.) V.1-5, 1967-72. For publisher information, see IJCBDX

IZSBAI Izvestiya Sibirskogo Otdeleniya Akademii Nauk S.S.S.R., Seriya Biologomeditsinskikh Nauk. (Novosibirsk, U.S.S.R) 1963-68. For publisher information, see ISBNBN

IZVIAK Internationale Zeitschrift fuer Vitaminforschung. International Review of Vitamin Research. (Bern, Switzerland) V.19-40, 1947-70. For publisher information, see IJVNAP

JAADDB Journal of the American Academy of Dermatology. (C. V. Mosby Co., 11830 Westline Industrial Dr., Saint Louis, MO 63141) 1979-

JACBBD Journal of Applied Chemistry and Biotechnology. (London, England) V.21-28, 1971-78.

JACHDX Journal of Antimicrobial Chemotherapy. (Academic Press, 24-28 Oval Rd., London NW1 7DX, England) V.1- 1975-

JACIBY Journal of Allergy and Clinical Immunology. (C.V. Mosby Co., 11830 Westline Industrial Dr., St. Louis, MO 63141) V.48- 1971-

JACSAT Journal of the American Chemical Society. (American Chemical Society Publications, 1155 16th St., N.W., Washington, DC 20036) V.1- 1879-

JACTDZ Journal of the American College of Toxicology. (Mary Ann Liebert, Inc., 500 East 85th St., New York, NY 10028) V.1- 1982-

JADSAY Journal of the American Dental Association. (American Dental Assoc., 211 E. Chicago Ave., Chicago, IL 60611) V.9- 1922-

JAFCAU Journal of Agricultural and Food Chemistry. (American Chemical Society Publications, 1155 16th St., N.W., Washington, DC 20036) V.1- 1953-

JAGRAC Journal of Agricultural Research. (Washington, DC) V.1-78, 1913-49. Discontinued.

JAGSAF Journal of the American Geriatrics Society. (American Geriatrics Society, Inc., 10 Columbus Circle, New York, NY 10019) V.1- 1953-

JAIHAQ Journal of the American Institute of Homeopathy. (Board of Trustees of the American Institute of Homeopathy, 910 17th St., N.W., Suite 428, Washington, DC 20006) V.1- 1909-

JAINAA Journal of the Anatomical Society of India. (Dept. of Anatomy, Medical College, Aurangabad MS, India) V.1- 1952-

JAJAAA Journal of Antibiotics, Series A. (Tokyo, Japan) V.6-20, 1953-67. For publisher information, see JANTAJ

JAJBAD Journal of Antibiotics, Series B. (Tokyo, Japan) V.6-20, 1953-67. For publisher information, see JJANAX

JAMAAP JAMA, Journal of the American Medical Association. (American Medical Association, 535 N. Dearborn St., Chicago, IL 60610) V.1- 1883-

JAMPA2 Journal of Animal Morphology and Physiology. (Society of Animal Morphologists and Physiologists, Dept. of Zoology, Faculty of Science, M.S. Univ. of Baroda, Baroda 2, India) V.1- 1954-

JANCA2 Journal of the Association of Official Analytical Chemists. (Association of Official Analytical Chemists., Box 540, Benjamin Franklin Station, Washington, DC 20044) V.49- 1966-

JANSAG Journal of Animal Science. (American Society of Animal Science, 309 West Clark Street, Champaign, IL 61820) V.1- 1942-

JANTAJ Journal of Antibiotics. (Japan Antibiotics Research Association, 2-20-8 Kamiosaki, Shinagawa-ku, Tokyo, Japan) V.2-5, 1948-52; V.21- 1968-

JAOCA7 Journal of the American Oil Chemists' Society. (American Oil Chemists' Society, 508 South 6th St., Champaign, IL 61820) V.24-56, 1947-79.

JAPHAR Journal of Anatomy and Physiology. (London, England) V.1-50, 1916. For publisher information, see JOANAY

JAPMA8 Journal of the American Pharmaceutical Assoc., Scientific Edition. (Washington, DC) V.29-49, 1940-60. For publisher information, see JPMSAE

JAPRAN Journal of Small Animal Practice. (Blackwell Scientific Publications Ltd., POB 88, Oxford, UK) V.1- 1960-

JAPYAA Journal of Applied Physiology. (American Physiological Society, 9650 Rockville Pike, Bethesda, MD 20014) V.1- 1948-

JASIAB Journal of Agricultural Science. (Cambridge University Press, London or Cambridge University Press, New York) V.1- 1905-

JATOD3 Journal of Analytical Toxicology. (Preston Publications, P.O. Box 48312, Niles, IL 60648) V.1- 1977-

JAVMA4 Journal of the American Veterinary Medical Association. (American Veterinary Medical Assoc., 600 S. Michigan Ave, Chicago, IL 60605) V.48- 1915-

JBCHA3 Journal of Biological Chemistry. (American Society of Biological Chemists, Inc., 428 E. Preston St., Baltimore, MD 21202) V.1- 1905-

JBJSA3 Journal of Bone and Joint Surgery. American Volume. (10 Shattuck St., Boston, MA 02115) V.30- 1948-

JBJSB4 Journal of Bone and Joint Surgery. (Boston, MA) V.20-45, 1922-47. For publisher information, see JBJSA3

JBMRBG Journal of Biomedical Materials Research. (J. Wiley and Sons, 605 3rd Ave., New York, NY 10016) V.1- 1967-

JBSUAK Journal of Bone and Joint Surgery, British Volume. (Journal of Bone and Joint Surgery Inc., 10 Shattuck St., Boston, MA 02115) V.30- 1948-

JCBSB5 Journal of the Chinese Biochemical Society. (Chinese Biochemical Society, c/o Institute of Biological Chemistry, Academia Sinica, POB 23-106, Taipei, Taiwan) V.1- 1972-

JCCPAY Journal of Cellular and Comparative Physiology. (Philadelphia, PA) V.1-66, 1932-65. For publisher information, see JCLLAX

JCEMAZ Journal of Clinical Endocrinology and Metabolism. (Williams and Wilkins Co., 428 East Preston St., Baltimore, MD 21202) V.12- 1952-

JCENA4 Journal of Clinical Endocrinology. (Springfield, IL) V.1-11, 1941-51. For publisher information, see JCEMAZ

JCGADC Journal of Clinical Gastroenterology. (Raven Press, 1185 Avenue of the Americas, New York, NY 10036)

JCGBDF Journal of Craniofacial Genetics and Developmental Biology. (Alan R. Liss, Inc., 150 Fifth Ave., New York, NY 10011) V.1- 1980-

JCGEDO Journal of Cytology and Genetics. (Society of Cytologists and Geneticists, Treasurer, Professor M. S. Chennaveeraiah, Karnatak University, Department of Botany, Dharwar 580003, India) V.1- 1966-

JCHAAE Journal of Chemotherapy and Advanced Therapeutics. (Philadephia, PA) V.3,1926; V.11(2-15), 1934-39.

JCHOAM Journal of Chemotherapy. (Philadelphia, PA) V.4-11(1), 1927-34.

JCHODP Journal of Clinical Hematology and Oncology. (Wadley Institutes of Molecular Medicine, 9000 Harry Hines Blvd., Dallas, Texas 75235) V.5, No.3- 1975-

JCINAO Journal of Clinical Investigation. (Rockefeller University Press, 1230 York Avenue, New York, NY 10021) V.1- 1924-

JCLBA3 Journal of Cell Biology. (Rockefeller University Press, 1230 York Avenue, New York, NY 10021) V.12- 1962-

JCLLAX Journal of Cellular Physiology. (Alan R. Liss, Inc., 150 Fifth Avenue, New York, NY 10011) V.67- 1966-

JCLPDE Journal of Clinical Psychiatry. (Physician Postgraduate Press, Inc., POB 24008, Memphis TN 38124) V.39- 1978-

JCNDBK Journal of Clinical Pharmacology and New Drugs. (Stamford, Connecticut) V.11-13, No. 4, 1971-3, For publisher information, see JCPCBR

JCPAAK Journal of Clinical Pathology. (British Medical Association, Tavistock Sq., London WC1H 9JR, England) V.1- 1947-

JCPBAN Journal de Chimie Physique et de Physicochimie Biologique. (Stechert Macmillan, Inc., 7250 Westfield Ave., Pennsauken, NJ 08110) V.1- 1903-

JCPCBR Journal of Clinical Pharmacology. (Hall Associates, P.O. Box 482, Stamford, CN 06904) V.13- 1973-

JCPCDT Journal of Cardiovascular Pharmacology. (Raven Press, 1140 Ave. of the Americas, New York, NY 10036) V.1- 1979-

JCPHB8 Journal of Clinical Pharmacology and Journal of New Drugs. (Albany, NY) V.7-10, 1967-70. For publisher information, see JCPCBR

JCPPAV Journal of Comparative and Physiological Psychology. (American Psychological Association, 1200 17th St., N.W., Washington, DC 20036) V.40- 1947-

JCPRB4 Journal of the Chemical Society, Perkin Transactions 1. (Chemical Society, Publications Sales Office, Burlington House, London W1V 0BN, England) No.1- 1972-

JCPTA9 Journal of Comparative Pathology and Therapeutics. (Liverpool, England) V.1-74, 1883-1964. For publisher information, see JCVPAR

JCPYDR Journal of Clinical Pyschopharmacology. (Williams & Wilkins Co., 428 E. Preston St., Baltimore, MD 21202) V.1- 1981-

JCREA8 Journal of Cancer Research. (Baltimore, MD) V.1-14, 1916-30. For publisher information, see CNREA8

JCROD7 Journal of Cancer Research and Clinical Oncology. (Springer-Verlag, Heidelberger Pl. 3, D-1 Berlin 33, Germany) V.93- 1979-

JCSOA9 Journal of the Chemical Society. (London, England) 1926-65. For publisher information, see JCPRB4

JCTODH Journal of Combustion Toxicology. (Technomic Publishing Co., 265 Post Rd. W., Westport, CT 06880) V.3, no.1- 1976-

JCUPBN Journal of Cutaneous Pathology. (Munksgaard, 35 Noerre Soegade, DK-1370 Copenhagen K, Denmark) V.1- 1974-

JCVPAR Journal of Comparative Pathology. (Academic Press Inc., Ltd., 24-28 Oval Rd., London NW1 7DX, England) V.75- 1965-

JDGRAX Journal of Drug Research. (Drug Research and Control Center, 6, Abou Hazem St., Pyramids Ave., POB 29, Cairo, Egypt) V.2- 1969-

JDREAF Journal of Dental Research. (American Association for Dental Research, 734 Fifteenth St., NW, Suite 809, Washington, D.C. 20005) V.1- 1919-

JDSCAE Journal of Dairy Science. (American Dairy Science Association, 309 W. Clark St., Champaign, IL 61820). V.1- 1917-

JEBIAM Journal of Experimental Biology. (P.O. Box 32, Commerce Way, Colchester CO2 8HP. Essex. U.K.) V.7- 1930-

JEEMAF Journal of Embryology and Experimental Morphology. (P.O. Box 32, Commerce Way, Colchester CO2 8HP. Essex. U.K.) V.1- 1953-

JEENAI Journal of Economic Entomology. (Entomological Society of America, 4603 Calvert Rd., College Park, MD 21201) V.1- 1908-

JEGPAY Journal of the Egyptian Public Health Association. (Egyptian Public Health Association., Shousha Bldg., Bloc. A, 31 Sharia 26 July, App. 116, Cairo, Egypt) V.1- 1926-

JELJA7 Journal of Electron Microscopy. (Japanese Society of Electron Microscopy, Japan Academy Societies Center, 2-4-16 Yayoi, Bunkyo-ku, Tokyo 113, Japan) V.1- 1953-

JEMAAJ Journal of the Egyptian Medical Association. (The Egyptian MedicaL Assoc., 42 Sharia Kasr-el-aini, Cairo, Egypt) V.11- 1928-

JEMEAV Journal of Experimental Medicine. (Rockefeller University Press, 1230 York Avenue, New York, NY 10021) V.1- 1896-

JEPOEC Journal of Environmental Pathology, Toxicology and Oncology. (Chem-Orbital, POB 134, Park Forest, IL 60466) V.5(4)- 1984-

JEPTDQ Journal of Environmental Pathology and Toxicology. (Park Forest South, IL) V.1-5, 1977-81.

JESEDU Journal of Environmental Science and Health, Part A: Environmental Science and Engineering. (Marcel Dekker, POB 11305, Church St. Station, New York, NY 10249) V.A11- 1976-

JETHDA Journal of Ethnopharmacology. (Elsevier Sequoia SA, POB 851, CH-1001 Lausanne 1, Switzerland) V.1- 1979-

JETOAS Journal Europeen de Toxicologie. (Paris, France) V.1-6, 1968-72. For publisher information, see TOERD9

JETPEZ Journal of Engineering for Gas Turbines and Power. (American Society of Mechanical Engineers. Order Dept., POB 2300, Fairfield, NJ 07007) V.106- 1984-

JEZOAO Journal of Experimental Zoology (Alan. R. Liss, Inc., 150 Fifth Avenue, New York, NY 10011) V.1- 1904-

JFALAX Journal of the Faculty of Agriculture, Tottori University. (Maruzen Co., Ltd., Export Dep., P.O. Box 5050, Tokyo Int., 100- 31 Tokyo, Japan) V.1- 1951-

JFDSAZ Journal of Food Science. (Institute of Food Technologists, Subscrip. Dept., Suite 2120, 221 N. La Salle St., Chicago, IL 60601) V.26- 1961-

JFHZA8 Journal of the Faculty of Science, Hokkaido University, Series 6: Zoology. (Hokkaido University, Faculty of Science, Nishi-8 chome, Kita-10jo, Kita-ku, Sapporo 060, Japan) V.1- 1930-

JFIBA9 Journal of Fish Biology. (Academic Press, 24-28 Oval Rd., London NW1 7DX, England) V.1- 1969-

JFLCAR Journal of Fluorine Chemistry. (Elsevier Sequoia SA, P.O. Box 851, CH 1001, Lausanne 1, Switzerland) V.1- 1971-

JFMAAQ Journal of the Florida Medical Association. (Florida Medical Association., POB 2411, Jacksonville, FL 32203) V.1- 1914-

JFSCAS Journal of Forensic Sciences. (American Society for Testing and Materials, 1916 Race St., Philadelphia, PA 19103) V.1- 1956-

JFSTAB Journal of Food Science and Technology. (Assoc. of Food Scientists and Technologists, Exec Secy., Central Food Technological Research Inst., Mysore 570013, India) V.1- 1964-

JGAMA9 Journal of General and Applied Microbiology. (Microbiology Research Foundation, c/o Japan Academic Societies Center Bldg., 4-16, Tokyo 113, Japan) V.1- 1955-

JGHUAY Journal de Genetique Humaine. (Editeurs Medecine et Hygiene, 78, ave de la Rosaraie, 1211 Geneva 4, Switzerland) V.1- 1952-

JGMIAN Journal of General Microbiology (P.O. Box 32, Commerce Way, Colchester CO2 8HP, UK) V.1- 1947-

JGOBAC Journal de Gynecologie Obstetrique et Biologie de la Reproduction. (Masson et Cie, Editeurs, 120 Blvd. Saint-Germain, P-75280, Paris 06, France) V.1- 1972-

JHCYAS Journal of Histochemistry and Cytochemistry. (Williams and Wilkins Co., 428 E. Preston St., Baltimore, MD 21202) V.1- 1953-

JHEMA2 Journal of Hygiene, Epidemiology, Microbiology and Immunology. (Avicenum, Zdravotnicke Nakladatelstvi, Malostranske namesti 28, Prague 1, Czechoslovakia) V.1- 1957-

JHHBAI Bulletin of the Johns Hopkins Hospital. (Baltimore, MD) V.1-119, 1889-1966. For publisher information, see JHMJAX

JHMAD9 Journal of Hazardous Materials. (Elsevier Scientific, P.O. Box 211, Amsterdam C. Netherlands) V.1- 1977-

JHMJAX Johns Hopkins Medical Journal. (Journals Dept., Johns Hopkins Univ. Press, Baltimore, MD 21218) V.120-151, 1967-82. Discontinued.

JICSAH Journal of the Indian Chemical Society. (Indian Chemical Society, 92, Acharya Prafulla Chandra Rd., Calcutta 700009, India) V.5- 1928-

JIDEAE Journal of Investigative Dermatology. (Williams and Wilkins Co., 428 E. Preston St., Baltimore, MD 21202) V.1- 1938-

JIDHAN Journal of Industrial Hygiene. (Baltimore, MD/New York, NY) V.1-17, 1919-35. For publisher information, see AEHLAU

JIDIAQ Journal of Infectious Diseases. (University of Chicago Press, 5801 S. Ellis Ave., Chicago, IL 60637) V.1- 1904-

JIDOAA Jikken Dobutsu. Experimental Animals. (Nippon Jikken Dobutsu Kenkyukai, c/o Tokyo Daigaku Ikngaku Kenkyusho, 6-1, Shiroganedai, Minato-ku, Tokyo 108, Japan) V.1- 1952-

JIDXA3 Journal of the Indiana State Medical Association. (3935 N. Meridian, Indianapolis, IN 46208) V.1- 1908-

JIDZA9 Jinrui Idengaku Zasshi. Journal of Human Genetics. (Nippon Jinrui Iden Gakkai, c/o Tokyo Ika Shika Daigaku, 1-5-45 Yushima, Bunkyo-ku, Tokyo 113, Japan) V.1- 1956-

JIHTAB Journal of Industrial Hygiene and Toxicology. (Baltimore, MD/New York, NY) V.18-31, 1936-49. For publisher information, see AEHLAU

JIISAD Journal of Indian Institute of Science. (The Institute, Bangalore 560012, India) V.42- 1960-

JIMAAD Journal of the Indian Medical Association. (Indian Medical Assoc., 53, Creek Row, Calcutta 700014, India) V.1- 1931-

JIMMBG Journal of Immunological Methods. (Elsevier North Holland Inc., 52 Vanderbilt Ave., New York, NY 10017) V.1- 1971-

JIMRBV Journal of International Medical Research. (Secy., Cambridge Medical Publications, Ltd., 435/437 Wellingborough Rd., Northampton NN1 4EZ, England) V.1- 1972-

JIMSAX Journal of the Irish Medical Association. (Irish Medical Association., I.M.A. House, 10 Fitzwilliam Pl., Dublin, Ireland) V.28- 1951-

JINCAO Journal of Inorganic Nuclear Chemistry. (Pergamon Press Ltd., Headington Hill Hall, Oxford OX3 0BW, England) V.1- 1955-

JISMAB Journal of the Iowa State Medical Society. (Iowa Medical Society, 1001 Grand Ave., West Des Moines, IA 50265) V.1- 1911-

JJANAX Japanese Journal of Antibiotics. (Japan Antibiotics Research Assoc., 2-20-8 Kamiosaki, Shinagawa-ku, Tokyo, Japan) V.21- 1968-

JJATDK JAT, Journal of Applied Toxicology. (Heyden and Son, Inc., 247 S. 41st St., Philadelphia, PA 19104) V.1- 1981-

JJCOAC Japanese Journal of Clinical Oncology. (Japanese Journal of Clinical Oncology, Editorial Committee, c/o National Cancer Center, Tsukiji, Tokyo, Japan) V.1- 1971-

JJCREP Japanese Journal of Cancer Research (Gann). (Elsevier Science Publishers B.V., POB 211, 1000 AE Amsterdam, Netherlands) V.76- 1985-

JJEMAG Japanese Journal of Experimental Medicine. (Editorial Office, The Institute of Medical Science, University of Tokyo, Shirokanedai, Minatoku, Tokyo, Japan) V.7- 1928-

JJIND8 JNCI, Journal of the National Cancer Institute. (U.S. Government Printing Office, Superintendent of Documents, Washington, DC 20402) V.61- 1978-

JJMBAN Japanese Journal of Microbiology. (Tokyo, Japan) V.1-20, 1957-76.

JJMCAQ Japanese Journal of Medical Science and Biology. (National Institute of Health, 2-chome, Kamiosaki, Shinagawa-ku, Tokyo 141, Japan) V.5- 1952-

JJMDAT Japanese Journal of Medicine. (Nankodo Co., Ltd., POB 5272, Tokyo International 100-31, Japan) V.1- 1962-

JJPAAZ Japanese Journal of Pharmacology. (Nippon Yakuri Gakkai, c/o Kyoto Daigaku Igakubu Yakurigaku Kyoshitu Sakyo-ku, Kyoto 606, Japan) V.1- 1951-

JJPAD4 JPMA, the Journal of the Pakistan Medical Association. (Pakistan Medical Association, P.M.A. House, Garden Rd., Karachi 3, Pakistan) V.24- 1974-

JJPHAM Japanese Journal of Physiology. (Business Center for Academic Societies Japan, 4-16, Yayoc 2-chome, Bunkyo-ku, Tokyo 113, Japan) V.1- 1950-

JKXXAF Japanese Kokai Tokkyo Koho Patents. (U.S. Patent Office, Science Library, 2021 Jefferson Davis Highway, Arlington, VA 22202)

JLACBF Justus Leibigs Annalen der Chemie. (Verlag Chemie GmbH,

Postfach 1260/1280, D-6940 Weinheim, Fed. Rep. Germany) V.33-1840-

JLCMAK Journal of Laboratory and Clinical Medicine. (C.V. Mosby Co., 11830 Westline Industrial Dr., St. Louis, MO 63141) V.1- 1915-

JLSMAW Journal of the Louisiana State Medical Society. (New Orleans, LA) V.105-134, 1953-82.

JMCMAR Journal of Medicinal Chemistry. (American Chemical Society Pub., 1155 16th St., N.W., Washington, DC 20036) V.6- 1963-

JMEJAS Jikeikai Medical Journal. (Pharmacological Institute, The Jikei University School of Medicine, Minato-Ku, Tokyo, Japan) V.1- 1954-

JMENA6 Journal of Medical Entomology. (Bishop Museum, Entomology Dept., POB 6037, Honolulu, Hawaii 96818) V.1- 1964-

JMFTAT Journal of Milk and Food Technology. (Managing Editor, P.O. Box 437, Shelbyville, IN 46176) V.10- 1947-

JMGZAI Japan Medical Gazette. (Tokyo, Japan) V.1-18(?), 1964-81. Discontinued.

JMIABM Journal of Medicine and International Medical Abstracts and Reviews. (Hemangini Publications, P.O. Box 16705, Calcutta, India) V.19-31, 1956-66(?)

JMICAR Journal of Microscopy. (Blackwell Scientific Publications, Ltd., Osney Mead, Oxford OX2 OEL, England) V.89- 1969-

JMIMDQ Journal of Microbiological Methods. (Elsevier Science Publishers B.V., POB 211, 1000 AE Amsterdam, Netherlands) V.1- 1983-

JMJOAY Japanese Medical Journal. (Tokyo, Japan) V.1-4, 1948-51. For publisher information, see JJMCAQ

JMMIAV Journal of Medical Microbiology. (Longman Group Journal Division, 4th Ave., Harlow Essex CM20 2JE, England) V.1- 1968-

JMOBAK Journal of Molecular Biology. (Academic Press, 111 Fifth Ave., New York, NY 10003) V.1- 1959-

JMPCAS Journal of Medicinal and Pharmaceutical Chemistry. (Washington, DC) V.1-5, 1959-62. For publisher information, see JMCMAR

JMSCA9 Journal of Mental Science. (London, England) V.4-108, 1857-1962. For publisher information, see BJPYAJ

JMSHAO Journal of the Mount Sinai Hospital. (New York, NY) V.1-36, 1934-69. For publisher information, see MSJMAZ

JMSTBR Journal de Medecine de Strasbourg. (Expansion Scientifique Francaise, 15 rue St. Benoit, Paris 6, France) V.1- 1970-

JMSUAT Journal of the Madras Agricultural Students' Union. (Madras Agricultural Journal, Tamil Nadu Agricultural University Campus, Coimbatore 641003, India) V.1-16, 1912-28

JMTHBU Journal of the Medical Association of Thailand. (Medical Assoc. of Thailand, New Pejburi Rd., Bangkok 10, Thailand) V.1- 1918-

JMXSAE Journal of Maxillofacial Surgery. (Georg Thieme Verlag, Postfach 732, Herdweg 63, D-7000 Stuttgart 1, Fed. Rep. Germany)

JNAPAX Journal of Applied Nutrition. (International College of Applied Nutrition, Box 386, La Habra, CA 90631) V.6- 1953-

JNCIAM Journal of the National Cancer Institute. (Washington, DC) V.1-60, No.6, 1940-78. For publisher information, see JJIND8

JNCSAI Journal of Cell Science. (P.O. Box 32, Commerce Way, Colchester CO2 8HP. Exxes. U.K.) V.1- 1966-

JNDRAK Journal of New Drugs. (Albany, NY) V.1-6, 1961-66. For publisher information, see JCPCBR

JNEUAY Journal of Neuropsychiatry. (Chicago, IL) V.1-5, 1959-64. For publisher information, see BENPBG

JNEUBZ Journal of Neurobiology. (John Wiley and Sons, Inc., 605 3rd Ave., New York, NY 10016) V.1- 1969-

JNMAAE Journal of the National Medical Association. (Appelton and Lange, 25 Van Zant St., E. Norwalk, CT 06855) V.1- 1909-

JNMDAN Journal of Nervous and Mental Disease. (Williams and Wilkins Co., 428 E. Preston St., Baltimore, MD 21202) V.3- 1876-

JNMDBO Journal of Medicine. (PJD Publications, POB 966, Westbury, NY 11590) V.1- 1970-

JNNPAU Journal of Neurology, Neurosurgery and Psychiatry. (British Medical Association, Tavistock Square, London WC1H 9JR, UK) V.7- 1944-

JNPHAG Journal de Pharmacologie. (Masson Publishing USA, Inc. (Journals), 211 E.43rd St., Suite 1306, New York, NY 10017) V.1- 1970-

JNRYA9 Journal of Neurology. (Springer-Verlag New York, Inc., Service Center, 44 Hartz Way, Secaucus, NJ 07094) V.207- 1974-

JNSCAG Journal of the Neurological Sciences. (Elsevier Publishing Co., P.O. Box 211, Amsterdam C, Netherlands) No.1- 1964-

JNSVA5 Journal of Nutritional Science and Vitaminology. (Business Center Acad. Soc., Japan) V.19- 1973-

JOALAS Journal of Allergy, Including Allergy Abstracts. (St. Louis, MO) V.1-47, 1929-71. For publisher information, see JACIBY

JOANAY Journal of Anatomy. (Cambridge University Press, The Pitt Building, Trumpington Street, Cambridge CB2 1RP, UK) V.51- 1916-

JOAND3 Journal of Andrology. (Lippincott/Harper, Journal Fulfillment Dept., 2350 Virginia Ave., Hagerstown, MD 21740) V.1- 1980-

JOBAAY Journal of Bacteriology. (American Society for Microbiology, 1913 I St., N.W., Washington, DC 20006) V.1- 1916-

JOBIAO Journal of Biochemistry (Tokyo). (Japanese Biochemical Society, Gakkai Senta Bldg., 2-4-16 Yayoi, Bunkyo-Ku, Tokyo 113, Japan or Japan Publication USA or Maruzen) V.1- 1922-

JOBSDN Journal of Bio. Sciences. (Indian Academy of Sciences, POB 8005, Bangalore 560080, India) V.1- 1979-

JOCDAE Journal of Chronic Diseases. (Pergamon Press, Maxwell House, Fairview Park, Elmsford, NY 10523) V.1- 1955-

JOCEAH Journal of Organic Chemistry. (American Chemical Society Pub., 1155 16th St., N.W., Washington, DC 20036) V.1- 1936-

JOCMA7 Journal of Occupational Medicine. (American Occupational Medical Association, 150 N. Wacker Dr., Chicago, IL 60606) V.1- 1959-

JOCRAM Journal of Chromatography. (Elsevier Pub. Co., P.O. Box 211, Amsterdam C, Netherlands) V.1- 1958-

JODUA2 Journal of Osaka Dental University. (Osaka Dental Univ., 1-47, Kyobashi Higashi-Ku Osaka 540, Japan) V.1- 1967-

JOENAK Journal of Endocrinology. (Biochemical Society Publications, P.O. Box 32, Commerce Way, Whitehall Industrial Estate, Colchester CO2 8HP, Essex, England) V.1- 1939-

JOETD7 Journal of Ethnopharmacology. (Elsevier Sequoia SA, POB 1304, CH-1001, Lausanne, Switzerland) V.1- 1979-

JOGBAS Journal of Obstetrics and Gynaecology of the British Commonwealth. (London, England) V.68-81, 1961-74. For publisher information, see BJOGAS

JOGNAU Journal of Genetics. (Asit Kr. Bhattacharyya, M. A., 18/1, Barrackpore Trunk Rd, Belghoria, India) V.1- 1910-

JOHEA8 Journal of Heredity. (American Genetic Association, 818 Eighteenth St., N.W. Washington, D.C. 20006) V.5- 1914-

JOHYAY Journal of Hygiene. (Cambridge University Press, P.O. Box 92, Bentley House, 200 Euston Rd., London NW1 2DB, England) V.1- 1901-

JOIMA3 Journal of Immunology. (Williams and Wilkins Co., 428 E. Preston St., Baltimore, MD 21202) V.1- 1916-

JOIMD6 Journal of Immunopharmacology. (Marcel Dekker, POB 11305, Church St. Station, New York, NY 10249) V.1- 1978/79-

JOMAAL Journal of Mammalogy. (The American Museum of Natural History, Central Park W at 79th St., New York, NY 10024) V.1- 1919-

JONRA9 Journal of Neurochemistry. (Pergamon Press Ltd., Headington Hill Hall, Oxford OX3 0BW, England) V.1- 1956-

JONUAI Journal of Nutrition. (Journal of Nutrition, Subscription Dept., 9650 Rockville Pike, Bethesda, MD 20014) V.1- 1928-

JOPAA2 Journal of Parasitology. (American Society of Parasitologists, 1041 New Hampshire St., Lawrence, KS 60044) V.1- 1914-

JOPDAB Journal of Pediatrics. (C.V. Mosby Co., 11830 Westline Industrial Dr., St. Louis, MO 63141) V.1- 1932-

JOPHAN Journal de Physiologie. (Masson et Cie, Editeurs, 120 Blvd. Saint- Germain, P-75280, Paris 06, France) V.39- 1946-

JOPHBO Journal of Oral Pathology. (Munksgaard, 35 Norre Sogade, DK 1370 Copenhagen K, Denmark) V.1- 1972-

JOPHDQ Journal of Pharmacobio-Dynamics. (Pharmaceutical Society of Japan, 12-15-501, Shibuya 2-chome, Shibuya-Ku, Tokyo 150, Japan) V.1- 1978-

JOPRAJ Journal of Periodontology. (American Academy of Periodontology 211 E. Chicago Ave., Chicago, Illinois 60611) V.1- 1930-

JOPSAM Journal of Psychology. (Journal Press, 2 Commercial St., Provincetown, MA 02657) V.1-1935/36-

JORCAI Journal of Organometallic Chemistry. (Elsevier Sequoia SA, POB 851, CH-1001 Lausanne l, Switzerland) V.1- 1963-

JOSCDQ Journal of SCCJ (Society of Cosmetic Chemists of Japan). (Nippon Keshohin Gijutsushakai, c/o Shiseido, 7-5-5 Ginza, Chuoku, Tokyo, 104, Japan) V.10- 1976-

JOSUA9 Journal of Oral Surgery. (American Dental Association, 211 E. Chicago Ave., Chicago, IL 60611) V.1-16, 1943-58; V.23- 1965-

JOTPAX Journal of Oral Therapeutics and Pharmacology. (Baltimore, MD) V.1-4, 1964-68. Discontinued.

JOUOD4 Journal of UOEH (University of Occupational Environmental Health). (University of Occupational and Environmental Health, 1-1 Iseigaoka, Yahata-nishi-ku, Kitakyushu, 807, Japan) V.1- 1979-

JOURAA Journal of Urology. (Williams and Wilkins Co., 428 E. Preston St., Baltimore, MD 21202) V.1- 1917-

JOVIAM Journal of Virology. (American Society for Microbiology, 1913 I St., N.W., Washington, DC 20006) V.1- 1967-

JPBAA7 Journal of Pathology and Bacteriology. (London, England) V.1-96, 1892-1968. For publisher information, see JPTLAS

JPBEAJ Journal de Pharmacie de Belgique. (Masson et Cie, Editeurs, 120 Blvd. Saint-Germain, P-75280, Paris 06, France) V.1- 1945/46-

JPCAAC Journal of the Air Pollution Control Association. (Air Pollution Control Association., 4400 5th Ave., Pittsburgh, PA 15213) V.5- 1955-

JPCEAO Journal fuer Praktische Chemie. (Johann Ambrosius Barth Verlag, Postfach 109, DDR-701, Leipzig, Ger. Dem. Rep.) V.1- 1834- (Several new series, but continuous vol. nos. also used)

JPDSA3 Journal of Pediatric Surgery. (Grune and Stratton, Inc., 111 Fifth Ave., New York, NY 10003) V.1- 1966-

JPEMAO Journal of Perinatal Medicine. (Walter de Gruyter, Inc., 200 Sawmill River Rd., Hawthorne 10532) V.1- 1973

JPETAB Journal of Pharmacology and Experimental Therapeutics. (Williams and Wilkins Co., 428 E. Preston St., Baltimore, MD 21202) V.1- 1909/10-

JPFCD2 Journal of Environmental Science and Health, Part B: Pesticides, Food Contaminants, and Agricultural Wastes. (Marcel Dekker, POB 11305, Church St. Station, New York, NY 10249) V.B11- 1976-

JPHAA3 Journal of the American Pharmaceutical Association. (American Pharmaceutical Assoc., 2215 Constitution Ave, N.W., Washington, DC 20037) V.1-28, 1912-39; New Series V.1-17, 1961-77.

JPHYA7 Journal of Physiology. (Cambridge University Press, P.O. Box 92, Bentley House, 200 Euston Rd., London NW1 2DB, England) V.1- 1878-

JPMRAB Japanese Journal of Medical Sciences, Part IV: Pharmacology. (Tokyo, Japan) V.1-16, 1926-43. For publisher information, see JJPAAZ

JPMSAE Journal of Pharmaceutical Sciences. (American Pharmaceutical Assoc., 2215 Constitution Ave., N.W., Washington, DC 20037) V.50- 1961-

JPPGAR Journal de Physiologie et de Pathologie Generale. (Paris, France) V.1-38, 1899-1945. For publisher information, see JOPHAN

JPPMAB Journal of Pharmacy and Pharmacology. (Pharmaceutical Society of Great Britain, 1 Lambeth High Street, London SE1 5JN, England) V.1- 1949-

JPROAR Journal of Protozoology. (Society of Protozoologists, P.O. Box 368, Lawrence, KS 66044) V.1- 1954-

JPRSB6 Journal of Periodontal Research, Supplement. (Munksgaard, 35 Noerre Soegade, DK-1370 Copenhagen K, Denmark) No. 1- 1967-

JPTLAS Journal of Pathology. (Longman Group Ltd., Subscriptions (Journals Department, Fourth Avenue, Harlow, Essex, CM19 5AA, UK) V.97- 1969-

JPYRA3 Journal of Psychiatric Research. (Pergamon Press Ltd., Headington Hill Hall, Oxford OX3 0BW, England) V.1- 1961-

JRARAX Journal of Radiation Research. (c/o Tokai Daigaku Igakubu Bunshi Seibutsugaku Kyoshitsu, Tenbodai, Isehara, 259-11, Japan) V.1- 1960-

JRBED2 Journal of Reproductive Biology and Comparative Endocrinology. (P.G. Institute of Basic Medical Sciences, Dept. of Endocrinology, Taramani, 600 113, India) V.1- 1981

JRFSAR Journal of Reproduction and Fertility, Supplement. (Journal of Reproduction and Fertility Ltd., 22 New Market Rd., Cambridge CB5 8D7, England) No. 1- 1966-

JRHUA9 Journal of Rheumatology. (Business Office, Suite 420, 920 Yonge St., Toronto, Ontario M4W 3J7 Canada) V.1- 1974-

JRIHDC Journal of Research in Indian Medicine Yoga and Homoeopathy. (Banaras Hindu University, Varanasi-5, India) V.11- 1976-

JRIMAO Journal of Research in Indian Medicine. (Varanasi, India) V.1-10, 1966-75. For publisher information, see JRIHDC

JRMSAS Journal of the Royal Microscopical Society. (London, England) V.47-88, 1927-68. For publisher information, see JMICAR

JRPFA4 Journal of Reproduction and Fertility. (Journal of Reproduction and Fertility Ltd., 22 New Market Rd., Cambridge CB5 8D7, England) V.1- 1960-

JRPMAP Journal of Reproductive Medicine. (Box 56, 5841 Maryland Ave., Chicago, IL 60637) V.3- 1969-

JRSMD9 Journal of the Royal Society of Medicine. (Academic Press Inc., Ltd., 24-28 Oval Road, London NW1 7DX, England) V.71- 1978-

JSALDP Journal of Studies on Alcohol. (Journal of Studies on Alcohol, Rutgers University, New Brunswick, NJ 08903) V.36- 1975-

JSCCA5 Journal of the Society of Cosmetic Chemists. (Society of Cosmetic Chemists, 50 E. 41st St., New York, NY 10017) V.1- 1947-

JSDKD3 Josai Shika Daiguku Kiyo. Bulletin of the Josai Dental University. (Josai Shika Daigaku, 1-1, Keyakidai, Sakato, Saitama-Ken 350-02, Japan) Number 1- 1972-

JSFAAE Journal of the Science of Food and Agriculture. (Society of Chemical Industry, 14 Belgrave Sq., London SW1X 8PS, England) V.1- 1950-

JSGRA2 Journal of Surgical Research. (Academic Press, 111 Fifth Ave., New York, NY 10003) V.1- 1961-

JSICAZ Journal of Scientific and Industrial Research, Section C: Biological Sciences. (New Delhi, India) V.14-21, 1955-62. For publisher information, see IJEBA6

JSIRAC Journal of Scientific and Industrial Research. (Hillside Rd., New Delhi 110012, India) V.1-4, 1942-45; V.22- 1963-

JSOMBS Journal of the Society of Occupational Medicine. (John Wright and Sons, 42-44 Triangle W., Bristol BS8 1EX, England)

JSONAU Journal of Surgical Oncology. (Alan R. Liss, Inc., 150 5th Ave., New York, NY 10011) V.1- 1969-

JSONDX Journal of Soviet Oncology. (New York) V. 1-5, 1980-84.

JSOOAX Journal of the Chemical Society, Section C: Organic. (London, England) No.1-24, 1966-71. For publisher information, see JCPRB4

JSPMAW Journal of Supramolecular Structure. (Alan. R. Liss, Inc., 150 Fifth Ave., New York, NY 10011) V.1- 1972-

JSTBBK Journal of Steroid Biochemistry. (Pergamon Press Ltd., Headington Hill Hall, Oxford OX3 0BW, England) V.1- 1969-

JSXRAJ Journal of Sex Research. (Society for the Scientific Study of Sex, Inc., Mt. Royal and Guilford Aves., Baltimore, MD 21202) V.1- 1965-

JTASAG Journal of the Tennessee Academy of Science. (Tennessee Academy of Science, Box 153, George Peabody College, Nashville, TN 37203) V.1- 1926-

JTBIDS Journal of Thermal Biology. (Pergamon Press Ltd., Headington Hill Hall, Oxford OX3 0BW, England) V.1- 1975-

JTCEEM Journal de Toxicologie Clinique et Experimentale. (SPPIF, B. P. 22, F-41353 Vineuil, France) V.5- 1985-

JTCSAQ Journal of Thoracic and Cardiovascular Surgery. (C.V. Mosby Co., 11830 Westline Industrial Dr., St. Louis, MO 63141) V.38- 1959-

JTCTDW Journal of Toxicology, Clinical Toxicology. (Marcel Dekker, POB 11305, Church St. Station, New York, NY 10249) V.19- 1982-

JTEHD6 Journal of Toxicology and Environmental Health. (Hemisphere Publ., 1025 Vermont Ave., N.W., Washington, DC 20005) V.1- 1975/76-

JTMHA9 Journal of Tropical Medicine and Hygiene. (Staples and Staples Ltd., 94-96 Wigmore St., London, England) V.10- 1907-

JTOTDO Journal of Toxicology, Cutaneous and Ocular Toxicology. (Marcel Dekker, POB 11305, Church St. Station, New York, NY 10249) V.1- 1982-

JTSCDR Journal of Toxicological Sciences. (Editorial Office, Higashi Nippon Gakuen Univ., 7F Fuji Bldg., Kita 3, Nishi 3, Sapporo 060, Japan) V.1- 1976-

JUIZAG Juntendo Igaku. Juntendo Medicine. (Juntendo Igakkai, 2-1-1, Hongo, Bunkyo-ku, Tokyo, 113, Japan) V.1- 1955-

JULRA7 Journal of Ultrastructure Research. (Academic Press, 111 Fifth Ave., New York, NY 10003) V.1- 1975-

JVPTD9 Journal of Veterinary Pharmacology and Therapeutics. (Blackwell Scientific Publications Ltd., POB 88, Oxford, UK) V.1- 1978-

JWMAA9 Journal of Wildlife Management. (Wildlife Society, Suite 611, 7101 Wisconsin Ave., Washington, DC 20014) V.1- 1937-

JZKEDZ Jitchuken, Zenrinsho Kenkyuho. Central Institute for Experimental Animals, Preclinical Reports. (The Institute, 1433 Nogawa, Takatsu- Ku, Kawasaki 211, Japan) V.1- 1974/75-

KorCJ# Personnal Communication to NIOSH, from C.J. Korpics, Sherwin-Williams Chemicals, 1310 Expressway Drive, Toledo, OH 43608, August 22, 1974

KAEMAW Kitasato Archives of Experimental Medicine. (Kitasato Institute, 5-9-1 Shirokane, Minato-ku, Tokyo 108, Japan) V.1- 1917-

KAIZAN Kaibogaku Zasshi. Journal of Anatomy. (Nihon Kaibo Gakkai, c/o Tokyo Daigaku Igakubu Kaibogaku Kyoshitsu, 7-3-1, Hongo, Bunkyo-ku, Tokyo, Japan) V.1- 1928-

KAKTAF Kagaku to Kogyo (Tokyo). Chemistry and Chemical Industry. (Nippon Kagakkai, 1-5 Kanda-Surugadai, Chiyoda-ku, Tokyo 101, Japan) V.1- 1948-

KAMJDW Kawasaki Medical Journal. (Kawasaki Medical School, Kurashiki 701-01, Japan) V.1- 1975-

KASEAA Kagaku To Seibutsu. Chemistry and Biology. (Tokyo Daigaku Shuppan-kai, 7-3-1, Hongo, Bunkyo-ku, Tokyo, Japan) V.1- 1962-

KASHAJ Kanko Shikiso. Medical Researches for Photosensitizing Dyes. (Kanko Shikiso Kenkyukai, c/o Pathological Institute, Faculty of Medicine, Kyoto Univ., Kyoto, Japan) No. 1- 1949-

KBAMAJ Kosmicheskaya Biologiya I Aviakosmicheskaya Meditsina Space Biology and Aerospace Medicine. (Mezhdunarodnaya Kniga, Kuznetskii Most 18, Moscow G-200, USSR) V.8- 1974-

KBMEAL Kosmicheskaya Biologiya i Meditsina. Space Biology and Medicine. (Moscow, USSR) V.1-7, 1967-73. For publisher information, see KBAMAJ

KCRZAE Kauchuk i Rezina. (v/o Mezhdunarodnaya Kniga, Kuznetskii Most 18, Moscow G-200, USSR) V.11- 1937-

KDIKAX Kobe Daigaku Igakubu Kiyo. (Kobe Daigaku Igakkai, Kusumoki-cho, Ikuta-ku, Kobe 650, Japan) V.30- 1968-

KDIZAA Kagoshima Daigaku Igaku Zasshi. Medical Journal of Kagoshima University. (Kagoshima Daigaku Igakkai, 1208-1, Usuki-cho, Kagoshima 890, Japan) V.1- 1945-

KDPU** Beitrag zur Toxikologischen Wirkung Technischer Losungsmittel, Otto Klimmer Dissertation. (Pharmakologischen Institut der Universitat Wurzburg, Germany, 1937)

KDYIA5 Kidney International. (Springer-Verlag New York, Inc., Service Center, 44 Hartz Way, Secaucus, NJ 07094) V.1- 1972-

KEKHA7 Koshu Eiseiin Kenkyu Hokoku. Bulletin of the Institute of Public Health. (Kokuritsu Koshu Eiseiin, 4-6-1 Shirokanedai Minato-ku, Tokyo, 108, Japan) V.1- 1951-

KEKHB8 Kanagawa-ken Eisei Kenkyusho Kenkyu Hokoku. Bulletin of Kanagawa Prefectural Public Health Laboratories. (Kanagawa Prefectural Public Health Laboratories, 52-2, Nakao-cho, Asahi-ku, Yokohama 221, Japan) No.1- 1971-

KFIZAO Kyoto-furitsu Ika Daigaku Zasshi. Journal of the Kyoto Prefectural School of Medicine. (Kyoto-furitsu Ika Daigaku Igakkai, Hirokoji, Kawaramachidori, Kamiyoku, Kyoto, Japan) V.1- 1927-

KFMBA6 Kaiser Foundation Medical Bulletin. (Oakland, CA) V.1-7, 1953-59. Discontinued.

KGGZAL Koku Geka Gakkai Zasshi. Oral Surgery. (Tokyo, Japan)

KHFKDF K'o Hsueh Fa Chan Yueh K'an. Progress in Sciences. (Kuo Chia K'o Hsueh Wei Yuan Hui, 2 Canton St., Taipei 107, Taiwan) V.1- 1973-

KHFZAN Khimiko-Farmatsevticheskii Zhurnal. Chemical Pharmaceutical Journal. (v/o Mezhdunarodnaya Kniga, Kuznetskii Most 18, Moscow G-200, U.S.S.R.) V.1- 1967-

KHKKA3 Kachiku Hanshoku Kenkyu Kaishi. Journal of the Society of Animal Reproduction. (Tokyo, Japan) V.1-22, 1955-77. For publisher information, see KHZADH

KHTPAT Kexue Tongbao (Chinese Edition). Science Bulletin. (Guoji Shudian, POB 2820, Beijing, Peop. Rep. China) V.1-17, 1950-66; V.18- 1973-

KHZADH Kachiku Hanshokugaku Zasshi. Journal of Animal Reproduction. (Japanese Society of Animal Reproduction, c/o Third Research Division, National Institute of Animal Health, Kodarira Tokyo-to 187, Japan) V.23- 1977-

KHZDAN Khigiena i Zdraveopazvane. (Hemus, blvd Russki 6, Sofia, Bulgaria) V.9- 1966-

KHZHAZ Khimiya i Zhizn. Chemistry and Life. (v/o Mezhdunarodnaya Kniga, Kuznetskii Most 18, Moscow G-200, U.S.S.R.) V.1- 1965-

KIDZAK Kansai Ika Daigaku Zasshi. Journal of the Kansai Medical School. (Kansai Ika Daigaku Igakkai, 1, Fumizono-cho, Moriguchi 570, Osaka, Japan) V.8- 1956-

KIHSDM Guangxi Yixue. Kuanghsi Medicine. (Kuang-hsi I Hsueh K'o Hsueh Ch'ing Pao Yen Chin So, 19 T'ien T'ao Lu, Nan-ning, Kuang-hsi, Peop. Rep. China) V.?- 1979-

KIKNAJ Kokuritsu Idengaku Kenkyusho Nempo. Annual Report of the National Institute of Genetics. (Kokuritsu Idengaku Kenkyusho, 1111 Yata, Mishima, Shizuoka-ken 411, Japan) No. 1- 1949-

KIZAAL Kurume Igakkai Zasshi. Journal of the Kurume Medical Association. (Kurume Igakkai, c/o Kurume Daigaku Igakubu, 67, Asahi-machi, Kurume 830, Japan) V.9- 1946-

KIZSB8 Kyorin Igakukai Zasshi. Journal of the Kyorin Medical Society. (6-20-2 Shinkawa, Mitaka City, Tokyo, Japan) V.1- 1970-

KJMDA6 Kobe Journal of the Medical Sciences. (Kobe Daigaku Igakubu, Editorial Board, Kobe, Japan) V.1- 1951-

KJMEA9 Keio Journal of Medicine. (Keio University, School of Medicine, 35 Shinano-machi, Shinjuku-Ku, Tokyo 160, Japan) V.1- 1952-

KJMSAH Kyushu Journal of Medical Science. (Fukuoka, Japan) V.6-15, 1955-64. Discontinued.

KLWOAZ Klinische Wochenschrift. (Springer-Verlag, Heidelberger Pl. 3, D-1 Berlin 33, Germany) V.1- 1922-

KMAUAI Klinische Monatsblaetter fuer Augenheilkunde. (Ferdinand Enke Verlag, Hasenbergsteige 3, 7000 Stuttgart 1 Germany) V.1- 1863-

KMBAAV Kaiser Foundation Medical Bulletin, Abstract Issue. (Oakland, CA) No.1-5, 1960-65. Discontinued.

KNZOAU Kanzo. Liver. (Nippon Kanzo Gakkai, c/o Toyo Bunko, 28-21, 2-chome, Hon Komagome, Bunkyo-ku, Tokyo 113, Japan) V.1- 1960-

KOBUA3 Kobunshi. High Polymers. (Kobunshi Gakkai, 5-12-8, Ginzai, Chuo-Ku, Tokyo 104, Japan) V.1- 1952-

KODAK* Kodak Company Reports. (343 State St., Rochester, NY 14650)

KOKABN Kobunshi Kako. Polymer Applications. (Kobunshi Kankokai, Chiekoin Maruta-machi Kundaru, Kamigyo-ku, Kyoto, 602, Japan) V.13- 1964-

KONODE Kongetsu No Noyaku. Agricultural Chemicals Monthly. (Kagoku Kogyo Nipposha, 3-19-16 Shibaura, Minato-ku, Tokyo 108, Japan) V.1- 1953(?)-

KOTTAM Kogai To Taisaku. Journal of Environmental Pollution Control. (Kogai Taisaku Gijutsu Dokokai, Shuwa-Akasaka Bldg., 9-1-244, Akasaka, Minato-ku, Tokyo, Japan) V.1- 1965-

KPJBAR Kalikasan. The Philippine Journal of Biology. (National Publishing Cooperative Inc., 2nd Fl., Santander Bldg., 20 M Hemady St. Cor Aurora Blvd., Quezon City, Philippines) V.1- 1972-

KRANAW Krankheitsforschung. (Leipzig, E. Germany) V.1-9, 1925-32. Discontinued.

KREBAG Krebsarzt. Zeitschrift fuer Erforschung und Bakaempfung der Krebskrankheit. (Vienna, Austria) V.1-24, 1946-69. For publisher information, see OZEBAE

KRKRDT Kriobiologiya i Kriomeditsina. (Izdatel'stvo Naukova Dumka, ul Repina 3, Kiev, USSR) No.1- 1975-

KRMJAC Kurume Medical Journal. (Kurume Igakkai, c/o Kurume Daigaku Igakubu, 67, Asahi-machi, Kurume, Japan) V.1- 1954-

KSGZA3 Kyushu Shika Gakkai Zasshi. Journal of the Kyushu Dental Society. (c/o Kyushu Shika Daigaku, 2-6-1 Manazuru, Kokurakita-ku, Kitakyushu, Japan) V.1- 1933-35; 1939-40; 1951-

KSKZAN Khimiya v Sel'skom Khozyaistve. Chemistry in Agriculture. (v/o Mezhdunarodnaya Kniga, Kuznetskii Most 18, Moscow G-200, U.S.S.R.) No.1- 1963-

KSRNAM Kiso to Rinsho. Clinical Report. (Yubunsha Co., Ltd., 1-5, Kanda Suda-Cho, Chiyoda-ku, KS Bldg., Tokyo 101, Japan) V.1- 1960-

KTHYAC Kuo Li Tai-Wan Ta Hsueh I Hsueh Yuan Yen Chiu Pao Kao. The memoirs of the College of Medicine of the National Taiwan University. (Kuo Li Tai-Wan Ta Hsueh I Hsueh Yuan, Taipei, Taiwan) 1947-

KTUNAA K'at'ollik Taehak Uihakpu Nonmunjip. Journal of Catholic Medical College. (Catholic Medical College, Seoul, South Korea) V.1- 1957-

KTUWDD Koryo Taehakkyo Uikwa Taehak Nonmunjip. Collection of Papers of the Medical School, Korea University.

KUIZAR Kumamoto Igakkai Zasshi. Journal of the Kumamoto Medical Society. (Kumamoto Igakkai, C/o Kumamoto Daigaku Igakubu, 2-1, 2-chome Honjo-machi, Kumamoto, Japan) V.1- 1925-

KUMJAX Kumamoto Medical Journal. (Kumamoto Daigaku Igakubu, Library, Kumamoto, Japan) V.1- 1938-

KYDKAJ Kyoritsu Yakka Daigaku Kenkyu Nempo. Annual Report of the Kyoritsu College of Pharmacy. (Kyoritsu Yakka Daigaku, 1-5-30 Shibakoen Minato-ku, Tokyo, Japan) V.1- 1955-

LacHB # Personal Communication from Mr. H.B. Lackey, Chemical Products Div., Crown Zellerbach, Camas, Washington 98607, to Dr. H.E. Christensen, NIOSH, Rockville, MD 20852, June 9, 1978

LilPW # Personal Communication from P.W. Lilley, Sun Company, Inc., P.O. Box 1135, Marcus Hook, PA 19061, to Richard Lewis, NIOSH, November 1, 1983

LitL ## Personal Communication from Larry Little, Occidental Chemical Corp., Hooker Chemical Center, 360 Rainbow Blvd. South, Box 728, Niagara Falls, NY 14302, to Doris Sweet, NIOSH, Cincinnati, OH 45226, May 6, 1985

LAACAR Laboratory Animal Care. (Joliet, IL) V.?-20, ?-1970. For publisher information, see LBASAE

LACHDL Liebigs Annalen der Chemie. (Verlag Chemie International, Inc. 175 Fifth Ave., New York, NY 10010) No. 1- 1979-

LAINAW Laboratory Investigation. (Williams and Wilkins Co., 428 E. Preston St., Baltimore, MD 21202) V.1- 1952-

LAKAA3 Laekartidningen. Medical News. (Swedish Medical Assoc., POB 5610, 11486 Stockholm 5, Sweden) V.62- 1965-

LAMEDS LARC Medical. (19 Bis, Rue d' Inkermann, 59000 Lille, France) V.1- 1981-

LANCAO Lancet. (7 Adam St., London WC2N 6AD, England) V.1- 1823-

LAPPA5 Lavori dell'Istituto di Anatomia e Istologia Patologica della Universita Degli Studi Perugia. (Istituto di Anatomia e Isologia Patologica, Caselle Postale 327, 06100 Perugia, Italy) V.1- 1939-

LARYA8 Laryngoscope. (222 Pine Lake Rd., Collinsville, IL 62234) V.1- 1896-

LBANAX Laboratory Animals. (Biochemical Society Book Depot, POB 32, Commerce Way, Colchester Essex CO2 8HP, England) V.1- 1967-

LBASAE Laboratory Animal Science. (American Association for Laboratory Animal Science, 210 N. Hammes Ave., Suite 205, Joliet, IL 60435) V.21- 1971-

LBWTAP Lebensmittel-Wissenschaft and Technologie. (Forster-Verlag, Postfach 295, CH-8033 Zurich, Switzerland) V.1- 1968-

LDBU** Langer Dissertation. (Breslow, 1932)

LDTU** Narkoseversuche mit Hoheren Alkoholen und Stickstoffderivaten, Fritz Leube Dissertation. (Pharmakologischen Institut der Universitat Tubingen, Germany, 1931)

LEREDD Leukemia Research. (Pergamon Press Ltd., Headington Hill Hall, Oxford OX3 OEW, England) V.1- 1977-

LGMDAU Liege Medical. (Brussels 18, Belgium) 1907-19. For publisher information, see SCALA9

LIFSAK Life Sciences. (Pergamon Press, Maxwell House, Fairview Park, Elmsford, NY 10523) V.1-8, 1962-69; V.14- 1974-

LITRC* Literature Research Co. Translation. (Annandale, VA)

LLOYA2 Lloydia. (Lloyd Library and Museum, 917 Plum St., Cincinnati, OH 45202) V.1- 1938-

LMSED6 L.E.R.S. Monograph Series. Laboratories d'Etudes et de Recherches Synthelabo Monograph Series. (Raven Press, 1185 Ave. of Americas, New York, NY 10036) V.1- 1983-

LONZA# Personal Communication from LONZA Ltd., CH-4002, Basel, Switzerland, to NIOSH, Cincinnati, OH 45226

LPDSAP Lipids. (American Oil Chemists' Society, 508 South Sixth St., Champaign, IL 61820) V.1- 1966-

LPPTAK Labo-Pharma-Problemes et Techniques (19, rue Louis-le-Grand, Paris, France) V.12- 1965-

LROTDW Laryngo-Rhinologie und Otolaryngologie. (Georg Thieme Verlag, Postfach 732, Herdweg 63, 7000 Stuttgart, Germany) V.54- 1975-

LSBGAY Life Sciences, Part 2: Biochemistry, General and Molecular Biology. (New York, NY) V.9-13, 1970-73. For publisher information, see LIFSAK

LSPPAT Life Sciences, Part 1: Physiology and Pharmacology. (New York, NY) V.9-13, 1970-73. For publisher information, see LIFSAK

LYPHAD Lyon Pharmaceutique. (Edition Paul Chatelain, 63 rue de la Republique, 69 Lyons 2, France) V.1- 1950-

LZAVAL Latvijas P.S.R. Zinatnu Akademijas Vestis. Proceedings of the Academy of Sciences of the Latvian S.S.R. (v/o Mezhdunarodnaya Kniga, Kuznetskii Most 18, Moscow G-200, U.S.S.R.) No.1- 1947-

MahWM# Personal Communication from W.M. Mahlburg, Hopkins Agricultural Chemical Co., P.O. Box 7532, Madison, WI 53707 to NIOSH, Cincinnati, OH 45226, November 16, 1982

MarJV# Personal Communication from Josef V. Marhold, VUOS, 539-18, Pardubice, Czechoslavakia, to the Editor of RTECS, Cincinnati, OH, March 29, 1977

MccSB # Personal Communication from Susan B. McCollister, Dow

Chemical U.S.A., Midland, MI 48640, to NIOSH, Cincinnati, OH 45226, June 15, 1984

MosJN # Personal Communication from J. N. Moss, Toxicology Department, Rohm and Haas Co., Spring House, PA 19477 to R. J. Lewis, Sr., NIOSH, Cincinnati, OH 45226, August 15, 1979

MACHAT Marseille Chirurgical. (Paris, France) V.1-24, 1949-72(?). For publisher information, see MDTMBF

MACPAJ Mayo Clinic Proceedings. (Mayo Foundation, Plummer Bldg., Mayo Clinic, Rochester, MN 55901) V.39- 1964-

MADCAJ Medical Annals of the District of Columbia. (Washington, DC) V.1-43, 1932-74. Discontinued.

MAEPBU Methods and Achievements in Experimental Pathology. (Yearbook Medical Publishers, 35 E. Wacker Dr., Chicago, IL 60601) V.1- 1966-

MAGDA3 Mechanisms of Ageing and Development. (Elsevier Sequoia SA, POB 851, Ch 1001, Lausanne 1, Switzerland) V.1- 1972-

MAGJAL Malaysian Agricultural Journal. (Ministry of Agriculture and Fisheries, Business Mgr., Kuala Lumpur, Malaysia) V.45- 1965-

MAIKD3 Maikotokishin (Tokyo). Mycotoxin. (Maikotokishin Kenkyukai, c/o Tokyo Rika Daigaku Yakugakubu, 12 Funagawaramachi, Ichigaya, Shinjuku-ku, Tokyo 162, Japan) No.1- 1975-

MAIZAB Manshu Igaku Zasshi. Manthuria Medical Journal. (Dairen/Shimmeicho, South Manchuria) V.1-33, 1923-40. Discontinued.

MAMDAX Marseille Medical. (Marseille, France) V.6-112, 1869-1975. For publisher information, see MDTMBF

MAMEA2 Maroc Medical. (Casablanca, Morocco) V.1-55, 1921-75. Discontinued.

MASCAQ Medical Arts and Sciences. (Loma Linda, CA) V.1-28, 1947-74. Discontinued.

MASODV Masui to Sosei. Anesthesia and Resuscitation. (Hiroshima Masui Igakkai Hiroshima Daigaku Igakubu Masuigaku Kyoshitsu, 1-2-3 Kasumi, Hiroshima 734, Hiroshima, Japan) V.6- 1970-

MBLED7 Marine Biology Letter. (Elsevier Biomedical Press, 52 Vanderbilt Ave., New York, NY 10017) V.1- 1980-

MCBIA7 Microbios. (Faculty Press, 88 Regent St., Cambridge, England) V.1- 1969-

MCEBD4 Molecular and Cellular Biology. (American Society for Microbiology, 1913 I St., NW, Washington, DC 20006) V.1- 1981-

MDACAP Medicamentos de Actualidad. (Medicamentos de Actualidad, Apartado de Correos 540, Barcelona, Spain) V.1- 1965-

MDBYAS Medical Biology (Editorial Office, Runeberginkatu 47A, SF-00260 Helsinki 26, Finland) V.52- 1974-

MDCHAG Medicinal Chemistry: A Series of Monographs. (Academic Press, 111 5th Ave., New York, NY 10003) V.1- 1963-

MDMIAZ Medycyna Doswiadczalna i Mikrobiologia. (Ars Polona-RUCH, POB 1001, 1, P-00068 Warsaw, 1, Poland) V.1- 1949- For English Translation, see EXMMAV

MDNKAC Miyazaki Daigaku Nogakubu, Kenkyu Hokoku. Bulletin of the Faculty of Agriculture, University of Miyazaki. (Miyazaki Daigaku Nogakubu, 100, Funazuka-cho, Miyazaki 880, Japan) V.1- 1955-

MDPGAA Mitteilungen der Deutschen Pharmazeutischen Gesellschaft. (Weinheim Ger.) V.1-21, 1924-44; V.22-41, 1952-71. For publisher information, see ARPMAS

MDREP* U. S. Army, Chemical Corps Medical Division Reports. (Army Chemical Center, MD)

MDSR** U. S. Army, Chemical Corps Medical Division Special Report. (Army Chemical Center, MD)

MDTMBF Mediterranee Medicale. (Sud-Regie, 58 Ave de la Marne, 92600 Asnieres, France) V.1- 1973-

MDWTAG Medycyna Weterynaryjna. Veterinary Medicine. (Ars Polona-RUCH, POB 1001, P-00068 Warsaw, Poland) V.1- 1945-

MDZEAK Medizin und Ernaehrung. (Stuttgart, Germany) V.1-13, 1959-72. Discontinued.

MECHAN Medicinal Chemistry, A Series of Reviews. (New York, NY) V.1-6, 1951-63. Discontinued.

MEDIAV Medicine. (Williams and Wilkins, 428 E. Preston St., Baltimore, MD 21202) V.1- 1922-

MEHYDY Medical Hypotheses. (Churchill Livingstone Inc., 19 W. 44 St., New York, NY 10036) V.1- 1975-

MEIEDD Merck Index. (Merck and Co., Inc., Rahway, NJ 07065) 10th ed.- 1983- Previous eds. had individual CODENS

MEKLA7 Medizinische Klinik. (Urban and Schwarzenberg, Pettenkoferst 18, D-8000 Munich 15, Germany) V.1- 1904-

MELAAD Medicina del Lavoro. Industrial Medicine. (Via S. Barnaba, 8 Milan, Italy) V.16- 1925-

MEMOAQ Medizinische Monatsschrift. (Wissenschaftliche Verlagsgesellschaft mbH, Postfach 40, 7000 Stuttgart 1, Germany) V.1- 1947-

MEPAAX Medycyna Pracy. Industrial Medicine. (ARs-Polona-RUSH, POB 1001, 00-068 Warsaw 1, Poland) V.1- 1950-

MEPHDN Medical Pharmacy. (Daiichi Seiyaku K.K., 3-14-10 Nihonbashi, Chuo-ku, Tokyo 103, Japan) V.1- 1966-

MERAA9 Meditsinskaya Radiologiya. (Akademiya Meditsinskikh Nauk SSSR, ul Solyanka 14, 109801 Moscow, USSR) V.1- 1956-

METAAJ Metabolism, Clinical and Experimental. (Grune and Stratton, Inc., 111 Fifth Ave., New York, NY 10003) V.1- 1952-

METRA2 Medecine Tropicale (Marseille). (Parc du Pharo, 13007 Marseille, France) V.1- 1941-

MEXPAG Medicina Experimentalis. (Basel, Switzerland) V.1-11, 1959-64; V.18-19, 1968-69. For publisher information, see JNMDBO

MFEPDX Methods and Findings in Experimental and Clinical Pharmacology. (Methods and Findings, Sub. Dept., Apartado Correos 1179, Barcelona, Spain) V.1- 1979-

MFLRA3 Mededelingen van de Faculteit Landbouwwetenschappen, Rijksuniversiteit Gent. Communications of the Faculty of Agricultural Sciences, State University of Ghent. (Bibliotheek, Faculteit Labdbouwwettenschappen, Rijksuniversiteit Gent, Coupure 533, B-9000, Ghent, Belgium) V.35- 1970-

MGBOAJ Magyar Tudomanyos Akademia Biologiai es Orvosi Tudomanyok Osztalyanak Kozlemenyei. Publications of the Biological and Medical Science Department of the Hungarian Academy of Science. (Budapest, Hungary) V.5-14, 1954-63. For publisher information, see ORVOBF

MGBUA3 Microbial Genetics Bulletin. (Ohio State University, College of Biological Science, Dept. of Microbiology, 484 W. 12th Ave., Columbus, Ohio 43210) Number 1- 1950-

MGGEAE Molecular and General Genetics. (Springer-Verlag, Heidelberger Pl. 3, D-1 Berlin 33, Germany) V.99- 1967-

MGLFDB Monographs of the Giovanni Lorenzini Foundation. (John Wiley and Sons, Inc., 605 Third Ave., New York, NY 10158) V.1- 1978-

MGLHAE Mitteilungen Aus Dem Gebiete Der Lebensmitteluntersuchung und Hygiene. (Eidgenoessiche Drucksachen und Materialzentrale, 3000 Bern, Switzerland) V.1- 1910-

MGONAD Magyar Onkologia. Hungarian Onkology. (Kultura, P.O. Box 149, H-1389 Budapest, Hungary) V.1- 1957-

MIANAP Minerva Anestesiologica. (Edizioni Minerva Medica, Casella Postale 491, I-10126 Turin, Italy) V.19- 1953-

MIBLAO Microbiology (Moscow). (Plenum Publishing Corp., 233 Spring St., New York, NY 10013) V.26- 1957-

MIBUBI Mikrobiyoloji Bulteni. (Dr. Ismail Gurer, Gulhane Askeri Tip Akademisi Intaniye Klinigi Profesoru, Etlik, Ankara, Turkey) V.1- 1967-

MIDAD4 NIDA Research Monograph. (National Institute on Drug Abuse, Division of Research, 5600 Fishers Lane, Rockville, MD 20857) No.1- 1975-

MIGIA6 Minerva Ginecologia. (Edizioni Minerva Medica, Casella Postale 491, I-10100 Turin, Italy) V.1- 1949-

MIFAAB Minerva Farmaceutica. (Turin, Italy) V.1- 13, 1952-64 Discontinues

MIIGAA Mie Igaku. Mie Medica Science. (Mie Igakkai, Mie-kenritsu Daigaku Igakubu, 11, Otani-machi, Tsu, Mie, Japan) V.1- 1957-

MIIMDV Microbiology and Immunology. (Center for Academic Publications Japan, 4-16, Yayoi 2-chome, Bunkyo-ku, Tokyo 113, Japan) V.21- 1977-

MIKBA5 Mikrobiologiya. (v/o Mezhdunarodnaya Kniga, Kuznetskii Most 18, Moscow G-200, USSR) V.1- 1932-

MIKSAM Mikroskopie. (Verlag Georg Fromme und Co., Spengergasse 39, A 1051 Vienna 5, Austria) V.1- 1946-

MILEDM Microbios Letters. (Faculty Press, 88 Regent St., Cambridge, England) V.1- 1976-

MIMDAL Minnesota Medicine. (375 Jackson St., St. Paul, MN 55101) V.1- 1918-

MIMEAO Minerva Medica. (Edizioni Minerva Medica, Casella Postale 491, Turin, Italy) V.1- 1909-

MIPEA5 Minerva Pediatrica. (Edizioni Minerva Medica, Cassella Postale 491, I-10126 Turin, Italy) V.1- 1949-

MIVRA6 Microvascular Research. (Academic Press, 111 Fifth Ave., New York, NY 10003) V.1- 1968-

MJAUAJ Medical Journal of Australia. (P.O. Box 116, Glebe 2037, NSW 2037, Australia) V.1- 1914-

MJDHDW Mukogawa Joshi Daigaku Kiyo, Yakugaku Hen. Bulletin of Mukogawa Women's College, Food Science Series. (Mukogawa Joshi Daigaku, 6-46, Ikebiraki-cho, Nishinomiya 663, Japan) No.19- 1972-

MJOUAL Medical Journal of Osaka University. (The University, 33, Joan-cho, Kita-Ku, Osaka, Japan) V.1- 1949-

MLBRBU Molecular Biology Reports. (Dr. W. Junk bv Publishers, POB 13713, 2501 ES The Hague, Netherlands) V.1- 1973-

MLDCAS Medecine Legale et Dommage Corporel. (Paris, France) V.1-7, 1968-74. Discontinued.

MLSR** U. S. Army, Chemical Corps Medical Laboratories Special Reports. (Army Chemical Center, MD)

MMAPAP Mycopathologia et Mycologia Applicata. (The Hague, Netherlands) V.5-54, No.4, 1950-74, For publisher information, see MYCPAH

MMDPA6 Materia Medica Polona (English Edition). (Ars Polona-RUCH, P.O. Box 1001, P-00 068 Warsaw, 1, Poland) V.1- 1969-

MMEDA9 Military Medicine. (Association of Military Surgeons of the United States, Box 104, Kensington, MD 20795) V.116- 1955-

MMIYAO Medical Microbiology and Immunology. (Springer-Verlag, Heidelberger, Pl. 3, D-1 Berlin 33, Germany) V.157- 1971-

MMJJAI Mie Medical Journal. (Mie Medical Society, Mie Prefectual Univ., School of Medicine, Tsu, Japan) V.3- 1952-

MMWOAU Muenchener Medizinische Wochenscrift. (Munich, Fed. Rep. Ger.) V.33-115, 1886-1973.

MMWR** MMWR. Morbidity and Mortality Weekly Report. (Centers for Disease Control, Atlanta, GA 30333) For RTECS citation, only V.34(Suppl 1S), 1985 is used. V.25(10)- 1976-

MOLAAF Monatsschrift fuer Ohrenheilkunde und Laryngo-rhinologie. (Vienna, Austria) V.1-108, 1867-1974. For publisher information, see LROTDW

MOLBBJ Molecular Biology. English of Molekuliarnaia Biologiia. (Plenum Publishing Corp., 233 Spring St., New York, NY 10013) V.1- 1967-

MONS** Monsanto Co. Toxicity Information. (Monsanto Industrial Chemicals Co., Bancroft Bldg., Suite 204, 3411 Silverside Rd., Wilmington, DE 19810)

MOPMA3 Molecular Pharmacology. (The American Society for Pharmacology and Experimental Therapeutics, 9650 Rockville Pike, Bethesda, MD 20014) V.1- 1965-

MPHEAE Medicina et Pharmacologia Experimentalis. (Basel, Switzerland) V.12-17, 1965-67. For publisher information, see PHMGBN

MPPBAB Meditsinskaya Parazitologiya i Parazitarnye Bolezni. Medical Parasitology and Parasitic Diseases. (v/o Mezhdunarodnaya Kniga, Kuznetskii Most 18, Moscow G-200, U.S.S.R.) V.1- 1932-

MPPPBK Modern Problems of Pharmacopsychiatry. (S. Karger AG, Postfach, CH-4009, Basel, Switzerland) V.1- 1968-

MRBUDF MARDI Research Bulletin. (MARDI, Secy., Publication Committee, POB 208, Serdang, Selangor, Malaysia) V.1- 1973-

MRCDAY Medical Record (1866-1922). (New York, NY) V.1-101, 1866-1922. For publisher information, see CLMEA3

MRCSAB Medical Research Council, Special Report Series. (Her Majesty's Stationery Office, P.O. Box 569, London SE1 9NH, England) SRS1- 1915-

MRLAB3 Mededelingen Rijksfaculteit Landbouwwetenschappen, Gent. Communications of the State University of Agricultural Sciences, Ghent. (Ghent, Belgium) V.31-34, 1966-69. For publisher information, see MFLRA3

MRLEDH Mutation Research Letters. (Elsevier/North-Holland Biomedical Press, POB 211, 1000 AE Amsterdam, Netherlands)

MRLR** U. S. Army, Chemical Corps Medical Laboratories Research Reports. (Army Chemical Center, Edgewood Arsenal, MD)

MSERDS Microbiology Series. (Marcel Dekker, Inc., POB 11305, Church St. Station, New York, NY 10249) V.1- 1973-

MSJMAZ Mount Sinai Journal of Medicine, New York. (The Annenberg Building, Room 10-35, 5th Ave. and 100th St., New York, NY 10029) V.37- 1970-

MSKNA9 Meiji Seika Kenkyu Nempo. (Meiji Seika K.K., Morooka-cho, Kohoku-ku, Yokohama 222, Japan) No.1- 1959-

MTENBL Modern Trends in Endocrinology. (London, England) V.1-4, 1958-72. Discontinued.

MTHEA8 Monographs on Therapy. (The Squibb Institute for Medical Research, New Brunswick, NJ)

MUREAV Mutation Research. (Elsevier/North Holland Biomedical Press, P.O. Box 211, 1000 AE Amsterdam, Netherlands) V.1- 1964-

MUTAEX Mutagenesis. (IRL Press Ltd. 1911 Jefferson Davis Highway, Suite 907, Arlington, VA 22202) V.1- 1986-

MVCRB3 "Fluorescent Whitening Agents, Proceedings of A Symposium." MVC-Report, Miljoevardscentrum, Stockholm Center for Environmental Sciences, No. 2, 1973

MVMZA8 Monatshefte fuer Veterinaermedizin. (VEB Gustav Fischer Verlag, Postfach 176, Villengang 2, 69 Jena, E. Germany) V.1- 1946-

MYCPAH Mycopathologia. (Dr. W. Junk bv Publishers, POB 13713, 2501 ES The Hague, Netherlands) V.1- 1938-

MZHUDX Mikrobiologicheskii Zhurnal (Kiev). Journal of Microbiology. (v/o Mezhdunaronaya Kniga, Kuznetskii Most 18, Moscow G-200, U.S.S.R.) V.40- 1978-

MZUZA8 Meditsinskii Zhurnal Uzbekistana (v/o Mezhdunarodnaya Kniga, Kuznetskii Most 18, Moscow G-200, U.S.S.R.) No.1- 1957-

NAGZAC Nagasaki Igakkai Zasshi. Journal of Nagasaki Medical Association. (Nagasaki Igakkai, c/o Nagasaki Daigaku Igakubu, 12-4 Sakamoto-machi, Nagasaki 852, Japan) V.1- 1923-

NAHRAR Nahrung. Chemistry, Biochemistry, Microbiology, Technology, Nutrition. (Akademie-Verlag GmBH, Leipziger St. 3-4, 108 Berlin, Ger. Dem. Rep.) V.1- 1957-

NAIZAM Nara Igaku Zasshi. Journal of the Nara Medical Association. (Nara Kenritsu Ika Daigaku, Kashihara, Nara, Japan) V.1- 1950-

NALSDJ NATO Advanced Study Institute Series, Series A: Life Sciences. (Plenum Pub. Corp., 233 Spring St., New York, NY 10013) V.53- 1983-

NAREA4 Nutrition Abstracts and Reviews. (Central Sales Branch, Commonwealth Agricultrual Bureaux, Farnham Royal, Slough SL2 3BN, England) V.1- 1931-

NARHAD Nucleic Acids Research. (Information Retrieval Inc., 1911 Jefferson Davis Highway, Arlington, VA 22202) V.1- 1974-

NASDA6 Nagoya Shiritsu Daigaku Igakkai Zasshi. Journal of the Nagoya City University Medical Association. (The University, Nagoya, Japan) V.1- 1950-

NASLDX National Academy Science Letters (India). (National Academy of Sciences, India, 5-Lajpat Rai Rd., Allahabad 211002, India) V.1- 1978-

NASSDK NATO Advanced Study Institute Series, Series A: Life Sciences. (New York, NY) V.1-52, 1975-82.

NATUAS Nature. (Macmillan Journals Ltd., Brunel Rd., Basingstoke RG21 2XS, UK) V.1- 1869-

NATWAY Naturwissenschaften. (Springer-Verlag, Heidelberger Platz 3, D-1000 Berlin 33, Federal Republic of Germany) V.1- 1913-

NCDREP New Cardiovascular Drugs. (Ravencrest, 1185 Avenue of the Americas, New Yoty, NY 10036) 1985-

NCIAL* Progress Report Submitted to the National Cancer Institute by Arthur D. Little, Inc. (15 Acorn Park, Cambridge, MA 02140)

NCIBR* Progress Report for Contract No. NIH-NCI-E-68-1311, Submitted to the National Cancer Institute by Bio-Research Consultants, Inc. (9 Commercial Ave., Cambridge, MA 02141)

NCICP* Progress Report Submitted to the National Cancer Institute by Charles Pfizer and Company.

NCIHL* Progress Report Submitted to the National Cancer Institute by Hazelton Laboratories, Inc.

NCIIR* Progress Report for Contract No. N01-CP-12338, Submitted to the National Cancer Institute by IIT Research Institute. (Chicago, IL)

NCILB* Progress Report for Contract No. NIH-NCI-E-C-72-3252, Submitted to the National Cancer Institute by Litton Bionetics, Inc. (Bethesda, MD)

NCIMAV National Cancer Institute, Monograph. (U. S. Government Printing Office, Supt. of Doc., Washington, DC 20402) No.1- 1959-

NCIMR* Progress Report Submitted to the National Cancer Institute by Mason Research Institute. (Worcester, MA)

NCINS* National Cancer Institute Report. (Bethesda, MD 20014)

NCIRI* Progress Report Submitted to the National Cancer Insitute by Piason Research Institute.

NCISA* Progress Report for Contract No. PH-43-63-1132, Submitted to the National Cancer Institute by Scientific Associates, Inc. (6200 S. Lindberg Blvd., St. Louis, MO 63123)

NCISP* National Cancer Institute Screening Program Data Summary, Developmental Therapeutics Program, Bethesda, MD 20205

NCISS* Progress Report Submitted to the National Cancer Institute by South Shore Analytical and Research Laboratory.

NCITR* National Cancer Institute Carcinogenesis Technical Report Series. (Bethesda, MD 20014) No. 0-205. For publisher information, see NTPTR*

NCIUS* Progress Report for Contract NO. PH-43-64-886, Submitted to the National Cancer Institute by the Institute of Chemical Biology, University of San Francisco. (San Francisco, CA 94117)

NCNSA6 National Academy of Sciences, National Research Council, Chemical- Biological Coordination Center, Review. (Washington, DC)

NCPBBY National Clearinghouse for Poison Control Centers, Bulletin. U.S. Department of Health, Education, and Welfare (Washington, DC)

NDADD8 New Drugs Annual: Cardiovascular Drugs. (New York, NY) V.1-2 1983-84 For publisher information see NCDREP

NDKIA2 Kankyo Igaku Kenkyusho Nenpo (Nagoya Daigaku). Annual Report of the Research Institute of Environmental Medicine, Nagoya University. (The University, Furo-cho, Chikosa-Ku, Nagoya, Japan) V.1- 1947/48(Pub. 1949)-

NDRC** National Defense Research Committee, Office of Scientific Research and Development, Progress Report.

NEACA9 News Edition, American Chemical Society, (Easton, PA) V.18-19, 1940-41. For publisher information, see CENEAR

NEAGDO Neurobiology of Aging. (ANKHO International Inc., POB 426, Fayetteville, NY, 13066) V.1- 1980-

NEJMAG New England Journal of Medicine. (Massachusetts Medical Society, 10 Shattuck St., Boston, MA 02115) V.198- 1928-

NEMAAT Nematologica. (E.J. Brill, Publishers, Oude Rijn 33-35 Leiden, Netherlands) V.1- 1956-

NEOLA4 Neoplasma. (Karger-Libri AG, Scientific Booksellers, Arnold-Boecklin-Strasse 25, CH-4000 Basel 11, Switzerland) V.4-1957-

NEPHBW Neuropharmacology. (Pergamon Press, Headington Hill Hall, Oxford OX3 OBW, England) V.9- 1970-

NEPSBV Neuropsychopharmacology, Proceedings of the Meeting of the Collogium Internationale, Neuropsychopharmacologicum. (Excerpta Medica Foundation, P.O. Box 1126, Amsterdam, Netherlands) V.1- 1959-

NEREDZ Neurochemical Research. (Plenum Publishing Corp., 233 Spring St., New York, NY 10013) V.1- 1976-

NETOD7 Neurobehavioral Toxicology. (ANKHO International, Inc., P.O. Box 426, Fayetteville, NY 13066) V.1-2, 1979-80, For publisher information, See NTOTDY

NEURAI Neurology. (Modern Medicine Publications, Inc., 757 Third Avenue, New York, NY 10017) V.1- 1951-

NEZAAQ Nippon Eiseigaku Zasshi. Japanese Journal of Hygiene. (Nippon Eisei Gakkai, c/o Kyoto Daigaku Igakubu Koshu Eiseigaku Kyoshita, Yoshida Konoe-cho, Sakyo-ku, Kyoto, Japan) V.1- 1946-

NEZTAF New Zealand Veterinary Journal. (New Zealand Veterinary Assoc., Massey Univ., Palmerston North, N.Z.) V.1- 1952-

NFGZAD Nippon Funin Gakkai Zasshi. Japanese Journal of Fertility and Sterility. (Nippon Funin Gakkai, 1-1 Sadohara-cho, Ichigaya, Shinjuku-ku, Toyko 162, Japan) V.1- 1956-

NGCJAK Nippon Gan Chiryo Gakkai-shi. Journal of Japan Society for Cancer Therapy. (Nihon Gan Chiryo Gakkai, Kyoto, Japan) V.1- 1966-

NGGKED Nippon Gan Gakkai Sokai Kiji. Proceedings of the Annual Meeting of the Japanese Cancer Association. (Japanese Cancer Association, Tokyo, Japan) V.1- 1956-

NGGZAK Nippon Geka Gakkai Zasshi. Journal of the Japanese Surgical Society. (Nippon Geka Gakkai, 2-3-10 Koraku, Bunkyo-ku, Tokyo, 112, Japan) V.8- 1908-

NGZAA6 Nippon Ganka Gakkai Zasshi. (Nippon Ganka Gakkai, 2-4-11-402 Sarugaku-cho, Chiyoda-ku, Tokyo 101, Japan) V.1- 1897-

NHOZAX Nippon Hoigaku Zasshi. Japanese Journal of Legal Medicine. (Nippon Hoi Gakkai, c/o Tokyo Daigaku Igakubu Hoigaku Kyoshitsu, 7-3-1, Hongo, Bunkyo-ku, Tokyo, Japan) V.1- 1944-

NHTIA7 Nordisk Hygienisk Tidskrift. Scandinavian Journal of Hygiene. (Prof. Gideon Gehardsson, Sekreterare i Foereningen foer Omgivningshygiene, Svenska Arbetsgivarefoereningen, Box 16120, 10323 Stockholm 16, Sweden) V.1- 1920-

NIAND5 NIPH Annals. (National Institute of Public Health, Postuttak, Oslo, 1, Norway) V.1- 1978-

NIBKAW Nippon Byori Gakkai Kaishi. Journal of the Japanese Pathological Society. (c/o Tokyo Daigaku Igakubu Byorigaku Kyoshitsu, 7-3-1 Hongo, Bunkyo-ku, Tokyo 113, Japan) V.1- 1911-

NICHAS Nichidai Igaku Zasshi. (Nihon Daigaku Igakkai, 30, Oyaguchi kami-machi, Itabashi-ku, Tokyo 173, Japan) V.1- 1937-

NICKA3 Nippon Chikusan Gakkai Ho. Journal of the Japanese Society of Zootechnical Science. (Nippon Chikusan Gakkai, 201 Nagatani Koporasu, 2-9-4 Ikenohata, Taito-ku, Tokyo 110, Japan) V.1- 1924-

NIGHAE Archiv Fuer Japanische Chirurgie. (Nippon Geka Hokan Henshushitsu, c/o Kyoto Daigaku Igakubu Geka Seikei Geka Kyoshitsu, 54 Kawara- machi, Shogoin, Sakyo-ku, Kyoto 606, Japan) V.1- 1924-

NIGZAY Niigata Igakkai Zasshi. Niigata Medical Journal. (Niigata Daigaku, 1 Asahi-Machi, Niigata, Japan) V.60- 1946-

NIHBAZ National Institutes of Health, Bulletin. (Bethesda, MD)

NIIHAO Nichidoku Iho. Japanese-German Medical Reports. (Nihon Schering Co., Ltd., 6-64, Nishimiyahara, 2-chome, Yodogawa-ku, Osaka 532, Japan) V.1- 1956-

NIIRDN "Drugs in Japan. Ethical Drugs, 6th Edition 1982," Edited by Japan Pharmaceutical Information Center. (Yakugyo Jiho Co., Ltd., Tokyo, Japan)

NIOSH* National Institute for Occupational Safety and Health, U. S. Dept. of Health, Education, and Welfare, Reports and Memoranda.

NIPAA4 Nippon Shokakibyo Gakkai Zasshi. Journal of the Japanese

Society of Gastroenterology. (Nippon Shokakibyo Gakkai, 4-12 7-Chome, Ginza, Chuo-Ku, Tokyo 104, Japan) V.1- 1899-

NIPDAD Nihon Daigaku No-Juigakubu Gakujutsu Kenkyu Hokoku. Research Reports of the College of Agriculture and Veterinary Medicine, Nihon Univ. (The Univ., 34-1 Shimouma, 3-chome, Setagaya-ku, Tokyo 154, Japan) No.1- 1953-

NIPOAC Nippon Shonika Gakkai Zasshi. (Nippon Shonika Gakkai, 1-29-8 Shinjuku, Shinjuku-Ku, Tokyo 160, Japan) V.55- 1951-

NISFAY Nippon Sanka Fujinka Gakkai Zasshi. Journal of Japanese Obstetrics and Gynecology. (Nippon Sanka Fujinka Gakkai, c/o Hoken Kaikan Building., 1-1 Sadohara-cho, Ichigaya, Shinjuku-ku, Tokyo 162, Japan) V.1- 1949-

NISIA9 Nippon Shika Ishikai Zasshi. Journal of the Japan Dental Association. (1-20 Kudan.-kita, 4-Chome, Chiyoda-ku, Tokyo, Japan) V.1- 1948-

NISZAQ Nippon Sanshigaku Zasshi. (Nippon Sanshi Gakkai, c/o Sanshi Shikenjo, 3-55-30 Wada, Suginami-ku, Tokyo 166, Japan) V.1- 1930-

NIVAAY Notiziario dell'Istituto Vaccinogeno Antitubercolare. Bulletin of the Institute for Antitubercular Vaccinogens. (l'Instituto, Via Clericetti, 45 Milan, Italy) V.1-11, 1951-61.

NJGKBV Snake. (Japan Snake Institute, Hon-machi, Yaduzuka, Nitta-gun, Gunma-ken, Japan) V.1- 1969-

NJMSAG Nagoya Journal of Medical Science. (Nagoya University School of Medicine, 65 Tsuruma-cho, Showa-ku, Nagoya 466, Japan) V.2- 1927-

NJUZA9 Japanese Journal of Veterinary Science. (Nippon Jui Gakkai, 1-37-20, Yoyogi, Shibuya-ku, Tokyo 151, Japan) V.1- 1939-

NKEZA4 Nippon Koshu Eisei Zasshi. Japanese Journal of Public Health. (Nippon Koshu Eisei Gakkai, 1-29-8 Shinjuku, Shinjuku-ku, Tokyo 160, Japan) V.1- 1954-

NKGZAE Nippon Ketsueki Gakkai Zasshi. Journal of Japan Haematological Society. (Kyoto Univ. Hospital, Faculty of Medicine, Kyoto, Japan) V.1- 1937-

NKOGAV Nippon Kokuka Gakkai Zasshi. Journal of the Japanese Stomatological Society. (7-3-1, Hongo, Bunkyo-Ku, Tokyo, Japan) V.1- 1952-

NKRZAZ Chemotherapy (Tokyo). (Nippon Kagaku Ryoho Gakkai, 2-20-8 Kamiosaki, Shinagawa-Ku, Tokyo, 141, Japan) V.1- 1953-

NMJOAA Nagoya Medical Journal. (Nagoya City University Medical School, 2-38 Nagarekawa-machi, Naka-ku, Nagoya 460, Japan) V.1- 1953-

NNAPBA Naunyn-Schmiedebergs Archiv fuer Pharmakologie. (Berlin, Germany) V.264-271, 1969-71. For publisher information, see NSAPCC

NNBYA7 Nature: New Biology. (Macmillan Journals Ltd., Houndmills Estate, Basingstoke, Hants RG21 2XS, England) V.229-246, 1971-73.

NNGAAS Nippon Naika Gakkai Zasshi. Journal of the Japanese Society of Internal Medicine. (Nippon Naika Gakkai, 33-5, 3-Chome, Hongo, Bunkyo-ku, Tokyo 13) V.1- 1913-

NNGADV Nippon Noyaku Gakkaishi. (Pesticide Science Society of Japan, 43-11, 1-Chome, Komagome, Toshima-ku, Tokyo 170, Japan) V.1- 1976-

NNGZAZ Nippon Naibumpi Gakkai Zasshi. Journal of the Japan Endocrine Society. (Nippon Naibumpi Gakkai, c/o Kyoto-Furitsu Ika Daigaku, Kojinbashi Nishizume-Sagaru, Kamigyo-ku, Kyoto 602, Japan) V.1- 1925-

NOALA4 Nova Acta Leopoldina. (Johann Ambrosius Barth Verlag, Postfach 109, Salomonst 18b, 701 Leipzig, E. Germany) V.1- 1932-

NOMDA6 Northwest Medicine. (Seattle, WA) V.1-73, 1903-73. Discontinued.

NOSYBW Nobel Symposium. (Plenum Publishing Corp., 233 Spring St., New York 10013) No.1- 1965-

NOVTAV Nordisk Veterinaermedicin. (Danske Dyrlaege Forening, Alhambravej 15, DK1826 Copenhagen V, Denmark) V.1- 1949-

NPIRI* Raw Material Data Handbook, V.1 Organic Solvents, 1974. (National Association of Printing Ink Research Institute, Francis McDonald Sinclair Memorial Laboratory, Lehigh University, Bethlehem, PA 18015)

NPMDAD Nouvelle Presse Medicale. (Masson et Cie, Editeurs, 120 Blvd. Saint-Germain, P-75280, Paris 06, France) V.1- 1972-

NPRNAY Nephron. (S. Karger Publishers, Inc., 150 Fifth Ave., Suite 1105, New York, NY 10011) V.1- 1964-

NRSCDN Neuroscience. (Pergamon Press Ltd., Headington Hill Hall, Oxford OX3 0BW, England) V.1- 1976-

NRTTA8 NRC Technical Translation. (The National Research Council of Canada, Ottawa, Ontario K1A 0R6, Canada) No.1- 1949-

NRTXDN Neurotoxicology. (Pathotox Publishers, Inc., 2405 Bond St., Park Forest South, IL 60464) V.1- 1979-

NSABA2 Nuclear Science Abstracts. (U. S. Atomic Energy Commission, Div. of Tech. Inform: Ext., P.O. Box 62, Oak Ridge, TN 37830) V.1- 1948

NSAPCC Naunyn-Schmiedeberg's Archives of Pharmacology. (Springer Verlag, Heidelberger, Pl. 3, D-1 Berlin 33, Germany) V.272- 1972-

NSMZDZ Nippon Shika Masui Gakkai Zasshi. (Tokyo Ikashika Daigaku Shika Masuigaku Kyoshitsu, 1-5-45 Yushima, Bunkyo-Ku, Tokyo 113, Japan) V.1- 1973-

NTIMBF Nauchnye Trudy, Irkutskii Gosudarstvennyi Meditsinskii Institut. Scientific Works, Irkutsk State Medical Institute. (Irkutskii Gosudarstvennyi Meditsinskii Institut, Irkutsk, USSR) No.80- 1967-

NTIS** National Technical Information Service. (Springfield, VA 22161) (Formerly U. S. Clearinghouse for Scientific and Technical Information)

NTOTDY Neurobehavioral Toxicology and Teratology. (ANKHO International Inc., P.O. Box 426, Fayetteville, NY 13066) V.3- 1981-

NTPTB* NTP Technical Bulletin. (National Toxicology Program, Landow Bldg. 3A-06, 7910 Woodmont Ave., Bethesda, Maryland 20205)

NTPTR* National Toxicology Program Technical Report Series. (Research Triangle Park, NC 27709) No.206-

NUCADQ Nutrition and Cancer. (Franklin Institute Press, POB 2266, Phildelphia, PA 19103) V.1- 1978-

NULSAK Nucleus (Calcutta). (Dr. A.K. Sharma, c/o Cytogenetics Laboratory, Department of Botany, University of Calcutta, 35 Ballygunge Circular Rd., Calcutta 700 019, India) V.1- 1958-

NUMEBI Nutrition and Metabolism. (S. Karger AG, Postfach CH-4009 Basel, Switzerland) V.12- 1970-

NUNDAJ Neuroendocrinology. (S. Karger AG, Postfach CH-4009 Basel, Switzerland) V.1- 1965/66-

NUPOBT Neuropatologia Polska. (Ars-Polona-RUCH, POB 1001, 00-068 Warsaw 1, Poland) V.1- 1963-

NUREA8 Nutrition Reviews. (Nitrutionn Foundation, c/o Dr. R.E. Olson, ed., Dept. of Biochemistry, St. Louis Univ. School of Medicine, St. Louis, MO 63104) V.1- 1942-

NURIBL Nutrition Reports International. (Geron-X, Inc., POB 1108. Los Altos, CA 94022) V.1- 1970-

NVTMAF Nordisk Veterinaermedicin Danske Udgave. (Copenhagen, Denmark)

NWSCAL New Scientist. (IPC Magazines Ltd., Tower House, Southampton, St., London WC2E 9QX, England) V.1- 1956-

NYJMAK New York Journal of Medicine and the Collateral Sciences. (New York, NY)

NYKGA7 Noyaku Kagaku. Pesticide Science. (Pesticide Science Society of Japan, 43-11, 1-Chome, Komagome, Toshima-ku, Tokyo 170, Japan) V.1-3, 1973-76.

NYKZAU Nippon Yakurigaku Zasshi. Japanese Journal of Pharmacology. (Nippon Yakuri Gakkai, 2-4-16, Yayoi, Bunkyo-Ku, Tokyo 113, Japan) V.40- 1944-

NYSJAM New York State Journal of Medicine. (Medical Society of the State of New York, 420 Lakeville Rd., Lake Success, NY 11040) V.1- 1901-

NYZZA3 Nippon Yakuzaishikai Zasshi. Journal of the Japan Pharmaceutical Association. (12-15-701, Shibuya 2-Chome, Shibuya-ku, Tokyo, Japan) V.1- 1949-

NZJEA3 New Zealand Journal of Experimental Agriculture. (New Zealand Dept. of Scientific and Industrial Research, Publications Officer, Science Information Div., POB 9741, Wellington, New Zealand) V.1- 1973-

NZMJAX New Zealand Medical Journal. (Otago Daily Times and Witness Newspapers, P.O. Box 181, Dunedin C1, New Zealand) V.1- 1900-

OBGNAS Obstetrics and Gynecology. (Elsevier/North Holland, Inc., 52 Vanderbilt Avenue, New York, NY 10017) V.1- 1953-

OCHRAI Occupational Health Review. (Ottowa, Canada) V.4-22, 1953-71. Discontinued.

OCMEA4 Occupational Medicine. (Chicago, IL) V.1-5, 1946-48. For publisher information, see AEHLAU

OCMJAJ Osaka City Medical Journal. (Osaka City Medical Center, 1-4-54, Asahimachi, Abenoku, Osaka 545, Japan) V.1- 1954-

OCPRAO Ochrona Pracy. Industrial Safety. (Ars Polona-RUCH, POB 1001, 00-068 Warsaw 1,Poland) V.7- 1952(?)-

OCRAAH Organometallic Chemistry Reviews, Section A: Subject Reviews. (Lausanne, Switzerland) V.3-8, 1968-72. For publisher information, see JORCAI

ODFU** Vergleichende pharmakologische Prufung Zweier neuer Lokalanas- thetika: Perkain und Pantokain, Wilhelm Ost Dissertation. (Pharmakologischen Institut der Universitat zu Frankfurt am Main, Germany, 1931)

ODIZAK Osaka Daigaku Igaku Zasshi. Journal of the Osaka University Faculty of Medicine. (Osaka Daigaku Igakkai, 33 Joancho, Kitaku, Osaka, Japan) V.1- 1949-

OEKSDJ Osaka-furitsu Koshu Eisei Kenkyusho Kenkyu Hokoku, Shokuhin Eisei Hen. (Osaka-furitsu Koshu Eisei Kenkyusho, 1-3-69 Nakamichi, Higashinari-ku, Osaka 537, Japan) No.1- 1970-

OFAJAE Okajimas Folia Anatomica Japonica. (Japan Publications Trading Co., 175 5th Ave., New York, NY 10010) V.14- 1936-

OGSUA8 Obstetrical and Gynecological Survey. (Williams and Wilkins, 428 E. Preston St., Baltimore, MD 21202) V.1- 1946-

OHSLAM Occupational Health and Safety Letter. (Environews, Inc., 1097 National Press Bldg. Washington, DC 20045) V.1- 1970-

OIGZDE Osaka-shi Igakkai Zasshi. Journal of Osaka City Medical Association. (Osaka-shi Igakkai, c/o Osaka-shiritsu Daigaku Igakubu, 1-4-54 Asahi-cho, Abeno-ku, Osaka, 545, Japan) V.24- 1975-

OIGZSE Osaka-shi Igakkai Zasshi. Journal of Osaka City Medical Association. (Osaka-shi Igakkai, c/o Osaka-shiritru Daigaku Igakubu, 1-4-54 Asahi-cho, Abeno-ku, Osaka, 545, Japan) V.24- 1975-

OIZAAV Osaka Igaku Zasshi. (Osaka, Japan)

OJVRAZ Onderstepoort Journal of Veterinary Research. (Div. of Agricultural Inform., Dept. of Agricultural Tech. Ser., Private Bag X-144, Pretoria, S. Africa) V.25- 1951-

OJVSA4 Onderstepoort Journal of Veterinary Science and Animal Industry. (Pretoria, S. Africa) V.1-24, 1933-50. For publisher information, see OJVRAZ

OKEHDW Osaka-furitsu Koshu Eisei Kenkyusho Kenkyu Hokoku, Koshu Eisei Hen. Research Reports of the Osaka Prefectural Institute of Public Health, Public Health Section. (1-3-69 Nakamichi, Higashinari-ku, Osaka, 537, Japan) No.3- 1966-

OMCDS* Olin Chemicals Data Sheet. (Industrial Development, 745 5th Ave., New York, NY 10022)

ONCOAR Oncologia. (Basel, Switzerland) V.1-20, 1948-66. For publisher information, see ONCOBS

ONCOBS Oncology. (S. Karger AG, Postfach CH-4009 Basel, Switzerland) V.21- 1967-

ONCODU Revista de Chirurgui, Oncologie, Radiologie, ORL, Oftalmologie, Stomatologie, Seria: Oncologia.(Rompresfilatelia, ILEXIM, POB 136-137, Bucharest, Romania) V.13, No.4- 1974-

ONKOD2 Onkologie. (S. Karger AG, Postfach CH-4009 Basel, Switzerland) V.1- 1978-

OPHTAD Ophtalmologica. (S. Karger AG, Postfach CH-4009 Basel, Switzerland) V.96- 1978-

ORVOBF Orvostudomany. (Akademiai Kiado, P.O. Box 24, Budapest 502, Hungary) V.19- 1968-

OSDIAF Osaka Shiritsu Daigaku Igaku Zasshi. Journal of the Osaka City Medical Center. (Osaka-Shiritsu Daigaku Igakubu, 1-4-54, Asahi-cho, Abeno-Ku, Osaka, Japan) V.4-23, 1955-74.

OSMJAT Ohio State Medical Journal. (Ohio State Medical Assoc., 600 S. High St., Suite 500, Columbus, OH 43215) V.1- 1905-

OSOMAE Oral Surgery, Oral Medicine and Oral Pathology. (C.V. Mosby Co., 11830 Westline Industrial Dr., St Louis, Mo. 63141) V.1- 1948-

OYYAA2 Oyo Yakuri. Pharmacometrics. (Oyo Yakuri Kenkyukai, Tohoku Daigaku, Kitayobancho, Sendai 980, Japan) V.1- 1967-

OZEBAE Oesterreichische Zeitschrift fuer Erforschung und Bekaempfung der Krebskrankheit. (Verlag Adolf Holzhausens Nachfolger, Kandlgasse 19/21, 1070 Vienna, Austria) V.25-28, 1970-73.

PetKP# Personal Communication from Dr. K.P. Petersen, DAK Labs., 59 Lergravsvej, DK-2300, Copenhagen, Demark, to B. Jones, Tracor Jitco, Inc., December 22, 1977

PreF## Personal Communication to Fran Preuhs, Neville Chemical Co., Neville Island, Pittsburgh, PA 15225, from Dr. Alan E. Friedman, Tracor Jitco, Inc., May 17,1979

PAACA3 Proceedings of the American Association for Cancer Research. (Waverly Press, 428 E. Preston St., Baltimore, MD 21202) V.1- 1954-

PAASAH Publication, American Association for the Advancement of Science. (AAAS, 1515 Massachusetts Ave., N.W., Washington, DC 20005) No.1-94, 1934-73. Discontinued.

PABIAQ Pathologie et Biologie. (Paris, France) V.1-16, 1953-68. For publisher information, see PTBIAN

PACHAS Pure and Applied Chemistry. (Butterworth and Co., Ltd., Borough Green, Sevenoaks, Kent TN15 8PH, England) V.1- 1960-

PAFEAY Patologicheskaya Fiziologiya i Eksperimental'naya Terapiya. (v/o Mezhdunarodnaya Kniga, Kuznetskii Most 18, Moscow G-200, USSR) V.1- 1957-

PAHEAA Pharmaceutica Acta Helvetiae. (Case Postale 210, CH 1211 Geneva 1, Switzerland) V.1- 1926-

PAMIAD Pathologia et Microbiologia. V.23-43, 1960-75. For publisher information, see ECEBDI

PANSCG PANS. (Centre for Overseas Pest Research, PANS Office, College House, Wright's Lane, London W8 5SJ, England) V.15- 1969-

PAPOAC Patologia Polska. (Ars-Polona-RUSH, POB 1001, 00-068 Warsaw 1, Poland) V.1- 1950-

PAREAQ Pharmacological Reviews. (Williams and Wilkins, 428 E. Preston St., Baltimore, MD 21202) V.1- 1949-

PARPDS Pathology, Research and Practice. (Gustav Fisher Verlag, Postfach 72 01 43, D-7000 Stuttgart 70, Federal Republic of Germany) V.162- 1978-

PARWAC Polskie Archiwum Weterynaryjne. Polish Archives of Veterinary Medicine. (Panstwowe Wydawnictwo Naukowe, POB 391, P-00251 Warsaw, Poland) V.1- 1951-

PATHAB Pathologica. (Via Alessandro Volta, 8 Casella Postale 894, 16128 Genoa, Italy) V.1- 1908-

PAVEAC Pathologia Veterinaria. (Basel, Switzerland) V.1-7, 1964-70. For publisher information, see VTPHAK

PAWIAT Praktische Anaesthesie, Wiederbelebung und Intensivtherapie. (Georg Thieme Verlag, Postfach 732, Herdweg 63, D-70000, Stuttgart 1, Fed. Rep. Ger.) V.9- 1974-

PBBHAU Pharmacology, Biochemistry and Behavior. (ANKHO International Inc., P.O. Box 426, Fayetteville, NY 13066) V.1- 1973-

PBFMAV Problemi na Farmatsiyata. (Durzhavno Izdatel'stvo Meditsina i Fizkultura, Pl. Slaveikov II, Sofia, Bulgaria) V.1- 1973-

PBPHAW Progress in Biochemical Pharmacology. (S. Karger AG, Postfach CH-4009 Basel, Switzerland) V.1- 1965-

PBPSDY "Pharmacological and Biochemical Properties of Drug Substances," Morton E. Goldberg, ed., Washington, D.C., American Pharmaceutical Assoc., V.1- 1977-

PBVMA9 Problemi na Vutreshnata Meditsina. Problems of Internal Medicine. (Durzhavno Izdatel'stvo Meditsina i Fizkultura, pl. Slaveikov 11, Sofia, Bulgaria) V.1- 1973-

PCBPBS Pesticide Biochemistry and Physiology. (Academic Press, 111 5th Ave., New York, NY 10003) V.1- 1971-

PCBRD2 Progress in Clinical and Biological Research. (Allan R. Liss, Inc., 150 5th Ave., New York, NY 10011) V.1- 1975-

PCCRA4 Proceedings of the Canadian Cancer Research Conference. (Univ. of Toronto Press, Toronto, M55 1A6, Ontario, Canada) V.1- 1954-

PCFTDS Progress in Chemical Fibrinolysis and Thrombolysis. (Raven Press, 1140 Ave. of the Americas, New York, NY 10036) V.1- 1975-

PCIPDV Beijing Yixueyuan Xuebao. Journal of Peking Medical College. (Beijing Yixueyuan, Beijiaohaidianqu, Beijing, Peop. Rep. China) Beginning history not known.

PCJOAU Pharmaceutical Chemistry Journal. English Translation of KHFZAN. (Plenum Publishing Corp., 233 Spring St., New York, NY 10013) No.1- 1967-

PCLCA4 Progress in Clinical Cancer. (Grune and Stratton Inc., 111 5th Ave., New York, NY 10003) V.1- 1965-

PCMBA6 Proceedings of the International Congress for Microbiology. (National Research Council, Division of Biosciences, Sussex Dr., Ottawa 2, Ontario, Canada) V.1- 1930-

PCNAA8 Pediatric Clinics of North America. (W.B. Saunders, W. Washington Sq., Philadelphia, PA 19105) V.1- 1954-

PCOC** Pesticide Chemicals Official Compendium, Association of the American Pesticide Control Officials, Inc. (Topeka, Kansas, 1966)

PDHU** Uber die Wirksamkeit einiger aliphatischaromatischer Amine, einiger Isochinolin derivate, sowie eines Chinolinabkommlings und einer Barbitursaure auf die Erreger der Schlafkrankheit bei Tieren, Adolf Puls Dissertation. (University of Hamburg, Germany, 1936)

PDTNBH Paediatrician. (S. Karger AG, Postfach CH-4009 Basel, Switzerland) V.1- 1972-

PEDIAU Pediatrics. (P.O. Box 1034, Evanston, IL 60204) V.1- 1948-

PEGTAA Problemy Endokrinologii i Gormonoterapii. Problems of Endocrinology and Hormone Therapy. (Moscow, USSR) V.1-12, 1955-66. For publisher information, see PROEAS

PEHPB4 Pei I Hsueh Pao. (Taipei Medical College, 250 Wu Shing St., Taipei, Taiwan) V.1- 1961-

PEMNDP Pesticide Manual. (British Crop Protection Council, 20 Bridgport Rd., Thornton Heath CR4 7QG, UK) V.1- 1968-

PENDAV Polish Endocrinology. English translation of Endokrynologia Polska. (Springfield, Virginia) V.13-23, 1962-72. Discontinued.

PENNS* Pennsalt Chemicals Corporation, Technical Div., New Products. (Philadelphia, PA)

PEREBL Pediatric Research. (Williams and Wilkins Co., 428 E. Preston St., Baltimore, MD 21202) V.1- 1967-

PESTC* Pesticide and Toxic Chemical News. (Food Chemical News, Inc., 400 Wyatt Bldg., 777 14th St., N.W. Washington, DC 20005) V.1- 1972-

PESTD5 Proceedings of the European Society of Toxicology. (Amsterdam, Netherlands) V.16-18, 1975-77, Discontinued.

PEXTAR Progress in Experimental Tumor Research. (S. Karger AG, Postfach CH-4009 Basel, Switzerland) V.1- 1960-

PFLABK Pfluegers Archiv. European Journal of Physiology. (Springer-Verlag, Heidelberger, Pl. 3, 1 Berlin 33, Germany) V.302- 1968-

PGMJAO Postgraduate Medical Journal. (Blackwell Scientific Publications, Osney Mead, Oxford OX2 OEL, England) V.1- 1925-

PGPKA8 Problemy Gematologii i Perelivaniia Krovi. Problems of Hematology and Blood Transfusion. (v/o Mezhdunarodnaya Kniga, Kuznetskii Most 18, Moscow G-200, U.S.S.R.) V.1- 1956-

PGTCA4 Pigment Cell. (S. Karger AG, Arnold-Boecklin St., 25 CH-4011 Basel, Switzerland) V.1- 1973-

PHABDI Proceedings of the Hungarian Annual Meeting for Biochemistry. (Magyar Kemikusok Egyesulete, Anker Koz 1, 1061 Budapest, Hungary) V.1- 1961-

PHARAT Pharmazie. (VEB Verlag Volk und Gesundheit, Neue Gruenstr 18, 102 Berlin, E. Germany) V.1- 1946-

PHBHA4 Physiology and Behavior. (Pergamon Press Inc., Maxwell House, Fairview Park, Elmsford, NY 10523) V.1- 1966-

PHBTH* "Pharmacology: Basis of Therapy," Goodman, L.S., ed., 3rd Ed. New York, Macmillan, 1967

PHBUA9 Pharmaceutical Bulletin. (Tokyo, Japan) V.1-5, 1953-57. For publisher information, see CPBTAL

PHCBAP Photochemistry and Photobiology. (Pergamon Press Ltd., Headington Hill Hall, Oxford OX3 0BW, England) V.1- 1962-

PHINDQ Pharmacy International. (Elsevier Science Publications Co.,Inc., 52 Vanderbilt Ave., New York, NY 10017) V.1- 1990-

PHJOAV Pharmaceutical Journal. (Pharmaceutical Press, 17 Bloomsbury Sq., London WC1A 2NN, England) V.131- 1933-

PHMCAA Pharmacologist. (American Society for Pharmacology and Experimental Therapeutics, 9650 Rockville Pike, Bethesda, MD 20014) V.1- 1959-

PHMGBN Pharmacology: International Journal of Experimental and Clinical Pharmacology. (S. Karger AG, Postfach CH-4009 Basel, Switzerland) V.1- 1968-

PHPHA6 Phytiatrie-Phytopharmacie. (Societe Francaise de Phytiatrie et de France) V.1- 1952- Phytopharmacie, Etoile de Choisy, Route de St-Cyr, 78000 Versailles,

PHREA7 Physiological Reviews. (9650 Rockville Pike, Bethesda, MD 20014) V.1- 1921-

PHRPA6 Public Health Reports. (U.S. Government Printing Office, Supt. of Doc., Washington, DC 20402) V.1- 1878-

PHTHDT Pharmacology and Therapeutics. (Pergamon Press Ltd., Headington Hall, Oxford OX3 0BW, England) V.4- 1979-

PHTXA6 Pharmacology and Toxicology. Translation of FATOAO. (New York, NY) V.20-22, 1957-59. Discontinued.

PHYTAJ Phytopathology. (Phytopathological Society, 3340 Pilot Knob Rd., St. Paul, MN 55121) V.1- 1911-

PHZOA9 Physiological Zoology. (University of Chicago Press, 5801 S. Ellis Ave., Chicago, IL 60637) V.1- 1928-

PIAIA9 Proceedings of the Iowa Academy of Science. (Iowa Academy of Science, Univ. of Northern Iowa, Cedar Falls, IA 50613) V.1- 1887/93-

PIATA8 Proceedings of the Imperial Academy of Tokyo. (Tokyo, Japan) V.1-21, 1912-45. For publisher information, see PJACAW

PICFEE Proceedings of the International Conference of the International Planned Parenthood Federation. (IPPF International Office, 18-20 Lower Regent St., London SW1Y 4PW, UK)

PISCAD Proceedings of the Indian Science Congress. (Indian Science Congress Association, 14, Dr. Biresh Guha St., Calcutta, 700017, India) 1st- 1914-

PISDDJ Pacific Information Service on Street Drugs. (J.K. Brown, School of Pharmacy, Univ. of the Pacific, Stockton, CA 95211) V.1- 1972(?)-

PIXXD2 PCT (Patent Cooperation Treaty) International Application. (U. S. Patent and Trademark Office, Foreign Patents, Washington, DC 20231)

PJABDW Proceedings of the Japan Academy, Series B: Physical and Biological Sciences. (Maruzen Co., Ltd., P.O. Box 5050, Tokyo International 100-31, Japan) V.53- 1977-

PJACAW Proceedings of the Japan Academy. (Tokyo, Japan) V.21- 53, 1945-77. For publisher information, see PJABDW

PJPHEO Pakistan Journal of Pharmacology. (c/o Univ. of Karachi, Faculty of Pharmacy, Karachi, 32, Pakistan) V.1- 1984-

PJPPAA Polish Journal of Pharmacology and Pharmacy. (ARS-Polona-Rush, POB 1001, 00-068 Warsaw 1, Poland) V.25- 1973-

PJSCAK Philippine Journal of Science. (National Science Development Board, POB 774, Manila, Philippines) V.14- 1919-

PLCHB4 Physiological Chemistry and Physics. (U.S. Government Printing Office, Superintendent of Documents, Washington, DC 20402) V.1- 1969-

PLENBW Pollution Engineering. (1301 S. Grove Ave., Barrington, IL 60010) V.1- 1969-

PLIRDW Progress in Lipid Research. (Pergamon Press Ltd., Headington Hill Hall, Oxford OX3 OBW, England) V.17- 1978-

PLMEAA Planta Medica. (Hippokrates-Verlag GmbH, Neckarstr 121, 7 Stuttgart, Germany) V.1- 1953-

PLMEDD Prostaglandins, Leukotrienes and Medicine. (Longman Inc., 19 W. 44th St., New York, NY 10036) V.8- 1982-

PLMJAP Pahlavi Medical Journal. (Editor, Pahlavi Medical Journal, School of Medicine, Pahlavi University, Shiraz, Iran) V.1- 1970-

PLPSAX Physiological Psychology. (Psychonomic Society, Inc., 1108 W. 34th Ave., Austin, TX 78705) V.1- 1973-

PLRCAT Pharmacological Research Communications. (Academic Press, 111 5th Ave., New York, NY 10003) V.1- 1969-

PMARAU Prensa Medica Argentina. (Junin 845, Buenos Aires, Argentina) V.1- 1914-

PMAYAH Pharmaceutical Manufacturers Association Yearbook. (1155 15th St., N.W., Washington, DC 20005) V.1- 1959/60-

PMDCAY Progress in Medical Chemistry. (American Elsevier Publishing Co., 52 Vanderbilt Ave., New York, NY 10017) V.1- 1961-

PMDFA9 Pakistan Medical Forum. (Karachi, Pakistan) V.1-8, 1966-73 Discontinued

PMESAJ Postgraduate Medical Journal, Supplement. (Blackwell Scientific Publications, Osney Mead, Oxford OX2 OEL, England) V.1- 1949-

PMHBAH Polish Medical Sciences and History Bulletin. (Chicago, IL) V.1-15, 1956-76. Discontinued.

PMJMAQ Proceedings of the Annual Meeting of the New Jersey Mosquito Extermination Association. (New Brunswick, NJ) V.1-61, 1914-74.

PMMDAE Panminerva Medica. (Edizioni Minerva Medica, Casella Postale 491, Turin, Italy) V.1- 1959-

PMRSDJ Progress in Mutation Research. (Elsevier North Holland, Inc., 52 Vanderbilt Ave., New York, NY 10017) V.1- 1981-

PMSBA4 Progress in Molecular and Subcellular Biology. (Springer-Verlag, Heidelberger Pl. 3, D-1000 Berlin 33, Federal Republic of Germany) V.1- 1969-

PNASA6 Proceedings of the National Academy of Sciences of the United States of America. (The Academy, Printing and Publishing Office, 2101 Constitution Ave., Washington, DC 20418) V.1- 1915-

PNCCA2 Proceedings, National Cancer Conference. (Philadelphia, PA) V.1-7, 1949-72, For publisher information, see CANCAR

POASAD Proceedings of the Oklahoma Academy of Science. (Oklahoma Academy of Science, c/o James F. Lowell, Executive Secretary-Treasurer, Southwestern Oklahoma State University, Weatherford, Oklahoma 73096) V.1- 1910/1920-

POKLA8 Problemi na Onkologiyata. Problems of Oncology. (Durzhavno Izdatelstvo Meditsina i Fizkultura, Pl. Slaveikov 11, Sofia, Bulgaria) V.1- 1973-

POLEAQ Polski Tygodnik Lekarski. Polish Medical Weekly. (Ars Polona-RUCH, POB 1001, P-00 068 Warsaw, 1, Poland) V.1- 1946-

POLMAG Polymer. The Chemistry, Physics and Technology of High Polymers. (IPC Science and Technology Press, POB 63, Guildford, Surrey GU2 5BH, England) V.1- 1960-

POMDAS Postgraduate Medicine. (McGraw-Hill, Inc., Distribution Center, Princeton Rd., Hightstown, NJ 08520) V.1- 1947-

POMJAC Polish Medical Journal. (Warsaw, Poland) V.1-11, 1962-72. Discontinued.

PONCAU Problems of Oncology. Translation of VOONAW. (New York, NY) V.3-7, 1957-61. Discontinued.

POSCAL Poultry Science. (Poultry Science Assoc., Texas A and M University, College Station, TX 77843) V.1- 1921-

PPASAK Proceedings of the Pennsylvania Academy of Science. (Pennsylvania Academy of Science, c/o Stanley Zagorski, Tr., Gannon College, Perry Sq., Erie, PA 16501) V.1- 1924/1926-

PPGDS* PPG Industries, Inc., Material Safety Data Sheet. (PPG Industries, Inc., Chemicals Group, One Gateway Center, Pittsburgh, PA 15222)

PPHAD4 Pediatric Pharmacology. (Alan R. Liss, Inc., 150 5th Ave., New York, NY 10011) V.1- 1980-

PPRPAS Produits and Problemes Pharmaceutiques. (Paris, France) V.17-28, 1962-73. Suspended.

PPTCBY Proceedings of the International Symposium of the Princess Takamatsu Cancer Research Fund. (Japan Scientific Societies Press, 2-10, Hongo 6-chome, Bunkyo-ku, Tokyo 113, Japan) (1st)-1970(Pub. 1971)-

PRACAK Practitioner. (5 Bentinck St., London, England) V.1- 1868-

PRAEAQ Progress in Aeronautical Sciences. (Pergamon Press, Maxwell House, Fairview Park, Elmsford, NY 10523) V.1-10, 1961-70. Name changed to Progress in Aerospace Science

PRAXAF Praxis. (Hallwag, AG, Nordring 4, 3001, Bern, Switzerland) V.12- 1923-

PRBIDC Progress in Reproductive Biology. (S. Karger AG, Arnold-Boecklin Str. 25, CH-4011, Basel, Switzerland) V.1- 1976-

PREBA3 Proceedings of the Royal Society of Edinburgh, Section B. (Royal Society of Edinburgh, 22 George St., Edinburgh, Scotland) V.61- 1943-

PRGLBA Prostaglandins. (Geron-X, Inc., P.O. Box 1108, Los Altos, California 94022) V.1- 1972-

PRKHDK Problemi na Khigienata. Problems in Hygiene. (Durzhavno Izdatel'stvo Meditsina i Zizkultura, Pl. Slaveikov 11, Sofia, Bulgaria) V.1- 1975-

PRLBA4 Proceedings of the Royal Society of London, Series B, Biological Sciences. (The Society, 6 Carlton House Terrace, London SW1Y 5AG, England) V.76- 1905-

PRLFAG Pracovni Lekarstvi. (PNS-Administrace Odborneho Tisku, Jindrisska 14, C2-125 05 Prague 1, Czechoslovakia) V.1- 1949-

PRMEAI Presse Medicale. (Paris, France) V.1-79, 1893-1971. For publisher information, see NPMDAD

PRNPAM Progress in Neuropathology. (Grune and Stratton, 111 5th Ave., New York, NY 10003) V.1- 1971-

PROEAS Problemy Endokrinologii. (v/o Mezhdunarodnaya Kniga, Kuznetskii Most 18, Moscow G-200, U.S.S.R.) V.1-6, 1936-41; V.13- 1967-

PROMDL Prostaglandins and Medicine. (The Longman Group Ltd., Journals Div., 43-45 Annandale St., Edinburgh EH47 4AT, Scotland) V.1- 1978(?)-

PROTA* "Problemes de Toxicologie Alimentaire," Truhaut, R., Paris, France, L'evolution Pharmaceutique, (1955?)

PRPHA8 Produits Pharmaceutiques. (Paris, France) V.1-16, 1946-61. For publisher information, see PPRPAS

PRSEAE Proceedings of the Royal Society of Edinburgh. (Royal Society of Edinburgh, 22 George St., Edinburgh EH2 2PQ, Scotland) V.1-60, 1832-1940.

PRSMA4 Proceedings of the Royal Society of Medicine. (Grune and Stratton Inc., 111 5th Ave., New York, NY 10003) V.1- 1907-

PRTODI "Progress in Toxicology," Zbinden, Gerhard, New York, Springer- Verlag, 1973

PRTUAX Problemy Tuberkuleza. Problems of Tuberculosis. (v/o Mezhdunarodnaya Kniga, Kuznetskii Most 18, Moscow G-200, USSR) V.1- 1923-

PSCBAY Psychopharmacology Service Center, Bulletin. (Bethesda, MD) V.1-3, 1961-65. For publisher information, see PSYBB9

PSCHDL Psychopharmacology (Berlin). (Springer-Verlag New York, Inc., Service Center, 44 Hartz Way, Secaucus, NJ 07094) V.47- 1976-

PSDAA2 Proceedings of the South Dakota Academy of Science. (Univ. of South Dakota, Vermillion, SD 57069) V.1- 1916-

PSDTAP Proceedings of the European Society for the Study of Drug Toxicity. (Princeton, NJ 08540) V.1-15, 1963-74. For publisher information, see PESTD5

PSEBAA Proceedings of the Society for Experimental Biology and Medicine. (Academic Press, 111 5th Ave., New York, NY 10003) V.1- 1903/04-

PSIRAA Pakistan Journal of Scientific and Industrial Research. (Senior Scientific Officer, Pakistan Council of Scientific and Industrial Research, 39, Garden Rd., Karachi, Pakistan) V.1- 1958-

PSMMAF Proceedings of the Staff Meetings of the Mayo Clinic. (Rochester, MN) V.1-38, 1926-63. For publisher information, see MACPAJ

PSSCBG Pesticide Science. (Blackwell Scientific Publications Ltd., Osney Mead, Oxford OX2 OEL, England) V.1- 1970-

PSSID2 Pergamon Series on Environmental Science. (Pergamon Press Ltd., Headington Hill Hall, Oxford OX3 0BW, England) V.1- 1978-

PSSYDG Proceedings of the Serono Symposia. (Academic Press Inc. Ltd., 24-28 Oval Rd., London NW1 7DX, England) V.1- 1973-

PSTDAN Pesticides. (Colour Publications Pvt. Ltd., 126-A Dhuruwadi, Off Dr. Nariman Rd., Bombay 400-025, India) V.1- 1967-

PSTFDW Proceedings of the Symposium on Textile Flammability. (LeBlanc Research Corp., 5454 Post Rd., East Greenwich, RI 02818) V.1- 1973-

PSTGAW Proceedings of the Scientific Section of the Toilet Goods Association. (The Toilet Goods Assoc., Inc., 1625 I St., N.W., Washington, DC 20006) No.1-48, 1944-67. Discontinued.

PSYBB9 Psychopharmacology Bulletin. (U.S. Government Printing Office, Supt. of Doc., Washington, DC 20402) V.3- 1966-

PSYPAG Psychopharmacologia (Berlin). (Berlin, Ger.) V.1-46, 1959-76. For publisher information, see PSCHDL

PTBIAN Pathologie-Biologie. (Expansion Scientifique Francaise, 15 rue St. Benoit, Paris 6, France) V.17- 1969-

PTEUA6 Pathologia Europaea. (Presses Academiques Europeennes, 98, Chaussee de Charleroi, Brussels, Belgium) V.1- 1966-

PTLGAX Pathology. (Royal College of Pathologists of Australia, 82 Windmill St., Sydney, NSW 2000, Australia) V.1- 1969-

PTPAD4 Pharmacology and Therapeutics, Part A: Chemotherapy, Toxicology and Metabolic Inhibitors. (Oxford, England) V.1-2, 1976-78. For publisher information, see PHTHDT

PTPBD7 Pharmacology and Therapeutics, Part B: General and Systematic Pharmacology. (Oxford, England) V.1-3, 1975-78. For publisher information, see PHTHDT

PTRBAE Philosophical Transactions of the Royal Society of London, Series B: Biological Sciences. (Royal Society, 6 Carlton House Terrace, London SW1Y 5AG, England) V.178- 1887-

PTRMAD Philosophical Transactions of the Royal Society of London, Series A: Mathematical and Physical Sciences. (Royal Society of London, 6 Carlton House Terrace, London SW1Y 5AG, England) V.178- 1887-

PTRSAV Philosophical Transactions of the Royal Society of London. (London, England) V.1-177, 1665-1886. For publisher information, see PTRMAD

PUMTAG Trace Substances in Environmental Health. Proceedings of University of Missouri's Annual Conference on Trace Substances in Environmental Health. (Environmental Trace Substances Research Center, Univ. of Missouri, Columbia, MO 65201) V.1- 1967-

PUOMA5 Proceedings of the University of Otago Medical School. (Otago Medical School Research Society, Box 913, Dunedin, New Zealand) V.1- 1922-

PVPRAW Paint and Varnish Production. (Palmerton Publishing Co., 101 W. 31 St., New York, NY 10001) V.29- 1949-

PVTMA3 Preventive Medicine. (Academic Press, 111 Fifth Ave., New York, NY 10003) V.1- 1972-

PWPSA8 Proceedings of the Western Pharmacology Society. (Univ.

of California Dept. of Pharmacology, Los Angeles, CA 94122) V.1- 1958-

PYRTAZ Psychological Reports. (Southern Universities Press, Baton Rouge, LA 70813) V.1- 1952-

PYTCAS Phytochemistry. An International Journal of Plant Biochemistry. (Pergamon Press Ltd., Headington Hill Hall, Oxford OX3 OEW, England) V.1- 1961-

QJDRAZ Quarterly Journal of Crude Drug Research. (Lisse, Netherlands) V.1- 19,-1961-81 For publisher information see IJCREE

QJMEA7 Quarterly Journal of Medicine. (Oxford University Press, Press Road, Neasden, London NW 10 0DD, England) V.1- 1932-

QJPAAA Quarterly Journal of Pharmacy and Allied Sciences. (London, England) V.1, 1928. For publisher information, see JPPMAB

QJPPAL Quarterly Journal of Pharmacy and Pharmacology. (London, England) V.2-21, 1929-48. For publisher information, see JPPMAB

QJSAAP Quarterly Journal of Studies on Alcohol. (New Brunswick, NJ) V.1-28, 1940-67. For publisher information, see QJSOAX

QJSOAX Quarterly Journal of Studies on Alcohol, Part A: Originals. (Rutgers State University, New Brunswick, NJ 08903) V.29- 1968-

QUNUAZ Quaderni della Nutrizione. (Moore-Cottrell Subscription Agencies Inc., North Cohocton, NY 14868) V.1- 1934-

QURBAW Quarterly Reviews of Biophysics. (Cambridge University Press, P.O. Box 92, Bentley House, 200 Euston Rd., London NW1 2DB, England) V.1- 1968-

RalRL # Personnel Communication to NIOSH from Robert L. Raleigh, M.D., Assistant Director of Health and Safety Lab., Eastman Kodak Company, Rochester, NY

RABIDH Revista de Igiena, Bacteriologie, Virusologie, Parazitologie, Epidemiologie, Pneumoftiziologie, Seria: Igiena. (Editura Medicala, Str.13 Decembrie 14, Bucharest, Romania) V.23- 1974-

RADLAX Radiology. (Radiological Society of North America, 20th and Northampton Sts., Easton, PA 18042) V.1- 1923-

RADOA8 Radiobiologiya. (v/o Mezhdunarodnaya Kniga, Kuznetskii Most 18, Moscow G-200, USSR) V.1- 1961

RAEHDT Revista de la Asociacion Espanola de Farmaceuticos de Hospitales. (c/o Camelias 86-1, Vigo, Spain) 1977-

RAMAAB Revista de la Asociacion Medica Argentina. (Santa Fe 1171, Buenos Aires, Argentina) V.1- 1915-

RAREAE Radiation Research. (Academic Press, 111 Fifth Ave., New York, NY 10003) V.1- 1954-

RARIAQ Record of Agricultural Research. (Department of Agricultural for Northern Ireland, Dundonald House, Upper Newtonards Rd., Belfast BT4 3SB, N. Ireland) V.13- 1963-

RARSAM Radiation Research, Supplement. (Academic Press, 111 5th Ave., New York, NY 10003) No.1- 1959-

RBBIAL Revista Brasileira de Biologia. (Caixa Postal 1587, ZC-00 Rio de Janeiro, Brazil) V.1- 1941-

RBPMAZ Revue Belge de Pathologie et de Medecine Experimentale. (Brussels, Belgium) V.18-31, 1947-65. For publisher information, see PTEUA6

RBPMB2 Revista Brasileira de Pesquisas Medicas e Biologicas. (Editora Medico -Biologica Brasileira, Rua Pedrosa de Alvarenga, 1255, Sao, Paulo, Brazil) V.1- 1968-

RCBIAS Revue Canadienne de Biologie. (Les Presses de l'Universite de Montreal, P.O. Box 6128, 101 Montreal 3, Quebec, Canada) V.1- 1942-

RCCRDT Revista do Centro de Ciencias Rurais (Universidade Federal Santa Maria). Review of the Center for Rural Sciences. (Universidade Federal de Santa Maria, Centro de Ciencias Rurais, 97.100 Santa Maria, Brazil) V.1- 1971-

RCHMA2 Revista Chilena de Higiene y Medicina Preventiva. (Santiago, Chile) V.1-15, 1937-53. Discontinued.

RCIBAS Research Council of Israel, Bulletin. (Israel Scientific Press, P.O.B. 801, Jerusalem King George Ave. 33, Jerusalem) V.1- 1951-

RCOCB8 Research Communications in Chemical Pathology and

Pharmacology. (PJD Publications, P.O. Box 966, Westbury, NY 11590) V.1- 1970-

RCPBDC Research Communications in Psychology, Psychiatry and Behavior. (PJD Publications, P.O. Box 966, Westbury, NY 11590) V.1- 1976-

RCPRAN Record of Chemical Progress. (Detroit, MI) V.1-32, 1939-71. Discontinued.

RCRVAB Russian Chemical Reviews. (Chemical Society, Publications Sales Office, Burlington House, London W1V 0BN, England) V.29- 1960-

RCSADO Research Communications in Substances Abuse. (PJD Publications Ltd., Box 966, Westbury, NY 11590) V.1- 1980-

RCTEA4 Rubber Chemistry and Technology. (Div. of Rubber Chemistry, American Chemical Society, University of Akron, Akron, OH 44325) V.1- 1928-

RCUFAC Revista Cubana de Farmacia. (Centro Nacional de Informacion de Ciencias Medicas, Calle 23, No. 177, Havana, Cuba) V.1- 1967-

RDBGAT Radiobiologia, Radiotherapia. (VEB Verlag Volk und Gesundheit, Neue Gruenstr. 18, DDR-102 Berlin, Germany) V.1- 1960-

RDCNBM Reproduction. (Calle Puerto de Bermeo, 11 Madrid 34, Spain) V.1- 1974-

RDMIDP Quarterly Reviews on Drug Metabolism and Drug Interactions. (Freund Pub. House, Suite 500, Chestham House, 150 Regent St., London W1R 5FA, UK) V.3- 1980-

RDMU** Vergleichende Untersuchungen uber Genatropin und Atropin, Ulrich Rathscheck Dissertation. (Pharmakologischen Institut der Westfalichen Wilhelms-Universitat Munster/Westf., Germany, 1934)

RDWU** Beitrage zur Pharmakologie des Berylliums, Ursula Richter Dissertation. (Pharmakologischen Institut der Universitat Wurzburg, Germany, 1930)

REANBJ Revista Espanola de Anestesiologia y Reanimacion. (Sociedad Espanola de Anestesiologia y Reanimacion, Mallorca, 189 Barcellona 36, Spain) V.1- 1954-

REBPAT Radiation and Environmental Biophysics. (Springer-Verlag New York, Inc., Service Center, 44 Hartz Way, Secaucus, NJ 07094) V.11- 1974-

RECYAR Revue Roumaine d'Embryologie et de Cytologie, Serie d'Embryologie. (Bucharest) V.1-8, 1964-71.

REDH** J.D. Riedel-E. de Haen A.-G., Laboratory. (Berlin, Germany)

REEBB3 Revue Europeenne d'Etudes Cliniques et Biologiques. European Journal of Clinical and Biological Research. (Paris, France) V.15-17, 1970-72. For publisher information, see BIMDB3

REMBA8 Revista Ecuatoriana de Medicina y Ciencias Biologicas. (Facultad de Ciencias Medicas, Quito, Ecuador) V.1- 1963-

REMVAY Revue d'Elevage et de Medecine Veterinaire des Pays Tropicaux. (Editions Vigot Freres, 23 rue de l'Ecole-de-Medecine, 75006 Paris 6, France) V.1- 1947-

REONBL Revista Espanola de Oncologia. (Instituto Nacional de Oncologia del Cancer, Ciudad Universitaria, Madrid 3, Spain) V.1- 1952-

REPMBN Revue d'Epidemiologie, Medecine Sociale et Sante Publique. (Masson et Cie, Editeurs, 120 Blvd. Saint-Germain, P-75280, Paris 06, France) V.1- 1953-

REPTED Reproductive Toxicology. (Pergamon Press Inc., Maxwell House, Fairview Park, Elmsford, NY 10523) V.1- 1987

RETOAE Research Today. (Eli Lilly Co., Indianapolis, IN) V.1-16, 1944-60. Discontinued.

REXMAS Research in Experimental Medicine. (Springer-Verlag, Heidelberger, Pl. 3, D-1 Berlin 33, Germany) V.157- 1972-

RFCTAJ Rassegna di Fisiopatologia Clinica e Terapeutica. Review of Clinical and Therapeutic Physiopathology. (Rome, Italy) V.9-42, 1937-70. Discontinued.

RFECAC Revue Francaise d'Etudes Cliniques et Biologiques. (Paris, France) V.1-14, 1956-69. For publisher information, see BIMDB3

RFGOAO Revue Francaise de Gynecologie et d'Obstetrique. (Masson Publishing USA, Inc., 14 E. 60th St., New York, NY 10022) V.20-1920-

RHPC** Rohm and Haas Company Petroleum Chemicals. (Philadelphia, PA 19105)

RIALA6 Revista do Instituto Adolfo Lutz. Review of the Adolfo Lutz Institute. (Biblioteca do Instituto Adoilfo Lutz, Caixa Postal, 2027, 01000 Sao Paulo, Brazil) V.1- 1941-

RIGAA3 Rinsho Ganka. (Igaku Shoin, 5-24-3 Hongo, Bunkyo-ku, Tokyo 113-91, Japan) V.1- 1947-

RIHYAC Rinsho Hinyokika. Clinical Urology. (Igakushoin Medical Publishers, Inc., 1140 Avenue of the Americas, New York, NY 10036) V.21- 1967-

RIMAAX Rivista di Malariologia. (Rome, Italy) V.5-46, 1926-67. Discontinued.

RISCAZ Ricerca Scientifica. Scientific Research. (Rome, Italy) V.18-30, 1948-60; V.36-46, 1966-76. Suspended.

RISSAF Rendiconti Istituto Superiore di Sanita (Italian Edition). (Rome, Italy) V.4-27, 1941-64.

RJARAV Rhodesian Journal of Agricultural Research. (Rhodesian Ministry of Agriculture, P.O. Box 8108, Causeway, Salisbury, Rhodesia) V.1- 1963-

RKGEDW Rinsho Kyobu Geka. Japanese Annals of Thoracic Surgery.

RMCHAW Revista Medica de Chile. (Sociedad Medica de Santiago, Esmeralda 678, Casilla 23-d, Santiago, Chile) V.1- 1872-

RMEMDQ Revue Roumaine de Morphologie, d'Embryologie et de Physiologie, Serie Morphologie et Embryologie. (ILEXIM, POB 136-137, Bucharest, Romania) V.21- 1975-

RMLIAC Revue Medicale de Liege. (Hopital de Baviere, Universite de Liege, Pl. du XX Aout 7, Liege, Belgium) V.1- 1946-

RMMJAK Rocky Mountain Medical Journal. (Denver, Co) V.35-76, 1938-79.

RMNIBN Revista Medico-Chirurgicala. (Societatea de Medici si Naturalisti, Bulevardul Independentei, Iasi, Romania) V.35- 1924-

RMSRA6 Revue Medicale de la Suisse Romande. (Societe Medicale de La Suisse Romande, 2 Bellefontaine, 1000 Lausanne, Switzerland) V.1- 1881-

RMVEAG Recueil de Medecine Veterinaire. (Editions Vigot Freres, 23 rue de l'Ecole-de-Medecine, Paris 6, France) V.1- 1824-

ROHM** Rohm and Haas Company Data Sheets, (Philadelphia, PA 19105)

RPCMBZ Revue de Pathologie Comparee et de Medecine Experimentale. (Editions Medicales et Scientifiques, Editeurs, Boite Postale 100, Paris 17, France) V.5- 1968-

RPHRA6 Recent Progress in Hormone Research. Proceedings of the Laurentian Hormone Conference. (Academic Press, 111 5th Ave., New York, NY 10003) V.1- 1947-

RPOBAR Research Progress in Organic Biological and Medicinal Chemistry. (New York) V.1-3, 1964-72. Discontinued.

RPPCAZ Revue De Pathologie Generale Et De Physiologie Clinique. (Paris, France) V.56-64, 1956-64. For publisher information, see RPCMBZ

RPTOAN Russian Pharmacology and Toxicology. Translation of FATOAO. (Euromed Publications, 97 Moore Park Rd., London SW6 2DA, England) V.30- 1967-

RPZHAW Roczniki Panstwowego Zakladu Higieny. (Ars Polona-RUSH, POB 1001, 00-068 Warsaw 1, Poland) V.1- 1950-

RRBCAD Revue Roumaine de Biochimie. (Rompresfilatelia, POB 2001, Calea Grivitei 64-66, Bucharest, Romania) V.1- 1964-

RRCRBU Recent Results in Cancer Research. (Springer Verlag New York, Inc., Service Center, 44 Hartz Way, Secaucus, NJ07094) V.1- 1965-

RRENAR Revue Roumaine d'Endocrinologie. (Bucharest, Romania) V.1-11, 1964-74. For publisher information, see RRENDU

RRENDU Revue Roumaine de Medecine, Serei Endocrinologie. (ILEXIM, POB 136-137, R-70116 Bucharest, Romania) V.13- 1975-

RRESA8 Rastitel'nye Resursy. Plant Resources. (v/o Mezhdunarodnaya Kniga, Kuznetskii Most 18, Moscow G-200, USSR) V.1- 1965-

RREVAH Residue Reviews. (Springer Verlag New York, Inc., Service Center, 44 Hartz Way, Secaucus, NJ 07094) V.1- 1962-

RSABAC Revista de la Sociedad Argentina de Biologia. (Associacion Medica Argentina, Santa Fe 1171, Buenos Aires, Argentina) V.1- 1925-

RSPSA2 Rassegna di Studi Psichiatrici. Review of Psychiatric Studies. (Ospedale Psichiatrico, Calata Capodichino. 232, 80141 Naples, Italy) V.1- 1911-

RSTEBF Research Steroids. (Pergamon Press, Headington Hill Hall, Oxford OX3 OEW, England) V.1- 1964-

RSTUDV Rivista di Scienza e Technologia degli Alimenti e di Nutrizione Umana. Review of Science and Technology of Food and Human Nutrition. (Bologna, Italy) V.5-6, 1975-76. Discontinued

RTCPA3 Recueil des Travaux Chimiques des Pays-Bas. Journal of the Royal Netherlands Chemical Society. (Koninklijke Nederlandse Chemische Vereniging, Burnierstr., p.o. box 90613, 2509 LP The Hague, Netherlands) V.64- 1965-

RTOPDW Regulatory Toxicology and Pharmacology. (Academic Press. Inc., 111 Fifth Ave., New York, NY 10003) V.1- 1981-

RTPCAT Rassegna di Terapia e Patologia Clinica. (Rome, Italy) V.1-8, 1929-36. For Publisher Information, see RFCTAJ

RVFTBB Rivista di Farmacologia e Terapia. (Prof. William Ferrari, Via G. Campi 287, 41100 Modena, Italy) V.1- 1970-

RVMVAH Revue de Medecine Veterinaire (Toulouse). (Ecole Nationale Veterinaire de Toulouse, 23 Chemin des Capelles, 31076 Toulouse, France) V.89, No.2- 1937-

RVTSA9 Research in Veterinary Science. (Blackwell Scientific Publications, Ltd., Osney Mead, Oxford OX2 OEL, England) V.1- 1960-

RZOVBM Rivista di Zootecnia e Veterinaria. Zootechnical and Veterinary Review. (Rivista di Zootecnia e Veterinaria, Via Viotti 3/5, 20133 Milan, Italy) Number 1- 1973-

SchH## Personal Communication to NIOSH from Dr. H. Schweinfurth, Schering Aktiengesellschaft, Postfach 65-03-11, D-1000 Berlin 65, Germany, on December 2, 1982

SchP## Personal Communication from Dr. P. Schmitz, Bayer AG, 5090 Leverkusen, Bayerwerk, Fed. Rep. Ger., to NIOSH, April 4, 1986

SheCW# Personal Communication from Dr. C.W. Sheu, Genetic Toxicology Branch, FDA, to H. Lau, Tracor Jitco, March 25, 1977

SinJF# Personal Communication from J.F. Sina, Merck Institute for Therapeutic Research, West Point, PA 19486, to the Editor of RTECS, Cincinnati, OH, on October 26, 1982

SpiEW# Personal Communication to Roger Tatken, NIOSH, from Dr. E.W. Spitz, E.P.A. on February 13, 1980

StoGD# Personal Communication to NIOSH from Fr. Gary D. Stoner, Dept. of Community Medicine, School of Medicine, University of California, La Jolla, CA 92037, May 25, 1975

SAAMDZ Substance and Alcohol Actions/Misuse. (Pergamon Press Inc., Maxwell House, Fairview Park, Elmsford, NY 10523) V.1- 1980-

SABOA9 Sabouraudia. (Longman Group Ltd., Journals Div., 4th Ave., Harlow, Essex CM20 2JE, England) V.1- 1961-

SACAB7 South African Cancer Bulletin. (National Cancer Association of South Africa, 9 Jubillee Rd., Parktown, Johannesburg, South Africa) V.1- 1957-

SADEAN Saglik Dergisi. (Turkiye Cumhuriheti esaglik ve Sosyal Yardim Bakanlinca Yayinlanir, Ankara, Turkey) V.19- 1945-

SAIGAK Saishin Igaku. Modern Medicine. (Saishin Igakusha, Senchuri Bldg., 3-6-1 Hirano-machi, Higashi-ku, Osaka 541, Japan) V.19- 1945-

SAIGBL Sangyo Igaku. Japanese Journal of Industrial Health. (Japan Association of Industrial Health, c/o Public Health Building, 78 Shinjuku 1-29-8, Shinjuku-Ku, Tokyo, Japan) V.1- 1959-

SAJSAR South African Journal of Science. (South African Association for the Advancement of Science, POB 61019, Marshalltown, Transvaal 2107, South Africa) V.6- 1909-

SAKNAH Soobshcheniia Akademii Nauk Gruzinskoi S.S.R. (v/o Mezhdunarodnaya Kniga, Kuznetskii Most 18, Moscow G-200, U.S.S.R.) V.2- 1941-

SAMJAF South African Medical Journal. (Medical Association of South Africa, Secy., P.O. Box 643, Cape Town, S. Africa) V.6- 1932-

SAPHAO Skandinavisches Archiv fuer Physiologic. (Karolinska Institutet, Editorial Office, Stockholm, Sweden) V.1-83, 1899-1940. Superseded by APSCAX

SATHAA Schweizer Archiv fuer Tierheilkunde. (Orell Fuessli Graphische Betriebe AG, Postfach 1459, 8036 Zurich, Switzerland) V.1- 1859-

SAVEAB Sammlung von Vergiftungsfaellen. (Berlin, Germany) V.1-13, 1930-1944; V.14, 1952-54. For publisher information, see ARTODN

SAZTA8 Schweizerische Apotheker-Zeitung. (2 rue Voltaire, 1211 Geneva 1, Switzerland) V.52- 1914-

SBJODP Science of Biology Journal. (POB 2245, Springfield, Ill. 62705) V.1- 1975-

SBLEA2 Sbornik Lekarsky. (PNS-Ustredni Expedice Tisku, Jindriska 14, Prague 1, Czechoslovakia) V.1- 1887-

SCALA9 Scalpel. (Brussels, Belgium) 1920-V.124, 1971. Discontinued.

SCAMAC Scientific American. (Scientific American, Inc., 415 Madison Ave., New York, NY 10017) V.1- 1859-

SCCSC8 Soap, Cosmetics, Chemical Specialties. (MacNair-Dorland, 101 W. 31st St., New York, NY 10001) V.47- 1971-

SCCUR* Shell Chemical Company. Unpublished Report. (2401 Crow Canyon Rd., San Romon, CA 94583)

SCGTDW Somatic Cell Genetics. (New York, NY) V.1-9, 1975-83.

SCHSAV Soap and Chemical Specialties. (New York, NY) V.30-47, 1954-71. For publisher information, see SCCSC8

SCIEAS Science. (American Assoc. for the Advancement of Science, 1515 Massachusetts Ave., NW, Washington, DC 20005) V.1- 1895-

SCINAL Science and Culture. (Indian Science News Association, 92 Acharya Prafulla Chandra Road, Calcutta 700009, India) V.1- 1935-

SCJUAD Science Journal. (London, England) V.1-7, 1965-71. For publisher information, see NWSCAL

SCMBE9 Spokane County Medical Society Bulletin. (Spokane, WA) 1929-81.

SCMGDN Somatic Cell and Molecular Genetics. (Plenum Publishing Corp., 233 Spring St., New York, NY 10013) V.10- 1984-

SCNEBK Science News. (Science Service, Inc., 1719 N. St., N.W., Washington, DC 20036) V.1- 1921-

SCPHA4 Scientia Pharmaceutica. (Oesterreichische Apotheker- Verlagsgesellschaft MBH, Spitalgasse 31, 1094 Vienna 9, Austria) V.1- 1930-

SCYYDZ Shengzhi Yu Biyun. Reproduction and Contraception. (China International Book Trading Corp., POB 2820, Beijing, Peop. Rep. China) 1980-

SDGTB3 Seminars in Drug Treatment. (Grune and Stratton, 111 5th Ave., New York, NY 10003) V.1- 1971-

SDLU** Ein Beitrag zum Kapitel der anorganischen Arsenvergiftung, Kurt Schaffer Dissertation. (Medizinischen Klinik Lindenburg der Universitat Koln, Germany, 1937)

SDMEAL South Dakota Journal of Medicine. (South Dakota Medical Association, 608 West Ave. N., Sioux Falls, SD 57104) V.18- 1965-

SDMU** Zur Pharmakologie einiger Triazoliumverbindungen I, Erhart Schulze Dissertation. (Pharmakologischen Institut der Martin Luther- Universitat Halle-Wittenberg, Germany, 1936)

SDSTBT Scientific and Technical Report - Soap and Detergent Association. (Soap and Detergent Assoc., 485 Madison Ave., New York, NY 10022) No.1- 1965-

SDUU** Ueber die Beeinflussung der Salvarsantoxizitat im

Tierversuch, Jurgen Steudemann Dissertation. (University of Hamburg, Germany, 1934)

SEIJBO Senten Ijo. Congenital Anomalies. (Nihon Senten Ijo Gakkai, Kyoto 606, Japan) V.1- 1960-

SEMEAS Semana Medica. (Sociedad Argentina de Gastroenterologica, Santa Fe 1171, Buenos Aires, Argentina) V.1- 1894-

SFCRAO Symposium on Fundamental Cancer Research (Williams and Wilkins Co., 428 E. Preston St., Baltimore, Md. 21202) V.1- 1947-

SFTIAE Svensk Farmaceutisk Tidskrift. Swedish Journal of Pharmacy. (Apotekarsocieteten, Upplandsgatan 6A, Stockholm C, Sweden) V.1- 1897-

SGOBA9 Surgery, Gynecology and Obstetrics. (Franklin H. Martin Memorial Foundation, 55 E. Erie St., Chicago, IL 60611) V.1- 1905-

SHBOAO Shokubutsu Boeki. (Nippon Shokubutsu Boeki Kyokai, 43-11, Komagome 1-chome, Toshima-Ku, Tokyo, Japan) V.1- 1947-

SHELL* Shell Chemical Company, Technical Data Bulletin. (Agricultural Div., Shell Chemical Co., 2401 Crow Canyon Rd., San Ramon, CA 94583)

SHGKA3 Shika Gakuho. Journal of Dentistry. (Tokyo Shika Daigaku Gakkai, 9-18, 2-chome, Sanzaki-cho, Chiyuoda-ku, Tokyo, Japan) V.1- 1895-

SHHUE8 Shiyou Huagong. Petrochemical Technology. (Beijing Huagon Yanjiuyan, Beikou, Hepingjie, Beijing, Peop. Rep. China)

SHIGAZ Shigaku. Ondotology. (Nippon Shika Daigaku Shigakkai, 1-9-20 Fujimi, Chiyodaku, Tokyo 102, Japan) V.38- 1949-

SHIYDO Shaanxi Xinyiyao. Shaanxi New Medical. (Shanxi Renmin Chubanshe, Xi'an, Shaanxi, Peop. Rep. China) V.1- 1972-

SHKKAN Shika Kiso Igakkai Zasshi. Journal of the Japanese Association for Basic Dentistry. (Shika Kiso Igakkai, Nihon Daigaku Shigakubu, 1-8 Surugadai, Kanda, Chiyoda-ku, Tokyo 101, Japan) V.1- 1959-

SHNSAS Shinryo to Shinyaku. Medical Consultation and New Remedies. (Iji Shuppan Sha, 1-2-8 Shinkawa, Chuo-Ku, Tokyo 104, Japan) V.1- 1964-

SHYCD4 Shengwu Huaxue Yu Shengwu Wuli Jinzhan. Progress in Biochemistry and Biophysics. (Kexue Chubanshe, 137 Zhaoyangmennei Dajie, Beijing, Peop. Rep. China) No.1- 1974(?)-

SHZAAY Shoyakugaku Zasshi. (Kyoto Daigaku Yakugakubu, Yoshida Shimoadachi-machi, Sakyo-Ku, Kyoto, Japan) V.6- 1952-

SIGAAE Shika Igaku. Odontology. (Osaka Shika Gakkai, 1 Kyobashi, Higashi-Ku, Osaka, Japan) V.1- 1930-

SIGZAL Showa Igakkai Zashi. Journal of the Showa Medical Association. (Showa Daigaku, Showa Igakki, 1-5-8, Hatanodai, Shinagawa-ku, Tokyo 142, Japan) V.1- 1939-

SIHSD8 Shang-hai I Hsueh. (Shanghi, China) V.1- 1978(?)-

SIIPD4 Shanghai Diyi Yixueyuan Xuebao. Journal of the Shanghai First Medical College. (Bejing, Pep. Rep China) 1956- V.12 1985.

SIZSAR Sapporo Igaku Zasshi. Sapporo Medical Journal. (Sapporo Igaku Daigaku, Nishi-17-Chome, Minami-1-jo, Chuo-ku Sapporo 060, Japan) V.3- 1952-

SJCLAY Scandinavian Journal of Clinical and Laboratory Investigation. (Universitetsforlaget, PO Box 142, Boston, MA 02113) V.1- 1949-

SJDBA9 Soviet Journal Developmental Biology (English Translation). (Plenum Publishing Corp., 233 Spring St., New York, NY 10013) V.1- 1970-

SJDRAN Scandinavian Journal of Dental Research. (Munksgaard, 6, Norregade, DK-1165, Copenhagen K, Denmark) V.78- 1970-

SJGRA4 Scandinavian Journal of Gastroenterology. (Universitetsforlaget, PO Box 142, Boston, MA 02113) V.1- 1966-

SJHAAQ Scandinavian Journal of Haematology. (Munksgaard, 35 Norre Sogade, DK 1370 Copenhagen K, Denmark) V.1- 1964-

SJHSBD Scandinavian Journal of Haematology. Supplementum. (Munksgaard, Noerre Soegade, DK-1370 Copenhagen K, Denmark) No.1- 1964-

SJRDAH Scandinavian Journal of Respiratory Diseases. (Munksgaard, 6 Norregade, DK-1165 Copenhagen K, Denmark) V.47- 1966-

SJRHAT Scandinavian Journal of Rheumatology. (Almqvist and Wiksell, POB 45150, S-10430 Stockholm, Sweden) V.1- 1972-

SJUNAS Scandinavian Journal of Urology and Nephrology. (Almqvist and Wiksell Periodical Co., POB 62, 26 Gamla Brogatan, S-101-20 Stockholm, Sweden) V.1- 1967-

SKEZAP Shokuhin Eiseigaku Zasshi. Journal of the Food Hygiene Society of Japan. (Nippon Shokuhin Eisei Gakkai, c/o Kokuritsu Eisei Shikenjo, 18-1, Kamiyoga 1-chome, Setagaya-Ku, Tokyo, Japan) V.1- 1960-

SKIZAB Shikoku Igaku Zasshi. Shikoku Medical Journal. (Tokushima Daigaku Igakubu, Kuramoto-cho, Tokushima, Japan) V.1- 1950-

SKKNAJ Sankyo Kenkyusho Nempo. Annual Report of Sankyo Research Laboratories (Sankyo K.K. Chuo Kenkyusho, 1-2-58 Hiromachi, Shinagawa-ku, Tokyo 140, Japan) No.15- 1963-

SKNEA7 Shionogi Kenkyusho Nempo. Annual Report of Shionogi Research Laboratory. (Shionogi Seiyaku K. K. Kenkyusho, 2-47 Sagisukami, Fukushima, Osaka, Japan) No.1- 1951-

SMDJAK Scottish Medical Journal. (Longman Group Journal Division, 4th Ave., Harlow, Essex CM20 2JE, England) V.1- 1956-

SMEZA5 Sudebno-Meditsinskaya Ekspertiza. Forensic Medical Examination. (v/o Mezhdunarodnaya Kniga, Kuznetskii Most 18, Moscow G-200, U.S.S.R.) V.1- 1958-

SMJOAV Southern Medical Journal. (Southern Medical Association, 2601 Highland Ave., Birmingham, AL 35205) V.1- 1908-

SMSJAR Scottish Medical and Surgical Journal. (Edinburgh, Scotland) For publisher information, see SMDJAK

SMWOAS Schweizerische Medizinische Wochenschrift. (Schwabe and Co., Steintorst 13, 4000 Basel 10, Switzerland) V.50- 1920-

SNSHBT Senshokutai. Chromosome. (Senshokutai Gakkai, Kaokusai Kirisutokyo Daigaku, Mitaka, Tokyo-to, Japan) No. 1- 1946- Adopted new numbering in 1976.

SOGEBZ Soviet Genetics. Translation of GNKAA5. (Plenum Publishing Corp., 233 Spring St., New York, NY 10013). V.2- 1966-

SOLGAV Seminars in Oncology. (Grune and Stratton Inc., 111 Fifth Ave., New York, NY 10003) V.1- 1974-

SOMEAU Sovetskaia Meditsina. (v/o Mezhdunarodnaya Kniga, Kuznetskii Most 18, Moscow G-200, U.S.S.R.) V.1- 1937-

SOVEA7 Southwestern Veterinarian. (College of Veterinary Medicine, Texas A and M University, College Station, TX 77843) V.1- 1948-

SPCOAH Soap, Perfumery and Cosmetics. (United Trade Press, Ltd., 33 Bowling Green Lane, London EC1R0DA, England) V.8, No.7- 1935-

SPEADM Special Publication of the Entomological Society of America. (4603 Calvert Rd., College Park, MD 20740) (Note that although the Registry cites data from the 1974 edition, this reference has been superseded by the 1978 edition.)

SPERA2 Sperimentale. (Universita degli Studi Piazza San Marco, 4, 50121 Florence, Italy) V.50- 1896-

SRCHAA Search. (Science House, 157 Gloucester St., Sydney, N.S.W. 2000, Australia) V.1- 1970-

SRTCDF Scientific Report. (Dr. C.R. Krishna Murti, Director, Industrial Toxicology Research Centre, Luchnow-226001, India) History unknown

SSBSEF Scientia Sinica, Series B: Chemical, Biological, Agricultural, Medical, and Earth Sciences (English Edition). (Scientific and Technical Books Service Ltd., POB 197, London WC2N 4DE, England) V.25- 1982-

SSCHAH Soap and Sanitary Chemicals. (New York, NY) V.14-30, 1938-54. For publisher information, see SCCSC8

SSCMBX Scripta Scientifica Medica. (Durzhavno Izdatelstvo Meditsina i Fizkultura, P1. Slaveikov II, Sofia, Bulgaria) V.4(2)- 1965-

SSEIBV Sumitomo Sangyo Eisie. Sumitomo Industrial Health. (Sumitomo Byoin Sangyo Eisei Kenkyushitsu, 5-2-2, Nakanoshima, Kita-ku, Osaka, 530, Japan) No.1- 1965-

SSHZAS Seishin Shinkeigaku Zasshi. (Nippon Seishin Shinkei Gakkai, c/o Toyo Bunko, 2-28-21, Honkomagome, Bunkyo-Ku, Tokyo 113, Japan) V.39- 1935-

SSINAV Scientia Sinica. (Peking, Peop. Rep. China) V.3-24, 1954-81. For publisher information, see SSBSEF

SSSEAK Studi Sassaresi, Sezione 2. (Societa Sassarese di Scienze Mediche e Naturali, Viale S, Pietro 12, I-07100 Sassari, Italy) V.42- 1964-

SSZBAC Sbornik Vysoke Skoly Zemedelske v Brne, Rada B: Spisy Veterinarni. (Brno, Czechoslovakia) 1953-66. For publisher information, see AUAAB7

STBIBN Studia Biophysica. (Akademie-Verlag GmbH, Liepziger Str. 3-4, DDR-108 Berlin, German Democratic Republic) V.1- 1966-

STCC** Stauffer Chemical Company, Product Safety Information Sheets. (Stauffer Chemical Company, Agricultural Chemical Division, Westport, Connecticut 06880)

STEDAM Steroids. (Holden-Day Inc., 500 Sansome St., San Francisco, CA 94111) V.1- 1963-

STEVA8 Science of the Total Environment. (Elsevier Scientific Publishing Co., P.O. Box 211, Amsterdam C, Netherlands) V.1- 1972-

STMEAV Strasbourg Medical. (Strasbourg, France) V.1-20, 1950-69. For publisher information, see JMSTBR

STOAAT Stomatologiia (Moscow). (v/o Mezhdunarodnaya Kniga, Kuznetskii Most 18, Moscow G-200, USSR) V.1- 1923-

STRAAA Strahlentherapie. (Urban and Schwarzenberg, Pattenkoferst 18, D-8000 Munich 15, Germany) V.1- 1912-

STRHAV Staub-Reinhaltung der Luft. (VDI-Verlag GmbH, Postfach 1139, D-4000 Duesseldorf 1, Fed. Rep. Germany) V.26- 1966-

STSOAN Sbornik Nauchnykh Trudov, Severo-Osetinskii Gosudarstvennyi Meditsinskii Institut. (The Institute, Tashkent, U.S.S.R.) No.1- 1948-

SUFOAX Surgical Forum. (American College of Surgeons, 55E. Erie St., Chicago, IL 60611) V.1- 1950-

SUNCO* Sun Company Material Safety Data Sheets. (1608 Walnut St., Philadelphia, PA 19103)

SURGAZ Surgery. (C.V. Mosby Co., 11830 Westline Industrial Dr., St. Louis, MO 63141) V.1- 1937-

SURR** Szechuan University Research Report, Natural Science. (Szechuan University, Szechuan, China) Number 2/3- 1978-

SVLKAO Sbornik Vedeckych Praci Lekarske Fakulty Univerzity Karlovy v Hradci Kralove. Collection of Scientific Works of the Charles University Faculty of Medicine in Hradec Kralove. (Univerzita Karlova, Ovocny Trh. 3, CS-116-36 Prague 1, Czechoslovakia) V.1- 1958-

SWEHDO Scandinavian Journal of Work, Environment and Health. (Haartmaninkatu 1, FIN-00290 Helsinki 29, Finland) V.1 1975-

SYHJAM Saengyak Hakhoechi. Journal of the Society of Pharmacognosy. (c/o Natural Products Research Institute, Seoul National University, 28 Yunkeon-Dong, Chong-ro-ku, Seoul 110, S. Korea) V.9- 1978-

SYSWAE Shiyan Shengwa Xuebao. Journal of Experimental Biology. (Guozi Shudian, China Publications Center, P.O. Box 399, Peking, Peop. Rep. China) V.1- 1953- (Suspended 1966-77)

SYXUD2 Shandong Yixueyuan Xuebao. Journal of Shandong Medical College (China International Book Trading Corp., POB 2820, Beijing, Peop. Rep. China)

SYXUE3 Shenyang Yaoxueyuan Xuebao. Journal of Shenyang College of Pharmacy. (Shenyang Yaoxueyuan, 7, 2-duan Wenhualu, Shenhequ, Shenyang, Peop. Rep. China) V.1- 1984-

SZAPAC Schweizerische Zietschrift fuer Allgemeine Pathologie und Bakteriologie. (White Plains, NY) V.1-3,1938-40; V.13-22, 1950-1959. For publisher information, see ECEBDI

SZPMAA Sozial-und Praeventivmedizin. (Art. Institut Orell Fuessli AG, Dietzingerstrasse 3, 8003 Zurich, Switzerland) V.19- 1974-

SZTPA5 Schweizerische Zeitschrift fuer Tuberkulose und Pneumonologie. (Basel, Switzerland) V.1- 18, 1944-1961

TABIA2 Tabulae Biologicae. (The Hague, Netherlands) V.1-22, 1925-63. Discontinued.

TAFSAI Transactions of the American Fisheries Society. (American Fisheries Society, 5410 Grosvenor Lane, Bethesda, Md 20014) V.1- 1870-

TAGTBR Trudy Azerbaidzhanskogo Nauchno-Issledovatel'skogo Instituta Gigieny Truda i Professional'nykh Zabolevanii. Transactions of the Azerbaidzhan Scientific Research Institute of Labor Hygiene and Occupational Diseases. (The Institute, Baku, U.S.S.R.) No.1- 1966-

TAKHAA Takeda Kenkyusho Ho. Journal of the Takeda Research Laboratories. (Takeda Yakuhin Kogyo K. K., 4-54 Juso-nishino-cho, Higashi Yodogawa-Ku, Osaka 532, Japan) V.29- 1970-

TAPPAP Tappi. (Tappi, One Dunwoody Park, Atlanta, GA 30341) V.32- 1949-

TCMUD8 Teratogenesis, Carcinogenesis, and Mutagenesis. (Alan R. Liss, Inc., 150 Fifth Ave., New York, NY 10011) V.1- 1980-

TCMUE9 Topics in Chemical Mutagenesis. (Plenum Publishing Corp., 233 Spring Street, New York, NY 10013) V.1- 1984-

TDBU** Ueber die Wirkung des Braunsteins, Paul Thissen Dissertation. (Pharmakologischen Institut der Tierarztlichen Hochschule zu Berlin, Germany, 1933)

TDKNAF Takeda Kenkyusho Nempo. Annual Report of the Takeda Research Laboratories. (Osaka, Japan) V.1-28, 1936-69. For publisher information, see TAKHAA

TEARAI Terapevticheskii Arkhiv. Archives of Therapeutics. v/o Mezhdunarodnaya Kniga, Kuznetskii Most 18, Moscow G-200, USSR) V.1-19, 1923-41; V.20- 1948-

TECSDY Toxicological and Environmental Chemistry. (Gordon and Breach Science Pub. Inc., 1 Park Ave., New York, NY 10016) V.3(3/4)- 1981-

TeiD## Personal Communication to NIOSH from Dian Teigler, Director of Biological Research, Lee Pharmaceuticals, 1444 Santa Anita Ave., South El Monte, CA 91733, June 16, 1975

TELEAY Tetrahedron Letters. (Pergamon Press Ltd., Headington Hill Hall, Oxford OX3 OBW, England) 1959-

TFAAA8 Trudy Sektora Fiziologii, Akademiia Nauk Azerbaidzhanskoi, SSR. Transactions of the Dept. of Physiology, Academy of Sciences of the Azerbaidzhan, SSR. (Baku, USSR) V.1-9, 1956-67. For publisher information, see TFZABW

TFAKA4 Trudy Instituta Fiziologii, Akademiya Nauk Kazakhshoi SSR. Transactions of the Institute of Physiology, Academy of Sciences of the Kazakh. SSR. (The Academy, Alma-Ata, USSR) V.1- 1955-

TFZABW Trudy Instituta Fiziologii, Akademiya Nauk Azerbaidzhanskoi SSR. Transactions of the Institute of Physiology, Academy of Sciences of the Azerbaidzhan SSR. (The Academy, Institut Fiziologii, Baku, USSR) V.10- 1967-

TGANAK Tsitologiya i Genetika. (v/o Mezhdunarodnaya Kniga, Kuznetskii Most 18, Moscow G-200, USSR) V.1- 1967-

TGMEAJ Tropical and Geographical Medicine. (DeErven F. Bohn NV, Box 66, Harlem, Netherlands) V.10- 1958-

TGNCDL "Handbook of Organic Industrial Solvents," 2nd ed., Chicago, IL, National Association of Mutual Casualty Companies, 1961

THAGA6 Theoretical and Applied Genetics. (Spinger Verlag New York, Inc., Service Center, 44 Hartz Way, Secaucus, NJ 07094) V.38- 1968-

THBRAA Thrombosis Research. (Pergamon Press, Maxwell House, Fairview Park, Elmsford, NY 10523) V.1- 1972-

THERAP Therapie. (Doin, Editeurs, 8 Place de l'Odeon, Paris 6, France) V.1- 1946-

THEWA6 Therapiewoche. (G. Braun GmbH, Postfach 1709, 7500 Karlsruhe 1, Fed. Rep. Germany) V.1- 1950-

THGEAU Therapie der Gegenwart. (Urban and Schwarzenberg,

Pettenkoferstr. 18, D-8000 Munich 2, Federal Republic of Germany) V.36-85, 1895-1944; V.86- 1946-

THGNBO Theriogenology. (Geron-X, Inc., P.O. Box 1108, Los Altos, CA 94022) V.1- 1974-

THHUAF Therapia Hungarica. (Akademiai Kiado, POB 24, Budapest 502, Hungary) V.1- 1953-

THMOAM Therapeutische Monatshefte. (Berlin, Germany) V.1-33, 1887-1919. For publisher information, see KLWOAZ

THORA7 Thorax. (British Medical Association, Travistock Square, London WC1H 9JR, England) V.1- 1946-

THUMAM Therapeutische Umschau, Revue Therapeutique. (Hans Huber, Laenggstrasse 76, 3000 Bern 9, Switzerland) V.1- 1944-

TIDZAH Tokyo Ika Daigaku Zasshi. Journal of Tokyo Medical College. (Tokyo Ika Daigaku Igakkai, 1-412, Higashi Okubo, Shinjuku-ku, Tokyo, Japan) 1918-

TIEUA7 Tieraerztliche Umschau. (Terra-Verlag, Neuhauser. Str. 21, 7750 Constance, Germany) V.1- 1946-

TIHHAH Tai-wan I Hsueh Hui Tsa Chih. Journal of the Formosan Medical Association. (Tai-wan I Hsueh Hui, 1, Sect.1 Jen-ai Road, Taipei, Taiwan) V.45- 1946-

TIUSAD Tin and Its Uses. (Tin Research Institute, 483 W. 6th Ave., Columbus, OH 43201) No.1- 1939-

TIVSAI Trudy, Vsesoyuznyi Nauchno-Issledovatel'skii Institut Veterinarnoi Sanitarii. Transactions, All-Union. Scientific Research Institute of Veterinary Sanitation. (The Institute, Moscow, U.S.S.R.) V.11- 1957-

TIYADG Tianjin Yiyao. Tianjin Medical Journal. (Guoji Shudian, China Publications Center, POB 399, Peking, China) V.1- 1973-

TJADAB Teratology, A Journal of Abnormal Development. (Wistar Institute Press, 3631 Spruce St., Philadelphia, PA 19104) V.1- 1968-

TJEMAO Tohoku Journal of Experimental Medicine. (Maruzen Co., Export Dept., P.O. Box 5050, Tokyo Int., 100-31 Tokyo, Japan) V.1- 1920-

TJEMDR Tokai Journal of Experimental and Clinical Medicine. (Tokai University Press, Shinjuku Tokai Bldg., 3-27-4 Shinjuku-Ku, Tokyo 160, Japan) V.1- 1976-

TJIDAH Tokyo Jikeikai Ika Daigaku Zasshi. Tokyo Jikeikai Medical Journal. (3-25-8, Nishi Shinbashi, Minato-ku, Tokyo, Japan) V.66- 1951-

TJIZAF Tokyo Joshi Ika Daigaku Zasshi. Journal of Tokyo Women's Medical College. (Society of Tokyo Women's Medical College, c/o Tokyo Joshi Ika Daigaku Toshokan, 10, Kawada-cho, Shinjuku-ku, Tokyo 162, Japan.) V.1- 1931-

TJSGA8 Tijdschrift Voor Sociale Geneeskunde. (B.V. Uitge-versmaatschappij. Reflex, Mathenesserlaan 310, Rotterdam 3003, Netherlands) V.1- 1923-

TJXMAH Tokushima Journal of Experimental Medicine. (Tokushima Diagaku Igakubu, 3, Kuramoto-cho, Tokushima, Japan) V.1- 1954-

TKIZAM Tokyo Igaku Zasshi. Tokyo Journal of Medical Sciences. (Tokyo, Japan) V.59-76, 1951-68.

TKORAS Trudy Kazakhskogo Nauchno-Issledovatel'skogo Instituta Onkologii i Radiologii. (The Institute, Alma-Ata, U.S.S.R.) V.1- 1965-

TMPBAX Trudy Moskovskogo Obshchestva Ispytatelei Prirody. Transactions of the Moscow Society of Naturalists. (Moskovskoe Obshchestvo Ispytaltelei, Prirody, Moscow, USSR) V.1- 1951-

TMPRAD Tropenmedizin und Parasitologie. (Georg Thieme Verlag, Postfach 732, Herdweg 63, 7000 Stuttgart, Germany) V.25- 1974-

TNEOAO Transactions of the New England Obstetrical and Gynecological Society. (Publisher unknown)

TNICS* Toxicology of New Industrial Chemical Substances. (U.S.S.R. Academy of Medical Sciences, Moscow-Meditsina)

TNNPA4 Trudy, Nauchno-Issledovatel'skii Institut Neftekhimicheskikh Proizvodstv. Transactions, Scientific Research Institute of the Manufacture of Petrochemicals. (The Institute, Moscow, U.S.S.R.) No.1- 1969-

TOANDB Toxicology Annual. (Marcel Dekker, POB 11305, Church St. Station, New York, NY 10249) 1974/75-

TobJS# Personal Communication to the Editor, Toxic Substances List from Dr. J.S. Tobin, Director, Health and Safety, FMC Corporation, New York, NY 14103, November 9, 1973

TOERD9 Toxicological European Research. (Editions Ouranos, 12 bis, rue Jean-Jaures, 92807 Puteaux, France) V.1- 1978-

TOFOD5 Tokishikoroji Foramu. Toxicology Forum. (Saiensu Foramu, c/o Kida Bldg., 1-2-13 Yushima, Bunkyo-ku, Tokyo, 113, Japan) V.6- 1983-

TOHJAV Tongmul Hakhoe Chi. Korean Journal of Zoology. (Zoological Society of Korea, c/o Department of Zoology, College of Natural Science, Seoul National University, Seoul 151, South Korea) V.1- 1958-

TOIZAG Toho Igakkai Zasshi. Journal of Medical Society of Toho University. (Toho Daigaku Igakubu Igakkai, 5-21-16, Omori, Otasku, Tokyo, Japan) V.1- 1954-

TOLED5 Toxicology Letters. (Elsevier Scientific Publishing Co., P.O. Box 211, Amsterdam, Netherlands) V.1- 1977-

TOPADD Toxicologic Pathology. (E.I. du Pont de Nemours Co., Inc., Elkton Rd., Newark, DE 19711) V.6(3/4)- 1978-

TORAAK El Torax. (Montevideo, Uruguay) V.1- 1952-

TOSUAH Transactions of the Ophthalmological Societies of the United Kingdom. (Churchill-Livingstone Inc., 19 W. 44th St., New York, NY 10036) V.1- 1880-

TOXIA6 Toxicon. (Pergamon Press, Headington Hill Hall, Oxford OX3 OBW, England) V.1- 1962-

TOXID9 Toxicologist. (Society of Toxicology, Inc., 475 Wolf Ledge Parkway, Akron, OH 44311) V.1- 1981-

TPHSDY Trends in Pharmacological Sciences. (Elsevien North Holland, Inc., 52 Vanderbilt Ave., New York, NY 10017) V.1- 1979-

TPKVAL Toksikologiya Novykh Promyshlennykh Khimicheskikh Veshchestv. Toxicology of New Industrial Chemical Sciences. (Akademiya Meditsinskikh Nauk S.S.R., Moscow, U.S.S.R.) No.1- 1961-

TRBMAV Texas Reports on Biology and Medicine. (Texas Reports, University of Texas Medical Branch, Galveston, TX 77550) V.1- 1943-

TRENAF Kenkyu Nenpo - Tokyo-toritsu Eisei Kenkyusho. Annual Report of Tokyo Metropolitan Research Laboratory of Public Health. (24-1, 3 Chome, Hyakunin-cho, Shin-Juku-Ku, Tokyo, Japan) V.1- 1949/50-

TRIPA7 Tin Research Institute, Publication. (The Institute, Fraser Rd., Greenford UB6 7AQ, Middlesex, England) No.1- 1934-

TRPLAU Transplantation. (Williams and Wilkins, Co., 428 E. Preston St., Baltimore, MD 21202) V.1- 1963-

TRSEAO Transactions of the Royal Society of Edinburgh. (Edinburgh, Scotland) V.1-70, 1783-1979.

TRSTAZ Transactions of the Royal Society of Tropical Medicine and Hygiene. (The Society, Manson House, 26 Portland Pl., London W1N 4EY, England) V.14- 1920-

TSCAT* Office of Toxic Substances Report. (U. S. Environmental Protection Agency, Office of Toxics Substances, 401 Mst SW, Washington, DC 20460)

TSITAQ Tsitologiya. Cytology. (v/o Mezhdunarodnaya Kniga, Kuznetskii Most 18, Moscow G-200, USSR) V.1 1959-

TUBEAS Tubercle. (Williams and Wilkins, 428 E. Preston Street, Baltimore, MD 21202) V.1- 1919-

TUMOAB Tumori. (Casa Editrice Ambrosiana, Via G. Frua 6, 20146 Milan, Italy) V.1- 1911-

TVLAA5 Trudy Vsesoyuznogo Nauchno-Issledovatel'skogo Instituta Lekarstvennykh Rastenii. Transactions of the All-Union Scientific Research Institute of Medicinal Plants. (Vsesoyuznyi Nauchno-Issledovatel'skii Institut Lekarstvennykh Rastenii, Bittsa, U.S.S.R.) No.5- 1936-

TWHPA3 Tung Wu Hsueh Pao. Journal of Zoology. (Guozi Shudian, China Publications Center, POB 399, Peking, Peop. R. China) V.1- 1949-

TXAPA9 Toxicology and Applied Pharmacology. (Academic Press, 111 5th Ave., New York, NY 10003) V.1- 1959-

TXCYAC Toxicology. (Elsevier/North-Holland Scientific Publishers Ltd., 52 Vanderbilt Ave., New York, NY 10017) V.1- 1973-

TXMDAX Texas Medicine. (Texas Medical Assoc., 1905 N. Lamar Blvd., Austin, TX 78705) V.60- 1964-

TYKNAQ Tohoku Yakka Daigaku Kenkyu Nempo. Annual Report of the Tohoku College of Pharmacy. (77, Odawara Minamikotaku, Hara-machi, Sendai, Japan) No.10- 1963-

UsuT## Personal Communication from Dr. Tetsuo Usui, Teilcoku Hormone Mfg. Co. Ltd., 1604 Shimosakunobe Takatsu-ku, Kawasaki, Japan, to Jenny Fang, Tracor Jitco, Inc., June 29, 1979

UBZHAZ Ukrains'kii Biokhimichnii Zhurnal (1946-77). Ukranian Biochemical Journal. (Kiev, USSR) V.18-49, 1946-77. For publisher information, see UBZHD4

UBZHD4 Ukrainskii Biokhimicheskii Zhurnal. Ukranian Biochemical Journal. U/0 Mezhdunarodnaya Kniga, Kuznetskii Most 18, (Moscow G-200, USSR) V.50- 1978-

UCDS** Union Carbide Data Sheet. (Industrial Medicine and Toxicology Dept., Union Carbide Corp., 270 Park Ave., New York, NY 10017)

UCMH** Union Carbide Manual Hazard Sheet. (Industrial Medicine and Toxicology Dept., Union Carbide Corp., 270 Park Ave., New York, NY 10017)

UCPHAQ University of California, Publications in Pharmacology. (Berkeley, CA) V.1-3, 1938-57. Discontinued.

UCREAR Urologic and Cutaneous Review. (St. Louis, MO) V.17-56, 1913-52. Discontinued.

UCRR** Union Carbide Research Report. (Union Carbide Corp., Old Ridgebury Rd., Danbury, CT 06817)

UDHU** Zur Frage der Entgiftung des Thalliums, Gerhart Urban Dissertation. (Pharmakologischen Institut der Martin Luther-Universitat Halle- Wittenberg, Germany, 1936)

UGLAAD Ugeskrift for Laeger. (Almindelige Danske Laegeforening, Kristianiagade 12A, 2100 Copenhagen, Denmark) V.1- 1839-

UICMAI UICC Monograph Series. (Springer Verlag New York, Inc., Service Center, 44 Hartz Way, Secaucus, NJ 07094)

UJMSAP Upsala Journal of Medical Sciences. (Almqvist and Wiksell, Periodical Co., POB 62, 26 Gamla Brogatan, S-101, 20 Stockholm, Sweden) V.77- 1972-

UMCAAA Union Mediale du Canada. (Union Medical du Canada, 1440 ouest rue Sainte-Catherine, Montreal H3G 2P9, Quebec, Canada) V.1- 1872-

UMGEA8 Uchenye Zapiski, Moskovskii Nauchno-Issledovatel'skii Institut Gigieny imeni F.F. Erismana. Scientific Papers, Moscow F.F. Erisman Scientific Research Institute of Hygiene. (The Institute, Moscow, USSR) No.1- 1959-

UMJOAJ Ulster Medical Journal. (Ulster Medical Society, Belfast City Hospital, Belfast BT9 7AD, N. Ireland) V.1- 1932-

UNDRA5 Unlisted Drugs. (Pharmaco-Medical Documentation, Inc., Box 401, 205 Main St., Chatham, NJ 07928) V.1- 1949-

UPJOH* "Compounds Available for Fundamental Research, Volume II-6, Antibiotics, A Program of Upjohn Company Research Laboratory." (Kalamazoo, MI 49001)

URGABW Urologe Ausgabe A. (Springer-Verlag, Heidelberger Pl.3, D-1000 Berlin 33, Fed. Rep. Ger.) V.9- 1970-

URLRA5 Urological Research. (Springer Verlag New York, Inc., Service Center, 44 Hartz Way, Secaucus, NJ 07094) V.1- 1973-

UROTAQ Urologia. (Libreria Editrice Canova, Via Panciera, 3b, 31100 Treviso, Italy) V.1- 1934-

USBCC* U. S. Borax and Chemical Company Reports. (New York, NY)

USXXAM United States Patent Document. (Commissioner of Patents and Trademarks, Washington, DC 20231)

UZBZAZ Uzbekskii Biologicheskii Zhurnal. (v/o Mezhdunarodnaya Kniga, Kuznetskii Most 18, Moscow G-200, USSR) No. 1- 1958-

VAAPB7 Virchows Archiv, Abteilung A: Pathologische Anatomie.

(Berlin, Germany) V.344-361, 1968-73. For publisher information, see VAPHDQ

VAAZA2 Virchows Archiv, Abteilung B: Zellpathologie, Cell Pathology. (Springer-Verlag, Heidelberger Pl. 3, D-1 Berlin 33, Germany) V.1- 1968-

VABBA3 Vestsi Akademii Navuk Belaruskai SSR, Seryya Biyalagichnykh Navuk. Proceedings of the Academy of Sciences of the Belorussian SSR, Biological Sciences Series. (v/o Mezhdunarodnaya Kniga, 12100 Moscow, USSR) No. 1- 1956-

VAMNAQ Vestnik Akademii Meditsinskikh Nauk S.S.S.R. Journal of the Academy of Medical Sciences of the U.S.S.R. (v/o Mezhdunarodnaya Kniga, Kuznetskii Most 18, Moscow G-200, U.S.S.R.) V.1- 1946-

VAPHDQ Virchows Archiv, Abteilung A: Pathological Anatomy and Histology. (Springer-Verlag, Heidelberger pl.3, D-1 Berlin. 33, Germany) V.362- 1974-

VCTDC* Vanderbilt, R.T., Co., Technical Data Sheet. (230 Park Ave., New York, NY 10017)

VDGIA2 Verhandlungen der Deutschen Gesellschaft fuer Innere Medizin. (J.F. Bergmann Verlag, Trogerstrasse 56, 8000 Munich 80, Federal Republic of Germany) V.33-52, 1921-40; V.54- 1948-

VDGPAN Verhandlungen der Deutschen Gesellschaft fuer Pathologie. (Gustav Fischer Verlag, Postfach 53, Wollgrasweg 49, 7000 Stuttgart-Hohenheim, Germany) V.1- 1898-

VDVEAV Vestnik Dermatologii i Venerologii. Journal of Dermatology and Venereology. (v/o Mezhdunarodnaya Kniga, Kuznetskii Most 18, Moscow G-200, U.S.S.R.) No.1- 1932-

VEARA6 Veterinarski Arhiv. (Jugoslovenska Knjiga, P.O. Box 36, Terazije 27, 11001, Belgrade, Yugoslavia) V.1- 1931-

VELPB* Velsicol Chemical Corporation, Product Bulletin. (341 E. Ohio St., Chicago, IL 60611)

VEMJA8 Veterinary Medical Journal. (Cairo Univ., Faculty of Veterinary Medicine, Giza, Egypt) V.1- 1953-

VETNAL Veterinariya. (v/o Mezhdunarodnaya Kniga, Kuznetskii Most 18, Moscow G-200, U.S.S.R.) V.1- 1924-

VETODR Veterinary Toxicology. (Manhatten, KS) V.15-18, 1973-76. For publisher information, see VHTODE

VETRAX Veterinary Record. (7 Mansfield St., London, W1M 0AT, England) V.1- 1888-

VHAGAS Verhandlungen der Anatomischen Gesellschaft. (VEB Gustav Fischer, Postfach 176, DDR-69 Jena, E. Germany) V.1- 1887-

VHTODE Veterinary and Human Toxicology. (American College of Veterinary Toxicologists, Office of the Secretary-Treasurer, Comparative Toxicology Laboratory, Kansas State University, Manhatten, Kansas 66506) V.19- 1977-

VIHOAQ Vitamins and Hormones. (Academic Press, 111 5th Ave., New York, NY 10003) V.1- 1943-

VINIT* Vsesoyuznyi Institut Nauchnoi i Tekhnicheskoi Informatsii (VINITI). All-Union Institute of Scientific and Technical Information. (Moscow, USSR)

VIRLAX Virology. (Academic Press, 111 Fifth Ave. New York, NY 10003) V.1- 1955-

VKMGA7 Voprosy Kommunal'noi Gigieny. Problems of Communal Hygiene. (Kiev, U.S.S.R.) V.1-6, 1963-66. For publisher information, see GNAMAP

VMDNAV Veterinarno-Meditsinski Nauki. (Hemus, Blvd. Russki 6, Sofia, Bulgaria) V.1- 1964-

VMEZA4 Voenno-Meditsinskii Zhurnal. Journal of Military Medicine. (v/o Mezhdunarodnaya Kniga, "Kuznetskii Most 18, Moscow G-200, USSR) 1944-

VOONAW Voprosy Onkologii. Problems of Onkology. (v/o Mezhdunarodnaya Kniga, Kuznetskii Most 18, Moscow G-200, U.S.S.R.) V.1- 1955-

VPITAR Voprosy Pitaniya. Problems of Nutrition. (v/o Mezhdunarodnaya Kniga, Kuznetskii Most 18, Moscow G-200, U.S.S.R.) V.1- 1932-

VRDEA5 Vrachebnoe Delo. Medical Profession. (v/o Mezhdunarodnaya Kniga, Kuznetskii Most 18, Moscow G-200, U.S.S.R.) No.1- 1918-

VRRAAT Vestnik Rentgenologii i Radiologii. Bulletin of Roentgenology and Radiology. (v/o Mezhdunarodnaya Kniga, Kuznetskii Most 18, Moscow G-200, U.S.S.R.) V.1- 1920-

VTPHAK Veterinary Pathology. (S. Karger AG, Postfach CH-4009 Basel, Switzerland) V.8- 1971-

VVIRAT Voprosy Virusologii. (v/o Mezhdunarodnaya Kniga, Kuznetskii Most 18, Moscow G-200, USSR) V.1- 1956-

WolMA# Personal Communication to Henry Lau, Tracor Jitco, Inc., from Mark A. Wolf, Dow Chemical Co., Midland, MI 48640

WADTAA Wright Air Development Center, Technical Report, U. S. Air Force. (Wright-Patterson Air Force Base, OH 45433)

WATRAG Water Research. (Pergamon Press Ltd., Headington Hill Hall, Oxford OX3 0BW, England) V.1- 1967-

WBEMAZ Wissenschaftliche Berichte E. Merck. (E. Merch AG, Darmstadt, Germany) No.1-6, 1960-71. Discontinued.

WDHU** Zur Pharmakologie des Antimons, Max Wieland Dissertation. (Pharmakologischen Institut der Hansischen Universitat, Hamburg, Germany, 1937)

WDMBAM Wadley Medical Bulletin. (Dallas, TX) V.1-5, 1970-75. For publisher information, see JCHODP

WDMU** Vergleichend-pharmakologische Untersuchungen uber die haemolytische Wirkung und die Toxizitat der Lokalanaesthetika, Martin Walther Dissertation. (Pharmakologischen Institut der Martin Luther- Universitat Halle-Wittenburg, Germany, 1936)

WDWU** Uber Komplexverbindungen von Pyramidon mit Rhodan-Ammonium, Rhodan-Natrium und Rhodan-Strontium, Walter Wittmann Dissertation. (Pharmakologischen Institut der Westfalischen Wilhems-Universitat zu Munster, Germany, 1933)

WEHRBJ Work, Environment, Health. (Helsinki, Finland) V.1-11, 1962-74. For publisher information, see SWEHDO

WHOTAC World Health Organization, Technical Report Series. (1211 Geneva, 27, Switzerland) No.1- 1950-

WILEAR Wiadomosci Lekarskie. (Ars Polona-RUSH, POB 1001, P-00068 Warsaw 1, Poland) V.1- 1948-

WIMJAD West Indian Medical Journal. (West Indian Medical Journal, University of the West Indies, Faculty of Medicine, Kingston 7, Jamaica) V.1- 1951-

WIPAAZ Wiadomosci Parazytologiczne. Parasitological News. (Ars Polona-RUCH, PB01001, 00-068 Warsaw 1, Poland) V.1- 1955-

WJMDA2 Western Journal of Medicine. (California Medical Assoc., 731 Market St., San Francisco, CA 94103) V.120- 1974-

WJSUDI World Journal of Surgery. (Springer-Verlag New York, Inc. Service Center, 44 Hartz Way, Secaucus, NJ 07094) V.1- 1978-

WKMRAH Wakayama Medical Report. (Wakayama Medical College Library, 9,9-bancho, Wakayama, 640, Japan) V.1- 1953-

WKWOAO Wiener Klinische Wochenschrift. (Springer-Verlag, Postfach 367, A-1011 Vienna, Austria) V.1- 1888-

WMHMAP Wissenschaftliche Zeitschrift der Martin-Luther Universitaet, Halle- Wittenberg, Mathematisch-Naturwissenschaftliche Reihe. (Deutsche Buch-Export und-Import GmbH, Leninstr. 16, 701 Leipzig, E. Germany) V.1- 1951-

WMWOA4 Wiener Medizinische Wochenschrift. (Brueder Hollinek, Gallgasse 40A, A-1130 Vienna, Austria) V.1- 1851-

WQCHM* "Water Quality Characteristics of Hazardous Materials, " W. Hann, and P.A. Jensen, Environmental Engineering Division, Civil Engineering Department, Texas A and M University, Volumes 1-4, 1974

WRABDT Wilhelm Roux's Archives of Developmental Biology. (Springer-Verlag New York, Inc., Service Center, 44 Hartz Way, Secaucus, New Jersey 07094) V.177- 1975-

WRPCA2 World Review of Pest Control. (London, England) V.1-10, 1962-71. Discontinued.

WTMOA3 Wiener Tieraerztliche Monatsschrift. (Ferdinand Berger and Soehne OHG, Wiene Str. 21-23, A-3580 Horn, Austria) V.1-1914-

WYCCO* Wyandotte Chemicals Corporation Reports. (Wyandotte, MI)

WZERDH Wissenschaftliche Zeitschrift der Ernst-Moritz-Arndt-Universitaet Greifswald, Medizinische Reihe. (Buchexport, Postfach 160, Leipzig DDR-7010, Ger. Dem. Rep.) V.21- 1972-

XAWPA2 U.S. Army Chemical Warfare Laboratories. (Army Chemical Center, MD)

XENOBH Xenobiotica. (Taylor and Francis Ltd., 4 John St., London WC1N 2ET, England) V.1- 1971-

XEURAQ U. S. Atomic Energy Commission, University of Rochester, Research and Development Reports. (Rochester, NY)

XHWBA9 U. S. Department of Health, Education and Welfare, Office of Education, Bulletin. (Washington, DC)

XIMMAX U. S. Department of the Interior, Bureau of Mines, Monographs. (Washington, DC)

XPHBAO U.S. Public Health Service, Public Health Bulletin. (Washington, DC)

XPHCI* U.S. Public Health Service, Current Intelligence Bulletin. (National Institute for Occupational Safety and Health, 5600 Fishers Lane, Rockville, MD 20857)

XPHPAW U.S. Public Health Service Publication. (U.S. Government Printing Office, Supt. of Doc., Washington, DC 20402) No.1- 1950-

YACHDS Yakuri to Chiryo. Pharmacology and Therapeutics. (Raifu Saiensu Pub. Co., 5-3-3 Yaesu, Chuo-ku, Tokyo 104, Japan) V.1-1973-

YAHOA3 Yakhak Hoeji. Journal of the Pharmaceutical Society of Korea. (Seoul Taehakkyo Yakhak Taehak, 28 YonGondong, Chongnogu, Seoul, S. Korea) V.1- 1956(?)-

YAKUD5 Gekkan Yakuji. Pharmaceuticals Monthly. (Yakugyo Jihosha, Inaoka Bldg., 2-36 Jinbo-cho, Kandu, Chiyoda-ku, Tokyo 101, Japan) V.1- 1959-

YCWCDP Yun-nan Chih Wu Yen Chiu. Yun-nan Botanical Research. (Yun-nan Jen Min C'hu Pan She, 100 Shu Lin Chieh, Kunming, Peop. Rep. China) V.1- 1979-

YFZADL Yaowu Fenxi Zazhi. Journal of Pharmaceutical Analysis. (Guoji Shudian, POB 2820, Beijing, Peop. Rep. China) V.1- 1981-

YHHPAL Yaoxue Xuebao. Acta Pharmaceutica Sinica. Pharmaceutical Journal. (China International Book Trading Corp., POB 2820, Beijing, Peop. Rep. China) V.1- 1953- (Suspended 1966-78)

YHTPAD Yaoxue Tongbao. Bulletin of Pharmacology. (China International Book Trading Corp., POB 2820, Beijing, Peop. Rep. China) V.1-12, 1953-66; V.13- 1978-

YIGODN Yiyao Gongye. Pharmaceutical Industry. (c/o Shanghai Yiyao Gongye Yanjiuyuan, 1320 Beijing-Xilu, Shanghai 200040, Pep. Rep. China) Begining history not known.

YIKUAO Yamaguchi Igaku. (Yamaguchi Daigaku Igakkai, Kogushi Ube, Yamaguchi-Ken 755, Japan) V.1- 1953-

YJBMAU Yale Journal of Biology and Medicine. (The Yale Journal of Biology and Medicine, Inc., 333 Cedar Street, New Haven, Connecticut 06510) V.1- 1928-

YKGKAM Yukagaku. Journal of Japan Oil Chemists' Society. (Nippon Yukagaku Kyokai, c/o Yushi Kogyo Kaigan, 3-13-11 Nihonbashi Chuo-Ku, Tokyo, Japan) V.5- 1956-

YKIGAK Yokohama Igaku. (Yokohama-Shiritsu Daigaku Igakkai, 2-33 Urafune-cho, Minami-Ku, Yokohama 232, Japan) V.1- 1948-

YKKKA8 Yakugaku Kenkyu. Japanese Journal of Pharmacy and Chemistry. (Osaka, Japan) V.1-39, 1929-68. Discontinued.

YKKZAJ Yakugaku Zasshi. Journal of Pharmacy. (Nippon Yakugakkai, 12-15-501, Shibuya 2-chome, Shibuya-ku, Tokyo 150, Japan) No.1- 1881-

YKRYAH Yakubutsu Ryoho. Medicinal Treatment. (Iji Nipposha, Kamiya Bldg., 4-11-7 Hatchobori, Chuo-Ku, Tokyo 104, Japan) V.1- 1968-

YKYUA6 Yakkyoku. Pharmacy. (Nanzando, 4-1-11, Yushima, Bunkyo-ku, Tokyo, Japan) V.1- 1950-

YMBUA7 Yokohama Medical Bulletin. (Npg. Yokohama Ika Daigaku, Urafune-cho, Minato-ku, Yokohama, Japan) V.1- 1950-

YOIZA3 Yonago Igaku Zasshi. Journal of the Yonago Medical Association. (c/o Tottori Daigaku Igakubu, Nishi-machi, Yonago, Japan) V.1- 1948-

YOMJA9 Yonsei Medical Journal. (Yonsei University College of Medicine, P.O. Box 71, Seoul, S. Korea) V.1- 1960-

YRTMA6 Yonsei Reports on Tropical Medicine. (Yonsei University, Institute of Tropical Medicine, Box 17, Seoul, South Korea) V.1- 1970-

ZolH## Personal Communication from H. Zollner, Institut fur Biochemie, Der Universitat, Gras, to H. Lau, Tracor Jitco, Inc., October 23, 1975

ZAARAM Zentralblatt fuer Arbeitsmedizin und Arbeitsschutz. (Dr. Dietrich Steinkopff Verlag, Saalbaustr 12,6100 Darmstadt, Germany) V.1- 1951-

ZACCAL Zacchia. Archivio di Medicina Legale, Sociale e Criminologica. Zacchia. Archives of Forensic, Social and Criminological Medicine. (Zacchia, Viale Regina Elena 336, Rome 00161, Italy) V.1- 1921-

ZAENAU Zeitschrift fuer Anatomie undd Entwicklungsgeschichte. (Berlin, Germany) V.60-145, 1921-74.

ZANZA9 Zeitschrift Angewandte Zoologie. (Duncker und Humblot, Postfach 410329, D-1000 Berlin 41, Fed. Rep. Germany) V.41- 1954-

ZAPOAK Zeitschrift fuer Allgemeine Mikrobiologie. Morphologie, Physiologie, Genetik und Oekologie der Mikroorganismen. (Akademie-Verlag GmbH, Liepziger Str. 3-4, DDR-108 Berlin, German Democratic Republic) V.1- 1960-

ZAPPAN Zentralblatt fuer Allgemeine Pathologie und Pathologische Anatomie. (VEB Gustav Fischer Verlag, Postfach 176, Villengang 2, 69 Jena, E. Germany) V.1- 1890-

ZBPHA6 Zentralblatt fuer Bakteriologie, Parasitenkunde, Infektionskrankheiten und Hygiene, Abteilung I: Medizinisch-Hygienische Bakteriologie, Virusforschung und Parasitologie, Originale. (Stuttgart, Federal Republic of Germany) V.31-151, 1902-45; V.152-216, 1947-71. For publisher information, see ZMMPAO

ZBPIA9 Zentralblatt fuer Bakteriologie, Parasitenkunde, Infektionskrankheiten und Hygiene, Abteilung II. Naturwissenschattliche: Allgemeine, Landwirtschattliche und Technische Miknobiologie. (VEB Gustav Fischer Verlag, Postfach 176, DDR-69 Jena, Ger. Dem. Rep.) V.107-132, 1952-77.

ZBPMAL Zentralblatt fuer Pharmazie. (Nurenburg, Germany) V.1-29, 1904-33. Discontinued.

ZDBEA9 Zdravookhranenie Belorussii. Public Health Service of White Russia. (v/o Mezhdunarodnaya Kniga, Kuznetskii Most 18, Moscow G-200, USSR) V.1- 1955-

ZDKAA8 Zdravookhranenie Kazakhstana. Public Health of Kazakhstan. (v/o Mezhdunarodnaya Kniga, Kuznetskii Most 18, Moscow G-200, U.S.S.R.) V.1- 1941-

ZDTUAB Zdravookhranenie Turkmenistana. Public Health of Turkmenistan. (v/o Mezhdunarodnaya Kniga, 113095 Moscow, USSR) V.1- 1957-

ZDVEA7 Zdravstveni Vestnik. Medical Journal. (Jugoslovenska Knjiga, POB 36, Terazije 27, YU-11001 Belgrade, Yugoslavia) V.15- 1946-

ZDVKAP Zdravookhranenie. Public Health. (v/o Mezhdunarodnaya Kniga, Kuznetskii Most 18, Moscow G-200, USSR) V.1- 1958-

ZEBLA3 Zeitschrift fuer Biologie. (Munich, Germany) V.1-116, 1865-1971. Discontinued.

ZECHAU Zentralblatt fuer Chirurgie. (Johann Ambrosius Barth Verlag, Salomonstr. 18b, 7O1 Leipzig, E. Germany) V.1- 1874-

ZEGYAX Zentralblatt fuer Gynaekologie. (Johann Ambrosius Barth Verlag, Postfach 109, DDR-7010, Leipzig, Ger. Dem. Rep.) V.1- 1877-

ZEKBAI Zeitschrift fuer Krebsforschung. (Berlin, Germany) V.1-75, 1903-71. For publisher information, see JCROD7

ZEKIA5 Zeitschrift fuer Kinderheilkunde. (Springer-Verlag, Heidelberger, Pl. 3, D-1 Berlin 33, Germany) V.1- 1910-

ZENBAX Zeitschrift fuer Naturforschung, Ausgabe B, Chemie, Biochemie, Biophysik, Biologie und Verwandten Gebieten. (Verlag der Zeitschrift fuer Naturforschung, Postfach 2645, D-7400, Tuebingen, Federal Republic of Germany) V.2- 1947-

ZEPHAR Zentralblatt fuer Physiologie. (Leipzig, E. Germany) V.1-34, 1887-1921. For publisher information, see PFLABK

ZEPRAN Zeitschrift fuer Praeventivmedizin. (Zurich, Switzerland) V.1-13, 1956-68. For publisher information, see SZPMAA

ZEPTAT Zeitscrift fuer Experimentelle Pathologie und Therapie. (Berlin, Germany) V.1-22, 1905-21. For publisher information, see REXMAS

ZERNAL Zeitschrift fuer Ernaehrungswissenschaft. (Steinkopff Verlag, Postfach 1008, 6100 Darmstadt, Germany) V.1- 1960-

ZEVBA5 Zeitschrift fuer Verebungslehre. (Springer-Verlag, Heidelberger Pl.3, D-1 Berlin 33, Germany) V.89-98, 1958-66.

ZEXCAI Zeitschrift fuer Experimentelle Chirurgie. (VEB Verlag Volk und Gesundheit, Neue Gruenstr 18, 102 Berlin, E. Germany) V.1- 1968-

ZGASAX Zeitscrift fuer Gastroenterologie. (Karl Demeter Verlag und Anzeigen-Verwaltung, Wuermstr 13, 8032 Graefelfing/Munich, Federal Republic of Germany) V.1- 1963-

ZGEMAZ Zeitschrift fuer die Gesamte Experimentelle Medizin. (Berlin, Germany) V.1-139, 1913-65. For publisher information, see REXMAS

ZGHEAQ Zentralblatt fuer die Gesamte Hygiene mit Einschluss der Bakteriologie und Immunitaetstehre. (Berlin, Germany) V.20-52, 1929-44. Discontinued.

ZGHGAW Zentralblatt fuer die Gesamte Hygiene und Ihre Grenzgebiete. (Berlin, Germany) V.1-19, 1922-29. Discontinued

ZGIMAL Zeitschrift fuer die Gesamte Innere Medizin und Ihre Grenzgebiete. (VEB Georg Thieme, Hainst 17/19, Postfach 946, 701 Leipzig, E. Germany) V.1- 1946-

ZGMOBE Zeitschrift fuer Gerichtliche Medizin. (Vienna, Austria)

ZGSHAM Zeitschrift fuer Gesundheitstechnik und Staedtehygiene. (Berlin, Germany) V.21-27, 1929-35. For publisher information, see ZANZA9

ZHINAV Zeitschrift fuer Hygiene und Infektionskrankheiten. (Berlin, Germany) V.11-151, 1892-1965. For publisher information, see MMIYAO

ZHPMAT Zentralblatt fuer Bakteriologie, Parasitenkunde, Infektionskrankranheiten und Hygiene, Abteilung 1: Originale, Reihe B: Hygiene, Praeventive Medizin. (Gustav Fischer Verlag, Postfach 72-01-43, D-7000 Stuttgart 70, Fed. Rep. Germany) V.155- 1971-

ZHYGAM Zeitschrift fuer die Gesamte Hygiene und Ihre Grenzgebiete. (VEB Georg Thieme, Hainst 17/19, Postfach 946, 701 Leipzig, E. Germany) V.1- 1955-

ZIEKBA Zeitschrift fuer Immunitaetsforschung, Experimentelle und Klinische Immunologie. (Fischer, Gustav, Verlag, Postfach 53, Wollgrasweg 49, 7000 Stuttgart-Hohenheim, Germany) V.141- 1970-

ZIETA2 Zeitschrift fuer Immunitaetsforschung und Experimentelle Therapie. (Stuttgart, Germany) 1924-V.124, 1962. For publisher information, see ZIEKBA

ZKCGAZ Zeitschrift fur Kinderchirurgie und Grenzgebiete. (Hippokrates Verlag, Postfach. 593, D-7000 Stuttgart 1, Fed. Rep. Germany) V.1- 1957(?)-

ZKKOBW Zeitschrift fuer Krebsforschung und Klinische Onkologie. (Berlin, Germany) V.76-92, 1971-78. For publisher information, see JCROD7

ZKMAAX Zhurnal Eksperimental'noi i Klinicheskoi Meditsiny. (v/o Mezhdunarodnaya Kniga, 121200 Moscow, USSR) V.2- 1962-

ZKMEAB Zeitschrift fuer Klin Medicine. (Berlin, Germany) V.1-158, 1879-1965. For publisher information, see EJCIB8

ZLUFAR Zeitschrift fuer Lebensmittel-Untersuchung und-Forschung. (Springer- Verlag, Heidelberger, Pl. 3, D-1 Berlin 33, Germany) V.86- 1943-

ZMEIAV Zhurnal Mikrobiologii, Epidemiologii i Immunobiologii.

(v/o Mezhdunarodnaya Kniga, Kuznetskii Most 18, Moscow G-200, U.S.S.R.) V.14- 1935-

ZMMPAO Zentralblatt fuer Bakteriologie, Parasitenkunde, Infektionskrankh eiten und Hygiene, Abteilung 1: Originale, Reihe A: Medizinische Mikrobiologie und Parasitologie. (Gustav Fischer Verlag, Postfach 72-01-43, D-7000 Stuttgart 70, Fed. Rep. Germany) V.217- 1971-

ZMOAAN Zeitschrift fuer Morphologie und Anthropologie. (Schweizerbart Verlagsbuchhandlung, Johannesst 3A, D-7000, Stuttgart 1, Germany) V.1- 1899-

ZMWIAJ Zentralblatt fuer die Medizinischen Wissenschaften. (Berlin, Germany)

ZNCBDA Zeitschrift fuer Naturforschung, Section C: Biosciences. (Verlag der Zeitschrift fuer Naturforschung, Postfach 2645, D-7400 Tuebingen, Federal Republic of Germany) V.29C- 1974-

ZNTFA2 Zeitschrift fuer Naturforschung. (Wiesbaden, Federal Republic of Germany) V.1, No.1-12 1946. For publisher information, See ZENBAX

ZOBIAU Zhurnal Obshchei Biologii. Journal of General Biology. (v/o Mezhdunarodnaya Kniga, Kuznetskii Most 18, Moscow G-200, U.S.S.R.) V.1- 1940-

ZPPLBF Zentralblatt fuer Pharmazie, Pharmakotherapie und Laboratoriums- diagnostik. (VEB Verlag Volk und Gesundheit, Neue Gruenstr 18, 102 Berlin, E. Germany) V.109- 1970-

ZRMDAN Zeitschrift fuer Rechtsmedizin. (Springer-Verlag, Heidelberger, Pl. 3, D-1 Berlin 33, Germany) V.67- 1970-

ZSNUAI Zeitschrift fuer Neurologie. (Berlin) V.198-206, 1970-74.

ZSRSBX Zashchita Rastenii (Moscow). Plant Protection. (v/o Mezhdunarodnaya Kniga, Kuznetskii Most 18, Moscow G-200, U.S.S.R.) V.11- 1966-

ZTMPA5 Zeitschrift fuer Tropenmedizin und Parasitologie. (Stuttgart, Germany) V.1-24, 1949-73. For publisher information, see TMPRAD

ZUBEAQ Zeitschrift fuer Unfallmedizin und Berufskrankheiten. (Revue de Medecine des Accidents et des Maladies Professionelles. (Verlag Berichthaus, Zwingliplatz 3, Postfach, CH-8022, Zurich, Switzerland) V.1- 1907-

ZVKOA6 Zhurnal Vsesoyuznogo Khimicheskogo Obshchestva Imeni D.I. Mendeleeva. Journal of the D.I. Mendeleeva All-Union Chemical Society. (v/oMezhdunarodnaya Kiga, Kuznetskii Most 18, Moscow G-200, U.S.S.R.) V.5- 1960-

ZVRAAX Zentralblatt fuer Veterinaermedizin, Reihe A. (Paul Parey, Linderst 44-47, 1000 Berlin 61, Germany) V.10- 1963-

ZYDXDM Zhejiang Yike Daxue Xuebao. Journal of Zhejiang Medical University. (Zhejiang Yike Daxue, Yan'an Lu, Hanzhou, Zhejiang, Peop. Rep. China) V.1- 1972-

ZZACAG Zeitschrift fuer Zellforschung und Mikroskopische Anatomie. (Berlin, Fed. Rep. Ger.) V.1-147, 1924-73. For publisher information, see CTSRCS

11FYAN "Fluorine Chemistry, Volume III. Biological Effects of Organic Fluorides," J.H. Simons, ed., New York, Academic Press, 1963

13BYAH "Morphological Precursors of Cancer," ed. Lucio Severi, Proceedings of an International Conference, Universita Degli Studi, Perugia, Italy, June 26-30, 1961, Perugia, Univ. of Perugia, 1962

13OPAL "Control of Ovulation," Proceedings of the Conference, Massachusetts Inst. of Technology, Edicott House, Dedham, MA, Feb. 26-28, 1960, C.A. Villee, ed., Oxford, Pergamon Press, 1961

13UZAD "Toksikologiia Redkikh Metablov," Ikh Soedin, 1963, U.S.S.R.

13ZGAF "Gigiena i Toksikologiya Novykh Pestitsidov i Klinika Otravlenii, Doklady Vsesoyuznoi Konferentsii," Komiteta po-Izucheniiu i Reglamenta-tsii Indokhimikatov Glavnoi Gosudarstvennoi Sanitarnoi Inspektsii U.S.S.R., 1962

14CYAT "Industrial Hygiene and Toxicology," 2nd rev. ed., Patty, F.A., ed., New York, Interscience Publishers, 1958-63

14FHAR "Venomous and Poisonous Animals and Noxious Plants of the Pacific Region, "Keegan, H.L., and W.V. Macfarlane, eds., NY, Pergamon Press Inc., 1963

14JTAF "Mycotoxins in Foodstuffs," Proceedings of the Symposium held at the Massachusetts Institute of Technology, Mar. 18-19, 1964, Wogan, G.N., ed., Cambridge, MA, MIT Press, 1965

14KTAK "Boron, Metallo-Boron Compounds and Boranes," R.M. Adams, New York, Wiley, 1964

14XBAV "International Congress of Chemotherapy," Proceedings of the 3rd Congress, July 22-27, 1963, Stuttgart, Thieme, 1964

14ZYA8 "Induction of Mutations and the Mutation Process," ed. Jiri Veleminsky, Proceedings of a Symposium held in Prague, Czechoslavok, 26-28, 1963, Prague, Publishing House of the Czechoslavok Academy of Science, 1965

15ARA5 "Voprosy Promyshlennoi i Sel'skokhozyai stvennoi Toksikologii," 1964

15MBAH "Proceedings of the International Symposium on Chemotherapy of Cancer," Lugano, Switzerland, April 28-May 1,1964, Plattner, P.A., ed., New York, American Elsevier Publishing Co., 1964

15QWAW "Agents Affecting Fertility, a Symposium," 1964, Austin, C.R., and J.S. Perry, eds., Boston, Little, Brown and Co., 1965

15SWA8 "Kirk-Othmer Encyclopedia of Chemical Technology, 2nd Ed. 1963-70," A. Standen, et al., eds., New York, Wiley, 1971

15TRAW "Preimplantation Stages of Pregnancy, Ciba Foundation Symposium, London, 1965," Wolstenholme, G.E.W., and M. O'Connor, eds., Boston, Little, Brown and Co., 1965

16KXAC "Farmakologiya i Farmakoterapiya Alkaloidov i Glikozidov," Pharmacology and Pharmacotherapy of Alkaloids and Glycosides, Kamilov, I.K., ed., 1966. Available from Chemical Abstracts Service, PO Box 3012, Columbus, OH 43210

17QLAD "Ratsionl'noe Pitanie," (Rational Nutrition) P.D. Leshchenko, ed., USSR, 1967

18IDAO "Nobel - Gas Compounds," A.J. Finkel, et al., Lamont, Illinois, Argonne National Laboratory, 1963

18JKAG "Proceedings of the Fifth British Insecticide and Fungicide Conference," Brighton, England, November 6-9, 1961

19DDA6 "Animal Toxins, a Collection of Papers Presented at the International Symposium on Animal Toxins, 1st, 1966," Russell, F.E., and P.R. Saunders, eds., NY, Pergamon Press Inc., 1967

19FKA3 "Biochemistry of Some Foodborne Microbial Toxins," Symposium on Microbial Toxins, New York, Cambridge, MIT Press, 1967

19UQAS "Khimicheskie Paktory Vneshnei Sredy I Ikh Gigienicheskoe Znachenie, Materialy Nauchnoi Konferentsii," 2nd, Pervyi Moskovskii Meditsinskii Institut, Moscow, June 22, 1965

20CEAM "Farmakologiya Alkaloidov i Glikozidov," Pharmacology of Alkaloids and Glycosides, Kamilov, I.K., ed., 1967. Available from Chemical Abstracts Service, PO Box 3012, Columbus, OH 43210

20MAAQ "Toxicity of Anesthetics," Proceedings of a Research Symposium, Seattle, Washington, May 12-13, 1967. Baltimore, Maryland, Williams and Wilkins Co., 1968

20MYAU "Mechanism of Mutation and Inducing Factors," Proceedings of a Symposium, Prague, Czechoslovakia, Aug. 9-11, 1965. New York, Academic Press, 1966

20PKA3 "Novye Dannye Po Toksikologii Redkikh Metalov Ikh Soedinenii," 1967

20ZJAG "Toksikologiia Novykh Khimicheskikh Veshchesty, Vnedriaemykh V," Rezinovuiu I Shinnaiu Promyshlennost, 1968

21ACAB "Imifos," S.A. Giller, ed., Izd, Riga, U.S.S.R., 1968

21NDAB "Symposium on Carbenoxolone Sodium," Robson, J.M., and F.M. Sullivan, eds., London, Butterworth, 1968

21PFAR "Conferentia Hungarica pro Therapia et Investigatione in Pharmacologia," Societas Pharmacologyica Hungarica, 4th, Budapest, Oct. 3-8, 1966, Akad. Kindo, 1968

22HAAD "Microbial Toxins, 1970-72," Ajil, S.J., et al., eds., New York, Academic Press, 1970-72

22XWAN "Chemical Mutagenesis in Mammals and Man," F. Vogel and G. Roehrborn, eds., Berlin, Springer-Verlag, 1970

23EIAT "Venomous Animals and Their Venoms," Buecherl, W., and E.E. Buckley, eds., New York, Academic Press, Inc., 3 vols., 1968-71

23HZAR "Carcinoma of the Colon and Antecedent Epithelium," W.J. Burdette, Springfield, Illinois, C.C. Thomas, 1970

24KLA7 "International Encyclopedia of Pharmacology and Therapeutics," New York, Pergamon Press, 1966-

24UTAD "Onkologiya, Informatsionen Byuletin, Prilozhenie." Natsionalen Kongres po Onkologiya, Sbornik Dokladi, 1st, Sofia, Bulgaria, October 22-24, 1969

24ZIA5 "International Council of Scientific Unions, International Upper Mantl Project, Programs and International Recommendation," 1960-63

25NJAN "International Cancer Congress, Abstracts, 10th," Houston, Texas, May 22-29, 1970, Cumley, R.W. and J.E. McCay, eds., Chicago, IL., Year Book Medical Publishers, Inc., 1970

26QZAP "Advances in Antimicrobial and Antineoplastic Chemotherapy, Progress in Research and Clinical Application," 7th Proceedings of the International Congress of Chemotherapy, Hejzlar, M. et al., eds., Prague, Aug. 23-28, 1971, 3 Vols., Prague, Czecholovakia, University Press, 1972

26RAAN "Benzodinzenines," S. Garattini, ed., New York, Raven Press, 1973

26UYA8 "Psoriasis," Proceedings of the International Symposium, Stanford, Calif., July 7-10, 1971, Farber, E.M. and A.J. Cox, eds., Stanford, Calif., Stanford University Press, 1971

26UZAB "Pesticides Symposia," collection of papers presented at the Sixth and Seventh Inter-American Conferences on Toxicology and Occupational Medicine, Miami, Florida, Univ. of Miami, School of Medicine, 1968-70

27CWAL "Medical Primatology," Selected Papers from the 3rd Conference on Experimental Medicine and Surgery in Primates, Lyons, France, June 21-23, 1972, Goldsmith, E.I. and J. Moor-Jankowski, eds., White Plains, New York, Phiebig, 1972

27GLA8 "Toksikologiya i Gigiena Produktov Neftekhimii i Neftekhimicheskikh Proczvodstv, Vsesoyuznaya Konferentsiya," Doklady, U.S.S.R., 1971. Toxicology and Hygiene of the Products of Petroleum Chemistry and Petrochemical Productions, All-Union Conference Reports. (Yaroslavskii Meditsinskii Institut, Yaroslavl, U.S.S.R.)

27ZIAQ "Drug Dosages in Laboratory Animals - A Handbook," C.D. Barnes and L.G. Eltherington, Berkeley, Univ. of California Press, 1965, 1973

27ZJAT "Tranquillizing and Antidepressive Drugs," W.M. Benson and B.C. Schille, Springfield, Illinois, C.C. Thomas, 1962

27ZKAW "Isolation and Identifications of Drugs in Pharmaceuticals, Body Fluids and Post-Mortem Material," E.G.C. Clarke, ed., London, Pharmaceutical Press, 1969

27ZMA4 "Psychopharmacological Agents, Volume 2," M. Gorden, ed., New York, Academic Press, 1967

27ZOAA "Compendium of Psychopharmacology," W. Poldinger, Basle, Switzerland, Hoffmann-La Roche, 1967

27ZPAD "Psychochemotherapy - The Physicians' Manual," E. Remmen, et al., Los Angeles, Western Medical Publications, 1962

27ZQAG "Psychotropic Drugs and Related Compounds," E. Usdin, and D.H. Efron, 2nd Ed., Washington, DC, 1972

27ZTAP "Clinical Toxicology of Commercial Products - Acute Poisoning," Gleason, et al., 3rd Ed., Baltimore, Williams, and Wilkins, 1969. (Note that although the Registry cites data from the 1969 edition, this reference has been superseded by the 4th edition, 1976.)

27ZWAY "Heffter's Handbuch der Experimentelle Pharmakologie."

27ZXA3 "Handbook of Poisoning: Diagnosis and Treatment," R.H. Dreisbach, Los Altos, California, Lange Medical Publications

27ZYA6 "Handbook of Toxicology of Pesticides to Wildlife," U. S. Department of Interior, Washington, DC

27ZZA9 "Manual of Pharmacology and Its Applications to Therapeutics and Toxicology," T. Sollman, 8th Ed., Philadelphia, W.B. Saunders, 1957

28FTAY "Environmental Toxicology of Pesticides, Proceedings of a United States-Japan Seminar, "Oiso, Japan, October 1971. (Pub. 1972). New York, Academic, 1972

28QFAD "Laboratory Animal in Drug Testing," Symposium of the 5th International Committee on Laboratory Animals, Hanover, Sept. 19-21, 1972, Speigel, A. ed., Stuttgart, Germany, Fischer, 1973

28ZAA9 "Manual of Pharmacology and Its Applications to Therapeutics and Toxicology," T. Sollman, 7th Ed., Philadelphia, W.B. Saunders, 1942

28ZDAI "Psychotropic Drugs," S. Garattini, ed., Amsterdam, Elsevier Press, 1957

28ZEAL "Pesticide Index," E.H. Frear, ed., State College, PA, College Science Publications, 1969 (Note that although the Registry cites data from the 1969 edition, this reference has been superseded by the 5th edition, 1976.)

28ZFAO "Pharmacology and Toxicology of Uranium Compounds," Voegtlin and Hodge, New York, McGraw Hill, 1949

28ZHAU "Proceedings of the American Society for Pharmacology and Experimental Therapeutics."

28ZLA8 "Toxicity of Industrial Metals," E. Browning, London, Butterworths, 1961

28ZNAE "Public Health Reports, Supplement." (Baltimore, MD/ Washington, DC)

28ZOAH "Chemicals in War," A.M. Prentiss, 1937

28ZPAK "Sbornik Vysledku Toxixologickeho Vysetreni Latek A Pripravku," J.V. Marhold, Institut Pro Vychovu Vedoucicn Pracovniku Chemickeho Prumyclu Praha, Czechoslovakia, 1972

28ZRAQ "Toxicology and Biochemistry of Aromatic Hydrocarbons," H. Gerarde, New York, Elsevier, 1960

28ZSAT "Drugs in Research," P. DeHaen, New York, Paul DeHaen, 1964

29QHAQ "Principles of Medicinal Chemistry," William O. Foye, eds., Philadelphia, Lea and Febiger, 1974

29QKAZ "Toxins of Animal and Plant Origin," A. de Vries and E. Kochva, eds., New York, Gordon and Breach Science Publishers, 1971

29ZQAU "Electrical and Drug Treatments in Psychiatry," A.S. Paterson, New York, Elsevier, 1963

29ZRAX "Pharmacology of Oriental Plants," K.K. Chen and B. Mukerji, Oxford Pergamon Press, 1965

29ZSA2 "Trifluoperazine: Further Clinical and Laboratory Studies," J.E. Moyer, ed., Philadelphia, Lea and Febiger, 1959

29ZTA5 "Neuro-Psychopharmaca in Klinik und Praxis," D. Wandreg and V. Leuther, Stuttgart, F.K. Schattauer, 1965

29ZUA8 "Toxicity and Metabolism of Industrial Solvents," E. Browning, New York, Elsevier, 1965

29ZVAB "Handbook of Analytical Toxicology," Irvings Sunshine, ed., Cleveland, Ohio, Chemical Rubber Company, 1969

29ZWAE "Practical Toxicology of Plastics," R. Lefaux, Cleveland, Ohio, Chemical Rubber Company, 1968

30FDAB "Neuropoisons: Their Pathophysiological Actions, 1971-1974," Simpson, L.L., and D.R. Curtis, eds., New York, Plenum Pub. Corp., 1974

30ZDA9 "Chemistry of Pesticides." N.N. Melnikov, New York, Springer-Verlag, 1971

30ZFAF "Toxic Aliphatic Fluorine Compounds," F.L.M. Pattison, London, Elsevier, 1959

30ZIAO "Vanadium Toxicology and Biological Significance," T.G. Faulkner- Hudson, New York, Elsevier Publishing Company, 1964

31BYAP "Experimental Lung Cancer: Carcinogenesis Bioassays, International Symposium, 1974," New York, Springer, 1974

31OGA2 "Alcohol and Aldehyde Metabolizing Systems," Papers presented at the International Symposium. Stockholm, July 9-11, 1973. New York, Academic, 1974

31TFAO "Progress in Chemotherapy," Proceedings of the 8th International Congress on Chemotherapy, Athens, Greece, 1973, Daikos, G.K. ed., Athens, Hill. Society of Chemotherapy, 1974

31ZOAD "Pesticide Manual," Worcesteshire, England, British Crop Protection Council, 1968

31ZPAG "Synthetic Analgesics," Oxford, Pergamon Press, 1966

31ZTAS "Alcohols: Their Chemistry, Properties and Manufacture," J.A. Monick, New York, Reinhold Book, 1968

31ZUAV "Investigations in the Field of Organogermanium Chemistry," Rijkens, F., and G.J.M. Van Der Kerk, Utrecht, Netherlands, Schotanus & Jens Utrecht N.V., 1964

32OAAP "Pesticide and the Environment: a Continuing Controversy," Papers Presented and Papers Reviewed at the 8th Inter-American Conference on Toxicology and Occupational Medicine, Miami, Florida, July 1973, Deichmann, W.B., ed., New York, Stratton, 1973

32RWA4 "Free Radicals in Biology," William A. Pryor, ed., New York, Academic, 1976

32XPAD "Teratology," C.L. Berry, ed., New York, Springer-Verlag, 1975

32YLA6 "Teratomas and Differentiation, a Symposium," Sherman, M.I. ed. Roche Inst. of Molecular Biology, Nutley, New Jersey, May 19-21, 1975, New York, Academic, 1975

32YWA5 "Radiation Research, Biomedical, Chemical, and Physical Perspectives, Proceedings of the International Congress of Radiation Research, 5th, Seattle, July 14-20, 1974 (Pub 1975)," O.F. Nygaard, et al., ed., New York, Academic, 1975

32ZCAI "Ganglion-Blocking and Ganglion-Stimulation Agents," D.A. Kharkevich, Moscow, First Moscow Medical Institute, 1967

32ZDAL "Aldrin Dieldrin Endrin and Telodrin: An Epidemiological and Toxicological Study of Long-Term Occupational Exposure," K.W. Jager, New York, Elsevier, 1970

32ZWAA "Handbook of Poisoning: Diagnosis and Treatment," R.H. Dreishach, 8th Ed., Los Altos, California, Lange Medical Publications, 1974

32ZXAD "Transactions of the 36th Annual Meeting of the American Conference of Governmental Industrial Hygienists," Miami Beach, Fla., May 12-17, 1974

33AQAA "Biology of Radiation Carcinogenesis," J. M. Yuhas, et al., ed., New York, Raven Press, 1976

33GVAL "Catharanthus Alkaloids," William I. Taylor, and Norman R. Farnsworth, eds., New York, Marcel Dekker, 1975

33IUAS "Mycotoxins," Purchase, I.F.H. ed., Amsterdam, Elsevier, 1974

33JOAF "Sublethal Effects of Toxic Chemicals on Aquatic Animals (Proceedings)," J.H. Koeman, and J.J. Strik, New York, Elsevier, 1975

33NFA8 "Animal Models in Dermatology," H. Maibach, ed., New York, Churchill Livingston, 1975

34LXAP "Insecticide Biochemistry and Physiology," C.F. Wilkinson, ed., New York, Plenum, 1976

34TRAD "Food-Drugs from the Sea," Proceedings of the Conference, 4th, University of Puerto Rico, Mayaguez, PR, November 17-21, 1974, Webber, H.H., and G.D. Ruggieri, eds., Washington, DC, Marine Technology Society, 1976

34ZHAD "Symposium on Mycotoxins in Human Health," I.F.H. Purchase, London, Macmillan, 1971

34ZIAG "Toxicology of Drugs and Chemicals," W.B. Deichmann, New York, Academic Press, 1969

34ZRA9 "Proceedings Quadrennial Conference on Cancer," L. Severi, ed., Perugia, Italy, University of Perugia, 1966

35FUAR "Animal, Plant, and Microbial Toxins," Akira Ohsaka, et al., eds., New York, Plenum Press, 1976

35NLA6 "Trace Elements in Human Health and Disease," Vol. 1 and 2, A.S. Prasad, et al., ed., New York, Academic Press, 1976

35ORAT Gigiena Truda i Sostoyanie Spetsificheskikh Funktsii u Rabotnits Neftekhimicheskoi i Khimicheskoi Promyshlennosti. Industrial Hygiene and Condition of Specific Functions of Working Women in the Petrochemical and Chemical Industry. (Sverdlovskii Nauchno - Issledovatel'skii Institut Okhrany Materinstva i Mladeuchestva Minzdrava RSFSR, Sverdiovsk, U.S.S.R.) 1974

35WYAM "In Vitro Metabolic Activation in Mutagenic Testing," Proceedings of the Symposium on the Role of Metabolic Activation in Producing Mutagenic and Carcinogenic Environmental Chemicals, Research Triangle Park, N.C., Feb. 9-11, 1976, De Serres, F.J. et al., eds., New York, Elsevier North Holland, 1976

35ZRAG "Medicinal Chemistry IV," J. Maas, ed., Proceedings of the 4th International Symposium on Medicinal Chemistry, Noordwijkerhout, The Netherlands, September 9-13, 1974. New York, Elsevier, 1974

36DLAV "Clinical and Pharmacological Aspects of Sodium Valproate (Epilim) in the Treatment of Epilepsy," Proceedings of a Symposium, Nottingham, England, Sept. 23-24, 1975, Legg, N.J. et al., eds., Kent, England, MCS Consultants, 1976

36PYAS "Pharmacology of Steroid Contraceptive Drugs," Garattini, S. and H.W. Berendes, eds., New York, Raven Press, 1977. (Monograph Mario Negri Inst. Pharm. Res.)

36SBA8 "Antifungal Compounds," M.R. Siegel and H.D. Sisler, eds., 2 vols., New York, Marcel Dekker, 1977

36THAV "Management of the Poisoned Patient," Rumack, B.H., and A.R. Temple, eds., Princeton, NJ, Science Press, 1977

36URAW "Microsomes and Drug Oxidations." Proceedings of the 2nd International Symposium on Microsomes, Berline, 21-24 July, 1976. Oxford, Pergamon Press, 1977

36YFAG "Biological Reactive Intermediates, Formation, Toxicity and Inactivation," Proceedings of the International Conference on Active Intermediates, Formation, Toxicity and Inactivation, University of Turku, Turku, Finland, July 26-27, 1975. New York, Plenum, 1977.

37ASAA "Kirk-Othmer Encyclopedia of Chemical Technology, 3rd Edition," Grayson, M. and D. Eckroth, eds., New York, Wiley, 1978

37ATAD "Prevention and Detection of Cancer," Proceedings of the 3rd International Symposium on Detection and Prevention of Cancer, New York, Marcel Dekker, 1977

37KXA7 "Water Chlorination: Environmental Impact and Health Effects," Proceedings of the Conferences on the Environment, Ann Arbor, MI, Ann Arbor Science Pub. Inc., 1975-

37OBAT "Peptides," Goodman, M., and Meienhafer, J., eds., 5th American Peptide Symposium, Univ. Calif., June 20-24, 1977, New York, Wiley, 1978

37QLAZ "Current Approaches in Toxicology," Ballantyne, B., ed., Bristol, UK, John Wright & Sons, Ltd., 1977

37XLA2 "Current Chemotherapy," Proceedings of the 10th International Congress of Chemotherapy, Zurich, Sept. 18-23, 1977, W. Siegenthaler and R. Luethy, eds., 2 vols., Washington, DC, Am. Soc. Microbiol., 1978

38KLAC "Structure-Activity Relationship among the Semisynthetic Antibiotics." Perlman, D., ed., New York, Academic Press, 1977

38MKAJ "Patty's Industrial Hygiene and Toxicology," 3rd rev. ed., Clayton, G.D., and F.E. Clayton, eds., New York, John Wiley & Sons, Inc., 1978-82. Vol. 3 originally pub. in 1979; pub. as 2nd rev. ed. in 1985.

39SDAH "Perspective Toxinology," A. W. Bernheimer, ed., New York, Wiley, 1977

39TKAR "Handbook of Cancer Immunology." Waters, Harold, ed., New York, Garland STPM Press, 1978-81

40RMA7 "Health and Sugar Substitutes," Proceedings of the ERGOB Conference on Sugar Substitutes, Geneva, 1978, Guggenheim, B., ed., Basel, Switzerland, S. Karger AG, 1979

40WDA5 "Bleomycin: Current Status and New Development," Papers Presented in a Symposium, Oakland, Calif., 1977, Carter, S.K., et al., eds., New York, Academic Press, 1978

40QBA3 "Proceedings of the International Congress on Toxicology, Toxicology as a Predictive Science, 1st, Toronto, 1977," Plaa, G.L., and W.A. Duncan, eds., New York, Academic Press, Inc., 1978

40YJAX "Evaluation of Embryotoxicity, Mutagenicity and Carcinogenicity Risks In New Drugs," Proceedings of the Symposium on Toxicological Testing for Safety of New Drugs, 3rd, Prague, Apr. 6-8, 1976, Benesova, O., et al., eds., Prague, Czechoslovakia, Univerzita Karlova, 1979

41CIAR "Effects of Poisonous Plants on Livestock," Proceedings of a Joint U.S. - Australian Symposium on Poisonous Plants, Logan, Utah, 1978, Keeler, R.F. et al. eds., New York, Academic Press, 1978

41HTAH "Aktual'nye Problemy Gigieny Truda. Current Problems of Labor Hygiene," Tarasenko, N.Y., ed., Moscow, Pervyi Moskovskii Meditsinskii Inst., 1978

41KEAL "Toxicology, Biochemistry and Pathology of Mycotoxins", Uraguchi, K. and M. Yamazaki, eds., New York, Wiley, 1978

41PDA9 "Kadmium-Symposium," Jena, 1977, Bolck F., ed., Jena, Ger. Dem. Rep., Friedrich-Schiller Universitaet, 1979

41UKAL "Mutagen-induced Chromosome Damage in Man," A.C. Stevenson and L. Evans eds., Edinburgh, Scotland, Edinburgh Univ. Press, 1978

42GEA8 "Environmental Carcinogenesis," P. Emmelot and E. Kriek, eds. Amsterdam, Elsevier, 1979

43FLAV "Burger's Medicinal Chemistry," Wolff, M.E., ed., 4th ed., New York, John Wiley and Sons, 1981

43GRAK "Dusts and Disease," Proceedings of the Conference on Occupational Exposures to Fibrous and Particulate Dust and Their Extension into the Environment, 1977, Lemen, R., and J.M. Dement, eds., Park Forest South, IL, Pathotox Publishers, 1979

43MKAT "Current Chemotherapy and Infectious Diseases," Proceedings of the 11th International Congress of Chemotherapy and the 19th Interscience Conference on Antimicrobial Agents and Chemotherapy, Boston, MA, 1-5 Oct., 1979, American Society for Microbiology, Washington, DC, 1980

43QNAO "Cadmium in the Environment," Nriagu, J.O., ed., 2 vols., New York, Wiley, 1980-81

43VMAC "Microsomes, Drug Oxidations, and Chemical Carcinogenesis," International Symposium on Microsomes and Drug Oxidations, 4th, 1979, Coon, M.J., et al., eds., New York, Academic Press, 1980

43VSAU "Phenytoin-Induced Teratology and Gingival Pathology," Hassell, T.M., et al., eds. New York, Raven Press, 1980

43XWAI "Brain Tumors and Chemical Injuries to the Central Nervous System," Proceedings of the International Neuropathological Symposium, Warsaw, Sep. 23-26, 1976, Mossakowski, M.J., ed. Warsaw, Panstwowe Wydawnictwo Naukowe, 1978

43ZRAD "Hemoperfusion: Kidney and Liver Support and Detoxification," Proceedings of the International Symposium, Haifa, 1979, Sideman, S., and T.M.S. Chang, eds., Washington, DC, Hemisphere Pub. Corp., 1980

44LQAF "International Sulprostone Symposium, (Papers)," Vienna, 1978, Friebel, K., et al., eds., Berlin, Fed. Rep. Ger., Schering AG, 1979

44LVAU "Peptides: Structure and Biological Function," Proceedings of the American Peptide Symposium, 6th, 1979, Cross, E. and J. Meienhofer, eds., Pierce Chemical Co., 1979

45ICAX "Nickel Toxicology," 2nd International Conference on Nickel Toxicology. Swansea, U.K., 3-5 Sept., 1980. London, Academic Press, 1980

45JVAR "Research Frontiers in Fertility Regulation," Proceedings of an International Workshop, 1980, Zatuchni, G.I., et al., eds., Hagerstown, MD, Harper and Row Medical Dep., 1980

45KQAH "Health Effects Investigation of Oil Shale Development," Meeting, Gatlinburg, Tenn. USA, June 23-24, 1980, Editors: Griest, W.H., M.R. Guerin and D.L. Coffins, Ann Arbor Science Publishers, Inc., 1980

45OHAA "Short-Term Test Systems for Detecting Carcinogens," Proceedings of the Symposium, 1978, Norpoth, K.H., and R.G. Garner, eds., Berlin, Springer-Verlag, 1980

45OJAG "Radiation Sensitizers: Their Use in the Clinical Management of Cancer," Proceedings of the Conference, Key Biscayne, Fla., Oct. 3-6, 1979, Brady, L.W., ed. New York, Masson Publishing USA, Inc., 1980

45REAG "Teratology of the Limbs," Symposium on Prenatal Development, 4th, 1980, Merker, H.J., et al., eds., Berlin, Walter de Gruyter and Co., 1980

45VOAW "Inflammation: Mechanisms and Treatment," Proceedings of the International Meeting, 4th, 1980, Willoughby, D.A., and J.P. Giroud, eds., Baltimore, MD, University Park Press, 1980

45ZJA3 "Nitrosoureas: Current Status and New Development," Prestayko, A.W. et al., eds., New York, Academic Press, 1981

46GFA5 "Safety Problems Related to Chloramphenicol and Thiamphenicol Therapy," Najean, Y., et al., eds., New York, Raven Press, 1981

46IEAC "Current Research Trends in Prenatal Craniofacial Development," Proceedings of an International Conference, 1980, Pratt, R.M., and R.L. Christiansen, eds., New York, Elsevier North Holland, 1980

46IZA7 "Retinoids, Advances in Basic Research and Therapy," Proceedings of the International Dermatology Symposium, Berlin, 1980, Orfanos, C.E., et al., eds., New York, Springer-Verlag, 1981

46OJAN "Chemical Analysis and Biological Fate: Polynuclear Aromatic Hydrocarbons," 5th International Symposium, Battelle's Labs., Oct. 1980. Columbus, Ohio, Battelle Press, 1981

47JMAE "Cytogenetic Assays of Environmental Mutagens," Hsu, T.C., ed., Totowa, NJ, Allanheld, Osmun and Co., 1982

47YKAF "Sporulation and Germination," Proceedings of the Eighth International Spore Conference, 1980, Washington, DC, American Society of Microbiology, 1981

48ECAY "Spurenelement-Symposium: Nickel, 3rd," Jena, Anke, M., et al., eds., Jena, Ger. Dem. Rep., Friedrich-Schiller-Universitaet, 1980

48HGAR "Current Chemotherapy and Immunotherapy," Proceedings of the International Congress of Chemotherapy, 12th, 1981, Periti, P., and G. Gialdroni-Grassi, eds., Washington, DC, American Society for Microbiology, 1982

48NTAS "Advances in Genetics, Development, and Evolution of Drosophila," Proceedings of the European Drosophila Research Conference, 7th, 1981, Lakovaara, S., ed., New York, Plenum Publishing Corp., 1982

48THAM "Proceedings of the International Symposium on Cyclodextrins," 1st, 1981, Szejtli, J., ed., Hingham, MA, Kluwer Boston Inc., 1982

49DWAS "Application of Behavioral Pharmacology in Toxicology," Zhinden, G., et al., eds. New York, Raven Press, 1983

49DXAV "Health Evaluation of Heavy Metals in Infant Formula and Junior Food," Proceedings of the Symposium, 1981, Schmidt, E.H.F., and A.G. Hildebrandt, eds., Berlin, Springer-Verlag, 1983

49RQAC "Chemistry and Biology of Hydroxamic Acids," Proceedings of the International Symposium, 1st, Dayton, 1981, Kehl, H., ed., Basel, Switzerland, S. Karger AG, 1982

49VWAG "Female Transcervical Sterilization," Proceedings of an International Workshop on Non-Surgical Methods for Female Tubal Occlusion, Chicago, June 22-24, 1982, Zatuchni, G.I., et al., eds. New York, Harper and Row, 1983

49WEAZ "Pesticide Chemistry: Human Welfare and the Environment," Proceedings of the International Congress of Pesticide Chemistry, 5th, Kyoto, Aug. 29-Sep. 4, 1982, Miyamoto, J. and Kearney, P.C., eds. Oxford, England, Pergamon Press Ltd., 1983

50EXAK "Formaldehyde Toxicity," Conference, 1980, Gibson, J.E., ed., Washington, DC, Hemisphere Publishing Corp., 1983

50EYAN "Structure-Activity Correlation as a Predictive Tool in Toxicology," Papers from a Symposium, 1981, Golberg, L., ed., Washington, DC, Hemisphere Publishing Corp., 1983

50NNAZ "Polynuclear Aromatic Hydrocarbons," International Symposium, 7th, 1982, Cooke, M., and A.J. Dennis, eds., Columbus, OH, Battelle Press, 1983

50RXAH "Formaldehyde: Toxicology, Epidemiology, Mechanisms," Papers of a Meeting, 2nd, 1982, Clary, J.J., ed., New York, Marcel Dekker, 1983

50WKA3 "Reproduction and Developmental Toxicity of Metals," Proceedings of a Joint Meeting, Rochester, 1982, Clarkson, T.W., et al., eds., New York, Plenum Publishing Corp., 1983

51UDAB "Heavy Metals in the Environment, International Conference, 4th, 1983," Heidelberg, S., 2 vols., Edinburgh, UK, CEP Consultants Ltd., 1983

51ZKAW "The Retinoids, Vol.2," Sporn, M.B., et al., eds., New York, Academic Press, Inc., 1984

52EDA6 Nitrendipine Present Int Workshop Nitrendipine 1st, New York, 1982, "Scriabine, A., et. al. eds., Baltimore, Urban & Schwarzenberg, 1984

52GUAX "Anorganische Stoffe in der Toxikologie und Kriminalistik, Symposium, Mosbach, Fed. Rep. Ger., 1983," Heppenheim, Fed. Rep. Ger., Verlag Dr. Dieter Helm, 1983

52MLA2 "Toxicology of Petroleum Hydrocarbons, Proceedings of the Symposium, 1st, 1982," MacFarland, H.N., et al., eds., Washington, DC, American Petroleum Institute, 1983

52OLAC "Veterinarnaya Entomologiya i Akarologiya," Veterinary Entomology and Acarology, Nepoklonov, A.A., ed., Moscow, Izdatel'stovo Kolos, 1983

53HJAC "Mouse Liver Neoplasia: Current Perspectives," Popp, J.A., ed., New York, Hemisphere Pub. Corp., 1984

55CXA9 "Trichothecenes and Other Mycotoxins, Proceedings of the International Mycotoxin Symposium, 1984," Lacey, J., ed., Chichester, UK, John Wiley and Sons Ltd., 1985

55DXAE "Progress in Nickel Toxicology, Proceedings of the International Conference on Nickel Metabolism and Toxicology, 3rd, 1984," Brown, S.S., and F.W. Sunderman, Jr., eds., Oxford, UK, Blackwell Scientific Pub. Ltd., 1985

85ADA6 "Carcinogenicity Tests of Oral Contraceptives," Great Britain Committee on Safety of Medicines, London, H.M. Stationery Off., 1972

85AEA9 "Environmental Hazards of Metals," I.T. Brakhnova, trans. by J.H. Slep, New York, Consultants Bureau, 1975

85AFAC "Ecological Toxicology Research: Effects of Heavy Metal and Organohalogen Compounds," A.D. McIntyre and C.F. Mills, eds., New York, Plenum Press, 1975

85AGAF "Effects and Dose-response Relationships of Toxic Metals," G.F. Nordberg, ed., Proceedings from an International Meeting Organized by the Subcommittee on the Toxicology of Metals of the Permanent Commission and International Association on Occupational Health, Tokyo, November 18-23, 1974, New York, Elsevier Scientific, 1976

85AHAI "Clinical Aspects of the Teratogenicity of Drugs," H. Nishimura, and T. Tanimum, New York, Elsevier, 1976

85AIAL "Toxicity of Pure Foods," E.M. Boyd, Cleveland, Ohio, CRC Press, 1973

85AJAO "Handbook of Vitamins and Hormones," R.J. Kutsky, New York, V.N. Reinhold, 1973

85AKAR "Toxicology of Insecticides," F. Matsumura, New York, Plenum Press, 1975

85ALAU "Carbamate Insecticides: Chemistry, Biochemistry, and Toxicology," R.J. Kuhr, and H.W. Dorough, Cleveland, Ohio, CRC Press, 1976

85AMAX "Industrial Toxicology," A. Hamilton and H.L. Hardy, 3rd Ed., Acton, Mass., Publishing Sciences Group, 1974

85ANA2 "Adverse Effects of Environmental Chemicals and Psychotropic Drugs, Volume 2," M. Horvath, ed., New York, Elsevier, 1976

85ARAE "Agricultural Chemicals, Books I, II, III, and IV," W.T. Thomson, Fresno, CA, Thomson Publications, 1976/77 revision

85CVA2 "Oncology 1970," Proceedings of the Tenth International Cancer Congress, Chicago, Year Book Medical Publishers, 1971

85CYAB "Chemistry of Industrial Toxicology," H.B. Elkins, 2nd Ed., New York, J. Wiley, 1959

85CZAE "Survey of Chemical and Biological Warfare," J.Cookson and J. Nottingham, New York, Monthly Review Press, 1969

85DAAC "Bladder Cancer, A Symposium," K.F. Lampe et al., eds., Fifth Inter- American Conference on Toxicology and Occupational Medicine, University of Miami, School of Medicine, Coral Gables, Florida, Aesculapius Pub., 1966

85DCAI "Poisoning; Toxicology, Symptoms, Treatments," J.M. Arena, 2nd Ed., Springfield, Illinois, C. C. Thomas, 1970

85DEAO "Occupational Medicine and Industrial Hygiene," R.T. Johnstone, St. Louis, C.V. Mosby, 1948

85DFAR "Report of the Enquiry into the Medical and Toxicological Aspects of CS (Orthochlorobenzylidene Malononitrile)," London, Her Majesty's Secretary of State for the Home Department, 1971

85DGAU "Catalog of Teratogenic Agents," T.H. Shepard, 2nd Ed., Baltimore, Maryland, Johns Hopkins University Press, 1976

85DHAX "Medical and Biologic Effects of Environmental Pollutants Series." Washington, DC, National Academy of Sciences, 1972-77

85DIA2 "CTFA Cosmetic Ingredient Dictionary," N.F. Estrin, 2nd Ed., Washington, DC, Cosmetic, Toiletry and Fragrance Association, 1977

85DJA5 "Malformations Congenitales des Mammiferes," H. Tuchmann-Duplessis, Paris, Masson et Cie, 1971

85DKA8 "Cutaneous Toxicity," V.A. Drill and P. Lazar, eds., New York, Academic Press, 1977

85DLAB "Studies on Chemical Carcinogenesis by Diels-Alder Adducts of Carcinogenic Aromatic Hydrocarbons," R.H. Earhart, Jr., Ann Arbor, Michigan, Xerox Univ. Microfilms, 1975

85DMAE "Drugs as Teratogens," J.L. Schardein, Cleveland, Ohio, CRC Press, 1976

85DNAH "Toxicologie Clinique et Analytique," J.P. Frejaville, and R. Bourdon, 2nd Ed., Paris, France, Flammarion Medecine-Sciences, 1975

85DPAN "Wirksubstanzen der Pflanzenschutz und Schadlingsbekampfungsmittel," Werner Perkow, Berlin, Verlag Paul Parey, 1971-1976

85DQAQ "Organophosphorus Pesticides," R. Derache, New York, Pergamon Press, 1977

85DUA4 "Chemical Tumour Problems," W. Nakahara, ed., Tokyo, Japanese Society for the Promotion of Science, 1970

85DVA7 Sonderdruck aus Bericht Uber den. VIII. Internationalen Kongress fuer Unfallmedizin und Berufskrankheiten. Frankfurt A.M. 26-30 September 1938, Leipzig, Georg Thieme, 1938

85DWAA International Congress on Occupational Health, Helsinki, 1957

85DXAD "Pediatric Nutrition in Developmental Disorders," S. Palmer and S. Ekvall, eds., Springfield, Ill., C.C. Thomas, 1978

85DYAG "Fourth National Medicinal Chemistry Symposium of the American Chemical Society, June 17-19, 1954, Syracuse, NY," Washington, DC, American Chemical Society, 1954

85DZAJ "Principles of Insect Chemosterilization," G.C. Labrecque and C.N. Smith, eds., New York, Appleton-Century-Crofts, 1968

85EAAH "Cancer," R.W. Raven, ed., London, Butterworth and Co., 1958

85EBAK "Noyaku Chudoku, Pesticide Poisoning," K. Hiraki and K. Hyodo, Tokyo, Igaku Shoin, 1969

85ECAN "Metal Toxicity in Mammals, Vol. 2: Chemical Toxicity of Metals and Metalloids," B. Venugopal and T.D. Luckey, New York, Plenum Press, 1978

85EGD4 "Toxins: Animal, Plant and Microbial. Proceedings of the International Symposium," P. Rosenberg, ed., New York, Pergamon Press, 1978

85EHD7 "Toxic Eye Hazards," Publication No. 494, National Society for the Prevention of Blindness, New York, The Society, 1949

85EIDA "A Survey of Cardiac Glycosides and Genins," J. H. Hoch, Charleston, Univ. of South Carolina, 1961

85EJDD "Chemical Induction of Cancer, Structural Bases and Biological Mechanisms," J. C. Arcos and M. F. Argus, New York, Academic, 1974

85EKDG "Etude Experimentale de la Glycoregulation Gravidique et de l'Action Teratogene des Perturbations du Metabolisme Glucidique," R. Meyer, Paris, Masson et Cie, 1961

85ELDJ "Die Herzwirksamen Glykoside," G. Baumgarten and W. Forster, Leipzig, VEB Georg Thieme, 1963

85ERAY "Antibiotics: Origin, Nature, and Properties," T. Korzyoski, Z. Kowszyk-Gindifer, and W. Kurylowicz, eds., Washington DC, American Society for Microbiology, 1978

85ESA3 "Merck Index; an Encylopedia of Chemicals and Drugs," 9th Ed. Rahway, N.J., Merck and Co., 1976

85FZAT "Index of Antibiotics from Actinomycetes," Umezawa, H. et al., eds., Tokyo, University of Tokyo Press, 1967

85GDA2 "CRC Handbook of Antibiotic Compounds, Volumes 1-9," Berdy, Janos. Boca Raton, Florida, CRC Press, 1980

85GLAQ "A Survey of Antimalarial Drugs, 1941-1945," Wiselogle, F.Y., 2 vols., Ann Arbor, MI, J.W. Edwards, 1946

85GMAT "Toxicometric Parameters of Industrial Toxic Chemicals Under Single Exposure," Izmerov, N.F., et al. Moscow, Centre of International Projects, GKNT, 1982

85GNAW "Pharmacodynamie de l'Acide Dipropylacetique (ou Propyl-2-Pentanoique) et de ses Amides," Carraz, G., Imprimerie Eymond, 1968

85GRAA "The Control of Fertility," Pincus, G., New York, Academic Press, 1965

85GUAJ "Antifertility Compounds in the Male and Female," Jackson, H., Springfield, IL, Charles C. Thomas, 1966

85GYAZ "Pflanzenschutz-und Schaedlingsbekaempfungsmittel: Abriss einer Toxikologie und Therapie von Vergiftungen," 2nd ed., Klimmer, O.R., Hattingen, Fed. Rep. Ger., Hundt-Verlag, 1971

85HBAZ "Environmental Variables in Oral Disease;" A Symposium Presented at the Montreal Meeting of the American Association for the Advancement of Science, 1964, Kreshover, S.J., et al., eds., Washington, DC, American Association for the Advancement of Science, 1966

85INA8 "Documentation of the Threshold Limit Values and Biological Exposure Indices," 5th ed., Cincinnati, Ohio, American Conference of Governmental Industrial Hygienists, Inc., 1986

85IPAE "Modern Pharmaceuticals of Japan, V," Tokyo, Japan Pharmaceutical, Medical and Dental Supply Exporters' Assoc., 1975

85IVAW "Possible Long-Term Health Effects of Short-Term Exposure to Chemical Agents," National Research Council, 3 vols., Washington, DC, National Academy Press, 1982-85

85IXA4 "Structure et Activite Pharmacodyanmique des Medicaments du Systeme Nerveux Vegetatif," Bovet, D., and F. Bovet-Nitti, New York, S. Karger, 1948

85JCAE "Prehled Prumyslove Toxikologie; Organicke Latky," Marhold, J., Prague, Czechoslovakia, Avicenum, 1986

85JDAH "Organophosphorus Pesticides: Organic and Biological Chemistry," Eto, M., Cleveland, OH, CRC Press, Inc., 1974

85JFAN "Agrochemicals Handbook," with updates, Hartley, D., and H. Kidd, eds., Nottingham, Royal Soc. of Chemistry, 1983-86

A Guide to Using This Book

Entry Number – Entries are indexed in order by this alphanumeric code.
See Introduction: paragraph 1, p. xiii.

Entry Name – A complete entry name and synonym cross-index is located in Section 2.
See Introduction: paragraph 2, p. xiii.

DOT: – The four digit hazard code assigned by the U.S. DOT.
See Introduction: paragraph 5 p. xiii.

mf: – The molecular formula
mw: – The molecular weight
See Introduction: paragraphs 6, and 7, p. xiv.

PROP: – Physical properties including solubility and flammability data. May contain a definition of the entry.
See Introduction: paragraph 9, p. xiv.

SYNS: – Synonyms for the entry. A complete synonym cross-index is located in Section 2.
See Introduction: paragraph 10, p. xiv.

Toxicity Data: – Data for skin and eye irritation, mutation, teratogenic, reproductive, carcinogenic, human, and acute lethal effects.
See Introduction: paragraphs 11, 12, 13, 14, and 15, p. xiv-xxiv.

Consensus Reports: – Supply additional information to enable the reader to make knowledgeable evaluations of potential chemical hazards.
See Introduction: paragraph 17, p. xxiv.

Standards and Recommendations: Here are listed the OSHA PEL, ACGIH TLV, DFG MAK, and NIOSH REL workplace air levels. U.S. DOT classification and labels are also listed.
See Introduction: paragraph 18, p. xxv.